35th Annual USSD Conference 2015

Managing Risk and Uncertainty – Ensuring a Sustainable Future for Dams, Levees and Reservoirs

Louisville, Kentucky, USA
13-17 April 2015

Volume 1 of 2

ISBN: 978-1-5108-2941-1

Printed from e-media with permission by:

Curran Associates, Inc.
57 Morehouse Lane
Red Hook, NY 12571

Some format issues inherent in the e-media version may also appear in this print version.

Printed by Curran Associates, Inc. (2016)

For permission requests, please contact United States Society on Dams (USSD) at the address below.

United States Society on Dams (USSD)
1616 Seventeenth Street #483
Denver, CO 80202
USA

Phone: 303-628-5430
Fax: 303-628-5431

gene@ussdams.org

Additional copies of this publication are available from:

Curran Associates, Inc.
57 Morehouse Lane
Red Hook, NY 12571 USA
Phone: 845-758-0400
Fax: 845-758-2633
Email: curran@proceedings.com
Web: www.proceedings.com

USSD

Managing Risk and Uncertainty — Ensuring a Sustainable Future for Dams, Levees and Reservoirs

35th Annual USSD Conference
Louisville, Kentucky, April 13-17, 2015

Hosted by Corps of Engineers

On the Cover

Aerial view of the Cannelton Hydroelectric Project — one of American Municipal Power's four run-of-river hydroelectric projects currently under construction on the Ohio River. The powerhouse is a three-unit, single monolith, reinforced concrete, combination dam/hydroelectric powerhouse founded on alluvial river deposits and designed to be submerged in floods. The powerhouse is equipped with three horizontal bulb turbines each with a capacity of 28 MW and is capable of generating a total of 84 MW at a gross head of 25 feet. The powerhouse is provided with an exterior-type 175-ton capacity permanent bridge crane that operates on concrete runway frames above the roof in a direction parallel to flow and is an unique application for hydroelectric projects on the Ohio River, first implemented at the Belleville Hydroelectric Project.

U.S. Society on Dams

Vision

A world class organization dedicated to advancing the role of dam and levee systems and building the community of practice.

Mission

USSD, as the United States member of the International Commission on Large Dams, is dedicated to:

ADVOCATE:	Champion the role of dam and levee systems in society.
EDUCATE:	Be the premier source for technical information about dam and levee systems.
COLLABORATE:	Build networks and relationships to strengthen the community of practice.
CULTIVATE:	Nurture the growth of the community of practice.

FOREWORD

Included in these Proceedings are papers presented during the **35th USSD Annual Meeting and Conference**, held April 13-17, 2015, in Louisville, Kentucky. A separate book includes abstracts for each paper.

The theme of the 35th USSD Conference was ***Managing Risk and Uncertainty — Ensuring a Sustainable Future for Dams, Levees and Reservoirs.*** The Conference technical program was organized by several USSD Committees under the leadership of **B. Alex Grenoble**, HDR. The theme highlights that, while dams and levees are critical components of the nation's infrastructure, there are always risks and uncertainties associated with ownership and operation of these facilities. Recognizing and effectively managing these risks and uncertainties is critical to ensuring that these structures will continue to provide the societal benefits that we depend on without posing an unacceptable hazard to life, property or the environment.

Technical Sessions during the Conference focused on best practices and new technologies related to concrete dams, embankment dams and levees; dam safety and security; construction and rehabilitation; hydraulics; foundations, monitoring; and seismic and environmental issues.

The papers in the Proceedings were selected from abstracts submitted in response to a Call for Papers, and include both oral and poster presentations. Authors are specialists with broad experience from government agencies, utilities, academia, water districts, consulting firms and private industry.

The Conference Organizing Committee extends thanks and appreciation to the **Corps of Engineers**, Host of the 35th Annual Meeting and Conference.

Special thanks are also extended to the Committee Members who selected the abstracts and reviewed the technical papers, and to the authors who prepared the papers included in the Proceedings.

CONTENTS

Plenary Session

Managing Risk and Uncertainty

Construction

Embankment Dams

Seismic

Hydrology and Hydraulics I

Concrete Dams

Levees

Tailings and Ash Impoundments

Dam Safety and Security

Monitoring

Foundations

United States Society on Dams

USSD

Managing Risk and Uncertainty — Ensuring a Sustainable Future for Dams, Levees and Reservoirs

35th Annual USSD Conference
Louisville, Kentucky, April 13-17, 2015

Unlocking Funding for Navigation Improvements
By formulating two procurement packages, we enabled the U.S. Army Corps of Engineers St. Paul District's first-ever design-build delivery. Expedited scheduling qualified Lock and Dam 3 for federal funding, keeping waters open for the 7 million tons of cargo passing through the lock annually.
This is where great begins.
hdrinc.com

SAFETY ASSESSMENT ON THE HEIGHTENING PROJECT OF DANJIANGKOU DAM

Y. Liu[1]
G. X. Zhang[2]
L. Zhang[3]
Q.J. Zhou[4]

INTRODUCTION

The Danjiangkou Reservoir is a water source of the middle route of China's South-to-North Water Diversion Project. To meet the requirements on the quantity of water diversion of the middle route, the Danjiangkou Project should be heightened and thickened on the basis of early construction. The dam of the early project is 162.0 m in elevation, with a normal storage water level of 157.0 m. According to the plan, after heightening, the height of the dam will be risen to 176.6 m, while the normal water level will be elevated to 170.0 m. Besides, the maximum reservoir capacity will be increased from 208.9 billion m^3 to 339.1 billion m^3.

In the early construction, the water retaining structure, with a total length of 2.494 km, consists of the concrete gravity dam in the riverbed and the earth and rockfill dam along the banks. The dam is 162.0 m in elevation, with a maximum height of 97.0 m. The length of the concrete dam is 1141 m while the length of the earth and rockfill dams along the left and right banks are 1223 and 130 m, respectively.

The early stage of Danjiangkou water control project began in Sept. 1958 and was completed in Feb. 1974. Many accidents due to poor quality occured in the concrete dam poured at the early stage, such as lack-of-foundation, cold joint and unqualified strength. A large number of cracks appeared on the concrete which were poured simultaneously, with a total number up to 2463 according to the statistics. The cracks were then reinforced at that time.

Since its first operation in 1972, the Danjiangkou water control reservoir showed a minimum water level of 131.28 m in Apr. 1979 and a maximum level as high as 160.07 m on Oct. 7th, 1983. The dam has undergone several high water levels after 1983. For Danjiangkou Dam in which several defects and cracks exist, safety assessments have been conducted many times, and the conclusion is that Danjiangkou Dam constructed at the early stage satisfies the requirements on safe operation.

---

[1] Deputy Director, Department of Structures and Materials, China Institute of Water Resources and Hydropower Research, Beijing, China, liuyi@iwhr.com
[2] Director, Department of Structures and Materials, China Institute of Water Resources and Hydropower Research, Beijing, China, zhanggx@iwhr.com
[3] Senior Engineer, Department of Structures and Materials, China Institute of Water Resources and Hydropower Research, Beijing, China, zhangl@iwhr.com
[4] Senior Engineer, Department of Structures and Materials, China Institute of Water Resources and Hydropower Research, Beijing, China, zhouqj@iwhr.com

After the heightening project of dam begun, we have carried out a comprehensive inspection on the cracks of dam. Totally 3169 cracks were found by Oct. 2011, of which 110 were III-type cracks and 166 were IV-type. Particularly, some cracks spanned over several sections of dam, and some cracks were located on the dam surfaces in both upstream and downstream. With the heightening of dam, the cracks in some sections of dam extended from the early construction to the concrete poured later.

Heightening and thickening are needed for the concrete dam with so many cracks and defects. Disconnection exists in the interface between old and new concrete surfaces due to the effects of temperature load. Therefore, the method of safety assessment of the heightening project is the major concern.

The cracks and defects have weakened the local strength of the concrete dam. Moreover, the hydraulic fractures have been generated. Given the disconnection between the old and new concrete surfaces, the old and new dams cannot bear the load as a whole structure. All these factors have threatened the safety of the dam. According to the Design Specifications for Concrete Gravity Dam (SL319-2005) [1], the safety of dam depends mainly on stability and stress. The stability safety factor of the dam is calculated using rigid body limit equilibrium method, while the stress is calculated by the method of material mechanics. However, these calculation models are quite simple, with no considerations on the effects induced by the cracks and defects inside the dam or the interface between old and new concrete surfaces. As a consequence, new method and criteria used for assessing the safety condition of the heightening project of Danjiangkou Dam are required.

In the present work, based on the whole-process simulations on the whole dam and engineering analogy method, we proposed a new method for the assessment of the safety condition of concrete dams. Using this method, the long-term stability of the cracks and the effects of cracks, defects and the interface between old and new concrete surfaces on the safety of dam were studied. Moreover, the concept of relative overload safety was introduced to assess the relative safety condition of dam. Thus, the difficulties in safety assessment on Danjiangkou Dam can be resolved.

SAFETY ASSESSMENT METHOD FOR THE CONCRETE DAM WITH DEFECTS TAKEN INTO ACCOUNT

Using finite element method (FEM), the interfaces between old and new concrete surfaces were simulated with the adoption of joint model; the local defects were simulated using parameter weakening method, by which the actual working condition of the dam could be simulated. The safety margin of the dam was studied using the nonlinear analysis of overload. The result was then compared with that of the dam with no defects. Accordingly, the effects of cracks and defects on the safety of dam and the relative safety of the dam were obtained.

Finite Element Method (FEM)

In FEM, four types of actual factors are considered: actual material, actual structure, actual loads, and actual process. Actual material parameters refer to hydration heat, time-dependent modulus, autogenous volume change and creep of the hydraulic concrete, and parameters for foundation on the basis of back analysis. The actual structure includes transverse joints in the dam, faults, defects in the foundation, and the waterproof-drainage system. Actual loads comprise the gravity load considering the construction process, water load that includes seepage load, and temperature load that includes the nonlinear temperature difference inside the dam. The actual process stands for the actual construction and impounding processes.

These four actual factors are dependent on one another. Real stress states in a dam are produced by the coupling effects of these four actual factors. However, the consideration of these factors is oversimplified in current design codes and conventional safety assessment methods. Hence, an integrated analytical approach is proposed in this work to simulate this coupling effect in finite element analysis.

The unsteady temperature field, unsteady seepage field, and concrete creep stress field computations have to be conducted for the actual working performance assessment. The unsteady temperature field analysis is governed by Eq. [1].

$$\lambda\left(\frac{\partial^2 T}{\partial x^2}+\frac{\partial^2 T}{\partial y^2}+\frac{\partial^2 T}{\partial z^2}\right)+c\rho\left(\frac{\partial \theta}{\partial \tau}-\frac{\partial T}{\partial \tau}\right)=0 \qquad [1]$$

where the input parameters include temperature $T=\overline{T}$, boundary heat flux $-\lambda\frac{\partial T}{\partial n}=q$, and third boundary conditions $-\lambda\frac{\partial T}{\partial n}=\beta(T-T_a)$, where τ is the concrete age, λ is the heat conductivity coefficient, ρ is the density, c is the specific heat, $a=\lambda/c\rho$ denotes the thermal diffusivity, θ is the adiabatic temperature rise, T_a is the environment temperature (which becomes the adiabatic wall temperature for forced convective conditions), and β is the heat emission factor.

The unsteady seepage field is computed according to Eq. [2].

$$\frac{\partial}{\partial x}\left(k_x\frac{\partial h}{\partial x}\right)+\frac{\partial}{\partial y}\left(k_y\frac{\partial h}{\partial y}\right)+\frac{\partial}{\partial z}\left(k_z\frac{\partial h}{\partial z}\right)=0 \qquad [2]$$

where h is the water head, and k_x, k_y, k_z represent the permeability coefficients of three directions. The initial condition is set as $h|_{t=0}=h_0(x,y,z,t)$, the water head boundary is set as $h|_{\Gamma_1}=f_1(x,y,z,t)$, and the flux boundary is set as $k_n\frac{\partial h}{\partial n}|_{\Gamma_2}=f_2(x,y,z,t)$.

Taking concrete as a visco-elastic material, we should consider the time-dependent effects of Young's modulus. The displacement/stress state of the dam-foundation system in every step can be solved by using the incremental finite element method. The governing equation is as follows:

$$[K]\{\Delta\delta_n\}=\{\Delta P_n\}^L+\{\Delta P_n\}^C+\{\Delta P_n\}^T+\{\Delta P_n\}^0+\{\Delta P_n\}^S \qquad [3]$$

where $[K]$ is the global stiffness matrix, $\{\Delta\delta_n\}$ is the nodal displacement increment, $\{\Delta P_n\}^L$ is the nodal load increment attributed to external load, $\{\Delta P_n\}^C$ is the nodal load increment attributed to creep strain, $\{\Delta P_n\}^T$ is the nodal load increment attributed to temperature change, $\{\Delta P_n\}^0$ is the nodal load increment attributed to the volume change of concrete, and $\{\Delta P_n\}^S$ is the nodal load increment attributed to seepage load. Detailed derivations of the calculation formula are explained by Zhu et al [2,3] .

Calculation Model of Joint Elements

The definition of joint element. Joint element refers to crack element and transverse joint element. The two types of joints are different in their constitutive relations, but same in the definition of displacement function and stiffness matrix. For cracks on spatial surface, they can be defined by 8~20 joint-varying thin layer element, as shown in Fig. 1.

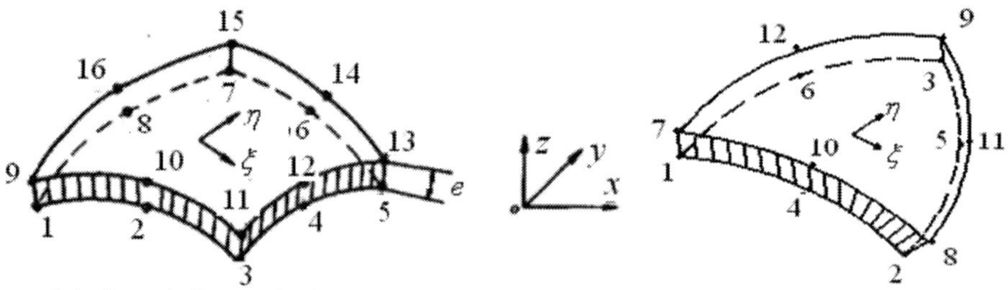

(a) Quadrilateral element (b) Triangle element

Figure 1. Joint element on curved surface.

The deformation at any point of the joint surface is:

$$\begin{bmatrix}\Delta u\\ \Delta v\\ \Delta w\end{bmatrix}=\sum_{i=1}^{N}[N_i]\begin{bmatrix}u_i\\ v_i\\ w_i\end{bmatrix} \qquad [4]$$

In the equation, $[N_i]$ refers to shape function. The stiffness matrix of the joint element is:

$$[k]^e = \int_{-1}^{1}\int_{-1}^{1}[N]^T[L]^T[D][L][N]d\xi d\eta \qquad [5]$$

In the equation, $[L]$ is coordinate transformation matrix and $[D]$ elastic matrix.

When joints yield or open, and the normal force or shearing strength on joint surface needs to be released, the non-balance force of the releasing stress should be calculated according to the equation below.

$$[P]^{\varepsilon}_{\sigma_0} = -\int_{-1}^{1}\int_{-1}^{1}[N]^T[L]^T[\sigma_0^{'}]|J|d\xi d\eta \qquad [6]$$

Local joint opening or yielding can be defined by Gauss point, and can be reflected through adjustment of the normal and tangential stiffness of the integral point in Eq. [5] and Eq. [6].

<u>The elastic coefficient matrix of joint element.</u> Matrix $[D]$ in the general definition equation [5] is the elastic matrix of joint element. By changing the value of the matrix, different joints and their state can be simulated. It is defined by the following equation:

$$[D] = \begin{bmatrix} K_s & & \\ & K_s & \\ & & K_n \end{bmatrix} = \frac{1}{e}\begin{bmatrix} G_s & & \\ & G_s & \\ & & G_n \end{bmatrix} \qquad [7]$$

e is element thickness, G_s and G_n refer to shear modulus of the interior materials of joint elements along the tangential direction and elastic modulus along the normal direction respectively. When the thickness of joint element is 0, the computation equation is:

$$[D] = \begin{bmatrix} \lambda_s & 0 & 0 \\ 0 & \lambda_s & 0 \\ 0 & 0 & \lambda_n \end{bmatrix} \qquad [8]$$

λ_s and λ_n are the normal and tangential stiffness of the joint surface. Their values are adopted according to the test results, if there are no test values adopt a bigger value so as to reflect the deformation of joint surface and avoid error.

<u>Simulation of joint failure.</u> Joint state depends on tensile strength and shearing strength. When the stress is lower than failure strength, the joint is in the state of cohesion, and the load transferring mechanism is consistent with that without joint; when the stress reaches the failure strength, the joint fails. The strength criterion of joints is as follows:

① tensile failure: $\sigma_n > \sigma_c$

② shearing failure

compress shearing: $\tau > c + \sigma_n tg\phi$

tensile shearing: $\tau > c$

σ_c is the tensile strength of joints, c is the cohesion, ϕ is the friction angle.

Open and close of joints and simulation of sliding. In operation period, dam will be influenced by the periodic variation of temperature and water level. Thus, the state of joints constantly changes because of repeated loading. For example, some joints open in winter and close in summer, while others are on the contrary. The same joint may open and close repeatedly under the influence of temperature and water level change. Moreover, the open and close of many joints have mutual actions with each other. Meanwhile, some joints may exist reciprocating shearing sliding. Therefore, the open-close and sliding of multiple joints is a non-linear problem. Shi Genhua [4] considers it as a high-dimension non-linear problem, which can be solved by the non-linear iterative method.

With reference to the non-linear iterative method proposed by Shi Genhua, this paper aims to solve the problems of open-close and sliding of joints. Since the change of temperature and water level is simulated, the incremental method is adopted to simulate loads, and the total increment method to simulate the state of open and close and sliding. The change of state and the calculation method of joints between Step *i* and Step *i+1* are shown as follows:

a) From close to close, without sliding yield: no treatment of the stiffness matrix, not to release non-balance force;

b) From close to close, with yield sliding, i.e. the shearing strength $|\tau| > c + \sigma_n \tan\phi$.

Stiffness coefficient: $K_n^i \to K_n^{i+1}$, $K_s^i \to K_s^{i+1} = 0$.

c) From close to open, stiffness coefficient: $K_n^i \to K_n^{i+1} = 0$, $K_s^i \to K_s^{i+1} = 0$. And release all the stress on joint face in Step *i*.

d) From open to close, stiffness coefficient: $K_n^i = 0 \to K_n^{i+1}$, $K_s^i = 0 \to K_s^{i+1}$.

In the above step, the normal opening is " + ". And in the present step, it is " - ". It is of necessity to calculate the tangential penetrating proportion in order to determine the tangential force for complementation. Stress needs to be complemented is:

$$[\sigma']^{\varepsilon} = [D] \begin{bmatrix} \delta'_x \\ \delta'_y \\ \delta'_n \end{bmatrix} \qquad [9]$$

δ_n' is penetration amount in the normal direction. δ_x' and δ_y' refer to the displacement in the direction of local coordinate x' and y' after penetration of the contact point. Direction of local coordinate x' and y' after penetration of the contact point.

Within one loading time step, all joint elements undergo repeated alternation in accordance with state a) ~ d) until convergence is obtained, then move on to the next loading time step.

Calculation Method of the Overload Safety

When the rock-dam concrete of the foundation of the dam are loaded to a certain degree, i.e., the stress exceed the strength, the material will yield. Therefore, the material becomes nonlinear, and the elasto-plastic finite element model should be used. The incremental method was combined with the iterative method. That is, the load was increased using the incremental method, while the nonlinearity of the material in each incremental load step was described using the iterative method.

Drucker-Prager yield criterion and Mohr Coulomb yield criterion were adopted for the basement and concrete, respectively. As we all know, both concrete and rocks are brittle materials. Upon failure, their strength will be reduced or even disappear. For instance, the tensile strength of the concrete which are split under tension is zero. Besides, their cohesive force and friction coefficient will decrease in a similar manner. Accordingly, the simulation results using the ideal elasto-plastic model will overestimate the bearing capacity of the structure to a certain extent, which is detrimental to the safety design of the structure. To avoid the overestimation of safety, the brittle failure model was introduced to solve the problem that the strength was reduced after the material yielded.

Currently, the most common index in the assessment of the structural safety of concrete dam is the overall safety . By the causes of failure, the overall safety can be classified into overload safety and strength reserve related safety. The safety coefficient can be calculated as follows. Using nonlinear finite element method, we performed calculations on each section of dam by incremental loading until the dam was destroyed. The inflection point on the relationship curve between displacement and overload factor can be regarded as the failure criterion, while the overload factor at the inflection point is the safety coefficient. According to the loading method, several types of safety coefficients can be acquired, as described in the following.

1) Water density overloading method, with the safety coefficient of K_5

$$K_5 = \frac{\gamma_u}{\gamma} \qquad [10]$$

where γ_u denotes the water density when the dam body and foundation fail, while γ denotes the water density under normal conditions.

2) Water-level overloading method, with the safety coefficient ofK_6

$$K_6 = \frac{P_u}{P} \quad [11]$$

where P_u denotes the hydraulic pressure of dam heel when the dam body and foundation failure, while P denotes the hydraulic pressure of dam heel under normal conditions.

3) Strength reserve coefficient method (or strength-reduction method), with the safety coefficient of K_8.

Various strength parameters of the concrete in the dam all decrease to $1/\lambda$ times that of the initial values, until the dam fails. Thus, the safety parameter can be calculated as

$$K_8 = \lambda_u \quad [12]$$

where λ_u is the value of λ when the dam fails.

Relative Safety Based on Engineering Analogy Method

As far as the heightening project of the dam with defects is concerned, the evaluation of safety is still a challenge. In our study, an engineering analogy method was introduced to assess the safety of the dam heightening project. To explain in detail, the safety of the old dam, the dam after heightened with defects taken into account and the dam after heightened without considerations of the defects were calculated under the same calculation conditions. Thus, the relative safety between the heightened dam and the old dam and the influence of the effect of defects on the safety of dam were evaluated.

SAFETY ASSESSMENT ON DANJIANGKOU DAM

For the typical sections of dam, the finite element calculation models with the considerations of defects, cracks and the interfaces between old and new concrete surfaces were constructed based on the actual dam size and the detection results of defects and cracks. The calculation parameters were selected by reference to the testing results for concrete specimens, and the whole process of the construction of dam and water impounding were simulated. The working status of the dam under normal conditions could be calculated. By analyzing the safety coefficient of the concrete where defects are located, as well as the stability of cracks and the interface between old and new concrete surfaces, we calculated the safety coefficients of undamaged dams and the dams with cracks, defects and interfaces between old and new concrete surfaces using the overload and strength-reduction methods, respectively. Finally, the actual safety of the dam could be assessed.

Calculation model and the related parameters

According to the concrete pouring data in the early stage of Danjiangkou Dam, 11 low strength regions were identified along the 21-st section of the dam. The III-type and IV-type cracks mainly existed at the dam crest, both being transverse cracks on the top of gate pier. The upstream cracks were mainly horizontal cracks, which were repaired during the operating period. The 21-st section of the dam was selected as the typical section in the safety analyses and assessment of the dam. Fig. 2 shows the calculation model of the 21-st section of dam, in which a total of 84952 elements and 98693 nodes were included. The number of joints elements on the interface and crack surfaces was 1924 and 1738, respectively. The calculations parameters are listed in Table 1. During the calculation, dead weight, hydraulic pressure, temperature load and uplift pressure were considered. The hydraulic pressure was taken into account when the joints were flooded.

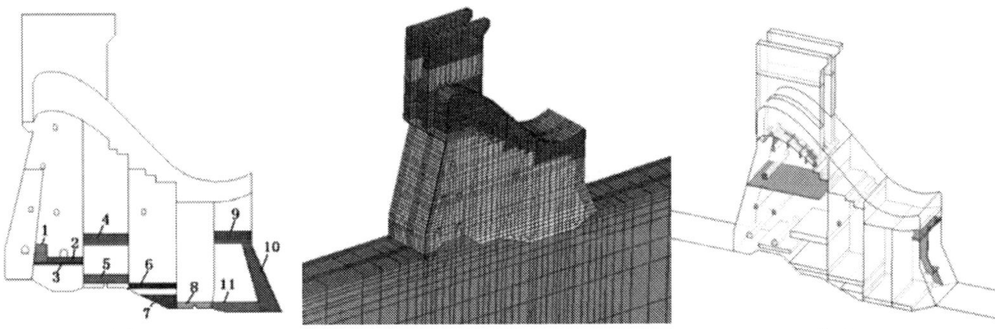

(a) (b) (c)

(a) Distribution of zones of low strength concrete (b) Grid maps in the calculation (c) Position of cracks

Figure 2. Calculation model of the 21-st section of dam.

Table1. Parameters in the calculation.

Material	Elastic modulus (GPa)	Density (kg/m3)	Poisson's ratio	Friction coefficient	Cohesive force (Mpa)
Basement	21.0	2700	0.22	1.0	1.0
Old concrete	35.0	2450	0.167	1.0	1.0
New concrete	36.5	2450	0.167	1.0	1.2
Low strength concrete	5.0	2450	0.167	0.8	0.6
Joints between the layers	/	/	/	0.8	0.6

Calculations and Analyses of the Dam Under Normal Working Conditions

When the dam is under normal conditions, the following three questions are of our concern: (i) how the low strength concrete defects affect its working condition; (ii) whether the cracks will extend after heightening; and (ii) what is the status of the interfaces between old and new concrete surfaces. In this section, using FEM, the working conditions during the past 50 years after the dam was poured was calculated. In-depth studies were performed on the above three questions.

1) Effects of low strength concrete defects on the working condition of dam

Two different working conditions, namely, the dam with and without low strength concrete, were considered. The stress at the dam heel and dam site were calculated and compared. In addition, the stress in the zones of low strength concrete was calculated. Table 2 lists the calculation results under two different working conditions. It can be seen that compared with the working condition of the dam without low strength concrete, for the dam with low strength concrete, the principal tensile stress at the dam heel increases from 0.9776 Mpa to 1.030 MPa; the tensile stress region of the foundation plane is still less than 7%. The principal compressive stress at the dam site increases from 4.185 Mpa to 4.978 Mpa; and the coefficient of safety against pressure is above 4.0. In the zone of low strength concrete, no tensile stress exists, while the maximum compressive stress is 1.944 MPa. According to the experimental results, the minimum compressive strength of the low strength concrete is 5.3 Mpa and the minimum safety factor is 2.73, which is less than the value of 4.0 as required in the specifications. However, the overall safety of the dam is unaffected.

Table 2. Calculation Results Under the Two Working Conditions (Unit: Mpa).

Working condition	Tensile stress at the heel of dam First principal stress	At the heel of dam Vertical stress	Compressive stress at the heel of dam Third principal stress	The maximum third principal stress of the low strength concrete	The location where the maximum third principal stress appears
Without considerations of the low strength concrete	0.979	-0.264	-4.185	无	无
With considerations of the low strength concrete	1.026	-0.245	-4.978	-1.944	8号区

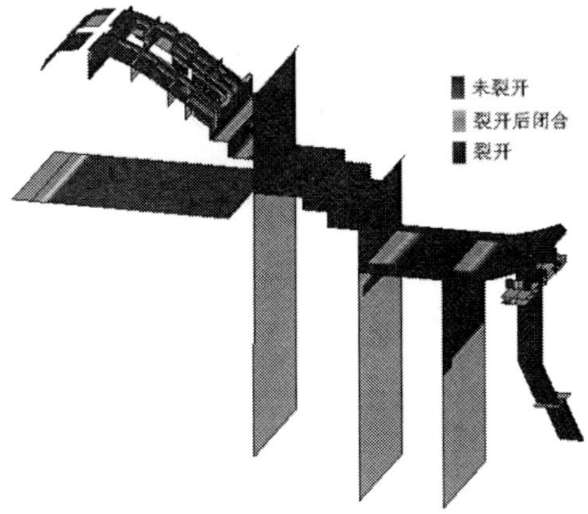

Figure 3. Cracks and interfaces between old and new concrete surfaces under normal working conditions

2) Crack stability and interface between old and new concrete surfaces in the dam after heightening

Figure 3 displays the cracks and interfaces between old and new concrete surfaces in the dam after a long-term service. It can be seen that after repair, the horizontal cracks are closed after opening within a certain range on the surface. This is mainly due to the effects of hydraulic pressure, dead weight and temperature load. In contrast, the internal cracks never open again. Because of external temperature, the horizontal cracks and the interfaces between old and new concrete surfaces near the gate pier remain opened.

Effects of cracks on the safety of the dam

According to the three calculation methods regarding safety as described in Eq. [10]-[12], it is assumed that the interfaces between old and new concrete surfaces were closed. We performed the calculations under four different conditions, namely, the old dam, the dam after heightening, the joint-free dam after heightening and the heightened dam with flooded total head. Finally, the calculation results under 12 different working conditions were obtained, as listed in Table 3. Fig. 4 shows the failure modes under overload and strength-reduction conditions. For the 21-st section of the dam with cracks after heightening, the safety decreases compared in both joint-free dam and original dam. Under waterhead overloading mode, the horizontal cracks in the upstream with the length of 113 m can be regarded as the control section. Its safety was weakened by the flooding of upstream cracks, showing a minimum safety of 2.00.

Table 3. A summary sheet of various safety parameters of the 21-st section of dam under different working conditions

Number of working condition	Overloading mode	Flooding in the upstream cracks	Research object	Safety coefficient
1-1	Water density overloading	No	The dam after heightening	5.50
1-2			The joint-free dam after heightening	6.50
1-3			Old dam	9.00
1-4		Total head	The dam after heightening	5.50
1-5	Waterhead overloading	No	The dam after heightening	2.20
1-6			The joint-free dam after heightening	2.76
1-7			Old dam	2.70
1-8		Total head	The dam after heightening	2.00
1-9	Strength-reduction method	No	The dam after heightening	3.90
1-10			The joint-free dam after heightening	4.00
1-11			Old dam	5.50
1-12		Total head	The dam after heightening	3.90

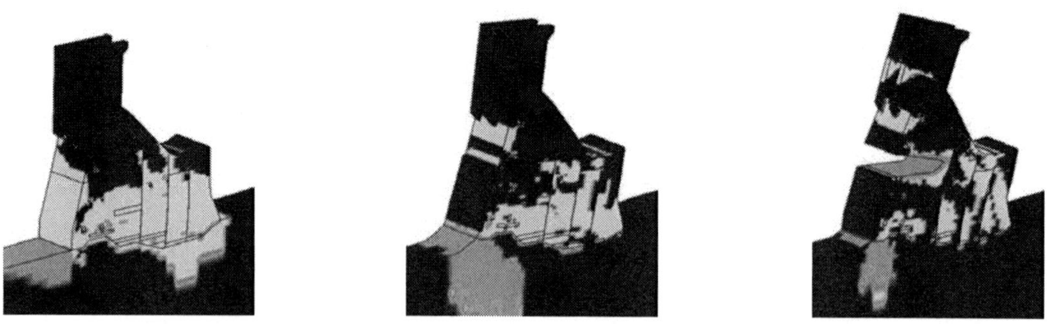

(a) Strength-reduction method (b) Water density overloading (c) Waterhead overloading

Figure 4. Failure modes under different loading conditions

Effects of the interfaces between old and new concrete surfaces on the safety of dam

Three calculation methods on safety described in Eq. [10]-[12] are used. With the cracks appearing at the early stage of project taken into account, the safeties were calculated under three different conditions, that is, the interface between old and new concrete surfaces is closed, the interface is opened, and the interface is opened with flooding of total head. Table 4 lists the calculation results under 9 different conditions. Figure 4 shows the failure modes under overloading and strength-reduction conditions. For the 21-st section of the dam after heightening, if no flooding exists, whether the interface is opened or closed has almost no effects on the safety of the dam. However, when the interface is flooded, the safety of the dam is significantly affected, with the minimum safety of 1.20. Consequently, the upstream face of the interface between old and new concrete surfaces should be sealed and the drainage should be ensured. This is important to avoid the working condition in which the interfaces bear the hydraulic pressure.

Table 4. A Summary Sheet of Various Safety Parameters of the 21st Section of Dam Under Different Working Conditions.

Number of working condition	Mode of overloading	Research object	Safety coefficient
2-1	Water density overloading	The dam with undamaged interface after heightening	6.50
2-2		The dam with opened interface after heightening	6.00
2-3		The dam with flooded interface after heightening	2.50
2-4	Waterhead overloading	The dam with undamaged interface after heightening	2.30
2-5		The dam with opened interface after heightening	2.20
2-6		The dam with flooded interface after heightening	1.20
2-7	Strength-reduction method	The dam with undamaged interface after heightening	4.00
2-8		The dam with opened interface after heightening	4.00
2-9		The dam with flooded interface after heightening	2.00

CONCLUSIONS

The safety assessment of defective dam heightening project is a great challenge in dam engineering field. The defects, cracks and the interface of old and new concrete surfaces have a major impact on the safety of the dam. However, it is difficult to make quantitative assessment of its safety status with the existing conventional methods. This research puts forward a relative safety assessment method based on simulation analysis and engineering analogy. A FEM model is built that considers defects, cracks and the interface of old and new concrete surfaces. The calculation parameters are selected by referring to the test results for concrete samples to simulate the whole process of dam construction and water storage. The data of the working status of the dam under normal conditions is obtained. We analyze the safety factor of concrete in the defective part, crack stability and the status of the interface between old and new concrete surfaces. The overloading and strength-reduction method are used to calculate the safety factor of pre-heightening dam, post-heightening intact dam, the post-heightening dam with cracks and defects, and the interface between old and new concrete surfaces. Finally the relative safety of the dam is evaluated. Using this method, a comparison is made on the safety of dam before and after heightening. The influence of crack defects on the safety of the dam is also evaluated.

The method was later applied for safety assessment of the heightening project of Danjiangkou Dam. The relative safeties of the dam body before and after heightening were calculated. The safeties of post-heightening dam body with cracks, flooding of cracks, opening of the interface between old and new concrete surfaces and the opening and flooding of the interface are also obtained by this method. If the cracks and the opening of interface between old and new concrete surfaces are not considered, then after the dam is heightened, the safety value is still large despite a slight decline. After the opening of cracks and the interface between old and new concrete surfaces, the safety further decreases but to a limited extent. Besides, the flooding of the interface between old and new concrete surfaces has a major impact on the safety of the dam body. The waterhead overload safety is only 1.20. Therefore, proper sealing and drainage of the upstream face of the interface between old and new concrete surfaces are required to avoid the condition that the interface bears the hydraulic pressure.

REFERENCES

[1] Design Specification for Concrete Gravity Dams (SL319-2005)

[2] ZHU Bofang, Thermal stresses and temperature control of mass concrete, Tsinghua University Press, 2014

[3] Zhu Bofang, Principle and Application of Finite Element Method (Second Edition), China Water & Power Press, 1998

[4] Shi Genhua, Numerical manifold method and discontinuous deformation analysis [M]. Translated by Pei Juemin. Tsinghua University Press. 1997

LESSONS LEARNED FROM RECLAMATION'S TWO DECADES OF RISK ASSESSMENT EXPERIENCE

William Engemoen[1]
Daniel Osmun[2]
William Fiedler[3]

ABSTRACT

In the mid-1990s, Reclamation began transitioning from a deterministic approach to dam safety evaluations to a probabilistic approach involving evaluation of potential failure modes and quantitative estimation of the risks of dam failure. Over these past two decades, risk approaches within Reclamation have been developed and have evolved, and in fact are still evolving today. This paper briefly outlines the development of quantitative risk analysis and risk-informed decision processes within Reclamation and discusses the benefits of analyzing the safety of dams using a risk perspective. In addition, key changes in methodology over the years are highlighted. There is little question that the risk methodology has evolved, and in some cases significantly. In large part, this is due to the learning involved in adopting a new process, and adjusting or modifying the process based on the experiences gained from applying the methodology on a wide variety of projects. Many lessons learned along this 20-year journey in risk assessment are presented, with the goal of sharing difficulties and successes to possibly ease the path for newer organizations or individuals just becoming involved in quantitative risk analysis.

INTRODUCTION

Prior to the mid to late 1970s, there was no formal dam safety program within the Bureau of Reclamation (Reclamation). Up to that time, Reclamation dam engineers in Denver were primarily engaged in the design of new dams (or engaged in planning studies for new dams). Periodic Review of Operation and Maintenance (ROM) examinations were conducted on existing Reclamation dams, with an obvious effort to identify any performance issues.

Several events in the United States during the 1970s led to the development of a dam safety program within Reclamation, including:

- The 1972 failure of the Buffalo Creek coal waste impoundment dam in West Virginia from slope failure during a rain event, resulting in 125 deaths.
- The 1972 failure of Canyon Lake Dam in South Dakota from overtopping during the Rapid City flood in June of that year, resulting in 33 deaths.

---

[1] Lead Civil Engineer – Geotechnical Services Division, Bureau of Reclamation, P.O. Box 25007, Denver, Colorado, wengemoen@usbr.gov
[2] Lead Civil Engineer – Geotechnical Services Division, Bureau of Reclamation, P.O. Box 25007, Denver, Colorado, dosmun@usbr.gov
[3] Lead Civil Engineer – Geotechnical Services Division, Bureau of Reclamation, P.O. Box 25007, Denver, Colorado, wfiedler@usbr.gov

- The 1972 National Dam Inspection Act, which authorized the U.S. Army Corps of Engineers to carry out a national program of dam inspections for the purpose of protecting human life and property. Although lack of federal funding led to no inspections until 1978, work did begin on the National Inventory of Dams.
- The 1976 failure of Bear Wallow Dam in North Carolina, apparently from flood overtopping, resulting in 4 deaths.
- Of most relevance to federal dam safety, the 1976 failure of Reclamation's Teton Dam in Idaho from internal erosion during initial reservoir filling, resulting in 11 deaths.
- The 1977 failure of Laurel Run Dam near Johnstown, Pennsylvania from flood overtopping, resulting in 40 deaths.
- The 1977 failure of Kelley Barnes Dam in Georgia from slope failure during a flood, resulting in 39 deaths.

These dam failures, as well as preceding ones in the 1960s, led to the passage of the Reclamation Safety of Dams Act of 1978 (Public Law 95-578, 1978). This law authorized the Dam Safety Program within Reclamation and authorized the Secretary of the Interior to perform necessary modifications to ensure dams were structurally competent to ensure public safety. This act made no specific mention of "risk."

However, in the following year of 1979, the Federal Guidelines for Dam Safety (FEMA 2004) were developed by a committee of federal agency representatives commissioned by the President of the United States, with the intent to promote prudent and reasonable dam safety practices among federal agencies. In these Federal Guidelines, risk-based analysis was recognized as a new approach for assessing dam safety, and development in this area was encouraged. Key language within the Guidelines included:

> *"Risk-based analytical techniques and methodologies are a relatively recent addition to the tools available for assessing dam safety. With further refinement and improvement, risk-based analyses will probably gain wider acceptance in the engineering profession and realize potential as a major aid to decision-making in the interest of public safety.....agencies should be encouraged to conduct research to refine and improve the techniques and to develop the methodologies and base of expertise necessary to apply them to dam safety evaluations.....On existing dams, a risk-based analysis should be considered in establishing priorities for examining and rehabilitating the dams, or for improving their safety."*

> *"The agencies should individually and cooperatively support research and development of risk-based analysis and methodologies as related to the safety of dams. This research should be directed specifically to the fields of hydrology, earthquake hazard and the potential for dam failure. Existing agency work in these fields should be continued and expanded more specifically into developing risk concepts useful in evaluating safety issues."*

EARLY RECLAMATION DAM SAFETY PROGRAM

In response to the failure of Teton Dam, the 1978 Reclamation Safety of Dams Act, and the 1979 Federal Guidelines, Reclamation began to establish procedures to analyze data and assess the condition and risk of its structures. Early emphasis tended to be more on seismic and flood issues and not on static issues. In accordance with standard dam engineering practice at the time, the initial focus was on deterministic, not probabilistic, evaluations.

Reclamation began dam safety evaluations of their dams in January 1979, under the newly developed Safety Evaluation of Existing Structures (SEED) process. Three-person teams spent 12 weeks inspecting dams and developing reports that were termed SEED Evaluation Reports. These reports discussed the design and construction, foundation conditions, loadings, performance, and identified potential deficiencies. Typical Safety of Dam (SOD) recommendations from these studies were to conduct updated evaluations of extreme loadings, evaluate performance under different loadings, document static stability, etc.

Subsequent actions resulting from the SEED evaluation reports led to flood and earthquake loading studies, analyses of structural performance under various loads – each study in response to the SOD recommendations made in the evaluation report. These analyses were strictly deterministic, and assessed static factors of safety and whether the dams could withstand the Maximum Credible Earthquake or Probable Maximum Flood. These studies were combined into one report termed the SEED Analysis Report. It is interesting to note that only limited evaluations of the safety of an embankment dam against seepage concerns were performed, yet the cause of Reclamation's only dam failure (Teton Dam) was internal erosion. In large part, this was likely due to the fact that there were no widely accepted deterministic evaluations of safety factors against an internal erosion failure mode; studies were generally limited to evaluation of filter compatibility and perhaps seepage analyses.

After the initial SEED examination (and evaluation), periodic SEED examinations followed. However, there was not a specific, mandated frequency for exams and evaluation and analysis reports. "Intermediate SEED Examinations" were conducted at intervals ranging from 2 to 6 years, in general, and it varied from dam to dam. Dams with seepage concerns were often subject to annual "Special SEED Examinations."

If a SEED Analysis Report resulted in new SOD recommendations, or remaining incomplete SOD recommendations from the SEED Evaluation, future actions were addressed in a Modification Decision Analysis (MDA) study. Essentially, the purpose of this level of study was to verify whether or not dam safety deficiencies existed that would need to be addressed by a dam modification. (Subsequently, this MDA process became known as the current Issue Evaluation (IE) process.) In early studies, a common focus was on developing modifications to address overtopping from the Probable Maximum Flood.

TRANSITION TO CURRENT DAM SAFETY PROGRAM

The SEED exams and MDA evaluations continued from the late 1970s to the early/mid 1990s. In or around 1995, the Comprehensive Facility Review (CFR) process was piloted on a few dams, and put in place shortly thereafter. These CFRs were set to occur on a 6-year cycle. The original CFRs (up to year 2000) featured 3 separate reports; Report of Findings, Examination Report, and Performance Parameters. The Performance Parameters introduced the concept of discussing dam safety deficiencies in terms of "Potential Failure Modes." Various ways that an existing dam could potentially fail were discussed and qualitatively evaluated for the likelihood of occurrence.

During this same time, Reclamation began moving from a deterministic approach to dam safety to a probabilistic approach. Reclamation had begun a gradual development of risk analysis methodology for dams since the release of the 1979 Federal Guidelines. However, progress was slow since there was little precedent/methodology in the United States or even world-wide. Implementation of comprehensive risk analysis procedures within Reclamation began in 1994 (Nambe Falls Dam), followed in the later 1990s by Issue Evaluation risk analyses of Bradbury, Salmon Lake, Conconully, Willow Creek, Pueblo, and Horsetooth Dams.

Between 1996 and 1998, quantitative risk analysis of potential failure modes was made a part of the CFR evaluation as well. During this same timeframe, Reclamation's "Risk Cadre" was formed, with 5 technical staff pulled from the line organization and dedicated almost solely to the development of risk methodology and procedures. From that point on, deterministic studies were de-emphasized in assessing dam safety deficiencies, with probabilistic approaches becoming more standard. Through the use of "toolboxes" (documents dedicated to specific methodologies) developed by the Risk Cadre and on-the-job training or exposure to risk analysis, the agency transitioned into the new process and procedures.

RECOGNIZED BENEFITS OF RISK ANALYSIS

In addition to being recommended in the Federal Guidelines, several benefits of incorporating risk analysis and risk-informed dam safety decision making appealed to Reclamation. In simplest terms, there are two components of a risk analysis - the development of potential failure modes (PFM) and quantitative estimation of risk. For the first component of developing PFMs, all relevant data and records are reviewed to gain a good understanding of the site. This is then followed by a brainstorming session to develop a list of plausible mechanisms by which a given structure might failure. Each potential failure mode is carefully considered, and those deemed to be most credible (or pose the greatest risk) are retained for additional consideration. Each credible PFM is carefully described in detail, ideally from the initiation phase all the way to dam breach. In addition, the PFM is sketched out to visually illustrate the location and progression of the mechanism. This potential failure mode evaluation:

- Requires detailed knowledge of the dam being evaluated (and ideally an understanding of relevant dam case histories)

- Leads to understanding of failure mechanisms and vulnerabilities
- Improves understanding of monitoring requirements

The next step (and second component) involves quantitatively estimating the probability that a given potential failure mode will lead to dam failure. Each credible potential failure mode is broken into a series of events (consistent with the description of the failure mode) required for that PFM to lead to dam failure. These steps are typically considered to constitute an "event tree." Building an event tree further reinforces the steps and considerations required for dam failure. Such a quantitative risk analysis:

- Extends the potential failure mode identification by detailing the various events and sequence of events required for dam failure
- Identifies areas of uncertainty where additional data collection or analyses would improve understanding of the loading condition or the performance of the dam
- Identifies areas of concern where temporary or permanent risk reduction activities are warranted, based on the overall risk estimate or insights gained about a PFM of concern
- Aids in prioritization of dam safety activities and overall dam safety program management

Within Reclamation, the risk analysis process is used by teams evaluating dam safety issues to:

- Gain a better understanding of what can cause the dam to fail (vulnerabilities, events necessary for failure)
- Quantify the engineering judgments (while also building the case to support the risk estimates)
- Identify the need for additional actions, which may include investigations or studies to reduce uncertainties in the risk analysis, better define risks, or design dam safety modifications to address a confirmed risk

Decision makers use the results of risk analyses to:

- Identify the source(s) of greatest risks and means to address the risks
- Determine a prudent course of action to address the risks
- Develop means to reduce uncertainty as needed
- Determine overall priority of dam safety actions within Reclamation
- Determine where the available resources can be most efficiently applied to reduce Reclamation's risk exposure
- Provide consistent and defensible dam safety decisions
- Provide a convincing means to explain to stakeholders why actions are being taken (or not) and funds expended

EVOLUTION OF RECLAMATION RISK METHODOLOGIES

In the 20 or so years since Reclamation embarked upon a risk-informed approach to dam safety, the state of practice in geotechnical and structural engineering has advanced. New interpretations and analysis processes continue to evolve in such areas as the development of probabilistic loadings (both hydrologic and seismic), evaluation of liquefaction and strength loss in soils, dynamic analyses of embankment and concrete dams (and appurtenant structures), among others. In addition, there is evolving research and studies related to dam safety; including the areas of internal erosion of soils, dam breach inundation, and potential life loss resulting from dam failure.

Given the constantly changing state of practice, the risk methodology at Reclamation has evolved to incorporate the latest standards of practice and processes for engineering analyses. Perhaps the three main changes to the Reclamation risk processes have involved the estimation of the probability of internal erosion, the estimation of potential life loss in the event of dam failure, and the portrayal of risks and decision-making guidelines.

Estimation of Internal Erosion Risks

In the mid to late 1990s and early 2000s, Reclamation's estimates of the probability of various internal erosion potential failure modes were based on work by the University of New South Wales (UNSW) in Australia (Foster et al. 1998). Robin Fell, Mark Foster, and others at UNSW researched a large number of world-wide dam failures and incidents and developed a simplified procedure to estimate failure probabilities for different types of internal erosion failure modes at different types of dams. Their method used case histories and total dam-years of operation to develop a statistical base rate frequency which could then be adjusted for site-specific features at the dam being evaluated. This process was used for most Reclamation dams during this early period of risk analysis.

In the mid-2000s the simplified UNSW approach was superseded by an event tree approach that included 7 or 8 events necessary to lead to an internal erosion dam failure. The event tree approach continues to be used today in Reclamation dam safety studies. This process was tailored after additional UNSW research that pioneered event trees for internal erosion. Perhaps the key event in an internal erosion event tree deals with estimating the probability that internal erosion will *initiate*, and the UNSW researchers provided an empirically-based approach to estimate that probability.

In 2009, Reclamation researched their own inventory of dams to determine the total number of dam-years of operation and the number of internal erosion incidents observed in Reclamation dams (Engemoen and Redlinger 2009, Engemoen 2011). Since then, the key event of initiation of internal erosion has been estimated using a base rate frequency developed specifically from this study of Reclamation incidents of internal erosion and including the interpretation of whether site conditions at a given dam being evaluated are better or worse than the average condition at Reclamation dams.

Estimate of Life Loss from Dam Failure

The earliest estimates of life loss from dam failure were typically based on equations developed by DeKay and McClelland (DeKay 1993) and Reclamation's Brown and Graham (Brown 1988). Whereas Brown and Graham related fatality rate to the population at risk (PAR) and amount of warning time, DeKay and McClelland related fatalities to the PAR and the "lethality" of the flooding. These equations were the primary tools used by Reclamation to estimate life loss up to 1999.

In September of 1999, Reclamation published DSO-99-06 (Reclamation 1999), a Dam Safety Office report entitled "A Procedure for Estimating Loss of Life Caused by Dam Failure." DSO-99-06 was based on the analysis of dam failures, flash floods and other floods located primarily in the United States, and provided suggested empirically-based fatality rates to be applied to downstream populations subjected to dam breach flows, considering the warning time, flood severity, and flood severity understanding. This empirical procedure was used by Reclamation for 14 years.

In 2014, Reclamation issued the Reclamation Consequence Estimating Methodology (RCEM), which is now the methodology used to estimate life loss that could result from failure of a Reclamation dam (Reclamation 2014). Like DSO-99-06, RCEM is rooted in empirical interpretation of dam failure and flood event case histories. However, additional case histories were investigated for RCEM to expand the database that forms the basis for the empirical approach by more than 50 percent. Rather than a tabular approach used in DSO-99-06, RCEM graphically relates fatality rates to flood severity (as defined by depth times velocity) and warning time. In addition, RCEM includes overall and suggested bounds of fatality rates, and encourages the portrayal of life loss as a range, as case histories illustrate that a large range of fatalities can occur during floods of similar severity.

Portrayal of Risk

The evolution of risk analysis at Reclamation has not been limited to the risk methodology, but also includes the development of decision making guidelines, termed Public Protection Guidelines. Prior to 2003, Reclamation compared estimated risks to Tier 1 and Tier 2 guidelines, shown on Figure 1 (Reclamation 1997). Tier 2 was the annual probability of failure, plotted for each loading category, and any category of risk that exceeded $1x10^{-4}$ indicated increasing justification to reduce the probability of failure. Tier 1 was the annual failure probability multiplied by the estimated life loss, also plotted for each loading category. Any Tier 1 risk that exceeded $1x10^{-3}$ provided a strong justification to reduce long term risk, while any Tier 1 risk that exceeded $1x10^{-2}$ provided a strong justification to reduce short term risk.

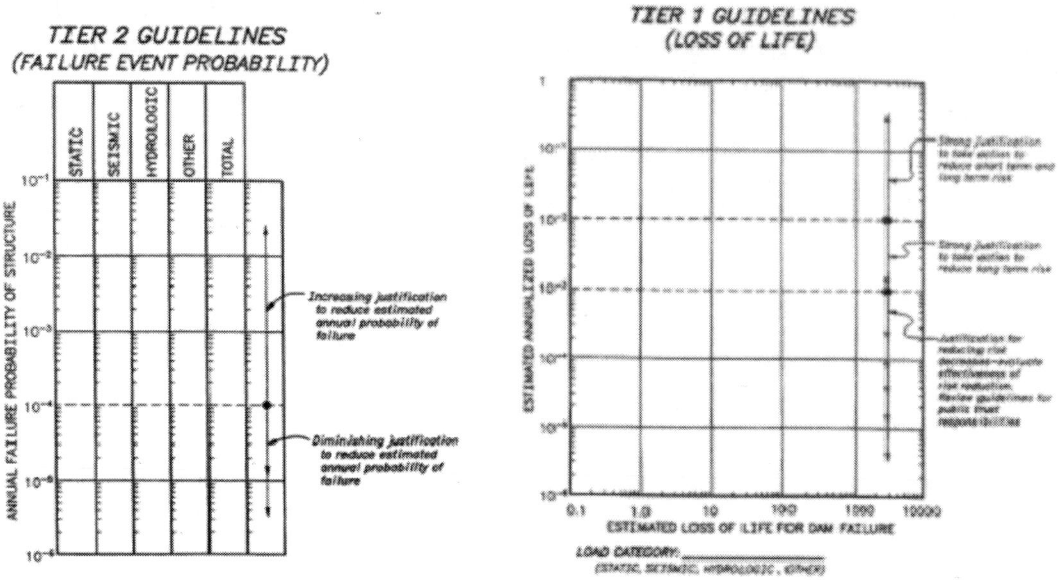

Figure 1. Reclamation's 1997 Public Protection Guidelines

Reclamation's Public Protection Guidelines were revised in 2003 (Reclamation 2003). Key changes involved changing terminology and the replacing the two charts in Figure 1 with a single f-N chart shown in Figure 2. With this revision, Annual Probability of Failure replaced Tier 2 as one measure of risk, and Estimated Risk replaced Tier 1 as the other measure of risk. Risks at a given dam were typically evaluated separately by load category.

The latest revision of Reclamation's Public Protection Guidelines occurred in 2011 (Reclamation 2011). These guidelines also updated terminology, the title of the two risk measures, the f-N chart, but added an "As Low as Reasonably Practicable" (ALARP) zone to the f-N chart as well. The current 2011 chart is depicted on Figure 3. The two measures of risk are now titled Annualized Failure Probability and Annualized Life Loss. The total risk, or sum of risks from all PFMs under all loading categories, is now considered as opposed to separate risks from the three loading categories.

As Reclamation has gained experience with risk analysis practices, the way risks are communicated and displayed has changed as well. Essentially, the goal is to portray risks in a manner that can be readily understood, as well as to provide thresholds for increasing justification to take additional actions to ensure public safety.

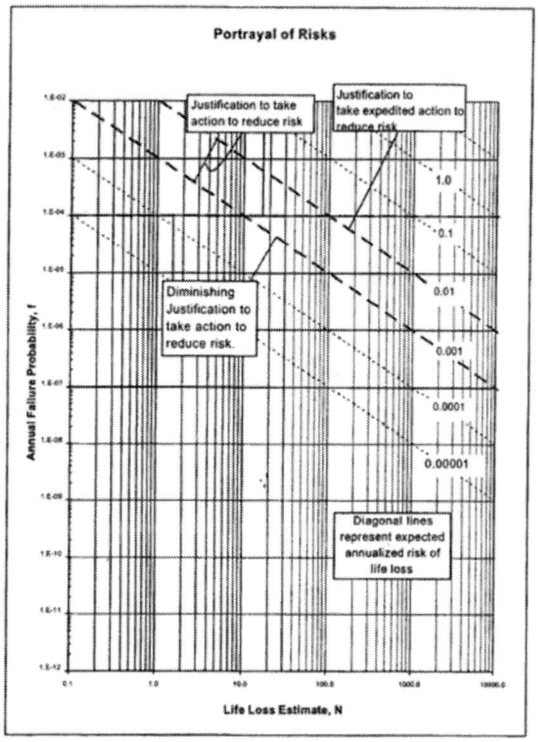

Figure 2. Reclamation's 2003 Public Protection Guidelines

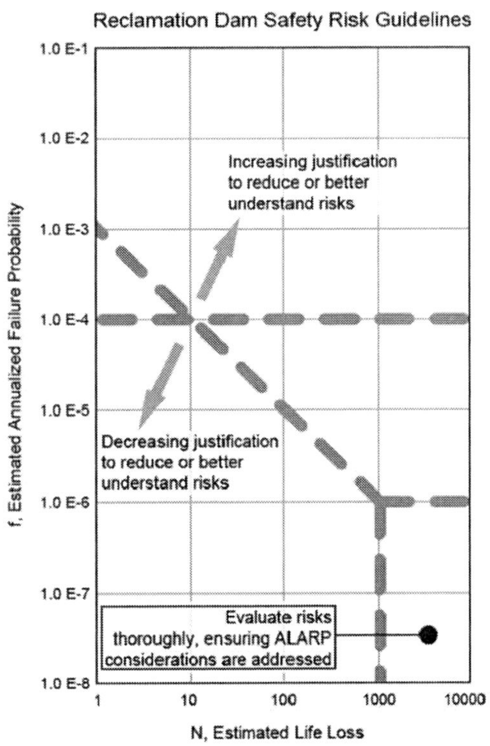

Figure 3. Reclamation's 2011 Public Protection Guidelines

LESSONS LEARNED IN APPLYING RISK ANALYSIS

Adopting a probabilistic, risk-informed approach to dam safety, after many years or decades of deterministic approaches, did not occur seamlessly or easily within Reclamation. It is a process that takes time and involves a cultural change to the way the organization and staff approach dam safety evaluations and decisions. Progress can be slow, and there may be some mis-steps along the way. As Reclamation has progressed toward a dam safety program utilizing risk analysis and probabilistic methods, many lessons have been learned over the course of 20 years. Some of the key lessons are detailed below.

Emphasize Potential Failure Mode Analysis

The risk analysis process should be fundamentally founded in the evaluation of potential failure modes, sometimes referred to as Potential Failure Mode Analysis (PFMA). Thinking about how a given dam might fail facilitates a transition toward a probabilistic approach, in that less thought is given to factors of safety and deterministic loads such as a Maximum Credible Earthquake or Probable Maximum Flood and expanded thought given to other failure mechanisms. To properly conduct a PFMA, teams must thoroughly digest all relevant data and available records to gain an understanding of the strengths and unique vulnerabilities of a given structure. A major advantage of this process is that it focuses on the entire failure process, and avoids such simplifying terms as "slope failure," or "failure by piping," or "failure by liquefaction." Teams are forced to dwell on many specifics such as probable location, possible weak links, likelihood of available freeboard, and many other variables that must combine to lead to failure. By considering the many aspects to a failure, teams become acutely aware of the likely and unlikely factors that help determine whether a potential failure mode is likely to lead to a dam failure. Once potential failure modes are developed and discussed, the development of each potential failure process into an event tree with probability estimates for each event is a logical and practical extension of this process.

Benefit of Dedicated Positions for Developing and Implementing Methodology

As Reclamation began to implement widespread PFMA and risk procedures into the dam safety program, a Risk Cadre comprised of 5 selected individuals was established to work solely on the development of risk analysis processes and guidelines, and facilitate team risk analyses. These were not part-time or collateral duty positions, but fully dedicated to these tasks. There is little doubt that dedicated positions enabled Reclamation to develop practices more quickly and efficiently than if this work was simply a part-time endeavor performed intermittently with other project work. The use of multiple people in a cadre also led to a diversity of opinion where various options and processes could be discussed.

Just Do It!

Although Reclamation utilized their Risk Cadre to establish processes and methodologies, much of the development occurred by forming risk analysis teams for given dams and simply working through the process of estimating risks. Detailed manuals were not developed first, but rather were developed later based on the lessons learned after many teams worked through the risk analysis process. This accomplished two things simultaneously: 1) the accomplishment of project work (i.e. the estimation of dam safety risks at a given dam); and 2) the trialing and subsequent development of sound, practical risk analysis methodologies and procedures that could reasonably be accomplished by teams with initial limited experience in risk analysis.

The Importance of Training

Initially, Reclamation relied primarily on "on the job" participation in risk analyses as the primary means of teaching the application of risk analysis practices and procedures to its staff. This method of training is still utilized today, and there is little doubt of the benefit of "learning by doing." However, other forms of training have proven to be beneficial as well. Reclamation developed a series of "Best Practices" chapters that detailed the various components of risk analysis and decision-making. Reclamation was joined by the U.S. Army Corps of Engineers in expanding and updating these best practices (Bureau of Reclamation and U.S. Army Corps of Engineers 2012), and the two agencies conduct joint trainings at least annually. In addition, Reclamation developed a program of monthly case histories, which present key dam failures or incidents, as well as studies of ongoing dams with interesting potential failure modes and risks. This effort has expanded to now include other Federal agencies presenting case histories as well, including the U.S. Army Corps of Engineers, the Federal Energy Regulatory Commission, and the Tennessee Valley Authority. Reclamation also presents a monthly technical training session to their staff involved in dam safety evaluations, with topics that cover the entire range of components that comprise dam safety studies from development of loadings through structural response to potential consequences of failure. These case history and technical training sessions have helped instill in both new and experienced staff members a better understanding of the different ways in which dams have (or could have) failed (i.e. enhanced the understanding of various potential failure modes).

A Team Approach is Critical

A risk analysis involves the elicitation of probability estimates from a team of individuals familiar with the dam and dam safety concepts. To ensure a diversity of opinion and thorough discussion of each event in a failure process, it is critical that risk analyses be conducted with a team rather than an individual. Whereas an individual may be focused on a particular opinion or viewpoint, a team is likely to bring different perspectives that need to be considered when evaluating complicated failure processes. A key consideration is to ensure that team members have the right level of expertise for the issue being evaluated; the value of a risk analysis is dependent on the knowledge, experience, and abilities of the team. Reclamation utilizes a facilitator for each team risk analysis. The facilitator is well experienced with the risk process and dam safety evaluations, and serves to impart knowledge of the risk process to team members, ensure that potential failure modes receive a thorough and fair evaluation, as well as to keep the process flowing logically and smoothly.

Involve the Entire Team

Given the sometimes difficult transition from deterministic to probabilistic analyses, and a natural discomfort of making estimates of probability that is typical for newcomers to risk analysis, it can be difficult to engage each and every team member assigned to a risk analysis. A key role of the risk analysis facilitator is to ensure that each team member be given the opportunity to offer both opinions and insights as well as to suggest a probability estimate for the event being discussed. As discussed above, in order to get a diversity of opinion, each team member's input is important. Patience, as well as the understanding of the need to involve all team members, is a key attribute of a risk analysis facilitator.

Avoid Tendency toward Conservatism

Particularly in the early phases of implementing risk analysis procedures, there appears to be a tendency toward over-conservatism in the estimates of the probability of events. In part this may be due to unfamiliarity or discomfort with a probabilistic process. Teams should be encouraged to carefully consider their estimates, and attempt to avoid any bias regarding the new process. Team discussions toward consensus values should help in obtaining more reasonable estimates than might be offered by a single individual.

When probability ranges are assigned to an event (e.g. a reasonable low of 0.01 and a reasonable high of 0.5), care needs to be given to the probability distribution assigned to this range. If a uniform distribution is assigned, the resulting mean would be 0.25, which is only half the reasonable high, but 25 times higher than the reasonable low. It is thus critical that risk teams recognize that the reasonable high estimate for an event can unduly weight the "best estimate."

In addition, experience appears to indicate that when a threat is not clearly understood or perceived, people tend to judge it to be more critical or dangerous than it may be. Hence, a key in avoiding conservatism in risk estimates to take the time necessary to thoroughly flesh out, discuss, and describe each PFM (and each step in the event tree) in detail. Each risk estimator must clearly understand the entire failure process in order to make reasonable probability estimates.

Expect Uncertainty

Predicting the performance of dams under unusual loadings entails a great deal of uncertainty. It is very difficult to develop probabilistic loads for extreme earthquake and flood events. A 1,000-year loading, either a flood or an earthquake, could take several different forms. In other words, such a flood might involve different combinations of peak flows and flood volumes, while such an earthquake could result in several different combinations of magnitude and ground motions. Given the relatively limited record of historical floods and earthquakes, the prediction of remote loadings has a severely limited basis in past events. In addition to the uncertainties associated with defining the probability of remote loadings, additional difficulties are posed when predicting a dam's response to an extreme load, particularly in the case of seismic loads. There are only a very small handful of dam failures due to earthquakes, as well as relatively few documented and detailed cases of dams that successfully withstood extreme seismic loads. Hence, the calibration of dams to existing case histories is limited.

Similarly, dam performance under normal operations is very difficult to predict. The major potential failure mode of concern, particularly for embankment dams, is internal erosion. Multiple studies have shown that an internal erosion failure mode can occur suddenly, and after decades of successful performance and no previous evidence of a problem. Predicting the performance of an existing dam after many years or decades of satisfactory performance is bound to have uncertainty.

Finally, the number of lives lost in the event of dam failure can be a function of a great deal of variations in flood severity, warning time, evacuation capability, and simply human nature (response to real or perceived threats). A study of case histories of flooding events has shown that the fatality rate (number of flooding fatalities for a given number of individuals within the flood inundation plain) can vary by several orders of magnitude for floods of similar depth and velocity.

None of these uncertainties is meant to dissuade the use of risk analysis; rather, it is just to acknowledge that "precise" risk estimates are an unrealistic expectation. Risk estimates should be given as ranges. Although a mean or "best estimate" can be assigned, it should be recognized that the potential result may be best defined as a band of values within a probability range that may span an order of magnitude or more.

Emphasize Sensitivity Studies

Sensitivity studies provide a useful means of understanding the potential uncertainty in estimated failure probabilities. Sensitivity studies refer to varying parameters in analyses or varying probability estimates for key events in an event tree, in both cases to determine the relative effect on the final numbers. Reclamation routinely employs sensitivity studies in all phases of dam safety engineering and risk analyses. For example, in the engineering analyses that provide insights into the likelihood of event, blow counts will be varied to determine the impact on liquefaction potential, material strengths will be varied to determine the impact on stability or deformations, and different initial reservoir elevations will be assumed in flood routings. The resulting range of results can provide useful, if not critical, insights into how performance may be affected by variations in key parameters. These insights are then used to help assign reasonable bounds for the probability of various events in a failure event tree.

It's The Process, Not The Final Number

As demonstrated above, it is unreasonable to expect that a probability of failure estimated during a quantitative risk analysis is an accurate number with a high level of precision. Rather, it is a general value that represents in a relative sense whether the risks posed by a potential failure mode are relatively high, relatively low, or near an accepted "threshold" value. These risk estimates are important for technical teams and decision makers in determining whether additional studies or corrective actions are justified, or whether normal but prudent actions are satisfactory at a given dam.

Apart from the resulting "number," the most helpful result from a risk analysis is the improved understanding of a dam's strengths and vulnerabilities that result from the detailed look at multiple potential failure modes. It is this enhanced understanding of a dam's potential performance that greatly aids Reclamation's understanding of potential risks during normal operations and potential unusual loadings, and helps target additional explorations, studies, and monitoring to best ensure a thoughtful and prudent continued operation of each dam. ***Emphasizing and re-emphasizing this benefit can also help teams get comfortable with risk analysis and see the value in the process.***

To emphasize this philosophy, Reclamation's Public Protection Guidelines include the following statements:

> *"Reclamation recognized that risk analysis procedures, although quantitative, do not provide precise numerical results. Thus, Reclamation wanted the nature of the risk evaluation to be advisory, not prescriptive, such that site specific considerations, good logic, and all relevant external factors could be applied in decision making, rather than reliance on a 'cookbook' numeric criteria approach."*

> *"The numbers, while important, are less important than understanding and clearly documenting what the major risk contributors are and why."*

Organizations can become overly attached to the importance of risk estimates, to the point where decisions are driven more by risk numbers than engineering judgment and the consideration of many other factors such as economics, environmental concerns, socio-political implications, and public perceptions. Dam safety decisions should be based on many of these factors, and not solely on risk. "Risk-based" decisions might be considered to be based primarily or solely on risk, while "risk-informed" decisions are based on more than just the estimated risks, but other key factors and the strength of the engineering case built to support a conclusion of low or high risk at a dam. Reclamation prefers the use of these "risk-informed" decisions.

Continue to Evolve

Risk analysis is still a relatively new endeavor to Reclamation, and to the dam safety profession as a whole. As such, the state of the practice continues to improve.

Individuals and organizations need to be flexible in their guidance and procedures, and be willing to move forward with different approaches as risk analysis practice improves. This means guidance and methodology documents will need to be periodically updated, and training needs to continue well into the future.

SUMMARY

Reclamation has found risk analysis to be an excellent tool for helping evaluate dam safety issues and setting priorities for additional explorations, studies, or corrective actions within our inventory of dams. In addition, it has greatly facilitated the discussion of the need for further actions with Reclamation water users and other project stakeholders.

It takes time and patience to establish a change from deterministic analyses to probabilistic analysis and quantitative risk analysis. The use of a team dedicated solely to the implementation of risk analysis helped Reclamation develop methodologies and procedures that were then taught to teams evaluating a given dam. No initial methodology or procedure will be perfect, and the path forward has evolved, and will continue to do so, as the practice of risk analysis becomes more and more accepted in the industry.

REFERENCES

Brown, C.J., and Graham, W.J., "Assessing the Threat to Life from Dam Failure," Water Resources Bulletin, Volume 24, No. 6, December 1988.

Bureau of Reclamation, "A Procedure for Estimating Loss of Life Caused by Dam Failure," Dam Safety Office Report DSO-99-06, Denver, Colorado, September 1999.

Bureau of Reclamation, "RCEM – Reclamation Consequence Estimating Methodology – Interim Guidelines for Estimating Life Loss for Dam Safety Risk Analysis," Denver, Colorado, February 2014.

Bureau of Reclamation, "Guidelines for Achieving Public Protection in Dam Safety Decision Making," Denver, Colorado, April 4, 1997.

Bureau of Reclamation, "Guidelines for Achieving Public Protection in Dam Safety Decisionmaking," Denver, Colorado, June 15, 2003.

Bureau of Reclamation, "Interim Dam Safety Public Protection Guidelines – A Risk Framework to Support Dam Safety Decision-Making," Denver, Colorado, August 2011.

Bureau of Reclamation and U.S. Army Corps of Engineers, "Best Practices in Dam and Levee Safety Risk Analysis," December 2012.

DeKay, M.L., and McClelland, G.H., "Predicting Loss of Life in Cases of Dam Failure and Flash Flood," Risk Analysis, Vol. 13, No, 2, 1993.

Engemoen, W. O., and Redlinger, C. G. (2009). "Internal erosion incidents at Bureau of Reclamation dams." Proc., 29th USSD Annual Meeting and Conference, United States Society on Dams, Nashville, TN, 731-745.

Engemoen, W. O. (2011). "Bureau of Reclamation experiences with internal erosion incidents." Proc., Internal Erosion in Embankment Dams and Their Foundations, European Working Group of ICOLD, Brno, Czech Republic, 11-18.

FEMA (2004 reprint), "Federal Guidelines for Dam Safety" Prepared by the Ad Hoc Interagency Committee on Dam Safety, Federal Coordinating Council for Science Engineering and Technology, Washington, D.C., June 25, 1979. (Reprinted by US Departement of Homeland Security, Federal Emergency Management Agency, April 2004)

Foster, M. A., Spannagle, M. and Fell, R. (1998). "Report on the analysis of embankment dam incidents." *UNICIV Report No. R-374*, University of New South Wales, Sydney, Australia.

Public Law 95-578, "The Reclamation Safety of Dams Act of 1978," 1978.

DAM SAFETY PORTFOLIO RISK MANAGEMENT: PRIVATE SECTOR VS. GOVERNMENT

George A. Kelley, P.E., SECB, F.ASCE[1]
David B. Paul, P.E.[2]
Robert C. Bachus, Ph.D. P.E., D.GE[3]
Glenn J. Rix, Ph.D.[4]

ABSTRACT

In the United States, modern dam safety evaluations for government-owned facilities are made using formalized risk-informed decision protocols. Regulatory agencies (e.g., the Nuclear Regulatory Commission (NRC), Environmental Protection Agency (USEPA)) dictate facility-specific risk assessment requirements that relate to dam safety. It is recognized that many other public and certain private entities, (e.g., electric power generating companies, port authorities, water management organizations, insurance companies) currently are not required to adopt a risk-based approach in making decisions that impact the safe operation of their assets, but have voluntarily embraced risk-informed approaches. Many organizations are considering the advantages and disadvantages of adopting risk management principles and risk-informed decision processes to help manage their assets and operations. While most organizations may consider the consequences to downstream communities from a dam breach, some may not formally consider the consequences to upstream stakeholders (e.g., loss of recreation, water supply, and/or flood storage). The purpose of this paper and the accompanying conference presentation is to introduce the challenges that are facing one federal agency, the Tennessee Valley Authority (TVA), as it considers the voluntary transition towards a risk-informed decision making (RIDM) strategy in evaluating dam safety and for managing other water impoundment assets (e.g., coal combustion residual (CCR) waste handling facilities, nuclear cooling ponds, etc.). The goal is to assist government agencies and private organizations that are not currently mandated to adopt risk-informed approaches, yet are currently considering the implementation of risk-informed decision processes in the management of their assets.

INTRODUCTION

While modern dam safety evaluations in the United States are made using formalized risk-informed approaches for government-owned facilities, the general use and recognition of the value of incorporating risk-informed methods are not yet common practice among many geo-professionals and private business organizations. Furthermore, the approach taken by some agencies and organizations is tailored to the authority,

[1] Senior Program Manager, Dam Safety Risk Program, Tennessee Valley Authority, 1101 Market Street, Chattanooga, TN, 37402, gakelleo@tva.gov
[2] Lead Civil Engineer, U.S. Army Corps of Engineers, Risk Management Center, 12596 West Bayaud Ave., Suite 400, Lakewood, CO, 80228, david.b.paul@usace.army.mil
[3] Principal, Geosyntec Consultants, 1255 Roberts Blvd NW, Kennesaw, GA, 30144, rbachus@geosyntec.com
[4] Principal, Geosyntec Consultants, 1255 Roberts Blvd NW, Kennesaw, GA, 30144, grix@geosyntec.com

mission, or management practices of the specific agency. It is also recognized that some government agencies (e.g., the Nuclear Regulatory Commission (NRC), Environmental Protection Agency (USEPA), etc.) dictate facility-specific risk assessment requirements that relate to "safety," including structural integrity, dam safety, and protection of human health. Some of these tailored approaches may not be immediately transferable to private industry, yet it is recognized that many non-regulated public and certain private entities (e.g., electric power generating companies, port authorities, water management organizations, or insurance companies) have benefited and may continue to benefit from a risk-informed approach in developing strategies that ensure the safe operation of their assets.

Many organizations, including the Tennessee Valley Authority (TVA), are considering the advantages and disadvantages of adopting risk management principles and risk-informed decision protocols to help manage their assets and operations because of their "appearance" within the dam safety community. In addition to current practices that focus on dam safety, the lessons, experiences, and practices from other engineering disciplines (e.g., hydraulics, seismicity, nuclear, etc.) may introduce recommendations that may be beneficially considered in the risk assessment for a portfolio of assets. Most organizations that assess dam safety consider the consequences to downstream communities from a dam breach, but may not necessarily consider the consequences to upstream stakeholders (e.g., loss of recreation use, water supply, flood storage and other benefits). While new guidelines for assessing risks and comparing these risks to tolerable risks guidelines are anticipated from the Federal Emergency Management Agency (FEMA), the details are as yet unknown. The publication date of these new guidelines has yet to be established.

This paper introduces the challenges and opportunities that are facing TVA as it considers the voluntary transition towards a risk-informed approach for evaluating dam safety and managing its other water impoundment assets (e.g., coal combustion residual (CCR) waste handling facilities, etc.). As such, a brief review of TVA's history and mission is presented along with the history of risk assessment procedures for dam safety. This is followed by an introduction to risks that extend beyond life safety for the portfolio of assets owned by TVA. The final section of the paper introduces the strategy that is being considered by TVA in its development of a risk-informed protocol that can be implemented across the entire TVA system.

HYDROELECTRIC POWER GENERATION AND THE EVOLUTION OF TVA

The safety of populations living downstream of dams is the responsibility of dam owners. Within the U.S., both the state and federal governments play a role in providing the requisite level of safety. These roles vary based on the historical development of dams within the country and the changing and evolving tolerance of society for the potential consequences of their failure. This section addresses the historical development of dams with a perspective of the role of TVA, while the following section addresses history regarding the use of risk assessment to evaluate dam safety.

The invention of the light bulb by Thomas Edison in 1879 symbolizes the initial surge in demand and investment in electric power, which preceded the financial corruption of the electric utility industry within 50 years. These concerns within the properties and natural resources of the country led to establishment of the Federal Power Commission in 1920. Later renamed the Federal Energy Regulatory Commission in 1977, the role of this federal agency was primarily to regulate the development and secondarily to regulate the ownership of dams and the power they produce. The ruling in Shoshone Hydro Development vs. U.S. Bureau of Reclamation (USBR) in 1929 provided the primary precedent for the development of hydroelectric power to address the growing demand for electricity. The time period of 1930 to 1980 saw the country's greatest investment in dams as energy infrastructure. Dam engineering advanced and more major dams were constructed than during any other time in U.S. history.

This time period also saw the birth and growth of the Tennessee Valley Authority which was in part founded on the ideas and concepts of comprehensive multipurpose water resources development. TVA was established by an act of the 73rd Congress in the TVA Act of 1933. This was a time period in our country's history when energy monopolies were forming and there was political discontent regarding the belief that these enterprises would benefit the public, particularly the public in the then largely rural southern states. Though originally established in the same manner as other agencies (i.e., fully funded by appropriations from Congress), TVA evolved into a public- or government-owned corporation whose legacy and assets are owned by the federal government, but whose mission and management have transformed it in the crucible of time and technology into an organization that resembles private energy companies. Specifically, the TVA is mandated to operate as a profitable business entity that must be able to sell electricity to its stakeholders at a rate commensurate with private-industry energy providers or to function as a "yardstick" within the industry. During its development through the last 80 years, the TVA has amassed significant energy assets within the Tennessee River Valley that includes an integrated system of 29 hydroelectric dams, 11 fossil-fired power plants, and 3 nuclear facilities that operate as three independent but centrally managed and coordinated business units all of which rely on the river.

HISTORY OF RISK ASSESSMENT FOR DAM SAFETY

Private and federal dam developers were primarily focused on the benefits of river operations and providing abundant electric power until the failure of Teton Dam in Idaho in 1976 and Kelly Barnes Dam in Georgia in 1977. It is well established that tragedy and public outcry often moves the government to act on issues that affect the public safety, health and welfare, thanks to political circumstances at the time; providence prevailed for the dam industry. In the aftermath of these and other dam failures, states enacted dam safety legislation and provided regulatory measures to ensure the safety of dams within their jurisdiction. Specifically, through a presidential executive memorandum dated 23 April 1977, President Carter ordered a review of the management of federal dams. The newly formed Federal Emergency Management Agency (FEMA) was requested to coordinate efforts to improve dam safety for all government-owned water-retaining structures, particularly those facilities located immediately above populations at risk. In

1979, FEMA issued the first *Federal Guidelines for Dam Safety* (FEMA Guidelines). Although standards-based, deterministic design and analysis procedures were the norm at the time, the following text from this document indicates recognition for the potential of a risk-based analysis approach as a new tool for decisions on prioritization of dam safety issues and remediation for existing dams.

> *"Risk-based analytical techniques and methodologies are a relatively recent addition to the tools available for assessing dam safety. With further refinement and improvement, risk-based analyses will probably gain wider acceptance in the engineering profession and realize potential as a major aid to decision-making in the interest of public safety."*

In the context of risk-based analyses for dam safety, risk is generally defined as:

$$\text{Risk of Consequence} = P[\text{Load}] \times P[\text{Failure} \mid \text{Load}] \times M[\text{Consequences} \mid \text{Failure}]$$

Where P[Load] is the probability or likelihood that a particular loading condition will occur, P[Failure | Load] is the probability or likelihood that the dam will perform adversely or fail as a result of that loading condition, and M[Consequences | Failure] is the magnitude of induce specific consequences resulting from the performance or failure of the structure. The loading conditions include both normal operating loads as well as extreme events such as flooding and earthquakes. Risks are determined in the context of potential failure modes which should include piping and internal erosion, overtopping, earthquake-induced deformations, and operational failures among others. Consequences include loss of life and economic losses due to physical damage to infrastructure, loss of use and environmental damages.

In the late 1970s and early 1980s, USBR used a risk-based approach at Jackson Lake Dam in Wyoming to evaluate the probability of overtopping due to a liquefaction-induced failure of the foundation and embankment soils. An example of the outcome of these analyses is shown in Figure 1. In this same time period, several university research groups developed risk-based approaches for dam safety evaluation (e.g., Howell et al., 1980; Vanmarcke and Bohnenblust, 1982; and McCann et al., 1983). However, most dam safety evaluations that ensued following the Federal Guidelines, including the Safety Evaluation of Existing Dams (SEED) and Comprehensive Facility Reviews (CFRs), remained largely deterministic and based on required minimum factors of safety for various loading conditions.

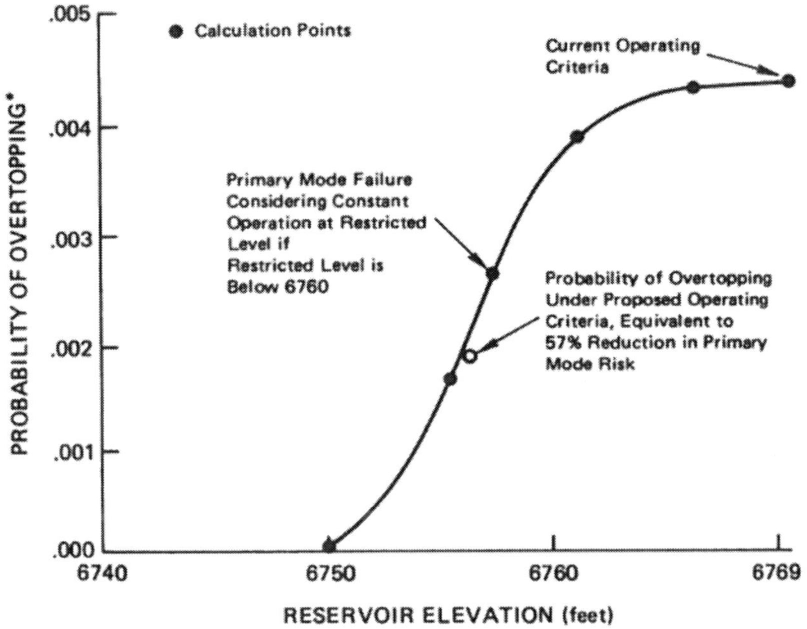

Figure 1. Example of Early Risk Analysis for Dam Safety Evaluation (from National Research Council, 1983)

In the late 1990s, USBR (1997) developed Public Protection Guidelines to address societal risk (Figure 2a) and individual risk (Figure 2b). s used for formal risk assessments on dams familiar to most readers. Shortly thereafter, USBR These figures represent the predecessors of modern-day f-N chart (1998) formed a team of five individuals from their Technical Service Center (referenced as the Risk Cadre or Risk Methodology Team) to further the development and implementation of risk analysis processes for dam safety.

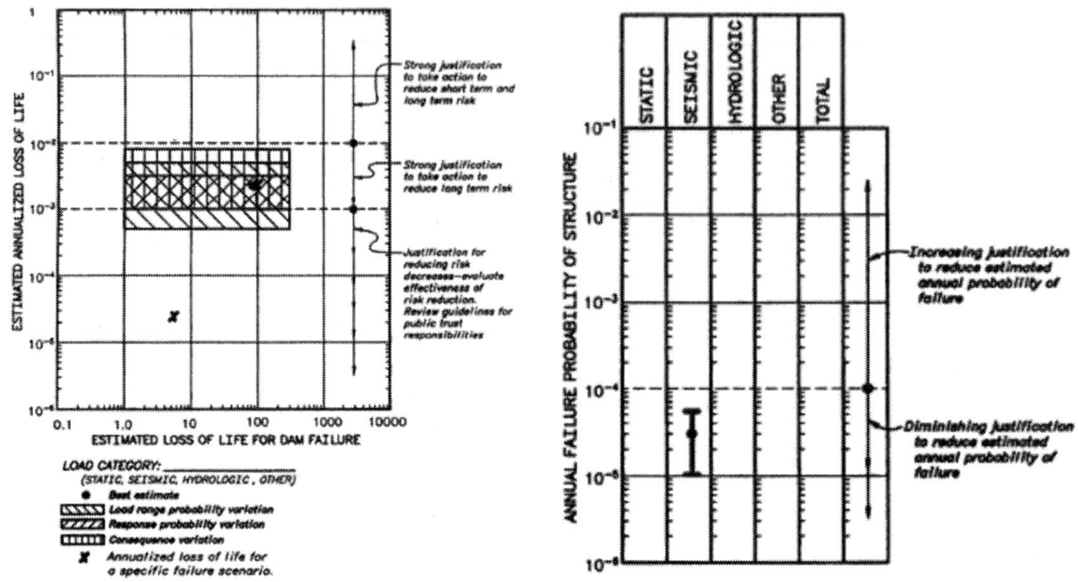

Figure 2. Acceptable Risk Criteria for Societal (a) and Individual (b) Risks (from USBR, 1997)

Following the levee failures in New Orleans as a result of Hurricane Katrina in 2005, the U.S. Army Corps of Engineers (USACE) was motivated to implement its Screening for Portfolio Risk Analysis (SPRA) program, which categorized dams among five Dam Safety Action Classification (DSAC) designations based on relative risks regarding dam safety. In 2009, USACE formed their first Risk Management Center (RMC) in Denver to, among other objectives, "*maintain and develop risk competencies.*"

Currently, many agencies (e.g., USACE, USBR, FEMA, FERC, etc.) use the risk assessment process to help inform decisions. There are, however, no uniform guidelines or procedures for performing these assessments. The Federal Energy Regulatory Commission (FERC), Division of Dam Safety and Inspections (D2SI) is currently developing a risk-informed decision making (RIDM[5]) methodology and guidelines as part of FERC's 2009-2014 Strategic Plan. As mentioned previously, FEMA's *Federal Guidelines for Dam Safety Risk Management* (publication pending) is anticipated to establish uniform guidelines that can be followed by all agencies (and other voluntary stakeholders). While RIDM methodologies are now well-recognized and widely practiced as the (albeit non-standardized) "norm" for dams under federal jurisdiction, adoption by many state and most private industry practice has been slow. However, momentum is building. It is hoped that the upcoming FEMA guidelines will help unify the RIDM efforts and provide the necessary impetus for state agencies and private dam owners to adopt RIDM practices.

[5] RIDM is used in the remainder of the paper in the broad sense of any approach to dam safety decision making that is based on principles of risk.

TVA is well aware of these efforts by other federal agencies in the U.S. that have adopted or are developing risk-based approaches to dam safety and has learned much from USBR, USACE, and FERC in pursuing (and hopefully developing) standard RIDM protocols. TVA also realizes that there is significant interest and progress in the international community (e.g., Canada, Netherlands, Australia) regarding the development of risk-based programs and is studying the approaches used by these organizations, particularly those that extend beyond "conventional" dam safety, in considering options that may be applicable.

The federal partnership, within their development of the Federal Guidelines for Dam Safety, clearly attributes the risks of dam failure in the downstream direction. TVA notes that a characteristic of the RIDM approaches developed by USBR, USACE, and FERC is that the consequences of a dam failure are focused on the potential loss of life associated with flooding downstream, but also consider downstream economic impacts. Thus, in most cases, the criteria for tolerable risk are expressed in terms of annualized life loss for societal risk and annualized failure probability for individual risk (i.e., the risk to the most exposed individual). These criteria are expressed in the form of f-N or F-N charts like that shown in Figure 2. However, TVA also understands that many privately owned organizations and other federal agencies have mandates that go beyond minimizing life loss and protecting the public. These organizations and agencies typically have corporate risk management strategies also aimed at protecting financial assets (i.e., avoiding economic losses due to physical damage and loss of use); protecting the environment; protecting their reputation in the eyes of the public; assuring business continuity after an extreme event; and generally protecting and improving shareholder value. For them, RIDM provides a framework for making educated business decisions to achieve multiple objectives. TVA hopes that it can learn and apply lessons that are gleaned by reviewing the strategies from these organizations.

PORTFOLIO OF RISKS AND CHALLENGES FACING TVA

TVA has historically incorporated (and currently incorporates) the conventional, standards-based approach to dam safety employed by many agencies that own or regulate dams. An example of the standards employed by TVA for embankment dams is presented in Table 1.

Table 1. Conventional Deterministic-Based Stability Criteria Used by TVA

Loading Condition	Minimum Factor of Safety[5]	Slope to be Analyzed
Case I - Construction condition [1]	1.3	Upstream and downstream
Case II - Operation	1.5	Upstream and downstream
Case III – Probable Maximum Flood[2]	1.1	Upstream and downstream
Case IVA - Operational Drawdown [3][4]	1.2	Upstream and downstream
Case IVB - Sudden drawdown from flood wave passage [3][4]	1.1	Upstream and downstream
Case IVC - Sudden drawdown – emergency drawdown[3]	1.0	Upstream
Case V - Earthquake (for steady seepage conditions with seismic loading using a pseudo static lateral force coefficient)	1.1[6]	Upstream and downstream
Case V - Earthquake Deformation	< 2 ft <u>and</u> <1/2 the thickness of filter layers [7]	Upstream and downstream
Case V - Post-Earthquake	1.1 [6][8]	Upstream and downstream

Minimum calculated factors of safety or allowable deformations are defined for a variety of loading conditions that include normal operations as well as extreme events (e.g., the probable maximum flood (PMF), the maximum considered earthquake (MCE), etc.). As with most projects that incorporate this deterministic approach, the implication is that a particular dam is considered "safe" if these conditions are satisfied. This approach is based in part on a philosophy that dam failures are unacceptable and that achieving an acceptable calculated factor of safety for various conditions is tantamount to a "safe" dam. Although the goal of no dam failures is laudable, lessons learned from a study of dam performance shows that it is not realistic. For example, failures of concrete and embankment dams occur at a frequency of approximately 0.8×10^{-4} per year (Foster et al., 1998; Douglas et al., 1998) despite the results of calculations that indicate adequate safety. Furthermore, this approach of selecting an analysis condition, performing a calculation, and checking the box marked "Adequate FS" often leads to a false sense of security, and the limited availability of resources makes it impossible to address all potential dam safety issues and render all dams "safe." In the case of TVA, the agency has spent the last several years reviewing all of the (mostly aging) dams in its portfolio that were designed following this conventional approach with a mission to assess performance under <u>current</u> design criteria. As a result, TVA has found that most facilities require unplanned maintenance, while several facilities require significant additional analyses, assessments, and rehabilitation. Risk-based methods that were not available at the time of original construction offer a more rational approach to dam safety. More significantly, however, a RIDM approach would help direct repair and rehabilitation priorities, as the process could help in the allocation of limited resources to address multiple demands.

Most federally owned dams are currently managed utilizing RIDM principles and owners of many dams that are federally regulated by the FERC are being given the option of managing their facilities utilizing this methodology on a case-by-case basis. Although TVA is a federal agency and therefore self-regulating with regard to dam safety, it is not required to adopt this strategy. However, as a partnering member of the Interagency Committee on Dam Safety and the National Dam Safety Review Board, TVA recognizes the advantages that an RIDM framework offers. In fact, the senior author is tasked internally within TVA with the initial development and implementation of a risk-based model that recognizes the many unique features of TVA. The unique circumstances that face TVA as it develops a RIDM process and procedure for its dam safety program include the following considerations with regard to governance, framework and implementation:

- TVA's mission extends beyond ensuring public safety. While dam safety risk management is traditionally and historically an issue of life and/or public safety within the dam designer/owner community and these remain the foremost objectives for TVA, TVA's mission also includes providing flood protection, electrical power, water supply, navigation, recreation, and economic development. To explicitly estimate an annualized risk, it is necessary to quantify the potential consequences associated with failure of a dam on each of these benefits, recognizing that there is considerable uncertainty and variability in our ability to quantify these consequences. TVA must also identify tolerable risk guidelines for individuals, society, the environment, economic losses, and other factors to inform decisions regarding the implementation of risk reduction measures.

- TVA has a portfolio of spatially correlated assets. Most TVA dams are constructed and operated in a linear (or series) system of energy assets that are mostly located within a well-defined river basin. As such, significant hydrologic and/or seismic events have the potential to impact multiple facilities essentially concurrently. Furthermore, the failure of a single dam may pose significant risks to other assets that are located both upstream and downstream of the failed dam and carry their own separate safety concerns. In most cases, these assets are also owned and managed by TVA.

- Certain facilities included in TVA's portfolio of dams (i.e., coal tailings and coal combustion residual structures) have a relatively low risk of life loss, yet the environmental and economic consequences may be high and must to be considered in risk estimates. As with the other societal benefits identified above, there is considerable uncertainty and variability regarding the ability to quantify these consequences. Additionally, assessing how these risks impact the annualized enterprise risk exposure is a significant challenge.

- TVA's risk-based dam safety program must be aligned with the TVA enterprise risk management program and implemented under the current austere business environment with TVA. For a risk-informed dam safety program to succeed at TVA, there must be an overarching approach to risk management that is accepted and applied within and across the organization.

- TVA is a large, diverse organization; developing and implementing an RIDM dam safety program throughout the entire organization will be a significant challenge. Based on experience, the authors recognize that various organizations have tried diligently to train personnel to adopt new processes. The proverbial "change is difficult" will be tested because the question "What is wrong with our current system?" will undoubtedly be asked.

- It will be a challenge to actively integrate risk-informed thinking into TVA's dam safety program while meeting current day-to-day organizational and operational demands. The RIDM approach requires that technical practitioners and project managers within TVA essentially re-think the manner in which they assess potential problems at facilities. Rather than just considering a calculated factor of safety for a specific mode of failure, individuals have to think in terms of risk and the opportunities to reduce them. Specifically, this means that for each facility, personnel must assess different and unique potential failure modes, estimate the likelihood of the specific failure, and the consequence of that specific mode of failure as well as the range of uncertainty surrounding these estimates. Each of these components are integral to the risk assessment process.

- It will be necessary to develop guidelines that can be used within TVA and by its consultants to consistently estimate and assess the wide range of risks at TVA facilities. This, in and of itself, is no minor undertaking because it will require changes in the TVA practices with consultants who's services are retained and who will likely be quick to offer alternative techniques and procedures. The challenges related to this stage will be in crafting the technical components of this guidance to achieve efficacy and consistency of the process. Many of the risks that need to be considered are not the traditional risks that are common to current dam safety programs.

- Program leadership which includes resident abilities in risk best practices within TVA will infuse authenticity and ownership to the dam safety risk management process. The decision making process is a governmental/ownership type process to which the risk experience in-house is key.

INITIAL STRATEGY FOR DEVELOPMENT AND IMPLEMENTATION BY TVA

The previous section presented and discussed the unique circumstances and challenges that face TVA in its efforts to implement an RIDM approach to dam safety within TVA. The activities needed to address these challenges are in their infancy and they will take time to fully develop and implement. This section identifies components of the strategy that are being considered for addressing the challenges identified above.

Universal Adoption: The successful implementation within a large and diverse organization such as TVA will requires a large-scale commitment on the part of senior management. Management acceptance and insistence on adoption at all levels of the organization will be paramount to success of the program. The authors suggest that the best strategy for garnering this support is to prepare documentation and provide a presentation to senior management that highlights the following:

- Upper Management Organization – As a TVA employee, the senior author is most aware of the organization of TVA and the critical role of senior management in making strategic decisions from the organization. It will be important to fully understand this structure to gain credibility. Both the first and second author have a diverse experience regarding the operations of large government organizations and presentations of risk concepts to senior decision makers.

- Diversity of the TVA Organization – Despite its relatively straightforward mission as described previously, TVA is a very diverse organization with different day-to-day decision drivers. The authors need to highlight and acknowledge this diversity and then demonstrate how the current system that is not "broken" can be improved systematically by adoption of a RIDM approach.

- Examples of Assets – One way of demonstrating the diversity of the organization is to highlight the diversity of the assets within the TVA portfolio. Therefore, it will be necessary for the authors to provide examples of the varied assets and how a risk-based approach can beneficially impact the operations of asset, particularly when the approach is adopted system-wide.

- Examples of Beneficial Change – The changes that are proposed by the authors are not insignificant and while any change is difficult, the transition to a RIDM approach represents a major shift. The old adage that "the proof is in the pudding" is an absolute reality and necessity in this case. It will be necessary to demonstrate to management that risk-based approaches as proposed to TVA are not experimental, but rather that they have a history of demonstrative success of providing value. To the extent that these success examples can be extended to private organizations and beyond dam safety itself, the better the likelihood for adoption.

- Training – The authors believe that it is important to present senior management with the strategy for a training program. The transition to a system-wide RIDM approach across TVA will be difficult and senior management needs to see that the necessary steps are developed to provide adequate training to the appropriate personnel.

- Metric of Success and Timetable – While the authors believe that senior management buy-in is the most important, presenting the executives with a realistic "path forward" in terms of a specific schedule for implementation is also critical. This schedule needs to include the appropriate strategy-development meetings with TVA personnel regarding TVA's risk tolerance, follow-on presentations to senior management, development of guidelines and training materials, and a phased implementation of the RIDM approach across the TVA business sectors.

Integrated Risk-Informed Thinking: The concept of risk must be integrated into the current dam safety procedures and policies:

- Contrast Conventional and RIDM Approaches – The authors have found that it is difficult to get professionals to understand and contrast the RIDM approach to conventional deterministic approaches. Therefore, it will be necessary to demonstrate through a series of case history examples the differences between deterministic and probabilistic approaches. These case histories must be for the type of projects that the personnel understand.

- Example of the Alternate RIDM Thinking – It would be beneficial to identify a typical TVA project that can be used as an example application of the RIDM approach. Ideally, this would be a dam project where a conventional analysis has indicated the need for rehabilitation. Citing current project-specific examples will help in developing acceptance of the RIDM process. Concepts of event and fault trees will be introduced to help explain this alternative process.

- Potential Failure Mode Analysis (PFMA) – Given the similarities between many TVA facilities, the authors propose to develop tables that identify the potential failure modes that have to be considered in the risk-based assessment. The process of performing a potential failure mode analysis (PFMA) will be described and an example PFMA will be presented.

- Likelihood of Occurrence – Using references developed by USACE and other agencies, the concept of assessing the relative likelihood of occurrence will be described. Guidelines will be presented for providing a quantitative as well as a qualitative assessment. For this specific task, concepts of statistics and probability will need to be explained and demonstrated using dam- and TVA-specific example whenever available .

- Consequences- As described previously, the RIDM approach for TVA will have to consider the "non-traditional" consequences of failure or unacceptable performance, including the financial and societal consequences (e.g., loss of recreation value, water supply, flood storage, etc.). This specific step will require considerable effort to compile information and obtain buy-in and consensus from TVA management.

- Training – All of the information presented in this section will have to be included in training sessions that are developed for TVA personnel. These sessions will include hands-on PFMA examples, development of event trees, and performing a qualitative and quantitative risk assessment. The authors recognize that different materials will need to be developed for TVA design personnel, TVA contractors, and TVA operational personnel. For example, operational personnel should be trained to integrate risk concepts into regular facility inspections and the development of emergency action plans.

- Metric for Success and Schedule – The authors will develop a schedule for the training program that includes sufficient time for preparation of training materials and for training of the various TVA personnel. Milestones will be included in the schedule to help establish quantitative measures for gauging the success of the RIDM initiative.

Development of Guidelines: Assuming that upper management endorses the RIDM approach and supports its implementation, it will be necessary to develop implementation guidelines to commence the transition to risk-informed decision making.

- Conventional Guidelines – As part of this effort, the authors propose to compile guidelines prepared by USACE, USBR, FERC, and others regarding guidelines for implementing a RIDM approach. To the extent possible, guidelines will be identified for addressing the non-traditional consequences.

- Proposed Guidelines – Because of the size and diversity of TVA, as well as TVA's dependence on consultants and contractors, it will be necessary to propose and implement agency-specific guidelines for the data compilation, analyses, and interpretation of results for application of a RIDM approach. The authors believe that one of the keys to success for this evolution of thinking will be the development of consistent guidelines that can be introduced and implemented across the entire TVA organization.

- Scalable – The guidelines (and associated tools) that are developed should be (i) based on a fundamental principles; (ii) implementable to the point that TVA can consistently meet the range of problems at hand; and (iii) scalable to address the evolution of the risk management program.

- Training – As described in the previous section, training is a critical step to assure success of the practice evolution to a RIDM approach. The proposed guidelines described above will play a critical role in the training phase. The goal of these efforts will be to have TVA personnel and its consultants comfortable with working with risk-based strategies.

- Metric for Success and Schedule – As with the previous two topics, the authors believe it necessary to establish a schedule for development and implementation. Measures of success will need to be developed and will be proposed as part of this final stage.

CONCLUSIONS

TVA has made a voluntary commitment to develop a RIDM process that can be integrated across TVA for its portfolio of assets. The purpose of this paper is to identify the challenges that are faced by any organization as large and diverse as TVA. The paper also identifies strategies directed towards this significant effort. Assuming that upper management endorses the RIDM approach and supports its implementation, it will be necessary to develop implementation and training programs to commence the transition to risk-informed decision making. The authors solicit input from the USSD members and participants at this conference to comment, critique, and (most important) provide alternative suggestions, particularly if the suggestions have been successfully at other organizations.

REFERENCES

Vanmarcke, E. H., and Bohnenblust, H. (1982). *Risk-Based Decision Analysis for Dam Safety*, Massachusetts Institute of Technology, Department of Civil Engineering Research Report R82-11.

Howell, J. C., Bowles, D. S., Anderson, L. R., and Canfield, R. V. (1980). *Risk Analysis of Earth Dams*, Department of Civil and Environmental Engineering, Utah State University.

McCann, M. W., Jr., Shah, H. C., and Franzini, J. B., (1983). *Application of Risk Analysis to the Assessment of Dam Safety*, Department of Civil Engineering Stanford University.

U.S. Bureau of Reclamation. (1997) *Guidelines for Achieving Public Protection in Dam Safety Decision Making*. Interim Guidelines.

National Research Council, (1983), *Safety of Existing Dams; Evaluation and Improvement,* Committee on the Safety of Existing Dams, Water Science and Technology Board, Commission on Engineering and Technical Systems.

THE FERC RIDM . . . WHAT'S IN IT FOR ME?

Guy S. Lund, P.E. [1]
Bill Christman, P.E., [2]
Gene Yow, P.E., [3]

ABSTRACT

The Federal Energy Regulatory Commission (FERC) began to publish guidelines for dam safety assessment using Risk Informed Decision Making (RIDM) in 2014. Risk analysis has been the subject of several presentations, most of which have discussed risk analysis from the regulatory and engineering perspective. This paper focuses on the owner's perspective – specifically, why was risk analysis considered appropriate for these particular applications, and what are some of the benefits that have resulted from these studies.

The Chelan County Public Utility District (Chelan PUD) owns and operates three large hydroelectric projects (Projects) on the Columbia and Chelan Rivers (Rock Island, Rocky Reach, and Lake Chelan) in central Washington with a combined generating capacity of approximately 2000 MW. Chelan PUD has made diligent efforts to maintain reliable operations and public safety, and has invested significant resources to address the safety due to the PMF, which until recently was considered the principal dam safety concern for the Chelan PUD's Projects.

Recent advances in earthquake engineering have shown that the seismic hazard for central Washington is significantly greater than was historically realized, and now the seismic loads have become the principal safety issue for the Chelan PUD Projects. Using traditional dam deterministic safety methods and guidelines would require the Chelan PUD to evaluate the projects for revised maximum credible earthquake (MCE) seismic hazards, and if necessary, perform modification to enhance the safety of the projects to current standards.

In cooperation with the FERC, a Probabilistic Seismic Hazard Analysis (PSHA) was performed for central Washington. The PSHA estimates the potential seismic hazard based on return frequency. The PSHA was then used to perform a pilot seismic risk assessment of the Chelan PUD projects. The initial results from the pilot risk studies have been very encouraging for the Chelan PUD, regarding the current safety of the projects and the long-term reliability of the hydroelectric resource for their constituents.

---

[1] Guy S. Lund, P.E., AECOM, 8181 E. Tufts Ave, Denver, CO 80237; (303) 740-3902, guy.lund@aecom.com
[2] Bill Christman, P.E., Chelan PUD, 327 N. Wenatchee Ave, Wenatchee, WA 98807, (509) 661-4283, bill.christman@chelanpud.org
[3] M. Gene Yow, P.E., Chelan PUD, 327 N. Wenatchee Ave, Wenatchee, WA 98807, (509) 661-4305, gene.yow@chelanpud.org

PROJECT BACKGROUND

Rock Island Project

Rock Island Hydroelectric Project is located on the Columbia River 12 miles downstream of Wenatchee, Washington, and is regulated under the Federal Energy Regulatory Commission (FERC). The principal structures include the north abutment gravity wall, gated spillways, and powerhouse I and II, as shown in Figure 1.

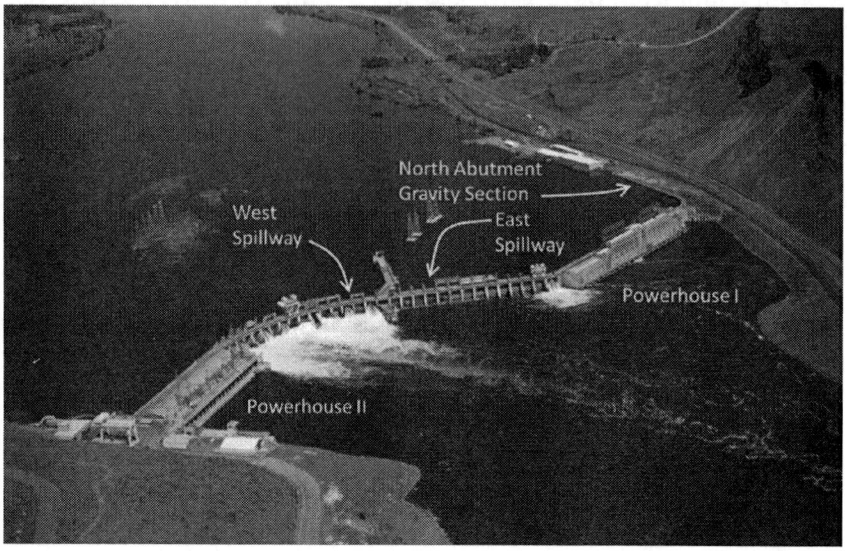

Figure 1. Aerial view of Rock Island Dam

The North Abutment Gravity Section is part of the main dam and is located on the left abutment. The section is a 590-foot-long concrete gravity dam section with a height varying from 20 to 75 feet. The gravity section was post-tensioned in the 1970s to provide additional stability for the current headwater elevations.

The 1184-foot-long spillway is located in the central portion of the dam, and consists of the East (left) Spillway and the West (right) Spillway. There are 31 spillway bays, which are each controlled by a vertical lift gate.

Powerhouse No. 1 is the original powerhouse and is located adjacent to the North Abutment Gravity Section. Powerhouse No. 2 was constructed in the 1970s and is located adjacent to the south (right) abutment. The generating capacity for the project is 648 MW.

The Project operates as a run-of-river project. Discharges are primarily regulated due to requirements set forth by other storage reservoirs along the Columbia River.

Rocky Reach Hydroelectric Project

The Rocky Reach Hydroelectric Project is located on the Columbia River, approximately 7 miles upstream from Wenatchee, Washington, and is shown on Figure 2. The primary purpose of the Project is power production with a total generating capacity is over 1300 MW.

The Project consists of several features, including a forebay wall on the west (right) abutment, powerhouse service bay, powerhouse, center dam, spillway, east (left) abutment retaining wall, and east embankment seepage cutoff. The spillway contains twelve, 50-foot-wide gated bays that are separated by 10-foot-wide piers. Each spillway bay consists of a mass-concrete ogee section. The spillway piers extend above the ogee crest and provide support for the spillway access bridge, radial gate trunnion anchorage, and gate hoists. The radial gates have recently been modified to enhance structural capacity due to potential trunnion friction.

Figure 2. Aerial view of Rocky Reach Dam

The left side of the spillway structure is connected to a non-overflow gravity section, called the east abutment retaining wall, which extends into east embankment section that forms the water barrier on the left abutment.

Lake Chelan

The Lake Chelan Hydroelectric Project (Project) is located near the town of Chelan, Washington, and the main purpose is power production. The first dam was constructed in 1892 for the purpose of furnishing water and electricity to the town of Chelan. The current dam was constructed from 1926 to 1928. The dam and surge tank are shown in Figure 3.

Figure 4. Aerial view of Lake Chelan Dam and the surge tank.

The Project consists of a main dam, spillway, power tunnel and penstock, powerhouse, and switchyard. The main dam consists of concrete gravity structures abutting both ends of a central gated spillway. The spillway section contains eight, 20-foot-wide bays with 14-foot-high tainter gates. A stilling basin downstream of the gate structure is created by two baffles which dissipate energy through two dentated tumble bays.

Reservoir releases are through a low level outlet works at the main dam (for river habitat enhancement), the spillways (for headwater control), or the power tunnel to the powerhouse (for power generation). The powerhouse is connected to the reservoir with an 11,600-foot-long power tunnel. The power tunnel is 14 feet in diameter, with a total vertical drop of about 400 feet. It is steel-and-concrete lined for approximately 1,000 feet upstream from the powerhouse and concrete-lined for the remainder of its length. A 14-foot-diameter butterfly valve is located in the upstream end of the tunnel near the toe of the dam. A vertical riser in the steel-lined section connects to a differential-type steel surge tank 113 feet high and 45 feet in diameter. At the lower end of the steel-lined section, a wye branch divides the flow into two, nine-foot-diameter penstocks.

The powerhouse is located near the Columbia River and consists of two vertical-shaft, Francis-type turbines with a total generating capacity of approximately 60 MW.

SUMMARY FROM THE CHELAN PUD PILOT RISK SEISMIC EVALUATION

The methodology used to perform the risk analysis for the selected potential failure modes (PFMs) is summarized below:

- Develop probabilistic Seismic hazard curves, as shown in Figure 4.
- Identify the critical failure modes (PFMs) for the project related to the seismic event.

- Perform analyses of the critical parameters related to the PFMs and develop a fragility curve for the system, as shown in Figure 5.

- Perform inundation studies to identify the population at risk from flooding, and estimate the range of potential life loss (PLL) due to the assumed PFM. Compute the risk and plot on the probability of failure (f) versus potential consequence (N) charts, as shown on Figure 6.

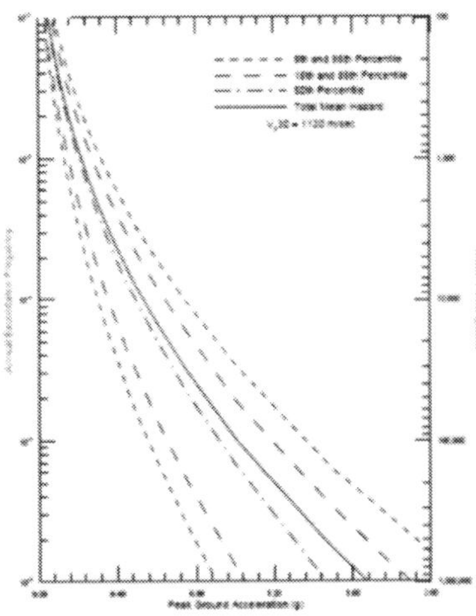

Figure 5. Probabilistic Seismic Hazard Analysis curve.

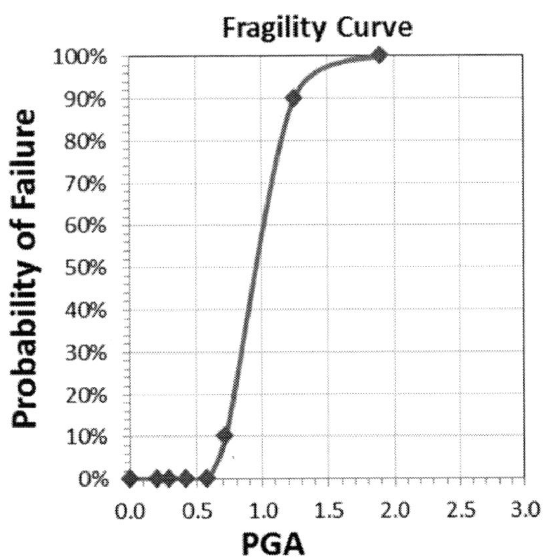

Figure 6. Spillway piers fragility curve, Rocky Reach Project.

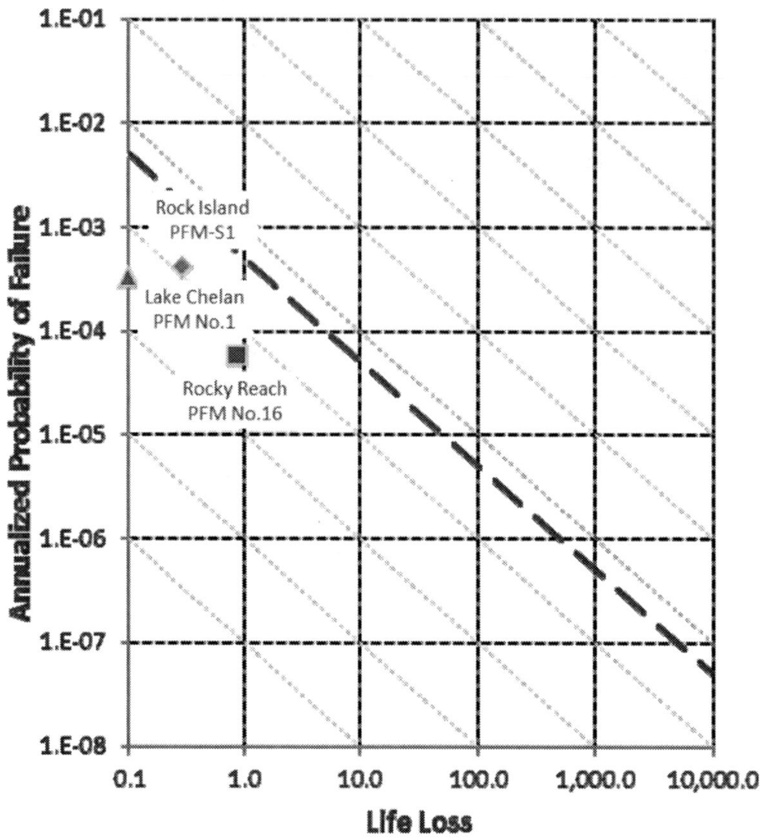

WHAT ARE THE BENEFITS OF RIDM?

The Chelan PUD has invested into the RIDM process for the following reasons:

- **Better Understanding of Risk.** The traditional regulatory philosophy evaluated project safety based on deterministic loads for normal operation, floods, and earthquakes. Through the process owners such as the Chelan PUD were not naïve to the project risks; it is evident that the traditional dam safety process over simplified the risk due to potential failure. Chelan PUD appreciates that a better understanding of the potential risks results in better informed decisions regarding safety, reducing risk, and informed decisions for capital investments for long-term operability.

- **Understand need to Monitor for Change.** The Folsom Dam radial gate failure showed that systems are always undergoing change; therefore, it is important to monitor for change in behavior. For example, the trunnions at Folsom Dam when installed most likely did not have any significant frictional resistance. However, with time, corrosion in the trunnion assembly resulted in a frictional resistance that increased the load on the gate structural members, and eventually resulted in a failure. The RIDM process includes evaluation of the long-term structural and equipment behavior, which includes how the potential for deterioration will affect the probability of failure and how downstream development may affect the potential for adverse failure consequences. Thus, through diligent use of the RIDM processes, the operators and engineers develop an understanding of potential issues that can develop with time, how these issues may increase the project risk, and should result in enhanced methods to monitor for change in behavior.

- **Useful Prioritization Tool.** The Chelan PUD operates three large hydroelectric projects. The organization has always had adequate resources to address dam safety concerns; however, in the present economic age, there is not necessarily unlimited capital to address all potential safety deficiencies. Therefore, it is necessary to develop a method that helps engineers prioritize safety deficiency repairs and effectively communicate decisions to stakeholders.

- **As Low as Reasonably Practical (ALARP).** The ultimate desire of the Chelan PUD is to provide proper stewardship of the available resource at the lowest practicable risk. The RIDM process has helped to identify areas in which the potential risk can be further reduced for reasonable costs. For example, Chelan PUD recently collaborated with a wide variety of stakeholders to update the seismic load in the broader area of the Mid-Columbia River. The update was performed in both a probabilistic and deterministic manner. The probabilistic results provided higher loads than the deterministic results for longer return periods. Therefore, the probabilistic results are being used to inform the evaluation of Project facilities which present a higher risk to public safety and operational resiliency (e.g. concrete dam structures, spill gates) and the deterministic results are being used to inform the evaluation of myriad facilities which have little potential effect upon public safety (e.g. transmission structures).

- **Work Effectively with Regulator.** In the traditional dam safety environment if a project was determined to be deficient, the regulatory agency typically required that the deficiency be repaired in a deterministic manner. This can place the owner in a difficult position; especially if the required repairs will have substantial cost or other impacts, particularly when the risk is not well understood or perceived to be very minor. For owners with limited resources, investing too much resource into a singular venture may take time, resources, and attention away from other important and urgent endeavors; particularly if the "other endeavors" are not well understood. The RIDM process provides options for the owners and Regulators. For example, it may be determined that the greatest risk is due to a lesser flood, primarily because the lesser flood is a much more frequent event, and has a greater probability of occurrence. The total risk may be significantly reduced by performing less costly remediation for the lesser flooding event. Interim risk reduction, is not the end of project safety, but a step the owner can implement to reduce risk while planning for more significant modifications and repairs to the project.

Better Understanding of Risk

Certainly one of the benefits from the RIDM process is a better understanding of the potential risks due to failure. Risk is simply defined as the product of *"probability of failure"* and *"consequences"*, as shown in the Equation 1. Although traditional dam safety programs have evaluated downstream inundation (i.e., consequences) to estimate the project hazard for use in the emergency action plans, these evaluations did not address the size of potential life loss. For example, the potential life loss of an unexpected fisherman 25 miles below the dam would make the project a HIGH hazard, the same as Lower San Fernando with nearly 100,000 people in the inundation zone within 5 minutes of the dam. Similarly, evaluating the "probability of failure" was rarely performed, if ever. Therefore, in today's context the understanding of dam safety risk was typically limited, or overly simplified.

$$Risk = (failure\ probability) \times (consequences) \tag{1}$$

Society evaluates risk practically every day in many different areas; such as home, business, finances, et cetera. Some examples:

- When developing a new business plan, it is common to research the marketing potential, product need, financial benefits, investment cost, and possible losses. Together, each of these areas layout the likelihood that the business can be successful; similar to probability of failure. A business will also invest wisely, and not put all the eggs in one basket, because the consequence of failure could result in a new sign on the door – going out of business.

- Financial investments must address the question regarding the level of risk for the investment. A financial investment with significant risk may have high payoff, but it can also result in large losses. Since no one was performing an economic consequence assessment due to a potential dam breach, the economic losses were largely ignored in the dam safety evaluations.

- Risk is also evaluated in everyday situations, such as cooking. When cooking chicken for dinner, most understand the importance of thoroughly cooking the meat. The risk of food poisoning from undercooked chicken cannot be ignored and, with better understanding of food-borne illness, measures to reduce risk are more often adequately employed.

There are hundreds of examples which illustrate how everyone uses risk to make decisions on an everyday basis. However, careful, pragmatic, and systematic evaluation of risk is relatively new to the dam safety environment.

RIDM and Consequences

Unfortunately, historically dam safety programs have categorized projected failure based solely on downstream consequences, as summarized in the FERC *"Engineering Guidelines for the Evaluation of Hydropwer Projects"*.

- LOW hazard projects are generally located in rural or agricultural areas where failure may damage outbuildings, limited agricultural land, or local roads. Low hazard potential dams have a small storage capacity, the release of which would be confined to the river channel in the event of a failure and therefore would represent no danger to human life.

- SIGNIFICANT hazard projects are usually located in predominately rural or agricultural areas where failure may damage isolated homes, secondary highways or minor railroads; cause interruption of use or service of relatively important public utilities; or cause some incremental flooding of structures with possible danger to human life.

- HIGH hazard projects are those located where failure may cause serious damage to homes, agricultural, industrial and commercial facilities, important public utilities, main highways, or railroads, and there would be danger to human life.

The hazard classification is then used to establish the design loads. For example, for LOW hazard structures, the inflow design flood (IDF) may be selected primarily to protect against loss of the dam and its benefits should a failure occur. That may be a flood event that corresponds to a 100-year return frequency. On the other hand, for some Significant and most HIGH hazard projects, IDF may be the probable maximum flood (i.e., PMF). In some cases, a similar approach was used to establish the design earthquake loads. However, the evaluation of consequences did not go much further than establishing the hazard classification. Consequently, dams might have been remediated sufficient to withstand the PMF, but operators might not have known the risks of lesser flood releases endangering persons within the downstream inundation zone (e.g. ephemeral recreational use), or risks to persons exposed to the upstream flow-influence area. An adequate risk-informed operational approach would obviate hazards due to all flow releases and all potential influence zones while ensuring the dam itself can withstand the appropriate extreme load.

The risk process is really just an extension of this older hazard classification system. Before, any dam that had at least one (1) life loss, was grouped in the HIGH hazard category. The RIDM process now takes the HIGH category and further evaluates it so that the consequences are better understood. For example, would the flood wave due to a potential breach result in less than 1, 10, or 100 lives lost. It is important to note, that RIDM process does not suggest that project safety is dependent solely upon the consequences. Rather, for example, if an owner is facing a situation where two projects require safety enhancements, the decision as to which project is given highest priority may be based on the level of consequences and the speed with which the deficiency can be remediated. For example, priority may be given to Project A, which has an estimated consequence of 100 lives lost, compared to Project B with an estimated consequence of one life loss. Repair of the project with lower consequences would still be required, but might be conducted after completion of modifications to the first project; or, the second remediation might be performed first if the lesser life lost potential is more immediate or compelling.

Furthermore, consequences are not always measured with respect to potential life loss. If an owner's primary purpose is water supply and there are little-to-no downstream consequences, then the economic cost due to potential loss of the water resource may be a better standard to evaluate consequences. In many cases, once the owner understands the potential economic consequences, it has resulted in self-imposed standards that exceed the levels published in typical dam safety guidelines.

RIDM and Failure Probability

The traditional dam safety programs require that the project has adequate safety for predefined loading conditions (i.e., normal, flooding, and earthquake), and the safety has typically been evaluated based on acceptable factors of safety. It has long been understood that there are inconsistencies in the methods combining deterministic loads and factors of safety. The following bullets state accepted dam safety practices (taken from 1977 Reclamation criteria) and can be used to explain this issue more clearly.

- The minimum allowable factors of safety commonly used to assess the safety against concrete dam sliding stability are 3, 2, and greater than 1, for the usual, unusual (PMF), and extreme (seismic MCE) loads, respectively.

- The extreme load consists of the maximum credible earthquake (MCE) used for the earthquake loading is defined as the largest magnitude due to an active seismic source. Historically, active seismic sources have been associated with movement within the last 10,000 years; in other words, the MCE corresponds to a 10,000 year return frequency.

- The unusual load is defined uses the reservoir levels associated with the probable maximum flood. The PMF is the largest flood due to meteorological and hydrological characteristics. While the PMF event does not correspond to a specific return period, some agencies have commonly associated the PMF event to correspond to 100,000 to 1,000,000-year return frequencies.

The inconsistency is shown in Table 1, and shows that the unusual load is a less frequent event, than the extreme load. Even though the unusual flood load is less frequent, it requires a greater factor of safety than the extreme load. Of course, this is an over simplification to illustrate a point, that there are inconsistencies within the typical standards for the different required loading combinations.

Table 1. Summary of Typical Dam Safety Guidelines

Load Description	Load Type	Minimum Required Factor of Safety	Estimated Return Frequency
Usual	Static	3.0	1
Unusual	Static Flood	2.0	100,000
Extreme	Dynamic Earthquake	>1	10,000

So for an example, assume that a project owner/operator has found that their dam does not have adequate safety for both the flood and extreme seismic load. The traditional dam safety guidelines would require that the project be modified so that it has adequate safety for both the unusual flood and extreme seismic load. In other words, the deficiencies were given equal priority. Yet, as shown in Table 1, the probability of the flood load was much less than the probability of the earthquake load.

What happens if the owner does not have the funds to repair the project for both loads? How do they justify which repair measure should be performed first?

To illustrate this point, assume that the project above results in a potential life loss of 5 for a flooding event, and 20 for an earthquake event (note, that even though inundation due to a flood would be more significant, there is warning time due to the hydrologic (rain) event that prepares the population, operator, and emergency responders. A seismic breach is typically without warning, and thus, it is not unusual for the seismic failure to result in greater potential life loss).

There is also the issue that dams with small PLL are at relatively less risk than project with large PLL. Thus, assessing the safety of the project with the same factor goes against the concept that society is increasingly adverse to accidents with larger life loss.

Table 2. Example Problem Risk Evaluation

Load Description	Estimated Return Frequency	Estimated Probability of Failure	Estimate Potential Life Loss	Estimated Risk
Unusual	1/100,000	50%	5	2.5×10^{-5}
Extreme	1/10,000	50%	20	1.0×10^{-3}

The table shows that the highest risk is due to the earthquake event, thus, the owner would probably invest resources to repair the project for the seismic loads. It is important to note, that the RIDM process is not indicating that remediation of the less probable event is not necessary. Rather, it may simply make sense to modify the project associated with a more frequent loading event first.

The RIDM process provides a level playing field, and allows the engineers to weigh the frequency of the load (i.e., 100,000 year flood, or 10,000 year earthquake) in the estimate of probability of failure.

Understand Need to Monitor for Change

Within the engineering community there is an elephant in the room, and that elephant is changing conditions. The radial gate failure at Folsom Dam demonstrated how changing operating conditions can impact project safety. At other Projects, changing downstream consequences may affect the level of appropriate system robustness. The fact is that systems undergo change, usually in the form of deterioration due to age, weather, use, etc. The effect change has on the ability of a system to safely support loads can be significant.

At Folsom Dam, weather and age led to the corrosion on the trunnion pin. The corrosion increased the load required to turn the trunnion and open the gate. The increase load resulted in additional load being applied to the radial gate members. Deterioration had developed in some gate connections due to weathering (corrosion) and after several decades these connections degraded strength. In the end, change in the structure due to weathering resulted in the failure of a connection, which led to buckling of critical support members, and failure of the gate, which life-safety and economic consequences.

Change happens, and it impacts the safety of our systems. The RIDM process considers the potential for change in the on-going safety evaluation of any project.

Chelan PUD benefited because of a better understanding of how potential change would affect the radial gates at Rocky Reach Dam. The large radial gates measure 50-foot-wide by 58 feet tall, and were designed without considering the potential for trunnion friction. An early analysis showed that the critical gate members would not have adequate capacity to support the load if trunnion friction increased. Although the current capacity of the gate was considered adequate for the deterministic standards, if the changes were to continue then the gate would have been more susceptible to failure. In addition, failure of the gate would mean loss of power generation, the primary business for Chelan PUD.

The downstream consequences due to a potential gate failure were minimal, because the Chelan PUD has prevented development within the downstream inundation zone. However, the potential for lost generation and reputational impact (i.e. consequences) due a radial gate failure were considered significant. Therefore, based on the consequences, none of which were related to potential life loss, the Chelan PUD implemented modifications to the critical gate members to be able to support increase loads due to trunnion friction and now regularly monitors the gate(s) structural and mechanical performance.

Useful Prioritization Tool

Originally risk analysis was intended to be used as a prioritization tool.
The Chelan PUD would like to use the RIDM process to better understand and reduce risk, both with regard to overall structural robustness and for systematically prioritizing mechanical interventions and maintenance. The deterministic guidelines typically governed performance of the basic dam structure. In the deterministic world, we have

too many persons with their own ideas of where rehabilitation or remediation needs to take place. With a RIDM process and clear criteria (e.g. tolerable risk on the f-N chart) we can work more effectively both internally with the FERC and with the public to show why we need to do what we believe is appropriate to reduce risk against a clear criteria, rather than do work first on something that isn't risky, but looks to some decision maker like it must be risky.

As Low as Reasonably Practical (ALARP)

The practice of evaluating dam safety through risk assessments has established tolerable risk definitions (generally as a line on an F-N chart), with the understanding the ultimate goal is to reduce risk to a state that is as low as reasonably practicable (ALARP). The RIDM process can be useful to inform Project owners/operators of where risk, and risk reduction measures might exist, and thusly highlight best available options for reducing Project risk. For example, if a Project is judged to currently satisfy tolerable risk criteria the RIDM process on concert with the ALARP principle may illustrate how a small investment would significantly reduce the project risk significantly below the tolerable risk metric. At Lake Chelan, the penstock surge tank is over 100-years old and the RIDM evaluation showed that the potential life loss is insignificant; however, the economic cost due to lost generation if a failure were to occur is substantial. The updated seismic loads, presented in a Probabilistic manner explicate that for relatively minor cost the probability of failure given a significant seismic event can be substantially reduced.

Consequently, Chelan PUD is planning to strengthen the surge tank structure, although there are no life loss consequences. The Chelan PUD considered the remediation expense a responsible alternative in seeking to reduce the risk of failure to less than the tolerable risk criteria as an affordable improvement in operation risk reduction.

Work Effectively with Regulator

The RIDM process provides more opportunity to work with the regulator to develop proper repairs, if necessary. Whereas, in the traditional dam safety program if the structure is found to be deficient for the maximum loads, the guidelines would indicate that repair for the deficiency is remotely necessary. However, if reducing the overall project risk is the issue, then there may be options to accomplish that apart from expensive construction repairs.

The Chelan PUD sees the RIDM process as a way to work even more effectively with the FERC Regulator and other stakeholders. For example, the spillway at Lake Chelan does not have adequate capacity for the PMF discharges, and as a result the dam would be expected to overtop under that extreme event. However, considering the infrequency of the event, and several monitoring programs that provide the operator several days of warning time regarding potential flood flows, and the unoccupied inundation zone, it was concluded that modifications to the structure were not necessary. By not performing expensive repairs to the Lake Chelan Project, the District was able to maintain satisfied stakeholders while concentrating upon important concerns at other projects.

CONCLUSIONS

This Chelan PUD operates three large hydroelectric projects on the Columbia and Chelan Rivers (Rock Island, Rocky Reach, and Lake Chelan) in central Washington and is currently participating in pilot risk informed decision making (RIDM) studies with the FERC. The initial findings from these studies have been very encouraging and have been compared to deterministic results to ensure the risk is adequately low. Through the process it is clear that RIDM will help dam owners and operators have a better understanding of project risk, understand the impacts changing conditions will have on the safety of the projects, be able to better prioritize safety investments into the projects, and work effectively with the regulators to ensure public safety.

REFERENCES

Federal Energy Regulatory Commission (FERC), "Engineering Guidelines for the Evaluation of Hydropower Projects", 1991

U.S. Department of the Interior, Bureau of Reclamation, *"A Procedure for Estimating Loss of Life Caused by Dam Failure"*, DSO-99-06, September 1999.

URS Corporation, "Site-Specific Seismic hazard Analyses for Lake Chelan, Rock Island, and Rocky Reach Dams, Central Washington," December 2012.

RISK PRIORITIZATION AND MITIGATION EVALUATION FOR A PORTFOLIO OF DAMS IN KENTUCKY

Michelle Meehan, PE [1]
Jonathan Keeling, PE, CFM[2]

ABSTRACT

As part of Kentucky Division of Water's (KDOW) Dam Safety Hazard Mitigation Program, Stantec used risk-informed methods to prioritize actions across a portfolio of 78 dams and identify potential hazard mitigation options for each dam.

To prioritize the dams, several methods were reviewed and compared. A tool created by the Natural Resources Conservation Service (NRCS) and based on the U.S. Bureau of Reclamation's "Risk Based Profile System" was selected. This spreadsheet-based method allowed for the determination of relative risk of the dams using limited information from historical files, results from a simplified dam breach analyses, notes from a recent field reconnaissance and publicly available information. This spreadsheet determines an overall Total Failure Index (which indicates the relative probability of failure) and a Total Risk Index (which indicates the relative risk to the downstream population if a failure occurs). The results were used to create a ranking for the portfolio of dams and to categorize them based on potential hazard mitigation techniques that would be applicable.

This paper will highlight a way to take a portfolio of dams and prioritize those dams using risk-informed methods (with or without complete detailed construction or design information). Examples of flood risk mitigation alternatives explored at these dams, including conservation easements, property buyout programs, and dam upgrades/modifications will also be reviewed, as well as how these potential mitigation alternatives are linked with the risk prioritization process. Finally, case studies of dams in the portfolio where mitigation strategies have been implemented will be highlighted.

INTRODUCTION

The Kentucky Division of Water Dam Safety Program has regulatory authority/responsibility for approximately 1,000 dams across the state of Kentucky. This inventory includes state, public and privately owned dams of various hazard classifications (low, moderate and high). KDOW has developed a Dam Safety Hazard Mitigation Plan (Plan) for an inventory of approximately 200 state and locally owned dams. As a part of this plan the Project Team (Stantec, in conjunction with KDOW), reviewed, tested and in some cases developed best practices to help quantify, communicate and mitigate current risks associated with dams, with the eventual goal of helping avoid future risks posed by dams.

---

[1] Water Resources Engineer, Stantec, 1409 North Forbes Road, Lexington, KY 40511, michelle.meehan@stantec.com
[2] Senior Associate, Stantec, 1409 North Forbes Road, Lexington, KY 40511, jon.keeling@stantec.com

A key component of the project was evaluating risks posed by dams and potential mitigation options to reduce risk. To that end, Stantec reviewed various risk prioritization techniques available to perform an overall preliminary risk prioritization of 78 of the 200 dams identified in the Hazard Mitigation Plan. Once the risk prioritization process was complete, mitigation options were identified for each of the dams and will be used for future consideration of fund allocation by KDOW.

COMPARISON OF RISK PRIORITIZATION METHODS

For this project, KDOW desired a method that was objective, that could produce reasonable results for the dams regardless of gaps in historical data, and that would be able to provide an overall ranked list of the dams. The amount of historical data available for each dam varied, with some dams having detailed as-built drawings and geotechnical data while other dams had no drawings or geotechnical information available. Therefore, a method that would produce results that were not skewed based on the information available for the dam were preferred. Three methods for risk prioritization of dams were reviewed including the FEMA Risk Prioritization Tool, the NRCS Risk Assessment Spreadsheet tool, a risk indexing method used by the Ohio Department of Natural resources (ODNR) and the US Army Corps of Engineers (USACE) Risk Analysis method.

Ultimately, the NRCS Risk Assessment Spreadsheet was used as the prioritization method for the project. A summary of the methods evaluated is provided below.

FEMA Risk Prioritization Method

The FEMA Risk Prioritization Tool is a spreadsheet based risk prioritization method. The purpose is to compare risks within a large inventory of dams and provide an understanding of key contributors to risk and how it can be quantified. The FEMA tool anticipates that only the minimal information (National Inventory of Dams data, Phase 1 report, or a site inspection report) is available for most of the dams within the inventory. Although many more details from dam sections, analyses, and inspections can be entered, the tool can perform conservative evaluations based on minimal input. A drawback to using this method for the KDOW project was that answers to questions about the dam within the tool are more subjective in nature. It is up to the user to determine what they think is the probability of failure for each element based on the information they have available. Due to this subjectivity, the evaluation of the risk associated with each element may vary between users which can lead to inconsistencies when trying to economically prioritize dams at a "screening" level.

NRCS Risk Assessment Spreadsheet

The Natural Resources Conservation Service has created a spreadsheet tool that is designed to evaluate dams to assess risk. The spreadsheet is based on the U.S. Bureau of Reclamation's "Risk Based Profile System" developed for dam safety prioritization. This spreadsheet provides an overall "Total Failure Index" and "Total Risk Index" that are based on the potential for failure and the consequences of a failure. These values can be used to rank and compare the dam with other dams to prioritize funding and future studies.

ODNR Risk Indexing Method

The Ohio Department of Natural Resources had conducted a prioritization for their dams for enforcement by using a risk indexing method. This method is somewhat similar to the NRCS Risk Assessment Spreadsheet tool; however it is simpler and more subjective. With this method, the result is a risk index for a sunny day failure and a rainy day failure. For each failure, a total of 100 points is available (with a higher point value indicating a higher likelihood of failure). One of the drawbacks of this method for application in the KDOW project was that it was unclear how the points were assigned to the different safety categories (i.e. seepage safety carries more weight than stability safety). In addition, while the method does not require much time or effort; the indexing process is subjective. Because of the subjectivity, the ranking of a dam could be skewed based on the user that was evaluating the dam.

USACE Risk Analysis Method

The USACE has recently transitioned their dam safety program from a standards-based approach to a dam safety portfolio risk management approach. The USACE and US Bureau of Reclamation (USBR) have jointly created a best practices manual as a guide to their risk assessment process. This process reviews considers failure modes similar to those analyzed within the NRCS Risk Assessment Spreadsheet (seismic, hydrologic, static). For this risk analysis, the user would also assign the likelihood of these failure methods, the consequences and determine a level of confidence (based on the information available).

Based on the evaluation of the four methods, the NRCS Risk Assessment Spreadsheet was utilized for the project because it was the most objective of the three methods reviewed, provided the ability to rank the dams based on the scores assigned in the spreadsheet and did not require detailed construction or design data. The methodology that was used to create the NRCS Risk Assessment Spreadsheet is very similar to the USACE method of risk analysis. The spreadsheet has the advantage of offering a systematic approach to evaluating a set group of failure modes based on a predetermined set of criteria, which was preferred in the case of the KDOW project.

RISK PRIORITIZATION PROCESS USING NRCS SPREADSHEET

The NRCS risk assessment spreadsheet consists of a series of questions about the dam. The questions are answered by the user based on available information for the dam and the spreadsheet calculates a Total Failure Index (based on the condition of the dam and its likelihood of failure) and a Total Risk Index (based on the consequences of a failure and the likelihood of failure). The process is relatively objective, which is important in order to be able to compare the dams to one another. The spreadsheet can be applied to earthen dams as well as concrete dams.

The spreadsheet contains five tabs: static, hydrologic, seismic, risk and consequences. Each of the tabs and their calculations is discussed below.

Static Failure Index

The static tab considers the risk of a sunny-day failure due to the condition of the dam including the condition of the principal spillway, past reservoir filling history, seepage and deformation, foundation geology, and the design, construction and monitoring of the embankment. A series of yes or no questions is answered by the user utilizing available sources of information including dam specific information such as the last inspection report and as-built drawings, as well as publicly available data such as geologic quadrangle maps and soil surveys. The answers to the yes or no questions result in a point value (i.e. a "no" answer for a question may give a point value of 10). The points are summed to determine the Static Failure Index.

Hydrologic Failure Index

The hydrologic tab considers the risk of a failure during a storm event based on the hydrologic capacity of the dam, and the geometric configuration of the spillways. The user enters values for each field based on available data (including previous hydrologic & hydraulic analyses) and an overall Hydrologic Failure Index is computed.

Seismic Failure Index

The seismic tab considers the proximity of the dam to seismic zones and the potential for liquefaction of the dam foundation. The user enters values based on as-built drawings or publicly available datasets such as seismic zones and geologic quadrangle maps. For this tab a Seismic Failure Index is computed.

Total Risk Index and Total Failure Index

The risk tab calculates the Total Failure Index and the Total Risk Index. The Total Failure Index is the sum of the three failure indices (static, hydrologic, and seismic). The maximum amount of points for both the static and hydrologic failure indices is 300. The maximum for the Seismic Failure Index is 100. Therefore, the maximum Total Failure Index that can be calculated is 700. A higher risk index number corresponds with a higher risk of failure. For the Total Risk Index the user enters the estimated population at risk and a fatality rate (based on warning time, the community's understanding of evacuation procedures, and the depth and velocity of a potential breach). The failure index for each scenario (static, hydrologic, and seismic) is multiplied by the fatality rate and the population at risk to come up with a risk index. These risk indices are summed to compute the Total Risk Index. Because the failure index is multiplied by the population at risk to compute the risk index, there is no maximum value for the Total Risk Index.

$$\text{Total Failure Index} = \text{Static Failure Index} + \text{Hydrologic Failure Index} + \text{Seismic Failure Index} \quad (1)$$

$$\text{Total Risk Index} = \text{Total Failure Index} * \text{Fatality Rate} * \text{Population At Risk} \quad (2)$$

The final tab is a consequences tab which summarizes the risk indices and gives an overview of the dam being considered. Most of the data on this tab is not considered in the computation of the risk indices, but can be used to compare the different failure impacts of two dams with similar risk indices. This information can help the end user determine which dam is more of a priority in terms of funding or further analysis.

One concern with using the NRCS spreadsheet was that some of the questions could not be answered with certainty without as-built or geotechnical information. In total, seven items within the spreadsheet required this information. Based on discussions with NRCS who had used this method previously, their process was to answer the questions conservatively when the information was unknown. This would result in an additional score of up to 55 points added to a dam with limited geotechnical or as-built data. Because there are a total of 700 points available, it was not envisioned that this methodology would skew the results significantly.

The data available for use in filling out the spreadsheets consisted of:

- KDOW Dam Safety Inspection Reports (available for all dams)
- USACE Phase 1 Inspection Reports (available for some dams)
- Hydrologic and Hydraulic (H&H) Analyses (available for most dams)
- As-Built Drawings (available for some dams)
- Geotechnical Data (available for some dams)
- Simplified Dam Breach Analysis & Inundation Mapping (available for all dams – performed as part of this project)
- NID Database, KDOW dam database (available for all dams)
- Peak Ground Acceleration map for Kentucky (published by Kentucky Geological Survey)
- Soil Survey, Geologic Quadrangles (US Geological Survey)
- Karst Potential Classification Shapefile (Kentucky Geological Survey)

Adjustments to Hydrologic Failure Index

Another concern with using the NRCS spreadsheet was that it does not specifically address the ability of the dam to pass its design storm in accordance with KDOW dam safety regulations. Rather, the spreadsheet looks at what design rainfall a dam can pass regardless of its design. Therefore, a low hazard dam would score higher on the Total Failure Index than a high hazard dam, even if they were both designed to KDOW regulatory standards, because a low hazard dam is not required to pass as large a rainfall amount. While this is logical from a hydrologic failure perspective (a dam designed to pass less rainfall has a greater chance of overtopping), KDOW also wanted the overall ranking to reflect whether the dams were hydraulically deficient. Therefore, the hydrologic failure index and Total Failure Index were adjusted to reflect this as shown in Equations 3 & 4.

$$\text{Adjusted Hydrologic Failure Index}=\frac{\text{Hydrologic Failure Index}}{\text{\% of Design Storm Passed}} \quad (3)$$

$$\text{Adjusted Total Failure Index}= \text{Static Failure Index} + \text{Adjusted Hydrologic Failure Index} + \text{Seismic Failure Index} \quad (4)$$

This adjustment allows the Adjusted Total Failure Index to be higher for dams that pass a smaller percentage of the regulatory design storm.

RESULTS

The NRCS spreadsheet was completed for 78 dams. For each dam both an Adjusted Total Failure Index and Total Risk Index was calculated. The dams were sorted by these values in an overall table. The results were also presented graphically, plotting Adjusted Total Failure Index versus Total Risk Index (see Figure 1).

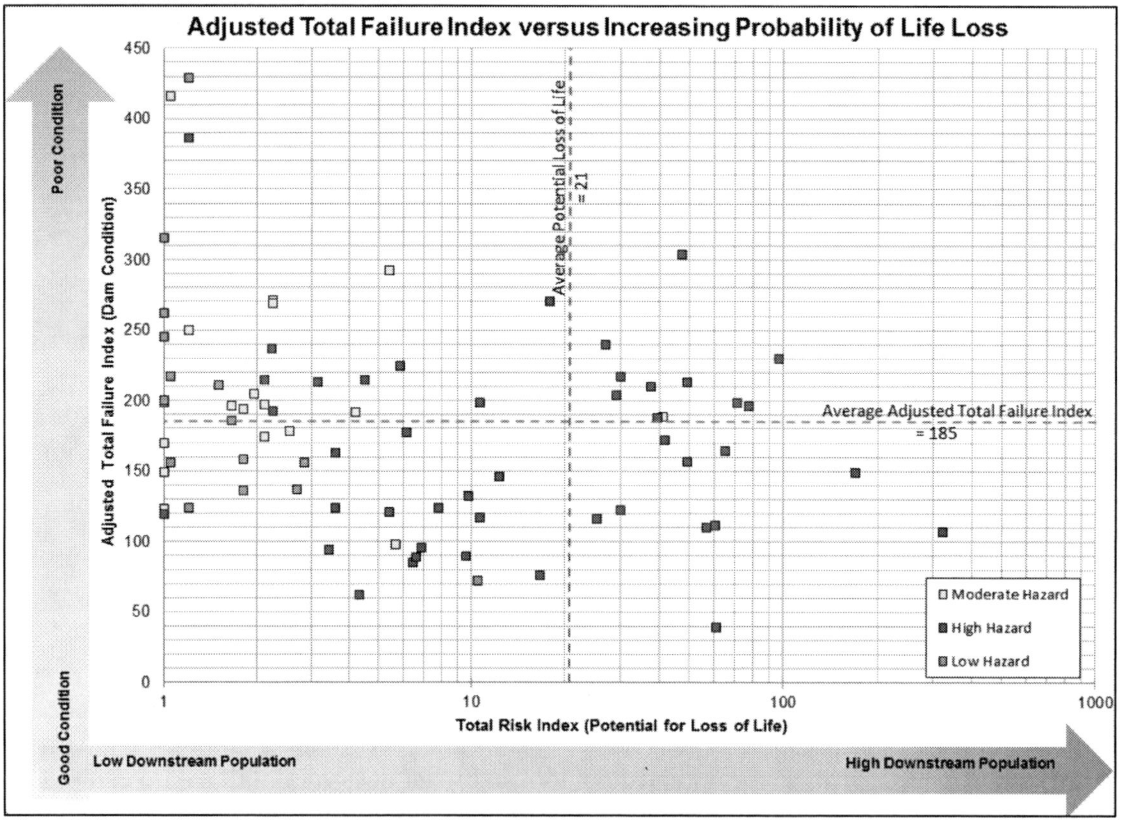

Figure 1. Adjusted Total Failure Index vs. Total Risk Index

The Adjusted Total Failure Index (plotted on the y-axis) is an indication of the condition of the dam and the Total Risk Index (plotted on the x-axis) is an indication of the consequences of a failure. The graphical results were then utilized as a first pass to assess potential mitigation options for each dam.

MITIGATION OPTION EVALUATION

General mitigation options can be identified for each dam based on where the dam appears spatially on the graph. Mitigation options considered include:

- Structural upgrades or repairs to dam and/or spillways
- Acquisition of easements in downstream inundation areas
- Adoption of higher standards in local regulations for development in the inundation areas
- Buy Outs or Relocations for impacted structures
- Berms, Floodwalls or Floodproofing
- Warning Systems
- Emergency Action Plans (EAPs)

Potential mitigation options were assessed based on the location of the dam on the graph. This evaluation is shown in Figure 2. The quadrants shown on the graph were based on average values of the risk screening analysis for the 78 dams. It should be noted that the dams were reviewed on a case-by-case basis and mitigation options were not identified solely based on the dam's location on the graph. Rather, the graphical representation was meant to provide a high-level overview of potential mitigation options as described below.

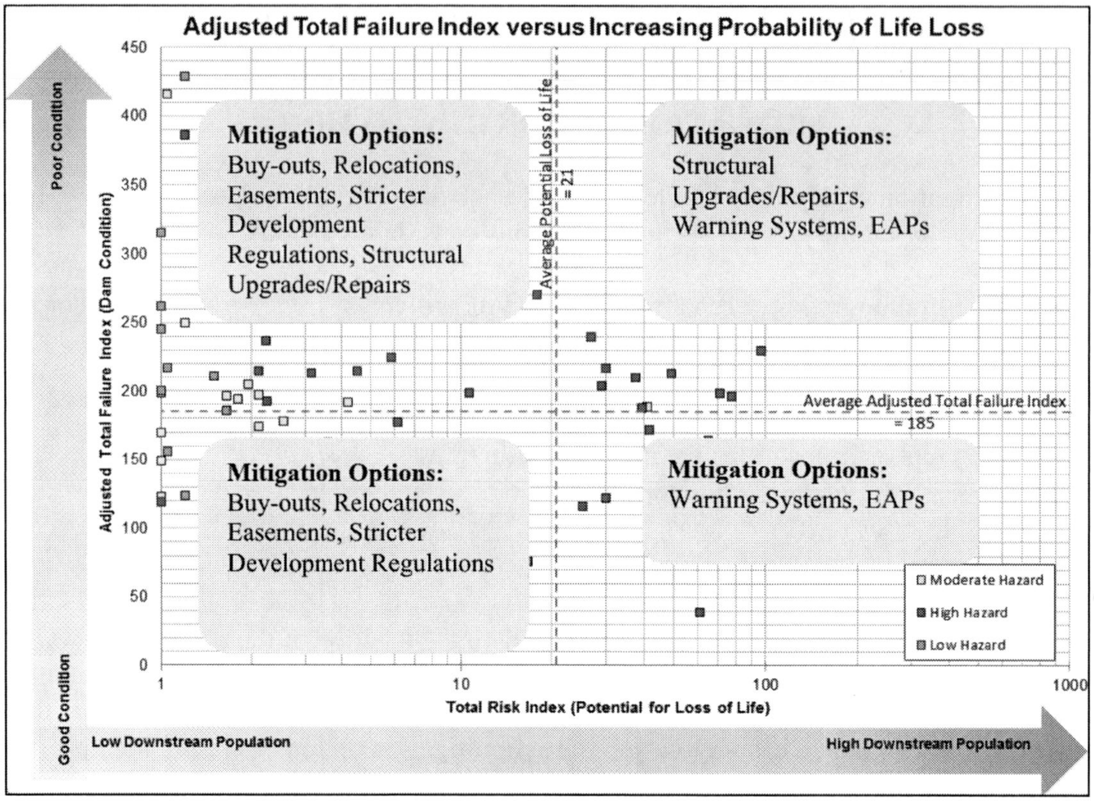

Figure 2. Mitigation Options

A high Adjusted Total Failure Index is indicative of problems with the condition of the dam. For these dams, the mitigation options vary depending on the downstream population. In the case of a larger population downstream, repairs to the dam should be pursued to lessen the likelihood of a dam failure. In addition, EAPs and warning systems could also be pursued to reduce the risk to the downstream population in the event a failure was to occur. For dams with a high Adjusted Total Failure Index but a low Total Risk Index (small population downstream), mitigation options such as buy-outs or structure relocation in combination with easements to protect the downstream inundation area from development can reduce the risk of a dam failure impacting human life. In some cases, the cost associated with removing or preventing the consequences of a dam failure may be significantly less than the cost to make repairs or upgrades to the dam to try and prevent a dam failure. For dams with a high downstream population, but a low Adjusted Total Failure Index, the mitigation options are focused on reducing the risk to the downstream communities in the event of a failure.

CASE STUDIES

For four of the 78 dams, mitigation options have been pursued or are being pursued to reduce risk. These dams are described individually below along with the graphical change (or expected change) in the dam scores after completion of the projects.

General Butler Dam

This dam is a low hazard dam with no population downstream, however there are structural problems with the spillway which cause it to have a high Total Failure Index. KDOW is currently pursuing repairs to the spillway which will lessen the Total Failure Index and reduce the risk of a failure.

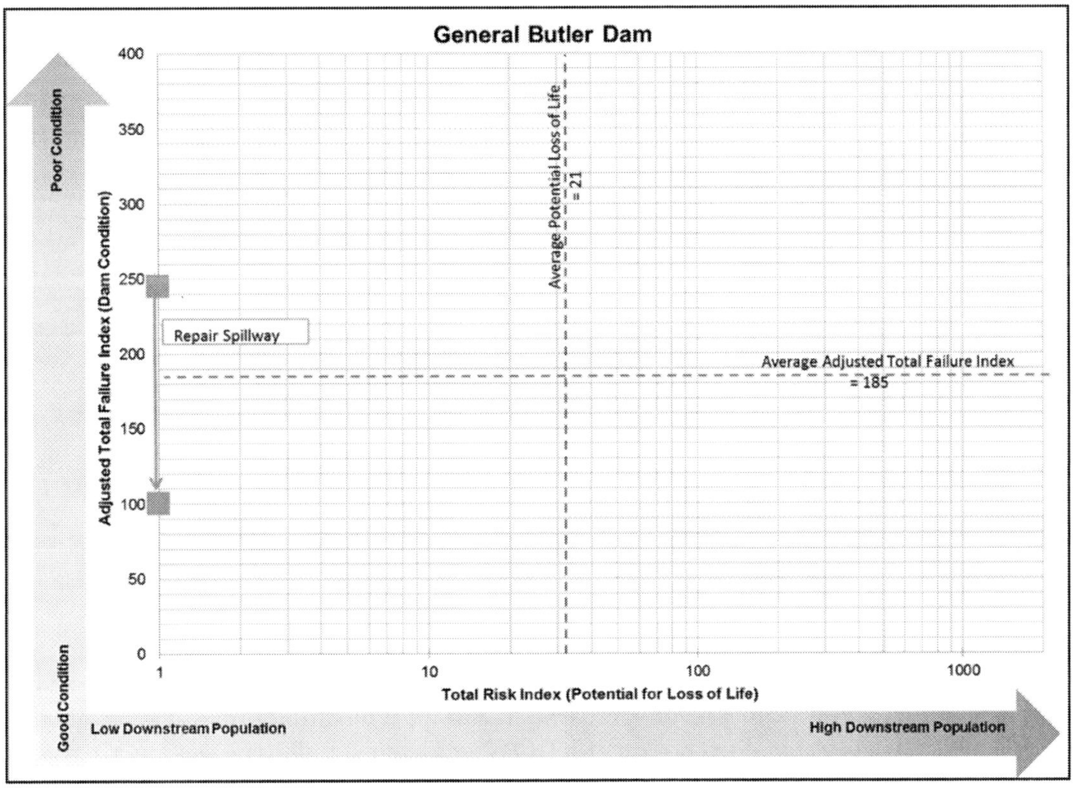

Figure 3a. Improvements to General Butler Dam

Bullock Pen Dam

This dam is a high hazard dam with seepage and low spillway capacity that does not meet KDOW's design standards. KDOW is pursuing an RCC overlay that will address the spillway capacity. Once constructed, the Adjusted Total Failure Index will decrease indicating a reduction in failure potential. In addition, an EAP is being prepared for the dam, which can increase the efficiency with which the dam owner and emergency responders are able to handle an emergency situation at the dam and evacuate potentially affected populations downstream. This increased efficiency during an emergency could result in a reduction of lives lost.

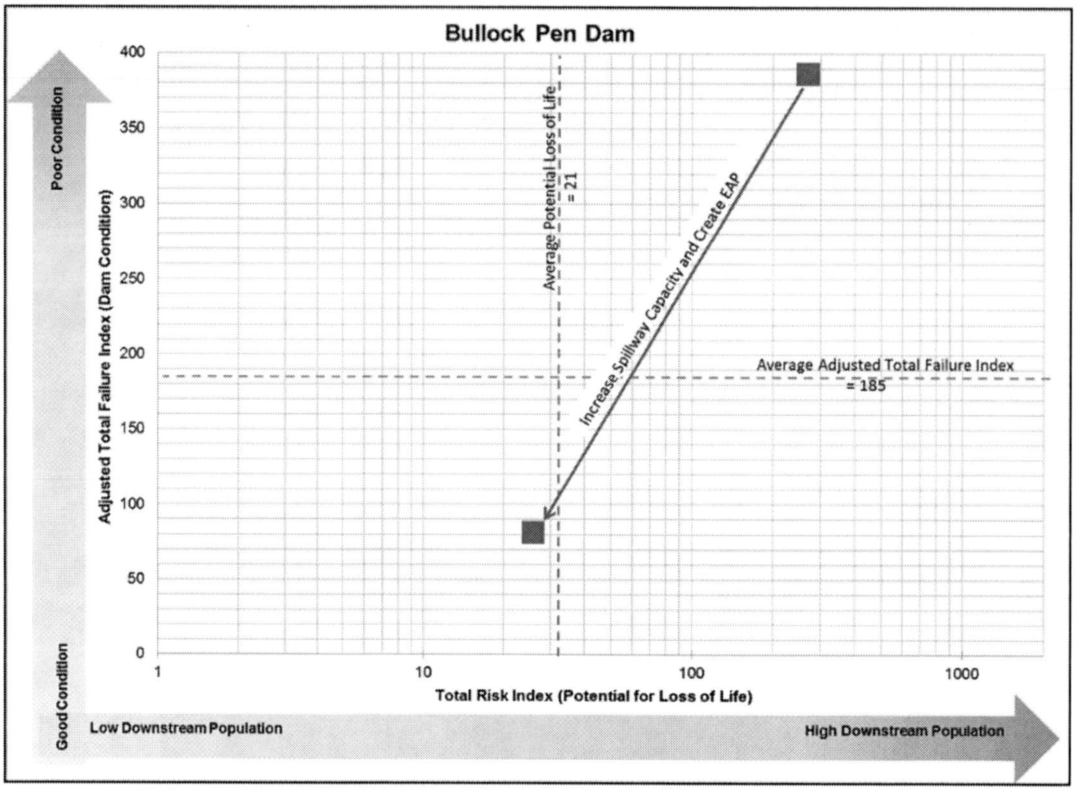

Figure 3b. Improvements to Bullock Pen Dam

Beech Creek Dam

Similar to Bullock Pen Dam, Beech Creek Dam is a high hazard dam with seepage and low spillway capacity that does not meet KDOW's design standards. Design alternatives are being pursued to address these issues. In addition, an EAP is being prepared for the dam. These activities would lessen the Adjusted Total Failure Index and Total Risk Index scores for the dam.

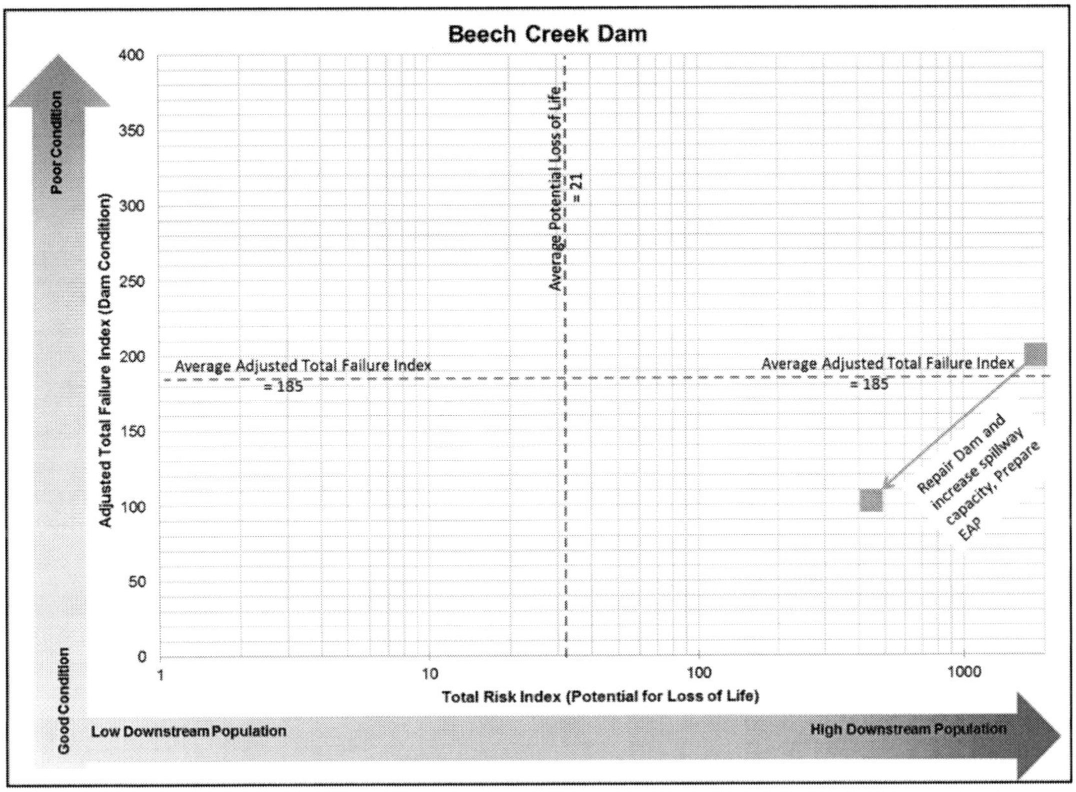

Figure 3c. Improvements to Beech Creek Dam

Spurlington Lake Dam

Spurlington Lake Dam was initially a high hazard dam with one residence identified as possibly in the downstream inundation area. A detailed H&H analysis was performed to define the downstream inundation limits. Based on this analysis, one home was purchased. In addition, these results were used to pursue purchasing conservation easements along the inundation zone to protect the area from future development. This reduces the Total Risk Index score for the dam. The adjusted total Failure Index is reduced because the dam is not required to pass as much rainfall due to the lower hazard classification.

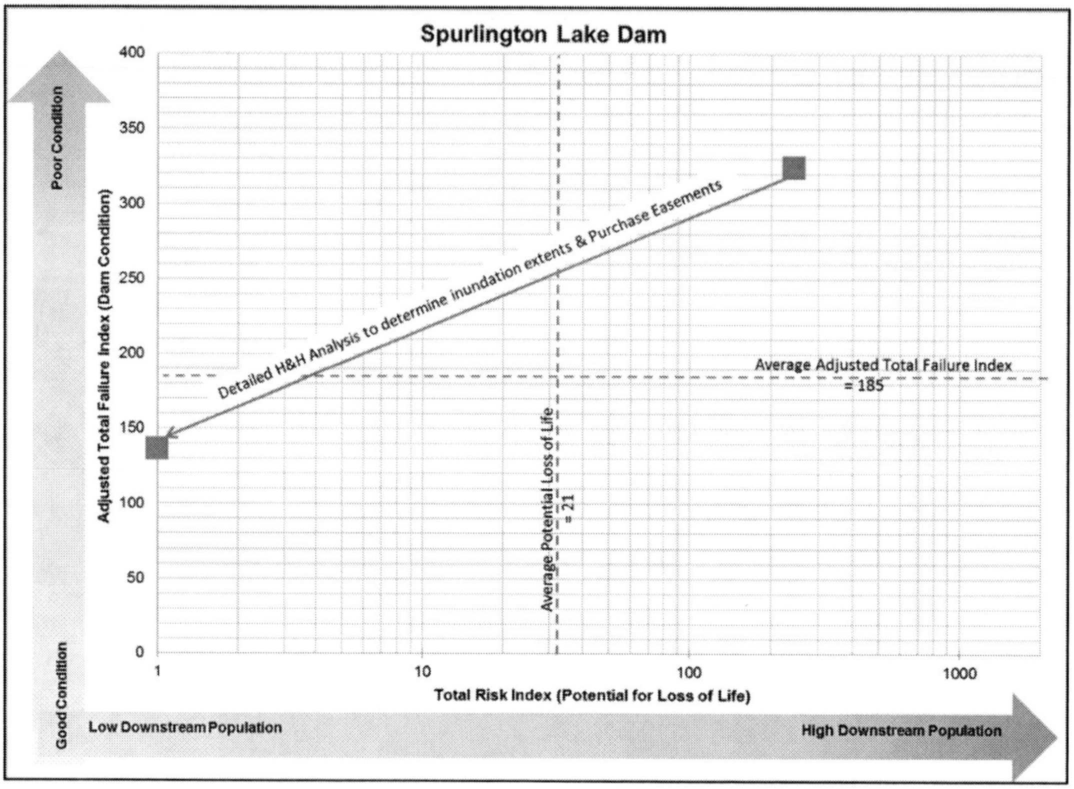

Figure 3d. Improvements to Spurlington Lake Dam

As shown in Figures 3a – 3d, the improvements performed to date (or those that will be performed) reduce the risk of failure of the dam and/or reduce the risk to the downstream population. Viewing the mitigation projects in this manner shows visually how the improvements affect the dam's risk and also allow the reduction in risk to be quantified.

CONCLUSIONS

The NRCS spreadsheet method provides an objective method to evaluate and rank a suite of dams using engineering records as well as publicly available data. For the KDOW project, 78 of their high priority dams were evaluated using this method and the resulting scores were used to begin analyzing mitigation options for groups of dams. The mitigation options identified focus on risk reduction (both risk of dam failure and risk to downstream population). The outcome of this study has been used by KDOW for planning and fund allocation purposes. In some instances, mitigation options have been completed or are being pursued. These case studies can be graphically illustrated using revised scoring for the dams to show the effect of these improvements.

REFERENCES

Federal Emergency Management Agency (FEMA), "Risk Prioritization Tool for Dams." June 1, 2008.

Natural Resources Conservation and Management (NRCS), "Evaluation of Potential Rehabilitation Projects." October 22, 2001.

Natural Resources Conservation and Management (NRCS). "Guidance for Completion of 'Evaluation of Potential Rehabilitation Projects'." February 19, 2010.

Ohio Department of Natural Resources (ODNR), "Dam Safety Enforcement Program." December 10, 2004.

Stateler Jay N., Anevski, John, Oellermann, Doug, and Brown, Michael. "Consequence Ratings: A Streamlined Method for Developing Loss of Life Estimates."

United States Department of the Interior, Bureau of Reclamation (USBR), "Acer Technical Memorandum No. 11 - Downstream Hazard Classification Guidelines." 1998.

United States Army Corps of Engineers (USACE). "Best Practices in Dam and Levee Safety Risk Analysis." December 3, 2012.

THE ROLE OF EMPIRICISM IN QUANTITATIVE RISK ANALYSIS

William Engemoen[1]
William Fiedler[2]
Daniel Osmun[3]

ABSTRACT

When undertaking quantitative risk analysis as a means of evaluating dam or levee safety, it is easy to get deeply involved in, and perhaps frustrated with, the theory of statistics and probability as well as uncertainty and confidence in complex engineering analyses and resulting estimates of probability of structural response. This is particularly so because typical dam safety evaluations involve predicted performance under extreme loadings such as previously unexperienced floods or earthquakes, or prediction of failure during normal operations when a facility may have operated relatively satisfactorily for decades. Tying estimates of probability developed during a risk analysis to empirical performance whenever possible is a critical consideration to help ground, or form an improved basis for, risk estimates. An empirical basis will not only help calibrate probability estimates of predicted behavior with actual case histories, but will lend credibility to the risk analysis process as well. Within Reclamation, empirical evidence is used in several aspects of the risk analysis process, including the three key components of the risk equation – loadings, response to loadings, and consequences of failure. This paper will discuss the use and advantages of empiricism in these aspects of risk analysis.

INTRODUCTION

Quantitative risk analysis is becoming more widespread as a tool for evaluating dam safety issues. For some, there is a perception that risk analysis is solely or largely based on the theory of probability, and involves a lot of complicated terms and principles. At times, the perceived need to make a quantitative risk analysis as detailed and accurate as possible (and thereby perhaps more defensible) results in an exceedingly lengthy and sometimes overwhelming process. Organizations and individuals need to be reminded that there is no one, absolute answer from a risk analysis. Rather, there is a probable range of likely outcomes as opposed to a single value. The true value gained from a risk analysis is likely more related to the insights and knowledge gained from the thorough evaluation of the failure process that occurs during a risk analysis, rather than the resulting value of failure probability. For this reason, overly complex risk analyses (particularly if they require large expenditures of money and time) may not offer any more value than more simple risk analyses.

---

[1] Lead Civil Engineer – Geotechnical Services Division, Bureau of Reclamation, P.O. Box 25007, Denver, Colorado, wengemoen@usbr.gov
[2] Lead Civil Engineer – Geotechnical Services Division, Bureau of Reclamation, P.O. Box 25007, Denver, Colorado, wfiedler@usbr.gov
[3] Lead Civil Engineer – Geotechnical Services Division, Bureau of Reclamation, P.O. Box 25007, Denver, Colorado, dosmun@usbr.gov

A complicated risk analysis with complex probability distributions and several statistical/theoretical adjustments, and based on elaborate numerical models using parameters based on theory and research, may prove difficult for decision makers to comprehend and difficult for technical teams to fully understand and defend. For risk analysis to be an effective tool for both evaluating dam safety issues and making decisions on future actions, it is wise to adopt the simplest and most straight-forward approach and analysis. Two quotes from Ralph Peck (Norwegian Geotechnical Institute 2000) illustrate this opinion:

> *"Simple calculations based on a range of variables are better than elaborate ones based on limited input."*

> *"We should be on guard not to ascribe to elaborate analytical routines a reliability they do not possess."*

In the authors' opinion, one consideration in keeping analyses simple, yet defensible and robust, is to use observations from case histories (actual performance) whenever possible. Reclamation has incorporated empirical observations into several aspects of risk analysis, including all three main components of risk – loading, structural response, and consequences.

EMPIRICISM IN PROBABILITY OF LOADING

Potential failure modes and quantitative risk analyses for dams usually involve three separate loading conditions – static (or normal operations), seismic (earthquake) loadings, and hydrologic (flood) loadings.

Static Loadings

The static loading condition imposed upon a dam during normal operations is the typical reservoir operation level, which generally fluctuates during the course of a year. Thus, for existing dams, the probability of reservoir loading can be determined from a historical record of reservoir operations at the dam being evaluated. An example reservoir operations curve at a Reclamation reservoir is shown in Figure 1. This loading is thus totally empirical, based solely on observed behavior. Reservoir levels that exceed normal operations will typically be due to a flood with a given return period, and those levels would be considered hydrologic loadings.

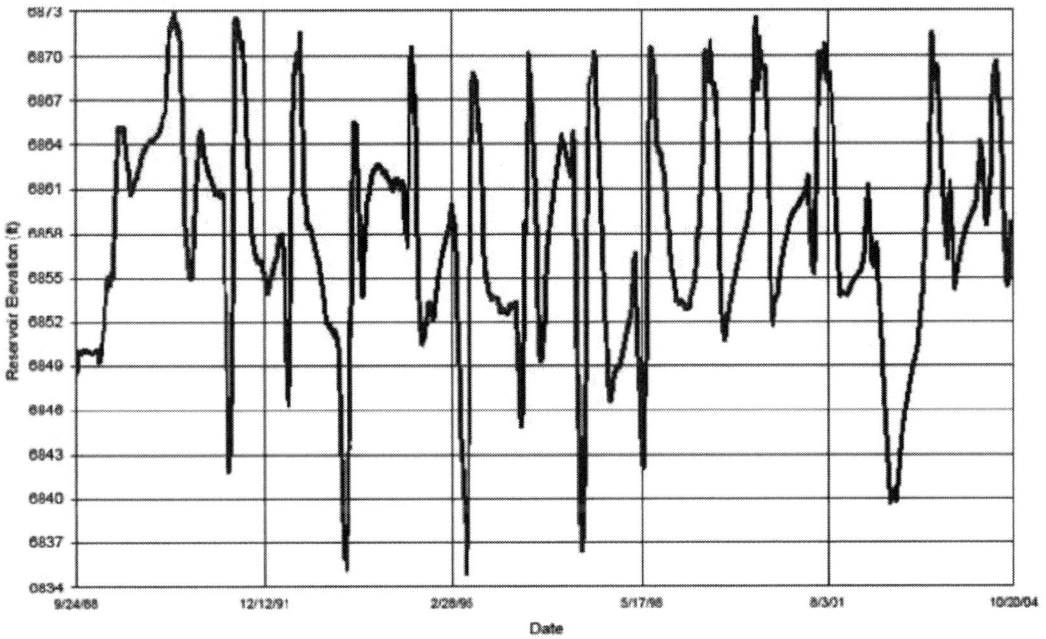

Figure 1. Reservoir Loading Curve

Hydrologic Loadings

Hydrologic loadings for risk analysis must often be projected to extreme events with return periods of thousands of years or more. However, Reclamation uses at least two sources of empirical evidence to help develop hydrologic hazard curves which portray peak flood flows (and sometimes flood volumes) with a return period. An example hazard curve is shown in Figure 2. Stream gage records exist at most dams and reservoirs, on a stream or river upstream from a Reclamation dam, or perhaps in a nearby, similar basin. Often, these records of measured stream flows go back 100 years or more. Thus, the early portion of a hazard curve, generally up to a 100-year return period, can be obtained by stream gage records.

Rather than simply extending the frequency curve on the trend of the stream gage portion, Reclamation utilizes streamflow and rainfall modeling extrapolation and modeling techniques, but also incorporates paleoflood hydrology to help identify the approximate return period of past flooding events in a drainage area being considered. Paleoflood hydrology is the study of past or ancient flood events which occurred before the time of human observation or direct measurement by modern procedures. River geomorphology and soils/stratigraphy adjacent to the stream are studied to provide information on past flood events, in addition to the gathering of evidence of past floods and streamflow to estimate ages of these events by such methods as carbon dating. The advantage of paleoflood data is that it is often possible to gain information about past floods that may be 10 to 100 times older than the stream gage records.

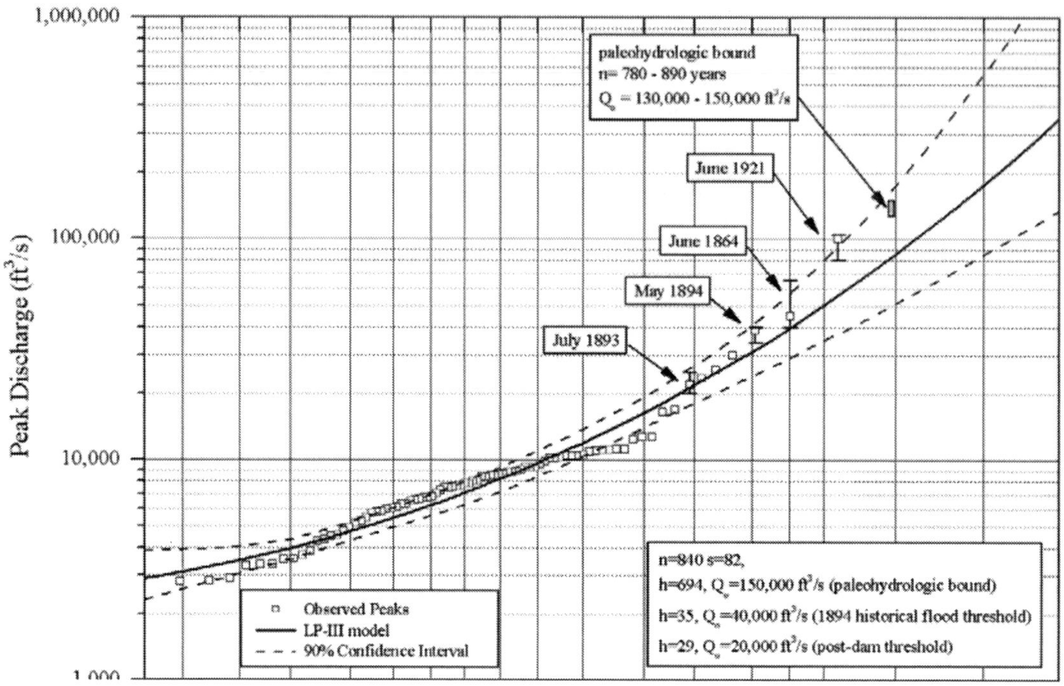

Figure 2. Hydrologic Hazard Curve

Seismic Loadings

Extreme seismic loadings are perhaps the most difficult to characterize with empirical data, primarily due to the relatively limited amount of site-specific measurements of earthquake events. Rarely is seismic strong motion instrumentation located at a dam that experiences a very strong earthquake loading that would be of interest to dam safety evaluations. Although numerous large earthquakes occur each year, peak horizontal ground motions and duration of shaking at a given site can often only be estimated due to a lack of strong motion instruments at the specific site of interest. Hence the development of seismic hazard curves, such as shown in Figure 3, depends to a large extent on the identification and characterization of earthquake sources, the predicted ground motions from computer codes that incorporate slip rates and estimated magnitudes for the various sources, and the application of attenuation relationships.

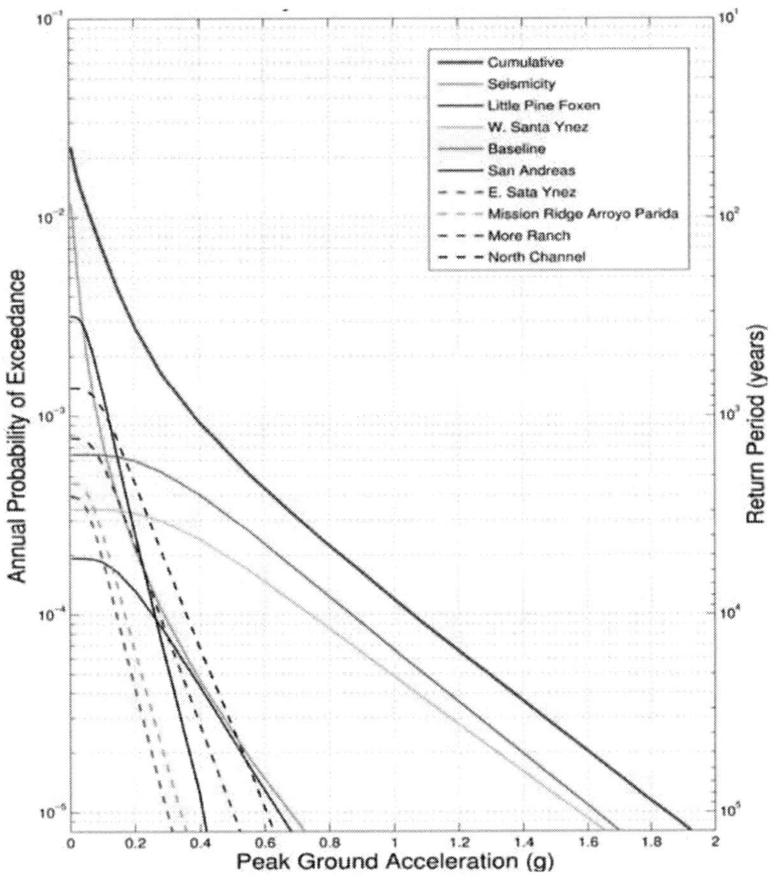

Figure 3. Seismic Hazard Curve

EMPIRICISM IN PROBABILITY OF STRUCTURAL RESPONSE

There are numerous aspects involved in estimating the probability of structural response to a loading, including deterministic and probabilistic engineering analyses that provide insights and guidance in estimating anticipated performance. Some of these key building blocks involve empirical procedures. A discussion of structural response is categorized below by loading conditions.

Static Potential Failure Modes

In Reclamation dam safety evaluations, inevitably the static potential failure modes of most concern for embankment dams deal with internal erosion. Static stability issues are rare in an inventory of existing dams, since the structures have typically experienced decades of normal loads. On the other hand, internal erosion incidents or failures have been reported at dams with many decades of successful performance. Internal erosion accounts for approximately half of all the historic failures of embankment dams (Foster et al. 1998).

There are no widely accepted deterministic analyses to estimate the safety factor against internal erosion in an embankment dam or its foundation. Risk analysis provides a means to consider the various steps that would have to occur to lead to dam failure by an internal erosion process. Event trees to evaluate internal erosion potential are presented elsewhere (Bureau of Reclamation and U.S. Army Corps of Engineers 2012). Arguably the critical event in the internal erosion process, and the most difficult to assign a probability to, is the *initiation of internal erosion*. Initiation is typically judged to have a relatively low probability of occurring; is based on a number of variables including presence of a concentrated seepage path, hydraulic gradients, and soil erodibility; and is thus difficult to estimate. Estimated historic rates of internal erosion initiation, or the number of observed incidents over a given number of years of operation, can provide risk teams with a relative range or average value for various types of internal erosion.

To investigate the frequency of internal erosion incidents within the Reclamation inventory, dam safety reports were reviewed with a focus on past performance. From this document review, supplemented with personal experience of senior staff, a lengthy list of incidents of internal erosion at Reclamation facilities was developed (Engemoen and Redlinger 2009, Engemoen 2011). There have been a total of 99 known incidents including one failure. An estimate of the historical rate of initiation of internal erosion in a Reclamation embankment dam can be obtained by dividing the number of incidents and failures by the number of dam-years of operation. The total number of dam-years was obtained by identifying each of Reclamation's approximately 220 major embankments, determining their present age, and summing all ages. In this manner, the number of dam-years of operation at Reclamation facilities was estimated at approximately 13,000.

Using the number of incidents and dam-years, the following table presents the number of incidents and the estimated historical rate at which erosion has initiated. The incidents and historical rates of initiation are portrayed two ways – based on a general category (or location) of internal erosion, and based on the type of internal erosion mechanism that was believed to have occurred (backward erosion piping, scour, etc.). In addition, the historical frequency is presented as a range, which is interpreted as the "best estimate" range of likelihood. The actual range could vary to a much greater extent depending on whether site conditions are better or worse than the average condition at a Reclamation dam.

Table 1. Number of Incidents and Best Estimate Frequencies of Initiation of Internal Erosion at Reclamation Dams

	Backward Piping	**Internal Migration**	**Scour**	**Suffusion/Suffosion**	**FREQUENCY**
Embankment	3	1	2	1	**$3x10^{-4}$ to $1x10^{-3}$**
Foundation	8	21	25	16	**$2x10^{-3}$ to $1x10^{-2}$**
Embankment into Foundation	0	0	6	0	**$2x10^{-4}$ to $1x10^{-3}$**
Into/Along Conduit	1	4	0	0	**$4x10^{-4}$ to $1x10^{-3}$**
Into Drain	4	7	0	0	**$5x10^{-4}$ to $2x10^{-3}$**
FREQUENCY	**$5x10^{-4}$ to $2.5x10^{-3}$**	**$1x10^{-3}$ to $5x10^{-3}$**	**$1x10^{-3}$ to $5x10^{-3}$**	**$6x10^{-4}$ to $2.5x10^{-3}$**	

Obviously, these historical estimates must be used with caution based on the general method in which they are developed; the potential for initiation of internal erosion at a given dam must be carefully considered in terms of site-specific factors. Nonetheless, average historical values can help ensure the team has an empirical reference point to help guide their probability estimates.

Hydrologic Potential Failure Modes

Hydrologic failure modes also account for nearly half of all embankment dam failures, with the majority of failures attributed to erosion from flood overtopping. In addition to actual performance of existing dams, there is great deal of research and large scale experiments that have been conducted to observe the performance of earth dam slopes subjected to various depths and durations of overtopping flows. The case histories and experiments provide Reclamation risk analysis teams with valuable information for estimating the ability of a dam to withstand various depths and durations of overtopping flows that are predicted by flood routings. In general, embankment dams have been shown to not survive long in overtopping events, and the performance is even worse if concentrated flows (higher velocities and greater depths) occur in a low spot in the crest (e.g. where crest camber ends near the dam abutments). Thus, it is not unusual for Reclamation teams to interpret past performance as generally poor, and estimate a relatively high probability of continuing crest erosion at an embankment dam subjected to overtopping. A lower likelihood of an erosional failure may be estimated for the more unusual cases of shallow or short duration of overtopping, paved crests with thick rockfill shells, and uniform or sheet flow overtopping as opposed to concentrated flows.

One contributing cause of flood overtopping can be the failure of spillway gates to operate. Given the many dam-years of operation of spillway gates both within Reclamation's inventory and worldwide, this appears to present an opportunity to develop base-rate frequencies for gate operation difficulties. Although not yet developed to a point that it can be used in quantitative risk analysis, a database of Reclamation gate performance is being developed with the intent to help guide risk teams in the assessment of the probability of gate operational problems during flood events.

Seismic Potential Failure Modes

Earthquake loading has accounted for less than 2 percent of all embankment dam failures. As such, there are very few case histories or empirical data on which to base structural performance of dams during seismic loading. However, some of the deterministic engineering analyses which form a basis for probability estimates for various seismic potential failure modes have empirical foundations. For example, a typical seismic potential failure mode involves foundation liquefaction that leads to slope failure, crest loss, and overtopping of an embankment dam. Reclamation's engineering evaluation of the potential for foundation liquefaction is based on curves of SPT (or similar) penetration resistance versus cyclic stress ratio (or measure of earthquake energy). These curves were based on penetration resistances from sites that had observed instances of liquefaction or no liquefaction. Unfortunately, these sites were level ground, with no data points beneath dam foundations.

Given the very limited number of dams subjected to large earthquakes, there is also very little empirical information on the amount of deformation and crest loss that can occur during earthquake shaking. Reclamation risk teams do consider that performance of such dams as Lower San Fernando Dam, Sheffield Dam, Austrian Dam, and others, but this is an admittedly small set of data. A Reclamation research effort that holds promise is a planned collection of data (including measured site-specific ground motions and embankment and foundation material properties) at a number of dams (both embankment and concrete dams) in such areas as Chile and Japan that have experienced significant earthquake shaking during large earthquakes in the past few years. It is planned that these data, as well as reported dam performance (settlement, slope movement, cracking, etc.), will be used in current state of the practice analyses to see how well predictions (without knowing the actual results) match observed behavior. The use of these types of empirical observations (particularly for non-failures) should prove highly informative and useful in attempting to calibrate numerical models with actual performance.

EMPIRICISM IN PROBABILITY OF CONSEQUENCES

Within the Reclamation dam safety program, the consequences of dam failure are primarily focused on life loss. The estimation of life loss that might occur as a result of dam failure is a difficult task, involving considerable uncertainty. Since the 1990s when Reclamation began utilizing quantitative risk analysis as a tool to evaluate dam safety, life loss estimation has been based on an empirical approach that utilized data from past flooding events. For 15 years, "A Procedure for Estimating Loss of Life Caused by Dam Failure," or report DSO-99-06 (Bureau of Reclamation 1999) was the approach used to estimate life loss. In 2014, Reclamation developed the Reclamation Consequence Estimating Methodology (RCEM) (Bureau of Reclamation 2014), which can be viewed as an expansion and improvement of DSO-99-06. Both procedures are based on a study of dam failures, flash floods, and other flooding events that occurred in both the United States and worldwide. RCEM includes an additional number of case histories, such that the current database of floods contains approximately 80 data points.

Analysis of the case histories indicates that flood severity (a measure of the lethality of the flood) and warning time are the factors that most influence the fatality rate for a flood. Flood severity is estimated as the depth times the velocity of the flood flows (DV). In developing RCEM, the numerical measure of flood severity, DV, was estimated for each case history event using available documentation and engineering judgment. The data were then plotted graphically. Two charts were developed for selecting fatality rates using the graphical approach; Fatality Rate vs. DV for Little or No Warning (Figure 4), and Fatality Rate vs. DV for Adequate Warning (Figure 5).

Each chart includes dashed lines that represent "suggested" and "overall" limits for fatality rates over the full range of DV values. The suggested limits were selected based on the most representative case history data points for each warning time scenario. Cases with questionable data were given less influence on the suggested range. The overall limits are intended to represent the upper and lower bounds of fatality rates, between which nearly all case history data falls.

As reflected in the figures, there is a large uncertainty inherent in the estimation of life loss resulting from dam failure (particularly for cases of adequate warning), in part due to large possible variations in the development and progression of breach flows, as well as numerous potential ways that the downstream public receives warning (if any) and the manner in which they respond to warning. The study of flooding case histories reinforces the finding that there are a wide range of possible outcomes from dam failure, ranging from no fatalities to thousands of lives lost. The use of such empirical findings and the resulting procedure based on these data are intended to reflect the variability associated with life loss, as well as encourage the use of judgment in considering the many variables associated with estimating life loss. Lessons learned from the case histories show that a wide range of fatality rates are possible, and thus a range of life loss should be portrayed rather than single point values.

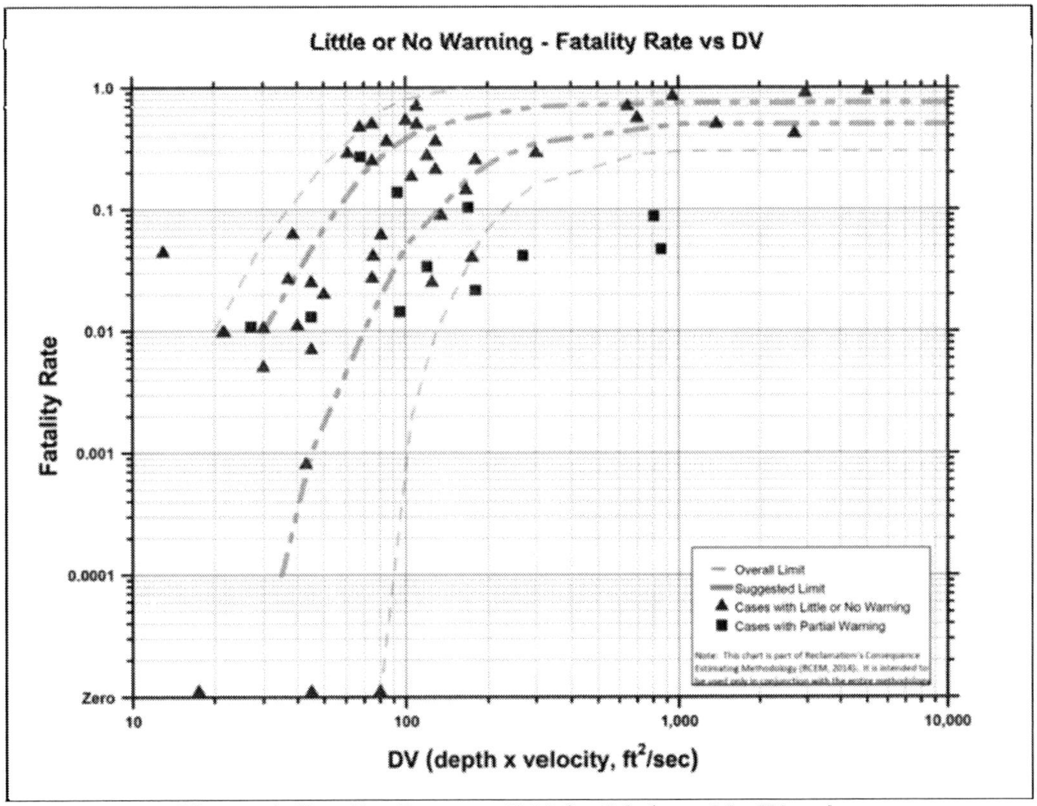

Figure 4. Fatality Rate vs. DV for Little or No Warning

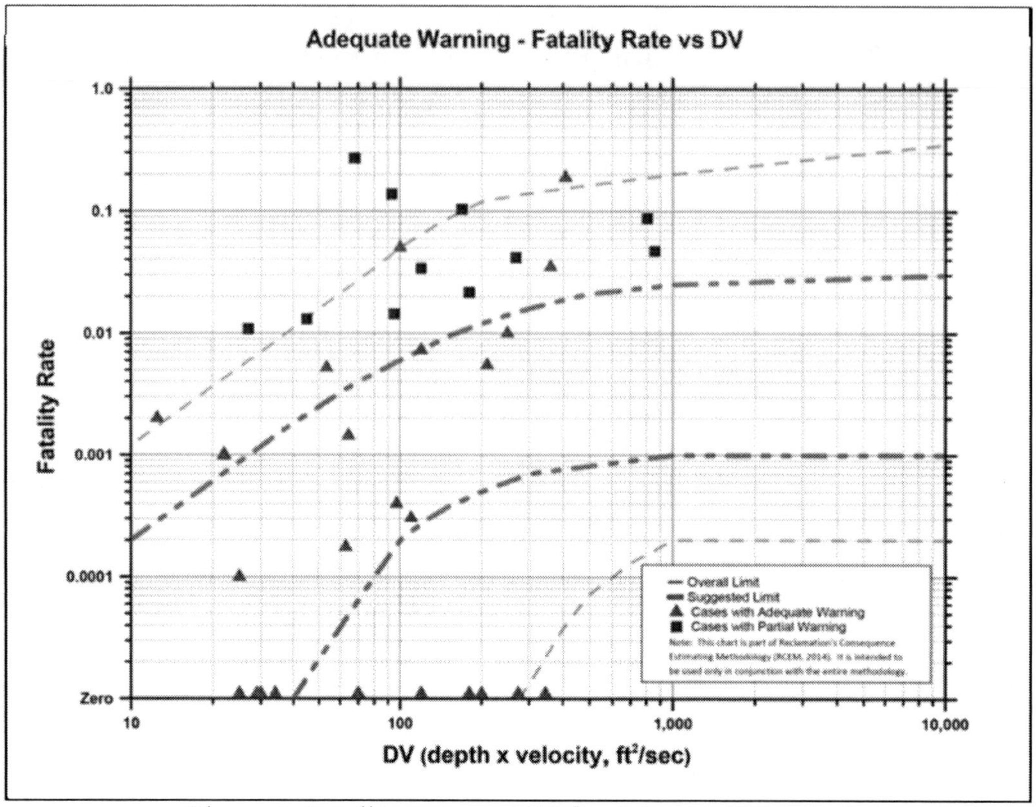

Figure 5. Fatality Rate vs. DV for Adequate Warning

SUMMARY

Quantitative risk analysis has proven to be a valuable tool to estimate the risks of potential failure modes at Reclamation dams and to provide a means of prioritizing additional actions within the inventory. In order to keep the risk analysis process from being overly based on numerical models and probability theory, Reclamation considers or utilizes empirical observations and data in as many steps of the analysis process as practicable. Tying probability estimates of future events to past observations and case histories helps risk teams determine whether estimates are realistic, grounds the estimates in an empirical rather than purely numerical basis, and thus improves the credibility of risk estimates.

REFERENCES

Bureau of Reclamation, "A Procedure for Estimating Loss of Life Caused by Dam Failure," Dam Safety Office Report DSO-99-06, Denver, Colorado, September 1999.

Bureau of Reclamation, "RCEM – Reclamation Consequence Estimating Methodology – Interim Guidelines for Estimating Life Loss for Dam Safety Risk Analysis," Denver, Colorado, February 2014.

Bureau of Reclamation and U.S. Army Corps of Engineers, "Best Practices in Dam and Levee Safety Risk Analysis," December 2012.

Engemoen, W. O., and Redlinger, C. G. (2009). "Internal erosion incidents at Bureau of Reclamation dams." Proc., 29th USSD Annual Meeting and Conference, United States Society on Dams, Nashville, TN, 731-745.

Engemoen, W. O. (2011). "Bureau of Reclamation experiences with internal erosion incidents." Proc., Internal Erosion in Embankment Dams and Their Foundations, European Working Group of ICOLD, Brno, Czech Republic, 11-18.

Foster, M. A., Spannagle, M. and Fell, R. (1998). "Report on the analysis of embankment dam incidents." *UNICIV Report No. R-374*, University of New South Wales, Sydney, Australia.

Norwegian Geotechnical Institute Publication 207, "Ralph Peck – Engineer – Educator – A Man of Judgment," Oslo, Norway, 2000.

TRANSITIONING TO RISK-INFORMED DAM AND LEVEE DESIGN STANDARDS — CORPS OF ENGINEERS STRATEGIES FOR MAJOR REVISIONS TO ITS DESIGN PHILOSOPHIES

Nathan J. Snorteland, M.S., P.E.[1]
David A. Margo, M.S., P.E.[2]
Jeffrey A. Schaefer, PhD, P.E, P.G.[3]
Mark W. Pabst, M.S., P.E.[4]
Scott E. Shewbridge, PhD, P.E., CEM[5]
Richard Varuso, PhD, P.E.[6]

ABSTRACT

The Corps of Engineers is currently revising many of its regulations and manuals related to dams and levees. As part of these revisions, the Corps has approached the changes to the existing standards by integrating the lessons it has learned over the last 10 years of integrating risk into its corporate decision-making. Over the next 5 years, the Corps will be explicitly considering failure modes, probability of failure or reliability, and consequences into many of the standards. This philosophical change will affect standards related to seismology, hydrology, geotechnical engineering, structural engineering, and hydraulic engineering. This paper will discuss the corporate strategy used to underpin all design standards, planned changes to some of the critical manuals, and some examples of how the Corps anticipates the manuals might be used.

INTRODUCTION

Engineering organizations, private consultants, and government agencies have been using regulations, manuals, and guidance published by the Corps of Engineers for nearly 75 years. The guidance currently published aggregates many of the lessons learned by the profession from their experience observing the performance of dams and levees worldwide. The approach taken by our predecessors, to pass that knowledge to future generations, has led to an improvement in the design and construction processes over the course of the last 75 years. Some fundamentals of guidance documents in the Corps have not changed since the 1940's. Despite this, the profession continues to learn. The small

[1] Director, Corps of Engineers Risk Management Center, 12596 West Bayaud Ave., Suite 400, Lakewood, CO 80228, nathan.snorteland@usace.army.mil.
[2] Lead Civil Engineer, Corps of Engineers Risk Management Center, 1000 Liberty Avenue, Pittsburgh, PA 15222, david.a.margo@usace.army.mil.
[3] Lead Civil Engineer, Corps of Engineers Risk Management Center, 600 Martin Luther King Jr Place, Louisville, KY 40202, jeffrey.a.schaefer@usace.army.
[4] Senior Geotechnical Engineer, Corps of Engineers Risk Management Center, 12596 West Bayaud Ave., Suite 400, Lakewood, CO 80228, mark.w.pabst@usace.army.mil.
[5] Senior Geotechnical Engineer, Corps of Engineers Risk Management Center, 12596 West Bayaud Ave., Suite 400, Lakewood, CO 80228, scott.e.shewbridge@usace.army.mil.
[6] Senior Program Manager, Corps of Engineers Risk Management Center, P. O. Box 60267, New Orleans, LA 70160, richard.j.varuso@usace.army.mil.

number of fundamental changes is a tribute to the robustness of the approaches taken during the development of those documents.

At the same time, the profession has experienced a number of structural failures including the levee system in New Orleans, the failure of Teton Dam, the failure of Taum Sauk Dam, and the failure of Swift No. 2 Dam.

Teton Dam Failure – Interface between structures and foundations not adequately understood or addressed.

Swift No. 2 Dam Failure – Design did not consider interface between embankment and foundation.

Taum Sauk Dam Failure – interaction between instrumentation and operations not adequately considered.

New Orleans I-Wall Failures – Original design criteria did not consider the failure mode that led to the eventual failure of these structures.

There are also structural failures that are not documented due to litigation that provide information that is valuable to the profession, but may be difficult to obtain. Many projects have also performed in ways that are less than ideal. Flood fighting and intervention activities on many levee systems have provided valuable insights into the performance of structures during floods. The profession learns much from the numerous structures that performed differently than expected.
And finally, there are natural processes that no guidance currently adequately addresses. This includes erosion through karst geology, flooding caused by several storms less than

the Probable Maximum Flood (PMF), and interactions between stiff and ductile materials during earthquakes (for instance spillways on earthen embankments).

PHILOSOPHY BEHIND THE REVISIONS

Given the large amount of information available, some changes to the approach taken in several critical areas are warranted. Revisions are routinely needed to keep up with advancements in the state of the practice which includes our understanding of system behavior and technological advancements for monitoring and analyzing system performance. Technology advances and obvious mistakes are to be addressed. Some approaches have proven to be incorrect or inefficient. The Corps is currently revising more than 50 guidance documents to address these items. In addition, there are five items that have been identified as fundamental improvements to the philosophies that underpin all guidance. Some of these are a direct result of the Corps moving towards a risk-informed decision process; some are as a result of lessons learned from managing a large dam and levee portfolio; while others are intended to increase consistency in approaches across governmental agencies.

The Corps is adopting some fundamental tenets for inclusion into all relevant dam and levee documents. Each document will include a requirement to:

- Analyze potential failure modes;
- Evaluate robustness, redundancy, and resilience;
- Incorporate risk or reliability concepts;
- Explicitly address monitoring and intervention during operation; and
- Examine potential intersections with other guidance documents.

The Corps is challenged to keep a connection with the provenance that exists in the current guidance. In many respects, these manuals have guided the construction of a robust nation-wide system that has performed well so far – with some notable exceptions. The Corps does not want to abandon the positive aspects of the existing guidance. However, it must be recognized that improvements are needed to address lessons learned from those instances where structures did not perform as intended even when designed in accordance with that guidance.

<u>Potential Failure Modes</u>

One of the most significant advancements in the dam and levee industries over the last 20 years has been the use of Potential Failure Mode Analysis (PFMA). This approach has led to more careful consideration of the weaknesses of engineered structures and their interaction with the natural environment. Each manual will include a procedure of how to use PFMA during the design process. The objective of this effort is to ensure potentially unique characteristics of the structure are still evaluated given that the guidance cannot explicitly address all possible scenarios.

Evaluating Robustness, Redundancy, and Resilience

Commonly referred to as the "3 R's", each manual will discuss robustness, redundancy, and resilience at least at the fundamental level. They may be either requirements or considerations. The objective is to ensure designs consider how a structure might fail. The goal is to design structures that perform well over a large range of loading and conditions, prevent fragile or brittle failures, and increase the reliability of structures by providing primary and backup features to prevent failure modes from developing.

Incorporate Risk and Reliability

Currently, most Corps design manuals inconsistently, if at all. Many of these manuals are used to support risk assessments, so explicitly considering risk and reliability will improve those processes. One of the weaknesses in the existing procedures is that the variability of input parameters is not addressed with the guidance. Designers used what was assumed to be conservative assumptions, but that conservatism was not as conservative as assumed because the natural variability is so high. Conversely, simply following guidance and choosing a combination of conservative assumptions has led to designs that are less efficient than optimal. Several risk and reliability methods are available to compensate for this issue, and each manual will address this in a slightly different way, but incorporating these concepts into the design process will improve our ability to develop sound designs.

Explicitly Address Monitoring and Intervention

Designers strive to achieve economical designs that perform as intended when loaded. Current guidance typically addresses many of the uncertainties and unknowns by increasing the robustness of the design. Despite this approach, structures still occasionally exhibit signs of distress when loaded such as settlement, movement, or seepage. Each year, the Corps "flood fights" a variety of structures in its inventory based on observed performance. Not all of this intervention is done because failure is imminent – it is part of sound practice and may even be part of the operational expectations from design, such as for truncated seepage berms. Some of these behaviors may be benign while others may indicate a developing failure mode. In either case, intervention is part of sound practice even though it has not often been explicitly considered in the design phase of a project. Where applicable, each guidance document will discuss monitoring and inspection requirements, expected performance, potential distress signals, and intervention strategies. In some cases more robust designs will be justified and in other cases an intervention strategy will be adopted. This new paradigm will allow the designer to explicitly consider intervention as a strategy for managing risk when it is economical and effective to do so.

Examining the Intersection Between Different Guidance or Between Separate Structures

One weakness that has been observed in the existing guidance involves the interface between disciplines because most Corps manuals are currently organized by discipline (e.g. geotechnical, structural, hydraulic). Examples of this include items like spillway drains which are included in structural guidance to reduce uplift under slabs. This can be problematic because the geotechnical aspects of these drains can easily be overlooked resulting in the introduction of erosion failure modes into critical areas. Another example is the Hurricane Katrina disaster where a soil-structure interaction problem contributed to several I-Wall failures. The Corps has since issued several guidance documents, including ETL 1110-2-575, Evaluation of I-Walls, to account for failure modes involving sheet pile walls and soil. The Corps recognizes that different project features may be designed by different disciplines with different objectives. The intent is to explicitly identify those areas that require a multi discipline perspective and highlight where additional consideration is warranted.

GUIDANCE RECENTLY UPDATED

Drilling in Embankment Dams – ER 1110-2-1807

This document was significantly revised in 2014. The primary purpose of this document is to prevent damage to embankments and their foundations from hydraulic fracturing, erosion, filter/drain contamination, heave, or other mechanisms during drilling operations, sampling, in-situ testing, grouting, instrumentation installation, borehole completion, and borehole abandonment. The Corps has hydraulically fractured several structures, and has included guidance to avoid doing so in the future. The potential for erosion and clogging of drainage features in a dam are discussed. Guidance is provided on drilling procedures to minimize the risk of damaging embankments along with requirements for borehole completion. The regulation specifies minimum experience requirements for drill rig operators and drilling program oversight personnel. The regulation requires Districts to develop a drilling plan that includes the objectives and justifications of the drilling program, lists the exploration team and their experience, summarizes the review of existing relevant information, presents essential geologic and engineering drawings, summarizes the drilling scope and methodology, and a risk evaluation for the potential risk of hydraulic fracture, erosion, contamination of drainage features, heave, or any other damage. In general, all drilling and investigation should be targeted to obtain information related to a specific failure mode identified from a Potential Failure Mode Analysis (PFMA). For dams, the justification for drilling must include an approved recommendation from a risk assessment performed in support of the Dam Safety risk management process described in ER 1110-2-1156 *Safety of Dams - Policy and Procedures*. If the structure has not had a PFMA, a thorough evaluation similar to the PFMA process must be performed and presented in the drilling program plan to show that the drilling is justified.

Policy and Procedural Guidance for Processing Requests to Alter U.S. Army Corps of Engineers Civil Works Projects Pursuant to 33 USC 408 (modifying levees or adding hydropower to Corps projects) – EC 1165-2-216

EC 1165-2-216 was significantly revised in 2014. The overall objective was to create a policy to improve consistency, flexibility, decision-level clarity, federal action clarity, and coordination. The purposes of the changes were to:

- Improve consistency in the way USACE considers, processes, and documents decisions for requests for alterations to Civil Works projects;
- Create a process that is applicable to all types of Civil Works projects;
- Be transparent on what information is required; and
- Create a process that can be tailored by districts to the appropriate scope, scale, and complexity of a proposed alteration.

Consistency is improved by including all types of Civil Works projects and establishing step-by-step procedures and decision-making criteria. Flexibility is improved by allowing for scalability based on scope and level of complexity for applications. Decision-level clarity is improved by providing criteria to determine when a Section 408 request requires HQUSACE review and decision. Federal action clarity is improved by clarifying environmental compliance requirements. Coordination is improved by encouraging coordination throughout process, requiring that requests be endorsed by the non-federal sponsors to ensure concurrence, and establishing a Section 408 coordinator in each District.

Where a project or a proposed modification poses a significant threat to human life, the permit application will be reviewed by the Risk Management Center, HQUSACE, and a Type II Independent External Peer Review panel. The permit will be reviewed to ensure that the proposed work does not increase risk to the protected area or to the structure itself.

EM 1110-2-3506, Grouting Technology

The update of EM 1110-2-3506 was revised in 2014. The manual provides detailed and comprehensive information on state-of-the-art technology, practices, procedures, materials, and equipment for use in planning and executing a grouting project. The manual is primarily focused on dams, levees, foundations, and appurtenant structures. Different grouting applications are discussed, including preconstruction foundation treatment and post-construction remedial measures. Methods of grouting that have proven to be effective are described, and various types of grout, their proportions, and their best-suited uses are addressed. The manual discusses grouts composed primarily of cementitious suspensions and additives, although other types of grout are mentioned.

ETL 1110-2-575, Evaluation of I-Walls

ETL 1110-2-575 was published in 2011 as a result of the failure of I-walls during Hurricane Katrina. The Corps no longer advocates constructing I-walls for new projects.

The manual provides updated technical criteria and guidance for evaluation of existing I-walls. General guidance for performing I-wall evaluation is provided along with detailed updates to existing guidance that focus on three I-wall performance issues:

- The flood-side gap that was discovered at I-walls in New Orleans after Hurricane Katrina;
- Rapid-flood-loading undrained strength considerations for under-, normally- and slightly overconsolidated fine-grained soils;
- The rotational stability failure mode was found to be critical for most I-walls evaluated;
- And criteria for consideration of deflections.

The manual provides guidance for evaluating the critical failure mode consisting of additional water pressure introduced when the wall deflects creating an opening on the water side of the I-wall.

GUIDANCE BEING UPDATED

Levee Design Manual – ER 1110-2-1913

The Levee Design Manual is being revised in coordination with the Corps' development of the Safety of Levees – Policy and Procedures guidance document. This will be a significant revision to accomplish the following objectives:

- Unify design practices that vary across USACE;
- Incorporate failure modes evaluation into design, analysis, and construction;
- Enhance evaluation of subsurface conditions;
- Include a discussion of construction practices;
- Integrate operations, maintenance, and flood fighting into the guidance; and
- The design process will be iterative starting with standards-based design, which will then be evaluated with risk and reliability methods to achieve life safety objectives.

The most significant changes to this document will be how seepage and internal erosion are characterized and how stability is evaluated. The document will also address the issue of levees that protect large population centers or other critical infrastructure. The Corps is planning to integrate design practices with the Corps life safety-based decision processes. An example of this is shown in figure below.

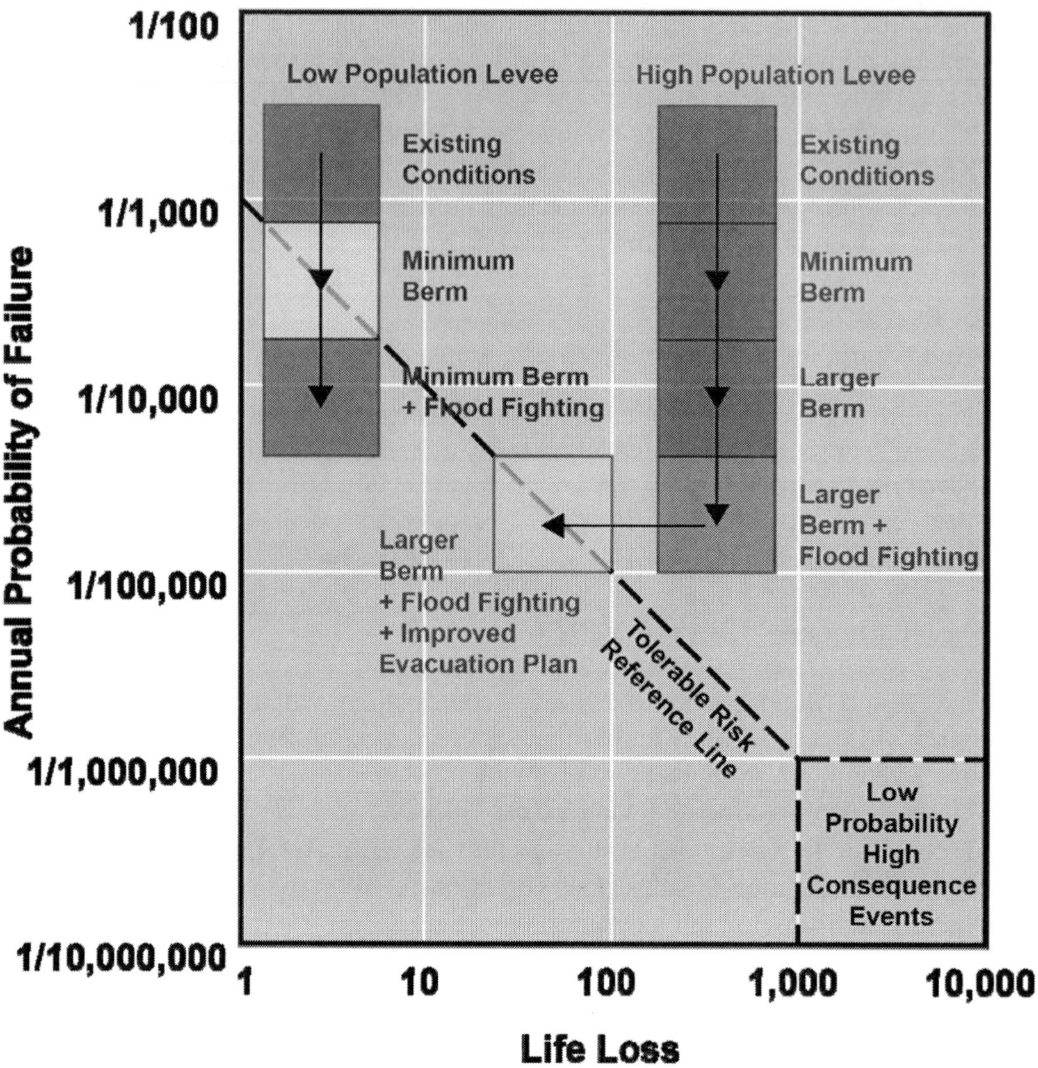

The Corps is planning to release this document for public comment and hold industry workshops in 2015.

<u>**Seepage Analysis Control for Dams – EM 1110-2-1901**</u>

The current manual follows a traditional approach to evaluating safety with respect to seepage, changes in head, gradients, and seepage quantities. Analysis methods are presented for seepage conditions typically encountered in embankment dams such as high exit gradients downstream of a dam or high hydrostatic pressures beneath confining layers in the foundation. Currently, factor of safety calculations are provided for these conditions and if a threshold criteria is not met the structure is considered to not satisfy safety criteria. This is transitioning to an approach that considers performance and includes considerations of how these conditions might lead to failure of the structure. In practice, users have trended towards assuming "worst case" conditions, such as the greatest reservoir loading, poorest soil conditions, simplifying geologic interpretation, etc. This has resulted in actions that may be overly conservative – but also in situations where combinations of conditions are unconservative.

The revised manual will focus on methods and procedures intended to be used in a potential failure modes approach coupled with a probabilistic assessment of each failure mode, a process which is detailed in the USBR/USACE *Best Practices in Dam and Levee Risk Analysis*[7]. Site specific failures modes are proposed for common internal erosion processes associated with embankment dams. While many failure modes may be considered, only those that result in failure, as defined by breach of the embankment and loss of the reservoir, are developed.

Development of these credible failure modes typically consists of defining an event tree that represents the internal erosion process where the failure mode initiates and progresses until loss of reservoir occurs. As part of that evaluation, the 'factor of safety' calculations described above are used as one consideration among many to assess the probability of failure, as opposed to a sole criterion. Thus even if a particular design standard is not met, such as uplift on a thin blanket downstream of an embankment, if the other events on the event tree indicate that failure is unlikely, such as high creep ratio or low likelihood of initiation of backwards erosion due to low horizontal gradients, the structure may still be seen to have relatively high reliability.

The probability of failure and associated risk (evaluated in accordance with USBR/USACE Best Practices) are then compared to tolerable risk guidelines to determine if a given failure modes exceeds acceptable risk levels defined in ER 1110-2-1156 *Safety of Dams – Policy and Procedures*. When risks are found to be unacceptable, a range of alternatives is examined and the level of risk reduction of each is estimated and evaluated. Risk reduction estimates are made using the same methodology that was used in the initial – or existing condition – risk assessment. Alternatives are evaluated by the amount of risk reduction they provide and consideration is required as to the desired reduction level as described in ER 1110-2-1156.

Levee System Evaluations – (Accreditation for the National Flood Insurance Program)

Both the U.S. Army Corps of Engineers Levee Safety Program and the Federal Emergency Management Agency (FEMA) National Flood Insurance Program (NFIP) are concerned with life safety, reduction of property damage due to floods and communicating flood hazards and risks. It has been recognized by both organizations that it would be beneficial to have a common set of standards related to the evaluation of levee systems that are more risk-informed and communicate these concepts to local sponsors. The goals of both agencies in this regard has been summarized in the Flood Protection Structure Accreditation Task Force report, prepared with input from both agencies and dated November 2013.

---

[7] Best Practices in Dam and Levee Risk Analysis, http://www.usbr.gov/ssle/damsafety/Risk/methodology.html

USACE is in the process of updating its Engineering Circular (EC) 1110-2-6067, Levee System Evaluations for the National Flood Insurance Program (NFIP), which expired in 2012, to provide guidance on utilizing risk assessment information associated with levee systems when making NFIP accreditation recommendations to FEMA. This updated document will consider concepts such as:

1. Adopt a risk-informed framework for levee evaluations for FEMA's NFIP;
2. Require flood warning, preparedness, and evacuation plans as accreditation criteria;
3. Require that the scenario for an overtopping event and the associated risk reduction measures to mitigate for such event be analyzed and included in the accreditation package; and/or
4. Strengthen floodplain management measures for leveed areas

The USACE plans to adopt a risk informed approach for levee evaluation decisions. The initial step for implementing this modification will be to specifically include a probabilistic assessment of levee performance. The NFIP regulations defining accreditation requirements, 44 CFR 65.10, currently require detailed deterministic engineering analysis. This intent of these requirements can be satisfied using a risk assessment which integrates the analytical methods of traditional engineering, the critical thinking of risk analysis, and professional judgment of engineers, review boards, and decision makers. The Levee System Evaluation regulation will focus on how to develop and use the results of a risk assessment to support levee evaluation decisions for NFIP purposes. The information available from a risk assessment will also help communities and sponsors target investments and risk reduction activities where they are most needed.

Levee Safety Program – Policy and Procedures – EC 1110-2-6072

The Corps is finalizing a companion document to ER 1110-2-1156 (Safety of Dams – Policy and Procedures) dedicated to levee safety. This document prescribes the guiding principles, policy, organization, responsibilities, and procedures for implementation of the Corps Levee Safety Program. The Levee Safety Program is transitioning to managing the portfolio of levees using a risk-informed process with this document as the foundation. The program scope includes inspections and assessments leading to recommendations to ensure levee systems will provide benefit to the Nation by managing risk associated with the USACE levee safety portfolio. The portfolio of levee systems covered by this document includes those that are: 1) USACE operated and maintained; 2) Federally authorized - local sponsor operated and maintained; and 3) Non-Federal levee systems in the Rehabilitation Program (RP). The goal of the levee safety program guidance is to ensure that all levee systems are designed, constructed, operated, maintained, inspected, and rehabilitated to achieve tolerable risk while meeting essential USACE guidelines. The Corps expects to release the guidance for public comment in 2015.

Hydrology and Hydraulics Guidance

Historically, hydrology and hydraulic guidance focused on establishing a single design event with its associated design parameters (e.g. a maximum water surface elevation for a levee or a maximum discharge for a spillway). In 2011, dams and levees throughout the Mississippi River Basin were significantly challenged by flooding which was less than the design magnitude but had a duration that was many months longer than anticipated during design. To support the use of potential failure modes during the design process, the focus of hydrology and hydraulic guidance is shifting to a more comprehensive look at a suite of plausible flood loading scenarios with several combinations of magnitude and duration. For example, explicit consideration for plausible combinations of loading scenarios will be included in an update to EM 1110-2-1619, Risk Assessment for Flood Risk Management Studies.

Risk and reliability concepts are being included in a revision to ER 1110-8-2, Inflow Design Floods for Dams and Reservoirs. Design floods for dams are obtained from a combination of reasonable worst-case hydrologic events (e.g. PMP) and parameters (e.g. wet antecedent conditions). Numerous decisions and assumptions are made by the analyst throughout the process without much consideration given to the plausibility of various combinations. Questions such as "Can a record snowpack precede an extreme rainfall?" or "Can a wind direction along the maximum fetch occur during the weather pattern necessary to produce an extreme rainfall?" seldom get asked. Updates to the guidance will aim to provide more transparency to the process by explicitly considering uncertainties in the estimate and the sensitivities associated with key parameters and assumptions. The goal is to develop a combination of parameters for the design flood that is both plausible and congruent.

Interim risk reduction measures (IRRM) have been adopted by USACE to manage existing risks over the near term while remedial actions are being formulated, designed, and constructed over the long term. One of the more common IRRM involves reducing dam safety risk by reducing the water level in a reservoir through a modification to the operating plan. A reduction in the water level may reduce the chance of breach by reducing the load, may reduce the potential consequences if a breach should occur, and may provide additional time and opportunities for intervention. Even though USACE does have existing policies for evaluating and approving these water control plan modifications, there are some unique challenges with IRRM because they are usually of larger magnitude, longer duration, and bigger impact than most typical modifications. ER 1110-2-240, Water Control Management, is currently being revised to clarify requirements and implement a risk informed approach to evaluate the benefits and impacts of changes to a water control plan as part of an IRRM.

A location for levee system overtopping can be selected by provided an increment of height along portions of a levee. This increment of height, called superiority, can provide resiliency by establishing an overtopping location that minimizes residual risk based on the combination of potential consequences and anticipated overtopping performance. Revisions to ETL 1110-2-299, Overtopping of Flood Control Levees and Floodwalls, will include procedures for estimating the increment of superiority required to achieve a predictable overtopping location under a variety of flood loading conditions. Guidance will also be developed to make a risk informed decision on the need and justification for other resiliency measures which could include providing erosion protection at the overtopping location. The goal of these measures could range from preventing a breach in a high consequence leveed area to slowing the progression to a breach in a leveed area where the additional time would provide adequate opportunity for evacuation. These resiliency decisions will be primarily driven by life safety considerations because resiliency measures typically cannot be justified economically as the added costs are significantly greater than the reduction in derived benefits. The guidance update will also need to address the fact that the overtopping design section (i.e. the low spot in the levee) is usually the first location where sand bags are placed during a flood to prevent overtopping.

Hydrology and hydraulic guidance defines the water loadings that are used for the structural and geotechnical design and analysis of dams and levees. Current guidance that defines these design loadings is inconsistent, occasionally contradictory, and at times confusing. Load cases for the other Corps manuals are defined using either a probabilistic approach (e.g. EM 1110-2-2100, Stability Analysis of Concrete Structures) or a deterministic approach (e.g. EM 1110-2-1913, Levee Design Manual). Load cases can also be defined based on a best or median estimate (EM 1110-2-2100, Stability Analysis of Concrete Structures) or an estimate with a specified level of assurance or confidence (EC 1110-2-6067, USACE Process for the National Flood Insurance Program Levee System Evaluation). The intent is to improve consistency in the definition of hydraulic design loads across the various guidance documents to the extent practicable while recognizing that differing objectives may occasionally result in differing load definitions for some documents.

Seepage Control Barriers for Embankments

The Corps is developing an Engineering Manual to address designing foundation treatments using vertical cutoff-type barriers for dams and levees – primarily cutoff walls. It does not discuss horizontal seepage barriers or other upstream sealing alternatives. It replaces elements of Chapter 9 of the existing EM 1110-2-1901, Seepage Analysis Control for Dams. It will discuss dam safety issues, geologic consideration, engineering considerations, and types of construction. Case histories for each type of wall will be included. A comprehensive summary table of projects for each type of cutoff wall has been compiled as well as a detailed list of terminology.

THE FUTURE

The dam and levee safety industries are transitioning to risk management as the principle that underpins credible and consistent decision-making. Combined with integrating lessons learned from incidents and failures, that transition is already causing risk and reliability concepts to be included in guidance documents. This will likely continue as potential failure modes and risk concepts become part of standard practices across the profession.

Existing guidance was written with lessons learned from past incidents and failures in mind. The transition to failure mode analysis and assuring designs explicitly address failure modes is a natural progression as the science, technology, and our understanding of nature advance and will continue into the future. The use of probabilistic methods to address the variability in design parameters is being promoted, while using deterministic methodology to supplement the probabilistic approach.

By contrast, the transition to a fully risk-based design philosophy has some limits, as the drawbacks for having criteria that become too flexible undermine the existence of guidance itself. It's not clear that a complete transition to solely risk-based design is possible and it's less clear whether that transition is even desirable. The transition to a fully risk-based design philosophy has some challenges, with one of the perceived drawbacks being the potential to undermine minimum essential guidelines resulting in acceptance of substandard dams and levees in areas with relatively low consequences and low risk. There are concepts embedded in tolerable risk that can support design decisions under various decision scenarios. For instance, more attention to detail and more robust designs can be justified when societal risks (both probability of failure and associated consequences) are high. Individual risk (typically probability of failure) concepts can also be used to ensure that all levees achieve minimum essential performance objectives. The cost effectiveness of risk reduction measures will be a consideration in all cases to achieve a balance between reducing risk and being prudent stewards of taxpayer dollars.

In the future, Corps guidance will be structured to improve the balance between the prescriptive content needed for consistency and the flexible content needed to support a risk informed decision process. The likely path forward appears to be risk concepts (i.e. PFMA, resiliency, risk analysis, intervention, and multi discipline perspective) informed by minimum design standards with an allowance for more robust designs where needed to ensure risks are reduced as low as practicable.

EVALUATION TOOL D-SMART FOR RISK-BASED DAM SAFETY MANAGEMENT

Jeong-Yeul Lim, PhD[1]
Nam-Ryong Kim, PhD[2]

ABSTRACT

The purpose of this study is to develop a dam safety management assessment tool based on risk assessment for the dam safety management paradigm shift. As the purpose of the dam safety management assessment tool is to develop the next generation dam safety management decision-making support system. Traditionally dam is considering non-risk structure, however recently natural disasters and the risk components are increasing. So, there has been increased interest for the dam vulnerability. Accordingly, the risk components of the dam are carried out systematically organized and analyzed. Directly on structures major load, such as floods and earthquakes, the probability of failure and the system response probability for the theoretical and analytical method are described. Risk analysis and assessment is to develop for the risk analysis model based potential failure mode can occur in the structure. The development of the model, the probability value of the load is applied to initial input parameter. Risk analysis and assessment is carried out using system response probability of the structure. Therefore in this study is to develop hydrological and geotechnical analysis module and risk analysis and assessment tool for the development of the dam safety management decision-making support tool D-SMART. Analysis module is composed of two parts. One is the initial load probability result part to the hydrological and an earthquake characteristic, the other is system response probability result part. Risk analysis and assessment tool is composed of the based event tree analysis method. In this year, dam risk analysis and assessment tool is developed D-SMART program, including analysis hydrological module, geotechnical module and consequence module. D-SMART is program of risk analysis and assessment utilizes technique for dam safety evaluation. In the future, Dam risk assessment carried out step by step in K-water plans and case study S dam.

INTRODUCTION

Severe disasters are increasing in the world due to global climate change. For example, probable maximum flood (PMF) level has been increased due to heavier rainfall. Strong earthquake is another major disaster in the world and it induces many casualties and economic loss. 2011 Tohoku earthquake in east Japan and its tsunami caused more than 24,000 casualties and its effect in the economy is uncountable. When major social infrastructure, such as large dams, is subjected to these kinds of natural disasters, the main factors threatening the safety of the structures are the consequences of climate change, aging, and/or deterioration.

[1] K-water Institute, 125 Yuseongdaero 1689beon-gil, Yuseong, Daejeon, Korea, passion@kwater.or.kr
[2] K-water Institute, 125 Yuseongdaero 1689beon-gil, Yuseong, Daejeon, Korea, namryong@kwater.or.kr

The state-of-the-art concept of safety management for dams and facility structures is changing from evaluation of its safety to evaluation of vulnerability and risk management. In countries such as the United States, Australia and the Netherlands, risk analysis and assessment is adopted as new safety management paradigm to assess the structural safety in terms of engineering or economic aspects. It is also a good issue which is actively discussed in major experts groups such as International Commission on Large Dams (ICOLD) or other international societies and the cases of risk analysis and assessment around the world are frequently adopted. Awareness of risk analysis and assessment in dam safety is also growing in Korea and systematic tools for risk based assessment are now under development for advanced safety management of dams.

This paper explains the development of risk analysis, assessment and management system for dams as a result of research for advanced dam safety management program. It is utilized not only for evaluation of dam safety but for management under new climate change. Introduction of risk analysis and assessment in dam safety management is necessary and tools for risk analysis specific to Korean being developed. A new risk assessment system for dams has been developed to integrate the most important inputs and analysis resulting from an international joint research program. It is used were to demonstrate applicability of this research.

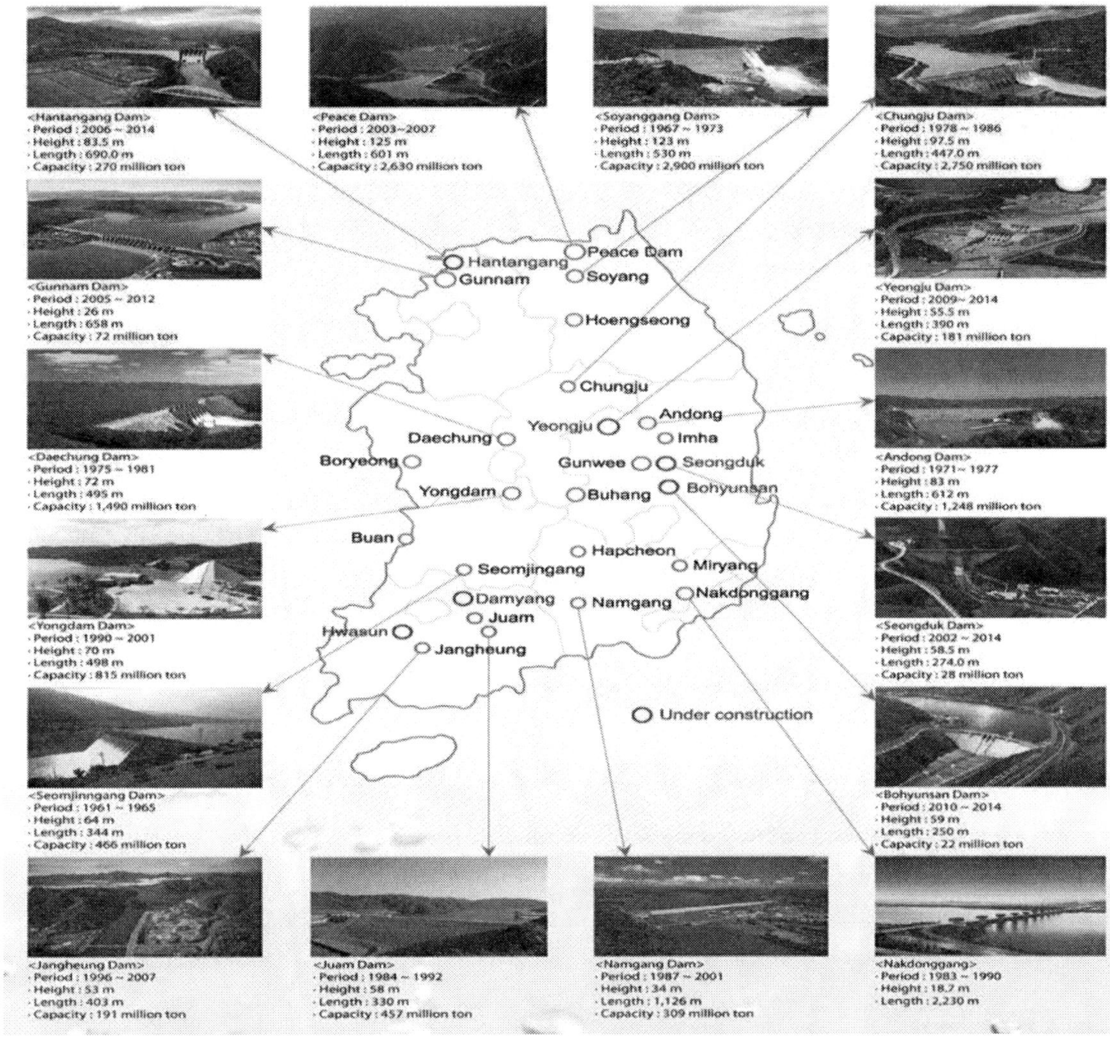

Figure 1. Korea Water Resources Corporation (K-water) Portfolio of Multi-Purpose Dams.

RISK ANALYSIS AND ASSESSMENT

Risk in dam safety can be defined as the level of danger related to or caused by a dam. Figure 2 shows examples of damage or failures. These are also good examples of the aspects which were mainly dealt with in risk analysis and assessment. The items for risk analysis and assessment include not only simple structural safety but also non-structural conditions, so the risk is defined by an entire dam and reservoir system.

The risk management of dams does not define the mode of failure from a singular condition, but it is to analyze and to evaluate failure scenario and consequent level of damages using a probability based approach assessing factors affecting the performance of dams. The key factors in risk analysis are to determine the major type of disaster, the vulnerable areas, and potential failure modes. A single major factor was simply considered in risk analysis of dam structure in the past, however a combination of various possible disaster scenarios are utilized in the state-of-the-art risk analysis method. Risk analysis is related with both awareness and forecast of risk.

Figure 2. Dam Failure due to Flood and Earthquake

Embankment dams typically cannot withstand any significant amount of overtopping, due to limited erosion resistance of the soil material used in their construction. Concrete dams are generally perceived to be more resistant to overtopping failure, due to the durability of the dam itself as well as the erosion resistance provided by a rock foundation. However, weak and fractured rock may be susceptible to significant erosion during overtopping flows, and if foundation support is lost due to overtopping erosion, the dam could be lost.

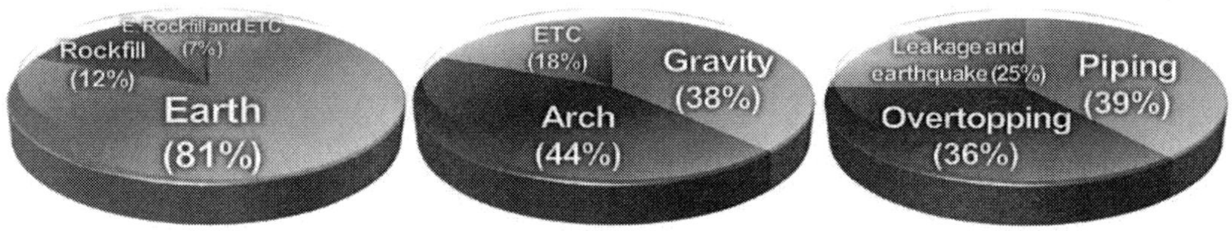

(a) (b) (c)

Figure 3. Investigation of Main Cause of Failure (a:fill type, b:concrete type, c:main cause)

For much of the last decade, USBR has used a standard event tree and methodology to estimate risks of internal erosion quantitatively. Recently, a new methodology reference document has been developed for assessing internal erosion risks (Fell et al, 2008). Both methodologies use similar event trees for formulating risk estimates.

The methodology of risk analysis can be classified as qualitative and quantitative approaches and their characteristics are listed in Table 1. These approaches must be utilized appropriately considering their conditions and objectives, and they are applied together since they are closely related or cannot applied alone in most cases.

Table 1. Methodology of Risk Analysis

Method	Characteristic	Objective
Qualitative Approach	Base of Empirical (FMEA, FMECA)	Risk Factor and Consequence Estimation
Quantitative Approach	Monte Carlo Simulation (ETA, FTA, PRA)	Numerical Analysis Method of Modelling

Figure 4 introduces the methodologies for risk assessment for existing dams. It is similar to Table 1, however, it provides more detail about qualitative and quantitative approaches and their use and objectives.

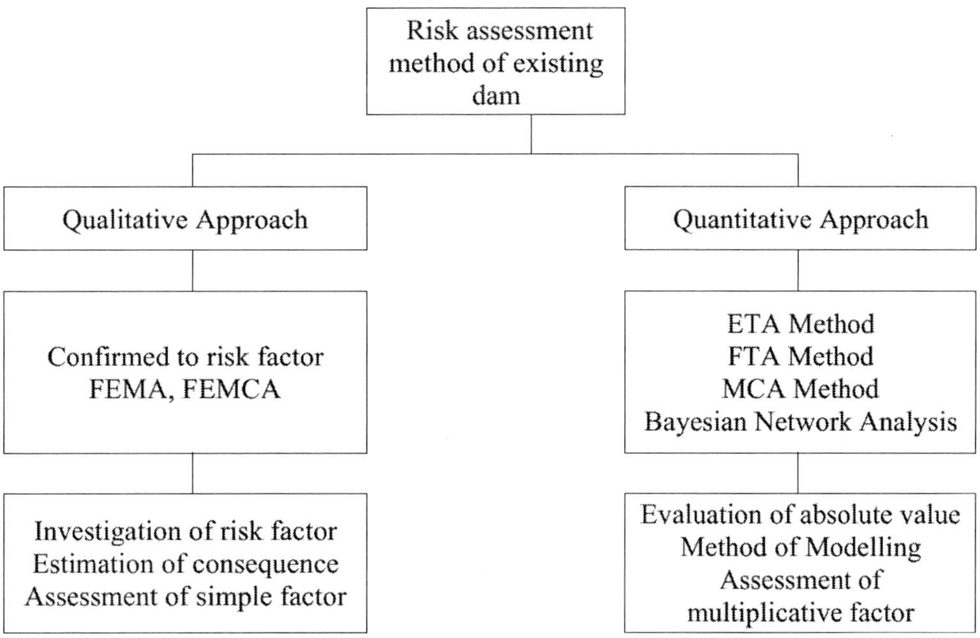

Figure 4. Risk Assessment Method of Existing Dam

Figure 5 summarizes risk analysis qualitative traditional diagnostic safety evaluation of an existing dam. The concept of the traditional routine safety evaluation side is explained by the scopes, objectives, and result of analysis. The items included in risk analysis and assessment is listed as the objectives of approach and analysis, and the results. So the difference between these two methodologies is below can be compared.

- Diagnostic safety evaluation
  - Individual safety assessment
  - Safety under each condition
  - Evaluation restricted to dam structure itself
  - Difficult to estimate the consequences of failure

Evaluation Tool D-SMART

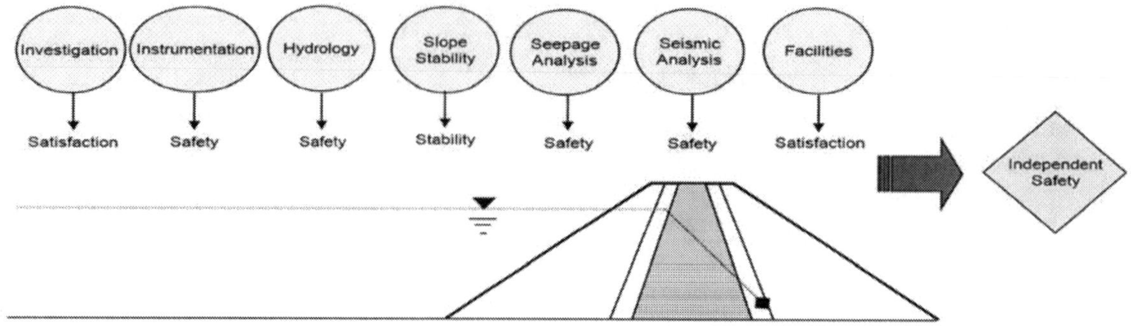

- Dam risk analysis and assessment
  - Safety evaluation considering multiple factors
  - Safety assessment considering correlations between each evaluation items
  - Evaluating entire system related with the target dam
  - Provide detailed consequences of dam failure or disaster
  - Directly supports decisions for dam safety countermeasures

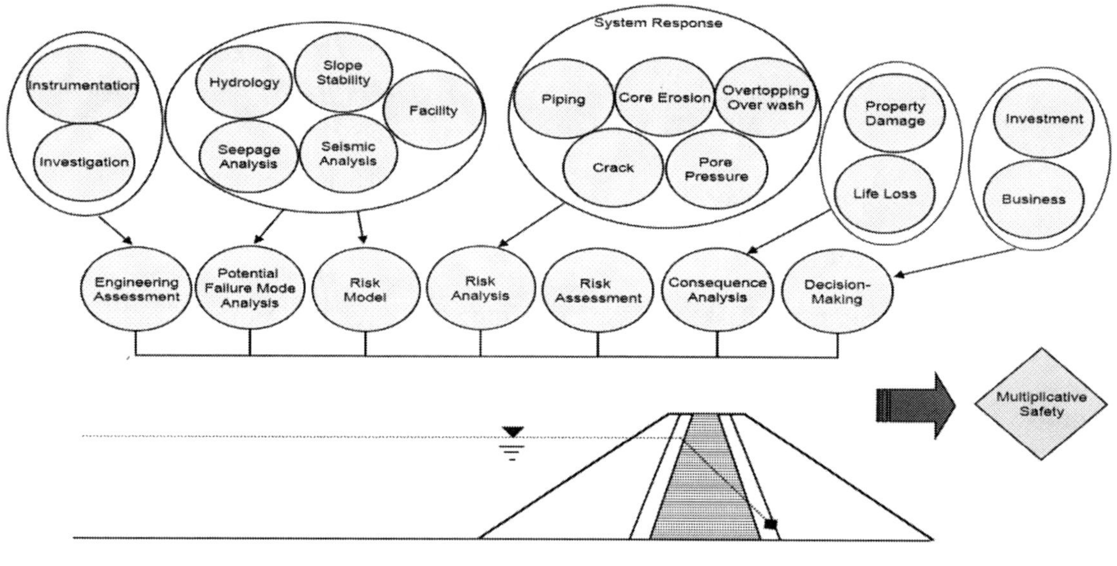

Figure 5. Comparison with between Diagnosis and Risk Analysis

Figure 6 shows the general process of risk analysis and assessment of dam safety. Generally, risk assessment is in two steps. One is the risk definition and to find the failure modes, the other is the development of the risk model and risk assessment. Steps of the risk assessment consist of 7 phases which can be adjusted. And then recommended for the risk reduction plan.

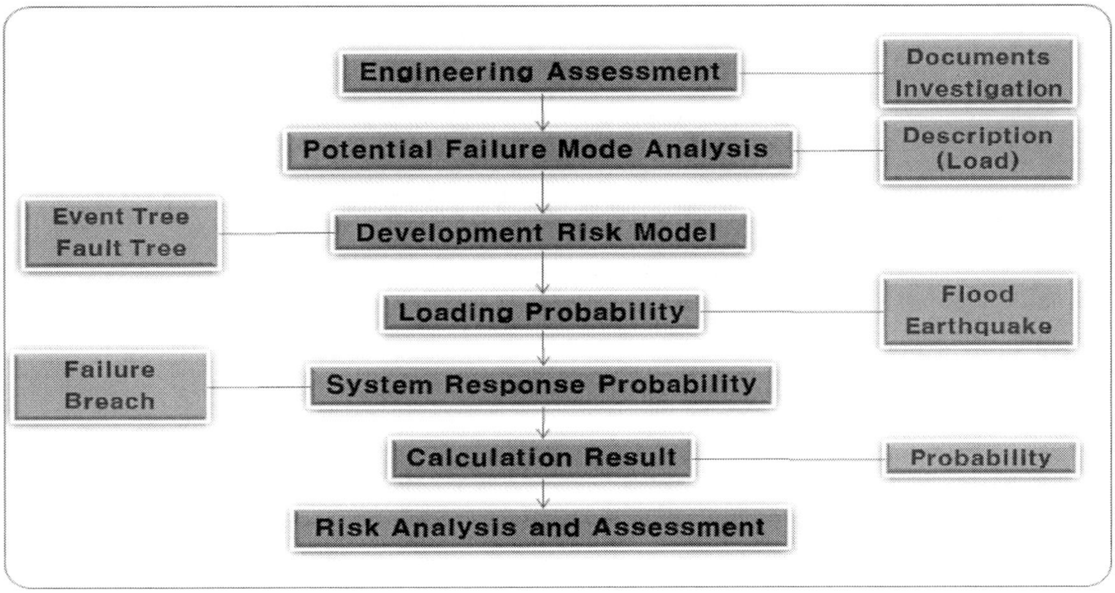

Figure 6. Process of Risk Assessment

DEVELOPMENT OF D-SMART

Concept

D-SMART uses "event trees" as the basic tool for estimating dam safety risks. Each identified potential failure mode is "decomposed" into a sequence of component events and conditions that all must occur for the failure to develop. The event tree depicts the sequence or progression of events and conditions in a branching fashion, with one or more paths leading to failure of the dam and the rest leading to no failure. Event trees consist of a series of linked nodes and branches. Each node represents an uncertain event (a crack forms in the embankment) or condition (existence of adversely oriented joint planes). Each branch represents one possible outcome of the event or condition, and it is assigned a probability. The event tree is constructed from left to right, usually starting with an initiating event (earthquake, flood, filling the reservoir for another water year) and proceeding through potential failure mode development to a dam breach and its consequences. Branch probabilities at each node must sum exactly to 1.0. They are estimated by the judgment of knowledgeable people, guided when possible by the numerical modeling, statistical analysis, case histories, and theory. In most risk analyses related to dam safety, there are certain questions that cannot be addressed except by judgment and degree-of-belief probabilities. When estimating likelihood for each branch in an event tree, it is assumed the even represented by the previous branch has occurred (regardless of it likelihood).

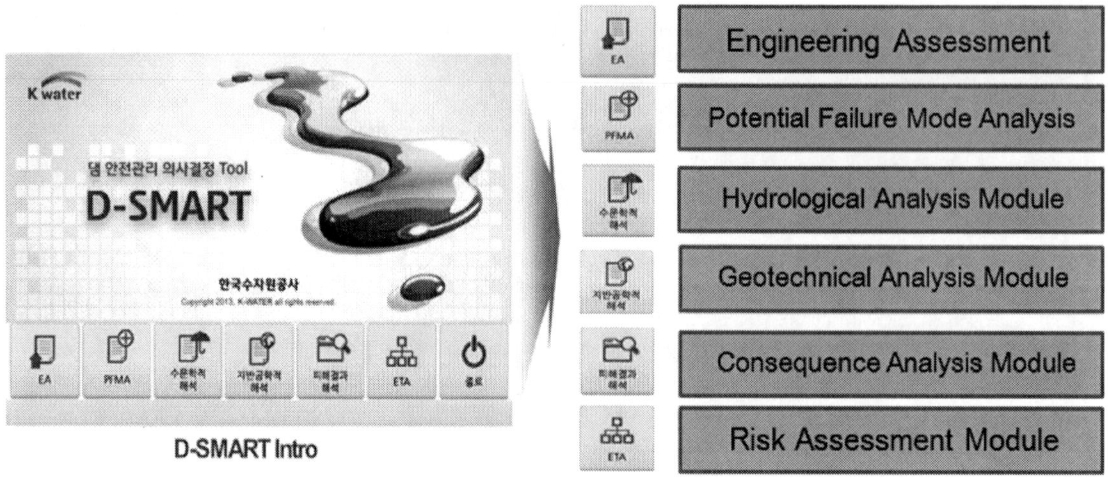

Figure 7. D-SMART Main

Figure 7 shows the Evaluation Tool D-SMART for dam safety management. D-SMART have 6 sub-module. Each module is described of as follows. Engineering assessment (EA), Potential failure mode analysis (PFMA), Hydrological analysis module, Geotechnical analysis module, Consequence analysis module and Risk assessment module.

Engineering Assessment

Engineering Assessment is consist of 3 items and 84 subsection. 3 items are flood, earthquake and normal operation. Figure 8 shows engineering assessment windows. Engineering assessments indicate whether individual dams are expected to meet current good dam safety practice in Korea. This approach, was developed by RAC. The assessment factors are grouped by flood, earthquake and flood-internal types of initiating events or loading, and by various dam systems and subsystems. An Engineering assessment rating is assigned to each assessment factor.

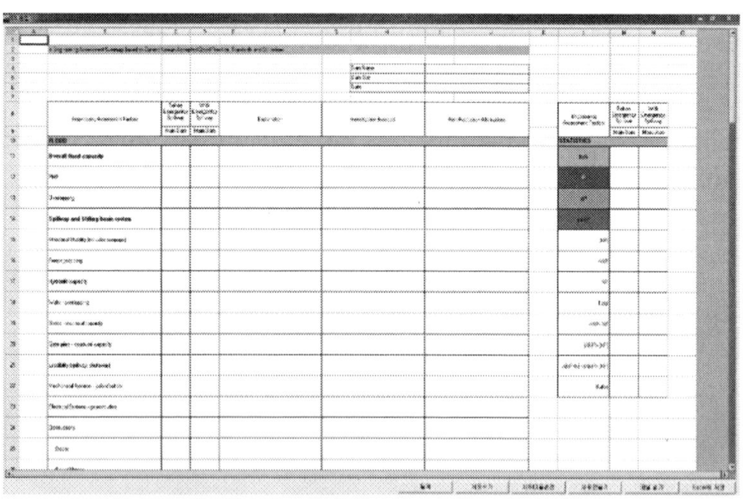

Figure 8. Engineering Assessment window

Potential Failure Mode Analysis

Generally, potential failure mode analysis (PFMA) follows the steps below. The failure modes decided engineering assessment of the dam condition.

- Collect all available information on the dam (refer to Engineering assessment)
- Identify failure mode (what, how and why)
- Investigate dam site (finding potential vulnerabilities)
- Confirm an engineering assessment against dam condition

Potential failure mode analysis results indicated significant and credible potential failure modes. Figure 9 shows the potential failure mode analysis windows. PFMA is described the critical failure scenario for flood and earthquake. Credible failure modes are physically possible even though they may be extremely unlikely to occur. Significant failure modes varies with the purpose of the risk assessment but in judging significance the cumulative effects of multiple failure mode should be considered.

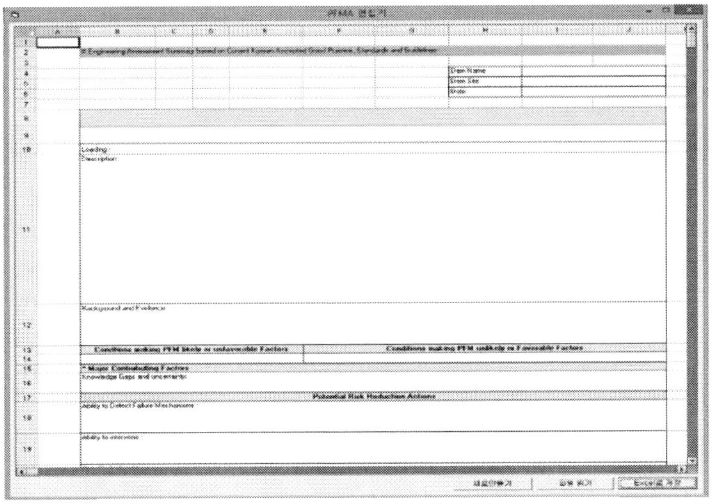

Figure 9. Potential Failure Mode Analysis

Hydrological Analysis Module

Hydrologic risk assessment method is suggested and it analyzes the risk considering three major factors. The factors are the rainfall simulation method considering parametric uncertainties, the rainfall-runoff ensemble scenario considering the uncertainty in rainfall-runoff model, and evaluating the probability of dam failure determined by gate operability. So the suggested method analyze considering these three factors. Uncertainty analysis associated with single rainfall-runoff model is provided and further discuss the use of the uncertainty analysis model for dam risk analysis. Floods associated with heavy convective rainfall may be connected with a relatively localized area of the watershed so that rainfall modeling using a single-site simulator for an entire watershed may be concluded in an underestimation or overestimation of design rainfall which leads to an unreliable estimation of design floods. For these reasons, the multisite rainfall simulation model for weather networks would require preserving spatial coherence. Figure 10 show the hydrological analysis module for estimate flood loading probabilities.

Hydrological analysis module is consist of 3 steps.

- Step 1 : Rainfall simulation uncertainty
- Step 2 : Calculate HEC-1
- Step 3 : Calculate HEC-5

This results is provided in the event tree analysis model for risk assessment.

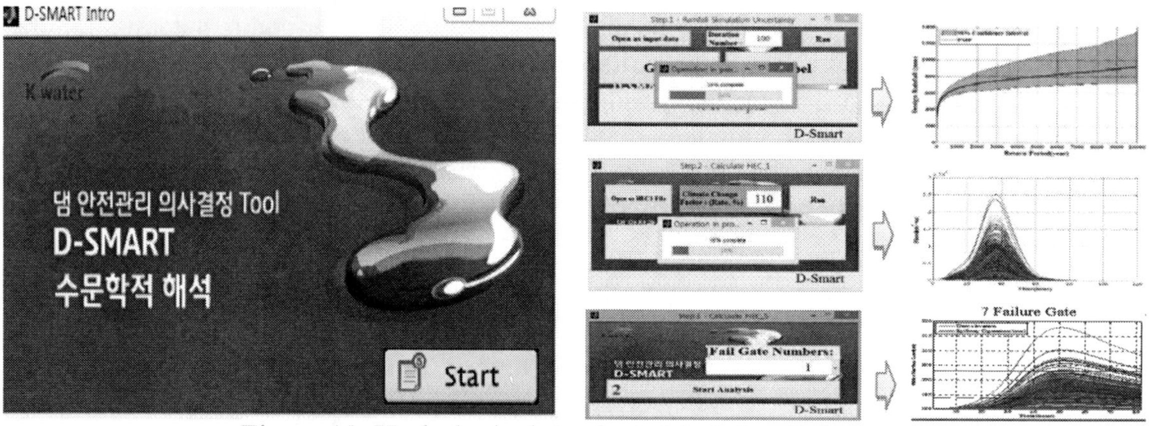

Figure 10. Hydrological Analysis Module windows

<u>Geotechnical Analysis Module</u>

To define geotechnical failure mechanism, material properties of geo-materials of dam must be configured. Major geotechnical failure modes of dam are internal erosion in embankment due to piping, embankment failure due to increase of pore pressure in soil material, and failure due to seismic event. The failure in embankment drives to failure of entire dam systems.

Internal erosion in embankment dams is not a completely understood phenomenon. Dam failures, accidents and deterioration of dams because of ageing effects do take place as a result of internal erosion. Even if dams generally should be considered as safe constructions, a large part of the present knowledge about problems associated with internal erosion is a result of studies of former dam incidents. This circumstance is almost inevitable since internal erosion processes are very complex and take place inside the embankment or foundation making the erosion transparency limited until it has progressed enough to be visible or detected by measurements.

According to Fell *et al.* (2005), in their excellent textbook about geotechnical engineering of dams, four conditions m ust exist for internal erosion and piping to occur. These are: "1) there must be a seepage flow path and a source of water; 2) there must be ero dible material within the flow path and this material must be ca rried by the seepage flow; 3) there must be an unprotected exit, from which the eroded material may escape and 4) for a pipe to form, the m aterial being piped, or th e material directly above, m ust be able to form and support 'roof' for the pipe".

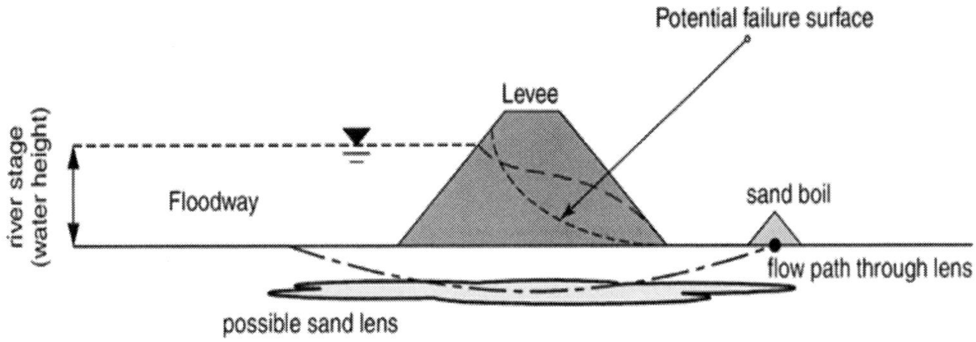

In geotechnical risk assessment, major cause of breach and the location associated with internal erosion and slope failure must be identified first, and probability of system failure must be evaluated later from simulation of embankment. Finally, appropriateness of presumed failure path from the analysis must be validated by probabilistic method.

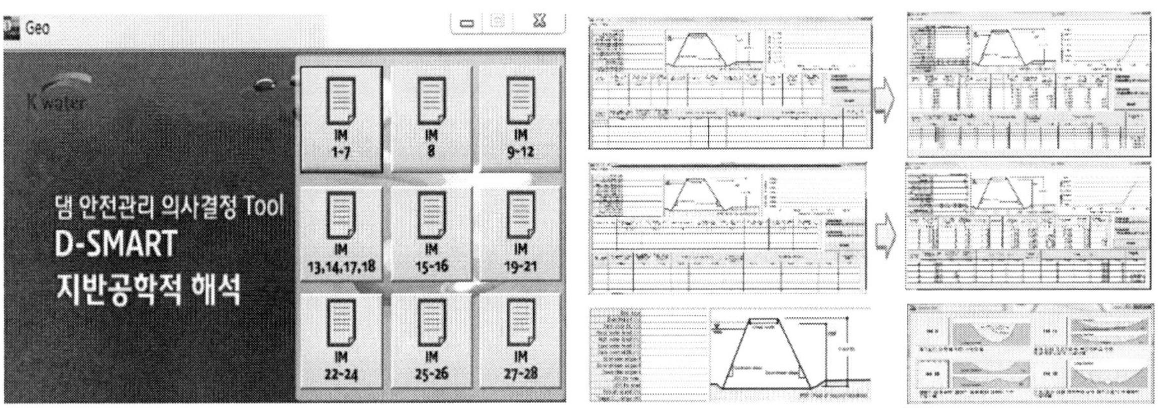

Figure 11. Geotechnical Analysis Module Windows

<u>Consequences Analysis Module</u>

Dams generally collapse either in a moment or progressively, and the main cause of failure and its mechanisms vary according to the type and the material composition of dams. Typical geometric characteristics of dam breach has not been clearly defined, however, various research suggest empirical method to define these terms based on number of cases previously investigated. MacDonald and Langridge-Monopolis (1984) investigated major dam breach cases in the history, and concluded that overflow through the breached zone forms 2:1 slope in side wall of the channel in early stage of flow, and finally makes trapezoidal channel in the dam as the flow continues. Houston (1985) investigated the cases of soundly constructed embankment dams with good material, and found that the breached channel forms a trapezoid with 1:1 slope side walls

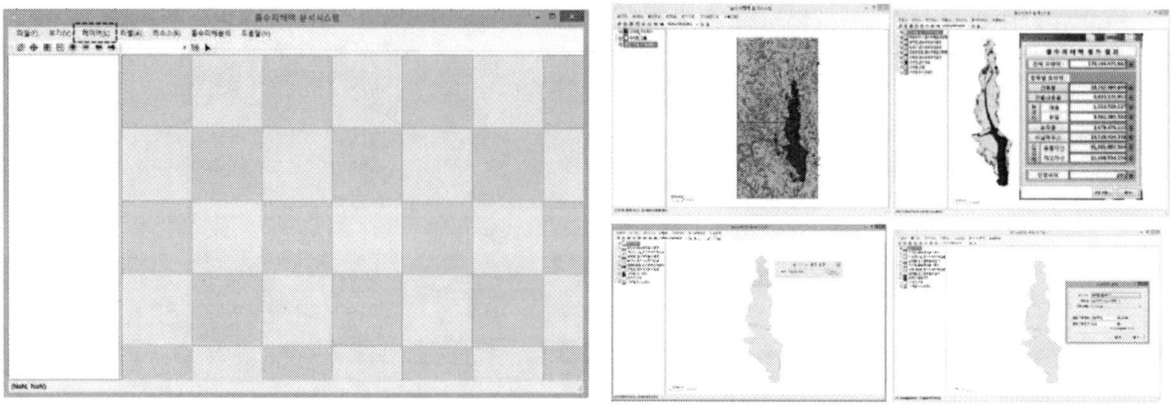

Figure 12. Consequences Analysis Module Windows

In Korea, multi-dimensional flood damage assessment (MD-FDA) method developed in 2004 is used for quantitative flood loss estimation. This method has been developed by Korean Ministry of Land, Infrastructure and Transport to complement the limitation of previous method which uses general regression analysis.

The damages for this assessment are categorized as seven items which are buildings, the properties in the buildings, farms, its plantations, business, storage, and public utilities, and it is separately analyzed for each administrative district. Furthermore, the reliability of this analysis could be improved by considering. The methods for casualties and damages assessments are developed using Duramax-Xr engine which is as GIS-based module. The damage assessment requires a catchment basin map, inundation map, etc. and the casualty assessment used population density for each district provided by Statistics Korea (national bureau of statistics in Korea).

Risk Assessment Tool

The purpose of ETA is to evaluate every result coming from initial events. It can be generally applied to not only entire system, but sub-systems, softwares, procedure, the human error, and etc. Safety and/or risk issues regarding dam construction can be diagnosed if ETA is applied in planning and design stage. ETA is a kind of binary tree structure to evaluate various decision making route, and its first application was risk assessment a nuclear power plant (WASH-1400, Nuclear Power Plant Safety Study, circa) in order to improve previous fault tree analysis (FTA). ETA is regarded better than FTA in terms of easy model construction and analysis.

Event Tree Analysis (ETA) is commonly used in dam safety risk analysis. Event trees are tools used to aid in understanding, analyzing, and communicating dam safety risks and in making decisions. Like all other tools they are imperfect, and their potential value depends on the skills of their user ; and in dam safety risk assessment this requires suitably qualified professional engineers.

Event trees can be used to obtain quantitative estimates of the probability of dam failure and its associated consequences. This can be done for an existing dam or for various risk reduction measures. Event trees can also serve as qualitative or diagrammatic representations of failure modes, their associated outcomes (system effects), and their consequences for various types of initiating events.

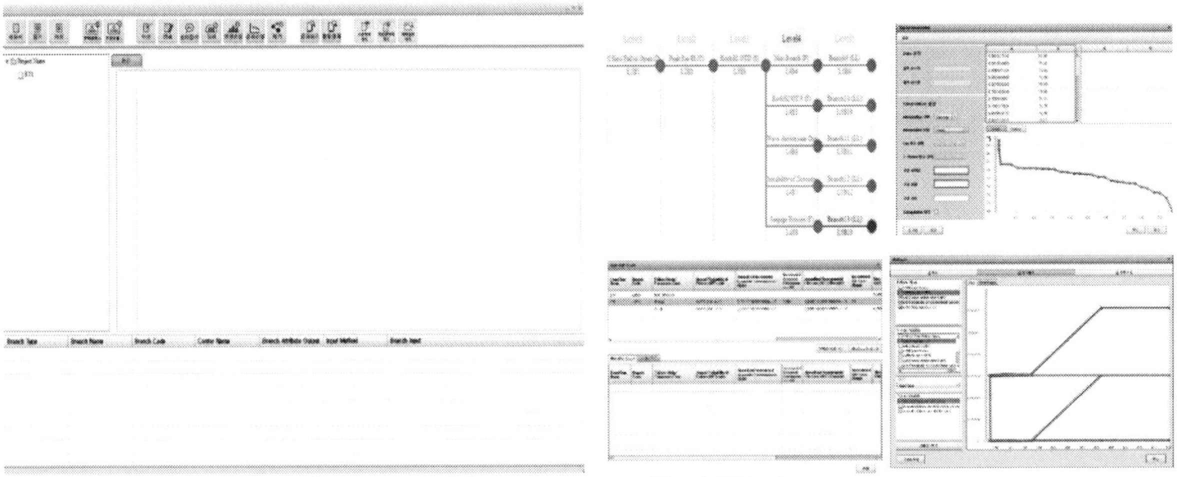

Figure 13. Risk Assessment Tool Windows

CASE STUDY OF RISK ASSESSMENT OF DAM

Korea has approximately 18,000 dams. There are 18 multipurpose dams, 2 flood control dams and the rest are agriculture water dams.

There was no risk analysis and assessment program for dam safety in Korea before this study. Therefore, one dam was chosen for a pilot study to develop the analysis and assessment process so it can be used to develop a risk based dam safety management and decision making system.

Information for the pilot dams is summarized as follow. 123m high zone type rockfill dam(S dam). Figure 14 show the pilot study dam view and information.

- Type: Centre core rock-fill dam
- Height: 123 m
- Length: 530 m
- Crest Level: 203.0 (EL. m)
- Catchment Area: 2,703 km^2
- Gross Storage Capacity: 2,900 million m^3
- Flood Water Level: 198.0 (EL. m)
- Normal High Water Level: 193.5 (EL. m)

Figure 14. View of a S-Dam in Korea

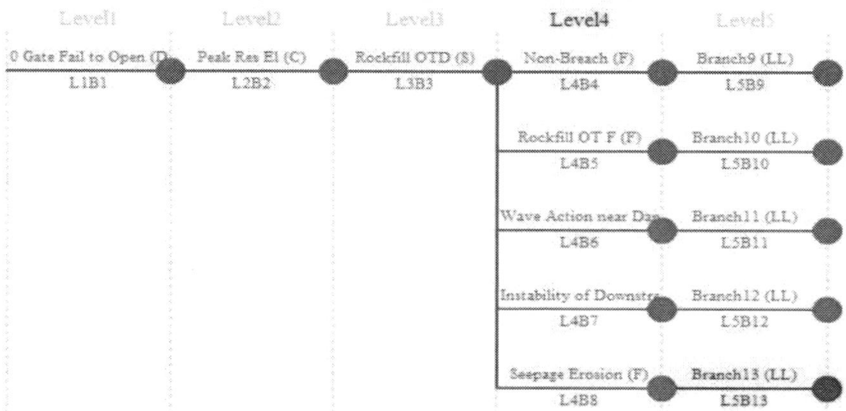

Figure 15. Risk analysis Model (Event Tree Analysis)

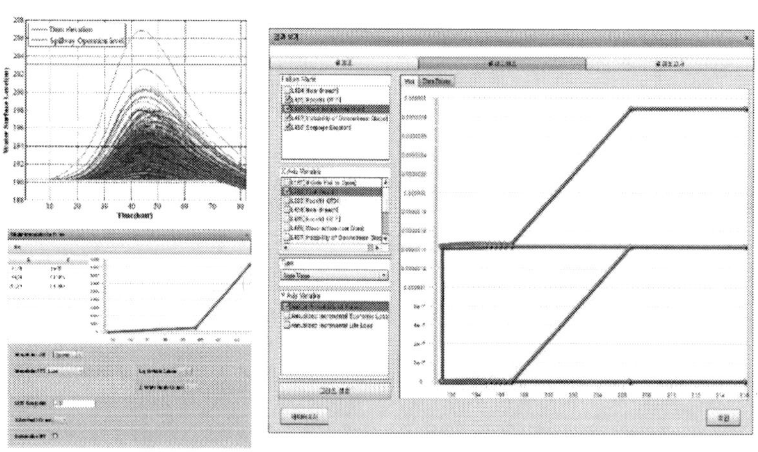

Figure 16. Results of the risk analysis

CONCLUSION

Risk assessment is an essential component of dam safety, and statistical analysis plays central role in dam risk assessment. Dam risk analysis relies on a series of probabilistic analyses of hydrologic, geotechnical, structural models and their associated inputs. However, it may be necessary to employ subjective information from the expert opinion in estimating loading probabilities and system response probabilities due to a lack of available data for some risk factors. In 2011 year, risk analysis and assessment for one pilot dam in Korea was undertaken for the first time. And then, development of evaluation tool D-SMART for dam safety management in Korea. D-SMART is provide total solution for the risk analysis and assessment of dam safety. For future, risk-based dam safety management system for dam safety will be improve risk analysis module for flood and earthquake.

REFERENCES

Korea Water Resource Corporation (2013), "Development of next generation dam safety management decision-making system", report.

Lim J.Y (2013), "Case study of risk assessment for dam safety management decision-making in Korea", Proceeding of the Australian Committee on Large Dams Conference.

USACE 2009, "Internal Erosion Toolbox-A Method for Estimating Probabilities of Failure of Embankment Dams due to Internal Erosion"

USBR 2003, "Dam Safety Risk Analysis Methodology", Ver.3.3.1

USBR 2001, "Risk Based Profiling System"

USACE 2011, "Safety of Dams-Policy and Procedure", ER 1110-2 1156

H.Mattsson, J.G.I Heiistrom & T.S.Lundstrom, 2008 "On Internal Erosion In Embankment Dams", Research Report.

L.M Zhang, Y.Xu & J.S.Jia, 2009 "Analysis of earth dam failure : A database approach", Georisk, Vol.3, No.3.

SITE CHARACTERIZATION OF DAM INFRASTRUCTURE FOR ENGINEERING RISK ASSESSMENTS USING A GIS

Meghann Wygonik[1]
Christopher Kelly[2]
William Walker[3]
Baron Worsham[4]

ABSTRACT

As the U.S. Army Corps of Engineers (USACE) adapts to meet the challenges associated with an aging dam infrastructure portfolio, risk informed decisions have become more prevalent. In order to determine the level of risk associated with the USACE flood risk management and navigation dam portfolio, the Risk Management Center has implemented several categories of risk assessments. These risk assessments are extremely data intensive activities requiring a detailed review of dam performance and documentation. Due to the large number of files associated with the life cycle of a dam, the quality of a risk assessment is generally a function of the ability to identify, relate, and understand a large amount of data and information in a relatively short period of time.

In order to aid the risk assessment process, research was conducted to investigate dam safety uses of a Geographic Information System (GIS). A site characterization schema was developed that captured the pertinent information requirements needed for a risk assessment. The schema was then used to develop GIS prototypes for a flood risk management dam and a navigation dam. The benefits of the site characterization GIS prototypes included spatially associating existing information in a dynamic environment; access to information and documents stored within existing data sources; and the ability to produce figures and diagrams that present complex relationships in a professional manner. Overall, the prototypes proved to be an interactive and information-rich resource for the risk assessment process. Feedback was captured and incorporated into recommendations for improvement and expansion.

INTRODUCTION

In 2009 the USACE Institute for Water Resources, Risk Management Center was established and tasked with assessing and managing the risk associated with the USACE portfolio of flood risk management and navigation dams (USACE, 2013). In order to accomplish this directive, the Risk Management Center has implemented several

[1] Engineer, US Army Corps of Engineers, Pittsburgh District, Pittsburgh, PA 15222, Meghann.M.Wygonik@usace.army.mil
[2] Engineer, US Army Corps of Engineers, Dam Safety Modification Mandatory Center of Expertise, Pittsburgh, PA 15222, Christopher.J.Kelly2@usace.army.mil
[3] Engineer, US Army Corps of Engineers, Nashville District, Nashville, TN 37209, William.G.Walker@usace.army.mil
[4] Engineer, US Army Corps of Engineers, Nashville District, Nashville, TN 37209, Baron.M.Worsham@usace.army.mil

categories of risk assessments (USACE, 2014). These risk assessments are extremely data intensive activities requiring a detailed review of the documentation and performance of the dam (Scott, 2010). Depending on the level of effort associated with the risk assessment, engineers have a few days during a Semi-Quantitative Risk Assessment to a few months during an Issue Evaluation Study to assess the level of risk.

Prior to a risk assessment, documentation from the life cycle of the dam is compiled and uploaded to a central dam safety documentation repository. This effort aims to capture all the pertinent design documents, engineering analyses, and known information about the dam for use during the risk assessment. This effort often results in a large number of files that the risk assessment team must then sort through in order to identify and integrate the data and information needed to perform the risk assessment (Jung, 2013). As a result, the quality of the risk assessment is generally a function of the ability to identify, relate, and understand a large amount of data and information associated with the dam in a relatively short period of time.

In an effort to leverage technology to aid risk assessments, research was conducted to investigate the use of a Geographic Information System (GIS) for dam safety purposes. The research focused on maximizing the use of existing USACE databases and licensed software to create an interactive and information-rich resource. The resource was intended to be expandable, scalable, and capable of serving as a template for easy replication. Development criteria included the ability to provide single point access to an organized structure of pertinent dam safety information; integrate existing data and information pertaining to the dam and historical performance; and create hyperlinks to established USACE data sources.

SITE CHARACTERIZATION GIS DEVELOPMENT

Site Characterization Schema

An extensive review of USACE risk assessment and site characterization documentation was performed to gain an understanding of the data needs of risk assessments and the current state of site characterization information. Data requirements were extracted from sources including USACE dam safety documentation, engineering regulations, external research reports, and discussions with USACE engineers. The data requirements were compiled into a list of pertinent information for risk assessments and organized by physical feature or engineering discipline. From this list, a site characterization schema was developed that contained 115 site characterization feature classes grouped by physical feature or engineering discipline within feature datasets. Table 1 shows the distribution of site characterization feature classes by feature dataset grouping. When possible, pertinent information requirements were kept at the most generic feature description to simplify the schema structure. For example, data requirements for ball valves, butterfly valves, check valves, and gate valves were generalized and complied into a single valve feature class.

Table 1. Site Characterization Schema Summary Table

Feature Dataset	Number of Feature Classes
General Information	6
Dam Features	26
Lock Features	5
Electrical & Mechanical Features	14
Site Features	25
Dam Safety	17
Geotechnics	11
Hydraulics & Hydrology	3
Consequences	3
Base Maps	5

Attribute Tables. Attribute tables for the site characterization feature classes were developed to contain a minimum amount of information that would be relevant and useful during risk assessments. At a minimum, each feature class contained six attributes; the National Inventory of Dams (NID) Number, the Feature ID, a field for condition description or link to inspection ratings, a link to associated reports, a link to associated drawings, and a field for notes. Table 2 is an example of a site characterization attribute table for a dam gate. Nine additional attributes, such as associated hydraulic feature, type of gate, pertinent elevations, and discharge capacity, were added to the six basic attributes to describe the pertinent information requirements for various gates present on dams. On average, each feature class required ten attributes to capture the pertinent information requirements.

Table 2. Site Characterization Schema Table: Dam Gate

Dam Gate	
Attribute	**Data Type**
Dam NID Number	text
Gate ID	text
Hydraulic Feature	text
Type	text
Function	text
Height_feet	double
Width_feet	double
Top of Gate Elevation_feet	double
Bottom of Gate Elevation_feet	double
Elevation Datum	text
Maximum Discharge_cfs	double
Condition	text
Reports	text
Drawings	text
Notes	text

Hydraulic Feature
Spillway
Conduit

Type
Tainter/Radial
Vertical Lift
Roller
Slide/Sluice

Function
Operational
Emergency

When applicable, attribute tables were developed based on information readily available in existing databases. For example, the general information feature class attribute table was developed based on the USACE Dam Safety Program Management Tool database. Geologic and seismic feature attribute tables were developed based on the information readily and freely available from the United States Geologic Survey. Consequence feature classes were developed based on products from the USACE Modeling, Mapping, and Consequence Center of Expertise.

Field Properties. The site characterization feature classes were created with simplified field properties for consistency. Attributes that were of known value, such as the NID Number, were given specific length and null requirements. Attributes with a number data format were given a data type of double. Attributes with a data format of date were given a data type of text to allow for different date formats, such as a month-year formatted compared to a day-month-year format. Table 3 shows a summary of the field properties used when creating the feature classes.

Table 3. Attribute Table Field Properties

		Field Properties		
Attribute	**Data Format**	**Data Type**	**Length**	**Null / Not Null**
NID Number	text	text	8	not null
Feature ID	text	text	50	not null
Attribute (text)	text	text	250	null
Attribute (number)	number	double	N/A	null
Attribute (date)	date	text	100	null

Symbology. Accompanying layer files were created for each site characterization feature class to capture the symbology. Preference was given to existing GIS symbology templates. When existing GIS symbology templates were unavailable, the symbology was developed based on basic shapes and colors that created unique symbols.

Geospatial Standards. The geospatial standard used by the Department of Defense is the Spatial Data Standards for Facilities, Infrastructure, and Environment (SDSFIE) (USACE 2011). Basic SDSFIE requirements were incorporated into the site characterization schema. Future efforts should continue to pursue the processes and requirements to become SDSFIE compliant.

Database Connections. When applicable, database connections were established to maximize the utilization of data and information stored in existing databases while minimizing data duplication. Common databases used by the USACE include ESRI's geodatabase for geospatial information, Bentley's ProjectWise for document management, and the Dam Safety Program Management Tool (DSPMT) for general information. Discipline specific databases include Bentley's gINT for geologic data, WinIDP for instrumentation data, and the Corps Water Management System (CWMS) for pool and precipitation data, among others. Additional databases that contain beneficial information for risk assessments include the USACE Resident Management System (RMS) for construction data, the Asset Management Database for operational condition assessment data, and the Mapping Modeling Consequences (MMC) oracle spatial database for inundation maps and consequence data. Figure 1 shows some of the different types of information and the associated databases used during risk assessments.

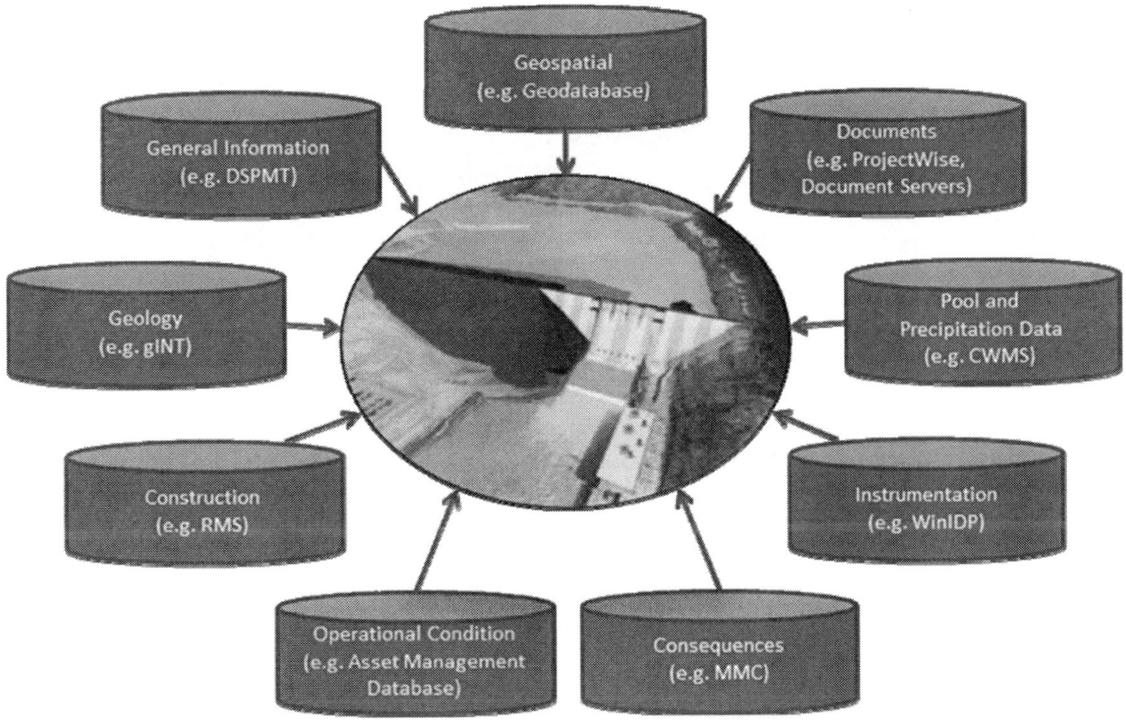

Figure 1. Databases Connections

Flood Risk Management Dam Prototype

A site characterization GIS was developed for a flood risk management dam with both concrete and embankment sections within ESRI's ArcMap. A second generation, three dimensional prototype was developed within ESRI's ArcGlobe. The two dimensional and three dimensional site characterization GIS prototypes are shown in Figure 2 and 3, respectively. The two dimensional features were digitized from georeferenced as-built drawings, while the three dimensional features were constructed from as-built drawings within AutoCAD 2013. The three dimensional AutoCAD solids were imported to ArcGlobe as multipatch features via ESRI's Data Interoperability Extension. A combination of inspection reports, original design documents, and as-built drawings were used to aggregate attribute data for the dam features and site features. Dam safety features were digitized and attributed according to as-built drawings and tabular data. Aerial photography and LiDAR elevation data were supplied by the host USACE district. Geologic information was downloaded from the United States Geological Survey. When applicable, information was extracted from existing USACE databases to attribute the features, such as the Dam Safety Program Management Tool for general information, WinIDP for instrumentation data, and the Mapping, Modeling, and Consequence database for inundation mapping and associated consequence information. Additional data and information associated with features not captured directly within the attribute tables were included by creating hyperlinks to documents or folders on district servers or within Bentley's ProjectWise.

Figure 2. Two Dimensional Prototype Figure 3. Three Dimensional Prototype

The flood risk management site characterization GIS was developed for use during a Periodic Risk Assessment. Due to scheduling changes, the risk assessment was postponed. Therefore, the flood risk management site characterization GIS prototype has not been field tested. However, the prototype has undergone a series of demonstrations to a wide audience highlighting various features of the prototype. Comments from engineers during the demonstrations were documented and analyzed to determine the benefits and shortcomings of the prototype during the risk assessment process.

Navigation Dam Prototype

A site characterization GIS was also prototyped for a navigation dam with a lock chamber, gated dam, embankment overflow section, and dike section within ESRI's ArcMap and ArcGlobe. An existing three dimensional CAD file within Bentley's Microstation was provided from the host district. The three dimensional Microstation solids were imported to ArcGlobe as multipatch features via ESRI's Data Interoperability Extension. A combination of inspection reports, district documentation, and as-built drawings were used to digitize and aggregate attribute data for the dam, lock, and site features. Subsurface data and information was digitized and attributed from as-built drawings and district files. Aerial photography and LiDAR elevation data were provided by the host district. Information was extracted or linked to the Dam Safety Program Management Tool database for the general information feature, the host district's spatial database engine for the distress indicator features, and the Mapping, Modeling, and Consequence database for inundation mapping and associated consequence information. Additional data and information not included within the feature attribute tables were included by creating hyperlinks to supporting documents or district folders.

The navigation site characterization GIS, shown in Figure 4, was developed for use during a Semi-Quantitative Risk Assessment and Dam Safety Modification Study. The site characterization GIS could potentially follow the dam through construction and the Post Implementation Evaluation if the dam was to progress through the USACE Portfolio Risk Management Process (USACE, 2014). The prototype was utilized during the Semi-Quantitative Risk Assessment in August 2014 and the Dam Safety Modification Study from the fall of 2014 through 2015. Utilization of the GIS prototype during the risk assessments and comments from risk assessment team were documented. The feedback was compiled and analyzed to determine the benefits and shortcomings of the prototype during the risk assessment process.

Figure 4. Navigation Site Characterization GIS

RESULTS

Overall, the prototypes were well received by the engineers. Comments from engineers were captured on a wide range of functions and features of the prototypes and were used to formulate the results. Evaluation of the comments revealed three main trends for prototype benefits and three main trends for prototype shortcomings.

Prototype Benefits

The majority of engineers saw the greatest benefit in spatially relating performance data to dam features and the surrounding environment. Other benefits included the ability to link dam features to documents, generate report diagrams and figures, and use the site characterization GIS as a communication tool. Of the many benefits identified by the engineers, the three main trends were the value of spatial association, improved access to information stored in databases, and ability to publish maps.

Spatial Association. A large benefit of the site characterization GIS was the ability to spatially relate physical features of the dam with historic and current dam performance. For example, during a recent periodic inspection for the flood risk management dam, four wet spots were observed near the toe of the embankment. These wet spots were digitized and attributed according to information located within periodic inspection reports and presented to engineers during demonstrations. Using the identify and hyperlink functions embedded in the GIS, engineers were able to interrogate the prototype for dam safety information such as instrumentation plots, instrument properties, and photographs in the area surrounding the wet spots, as shown in Figure 5. Subsurface information was accessible within the prototype via borings logs and geologic cross sections near the area of interest, as shown in Figure 6. A published geologic map from the United States Geologic Survey provided a more regional perspective to the subsurface soils and rock present at the dam site. As the interrogation of the prototype became more detailed, engineers were able to select individual features from the map, or query features based on properties within the attribute table, such as a query based on the piezometer tip elevation.

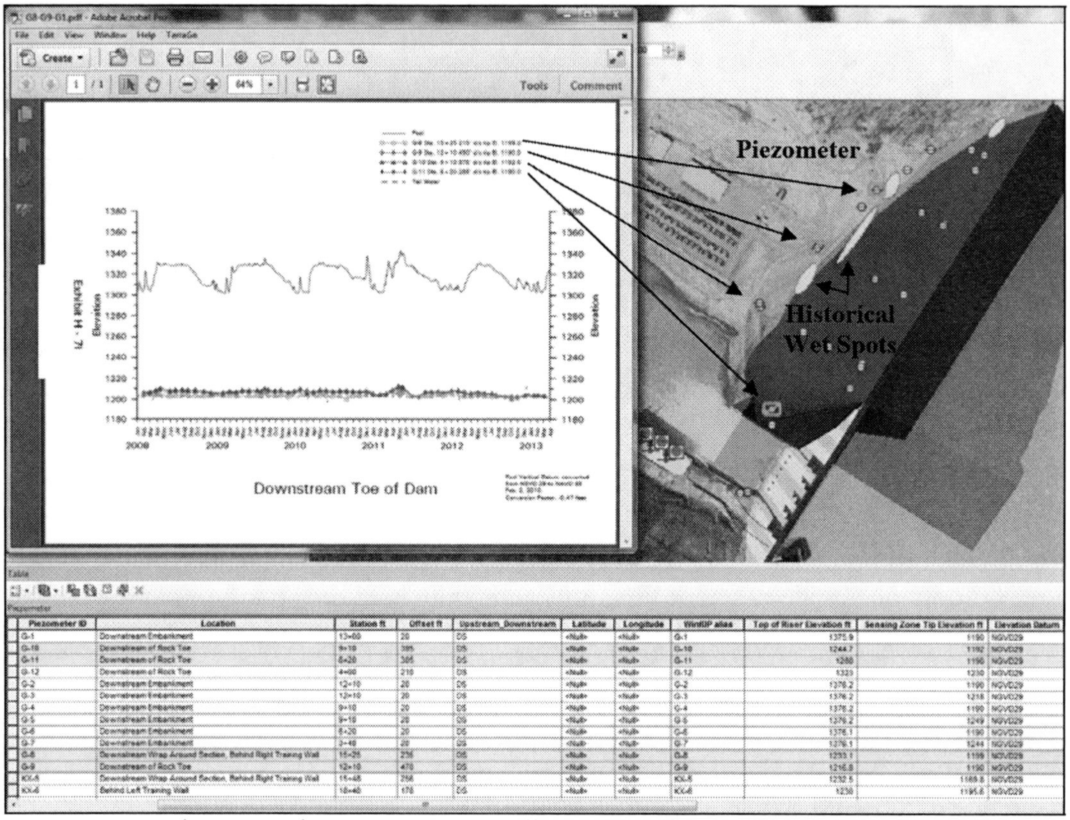

Figure 5. Piezometer Data Adjacent to Historical Wet Spots

Additional benefits were noted in the ability to visualize a combination of feature from different times in the dam's history, such as the original river channel and temporary construction features, in relation to dam features of interest, such as the conduit, the abutments, or the concrete-embankment interface. Incorporating different features from the dam's history into a common visual allowed the engineers to better understand how features of the dam may or may not be related to the dam's historic, current, or future performance.

Database Connectivity. Database connections proved to be beneficial by making existing data and information readily available to the engineers during the risk assessment. One example of the efficiencies experienced from database connections is highlighted within the consequences evaluation portion of the risk assessment. A web map service connection was established with the Mapping Modeling and Consequence oracle spatial database which allowed the engineers to view the inundation mapping, time-to-rise estimates, and distance markers from the dam to points of interest downstream within the GIS prototype, as shown in Figure 7. Additional consequence information was accessed via hyperlinks to the Mapping Modeling and Consequence reports and working files. Additionally, the database connections limited data duplication and ensured the most up to date information was presented within the GIS prototype.

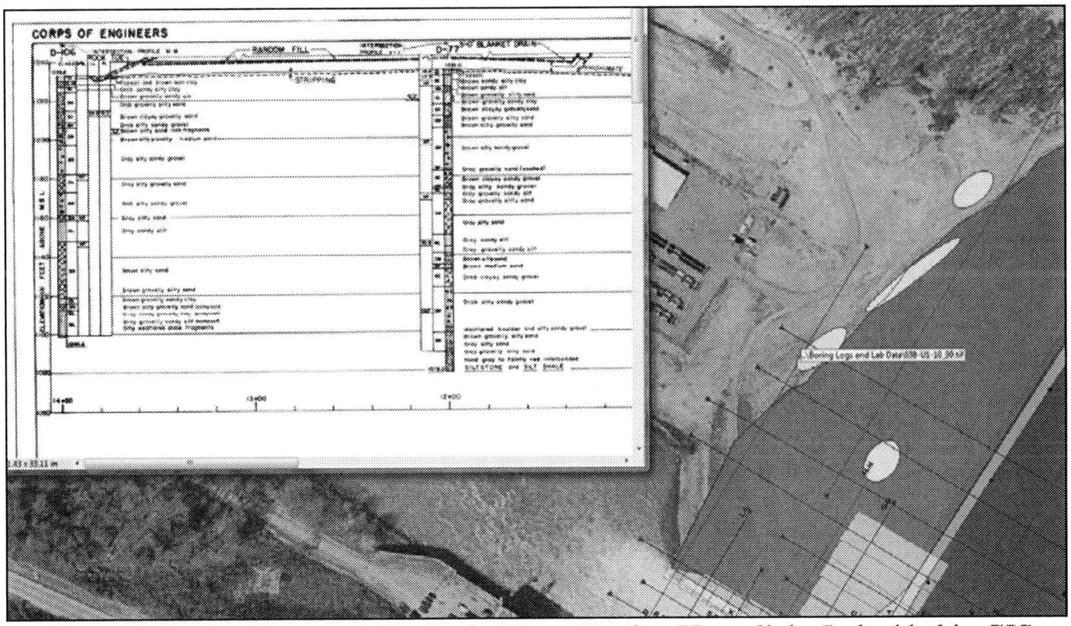

Figure 6. Boring Logs and Geologic Cross Section Hyperlinks Imbedded in GIS

Figure 7. Inundation Mapping for a Modeled Dam Break Scenario

Published Maps. The site characterization prototypes were also used to generate report figures to more effectively communicate areas of interest or diagram potential failure modes. The published maps presented the information in a professional manner, with figures to scale, annotations, and reference images. Engineers expressed this as beneficial because of the time saved during report writing and the ability to communicate complex relationships using one graphic. Figure 8 is an example report figure from the navigation dam prototype that shows the location of loose barges during a marine accident in 2011 in relationship to the scour surveys which was used during a credible potential failure mode evaluation.

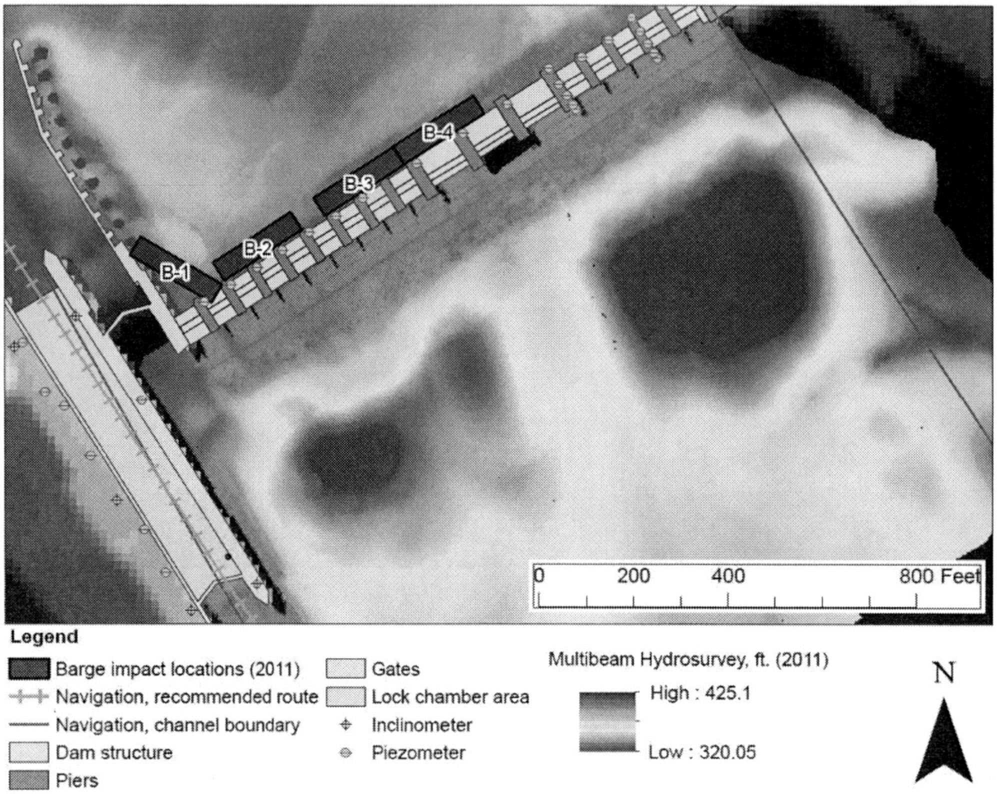

Figure 8. Published Map showing 2011 Marine Accident

Prototype Shortcomings

While the prototypes had multiple benefits, the engineers also identified shortcomings of the prototypes. Evaluation of the engineers' comments revealed three main trends. They were unintuitive discipline specific characteristic feature classes, complex structure of database connections, and the inability to generate custom cross sections.

Discipline Specific Characteristic Feature Classes. The prototypes were developed with discipline specific characteristic feature classes for each engineering discipline to store information that was not directly related to a physical feature. The discipline specific characteristic feature classes were categorized by discipline and included a wide range of data from historic and design information to current institutional and operational knowledge of the dam. While these tables attempted to capture the discipline specific information needed for risk assessments, the information was not easily conveyed in this format. Accessing the information was described as not intuitive and did not contain the level of detail needed to inform engineers during a risk assessment. For example, hydraulic and hydrology information needed to evaluate an overtopping failure mode requires the probable maximum flood hydrograph and pertinent dam elevations. The probable maximum flood hydrograph is not directly associated with a physical feature on the dam. Therefore, the probable maximum flood hydrograph was presented within the prototypes via a hyperlink to the source document in the H&H Curves feature. Metadata associated with the hydrograph and pertinent dam elevations were contained in the H&H Characteristic and Reservoir Data features, respectively. As a result, the engineers were not able to intuitively locate the data and information needed to evaluate an overtopping failure mode.

Database Connections. While the prototypes provided ready access to data stored in databases, the database connections did not offer any additional information or data not already provided by the host database. Additionally, setting up the database connections can be tedious requiring join and relate functions to be established between numerous tables, as was the case with the Dam Safety Program Management Tool database connection. Due to the size and structure of some databases, multiple join and relate functions, including one-to-one, one-to-many, and many-to many join and relate functions, could need to be established to read in the attribute data. For databases with a more complex structure, a onetime data export may be more efficient. However, the onetime export is not connected to the database, so future updates to the database would not be reflected in the attribute table.

3D Cross Section Generation. While the prototypes were able to provide professional drawings in plan view using the layout view, the site characterization GIS was not able to create custom cross sections from the 3D multipatch features. Engineers identified the ability to create custom cross sections by drawing a line across 3D multipatch features within the site characterization GIS as a large benefit. Although this ability is not currently available in the site characterization GIS, it could be researched during future efforts.

CONCLUSION

As the USACE dam safety program continues to manage the risk associated with an aging portfolio, a site characterization GIS can help engineers to better understand the features and performance of a dam which contribute to risk. The research showed that the site characterization prototypes were able to provide a platform for data collection, organization, and presentation on a wide range of information from detailed feature attributes to institutional knowledge. Existing USACE systems and data sources were used to create an interactive and information-rich resource for dam safety risk assessments.

The three main benefits of the site characterization GIS were spatial association of dam features and performance, access to data and information stored in existing databases, and the ability to present information and complex relationships in a comprehensive and professional manner. However, the research also revealed areas for improvement. The prototypes showed that information not directly related to physical features was not intuitive to locate and was not comprehensive. Additionally, database connections were complex and the GIS was not able to create custom multipatch cross sections.

The research also showed that there is much potential for expansion and further refinement. Beyond the risk assessment process, a site characterization GIS could also add benefit to construction activities, Post Implementation Evaluations, Periodic Inspections, and Operational Condition Assessments. Pilot studies could be conducted to investigate the potential uses of ArcGIS Online for risk assessments, research potential ways to create 3D multipatch cross sections, and refine the discipline specific characteristic feature classes. Pilot studies could also continue researching spatial standards, expanding and developing database connections, and refining workflows associated with assembly and utilization during risk assessments.

ACKNOWLEDGEMENTS

The authors would like to thank all the USACE dam safety professionals that provided guidance, support, and input on this research.

REFERENCES

Jung, I., Fraizer, J., Akinci, B., Berges, M., Ergan, S., Garrett, J. H., and Kelly, C., "Towards Dam Information Modeling: Best Practices Learned From the AEC/FM Domain", 2013 International Commission on Large Dams (ICOLD), Seattle, WA, 2013.

Scott, G., "Shedding some light on this thing called Risk Assessment Part 1 – Risk Analysis Basics", *The Journal of Dam Safety* 8.1(2010): 20-26.

U.S. Army Corps of Engineers. ER 10-1-55, Roles and Responsibly Risk Management Center, 30 June 2013.

U.S. Army Corps of Engineers. ER 1110-2-1156, Safety of Dams – Policy and Procedures, 31 March 2014.

U.S. Army Corps of Engineers. "Spatial Data Standards for Facilities, Infrastructure, and Environment (SDSFIE) 3.0 Implementation Guidance". 2 December 2011. http://www.sdsfieonline.org/Documents/USACE%20SDSFIE%20Implementation%20Plan%20FINAL.pdf.

DESIGN AND CONSTRUCTION OF A SECANT-PILE CUTOFF WALL IN VARIABLE GROUND CONDITIONS

Erik Newman, PhD, PE[1]
Michael Forrest, PE[2]
Kylan Kegel, PE[3]
Kenneth Pattermann, PE, GE[4]

ABSTRACT

The Folsom Dam Joint Federal Project is being undertaken to increase the spillway capacity of Folsom Reservoir in order to meet the demands imposed by an updated assessment of the probable maximum flood. The auxiliary spillway consists of an approach channel, a control structure, and a spillway chute and stilling basin that discharge spillway flows into the American River downstream of the existing spillway. Staged construction allowing the gated control structure, spillway chute and stilling basin to be completed without impacting operation of Folsom Reservoir was accomplished by leaving a rock plug between the reservoir and the gated control structure.

A design was selected in which a temporary embankment will be constructed on the upstream side of the rock plug to form a cofferdam together with the upstream portion of the rock plug. A secant-pile seepage cutoff wall was constructed through the embankment portion of the cofferdam and into the underlying weathered quartz diorite to reduce seepage through pervious soil and decomposed rock into downstream work areas.

The secant-pile cutoff wall is 1170 feet long with a maximum depth of 102 feet, totaling about 98,000 square feet, with a minimum specified thickness of 12 inches. The pile shafts were required to be cased through fill and native soil with concrete placement occurring ahead of casing withdrawal. Shaft drilling challenges included joint-controlled weathering present at the site, involving unweathered granodiorite core stones surrounded by decomposed granite.

The specification required that core borings and water pressure testing be performed by the contractor every 50 feet along the wall alignment to better define the conditions before production began. QC procedures during construction included verticality and diameter surveys of each shaft, a review of foundation materials at the base of the shafts, as well as monitoring placed concrete quantities for comparison with the theoretical quantities. Drilling and concrete placement began in early March 2014 and the wall was completed on August 1, 2014.

---

[1] Project Engineer, URS Corporation, 1333 Broadway, Suite 800, Oakland, CA 94612, (510) 874-3296, erik.newman@urs.com
[2] Project Manager, URS Corporation, 1333 Broadway, Suite 800, Oakland, CA 94612, (510) 874-3012, michael.forrest@urs.com
[3] Technical Lead, US Army Corps of Engineers Sacramento District, CESPK-JFPO, 1325 J Street, Sacramento, CA 95814-2922, (916) 557-7775, Kylan.A.Kegel@usace.army.mil
[4] Geotechnical Engineer, US Army Corps of Engineers Sacramento District, CESPK-JFPO, 1325 J Street, Sacramento, CA 95814-2922, Kenneth.R.Pattermann@usace.army.mil

---

INTRODUCTION

The Folsom Dam Joint Federal Project (JFP) is being undertaken to increase the spillway capacity of Folsom Reservoir in order to meet the demands imposed by the updated assessment of the probable maximum flood, and to allow more control of releases for other flood events. The auxiliary spillway consists of an approach channel, a control structure, and a spillway chute and stilling basin that discharge spillway flows into the American River downstream of the existing spillway (see Figure 1). Staged construction will allow the gated control structure, spillway chute and stilling basin to be completed without impacting operation of the existing Folsom Dam by keeping intact an existing rock plug, including a seepage cutoff wall, between the reservoir and the control structure until the control structure is operational. All elevations in this paper are referenced to the North American Vertical Datum of 1988 (NAVD88).

The work discussed in this paper is part of Phase IV of the project, which continues from Phases I through III. Phases I through III, which included excavation of the new spillway chute and control structure foundation, began in 2008. Excavation for the control structure foundation was completed in May 2012. The excavation for the control structure foundation stopped about 100 feet upstream of the control structure, terminating with a temporary 120-foot-high face. This face created a rock plug retaining Folsom Lake (see Figure 2) with a minimum width of about 200 feet at the high lake level elevation and a width of over 1000 feet at the channel invert elevation. This rock plug met permanent dam safety design criteria, and performed well with only minor seepage observed through joints in moderately weathered or better rock. However, the Phase IV excavation reduces the width of this rock plug, and a cutoff wall is necessary to meet permanent dam safety criteria during Phase IV.

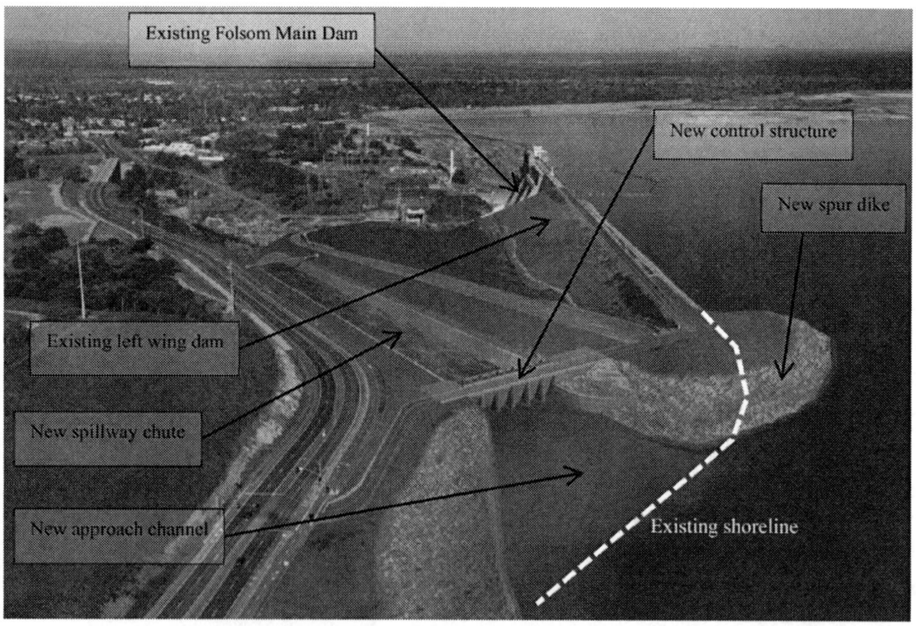

Figure 1. Rendering of completed Folsom Auxiliary Spillway

The Phase IV rock plug and temporary embankment discussed below were designed to maximize in-the-dry excavation of the remainder of the approach channel without impacting operation of Folsom Lake and to minimize interference with the construction of the control structure by the Phase III contractor, whose schedule overlaps with the start of Phase IV excavation.

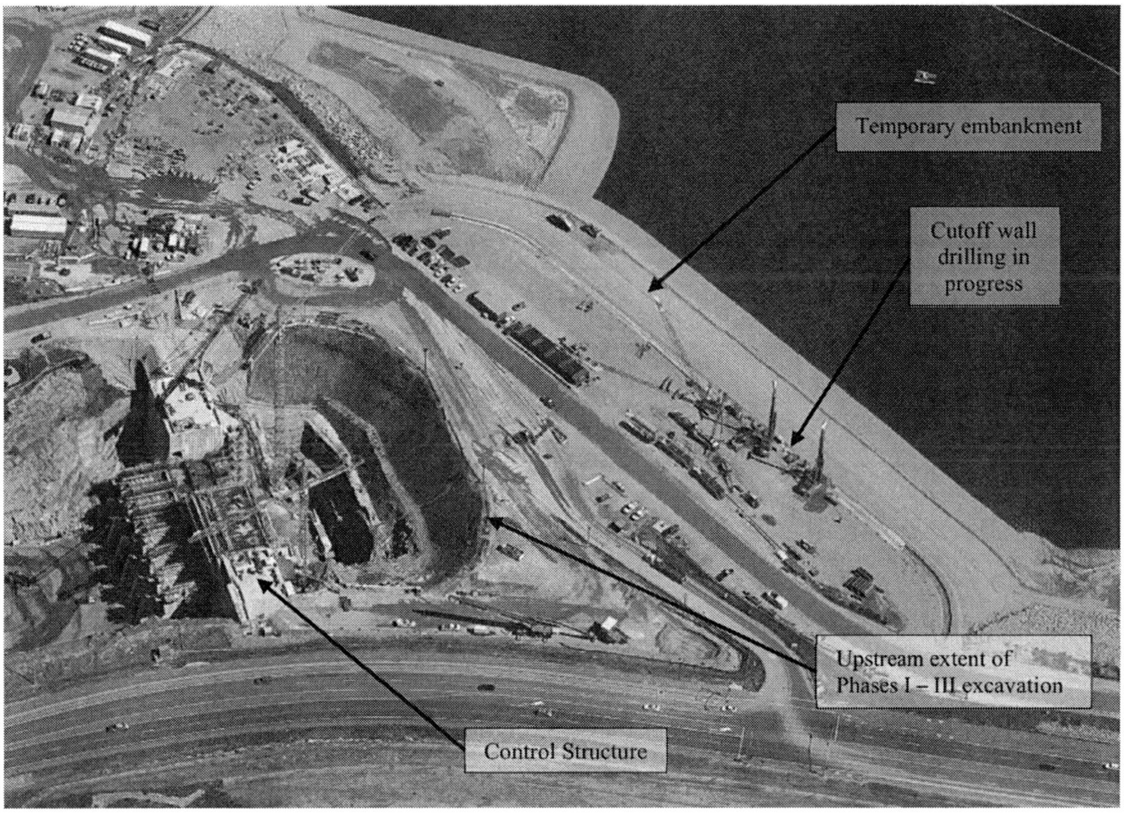

Figure 2. Overview of project site, March 2014

MATERIAL CHARACTERIZATION

Site Geology and Subsurface Conditions

The Folsom Dam JFP Auxiliary Spillway is located in the western foothills in the northern portion of the Sierra Nevada physiographic province. The predominant rock type in the project area is described as a light gray and white speckled, medium-to-coarse-grained, strong to very strong quartz diorite (the terms "granodiorite" and "granite" are used interchangeably in this paper instead of quartz diorite, which is the rock classification used by the USACE). Mapping from the Phase III excavation, supplemented with core borings drilled for Phase IV design identified four major joint sets striking in different directions, with dips varying from 27 to 81 degrees. Three of the four joint sets daylight on the face of the rock plug. There are also a significant number of randomly-oriented joints.

Secant-Pile Cutoff Wall

The granodiorite bedrock exhibits variable weathering, fracture spacing and aperture, strength, and associated seismic P-wave velocities. In a very broad sense, the weathering grade decreases with depth, except where shears or joints are present and the weathering extends deeper. This type of weathering along joint surfaces, where the geometry of the joints creates joint-bounded rock masses in the subsurface, has led to the formation of resistant, hard unweathered "core stones" surrounded by highly weathered to decomposed rock.

In some locations, weathering grade changes gradually, often with alternating zones of more and less weathered rock, usually associated with joints or shears. In other places the transition is more abrupt, occurring over less than a 1-foot-wide zone, and occasionally skipping a weathering grade, i.e. decomposed rock abruptly becoming moderately or slightly weathered. Decomposed to highly weathered rock typically has P-wave velocities below 9,000 feet/sec and moderately weathered rock has P-wave velocities between 9,000 and 14,000 feet/sec. Slightly weathered and fresh rock have P-wave velocities above 14,000 feet/sec. The unconfined compressive strength (UCS) of decomposed and highly weathered rock at the site is on the order of 1000 psi or less, moderately weathered rock is up to about 5000 psi, and slightly weathered or better rock had UCS values as high as 20,000 psi.

Overlying the decomposed granite (DG) are natural soils (alluvium, colluvium, and residual soil) and artificial fill placed during construction of Folsom Dam in the 1950s. The natural soils are generally less than 10 feet thick. The thickness of the existing fill overlying the natural soil ranges from 5 to 45 feet, but is generally less than 20 feet.

Engineering Properties for Analysis

The hydraulic conductivities of the rock materials for use in seepage analyses were selected on the basis of results from in-situ testing performed during the field explorations for Phase IV. The hydraulic conductivity of the soil materials (alluvium/colluvium) was selected based on the USCS soil classification types corresponding to these materials (i.e., SM, SC-SP, SC-SM), while the hydraulic conductivity of the original fill materials was obtained from Phase III characterization studies.

The hydraulic conductivity values assigned to each material are summarized in Table 1. To consider the effect of discontinuities in the seepage analysis, the hydraulic conductivities in the direction of the discontinuities are assumed to be 10 times greater than the rock mass as a whole. The preferred orientation for seepage was assumed to coincide with Joint Set D (see Newman et al. 2014), which presents an apparent dip of 14° along the rock plug design section, dipping in the downstream direction. Both isotropic and anisotropic hydraulic properties were assigned to the decomposed granodiorite and weathered granodiorite; only isotropic properties were used for the other materials. Properties for the cutoff wall were developed before the type of wall was selected and are therefore conservative for a concrete wall.

The cutoff wall is located in a temporary fill and rock plug within the approach channel and provides seepage control through the rock plug between the excavation upstream of the control structure and Folsom Lake. The 1,170-foot-long cutoff wall consists of a concrete wall installed across the width of the future approach channel as shown on Figure 3. The cutoff wall alignment shown on Figure 3 corresponds to a configuration that maximizes in-the-dry excavation for the approach channel. The cutoff wall depth was expected to vary between approximately 25 feet at the north abutment tie-in to a maximum of 100 feet. The depth was intended to penetrate through fill and DG, extending at least five feet into highly weathered (or better) rock. This requirement was developed with input from stakeholders and was required for long-term dam safety criteria to be met during construction.

Table 1. Hydraulic Conductivities

Material	Isotropic Condition K (cm/sec)	Anisotropic Condition			
		K_x (cm/sec)	K_y (cm/sec)	K_y/K_x	Direction (°) *
Existing fill	5×10^{-2}	5×10^{-2}	5×10^{-2}	1.0	N/A
Colluvium/Alluvium	1×10^{-3}	1×10^{-3}	1×10^{-3}	1.0	N/A
Temporary embankment – rockfill	10	10	10	1.0	N/A
Temporary embankment - earthfill	1×10^{-3}	1×10^{-3}	1×10^{-3}	1.0	N/A
Decomposed granodiorite	1×10^{-4}	1×10^{-3}	1×10^{-4}	0.1	-14
Highly weathered granodiorite	3×10^{-5}	3×10^{-4}	3×10^{-5}	0.1	-14
Moderately weathered granodiorite	3×10^{-5}	3×10^{-4}	3×10^{-5}	0.1	-14
Cutoff wall	1×10^{-6}	1×10^{-6}	1×10^{-6}	1.0	N/A

* Measured as the angle between the horizon and the preferential flow direction (dipping in the downstream direction).

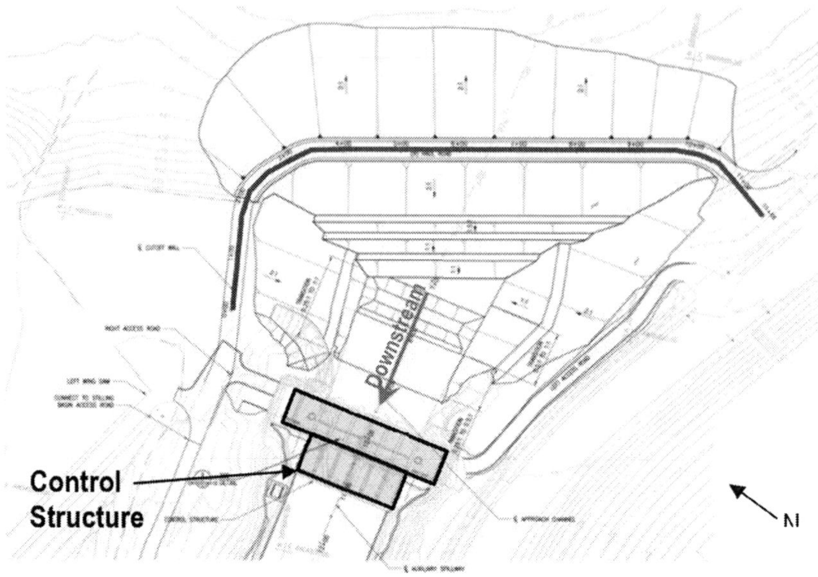

Figure 3. Cutoff wall alignment in temporary embankment

During design, the wall alignment was shifted 90 to 100 feet upstream with the addition of a temporary fill placed over the existing reservoir shoreline. This resulted in more in-the-dry excavation and less in-the-wet excavation, reducing cost and schedule. This change also reduced the amount of wall installation through the existing loose fill containing hard boulders because loose fill with boulders would pose significant constructability challenges. Additionally, the realignment results in less interference with the control structure contractor because the cutoff wall work is farther away from their work area.

Temporary Embankment

The temporary embankment was constructed of materials referred to as Zone 1 and Zone 2. Both are crushed rock obtained from on-site and nearby off-site sources. The typical temporary embankment and cutoff wall geometry are shown in section on Figure 4 together with the soil and rock profiles. Zone 1 is finer, with a uniform gradation between one inch and three inches, intended to make drilling of the cutoff wall piles easier. Zone 1 was initially designed as an earthfill. Zone 2 is coarser, with a uniform gradation between one and six inches. It was anticipated that the lower portion of Zone 2 might be dumped underwater if the lake level was high at the start of construction. Filter compatibility between the Zone 2 and Zone 1 materials was also checked to confirm that the gradation of the Zone 1 material would prevent material in Zone 2 from piping downstream during normal seepage conditions, and to confirm that the gradation of the Zone 2 material would prevent material in Zone 1 from piping upstream under rapid drawdown conditions.

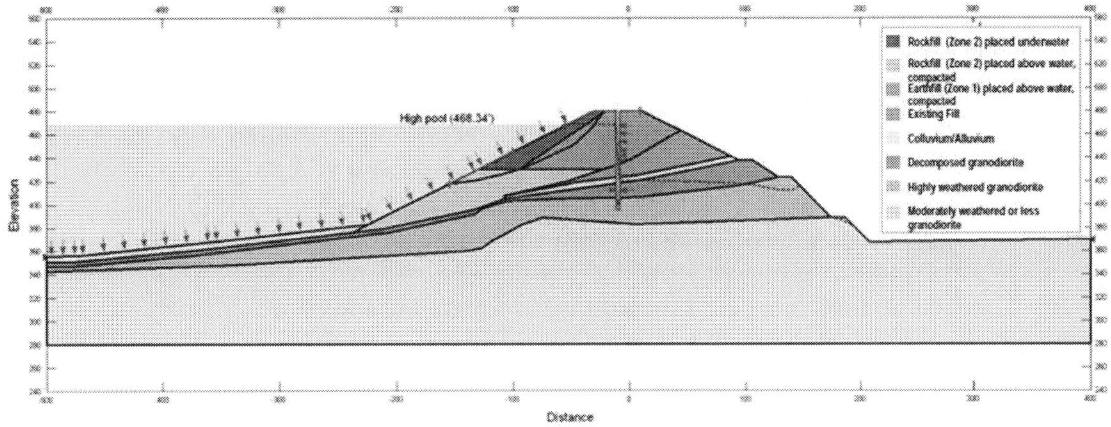

Figure 4. Typical temporary embankment and cutoff wall section

Cutoff wall

Two types of cutoff walls were considered: a secant pile wall and a slurry panel wall. Specialist contractors were contacted to gain information on the preferred cutoff wall type, constructability issues in ground with boulders, and productivity and cost information.

Construction of a slurry wall through loose fill with boulders would involve the potential for slurry loss, resulting in caving. Also, some shifting/settling of boulders may occur during drilling activities. Slurry loss could be controlled by pre-treatment, such as by grouting, along the sides of the cutoff wall, which would add to cost and schedule. Due to this potential for slurry loss and caving, a panel wall, or any wall system relying on slurry for ground support, was discounted as a potential option for seepage control. As such, a secant pile wall with casing was adopted. A secant wall would be cased through the existing fill, so drill fluid loss would not be an issue for ground support.

The secant pile shaft diameters were to be selected by the contractor, with the diameter required to be greater than three feet. Vertical tolerance was required to be within 0.5% of vertical. Column continuity was required to achieve a 12-inch minimum overlap (perpendicular to the plane of the cutoff wall) at any point between the primary and secondary shafts. A downhole survey of each shaft was required to verify verticality, diameter, and overlap.

Planned Instrumentation and Monitoring

Geotechnical monitoring of the rock plug was required to confirm that the temporary embankment and downstream slope of the rock plug perform in accordance with design expectations. Because the embankment and rock plug are temporary features, the instrumentation will only need to be sufficient for a limited period of time (up to 18 months). As such, a limited instrumentation plan was developed that included vibrating wire piezometers, survey monuments, and inclinometers, all located at the crest of the temporary embankment. The piezometers were installed both at the wall tip elevation and about 30 feet above the wall tip elevation to confirm the water levels calculated from the seepage analysis described below. Inclinometer casings extend below the final channel invert elevation.

The instrumentation will be monitored by the contractor and instrumentation readings will be taken weekly to show that the temporary embankment and plug are performing in accordance with design expectations. Seepage and erosion monitoring will also be performed where seepage is observed in the excavation.

An instrumentation and inspection manual was prepared to indicate frequency of instrument readings, data recording, reporting, and evaluation protocols. Threshold values for the instruments were also included in the manual.

ENGINEERING ANALYSIS

Steady-state seepage analyses were performed using the computer program SEEP/W (Geo-Slope 2007). SEEP/W is a two-dimensional finite element model and was used to estimate the phreatic surface, hydraulic gradients, and flow quantities for the proposed cutoff wall and plug configuration. It was assumed that the materials can be modeled according to Darcy's Law. The section used for the seepage analyses is shown on Figure 4.

Boundary Conditions

The following boundary conditions were assumed for the seepage analyses:

- The lake level was assigned as a constant head boundary condition to all surface nodes on the upstream side of the plug. Water levels corresponding to the high water reservoir level (El. 468.34 feet) and the low water reservoir level (El. 428.34 feet) were used.
- The water level at the downstream end of the models was assumed to be at the approach channel invert (El. 362.34 feet). This constant head boundary condition was applied to the nodes on the downstream border.
- A "potential seepage face" boundary condition was applied to the downstream slope.
- A "no flow" boundary condition was applied at the bottom of the model (elevation 240 feet was used as the bottom of the model).
- The cutoff wall was not applied as a boundary condition but was modeled explicitly using the properties described above.

Results

Results for the worst-case analysis including the high reservoir level and the effect of anisotropy in the rock are shown in Figure 5. The calculated phreatic surfaces show sharp drops at the cutoff wall for all analyzed cases. Estimates of unit flow rates for the two reservoir levels considered are summarized in Table 2. The estimated magnitude of seepage flow for the different conditions considered varies between 10 and 85 gallons per minute (gpm). These flow rates were used to specify the required capacity of the excavation dewatering system. While rock joints were not modeled explicitly, they were accounted for in the measured hydraulic conductivities and the assumed degree of anisotropy, and it is anticipated that seepage will occur through joints in the moderately-weathered or better rock below the cutoff wall tip elevation.

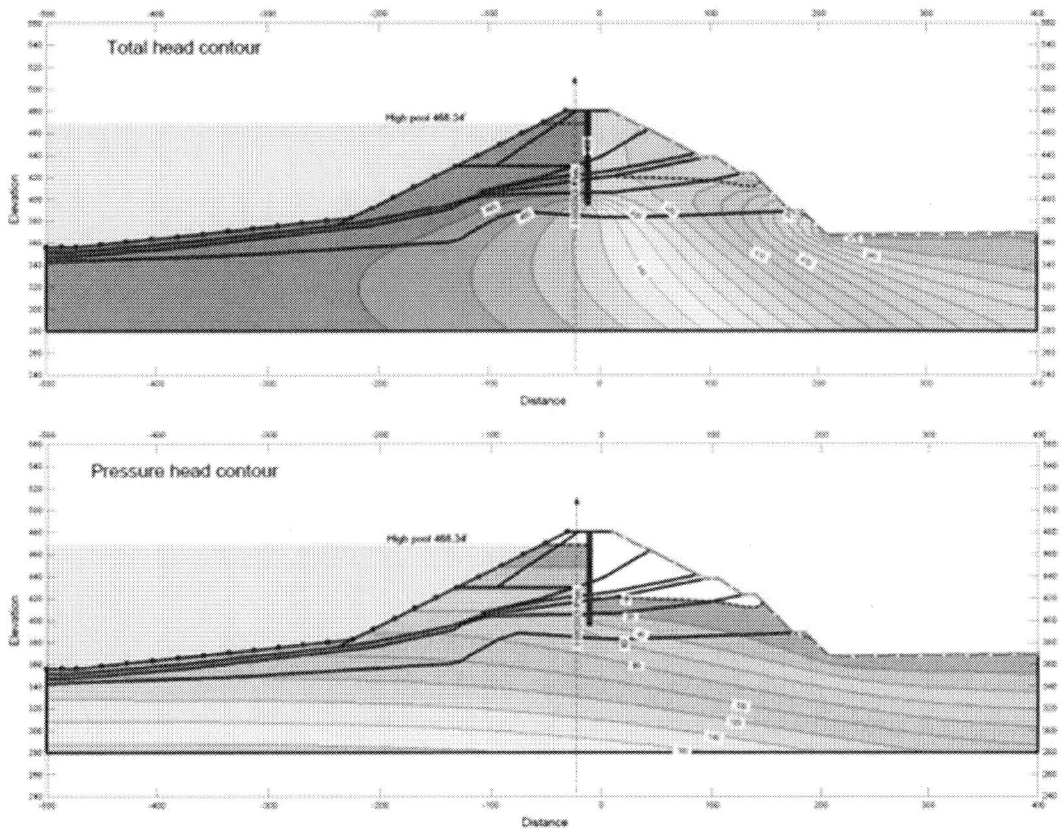

Figure 5. Seepage analysis results

Pore pressures calculated from the seepage analyses were used in the stability analysis of the rock plug and temporary embankment slopes and in the design of rock reinforcement and drainage measures for the downstream slope of the rock plug, as required by the seepage shown to be daylighting on the face of the excavation in Figure 5. Note that this phreatic line on the downstream face of the rock plug is an artifact of the analysis; seepage near the downstream face of the rock plug is likely to be concentrated in joints.

CONSTRUCTION

The design was required to make a number of assumptions to accommodate uncertainty in the lake level because reservoir operations were not allowed to be modified for construction. However, unusually low lake levels during the winter of 2013-2014 allowed all of the excavation at the upstream end of the approach channel (upstream of the temporary embankment) to be completed in-the-dry and ahead of schedule. This excavation had originally been anticipated to occur in the wet and at the end of the project, so this change represents a significant gain in the schedule.

Secant-Pile Cutoff Wall

Table 2. Summary of calculated seepage flow rates

Reservoir Water Elevation (ft)	Seepage Condition	Total unit flow rate (ft^3/sec per ft)	Total flow rate (gpm)*
468.34	Isotropic	6.1×10^{-5}	20
	Anisotropic	2.6×10^{-4}	85
428.34	Isotropic	3.0×10^{-5}	10
	Anisotropic	1.5×10^{-4}	50

* based on an approach channel width of 700 feet.

The low lake level also allowed the contractor to excavate and replace all of the existing loose, bouldery fill (to a maximum depth of about 20 feet below the pre-construction ground surface) that was present along the cutoff wall alignment before construction of the temporary embankment. The replacement of this existing fill reduced much of the potential for delays associated with difficult drilling conditions.

Construction Exploration Program and Update of Design Wall Profile

The design exploration program included ten exploratory core borings drilled within 100 feet of the 1170-foot-long cutoff wall alignment, with only four of them less than 30 feet from the alignment. The preliminary design cutoff wall profile that was the basis for bids was developed from these design exploratory borings. This was accomplished by interpolating an idealized surface representing the contact between DG and highly-weathered (HW) granite. The cutoff wall was designed to penetrate five feet below the HW granite surface. However, as described above, weathering at the site is joint-controlled, and thus the depth of the DG/HW contact interpreted from the widely-spaced design borings was only an approximation.

The core drilling and water pressure testing program required by the contract was begun by the Phase IV contractor (Kiewit Infrastructure West Co.) in December 2013 and completed in January 2014. Core borings were drilled every 50 feet along the planned cutoff wall alignment (starting at Station 1+00, see Figure 9) to supplement the design borings. Borings were logged by USACE geologists and all rock core was boxed for later inspection. Water pressure testing was performed in each of the exploratory borings with the goal of confirming that the hydraulic conductivity of the material near the wall tip elevation was consistent with the properties assumed during design. Results from the water pressure testing were interpreted following the recommendations of Houlsby (1976) and Quiñones-Rozo (2010).

After completion of the borings and water pressure testing, logs, core photos, and water pressure test data were provided to the designer to refine the preliminary cutoff wall profile. Field visits were also made by engineers and geologists to observe the drilling and water pressure testing, and to inspect the core as borings were completed. The construction core borings showed depths to the DG/HW contact in general agreement with the design borings, but with some significant local variation. They also showed areas where slightly weathered to unweathered core stones were present above DG or HW material, consistent with observations from the design borings. These core stones were of particular interest because, while not as difficult to drill through as the hard boulders encountered in the loose existing fill, they were still likely to present drilling difficulties during installation of the cutoff wall. There were also areas where highly weathered or DG seams were encountered at depths greater than the preliminary cutoff wall depth.

A number of factors were considered in selecting the final target cutoff wall depth. In general, the target wall tip elevation is five feet below the interpreted bottom of DG in each exploratory boring, with the following exceptions:

1. The target wall tip elevation is not a strict requirement; drilling may be stopped at a higher elevation if hard drilling conditions are reached above, but close to, the target elevation.
2. If the bottom of DG is more than five feet above the preliminary design line, then the target cutoff wall tip elevation is set at the original design line unless water pressure testing, inspection of the core, or consideration of overall geometry show that hydraulic conductivities in the material above the original design line are low and that the potential for piping through the rock plug is low. Note that the target tip elevation may not be reached at every location if hard drilling conditions are encountered as noted in exception 1.
3. If the bottom of DG is below the preliminary design line then the target depth may be limited to the preliminary design line if inspection of the core and consideration of overall geometry suggest that DG present below the wall tip elevation is not subject to erosion and piping through the rock plug based on its gradation.

Consideration of the overall geometry is mainly relevant at the south end of the alignment (see Figure 2). South of that point, seepage passing beneath the wall has to flow through a greater distance of soil and rock before it reaches the excavation face, which reduces gradients and the potential for erosion to acceptable levels. At the north end of the alignment, the cutoff wall is closer to the excavation and, therefore, there is no increase in distance that can be counted on to reduce the potential for piping through pervious DG.

The design intent was not for the cutoff wall to be drilled through a significant amount of moderately weathered or better rock to cut off relatively thin zones of DG at great depths. Inspection of core in the borings where this condition is present showed that the DG at depth, after being broken down by hand, has the appearance of a coarse sand or fine gravel with little to no fines. As such, for piping to occur, flow velocities would have to be high and the flow would have to reach a relatively wide, open joint which would have to be persistent enough to daylight on the downstream face of the rock plug. Even if flow velocities are high enough to erode DG at depth, a joint that is neither persistent nor wide enough would trap the eroded particles and stop the incipient piping. The likelihood of piping occurring in this situation decreases with depth because the distance from the cutoff wall to the face of the excavation increases.

Secant-Pile Cutoff Wall

Based on this, it was determined that drilling of the cutoff wall shafts may be halted after the drilling has advanced two feet into hard drilling conditions. The criteria used to define hard drilling would be based on rate of progress or on some other measure of drilling resistance, to be developed with input from the contractor. The intent was that the contractor not spend a great deal of time advancing the shafts unnecessarily into moderately weathered or better rock.

The designer was notified if hard drilling conditions were encountered more than 10 feet above the target wall tip elevation in production shafts. A shallower wall depth at those locations may be acceptable depending on the conditions, or additional measures may be required to probe ahead of the current depth and determine how deep the hard rock might extend below the current depth. The worst-case condition in this scenario is where drilling stops after encountering a core stone at a relatively high elevation with DG still present below it.

Cutoff Wall Drilling

Construction of the cutoff wall began with the construction of an unreinforced concrete guide wall in March 2014; drilling was completed in July 2014. The cutoff wall subcontractor was Malcolm Drilling.

The wall was constructed with the drilling equipment situated approximately perpendicular to the centerline of the wall (equipment straddled the centerline on another secant wall on the project) . Two rigs (Bauer BG 40) were used at the beginning and a 3rd rig (Liebherr LB44) was brought in for a short period of time to get the wall completion date back on schedule. The general layout of the drilling operation is visible on Figure 2. Shaft diameters were generally 3.3 ft, with spacings of 27 to 30 inches on-center. Diameters were increased to 3.9 ft spaced 35 inches on-center over a limited reach where the wall depth was about 100 feet.

The specifications originally required that the guide wall be reinforced, but the contractor was allowed to build it without reinforcement so that any supplemental shafts did not have to be drilled through reinforcing bars. The unreinforced guide wall experienced severe cracking in some areas and made it difficult to maintain the centerline of the wall. Measuring the verticality of the shafts was a critical specification requirement (<0.5%) in calculating the minimum overlap. The plans and specifications required a minimum overlap of 12-inches as measured perpendicular to the wall.

During the test section, the contractor tried several methods to accurately determine the verticality of the shafts. These included the SoniCaliper (manufactured by Loadtest) and the PRAD TLK system (manufactured by Lutz). The contractor also attempted to use an inclinometer down the shafts, but it became apparent that the amount of time required to conduct this procedure would have been prohibitive. The contractor also used a smart level to check the verticality of the steel casings as they were being added to the shafts.

The SoniCaliper was tried several times at the beginning of the construction of the wall, but did not work well. Occasionally, there were abrupt changes measured in the casing alignment which seemed unusual for rigid steel shafts. It was suspected that the errors were due to the fact that the holes (or shafts) were dry. SoniCaliper data sheets indicate: "The SoniCaliper is a profiling sonar device, specially adapted to function in drilling fluids. Because the properties of drilling fluids vary widely, a calibration must be performed for each shaft to determine fluid wavespeed."

The Lutz system (PRAD sensor) was used and placed right above the auger as shown in Figure 6. The PRAD sensor measures the deviation and the bottom pressure and uses technologies based on accelerometers and gyroscopes. The VR28 (Lutz) was also used to measure rotation and speed of the Kelly. The detection element creates a high frequency electromagnetic field at the tip of the sensor and its amplitude is dampened by the presence of a nearby metal mass. This system seemed to function well and provided shaft deviation measurements both perpendicular and parallel to the wall centerline every 5 feet in depth.

Figure 6. Lutz PRAD above auger.

The Phase IV prime contractor was responsible for quality control (QC) for the wall construction. In this capacity, the contractor performed all measurements of depth, inclination, and overlap, as well as ensuring that piles conformed to specified tolerances and that ultimately a functional seepage barrier was constructed (by calculating overlap). The responsibility of USACE is that of quality assurance (QA) and engineering during construction. QA generally consists of overseeing the contractor's QC procedures and spot-checking specific aspects of the work. For the Phase IV cutoff wall, these items included checking depth and hole cleanliness for every pile, observing excavation and foundation conditions at depth, monitoring groundwater levels, a nearly continuous presence of USACE QA personnel during the drilling and concrete operations, and involvement in ongoing discussions related to challenges presented during the work.

Several challenges were encountered during the construction of the cutoff wall. First and foremost was the presence of unweathered core stones and variable elevation of hard bedrock near the design tip of the cutoff wall. The contractor made the decision to employ one Liebherr LB44 drill rig and two Bauer BG-40 rigs with soil augers, rock augers, and core barrels to execute the work. This equipment was selected in spite of the fact that the contract documents indicate that higher-strength core stones (up to 20,000 psi unconfined compressive strength) would likely be encountered. These tools were efficient in drilling through material up to around 4,000 psi in strength, but productivity slowed significantly in material of higher strength (in some cases, from one foot per minute to one foot per hour or less).

As the work progressed, a system was developed between USACE and the contractor to identify drilling conditions that indicated that competent rock (and not a core stone) was encountered above the design line. A USACE point-of-contact was contacted on each occasion, and based on criteria related to the depth of the specific pile in relation to material encountered at a similar depth in adjacent piles, a decision would be made whether to continue drilling through the hard material or to stop drilling at the current depth. Overall, this iterative and intensive process saved the contractor many hours of work in drilling through material ultimately deemed to be massive bedrock, and associated equipment damage.

Figure 7 shows the as-built wall profile compared to both the preliminary design profile and the final target profile developed based the core drilling and water testing program. Note that the final target wall depth was shallower than the design profile at the south end of the wall (past Station 10+50) because it was determined based on the criteria discussed above that the DG present at depth did not represent a seepage and piping risk. Also note the high spot in the wall between Station 0+00 and 1+00 and between Station 3+00 and 3+50. The wall was shortened in these areas based on observations from some initial shafts that were drilled to the design depths and encountered slightly weathered to unweathered rock with no DG.

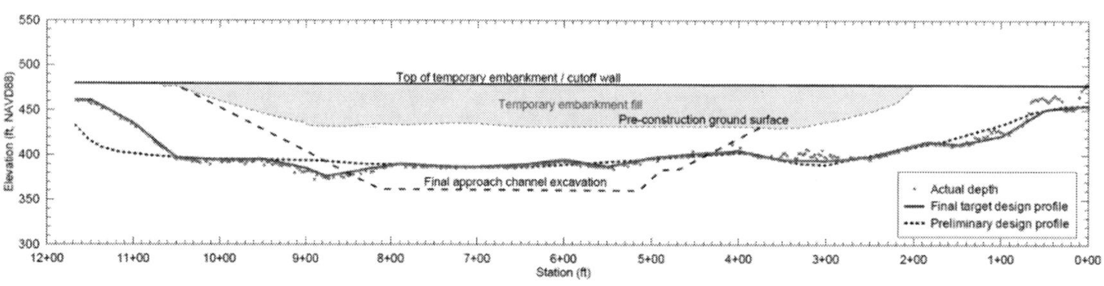

Figure 7. Comparison between design and as-built cutoff wall profiles

Performance

At the time of writing, the Phase IV excavation has advanced to only about 50% of the height of the rock plug, and the lake level remains very low (at or below the wall tip elevation). As such, only limited conclusions can be drawn about performance. Initial piezometer readings show groundwater within the rock plug to be slightly higher than the reservoir elevation, and falling over time. This appears to be due to trapped water within the rock mass or existing fill. Inclinometers show negligible movements into the excavation with no indications of shearing concentrated on a slip surface. Survey monuments also show negligible movements.

CONCLUSIONS

The rock plug, temporary embankment, and cutoff wall were designed to meet permanent dam safety criteria taking into account geologic conditions at the site and construction-related considerations. Seepage analyses showed acceptable conditions. Exploratory data collected during construction allowed the preliminary cutoff wall design profile to be refined and procedures developed to speed up coordination between the contractor, the USACE, and URS to minimize the impact of difficult drilling conditions on the construction schedule. The timely sharing of engineering data between the contractor and the design engineers and geologists made it easier to make decisions quickly. Low lake levels during construction allowed the project to save cost and schedule when possible. The variable geologic conditions encountered during cutoff wall construction required efficient and timely collaboration between all interested parties.

REFERENCES

Geo-Slope International Ltd. (2007). Seepage Modeling with SEEP/W 2007: An Engineering Methodology. Second Edition.

Houlsby, A. (1976). Routine Interpretation of the Lugeon Water-Test. Q. J. Engineering Geology, Vol. 9, pp. 303-313.

Newman, E., M. Forrest, C. Jaramillo, K. Kegel, J. Wisniewski (2014). Rock Mechanics Considerations in the Design of Folsom Auxiliary Spillway Approach Channel, 48th Annual US Rock Mechanics Symposium, paper 14-7629.

Quiñones-Rozo, C. (2010). Lugeon Test Interpretation, Revised, USSD Annual Conference Proceedings, pp. 405-414.

HYBRID TRD AND JET GROUT CUTOFF WALL AT LAKE MANATEE DAM

James D. Hussin, PE[1]
Glen R. Andersen, PE, ScD[2]
Dennis Boehm[3]
Curtis Cook, PE[4]

ABSTRACT

Lake Manatee Dam is nearly 50 years old and consists of a pervious outer shell with a central clayey core. In early 2014, significant erosion observed during annual inspections and supplemental subsurface investigations revealed a compromised dam core with areas of permeable soils, allowing excessive seepage through and under the dam, which threatened major flooding and depletion of the water supply if left untreated. Priority remediation goals implemented included reestablishing seepage control with a new cutoff wall within the existing dam embankment and beneath the service spillway. To achieve this, a hybrid cutoff wall was constructed in the critical areas of the spillway and adjacent embankment. The cutoff wall consisted of a 95 to 105-foot deep Trench Cutting and Remixing Deep (TRD) cutoff wall through the earth dam extending to about 1,000 feet south and 2,000 feet north of the spillway. Beneath the 300-foot long spillway, 105-foot-deep jet grouted cutoff wall was constructed.

This paper describes the design, construction, quality control, and verification testing of the composite seepage cutoff wall.

INTRODUCTION

Manatee County runs and operates Lake Manatee Dam in Manatee County, Florida. This dam was designed and constructed between 1965 and 1967. It is an in-stream reservoir on the Manatee River approximately 20 miles upstream of its discharge point into Tampa Bay. Its primary function is a freshwater supply for users in Manatee and Sarasota Counties.

The dam consists of an approximately 4,700-foot-long zoned-earth embankment. Based upon NGVD29 vertical datum, the present elevation of the crest of the dam ranges from approximately 51 to 55 feet. An aerial view of the dam and spillway structure is presented in Figure 1. The principal spillway (called the service spillway) is located near the southern end of the dam and just south of the original Manatee River channel.

---

[1]Hayward Baker Inc., 6850 Benjamin Road, Tampa, Florida, USA, 813-884-3441, jdhussin@haywardbaker.com
[2]Amec Foster Wheeler plc, 2000 E. Edgewood Drive, Suite 215, Lakeland, Florida, 863-899-4130, glen.andersen@amecfw.com
[3]Hayward Baker Inc., 509 Sam Houston Parkway E., Suite 300, Houston, Texas, USA, 281-668-1870, dwboehm@haywardbaker.com
[4]Hayward Baker Inc., 6850 Benjamin Road, Tampa, Florida, USA, 813-884-3441, crcook@haywardbaker.com

Figure 1. Aerial of Lake Manatee Dam (courtesy of Manatee County).

The embankment is comprised of a pervious outer shell with a central clayey core. This clayey core is 10 feet wide at the top and increases to 16 feet wide near the bottom, where it is keyed into a relatively impermeable stratum. Refer to Figure 2.

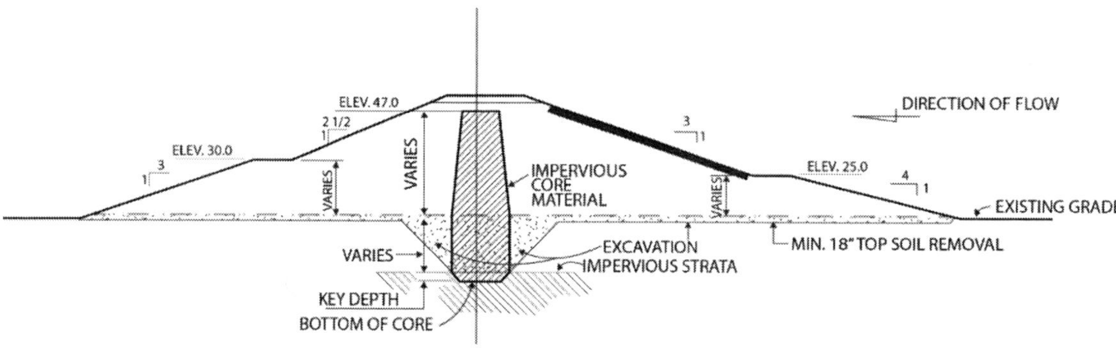

Figure 2. Original Embankment Cross Section.

Shortly after going into service, the embankment began exhibiting signs consistent with internal erosion and piping. Evidence of this particular failure mechanism was found near the spillway and at other locations along the embankment by the U.S. Army Corps of Engineers, Jacksonville District during the first official inspection of the dam in 1978. They reported evidence of sinkholes, loss of embankment surface material and vortices in the reservoir during low water conditions (as told by the dam operators).

A thorough review and integration of the historical inspection data in early 2014 indicated that the dam was in a severely distressed state and that it could experience a breach following a prolonged high flow rate release caused by heavy and constant rain. To prevent such a breach, an emergency repair was initiated to construct a seepage cutoff wall. This emergency repair was called Phase I and included the reestablishment of the dam seepage cutoff function in the critical area beneath the spillway and approximately 1,000 feet south and 2,000 feet north of the spillway. This emergency phase of the work was required to be completed before the start of the impending rainy season.

A Phase II non-emergency repair will involve the filling of an interconnected void system in and around the spillway itself and the reestablishment of the under seepage control system beneath the service spillway.

SITE AND SUBSURFACE INVESTIGATIONS

Various subsurface investigations have been performed over the years at Lake Manatee Dam. These have included the original subsurface exploration, a subsurface investigation in 1979 during a Phase II Inspection, a subsurface investigation in the 1980s, and subsurface investigations in response to erosion and piping concerns that arose during annual inspections from 1998 to 2011. Of concern were the causes for continuing loss of material adjacent to the spillway and movements of the upstream approach walls and downstream training walls. Each of these subsurface explorations found evidence consistent with an internal erosion and piping failure mechanism in the embankment and foundation soils. This evidence included loss of circulation of the drilling fluid and soils with Standard Penetration Test (SPT) N values of less than 1 (weight-of-hammer) in the embankment core and foundation soils.

During a supplemental inspection in 2013, verification was performed of the presence of the very loose soils in the embankment and surrounding the service spillway and the presence of a large void system under the downstream stilling basin and footings for the downstream training walls that had been reported by others. Stick logs of the SPT borings for the subsurface exploration are shown in Figure 3. Also shown in this figure is an indication of the location of higher and lower permeability layers (yellow zones indicating higher permeability and green zones indicating lower permeability).

According to Figure 3, there is a relatively high permeability layer beneath the original seepage cutoff layer. Given that both the bottom of the dam's clay core and cutoff layer exhibited signs of internal erosion, and a high permeability layer is located immediately beneath, the decision was made to target the deeper lower permeability layer as the key-in layer for the new seepage cutoff wall.

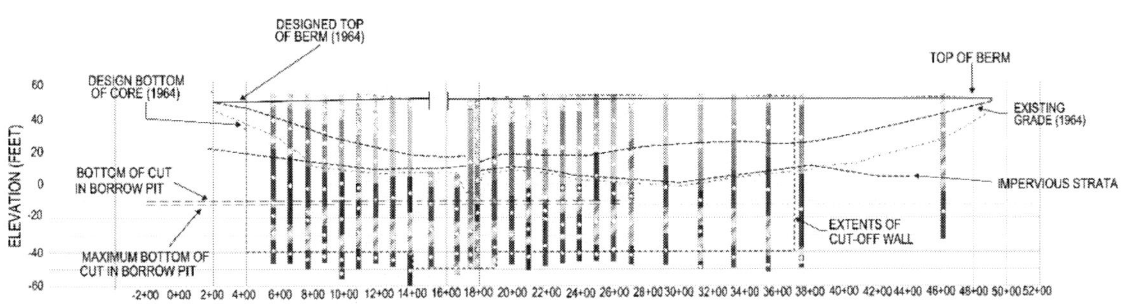

Figure 3. Subsurface profile along dam centerline from SPT borings.

HISTORY OF DAM ISSUES

Lake Manatee Dam has exhibited signs of active internal erosion and piping failure mechanism since the very first inspection by the U.S. Army Corps of Engineers in 1978. Details of the analytical approach used to evaluate the historical inspection data and specific details of this data will not be presented herein. Figure 4 summarizes the general locations of visual observations in the vicinity of the service spillway that are consistent with internal erosion and piping failure mechanism. The visual evidence found at the locations indicated on Figure 4 includes vortices in the reservoir; sinkholes; loose and raveled soils; loss of surficial soils; the formation of a sediment island in the downstream river channel; and loss and undermining of riverbank erosion protection materials.

Also shown on Figure 4 is the approximate head in Lake Manatee Reservoir and in the Manatee River. These drive seepage through the embankment and foundation, and are considered to be largely responsible for activating the internal erosion and piping failure mechanism. Dashed lines with arrowheads are also drawn to indicate the shortest seepage paths for water coming from the reservoir to the Manatee River. It is clear that the evidence of internal erosion and piping contained in the prior inspection reports coincides with the shortest seepage paths and hence with the highest seepage gradients expected at the dam.

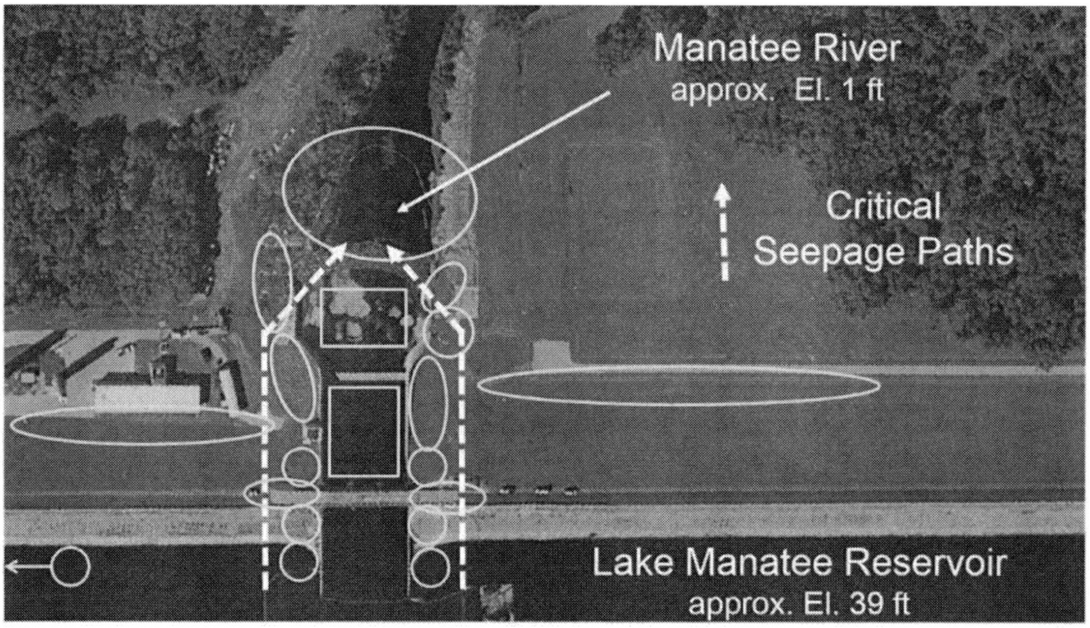

Figure 4. Spillway area with evidence of internal erosion and piping failure mechanism reoccurring throughout dam's history.

EVALUATION AND REMEDIATION DESIGN

The greatest concern for the stability of Lake Manatee Dam was the potential for a rapid drawdown condition resulting in a large sinkhole at or near the embankment core (such as occurred in 2009) and a breach of the dam immediately following a prolonged and high-flow rate release through the service spillway. To prevent this, a deep cutoff wall was recommended. This wall was designed to cutoff the larger flows that were expected to be occurring in the previously identified higher permeability layer immediately beneath the original key-in layer and that were creating the high pore water pressures that triggered the internal erosion mechanism. Figure 5 shows a cross section for the embankment with the seepage cutoff wall placed on the upstream edge of the original impervious core. Figure 6 demonstrates the alignment of the cutoff wall.

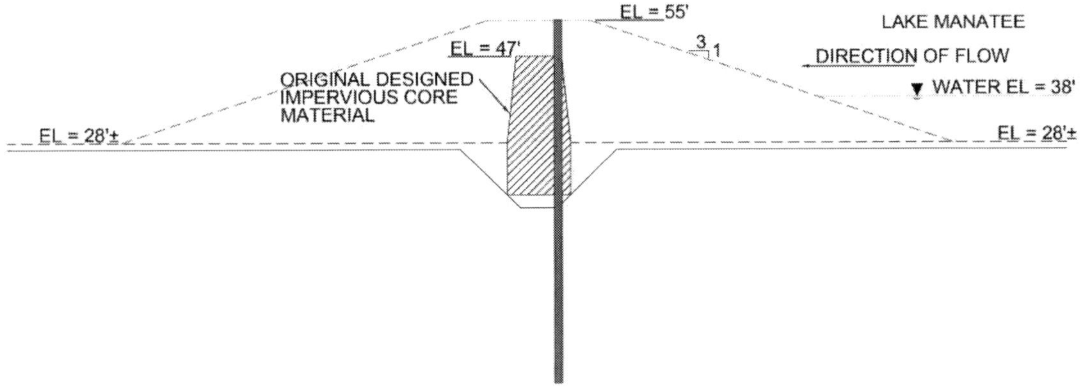

Figure 5. Cross section showing the approximate position of the seepage cutoff wall.

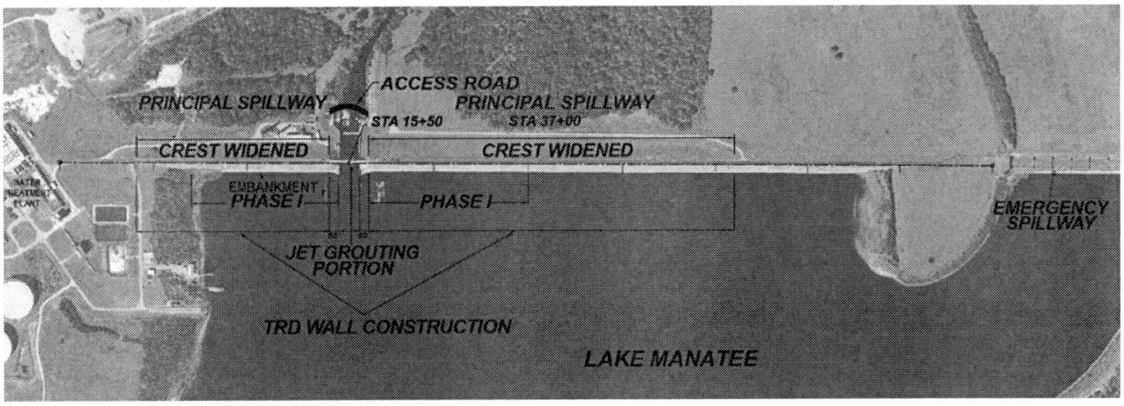

Figure 6. Alignment of cutoff wall.

REMEDIAL CONSTRUCTION

The cutoff wall was constructed by two methods: the jet grouting method and the TRD method. Jet grouting was constructed beneath the spillway and the TRD through the adjacent earth dam.

TRD Method

TRD soil mix walls are mixed-in-place walls that restrict groundwater flow and/or support excavations using a specialized vertical cutter post mounted on a base crawler machine. The vertical cutter post, resembling a large chain saw, is inserted vertically in segments by the crawler machine until the design depth of the wall is reached. The crawler machine then advances along the wall alignment while the cutter post cuts and mixes the in situ soil with cement-based binder slurry injected from ports on the post. The vertical mixing action blends the entire soil profile eliminating any stratification and creates a soilcrete (soil and cement) wall with a high degree of homogeneity and low permeability (Figure 7). This technique has been performed to construct cut off walls as deep as 160 feet in soils and soft rock. In hard rock, predrilling ahead of the machine is required to break up the rock prior to mixing.

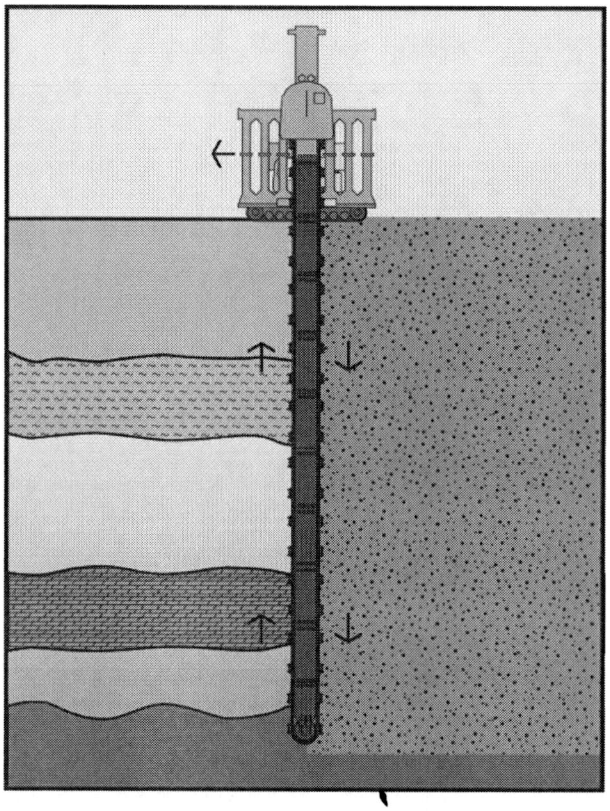

Figure 7. TRD wall method (courtesy of Hayward Baker Inc.).

The TRD post advancement rate and cutter chain speed were adjusted by the operator throughout wall construction, dependent upon the soil conditions encountered and the consistency of the soilcrete created.

The TRD wall alignment (Figure 8) was located near the upstream edge of the dam's core. The TRD equipment was configured to construct a 22-inch-wide wall. To help assure that no grout slurry entered the reservoir, a 3-foot-deep trench was excavated along the alignment ahead of the TRD rig to contain the excess soilcrete produced during wall construction. A backhoe continuously bailed the trench, placing the excess material in dump trucks for transport to an on-site storage area. The excess material was later used as fill to widen selected sections of the dam.

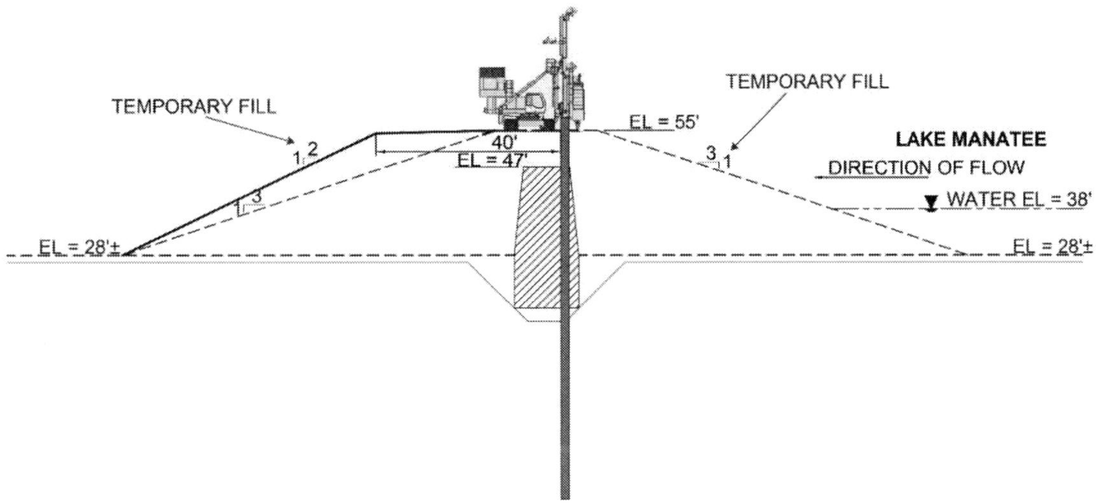

Figure 8. TRD wall alignment and fill placement for equipment access.

A pre-production test section consisting of a 40-foot-long cutoff wall was constructed to full depth utilizing the production methods, verifying the procedures and equipment would meet or exceed the project specifications. The geotechnical contractor utilized a 5 component mix which included water, Portland Cement Type I-II, bentonite, slag cement, and in situ soils to meet the design requirements. This mix produced the specified low permeability soilcrete wall. The TRD wall construction at Lake Manatee Dam is shown in Figure 9.

Figure 9. TRD wall construction at Lake Manatee Dam.

Hybrid TRD and Jet Grout Cutoff Wall

Jet Grouting Method

Jet grouting is a grouting technique that creates in situ geometries of soilcrete using a grouting monitor attached to the end of a drill stem (Figure 10). The grout is injected at high velocities into the in situ soils during rotation and withdrawal of the monitor. The energy erodes the soil and blends it with the grout slurry, leaving behind soilcrete.

Figure 10. Illustration of the jet grouted wall method (courtesy of Hayward Baker Inc.).

Jet grouting is effective across a wide range of soil types, including silts and most clays. Because it is an erosion-based system, soil erodibility plays a major role in predicting geometry, quality, and production. Since the geometry and physical properties of the soilcrete are engineered, the properties of the soilcrete are readily predictable.

The jet grout monitor was advanced to the maximum treatment depth, at which time high velocity grout jets were initiated from ports in the side of the monitor. The grout slurry injected from the jets eroded and mixed the in situ soil as the drill stem and jet grout monitor were rotated and raised.

The jet grouted portion of the wall generally consisted of 4-foot-diameter columns constructed on a 3-foot center-to-center spacing (Figure 11). Near the north and south edge of the stilling basin slab, the approach wall foundations restricted the drilling locations so 8-foot-diameter columns were constructed on approximately 6.5-foot centers at these locations. To enable safe and stable access to the sloped face of the dam on each side of the spillway, a temporary embankment was built utilizing borrowed material from a nearby source. The temporary embankment was held in place by retaining piles that were installed prior to placement of the embankment. Within the stilling basin, jet grouting was performed from a floating barge as shown in Figure 12. The composite TRD and jet grout cutoff wall near the spillway can be seen in Figure 13.

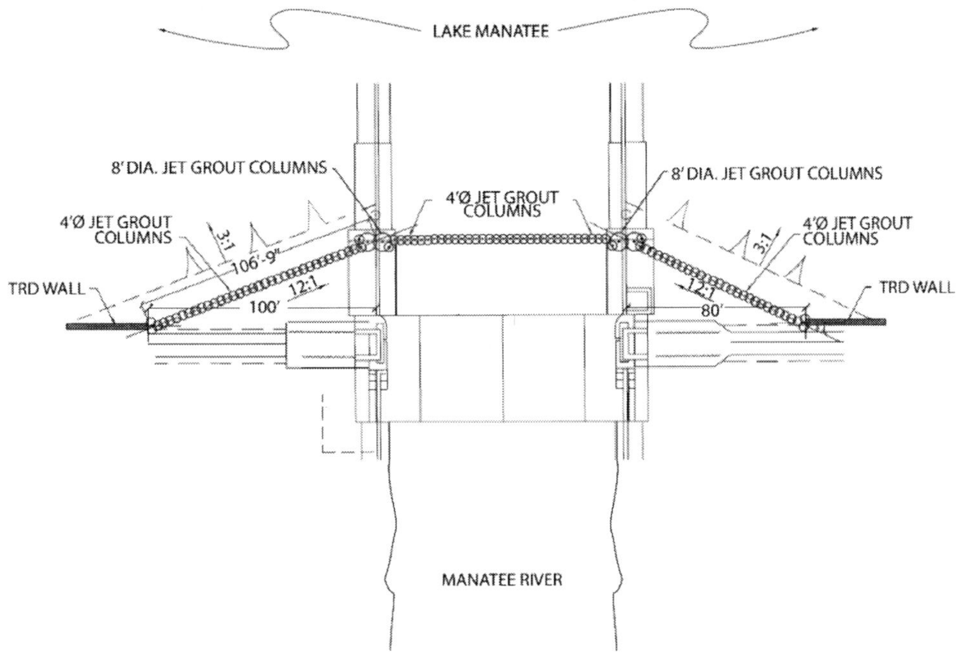

Figure 11. Plan view of jet grout wall connecting TRD walls on either end of spillway.

Figure 12. Jet grout cutoff wall construction through standpipes fastened to spillway slab.

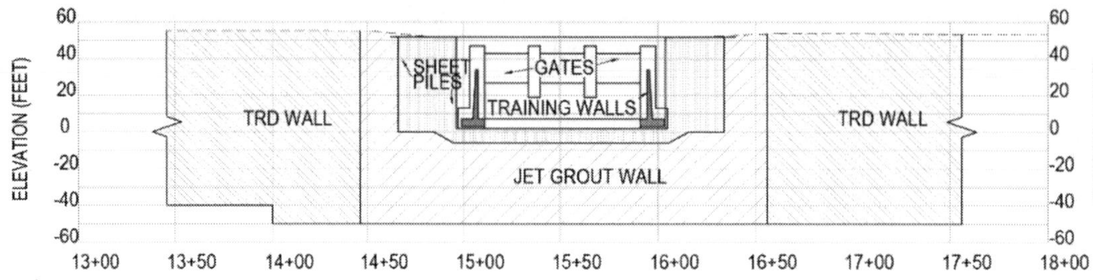

Figure 13. Profile of the composite TRD and jet grout cutoff wall near the spillway.

CONSTRUCTION QUALITY CONTROL

Prior to production work, samples of soils to be treated were retrieved for both soil mixing and TRD laboratory bench scale testing. The samples were mixed in a laboratory with varying binder contents and types to determine the mix design that would be used to achieve the required performance. Because of the urgent nature of the work, initial mix designs were based on the contractor's experience and then modified as the laboratory results were available.

The field TRD quality control consisted of manual measuring, sampling and testing, as well as a computerized data acquisition (DAQ) system, which monitored sensors and instrumentation located in the equipment and grout line. The depth of wall construction was verified by measuring and documenting the length of each section of cutter post during insertion and extraction operations and the length of the post above the ground surface during wall construction. The wall width was controlled by the width of the cutter plates, which were measured and documented during cutter post insertion. The maximum plate width and wall width was ~21.7 inches. In addition, the wall verticality was measured and recorded with a digital level during the initial insertion. Cutter post verticality monitoring utilized in-post inclinometers with data shown in real-time on a proprietary DAQ in-cab monitor (Fig. 14). The allowable inclination of the post was 2°.

Specific gravity of the binder slurry was measured a minimum of 3 times per shift with a mud balance in general accordance with ASTM D2419. Specific gravity was required to stay within 3% of the mix design value. The TRD wall required a permeability of $1x10^{-6}$ cm/sec. Wet grab samples cast into cylinder molds and later tested in a laboratory confirmed the cutoff wall met the design compressive strength of 50 psi at 28 days and permeability of $1x10^{-6}$ cm/sec.

Figure 14. TRD in-cab DAQ screens for real-time operator monitoring of post inclination and grout specific gravity, volume, and temperature as well as other equipment parameters.

The jet grouting rig was also fully instrumented with an on-board DAQ system to monitor specific parameters including RPM, depth, and lift rate; water and grout parameters including flow rate, pressure, specific gravity, and volume; and air pressure and flow. Data from the DAQ system were recorded and displayed alongside specified target values on an in-cab monitor (Figure 15). The operator monitored all parameters, allowing for real-time adjustments during construction to adhere to specifications. Similar to the TRD quality control, samples were retrieved from the wet jet grout material and cast into cylinders for specific strength and permeability testing with similar requirements being met.

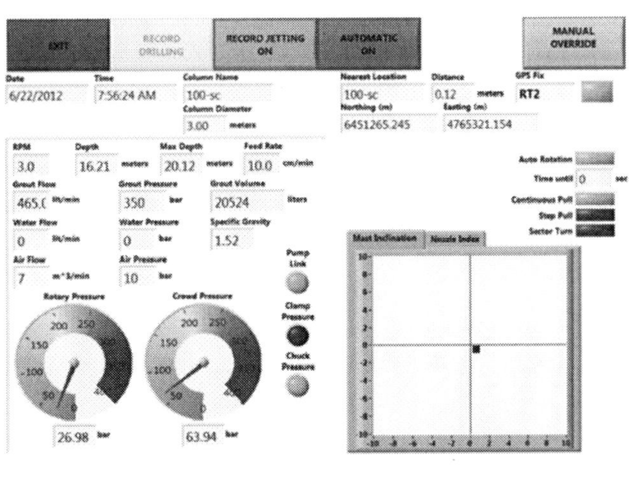

Figure 15. Jet grout DAQ screens for real-time operator QC monitoring.

POST REMEDIATION EVALUATION

Based upon preliminary results from a series of nested piezometers both upstream and downstream of the new seepage cutoff wall, the cutoff wall appears to be functioning properly with a small portion of the seepage from the reservoir passing beneath the wall but with a majority of the seepage flowing around the northern and southern ends of the TRD wall. Relatively small amounts of seepage appear to be passing through the jet grout wall in the general vicinity of the service spillway ranging from approximately 0.2 gallons per minute to about 2 gallons per minute.

These piezometers also indicate that there may be the presence of a regional artesian pressure condition in the high permeability layer just beneath the original seepage cutoff layer. However, a detailed evaluation of the performance of the wall will be presented in a later paper.

CONCLUSIONS

The nearly 50-year-old Lake Manatee Dam required emergency remediation to address an internal erosion and piping failure mechanism. Remediation consisted of a 95 to 105-foot deep cutoff wall constructed in the critical areas of the spillway and adjacent embankment. The TRD and jet grout methods proved effective in constructing an emergency remedial hybrid cutoff wall through an existing compromised earth dam and spillway.

AKNOWLEDGEMENTS

The authors of this paper acknowledge the efforts and contributions made to the formulation, approval, and implementation of the engineering and construction solutions presented in this paper by the staffs of the Owner (Manatee County), Designer (Amec Foster Wheeler plc), and the Contractor (Hayward Baker Inc.). The professional teamwork by all members enabled the successful completion of this challenging dam remediation program.

DOUBLE ROW GROUT CURTAIN DESIGN AND CONSTRUCTION FOR CEDAR CREEK DAM IN MALAKOFF, TEXAS

Daniel P. Bole[1]
Marc T. Miller, P.E.[2]
Billy L. Fisher[3]

ABSTRACT

This case history presents the design and construction of a grout curtain, trench drain and collector system, and quality control and verification testing for the Tarrant Regional Water District's Cedar Creek Dam (Joe B. Hogsett Dam) project on Cedar Creek Reservoir in Malakoff, Texas. The demand to maintain fresh water supplies required plans to design, build, and maintain operations of a raw water infrastructure to tap into existing resources. Since the original construction of the reservoir embankment, seepage related issues have slowly developed that have resulted in wet and marshy areas that are difficult to maintain, and may indicate a possible reduction in dam safety. Measures installed to limit these issues, such as relief wells, are now aged and difficult to maintain. The modifications for the project includes a series of contracts to address these issues, and this discussion deals specifically with a grout curtain used to reduce seepage through a cap of terrace deposits around the cutoff and through the left abutment.

For this project, two significant work phases included the construction of a 1,400- foot double row grout curtain along the left abutment area and approximately 1,300 feet of a trench drain and collector system.

The grout holes were spaced 10 feet on center and battered at 15° from vertical to an average depth of 80 feet. The design required work in two different grout zones. The upper grout zone was composed of terrace deposits typically consisting of sands and gravels, while the lower grout zone consisted primarily of fractured shale.

Since the upper grout zone would be unstable during drilling, the geotechnical contractor utilized a ported PVC pipe to maintain an open hole. Following completion of the upper grout zone, drilling in the rock portion allowed for final grouting. This required a suite of balanced stable grout mixes. In addition, all grouting operations utilized the geotechnical contractor's proprietary iGrout™ system, allowing the owners and engineers to monitor the operations and progress remotely. Permeability tests completed along the centerline confirmed achievement of the performance requirements.

[1]Hayward Baker Inc., 11575 Wadsworth Blvd., Broomfield, Colorado, USA, 303-469-1136, dpbole@haywardbaker.com.
[2]Freese & Nichols Inc., 4055 International Plaza, Suite 200, Fort Worth, Texas, USA, 817-735-7525, Marc.Miller@freese.com
[3]Hayward Baker Inc., 661 Shahan Prairie Road, Little Elm, Texas, USA, 972-294-5000 blfisher@haywardbaker.com

The final stage of construction included installation of a drainage trench and collector system to handle additional water. The trench consists of two short legs and one long leg forming a "Y." Excavating the trench into a permeable sand layer allowed for transport of water into the trench drain and collection in a controlled manner. The flow will be monitored through a weir along the drain alignment.

INTRODUCTION

The Tarrant Regional Water District's (TRWD) Cedar Creek Dam, built from 1961 to 1965, includes a 7,700-foot-long, 91-foot-high main embankment. In addition, the levee, which extends to the northwest from the main embankment, is 9,800 feet long with a maximum height of 35 feet.

The embankment (Figures 1 & 2) includes a zoned earth fill dam with an open-cut cutoff trench. The internal drainage system consists of a blanket-type sand drain under portions of the downstream shell and numerous relief wells. The upstream slope protection consists of 30 inches of rock riprap at higher levels and 24 inches of rock at lower levels of the dam face.

Figure 1. Aerial view of the Cedar Creek Dam main embankment, located on Cedar Creek Lake in Malakoff, Texas.

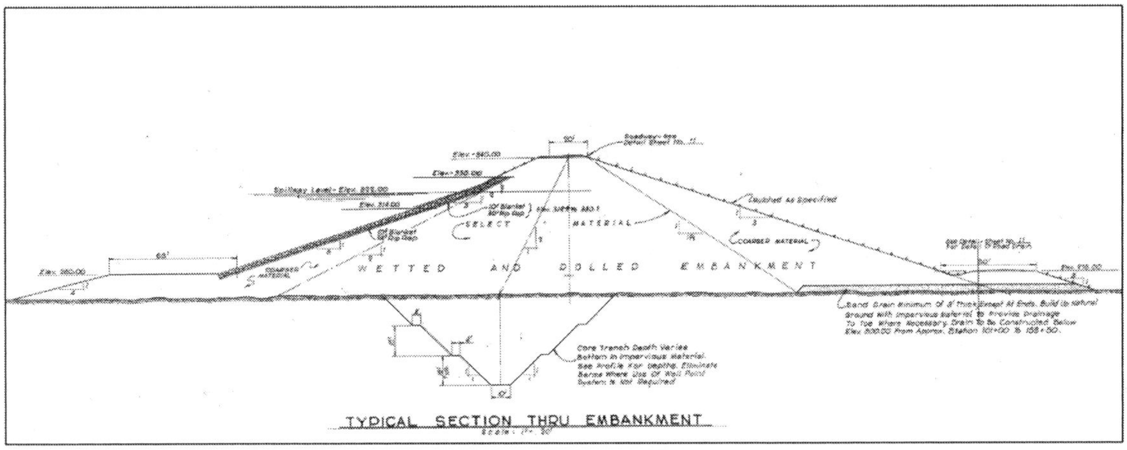

Figure 2. Typical cross-section of Cedar Creek Dam.

DESIGN

The Owner authorized a study of the project based on on-going seepage and maintenance related issues that made many areas difficult to mow and inspect, and hindered access to relief wells. Because of this, many of these areas have developed brush and tree stands that further complicate inspection activities. A comprehensive evaluation of the Cedar Creek Dam included the review of historical data, geophysical testing of the embankment and foundation, 25 soil borings with laboratory testing, sonar investigations of the spillway basin and outlet conduit, additional field inspections, and clearing and excavation to identify the potentially damaged internal drainage and relief well systems. During the course of the study, terrace deposits (Qt) were identified at the left abutment, as noted on Figure 3.

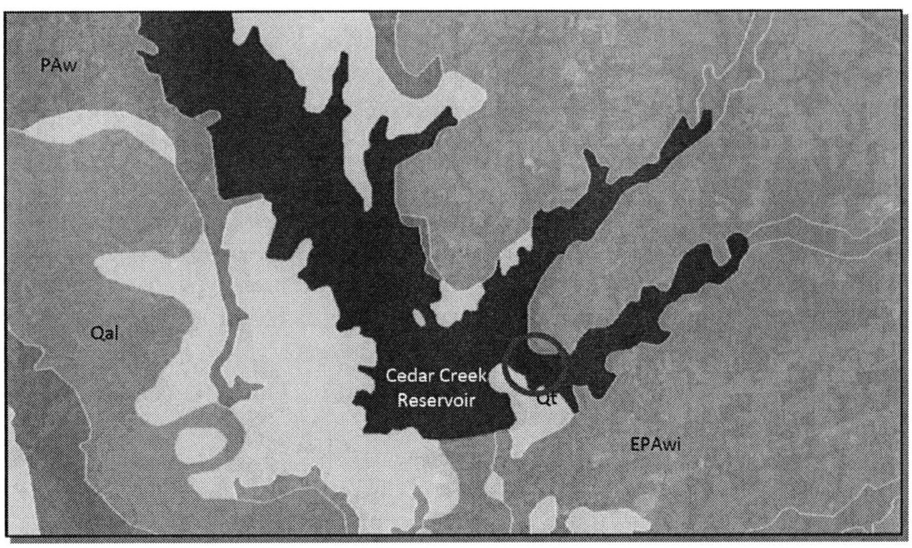

Figure 3. Geologic Atlas Map with Left Abutment circled.
(USGS National Geologic Map Database Catalog, 2013)

These terrace deposits included a wide variety of material types, but were relatively clean and coarse between Elev. 270 and 285, as shown in Figure 4.

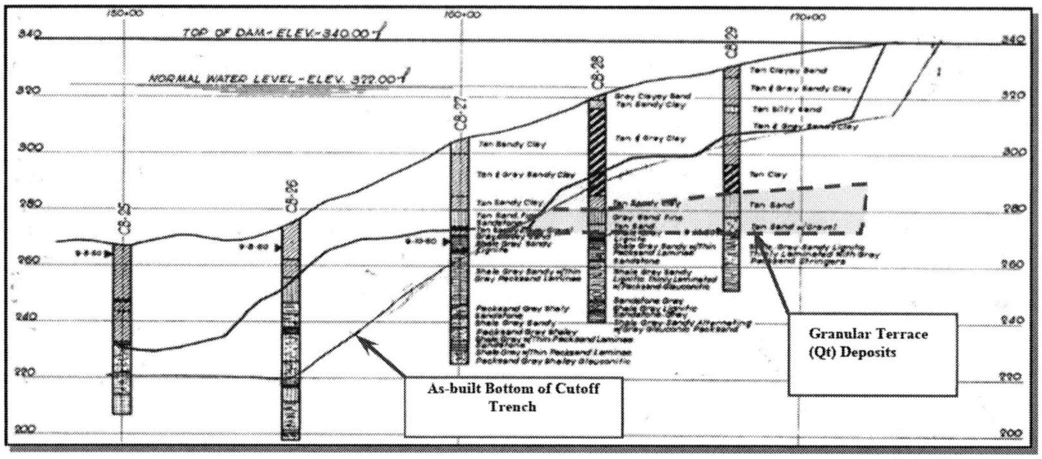

Figure 4. Centerline Cross Section at Left End with Qt Deposits Noted.

Investigations revealed the fill embankment generally consisted of clay and sandy clay underlain by an alluvium formation containing sand, clayey sand, gravel, and clay with sand. Below the alluvium formation, the soils primarily consisted of weakly cemented sandstone and shale of the Wilcox Group.

Because it was suspected that these deposits were exposed on both the upstream and downstream sides of the left abutment, these deposits allowed for through seepage around the left abutment of the dam. Because of the length of the seepage path, internal erosion was not considered a probable failure mode, but the seepage contributed to wet surface conditions at the left groin and surrounding downstream area. The seepage is likely intensified by the termination of the dam cutoff at about Station 155+50 that occurred due to constructability issues related to the excavation depth. This resulted in a section that likely exposed the base of the dam embankment to a window of granular terrace deposits between Station 162+00 to 164+00. Because the source of this seepage differs from the known underseepage issue, the left abutment improvements and the main embankment improvements were divided into separate improvement contracts.

Various cutoff options were explored and the project was identified as a candidate for injection curtain grouting because of the relatively discrete zone of coarse granular terrace deposits in the abutment. The injection grouting also allowed the creation of a lower permeability zone to establish the cutoff along the full length of the dam and to redirect the terrace deposits seepage further downstream of the dam to reduce exit gradients and provide a filtered exit along a collector trench.

Development of the grout curtain geometry was achieved in part with the use of SEEP/W, GeoStudio 2012 developed by GEO-SLOPE International Ltd. The "plan view" model option was used to approximate seepage behavior through the confined granular deposits. The analysis was performed to study flow paths and subsurface pressures within the mapped terrace deposits for the proposed improvements. The results indicate that a permeability reduction of two orders of magnitude will result in approximately 80 to 90% reduction in seepage quantities, and that the resulting pressure regime significantly reduces horizontal gradients and reduces the potential for internal erosion.

The selected grout curtain geometry extended from dam centerline Station 160+00 to 167+00 to establish an improved cutoff and reduce seepage through windows remaining from the original construction. The grout curtain then turned 90° downstream for a distance of 700 feet to direct seepage east and downstream of the dam and improve maintenance in the valley area around the left abutment. The design injection depth was selected to target the granular terrace deposits, but was extended 10 feet into the underlying Wilcox Group to address potential permeable or jointed zones in the shale and sandstone material. Refer to Figures 5 and 6 for plan and cross section views.

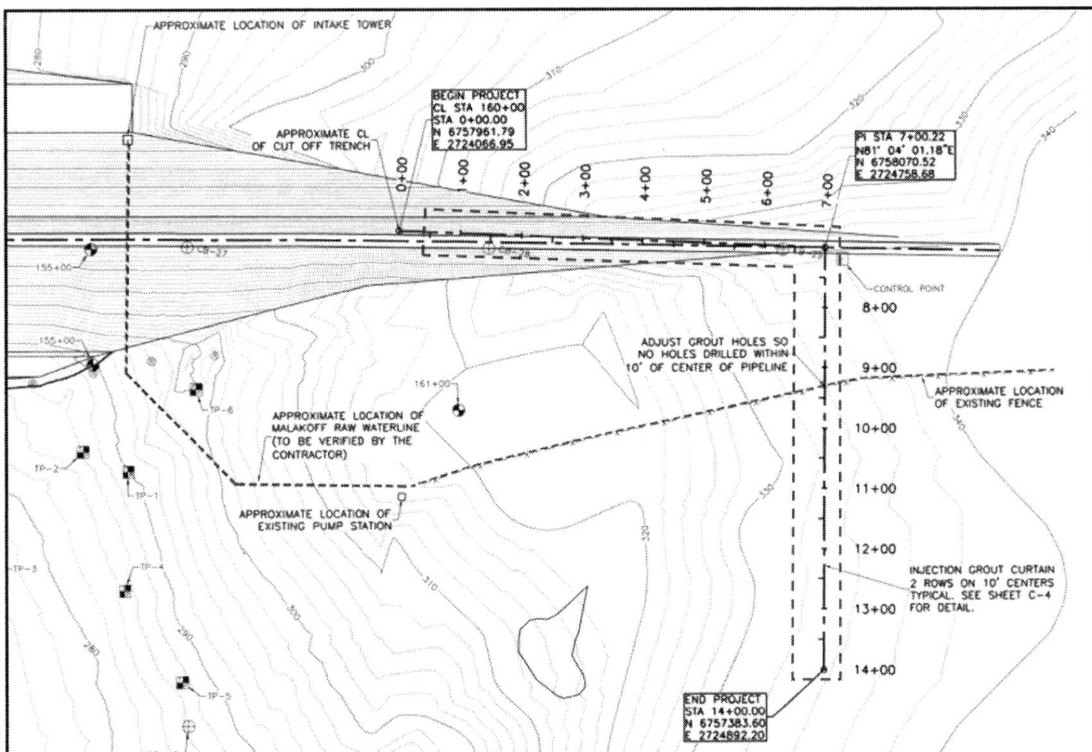

Figure 5. Plan View of Grout Curtain Alignment.

Double Row Grout Curtain

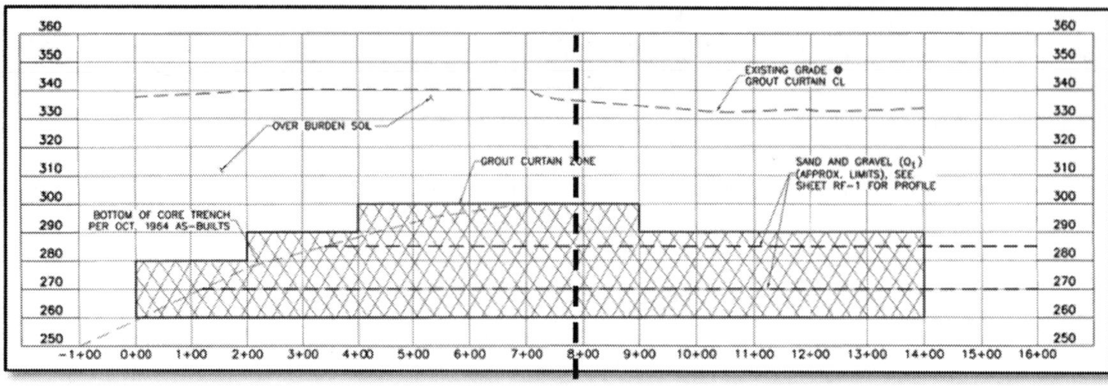

Figure 6. Cross Section of Grout Curtain Injection Zone.

The analysis indicated about 20 to 40 gpm of seepage could potentially pass through the grout curtain and daylight at the dam toe. A secondary system downstream of the grout curtain was included for the filtration and collection of this seepage. Although this system (Figure 7) was not critical for stability, it was desirable to the owner as it would reduce the potential for wet downstream areas and improve access for maintenance and inspection. The collector was divided into two sections to collect seepage from the left groin (Line B) and parallel to the abutment topography (Line A) and discharge into a vault with a weir system. A third line (Line C) was then used to divert the seepage into an existing outfall channel for the relief well system. These perforated collection lines were placed in trenches that included a gravel pipe backfill with a fine sand filter layer. The location of the lines were selected so that the depth of the trench would intercept the granular terrace deposits.

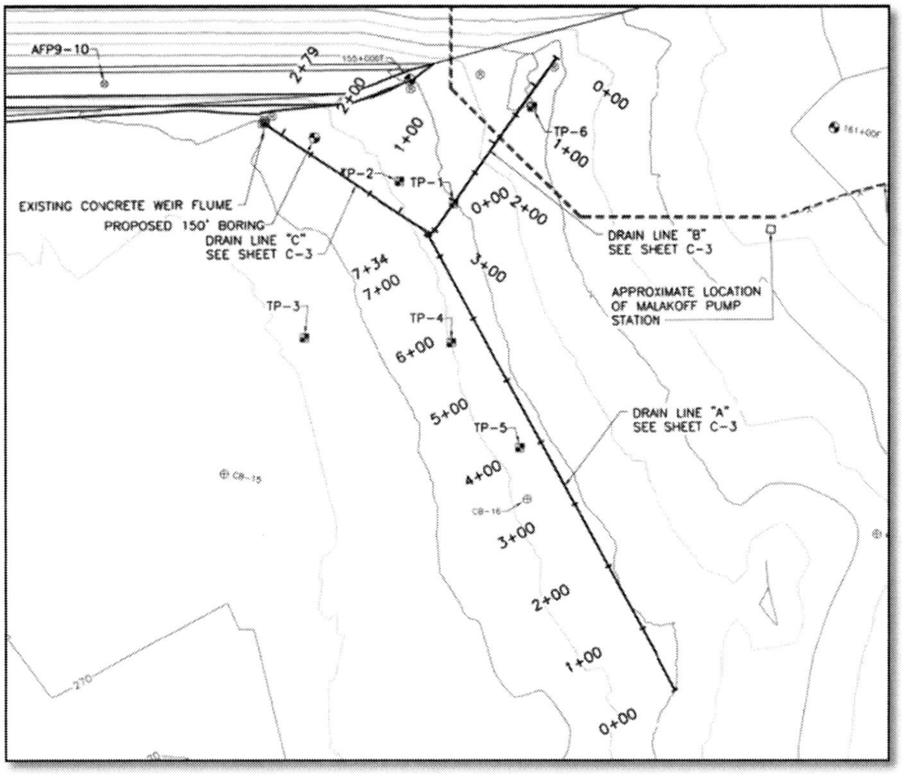

Figure 7. Centerline Cross Section at Left End with Qt Deposits Noted.

CONSTRUCTION

Phase 1

Drilling. Grout hole drilling was performed utilizing two techniques: a rotary-duplex method for the drilling of the embankment overburden, and a single rotary method for the foundation rock. All drilling was completed utilizing water as the flushing fluid to remove the cuttings. The duplex drilling (Figure 8) through the embankment involved the advancement of the outer steel casing to the required minimum 6-inch embedment into the rock beneath the earth embankment. After the drill tooling was advanced to the proper embedment, the inner tooling was withdrawn and a 4-inch-diameter flush-threaded PVC casing was installed inside the steel casing to mitigate potential stability concerns and to protect the embankment during future grouting related activities. The annulus between the steel and PVC casings was then grouted utilizing a bentonite-cement grout. Following extraction of the steel casing, the PVC casing was re-grouted in-place to ensure the casing was properly sealed. The PVC consisted of two different sections of pipe to accommodate the grout curtain design: the upper portion was a plain pipe and the lower portion placed through the intended grout zone was a PVC sleeve-port pipe. The PVC sleeve-port pipe provided grouting access to the sands and gravels that were schedule for treatment by the designated grout zone. After grouting of the upper zone was completed, the rock portion below the PVC casing was typically drilled in a single stage using a roller-bit with a water-flush. The average drilled rock depth was approximately 8 feet.

Figure 8. Rotary duplex drilling on Cedar Creek Dam (2014).

Grouting. The curtain grouting was completed in two separate phases. The upper grout zone, sands, and gravels treatment utilized the ported pipe and was performed in an upstage, bottom-up manner. The lower grout zone consisted of a shale bedrock and was completed utilizing a combination of upstage and downstage treatments, pending hole stability. The stage length was varied but typically did not exceed a length of 10 feet. The grout curtain design was a double row approach that was comprised of four hole series: A-line primary, A-line secondary, B-line primary, and B-line secondary. The grouting operation would move in a particular order starting with A-line primary, then completing the A-secondary series, following by B-line primary and secondary. The sequencing is critical for consistent and accurate evaluation of the grouting operations.

Grouting was completed using balanced stable grout mixes and performed through inflatable packers that were sized to match the drill hole diameter. Each grouting mix was batched on-site using a high-shear colloidal mixer. The grout was then sent to the hole location through 1-inch high pressure grouting lines arranged in full circulation to allow for proper controls of all pressures imposed by the soil and rock formations. An inflatable packer was lowered into the hole to the appropriate depth and then inflated. The connection between the grout manifold and the packer was a hydraulic hose with the ability to be pulled or reeled up to each depth required. After the packer inflation, the desired grout mix was injected into the formation and specific grouting criterion was utilized for the proper closure of each stage. When upstage grouting, the packer was pulled up and inflated to begin grouting on the next stage immediately following the completion of grouting the previous stage. When downstage grouting, the packer was removed and operations continued to the next scheduled hole.

Refusal Criteria. The GIN, or Grouting Intensity Number, was the refusal criterion for terminating a grout stage. Following refusal of the first stage, the packer was pulled and the grout was allowed to reach initial gel time prior to the washing of the ported PVC pipe. However, if grout refusal was not established from the previous requirements, the hole was rested for 4 hours, or until initial gel time completion, and then flushed with 5 to 8 cubic feet of water. Following the water flush, the hole was grouted in accordance to the general procedures.

The refusal criterion for this project was based on closure confirmation on the GIN method, or limiting envelop. The required GIN for this project was established as 150. The following equation was utilized for calculating the GIN (Deere Lombardi 1993):

$$GIN = p*V$$

where:

GIN: constant, specific energy
p: pressure (psi)
V: grout volume (ft^3)/length of stage (ft)

The refusal criteria for this project remained the same throughout the entire grouting process; however, the approach to achieving the required closure varied by the engineer and contractor to meet desired seepage reductions while still maintaining cost considerations of the owner. Data acquisition systems were utilized during all pressure testing and grouting operations. The system collected information in real-time and provided insight to the engineer and contractor during all phases of the operation. The results were uploaded to a file transfer system nightly for the engineers and owner, allowing review while on- or offsite.

During the early stages of the project, the grouting was monitored closely for injection rate and to determine the most suitable grout mix for the successful completion of the project. While utilizing the GIN method it is important to note that it is preferable to utilize a single thick mix to complete the grouting. This differs from other grouting approaches that utilize a mix-thickening sequence that starts with a thin, flowable mix and thickens to a much more viscous grout mix with less flowable characteristics. The geology at the Cedar Creek Dam site varied from porous gravel zones to less permeable sand layers to bedded shales/sandstones. It was difficult to determine prior to grouting how thin or thick the mix should be for optimal treatment. To assist in the decision, a suite of grout mixes was submitted for approval on the project with varying viscosities. All grout mixes were balanced and stable mixes, yielding a bleed of less than 3%.

The grouting operation was initially completed with the thinnest mix, A-mix, to make sure small fractures and low porosity zones were not missed as a result of the mix thickness. The A-mix produced positive results and consistently established the GIN value set in the project specifications. The data was monitored continuously with the iGrout system and submitted daily for review. A small, less critical section of the grout curtain was selected for treatment of all four hole series to be completed for analysis and to determine if any adjustments to the grouting program were necessary. The initial grouting utilized a high threshold for grout take before intermittent grouting was employed. Stopping of a stage that encountered back pressure was not practiced due to fear of losing a stage, or inadequate treatment of the groutable zone. The results of the completed section were graphed for analysis (Figures 9 & 10).

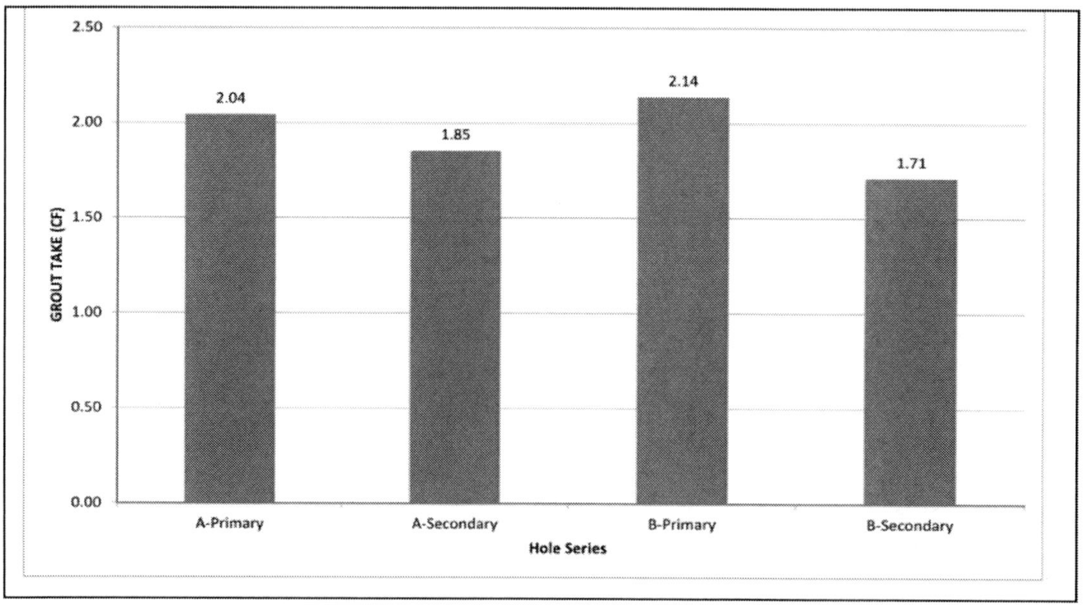

Figure 9. Grout takes by series during early grouting.

The grout takes are presented as the grout volume per foot of borehole. The results indicated that grout takes reduced from primary to secondary, however, the change was not significant enough to leave the grouting program unchanged. The project team reviewed the data and procedures and adjusted the grouting approach to limit "excessive" grouting and provide a more targeted grout curtain. The first adjustment was thickening the mix from the thinnest mix to a medium viscosity mix. The early phases of grouting and pressure testing indicated that the grout zone could be grouted with a medium viscosity mix. The initial pressure testing indicated levels of high permeability with Lugeon values of 20 or greater. A Lugeon is defined as: 1 Lugeon = 1 L/m/bar at a test pressure of 10 bars *(Bruce 2003)*. The Lugeon value of 10 or greater warrants grouting for seepage mitigation *(Houlsby 1990)*. The second adjustment was a more rigid intermittent grouting program with a lower injection quantity. After the program changes, the geotechnical contractor monitored the data and plotted it regularly to evaluate the effectiveness of the changes.

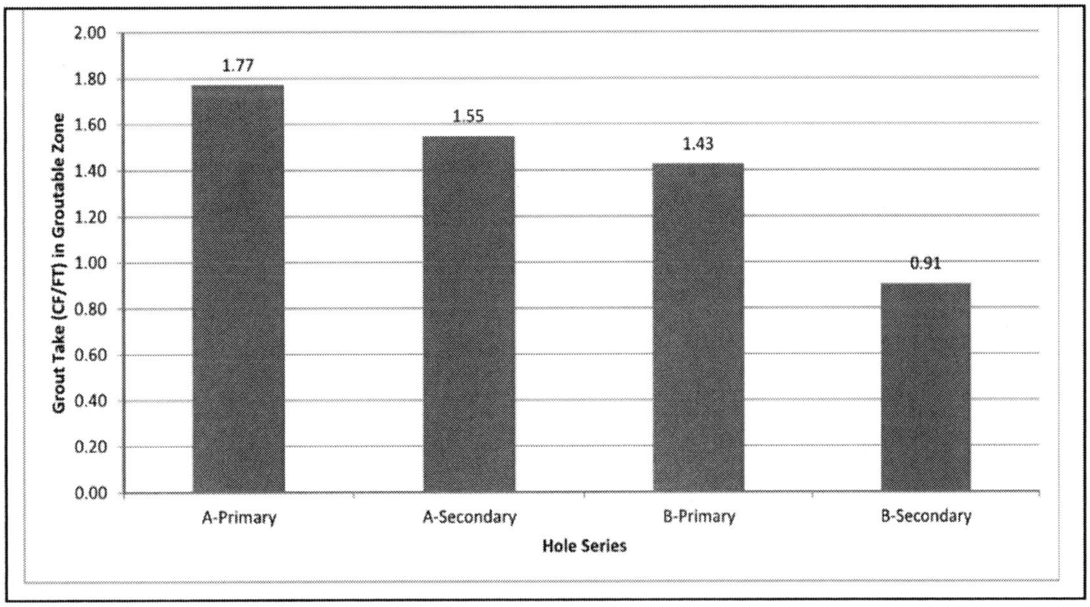

Figure 10. Bar chart demonstrating grout takes by series following changes to grout program.

The updated grout analysis indicated a very steady reduction in grout take by order of hole series. The grout take per linear foot was reduced by 50% from A-line primary to the final B-line secondary series and grout takes decreased considerably for each series. The results provided the project team with a high level of confidence in the effectiveness of the grouting program. To confirm the analysis of the data, verification holes were drilled along the centerline of the grout curtain, in between the rows, and pressure tested to evaluate the post-treatment permeability.

The verification test required the drilling of a hole in the center of the grout curtain in an area where all four series were completed. It should be noted that coring was specifically eliminated due to the nature of grouted gravel zones and the recovery of such material. The hole was drilled into the grout zone and terminated prior to exiting the bottom of the grout curtain. The packer was inflated at the top of the grout zone for a water pressure test. The geotechnical contractor calculated the pressure utilized for the test by taking the maximum pressure on the grout curtain at maximum pool height and multiplying it by 2. The results of the verification test indicated a Lugeon value of less than 5. The results indicated a low Lugeon value, or very tight zone, that is difficult to grout. The low Lugeon values indicated that additional series of holes would not be required.

Phase 2

Drainage Trench Installation. Following the completion of the grout curtain, a collector trench drain was installed through permeable sands downstream of the dam embankment. The trench consists of two short legs and one long leg forming a "Y." The total length of the trench is 1,300 feet, which includes 3 weir monitors within a collector drain for measurement of seepage through the grout curtain and embankment along the left abutment. The trench was installed into a permeable sand layer to allow transport of water into the trench drain and collection in a controlled manner. The vaults and weir design were established based on the design goals for a total seepage of 20 to 40 gpm.

Installation of the collector drain (Figure 11) took approximately 4 weeks from start to finish. The excavation was partially completed in a running sand that created some difficulty with maintaining an open excavation. A combination of dewatering and trench boxes were utilized for the completion of the work. The collector drain was composed of SDR-135 PVC perforated pipe that was surrounded by 1 inch crushed limestone number 4 stone aggregate, wrapped in a geotextile filter fabric, and placed on a clean filter sand to allow for the flow of water into the pipe. Once the pipe was installed, a vault was formed and poured to allow the placement of vibrating-wire weir monitors.

Figure 11. Trench collector installation August 2014.

DATA ACQUISITION (DAQ)

The geotechnical contractor employed a proprietary DAQ system to monitor all pressure testing and grouting operations in real-time. The grout technician simultaneously controlled up to 3 water pressure testing or grouting operations from a single computer screen. The system automatically computed target pressures for each grouting stage based on the project specifications, and recorded various monitoring parameters, including injection pressure, flow rate, and volume. The utilization of iGrout allowed for real-time observation of all pressure testing and grouting operations. The data were used for grouting analysis to determine the grouting effectiveness and to allow for on-the-fly decision making regarding the proper completion of each stage. The data were displayed on a desktop computer screen for a grout technician (Figure 12) and engineer to monitor progress. In addition, pressure testing and grouting plots (Figure 13) were created and submitted daily for each stage. The water testing and grouting data were synced with a CAD interface and plotted automatically to produce updated "as-built" drawings.

Figure 12. iGrout Technician at Cedar Creek Dam.

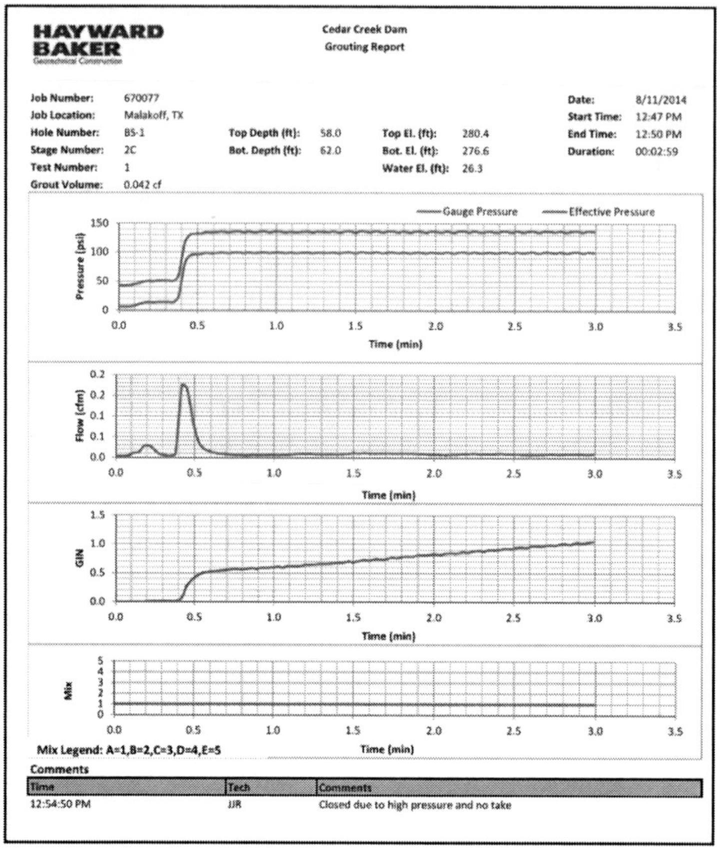

Figure 13. Individual grouting reports included all grouting parameters.

CONCLUSIONS

The project achieved the goals set forth in the project specifications. The GIN methodology proved effective for this geology, however, it should be noted that success was not achieved blindly; it required constant evaluation and targeted modifications to the approach based on information gained during construction. The data acquisition systems enabled the project team to evaluate the effectiveness of the grouting program and modify when necessary. The recorded information provided the opportunity for the proper interpretation of the ground conditions. Throughout the entire project, a high level of communication between the owner, engineer, and contractor ensured that any changes were agreed upon, and that the time required to implement the changes was minimal.

The post-treatment seepage results indicate that the system is effective. Observations of the left groin indicate that the surface conditions are significantly improved, and areas with standing water at the start of construction are now dry. Initial seepage observations of the drainage collection lines indicated flows of about 30 gpm, which agrees closely with the seepage quantity predicted during design. The weir and instrumentation will be used to automatically monitor seepage on a daily basis, and trends will be developed to relate seepage levels to the reservoir level.

REFERENCES

Lombardi, G., and De ere, D.U., 1993, " *Grouting Design and Control Using the GIN Principle*," International Water Power and Dam Construction, June 1993

Houlsby, A.C., 1990, " *Construction and Design of Cement Grouting*," John Wiley and Sons.

Hayward Baker Incorporated, 2012, Promotional Materials, I-Grout Com puter Aided Data Acquisition and Monitoring System.

Weaver, K.D., and Bruce, D.A., 2007, "*Dam Foundation Grouting*," ASCE Press

USGS National Geologic Map Database Catalog, 2013

GROUTING UNDERWATER JOINTS IN CONCRETE INTAKE STRUCTURE

Michael McCaffrey[1]
Mark Gross[2]
Tony Plizga[3]
Florijon Dhimitri[4]

ABSTRACT

Grouting was used to seal underwater open joints and cracks in the concrete intake structure of Narrows Dam. Narrows Dam is a 1917 constructed concrete gravity structure 200 feet high located on the Yadkin River near Badin, North Carolina. It is owned and operated by Alcoa Power Generating Inc., Yadkin Division.

Most of the seepage in the lower dam gallery was coming directly from the intake's vertical inter-gallery drains that are in the middle of each pier between the four intake bays. Seepage exiting the inter-gallery drains has been a concern over the years with an increasing trend since the 1990s, with flows of up to 900 gpm for the intake. In 1986 a cement grouting program in the piers between bays had limited success because high seepage rates in the joints and cracks made placement and containment of the grout difficult.

Inter-gallery drain seepage reduced immediately when two of the four intake bays were dewatered in 2007 for maintenance, indicating a short and direct path from the intake bay to the inter-gallery drains. Dewatered bay inspections in 2010 and 2011 found evidence of cross-flow seepage from the adjacent watered bay through cracks and deteriorated construction joints in the concrete pier between intake bays. Intake seepage was reduced in 2008 to 200 gpm after divers placed temporary seals on the surface of the joints in Bay 3.

A permanent repair was needed before the temporary seals deteriorated and allowed seepage to increase. Design and construction of a permanent grouting repair is discussed, including removing deteriorated temporary surface patching, joint and crack surface sealing, and injection grouting into the joints and cracks behind the surface sealing. All sealing and grouting work was done by divers in watered bays.

---

[1] Principal Engineer, Parsons Brinckerhoff, 75 Arlington St. Boston, MA 02116, (617) 960-5041, McCaffrey@pbworld.com
[2] Vice President of Hydro Operations, APGI, Yadkin Division, 293 NC 740 Hwy, PO Box 576, Badin, NC 28009, (704) 422-5774 Mark.gross@alcoa.com
[3] Senior Principal Engineer, Parsons Brinckerhoff, 75 Arlington Street, Boston, MA 02116, (617) 960-4972, Plizga@pbworld.com
[4] Project Engineer, Parsons Brinckerhoff, 75 Arlington St. Boston, MA 02116, (617) 960-4847, Dhimitri@pbworld.com

INTRODUCTION

Seepage from the three inter-gallery drains in the intake (intake drains) that are between the four intake bays account for nearly all of the seepage in the lower gallery of Narrows Dam. The amount of seepage has been a significant concern since the 1990's. The assessment of past approaches to reducing seepage through the intake drains and recent evaluation of the intake bays were used to develop a design for a permanent seal to reduce seepage between the intake bays through concrete joints and cracks, reduce the seepage in the intake drains, and prevent the further deterioration of the intake piers.

DAM

Narrows Dam shown in Figure 1 consists of a right non-overflow gravity section, two gated spillway sections, an integral intake section with four 15-ft-diameter steel penstocks, and a 110 MW downstream powerhouse. The dam is a curved concrete gravity structure 1,144 feet long and 200 feet high and has upper and lower level inspection galleries with 18-inch diameter vertical inter-gallery drains extending between the two gallery levels.

Figure 1. Narrows Dam with four penstocks exiting the intake section.

The intake structure has four intake bays, one for each penstock. Each bay consists of an intake enclosure with trash racks upstream of the intake gates and the intake transition area between the intake gates and penstock, as shown in Figures 2 and 3. A vertical construction joint parallel to the intake gate separates the enclosure from the transition area, and a vertical construction joint at the center of each pier separates each bay. The intake is constructed of concrete in approximately 5-foot thick lifts. Each bay has two vertical lift gates to control flow through 20-foot by 20-foot opening into the transition area. The transition area transitions from the 20-foot by 20-foot square intake opening into a round 15-foot diameter steel lined penstock 15 feet downstream of the intake gates. Three drainage and inspection galleries extend through the intake. They include the upper inspection gallery and two lower inspection galleries. Three 18-inch diameter intake drains connect the upper and lower inspection galleries, as shown in Figures 2 and 3. Each intake drain is a non-sleeved formed hole located at the center of the pier between bay transition areas. The three intake drains contribute the majority of the seepage that is measured in the lower gallery.

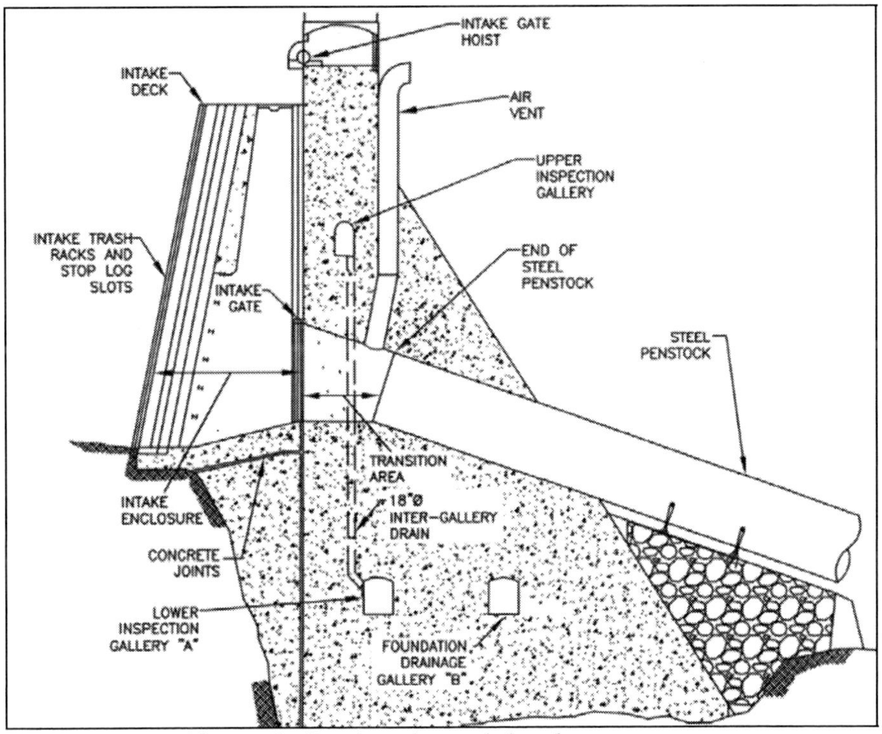

Figure 2. Section through intake structure.

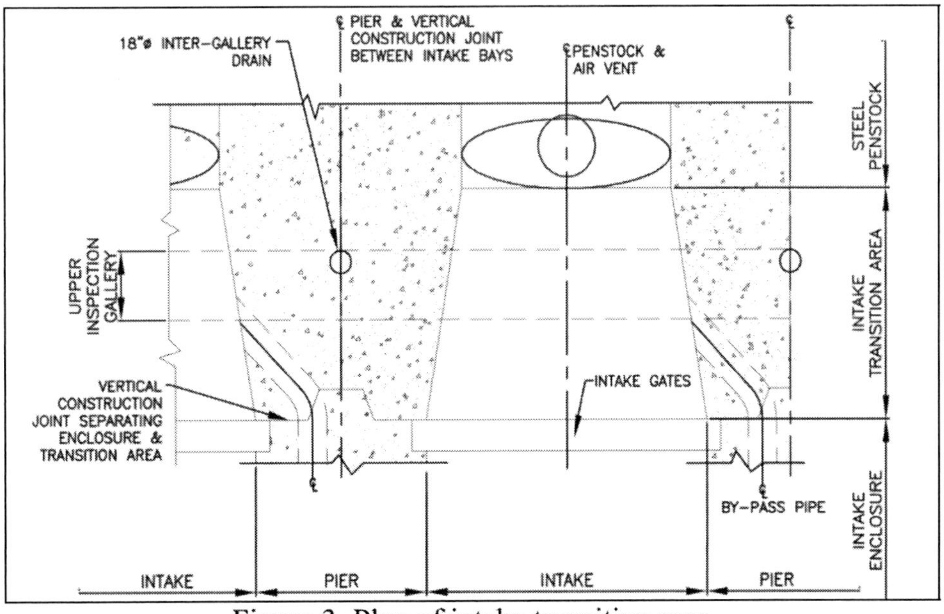

Figure 3. Plan of intake transition area.

Grouting Underwater Joints

SEEPAGE HISTORY

A cement grouting program was performed in 1986 by others to reduce intake seepage. Eight holes were drilled in the concrete piers from the intake deck. Holes were angled downstream across the vertical construction joint that separates the enclosure from the transition area. In addition, surface repairs were carried out within Bays 3 and 4, including removal of spalled and cracked concrete, and placement of new concrete patches in scour voids along the edge of the steel penstock. Intake seepage was reduced about 50%. During the course of the grouting work, two of the four bays remained watered. Seepage rates through the joints and cracks of the piers appeared to wash out the injected grout before it could set, even when both adjacent bays were dewatered.

Seepage through the intake drains has been a concern over the years with an increasing trend in the 1990s and early 2000s. Seepage has historically increased in the warm months and decreased in the cold months, indicating thermal cycles open and close joints. These cycles have progressively increase seepage over the years. From 1995 to 2001, the maximum seepage increased from 300 gpm to 635 gpm. In June 2002, slots were cut in the concrete spillway deck slab to provide expansion joints to reduce the thermal stresses in the main spillway piers which were deflecting, creating Tainter gate operation problems. The spillway deck slab appeared to provide a horizontal jacking force towards the intake, possibly opening horizontal construction joints in the intake. A substantial decrease in intake seepage was observed after the deck slab slot cut, indicating a correlation to the reduction of an apparent jacking force. However, in 2007 the intake seepage increased to a maximum of 900 gpm.

Temporary surface repairs were made in 2008 to the horizontal construction joints in the transition area in Bay 3. Divers entered the bay, cleaned open joint surfaces, and hand packed epoxy patching compound to seal them. The repair decreased total seepage from approximately 500 gpm to approximately 200 gpm. Seepage since the 2008 repair has remained in this range, but a permanent repair was needed to ensure that the seepage rates did not increase to the pre 2008 rates.

ENGINEERING ASSESSMENT

In September 2007, Bay 2 was dewatered resulting in a 15% decrease in total seepage from the intake drains and in October 2007, Bays 2 and 3 were dewatered resulting in an 84% decrease in total seepage. Seepage reduction occurred immediately after dewatering, indicating a short and direct seepage path from the transition area to the intake drains.

An internal dewatered inspection of Bay 2 in November 2007 found cross-flow seepage from the adjacent bays through open construction joints in each of the three concrete piers between transition areas. A detailed internal inspection of the four intake bays was performed in 2010 and 2011 to document the conditions. The inspected bay was dewatered and the other three remained watered. During this inspection, the construction joints and cracks (joints), seeping joints, and spalls were identified, as shown in Figures 4 through 6.

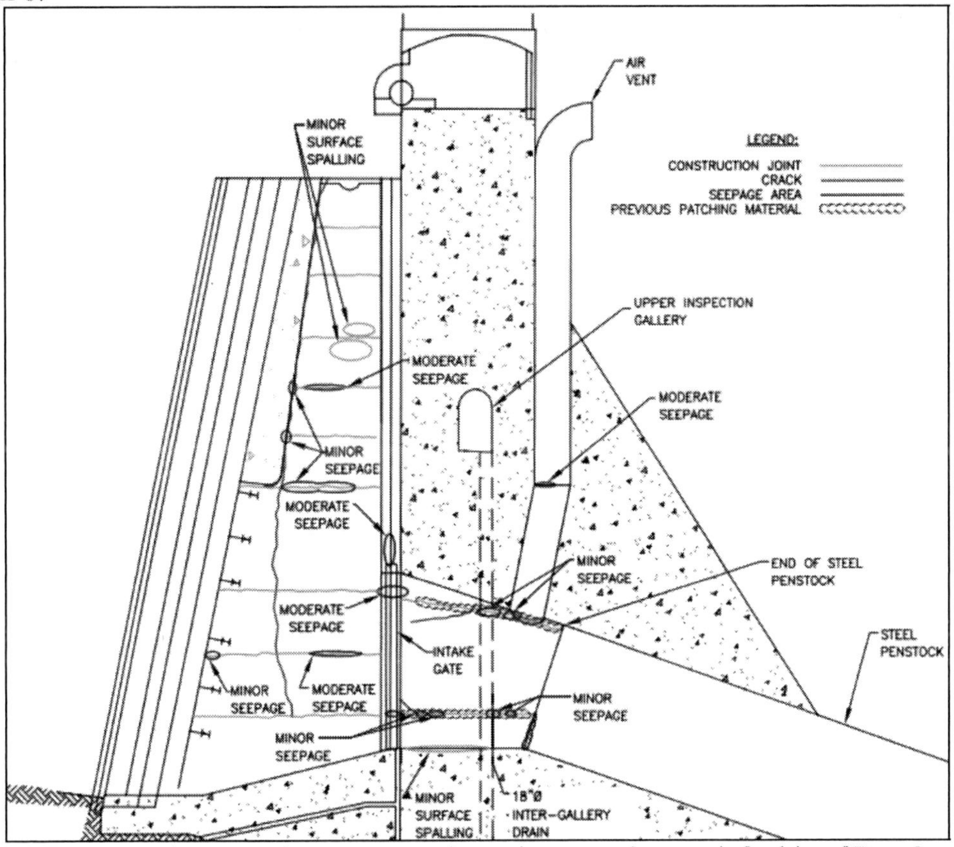

Figure 4. Joints, cracks, seeps, and previous patches on left side of Bay 3.

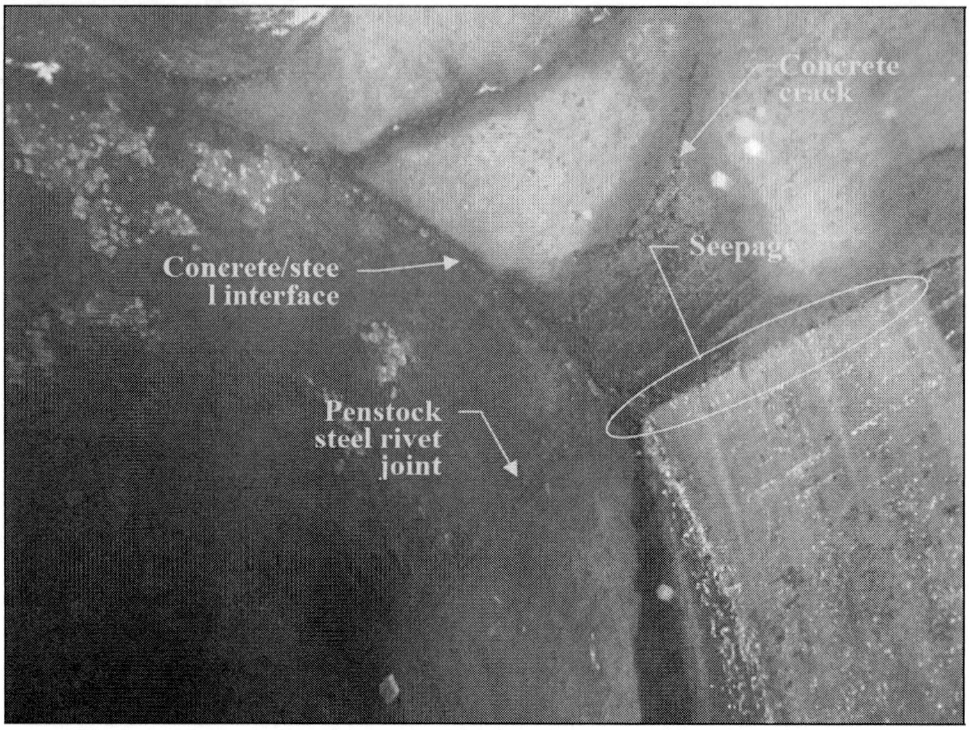

Figure 5. Bay 4 transition area during dewatered intake inspection.

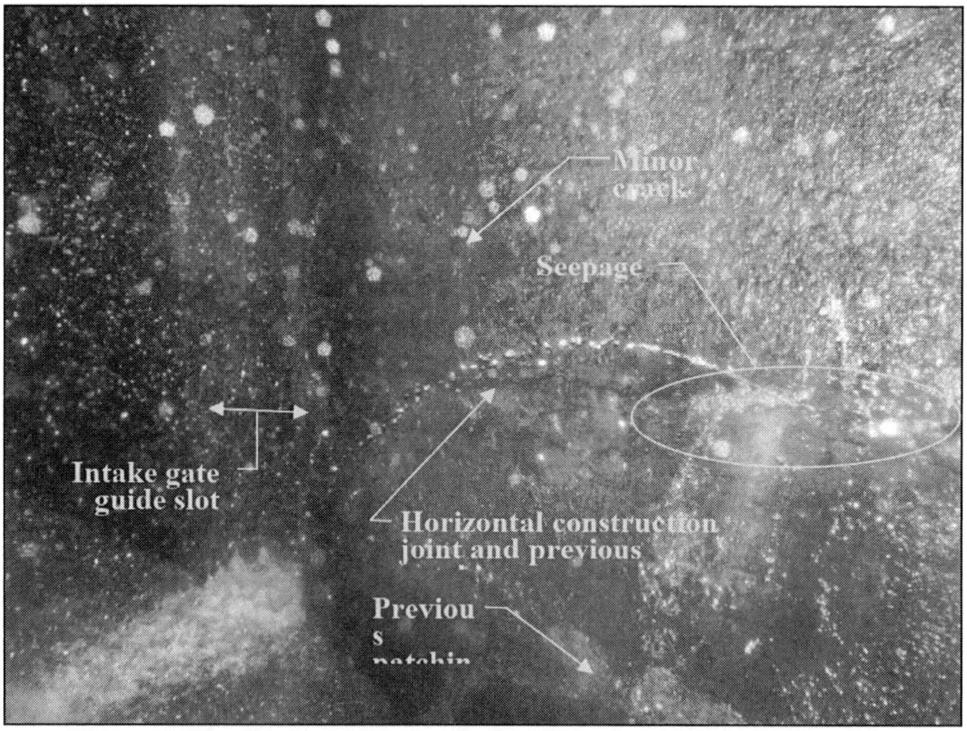

Figure 6. Bay 3 transition area during dewatered intake inspection.

Most of the seepage into the intake drains appeared to be coming from open construction joints, based on flashlight and borescope inspections. The inspection of the transition area in Bay 3 found that the 2008 patching repairs were mostly still effective, but some repairs were now deteriorated and seeping. This supported the seepage measurement data indicating the 2008 repair remained mostly effective.

In a dewatered bay, when the adjacent bay is watered, seepage flows from the watered bay through joints in the concrete pier into the dewatered bay, and also into the intake drain between bays. When both adjacent bays are watered, seepage only flows from the bay to the intake drain between bays. Therefore, a watered inspection will find different seepage locations in a pier than a dewatered inspection.

Based on the dewatered inspections of 2010 and 2011, open seepage paths in the piers between transition areas that were behind the temporary seals required grouting to provide a more permanent seal. A design was developed to seal and grout the joints, and repair the spalls.

GROUTING REPAIR

General

Parsons Brinkerhoff (PB) was the design engineer and also provided field engineering services during the grouting repair effort. The repair work and diving were performed by Glenn Underwater Services (GUS), and the construction manager for the project was Fluor Global Services (FGS). Repair work w as performed in accordance with the PB project plans and specifications, which included cleaning and sealing concrete joints, injection grouting in seeping joints, and repairing concrete spalls. Repair work was performed in the transition area between the head gate and the steel penstock. It also included cleaning and sealing concrete joints in the lower 35 feet of the head gate guide slots, and in the bottom 15 feet of the headwall above the intake opening.

The grouting repair work was completed for the pier between Bays 3 and 4 (Pier 3 / 4) in 2013 during Phase 1 and for the pier between Bays 2 and 3 (Pier 2 / 3) in 2014 during Phase 2. The repair work took about 3 weeks during each phase with a diving crew of about 12. Diving work was performed in about 40 to 60 feet of water depth, and visibility under water was about 3 feet.

As a safety precaution when divers were in the bay, generation was shut down in the bay being repaired as well as the adjacent bays, thus the intake bay remained watered throughout the duration of the pier repairs to minimize generation shut down. In addition, before the start of every morning dive shift, a remote operated vehicle (ROV) was deployed in the intake areas to test for potential water currents, such as from leaks in the wicket gates or significant seepage in joints. None were found by the ROV. The ROV was also used throughout the repair effort to monitor diver safety, work progress, and look for grout leaks.

Figure 7. Diving operation equipment on intake deck.

Joint Preparation

The concrete at the joint surface was cleaned to provide a sound and rough surface for the two-part epoxy surface sealant to adhere. A high-pressure water jet was used to remove contaminants and loose material, such as organics and poor concrete. Cleaning extended on the concrete surface several inches beyond the joint, and into the joint opening. Where previous surface sealant covered the joint, a chipping hammer was used to remove any loose or weak sealant.

After the joints were cleaned, divers used dye along the length of each joint to find locations where seepage was entering the joint. Limits of these seeps and those previously identified in the 2010 and 2011 dewatered inspections were marked on the concrete. Seeping joints required injection grouting. A backer rod was inserted into the more open seeping joints to prevent the surface sealant from being sucked into the joint by the seepage pressure. A borehole camera was used to identify seepage locations entering the intake drain.

Grout injection holes were drilled at 1-foot intervals 6 inches above or below the joint to intersect the joint at 45 degrees. Each grout injection hole and the adjacent seeping joint were thoroughly washed by injecting water to remove drill cuttings and loose particles, starting with the most upstream hole of each joint. Divers installed a grout injection port, shown in Figure 8, into each drilled hole and tightened the gasket to make a pressure resistant seal.

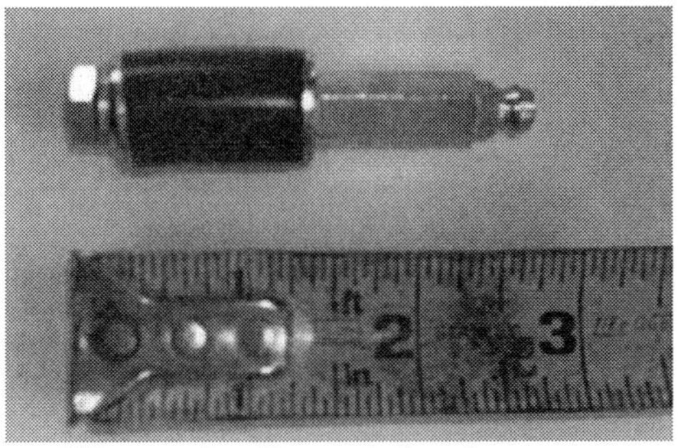

Figure 8. Grout injection port.

All cleaned joint surfaces were sealed with a quick setting underwater epoxy surface sealant, which forms a barrier against water entering into open joints from the transition area and flowing to the intake drains. Surface sealant was inserted into the joints by hand until it could no longer be pushed in (refusal), and overlapped the surrounding concrete at least 2 inches to provide an adequate seal to prevent grout from exiting during grout injection. Surface sealant was hand finished to provide a smooth final surface without protrusions from the existing concrete surface, and was allowed to cure at least 12 hours before starting injection grouting.

Injection Grouting

Grout injection ports were secured in place and the epoxy surface sealant was applied and allowed to cure. The grout used was a low viscosity hydrophobic polyurethane resin. Cured grout formed a flexible closed cell foam with excellent elastic, tensile, and adhesion properties. The grout expanded up to 27 times its liquid volume when it reacted with water, as shown in Figure 9. The elastic properties of the polyurethane grout allow it to expand and contract with the normal thermal cycles of the concrete dam as the joints open and close, allowing the seal to be maintained during future years of normal cyclical dam movements.

Figure 9. Grout sample after expanding and curing inside a small container.

A 3-foot long packer was inflated in the 18-inch diameter intake drain to prevent grout from exiting a joint that extended from the transition zone face back to the drain. The packer had a center drain hole to allow seepage entering the intake drain from above the packer to pass and drain to the lower gallery. A borehole camera was used in the drain to monitor grout leaks above and below the packer.

Grout was injected through the injection port. A water heated shroud covered the grout injection line to keep the liquid grout warm as it passed through the cold reservoir water. Grout injection began at the most downstream injection hole, and moved upstream along each joint. During this process, the ROV was present to observe any grout exiting the joints in the transition area between the intake gates and the steel penstock, gate guide slot, and the headwall. A borehole camera was stationed inside the by-pass pipe to monitor for any potential grout exiting into it or the gate valve. In addition, dam personnel were stationed in the lower gallery to observe any grout exiting the intake drain. When grout was observed leaking through or around the surface sealant or through pin holes in the concrete, a cloth was jammed into the leak to stop the grout exit. Grout that leaked out was collected and disposed.

Some grout entered the intake drain, but was washed down the drain with water pumped from the reservoir. Water was pumped when grout was observed exiting the bottom of the intake drain by personnel in the lower gallery and at regular intervals when personnel were not stationed in the lower gallery. Flushed water and grout was washed down the lower gallery and allowed to flow into the tailrace where a floating containment boom prevented the floating foam grout chunks from exiting uncontrolled beyond the dam toe.

Grout injection was stopped in a grout hole once: 1) grout pump refusal was reached indicating the joint had been filled with grout up to the 1,300 psi limiting pressure set for the pump that had a rated capacity of 3,000 psi, 2) a volume of grout had been injected that far exceeded the expected volume of the joint, 3) grout was observed exiting the intake drain indicating the joint between the bay and drain had filled with grout, 4) grout was observed exiting the downstream face of the dam, at locations of previously identified seepage.

When pump refusal was not reached after pumping a reasonable volume of grout, a thorough inspection was made looking for evidence of grout exiting any adjacent joints in the bay, by-pass pipe, intake drain, and downstream face of the dam. If the desired result was not obtained, grouting in the injection hole was discontinued, and grouting in the next upstream injection hole commenced. The refusal grout pressure was developed based on past experience with a similar pump, grout, joints and structure.
About 25 gallons of liquid grout was injected into the pier between Bays 3 and 4, and about 35 gallons of liquid grout was injected into the pier between Bays 2 and 3. Some of this grout exited in the intake drain and ended up in the lower gallery. Some of the grout exited the downstream face of the intake structure at locations of previously observed seepage. A small amount of grout was lost in the grout hoses and equipment. However, most of the grout sealed the concrete voids associated with joints between bays of the intake.

After grout injection, the grout ports were removed and the top of the injection holes were filled with surface sealant to protect the polyurethane grout.

Spall Repair

Concrete spalls at the upstream end of the steel penstocks in Bays 3 and 4 were cleaned by pressure washing and chipping where needed to prepare for repairs. Most spalls enlarged and deepened during this cleaning process. Spalls greater than 4 inches deep were repaired by saw cutting the perimeter, installing epoxy anchored dowels, and filling the spall with cement grout using form boards that were mounted flush to the circular shaped concrete surface. Shallower spalls were filled with surface sealant.

Seepage Monitoring

Seepage monitoring continues in the lower gallery, measuring seepage discharge from the three intake drains. Seepage for the entire intake after the repairs was about 150 gpm, which is a slight reduction from before the repairs, indicating new seepage paths were permanently sealed. Seepage temporarily increased during the repairs when existing deteriorated sealant patches were removed from joints. Seepage reduced after new surface sealant was re-applied to the areas of removed patches.

RESULTS AND CONCLUSIONS

This repair effort sealed many open and seeping joints in intake Bays 4, 3 and the left side of 2. Repairs in Bay 1 and the right side of Bay 2 have not yet been performed. The repairs met the goals of the project, to provide a more permanent joint seal to prevent further seepage increases and prevent further deterioration of open and seeping joints. The repairs are also expected to reduce future seepage into the intake drains.

The polyurethane grout injected into the joints reacted with water that filled the joints and expanded inside the joint to penetrate deeper into the joint. It cured to a flexible closed cell foam which sealed open joints. The grout's expansive and contractive properties maintain a water tight seal during annual thermal cycles of the dam as the dam expands and contracts, which cause joints to open and close responding to these thermal cycles.

Many joints in the intake were not grouted, such as in the piers and enclosure area upstream of the gates, and joints that are above or below the intake bays. For this reason, some seepage is expected to continue in the intake. But all joints in the transition area of the bay were sealed and the seeping joints were grouted. This area was the closest to the intake drain, and thus open and seeping joints with direct and short seepage paths to the intake drain have been sealed.

Grout observed exiting the downstream face of the intake indicate that there were direct seepage paths through the concrete intake structure, of which some are now filled or partially filled with grout. After the grouting repair, some of the previous seepage locations on the downstream face no longer exhibit seepage.

Intake head gates were closed in various combinations after the repairs to check the differences in seepage from the intake transition area that are measured in the lower gallery. Seepage was 50 gpm when the head gates for all four bays were closed and the bays dewatered, compared with 150 gpm when the gates were open and bays watered. Therefore about 50 gpm of the current 150 gpm seepage is coming from areas that were not repaired as part of this effort, such as the concrete piers upstream of the head gates and the intake structure headwall itself. Additional seepage of about 50 gpm is coming from the portions of bays that have not yet been repaired.

LEAK SEALING OF UNDERWATER JOINT
ON UPSTREAM FACE OF A CONCRETE GRAVITY DAM

Kevin Finn[1]
Jeff Auser[2]
Matthew Johnson[3]
Paul Shiers[4]
Michael McCaffrey[5]

ABSTRACT

Minor seepage of about 30 gpm was historically observed in the gallery of Cheoah Dam, exiting from the bottom of an 18-inch diameter vertical intergallery drain. A significant increase in seepage from the same intergallery drain to around 1,000 gpm occurred during 2011. Cheoah Dam is located on the Little Tennessee River in North Carolina and is a 229 ft-high arched concrete gravity dam constructed in 1919. The dam was recently acquired and is operated by Brookfield Smoky Mountain Hydropower, LLC.

The seepage had progressively increased steadily to about 1,900 gpm by mid 2012. During this time the source of the leak was investigated and determined to be a deteriorated horizontal construction joint on the upstream face of the dam at a depth of 160 feet below reservoir water surface. Aspects complicating the investigation and eventual repairs included significant seepage velocities entering the joint at this depth and a debris pile that included tree limbs piled up against the dam face at the location of the joint, covering part of the joint. To reduce the leakage, several temporary repairs such as placing different combinations of cinders, stones and other materials in the leaking construction joint were implemented over a 10 month period. These repairs had limited, temporary success. A long term repair program based on use of a packer and grouting was implemented, along with the repair of other minor leaking joints. This resulted in the total seepage being reduced to pre-2011 levels.

This paper documents the history, various methods of investigations and repair options, and the temporary and long term repairs to seal the joint leak.

---

[1] Senior Engineer, Parsons Brinckerhoff, 75 Arlington St., Boston, MA 02116, (617) 960-5031, Finnk@pbworld.com
[2] Chief Dam Safety Engineer, Brookfield Renewable Power, 75 State Street, Suite 2701 Boston, MA 02109, (857) 313-7716, Jeff.Auser@brookfieldrenewable.com
[3] Director of Asset Management, Brookfield Renewable Power, New York East Regional Operating Center, 399 Big Bay Road Queensbury, NY 12804, (518) 743-2081, Matthew.Johnson@brookfieldrenewable.com
[4] Vice President, Parsons Brinckerhoff, 75 Arlington St., Boston, MA 02116, (617) 960-4990, Shiers@pbworld.com
[5] Principal Engineer, Parsons Brinckerhoff, 75 Arlington St. Boston, MA 02116, (617) 960-5041, McCaffrey@pbworld.com

CHEOAH DAM

The Brookfield Smoky Mountain Hydro Project is located on the Little Tennessee and Cheoah Rivers in Graham and Swain Counties, North Carolina and Blount and Monroe Counties, Tennessee. The Project, which consists of four hydropower stations, is owned and operated by Brookfield Smoky Mountain Hydropower LLC (Brookfield).

The Cheoah Dam is located on the Little Tennessee River, just upstream of the mouth of the Cheoah River in Graham and Swain Counties, North Carolina.

Figure 1. Cheoah Dam, looking from downstream toward upstream.

The dam is a curved concrete gravity structure approximately 750 feet long with a maximum height of about 230 feet. Completed in 1919, the dam includes an integral intake section, a 600-ft-long Tainter gate spillway section, and a short gravity section at either end of the gate-controlled spillway. Nineteen Tainter gates, each 25 ft wide by 19 ft high are installed atop the dam to release surplus water during storm or flooding events.

The station has a usable storage capacity of approximately 4,200 acre-feet at full pond elevation, with a reservoir surface area of approximately 644 acres. Cheoah Station is generally operated in run-of-river mode, where the average daily discharge is approximately equal to the average daily inflow, with daily fluctuations occurring to meet system operating demands.

DAM DRAINAGE SYSTEMS

The spillway section includes three drainage and inspection galleries. The upper inspection gallery has a constant floor elevation about 30 feet below the crest of the gated spillway. The lower inspection gallery has a floor elevation that generally follows the contour of the base of the dam. At the central (deepest) portion of the dam, the lower gallery is 140 feet below the upper gallery. A foundation drainage gallery is located adjacent to and approximately 15 feet downstream of the lower inspection gallery at the deepest portion of the dam, with similar floor elevations.

A series of eighteen inch diameter vertical drains, 12.5 feet on center, connect the upper inspection gallery and the lower inspection and grouting gallery. These drains were cast as a formed hole during original construction to intercept water which may potentially leak through the dam at horizontal construction joints, and create unfavorable uplift pressures at the construction joints, and route the water through a discharge passage in the lower gallery. An elevation view of the dam, showing the inspection galleries and intergallery drains, is shown in Figure 2.

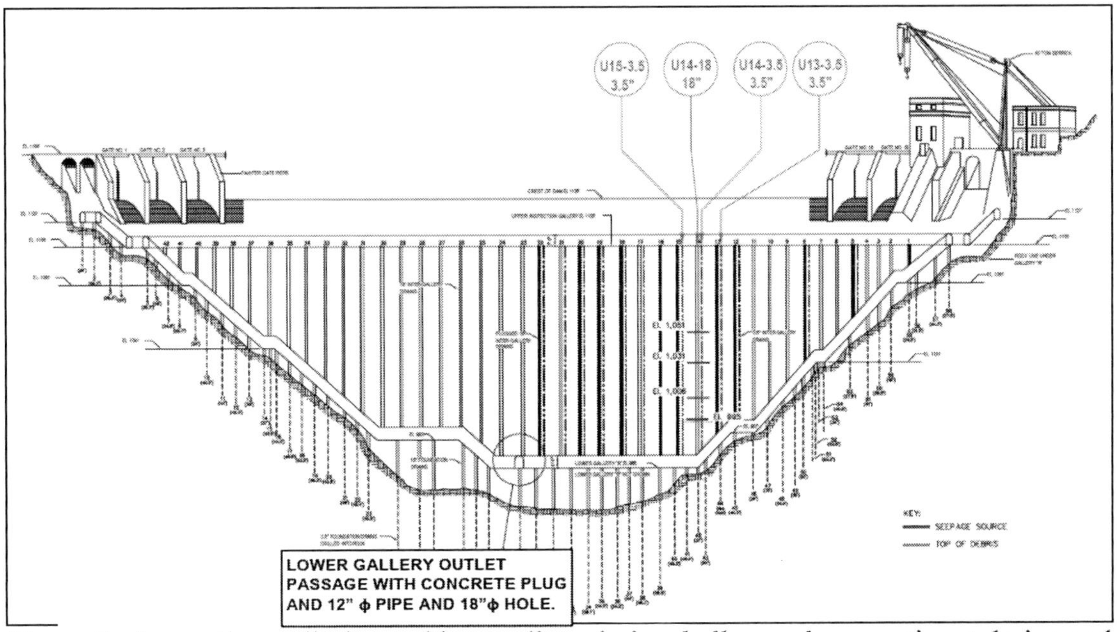

Figure 2. Inspection galleries and intergallery drains; balloons show pertinent drains and their diameter.

MONITORING OF SEEPAGE FLOWS: JUNE 2005 TO OCTOBER 2012

During the 2005-2007 annual in-house inspections conducted in the presence of the FERC, a minor seep of about 30 gpm was observed exiting from the bottom of one of the intergallery drains (U14-18) in the lower gallery. Because there was minimal water observed entering this drain at the upper gallery, it was surmised that this drain was performing as intended, and was intercepting water that was passing along a construction joint somewhere between the upper and lower galleries.

A slight increase in seepage at this drain was noted during the in-house inspection in 2008. The flow remained relatively constant through 2010, until a significant increase in seepage to an estimated 1,000 gpm was noted in 2011. The steady increase did not signify a dam safety issue; the drain was operating as intended, to allow a more favorable route for water leaking along horizontal joints to be discharged, so that an unfavorable pressurized joint condition did not develop. Although the exact cause of the abrupt change in seepage was not identified, it was likely due to continued concrete deterioration and water scour along the surfaces of the leaking joint, which opened a larger path for water. There were no dam deformations that could explain the joint opening, and the joint opening did not extend beyond one of the vertical block joints, with the other hidden by a debris pile, described below. The change in flow was documented by visual estimates of flow entering the lower gallery and photographs, as exhibited in Figure 3.

Figure 3. Change in flow at Drain U14-18 from May 2006 (left) to November 2011.

Early seepage measurements were obtained using a stopwatch to determine the time to fill a 55 gallon drum, and converting the fill time to gallons per minute. By late 2012, the seepage had increased to approximately 1,900 gpm; at these high flows, the bucket and stopwatch method was no longer adequate for measuring the flow rate through the intergallery drain. In April 2012 a sharp crested weir was installed in the lower gallery downstream of the drain to facilitate the measurement of inflows, as shown in Figure 4.

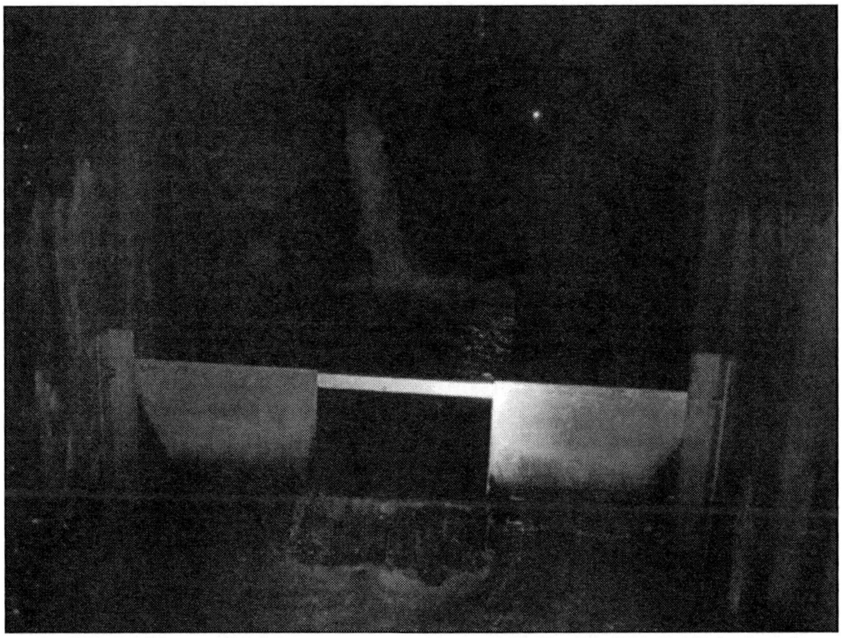

Figure 4. Weir for seepage measurement, with Drain U14-18 in background.

By measuring flow depths on the up- and downstream sides of the weir, inflows to the lower gallery were estimated. These measurements are summarized in Figure 5.

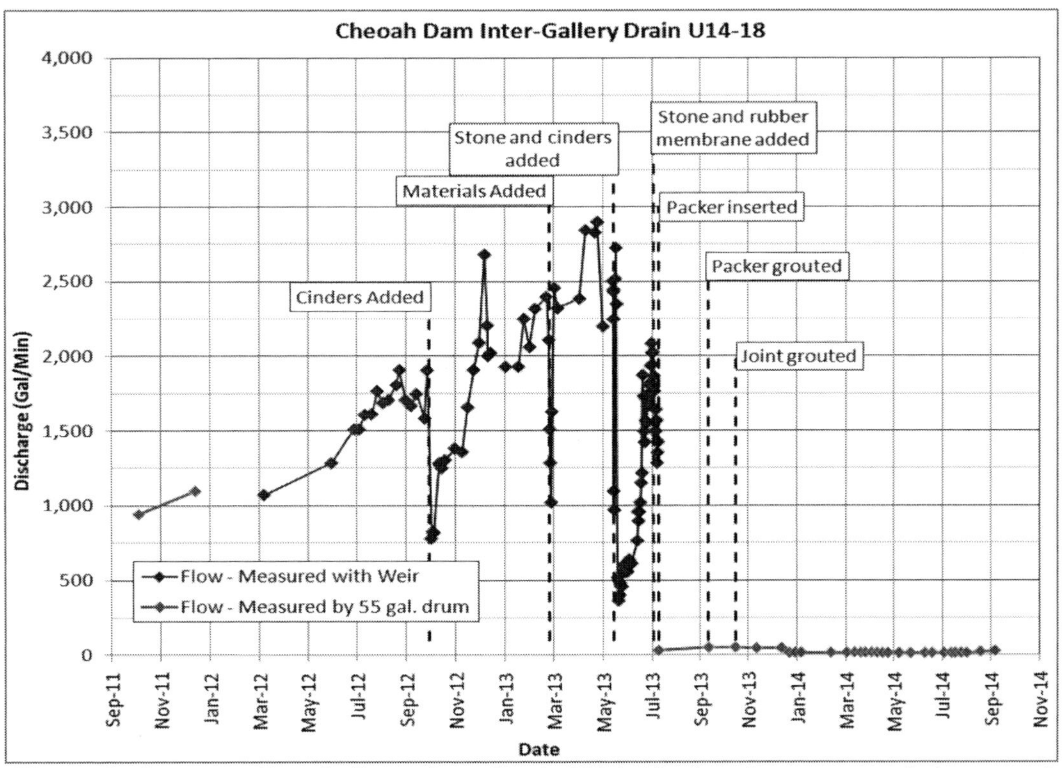

Figure 5. Lower gallery inflows from Drain U14-18, showing post-repair improved conditions.

SHORT-TERM SEEPAGE REPAIRS: OCTOBER 2012 TO JULY 2013

The source of the seepage was identified as an open construction joint, which allowed reservoir water to infiltrate the dam. The relatively sudden increase in seepage was likely due to increased scour of the joint surfaces caused by increased seepage velocity.

Beginning in October 2012, several temporary repairs were implemented in an effort to reduce the seepage. These included the placement of cinders, stones and other materials into the leaking construction joint and on the debris pile at the upstream face of the dam. The "debris pile" consisted of sediment, tree branches and other debris which had built up on the reservoir bottom over time against the face of the dam. Although these repairs were successful for the short term, as the smaller materials temporarily created a patch on the upstream face of the dam or in the joint, the repairs had limited, temporary success. With time, the materials either washed through the open joint, or sloughed off of the debris pile, allowing the water path to reopen.

Figure 6. Cinder delivery system, to introduce material to leaking joint.

These efforts reduced the seepage to 775 gpm. Additional cinders poured through a 6-inch diameter hose in November 2012 further reduced the seepage by about 30 gpm. However, cinders were subsequently observed on the floor of the lower gallery, trapped by the weir, with a maximum particle size of 1 inch and averaging about ¼ inch.

By March 2013, about four months after the initial repair effort, the seepage had increased to 2,400 gpm. To seal the joint, cinders, cinder-filled socks, oakum strands, and oakum rope knotted balls with expansive caulk were inserted into the joint from the reservoir side.

Figure 7. Cinders (up to 2 inch diameter) used for temporary joint sealing.

This repair effort had similar success; the short term benefit reduced the measured flow to about half of the 2,400 gpm rate, but the flow had increased to 3,025 gpm by June 2013. To reduce this large flow, additional materials (1 to 3 inch stones, cinders, peastone) were inserted at the joint opening, or were placed on the debris pile high enough to cover the joint opening. By mid-June, these efforts had reduced the seepage to 360 gpm.

Because of the increased flow into the lower gallery, the existing horizontal 12-inch diameter drain pipe from the lower gallery to the downstream face of the dam was no longer sufficient to adequately pass the water out of the lower gallery of the dam. In July 2013, a new 18-inch diameter horizontal exit drain was drilled through a concrete plug at the downstream end of the lower transverse inspection gallery, parallel to the 12-inch drain pipe. This drain hole, together with the existing 12-inch diameter exit pipe, reduced the depth of flooding in the lower gallery to 12 inches or less.

UNDERWATER INVESTIGATIONS: AUGUST 2011 TO NOVEMBER 2013

The Owner and PB felt confident that these temporary repair efforts were targeting the primary source of water entering the galleries due to a series of underwater investigations undertaken between 2011 and 2013. While the temporary repairs were being performed, these investigations were undertaken in an effort to evaluate the leaking joints and develop a long term repair program. These investigations included remotely operated vehicle (ROV) surveys, dye tests, drain hole borescope video inspections and a bathymetric survey.

An ROV with video camera was operated by Glenn Underwater Services (GUS) of Charlotte, NC to inspect the joints on the upstream face of the dam. The ROV was the primary means of obtaining video and photographic evidence of the state of deterioration of the joints on the upstream face. In some instances, such as when cinders or peastone were being placed, or flow into the joint was too strong for the ROV to operate without the risk of being sucked in, a frame-mounted camera was lowered by ropes from the dam deck to the inspection areas.

The ROV inspections were successful at identifying the size and location of the leaking joints, and the debris pile upstream of the dam that obscured portions of the primary leaking joint. Two vertical construction joints, about 50 ft apart forming the sides of a concrete block, were found in the vicinity of the leaking joints. The leaking joints began at or near the right (facing downstream) vertical construction joint. The most prominent joint began at the vertical construction joint, continued left about 12 feet, and disappeared behind the debris pile. The joint opening was about 2 inches at the face, and continued into the dam farther than the ROV video could observe, possibly more than 1 or 2 feet.

One principal finding of the ROV survey was that the primary leaking joint was partially obscured by the debris pile upstream of the dam. Although the introduction of cinders and other small sized materials helped to seal exposed portions of the joint, they had the unintended effect of causing the debris pile to settle. As the pile settled, additional portions of the joint became exposed, causing the seepage into the gallery to increase.

Dye was injected into the water in front of the ROV camera during inspections to determine if reservoir water was seeping into a joint, and if so, the location and relative velocity. Seepage velocity was highest into the lower, partially obscured joint, with a significantly lower flow rate observed at the upper joints. The highest seepage velocity into the lowest joint occurred where the joint intersected the debris pile (see Figure 8).

Figure 8. Dye test at the lower horizontal joint, showing significant flow into the 2 inch open joint.

Borescope video inspections were performed in intergallery drain U14-18 in August 2011, and during June 2013 in U14-18 and in the three 3.5 inch intergallery drains adjacent to U14-18, to identify the elevations at which flow was entering the drain. A joint with significant seepage was found in U14-18 at an elevation which corresponded to the significant leaking joint at the upstream face, with minor seepage identified at the elevation of the other minor upstream leaking joints. The significant joint also communicated with the drain to the right of U14-18, although the flow into this drain was significantly lower.

A bathymetry survey was performed in July 2013 on the debris pile, to calculate the volume of material to remove to expose the remainder of the leaking portion of the primary leaking joint. The survey indicated a greater debris volume than previous ROV inspections. This increased volume, and the corresponding cost to dredge and remove this material, necessitated a leak repair that did not require debris removal.

EVALUATION OF LONG TERM REPAIR OPTIONS

While these investigations were ongoing, and as it became evident that introducing cinders upstream of the joint provided only a short term solution, possible long term repair options were evaluated.

The principal concern for these repair options was to protect the safety of the work crew that would institute the repairs. As such, numerous methodologies were reviewed, to select a long term repair that was acceptable to all parties and which would ensure the long term stability of the dam.

In all, fourteen options were developed and evaluated by the diving contractor and by Parsons Brinckerhoff. Among these options, 5 options did not use grout to seal the joint, while the remaining options did. For each option, the detailed steps needed to complete the repair, with a description of key issues specific to each, were developed. Each option was eventually assigned an overall rating, based on the likelihood of success, the estimated cost, and the estimated longevity of the repair.

In June 2013 a leak repair option was mutually chosen by the diving contractor, Parsons Brinckerhoff and the Owner. This option, which required the use of divers at the 160-ft depth of the leaking joints, included installing a packer inside the U14-18 drain at the primary leaking joint to temporarily stop the seepage through the joint to ensure diver safety. To ensure a complete repair, the debris in the exposed joint would be removed, the joint would be cleaned, a membrane would be bolted across the joint opening to create a seal, and the joint would be filled with nonshrink grout to stop the inflow of water.

To improve safety for the divers, the water-inflatable packer in the intergallery drain was filled with grout to prevent a rupture of the packer bladder. Such a rupture could have caused a sudden and large seepage flow into the joint, putting a diver working near the joint at risk of being drawn to the joint opening and held fast.

LONG TERM REPAIR PROGRAM

Long term repairs began in August 2013 with the insertion of the inflatable packer in U14-18 across the primary seeping joint. Based on field testing of the diameter of the drain, a custom inflatable packer 14 inches in diameter by 36 inches long was built by Plug-It Products of Lodi, California to be installed in the drain (see Figure 9).

Figure 9. Custom inflatable packer.

In August 2013 the water-inflatable packer was installed in U14-18 across the primary seeping joint, successfully stopping flow into the drain. Although flow was stopped, the joint remained pressurized with 160 ft of head, which was unacceptable for a long term repair. A second inflatable packer was installed in the adjacent 3.5 inch drain at a similar elevation, to stop the minor seepage flow that had been discovered in the June borescope investigation. Had this packer not been installed, grout injected into the joint under pressure would have migrated to the nearby drain and washed out of the joint. By blocking this communication, a fully pressurized condition could be developed.

In October 2013 the water in the bladder of the U14-18 packer was replaced with grout. Grout was injected into the inflation line of the packer, which pushed the bladder water out the outlet line. Grout was injected with a high enough back pressure on the outlet line (minimum 100 psi) to keep the bladder fully inflated against the water pressure in the joint, thereby ensuring a watertight seal. A 6 inch diameter pipe sleeve through the packer allowed seepage entering U14-18 from joints above the packer to flow past the packer, preventing a backup of water into the upper gallery. It was important to include this sleeve; the packer was designed to be grouted in place, and left in the drain permanently at the conclusion of the remediation effort.

After the joint was cleaned with a pressure washer, a stainless steel plate, 12 ft long, 1 ft high, and ¼ inch thick was installed by divers on the upstream face across the exposed joint. 5/8 inch diameter anchor bolt holes were pre-drilled 9 inches on center along the top and bottom perimeter of the plate. ½ inch stainless steel Hilti® anchors were then used to fasten the plate to the concrete, using both mechanical and epoxy bonds. A rubber gasket provided a watertight seal between the plate and the concrete dam face.

During stainless steel plate fabrication, twelve 2-inch diameter grout ports were welded to the plate at 1 ft on center spacing along the plate centerline. A ball valve on each grout port allowed it to be opened or closed. The port farthest to the right extended about 3 inches to the right of the vertical construction joint, outside of the limits of the observed leaking joint. A 1-inch outside diameter, 5 ft long tremie pipe was inserted by the divers into each grout port, extending horizontally into the joint as deep as the pipe could be pushed. This depth, measured at each grout port, varied from 10 inches to 4.5 ft, as shown in Figure 10.

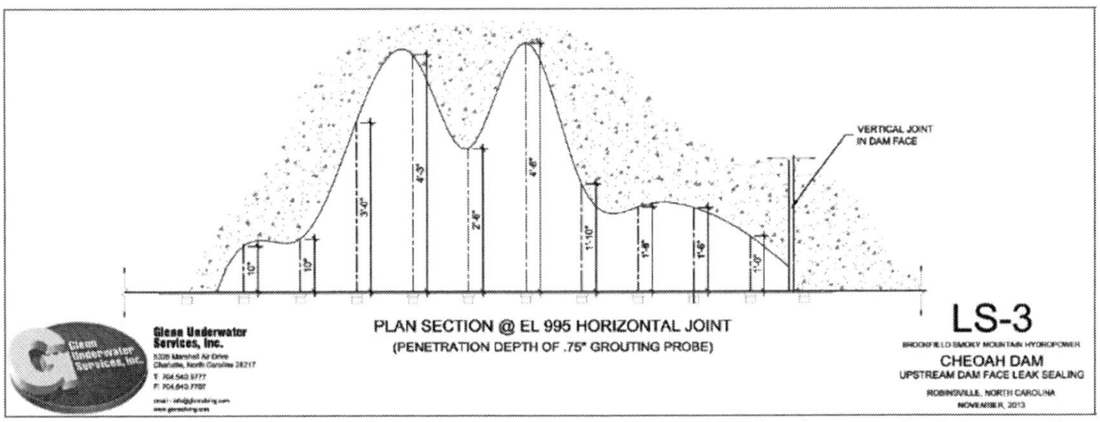

Figure 10. Plan of ten grout injection tremie pipe insertion depths into the joint; twelve ports are identified as rectangles.

Grout mixes were shop tested during the weeks prior to grouting using three base options, with the final mix selected for its pre-cure properties. For a successful repair, it was important that the grout gel quickly without running out of the joint where the debris pile covered the joint. This was achieved with a rapid thixotropic set time of less than 10 minutes. However, the grout had to be runny enough to be pumped more than 200 ft through a hose and injected into the joint through the tremie pipe. In addition, the grout had to achieve a 28-day strength of at least 1,000 psi to ensure durability. Bentonite was used to prevent the grout from shrinking; otherwise, a crack could form between the grout and the top or bottom surface of the joint, allowing water to re-enter the intergallery drains.

The joints were grouted on November 13, 2013. The grout was mixed in 2-bag batches in a colloidal mixer located on the dam crest. After mixing, it was distributed through a hose to the pump hopper, and pumped from the dam crest to the underwater tremie pipe.

Grout was fist injected through the tremie pipe into the deepest section of the joint (the middle port), with all other grout ports closed except Port 1 (the farthest left), and Port 12 (the farthest right), to allow water in the joint to be displaced by the grout. After 125 gallons of grout had been injected, it began seeping out at several locations, including

Port 1, several upper anchor bolts between Ports 1 and 3, and behind the top of the plate between Ports 1 and 5 (see Figure 11).

Figure 11. Grout exiting above the plate at the conclusion of grouting

A total of 230 gallons of grout were injected into Port 6, with a maximum pressure of 90 psi. It is estimated that several gallons of grout exited from the three abovementioned locations. The tremie pipe was inserted into Port 8 with all ports open, and after injecting 10 gallons of grout, it was observed exiting from Ports 7, 9, 10 and 11. The grouting process was then repeated with all valves closed, with 4 gallons injected into Port 3, and 78 gallons injected into Port 1. When the tremie pipe was finally removed, approximately 1.5 CY of grout had been injected into the joint. Because the debris pile obscured part of the joint opening, it was impossible to determine the total volume of grout exiting the joint. However, about 20 gallons of grout accumulated on top of the debris pile during the injection at Port 1.

During the joint grouting, six-inch-tall grout cylinders were obtained. The unconfined compressive strength of the grout was acceptable at over 3,000 psi at 28 days. 7 day and 14 day tests were also completed, with strengths in excess of 2,100 psi and 2,600 psi, respectively.

The three additional leaking horizontal lift joints were also repaired. Although their total contribution to the seepage measured at U14-18 was a small portion of the total flow, the estimated 40 to 50 gpm was not insignificant. The seeping joint openings at the dam face were small, typically 1/4 inch or less, although some locations were up to 6 inches high at the face, narrowing to 1/8 inch a few inches into the joint.

SUMMARY AND CONCLUSIONS

In July 2013, shortly before the 18 inch diameter horizontal exit drain was drilled, seepage flows in excess of 2,000 gpm were entering the lower gallery from U14-18, with flow depths in the gallery averaging 1.5 to 2 ft. The U14-18 intergallery drain was functioning exactly as intended – it provided a more favorable route for water entering the dam than through horizontal construction joints, routing the water to an unpressurized drain so that uplift pressures did not extend downstream of the drain. However, the volume of flow inhibited access to the lower gallery, and created an unsafe environment for maintenance staff.

Through an extensive underwater and remote camera inspection program, the extent and cause of the primary seepage was determined, and isolated to a few open horizontal construction joints at the upstream face. A repair program was developed jointly by the diving contractor, Parsons Brinckerhoff and the Owner, to both ensure that the leakage into the dam was significantly reduced, and that the repairs were completed in a manner that mitigated the exposure of the divers to potentially unsafe working conditions. As a result of this program, the seepage measured at U14-18 has been significantly reduced, while maintaining the integrity of the intergallery drain system. This reduction was achieved in a safe and cost effective manner, and is applicable to other concrete gravity dams where lowering of the reservoir or the use of a bulkhead to investigate observed seepage and perform remediation measures are not feasible.

DEEP WATER AT HORSE MESA DAM — DESIGN OF REPAIRS

George V. Sabol, PhD, PE[1]
Roger Baker, PE[2]
Michael Langen, PE[3]

ABSTRACT

On 12 June 2012, a 50 ton concrete guide valve inside the Horse Mesa Dam pump storage intake failed, passed through the steel penstock, and came to rest in the scroll case of the Unit 4 turbine and knocked that unit out of peak-load generating capacity. Horse Mesa Dam is a 300-foot high concrete arch dam on the Salt River in Arizona. It was designed and constructed by the U.S. Bureau of Reclamation in 1924, including hydrogenerating Units 1-3 with a combined capacity of 32 MW. In 1971, pump-storage was added to the dam with a capacity of 97 MW in Unit 4. Getting Unit No. 4 back on line and repairing the intake quickly and correctly was paramount to SRP. Since the intake structure is more than 160 feet under water on the upstream side of the dam, a major decision was whether to perform the repair by lowering Apache Lake about 175 feet or performing underwater construction. Stantec embarked on concept design, constructability, and cost estimating for both methods of construction and hired Kiewit Infrastructure West Company for construction in the "dry" and Global Diving & Salvage Inc. (Global) for underwater construction. Within months of the failure, both concepts were evaluated and SPR decided on underwater repair of the intake.

Stantec analyzed the flow, using Computation Fluid Dynamics, transients and modal analysis from "Do Nothing" (other than remove the concrete debris from the intake) to sophisticated vane arrangements to improve hydraulic performance for both generation and pumping. Nineteen alternatives were developed and evaluated. SRP rejected any alternative that would leave the remaining vane in place without structural enhancement. The adopted design consists of two vertical vanes and nine horizontal vanes, which were not used in the original configuration. The vanes were fabricated of steel and built by SRP in its shop. The design was bid by SRP in April 2013, the contractor, Global, mobilized in May 2013, and Unit 4 was back on-line in August 2013, 13 months after failure.

PROJECT BACKGROUND

The Salt River Project (SRP) is located near Phoenix, Arizona, includes a water service area of about 375 square miles and an electric service area covering 2,900 square miles spanning portions of Maricopa, Gila and Pinal Counties in central Arizona. Water

[1] Principal, Stantec Consulting Services Inc., 8211 South 48th Street, Phoenix, AZ 85044, gsabol@stantec.com
[2] Principal Engineer, SRP, PO Box 52025, M-S 52025, Phoenix, AZ 85072, roger.baker@srpnet.com
[3] Vice President, Global Diving and Salvage, Inc., 3840 West Marginal Way SW, Seattle, WA 98106, mlangen@gdiving.com

supplies from the Salt and Verde Rivers are controlled by six storage dams, as illustrated in Figure 1.

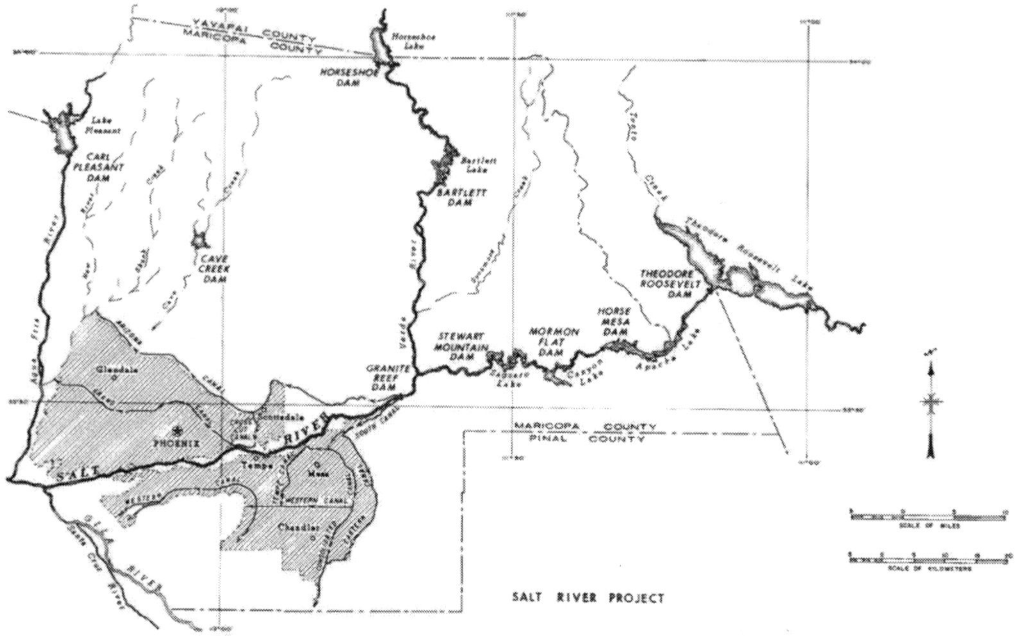

Figure 1. Location of Horse Mesa Dam

The SRP power system includes eight hydroelectric units at four dam sites on the Salt River: Theodore Roosevelt Dam, Horse Mesa Dam, Mormon Flat Dam, and Stewart Mountain Dam.

Located 65 miles northeast of Phoenix, Horse Mesa Dam (see Figure 2) is a concrete thin-arch structure 305 feet high and 66 feet long. Constructed by the Salt River Valley Water Users' Association between 1924 and 1927, the dam contains 162,000 cubic yards of concrete, and forms a 245,138 acre-foot reservoir called Apache Lake.

Figure 2. Horse Mesa Dam from Salt River Canyon

The dam has three conventional hydroelectric generating units, Units 1-3, rated at a total of 32MW installed during original construction, and one pumped storage unit, Unit 4, rated at 114MW, added in 1972 and upgraded in 2002. The intake during construction is shown in Figure 3.

Figure 3. Intake during construction

On 12 June 2012, while Unit 4 was undergoing a normal generation startup, site operators noticed a loud rumbling in the penstock and immediately shut the unit down. After proper clearances were taken, the scroll case was unwatered and an inspection performed, where large sections of reinforced concrete were found lodged in the bullnose of the scroll case. Further investigation determined that failure of one of two 26 ft x 24 ft x 1 ft guide vanes located within the intake structure upstream of the dam under 165 feet of water was the likely source of the concrete in the scroll case. A photo of the intake guide vanes during construction is provided in Figure 4 (the failed vane is on the left), and an illustration of the failure is shown in Figure 5.

Figure 4. Intake vanes during construction

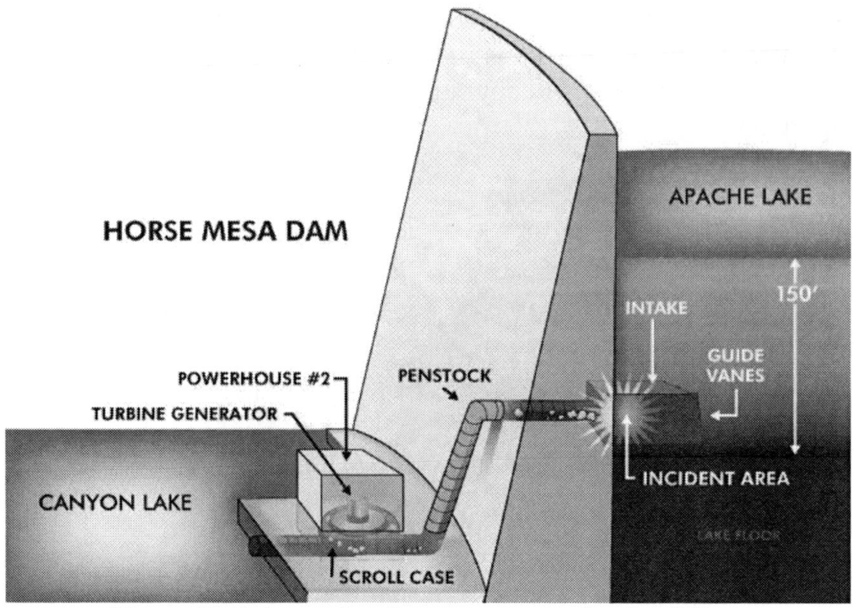

Figure 5. Schematic of pump-storage intake failure

The next step undertaken was to perform an underwater inspection of the intake structure. SRP contracted with Underwater Engineering Services Inc. (UESI) to mobilize a crane, barge, and diving operations to remove intake trashracks and perform a detailed inspection. Once the six 10 ft x 13 ft trashrack panels were removed, a remotely operated vehicle was used to survey the interior of the intake structure to verify that it was safe for divers to enter. Divers then performed a detailed inspection which verified one of the guide vanes had failed and that the remainder of the structure was not damaged.

Concurrent with the intake inspection, removal of the concrete from the scroll case was initiated, which proved to be no simple task. Due to the precarious position of some of the concrete, an engineered procedure for shoring and removal was developed. Further challenges included working in a confined space and the fact that all the concrete needed to be reduced to sizes manageable by hand and would fit through the 36 inch diameter scroll case mandoor. Subsequent to the removal of the concrete, SRP contracted with Alstom Power to provide a detailed turbine inspection, which revealed the damage to the distributor was limited to coating damage and minor impingement damage to stayvanes, wicket gates, and the runner.

UNIT 4 REPAIR DESCRIPTION

Repairs and modifications were required to rehabilitate the Unit 4 intake structure and restore normal pumped-storage operations to generate electrical power. Stantec was contracted by SRP to evaluate, compare, and recommend repair modification options for the Unit 4 intake structure. After several workshops and alternative evaluations for the repair modification, Stantec recommended an option to remove the remaining concrete North Vane and replace the original two vertical concrete vanes with two new steel vertical vanes (in similar locations). However, the recommended modifications also required the three vertical bay areas between the new vertical vanes and intake walls would have three additional levels of steel horizontal vanes, all of which would be installed underwater.

Design of the repair vanes involves extensive analyses. There were four major areas of analysis required for the hydraulic and structural designs of the new vertical and horizontal vane modifications. The first aspect of analysis performed by Stantec was Computational Fluid Dynamics (CFD) modeling. The CFD modeling evaluated vane modification configurations and locations to improve the hydraulics and flow velocity distribution during pumping and power generation operations. The second analysis was the hydraulic transient that was completed to estimate the maximum hydraulic pressure fluctuation loadings that could occur during system pumping and power generation operations. The third aspect was the load frequency modal analysis performed to evaluate potential system vibrations and natural frequencies for design of the vane modifications. The critical modal frequency loadings and system responses were determined to minimize adverse effects and avoid potential future fatigue failures. The final analyses were for the structural design. The structural analyses considered the maximum hydraulic pressure and structural loadings provided from the other analyses; including the results from the CFD, modal, and hydraulic transient analyses.

COMPUTATIONAL FLUID DYNAMICS (CFD) MODELING

CFD modeling was performed to develop a replacement vane configuration that provides satisfactory hydraulic performance. The model results were also used to estimate the steady-state hydrodynamic loads on the vanes. The modeling was performed for both the generating and pumping modes. The CFD analyses were conducted in three phases.

- Phase 1 – This was modeling of the existing (1970 design) with either the North Vane only, or without either the North or South Vanes.
- Phase 2 – A more advanced phase to develop a CFD model that included the trashrack. This phase confirmed that a new vane configuration was needed to provide acceptable hydraulic conditions including more uniform velocity through the trashrack.
- Phase 3 – Various vane configurations were evaluated to develop the selected vane concept. Phase 3 also resulted in the estimation of the steady-state hydrodynamic loads on the vanes, trashrack and intake structure.

Phase 1 demonstrated the following for the pumping mode:

- Without the South Vane, the velocity distribution at the trashrack is unacceptably non-uniform, and a flow recirculation zone exists behind the North Vane which could cause undesirable vane vibrations.
- Forces on the North Vane increase by almost 90 percent.
- Adverse hydraulic conditions exist with no vanes.

Phase 1 demonstrated the following for the generating mode:

- Velocity distributions are only slightly impacted whether operated with the North Vane only or no vane.

- Forces on the North Vane increase by more than 50 percent.

Phase 1 also concluded that:

- Reservoir water surface, in the normal operating range, does not have significant impact on forces on the vane or intake, or on hydraulic performance.
- Vane configuration has a significant impact at the trashrack during pumping mode.
- Conditions at the scroll case are not impacted by vane configuration.

Phase 2 modeling included the details of the existing trashrack. Results included forces on the trashrack and the internal surfaces of the intake structure. Four scenarios were modeled: pumping at 5,300 cfs and generating at 6,414 cfs for the original two vane configuration, and for the no vane configuration. Phase 2 demonstrated that the intake should not be operated without vanes, and it was concluded that a new vane configuration was needed to achieve acceptable hydraulic conditions in the intake.

During Phase 3, eight vane layouts were modeled and evaluated; each concept having two vertical vanes as per the original 1970 design, and three sets of horizontal vanes at selected locations and inclination angles. The preferred configuration, illustrated in Figure 6, significantly improved the hydraulics of the intake structure. For the pumping mode, the velocity distribution at the trashrack is much more uniform with the previous low velocity zone near the floor eliminated. The maximum velocity at the trashrack was reduced from near 15 ft/sec to about 12 ft/sec while maintaining full pumping capacity. For the generating mode, the velocity through the vane area of the intake increased about 7 percent due to the reduced flow area from the incorporation of the horizontal vanes. However, generating capacity was improved due to enhanced hydraulics. The velocity distribution at the scroll case is nearly identical to the 1970 baseline condition. An example of CFD output for the pumping mode is shown in Figure 7.

The CFD modeling of the selected concept generated Excel files for pressure and viscous (shear) pressures acting on the vane surfaces, concrete intake surfaces, trashrack piers, and the trashrack. Those data were used in the structural analyses.

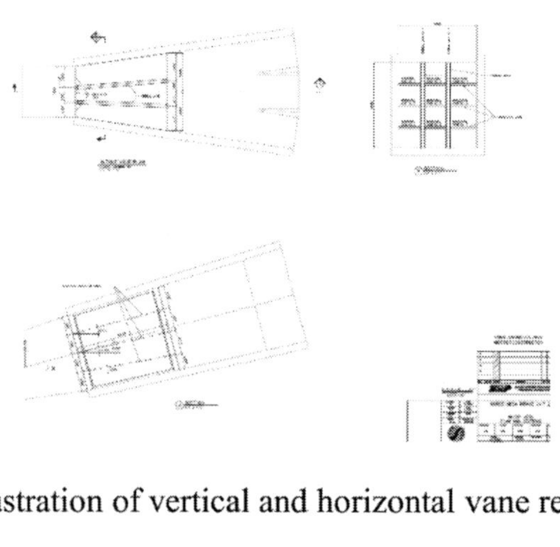

Figure 6. Illustration of vertical and horizontal vane replacement

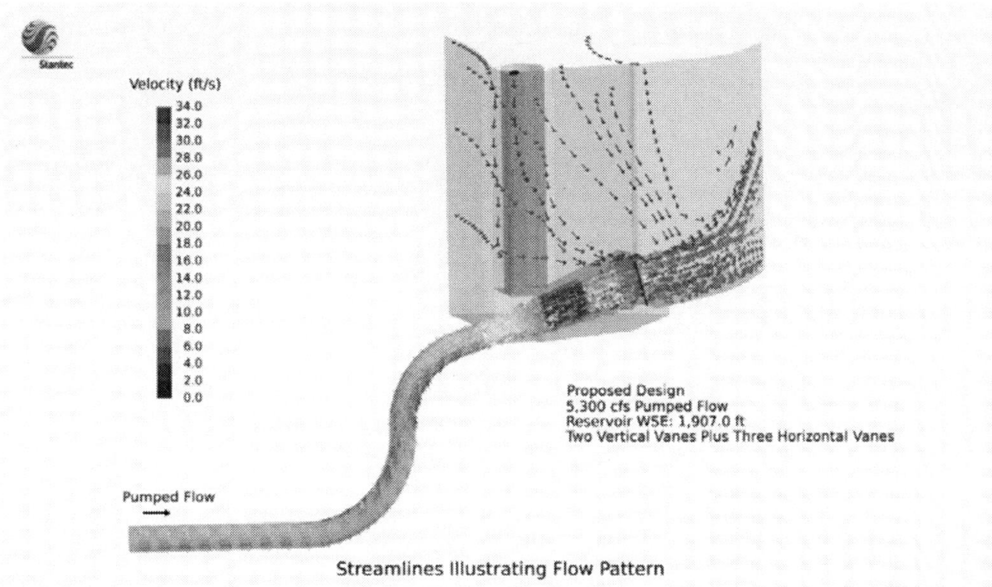

Figure 7. CFD model of pumping

TRANSIENT ANALYSIS

A transient analysis, often called water hammer, was performed for the selected vane configuration. The purpose of the analysis was to estimate the loads that are transmitted to the new vanes and the intake structure during pumping start-up and shut-down operations when in pumping mode as well as turbine load acceptance and full load rejection operations when in generating mode. From those analyses, the minimum and maximum hydraulic transient pressure fluctuations were calculated and input to the structural analyses.

The design concept for the vanes consist of two vertical vanes and three sets of horizontal vanes. Therefore, from a hydraulic analysis perspective, the intake at the vanes consists of 12 "parallel" flow conduits. For the transient analyses, each conduit was modeled as an equivalent flow area with discharges, velocities and hydraulic parameters as determined by the CFD analyses. The transient analyses of the 12 parallel conduits (multiple-path method) allowed for estimation of the differential surge pressures between adjacent conduits.

The transient model was calibrated to the 2002 configuration using data from the 2004 Commissioning Report. The following test results were duplicated with the model:

- Pump start-up
- Pump shut-down
- 60 MW emergency trip

The model calibration provided a means to estimate hydraulic parameters to the model that greatly improved the model reliability. Model results and 2004 Commissioning Test results are summarized as:

	Scroll Case Pressure, in kPa	
	2004 Commissioning Test	Model Result
Pump start-up	940	960
Pump shut-down	985	1018
60 MW emergency trip	982	960

The 2-vane model very closely approximated recorded pressures from the 2004 testing and there is high confidence in the pressure loadings to the vanes and structures from the transient analyses of the modified vane arrangement.

The transient analysis was performed for four conditions:

- Condition 1 - Pump start-up at 5,300 cfs
- Condition 2 - Pump shut-down at 5,300 cfs
- Condition 3 - Generating, full load rejection at 6,397 cfs
- Condition 4 - Generating, load acceptance at 6,397 cfs

From the analyses, Condition 2 – Pump shut-down, produces the governing transient pressures for the structural analyses. Pressures on the intake structure surfaces and on the vertical and horizontal vane surfaces are provided below:

Recommended Transient Differential Pressure Loadings

Scenario	At Shut-Down		Post Shut-Down	
	(psi)	(psf)	(psi)	(psf)
Intake Box Section (External Loading)[1,3]	13.1	1,900	17	2,500
Intake Box Section (internal Loading)[2,3]	6.5	1,000	7	1,000
Verticle Vanes[4]	4.1	600	-	-
Horizontal Vanes[5]	1.7	250	-	-

Notes:

1. Pressing inward on box section
2. Pressing outward on box section
3. Use whichever loading is greater
4. Applied normal in either horizontal direction on exterior plates
5. Applied normal in either vertical direction on exterior plates

MODAL ANALYSIS

A modal analysis of the Unit 4 pump-storage facility was performed to assess how various parts (draft tube, wicket gates, penstock, vanes, and trashrack) react to specific forcing functions. The purpose of the forced vibration analysis was to identify the frequencies that may be transmitted from the pump/turbine to the new vanes and replacement trashrack. Data from the 2004 Commissioning Test was used to obtain the pressure fluctuations at the pump/turbine.

Significant results of the modal analysis are:

- At a wicket gate opening of 65.3 percent and power output of 86 MW, a frequency of about 12 Hz develops. That frequency is synchronous with the draft tube pressure fluctuations from twin vortices in the draft tube, and those vortices are exciting the penstock to resonance. This explains the reported "rumbling" that can be heard throughout the power plant.
- The natural frequency of the new steel vanes is increased to over 452 Hz when filled with concrete grout. That frequency is well above any resonance with oscillations coming from the pump/turbine.
- The trashrack was not subjected to any resonant conditions while it was in service from 1972 through 2012.
- There is no evidence from either modal test or nondestructive test (NDT) of the trashrack that indicates that the trashrack was subjected to vibrations during either the pumping or generating operations, or from vortex shedding.
- Modifications were made to the trashrack design to enhance structural integrity and improve performance of some welds that had cracked. It was determined that the trashrack would not be subjected to excitations from oscillations originating during pumping or generating operations.

STRUCTURAL ANALYSES OF THE REPLACEMENT VANES

The structural analyses of the steel replacement vertical vanes and the newly incorporated horizontal vanes was conducted concurrently with the design of the vanes. Analyses were performed of the major components of the vane system including:

- Welded frame
- Skin plate
- Connection details and fasteners to the concrete intake walls, floor slab, and roof slab, including the grout plate.
- Clips welded to the vertical vanes and connected by 1-inch bolts to the bottom of the horizontal vanes.
- Bolted connections (3/4 inch and 1 inch) of the horizontal vanes to the vertical vanes
- Vane in-fill grout.

During the alternatives analysis for the project, a basic concept for the fabricated steel horizontal and vertical vanes was developed based on assumptions for the surface pressures on the vanes. The surface pressures were estimated based on the initial (Phase 1 and 2) CFD analyses. Simple beam equations and hand analyses were used to estimate the beam sizes and skin plate thickness. No analysis of the connections was performed at that time. The concept indicated that the steel vanes would be connected to the concrete walls and slabs with embedded anchors. The concept indicated a basic bolted vane-to-vane connection. Slotted holes were provided in the vane-to-vane connection and a grout pad was provided in the vane-to-concrete connection.

For the final design, steady-state loads were provided from the Phase 3 CFD models for both pumping and generating modes. In addition, hydraulic transient loads during start-up and shut-down of the pump-turbine were applied as uniform differential pressures to the surfaces of the intake structure and vanes in conjunction with steady-state loads from the CFD modeling. Dead loads were also applied to the structural elements as appropriate.

The initial concept was for the steel vanes to be a grid of steel beams with a skin plate welded to those members. The vanes were to be hollow and filled with water. That resulted in two adverse considerations: First, the skin plates would be subjected to fluctuations of positive and negative pressures that would need to be reacted by relatively thick steel plate. Second, modal analysis indicated that the unsupported skin plate may resonate with other forcing frequencies in the intake system, and that could lead to excessive vibration and potential failure. To alleviate those structural concerns, the design includes filling the steel vanes with a cement grout. That grout also provides corrosion protection to the otherwise unprotected interior surfaces of the steel vanes.

The analysis was initially conducted with a two-dimensional, planar rigid frame structural model. That resulted in the preliminary sizes of the beams in the vanes. The horizontal and vertical wide flange beams were modeled as frame members rigidly connected at the joints and to the existing concrete walls and slabs. The results from the frame analysis were used to prepare preliminary typical vane-to-vane and vane-to-concrete connection designs. The frame analysis was used to evaluate 16 load combinations and determine the critical loads for design.

A three-dimensional finite element model of the intake structure and vanes was used to more accurately analyze the stresses and connection loads in the intake structure and vanes for the varying pressures from the combined CFD and transient analyses. The critical load combinations identified from the two-dimensional frame analysis were evaluated using the three-dimensional model to refine the design of the vanes and develop the loads for further detailed analysis and design of the connections.

The three-dimensional model also provided more detailed information for the loads on the vane-to-vane and vane-to-concrete connections for the various load combinations. The walls and top slab of the intake structure were anticipated to behave as two-way slabs and transfer additional load to the east end of the vanes. This two-way bending behavior of the concrete walls and top slab increased the loads on all of the connections at the east end of the vanes and required further analysis and modification of the connections to handle the higher loads.

The axial loads and moments in the beams from the frame analysis were used to analyze and design typical vane-to-vane and vane-to-concrete connections. The tension and compression forces indicated that the vane-to-vane connection which were to be slotted to provide adjustability for construction, would need to be designed as "slip-critical" and the bolts would need to be torqued to provide the shear capacity required to resist the loads. It also was determined that both tension and compression fitting would be required to provide a load path through the beams in the vertical vanes. Designs for the vane-to-concrete connections were developed using the Hilti anchor design program based on the axial loads, moments and shears at the ends of the beams from the two-dimensional frame analysis.

The three-dimensional model showed that the two-way bending of the walls and roof slab transferred additional load to the east end of the vanes, which increases the loads on the vane-to-concrete connections. The loads on the vane-to-concrete connections were used to modify the connection designs for the higher loads from the transient pressures on the walls and roof slab of the intake structure. This required additional anchors at closer spacing at the east end of the vanes to distribute and support the high tension and compression forces and shears along the edge of the vanes. Some additional anchors were provided beyond the ends of the vanes by extending the edge beam and support angles. Additional anchors also were required beyond the west end of the vanes to distribute and support the higher loads at those locations.

The skin plates were analyzed as shell elements between the beams. The results dictated skin plate thickness of 1/4 inch with stud anchors spaced at 9 inches on center in both directions. The skin plate is welded to the flanges of the beams with full penetration groove welds.

The frame members are W6x15 beams in the horizontal vanes and W12x16 beams in the vertical vanes. The angles at each side of the top and bottom of the vertical vanes are 8x8x1, and those are used for the vane-to-concrete connection. The angles at the top side and bottom side of the horizontal vanes are 6x6x1, and those are used for the vane-to-concrete connection.

The dimensional basis of design was a sonar survey. The sonar data was accurate within a 0.1 foot. In order to allow for that variability and to improve underwater constructability, the following was incorporated in the design:

1. Structural grout pads were installed at the interface between the vanes and the intake concrete surfaces (intake roof and floor for the vertical vanes, and both walls for the horizontal vanes). The grout pads also filled the voids between the angles at the ends of the vanes, account for variability in the concrete surfaces, and provided up to a maximum of 2 inches in dimensional tolerance for the vanes.
2. The vertical vanes were designed to accommodate structural grout pads between the ends of the vanes and the concrete surfaces at both the floor and the roof. The design gap for the vertical vanes was not to exceed 2 inches at the top and bottom. Due to dimensional uncertainty of the intake, and to avoid vertical vanes that were too high, the vanes were designed to be 4 inches short of the sonar measurement data.
3. The horizontal vanes were designed with slotted holes at the connections of the horizontal vanes to the vertical vanes. The slots provided 3/4 inch adjustment in each direction. Additionally, the maximum gap between the intake walls and horizontal vanes was set at 1 inch.

SUMMARY OF DESIGN OF THE REPAIR

- Repair of the intake considered drawing down the reservoir 170 feet in order to perform the construction in the "dry" and by underwater construction.
- Underwater construction was selected due to environmental, operational, constructability and other issues.
- There was considerable design uncertainty due to inability to obtain all of the necessary dimensional data for the installation of the replacement vane elements.
- Sonar survey data of the inside of the intake was obtained. That was complicated by the presence of the North Vane. Sonar data accuracy was to be ± 0.1 foot.
- The concrete vanes were replaced with steel vanes that were fabricated in a shop and installed by divers.
- The two concrete vanes were designed to be replaced by the steel vanes in the same location and configuration.
- In addition, three sets of steel horizontal vanes were added.
- The vanes were attached to the walls, roof and floor of the existing structure by concrete anchors.
- The horizontal and vertical vanes were assembled underwater with bolted connections.
- Due to dimensional uncertainty and to aid constructability the horizontal and vertical vanes were designed to be shorter than data indicated. The gap at the top and bottom is filled by a grout cap.
- The horizontal to vertical vane connections used slotted holes to account for dimensional uncertainty. Those connections are "slip critical" and required specified torque to carry the loads.
- Loads were determined by computational fluid dynamics (CFD) and transient analyses for both pumping and generating modes of operation.
- Modal analyses were performed to avoid resonant frequency excitation of the various elements of the intake system.
- The vanes were filled with concrete grout after underwater installation. That grout provided structural integrity to the thin (1/4 inch steel plate) vane walls, provided internal corrosion protection and avoided low frequency excitation of the skin plate and potential fatigue failure.
- "Fit" of the vanes during construction was complicated due to inaccuracies in the best determined dimensions from design drawings and sonar measurements. The design had adequate robustness so that changes could be incorporated during construction to account for dimensional uncertainty.

ACKNOWLEDGEMENT

The Horse Mesa Dam Unit 4 hydrogenarating plant was back online with 14 months of the failure. That success is attributed to the team effort of SRP's engineers, SRP's MCM shop that fabricated the steel vanes, Global Diving and Salvation, Inc's engineers and construction crews, and to Stantec's engineers and designers collaborating in five offices in Arizona, Colorado, Kentucky and Washington. The success of this project is due to the diverse talents, expertise, collaborative spirit, enthusiasm, dedication and good engineering judgment of many at SRP, Global and Stantec. It is a project that all will remember and recall with a sense of accomplishment under the most difficult conditions.

DEEP WATER CONSTRUCTION AT HORSE MESA DAM

Michael Langen[1]
George V. Sabol, PhD, PE[2]
Roger E. Baker, PE[3]

ABSTRACT

On 12 June 2012, a 50 ton concrete guide vane inside the Horse Mesa Dam pump storage intake failed, passed through the steel penstock and came to rest in the scroll case of the Unit 4 turbine, knocking that unit out of peak-load generating capacity. Horse Mesa Dam is a 300-foot high concrete arch dam on the Salt River in Arizona, operated and maintained by Salt River Project (SRP). It was designed and constructed by the U.S. Bureau of Reclamation in 1924, including hydro generating Units 1-3 with a combined capacity of 32 MW. In 1971, pump-storage was added to the dam with an original capacity of 97 MW in Unit 4. Getting Unit 4 back on line and repairing the intake quickly and correctly was paramount to SRP. Stantec was contracted by SRP to design the repairs to the Horse Mesa Dam intake. Since the intake structure is more than 160 feet under water on the upstream side of the dam, a major decision was whether to perform the repair by lowering Apache Lake about 175 feet or performing underwater construction. Stantec embarked on concept design, constructability, and cost estimating for both methods of construction and hired Kiewit Infrastructure West Company to provide technical input on construction in the "dry" and Global Diving & Salvage Inc. (Global) for underwater construction. Within months of the failure, both concepts were evaluated and SRP decided on underwater repair of the intake.

Stantec proceeded with a design consisting of two vertical vanes, each approximately 25 feet high by 25 feet long and nine horizontal vanes, which were not used in the original configuration. In order to expedite construction, SRP made the decision to self perform the fabrication of the vanes utilizing their in-house Mechanical Construction and Maintenance Group (MC&M) to fabricate all of the components. The installation of the guide vanes was issued for bid by SRP in April, 2013 and Global was selected to perform the construction. By May, 2013 Global had mobilized to the site and commenced repairs on Unit 4. All together over 100,000 pounds of steel was installed using 4,000 pounds of nuts, bolts and concrete fasteners. Upon assembly the steel vanes were filled with 80 cubic yards of concrete grout. Unit 4 was back on-line in August 2013, 13 months after failure.

The Horse Mesa Dam project provides an excellent case study in which to examine a number of considerations to be taken into account when undertaking an underwater construction project.

[1] Vice President, Global Diving & Salvage, Inc., Seattle, WA
[2] Principal, Stantec Consulting Services Inc, Phoenix, AZ
[3] Principal Engineer, Salt River Project, Phoenix, AZ

INTRODUCTION

The Salt River Project (SRP) located near Phoenix, Arizona, includes a water service area of about 375 square miles and an electric service area covering 2,900 square miles spanning portions of Maricopa, Gila and Pinal Counties in central Arizona. The SRP power system includes eight hydroelectric units at four dam sites on the Salt River: Theodore Roosevelt Dam, Horse Mesa Dam, Mormon Flat Dam, and Stewart Mountain Dam. Located 65 miles northeast of Phoenix, Horse Mesa Dam is a concrete thin-arch structure 305 feet high and 66 feet long. Constructed by the Salt River Valley Water Users' Association between 1924 and 1927, the dam contains 162,000 cubic yards of concrete, and forms a 245,138 acre-foot reservoir called Apache Lake.

The dam has three conventional hydroelectric generating units, Units 1, 2 and 3, rated at a total of 32MW installed during original construction, and one pumped storage unit, Unit 4, added in 1972. Rated at 96MW, it was upgraded to 114 MW in 2002. Water enters the unit from Apache Lake through a concrete intake structure that was originally constructed with two vertical concrete guide vanes running the length of the intake. These guide vanes were designed to provide improved hydrodynamic flow during the pumping cycle. On June 12, 2012, one of the 50 ton concrete guide vanes failed, passed through the steel penstock and came to rest in the scroll case of the Unit 4 turbine, knocking that unit out of peak-load generating capacity. Getting Unit 4 back on line and repairing the intake quickly and correctly was paramount to SRP.

Figure 1. Unit 4 intake structure during original construction

Stantec Engineering was contracted by SRP to design the repairs to the Horse Mesa Dam intake. They embarked on concept design, constructability, and cost estimating for various repair methodologies. Since the intake structure is more than 160 feet under water on the upstream side of the dam, a major decision was whether to perform the repair by lowering Apache Lake about 175 feet or performing underwater construction. Stantec hired Kiewit Infrastructure West Company to provide expertise for construction in the "dry" and Global Diving & Salvage Inc. (Global) for underwater construction. Within months of the failure, all concepts were evaluated and SRP had decided on underwater repair of the intake.

SELECTION OF UNDERWATER VERSUS DRY CONSTRUCTION FOR RETROFIT AND REHABILITATION WORK ON HYDRO FACILITIES

To evaluate repair options, SRP and Stantec hosted a series of workshops attended by a multi-disciplinary team including outside contractors and outside consultants, as well as representatives from a number of departments across Salt River Project. Seventeen options were considered, ranging from "do nothing" to eight different vane configurations of increasing sophistication. Conceptual designs of each of the configurations were developed and each evaluated for reliability versus risk, constructability, operational impacts and cost using a weighted decision matrix. In addition each option was also evaluated for the relative risk, costs and benefits of diving vs. drawing down the reservoir.

The logistics of accessing and supporting the dry construction with the reservoir in a drawn down condition was a major factor in the feasibility and costs of performing the work in the dry. Additionally, draining the lake would have significantly impacted the Owner's water operations; SRP delivers up to 1 million acre-feet annually to municipal, industrial, agricultural and residential users within metropolitan Phoenix. The move also would have adversely affected the many recreational and commercial users at Apache Lake. Performing the work using underwater construction methods allowed the selected repair option to be installed with minimal impact on facility operations while maintaining lake levels required to support activities of the recreational and commercial stakeholders.

Utilizing the results from the workshops, the recommended option was to remove the remaining concrete vane and replace the original two vertical concrete vanes with two new steel vertical vanes (in similar locations). The recommended modifications also required that the three vertical bay areas between the new vertical vanes and intake walls would have three additional levels of steel horizontal vanes, all of which would be installed underwater. With the repair option selected Stantec then proceeded to finalize design of the repair. For further information on the design efforts the reader is referenced to ***Deep Water at Horse Mesa Dam – Design of Repairs*** by Dr. George Sabol also included in these proceedings.

ADVANTAGES OF USING SATURATION DIVING AS A METHOD FOR UNDERWATER CONSTRUCTION

Divers working at depth are exposed to pressures greater than standard atmospheric conditions. When diving, a diver's body absorbs gases in proportion to the surrounding pressure. As a diver surfaces, pressure decreases and the excess gas is released. By ascending slowly, gas is absorbed into the bloodstream, travels to the lungs and is exhaled safely. Ascending too fast can cause gas to come out of solution and form dangerous bubbles in the bloodstream — bubbles that can cause decompression sickness, also known as "the bends."

Saturation diving reduces the risk of decompression sickness and greatly increases the efficiency of working in deeper water.

In saturation diving, divers breathe a blend of oxygen and helium in a pressurized environment that is maintained for extended periods; they are transported to and from the underwater work site inside a pressurized diving bell. Divers are decompressed to surface pressure only once, at the end of their "tour of duty." Limiting decompressions reduces the risk of decompression sickness and increases the time available for working.

Figure 2. Saturation diving system assembled onto a modular barge

Saturation diving can be performed in a range of water depths from relatively shallow (120 to 150 feet) to in excess of 1,000 feet of seawater. It becomes most cost effective when executing work in deeper waters or where labor intensive tasks and / or skilled workmanship is required.

The depth and complexity of the tasks required for the repair of Unit 4 made this project an excellent candidate for the use of saturation diving techniques. The method would have no impact on water operations and offered the earliest return-to-service date for completing the repairs. Additionally, a long needed repair on the dam's other three intake structures could be accomplished at the same time. Units 1, 2, and 3 are also located on the Apache Lake side of the dam, approximately 100 feet deeper than Unit 4. The bulkheads for these units did not seal adequately to the penstock intakes due to deteriorated concrete on the intake opening face, making it difficult to isolate the units downstream of the reservoir. The proposed repair involved the installation of new stainless steel bulkhead sealing surfaces on each of the three unit intakes. This would be accomplished immediately following startup of the repaired Unit 4.

UNDERWATER CONSTRUCTION OPERATIONS

There are a full range of capabilities that can be brought to bear on construction projects performed below the water. With proper planning, preparation and design almost all operations that can be performed above the water can also be performed underwater. Underwater construction also has limitations which must be taken into account when designing a repair or new construction. An examination of the underwater construction operations performed at Horse Mesa Dam will illustrate many of these capabilities as well as demonstrate methodologies that can be used to mitigate inherent limitations.

Access to Horse Mesa Dam was limited, all equipment and materials had to be brought in via Apache Lake. Diving work was performed from an 80 x 80 foot sectional barge assembly. A second 40 x 40 foot materials barge was used to support the work and transport materials to and from the worksite. Primitive roads leading into Apache Lake limited truck and crane access so all equipment was received and assembled at a location 15 miles upstream from the dam on Apache Lake. During a 10-day period, Global Diving brought in 29 truckloads of equipment. As the trucks arrived and unloaded, crews assembled the barges, installed diving and ancillary equipment (including a 50 ton hydraulic crane) and commissioned the diving systems. Once everything was readied for operation, the entire flotilla was pushed down river to Horse Mesa Dam.

Smaller transport vehicles could make it along the road as far as Apache Lake Marina, located approximately seven miles upstream from the dam. Once initial mobilization was complete, a staging area was established at this location with a 90 ton RT (rough terrain) hydraulic crane set up next to the water's edge. This crane would remain throughout the project to handle material and equipment in support of the diving operations. The material handling barge and a tug were used to transport material and equipment from the marina to the dam. Fabrication of all steel replacement components was performed by SRP's Mechanical Construction and Maintenance Group (MC&M) in their shop in Tempe and trucked to the marina staging site. Over 100,000 pounds of steel and 4,000 pounds of bolts, nuts, washers and concrete anchors passed through the marina to be installed at the dam.

North Vane Demolition

Figure 3. Section of the North Vane being removed from the reservoir.

The first order of business for the divers was to remove all leftover pieces of the failed vane. They also dismantled and removed the remaining guide vane. In order to remove the north vane safely and efficiently from inside the intake structure the divers used diamond wire saws to cut the remaining vane into a number of individual pieces. Because there was no direct vertical access to the work areas inside of the intake, a variety of lifting devices and rigging was used during the course of the work inside of Unit 4. For handling the concrete during the demolition phase, barge mounted winch wires were used to "yard and stay" the cut blocks from inside the intake to the retrieval location. Haul back lines moved materials and equipment along the length of the intake structure. Divers installed pad eyes and snatch blocks as needed to facilitate this rigging. Additional rigging was installed as the work progressed from demolition to installation. The demolition included use of hydraulic core drills to establish cut lines and rigging points. False work was installed to maintain stability of the vane as divers cut it into sections. Each block was hauled out of the intake, then lifted by crane to the surface and loaded onto the materials barge. The sections were turned over to SRP who reassembled them in an adjacent lay down yard for forensic investigation. Any remaining concrete from the north or south vane was removed by chipping and / or grinding so that all surfaces were flush with the floor and ceiling. With the concrete vanes removed the new vane assemblies could now be put into place.

Installation of Vertical Vanes

The vanes were designed to be attached to the walls, roof and floor of the existing structure using epoxy anchors. Bolted connections were assembled underwater to secure the horizontal to the vertical vanes. Due to dimensional uncertainty and to aid constructability the horizontal and vertical vanes were designed to allow a variable gap between the vane and the existing concrete. Post installation the gap was formed and then filled with a high strength grout. The two vertical vanes were each delivered to the site in three segments. The vane segments were erected on the material barge and bolted together. When fully assembled each vane measured approximately 25 feet in length and varied in height from 25 feet on the upstream end to approximately 20 feet on the downstream end.

Figure 4. Vertical vane fully assembled and prepped for installation, being lowered into the reservoir.

During assembly, several field modifications were made to the vanes to facilitate installation. Modifications included the following.

- Lifting frame bolted at each vertical joint for rigging and handling the vane during installation.
- Clip plates used to stabilize each vertical joint.
- Six sets of roller assemblies. Each segment of vane had its own set of roller assemblies, one on each side. The roller assemblies were mounted at the

appropriate elevation to maintain the specified spacing between the intake floor and bottom of the vane.

- Flotation tanks attached along the top of the vane on each side; buoyancy of the tanks assisted in keeping the vanes vertical, preventing them from tipping over during installation.
- Leveling / jacking bolts installed on the base of the vanes to plumb up the vane and maintain the specified floor clearance
- Grout ports in both bottom and top angles and the body of vertical vanes.

While the vertical vanes were being assembled on the surface, divers were busy surveying and laying out vane locations inside the intake. Based on the layouts, a purpose built stop was mounted to the floor of the intake at the final location of the downstream end of each vane. In addition guide rails were installed onto the floor to guide the vane into the vane pocket stops and provide correct orientation of the vane in the structure. Divers also installed a porch like structure extending past the intake entrance to support the vane as it was lowered from the surface and then brought into the structure. The interior geometry of the intake is complicated, the entire structure tapering as it proceeds downstream from the intake entrance to the interface with the penstock Walls tapering at different angles than the floor and ceiling. In addition the entire structure is angled downward at approximately fifteen degrees from the horizontal.

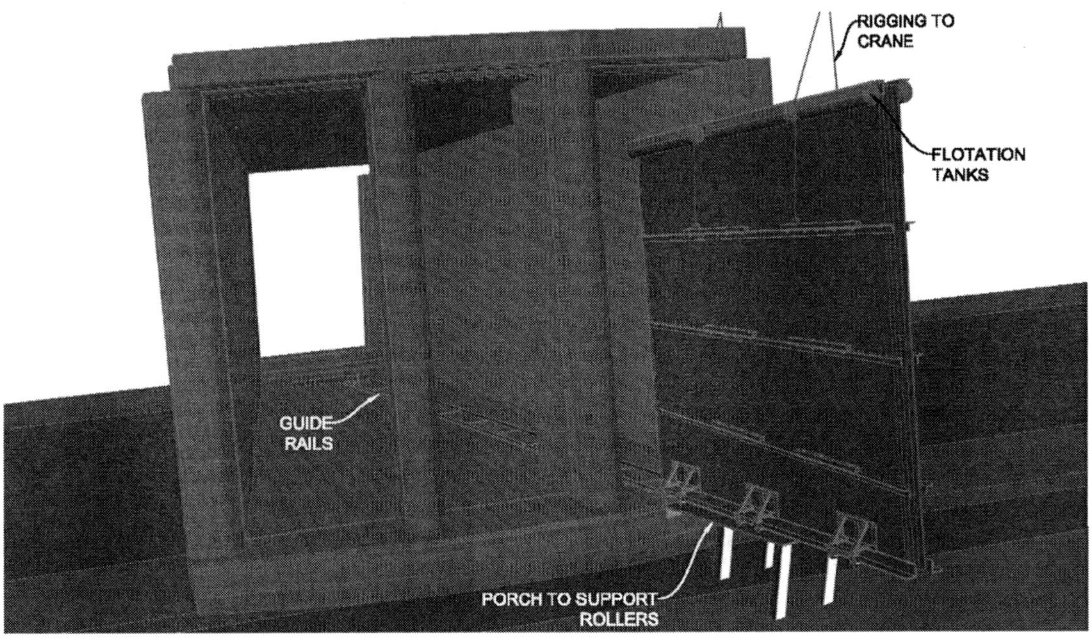

Figure 5. Vertical vane being positioned onto the guide rails in front of intake.

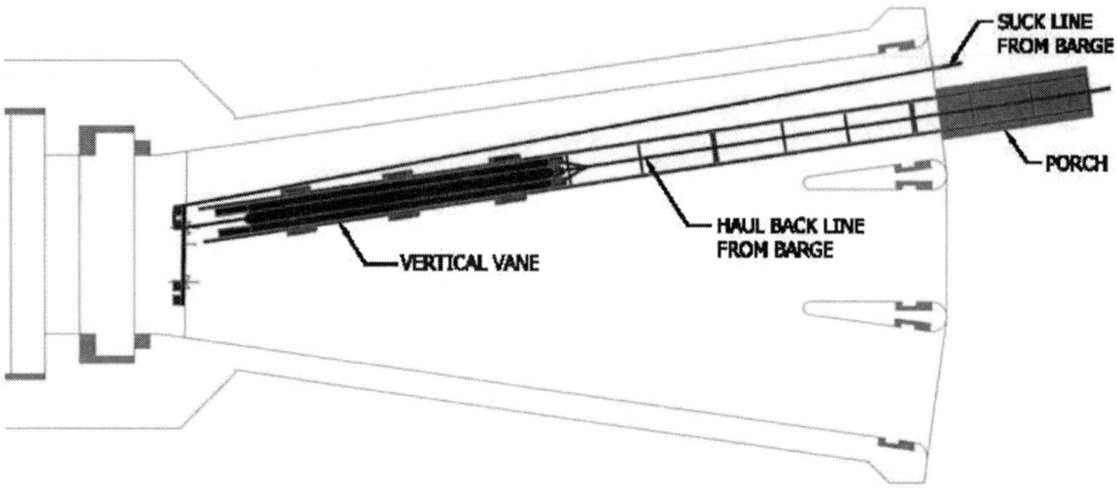

Figure 6. Schematic depicting rigging of vertical vane during installation.

The complex geometry of the intake, coupled with a lack of definitive as built dimensional data resulted in some misalignment in the final location of the vanes. This resulted in the installation of the vertical vanes becoming an iterative process to find the best fit location. Go-no-go gauges mimicking the as built dimensions of the horizontal vanes were utilized to verify fit and alignment during positioning of the vertical vanes. A comprehensive survey was performed and revised positions of the vanes were modeled in AutoCAD to check for fit and potential interferences. Working with the Owner's representatives and the design engineers, positioning was verified as acceptable within the design criteria. Additional required modifications were identified and agreed on.

Once the vane was determined to be in its desired location, the diver set the leveling / jacking bolts and installed temporary anchors to secure the vane in place. The roller assemblies and flotation tanks were removed and returned to the surface for mounting to the second vane. The rigging and guide rails were repositioned to accommodate the second vane and that vane installed using the best fit criteria and go-no-go gauges. Once in position the leveling / jacking bolts and temporary anchors were installed to secure the vane in place. The roller assemblies and flotation tanks were removed and returned to the surface. Divers then used a purpose built core drill jig on a track system to drill out the epoxy anchor holes on top and bottom of each vane. Depth of hole were recorded and verified to ensure the required minimum embedment into the existing concrete. The gap between the vertical vanes and intake structure floor and ceiling was filled with a high-strength grout. Divers anchored and secured steel forms to contain the grout.

Installation of Horizontal Vanes

An additional rigging fixture was added for installation of the horizontal vanes. This fixture, installed in the downstream end of the intake structure, was designed to accommodate a set of snatch blocks mounted in the center of each vane "slot". A barge mounted winch wire ran down into the structure, through both snatch blocks, back up and dead headed at the barge.

A purpose built picking frame, incorporating a snatch block at each of the four corners of the frame, through which the above described wire was reeved. The frame was attached to the horizontal vane to be deployed. By coming up or down on the winch wire direction the elevation of the vane could be controlled when in the structure. The previously installed hold back line was attached to the frame for lowering and upstream / downstream positioning of the vane.

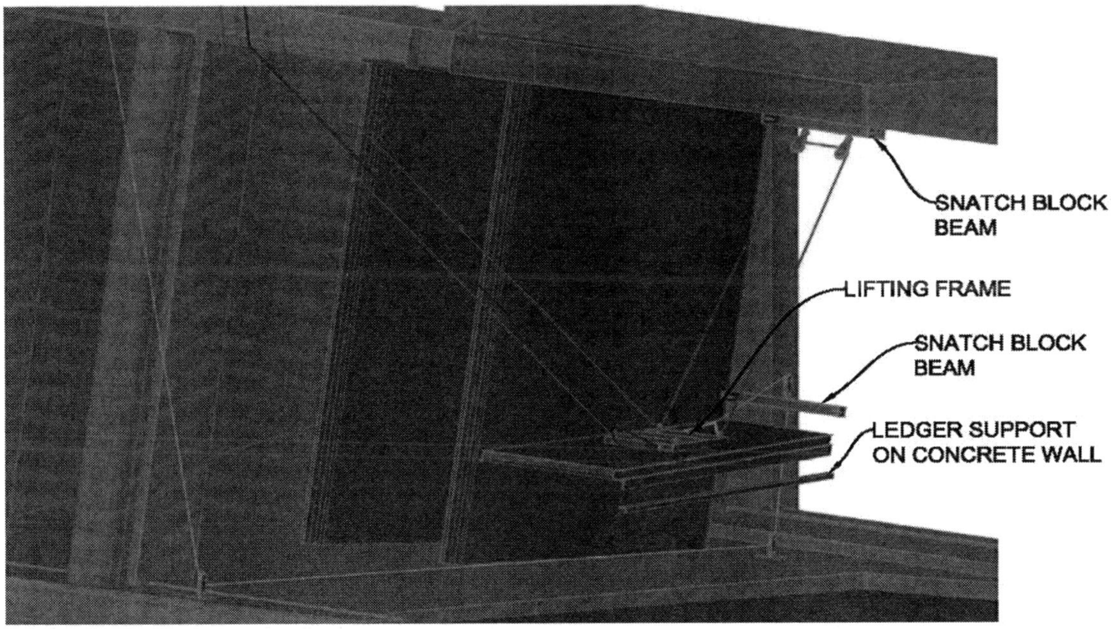

Figure 7. Illustration depicting installation rigging for one of the horizontal vanes. Winch wire used to control elevation is depicted in red, hold back line is blue.

Six angle ledgers were anchored to the north and south concrete walls to support the horizontal vanes as they were positioned at each elevation. An alignment jig was designed to ensure that the ledgers were installed at the same elevation and angle as the connection brackets welded to the vertical vanes. These ledgers were later used as part of the grout forms for the horizontal vane to concrete wall connections.
Each horizontal vane was generally installed as follows.

- Install picking frame onto selected Horizontal Vane
- Use crane to pick the vane off of the material barge and swing it around in front of the dive barge over the water. Install the winch wire into the snatch blocks on the frame and snug up the wire. Connect the hold back winch wire and snug up the line.
- Transfer load to haul back line and disconnect the crane.
- Lower the vane down the wires using the hold back line until vane is close to the opening of the intake structure.
- Diver hooks up suck line to the frame.
- Diver directs top side crew to operate the winches as needed and pull the vane into the structure and set it down on the first set of angle ledgers.

- Use suck line and haul back line to position vane. With vane in the correct position diver installs 1-inch bolts into the brackets on the vertical vane.
- Detach the picking frame and pull it back up to the surface for mounting onto the next horizontal vane to be installed. Diver moves the set of snatch blocks and wire to the next slot on the rigging fixture.

Vanes which abutted the intake structure walls were secured to the structure using epoxy anchors. Divers used a purpose built core drill jig on a track system to drill out the epoxy anchor holes on top and bottom of each vane. Depth of hole was recorded and verified to ensure the required minimum embedment into the existing concrete. The gap between the horizontal vanes and intake structure walls was filled with a high-strength grout. Divers anchored and secured steel forms to contain the grout. Vanes which abutted the vertical vanes were designed to be attached using bolted connections. However, misalignments between some of the brackets resulted in the need to modify connections between the vanes at some locations. Working collaboratively with the design engineers and the Owner, allowed for in-stride design modifications to be put in place while minimizing impact on the project schedule. Modification of the connection brackets (enlarging bolt holes, installation of additional washer plates, etc.) or in some cases underwater welding was utilized to ensure the connection was adequate for the specific loading condition of the bracket in question.

In order to provide structural integrity to the thin (1/4 inch steel plate) vane walls, provide internal corrosion protection and avoid potential fatigue failure caused by low frequency excitation of the skin plate, the vanes were specified to be filled with a cementitous grout. The grout was placed into the vertical vanes in a series of eight foot lifts. Starting from the lowermost elevation, after each level of horizontal vanes were installed and secured in place, grout was pumped into the vertical and horizontal vanes through preinstalled grout ports. A final lift of infill grout was required to completely fill the vertical vanes.

Following completion of the vane installation diver's performed cleanup activities including the following.

- Removal of all grout ports and hardware, plug with pipe plugs and grind flush.
- Strip all grout forms, remove all rigging and blocks used for installation, cut and grind down flush all extra anchor bolts that were used during construction.
- Tighten all fasteners to engineer specifications.
- Coat all exposed concrete anchors, bolted connections, exposed metal and any areas where coating has been compromised with a two-part epoxy coating (splash zone).

Divers then performed a final inspection of the installation and installed new trash racks on the intake entrance.

Figure 8. Lower level unit penstock intakes during construction of the dam.

With the work on Unit 4 complete, divers then focused their attention to the repairs needed on Units 1 through 3. Bulkheads used to isolate the units downstream of the reservoir did not seal adequately to the penstock intakes due to deteriorated concrete on the intake opening face. In addition, the bulkhead guide slots had a number of spalls and areas of exposed rebar which required repair. While the work was not quite as involved, there was some adjustment as the intakes for the original three units are about 100 feet deeper than the Unit 4 intake. Divers installed a 12-foot by 14-foot by 1-inch thick stainless steel bulkhead seal plate around the bell mouth of each intake, fastened the seal plate using concrete expansion anchors and placed a grout pad between the seal plate and the existing concrete. Then a combination of saturation and surface divers repaired identified spalls and damaged areas along the bulkhead guide slots.

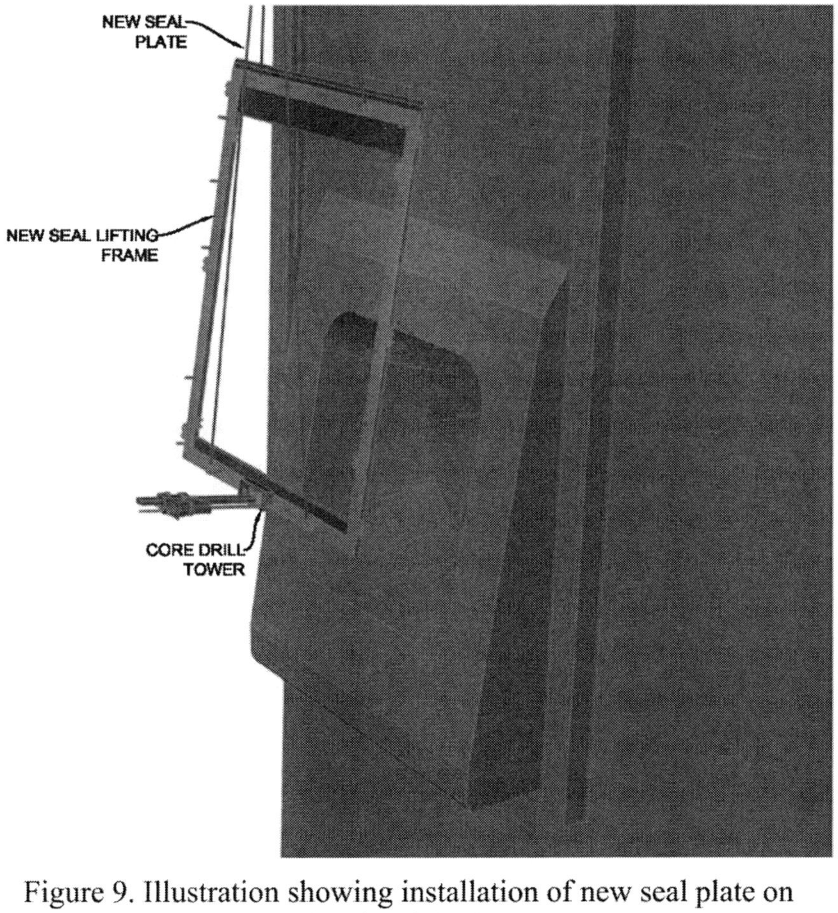

Figure 9. Illustration showing installation of new seal plate on low level intake.

CONCLUSION

There are a wide range of capabilities to facilitate underwater construction projects. Reliability versus risk, constructability, operational impacts and cost must be evaluated when determining the best method for executing infrastructure repair. With proper planning, preparation and design almost all operations that can be performed above the water can also be performed underwater. Limitations posed by performing work underwater can often be mitigated through pre-job planning and design of specialty tools and equipment. Unknown conditions that may exist below the surface will require collaborative efforts between contractor, engineer and owner in developing risk management plans and facilitating efficient solutions.

ACKNOWLEDGEMENTS

The Horse Mesa Dam Unit 4 hydro generating plant was back online within 14 months of the failure of the intake vane. This success can be attributed to the team effort of SRP's engineers, SRP's MC&M shop, Global Diving and Salvage, Inc's engineers and construction crews and to Stantec's engineers and designers. I would like to extend a special recognition to those members of the team who spent many, many long days in the field, under extremely challenging conditions to bring this project to a safe and successful conclusion.

EVALUATING THE MECHANICS OF PIPING EROSION INITIATION THROUGH LABORATORY MODELING

Richard Keizer[1]
John Rice[2]

ABSTRACT

Internal erosion is one of the predominant mechanisms responsible for incidents and failures of dams and levees. Piping (or Backward Erosion Piping) represents one of the most common mechanisms of internal erosion incidents in dams and levees. Current geotechnical engineering practice for the assessment of piping potential consists of comparing expected exit gradients with the critical gradient of the soil at the seepage exit point. This method was developed to model the heave mechanism and does not fully model the mechanisms associated with piping. The critical gradient is generally considered as the ratio of soil buoyant unit weight and the unit weight of water; suggesting that the critical gradient only depends on the void ratio and specific gravity of the solids. However, in the field and in research it has been observed that piping can initiate at average gradients much lower than unity due to concentrations in flow and non-vertical exit faces. Therefore, there is a need for deeper understanding of the granular scale mechanisms of the piping erosion process.

This paper presents the results of a laboratory study to assess the effects that inclination of the exit face and concentration of flow have on the potential for initiation of piping and the piping mechanisms. Using a laboratory device designed and constructed specifically for this study, the critical gradients needed to initiate piping in a variety of sandy soils were measured over a range of exit face inclinations. The tests results were observed to assess the effects the inclination has on the initiation mechanisms for piping as well as the critical gradients for piping. Tests were also performed with a constricted exit and a riser tube on the device outlet to look at the effects of flow concentration on the initiation of piping. With the flow concentration the behavior becomes a complex three-dimensional mechanism with mechanisms of fluidization and transverse piping coming into play.

INTRODUCTION

Internal erosion is the erosion of soil or rock due to forces imposed by subsurface water flow or seepage. Internal erosion mechanisms take several forms, including: heave, piping, concentrated leak erosion, contact erosion, and suffosion (ICOLD 2012). The effects of these various internal erosion mechanisms have been reported to account for approximately half of dam and levee failures and incidents throughout the world (Foster et. al. 2000, Richards and Reddy, 2007).

In the early 1900s Terzaghi derived a method for assessing the heave mechanism (Terzaghi 1922, Terzaghi and Peck 1948). Terzaghi's work compared the vertical the

[1] Staff Geotechnical Engineer, GEI Consultants, Sacramento, CA, rock3thekeiz@yahoo.com
[2] Assistant Professor Utah State University, UMC 4110, Logan, UT, john.rice@usu.edu

hydraulic gradient at the seepage exit to the critical gradient, i_{cr}, needed to initiate erosion in the affected soil. Terzaghi defined the critical gradient is a function of soil buoyant unit weight, γ' , through the following equation,

$$i_{cr} = \gamma' / \gamma_w \tag{1}$$

where γ_W is the unit weight of water.

Understanding the limits of this derivation, Terzaghi differentiated between the *heave* mechanism and the mechanics of *subsurface erosion*, which Terzaghi claimed "defy theoretical approach" (Terzaghi and Peck, 1948). While Terzaghi made it clear that his derivation was intended to model the heave mechanism, the critical gradient as described by Equation 1 has often been used to assess piping potential. For example, the hydraulic gradient across a low-permeability "blanket" layer is often calculated to assess the potential for piping in the underlying sand layer in Figure 1 (sand boil formation). However, once a crack has formed through the blanket layer the seepage flow is concentrated into crack and the soil particles are moved individually through the crack to the ground surface forming a boil, a mechanism commonly referred to as piping.

Based on the above discussion, it is apparent that greater understanding of the piping mechanism is needed. This paper presents the results of a laboratory study of the piping mechanism and the soil parameters that affect the hydraulic conditions needed to initiate piping. The results are presented with respect to the various stages of the initiation of piping that were observed in the laboratory models.

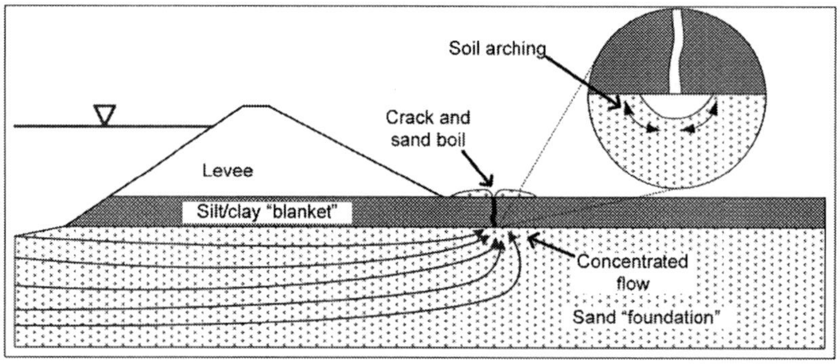

Figure 1. Sand boil formation in a levee foundation.

In previous papers (Swainston-Fleshman and Rice 2014a, 2014b, Fleshman and Rice, 2013) the results of tests performed with a vertically-oriented (horizontal exit face), 2-inch diameter sampler were presented.

This study was designed to evaluate the mechanisms associated with the initiation of piping in sandy soils through observation of soil behavior in simple laboratory models. In portions of this study presented in previous papers (Swainston-Fleshman and Rice 2014a, 2014b, Fleshman and Rice, 2013) the results of laboratory tests modeling a horizontal exit face were presented. This paper presents the results of latter portions of this study that investigate the effects of inclination of the exit face as well as effects of the sample diameter. The models were conducted using a laboratory apparatus designed and constructed at the Earth Structures Laboratory at Utah State University. The device forces a hydraulic gradient through a soil sample to produce a flow regime without converging or diverging flows. Thus, the hydraulic pressures and gradients can easily be assessed throughout the test and the critical hydraulic conditions necessary to initiate the piping process can be assessed. The study looked at various gradations of sand, glass beads, and zirconium oxide beads so that the effects of various soil parameters on piping initiation and progression can be assessed.

The primary objectives of the study were to provide understanding of the mechanics of the piping phenomenon and assess the effects basic soil properties, exit face inclination, and the size of the free exit face (sample diameter) have on the critical hydraulic conditions needed to initiate piping. Eventually, the study and continued research are expected to provide sufficient information to develop practical solutions for assessing piping potential in sandy soils that account for soil properties, complex flow conditions, and exit face conditions.

TESTING APPARATUS

A schematic illustration of the apparatus used in this study is presented in Figure 2. The device is designed to apply a uniform hydraulic gradient through a soil sample so that the mechanisms of piping initiation can be observed in a controlled and measurable seepage regime. A brief description of the device is provided below. Detailed descriptions of the device and testing procedure are presented in Fleshman (2012) and Fleshman and Rice (2013).

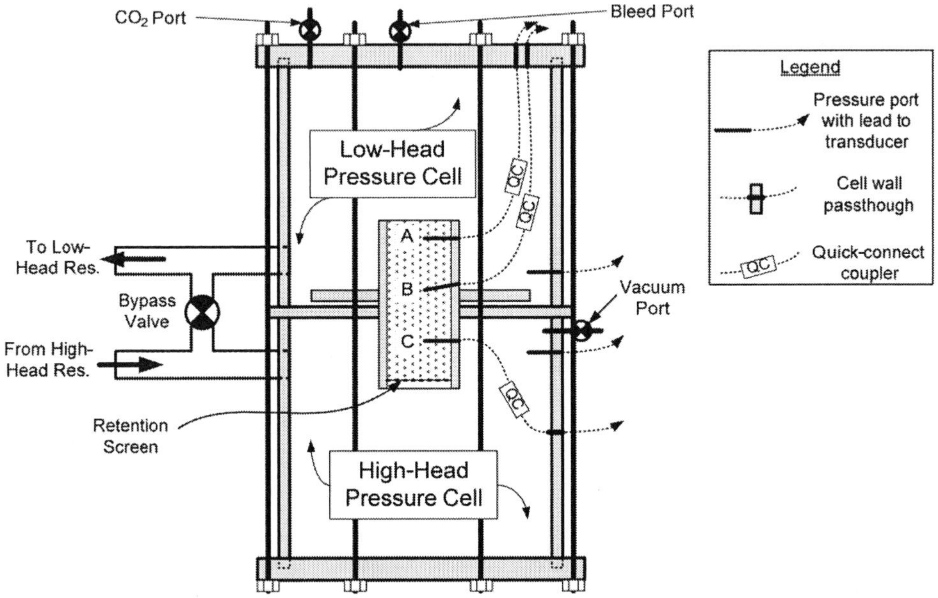

Figure 2. Schematic illustration of the pressure cells, sample holder, and pore pressure ports.

The soil sample holders used in the study are is a 5-inch (13 cm) long cylinder-shaped Plexiglas molds with screens at the base to retain the soil while allowing water to flow freely through to the soil sample. The cylinder is sealed between two enclosed pressure cells forcing all flow through the soil sample. The hydraulic heads of both pressure cells are controlled by the high-head and low head reservoirs. The head in the high head reservoir is attached to the high head pressure cell and can be adjusted to change the differential head across the sample. To assist in the deairing of samples, the reservoirs were modified to allow pressuring the system with 15 psi (103 kPa) back pressure while maintaining a precise small differential head (ranging from 0 to 0.8 psi (0 to 5 kPa)) between the cells. Tests reported in previous papers (Swainston-Fleshman and Rice 2014a, 2014b, Fleshman and Rice, 2013) were run using a 2-inch (5 cm) diameter sampler. Tests run for the tests conducted for the latter portion of the study were run using a 4-inch (10 cm) diameter sampler.

Both sample holders are instrumented to measure the pore pressures within the soil sample. Three ports are located at three elevations within the soil sample to measure pore pressures at ¾ of an inch (1.9 cm), 2-¼ inches (5.7 cm), and 3-¾ inches (9.5 cm) down from the top of the sampler. The ports consist of 1/8-inch (3mm) tubes with filtered ends to prevent soil inflow and are denoted A, B, and C, respectively, as can be seen in Figure 2. The port tubes are connected to differential pressure transducers installed between the port and the low-head pressure cell. The total differential head between the pressure cells (and thus the total differential head across the sample) is also measured. Flow through the device is measured with a magnetic-flux flow meter. Figure 3 shows the device inclined to perform tests with the exit face inclined.

Figure 3. Photo of device inclined to perform tests with inclined exit faces

An automated instrumentation and data collection system automatically collects the differential pressure and flow data throughout the test. The data collected is recorded on a data logger and can be displayed in real time during the test on a dedicated computer screen. A video taken of each test is linked to the data through a digital counter that visually displays the time since the start of data collection in the view window of the video. This system allows observations of soil behavior made in the video to be linked precisely with the collected data.

TESTING PROCEDURE

The soil sample is prepared by using a dry-raining and tapping technique. The sand is placed into the soil holder in small lifts and the sampler is tapped on the side to densify the sand. This technique results in a narrow range of densities for each soil type when duplicate experiments are run. The 5-inch (13-cm) tall soil holder is filled over capacity and the top is struck off to produce a level surface with the top. The soil sample is then weighed and the density of the soil is calculated and recorded. The soil sample holder is then placed and sealed between the pressure cells and the pore pressure sensor connections are connected before the top plate is placed.

The valves are then closed to completely seal off the pressure cells. The sample is then saturated through a process consisting of 1) flushing with CO_2, 2) applying a vacuum to the cells, 3) saturating with de-aired water, and 4) applying 15 psi (103 kPa) back pressure to both cells. Each test starts with zero differential head across the sample and proceeds by incrementally increasing the differential head across the soil sample and thus the gradient in the soil. Once movement is detected in the sample, the increasing head is halted to allow the sample to reach an equilibrium condition with the applied gradient. The differential head is again decreased once no additional soil movement is observed. The test progresses until complete failure of the sample. Observations are made during the test of the intermediate stages of failure progression.

MATERIALS TESTED

Tests were performed on a variety of sandy soils, glass beads, and zirconium oxide beads. A summary of key properties of the materials tested is presented in Table 1. Ottawa 20-30 and Graded sands conforming to ASTM C778-03 (well rounded silica sands) were tested. To investigate the effect of grain shape, specimens of angular silica sand and zirconium oxide beads were prepared to the same gradations as the Ottawa 20-30 and Graded sands. Specimens of a uniform fine-grained (No 100 sieve) garnet sand were also tested. The garnet sand and zirconium oxide have much higher specific gravities than the quartz sands (4.0 and 3.8 for garnet and zirconium oxide, respectively, verses 2.6 for quartz).

Table 1. Properties and characteristics of the soils tested

Soil Type	Sphericity	Roundness	Specific Gravity	Average Grain Size (D_{50} mm)	Coefficient of Uniformity	Internal Friction Angle$_{т}$	Min & Max Void Ratio$_{\Psi}$
Graded Ottawa Sand	0.9	0.7	2.65	0.373	1.76	38°	**0.475 - 0.719**
20-30 Ottawa Sand	0.9	0.7	2.66	0.574	1.51	38°	**0.493 - 0.707**
Garnet Sand	0.3	0.1 - 0.3	4.05	0.155	1.25	44°	**0.960 - 1.537**
Graded Anglular Sand	0.5 - 0.7	0.3	2.63	0.305	2.18	44°	**0.633 - 1.145**
Graded Zirconium Beads	1.0	1.0	3.84	0.373	1.57	35°	*0.455 - 0.596*
Uniform Zirconium Beads	1.0	1.0	3.81	0.312	1.33	33°	*0.515 - 0.670*
Uniform Glass Beads	1.0	1.0	2.49	0.503	1.20	28°	*0.522 - 0.651*

т - Internal Friction Angle measured using direct shear testing

Ψ - Min & Max Void Ratio measured by **ASTM** and *Modified* Relative Density tests

OBSERVED BEHAVIOR

The video for each test was carefully inspected to observe stages in the development of the failure. While the general soil behavior varied between tests, four stages of piping development were identified to describe the progression of the failure: 1) first visible movement, 2) boil formation, 3) heave, and 4) total failure. Details of these stages and interpretations of the mechanisms occurring during each stage are presented below. Further evidence supporting these interpretations will be presented in discussions of data interpretation later in this paper.

The first stage is defined by a slight heave or rocking movement of individual sand grains along the exit face. This movement is difficult to detect and often requires repeated viewing of the portion of the video where movement first occurs. This stage is attributed to the sand grains along the free face of the sample reaching a state of incipient motion as the forces imposed by the seeping water approach the magnitude of the forces retaining the soil grains. While movement is occurring with the individual grains, there is little overall movement heaving of the exit face or displacement of individual grains.

The second stage is formation of one or several sand boils on the exit face. The boils develop when a preferential seepage pathway forms through the upper portion of the sample as a result of the loosening and rearrangement of the grains during progressive heaving. The preferential seepage path relieves pressure from within the sample much like a miniature relief well; allowing the sample to reach equilibrium without further heaving.

The third Stage, heave movement, occurs when any significant portion of the soil is displaced off of the cross-sectional area of the sample exit face due to progressive heaving of the soil. In other words, the soil has heaved or sloughed off of the cross-sectional area. Fig. 4 shows heave movement in the form of sloughing of the soil exit face.

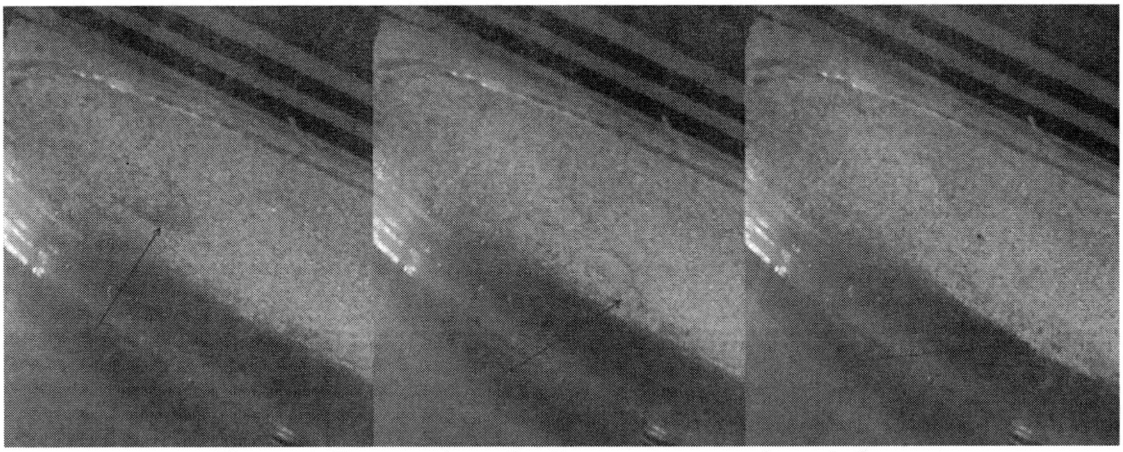

Figure 4. Progression of soil movement on the exit face.

Total failure is defined as the largest gradient before the sample fails. Because the exact gradient can only be estimated during the test, the data is analyzed after the test is complete and the maximum gradient is recorded. Gradient corrections are made to account for the decreased length of the sampler that occurs due to sloughing of the exit face prior to either boil formation and/or total failure.

MECHANICS OF PIPING INITIATION

As discussed in previous papers (Swainston-Fleshman and Rice 2014a, 2014b, Fleshman and Rice, 2013), the sequence of initiation of the piping process observed in the laboratory tests consists of four observable stages: 1) first visible movement, 2) heave progression, 3) boil formation, and 4) total heave. The mechanics of this progression can be explained as the soil structure reacting to the increasing hydraulic gradient and flow by adjusting its structure with a zone of decreased density in conjunction with preferred seepage channels (sand boils). These mechanisms will be examined by considering the interaction of the various forces acting on the soil grains as the soil structure reacts to the increasing hydraulic gradient.

Figure 5 depicts a grain at the exit face of a test sample. As a vertical hydraulic gradient is imposed on the sample, the forces acting on the grain include: 1) the weight of the grain (W), 2) the buoyant force (B), 3) the seepage force (S), 4) normal (N) and shear (F) forces at intergrain contacts, and 5) the viscous shear forces from the seeping water (S_V). The weight and buoyant forces remain constant throughout the test. The intergranular forces will vary as the other forces on the grain change; increasing or decreasing depending on the orientation of the grain contact. The seepage force is due to the hydraulic gradient across the grain and is a result of higher heads on the upstream side of the grain. The magnitude of the seepage forces are a function of the size of the interstitial voids and the velocity of the water seeping through the voids. As water seeps through a void, the water flows at a maximum velocity at the center of the void and decreases to near zero adjacent to the void. According to Navier-Stokes theory, the magnitude of the viscous shear imposed on the grain increases not only as the average velocity through the void increases but also as the distance from the maximum velocity to the grain surface decreases. Thus, as the size of the void increases, not only does the velocity of the seeping water decrease but, given a larger void, the shear transfer from the water to the grain decreases due to the greater distance to the maximum flow velocity.

Initiation of the initial movement of piping development (Stage 1) occurs when the seepage force and the viscous shear forces approach the magnitude of the retaining forces (buoyant weight and intergranular forces) on the soil grains at the exit face. At this point the surface grains are in a state of incipient movement and begin to move with increasing hydraulic gradient. However, their movement increases the size of the surrounding interstitial voids, allowing the velocity the decrease of both the seepage water as well as the hydraulic gradient across the grain. As a result, the viscous shear forces and seepage forces decrease and the grains reach a state of equilibrium after a small movement or rotation. As the hydraulic gradient across the sample is again increased, the surficial grains move until a maximum void ratio is achieved. Additional increases in the gradient reduces the downward force of the upper grains on the next layer of underlying grains allowing them to loosen until they have reached a state of equilibrium. This process continues with increasing gradient across the sample as the loosened zone increases in thickness and the exit face of the sample progressively heaves (the second stage).

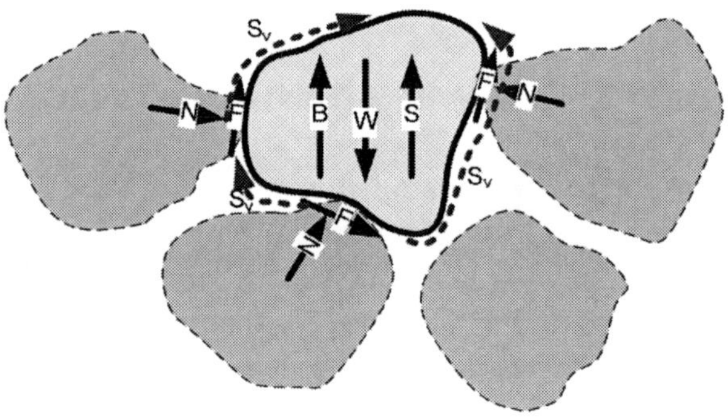

Figure 5. Plot of raw pore pressure data for test on graded Ottawa sand.

In some of the tests, the heave progression described above is temporarily interrupted by the formation of a sand boil (the third stage). In the tests performed, sand boils form when there is a random alignment of large interstitial voids in the near-surface soil structure, thus forming a preferential, low-resistance pathway for seeping water. The pathway acts similar to a relief well, allowing water from within the sample to escape. With the water flowing out the preferential pathway pressure from increased gradients is relieved without further progression of the heave. In graded soils the pathway may be large enough to allow the finer portion of the gradation to be removed from the soil matrix and deposited at the surface in the sand boil. The removal of the fine grains increases the hydraulic conductivity of the soil surrounding the preferential pathway; thus, increasing the effectiveness of the pathway in providing drainage to the interior of the sample. Sand boils were observed to collapse with increased hydraulic gradients. This collapse is likely due to their drainage capacity being exceeded and the continuation of heave progression disturbing the alignment of voids in the preferential pathway.

As long as the heaved zone remains effectively on top of the sample, the process of progressively achieving equilibrium can be maintained as the remaining downward force of the upper grains continues to act on the lower grains. However, when the sides of the heaved mound exceed a stable inclination the sloughing of the sides of the heave mound will occur. This sloughing removes the overburden of the loosened zone, and prevents equilibrium from being achieved. At this point, the rapid progression of total heave (fourth stage) rapidly occurs as observed in the tests.

APPLICATION TO FIELD CONDITIONS

The observed behavior of the soils in the laboratory models can be translated to expected behavior in the field. Consider the configuration in Figure 6a, consisting of a low permeability layer of silt and clay (often referred to as a "blanket" layer) overlying a layer of higher permeability sand. An irregularly shaped defect exists in the blanket layer. While the defect is admittedly an odd shape to be found in nature, the shape is convenient for the purposes of this discussion. The seepage flow in the model is from left to right, as would be imposed by the flow conditions depicted in Figure 1. As the water rises on the left side of Figure 1, the differential head imposes increasing hydraulic gradients at the base of the defect in Figure 6a. When the gradients reach a critical level, the Initial Movement Stage identified in the laboratory models will occur at the base of the defect. Increasing differential head will increase the gradients and the Progressive Heave Stage will occur, forming a zone of loosened soil below the defect and filling the lower portion of the defect with similarly loosened soil, as depicted in Figure 6b. The Boil Formation Stage may also occur intermittently with the Progressive Heave Stage, as was observed in some of the laboratory models. The loosened zone around the base of the defect, having higher hydraulic conductivity than the surrounding soil, will allow more flow to enter the defect. This flow increase, combined with increased flow due to increasing differential head, may increase the flow velocity in the upper portion of the defect to the point where detached soil particles are able to be carried to the ground surface in fluid suspension (i.e. Stoke's Law phenomenon) as shown in Figure 6c. Once out of the defect, the sand particles are deposited on the ground surface forming a sand boil. With the removal of sand particles from the flow path the loosened zone can increase in size and progress toward the source of the seepage, further decreasing the seepage resistance and progression toward a piping failure.

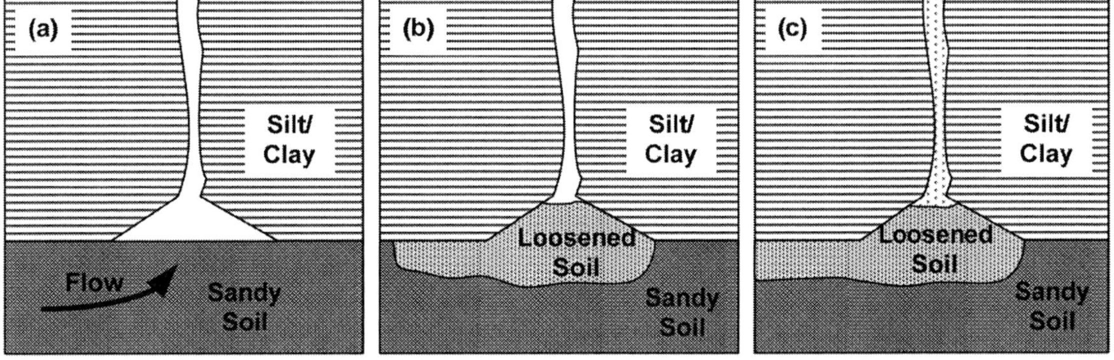

Figure 5. Initiation of piping mechanisms in a field condition.

The mechanisms described above can also be used to explain a phenomenon often observed in levees; the reduced resistance to piping over successive flood events. As the loosened zone depicted in Figure 6c progresses toward the source of flow, the loosened soils provide a zone of increased hydraulic conductivity and reduced seepage resistance. This zone provides a preferred seepage pathway to the defect allowing piping to initiate at a lower differential head. Over successive flood events the loosened zone progressively advances toward the seepage source, incrementally increasing the piping risk.

RESULTS OF TESTING

In previous papers (Swainston-Fleshman and Rice 2014a, 2014b, Fleshman and Rice, 2013) the results of tests performed using only the 2-inch diameter sampler in a vertical orientation were presented. The tests were performed on various sandy soils as well as gradations of glass and zirconium oxide beads. Comparison of the results was used to assess the effects of various soil parameters on the critical hydraulic conditions necessary to initiate piping in this configuration.

The current study compares the results presented in the previous study with tests performed using the 4-inch diameter sample holder in both vertical and inclined positions. The purpose of these tests is to model behavior that may be observed in the field where the opening or defect (such as that depicted in Figure 6) is larger than in the previous tests. The inclined tests are intended to model piping initiation along inclined exit faces or at the head of a progressing pipe (i.e. the left side of the loosened soil in Figure 6b).

EFFECT OF SAMPLE INCLINATION

Sample of the soils presented in Table 1 were tested in the 4-inch sample holder under exit face inclinations ranging from horizontal to the angle at which the soil fails with no seepage flow. Figures 7 through 10 present plots for graded Ottawa Sand of the gradients under which the four stages of failure were observed: 1) first visible movement (i_m), 2) boil formation (i_b), 3) heave (i_h), and 4) total failure (i_f).

Figure 7 is a plot of gradients needed to cause initial movement of soils, i_m. These results indicate that at small inclination angles (up to 25 percent of the angle of repose) the gradient is about the same as that for the horizontal exit face. However, at angles greater than 50 percent of the angle of repose the magnitude of the gradient to cause initial movement decreased rapidly, approaching zero for the failure angle.

Figure 8 plots gradients needed to cause sand boils versus the angle of inclination, i_b. Figure 8a presents the gradients calculated using the full length of the sample holder while Figure 8b presents the gradient that have been corrected to account for sloughing of the exit face prior to boil formation. With the gradients corrected (Figure 8b) the relationship between gradient and inclination angle is nearly linear.

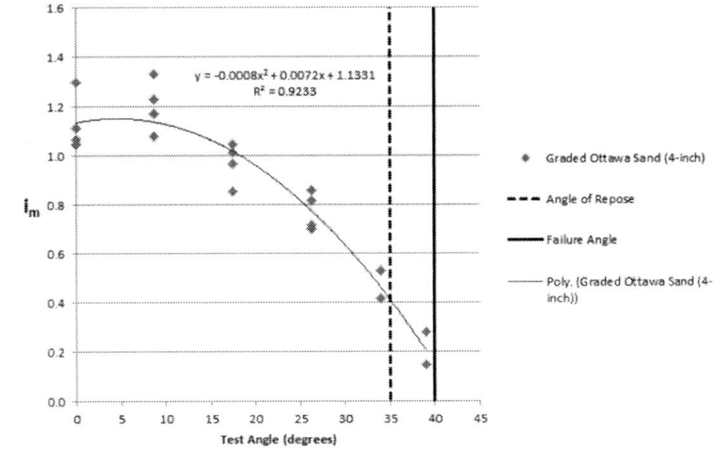

Figure 7. Plot of gradient required to cause initial movement versus angle of inclination for Graded Ottawa Sand

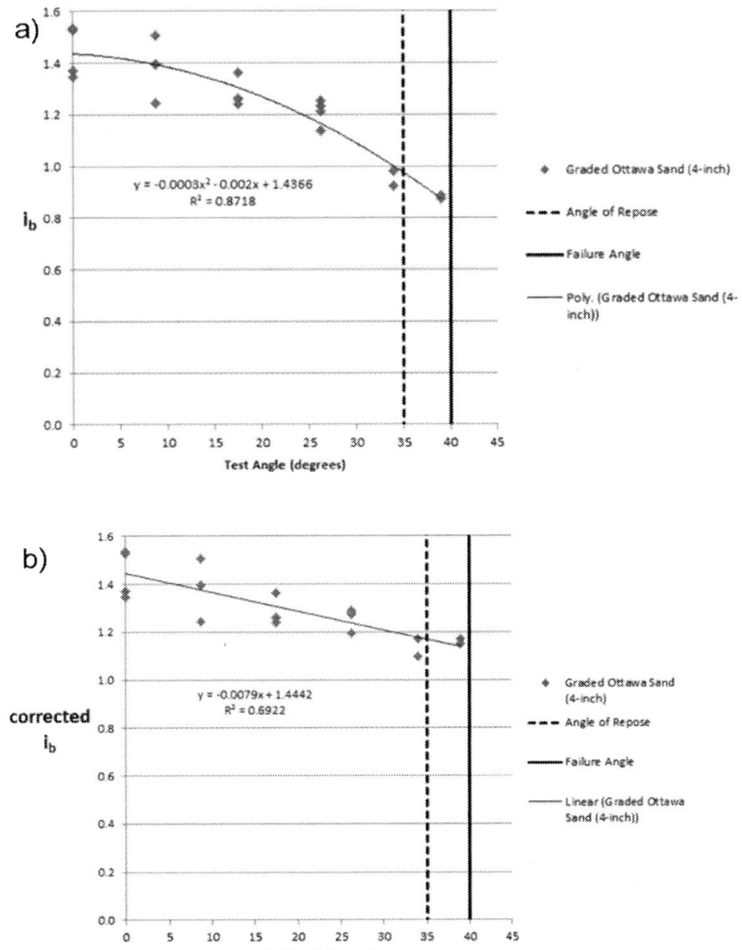

Figure 8. Plot of gradient required to cause sand boils versus angle of inclination for Graded Ottawa Sand: a) uncorrected gradient, b) corrected gradient

Figure 9 is a plot of gradients needed to cause sufficient heave or sloughing of soils so that the soil extends beyond the edges of the sampler, i_h. Similar to the plot of i_m (Figure 7) there is little effect for angles less than 25 percent of the angle of repose but a marked decrease in gradient with increasing inclination.

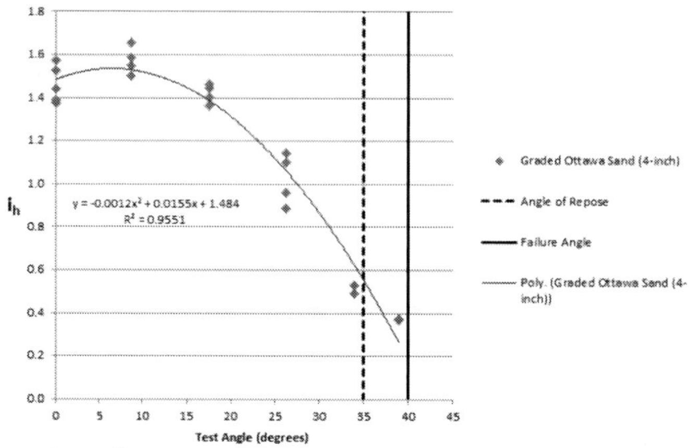

Figure 9. Plot of gradient required to cause heave or sloughing versus angle of inclination for Graded Ottawa Sand

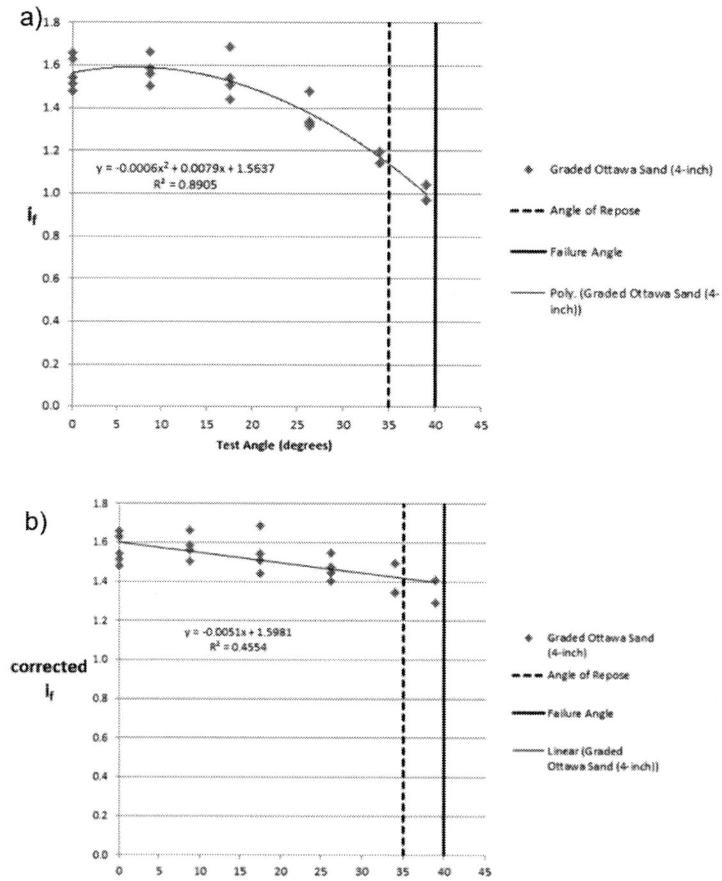

Figure 10. Plot of gradient required to total failure versus angle of inclination for Graded Ottawa Sand: a) uncorrected gradient, b) corrected gradient

Figure 10 plots gradients needed to cause total failure versus the angle of inclination, i_f. Figure 10a presents the gradients calculated using the full length of the sample holder while Figure 10b presents the gradient that have been corrected to account for sloughing of the exit face prior to boil formation. With the gradients corrected (Figure 10b) the relationship between gradient and inclination angle is nearly linear.

EFFECT OF SAMPLER SIZE

Figures 11 and 12 present comparisons of the tests results from tests on all of the soil types tested run using the 2-inch diameter sample holder and the 4-inch diameter sample holder. Figure 11 presents tests run with a horizontal exit face while Figure 12 presented tests run with an exit face inclination of 75 percent of the angle of repose. Also presented in both figures are theoretical critical gradients calculated using Equation1 for each soil.

Inspection of Figure 11 indicates that the sample holder size has different effects on the various stages of the piping initiation. In Figure 11a the sample holder size has a small effect on gradients for initial movement in the rounded and graded samples and almost no effect on the remaining samples. This is likely due to this stage being an intergranular mechanism where interaction with the sides of the sampler were of little importance. In Figure 11b, the sampler holder size had a somewhat greater effect on the results. This is likely due to more of the soil mass being involved with this phenomenon. Figures 11c and 11d indicate the sampler size has even greater effect on the heave and total failure stages. This is likely due to these phenomena involving the entire soil mass so that bridging and interaction with the sides of the sampler were of more importance.

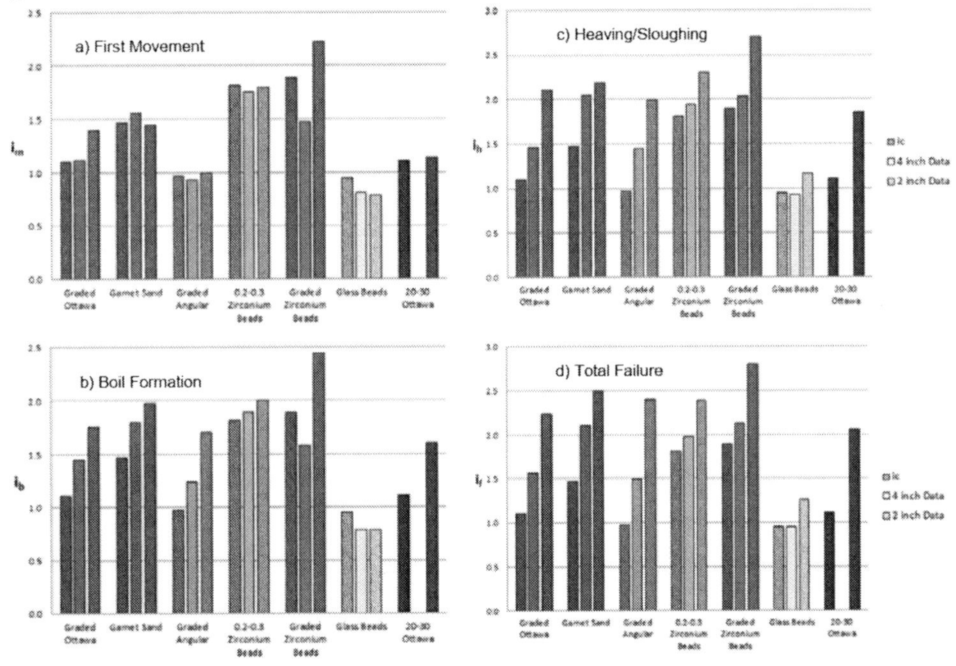

Figure 11. Comparison of observed gradients in tests with **horizontal exit face** between 2-inch and 4-inch diameter sample holders for : a) first visible movement (i_m), b) boil formation (i_b), c) heave (i_h), and d) total failure (i_f).

In Figure 12 (presenting tests performed at 75 percent of the angle of inclination), while the resulting gradients are lower than those in the horizontal exit face case, The trends and effects of sampler size on the various stages are similar. In general, for both inclinations reported in Figures 11 and 12, the effect of increasing sampler size is to decrease the gradients needed to initiate piping. It is expected that with an infinite sampler size, total failure of the sample would occur at a gradient similar to that predicted by Equation 1, which does not account for side friction or soil arching.

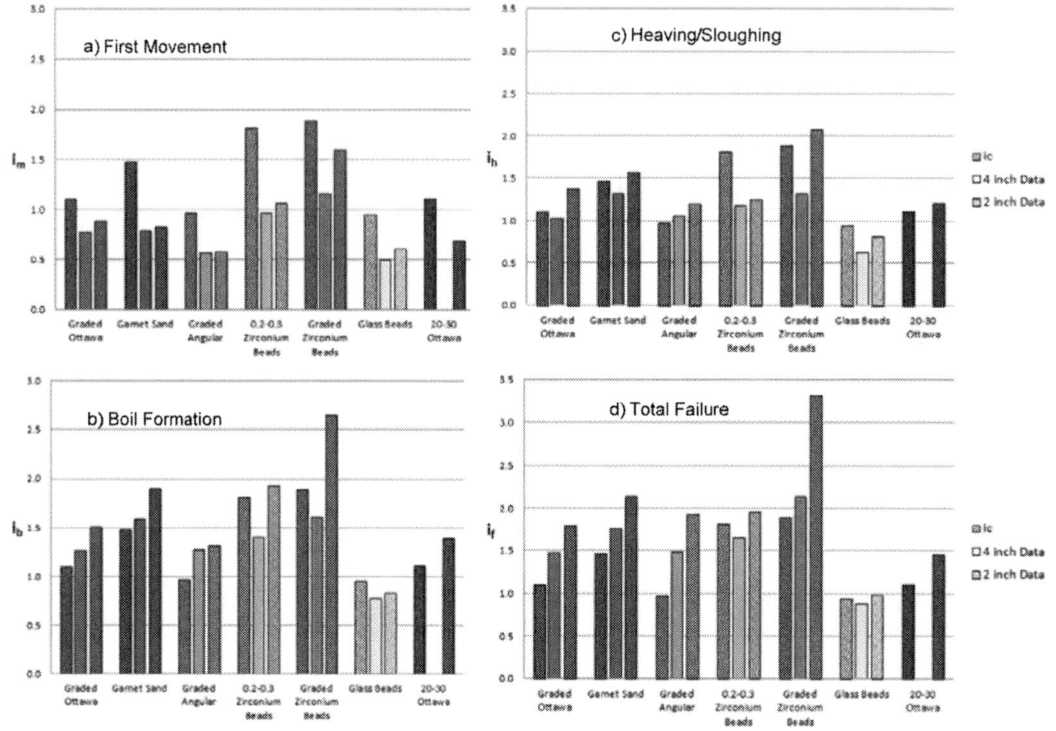

Figure 12. Comparison of observed gradients in tests at **75% if the angle of repose** between 2-inch and 4-inch diameter sample holders for : a) first visible movement (i_m), b) boil formation (i_b), c) heave (i_h), and d) total failure (i_f).

CONCLUSIONS

The paper presents a laboratory modeling program designed to identify and study the mechanisms of the initiation of piping erosion in sandy soils. The modeling was performed using a laboratory apparatus designed to measure critical hydraulic conditions for the initiation of piping in sandy soils and observe the mechanisms associated with the initiation of piping. Various sandy soils, glass beads and zirconium oxide beads were tested in the research in order to assess the effects of various soil properties on the critical hydraulic conditions needed to initiate piping. Tests were performed using a 2-inch diameter and a 4-inch diameter sample holder and were performed over a range of inclinations of the exit face.

A hypothetical example of how the stages of piping initiation observed in the laboratory models can take place in a field condition. This example shows how the observed laboratory behavior can be formulated into a new model for the initiation and progression of piping in earth structures such as dams and levees.

The test results indicated that the inclination of the exit face has a large effect on the initial movement and heave phases of piping initiation and a much less pronounced effect on the boil formation and total failure mechanisms. The size of the sampler was found to have a small effect on the first movement and boil formation mechanisms. This is attributed to these mechanisms being more intergranular phenomena, where interaction with the sides of the sampler is less of an issue. The effect of sampler size was much more pronounced with the heave and total failure stages since these stages incorporate movement of the entire soil mass and thus are influenced more by bridging and side friction.

ACKNOWLEDGEMENTS

This material is based upon work supported by the National Science Foundation (NSF) under Grant CMMI 1131518. Any opinions, findings, and conclusions or recommendations expressed in this material are those of the authors and not necessarily the views of NSF

REFERENCES

Fleshman, Mandie, 2012, "Laboratory Modeling of Critical Hydraulic Conditions for the Initiation of Piping", Master's Thesis, Utah State University, 2013.

Fleshman, M. and Rice, J., 2013, "Constant Gradient Piping Test Apparatus for Evaluation of Critical Hydraulic Conditions for the Initiation of Piping", *ASTM Geotechncial Testing Journal*, 36(6), pp 1-14.

Foster, M., Fell, R., and Spannagel, M., 2000, "Statistics of embankment dam failures and accidents," *Canadian Geotechnical Journal*, 37(5), pp. 1000-1024.

ICOLD (International Committee on Large Dams), 2012, "Internal Erosion of Existing Dams, Levees, and Dikes, and their Foundations", *ICOLD Bulletin IXX*, Draft of May 21, 2012.

Richards, K.S. and Reddy, K.R., 2007, "Critical appraisal of piping phenomena in earth dams," *Bulletin of Engineering Geology and the Environment*, 66(4), pp. 381-402.

Swainston-Fleshman, M., and Rice, J. (2014a). *Evaluating the Mechanics of Piping Erosion Initiation through Laboratory Modeling.* Proceedings of USSD Annual Conference, April 7-11, 2014. San Francisco, CA.

Swainston-Fleshman, M., and Rice, J. (2014b). *Laboratory Modeling of the Mechanisms of Piping Erosion Initiation.* ASCE Journal of Geotechnical and Geoenvironmental Engineering, 140-6.

Terzaghi, K., 1922, "Der Grundbruch an Stauwerken und Seine Verhutung [The Failure of Dams and Its Prevention]," *Die Wasserkraft*, Vol. 17, pp. 445-449.

Terzaghi, K., 1948, *Theoretical Soil Mechanics*. Wiley, New York, 510 p.

Terzaghi, K., and Peck R. B. (1948). *Theoretical Soil Mechanics in Engineering Practice*. Wiley, New York.

EXAMINATION OF THREE-DIMENSIONAL EFFECTS OF INTERNAL EROSION (IE) AND PIPING PROCESSES IN SOIL

Scott Anderson, P.E.[1]
Keith Ferguson, P.E.[2]

ABSTRACT

Internal erosion and piping is one of the primary causes of failure of embankment dams. However evaluating seepage conditions that could lead to the development of this potential failure mode is a difficult task and requires understanding of the processes associated with soil erosion, extensive experience, and keen engineering insight and judgment. Two dimensional seepage models, while providing valuable information, cannot adequately model backward erosion and piping phenomenon. This paper takes a look at the current understanding of internal erosion (IE) processes and mechanisms, and discusses the importance of three dimensional effects on the assessment of these processes in soils susceptible to erosion. The paper includes a re-examination of the laboratory testing and three dimensional modeling research conducted by the University of Florida on defect propagation that became the basis of Schmertmann's methodology for estimating factor of safety against piping (Schmertmann, 2000). The previous numerical models are updated using modern 3-D seepage models to take a closer look at seepage gradients and flow concentration that may develop at, and near the active face (upstream end) of the defect. In addition, a 3-D model of a layered system is presented and compared to the 2-D modeling results and observations of a seepage defect that developed at Isabella dam (CA). Results from these two and three dimensional seepage models will be presented and compared, providing practical guidance on the evaluation of erosion/piping failure modes in embankment dams. Three dimensional effects at the entrances to defects in outlet conduits, bedrock openings and at contacts with coarse materials that cannot filter finer erodible soils are also discussed.

INTRODUCTION

The process and mechanisms of internal erosion are a key area of emphasis within the dam and levee engineering community in the US and around the world. The ongoing research and documentation that is occurring are summarized in the following sections to establish a framework for the later discussion of the importance of 3-D effects when assessing potential internal erosion failure mechanisms, particularly in foundation soils immediately below the base of an embankment dam.

---

[1] National Technology Leader for Numerical Modeling , HDR Engineering, Inc., 1670 Broadway, Suite 3400, Denver, CO 80202, 303-764-1520, scott.t.anderson@hdrinc.com
[2] National Dam and Hydraulic Structures Practice Leader, HDR Engineering, Inc., 1670 Broadway, Suite 3400, Denver, CO 80202, 303-764-1520, keith.ferguson@hdrinc.com

THE IMPORTANCE OF INTERNAL EROSION FAILURE MODES

ICOLD (Bulletin 164, 2013), and Chapter 26 of the USBR/USACE Best Practices for Dam and Levee Safety Risk Analysis documents provide excellent summaries of both internal erosion failures and incidents associated with embankment dams. Much of these summaries are derived from the work of Foster and Fell documented in a variety of references beginning in about 1998, as well as Engemoen (et. al.) at the U.S. Bureau of Reclamation beginning in about 2009. A few excerpts from these documents indicating the relative importance of IE to other potential failure modes are provided in Tables 1 and 2, below.

As can be seen in Table 1, 46 percent of all embankment dam failures can be attributed to internal erosion processes. Table 2 further indicates that 67 percent of internal erosion failures occur within the embankment dam and the remaining 33 percent occur within the foundation. Nearly one half of the failures within the embankment were associated with conduits which penetrate, or walls which support the embankment.

Table 1. Overall Statistics of Embankment Dam Failures

Failure Mechanism	Erosion		Embankment Sliding	
Mode of failure	External erosion (overtopping)	Internal erosion (IE)	Static Instability	Seismic instability
% over the world	48%	46%	4%	2%
% over the world	94%		6%	

Table 2. Further Statistics of Dam Failures through the Embankment or Foundation

No. of Cases		% of Failures		Average Probability of	
All Failures	Failures in Operation	All Failures	Failures in Operation	All Failures	Failures in Operation
Internal Erosion through the Embankment					
39	38 30		33	3.5E-03	3.5E-03
Internal Erosion through the Foundation					
19	18 15		15	1.5E-03	1.5E-03

Both Redlinger and Engemoen, on behalf of the U.S. Bureau of Reclamation have recently carried out detailed analyses of internal erosion incidents at 220 of Reclamations dams (ICOLD, 2013). It is important to note that this portfolio of dams generally represent 1) a subset of the entire US portfolio designed and constructed within a single agency and their engineering practice guidelines and 2) relatively large structures that are usually filled to 'normal "high" pool levels' (within 7 to 12 feet of the dam crest) at the beginning of each year. ICOLD notes that the definition of "incident" in this assessment is not the same as the ICOLD definition. The Reclamation findings are as follows:

- 99 of the 220 embankment dams had experienced incidents (45%)

- Of these 99 incidents, 1 was a failure, 53 involved definite particle transport, and the remaining 45 were excessive seepage or sand boils.
- Of these 99 incidents, 9 occurred in the embankment, 70 in the foundation, 6 from the embankment into a conduit, 5 were into or along a conduit, and 11 were into a toe or under drain.

INTERNAL EROSION MECHANISMS

The following is a summary of internal erosion mechanisms that are currently in use in the dam and levee engineering community of practice within the US and internationally (USBR/USACE Best Practices, Chapter 26, 2014; ICOLD, bulletin 164, 2013; Authors'experience).

- ***Piping*** (Backward erosion piping (BEP)) – Occurs when soil erosion begins at an unfiltered seepage exit and continues its way upstream through the embankment, foundation, or both, forming a continuous pathway toward the reservoir . Two primary physical "erosion" *processes* are present in this overall mechanism; 1) adequate seepage gradients and flow quantities (energy) for **detachment** particles within the active erosion front (pipe head), and 2) once a particle is detached, adequate flow quantity and velocity in open "pipes" to **transport** eroded soils through the forming defect. In addition to these erosion processes several other conditions must be present for this overall mechanism to continue (progress) toward and open into a reservoir:

    1) a continuous flaw or pathway for the "pipe" to develop. Note that different types of flaws can connect to form one continuous flaw,
    2) relatively permeable and erodible material within the flaw/flow path (fine sand, silty sands, and sandy silts), and
    3) conditions allowing for the formation of a stable and open "pipe" downstream of the active front. Typically, this is described as a "roof".

    In general, recent publications suggest that this mechanism is controlled primarily by horizontal gradients and seepage flows. As will be subsequently demonstrated, idealized 3-D seepage models suggest that within the active seepage "front" within a developing "pipe" (typically two or more diameters from the active "face" or pipehead), the gradient and velocity vectors are highly variable ranging from horizontal to vertical and both component likely influence the erosion process. This confirms some of the original observations by Townsend (1981) and Schmertmann (2000) and suggests that **detachment** is a highly complex process with important contributions from vertical gradients typically associated with a "heave" or "quick" condition described below, horizontal gradients, and the actual quantity of flow entering the pipe at that quantities ability to transport the detached particles from the active erosion region.

- ***Stoping*** (Internal migration) erosion – Particle movement is driven by gravity in materials that are not capable of sustaining a roof or pipe. It is characterized by the formation of localized voids that successively grow and collapse leading to the formation of sinkholes, and localized over-steepening of embankment slopes leading to progressive instability.

- ***Heave or Quick Condition*** (sand boils) – The original pioneering work on this issue was performed by Terzaghi at Robert College, Bebek, Turkey beginning in about 1918. Terzaghi's experiments showed that as upward hydraulic forces through a column of sand were increased, "the quantity of percolation would increase proportionally (in accordance with Darcy's Law) until the upward hydraulic force became equal to the submerged weight of the column of sand. Then the flow would immediately and dramatically increase while the sand underwent a complete change in its fabric, acquiring a very loose grain packing. In other words, the structure of the sand had rearranged itself automatically into a looser and much more permeable fabric in order to reestablish hydraulic equilibrium between the weight of the sand and the now much smaller force of the flowing water." (Goodman, 1999). Terzaghi subsequently constructed model dams, of various shapes, and observed the effects of gradually raising the water level stored behind them. As the water height increased, a critical height would be reached at which sand on the submerged surface at the downstream toe of the dam would appear to "boil" as it suffered erosion of fine particles; shortly afterwards, a tunnel would be eroded completely under the dam, and the reservoir would summarily wash through it – a classical failure by "piping". This critical height was easily estimated from the hydraulic flux and permeability at the worst point along the seepage path, which was usually at the downstream toe of the dam.

- ***Uplift and Blowout*** – This mechanism can occur if the embankment or a confining, low permeability foundation layer overlies a more permeable layer under and at the downstream toe of an embankment. Under this condition relatively high (artesian) water pressures may develop at the base of the embankment/confining layer. When these pressure become higher than the submerged weight of the confining layer materials the confining layer is "lifted" upwards resulting in the formation of cracks providing an unfiltered pathway for the erosion (blowout) of the higher permeability soils below the confining layer.

- ***Concentrated Leak Erosion*** – Involves and erosion mechanism along the edges or perimeter of an opening (crack or pipe) in which concentrated leakage is occurring. This is sometimes referred to as "scour" and is related to **contact erosion** (ICOLD, 2013, USBR/USACE, 2012). The primary erosion mechanism at work is associated with the removal of soil at the exposed face or surface of the defect due to concentrated water flow in the defect. This type of erosion mechanism has typically be the subject of research associated with river mechanics and scour and authors such as Hjulstrom, 1935; Ishbash, 1936; Briaud, 2007, and others provide excellent research results that can inform our understanding of internal erosion when the primary erosion mechanism is concentrated flow within a defect. This mechanism is at work during the gross enlargement stage once a continuous pipe or defect (crack) has reach an open source of water such as a highly permeable shell, open joint, or the reservoir.

- ***Suffusion*** – Selective erosion of finer particles from the matrix of coarser particles, leaving behind a soil skeleton formed by the coarser particles (ICOLD, 2013). Because the coarser particles carry stresses within the soil, little to no loss of matrix integrity or change in volume occurs. It is interesting to note that this phenomenon was first referred to as "scour" by Terzaghi when evaluating the left abutment seepage at Cleveland Dam in British Columbia (Goodman, 1999). Later, Sherard referred to this phenomenon "internally unstable gradings" (Sherard, 1979). Kezdi (1979) introduced the concept of suffusion as an erosion mechanism and Kenney and Lau (e.g. 1985) furthered the concept of internal stability by introducing a method for shape analysis of gradings. Skepton and Brogan (1994) confirmed the Kenney-Lau method with piping tests and showed that internally unstable grading seroded at significantly lower gradients than would be expected from classical piping theory of Tezaghi. An excellent summary of the Kenney-Lau approach has been provided by Ronnqvist and Viklander (2014).

- ***Suffosion*** – Similar to suffusion except effective stress is carried by the finer fraction and movement of these fines results in a reduction in the total volume and collapse of the soil matrix.

INTERNAL EROSION STAGES

The following is a summary of the stages of internal erosion that are generally used in the US and throughout the world when performing potential failure modes and risk assessment activities for embankment dams (ICOLD bulletin 164, 2013, Authors experience). Some additional clarifications of the stages are presented. The Author's note that these stages may involve very different considerations and factors for different potential failure pathways through a dam verses through the dam foundation. Much of the literature today provides a somewhat ambiguous differentiation between the "continuation" and "progression" stages. The Authors offer a more definitive definition based on the specific erosion mechanisms at work in the overall erosion process.

- ***Initiation*** – Initiation is the time and location where erosion begins at a point of seepage discharge:
    1) from an unfiltered defect, geologic anomaly, or crack along the downstream toe of embankments,
    2) at defects in conduits, drain systems, and along structures;
    3) at location(s) of filter incompatibility at zone contacts in embankments, within foundations, and at the contact between the embankment and foundation;
    4) at location(s) of concentrated leakage discharge from an embankment/foundation crack due to differential settlement or hydraulic fracturing (Sherard, 1985; McCook and Grotrian, 2010; Ferguson et al, 2013); and

5) from the embankment into foundation defects such as open joints, and karst features.

- ***Continuation*** – the portion of the piping and erosion mechanism where internal erosion is governed by **detachment** and **transport** processes. Seepage into developing defects and the corresponding detachment and transport of soil during this stage is controlled by the location and characteristics of the flaw (pathway dimensions, coefficient of uniformity (C_u) and permeability of the soil surrounding the defect, ability to form a roof, etc.). During this stage, developing pipes remain relatively small and difficult to detect unless other mechanisms such as stoping and sinkhole development occur as part of the IE process (e.g. Ferguson, et. al., 2014).

- ***Progression*** – the portion of the piping and erosion process where significant gross enlargement occurs and erosion is governed by **concentrated leak erosion** (also referred to as "**scour**") processes. The rate of erosion is controlled by the erodibility of the embankment/foundation soils at the surface of the defect when exposed to flowing water within the defect. Other considerations related to the progression stage include the ability of 1) embankment zoning (coarser shell materials) or conditions within a foundation defect (e.g. the walls of a joint or karst feature) that could limit or terminate the growth of the defect, 2) material characteristics (dispersion), or 3) the potential for embankment or foundation material to collapse, swell, or otherwise self-heal (such as development of a natural filter) along the erosion pathway.

- ***Breach formation*** – The gross enlargement process is completed during this stage resulting in the uncontrolled release of the reservoir.

FACTORS INFLUENCING THE INTERNAL EROSION (PIPING) PROCESS

The following is a summary of the factors influencing the internal erosion (piping) process as discussed by a number of noted researchers including Townsend, Schmertmann, USBR, Fell et al (2008), ICOLD bulletin 164, Hjulstrom (1935, 1956), Briaud (2007).

- Particle **Detachment** (at unfiltered discharge)
  - ✓ The primary factors influencing detachment include 1) particle size and size distribution (coefficient of uniformity), 2) seepage gradients/velocity/energy at the active front, and 3) soil density/stress state.
  - ✓ Townsend et. al. (1988) observed the following from the University of Florida piping flume tests: "Piping begins to form with the displacement of several sand particles at the tip of the initial pipe. The particles slide into the newly created channel, and when finally washed away, several more new particles became displaced as a retrogressive slide. A continuous landslide and erosion process exists which propagates the pipe. However for sand with large grain sizes or large coefficients of uniformity, C_u, insufficient water velocities exist

to erode the soil particles after they slide into the initial pipe and deposition may occur downstream in the pipe. Because of this deposition, the piping may be retarded or even ceased by self-healing."

- Particle **Transport** and Deposition (sedimentation)
  - ✓ Particle size, size distribution, and specific gravity
  - ✓ Flow velocity in the defect, depth, and velocity variations within system of developing defects
  - ✓ Other factors contributing to a transition from erosion to deposition within a defect
- **Concentrated Leak Erosion (Scour)** – gross enlargement processes
  - ✓ Particle size and specific gravity, shear stress, flow velocity in the defect

KEY FACTORS IN ASSESSING DAM SAFETY RISKS ASSOCIATED WITH INTERNAL EROSION

Based on the authors experience investigation and assessing seepage failure modes (both incidents and actual failures), the following is a summary of the most significant considerations associated with each stage of the seepage failure mode process:

- Initiation
  - ✓ The potential for an unfiltered natural or man-made defect that would result in the concentration of seepage gradients and flows that could cause initiation of particle migration
- Continuation and Progression/gross enlargement
  - ✓ The potential for a continuous pathway (layer, crack, hydraulic fracture or other natural or man-made defect from the point of initiation to the reservoir or water source
  - ✓ The permeability of the soils and ability to generate adequate seepage quantity and velocity (energy) to **detach** ***and*** **transport** the full range of soil particle sizes along the pathway
  - ✓ The erodibility of the soils along the pathway (coefficient of uniformity, C_u; plasticity, density/specific gravity)
  - ✓ Conditions that promote the stability of the piping feature (roof or pipe channel)
  - ✓ Conditions that could arrest the advancement of the piping/erosion feature such as changes in material type along the pathway, collapse, and development of a natural filter. Natural filters may develop in the presence of more widely graded soils and sufficient energy only to transport a portion of the gradation, leaving coarse materials deposited in the pipe that eventually cause a natural filter to develop.
- 2-D verses 3-D influences
  - ✓ Once initiation has occurred, seepage conditions at the advancing face (head) of the pipe change dramatically with significant concentration of seepage gradients, flow quantities and velocity.
  - ✓ If initiation is suspected, consideration of 3-D seepage influences on the fundamental factors of detachment and transport become the primary

consideration in estimating the likelihood that the continuation process will advance all the way to the reservoir. Methods of assessment that rely on global or point gradients using 2D flow nets or experimental test results are likely not conservative and hence may not be suitable in estimating risk if an initiating event has occurred or is suspected.

USING NUMERICAL MODELS TO INFORM OUR UNDERSTANDING OF INTERNAL EROSION PROCESSES

Two-dimensional (2D) versus Three-dimensional (3D) Modeling

Pioneering work on internal erosion process including the development of laboratory test equipment and procedures that allow for the observation of internal erosion processes has been performed at the University of Florida, and the Technical University in Delft. The results of this research were first published in 1981 by Townsend, (et. al.), and in 1988 by Sellmeyer. The work at the University of Florida was triggered by the Martin Plant Dike failure (Florida Power and Light Company), and the authors noted the following as one of the primary basis for their research:

> "The theory (*summarized in the report*) is based upon the observation that the horizontal seepage velocities of water flowing through soils are insufficient by one or more orders of magnitude to cause erosion; e.g. piping. Consequently, an additional factor must be present. It is theorized that a significant vertical gradient exists causing the soil grains to become quick; and thus the seepage velocity required for scour approaches zero, making possible the progressive erosion of the pipe. Intuitively, the vertical gradients will become concentrated at the head of the advancing pipe thereby contributing to the pipe's advance. Dr. Schmertmann quantified the vertical gradients at a pipe tip based upon sketching of flow nets, and related these vertical gradients to the average gradient via a concentration factor, i.e. $i_{(vert)} = c \times i_{(ave)}$, where $i_{(vert)}$ = perpendicular (vertical) gradient at pipe tip, C = concentration factor, and $i_{(ave)}$ = average hydraulic gradient (total head difference/total linear flow distance). The development of these correction factors to account for **three-dimensional** flow into an advancing pipe was limited by flow net sketching, and refinement became an objective of this project... Specifically, the concentration factor values were to be refined using numerical programs and extended for a variety of conditions. Physical model tests were to be performed to verify theoretical concepts concerning the mechanisms of piping and to provide physical data for verifying concentration factor values."

On the other hand, Sellmeyer (1988, also spelled Sellmeijer) notes the following as one of the fundamental bases for his initial research:

> 'The whole mechanism (*piping as a form of seepage erosion*) is quite complex and use is made of branches of soil mechanics (both continuum

> approaches and particulate aspects), groundwater flow and hydraulics. A certain amount of simplification has to be introduced so as to make the problem suitable for mathematical analysis. Inspiration is drawn from simple visual tests...The model is essentially **two-dimensional**. It is believed that this does not seriously affect the validity of the results. The design rules obtained allow for a great variety of geotechnical conditions, The analysis gives insight into the safety factor of the design."

Recent small-scale experiments have been performed to validate the Sellmeijer's model for multi-layer aquifers (V. van Beek, et al, 2012). This study demonstrated that 2D numerical groundwater flow models in which the Sellmeijer numerical model has been implemented (MSEEP) accurately calculated critical heads that were experimentally obtained in small-scale experiments where discharge occurs to a continuous defect (perpendicular to the direction of flow through the models) such as an open channel on the landside of a levee or dike. However, the critical head to initiate erosion into a circular exit point (sand boil) through a confining cover could not be predicted by the 2D numerical program. Subsequent 3D modeling by Vandenboer, et. al. (2014) demonstrated the inherent 3D nature of the piping phenomenon. The authors suggest the superiority of 3D numerical model computations compared to 2D approaches. These authors further suggest that 3D numerical results enable a better understanding of the complex physical mechanisms involved in backward erosion piping and thus can lead to a significant improvement in the safety assessments of water retaining structures.

Ferguson presented information on the 3-D nature of the seepage equipotential and flow lines surrounding the Isabella Auxiliary Dam, CA discharge conduit in 2010 (Ferguson, 2010) based on interpretations of instrumentation readings. The instrumentation readings, along with other observations, and detection of suspected initiation locations in a foundation drain pipe and through defects in the canal lining immediately downstream of the outlet works discharge at Isabella Auxiliary dam suggested that the failure mode process at Isabella Auxiliary Dam had initiated and was in an advanced continuation stage in an erodible silty sand foundation layer adjacent to the outlet conduit excavation. In 2012, Ferguson, et al (2012) presented the results of 2D seepage modeling of a hypothetical defect within an idealized confined and layered foundation built from the observed conditions at Isabella Dam. Results of 2D seepage modeling suggested that exit gradients at all possible points of an active seepage front within a model defect (including the point of initiation) were likely sufficient to result in the continuation of a piping erosion process beneath the dam, particularly when considering how gradients concentrate around the advancing front (head) of a seepage defect. However, the estimated seepage quantities into the model piping defect (0.001 to 0.01 gallons per minute (gpm) per foot) were not consistent with the observed flow discharging from a drain pipe where the defect initiated (>1 to as much as 7 gpm). Specifically, the estimated seepage quantities for various reservoir levels from the 2D model were likely insufficient to effectively transport detached particles from the active front of the defect. The observed seepage was up to at least 100 times larger than the seepage estimated by the 2D model. At that time, Ferguson et. al., cautioned that the 2D model was a simplification and acknowledged that the actual development of the piping or erosion feature, and the corresponding regime of water pressures (equipotential lines), gradients and flow quantities that would develop in the vicinity of the defect is a three dimensional (3D) problem. Hence, "the actual gradients and flow quantities into the active front and upstream portion of a pipe or erosion defect could be substantially higher than the 2D model results indicate."

REVIEW OF SCHMERTMAN RESEARCH AND FACTOR OF SAFETY METHOD

In response to the October 30, 1979 failure of a portion of the embankment containing a cooling water reservoir at a Florida Power and Light Company power plant, Dr. Schmertmann, a member of a Board of Consultants convened to investigate the cause of failure, developed a theory to assist in analyzing the failure and to investigate the safety of other earth embankments (Schmertmann, 1980). At that time, Schmertmann had noted that only empirical methods and engineering judgment were available for the design of dams/ or to evaluate foundation materials and their resistance to piping. Subsequent to the development of the theory, a research program was initiated at the University of Florida (UF), to help verify and refine Schmertmann's piping theory.

The seepage research work at the UF first published in 1981 by Townsend et. al. had, as a principal component of the work, the development and use of a 3-D finite element seepage program to examine gradient concentration factors parallel and perpendicular to an internal erosion "pipe" for various pipe penetrations and boundary conditions. The other primary component of the research work included the development and evaluation of an experimental flume to identify the average gradients required to initiate piping, and to observe the piping process.

A 3-D finite element program for seepage written by the US Waterways Experiment Station (WES) was used (Wong, 1981; Tracy, no date) as the basis for the UF research work. The program identified by "SEEP3D" could be used to solve both transient and steady state seepage problems. The UF research used only the steady state formulation. The program at that time had the limitation of permitting a maximum of only 1000 nodal points in the finite element grid. This proved a severe limitation and required some undesirable approximations in interpreting the results (Wong, 1981).

In later advancing his initial theory for evaluating the potential for piping and internal erosion, Schmertmann performed a comprehensive review of the 3-D modeling and laboratory testing procedures and observations at UF, as well as related research performed at the Technical University at Delft (Sellmeyer, 1988). Some important observations from the Schmertman paper (2000) are as follows:

- The term "piping" is defined as the erosion of soil, due to concentrated internal seepage, within a mass of soil, and "scouring" as the erosion along a soil surface due to water flowing along that surface, without the need for seepage through the soil to that surface. He further notes the importance of both seepage gradient and "roof" support conditions to the "piping" erosion process.

- 3D seepage concentration is required to cause "piping" or for internal erosion to occur. Under normal conditions in a dam foundation, the gradients are not sufficient to cause the initiation or "progression" of piping.
- Both 2D and 3D flow nets were used to understand the piping phenomenon and inform the factor of safety method presented in the paper. The 3D computer studies by Wong (Townsend, 1981) are noted but it appears that the majority of the assessment method relied on the combining longitudinal, transverse and horizontal section 2D flow nets to estimate 3D conditions.
- The proposed method in the paper uses an average or global factor of safety as well as the point-by-point variation of the factor of safety vs. piping along a potential pipe path. The global gradient needed for piping to continue in the various experiments at the UF and in Delft varied with the percent downstream-upstream advance of the pipe and reached a maximum at a penetration of 30 to 50 percent in the Delft experiments, and about 20% in the UF tests. This experimental observation is illustrated in the figures below taken from Chapter 26 of the USBR/USACE Best Practices for Risk Analysis. Note the reference to a summary of this phenomenon by van Beek taken from an article written in 2010. This information appears in work published by various Delft researchers including Sellmeijer from 1981 to 1993 and was reviewed in detail by Schmertmann (2000).

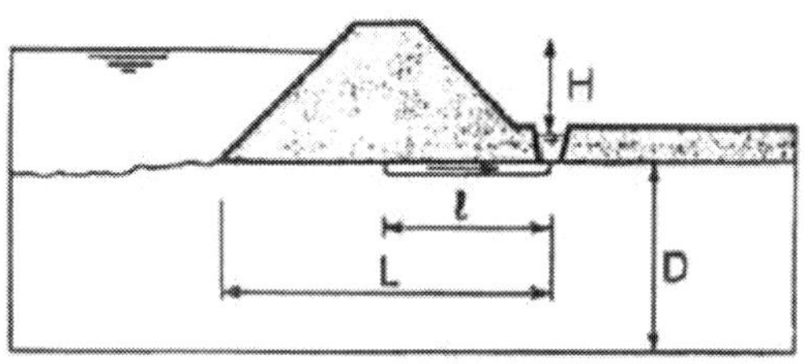

Figure 26-17. Geometry of Delft Backward Erosion Model (Weijers and Sellmeijer, 1993)

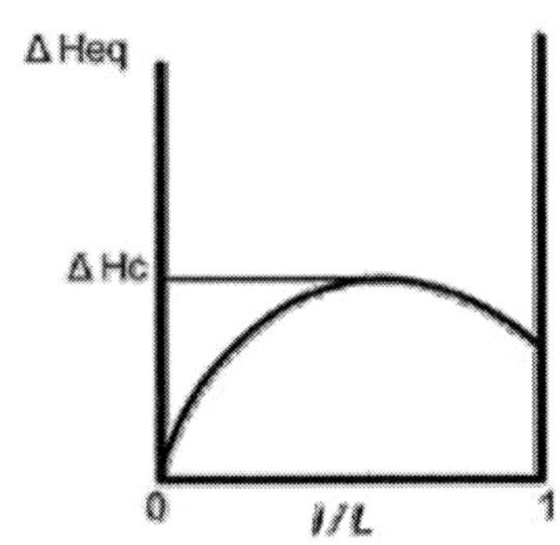

Figure 26-18. Hydraulic Head as a Function of the Ratio between Pipe Length and Seepage Length (van Beek et al., 2010)

Figures 1 and 2. Taken from Chapter 26 of the USBR/USACE *Best Practices for Risk Analysis*

Schmertmann believed that this pattern could be explained from the flow net pattern of equipotential line concentrations associated with the dimensions of the testing equipment and not as a result of a fundamental aspect of the piping phenomenon itself. To utilize the experimental results however, required the use of a horizontal gradient correction factor (C_{Gt}) to relate the experimental results to an average gradient calculated using flow nets. In other words, it is necessary to multiply the average or global gradient required to sustain the piping process in the lab tests by the horizontal gradient correction factor to estimate the global gradient that is required for sustaining the piping process that would be compared to flow net calculations:

$$i_{pmt} = i_{pmte} \times C_{Gt} \qquad (1)$$

Where: i_{pmt} = the maximum (m) value of the Global (or average gradient) in a flume test (t) required to sustain downstream to upstream piping (p). It represents the minimum Global gradient needed to complete the piping process from the point of initiation to the reservoir

i_{pmte} = the maximum (m) value of the global gradient required to sustain piping in lab experiments. Note that multiplying this experimental value by the correction factor, C_{Gt} give a constant value of i_{pmt} at any L.

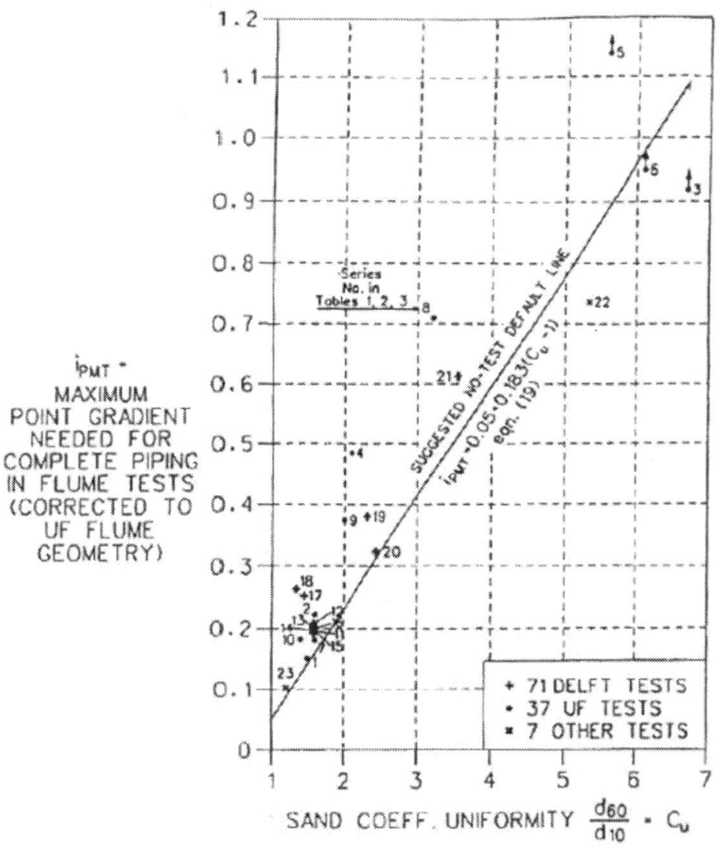

Figure 3. Summary of all the UF and Delft experimental results with the C_{Gt} correction applied

Figure 3 shows a summary of all the UF and Delft experimental results with the C_{Gt} correction applied. Note the limited range of C_u values in the laboratory experiment samples. Due to concerns about the samples, the relationship shown in this figure should be used with caution and perhaps only for soils with C_u values that are less than 3.2 (a very narrow range that is seldom encountered in nature).

- The paper summarizes the "meandering" nature of the piping phenomenon noting that "the piping channel does not advance in a downstream to upstream straight line. Rather, it advances in a meandering, braded system of channels, sometimes with oxbows, and often with one path clogging and stopping progress followed by a restart of another path". This is consistent with observations in the Delft research that has noted that "the meandering of the pipes is due to the search for weakness in the granular structure of the sand. It is associated with the non-homogeneity of the sand properties and is therefore not a essential but an added feature of the piping problem."
- Schmertmann notes that the flow velocity needed for the piping process is substantially less that the flow velocity required in a concentrated leak to cause "scour" (concentrated leak erosion).
- The paper summarizes the lack of experimental evidence that initial effective stress state influences the piping pathway but goes on to suggest that the "micro" condition of very low effective stresses along its boundaries (the water/sand erosion interface), may control the piping behavior irrespective of the prior "macro" effective stress. Macro effective stresses may have some effect on the piping process such as changing the cross-sectional area or slope of the pipe, but do not appear to change whether or not piping will occur.
- Based on the University of Florida experimental results, the influence of a developing pipe on distortion of seepage flow patterns and corresponding water pressures in other portions of the dam were postulated to be relatively small. Specifically, in a homogeneous sand, Schmertmann suggests that the influence of a developing pipe extends no more that 80 radii in any direction from a pipe channel and a half-circular cross-section of the pipe. Put into context, for a 12.5mm pipe radius, the local influence of the pipe would not extend a horizontal or vertical distance of more than 1 meter from the pipe. Of course, anisotropy could have a significant impact on this observation.

Considering all of the factors summarized in the paper, including those noted above, Schmertmann summarized two simplifying mechanisms as primary considerations in the development of his Factor of Safety methodology:

<u>Mechanism 1</u> – Critical Horizontal Gradients and Regressive Slope Failures
<u>Mechanism 2</u> – Critical Vertical Gradients and Liquefaction

Both of the simplifying mechanisms considered by Schmertmann involve very low "micro" effective stresses at the active front of the pipe (pipehead). Both were observed to have the potential to capture the essence of the piping problem. Schmertmann chose to use the critical vertical gradient method for a number of reasons including:

a. The ease at understanding this mechanism
b. Ability to visualize and quantify based on using flownets.
c. Mechanism seemed to match, at least qualitatively with testing experience.

Careful study of the 3D pipehead gradients determined by Wong (Townsend, 1981) as well as 2D flownet approximations suggested that:

$$i_{he3} = 1.33 \text{ x } i_{ve3} \tag{2}$$

Where : i_{he3} = the 3D horizontal gradient at the pipehead
i_{ve3} = the 3D vertical gradient at the pipehead

The point-by-point methodology in the paper provides an estimate of a factor of safety (F_{px}) verses piping along any postulated failure pathway (x):

$$F_{px} = (i_p/i)_{xf} \tag{3}$$

Where: i_{pxf} = the gradient at a specific point (x) along a potential piping path in the field (f) that would cause piping (p)
i_{xf} = the estimated actual gradient in the field (f) at the sample point (x) from a 2D flow net

The calculated factor of utilizes the I_{pmt} from laboratory tests and the application of a range of lab-to-field correction factors. A total of eleven correction factors are described. Through a number of mathematical refinements, the following equation is the final model for estimating the factor of safety along any hypothetical piping pathway of concern.

$$F_{px} = [(C_D C_L C_S C_K C_Z C_\gamma) i_{pmt}(C_\alpha)/(C_R \text{ x } i_f)]_x \tag{4}$$

The USBR/USACE Best Practice for Risk Assessment (2012), Chapter 26 notes that Schmertmann (2000) contained several errors which have been correct in the equation below to estimate the critical gradient for particle transport (i.e. to advance a pipe to the reservoir, i_{adv}).

$$i_{adv} = (i_{pmt})_{corrected} = [(C_D)(C_L)(C_S)(C_K)(C_Z)(C_\gamma)(C_\alpha)/C_R](i_{pmt}) \tag{5}$$

The letter C in these equations denotes a "Correction Factor" with each subscript identifying the factor or variable that can be applied when field conditions vary from the laboratory conditions:

Where:
D = the Depth/Length (D/L) factor
L = the Length factor
S = the Grain Size factor
K = the Anisotropic Permeability factor
Z = the (high permeability) underlayer factor

γ =	the Density factor
α =	the Pipe Inclination adjustment
R =	the Convergent/Divergent flow factor

In general, while being highly informative about the IE process, there are a number of important observations and considerations related to the use of the model presented by Schmertmann;

a. The Factor of Safety methodology cannot be easily adopted in todays risk informed decision making process where an engineer and/or organization is faced with making best estimates of the probability of failure associated with different potential seepage failure mode pathways, and also consider uncertainty in the range of critical parameters used in the analysis.
b. Further, the method is very difficult to apply in complex geologic environments where the continuity of potential erosion pathways, and the variability of material properties and dimensions along that pathway can be substantial and introduce significant uncertainties into the assessment process. The method is severely limited by two important factors: 1) the limited range of materials considered in the laboratory tests, and 2) the limitations associated with the 3D modeling that was performed (number nodes, the graphical interface for presentation and interpretation of results, and the number of study cases analyzed).
c. The method does not explicitly consider the conditions that develop at the active face of the piping feature once initiation has occurred. These conditions, and the primary detachment and transport mechanisms associated with the horizontal and vertical gradient, velocity vectors, and flow quantities at the active front are only implicitly considered through the i_{pmt} parameter.

RESULTS OF THREE-DIMENSIONAL COMPUTER MODELING

Model Configuration

To further examine and evaluate the influence of 3-D seepage conditions in the formation of a pipe during the internal erosion process, the authors have developed a 3-D seepage model as shown on Figure 4. This model is similar to the 3-D seepage model created as part of the UF research (Townsend, et al, 1981). The results from this model that are presented in this paper are considered introductory and will serve as the starting point for future evaluations. It should be noted that the dimensions shown on Figure 4 are in feet. In general, the Y dimension of the model represents conditions from the upstream (Y = 0) to a downstream discharge (Y = 100) location. The Z dimension represents the elevation above the bottom of the model. An upper confining layer is located at Z = 100 and the bottom of the model is at Z = 0 feet.

Note that the bottom of the model was set as an impermeable boundary for an initial set of study cases. A lower confining layer boundary was later studied at Z = 85 feet as discussed further below. The X dimension represents the model horizontal dimension and was set with a total width of 200 feet, well in excess of any anticipated 3-D influence of a piping feature based on the results of previous physical and computer modeling investigations. The piping feature included in the model extends from the downstream end at Y = 100, X = 0; and Z = 100, along the upper left hand edge (looking downstream). The model is axis-symmetric along the X = 0 boundary. Hence the estimated gradients represent conditions along half of the piping feature but would be expected similar for the other half of the pipe feature not included in the model. However, estimated seepage flow into the pipe in the axis-symmetric model represents only half of what would actually occur. A net average gradient of 1 was applied in the model (similar to the models at UF). Establishing conditions for an applied average gradient of 1 provides a number of numerical conveniences. Perhaps the most important convenience is that actual gradients and flow quantities for conditions where the average gradient is less than 1.0 can be easily obtained by simply multiplying the computer model gradients and/or flow quantities presented in this paper by the average gradient that exists under specific field condition in question. For the Isabella Dam example that will be subsequently discussed, the average gradient is on the order of 0.1 to 0.2, depending on the reservoir level. The computer results can hence, be multiplied by 0.2 to estimate the actual gradient and seepage flow conditions that would ideally occur in the field.

The UF model was limited to 1,000 nodes, resulting in a maximum of 729 elements. As previously noted, the boundary conditions of the UF model were set to generate an average gradient of unity (i_{ave} = 1.0), achieved by applying a head of 100 (feet) across a model horizontal dimension L of 100 (feet). The pipe was represented by a triangular shaped void in the mesh. The void "radius" was the length of the sides of the triangle. The radii of 0.005 x L (0.5 feet or 6 inches) and 0.01 x L (1 foot) were evaluated as one of the variables explored in their study. A considerable portion of the study was performed to evaluate the influence in the model dimensions on the estimates of seepage gradients entering the void. For the current modeling effort, the final dimensions of the UF model were adopted.

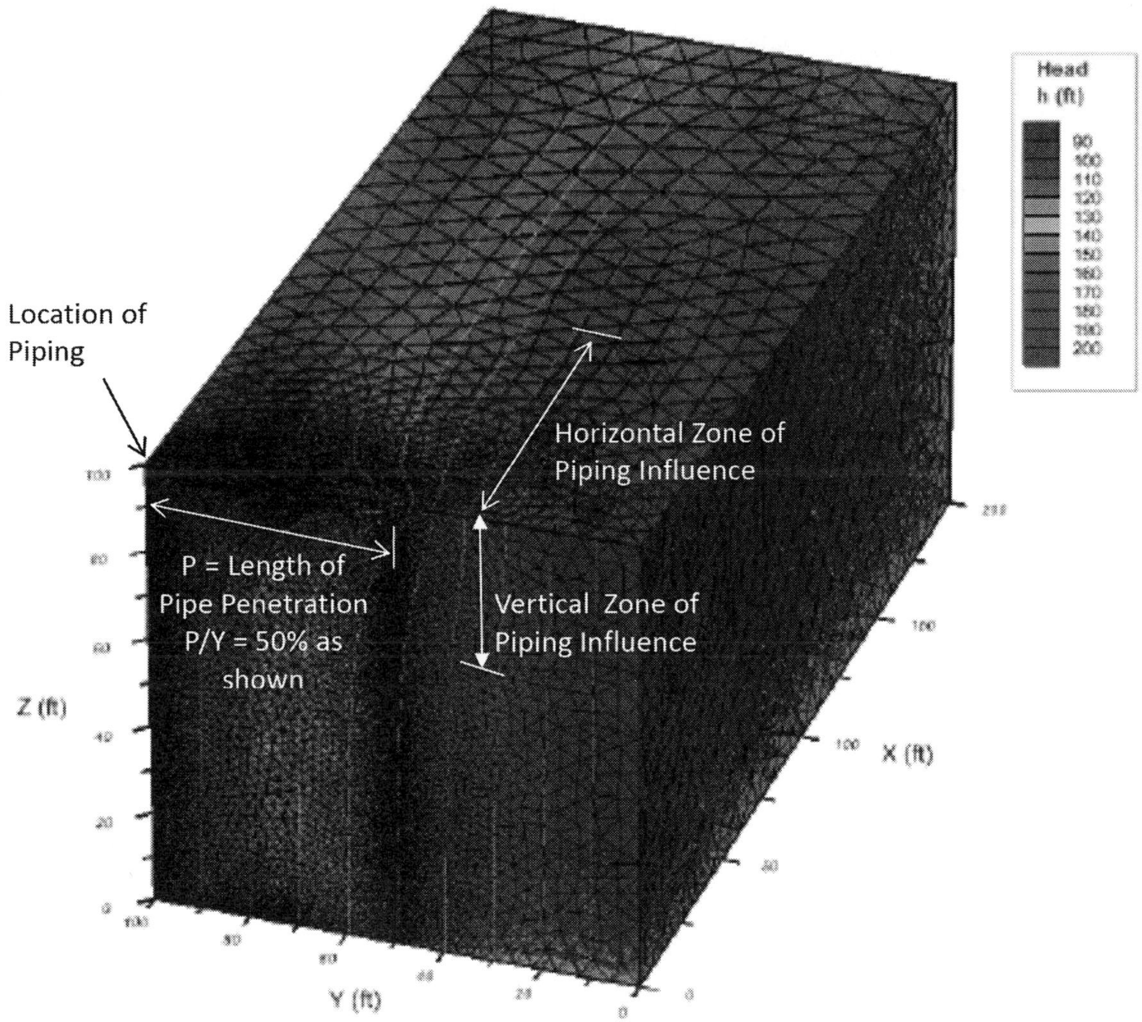

Figure 4. 3-D Seepage Model Configuration

The current 3-D model was prepared using the software program SVFlux 2009 by Soilvision Software. This software can be used for 1-, 2-, and 3-D finite element modeling of seepage. The program uses hp-adaptive meshing to generate the mesh on geometry that the user provides. The adaptive meshing algorithm allows the program to refine the mesh in regions of high gradients in order to more accurately estimate the gradients without the user having to iterate through mesh refinements. Unfortunately the exact recreation of the UF triangular pipe geometry was not possible due to the geometric limitations of the SVFlux software. The pipe was modeled as a rectangular pipe, with a square profile. The "radius" reported in the current model results is equal to the length of the sides of the square.

The differences observed between the UF and the current research studies is likely a result of the differences in geometry of the pipe and the density of the mesh that could be

achieved in the current study as compared to the mesh limitations inherent to the software used in the original study. Based on our comparison of the two different study results, we believe that the gradients reported in the UF study were less that what actually would occur due to averaging across larger elements. The model for the current study uses approximately 70,000 nodes and 50,000 elements. Mesh density in the current study was controlled by the software and parameters controlled by the software which were optimized to assure convergence of the finite element solution.
The current study looked at the influence of the following factors on estimated gradients:

a. radius of the pipe, R,
b. depth of the permeable layer beneath the upper confining layer in the model, D,
c. penetration of the pipe, P, and
d. anisotropy of the upper permeable layer of material.

Three radii were considered for basic model, R = 0.0050 x L (0.5 feet), 0.0075 x L (0.75 feet) and 0.0100 x L (1 foot); the basic model comprised of the full depth permeable layer (D = Z = 100 feet) and L = Y = 100 feet. For the remaining models only two radii were considered; 0.0050 x L (0.5 feet) and 0.0100 x L (1 foot). In addition to the depth of the permeable layer in the model being set at 100 feet, an analysis was also conducted with the depth of the permeable layer limited to the upper 15 feet immediately below the top confining layer of the model. The penetration of the pipe was varied from 5 percent (5 feet) to 95 percent (95 feet) of the model length Y of 100 feet, generally in steps of 10 percent (10 feet). A single permeable material was used to reduce the complexity of the model. A hydraulic conductivity of 2.8 feet per day (feet/day) (1.0 x 10^{-3} centimeters/second) with a saturated volumetric water content of 0.3 was used for the permeable layer. The anisotropy of the material was set to $k_h/k_v = 1$ for the basic model, with a variation of $k_h/k_v = 4$ in subsequent study cases. A head of 200 feet was applied uniformly along the upstream face of the model (Y = 100 feet) and a head of 100 feet was applied uniformly along the downstream face of the model (Y = 0 feet) to generate an average gradient of 1.0 in the model. Review boundaries were applied along the faces of the pipe to allow water to flow out of the model, while the remaining boundaries were input as no (zero) flow boundaries

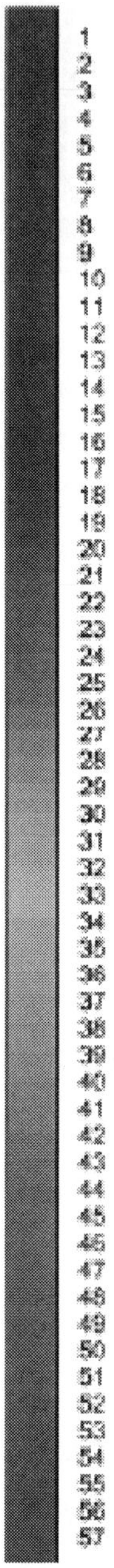

Figure 5. Gradient Scale

The results for an equivalent 2-dimensional models (at X = 0) were compared with the basic 3-D model results, using the same dimensions and properties incorporated in the 3D models.

Modelling Results

The first use of our model results was an examination of the direction and relative magnitude of gradient and flow vectors in the vicinity of the active erosion face (pipehead) as well as in the region between the pipehead and the downstream point of initiation. The direction and magnitude of these vectors in the immediate vicinity of the pipehead will help to inform our understanding of the **detachment** erosion mechanism, whereas, examination of the magnitude of these vectors between the pipehead and the discharge location help to inform our understanding of the **transport** erosion mechanism.

Example results showing gradient contours (see Figure 5 for gradient color scale definition) and velocity (flux) vectors at the X = 0 axis-symmetric boundary of the model with the pipe penetration at 50 feet (50% of the model length along the Y axis) are shown on Figure 6. Gradient contours and velocity (flux) vectors along a cross-section perpendicular to the pipe feature at the pipehead are shown on Figure 7 for the same pipe penetration of 50 feet. The pipe radius in these results is 0.5 feet (6-inches).

Examination of the results illustrated on these example figures and other 3-D modeling results lead to the following conclusions:

- Gradients at, and in the immediate vicinity of the pipehead and along the length of the pipe are very high; e.g. > 40 in the pipehead region to 25 to 30 at the downstream end of the pipe). If an average gradient correction factor of 0.15 to 0.2 is applied to the results, simulating the kind of conditions that would exist in the foundation of most dams and levees under normal maximum hydraulic loading conditions, the estimated 3-D gradients would be at least 6 to 8 in the pipehead region and 4 to 6 along the remainder of the pipe). Hence the model results suggest that the 3-D influence of the pipe reinforces that the gradients are sufficient from the time of initiation through full penetration of the pipe to **detach** a range of soil particles during the piping process.
- The direction of the gradient vectors (the same as velocity/flux vectors) at the pipehead is highly variable ranging from horizontal at the top of the pipe adjacent to the "confining" top layer, to vertical along the base of the pipe along the pipe axis. Vertical gradients appear to range from greater than 20% higher to approximately equal to the horizontal gradients in the pipehead area.

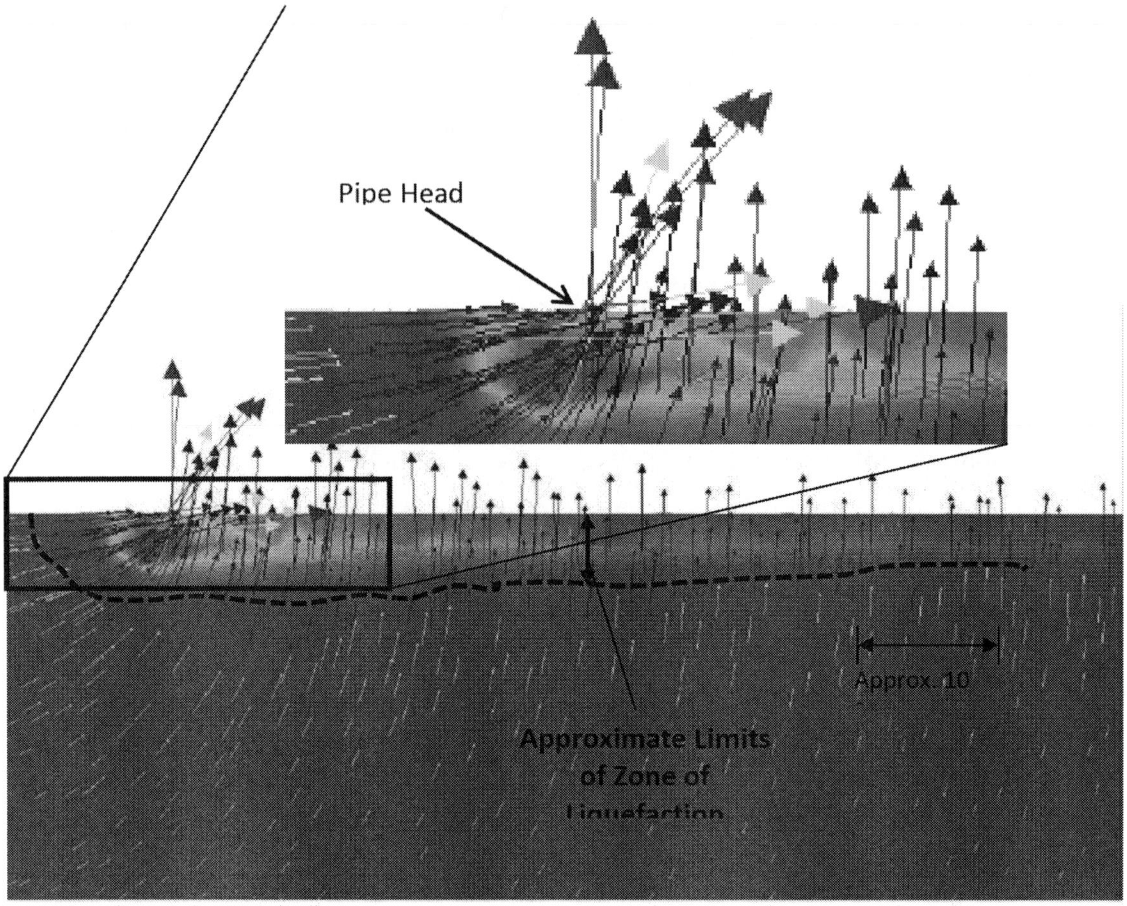

Figure 6. Example Gradient/Flux Vectors Entering Pipe Head Parallel to Pipe (X = 0); Pipe Head at 50% Penetration (Y = 50 Feet)

- The gradient and velocity (flux) vectors along the axis of the pipe become essentially vertical within about 10 to 20 pipe radii downstream from the pipe head.
- The horizontal zone of 3-D influence of the erosion feature is a maximum of about 80 radii similar to the findings by Schmertmann.
- The vertical zone of 3-D influence of the erosion feature is approximately comparable to the depth of pipe penetration when the permeable layer is the full height of the model. The vertical zone of 3-D influence is essentially the entire layer when the layer depth is less than the amount of pipe penetration.

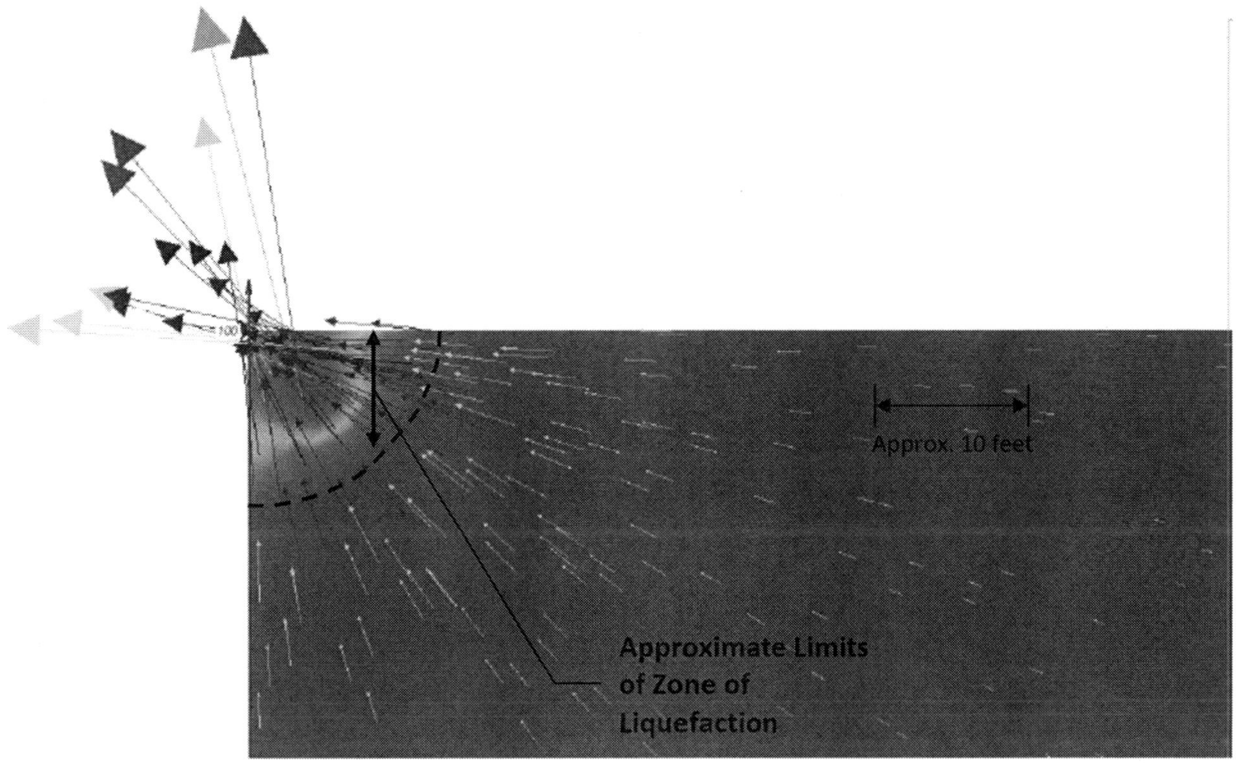

Figure 7. Example of Gradient/Flux Vectors Entering Pipe Head Area Perpendicular to Pipe; 50% Pipe Penetration (Y = 50 Feet)

- Vertical gradients are substantially higher than 1.0 for more than 70 to 80% of the length of the pipe even when a correction factor of 0.2 is applied to the computer model results. These model results suggest that a zone of disturbance (liquefaction) around the perimeter of the pipe develops as postulated by Schmertmann. This zone of disturbance may extend as much as 10 to 20 radii around the pipe. For a 3-inch radius, the zone of disturbance may range for a radius of 30 to 60 inches. For a 6-inch radius, the zone may range for a radius ranging from 5 to 10 feet. The result of this occurrence suggests that upward (near vertical) gradients and velocity vectors would have a significant influence on the **transport** of particles in the pipe once **detachment** has occurred in the pipehead area. This observation is consistent with Schmertmann's observation noted above: "the flow velocity needed for the piping process is substantially less that the flow velocity required in a concentrated leak to cause "scour" (concentrated leak erosion)" such as those presented by Briaud (2007). However, as vertical gradients and upward flows decrease near the downstream end of the pipe, there may be a heightened potential for deposition to influence the pipe development process.

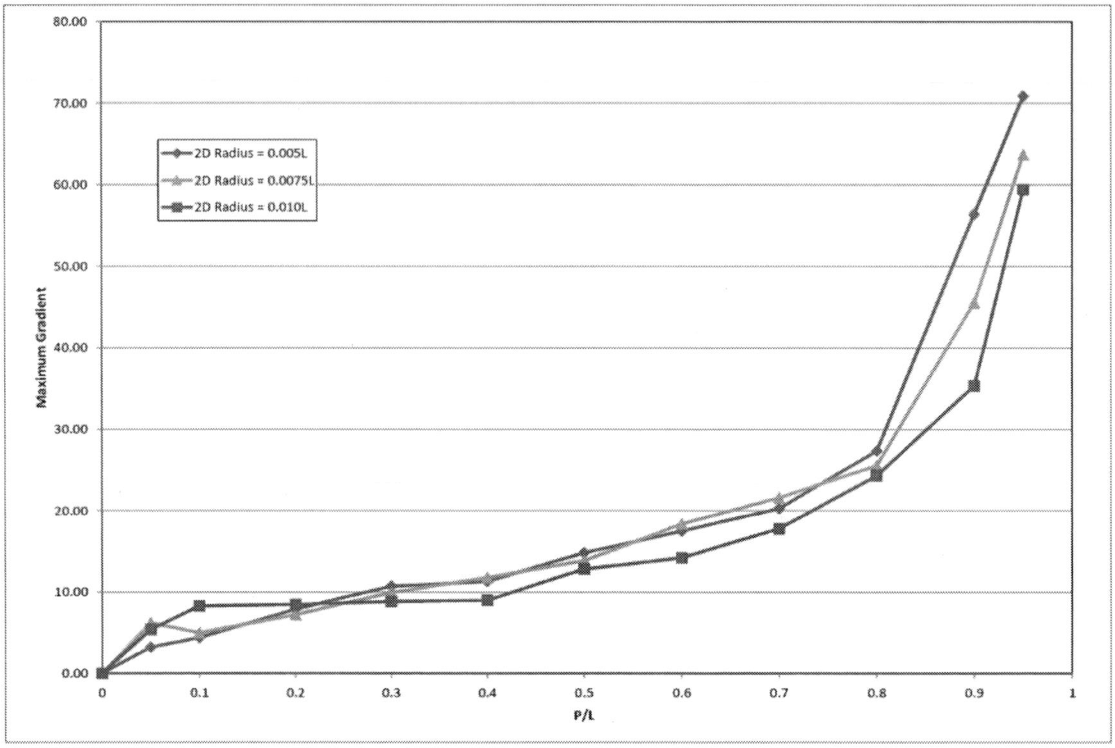

Figure 8. Summary of Estimated 2-D Gradients at the Pipehead as a Function of Pipe Penetration into the Model

Figures 8 through 10 contain a summary of the estimated 2-D verses 3-D gradients at the pipehead, and total flux from the discharge end of the piping feature as a function of the piping penetration D/L (where D is the penetration length and L is the total length (100 feet) of the model along the y-axis shown on Figure 5). The computer program SVFlux provides both the 3-D and 2-D results for these comparisons. Note that the results shown on Figure 10 have not been corrected for the axis-symmetric condition of the model. Actual seepage values shown on this figure are 2 times greater.

A number of the study case results are included on these Figures representing 1) various pipe diameters as a function of the y-axis length (L), 2) for permeable layer depths of 15 feet (labeled as shallow) and 100 feet (no label other than the pipe radius) and for a shallow layer when the anisotropy ratio is changed from 1 to 4 (k_h/k_v).

Table 3 contains a summary of the average differences in the estimated gradients between the 2-D and 3-D models for the case of:

- ✓ R = 0.005L (6-inches)
- ✓ Shallow permeable layer (15 feet below the upper confining layer at A = 100 feet)
- ✓ Anisotropy Ratio of 1

The results shown in Table 3 can be roughly compared to the results of previous 2-D evaluations of seepage conditions along the Borel Conduit beneath the Isabella Auxiliary Dam (Ferguson, 2012) by comparing the results in column 3 with the noted gradients summarized in the comment column.

Table 3. Comparison of 2-D verses 3-D Gradients at Pipehead

P/L	Est. Max.2-D Gradient in Model	Est. Max 2-D Gradient with Correction Factor of 0.2 applied	Est. Max 3-D Gradient in the Model[1]	Est. Max 3-D Gradient with Correction Factor of 0.2 applied[2]	Ratio of Corrected 3-D to 2-D Gradients	Comments
0	0.00	0.0	8.14	1.6	> 16	
0.05	3.24	0.6	12.56	2.5	3.88	
0.1	4.41	0.9	20.70	4.1	4.69	(2-D i=0.4)[3]
0.2	7.90	1.6	32.38	6.5	4.10	
0.3	10.71	2.1	43.66	8.7	4.08	(2-D i=0.5)[3]
0.4	11.32	2.3	67.98	13.6	6.01	Point of maximum 3-D effect
0.5	14.82	3.0	66.69	13.3	4.50	(2-D i=0.8)[3]
0.6	17.50	3.5	83.65	16.7	4.78	
0.7	20.27	4.1	116.60	23.3	5.75	3-D gradients become more erratic at this penetration
0.8	27.34	5.5	129.70	28.5	5.22	(2-D i=0.9)[3]
0.9	56.43	11.3	162.10	32.4	2.87	3-D effect diminishes at 90%
0.95	70.90	14.2	199.40	39.9	2.81	

Notes:
[1] The gradients shown in this column represent the Model estimated concentration factors based on an applied average gradient, $i_{ave} = 1.0$ across the y-axis.
[2] The gradients shown in this column represent an approximate concentration factor that would be expected for a model with an applied average gradient, $i_{ave} = 0.2$ applied across the y-axis.
[3] These are the estimated 2-D gradients reported for Isabella Auxiliary Dam in Ferguson, 2012 estimated with Seep/W for comparison with the SVFlux results

The authors attribute the difference between the 2-D estimated gradients with the two different computer programs to two primary factors: 1) the estimated maximum gradients from the SVFlux 2-D analyses in this table are typically at the lower corner of the pipehead and include both x and y components, and 2) the SVFlux adaptive meshing generated a much finer mesh at the pipehead and hence higher 2-D estimated gradients. This comparison hints at the uncertainties associated with 2-D seepage analyses methods.

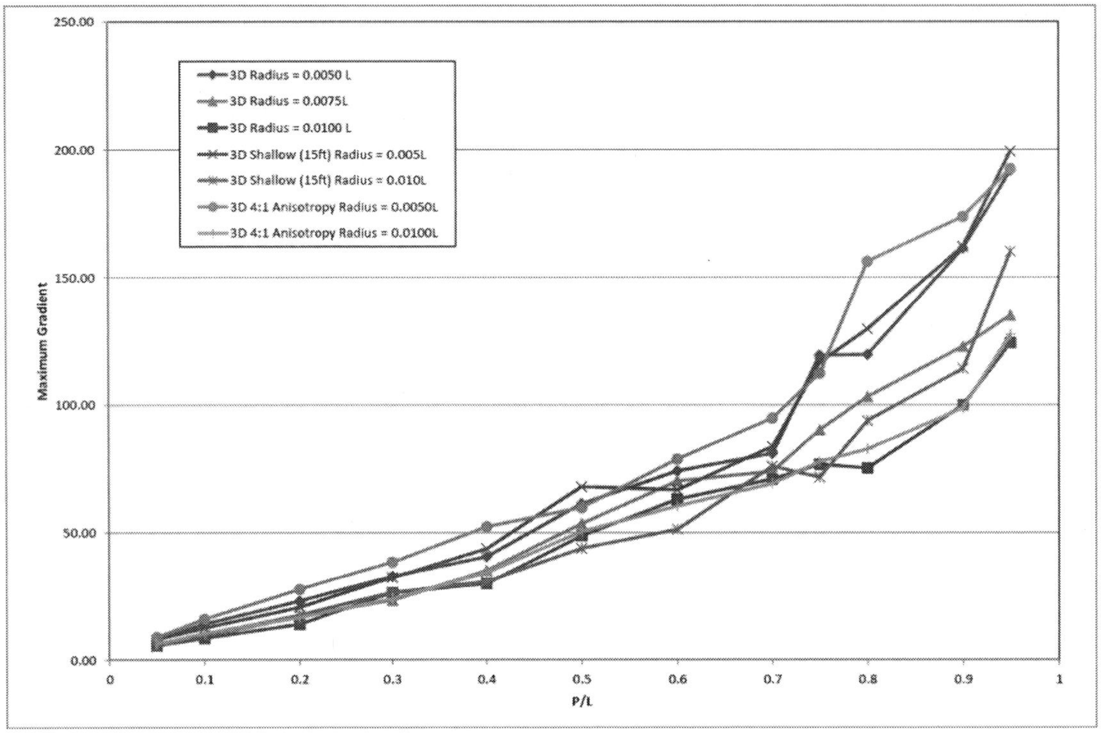

Figure 9. Summary of Estimate 3-D Gradients at the Pipehead as a Function of Pipe Penetration into the Model

Table 4 contains a summary of the estimated flux (flow) through all surfaces of the pipe for both the 2-D and 3-D conditions for the case of:

- ✓ R = 0.005L (6-inches)
- ✓ Shallow permeable layer (15 feet below the upper confining layer at A = 100 feet)
- ✓ Anisotropy Ratio of 1

Note that the results in Table 4 with all correction factors applied, indicated that the total flux discharging from the pipe is about 9 to 17 times greater in the 3-D model verses what is predicted in the SVFlux 2-D model.

The corrected 3-D model results in Table 4 have been set to approximately simulate the maximum average gradient conditions in the foundation of Isabella Auxiliary Dam. The maximum flow quantities observed discharging from a defective toe drain pipe at the Isabella Auxiliary Dam where piping had initiated and was in the process of continuing toward the reservoir ranged from 0 to as much as 8 gpm depending on reservoir level as shown in Figure 11. These observations are generally consistent with the 3-D seepage model results.

Table 4. Summary of Estimated 2-D and 3-D Flux (flow quantity) Values

P/L	Est. Max.2-D Flux in Model (gpm)	Est. Max 2-D Flux with Correction Factor of 0.2 applied[2] (gpm)	Est. Max 3-D Flux in the Model (gpm)	Est. Max 3-D Flux with Correction Factor of 0.4 applied[1] (gpm)	Ratio of Corrected 3-D to 2-D Flux	Comments
0	0	0	0.16	0.07	> 0.3	
0.05	0.16	0.04	0.48	0.19	6.20	
0.10	0.23	0.08	1.39	0.55	12.24	
0.2	0.40	0.14	2.67	1.07	13.44	
0.3	0.57	0.20	4.74	1.90	16.71	
0.4	0.73	0.27	5.92	2.37	16.26	
0.5	0.94	0.35	7.84	3.14	16.64	
0.6	1.17	0.44	10.10	4.04	17.28	Point of max 3-D effect
0.7	1.47	0.55	11.62	4.65	15.81	
0.8	1.89	0.72	14.91	5.96	15.77	
0.9	2.61	0.99	17.03	6.81	13.03	
0.95	3.41	1.28	19.69	7.88	11.57	

Notes: [1] This is a net correction factor which is the product of the average gradient correction of 100/500 = 0.2 x 2. The second factor of 2 is applied to compute the total flux from the axis-symmetric model that includes only ½ of the piping defect.

[2] This is corrected gradient estimated by applying the average gradient correction of 0.2 to the results of the pipe with a radius of 1 foot.

CONCLUSIONS

This paper has presented a review of the primary considerations related to internal erosion and the development of piping features that can lead to the development of a seepage failure mode. Three-dimensional influences have been long been suspected to be a significant factor in the internal erosion process but until recently, sufficient computer modeling tools were not widely available to allow for a deeper examination to inform our understanding of these 3-D effects. A state-of-the-art 3-D model using the computer program SVFlux has been developed and presented as part of this paper. The model replicates the initial work completed at the University of Florida but with substantially enhanced meshing and graphical output capabilities.

The initial results of the 3-D modeling work confirm many of the insights presented by Schmertmann and allow for an in-depth examination of the internal erosion process. While much still remains to be learned, the following general conclusions can be made.

1. The potential for unfiltered geologic and/or man-made defects are a primary consideration in the assessment of "initiation". The 3-D model results suggest that the concentration of gradients and flows around defects is significant and likely provides the energy required to initiate the erosion process. Finding such defects is a critical first step to the assessment of potential failure modes, estimating the probability of initiation, and in the overall seepage safety of a structure.

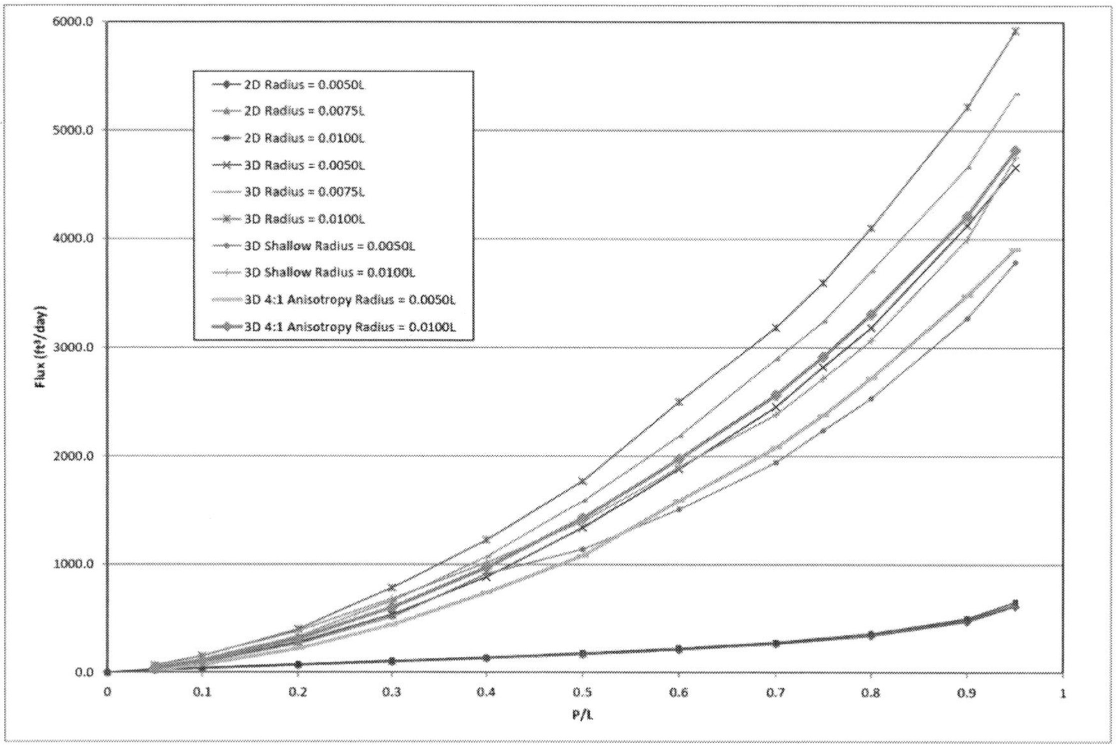

Figure 10. Summary of Estimated 2-D and 3-D Total Flux into the Pipe as a Function of Pipe Penetration into the Model

2. Once initiated, gradients at the active front (pipehead) will be sufficient for the entire length of a potential erosion pathway to sustain an erosion process. However, there are a number of other important considerations in assessing failure mode potential. Results of the 3-D analyses confirm that these key considerations for the estimation of the potential for a piping failure are:

- ✓ The potential for a continuous pathway (layer, crack, hydraulic fracture or other natural or man-made defect from the point of initiation to the reservoir or water source
- ✓ The permeability of the soils and ability to generate adequate seepage quantity and velocity (energy) to **detach** *and* **transport** the full range of soil particle sizes along the pathway

✓ The erodibility of the soils along the pathway (coefficient of uniformity, C_u; plasticity, density/specific gravity)

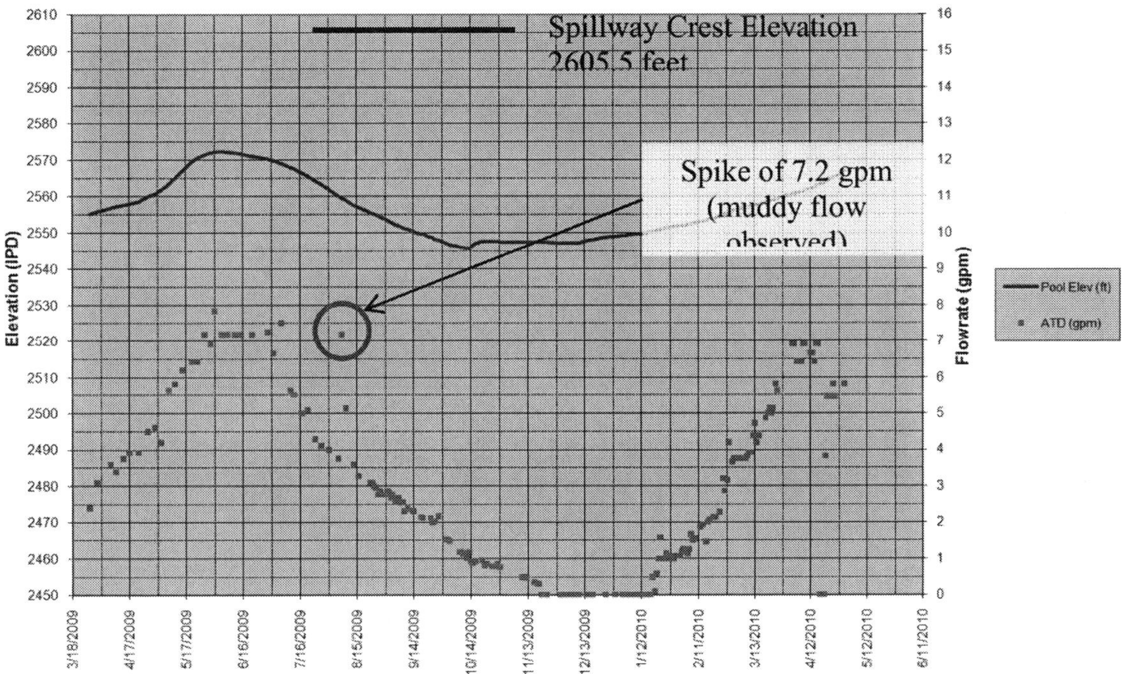

Figure 11. Plot of Seepage Discharging from Defective Isabella Auxiliary Dam Drain Pipe

✓ Conditions that promote the stability of the piping feature (roof or pipe channel)

✓ Conditions that could arrest the advancement of the piping/erosion feature such as changes in material type along the pathway, collapse, and development of a natural filter. Natural filters may develop in the presence of more widely graded soils and sufficient energy only to transport a portion of the gradation, leaving coarse materials deposited in the pipe that eventually cause a natural filter to develop.

3. Transport of detached particles will be significantly influenced by vertical seepage gradients for the majority (75 to 80%) of the pipe length that has developed.

4. A significant zone of disturbance or liquefaction likely develops in the region surrounding a pipe. This phenomenon could cause a very rapid gross enlargement process once the pipe feature breaks through to a water source. This zone may also contribute to localized embankment and/or foundation instability such as observed at the downstream toe of Big Bay dam at the onset of the gross enlargement process.

REFERENCES

Briaud, J-L, Case Histories in Soil and Rock Erosion, 9th Peck Lecture, February 12, 2007.

Donahue, C., Critical Gradient for Internal Erosion in Earthen Dams: A Comparative Analysis of Two Predictive Methodologies, Master of Engineering Thesis, Massachusetts Institute of Technology, June 2013.

Federal Emergency Management Agency (FEMA), Technical Manual: Conduits through Embankment Dams, FEMA 484CD, September, 2005.

Fell, R., Wan, C. F., Cyganiewicz, J., and M Foster, Time for Development of Internal Erosion and Piping in Embankment Dams, Journal of Geotechnical and Geoenvironmental Engineering, ASCE, April 2003, pp 307- 314.

Ferguson, K.A., Investigation and Evaluation of Seepage Conditions Around Outlet Conduits – Two Case Histories, Presentation at the Ohio River Valley Soil Seminar (ORVSS), Louisville, KY, October 20, 2010.

Ferguson, K.A., Investigation and Evaluation of Seepage Conditions and Potential Failure Modes Around Outlet Conduits, Proceedings of the 2012 Annual Conference, United States Society on Dams (USSD), New Orleans, LA, April, 2012.

Ferguson, K.A., Soudkhah, M., and E. Sossenkina, Potential for Cracking around Outlet Conduits and Its Impact on Seepage Safety, Proceedings of the 2013 Annual Conference, United States Society on Dams (USSD), Phoenix, AZ, February, 2013.

Ferguson, K.A., Anderson, S., and E. Sossenkina, Reexamination of the 2004 Failure of Big Bay Dam, Mississippi, Proceedings of the 2014 Annual Conference, United States Society on Dams (USSD), San Francisco, CA, April 2014.

Goodman, R. E., Karl Terzaghi, The Engineer as Artist, ASCE press, 1999, page 266.

Halpin, E., and K. Ferguson, US Army Corps of Engineers Dam Safety Program Status and Lessons in Transitioning to Risk Informed Approaches, ASDSO Journal of Dam Safety, Vol 2, Spring 2007, pp 24-36.

Hjulstrom, F., Doctoral Thesis, "Studies of the Morphological Activity of Rivers as Illustrated by the River Fyris", Bulletin Mineral Geol. Inst. University of Uppsala, 25: 221-528, 1935.

International Commission on Large Dams (ICOLD), Internal Erosion of Existing Dams, Levees and Dikes, and Their Foundation, Bulletin 164, Volume 1: Internal Erosion Processes and Engineering Assessment, 2013.

Ishbash, S. V. 2nd Conference on Large Dams, Vol, V, 1936.

Kenney, T.C., and Lau, D. "Internal Stability of Granular Filters, Canadian Geotechnical Journal, 22, 215-225, 1985.

Kezdi, A., Soil Physics-Selected Topics. Elsevier Scientific Publishing Co., Amsterdam, 160p., 1979.

McCook, D.K., and K.O. Grotrian, Using SIGMA/W to Predict Hydraulic Fracture in an Earthen Embankment, Proceeding of the Annual Conference, Association of State Dam Safety Officials, Seattle, WA, 2010.

Perzlmaier, S., "Hydraulic Criteria Focusing Critical Gradient and Flow Velocity (cohesionless soils)",Annual Meeting of the European Working Group on Internal Erosion in Embankment Dams, Stockholm, Sweden, September, 2006.

Ronnqvist, H., and P. Viklander, On the Kenney-Lau Approach to Internal Stability Evaluation of Soils, Geomaterials, 4, 129-140, 2014.

Schmertmann, J.H., "The Non-filter Factor of Safety Against Piping Through Sand", ASCE Geotechnical Special Publication No. 111, Judgment and Innovation, F. Silva and E Kavazanjian, eds., ASCE, Reston, VA, 65-132, 2000.

Sellmeyer, J.B., On the Mechanism of Piping Under Impervious Structures, *Doctorate Dissertation,* Technical University Delft, 1988.

Sherard, J.L., "Study of Piping Failures and Erosion Damage from Rain in Clay Dams in Oklahoma and Mississippi," U.S. Department of Agriculture, Soil Conservation Service, Washington, D.C., March 1972.

Sherard, J.L., Decker, R.S., and N.L. Ryker, "Piping in Earth Dams of Dispersive Clays," Proceedings of a Specialty Conference on Performance Earth and Earth-Supported Structures, ASCE, Vol. 1, part 1, pp. 584-626, 1972.

Sherard, J.L., "Sinkholes in Dams of Coarse, Broadly Graded Soils", Proceedings of the 13th ICOLD Congress, India, Vol. 2, 25-35, 1979.

Sherard, J.L., Hydraulic Fracturing in Embankment Dams, American Society of Civil Engineers (ASCE) Journal of Geotechnical Engineering, October, 1986 (First presented at the ASCE Spring Convention held in Denver, CO, 1985).

Skempton, A.W. and J.M. Brogan, Experiments on Piping in Sandy Gravels, Geotechnique, 44, pp 449-460, 1994.

Townsend, F.C., Schmertmann, J.H., Logan, T.J., Pietrus, T.J., and Y.W. Wong, An Analytical and Experimental Investigation of a Quantitative Theory for Piping in Sand, Final Report, University of Florida, 1981.

Townsend, F.C., Bloomquist, D., Shiau, J-M, Martinex, R. and H. Rubin, Analytical and Experimental Evaluation of Piping and Filter Design for Sands, Report to the U.S.

Bureau of Reclamation, University of Florida, Depar. Of Civil Engineering, January, 1988.

Tracy, F.T., "Three-Dimensional Finite Element Program for Steady-State and Transient Seepage Problems", U.S. Army Waterways Experimental Station, Automatic Data Processing Center, Vicksburg, Mississippi.

U.S. Army Corps of Engineers (Corps), ER 1110-2-1156, Safety of Dams – Policy and Procedures, October 28, 2011.

U.S. Bureau of Reclamation and the U.S. Army Corps of Engineers, Best Practices in Dam and Levee Safety Risk Analysis, 3 December 2012.

van Beek, V., Yao, Q. Van, M., and F. Barends, "Validation of Sellmeijer's model for backward piping under kikes on multiple sand layers, ICSE6, Paris, France, pp 543 – 550, August 27-31, 2012.

Vandenboer, K., van Beek, V., and A. Bezuijen, 3D finite element method (FEM simulation of groundwater flow during backward erosion piping, Front. Struct. Civil Engineering, Vol 8(2), pp 160-162, 2014.

Wong, Y.W., Three Dimensional Finite Element Analysis of a Quantitative Piping Theory, Report Presented to the Department of Civil Engineering at the University of Florida in Partial Fulfillment of the Requirements for the Degree of Master of Engineering, August, 1981.

EVALUATION OF IMPACT OF DENSITY ON FILTERING PROPERTIES OF NARROW VERTICAL FILTERS

Edwin R. Friend, P.E., P.G.[1]
Kerry Repola[2]
Robert J. Huzjak, P.E.[3]
Chris Wienecke[4]

ABSTRACT

Filters are necessary in embankment dams to provide protection from internal erosion. Design and performance criteria for typical filters (chimney drains, blanket drains, toe drains, conduit collars, etc.) is well established and the authors are advocating following the standard of practice for these facilities. However, it can be difficult to appropriately compact filter materials in deep vertical filter trenches. Examples of deep vertical filter trenches include center filters installed in dry-earth dams with unsupported narrow excavations and in downstream filters installed in embankment dams and foundations with bio-polymer slurry supported excavation. Based on the authors' research, very limited information has been published related to the impact of filter material density on the performance of the filtering characteristics.

Extensive laboratory testing was performed by Sherard, Dunnigan, and Talbot (ASCE, 1984) to identify criteria for filter design for clay and silt soils in embankment dams. These analyses consisted of a test apparatus that contained a compacted specimen of the base soil of interest. This specimen was created with a defect simulating a crack or other opening. Beneath the soil specimen was sand carefully graded and compacted to serve as a filter material, which prevented the loss of the base soil with water running through the defect at a pressure of 40 pounds per square inch.

The Authors performed filter tests to evaluate the impact of filter density on filter performance using a specific clayey base soil utilizing methods that generally replicated the method by Sherard, Dunnigan, and Talbot (ASCE, 1984).

INTRODUCTION

Narrow vertical filters for dams have become more prevalent in recent years including installation of central filters in dry earth dams in the southwestern United States and the use of bio-polymers supported trenches to install filters at the toes of dams. Both of these installation methods allow for limited compaction of the filter materials. The Authors are

[1] Principal, Project Manager, RJH Consultants, Inc., 9800 Mt Pyramid Court, Suite 330, Englewood, CO 80015, efriend@rjh-consultants.com
[2] Laboratory Manager, Advanced Terra Testing, 833 Parfet St, Lakewood, CO 80215, krepola@terratesting.com
[3] President, RJH Consultants, Inc., 9800 Mt Pyramid Court, Suite 330, Englewood, CO 80015, rhuzjak@rjh-consultants.com
[4] Laboratory Owner, Advanced Terra Testing, 833 Parfet St, Lakewood, CO 80215, cwienecke@terratesting.com

not proposing a change to the current standard of practice which recommends compacting filters (FEMA, 2011). However, for the above conditions and technologies, compaction and consolidation of a filter can be difficult.

The goal of the testing was to identify if the density of the filter material impacts the filter properties related to the retention of a particular base soil. This paper will present the results of filter testing performed for a specific clayey base soil using methods that generally replicated the method by Sherard, Dunnigan, and Talbot (ASCE, 1984).

PREVIOUS RESEARCH AND CURRENT DESIGN PRACTICE

The basis for current filter design criteria is primarily rooted in work performed by Sherard, Dunnigan, and Talbot (ASCE, 1984 and 1984A). The guidelines proposed by Sherard, Dunnigan, and Talbot resulted in a factor of safety of at least 2 for clayey base soils with at least 85 percent fines (passing the No. 200 sieve). It should be noted that this testing showed a wide scatter for the boundary between success and failure for clay soils, but the criteria for filter design was set so the factor of safety was at least 2 for the most critical soils. This work used filters that were compacted in a saturated condition on a vibrating table with a 10 kg. surcharge weight. Filter densities were not measured, but likely exceeded 70% relative density. Subsequent testing to verify this work found that relative densities of about 60% produced similar results (personal communication, Talbot, 2014). The Federal Emergency Management Agency (FEMA) guidelines for Filters for Embankments Dams (FEMA, 2011) recommends compacting filters to at least 70 percent relative density according to ASTM D4253 and ASTM D4254 (ASTM, 2014).

The Authors only identified one reference where a similar type of filter test was performed to evaluate the impact of density of a filter on the ability to retain particles. Research by Park (Park, et. al, 2004) performed a modified type of filter test that included a crack along the side of the test apparatus. Water under pressure was applied to the base soil and filter material. Tests were performed with filter materials compacted between 18 percent and 70 percent relative density and all tests resulted in successful retention of the base soil. The filter design criteria used by Park et. al. was United States Bureau of Reclamation Standards (USBR, 1994), which are similar to FEMA criteria (FEMA, 2011).

TEST METHODLOGY

Base Soil and Filter Material Properties

The selected base soil was classified as a Category 1 soil according FEMA because it contained more than 85 percent fines. The particle size for the $D_{85}B$ ($D_{85}B$ = the particle size of the base soil where 85 percent by weight of soil particles are finer) for the selected base soil was 0.063mm. The liquid limit was 48 and the plasticity index was 36. Based on a crumb test (USBR, 1990), the soil was classified as non dispersive. Using the FEMA (FEMA, 2011) guidelines a filter design was performed and the maximum $D_{15}F$

was calculated to be 0.5 mm. The base soil and required filter gradation using FEMA guidelines are shown on Figure 1.

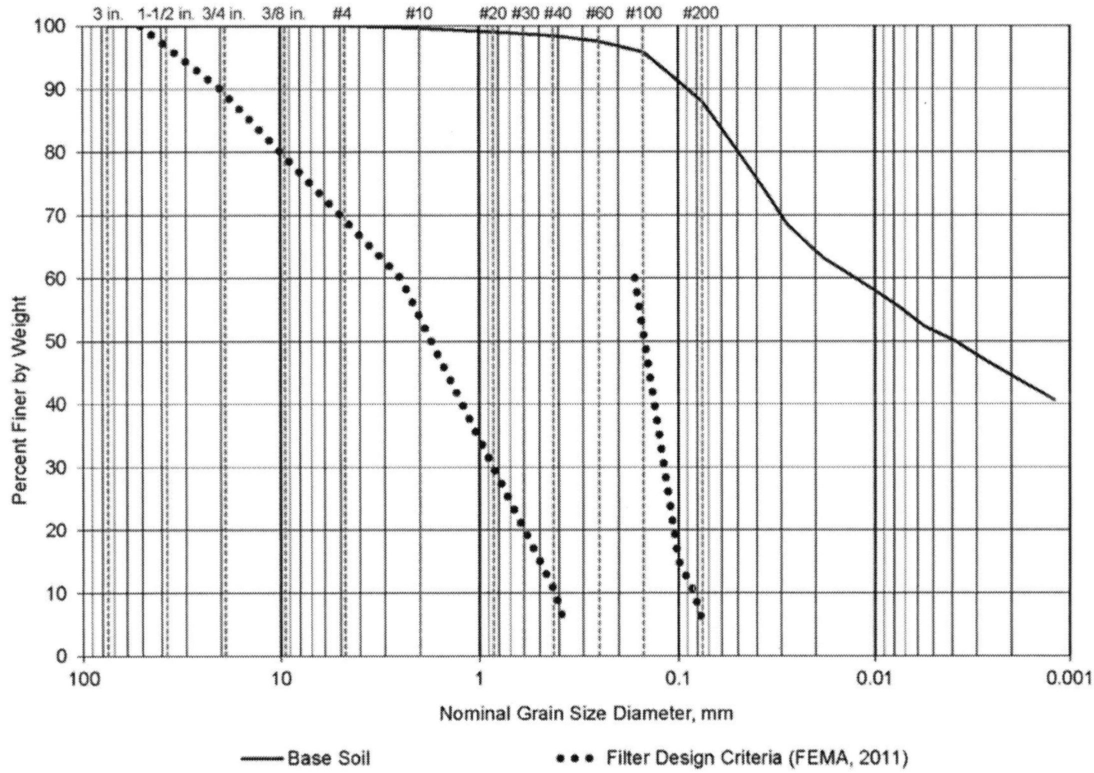

Figure 1. Base Soil and Filter Design Gradations

Filter materials were developed using subrounded to rounded quartz well screen sands. Well screen sands are very poorly graded and therefore mathematical estimates could be performed to developed ratios using various sands to prepare a filter material that would have the desired $D_{15}F$ ($D_{15}F$ = the particle size of the filter where 15 percent by weight of soil particles are finer).

After completion of successful filter tests, grain size tests were performed on the base soil to confirm the material was similar to the initially evaluated base soil. Grain size tests and relative densities were performed on the filter material to identify the actual $D_{15}F$ and relative density for each specific test.

Test Apparatus and Sample Preparation

The tests performed used an apparatus similar to the apparatus used by Sherard, Dunnigan, and Talbot, as shown in Figure 2.

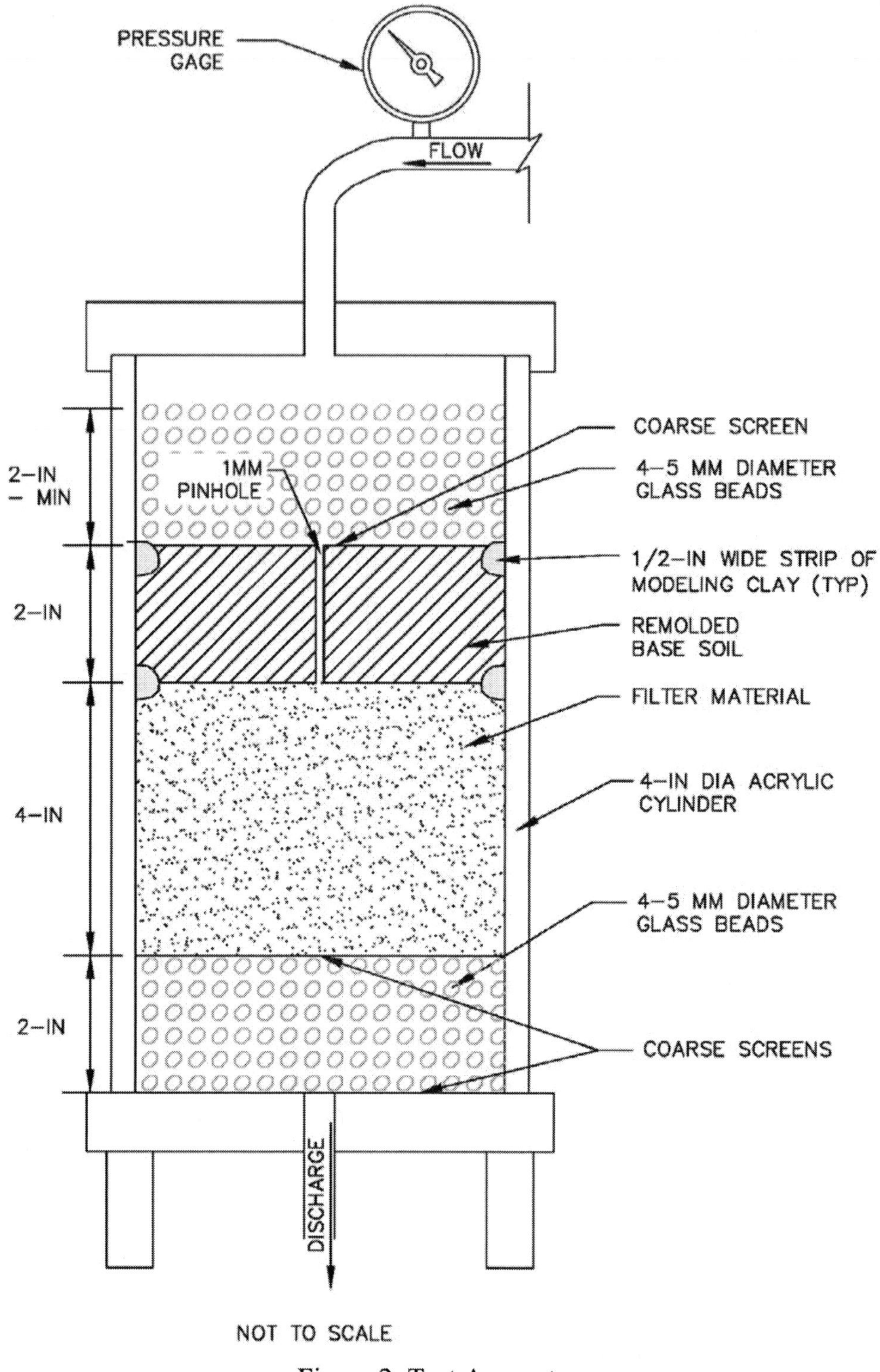

Figure 2. Test Apparatus

The sample was remolded within the test apparatus with the test apparatus placed vertically as follows:

1. Blocks were inserted into the apparatus that set on a solid surface.
2. The base soil was placed over the blocks in two, one-inch lifts and were lightly compacted with minimal effort. The contact between the first and second lift was scored prior to placement of the second lift.
3. Modeling clay was placed around the inside of the cylinder wall at the top and bottom of the base soil.
4. The apparatus was turned 180 degrees and blocks were inserted into the opposite side of the apparatus to maintain the position of the base soil in the apparatus. The blocks were set on a solid surface.
5. Filter material was placed into the cylinder. Two methods of placement were used to control density of the filter material:
   a. For dense samples – filter material was placed into the apparatus while rodding the filter material and striking the outside of the test apparatus with a rubber mallet.
   b. For low density samples – filter material was placed with methods similar to ASTM D4253 (ASTM, 2014) for placement at minimum relative density.
6. A 1-mm diameter pin hole was introduced into the base soil and extended into the filter material
7. Coarse screens were placed between the glass beads and the base soil and between the glass beads and the bottom of the apparatus so that the glass beads would not plug the pin hole or discharge outlet. A coarse screen was also included between the filter material and the glass beads to aid in disassembly of the test apparatus.
8. Glass beads were placed above the base soil and below the filter material. Care was taken to not modify the density of the filter material during assembly.

General Test Results

Tests were performed with the cylinder orientated vertically. Water was supplied to the apparatus at 40 pounds per square inch (psi). A valve was opened to allow the pressurized water to enter the apparatus. Discharge water was collected in glass beakers to visually evaluate turbidity. A surge of discolored water that would progressively become more turbid and cloudy was observed within about the first minute of all tests.

For successful tests, after the initial surge of turbid water, either flow completely stopped or the flow greatly diminished and eventually would become clear with no observable fines. A successful test was usually identified within the first one to five minutes. Upon disassembly of the test apparatus there was very limited widening of the pin hole and no other cracks or pipes were identified in the base soil. It is the opinion of the authors that the flow either completely stopped or was greatly diminished because the filter performed properly and retained the fines, which created a filter cake that reduced or stopped flow.

For unsuccessful tests, the water flow would generally continue at about the same rate throughout the test. Varying degrees of discolored water would continue throughout the duration of these tests and between six and seven liters of water flowed through the apparatus for each unsuccessful test. At the end of several tests (after six to seven liters of flow), the water was mostly clear, but fines were observable in the bottom of the beakers about 10 to 15 minutes after completion of the tests. It appears that for this particular clay, when the pipe was of sufficient size to convey the seepage allowed by the filter, erosion was significantly reduced, but not halted. Upon disassembly of the test apparatus, generally the pin hole had eroded (widened) and other cracks or pipes had formed during testing.

Care needs to be taken with assembly of the test apparatus and particularly with the placement width of the modeling clay. Based on observations, the void space is larger at the contact between the filter material and the side of the test apparatus. Several failed tests occurred because the modeling clay was not adequately adhered to the side of the test apparatus or the width of the modeling clay was to narrow. This condition resulted in fines bypassing the filter material and flowing along the side of the test apparatus where the void space was larger and not through the center of the filter material.

Variation of Density

After the test apparatus was assembled with the base soil and filter material, several tests were performed with a compacted sample of filter material to identify the $D_{15}F$ required to retain the base soil. The $D_{15}F$ was varied by about 0.1mm to 0.3mm until the successful $D_{15}F$ was identified. Initial filter material gradations were prepared based on maximum $D_{15}F$ for a successful test identified by Sherard, Dunnigan, and Talbot (ASCE, 1984). After a successful test was performed for compacted filter material, the test was repeated with low density filter material and the $D_{15}F$ was varied by about 0.1 mm until a successful test was performed with the low density filter material.

RESULTS

The successful $D_{15}F$ for the compacted filter material with a relative density of 55 percent was 1.1 mm, which results in a factor of safety of about 2.2 when compared to the maximum $D_{15}F$ according to FEMA criteria. This is similar to factors of safeties identified by Sherard, Dunnigan, and Talbot (ASCE, 1984) for Category 1 base soils. The successful $D_{15}F$ for the low density filter material with a relative density of 17 percent was 0.7 mm, which results in a factor of safety of about 1.4 when compared to the maximum $D_{15}F$ according to FEMA criteria. The base soil gradation, required filter gradation using FEMA guidelines, the filter material gradations, and respective $D_{15}Fs$ are show on Figure 3.

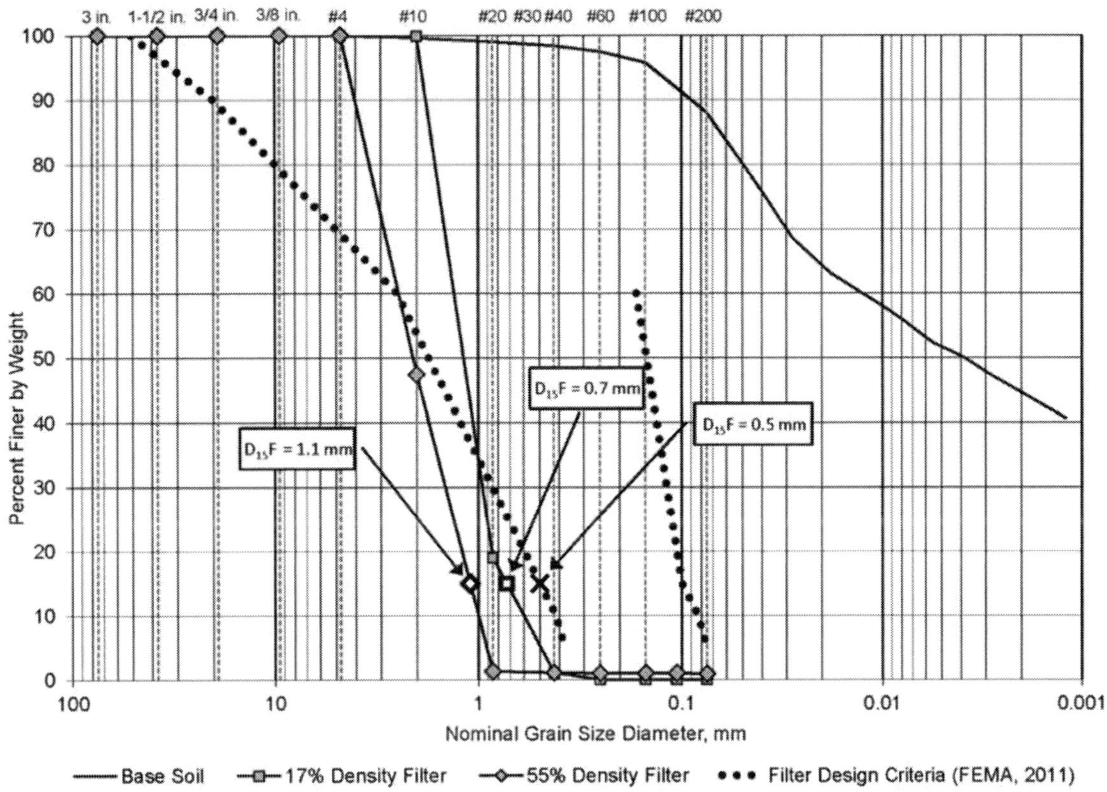

Figure 3. Filter Testing Gradations

CONCLUSIONS

The Authors reiterate that the standard practice of compacting filters should be adhered to when possible. The purpose of this paper was to present a test method to evaluate the impact of density on the filter material properties (ability to retain the base soil) for a specific base soil and is not intended to be design guidance. Based on the results of the tests performed we offer the following conclusions that pertain to the tested materials:

- Density does impact filter material's ability to retain the base soil.
- Low density filter materials would meet the intent of the FEMA guidelines, but with a lower factor of safety.
- The factor of safety of the $D_{15}F$ at failure versus the design guideline $D_{15}F$ was reduced from 2.2 to 1.4.
- Erosion of the tested base soil continued throughout the unsuccessful tests, however erosion appeared to significantly reduce after a sufficient pipe size was developed to convey the flow accepted by the filter material. It should be noted that this particular clay soil has a high plasticity index (near the upper limit for a liquid limit of 48). This would result in a rather high resistance to erosion in the simulated crack (pin hole). Silt or clay soils with lower plasticity indexes would likely erode more in unsuccessful tests and test durations would likely be shorter.

RECOMMENDATIONS

For situations where compaction cannot be performed, as in narrow vertical trenches and bio-polymers supported trenches we recommend performing a filter test on the site specific base soils and intended filters to confirm the design guidelines are met. As shown in Sherard, Dunnigan, and Talbot (ASCE, 1984), the factor of safety for filters varies widely for Category 1 soils and a filter test would provide site specific test data to confirm impact on density to specific base soils.

Proper setup of the test apparatus is critical to test accuracy, particularly the modeling clay separating the base soil and filter material.

The duration of the filter tests needs to be sufficient to confirm the lack of fines in the discharge water. Several unsuccessful tests had initial appearance of successful tests, but upon review, a small quantity of fines were visible in the discharge water.

Gradations, and relative densities should be performed on each successful test to obtain actual $D_{15}F$ and relative densities.

ACKNOWLEDGEMENTS

The Authors would like to acknowledge Mr. James Talbot, P.E., for his invaluable guidance, suggestions, and review of our test apparatus, test procedures, and this paper.

REFERENCES

ASTM International, Annual Book of Standards, Volume 04.08, 2014

FEMA, Filters for Embankment Dams, Best Practices for Design and Construction, 2011.

Park, Youngjin; Brandon, Thomas L.; Duncan, Michael J.. Investigation of the Ability of Filters to Stop Erosion through Cracks in Dams. Virginia Tech, Department of Civil and Environmental Engineering, 2004.

Sherard, James L; Dunnigan, Lorn, P.; and Talbot, James R.. Filters for Silts and Clay, Journal of Geotechnical Engineering, Volume 110, pp 701-718, ASCE, 1984

Sherard, James L; and Dunnigan, Lorn, P.. Critical Filters for Impervious Soils, Journal of Geotechnical Engineering, Volume 115, pp 927-947, ASCE, 1984A

Talbot, James, R.. Personal communications, July, 2014.

U.S. Bureau of Reclamation. Earth Manual, Part 2, 1990.

U. S. Bureau of Reclamation. Design Standards No. 13: Embankment Dams. Chapter 5 – Protective Filters, United States Department of the Interior Bureau of Reclamation, Technical Service Center, Denver, CO, 1994.

MIX DESIGN AND CONSTRUCTION OF ASPHALT CONCRETE FACE SLAB IN FROZEN REGION

Lu Yihui[1]
Feng Minghui
Xia Shifa
Hao Jutao
Liu Zenghong

ABSTRACT

In cold regions, the asphalt concrete face slab is superior to the conventional reinforced concrete face slab for anti-seepage. It is a critical decision for a successful project to choose the right material of face slab in the frozen region. What was shown in this paper is that the performance of asphalt concrete is excellent for resisting frozen-fracture in very low temperature environment as face slab. The Asphalt Concrete face slab is recommended in the frozen region. A sample work is given to shown how to design the mixture by adjusting the original bitumen to suit to the special environment. In addition, construction rules are explained to share the experiences of how to pave the asphalt face slab at very cold weather.

INTRODUCTION

The technology of asphalt concrete face slab has been used in dams abroad since 1930s. In China, the technology was first adopted in 1970s in Shiertaizi dam in Liaoning Province, and in Zhengcha dam in Sh'anxi Province. The development of technology of asphalt concrete face slab in China has experienced three stages: conscious exploration, cooperation with foreign parties, and creative development on self-reliance. During the stage of cooperation with foreign parties, the Tianhuangping, Zhanghewan and Xilongchi Upper Pond projects of asphalt concrete face slab were constructed. In the stage of creative development on self-reliance, the Baoquan and Huhehot projects were accomplished independently. In China, with the improvement of the quality of asphalt and the construction technology, the projects of asphalt concrete have been successfully constructed not only in the south area with warm climate, but also in the north area where it is very cold. In cold regions, because of the low air temperature, cracks are prone to occur at conventional reinforced concrete face slabs; and the waterproof seals of the joints often cease to be effective, resulting in bad anti-seepage effect. Therefore, asphalt concrete face slab is chosen first in cold regions for its no joints and good performance in resisting frozen-fracture, though the unit price of asphalt concrete is higher. For instance, the Xilongchi Upper Pond project with the extremely low temperature of -34.5℃ and the Baoquan Upper Pond project with the lowest temperature of -28.0℃, of which the construction started in 2005 and 2006 respectively, both adopted the asphalt concrete face slab as the anti-seepage structure of the Upper Ponds.

[1] China Institute of Water Resources and Hydropower Research (all authors)

The mix proportion design and on-site construction of asphalt concrete face slab in cold regions are distinctly different from those in warm areas. Based on the Huhehot Upper Pond project of asphalt concrete face slab, this paper introduces the critical technology of mix proportion design and construction of asphalt concrete in cold regions, which can be referred by similar projects.

INTRODUCTION TO THE HUHEHOT UPPER POND PROJECT OF ASPHALT CONCRETE FACE SLAB

The Huhehot pumped storage power station is located in the area of Daqingshan Mountains, which is about 20km from the center of Huhehot City. The installed capacity of the station is 1200MW, totally 4 generators with each capacity of 300MW. After finishing construction, the station will be connected to the power grid of West Inner Mongolia to take charge of peak and valley regulation, frequency and phase modification, standby for power accident and so on.

The average air temperature at Upper Pond is 1.1℃ with the highest temperature of 35.1℃ and the lowest of -41.8℃. The Upper Pond is formed by digging and filling the ground with the width of the crest of dam being 10m. The altitude of the bottom is 1900m while that of the crest is 1943.0m. The normal water level and dead water level are 1940m and 1903m, respectively. The asphalt concrete face slab with slope of 1:1.75 is adopted as the anti-seepage structure of the Upper Pond. The total area of slabs is 244,800 m^2, including 101,100 m^2 for the bottom and 143,700 m^2 for the incline.

The asphalt concrete face slab is designed as a simplified section. The macadam bedding with a thickness of 60cm is paved under the slab for water discharge. The asphalt concrete slab of totally 18.2cm thick includes leveling layer, impermeable layer and seal layer, which are 8cm, 10cm and 2mm thick respectively. In addition, in some positions of incline where deformation may be large, such as the intersection part of the bottom and the slope, which is designed as an ogee shape, the joint for slab and wave protection wall and the boundary line of digging area and filling area, a thickened impermeable layer is designed with an additional thickness of 5cm.

Photo 1. Upper Pond of Huhehot Pumped Storage Power Station

The Huhehot Upper Pond project of asphalt concrete face slab is designed and constructed by Chinese companies on self-reliance. The Beijing IWHR-KHL Co., Ltd (abbreviated as KHL) took charge of the mix proportion design and the production of asphalt concrete. The construction of the whole project was accomplished jointly by KHL and others.

MIX DESIGN FOR ASPHALT CONCRETE OF IMPERMEABLE LAYER

The Huhehot pumped storage power station is located in high cold regions. For the Upper Pond, the extremely low temperature is -41.8℃ while the highest temperature is 35.1℃, resulting in the maximum temperature difference of 77℃. According to the practice of Baoquan, Zhanghewan and Xilongchi projects, the current quality of asphalt in China can meet the requirements for anti-seepage, deformation, slope flow, water stability, aging resistance of asphalt concrete. However, the extremely low temperature in the Huhehaot Upper Pond is -41.8℃, the lowest temperature in similar projects in China and 7.3℃ lower than the lowest temperature of the Xilongchi project, -34.5℃. Therefore, the performance for resisting frozen-fracture of asphalt concrete face slab is extremely important for the Huhehot Upper Pond project.

The performance for resisting frozen-fracture of asphalt concrete of impermeable layer is determined mainly by the asphalt quality and mix proportion design. In order to solve the problem of frozen-fracture of asphalt concrete face slab, KHL executed performance test and inspection analysis of the main raw materials including coarse aggregates, fine aggregates, packing materials and so on, and gave the positive evaluation to these raw materials. At the same time, KHL also carried out the investigation for the newest asphalt products, and executed the comprehensive performance tests for the 13 sorts of asphalt. Besides, the physical and mechanical performance tests were carried out. Especially a large amount of frozen-fracture tests were conducted to select the appropriate asphalt and mix proportion to meet the requirement on resisting cracks of slabs in the temperature of lower than -43℃. Finally, the most favorable asphalt and mix proportion were chosen. The modified asphalt is adopted in the Huhehot Upper Pond project of asphalt concrete face slab, which is distinctly different from other project in China, as shown in Table 1.

Table 1. Comparison of the performance of modified asphalt used in the Huhehot and Xilongchi projects

No.	Item	Unit	Test values in Xilongchi[2]	Test values in Huhehot
1	Penetration(25°C, 100g, 5s)	1/10mm	98	118
2	Ductility in low temperature	Cm	64	80
3	Ductility(15°C, 5cm/min)	Cm	103	85
4	Softening point	°C	82	65

No.	Item	Unit	Test values in Xilongchi[2]	Test values in Huhehot
5	Ratio for penetration	%	93	109
6	Brittle point	°C	-21	-28

It is worth noting that the test values of frozen-fracture of asphalt concrete on site in Xilongchi project with the average value of -39℃ are higher than those of Huhehot project with the average value of -44.7℃. However, it is indicated in Table 1 that the brittle point of the modified asphalt used in the Xilongchi project is lower than that used in the Huhehot project. This is due to that the frozen-fracture temperature of asphalt concrete is not only related to the brittle point of asphalt, but also influenced by the mix proportion. Ordinarily, the softening point of asphalt rises after aging; however, the modified asphalt adopted in the Xilongchi project and Huhehot project show the reverse results that the softening point decreases after aging. The softening point of the modified asphalt used in the Xilongchi and Huhehot project decreased 5℃ and 4.55℃, respectively. This phenomena need to be studied further.

After the indoor experiment, usability test and production test, the final mix proportion for the impermeable layer of the Xilongchi and Huhehot projects are shown in Table 2 and Table 3 respectively.

Table 2. The mix proportion for impermeable layer of Xilongchi project

Mineral aggregate gradation										
Sieve Diameter (mm)	16	13.2	9.5	4.75	2.36	1.18	0.6	0.3	0.15	0.075
Pass percent(%)	100	95.6	86.2	73.5	57.6	42.4	30.7	19.3	13.9	12
Asphalt content(the percentage of the asphalt weight in the gross weight of asphalt and mineral aggregate):7.8%										
Fiber content(the percentage of the fiber weight in the gross weight of asphalt and mineral aggregate):0.2%										

Table 3. The mix proportion for impermeable layer of Huhehot project

Mineral aggregate gradation										
Sieve Diameter (mm)	16	13.2	9.5	4.75	2.36	1.18	0.6	0.3	0.15	0.075
Pass percent(%)	100	96.6	89.3	65.4	50	37.2	27.2	17.6	14.2	11
Asphalt content(the percentage of the asphalt weight in the gross weight of asphalt and mineral aggregate):7.3%										

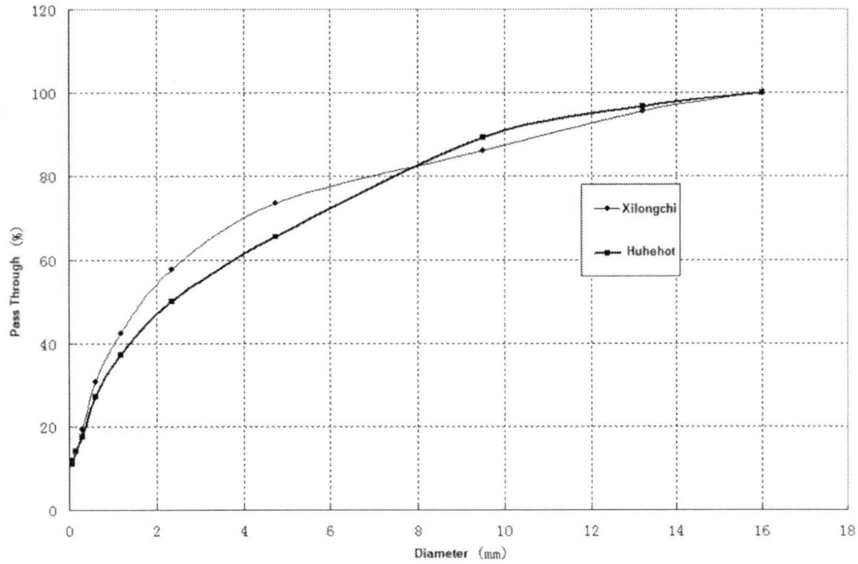

Figure 1. Curves of the mix proportion for impermeable layer of Xilongchi and Huhehot project

CONSTRUCTION OF ASPHALT CONCRETE FACE SLAB

Mixing of Asphalt Mixture

The quality of asphalt concrete depends on the mixture first. The precision of mixing and temperature of the mixture must meet the design requirements. The mixer used in the Huhehot project has 6 bins for cold aggregates and 5 bins for hot aggregates. The mixing is totally automatic with the precision as follows: weighting error of asphalt is ±0.2%, weighing error of mineral powder is ±0.3%, weighing error of aggregates is ±0.5%. The production capacity of the mixer is 120~160t/h. The temperature standard for asphalt, aggregates and mixture during the mixing is shown in Table 4.

Photo 2. Asphalt mixer used in the Huhehot project.

Table 4. Temperature standard for asphalt mixer in Huhehot project

Project	Asphalt concrete of impermeable layer (°C)	asphalt concrete of leveling layer (°C)
Asphalt	160~180	150~170
Aggregates	180~200	170~190
Asphalt Mixture	160~180	150~170

Paving and Rolling of Asphalt Concrete

Asphalt concrete is used in the whole basin for anti-seepage in the Upper Pond of Huhehot pumped storage power station. Therefore, the construction area can be divided into the flat area on bottom and the slope area on incline. The construction on slope is the key point and the difficulty of the Huhehot project. Its process is described in the following:

1. Construction Technology and Control of Paving on Slope

Asphalt concrete on slope is paved in a vertical way from bottom to top. In order to reduce construction joints, and increase anti-seepage performance and integrity of face slabs, the paving width must be as large as possible. The paving width of the paver used in the Huhehot project is 2.5-5m; and the paving width is actually controlled as 4.25m. The standard for paving thickness of each layer is shown in Table 5. The rolling temperature and number of rolling are shown in Table 6.

Table 5. Standard for paving and rolled thickness of each layer

Project	**Paving**	**Rolled thickness（cm）**
Level layer	≥9.2	≥8
Impermeable layer	≥11.2	≥10
Thickening layer	≥6	≥5

Table 6. Standard for construction temperature and process on slope paving

Project	**Rolling process**	**Temperature (℃)**	**Number of rolling**	**Way of rolling**
Level layer	Initial rolling	>130	1	Quiet rolling
	Double rolling	>105	3	Vibration rolling when forward, Quiet rolling when backward
	Final rolling	>90	1	Quiet rolling
Impermeable layer	Initial rolling	>140	1	Quiet rolling
	Double rolling	>115	3	Vibration rolling when forward, Quiet rolling when backward
	Final rolling	>90	1	Quiet rolling

The control of rolling temperature is the key point to ensure the quality of asphalt concrete. Asphalt concrete must be rolled when its temperature is higher than required to make the loose mixture dense. Thus it can resist the follow-up heavy load and reduce the heat loss. If the rolling temperature is too high, cracks will appear. If the rolling temperature is too low, the mixture could not be fully compacted which is bad for anti-seepage performance and durability. After the final rolling, quiet rolling is used to eliminate the wheel tracks and make the face smooth.

Photo 3. Site view of paving and rolling of asphalt concrete face slab at the Upper Pond of Huhehot pumped storage station.

2. Treatment of Joints

Modified asphalt was adapted in the Huhehot Upper Pond project of asphalt concrete face slab. Compared with other constructed projects in China, the modified asphalt used in the

Huhehot project has a high viscosity and need a higher construction temperature; and hence the construction is more difficult. As can be seen from the usability experiment, the compaction of impermeable layer was difficult. Most porosities of asphalt concrete measured were close to the upper limit. Some porosities of concrete on joints were even higher than the upper limit. Therefore, paving, rolling and joints treatment must be further studied and well organized to ensure that the porosity of impermeable layer is below 3%, especially for the treatment of joints. Joints are divided into hot joints and cold joints.

Treatment of Hot Joints. Hot joints are those joints whose adjoining strip is compacted more than 90% and the temperature is higher than100℃ which is suitable for rolling. The treatment of hot joints is to use the paver to make the side a 45° slope, and then paving the next strip and rolling the both side.

Treatment of Cold Joints. First, joints must be coated with cold asphalt. The paving of next strip must wait for the gasoline in asphalt completely evaporated. When paving the new strip, infrared heater must be used to heat the joint to make the joint smooth. The heating temperature should be controlled between 100-130℃, and the heating depth should not be less than 7cm. The heating temperature and heating time must be well controlled to prevent the aging of asphalt.

Before rolling the new strip, use vibration roller to roll the joint on both sides. Roller must cover at least 10cm on the new strip, and the temperature 7cm below joint surface must be higher than 60℃. The joint should be rolled at least for twice, using vibration rolling when forward and quiet rolling when backward.

Construction joint is the weakness of asphalt concrete. During the construction, the cold joints must be well recorded. After the construction of cold joints, the quality of joints must be tested according to specifications and technical requirements. If the joints are unqualified, they must be treated. The procedure is to use infrared heater to heat the joint, and then tamp it.

3. Construction at the Corner

There is a fan area at the corner on the basin slope. In the past time fixed width paver must cooperate with manual paving to complete the construction at the corner. This makes difficult to control the quality. The paver used in the Huhehot project is a variable width paver which can change pave width from 2.5m to 5m. This can avoid the manual paving area to ensure the construction quality.

Construction of Leveling-Layer Asphalt Concrete at Low Temperature

The Huhehot project is located in a severe cold region, with high altitude and large wind in the whole year. If the construction criterion (daily mean temperature higher than 5℃, wind force lower than level 4) is totally obeyed, the construction time will be very short. Therefore, construction at low temperature should be explored. For safety reason, construction of leveling-layer asphalt concrete at low temperature was only tried from mid-September to the end of October in 2012.

Low Temperature Reformation of Mixer. The daily mean temperature in Huhehot project is most lower than 5℃ and the night temperature is lower than 0℃ after mid-September. Under this condition, in order to ensure the quality and production capacity of asphalt concrete, the mixer was reformed. The reformation including: water proof and freeze proof of aggregates, heat preservation of pneumatic system (building air compressor room and heating device) , change fuel oil, raising the temperature of hot aggregates and the mixture and so on. 25000t mixtures were produced from mid-September to the end of October in 2012, with the qualification rate of 100%.

Paving and Rolling On Site. When paving at low temperature condition, the weather should be concerned, construction was forbidden when there was large wind, rain and snow. Asphalt concrete must be constructed at warm time during daytime, construction at night was forbidden. The paving time was better from 9am. to 4pm. The construction had 5 principles: quickly unloading, quickly transporting, quickly feeding, quickly paving, quickly rolling. Detailed steps are as follows.

1. Temperature control. The temperature of paving and rolling must take the upper limit of the standard. Paving temperature of leveling layer should not be lower than 155℃, and the initial rolling temperature could not be lower than 145℃. Because of the low temperature and large wind, the heat of asphalt concrete lost quickly during paving and rolling. So the initial rolling should be rolled as soon as possible after asphalt concrete was paved to reduce the heat loss.

2. Control of feeding car. The load and speed of the car must be controlled during the low temperature construction. The amount of the mixture in the feeding car must be well controlled, made sure the mixture was feeding equally (8t each time). The feeding speed should be as quickly as possible on the premise of safety.

4. Control of the paver. Paver with high compaction capacity should be used when construction at low temperature. Before paving, the iron plate should be heated and the compaction power should be adjust to make sure the compactness was higher than 85% after paving. Paving speed should be 1.5m/min~2m/min at the lower half of the strip. It could be properly increased at the upper half if the compactness could be ensured. At the lower half of the strip, the mixture could not be rolled in time, so carpet or cotton quilt was used to cover this part to ensure heat-preservation and reduce heat loss.

4. Control of rolling. Rolling is the most important point during the low temperature construction. Initial rolling temperature and final rolling temperature must be ensured. Initial rolling temperature must take the upper limit of the standard. In case of low temperature construction, the compactness of asphalt concrete after paved should be higher than 85%, and wrinkles would not occur during rolling. Vibration should be used in initial rolling during low temperature. Initial rolling should be rolled as soon as possible after asphalt concrete was paved (use 3T double wheels vibration roller). As for

the asphalt concrete paved where the roller couldn't reach in time because of the disturbing of paver, carpet or cotton quilt with enough thickness should be covered to decrease the velocity of heat loss and ensure the initial temperature of rolling.

The area of leveling layer constructed at low temperature in Huhehot project is about $50000m^2$, after taking the upper measures, testing results showed the qualification rate of the leveling layer was 95%. The unqualified area was removed and re-paved in 2013.

CONCLUSIONS

1. The development of asphalt concrete face slab technology has experienced the stages of conscious exploration and cooperation with foreign parties; and it is now in the stage of creative development on self-reliance. In cold regions, the anti-seepage performance of asphalt concrete face slab is superior to that of traditional reinforced concrete face slab owing to its properties of high crack resistance under low temperature, seams free and high adaptability to deformations.

2. The low-temperature crack resistance of asphalt concrete in cold regions is the key point to the success of the projects; and it can be improved through asphalt types and mix proportion optimization. In China, the mix proportion technologies of asphalt concrete have been mature; yet the properties of modified asphalt should be further studied.

3. In cold regions, modified asphalt is usually adopted at impermeable layer of face slab. However, its high viscosity and high dosage can result in the difficulties in compaction during construction process. Based on the experiences of the Huhehot project, it is difficult to control the porosity of joints between strips of impermeable layer. To meet the design requirement, perfect construction technology and elaborate treatment are necessary. The construction technologies of rolling equipment after asphalt paver should be further studied, which means rolling the joints after paving and consequently improve the quality of the joints.

4. In cold regions, the construction period is shortened if asphalt concrete is paved strictly based on the requirement of specifications. Therefore, construction technology of leveling layer under low temperature was explored in the Huhehot project. Such measures as low-temperature reformation of mixer, heat preservation with quilts and quick rolling to ensure the initial rolling temperature of asphalt concrete were adopted during the construction process, which ensured the quality of mixtures and asphalt concrete after rolling compaction. The test results of the Huhehot project indicate that the qualification rate of on-site paving and rolling quality is higher than 95%, which proves the practicability of above measures. However, the construction technologies of impermeable layers under low temperature should be further studied.

REVISITING A FORGOTTEN DAM BUILDING TECHNOLOGY: CONCRETE CORE WALLS IN EMBANKMENT DAMS

Richard R. Davidson, P.E.[1]
Kenneth B. Hansen, P.E.[2]

ABSTRACT

Early in the twentieth century, placing concrete core walls within embankment dams was a popular construction technique for small to medium height dams. It became in vogue as a replacement for the popular British dam construction technology of puddle clay core dams which were used between the 1860's and 1920's. It avoided the many problems with semi-hydraulic / manned placement methods of the puddle clay cores within narrow trenches. However, after the mid 1930's this concrete core wall construction fell out of favor because of the improvements made in embankment compaction methods and the difficulties in building reinforced concrete core walls to more significant heights.

Today concrete core wall embankment dams are now reaching an age where their continued performance is being questioned. This dam building technology has become extinct and is unknown to the last few generations of dam engineers. Therefore, it is relevant to re-examine this dam building technology in a modern context and work on answering the following questions. How have these dams performed after almost a century of service? Are there unanticipated performance features that have produced positive results when subjected to extreme flood and seismic events? Does the concrete provide enhanced performance over time? What role does steel reinforcement play it the performance of the core wall? Are there lessons here that can be applied to the more common concrete cutoff wall solutions being applied to embankment dams with seepage problems?

This paper will examine these questions with a number of illustrative case histories to provide a retrospective illumination of this forgotten dam building technology.

INTRODUCTION

Embankment dams with concrete core walls early in the twentieth century were a popular dam building methodology that overcame a number of problems with the puddle clay core embankments adopted from British dam engineers. However, as embankment construction methods improved in the 1940's, concrete core walls virtually disappeared from use because of their high cost and difficulty in achieving the heights needed with more modern dams. However, those dams that survived this early era of dam building have performed surprising well in spite of the inherently challenging concept of placing a rigid diaphragm within a deformable earth structure. This dam section can provide

[1] Senior Principal and Vice President, URS Corporation, Denver, Colorado, richard.d.davidson@aecom.com.
[2] Consulting Engineer, Englewood Colorado

advantages, but there are height limitations where they are most suitable. Many of these dams are approaching 100 years of age or more.

EVOLUTION OF CONCRETE CORE WALL TECHNOLOGY

The concrete core wall era began in about 1865 with Mill River Dam in Massachusetts. The first core wall dam in Australia was Ness Gully Dam also called Korumburra No. 1 built in 1885 in Victoria. This dam was built by the State Rivers and Water Supply Commission of Victoria for water supply for this farming region. Other early projects include the 130-ft high Lower Otay Dam in California built in 1897 and a 72-ft high dam for the Anaconda Mine in Montana built in 1898.

Concrete core walls were generally keyed into the foundation bedrock and extended up into the embankment to various levels. They were generally tapered from a wide base to a narrow top, perhaps only 12 inches wide. Some were reinforced with steel cages. A typical section is shown in Figure 1. Initially they were often built with a puddle clay core which may have been extended to the top of the dam. Puddle clay cores originated in Great Britain, of which Australia was a colony at the time. The puddle clay was a narrow zone of selected clay processed to a putty-like consistency and compacted by animals such as sheep (the origin of sheepsfoot rollers) or placed hydraulically in puddle ponds. The exterior (more random) embankment shells were generally not particularly well compacted and were often placed by horse or oxen drawn drays or small wagons which transported and compacted the shell material. Core walls controlled seepage through the dams and allowed more flexibility with the exterior shell materials, especially if there was little clay borrow available. Later toe drains were incorporated in the section to collect seepage. However, many of the unreinforced concrete core walls cracked with the deforming, poorly compacted earthfill embankment, leading to seepage and piping problems downstream.

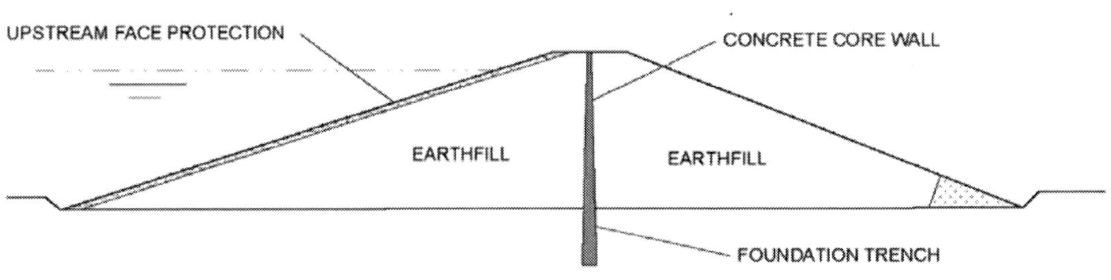

Figure 1. Typical Concrete Core Wall Configuration Within an Earthfill Embankment (Cole, 2000)

Photo 1. Lance Creek Dam, Victoria

Lance Creek Dam is a typical 69-ft high embankment with a concrete core wall built in 1911 and raised in the 1960's. Lance Creek underwent a seismic upgrade with deep soil mixing columns upstream and downstream in 2003.

Crane Valley Dam in California built in 1910 had a concrete core wall and the embankment shells were placed using hydraulic fill methods.

Photo 2. Crane Valley Dam Hydraulic Fill Placement (Kollgaard et al, 1988)

Concrete core walls continued to be utilized up until the 1960's, but the most significant structures were built in the 1910's to 1930's. Table 1 provides an incomplete list of core wall embankment dams built primarily in the USA and Australia. This list is biased toward failures since these were highlighted in the historical paper by Middlebrooks (1953) on unsatisfactory performance of earth dams.

TABLE 1. Embankment Dams with Concrete Core Walls

Dam	Country	State	Year of Construction	Height, ft	Length, ft	Characteristics of Core Wall		Comment
						Material	Thickness: top-bottom,ft	
Mill River	USA	Massachusettes	1865	43		Concrete		Failed in 1874 from seepage
Korumburra No 1	Australia	Victoria	1885	49	682	Reinforced Concrete		Clay core, Ness Gully, South Gippsland Water
Walden Pond Outlet	USA	Massachusettes	1890	43		Concrete	4.7-9.5	Significant Leakage, Lynn MA
Lower Otay	USA	California	1897	130		Concrete		Failed in 1916 from overtopping. Replaced by 119 ft high Savage Dam in 1919.
Anaconda	USA	Montana	1898	72		Concrete		Failed in 1938 from seepage slide
Hales Bar	USA	Tennessee	1913	85	500	Concrete		Karst underseepage secant pile cutoff wall in 1940-1944, dam removed in 1968
Crane Valley	USA	California	1910	145	1880	Concrete	2-7.8	Hydraulic fill, Bass Lake, PG&E. Reinforced portion of concrete core wall during seismic retrofit in 2013
Dells and Hatfield	USA	Wisconsin	1908	34		Concrete		Failed in 1911 due to overtopping. Replaced with 195 ft high Hatfield Dam in 1912
Korweinguboora	Australia	Victoria	1910	39	2043	Concrete		Clay core, Water Supply Dept
Lance Creek	Australia	Victoria	1911	69	1574	Concrete		South Gippsland Water, puddle clay
Hornell	USA	New York	1912			Concrete		Failed in 1912 due to seepage
Terrace	USA	Colorado	1912	55	605	Concrete		Hydraulic Fill
Sorpe	Germany	North Rhine	1935	226	2300	Concrete		Bombed in WWII causing seepage problems. Rockfill with clay core
Ashokan Dam & Dike	USA	New York	1912	252	4650	Concrete	4-7	NYCDEP
Oakley (Goose Creek)	USA		1913	145	1050	Reinforced concrete	1-3	
Hebgen	USA	Montana	1914	82	718	Concrete	3-16	Settlement from 1959 M7.5 Earthquake
Beargamil	Australia	NSW	1914	53	1014	Concrete		NSW Public Works Dept
Moorabool	Australia	Victoria	1915	59	2440	Concrete		Clay core, JS Sherland
Harvey	Australia	WA	1916	79	1096	Concrete		WA Public Works Dept
Melton	Australia	Victoria	1916	35	1095	Concrete		Rockfill and Puddle clay ore, SRWSC
Pisgah	USA	Minnesota	1918	40		Concrete		Otter Tail Power Co.

TABLE 1. Embankment Dams with Concrete Core Walls

Dam	Country	State	Year of Construction	Height, ft	Length, ft	Characteristics of Core Wall: Material	Characteristics of Core Wall: Thickness: top-bottom,ft	Comment
Catawba	USA	North Carolina	1919	171	3360	Concrete		Semi-hydraulic Fill, Duke Energy
Tieton	USA	Washington	1924	235	920	Reinforced Concrete	5	USBR - Wall is 319 ft high
Tourist Park	USA	Michigan	1924	25	1500	Reinforced Concrete	1 - 1.9	Failure in 2003 from overtopping
Ued Kabvir	Tunis		1925	115		Reinforced Concrete	2.0-9.0	Damaged in 1925
Piedmont #1	USA	California	1903	96	342			East Bay MUD, Conduit failure
Waranga Basin	Australia	Victoria	1908, 1925	40	5000	Reinforced Concrete	0.5-1.5	Goulbum - Murray Water, Core wall stopped piping incidents.
Cascade No 2	Australia	NSW	1926	85	420	Concrete		NSW Public Works Dept
Balsam	USA	New Hampshire	1927	60		Concrete		Failure in 1929 due to erosion from spillway
Connolly	Australia	Queensland	1927	22	476	Concrete		Rockfill, QWRC
Sugarloaf	Australia	Victoria	1927	46	3050	Concrete		Rockfill and earth core, Slump in 1929, SRWSC
Apishapa	USA	Colorado	1920	115	600	Concrete		Piping through settlement cracks in 1923
St Mary's	USA	California	1928	55	320	Concrete		Embankment crest cracking during 1955 earthquake
O'Shannassy	Australia	Victoria	1928	116	741	Concrete		Clay core, Melbourne Water
Spring Gully	Australia	Victoria	1929	79	1679	Concrete		Clay Core
Mangamahoe	New Zealand		1931	79		Reinforced Concrete		Trust Power
Silvan	Australia	Victoria	1931	141	2112	Concrete		Melbourne Water
Little Bear Valley (Arrowhead)	USA	California	1922	190	720	Concrete	3 - 20	Hydraulic fill, cracking repaired
Priest	USA		1931	144		Concrete		
Ashokan Dikes	USA	New York	1931	98		Concrete		NYCDEP
Kahl	Germany		1935	116		Reinforced Concrete	4-8	
Hume	Australia	Victoria	1936	128	4264	Concrete		Clay core, State Rivers & WS Commission
Cascade No 3	Australia	NSW	1938	100	810	Concrete		
Lake Endeavour	Australia	NSW	1939	69	656	Concrete		NSW Public Works Dept

TABLE 1. Embankment Dams with Concrete Core Walls

Dam	Country	State	Year of Construction	Height, ft	Length, ft	Characteristics of Core Wall: Material	Characteristics of Core Wall: Thickness: top-bottom,ft	Comment
Horspranget	Sweden		1951	170		Reinforced Concrete	1.3-16.4	
Rocky Creek	Australia	NSW	1952	92	669	Concrete		Gutteridge Haskins & Davey
Cochrane	Australia	NSW	1958	95	652	Concrete		Blair & Stucky
Bockhartzee	Austria		1982	100		Reinforced Concrete	2	
Baglihar Cofferdam	India	Kashmir	2002	59, 131	285	Concrete	13	
VanArsdale #1 (Cape Horn)	USA	California	1907	59	237			First combined concrete gravity & earth Dam
Delhi	USA	Iowa	1927	40	550	Reinforced Concrete		Failed due to overtopping in 2010
Mammoth	USA	Utah	1920's	100		Concrete	2	Puddle Clay, Buttress added to core wall
Pleasant Valley	USA	Washington	1920's	43		Concrete	1 - 3	12 ft puddle clay core

The failures are quite instructive about the early problems with this method of construction. Many of the failures were related to overtopping of the earth embankments, while there were a few seepage and piping related problems through cracks in the unreinforced core walls. Piedmont No. 1 in California suffered enough embankment settlement to shear off the outlet pipe where it passed through the concrte core wall. There were, of course, many other core wall embankment dams built in this era that have performed satisfactorally.

Concrete core wall dams were expensive to build, both for the concrete itself and for the restrictions imposed on the placement of the supporting earthfill. Early on, it was not fully recognized that the need for compaction of the supporting earthfill was more important than for the ductile puddle core construction, hence the more rigid concrete core walls often experienced significant settlement-induced cracking, and relatively large lateral movement during reservoir filling and over time. Improvements in the design and compaction of earthfill embankments eventually made concrete core walls unnecessary when clay core materials were available.

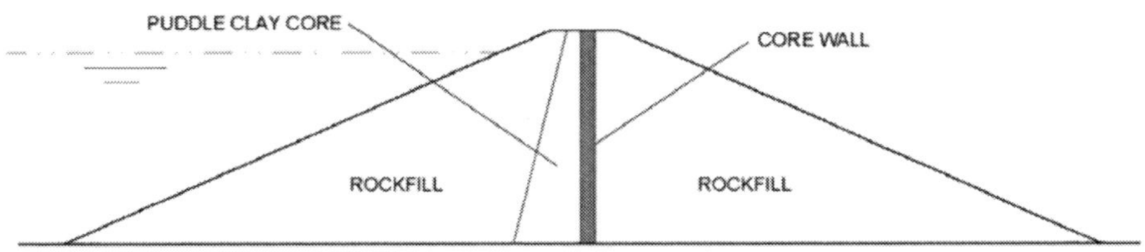

Figure 2. Typical Configuration of Concrete Core Wall in a Rockfill Embankment (Cole, 2000)

Concrete core walls were also used in rockfill dams as shown in Figure 2. The first case in Australia was Melton Dam in Victoria built in 1916 which also has a puddle clay core upstream of the concrete core wall. This technology was used before filters were adopted to reduce seepage and piping of the core material into the surrounding rockfill. However, the core wall did not provide protection upstream and the Sugarloaf Dam (predecessor of Eildon Dam) built in 1927 suffered substantial upstream subsidence on drawdown (Figure 3) and had to be supplemented with a rockfill buttress in 1929 to avoid further movement. Interestingly, the core wall prevented complete failure of the dam.

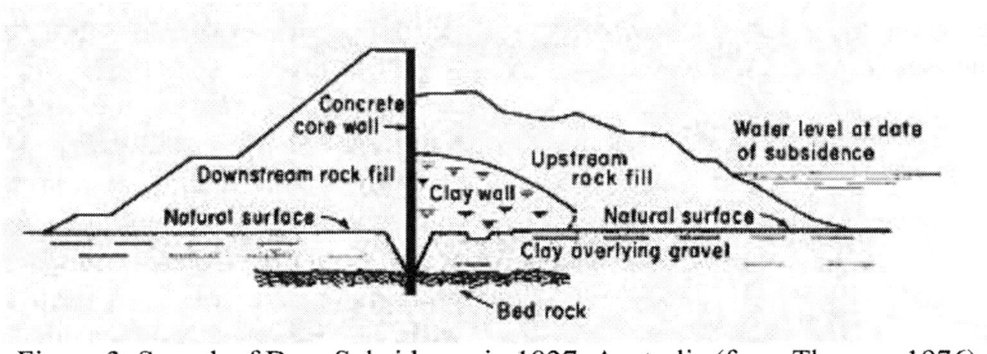

Figure 3. Sugarloaf Dam Subsidence in 1927, Australia (from Thomas, 1976)

Concrete core wall dams were built extensively in the United States in the 1920's. Most were relatively small structures less than 100 ft high, but a few much larger structures were built including Tieton Dam (1924) on the Yakima River in Washington by the Bureau of Reclamation. According to Yagin et al (1998), 41 concrete core wall embankment dams were built in the USA, with 6 in Germany and 4 in Sweden. Cole (1999) identifies another 20 concrete core wall dams built in Australia.

CONCRETE CORE WALL EMBANKMENT DAM PRECEDENTS

Tourist Park Dam, Michigan

Tourist Park is a 25-ft high embankment dam completed in 1924 with a concrete core wall for most of its length. The dam owned by the Marquette Board of Light & Power is located on the Dead River at Marquette in the Upper Peninsula of Michigan about one mile upstream from Lake Superior.

On 15 May 2003, flow from the failure of the upstream Silver Lake Dam caused the failure of Tourist Park Dam due to overtopping. The flow over the dam was estimated at 7,500 cfs which produced about a 1-ft depth of flow over the embankment. The concrete core wall was intended to extend to a rock knob at the right abutment; however the concrete core wall was not completed as planned probably due to difficulty in properly constructing the wall where the rock foundation dipped sharply downward.

Photo 3. Overtopping of Tourist Park Dam in 2003 Silver Lake Event

The silty sand embankment material on both sides of the concrete core wall was washed away due to the overtopping flow. There was no movement or damage to the reinforced concrete core wall. It is unlikely that any of the reinforcing bars from the wall extended into the foundation rock. The height of the concrete core wall at its exposed right end was estimated at about 12 ft. It is 1-ft wide at the top and battered at a 0.04H/1.0V slope on both sides. This means the thickness at the base at this location is slightly less than 2 feet.

It is the opinion of the authors, both of whom served on the Board of Consultants for reconstruction of the dam, that had the concrete core wall been extended as planned to the right rock abutment, the dam would likely not have failed. Thus, the concrete core wall would have served not only as a seepage control barrier, but also as a positive dam safety overtopping feature for the dam.

Reconstruction of the dam included extending the reinforced concrete core wall socketed into the rock foundation to a new 60-ft wide mass concrete auxiliary spillway located in the dam breach area. The reconstruction work was completed in 2013.

Photo 4. Remaining Concrete Core Wall at Tourist Park Dam After Failure

Hebgen Lake Dam, Montana

Hebgen Lake Dam (Sherard et al., 1963) located on the Madison River in southwest Montana close to Yellowstone National Park is an 82-ft high earth dam (Figure 4) which was subjected to a high intensity earthquake on 17 August 1959. The dam which was completed in 1914 included a concrete core wall that extended through the embankment and alluvium to a rock foundation, except where it was founded on alluvium near the right abutment near the spillway. The concrete core wall is unreinforced and varies in thickness from 16 ft at the bottom to 3 ft at the top. The core wall was terminated about 6 ft below the embankment crest when the dam was completed.

The dam was subjected to the most severe earthquake recorded in the Rocky Mountain region, estimated to be between M7.3 to M7.6 on the Richter scale. The quake's intensity was most likely on the high side of this magnitude. One report from the California Seismological Center at Berkeley, California had the quake's magnitude measured at M7.5 to 7.75. Twenty-two fatalities and $11 million in damage (1959 dollars) were attributed to this earthquake in this sparsely populated area of Montana.

Seiche waves caused by a landslide into the reservoir caused the dam to be overtopped on four occasions for about 10 minutes each time. The maximum height of the overtopping was more than 3 ft. The extreme seismic shaking caused the dam crest to settle nearly 6 ft in relation to the core wall on the upstream side and about 4 ft on the downstream side. The dam's settlement and washing away of some embankment material left the concrete core wall exposed on both sides as shown in the Photo 5.

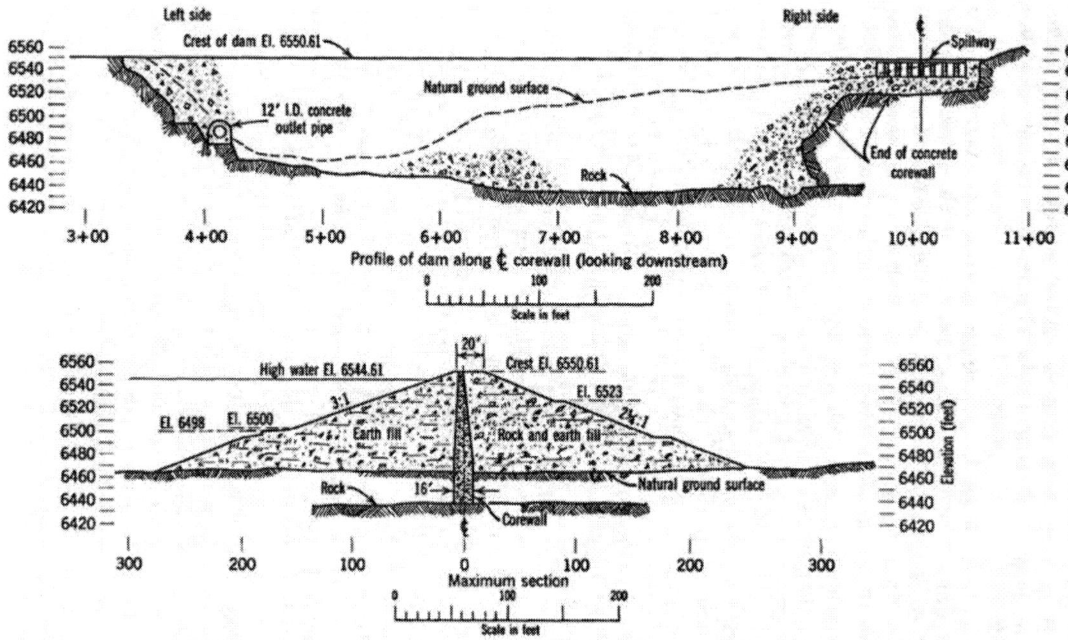

Figure 4. Hebgen Lake Dam (From Sherard et al. 1963)

The bedrock underlying the dam settled 9.7 ft uniformly. After the earthquake, levels taken on the top of the concrete core wall indicated the crest of the portion of the wall founded on bedrock remained no more than 0.1 ft from perfectly level. From the right abutment to a point about 400 ft toward the dam's center, the core wall deflected downstream 0.8 ft in the shape of an arch. From that point to the spillway on the left abutment, a distance of about 220 ft, the arch continued on with a maximum deviation of 2.8 ft from its original center line.

Photo 5. Exposed Core Wall and Cracking at Hebgen Dam (Sherard et. al, 1963)

Following the earthquake, an inspection of the concrete wall found 16 generally vertical cracks. Most of the cracks were hairline, but three cracks were 2 to 3-1/2 inches wide. These wide cracks were located near the right abutment where the core wall was not founded on rock. The wide cracks caused the length of the concrete core wall to elongate about 7 inches. The core wall was repaired by drilling 365 ft of holes in the concrete to intersect the visible cracks and pumping a water-cement grout into the cracks. The repairs required 412 bags of cement (19.4 tons).

Jim Sherard who visited the site soon after the event noted that the dam did not fail due to a "fortunate array of conditions". They included slightly cohesive well graded sand and gravel embankment materials which resisted erosion and the development of high pore pressures. However, in the opinion of the authors, the concrete core wall was a key element in preventing failure of the dam.

Walden Pond Outlet Dam, Massachusetts

The Walden Pond Outlet Dam forms one of four interconnected reservoirs that supplies about 11 million gallons of water a day to Lynn, Massachusetts. The 43–ft high earth dam has an unreinforced concrete core wall, built in 1890 and raised in 1905. The crest width is quite wide at 45 ft. The core wall extends 25 ft below the base of the dam and was built between two rows of steel sheetpiling. According to ENR (1908), the core wall developed leakage in 1908 of between 0.4 to 0.7 cfs through cracks in the wall. The wall was repaired and has functioned adequately ever since. The leakage led one ENR author to conclude the "unworthiness of concrete or masonry core walls."

Baglihar Cofferdam, India

The Baglihar hydroelectric project (Du, 2009) is located on the Chenab River in the States of Jammu and Kashmir in India. In order to provide seepage control for the rockfill upstream cofferdam in the Himalayan Mountains where clay is not available, the design engineers decided to use a 13-ft thick unreinforced concrete core wall.

This first stage cofferdam was 59 ft high and was founded on quartzite rock, except in the main course of the river where it sits on a concrete plinth. The concrete core wall for the 285 ft long cofferdam has a straight axis and ties into rock abutments.

Concrete for the wall using 1-1/2 inch maximum size aggregate was designed to achieve a minimum 28-day compressive strength of 2150 psi. Two transverse (contraction) joints divided the wall into three sections. From the right end, the lengths of the sections were 70 ft, 78 ft and 135 ft. Two lines of waterstops were placed in each transverse joint.

Construction of this Stage 1 cofferdam which was designed to be overtopped was completed in April 2002. In order for the cofferdam to achieve resistance to overtopping, cubical rock filled steel crates were placed in stair-stepped fashion on the downstream slope of the rockfill section. The crates which measured 4.9 ft on each side were covered on top with a 12-inch layer of concrete. These rock-filled crates could be called "gabions".

Soon after completion of the cofferdam, the rainy season started and the cofferdam was overtopped for four months from 6 May to 5 September 2002. After this long overtopping duration, all downstream support for the concrete core wall had been washed away as shown in Photo 6. This included the concrete capped rock-filled crates as well as the rockfill section. In addition, the river had been severely scoured downstream of the concrete core wall. The concrete core wall is completely exposed on its downstream face.

Upon inspection of the concrete wall, it was found the transverse joints were firmly closed on the upstream face, but open downstream from 0.4 to 1.0 inches, indicating some movement of the crest downstream. There were also a construction lift joint adjacent to the right side contraction joint and a construction lift joint near the bottom of the central section that had opened. The latter opening created leakage on the order of 14 to 18 cfs. The interface between the concrete wall and the foundation rock also opened. In this lower area, the upstream was in tension and downstream was in compression, or opposite of the condition at the crest. This would be the case for a cantilever wall loaded from upstream.

Following the four months of overtopping during the 2002 rainy season, the cofferdam was not remediated or rebuilt. It then survived four more months of overtopping in 2003. In the third year it was rehabilitated and raised to its planned 131-ft height.

A gravity dam analysiswould conclude that the concrete core wall should have failed. With a back analysis using a finite element model, Du (2009) concluded there was some arching action in the straight gravity dam in which some of the load was taken into the abutments. In this case study, the concrete core wall not only provided seepage control for the rockfill cofferdam, but also prevented the structure from failing when overtopped for four month periods in two consecutive years.

Photo 6. Baglihar Cofferdam after Flood Season Showing Exposed Core Wall (Du, 2009)

Lake Delhi Dam, Iowa

Delhi Dam on the Maquoketa River in Iowa breached during a flood in July 2010 with a peak outflow of about 69,000 cfs. The results of a failure investigation were reported by Fieldler et. al. (2011). The dam was built between 1922 and 1929 for hydropower generation. The dam section included a 425-ft long, 43-ft high embankment adjacent to the right abutment comprised of low plasticity sandy clay and a central core wall founded on sheetpiles which extended to rock. The core wall ended 6 ft below the dam crest. The crest width was 25 ft, with embankment slopes of 3:1 upstream and 2:1 downstream.

The failure investigation panel postulated that differential settlement between the relatively rigid core wall and the deformable embankment may have caused transverse cracking above the top of the wall which created seepage pathways in the upper portion of the embankment. During the July flood, one of the spillway gates could not be opened fully and the reservoir rose above the core wall. Within 8 hours vortices in the pool and sinkholes upstream were observed, and then cloudy seepage emanated from the downstream face along with settlement of the embankment, suggesting that internal erosion had initiated. Within about 7 hours from observing seepage downstream, the full breach initiated with erosion of the downstream embankment from overtopping flows and toppling of sections of the core wall until a breach width of 235 ft developed. Fieldler et al. (2011) suggested that "it is likely that the concrete core wall slowed down the rate at which the embankment was breached."

Photo 7. Delhi Dam Failure with Core Wall Still Intact, 24 July 2010

Not continuing the core wall to the crest and gate operation problems were important factors in this dam failure. Correcting both of these flaws as well as adding filter protection would have improved the performance of Delhi Dam in the 2010 flood.

Ness Gully Dam, Australia

Ness Gully Dam or Kurumburra No. 1 is the first concrete core wall dam built in Australia. The dam is 49 ft high with a homogenous earth fill section and a rockfill toe drain added about 30 years ago. The concrete core wall was excavated into bedrock and reinforced with rectangular steel bars. The core wall extends to the crest of the dam, as shown in Photo 7.

The dam is still operated by South Gippsland Water for water supply and has been in continuous operation since 1895. The wall has a noticeable downstream bow caused by embankment deformation over the years, but continues to perform well. A small amount of through seepage is picked up by the toe drain.

Photo 8. Ness Gully Dam in 2001 Showing Exposed Core Wall on the Crest

Waranga Basin, Australia

Waranga Basin is an important regulating basin for the rural irrigation system in Victoria. The 40-ft high dam was originally built in 1908 as a puddle clay core embankment and was raised in 1919 creating an unusual step in the puddle section. In 1922 five significant leaks developed through the puddle clay. In an emergency response, a timber sheet pile wall was installed downstream to stop the leakage successfully.

General Sir John Monash, engineer and army commander from World War I, and founder of Monash University was called in to assess the cause of the near failure. He concluded that internal erosion and piping was caused by yabbie (crayfish) holes through the core. He suggested that a concrete core wall be built along a downstream bench at the location of the temporary sheet pile wall. This 20-ft high, 18- to 6-inch wide reinforced concrete wall was built in 1925 as shown in Photo 9. The fill around the wall was poorly compacted with horse drawn drays (small carts) and deformation of the wall is noticeable during visual nspection. However, the wall and dam have performed well ever since without any further piping through the puddle clay core.

Photo 9. Construction of Concrete Core Wall at Waranga Basin, Australia, 1925

Hume Dam, Australia

Hume Dam is the most important dam in Australia. It is located on the River Murray just above the towns of Albury, New South Wales and Wadonga, Victoria. It provides essential water supply for agricultural needs in four states and as such represents the center point in the Murray-Darling Basin system. This iconic dam was constructed in the 1920's and has experienced a number of major dam safety upgrades over the decades. Most recently, beginning in 1996 after a serious incident at the southern junction between the main embankment concrete core wall and the concrete spillway, the dam was upgraded with structural and geotechnical enhancements to address various static and extreme loading conditions.

The main embankments were constructed with concrete core walls through each of the embankments (Photo 10). On the 128-ft high South main embankment the core wall was supplemented with a blister gallery that collected drainage water from a cobble drain installed along the base of the wall. A large concrete tower was built adjacent to the mass concrete gated spillway and the core wall was extended into each tower with an enlarged tongue and groove section surrounded by a bituminous seal. Over the years, the deformation of the compressible embankment caused some cracking at the junction between the core wall and the tower, and the nib cracked through. A grouting effort was implemented in 1996 to seal the leakage but unexpectedly large movement during grouting revealed that the embankment at the junction was far less stable than expected. This led to an emergency drawdown of the reservoir. Once the extent of the defects at the junction was understood, a major remedial works program was implemented including construction of downstream buttress, anchoring of the southern training wall, installation of a plastic concrete and soil bentonite cutoff wall upstream and installation of a row of filter columns adjacent to the distressed junction area.

Subsequent risk assessments revealed that the piping risk remained too high, so the filter column treatment was expanded in 2011. A major concrete buttress was also added in 2013 to address seismic concerns with the Southern training wall. In spite of the overall dam safety concerns, the concrete core wall has generally performed well over it's nearly 100-year life, except for the inherent weakness of a rigid barrier within a deformable poorly compacted embankment connecting to the rigid spillway structure. Manhole inspection along the full depth of the core wall reveals that the concrete is in excellent condition.

Photo 10. Core Wall and South Embankment Construction at Hume Dam, Australia, Early 1930's

Tieton Dam, Washington USA

Tieton Dam, part of the Yakima Storage Project in Washington, was completed by the Bureau of Reclamation in 1924. The embankment section includes both a central concrete core wall and puddle clay core upstream of the core wall. The surrounding embankment is comprised of sands, gravels, cobbles and boulders, compacted and graded with water jets. The downstream face is protected with a thick zone of dumped rockfill.

The core wall was comprised of unreinforced mass concrete below original ground, and was a 1:100 battered reinforced concrete wall above ground. The puddle clay was sluiced and ponded against the core wall and was comprised of sandy silt with some clay. The 5-ft thick core wall was embedded in the bedrock at least 10 ft. The maximum height of the core wall was 319 ft as shown in Figure 5, rivaling the cutoff wall at Wolf Creek Dam in Kentucky built 50 years later.

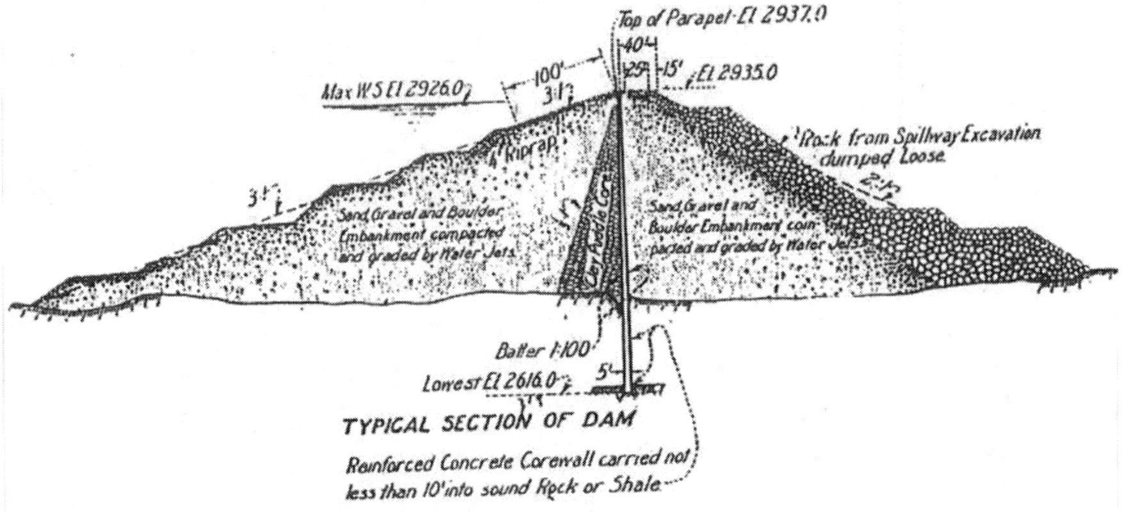

Figure 5. Tieton Dam Section (Courtesy USBR)

The only major issue with the dam was observed deterioration of the upper 5-ft portion of the concrete core wall probably due to freeze-thaw cycles on the non air-entrained concrete. The concrete was repaired by excavation, saw cut removal and replacement in 2001. The core wall was observed to have deflected 1.8 ft downstream over the years with no detrimental effect on cracking or seepage. Inspection wells revealed no significant cracking or deterioration of the concrete with depth.

Photo 11. Tieton Dam (Kollgaard, 1998)

Several other USBR Yakima river dams have concrete core walls for foundation cut offs.

DEFORMATION PERFORMANCE OF CONCRETE CORE WALLS

The deformation performance of a core wall is dominated by the large differential water pressures across the wall and tendency for long term deflection due to the poorly compacted nature of the downstream fill zone that typifies the era of these dams. There are some cases in Australia where the core wall has undergone very large deflections (up to 3 ft of downstream deflection in a couple of the Victorian dams). Lake Endeavour Dam in NSW (Table 1) has a banana shape along the top of the core wall.

However, concrete core walls within embankment dams can sustain relatively large deformation without being damaged if they are reinforced. Several case histories have demonstrated this resilient behavior. The almost 80 years of performance of Sorpe Dam in Germany, even after being bombed by the Allies attests to this resilience. Decades of good performance of permanent reinforced concrete diaphragm slurry walls for basement excavations confirm this excellent long term behavior. However, at the interface between the deformable embankment and a rigid spillway structure or rock abutment there can be issues with incompatible deformation behavior and cracking. Un-reinforced core walls can be more problematic with cracking within poorly compacted embankments. These cracks have led to excessive leakage and in some cases seepage and piping failures.

SEEPAGE PERFORMANCE OF CONCRETE CORE WALLS

Concrete core walls provide an effective seepage control measure by virtue of their low permeability and high erosion resistance. Problems have occurred with walls that do not extend sufficiently into the dam abutment rock or foundation rock. In this case high seepage gradients can develop at the edge or base of the wall. If the wall is founded in erodible soils then this can cause internal erosion at those locations. When cracks develop in the wall, the concrete provides erosion resistance to prevent crack enlargement. At Tourist Park an additional defensive design feature was added downstream of the wall with a chimney filter which provided piping protection for the highly erodible embankment backfill. A similar design philosophy was implemented twenty years before on the Manasquan Dam in New Jersey where a chimney filter was provided downstream of the soil bentonite cutoff wall.

OVERTOPPING PERFORMANCE OF CONCRETE CORE WALLS

The excellent performance of the Tourist Park, Baglihar and Hebgen Lake dams have demonstrated the excellent overtopping resistance of concrete core walls. The overtopping failure a Lake Delhi was slowed by the core wall. Reinforcement allows the wall to remain free standing to a limited height when portions of the embankment fill have been removed by overtopping erosion.

However, there are cases of unreinforced concrete core walls that have failed during overtopping events such as Lower Otay and Dells Dams in Table 1, among others cited in Middlebrooks (1953).

Photo 12. Exposed Core Wall at Tourist Park Dam

LONG TERM PERFORMANCE OF CONCRETE IN THE CORE WALLS

Nearly all of the concrete core walls were constructed prior to the development of air-entrainment in the mid 1930's which provides enhanced freeze-thaw resistance of the exposed concrete. However, because the concrete wall is protected from weathering on both sides by earthfill, its durability has been very good. The only exceptions have been near the dam crest where the concrete is exposed.

Unless the core wall is reinforced to help control crack width and provide some resistance to bending, the concrete has invariably cracked due to a combination of drying and thermal related shrinkage. In a number of cases, cracks have developed to sufficient width to pass water, thus reducing the effectiveness of the core wall as a seepage barrier.

Where the core walls have been cored in recent years, the strength of the concrete has generally showed excellent long term strength gain and little degredation in properties.

SUMMARY OF DAM SAFETY LESSONS

Reinforced concrete core walls within soil embankment dams have generally performed well over many decades. They can sustain some amount of overtopping and have a high resistance to erosion. The history of performance of concrete core walls in rockfill dams is more problematic when incompatible puddle clay cores were added to the section for seepage control redundancy. The filter incompatibility can lead to loss of material upstream during drawdown events as was the case for Sugarloaf dam in Australia. Adding filter zone protection as is done with concrete face rockfill dams is a prudent design approach.

Concrete core walls fell out of favor because of cost and difficulties in building the high walls required for modern dams. There was also concerns about placing a relatively rigid seepage control element within a deformable embankment. The key question is how much deformation can a concrete core wall sustain without losing its structural integrity. Core wall dams have even sustained large earthquakes with relatively minor damage. For those dams that survived the early years, they have in fact performed well over the decades and there is no reason to suspect they will deteriorate any more than any other concrete dam. Surficial deterioration in an exposed environment can be a concern as was the case at Tieton dam in Washington.

LESSONS FOR FUTURE CUTOFF WALLS

Most modern cutoff walls for dams are constructed without reinforcement. For embankments not expected to undergo significant deformation this is probably not a concern. But, for deforming embankments or those that could be overtopped such as temporary cofferdams, reinforcement should be considered in the design. It does add cost but provides significant ductility and resiliency to the wall.

CONCLUSIONS

Concrete core walls within embankment dams is a forgotten dam building technology. However, those dams built in this way in the early twentieth century that survived the early years of operation have generally performed well. The technology was overcome by significant improvements in embankment compaction techniques as dam heights rose during the last century.

- Concrete core wall embankments do provide specific advantages over other dam technologies;
- Core walls have a very low permeability and are highly resistant to erosion;
- Core walls can be overtopped to some extent without losing their integrity;
- Reinforced concrete core walls can sustain significant deformation and some loss of embankment section; and
- Providing a filter zone downstream is a prudent defensive design feature.

Risks with concrete core wall embankment dams are embankment deformation-induced cracking that can lead to seepage problems and limited ability to sustain overtopping erosion of the surrounding embankment. The practice of stopping the core wall well below crest level is an inherent flaw during extreme flood events because differential settlement-induced cracking above and adjacent to the rigid wall in the poorly compacted earthfill can lead to internal erosion and piping.

Although embankment core wall dams are unlikely to enjoy a resurgence in popularity, they do provide some important precedent for modern dam cutoff walls. Perhaps the added resiliency of reinforcement will be considered more seriously for future dam cutoff walls.

Many embankments with concrete core walls are approaching a century of age. Aside from surface concrete deterioration and deformation-induced cracking, there appears to be no reason to expect these dams to not continue to perform adequately.

REFERENCES

Cole, B., Editor (2000) *Dam Technology in Australia 1850 – 1999,* ANCOLD.

Du, Chongjiang (2005) "Arch action in concrete dams with straight axis", Dam Engineering Vol. XX, Issue 2.

Fieldler, W, W King, N. Schwanz, J. Garton and L. McDaniel (2011) "Dam Safety: What Happened to Lake Delhi Dam", Journal of Dam Safety, Issue 9.2 Summer, ASDSO

Kollgaard, E. B. and W. L. Chadwick, Editors (1988) *Developments of Dam Engineering in the United States,* Pergamon Press, New York.

Middlebrooks, T. A. (1953) "Earth Dam Practice in the United States", Transactions, ASCE.

Paul, D. B., R. Davidson and N. Cavalli, Editors (1991) *Slurry Walls Design Construction and Quality Control,* ASTM STP 1129

Rice, J, J. M. Duncan and R. Davidson (2007) "Identification of Failure Mechanisms Associated with Seepage Barriers in Dams", ASCE GeoDenver 2007, Denver Colorado February

Sherard, J, R. J. Woodward, S. F. Giziensky and W. A. Clevenger (1963) *Earth and Earth-Rock Dams*, New York.

Thomas, H. H. (1976) *The Engineering of Large Dams*, John Wiley, New York.

URS Australia (2000) Waranga Basin Design Review, Report prepared for Goulburn-Murray Water, October.

USSD (2013) *Dams of the United States, A Pictorial Display of Landmark Dams.*

Yagin, V. P., I. A. Davydov, V. V. Mik, and T. V. Leimann (1998) "Earth Dams with Concrete and Reinforced-Concrete Core Walls", Hydromechanical Construction, Vol. 32, No. 2.

RISK ASSESSMENT: BALANCING ACT TO ACHIEVE LONG-TERM STABILITY FOR SAFETY OF AGING DAMS

William J. Friers, P.E.[1]
Kapila S. Pathirage, P.E.[2]
Chris E. Wheland[3]

ABSTRACT

This paper presents the comprehensive approach for assessing the condition, evaluating potential risks, and managing the risks taken by one city to extend the useful life of a 110-year-old dam. The city of Troy, New York (City), owns, operates, and maintains the Tomhannock Reservoir. The reservoir is a top priority, serving as the city's principal water source, providing the community with an average of 17 million gallons of water per day. As with all reservoirs, Tomhannock was constructed to reduce the risks of water shortage. The City understands that its construction, however, created a new risk of dam failure. While this risk is characteristically of low probability, it can be of high consequence, with a potential for life loss in addition to economic and environmental damages and social disruption, thus making the integrity of the dam a top priority. CDM Smith was engaged perform a safety inspection for the Tomhannock Dam in the summer of 2010. The safety inspection found areas of seepage on the downstream slope of the embankment, deterioration of the primary spillway's abutment walls, and apparent leaking in 5-foot-diameter steel low-level outlet conduit running through the embankment. Based on the safety inspection findings and the essential nature of the dam/reservoir the City moved forward, engaging CDM Smith to perform an Engineering Assessment (EA) of the dam. Recommendations coming out of the EA, based on geotechnical investigations and analyses, included construction of an earth berm on the downstream slope of the embankment, ground modification of weak soils along the toe of the dam, and rehabilitation of the leaking low-level outlet conduit, situated near the center of the embankment. All recommendations were accepted by the City. CDM Smith's rehabilitation design for the downstream embankment slope included construction of a buttress berm and shallow soil mixing to stabilize the toe of the embankment and structural refurbishment of the damaged low-level outlet conduit using a centrifugally cast, ultra-high-strength cementitious liner system. The spillway's left and right abutment walls were replaced with new reinforced concrete retaining walls. Construction was completed in November 2013.

---

[1] William J. Friers, P.E. CDM Smith, 11 British American Boulevard, Suite 200, Latham, NY 12110, 518.782.4513, frierswj@cdmsmith.com

[2] Kapila S. Pathirage, P.E. CDM Smith, Raritan Plaza I, 110 Fieldcrest Avenue, 6th Floor, Edison, NJ 08837, 732.590.4625, pathirageks@cdmsmith.com

[3] Chris Wheland. Department of Public Utilities, City of Troy, NY, 25 Water Plant Road, Troy, NY 12182, 518.237.0865, Chris.Wheland@troyny.gov

INTRODUCTION

Concerned with the public safety of New York dams, most of which are more than 50 years old, New York State Department of Environmental Conservation (NYSDEC) updated its regulations in August 2009 to focus on maintaining these structures in a safe manner and to enhance its role in helping owners remain in compliance. NYSDEC's amended regulations, 6 NYCRR Part 673, mandate periodic Engineering Assessments for all Intermediate and High Hazard Dams. The city of Troy, New York (City), owns, operates, and maintains nine dams; NYSDEC classifies the city's Tomhannock Reservoir Dam, Wright Lake Dam, and Bradley Lake Dam as Class C High-Hazard structures. Tomhannock Reservoir serves as the city's principal water source, providing the community with an average of 17 million gallons of water per day. The city's Department of Public Utilities (DPU) is responsible for the operation and maintenance of Tomhannock Reservoir and Dam. Tomhannock Reservoir, a man-made reservoir 6 ½ miles northeast of the city is 5 ½ miles long and holds 12.3 billion gallons when full. While the risk of failure is characteristically of low probability, a failure of the Tomhannock Reservoir Dam may result in significant damage to the houses and structures downstream of the dam with potential for personal injury or loss of life. Failure of the Tomhannock Reservoir Dam would also eliminate the City's water supply, causing economic hardship; putting the city's fire-fighting capabilities at risk; and creating overall social disruption. Accordingly, the City has made the integrity of the dam a top priority.

This paper presents the comprehensive approach for assessing the condition, evaluating potential risks, and managing the risks taken by the City to extend the useful life of the 110-year-old Tomhannock Reservoir Dam.

BACKGROUND

History

"By an act of legislature, passed on March 18, 1791, the town of Troy was erected from a part of the town of Rensselaerswyck. On Monday, April 4 a town meeting was held at Ashley's tavern, and the first town officers elected"[4]. An act incorporating the Troy Water Works Commission (Commission) was passed on April 18, 1829. In 1833, the Commission initiated the construction of a series of reservoirs on the Piscawen Kill. The aggregate capacity of the several reservoirs was slightly more than 1 million gallons. In 1843 and 1853, other reservoirs were constructed along the Piscawen Kill. Periodic shortages starting around 1860 made it necessary to supplement supply by pumping water from the Hudson River. Several additional reservoirs were constructed between 1868 and 1880 to meet the city's growing demand for water (Figure 1). Periodic shortages continued and in 1897 a study was authorized by the Commission to investigate alternatives to address the long-term water demands of the city. The study recommended the Tomhannock Creek be developed to supply to Commission's service area. In 1899 the Hudson River water was found unfit for human consumption. On December 31, 1899 the Commission's duties were transferred to the newly created Department of Public

[4] Arthur James Weise, The City of Troy and its vicinity (Astor, Lennox and Tilden 1886), 314

Utilities. Construction of the Tomhannock Reservoir Dam (Dam) began in 1900, concluding in 1905 as an earth-filled dam with a vertical concrete core wall along the centerline of the dam.

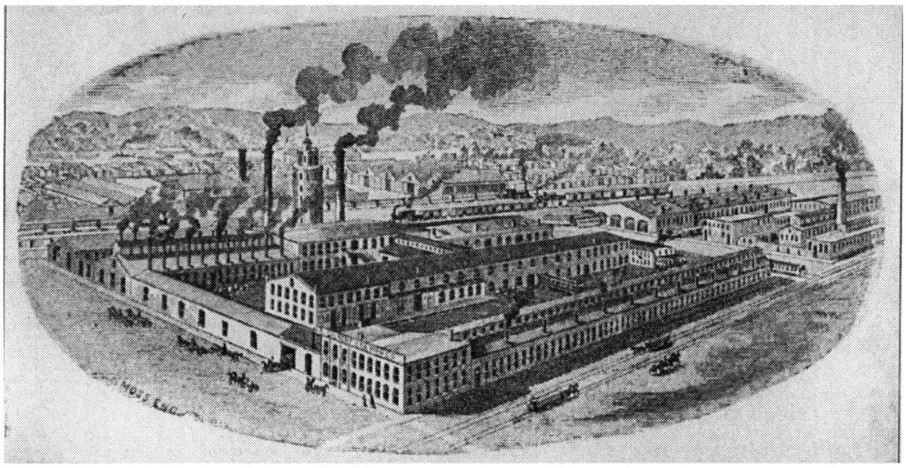

Figure 1. Troy NY, View of Gilbert Car Manufacturing Company. Circa 1890

Today the Tomhannock Reservoir serves as the city's principal water source, providing the community (Figure 2) with an average of 17 million gallons of water per day.

Figure 2. Troy, NY (2014)

Description of Project

A low-level outlet conduit consisting of a 5-foot-diameter steel pipe encased in concrete is situated near the center of the embankment. A gatehouse at the top of the embankment (upper gatehouse) controls three 1'-6" x 4'-6" sluice gates at the entrance to the outlet. A second gatehouse located at the toe of the downstream slope controls the outlet of the drain line through four 30-inch-diameter gate valves that are manifolded into the drain line. The principal spillway (Figure 5) is located approximately 1,000 feet south of the main embankment. The spillway consists of a 300-foot-long, broad-crested weir with an ogee-shaped spillway face (upper spillway). This spillway is approximately seven feet high and discharges to a concrete apron below followed by a second ogee-shaped spillway (lower spillway) with a height of approximately 11 feet. The water intake for the City's water system is located remotely from the dam and spillway site and has no effect on dam safety.

Figure 5. Principal Spillway (2010)

REGULATORY COMPLIANCE

NYSDEC updated its regulations in August 2009 to enhance its role in helping owners remain in compliance with dam safety requirements while emphasizing maintenance and safe operation of dams. The revised NYCRR Part 673 regulations state that NY dam owners must:

- Prepare and implement an Inspection and Maintenance (I&M) Plan by August 19, 2010;

- Have an Emergency Action Plan (EAP) prepared and submitted to NYSDEC by August 19, 2010 for Class C (High-Hazard) dams, and August 19, 2011 for Class B (Intermediate-Hazard) dams;

- Submit an Annual Certification to NYSDEC by January 31 of each year, beginning in 2011;

- Engage a NY-licensed Professional Engineer to conduct safety inspections at regular intervals (every 2 years for Class C dams and every 4 years for Class B dams); and

- Engage a NY-licensed Professional Engineer to conduct an Engineering Assessment by August 19, 2012, for large Class C dams; August 19, 2014, for small Class C dams; and August 19, 2015, for Class B dams.

DPU engaged CDM Smith in 2010 to support their efforts to comply with the newly promulgated NYSDEC regulations. Work performed included preparation of an I&M Plan and EAP for the City's high-hazard dams including Wright and Bradley Lake Dams and the Tomhannock Reservoir Dam, and the inspection of the three dams. I&M Plans were prepared by CDM Smith based upon interviews with DPU Chief Water Plant Operator, Neil Bonesteel, P.E. and Assistant Operations Manager, Chris Wheland. The Plans were completed in advance of the NYSDEC August 2010 deadline. CDM Smith then performed the dam-break analyses and inundation mapping for the EAP's and the Safety Inspection Reports, for the City's August 2011 submittal to NYSDEC.

The Tomhannock Reservoir Dam was inspected May, 2010 by CDM Smith engineers, accompanied by Mr. Wheland. The purpose of the inspection was to observe and evaluate the present condition of the dam and appurtenant structures in accordance with 6 NYCRR Part 673 to provide information that will assist in both prioritizing dam repair needs, if any, and planning/conducting maintenance and operation. Underwater areas were not inspected. The condition of the dam was judged by CDM Smith in accordance with NYSDEC regulations and was found to be in a "Deficiently Maintained" Condition. The safety inspection found seepage and wet areas on the downstream slope of the embankment and deterioration of the primary spillway's abutment walls. In addition, the 5-foot-diameter steel conduit running through the embankment was apparently leaking. Other deficiencies included:

1. Heavy vegetation and small trees were observed on the upstream slope. Riprap armoring of the upstream slope had been eroded away in some locations.
2. Heavy vegetation and several woodchuck burrows were observed on the downstream slope.
3. Only two of the three inlet control gates located at the upper gate house functioned properly. One of the four 30-inch valves located at the lower gatehouse was leaking.
4. Concrete at construction joints on the upper and lower spillways was eroded and deteriorated from around the joints.
5. Deterioration and cracks were observed on both spillway abutment walls. Significant horizontal deflection was observed in the right abutment wall, and significant erosion was occurring at toe of the left abutment wall.
6. The surface of the concrete apron between the stilling basin of the upper spillway, and the crest of the lower spillway was severely deteriorated and vegetation/trees were observed.
7. Severe erosion was observed at the toe of the lower spillway. The erodible spillway channel section was covered with trees and heavy growth of vegetation.

After the safety inspection, CDM Smith engineers met with DPU personnel to discuss the assessed condition of the dam and review recommendations contained in the Safety Inspection Report. DPU responded in a proactive manner. They developed a forward-looking and aggressive plan to address the identified deficiencies and investigate and evaluate the condition of the dam.

MANAGING RISKS

DPU's plan to improve the condition of the dam was well thought out; founded on logic and practicality. Resources were available to perform vegetation and rodent control on the downstream slope of the dam. Accordingly, DPU cleared the dense vegetation (Figure 6) from the embankment slopes, and removed the woodchucks and backfilled burrows. When the work was completed, they had achieved their first risk-reduction undertaking.

Figure 6. Removal of Vegetation from Downstream Slope (2011)

Then, interested in investigating the nature of the observed seepage and knowing the dam's record drawings provided insufficient detail for an engineering assessment, DPU authorized CDM Smith to complete an Engineering Assessment (EA) of the dam in early 2011—nearly a year before regulations required. Their logic: you need to know what level you're dealing with, before you can begin to move forward to reduce your exposure to risk.

A detailed hydrologic and hydraulic analysis was performed to determine the appropriate spillway design flood (SDF) for the Tomhannock Reservoir. According to the NYSDEC regulations, the design storm event is the one-half Probable Maximum Flood (1/2 PMF) event. The contributing watershed area was determined to be 67.0 square miles including the 1,740-acre normal reservoir water surface. The results from the analysis show that the peak ½ PMF inflow into the Tomhannock Reservoir would be 42,500 cfs and that the routed SDF discharge would be 36,000 cfs. This analysis is for an initial reservoir level at the spillway crest elevation (El. 390.4) at the start of the ½ PMF analyses. The corresponding maximum reservoir level is El. 401.7, 0.3 feet below the crest of the dam.

The assessment of the Tomhannock Reservoir Dam, spillway, and appurtenant works was based upon a dam safety inspections performed by CDM Smith; review of previous reports, the Owner's and NYSDECs project files, and available drawings; results of a subsurface exploration program (Figure 7), including test borings, field and laboratory testing; and engineering analyses of the structures.

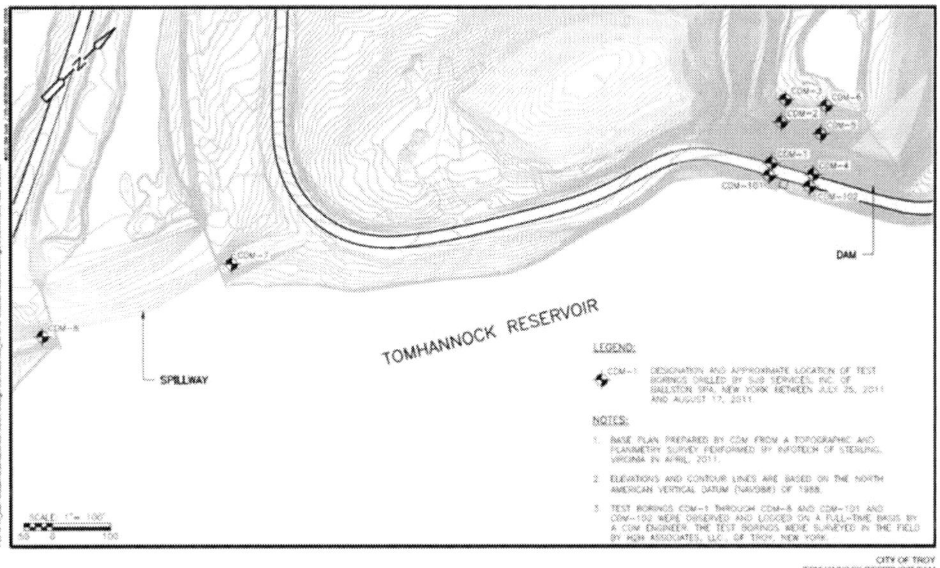

Figure 7. Subsurface Exploration Location Plan

The test borings were advanced using 4 ¼ -inch-inside-diameter hollow stem continuous flight augers (HSA) to depths between 25.0 and 82.0 feet below the existing ground surface. Split spoon samples were collected continuously for the first 16 feet in borings CDM-3, CDM-6, CDM-7 and CDM-8, and for the first 20 feet in borings CDM-1, CDM-2, CDM-4, CDM-5, CDM-101 and CDM-102. Subsurface soil conditions were interpreted from the test borings conducted as part of the study along with CDM Smith's understanding of the local geology. Test borings generally encountered fill underlain by silt and clay, glacial till, and bedrock (Table 1). A notable exception was found at the toe of the dam. Boring logs for CDM 3 and CDM 6 recorded blow-counts ranging from "weight-of-hammer" (WOH) to 2, between El. 341' and 332'.

Table 1. Summary of Subsurface Explorations

Boring No.	Location	Approx. Ground Surface	Exploration Depth (ft)	Stratum Thickness (ft)					
				Topsoil	Fill	Silt & Clay	Sand & Gravel	Clay & Silt Fill	Till
CDM-1	Downstream Crest Edge	402.0	77.0	2.0	60.5	NE	NE	NE	>14.5
CDM-2	Downstream Mid-Slope	364.3	47.0	NE	32.5	NE	NE	NE	>14.5
CDM-3	Downstream Toe	346.1	30.0	0.3	11.7	NE	NE	NE	>18
CDM-4	Downstream Crest Edge	401.1	77.0	0.3	67.2	NE	NE	NE	>9.5
CDM-5	Downstream Mid-Slope	365.0	47.0	0.3	32.2	NE	NE	NE	>14.5
CDM-6	Downstream Toe	341.8	25.0	0.3	10.7	NE	NE	NE	>14.0
CDM-7	Spillway Abutment	402.2	65.0	0.3	15.6				>12.5
CDM-8	Spillway Abutment	400.7	43.5	0.3	17.2	9.0	NE	NE	12.0

Representative soil samples from each split spoon were collected and stored in jars for subsequent review and laboratory testing (Table 2). Moisture content tests, grain size analyses, hydrometer analyses, Atterberg limits tests, and specific gravity tests were performed at the SJB Services, Inc. Testing Laboratory in Ballston Spa, New York on soil samples collected in the test borings. Consolidated undrained (CU) triaxial tests were performed at the SJB Services, Inc. Testing Laboratory in Buffalo, New York on undisturbed soil samples collected in the test borings.

Table 2. Summary of Geotechnical Laboratory Index Test Results

Subsurface Exploration No.	Sample No.	Sample Depth	Moisture Content (%)	Stratum	Grain Size Analyses (%)					Specific Gravity	USCS Class.
					Gravel	Sand	Silt	Clay	Fines		
CDM-1	S-2	2-4	-	Fill	15.5	69.8	10.5	4.2	14.7	-	SM
CDM-1	S-9	16-18	-	Fill	-	-	-	-	-	2.35	-
CDM-1	S-13	30-32	9.5	Fill	26.4	52.0	21.6		21.6	-	SM
CDM-1	S-17	50-52	-	Fill	43.3	29.5	21.0	6.2	27.2	-	GM
CDM-1	S-20	65-67	-	Till	34.7	32.4	22.1	10.8	32.9	2.26	-
CDM-2	S-3	4-6	-	Fill	17.5	46.7	22.7	13.1	35.8	-	SM
CDM-2	S-15	40-42	-	Till	32.0	38.1	16.2	13.7	29.9	2.58	-
CDM-3	S-4	6-8	-	Fill	33.0	36.5	20.4	10.1	30.5	-	SM
CDM-3	S-5	8-10	-	Fill	39.9	33.7	20.8	5.6	26.4	-	GM
CDM-3	S-8	14-16	-	Till	37.4	41.4	14.4	6.8	21.2	-	SM
CDM-4	S-3	4-6	-	Fill	38.7	41.8	12.1	7.4	19.5	-	SM
CDM-4	S-11	20-22	-	Fill	-	-	-	-	-	2.52	-
CDM-4	S-14	35-37	-	Fill	30.6	42.7	19.1	7.6	26.7	-	SM
CDM-4	S-21	70-72	-	Till	28.9	51.4	13.4	6.3	19.7	-	SM

The seepage and slope stability analyses were conducted using finite element software on the existing geometry of the dam and soil properties evaluated from the data collected from the subsurface exploration and laboratory testing programs. The spillway stability evaluation was performed using the existing spillway cross-section obtained from as-built drawings and recent and historical test boring information. The hydrologic and hydraulic analyses were conducted using stage-storage curve and precipitation data to evaluate the various reservoir pool elevations and volumes. The water loading conditions as evaluated from the hydrologic and hydraulic analyses were used in the seepage and slope stability analyses of the dam.

Seepage and stability analyses were performed to evaluate the condition of the existing earthen dam based upon the guidelines outlined in the U. S. Army Corps of Engineers engineering manual EM 1110-2-1902, Slope Stability, as prescribed by NYSDEC Guidelines for Design of Dams, paragraph 9.2. Design cross-sections (Figure 8) were developed for the earthen dam using the information obtained from the subsurface investigations. The overall geometry of the dam for the design cross-sections was based historical drawings and confirmed by survey data obtained from recent aerial surveys performed by the City under a related project.

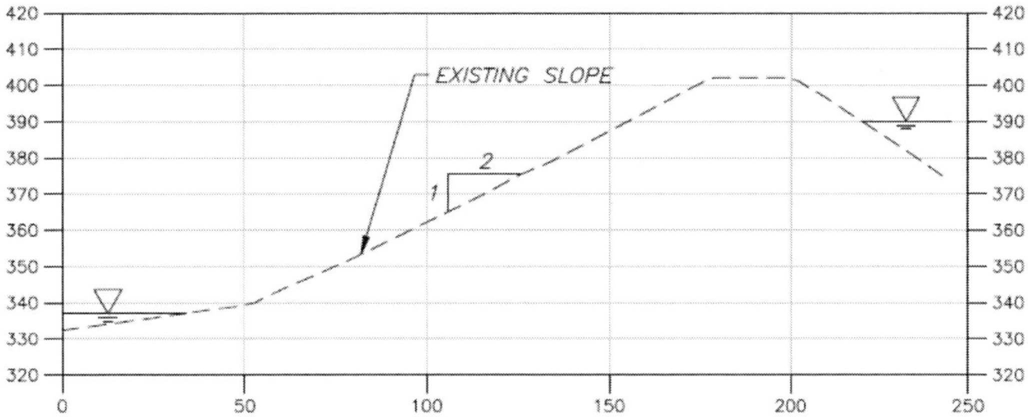

Figure 8. Embankment Dam Cross-Section

The results of the analyses showed the existing downstream slope did not meet the minimum factor of safety requirements for maximum storage pool and maximum surcharge pool steady-state conditions (Figure 9).

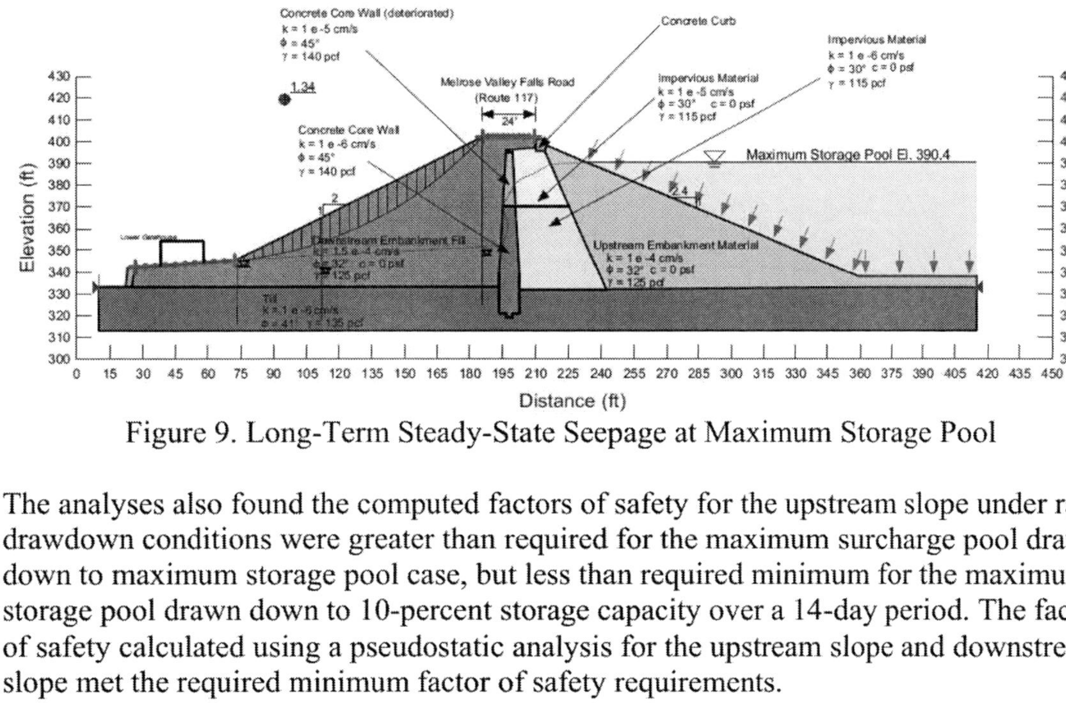

Figure 9. Long-Term Steady-State Seepage at Maximum Storage Pool

The analyses also found the computed factors of safety for the upstream slope under rapid drawdown conditions were greater than required for the maximum surcharge pool drawn down to maximum storage pool case, but less than required minimum for the maximum storage pool drawn down to 10-percent storage capacity over a 14-day period. The factors of safety calculated using a pseudostatic analysis for the upstream slope and downstream slope met the required minimum factor of safety requirements.

Stability analyses of the spillway were performed for the four loading conditions outlined in NYSDEC Division of Water – Guidelines for Design of Dams, dated January 1985 and revised January 1989. The design cross-section (Figure 10) was developed for the spillway using the information from drawings provided by DPU. The water loading conditions as evaluated from the hydrologic and hydraulic analyses were used in the stability analyses of the concrete spillway. The calculations analyzed the structure for overturning, sliding, concrete stresses, and base pressures for the four loading conditions. Loading conditions considered included:

- Case 1, Normal Pool;
- Case 2, Normal Pool plus an ice load of 5,000 pounds per linear foot;
- Case 3, Spillway Design Flood; and
- Case 4, Seismic

Table 3. Summary of Spillway Stability Analyses

Loading Condition	ΣFV (k/ft)	ΣFH (k/ft)	ΣMR (k-ft/ft)	ΣMO (k-ft/ft)	ΣMR / ΣMO	Location from Toe	Sliding Factor of Safety	
							Calculated	Required
Case 1, Normal Pool	31.2	8.8	1136.8	446.3	2.55	22.1	2.05	1.50
Case 2, Normal Pool with Ice	31.2	13.8	1136.8	498.3	2.28	20.1	1.31	1.25
Case 3, Spillway Design Flood	24.2	7.0	1510.4	995.3	1.51	19.7	1.37	1.25
Case 4, Seismic	31.2	14.2	1136.8	476.8	2.38	21.1	1.30	1.00

The computed factors of safety (Table 3) indicated the concrete spillway structure was stable under all required loading conditions based on NYSDEC criteria.

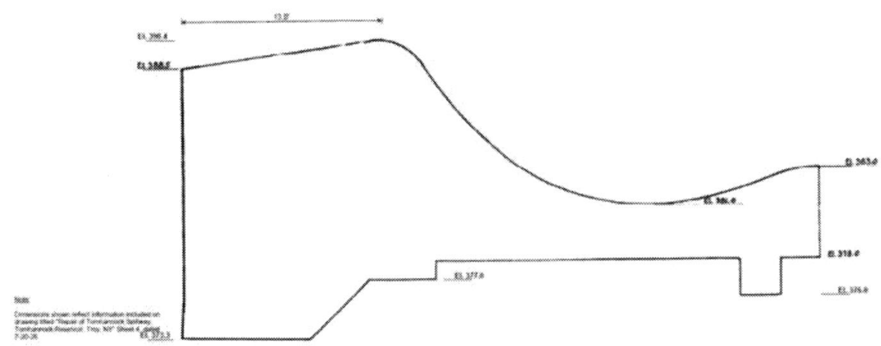

Figure 10. Spillway Cross-Section

RISK MANAGEMENT/REHABILITATION

The Engineering Assessment identified a scope of work to bring the Tomhannock Reservoir Dam into compliance with NYSDEC regulatory requirements. Recommended remedial work, requiring engineering support, included:

- Increase the factor of safety of downstream slope of the embankment under maximum storage pool and maximum surcharge pool steady-state conditions
- Remediate seepage on the downstream slope of the embankment
- Remediate the "soft soil condition" at the toe of the dam, between El 341' and 332'.

- Increase the low-level outlet capacity to meet NYSDEC emergency draw-down requirements.
- Repair leak(s) in the 5-foot-diameter steel conduit running through the embankment.
- Increase the factor of safety for the upstream slope under rapid drawdown conditions.
- Remove and reconstruct the spillway's left and right abutment walls.

The Engineering Assessment provided DPU the information they needed to begin efforts to reduce your exposure to risk. After consultation with CDM Smith engineers, DPU developed a logical approach to address the observed conditions, with due consideration given to resource availability. The first priority was the 5-foot-diameter conduit—to relieve internal pipe pressures, gate valves were opened at the downstream outlet, then a qualified diver was brought in to close the inoperable valve. This effort resulted in significant reduction of seepage and reduction of associated risks.

In August 2012, DPU authorized CDM Smith to perform the design and develop construction documents for the rehabilitation of the Tomhannock Reservoir Dam. The authorized scope of work included engineering services for the above recommended remedial work. DPU committed to a future study of alternatives to increase the dam's low-level outlet (LLO) capacity to meet NYSDEC emergency draw-down requirements and elected to delay work that would address recommendations to increase the factor of safety for the upstream slope under rapid drawdown conditions. DPU's decision was based on their evaluation of the risks associated with the LLO capacity and the upstream slope stability. Remediation of the downstream slope; the soft soil condition at the toe of the dam; repair of the 5-foot diameter pipe brings the dam; and replacing the spillway abutment walls address potential failure modes (PFM) and brings the dam into compliance with NYSDEC regulations. With the PFM's addressed and increased factors of safety (compliant with DEC regulations), the projected need for emergency draw-down capacity is significantly reduced. The reduced likelihood of an emergency reservoir drawdown also reduces the potential for upstream slope failure that can only occur during emergency drawdown.

CDM Smith began design of the Tomhannock Reservoir Dam in 2012. Rehabilitation of the downstream embankment slope included construction of a buttress berm and shallow soil mixing to stabilize the toe of the embankment (Figure 11). Plans and specifications were completed in December 2012 and the work was bid in January 2013. Construction began in June 2013, after spring runoff. Construction was completed November 2013 (Figure 12).

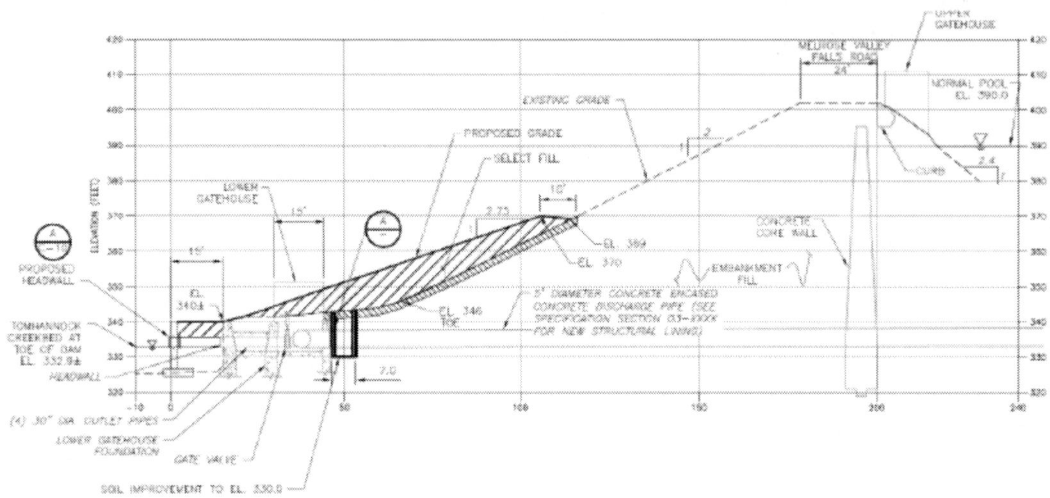

Figure 11. Embankment Stabilization - Buttress Berm

Figure 12. Downstream Slope Stabilization - Buttress Berm (2014)

CONCLUSION

Whether a dam was constructed in 1904 or 2004, failure or misoperation could threaten public safety, and local or regional economics. Assuring the safety of dams is a cooperative effort between the owner, consultant, and regulating agency with the most important role being that of the owner. The owner is the one who sees the dam regularly and through surveillance and monitoring is monitoring and evaluating the health of the structure.

The use of risk analysis to evaluate proposals for the major rehabilitation of the Tomhannock Reservoir Dam allowed the City of Troy, NY to implement a rehabilitation program that addressed dam safety issues while adhering to a finite budget. Risk analysis requires adequate information to evaluate alternatives. If an Owner has a detailed engineering analysis with recommendations for rehabilitation, and practical insight into reservoir operations, he can, based on risk analysis, establish a prioritized plan of remedial action with a level of confidence.

Owners of dams should budget for costs of operating and maintaining their dams. The City of Troy's DPU was proactive in their management of the Tomhannock Reservoir Dam safety. Lessons learned from this project include:

- Periodic dam safety inspections identify not only deficiencies, but also serve to remind owners of their obligation to properly maintain their dams.
- DPU was committed to dam safety and worked in cooperation with CDM Smith and NYSDEC officials to advance the rehabilitation of the Tomhannock Reservoir's dam safety program.
- A proactive dam safety program significantly reduces owner risks/liability and leads to an increased ability to sleep at night.

PRAGMATIC DESIGN OF AN EMBANKMENT DAM: BURY DEADMEN AND FORGET ABOUT IT (FORGETABOUTIT)

Michael S. Quinn[1]
Gary Dale[2]

ABSTRACT

Investigation of Taylor Pond Dam located within New York's Adirondack Park found the embankment upstream and downstream of a concrete corewall was comprised of vibration sensitive loose sands raising the concern for liquefaction during the design earthquake. A liquefaction analysis was performed for the embankment using the procedures outlined in the 2005 publication, "Earthquake Analyses and Design of Dams" (Federal Emergency Management Agency - FEMA), and the 2008 publication, "Soil Liquefaction during Earthquakes" (Idriss & Boulanger).

The analysis involved evaluating the embankment and foundation to determine if the design earthquake and associated peak ground acceleration (PGA) will induce cyclic shear stresses that exceed the in-situ cyclic resistance of the dam embankment soils, potentially triggering liquefaction. The embankment soil strengths were estimated for the potentially liquefiable zones using SPT/CPT data, and residual shear strength equations developed by Idriss & Boulanger. Residual soil strength properties were assigned to defined stratigraphic zones within critical sections to perform a post-seismic stability analysis. The analysis concluded that downstream embankment stability is maintained while the upstream embankment would undergo significant sloughing failure. The upstream sloughing alone would not breach the existing dam, however when combined with raising the dam crest, which was required to prevent overtopping during the spillway design flood (SDF), a mode of failure was identified where increased lateral earth pressures would topple the corewall.

The need for raising the concrete corewall and downstream embankment to accommodate the SDF without overtopping the dam resulted in embankment stability conditions that required a dam modification. Following a potential upstream embankment slough during the design earthquake, the design solution was to construct an anchor (dead-men) system to stabilize a cantilevered corewall height of approximately 12.5 feet, which will support the downstream embankment. The selection of a deadman system to provide seismic stability of the embankment at a new crest height simplified the dam rehabilitation, and significantly reduced construction costs relative to alternatives that had included spillway expansion and ground improvement methods.

[1] Senior Dam Safety and Geotechnical Engineer – CHA Consulting, Inc., mquinn@chacompanies.com
[2] Senior Dam Safety and Geotechnical Engineer - CHA Consulting, Inc., gdale@chacompanies.com

SITE DESCRIPTION

The Taylor Pond Dam is an earthen fill dam with a concrete corewall built in 1924 that impounds Taylor Pond (normal pool volume of 31,242 acre-feet) and is owned by New York State Department of Environmental Conservation (NYSDEC). The dam is approximately 400 feet long and 25 feet in height, with downstream slopes of approximately 2H:1V (horizontal to vertical) and upstream slopes on the order of 3H:1V to 3.5H:1V. The upstream face of the dam is armored with rip-rap (from the crest an unknown distance below the normal pool level) to protect against wave erosion. The width of the embankment crest, at approximately elevation 1,394 feet varies from 6 to 9 feet due to shoreline erosion. Prior to rehabilitation, the concrete core wall at several locations along the crest of the dam was exposed revealing a top of wall thickness of 1.5 feet. Based on test pit exploration and construction documentation the core wall has a tapered section that extends to a depth of approximately 38 feet at a batter of approximately 15V:1H (Figure 1).

Figure 1. Taylor Pond Dam – Beginning of Rehabilitation Work

Approximately 12 to 14 feet below the crest of the dam the downstream slope was benched. This bench area was approximately 20 feet wide and 8 to 10 feet above the toe. The benched slope provided the primary access from the left to the right side of the dam. At the downstream toe the dam is approximately 140 feet wide between earthen abutment slopes.

The existing outlet works consisted of a gatehouse set on a concrete riser structure positioned near the center of the dam and extending through embankment to the soil foundation (Figure 2). The gatehouse housed a manual actuator that was used to operate a single 4-foot diameter gate seated against the downstream wall of the riser structure. The configuration provides flow control within a flooded riser structure that is connected to an upstream and downstream 4-foot by 4-foot concrete box culvert. Within the riser structure the invert elevation of the upstream and downstream segments of the box culvert is 1,365.0 feet. Past operations included a stop log system within the riser structure to regulate the pool level at Taylor Pond and to provide access to the slide gate.

Figure 2. Taylor Pond Dam Crest Condition

Base and storm flows from Taylor Pond pass over a concrete outlet weir at elevation 1,390.5 and flow is conveyed through a 40-foot wide primary spillway excavated within the left abutment of the dam (Figure 2 & 3). The first 520 feet of the spillway channel is sloped at 0.5 percent to a channel grade control structure. Beyond the grade control structure the spillway channel steepens to a 6.5 percent grade over a distance of 230 feet discharging into a beaver pond below the dam at elevation 1,373 feet.

Figure 3. Taylor Pond Primary Spillway

LIQUIFACTION AND SEISMIC STABILITY APPROACH

In 2007 an Engineering Evaluation (Phase 2) report assessing the stability of the earthen embankment was completed. The assessment included obtaining standard penetration test (SPT) data from a subsurface investigation, assessing other observed deficiencies and evaluating the hydraulic capacity of the spillway. The primary concern of this initial engineering assessment was the factors of safety for liquefaction and slope stability during the post-seismic event where the computed factors of safety were found to be less than the target factors of safety per the publication "Earthquake Analyses and Design of Dams" (Federal Emergency Management Agency, May 2005). To address liquefaction several ground improvement methods were recommended at the conceptual level prior to the owner entering a planning phase to fund a dam rehabilitation project at Taylor Pond Dam.

In March 2012 the consulting engineer, NYSDEC and New York State Office of General Services (OGS) agreed to an additional subsurface investigation of the embankment using cone penetration tests (CPT) to confirm the results of the 2006 SPT data. The intent of the new CPT data was to provide additional data for stability and liquefaction analysis in an effort to better define the scope of rehabilitation work needed at Taylor Pond Dam. Stratigraphic sections (critical sections) through center and adjacent to the right abutment of the dam were developed to identify and define zones of the embankment or foundation where soils are potentially liquefiable.

A re-evaluation of the liquefaction potential was performed using the CPT data and a probabilistic seismic hazard assessment (PSHA). The PSHA provided an estimate of the earthquake magnitude and the effective peak ground acceleration at the site for an event equivalent to a recurrence interval of 2,500 years. The consulting engineer performed a liquefaction analysis of the Taylor Pond Dam embankment and foundation soils in general accordance with the procedures outlined in the publication "Earthquake Analyses and Design of Dams" (Federal Emergency Management Agency, May 2005), and the publication "Soil Liquefaction During Earthquakes" (Idriss, I.M. & Boulanger, R.W., Earthquake Engineering Research Institute, 2008). The analysis involved evaluating the embankment fill and soil foundation to determine if the selected design earthquake and associated peak ground acceleration (PGA) will induce cyclic shear stresses that exceed the in-situ cyclic resistance of the dam, thereby resulting in potential liquefaction of the embankment or foundation. Once potentially liquefiable zones were defined, the analysis estimated the residual shear strength using CPT data within these zones. CHA then ran a post-seismic stability analysis of critical sections through the dam (Sections A & B Figure 4) with the computed residual shear strength properties and the defined geometry of liquefiable zones.

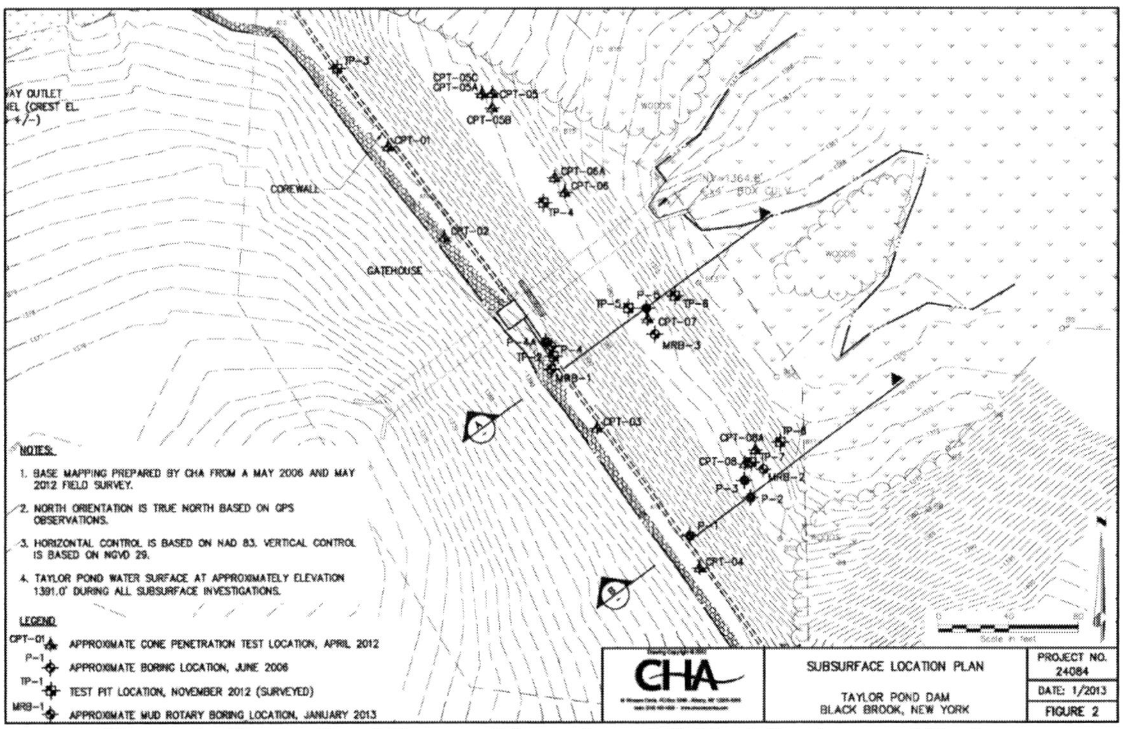

Figure 4. Taylor Pond Dam and Location of Critical Sections and Borings

Following third party and regulatory review questions arising from the very low tip and sleeve resistance values from the CPT investigation, the design team concluded that sufficient uncertainty remained regarding the influence of the concrete corewall and the proximity of grouted boreholes had on the CPT data. This conclusion resulted in the recommendation to complete a supplemental drilling investigation to reduce uncertainties related to the CPT data. The supplemental investigation work consisted of using mud-rotary boring (MRB) methods and advancing three boreholes that would fill SPT data gaps in the 2006 subsurface investigation. The supplemental investigation also provided N-values of the upstream embankment soils using SPT. The supplemental investigation successfully provided the additional data needed to complete the liquefaction evaluation and post-seismic stability analysis along the two critical sections through the dam.

LIQUEFACTION TRIGGERING POTENTIAL

The liquefaction analysis utilized the SPT data (specifically the N-values) obtained from the boring investigations, average soil unit weight (obtained from in-place testing of the dam embankment in test pits) and the phreatic surface across the dam to identify the soils with the potential to liquefy (i.e., sand-like behavior). Soils within the dam embankment and foundation deemed susceptible to liquefaction met the following conditions:

- soils in a loose to very loose state as estimated from SPT values;
- soils with PI value less than 7; and
- soils that are saturated or below the phreatic surface.

The observed loose and very loose silty sand soils above the phreatic surface are not subject to liquefaction. Table 1 provides a summary of uncorrected N-values obtained

Table 1. Summary of N-Values (uncorrected) through Taylor Pond Dam

Depth of Sample (feet)	Crest Boring U/S of Corewall	Data Used to Develop Section B-B' Along Right Abutment: Crest Boring D/S of Corewall	Bench Borings		Data Used to Develop Section A-A' Along LLO Conduit: Crest Boring D/S of Corewall	Bench Borings	
	MRB-1	P-1	P-2	MRB-2	P-4	P-5	MRB-3
1	13	2	4	6	4	3	20
3	8	6	10	14	7		36
5	9	2	11	3	3	13	24
7	2	3	**3**	6	3		16
9	9	2	**1**	13	2		21
11	**1**	4	12	**3**	3	**6**	7
13	**2**	9		48	7		21
15	**2**	3	27	59	34	27	12
17	**3**	**1**			11		14
19	**1**	**1**			4		49
21	**2**	**1**	90	50	3	90	50
23	**7**	8		50	5		35
25	**5**	10			13		50
27	56	4			14		
29	56	Concrete			9		
31	56	Concrete			15		
33	56	Concrete			6		
35	56	50			24		
37	56	50			5		
39	56	50					

Note: N-values in bold and underlined represent zones that are liquefiable during a 2,500-year return frequency seismic event.

during both the 2006 and 2013 subsurface investigations. In addition to the embankment soil data, the PGA at the site was assessed using the USGS Java Ground Motion

Parameter Calculator. An earthquake magnitude was determined using the 2008 PSHA Interactive Deaggregation website for the 0.2 second period (PGA). Both the magnitude and PGA were estimated assuming a 2 percent Probability of Exceedance in 50 Years (2,500-year event) resulting in a mean magnitude of 5.86 and a PGA for a Site Class D of 0.17g. Using a speadsheet analysis, the soil and earthquake data was inputted to compute vertical and effective stresses within the dam embankment which provided the input to compute cyclic stress ratios (CSR) using the equation below.

$$CSR = 0.65(\sigma_{vc}/\sigma'_{vc})*(a_{max}/g)*r_d$$

σ_{vc} = vertical stress at consolidation
σ'_{vc} = vertical effective stress at consolidation
a_{max} = maximum horizontal ground surface acceleration
r_d = shear stress reduction coefficient

Using the N-values from the borehole SPTs, the equivalent clean sand SPT values ($N_{(1)60cs}$) can be estimated to calculate cyclic resistance ratios (CRR) using the equation:

$$CRR = \exp(N_{(1)60cs}/14.1 + (N_{(1)60cs}/126)^{\wedge 2} - (N_{(1)60cs}/23.6)^{\wedge 3} + (N_{(1)60cs}/25.4)^{\wedge 4} - 2.8)$$

$N_{(1)60cs}$ = equivalent clean sand $N_{(1)60}$ for computing CRR

A factor of safety is computed by dividing the CRR by the CSR value to determine the potential for triggering liquefaction. Liquefaction triggering is assumed if the factors of safety are less than 1.1 which follows Idriss & Boulanger. Using these results, the zones of triggered liquefaction within the dam embankment were defined and these zones are assumed to undergo a reduction in shear strength equivalent to the residual shear strength of the embankment and foundation soils.

Table 2. Spreadsheet Analysis Summary for Computation of Liquefaction FOS Based on the Ratio of CRR to CSR

Input Parameters: Borehole MRB-1 Upstream Embankment								
Peak ground accel (g) =				0.173				
Earthquake magnitude, M =				5.86				
Water table depth (ft) =				2				
Phi Angle φ =				32				
Slope angle degrees =				26.6	(2H:1V)			
Sample #	Depth (ft)	Soil Type	N	(N1)60-CS	r_d	CSR	CRR for M=7.5 @ σ_{vc}'=1 atm	Factor of Safety
1	1	Emb. Fill	13	21.6	1.00	0.11	0.237	2.00
2	3	Emb. Fill	8	13.6	0.99	0.14	0.147	1.83
3	5	Emb. Fill	9	16.2	0.99	0.16	0.172	1.81
4	7	Emb. Fill	2	4.3	0.99	0.17	0.067	0.65
5	9	Emb. Fill	9	17.1	0.98	0.18	0.182	1.70
6	11	Emb. Fill	1	2.6	0.98	0.19	0.056	0.49
7	13	Emb. Fill	2	4.1	0.97	0.19	0.065	0.56
8	15	Emb. Fill	2	8.9	0.97	0.19	0.104	0.88
9	17	Emb. Fill	3	5.6	0.96	0.19	0.077	0.64
10	19	Emb. Fill	1	2.4	0.96	0.19	0.055	0.45
11	21	Emb. Fill	2	3.7	0.95	0.19	0.063	0.51
12	23	Emb. Fill	7	10.5	0.95	0.19	0.117	0.95
13	25	Emb. Fill	5	9.3	0.94	0.19	0.107	0.86
14	27	Till	56	NA	0.93	0.19	NA	NA
15	29	Till	56	NA	0.93	0.19	NA	NA
16	31	Till	56	NA	0.92	0.19	NA	NA
17	33	Till	56	NA	0.90	0.19	NA	NA
18	35	Till	56	NA	0.89	0.19	NA	NA
19	37	Till	56	NA	0.88	0.18	NA	NA
20	39	Till	56	NA	0.86	0.18	NA	NA

Table 2 presents the summary of the results from boring MRB-1 located within the upstream embankment and foundation soils. These results show a zone, extending from a depth of 7 feet to a depth of 25 feet, which is subject to liquefaction in the event of the design earthquake. This same analysis was performed for five other borings (i.e., borings P-4, P-5, MRB-3, MRB-2 and P-1) along the two critical sections depicted on Figure 4.

Following the identification of liquefiable zones within the upstream and downstream dam embankment the next step of the analysis consisted of computing residual shear strengths within the liquefiable zones and then using these values to complete a post–earthquake stability analysis.

POST-EARTHQUAKE SLOPE STABILITY

Using the liquefaction-potential results, Idriss & Boulanger presents a computation methodology to compute the residual shear strength for each SPT interval identified as a liquefiable soil. Idriss & Boulanger developed the expression below for computing residual shear strength of liquefiable soils using equivalent clean-sand SPT-corrected N-values.

$$S_r = [\exp(N_{(1)60cs\text{-}Sr}/16 + ((N_{(1)60cs\text{-}Sr} - 16)/21.2)^{\wedge 3} - 3.0) * (1 + \exp(N_{(1)60cs\text{-}Sr}/2.4 - 6.6))] * \sigma'_{vo}$$

S_r = residual shear strength
$N_{(1)60cs\text{-}Sr}$ = equivalent clean sand $N_{(1)60}$ for computing residual shear strength
σ'_{vo} = vertical effective stress overburden stress

Presented graphically, the residual shear strength and vertical effective overburden stress ratio may be estimated from Figure 5. It is important to note that the above equation is represented by the upper curve in Figure 5 where void redistribution effects are expected

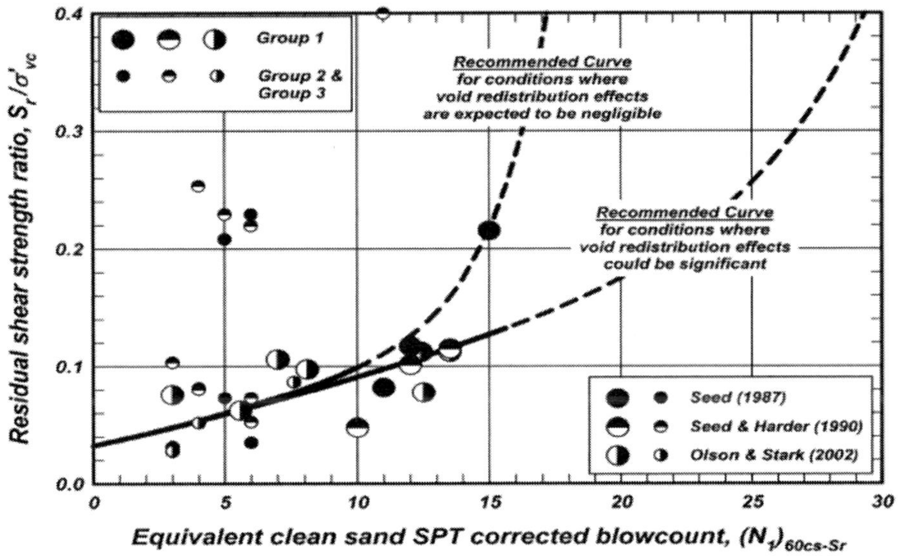

Figure 5. Normalized Residual Shear Strength Ratio of Liquefied Sand vs. Equivalent Clean-Sand SPT Corrected Blow Count (Idriss & Boulanger, 2008)

to be negligible. Within the Taylor Pond Dam embankment the potentially liquefiable soils underlie relatively free draining embankment soils and therefore the undrained shear resistance would increase rapidly as the $N_{(1)60cs\text{-}Sr}$ value approaches 15 (Idriss & Boulanger).

Using the above residual shear equation, the residual shear strengths were computed for soils within the defined embankment intervals where the ratio of CRR to CRS was less than 1.1 (i.e., FOS). For soils above the 1.1 FOS, an undrained shear strength was assigned. For the post-earthquake slope stability analysis the downstream embankment soils were found to have limited zones of liquefiable soils requiring an assignment of residual shear strength. Specifically, the section analyzed along the right abutment (Section B) has a 2 to 4 foot zone. Loose downstream embankment soils along the centerline of the dam were found not to liquefy using the SPT N-values obtained during the mud rotary drilling investigation. For the upstream embankment soils observed during the mud rotary investigation, it was confirmed that these soils become very loose within 6 to 8 feet of the lake bottom surface (elev. 1,388 to 1,386 feet). The weak zone of upstream embankment soil extends to a depth of 25 feet, and for these soils a residual shear strength value was calculated and assigned for post-earthquake stability analysis. The calculations for residual shear strength at MRB-1 upstream embankment are provided in Table 3.

Table 3. Summary of Results for Residual Shear Strengths at Borehole MRB-1

Input Parameters: Borehole MRB-1 Upstream Embankment					
Phi Angle φ =				32	
Sample #	Depth	σ_{vc}'	(N1)60-CS	Factor of Safety	Residual Strength (psf)
	(ft)				
1	1	6	21.6	2.00	71.88
2	3	14	13.6	1.83	184.12
3	5	20	16.2	1.81	264.82
4	7	26	4.3	0.65	30.60
5	9	33	17.1	1.70	426.23
6	11	39	2.6	0.49	36.99
7	13	45	4.1	0.56	50.82
8	15	51	8.9	0.88	94.63
9	17	57	5.6	0.64	76.32
10	19	64	2.4	0.45	58.97
11	21	70	3.7	0.51	75.93
12	23	76	10.5	0.95	165.13
13	25	82	9.3	0.86	157.05
14	27	88	NA	NA	*Undrained*

A computer-aided embankment slope stability analysis was performed using SLIDE™ Version 6.0, a 2D Limit Equilibrium Slope Stability Analysis software program developed by Rocscience, Inc. Using input data, including dam geometry, embankment and foundation soils, phreatic conditions and Taylor Pond water levels the factor of safety of soil slopes using Bishop's Modified Method were computed. The stability analyses performed for the embankment post-earthquake analysis and the computed factors of safety include:

- Downstream centerline section (Section A) using strength properties interpreted from SPT at boring locations P-4 and MRB-3; the computed FOS is 1.40.
- Upstream centerline section (Section A) using strength properties interpreted from SPT at boring location MRB-1; the computed FOS is 0.9.
- Downstream right section (Section B) using strength properties interpreted from SPT at boring locations P-1 and MRB-2; the computed FOS is 1.05.
- Upstream right section (Section B) using strength properties interpreted from SPT at boring location MRB-1; the computed FOS is 0.89.

Figures 6 and 7 present the screen shots of the output graphics from the SLIDE™ slope stability model. The results show the embankment geometry configured following the proposed modifications, which include raising the dam embankment and corewall and the inclusion of a trench drain near the toe of the dam. The results show that with the modification the downstream embankment meets the criteria for post-seismic stability (FOS greater than 1.0) and that the upstream embankment does not.

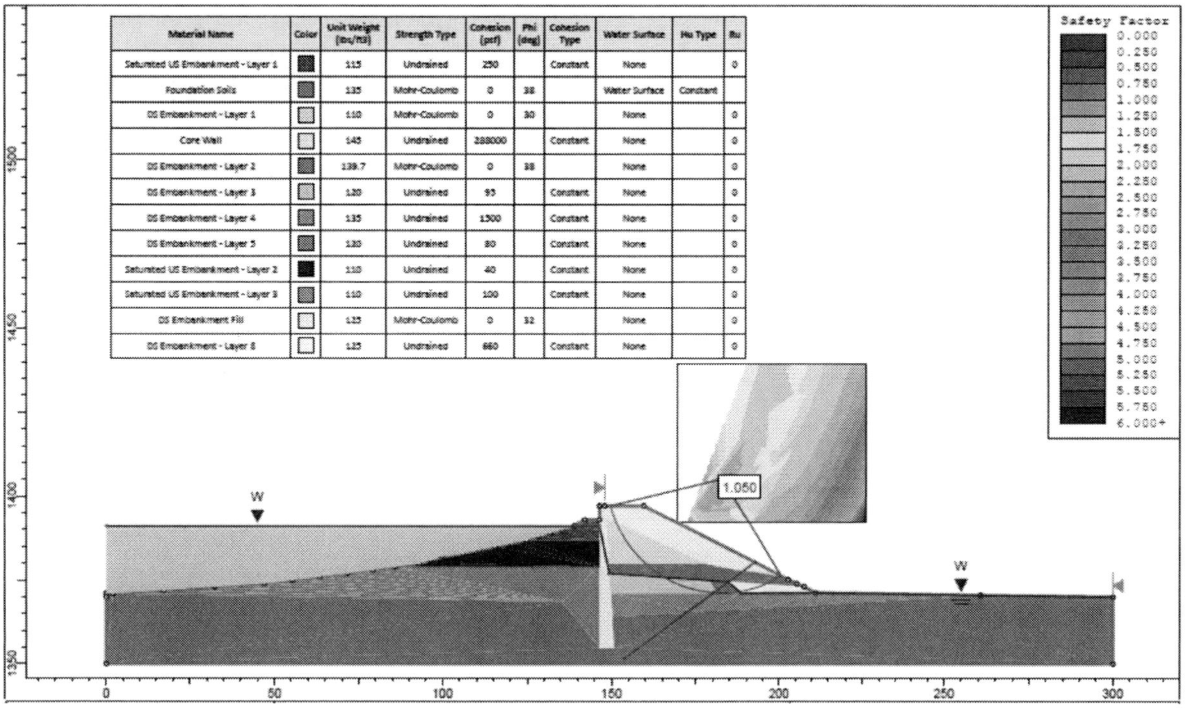

Material Name	Color	Unit Weight (lbs/ft3)	Strength Type	Cohesion (psf)	Phi (deg)	Cohesion Type	Water Surface	Hu Type	Ru
Saturated US Embankment - Layer 1		115	Undrained	250		Constant	None		0
Foundation Soils		135	Mohr-Coulomb	0	38		Water Surface	Constant	
DS Embankment - Layer 1		110	Mohr-Coulomb	0	30		None		0
Core Wall		145	Undrained	288000		Constant	None		0
DS Embankment - Layer 2		139.7	Mohr-Coulomb	0	38		None		0
DS Embankment - Layer 3		120	Undrained	93		Constant	None		0
DS Embankment - Layer 4		135	Undrained	1500		Constant	None		0
DS Embankment - Layer 5		120	Undrained	80		Constant	None		0
Saturated US Embankment - Layer 2		110	Undrained	40		Constant	None		0
Saturated US Embankment - Layer 3		110	Undrained	100		Constant	None		0
DS Embankment Fill		125	Mohr-Coulomb	0	32		None		0
DS Embankment - Layer 8		125	Undrained	660		Constant	None		0

Figure 6. Post-Seismic Stability Analysis of Downstream Embankment – Section B

As discussed, the post-seismic analysis is based on an assumed seismic return frequency of 2,500 years. The post-seismic analysis results indicate that the downstream embankment would remain stable with potentially minor ground deformation in the form of settlement over the relatively thin critical layers near the base of the embankment. For the upstream dam embankment, the amount of settlement or slumping is estimated to be severe given a significant portion of the embankment height may liquefy and the soil shear strengths decrease to residual shear strength values.

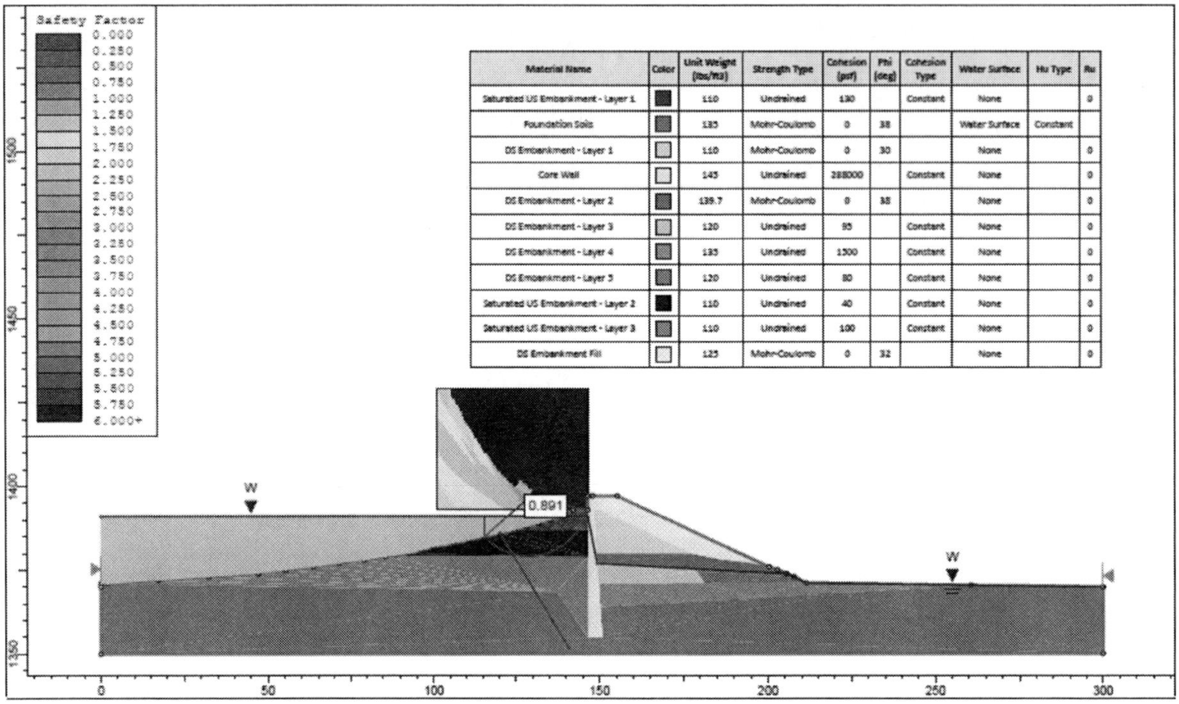

Material Name	Color	Unit Weight (lbs/ft3)	Strength Type	Cohesion (psf)	Phi (deg)	Cohesion Type	Water Surface	Hu Type	Ru
Saturated US Embankment - Layer 1		110	Undrained	130		Constant	None		0
Foundation Soils		135	Mohr-Coulomb	0	38		Water Surface	Constant	
DS Embankment - Layer 1		110	Mohr-Coulomb	0	30		None		0
Core Wall		145	Undrained	288000		Constant	None		0
DS Embankment - Layer 2		139.7	Mohr-Coulomb	0	38		None		0
DS Embankment - Layer 3		120	Undrained	95		Constant	None		0
DS Embankment - Layer 4		135	Undrained	1300		Constant	None		0
DS Embankment - Layer 5		120	Undrained	80		Constant	None		0
Saturated US Embankment - Layer 2		110	Undrained	40		Constant	None		0
Saturated US Embankment - Layer 3		110	Undrained	100		Constant	None		0
DS Embankment Fill		125	Mohr-Coulomb	0	32		None		0

Figure 7. Post-Seismic Stability Analysis of Upstream Embankment – Section B

The post-seismic deformation evaluation of the upstream embankment slope considered the bathometric lake floor grades and conservatively estimated that slumping soils will laterally spread to reduce the existing embankment slope of 3H:1V to 7H:1V. The associated embankment crest deformation was checked using a simplified post-earthquake screening parametric analysis as developed by FMSM Engineers (2007) for the USACE, and applying a factor of safety of approximately 3.0. Using this approach the estimated crest deformation would result in a failure surface scarp that would expose approximately 8 feet of the upstream face of the concrete corewall, but more importantly the deformation would not cause a loss in freeboard height. Structural analysis of the corewall concluded that this degree of exposure from the slumping upstream (submerged) embankment would not topple the wall, understanding that normal pool hydraulic head would be present following the seismic event. At the original embankment crest height (elev. 1,394 feet) the downstream embankment soils would be retained by the corewall and remain stable, preventing the uncontrolled release of Taylor Pond.

HYDRAULIC AND HYDROLOGIC COMPLICATIONS

As required for all engineering assessments in New York State, a hazard class evaluation is performed to confirm the existing dam safety hazard classification. For Taylor Pond Dam the evaluation concluded that the hazard class should be raised from an intermediate (Class B) hazard dam to a high (Class C) hazard dam. The change in hazard class required an increase in the spillway design flood (SDF) from 150 percent of the 100-year storm to 50 percent of the probable maximum flood (PMF) without overtopping the earthen dam embankment. To achieve this several conceptual designs were evaluated that involved primary spillway modifications or raising the dam embankment. Based on the conceptual designs and associated costs, the option to raise the dam embankment crest elevation to 1,398.5 feet with no primary spillway channel modifications was selected. This option was preferred since the relative ease of raising the embankment and corewall approximately 6.5 feet provides the necessary head and freeboard during the SDF to pass the peak outflows through the existing primary spillway without modification to the existing spillway. Eliminating the need to modify the 750 foot long primary spillway significantly reduced environmental impacts and structural spillway modifications which would have significantly increased the project construction and permitting costs.

The choice to raise the embankment crest, however, brought back into question the post-seismic stability of the dam following a design earthquake event. Raising the concrete corewall and downstream embankment, which was needed to buttress the corewall during the SDF hydraulic load, was found to create sufficient unbalanced load under the design earthquake stability scenario. The post-seismic stability scenario assumes a normal pool level and the slumping failure of the upstream embankment, and this combination would result in significantly unbalanced loading on the corewall that would potentially topple the corewall. To mitigate the unbalanced loading several methods of downstream embankment modifications were considered including:

- Ground improvement by densification methods;
- Ground improvement by soil mixing methods;
- Earth reinforcement using geogrid;
- Reinforcing the corewall and extension with post-tensioned reinforcement anchors; and
- Corewall support by use of tie-backs anchored to deadman anchor blocks buried within the downstream embankment.

Each stabilization method had to ensure that the dam embankment is stable under all the NYSDEC required load cases with seismic loading being the most critical at Taylor Pond dam. Again a conceptual design review and cost evaluation was performed and based on relative ease of construction and low cost the dead-men anchor design was selected.

COREWALL AND EMBANKMENT MODIFICATION

The embankment modification at Taylor Pond Dam effectively raised the existing mid-slope bench on the downstream embankment from the current elevation of approximately 1,379 feet to the new crest elevation 1,397.5 feet with the placement of fill. The new fill buttresses the corewall during the spillway design flood, but must be retained by the corewall during normal, and more critically, following the design earthquake event. As shown in Figure 8 the concrete corewall extension to elevation 1,398.5 feet creates a cantilevered wall section of approximately 4.5 feet above the crest of the upstream embankment that is loaded by approximately 3.5 feet of downstream embankment during normal pool conditions. The cantilever length is estimated to increase to 12.5 following the slumping of the upstream embankment subsequent to the occurrence of a design earthquake event.

Given the existing corewall is lightly reinforced and not designed to serve as an earth retaining structure it was necessary to reduce the loading of the wall in the event of an upstream embankment slump. The extension of the concrete corewall and raising of the downstream embankment required the design and installation of a dead-man and tie-back anchor system to stabilize the existing upper portion (approximately 7 feet) plus 5-foot extension of the corewall (Figure 8).

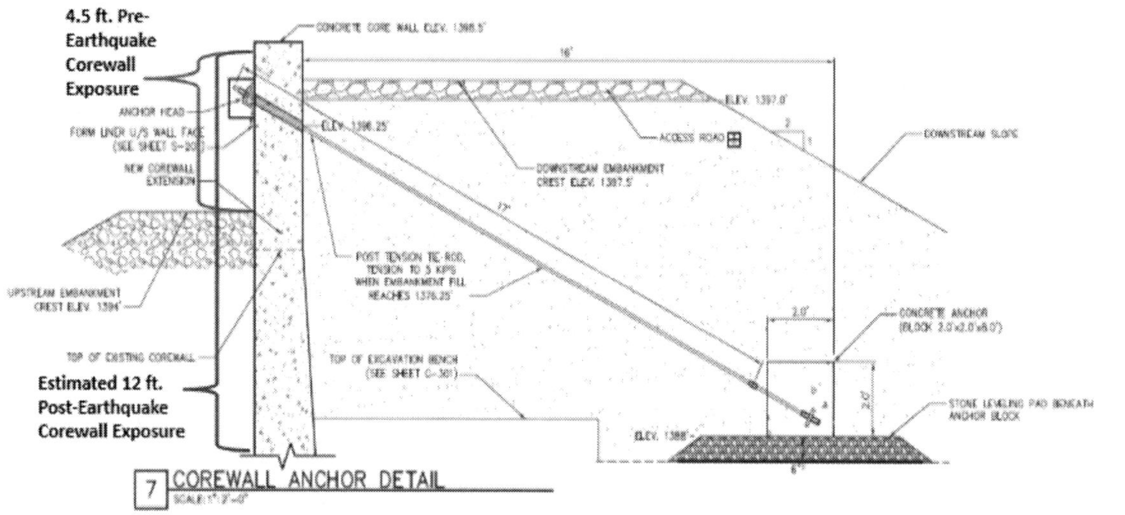

Figure 8. Corewall Raise and Dead-Men Tie-Back Anchor System Design

The dead-man and tie-back anchor system consists of 17-foot long tie-back tension elements seated on the concrete corewall extension and linked back to concrete dead-men buried within new compacted embankment fill. The 2 feet by 2 feet by 8 feet long dead-men anchors are spaced on 15-foot centers along the 400-foot long crest of the dam. The dead-man and tie-back anchor system was designed to provide 18 kips of resistance which is sufficient to mitigate the toppling of the corewall in the event of slumping of the upstream embankment.

Construction of the embankment rehabilitation (Figure 9) included:

- Subgrade preparation (benching to receive fill) across the downstream embankment slope above elevation 1,379 feet.
- Embankment soil placement and compaction to raise the downstream embankment to elevation 1,397.5 feet, creating a new crest across the dam for light vehicle access.
- A reinforced concrete corewall extension raising the existing corewall from elevation 1,393.5 feet to 1,398.5 feet.

- Placement of concrete dead-men with tie-backs to the dam embankment concrete corewall.
- Grading and replacing rip rap across the upstream embankment crest to re-establish a uniform shoreline upstream of the corewall.

Figure 9. Construction of Dead-Men Tie-Back Anchor System

In addition to the embankment crest modification the proposed improvements include installing two trench drains within the left and right groins (i.e., abutment contacts) downstream of the corewall (where significant seepage was historically observed), and construction of a toe-of-slope blanket drain. These additional drainage features provide and maintain stable seepage conditions within the lower portion of the dam embankment. The trench drains include V-notch seepage flow monitoring weir boxes to monitor seepage discharge through the embankment soils.

Inadvertent validation of the seismic design sensitivity was confirmed during construction when heavy equipment induced vibrations had to be controlled prior to the completion of the downstream embankment soil drainage improvements. Near the onset of the project, while performing clearing and grubbing, the contractor developed sufficient ground motion with a tracked excavator to cause a localized area of liquefaction to develop near the base of the right embankment-abutment contact (Figure 10).

In response to the development of longitudinal fractures in the dam embankment soils the Engineer directed the contractor to temporarily stabilize the base of the dam embankment with a stone fill toe buttress. Subsequently with the trench drain installion the embankment and abutment soil porewater pressures dissipated and the soils strengthened allowing the embankment raise and corewall extension to be completed without further stability concerns.

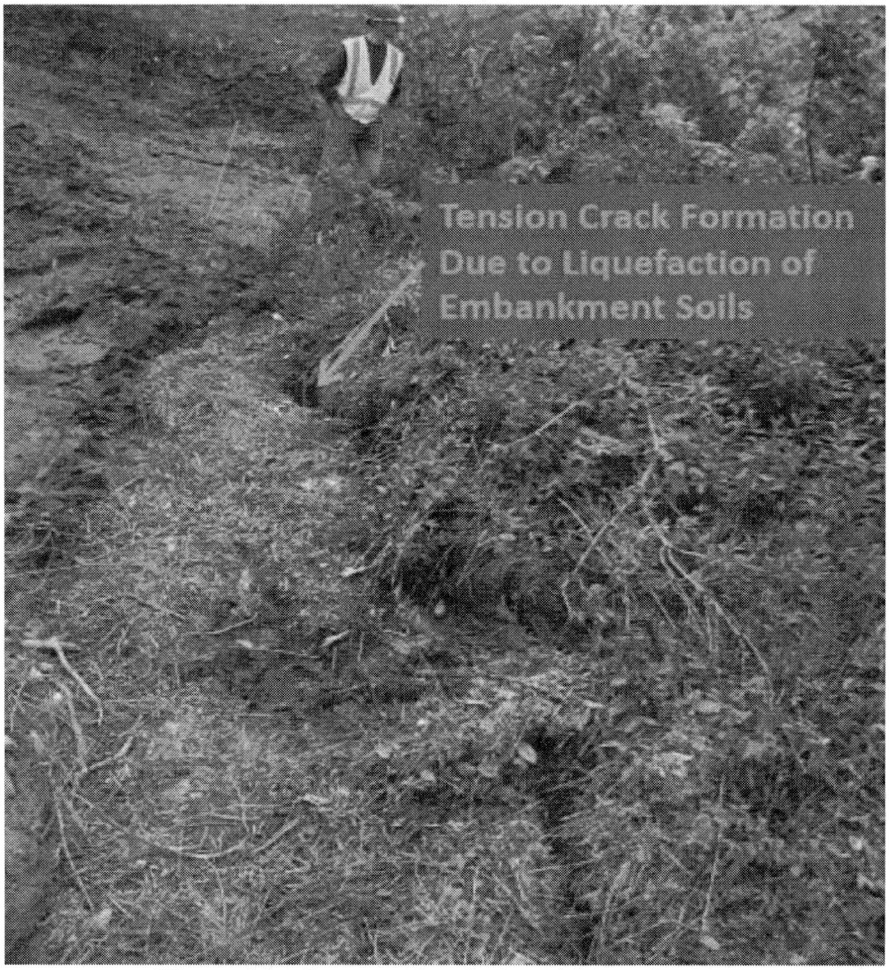

Figure 10. Construction Induced Liquefaction

SUMMARY

The rehabilitation and improvements at Taylor Pond Dam were necessary to bring the dam into compliance with the current NYSDEC dam safety regulations. To achieve regulatory compliance the modifications presented in this paper were designed to provide safe passage of peak outflows during the SDF while maintaining the stability of the dam embankment in the event of a design earthquake generating a magnitude and peak ground acceleration with a recurrence interval of 2,500 years. In addition to the seismic loading, the static load cases required for analysis by the NYSDEC were also evaluated and found to meet the target FOS for embankment stability.

The design solution acknowledges the potential failure (slump) of the upstream embankment, but recognizes the very low probability of occurrence. A dead-men tie-back anchor system is a low cost solution to mitigate the potential mode of failure, whereby the loss of upstream embankment would result in the potential toppling of the concrete corewall, followed by the loss of freeboard and the breach of the downstream embankment soils. The construction of the crest raise per the design at Taylor Pond Dam

Figure 11. Completed Rehabilitation of the Taylor Pond Dam

will preserve the freeboard, as well as, the downstream embankment and concrete corewall integrity. The dam rehabilitation design will insure an effective impoundment during the design earthquake event and anticipated liquefaction of the upstream embankment, while preventing overtopping of the dam at the corewall during the SDF event.

REFERENCES

Das, Braja M., Fundamentals of Geotechnical Engineering, Fourth Edition, 2005, 2008 and 2013.

Federal Guidelines for Dam Safety: Earthquake Analyses and Design of Dams, FEMA 65, printed May 2005.

Finn, W. D. L., Ledbetter, R. H., and Marcuson III, W. F., "Seismic Deformations in Embankments and Slopes," Proceedings, Symposium on Developments in Geotechnical Engineering (From Harvard to New Delhi, 1936-1994), Bangkok, Thailand, pp. 233-264 January 12-16, 1994.

FMSM Engineers (2007), A Simplified Tool for Assessing the Deformation of Embankment Dams and Levees on Liquefied Soils, 27. Methods to Evaluate Seismic Risks for Embankments November, 2012.

I. M. Idriss and R. W. Boulanger, Earthquake Engineering Research Institute MNO-12, *Soil Liquefaction during Earthquakes*, 2008.

New York State Department of Environmental Conservation Guidelines for the Design of Dams, NYSDEC Division of Dam Safety, January 1989.

Rocscience Inc.; Computer Program Slide Version 6.0, 2D Limit Equilibrium Slope Stability Analysis, July 2010.

U.S. Department of the Interior | U.S. Geological Survey, U.S. Seismic Design Maps web application, March 12, 2014 16:51:43 UTC

GENERALIZED SCALING PRINCIPLES FOR PHYSICAL MODELING OF EMBANKMENT DAMS IN EVALUATION OF SEISMIC BEHAVIOR

Nam-Ryong Kim, PhD[1]
Jeong-Yeul Lim, PhD[2]
Ik-Soo Ha, PhD[3]

ABSTRACT

Reduced scale physical modeling is often carried out in order to evaluate or verify performance of large geotechnical systems such as embankment dams during design or management stage. The modeling using geotechnical centrifuge can provide correct scaling of the self-weight stresses so that the strength or deformation characteristics of materials as well as the global behavior of the system can be considered more realistic. Generally, reduced testing model is accelerated within a centrifuge up to the acceleration corresponding to modeling scale. However it is sometimes difficult to follow this principle due to limitation of testing equipment to be utilized for large prototype structure. Especially, this problem is more frequently raised in seismic test due to limited platform area and allowable centrifugal acceleration level of earthquake simulator so appropriate modeling strategies is required. This paper introduces generalized scaling principles used for seismic test of large earth structures using geotechnical centrifuge. The scaling principles have to be applied to cases which the acceleration level for centrifuge test according to the model scale cannot be achieved due to limitations of testing equipment. These can be derived from basic soil mechanics and validated by comparing with previously suggested scaling factors. A series of tests using geotechnical centrifuge and earthquake simulator are carried out to verify the principles varying acceleration level, loading frequency and other major factors. Finally, the experimental design strategies for appropriate physical model test considering these scaling principles will be discussed.

INTRODUCTION

In last decades, there were several cases which strong earthquake ground motion caused partial or even critical damages in embankment dams in the world and these events highlights the importance of seismic design and evaluation of seismic response of the systems. Damages in the embankment by strong earthquake such as excessive settlement, movement or failure of slopes, liquefaction can deteriorate the safety performance of entire system and sometimes drive into tragic consequences. Therefore, the seismic response and performance of dams subjected to strong earthquake motion must be evaluated to prevent critical consequences, especially for the dams located in severe seismic zone and countermeasures must be prepared.

---

[1] K-water Institute, 125 Yuseongdaero 1689beon-gil, Yuseong, Daejeon, Korea, namryong@kwater.or.kr
[2] K-water Institute, 125 Yuseongdaero 1689beon-gil, Yuseong, Daejeon, Korea, passion@kwater.or.kr
[3] Kyungnam University, 7 Kyungnamdaehak-ro, Masanhappo, Changwon, Gyeongsangnamdo, Korea, geotech@kyungnam.ac.kr

Various methods are employed in order to evaluate safety performance or validate the design parameters of large earth structures such as embankment dams. Numerical methods such as finite element analysis are widely adopted in engineering practice for these purposes since the engineers can easily model entire structure and simulate various loading condition required, and seismic response and behavior of entire system can be also calculated by time integration algorithm. While we are able to model complicated structural and material characteristics, and to simulate various and more realistic loading conditions by numerical simulation with great advances of technology, reduced scale physical modeling is still useful and conducted in both research and practice. It is carried out in order to simulate complex model such as boundary conditions or effect of interfaces between different materials, and to consider more precise material characteristics, validate the result of numerical simulation.

Physical modeling using a centrifuge is widely applied in most of geotechnical engineering field, and this method provides appropriate scaling of self-weight stress conditions in large geotechnical systems (Gaudin *et al.* 2009; Schofield 1980; Taylor 1995). When the reduced scale model is accelerated within a centrifuge, the self-weight of the soil can be raised to the field conditions such that the entire behavior of the model can be similar to full scale geotechnical systems; hence it can reliably represent performance behavior of prototype structures. Based on this advantage, physical modeling technology using a centrifuge has been applied in various geotechnical engineering fields, especially in earthquake related problems.

There have been several cases of physical modeling of embankment dams to evaluate seismic response characteristics such as amplification, deformation and/or settlement, liquefaction susceptibility, slope stability, etc. With development of in-flight shaking table for centrifuge, various seismic tests for embankment dams were conducted by number of researchers, especially for the embankments sitting on the liquefiable foundation (Ng *et al.*, 2004; Sharp & Adalier, 2006; Peiris *et al.*, 2008). More recently, seismic response and amplification characteristics of earth core rock-fill dam and concrete faced rock-fill dams were investigated by a series of centrifuge tests (Kim *et al.*, 2011).

The prototype dams of previous research are relatively small. In reality, failure of large dam has bigger risk to public safety so most of cases which geotechnical engineers are asked to evaluate the seismic response and safety by physical modeling deal with large dams. However, the size of reduced scale model which can be mounted to in-flight earthquake simulator for centrifuge the level of centrifugal acceleration which the simulator can operate are limited, hence special modeling strategy is required to model and simulate large dam for seismic safety evaluation.

In this paper, general principles of geotechnical centrifuge technology as well as the principles for seismic test of large embankment dams are introduced. The scaling factors are derived from basic soil mechanics and validated. Seismic test using geotechnical centrifuge and earthquake simulator are carried out to verify the principles varying test and loading conditions. Finally, the experimental design strategies for appropriate physical model test considering these scaling principles are suggested.

CENTRIFUGE MODELING OF DAMS — PRINCIPLES AND LIMITATIONS

Principles of Geotechnical Centrifuge Modeling

Scaled model for centrifuge test is generally built using the same material following the same geometrical arrangement, and the experiments must be designed based on similarity laws derived from the fundamental equations governing the phenomena to be investigated. The basic scaling law for the geotechnical centrifuge testing is derived from the need to ensure stress similarity between the model and corresponding prototype. When an acceleration of N times Earth gravity (g) is applied to a material of density ρ, the vertical stress, σ_v at depth h_m in the model is given by:

$$\sigma_{vm} = \rho \cdot Ng \cdot h_m \qquad (1)$$

In the full scale prototype, indicated by the subscript p, the vertical stress is expressed by:

$$\sigma_{vp} = \rho \cdot g \cdot h_p \qquad (2)$$

From the fundamental idea of the centrifuge modeling, the vertical stress in the model and the prototype is identical hence the scale factor for linear dimensions is 1:N. Basic scaling factors for physical quantities in centrifuge testing are derived based on dimensional analysis using this linear dimension scale. Since the mechanical properties of geomaterials used for general centrifuge models are identical with the prototype materials, those physical quantities can be easily derived. The scaling factors for basic quantities in centrifuge modeling are listed in Table 1. Most of the scaling factors for the modeling can be derived from these bases. Details for the centrifuge modeling principles can be found in Taylor (1995).

Table 1. Basic scaling factors for Ng geotechnical model tests

Quantity	Scaling factor	Quantity	Scaling factor
Stress, stiffness	1	Force	N^{-2}
Density	1	Mass	N^{-3}
Length	N^{-2}	Diffusion time	N^{-2}
Gravity	N	Wave velocity	1
Strain	1	Acceleration	N

Scaled Modeling of Large Dams — Limitations

Let us assume that we are carrying out a seismic test using a geotechnical centrifuge for a prototype embankment dam which is 100m high and 300m long in up- to down-stream direction and applying 0.4g peak ground acceleration (PGA) motion. If we can build a model in a 1.5m long and 0.75m high container, which is almost the biggest container in the world for an in-flight shaking table, the scale is 1:200 so we need to run the centrifuge up to 200g and simulate 80g ground motion to simulate appropriate stress and loading conditions according to basic scaling principles. However, most of large geotechnical centrifuges are seldom operated to such a high acceleration level and there are only a few shaking tables with such a large platform area. Furthermore, most of in-flight shaking tables can be operated lower than 50g centrifugal acceleration conditions (Iai *et al.*, 2005), and maximum shaking acceleration level is generally smaller than 40g. Therefore, it is not possible to carry out physical model test of such a large embankment dam based on basic scaling principles for centrifuge modeling, while there are needs and requirements from engineering practice to assess seismic safety of large dams subjected to strong earthquake ground motions.

If we test the model despite of these limited conditions, we generally operate the centrifuge equipment up to maximum allowable acceleration level for shaking table operation. If the centrifugal acceleration applied to the model above is 50g for example, the model can be regarded as simulating 25m high embankment dam so it is a completely different case from original prototype structure hence the response will be also different. Therefore it is difficult to directly estimate the seismic response or safety conditions of prototype structure in these kinds of limitations so we need to find and validate alternative modeling strategy based on solid dimensional analysis. A special scaling principle for large structures for seismic centrifuge tests is suggested. This principle mainly focuses on the dynamic response characteristics of entire structure and is compared with previous research.

SCALING OF LARGE PROTOTYPE STRUCTURES FOR SEISMIC TEST

New Scaling Considering Dynamic Response Characteristics

In cases of large earth structures where ordinary scaling principles for centrifuge test cannot be applied due their scale, the dynamic response characteristics of the systems have to be considered so that appropriate modeling strategy can be provided. Considering the limitations of scale and acceleration, we can review two modeling cases, 1:N_0 scale model tested at N_0 times gravity condition as a general centrifuge modeling case, and the same model tested at kN_0 times gravity condition where the testing equipment has limited capacity. Assuming that the height of the model is h_m, the prototypes represented by each case can be expressed as Equations (3) and (4)

$$H_p = kN\,h_m \tag{3}$$

$$H_{vp} = N\,h_m \tag{4}$$

where H_p is the height of prototype structure, and H_{vp} is the height of a virtual prototype structure which is simulated by Ng acceleration condition. For instance, if we consider 1:200 scaled model of 100m high embankment dam is prepared and tested at 200g and 50g conditions, N becomes 50 and k becomes 4, hence these two cases represent 100m and 25m high dams respectively. Generally, the natural period of an earth system can be expressed as

$$T_n = c\frac{H}{V_S} \tag{5}$$

In this equation, 1st order coefficient c for semi-infinite horizontal soil deposit is 4, and it is 2 to 2.6 for embankment dam (Gazetas, 1987; Bray & Travasarou, 2005). Shear wave velocity V_S of the geomaterial for embankment dam is highly affected by confining stress and can be expressed as

$$V_S = \alpha\sigma_m{}^{\beta} \tag{6}$$

where α is V_S at reference confining pressure, and stress exponent β is generally from 0.2 to 0.3 (Cho et al., 2006). The effective stress in a small scale model which is accelerated in the centrifuge is proportional acceleration level, the stress level for Ng and kNg model can be expressed as

$$\sigma_{m,kN} = k\sigma_{m,N} \tag{7}$$

The difference of confining pressure induces the difference of shear wave velocity. Hence, equation (7) can be rewritten using equation (6), and the natural period or frequency of embankment can be expressed as

$$\overline{V}_{S,kN} = k^{\beta}\overline{V}_{S,N} \tag{8}$$

$$T_{kN} = k^{-\beta}T_N \text{ or } f_{n,kN} = k^{\beta}f_{n,N} \tag{9}$$

Therefore, we can understand that if the acceleration increases k times, the fundamental natural frequency of the model increases as k^{β}. In general, the fundamental period of prototype structures are proportional to centrifugal acceleration, the period of the prototype can be derived from the testing result at Ng acceleration condition as following

$$T_p = kNT_{kN} = k^{1-\beta}NT_N \tag{10}$$

Therefore, the fundamental period of 1:kN scaled model tested at Ng centrifugal acceleration condition is $1/k^{1-\beta}N$ times of the fundamental period of prototype structure.

Generalized Scaling Relations

This approach can be compared with "generalized scaling relations" for seismic test of scaled models. Iai et al. (2005) derived scaling factors for similar cases that ordinary scaling principles for centrifuge test cannot be directly applied because of either the testing machine or the largeness of prototype structure, by combining scaling principles for $1g$ shaking table tests and Ng dynamic centrifuge tests. In their method, if we define the scale between virtual intermediate prototype and model as η (= N) and that between virtual prototype and full scale prototype as μ (= k), every scale factor for physical quantities in a test is derived by combination of centrifuge scaling factors by η and $1g$ shaking table test scaling factors by μ.

In general, the scaling factor for frequency used for scaling of earthquake motion in 1g shaking table test is defined as

$$\mu_f = (\mu\mu_\varepsilon)^{-0.5} \tag{11}$$

where μ_ε is scaling factor for strain. In general cases, it is used as $\mu^{0.5}$ and it can be expressed as $\mu^{2\beta}$ more precisely (Iai, 1989; Iai & Sugano, 1999). The scaling factor of frequency for centrifuge test is

$$\eta_f = \eta^{-1} \tag{12}$$

hence the generalized scaling factor for frequency or period can be expressed as

$$f_{\mathrm{GS}} = (\mu_f \eta_f) f_{\eta g} = (\mu^{-0.75}\eta^{-1}) f_{\eta g} \tag{13}$$

$$T_{\mathrm{GS}} = \mu^{0.75}\eta T_{\eta g} \tag{14}$$

The factor for period in Equation (14) is exactly same as Equation (10) when β is 0.25 which is generally adopted value for most of granular material.

VERIFICATION OF NEW SCALING APPROACHES — PHYSICAL MODELING

Geotechnical centrifuge

The test was carried out using the geotechnical centrifuge at K-water Institute and it is one of the largest centrifuges in the world. The platform radius of this centrifuge is 8.0m and the payload dimension is 2m×2m×2m. This large equipment provides advantages in physical modeling; relatively uniform acceleration field can be achieved from large platform radius and large testing model can reduce errors or uncertainties from testing results. The centrifuge can accelerate maximum 8,000kg payload up to 100g and the maximum acceleration is 150g with reduced payload mass. The centrifuge and in-flight earthquake simulator are shown in Figure 1.

The earthquake simulator is used to generate base shaking motion in the model during centrifuge operation. The payload dimension is 1.8m(L)×0.8m(W) and it allows maximum 1.44m long testing model to be equipped for seismic test which is the biggest as centrifuge test. The shaking table can be operated under maximum 100g centrifugal acceleration and the maximum horizontal shaking acceleration is 22g for 1,500kg payload mass. It can simulate not only sinusoidal shaking motions but recorded earthquake motions, and the operational frequency range for reliable response is from 20 to 350Hz.

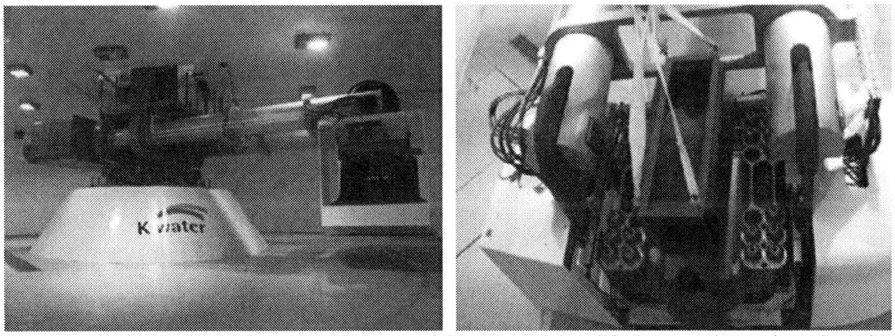

Figure 1. K-water geotechnical centrifuge and in-flight earthquake simulator

Testing Model and Procedure

Figure 2 shows the design of embankment dam model used to verify the applicability of the scaling principles. The simplified model was designed as a uniform earth structure with 1:1.4 slopes on both sides. The length and the height of the model are 1440mm and 496mm, and they were determined considering the size of container. Because of the purpose of these tests, the model was specially designed to have fundamental natural frequency within the range that the earthquake simulators can generate sinusoidal excitation, while the global seismic response of the embankment is similar with those from shear-beam model approach by Dakoulas & Gazetas (1985). Hence the geometry of the model is simplified as much as possible and the slope had to be steep to have flexible behavior.

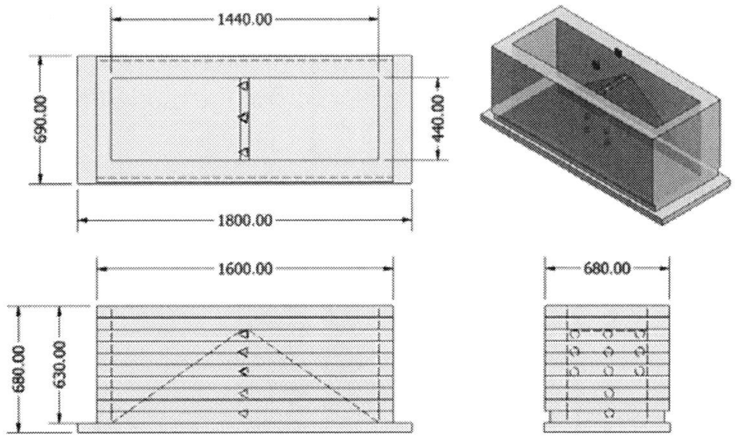

Figure 2. Simplified embankment dam model for seismic tests

Generalized Scaling Principles

The model was constructed using gravelly sand material. The maximum grain size is limited to 15mm and the material is classified as well-graded sand (SW). Thirteen accelerometers were installed in the center of the model to measure the response at each depth, $0H$, $0.25H$, $0.5H$, $0.75H$, and $1H$, so the frequency response subjected to an input motion could be calculated from the measured data.

After the model is prepared, it was installed in the shaking table as shown in Figure 3. The seismic tests were conducted under 15 and 30g centrifugal acceleration conditions hence k, N and β in Equation (9) are 2, 15 and 0.25 respectively. Various input motions such as sinusoids varying frequency, band limited white noise and Ormsby wavelet were applied in the base and the peak ground acceleration (PGA) was limited to 0.2g in prototype scale; $3g_{shaking}$ at $15g_{centrifuge}$ and $6g_{shaking}$ at $30g_{centrifuge}$.

There are various methods to evaluate response characteristics of dynamic system and it is generally called "system identification". In this study, the frequency characteristics of embankment model were calculated in frequency domain by comparing the crest response subjected to base motion which are both measured from each test.

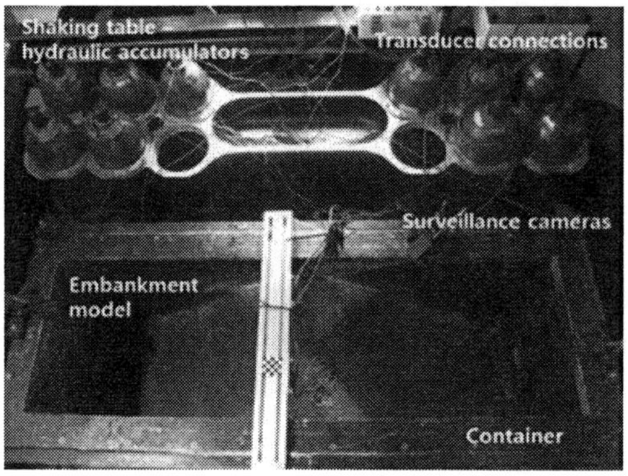

Figure 3. Dynamic centrifuge testing setup

Results and Discussions

Figure 4 shows typical acceleration signals measured at the model. In this result, Ormsby wavelet signal which has equivalent frequency amplitudes from 100 to 250Hz has been applied at the base (Z=0) and it could be observed that the motion is amplified as it goes up to crest.

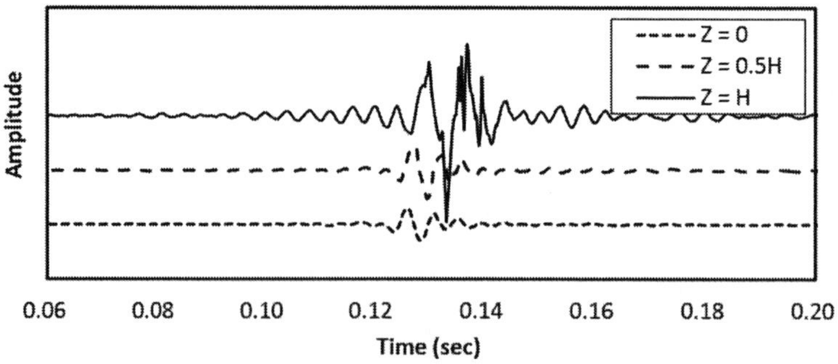

Figure 4. An example of test result; amplification from base motion to crest response

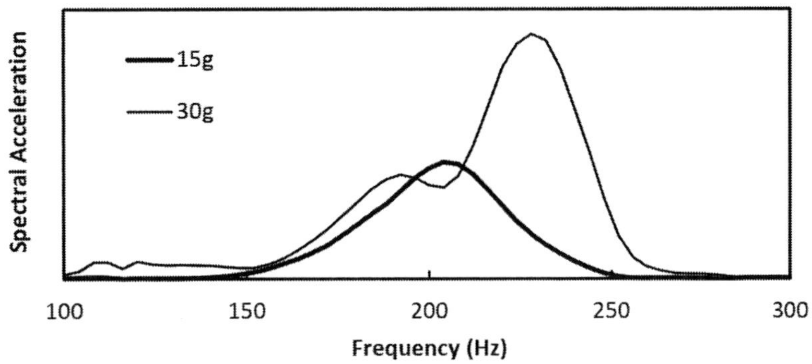

Figure 5. Frequency response of the crest motion at each g-level

The power spectrum of acceleration history measured at the crest can be regarded as the frequency response due to the frequency characteristic of input signal hence the fundamental mode natural frequency of the embankment at each g-level could be easily calculated and they are shown in Figure 5. In these results, the 1st mod fundamental frequencies (f_r) calculated from 15 and 30g conditions are 204 and 228Hz respectively. From Equation (9), we can calculate the theoretical ratio between these two frequencies since k, the ratio of centrifugal accelerations is 2. Table 2 summarizes the key results from experiments and the error between these two frequencies is about 6.4%. When we assume that the model represents a 1/200 scale of virtual prototype embankment dam, the natural period of the prototype is about 0.51 to 0.55 seconds which are reasonable value for a 100m high embankment.

Table 2. The results of experiments and natural frequency estimation of prototype dam

Test condition ($g_{centrifuge}$)	Frequency (Hz)	Frequency ratio (general. scale)	Experimental ratio	$T_{Prototype}$ (sec)
15	204	$2^{0.25} = 1.189$	1.118 (6.4% error)	0.51
30	228			0.55

CONCLUSION

The strategy and principles for physical modeling of large embankment dams for experimental evaluation of seismic performance evaluation are introduced. The scaling principles are derived using stress-stiffness relations of geomaterials and validated by a series of centrifuge tests. Even though appropriate modeling for dynamic centrifuge test of large embankment dam is not available due to the limitations of testing equipment, the test can be designed by two stage modeling method so called generalize scaling relations and the behavior of prototype dam can be estimated. To validate this modeling method, a simplified embankment dam model was prepared and seismic tests using geotechnical centrifuge have been conducted at two different acceleration states. The frequency responses were similar with general behavior of embankment dams, and the dominant frequencies at each acceleration level validated the applicability of the method.

REFERENCES

Bray, J.D. and Travasarou, T. (2007) Simplified Procedure for Estimating Earthquake-Induced Deviatoric Slope Displacements, J Geotech Geoenviron, Vol.133, No.4, pp. 381-392.

Cho, G.C., Doods, J., and Santamarina, J.C. (2006) Particle shape effects on packing density, stiffness, and strenth: natural and crushed sands, J Geotech Geoenviron, Vol.132, No.5, pp. 591-602.

Dakoulas, P. and Gazetas, G (1985) A class of inhomogeneous shear models for seismic response of dams and embankments, Soil Dyn Earthq Eng, Vol.4, No.4, pp. 166-182.

Gaudin, C., White, D. J., Boylan, N., Breen, J., Brown, T., De Catania, S., and Hortin, P. (2009) A wireless high-speed data acquisition system for geotechnical centrifuge model testing, Meas Sci Technol, Vol. 20, No. 9, pp. 1–11.

Gazetas, G. (1987) Seismic response of earth dams: some recent developments, Soil Dyn Earthq Eng, Vol.6, No.1, pp. 2-47.

Iai, S. (1989) Similitude for shaking table tests on soil-structure-fluid model in 1g gravitational field, Soils Found, Vol.29, No.1, pp. 105-118.

Iai, S. and Sugano, T. (1999) Soil-structure interaction studies through shaking-table tests, Proc. 2nd Int. Conf. on Earthquake Geotech. Engrg, Lisbon, pp. 927-940.

Iai, S., Tobita, T. and Nakahara, T. (2005) Generalized scaling relations for dynamic centrifuge tests, Géotechnique, Vol.55, No.5, pp. 355-362.

Kim, M. K., Lee, S. H., Choo, Y. W., and Kim, D. S. (2011) Seismic behaviors of earth-core and concrete-faced rockfill dams by dynamic centrifuge tests, Soil Dyn Earthq Eng, Vol. 31, No. 11, pp. 1579–1593.

Ng, C.W.W., Li, X.S., Van Laak, P.A. and Hou, D.Y.J. (2004) Centrifuge modeling of loose fill embankment subjected to uni-axial and bi-axial earthquake, Soil Dyn Earthq Eng, Vol. 24, No. 4, pp. 305–318.

Peiris, L.M.N., Madabhush, S.P.G. and Schofield, A.N. Schofield (2008) Centrifuge modeling of rock-fill embankments on deep loose saturated sand deposits subjected to earthquakes, J Geotech Geoenviron, Vol. 134, No. 9, pp. 1364-1374.

Schofield, A. N. (1980) Cambridge geotechnical centrifuge operation, Géotechnique, Vol. 20, No. 3, pp. 227–268.

Seed, H.B. and Idriss, I.M. (1970) Soil moduli and damping factors for dynamic response analyses, Report No. EERC 70-10, Earthquake Engineering Research Center, Univ. of California Berkeley, California, pp. 1-43.

Sharp, M.K. and Adalier, K. (2006) Seismic response of earth dam with varying depth of liquefiable foundation layer, Soil Dyn Earthq Eng, Vol. 26, No. 11, pp. 1028-1037.

Taylor, R. N. (1995) Geotechnical centrifuge technology, Blackie Academic, London, UK.

WATER DELIVERY AFTER A LARGE SAN ANDREAS EARTHQUAKE AN UNCONVENTIONAL RETROFIT APPROACH

Michaele Monaghan, PE, LEED® AP[1]
Gregory Reichert, PE[2]
Dustin Jolley[3]
Leroy Gullette, PE[4]
Tasso Mavroudis, PE[5]

ABSTRACT

The San Francisco Public Utilities Commission (SFPUC) is completing a multi-billion dollar Water System Improvement Program (WSIP) to increase the seismic reliability of SFPUC's water system. The primary level of service goal is to maintain/restore, within 24 hours, potable water supply to the San Francisco Bay Area. One of the major projects of the program is the Crystal Springs – San Andreas Transmission System Upgrade Project. It involves improvements to reservoirs that serve as the only emergency water supply for 1.4 million people on the San Francisco peninsula. The San Andreas Reservoir is the last available source for water delivery. This paper focuses on the rehabilitation of the San Andreas Outlet Structures (SAOS2 & SAOS3).

The structures are reinforced concrete shafts embedded in the reservoir bank, located within 100 feet of the San Andreas Fault. Built in the 1920's (SAOS2) and the 1930's (SAOS3), as originally constructed the shaft were 12-foot inside diameter (ID), roughly 60-feet deep (SAOS2); and 14-foot ID, roughly 70-feet deep (SAOS3). FLAC software was used to develop seismic loading representing the non-linear soil-structure interaction at the shaft. The results indicated that the reservoir bank would fail such that the point where the shaft daylights would move further from the tower base and displace approximately 3-feet. The thickness of any new internal liner was limited to 1-foot in order to maintain access to the mechanical equipment within the shaft. Conventional structural design considerations could not accommodate the seismic loading given these space limitations. Therefore, the tower was designed to accommodate a controlled failure. The design included a "hinge" at the base allowing rotation of the shaft and expansion joints along the shaft to accommodate the lengthening of the structure. Mechanical systems within the tower were designed to follow the rotation and expansion of the shaft. The completed design allows both structures to remain stable and operational after an extreme event on the San Andreas Fault.

---

[1] Sr. Engineer – Hydraulic Structures, AECOM, 1333 Broadway, Suite 800, Oakland, CA 94612, (510) 874-3223, Michaele.monaghan@aecom.com
[2] Sr. Project Manager, AECOM, 1333 Broadway, Suite 800, Oakland, CA 94612, (510) 874-3090, gregory.reichert@aecom.com
[3] Project Engineer, AECOM, 1333 Broadway, Suite 800, Oakland, CA 94612, (510) 874-1788, dustin.jolley@aecom.com
[4] Project Engineer, San Francisco Public Utilities Commission, 525 Golden Gate Ave., San Francisco, CA 94102, (415) 551-4807, lgullette@sfwater.org
[5] Sr. Project Manager, San Francisco Public Utilities Commission, 525 Golden Gate Ave., San Francisco, CA 94102, (415) 554-1809, amavroudis@sfwater.org

INTRODUCTION

The Crystal Springs – San Andreas Transmission System Upgrade Project (CSSATSUP), one of 81 WSIP projects, is spread out over 8 miles of environmentally sensitive San Francisco Peninsula Watershed and has seven different project jobsites which cross or are within close proximity to the San Andreas Fault (SAF) as well as other minor fault zones. The Crystal Springs - San Andreas Transmission System (CSSATS) transmits water between three reservoirs including the Upper Crystal Springs Reservoir (UCSR), Lower Crystal Springs Reservoir (LCSR), and the San Andreas Reservoir (SAR) which in turn supply source water to the Harry Tracy Water Treatment Plant (HTWTP) for use as both supplemental supply during peak demands and emergency drinking water. In the event of a seismic catastrophe where East Bay and Hetch Hetchy supply is lost, the peninsula watershed reservoirs would serve as the only emergency water supply for 1.4 Million people on the San Francisco Peninsula. The CSSATSUP includes seismic retrofit upgrades to a submerged culvert connecting two reservoirs that crosses the SAF, four reservoir outlet structures within 1000 feet of the SAF, and rehabilitation of six-miles of 60-inch diameter steel pipeline running parallel to the SAF, as well as construction of a new electrical substation and new pump station.

The Project included improvements to the seven conveyance facilities listed below:

- Upper Crystal Springs Dam (UCSD)
- Crystal Springs Outlet Structures 1 & 2 (CSOS1 & CSOS2)
- Crystal Springs Pump Station (CSPS)
- Crystal Springs / San Andreas Pipeline (CSSAPL)
- San Andreas Outlet Structures 2 & 3 (SAOS2 & SAOS 3)

This paper focuses primarily on the SAOS facilities as they represent the most critical vulnerability of the system.

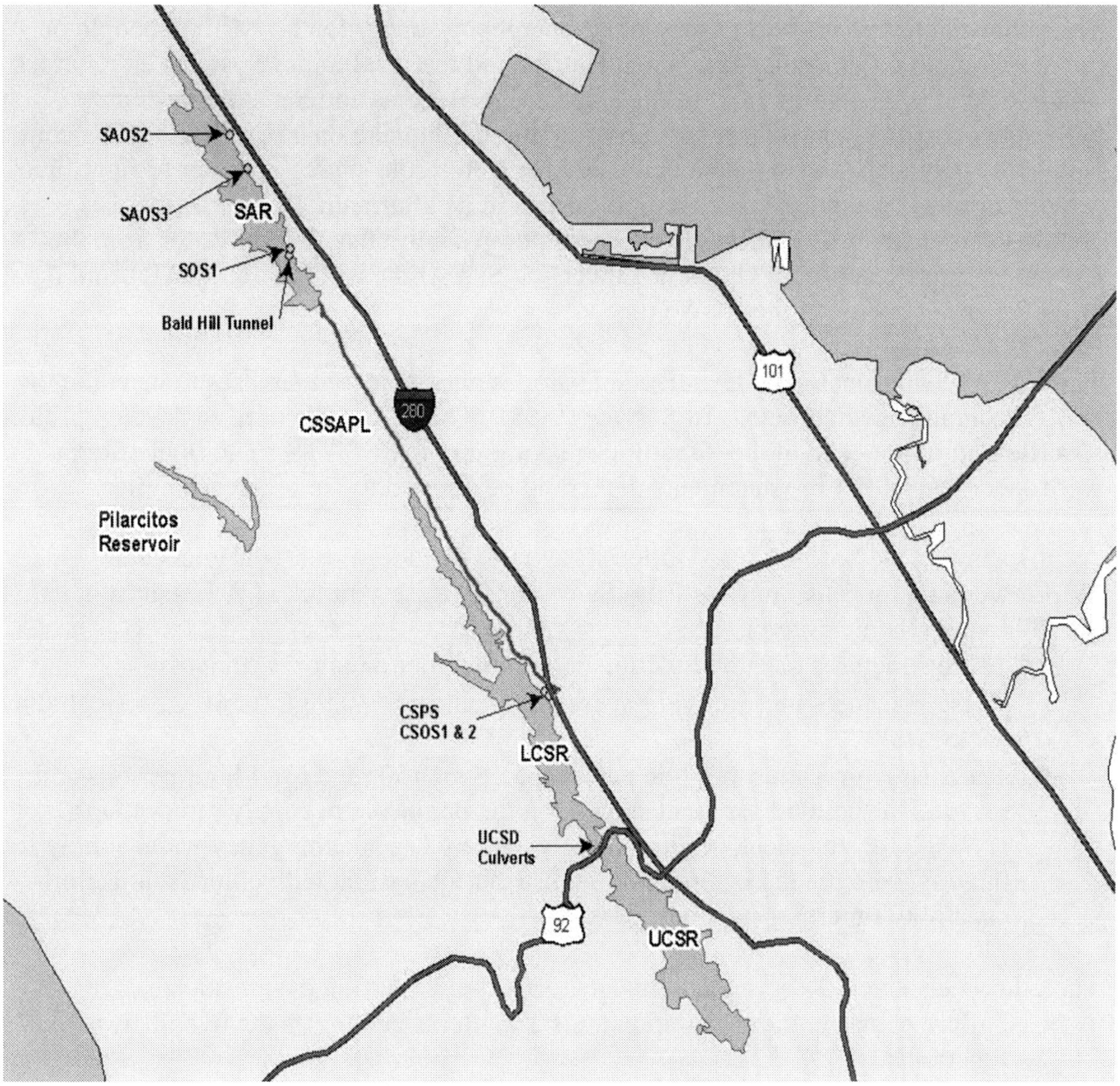

Figure 1. CSSATUP Area Map

SYSTEM OPERATIONAL CONSTRAINTS

The Hetch Hetchy Water System (owned and operated by the City and County of San Francisco's Public Utilities Commission) provides water to 2.4 million people in San Francisco, Santa Clara, Alameda and San Mateo counties. Eighty-five percent of the water comes from Sierra Nevada snowmelt and is stored in Hetch Hetchy reservoir located in Yosemite National Park. The remaining 15% of the SFPUC's supply comes from local catchment areas and is stored in reservoirs in the East Bay's Sunol Valley and in the Peninsula Watersheds. Two of these three water sources must be fully operational to meet the overall system water demand. More importantly, the local reservoirs are likely the only water source able to provide water immediately following the next big earthquake. USGS predicts a major seismic event will occur in the San Francisco Bay area within the next 20 years.

The peninsula reservoirs serve as the only emergency supply for 1.4 Million people on the San Francisco Peninsula. The Hetch Hetchy and the local Sunol Systems either cross or are located in close proximity to the Hayward, Calaveras and San Andreas Faults before reaching the peninsula reservoirs. A major earthquake on any of these faults could isolate the peninsula leaving the local reservoirs as the sole source of water to fight fires, provide drinking water, and meet sanitary needs in its aftermath. The CSSATS is comprised of a series of tunnels, pipes, and a pump station that deliver water from the Crystal Springs Reservoirs (Upper & Lower), to San Andreas Reservoir (SAR), and then into the water system via the HTWTP.

The CSSATSU Project includes seismic upgrades and capacity increases to meet the WSIP Program elevated Levels of Service (LOS). The LOS were used to develop project specific criteria for each of the 7 facilities making up the CSSATS. The final criteria were specific quantitative statements that could be applied to the project facilities.

As an example, the San Andreas Outlet Structures (including reservoir inlets, the shaft/tower structures and tunnels from San Andreas Reservoir to HTWTP) must meet the following seismic criteria:

- These structures must be functional 24 hours after a major earthquake.
- These structures must provide average capacity of 140 mgd for 30 days (including backwash)
- These structures must provide peak capacity of 150 mgd for 4 hour periods and serve as the conduit for local storage on the peninsula to supply raw water to HTWTP for treatment, or alternatively, to serve as an emergency water supply fed directly from the reservoirs into the distribution system with chlorine injection capability for emergency purposes.

The criteria also included general concepts to be applied during design such as:

- Failure of an individual component is acceptable as long as the overall system is functional.
- Under a major earthquake on the Calaveras or Hayward Fault, the system should suffer minimal damage and remain fully functional.
- Under a major earthquake on the San Andreas Fault, the system may suffer some damage but must remain functional.

Authorities having jurisdiction over some or all of our facilities imposed additional criteria.

The California Department of Water Resources - Division of Safety of Dams (DSOD) has jurisdiction over approximately 1,250 state regulated dams. An integral part of their program is ensuring that outlet works have sufficient capacity to draw down the reservoir in case of an emergency. For dams that impound more than 5,000 acre-feet such as San Andreas, DSOD establishes specific criterion after considering factors such as dam design and construction, the dam's overall condition, type of dam, downstream hazard, and seismic hazard. The criterion established for San Andreas is 10% of the reservoir head (measured from the upstream toe to the maximum water surface elevation) in 7 to 10 days, and the full reservoir capacity in 90 to 120 days.

Finally, designs had to consider operations of the system during construction. The Quality Control Inspection Program (QCIP) scheduled all required shutdowns at the overall program level, a coordination effort that included 81 projects across the system. Each project had to request shutdown windows early on in the planning for the full duration of the project to ensure the system could continue to reliably deliver water.

The CSSATS includes the facilities necessary to move water from Upper Crystal Springs Reservoir to the raw water pump station of the HTWTP. The facilities include the UCSD Discharge Culverts, Lower Crystal Springs Dam Outlet facilities, Crystal Springs Pump Station, Crystal Springs/San Andreas Pipeline, and the San Andreas Outlet facilities. This project includes hydraulic and seismic improvements to these facilities in order to reliably supply 140 mgd of raw water to the HTWTP and to serve as the sole emergency supply for the San Francisco Peninsula.

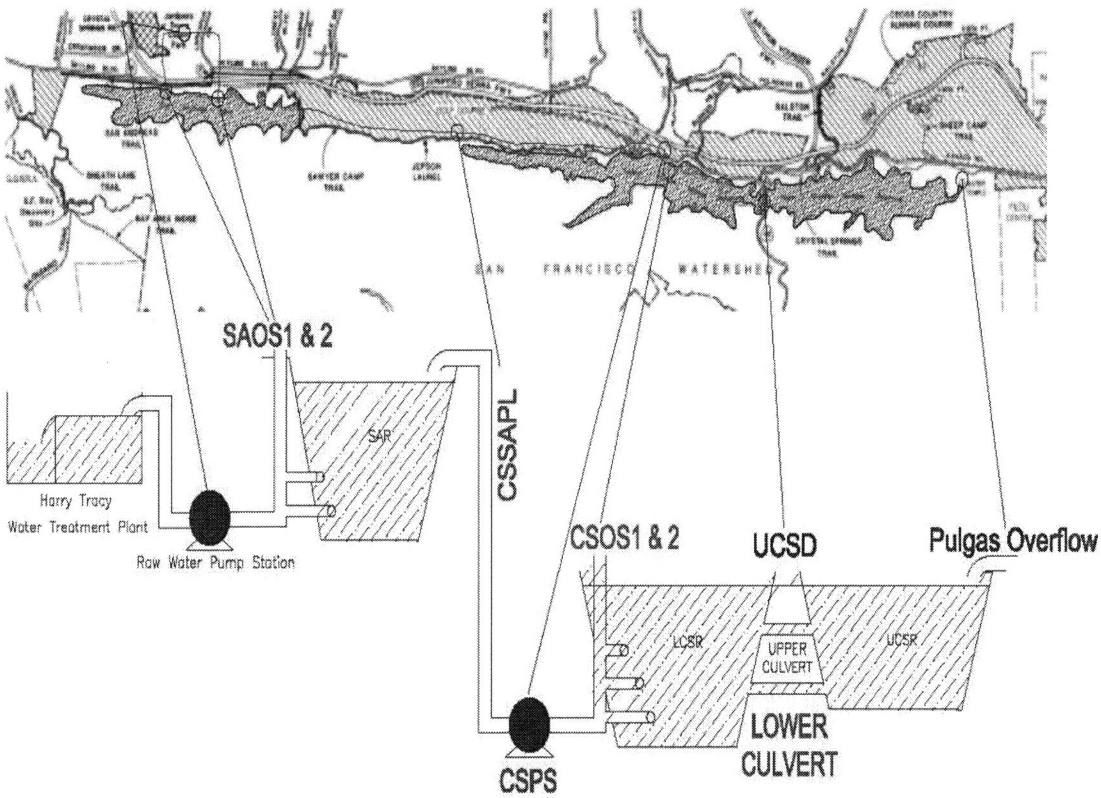

Figure 2. CSSATS Conveyance Overview

STRUCTURAL DESIGN CONSIDERATIONS

The seismic retrofitting of the historic CSSATS presented a unique challenge for the SFPUC and its engineers due to the close proximity to the San Andreas Fault. The San Andreas Fault, a right-lateral strike-slip fault, runs in an approximate North-South orientation and is the cause of the valley formation which is currently used for water storage (i.e., Crystal Springs and San Andreas Reservoirs). The fault line in some areas actually transects the project and runs directly through the Eastern abutment of UCSD lower culvert and the within 100 feet of both SAOS outlet structures.

The seismic criteria was divided into two categories, "Seismic Capacity" and "Functional Capacity" and defined per review of the structural code (IBC) and functional requirements of the system. The following is a list of requirements for each scenario:

San Andreas Fault Event:

Seismic Capacity – the ability for structures to survive

- Structure does not collapse but may have damage beyond economical repair

Functional Capacity – the ability to operate the water system

- Mechanical, Electrical, and Pipeline failures are operable within thirty days (24 hours for SAOS2 and SAOS3)
- Direct Delivery System is immediately operable as a fail-safe during specific system failures, (for use during a boil water order).

In applying the above criteria it is necessary to consider the operational strategy of the system. From a system standpoint, the water stored on the peninsula should be drawn down in excess of normal levels only under extreme conditions and as a last resort. Interpretation of the operational strategy identifies the peninsula reservoirs as key to providing operational reliability and earthquake preparedness. Under any scenario, restoring supply from the East side of San Francisco Bay (East Bay) and minimizing the draw-down of the peninsula reservoirs should be a priority. This would result in requiring HTWTP to meet a maximum rate (153 MGD) only for the duration necessary to restore supply from the East Bay.

Seismic Criteria Applied to San Andreas Outlet Structures:

The outlet facilities from San Andreas Reservoir to HTWTP and the direct connections to the system are of the highest priority to the SFPUC. Failure of these facilities would result in not meeting the levels of service under any of the scenarios stated above. Furthermore, direct delivery of water from SAR to the distribution system would not be achievable. System reliability will be provided by redundancy of outlet facilities.

SAOS2 and 3 deliver raw water from the SAR to the Harry Tracy Water Treatment Plant (HTWTP). Both outlet structures consist of shafts excavated on shore with adit tunnels connecting to the reservoir and outlet tunnels connecting to pipelines. Water is conveyed in closed conduits such that the towers themselves are dry. In addition to delivering to the HTWTP, the outlet structures also have the ability to deliver raw water directly into the San Francisco water supply through existing connections to San Andreas Pipelines No. 2 and 3. The San Andreas Fault is located within 100-feet to the west of the SAOS structures. The fault also has undefined splays that are unpredictable for future rupture.

After a major seismic event on the San Andreas Fault, the supply water from the SAR must be available to the HTWTP within 24 hours. In the event of an extreme emergency, in which HTWTP has failed, either SAOS 2 or SAOS 3 must remain operational to deliver water directly to the system as a fail-safe option. The design criteria forced the design team to produce a unique solution to the seismic problem.

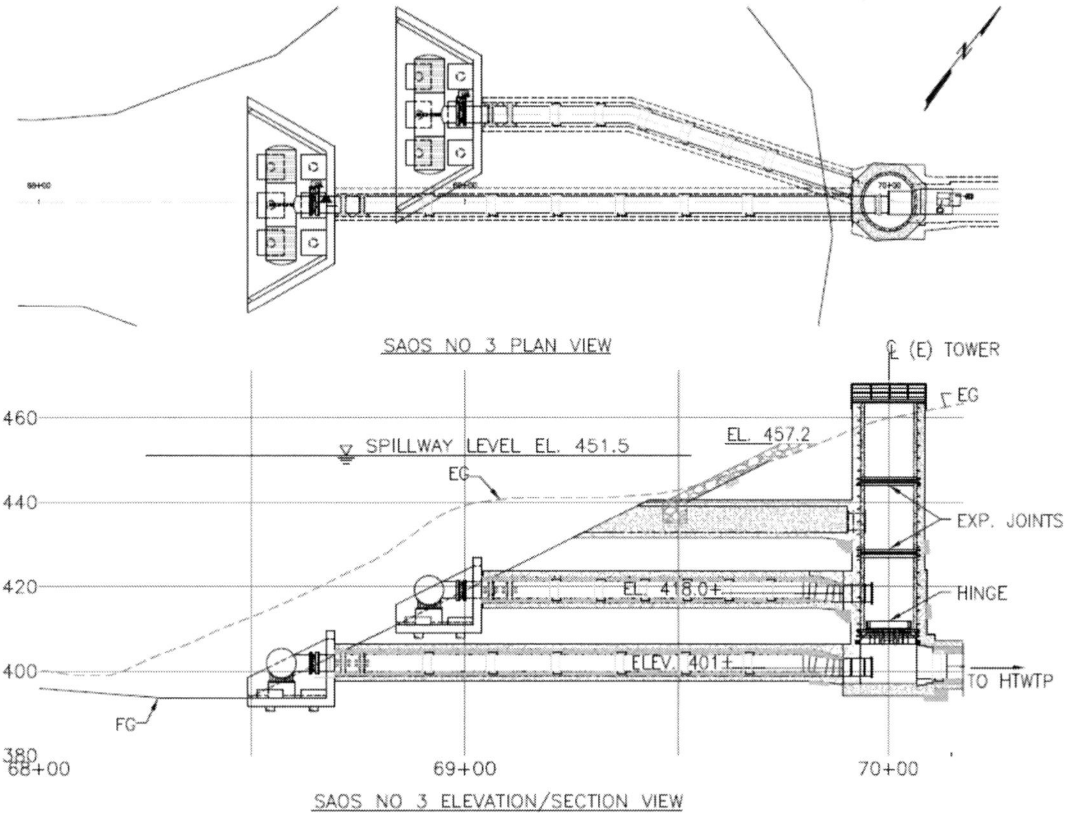

Figure 3. SAOS3 Tower Seismic Upgrade Section

Because the two intake towers/shafts are embedded in the reservoir bank the design required consideration of the soil structure interaction. For this reason, the towers/shafts were modelled in FLAC (Fast Lagrangian Analysis of Continua; an explicit finite difference modelling software).

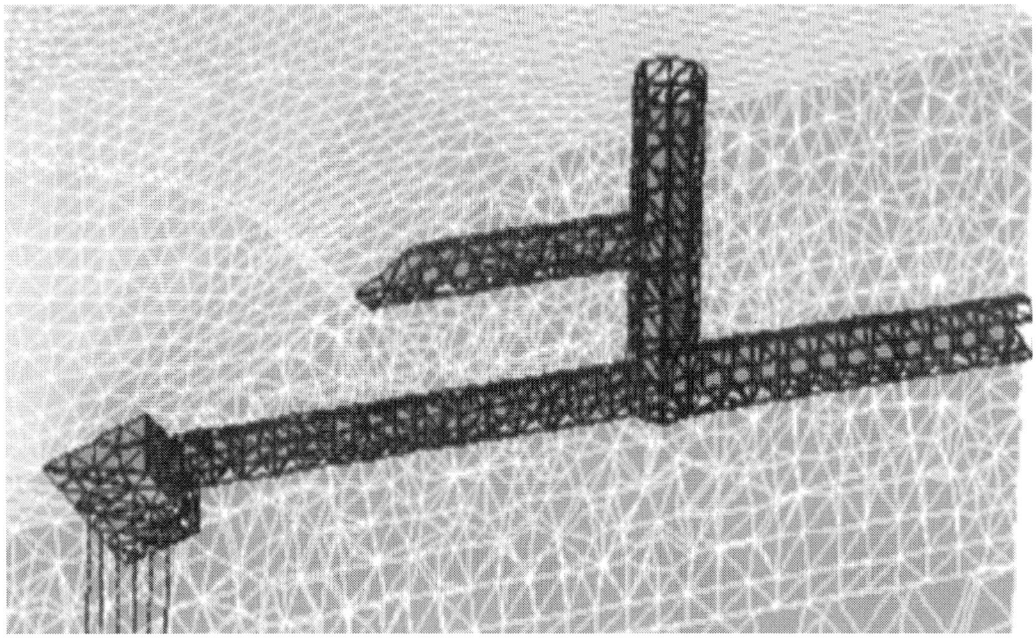

Figure 4. SAOS2 Tower FLAC Model Mesh

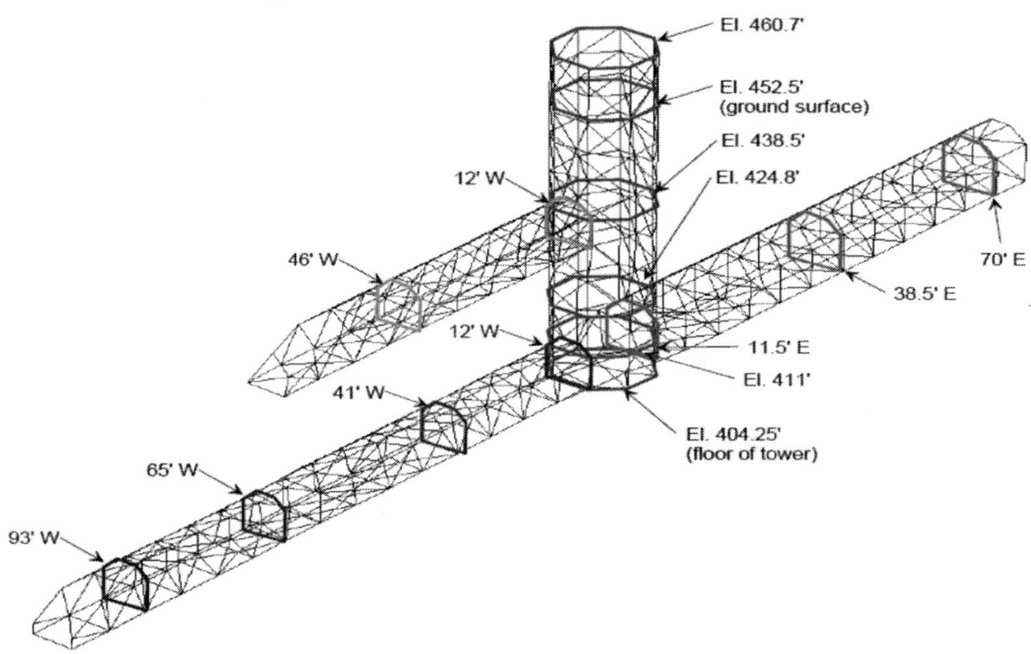

Figure 5. Critical Areas of FLAC Model

The FLAC Model provided deflections and rotations at critical points in the structure. The deflections and rotations obtained from FLAC were then input to a SAP (Structural Analysis Program; a finite element modelling system) model for structural analysis. The SAP model was configured as a "stick" model with actual sections of the tower modelled through the SAP program SECTION DESIGNER module which allows the modeler to include section geometry and material properties of the various tower cross sections.

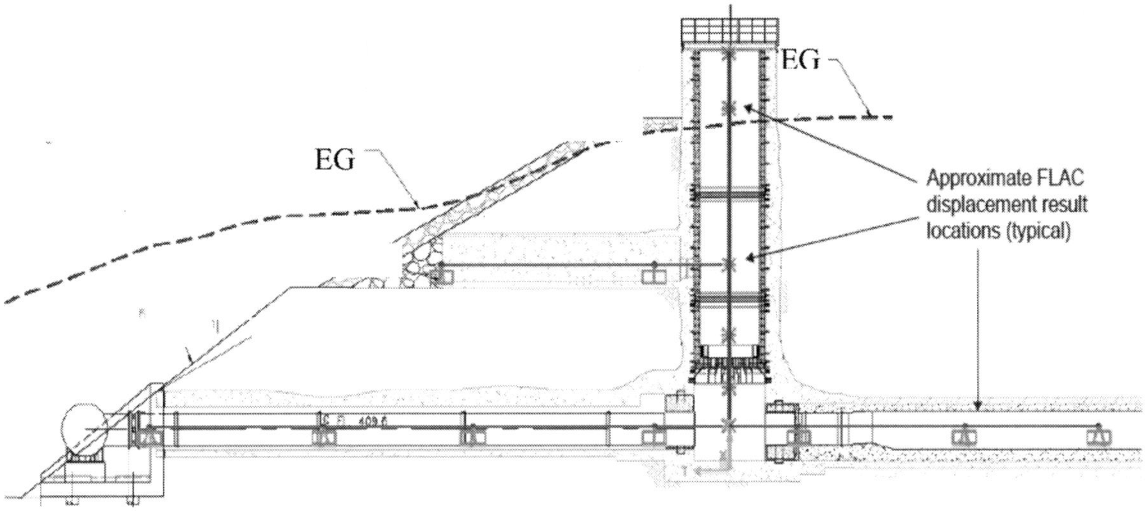

Figure 6. SAP Model Overlain by Actual Structure

Figure 7. Extruded SAP Model

Conventional structural design considerations could not accommodate the seismic loading and space limitations. Analysis results for the loading (moment) at the base of the towers/shafts indicated that the seismic loads would be significantly larger than could be adequately reinforced against. Additionally, analysis models indicated that the natural ground could slough up to 3 feet with the final location of the top of the shaft moving towards the reservoir, with the top of the towers/shafts on the ground surface further from the base. For these reasons, the tower was designed to fail in a controlled manner. The FLAC model was adjusted to allow for rotation and extension at critical points and re-analyzed. Again, the resulting rotations and displacements were input to the SAP model for final design of the tower liner.

To achieve the goal of allowing the tower to rotate, a physical hinge was designed for the base section of both towers just above the lower adit/tunnels to allow the tower to break and rotate at a preferred location.

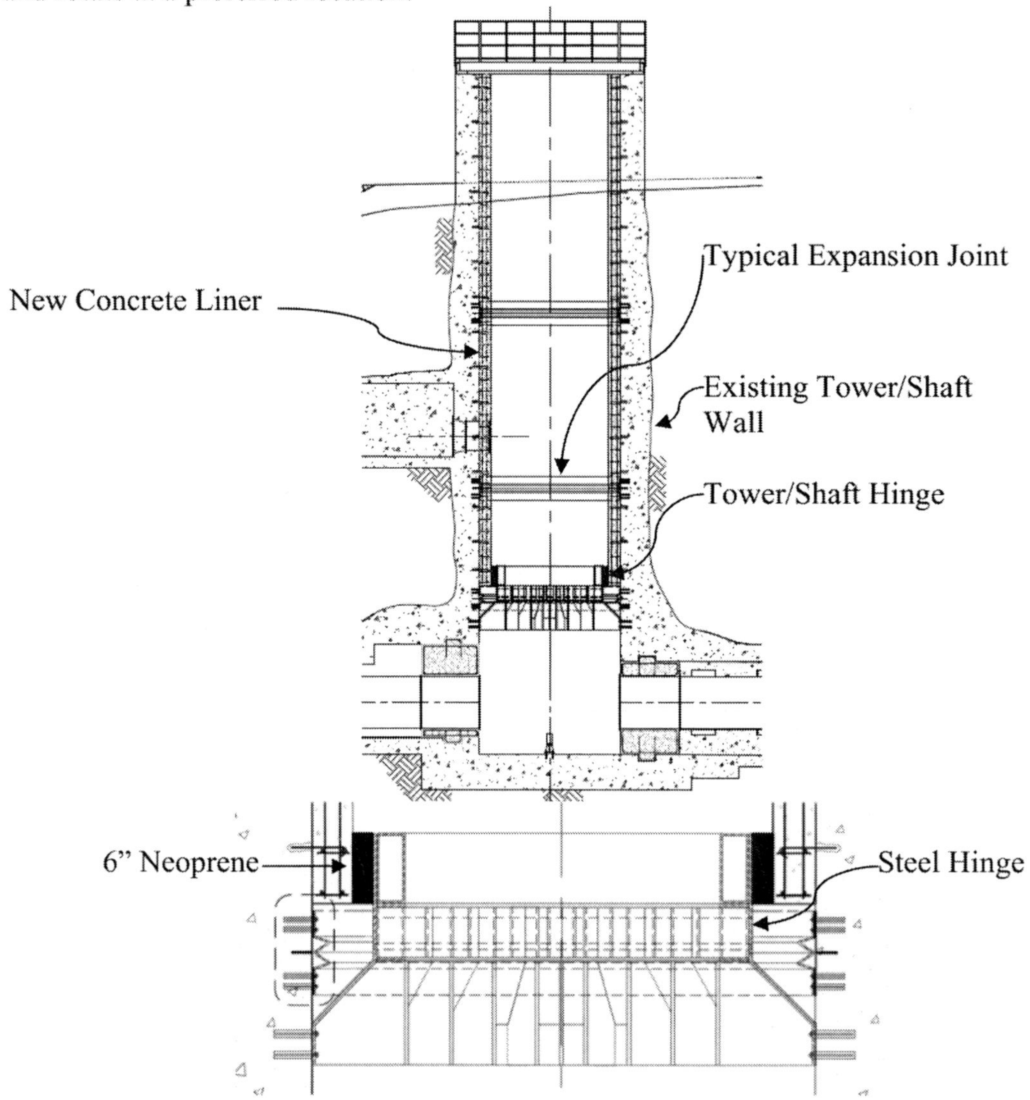

Figure 8. SAOS2 Tower Hinge Details

To produce the hinge, the existing tower structure was "weakened" by saw-cutting a 6-inch deep notch in the interior side of the existing tower wall. The saw-cutting in this location forms a weakened plane in the concrete and interrupts steel reinforcement so that "failure" is focused at this point. The saw-cut location is covered by a sheet metal "bellows" that will allow the structure to expand and contract as the tower rotates. This bellows is backed by neoprene at the anchorage points to inhibit water inflow after a large seismic event and avoid flooding the tower.

The hinge is designed to be anchored into the stronger stable base of the tower just above the tunnel/adit connections while allowing rotation of the tower above the weakened plane. The upper ring of the hinge is sized to be 6-inches smaller than the internal radius of the tower with the new liner in place. In addition, the thickness of any new internal liner was limited to one-foot in order to maintain access to the mechanical equipment within the shaft. This void is filled with a 6" neoprene pad to permit the rotation of the tower. This neoprene pad also provides damping of the system to diffuse the seismic energy at this location.

To address the lengthening of the towers/shafts, expansion joints were designed in the tower to enable the elongation of the tower structure due to lateral spreading of the bank. These are formed in much the same way as the tower hinge in that key locations adjacent to the adits were intentionally weakened and steel bellows were anchored to either side of the weakened planes. The bellows will allow the tower to "stretch" and still remain intact. As with the hinge, the bellows are anchored to the tower liner with neoprene backing. Additionally, the new tower liner reinforcement is connected above and below the expansion joint with smooth dowels greased and inserted into PVC sleeves and tied with stirrups to the liner reinforcing, thereby maintaining the integrity of the new tower liner.

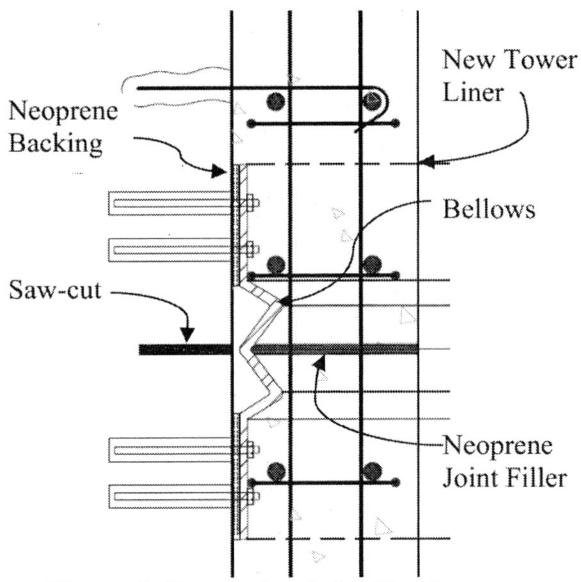

Figure 9. Expansion Joint Section

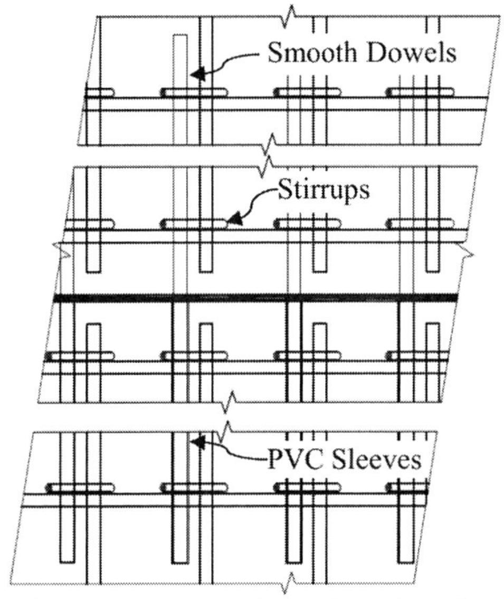

Figure 10. Expansion Joint Elevation

CONVEYANCE PIPE DESIGN CONSIDERATIONS

A new steel pipeline was installed at the center of each concrete lined adit tunnel and encased in low strength concrete. The reduced capacity through the adit was offset by an increased capacity in the tunnel where the existing pipe was replaced with a new larger diameter steel pipe. The hydraulic head loss is slightly less than before. The new pipe was installed with several mechanical expansion joints to allow for movement (articulation, expansion and contraction) during a seismic event and the annular space was filled with cellular concrete. The length of adit tunnel extending out into the reservoir toward and/or crossing the fault line was cut back and removed so that a new terminal structure could be installed outside the immediate known fault rupture zone. If a seismic event resulted in the failure of the reservoir bank/shoreline near and/or above the adit structures, the pipeline terminal structure would be pull away from the structure while the pipeline expansion joints would slip allowing for displacement. A failure of the concrete tunnel with displacement perpendicular to the tunnel length would crush the low strength cellular concrete between the pipe walls but mitigate the potential for pipe wall damage.

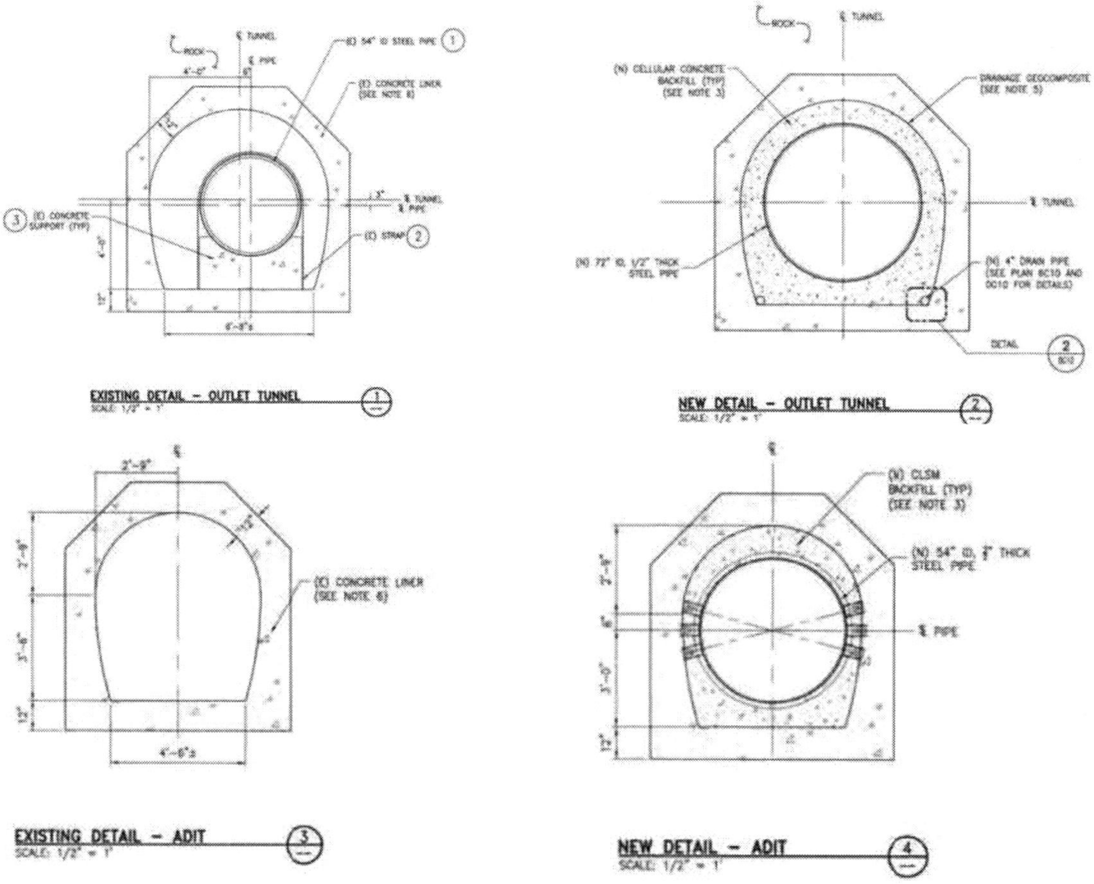

Figure 11. SAOS2 Adit and Tunnel Seismic Upgrade Sections

Pipes and appurtenances inside the shafts, adits and tunnels were structurally braced and fitted with articulating features to move in sync with the shaft structural design, with the intent of maintaining flow immediately after a seismic event. Mechanical joints were strategically placed in locations along all of the pipe alignments to provide the required articulation.

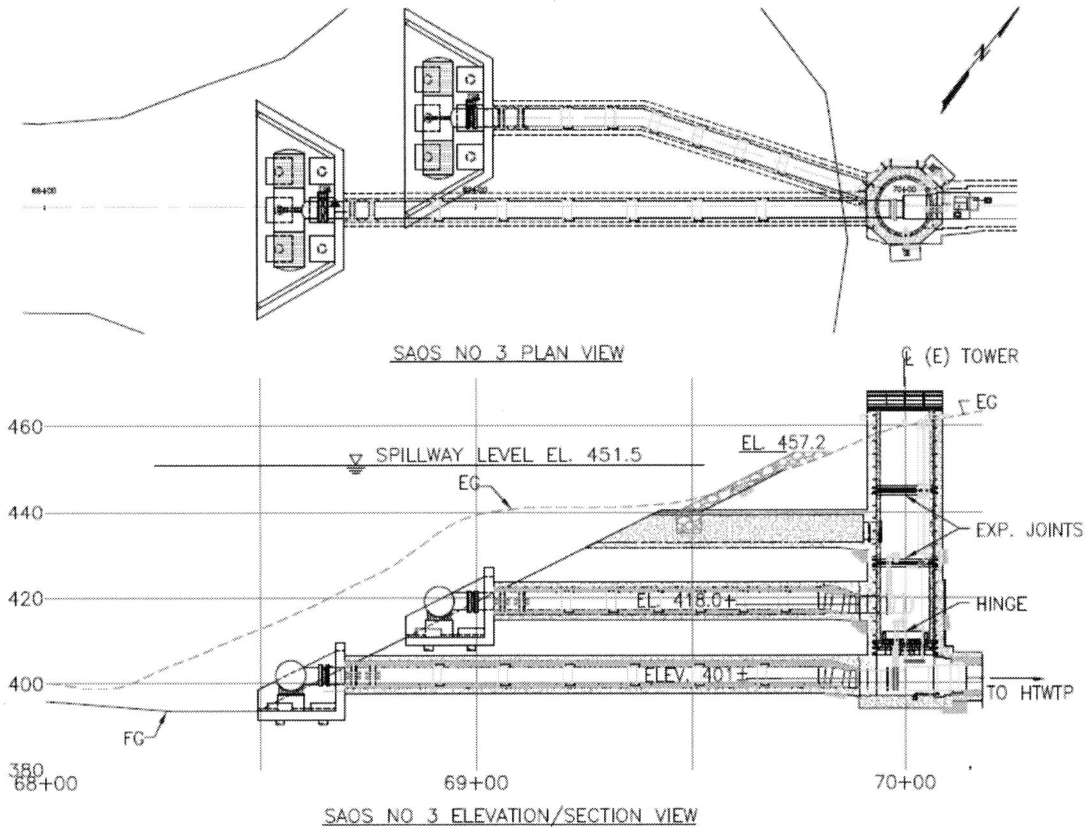

Figure 12. SAOS3 adit, shaft and connection to tunnel; general arrangement and articulating joint layout

CONCLUSIONS

Final design and analysis of both structures indicates that the towers/shafts, even if permanently deformed, will remain stable and functional following a large San Andreas earthquake. The design permits the SFPUC to meet their post-earthquake goals for water delivery to the City and County of San Francisco.

SEISMIC RISK TO FLOOD-CONTROL STRUCTURES AND AGRICULTURE IN THE NEW MADRID SEISMIC ZONE OF ARKANSAS AND MISSOURI

Christopher A. King, P.G.[1]

ABSTRACT

Over the last decade, engineers and scientists published various assessments regarding the impact of large magnitude seismic events on infrastructure in the New Madrid Seismic Zone (NMSZ). However, there is little mention concerning potential effects on flood control structures and agriculture in the relatively flat and largely rural Mississippi River Delta region. The counties most at risk in northeast Arkansas and southeast Missouri produce over $3 billion worth of row crops annually - predominantly corn, soybeans, and rice - making agriculture the largest part of the regional economy. This level of production would never have been possible without numerous earthen dams, levees, and drainage canals constructed over the last century with assistance from the U.S. Department of Agriculture's Natural Resources Conservation Service (formerly SCS) and U.S. Army Corps of Engineers. The northern Mississippi Delta region leads the nation in rice production due to favorable climate, soils, and a productive alluvial aquifer. Unfortunately, this region is at the heart of the NMSZ, where even a 6.0 Mw quake results in widespread liquefaction because of susceptible soil conditions.

Economic devastation resulting from compromised flood control structures and ruined fields is a real threat. There is a 25 – 40% chance that this region will experience a magnitude 6.0+ seismic event over the next 50 years. A Mw 6.7 tremblor struck Charleston, Missouri in 1895, and recent estimates suggest an event of this magnitude could generate liquefaction over more than 14,000 square miles.

INTRODUCTION

Covering parts of Arkansas, Illinois, Kentucky, Missouri, and Tennessee, the New Madrid Seismic Zone (NMSZ) is the most active earthquake region in the United States east of the Rocky Mountains, with more than 200 earthquakes each year. Fifteen earthquakes with estimated magnitudes of 6.5 to 8.0 occurred in the NMSZ between December 16, 1811 and March 16, 1812, along with thousands of smaller earthquakes. There is a 7 – 10% chance for a Mw 7.0 or greater quake occurring in the NMSZ during the next 50 years, and about a 25 – 40% probability of a 6.0 or greater event happening during this time (USGS, 2009).

Major events of 1811 – 1812 were located in northeast Arkansas and southeast Missouri, at the northern end of the Mississippi Embayment (Figures 1-3). Ground failures such as lateral spreading and widespread subsidence of the land surface resulted from soil liquefaction across the Mississippi River flood plain and its tributaries (USGS, 2009).

[1] USDA Natural Resources Conservation Service, 700 W. Capitol Ave., Room 3416 Federal Building, Little Rock, Arkansas, 72201 E-mail: chris.king@ar.usda.gov

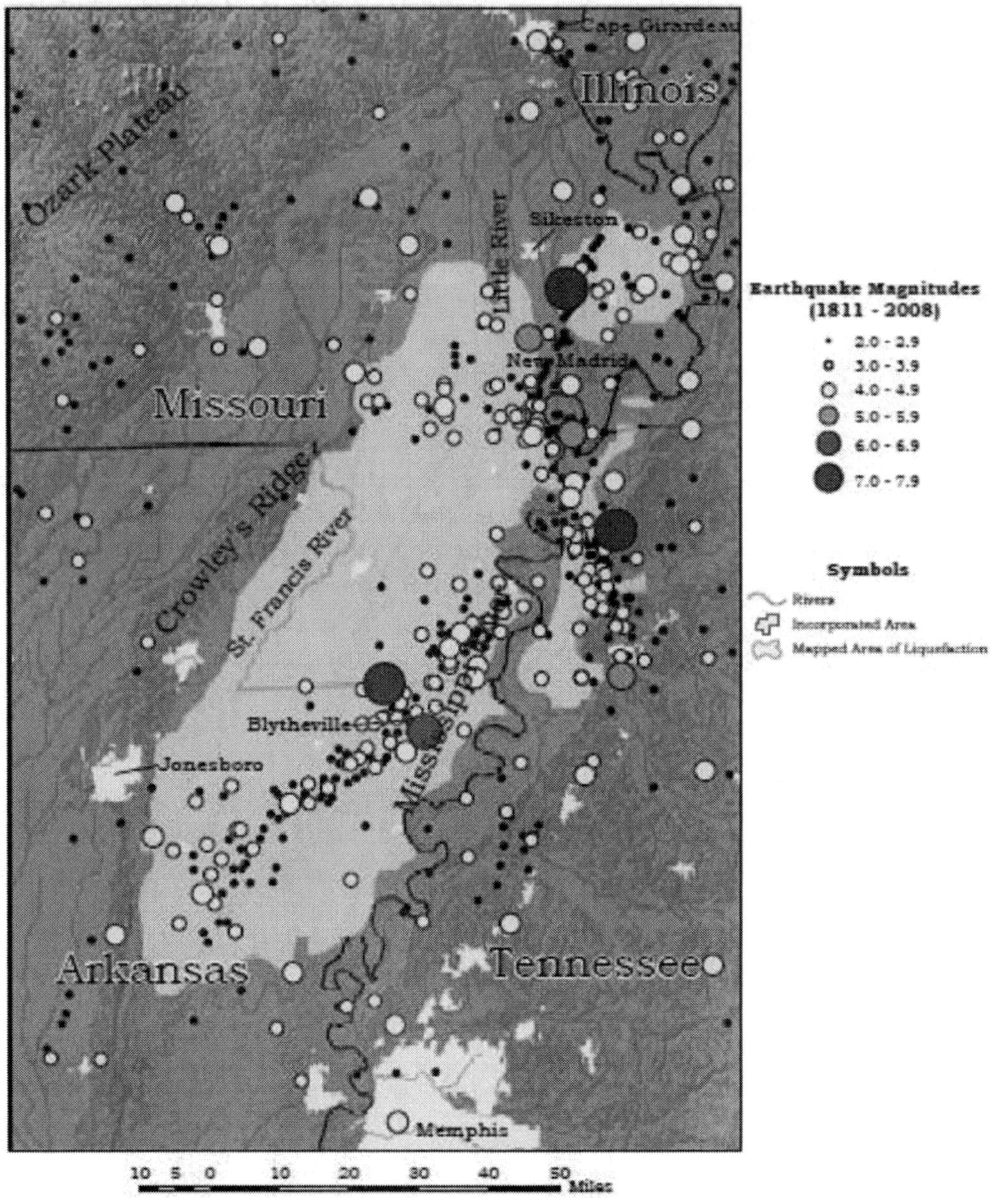

Figure 1. Area incorporating the New Madrid Seismic Zone, showing locations of selected earthquakes from the CEUS Project Database (http://www.ceus-ssc.com/Report/Downloads.html). Blue shaded area is region where extensive sand blows and lateral spreading have been mapped (Crone and Wheeler, 2000).

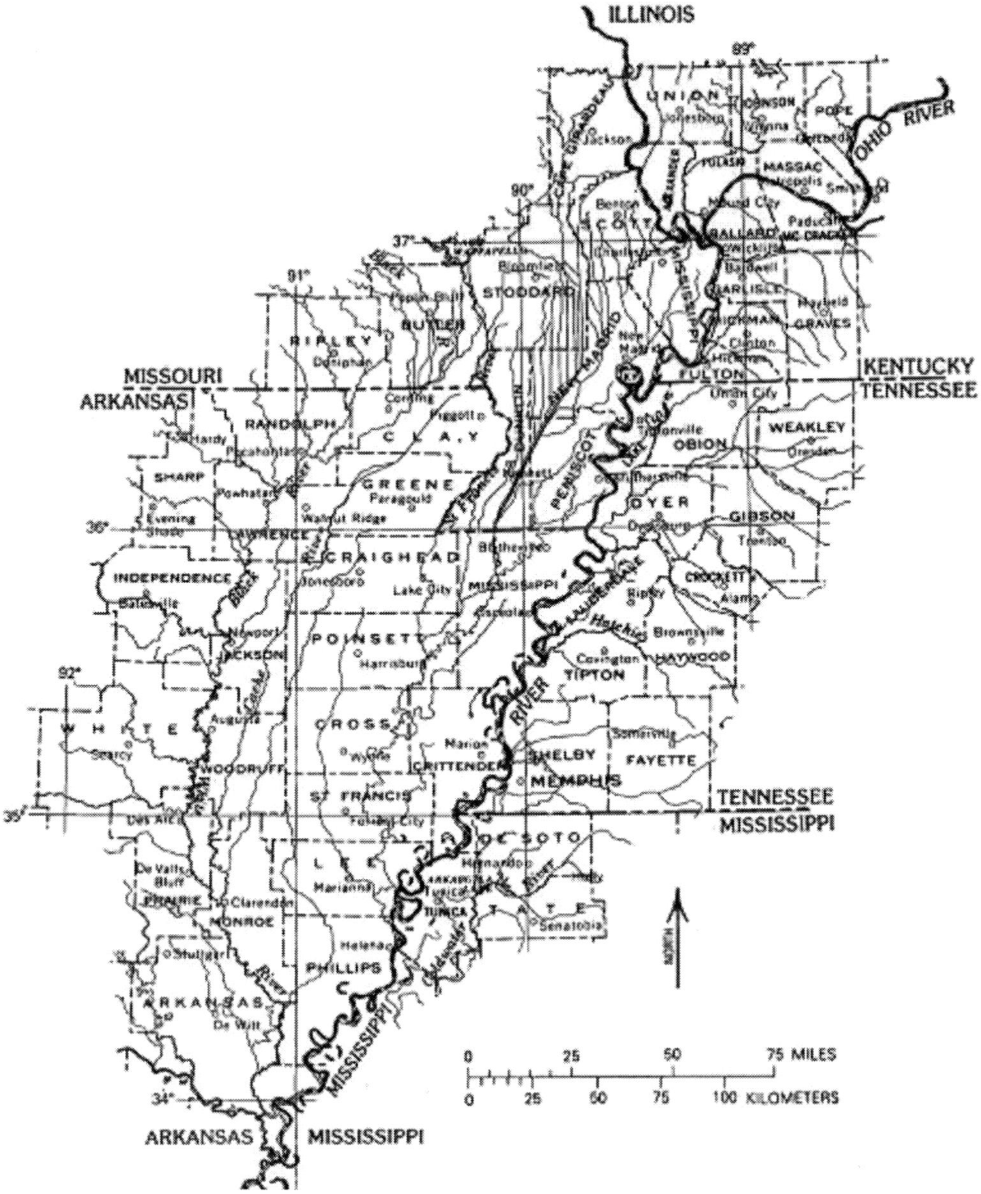

Figure 2. Region encompassing the New Madrid Seismic Zone, showing counties, major streams, and towns (Obermeier, 1989).

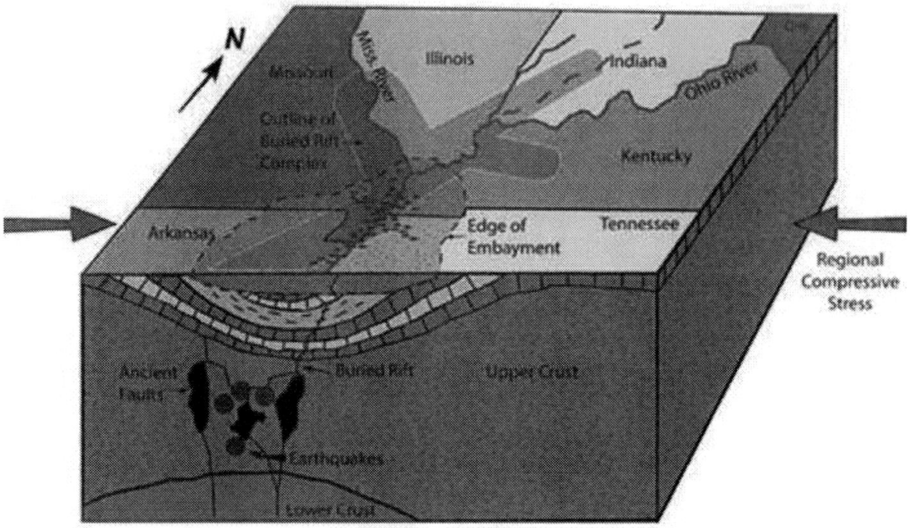

Figure 3. Block diagram of the New Madrid Seismic Zone, showing northern Mississippi Embayment and outline of Reelfoot Rift (http://gsa.state.al.us/gsa/geologichazards/NewMadrid.htm).

When the first principal quake struck during the winter of 1811 - 1812, the New Madrid region was sparsely populated. An early chronicler described the earthquake's effect as the ground moving like waves on the land, when suddenly the earth would burst, sending up huge volumes of water and sand, leaving chasms where the earth had burst open (Hendricks, 2011). Near the St. Francis River in northeast Arkansas, vast tracks of land sank as far as fifty feet into the earth or disappeared into the river. These "sunk lands" of northeast Arkansas cover parts of Craighead, Mississippi, and Poinsett counties. After 1812, government surveys described ninety percent of the region as a hopeless, permanent mosquito and malaria-infested swamp (Jackson, 2014 and Blackwell, 2007). However, early residents realized that if the land was cleared of timber and drained, the rich alluvial soil would be productive for a variety of crops, especially cotton. Reclaiming the swamp and overflow lands of the Mississippi River Delta required building levees and other structures to mitigate frequent floods. In the early 1900's wealthy landowners established numerous local watershed districts to build canals and levees. By the 1980's watershed districts constructed a series of dams in highlands bordering the Delta.

The flat, low-lying northern Mississippi River Delta region is now a major agricultural production area, due in part to a shallow and prolific alluvial aquifer along with rich alluvial soils. The seven southeastern-most counties of Missouri (the "Bootheel") produce one third of the state's agricultural income. In 2012, the counties comprising northeast Arkansas produced $2 billion in corn, soybeans, and rice alone (USDA, 2012).

The same conditions that make this area fertile ground also make it highly susceptible to liquefaction. Today, a repeat event similar to the 1811-12 earthquakes would cause extensive damage to flood control structures and the land itself, disrupting farming for months to years. Geysers of sandy soil and water known as sand blows would erupt tens of feet into the air, each depositing a blanket of infertile sand sufficient to cover an acre of cropland. Fields leveled for irrigation would subside, and drainage ditches would be blocked (MO DNR, 2014). Failure of levees and streambanks would contribute to widespread flooding (USGS, 2009). Even a much smaller event similar to the 6.3 Mw quake that struck near Lepanto, Arkansas in 1843 has the potential to induce liquefaction over a vast area.

LEVEES AND DRAINAGE DISTRICTS

The rich agricultural lands of northeast Arkansas and southeast Missouri experienced frequent flooding on the east by the Mississippi and St. Francis Rivers, while on the west floodwaters draining the Ozark highlands ensured that the low-lying region remained wet year-round. Situated in the center of this area, a geographic prominence known as Crowley's Ridge was the only high ground not subject to inundation.

Following the New Madrid Earthquakes of 1811-1812, drainage problems intensified. Caved banks and felled trees along the St. Francis and other rivers created vast rafts, which inundated large portions of the area. Beginning six miles north of Marked Tree in Poinsett County, Lake St. Francis continued southward for twenty-four miles, stretching to twelve miles in width, and reached depths of 15 ft. Another formidable challenge to travel and development was the Great Swamp, a nearly two-million acre region of marshland in southeastern Missouri. Situated at the northern end of Crowley's Ridge, this marshland was also known as Dark Cypress and the Missouri Glades. According to historians, this marshland was the largest wetland in America (Blackwell, 2007).

With land in the region selling for pennies on the acre, prominent planters came together to form drainage districts. Organization of drainage districts required wealthy landowners to petition county courts to place a lien on the lands through a court order. This order ensured all landowners would pay improvement taxes. Money collected from the taxes paid the principle; this and interest on bonds issued by the drainage district, along with proceeds from the bond sales, were used to build levees and drainage canals. Often the drainage districts received matching funds from the federal government.

In 1907, prominent southeast Missouri planters formed the Little River Drainage District (LRDD) to tackle the challenge of draining the Great Swamp. By 1928, the district had built nearly 1,000 miles of ditches, more than 300 miles of levees and drained 1.2 million acres of land in what was the world's largest drainage project of its time. The project moved more earth than was excavated during construction of the Panama Canal. Before 1907, swamp covered over 90 percent of the land administered by the LRDD. Nearly 96 percent of this territory was cleared by 1928 (Lewis, 2012).

The LRRD work resulted in two significant changes. First, the project diverted Ozark highlands water that formerly flowed into the New Madrid hinterland into more direct routes to the Mississippi. Second, the district replaced irregular creeping fingers of the area's myriad small streams with straight drainage ditches at right angles to each other, roughly one mile apart, all ultimately draining into Big Lake just west of Blytheville,

Arkansas. The main ditches are 200 feet wide with average water depths of 8 feet. Smaller ditches are around 30 feet wide. This water is picked up by watershed districts in Arkansas, where it makes its way to the Mississippi River in Helena, Arkansas, over 150 miles downstream (Figure 4).

Today the LRDD extends from the Headwaters Diversion Channel just south of Cape Girardeau to the Arkansas border and includes parts of seven counties – Cape Girardeau, Bollinger, Dunklin, New Madrid, Pemiscot, Scott, and Stoddard. The total leveed area is 1,040 square miles, encompassing a population of approximately 33,000 people. The Headwaters Diversion Channel is a crucial component within the LRDD, running 40 miles from around Greenbrier in Stoddard County to its outlet into the Mississippi River south of Cape Girardeau. Bounded by large levees, the channel collects water from the St. Francis, Castor, and Whitewater rivers, as well as four large streams, and redirects the flow into the Mississippi River. Three water detention basins store floodwater to prevent the southern levee of the Diversion Channel from overtopping (Blackwell, 2007).

Organized levee construction in northeastern Arkansas began in 1893 with the formation of the St. Francis Levee District, chartered by the State legislature to drain the sunken lands. Relying on an inconstant system of levees, the district was unable to control a disastrous flood in 1897, and again in 1903, 1912, 1913, 1915, and 1916. With the accumulating damage, the U.S. Congress finally responded, passing the Ransdell-Humphrey's Flood Control Act of 1917. Although insufficient, it provided $45 million in cost-share for local drainage districts within the Mississippi River Valley.

Efforts to control the Mississippi and its tributaries upstream only intensified the severity of floods in unprotected areas downstream. In 1917, U.S. Senator Thaddeus R. Carraway warned the "prominent men in eastern Arkansas" that Missouri intended to divert the overflow from the Ozarks and "turn the entire column loose on Arkansas." Caraway counseled Arkansans to seek an injunction against the Missouri improvement districts until the United States government canalized or dredged the St. Francis River to control the increased volume of water. This sparked new interest in flood prevention for the region. In 1917 one of the Country's largest districts, Drainage District Number Seven, was formed in Poinsett County

To stabilize and control the tortuous St. Francis, District No. 7 planned to strictly regulate the river's flow north of Marked Tree and divert overflow water to the Steep Gut Floodway. This one-mile wide artificial, 20,000-acre channel extends from the foot of Lake St. Francis to the Poinsett – Cross county line, where the upper end of the St. Francis Floodway begins. Designated as navigable to Wappapello, Missouri, the St. Francis River carried a considerable traffic of lumber and boats. The Steep Gut Floodway, lock, and sluiceway were completed in 1926, just prior to the Flood of 1927, which destroyed many of the Drainage District's levees (AHPP, 2014; Mitchel, 2014).

Figure 4. Mississippi River and Tributaries flood control projects. MVM 2: Commerce - St. Francis River System Levee; MVM 10: West Bank St. Francis Floodway System; MVM 11: St. Francis East to Big Lake West Levee System; MVM 12: Big Lake and St. Francis Floodway East System; MVM 14: Elk Chute Levee System; MVM 21: Little River Drainage District Levee System (USACE, 2014).

U.S. Army Corps of Engineers Flood Control Projects

The great Mississippi River flood of 1927 covered nearly 26,000 square miles and affected eleven states. During this event the Mississippi flowed behind New Madrid down the St. Francis basin, rejoining its usual bed just north of Helena, Arkansas (Bolton Valencius, 2013). In response to the previous year's devastating overflow, Congress approved the Flood Control Act of 1928, which formalized flood control in the Mississippi Valley. Promoted by Army Corps of Engineers Chief Edgar Jadwin, it rejected a plan to convert the St. Francis Basin into a reservoir, instead recommending raising levees in the basin an average of three-and-a-half feet above their current grade and increasing the width of the Steep Gut Floodway to two miles.

Jadwin's plan called for constructing the Birds Point – New Madrid Floodway in southeast Missouri to reduce flooding at and above Cairo, Illinois during a major flood. Built by the Corps between 1929 and 1932, this floodway is between 3 and 15 miles wide. It includes the 55 mile long Upper St. Francis Frontline Levee, which provides protection to a 205 square mile agricultural area within the floodway during periods when it is not in operation. The floodway diverts 550,000 cubic feet per second from the Mississippi River during the project design flood. At this flow level the Mississippi will drop 7 feet at Cairo. The floodway activates through a controlled destruction of the levee, either with explosives or by overtopping (AHPP, 2014; USACE, 2014).

With assistance from the Corps, by 1934 District No. 7 completed improvements to the Steep Gut floodway allowing twice the previous flow at the same stage; however, serious erosion issues and levee collapse at the outlet end of sluiceway dogged the project, culminating in a massive winter flood in 1937. To save the town of Cairo, Illinois, the Army Corps of Engineers dynamited the levees on the Mississippi Valley floodway, flooding much of the Bootheel (Bolton Valencius, 2013). In 1938 the Flood Control Act was amended to include over $21 million for the St. Francis River Basin Flood Control Project in Missouri and Arkansas, designed to protect one million acres in Arkansas, as well as repair existing levees and construct new ones along the St. Francis and Little Rivers. Later that year, as the Corps and drainage district personnel worked to repair the sluiceway, high waters crevassed the levee and washed out a ninety-foot gap, damaging it beyond repair.

Providing an excellent example of the challenges designers face when building flood control structures in the NMSZ, investigations by the Corps of Engineers determined the sluiceway and levee failed due to underlying strata of fine sands, which tended to become "quick" when saturated. Out of concern for the safety of the levee and any replacement structure on the uncertain foundation, the Corps of Engineers decided to dam the levee gap, and instead of passing water under or through the dam, siphon water over it. In seven months, the Corp's Memphis District had designed and installed the Marked Tree Siphons. Dedicated in 1939, Drainage District Number Seven still operates and maintains the siphons (AHPP, 2014).

The Corps designed the Marked Tree Siphons to take water out of St. Francis Lake (a.k.a. the Sunken Lands), hoist it over the levee, and discharge it into the St. Francis River, thereby maintaining structural integrity of the levee. There are no water pumps and no moving internal valves. Once a vacuum pump primes the siphons, operators can constrict the flow by opening an air valve, which introduces a bubble into the top of the arch. Once primed, the siphons can run indefinitely powered only by gravity. The siphons preserve the original channel of the St. Francis River, provide for emergency relief of the floodway, and allow farming of lands around the lake. The Marked Tree Siphons were essential to the St. Francis River Basin Flood Control Project and a major component in the Army Corps of Engineers flood control plan for the Mississippi River Valley. Built of three nine-foot diameter, 228-foot long steel tubes, the Marked Tree Siphons are a unique engineering achievement and the only siphons of their type in the United States (AHPP, 2014; Mitchel, 2014).

As remarkable as the siphons are, perhaps the most impressive levee in the NMSZ is the 277 mile long Lower St. Francis Levee, which begins at New Madrid and ends near the mouth of the St. Francis River, about six miles north of Helena, Arkansas. This structure protects 3,760 square miles, including portions of New Madrid and Pemiscot counties in Missouri, and Mississippi, Craighead, Poinsett, Crittenden, St. Francis, and Lee counties in Arkansas. This highly productive agricultural area has not been flooded since 1927. A break in the upper reaches of this levee would create a major disaster during an extreme flood. With a total population of 280,000 in this area, estimated property damages of $11 billion could occur.

Today, main stem levees on the west bank of the Mississippi River begin just south of Cape Girardeau, and except for gaps at points where tributaries enter the Mississippi, extend continuously to Venice, Louisiana. On the east bank, the levees are not continuous because of high bluff lines near the riverbank (USACE, 2014).

Effects of Drainage Projects

Regrettably, early unregulated drainage projects to create farmlands within the Mississippi Delta region destroyed enormous swaths of wetlands and forests. But with increased awareness of environmental consequences in the 1960's and 70's, numerous laws and programs were created to prevent future wetland loss, such as Section 404 of the Clean Water Act and the National Environmental Policy Act. Since 1977, when President Carter issued the Executive Order on Wetland Protection, direct Federal action to drain wetlands has been limited. The Wetlands Reserve Program administered by the USDA Natural Resources Conservation Service provides financial incentives for landowners to protect wetlands on their property.

Although some of the nation's most economically depressed rural counties are located in the Delta, modern drainage techniques have allowed towns in the region to prosper, turned waterlogged bottomlands into tillable soil, and eliminated malaria. Large farms remain the dominant land use in southeast Missouri and northeast Arkansas. Prior to adequate levees only about ten percent of the land in Mississippi County, Arkansas was capable of cultivation, with the rest consisting of impassable swampland. Through a widespread network of drainage ditches, levees, and innovations such as the Marked Tree siphons, the northern Mississippi Delta region has become more habitable. These structures have maintained the St. Francis River in its natural channel, as well as protected agricultural resources and permitted expanding settlement in the area.

PUBLIC LAW 566

Following World War II there was a great deal of rethinking on the issue of levees and drainage districts. Lessons learned during the Dust Bowl and catastrophic floods in the lower Mississippi River Valley during the 1920's and 30's led to a more holistic approach to controlling soil erosion and floods.

The Watershed Protection and Flood Control Act of 1954 (commonly referred to as PL-566) authorized the Soil Conservation Service (now NRCS) of the U.S. Department of Agriculture to provide planning assistance and construction funding for flood control projects. The program provides technical and financial assistance (cost sharing) to local sponsors (normally watershed districts or local governments) to develop and implement watershed protection, flood prevention, and agricultural water management.

Included under agricultural water management are irrigation, rural water supplies, agricultural drainage, and measures for water conservation and water quality. Non-agricultural water management activities include measures to improve fish and wildlife habitat and water-based recreation. Practices normally included in project plans are land treatment, nonstructural measures, and structural measures. Structural measures can include floodwater-retarding structures and channel work.

The PL-566 program bridges the gap between on-farm conservation practices installed by individual landowners and the large projects installed on major rivers by the Corps of Engineers and the Bureau of Reclamation. The Corps is chiefly responsible for large-scale projects, while the NRCS undertakes smaller-scale projects. Restrictions on projects include the following: the size of the watershed must be 250,000 acres or less; no single structure may provide more than 12,500 acre-feet of floodwater retention; and no single dam may provide more than 25,000 acre-feet of total capacity. Congress must approve projects with costs greater than $5 million or structures with total storage greater than 25,000 acre-feet. The Federal cost-sharing rate for NRCS flood control projects has been 100 percent, while single-purpose projects without flood control benefits, e.g., those involving only agricultural benefits, receive only 50 percent Federal support. However, a project that involves both flood control and agricultural benefits is eligible to receive 75 percent Federal financing of the total project costs even though agricultural drainage is the primary purpose. NRCS only provides financial and technical assistance if the loss of wetlands incidental to project purposes is mitigated.

From the 1940s through the 1970s, local communities using NRCS assistance constructed more than 11,800 dams in 47 states. These watershed management projects provide an estimated $2.2 billion in annual benefits in reduced flooding and erosion damages, and improved recreation, water supplies, and wildlife habitat for roughly 47 million people.

These reservoirs prevent more severe downstream flooding by providing a minimum 10-day residence time for surface flow contributions to their storage in the upper reaches of watersheds. PL-566s also attenuate sediment and nutrient pollution carried with surface runoff by storing these constituents in the upper basin, allowing for dissolved nutrients and sediments to settle out of suspension to the bottom of the reservoirs where they can be stored (Prochnow et al, 2007).

The potential failure of dams located upstream from towns and cities creates a risk for loss of life and property, particularly in seismically active areas. Many NRCS-assisted dams begin with a low or significant hazard classification, as most of these dams protect rural agricultural areas from flooding. Although only a handful of small watershed dams built under the authority of PL-566 exist in the Missouri Bootheel, there are 90 structures in northeast Arkansas, with an average maximum storage capacity of 1,700 acre-feet (Figure 5).

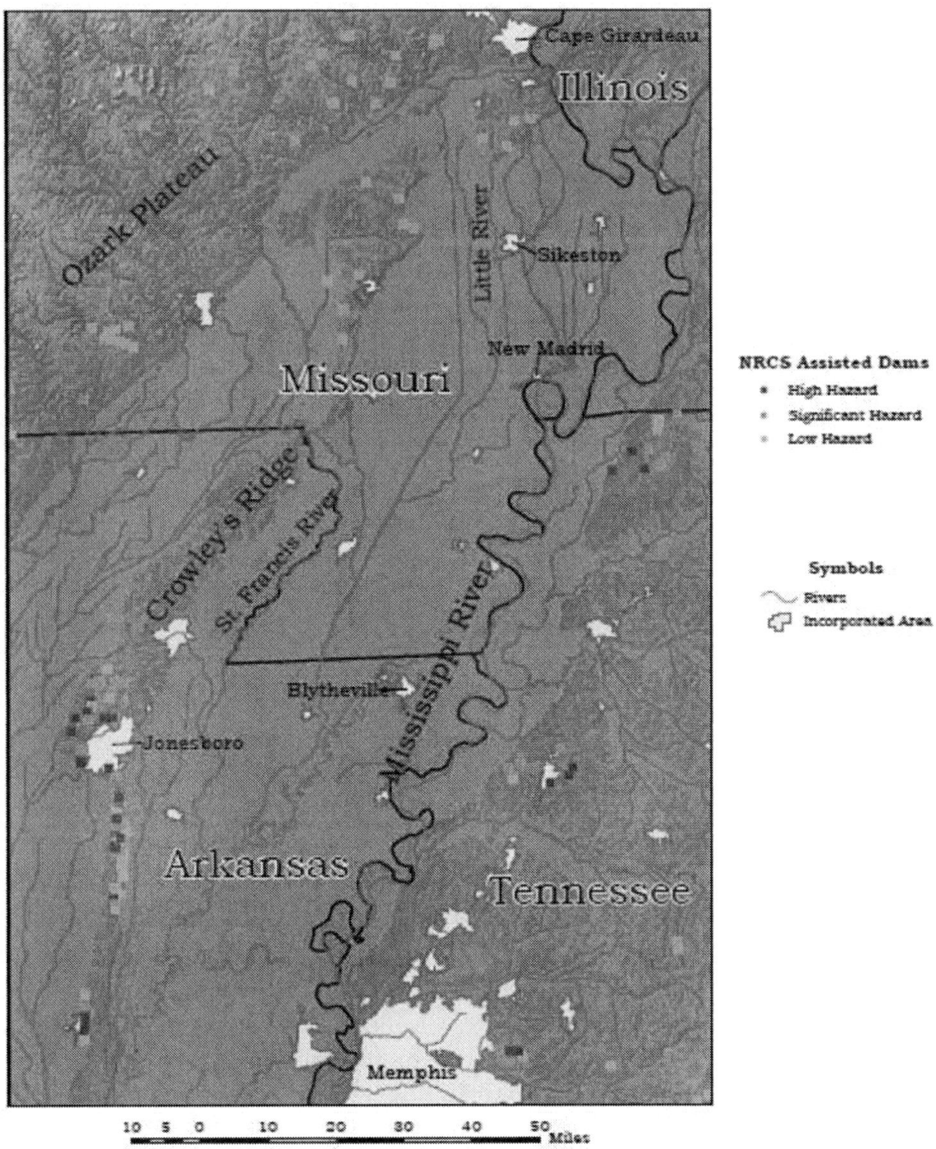

Figure 5. USDA-NRCS assisted PL-566 watershed dams in northeast Arkansas and southeast Missouri.

Twenty-six of these structures are now high hazard because development has occurred downstream in areas that would flood if the dam failed. High hazard potential dams are those where failure or mis-operation will probably cause loss of life and serious damage to important structures, according to the 2004 Federal guidelines for dam safety published by the Federal Emergency Management Agency.

With a maximum storage capacity of 6,400 acre-feet, the largest high-hazard dam in northeast Arkansas is Big Creek Site 2, which impounds Lake Frierson near the Greene – Craighead County line. This structure is located roughly six miles north of the Jonesboro metropolitan area. At a 2010 population of 120,000, this is the largest metropolitan area in the NMSZ of northeast Arkansas and southeast Missouri.

Each high hazard dam has an Emergency Action Plan (EAP), a formal document that identifies potential emergency conditions at a dam and specifies actions for operators to follow in order to minimize loss of life and property damage. Most EAPs include a breach inundation map outlining the areas that could be flooded should a total dam failure occur. The inundation map serves as a general evacuation map, beginning with a breach routing analysis and floodplain surveys. The inundation map for Big Creek Site 2 shows a worst-case scenario assuming the lake is at full stage and a sudden catastrophic failure were to occur (Figure 6).

REGIONAL GEOLOGY

Northeast Arkansas and Southeast Missouri are located within the Upper Mississippi Embayment, a southward plunging trough that extends from southern Illinois to the Gulf of Mexico. Embayment deposits are composed of unconsolidated sediments from the post-Paleozoic period, ranging in thickness from a few feet along the edges of the embayment to more than 3,000 feet along the axis of the embayment (Figure 3). The Mississippi Alluvial Plain consists of the youngest and uppermost sediments in the embayment, underlain by about 100 to 150 feet of Quaternary alluvium. The basal Quaternary is also known as the Mississippi River Valley Alluvial Aquifer (MRVAA), comprised of gravel and/or coarse to medium sand overlain by fine sand which generally grades to silt or clay within 20 to 30 feet of the surface. The underlying Tertiary deposits are unconsolidated clays and sands.

Crowley's Ridge, a geomorphic prominence, rises roughly 200 feet above the alluvial plain trending northeastward, roughly parallel to the axis of the embayment. The ridge is an erosional remnant separating the floodplain of the present Mississippi River from that of its former course west of the ridge. Containing Quaternary clay, silt, and sand with lenses of gravel, Crowley's Ridge is capped with Pleistocene loess, which can reach thicknesses of over 25 feet. The ridge is an erosional remnant, containing older Tertiary alluvial deposits at depth. Crowley's Ridge and the Ozark Highlands form the headward reaches of many tributaries to the St. Francis and other rivers.

The foundation strata of most PL566 dams in northeast Arkansas consist of Mississippi Alluvial Plain terrace deposits and channel fill sediments derived from Crowley's Ridge. These deposits include various sequences of gravelly sands, sands, silts, and clays with assorted mixtures. The layers more often resemble discontinuous lenses with various thicknesses (King, 2010).

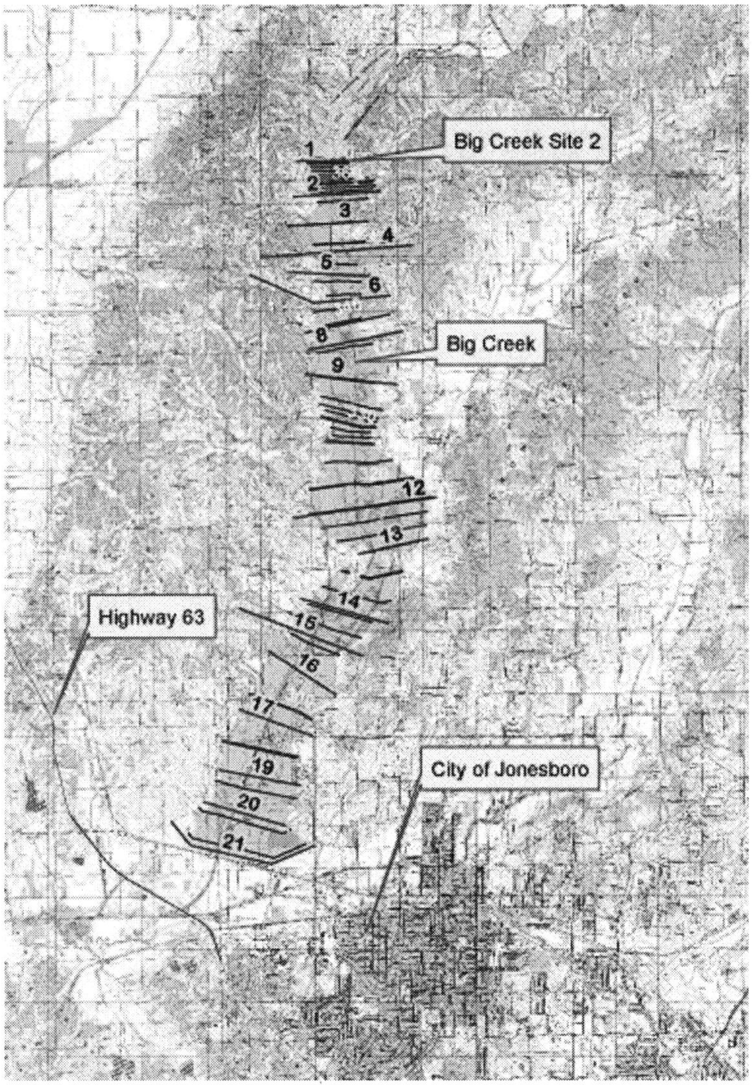

Figure 6. Breach inundation map for Big Creek Site 2, a USDA-NRCS sponsored watershed dam located 6 miles north of Jonesboro, Arkansas. Numbered lines are survey profile lines across the Big Creek floodplain. It is important to note that the figure is only an approximation of the effects should a breach occur.

The New Madrid Seismic Zone contains reactivated faults that originally formed when North America began to split or rift apart during the breakup of the supercontinent Rodinia some 750 million years ago. The rift failed, but remained as a scar or zone of weakness.

As the Atlantic Ocean was opening during the Mesozoic Era (about 200 million years ago), rifting was once again re-activated and intrusive igneous rocks were emplaced. Again, the rifting failed and the continent remained intact, although with a significant zone of weakness. This rift, commonly known as the Reelfoot Rift, coincides with the northernmost portion of the Mississippi embayment.

A regional stress field with maximum compressive stress oriented approximately east-northeast-west-southwest controls contemporary seismicity and current deformation in the New Madrid region. Ancient faults are now active primarily as strike-slip faults. The modern seismicity is concentrated into two major trends that form a zigzag pattern with an overall northeasterly trend. Most of the seismicity is located from three to 15 miles beneath the Earth's surface. Depths of igneous rock bodies closely correspond to the depth of seismic activity

Along the linear, northeasterly trending seismicity zones faults have near-vertical dips. Faults in the cross-rift trend are oriented NW-SE and have dips of about 31 – 48 degrees, with a main dip direction to the southwest (Crone and Wheeler, 2000).

SEISMIC EFFECTS

Dams must withstand severe earthquakes with minimal deformation to prevent overtopping and protect life and property. Earthen dams may slump or deform under severe earthquake shaking. A dam's ability to withstand an earthquake is termed seismic stability.

Soils under earthquake shaking generally exhibit softening of stiffness or a loss of strength, which may cause settlement of structures, slope failures, or loss of bearing capacity. Known as liquefaction, this phenomenon is especially associated with saturated cohesionless soils. The crucial question in assessing the dynamic stability of a dam or levee is whether the embankment or its foundation is prone to liquefaction or not (Ozkan, 1998). While there are other earthquake-induced phenomena the engineer needs to consider, their importance pales in comparison to that of liquefaction.

Exactly how an earthen embankment, such as a levee or dam, will perform during a large-magnitude seismic event is difficult to predict, as it depends on the foundation conditions, embankment materials, orientation of the structure, magnitude, distance, and depth of the seismic source, and many other variables (Marcuson et al, 2007). PL-566 structures have been through thorough geotechnical investigations prior to construction, including drilling and sampling to look for potentially liquefiable sediments, faults, groundwater conditions, and the stability of foundation materials.

PL-566 watershed dams such as Big Creek Site 2 would likely suffer damage should a strong to major (6.0 – 7.9 Mw) seismic event occur within the NMSZ. Sudden, catastrophic failure is unlikely, however. Research by Foster et al (2000) notes 136 complete dam failures out of 11,192 documented dams worldwide, with only 1.5 percent of failures attributed to earthquakes. These dam failures were tailings or hydraulic fill dams, or relatively small earthfill embankments of older and possibly inadequate design and construction.

Given the lack of case history performance, prediction of embankment behavior under earthquake loading often has significant uncertainty. Despite few historical cases, engineers consider seismic failures as a legitimate threat, with capability of sudden loss to a reservoir (USBOR and USACE, 2012).

Thick alluvial sediments and a shallow water table present throughout the Mississippi River Delta region, combined with a high seismicity risk, result in elevated liquefaction potential for watershed structures within the NMSZ.

There are three general requirements for liquefaction to occur: 1 – cohesionless, silty to sandy soil, 2 – fault capable of generated an earthquake of magnitude 5.0 or greater, and 3 – depth to groundwater should be less than 50 feet (Lund et al, 2008). East of Crowley's Ridge water levels in the MRVAA are often within 20 feet of the land surface (Ausbrooks and Prior, 2008). Limited recharge and high pumping rates for irrigation purposes keep water levels deeper west of the ridge, but still commonly less than 50 feet from the land surface (Figure 7). The average depth to water in the MRVAA of southeast Missouri was from ten to eighteen feet in 2012 (Henggeler, 2013). Localized areas with shallower, saturated sand & silt deposits exist throughout the alluvial plains along numerous streams and abandoned river channels.

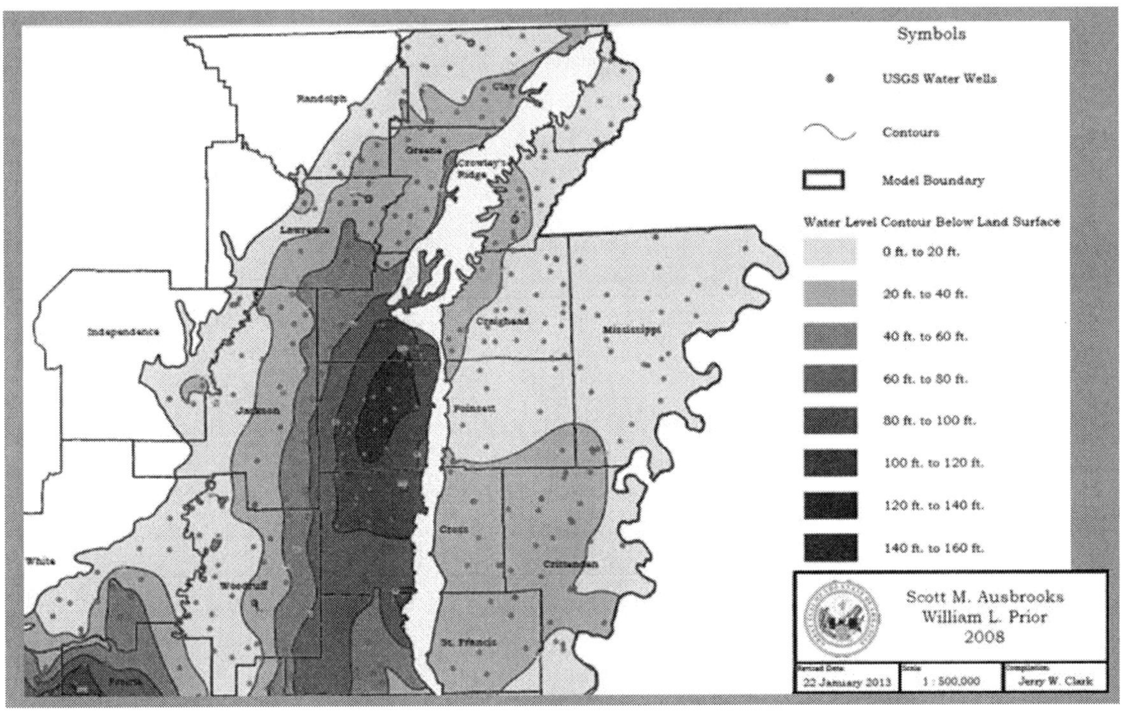

Figure 7. Section of map showing depth to groundwater in the Mississippi River Valley Alluvial Aquifer of northeast Arkansas (Ausbrooks and Prior, 2008).

The occurrence of liquefaction depends on many factors, such as earthquake magnitude, shaking duration, peak ground motion (acceleration/velocity), depth to groundwater table, basin structures, site effects and liquefaction susceptibility of sediments. The occurrence of liquefaction is difficult to predict on a theoretical basis; empirical approaches are adopted as a rule in assessing the liquefaction potential of an area (Wang et al, 2006).

At specific sites of engineering importance, the assessment method most often used is the standard penetration test. Field tests such as standard penetration tests (SPT) and cone penetration tests (CPT) help predict whether liquefaction is likely or not. The measured SPT blow counts, however, are subject to some corrections for overburden pressure, initial static shear stress and effective confining pressure. SPT value and cyclic stress ratio are not the only factors affecting liquefaction potential, where other factors such as fines content, gravel content, cementation, age, fabric, thickness of the liquefied layer, thickness of overlying non-liquefied layer, impeded seepage, and topography may affect this potential as well as the extent of it (Ozkan, 1998).

On a broader scale, empirical relations between the magnitude of ground shaking and liquefaction may be combined with numerical simulations of ground shaking during an earthquake to evaluate the liquefaction potential of a large area. However, this approach requires extensive information about sediment properties and subsurface structure of the area of interest, and both are often unavailable.

In areas where such data are not available, a simpler approach is available to set some limits to the expected extent of liquefaction during potential earthquakes. Field observations show that for earthquakes of a given magnitude, the occurrence of liquefaction is confined within a particular distance from the earthquake focus, defined as the liquefaction limit, beyond which liquefaction may not be expected (Papadopoulous and Lefkopulos, 1993; Wang et al, 2006). The liquefied sites at the liquefaction limit are likely to be those with the optimal conditions for liquefaction, that is, saturated soils with high liquefaction susceptibility. Thus, the application of the liquefaction limit to an area without prior information on its liquefaction susceptibility merely represents an estimate of the maximum likelihood of liquefaction occurrence during a potential earthquake. Sites at closer distances may include less optimal conditions but receive greater seismic input.

One of the most important problems in the study of soil liquefaction is determining how the maximum distance at which liquefaction can occur from a seismic source increases with earthquake magnitude. Several researchers have examined and mapped damage suspected to be the result of liquefaction following large-scale seismic events in various locations throughout the world (Papadopoulous and Lefkopulos, 1993; Wang et al, 2006). By noting the distance of damage from the epicenter of numerous events, liquefaction curves plotting moment magnitude (Mw) and distance help estimate the radius of potential liquefaction from various Mw events (Figure 8).

In 1843, a magnitude 6.3 earthquake occurred near the town of Lepanto, Arkansas. Utilizing the most recent (and more conservative) liquefaction curve developed by Wang and others, this event could have conceivably induced liquefaction over almost 7,000 square miles. For comparison, a 7.5 Mw event at this same location could produce liquefaction over 75,000 square miles (Figure 9).

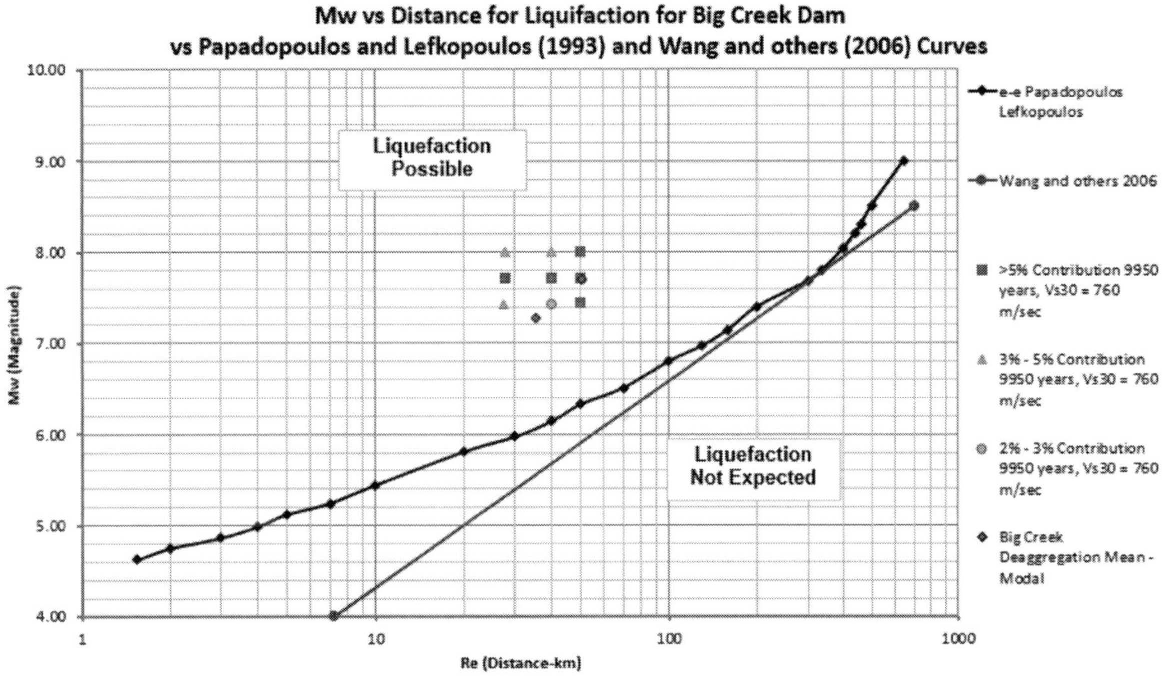

Figure 8. Magnitude vs. distance curves for assessing liquefaction potential at Big Creek Site 2 dam site, based on contributions from known seismic events in the NMSZ. Note that all points plot well above the liquefaction curves.

Sand Blows and Lateral Spreading

The extent of sand blows and lateral spreading during the 1811-1812 events is well documented, most notably between Crowley's Ridge and the Mississippi River (Fuller, 1912; Obermeier, 1989; Crone and Wheeler, 2000; Tuttle, 2005).

Earthquake-induced liquefaction temporarily builds up water pressure in the pores between sand and/or silt grains as seismic waves pass through the sediment. If the pore-water pressure increases to the point that it equals the weight of the overlying soil, the sediment liquefies and behaves as a fluid. The resulting slurry of water and sediment tends to flow towards the ground surface along cracks and other weaknesses. Overlying soil "floating" on liquefied sediments moves down even gentle slopes, causing fissuring with lateral and vertical displacements. This type of landslide known as lateral spreading is commonly responsible for damage to infrastructure (bridges, roads, buildings) during major earthquakes (Tuttle, 2011). Lateral spreading may occur on very shallow slopes with angles of only one or two degrees from the horizontal (Bethapudi, 2008).

The most obvious effects of the 1811-1812 earthquakes are large sandy deposits, known as sand blows, resulting from eruption of water and sand to the ground surface. Some of these eruptions reached tens of feet into the air. Sand blow deposits exist over an extremely large area of about 4,000 square miles (Figure 1). Sand blows have circular, elliptical, and linear shapes that often exceed a hundred feet in width and hundreds of feet in length. Viewed in cross-section, sand blows commonly take the form of large lenses 3 to 6 feet in thickness. Some sand blows pre-date 1811 as the result of prehistoric New Madrid earthquakes.

Studies of paleoliquefaction features provide estimates as to the magnitude, timing, and locations (all with inherent uncertainties) of moderate-to-large paleoearthquakes that occurred in the NMSZ. Tuttle (2005) found features formed during paleoearthquakes dated to around 1810 AD, 1450 AD, 900 AD, 300 AD, and 1100 BC.

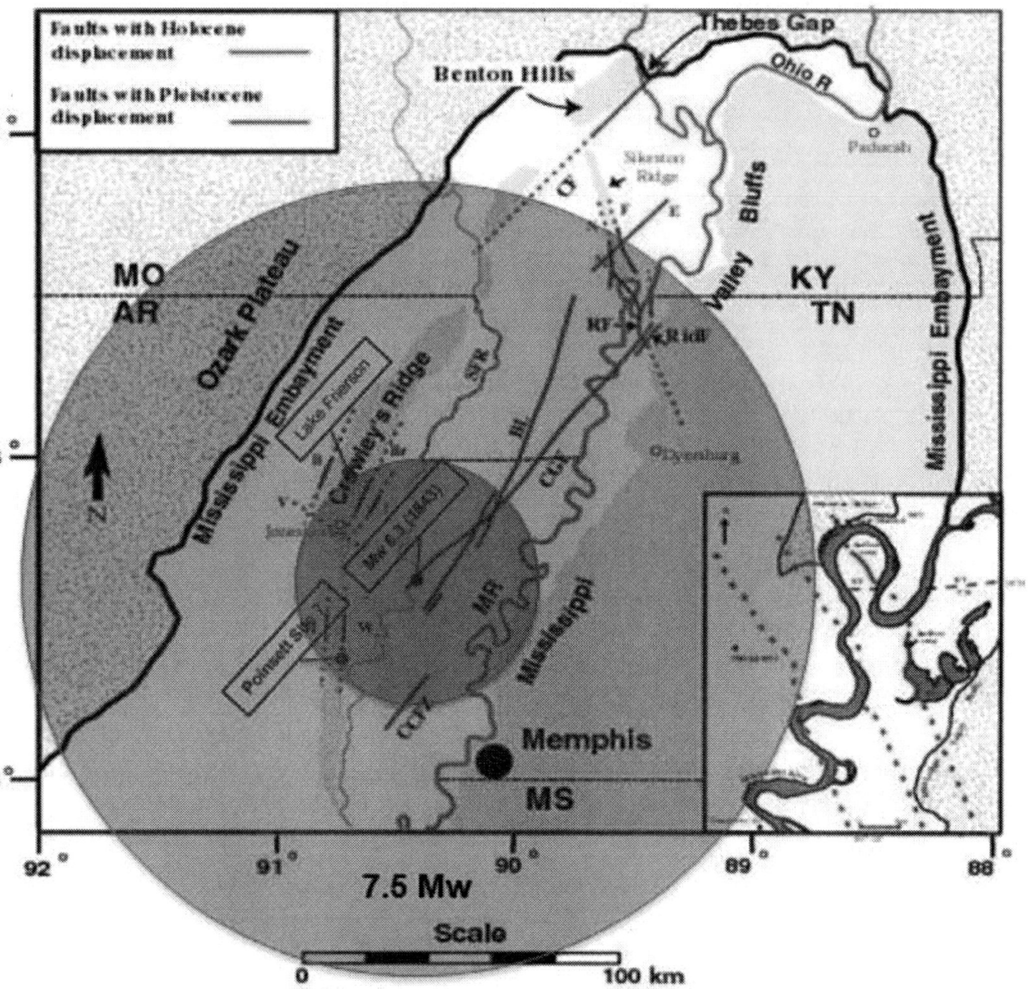

Figure 9. Extent of potential liquefaction for the Mw 6.3 earthquake at Lepanto, Arkansas in 1863 (red circle). Larger blue circle represents potential extent of liquefaction should a 7.5 Mw event happen at same location. Modified from Van Arsdale and Ellis (2002).

A much wider area of Arkansas is at risk for earthquake activity than originally believed, according to a revised USGS seismic hazards map released in July 2014 (Figure 10), and the next big earthquake may not be within the traditional boundary of the NMSZ. The 2014 USGS map accounts for paleoseismic evidence indicating at least two major earthquakes centered in Lee County, Arkansas. In 2011 geologists discovered ancient sand blows south of Marianna dated to between 4,800 and 10,000 years ago (Heard, 2014).

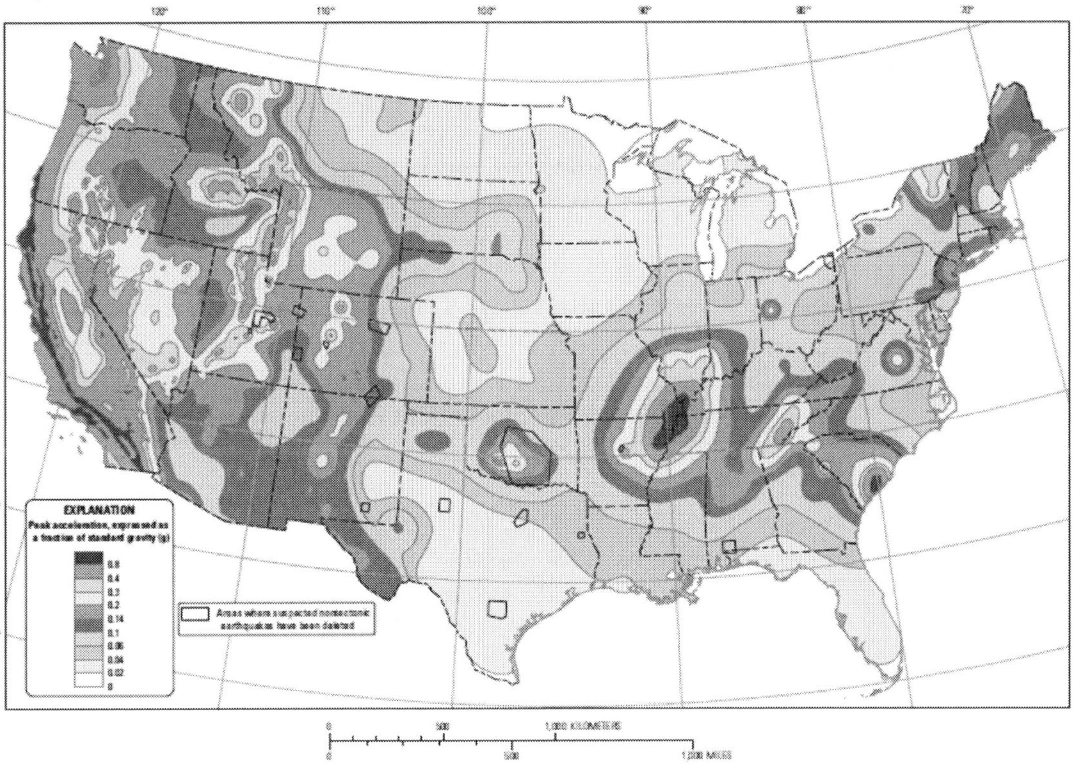

Two-percent probability of exceedance in 50 years map of peak ground acceleration

Figure 10. 2014 U.S. Geological Survey National Seismic Hazard Map for the Conterminous United States.

<u>Differential Settlement and Cracking</u>

Recent publications mention several predominant earthquake effects on embankment dams (Fell, 2005; USBOR & USACE, 2012). Liquefaction and loss of shear strength due to increased pore pressures in the embankment or foundation may result in settlement and cracking, particularly near the dam crest. Cracks can also form by differential movement between the embankment, abutments, and spillway structures leading to cracks. Variations in foundation conditions and zonation of dam materials, such as a compacted clay core can lead to differential settlement, and water entering cracks can result in internal erosion over time. Damage to outlet works (i.e. principal spillway) through the embankment can also lead to leakage and potential internal erosion of the embankment. As with embankment dams, the conditions that define a levee failure include piping, overtopping, cracking, slumping and excessive settlement (Rosidi, 2007).

A reduction in freeboard due to settlement or instability of the upstream and downstream slopes may result in overtopping of the dam. Similarly, seiches from a large tectonic movement in the reservoir basin or waves from earthquake-induced landslides in the reservoir may result in overtopping of the dam. Overtopping of NRCS watershed dams from seiches is unlikely in the NMSZ, due to limited topographic relief and small size of the reservoirs. NRCS dams have auxiliary spillways that are at least 3 feet below the dam crest to reduce the likelihood of overtopping.

Structural damage can result from differential movements on faults in the dam foundation. This is not likely to happen on NRCS watershed dams, as part of the initial dam site geotechnical investigation involves identifying any potentially active fault planes in the project area to ensure the structure is not at risk.

Seed et al (1978) presented a review of earth dam performance during earthquakes. They concluded that earth dams perform well, with no harmful effects, when built with reasonable slopes on good foundations under moderate ground shaking (up to 0.2g from magnitude 6.5 – 7 earthquakes). The authors also noted that dams constructed of clay soils on clay or on rock foundations withstand extremely strong ground shaking ranging from 0.35 to 0.8g from a magnitude 8.25 earthquake with no apparent damage.

Pore pressure buildup during ground shaking in saturated, loose and medium dense granular materials affects the resistance, and consequently the stability of the embankment under cyclic loading, during and following the seismic event. A well-known event due to the buildup of pore pressure during an earthquake is the Sheffield Dam failure. This 25 foot high, 700 foot long embankment was composed of silty sand and sandy silt, with an upstream slope faced with a 3-foot thick clay blank which extended as much as 42 feet down into the foundation to serve as a cutoff. A 6.3 Mw earthquake centered about seven miles away rattled the dam site, which experienced an estimated peak acceleration of 0.2g. A pseudo-static analysis using a seismic coefficient of 0.1 for the respective magnitude and distance to the site yielded a factor of safety of 1.2 for the embankment. Despite this, the dam failed due to liquefaction or loss of strength in a loose sand layer near the base of the embankment (Ozkan, 1998).

Internal Erosion

The formation of cracks and resultant internal erosion are of particular concern in dams and levees throughout the NMSZ, as soils in the Mississippi River Alluvial Plain often contain a dispersive clay fraction. Dispersive soils are highly prone to erosion when in contact with water, often leading to tunnel and gully formation. Structures previously constructed by NRCS near Crowley's Ridge also have significant occurrences of dispersive clays (McCook, 2003).

Seed et al (1978) noted that extensive damage or failure of dams typically occurs in the 24 hours following an earthquake, suggesting that piping through cracks formed during earthquake shaking may be responsible. Associated problems range from the complete breaching of flood retarding embankments to the severe surface erosion of earth fills.

Failures due to internal erosion through cracks or other openings usually take the form of an irregularly shaped tunnel through the fill. Water flowing through the crack rapidly enlarges the crack, and the accompanying failure is sudden and dramatic. In other cases, the erosion results in an unusual appearance referred to as "jugging" when water infiltrates surface cracks. Erosion of drying cracks from the bottom upwards results in the "jugging" pattern. A tunneling action initiates as easily eroded clay particles are carried laterally to the slope surface from the bottom of the vertical crack (Figure 11). The moist strength of overlying soils results in the jug shaped holes often observed. Interconnected holes and rills provide a path for breaching of the embankment at high water levels (SCS, 1991).

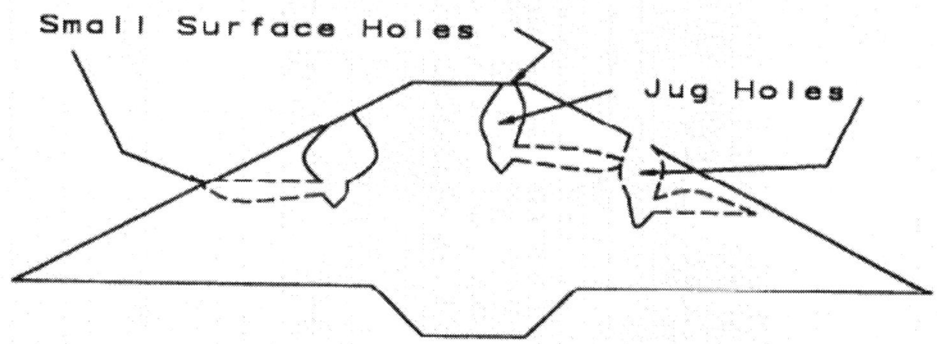

Figure 11. Cross-section illustrating jug hole features on an embankment built with dispersive soils (SCS, 1991).

Rainfall and runoff on exposed slopes of dispersive clay soils can also cause severe erosion, exacerbating damage to watershed structures following seismic events. The erosion often results in severe rilling and gullying on the slopes. Erosion of cut slopes in constructed drainage channels is another common problem associated with dispersive clays. The eroded slopes have high roughness that hampers maintenance. Sloughing of the banks and enlargement of the excavated channel can also be a problem (SCS, 1991).

Defensive Measures to Prevent Internal Erosion

The most effective design measure for preventing internal erosion of earth fills that impound water is a sand chimney filter. Eroding water flowing through a crack in an embankment carries particles of the eroding soil to the sand filter interface. Rapid plugging of the flow results as particles build up on the face of the filter. Engineers frequently design sand chimney filters as a vertical zone in the embankment. This zone is generally about 2 to 3 feet wide and extends below any excavation. The chimney filter zone normally continues upwards in a fill to the elevation of the planned maximum water storage height (SCS, 1991).

NRCS watershed dams built after 1970 in the Mississippi Delta region commonly include chimney filters. Archived design drawings reveal that ten of the twenty-six NRCS sponsored dams currently classified as high hazard in northeast Arkansas contain chimney filters. Only two of 136 dam failures documented worldwide failed by piping through the embankment despite a chimney filter being present. In both cases failure is likely due to piping through dispersive backfill materials around the outlet conduit (Foster et al, 2000).

SEISMIC ANALYSIS

A revival of interest by project sponsors in the Big Slough watershed resulted in preliminary geologic investigations for seven proposed embankment dams in 2002. Big Slough creek is a tributary to the St. Francis River, with headwaters on Crowley's Ridge in Clay County, Arkansas. NRCS also performed detailed follow-up investigations at two of these sites in 2009. NRCS performed a geotechnical investigation in 2007 at the proposed Poinsett Site 7 dam site, about 20 miles south of Jonesboro. This site is located on the Mississippi River Alluvial Plain near the base of Crowley's Ridge. Samples from these investigations underwent soil mechanics testing and associated seismic analyses. These investigations provide a recent, realistic assessment as to the general foundation characteristics and seismic risk to structures in the NMSZ.

NRCS designs called for a height of about 30 feet for the Poinsett Site 7 and Big Slough dam sites. The potential for large earthquakes influenced their design. United States Geological Survey (USGS) web sites provided information to predict peak horizontal acceleration for various degrees of conservatism.

The USGS Earthquake Hazards Program provides information via an online database to predict the peak horizontal acceleration for earthquakes that have a fairly frequent predicted occurrence - such as 10 percent probability of exceedance in 50 years (475 year return period), to a less frequent prediction, such as an earthquake with about a 5,000 year return period (1% probability of exceedance in 50 years).

Using the coordinates for Piggott Arkansas of 36.4 latitude and -90.2 longitude as a representative for seismic risk in the Big Slough Watershed, the maximum peak horizontal acceleration value obtained from the USGS web site for an earthquake with a 1% probability of exceedance in 50 years (~5,000-year return period) is about 130% of gravity. This extremely high value is attributable to the proximity of these sites to the New Madrid seismic source zone. Special problems on these sites include the possibility that embankment soils will be highly dispersive clays with very low resistance to internal erosion, and the seismicity of the area (McCook, 2003).

Alluvial and residual soils at the sites are predominantly low to medium plasticity clays classifying as CL. Soils in the Big Slough watershed have liquid limit (LL) values in the range of 30 – 40 and plasticity index (PI) values in the range of 9 to 20. Some sandy and silty horizons also occur, usually at depth in either the flood plain or abutments.

Very little groundwater was encountered in abutment borings at the Big Slough sites, however groundwater was found at less than 20 feet from the land surface in some borings on the flood plains, most often in the vicinity of the existing stream channel. Investigations concluded that seepage would not be a problem at the sites. A minor cutoff trench of normal dimensions was proposed.

The probable saturated consistency of the soil types was determined utilizing the liquidity index. A liquidity index of 1.0 indicates that the saturated water content of the soil is at the liquid limit. A liquidity index of zero indicates that the saturated water content of the soil is at its plastic limit. Values are in the range of 0.24 – 0.6, reflecting a medium to stiff consistency. In regards to shear strength, the undrained strengths of the samples were high to very high, reflecting the low liquidity index of the soils. These results are typical of clay soils that have been overconsolidated, probably caused by extensive desiccation of the soil profiles subsequent to deposition. Testing by the NRCS Soil Mechanics Laboratory shows a common preconsolidation load of at least 8,500 psf.

Based on these findings, foundation settlement at the Big Slough sites should be minimal because the foundation soils are overconsolidated. Depending on depth of alluvium, analyses suggest a total settlement in the foundations of these 30-foot high dams at no more than 0.2 to 0.4 foot.

Engineers used the "Makdisi and Seed" method (1978) to predict potential embankment slumping under a design earthquake for Big Slough dam sites. In this procedure, you first compute a ratio between the yield acceleration to the design earthquake. Using a design earthquake of 1.27g, the ratio becomes 0.3/1.27 = 0.24. The chart associated with this procedure predicts a crest displacement of about 16 feet. However, the reference article notes that predicted values of displacement are probably not realistic for low values of the ratio of yield acceleration to peak acceleration, and field experience indicates that well-compacted cohesive embankments on non-liquefiable foundations experience few problems even when subjected to very large earthquakes (McCook, 2003).

Due to the large freeboard likely present at these dams when an earthquake might strike, along with the character of the dam foundation, the NRCS National Design Center recommended planning for 3H:1V side slopes with a chimney filter to provide sufficient defensive design against seismicity. The presence of dispersive clays and high seismicity of the sites warranted incorporation of substantial chimney filters in each dam. Recommended dimensions for chimney filters were 5 feet in width, extending upward into the embankment to at least the crest of the principal spillway riser, and about 4 feet below the flood plain elevation. Engineering design reviews determined that a zoned fill configuration is not necessary for most sites, but to prevent severe surface erosion of the embankment some special measures are required.

Funding shortfalls prevented the construction of Big Slough dam sites. Poinsett Site 7 underwent several geologic investigations, but substantial seismic risk and the design modifications required to obtain seismic stability kept this site from construction as well.

NRCS designers extracted data from the USGS Earthquake Hazards Program in 2006 to determine potential earthquake loading at Poinsett Site 7. This program presents earthquake data in a format limited to 2.13g. At Poinsett Site 7 the estimated peak ground acceleration (pga) with an annual exceedance probability of $2x10^{-4}$ would slightly exceed this value. Following preliminary seismic stability evaluations it became clear that removal or treatment of liquefiable foundation materials and other design measures would exceed available funds for this project.

Seismic Analysis for Levees

Engineers assess seismic risk for levees through the probability that the levee will experience one or more failures during a future earthquake. The likelihood of a levee experiencing failure is a function of seismic-induced slope deformation. The expected levee damage from future earthquakes is calculated by integrating the probabilities of failure estimated for the various conditions associated with levee and site soil conditions, subsurface and surface water levels and earthquake ground motion and liquefaction scenarios (Rosidi, 2007).

Depending on its length and the propagation speeds of the seismic waves, earthquake ground motion may vary significantly along the levee, which is a challenge on levees in northeast Arkansas and southeast Missouri that can exceed 100 miles in length. For a major earthquake in the NMSZ, the dispersion of surface waves, combined with their low attenuation, would result in large amplitude, long duration sinusoidal type of motion with periods in the same range as the natural periods of tall buildings (Nuttli, 1974).

DAMAGE POTENTIAL

In 2009, the USGS issued a fact sheet emphasizing that according to their estimates, over the next 50 years the NMSZ has about a 7 – 10 percent chance of having an earthquake roughly the same size of the major 1811 – 1812 quakes. The chance of a magnitude 6 or larger earthquake in the next half century is much greater – 25 to 40 percent. Consequences of even a seemingly moderate earthquake sequence include impassable roads, extensive flooding, levee failure, and disruption to telecommunications and fossil fuel pipelines.

A 2009 FEMA-funded report underscored these seismic concerns. Released by the University of Illinois Mid-America Earthquake Center, the report estimates damage from a magnitude 7.7 New Madrid earthquake. The authors concluded that direct losses could reach $300 billion, with indirect losses at least twice that. Tennessee, Arkansas, and Missouri would suffer the most, with damage reaching into Illinois and Kentucky. The report estimates damage to about 715,000 buildings, with major destruction to infrastructure like roads and bridges, hampering services from emergency personnel. Such an earthquake could kill over 3,000 people and injure 86,000; leaving 2.6 million households without power, and 215,000 individuals seeking shelter. The report predicts damage to 20 levees in Craighead, Greene, Mississippi, and Poinsett Counties of Arkansas. This scenario suggests that Poinsett County would have 55 dams damaged (curiously, the National Inventory of Dams database only lists 30 dams in Poinsett County). The report does not mention potential damage to dams in neighboring counties, even though they also have a significant number that would surely experience damage as well.

Severe shaking and ground deformation would lead to numerous damaged facilities in southeast Missouri. Several dams in Pemiscot County along with a few dams in Mississippi County are likely to incur substantial damage. This includes cracking of the dam structure and measurable deformation of the ground beneath the dam. It is also possible for the earth dams in these counties to experience very intense shaking, resulting in some release of water in the reservoirs behind dams. Levees are likely to incur damage at slightly lower levels of ground shaking than dams. As a result, most levees in Butler, Dunklin, Pemiscot, New Madrid, Mississippi, Scott, and Cape Girardeau Counties will incur large cracks along the length and width of the levee, as well as some displacement of soil beneath the levee. If cracking is severe enough, some release of water is possible as long as there is standing water behind the levee (Mid-America Earthquake Center, 2009).

If the future resembles the past, repeated shakes equal to or greater in intensity will likely occur several months after an initial jolt, as repairs are underway. The initial December 16, 1811 earthquake centered near Blytheville, Arkansas was an estimated 7.5 - 7.7 magnitude event, followed months later by three other major events, as well as many others in the strong to moderate category. It is hard enough to plan for one earthquake, but no one has any real sense of how to plan for three or four in quick succession (Bolton Vallencius, 2013).

It is important to note that incidence of failure or damage to well-engineered dams is uncommon, even under strong ground shaking (Ozkan, 1998). We have only limited ability to predict the extent of internal damage when earth dams experience seismic shaking or fault movement in the foundation. Research is lacking regarding the effects of internal damage and cracking on defensive measures installed into the dam to control piping. Today's remediation practice requires considering the possibility that some of these defenses might be breached (Marcuson et al, 2007).

Severe damage to drainage canals, levees, and dams in the NMSZ would result in a loss of the ability to attenuate floodwaters, destroying the agricultural production capacity that is the primary economic engine of the region. Northeast Arkansas and southeast Missouri produce over $3 billion worth of row crops annually - predominantly corn, soybeans, rice, and cotton. Nearly 70 percent of U.S. rice production is concentrated in the western half of the Mississippi Delta, stretching from Louisiana to the Missouri Bootheel. According to *Arkansas Rice Facts*, Arkansas ranks first among the six major rice-producing states, accounting for approximately 48 percent of U.S. rice production. Poinsett, Cross, and Lawrence are among the top five rice-producing counties in Arkansas. Butler County produces roughly half of the rice in Missouri.

Rice farmers use GPS and laser systems to precision-grade large tracts of land so that they are perfectly flat or tilted slightly to one side or another in order to drain smoothly. Landowners completely scape off the top layer of soil, grade the remaining layers to the desired angle, and then respread the surface. Known as land leveling, this practice allows for efficient cultivation of rice. Water that runs off is captured in ditches bordering the fields, and directed or pumped into irrigation storage reservoirs resembling lakes. In the

Mississippi River Delta region of Arkansas alone, NRCS provided over $66 million in incentives from 2009 – 2013 to assist farmers with implementing irrigation practices such as these to conserve water use and reduce runoff from fields.

The 1811 – 12 earthquakes caused many types of ground failures including landslides along the Mississippi River bluffs bordering the eastern floodplain from Mississippi to Kentucky. Ground failures also included lateral spreading and ground subsidence by soil liquefaction across the Mississippi River flood plain and along tributaries to the Mississippi over 5,500 square miles. Large fissures formed in the soft alluvial ground that were too wide to be crossed on horseback. The earthquake made much of this land unusable for the subsistence type agriculture of that day because of the deep fissures and land subsidence, which converted good fields to swamps, and numerous sand blows that covered the ground with sand and mud. Heavy damage inflicted on the land by these earthquakes led Congress to pass the first disaster relief act in 1815, providing landowners of ravaged ground with an equal amount of land in unaffected regions (Nuttli, 1974).

Today, a repeat event will produce similar effects in northeast Arkansas, southeastern Missouri, western Tennessee and Kentucky, and southern Illinois. Roadways in the Mississippi Delta of Arkansas and Missouri could become impassable because of bridge failures and fissuring of road surfaces. Venting of large quantities of water, sand, and mud because of liquefaction could flood fields and roads, wreaking havoc on agricultural activities. Failure of levees and streambanks, especially during high water, would contribute to flooding (USGS, 2009).

NRCS has provided financial and technical support to construct numerous secondary drainage canals, which feed into larger ones built by watershed districts with assistance from the U.S. Army Corps of Engineers. Looking at aerial imagery the scale of this drainage network is readily apparent (Figure 12). Highways and rural roads cross these canals every few thousand feet across much of the NMSZ. Each crossing is a potential weak point where road collapse is likely during an earthquake of sufficient magnitude to induce liquefaction. This collapse would effectively strand rural residents on their farms, and debris from bridges and culverts would obstruct ditches and canals, compromising their ability to drain fields.

Even though a 7.0+ is a possibility in the next 50 years, it is far more likely that a 6.0 would hit the region, and that would still be a major problem. A 6.0 magnitude earthquake in the NMSZ would affect 10 times more area than a 6.0 in California. This is because the earth's crust is denser in the central U.S., and the basement rocks do not dissipate the energy as much as they do in California (Heard, 2014).

The NMSZ appears to be about 30 years overdue for a magnitude 6.0+ quake, as the last quake of this size, a magnitude 6.7 occurred at Charleston, Missouri in 1895. A magnitude 6.3 quake near Lepanto, Arkansas in 1843 was the next prior quake of this magnitude (MDNR, 2014). Using the distance versus liquefaction curve of Wang et al (2006), a 6.0 magnitude earthquake could induce liquefaction over 3,600 square miles.

Figure 12. Map showing area just east of Jonesboro city limits. Note the large number of drainage canals and road crossings. This is the typical situation throughout northeast Arkansas and southeast Missouri. The large St. Francis Sunken Lands Ditch is labeled in blue and can be seen on the far right.

When assessing the potential seismic effects on watershed structures it is not sufficient to determine that materials are potentially liquefiable; one must estimate the amount of permanent deformation a dam is likely to experience and determine whether that deformation is tolerable, given such conditions as filter zone and core geometry, allowable freeboard and downstream development. This is goal of seismic stability analyses.

If predicted deformations to an embankment are small and tolerable, the engineer's job is finished. On the other hand, if estimated deformations are intolerable the engineer must consider seismic remediation. Seismic retrofits of structures designed prior to 1960 and built on potentially liquefiable alluvial material are a key concern for geotechnical engineers. In order to rehabilitate a deficient embankment to prevent potential seismic instability, one must either change the engineering properties of the dam and/or foundation, modify the geometry of the existing dam, or both (Marcuson et al, 2007).

In the NMSZ of Arkansas and Missouri, six PL566 dams are pre-1960 vintage, three of which are high hazard. Mississippi River and tributary levees were designed in the early 1950's (USACE, 2014). Securing adequate funds to retrofit all of these structures so they withstand a large magnitude seismic event is not likely. Owners and operators of dams and levees need to prioritize and assess structures in areas posing the greatest risk to life and property should a failure occur.

CONCLUSION

Watersheds across the NMSZ contain multiple PL-566 structures. Multiple structures in the upper reaches of a watershed offer superior safety over building one large structure on a major stream. Substantial flood protection would still be available within the Poinsett, Big Creek, and Caney Creek watersheds even if one PL-566 structure fails. However, many of these structures are nearing the end of their design life. Maintaining existing small watershed structures is critical. They provide additional insurance to levees in retarding floodwaters that would otherwise inundate vast tracts of farmland within the northern Mississippi Delta.

The degradation of PL566s should be of great concern to environmental planners, because most of the dams have surpassed or are nearing the end of their design life. Many may already be losing much of original storage potential due to sedimentation. On a watershed scale, small upland reservoirs act as positive ecological features in that they store significant amounts of nutrients that can cause even more environmental problems when concentrated downstream (Prochnow et al, 2007).

NRCS' Watershed Rehabilitation Program addresses the importance of these watershed structures and their need for repairs. The 2014 Farm Bill includes a $262 million investment to rehabilitate dams. This represents a 21-fold increase in funding over the previous Farm Bill. The Watershed Rehabilitation Program provides for safety assessments of 500 dam sites. Arkansas alone will receive $780,000 to assess 39 dam sites.

Seismic stability evaluations of PL-566 structures are eligible for funding from the Watershed Rehabilitation Program. These evaluations look at how a dam will perform during a large earthquake, and examine the effects of seismic hazards such as shaking and surface fault rupture.

If a large Mw 7.5 quake hit, in additional to the commonly discussed destruction of infrastructure and utilities, there would be considerable damage to dams, levees, and canals. Differential settlement with resultant cracks leading to internal erosion and formation of seepage pathways through embankments would be common. Dams would experience toppled principal spillway risers. Bridges crossing drainage ditches and canals would collapse, resulting in blockages that grow over time by ensnaring woody debris. The large-scale agriculture of today is impossible with widespread flooding, subsidence of the land surface, and thick sand blow deposits

It has taken over a century to achieve the current level of agricultural productivity within the NMSZ. A magnitude 6.0+ earthquake takes only minutes to wipe out much of this work. There is a 25 – 40% chance this could happen within the next 50 years.

Recommendations

Dams upstream of towns and cities create a high risk potential for loss of life and property, particularly in seismically-active areas. This results in additional responsibility for project sponsors. Sponsors should re-evaluate the hazard class of their dams, as many are no longer low-hazard structures as originally designated. Urban encroachment and downstream development will necessitate a higher hazard classification.

Inundation maps show areas of potential flooding in the event of sudden or total failure of flood control structures. Owners should develop these maps and share them with local governments while developing emergency procedures for the evacuation and control of areas in the event of a failure.

Dam owners should inspect high hazard dams at least annually, monitor dams during heavy rainfall events, examine dams for damage following such events, and place high priority on operation and maintenance. Repairs need to occur in a timely manner.

Watershed districts and local governments that own and operate all types of flood control structures should not hesitate to contact their local USDA-NRCS or U.S. Army Corps of Engineers representatives for assistance with inspections, seismic stability assessments, and emergency action plans.

ACKNOWLEDGEMENTS

I would like to express my gratitude to the assistance provided by staff of the Arkansas Geological Survey, especially Brian Kehner with preparation of certain graphics. I would also like to thank NRCS Geologist Ana Vargo for support with liquefaction curve issues, NRCS National Design Engineer Steve Durgin for editing suggestions, and my wife, Katrina, for assistance with proofreading.

REFERENCES

Arkansas Historic Preservation Program, 2014. Marked Tree Lock and Siphons, Marked Tree, Poinsett County. Accessed at http://www.arkansaspreservation.com/historic-properties/_search_nomination_popup.aspx?id=1819

Ausbrooks, Scott, and Prior, William, 2008. Depth to Groundwater in the Mississippi River Valley Alluvial Aquifer in Eastern Arkansas. Published by the Arkansas Geological Survey. Updated January 2013.

Bethapudi, Raghudeep, 2008. Liquefaction Induced Lateral Spreading in Large-scale Shake Testing. M.S. thesis in Civil Engineering, State University of New York at Buffalo, 234 p.

Blackwell, Sam, 2007. A landscape transformed by the Little River Drainage District. Accessed at http://medialab.semissourian.com/story/1288730.html on 7/24/2014.

Bolton Valencius, Conevery, 2013. *The Lost History of the New Madrid Earthquakes.* The University of Chicago Press, 472 p.

Crone, Anthony, and Wheeler, Russell, 2000. Data for Quaternary Faults, Liquefaction Features, and Possible Tectonic Features in the Central and Easter United States, East of the Rocky Mountain Front. USGS Open-File Report 00-260.

Fell, Robin, MacGregor, P., Stapledon, D., and Bell, G., 2005. *Geotechnical Engineering of Dams*. CRC Press, 912 p.

Fuller, M., 1912. The New Madrid Earthquake. United States Geological Survey Bulletin 494.

Heard, Kenneth, 2014. Survey places more of state at quake risk. Accessed at http://www.arkansasonline.com/news/2014/jul/21/survey-places-more-of-state-at-quake-ri/ on 7/21/2014.

Hendriks, Nancy, 2011. Sunken Lands. *The Encyclopedia of Arkansas History & Culture.* Accessed at http://www.encyclopediaofarkansas.net/encyclopedia/entry-detail.aspx?entryID=2222 on 9/15/2014.

Hendriks, Nancy, 2014. Sunken Lands. *Bulletin of the Hot Springs Geology Club.* Vol. 14 No. 6 Issue 76.

Henggeler, Joseph C., 2013. Irrigation Systems, Wells, and Pumps of the Mississippi River Alluvium Aquifer of Southeast Missouri. Unpublished report distributed during "Irrigation Water Well Drilling Workshop" at University of Missouri Agricultural Extension Office in Portageville, Missouri on April 29, 2013.

Jackson, Donna Brewer, Levees and Drainage Districts. *The Encyclopedia of Arkansas History & Culture.* Accessed at http://www.encyclopediaofarkansas.net/encyclopedia/entry-detail.aspx?entryID=1165 on 9/15/2014.

King, Christopher A., 2010. Geologic Investigation for Big Slough Watershed Floodwater Retarding Structure Site No. 1. USDA-NRCS project report.

King, Christopher A., 2010. Geologic Investigation for Big Slough Watershed Floodwater Retarding Structure Site No. 7. USDA-NRCS project report.

Lewis, Keith, 2012. Work on the Little River Drainage District began 100 years ago today. Accessed at http://www.semissourian.com/story/1900459.html on 7/24/2014.

Lund, W.R., Knudsen, T.R., Vice, G.S. and Shaw, L.M., 2008. Geologic Hazards and Adverse Construction Conditions: Liquefaction-Hazard Map for the St. George-Hurricane Metropolitan Area. Utah Geological Survey Special Study 127, plate 2.

Maksidi, F.I. and Seed, H.B., 1978. Simplified procedure for estimating dam and embankment deformations. *J. Geotechnical Engrg. Div., ASCE,* 104(GT7), 849 – 867.

Marcuson, W.F., Hynes, M.E., and Franklin, A.G., 2007. Seismic Design and Analysis of Embankment Dams: The State of Practice. Paper published in association with The

Donald M. Burmister Lecture at the Department of Civil Engineering and Engineering Mechanics at Columbia University.

McCook, D., 2003. Soil Mechanics Report for Big Slough Creek, Sites 1, 2, 3, 6, 7, 10, & 11. USDA-NRCS National Design, Construction, and Soil Mechanics Center.

Mid-America Earthquake Center, 2009. Impact of New Madrid Seismic Zone Earthquakes on the Central USA – Volume I & II. MAE Center Report No. 09-03. Accessed at https://www.ideals.illinois.edu/handle/2142/14810

Mississippi River Commission, 2008. The Mississippi River & Tributaries Project: Backwater Areas. Information Paper dated April 2008.

Missouri Department of Natural Resources, *Facts about the New Madrid Seismic Zone*. Accessed at http://www.dnr.mo.gov/geology/geosrv/geores/techbulletin1.htm on 6/30/2014.

Missouri Department of Natural Resources, Missouri Geological Survey Fact Sheet Number 26, *The New Madrid Seismic Zone*. February, 2014.

Mitchell, Steven, 2014. Marked Tree Siphons. *The Encyclopedia of Arkansas History & Culture*. Accessed at http://www.encyclopediaofarkansas.net/encyclopedia/entry-detail.aspx?entryID=3999 on 9/15/2014.

Natural Resources Conservation Service, 2005. TR 60 - Earth Dams and Reservoirs.

Nuttli, Otto, 1974. Earthquake Information Bulletin, Volume 6, No. 2

Obermeier, S. F., 1989, The New Madrid Earthquakes: An engineering-geologic interpretation of relict liquefaction features, US Geological Survey, Professional Paper 1336-B, 114 p.

Ozkan, M.Y., 1998. A review of considerations on seismic safety of embankments and earth and rock-fill dams. *Soil Dynamics and Earthquake Engineering*. 17 (1998), 439 – 458.

Papadopoulos, G.A. and Lefkopoulos, G., 1993. Magnitude-distance relations for liquefaction in soil from earthquakes. *Bulletin of the Seismological Society of America*, 83(3), pp. 925-938.

Prochnow, S.J., White, J.D., Scott, T., and Filstrup, C., 2007. Small Reservoir Impact on Simulated Watershed-Scale Nutrient Yield. *Research Letters in Ecology*. Volume 2007, Article ID 12571, 4 pp.

Rosidi, Dario, 2007. Seismic Risk Assessment of Levees. *Civil Engineering Dimension, Vol. 9, No. 2, 57 – 63.*

Seed, H.B., Maksidi, F.I, and De Alba, P., 1978. Performance of earth dams during earthquakes. *J. Geotechnical Division, ASCE*, 104(GT), 967 – 995.

Soil Conservation Service, United States Department of Agriculture. *Soil Mechanics Note 13 – Dispersive Clays*. February, 1991.

Tuttle, M., 2005. Search for and Study of Sand Blows at Distant Sites Resulting from Prehistoric and Historic New Madrid Earthquakes: Final Technical Report for U.S. Geological Survey Award No. 1434-02HQGR0097.

Tuttle, M., 2011. Sand Blows and the New Madrid Earthquakes of 1811 – 1812. Spring 2010 Special Edition of The Earth Scientist, Volume XXVII Issue 1.

U.S. Army Corps of Engineers, 2014. Mississippi River and Tributaries Project – Levee System Evaluation Report for the National Flood Insurance Program.

U.S. Bureau of Reclamation and U.S. Army Corps of Engineers, 2012. Best Practices in Dam and Levee Safety Risk Analysis – Methods to Evaluate Seismic Risks for Embankments. Accessed at http://www.usbr.gov/ssle/damsafety/Risk/BestPractices/27-SeismicEmbankmentPP20121121.pdf on 10-10-2014.

U.S. Department of Agriculture, 2012. Census of Agriculture – County Data. Available online at http://www.agcensus.usda.gov/Publications/2012/

U.S. Geological Survey, Fact Sheet 2009-3071, *Earthquake Hazard in the New Madrid Seismic Zone Remains a Concern*. August, 2009.

U.S. Geological Survey, 2011. NMSZ Expert Panel Report to NEPEC. Available online: http://earthquake.usgs.gov/aboutus/nepec/reports/NEPEC_NMSZ_expert_panel_report.pdf

Van Arsdale, Roy, and Ellis, Michael, 2002. Final Report for SG-4 - Characterization of Active Faults in the New Madrid Seismic Zone. Center for Earthquake Research and Information, University of Memphis. Available online: http://mae.cee.illinois.edu/publications/reports/Report04-02.pdf

Wang, C., Wong, A., Dreger, D., and Manga, M., 2006. Liquefaction Limit during Earthquakes and Underground Explosions: Implications on Ground-Motion Attenuation. *Bulletin of the Seismological Society of America*, Vol. 96, No. 1, pp. 355–363. February, 2006.

ANALYSIS OF SEISMIC DEFORMATIONS AND CRACKING FOR EMBANKMENTS OF THE TEKAPO CANAL

Lelio Mejia[1]
Ethan Dawson[2]

ABSTRACT

The 26 km long Tekapo Canal is a major component of the Upper Waitaki Power Scheme in the Mackenzie Basin in New Zealand's South Island. The canal, commissioned in 1977, conveys water from a power plant at Lake Tekapo to a power plant at Lake Pukaki. To support a re-lining and repair works project along sections of the canal, seismic deformation analyses were performed. Earthquake-induced settlements and deformations for three critical embankment sections were estimated. Two dimensional, nonlinear, dynamic numerical analyses were performed with the computer code FLAC. Analyses were performed for the Maximum Design Earthquake (MDE), the Serviceability Level Earthquake (SLE) and for aftershocks of various magnitudes. A critical feature in the repairs and upgrades was a geosynthetic liner to be placed along portions of the canal. Seismic performance of this liner would be affected by cracking in the underlying embankment. Crack size estimates (width and depth) were developed based on evaluation of the computed deformations and the empirical correlations of Fong and Bennett (1995) and Pells and Fell (2002). Calculated deformations were generally small, and indicative of adequate seismic stability. Recommended design crack widths for the MDE ranged from 20 to 50 mm, while recommended design crack depths for the MDE ranged from 1.0 to 2.5 m.

INTRODUCTION

The 26 km long Tekapo Canal is a major component of the Upper Waitaki Power Scheme in the Mackenzie Basin in New Zealand's South Island. The canal, commissioned in 1977, conveys water from a power plant at Lake Tekapo to a power plant at Lake Pukaki. To support a re-lining and repair works project along sections of the canal, seismic deformation analyses were performed. Earthquake-induced settlements and deformations for three critical embankment sections were estimated through two-dimensional, nonlinear, dynamic numerical analyses performed with the computer code FLAC. Analyses were performed for the Maximum Design Earthquake (MDE), the Serviceability Level Earthquake (SLE) and for aftershocks of various magnitudes. Acceleration time-histories for these earthquake events were developed from the seismic hazard analysis for the project as described by Somerville (2013).

A critical feature in the repairs and upgrades was a geosynthetic liner to be placed along portions of the canal, Seismic performance of this liner would be affected by cracking in the underlying embankment. Crack size estimates (width and depth) were developed

[1] URS Corporation, Oakland, California ,USA, lelio.mejia@urs.com
[2] URS Corporation, Los Angeles, California, USA, ethan.dawson@urs.com

based on evaluation of the computed deformations and the empirical correlations of Fong and Bennett (1995) and Pells and Fell (2002).

A map of the Tekapo Canal is shown in Figure 1. Seismic deformation analyses were performed to estimate earthquake shaking induced settlements and deformations for three embankment sections along critical segments of the Tekapo Canal. These sections were selected as representative of the conditions at canal stations with the most severe embankment heights and slopes along those canal segments. Such stations would be expected to undergo seismic deformations near the upper end of the range that might materialize along the canal after an earthquake. Analyses were performed for:

- Patterson Terrace, km 5.9
- Irishman's Creek, km 13.1
- Maryburn Fill, km 18.1

The critical embankment sections analyzed are illustrated in Figures 2 through 4, which show the FLAC numerical meshes for these sections.

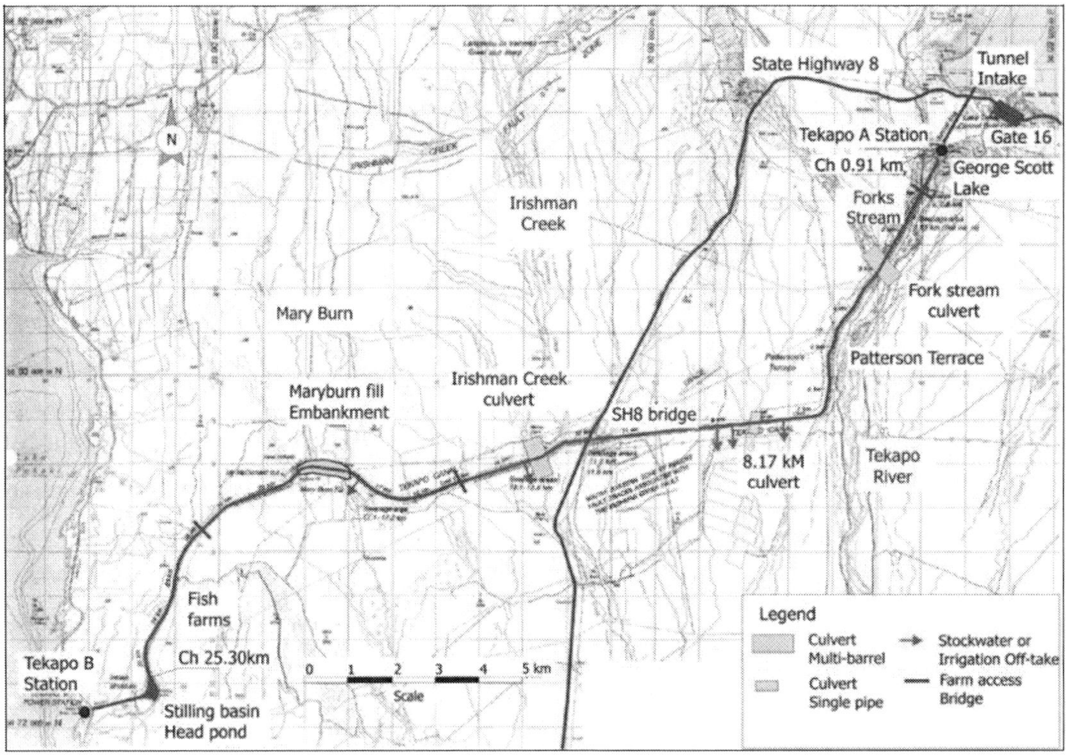

Figure 1. Tekapo Canal Map.

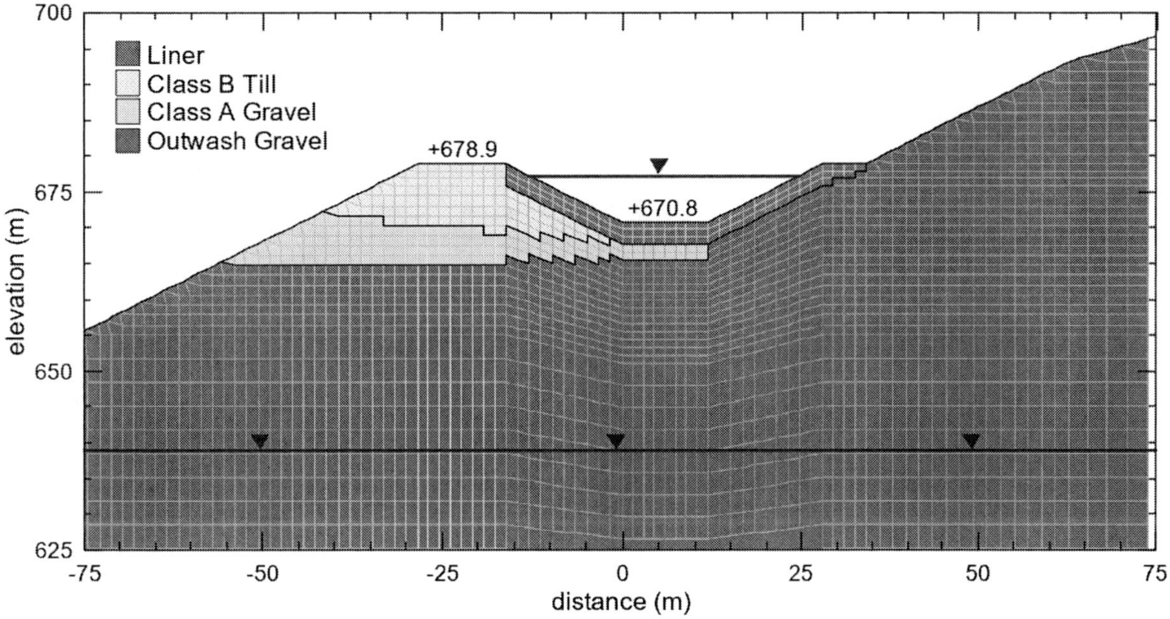

Figure 2. FLAC numerical mesh for Patterson Terrace section (km 5.9).

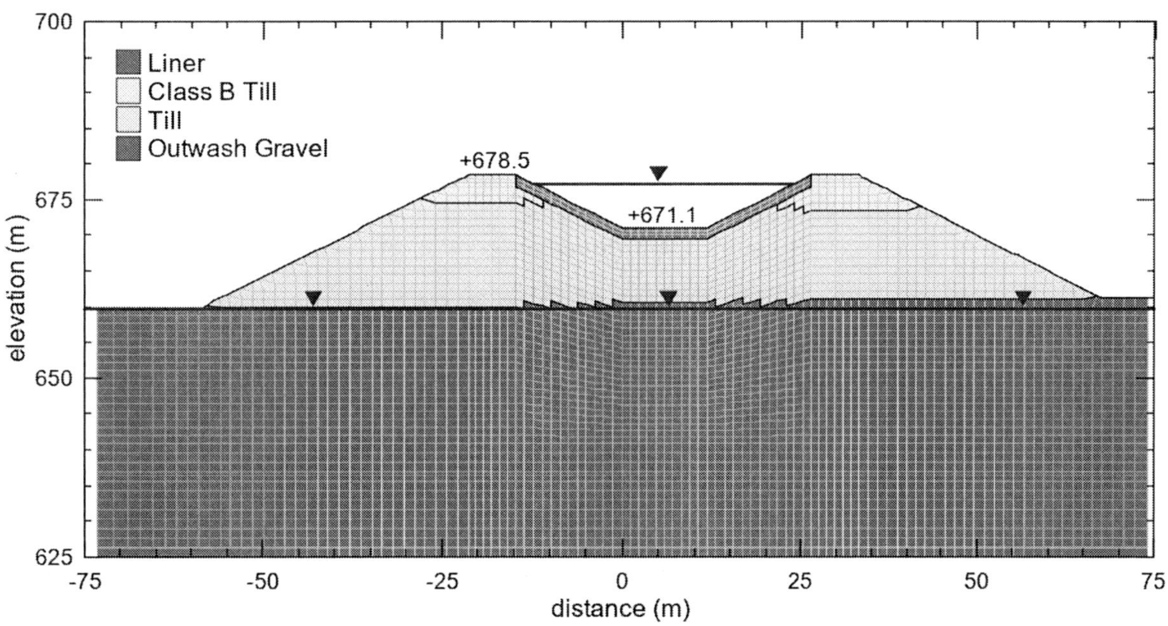

Figure 3. FLAC numerical mesh for Irishman's Creek section (km 13.1).

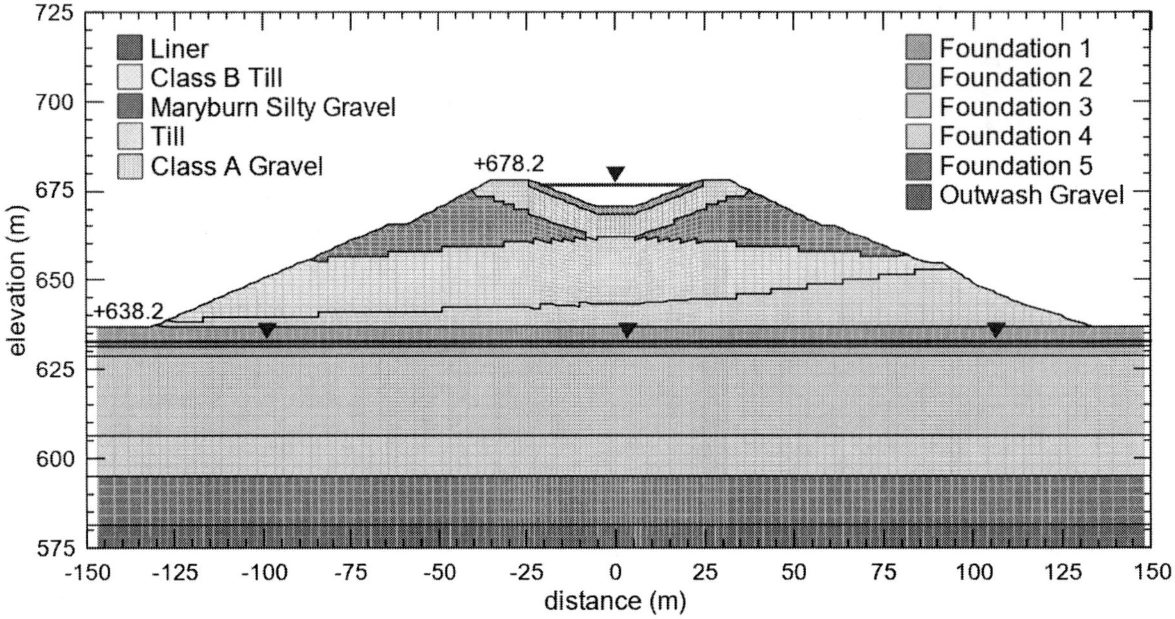

Figure 4. FLAC numerical mesh for Maryburn Fill section (km 18.1).

MATERIAL PROPERTIES

Table 1 presents the key properties for seismic analysis of the Maryburn embankment, including the unit weight, strength parameters and dynamic stiffness parameters selected for each material zone in the embankment model. Material properties for the Patterson Terrace and Irishman's Creek models were similar, although not identical.
Material properties were developed based on information presented in reports by Genesis Energy (2011), DamWatch (2010a), and DamWatch (2010b), and based on engineering judgment and experience with similar materials. In selecting the properties for the analyses, it was noted that there was a change in specifications for construction of the embankment fills known to have been implemented at approximately km 13.

The unit weights and strength parameters for the Maryburn and Irishman's Creek models were based primarily on the parameters in the Genesis Energy (2011) report. The strength parameters for the Patterson Terrace model were partly based on those shown in the same report.

The dynamic stiffness parameters were based on interpretation of seismic wave velocity surveys presented by DamWatch (2010b). Although the data were obtained at sites other than those selected for analysis, they represent materials from the same geologic origin and placed under similar conditions as those present at the embankment sites used for the analyses. Thus, the data were used to estimate the dynamic parameters for the analysis models based on experience with similar materials elsewhere in the Waitaki River valley. The resulting parameters are also consistent with published information (e.g. Seed et al., 1984).

Table 1. Material Properties for Maryburn Fill Model

Material	Unit Weight (kN/m^3)	Strength		Dynamic Stiffness	
		Cohesion C (kPa)	Friction φ (°)	K_{2max}	Vs (m/s)
Liner	21.5	10	37	250	-
Class B Till	21.5	10	37	300	-
Maryburn Silty Gravel	20.6	0	38	150	-
Till	21.5	10	37	300	-
Class A Gravel	21.5	0	40	150	-
Foundation 1	19.7	0	37	80	-
Foundation 2	21.5	0	32	80	-
Foundation 3	18.7	0	30	80	-
Foundation 4	20.6	0	34	80	-
Foundation 5	22.6	0	40	100	-
Outwash Gravels	22.6	Elastic		-	650

TECHNICAL APPROACH

Two-dimensional, nonlinear, dynamic numerical analyses were performed with the computer code FLAC, Version. 6.0 (Itasca 2008). Soils were modelled with a linear-elastic, perfectly-plastic Mohr-Coulomb constitutive model. In this model, the soil shear strength is defined in terms of a friction angle and cohesion. For stress states below the shear strength, the behavior of the model is linear-elastic.

The initial in situ stress state prior to earthquake shaking was computed by simple gravity turn-on analyses. The dynamic shear modulus was modelled as a function of the initial mean effective stress using the relation

$$G_{max} = 1000\, K_{2max}(\sigma'_m)^{0.5} \quad (1)$$

where σ'_m is the mean effective stress in units of pounds per square foot (psf) and G_{max} is also expressed in psf.

Dynamic shaking was simulated by applying the design acceleration-time histories (both horizontal and vertical) to the base of the model through a compliant base. This non-reflective boundary absorbs downward propagating shear and compression waves, eliminating non-physical reflections off the base of the numerical mesh. The input time histories (Somerville, 2013) were developed for a site condition corresponding to a Vs30 of 650 m/sec, the shear wave velocity of the Outwash Gravels at the base of the embankment numerical models. Thus, the time histories were developed to be used directly in the embankment analyses, without modification. The time histories were input into FLAC models as discussed in detail by Mejia and Dawson (2006).

In order to minimize boundary disturbances, the left and right sides of the mesh were equipped with free-field boundaries simulating level ground conditions. A small value of Rayleigh damping (2%) was used to provide damping at small strains. At large strains, beyond the yield strain, the Mohr-Coulomb model automatically produces hysteretic damping. Directivity and basin effects were considered in the seismic deformation analyses to the extent that such effects were incorporated in the acceleration time histories provided for input into the analyses. The effects of foundation geometry on ground motion amplification were directly included in the analyses by including the foundation geometry in the analysis models.

The numerical meshes used in the analyses are shown in Figures 2 through 4. The analyses were performed for the post-upgrade condition which assumes a geosynthetic liner is in place so that the embankment materials are not saturated. Native soils were modelled to be saturated below the phreatic surface indicated on the figures. Loading due to water in the canal was simulated by applying a hydrostatic boundary load acting on the canal bottom and sides. Note that the geosynthetic liner itself was not modelled because of its small mass and stiffness relative to the earth liner and canal embankments.

ANALYSIS RESULTS

Details of the analysis results are presented for the Maryburn embankment, while summary tables are provided for the Patterson Terrace and Irishman's Creek analyses. Dynamic deformation analyses were performed for each of 7 time histories developed for the Maximum Design Earthquake (MDE), the Serviceability Level Earthquake (SLE), and aftershocks of magnitudes M6, M5.5 and M5. That number of time histories was judged sufficient to reasonably represent the range of variability in embankment response that might be associated with the MDE and SLE. The development of the time histories is discussed by Somerville (2013). Figures 5 and 6 show time histories of computed horizontal and vertical accelerations at the left and right embankment crests, along with the acceleration input at the base of the model for MDE time history No. 4. The settlements and displacements at the center of the embankment crests were selected as a representative index of the overall embankment deformation and are listed in Table 2 for the seven time histories used to represent the MDE. Also listed in Table 2 are the normalized crest settlements: i.e. the crest settlement divided by the height of the embankment.

Summaries of the computed crest settlements for the three embankment sections analyzed are provided in Tables 3 and 4, which list the average computed settlements for the MDE, SLE and aftershock scenarios.

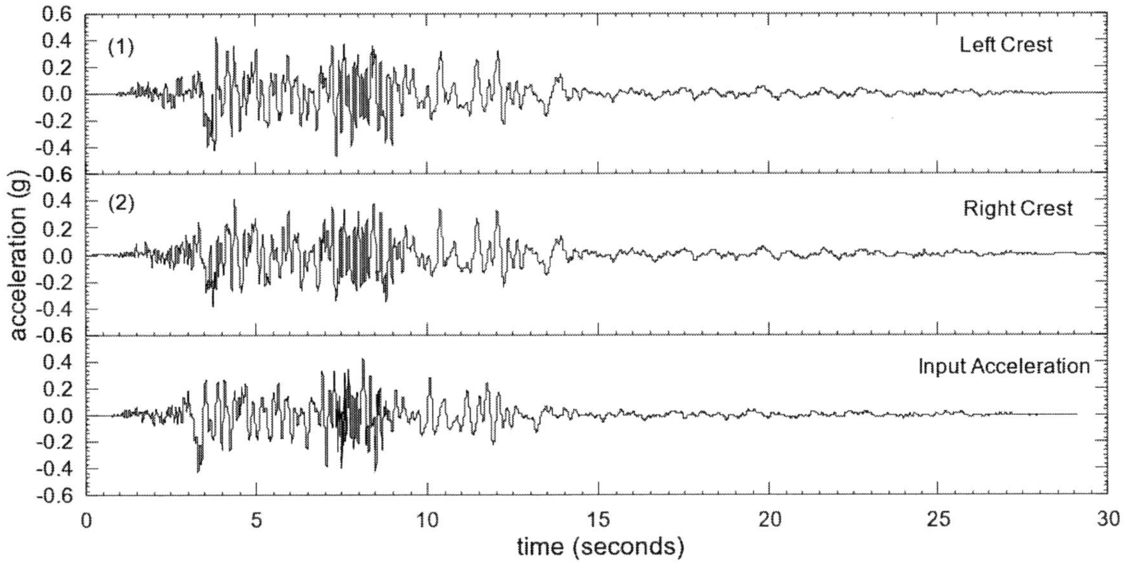

Figure 5. Maryburn Fill computed horizontal accelerations: MDE Time History No. 4.

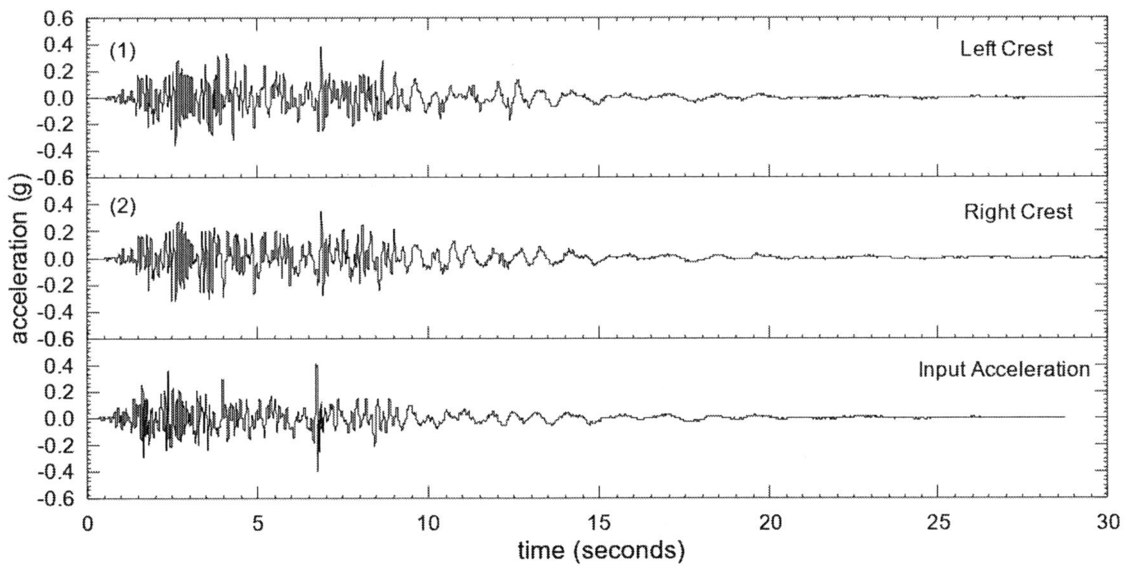

Figure 6. Maryburn Fill computed vertical accelerations: MDE Time History No. 4.

Table 2. Computed Crest Settlements, Maryburn Fill, MDE (2,500 ARI).

Time History		Crest Settlement (cm)		Normalised Crest Settlement (%)	
no.	name	left	right	left	right
1	LP-BRN-090	7.8	7.1	0.19	0.17
2	LP-CAP-000	7.2	6.0	0.18	0.15
3	LP-LPGC-000	6.6	5.9	0.16	0.14
4	NR-655-022	7.3	7.1	0.18	0.17
5	NR-LDM-064	5.9	6.0	0.14	0.15
6	NR-RRS-228	5.7	7.2	0.14	0.18
7	NR-SCE-018	6.6	5.9	0.16	0.14
Average		6.7	6.5	0.16	0.16

Table 3. Summary of Computed MDE and SLE Crest Settlements for the Three Sites.

		MDE$^{(1)}$		SLE$^{(2)}$	
Embankment	Embankment Height (m)	Crest Settlement (cm)	Normalized Crest Settlement (%)	Crest Settlement (cm)	Normalized Crest Settlement (%)
Irishman's Creek	19	8	0.42	0.5	0.03
Maryburn Fill	41	7	0.16	1.5	0.04
Patterson Terrace	15	2	0.13	0.2	0.01

Notes:
(1) The Annual Return Interval (ARI) of the MDE for the Irishman's Creek site is 10,000 years, while for the Maryburn Fill and Patterson Terrace sites it is 2500 years.
(2) The ARI of the SLE is 500 years for all three sites

Table 4. Summary of Computed Aftershock Crest Settlements for the Three Sites.

Embankment	Embankment Height (m)	M6 Aftershock Crest Settlement (cm)	M6 Aftershock Normalized Crest Settlement (%)	M5 Aftershock Crest Settlement (cm)	M5 Aftershock Normalized Crest Settlement (%)
Irishman's Creek	19	1.4	0.07	0.3	0.020
Maryburn Fill	41	4.1	0.01	0.5	0.010
Patterson Terrace	15	0.6	0.04	0.01	0.001

EMPIRICAL CRACK SIZE ESTIMATES

Crack sizes (width and depth) were estimated using the empirical correlations of Fong and Bennett (1995) and Pells and Fell (2002). In these studies, a large number of case histories of dams and embankments subjected to earthquake shaking were reviewed. Empirical correlations were then developed for crack width and depth as a function of the normalized crest settlement (the crest settlement divided by the height of the embankment). These empirical correlations were applied to Tekapo Canal embankments using the computed MDE deformations for each analysis section. Estimated crack depths and maximum widths for the MDE are listed in Table 5.

Table 5. Empirical Crack Estimates for MDE.

Embankment	Height (m)	Estimated Crest Settlement (cm)	Normalized Crest Settlement (%)	Fong and Bennett (1995) Crack Depth (m)	Fong and Bennett (1995) Crack Width (mm)	Pells and Fell (2002) Crack Depth* (m)	Pells and Fell (2002) Crack Width (mm)
Irishman's Creek	19	8	0.42%	1 - 2	30 - 40	1.2 - 3.2	30 -80
Maryburn Fill	41	7	0.17%	0.9 - 1.2	20 -30	0.4 - 1.2	10 - 30
Patterson Terrace	15	2	0.13%	0.3 - 0.6	10 - 20	0.4 - 1.2	10 - 30

* Estimated crack depth per Pells and Fell (2002) is obtained as 40 times crack width

CONCLUSIONS AND RECOMMENDATIONS

Seismic deformation analyses were performed to estimate earthquake shaking induced settlements and deformations for three embankment sections representative of the most severe conditions along critical segments of the Tekapo Canal. Analyses were performed for the Maximum Design Earthquake (MDE), the Serviceability Level Earthquake (SLE) and for aftershocks of various magnitudes.

Based on the analyses the following conclusions and observations were made:

1. Calculated deformations were generally small (see Table 3) and indicative of adequate seismic stability
2. Calculated deformations are consistent with observations of well-built dams and embankments during past earthquakes
3. Embankment crest settlements (see Table 3) indicate small freeboard loss
4. Seismic deformations could lead to cracking near the embankment crests with occasional cracks to depths of up to about a few meters and widths of up to a few tens of millimetres
5. Deformation patterns suggest potential for minor longitudinal cracking near the water line or crest edges with occasional cracks to depths of up to about a few meters and widths of up to a few tens of millimetres.
6. Crest settlements may also be associated with minor transverse cracking near the crest
7. Deformations at the canal invert are very small in all cases
8. Data presented by Fong and Bennett (1995) and Pells and Fell (2002) suggest relatively small crack depths and widths as shown in Table 5

Recommended MDE cracks sizes for design are listed in Table 6. These are based on the analyses assumptions and results, the estimated degree of applicability of empirical crack size estimates, and engineering judgment. Generally, to accommodate a given zone of tensile strain in an embankment, it may be possible to have fewer wider cracks at larger spacings, or more frequent narrower cracks at smaller spacings. The recommended crack sizes assume widely spaced, essentially isolated cracks. They would be representative of the first scenario described above.

It was recommended that the design crack sizes for the SLE and aftershocks be scaled from the recommended sizes for the MDE, in proportion to the calculated crest settlements.

Table 6. Recommended design crack widths and depths for MDE.

Embankment	Recommended Crack Depth (m)	Recommended Crack Width (mm)
Irishman's Creek	2.5	50
Maryburn Fill	1.5	30
Patterson Terrace	1	20

REFERENCES

Damwatch 2010a. *Tekapo Canal Condition Assesment and Mitigation Concepts* Report, for Meridian Energy. 02, August, 2010.

Damwatch 2010b. *Tekapo Canal Safety Risk Management Project Geophysical Investigation Report,* for Meridian Energy. October, 2010.

Fong, F.C. and Bennett, W.J. 1995 Transverse Cracking on Embankment Dams Due to Earthquakes *presented at the 1995 ASDSO Western Regional Conference, Red Lodge, Montana.*

Genesis Energy, 2011. *Tekapo Canal Options Study*, August, 2011.

Itasca Consulting Group, 2008. FLAC, *Fast Lagrangian Analysis of Continua, Version 6.0.* Minneapolis, Minnesota, USA.

Mejia, L.H. and Dawson, E.M. 2006. Earthquake Deconvolution for FLAC, *in FLAC and Numerical Modeling in Geomechanics (Proceedings of the 4th International FLAC Symposium, Madrid, Spain, May 2006)*, pp. 211-219. P. Varona & R. Hart, eds. Minneapolis, Minnesota: Itasca Consulting Group Inc.

Pells, S. and Fell, R. 2002. Damage and Cracking of Embankment Dams by Earthquakes, and the Implications for Internal Erosion and Piping. *UNICIV Report No. R-408, ISBN 85841 375 2. School of Civil and Environmental Engineering, The University of New South Wales.*

Somerville, P. 2013, Development of Design Ground Motions for the Tekapo Canal Project. *Proceedings, NZSOLD/ANCOLD 2013 Conference, Rotorua, New Zealand, 13-15 November.*

DAM FAILURE-NUCLEAR PLANT SEISMIC PRA MODEL — MODELING CORRELATIONS AND UNCERTAINTY

Martin W. McCann, Jr.[1]
Greg Paxson[2]
Raymond Schneider[3]
Andrea Maioli[4]

ABSTRACT

The events that occurred as a result of the Tohoku earthquake in 2011 provided one example of a combined seismic-flood event. In this case an off-shore seismic event created a tsunami that inundated the Fukushima Daiichi nuclear power plant site. While the seismic event was felt at the site, the robust seismic design protected the plant from seismic related damage. In this paper we consider an analogous, but hypothetical scenario and describe an approach to evaluate the seismic risk to a nuclear power plant (NPP) that is located downstream of a dam whose failure could result in flooding onsite. The upstream dam and NPP are co-located with respect to the location of future earthquakes and as a result both facilities could experience strong motion during the same event. As a consequence the NPP seismic risk model must consider the simultaneous occurrence and correlation of earthquake ground motions at the two facilities and the effects of flooding in the event of dam failure and NPP flooding. The seismic risk analysis models the NPP and dam performance as a result of the earthquake, the probabilistic character of uncontrolled release that could occur as a result of a breach of the dam and inundation at the NPP, and the consequences of the inundation at a potentially seismically damaged plant.

---

[1] Jack R. Benjamin & Associates, Inc., 530 Oak Grove Avenue, Ste. 202, Menlo Park, CA 94025
mccann@jbaslc.best.vwh.net
[2]- Schnabel Engineering, 1380 Wilmington Pike, Suite 100, West Chester, PA 19380
GPAXSON@schnabel-eng.com
[3] Westinghouse Electric Company, LLC., 20 International Drive, Windsor, CT,
schneire@westinghouse.com
[4] Westinghouse Electric Company, LLC., 1000 Westinghouse Drive, Cranberry Township, PA,
maiolia@westinghouse.com

INTRODUCTION

A unique problem in seismic risk analysis for infrastructure systems is the potential that multiple systems or geographically distributed elements of the same system may be adversely affected by an earthquake, leading to a cascade of impacts. One such example is the case of a dam located upstream of a NPP whose failure could result in flooding at the plant. Both facilities may experience ground motions that may damage one or both facilities. From the perspective of NPP safety, core damage may be the direct result of seismically initiated damage, flooding that occurs in the event of upstream dam failure, or as a result of the combined effect of seismic damage to plant structures, systems and components (SSCs) that compromises the capacity of the plant to subsequent onsite flooding. The subject of this paper is to: a) describe the seismic risk modeling issues associated with this inter-system risk issue, b) describe a methodology to evaluate the NPP and dam seismic risk problem, and 3) to present results of a parametric analysis that illustrates the relative importance of key factors in evaluating this problem.

The post-earthquake response of a NPP to a potential flood threat could be complicated by a number of factors. For instance, the damage caused by the earthquake might compromise flood protection systems (e.g., building exterior walls, building penetrations, off-site and/or on-site emergency power sources) or the ability of plant personnel to carry out emergency operations (i.e., install temporary flood barriers.

A full-scope seismic risk analysis for the NPP-upstream dam system can be particularly complex. It would entail a seismic risk analysis for the NPP, a similar analysis for the dam, including the analysis of the timing and magnitude of downstream flooding that occurs in the event of a dam breach, an assessment of the vulnerability of the NPP to external flooding, and a seismic hazard and risk quantification analysis that accounts for the effects of ground motion correlations and dependencies that affect the performance of each facility.

In this paper we focus on limited parts of the NPP-dam seismic risk problem. The first is the issue of modeling the performance of facilities that are co-located with respect to an earthquake event. The assessment of this problem requires that sources of dependence/correlation associated with the potential earthquake ground motions be modeled explicitly. The second part of the problem concerns the uncertainties (aleatory and epistemic) associated with estimating the timing and depth of flooding that could occur at the NPP; the probabilistic aspects of dam breach floods. Elements of the seismic risk analysis that will not be addressed in this paper include the evaluation and quantification of epistemic uncertainties in the seismic fragility of SCCs at the NPP and the dam. In addition, post-earthquake operational aspects of the dam (e.g., operation of gates, low-level outlets) and at the NPP are not modeled.

PLANT AND DAM LOCATION AND CHARACTERISTICS

Plant and Dam Location

In this problem the NPP is located on a river that provides plant service water. Upstream of the plant, approximately 40 river miles is a water supply dam. As part of the plant design basis river flooding and potential dam failure were considered in the context of the deterministic standards applicable at the time.

Plant Characteristics

The hypothetical plant of interest is a pressurized water reactor (PWR) situated on a river which serves as an ultimate heat sink. Mitigation of external events involves maneuvering the plant to a shutdown condition and removing decay heat via either the use of the residual heat removal (RHR) system or by way of steam generator (SG) heat removal using the plant's auxiliary feedwater (AFW) system. In addition it is further assumed that a post-Fukushima FLEX system for responding to extended loss of AC power (ELAP) events is also available. As is expected for typical plant designs, the pump rooms housing permanently installed decay heat removal pumps are designed to be water-proof (primarily to protect from internal floods). Small capacity sump pumps are located in the pump rooms to accommodate minor system and groundwater leakage. External building flood protection is provided by permanent flood barriers, and flood resistant penetrations. Recently FLEX equipment has been designed and procured to back-up existing permanent in-plant systems. The plant FLEX strategy is to store portable low capacity heat removal pumps, a portable diesel generator and a fuel oil tank at approximately elevation 177 feet msl of the auxiliary building. Water for injection into the SG is to be taken from the RWST located at the lower level of the building well above the potential dam flood scenario. Should it be needed, the RWST may be replenished by a large inventory of water located on a hill near the site boundary. Following notification of an ELAP condition, the FLEX system is aligned and portable diesel fuel tank is filled. This overall strategy provides multiple layers of defense in depth. For simplicity all key decay heat removal systems of concern are located in the Auxiliary Building. The intake structure and ultimate heat sink are not shown but are assumed to not be impacted by the seismic event. The mitigation system layout is illustrated in Figure 1.

Dam Characteristics

The dam is a zoned earth embankment with a maximum height of 150-ft and the maximum depth of water at normal pool is about 140-ft. The reservoir has a volume of 58,000 acre-ft (19 billion gallons) at normal pool. Typically, the reservoir is at or above normal pool about 75% of the year, within 5-ft of normal pool about 90% of the year, and within 10-ft of normal for more than 95% of the year.

SEISMIC HAZARD

Seismic Source

For purposes of this analysis a single seismic source is considered that is the host for both the dam and the NPP. The annual rate of occurrence of earthquakes of magnitude (**M**) 5 and greater is 10^{-2}. The maximum magnitude of earthquakes that can occur in the source is 7.2. Figure 2 shows the seismic source and the location of the dam and NPP.

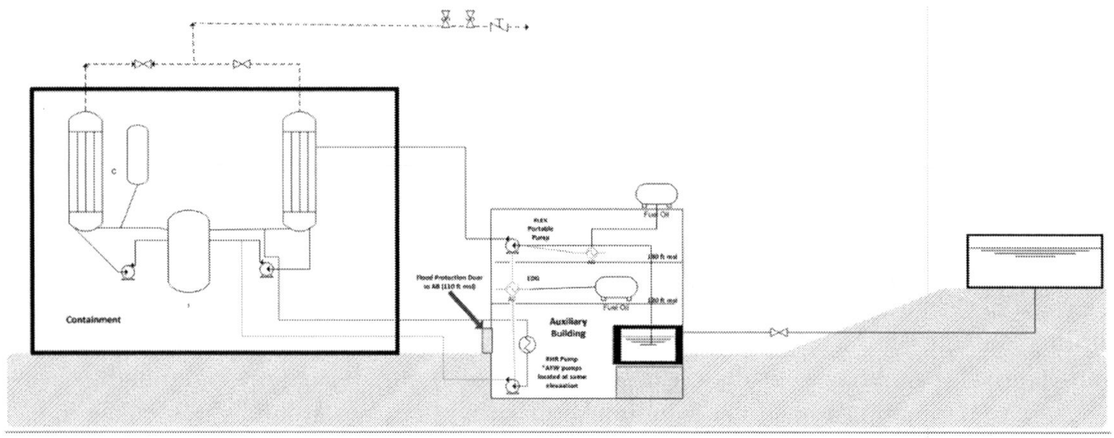

Figure 1. Illustration of the NPP layout.

Ground Motion Correlation

Empirical evaluation of ground motion suggests the residuals with respect to the mean log or best estimate can be represented by,

$$\varepsilon_{ij} = \tau_i + \varsigma_{ij} \tag{1}$$

where, ε_{ij} = deviation for the jth strong motion recording (at an individual recording station) due to the ith earthquake; τ_i = event term for the ith earthquake; ς_{ij} = intra-event variability for the jth strong motion recording (at an individual recording station) due to the ith earthquake.

The total logarithmic (natural log) standard deviation of ground motion residuals (of the ε_{ij}'s) is approximately 0.60 which produces a +/- two standard deviation range corresponding to a factor of approximately 11. This variability corresponds to the randomness or aleatory uncertainty in earthquake ground motions. The logarithmic standard deviation of the event terms (the τ_i terms in equation 1) is approximately 0.31 and the intra-event variability is approximately 0.51 (Chou and Youngs, 2008).

The inter-event variability, τ_i, is a source of parametric aleatory uncertainty that quantifies a random, but systematic difference between earthquakes of the same magnitude. The inter-event term is attributable to random differences between earthquakes (i.e., stress drop), where the ground motions are earthquake systematically different from the average trend across all earthquakes of the same magnitude.

In terms of evaluating the seismic risk for multiple facilities, the tau effect says that ground motions at all sites may be systematically higher or lower for a given earthquake. The last source of correlation addressed is associated with the random field of ground motions that occurs during an earthquake. Empirical studies have evaluated the correlation between sites as a function of their separation distance. Figure 3 shows a model for the correlation coefficient for ground motions estimated by Boore, et al. (2003). For spatially co-located systems with relatively small separation distances (<5km), the correlation coefficient is greater than 0.5 indicating that if the ground motions at nearby sites are also high. Similarly, if motions are low, they are likely low at nearby sites.

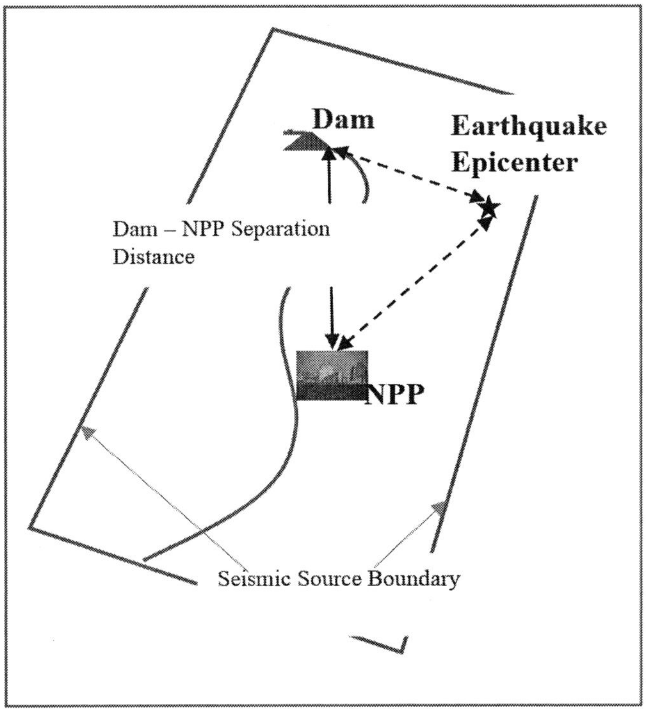

Figure 2. Seismic source considered in this analysis and the location of the NPP and dam.

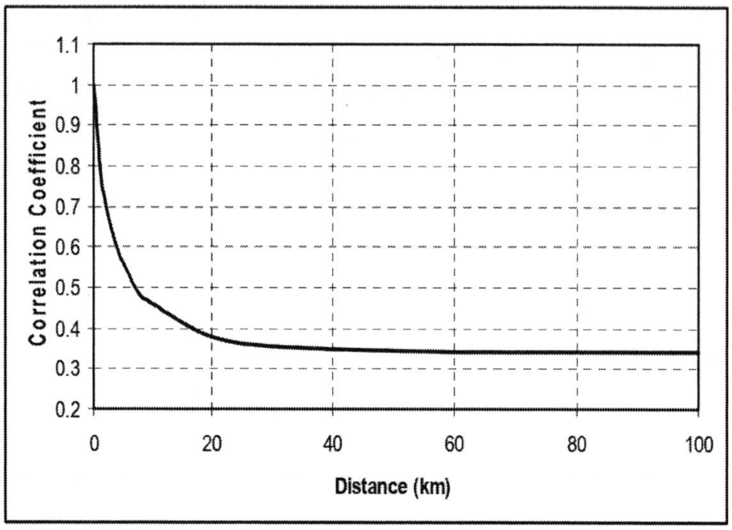

Figure 3. Correlation coefficient for ground motions as a function of site separation distance.

SEISMIC RISK MODEL

Seismic probabilistic risk analysis (SPRA) for NPPs has been performed for over 30 years. As such the methodology for their implementation is reasonably well established. However, the evaluation of plant performance in the context of simultaneous (or near simultaneous) exposure to dual external hazards is unique. In this case, the plant has three unique damage states:

1. Seismic induced core damage (CD_S)
2. Flood induced core damage (resulting from a dam failure) (CD_FLD)
3. Joint seismic-flood induced core damage (CD_S_FLD)

The first of these states is the typical product of a NPP SPRA. The second state, CD_FLD, is the equivalent result if an external flood risk analysis were performed for the plant. In this case it represents the core damage that results from a seismically-initiated dam failure that floods the plant site. While external flood risk studies are not as common, conceptually they are the equivalent of SPRAs. The last state, CD_S_FLD, is unique and represents the performance of the plant when it is exposed to the effects of earthquake ground motions, followed by the flooding that could occur as a result of an upstream dam failure.

Plant Seismic Risk

NPP SPRAs evaluate the vulnerability of equipment items and critical structures to ground motions. A seismic event is likely to cause a loss of off-site power and, depending on the level of ground motion, it may damage permanent safety related equipment needed for safe shutdown. In addition, damage to structures my compromise the availability of some equipment items or their flood protection (below grade cracking of walls, damage to doorways, etc.).

Plant Flood Risk

A plant flood risk model is similar in principle to a seismic model in terms of its purpose and general elements; plant logic model, estimates of structure and component fragility, and operator actions. In general a plant flood response model will build upon the internal events and internal flooding model; the flood response logic for the plant is simpler (than the seismic model) in that there are fewer SSCs that are exposed to the flood hazard. At the same time however, the seismic equipment list may have to be augmented with other components such as external flood barriers, structures or equipment items that are not considered in the seismic PRA.

There are differences between seismic and flood models. For one, in a flood event, the hazard does not challenge the entire plant, thus limiting the complexity of the modeling that must be done. For example, floods cannot cause spurious system actuation or fail system piping. A second difference is the fact that for flood events there is often some warning that is available, possibly on both the timing and the level of flooding that may occur. This is often the case for major riverine floods events as well as for storm surge associated with hurricanes. As a consequence there could be forewarning of a beyond design basis event and adequate time to implement flood mitigation actions to protect the plant.

Depending on the circumstances this may not be the case in the event of a seismically initiated upstream dam failure, where the forewarning is a function of the failure mode of the dam, the details of the breach formation, and conditions in the river at the time of the event, etc. As a consequence, the forewarning may be limited to the travel time of the flood, or less (depending on communications at the time). In the current hypothetical illustration the plant would have about 6 to 25 hours before the initial arrival of the flood to prepare the plant.

Plant Seismic-Flood Risk

The third core damage scenario (CD_S_FLD) involves a seismic event which can damage the plant (though not cause core damage) and cause a breach of the upstream dam. As a consequence of the plant seismic damage, the response to the flood event may be complex. Complicating factors may include early loss of offsite power, reduced access to plant locations, potential distortions to frames used for temporary barrier installations, etc. It may also complicate post-event emergency response. The occurrence of a seismic event in the region will provide plant operators a form of early warning of a potential dam failure, and an indication that plant actions that may be taken for flood preparation are taken expeditiously. These activities may include installation of flood barriers, deployment of flood mitigation equipment to protected staging areas. In this illustration, the plant will respond to the flood by installing flood barriers at the auxiliary building and intake structures and place the plant on shutdown cooling. While not directly considered in this discussion, simultaneous implementation of on-site flood preparation and post-seismic actions could pose unexpected conflicts further complicating implementation of protective actions.

There is a potential that FLEX strategies cannot be implemented in time. Depending on the level of ground motion at the dam, embankment instability may lead to a relatively sudden breach and uncontrolled release of the reservoir. Alternatively, less severe damage to the embankment may result in a delayed breach if the reservoir cannot be lowered in time.

A complicating effect of the seismic event is that fact that the plant may survive the event, but at the same time experience damage to structures and selected equipment items that compromises the response to flooding. For example, damage to structures may deform doorway frames complicating installation of flood barriers and create leak paths, crack exterior walls creating a leak path into the building, and potentially expanding small cracks that may exist in the foundations of structures containing flood mitigation equipment. Furthermore, following a seismic event there is a likelihood that flood preparatory actions would need to take place without the benefit of off-site power.

Strategies for addressing leakage into buildings typically involve operation of sump pumps. However, power for installed sump pumps are normally assigned to non-safety buses which would not be available following a loss of offsite power. The relatively larger failure potential reflects the consequences of the seismic event on the flood hazard may not be recognized until flood waters arrive on site. Success of flood mitigation under these circumstances is significantly aided by FLEX related equipment which is deployed well above potential flood elevations.

Dam Seismic Risk Model

Depending on the layout, design and operating characteristics of a dam, there are a number of different seismic failure modes that could lead to a breach and uncontrolled release of the reservoir (URR). Like a NPP, a dam is a system of structures (natural and man-made), systems (reservoir retention and control systems such as gates and valves, structures, powerhouse), and components (mechanical, electrical and electronic), operators, and operating procedures. There are various combinations of events that can lead to URR as a result of a seismic event. Potential major seismic failure modes include:

- Embankment slope instability leading to loss of freeboard, overtopping, and erosion of the embankment.
- Embankment settlement and cracking, resulting in seepage, that evolves to internal erosion of embankment materials, and collapse of the dam crest. This failure mode may be coupled with an inability to operate gates and/or low level outlets.
- Embankment overtopping as a result of seiche or landslides into the reservoir.
- Foundation instability and loss of bearing capacity leading to embankment sliding or settlement.

- Failure of dam gates resulting in controlled release (e.g., Folsom Gate failure in 1995).

Therefore, much like a NPP an SPRA model would be developed to describe the events that may result in URR, and which sequences pose a flood hazard to the NPP.

NPP-Dam Seismic and Flood Risk Model

To model the performance of the dam and the NPP during a seismic and dam break flood event, an event tree model was developed. The combined seismic-flood model can become complicated, however for purposes of this demonstration assessment a simplified dam and NPP model is used. In addition, a limited number of the elements of the NPP that relate to flood protection and vulnerability are modeled. Figure 4 illustrates the NPP-dam seismic event tree that was developed for this analysis. The top events in the event tree are summarized in Table 1.

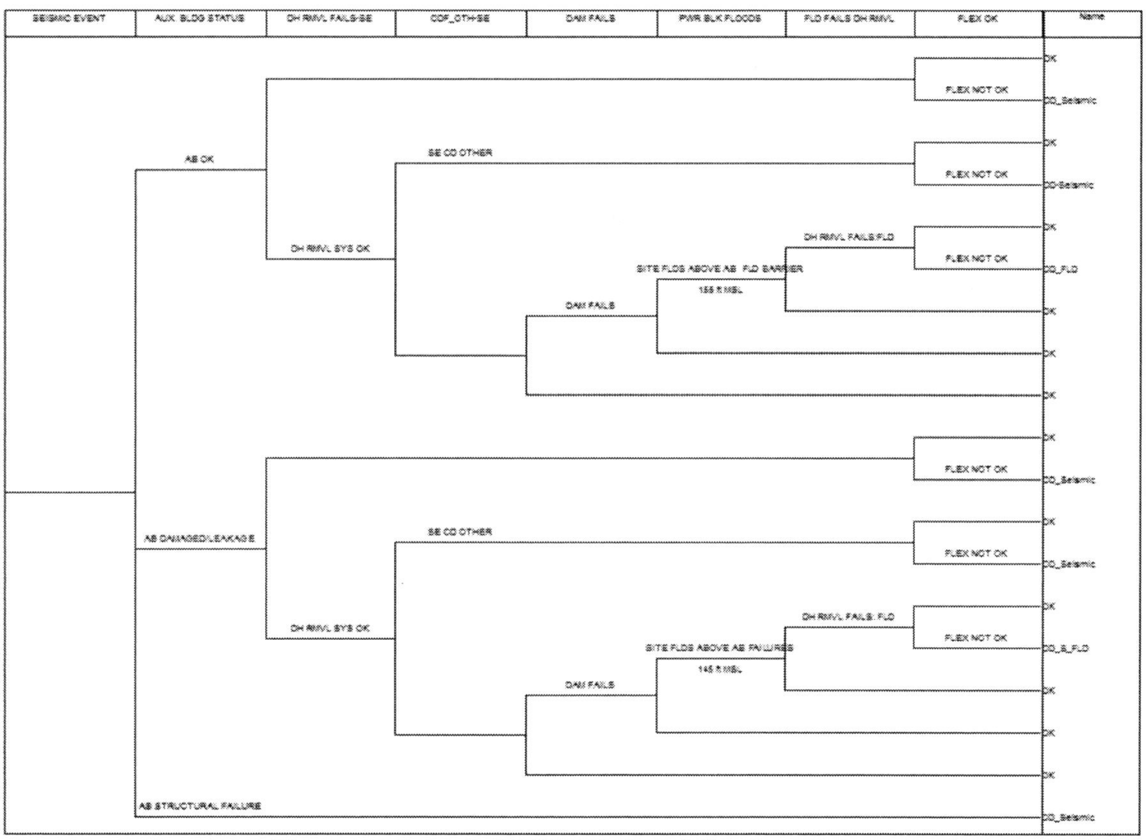

Figure 4. Seismic NPP and dam failure event tree.

Table 1. Event Tree Top Events

Top Event	Description
Seismic Event (SE)	Seismic initiating event
Auxiliary Building Status (AB)	Models the state of the Auxiliary Building (AB); there are three possible outcomes: • AB OK: AB structure not damaged by the he SE • AB Damaged/Leakage – SE causes cracking, loosening of penetrations or seal failures sufficient to create leak paths to the building interior. • AB Failure: AB fails rendering equipment interior to the building inoperable
DH REMVL Fails-SE	Seismic event causes failure of mitigation equipment required for decay heat removal. Damage assumed to be sufficient to fail all installed decay heat removal capability. Without timely implementation of FLEX strategies, core damage occurs.
CDF_OTH-SE	A potential core damage event, not directly associated with the loss of decay heat, occurs. These events include induced LOCAs. These events may not be associated with a loss of power but potential for FLEX inventory makeup systems may mitigate the consequences of a class of these scenarios.
Dam Fails	The seismic event causes a failure of the upstream dam
Site Power Block Flooded	Dam failure results in flooding at the plant. Without seismic damage the AB is assumed to be protected from the dam flood (e.g. event included in design basis). Floods at the site above the level where the AB leakages/failures occur can propagate into the AB and challenge the operability of key flood mitigation equipment.
FLD Fails DH Removal Systems	Following an AB flood this event asks whether the water infiltration into the building floods out mitigation equipment. The flood fails all installed decay heat removal equipment.
FLEX OK	FLEX system mitigates the loss of decay heat removal.

The NPP seismic-dam failure event tree initially addresses the state of the plant as a result of the seismic event. These top events consider; did seismic CD occur, if not, what is the state of structures and critical safety functions that will be required to deal with the flooding in the event the dam fails (e.g., decay heat removal). For example, in an SPRA for a NPP there are typically singleton cutsets involving the failure of major structures such as the auxiliary building. If the auxiliary building fails there is CD. The seismic event may produce a level of damage to the auxiliary building that does not constitute failure, but does compromise the flood protection function of the building. For instance, damage may occur that results in cracking of walls above and below grade, damage to doorways, etc. that could allow flooding in the building to occur.

The next part of the event tree addresses whether the dam has failed as a result of the seismic event and if failure occurs the level of the flood hazard at the NPP.

The final part of the event tree address the NPP performance given the occurrence of seismic and flood damage.

DAM BREACH AND INUNDATION ANALYSIS

When a dam breach occurs the outflow and the inundation downstream depends on factors/parameters related to the details of the breach and the hydraulic characteristics of the downstream channel (Xu and Zhang, 2009). With respect to the breach, these include:

- Dam failure mode (e.g., overtopping from a flood event, piping),
- Location of the breach along the embankment,
- Breach geometry, including the width, side slopes, and bottom elevation,
- Erodibility of the embankment materials,
- Time to failure (time it takes for the full breach to develop), and
- Reservoir volume at the time of the breach.

In terms of the estimate of the inundation that occurs downstream, factors that influence the travel time, depth of inundation, velocity of flow, and the hazard to the NPP include:

- Pre-existing flow in the river at the time of the earthquake (varies with the time of the year),
- Channel and overbank characteristics (slope, geometry, roughness),
- Changes in channel and overbank conditions during the year (winter vs summer),
- Interaction between main and side channel flows,
- Tidal effects (if applicable),
- Effects of obstructions,
- Debris content,
- River icing, and
- Topography.

These factors, each of which is specific to the river channel and overbank conditions contribute to aleatory and epistemic uncertainties in predicting flood stages (Freeman, et al., 1995). Estimates of the aleatory uncertainty in peak flood elevation levels can be as much as two feet (Freeman, et al., 1995). For streams or rivers that are not gaged (flows and flood stage data is not collected) the aleatory uncertainty may be greater. In addition there will be epistemic uncertainties that will impact the estimate of flood levels.

The analysis of flooding that could occur at the downstream NPP is carried out in the following steps:

1. The outflow hydrograph (a time series of flow rate) from a dam breach is estimated based on the characteristics of the dam, reservoir volume, failure mode, and the breach.
2. The breach outflow hydrograph is then routed downstream using a model of the river channel and overbank areas.

In this analysis the U.S. Army Corps of Engineers HEC-RAS model (HEC, 2010) is used to model the dam breach outflow and to route the flood downstream.

The flood stage at the NPP site can be represented by the following relationship:

$$S = g(RV, ER, BR, FM, CH, Q_R) + \varepsilon \quad (2)$$

where,

$g()$ = represents the dam breach and flood routing model
RV = reservoir volume
BR = breach characteristics (time to failure, failure mode, breach geometry, etc.)
ER = embankment erodibility
FM = dam failure mode
CH = downstream channel characteristics (slope, Mannings n, etc.)
Q_R = river flow at the time of the dam failure
ε = random term that models the variability in flood stages (Freeman, et al., 1995)

There are a number of sources of aleatory and epistemic uncertainty associated with estimating the flooding at the NPP; estimating the parameters of the dam breach, river channel response, the estimated breach outflow, the estimated flood stage, etc.

To model the epistemic uncertainty in the estimate of the breach characteristics (breach bottom width and time to failure) a logic tree is constructed. The logic tree includes alternative breach models and the uncertainty in the estimates of the model parameters. Two empirical models are used in this analysis; Froehlich (2008) and Xu and Zhang (2009). Table 2 summarizes the dam breach models that are used in the analysis. The standard deviation corresponds to the aleatory variability in the parameter (e.g., breach bottom width) estimate. Each model is assigned a probability weight as shown in Table 2. To represent the uncertainty in the model parameters, three estimates of the mean value of each parameter are considered that correspond to the 0.05, 0.50 and 0.95 percentile estimates. Each parameter is modeled using a truncated normal distribution that is defined by its mean and standard deviation. Figure 5 shows the logic tree for modeling the uncertainty in the breach parameters.

Table 2. Breach Models Used

Model	Weight	Bottom Breach Width (ft)[1]	Time to Failure (Hrs)[1]
Froehlich (2008)	0.40	Mean: 230.3 SD[2]: 162.8	Mean: 1.1 SD: 0.34
Xu and Zhang (2009)	0.60	Mean: 137.2 SD: 59.6	Mean: 2.33 SD: 1.35

[1]Best Estimate of the Mean; [2]SD = Standard Deviation

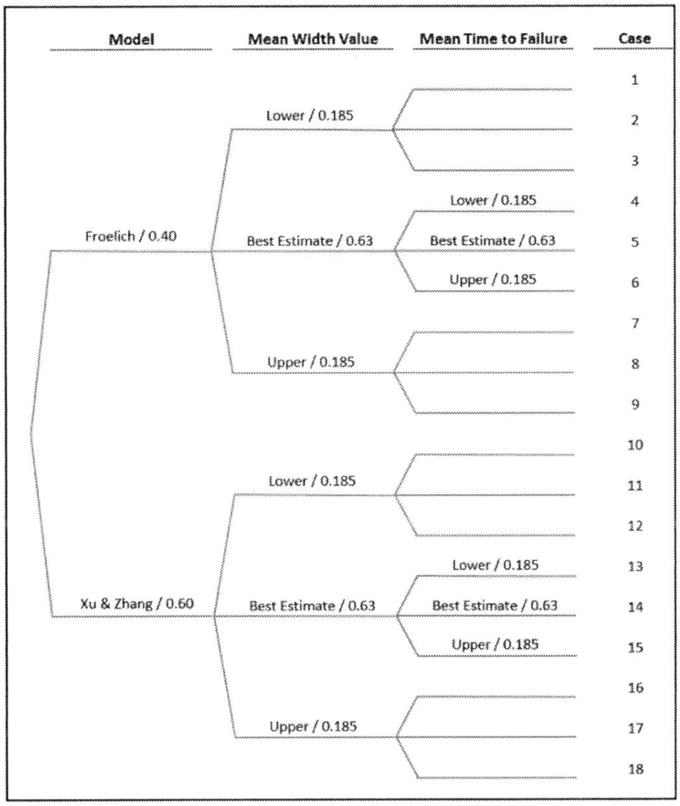

Figure 5. Logic tree for modeling the uncertainty in breach parameters.

Each branch in the logic tree corresponds to an alternative breach model (an alternative prediction of the breach characteristics). For each logic tree branch, a random breach analysis is performed by Monte Carlo simulation using the parameters defined on the branch. The Monte Carlo simulation was performed using the MCBreach external controller for HEC-RAS, developed by WEST Consultants (Goodell, 2013). The result of the simulation is a series of dam breach outflow hydrographs that quantifies the aleatory variability in breach outflow. The complete set of branches in the logic tree models the epistemic uncertainty in the estimate of the breach outflow. The aleatory and epistemic uncertainty in the estimated flood stage at the NPP is determined by routing the outflow hydrographs downstream to the plant site. Once this routing is completed, a probability distribution on the peak flood stage can be determined through the use of equation 2, where ε has a zero mean and a standard deviation of 1.0 feet (Freeman, et al., 1995).

The following lists the sources of aleatory and epistemic uncertainty considered in this analysis in estimating the flooding at the NPP.

Element of the Analysis	Sources of Uncertainty	
	Epistemic	Aleatory
Dam Breach Geometry		
a. Breach Width	X	X
b. Time to Failure	X	X
c. All other parameters at fixed values (*RV, FM, CH,* Q_R)	--	--
Flood Routing	Not evaluated	X

RISK QUANTIFICATION

The risk quantification is carried out in a series of steps:

1. Monte Carlo simulations are performed for each breach case shown in the logic tree in Figure 5 to estimate the probability distribution on the breach outflow at the dam.
2. The breach outflow hydrographs are routed downstream to determine the distribution of peak flood depths at the NPP, including the variability in the flood routing.
3. The seismic part of the event tree is quantified to determine the frequency of occurrence of the event tree sequences, involving the seismic performance of the plant and the dam.
4. The seismic sequence frequencies are combined with the flood hazard curves (from Step 2) and the flood fragility estimates to determine the CD frequency from seismic, dam break floods and combined events.

RESULTS

This section presents results of the different elements of the analysis. These include:

- Reviewing the results of the dam breach and flood routing analysis.
- Estimate of the frequency of seismically initiate failure of the dam and NPP (as standalone structures).
- Frequency of NPP core damage due to the seismic event or the flooding as a result of dam failure.
- Probabilistic flood hazard curves due to dam failure.

Dam Breach and Flood Routing Results

For each branch in the logic tree 1,000 simulations were performed using the MCBreach model to estimate the aleatory uncertainty in the breach outflow hydrograph (a total of 18,000 simulations; 18 logic tree branches x 1,000 simulations per branch). Figure 6 illustrates the variation in the breach outflow hydrographs for Case 1 in the dam breach logic tree. In this case the peak discharge from the lowest to highest estimates varies by more than a factor of 3 (520,257 to 1,700,684 cubic feet per second (cfs)). In addition, the results in Figure 6 illustrate the variation in the shape of the hydrograph and the timing of the peak outflow.

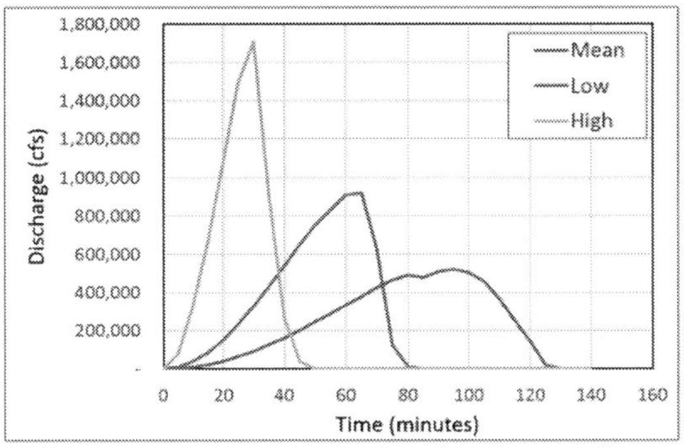

Figure 6. Case 1variation in the outflow hydrographs from the dam breach; low, mean, and high estimates of the peak outflow discharge from the simulations.

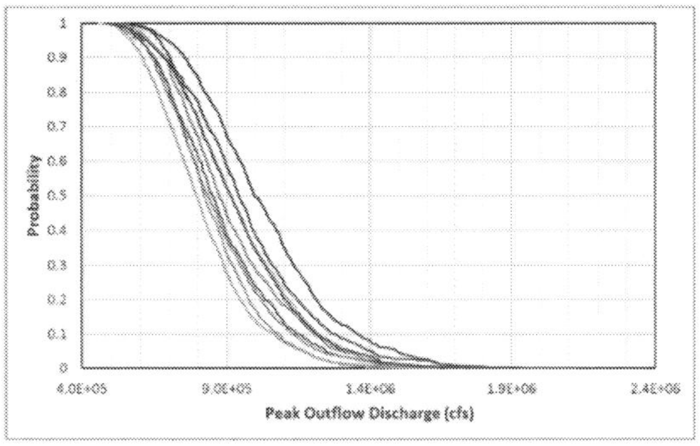

Figure 7. Cumulative distribution of peak outflows for the dam breach logic tree branches using the Froehlich model.

Figure 7 shows the probability distributions on the peak outflow discharge for the 9 logic tree branches corresponding to the Froehlich model. Each probability distribution quantifies an estimate of the aleatory uncertainty in breach outflow (due to the randomness in breach characteristics) and the different distributions capture the epistemic uncertainty in the estimate of the breach outflows.

The results of the dam breach analysis are input to a HEC RAS model of the river channel and overbank areas to estimate the flooding at the plant site. These calculation are carried out for each of the eighteen sets of outflow hydrographs. The result is a corresponding set (18 in total) of probability distributions on the peak flood elevations at the plant site. Figure 8 shows these results for the peak flood elevation at the plant. The figure on the top shows the results based on the Froehlich model, while the bottom figure is based on the Xu and Zhang model.

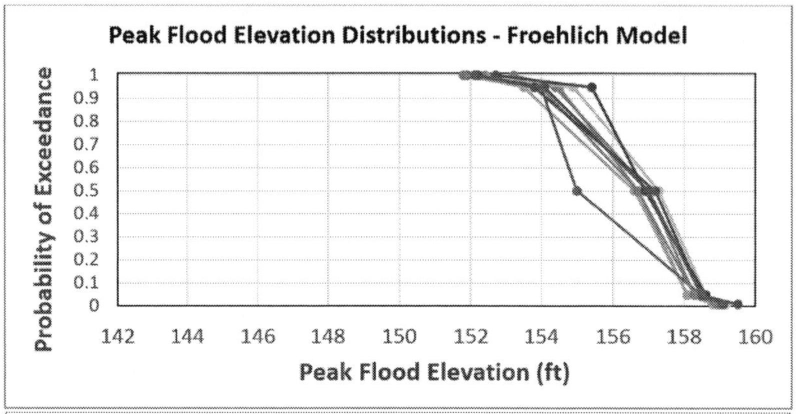

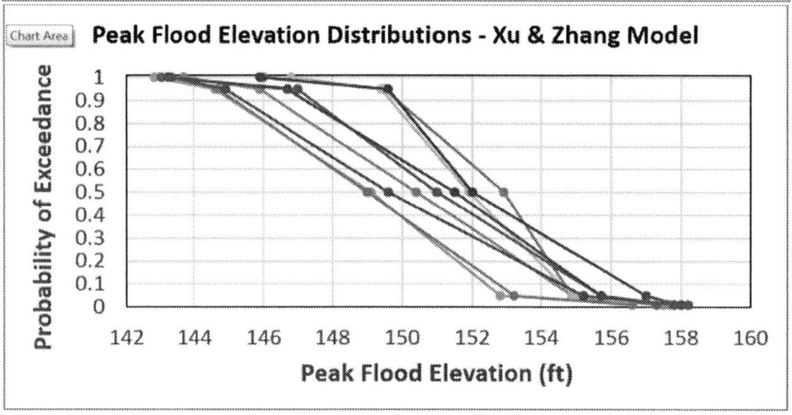

Figure 8. Cumulative distributions on the peak flood elevations at the plant for the Froehlich (top) and Xu and Zhang (bottom) models.

Observations relative to the peak flood elevation results in Figure 8 are:

- For both models there is considerable aleatory variability in the estimate of the peak flood levels that may occur at the plant. There is considerably more variability in the estimates based on the Xu and Zhang model which has a range of about 15 feet, while the range for the Froehlich model is about 7 feet.
- There is a limited difference in the location of the median of each distribution (a measure of the epistemic uncertainty in the model); for the Froehlich model the range is 155-157 feet and for Xu and Zhang it is 149-153 feet.

Given these models are both empirical and their datasets are similar, the differences in the two models represent the uncertainty in estimating the character of a dam breach and this uncertainty is particularly important with respect to estimate the level of flooding downstream.

Figures 9 and 10 present similar results for the timing of the flood event expressed in hours from the time the breach begins. Figure 9 shows the distributions on the time of initial arrival of the flood at the NPP site for the two models. Figure 10 presents the results for the time of arrival of the peak flood.

Here again there is considerable variability in the results and differences between the two models. For both the flood elevation results and the timing, a significant contributor to the differences in the two models is the estimate of the timing of breach formation. The Xu and Zhang model allows for longer breach formation times than the Froehlich model. As a result, the flood outflow is characterized by low peak flows and longer durations and delayed time of the flood peak. The net result is lower flood levels downstream.

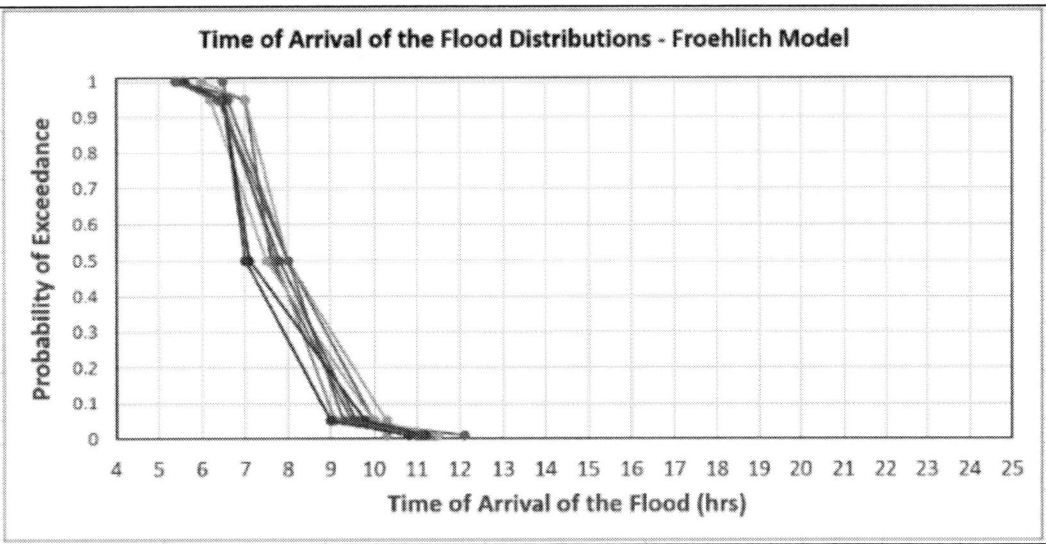

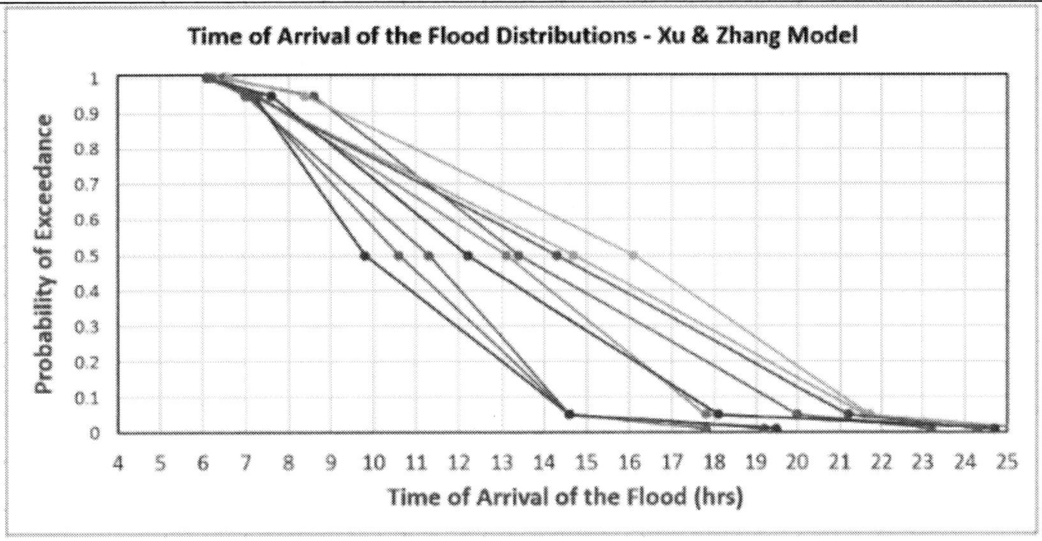

Figure 9. Cumulative distributions on the time of arrival of the flood at the plant for the Froehlich (top) and Xu and Zhang (bottom models.

For both models there is considerable variation (aleatory uncertainty) in the flood levels and the predicted timing of flooding. This can have important implications to the assessment of NPP risk as well.

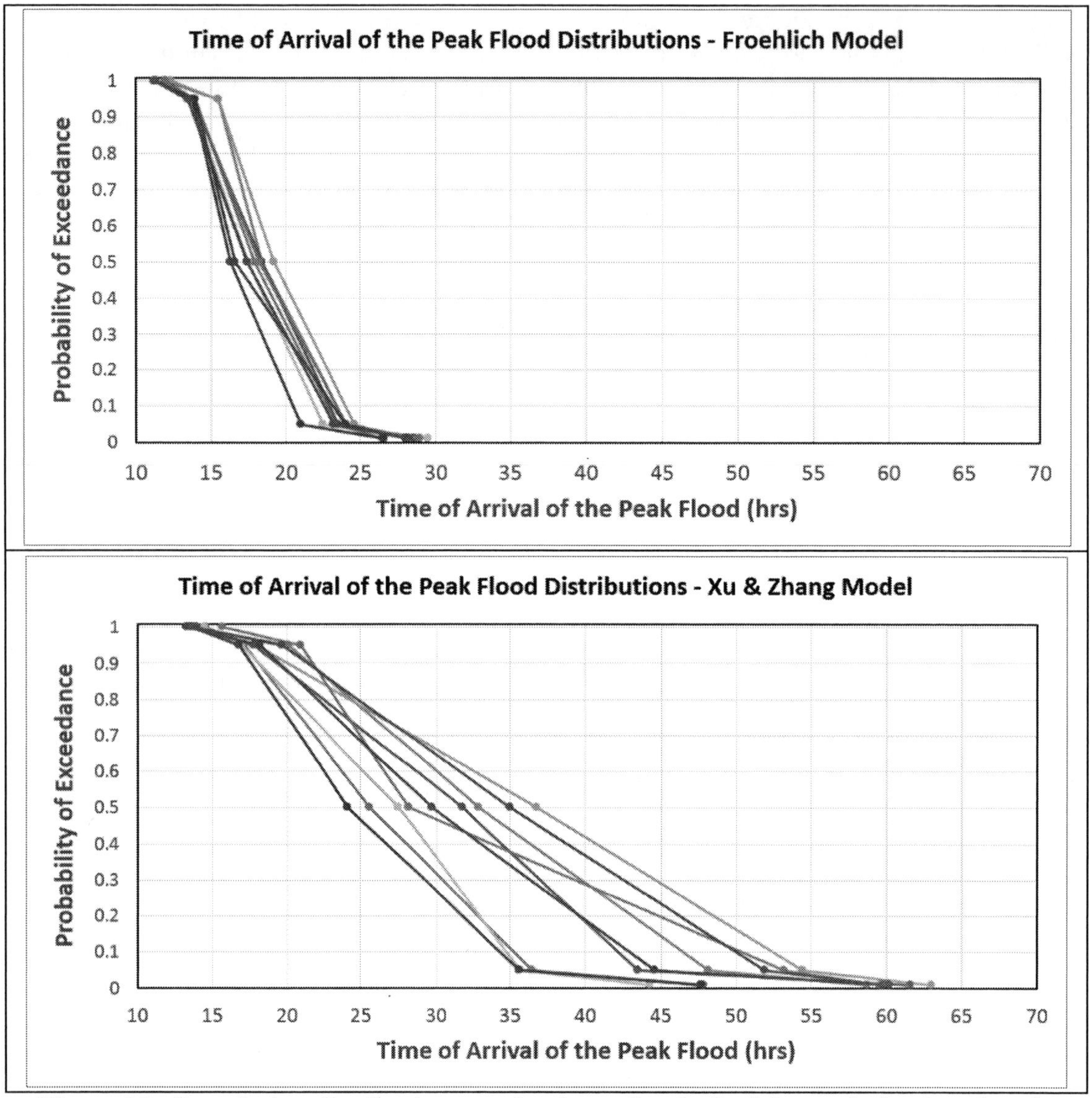

Figure 10. Cumulative distributions on the time of arrival of the peak flood at the plant for the Froehlich (top) and Xu and Zhang (bottom) models.

Dam and NPP Seismic Risk

As a frame of reference the frequency of failure of each facility can be evaluated individually for the seismic source and locations shown in Figure 2. The base case (labeled as Case 1 blow) median seismic capacity for the dam and NPP plant are 0.45 and 0.50g peak ground acceleration, respectively. The frequency of failure per year of each facility is 1.02×10^{-5} and 1.53×10^{-5} per year respectively.

Flood Hazard Curves at the NPP Site

Flood hazard curves at the NPP site due to dam failure are determined as the product of the frequency of dam failure and the conditional probability of exceedance distribution on peak flood elevation (see Figure 8). (Recall in this analysis the uncertainty in the seismic hazard and fragility estimates are not considered in the analysis.) Figure 11 shows the flood hazard curves taking into account the uncertainty in the dam breach characteristics (only) for each of the 18 dam breach and inundation cases considered. These curves quantify the uncertainty in the frequency that peak flood elevations will exceed different levels. The mean flood hazard curve (thick black line) is also shown in the figure.

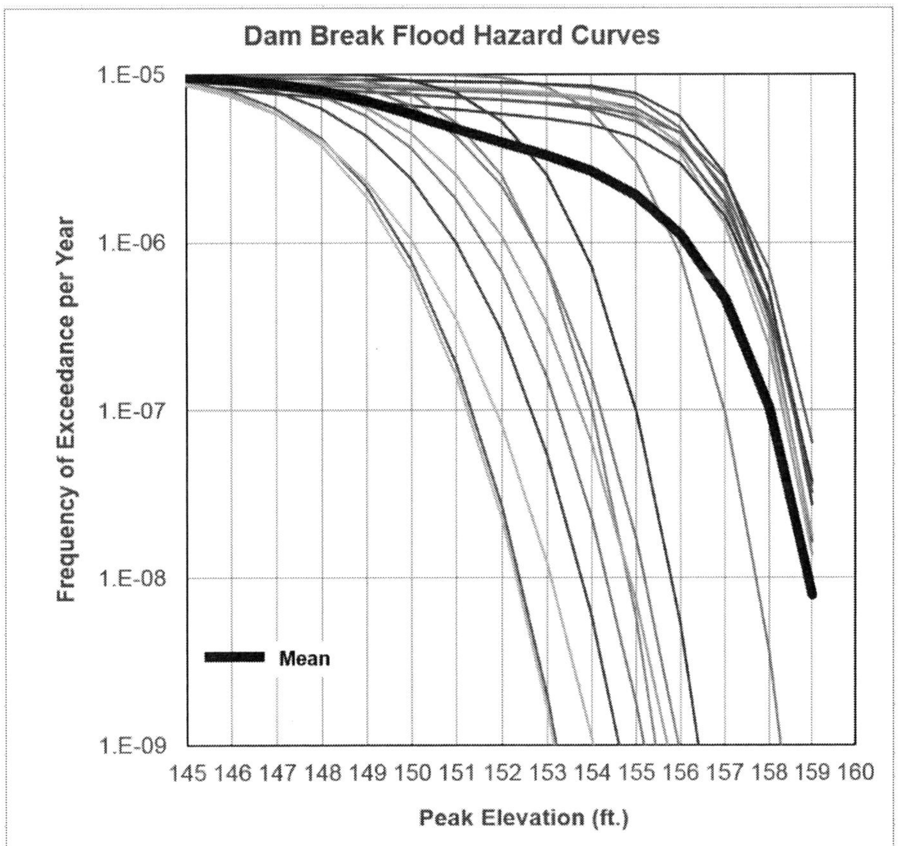

Figure 11. Flood hazard curves due to dam failure for the NPP site. The thick black line corresponds to the mean hazard curve.

In this analysis, plant grade is 153 feet. The mean frequency of occurrence of the plant site being flooded is 3.2×10^{-5} per year.

NPP Seismic-Flood Risk

The risk of core damage is evaluated in a series of risk calculations in which key parameters in the model are varied. The first parameter variation considered is the relative seismic capacity of the dam and the NPP. Three cases are considered; 1)the dam and NPP have similar seismic median capacity (the ground motion level at which there is a 50-50 chance of failure); 2) the dam's median seismic capacity is greater than the NPP (therefore the dam is less likely to fail); and 3) the dam capacity is less than the NPP.

The second parameter/attribute of the model that is considered is the FLEX system. Two cases are considered for FLEX; the first considers the FLEX system with the same seismic capacity as the NPP and the second cases considers the plant without the FLEX system available (a pre-Fukashima scenario). These cases are listed in Table 3.

Table 3. Seismic Risk Cases – Dam and Nuclear Power Plant

No.	Seismic Capacity	FLEX
1	Dam ~ NPP	Yes
2	Dam ~ NPP	No
3	Dam > NPP	Yes
4	Dam > NPP	No
5	Dam < NPP	Yes
6	Dam < NPP	No

Initially the event tree is quantified to determine the frequency of occurrence of event tree sequences involving damage to the plant and the dam (see the event tree in Figure 4). These calculations quantify the frequency that the seismic event and/or the flood event initiate damage to the plant. (The frequency of flooding and the potential failure to mitigate the flood at the plant are not considered.) These results are given in the first row in Table 4. The second set of results take into account the conditional probability of flooding that could damage the plant and the reliability of the plant response. The results for this case are given in the second row of Table 4.

Table 4. Mean Results Considering the Seismic Initiator, Chance of Flooding Exceeding Flood Barriers & Loss of Decay Heat Removal Due to the Flood

	Case					
Result	**1**	**2**	**3**	**4**	**5**	**6**
Seismic & Flood Initiators	1.56E-05	2.95E-05	1.54E-05	2.39E-05	1.59E-05	4.09E-05
CDF	1.53E-05	2.27E-05	1.53E-05	2.25E-05	1.53E-05	2.29E-05

When the FLEX system is available (Cases 1, 3, and 5) there is little impact of the upstream dam failure; the CDF is the same in all cases. Since the FLEX system is protected from the flooding that occurs onsite and has similar seismic capacity as the rest of the NPP, this is reasonable.

If the FLEX system is not available, there is a potentially moderate increase in the small impact on the frequency of seismic and/or flood damage to the plant. This impact is about a factor of 2 increase (see the results in the first row of Table 4).

Taking into account the probability of flood levels that will damage the NPP and the reliability of the plant response, the increase in the CDF is small and does not vary considerably from case-to-case.

SUMMARY

This paper explores the inter-relationship between the seismic and flood hazard for NPPs with upstream dams where seismic events can cause coincident floods. A case (dam-NPP arrangement) was considered in which the potential for dam failure was considered in the plant design and therefore was designed to have protection against the effects of flooding. In addition, the plant has additional reliability due to the presence of a FLEX system. Results of parameter evaluations to assess the relative contribution to NPP risk of different factors considered in the analysis are provided. For the case considered, the impact of a flooding due to a dam failure is at most moderate. This assessment is of course specific to this case and would be different for different circumstances. The example illustrates that while the earthquake may not directly lead to plant core damage, it could damage structures that protect equipment from the effects of flooding and possibly compromise emergency response capabilities required to address the potential flooding at the plant. The performance of the dam system during a seismic event and the uncertainty in estimating the breach characteristics can substantially impact the potential time and magnitude of the flooding that occurs downstream. In the event of a dam breach (immediate or delayed), the extent and timing of flooding at the plant may be highly variable due to such factors as the mode of dam failure, breach characteristics (location, size and timing), reservoir level and flow conditions in the river at the time of the breach, potential for debris, etc.

ACKNOWLEDGMENTS

The authors acknowledge Chris Goodell of WEST Consultants who provided the MCBreach code used to perform Monte Carlo simulations within HEC-RAS, varying breach parameters to estimate outflow hydrographs.

REFERENCES

Chiou, B.S.J. and R.R. Youngs, (2008). "An NGA Model for the Average Horizontal Component of Peak Ground Motion and Response Spectra," *Earthquake Spectra*, 24, 173-215.

Boore, D.M. J.F. Gibbs, W.B. Joyner, J.C. Tinsley, and D.J. Ponti, (2003). "Estimated Ground Motion from the 1994 Northridge, California, Earthquake at the Site of the Interstate 10 and La Cienega Boulevard Bridge Collapse," West Los Angeles, California, *Bulletin of the Seismological Society of America,* 93 (6), 2737–2751.

Xu, Y. and L.M. Zhang, (2009). "Breaching Parameters for Earth and Rockfill Dams," *ASCE, Journal of Geotechnical and Geoenvironmental Engineering*, pp. 1957-1970.

Freeman, G.E. R.R. Copeland, and M.A. Cowan, (1995). "Quantifying Stage-Discharge Uncertainty at Gaging Stations", Proceedings of the First International Conference on Water Resources Engineering, San Antonio, TX. American Society of Civil Engineers, New York, NY.

Hydrologic Engineering Center (HEC) (2010), "HEC –RAS," Version 4.1.0., U.S. Army Corps of Engineers, Davis, CA.

Goodell, C.R. (2013). "Moving Towards Risk-Informed Dam Breach Modeling," Dam Safety 2013, Association of State Dam Safety Officials, September.

Froehlich, D.C. (2008). "Dam Breach Parameters and their Uncertainties," *ASCE Journal of Hydraulic Engineering*, Vol. 134, No. 12.

LESSONS LEARNED FROM THE BLUESTONE DAM INFLOW DESIGN FLOOD UPDATE

D. Moses, PE, PH[1]
N. Koutsunis, EI[2]
K. Halstead, PE[3]
D. Margo, PE[4]

ABSTRACT

Bluestone Dam is a large concrete gravity dam designed during the late 1930's and constructed in the 1940's. Like many dams of the era, the inflow design flood (IDF) was based on the transposition of the largest observed storm of the region. The Bluestone Dam original IDF had a peak inflow of 430,000-ft^3/s.

During the 1980's, the National Weather Service (NWS) published improved estimates of Probable Maximum Precipitation (PMP). Subsequently, many existing projects were re-evaluated for hydrologic adequacy given this new understanding of the maximum precipitation over a given drainage area. The inflow design flood for Bluestone Dam was re-evaluated in 1982, including a site-specific PMP performed by the NWS, and the peak inflow resulting from the updated probable maximum precipitation was estimated at 1,086,000-ft^3/s; more than double the original inflow design flood peak.

A more recent analysis of the Bluestone Basin PMP performed in 2014 resulted in a 10-percent reduction of precipitation depth compared to prior estimates. Although the rainfall estimates were reduced, refinements to the rainfall-runoff modeling resulted in a peak inflow estimate at the dam site in the order of 1,444,000- ft^3/s to 1,661,000- ft^3/s, a 33- to 53-percent increase as compared to the 1982 study.

The limitations of the 1982 study that resulted in a lower peak inflow compared to the 2014 study include; the intensity and time-step of unit hydrographs derived from 6-hour precipitation data, the computational time-step of the numeric analysis and the use of a linear reach routing method. The results of this study indicate that the improvements in data and computational analysis, compared to 1980's and previous era IDFs, could significantly change the IDF for other large dams as well.

[1] Hydraulic Engineer, USACE, 502 8th Street, Huntington, WV 25701, dana.w.moses@usace.army.mil
[2] Hydraulic Engineer, USACE, 502 8th Street, Huntington, WV 25701, nicholas.a.koutsunis@usace.army.mil
[3] Lead Hydraulic Engineer, USACE, 502 8th Street, Huntington, WV 25701, kenneth.c.halstead@usace.army.mil
[4] Lead Civil Engineer, USACE, 1000 Liberty Ave, Pittsburg, PA 15222, david.a.margo@usace.army.mil

INTRODUCTION

Bluestone Dam is located along the New River in southern West Virginia and has a drainage area of approximately 4,600 square miles. The watershed is located in parts of North Carolina, Virginia, and West Virginia. The watershed lies within a highly dissected portion of the Allegheny Plateau that is characterized by high relief and narrow river valleys. Watershed elevations vary from around 1,300 feet near the dam site, to greater than 5,700 feet in the headwaters. The soils in the basin are primarily ultisols, alfisols, and inceptisols, with generally high runoff potential. The watershed is also characterized with generally shallow soil and exposed rock outcrops.

The precipitation in the watershed is often driven by convection storms and tropical systems in the summer months, and extra-tropical systems (i.e. frontal storms) in the spring and winter months. The area is also subject to Meso-Scale Convective Complexes (MCC), which can be enhanced by the orographic influences of the basin. In general, the largest floods in the basin have resulted from tropical systems in the summer/early fall and by extra-tropical systems in the winter.

Planning and design for the project started in the late 1930's with construction starting in 1942; but was halted during World War II. The project was completed in 1947, although the primary project purpose was modified to flood control, resulting in a decrease of the normal pool elevation from 1,490-feet to 1,410-feet. The Dam and New River are illustrated in Figure 1.

As part of the ongoing Dam Safety Modification Study (DSMS), a re-evaluation of the PMP and IDF was performed considering the additional availability of advanced computational methods, additional storm/flood data, and potential changes in policy and state-of-practice.

Figure 1. Bluestone Dam

BACKGROUND AND PREVIOUS ANALYSIS

Most dams designed prior to the 1980's were designed for an inflow design flood that was based on either a transposition of the largest regional storm or flood, or previous estimates of seasonal probable maximum precipitation depths. The IDF for Bluestone Dam, located along the New River in southern West Virginia, was developed using a transposition of a tropical storm that was centered near Altapass, NC. This storm was south of the New River basin; along the eastern slope of the Blue Ridge Mountains. The resulting IDF had a peak inflow of 430,000-ft^3/s and a peak reservoir stage of 1,523-feet.

In the early 1980's the NWS published two significant publications titled "Hydro-Meteorological Report 51 - Probable Maximum Precipitation Estimates, United States East of the 105th Meridian" and "Hydro-Meteorological Report 52 - Application of Probable Maximum Precipitation Estimates – United States East of the 105th Meridian". These publications were the primary sources for PMP data to produce IDFs for the region stretching from the Rocky Mountains to the Atlantic Ocean.

During the same time period there was a marked increase in attention, legislation, and funding related to Dam Safety as a response to several catastrophic dam failures. Considering The National Dam Safety Act (PL 92-367) of August 1978, the USACE initiated a comprehensive review of existing dams under the Dam Safety Assurance (DSA) Program. The DSA Program allowed for modification of existing infrastructure based on changes in state-of-the-art hydrologic analysis and seismic design criteria. The Bluestone Dam IDF was revisited in the early 1980's considering the availability of updated precipitation estimates, additional watershed hydrology data, and the availability of computer aided analysis.

The 1982 update utilized site-specific PMP data, developed by the NWS, since the Bluestone Basin is within a "stippled" region of the Appalachian Mountains. It is identified in HMR 51 that, "the generalized PMP estimates might be deficient because detailed terrain effects have not been evaluated". As part of this site-specific PMP analysis, the NWS concluded that the prototype storm mechanism that could create the maximum precipitation depths over the Bluestone Dam basin would be a tropical storm out of the south east. The study evaluated the elevation barrier potential and the broad scale and local scale impacts of the orography and it's impact on the basins PMP. The resulting PMP estimate, considering both limits to transposition based on elevation barriers and the affects of orography, resulted in a basin average 72-hour PMP depth estimate of approximately 20-inches; which is approximately 17-percent less than obtained using the generalized HMR 51 estimates.

A HEC-1 hydrologic model was utilized to perform the rainfall-runoff modeling. The initial-constant method was used to estimate precipitation losses, non-parametric Unit Hydrographs (UH) were used to transform rainfall into runoff and the Muskingum routing method translated flows from upstream sub-basins to the dam site. The basin was divided into 14 sub-basins that were then combined into 5-sub-areas for computational convenience. The hydrologic model was analyzed using a 6-hour computational time step. The rainfall-runoff information utilized for this update was developed in the Mid-1950's by a government contractor with storm events during the period ranging from 1940 to 1953. The precipitation and stream flow data in-which the UHs were derived, had a 6-hour time-step; therefore the UH data was also 6-hour.

The 1982 IDF analysis resulted in a peak inflow of 1,086,000-ft^3/s and a resulting reservoir stage of 1,555-feet. The peak inflow was more than twice the original IDF and the resulting stage overtops the dam by a significant amount. This resulted in several modifications to the existing dam including; providing fusible gates on previously unused penstocks, post-tension anchoring of the gravity concrete monolith sections, and construction of a penstock stilling basin as a means to dissipate energy downstream of the fusible penstocks. The addition of the fusible penstock spillway and stilling basin results in a peak reservoir stage of approximately 1,542-feet; which is a reduction of 13-feet in maximum reservoir stage over the pre-modification condition, assuming level-pool reservoir routing.

2014 UPDATE OF THE INFLOW DESIGN FLOOD

In 2014, as part of the ongoing Dam Safety Modification Study (DSMS), a re-evaluation of the IDF was performed considering the additional availability of advanced computational methods, additional storm/flood data, and potential changes in policy and state-of-practice. This re-evaluation of the IDF included another site-specific PMP review which recommended reducing the previous PMP depth by approximately 10-percent. Although the rainfall estimates were reduced, refinements to the prior rainfall-runoff modeling resulted in a peak inflow at the dam site in the range of 1,444,000- to 1,661,000- ft^3/s, depending on UH peaking. This represents a 33- to 53-percent increase over the 1982 analysis and more than 3 times the original IDF.

Revised Probable Maximum Precipitation

Given that both the PMP and resulting IDF estimates were more than 30-years old and with additional data and advancements in the science and understanding of extreme precipitation and flood events, the PMP and IDF were re-evaluated. Additional data includes the availability of hourly precipitation and stream flow gauges, additional large storms and floods, and the availability of gridded precipitation. Also since the completion of the 1982 studies, there has been large advancement in geospatial and hydrologic software and computational speeds.

Considering the factors listed above, a re-analysis of PMP estimates for the Bluestone Dam basin was performed by USACE. The re-analysis concluded that the 1982 site-specific PMP analysis potentially transposed storms that may not be plausible in the Bluestone Basin. Also, investigation into more recent storms revealed that it was highly unlikely that a tropical system prototype storm would produce the PMP within this watershed due to the terrain barriers surrounding the basin, and that it is very possible that a Meso-Scale Convective Complex would be the prototype PMP storm for the basin. The results of the 2014 PMP re-analysis was ultimately a 10-percent reduction in total

PMP estimate depths, as compared to the 1982 site-specific study. The average 72-hour PMP rainfall over the entire Bluestone Dam basin is currently estimated to be 17.9 inches.

Current USACE Policy on the Development of the Inflow Design Flood

The current guidance on the development of an IDF for USACE projects is contained in Engineering Regulation 1110-8-2 (FR) – Inflow Design Floods for Dams and Reservoirs. This policy outlines a list of assumptions and procedures that should be adhered to in the development of an inflow design flood. The assumptions and procedures of interest for this analysis is the assumption on rainfall losses, UH peaking, and hydrologic configuration of the basin.

The policy currently prescribes that "Rainfall to runoff conversion (UHs) and loss rates will be derived to correspond to patterns favorable for rapid concentrations of runoff from the drainage basin." The policy also states that "Reservoir inflow UHs for IDF determinations should be peaked 25 to 50 percent to account for the fact that UHs are usually derived from smaller floods." This policy is intended to address the limiting assumption of watershed linearity that is inherent in UH theory. For example, the time vs. discharge relationship (i.e. hydrograph shape) is scaled up or down directly proportional to the amount of rainfall excess without changing the shape of the UH. In reality, the shape can change based on many factors, most notably the effective rainfall and season (Chow, 1964). For the case of Bluestone Dam, the UH peaking was applied to the UHs at each sub-basin. Additional analysis is being performed to determine the appropriate magnitude of peaking for the IDF within the prescribed range. This analysis includes additional numerical watershed analysis utilizing 2-Dimensional hydraulic flow modeling.

The other policy aspect that is of importance is "Inflow design flood hydrographs will be computed as inflow into a full reservoir in order to allow for the effects of the reservoir in accelerating concentration of runoff under critical conditions". Simply put, this suggests that the hydrologic configuration (i.e. basin layout) should be that of the full reservoir pool (typically taken as top of flood control pool). This attempts to address the limitation of considering a static basin layout, when in reality, the basin response changes based on varying reservoir pool stages. Similar to this, the policy states "It is common practice to assume a zero travel time through the pool unless the pool is very long". This portion of policy is intended to leave discretion to the analyst to consider a dynamic reservoir routing, as opposed to a level pool routing. This accounts for the time and magnitude translation through the pool as well as the potential for wedge storage resulting from hydraulic gradient through the reservoir. Initial results suggest that the reservoir routing is sensitive to the consideration of a non-level pool routing and this method will be utilized in the final IDF routing.

Hydrologic Modeling Approach

The USACE HEC-HMS computer software was utilized for developing the rainfall-runoff analysis for the Bluestone Dam Basin. HEC-HMS allows the user a wide array of hydrologic modeling methods. The methods adopted for use in the Bluestone Dam hydrologic model are illustrated in Table 1.

Table 1. 2014 Hydrologic Modeling Methods

Modeling Method	Parameter	Description
Initial and Constant Loss Method	Initial Loss (inches)	Initial loss parameter that accounts for the moisture condition in the watershed at the beginning of the simulation – loss from canopy interception, surface storage, and infiltration.
	Constant Loss Rate (inches/hour)	Infiltration rate during saturated soil conditions.
	Percent Impervious Area	Impervious area directly connected to the channel network (no losses are computed).
Snyder Synthetic Unit Hydrograph Parameters	Standard Lag (hours)	The time from the center of mass of excess rainfall to the hydrograph peak.
	Peaking Coefficient	Dimensionless parameter affecting hydrograph shape.
Recession Baseflow	Initial Discharge (cfs)	Initial baseflow at the beginning of the simulation
	Initial Discharge per Area (cfs/mi^2)	Initial baseflow at the beginning of the simulation
	Recession Constant	Parameter used to control how fast the recession hydrograph trends to 0 cfs.
	Ratio to Peak	Parameter used to control when baseflow is "turned on" during a flood event. This parameter tells the program to start the baseflow recession once the receding limb of the direct runoff hydrograph reaches a certain threshold (the threshold is computed using the ratio of the flow to the peak flow).
Muskingum-Cunge Routing Method	Physical Basin Parameters	Reach Length (ft); Slope (ft/ft); Manning's n-values; 8-point Cross-Section

Sub-basin Delineation and UH Development

Similarly to the 1982 IDF study, the basin was broken down into manageable sub-basins based on the basin drainage configuration and the availability of observed stream flow data. This resulted in the basin being divided into 17 individual sub-basins for the purposes of UH development and one sub-basin representing the pool itself. The sub-basin configuration is illustrated in Figure 2. Synthetic UH parameters were then developed based on calibration to events for the available period when both hourly stream flow and precipitation was available.

The synthetic UHs were developed for multiple storm events and the worst reasonable set of parameters from the calibration events were adopted for use in the IDF analysis. To satisfy the policy requirement for UH peaking and to account for flood magnitudes many times larger than calibration events, the synthetic UH lag parameters were adjusted to produce a peak unit discharge that is 25- and 50-percent larger than the unadjusted synthetic UHs, while maintaining the same volume.

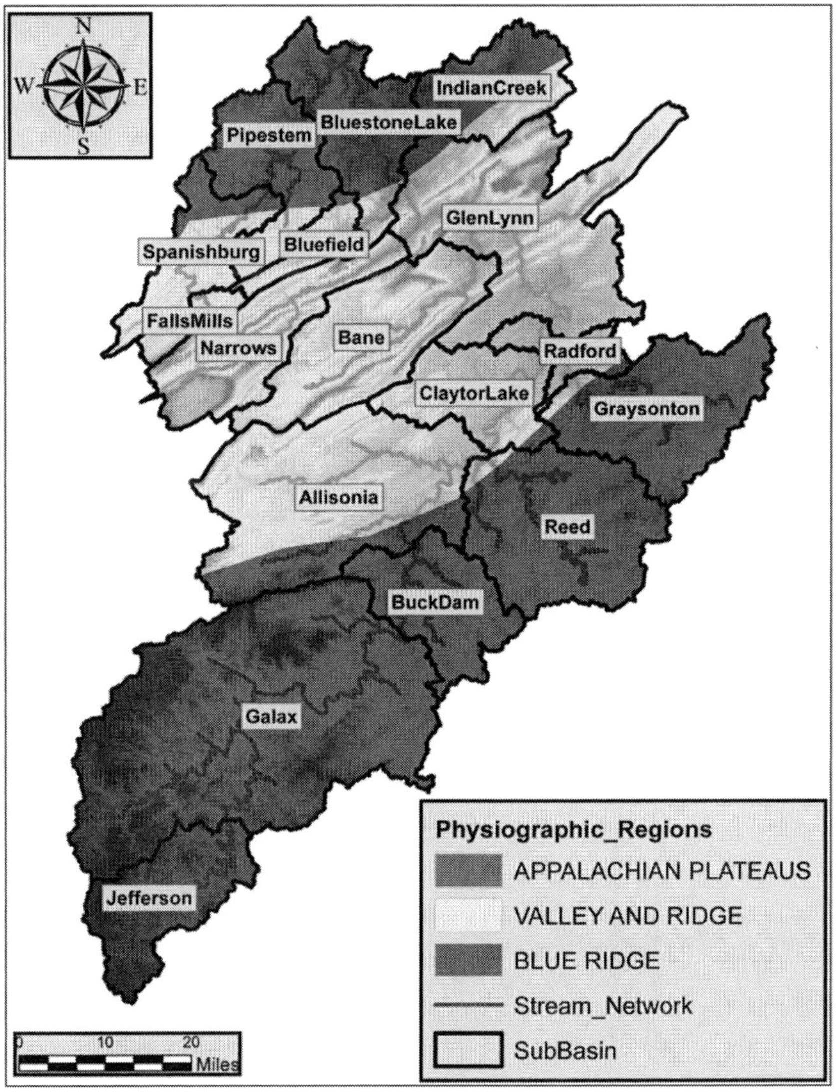

Figure 2. Hydrologic Model Sub-Basin Configuration

Hydrologic Routing

The Muskingum-Cunge reach routing method was utilized to translate the computed flows from upper portions of the basin to the dam site. This method is different than the Muskingum method used in the prior analysis. The Muskingum method must be calibrated to observed or modeled events and is not easily adjusted for flows that exceed calibration events; while the Muskingum-Cunge method explicitly accounts for the magnitude of discharge in each computational time-step. The Muskingum-Cunge parameters were calibrated to observed events for those flow magnitudes available; and supplemented with results from a steady flow hydraulic model for the magnitude of floods approaching the IDF.

Application of the PMP to the Bluestone Dam Basin

The updated PMP values were then applied to the updated hydrologic model to develop the IDF for the non-peaked UH's, along with UH's peaked 25- and 50-percent; resulting in peak reservoir inflows of 1,444,000-, 1,564,000-, and 1,661,000- ft^3/s; respectively. The results of this analysis are demonstrated in Figure 3. A sub-basin UH peaking of 50-percent was utilized in the 1982 study. The results of the updated IDF analysis are 33- to 53-percent greater than the 1982 IDF study, depending on UH peaking, although the PMP depths have been reduced by 10-percent.

Preliminary reservoir routing suggests that the resulting reservoir stage associated with the updated IDF is significantly above the existing dam crest given the existing condition of the dam, including the fusable penstock discharge. These results are considering a non-level pool routing assumption. When considering a level pool-routing assumption, as was done in the 1982 analysis, the resulting reservoir stage would be higher.

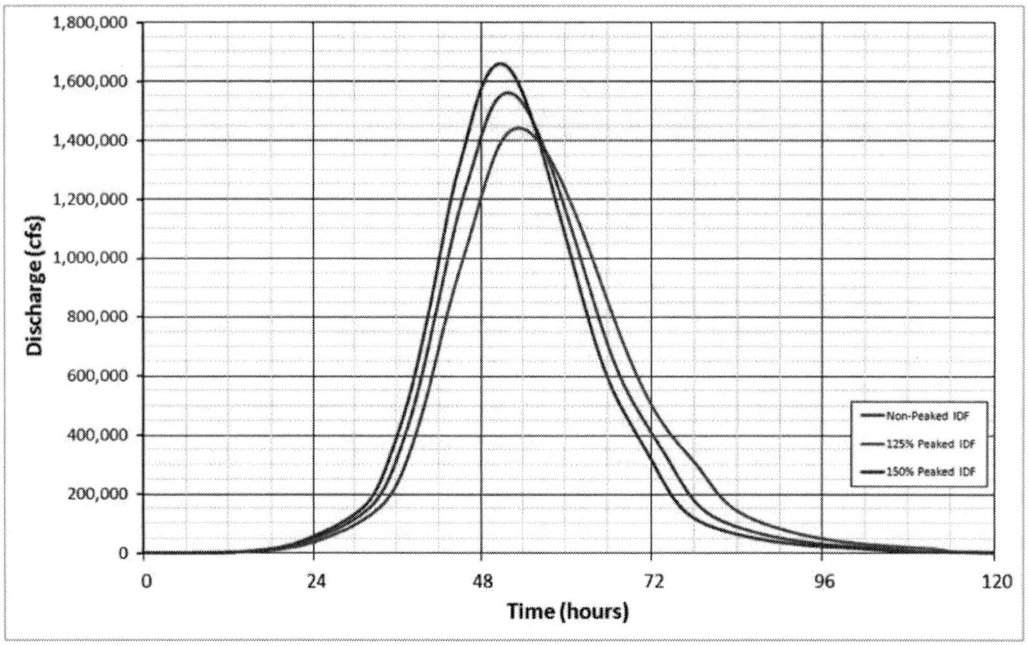

Figure 3. 2014 Inflow Design Flood Hydrographs

COMPARISION BETWEEN 1982 AND 2014 ANALYSIS

A comparison of the procedures, assumptions, and numerical methods applied between the 1982 and 2014 IDF analysis was performed. It was determined that the primary reasons for the significant increase in IDF peak discharge could be simplified to three drivers; sub-basins unit hydrographs, the computational time step, and reach routing method. The comparison of total inflow hydrographs are illustrated in Figure 4.

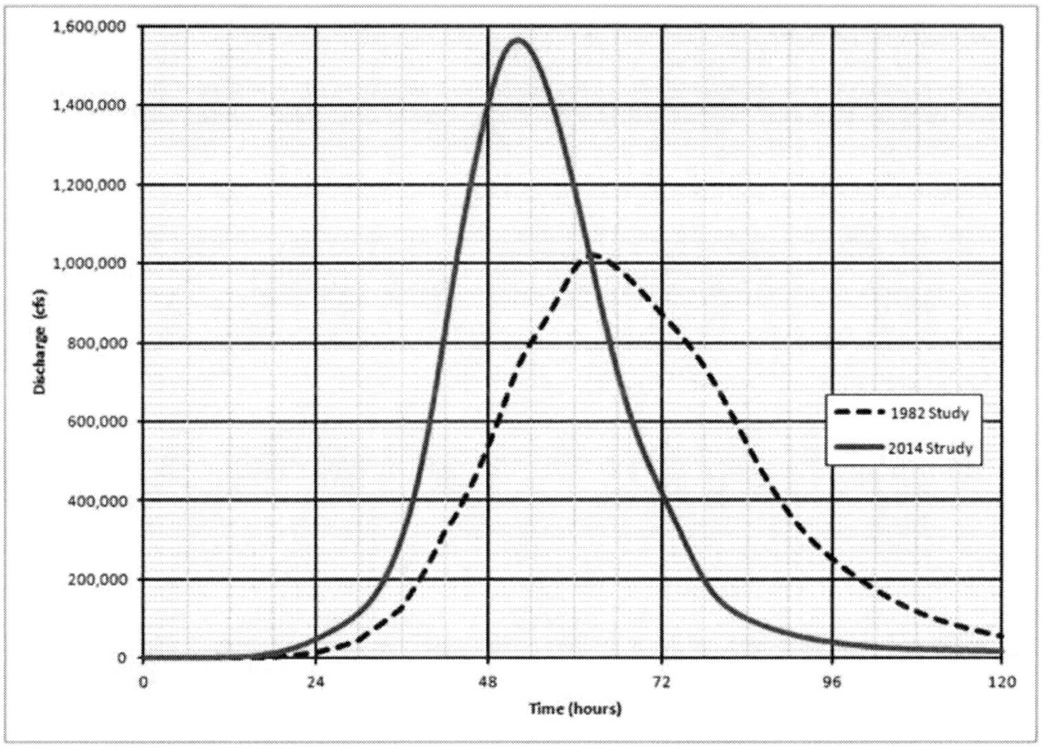

Figure 4. Comparison of 1982 and 2014 Inflow Design Flood Hydrographs

Unit Hydrographs

The sub-basin unit hydrographs from the 1982 analysis were taken from a 1950's hydrology study that included 6-hour unit hydrographs for various flood events from 1940 to 1953. These were based on 6-hour precipitation and stream flow data for the largest floods in each sub-basin during that period. There are limitations to using both the 6-hour time step and a short period of unit hydrograph development. The 6-hour time step results in poor hydrograph resolution of rapid response sub-basins, as the peak can often not be captured and the timing of the response is forced into 6-hour time ordinates.

The short period of record in-which unit hydrographs were derived for the 1982 study, resulted in some sub-basins UH's being derived from floods that are smaller than the flood of record within the sub-basin, considering that there is an additional 60-years of record. The magnitude of rainfall excess and the resulting runoff can have a significant effect on the shape of the unit hydrographs. For example, the Bane sub-basin's unit hydrograph for the 1982 study was based on a runoff event that had a peak discharge of approximately 10,000-ft^3/s, while the event used in the 2014 study had a peak discharge of over 20,000-ft^3/s. The UHs derived from these two events had a lag time of 15-hours and 7-hours, for the 1982 study and 2014 study; respectively. It is likely that the difference in the response times is both, the errors resulting from 6-hour data and the magnitude of the observed events used to derive the UHs.

Computational Time Step

The computational time step used for the 1982 analysis was 6-hour to coincide with the 6-hour unit hydrographs. The 2014 analysis utilized a 15-minute computational time-step to provide the appropriate resolution and minimize the numerical errors. Given the characteristics of the many of the individual sub-basins, a 6-hour time step can introduce numerical attenuation errors in both the magnitude and timing of discharges. An example is illustrated in Figure 5, of a unit rainfall (i.e. 1-inch of runoff over 1-hour) applied to identical synthetic UH parameters for varies computational time-steps. This example illustrates the increasing numerical error with coarser computational time step.

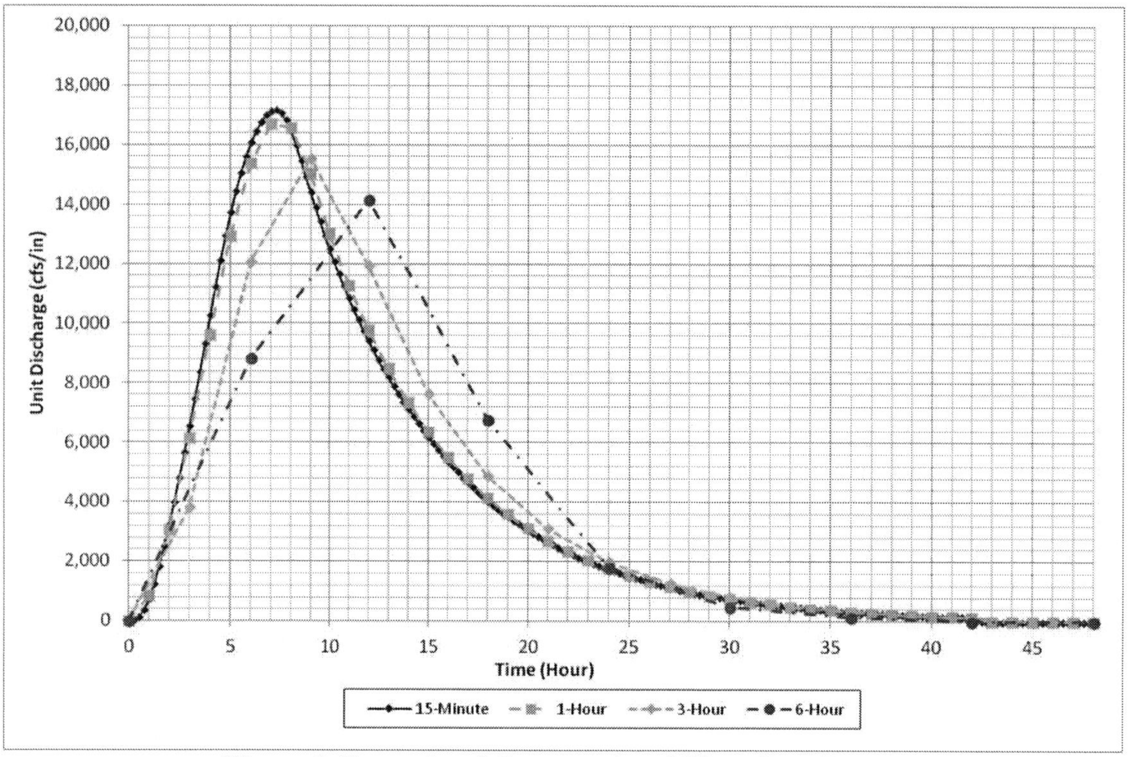

Figure 5. Example of Computational Time Step Errors

Hydrologic Routing

Lastly, the reach routing in the 1982 study was accomplished using the Muskingum routing method. This method is defined by two empirical parameters that are arrived at through calibration to observed floods. The Muskingum parameters in the 1982 analysis were developed based on normal to large floods that were an order of magnitude smaller than the flows that need to be represented in the IDF analysis. To overcome this limitation in the 2014 analysis, the Muskingum-Cunge method was used. This method solves the Muskingum parameters by computing the hydraulic characteristics of a representative cross-section and reach length; therefore computing the translations of time and magnitude each computational time step. The most notable difference between the two methods for this particular application is the time translation component of the reach routing. An example of the incorrect attenuation provided by the Muskingum method used in the 1982 analysis is provided in Figure 6. The reach routing was found to be the largest contributing factor to the increase in peak of the IDF between the 1982 and 2014 analysis.

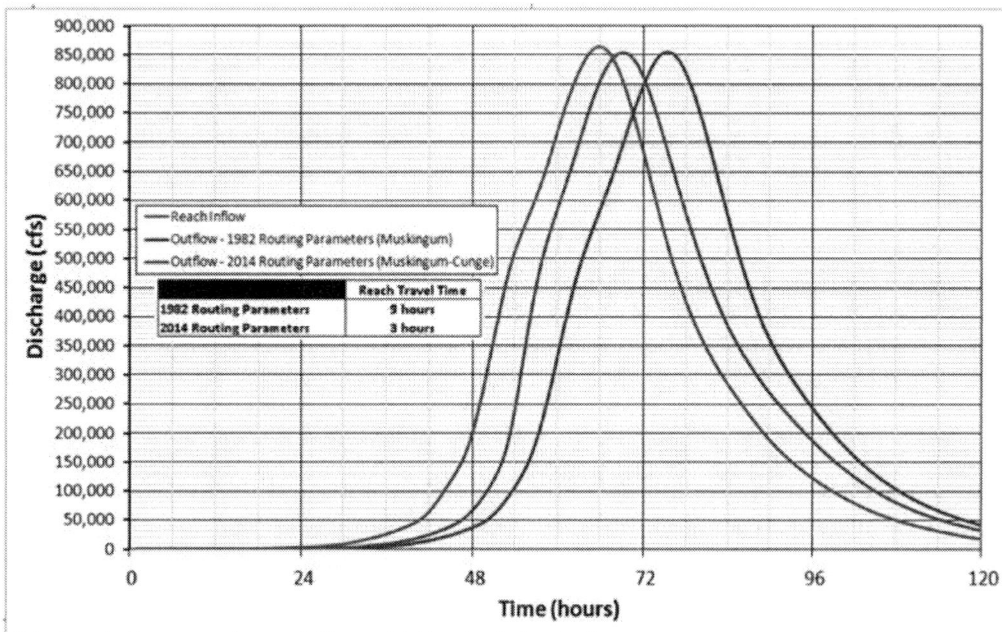

Figure 6. Example Reach Routing Comparison Between 1982 and 2014 Analysis

CONCLUSION

In conclusion, limitations in prior hydrologic modeling have resulted in an increase in peak discharge of the IDF in the range of 33- to 53-percent over prior estimates. There are several major lessons learned during the course of the Bluestone Dam IDF update. First, the unit hydrographs should be derived from the largest rainfall and/or runoff events within the basin in which they are being derived for, giving consideration for time-step and integrity of rainfall and stream flow data. If the rainfall and stream flow data is not at a time step that can adequately represent the runoff, the peak can often be omitted and errors can result from poor hydrograph definition. Previous IDF studies that do not utilize the largest storms/floods, in which data is available, should be revisited. Given the inherent non-linearity in the basin rainfall-runoff relationships, utilizing the largest floods within a basin with storm characteristics that are consistent with the prototype PMP storm and season will minimize these errors when applying linear-unit hydrograph theory. The use of distributed rainfall-runoff modeling that utilizes 2-dimensional unsteady flow analysis may be used in future studies to improve estimates of rainfall-runoff relationships for extreme storm events.

Second, the time-step in which numeric computations are performed can be an important aspect of developing a numeric hydrologic model. Given the computational ability of modern hydrologic software, this issue should not be widespread on current and future studies. Simple sensitivity analysis of computational time-step should be performed and a time-step that results in negligible difference in results should be obtained. The important lesson is that hydrologic analyses performed prior to the ease of advanced computational ability should be reviewed for potential errors resulting from a computational time-step that is to coarse for a given watershed.

Third, the reach routing method incorporated in a hydrologic model being applied to an IDF study should account for the non-linearity in reach routing. Often the flood of record within a basin or a specific stream reach is many times smaller than the expected IDF. Calibrating linear hydrologic routing methods to these floods can incorrectly attenuate predicted flows of higher magnitude, thereby reducing the peak discharge at the watershed outlet. Consideration should be given to non-linear hydrologic routing methods (e.g. Muskingum-Cunge) that can account for varying reach routing characteristics within a single simulation. It may also be necessary to perform hydraulic reach routing using an unsteady flow hydraulic model if the reach cannot reasonably be represented with hydrologic routing methods. The hydrologic reach routing can be compared to the unsteady flow hydraulic routing to determine the need for hydraulic routing.

Lastly, and perhaps the most valuable lesson learned from the Bluestone Dam IDF update, is that the understanding of the meteorology of extreme storms and prediction of the resulting extreme floods is uncertain and it is likely that any analysis to determine these is only a current estimate. The IDF for any high hazard structure should be periodically evaluated for any changes in the understanding of extreme meteorological events, changes in basin conditions, advancement in analysis methods, or changes in policy related to the IDF.

REFERENCES

Chow, VT. 1964, Handbook of Applied Hydrology – A Compendium of Water-Resources Technology, McGRAW-HILL BOOK COMPACY, New York.

USACE, 1982. Bluestone Dam – Spillway Design Flood, U.S. Army Corps of Engineers, Huntington District.

USACE, 1991. Inflow Design Floods for Dams and Reservoirs, Engineer Regulation 1110-8-2. U.S. Army Corps of Engineers, Washington, D.C.

USACE, 1994. Flood Runoff Analysis, Engineer Manual 1110-2-1417. U.S. Army Corps of Engineers, Washington, D.C.

USACE, 2014. Bluestone Basin PMP Review, U.S. Army Corps of Engineers, Omaha District

USACE, 2014. Development of the Inflow Design Flood for Bluestone Dam - DRAFT, U.S. Army Corps of Engineers, Huntington District.

USACE, 2014. Hydrologic Modeling System (HEC-HMS), U.S. Army Corps of Engineers, Hydrologic Engineering Center.

HYDRAULIC MODELING FOR COASTAL DAMS

Chad W. Cox, P.E.[1]
Kristina Ekholm, P.E.[2]
Bin Wang, P.E.[3]

ABSTRACT

In coastal areas there are many dams which are subject to both riverine and coastal flooding mechanisms. Such dams often form the tidal boundary between upstream freshwater rivers and downstream brackish areas influenced by tides, waves, and storm surge. These dams are sometimes located several miles upstream from the coast and were not designed or constructed to serve as a barrier against extreme surges. Most dam engineers are well versed in understanding and modeling water flowing downstream from the upgradient watershed. In a coastal environment however, water can also flow upstream towards the dam. This can result in increased tailwater elevations that can affect spillway hydraulics or in extreme conditions can even result in overtopping of the dam from the downstream direction. Hydraulic modeling for dams at tidal interfaces requires a specialized set of tools to supplement those typically used for inland riverine conditions. An understanding of tides and use of tide gage data is critical. Computer programs such as NOAA's SLOSH and the ADCIRC model developed by the USACE can be used to estimate the elevations of storm surges at coastlines and surges propagating upriver. Programs like SWAN can be used to model wave heights and run-up. Once coastal flood mechanisms are understood, the potential for combined effects on flooding must also be examined as many coastal storms can produce conditions which lead to an interaction between surge and riverine flooding. This paper will examine the issues present and tools available for hydraulic modeling at coastal dams.

INTRODUCTION

Dams that function as tidal barriers must be designed to withstand the tides and storm surges that they are subjected to from the saltwater side as well as riverine flood that approach from the freshwater site. These dams must not only function in the same manner as all other dams with respect to flooding from "upstream" sources, but they must also have appropriate structural and hydraulic capacity to accommodate flood mechanisms associated with high tides, extreme surges, and wave action acting from what would traditionally be called the "downstream" side of the dam. The combination of elevated water levels on both the saltwater and freshwater sides of a tidal barrier dam can result in loading combinations which are more extreme than either individual condition. Therefore an understanding of both traditional riverine hydrology and coastal flooding conditions is necessary when dealing with dams at the tidal boundary.

---

[1] GZA GeoEnvironmental Inc., 249 Vanderbilt Ave, Norwood, MA 02062, 781-278-5787, Chad.Cox@gza.com.
[2] GZA GeoEnvironmental Inc., 249 Vanderbilt Ave, Norwood, MA 02062, 781-278-5823, Kristina.Ekholm@gza.com
[3] GZA GeoEnvironmental Inc., 249 Vanderbilt Ave, Norwood, MA 02062, 781-278-5809, Bin.Wang@gza.com

The majority of dams are located in areas where coastal flooding is not a concern (other than the inland precipitation generated by the storm which may have originated off shore.) Dam engineers are therefore traditionally well versed in the theory behind riverine flooding which transforms rainfall into runoff and generally assumes antecedent water level conditions based on physical constraints such as spillway crest elevation. Coastal flooding is a specialized field which is more often the domain of ocean and marine engineers. Storm surge elevations on the ocean are driven by wind rather than rain, and the cyclical tide determines antecedent water conditions.

This paper will therefore seek to aid the dam engineer by: 1) defining several key concepts of tidal flooding; 2) identifying the conditions where dams and dam-like structures could be subject to the conditions of tidal flooding; 3) discussing potential failure mechanisms which could result from coastal and combined coastal/riverine flooding; 4) reviewing quantitative methods for analyzing coastal flood effects; and 5) briefly providing case studies of dams where these concepts have been applied.

KEY COASTAL FLOODING TERMS

Tides

Tide, also referred to as astronomical tide, is defined as "the periodic rise and fall of a body of water resulting from gravitational interactions between Sun, Moon, and Earth", according to NOAA. In general, most areas have two high and two low tides each day. This pattern is called a semi-daily or semidiurnal tide. Some areas such as the Gulf of Mexico only have one high and one low tide each day. This is called a diurnal tide.

Tidal datums are used as references to measure local water levels (NOAA, 2014a). Mean High Water (MHW) is the average of all the high water heights observed over the National Tidal Datum Epoch (NTDE). The National Tidal Epoch Datum is the specific 19 year period adopted by the National Ocean Service as the official time segment over which tide observations are taken and reduced to obtain mean values to tidal datums. Mean Low Water (MLW) is the average of all the low water heights observed over the NTDE. Mean Tide Level (MTL) is the arithmetic mean of mean high water and mean low water. Mean Range of Tide (MN) is the difference in height between MHW and MLW. Mean Higher High Water (MHHW) is the average of the higher high water height of each tidal day observed over the NTDE. Similarly, Mean Lower Low Water (MLLW) is the average of the lower low water height of each tidal day observed over the NTDE.

Amplitude of tides varies geographically. For example, a typical tide range along the north Atlantic coast is around 3 to 4 feet, whereas the highest tides in Alaska are up to 40 feet. Local hydrologic setting and coastal line geometry also play important roles on the tides. Generally speaking, tides fluctuate more significantly around water bodies with more local constraints.

Astronomical tides alone usually do not cause flooding events, because most manmade structures are located above the tidal influenced zone or protected to a level higher than the high tide. However, when other flooding events (e.g., storm surge or wind-generated waves) coincide with high tide, the adverse consequence can be significantly magnified.

Storm Surge

Storm surge is defined as the abnormal rise of water generated by a storm, over and above the predicted astronomical tides. Storm surge is produced by water being pushed toward the shore by the force of the winds moving cyclonically around the storm. Storm tide is the water level rise including the combination of the storm surge and the astronomical tide, as shown in Figure 1.

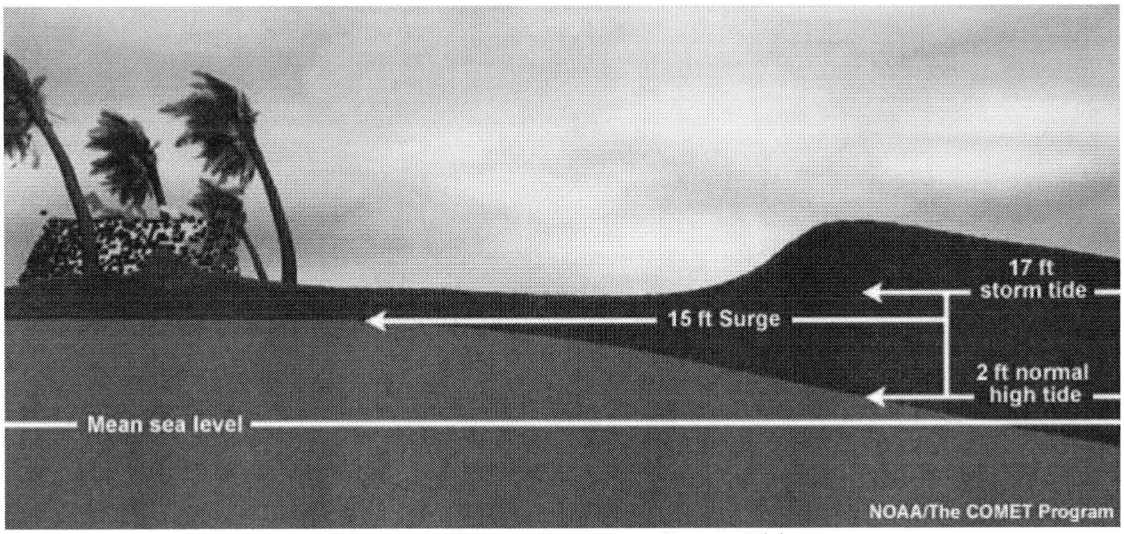

Figure 1. Storm Surge vs. Storm Tide

Surge caused by a specific storm is often approximated by subtracting predicted tide from the observed water level, assuming negligible hydrodynamic interactions between the two mechanisms. Figure 2 shows an example of predicted tide, observed storm tide and the difference between the two (i.e., storm surge) at the NOAA CO-OP gage (Station ID 8581750) at The Battery, NY, southern tip of the Manhattan Peninsula. The storm surge occurred around the vicinity of a semidiurnal high tide and the total observed water elevation reached 11 feet, NAVD88.

Both statistical and numerical modeling methods can be used to identify the high tide and surge elevations to which a coastal dam may be subjected, and to calculate the height of the wind waves that may act on the downstream side of the dam. Should the consequences of dam failure warrant it, combined event flooding (i.e. quantifying the effects of coincident riverine and coastal flooding) should be analyzed. Once the elevation of high tides, surge, wave action and combined event flooding is known, the resulting forces acting on a dam can be analyzed.

Wind-generated Waves

Waves are undulatory motion of a water surface. Significant wave height, average of the highest one-third of waves in a given period, is one of the key parameters for calculating wave setup and wave runup. Waves are essentially a form of energy transmission through water, due to some external energy source, such as a wind storm. Waves are generally of greatest concern on ocean coasts; however, inland water bodies such as natural lakes and even reservoirs are also subject to wind-generated waves under extreme storm conditions. The Great Lakes have been constantly monitored with data collection buoys. National Weather Service of NOAA has been performing lake wind/wave forecasting in North Dakota for smaller lakes (http://www.crh.noaa.gov/fgf/hydro/lake_fcst.php). Therefore wave action (setup and runup) has the potential to affect all dams, whether or not they are in a coastal setting.

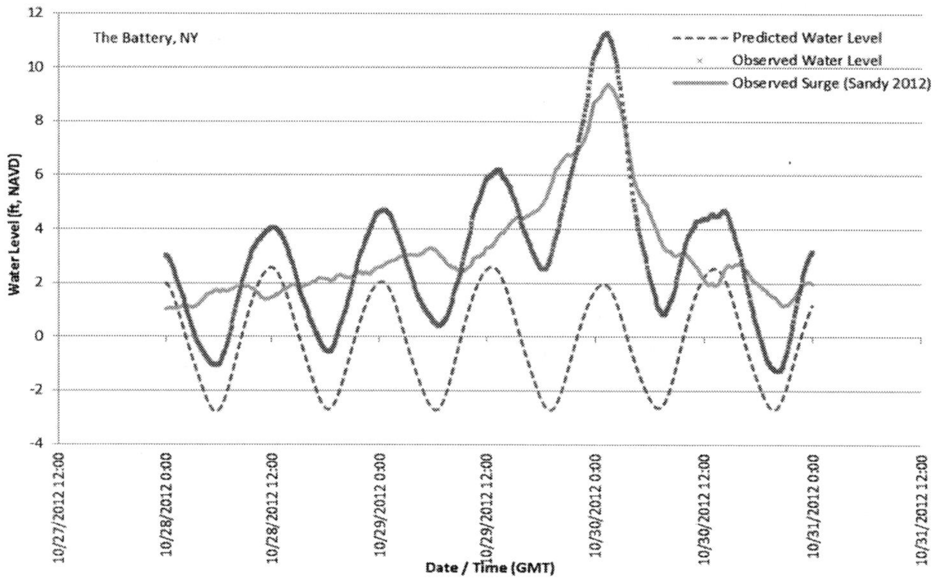

Figure 2. Tide (dash line), Storm Surge (orange line)) and Storm Tide (blue x) during Sandy 2012

Wave Setup

Wave setup is the increase in mean water level above the stillwater level due to momentum transfer to the water column by waves that are breaking or otherwise dissipating their energy (FEMA, 2005a).

Wave Runup

Wave runup refers to the height above the stillwater elevation (tide and surge) reached by the "swash" or water uprushing onto a slope (FEMA, 2005b). Runup is a complex phenomenon that is known to depend on the local water level, the incident wave conditions (height, period, steepness, direction), and the nature of the beach or structure being run up (e.g., slope, reflectivity, height, permeability, roughness).

Figure 3 illustrates the concepts of wave setup and wave runup, in relationship with waves and stillwater elevation.

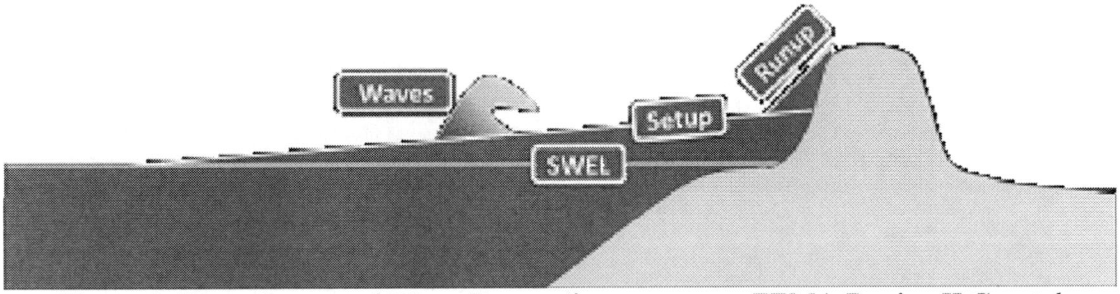

Figure 3. Wave Setup and Wave Runup (image source: FEMA Region II Coastal Analysis and Mapping)

COASTAL DAMS

The function of a dam is to retain water while the function of a floodgate, levee or dike is to prevent flow of water into specific areas. The focus of this paper is on dams; however, the concepts can be applied to floodgates, levees and dikes.

Coastal dams are frequently located immediately at the discharge point of a river or stream into a coastal water body. Coastal dams, can, however be located many miles upstream of the mouth. The criterion for a dam to be considered a coastal dam is that it be subject to the effects of coastal flooding mechanisms. Such dams can form the boundary between brackish, tidally-influenced water and upstream freshwater riverine impoundments. In some cases, it may not be apparent that a dam is subjected to coastal processes until a significant coastal tide or surge event occurs. An example of coastal dams is the Charles River New Dam located in Boston, MA, see Figure 4. The dam is located 3.5 rivers miles upstream of the confluence between Charles River and the Boston Harbor. The dam replaced an older structure further upstream. Both dams were constructed to create the Charles River Basin and protect the low-lying city areas from tidal flooding. The dam has three locks to permit the passage of boat into and out of the Charles River. Like many coastal dams, the Charles River Dam also includes a fish ladder to support the passage of anadromous fish into the freshwater portion of the river (USACE, 2014a). Figure 5 shows a photo of the dam taken by the U.S. Army Corps of Engineers (USACE).

FAILURE MECHANISMS

During a coastal flooding event cause by a hurricane, tropical storm, "nor'easter", or other such driver, water may flow upriver and cause flooding of areas many miles upstream of the coast. Water flowing upriver may have adverse effects on coastal dams. Dam failure mechanisms resulting from coastal flooding are similar to but different than the dam failure mechanisms associated with riverine dams and riverine flooding.

During a coastal flooding event, storm surge can cause the tailwater at a dam to rise or even result in water flowing from downstream to upstream. This rise in the tailwater can have a number of detrimental effects.

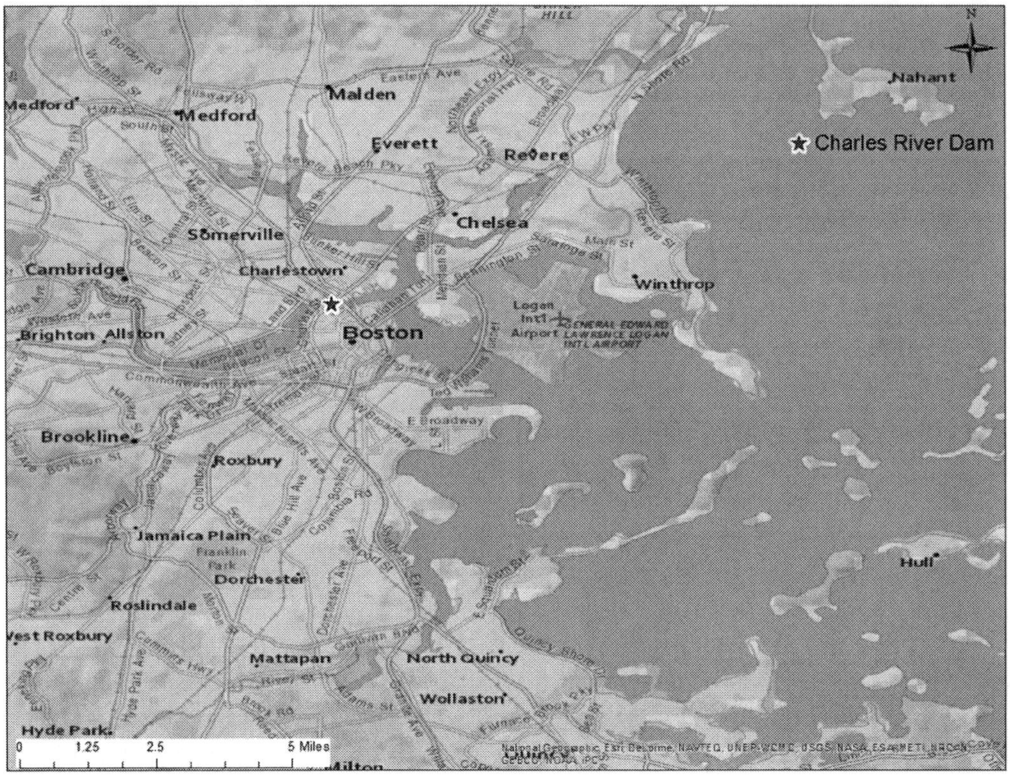

Figure 4. Locus Map of the Charles River Dam

Figure 5. Photo of the Charles River Dam (image source: USACE, 2014a)

Increased tailwater elevations caused by storm surge can reduce or even eliminate the typical capacity of a spillway or other outlet to discharge inflow coming from the upgradient watershed. Submergence of a spillway reduces the flow capacity which can in-turn increase the upstream peak water surface that a dam will experience as a result of a given rainfall-runoff event. This higher peak water surface elevation may result in overtopping of the dam from the upstream side. Dam overtopping from the upstream side can occur strictly due to the stillwater elevation or because of overwash from waves created by the high winds of a coastal storm. Overtopping may result in dam failure by causing erosion of an earthen structure or excessive loading of a concrete or masonry structure.

Overtopping of the dam can also occur from the downstream side. Extreme storm surge can backflow over a spillway or even rise above the top of the dam. Wind-generated waves propagating upstream from offshore can be very large and also overtop a dam and/or generate significant hydrodynamic loading against the dam or appurtenant structures. If the downstream slope an embankment dam is not sufficiently armored against wave action (through placement of riprap, etc.), scour and erosion due to waves can occur. Coastal wave action is heavily dependent on many factures including local bathymetry, alignment of the river, and coastal features such as islands and breakwaters.

Dams are subjected to uplift forces under their base. An uplift force is the result of the hydrostatic water pressure under the dam pushing up on the mass of the structure. During a coastal event, the tailwater at the dam may increase substantially. As the height of downstream water column increases, the uplift pressure distribution will change and the total uplift force increase. Increases in uplift at the structure/foundation interface or other joints can decrease the stability of a gravity structure, even though the higher tailwater may reduce the net horizontal driving force.

Predicting the timing (stage hydrograph) of increases in tailwater at a coastal dam can be complex. Storm surge is additive to the level of the tide. The peak water level can vary greatly depending on whether the greatest surge arrives at high or low tide. In addition, the highest surge may not be synchronized at all with the peak of the inflow hydrograph from upstream. If the upstream watershed is very large, there may be a lengthy lag time between the arrival of the storm which generates the rain and surge, and the arrival of significant river flows at the coastal dam.

Prediction of dam failure impacts resulting from a coastal storm can likewise be complex. If the dam is immediately upstream of a large estuary or open ocean, a dam failure may not significantly raise downstream water levels. This is because of the tremendous storage volume provided by such areas. Incremental loss of life and/or property damage from the dam failure may be mitigated because downstream areas may already be heavily affected by the coastal flooding. Loss of functionality (water supply, hydropower, etc.) associated with the failure of a coastal dam would be comparable to that of a similar dam located elsewhere. The major difference would be that areas upstream of a failed tidal barrier dam would again be subject to the upstream movement of brackish water.

QUANTITIATIVE METHODS FOR DEFINING COASTAL FLOODING HAZARDS

Quantitative methods such as statistical analysis or numerical modeling may be used to identify design elevations associated with coastal flooding mechanisms. This can be done either deterministically or probabilistically. The engineer will need to give consideration to a number of factors when selecting a method, including the consequences of failure, availability of data (tidal records, bathymetric data, etc.), and the time and budget available to perform the analysis.

Tide Data

Tide gage stations installed along the U.S. coast have been employed for years to record tidal fluctuations. Many "real-time" gages are currently being operated by NOAA's Center for Operational Oceanographic Products and Services. These gages, both active and historical, provide vital information to tidal analyses such as datums, harmonic constituents and tide predictions. Data points are taken at either 6-minute or hourly intervals.

Historical tide data can provide insights as to the controlling coastal flood mechanism for a given location. For example, the top water levels in the Gulf of Mexico were all caused by extreme tropical cyclones (i.e., hurricanes), whereas extratropical systems such as nor'easters are more likely to be responsible for the high water levels observed historically along the US north Atlantic coast.

Tide data can be analyzed with statistical tools similar to those used for stream flow. Historic tide elevations can be subjected to Log Pearson Type III analysis to estimate annual exceedance probabilities (AEP) for storm surges with relatively frequent events. For hydrodynamic modeling of extreme floods, observed historical tide data is often used to determine the antecedent water level required as the initial condition.

Long term sea level trends are being analyzed and updated. The global mean sea level rise was determined to be approximately 1.7 to 1.8 mm/yr (IPCC, 2007), but rates vary greatly by location. Observed rates are approximately 3 mm/yr in the mid-Atlantic region and up to 9.7 mm/yr along the northern central Gulf of Mexico (largely due to regional land subsidence).

Many studies suggest that sea level rise be viewed as an indicator of global climate change (Parker, 1992). USACE evaluated different sea level rise scenarios with varying climate change models (USACE, 2014b). Using the analysis results from this study, the predicted sea level rise for Boston, for example, could be as high as 5 feet in 2100, relative to the current conditions. Figure 6 shows three sea level trends for The Battery in New York City using 1-percent AEP base flood elevation of 14.8 feet (FEMA, 2013) at this location. Figure 6 indicates that a significant portion of lower Manhattan will be under water if the highest predicted scenario becomes a reality, due to sea level rise alone. The combined effect of sea level rise and climate change poses a risk to all coastal structures including coastal dams.

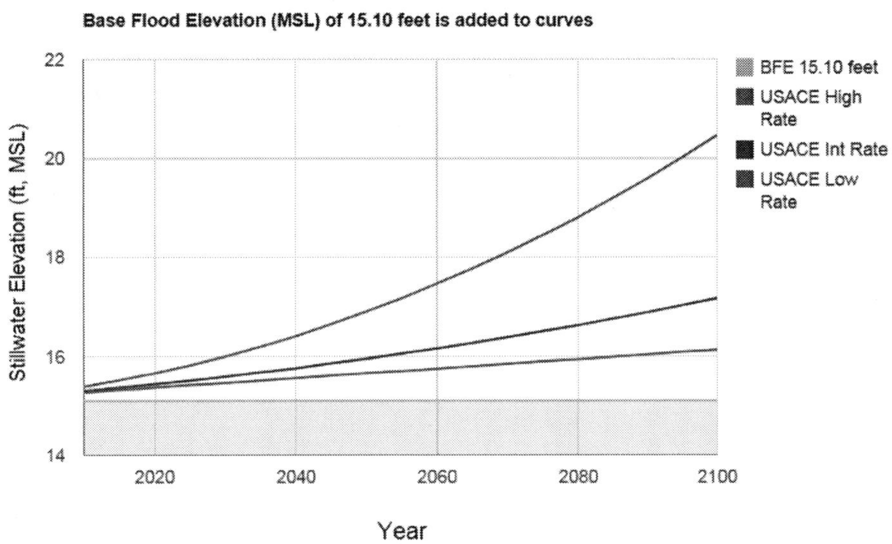

Figure 6: USACE-Projected Sea Level Rise Calculator for The Battery, New York

Review Flood Insurance Study and Flood Insurance Rate Map

The Federal Emergency Management Agency publishes Flood Insurance Studies (FISs) and Flood Insurance Rate Maps (FIRMs) for each county. The FIS is a compilation and presentation of flood risk data for specific watercourses, lakes and coastal flood hazard areas within a community. The FIS contains detailed flood elevation data in flood profiles and data tables. FEMA FISs and FIRMs can be downloaded free of charge at https://msc.fema.gov/portal.

As with any dam, review of the FEMA FIS and FIRM is a fast and cost effective method for obtaining basic hydrologic and hydraulic information for a coastal dam. Review of the FEMA FIS can give an engineer a general idea as to whether a dam is subject to coastal flooding mechanisms. The designer should review the methodology section to confirm that the analysis is up to date and uses appropriate methodologies. In some cases, parts of the analysis may have been performed many years before the publication date on the cover.

When reviewing the FIS the engineer should examine the flood profile. The profile will provide the 100 year and 500 year flood elevations calculated by FEMA. The profile may specifically indicate whether an area is tidally influenced. In some cases the profile does not contain information about the limits of tidal influences at the 500-year level.

The engineer should review the Flood Insurance Rate Map. This is an official map of a community on which FEMA has delineated the one percent annual chance (base) floodplain or Special Flood Hazard Area, the Base Flood Elevations (BFEs), and the risk premium zones applicable to the community. In coastal areas the map will identify the Coastal High Hazard Area/ V Zone (the area where wave heights are greater than three feet) and the Coastal A Zone (the area where wave heights are between 1.5 ft and 3 ft).

Many dams are outside of the Coastal V and Coastal A zones but are still subject to coastal flooding. An area with wave heights less than 1.5 feet and subject to the 100 year flood is classified to be within an A zone. Therefore examination of the maps alone will not indicate whether the flooding is associated with coastal events.

In areas subject to coastal flooding, it is important to know that the BFEs shown on the FIRMs may include wave heights on top of stillwater elevations. The designer should review the Transect Data table and identify the transect that is nearest to the dam (from the FIRM). This table typically provides the 10- year, 50-year, 100-year, and 500-year stillwater elevations at each transect. The stillwater elevation represents the projected elevation of floodwaters in the absence of waves. The table also contains the 100-year total water level, i.e. stillwater elevation plus wave setup.

It is noted that FEMA transects typically do not extend far upriver. Additionally, FEMA does not route storm surges upriver. FEMA provides the calculated surge and waves along the transect. The surge may attenuate or may amplify as it progresses upriver and analysis to better quantify the design elevation may be warranted.

Computer Programs

SLOSH. The Sea, Lake and Overland Surges from Hurricanes (SLOSH) model is a computerized numerical model developed by the National Weather Service (NWS) to estimate storm surge heights resulting from historical, hypothetical, or predicted hurricanes by taking into account the atmospheric pressure, size, forward speed, and track data. The SLOSH model is not publically available.

The SLOSH Display Program (SDP) was developed to display storm surge output (covering the entire U.S. Atlantic coast including the Gulf of Mexico and Caribbean Islands) and related information such as tides that was calculated through the SLOSH model. The SLOSH display program is publically available here: http://slosh.nws.noaa.gov/sloshPub/disclaim.php.

The designer can use the SDP to identify the Maximum Envelope of Water (MEOW) identified with a particular storm or identify the Maximum of Maximums (MOMs). The designer can click on the grid cell representing a structure (or the cell nearest to a structure) in order to identify the surge elevation (for a particular storm, or for all storms) at the cell. In addition, to identifying the elevations associated with these hypothetical storms, the program can be used to animate historic storms and identify the surge elevations associated with historic storms. Figure 7 shows an example of the SLOSH-computed MOM for the Louisiana coast, for Category 5 hurricanes.

The SLOSH program can also be run directly with custom storm inputs such as landfall location, track bearing, forward speed, and central pressure deficit. Based on these user-defined storm parameters, the model will predict storm surge height and inundation.

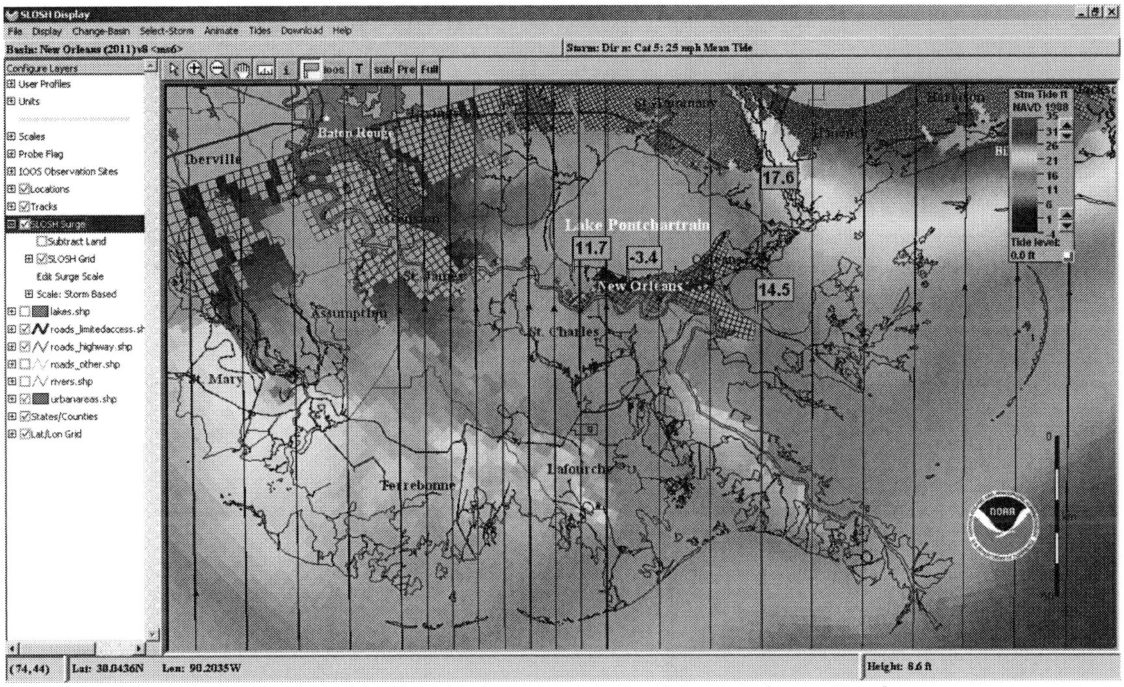

Figure 7. Slosh Display Program – New Orleans Basin

The limitation to using the SLOSH model is that the SLOSH grid typically does not extent far upstream from the coast. Dams that are located several miles upstream of the coast and on smaller rivers are not represented by the grid. This is important because surge may amplify upstream if the currents are perpendicular to the contact with the shore. Alternatively, the surge may attenuate if the current bends away from the shore. Coastal morphology may also affect the extent of rise of water. Additional analysis to better quantify the design elevation may be warranted at a dam upriver from the SLOSH grid.

<u>ADCIRC</u>. A two-dimensional, depth-integrated, barotropic time-dependent long wave, hydrodynamic circulation model, developed by the Coastal Hydraulics Laboratory (USACE) in collaboration with the University of North Carolina at Chapel Hill and the University of Texas at Austin. ADCIRC is a highly developed computer program for solving the equations of motion for a moving fluid on a rotating earth. These equations have been formulated using the traditional hydrostatic pressure and Boussinesq approximations and have been discretized in space using the finite element method and in time using the finite difference method. ADCIRC uses a non-structured, triangulated model grid with variable grid resolution. A number of large, regional-scale model grids were developed and validated by FEMA for various regions. For example, the development of the Region II model grid for the mid-Atlantic region was documented in FEMA publications (FEMA, 2011a, 2011b and 2012) The topography and bathymetry (elevation in meters, NAVD88) of the Region II grid in the vicinity of the New York City is shown in Figure 8, as an example of the model resolution. Similar resolutions are available for other coastal areas. Major river reaches have been incorporated into the FEMA grids. For coastal dams on less significant rivers, local refinement can be performed to improve the model resolution and achieve a higher accuracy of the coastal modeling results.

Figure 8. ADCIRD Model Region II Grid – Topography and Bathymetry

SWAN. The software is used to simulate waves nearshore, using wind as the primary forcing input. The model accounts for wave propagation in time and space, shoaling, refraction due to current and depth, frequency shifting due to stationary depth. The model also considers other factors such bottom friction and depth-induced wave-breaking. The program calculates wave characteristics such significant wave height, wave period as well as spectrum-related parameters. The software is developed and maintained by Delft University of Technology in Netherland (TU Delft).

CHAMP. Coastal Hazard Analysis Modeling Program (CHAMP), a FEMA-developed coastal modeling suite, includes three modules for three different applications:

1. Erosion. The erosion module performs erosion assessment for dunes, including eroded area and total volume.
2. WHAFIS. The FEMA-developed software name stands for wave height analysis for flood insurance studies. The program is a wave attenuation model and performs calculation of inland wave propagation associated with storm surge.
3. Runup. This model determines the height of wave uprush against a beach, a levee, or man-made structures.

Wave setup can be calculated by widely accepted empirical functions based on site topographic features and stillwater elevations. In storm surge-wave coupled models, outputs from this type of models also include wave setup. For example, SWAN calculates wave setup simultaneously with the wind-wave computations in ADCIRC/SWAN-coupled models.

CASE STUDIES

Charles River New Dam

The Charles River New Dam is a flood control structure on the Charles River in Boston, Massachusetts. It is located at the Boston Inner Harbor. The original Charles River Dam had simply used gravity sluicing to maintain the basin elevation. The dam had inadequate discharge capabilities. Runoff from major storms during the high tide cycle produced high basin elevations and flooding.

The Charles River New Dam was constructed in 1978 to replace the Charles River Dam. The purpose of the Charles River New Dam is to control the surface level of the Charles River and its tributaries and to prevent sea water from entering the Charles River Basin during high tides. The Charles River New Dam is an earthfill and concrete dam with stone slope protection and a top elevation of 12.5 ft above Mean Sea Level (NGVD29). The structure includes three navigation locks to pass commercial and recreational boats and a fish ladder. A key part of the operation of the dam is the pump station with six diesel pumps. Under typical operations, inflow to the basin is cyclically released via gravity through a service sluiceway. However, because high tide is above the normal level of the upstream basin, the sluice must be closed twice daily. In a flood condition at high tide, the only way to discharge river flow is for the dam operator, the Massachusetts Department of Conservation and Recreation, to pump water over the dam.

Design of the Charles River New Dam required an understanding of the watershed hydrology as well as coastal hydraulics. The project design assumes that the design conditions peak flood is coincident with the design high tide. The project designer (the Army Corps of Engineers (USACE)) developed a maximum tide elevation vs frequency curve for Boston Harbor based upon 200 years of tide data. USACE gave consideration to tide elevations associated with a hurricane approaching from the east or northeast but determined that they would be much larger than any tide recorded thus far and that such tides would be an extremely rare event. The USACE also evaluated wave heights for various fetch and wind speeds. The design went so far as to consider sea level rise at a rate of 0.6 feet per 100 years (USACE, 1971). Design of the Charles River New Dam and pump station and development of the operations plan required evaluation of coastal and inland systems.

Bargaintown Mill Pond Dam

The Bargaintown Mill Pond Dam is located at the headwaters of Patcong Creek in the Great Egg Harbor River Basin in Egg Harbor Township and Linwood City, New Jersey. The dam is located approximately 6 miles upriver of the Great Egg Harbor Bay. The dam impounds water for recreational purposes. It was reportedly constructed in 1718 and reconstructed in 1930, 1940, 1975 and 1997. The dam is an approximately eleven foot high earthen structure with a top of dam elevation of 10.7. Per New Jersey State Dam Safety Standards (N.J.A.C. 7:20) the spillway design storm for a Class II, Significant Hazard Dam is specified as the one half probable maximum precipitation (1/2 PMP).

The Bargaintown Mill Pond Dam did not have enough spillway capacity to convey the flood resulting from the ½ PMP.

An Inflow Design Flood (IDF) analysis was performed to evaluate the appropriateness of using the 100-year flood as the spillway design flood (SDF). The toe of the dam is at elevation 0, NAVD 88, the spillway is at elevation 6.4 and the top of dam is at elevation 10.7. High and low tide elevations at Great Egg Harbor Bay are 4.3 and 3.0 feet, respectively. Therefore peak flow from a hypothetical breach would be influenced by the tide condition at the time of breach. Breach analysis was performed during both high and low tide to provide consideration that potential incremental damage needed to be evaluated for both tide conditions. The low tide condition was more likely to produce a greater increase in water surface elevation due to the smaller footprint associated with the low tide but the high tide was potentially more likely to impact structures because the initial water surface elevation was closer to these structures. The tailwater condition also could have affected the outflow rate from estimated to result from a hypothetical dam failure. The result of the study indicated that a 100-year flood was appropriate as the SDF.

Lake Okeechobee

Lake Okeechobee is the largest freshwater lake in Florida, covering approximately 730 square miles. At its closest points, the dam is located approximately thirty five miles from the Atlantic Ocean and seventy miles from the Gulf of Mexico. Due to its location, it is considered to be an inland lake, not a coastal lake. Simply upon examination of the location of the lake, one might not consider overtopping as a result of storm surge as a potential failure mechanism for the structure. The lake is currently enclosed by the Herbert Hoover Dike which is up to thirty feet high in some locations. On September 16, 1928, the Lake Okeechobee Hurricane passed near Lake Okeechobee (NOAA, 2014b). The Hurricane produced a storm surge in the Lake itself of six to nine feet. This was an asymmetrical rise in water surface elevation caused by wind rather than rain. The rise was initially on the southern side of the lake but reversed to the north as the eye of the hurricane passed over the lake. The storm surge drove lake water to flow over its banks, failed the dike and inundated the surrounding area to water depths between ten and fifteen feet and killing approximately 1,500 people, primarily from drowning. The National Weather Service has run the SLOSH model in advance of significant storms (including Hurricanes Katrina and Rita) to predict storm surge. The storm surge and overtopping that occurred at Lake Okeechobee emphasizes the need to give consideration to coastal flooding mechanisms at dam sites in immediate proximity to the coast but also to dam sites that are not located at the coast or in direct connection with the ocean.

CONCLUSIONS

Dams located in coastal areas may be subject to hazards originating from both river flooding and coastal flooding. Dams at the tidal boundary require a higher level of scrutiny for potential flood mechanism and resulting impacts. A number of tools typically used by coastal engineers are available to the dam engineer for this purpose.

REFERENCES

FEMA, 2005a. Wave Setup, FEMA Coastal Flood Hazard Analysis and Mapping Guidelines Focused Study Report, Dean et al., February 2005.

FEMA, 2005b. Wave Runup and Overtopping, FEMA Coastal Flood Hazard Analysis and Mapping Guidelines Focused Study Report, Jones et al., February 2005.

FEMA, 2011a. "Region II Storm Surge Project, Mesh Development Report" – Draft Report, Federal Emergency Management Agency, Department of Homeland Security, Washington DC, September 2011.

FEMA, 2011b. "Region II Coastal Terrain Processing Methodology Documentation Report" – Draft Report, Federal Emergency Management Agency, Department of Homeland Security, Washington DC, September 2011.

FEMA, 2012. "Region II Storm Surge Analysis – Spatially Varying Nodal Attribute Parameters" – Draft Land Cover Data Report, Federal Emergency Management Agency, Department of Homeland Security, Washington DC, October 2012.

FEMA, 2013. City of New York, Bronx, Kings, New York, Queens and Richmond Counties, New York, Flood Insurance Study, Preliminary December 2013. FIS # 360497V000B, version number 1.0.0.0.

IPCC, 2007. Intergovernmental Panel on Climate Change, IPCC Fourth Assessment Report (http://www.ipcc.ch/publications_and_data/ar4/wg1/en/ch5s5-5-2.html).

NOAA, 2014a. Tides and Currents – About Tidal Datums, National Oceanographic and Atmospheric Administration, Available at http://tidesandcurrents.noaa.gov/datum_options.html.

NOAA, 2014b. Storm Surge and Coastal Inundation Event History, National Oceanographic and Atmospheric Administration, Available at http://www.stormsurge.noaa.gov/event_history_pre1930s.html.

Parker, 1992. Sea Level as An Indicator of Climate and Global Change, B.B. Parker, Marine Technology Society Journal, Vol.25, No.4, 1992 (condensed version available at http://tidesandcurrents.noaa.gov/sltrends/mtsparker.html).

USACE, 1971. Charles River Locks and Dam Design Memorandum No. 1 Hydrology and Hydraulics, U.S. Army Corps of Engineers, New England District, Waltham, Massachusetts, May 1971.

USACE, 2014a. Charles River Dam Local Protection Project, U.S. Army Corps of Engineers, New England District. Available at http://www.nae.usace.army.mil/Missions/CivilWorks/FloodRiskManagement/Massachusetts/CharlesRiverDam.aspx.

USACE, 2014b. North Atlantic Coast Comprehensive Study (NACCS), U.S. Army Corps of Engineers, Atlantic Division (Resources available at http://www.nad.usace.army.mil/CompStudy.aspx).

HYDROLOGIC INVESTIGATION FOR SADDLEBACK FLOOD RETARDING STRUCTURE

Bert Clemmens, PhD, PE, D.WRE[1]
Suzanne Monk[2]
Stephanie Gerlach[3]
Chuck Davis, PE[4]

ABSTRACT

A hydrologic and hydraulic investigation of the Saddleback Flood Retarding Structure (FRS), an earth fill dam, was conducted in 2013. Deficiencies in the structure prompted an evaluation of its hydraulic performance. The goals of this study were to (1) develop new hydrologic models for the Saddleback FRS using HEC-1 and (2) to route flood flows through Saddleback FRS using unsteady HEC-RAS. Two upstream structures discharge water into Saddleback FRS; Harquahala Floodway, which routes principal spillway flows from Harquahala FRS, and a canal overchute which receives water from an embankment that is protecting the Central Arizona Project (CAP) canal. HEC-RAS was used to model flows behind these embankments. For the CAP canal embankment and parts of Harquahala FRS, storage areas were used to model parts of the area behind these structures. Where the storage areas were not used, pilot channels were used to provide stable HEC-RAS calculations and to minimize volume errors. Inflow to these reservoirs was modeled based on storms with a 100-yr recurrence interval, as in the original design. Probable Maximum Precipitation (PMP) was estimated based on HMR-49 and from the new Arizona statewide standard (ADWR). The PMP storms were used to generate the Probable Maximum Flow (PMF) and 1/2 PMF inflow hydrographs. The water surface elevation in Saddleback FRS is above the top of the dam for all the HMR-49 PMF storms examined. The 72-hr Tropical PMF (ADWR methodology) did not overtop Saddleback FRS. The highest water surface elevation occurred for the Local 6-hr PMF when only the contributing area of Saddleback FRS (Saddleback Basin) was included in determining the PMP for both HMR-49 and ADWR PMPs. This project included a unique set of modeling approaches and was one of the first studies which used the new Arizona statewide PMP procedure.

INTRODUCTION

The Saddleback Flood Retarding Structure (FRS) is located in western Maricopa County about 75 miles west of downtown Phoenix. Saddleback FRS is an earth-fill dam operated and maintained by the Flood Control District of Maricopa County (District).

---

[1]WEST Consultants, Inc., 8950 South 52nd Street, Ste 210, Tempe, AZ 85284-1043; PH (480) 345-2155; FAX (480) 345-2156; email: bclemmens@westconsultants.com
[2] WEST Consultants, Inc., 8950 South 52nd Street, Ste 210, Tempe, AZ 85284-1043; PH (480) 345-2155; FAX (480) 345-2156; email: smonk@westconsultants.com
[3] Flood Control District of Maricopa County, 2801 W. Durango St., Phoenix, AZ 85009; PH (602) 506-1501; FAX (602) 506-4601; email: StephanieGerlach@mail.maricopa.gov
[4] WEST Consultants, Inc., 8950 South 52nd Street, Ste 210, Tempe, AZ 85284-1043; PH (480) 345-2155; FAX (480) 345-2156; email: cdavis@westconsultants.com

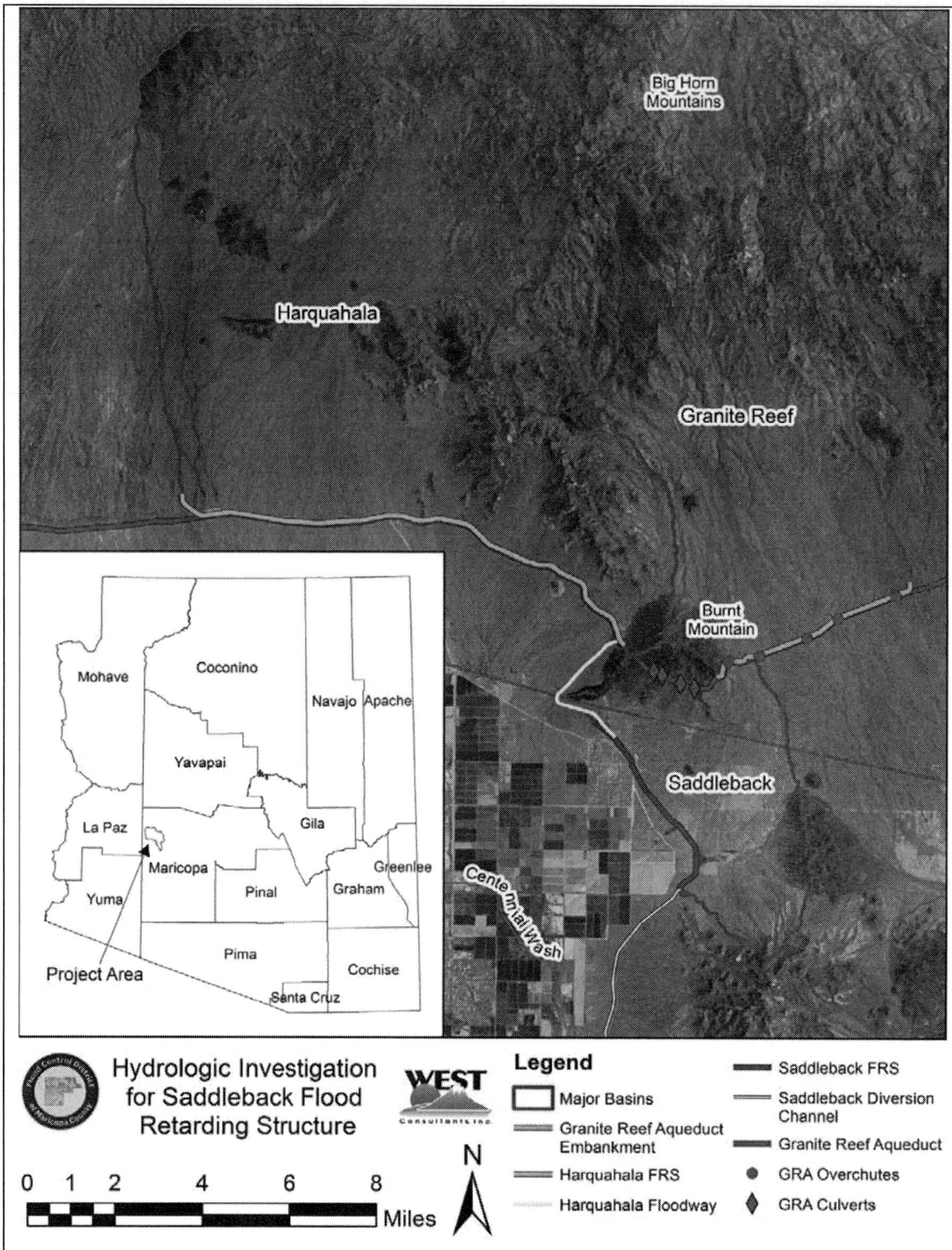

Figure 1. Saddleback FRS location map.

The structure was designed and constructed by the Natural Resources Conservation Service (NRCS), formerly the Soil Conservation Service (SCS), in the early 1980s. It was designed to protect farm land by detaining flood water from the upstream area and releasing those waters through the Saddleback Diversion Channel to Centennial Wash. Currently, Saddleback FRS is in need of repairs or, possibly, a full rehabilitation (Gannett Fleming 2011). The prior hydrologic studies included the original design (PCR Toups 1979) for the 100 yr storm and a dam break study conducted in 1991 (Carter Associates 1991) for the ½ PMF. Prior reports did not use the same drainage areas and the ½ PMF was no longer considered adequate by Arizona Department of Water Resources (ADWR) standards. Thus, new hydrology was needed to determine the requirements for the dam crest elevation or top of the dam. This study was somewhat comprehensive since alternatives to be evaluated later may include: decommissioning and breaching the FRS, converting the FRS into a levee, and rebuilding the FRS.

The plan for this study was to (1) develop new hydrologic models for Saddleback FRS using HEC-1 (HEC 1997) and (2) to route flood flows through Saddleback FRS using unsteady HEC-RAS. Harquahala FRS, as designed by SCS, discharges into the Harquahala Floodway, which discharges into Saddleback FRS. The Central Arizona Project (CAP) canal is just south of Harquahala FRS and is north of Saddleback FRS, as shown in Figure 1. Immediately north of Saddleback FRS, runoff from Burnt Mountain crosses under the CAP canal and discharges directly into Saddleback FRS, essentially unobstructed. Further to the east, an embankment protects the CAP canal from storm-water runoff (referred to as GRA for Granite Reef Aqueduct). Large overchutes allow water to leave the storage area behind this embankment and cross over the canal. The west-most overchute discharges into a stream channel that eventually enters Saddleback FRS. Much of the confusion regarding the drainage area of Saddleback FRS resulted from different estimates of the area behind this embankment that contributes water to Saddleback FRS. At low flows, the west-most overchute is isolated from the remaining drainage area by a road. At high flows, the water surface elevation is above the road surface allowing water to cross. For a PMF storm of uniform intensity over this watershed, water initially flows over this road from west to east. Later in the storm, the flow reverses and flows east to west over the road. This results from differences in the drainage path length for these different contributing areas.

RAINFALL INPUT

The project area is divided into three basins; specifically areas contributing directly to Harquahala FRS (105 ac), the GRA embankment (59 ac), and Saddleback FRS (22 ac), as shown in Figure 1. Rainfall amounts were determined for various storms for Saddleback Only (SO), Saddleback and Harquahala (SH), Saddleback and GRA (SG) and the entire drainage area. Rainfall data were assumed uniform over the area considered. Rainfall amounts were for storm shown in Table 1 for different combinations of basin. Storms for recurrence intervals of 100 yrs and less were determined from NOAA Atlas 14 (NOAA, 2006). Harquahala Only (HO) was evaluated for the 100-yr 6-hr and 100-yr 24 hr storms to evaluate the design storm. The District provided rainfall amounts for 500-yr 6-hr, 500-yr 24-hr, 1000-yr 6-hr, and 1000-yr 24-hr storms. HMR-49 (NOAA 1984) was used to determine rainfall amounts for the 6-hr local PMP and the 72-hr General PMP. The ADWR PMP Evaluation Tool (Arizona Department of Water Resources 2013) was used to develop the 6-hr Local ADWR PMP, and the 72-hr Tropical PMP. Other PMP storms had significantly lower rainfall totals and were not used. A total of 29 storms were evaluated, as shown in Table 1.

Table 1. Storms used to determine inflow into Saddleback FRS

Storm Event	Entire Area	Saddleback Basin Only	Harquahala Basin Only	Saddleback &Harquahala Basins	Saddleback & Granite Reef Aqueduct Basin
2-yr 6 hr		X			
2-yr 24 hr		X			
10-yr 6 hr		X			
10-yr 24 hr		X			
50-yr 6 hr		X			
50-yr 24 hr		X			
100-yr 6 hr		X	X	X	
100-yr 24 hr		X	X	X	
500-yr 6 hr	X				
500-yr 24 hr	X				
1000-yr 6 hr	X				
1000-yr 24 hr	X				
HMR-49 Local PMP 6-hr	X	X		X	X
HMR-49 General PMP 72-hr	X				
ADWR Local PMP 6-hr	X	X		X	X
ADWR Tropical PMP 72-hr	X	X		X	X

DDMSW AND HEC-1

The District uses a specially developed modeling system for generating input to HEC-1 (HEC 1997), known as Drainage Design Management System for Windows or DDMSW (FCDMC 2011). This system automatically determines sub basin properties from GIS data shape files such as Manning n, land use, infiltration parameters, etc. This system is not compatible with HEC-HMS. Manning n values and land use values (and associated parameters) were generated for individual sub basins from shape files developed for the project area. This area ranges from desert to foothills to mountainous. It is all essentially undeveloped land. Details can be found in WEST (2014).

DDMSW can only enter data up to a 100-yr storm. For low recurrence interval storms, the HEC-1 file generated by DDMSW has to be modified to input the rainfall amounts for these storms. HEC-1 was run with these new rainfall amounts and inflow into the storage area behind each FRS were saved as HEC-DSS (Data Storage System) files for input into HEC-RAS (HEC 2010). The number of sub-basins with input into these reservoirs were 8, 6, and 3; for Harquahala, GRA and Saddleback, respectively. For PMP rainfall events, the flow into these structures was the Probable Maximum Flood (PMF). Where ½ PMFs were required, the PMF flows in the DSS files were divided by two.

HEC-RAS

HEC-RAS was used to model flows behind these embankments. HEC-RAS requires a minimum flow to operate properly. Initial flows into Harquahala FRS resulted in large initial volume. A pilot channel was used so that the water surface elevations would be below the thalweg for the channel behind the dam. This reduced the volume error to acceptable levels. The emergency spillway is located at a high spot in the channel and the principal spillway is located at a low spot a short distance away. During some flow conditions, flow is supercritical between these two spillways. This caused computational difficulties. The area around the principal spillway was modeled as a storage area to avoid this issue. This allows critical flow at the storage area connection. This appeared to be a reasonable method for modeling this area of the channel. A diagram of the downstream portion of Harquahala FRS is shown in Figure 2.

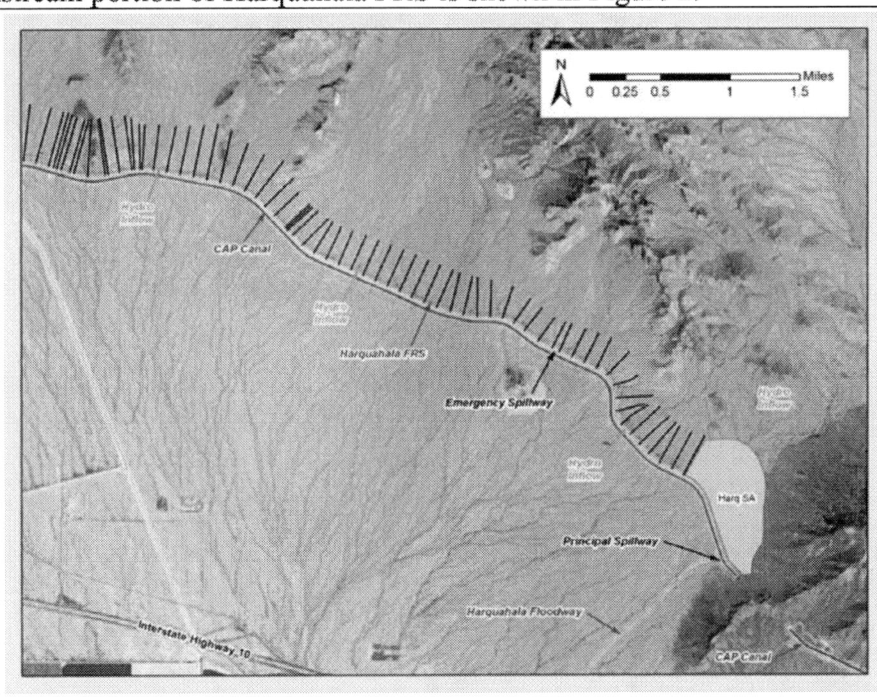

Figure 2. Cross section layout for Harquahala FRS (downstream).

Saddleback Flood Retarding Structure

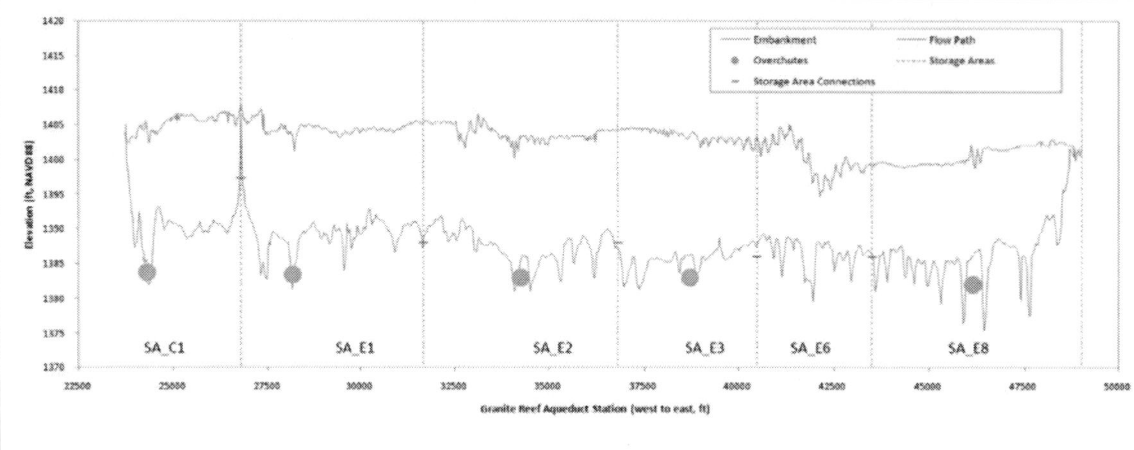

Figure 3. Profile along GRA embankment. (Storage basins locations are approximate).

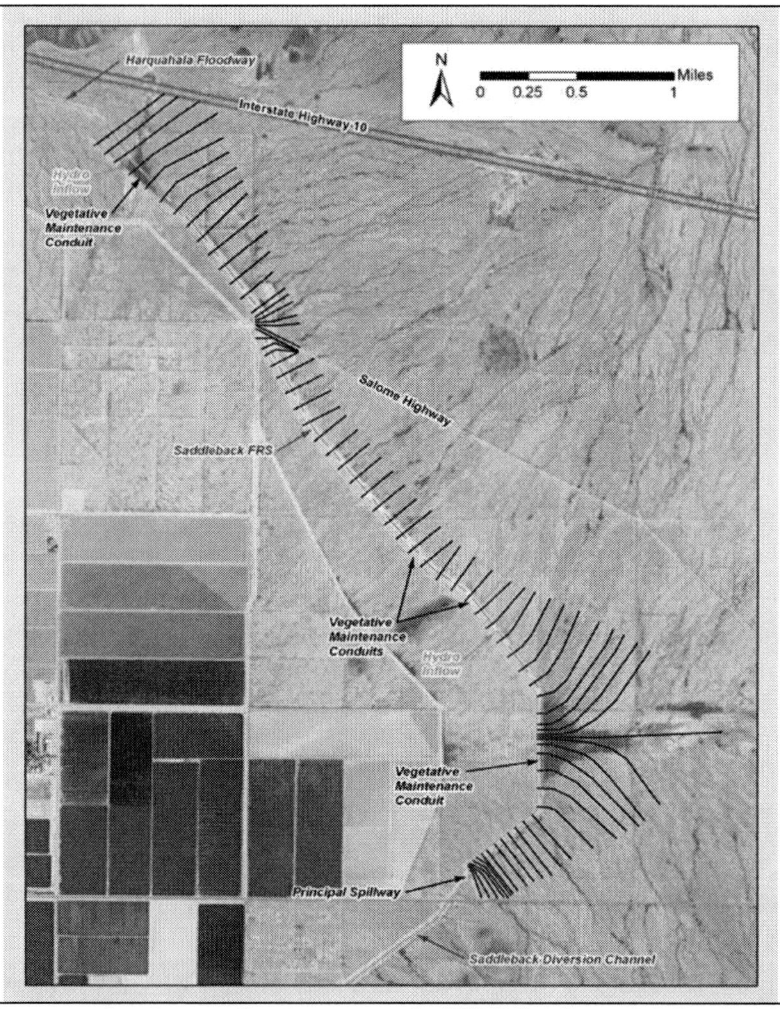

Figure 4. Cross section layout for the unsteady HEC-RAS model for Saddleback FRS.

For the GRA embankment, analysis of Saddleback FRS is only concerned with the flow through the west-most overchute. The direct contributing sub-basin is SA_C1. Since the area behind the embankment in this area is only connected to the rest of the area by flow over a road, storage areas seemed to be a good way to model this entire reservoir, which was divided into six storage areas, as shown in Figure 3.

Saddleback FRS has the same issue with volume errors as Harquahala FRS. Thus a pilot channel was used for this FRS. It was difficult to start at a low water level because the outlet into Saddleback diversion channel was not at the low point in the channel thalweg. So the outlet at the downstream end of the pilot channel had to be lower than the elevation of the principal spillway inlet. This channel also has culverts for low flows at a road crossing. Since pilot channels go through internal structures, a low flow channel had to be built into the culverts to accommodate flow through the pilot channel. The plan view for the cross sections for Saddleback FRS is shown in Figure 4. Eighty three separate HEC-RAS model runs were required to model the 29 rainfall events over the separate combinations of FRS and floodway models.

RESULTS

Rainfall

The rainfall amount for the HMR-49 General storm was only determined for the entire drainage area. It was 15.4 in over 6 hrs. The ADWR General storm resulted in a rainfall value that was less than the local storm rainfall value, it was not considered further. The local and tropical storm PMPs are shown in Table 2.

Table 2. Local Storm PMP Rainfall amounts for various contributing areas.

Drainage Basin	Basin	Total Area (mi^2)	HMR 49 Local Rainfall (inches)	ADWR Local Rainfall (inches)	ADWR Tropical Rainfall (inches)
Adjacent to Saddleback	S	22.1	12.7	11.0	12.2
Saddleback and GRA Embankment	S, G	81.3	10.2	9.9	12.9
Saddleback and Harquahala	S, H	127.4	9.3	9.5	12.3
Total Drainage Area	S, G, H	186.6	8.2	8.8	13.2

Peak Flows

The peak flows into Saddleback FRS are shown in Table 3 for the Saddleback Only drainage area. The labels with CS are the drainage sub basins. These tended to give the maximum water surface elevation behind the FRS.

Table 3. Saddleback FRS peak flows and time-to-peak (6-hour storms)

Storm	Peak Flow (cfs)			Time-to-Peak (hr)		
	CS05	CS06	CS09	CS05	CS06	CS09
2-yr, 6-hr	535	66	134	4.50	4.42	6.08
10-yr, 6-hr	1,513	898	967	4.42	4.5	4.25
50-yr, 6-hr	3,593	2,576	3,075	4.42	4.50	5.17
100-yr, 6-hr	4,614	3,371	4,329	4.42	4.50	5.08
HMR 49 Local PMP	38,163	27,960	52,529	3.58	3.67	4.00
ADWR Local PMP	34,013	25,187	54,697	3.42	3.50	3.75
ADWR Tropical PMP	1,978	1,281	3,359	42.08	42.08	42.08

Water Surface Elevations

Harquahala FRS. Analysis of the Harquahala FRS showed that water flows through the emergency spillway during the 100-yr 6-hr storm event. This is in contrast to design practice, which suggests that the Principal Spillway should handle the 100-yr storm without flow over the emergency spillway. The results are shown in Table 4. It was noted that constriction in flow occurs in an area behind Harquahala FRS, upstream from the emergency spillway. There are two constrictions, with an area of storage in between, as shown in Figure 5. Flow is restricted from passing through this area from both upstream and downstream.

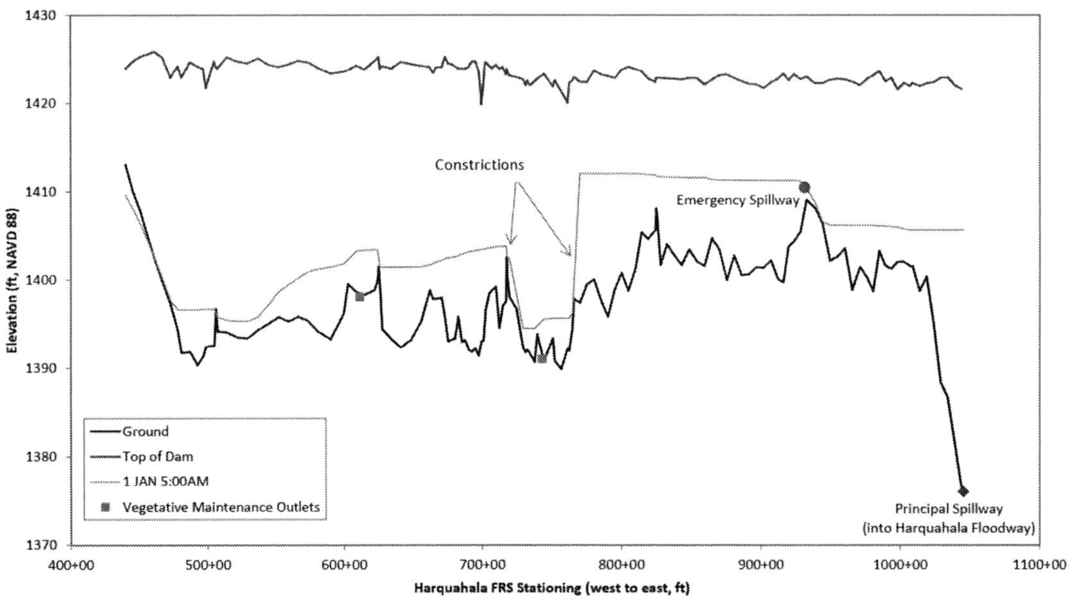

Figure 5. Choke point in Harquahala FRS. Note; constrictions are not apparent in 1-D ground elevations.

As a result, flows from the sub-basins near the emergency spillway are restricted in flowing upstream to fill the available storage. This causes the depth to be greater in the area of the emergency spillway. In HEC-RAS, flows go through critical depth for a short time at the constrictions on either side of the constricted area. Eventually the constriction reservoir area begins to fill and critical depth is drowned out. This is a difficult situation for a 1-D model, thus while interesting, these results have to be taken with a grain of salt. It is likely that the model did not allow sufficient flow through these restrictions. Thus the predicted water surface is likely a bit too high. Just a small increase in the flow at these constrictions would have resulted in no flow over the emergency spillway. A more accurate solution may have resulted if the constrictions had been modeled as in-line structures or if the area between the constrictions had been modeled as a storage area. The ground elevation at the constrictions could have been used to define the structure cross section, as was done for the GRA embankment storage areas. The crest of the emergency spillway is 1410.46 ft (NAVD88). Harquahala FRS was overtopped during the 6-hr PMP storms. These constrictions did not influence the water depth during the PMP events because they were quickly overtopped and drowned out.

Table 4. Results from analysis of storm inflow into Harquahala FRS.

Storm	Condition	Maximum Water Surface Elevation (ft)	Maximum Inflow (cfs)	Principal Spillway Flow (cfs)	Emergency Spillway Flow (cfs)
100-yr 6 hr	Harquahala only	1,412.32	32,445	353	8
100-yr 24 hr	Harquahala only	1,412.55	38,563	351	0

Saddleback FRS. The maximum water surface elevations and minimum freeboard in the Saddleback FRS for the various storm events are shown in Table 5. Based on these results, Saddleback FRS can handle all the storms up through the 1,000-year storm without overtopping. Even the ½ PMF from the worst case scenario PMP overtops Saddleback FRS. However, Saddleback FRS was overtopped for both the HMR-49 General Storm and the HMR-49 and ADWR Local Storms. The ADWR 72-hr tropical storm did not overtop Saddleback FRS. The Local Storm PMF (regardless of the contributing area) overtopped Saddleback significantly. The worst case was the Local Storm PMF for the Saddleback Basin only, where the dam was overtopped by more than 4 ft for both HMR-49 and ADWR storms. Water surface profiles for the maximum water level are shown in Figure 6.

The goal of this project was to determine whether or not Saddleback FRS overtops and to determine how high the water level gets assuming the FRS embankment is raised. Thus no flow is allowed over the embankment, as if the dam were higher. When overtopping occurred at the downstream end of Saddleback FRS (near river station 887), the rating curve for the outlet given by the District does not extend high enough. The rating curve was extrapolated. However, this extrapolation may not be accurate. The maximum water surface elevation was always at the upstream end (station 25943). However, the minimum freeboard sometimes occurred further downstream, where the top of the FRS is lower. This typically only occurred during the PMF events.

Table 5. Results from analysis of storm inflow into Saddleback FRS.

Storm	Condition	Maximum Water Surface Elevation (ft)	Minimum Freeboard (ft)	Station where Minimum Freeboard Occurs (ft)	Maximum Inflow (cfs)	Saddleback Maximum Outflow (cfs)
2-yr 6 hr	SO[1]	1,189.31	6.59	25943	716	15
2-yr 24 hr	SO	1,189.14	6.76	25943	1,144	19
10-yr 6 hr	SO	1,190.49	5.41	25943	3,228	88
10-yr 24 hr	SO	1,190.77	5.13	25943	5,503	169
50-yr 6 hr	SO	1,191.85	4.05	25943	8,481	453
50-yr 24 hr	SO	1,191.80	4.10	25943	10,364	488
100-yr 6 hr	SO	1,192.35	3.55	25943	10,961	613
100-yr 6 hr	S & H	1,192.49	3.41	25943	11,491	659
100-yr 24 hr	SO	1,192.23	3.67	25943	12,821	606
100-yr 24 hr	S & H	1,192.51	3.39	25943	14,308	682
500-yr 6 hr	S & H & G	1,193.67	2.23	25943	20,198	953
500-yr 24 hr	S & H & G	1,193.64	2.26	25943	23,941	941
1000-yr 6 hr	S & H & G	1,194.16	1.74	25943	24,048	1,039
1000-yr 24 hr	S & H & G	1,194.06	1.84	25943	27,471	1,010
PMF from HMR-49 PMP 6-hr Local	SO	1,199.71	-4.80	887	106,649	1,610
PMF from HMR-49 PMP 6-hr Local	S & H	1,196.68	-1.43	887	53,777	1,420
PMF from HMR-49 PMP 6-hr Local	S & G	1,197.41	-2.51	887	69,563	1,484
PMF from HMR-49 PMP 6-hr Local	S & H & G	1,196.10	-0.30	887	44,043	1,349
PMF from HMR-49 PMP 72-hr General	S & H & G	1,196.33	-1.45	887	13,503	1,421
PMF from ADWR PMP 6-hr Local	SO	1,199.20	-4.16	887	108,101	1,576
PMF from ADWR PMP 6-hr Local	S & H	1,198.14	-2.78	887	89,205	1,500
PMF from ADWR PMP 6-hr Local	S & G	1,198.57	-3.21	887	94,811	1,523
PMF from ADWR PMP 6-hr Local	S & H & G	1,197.91	-2.20	887	81,690	1,466
PMF from ADWR PMP 72-hr Tropical	SO	1,192.68	2.53	887	6,618	1,154
PMF from ADWR PMP 72-hr Tropical	S & H	1,192.76	2.76	887	6,201	1,137
PMF from ADWR PMP 72-hr Tropical	S & G	1,192.68	2.29	887	7,424	1,171
PMF from ADWR PMP 72-hr Tropical	S & H & G	1,192.65	2.97	887	5,908	1,122
½ PMF from HMR-49 PMP 6-hr local	SO	1,196.10	-0.20	25943	53,325	1,307
½ PMF from PMP 72-hr General	S & H & G	1,192.78	2.78	887	7,115	1,136
½ PMF from ADWR PMP 6-hr Local	SO	1,196.14	-0.24	25943	54,050	1,277

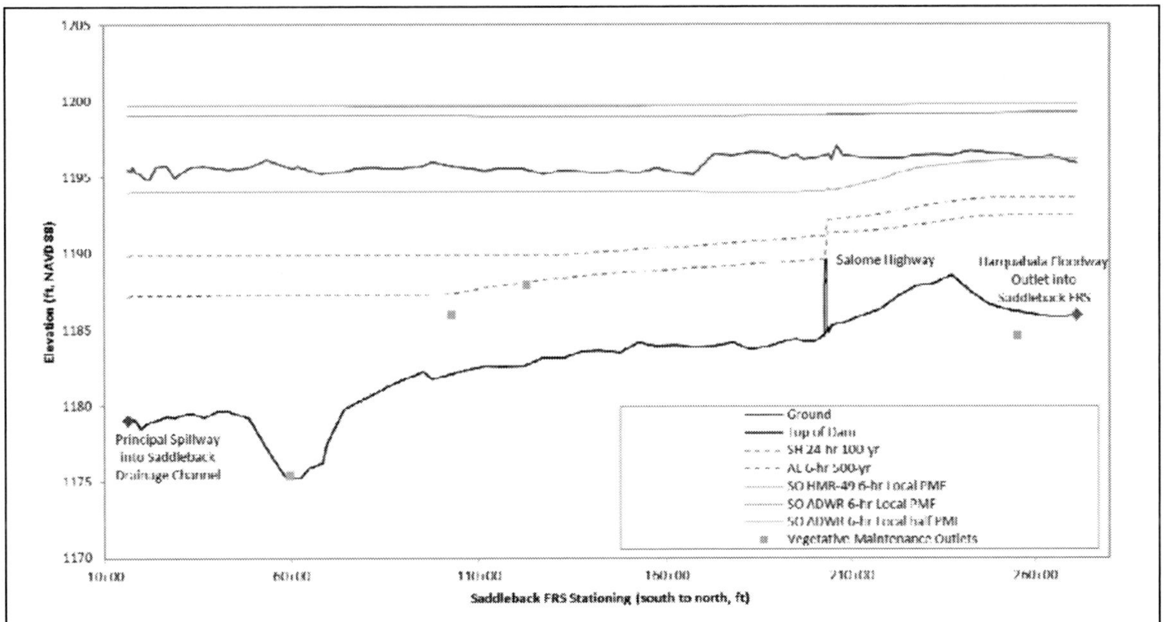

Figure 6. Water surface profiles for Saddleback FRS.

CONCLUSIONS

To our surprise for Harquahala FRS, the emergency spillway flowed during the 100-yr storms. Since it was not the intent of this project to examine Harquahala FRS and because there was uncertainty regarding whether or not this flow would occur. This was not examined further.

Saddleback FRS clearly overtops during ½ PMF and PMF events. The District will have to decide whether or not to rehabilitate this FRS. Saddleback FRS does not have an emergency spillway. It operates more like a channel than a dam, and thus converting it into a channel and making the FRS a levee makes sense. If it is abandoned altogether, the FRS would have to be breached in several locations, particularly upstream from Salome Highway and near the downstream end where large washes enter. The District has not decided which option they will take.

REFERENCES

ADWR. (2013). PMP Evaluation Tool Description and Usage. Phoenix: Arizona Department of Water Resources.

Carter Associates. (1991). Final Dambreak Analysis for the Harquahala and Saddleback Flood Retarding Structures. Phoenix, AZ, for Flood Control District of Maricopa County.

FCDMC. (2011). Drainage Design Manual for Maricopa County, Arizona: Hydrology. Flood Control District of Maricopa County, Phoenix, AZ.

Gannett Fleming. (2011). Saddleback Flood Retarding Structure (FRS) Failure Mode and Effects Analysis. Phoenix, AZ, for Flood Control District of Maricopa County.

HEC. (1997). HEC-1 Flood Hydrograph Package, v. 4.1. Hydrologic Engineering Center, US Army Corps of Engineers, Davis, CA.

HEC (2010, January). HEC-RAS River Analysis System, v. 4.1.0. Hydrologic Engineering Center, US Army Corps of Engineers, Davis, CA,.

NOAA (1984). Hydrometeorological Report No. 49: Probable Maximum Precipitation Estimates, Colorado River and Great Basin Drainages. National Weather Service, National Oceanic and Atmospheric Administration, Silver Spring, MD:

NOAA. (2006). NOAA Atlas 14: Precipitation-Frequency Atlas of the United States. National Weather Service, National Oceanic and Atmospheric Administration. Silver Spring, MD

PRC Toups. (1979). Saddleback Floodwater Retarding Structure, Harquahala Valley Watershed, Maricopa County, Arizona, Design Report. Phoenix, AZ: Volume I.

WEST. (2014). Hydrologic Investigation for Saddleback Flood Retarding Structure, Technical Support Data Notebook, WEST Consultants, Inc., Tempe, AZ, for Flood Control District of Maricopa County.

ADDRESSING OPERATIONAL RISKS AND UNCERTAINTIES FOR GATED SPILLWAYS

Greg Paxson, P.E.[1]
Robert Indri, P.E.[2]
Mark Landis, P.E., P.G.[3]

ABSTRACT

Gated spillways are commonly incorporated into dam design to provide flexibility with regard to reservoir operations and/or to pass large floods. Spillway gates present operation and maintenance challenges to dam owners. Gate operations during floods can require on-site personnel to react quickly, often during storms when flooding can limit access to the site or gates and compromise the availability of power. Additionally, the operation of gates requires decision making related to releasing large flows that could cause downstream damage. Interpretation of flooding potential has to be considered in the decision making process, since a delayed reaction could result in a dam safety emergency. Even when post-incident assessments validate that gate operations did not increase flooding, downstream landowners are often quick to condemn the dam owner for mis-operation of gates. In addition to operational challenges, dam owners are faced with costly ongoing maintenance to ensure that gates will function as designed when the need arises. This includes maintenance of structural and mechanical components (including hydraulics, electrical systems, and power feeds), training of operators, regular testing of the gate systems, and debris removal.

This paper discusses common structural, mechanical, electrical, instrumentation, and operational issues along with recommendations for regular maintenance of spillway gates. Hydraulically-efficient fixed crest labyrinth and piano key weirs are discussed as an option to reduce reliance on spillway gates during floods. Case histories include projects where gated spillways presented challenges to the dam owner and were partially or completely replaced with passive, fixed crest spillways or automated gates. Finally, a hypothetical example is presented to illustrate changes in risks through replacement of a gated spillway with a fixed crest labyrinth weir.

INTRODUCTION

Gated spillways are commonly incorporated into dams and are generally more typical for structures where:

- frequent regulation of the reservoir level is required

[1] Schnabel Engineering, 1380 Wilmington Pike, West Chester, PA 19382, 610-696-6066, gpaxson@schnabel-eng.com.
[2] Schnabel Engineering, 11A Oak Branch Drive, Greensboro, NC 27407, 336-274-9486, rindri@schnabel-eng.com
[3] Schnabel Engineering, 11A Oak Branch Drive, Greensboro, NC 27407, 336-274-9486, mlandis@schnabel-eng.com

- outflows need to be controlled for different flood events (e.g. flood control dams)
- significant discharge capacity is required to pass the design flood (e.g. dams on large watersheds), sometimes making a fixed crest spillway impractical

Spillway gates can be categorized as "underflow" gates, or "overflow" gates. Underflow gates are raised and flow is pressurized unless the gate is lifted to above the upstream water level, allowing free discharge over the fixed spillway crest. Overflow gates are lowered and flow passes over the gate until it is completely lowered to the fixed crest elevation. Common underflow gates include radial (or Tainter) gates and lift gates. Overflow or crest gates can be hinged at the fixed crest and hydraulically or pneumatically actuated. Other types of overflow gates include inflatable bladders (rubber dams) and Hydroplus Fusegates, a proprietary system designed such that the gate(s) tip during an extreme flood. Typically, radial gates are not designed for overflow; however, gates can be specially fabricated to facilitate overflow (e.g. Fontana Dam, NC – Photo 1).

Photo 1. Radial Gate designed to accommodate overflow, Fontana Dam, NC.

Gated spillways can provide significantly more discharge capacity in a given spillway footprint than conventional fixed crest weirs by effectively increasing the head over the fixed crest as the gates are operated. Additionally, gates can accommodate significant flood storage with little or no outflow until the gates are operated, making gated spillways very effective for flood risk reduction projects. However, spillway gates present operation and maintenance challenges to dam owners.

GATED SPILLWAY ISSUES AND RISKS

Spillway gates generally require more operation and maintenance than fixed crest spillways. The following is an excerpt from the US Army Corps of Engineers' *Hydraulic Design of Spillways* (USACE, 1990):

"The value of an uncontrolled fixed crest spillway in providing an extremely reliable operation and a very low cost maintenance facility is undeniable. [Various] considerations at many damsites may restrict the use of an uncontrolled fixed crest spillway."

USACE (1990) recommends that where possible, fixed crest spillways should be incorporated into designs, particularly when the watershed response time is short. In addition, USACE (1990) recommends installation of two or more gates to "satisfy safety concerns"; presumably to provide redundancy in case a spillway gate cannot be operated.

Common issues with gated systems are associated with their relatively complex mechanical systems (when compared to passive systems) and the user supplied input required to operate them.

Risks related to spillway gates can result from:

Unintended and uncontrolled releases from:

- accidental gate operation through failure of automation equipment or some other unexpected occurrence;
- structural failure of the gate from damaged components or from unexpected loadings (hydraulic, seismic, debris loads, vibration, etc.); and,
- inability to close gates due to damage, binding or debris obstruction.

Reduction in spillway capacity and increased potential for overtopping and failure of the dam from:

- inability to operate the gates during large floods; and,
- blockage of spillway gate openings from debris.

Flooding impacts can often result from the operational issues listed above and include:

- downstream flooding from un-intended and uncontrolled releases;
- upstream flooding from reduced spillway capacity due to inability to operate gates or blockage of the spillway gate openings; and,
- downstream flooding from failure of the dam due to reduced spillway capacity.

In addition to these issues with the gates and appurtenant works, effective operation of spillway gates commonly requires actions by the dam owner/operator, which introduces risks and uncertainties related to the human element. On-site personnel have to react quickly, often during storms when flooding can limit access to the site or gates and compromise the availability of power. Additionally, the operation of gates requires decision making related to releasing large flows that could cause downstream damage. Interpretation of flooding potential has to be considered in the decision making process, since a delayed reaction could result in a dam safety emergency.

Resources. There have been several publications summarizing historic gate failures and lessons learned (including Graham and Hilldale, 2001 and ASDSO/EPRI, 2000). Additionally, several publications provide best practices for gate operations and maintenance (including ASCE, 2012 and USSD, 2002) By following current best practices in gate operation, maintenance and training, risks related to gated spillways can be dramatically reduced, but not eliminated.

Dam or Gate Failure Examples

Poor maintenance, infrequent inspections, and operational issues can all lead to failure of either the gate or, in cases where gates cannot be effectively operated during extreme floods, overtopping and failure of the dam.

Using the National Performance of Dams Program (NPDP) Dam Incident Query tool (NPDP, 2014), eighteen "Gate Structural Failure" incidents are identified. Six of these incidents were related to failure of lifting systems, including hoisting mechanisms (1) and lifting chains or wire ropes (5). Three events were reportedly a result of concrete failure at piers that support gates. Only two events were reported as a gate failure, resulting in an uncontrolled release.

One of the more notable gate failures was at Folsom Dam in 1995 (Photo 2), where a Tainter gate failed during opening, reportedly as the result of trunnion pin friction. This failure generated significant interest in evaluating gated spillway systems by numerous agencies (ASDSO/EPRI, 2000).

Photo 2. Gate failure at Folsom Dam, 1995
(courtesy of Bureau of Reclamation)

The 2010 failure of Delhi Dam (Photo 3) has been partially attributed to inability to operate one of the spillway gates during an extreme flood event. Fiedler et al (2011) concluded that the failure was the result of internal erosion combined with overtopping of the embankment due to inadequate spillway capacity. The inability to operate one of the gates likely contributed to both failure modes as the internal erosion failure mode was believed to be initiated due to high reservoir levels.

Photo 3. Delhi Dam failure
(courtesy of Iowa Wing Civil Air Patrol)

Examples of Flooding Impacts Resulting from Gate Operations

In addition to the potentially catastrophic consequences related to failure of the dam or spillway gate(s), dam owners are faced with the challenge of operating the gate systems to consider potential upstream and downstream flooding impacts. These impacts may be real or merely public perception.

The 2014 operation of floodgates at two dams in India are notable recent examples of significant downstream impacts. On June 8, water was released from the Larij Hydropower Project and 24 students and a tour guide were carried away in the floodwaters; the bodies of 12 students have been recovered and the remaining 13 people are feared dead (NDTV, 2014). Less than two weeks later, ten boys were stranded in floodwaters reportedly caused by operation of gates at the Tenughat Dam (The Telegraph, 2014). The boys were rescued by divers following the closing of the gates. Both of these incidents have triggered inquiries with regard to the operations at the dams. The Larij inquiry report noted an increase in discharge from about 700 cfs to nearly 16,000 cfs over one hour and criticized the dam operators (NDTV, 2014):

> *"There are not standard operating procedures with the release of water...[Those] involved in powerhouse operation and those handling [dam] operations are not working in tandem..."*
>
> *"The warning system is also inadequate. All this constitutes a systemic failure..."*

A search for papers and articles related to the impacts of gate operations at dams results in numerous interesting and controversial case histories. As recent as July 2014, the US Army Corps of Engineers had to defend itself against criticisms related to releases at Coralville Reservoir in Iowa, which included charges from residents of both not releasing enough water prior to the flood and releasing too much water during the flood (O'Leary, 2014).

The Washington Suburban Sanitary Commission (WSSC) was criticized for causing flooding in downtown Laurel, Maryland by operations at Duckett Dam during a storm in April 2014, with claims from residents and politicians including:

> *"I don't have a problem with natural flooding that may take place. I have a problem with manufacturing a flood, and that's exactly what WSSC did"* (quote obtained from Shaver, 2014)
>
> *"I wonder about WSSC's mitigation plans. Can we release that water sooner? With all that technology at their disposal, you'd think they could avoid this."* (quote obtained from Pichaske, 2014)

The WSSC (Johnson, 2014) and many residents defended the operations, noting that the gate operations were necessitated due to a concern with the safety of the dam.

A major difference between these examples is that Coralville Dam is operated as a flood control structure while Duckett Dam impounds a water supply reservoir and was not intended to reduce downstream flooding during storms.

The Pompton Lakes Dam Floodgate Facility was developed to provide flood reduction benefits to upstream properties; however, there have been claims that operations at the dam have contributed to downstream flooding (Augenstein, 2012). An extensive study was performed to evaluate the impacts of gate operations on downstream flooding, with the finding that "the gates function as intended with no significant downstream impacts" (Miller et al, 2013).

There are also cases where dam operators are blamed for creating upstream flooding as the result of gate mis-operation. Lack of maintenance and the resulting inability to operate gates was alleged to cause flooding around Lady Bird Lake in Austin, Texas (Buchele, 2013).

Graham and Hilldale (2001) summarized several other notable incidents resulting from spillway gates opened intentionally during floods, including:

- Dominican Republic, 1998, Sabaneta Dam: During Hurricane George spillway gates were operated without notification of downstream residences. Flooding from the gate operation contributed to numerous deaths in the village downstream of the dam.

- Nigeria, 1999, Jebba and Shiriro Dams: Gates were operated during heavy seasonal rains, the gates were operated to prevent overtopping of the dams. The resultant flooding submerged 400 villages, killing hundreds and leaving 300,000 homeless.

- Mexico,1999, Penitas Dam: Spillway gates were operated to prevent overtopping of the dam resulting in heavy damage downstream of the dam. More than 400 deaths were blamed on the flood.

It is clear that whether or not the operation of gates during a flood contributes to downstream flooding, it is likely that the owner of the facility will be criticized for mis-operation of the structure.

Gate Operation Plans. Most projects with spillway gates have some form of operations plan to define the opening of gates during floods. These plans can range from "if the lake is rising, keep opening the gates" to complex plans that require monitoring of precipitation (forecasted and actual), lake levels, and upstream and downstream stream gage data.

While flood operations plans are prudent, the execution of these plans can be unrealistic or impracticable especially if the plan requires early operation of gates to increase flood storage prior to the actual rainfall event. For example, the authors have encountered one plan that required full opening of all spillway gates prior to the start of the storm event in order to lower the pool sufficiently to create additional flood storage to pass the design flood. Considering that opening of a single gate by only five percent causes downstream flooding of residences, it is unlikely that operators will follow the plan, since this would cause downstream flooding before the rain even starts. However, delaying these actions could put the dam at risk.

ADDRESSING RISKS RELATED TO SPILLWAY GATES

As presented in the examples herein, risks related to spillway gates can result from:

- inability to operate the gates during large floods, reducing spillway capacity and increasing the potential for overtopping and failure of the dam
- improper gate operations which cause downstream or upstream flooding (and create the potential for liability related to this flooding)
- spillway gates are opened accidentally through failure of automation equipment or some other unexpected occurrence
- spillway gates are opened intentionally during a major flood in accordance with a flood operating plan resulting in major downstream flooding
- spillway gates fail structurally
- debris blockage of spillway gate openings impedes flow, possibly leading to damage to the gates and/or potential overtopping and failure of the dam

To reduce these risks, actions can range from improved operation and maintenance and training, replacement of manual operations with systems that are programed for automated response in addition to manual and remote operation, to replacement of some or all of the gates with a more reliable passive, fixed crest spillway.

Maintenance and Modifications

Proper maintenance is critical to preventing failure of the gates and improving reliability of operations. This includes:

- periodic visual and hands-on structural inspection of the gate components;
- periodic gate performance testing;
- development of a maintenance manual and implementation of a preventive maintenance plan; and,
- providing a reliable power back-up system and maintaining this system.

Development of a written record of the gate design, components, modifications, inspections, operations, and maintenance is an important part of the gate life cycle and aids in the transition of key personnel responsible for gate operation and maintenance.

Gate rehabilitation could include structural modifications or reinforcement, replacement of lifting mechanisms, and automation of operations.

Effective Gate Operation Plans

An effective and well thought out gate operation plan can significantly reduce risk related to improper gate operations. The plan should include:

- gate operation procedures for normal and flood/emergency operations;
- identification of primary gate operations personnel and secondary responders for emergency gate operations;
- training on gate operations procedures for the primary and secondary responders

As with any critical plan, table top and full scale exercises should be conducted on a regular basis and include training for primary and secondary personnel. The effect of failure of different components of the operations plan should be analyzed, including loss of communications, road closures, un-availability of key personnel, power failure, etc. After exercises are conducted an appropriate amount of time should be allocated to conduct an after action review of the exercise to talk through failures and lessons learned, develop solutions, and incorporate the solutions into the plan and the next training cycle.

Dam owners need to balance downstream flooding impacts with gate operation for the safety of the project. Many older dams, designed and built when there was limited downstream development, could now impact downstream landowners with routine gate operations. As part of their gate operations plan, owners should evaluate the practicality of their plans in respect to upstream and downstream flooding impacts as well as dam safety. Every dam that includes gates for spillway capacity should have a well thought out operations plan. In developing the plan, an incremental damage assessment should be performed for various gate operation scenarios to reduce flooding impacts. For example, opening the gates "in advance of the storm" to preserve flood storage and increase spillway capacity for the peak of the event could present a higher risk to downstream residents than opening the gate after natural flooding begins to occur downstream. This is particularly true if the dam owner does not provide adequate warning with regard to gate operations.

Replacement

For projects where mechanical gates are significantly deteriorated, structurally or operationally deficient, or operations are difficult or present significant potential risks, replacement of some or all of the spillway gates may be justified. Options for replacement could include:

- Hydraulic or pneumatic controlled gates which utilize automated operations
- Fusegates, which do not rely on manual or automated systems for operation
- Fixed crest spillways which require little maintenance and no operation - Labyrinth and piano key weirs are hydraulically efficient solutions

The next section discusses specific examples related to alternatives to gated spillways, including case histories where traditional gated spillways were replaced.

ALTERNATIVES FOR GATED SPILLWAYS: EXAMPLES AND CASE HISTORIES

Numerous projects have included replacement of either deteriorated or antiquated gate systems.

Pneumatically actuated steel crest gates have been used to replace mechanically operated radial, flap, and slide gates for many projects (such as the patented bottom hinged spillway gates by Obermeyer Hydro, Inc.), which include automated operation to reduce the risk and uncertainty related to human error in manual operations. Similarly, rubber bladders have replaced mechanical systems.

Tipping Fusegates were selected in lieu of more traditional Tainter gates at Canton Lake Dam in Oklahoma (Hydroplus, 2014a) and Jindabyne Dam in Australia (Hydroplus, 2014b). This was reportedly based on initial and long term (maintenance) costs.

For several projects where gate operations presented challenges to the dam owner, fixed crest labyrinth and piano key weirs have been used to reduce routine maintenance and prevent operational issues.

The original design for Ute Dam in New Mexico included 27-ft high gates which were not installed as part of the 1962 construction. In the early 1980s, the US Bureau of Reclamation (Reclamation) evaluated options to increase reservoir storage by installing gates or identifying a more economical solution, which resulted in selection of a labyrinth spillway (Houston, 1982). In addition to being more cost effective to construct, this structure was presumably more reliable and considerably reduced operation and maintenance costs.

Examples of labyrinth spillways being used to replace gated structures include the Brazos Dam in Texas and the New London Dam in Minnesota. At Brazos Dam, the original drum gates did not maintain a reliable water level and were replaced in 1985 with hydraulically actuated leaf gates, which also had operational problems. In 2007, construction of a new labyrinth spillway was completed to replace the gated spillway (Vasquez et al, 2009). New London Dam also experienced problems with the original gate system and was replaced with a labyrinth spillway in 2010 (Minnesota DNR, 2014).

Piano key weirs are similar to labyrinth spillways, with a smaller foundation footprint that allows construction of the weir on gravity dams. At Malarce Dam in France (Photo 4), a piano key weir was installed to supplement the capacity of the existing gated spillway and increased total discharge capacity from about 140,000 to more than 160,000 cfs. In addition, the piano key weir was designed to discharge flood flows prior to needing to operate the gates. This provided twice as much time for the dam operator to get to the dam to open gates, which was very important given the rapid watershed response time (Laugier et al, 2014).

Photo 4. Piano key weir at Malarce Dam

Additional detailed project case histories follow.

Sugar Hollow Dam

Sugar Hollow Dam is a concrete gravity dam with a maximum height of about 80 feet, owned and operated by the Rivanna Water and Sewer Authority. The dam was constructed in 1947 to provide water supply for the City of Charlottesville, Virginia and surrounding communities. The 225-foot long ogee spillway was originally equipped with a system of eight, 25-foot wide by 5-foot high vertical lift gates. The gates were operated manually during floods using a hoisting system mounted to a rail that traversed a bridge over the spillway (Photo 5). This system was difficult to operate during floods and the bridge and other superstructure reduced hydraulic capacity. In addition, the configuration of the gates and superstructure made the spillway prone to clogging from logs and debris, further reducing capacity. Finally, the crest of the non-overflow sections of the dam was only five feet higher than the top of gates. Given the very steeply sloping watershed, the dam operator had to respond quickly during floods to operate the gates to prevent overtopping of the dam.

Photo 5. Sugar Hollow Dam: original gated spillway and hoisting system

The dam was found to have inadequate spillway capacity to pass the design flood without overtopping the non-overflow sections of the dam. Furthermore, stability analyses indicated the structure did not meet criteria for flood loading conditions. Rehabilitation of the dam was completed in 2000 and included installation of post tensioned rock anchors for stability, raising portions of the non-overflow sections to contain extreme flows to the central part of the dam, and replacement of the gate system with a five foot high inflatable bladder crest gate (Photo 6). The upgrading of the dam and design of the crest gate system is documented in (Campbell et al, 2000) and (Paxson et al, 1999).

Photo 6. Sugar Hollow Dam: inflatable bladder crest gate

With the removal of the piers and superstructure, the inflatable bladder crest gate provided better hydraulic performance and passage of debris than the original system. In addition, the bladder system was equipped with a computer system to automate operations, based on the reservoir level. Given the rapid hydrologic response of the watershed and the presence of several homes within the downstream floodplain, the designers were concerned with the potential for a sudden release from the gate system during operations. Therefore, various operating scenarios were evaluated under hypothetical flooding conditions to reduce the potential for surges in outflows. A spreadsheet program was developed to characterize outflow conditions for a wide range of operating conditions using a small computational time step.

The inflatable bladder system operates on water level control: the system attempts to maintain a specified constant water surface elevation throughout deflation. For this project, the target water surface elevation corresponded to one foot of flow over the inflated bladder (e.g. operations initiated with the reservoir at one foot above normal pool). Two methods of control ("fine" and "coarse") are commonly incorporated into the operating system. Each of these control modes samples water surface elevation at a discrete time interval (say every five minutes) and deflates the bladder incrementally in response to a water level above the target elevation. The target elevation for coarse control is higher than for fine control, and the deflation increment is typically larger for coarse control. The typical operating scenario used by the manufacturer was evaluated and found to result in surges in the flow from bladder deflation, as shown in Figure 1 (Paxson et al, 1999).

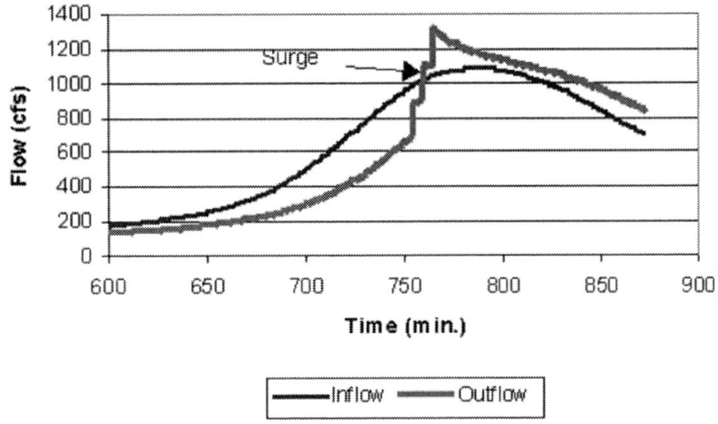

Figure 1. "Typical" gate operation model results – 2 year flood

Using the spreadsheet tool, an operating scheme was developed that would provide adequate hydraulic capacity while reducing the impacts of gate operations on downstream properties. As designed, operation (deflation) of the bladder will initiate for a storm between the 1- and 2-year events and would be fully deflated for a storm corresponding to roughly the estimated 100-year flood.

The replacement of the antiquated steel lift gates, piers, and superstructure with an inflatable bladder crest gate with an automated operating system successfully reduced risks related to inability to operate the gates during floods. The original gates required regular maintenance to ensure operability and manual operation was difficult during large floods. In addition, the replacement system is more suitable for passing debris, further reducing hydrologic risks. The automated gate operating system was programmed to consider site specific hydrologic and hydraulic conditions, to reduce the potential for sudden releases from the dam. The new gate system has performed without notable issues for nearly fifteen years.

Lake Townsend Dam

Lake Townsend provides the primary water supply for the City of Greensboro, North Carolina. The dam was constructed in 1967 and consisted of a 25-ft high concrete spillway flanked by earth embankments. The spillway was an ogee-shaped weir divided into nine, 25-ft wide bays and one 15-ft wide bay, with a total spillway width of 276-ft. Ten foot high steel vertical roller gates were located in each of bays and floods were passed by operating the gates as needed to maintain the reservoir level near the normal pool (top of gates).

By the late 1970s, cracking of the concrete spillway was observed and was eventually revealed to be the result of alkali-silica reactivity (ASR). Not only was the ASR a concern with regard to the integrity of the concrete structure, it presented potential issues related to gate operations, due to expansion and deterioration of the piers on which the gates were mounted. The staff at the dam reported vibration of the gates when they are overtopped by more than a few inches but also had concerns with vibration when the gates were operated and potential damage to the deteriorated piers. Based on these concerns, operations staff installed stops on the gates to prevent opening of the gates more than two feet.

In addition to the ASR concerns, the spillway had inadequate capacity to pass the design flood required by North Carolina dam safety regulations, even if the gates could be fully opened to allow unobstructed weir flow over the ogee.

Based on the structural and hydraulic concerns, the concrete gated spillway was replaced with a 7-cycle labyrinth spillway; two 20-ft high cycles were set with the crest at normal pool and the remaining five cycles are 21-ft high, with crest one foot above normal pool. The spillway has a total width of 307-ft, which includes a single bay with a low level and skimmer gate. The labyrinth spillway has capacity to pass about 80,000 cfs with the reservoir at top of dam, which is similar to the capacity of the original gated spillway, assuming full gate operation. Weir submergence from tailwater made it impractical to design the spillway to pass the design flood. Therefore, the embankments were armored to allow overtopping for extreme floods (Paxson et al, 2008).

A comparison of the discharge rating curves for the original gated spillway and the labyrinth weir is presented in Figure 2. The rating curve for the gated spillway assumed that the gates can be fully opened and that gates are operated as needed to maintain the reservoir within about three inches of the normal pool. As previously noted, the gated and labyrinth spillways have similar hypothetical discharge capacity with the pool at top of dam (EL 725.5); however, if operated as shown, outflows for a given storm event would be higher for the gated spillway. For the fixed crest labyrinth spillway, the pool levels would be higher for a given inflow, resulting in more flood attenuation. Although the dam was not designed to provide flood protection, this is considered an ancillary benefit of the new labyrinth spillway. It should be noted that there are no structures around Lake Townsend with elevations below the top of dam.

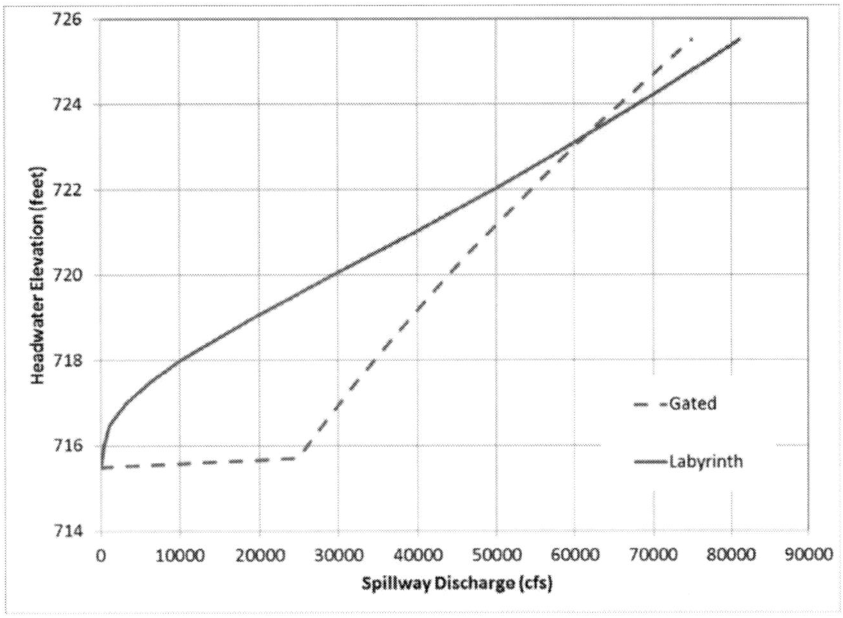

Figure 2. Lake Townsend discharge rating curves

For the Lake Townsend project, complete replacement of the gated spillway with a fixed crest labyrinth (Photo 7) reduced operational and hydrologic risks of the project. The need for complete replacement due to structural deterioration and concerns also made this a cost effective approach.

Photo 7. Lake Townsend Dam: Labyrinth spillway constructed downstream of original gated spillway

Linville Land Harbor Dam

Linville Land Harbor Dam, located in western North Carolina, is a high hazard dam owned and operated by the Linville Land Harbor Property Owners Association (POA). The dam was originally constructed in 1927, destroyed by floods in 1940, and was rebuilt in 1971. The concrete primary spillway, located at the left abutment, included five

Tainter gates and a 10-ft long fixed crest weir (Photo 8). Two gates were 16-ft wide and 12.1-ft high and three gates were 25-ft wide and 16.5-ft high. A concrete auxiliary spillway was located at the right abutment, with the crest set just above the normal pool.

Photo 8. Linville Land Harbor Dam: original spillways

Based on dam safety concerns related to significant deterioration of the spillway concrete (ASR) and gates, the State of North Carolina ordered the POA to lower the lake by about one foot until the structure was replaced or repaired. In addition, rigorous notification protocols were implemented so maintenance personnel could raise the gates in advance of floods to reduce stresses on the structure and provide hydraulic capacity.

Based on the significant deterioration of the structures, a complete replacement of the spillways was the preferred alternative. A fixed crest labyrinth spillway was found to be a cost effective replacement, eliminating the operation and maintenance issues related to the gates. A four cycle labyrinth weir (Photo 9) with a total width of 125-ft (included a single bay for low flow releases and drawdown) was found to have adequate discharge capacity to pass the design flood (about 25,000 cfs) without overtopping the embankment. If the original gated spillway could be operated such that all gates could be fully opened, the discharge capacity with the reservoir at top of dam would have been about 47,000 cfs, which is more than the required spillway design flood. However, based on past experience, it is unlikely that all of the gates could have been fully opened. The original auxiliary spillway had an estimated maximum capacity of about 8,000 cfs and if the gates could not be operated at all, the total capacity (flow overtopping gates plus auxiliary spillway) would have been about 16,500 cfs. However, there is a high probability that the gates would fail structurally under the high loads resulting from overtopping flow.

Photo 9. Linville Land Harbor Dam: labyrinth replacement spillway

Similar to the Lake Townsend project, the complete replacement of the gated spillway with a fixed crest labyrinth weir met regulatory criteria, reduced hydrologic and operational risks, and eliminated operation and maintenance costs related to the gates.

HYPOTHETICAL REPLACEMENT EXAMPLE

To illustrate potential reduction in risks through replacement of a gated spillway with a fixed crest spillway, a hypothetical example was developed. The hypothetical dam is a flood control structure located on a 250 square mile watershed. The dam includes an outlet works structure for smaller releases and a gated overflow spillway. Properties of the dam are presented in Table 1.

Table 1. Hypothetical dam - properties

Normal pool elevation	1000 ft
Gated spillway fixed crest elevation	1020 ft
Top of gates elevation	1060 ft
Top of dam elevation	1063 ft
Normal pool storage	44,000 acre-ft
Top of dam storage	137,400 acre-ft
Spillway gate width	40 ft
Number of gates	5
Total effective spillway width	200 ft

A hypothetical gate operations plan was developed and used to develop a discharge rating curve. The operations plan was designed to reduce outflows for more "frequent" storms but still allow passage of the PMF with three feet of freeboard (i.e. PMF peak stage EL 1060). HEC-HMS was used to perform rainfall and runoff calculations and route flows through the reservoir and dam. This is considered a simplified model since dynamic gate operations were not included in the model. Results of the analysis for several storm events are presented in Table 2.

Table 2. Hypothetical results – gates operated per design

Storm	Inflow (cfs)	Outflow (cfs)	Peak Stage (EL, ft)
10-year	14,660	624	1026.0
50-year	24,870	4,947	1032.2
100-year	30,030	7,306	1035.0
500-year	44,610	22,110	1037.5
1000-year	52,260	29,960	1038.8
PMF	174,200	153,300	1060.0

To evaluate potential risks resulting from mis-operation of the gates or possible inability to operate gate(s), two additional operational plans were considered and discharge rating curves developed. The first considered gate operations sooner than recommended (i.e. gates were opened when the reservoir reached a lower elevation). The results of this model are presented in Table 3.

Table 3. Hypothetical results – gates operated too early in storm*

Storm	Outflow (cfs)	Peak Stage (EL, ft)
10-year	2,432	1022.1
50-year	8,185	1026.8
100-year	11,450	1028.9
500-year	22,450	1033.7
1000-year	28,540	1036.2
PMF	152,600	1059.9

* Inflow same as Table 2

As expected, the outflows for the smaller storms are higher than if the gates are operated as designed (Table 4). Outflow for the 100-year flood is 57 percent higher than for the "as designed" case.

For the second case, the rating curve was developed to simulate one inoperable gate and opening of the other gates as needed to pass the flows. As anticipated, for this example the gated spillway had inadequate capacity to pass the PMF, resulting in about one foot of overtopping of the dam.

To address the potential risks related to gate operations, a labyrinth spillway was sized to replace the gates and provide similar flood attenuation, but still have capacity to pass the PMF with three feet of freeboard. To meet these requirements, a rating curve was developed for a three cycle labyrinth with a weir height of 30 feet, crest EL 1040, and a total spillway width of 216 feet. This spillway was modeled within HEC-HMS; results for the various storms are presented in Table 4.

Table 4. Hypothetical results – labyrinth spillway

Storm	Outflow (cfs)	Peak Stage (EL, ft)
10-year	50	1027.3
50-year	1,888	1040.2
100-year	6,549	1040.8
500-year	19,780	1042.5
1000-year	26,970	1043.5
PMF	161,200	1059.3

* Inflow same as Table 2

As shown, the labyrinth spillway results in similar or lower outflows for the return period storms but has adequate capacity to pass the PMF with the required freeboard. This would appear to be a reliable method for replacing the gated spillway.

The results for each of the scenarios are compared in Figures 3 and 4.

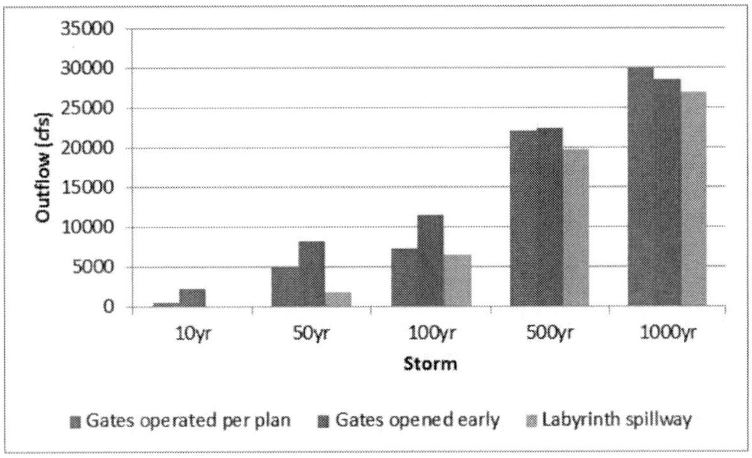

Figure 3. Hypothetical results – peak outflow

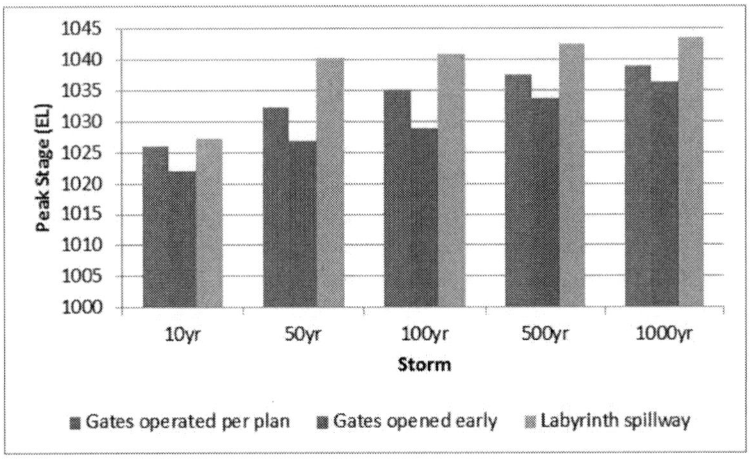

Figure 4. Hypothetical results – peak stage

Considering gate operations is considered critical to risk analysis. Figure 3 shows that operating the gates earlier than described in the plan would result in potential increased damages and loss of life for the smaller (higher probability) storm events (e.g. outflow for the 10- and 100-year events are 275 and 57 percent higher, respectively, than if the gates are operated per plan). The use of a fixed crest spillway to replace the gated structure would essentially eliminate the risks related to improper operation. However, for an embankment dam, it is possible that the labyrinth spillway could actually increase risks for geotechnical failure modes (e.g. internal erosion), since the higher peak stage would result in higher loadings for the more frequent (probable) events.

To be able to adequately consider risks, the probability of gate misoperation would need to be included in the event tree. Estimating the probability and type of misoperation presents a challenge in the risk analysis. Some insight with regard to this challenge could be gleaned through real-time emergency response testing with operations personnel.

It is important to note that this example is for a flood control dam, where even the misoperation of gates does not increase downstream consequences compared to the conditions before the dam was constructed. Another example could be evaluated for a gated spillway at a dam where flood control is not considered. For many of these structures, the gates are operated to maintain a given reservoir level through the storm (e.g. outflow = inflow). In this example, a misoperation of the gates could result in a significant increase in downstream damage over what would have occurred prior to the construction and operation of the dam.

CONCLUSIONS

Gated spillway systems present challenges to dam owners and if improperly maintained, can contribute to a dam incident or failure. In addition, the operation of gates needs to balance considerations related to downstream flooding resulting from operations and the potential for reduced spillway capacity if gates are not operated, which could lead to a dam failure.

Regular maintenance, vigilant inspections and testing of operations, and training of staff will significantly reduce the risks related to spillway gates presented herein. Development of a realistic and easy to follow flood operations plan is also critical.

Control systems also present operational risks. While manual control at the site is the most common means for operation, it also presents the highest risks (USSD, 2002). Automating the controls will reduce these risks; however, the owner needs to consider the potential for issues related to automated controls where the owner is not on site to observe operations. Remote automated controls should be combined with alarms that notify the owner of operations.

In cases where significant dam upgrades are required to address spillway structural, hydraulic, or other deficiencies, it may be possible to replace the gated spillway with a fixed crest weir without significantly increase project cost. Labyrinth and piano key weirs provide a potential solution to achieve desired hydraulic performance. This solution has been used successfully for several projects. While complete replacement of the gated spillway would eliminate the risks related to gate operations, it may not be practical for all projects, particularly where gates are essential to the project purpose (e.g. hydropower or flood control) or the required spillway capacity is so large that fixed crest spillway cannot meet the requirements. In these cases, one solution is to supplement or replace a portion of the gated spillway with a labyrinth or piano key weir. To reduce reliance on gate operation for more frequent floods, the fixed crest weir crest could be set such that this passive spillway flows before gate operations are needed. This reduces the risk related to increased downstream flooding (even if only perceived) from misoperation and also would provide more time for the dam owner to reach the site to operate gates during larger storms. It also provides more time for decision making during a flood event.

ACKNOWLEDGEMENTS

The authors would like to acknowledge the owners of the projects presented as case histories herein.

REFERENCES

American Society of Civil Engineers (ASCE) (2012). "Water Control Gates-Guidelines for Inspection and Evaluation". ASCE, Reston, VA.

ASDSO/EPRI (2000). "Association of State Dam Safety Officials (ASDSO)/ EPRI Spillway Gate Workshop: January5 & 6, 2000". EPRI, Palo Alto, CA, and the Association of State Dam Safety Officials, Lexington,KY.

Augenstein, Seth (2012). "Pompton Lakes dam served its purpose during Hurricane Irene, report finds". The Star Ledger, http://www.nj.com (April 12, 2012).

Buchele, Mose (2013). "Near-Catastrophe During Flooding Highlights Issues at Dam in Austin". https://stateimpact.npr.org (December 10, 2013).

Campbell, D., Harrison, J., and Kent, C. (1999). "Upgrading of Sugar Hollow Dam Is Enhanced by Team Approach," Dam Safety 1999, the 16th annual conference of the Association of State Dam Safety Officials, held Oct. 10-13, 2000 in St. Louis, Missouri.

Fiedler, W., King, W., Schwanz, N., Garton J., and McDaniel L. (2011). "Dam Safety: What Happened to Lake Delhi Dam?". Hydroworld, 30 (September), 8-15.

Graham, W. J., and Hilldale, R.C. (2001). "Spillway Gate Failure or Misoperation: Representative Case Histories". Report DSO-01-01, Bureau of Reclamation, Denver, CO.

Houston, Kathleen L. (1982). "Hydraulic Model Study of Ute Dam Labyrinth Spillway". Report GR-82-7. Bureau of Reclamation Report, Denver, CO.

Hydroplus (2014a). "Canton Dam". Information sheet provide to authors by email from Hasan Kocahan, Hydroplus representative, December 2014

Hydroplus (2014b). "Jindabayne Dam". Information sheet provide to authors by email from Hasan Kocahan, Hydroplus representative, December 2014

Johnson, Jerry N. (2014). "Duckett Dam was operated responsibly during heavy rainstorm [Commentary]". Baltimore Sun, http://baltimoresun.com (September 9, 2014).

Laugier, F., Vermeulen, J. and Lefebvre, V. (2014). "Overview of Piano Key Weirs experience developed at EDF during the past few years". Labyrinth and Piano Key Weirs II, PKW 2013, CRC Press, Taylor and Francis Group, Copyright 2014

Miller, A.C., Moyle, J.H., Ray, R., Cahill, B. and Mak, M. (2013). "Downstream Effects Resulting from the Operation of the Pompton Lake Dam Floodgates". ASDSO Journal of Dam Safety, (Vol.11, Iss.3, Art.2)

Minnesota DNR (2014). "New London Dam Reconstruction". http://www.dnr.state.mn (October 14, 2014).

National Performance of Dams Program (2014) query tool. http://ce-npdp-serv2.stanford.edu/DamDirectory/DamIncidentQuery/IncidentForm.jsp, Stanford University, Stanford, CA.

NDTV (2014). "Inquiry Report Blames Hydropower Project Authorities for Beas River Tragedy." www.ndtv.com (June 20, 2014)

O'Leary, Josh (2014). "Residents call for new dam outflow schedule". Iowa City Press-Citizen, http://www.press-citizen.com (September 15, 2014).

Paxson G., Harrison, J., and Campbell, D. (1999). "Hydraulic Assessment of Rubber Dam Spillway Crest Control," Dam Safety 1999, the 16th annual conference of the Association of State Dam Safety Officials, held Oct. 10-13, 2000 in St. Louis, Missouri.

Paxson, G., Crookston, B., Savage, B., Tullis, B., and Lux III, F. (2008). "The Hydraulic Design Toolbox: Theory and Modeling for the Lake Townsend Spillway Replacement Project." Dam Safety 2008, the 25th annual conference of the Association of State Dam Safety Officials, held Sept. 7-11, 2008 in Indian Wells, California.

Pichaske, Pete (2014). "WSSC under fire after another major flooding of Laurel". Baltimore Sun, http://baltimoresun.com (June 19, 2014).

Shaver, Katherine (2014). "Laurel officials, businesses say WSSC should pay for Duckett Dam's flood damage". Washington Post, http://www.washingtonpost.com (September 8, 2014).

United States Society on Dams (USSD) (2002). "Improving Reliability of Spillway Gates". USSD, Denver, CO.

U.S. Army Corp of Engineers (1990). "Hydraulic Design of Spillways". Engineering Manual 1110-2-1603, Washington, DC.

Vasquez, Victor M., Wolfhope, J.S. and Garrett, R. (2009). "An Elegant Solution: The Lake Brazos Labyrinth Weir". Civil Engineering Magazine, (Vol. 7, Iss.1), 48-55.

DEBRIS AND MAINTENANCE OF LABYRINTH SPILLWAYS

Brian M. Crookston, Ph.D., P.E.[1]
Dustin Mortensen, P.E.[2]
Tina Stanard, P.E.[2]
Blake P. Tullis, Ph.D.[3]
Victor M. Vasquez, P.E.[2]

ABSTRACT

Labyrinth spillways are passive control structures that are commonly implemented in dam projects (i.e., new or rehabilitation) due to their hydraulic performance, flexible design, and cost effectiveness.

The accumulation of sediment or floating debris can be problematic at spillways; sediment deposition or log jams can require cleaning and removal at regular intervals to eliminate flow impediments and maintain spillway performance. If neglected, debris or sediment could potentially reduce flow capacity and increase upstream flooding. Floating debris may be more likely to collect upstream of labyrinth weirs than linear weirs due to the reduction in flow depth and flow momentum per unit weir length.

This paper presents performance data and general information on the effects of debris and sediment on labyrinth spillways, including survey results of 65 labyrinth spillways in the USA and Portugal. In general, labyrinth weirs were found to be self-cleaning. In some instances, labyrinth weirs that replaced gated structures greatly improved existing maintenance issues with woody debris. Information is also presented regarding a labyrinth spillway with unique sediment and woody debris challenges.

The paper also presents the results of a hydraulic model study performed at the Utah Water Research Laboratory that investigated the impact of upstream sediment deposits on labyrinth hydraulic performance. In addition, a dam owners' experience with sedimentation at a labyrinth weir is discussed.

Finally, the effects of driftwood on a unique labyrinth weir geometry, called a piano key weir (PKW), are also presented. Physical modeling was performed at the Laboratory of Hydraulic Constructions (LCH) located in Lausanne, Switzerland. The results include observations regarding probabilities of debris 'passage' as a function of trunk size and flow depths. The impacts of debris accumulation on the discharge coefficient are also presented, as well as observations related to driftwood collection.

[1] Schnabel Engineering, Inc., 1380 Wilmington Pike Suite 100, West Chester, Pennsylvania, 19382, (610) 696-6066, bcrookston@gmail.com
[2] Freese and Nichols, Inc., 10431 Morado Circle Suite 300, Austin, Texas ,78759, (512) 617-3100, dgm@freese.com, ces@freese.com, vmv@freese.com
[3] Utah Water Research Laboratory, Utah State University, 8200 Old Main Hill, Logan, Utah, 84322-8200, (435) 797-3194, blake.tullis@usu.edu

INTRODUCTION

Spillway Deficiencies and Issues

A spillway is a safety device that may serve a variety of functions, such as: flood control, regulating the upstream water surface elevation, emergency drawdown, and an ecological bridge (e.g., fish passage). Spillways have been classified based upon type of control structure (active, passive, etc.), function (service, auxiliary, emergency), and according to geometry-specific details (e.g., ogee, side channel, shaft, siphon, stepped, labyrinth, etc.) (Khatsuria 2005). In many instances these hydraulic structures provide necessary support for communities throughout the globe.

However, no spillway is ageless. Construction of a new spillway or the rehabilitation of an existing spillway can be motivated by dam safety and owner concerns related to hazard creep, structural deterioration or damage, hydrologic revisions, or inadequate performance. Inadequate discharge capacity is a common spillway deficiency. Also, a cumbersome and expensive maintenance schedule may be required to ensure that a mechanical spillway successfully operates when needed. In some instances, spillway maintenance issues have justified the modification or even replacement of a spillway.

Labyrinth and Piano Key Weirs

A labyrinth weir (Crookston and Tullis 2013a) is a passive-control spillway with a unique geometric shape (Fig. 1). The non-linear nature of labyrinth weirs (folded back and forth in plan view) allows for significantly longer weir lengths than could be achieved using a traditional linear weir for a given spillway width. These spillways are used in a variety of applications (e.g., reservoirs, run-of-river). For example, a labyrinth spillway may be the only spillway operating during a storm, with the crest set at the normal pool elevation. Or, it may serve as an emergency spillway adjacent to spillway gates and engage near the peak of the inflow hydrograph. Labyrinth spillways are particularly well suited for instances where increased spillway capacity is needed or spillway footprint is restricted (Crookston et al. 2013). For a given spillway width, the water depth at the crest of a labyrinth will be less at a given discharge than for other linear free overfall structures due to the additional crest length (relative to a linear weir).

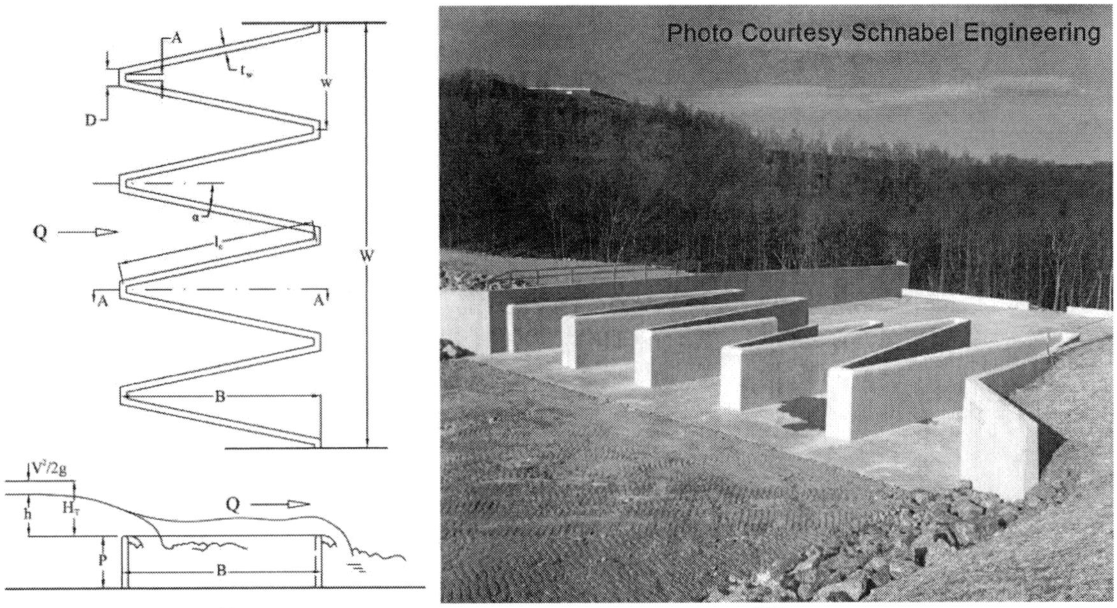

(A) (B)

Figure 1. Labyrinth weir geometric parameters (A) and a photo of a prototype example.

The U.S. Bureau of Reclamation constructed the first labyrinth weir in the USA at East Park Dam in California in 1910 (a 9-cycle, curved labyrinth weir). However, research and construction of additional labyrinth spillways did not begin in earnest until the 1970's. Following the publication of the Tullis et al. 1995 systematic labyrinth weir design method, an escalation in labyrinth weir construction occurred. It is estimated that about 100 labyrinth weirs exist today in the United States. Furthermore, there is a current annual trend of 10 or more labyrinth spillways in planning or design in the USA. More detailed information regarding the history of labyrinth weirs, published literature, and hydraulic design methods are presented in Crookston (2010), Crookston and Tullis (2013a,b), Crookston et al. (2013), and Falvey (2003).

A relatively new type of labyrinth weir is the piano key weir (PKW) (Fig. 2); the traditional labyrinth design was modified to improve hydraulic performance and decrease construction cost. The PKW is a rectangular labyrinth weir (in plan form) with cantilevered apexes and ramped floors; this geometry increases the allowable weir length for restricted footprint applications (e.g., top-of dam) (EDF-LNHE and University of Biskra, Algeria). A great deal of research has been performed regarding PKW hydraulic performance [Erpicum et al. (2011, 2013); Anderson and Tullis (2012a,b), (2013), Dabling and Tullis (2012)], and a number of structures have been constructed in reservoir or channel applications in Europe, Asia, and Australia.

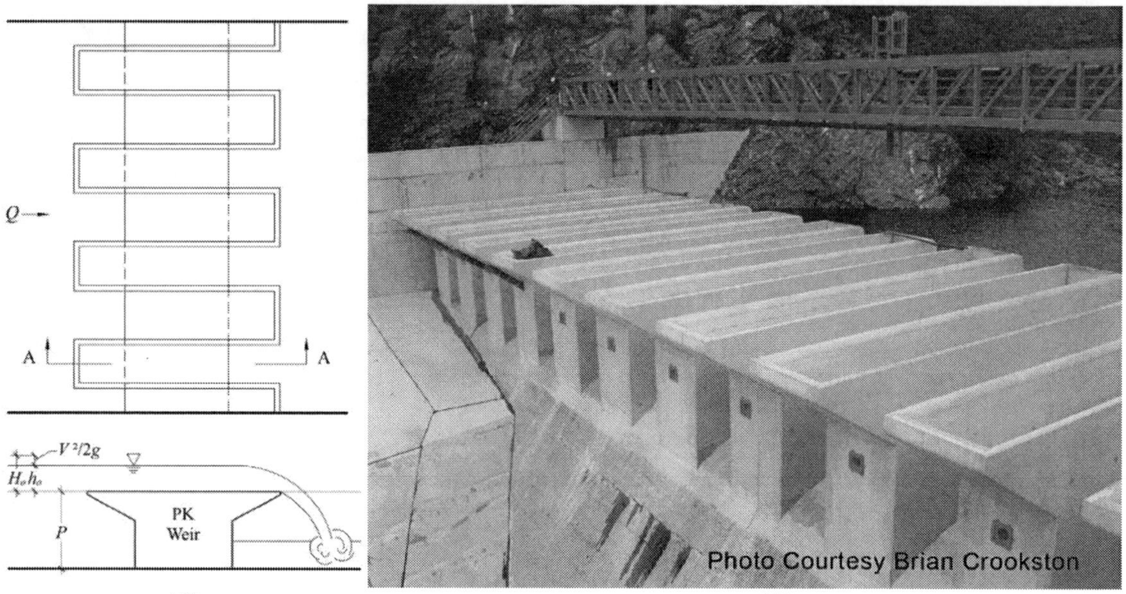

(5) (B)

Figure 2. PKW geometric detail and a photo of a PKW prototype structure.

The accumulation of sediment or floating debris can be problematic at spillways, particularly if there is a flow impediment. Weirs generally have a greater ability to pass floating debris than active-control spillways such as gates (Khatsuria 2005). However, due to the geometry of labyrinth spillways, branches and trees from forested upland areas can be transported to a spillway during flood events and form a blockage, which could potentially reduce spillway discharge capacity and increase upstream flooding. Furthermore, floating woody debris may be more likely to collect upstream of and atop labyrinth weirs than linear weirs due to the reduction in flow depth and flow momentum per unit weir length.

The potential for debris blockages and required maintenance to remove sediment or debris would be of interest to designers, operators, regulators, and owners. A few case studies have been conducted specific to labyrinth type weirs (PKW) and driftwood [Ouamane and Lempérièr (2006), Laugier (2007), Sogreah (2011)]. In addition, effects of sediment on labyrinth discharge capacity were investigated for the Boardman Labyrinth Spillway (Babb, 1976), and the Garland Canal, yet little field data and design information is available regarding the performance of these spillways with regards to floating debris and sediment.

This paper presents information on floating debris at labyrinth spillways, including a case study where sedimentation and debris were of concern. This information comes from the authors' experiences with specific projects, and information gathered from USA dam safety officials and colleagues in Europe. This paper also includes a summary of the results of recent laboratory studies conducted in the USA on labyrinth weirs and in Switzerland regarding the influence of woody debris on the discharge capacity of PKWs.

STUDY OVERVIEW

Labyrinth Spillways

Numerous owners and dam safety officials were contacted to compile their experiences regarding labyrinth weirs, woody debris, and maintenance. The locations of the labyrinth spillways in this dataset are presented in Fig. 3. The surveyed labyrinth spillways are subdivided into structures designed by Schnabel Engineering, Freese and Nichols, and by others. Fig. 3 does not present all labyrinth spillways in the USA nor surveyed labyrinths located in Portugal.

Figure 3. Labyrinth weir locations in the Unites States Surveyed for this current study.

The current study research focused on the following points:

- Occurrence of debris and type at labyrinth spillways
- How labyrinth spillways handle debris (self-cleaning, maintenance required, etc.)
- Any specific issues with blockages, log jams, etc.
- Typical maintenance performed
- For rehabilitation projects, changes in debris handling and related issues before and after the labyrinth spillway construction
- Considerations regarding potential debris or sediment issues during design
- Any other issues or observations

Debris and Maintenance of Labyrinth Spillways

Laboratory Investigations

Extensive physical modeling was performed during the design of the Lake Brazos labyrinth spillway. Sectional and full-width models of the dam were tested at the Utah Water Research Laboratory (Utah State University) in Logan, Utah. The sectional models, which featured a variety of weir geometries (wall angle, wall width, cycle width, cycle depth, apex shape, and crest shape), were tested at a ~1:10 scale (Froude similitude). The effects of tailwater submergence on discharge capacity were also explored. The effects of sediment build up on labyrinth weir discharge characteristics were also evaluated; the results are discussed later in this paper.

The full-width model (1:15 scale) was used to confirm hydraulic results from the sectional model and assist in evaluating the hydraulic requirements immediately upstream and downstream of the labyrinth weir. One use of the full-width model was to evaluate the ability of the labyrinth weir to pass anticipated sediment and debris. Debris was modeled with sagebrush and fabricated trees with dimensions and size distributions based on site-specific field observations of 'naturally occurring' floating debris. In general, the branching and volume were conservatively selected to evaluate extreme conditions and impacts on discharge capacity. Because of downstream controls, the labyrinth weir operated under submerged conditions at the larger design flows.

Physical modeling of three PKW geometries was conducted at the Laboratory of Hydraulic Constructions of the École Polytechique Fédérale de Lausanne (EPFL) in Lausanne, Switzerland. The principal investigators included Dr. Michael Pfister, Prof. Dr. Anton Schleiss (Director), and Prof. Dr. Blake Tullis. The 1:30 scale PKW tested featured weir heights ranging from 3 to 6 inches. A photograph of one of the models is presented in Fig. 4. The range of tested headwater ratios (upstream total head H to weir height P) were $0.1 \leq H/P \leq 0.5$.

Figure 4. Photo of one PKW geometry evaluated by Pfister et al. (2013).

Driftwood element dimensions and size distributions were selected based upon field data presented by Bezzola and Hegg (2007) to represent 'naturally occurring' floating debris. 10 driftwood test groups, comprised of up to 22 elements, were prepared and released sequentially upstream of the spillway. The volume of wood and the impact of each driftwood group on H were documented, resulting in a data set of 120 points.

RESULTS

Labyrinth Weir Debris Handling Performance

The size and nature of the woody debris and the non-linear weir geometry (e.g., labyrinth or PKWs) influence the spillway's potential as a debris collector. The debris handling performance of 75 labyrinth spillways is presented in Table 1 with key observations summarized as follows:

- Labyrinth spillways are generally self-cleaning.
- Large tree logs or stumps may require removal.
- Labyrinth weirs that replaced existing spillways were observed to significantly improve debris handling (5 structures).
- Field observations regarding debris loads or frequency of spillway operation are incomplete. It can be difficult to observe the debris handling performance of these spillways due to location and accessibility, as well as the infrequency of events generating large debris loads.

In general, surveyed owners expressed satisfaction with the debris handling characteristics of their labyrinth weirs and very little maintenance has been required. Photographs of several examples of labyrinth spillways with typical debris conditions are presented in Fig. 5. According to the survey responses, infrequent removal of large woody debris elements was only required for less than 7% of the labyrinth spillways. Management of large woody debris elements at Dog River dam was reported to simply involve pushing the occasional stump or log off the crest into the downstream channel when it became an aesthetic nuisance. Tongue River and Middle Creek have debris booms upstream to avoid debris collection/removal at the structure. Only one labyrinth, Lake Brazos Dam (unique debris conditions), reported significant debris loading and accumulation under normal conditions. The dam owner reported, however, that the labyrinth weir debris collection and management along with other maintenance issues were less problematic that with the gated spillway structure it replaced. The debris at Lake Brazos Dam is discussed in the results section.

Lake Townsend Dam and Linville Dam labyrinth spillways also replaced gated structures that required frequent maintenance to remove debris (Linville Dam shown in Fig. 6). Significant improvements have been reported, with required maintenance greatly reduced. When asked, the Water Supply Division Manager said the following about Lake Townsend, "It's been a dream. We have noticed the design does not allow much debris to build up – very little maintenance."

For some applications, where the labyrinth crest (auxiliary or emergency spillway) is located at an elevation above typical operating pool levels, a few owners expressed an eagerness for an occasional rainstorm to 'flush' the debris that accumulates at the labyrinth spillway during base flow conditions. One owner, who maintains normal pool at the crest of the labyrinth with little to no discharge passing over, reported an occasional jumping fish and/or climbing turtle that climb/jump over the crest and land on the dry, downstream side of the labyrinth. The resulting fish smell did impact park aesthetics near the structure.

Table 1. Labyrinth Weir Debris Handling Performance

Dam	Location	Cycles	Spilling Frequency	Debris Handling
Accord Pond Dam	MA	2		No issues
Agua Branca	Portugal	2		No issues
Alfaiates	Portugal	1		No issues
Alijó	Portugal	1		No issues
Alloway Lake Dam	NJ	5		No issues
Antelope Creek Channel	NE	3		No issues
Arcossó	Portugal	1		No issues
Calde	Portugal	1		No issues
Capital City Country Club Dam	GA	1		No issues
Concourse Lake Dam	VA	2		No issues
Cloe Dam	PA	7		No issues
Crystal Bridges (2 Labyrinths)	AR	2	Frequently	No issues
Dog River – raised	GA	8		Minor maintenance
DRA Detention Structure	GA	4		No issues
Elmendorf Lake Dam	TX	6	Frequently	No issues
Fort Miller Dam	NY	45		No issues
Glen Park Dam	NY	25	Frequently	No issues
Greystone Development Dam	GA	1		No issues
Hard Labor Creek Dam	GA	4		No issues
Hollis G Lathem Reservoir Dam	GA	3		No issues
Huntington Hills Dam	SC	3		No issues
Indian Run Dam	PA	2.5		No issues
Kauffman Dam	PA	5		No issues
Kings Falls Dam	NY	6		No issues
Lake Brazos Dam	TX	24	Constantly	Significant maintenance
Lake Natalie Dam	GA	4		No issues
Lake Paupacken Dam	PA	2		No issues
Lake Sovereign Dam	GA	2		No issues
Lake Townsend Dam	NC	7		No issues
Lake Upchurch Dam	NC	6		No issues
Leaser Lake Dam	PA	3		No issues
Linville Land Harbor Dam	NC	4		No issues
Little Blue Run Dam	PA	10		No issues
Lyman Run Dam	PA	8		No issues
Meacham Grove Stormwater Facility	IL	4	Frequently	Minor maintenance
Mercer Dam	OR	3	Frequently	No issues
Middle Creek Dam	MT	1	Yearly	No issues
New London Dam	MN	4		Minor maintenance
North Fork of Smith River Dam	MT	2	Yearly	No issues

Dam	Location	Cycles	Spilling Frequency	Debris Handling
Opossum Dam	PA	4		No issues
Pawnee Wetlands	NE	3		No issues
Pelican lake Dam	MN			Minor maintenance
Pine Lake Dam	GA	2		No issues
Pine Lake Dam	NY	3		No issues
Pine Run Dam	PA	4		No issues
Pisão	Portugal	1		No issues
Potato Lake Dam	MN			Minor maintenance
Pye Lake Dam	GA	3		No issues
Quincy Dam	CO	4	Seldom	no debris
Rapp Run Flood Control	PA	6		No issues
Roy F. Varner Reservoir Dam	GA	8		No issues
Ruby River	MT	3	Yearly	No issues
Sam Rayburn Dam	TX	16	Never	n/a
Santa Justa	Portugal	2		No issues
São Domingos	Portugal	2		No issues
Shawnee Dam	PA			No issues
South River No. 29	GA	3		No issues
South River No. 4	GA	2		No issues
Sun Mountain Dam	PA			No issues
Teja	Portugal	1		No issues
Tired Creek Dam	GA	3		No issues
Tongue River Dam	MT	4	Yearly	No issues
Treesdale Impoundment No. 1	PA			No issues
Unicoi State Park Dam	GA	3		No issues
Upper Dam – Rangeley	ME	4		No issues
Upper Norton Dam	VA	3		No issues
Upper Owl Creek Dam	PA	3		No issues
Valentine Mill Pond	NE	2	Frequently	No issues
Wanahoo Dam	NE	3		No issues
Waterford Lake Dam	NC	3		No issues
White Heron Dam	PA			No issues
Whitewater Creek Dam	GA	12		No issues
Yahoola Creek Dam	GA	5		No issues
Yellow River No. 3	GA	2		No issues
Zorinsky Basin No 1	NE			No issues

Figure 5. Examples of floating debris collected at labyrinth spillways
(photos courtesy Schnabel Engineering).

(A)

(B)

Figure 6. Examples of typical debris collections conditions at one reservoir where the gated structure (A) was replaced with a labyrinth spillway (B) (photos courtesy Schnabel Engineering).

Sediment Impacts on Labyrinth Weirs

Labyrinth weir discharge coefficients generally peak at a relatively low head and then decrease as *H/P* increases. Sediment buildup within the upstream cycles of a labyrinth weir would have the effect of reducing the effective wall height (*P*) and therefore increasing the *H/P* ratio, which could potentially decrease spillway discharge. Sedimentation was examined as part of the Lake Brazos Dam design and found to decrease discharge as predicted by the discharge coefficient curves and the H/P changes. None of the surveyed labyrinth weirs reported sediment deposits in the upstream cycles, despite the fact that several of the reservoirs, including Elmendorf and Valentine Mill, reported sedimentation problems in other parts of the reservoir.

As part of the Lake Brazos dam model study, Willmore (2004) conducted 4 model tests to examine the impacts uniform sediment deposition upstream of a labyrinth weir would have on H for a given discharge. The labyrinth weir geometric properties are summarized in Table 2. Sediment deposits were simulated with a horizontal false floor placed on the upstream of the weir (Fig. 7). The false floor elevation was adjusted to reduce the weir height by 20%, 40% and 50%.

Table 2. Sediment effects - labyrinth weir sectional model geometric information

Parameter	Value
α	8°
P	1.0 ft
$L_{c\text{-}cycle}$	11.5 ft
N	2.0
t_w	1.5 in
Crest Shape	Partial Ogee

Figure 7. Sediment simulation test with wall height reduced by 50%

Head-discharge rating curves were developed for each model for $0.1 \leq H/P \leq 0.7$. Total upstream head values for model tests that included the simulated sediment deposits are identified as H_{sed}. In Fig. 8, *Ps* is the reduced weir height due to sediment deposition. The H_{sed} to H ratio is presented in Fig. 8; sediment deposits that reduce the weir height by 20% or less were found to have a minor impact on the discharge rating. However, for sediment deposits that reduced the weir height by 40% and 50%, the increase in *H*, and hence upstream reservoir pool elevation, was more than 10% and 25% for about $0.1 \leq H/P \leq 0.6$. It should be noted that the applicability of these results are limited as the simulated sediment deposits were not actual, erodible sediments. For higher *H/P* ratios, the scour and removal of actual sediments at a labyrinth spillway might be expected if the sediment carrying capacity exceeds sediment inflow. Tests conducted in the laboratory using fine sand suggest the ability of the labyrinth to be self-cleaning; velocity profiling in the laboratory also found that flow velocities in the upstream cycles are relatively uniform, which suggests that the self-cleaning nature of the upstream cycles should also be relatively uniform. Bathymetric surveys of Lake Brazos Dam confirmed the structures' sediment self-cleaning ability.

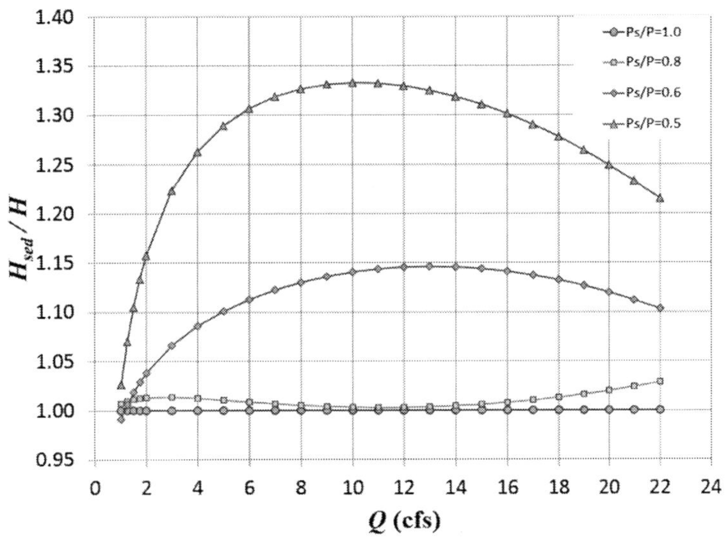

Figure 8. Influence of sediment deposits on 8° labyrinth weir discharge rating curve.

Lake Brazos Case Study

The Lake Brazos Dam was originally constructed as a gated spillway by the City of Waco in 1970. It is a run of the river dam and creates a recreational lake. The original drum gates could not maintain a constant reservoir level due to corrosion and debris issues. The drum gate system was replaced with leaf gates and hydraulic cylinders in 1985, but operational problems associated with debris and the gate operating mechanism persisted and often led to lowered lake levels and unappealing reservoir conditions. The continual problems with the gates led the City to replace the gates with a labyrinth weir.

One of the principle project challenges was to complete the spillway replacement without significantly affecting the upstream lake levels. The challenge led to a significant physical modeling effort that resulted in a highly customized design (Vasquez, 2006). Sediment and debris management represented two significant operation and maintenance design concerns. Since the dam is run-of-river, sediment can reach the dam rather than being dropped out when the river enters the lake. The design team did not know if the labyrinth weir would trap more sediment than the previous gated structure. In addition, the 46 river miles of the Brazos River, upstream of the dam site, carries a large debris load consisting of trees with root balls. Much of the debris was passed through the gated spillway although some of it would get stuck on the adjacent concrete-overlaid auxiliary embankment spillway (Fig. 9). The debris that passed over the gates often damaged the hydraulic cylinders located beneath the gates, which would require a lowering of the reservoir for repairs. As with the gated spillway, sediment and debris also had the potential to cause significant operations and maintenance challenges for the labyrinth weir. Therefore, sediment and debris impacts to the labyrinth were investigated as part of the physical model study.

Figure 9. Typical Debris on original Lake Brazos Dam

The results of the sedimentation modeling was presented previously. The impact of debris was modeled using two types of debris. The first was with model trees fabricated to replicate the type of tree debris seen at the site. The second method to model debris was with sagebrush (Fig. 10). The sagebrush increased the lake level by 3-inches for the 1-year storm event and by 1-foot for the 10-year event (42,000 cfs). By 58,000 cfs, the debris washed over the labyrinth. The fabricated trees had approximately the same impact on the lake level as the sagebrush and were washed over at approximately 35,000 cfs, a little less than the 5-year event (see Fig. 10).

Figure 10. Debris tests on the full-width model

After completion of the Lake Brazos dam labyrinth weir project in 2007, the City has monitored the dam for debris and sedimentation. The spillway passes an average daily flow of 1,000 cfs and debris collection at the dam is common (Fig. 11). The largest flood to date since the completion of the labyrinth weir is 30,000 cfs (< 5-year event), which was not large enough to cause the debris to pass over the weir.

The City has removed debris seven times (~ once per year on average) since the end of construction, resulting in a total expenditure of ~$300,000 (current cost is about $25k per cleanup). Since 2012, the City has retained a contractor for on-call services to assist with the debris removal on an as-needed basis. Prior to 2012, the removal was performed by the City and contracting or schedule difficulties resulted in the debris accumulating upstream of the dam for several months before it was removed.

The debris may have increased the river level slightly during storm events. There is a USGS flow gage downstream of the dam but not a reliable upstream lake level gage. The City has a lake level gage at the dam but operation is inconsistent and the accuracy is unknown. Data from this lake level gage is shown in Fig. 12 along with the design rating curves.

Figure 11. Debris on Lake Brazos Labyrinth Weir

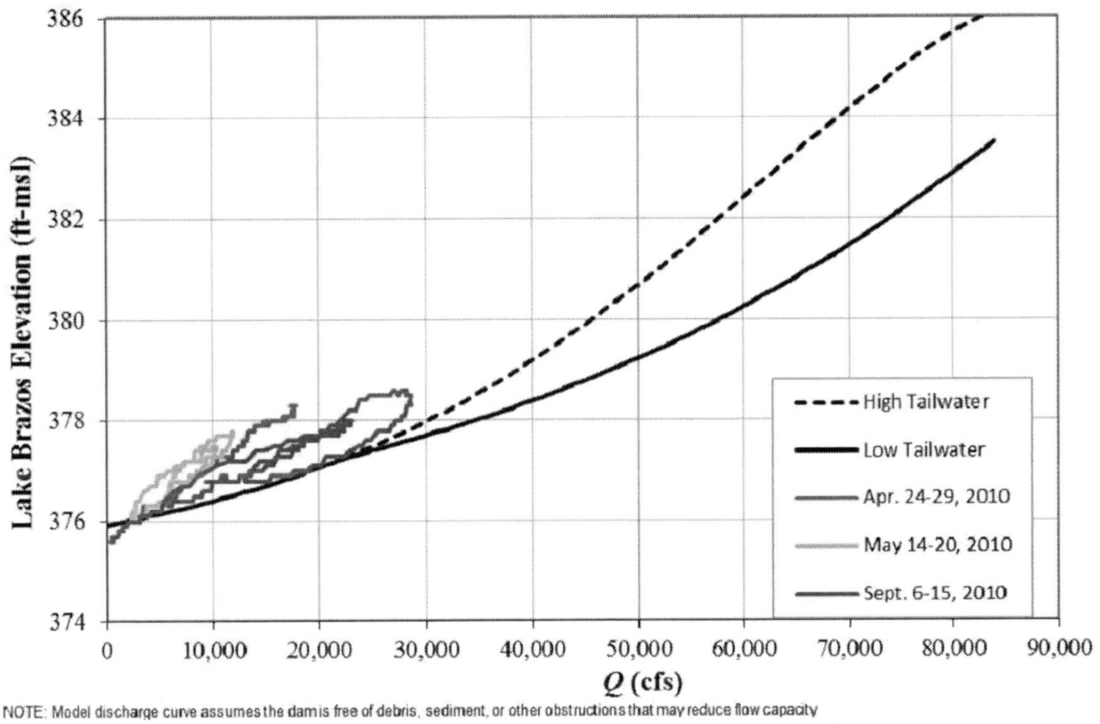

Figure 12. Lake Brazos Dam Rating Curve with Historical Performance

A bathymetric survey of Lake Brazos was performed in 2003 during the labyrinth weir design phase. Surveys were also performed in 2007 after the labyrinth was completed and five years later in 2013. The City removed floating debris prior to the 2013 survey to facilitate surveyor access. The most recent survey found significant debris accumulation on the river bottom upstream of the dam up to nine feet thick. Fig. 13 depicts the difference in elevation (or the thickness of the sediment/debris) from 2007 to 2013. As shown, the cycles were relatively free of sediment and the majority of accumulation is immediately upstream of the cycles along the original river channel. There does not appear to be accumulation inside the cycles; therefore, it is assumed the accumulation consists of woody debris, which floated to the dam and subsequently became waterlogged. The debris accumulation extends approximately 150 feet upstream of the dam.

Figure 13. Accumulated sediment/debris channel bottom layer thickness in Lake Brazos

Debris Blockage Sensitivity of Piano Key Weirs

As previously discussed, a PKW is a modified labyrinth weir with narrow rectangular cycles or keys. The narrow design of the PKW's inlet and outlet keys produce an increased potential for debris collection (e.g., the weir functions similarly to a trash rack for longer woody debris). If the catchment is forested, the potential effects of driftwood accumulation on the PKW head-discharge relation should be considered as part of the hydraulic design (Leite Ribeiro et al. 2012a, b).

Pfister et al. (2013) conducted systematic model testing of the effect of debris (i.e., wooden trunks and rootstock) on the PKW discharge efficiency for three different laboratory-scaled PKW geometries (see Fig. 14 and 15). The first test series determined the blocking probability of individual debris elements from seven different size classes of driftwood trunks; at a 1:30 model scale, the trunk diameters ranged from ~2.5 to 26 inches prototype scale. These tests found that the trunk diameter and upstream head represent the primary dimensional parameters influencing floating debris blockage probability. The second test series more closely approximated real-world conditions by investigating the nature of individual driftwood element accumulation on the PKWs and the effect on discharge efficiency.

For each individual debris batch, the number and sizes of the passing and trapped elements were noted along with the corresponding H. Without removing the debris trapped on the PKW from previous batches, subsequent debris batches were introduced upstream of the flume; this was repeated for all 10 debris batches. The effects of debris accumulation on the upstream head varied with the value of the no-debris upstream head reference conditions. At "lower" upstream head values, the cumulative debris tests produced increases in the debris-associated upstream head of up to 67%; higher upstream reference head values produced upstream head increases limited to approximately 16%.

Fig. 14 shows photographic examples of the debris accumulation tests for one of the PKW geometries tested following the addition of the 10 driftwood batches.

(a) (b)

Fig. 14 Model debris including (a) trunks, and (b) rootstocks

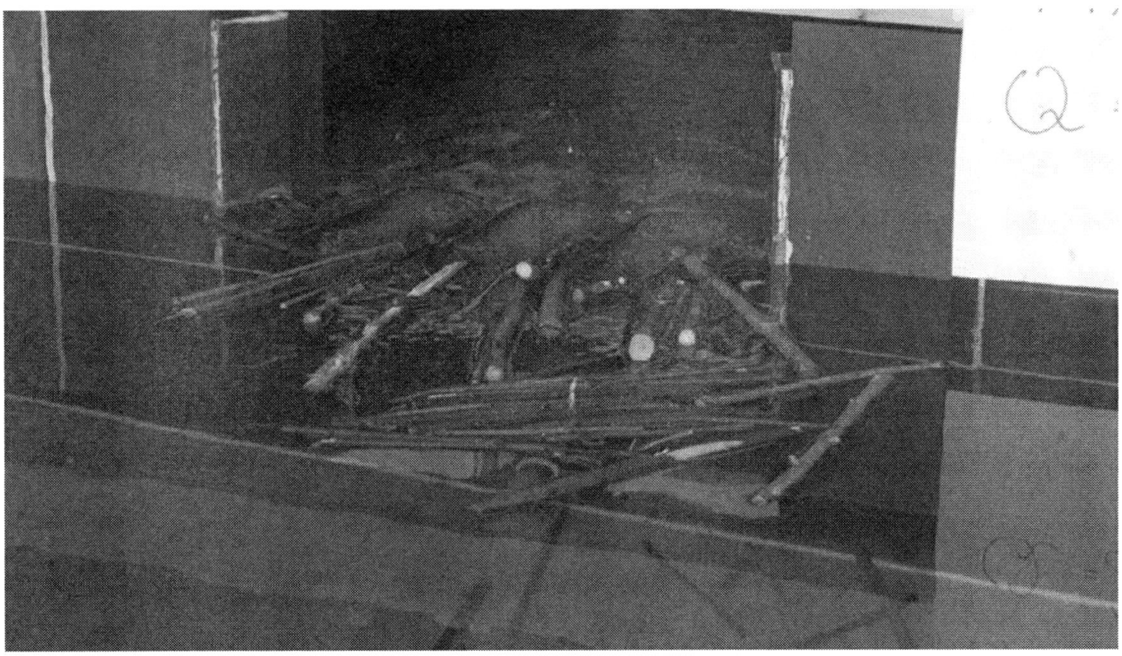

Figure. 15. Example of a blockage configuration following the addition of 10 debris batches.

It was found that despite the blockages, flow continues to pass under the floating debris and through the inlet cycles, with a maximum reduction in C_d of about 25% as a suggested rule of thumb or rough estimate for typical situations. Pfister et al. (2013)

Debris and Maintenance of Labyrinth Spillways

proposed Eq. 1 to estimate the new discharge coefficient impacted by woody debris, with a selection of results presented in Table 3.

$$\frac{C_{d-debris}}{C_d} = \left(\frac{H_{debris}}{H}\right)^{-3/2} \quad (1)$$

Table 3. Driftwood impact ratios – results sample

H_{debris}/H	$C_{d\text{-}debris}/C_d$
1.1	0.87
1.2 (typical cond.)	0.76
1.7 (meas. maximum)	0.45

In equation 1, C_d and H are the discharge coefficient and total upstream head, respectively, of a PKW without woody debris. $C_{d\text{-}debris}$ and H_{debris} are the discharge coefficient and total upstream head of a PKW with debris. For more detailed results, please see Pfister et al. (2013).

CONCLUSIONS

The increased weir length provided by a labyrinth weir facilitates an increase in discharge efficiency and lower upstream head for a given discharge relative to linear weirs of equal spillway width. Consequently, concern existed that this type of weir may have a higher potential as a debris collector. However, 74 of 75 surveyed dam owners/operators of labyrinth weirs in the USA and Portugal reported no significant debris accumulation or cleaning effort. Catchments that contribute significant amounts of large debris may not allow passage over the spillway; therefore, some maintenance is required. The results of a laboratory study on three PKW geometries (a type of labyrinth weir) in a reservoir scenario indicate a 16% increase in upstream head for higher flows, and a 67% upstream head increase for lower flows. Similar trends were also reported in a physical model study for a large labyrinth weir. Consequently, the potential impacts of debris on the discharge capacity of a labyrinth spillway during a large storm event may be less than expected. However, the design and maintenance of labyrinth spillways must certainly consider the effects of debris for situations where a heavy debris load is anticipated.

ACKNOWLEDGEMENTS

This study was supported by Freese and Nichols, Utah State University, and Schnabel Engineering. The authors are grateful to the many individuals who provided information regarding labyrinth spillways in the USA and abroad, with a special thanks to Prof. Jorge Matos (Universidade de Lisboa, Portugal) and António Pereira da Siva (Consultores de Enganharia e Ambiente).

REFERENCES

Anderson, R.M. and Tullis, B.P. (2013). "Piano Key Hydraulics and Labyrinth Weir Comparisons." J. Irrig. Drain., 139(3), 246-253.

Anderson, R.M. and Tullis, B.P. (2012). "Piano Key Weirs: Reservoir versus Channel Applications." J. Irrig. Drain. Eng., 138(8), 773-776.

Anderson, R.M. and Tullis, B.P. (2012). "Comparison of Piano Key and Rectangular Labyrinth Weir Hydraulics." J. Irrig. Drain. Eng., 138(6), 358-361.

Babb, A. (1976). "Hydraulic model study of the Boardman Reservoir Spillway." R.L. Albrook Hydraulic Laboratory, Washington State University, Pullman, WA.

Bezzola, G.R., Hegg, C., eds. (2007). *Ereignisanalyse Hochwasser* 2005 [2005 *Flood event analysis*]. Bundesamt für Umwelt & Eidgenössischen Forschungsanstalt für Wald, Schnee und Landschaft, Bern, Switzerland (in German).

Crookston, B.M. (2010). "Labyrinth weirs." Ph.D. Dissertation, Utah State University, Logan, UT.

Crookston, B.M., Paxson, G.S., and Campbell, D.B. (2013). "Effective Spillways: Harmonizing labyrinth weir hydraulic efficiency and project requirements" in Labyrinth and Piano Key Weirs II – PKW 2013. S. Erpicum, F. Laugier, M. Pfister, M. Pirotton, G Cicéro, A. Schleiss Editors. CRC Press, London, UK.

Crookston, B.M. and Tullis, B.P. (2013a). "Hydraulic design and analysis of labyrinth weirs. Part 1: Discharge relationships." *J. Irrig. Drain. Engr.,* 139(5) 363-370.

Crookston, B.M. and Tullis, B.P. (2013b). "Hydraulic design and analysis of labyrinth weirs. Part 2: Nappe Aeration, Instability, and Vibration." *J. Irrig. Drain. Engr.* 139(5) 371-377.

Dabling, M.R. and Tullis, B.P. (2012). "Piano Key Weir Submergence in Channel Applications." J. Hydraul. Eng., 138(7), 661-666.

Erpicum, S., Laugier, F., Pirotton, M., Boillat, J., Reverchon, B., and Schleiss, A. (2011). Labyrinth and Piano Key Weirs: PKW 2011. CRC Press, London, UK.

Erpicum, S., Laugier, F., Pfister, M., Pirotton, M., Cicéro, G., and Schleiss, A. (2013). Labyrinth and Piano Key Weirs II: PKW 2013. CRC Press, London, UK.

Falvey, H. (2003). Hydraulic design of labyrinth weirs. ASCE, Reston, VA.

Khatsuria, R. (2005). Hydraulics of spillways and energy dissipators. Marcel Dekker, NY.

Laugier, F. (2007). "Design and construction of the first piano-key weir (PKW) spillway at the Goulours Dam." *Hydropower and Dams,* Issue 5.

Leite Ribeiro, M., Pfister, M., Schleiss, A.J., Boillat, J.-L. (2012a). Hydraulic design of A-type Piano Key Weirs. *J. Hydraulic Res.* 50(4), 400–408.

Leite Ribeiro, M., Bieri, M., Boillat, J.-L., Schleiss, A.J., Singhal, G., Sharma, N. (2012b). Discharge Capacity of Piano Key Weirs. *J. Hydraulic Eng.* 138(2), 199–203.

Ouamane, A. and Lempérière, F. (2006). "Design of a new economic shape of weir. *Proc. Intl. Symp. Dams in the Societies of the 21st Century,* Barcelona, 463-470.

Pfister, M., Schleiss, A., and Tullis, B. (2013). "Effect of driftwood on hydraulic head of Piano Key Weirs" in Labyrinth and Piano Key Weirs II – PKW 2013. S. Erpicum, F. Laugier, M. Pfister, M. Pirotton, G Cicéro, A. Schleiss Editors. CRC Press, London, UK.

SOGREAH (2011). Modèle physique du barrage de Luzieres. Report No. 136 0 886, SOGREAH, France (in French).

Tullis, P., Amanian, N., and Waldron, D. (1995). "Design of labyrinth weir spillways." *J. of Hydr. Eng.,* 121(3), 247-255.

Vasquez, V.M., Wolfhope, J.S., Garrett, R., and Tullis, B., "Providing the reliability of a labyrinth weir at half the cost," Transactions of the International Committee on Large Dams, pp. 549–572, Barcelona, Spain, June 2006.

Willmore, C. (2004). "Hydraulic characteristics of labyrinth weirs." M.S. Report, Utah State University, Logan, UT.

DESIGN APPROACHES AND NUMERICAL MODELING OF A STEPPED SPILLWAY UNDER HIGH TAILWATER CONDITIONS

Laura Shearin-Feimster P.E.[1],
Brian M. Crookston, Ph.D., P.E.[2],
Stefan Felder, Ph.D[3]

ABSTRACT

Rehabilitation of existing embankment dams often includes addressing deficiencies in the spillways and outlet works. One common cost-effective solution is to armor the embankment using RCC, forming a stepped chute that increases spillway capacity while also offering energy dissipation. With the wealth of knowledge on the subject, it can be challenging to identify suitable methodologies for a specific rehabilitation project. Additional challenges arise when the design methodologies and research do not specifically account for site conditions and constraints that impact a specific spillway's hydraulics.

Schnabel recently completed an embankment dam rehabilitation project with a stepped chute subject to backwater effects. The design was based upon previous experimental studies identifying design parameters for embankment dam stepped spillways including flow depth, inception point of air entrainment position, and residual energy at the downstream end of the stepped chute. Moreover, CFD modeling of this specific stepped spillway geometry and hydraulic conditions was performed and the results were compared to the predicted results from published design methodologies.

This paper presents an overview of a unique rehabilitation project with high tailwater, a review of appropriate design methodologies, and a discussion regarding the applicability and limitations of the current design parameters to the described project, including tailwater effects. This paper also discusses the current challenges of modeling numerically stepped spillways. These insights should aid practitioners in the design of stepped chutes for embankment dams.

INTRODUCTION

The most common type of dam used today is the embankment dam. Insufficient spillway capacity, upstream and downstream development, and aging continue to justify rehabilitation and upgrading projects. One common cost-effective solution is to armor the embankment using RCC or conventional concrete, forming a secondary stepped spillway that increases spillway capacity while also offering energy dissipation (see

[1] Senior Engineer, Schnabel Engineering, 11A Oak Branch Drive, Greensboro, NC, USA 27407. Phone: (336) 274-9456, Email: lshearin@schnabel-eng.com
[2] Project Engineer, Schnabel Engineering, 1380 Wilmington Pike, Suite 100, West Chester, PA, USA 19382. Phone: (610) 696-6066, Email: bcrookston@gmail.com
[3] Lecturer, University of New South Wales, Water Research Laboratory, School of Civil and Environmental Engineering, Sydney NSW 2052 Australia. Phone: (+612) 9385-5562, Email: s.felder@unsw.edu.au

Figure 1), which is one advantage of stepped chutes. It is not uncommon to cover a stepped chute with topsoil and turf, where these surface treatments can be inexpensively replaced as required (below right).

Figure 1. Examples of embankment dams with stepped chutes (courtesy Schnabel)

The use of stepped chutes to rehabilitate embankment dams has encouraged detailed studies regarding the hydraulics of moderate-sloped stepped chutes during the past 12 years. Detailed research has been conducted throughout the world, including hydraulic laboratories located in Australia, Switzerland, Portugal, Germany, Japan, Italy, and the USA. A selection of published literature specific to moderate-sloped stepped spillways and the effects of steps on energy dissipation includes Peyras et al. (1992), Chanson (2002), Chanson and Toombes (2002), Ohtsu et al. (2004), Gonzalez and Chanson (2007), Felder and Chanson (2009, 2011, 2013), Meireles and Matos (2009), Hunt and Kadavy (2010), Bung (2011), Frizell and Svoboda (2012) and Hunt et al. (2014), For example, the multi-year research conducted at the USDA-ARS hydraulic laboratory was of stepped chutes at or near prototype scale (Hunt and Kadavy 2014).

The general flow characteristics of a moderate-sloped stepped chute are presented in Figure 2. As shown there is a non-aerated flow region and an aerated flow region, which begins downstream of the inception point, defined as the location where the boundary layer reaches the water surface; the length from the downstream edge of the crest to this location is L_i. Q is discharge, θ is the spillway slope, H_o is the total energy head at the crest to the point of interest, $V^2/2g$ is the velocity head component where V is the mean velocity and g is the acceleration constant of gravity. The approach depth is y_o. The step size is defined by the step height, h, and the step length, l. The equivalent clear water depth is y_{cw}. Regarding energy dissipation, H is residual energy and ΔH is total energy lost at a distance L from the downstream edge of the crest.

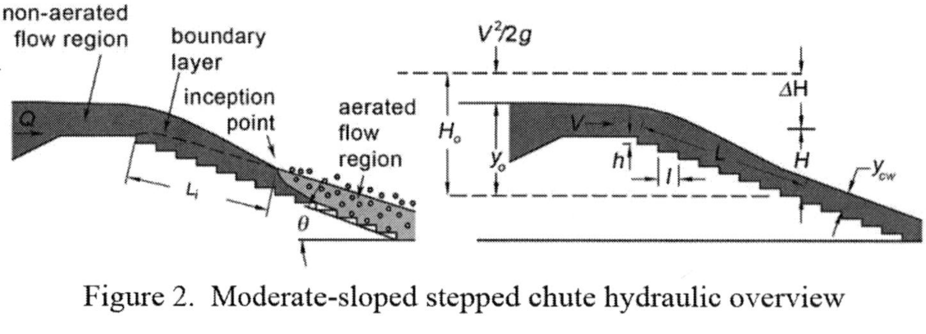

Figure 2. Moderate-sloped stepped chute hydraulic overview

Despite the wealth of research, there is still lack of consensus in the research community regarding several key aspects of stepped chute hydraulics, including the occurrence of uniform flow conditions, the occurrence of the state of equilibrium, and roughness height. Also, it can be challenging to identify and compare suitable design methodologies for a specific project that may include unique geometric or hydraulic components, including the inclusion of various weir types upstream of the stepped chute. Furthermore, the application of computational fluid dynamics (CFD) appears to produce favorable results in the clear-water (non-aerated) region, but significant challenges have been encountered when modeling the two-phase flow region.

Schnabel has encountered numerous sites where a stepped spillway has been the preferred option for rehabilitation. Many of these sites are estimated to have a high tailwater condition for large storm events. This paper presents an overview of a unique rehabilitation project with high tailwater, a review of appropriate design methodologies, and a discussion regarding the applicability and limitations of the current design parameters to the current project, including backwater effects. This paper also discusses the current challenges of modeling numerically stepped spillways. These insights should aid practitioners in the design of stepped chutes for embankment dams.

BASIS OF CASE STUDY

Fox Creek Multi-Purpose Structure Number 4 (Fox Creek 4) was used as a basis for examining the various methods noted herein. This particular site was selected due to its availability of site data and presence of high tailwater during flood flow conditions.

Background

Fox Creek 4 is located near Fleming, Kentucky, USA. It was originally designed by NRCS for the purpose of flood control and recreation. The dam was originally constructed in 1968 as a significant hazard structure. Later, the dam was considered to be high hazard based on downstream structures. In 2010, Schnabel was contracted to design the selected upgrade of a stepped auxiliary spillway over the embankment to meet the NRCS requirements for high hazard dams. Construction was completed in 2012; two photographs of the constructed spillway are presented in Figure 3.

Figure 3. Fox Creek 4 Spillway (photos courtesy Schnabel Engineering)

Site Data

For the purposes of this paper, Fox Creek 4's dimensions and some site characteristics were used to compare the cited published methods with the CFD model. This paper does not intend to analyze Fox Creek 4, but rather to form a basis of comparison of the various cited methods. The scenarios analyzed do not necessarily match site conditions used for design.

The RCC stepped spillway at Fox Creek 4 features an ogee crested weir upstream of the steps. The stepped chute is comprised of four 4-ft high steps, with the upper-most step at 3 ft and the final step at 5 ft. In addition, it is a 15° converging chute, with the top width of the spillway of 295 ft and a bottom-width of 263 ft. The chute length along the slope (*L*) is 64.4 ft. Additional pertinent project information is summarized in Table 1.

Table 1. Project Overview

Parameter	**Value**
Top of Dam Elevation	761.8 ft
Spillway Crest Elevation	757.5 ft
Spillway Width, Top	295 ft
Spillway Width, Bottom	263 ft
Spillway Slope	2.5H:1V (22°)
Basin Apron Elevation	728.0 ft, 722 ft

Two flow conditions were analyzed for energy dissipation for this study; the spillway design flood (SDF) that is the probable maximum flood (PMF) and about 2/3 of the PMF. When the spillway is passing the PMF and operating at full capacity, the peak flow rate was estimated to be 37,040 cfs (headwater elevation 767.3ft, tailwater elevation 748.4 ft). During the 2/3 PMF, the flow was estimated 24,705 cfs (headwater elevation 765 ft, tailwater elevation 744.3 ft).

STEPPED CHUTE RESIDUAL ENERGY PREDICTIVE METHODS

Four stepped chute energy dissipation predictive methods, summarized in Table 2, were selected for this study. Meireles and Matos (2009); Hunt and Kadavy (2010), Hunt et al. (2014); Gonzalez and Chanson (2007); and Felder and Chanson (2011, 2013, 2014) were used to calculate the energy loss along the stepped spillway. These methods base the energy dissipation capacity of the steps on air-water flow data. Note that the stepped chute for Fox Creek 4 is estimated to be too short for inception to occur for the 2/3PMF and PMF flow conditions. The clear-water flow regime, with a hydraulic jump on the steps followed by high tailwater in the stilling basin, is anticipated.

Table 2. Stepped Chute Energy Dissipation Predictive Methods

Reference	Slope	Flow Regime	d_c/h
Meireles and Matos (2009)	26.6°	Skimming	1.27 to 2.85
Hunt and Kadavy (2010), Hunt et al. (2014)	10° to 30°	Skimming	0.035 to 2.27
Gonzalez and Chanson (2007)	10° to 25°	Skimming	1.0 to 3.2
Felder and Chanson (2011, 2013, 2014) †	3.4° to 26.6°	Transition & Skimming	0.61 to 4

† energy prediction method based upon air-water flow data

Meireles and Matos

Meireles and Matos (2009) focused on using the inception point to evaluate energy dissipation. This experimental setup used a 2H:1V slope, which is similar to our case. This setup focused on the non-aerated flow regime, or the area upstream of the inception point. The researchers found the energy dissipation to be less than what had been found by others for uniform or fully developed flow conditions.

To calculate the location of the inception point, Meireles and Matos used a relationship developed by Chanson based on the roughness Froude number, F^* defined as

$$F^* = \frac{q}{\left(g\sin(\theta)k_s^{\ 3}\right)^{1/2}} \tag{1}$$

where k_s is the roughness height perpendicular to the pseudo-bottom, defined as

$$k_s = h\cos(\theta) \tag{2}$$

The location of the inception point, L_i, and the depth of flow at this location, d_i, are calculated using

$$L_i = 5.25 k_s F^{*0.95} \tag{3}$$

$$d_i = 0.28 k_s F^{*0.68} \tag{4}$$

where di is the flow depth at the inception point. The relationship between the clear-water depth d and the flow depth at the inception point was proposed as

$$\frac{d}{d_i} = 0.971 + 0.891 e^{-3.41(L/L_i)} \tag{5}$$

The resulting relation developed by Meireles and Matos to account for energy head loss along the stepped chute is presented as Eq. 6. As with the clear-water depth, head loss was found to be dependent on the normalized length, L/L_i.

$$\frac{\Delta H}{H_o} = 1 - \exp\left(-0.3\left(\frac{L}{L_i}\right)^{2.5}\right) \quad \text{(non-aerated flow region) (6)}$$

where $\Delta H = H_o - H$, the upstream total head at the crest of the stepped chute $H_{o}= H_{dam} + 1.5d_c$ with d_c the critical flow depth and the residual energy at any point along the non-aerated spillway:

$$H_o = y_{cw}\cos(\theta) + \alpha\left(\frac{V^2}{2g}\right) \quad (7)$$

In Eq. 7, y_{cw} is the equivalent clear water flow depth, α is the energy coefficient, V is mean velocity computed as $V = q/y_{cw}$.

Hunt and Kadavy

Hunt and Kadavy (2010) and Hunt et al. (2014) provide a very similar design approach design procedure that was developed for flat embankment slopes in the non-aerated flow region ($14° \leq \theta \leq 26.6°$). A portion of their method utilizes the computed inception point location to estimate energy dissipation. Hunt and Kadavy present multiple ways to calculate the location of the inception point, L_i, depending on specific site characteristics. For our case study, the following relation was utilized.

$$\frac{L_i}{k_s} = 5.19(F^*)^{0.89} \quad (8)$$

In our case study, the chute is not long enough for the flow to reach the inception point for the 2/3 PMF or PMF design points. However, Hunt and Kadavy (2010) found a linear trend for energy dissipation upstream of the inception point (clear water flow region):

$$\frac{\Delta H}{H_o} = 0.30\left(\frac{L}{L_i}\right) \quad \text{(non-aerated flow region) (9)}$$

Hunt et al. (2014) also utilize Eq. 7 to compute energy dissipation (see for additional information).

Gonzalez and Chanson

Gonzalez and Chanson (2007) discusses a design procedure based on an experimental setup tailored toward stepped chutes over embankment dams. The paper presents separate design procedures depending on whether aerated flow conditions are reached above the toe of the dam. The procedure is valid for slopes ranging from 10° to 25° and a critical depth to step height ratio (d_c/h) from 1.0 to 3.2. Due to the case study site conditions, the procedure for small embankment spillways with non-aerated flow conditions was used.

The ideal velocity for this condition is estimated using the Bernoulli equation

$$V_{max} = \sqrt{2g(H_o - d\cos(\theta))} \quad \text{(non-aerated flow region) (10)}$$

where d is the depth of flow at the base of the spillway. This equation is combined with the following relation

$$d = \frac{q}{V_{max}} + \frac{\delta}{N+1} \qquad \text{(non-aerated flow region) (11)}$$

where the velocity distribution exponent, N, is estimated to be approximately 5 for stepped chutes. The boundary layer in this relation, δ, is estimated using the following:

$$\frac{\delta}{L} = 0.0301\frac{1}{\sin(\theta)^{0.11}}\left(\frac{L}{h\cos(\theta)}\right)^{-0.17} \qquad \text{(non-aerated flow region) (12)}$$

From the unit flow rate and calculated depth at the spillway exit, the residual energy head can be calculated with Eq (7).

Felder and Chanson

Often stepped spillways are aerated when the chute is long enough to reach the inception point of air entrainment. In such a case, the design must be based upon air-water flow estimates of the residual energy of the spillway toe to accurately predict the energy to be dissipated in the stilling structure. Felder and Chanson (2014) developed a simple design guideline for slopes ranging from 8.9° and 26.6° based upon all available aerated flow data comprising both aerated transition and skimming flows. They identified characteristic slope groups for which the energy dissipation performance was similar and linked the residual energy to the flow rate via the critical flow depth at the upstream end. A embankment dam chute slope of 21.8° (2.5H:1V) is suggested to be the most efficient system for energy dissipation. The design guideline is valid for the aerated flow region downstream of the inception point and not for the non-aerated flow region. Based upon air-water flow data, the residual energy at the downstream end can be calculated using:

$$H = \int_0^{Y_{90}} (1-C)\cos(\theta)dy + \frac{q_w^{\,2}}{2g\left[\int_0^{Y_{90}} (1-C)dy\right]^2} \qquad \text{(aerated flow region) (13)}$$

where C is the void fraction and Y_{90} is the flow depth where the void fraction is 90%. [Note that the air-water flow Eq. (13) is equivalent to Eq. (7)].

For our case study, the recommended design guidance is a ratio of residual energy head to critical depth (H/d_c) of 3.37. Because the spillway in our case study is not of sufficient length for the occurrence of an air-water flow region downstream of the inception point for the flow rates of interest (PMF, 2/3 PMF), this method may not be directly applicable as far as estimating head losses for the design discharges. However, Felder and Chanson (2011) investigated stepped spillways with varying step heights, as is the case with this case study. Overall, their conclusions were that the dissipation was very similar when comparing uniform and non-uniform steps heights. It should be emphasized that, discharges less than 2/3 PMF (not included herein) on the present stepped spillway would be aerated resulting in nappe flow regime and also in transition flows. Transition flows can cause instable flows in particular in step geometries with non-uniform or unconventional step designs.

Method Comparison

Table 3 presents the computed results from the stepped chute methodologies discussed previously, as applied to Fox Creek 4. The calculations were performed separately for the main section of the spillway (the high section) and the section lowered to handle the principal spillway outflow (the low section). For the 2/3 PMF, the estimated energy head at the top of the slope is 36 ft for the high section and 42 feet for the low section. For the full spillway capacity, the estimated energy head at the top of the slope is 38.7 ft for the high section and 44.7 ft for the low section.

Table 3. Summary of calculated residual energy head at the base of the spillway

Method†	**Residual Energy Head† (ft)**			
	2/3 PMF		PMF	
	High Section	Low Section	High Section	Low Section
Meireles and Matos	24.6	24.6	33.3	35.2
Hunt et al.	24.8	24.8	30.0	32.4
Gonzalez and Chanson	29.3	29.3	33.6	37.3
Felder and Chanson	21.0 (aerated flows)		27.6 (aerated flows)	

†does not include tailwater effects

NUMERICAL MODELING

Although the aforementioned stepped chute methods were identified to be most applicable to Fox Creek 4, no method was fully applicable to Fox Creek 4 (clear water flow regime, varying step height, ogee weir upstream of chute, high tailwater, converging chute). Therefore, numerical modeling of Fox Creek 4 was performed to

- Evaluate spillway hydraulics for this specific geometry and site conditions
- Compare the numerical results to the previously cited stepped chute methods

Model Characteristics

The commercially available CFD solver, FLOW-3D (by Flow Science) was used to develop numerical models of Fox Creek 4. FLOW-3D solves the Reynolds-average Navier-Stokes equations (RANS) with modified algorithms to model flow past solid objects (e.g., weirs and steps), model the volume fraction of fluid in each discretized cell (VOF model), track the free surface, and model air void fraction in the flow. The RANS equations, with modified algorithms for continuity and momentum, are presented as Eqs. 14 and 15, respectively.

$$\frac{\partial}{\partial x}(uA_x)+\frac{\partial}{\partial y}(vA_y)+\frac{\partial}{\partial z}(wA_z)=0 \tag{14}$$

$$\frac{\partial U_i}{\partial t}+\frac{1}{V_F}\left(U_j A_j \frac{\partial u_i}{\partial x_j}\right)=-\frac{1}{P}\frac{\partial P}{\partial x_i}+g_i+f_i \tag{15}$$

Numerical modeling conducted for this study utilized the Renormalized Group (RNG) turbulence model. The Fractional Area/Volume Obstacle Representation (FAVOR) algorithm was used to embed the spillway and adjacent regions (drafted in Civil3D) in the computational mesh. Because air entrainment due to surface turbulence and a hydraulic jump is anticipated, simulations were performed with and without the Air Entrainment, Density Evaluation, and Drift-flux models. These three models are a parameterized resolution that simulates the volumetric fraction of entrained air in the numerical flow. An entrainment rate coefficient of 0.5 and surface tension coefficient of 0.005 were selected. Also, a drag coefficient of 0.5, average bubble radius of 1 mm, and the Richardson-Zaki coefficient of 1 were selected. Although a spatially-dependent bubble size distribution in the stepped chute would be expected, several bubble diameters were tested and the selected bubble diameter appears to be reasonable in this instance. Flow domain boundary conditions were specified (pressures, symmetry, walls, etc.) based upon hydrologic routings and a tailwater rating curve estimated using the USACE 1-dimensional HEC-RAS program.

No embedded mesh blocks were incorporated to minimized numerical interpolation and truncation at mesh block boundaries. The cross-sectional models were 1 cell wide perpendicular to the flow. Sectional models were more computationally expensive at 64 ft in width. Both models also included flux surfaces and history probes to record numerical results at locations of interest within the models (see Fig. 4). Simulation times were based upon the flow field and achieving solution convergence under steady-state conditions. Each simulation included multiple iterations to analyze runtime diagnostics and time-history plots. Final simulation characteristics are summarized in Table 4. By way of example, an isometric view of the section model (simulations 5 and 8) is presented in Figure 4.

Table 4. Numerical Simulations Test Matrix

Simulation		Event	Mesh Size (ft)	Total Cell Count ()	Simulation Time (s)	Computation Time (hr:min:s)	Physics†
1	X	2/3 PMF	1	11,562	120	00:06:32	*g*,*v*,T
2	X		0.5	45,120	120	00:37:42	*g*,*v*, T
3	X		0.25	180,480	120	07:38:50	*g*,*v*,T
4	X		0.5	45,120	120	00:50:57	*g*,*v*,T,AE,DE,DF
5	S		0.5	5,775,360	40	16:23:31	*g*,*v*,T,AE,DE,DF
6	X	PMF	0.5	45,120	120	00:42:22	*g*,*v*,T
7	X		0.5	45,120	120	01:06:28	*g*,*v*,T,AE,DE,DF
8	S		0.5	5,919,744	40	20:51:16	*g*,*v*,T,AE,DE,DF

†X = cross-section, S = section, *g* = gravity, ν = viscosity, T = turbulence closure method, AE = Air Entrainment, DF = Drift Flux, DE = Density Evaluation

As a reference for recorded computation times reported in Table 4, simulations were performed using a workstation with 32 cores @ 3.4 GHz (Xeon® processors) and 64 GB of RAM. As shown in Table 4 and Figure 5, three cell sizes for the meshed domain were simulated to refine the grid and investigate mesh convergence and solution independence.

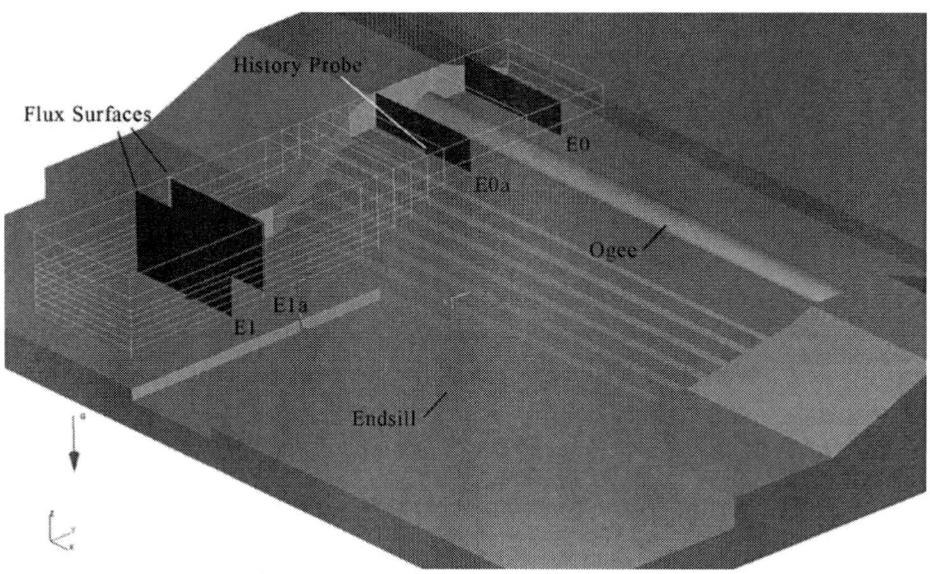

Figure 4. Sectional model overview

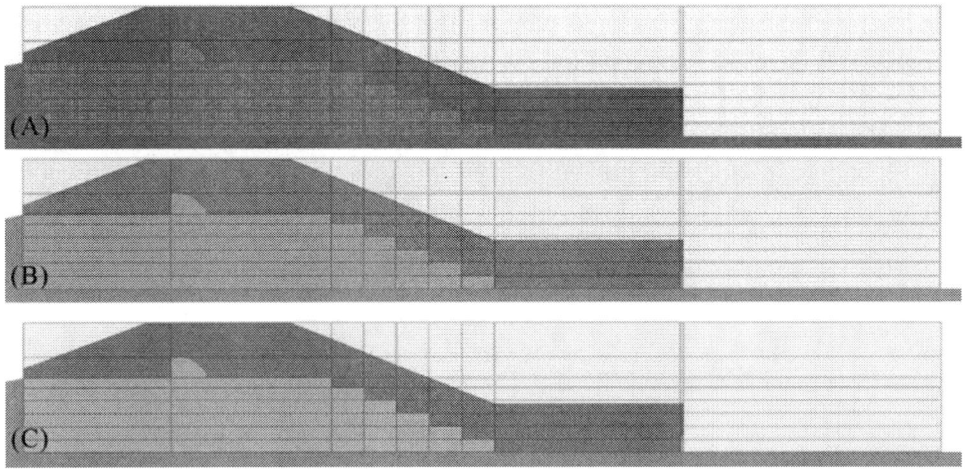

Figure 5. Cross-section models with mesh size at 1.0 ft (A), 0.5 ft (B), and 0.25 ft (C)

Due to the non-conforming hexahedral mesh in FLOW-3D, the solution is dependent upon discretization of the domain, specifically the selected mesh size. For the purposes of this study, and in view of the level of uncertainty in the hydrologic analysis used to develop the spillway design flood, a detailed mesh refinement study and quantification of numerical uncertainties were not performed. However, computed discharge at four locations during the simulation time (120 sec) for simulations 1, 2, and 3 are presented in Figure 6. Note that flux surface E1a was located at the endsill of the stilling basin and flux surface E1 downstream, locations in close proximity to the hydraulic jump.

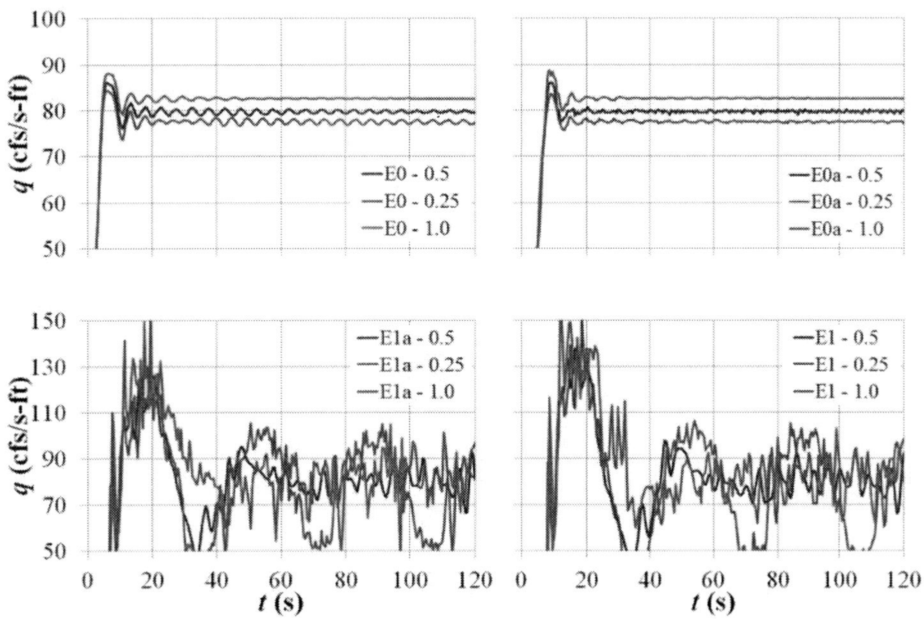

Figure 6. Discharge results for grids with 1 ft, 0.5 ft, and 0.25 ft cell size.

The percent difference in computed discharges, relative to the finest cell size of 0.25 ft is presented in Figure 7. As shown, discharges measured upstream of the stepped chute were approximately 7% greater and 3% greater for the 1 ft and 0.5 ft cell sizes. In addition, each step was represented with 8 cells in the vertical direction for the 0.5 ft cell size, which was viewed to be an adequate mesh resolution for this study.

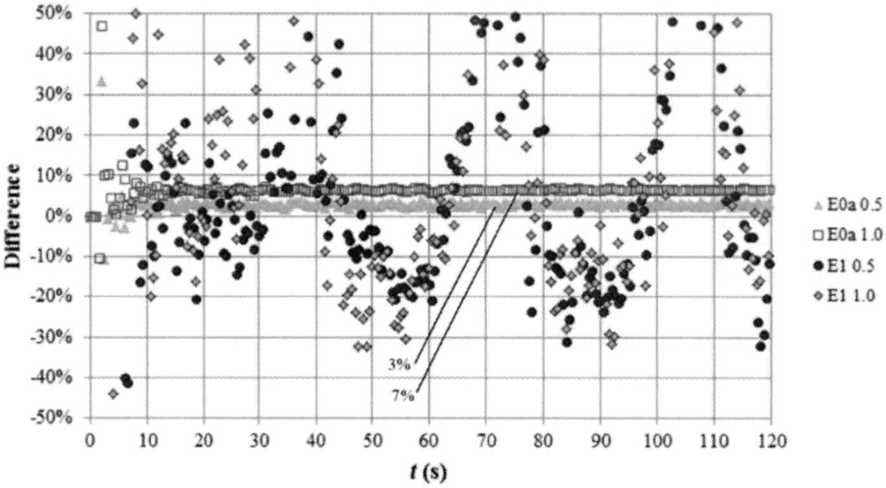

Figure 7. Percent difference relative to 0.25 ft cell size

Numerical Results

No field data or physical model experimental results exist for calibration of the CFD models. Therefore, the results have not been verified. Computed unit discharges are presented in Table 5. Also included is the ratio of the numerically computed q_{CFD} versus the original q estimated during the hydrologic analysis. Good agreement exists between the empirical estimate and the CFD results (within 3% to 5% for the finer meshes). Also, the sectional model results agreed well with the cross-section model results (3.5% and <1% for 0.5 ft cells).

Table 5. Computed unit discharges

Simulation	Event	q (cfs/ft)	q_{CFD}/q
1	2/3 PMF	82.6	0.99
2		79.7	0.95
3		77.5	0.93
4		79.8	0.95
5		80.3	0.96
6	PMF	121.5	0.97
7		121.5	0.97
8		121.4	0.97

Total energy was computed using CFD (flux surfaces) for all simulations for the higher section of the spillway. H_o, H, and ΔH are presented in Table 6. Total energy at the crest for the 0.5 ft and 0.25 ft cells, with and without air had computed differences less than 1%. However, differences of about 12% and 5% were computed for H for the 2/3 PMF and PMF, respectively. The numerical results estimate that about 50% of the total energy is dissipated by the stepped chute. As with discharge, energy dissipation results were similar between the sectional and cross-section models.

Table 6. Computed energy for Fox Creek 4 Stepped Spillway

Simulation	Event	Numerical Results			
		H_o (ft)	H (ft)	ΔH (ft)	$\Delta H / H_o$ ()
1	2/3 PMF	36.8	17.0	19.8	0.54
2		36.7	16.8	19.9	0.54
3		36.5	16.8	19.7	0.54
4		36.7	19.1	17.6	0.48
5		36.3	16.5	19.8	0.55
6	PMF	39.0	21.9	17.1	0.44
7		39.0	23.0	16.0	0.41
8		39.0	21.3	17.7	0.45

The numerical results would be expected to be a good approximation of flow rate and energy dissipation for Fox Creek 4; however, differences in energy dissipation between the numerical results and method predictions (see Table 2) exist. Therefore, a comparison of predicted H is shown in Table 7.

Table 7. Residual Energy Predictive Comparison

Method	2/3 PMF		PMF	
	H_{pred} (ft)	$H_{pred} / H_{CFD\text{-}Air}$	H_{pred} (ft)	$H_{pred} / H_{CFD\text{-}Air}$
CFD (No Air)	16.8	0.88	21.9	0.95
CFD (Air)	19.1	1.00	23.0	1.00
Meireles and Matos(No Air)	24.6	1.29	33.0	1.40
Hunt et al. (No Air)	24.8	1.30	30	1.30
Gonzalez and Chanson(No Air)	29.3	1.53	33.6	1.46
Felder and Chanson (Air)	21.0	1.1	27.6	1.2

The CFD results fall within the reported range for residual energy by Felder and Chanson. In general, it appears that the CFD results for H differ from the method predictions by about 15% to 46%. This discrepancy may be due, in part, to tailwater.

In addition to energy and unit discharge, the velocity flow field (in profile) was also extracted from the numerical results; views of simulations 4 and 7 (cross-section models) are presented in Figure 8. One notable feature of these simulations is the water surface profile where a hydraulic jump is expected. The cross-sectional models depict a plunging jet and do not appear to adequately resolve the profile of the hydraulic jump, despite there being close agreement in q and H. Figure 9 present a profile of simulation 5, which was a 65-ft wide simulation of the stepped chute. Note the differences in the water surface profile.

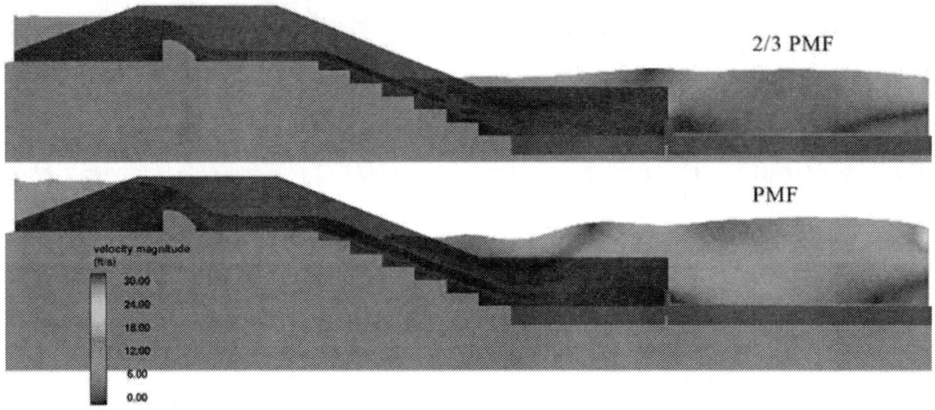

Figure 8. Velocity magnitudes for Fox Creek 4 stepped spillway

Figure 9. Water surface profile for simulation 5

As previously noted, several simulations were performed using air entrainment models to simulate air fraction (instead of a computationally more expensive 2 fluid approach). Figure 10 presents the volume fraction of entrained air for simulations 5 and 8. A range of physical model experimental results for the local void fraction C exists between stepped spillway research studies The numerical results for Fox Creek 4 are interesting with air concentrations near the surface generally $0.2 \leq C \leq 0.4$. However, additional model refinement and calibration is needed to better simulate the two-phase flow characteristics. Furthermore, the results shown in Fig. 10 indicate that the inception point of air entrainment occurs before the end of the spillway, which based on the design methods discussed herein, is viewed by the authors to be incorrect.

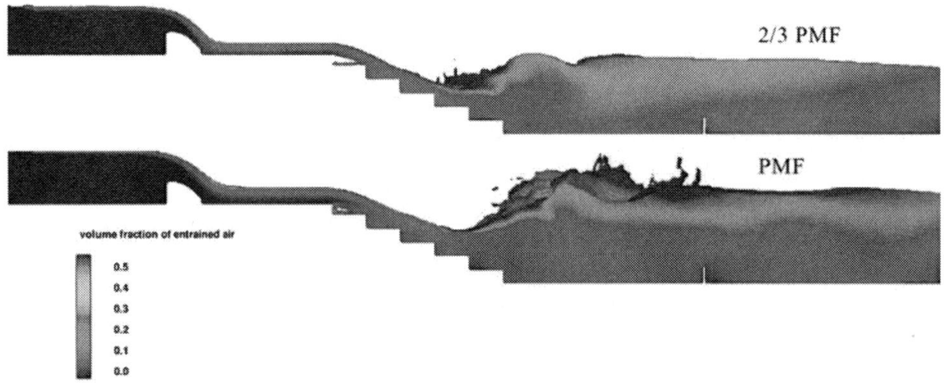

Figure 10. Volume fraction of entrained air.

DISCUSSION AND CONCLUSIONS

Numerical Model Limitations

Numerical modeling can provide important insights for spillway designers. For example, flow patterns are readily mapped and any problematic areas identified. Also, a great deal of information (e.g., depth, velocity, pressure, turbulence, etc.) can be computed across the entire domain and at each time step. And even though numerical models have been validated for stepped chutes, modelers and designers should be cognizant of the limitations of this tool.

One issue that can quickly arise is lengthy computation times. Although it may be possible to build a numerical model in less time than a physical model, a great deal of 'waiting' for results might occur. In this study, cross-sectional simulations were performed initially in order to more quickly analyze the results and make adjustments. A full-width numerical model was computationally expensive, so a portion of the model was simulated. Also, it was computationally intensive to simulate multiple flow rates; simulations for intermediate flows are planned. The authors wish to emphasize that modeling intermediate flows would also provide important insight for designers, as flow instabilities or hazardous conditions may exist for flow rates that are less than the SDF.

Another common issue is model calibration. It is always desirable to have accurate experimental data that can be used for numerical calibration. However, the numerical models used in this study were not verified. As previously noted, the numerical results in terms of unit discharge and energy dissipation performance appear reasonable and are of a design not explicitly accounted for in a published design method, yet without experimental results it was not possible to isolate the sources for reported differences in energy dissipation for this study. However the present numerical model also highlighted an issue with the modeling of aerated flows and was not able to predict accurately the aeration on the spillway and in particular in the hydraulic jump downstream. The velocity profiles on the spillway appear accurate, but seem questionable within the hydraulic jump. The authors readily acknowledge that project limitations exist and that at times it may be acceptable by owners/stake holders to increase design conservatism when it is unrealistic to perform a physical model study.

In general, it is challenging to accurately model shallow flows, unstable flows, and two-phase flows numerically. These challenges come from several sources, including turbulence model limitations, roughness, parametric model limitations, and size of the computational domain. As noted, estimation of residual energy and head loss was challenging and a critical design component. Also, the numerical model provided some information regarding wave and splash along the converging sidewall (see Fig. 11). However, resolution of the complex and intricate flow characteristics (air entrainment, surface splash and spray) was not captured numerically, as shown when compared with the photographs of a physical model in Fig. 12.

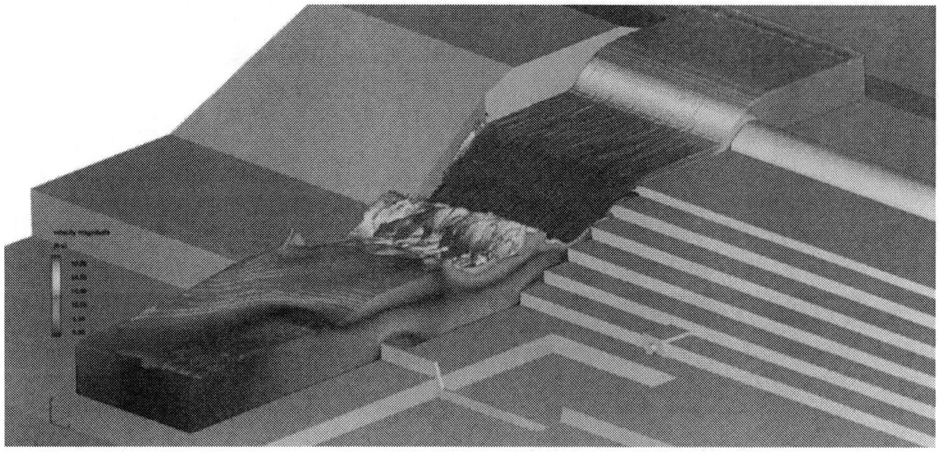

Figure 11. Numerical results 2/3 PMF

Figure 12. Flow characteristics of large scale physical model of a stepped chute at an embankment dam. Looking upstream (A), downstream (B) and inception point (C) (photos courtesy Sherry Hunt)

The use of numerical modeling for more complex step geometries must be considered carefully. It has been shown that flow instabilities can occur for transitional flow rates (i.e. flow rates lower than the design flow rates) which may result in unsafe operation of the spillway system (Felder and Chanson 2013). Examples were documented on pooled stepped spillways where instable jump waves can propagate down the spillway (Figure 13) or on non-uniform stepped spillways which may experience stronger instabilities and flow bulking compared to smooth flat stepped chutes. As a general guidance, whenever the design geometries of a hydraulic structure comprises non-conventional geometries, physical modeling (of adequate size scale to minimize scale effects) should be considered as an important design tool since the numerical modeling may not be able to predict complex flows accurately, and to calibrate a numerical model.

Figure 13. Flow instabilities on a pooled stepped spillway in transition flows showing an instationary jump waves propagating downstream (Felder and Chanson 2013)

In conclusion, Schnabel has encountered numerous sites where a stepped spillway has been the preferred option for rehabilitation. Many of these sites are estimated to have a high tailwater condition for large storm events. This paper presents an overview of a unique rehabilitation project with high tailwater, a brief review of several physical model-based design methodologies, and a discussion regarding the applicability and limitations of design parameters to the cited project, including tailwater effects. This paper also discusses some current challenges and limitations of numerical modeling for stepped spillways.

ACKNOWLEDGEMENTS

Special thanks to Professor Jorge Matos, PhD for personal communications on the subject. Also, the authors wish to thank Sherry Hunt, PhD for photographs of physical modeling conducted at the USDA Hydraulic Engineering Research Unit (Stillwater, Oklahoma, USA). This study was funded in part by Schnabel Engineering.

NOMENCLATURE

C	Air concentration
d	Equivalent clear-water depth (also shown as y_{cw})
d_c	Critical depth
d_c/h	Ratio of Critical depth to step height
d_i	Depth of flow at inception point
F^*	Froude number defined in terms of the roughness height
g	Acceleration constant of gravity
h	Step height
H	Residual energy head
$H_{CFD\text{-}Air}$	Residual energy head predicted by the CFD model
H_{dam}	Height of dam
H_o	Total energy head at crest
H_{pred}	Residual energy head predicted by a specified method
H/d_c	Ratio of residual energy head to critical depth
k_s	surface roughness
l	step length
L	Distance along slope from downstream edge to point of interest

L_i	Characteristic length from downstream edge to the surface inception point along slope
L/L_i	Normalized length
N	Velocity power law exponent
q	unit flow rate
Q	Total flow rate
q_{CFD}	CFD numerically computed unit flow rate
V	Average velocity at a specified location
V_o	Ideal velocity at chute exit
y_{cw}	Equivalent clear-water depth (also shown as d)
y_o	Approach depth
y_{90}	Characteristic depth when the air concentration is 90%
α	Energy coefficient
ΔH	Total energy head lost at a specified location
$\Delta H/H_o$	Relative energy loss
δ	Boundary layer thickness
θ	Spillway slope, degrees

REFERENCES

Bung, D. B. (2011). "Developing flow in skimming flow regime on embankment stepped spillways", *Journal of Hydraulic Research*, 49:5, pp. 639-648

Chanson, H. (2002). *The Hydraulics of Stepped Chutes and Spillways*. Balkema, Lisse, The Netherlands.

Chanson, H. and Toombes, L. (2002). "Experimental study of gas-liquid interfacial properties in a stepped cascade flow." *Envir. Fluid Mechanics,* Vol. 2, No. 3, pp. 241-263

Felder, S., and Chanson, H. (2014). "Aeration and air-water mass transfer on stepped chutes with embankment dam slopes." *Envir. Fluid Mechanics* Ahead of Print. (DOI 10.1007/s10652-14-9376-x)

Felder, S., and Chanson, H. (2013). "Aeration, flow instabilities, and residual energy on pooled stepped spillways of embankment dams." *J. Irrig. Drain Eng.*, 139(10), 880–887

Felder, S., and Chanson, H. (2011): "Energy dissipation down a stepped spillway with non-uniform step heights." *Jour. Hydr. Eng.*, ASCE, Vol. 137, No. 11, pp. 1543-1548.

Felder, S., and Chanson, H. (2009): "Energy dissipation, flow resistance and gas-liquid interfacial area in skimming flows on moderate-slope stepped spillways." *Environ. Fluid Mechanics*, Springer. Vol. 9, No. 4, pp. 427-441.

Frizell, K.W., and Svoboda, C.D. (2012). "Performance of type III stilling basins--Stepped spillway studies--Do stepped spillways affect traditional design parameters?"

Hydraulic Laboratory Rep. HL-2012-02, United States Dept. of the Interior, Bureau of Reclamation, Denver.

Gonzalez, C.A., and Chanson, H. (2007): "Hydraulic design of stepped spillways and downstream energy dissipators for embankment dams." *Dam Engineering*, Vol. 17, No. 4, pp. 223-244.

Hunt S.L., and Kadavy, K.C. (2014): "Flow depth and energy coefficient relationships for stepped spillways" *Hydraulic Structures and Society: Engineering Challenges and Extremes*, Proceedings of the 5th International Symposium on Hydraulic Structures, (2014), Brisbane, Australia.

Hunt S.L., and Kadavy, K.C. (2010): "Energy dissipation on flat-sloped stepped spillways: Part 1. Upstream of the inception point." Transactions of the ASABE, Vol. 53, No. 1, pp. 103-109.

Hunt S.L., Kadavy, K.C., and Hanson, G.J. (2014): "Simplistic design methods for moderate-sloped stepped chutes" *Jour. of Hydr. Eng.*, ASCE, Vol. 140, No. 12, pp. 04014062-1 to 04014062-15.

Meireles, I., and matos, J. (2009): "Skimming flow in the nonaerated region of stepped spillways over embankment dams." *Jour. of Hydr. Eng.*, ASCE, Vol. 135, No. 8, pp. 685-689.

ohtsu, I, Yasuda, Y., and Takahashi, M. (2004). "Flow characteristics of skimming flows in stepped channels." *Jour. of Hydr. Eng.*, ASCE, Vol. 130, No. 9, pp. 860-869.

Peyras, L., Royet, P., and Degoutte, G. (1992): "Flow and energy dissipation over stepped gabion weirs." *Jour. of Hydr. Eng.*, ASCE, Vol. 118, No. 5, pp. 707-717.

MODELING FLOODING FROM A CATASTROPHIC DIKE BREACH ALONG ELBE RIVER, GERMANY JUNE 2013

Robert Jüpner, Ph.D.[1]
André Assmann, Ph.D.[2]
Thilo Weichel Ph.D.[3]
Bernd Ettmer, Ph.D.[4]
Stefan Müller, M.Eng.[5]
Nageshwar Bhaskar, Ph.D.[6]

ABSTRACT

A failure of a dike during a catastrophic flood event usually leads to inundation of large areas. The size of the inundation area and water depths of such an event can be analyzed using two-dimensional (2D) hydraulic models together with a Digital Elevation Model (DEM) and detailed information about the dike breach itself, specifically, the geometry of the breach and discharge. Unfortunately information describing the physical extent of the dike breach is usually not available. Furthermore, the emergency management agencies usually need reliable information in a very short time frame to be able to take appropriate action like implementing evacuation measures.

During the recent Elbe flood disaster in June 2013, a catastrophic dike breach occurred near the village of Fischbeck in Saxony-Anhalt, Germany. As a result, nearly 500 m^3/s of flood water was released into relatively flat areas behind the dike through a 90 m breach. As a first step, an area-wide DEM with a high resolution was spatially aggregated to a practical 2D - grid level of 5 m x 5 m. Based on this data set, water levels as measured directly in the flooded zone, and the known slope of water surface in the main stream, different maps of the actual and potential flood extent were computed using readily available GIS computational tools. Additionally, a daily surveying flight was undertaken to monitor the inundation area and to receive real-time flood level information. As a result, daily maps illustrating current and future flood inundation areas in real time were created.

[1] Professor and Dean, Department of Civil Engineering, University of Kaiserslautern, Paul-Ehrlich-Straße 14, 67663 Kaiserslautern, Germany E-Mail: robert.juepner@bauing.uni-kl.de
[2] Executive board, geomer GmbH, Im Breitspiel 11 B, 69126 Heidelberg, Germany, E-Mail: aassm@geomer.de
[3] Senior Scientist, State authority for flood protection and water management Saxony-Anhalt, Halle, Germany, E-Mail: thilo.weichel@lhw.mlu.sachsen-anhalt.de
[4] Vice Dean, University of Applied Sciences Magdeburg-Stendal, Department of Water and Waste Management, 239114 Magdeburg Germany, E-Mail: bernd.ettmer@hs-magdeburg.de
[5] Scientist, Department of Water and Waste Management, 239114 Magdeburg Germany, E-Mail: Stefan.Mueller@hs-magdeburg.de
[6] Professor, Department of Civil and Environmental Engineering, University of Louisville, Louisville, Kentucky 40292, USA, E-Mail: bhaskar@louisville.edu

INTRODUCTION

In June 2013 the catchment areas of the Elbe and of the upper Danube rivers (refer to Figure 1(a)) were impacted by a devastating flooding. This flooding was caused by heavy rainfall in Central Europe, especially in the Southern and the Eastern parts of Germany between May 17 and June 2. The rainfall events in combination with high antecedent soil moisture led to the extreme flooding in Germany. As a result, the highest flood levels over a distance of more than 250 km (around 155 mi) along the central Elbe and of the lower Saale Rivers ever recorded (see Figure 1(b)) were exceeded in some locations (BfG, 2013; LHW, 2014).

Figure 1(a). Project Location — the catchment area of the river Elbe

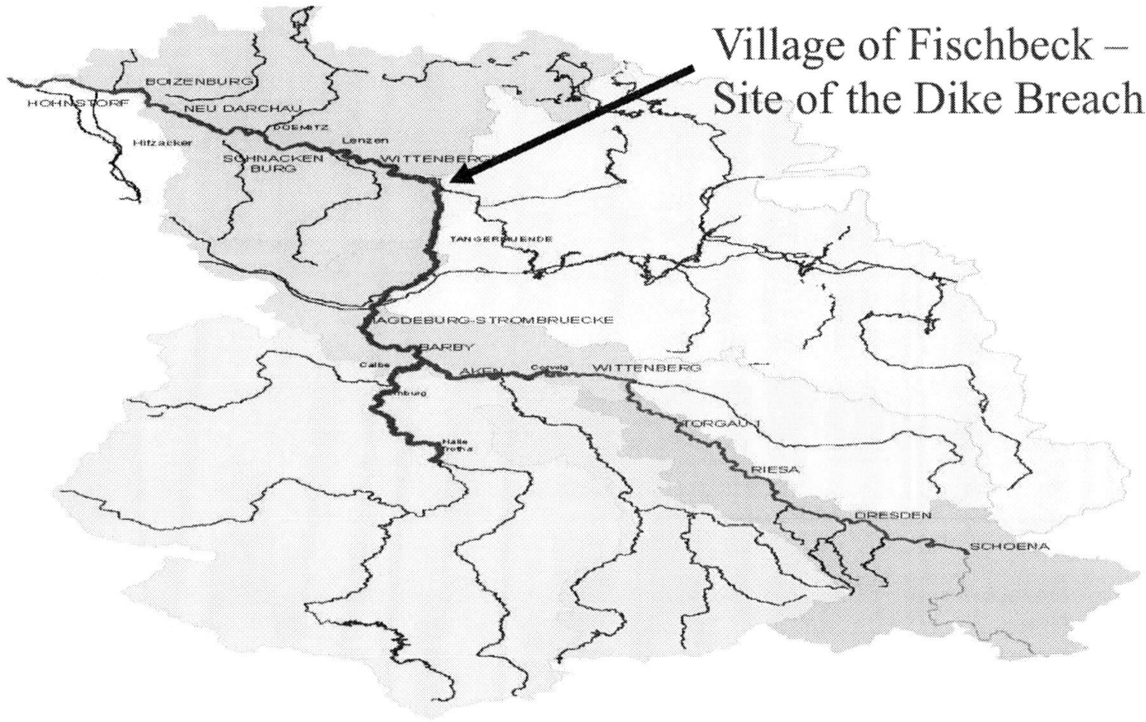

Figure 1(b). June 2013: Violations of HHW (highest known water level) since level institution in the course of federal waterways Elbe and Saale (red marking) (Source: German Federal Institute of Hydrology, 2013)

On June 10, 2013 at nearly midnight, a breach in the dike (referred to as a levee in the US) developed along the river Elbe close to the village of Fischbeck (Fig. 2) and expanded to a final width of 90 m. Although a specific cause of the dike failure was not identified, it appeared that it was presumably due to soil instability induced by water seepage and subsequent soil erosion that led to a slipping of both slopes of the dike (see Figure 2). Even though large efforts had been taken to protect the integrity of the dike, an overflow of the affected section could not be prevented (see Figure 3).

Figure 2. The right main dike of the Elbe close to the village of Fischbeck on June 9, around 10 hours before the crevasse occurred (Picture: Ellmann, 2013)

Figure 3. Dike breach close to the village of Fischbeck on June 10, around 10 hours after occurring (Picture: Jüpner, 2013)

At that time, the water level at the Elbe was measured to 8.36 m at Tangermuende gauging station. This level was 68 cm above the highest recorded water level in the river with nearly no remaining freeboard at the dike. The discharge to the flat areas (located behind the dike) through the dike breach was estimated to 500 m^3/s. After 9 days, initial efforts to close the breach with ships were partially successful by reducing the discharge through the crevasse considerably. The breach could finally be closed on June 20 after building an auxiliary dike (see Figure 4 below). Altogether a total of around 227 million cubic meters (m^3) of water was discharged from the Elbe through the breach to the flat areas behind the dike inundating a total area of around 150 km^2 between Elbe and Havel *(LHW, 2014)*.

Figure 4. Dike breach close to the village of Fischbeck on June 22, 12 days after occurring. The successful closing of the dike breach with the aid of ships and an auxiliary dike can easily be identified (Picture: Müller, 2013).

CONSEQUENCES OF THE CREVASSE

The area between Elbe and Havel rivers which was inundated by the crevasse can be characterized as an ancestral valley formed by the last glacial period and is relatively flat. The flood water was moving parallel to the river Elbe for roughly 30 km following the slope of the inland valley. The huge amounts of water reaching the areas behind the dike (more than 40 million m^3) within the first few days of the breach spread quickly.

Although the area impacted by the dike breach was not heavily populated a couple thousands of people who lived in the nearly 150 km^2 inundated area needed to be evacuated. This unpredictable flooding brought a lot of harm and high material damage to thousands of people. Only due to the high level of professional emergency management nobody got killed by this flooding disaster in the affected area.

ESTIMATION OF THE SPATIAL EXTENT AND THE DYNAMICS OF FLOODING IN THE AREA BETWEEN ELBE AND HAVEL RIVERS DURING THE FLOOD OF JUNE 2013

The flood 2013 along the Elbe downstream of the confluence with river Saale, was characterised by water levels never observed in the past. Table 1 shows the difference between the maximum water level during the flood in June 2013 and the highest water levels ever recorded up to that event. In addition to that, design flood levels for a 100- and 200-year flood event (LHW, 2014) are also displayed in the Table 1 for comparative purpose. Due to the German building codes, dikes are usually designed to a 100-year flood including freeboard (DIN 19712).

Table 1. Measured highest flood levels during the flooding of June 2013

Gauge	Measured highest flood level [m]	Difference to HHW [m]	HW_{100} [m]	HW_{200} [m]
Tangermünde	8.38	+ 0.70	8.00	8.28
Magdeburg-Strombrücke	7.47	+ 0.46	6.95	7.13
Barby	7.62	+ 0.61	7.06	7.24

In the case of any flood disaster the emergency management forces usually use a map that spatially illustrates flooding scenarios with the essential and relevant information such as water depth and spatial extent of the area inundated by the flood.

In all member states of the European Union, flood risk and flood hazard maps are available according to the European Flood Risk Management Directive (EU, 2007). They are created based on hydraulic models assuming a certain flood, for example, a 100-year-flood. These maps a very helpful instruments for emergency management forces, even if in the practice of water management these previously calculated flooding situations rarely occur if at all.

A crevasse usually creates a situation of extraordinary danger. Therefore the responsible emergency management forces need to have the following essential information to be able to react quick and appropriate:

- Spatial and temporal dimensions of the flooding in the area behind the dike.

- Main direction, intensity and maximum dimensions of the flood wave caused by the crevasse.

- Information on specific dangerous situations.

- Differences in height between the predicted maximum water levels and the height of the defense facilities like (auxiliary) dikes, and sandbag barriers.

- Additional information if evacuation routes are affected.

To provide the above information it is necessary to generate flood hazard maps (see Figure 5) that contain the relevant data on spatial and temporal dimensions of the inundated area, water levels, main direction of the flood wave, heights of dikes, routes of evacuation and other pertinent details (Patt & Jüpner, 2013).

The experiences from this flood event demonstrated once again that maps together with a plot of digital data (Juepner 2003) are the best kind of documents for any emergency management agency during a catastrophic flood event. Also in the real-time forecast of potential flood situations, the recording and monitoring of a flood event is to be made by using maps.

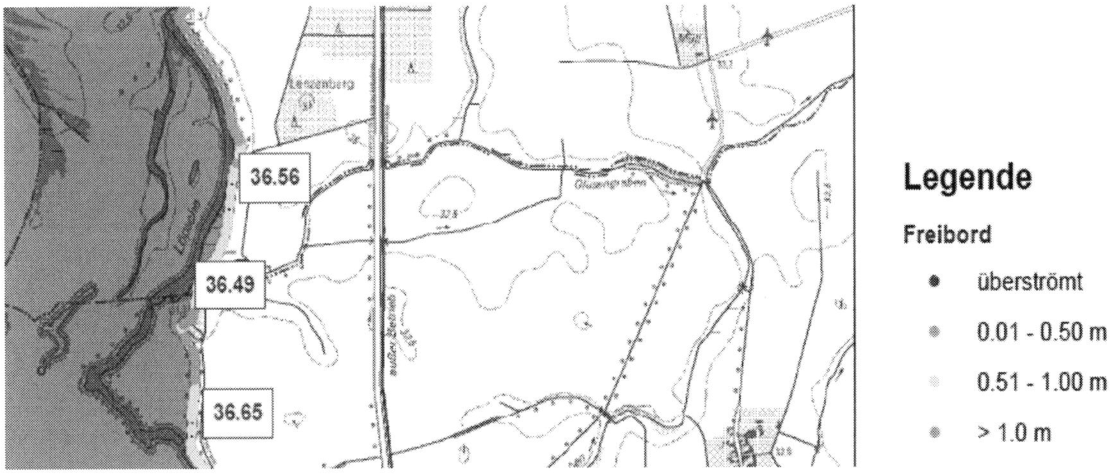

Figure 5. Map with a diagram of the remaining height of freeboard (grading in terms of color) for the right main dike of the Elbe during a flooding event with a return period of 200 years, inclusive the relevant heights of the crest of the dike (LHW 2013)

For flooding conditions involving a large-scale inundation of comparatively flat areas special applications, like the hydrodynamic simulation program FloodArea *(Assmann, 2013)*, have been developed. In this model the hydrodynamic part of the simulation program was integrated with the geographical information system software referred to as ArcGIS (ESRI, 2011). A similar model has been successfully applied to develop the "Elbe-Atlas" which contains the modeling of inundation areas due to crevasses *(Assmann et al., 2006)*.

One characteristic of the hydrodynamic model developed in the present study involves a raster based computational method that allows to immediately process existing topographic models at their full spatial resolution. In doing so, the different vectors of flow for each cell are determined in an iterative sequence using the calculated flow volumes from the eight neighboring cells. Due to the selected constant size of the cells this calculation can be carried out at high resolution and for large areas (a couple hundreds of millions of calculation nodes) and shows a very robust calculation performance. The model was optimized and tested for operational use in the context of the European research projects PREVIEW (FP6) and SAFER (FP7). The optimization mainly took place by the division of the processes of calculation to almost any number of calculation kernels, either at one single computer or in a cluster computers. The model validation happened amongst others in operational tests and in real claims of support, for example for the Hunza-River in Pakistan.

Based on hydrodynamic model described above, a simulation of the dike breach due to the flood of June 2013 along the Elbe River close to Fischbeck was carried out on behalf of the German Federal Institute of Hydrology (BfG) and showed promising results. Although, this modeling effort proved valuable for predicting flood levels, valuable time was lost providing editing data. To be able to apply these results for any future flood disaster management operations, the process of providing required data must be carried out in a more efficient and timely manner.

The problems associated with the operational use of the hydrodynamic model described in this study arise mainly from the availability of required data and computer capabilities. In addition, changes of the flooding situation in real time, for example, the width of the dike breach or water levels, need to be integrated in the simulation more efficiently. Especially suitable computer processors with enough capacities must be readily available in real time. Overall, the use of modeling techniques can certainly be valuable if the results of modeling processes can be used as a reliable base of information in operational decision processes.

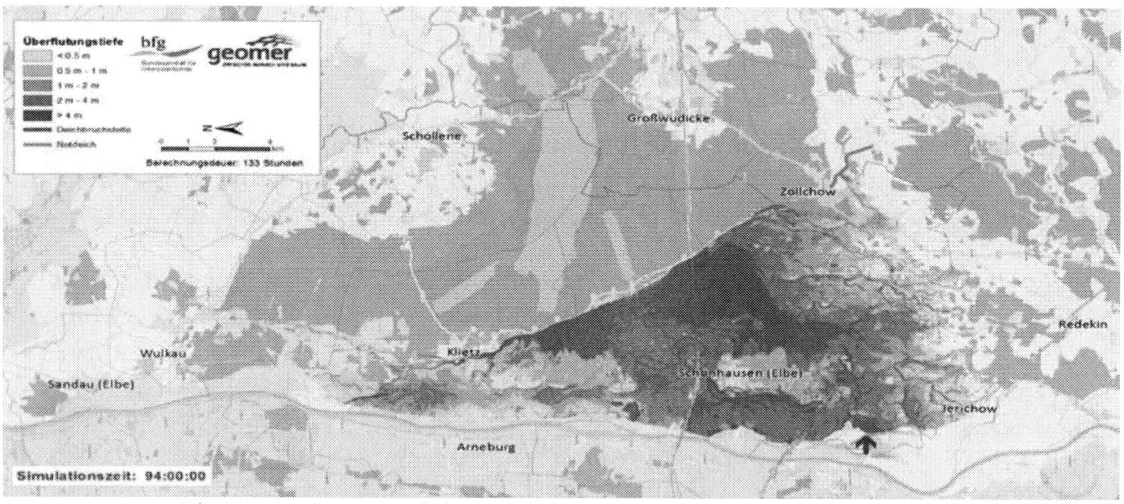

Figure 6. Extract of the modeling results of the Elbe crevasse

In a previous study, virtual scenarios of crevasses in the dyke were calculated in the context of flood control with the hydrodynamic simulation program FloodArea (Assmann et al, 2006) for the county of Stendal in 2008 and 2012. The aim was to identify the affected areas behind a dike prior to and after formation of a crevasse and to derive strategies of action and protection based on the modeling results. The hydraulic boundary conditions were defined for a worst case scenario. For instance, a dike breach of 100 m (around 330 ft) and a water level in the Elbe River of a flood event with a return period of 100 years plus an additional 0.5 m (around 1.6 ft) was assumed. In this way virtual crevasses for dike breaches at ten different places in the county of Stendal were simulated. The results of these simulations showed that the position of the dike breach is important for the dimensions of the flood wave behind the dike and that the manner the flood waters can spread in the area adjacent to breached dyke can differ significantly.

Each single crevasse, therefore, presents itself as a unique design case scenario. To completely cover alternative crevasse scenarios even more simulations must to be carried out.

In the specific case of the Fischbeck crevasse in June 2013 no previously calculated flooding scenarios were available for the emergency management forces. The calculated flooding scenarios did not match the situation that occurred in real time. Nevertheless, there was an urgent need for the entire emergency management to get an estimation of the size and the dynamics of the inundation as a result of the crevasse. In this context the usage of DEM-data proved to be extremely valuable and important basic data for further investigations and management of the flood in real time. The availability of an area-wide detailed digital elevation model (DEM) with a spatial resolution of 1m x 1m and vertical precision of 0.15m was used to derive information about the topography and the potential flooding situation. As a first step the original DEM (1m x 1m) was aggregated to a DEM5 with a spatial resolution of 5m x 5m to permit a better handling and performance. The deficit to a DEM with a resolution of 1m x 1m was neglected in this kind of usage. Finally, by applying the Geographic Information Systems (GIS) together with this topographic data, it was possible to visualise and compute information needed by the flood emergency management authorities (refer to Figure 7 below).

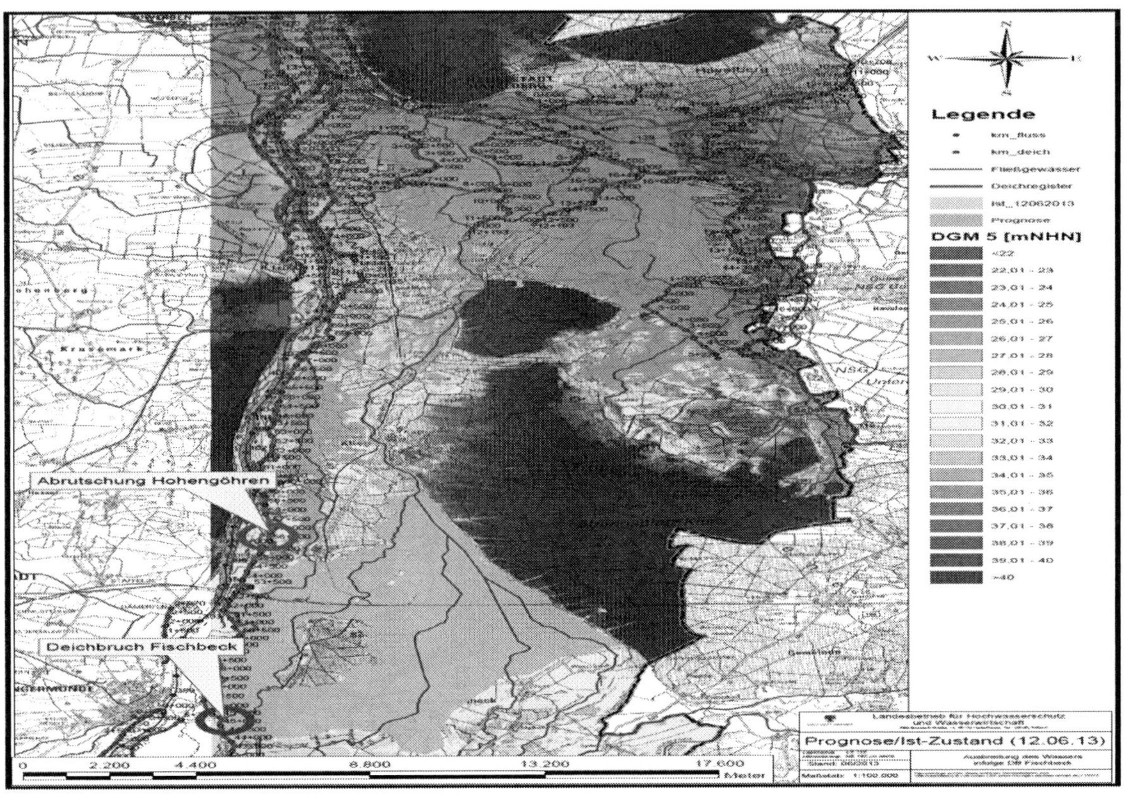

Figure 7. Map illustrating the existing inundated area (light blue) and forecasted inundated area (pink) on June, 12 (Juepner, Mueller and Weichel, 2013).

Local water levels in the fl ooded area were m easured at di fferent times after the dike break along the water-land-line. Considering the known slope (Figure 8/1) of the water surface (adopted from the already existing hydraulic modelling of the Elbe), the surveyed water levels were recalculated in an area wide GIS-datase t (Figure 8 /2). An additional new GIS raster dataset, which represented the theoretical water surface, was created with the same extent and spatial resolu tion of the DEM5. Because of the same spatial ex tent, the intersection of both raster da tasets could be realised (Figure 8 /3). In the next step the difference between DEM5 and the GIS-dataset of the theoretical water surface, bas ed on local measures, was calculated (Figure 8/4). The result of the intersection was a new GIS-dataset with potential inundation area and water depths based on a 5m x 5m resolution.

The results described above were drawn on a map and covered an area extending 30 km from the origin of the dike break (near Fisc hbeck) to the city of Havelberg (Figure 7). Based on this inform ation, the large-scale impact of the dike break flooding could be illustrated. The main advantage of this sim ple GIS-intersection over other m ethods is the relative short time to compute the results. In such extrem e situation, like the flood 2013, no detailed information about culverts, bridges or the stream network are necessary, if the primary interest is on the forecast of inundated areas by dike break.

Because of som e missing detailed information in the resu lts presented above an d to evaluate their accu racy, further inv estigation was carried o ut using daily observation flights over the flooded area. By using helicop ter an airborne based data acquisition was realized with this aerial reconnaissance. Beside aerial pictures, the delineation of the real flooded area was daily docum ented as well as video records by the on board technology of the Germ an Federal Police (Bundespolizei). The additional inform ation of the so called aerial reconnaissance was used to cr eate and update daily inundation m aps as instrument for the real-time recording and prediction for the emergency management.

Figure 8 illustrates a m ap of the actual flooded area on 12 th of June and the prediction of the further flooding, created by the GIS intersection. F igure 9 shows the flooded areas around the village of Hohengöhren using sim ulated water depths using GIS m odeling together with a measured reference point.

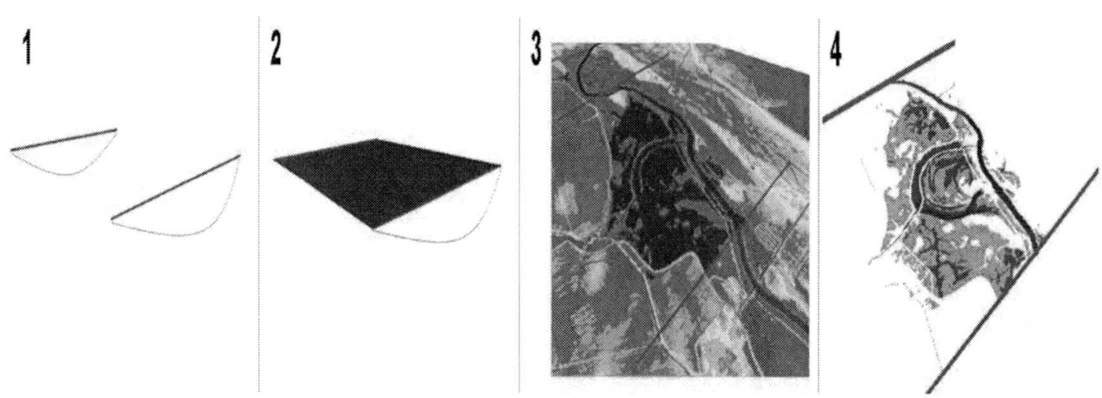

Figure 8. GIS calculation of flooding areas (Weichel, 2011)

Figure 9 below shows an example of a flooding situation in the village Hohengöhren, which is located around 5 km (around 3 mi) north of Fischbeck. The gradations in color stand for different depths of water.

Figure 9. Map of water depths calculated by the intersection of DEM5 and water surface based on local measurement

RECENT CATASTROPIC LEVEE (DIKE) BREACHES IN THE UNITED STATES

Over the past two decades some extreme flood events have occurred in the United States (refer to Table 2), mainly along the Mississippi River, that brought to limelight the inadequacies and deficiencies in an aging levee system built mostly by the US Army Corps of Engineers for the purpose of flood protection. What was deemed as an effective structural measure to reduce flood risk and damage has now become more of a liability by promoting a false sense of security for those communities living next to them and left them more vulnerable to future flood damage. This enhanced risk is even more so today because of physical impairment, inadequate and improper structural design of the levee system and lack of maintenance over the years. Furthermore, the shift in climatic patterns and regimes brought upon by global warming has resulted in more extreme and localized floods. The combined effect of all this caused some major breaches in the present levee system in the United States as observed during all the major floods listed in Table 2 below.

Table 2. Recent US Catastrophic Floods Floods in Chronological Over the Past Two Decades

Flood	**Period**	**Cause**	**Major Impacts and Damages**
The Great Mississippi Flood of 1993	May 1 – August 31, 1993	Wet Period during Fall of 1992 - above normal Snow Pack winter of 1993; heavy rainfall in the	Transporation disrupted; Levee Breaches; Damage to locks and dams.
New Orleans Flooding of 2005	August 29, 2005	Storm Surge and Flooding Caused by Hurricane Katrina	Over 50 failures of levees and floodwalls including three major levee breaches; More than 1800 lives lost; over 200 billion dollars in economic damages.
Midwest Flooding of 2008	June 2008	Heavy rainfall in upper midwest in parts of Indiana, Illinois, Iowa and Wisconsin	Major levee breaches in Des Moines and Cedar Rapids, Iowa forcing evavuations and causing extensive economic damage.
Mississippi River Flood of 2011	April –May 2011	Late winter snowmelt and heavy rainfall in early spring of 2011 and Tornado Super Outbreak of	Several breaches of levees in Missouri, Arkansas, Mississippi and Tennesse; Concerns about possible levee breaches in Baton

The vulnerability of the existing levees was clearly evident in 2005 when Hurricane Katrina came ashore the Gulf Coast near New Orleans, Louisiana. It caused major braches to a number of levees due to storm surge from the effects of Katrina. One of the main breaches occurred at 17th street in New Orleans as shown in Figure 10 below. The result was loss of over 1800 lives, evacuation of many communities and extensive property damage. The overall economic damage and disruption of lives was so extensive and noticeable, it prompted the Federal Government to pass Water Resources Development Act (WRDA) of 2007 (Public Law 110-114) – National Levee Safety Act of 2007. At the same time the Congress authorized the establishment of a National Committee on Levee Safety (NCLS) and charged it with developing a national levee safety policy. Section 9003 of Title IX of the WDRA act listed areas of concern to be addressed by the NCLS in a report to Congress.

The areas of concern identified by NCLS are: (1) Ensuring the protection of human life and property by levees through the development of technologically, economically, socially, and environmentally feasible programs and procedures for hazard reduction and mitigation relating to levees; (2) Encouraging use of the best available engineering policies and procedures for levee site investigation, design, construction, operation and maintenance, and emergency preparedness; (3) Encouraging the establishment and implementation of an effective national levee safety program that may be delegated to qualified States for implementation, including identification of incentives and disincentives for State levee safety programs; (4) Ensuring that levees are operated and maintained in accordance with appropriate and protective standards by conducting an inventory and inspection of levees; (5) Developing and supporting public education and awareness projects to increase public acceptance and support of State and national levee safety programs; (6) Building public awareness of the residual risks associated with living in levee protected areas; (7) Developing technical assistance materials for State and national levee safety programs (8) Developing methods to provide technical assistance relating to levee safety to non-federal entities; and (9) Developing technical assistance materials, seminars, and guidelines relating to the physical integrity of levees in the United States (adapted from the report by the National Committee on Levee Safety, 2009). Based on these the NCLS submitted 22 recommendations in a report to the Congress (NCLS, 2009).

Figure 10. Levee breach and resulting flooding at 17th street, New Orleans due to Hurricane Katrina in June 2005.

DISCUSSION AND CONCLUSION

The crevasse due to a breach in the right section of the main dike of the Elbe River close to Fischbeck caused a catastrophic event by flooding large areas behind the dike during the June flood of 2013. In the early stages of the flood more than 40 million m^3 of water was released each day through the dike breach into the area between Elbe and Havel Rivers. The waters spread quickly over a large area due to the flat landscape. Therefore the disaster action force worked under a lot of stress to save human life and to limit material damage. To do so information on the expected spatial extent of the flooding, the dynamics of the flooding, and the affected individual facilities like routes of evacuation or electric transformation stations became very important. The responsible water managers of the flood management staff of the LHW Saxony-Anhalt had to produce and provide this information to the disaster action force as quick as possible. This situation was new for all involved parties since it had never occurred before. Also there was no calculated "scenario of flooding" available. The use of the GIS based digital terrain model with selected measured local water levels seemed to be a promising approach to provide a quick estimation of the total flooding situation for the whole area with an acceptable accuracy and to repeat such procedures every 24 hours. Even though the DTM5 provides a spatially less accurate compared to the DTM1, the advantages of a better handling and a fast calculation justified its use in this situation where time was of the essence.

The hydrodynamic model used in this study could not have provided results with an acceptable accuracy for operational use without the simultaneous real-time monitoring with the help of helicopter surveying flights. Furthermore, the method required a large degree of improvisation and adaptation to the precise daily boundary conditions in real time during the disaster management operation. As a result important flooding situations could be estimated in advance with an acceptable accuracy. One particular problematic issue was the presence of many local facilities of flood protection like sandbag barriers. These facilities partially had a large influence on the local discharge dynamics and had to be included in the real-time monitoring and considered in the hydrodynamic model as they were being "discovered". Another problem was the collection of water level measurements. In the large evacuated flooding area all the relevant physical constraints also applied to the action force who wanted to obtain these important measurements. Only in due course of time the LHW succeeded in setting a reliable special flood level monitoring network. In this context the excellent cooperation with the German Federal Armed Forces and the German Federal Police needs to be emphasized. This led to obtaining needed results quickly and pragmatically. Furthermore it was hard to locate the boundaries of the inundated area with an acceptable accuracy just by air surveillance, especially in agricultural areas where the current spatial extent of the flooding were often hard to recognize. Even though development of a large crevasse in a dike along one of the largest rivers in Europe rarely occurs, the high number of (mostly smaller) crevasses during the flood events in the past shows the necessity to develop methodical procedures for event related calculation of the spatial extent and the dynamics of flood inundation and to derive practically useable approaches from this. The federal state of Saxony-Anhalt will immediately start acting in this direction according to their own statements.

In comparing the German flood experience along the Elbe River in year 2013 with the recent floods in the US it appears that in both countries levee or dike breaches occurred during major floods resulting from excessive rainfall. The underlying cause of the levee breach itself was not evident immediately in the Elbe river breach and with similar disasters in the US. In both cases structural failure due to soil erosion associated with the flood seems to be the underlying cause of the levee failure. The containment of the flooding due to a dike breach in real time is always an issue that needs to be seriously addressed if the damage is to be minimized. The use of ships to as an emergency measure to stop or contain the extent of flooding due to the dike breach along the Elbe River is a unique way the German flood managers decided to prevent the disaster from escalating further. Other ways to handle such a disaster in real time need to be closely examined wherever dikes or levees are vulnerable to breaches.

ACKNOWLEDGEMENTS

The present study, describing procedures used during the disaster operation in the middle section of the Elbe River during the flood of June 2013, would not have been possible without the assistance and cooperation of several organizations and individuals. First of all we would like to extend our sincere thanks to the State Service for Flood Protection and Water Management (LHW), especially the flood management staff, for providing most of the data used in this study and for the excellent and loyal cooperation we received from them during these difficult days during June 2013. Furthermore, the authors especially thank the helicopter pilots of the German Federal Police and State Police of Saxony-Anhalt for their loyal and dedicated support. Amongst the circle of many professional and voluntary aides, the authors would specially like to extend their appreciation to Mr. Hauptmann (Captain) Wolter, whose personal commitment was a great example to us.

REFERENCES

Assmann, A.: Grafe, M.; Runge, I., Thäger, F. (2006): Einflüsse des Berechnungsverfahrens und der Qualität der Grundlagendaten auf die Ermittlung überschwemmungsgefährdeter Gebiete. Hydrologie und Wasserbewirtschaftung, H. 1, pp 19 – 24.

Bundesanstalt für Gewässerkunde (BfG) (2013): Das Juni-Hochwasser des Jahres 2013 in Deutschland, BfG-1793, Koblenz.

ESRI (2011); ArcGIS Desktop: Release 10, Environmental Systems Research Institute, Redlands, CA.

Ettmer, B. (2011): Jahresbericht 2011 des Instituts für Wasserwirtschaft und Ökotechnologie (IWO), Magdeburg.

geomer GmbH (2013): FloodArea, ArcGIS-Erweiterung zur Berechnung von Überschwemmungsbereichen, Software-Handbuch, Heidelberg.

Jüpner, R. (2003): Erfahrungen aus dem operativen Hochwasserschutz: WASSER und ABFALL, 5/2003
Landesbetrieb für Hochwasserschutz und Wasserwirtschaft Sachsen-Anhalt (LHW) (2014): Bericht über das Hochwasser im Juni 2013, unveröffentlicht.

NCLS (2009): Recommendations for a National Levee Safety Program, a Report to Congress from the National Committee on Leveee Safety (Draft), January 15, 2009, Washington DC, USA pp 98.

Patt, H. and Jüpner, R. (2013): Hochwasser-Handbuch, 2. Auflage, Springer Berlin Heidelberg.

Weichel, T. (2011): Entwicklung eines Werkzeugs zur systematischen Bewertung der Grundlagen von Hochwassergefahrenkarten, Dissertation, Fachbereich Bauingenieurwesen der Technischen Universität Kaiserslautern, Shaker-Verlag, Aachen.

DAM REMOVAL AND RESTORATION METHODS: HEC-RAS MODELING OF THE FORMER CHAMBERSBURG RESERVOIR

Zachary R. Phillips[1]

ABSTRACT

Since the draining of the Chambersburg Reservoir and removal of the Birch Run Dam in 2005 the newly exposed channel and floodplain within the reservoir footprint has been rapidly eroding to establish a new equilibrium. As part of the removal project, 450 feet of channel within the newly carved dam breach were lined with limestone riprap. Perhaps due to channel armoring, the reach downstream of this area has shown significant bank erosion and instability. To determine the current hydraulic state of the site, the US Army Corps of Engineers Hydraulic Engineering Center – River Analysis System (HEC-RAS) was used to model the hydraulic impact of rip-rap armoring, and investigate possible impacts of wetland formation at the site. Results for modeling show elevated slopes, a larger bankfull geometry, and higher flow velocities in the rip rap lined reaches within the former embankment. The ability of the armored channel to channelize up to 10-yr flood discharges is also shown in modeling results. This is supporting evidence of susceptibility to scouring and bank erosion downstream of channel armoring. Additionally, channel hydraulics does not appear to be significantly controlled by secondary channel formation at the site, but further work may be necessary.

INTRODUCTION

The Chambersburg Reservoir (Adams County, PA) was drained in 2004 and the Birch Run Dam was removed from the Conococheague Creek in 2005. Since the completion of restoration work at the site in 2007, Shippensburg University Geography/Earth Science students have been monitoring the physical state of the site. Within the former area of inundation, GPS surveys have recorded the formation of a secondary channel and multiple channel avulsions at the site. Within the reach just downstream of the riprap armoring the channel has morphed into a braided pattern with large sandbar formations, many signs of bank failure, and flood-deposited sands along its banks, displaying that this reach is still morphologically active.

The Birch Run Dam was removed in 2005, after 72 years of service. It was a 700 foot long, 65-foot tall earthen dam with a concrete core wall. Johnsten et al. [2007] first documented the removal of the Chambersburg Reservoir, and Manuel [2009] established the first set of monitoring procedures to capture the site's geomorphic evolution. This study will focus on the riprap armoring in the vicinity of the dam, and the influence of wetland development on main channel hydraulics at the site. Figure 1 shows the geographic location of the former Chambersburg Reservoir amongst Adams County, Pennsylvania. The upstream watershed is located in the dense hardwood forests of the

[1] Master's Student, M.S. Geoenvironmental Studies, Geography/Earth Science Dept., Shippensburg University, PA 17257, zp9358@ship.edu

Michuax State Forest, and over half its drainage area flows into the up-shed Long Pine Run Reservoir.

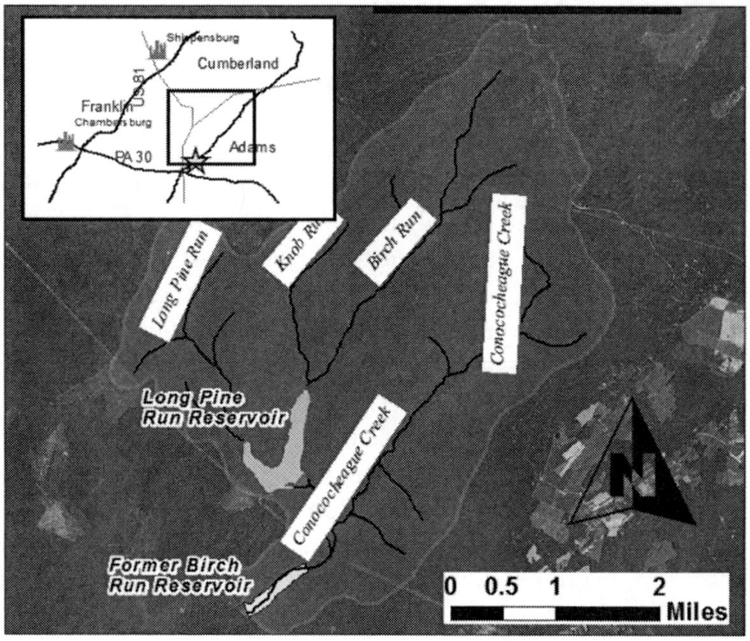

Figure 1. Map showing geographic location and watershed extent for the Chambersburg Reservoir study area. Star marks location of Chambersburg/Birch Run Reservoir

Cohorts at Shippensburg University in the Geography/Earth Science Department have been collecting data on the site every summer since 2008. Annually collected data includes main and secondary channel centerlines, Wolman pebble counts, stream cross-sections, and photo-station data. Throughout the monitoring period channels have meandered and jumped path, floods have deposited sands on downstream floodplains, a wetland has formed, and a connecting channel has joined the wetland to the main channel. So, this study is the first at the site to model the hydraulic effects of riprap on the reach immediately downstream and the influence of wetland formation on the site-wide hydraulics.

To begin connecting observations recorded in the field to numerical evidence, a HEC-RAS model was necessary to compute hydraulic characteristics for two designed experiments; one focusing on the riprap armored reach running through the leftover portion of the embankment, and a comparison of channel hydraulics for cross-sections either upstream or downstream of the wetland connector channel.

REVIEW OF LITERATURE

Dam Removal Examples in Pennsylvania

The Commonwealth of Pennsylvania has led the United States in numbers of dam removals for each of the last 11 years. The top three states for number of dam removals in

2013 were Pennsylvania (12), Oregon (8), and New Jersey (4) (American Rivers, 2013). Within the national spectrum, Pennsylvania accounted for 23.5% of the 51 total removals in the US for 2013. With Pennsylvania's large number of dam removals there exist various examples of removals for the state. American Rivers keeps a very impressive national database of dam removals through 2014.

In 2000, for approximately $5,000 dollars, the 5-foot tall by 75-foot wide Wild Lands Conservancy Dam was removed from the Little Lehigh Creek as part of a stream restoration demonstration project. The removal has improved substrate conditions for macroinvertebrate communities and the health of natural fisheries have also improved (American Rivers, 2007).

Another example of removals in Pennsylvania is the Hackenberg Dam. The Hackenberg Dam was removed in April of 2006, for a price tag of $6,000 dollars, and was the nearest upstream dam to Strodes Run's confluence with the Juniata River. The Hackenberg Dam was a small 4-foot by 35-foot concrete dam and was in danger of failure due to deterioration. The removal was conducted to relieve this danger, allow for fish passage, minimize erosion, and improve water quality (American Rivers, 2007).

Mill dams are very abundant in Pennsylvania, and the exact number is not known. One example of a mill dam removal is that of Sharrer's Mill Dam from the Conewago Creek. The dam was 6.5 feet tall and 260 feet long. During its operation the Conewago Creek breached the south bank of the dam and created a new creek. This obvious issue with the dam prompted for its removal in 2005 due to safety concern and erosional issues. The removal project removed the section of dam blocking the main channel, but left each abutment for stability.

Johnsten et al. [2007] and American Rivers [2007] reported on the removal of the 60-foot earthen Birch Run Dam being focused on in this study. The dam is considered one of medium height, and was considered the highest-risk dam in Pennsylvania before its draining in 2004. In comparison to other dams being removed in Pennsylvania, it is taller than most being removed. Before its removal, plans were proposed to refurbish the dam. These options to repair, replace, and lower the head level were priced from $16 million to $27 million. Although, the decision was made to breach and remove a portion of the dam to allow water to flow freely. The bill on the removal was $1.35 million, and it restored uninhibited flows to 4,000 feet of stream channel, for $337.50 per foot of channel (Johnsten et al., 2007).

RIPRAP ARMORING

Riprap is defined as a large, boulder or stone that is placed to protect earth surfaces from the forces of water waves or current (ASCE, 1962) and is heavily used in stream restoration when trying to protect channels from eroding. Riprap is a readily available material, thus is an economic selection as a natural method of erosion protection. Riprap systems are traditionally designed with the "Life-Cycle Approach," meaning that production, transport, installation, inspection, and maintenance of the riprap is taken into account prior to the start of a project (NCHRP, 2006). Each individual riprap boulder is designed to stay in place under peak flow scenarios, making it possible for reaches lined with riprap to be designed with unnaturally high channel slopes while still being seen as effective riprap design.

In designing effective riprap, the Isbash [1936] formula is used to compare gravitational and frictional forces that act upon each piece of riprap as water flows over it. The USACE criteria for effective riprap design states that riprap lining should extend to a downstream extent where erosional forces are no longer present and high enough on banks as to protect from flooding during maximum discharge events (USACE, 1980).

Lagasse et al. [2009] states that riprap can fail by being moved in any way, and that failures can occur even when riprap is "effectively" designed. With effectiveness in design based on staying in place and protecting banks in the areas of lining, the effects of riprap on reaches upstream and downstream are not the primary concern taken into account when evaluating a project.
For this project, channel hydraulics within the riprap lined reach were compared to natural channel reaches at the site. As seen in Figure 2, the riprap present within the area of the former embankment lines both the channel and bottom of the existing embankment so effects of the riprap are influential in both channel and floodplain hydraulics.

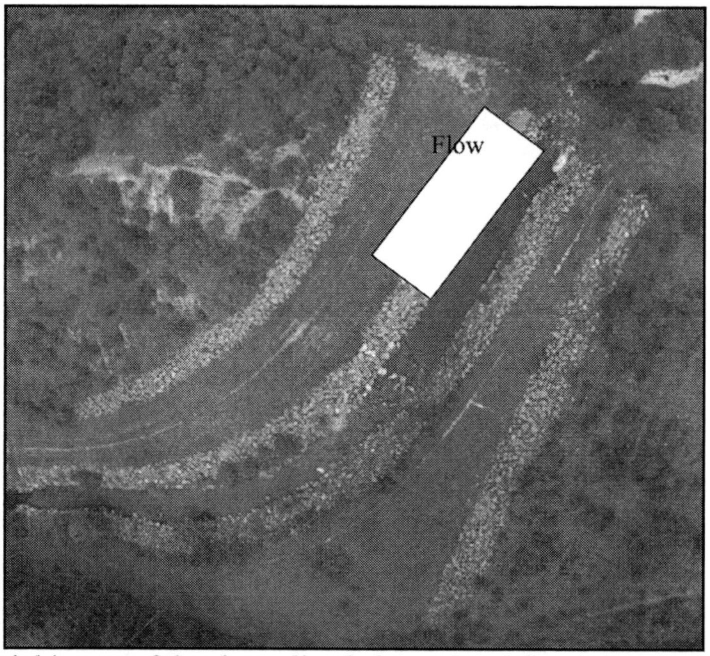

Figure 2. Aerial image of the riprap lined channel and floodplain in the area of the embankment removal. Flow is from top-right to bottom-left (Photo compliments of Dr. Scott Drzyzga).

UNITED STATES GEOLOGICAL SERVICE STREAMSTATS APPLICATION

The USGS StreamStats Application is an online GIS interface for delineating watersheds and estimating discharges for stream networks without gauging stations. Since geographic factors control runoff and channel formation, regression equations used are customized by state. For Pennsylvania, Roland and Stuckey [2008] developed a set of regional regression equations used to determine flow rates for the 2-, 5-, 10-, 50-, 100-, and 500-year discharge events. Additionally, Stuckey [2006] developed a set of low-flow, base-flow, and mean-flow regression equations for Pennsylvania. The use of this program allows for multiple flow magnitudes with known recurrence intervals to be estimated where gauging data may not be available.

US ARMY CORPS OF ENGINEERS HEC-RAS COMPUTER MODELING PROGRAM

HEC-RAS (or Hydraulic Engineering Center-River Analysis System) is a hydraulic modeling software package capable of modeling steady flow, unsteady flow, movable boundary sediment transport, and water quality for fluvial systems. The modeling program performs one-dimensional hydraulic calculations so that site-wide channel and floodplain hydraulics can be compared and visualized.

After running the model, cross-sectional results are ready for analysis. Each cross-section is broken up into either the "Left Overbank," "Channel," or Right Overbank." This allows for assessment of both floodplains and the channel as separate entities. Results that are output include a 3-D view of multiple cross-sections, a detailed summary for each cross-section, and a summary for the entire study area.

This study concentrated on the channel results for the following parameters; 1) Energy Gradient elevation and slope, 2) Top Width, 3) Flow Area, 4) Flow, 5) Channel Velocity and 6) Shear Stress. The comparisons of these numerical modeling results allows 'natural' channels at the site to be compared to the rip-rap armored reach, and for the impact of wetland/connector formation on site-wide channel hydraulics to be investigated.

METHODS

This study used the USGS StreamStats application and USACE HEC-RAS modeling platform to visualize and quantify the current hydraulic state of the Chambersburg Reservoir Birch Run Dam removal site under a series of flows. Modeling comparisons were conducted to display the effects of rip-rap armoring on local channel hydraulics and impact of wetland and connector channel formation on channel hydraulics across the site.

The methods used in this study were:

1. Flow data for HEC-RAS model input were estimated using the USGS StreamStats Application. The graphical user interface (GUI) was used to select a pour point for

the watershed to be delineated from. From there, the delineation tool uses elevation data, spatial hydrology data, and the flow estimation tool uses regional regression analysis to estimate discharge events for baseflow, 2-, 5-, 10-, 50-, 100-, and 200-year recurrence intervals.

2. The HEC-RAS model data was collected and assembled. A HEC-RAS model needs two types of data; geometric and flow data. Geometry data includes a stream "schematic" and cross-sections. For this project, a schematic was input, cross-sections were selected, and Manning's roughness coefficients were noted for each cross-section. Geometric data collection included 12 cross-sections for the riprap study, and 16 cross-sections for the site-wide comparison. Cross-sections were spaced throughout the study area so that important geomorphic features (i.e. pools, riffles, runs) were captured in the modeled stream reaches. The cross-section tool also allows for input of Manning's roughness data for each overbank floodplain section and channel at each cross-section. An example of the stream schematic and selected cross-sections for the site-wide study is seen in Figure 3.

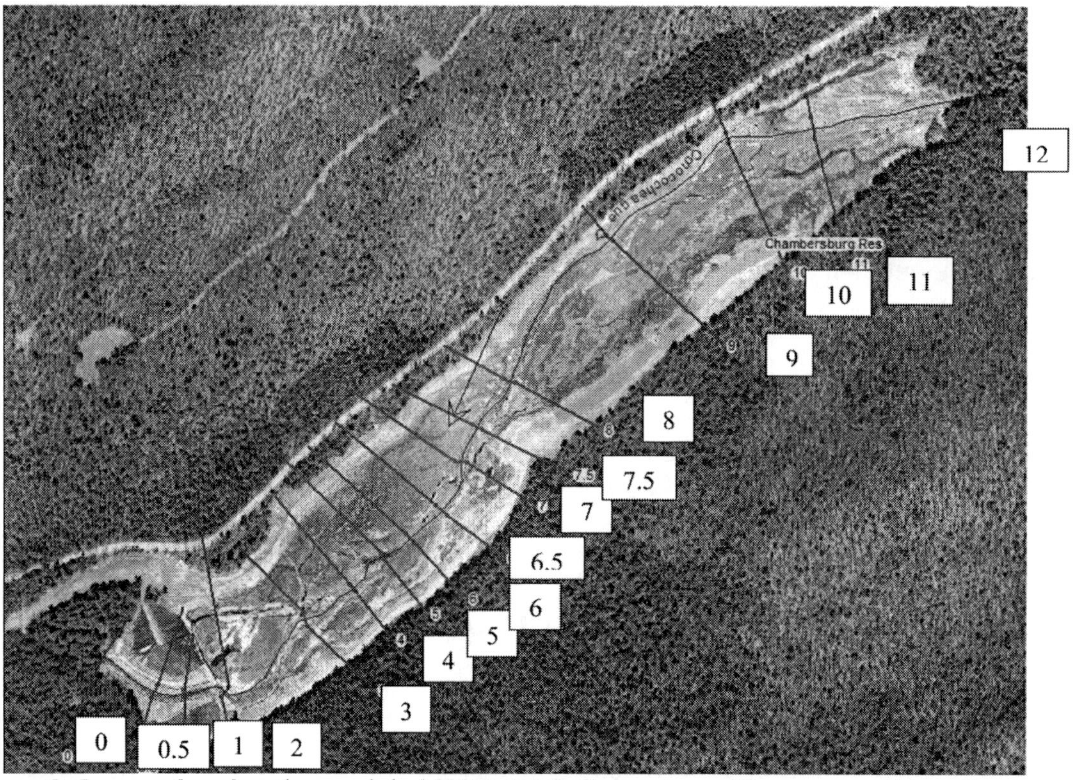

Figure 3. Image showing the modeled HEC-RAS Geometry with cross-sections chosen to capture significant geomorphic features across the site. The riprap lined reaches extend from cross-sections 0 to 2, and the wetland being investigated connects with the channel just downstream of cross-section 3.

For each cross-section; floodplain elevation profiles were extracted from a 1-m LiDAR-derived DEM, and channel bathymetry profiles were surveyed at 1-foot increments using a TOPCON ES-Series "Easy" Station. Data for Channel Roughness was estimated using the "Manning's 'n' Value Table" included in Chapter 3 of the HEC-RAS Hydraulic Reference Manual (USACE 2010). All surveying work was tied to the benchmark on top of the dam. Cross-sections were classified by their geomorphic location, i.e. whether they are located in a rip-rap lined reach, a reach within the legacy silt cap, deltaic reach, or "natural" reach. Classifications were formed using sediment data collected by Shippensburg University cohorts in the summer of 2008. This geomorphic classification map and the series of cross-sections are seen in Figure 4.

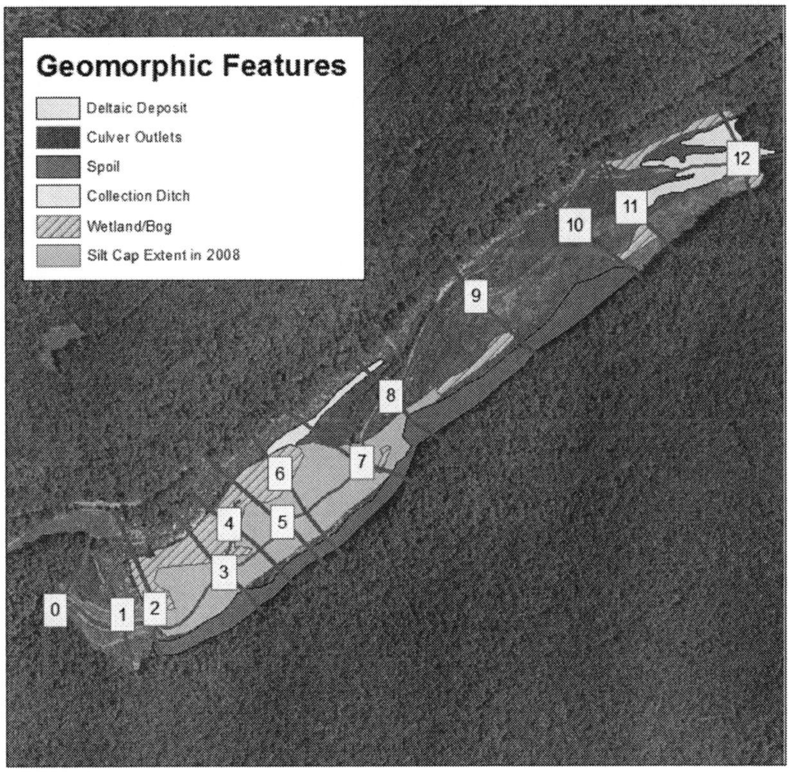

Figure 4. Geomorphic features within the former inundated area. For the tests, cross-sections were classified by their presence within a geomorphic feature. If the cross-section does not fall within one of the features, it was then classified as 'natural'

3. The model was run two separate times; once to acquire detailed results for riprap modeling, and once for the entire site so that wetland connection influences on channel geomorphology could be observed. Results were collected from the Cross-Section Output tables and exported to Microsoft Excel for analysis.

   Exported data were analyzed for each of the chosen tests for the study. For the model runs performed, the results displayed; 1) the impact of riprap armoring on local channel hydraulics, 2) the riprap's hydraulic impacts on adjacent reaches, 3) the effectiveness of the riprap armored reach in channelizing flood discharges higher than the bankfull (2-yr recurrence interval) discharge and 4) the effect of wetland formation (and the connecting of wetland and main channel via connector channel) on channel hydraulics.

For the comparison of the rip-rap armored reach at the site to 'natural' channel reaches, 12 more cross-sections were surveyed in the vicinity of the dam breach. These cross-sections are seen in Figure 5.

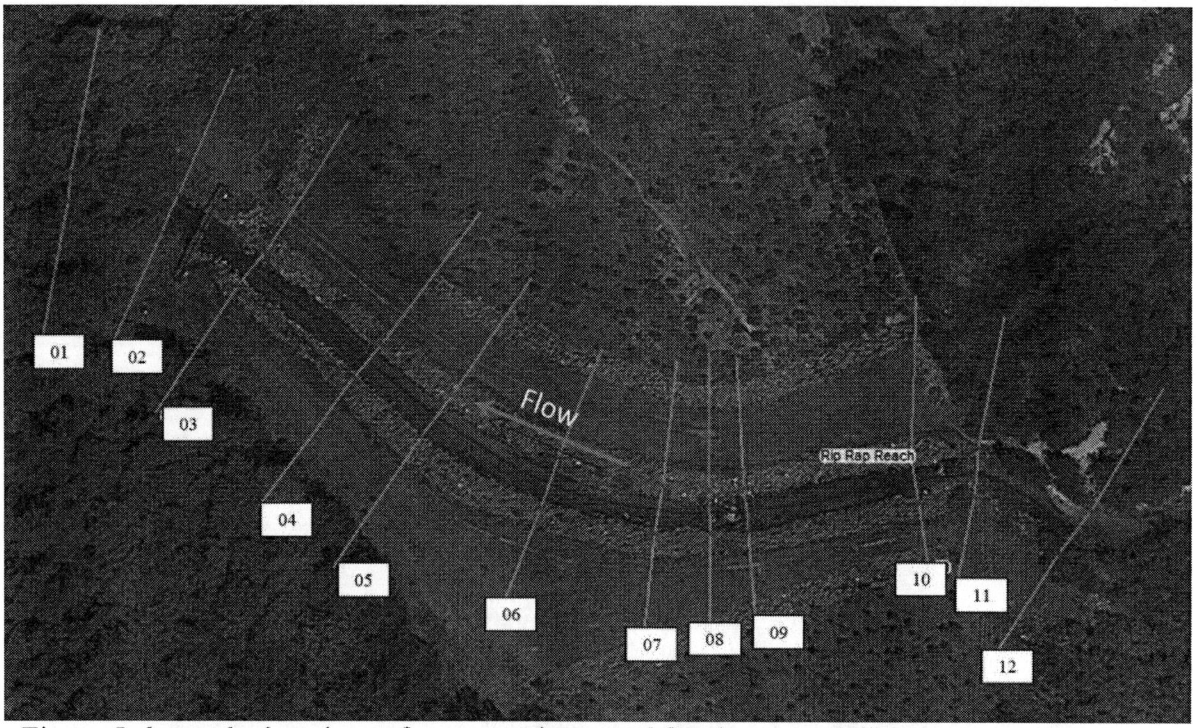

Figure 5 shows the locations of cross-sections used for riprap modeling. Cross-sections were placed to capture transitions in flow character upstream and downstream of the armored reach. Flow proceeds from cross-section 12, to cross-section 1.

Riprap lined cross sections (03 thru 10) were compared to 'natural' channel cross-sections upstream and downstream of the lined channel to observe the protected channel's ability to channelize floods, accelerate flow velocities, and control the energy gradient elevation and slope throughout the reach. The large bankfull volume, un-natural channel geometry and relatively steep slope of the riprap lined channel cause the deflection of flow energy upstream and downstream of the lined reach. The lined channel's inability to dissipate flow energy outside of its stream banks causes the need for flow energy to be dissipated either immediately downstream or upstream of the armored segment.

The hydraulic impacts of the riprap lined channel are seen in both visual and numerical modeling results. Both forms are included in the results section as a means to better connect the numerical model to the probable flow behavior. Numerical comparisons for; 1) Energy Gradient, 2) Top Width, 3) Flow Area, 4) Average Velocity, 5) Shear Force, and 6) Stream Power were used to perform the tests aforementioned. Visual results display the 2-yr and 10-yr discharges for general comparison. The 2-, 5-, 10-, 50-, and 100-year flow profiles were modeled in HEC-RAS once for the riprap lined reach geometry, and once for the site-wide model used to study wetland formation. To test the hydraulic impact of wetland and connector formation modeling results for cross-sections upstream of the connector were compared to those downstream. So, for this portion of the study, cross-sections downstream of cross-section 3 were compared to those upstream. Results for this portion of the modeling are presented numerically because visual modeling results did not obviously display any variation between upstream and downstream cross-sections.

RESULTS

HEC-RAS modeling was used to display the hydraulic impacts of rip-rap lining on the channel and floodplains and impact of wetland/connector formation on upstream and downstream cross-sections. Flow data estimated from the USGS StreamStats Application was found to be somewhat inaccurate. Errors range from 12% to 38%, and predictions are equivalent to up to 13 years of flow records (a tremendously low number when dealing with up to 100-year peak flows). Estimates for 10-yr baseflow, as well as weighted estimates for 2-yr peak, 5-yr peak, 10-yr peak, 50-yr peak, and 100-yr peak flood discharges are shown below in Table 1. Errors could possibly lead to errors in modeling bankfull discharge. Regardless, due to lack of flow data, discharge values from StreamStats were still used for HEC-RAS modeling.

Table 1. Results of USGS StreamStats application using nearby stream gages (USGS gage 01614090) and regression equations to estimate flow data for the undammed channel

Estimated Streamflow Statistics					
Flow Profile	Statistic	Flow (ft^3/s)	Flow (cms)	Prediction Error (%)	Equivalent yrs of record
1	10-yr Baseflow	4.25	0.12	12	
2	2-yr Peak Flood	152	4.30	28	4
3	5-yr Peak Flood	326	9.23	26	7
4	10-yr Peak Flood	496	14.05	28	10
5	50-yr Peak Flood	996	28.20	33	13
6	100-yr Peak Flood	1270	35.96	38	13

Each of the estimated flows was input into HEC-RAS modeling, and cross-sections were classified as 'riprap' or 'natural', and as being upstream or downstream of the wetland connector channel. To study the recent morphology of the site, HEC-RAS modeling results below were produced with the 2-yr flood discharge. Work by Williams [1978] suggests that the recurrence interval of effective discharge for 35 floodplains studied on streams in the US ranged anywhere from once every 1 to 2 years. For the Chambersburg Reservoir site, the effective discharge was estimated to be about two years. Thus, results for comparisons were produced with the 2-yr peak discharge HEC-RAS results, averaged individually for each cross-section.

Results for HEC-RAS modeling of riprap impacts are shown below in numerical form (Table 2). Modeling included that of baseflow, bankfull (2-yr), 5-yr, 10-yr, 50-yr and 100-yr discharges estimated from USGS StreamStats to indicate the lined channel's ability to channelize flooding.

To investigate bankfull hydraulics, the 2-yr interval was used to produce results seen in Table 2. Significantly, in Table 2, "Top Width" displays whether the water is fully contained within the channel. Larger values are seen in the 'natural' sections, meaning that flow is overtopping one or both of the banks at the 2-yr recurrence interval discharge. For example, Cross-sections 12, 11, 2, and 1 appear to have flow that has entered the floodplains. Diagrams summarizing the relationship between channel velocity, water surface and channel bottom elevation, and shear stress of results are seen in Figure 6a and b.

Table 2. Cross-section averaged HEC-RAS modeling results for a 2-yr peak discharge event. 'Natural' and riprap lined sections are noted in the "Class" column to allow for comparison.

Cross-section	Class	Min. Ch El.	W.S. Elev.	E.G. Elev.	E.G. Slope	Vel. Chnl.	Shear Chnl.	Flow Area	Top Width
	(type)	(ft)	(ft)	(ft)	(ft/ft)	(ft/s)	(lb/sq ft)	(sq ft)	(ft)
12	Natural	1039.07	1040.99	1041.39	1.155	4.06	88.18	34.76	39.21
11	Natural	1037.54	1041.07	1041.1	0.001	1.45	0.11	138.3	106.82
10	Riprap	1035.59	1040.91	1041.03	0.033	1.64	5.82	84.88	28.29
9	Riprap	1036.17	1038.06	1038.56	0.019	5.68	1.38	26.78	20.53
8	Riprap	1035.02	1037.45	1038.07	0.029	6.28	1.78	24.2	20.46
7	Riprap	1033.03	1036.19	1036.4	0.003	3.67	0.34	41.39	19.17
6	Riprap	1033.26	1035.85	1036.12	0.006	4.2	0.49	36.22	23.18
5	Riprap	1032.67	1035.58	1035.76	0.003	3.41	0.31	44.59	24.1
4	Riprap	1032.87	1035.05	1035.47	0.010	5.18	0.76	29.33	19.73
3	Riprap	1031.29	1033.86	1034.46	0.014	6.21	1.09	25.12	25.77
2	Natural	1030.39	1034.08	1034.14	0.001	2.06	0.18	80.65	52.2
1	Natural	1030.11	1034.02	1034.04	0.002	1.42	0.39	152.63	97.71

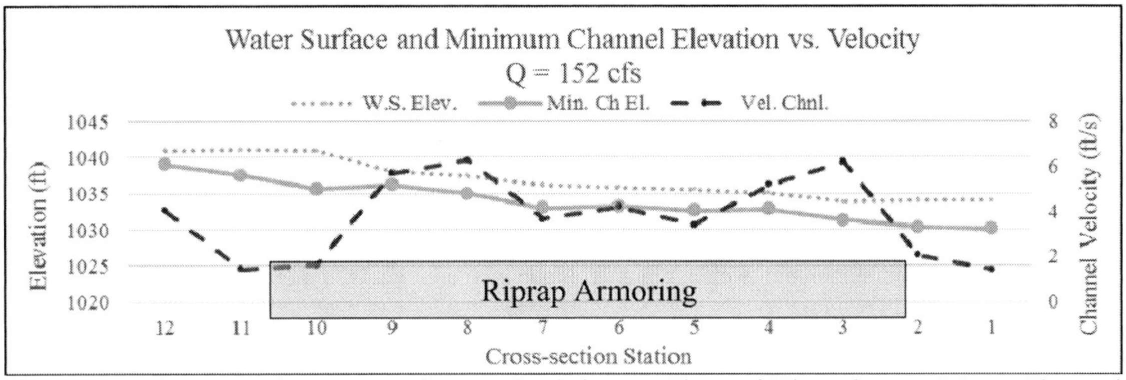

Figure 6a. Diagram of Water Surface and Minimum Channel Elevation vs. Mean Channel Velocity. Grey bar indicates the extent of the riprap lined reach.

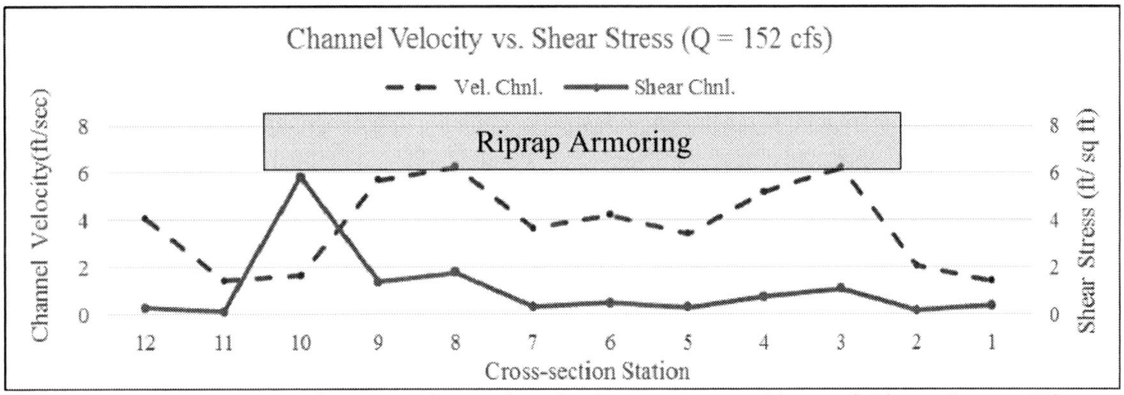

Figure 6b. Diagram of Mean Channel Velocity vs. Mean Channel Shear Stress. The extent of riprap lining in the channel is noted by the grey bar.

In modeling, average energy gradient slope was much higher in the riprap protected reaches (0.014) than in 'natural' reaches (0.001). Consequently, flow velocity within the channel is also much higher in riprap protected segments. The mean flow velocity for riprap armored segments is 4.53 ft. per second. For 'natural' segments, the mean flow velocity in the channel is 1.64 ft. per second. Shear stress is also higher within the riprap lined reaches. On average, an extra 1.27 lbs. per sq. ft. of friction is exerted on the channel bottom within riprap lined sections. Flow area and top width are also dramatically different. The flow area seen in riprap lined sections is 38% of the flow area averaged in 'natural' reaches. The average top widths in riprap lined and 'natural' reaches are 22.65 ft. and 73.99 ft., respectively. These results are viewable in Table 3.

Table 3. Averages for Energy Gradient Slope, Channel Velocity, Channel Shear Stress, Flow Area, and Top width for riprap lined cross-sections and 'natural' cross-sections. Results produced using the bankfull (2-yr recurrence) discharge.

	Riprap Averages	'Natural' Averages
E.G. Slope (ft/ft)	0.014	0.001
Vel. Chnl. (ft/sec)	4.53	1.64
Shear Chnl. (lb/ sq ft)	1.50	0.23
Flow Area (sq ft)	39.06	101.59
Top Width (ft)	22.65	73.99

Overall, the detailed riprap modeling revealed some interesting characteristics of the riprap lined reach. Using the estimated 2-yr flood discharge from USGS StreamStats, flow overtops the banks upstream and downstream of the armored reach. This result, seen in Figures 7a, b, and c, is best observed in visual modeling outputs. Massive channelization is also seen in Figures 7b and c. For the 2-yr recurrence flood discharge, flow is fully channelized between cross-sections 4 and 10. For the 10-yr discharge, flow is still channelized, but only between sections 4 and 9. This channelization of the 10-yr flood discharge causes a tremendous surplus of energy to be carried downstream, and dissipated along the unprotected banks of cross-sections 1 and 2.

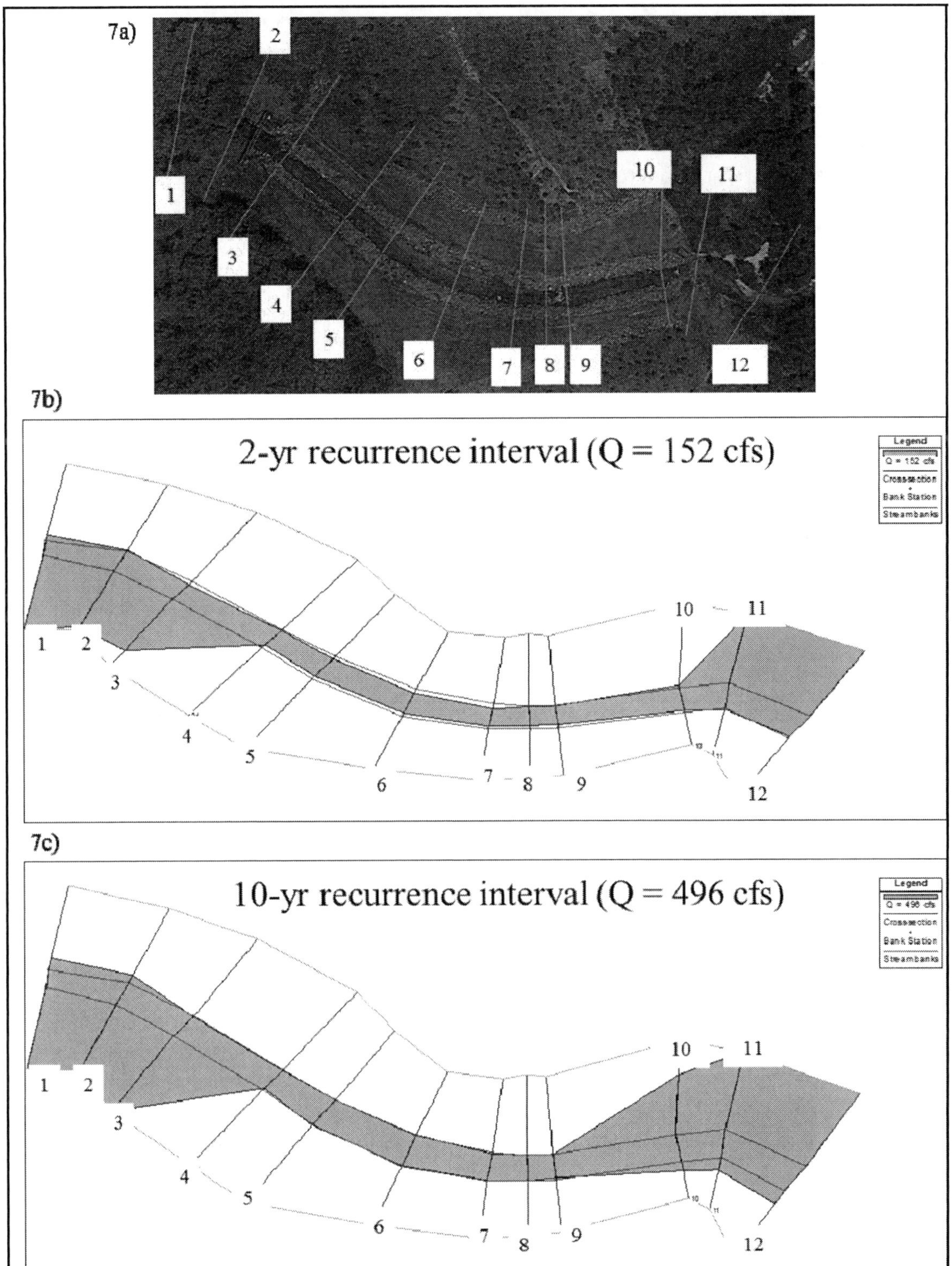

Figure 7a, b, and c. Figure 7a shows locations of cross-sections for riprap modeling. Figure 7b shows modeling results for the 2-yr recurrence discharge. Figure 7c displays results for modeling of the 10-yr flood discharge.

WETLAND CONNECTOR INFLUENCE

The influence of wetland connection was investigated by dividing the cross-sections into two groups for comparison. Cross-sections were classified as being either upstream or downstream of the secondary channel. In total, 11 cross-sections were upstream and 5 were downstream. Complete results are seen in Table 4.

Table 4. Results of HEC-RAS modeling for the 2-yr recurrence interval discharge (q = 152 cfs). Channels are classified as being upstream or downstream of the wetland connector, and are also classified by their geomorphic class.

Loc.	Above/Below Connector	Geomorph Class	W.S. Elev.	E.G. Elev	E.G. Slope	Vel. Chnl.	Shear Chnl.	Flow Area	Top Width
	(Up/ Downstream)		(ft)	(ft)	(ft/ft)	(ft/s)	(lb/sq ft)	(sq ft)	(ft)
12	Upstream	Deltaic	1086.40	1086.68	0.007	4.24	0.45	35.82	34.22
11	Upstream	'Natural'	1080.40	1080.54	0.013	3.13	0.42	53.42	143.88
10	Upstream	'Natural'	1076.16	1076.45	0.013	4.35	0.69	35.87	50.38
9	Upstream	'Natural'	1068.04	1068.44	0.015	5.09	1.10	29.88	24.51
8	Upstream	'Natural'	1059.09	1059.47	0.012	5.01	0.67	31.62	44.09
7.5	Upstream	Silt Cap	1056.73	1057.01	0.010	4.58	0.55	40.25	79.46
7	Upstream	Silt Cap	1054.29	1054.62	0.013	4.83	0.64	33.92	52.06
6.5	Upstream	Silt Cap	1051.93	1052.32	0.010	5.19	0.66	33.72	56.70
6	Upstream	Silt Cap	1050.10	1050.43	0.010	5.06	0.81	36.81	57.83
5	Upstream	Silt Cap	1048.41	1048.69	0.007	4.68	0.52	41.32	68.64
4	Upstream	Silt Cap	1047.06	1047.23	0.010	4.26	0.50	52.64	150.19
3	Downstream	Silt Cap	1043.64	1043.92	0.027	4.40	1.45	36.23	59.71
2	Downstream	Silt Cap	1040.13	1040.47	0.003	4.81	0.21	38.94	80.95
1	Downstream	Riprap	1038.63	1039.17	0.016	5.93	1.06	25.63	23.72
0.5	Downstream	Riprap	1036.69	1037.18	0.016	5.67	1.00	26.82	27.32
0	Downstream	'Natural'	1033.47	1033.76	0.007	4.46	0.55	44.26	54.11

The averages for 'Energy Gradient Slope,' 'Mean Channel Velocity,' 'Flow Area,' 'Shear Stress,' and 'Top Width' were calculated and compared. Because there are several different geomorphic classifications upstream and downstream of the connector channel, a separate comparison for those cross-sections falling only under the geomorphic classification of 'natural' and silt cap was also conducted.

Results for the comparison of cross-sections upstream and downstream of the wetland connector channel are viewable in Table 5. The comparison of cross-sections including all of the sections averaged 16 cross-sectional outputs. The comparison utilizing only cross-sections classified as 'natural' used 13 cross-sections. It is expected that the averages included in the comparison using only silt cap and 'natural' classes would be more significant due to their lack of impact from armoring. Thus, all of the riprap lined cross-sections and the deltaic cross-section were not used in this portion of the comparison.

Table 5. Summary of average hydraulic characteristics of cross-sections upstream and downstream of the wetland connector channel. Results have been separated into one comparison including all cross sections and, to deal with the issue of geomorphic control on channel shape, one including only the classifications of 'natural' and silt cap.

	Wetland Connector Impact Results (all cross-sections)		Wetland Connector Impact Results (silt cap and 'natural' cross-sections)	
	Downstream	**Upstream**	**Downstream**	**Upstream**
E.G. Slope (ft/ft)	0.014	0.011	0.012	0.011
Vel. Chnl. (ft/s)	5.05	4.58	4.56	4.62
Shear Chnl. (lb/sq ft)	0.85	0.64	0.74	0.66
Flow Area (sq ft)	34.38	38.66	39.81	38.95
Top Width (ft)	49.16	69.27	64.92	72.77

CONCLUSIONS

From the comparisons included in this study, which were direct comparisons of; 1) the "natural" channel forms at the site and those that are rip-rap armored, and 2) the hydraulic characteristics of cross-sections either upstream or downstream of the secondary channel that connects the main channel to the wetland at the site.

The comparison of "natural" and rip-rap armored reaches at the site revealed that the riprap reach has a slope that is 0.13 feet steeper per linear foot. This points to the riprap reach as having an unnaturally high slope. The flow area and top width are also useful in projecting how efficiently flow is channelized downstream. Within the riprap lined cross-sections, values for the flow area and top width were much smaller than those seen in the 'natural' reaches. In the riprap armored reach flow area and top width are 39.06 square feet and 22.65 feet, respectively. For the same values in the 'natural' cross-sections, flow area is 101.59 square feet and top width is 73.99 feet. This means that, for bankfull flow, the flow area for riprap armored channels is 38.5% of the same discharge's flow area in 'natural' channels. Similarly, the top width displays the same channelizing effect. Results seen in Table 2 show flow at its highest velocity as it exits the riprap lined reach. The exit velocity for flow leaving the riprap-lined reach is 4.76 ft. per second faster than the entrance velocity for the lined reach. This is significant because the reach immediately downstream is not rip-rap lined, and flow may readily scour sediment due to excessive velocities and shear stresses entering the wooded reach downstream. Additionally, the elevated velocities within the rip-rap lined reaches would also more actively transport sediment. Sediment being transported out of the riprap lined reach is capable of further clogging channels and burying ecosystems downstream, having a negative impact on the effectiveness of channel design.

Lastly, the impact of secondary channel formation on main channel hydraulics was studied by comparing cross-sections upstream and downstream of the connector channel.

These results showed that cross-sections with similar morphological classes were slightly steeper downstream, had a slightly higher average flow velocity, higher shear stress, and lower flow area and top width. There isn't any significant data that shows that hydraulics are being impacted across the site due to the secondary channel forming. But it is possible that the upstream channel slope is being controlled by the connector, due to the addition of sediment and disruption of flow at its confluence with the main channel.

Overall, the site seems to have the highest slopes and channel velocities within the rip-rap lined reach. The rip-rap is effective in protecting the channel and remaining dam embankment from scour, but its design also seems to accelerate water with its unnaturally high slopes and linear shape. The riprap is extremely successful in channelizing flood discharges due to its large bankfull area. The secondary channel does not seem to have any effect on channel hydraulics, but may be a control of channel slope as it adds sediment downstream. With the 2013 installation of a series of upstream pointing, V-shaped grade control log-weirs, further work will investigate the local and reach-wide impact of these structures.

ACKNOWLEDGEMENTS

I would like to thank the Shippensburg University Graduate Research Grant for aiding in the dissemination of project results. Special thanks goes out to my parents, I would never be here without your support. Thank you to Molly Moore and Caz Guzowski for the help with field surveying work. And finally, thank you to Drs. Scott Drzyzga, Christopher Woltemade, and Sean Cornell. Your guidance throughout my research has been all that I could ask for.

REFERENCES

American Society for Civil Engineering, 1962. "Nomenclature for Hydraulics," *Manuals and Reports on Engineering Practice*, No. 43, NY

Isbash, S.V., 1936. "Construction of dams by depositing rock in running water," *Transactions, Second Congress on Large Dams*, Vol. 5, p 123-13

Lagasse, P.F., P.E. Clopper, J.E. Pagán-Ortiz, L.W. Zevenbergen, L.A. Arneson, J.D. Schall, L.G. Girard, 2009. Bridge Scour and Stream Instability Countermeasures Experience, Selection and Design Guidance, *HEC-23 Volumes 1 and 2*, 3rd Ed.

Roland, M.A., and Stuckey, M.H., 2008, Regression equations for estimating flood flows at selected recurrence intervals for ungaged streams in Pennsylvania: *U.S. Geological Survey Scientific Investigations Report* 2008-5102, 57 p.

Stuckey, M.H., 2006, Low-flow, base-flow, and mean-flow regression equations for Pennsylvania streams: *U.S. Geological Survey Scientific Investigations Report* 2006-5130, 84 p.

USACE. 2010. HEC-RAS River Analysis System, Hydraulic Reference Manual (Version 4.1), USACE Institute for Water Resources, Davis, CA

Williams, G. P., 1978. "Bankfull discharge of rivers," *Water Resources Research*, 14(6), 1141-1154.

CONSTRUCTION FLOOD CASE HISTORIES

Daniel D. Mares[1]
David P. Keeney[2]

ABSTRACT

The purpose of this paper is to review case histories of significant floods that have occurred during construction of Bureau of Reclamation (Reclamation) dam modifications, the meteorological and hydrologic conditions that occurred, the design construction flood, the response of the cofferdams or design features to the loading and any operational changes in response to the flood conditions. The case histories are useful in reviewing the design floods selected, the cofferdam type, and in reinforcing the importance of evaluating risks during construction. The meteorological data (i.e. rain gauge observations, synoptic maps, NCAR/NCEP reanalysis, and radar) were reviewed and a summary of the meteorological conditions is presented.

INTRODUCTION

Flood conditions have occurred during construction at many of Reclamation's dams. As an example, in 1909, a large flood of record occurred during construction at one of Reclamation's dams. Inflows were 50 percent greater than any previous recordings. By the middle of June, with the main dam partially completed, the reservoir continued to fill. Based on the projected inflow estimates, it was concluded that the dike site was in danger of overtopping. Work to construct the dike was advertised, but all bids were rejected because the cost was considered too high. The use of Government forces was authorized and two 10-hour shifts were used to construct a small 4-foot-high "emergency" dike. The crisis ended on July 10 when the reservoir reached an elevation just 2 feet below the top of the emergency dike.

In the past, risks during construction have primarily focused on damage to the work in progress. The contractor was given the historical flood inflow hydrographs and related data, and the contractor designed the diversion capacity and cofferdam based on risks associated with damages to the work under construction. In the last 30 years, very few new dams have been built by Reclamation, but many dams have been modified to address dam safety concerns. The modification of existing dams actually presents different and, in some cases, more significant challenges than those encountered during the initial construction of projects due to the larger potential downstream consequences.

The case histories in this report include construction floods that occurred at Jackson Lake Dam in 1986, Theodore Roosevelt Dam in 1993, and Glendo Dam in 2010 and 2011. These case histories present in this paper are useful in reinforcing the need to evaluate the

[1] Civil Engineer, Waterways and Concrete Dams Group 2, Bureau of Reclamation, Technical Service Center, Denver, CO; 303-445-3252; dmares@usbr.gov.
[2] Meteorologist, Flood Hydrology and Consequences Group, Bureau of Reclamation, Technical Service Center, Denver, CO; 303-445-2533; dpkeeney@usbr.gov.

potential risks during construction, and to perform the studies necessary to determine the appropriate design flood and operational considerations. Reclamation has developed some considerations for evaluating risks during construction (e.g. Best Practices in Dam and Levee Safety Risk Analysis, 2012 and Appurtenant Structures for Dams, Spillway and Outlet Works, Design Standards 14, 2013) however, because of the responsibility associated with these decisions, each owner and experienced design engineer needs to evaluate each situation on a case by case basis.

Due to advances in meteorological forecasting capabilities, it is also important to understand the atmospheric and antecedent conditions that contributed to the flood conditions. This report includes selected meteorological data for each event.

THEODORE ROOSEVELT DAM MODIFICATION

Background and Description

Theodore Roosevelt Dam is a concrete, gravity-arch structure on the Salt River, located northeast of Phoenix, Arizona. The original dam was a cyclopean masonry structure completed in 1911. Theodore Roosevelt Dam was identified as requiring dam safety modifications in 1978.

Figure 1. Photograph of Theodore Roosevelt Dam under construction showing the cellular cofferdam, thrust block and spillway on left abutment. This photograph shows the restored construction site after the flood event.

The dam modification construction began in 1991 and was completed in 1996. The completed modifications to Theodore Roosevelt Dam increased the structural height of the dam by 77 feet with the construction of a mass concrete overlay on the downstream face and above the original dam crest. Modifications also included gated spillways (Figure 1) constructed on both abutments within massive thrust blocks, and a new river outlet works constructed through the left abutment for reservoir evacuation. The original 36 MW powerplant was modified for operation under higher reservoir heads and tailwater.

Theodore Roosevelt Dam Flooding Event: January 6 through 19, 1993

A series of high precipitation events occurred over central Arizona between January 6 and January 19, 1993. These events corresponded to record rainfall and high inflows into Theodore Roosevelt Dams during construction. During these events, a cellular cofferdam was in place and the left mass concrete thrust blocks and spillway were partially constructed at Theodore Roosevelt Dam.

Meteorological Setup for the Event

A long lasting El Nino phase of the El Nino Southern Oscillation (ENSO) was in place from 1990 until 1996. El Nino typically causes the jet stream to transport tropical moisture from near the Equator northward into the Desert Southwest. Even though only a weak El Nino was present during the winter of 1992-93, it was the impetus for the flooding event that occurred. In December, a series of Pacific storms, occurring at a rate of nearly one storm per week, caused saturated soil conditions across much of central Arizona and a slightly above average snowpack in the mountains above 6,000 feet.

This flooding event featured four main storms that caused the majority of the precipitation that occurred over the Theodore Roosevelt Dam watershed. All four storms are considered Atmospheric River events (narrow corridors of water vapor transport, usually a few thousand feet above the surface of the Earth). Atmospheric River events are characterized by intense, widespread rainfall that is often caused by the forced ascent of the water vapor band by a topographic boundary. On January 6, a large storm fueled by the subtropical jet stream moved onshore in California. The subtropical jet stream was able to carry ample amounts of moisture into the Desert Southwest of Arizona. As a result, the snow level rose above 8,000 feet from the initial 6,000 feet over much of Arizona which resulted in a rain-on-snow event for the first portion of this heavy rain event. The SNOTEL site near Workman Creek, Arizona (elevation 6,900) recorded that, during this timeframe, all of the previous snow melted due to the rain (e.g. NRCS, 2014). Given the antecedent conditions of saturated soils, the rain-on-snow produced abundant and rapid runoff. Many locations received between 2 to 4 inches of rain with the first storm. With the arrival of the second storm system around January 10, the snow level dropped back down below 7,000 feet and remained there through the rest of the event which caused the snowpack to begin building again. This storm system brought widespread rainfall amounts of 1 to 3 inches across much of Arizona below 7,000 feet. The third storm system arrived around January 13 and rainfall amounts generally less than 1.5 inches fell over the watersheds. The final system arrived around January 18 and lasted about 24 hours. It produced rainfall amounts in the 1- to 3-inch range. The last three storm systems caused the snowpack, especially at elevations above 9,000 feet, to become quite large, with point rainfall return periods calculated as high as 50 to 60 years (e.g. Bonnin, Martin, Lin, Parzybok, Yekta, and Riley, 2004).

It is common for Arizona to receive relatively large amounts of precipitation during the winter (December, January, and February). On average, Theodore Roosevelt Dam receives nearly one third of its annual precipitation during these three months. Only during the month of August in the Monsoon season does it receive more rain (e.g. House 1993). However, the January 1993 precipitation totals were between 300 to 500 percent of the normal amount of precipitation that occurs in an average January (e.g. NWS, 2014, Office of the State Climatologist, 1993, and USGS, 1993). At Theodore Roosevelt Dam, the 14 day rainfall total was 10.8 inches, which corresponds to a return period of about 200-years (e.g. Bonnin, Martin, Lin, Parzybok, Yekta, and Riley, 2004).

Diversion Plan and Performance of Cofferdam

The diversion plan for Theodore Roosevelt Dam consisted of staged construction, so that one spillway structure was available at all times to provide protection for up to a 25-year flood event; a cellular cofferdam to provide protection for the spillway and thrust block under construction; a reservoir drawdown of 21 feet from the crest of the cofferdam to provide about 341,000 acre-feet of storage space; and provisions for diversion flows through a diversion pipe line extending downstream about 530 feet from the river outlet works control structure. The 25-year design flood had a peak inflow of 122,000 ft^3/s and a 5 day volume of 360,000 acre-feet. The staged construction of the thrust blocks and spillways allowed releases from the existing right spillway gates until the left thrust block and new spillway were completed. The diversion pipeline capacity was 6,000 ft^3/s and permitted releases around the unwatered tailrace area during construction.

The cellular cofferdam consisted of three cells and two connecting arcs with two connecting concrete gravity walls on each end for closure. The cells were 37.6 feet in diameter and the arcs had a radius of 11.4 feet. The cellular cofferdam sheet piles were installed and the sand fill was capped with a one foot thick concrete slab reinforced with welded wire fabric. The concrete cap was provided to prevent erosion of the sand within the cellular cofferdam in the event that overtopping of the cofferdam structures occurred. The cellular cofferdam incorporated a couple of features that were noteworthy. Concrete toe buttresses were constructed on the downstream sides of the cellular cofferdams to provide support of the cells and provide some protection for overtopping. Remote monitoring of piezometric levels within the cells was also done during overtopping. The gravity walls were constructed with a two foot high parapet wall on the upstream side.

In late December 1992, a winter storm caused the reservoir elevation to rise to an elevation 28 below the cofferdam crest elevation. This storm produced about 2.3 inches of rain, which fell on the 5,762 mi^2 drainage basin. This additional rain elevated the reservoir levels and saturated the soils in the drainage basin. On January 7, with the reservoir elevation 24 feet below the cofferdam crest elevation, a second and more significant flood event occurred. The storm dumped about 6 inches of rain on the drainage basin over the next 5 days. Anticipating that the eight spillway gates on the right abutment would need to be opened, all equipment and materials were removed from the unwatered tailrace area downstream of the dam.

Then, on January 8, 1993, a peak inflow of 135,800 ft^3/s occurred which was the flood of record. The high inflows and rising reservoir forced the operation of the eight available existing spillway gates on the right abutment. The right spillway gates were opened with the reservoir at elevation 12 feet below the crest elevation of the cofferdam. On the left abutment, where the upstream cellular cofferdam had been constructed; the existing spillway had been removed; and mass concrete placements on the left abutment thrust block had already begun. The storm resulted in a maximum five day volume of 580,000 acre-feet (e.g. Hepler 1993). A subsequent storm caused the reservoir to rise to the highest reservoir elevation in the history of the dam on January 19.

Figure 2. Theodore Roosevelt Dam during 1993 flood showing partial construction of the left thrust block and overtopping of the cellular cofferdam.

Overtopping of the left abutment concrete-capped cellular cofferdam and left thrust block began on January 19 (see Figures 2, 3 and 4). The cofferdam overtopped for about two days and ended on about midnight on January 20 (e.g. Woodby, 1993). The reservoir reached the highest elevation since the original construction of the dam. The flow over the cofferdam and partially constructed spillway was estimated to be about 700 ft^3/s. The concrete-capped cofferdam performed fairly well for the approximately 1.5 foot depth of overtopping (see Figure 3). The cellular cofferdam sustained minimal increases in deflection and bulging, none of the sheet piles went out of interlock, and the cone-shaped wire mesh inserted in the weep holes helped to minimize sand loss during overtopping.

Figure 3. Aerial view of cellular cofferdams being overtopped.

Figure 4. Aerial view of Theodore Roosevelt during the 1993 flood event.

Releases from the right existing spillway gates were made again from February 12 until March 8, 1993 which continued to flood the work site. The flood caused a delay in construction and resulted in damages to the construction site. The dam construction was completed in 1996 (see Figure 5).

Figure 5. Aerial view of completed Theodore Roosevelt Dam.

JACKSON LAKE DAM MODIFICATION

Background and Description

Jackson Lake Dam is a composite earthfill embankment and concrete gravity structure on the Snake River in northwestern Wyoming. The dam was originally constructed as a temporary timber-crib structure between 1905 and 1907 to enlarge a natural lake. The temporary structure failed in 1910 and was replaced with a permanent dam. The new dam was completed in 1916.

Safety concerns were identified in the mid-1970s and the dam was modified between 1986 and 1989. In its current configuration, the dam consists of a concrete gravity section flanked by a zoned earthfill embankment on its left (north) end and a short earthfill embankment on its right (south) end.

Jackson Lake Dam Flooding Event: June 1 to 14, 1986

A flood event occurred at Jackson Lake Dam in Wyoming between June 1 and June 24, 1986, while the dam was being modified. This event corresponded to rapid snowmelt due to warm temperatures around Grand Teton National Park and the higher elevations just south of Yellowstone National Park.

Meteorological Setup for the Event

A large snowpack had setup during the winter of 1985 into the spring of 1986 over the headwaters of the Snake River in northwest Wyoming. This is in part due to the ENSO being in a La Nina phase (cooling waters off the coast of South America into the central

Pacific Ocean). In response to La Nina, the jet stream moved farther south than normal and funneled moisture over the Pacific Northwest which continued east into western Wyoming. The large snowpack persisted into June at elevations above 8,000 feet. The return period of the snowpack above 8,000 feet for the beginning of June was roughly 4 to 7 years (e.g. Cunnane, 1988).

The end of May and beginning of June 1986 featured a heat wave for portions of the inland northwest and northern Rocky Mountains. Many locations had high temperatures in the 90's, and a few even had 100 degree readings (e.g. WRC 2014). The higher elevations near Jackson Lake Dam had high temperatures in the 70's and 80's, with low temperatures staying mostly above freezing, which led to rapid melting of the large snowpack. A stationary boundary moved over northern Wyoming on June 4, which helped to abate the heat and bring isolated to scattered thunderstorms over the watershed. The stationary boundary caused a rain-on-snow condition above 8,000 feet, which lasted through June 15 in isolated locations at higher elevations. By June 24, the main threat of flooding had diminished because most of the snow had melted, even though isolated thunderstorms persisted. This flood event was mostly predicated by rapid snowmelt, rather than heavy rainfall (see Figure 6).

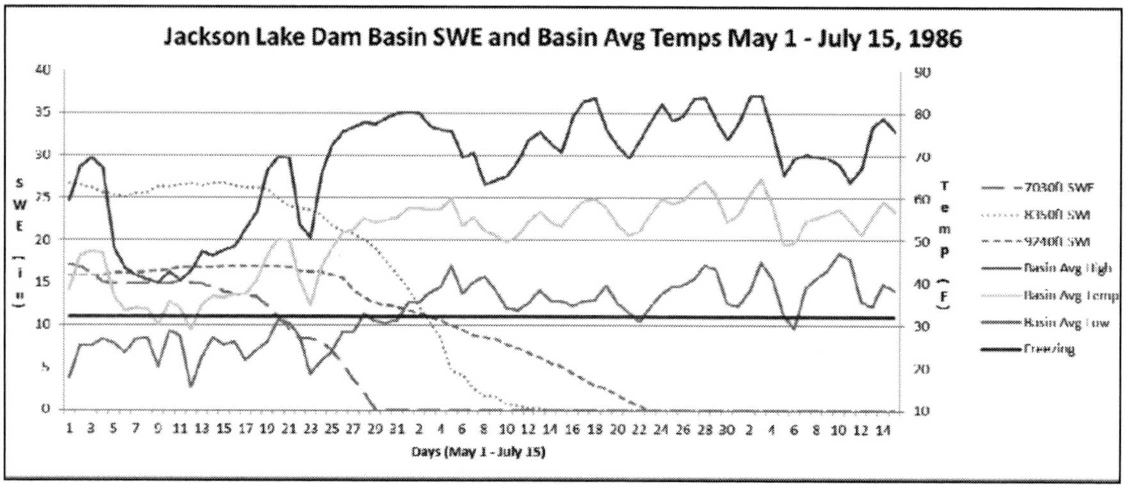

Figure 6. Graph showing the snowpack (light blue) in inches on the left axis for the May 1 to July 15 time period. High temperature, average temperature, low temperature, and 32 degrees are plotted on the right axis for the same time period. Notice how quickly the snowpack begins to melt once the average basin temperature rises above 32 degrees.

Hydrologic Loading, Diversion Plan, and Performance of Cofferdam during Construction

The diversion plan during construction was to have a cofferdam that could be installed in the wet, allow for staged construction of the left and right sides of the concrete gravity section, and allow for diversion of flows by utilizing the existing outlet works without restricting the approach channel area and the required discharge capacity. Due to the space limitations and high localized velocities, a cellular cofferdam was selected with embankment tie-in sections. The diversion plan with the cellular cofferdam provided definite advantages associated with the construction schedule during stage 2 construction. The contractor elected to use a double sheet pile wall to tie into the cellular cofferdam. The center cofferdam consisted of three 40-foot-diameter cells and two 11-foot-radius connecting arcs with a connecting concrete gravity walls for closure.

Flood routings were performed for both spring runoff and thunderstorm flood conditions to evaluate flood control operations, and determine specification requirements for the level of flood protection, operation of the existing outlet works, and height of the cofferdam. It was determined that dam operations needed to be similar to normal operations, except that the reservoir would be operated at a lower reservoir level during construction. The diversion plan included 200,000 acre-feet of flood storage to account for spring runoff conditions and to allow some flexibility in managing releases to allow downstream tributary flows to pass prior to increasing releases from Jackson Lake Dam. The outlet works discharge capacity needed to be at least 9,000 ft^3/s during the spring flood operating season between May 1 and July 10.

Spring runoff floods with return periods of 5 years to 100 years were routed. The flood routing for the 100-year spring runoff flood resulted in a freeboard of a little less than 5 feet. The results of the spring runoff flood routings indicated that the difference in maximum reservoir elevation between the 25-year flood and the 100-year flood was only 1 foot, which supported the decision to increase the flood protection level to a 100-year return period because the increased cost would be insignificant compared to the increased flood protection that would be provided.

The 100-year thunderstorm flood routings did not include any flood storage space prior to the flood since thunderstorms could occur after July 10 which is historically the end of the spring runoff time period. As a result, the 100-year thunderstorm flood controlled the design of the cofferdam crest elevation, and the estimated freeboard for this flood was 2.4 feet.

For diversion during Stage 1 of construction, the existing outlets were made available for spring runoff flood flows and the specifications required that the tie-in cofferdam, which would restrict the number of outlets available, could not be constructed until after July 10, and be removed by May 1 of the following year (see Figure 7). In addition, consideration would be given to breaching the tie-in cofferdam if more discharge capacity was needed after the July 10 date.

Figure 7. Aerial view of cellular cofferdam upstream of Jackson Lake Dam during the first stage of construction.

For diversion during Stage 2 of construction, 11 outlets were available, which satisfied the 9,000-ft^3/s discharge capacity requirement (see Figure 8).

Figure 8. Aerial view of cellular cofferdam upstream of Jackson Lake Dam during the second stage of construction.

Based on the peak inflow data, the maximum inflow on June 2, 1986 was estimated to have a return period of about 500 years. The highest reservoir water surface elevation reached during construction resulted in about 2.6 feet of freeboard remaining on the cofferdam which was about 2.5 feet higher in elevation than the flood routings had calculated for 100-year spring runoff design flood. The diversion plan worked well in allowing staged construction, while still providing protection for the flood of record. The cellular cofferdam was inspected during construction and performed as designed throughout construction, including during the 1986 flood.

GLENDO DAM MODIFICATION

Background and Description

Glendo Dam and Reservoir are located on the North Platte River in Southeast Wyoming. The reservoir is used for flood control, irrigation, and fish and wildlife enhancement. The construction of the dam, dikes, and appurtenances was completed in 1959. Glendo Dam is a zoned earthfill structure. Three dikes are located in the low areas between the hills of the reservoir's south shore. In 1989, a seepage berm was constructed on the downstream toe of the dikes.

The contract for the Glendo Dam modification was awarded in 2009, and the contractor had already begun some of the Safety of Dams construction work in 2010. In advance of construction, a risk during construction report and an interim construction Emergency Action Plan (EAP) were prepared. The interim construction EAP was prepared to supplement the existing EAP, to identify potential failure modes specific to construction and include the line of communication in the event of an emergency. In addition, decisionmakers participated in a conference call every 2 weeks to coordinate construction issues, which facilitated discussions and decisions related to concerns associated with changing conditions.

Meteorological Setup for the 2010 Flood Event

During the summer of 2010, the ENSO phase was transitioning from an El Nino phase (warm ocean temperatures off the coast of South America into the central Pacific Ocean) to a La Nina phase (cool ocean temperatures off the coast of South America into the central Pacific Ocean) (e.g. NOAA). The El Nino phase during the winter and spring allowed a large snowpack to build in the Rockies of Colorado and Wyoming and also allowed cooler than normal temperatures to continue into the summer. The cooler than normal temperatures led to a snowpack above 9,500 feet, which persisted into June. As the ENSO phase transitioned from El Nino to La Nina during the early summer, the jet stream (relatively strong winds concentrated within a narrow stream in the atmosphere typically above 30,000 feet) transitioned and funneled more Pacific moisture over the State of Wyoming. As such, the soil for the North Platte River basin was saturated when the high precipitation event occurred from June 10 to June 20 leading to substantial amounts of runoff, including a rain-on-snow component above 9,500 feet.

A series of upper level lows crossed northern Colorado and Wyoming starting on June 7 and continuing until nearly the end of the month. Due to these upper level lows, stratiform and convective precipitation occurred almost daily over some portion of the North Platte River basin west of Glendo Dam from June 7 until June 17. The heaviest precipitation that fell near Glendo Dam occurred approximately 25 miles south, near Wheatland, Wyoming, when 2.3 inches of rain fell within a 48-hour period from June 12 to 13. Given a 2.3-inch rainfall over a time period of 48 hours near Glendo Dam and a 2.5-inch rainfall over a 4-day period near Jeffrey City, Wyoming, the return period for the precipitation was estimated to be about 2 to10 years (e.g. Miller, 1964).

This event occurred during the normal storm seasonality timeframe, from June through August, when approximately 65 percent of all extreme storms occur in the Rocky Mountain region (e.g. Levish, England Jr, Klawon, O'Connell, 2003). This storm was unusual in that it lasted over a week while nearly 95 percent of all extreme storms in the Rocky Mountain region last less than a week.

Comparing 2010 and 2011 Snowpack and Flooding

While the 2010 winter snowpack was slightly above average, the 2011 winter produced an abundance of snowfall. In many cases it was one of the snowiest winter seasons on record across central Wyoming and northern Colorado. The Snow Water Equivalent in the 2011snowpack was estimated to be a 25- to 50-year event. By comparison, the 2010 Snow Water Equivalent was estimated to have a return period of 5 to 6 years (e.g. Miller, 1964).

The 2011 record snowpack was much larger than the 2010 snowpack; however, flooding occurred in 2010, and not 2011. In 2010, a series of upper level lows caused rain to fall on snow for much of the first half of June. However, in 2011 most of the major storm systems produced heavy rain either east or south of the North Platte River Basin upstream of Glendo Dam despite the Western United States being in an active storm pattern. The main difference in the flooding between 2010 and 2011 is that there was no rain-on-snow component. Because the rain in 2010 had a much larger aerial extent in the basins downstream of Glendo Dam than the rain in 2011, heavy downstream tributary flows and reduced irrigation demand required reduced releases from Glendo Dam, and this also contributed to higher reservoir levels. Without any major rain-on-snow events in 2011, the snow melted at a normal rate without the threat of flooding, as was seen in 2010.

Hydrologic Loading, Operations, and Response

In 2010, the snowpack in the basin above Glendo Dam was below average from December through early April. The forecasts were for a below-average inflow in April 2010. In June, however, the precipitation was widespread in the basin and extended downstream into the irrigation delivery area. Because of the June rains, there was very little irrigation demand. The timing of high flows coming down the North Platte also coincided with reservoirs on the Laramie River filling and spilling. The high flows from the Laramie River entered the North Platte River downstream of Glendo Dam. In an attempt to mitigate downstream flooding resulting from Glendo Dam releases, combined with Laramie River releases and the reduced irrigation demand, the U.S. Army Corps of Engineers directed that releases from Glendo Dam be reduced to allow the Laramie River flood peak to pass.

The combination of June rain storms in the basin above Glendo Dam with the above average runoff from the snowmelt and the reduced releases contributed to the high reservoir levels at Glendo Dam during construction in 2010. The reservoir reached the highest elevation since the 1989 construction of the dike seepage berms (see Figure 9). These high reservoir levels led to concerns at the dikes on June 20, 2010, when cloudy seepage was observed and instrumentation readings were abnormal and out of expected limits. Several wet spots were observed during this incident. On June 28, the most notable location of seepage was at the left groin of Dike 3, where seepage was observed bubbling and exiting at a point 4 feet lower than the reservoir elevation. The exit gradient produced sufficient velocity flow to cause incipient sand size particle movement. Water could be heard flowing under the cobbles on top of the seepage berm. The water level at inspection well IW-H was about 1 foot below the top of the cobbles, and water was observed entering the manhole through the joints in the riser pipes. Many other seepage locations were observed at the dikes, and their locations were surveyed by Reclamation. As a result, a risk team was formed and the hydrologic internal erosion failure modes at the dikes were evaluated. The understanding of the failure mode changed sufficiently to result in the removal of dike raise modifications from the contract. A corrective action study was initiated to evaluate alternatives to mitigate risks associated with internal erosion failure modes at the dikes.

Figure 9. Aerial view of the Glendo dikes during the 2010 flood.

In 2011, the largest snowpack on record was in the basin upstream of Glendo Dam. The snowpack was equivalent to 1.97 million acre-feet of runoff which was about 2.8 times the average of 700,000 acre-feet and significantly larger than the 1.5 million acre-feet that was the second highest year on record. Due to this large record snowpack, the reservoir was predicted to potentially reach an elevation 9 feet higher than the highest reservoir on record depending on the temperature conditions and the potential spring rains. The Wyoming Area Office began releasing water in March, about 2 months earlier than a normal water year, to create reservoir storage capacity in the system reservoirs with the key concern that a repeat of the 2010 water year with June rains and a fast snowmelt could produce high flows in the North Platte River.

Based on the projections for high reservoir conditions associated with the large snowpack, Reclamation implemented a temporary emergency modification to the 1989 seepage berms downstream of two of the dikes. The emergency modification to the dikes was completed using the contractor who was already working onsite on the auxiliary spillway and dam raise.

The scope of the modification and construction time period were limited due to the rate at which the reservoir was rising. The emergency modification consisted of the construction of an emergency berm which required removing approximately 2 to 3 feet of cobbles from the top and downstream slope of Dike 1 and Dike 3 in order to expose the sand and gravel zone of the 1989 seepage berms. The topsoil on a portion of the downstream slope of Dike 1 and Dike 3 was stripped. The excavated surface was lined with a sand filter zone that was 1.5 feet thick on horizontal surfaces and 3.5 feet thick on sloped surfaces (see Figure 10). The sand filter zone extended up the downstream slope of Dike 1 and Dike 3 to a height of about 6 to 9 feet. A gravel drain zone surrounded the perforated toe drain pipe downstream of the sand filter zone. The gravel drain extended to the same elevation up the downstream slope of Dike 1 and Dike 3 as the filter sand. Covering the sand filter and gravel drain and extending downstream was a zone of miscellaneous bermfill. The emergency berm had a width of about 45 feet.

Figure 10. View of the seepage berm modifications at the Glendo Dikes in 2011.

With the contractor working 7 days a week, the emergency repair work was completed in 5 weeks and was in place by the end of June 9. Fortunately, a cool spring with no rain-on-snow events allowed the snowpack to melt slowly through the summer months. In the summer of 2011, the reservoir eventually rose to an elevation that was much lower than was projected in May. The seepage berm modifications remained until they were removed as part of permanent Dike modifications in the summer of 2014.

SUMMARY AND CONCLUSIONS

Three case histories for dam modification have been presented which highlight the need for determining the appropriate level of protection for large floods during construction especially where the dam being modified is a significant or high hazard facility and consequences may extend beyond the value of the construction work and into the downstream population at risk and property. These case histories also highlight the usefulness of understanding potential meteorological conditions that could occur and the need for a plan that can be put in place to reduce risks associated with construction floods.

In the case of the Glendo Dam modification, risks during construction were evaluated, and an interim construction Emergency Action Plan (EAP) was prepared in advance of construction to supplement the existing EAP. In 2011, when it was realized that the snowpack could melt rapidly in combination with early summer rains, as had occurred the previous year, the National Weather Service (NWS) was used to provide real-time information on snowpack, Snow Water equivalent data, runoff data, and real-time future projections on atmospheric conditions. The reservoir operation plans were adapted to begin releasing water over a month and a half earlier than normal to create space in the upstream reservoirs for the eventual runoff from the large snowpack in the upper basin and the potential for rain-on-snow events. The NWS data supported the decision to perform a modification on the dikes during construction to reduce the risk associated with rising reservoir levels. With the longer lead time provided by the early March snowpack data, the decisions made in 2011 by the Wyoming Area Office to adapt the reservoir system operation was key to preventing downstream flooding conditions and high reservoir levels at Glendo Dam. The Glendo Dam case history also highlights a situation where a contractor already on site can assist in implementing emergency modifications.

The flood event during the Theodore Roosevelt Dam modification highlights the potential for rare hydrologic conditions to occur and the importance of an appropriately designed cofferdam, which could allow for overtopping. In this case, multiple back-to-back atmospheric river events brought rain to an area of saturated soils. The capability for the cofferdam to withstand overtopping without failing prevented a much larger and sudden discharge, as well as the benefit of attenuating the flood flow releases.

For the Jackson Dam modification, the key was the adequate design of a cofferdam/diversion system. Understanding the consequences of failure of a cofferdam highlights the need use risk based analysis approaches when defining the risks during construction and selecting the appropriate level of flood protection for the cofferdam/diversion system.

Based on these case histories, it is essential that the following be considered when determining the design level of flood protection and other requirements during construction, as well as the process for monitoring and responding to emergencies:

1. If necessary, perform a risk based analysis to determine the level of flood protection needed to protect the population at risk downstream of the dam, as well as to protect the work associated with construction. The risk based analysis may be useful in evaluating and selecting the construction design flood, the type of cofferdam/diversion plan, reservoir flood storage requirements, construction staging, work restriction requirements, and actions that may be needed in the event of an emergency. The evaluation of alternatives and operations during construction (including storage space and diversion release requirements) and the flood routing studies can be a significant design effort, and their importance should not be reduced or diminished due to design schedules and budgets. Consultant review board and management concurrence on the proposed level of flood protection is an important part of the design and decision process.

2. The specifications should provide the contractor with information necessary to design or to understand their role in the diversion and care of the inflows during construction. The specifications should include work restrictions, the design flood level of flood protection, and any construction staging requirements related to the times of the year when there is a reduced risk of flooding.

3. An interim or supplemental construction EAP should be developed that identifies potential failure modes during construction, as well as a plan to monitor and take appropriate actions if a large flood event develops. The supplemental construction EAP should define the roles, responsibilities, and line of communication. If an operations office that can monitor snowpack and runoff is not part of the EAP team, the tasks of monitoring rainfall and runoff, interpreting information produced by the NWS, and advising decisionmakers on the potential of changing conditions can be given to a meteorologist or hydrologist assigned to the construction support team. The NWS can be used to provide real-time information on precipitation, snowpack, and runoff data. In addition, the NWS provides forecasts and outlooks on atmospheric conditions. This information can be used by decisionmakers to reduce the risk associated with rising reservoir levels, and to adapt the reservoir operation plans to address changing conditions.

ACKNOWLEDGEMENTS

The research for this paper was sponsored by the Bureau of Reclamation's Dam Safety Technology Development. Many provided significant support and contributions to the report including the following: Lisa Krosley, Program Manager, Dam Safety Office, Reclamation; Victoria Sankovich Bahls, Meteorologist, Reclamation; John F. England,

Jr., Ph.D., Flood Hydrology Technical Specialist, Reclamation; Mahonri Williams, Chief, Water and Land Operations Division, Wyoming Area Office, Reclamation; Jeff Wormer, Geotechnical Engineer, Reclamation; Tom Hepler, Civil Engineer, Reclamation/Schnabel Engineering; and Hillery Venturini, Civil Engineer, Reclamation.

REFERENCES

Appurtenant Structures for Dams (Spillway and Outlet Works), Design Standards 14, Chapter 2: Hydrologic Considerations, Reclamation, Denver, CO, November 2013. http://www.usbr.gov/pmts/tech_services/engineering/design/DS14-2.pdf

Best Practices in Dam and Levee Safety Risk Analysis, a Joint Publication by Bureau of Reclamation, and U.S. Army Corps of Engineers, Denver, CO, December 3, 2012.

Bonnin, G.M., D. Martin, B. Lin, T. Parzybok, M. Yekta, and D. Riley, 2004, NOAA Atlas 14, Precipitation-Frequency Analysis of the United States – Volume 1 Version 5.0: Semiarid Southwest (Arizona, Southeast California, Nevada, New Mexico, Utah), National Weather Service, National Oceanic and Atmospheric Administration, U.S. Department of Commerce, Silver Spring, MD, pp 271.

Cunnane, C. 1988, *Methods and Merits of Regional Flood Frequency Analysis*, Journal of Hydrology, pp 296-90.

House, P.K., 1993, *The Arizona Floods of January and February 1993*, Arizona Geology, vol. 23, no. 2, Summer 1993, pp 5-9.

Hepler, Thomas E., 1993 Floods Test Large Dams in Arizona (Theodore Roosevelt Dam, Flooding Summary), USCOLD Newsletter, Issue No. 100, March 1993.

Levish, D.R., J.F. England Jr, J.E. Klawon, D.R.H. O'Connell, 2003, *Flood Hazard Analysis: Seminoe and Glendo Dams Kendrick Project and Pick-Sloan Missouri Basin Program Wyoming*. Flood Hazard Analysis, Bureau of Reclamation, Technical Services Center, Denver, Colorado, 2003.

Miller, J.F., 1964. *Technical paper No. 49 Two-to-Ten-Day Precipitation for Return Periods of 2 to 100 Years in the Contiguous United States,* U.S. Department of Agriculture, Office of Hydrology, US Weather Bureau, Washington, D.C., p. 29, p. 10.

National Oceanic and Atmospheric Administration, *El Nino/Southern Oscillation (ENSO) Technical Discussion*. National Climatic Data CenterAvailable at http://www.ncdc.noaa.gov/teleconnections/enso/enso-tech.php
Accessed April 15, 2014.

National Weather Service, 1993, Daily Weather Maps for December, January, and February, National Oceanic and Atmospheric Administration.

Natural Resources Conservation Service (NRCS), 2014, Website: http://www.wcc.nrcs.usda.gov/snow/ Accessed April 18, 2014.

Office of the State Climatologist, 1993, *Arizona Climate Summary*, Arizona State University, Department of Geography, vol 19, no. 7, 8 and 9.

United States Geological Survey (USGS, 1993), *Floods in Arizona, January 1993*, Department of the Interior, Water Resources Division, Tucson, Arizona. 1993.

Western Regional Climate Center (WRCC), 2014, Website: http://www.wrcc.dri.edu/summary/ Accessed April 15, 2014.

Woodby, Bob, Overtopping Performance of a Cellular Cofferdam, USCOLD Newsletter, Issue No. 102, November 1993.

UPDATE NO. 2 OF THE CALAVERAS DAM REPLACEMENT PROJECT

Michael Forrest[1]
John Roadifer[2]
Daniel Wade[3]
Susan Hou[4]
Tedman Lee[5]
Carman Ng[6]

ABSTRACT

Calaveras Dam is a major component of the San Francisco Public Utilities Commission (SFPUC) Regional Water System. Since 2001, in response to seismic stability concerns from the California Division of Safety of Dams about this 90-year-old hydraulic fill dam, the SFPUC has lowered water levels in Calaveras Reservoir to about 39 percent of its 96,850-acre-foot capacity. To restore reservoir capacity, a replacement dam and new appurtenant works were designed to remain functional after the design earthquake, which is a magnitude 7¼ maximum credible earthquake (MCE) on the Calaveras Fault located 0.3 miles from the dam. The MCE peak ground acceleration would be 1.1 g. The dam site also contains numerous landslides.

Construction of the replacement dam, which started in August 2011, is approximately 60% complete and in the fourth year of construction. Grouting of the right abutment is complete, the new intake tower and shaft are constructed, and excavations for the dam and spillway are nearing completion. The status of Calaveras Dam Replacement Project (CDRP) was presented at the USSD 2014 Conference. This paper provides an update of the project's status as well as engineering challenges that have arisen at this geologically complex site. Challenges and risks are being managed by early mitigation actions to avoid production delays and developing contingency plans to implement in response to triggers in the project's risk register.

INTRODUCTION

Construction of the CDRP began in September 2011. Excavation of the foundation for the spillway and dam is currently (October 2014) about 90 percent complete and is

[1] Vice President, URS Corporation, 1333 Broadway, Suite 800, Oakland, CA 94612, (510) 874-3012, michael.forrest@urs.com
[2] Sr. Project Manager, URS Corporation, 1333 Broadway, Suite 800, Oakland, CA 94612, (510) 874-1732, john.roadifer@urs.com
[3] Director, Water System Improvement Program, San Francisco Public Utilities Commission, 525 Golden Gate Ave., San Francisco, CA 94102, (415) 554-1853, dwade@sfwater.org
[4] Project Manager, San Francisco Public Utilities Commission, 525 Golden Gate Ave., San Francisco, CA 94102, (415) 551-4666, shou@sfwater.org
[5] Project Engineer, San Francisco Public Utilities Commission, 525 Golden Gate Ave., San Francisco, CA 94102, (415) 551-4866, tclee@sfwater.org
[6] Assistant Project Engineer, San Francisco Public Utilities Commission, 525 Golden Gate Ave., San Francisco, CA 94102, (415) 551-4867, cng@sfwater.org

anticipated to be essentially completed in November of 2014. The high cut in Observation Hill for the spillway, excavation for the spillway chute, and the excavation for the right abutment of the dam have been completed, except for a small area at the downstream side of the valley bottom. Excavation of the left abutment of the dam and the stilling basin for the spillway is ongoing. Installation of the two-row foundation grout curtain under the core of the new dam has been completed in the right abutment and is currently under construction at the top of the left abutment. The new intake shaft and tower have been completed and installation of miscellaneous metals and mechanical components in the tower and shaft are nearly completed. An overview of the project area as of November 2014 is shown on Figure 1.

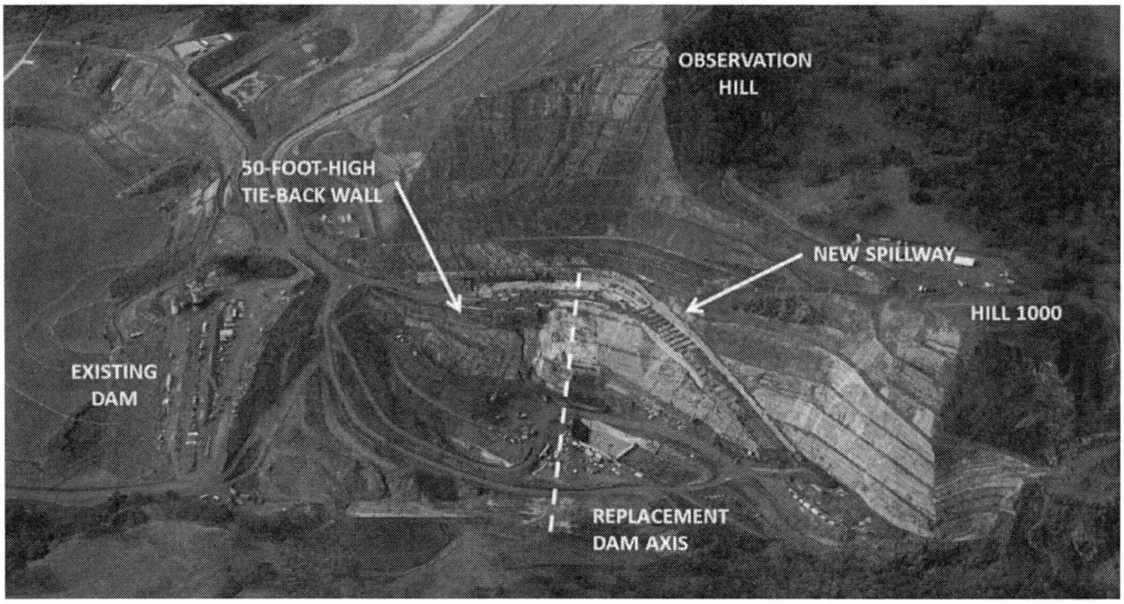

Figure 1. Aerial View of CDRP Site (November 2014).

Design challenges during 2013 at this geologically complex project site were previously described in Forrest, et al. (2014). Additional engineering challenges discussed in this paper that have presented as excavation and construction has progressed in 2014 include:

- Continuing characterizing and mitigating a large left abutment slide complex;
- Construction of a 50-foot-high tie-back wall to anchor a portion of the slide to create a platform for construction access;
- Stability analyses and reinforcement of the 400-foot-high stilling basin excavation in the Franciscan Complex;
- Protection of concrete structures against sulfate and acid attack from the geologic formation; and
- Construction of the spillway and controlling temperature.

CONTINUING CHARACTERIZING AND MITIGATING LARGE SLIDE COMPLEX IN LEFT ABUTMENT

Initial characterization of the previously unknown large slide complex in the left abutment has been described in Kelson, et al. (2014). The slide complex included Slide A, a late Pleistocene landslide located on the steep northeast facing slope of Observation Hill, and Slide B, a later landslide of large, coherent to partially articulated blocks of weathered material located on the eastern side of Observation Hill. The upper portion of Slide B likely includes re-mobilized parts of Slide A.

Mitigation for Slide A required flattening of the cut slope for the spillway from 1.3H:1V to 2H:1V (Forrest, et al., 2014). Mitigation for Slide B to date has required most of the slide material from the dam foundation, outlet pipe bench, and spillway foundations and has resulted in excavated grades that are lower than anticipated in the design. The over-excavation has required blocks of concrete backfill up to 35-feet-high below the spillway invert (see Figure 2) and will require concrete backfill up to 23 feet under portions of the outlet pipe bench downstream of the core trench and increased embankment fill quantities. As discussed later in this paper, a 50-foot-high tie-back wall was designed to maintain access and a staging area required for construction of the upper spillway.

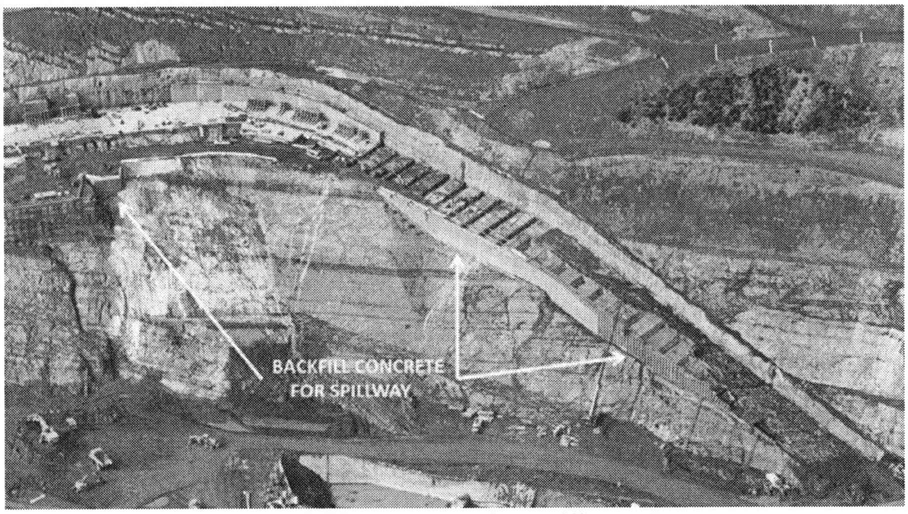

Figure 2. Backfill Concrete for Spillway Structures.

The complexity of the dam site has continued to require constant engineering geologic assessment of newly generated exposures in order to anticipate and be prepared for potential issues that could arise and develop feasible and timely construction recommendations. The basal surface of Slide B has been found to be an undulating surface that takes advantage of a series of tectonic discontinuities that strike generally subparallel to the valley and generally dip between 30 to 40 degrees towards the valley. As a result of that mapping, a portion of the basal slide surface in the upstream shell of the dam foundation was observed to be oriented more steeply than investigations in 2013 and early 2014 had indicated. A campaign of an additional 14 borings was initiated in September 2014 to characterize this area of the slide that had not been previously accessible in 2013 and 2014 due to steep terrain that has since become available with the advancing excavation. The results of these investigations indicate that the base of the landslide extends below portions of the pipe bench at the downstream toe of the dam and under a portion of the upstream core and upstream shell. After removal of the slide materials, backfill concrete similar to that used for the spillway discussed above, is being placed at these locations to build back up the pipe bench.

CONSTRUCTION OF 50-FOOT-HIGH TIE-BACK WALL

In order to provide contractor access to the upper spillway area, a portion of Slide B was stabilized in place using post-tensioned tie-backs. The design of this tied-back remnant considered crane loading and other construction live loads. Based on input from the contractor, the total weight of the crane and the maximum payload result in an average surcharge of 725 psf when distributed over the total footprint of the crane. A uniform construction live load of 250 psf was applied to the areas outside of the crane footprint.

Stability Analysis

Stability analyses of the crane pad tie-back wall were performed with the Slide B remnant left in place below the crane. The analysis includes the construction loads described above. The stabilizing pressure applied to the 1H:4V cut is that of a 140-kip post-tensioned anchor spaced on an eight-foot grid. Figure 3 shows that the factor of safety is about 1.6, which is greater than the required minimum of 1.5.

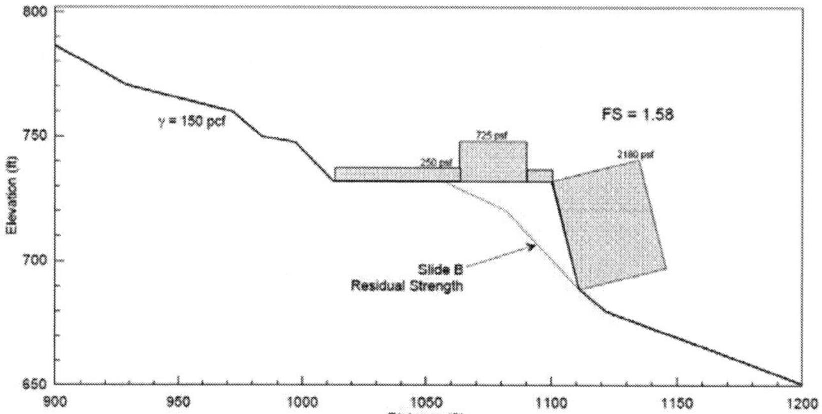

Figure 3. Stability Analysis of Tie-back Wall.

Tie-back Design

The tie-backs are four-strand, 270 ksi each, having a design load of 140 kips. Each tie-back was installed in a drill hole having a minimum diameter of 6 inches and a minimum bond length of 20 feet. To account for uncertainties in the location of the slide surface, an additional 10 feet was added to the unbonded length as determined from the estimated contours of the Slide B base. Required bond lengths were calculated using the design load and an average bond stress of 100 psi per Post-Tensioning Institute (PTI, 2006) recommendations for weathered sandstone. Performance testing of 10 percent of the tiebacks and proof testing of the remaining tiebacks was required during construction to verify the required design load was met. The 0.25(H):1(V) cut face was covered with a 4-inch-thick fiber-reinforced shotcrete layer supported by the tie-backs. Spot bolting was required to support localized weak areas or unfavorable wedges during excavation.

The required tie-backs are temporary in that they are not necessary after the end of construction when the slope will be buttressed by disposal fill placed between the existing and replacement dams. As such, only minimal corrosion protection is needed; the tendons were coated in corrosion-inhibiting compound by the manufacturer, encapsulated in grout during installation, and the exposed anchorage hardware coated in coal tar epoxy after post-tensioning. The completed tie-back wall is shown on Figure 4.

Figure 4. Completed Tie-back Wall.

STABILITY ANALYSES AND REINFORCEMENT OF 400-FOOT-HIGH STILLING BASIN EXCAVATION IN FRANCISCAN COMPLEX

Overall Slope Condition

The exposed face of the 400-foot-high stilling basin cut slope during excavation as of September 2014 is shown on Figure 5. The benches are generally at vertical intervals of 50 feet. The lowest visible bench (Bench G) in Figure 5 is at elevation 632. Slopes are 0.75H:1V between benches from the top of the excavation down to Bench G at elevation 632. Slope H at 0.5H:1V (in progress on Figure 5) will be below Bench G, then Bench H at elevation 582, then a 0.25H:1V slope to the floor of the stilling basin excavation at elevation 532.

The upper portion of the slope is Temblor sandstone and the lower portion of the slope is Franciscan. The Franciscan portion of the slope consists of a mélange shale matrix containing extensive quartz veins and blocks of hard greenstone, blueschist, and greywacke. The mélange shale is protected with bolts and shotcrete (visible in Figure 5) while two large greenstone and blueschist blocks are left untreated. The dimensions of the blocks range from inches to hundreds of feet.

Figure 5. Stilling Basin Cut Slope (March 6, 2014).

Shear Strength

Shear strength of the Franciscan melange cut slope was evaluated using the procedures of Marinos and Hoek (2000). The interpreted geological strength index (GSI) values for use in the shear strength evaluation were based on rock exposure conditions and of rock cores obtained during the design investigation phase. Figure 6 shows a selection of interpreted GSI values based on observed rock slope conditions at various locations in the slope. Shear strength parameters for the Franciscan mélange currently being evaluated using a block-in-matrix approach.

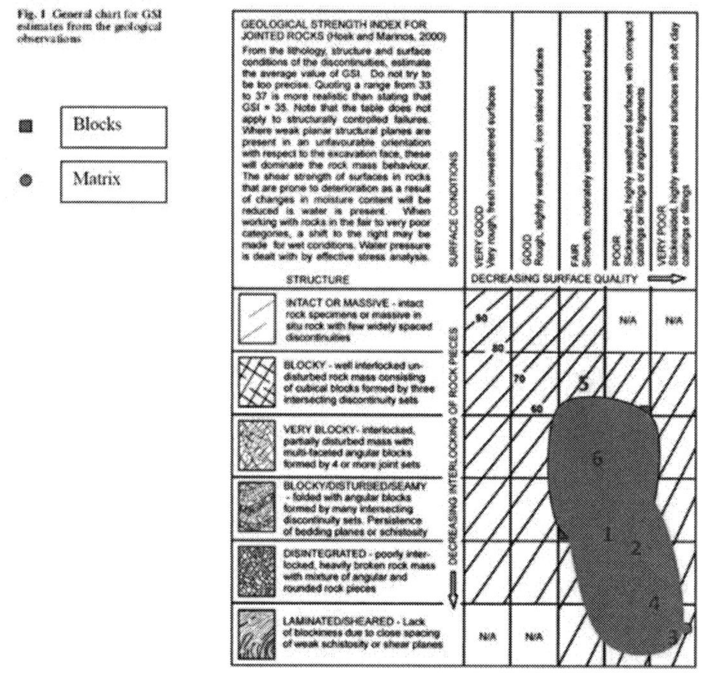

Figure 6. GSI Interpretation.

Stability Analysis

A number of variables affect the results described above and the most significant uncertainty is in the strength characterization. The results of the stability sensitivity analyses showing the range of GSI values are summarized in Table 1 below.

Table 1. Summary of Sensitivity Analysis Results

Analysis Case	**Factor of Safety**		
	One Bench	**Two Benches**	**Three Benches**
GSI = 45 No reinforcement	1.65	NA	1.63
GSI = 30 No reinforcement	1.15	1.22	1.29
GSI = 30 Planned reinforcement[1]	1.62	1.45	NA
GSI = 35	NA	1.68 (planned reinforcement)	1.39 (no reinforcement)
GSI = 30 Reduced bar spacing[2]	NA	1.63	NA
GSI = 30 Increased bar size[3]	NA	1.59	NA
GSI = 30 Planned reinforcement Long-term water level	1.60	1.44	NA

Note: NA = Not Applicable; [1] #8 bars, 12 feet on center, and 24 feet long; [2] #8 bars, 8 feet on center, and 24 feet long; [3] #11 bars, 12 feet on center, and 24 feet long.

The following is a summary of the stability analysis results described in Table 1:

- Without considering reinforcement, all slopes are stable (FS > 1.1).
- With the currently-planned reinforcement included (#8, Gr 75 bars, 24 feet long, 12 feet on-center), the factors of safety for the single-bench slip surfaces all meet stability requirements (FS > 1.5). The two-bench slip surface is slightly below the required FS with GSI = 30 (FS = 1.45, see Figure 7) but meets the requirement with GSI = 35. It also meets requirements with GSI = 30 with increased reinforcement: #8 bars, 8 feet on-center, and 24 feet long, or #11 bars, 12 feet on-center, and 24 feet long.
- The three-bench slip surface is large enough that it will be favorably influenced by 3D effects due to the locations of the large blocks in the slope or by a higher representative strength if the large blocks could be taken into account in the strength characterization. The stability of this slip surface will also be increased by the 24-foot rock bolts that are able to reach this surface.

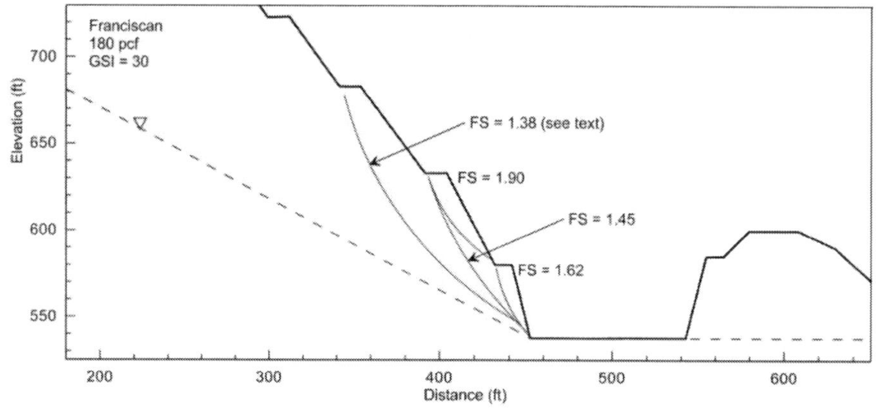

Figure 7. Stability Analysis using GSI = 30, with Reinforcement.

Slope Reinforcement, Drainage and Instrumentation

The following slope reinforcement was recommended based on the results of the stability analyses:

- For slip surfaces involving a single bench, it was concluded that the current slope reinforcement (#8 bars on a 12-foot grid, 24 feet long) is adequate.
- For the slip surfaces involving two benches, within the 0.25H:1V cut slope (below the lowest bench of the stilling basin cut), it was concluded that either #8 bars on an 8-foot grid, or #11 bars on a 12-foot grid. The lengths would remain 24 feet.
- For the slip surface involving three benches, the influence of the larger blocks within the slope is such that the conditions are acceptable with the planned 24-foot bolts as reinforcement.

To promote drainage of the hillside, 4-inch-diameter, sub-horizontal drains ("hydraugers") were drilled 80 feet into the hillside. The drains were cased with slotted PVC. The slope was instrumented with inclinometers and survey monuments to monitor for slope movements, which included after blasting. Piezometers were used to monitor groundwater levels. No adverse conditions have been observed.

PROTECTION OF CONCRETE STRUCTURES AGAINST SULFATE AND ACID ATTACK FROM GEOLOGIC FORMATION

During excavation of the left abutment Temblor Sandstone in 2013, two short-lived steam vents were encountered in the area above the new spillway. The vents disappeared over several days. The disturbance of bedrock probably allowed the steam to escape to the ground surface through cracks in the rock. The steam vents had a sulfurous odor, which led to the evaluation of the rock in the left abutment for potential sulfate attack.

Following runoff from rains in February 2014, orange iron stains were observed on the shotcrete that had been placed on the left wall of the spillway (see Figure 8). It was concluded that the iron stains resulted from acidic conditions in the rock slope above the spillway. Sulfates and acidic conditions can be deleterious to concrete structures.

Figure 8. Staining of Shotcrete on Left Spillway Wall.

Due to the potential deleterious effects on concrete noted above, Temblor Sandstone samples were obtained for laboratory corrosivity analysis from near or in the dam core trench and from the recently excavated spillway subgrade. The sulfate concentration and pH varied widely, from 60 to 3600 mg/kg and from 3.2 to 7.9, respectively.

Potential Deleterious Exposure of Concrete to Sulfates and Acid

Based on ACI 201.2R, the concentrations of sulfate between constitute "moderate" to "severe" level of exposure to concrete. Materials with a pH less than 6.0 are considered to be corrosive to concrete. Based on the laboratory test results and the ACI recommendations, protection measures were determined to be needed to mitigate the potential for sulfate and acid attack on concrete. The area of attack to be mitigated is at the interface between the rock and concrete.

Mitigation of Potential Sulfate and Acid Attack on Concrete

The project concrete uses Type II/V cement (low alkali) and incorporates Class F fly ash in the mixes. Water-cement ratios (w/c) for the concrete mixes are as follows: (a) 4,000 psi concrete, w/c = 0.40, (b) 3,000 psi concrete, w/c = 0.45, and (c) 2,000 psi concrete, w/c = 0.54. Based on the ACI provisions, the 4,000 psi concrete (with w/c = 0.40; e.g., spillway invert slabs and outlet conduit encasement) would be resistant to sulfate attack; 3,000 psi concrete may be vulnerable to sulfate attack; and 2,000 psi concrete would be vulnerable to attack.

Two methods are being implemented to isolate 2,000 and 3,000 psi concrete from the foundation rock to mitigate sulfate attack; steel fiber-reinforced shotcrete being used routinely as part of rock slope stabilization (specified 5,000 psi, w/c = 0.40, and contains Type II/V cement and Class F fly ash), or a 4,000 psi base concrete layer (with w/c = 0.40). Backfill concrete and dental concrete in the dam foundation was changed from 2,000 psi to 4,000 psi concrete. Asphaltic coatings were not considered for use on concrete structure foundations due to sliding stability concerns.

Concrete is not resistant to acid attack. Concrete on this site is potentially subject to incidental exposure to acid. However, historical performance of the 80-year-old concrete on this site has not suggested damage to concrete related to acid attack. Therefore, it was assumed that the exposure to acid was not significant. However, asphaltic spray-on coatings are being used to isolate the back of the cantilevered spillway stem walls of the spillway from potentially acidic runoff. The asphaltic membrane will be sprayed in two 1/16-inch thick layers (totaling 1/8-inch thick) at ambient temperatures.

CONSTRUCTION OF SPILLWAY AND CONTROLLING TEMPERATURE

The spillway structure at the CDRP requires a significant amount of concrete, some of which is 3,000 psi mass concrete. To mitigate potential sulfate exposure of concrete placed against the foundation rock as discussed above, it was decided that the approved 4,000 psi structural concrete or shotcrete mix would be placed against the exposed rock surfaces as mass concrete and the approved 3,000 psi mass concrete mix would be placed only against 4,000 psi concrete or shotcrete. The contract specifications for CDRP limit mass concrete placement temperature, lift height, and adjacent placement intervals. The specifications also provide specific guidance for temperature monitoring and curing.

An analysis was performed using ConcreteWorks (TxDOT, 2005) to compute the temperatures expected to occur in the spillway mass concrete. The analysis was used to evaluate both the expected greater temperatures in the initial 4,000 psi mass concrete placement against the foundation rock and the effects of placing the 3,000 psi mass concrete mix against the first lift. The analyses used mix design data and key assumptions based on requirements of the technical specifications and plans for concrete placement. The following specification requirements are critical to the analysis assumptions and interpretation of the results:

- First lift not to exceed 5 feet and subsequent lifts not to exceed 7.5 feet.
- Minimum of 4 days between placements of adjacent mass concrete blocks.
- Maximum decrease in concrete temperature in 24 hour period after end of protection period of 20° F.
- Maximum differential between surface of concrete and ambient temperature for removal of protection is 20° F.
- Maximum differential temperature between core of concrete and surface less than 35° F.
- Membrane curing of mass concrete surfaces after form removal and for a minimum of 4 days after placement.

Spillway Gravity Wall Mass Concrete Placement – First Lift

Key input for the thermal analysis of the first lift of the spillway gravity wall mass concrete placement is summarized as follows: (a) cement = 375 lb/yd^3, (b) Class F fly ash = 150 lb/yd^3, (c) water /cementitious ratio= 0.4, (d) concrete placement temperature = 50 °F, (e) substrate (concrete) Temperature = 57 °F, and (f) insulation blanket R-value = 5.67. The results of the analysis are shown below:

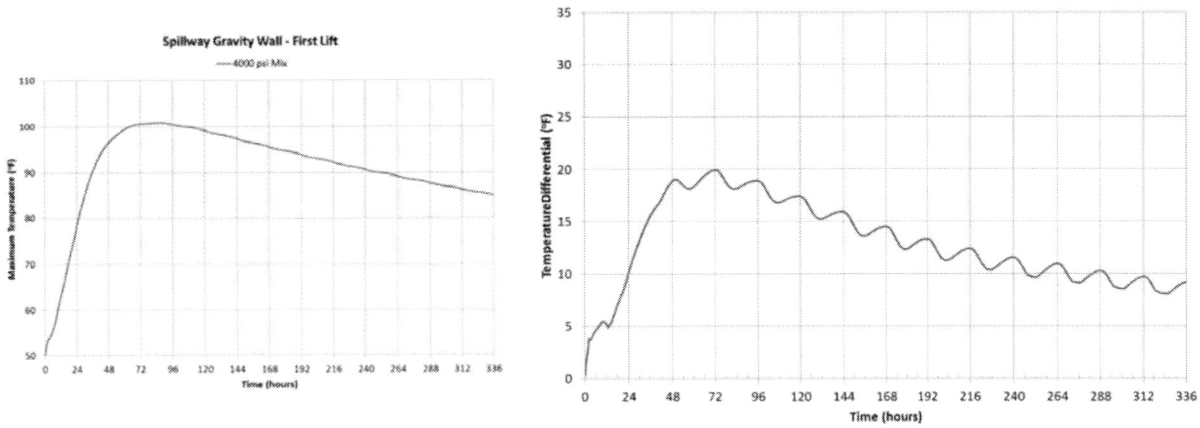

Figure 9. Thermal Analysis - Spillway Gravity Wall First Lift.

Figure 9 indicates that the maximum temperature of 100° F and the temperature differential of 19° F are well within the specified temperature requirements for a placement temperature of 50° F. Therefore, it is expected that the temperatures in the first lift of the spillway gravity wall mass concrete will be within specified values for the materials and construction methods proposed for the first lift of the mass concrete spillway gravity walls.

Spillway Gravity Wall Mass Concrete Placement – Upper Lifts

Key input for the thermal analysis of all but the first lift of the spillway gravity wall mass concrete placement is summarized as follows: (a) cement = 350 lb/yd^3, (b) Class F fly ash = 117 lb/yd^3, (c) water /cementitious ratio= 0.45, (d) concrete placement temperature = 50 °F, (e) substrate (concrete) Temperature = 103 °F, and (f) insulation blanket R-value = 5.67. This input is for the third lift above the foundation. Preliminary analyses of the first two lifts were run to calculate the substrate temperature. The results of the analysis are shown below:

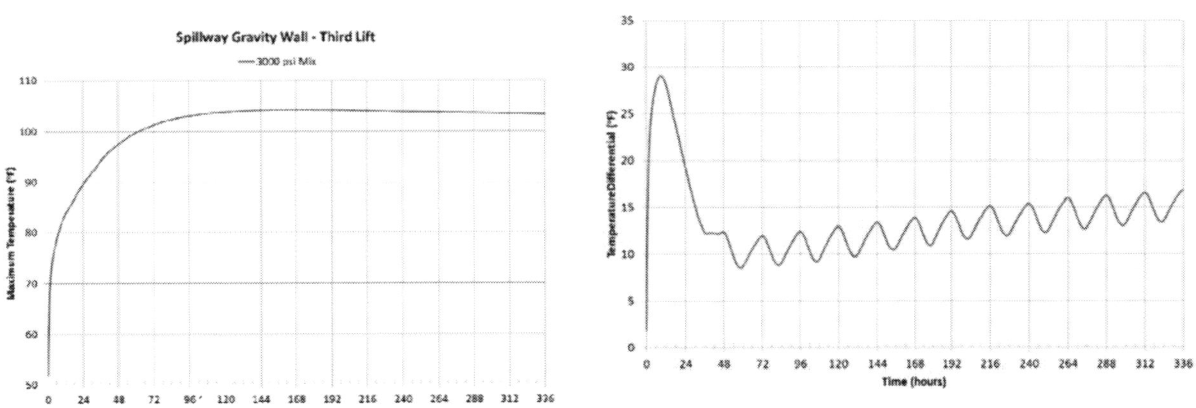

Figure 10. Thermal Analysis - Spillway Gravity Wall Third Lift.

Figure 10 indicates that the maximum temperature of 104° F and the temperature differential of 29° F are well within the specified temperature requirements for a placement temperature of 50° F. It is noted that the temperature differentials are increasing at the end of the 14-day analysis interval; however, the maximum temperature differential at this time is approximately 17° F, less than half of the allowable temperature differential. Additional insulation was applied consistent with the proposed thermal control plan for the spillway mass concrete placements, and demonstrated that this approach is effective in controlling increasing differential temperatures. Based on the results of the analysis for the spillway gravity wall mass concrete placements, it is expected that the temperatures in the upper lifts of the spillway gravity wall mass concrete will be within specified values for the materials and construction methods proposed for the mass concrete spillway gravity walls.

CONCLUSIONS

This paper presents an update of additional design challenges since the USSD 2014 Conference that have resulted from the complex geologic and geotechnical conditions encountered during excavation for the construction of the CDRP. Such complex geologic conditions described in this paper and in 2015 will require continuous monitoring of conditions as they are revealed during excavation. Adjustments to the design to accommodate such conditions will periodically be required.

The project is scheduled to be completed in 2018.

ACKNOWLEDGEMENTS

The contributions from Erik Newman (URS) on geotechnical topics and Scott Jones (URS) on concrete topics are gratefully acknowledged. In addition, the authors of this paper acknowledge the substantial efforts and contributions made to the formulation, approval, and implementation of the engineering solutions presented in this paper by the staffs of the Designer (URS Corporation and SFPUC), the San Francisco Planning Department, the Construction Manager (Black and Veatch), the Contractor (Dragados, Flatiron, and Sukut Joint Venture), the Calaveras Technical Advisory Panel, and the California Division of Safety of Dams.

REFERENCES

Forrest, M., Roadifer, J., Wade, D., Hou, S., and Tang, G. (2014). Update on the Calaveras Dam Replacement Project, Alameda County, California, "Proceedings of the 34th USSD Conference, San Francisco, California, April 2014.

Kelson, K., Respess, P., Newman, E., Wade, D., Hou, S., and Tang, G. (2014). Engineering Geologic Conditions at the Calaveras Dam Replacement Project, Alameda County, California, "Proceedings of the 34th USSD Conference, San Francisco, California, April 2014.

Marinos, P, and Hoek, E. (2000). *GSI A Geologically Friendly Tool for Rock Mass Strength Estimation*, Proceedings GeoEng2000, Melbourne, Technomic Publishers.

Post Tensioning Institute (PTI) (2006). Recommendations for Prestressed Rock and Soil Anchors, 6th Ed., Post Tensioning Institute, Pheonix, Arizona.

TxDOT. (2005). *Concrete Works Version 2.0.*

SOIL CEMENT ARMORING FOR DAMS: KEY PROJECT CONTROLS FOR SUCCESSFUL CONSTRUCTION

Don Roarabaugh, P.E.[1]
Dan Stare, P.E.[2]
Larry Corn, P.G.[3]

ABSTRACT

Soil cement and roller-compacted concrete are two construction materials frequently used to armor earth embankment dams. Soil cement is most often used for erosion protection on the reservoir side or upstream slope of embankments for protection from wave action while roller-compacted concrete is most often used for erosion protection on the downstream slope of embankments when there is potential for overtopping due to extreme storm events.

Guidance documents related to soil cement design and construction have been published by numerous agencies including the American Concrete Institute, International Commission on Large Dams, Portland Cement Association, U.S. Army Corps of Engineers, U.S. Bureau of Reclamation, and others. These documents provide significant value; however they may not distinguish between construction practices appropriate for low versus high hazard potential projects. For example, some soil cement guidance documents do not distinguish between practices appropriate for below grade channel slopes or pavement subgrades versus above grade high hazard potential dams.

This paper presents some key soil cement construction-phase project controls and techniques considered appropriate for high hazard potential and/or high value dams based on the authors' recent experiences. The Tampa Bay Reservoir Renovation Project completed in 2014, a 15.5 billion gallon above-grade reservoir in Florida which included approximately 550,000 cubic yards of soil cement armoring, is referenced for discussion and as a recent example of soil cement armoring. Project control recommendations are made for consideration on future high hazard potential and/or high value dam projects utilizing soil cement armoring for embankment protection.

INTRODUCTION

Soil cement slope protection for armoring the reservoir side of embankment dams was first tested by the U.S. Bureau of Reclamation (USBR) in 1951 at an experimental section constructed at the Bonny Dam in eastern Colorado. The test section was constructed separate from the dam and situated for maximum exposure to attack from the elements.

---

[1] Senior Project Engineer, Gannett Fleming, Inc., 207 Senate Ave, Camp Hill, PA 17011. droarabaugh@gfnet.com.
[2] Senior Geotechnical Engineer, Gannett Fleming, Inc., 207 Senate Ave, Camp Hill, PA 17011. dstare@gfnet.com.
[3] District Quality Manager, Kiewit Infrastructure South Co., 5100 W. Kennedy Blvd, Suite 250, Tampa, FL 33609. Larry.corn@kiewit.net.

After 10 years of monitoring the Bonny test section, the USBR added soil cement to their specifications as an alternative to riprap for upstream slope protection, primarily for potential use where riprap was not readily available [USBR 1984].

The first dam to utilize soil cement for upstream slope protection was Merritt Dam, a 126-foot high, 3,200-foot long dam completed by the USBR in 1963. Subsequently, the USBR used soil cement slope protection for Cheney Dam, an 84-foot high, 24,500-foot long dam completed in 1964, followed by five more embankments completed from 1966 to 1969 including two major earth dams, Glen Elder Dam and Starvation Dam. By 1984, three of these seven soil cement linings required repair, including Merritt and Cheney Dams, while the remaining four soil cement linings reportedly remained in excellent condition [USBR 1984].

While considerable progress in construction techniques and project controls has been made since the early uses of soil cement armoring for embankment dams, issues with recently constructed soil cement armored dam embankments emphasize the need for continued improvement with the technology and sharing of project experiences.

FUNDAMENTAL PROCESSES

Soil cement construction for dam embankment armoring can be broken down to two fundamental processes – producing the soil cement mix, and placing soil cement on the embankment. Acceptance criteria are instrumental in achieving success with the mix and the placement, and thus should be carefully established and factored into selection of project controls and construction techniques.

Success of the soil cement armoring, primarily the ability of the armor to withstand imposed loads and provide long-term durability, depend heavily on three key factors – compressive strength (as related to durability), in-place density, and bonding at lift joints if placed in multiple layers. Thus, satisfactorily achieving the three properties of strength, density, and lift joint bond, should be factored into each part of the process and be a focus of the project controls and construction techniques.

Figure 1, a photo of soil cement placement for the Tampa Bay Reservoir Renovation Project, shows the two fundamental elements of soil cement construction. Success with soil cement construction is primarily related to controlling the mix and controlling the placement. The next two sections delve into the soil cement mix and placing the soil cement armoring, including discussion of project controls and techniques to achieve the three key properties of strength, density, and lift joint bond.

Figure 1. Two Fundamental Processes for Successful Soil Cement Construction are Controlling the Mix and Controlling the Placement.

SOIL CEMENT MIX

Numerous articles and guidance documents have been written and published regarding designing and producing the soil cement mix. For completeness, many facets of the mix production process are briefly discussed herein. However, the focus of this section is on controlling the mix cement content using a statistically-based approach combined with statistically-based acceptance criteria similar to that used for conventional concrete to achieve the desired strength results.

Materials

Soil cement is most often composed of Portland cement, soil, and water, with the soil obtained from onsite or nearby sources. The reason the USBR investigated soil cement at the Bonny Reservoir test section in 1951 was to obtain the benefit of using onsite materials for embankment armoring in lieu of riprap, especially in areas where locally available riprap was difficult or impossible to obtain.

Evaluation and Selection of Soil Sources. Sandy soils are predominately used for production of soil cement for embankment armoring while cohesive soils are avoided. In addition, it is preferable to use onsite soils due to cost savings compared to hauling soil from offsite borrow areas. Prior to the start of any soil cement placement, investigations should be performed to evaluate the suitability of proposed soil sources for production of soil cement, and identify potentially detrimental soils that should be avoided.

Pre-production soil investigations performed during the mix design process are aimed at identifying borrow sources that react well with cement and produce the desired strength with the least amount of added cement. Borrow sources with undesirable characteristics, including organic matter, cohesive soils, and soils with low pH are identified and avoided to the extent practical.

Soil used for construction of the recently completed Tampa Bay Reservoir Renovation Project was a silty sand classified as SM or SP-SM based on the Unified Soil Classification System or A-2-4 based on the AASHTO Classification System. Onsite source evaluations were used to preclude undesirable soils to the extent practical which included soils with low pH, excessive organics, or excessive clay content.

Mix Proportions

Since the soil cement mix is most often composed solely of soil, cement, and water; and the water content is based on the optimum moisture for compaction using soil fill placement techniques, developing mix proportions for soil cement is focused almost entirely on selecting the cement content of the mix once the soil sources have been identified and evaluated. The cement content is expressed as a percent of the dry weight of the soil or as a percent of the dry density of the soil cement mix. Various approaches to selecting the appropriate cement content are discussed in the following sections.

Pre-production trial mix batching using a range of cement contents is commonly performed to determine the required cement content to meet project performance criteria. Soil cement specimens, commonly referred to as pills, are molded from the trial batches and tested for compressive strength and durability testing. If the soil sources are known to vary significantly, trial batching and testing should be performed using soils from multiple borrow areas, and/or using a blend of soils.

Typically the results of compressive strength and durability tests are correlated during the mix design phase, then compressive strength is monitored during construction to ensure that both strength and durability requirements are being met. Using compressive strength and durability criteria suggested in many guidance documents, it is common for the durability requirements to be exceeded when compressive strength criteria are met, thus compressive strength is often the dominant criteria of the two [USACE 2000].

A photo of pre-production soil cement pills prepared for the Tampa Bay Reservoir Renovation Project is presented in Figure 2. The pills were molded using soil from a borrow area under investigation and with a cement content of 12 percent. Pill 3B demonstrates a nominal loss of mass after completion of wet-dry durability testing, which was well within project criteria.

Figure 2. Soil Cement Pills. Pill on Left as Molded. Pill on Right Shows Nominal Loss of Mass at Completion of Wet-Dry Durability Testing.

Mix Design Criteria. It is common for the specified minimum compressive strength and maximum loss of mass following durability testing to be based primarily on performance of prior projects. Criteria presented in most guidance documents suggest a minimum compressive strength in the range of 600 to 750 psi at 7 days with further restrictions on the 28-day compressive strength in some cases [PCA 2007]. Many guidance documents reference USBR criteria since the agency performed some of the earliest research and completed the earliest projects using soil cement for armoring dam embankments.

The USBR criteria are often referred to as the "Bonny Criteria" since the basis of the criteria relates back to successful performance of the Bonny Reservoir test section. As noted by the USBR, however, it is important that project planners consider other factors including anticipated conditions and acceptable risk at each project site [USBR 2013].

Cement Content – Historical Soil Cement Approach. A common approach for establishing the cement content of the mix is to determine the cement content required to meet specified criteria based on trial batching and testing, then add one to two percent of additional cement to account for field variations. For example, if the specified strength is a minimum of 650 psi at 7 days, and trial batching test results indicate 650 psi is met with a cement content of 10 percent, then production might proceed with a cement content of 11 to 12 percent to account for variability.

Cement Content – Concrete Approach. Prior to discussing how the cement content can be controlled using a statistically based approach, it is necessary to understand the terminology. In particular, the term "minimum" as applied to the specified strength needs to be clearly understood. Is it the minimum for acceptance, or the minimum for selection of the mix design cement content? The difference is very significant.

The specified strength in the concrete industry, commonly annotated as f'_c, is not the minimum value for acceptance if ACI criteria are followed. Referring to Table 1, for concrete with a specified compressive strength of less than 5,000 psi, the mix design is selected to provide the higher strength of two options: 1) specified strength plus 1.34 times the sample standard deviation (ACI equation 5-1), or 2) specified strength plus 2.33 times the sample standard deviation minus 500 (ACI equation 5-2). ACI provides optional criteria when statistical data are not available, which could be compared to the soil cement approach of adding an arbitrary one to two percent of additional cement.

The added strength in the ACI equations, either $1.34S_s$ or $2.33S_s$-500, is at times referred to as an overdesign factor, similar to the added one to two percent of cement in the soil cement industry. The difference with the concrete approach, if the equations in Table 1 are used, is that the "overdesign factor" is statistically based, and more importantly, directly correlated to the acceptance criteria. If the average strength of the mix exactly meets ACI equation 5-1, there is a one in one hundred chance that the 3-test running average will fall below the specified strength. Further, if the average strength of the mix exactly meets ACI equation 5-2, there is a one in one hundred chance that any individual test will be more than 500 psi below the specified strength. The ACI acceptance criteria reflect these two positions: the 3-test running average must equal or exceed the specified strength, with no individual test more than 500 psi below the specified strength.

Based on the foregoing discussion, it is apparent that the specified strength for concrete, as referenced in the ACI structural concrete building code, is not the minimum value for acceptance, an important clarification for discussion of soil cement acceptance criteria.

Table 1. ACI 318 Criteria for Structural Concrete Mix Design (Copy of Table 5.3.2.1 from ACI 318-11) [ACI 2011]

TABLE 5.3.2.1 — REQUIRED AVERAGE COMPRESSIVE STRENGTH WHEN DATA ARE AVAILABLE TO ESTABLISH A SAMPLE STANDARD DEVIATION

Specified compressive strength, psi	Required average compressive strength, psi
$f'_c \leq 5000$	Use the larger value computed from Eq. (5-1) and (5-2) $f'_{cr} = f'_c + 1.34s_s$ (5-1) $f'_{cr} = f'_c + 2.33s_s - 500$ (5-2)
$f'_c > 5000$	Use the larger value computed from Eq. (5-1) and (5-3) $f'_{cr} = f'_c + 1.34s_s$ (5-1) $f'_{cr} = 0.90f'_c + 2.33s_s$ (5-3)

Assuming that concrete test results will follow a normal distribution, which is the typical assumption, it can be calculated that if concrete test results exactly satisfy ACI equation 5-1, approximately nine percent of all test results will be less than the specified strength [ACI 214], which would be acceptable following ACI structural concrete acceptance criteria. This is depicted graphically in Figure 3.

On the other hand, if the specified strength is taken as the minimum value for acceptance, then more cement will need to be added pushing the peak of the bell curve depicted in Figure 3 further to the right of f'_c and the result will be higher project costs and an increased cracking potential due to the higher average strength, higher cement content, and higher thermal stresses.

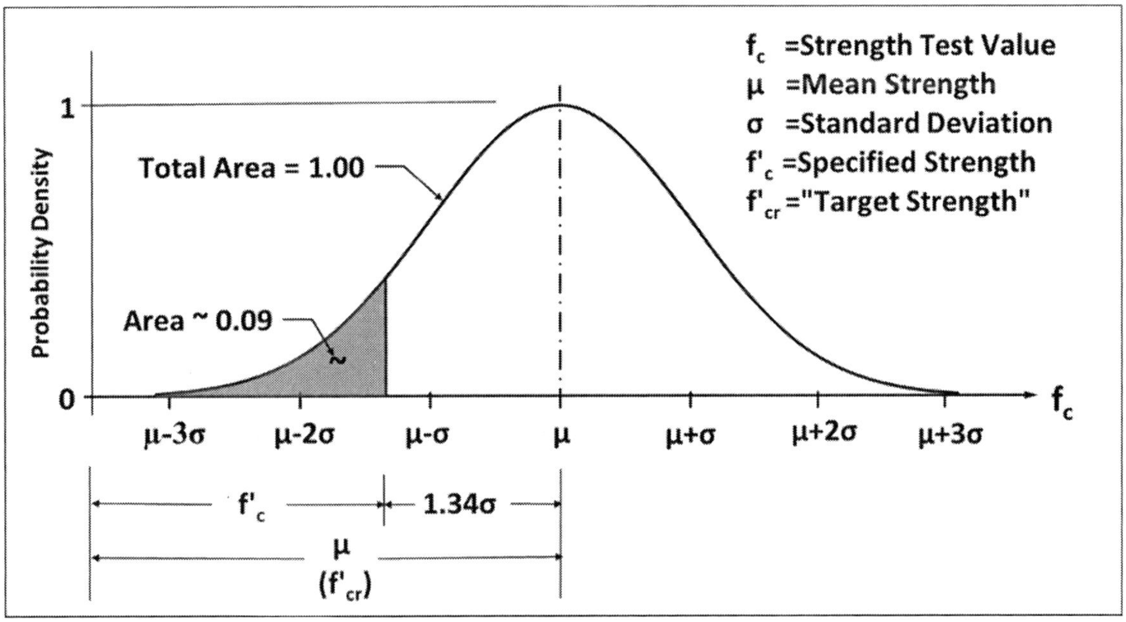

Figure 3. Normal Distribution of Strength Test Results and Relation to ACI Equation 5-1.

<u>Cement Content – Tampa Bay Reservoir Renovation Project Approach.</u> Selection of the cement content for the Tampa Bay Reservoir Renovation Project soil cement mix followed the historical soil cement approach for the start of production, which included trial batching and testing to determine the cement content needed to satisfy the specified criteria, then adding 1 to 2 percent additional cement to establish the starting production-phase cement content. After production-phase test data became available, the cement content was adjusted based on a statistical approach adopted from the concrete industry.

Although acceptance criteria are discussed in a later section of this paper, due to relating cement content control with acceptance criteria, some discussion is necessary in this section. The strength acceptance criteria adopted for the Tampa Bay Reservoir Renovation Project included the following: 1) the 5-test running average must be equal to or greater than the specified strength of 650 psi at 7 days, and 2) no individual test will fall below 500 psi. It is recognized that the 5-test running average is more "lenient" than the ACI code 3-test running average for structural concrete; however it is also noted that soil cement is intended to be a durable armoring in this case, not a load carrying beam or column. In addition, due to the use of in-situ soils in the mix as opposed to quarried and processed aggregates used in concrete mixes, and the potential higher variability associated with earlier test ages (specified strength at 7 days for soil cement as opposed to 28 days for concrete), consideration was given for additional variability in order to avoid forcing an unnecessarily high cement content.

Relating cement content to the acceptance criteria, the project adopted the approach that if either of the two aforementioned strength acceptance criteria were not met at any time, the cement content was increased by at least 0.5 percent. On the other hand, if the 30-test running average exceeded f'_{cr} by more than 50 psi, then the contractor could reduce the cement content by 0.5 percent. In this case, $f'_{cr}=f'_c+1.04S_s$, which is the equivalent one-in-one hundred chance provided by ACI equation 5-1 only revised for a 5-test running average in lieu of a 3-test running average.

The adopted strength acceptance criteria and related cement content adjustment controls were intended to provide the contractor with flexibility to achieve project goals beneficial to all parties by providing performance-based incentives for good quality control of soils and other factors leading to less strength variability in return for economies of reducing the cement content. In other words, achieve the best quality possible at the lowest cost; and provide the contractor with the flexibility to achieve this goal. Further, meeting the project strength and durability requirements with less cement included for overdesign has other beneficial project quality aspects.

<u>Cement Content – Summary.</u> It may be desirable to control the cement content for large dam embankment armoring projects using the historical soil cement approach to establish the initial cement content, combined with a statistical based approach adopted from structural concrete practices for ongoing cement content control during production. It is recognized that soil cement for embankment armoring is not structural concrete such as load carrying beams or columns, nor is it interior mass concrete such as placed for the interior of a gravity dam. However soil cement for embankment armoring does require concrete performance criteria, including durability, which is related to compressive strength; and further, soil cement should be treated with considerations appropriate for mass concrete, including balancing the probability of low-strength test results against undesirable consequences such as increased cracking potential at higher cement contents. Accordingly, adopting statistical based practices for soil cement mix control and strength acceptance from the concrete industry may be appropriate, particularly for high hazard potential and/or high value dam applications.

A full discussion of statistical based acceptance criteria, related mix design controls, and contractor incentives to optimize economies against project requirements is beyond the intended scope of this paper. Additional information can be found in concrete literature including some of the documents in the references section of this paper such as the reports prepared by ACI Committees 214, 301, and 318; and an article by Ernest K.

Schrader as published in Concrete International titled "Statistical Acceptance Criteria for Strength of Mass Concrete" [ACI 214] [ACI 301] [ACI 318] [Schrader 2007].

Centralized Soil Cement Plant

The first of three soil cement plants used for the Tampa Bay Reservoir Renovation Project and associated activities around the plant related to producing the soil cement mix are shown in the aerial photo presented in Figure 4. The plant and soil stockpile are discussed in more detail in this section.

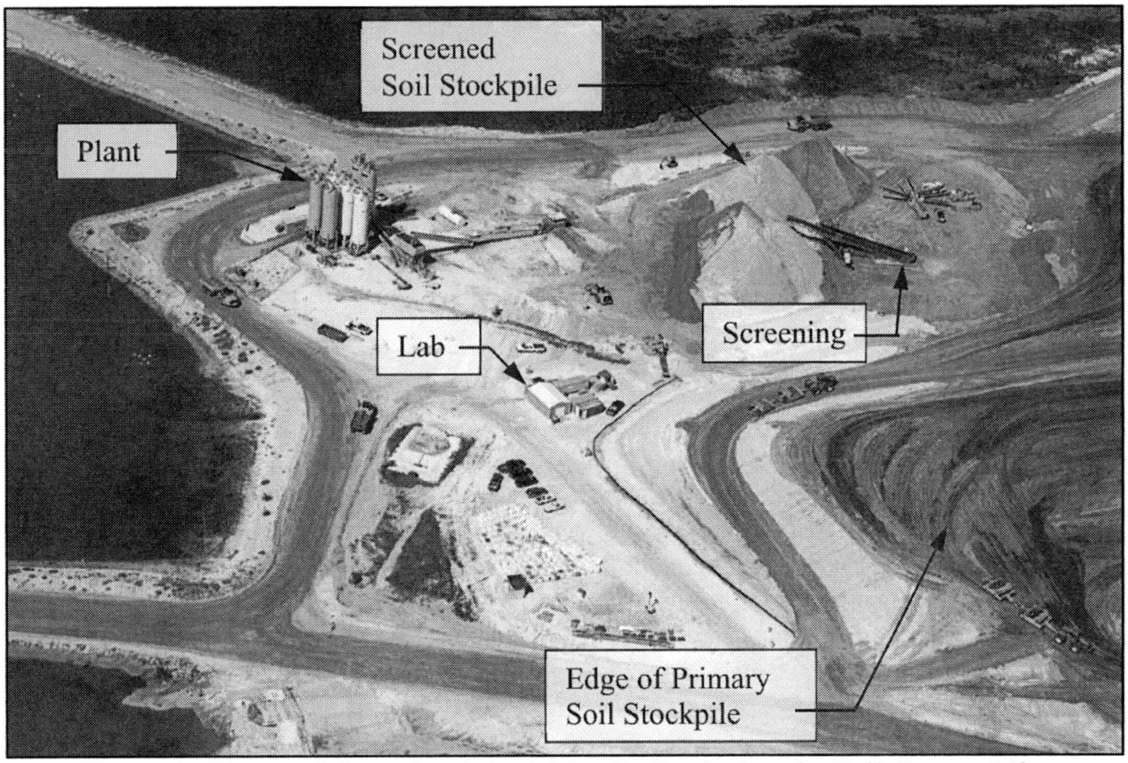

Figure 4. Plant and Associated Activities for Producing the Soil Cement Mix

Soil stockpile. An aerial view showing the footprint of the primary soil stockpile at the Tampa Bay Reservoir Renovation Project is shown in Figure 5. As discussed earlier, pre-production evaluation of soil borrow areas is a key element of the soil cement quality control process. After suitable borrow areas are identified, the potential for isolated concentrations of problematic soils can be further reduced by stockpiling the selected soils at the plant in horizontal layers and withdrawing by face loading as a means of blending the soils.

Figure 5. Footprint of Soil Stockpile at the Tampa Bay Reservoir Renovation Project

As shown in Figures 4 and 5, borrow sources selected for the soil cement mix at the Tampa Bay Reservoir Renovation Project were stockpiled near the plant. The stockpile was constructed in horizontal layers and material was withdrawn from the stockpile by face loading with front-end loaders. All stockpiled soil, approximately 1,000,000 tons, was then processed through ½-inch screens to proactively manage product quality. Screening operations and surge stockpiles ready for loading into the soil cement plant are shown in Figure 6.

Even with the screening efforts, clay agglomerations ("clay balls") would at times accumulate near the base of the screened surge piles as shown in Figure 7. The collections of clay balls were routinely removed to avoid concentrations in the mix, which could lead to low compressive strength test results.

Overall, the soil source management required a large effort on the part of the contractor; however this high level of proactive quality control assisted with reducing strength variability of the mix and thus allowed for a quality mix to be placed with a lower cement content than would have occurred without these focused efforts.

Figure 6. Screening Soil Can be an Effective Means of Reducing Product Variability.

Figure 7. Concentrations of Clay Balls in the Mix can be Problematic.

Batching and Mixing

For dam armoring, the materials are typically mixed in pugmill mixing plants. Inside the mixing chamber, rotating twin shafts equipped with paddles provide the shearing action needed for thorough mixing. Continuous mix pugmills are the most common type of plant used; however batch-type plants can also be very effective.

The three plants used to produce soil cement for the Tampa Bay Reservoir Renovation Project are shown in Figure 8. The plant on the right is a Coneco batch-type plant with a mixing chamber rated at 8 cubic yards. The two plants on the left are continuous mix type, an Accumix 600B and an Accumix 750XB, rated at 600 and 750 tons per hour, respectively. The plants were operated at production rates significantly lower than the rated capacity to achieve thorough mixing of the soil cement. Actual rates achieved during production were approximately one-third of the rated capacity for the continuous mix plants and about two-thirds of the rated capacity for the batch plant.

Figure 8. Two Continuous Mix Plants and One Batch Plant Used for Mixing Soil Cement at the Tampa Bay Reservoir Renovation Project.

SOIL CEMENT PLACEMENT

Producing a quality soil cement mix and placing the product are both very important processes for successful soil cement armoring. The best mix will not perform as intended if it is not well compacted. Both strength and durability will suffer. Additionally, if the mix is placed in multiple lifts, good bonding between successive lifts may be necessary depending on anticipated conditions. This is quite often the case for embankment armoring of large reservoirs where long fetch distances can lead to significant wave action. Without sufficient bonding, the wave loading may separate the soil cement lifts.

Damage that occurred to the early USBR dams armored with soil cement, as discussed in the introduction to this paper, was attributed to lack of bonding between the lifts combined with the severity of exposure. The soil cement armoring at Cheney Dam experienced damage during a severe windstorm, while the armoring at Merritt Dam was damaged during what was considered average wave action. The damage at both of these dams generally consisted of broken and displaced slabs and was attributed to uplift at the lift interfaces as a result of wave action. Poor bonding of the lifts was considered to be the root deficiency with the armoring [USBR 1984].

Lift Surface Care and Preparation

Soil cement placed in multiple lifts requires special care of the lift surfaces to promote bonding of successive lifts so the in-place armoring can perform as a single mass unit rather than discrete layers that can be separated by exposure to freeze-thaw conditions or wave action.

Cleaning. Lifts should be maintained in a clean condition and any laitance, debris, ponded water, or other potential contaminants that may inhibit bonding should be removed prior to placement of subsequent lifts. Cleaning immediately prior to placement of a bonding agent and subsequent lifts of soil cement is accomplished by blowing with compressed air, pressure washing, and/or sweeping with mechanically driven brooms.

Bonding Agent. If the joint between subsequent lifts will be exposed to conditions that may tend to separate the lifts, a bonding agent should be used. Historically, the bonding agent has been either dry cement or a cement slurry (grout) spread on the clean lift surface immediately in advance of soil cement placement for the next lift. Due to difficulties with controlling the application rate of dry cement, and potential for the dry cement to become airborne leading to loss of the bonding agent, environmental concerns, and/or health concerns, dry cement is not recommended. Cement slurry is applied in thin layers and thus is susceptible to drying from exposure. Therefore, the slurry should be applied immediately in advance of the soil cement.

The use of cement slurry bonding treatment between all soil cement lifts placed for dam embankment armoring is recommended in the current USBR guidance document published in late 2013 [USBR 2013]. Application of bonding slurry in advance of soil cement placement at the Tampa Bay Reservoir Renovation Project is shown in Figure 9.

Figure 9. Applying Bonding Slurry in Advance of Soil Cement Placement

Delivery and Spreading

The soil cement mix needs to be delivered and placed in a timely manner so it can be compacted prior to setting or drying from exposure. The mix also needs to be placed without segregation, although segregation is only a concern if there are coarse particles in the mix. Most soils used for soil cement, including the most desirable soils which are silty sands, are not highly susceptible to segregation concerns.

Soil cement can be delivered to the placement area with trucks, conveyors, or a combination of both. It is typically spread on the lift surface using dozers ranging in size from a D3 to a D6 equivalent. Paving machines, skid steers, and front-end wheel loaders have also been used successfully to spread soil cement.

Placement for the Tampa Bay Reservoir Renovation Project was accomplished with both truck and conveyor delivery, and most spreading was performed with dozers. The soil cement was placed in lifts ranging from 6 to 12 inches thick, after compaction, depending on the location on the project. Figure 10 shows delivery and spreading with a mobile conveyor and a spreader box mounted on the front of a D3 dozer. This delivery and spreading method was used for relatively narrow lifts on some parts of the project.

Figure 10. Conveyor and Dozer with Spreader Box Placing and Spreading Soil Cement

Compacting

Compaction, and achieving a dense in-place soil cement product, is considered to be one of the most important factors for successful soil cement armoring. Compaction of soil cement can be successfully accomplished using a variety of equipment, although smooth drum vibratory rollers are perhaps the most commonly used compaction equipment for sandy soils.

Adequacy of soil cement compaction is typically checked with a single-probe nuclear density gauge. The density test is considered to be one of the most important tests run on the project for two reasons: 1) as previously discussed, achieving adequate compaction is very important to the durability of the in-place product, and 2) the nuclear density test provides results immediately as opposed to compressive strength tests which may not be available for days. Thus, if the density test indicates a potential problem, it is known immediately and can be addressed prior to placement of additional soil cement.

Due to the variability of soils, and thus soil cement, compaction curves are a very important element in determining if the specified compaction is achieved. Techniques commonly used for control of soil fill placements should also be used for soil cement; specifically developing a family of curves representative of the soil sources (soil cement mixes), combined with frequent one-point compaction tests to ensure the correct compaction curve is being used at any given time for comparison with the nuclear density test results.

The nuclear density test result is meaningless if it is not compared to the proper target for the soil or soil cement mix being placed. Accordingly, and since density is so important to the durability of the in-place product, consideration should be given to performing compaction tests on a daily basis, and performing one-point tests every few hours during placement of soil cement armoring for a high hazard potential and/or high value dam.

A family of compaction curves developed for the Tampa Bay Reservoir Renovation Project for soil cement with a cement content of 11.5 percent is shown in Figure 11. As a result of the available borrow materials, diligent pre-production soil evaluations, and good soil stockpile management practices including blending and screening, the family of curves shown in Figure 11 is relatively tight with the maximum density ranging from approximately 122 to 123.5 pcf with a couple of outliers. The required in-place soil cement density was 98 percent of the maximum density.

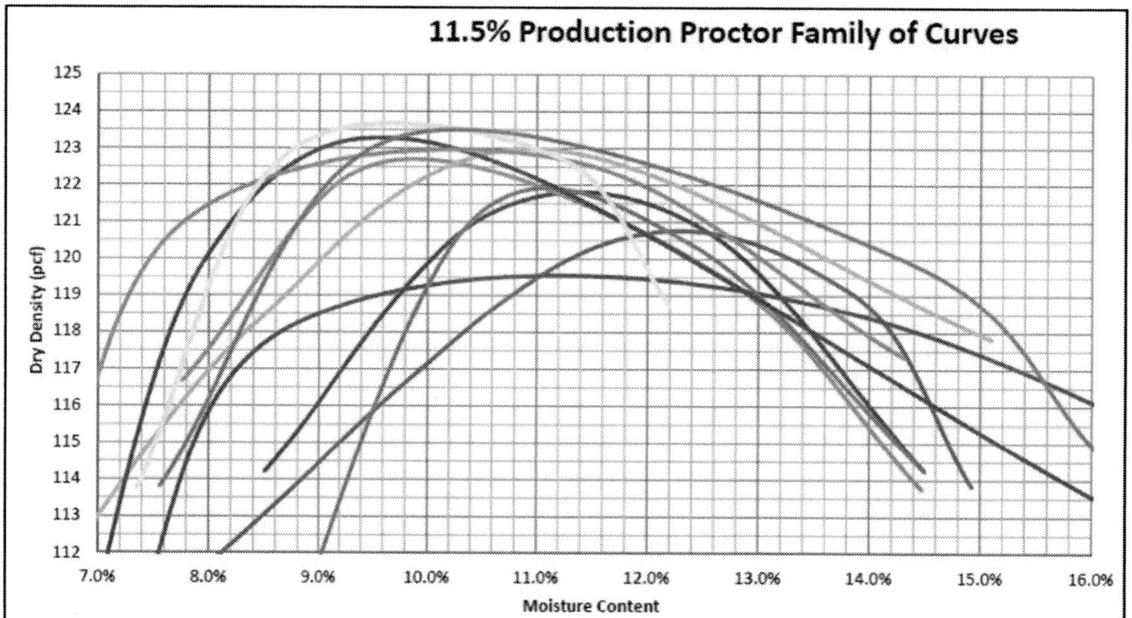

Figure 11. Example Family of Compaction Curves for Soil Cement with a Cement Content of 11.5 Percent

ACCEPTANCE CRITERIA

Acceptance criteria for soil cement armoring were discussed in prior sections of this paper where relevant to the topic at hand. A summary of acceptance criteria considerations is provided in this section.

Various criteria may be included in contract documents for acceptance; however, the focus should be on factors most important to achieving the primary purpose of soil cement armoring, which is protecting the dam embankment from loading conditions imposed by the environment. For soil cement armoring on the upstream embankment slope of high-hazard potential and/or high value dams, three key factors are compressive strength (as related to durability), density, and adequate bonding at lift joints.

Performance Based Criteria

To the extent practical, criteria established to achieve the desired in-place properties of the soil cement armoring should be performance based. The contractor is part of the project team for their construction expertise, equipment, personnel, and ability to construct the project, thus criteria should focus on defining the project requirements, not on telling the contractor how to do their task.

Density. As discussed previously, density is one of the key factors to achieving a quality in-place soil cement armor. Density criteria should define the percent compaction required, which is typically a percent of the maximum density based on lab compaction curves. Density criteria should not define compaction equipment, a practice often seen in specifications. Selection of equipment is in the realm of the contractor's expertise.

In addition to specifying the required density; due to the variability of soils, and thus the variable maximum density of soil cement, it is very important to continuously evaluate the density target. This is best evaluated by following the approach used for control of soil fills including developing a family of compaction curves, frequently re-evaluating the compaction curves, and performing frequent one-point compaction tests to ensure the correct compaction curve is being used for the target density at any given time.

Compressive Strength. As previously discussed in the mix proportion section, compressive strength acceptance criteria should follow a statistically based approach similar to that used for structural concrete. The acceptance criteria should be reflected in controls for the cement content of the mix. By allowing the contractor to adjust the cement content based on statistical analyses, in lieu of prescribing the mix cement content or prescribing the overdesign cement content, the contractor can apply good construction practices to achieve the specified properties in ways other than just adding more cement; such as carefully evaluating soil borrow areas, good control of the soil stockpile, and good control of the plant including accurate batching and thorough mixing.

The strength acceptance criteria adopted for soil cement near and above finished grade for the Tampa Bay Reservoir Renovation Project were the 5-test running average must equal or exceed the specified strength of 650 psi at 7 days with no individual test result below 500 psi. A sample of the project strength test results as related to the project acceptance criteria is presented in Figure 12. Note that all tests shown meet the acceptance criteria, even results below the specified strength.

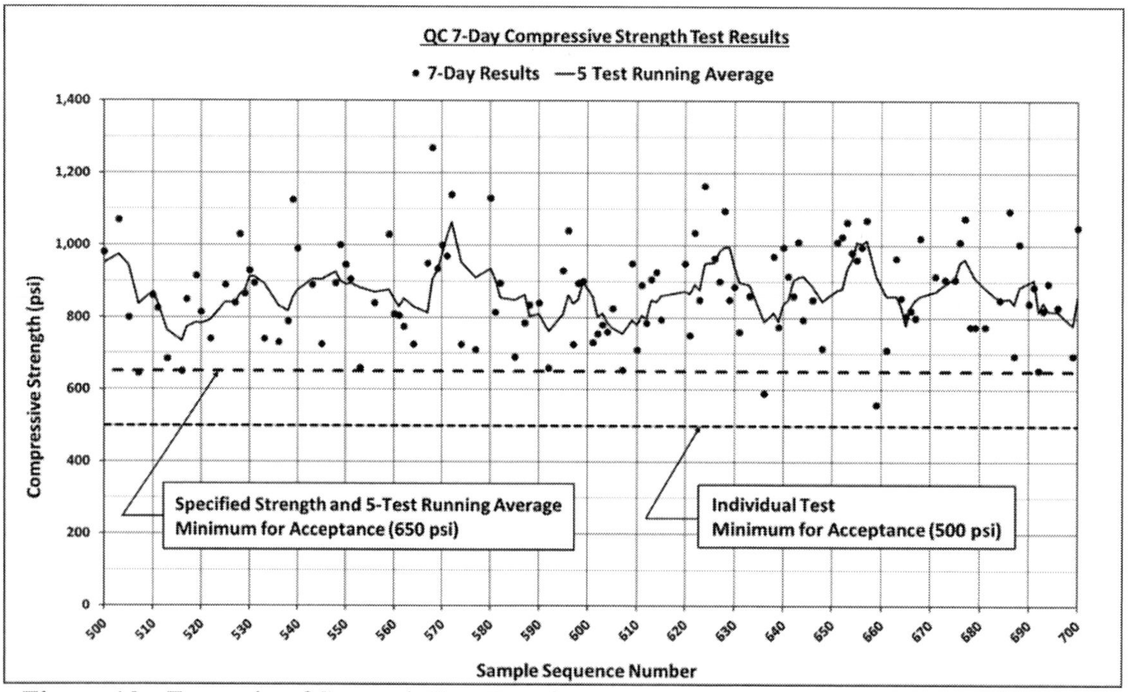

Figure 12. Example of Strength Test Results and Statistical Based Acceptance Criteria

If the compressive strength 5-test running average is below the acceptance criteria, steps should be taken to increase the average strength. If an individual test result falls below the acceptance criteria, procedures similar to those recommended in ACI guidelines for structural concrete should be followed including evaluation of cores and evaluation of the potential impact of the reported test result on the performance of the structure.

Prescriptive Criteria

Prescriptive project controls may be needed in addition to the performance based controls. These may include application of bonding slurry on lift joints, limiting the time from mixing to compaction to reduce the chance of setting or drying prior to compaction, and establishing procedures for hot and cold weather placements and limits for as-placed material temperatures.

Bonding of Lift Joints. Application of bonding slurry prior to placement of subsequent soil cement lifts should be prescribed, if bonding is considered necessary. This is a process that needs observation during construction to ensure project requirements are met, including adequate cleaning of the lift surface and placement of the bonding slurry immediately in advance of the soil cement.

Observation and Judgment During Construction

A subject that comes up on nearly all construction projects is what happens if a prescriptive project control is not met (or a performance-based target for that matter). For example, what if the specification calls for soil cement to be placed and compacted within 60 minutes of batching, and monitoring indicates it took 65 minutes? Should production be stopped and the material removed? With regard to the quality of the in-place product, it is noted that there is a significant difference between 60 minutes in the daytime sun and wind versus 60 minutes in cool damp night air.

As another example, what happens if the minimum strength for acceptance is 500 psi at 7 days, but a reported test result comes in at 490 psi? Should the soil cement lift in question, and perhaps 15 lifts that have been placed above it, be removed and replaced?

Removal and replacement is not always the best option in the opinion of the authors, and in some cases, is a significant waste of resources and cost. Further, due to the destructive nature of removal, and the break in continuity, the soil cement armoring after removal and replacement, even if performed in strict accordance with project controls, may be inferior to the armoring if other actions (or no action) were taken in lieu of removal.

It is the opinion of the authors that soil cement embankment armoring for high-hazard potential and/or high value dams should be observed and monitored by experienced personnel who understand the significance of the specified project controls. In addition, the design engineer should be consulted in cases of removal and replacement, unless it is clear that the product is unacceptable and qualified onsite personnel are in agreement. Appropriate project controls and acceptance criteria established beforehand should be combined with onsite observations and offsite consultations with qualified personnel who fully understand the issues at hand. The human element is part of project conception, design, construction, and post-construction inspection and monitoring. Is it correct to remove the human element from construction-phase project controls and acceptance?

CONCLUSIONS AND RECOMMENDATIONS

Soil cement involves construction procedures and issues relevant to placement of both soil fills and concrete. Comparisons to placement of soil fills include evaluation of borrow sources, control of the soil stockpile, and control of the target density using compaction curves. Comparisons to placement of concrete (both conventional and roller-compacted concrete) include compressive strength, durability, and bonding of lifts; plus other concrete factors such as set time and heat of hydration.

Following are some of the concepts presented in this paper as related to soil cement armoring for high-hazard potential and/or high value dams:

- Control of the soil cement mix and the soil cement placement are equally important to successful construction.

- Three key elements that should factor heavily into the project controls and construction techniques are compressive strength (as related to durability), density, and adequate bonding at the lift joints.
- The density target should be continuously evaluated due to the varying nature of soils. A family of compaction curves should be established for each mix design (cement content), and one-point compaction tests should be run routinely during placement to ensure the correct compaction curve is being used at all times.
- The specified compressive strength and durability requirements should be established by the design professional based on the application, and included in the project specifications.
- The cement content of the mix should be controlled by the contractor to meet the specified compressive strength and durability requirements. The starting cement content can be based on trial batching. After production phase test data become available, the data can be used for ongoing adjustments to the cement content.
- Compressive strength acceptance criteria should follow a statistically based approach similar to that used for concrete. The design professional should establish the acceptable frequency of low breaks and the minimum acceptable strength of individual test results based on factors including anticipated conditions and acceptable risk at the project site, durability testing at the mix design stage, and prior experience.
- Performance based project controls should be used to the extent practical to allow the contractor to select the construction equipment and methods. The contractor is part of the project team for their construction expertise, equipment, personnel, and ability to construct the project, thus criteria should focus on defining the project requirements, not on telling the contractor how to do their task.
- Bonding slurry (cement grout) should be used to bond lift joints when the armoring is composed of multiple lifts.

SUMMARY

Since construction of the experimental soil cement test section at Bonny Dam in 1951, soil cement has been used on numerous dam embankment armoring projects ranging from high-hazard potential dams with long fetch distances to small retention ponds. Soil cement is also used for projects other than dam applications such as parking lot and roadway subgrade improvements.

While considerable progress in construction techniques and project controls has been made since the early pioneering use of soil cement armoring for embankment dams by the USBR in the 1960s, issues with recently constructed projects emphasize the need for continued improvement with the technology and sharing of project experiences.

Based on the wide range of soil cement applications, it is readily apparent that project controls, construction techniques, and acceptance criteria suitable for some projects will not be applicable to all. This paper presented what are considered by the authors to be some of the key soil cement construction-phase project controls appropriate for embankment armoring of high hazard potential and/or high value dams.

REFERENCES

American Concrete Institute, ACI Committee 214. ACI 214R-11: *Guide to Evaluation of Strength Test Results of Concrete*. April 2011.

American Concrete Institute, ACI Committee 230. ACI 230.1R-09: *Report on Soil Cement*. July 2009.

American Concrete Institute, ACI Committee 301. ACI 301-10: *Specifications for Structural Concrete*. December 2010.

American Concrete Institute, ACI Committee 318. ACI 318-11: *Building Code Requirements for Structural Concrete*. May 2011.

International Commission on Large Dams, ICOLD. Bulletin 54: *Soil-Cement for Embankment Dams*. Report prepared for ICOLD by Kenneth D. Hansen. 1986.

Portland Cement Association, PCA. *Soil-Cement Guide for Water Resources Applications*. Report prepared for PCA by WEST Consultants. 2006.

Portland Cement Association, PCA. *Soil-Cement Laboratory Handbook*. 1992.

Schrader, Ernest K. *Statistical Acceptance Criteria for Strength of Mass Concrete*. Published in Concrete International. June 2007.

U.S. Army Corps of Engineers, USACE. Engineering Manual 1110-2-1913, Design and Construction of Levees; Appendix G: *Use of Soil Cement for Levee Protection*. April 2000.

U.S. Department of the Interior, Bureau of Reclamation, USBR. Design Standards No. 13, Embankment Dams; Chapter 17: *Soil-Cement Slope Protection*. August 2013.

U.S. Department of the Interior, Bureau of Reclamation, USBR. *Performance of Soil-Cement Dam Facings – 20-Year Report*. September 1984.

REHABILITATION OF THE CYLINDER GATE AT SRP'S HORSESHOE DAM

Gary Brady, PE[1]
Noel Guercio, PE[2]
Kevin Wolford[3]
Roger Baker, PE[4]

ABSTRACT

Horseshoe Dam is located on the Verde River in central Arizona and was constructed in 1945-1946 near the end of World War II by the Phelps-Dodge Corporation for the Salt River Valley Water Users' Association (known now as Salt River Project). The Salt River Project (SRP) Power and Water utility of Arizona has embarked most recently on a program of rehabilitation of the River Outlet Works (ROW) of Horseshoe Dam. The ROW is one component of the dam facility and consists of a 9-foot diameter, 126-foot high cylinder gate within a circular concrete intake tower. Releases through the ROW are controlled by the cylinder gate by raising and lowering it within a 4-foot vertical within the concrete tower.

A structural inspection of the ROW was completed in 2012. The structural inspection provided physical data and photography that was utilized to determine the best rehabilitation approach for the cylinder gate and associated facilities. The rehabilitation approach was completed to address the following:

- The gate and intake tower experienced severe vibration during operation, particularly at partial gate openings. There was concern about adverse impacts to the structural integrity of the tower and for operator safety while on the tower platform.

- The cylinder gate had several corroded, aged, or broken components.

This paper will present a brief history of Horseshoe Dam and the River Outlet Works facilities, existing condition inspection and findings by the structural team, and the proposed rehabilitation completed in 2013 and early 2014.

INTRODUCTION

Horseshoe Dam is located on the Verde River in central Arizona and was constructed during 1945-1946 by the Phelps-Dodge Corporation for the Salt River Valley Water Users' Association (known now as Salt River Project). The Salt River Project (SRP) Power and Water utility of Arizona has embarked most recently on a program of rehabilitation of the River Outlet Works (ROW) of Horseshoe Dam. The ROW consists

[1] Stantec Consulting Services Inc., 8211 S. 48th Street, Phoenix, AZ, (602) 438-2200, gary.brady@stantec.com.
[2] Stantec Consulting Services Inc., 8211 S. 48th Street, Phoenix, AZ, (602) 438-2200, noel.guercio@stantec.com.
[3] Salt River Project, 7050 E. University Drive, Mesa, AZ, (602) 236-6252, kevin.wolford@srpnet.com.
[4] Salt River Project, 7050 E. University Drive, Mesa, AZ, (602) 236-6078, roger.baker@srpnet.com.

of a 9-foot diameter, 126-foot high cylinder gate within a circular concrete intake tower, connected by a concrete draft tube to a 720-foot long, horseshoe shaped tunnel. Figure 1 shows the intake tower under construction in 1944.

Figure 1. Intake Tower Under Construction. (from SRP files)

The tunnel was driven through rock in the right abutment of the earthen embankment dam, and is fully concrete lined. The tunnel formerly terminated at a steel outlet nozzle. Figure 2 shows historic photos of the tunnel construction. Releases through the ROW are controlled by the cylinder gate within a range of 4-foot vertical movements within the intake tower.

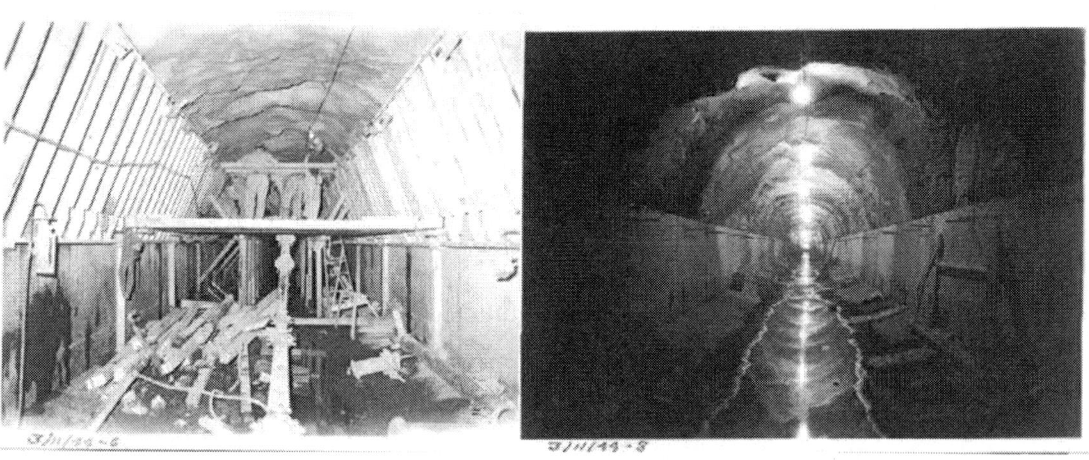

Figure 2. Construction Photos of ROW Tunnel. (from SRP files)

Because of the number of years that the cylinder gate was in operation and its visible deteriorating condition, Stantec Consulting Services Inc. was engaged to conduct a structural inspection and to produce an assessment report of the Horseshoe Dam intake tower and steel cylinder gate.

STRUCTURAL INSPECTION

The structural inspection was preformed April 24 – 26, 2012 utilizing rope access techniques allowing access to virtually the entire exposed surface area of the tower, gate and associated elements (see Figure 3). The exterior portion of the intake tower and cylinder gate was inspected only to the level of the water and sediment in the reservoir. Full height interior access, to within 2 feet of the bottom of the cylinder gate, was gained by the inspectors. Standing and incoming water was present within the gate even with the gate in the fully closed position, preventing the observation of the gate seal and seat in the draft tube.

Figure 3. Structural inspection engineers inspect and assess the outside of tower and inside of the cylinder gate in April 2012.

The structural inspection provided physical data and photography that was utilized to determine the best rehabilitation and betterment approach for the cylinder gate and associated facilities.

The rehabilitation and betterment approach was established to address the following:

- The cylinder gate is currently the single point of releases from Horseshoe Reservoir below the spillway crest elevation and therefore there is no means of controlling releases should the cylinder gate fail. SRP dams typically have two gates or valves on outlet works, an upstream isolation gate and a downstream flow control valve. Redundancy in flow control at Horseshoe Dam is desired by SRP.

- The seal at the bottom of the cylinder gate and the seat in the draft tube historically leaked excessively (possibly as much as 25 cfs). It was thought that this continuous leakage was due to a likely combination of normal wear, cavitation, and erosion by sediment laden water passing through the gate opening. This was exacerbated over the 65 years of service by using the cylinder gate for flow modulation, which most certainly exposed the gate to a significant amount of cavitation. In particular, the gate was operated in partial open positions, which caused excessively high velocities of water to pass between the gate seal and seat. Cylinder gates are normally operated as isolation gates, are operated either fully open or fully closed, and are not utilized for flow modulation. Although the gate seat is believed to be worn and in need of maintenance replacement and repair, it was discovered during the gate rehabilitation that the gate has not been seated properly for as long as 65 years and seating it properly during the rehabilitation reduced the leakage to almost zero.

- The gate and intake tower experienced severe vibration during operation, particularly at partial gate openings. SRP was concerned about adverse impacts to the structural integrity of the tower and for operator safety while on the tower platform.

- The cylinder gate had several corroded, aged, or broken components including:

    - Corroded steel cylinder, including much of its structural reinforcement components.

    - Aged cylinder gate counterweight wire rope.

    - Broken cylinder gate guides and rollers.

    - Concrete breaks and spalls in the tower.

    - Corroded and broken steel trashrack bars in the tower.

    - Broken or severely corroded access ladders and landings.

- There was some belief that the 126-foot high cylinder gate may be out of alignment with the gate seat in the draft tube.

- Operator access to the gate platform, as well as to the outlet, was poor and needed to be improved.

The gate is positioned over an orifice in the draft tube (Elevation 1907.5 feet) about 8 foot, 6 inch diameter. Opening of the gate is accomplished by raising the gate vertically with a hoist mounted in the top of the intake tower (Figure 4).

Figure 4. Cylinder gate hoist on intake tower. (Photograph by Rodney Tang, USBR)

When opened, water passes through the space between the bottom of the gate and the seat in the orifice. The gate is constructed of steel plate forming the 9-foot diameter cylinder and is stiffened internally by angles welded circumferentially to the cylinder and by cross-bracing (Figure 5). Historic photographs of the cylinder gate and draft tube from the 2000 USBR inspection report are shown in Figure 6.

Figure 5. Internal braces in the cylinder gate. (Photograph by Rodney Tang, USBR)

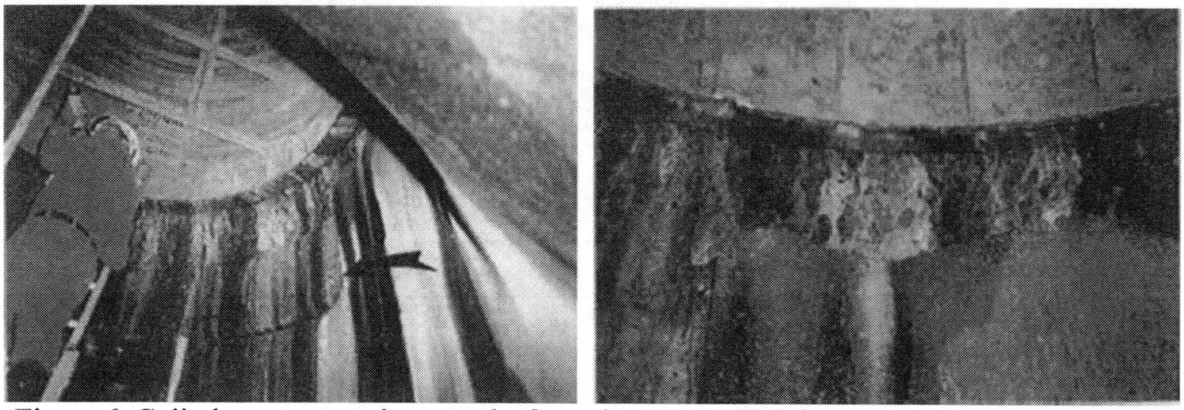

Figure 6. Cylinder gate seat photographs from the 2000 USBR inspection. (Photographs by Rodney Tang, USBR)

The condition of the cylinder gate, as assessed from limited observation from the deck at the top of the inlet tower, photographs, and reports, was that the gate is operable, but needs considerable rehabilitation for continued operation.

A sketch of the main components of the cylinder gate is shown in Figure 7.

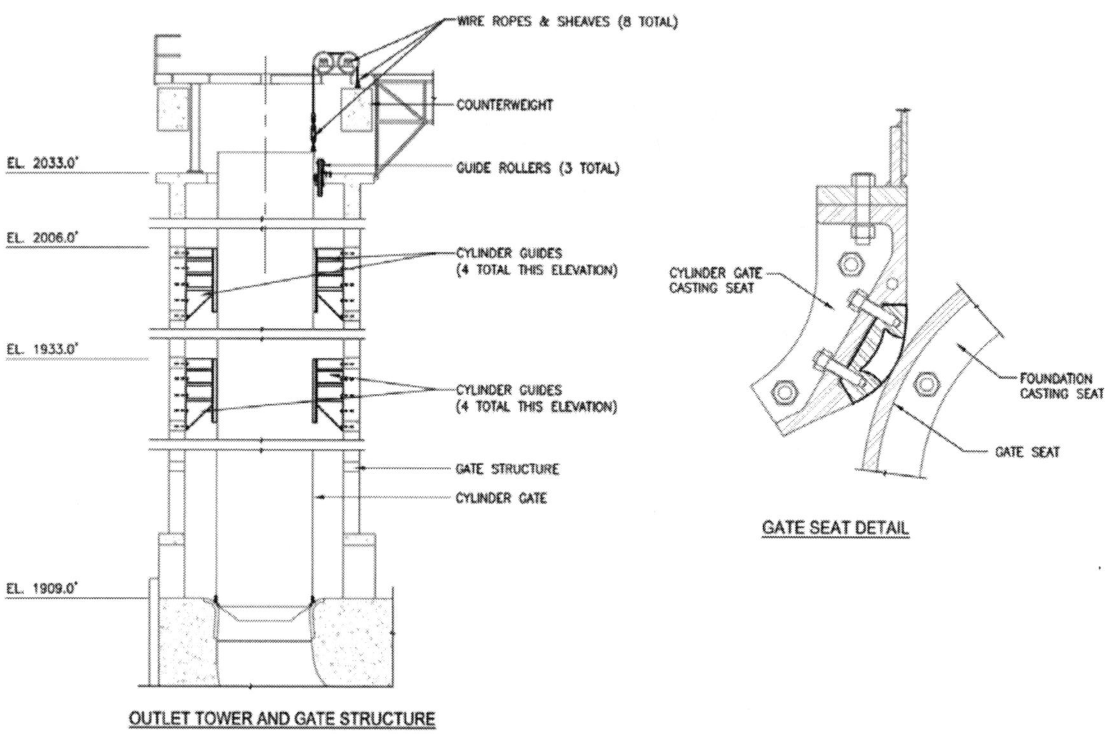

Figure 7. Sketch of the Intake Tower Components.

Rehabilitation recommendations and results for the components of the intake tower included the following:

Counterweight

A circular concrete counterweight (Figure 8) is connected to the cylinder gate by wire ropes passing over sheaves (Figure 9). The counterweight offsets 80,000 pounds of the 100,000 pound gate weight, thus allowing a much reduced hoist effort to raise the gate. The counterweight was found to be in satisfactory condition. The gate was installed with a maximum vertical travel of 4 feet. It appears that the current maximum travel may be about 3 1/2 feet or less. Restoring the 4 feet of gate opening is not necessary at this time.

Figure 8. Cylinder gate counterweight. (Photograph by Rodney Tang, USBR)

Wire Ropes Between the Cylinder and the Counterweight

Due to the age of the wire ropes, it was anticipated that these wire ropes were near the end of their safe and useable life. The replacement of the wire ropes that hoist the cylinder gate up and down was recommended. This was based upon the age of the existing wire ropes, rather than visible indication of corrosion or deteriorated condition. All wire ropes were replaced.

Figure 9. Wire ropes from the cylinder gate to the counterweight. (Photograph by Rodney Tang, USBR)

Guide Rollers

Three guide rollers (Figure 10) at spacings of 120° are positioned at elevation 2033 feet. Due to age and reports discussing the poor condition or missing guide rollers and slide guides lower down, it was anticipated that these rollers may not be functioning properly, and may be broken. New guide rollers that limited lateral movement and contribute to the excessive vibration when the gate was opened and closed were installed.

Figure 10. Guide Rollers. (Photograph by Rodney Tang, USBR)

Slide Guides

Sets of four slide guides each, (Figure 11) are located at about elevations 2006 and 1933 feet. The guides are located at spacings of 90° around the cylinder gate. Reports indicate that some of these slide guides may be broken or missing.

The lower guides were initially submerged. However, the reservoir water level dropped when the cylinder gate was fully opened and allowed for access to the lower guides. All the slide guides were repaired so they would function as original intended.

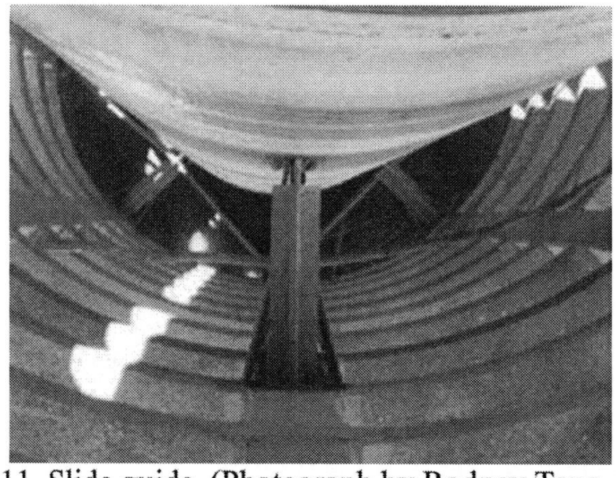

Figure 11. Slide guide. (Photograph by Rodney Tang, USBR)

Corrosion

The corrosion depicted in Figure 12 and discussions in the inspection reports indicated that there was considerable corrosion to steel members within the interior of the cylinder gate. Corrosion is seen on lateral cross bracing and on the angles that are welded circumferentially around the interior of the cylinder gate. In some places, pieces of the angles are missing. Similarly, there was found to be significant corrosion to the steel plate of the cylinder gate.

Review of the findings from a corrosion condition inspection and discussion with SRP staff determined the degree of rehabilitation and corrosion protection recommended for the cylinder gate. Corrosion protection was performed on the trashrack bars, the outside cylinder gate down to the reservoir level of about elevation 1950 feet and to the bottom of the cylinder on the interior. The corrosion protection approach generally consisted of the following approach:

- Cleaning, Decontamination, and Surface Preparation –
    - Low pressure (3,500 psi) waterblasting – Loose coatings, mill scale, and foreign matter was then removed with putty knives. No sandblasting was allowed.
    - Decontamination – Application of material known as Chlor-Rid, which removes ferrous and ferric salts. These salts can cause premature coating failure.
    - Surface Preparation – Low pressure waterblasting again prior to coating application
- Coating Application – Prime and apply surface epoxy paint

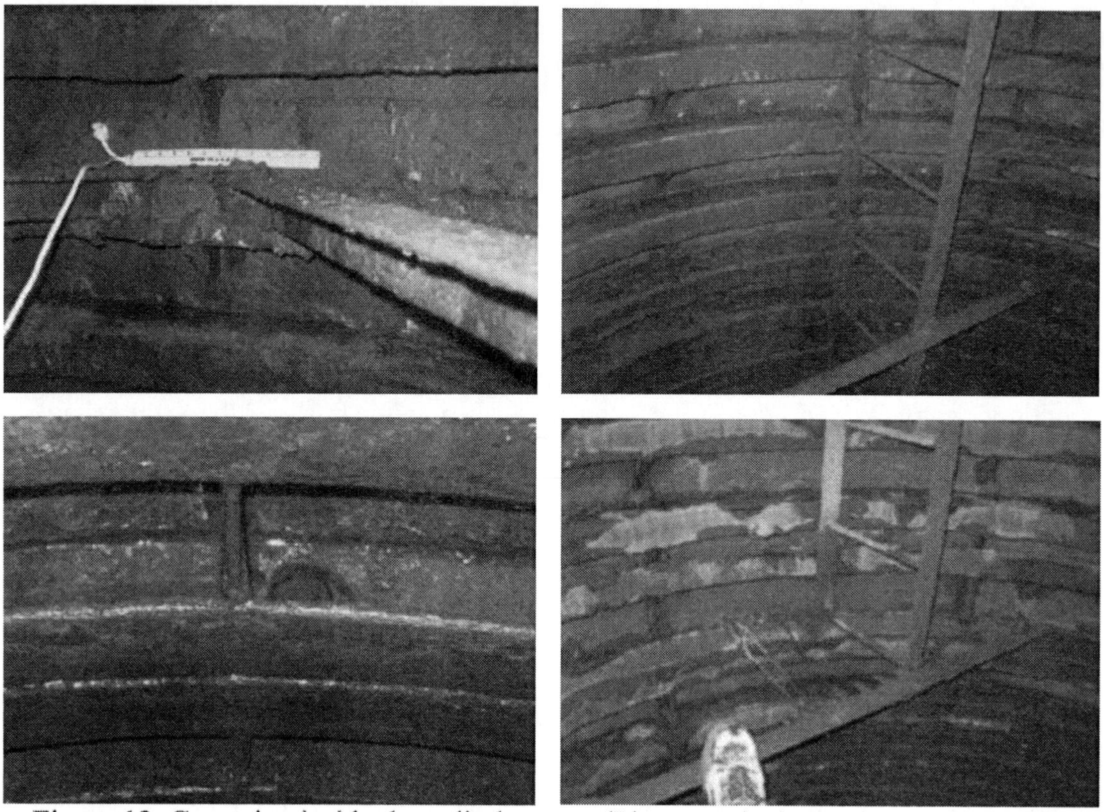

Figure 12. Corrosion inside the cylinder gate. (Photographs by Rodney Tang, USBR)

Seal Ring and Cylinder Gate Seal

Reports indicated that there was considerable leakage, perhaps up to 25 cfs, past the gate. The sealing parts of the gate consist of a gate seal mounted into a seal ring on the gate, and a gate seat mounted on the orifice. It was speculated that the leakage was caused by deterioration of either of these sealing components, and/or having the gate out of plumb. If the gate is out of plumb, the sealing may be incomplete over part of the circumference of the orifice. Photographs from the 2000 inspection are provided in Figure 7. During the removal and replacement of the cylinder gate cables in 2013, the counterweight was disconnected and the contractor's crane was used to lower the cylinder gate down with its full weight on the gate seat. Additionally, plumbing the cylinder gate as it was lowered improved its seating. Under these improved conditions, the cylinder gate achieved near complete seal with only minor leakage that allowed downstream tunnel inspection at a later date. Further rehabilitation of the seal ring and cylinder gate seal at this time was not considered necessary as a result.

Intake Tower Structure

The gate structure consists of a circular concrete tower about 124 feet tall, as shown in Figure 1. The tower is set on a concrete base and resembles lattice work as 5-foot high windows completely surround the tower. The windows allow flow to enter the space between the tower and the cylinder gate and flow down around the cylinder gate to the orifice at the top of the draft tube. The windows are each divided into three vertical segments by vertical trash bars, Figure 13.

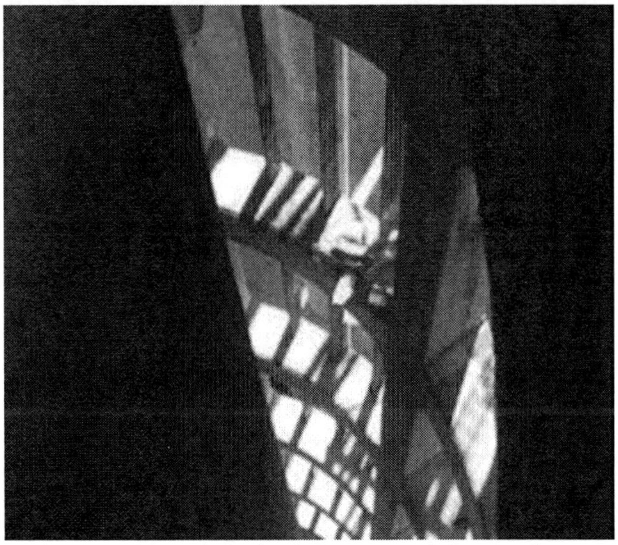

Figure 13. Intake tower trashrack. (Photograph by Rodney Tang, USBR)

The condition of the tower and trash bars appeared satisfactory, although spalling of the concrete tower was evident in many locations and a majority of the trash bars were severely corroded. All trash bars were rehabilitated. A concern noted is the lateral stability of the tower in the event of an earthquake or other lateral load. From construction photos (Figure 14 and 15), it is observed that no reinforcement appears to extend up from the foundation into the inlet tower, giving the appearance that the gate tower superstructure may be inadequately connected to the concrete foundation. Figure 15 shows a continuous, circular concrete wall on the foundation and illustrates reinforcement steel extending from the stem wall into the columns for the tower. However, it is doubtful from review of Figure 14 that the steel shown extends into the cylinder gate foundation. It is also unclear as to how the steel reinforcement shown would extend downward into the foundation of the tower. The tower lateral stability issue was not ultimately resolved for this assessment, as historic plans that could conclusively provide the structural design of the foundation and tower were not located.

Figure 14. 1944 photo of tower foundation. (from SRP files)

Figure 15. 1944 photo of tower stem wall. (from SRP files)

CONCLUSION

Design life for many facilities often exceeds and needs to exceed common expectations for engineering projects. The Horseshoe Dam Outlet Works cylinder gate, tower, and tunnel have been in operation for almost 70 years and the current maintenance rehabilitation effort is intended to extend required replacement for as much as another 30 years or more. Maintenance of the Horseshoe tower and cylinder gate facilities is a key in keeping the dam outlet works facility operational.

ACKNOWLEDGEMENTS

The authors would like to thank the staff and management of Salt River Project for their support and assistance throughout the project.

SEISMIC FRAGILITY FOR RISK ASSESSMENT OF CONCRETE GRAVITY DAMS

Yusof Ghanaat[1]
Robert C. Patev[2]
Anjana K. Chudgar[3]

ABSTRACT

This paper presents a quantitative methodology including examples for developing seismic fragilities for concrete gravity dams using non-linear analyses with Latin Hypercube Simulation (LHS). Potential failure modes are defined, and seismic fragility curves are developed for the global failure of the dam as well as failure of the piers. Both the aleatory and epistemic uncertainties are considered, and most significant variables influencing the seismic fragility are identified and used in the analysis. Values of input variables influencing the seismic fragility are generated using LHS and randomly selected to develop seismic fragilities. Nonlinear seismic analyses are performed for thirty simulations at several ground motion levels. The ground motion variable includes 30 sets of acceleration time histories obtained from the actual earthquake recordings and matched to the target response spectrum. At each ground motion level, the probability of failure is determined based on the fraction of the simulation that resulted in dam failure. The results indicate that the seismic fragility is defined reasonably well with the lognormal distribution and provide the median seismic fragility including randomness and uncertainty variability for the base sliding and the pier failure.

INTRODUCTION

Seismic risk assessment of dams is a critical component of a dam safety program. It allows estimating life, economic, and social and environmental loss from earthquakes and mitigating losses when earthquakes occur. Risk and uncertainty are intrinsic in water resource management activities. Uncertainty arises from the lack of knowledge and randomness of the loads that a dam will actually experience, the lack of perfect information about the manner in which the dam will respond to those loads, and limited information about what the resulting consequences would be. The aim of seismic risk assessment is to identify loading conditions, potential failure modes, and consequences, and how to estimate the probabilities for each event. The overall process consists of hazard definition, seismic fragility, consequence analysis, and integration of the three. Among these components, the seismic fragility which associates structural failure with the intensity of ground motion is addressed in this paper. The seismic fragility of a dam is defined as the conditional probability of its failure at a given earthquake ground motion intensity. Contemporary methods of risk assessment are based on subjective estimation of the probability of failure and often ignore or inadequately address uncertainties. Subjective estimate of the probability is a mental math that cannot be trusted. The seismic

[1] Quest Structures, Inc., 3 Altarinda Road, Suite 203, Orinda, CA 94563, yghanaat@QuestStructures.com
[2] National Risk Advisor, Risk Management Center, US Army Corps of Engineers, 696 Virginia Road, Concord, MA 01742, robert.patev@usace.army.m
[3] US Army Corps of Engineers, Headquarters, 441 G. Street, NW, Washington, DC 20314-1000

fragility described in this paper removes these limitations and formally incorporates both the aleatory variability and epistemic uncertainties in the analysis.

DAM DESCRIPTION

The dam sections employed in this study are geometrically similar to the Folsom Dam tallest overflow and non-overflow monoliths. The Folsom Main Concrete Dam consists of 28 individual monoliths, each 50 feet in width, and numbered 1 to 28, sequentially, right to left (Figure 1). The overflow monoliths in the river channel are numbered 12 to 20, with end Monoliths 12 and 20 being half-width monoliths. Figure 2 shows all nine piers and eight gates. Five gates on the right are service gates and three on the left are emergency gates. The tallest non-overflow (Figure 3) that rises 336.5 feet above the base El. 144 ft was studied previously (Ghanaat et.al, 2012b). The overflow section employed in this study is a replica of Monolith 14, the tallest overflow monolith. It is 332.25 feet tall rising from the base El. 144 to El. 476.25 ft at the highest pier point (Figure 4).

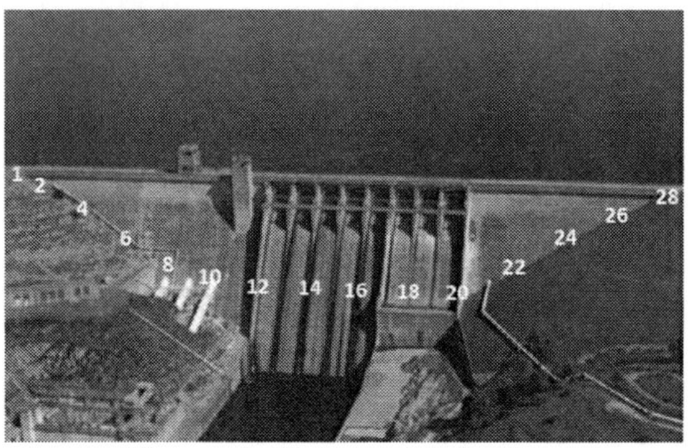

Figure 1. Folsom Main Dam.

Figure 2. Folsom Dam spillway monoliths.

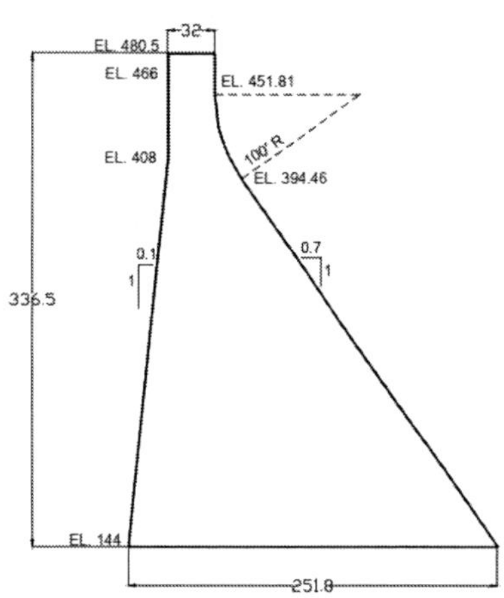

Figure 3. Non-overflow section.

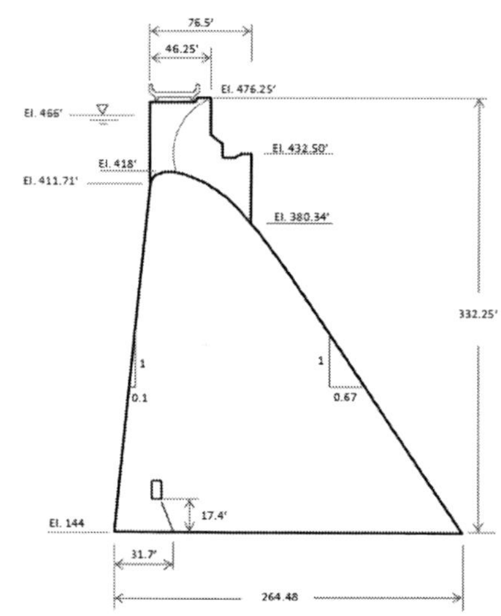

Figure 4. Overflow section.

The 8-foot-thick reinforced concrete piers are located in the middle of each overflow monolith. They are 46.25 ft wide at the top where they support a bridge and 76.5 ft at the bottom below the trunnion seat. The piers include one row of horizontal and vertical #8 bars at 1 foot on centers beneath the pier surfaces. Additional second and third rows of #8 bars also at one foot on centers are present in the nose area. The vertical reinforcing in the nose area penetrates some 38 feet into the spillway ogee.

SEISMIC FRAGILITY METHODOLOGY

Seismic fragility analysis of structures is a time-variant reliability problem for which a closed form solution does not exist and, therefore, simulation is inevitable. Full Monte Carlo Simulation (MCS) can produce most accurate results, but it costs more and requires an enormous amount of computational efforts when it involves nonlinear time-history analyses. The Separation-of-Variables Method (EPRI, 1994) is another costly procedure that requires a large number of nonlinear analyses. Application of this method in combination with nonlinear analysis would first require development and analysis of a model with median-centered properties and multiple input time histories to account for variability of the earthquake ground motion. The variability of seismic capacity due to one significant variable would necessitate the analysis using a model with the variable set to its 16% or 84% confidence level, while median values assigned to other variables. A repeat of such analyses is needed for each of the variables significant to seismic capacity and multiple ground motion input levels to determine the failure capacity. For example, 4,050 nonlinear analyses would be necessary to implement the Separation-of-Variables to develop seismic fragility involving 13 variables with 30 time-histories and five ground motion levels.

For most seismic fragility analyses, the pure random sampling of MCS may not be necessary, as long as the sampling reproduces distributions of the input parameters accurately. Latin Hypercube Simulation or LHS is one such option that satisfies this condition by providing a sampling method that appears random, but also reproduces the input distributions with much greater efficiency than MCS. As a more efficient alternative, we adopted LHS in combination with nonlinear analyses to develop seismic fragility for the gravity dam sections. As we will see later, only 300 nonlinear analyses would be necessary to implement the LHS for the above example, as opposed to 4,050.

Latin Hypercube Simulation

We develop the seismic fragility of concrete gravity dams based on the definition of failure that corresponds to the loss of water retention capability. Such a failure could occur during the earthquake ground shaking if cracking is significant and resistance to sliding and overturning is inadequate. Or it could happen in the post-earthquake condition due to increased pore pressures in the cracks, loss of cohesion and degradation of the friction coefficient as a result of seismic-induced sliding displacements.

Seismic fragility evaluations based on explicit nonlinear analysis are not a customary practice. A nonlinear response analysis is influenced by not only the various input parameters, but also by the characteristics of a particular earthquake record used as the seismic input. Thus, nonlinear analyses will be performed for multiple earthquake records to account for the effects of ground motion variability on the seismic response. Such nonlinear analyses are computation intensive; they should not only account for the formation and propagation of cracks, but also for the potential sliding and rocking responses that could follow. With nonlinear analysis the LHS (Ghanaat et al., 2011) is a more efficient method for the seismic fragility evaluation of the dam.

In the LHS Method, we first select the number of simulations or trials, N. We then develop N sets of earthquake time histories, each set consisting of the vertical and two horizontal components. The median spectral accelerations of the N sets of time histories should match the target ground response spectra over the frequency range significant to the response of the structure. N probability bins are defined, with each bin having an equal probability of occurrence. For each variable significant to seismic capacity of the structure or component, values are assigned to each of the N bins based upon the variable's probability distribution function (PDF). For the first trial, a value for each of the significant variables is randomly selected from the N equal probability bins. One set of earthquake time histories is also randomly selected from the N sets available. For the second trial, a value for each of the significant variables is randomly chosen from the remaining (N – 1) equal probability bins and one set of earthquake time histories is randomly chosen from the remaining (N – 1) sets. This process is repeated until the N trials are populated.

We perform seismic analyses for each of the N trials. At a given ground motion level, the probability of failure is determined based on the fraction of trials that result in dam failure (e.g., *n* trials resulting in failure out of N total). The probability of failure is determined for multiple ground motion levels. The seismic fragility is developed based on a fit to the failure data. This fragility is a composite of the randomness and uncertainty variables. The random variability is determined by performing analyses that vary the random variables but assigns median values to the variables associated with uncertainty (i.e., lack of knowledge).

In this study, we used a set of 30 trials to evaluate the seismic fragility of the overflow gravity section. We considered this quantity sufficient to obtain a reliable fragility without expending excessive computational effort. For the non-overflow section, we used a set of 10 trials, but for each simulation the ground motion level repeatedly increased until failure (Ghanaat et.al, 2011).

SEISMIC FRAGILITY INPUT VARIABLES

Earthquake Ground Motion – Aleatory Variability

The variability of earthquake ground motion can significantly affect the linear and particularly the nonlinear response of dams. In this study, the variability of earthquake ground motion was accounted for by using 30 sets of ground motion records for the overflow section and 10 sets for the previous study of the non-overflow section (Ghanaat et.al, 2011). We selected the records from previous earthquake events having magnitudes and source-to-site distances consistent with the maximum credible earthquake (MCE) at Folsom Dam site. The records represented different styles of faulting, pulse characteristics, pulse sequencing, duration, etc. The MCE is an M6.5 event at a source-to-site distance of 14 km to Folsom Dam (URS, 2008). The 84th percentile horizontal and vertical peak ground accelerations (PGA) for the MCE are 0.23g and 0.19g, respectively.

Earthquake time-history records were obtained from the Pacific Earthquake Engineering Research (PEER, 2005) Ground Motion Database. The records were selected searching the database for earthquake magnitudes in the range of M5.75 to M7.25, source-to-site distances up to 20.5 km, and strike-slip plus reverse/oblique fault type. Figure 5 compares horizontal and vertical target response spectra (thick black line) with response spectra of selected 30 sets of acceleration time histories. The geometric mean of the response spectra of time histories (thick red line) is in close agreement with the target response spectra in the period range of interest between 0.1 to 1 second.

To account for variability of time histories about the geomean, we first matched the selected records to the 84th percentile MCE response spectra, and then later during the nonlinear analyses scaled each of the two horizontal and vertical time histories by so-called "directionality factors" with respect to the geomean. A logarithmic standard deviation of 0.13 was used for variability of the two horizontal components and 0.18 for the vertical component.

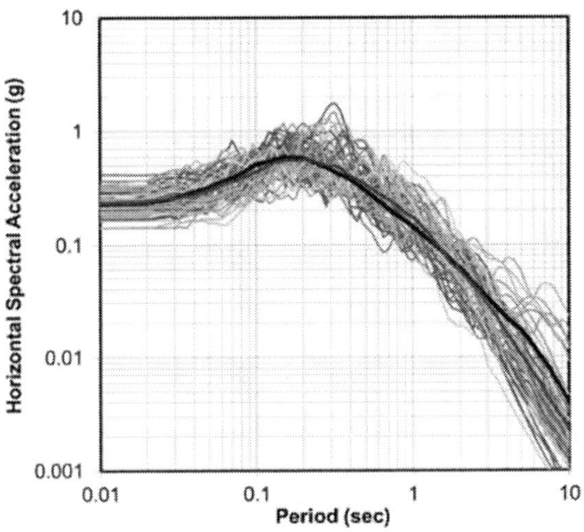

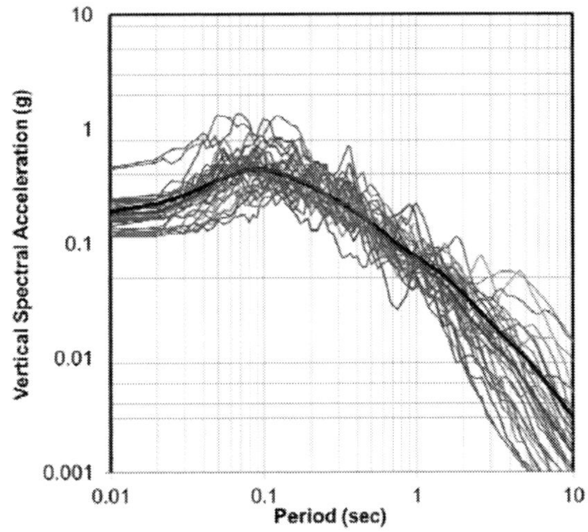

Figure 5. Comparison of Horizontal (left) and Vertical (right) Target Response Spectra with Response Spectra of Selected Time History Records.

Model Input Variables – Epistemic Uncertainty

Tables 1 and 2 summarize input variables we considered affecting potential failure modes and to be most significant to seismic fragility of the overflow and non-overflow sections, respectively. These variables are the only input parameters whose uncertainty or likelihood of occurrence is formally included in the analysis. All other input parameters are kept constant and represented by their median values. The likelihood of a variable having a particular value is determined from a PDF assigned to that variable. The PDF for a variable may be defined by statistical methods if sufficient data are available, or assumed based on experience and knowledge of low, best-estimate, and high values of the variable. We assumed lognormal PDF for variables for which adequate data were available and triangular for those with inadequate data (Tables 1 and 2). For each variable, 30 equal probability values for the overflow section were sampled from the variable's PDF using Latin Hypercube Sampling procedure. The complete definition of the PDFs and sampling of variables are provided in Ghanaat (2012a and 2014).

Table 1. Overflow Section.

Input Variables	**PDF**
EQ Ground Motion	THs
Concrete	
• Elastic Modulus, E_c	Lognormal
• Compressive Strength, f'_c	Lognormal
• Tensile Strength, f'_t	Lognormal
• Max Size Aggregate, A_s	Lognormal
Dam-Rock Contact	
• Rock Modulus, E_r	Lognormal
• Tensile Strength, f'_{it}	Lognormal
• Cohesion, c_i	Triangular
• Friction Angle, ϕ_i	Triangular
Pier Reinforcing Steel	
• Yield Strength, f_y	Triangular
• Strain-hardening Slope, E_{sh}	Lognormal
• Rupture Strain, ε_u	Lognormal
Drain Efficiency, D_{eff}	Triangular
Damping, ξ	Lognormal

Table 2. Non-overflow Section.

Input Variables	**PDF**
EQ Ground Motion	THs
Concrete	
• Elastic Modulus, E_c	Lognormal
Dam-Rock Contact	
• Rock Modulus, E_r	Lognormal
• Tensile Strength, f'_{it}	Lognormal
• Cohesion, c_i	Triangular
• Friction Angle, ϕ_i	Triangular
Lift Joint	
• Tensile Strength, f'_{lt}	Lognormal
• Cohesion, c_l	Triangular
• Friction Angle, ϕ_l	Triangular
Drain Efficiency	Triangular
Damping	Lognormal

FINITE-ELEMENT MODELS

Model Nonlinearity – Potential Failure Modes

The previous seismic analyses of Folsom Dam by the US Bureau of Reclamation indicated that failure mechanisms of the overflow monoliths would involve cracking and sliding at the base with shear and flexural failure of the piers. The base cracking starts from the upstream and continues to the downstream as the ground motion intensity increases. At increased ground motion levels, the monoliths fully crack at the base and slide on the foundation contact. The monoliths slide in a group of two or more on the foundation contact and also with respect to one another since the contraction joints are not keyed. The spillway piers suffer shear failure in the upstream-downstream direction and a moment or shear failure in the cross-stream direction. The base sliding of the monoliths is modeled using tiebreak contacts with tension-shear failure criteria. The vertical contraction joints with no shear keys are modeled using compression-only sliding contacts. Sliding contacts with zero friction resistance are employed at water surfaces with the dam, the gates and piers, and the foundation rock to allow free sliding against these surfaces.

Spillway piers are vulnerable to earthquake ground shaking and could fail in shear in the upstream-downstream direction and flexural and shear in the cross-stream direction.

Failure mechanisms of the reinforced concrete pier are modeled using nonlinear concrete and nonlinear reinforcing steel models to permit tensile and shear cracking of the concrete with yielding and rupturing of the reinforcing steel.

Nonlinear Models for Seismic Fragility

We evaluate the seismic fragility of the overflow section for a single monolith that is referred to as overflow section. A detailed model of the overflow section is developed to capture its nonlinear response while the two adjacent monoliths and the remainder of the dam are represented approximately to provide lateral support, interaction between monoliths, and connection to the abutments for the cross-stream seismic excitation. We developed and analyzed the model using the computer program LS-DYNA (LSTC, 2014). The adjacent monoliths were modeled to behave linearly, but, like the main overflow section, they could slide at the bottom and the sides when shear demands exceed the sliding shear resistance. The remainder of the dam merely provides a linear-elastic medium connection to the foundation and abutments. Figure 6 shows the overall finite-element model consisting of the dam, the foundation rock, and the impounded water. A foundation model with inertia, flexibility and damping effects was used to account for the dam-foundation interaction effects. The foundation model extends several dam heights in all three directions and includes non-reflecting boundary condition at the bottom and side surfaces. The complete dam-water interaction is represented by modeling the impounded water using 3D compressible fluid elements with non-reflecting boundary at the upstream extent of the fluid domain to include radiation damping.

Figures 7 and 8 provide more details on the finite-element discretization of the overflow section including the pier, gates, and connection to the adjacent dummy monoliths. The sloped portion of the overflow section was modeled using 3D solid element with linear concrete material properties (orange in Figure 7). The pier and ogee portion of the overflow section (pink in Figures 7 and 8), where the pier vertical rebars are anchored, are modeled using the Winfrith nonlinear concrete model (LSTC, 2014; Ghanaat, 2014). The pier reinforcing steel cage shown in Figure 8 is modeled using the "Plastic Kinematic" steel model (LSTC, 2014). The vertical and horizontal rebars in the steel cage were modeled individually and positioned at their exact locations. The tainter gate skin plates were modeled using linear shell, and beams, braces, and trunnion anchors linear beam elements.

Sliding contact surfaces were introduced between the overflow section and adjacent dummy monoliths and between the dummy monoliths and the remainder of the dam to allow for opening and sliding of the contraction joints. A friction angle of 45 degrees with zero cohesion were used for the contraction joints. The "tie-break" sliding contact surfaces were employed at the base of the overflow section and dummy monoliths. Initially, the "tie-break" contacts tie the overflow section and monoliths to the foundation rock, but the contact breaks when stresses exceed the contact tensile or shear strength. After the break, the "tie-break" contact behaves the same way as the traditional compression-only contacts where shear resistance is limited to frictional force and tensile resistance is zero. The sliding contact surfaces at the water interfaces with the gates and piers, as well as between the gates and piers use zero cohesion and friction.

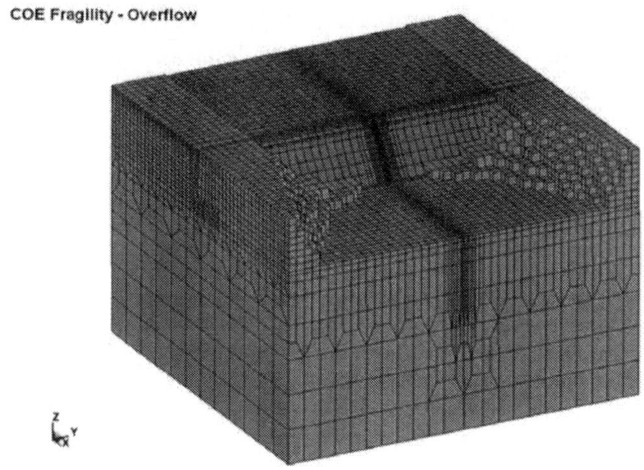

Figure 6. Overall FEM for Analysis of Overflow Section.

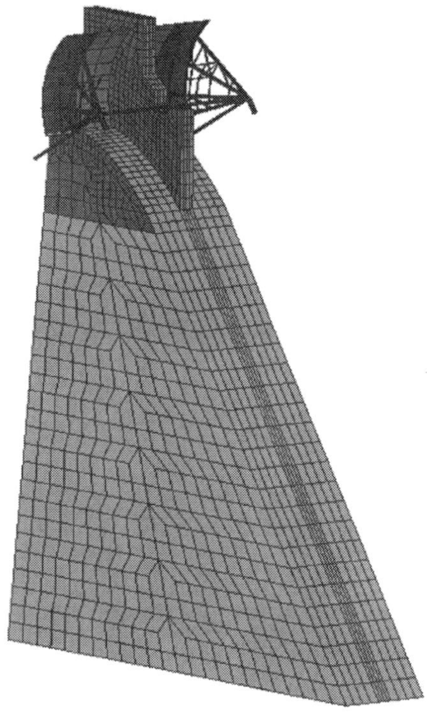

Figure 7. Main Overflow Section.

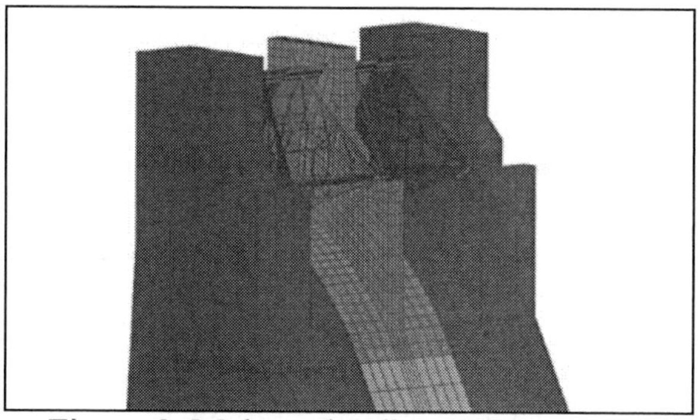

Figure 8. Main and Adjacent Monoliths.

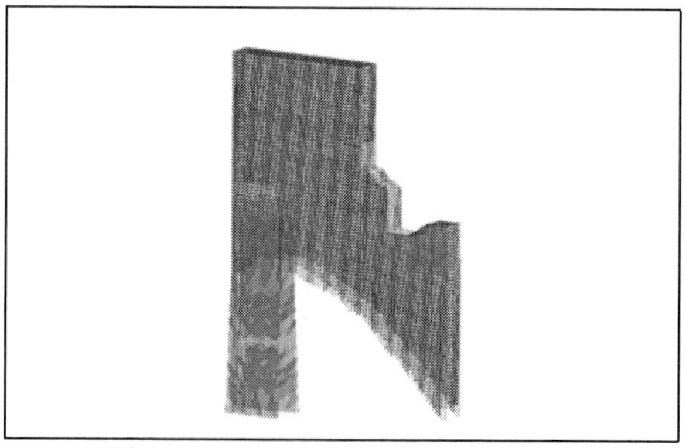

Figure 9. Pier Steel Rebar Cage.

NONLINEAR SEISMIC ANALYSIS

Generation of Traction Seismic Input

Because the LS-DYNA model includes non-reflecting boundaries, seismic input must consist of stress or force time histories rather than acceleration time histories. We converted the earthquake acceleration time histories to stress time histories following these steps:

1. Ground Acceleration at Depth. Performed deconvolution using the free-field acceleration time histories as the input to obtain acceleration time histories at the bottom surface of the LS-DYNA model located at a depth of 1,202 ft below the base of the dam.

2. Ground Velocity at Depth. Integrated the in-depth acceleration time histories from Step 1 to generate at-depth velocity time histories at the bottom of the LS-DYNA model.

3. Traction Time Histories at Depth. Converted the in-depth velocity time histories from Step-2 to shear (σ_s) and normal (σ_n) stress time histories using the following relationships.

$$\sigma_s = 2(\rho \; c_s) \; v_s \tag{1}$$

$$\sigma_n = 2(\rho \; c_n) \; v_s \tag{2}$$

where ρ is mass density of the foundation rock, c_s shear wave velocity of the foundation rock at the boundary, c_n compression wave velocity of the foundation rock at the boundary, v_s horizontal ground velocity at the boundary, and v_n vertical ground velocity at the boundary.

4. Verification of Boundary Tractions as Seismic Input. An LS-DYNA box-shaped foundation model without the dam was subjected to shear and normal seismic tractions. Horizontal and vertical accelerations at the top surface of the model were retrieved, and their response spectra were compared with the target response spectra. Differences were reconciled by adjusting the input seismic tractions and repeating the process until the response spectra of the ground surface accelerations matched closely with the target response spectra.

Nonlinear Analysis Considering All Variables

We first performed nonlinear seismic analyses including all input variables with the earthquake ground motion scaled to horizontal PGAs of 0.2g, 0.4g, 0.5g, 0.6g, and 0.8g following the strategy described previously. At each selected ground motion level, thirty analyses were performed with values of the variables randomly assigned to the 30 trial models. The probability of dam failure (base sliding) at each ground motion level was determined from the fraction of trials that resulted in dam failure. Table 3 summarizes the results of the nonlinear analyses for simulations with all variables. For example, the probability of dam failure at 0.6g ground motion level, which indicated 26 dam failures out of 30 trials, is 26/30 = 86.7%.

Figure 10 shows sliding displacement time histories for all thirty trials at the 0.6g ground motion level. The permanent sliding displacement for the failed 26 trials varied from 0.1 to 3 ft.

Table 3. Base Sliding Results Including All Variables.

	30 simulations at each ground motion level				
Ground Motion Level	0.2g	0.4g	0.5g	0.6g	0.8g
No. of Failure	6	18	23	26	28
Probability of failure	20.0%	60.0%	76.7%	86.7%	93.3%

Seismic Fragility for Risk Assessment

Nonlinear Analysis Considering Random Variables Only

We also performed nonlinear analyses varying only the random variables and assigning median values to all variables associated with the epistemic uncertainty (i.e., lack of knowledge). The only random variable considered was the earthquake ground motion input. All of the other variables (i.e., concrete, rock, and steel properties, drain efficiency, and damping) considered to be uncertainties.

We performed these nonlinear analyses with the earthquake ground motion scaled to the same horizontal PGAs of 0.2g, 0.4g, 0.5g, 0.6g, and 0.8g. Table 4 summarizes the analysis results, indicating that probabilities of sliding failure for ground motion levels in the range of 0.2g to 0.8g vary from 0.0 to 96.7 percent. These probabilities were calculated following the same procedure described earlier.

Table 4. Base Sliding Results Including Random Variables Only.

	30 simulations at each ground motion level			
Ground Motion Level	0.2g	0.4g	0.5g 0.6g 0.8g	
No. of Failure	0	25	29 29 29	
Probability of failure	0.0%	83.3%	96.7% 96.7% 96.7%	

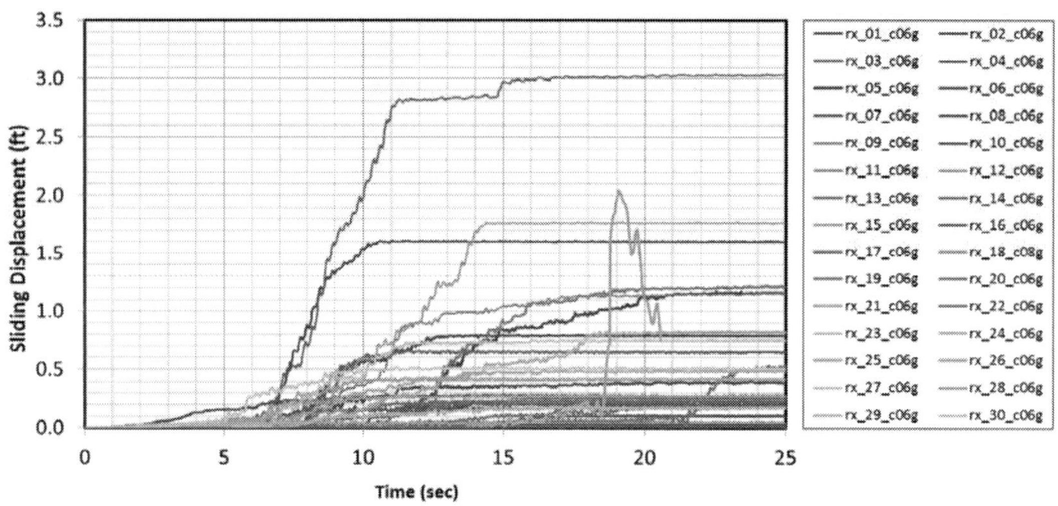

Figure 10. Dam sliding displacement histories for 0.6g and all variables

Nonlinear Response of Reinforced Concrete Pier

In this study, the spillway pier was included in the overflow section and modeled using nonlinear concrete and steel material models to determine its seismic fragility. Tables 5 and 6 summarize the nonlinear results for the two sets of 30 runs, one with all variables and another with random variables only. Overall, three types of nonlinear response behavior are observed: 1) no failure with minor cracking and no steel yielding, 2) failure with significant cracking and yielding but no steel rupturing and pier collapse, and 3) failure with steel rupturing and collapse. The simulations that did not fail showed modest concrete cracking with less than an inch of deformation and no steel yielding. The simulations that failed without collapse indicated significant cracking and with steel yielding but not rupture. Through section cracking and steel yielding were significant enough to generate large movements impacting the gates and the trunnion anchors. The most severe damage involved the through section cracking with steel yielding and rupturing which led to collapse of the entire or a portion of the pier. A few simulations at high ground motion levels and also at low ground motion level combined with low input parameters experienced such failure. Figure 11 displays pier major principal stresses with rebars yielding and rupturing for such failure. In this figure, red regions indicate severe cracking that covers the pier-ogee contact surface and most of the pier; steel rupturing occurs at the pier nose, the re-entrant corner near the trunnion, and next to the ogee.

Table 5. Pier Failure Results Including all Variables.

	30 simulations at each ground motion level			
Ground Motion Level	0.2g	0.4g	0.5g 0.6g 0.8g	
No. of Failure	8	13	18 21 24	
Probability of failure	26.7%	43.3%	60.0% 70.0% 80.0%	

Table 6. Pier Failure Results Including Random Variables Only.

	30 simulations at each ground motion level			
Ground Motion Level	0.2g	0.4g	0.5g 0.6g 0.8g	
No. of Failure	5	16	20 25 28	
Probability of failure	16.7%	53.3%	66.7% 83.3% 93.3%	

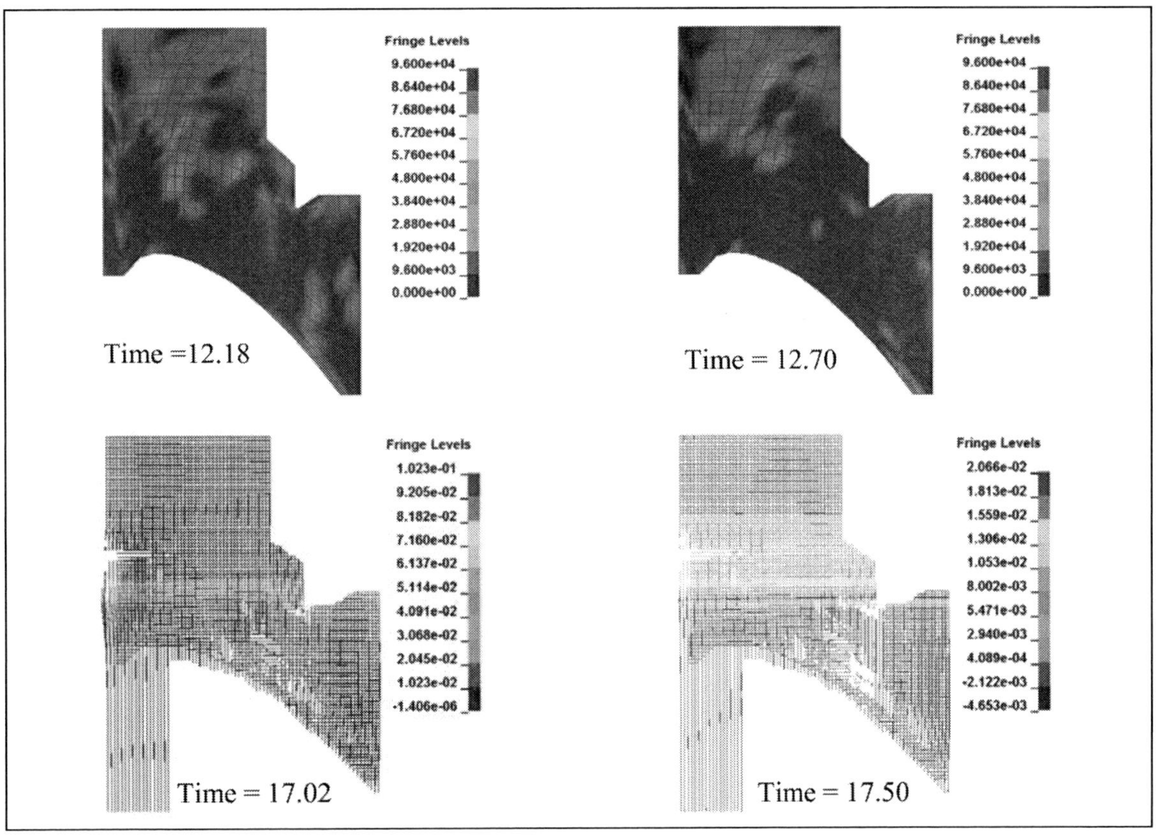

Figure 11. Concrete Major Principal Stresses (top) and Rebar Axial Strains (bottom).

SEISMIC FRAGILITY DEVELOPMENT

Composite Seismic Fragility

We developed the composite seismic fragility including both randomness and uncertainty from the nonlinear analysis results. Figure 12 summarizes the nonlinear analysis data used to determine the composite seismic fragility for the base sliding. For example, at the 0.4g ground motion level there is a 60% probability of dam failure, which is obtained from the fraction of trials failed at 0.4g ground motion level.

Assuming that the dam seismic fragility has a lognormal distribution, we estimated the parameters of the composite seismic fragility based on a least-squares fit between the natural logarithms of the failure PGAs and the number of standard normal variances associated with the probabilities of failure, as illustrated in Figure 12. We determined the median seismic fragility (PGA capacity) and composite logarithmic standard deviation to be 0.33g and 0.5773, respectively. The calculated coefficient of determination, R^2, is 0.9958, signifying that the lognormal distribution provides an excellent fit to the data. The composite seismic fragility and the failure data are plotted in Figure 14.

Failure Acc. (g) A	Probability of Failure	Standard Normal Variance	ln(A)
		X	Y
0.2	0.200	-0.842	-1.609
0.4	0.600	0.253	-0.916
0.5	0.767	0.728	-0.693
0.6	0.867	1.111	-0.511
0.8	0.933	1.501	-0.223

$y = 0.5773x - 1.1082$

$R^2 = 0.9958$

$A_m = \exp(-1.1082) = 0.33$

$\beta_c = 0.5773$

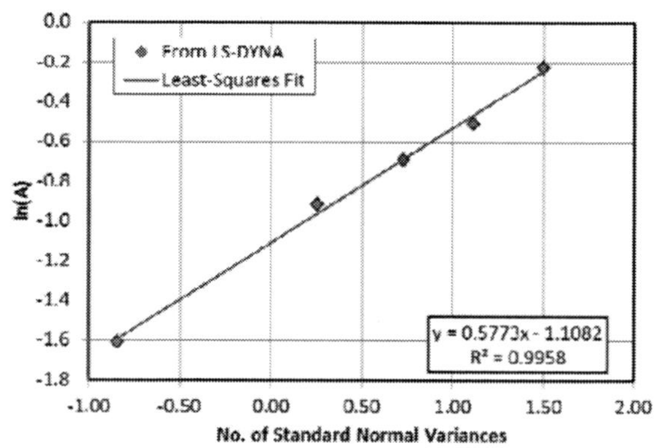

Figure 12. Least-Square Fit of Nonlinear Analysis Data for Composite Fragility.

Randomness Seismic Fragility

As shown in Figure 13, the probabilities of failure considering only random variability are 0%, 66.7%, 96.7%, 99.7%, and 100% at horizontal PGAs of 0.2g, 0.4g, 0.5g, 0.6g, and 0.8g, respectively. By applying least-squares fit to the nonlinear analysis data, we obtain a lognormal distribution for the randomness seismic fragility with the median capacity and randomness logarithmic standard of 0.35g and 0.2017, respectively. Figure 13 demonstrates the least-squares line fit to failure data and Figure 14 plots the randomness seismic fragility. The calculated coefficient of determination, R^2, is 0.9963, indicating that the lognormal distribution also provides an excellent fit to the random variability data.

Failure Acc. (g) A	Probability of Failure	Standard Normal Variance X	ln(A) Y
0.2	0.003	-2.713	-1.609
0.4	0.667	0.431	-0.916
0.5	0.967	1.834	-0.693
0.6	0.997	2.713	-0.511
0.8	1.000	3.403	-0.223

y = 0.2017x -1.0466

$R^2 = 0.9963$

A_m = exp (-1.0466) = 0.35

$\beta_r = 0.2017$

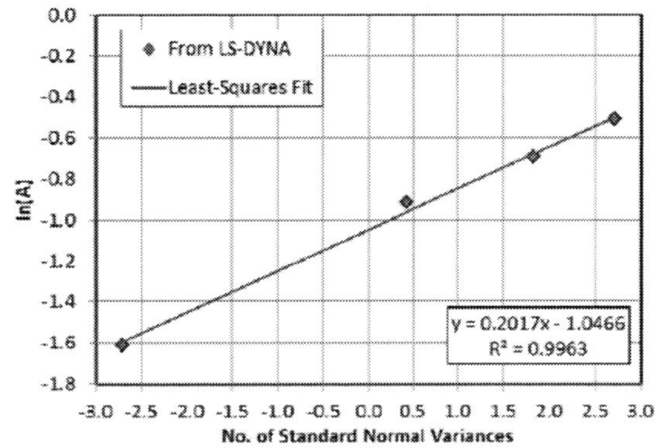

Figure 13. Least-Squares Fit of Nonlinear Analysis Data for Randomness Fragility.

Seismic Fragility Curves for Base Sliding

Based on the nonlinear analyses presented above, we conclude that the overflow section has median capacity, expressed in terms of the horizontal PGA, of A_m = 0.33g. We determine a logarithmic standard deviation for uncertainty of β_u =0.541 by extracting the logarithmic standard deviation for randomness of β_r = 0.2017 from the composite logarithmic standard deviation of β_c = 0.578.

$$\beta_u = \sqrt{\beta_c^2 - \beta_r^2} \tag{3}$$

Figure 14 summarizes the composite and randomness seismic fragility with the 5% and 95% confidence level curves.

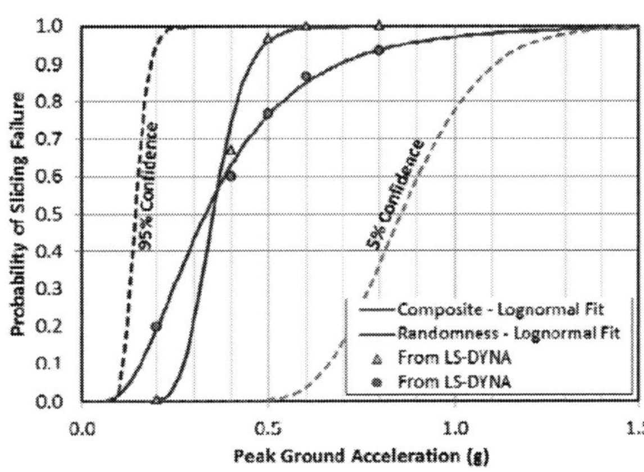

Figure 14. Dam seismic fragility for base sliding

Non-Exceedance Probability of Sliding Displacement

In addition to developing seismic fragility for the sliding of the dam, the nonlinear analysis results were also used to generate non-exceedance probability for magnitude of sliding displacements at various ground motion levels. For each ground motion level, first the permanent sliding displacements for the 30 simulations were reordered from low to high values, Then the non-exceedance probability (NEP) of the sliding displacement for each ground motion level was computed using Equation (4).

$$NEP = 100\% \left(\frac{n - 0.5}{N} \right) \tag{4}$$

Where $N = 30$ is the number of simulations and n is the ordered simulation number.

Assuming that the non-exceedance probability of sliding displacement has a lognormal distribution, we estimated the lognormal distribution parameters for the NEP using a least-squares fit to the data. Figure 15 compares the NEP sliding displacement functions with the failure data. The median (mu) and logarithmic standard deviation (beta) for each ground motion level are shown in the figure. These graphs provide very useful information for risk assessment. For example, the likelihood of sliding displacement not to exceed 1 foot is 99% at 0.2g, 94% at 0.4g, 87% at 0.5g, 79% at 0.6, and 58% at 0.8g.

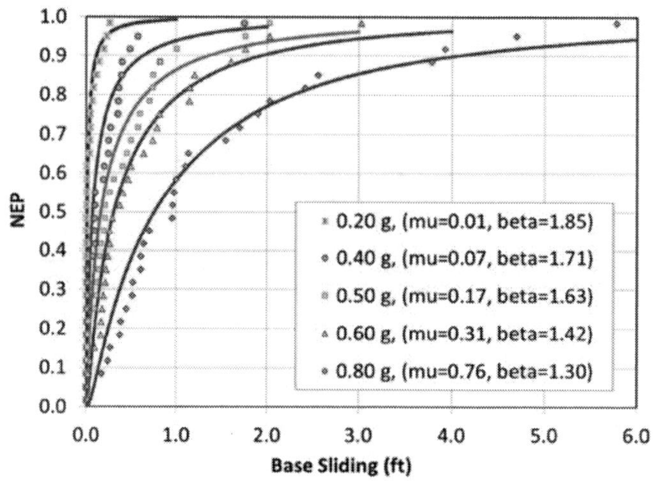

Figure 15. Non-exceedance probability of sliding displacement.

Seismic Fragility of Pier Failure

We also developed seismic fragility of the pier failure similar to the procedure used for the sliding of the overflow section. We first developed the composite seismic fragility including both randomness and uncertainty from the nonlinear analysis results of the pier described previously. Again, assuming that the pier seismic fragility has a lognormal distribution, the composite seismic fragility of the pier was obtained using a least-squares fit to the nonlinear analysis data. We determined the median PGA capacity and composite logarithmic standard deviation for the pier failure to be 0.39g and $\beta_c = 0.890$, respectively.

The pier seismic fragility was then developed including random variables only. Assuming a lognormal distribution, a least-squares fit to the nonlinear analysis data, resulted in the randomness seismic fragility having the median capacity and logarithmic standard of 0.36g and $\beta_r = 0.5552$, respectively.

Based on the above results, we conclude that the spillway pier has median capacity, expressed in terms of the horizontal PGA of $A_m = 0.39$g. We determ ined a logarithmic standard deviation for uncertainty of $\beta_u = 0.695$ by extracting the logarithmic standard deviation for random ness of $\beta_r = 0.555$ from the composite logarithm ic standard deviation of $\beta_c = 0.890$.

Figure 16 summarizes the pier composite a nd randomness seismic fragility with the 5% and 95% confidence level curves.

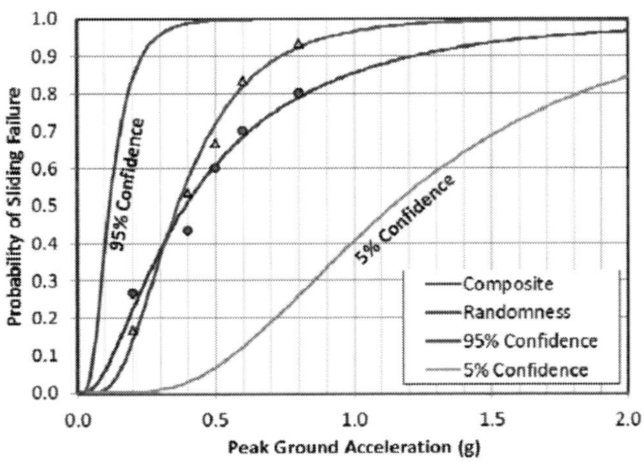

Figure 16. Pier Seismic Fragility Curves.

CONCLUSION

This study has proven that the nonlinear analysis in combination with Latin Hypercube Simulation provides a practical probabilistic method for developing seismic fragility for concrete gravity dams. The method uses advanced nonlinear analysis to capture nonlinear response behavior and potential failure modes of the dam and its components. The results show that the overflow section experiences sliding failure at the base of the monolith or severe damage and collapse of the pier, or both. The study included both the aleatory variability (randomness) and epistemic uncertainty (lack of knowledge) in developing the seismic fragility.

The results proved that a total of 30 simulations is sufficient to obtain a reliable fragility and that the seismic fragility is defined reasonably well with the lognormal distribution. The study indicates a median seismic fragility of 0.33g for sliding at the base of the example dam with a logarithmic standard deviation of $\beta_r = 0.202$ for randomness and $\beta_u = 0.541$for uncertainty. The seismic fragility for the pier failure was found to be 0.39g with a logarithmic standard deviation of $\beta_r = 0.555$ for randomness and $\beta_u = 0.695$ for uncertainty. These results show that the epistemic uncertainty is quite high for the base sliding and the pier failure, a factor that is often ignored in the contemporary risk assessment of dams. The use of median or best estimate parameters underestimates the risk and ignoring the aleatory variability of the ground motion could result in even higher underestimation of the risk.

We recommend such a study be repeated for two or more other gravity dams to confirm the applicability of the lognormal distribution and generate additional data on the aleatory variability and epistemic uncertainty. The results of the present study combined with the findings of the future studies can be used to establish a verified seismic fragility distribution with an average aleatory variability and epistemic uncertainty for application to other gravity dams. Such information would avoid the need for expensive simulations in the future. Instead, only a few deterministic analyses would be required to estimate the median capacity and then use the established fragility distribution and associated variability to generate seismic fragility curves at other confidence levels for risk assessment of the dam.

ACKNOWLEDGEMENTS

The authors acknowledge the financial support of the US Army Corps of Engineers, Headquarters, Washington D.C. The view of the authors does not purport to reflect the positions of the Department of the Army or the Department of Defense.

REFERENCES

EPRI (1994). Methodology for developing seismic fragilities. EPRI Report TR-103959, Palo Alto, CA.

Ghanaat, Y., Hashimoto, P.S., Olivier, Z., and Kennedy, R.P. (2011). "Seismic Fragility of Mühleberg Dam using nonlinear analysis with Latin Hypercube Simulation," *Proceedings of the 31st Annual USSD Conference*, San Diego, California, April 11-15, 2011.

Ghanaat, Y. (2012a). Seismic Fragility of Gravity Dams – Phase-I: Non-overflow Sections. Report submitted to US Army Corps of Engineers, May 7, 2012.

Ghanaat, Y., Patev, R.C., and Chudgar, A.K. (2012b). "Seismic Fragility Analysis of Concrete Gravity Dams," *Proceedings of the 15th World Conference on Earthquake Engineering*, Lisbon, Portugal, September 24-27, 2012.

Ghanaat, Y. (2014). Seismic Fragility of Gravity Dams – Phase-II: Overflow Section. Report submitted to US Army Corps of Engineers, December 16, 2014.

LSTC (2014), LS-DYNA, A general static and dynamic purpose finite-element code for large deformations static and dynamic analyses including soil-structure and fluid-structure interactions, Version LS971R7.1, Livermore, CA: Livermore Software Technology Corp., 2014.

PEER (2005), Pacific Earthquake Engineering Center. *PEER NGA Data Set*, Version 5.0, University of California, Berkeley.

URS (2008). Site-Specific Seismic Hazard Analyses and Development of Design Ground Motions, submitted to Sacramento District of USACE, September 29, 2008.

EVALUATING SEISMICALLY INDUCED HYDRODYNAMIC LOADS ON SPILLWAY GATES

Jerzy Salamon PE, PhD[1]

ABSTRACT

The determination of seismically induced hydrodynamic loads on spillway gates is a complex process that involves the seismic response of the dam, the response of the spillway gates, and the response of the reservoir. Several research studies have been conducted to determine the reservoir-dam and reservoir-spillway gate interaction during an earthquake, but there is not a comprehensive method that would provide satisfactory and accurate results without conducting sophisticated numerical analysis.

As a background, the dynamic reservoir loads developed during an earthquake are of importance in the design and evaluation of the spillway gates. The ground acceleration at the base of a dam during an earthquake can be considerably amplified at the top of the dam. Spillway gates will be subjected to this amplified acceleration, sometimes several times greater than that measured on rock at the abutment, depending on the response of the dam structure, location of the spillway gate, flexibility of the dam and the gate structure, actual water head on the gate, and whether the transverse, longitudinal or vertical acceleration is considered.

Dynamic stability analyses of spillway gates performed with the simplified methods could grossly miscalculate hydrodynamic loads on the gate structure. The result could be costly, where potentially unnecessary modifications to existing gates, over-design of new gates, or unreliable risk assessments based on inaccurate analyses of the gates are performed.

Standard methods for computing seismically induced reservoir loads on spillway gates are based on theories developed for concrete dams. In particular, the typical practice for computing hydrodynamic loads on dams relies on an analytical method developed by Westergaard. These solutions do not account for the flexibility of the dam nor the gate structure, amplifications of the ground motion acceleration up through the dam, the three dimensional effects, actual geometry of the gate skinplate, or the position of gates with respect to the face of the dam.

This paper presents the results of numerical analyses for rigid dams interacting with the reservoir, compares the results with the classical Westergaard solutions, and evaluates the spillway gate set back effect on the hydrodynamic loads induced during earthquakes.

---

[1]Technical Specialist, Waterways & Concrete Dams Group, Bureau of Reclamation, P.O. Box 25007, Denver, Colorado, 80225, USA, e-mail: jsalamon@usbr.gov

INTRODUCTION

Spillway gate structures in the closed position are exposed to a variety of loads including dead loads, hydrostatic pressure, seismic loads, ice loads, wave's effects from the reservoir and earthquake induced hydrodynamic reservoir loads. Some other loads are developed during gate operations.

Several research studies have been conducted to determine the reservoir-dam interaction during earthquakes but there is not a comprehensive method that would provide satisfactory and accurate results of hydrodynamic loads on spillway gates during seismic events. An overview of the analytical analysis methods used to determine hydrodynamic loads on dams and spillway gates was performed by Salamon (2011). A brief summary of the selected methods is provided below.

Hydrodynamic Loads on Dams

The first systematic analysis of the dam-reservoir interaction was performed by Westergaard (1931) in connection with the design of Hoover Dam by the Bureau of Reclamation (Reclamation). In the study of the earthquake response of a rigid dam with vertical upstream face, Westergaard developed an analytical solution for the hydrodynamic pressure distribution as a result of a harmonic horizontal motion of the dam. The approximate formula for the parabolic hydrodynamic pressure distribution over a vertical dam face, presented in a fundamental and simplistic form, has been used continuously by the industry in the analysis and design of concrete dams.

Zangar (1952) of Reclamation estimated experimentally the hydrodynamic pressures at the inclined upstream face of rigid dams using the electric analogy tray and then developed empirical formulas for various shapes of the dam upstream face.

Housner (1954) derived and solved an equation for fluid containers under earthquake loading incorporating a length of the reservoir in the formula. Housner made a distinction between an impulsive pressure that relates to the portion of the fluid that moves in coherence with the structure (the added mass), and the convective pressure that relates to effects like sloshing. Solutions for a rectangular, trapezoidal, segment, and stepped dam were provided as well as for flexible retaining walls.

Chwang (1978) and Chwang & Housner (1978) developed an analytical solution of seismic induced hydrodynamic forces for a rigid dam with an inclined upstream face of constant slope by potential-flow theory and momentum method, adopting von Karman's (1931) momentum-balance approach.

Effect of pressure wave absorption by sediment at the bottom of the reservoir was investigated by Chopra and Fenves (1983). The research results demonstrated that sediment in reservoirs plays an important role in assessing the real hydrodynamic reservoir pressure on the dam.

An extensive study was conducted at the University of California at Berkeley for earthquake analysis of concrete dams. The wave equation governing the motion of water in a reservoir was solved in the frequency domain using numerical methods and the solution was implemented in computer programs EAGD by Chopra (1983) and EADAP by Ghanaat (1989).

Lee and Tsai (1991) developed a closed-form solution for the analysis of the dam-reservoir system in the time domain. Studies were performed for the dam when the reservoir was both empty and full, considering the interaction between the fluid and the structure. The results of the investigation demonstrated that compressibility of the fluid and flexibility of the dam structure had greatest impact on interactive forces to both the structure and reservoir.

Several analyses were performed on a coupled dam-reservoir system composed of elastic and non-elastic dam material with compressible water, structural damping of the dam material and radiation damping of the water using numerical methods.

Hydrodynamic Loads on Spillway Gates

An experimental and Finite Element (FE) study was performed by Nakayama et al. (1991) for crest spillway gates set back from the vertical face of a rigid dam. As a result of the study, an analytical method was formulated for estimation of the seismic induced hydrodynamic loads on vertical spillway gates based on the simplified Westergaard solution.

Hydrodynamic loads on Folsom Dam spillway radial gates were analyzed by Reclamation. A methodology for determining the seismic induced hydrodynamic loads on spillway gates was summarized by Todd (2002). In the approach, a pseudo-static earthquake load was determined by including the effect of the "water added-mass" (reduced for a curved radial gate skinplate shape) and the earthquake related magnification factor for the dam.

Sasaki, Iwashita and Yamaguchi (2007) studied the basic characteristics of hydrodynamic pressure acting on spillway gates during an earthquake based on numerical analysis that considered vibrations of the dam body and the gate. They proposed a method for calculating hydrodynamic pressure during an earthquake in the seismic analysis of gates.

Versluis (2010) performed an extensive study on the hydrodynamic loads on large lock gates at the Technical University in Delft. This research was related to the design of the rolling gates for the new Panama Canal Locks. The Westergaard and Housner methods were implemented. Versluis concluded that the Housner formula for calculating hydrodynamic loads on the gates is more appropriate for the lock structure when the length of the reservoir is comparable with its depth. In the research a contribution of sloshing effects on the gates was also investigated.

PROBLEM STATEMENT FOR INTERACTION BETWEEN RESERVOIR AND SPILLWAY GATE

Analytical solutions obtained by Westergaard, Chwang, Housner, and others for a regular shape of the dam face would be impossible to derive for an irregular geometry of a dam with installed spillway gates. However, numerical methods can be used to solve the system of governing equations for the dam/gate system with complex boundary conditions. For the dam-gate-reservoir model shown in Figure 1, the problem for seismic induced motions can be formulated in a form of the Boundary Value Problem using Equations 1 through 5 below.

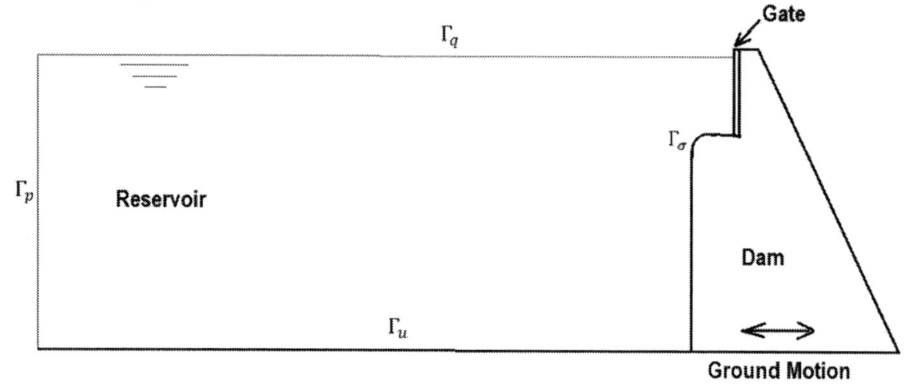

Figure 1. Model of the Dam-Gate-Reservoir System.

Problem Statement for Dam-Gate-Reservoir System

In the continuum theory of mechanics, the set of basic equations that includes equations of motions (Eq.1), geometric equations (Eq.2), and material constitutive law (Eq.3), together with boundary and initial conditions (Eq.4 and Eq.5), defines a mathematical formulation of the dynamic problem for linear elastic materials for fluid and concrete structure.

Equations of Motions. $\sigma_{ij,j} + \rho b_i = \rho \ddot{u}_i \qquad i,j,k = 1,2,3$ (Eq.1)

Geometric Equation. $\varepsilon_{ij} = \frac{1}{2}\left(u_{i,j} + u_{j,i} + u_{k,i}u_{k,j}\right)$ (Eq.2)

Constitutive Law for the Dam. $\sigma_{ij} = \frac{E}{1+\nu}\left(\varepsilon_{ij} + \frac{\nu}{1-2\nu}\varepsilon_{kk}\delta_{ij}\right)$ (Eq.3)

Constitutive Law for the Reservoir. $p = \frac{1}{3}\sigma_{kk} = K\,\varepsilon_{kk}$

where: $\underline{u}$, $\underline{\varepsilon}$, $\underline{\sigma}$, p are displacement, strain, stress tensors, and pressure
ρ, b, are mass density and body force
E, ν, K are elastic modulus and Poisson ratio of concrete, and bulk modulus of fluid.

Boundary Conditions. $\sigma_{ij}n_j = T_i$ on Γ_σ

$$\bar{q} = 0 \qquad on\ \Gamma_q$$

$$u_i = 0 \qquad on\ \Gamma_u \qquad \text{(Eq.4)}$$

$$p = \bar{p} \qquad on\ \Gamma_p$$

Γ_σ, Γ_q, Γ_u, Γ_p are parts of the boundary shown in Figure 1.

Initial Conditions.

$$u_i(x, t_0) = u_i^0(x) \qquad \dot{u}_i(x, t_0) = v_i^0(x) \qquad \ddot{u}_i(x, t_0) = a_i^0(x) \qquad \text{(Eq.5)}$$

Solution. Applying the standard procedure for derivation of the virtual power principle and considering geometric relationships between the displacements and strains, constitutive equations, and the initial conditions, the finite element discretization of the Galerkin variational formulation of the preceding equations results in the following system of second order differential equations:

$$\mathbf{M}_S\ddot{\mathbf{U}} + \mathbf{C}_S\dot{\mathbf{U}} + \mathbf{K}_S\mathbf{U} = \mathbf{F}_g + \mathbf{F}_p \qquad \text{(Eq.6)}$$

where: $\mathbf{M}_S$, $\mathbf{C}_S$, and $\mathbf{K}_S$, are the mass, damping and stiffness matrices, respectively

The unknown vector of nodal variable $\mathbf{U}$ represents the relative displacements at the nodes of the FE model of the dam. The forcing vector $\mathbf{F}_g = \mathbf{M}_s\ddot{\mathbf{U}}_g$ contains forces generated by the ground acceleration applied to the dam nodes. The vector $\mathbf{F}_P$ represents the hydrodynamic forces acting on the upstream face of the dam and it is related to the unknown vector of nodal pressures. To solve the problem in the time domain, an explicit approach has been implemented.

WESTERGAARD'S SOLUTIONS

Description of Westergaard's Approach

Exact Solution. The seismic motion of a straight rigid concrete gravity dam of height h with an infinite reservoir length, shown in Figure 1, was mathematically expressed by Westergaard in terms of the theory of elasticity of solids based on the formulation provided by Lamb (1924). Two equations of motion, a linear kinematic relation for small deformations, and an elastic constitutive equation without shearing stresses, together with the boundary conditions (stresses equal zero on the reservoir surface, vertical displacement equal zero at the bottom of the reservoir) described the two-dimensional physical model of the dam-reservoir system. The solution of the problem with horizontal and vertical motions of the water (plane strain) was given by Westergaard in the form of a stress (pressure) distribution in the reservoir by Eq.7.

$$\sigma = -\frac{8\alpha w h}{\pi^2} \cos\frac{2\pi t}{T} \sum_{1,3,5,..}^{n} \frac{1}{n^2 c_n} e^{-q_n} \sin\frac{n\pi y}{2h}$$

(Eq.7)

$$q_n = \frac{n\pi c_n x}{2h} \quad c_n = \sqrt{1 - \frac{16 w h^2}{n^2 g k T^2}}$$

According to Eq. 7, the maximum pressure p occurs when the dam is in the extreme positions (t = 0, T, 2T, etc.) during motion, so the maximum water pressure distribution at the upstream face of dam (for x =0, q_n = 0) could be expressed by Eq.8.

$$p = \frac{8\alpha w h}{\pi^2} \sum_{1,3,5,..}^{n} \frac{1}{n^2 c_n} \sin\frac{n\pi y}{2h}$$

(Eq.8)

$$c_n = \sqrt{1 - \frac{16 w h^2}{n^2 g k T^2}} = \sqrt{1 - \frac{0.71889}{n^2}\left(\frac{h \text{ sec.}}{1\,000\, T \text{ ft.}}\right)^2}$$

where: x, y = the axis of x is at the surface of the water directed upstream and the axis y is vertical downward (Figure 2),

w = weight of water per unit volume (w = 62.4 lb/ft^3),

g = acceleration due to gravity (g = 32.2 ft/sec^2)

α = maximum horizontal acceleration of foundation divided by g,

T = period of horizontal vibration of the foundation,

t = time,

k = bulk modulus of water (k = 300,000 lb/ft^2)

The solution expressed by Eq.7 and 8 was derived with the following assumptions:

- The dam upstream face is straight and vertical,
- The dam does not deform and is considered to be a rigid block,
- The reservoir is infinite in length,
- Dam sinusoidal oscillations are horizontal,
- Small motions are assumed during earthquake,
- The problem is defined in 2-D space,
- Period of free vibration of the dam, T_0, is significantly smaller than the period of vibration, T, of the earthquake (resonance is not expected),
- Non-dimensional horizontal acceleration of α = 0.1,

The effect of water compressibility was found to be small in the range of the frequencies that are supposed to occur in the oscillations due to earthquake.

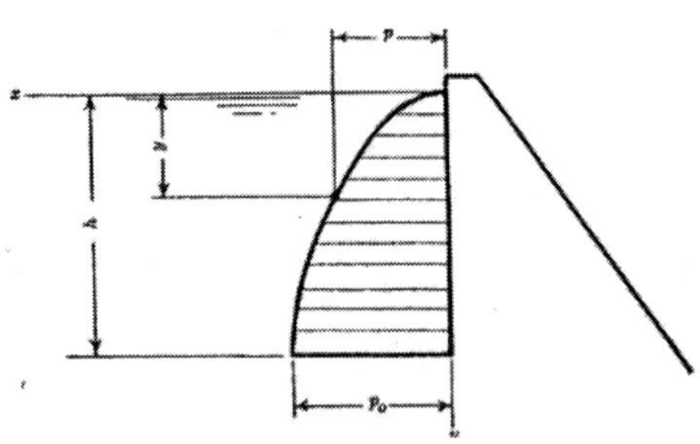

Figure 2. Pressure distribution on dam for the Westergaard's exact solution.

Approximate Solution. The hydrodynamic water pressure on the dam expressed by Eq. 8 was simplified by a parabola described by Eq. 9.

$$p = 0.875\ w\ \alpha\ (h\ y)^{0.5} \qquad \text{(Eq.9)}$$

This approximate Westergaard's formula (Eq. 9) is widely used by the industry in the preliminary calculations of the hydrodynamic pressure on dams.

Hydrodynamic Loads on Vertical Face of Rigid Dam

In this section Westergaard's approximate and exact solutions are compared with the FE analysis solution. For the purpose of this study a spreadsheet was developed to calculate the pressure distribution at the face of the dam for both Westergaard's formulas. The spreadsheet was calibrated with the results derived by Westergaard (1931). Figure 3 presents hydrodynamic pressure distribution for the reservoir depths of 200, 400, and 600 feet with the period of horizontal dam vibrations, *T*, between 0.33 and 4 seconds.

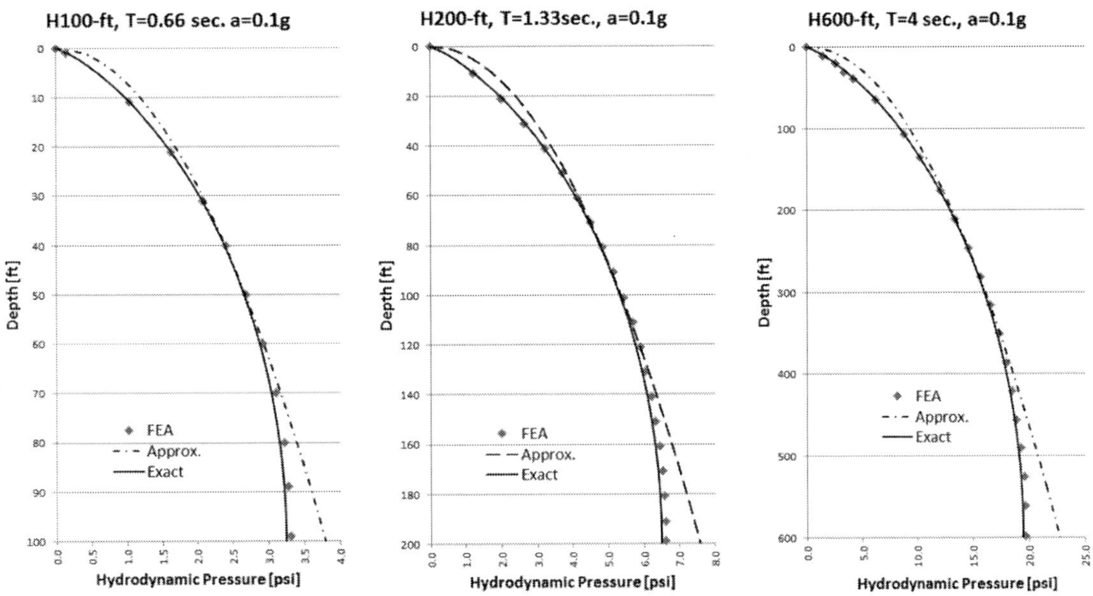

Figure 3. Comparison of hydrodynamic pressures calculated according to Westergaard's exact and approximate formulas and the FE results for ground acceleration of 0.1g.

Good agreement between the exact Westergaard's solution and the FE results for a rigid dam with vertical upstream face can be observed in Figure 3. In general the approximate Westergaard's formula (Eq.9) overestimates the hydrodynamic pressure at the upper part of a rigid dam when it is compared with the exact Westergaard's solution and the FE analysis results.

HYDRODYNAMIC LOADS ON SPILLWAY GATES

Seismically induced hydrodynamic pressure from the reservoir during an earthquake acts on the spillway gates in the same way it acts on the dam body. In this section, a basic study is performed to investigate the effect of the gate set back from the dam face. A FE analyses is performed for a rigid dam with vertical upstream face and the spillway gate located at various distances from the upstream face of the dam. Finally, an analytical method is proposed for estimation of the hydrodynamic loads on vertical spillway gates when the gates are set back from the dam face.

Gate Set-back Effect

Nakayama et al. (2007) performed an experimental study to determine hydrodynamic loads on spillway gates installed at a rigid dam with vertical upstream face. The results of the experiments supported by the FE analysis showed that during earthquakes, the spillway gates experience significantly lower loads from the reservoir when the gates are set back from the upstream face of the dam. They developed an empirical formula (Eq. 10) for the pressure distribution over the vertical face of the spillway gate based on the FE analysis results. The effect of the gate height and the set-back distance (Eq. 11 and Eq. 12) was incorporated in the study.

$$p_g(h) = p_{FEM,\beta=0}(h) \cdot C(h) \qquad \text{(Eq.10)}$$

$$C(h) = 1 - \frac{\beta h}{2h_1} \qquad (\alpha \leqq 0.3, \beta \leqq 0.7) \qquad \text{(Eq.11)}$$

$$C(h) = (1-\sqrt{\alpha})\exp\left(\frac{-1.4\beta h}{h_1}\right) + \sqrt{\alpha} \qquad \text{(Eq.12)}$$

where:
α is the ratio of the depth of the gate to the reservoir depth = h_1 / H
$\beta = d / h_1$ is the gate set-back ratio defined as a distance of the gate location to the face of the dam divided the gate height

Nakayama et al. developed also a simplified formula (Eq. 13) for the total hydrodynamic load on spillway gates that was based on the approximate Westergaard's solution (Eq.9).

$$P_g = {}^7/_{12}\, k\, w \sqrt{H}\ h_1^{1.5} \left(1 - {}^{3\beta}/_{10}\right) \qquad \alpha \leqq 0.3,\ \beta \leqq 0.7 \qquad \text{(Eq.13)}$$

FE ANALYSIS

For a cross-section of the spillway typical for Reclamation's gravity dams, a FE analysis was performed for a rigid dam with various gate heights (α), various gate set back distances (β) and a 5-foot-radius of the spillway approach channel (Figure 4). Distribution of the hydrodynamic pressure in the 200-foot-deep reservoir is shown in

Figure 4 for the gate to dam height ratio of 0.2, the ratio of the gate set back distance to the gate height of 0.5, and for ground acceleration of 0.1g.

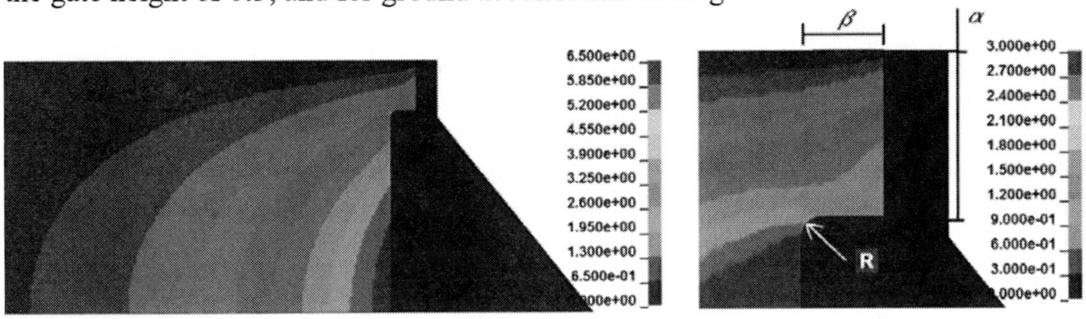

Figure 4. Hydrodynamic pressure [psi] for 200-feet deep reservoir with α=0.2, β=0.5 and ground acceleration of 0.1g.

Nakayama's Solutions vs FE Results

In this section, the solution proposed by Nakayama et al. (NOI) by Eq. 10 is compared with the FE analysis results obtained for the model shown in Figure 4.

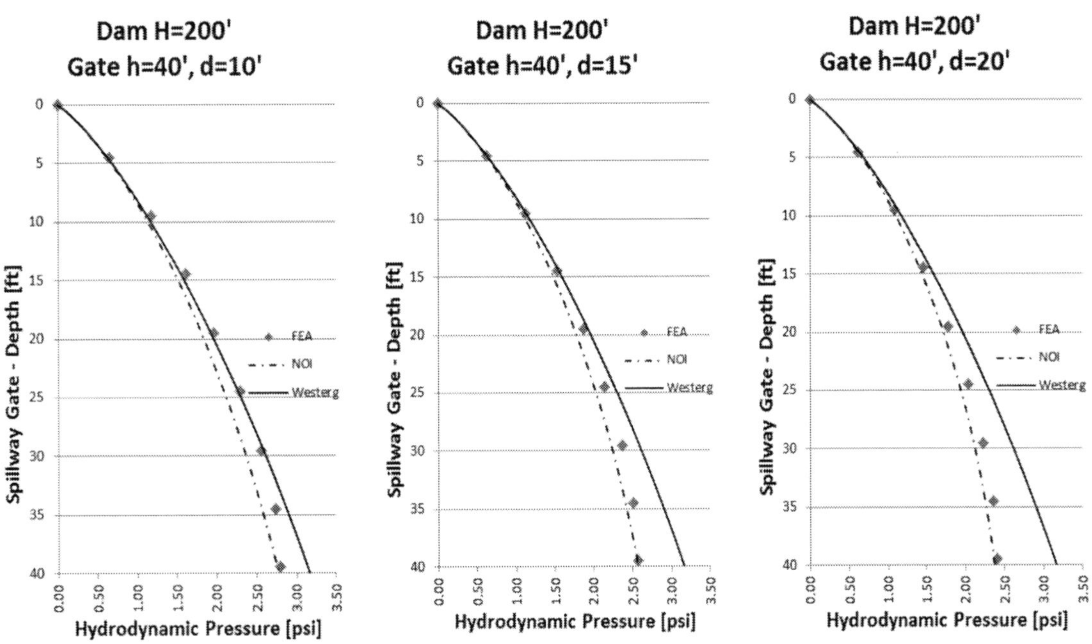

Figure 5. Comparison of hydrodynamic pressures [psi] for the gate set back distance at 10-, 15-, and 20-feet (β=0.25, 0.375, and 0.5, respectively), the 40-ft high spillway gate, 200-feet deep reservoir (α=0.2) and ground acceleration of 0.1g.

According to Figure 5, the hydrodynamic pressure on the gate decreases as the gate set back distance from the dam face increases. Nakayama et al. (NOI) solution shows lower pressure on the gate when compared with the FE analysis results.

HYDRODYNAMC LOADS ON SPILLWAY GATES BASED ON WESTERGAARD'S EXACT SOLUTION

A new approach for calculating hydrodynamic loads on spillway gates is proposed in this paper. The approach is based on the Westergaard's exact solution (Eq.7), and is calibrated with the FE analysis results performed for various gate heights and gate set-back distances and using Nakayama et al. equations (Eq. 10 and Eq. 11). The main advantage of the proposed method is simplicity and the fact that it does not require FE analysis of the dam to determine loads on vertical spillway gates.

For rigid dams with a vertical face, the total load on the vertical gates computed in the FE analysis is presented in Figure 6 as a function of the gate set back distance to gate height ratio. The Figure shows also the "New Curve" (Eq. 14) that is an approximation of the FE analysis results and the curve "NOI" developed by Nakayama et al. The new equation for the total hydrodynamic load on a vertical gate is:

$$P_{Total}(h) = P_{Westergaard} \cdot (1.1 - 0.45\beta) \qquad (\beta \leqq 0.7) \qquad \text{(Eq.14)}$$

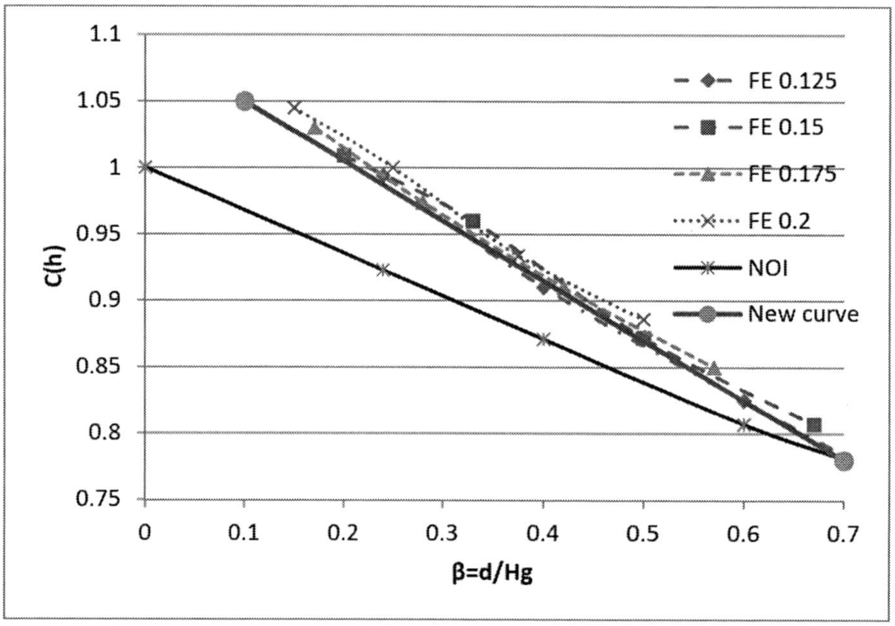

Figure 6. Calibration of "new curve" based on FE analysis results for total hydrodynamic load on vertical gate set back from the vertical face of a rigid dam.

CONCLUSION

Several conclusions can be made from the study on the hydrodynamic loads on spillway gates induced during earthquakes. First, the location of the spillway gate in respect to the dam face has significant influence on the pressure level developed at the gate skinplate. The study shows that pressure reduction is observed as the crest spillway gate is set back from the face of the gravity dam. Some other important factors under investigation by the author of this paper that may have impact on the gate loads during earthquake are: curvature and slope of the gate skinplate, geometry of the upstream dam face, and flexibility of the dam and the gate structure.

The results in Figure 3 show that the Westergaard's approximate equation (9), commonly used in the engineering practice, overestimates the hydrodynamic loads induced during earthquakes at the upper part of the rigid dam, where the crest spillways are usually installed.

In the presented study, a formula was developed to determine seismically induced hydrodynamic loads on vertical spillway gates installed at rigid dams with the vertical upstream face. The approach is based on the exact Westergaard's solution with vertical gate set back from the dam face. Implemented in an Excel spreadsheet, the formula can be utilized in the preliminary seismic analysis and design of the spillway gates installed at gravity dams.

ACKNOWLEDGEMENTS

The author would like to express his sincere thanks to Dam Safety Office at Reclamation for supporting the research on seismic induced hydrodynamic loads on spillway gates.

REFERENCES

Chopra A. K. (1967), *Hydrodynamic Pressures on Dams during Earthquakes*, Journal of the Engineering Mechanics Division, Proceedings of the ASCE, pp 205-223.

Chwang A. T., Housner G. (1977), *Hydrodynamic Pressures on sloping Dams during Earthquakes, Part 1 – Momentum Methods*, J. Fluid Mech. Vol. 87, Part 2, pp. 335-341.

Chwang A. T. (1978), *Hydrodynamic Pressures on sloping Dams during Earthquakes, Part 2 – Exact theory*, J. Fluid Mech. Vol. 87, Part 2, pp. 343-348.

Housner G. W. (1990), *The Momentum-balance Method in Earthquake Engineering*, Selected Earthquake Engineering papers of G.W. Housner, ASCE.

Nakayama Y., Ohmachi T., Inoue S. (1991), *Practical Evaluation of Hydrodynamic Pressures on Dam-Gates during Earthquake*, Proceeding of JSCE A, Vol. 64, No.4, pp. 959-969, 2008. (in Japanese).

Salamon J.W. (2011), *Seismic Induced Loads on Spillway Gates - Literature Review*, Bureau of Reclamation, DSO-11-06 Report, Denver, Colorado.

Sasaki T., Iwashita T., Yamaguchi Y. (2007), *Calculation Method of Hydrodynamic Pressure in Seismic Response Analysis of Gates*, Technical Report on the 39th Joint Meeting May 2007, Panel on Wind and Seismic Effects, Tsukuba, Japan.

Todd R.V. (2002), *Determining Earthquake Loading on Spillway Gates*, Hydro Review, Volume XXI, No.4, pp 92-99.

USACE (2003), U.S. Army Corp of Engineers, *EM 1110-2-6051 – Time-History Dynamic Analysis of Concrete Hydraulic Structures*, December 2003.

Westergaard H. M. (1931), *Water Pressures on Dams during Earthquakes*, ASCE Transactions, pp. 418-433, November 1931.

Von Karman T. (1931), *Discussion on: Water Pressures on Dams during Earthquakes by Westergaard*, ASCE Transactions, pp. 434-436, November 1931.

Zanger C. N., (1952), *Hydrodynamic pressures on Dams due to horizontal Earthquake Effects*, Bureau of Reclamation, Engineering Monograph No.11.

CRACK MECHANISM AND PREVENTION MEASURES OF CONCRETE IN THE INITIAL POURING BASED ON FIELD TEMPERATURE GRADIENT EXPERIMENT

Liu Youzhi[1]
Zhang Guoxin[2]
Wang Zhenhong[3]
Li Renjiang[4]

ABSTRACT

To study concrete cracking, a feedback analysis on variations of surface temperature gradient has been conducted for concrete at the intermission stage of layer pouring with a precise model and finite element simulation method combined with in-situ experiments of a dry hot valley dam. This paper presents the results of concrete stresses and their development under real concrete temperature gradient and explores the internal crack mechanism and possible hidden cracks of the concrete. A discussion on effective measures for prevention of such cracking risk, particularly for construction of valley dams in a dry hot season is also given. This work is of practical reference value for construction of concrete structure in dry hot valley.

INTRODUCTION

In China, some regions in Yunnan, Sichuan and the Yunnan - Guizhou Plateau belong to those regions with the dry and hot weather, where some river valleys such as Jinsha River, Yuan River, Nanpan River and Nu River belong to the dry and hot river valleys. Along with development of Chinese water conservancy construction, many concrete dams were built in these regions. The typical climate characteristics of dry and hot river valleys are high temperature, lack of rainfalls, the ambient temperature in summer being often up to above 40°C, and frequent short-term extreme climates. When damming in these regions, the temperature control and crack prevention are relatively difficult (particularly, damming in high temperature season); and the cracking risks such as concrete being lowered sharply and low temperature cooling strike due to rainstorms are relatively common [1-3]. In some actual projects, cracks often occur during the construction period (Figure 1) due to many factors except for some objective weather factors, which include one key factor, namely being lack of experiment data and systematic analysis on the temperature gradient and real stress of concrete at early age; on the other hand, the previous analysis on the development variation processes of concrete surface temperature gradient during the interval is not fine to really reflect the actual temperature variation process, let alone master the actual stress variation processes of early age concrete[4-7]. Thus, the forecast of cracking risks of early age concrete is also short of objective evaluation and how to rationally select the temperature control

[1] China Institute of Water Resources and Hydropower Research, Beijing, China 100038, youzl@iwhr.com
[2] China Institute of Water Resources and Hydropower Research, Beijing, China 100038
[3] China Institute of Water Resources and Hydropower Research, Beijing, China 100038
[4] China Three Gorges Corporation Yichang, Hubei, China 443000, li_renjiang@ctgpc.com

measures for prevention of such cracking risks is free of effective basis, so cracking may be difficultly avoided.

The evaluation of the actual stress in pouring concrete in high temperature season, especially the correct analysis on the mutual relation between the surface temperature and stress variation processes after just pouring concrete, is greatly of importance; and it is crucial to solve the current concrete surface crack prevention issue during the interval. As for concrete, pouring may give rise to the hydration reaction and the surface concrete may be influenced by the ambient temperature and hydration heat; the adiabatic temperature rise curves of indoor concrete laboratory experiments may basically reflect the actual heating processes; thus, the simulation of temperature fields is rational by using the indoor adiabatic temperature rise curves as the model in the traditional method for simulation and analysis. However, as for the stress fields, the similar method may not be used for simulation because the initial setting time is generally among 9~15 hours and concrete is plastic after pouring concrete, during which concrete temperature variation may not give rise to the stress variation; thus, the relation between the initial setting time and stress must be taken into account; especially, the early age concrete stress shall be estimated [8,9]. In previous work, such relation between the initial setting time and stress is always ignored in the traditional analysis method so that the analysis results may always be not consistent with the actual results[10-14].

The study was carried out in view of such issue. Based on the temperature gradient tests, the symmetrical study was carried out to the relations between those factors such as concrete initial setting time, bad time variation, and the variation rules of temperature and stress of concrete at early age under different climate conditions; the further analysis was carried out to possible cracking risk and safety hidden dangers of the surface concrete under the interval; in addition, the corresponding measures and methods for prevention of such cracking risks in high temperature season were put forward to provide references for the crack prevention in similar subsequent projects.

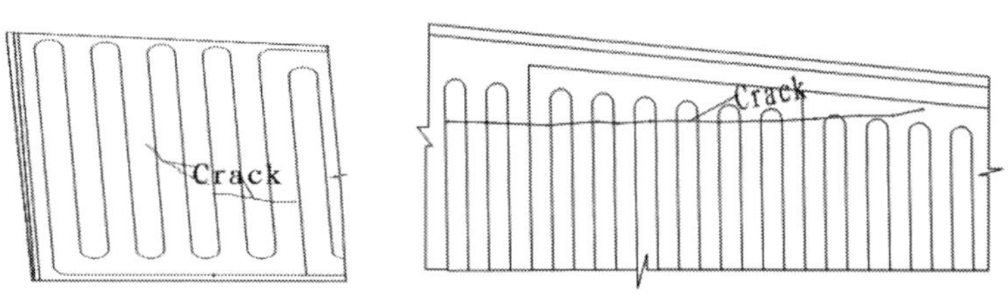

Figure 1. Schematic diagram of the crack occurred at the interval during the construction period.

FEEDBACK ANALYSIS ON OBSERVATION RESULTS OF TEMPERATURE GRADIENT TESTS AND TEMPERATURE FIELDS

Some high arc dam is located in a dry and heat river valley in China South - West Region, which was under the annual continuous construction, the maximum air temperature is often above 40°C in high temperature season and such region belongs to the typical dry and heat river valley climate. In accordance with the project crack prevention requirements, the comprehensive control temperature control measures such as low temperature pouring, flowing water cooling and surface heat preservation were taken for pouring such dam. For accurately obtaining the temperature gradient and the corresponding stress variation rules of the surface concrete at the interval during the dam construction period, the temperature gradient field tests were carried out. Such concrete lift is located in the elevation (361.8m - 364.8m) and is equipped with two layers of water pipes (1.5×1.5 layout). The cooling water temperature, water flowing rate, pouring temperature and ambient temperature are actual monitoring data in the field. Six thermometers are equipped along the elevation direction, whose specific positions are shown in Figure 2. The temperature gradient is monitored by a automatic temperature monitoring and recording device developed by our research group[5]. At the same time, a refined simulation model was established for simulation of temperature fields in other regions of such concrete lift, in which the grid size for the surrounding tube walls is in mm - level; and the calculation model is shown in Figure 3.

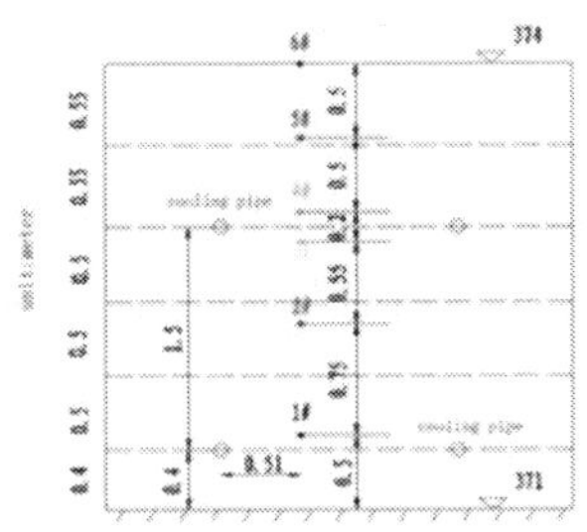

Figure 2. Schematic diagram of the temperature gradient observation layout points

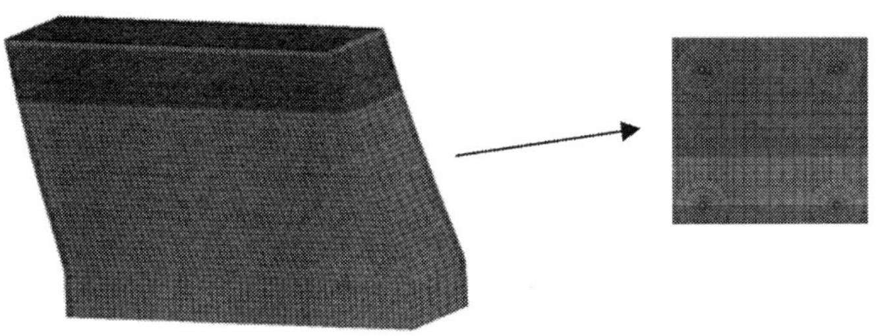

Figure 3. Simulation refined model

The inverting analysis was carried out to the temperature field of each lift of concrete for this dam; moreover, the refined feedback was carried out to the testing temperature variation processes. The interval of the concrete lift is 7 days; and the monitoring temperature was mainly from those data during such 7 testing days. As shown in monitoring data in Figure 4, the temperature variation rules for different locations along the elevation direction are obviously different. The locations (5# and 6#) near the surface are influenced by the surface ambient temperature, water curing and other boundary conditions; the temperature of 5# location was maximum after about 1 day and then it was basically maintained in the subsequent 3 days; on the other hand, the temperatures of those locations (1#~4#) relatively far away from the surface basically gradually rose along time. Figure 5 shows the obvious inside and outside temperature differences for concrete after 3~4 days. The comparison curves of the monitoring temperature values and the inversion temperature values for the typical locations are shown in Figure 6(a) to (d), from which it may be seen the feedback simulation results may relatively well reflect the actual temperature field distribution rules for this project; and the accuracy of simulation results of temperature and stress fields may provide the guarantee.

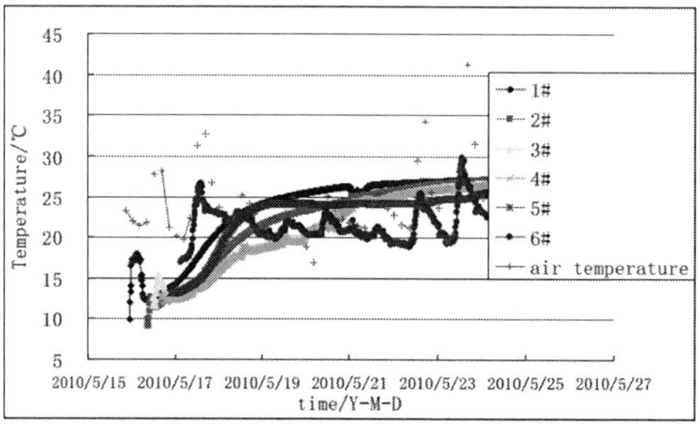

Figure 4. The monitoring values of actual temperatures of measurement locations.

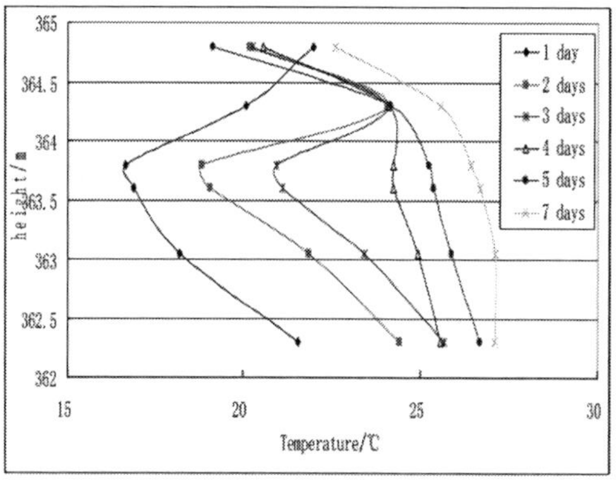

Figure 5. The measured temperature gradient distribution for different age.

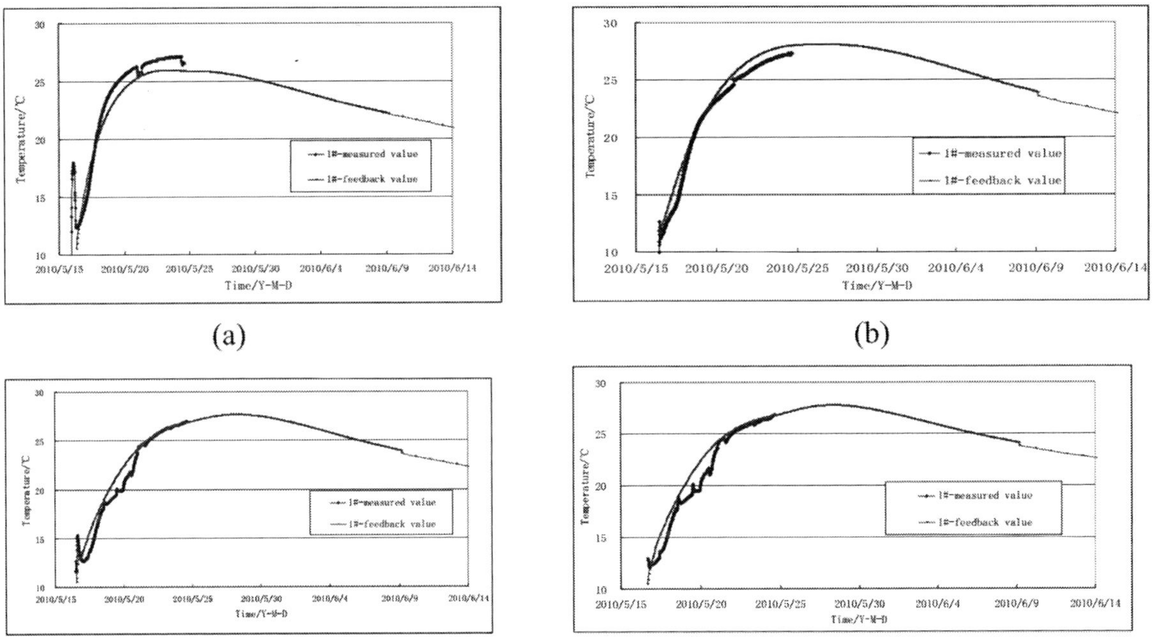

(a) (b)

(c) (d)

Figure 6. Comparison of the measured temperature results and feedback simulation results.

SIMULATION OF STRESS FIELDS BASED ON THE FIELD TEMPERATURE GRADIENT TESTS AND ANALYSIS ON THE CRACKING MECHANISM

Based on the field temperature gradient tests and the feedback analysis on temperature fields, the simulation and analysis were carried out to concrete temperature and stress fields; moreover, the possible cracking mechanisms were analyzed.

As shown the feedback temperature results in Figure 7(a), concrete temperature variation is independent on the initial setting time. After concrete pouring completed, the temperatures of inner locations were gradually increased along with the hydration reaction, and the inner temperature is more than the surface temperature after about 3 days; the peak temperature occurred at the age of about 5~6 days. The temperature of locations in different elevations changes variously with the flowing water cooling effect; and their temperatures are slightly different. In addition, the temperature of each surface location is greatly dependent on the ambient temperature; and the daytime and night temperature shows a trend of fluctuation.

With corresponding to the temperature processes, there is no inner stress in newly pouring concrete within the initial setting time (12 hours); and after such initial setting time, the inner stress in concrete changes along with the temperature variation. The inner stress is generally compressive when the temperature rises; on the other hand, the temperature of each surface location is remarkably dependent on the ambient temperature; the stress is compressive when the temperature rises and the stress is primarily tensile when the temperature drops. The main cause for there being tensile stress in the surface locations is the temperature drop process due to the temperature difference between day and night. As shown in Figure 7(b), the surface temperature is more than the inner temperature within 2 days after concrete pouring completed; on the other hand, the daily temperature drop process gives rise to the tensile stress in the surface. After the age of 3 days, the outer surface temperature is lowered below the inner concrete temperature; at this time, the superposed effect from the temperature difference between day and night and the difference between the inner and outer temperatures occurs. As shown in the comparison results between the surface concrete stress process and the allowable stress, concrete early age tensile stress is basically near or even more than the allowable tensile stress; such case shows that there may generally be relatively large cracking risk in the early age concrete surface. If the field prevention measure is not proper, cracks may possibly occur in the surface concrete.

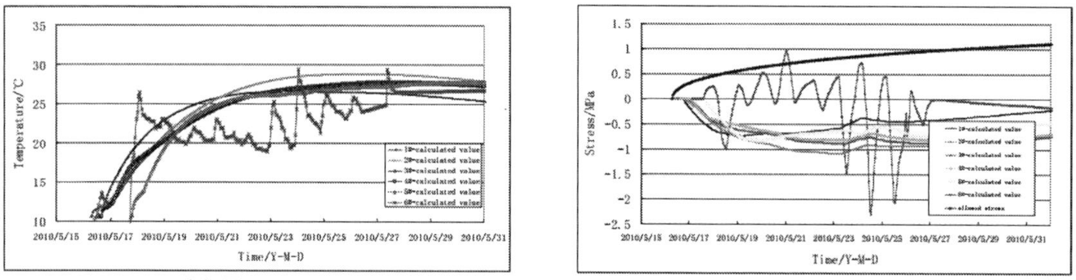

(a) Temperature (b) Stress

Figure 7. The feedback analysis temperature and stress process lines of the typical testing concrete locations (1# - 6# measurement locations)

STUDY ON INFLUENCING FACTORS OF EARLY AGE CONCRETE STRESS AND CRACK PREVENTION MEASURES

For further understanding those influencing factors of concrete temperature and stress in the surface at the interval under different boundary conditions and easily providing the basis for adopting the corresponding temperature control and crack prevention measures, comparison and analysis of several schemes were carried out and the main results are listed as follows:

Influencing of the Initial Setting Time for Concrete Stress

As shown in those results in Figure 8, the initial setting time may greatly influence the surface concrete stress; and the actual completed time for this concrete lift was about 3:00 in the morning. As shown in the measured ambient temperature, the subsequent temperature was always in the growth trend; if the initial setting time for concrete is not taken into account, the compressive stress occurs along with the growth of the ambient temperature; and the early compressive stress may be favorable for the crack prevention subsequently. On the other hand, there is no actually such reservation process for such compressive stress due to there being the initial setting time. After concrete has been initially set within 12 hours, concrete surface ambient temperature is up to the maximum value among one whole day and then it always drops; therefore, the initial tensile stress initially primarily occurs in concrete after its peak temperature but there is no any compressive stress reservation. Thus, if the initial setting time is not taken into account, the simulation results may possibly result in misled or weaken our understanding to the cracking risk of concrete at early age.

As shown in test results in Figure 9, different initial setting times, pouring completed time and the occurrence time for the daily maximum and minimum temperatures may comprehensively influence the development and variation rules of the surface stress in concrete at early age, and the optimization control shall be carried out by combining the actual conditions during the construction period.

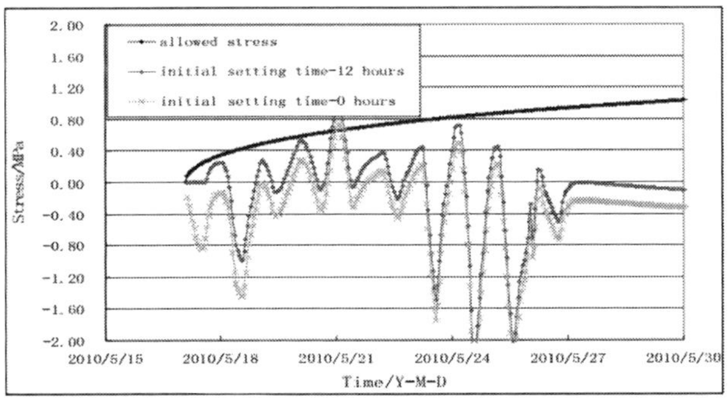

Figure 8. Stress compared when considering the concrete initial setting time or not

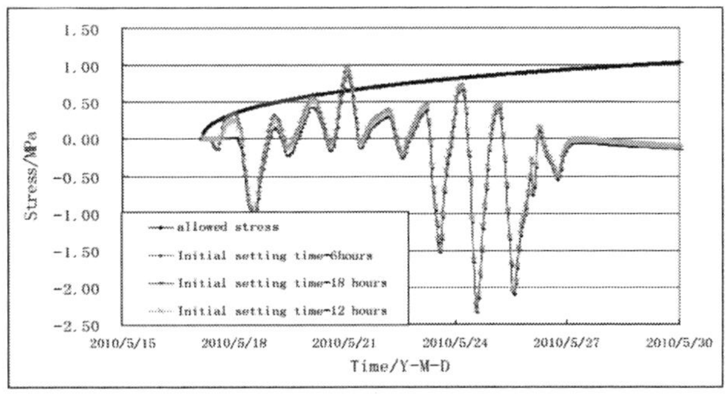

Figure 9. Initial setting time vs. concrete stress

Influencing of the pouring completed time for concrete stress

With assuming concrete initial setting time being 12 hours and the temperature difference between day and night being 10°C, the maximum and minimum temperatures occur in 3:00 o'clock and 15: 00 o'clock. Comparison and analysis were carried out to the influence of concrete temperature and stress when concrete pouring completed at the daily maximum and minimum temperatures respectively. As shown in Figure 10, if concrete pouring is completed under the daily minimum temperature, the initial setting for concrete is completed in the corresponding maximum temperature; and then the ambient temperature is lowered and the initial stress is the tensile stress; on the other hand, if concrete pouring is completed under the daily average maximum temperature, the concrete surface stress will first manifests for the compressive stress after the initial setting, and such case should be favorable for crack prevention. In contrast, if concrete lift completed under the maximum concrete temperature, it is obviously favorable for lowering the surface tensile stress and the cracking risk. Therefore, how to hold the pouring completed time to more favorable for concrete temperature control and crack prevention in accordance with the field pouring capacity and concrete initial setting time is very important.

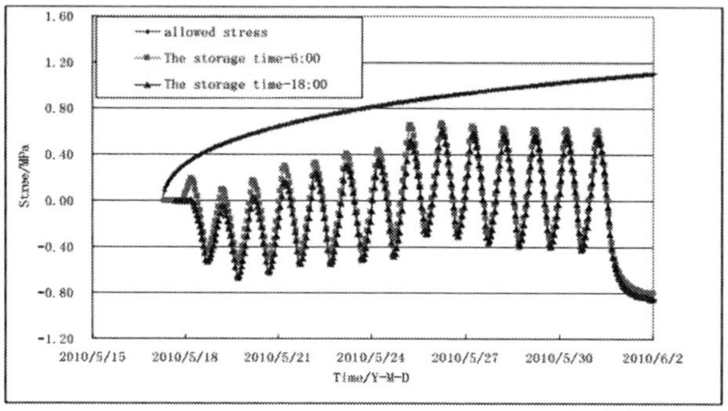

Figure10. Concrete stress vs. different pouring completed time

CRACK PREVENTION MEASURES AND SCHEME

By combining the above analysis, reasonable planning the concrete lift completed time is favorable for reducing the surface cracking risk, When concrete is initially set and the ambient temperature is the minimum value among one day, the optimal effect may be achieved; on the other hand, the pouring completed time of concrete may be deduced inversely in accordance with the initial setting time. In addition, the surface heat preservation and curing maintenance work must also be carried out to concrete at early age so as to lower the unfavorable effects due to the temperature difference between day and night or short-term temperature drop as can as possible; at the same time, the occurrence of local defect due to improper curing maintenance shall be avoided; thus, the cracking risk of concrete at early age at the interval may be effective prevented.

CONCLUSIONS AND SUGGESTIONS

(1) The tested field temperature gradient may be used to well reflect the distribution of temperature gradient of the surface concrete at the interval; thus it is favorable to reasonably evaluate the actual stress situations and risks of concrete at early age.

(2) As for the general concrete, once concrete is initially set under the comprehensive effect of the temperature difference between day and night and the inner and outer surface temperature difference, the relatively large cracking risks possibly occur; thus, the early surface curing and thermal insulation shall be well performed.

(3) The initial setting time must be taken into account for evaluation of the stress safety status of concrete at early age; otherwise, the judgment of cracking risks of concrete may possibly be weakened or misled.

(4) The best time for completion of initial setting of concrete is the time when the ambient temperature is the minimum value per day; and the corresponding inverse deduction may be carried out to the optimal beginning and closing placed times. As shown in the calculation results, the cracking risks of concrete at the early age may be effectively lowered through the rational optimization of the pouring completed time.

ACKNOWLEDGEMENTS

This work was financially supported by the National Basic Research Program of China (973 Program 2013CB036406 and 2013CB032904), the Twelfth-Five Science and Technology Support Project (2013BAB06B02) , IWHR and the Basin Water Cycle Simulation and Regulation of State Key Laboratory Special Research Foundation.

The authors declare that there is no conflict of interests regarding the publication of this article.

REFERENCES

[1] Zhang Guoxin, Liu Youzhi, Liu Yi et al. Cause analysis and temperature control and crack prevention measures on cracks in the super-high arch dam during the construction period [J]. Journal of Hydroelectric Engineering, 2010, Vol. 29(5): 45-50 (in Chinese).

[2] Zhang Guoxin, Ai Yongping, Liu Youzhi et al. Investigation on temperature control and crack prevention issues for the super-high arch dam during the construction period [J]. Journal of Hydroelectric Engineering, 2010, Vol. 29, No. 5: 125-130 (in Chinese).

[3] Zhang Guoxin, Liu Youzhi, Ma Xiaofang. Investigation on low temperature pouring cracking in a concrete dam [J], Water Resources and Hydropower Engineering, 2010, 41(7): 45~48 (in Chinese).

[4] Wang Jiachun, Yan Peiyu. Crack risk Analysis of concrete structure at early age [J]. Journal of Basic Science and Engineering, 2006, Vol.30, NO.2:33-39 (in Chinese).

[5] Jin nanguo, Jin Xianyu, Wu Weihe, Tian ye. Application of equivalent age method for early age concrete temperature crack control[J]. Journal of Zhejiang University(Engineering Science) 2008, Vol.39.NO.1:29-35 (in Chinese).

[6] Zhang Jun， Qi Kun； Zhang Minghua. nonlinear temperature field and stress Calculation of early age concrete panel[J].Low Temperature Architecture Technology, 2011,NO.33.No.6:22-26 (in Chinese).

[7] C. Providakis , E. Liarakos T-WiEYE: An early-age concrete strength development monitoring and miniaturized wireless impedance sensing system[J]. Procedia Engineering, 2011, Vol.10 , pp:484-489.

[8] Zhang Guoxin, LIU Yi,ZHANG Chong,et al. Dynamic model for concrete early-age response of elastic modulus and its application [J]. Water Resources and Hydropower Engineering, 2011.3(42) :62-66 (in Chinese).

[9] China Institute of Water resources and Hydropower Research, Concrete structure temperature distribution, temperature gradient, thermal insulation effect and sun heat radiation detection method[p]. Chinese invention patent, NO:ZL201010221450.5.2012-1-25 (in Chinese).

[10] J.Zreiki, F.Bouchelaghem, M.Chaouche Early-age behaviour of concrete in massive structures, experimentation and modeling[J]. Nuclear Engineering and Design, 2010, Vol.240 (10), pp.2643-2654.

[11] Miguel Azenha , Rui Faria. Denise Ferreira Identification of early-age concrete temperatures and strains: Monitoring and numerical simulation[J]. Cement and Concrete Composites, 2009, Vol.31 (6), pp.369-378.

[12] Huang Dahai； Yang Hu； Su Junan. Analysis of early age concrete surface protection time in winter[J]. Water Resources and Hydropower Engineering, 2008, Vol.39,No.9:34-39 (in Chinese).

[13] Wang Guojie, Zheng Jianlan， Site monitoring and analysis of stress related strain of early age concrete[J]. Engineering Mechanics, 2009 Vol.26. No.9,PP:61-67 (in Chinese).

[14] Zhou Yunchuan , Bai Liang , Yang Shengyuan , Chen Guting Simulation Analysis of Mass Concrete Temperature Field[J]. Procedia Earth and Planetary Science, 2012, Vol.5 , pp.5-12.

EVALUATING RELATIVE CONDITION OF REPEATED STRUCTURAL ELEMENTS WITH PERFORMANCE-BASED TESTING

Spenser Anderson[1]
Peter Orme[2]
Ziyad Duron, PhD[3]
Randy Nash, P.E.[4]
Kevin Marshall, P.E.[5]

ABSTRACT

Performance-Based Testing (PBT) is a new experimental technique developed to assess the performance of a structure by measuring its response to a high-amplitude short-duration impulsive load. PBT was recently conducted on Wanapum Dam, a large, power-generating dam whose spillway consists of several nearly identical monolith-pier elements, one of which was known to contain a large fracture. Testing is described in detail and several Performance Indicators are developed to assist with the determination of the relative health of similar structures. A Stiffness Indicator is defined as the ratio of the Shock Response Spectrum (SRS) of the applied force and the SRS of the top-of-pier displacement. An Amplification Indicator is defined as the ratio of the SRS of the top-of-pier acceleration and the SRS of the grout gallery acceleration. The ability of each of these indicators to reveal meaningful differences in condition across piers is evaluated, and the advantages and disadvantages of each technique are discussed. The Amplification Indicator is found to be most effective for highlighting the damaged pier. A companion paper details how the results of PBT can be applied to the validation of simple lumped-element or complex finite element numerical models.

INTRODUCTION

The presence of a large horizontal fracture in one of the spillway monoliths at Wanapum Dam prompted a failure investigation that ultimately resulted in a remediation and repair program. The program included the use of post-tension anchors to enhance the stability of the pier. Sequencing this repair strategy across the spillway was aided by evaluating relative pier condition to determine whether any of the piers present a higher risk for failure. If a pier or a group of piers was found to be of similar condition to the known damaged pier, a suitable repair sequence could be followed. A new testing technique is described that produced transient response behavior in the spillway which was used to develop Performance Indicators for evaluating relative pier condition. The technique employs an excitation source that delivers a high amplitude, short duration impulse load to the pier and induces high amplitude, broad frequency, transient responses in the spillway. Analysis techniques that are normally reserved for aerospace applications are

---

[1] De Pietro Engineering Fellow, Harvey Mudd College, Claremont, CA, 91711, sanderson@hmc.edu
[2] De Pietro Engineering Fellow, Harvey Mudd College, Claremont, CA, 91711, porme@hmc.edu
[3] Professor of Engineering, Harvey Mudd College, Claremont, CA, 91711, duron@hmc.edu
[4] Grant County Public Utility District Engineer, Beverly, WA 99321, rnash@gcpud.org
[5] Hydro Eng. Manager, Grant County Public Utility District, Beverly, WA 99321, kmarsha@gcpud.org

applied to the measured transient responses in the dam to produce a series of indicators for evaluating a pier's structural health or condition relative to the damaged pier.

PERFORMANCE-BASED TESTING ON WANAPUM DAM

Performance-Based Testing (PBT) is a relatively new technique for evaluating the performance of large civil infrastructure, with previous field investigations completed on two large suspension cable bridges and on two large concrete (one multiple arch, and one RCC arch) dams (Chiarito and Duron 2009, Duron et al 2014, Goldkamp et al 2014). The testing technique employs an excitation source capable of producing impulsive (or explosive) loading conditions characterized by large amplitude, broad-band frequency content (Hill et al 2012). The resulting response measurements are considered as impulse responses that contain inherent behavior characteristics of the structure. The shock-like nature of the response transients facilitates analysis techniques that produce indicators of performance or relative condition. The literature provides numerous references to structural health monitoring (SHM) techniques for large civil structures that show performance indicators based on changes in resonant frequency and damping are weakly correlated to damage states or condition (Doebling et al 1996). However, performance indicators that are based on high amplitude, broad-band frequency transient behavior have been shown to produce early warning of impending collapse in burning buildings (Duron 2011), and, when combined with shock-based analysis techniques (adapted from aerospace engineering applications), can produce indication of changing structural conditions.

Excitation Source – Cold Gas Thruster (CGT)

PBT employs an excitation device that delivers a high amplitude broad frequency load that mimics impulsive loading conditions. The device, originally developed for evaluating impact resistance of instrumentation inside aircraft, consists of a cylindrical metal chamber with a thin metal diaphragm sandwiched between a mated pair of converging and diverging nozzles. An inert gas (nitrogen gas was used at Wanapum Dam) is used to pressurize the chamber until the diaphragm ruptures or bursts, delivering a concentrated – high amplitude and short duration force pulse – to the pier. The device is referred to as a Cold Gas Thruster (CGT) and is shown installed on the top of a pier in Figure 1.

The CGT is instrumented with a load cell sandwiched between the back plate and the mounting bracket on the axis of the CGT, in order to capture the force pulse at firing. A high-g shock accelerometer is located near the axis of the CGT on the mounting bracket to record the acceleration of the CGT after firing. CGT force pulse and shock acceleration measurements acquired on Piers 11 and 4 are shown in Figure 2 through Figure 4, and were typical of the quality and type of measurements observed during testing.

Figure 1. CGT mounted to a bracket and bolted onto the top of a pier.

The portion of the force pulse that corresponds to initial diaphragm rupture (burst) is the first peak (see Figure 3), typically 3 ms in duration. Peak force levels were limited to 4000 lbf on Piers 7, 6, and 4, but were increased to 8000 lbf for other piers, and shock acceleration levels (see Figure 4) reached approximately 50 g (Pier 4) to 80 g (Pier 11). The high-g shock occurs over an extremely short period of time, contains high frequency content, and is localized primarily to the CGT mounting bracket.

Pier Response Instrumentation – Triaxial Accelerometer

Pier responses were acquired using servo-force balance accelerometers that were mounted in a triaxial configuration, acquiring responses in the stream, cross stream, and vertical directions. Two triaxial configurations were used – one on the top of the pier and another inside the gallery – on each pier. A photo of one of the triaxial configurations used at Wanapum Dam is shown in Figure 5.

At each measurement location, the mounting plate is simply placed on top of the concrete surface. Leveling screws are adjusted to ensure accelerometers are properly oriented in the stream, cross stream, and vertical directions. Instrument sensor sensitivities, amplification, filtering, and sampling rate (samples per second) are listed in Table 1.

Table 1. Instrumentation and Acquisition Parameters

Sensor	*Sensitivity (Nominal)*	*Gain*	*Filter Cutoff (hz)*	*Sampling Rate (sps)*
Load Cell	.11v/lbf	1	n/a	10,000
Shock Accel	49.53mv/g	0.1	1000	10,000
Pier Accel	1.2v/g or 11.8v/g	1 or 10	1000	10,000

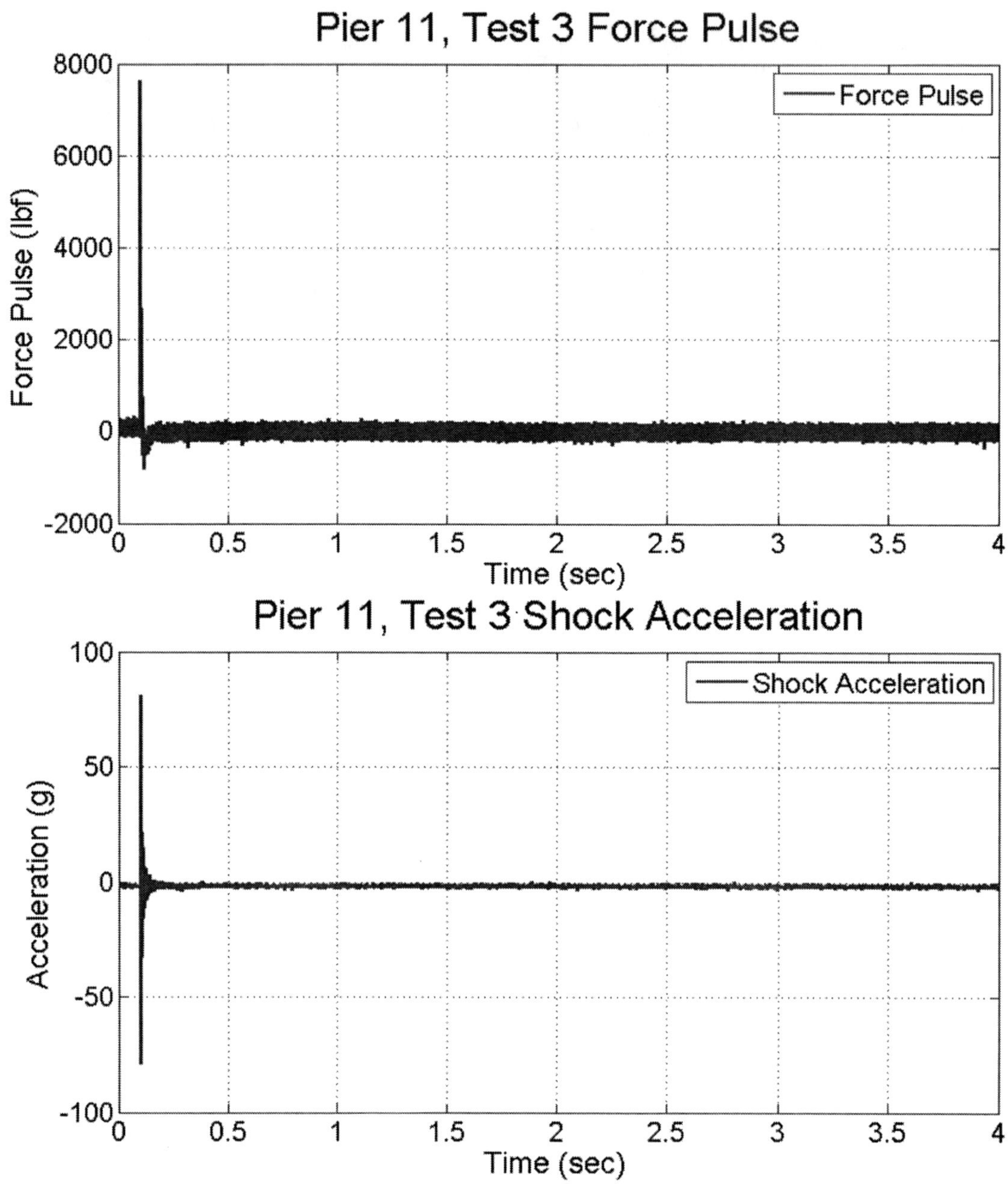

Figure 2. CGT force pulse and shock acceleration measurements on an undamaged pier.

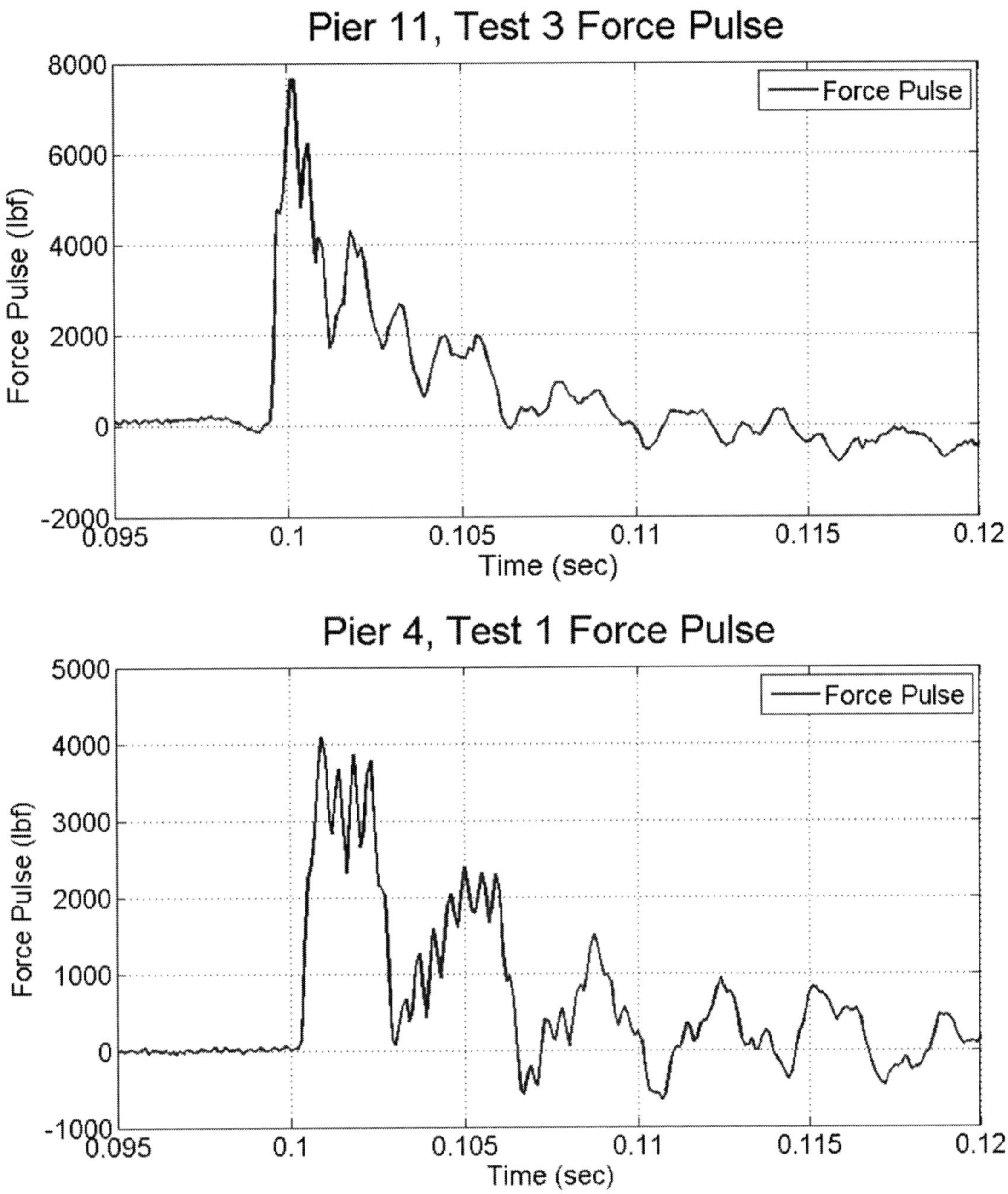

Figure 3. Comparison of CGT force pulses on Piers 11 (top) and Pier 4 (gallery) at diaphragm burst.

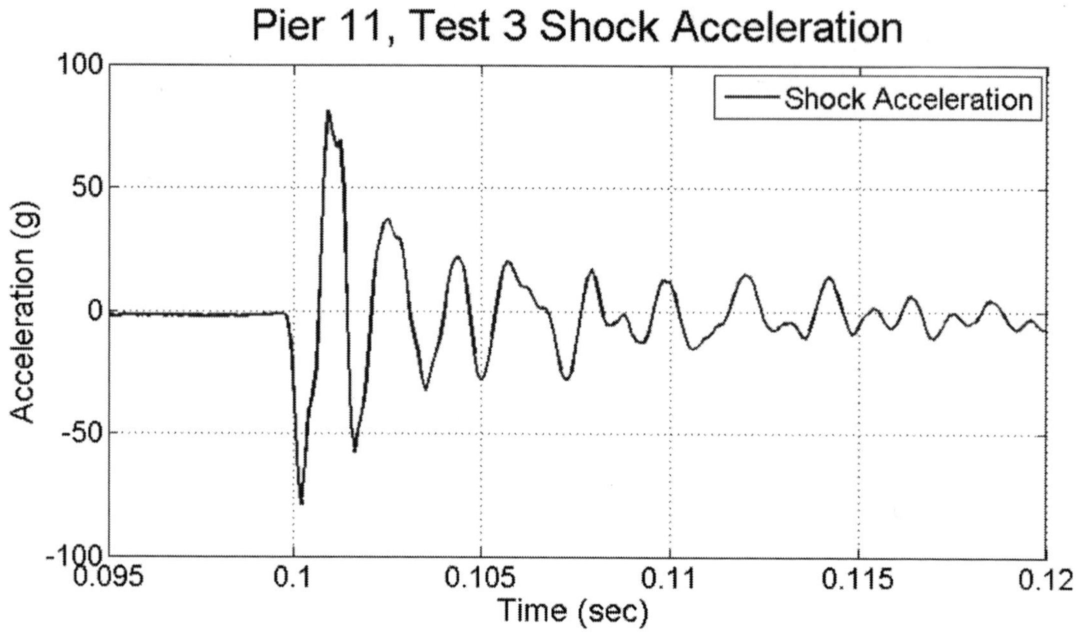

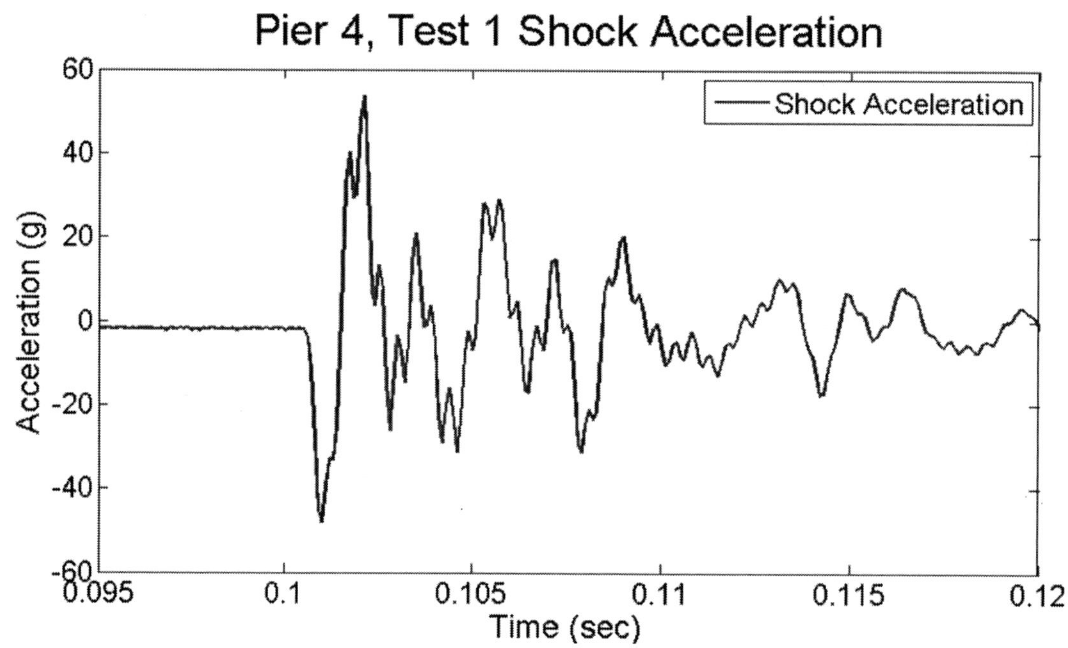

Figure 4. Comparison of CGT shock acceleration responses on Piers 11 (top) and Pier 4 (gallery) at diaphragm burst.

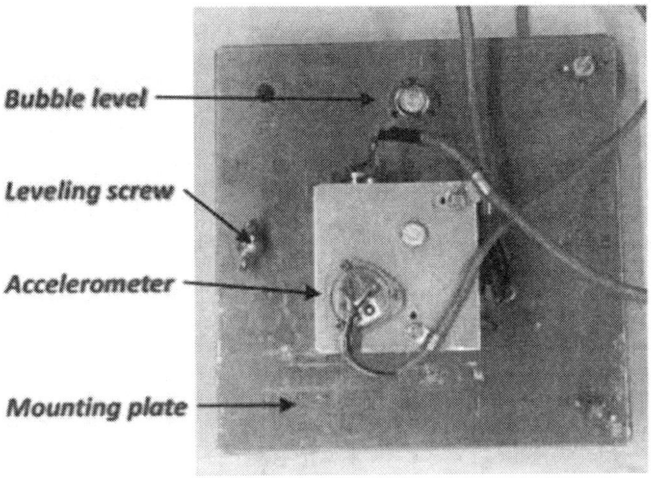

Figure 5. Typical triaxial acceleration configuration used for monitoring top and gallery measurement locations on a pier.

Measurement Layout and Description of Completed Tests

A schematic of the measurement layout for each pier tested is shown in Figure 6.

The CGT was mounted in the center of each pier with the nozzle pointing downstream, and triaxial accelerometers on the top of each pier were placed directly downstream of the CGT location. Gallery accelerometers were installed directly below the top-of-pier monitoring location. All 6 accelerometers recorded data simultaneously during each firing of the CGT.

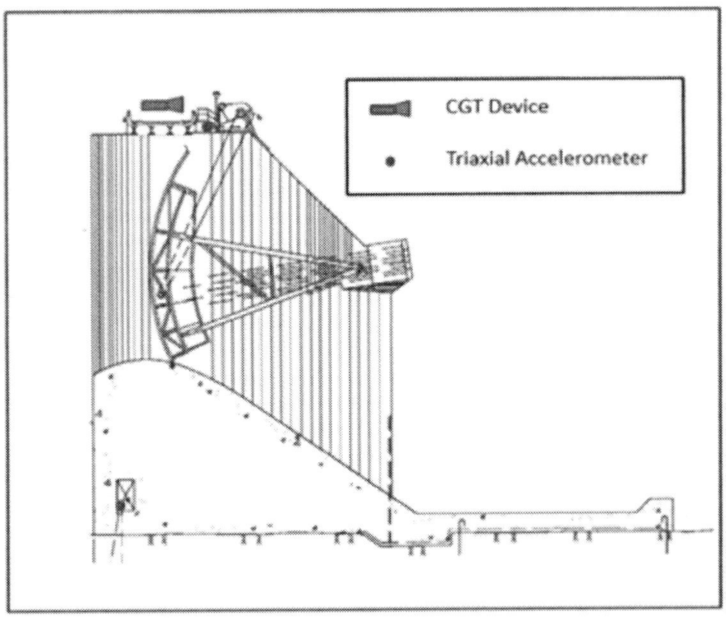

Figure 6. Measurement layout for each pier tested.

The damaged pier and three other piers were tested in March of 2014. In May of 2014, all piers were tested except for Pier 4, the pier with the large horizontal fracture, which was inaccessible due to construction work. A sample acceleration time record is shown in Figure 7, including the stream, cross stream, and vertical components recorded at the top of a pier and in the gallery beneath the pier.

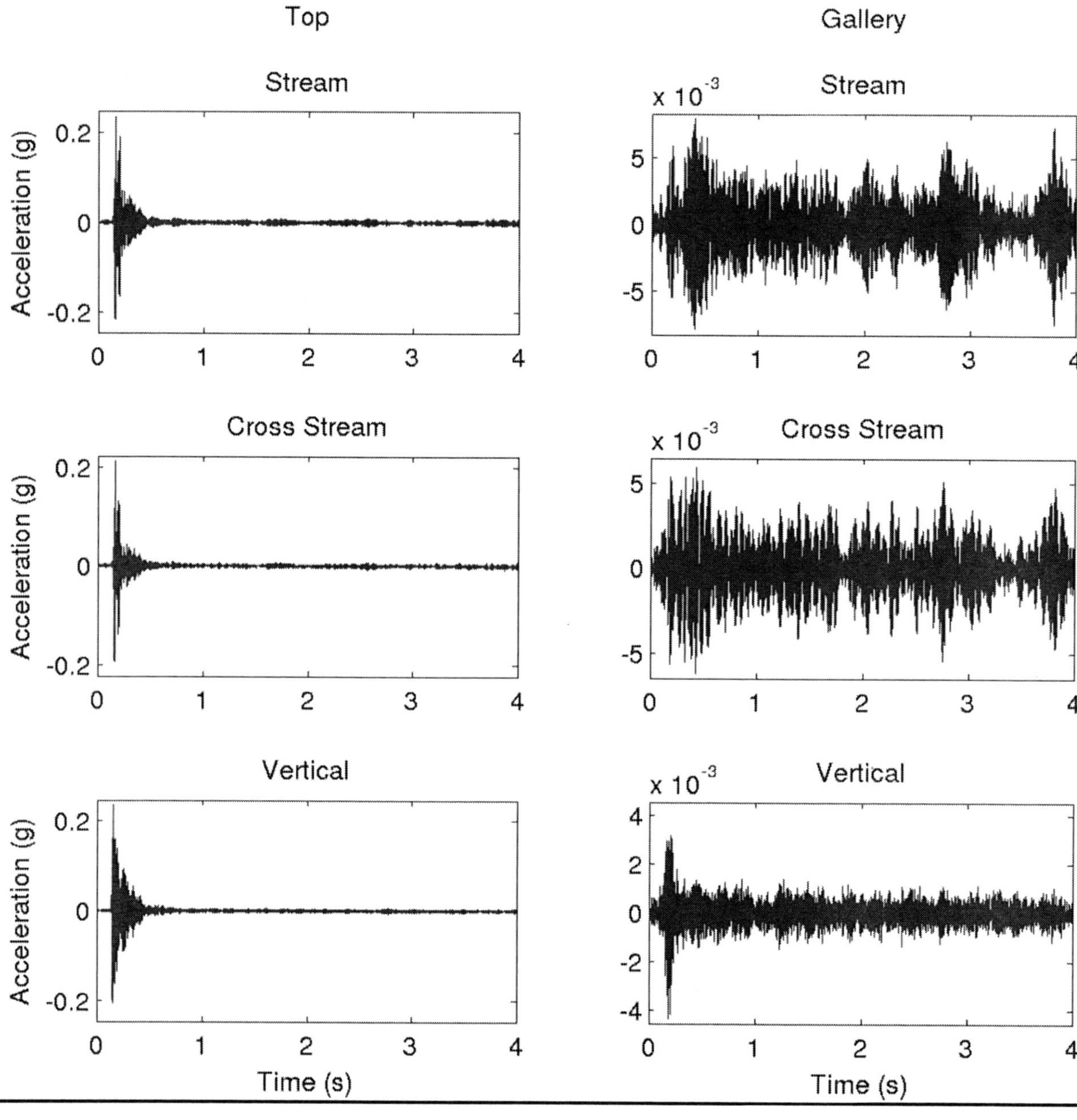

Figure 7. Sample acceleration time histories during a CGT blast. Left: Time histories recorded on the top of the pier. Right: Time histories recorded in the gallery beneath the pier.

EFFECT OF DAMAGE ON STRUCTURAL DYNAMICS

Damage to any civil structure can change the basic properties of the structure including mass and stiffness, and may even alter properties associated with the structure's ability to dissipate energy. The challenge for confirming that damaged conditions exist inside the structure rests with extracting relevant information that is influenced by the particular damage. Traditional field tests, developed in the 1960s, which employed steady state vibration generators, often produced response information that could not be tied to damaged conditions due to the low amplitude, band-limited excitations delivered to the structure. Locations inside structures where damage has occurred are characterized by relative motions near or around the damaged condition that are transient, containing higher frequency content due to motions that are influenced by the localized damaged conditions. As described below, PBT applied to the spillway targets this transient behavior, and corresponding Performance Indicators can provide effective means for highlighting relative condition between the spillway piers.

PERFORMANCE INDICATORS

The location and characteristics (i.e. time, frequency, decay) of the PBT responses prompt the calculation of Performance Indicators in order to highlight the relative condition between spillway piers. For example, amplification response in a pier can be estimated by the ratio of the measured responses acquired on the top and in the gallery of the pier. Pier stiffness can be estimated as the ratio of force applied to the pier and the corresponding displacement of the pier. Examples of Performance Indicators are provided below:

- Directional Amplification in each pier defined as the ratio of top to gallery response for each orientation (stream, cross stream, vertical) excited by the CGT (oriented in the stream direction). Large values would suggest increased response levels are present in the pier, indicating reduced strength in a particular orientation.
- Combined Amplification in each pier defined as the quadrature sum of the directional amplifications indicating overall reduced strength.
- Standard Error associated with the Combined Amplification for each pier. Wide variations and large values suggest deviations from expected normal pier behavior.
- Directional Stiffness in each pier defined as the ratio of the force delivered to the pier to the displacement of the (top) pier. Piers with large valued Directional Stiffness are considered stronger than their adjacent piers.
- Torsional Vulnerability defined as the ratio of the lateral response orientations (ie. stream and cross stream). Values near unity may suggest higher tendency for torsional behavior.
- Vertical Uplift indicating dominant vertical response defined as the ratio of vertical to minimum lateral response, where the minimum occurs in either the stream or cross stream direction.

- Performance Indicator (PI) Rank Order defined as the numerical ranking of each pier based on PI values.
- Overall Rank Ordering defined as the sum of the individual PI Rank Order values.

Two of these indicators, an Amplification Indicator and a Stiffness Indicator, are discussed in more detail in the following section.

<u>**Amplification Indicator**</u>

The Amplification Indicator is meant to determine the relative condition of each pier by examining frequency spectrum derived amplification (the ratio of the response at the top to the response in the gallery) in the pier. Specifically, it is defined as

$$Amplification\ Index(\omega) = \frac{SRS_{top}(\omega)}{SRS_{gallery}(\omega)}, \quad (2)$$

where the SRS is the shock response spectrum of the transient signal recorded in response to a CGT blast. This indicator can also be reduced to a single number that is a measure of the "size" of the indicator by integrating it, producing an Amplification Index. This is defined as

$$Amplification\ Index = -\int_{\omega_1}^{\omega_2} \frac{SRS_{top}(\omega)}{SRS_{gallery}(\omega)} d\omega \quad (3)$$

Three directional Amplification Indicators can be computed associated with the stream, cross stream, and vertical directions recorded in the piers.

<u>Experimental Results.</u> Directional Amplification Indices were calculated for several tests on each pier and were summed in quadrature to obtain the Combined Amplification indices shown in Figure 8.

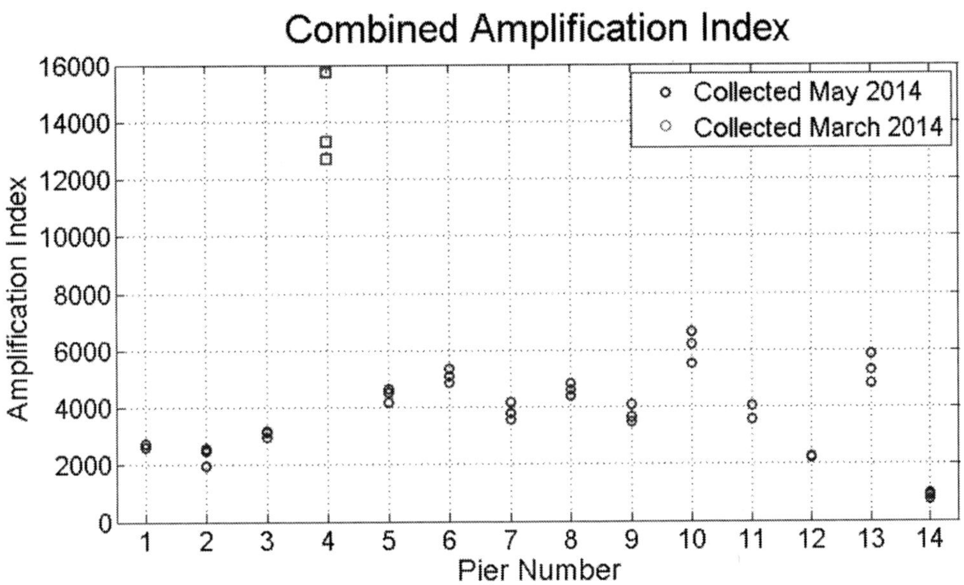

Figure 8. Amplification Indices for several tests on each pier.

It is apparent that the Amplification Index of Pier 4, which is known to have a large fracture, is much larger than that of the other piers. It is also seen that while there is variability across tests on a single pier, this variability is smaller than the variability between piers. The large index values associated with Pier 4 mask the grouping that is present across the remaining piers since closer examination highlights groupings of similar PI index values. For example, piers 5, 6, 7, 8, 9, and 11 all exhibit similar PI values and variations. Piers 2, 3 and 12 form another grouping, and piers 10 and 13 would form another. Although included in Figure 8, index values associated with monoliths 1 and 14 are not grouped along with the piers due to the structural differences in the monoliths. These groupings and their significance are discussed later in the paper.

Stiffness Indicator

The Stiffness Indicator is meant to determine the relative condition of each pier by examining a frequency spectrum derived stiffness of the pier. In the frequency domain, stiffness can be defined by the ratio of spectral load to displacement using a shock response spectrum (SRS) analysis technique. In the frequency domain, spectral displacement is obtained directly from the acceleration spectrum through division by $(j\omega)^2 = -\omega^2$. Specifically, the Stiffness Indicator is defined as

$$Stiffness\ Indicator = -\frac{SRS(\omega)_{load\ cell}}{\frac{SRS(\omega)_{accel}}{\omega^2}}, \quad (1)$$

where $SRS(\omega)_{load\ cell}$ is the shock response spectrum of the time history from the CGT load cell, $SRS(\omega)_{accel}$ is the shock response spectrum of an accelerometer time history.

Three directional stiffness indicators were computed using each of the stream, cross stream, and vertical response measurements.

The Stiffness Indicator can be reduced to a single number through integration producing a frequency-averaged index value. This produces a single number Stiffness Index defined as

$$Stiffness\ Index(\omega) = -\int_{\omega_1}^{\omega_2} - \frac{SRS(\omega)_{load\ cell}}{\frac{SRS(\omega)_{accel}}{\omega^2}}\, d\omega \tag{3}$$

Experimental Results. Stiffness Indices were calculated for several tests on each pier. The resulting values for all tests with adequate data quality are shown in Figure . These values were produced by calculating a Stiffness Index in all three directions in response to a stream load, and then adding the resulting indices in quadrature.

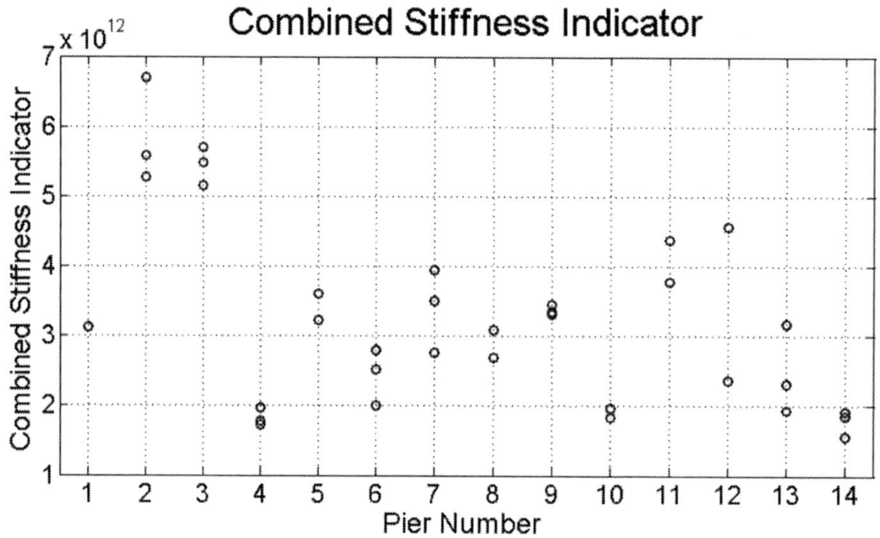

Figure 9. Stiffness Indices for several tests on each pier.

The known damaged condition in Pier 4 is believed to account for the relative low stiffness value indicated, and by comparison, this would suggest that Pier 10 has a comparatively low stiffness as well. The stiffness values associated with monoliths 1 and 14, and piers 2 and 13 are not considered comparable to the remaining piers due to their structural dissimilarities. Piers 5, 6, 7, 8, 9, and 11 indicate similar stiffness values, while Pier 3 demonstrates relatively higher stiffness. The index value for Pier 12 illustrates how not every Performance Indicator may produce repeatable findings, which is addressed below.

DISCUSSION

A good performance indicator should meet several criteria, including the ability to distinguish between damaged and undamaged conditions. This means that the values calculated for a damaged structure should be very different than those calculated for an undamaged or relatively healthy structure. Variations in estimates obtained from repeated testing should also be small, indicating confidence in both the testing and analysis techniques. A number of Performance Indicators were evaluated and reviewed for Wanapum Dam, and of those considered, the Directional and Combined Amplification Indicators were judged to provide the highest degree of confidence. Specifically, the tests on Pier 4 repeatedly produced index values outside the range of the remaining piers, which was expected due to its damaged condition. In addition, the tests on Pier 4 exhibited the highest amplification and lowest stiffness index values of any pier, which provided a sanity check for the PIs developed for the dam.

A second damaged condition was discovered by the dam owner in Pier 2 which was not highlighted or identified from the original PI analyses performed. In part, this resulted from the fact that the PIs developed for Wanapum Dam were focused on the large horizontal fracture already present in Pier 4. As a result, these PIs were developed to find similar fractures in adjacent piers. The problem with this approach is now obvious: it biases the development of appropriate PIs. The discovery of a second fracture in another pier with different orientation (i.e. vertical compared to horizontal) has motivated two significant improvements for future PBT and evaluations of large civil structures:

- Performance Based Testing on a structure will be completed by exciting the structure in each direction (stream, cross stream, and vertical) whenever possible. This will allow the full set of indicators to be computed using three directions of force and three direction of measured response.
- Performance Indicators for each structure will be determined according to known Potential Failure Modes (PFMs) for the structure. PFMs will be determined from potential failure mode analyses or risk assessment workshops, but will form the basis for appropriate PIs for highlighting relative conditions within the structure itself.

CONCLUSION

Performance-Based Testing was performed on each of the piers of Wanapum Dam, one of which was known to contain a large horizontal fracture. Several performance indicators were developed, and their ability to determine the relative condition of the piers was assessed by examining their ability to identify the known damage condition. It was found that the Amplification Indicator, the frequency-dependent ratio of the responses at the top and in the gallery of the pier, and its corresponding Amplification Index (the integral of the Amplification Indicator) was most successful. It clearly distinguished the damaged pier from piers that are not known to be damaged, and the variation in values calculated across tests on a single pier was small.

The inability to identify a second damaged state in the dam has prompted two significant improvements to PBT for future evaluations. The first is a modification to the testing procedure in which the excitation source delivers impulse loading in each of the three directions (two lateral, one vertical) to the structure. This will allow estimates of performance indicators to be acquired for principal and coupled behavioral characteristics in the structure. Equally important is the understanding that Performance Indicators for structures should not be based on known conditions, but, rather, they should be based on identified Potential Failure Modes (PFMs) of the structure. When applied in this manner, PIs can provide meaningful information on conditions that may lead to potentially catastrophic failure, and, thereby, can be used to develop short and long term mitigation strategies for the structure.

ACKNOWLEDGEMENTS

The testing completed at Wanapum Dam represents the first of its kind on a large concrete dam. The short schedule due to the current conditions in and around the dam necessitated extensive assistance from the engineering and support staff at the dam. The following personnel have contributed to this effort, and their contribution is gratefully acknowledged.

David Vedder, Electrical Engineer

Joe Powers, Electrician
Shane Lee, Electrician
Rob DeGroat, Electrician
Alex Davidson, Electrician
Chris Goe, Electrician

Tom Marty, Mechanic Foreman
Rick Collier, Mechanic
Brandon Horner, Mechanic
Andy Martin, Mechanic
William Gifford, Mechanic

Brad Martinez, Warehouse Foreman
Karen Mus, Warehouseman
Gary Gosvener, Warehouseman

Special recognition goes to Mr. Randy Nash for his technical and logistical support. This effort would not have been possible with out the technical leadership and support of Mr. Kevin Marshall.

REFERENCES

Chiarito, V., Z. Duron, and et.al, "Dyna mic Testing of the Cable Suspension System at the Waldo-Hancock Bridge," US Army Corps of Engineers, ER DC/GSL report, September 2009.

Doebling, S. W, C. R. Farrar, et.al, "Dam age Identification and H ealth Monitoring of Structural and Mechanical System s from Changes in Their Vibration Characteristics: A Literature Review," Los Alamos National Laboratory, report no. LA-13070-MS, 1996.

Duron, Z., K. Crane, and V. Chiarito, "Suspension Cable Response to Explosive Like Loading Conditions," US Army Corps of Engineers, ERDC/GSL report, January 2012.

Duron, Z., "Stability Monitoring of Burning Structures Based on Fire-induced Vibration Monitoring," Conference Proceedings of the Society for Experimental Mechanics, pp. 149-156, 2011.

Goldkamp, M. R., T. H. Savage, J. B. Edelman, P. J. Orme, Z. Duron, and M. Knarr, "Performance-Based Analysis of a Large Concrete Dam," San Francisco, California, 34th Annual USSD Conference, 2014.

Hill, N., K. Lownsbery, Z. Duron, and V. Chiarito, "Cold Gas Thruster Redesign and Characterization," US Army Corps of Engineers, ERDC/GSL report, January 2012.

ANALYTICAL AND NUMERICAL MODELING OF THE WANAPUM PIER-OGEE SPILLWAY

Peter Orme[1]
Spenser Anderson[2]
Ziyad Duron, PhD[3]
Randy Nash, P.E.[4]
Kevin Marshall, P.E.[5]

ABSTRACT

A failure investigation of Wanapum Dam, a large, power-generating dam, was initiated after the discovery of a significant horizontal fracture in one of the piers that comprise a portion of the spillway section of the dam. Performance-Based Testing (PBT) of the damaged structure and field condition assessment were conducted and are described in a companion paper. After the completion of PBT, lumped analytical systems modeling as well as finite element modeling techniques were used to explore the dynamic characteristics of the spillway section. Models were developed and tuned using measurements collected during PBT. The modeling procedures revealed important dynamic behavior that may contribute to root-cause for the damaged conditions in the spillway and that may be relevant for long-term remediation plans for the spillway. For example, analytical models were used to highlight the excitation and absorption characteristics in the spillway piers. Evidence of vibration absorption is shown in both the PBT and flow-induced responses measured in the field. This behavior could introduce dynamic loads at the monolith-pier interface, exacerbating pre-existing conditions that may have influenced the horizontal fracture. Validation of a finite element model against observed behavior is demonstrated based on resonant frequencies, spacing, and energy dissipation in the spillway.

INTRODUCTION

Performance Based Tests (PBT) conducted on Wanapum Dam form the basis for the analytical and numerical modeling techniques described herein. Observational evidence of large amplitude, broad-band frequency, transient decay behavior (Anderson et al. 2015) was used to motivate the development of lumped as well as finite element models of the dam's spillway. From each model, fundamental dynamic response characteristics are highlighted that point to conditions that may have contributed to fracture formation in one or more of the spillway piers. To contextualize the development of these models, physical elements of the spillway are described.

[1] De Pietro Engineering Fellow, Harvey Mudd College, Claremont, CA, 91711, porme@hmc.edu
[2] De Pietro Engineering Fellow, Harvey Mudd College, Claremont, CA, 91711, sanderson@hmc.edu
[3] Professor of Engineering, Harvey Mudd College, Claremont, CA, 91711, duron@hmc.edu
[4] Grant County Public Utility District Engineer, Beverly, WA 99321, rnash@gcpud.org
[5] Hydro Eng. Manager, Grant County Public Utility District, Beverly, WA 99321, Kmarsha@gcpud.org

Flow through the spillway at W anapum Dam is controlled through the operation of twelve large Tainter gates sandwiched between thirteen concrete piers. The piers are supported by concrete monoliths that include an ogee crest, as shown in Figure 1.

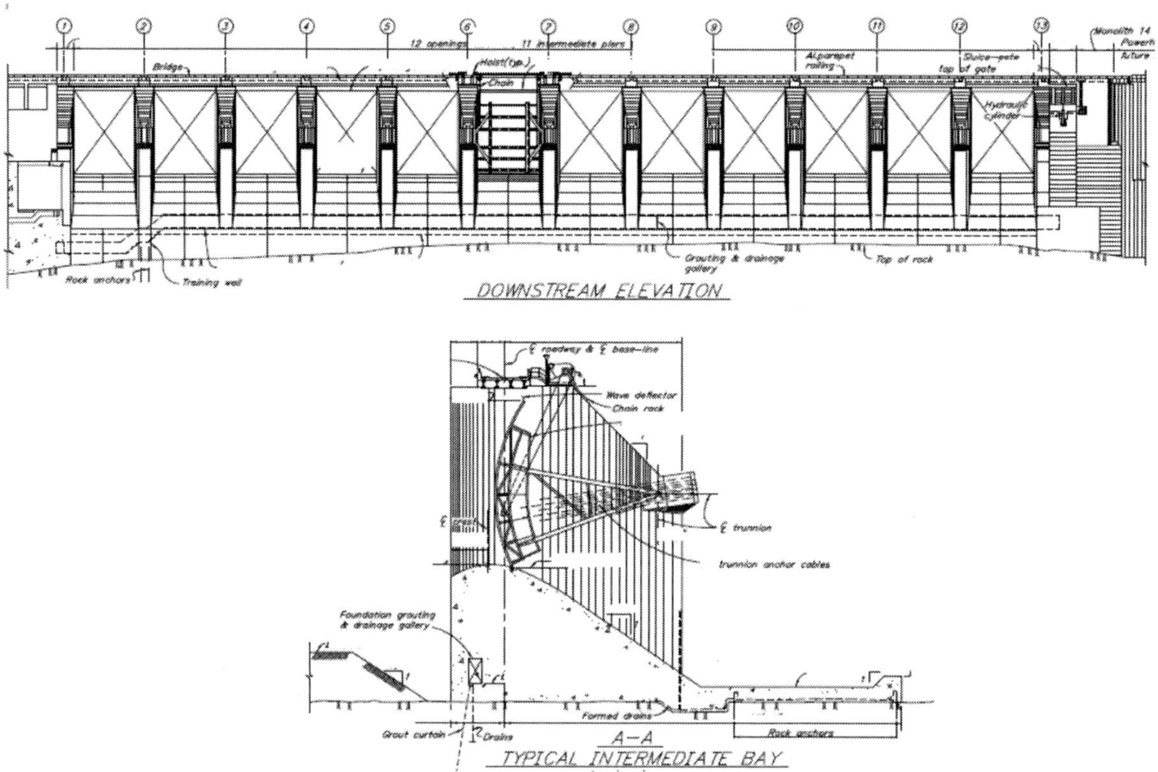

Figure 1. Spillway and pier configuration at Wanapum Dam

The repeated use of the same structural elements in the spillway's design contributes to the notion that the piers are structurally the same, and that a numerical model of a single monolith-pier configuration can be used to evaluate dynamic behavior across the entire spillway. However, experience with full-scale testing of large civil infrastructure and in the modeling of these systems suggest that the spillway is actually a coupled structural system whose components contribute to an integrated response behavior.

ANALYTICAL LUMPED ELEMENT MODELING

From a system modeling perspective, the spillway consists of a main system that can be characterized as a "gravity dam" defined by the monoliths. The monoliths are not keyed and each monolith supports one smaller coupled system, the pier. A simple lumped-element model can be used to describe a single monolith-pier configuration as indicated in Figure 2. In this representation, the main system mass (M) represents the monolith that is subject to base excitation, y, at a known frequency, f_y. A secondary lumped-element system can be attached to the main system, which, from a dynamics perspective, can be made to absorb certain vibrations in the main system. For the spillway, this secondary system represents the concrete pier attached to and supported by the monolith.

Vibration absorption of the main system results when the secondary system is tuned to the resonant frequency of the main system. To illustrate the relevance of this model to the spillway at Wanapum Dam, consider the following scenario.

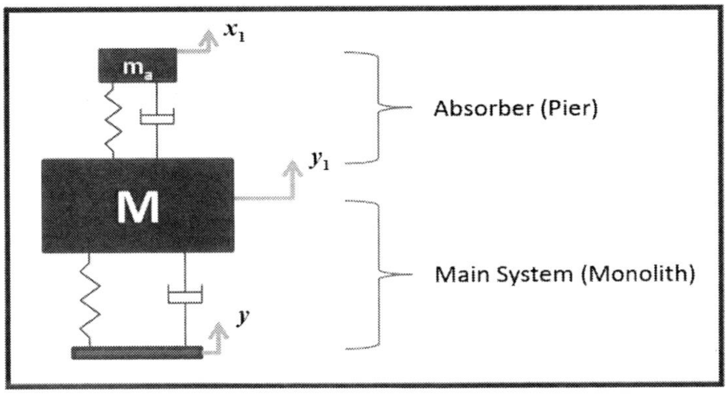

Figure 2. Lumped-element representation a monolith-pier configuration

A loading condition occurs which excites the spillway along the monolith-foundation interface. The frequency content of the base excitation extends to 20 Hz, and it is presumed that resonances in the monoliths as well as in the attached piers will be excited. When a pier resonance is near to, or coincides with a resonance in the supporting monolith, vibration absorption at that frequency in the monolith will result. To illustrate this the mass and stiffness were adjusted so that the fundamental mode for the pier and monolith are both 4.3 Hz[6] with a damping of 5% to produce the response curves in Figure 3. Notice the attenuation in the main system's response that occurs at 4.3 Hz, but at the cost of two new resonances as shown.

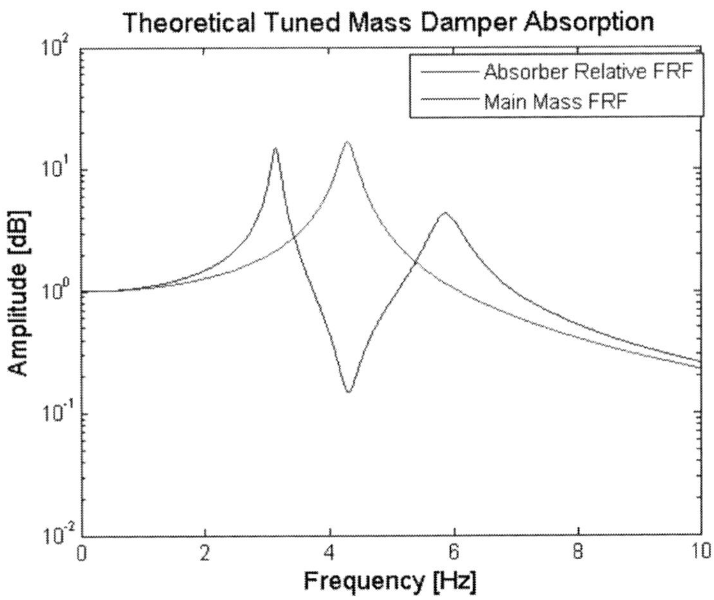

Figure 3. Tuned vibration absorber behavior

[6] This value is arbitrary for illustrating the behavior of the discrete model, but it also corresponds to known cross stream resonances in the piers found during PBT Phase II.

Extending this lumped-element model across the entire spillway results in the coupled model shown in Figure 4.

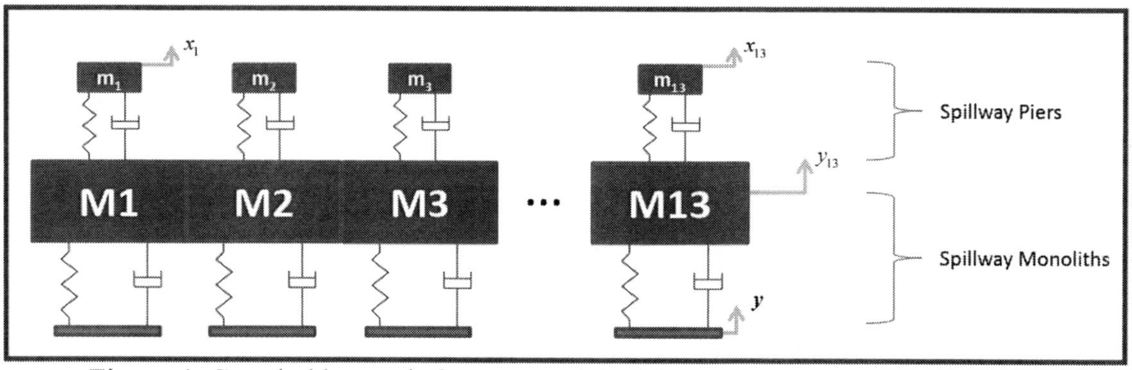

Figure 4. Coupled lumped-element model of the spillway at Wanapum Dam

Model parameters can be selected to represent each of the monolith-pier configurations in the spillway as uniform (i.e. same parameters for each configuration) or detuned (i.e. each configuration has a unique parameter set based on different material, geometric, or boundary conditions). Furthermore, the boundary interfaces (i.e. vertical contraction joints) between monoliths can be adjusted, but are assumed as rigid in Figure 4. Under these assumptions, all of the monolith responses (denoted as $y_{i,\ i=1\ to\ 13}$) are equal (and can be denoted as y_M) resulting in a 14 degree-of-freedom (dof) model. An expression can be derived for the frequency-dependent ratio of the monolithic response in the spillway (denoted as $Y_M(s)$) to the base excitation (denoted as $Y(s)$) as

$$\frac{Y_M(s)}{Y(s)} = \frac{s2\zeta\omega_n + \omega_n^2}{s^2 + 2\zeta\omega_n s + \omega_n^2 + \sum_{i=1}^{N_{Pier}} \left[\left(2\zeta_i\omega_i s + \omega_i^2\right)\left(\frac{M_i}{M}\right) \right] \left(1 - \frac{a_i}{b_i}\right)} \quad i = 1, 2, \ldots, N_{pier} \tag{1}$$

$$a_i = 2\zeta_i\omega_i s + \omega_i^2 \,, \; b_i = s^2 + 2\zeta_i\omega_i s + \omega_i^2 \,,$$

where N_{Pier} denotes the number of piers in the spillway (i.e. 13), and the mass ratios for each of the monolith-pier configurations are given by $\frac{M_i}{M}$[7]. The dynamic characteristics of each pier are represented by their individual natural frequencies and fractional damping (denoted as ω_i and ζ_i) pairs and since the monoliths are assumed to be rigidly connected to each other, a single parameter set (ω_n and ζ) can be used.

The analytical model can be used to evaluate spillway response behavior in any of the stream, cross stream and vertical directions as long as the appropriate resonances and damping estimates for each direction are employed. Damped natural pier frequencies were determined from PBT responses acquired in the cross stream direction and were found to vary from 3.2 Hz-5.3 Hz across the spillway, as listed in Table 1. It is important

[7] A mass ratio of 0.4 (pier to monolith mass ratio) was used, determined from numerical models of a single pier and monolith.

to note that Piers 1, 12, and 13 are dissimilar in construction to the rest of the piers and that one pier exists in known damaged conditions.

Table 1. The fundamental cross stream frequency of each pier as determined with PBT.

Pier Number	Fundamental Cross Stream Frequency (Hz)
1	3.2
2	3.97
3	3.51
4	4.73
5	5.34
6	3.83
7	4.43
8	4.35
9	5.23
10	4.12
11	4.45
12	4.95
13	4.27

Based on a frequency analysis of a finite element model of a single stand-alone monolith, the fundamental cross stream resonance of the monolith was found to be 26.5 Hz. Pier responses were examined to see if the vibration absorption frequency characteristics of the piers as shown by the analytical model were also evident in the field PBT results from the actual spillway. The comparison in Figure 5 was taken on Pier No. 6 and is offered as evidence of vibration absorption characteristics in the spillway.

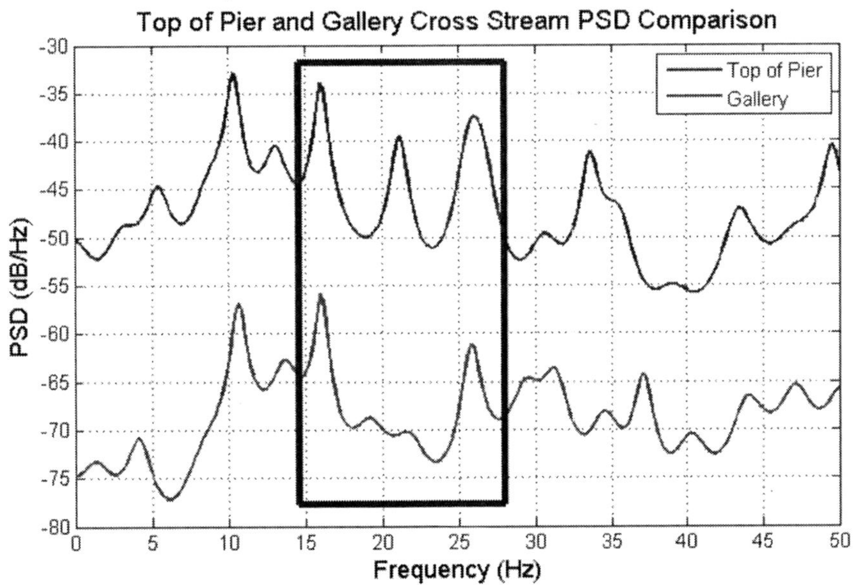

Figure 5. Top of pier and gallery cross stream Power Spectral Density comparison from Pier 6. Vibration absorption by the pier is observed at a resonance around 22 Hz

A rectangular box in Figure 5 identifies the frequency band where absorption effects of the pier on monolith response are presumed to exist. The comparison highlights the resonant peak in the pier at approximately 22 Hz (blue curve) that coincides with an attenuated response in the monolith (red curve). The split resonances that are associated with ideal vibration absorption (originally shown in Figure 3) are not as pronounced due to the distribution of (lower and higher order) resonances in both the pier and monolith. To focus on how the range of pier resonances can affect monolith behavior, the lumped-element model of the spillway was used to produce the result in Figure 6.

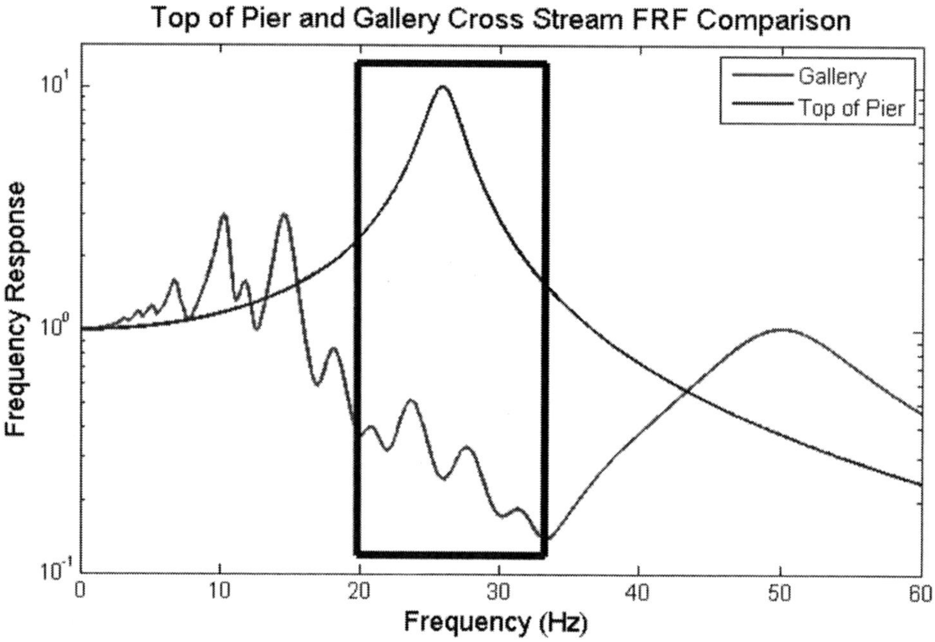

Figure 6. Analytical FRFs with one of thirteen piers tuned to the monolith resonance

The analysis illustrated in Figure **6** was performed by tuning the 13 piers in the model to resonances ranging from 3.3 Hz-30 Hz, with the monolith resonance at 26 Hz. The monolith response is attenuated in the vicinity of its fundamental resonance, but shows elevated response well below 26 Hz in this example. This is due to the presence of pier resonances that induce forced response behavior in the monolith. As many as seven pier resonances are estimated to exist below the fundamental monolith resonance that could, in effect, act as vibration generators that deliver dynamic loads to the monolith. The frequency response comparison below 15 Hz (i.e. "outside the box") in Pier No. 6 (Figure 5) is considered to be evidence of this forced response characteristic in the monolith.

Pier resonant behavior below the fundamental monolith resonance contributes to a dynamic loading condition that results in elevated, forced monolith response behavior. This condition may adversely affect pre-existing conditions inside the monolith. Pier resonant behavior aligned with fundamental monolith resonances act as vibration absorbers that attenuate the response of the monolith while exposing the pier-monolith "interface" to a dynamic load.

Spilling Induced Responses

The described vibration absorption behavior was observed in the flow induced motion recorded during spilling operations. As adjacent tainter gates are raised or lowered, the spillway piers exhibit increased or decreased magnitude vibrations, while the monoliths they rest on exhibit no comparable change. This effect is shown in Figure 7 below.

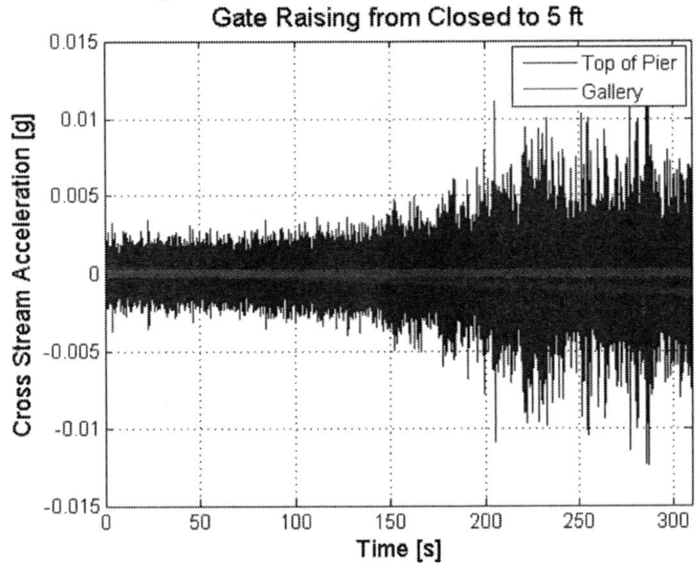

Figure 7. Effect of gate movement on spilling induced vibrations in pier and monolith

A Frequency Response Function of a pier with respect to the gallery can illustrate the relative response behavior as a function of frequency and spilling. Near the first expected monolith resonance in Figure 8, there is a 20 dB difference in the relative pier response when the gates are opened for spilling. The vibration absorption of the pier on its supporting monolith causes elevated response at the pier's resonance during spilling.

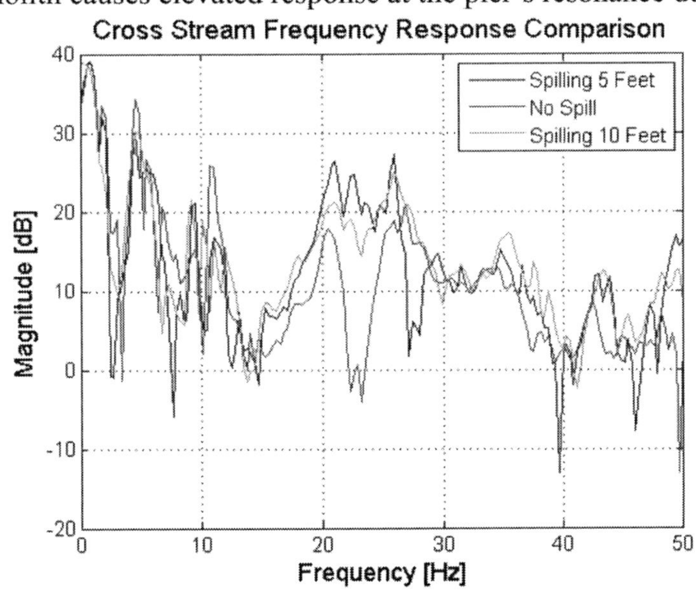

Figure 8. FRF of pier to gallery, indicating absorption at the first monolith resonance

FINITE ELEMENT MODELING

PBT responses can be used to validate finite element models for evaluating performance limits against design or hazard loading conditions (Edelman, et al. 2014). An advantage of using PBT responses over those acquired from traditional testing techniques is the ability to validate models using both frequency and time domain information. An illustration of this validation is given with a finite element model of a single pier.

Abaqus was used to develop a numerical model of a single representative, though not exact, pier, shown in Figure 9. The dimensions, geometry of the plan view, and the monolith-pier interface curved profile were approximated. The concrete was modeled as a homogeneous linear elastic material with Rayleigh damping, and the interface was fixed. No reservoir water was attached to the pier, and no nonlinear mechanisms were included in this model.

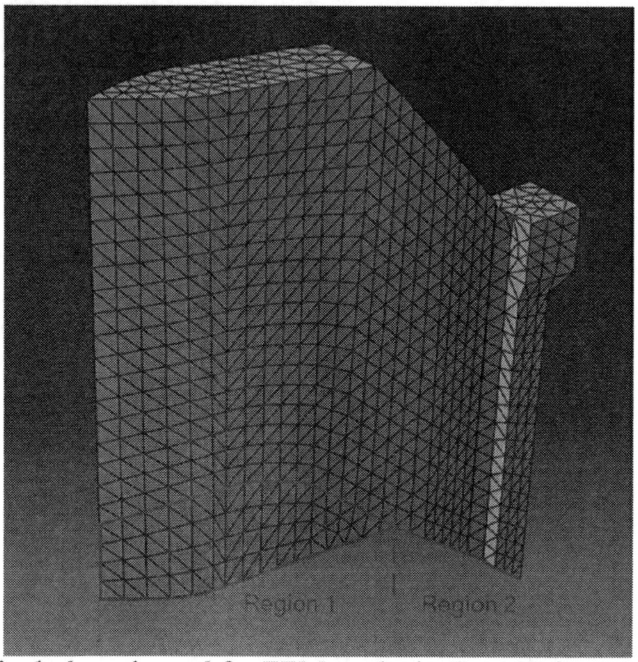

Figure 9. Tetrahedral mesh used for FEM analysis. Two regions whose boundary conditions can be varied independently are shown

Material properties for concrete (i.e. Young's Modulus, mass density, and Poisson's ratio) were varied to achieve a match between model and observed resonant frequencies. Rayleigh damping parameters, α and β, were selected to achieve an equivalent damping ratio of 5% at 4 Hz and 40 Hz using the relationship

$$\zeta = .5\left(\beta\omega + \frac{\alpha}{\omega}\right) \quad (2)$$

where ζ represents fractional damping and ω is the modal frequency in radians per second (Mills-Bria 2006). Model parameters that achieved a reasonable match with observed pier resonances are listed in Table 2.

Table 2. List of material properties used in numerical model

Model Parameter	Value Used
Young's Modulus	$3.6 * 10^6$ psi
Density	$2.15 * 10^{-4}\ lb_f s^2$
Poisson's Ratio	0.2
Alpha	2.28
Beta	$3.61 * 10^{-4}$

The comparison of resonant frequencies derived from field tests on structures acts as a reasonable first step toward model validation. For any model, it is always possible to achieve a good match with the fundamental resonance, and perhaps even with one of the higher resonances. However, it is usually the spacing between resonances that is difficult to reproduce in the absence of other indications of observed structural behavior. This is because resonant spacing is determined by the distribution of stiffness and mass in the actual structure which is sometimes difficult to reproduce in the model. Often inaccurate stiffness or boundary (interface) conditions in the model are to blame for improper resonant spacing. The spectral responses in Pier No. 6 (Figure 10) illustrate the typical spacing encountered during PBT in the spillway, and was used to evaluate the model's ability to reproduce pier resonant values and spacing, shown in Table 3.

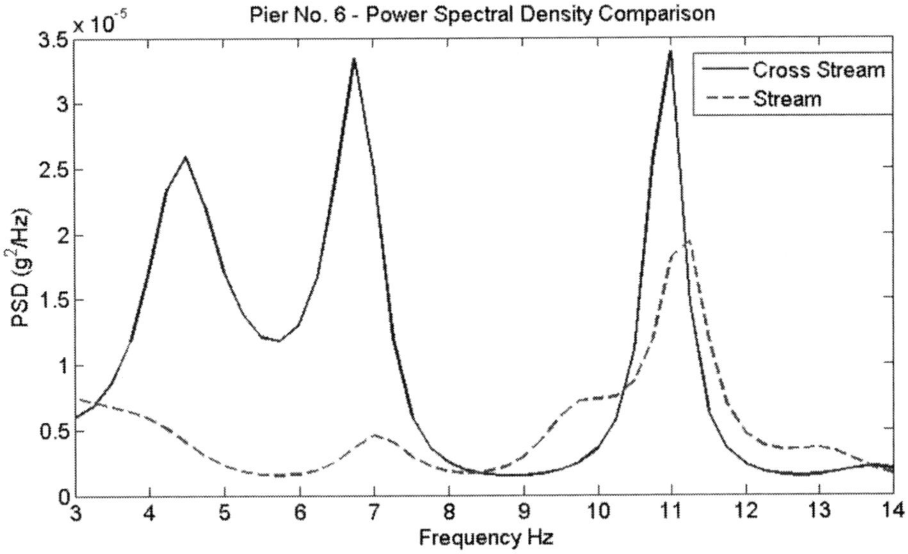

Figure 10. Pier 6 stream and cross stream PSD used to identify PBT resonant frequencies

Table 3. Comparison of measured and model-generated resonant frequencies

Resonance Number	PBT Data	Fixed Base	Free Toe End
1	4.5	4.36	4.31
2	7.0	8.42	7.73
3	10.5	14.82	11.38
4	11.0	15.28	11.78

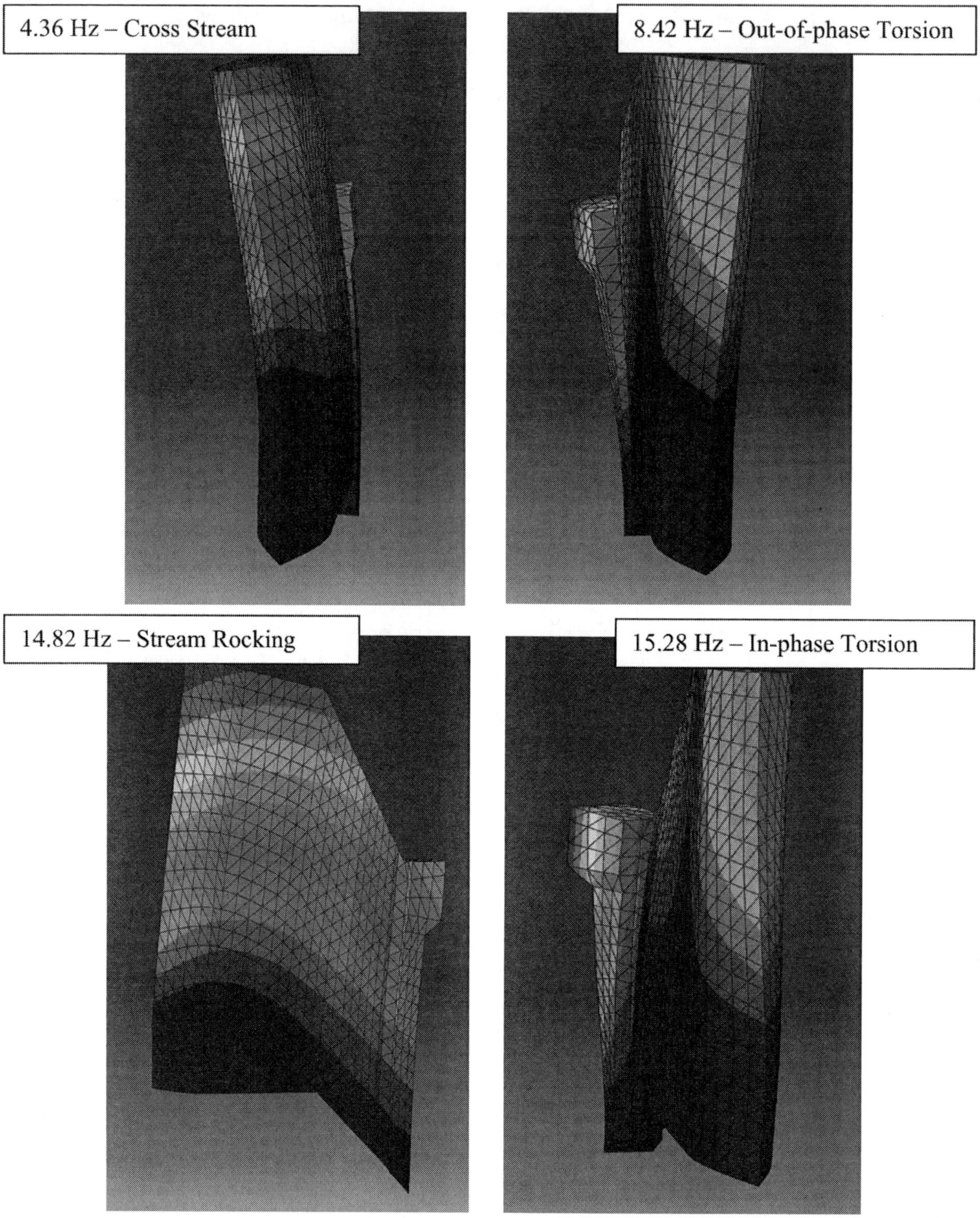

Figure 11. The first four mode shapes generated by the fixed base model

The comparison in Table 3 considers two different methods of representing the monolith-pier interface conditions. The interface location loosely represents the termination level of the rebar embedded in both the monolith and the pier. The first condition, fixed base, considers both Regions 1 and 2 from Figure 9 as fixed. While the first resonant frequency matches well, the spacing between the first and subsequent resonances does not match well with the measured results. However, when only Region 1 is fixed and the downstream interface, Region 2, is released (free toe end, mimicking flexible or soft interface conditions), a better agreement for both resonant frequencies and their spacing is achieved. The mode shapes corresponding to the fixed base model's first four resonant frequencies are shown in Figure 11. Construction records indicate a monolithic pour along the interface, though these results suggest measureable differences in fundamental pier resonances exist in the spillway and could be indicative of varying monolith-pier interface conditions.

Model Performance

A unique feature of PBT is the ability to use measured time history responses to evaluate a model's energy dissipation characteristics. This is made possible as a result of PBT's ability to induce impulse response behavior in a structure that has transient peak and decay behavior not otherwise available from traditional testing techniques. Contained within the transient responses are:

- The transition from forced to free vibration behavior (available as a result of the short duration, large magnitude excitation force pulse delivered to the structure),
- A broad frequency content that provides evidence of high, medium, and low frequency behavior in the structure, and
- Decay behavior that varies with both frequency and time.

PBT on a structure can be simulated by subjecting the numerical model to the measured force pulse applied to the structure. The measurement used is obtained from the load cell in the Cold Gas Thruster (CGT) and is applied to a node in the model coincident with the CGT attachment location on the top of the pier during testing. Acceleration time histories predicted by the model can be computed at any node in the model, allowing direct comparisons with actual acceleration responses acquired during PBT in the pier. The first step in this comparison examines the model's decay characteristics, specifically the time it takes for the transient induced by the CGT load to return to ambient (pre-load) levels. This is directly compared to observed decay behavior and is entirely based on the relative accuracy of the model's energy dissipation capabilities as compared to those in the real structure.

The remaining step in the model's validation is examining the model's frequency performance. In a typical seismic evaluation, an earthquake's spectral content is reported to 100Hz (a record's PGA is tied to this frequency) and there is typically no attempt to validate the model's performance at or near 100Hz. The broad band frequency content in PBT responses extends beyond 100Hz and is well suited for checking the model's frequency performance. To do this, the measured PBT responses are low-pass filtered until a cutoff frequency is found that produces a good match between peak responses. This cutoff frequency defines the bandwidth over which the model captures observed behavior. The predicted model responses in the stream, cross stream, and vertical directions at the node coincident with the measured location on the pier are shown in Figure 12.

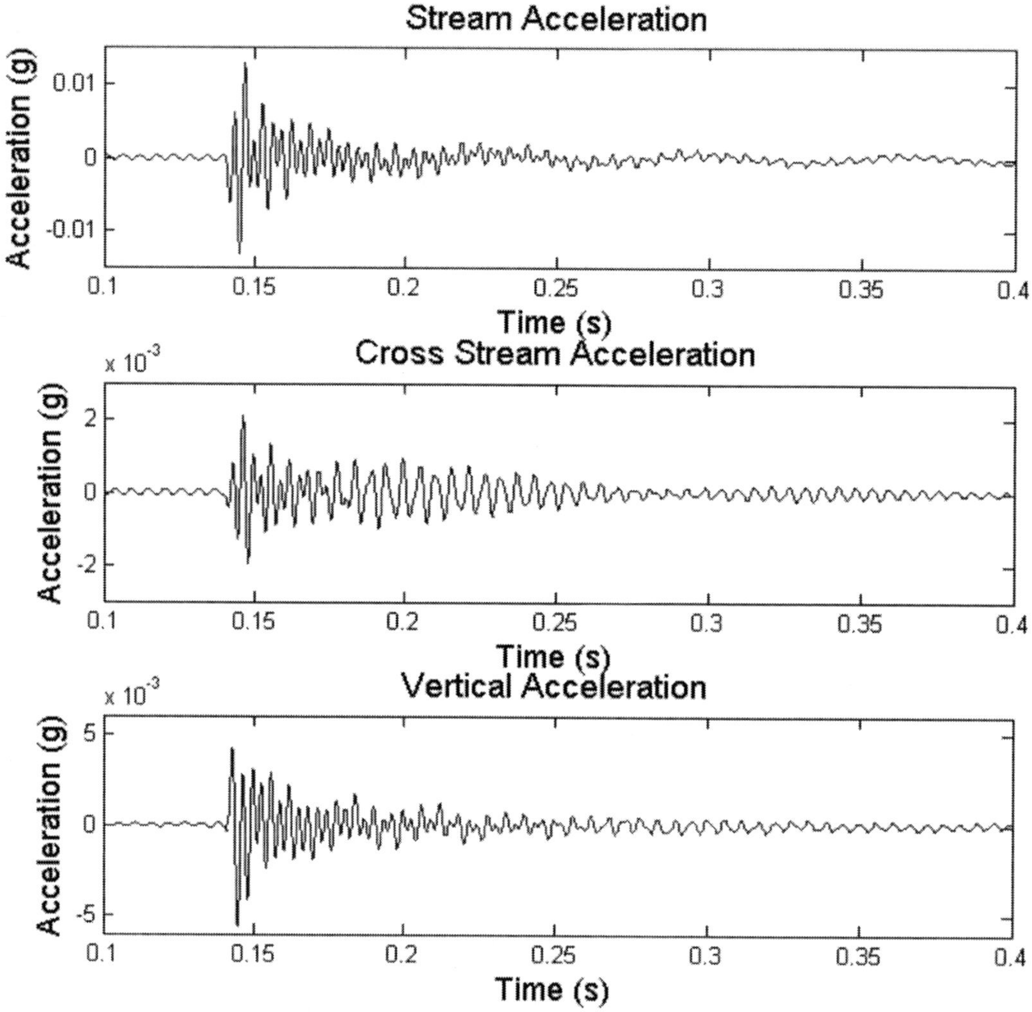

Figure 12. Stream, cross stream, and vertical responses from the model with the properties described in Table 2

An overlay of the stream direction model response and a PBT measurement from the same test as the force pulse is shown in Figure 13. The measured response has been low-pass filtered with an eighth-order Butterworth, zero-phase distortion digital filter with a 100 Hz cutoff. The comparison shown is considered sufficient to lend confidence to this model's performance up to 100 Hz. As more details (i.e. physical mechanisms) are included in the model, the comparison can be re-evaluated to ensure adequate model performance is maintained.

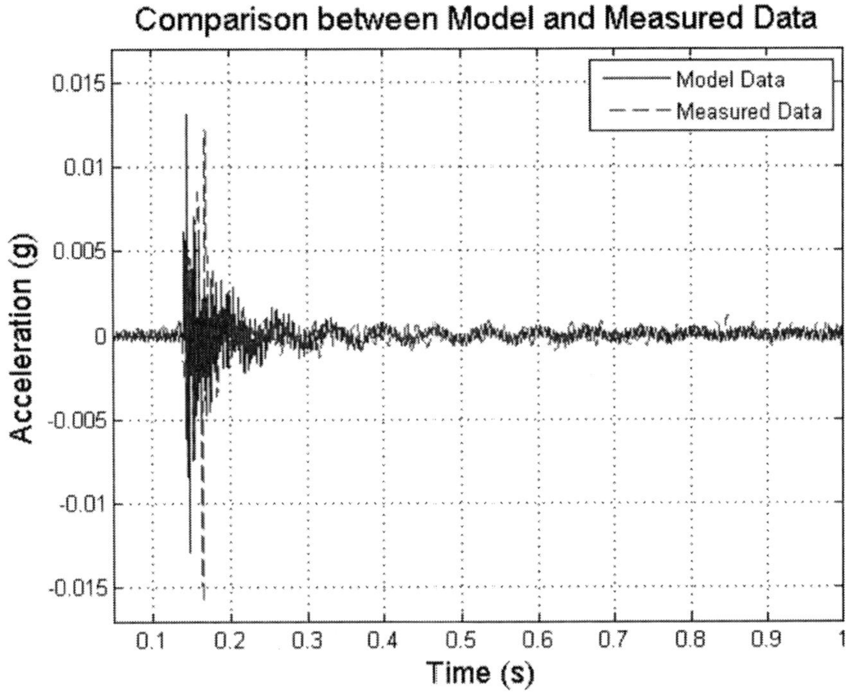

Figure 13. Comparison of measured PBT data filtered at 100 Hz and Model output data

CONCLUSION

This paper demonstrates the value Performance-Based Testing can add to understanding fundamental behavioral characteristics in a dam. When PBT measurements are used alongside analytical and numerical models, they can be used for validation of dynamic properties of the models. Demonstrations are given of both validation of a finite element model's impulse response and also corroboration of vibration absorption seen in the analytical model. The absorption behavior in the piers is offered as a potential contributor to horizontal fracture planes that developed in the Wanapum Spillway. In conjunction with the Performance Indicators derived from PBT measurements and described in the companion paper, PBT-informed modeling added significant value to the failure investigation at Wanapum Dam. As PBT on large structures is still at an early stage, further work can be done to refine and develop the process and the tools that can utilize the unique measurements it produces.

ACKNOWLEDGEMENTS

The testing completed at Wanapum Dam represents the first of its kind on a large concrete dam. The short schedule demanded by the current conditions in and around the dam necessitated extensive and full support of the engineering and support staff at the dam. The following personnel have contributed to this effort, and their support is gratefully acknowledged.

David Vedder, Electrical Engineer

Joe Powers, Electrician
Shane Lee, Electrician
Rob DeGroat, Electrician
Alex Davidson, Electrician
Chris Goe, Electrician

Tom Marty, Mechanic Foreman
Rick Collier, Mechanic
Brandon Horner, Mechanic
Andy Martin, Mechanic
William Gifford, Mechanic

Brad Martinez, Warehouse Foreman
Karen Mus, Warehouseman
Gary Gosvener, Warehouseman

Special recognition goes to Mr. Randy Nash for his technical and logistical support. This effort would not have been possible with out the technical leadership and support of Mr. Kevin Marshall.

REFERENCES

Anderson, S. et al. "Evaluating Relative Condition of Repeated Structural Elements with Performance-Based Testing". Louisville, KY: 35th Annual USSD Conference, 2015.

Edelman, J. B., et al. "Performance-Based Analysis of a Large Concrete Dam". San Francisco, CA: 34th Annual USSD Conference, 2014.

Mills-Bria, B., L. Nuss, D. O'Connel, and D. Harris. "4.0 Material Properties." *State-of practice for the Nonlinear Analysis of Concrete Dams at the Bureau of Reclamation*. Denver, CO: U.S. Dept. of the Interior, Bureau of Reclamation, 2006. 11-23.

WANAPUM FUTURE UNIT INTAKE DYNAMIC ANALYSIS FOR RISK INFORMED DECISION MAKING

James H. Rutherford, P.E.[1]
Kevin Marshall, P.E.[2]
Dave Mishalanie, P.E.[3]
Dan Curtis, P.Eng.[4]
Gurinderbir Sooch[5]

ABSTRACT

Grant County Public Utility District No. 2 (the District) engaged Hatch Associates Consultants Inc. (Hatch), to perform non-linear time history analysis of the Wanapum Future Unit Intakes (FUIs) using FLAC 2D as part of a risk informed review of the Wanapum Dam. The stability of the FUIs under earthquake loading relies on existing post-tensioned anchors to resist sliding and overturning.

In addition to reporting on our initial risk informed review of the FUIs, this paper presents some of the challenges faced during the analysis including how to model fully-grouted anchors that cross the foundation/rock interface, account for foundation rock roughness and interface geometry, and assess post-seismic stability. The results of Hatch's analysis are being incorporated into a prototype Risk Informed Decision Making (RIDM) process that is underway with participation of the District, FERC, and other engineering consultants.

INTRODUCTION

Grant County Public Utility District (the District) and Hatch conducted an evaluation of the seismic safety of the Wanapum Dam Future Unit Intake (FUI) Monoliths using a recent site-specific Probabilistic Seismic Hazard Analysis (PSHA) and Risk Informed Decision Making (RIDM) methods. The RIDM methods were established by the District and Chelan Public Utility District (Chelan PUD) and their consultants in collaboration with the Federal Energy Regulatory Commission (FERC). The results from our initial RIDM evaluation and screening level structural analysis indicate that the seismic performance of the Wanapum Future Unit Intake Monoliths are likely to meet prototype dam safety tolerable risk criterion established by the District and FERC specifically for this study. This project is considered to be a prototype risk evaluation because it is one of the initial trial runs conducted by a FERC licensee using RIDM methods to assess dam safety evaluations.

---

[1] Hatch Project Manager, Hatch Associates Consultants, Inc., Seattle, WA, jrutherford@hatchusa.com
[2] Engineering Manager, Grant County PUD No. 2, Beverly, WA, kmarsha@gcpud.org
[3] Project Manager, Grant County PUD No. 2, Beverly, WA, dmishal@gcpud.org
[4] Hatch Dam Analysis Group Lead, Hatch Associates Consultants, Inc., Niagara Falls, ON dcurtis@hatch.ca
[5] Structural Analyst, Hatch Project Engineer, Hatch Associates Consultants, Inc., Niagara Falls, ON gsooch@hatch.ca

The District co-sponsored a Probabilistic Seismic Hazard Assessment (PSHA) for the mid-Columbia region with Chelan and Douglas PUDs. The study was conducted by Jack Benjamin Associates, et al., in 2012. PSHA calculations were performed for three evaluation sites in the mid-Columbia region. The calculations at each site were performed for soft rock site conditions (average shear wave velocity of the top 30 meter of a site (Vs30) of 760 meter per second). The ground motion hazard was computed in terms of spectral acceleration (SA) at 12 spectral periods (0.01, 0.05, 0.075, 0.1, 0.15, 0.2, 0.3, 0.5, 1, 2, 3, and 4 sec) where the peak ground acceleration (PGA) corresponds to 0.01 sec period. Shannon & Wilson prepared site specific response spectra and time histories that are being used in the analysis of the Future Unit Intakes.

BACKGROUND

The Wanapum Hydroelectric Project is part of the Priest Rapids Development on the Columbia River just downstream of the town of Vantage on Interstate 90. Wanapum Dam consists of an earthen embankment on the left abutment, a gravity section with fish facilities, a powerhouse with ten generation units, six future unit intakes (FUI), a reinforced concrete spillway structure with twelve radial gates, right gravity section with fish facilities, and another earthen embankment on the right abutment. The FUIs are identified on as-built drawings as being "Intakes" 11 through 16. In this paper, FUIs are identified by numbers 11 through 16.

Figure 1 is a site plan and key map of the Wanapum Project showing project features including the FUIs. Each FUI is 90 feet wide, 68 feet long in the upstream/downstream direction and 162 feet high. Each FUI has four pier walls with three 20-foot-wide openings, referred to as bays. FUI 11 was previously modified to incorporate a fish passage structure. Photo 1 shows Wanapum Dam and the location of the FUIs. Photo 2 shows the FUIs as viewed from the left abutment.

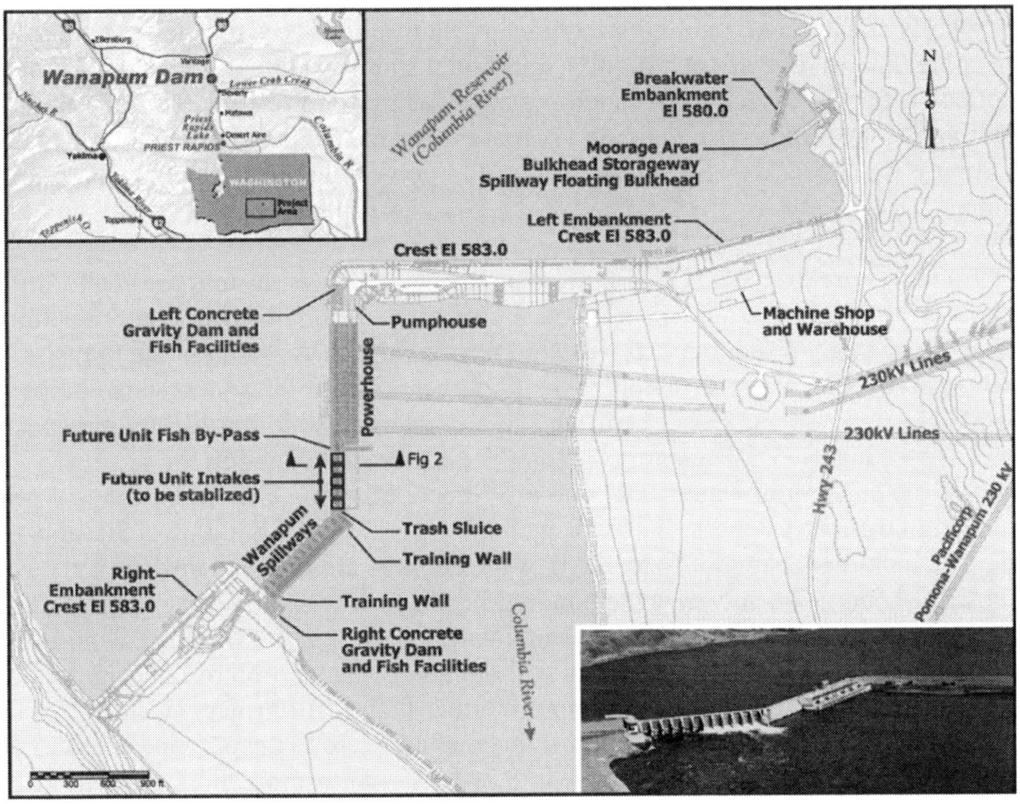

Figure 1. Wanapum Dam Location Map and Plan

Photo 1. Aerial View of Wanapum Dam from Downstream

As originally constructed, the FUI structures by themselves were not stable against overturning. At the time of construction, additional stabilization was provided by post-tensioned anchors (PT anchors) as a temporary measure for a few years prior to expanding the powerhouse and adding six more turbines and generators. The powerhouse was never expanded because the addition of the unit was not economically feasible.

The PT anchors are protected against corrosion by grout cover around the anchor in the drill hole. Although grout alone is not considered to be adequate corrosion protection for PT anchors by today's standards, it was common to rely on the grout alone to resist corrosion at the time of construction of the FUI anchors. The PT anchors are completely embedded in concrete and basalt bedrock and are not accessible for testing or inspection. Each of the FUIs has 13 PT anchors. Nine of the 13 are embedded in the upstream portion of the intake foundation slab, two are embedded in the end walls, and two are embedded in the interior piers. Each PT anchor consists of four bundles of 90 high-tensile strength ¼-inch-diameter solid wires. All of the anchors are inclined upstream six degrees. Without means to verify the condition of the anchors, the FERC directed the District to provide sufficient additional stabilization such that the structure would be stable under all static loading conditions without any consideration of the original PT anchors. The District initiated the Wanapum Future Unit Infill Project (Infill Project) to add mass to the structures to ensure that if the anchors were to become ineffective, the weight of the structure would be sufficient to prevent overturning and/or sliding instability under normal operations. (Rutherford, et al., 2011)

Photo 2. FUIs Viewed from the Left Abutment

Figure 2 shows the FUIs with stay-in-place stoplogs and concrete infill as completed in June 2010. This infill provides an overturning safety factor of 1.1 against overturning without including contribution of the existing PT anchors.

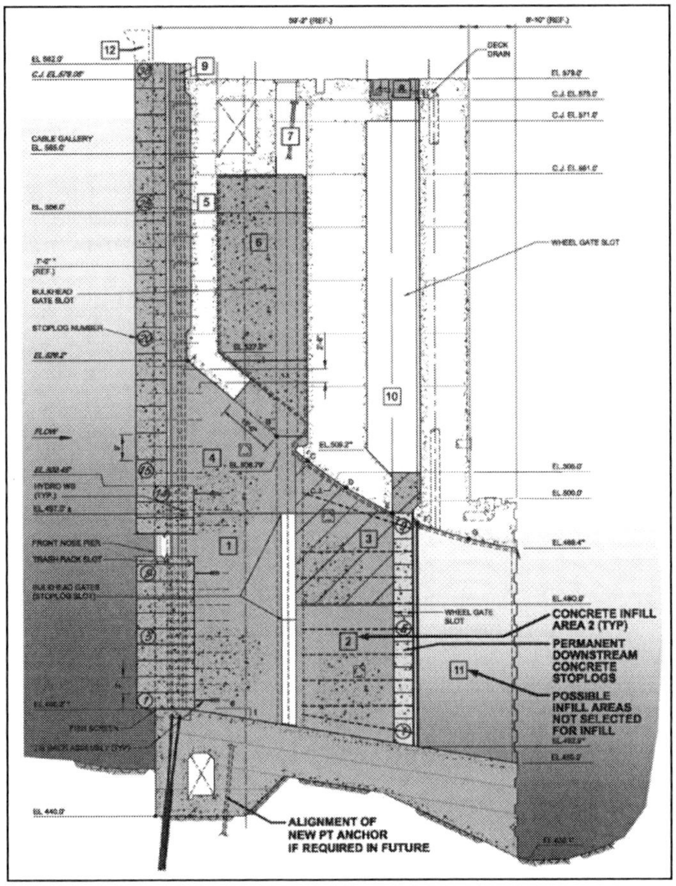

Figure 2. Section through FUI after Infill

ANALYSIS APPROACH

A response spectra analysis (RSA) was performed in 2009 using a linear elastic 2-D finite element model. The response spectra used were based on USGS recommendation scaled to a 0.2 g, 0.3g, and 0.41g PGA. The RSA results indicated that the tensile capacity of the existing post-tensioned (PT) anchors would be exceeded, even with a 0.2 g peak horizontal ground acceleration. It was concluded that, although these results revealed the existing PT anchors would become overstressed, the analysis by RSA would give conservative results since it does not account for non-linear effects, changes in the anchor to rock interface, and overall softening of the structure that would occur during the course of a severe seismic event. A preliminary 2D FLAC time history analysis (0.41pga) previously carried out in 2005 supported these conclusions since it showed that, for the infill case, the FUI would remain stable even if over 30% of the anchors were ineffective.

In advance of carrying out additional analysis as part of the RIDM process, we carefully defined the functional requirement for the FUI in the context of the entire project, established analysis boundary conditions, and reviewed the relevant failure modes.

Table 1 lists the functional requirements that were identified by the District, Jack Benjamin Associates, and Hatch. They are dependent upon river flow into the reservoir at the time of the earthquake, river flow coordination with other projects, upstream storage availability, post seismic damage to upstream and downstream facilities, and the post seismic condition of the Wanapum Dam and its components.

Table 1. FUI Structural System Functional Requirements

Category	Functional Requirement
Post-Seismic Access	Acceptably low risk of block movement that could prevent access to the spillway to allow local operation of the spillway gates that may be subject to loss of remote operating capability from the control room.
Control of Water	Acceptably low risk of an uncontrolled rapid release of the reservoir. Remain capable of retaining the reservoir to allow controlled discharge through the spillways and/or powerhouse.
Tolerable Risk	For seismic analysis, meet seismic tolerable risk of 50% as defined by USACE societal tolerable risk criteria.
Generation	Acceptably low risk to generation either due to loss of pool or need to draw down pool.

Table 2 lists the boundary conditions selected for this analysis. FUI earthquake performance has been judged primarily on maximum lateral shear displacement and cumulative strain. The maximum lateral shear displacement has been found to be the dominant performance metric. The upstream pool and downstream pool elevations were held at normal maximum and minimum elevations for this analysis. The use of average values would have reduced the overturning load on the FUIs under seismic loading however we elected to not account for this variable in the prototype analysis. Likewise, the drain efficiency was set at 75%. Based on data collected by Grant County, the drain efficiencies prior to infill were, on average, higher than this.

Table 2. FUI Structural System Boundary Conditions

Parameter	Assumed Value	Estimated Variability	Notes
Upstream Pool Level	El. 571.5	El.560 to El.571.5	Normal maximum operating level.
Tailwater Level	El. 481.5	El.481.5 to El.493	Normal minimum operating level.
Drain Efficiency	75%		Post EQ value influenced by extend of base cracking.
Effective Phi and Cohesion for Shear thru Intact Rock Knob (FUIs 12 and 13)	62 deg. 450 psi	- 2 to +4 deg	Estimated using Hoek-Brown Failure criterion.

Parameter	Assumed Value	Estimated Variability	Notes
Effective Phi for Rock – Foundation Contact (FUIs 14 -16)	56 deg.	- 2 to + 4 deg	Based on large scale waviness estimate.
PT Anchor Ultimate Strength	240 ksi	Not estimated	Original anchor design information and testing.
Anchor Grout Compressive Strength	2,500 psi	2500 to 3500 psi	Based on grout cube compressive test records.
PT Anchor Load	60% of Ultimate	60% to 70%	PT anchors were locked off at 70% and had assumed working stress of 60% of ultimate.
Initial PT Anchor Debond Length	8 ft	6 ft to 10 ft	Based on PT anchor prototype testing at site prior to FUI construction.
Base Shear Displacement Limit	1 in	1 in to 1.5 in	Performance Parameter: Professional judgment based on review of anchor shear displacement mechanisms.
Cumulative Strain Limit	3 %	Not estimated	Performance Parameter: Anchor wire was tested prior to construction. Stress/strain plot shows 3% as approximate limit.

For FUIs 12 and 13, the friction angle and the cohesion for the intact rock mass were estimated using the Hoek-Brown Failure Criterion. Input parameters for Roclab analysis (Rocscience, 2011) included a Geological Strength Index (GSI) of 50 based on construction photos that show a very blocky rock with up to four discontinuity sets. Although the discontinuities have a low persistence, incomplete drill logs described the basalt foundation rock as highly fractured with iron staining on joint surfaces. An intact uniaxial compressive strength of 14,500 psi was assumed although this is a low end of range for basalt as outlined in the literature (Peck, Hanson and Thornburn). A Hoek-Brown constant of 10 and a Disturbance Factor of 0 were assumed based on the fact that the dam was built in the 1960s and any poor quality and weathered rock would have been removed from the dam foundations as shown on the construction photos.

For FUIs 14 – 16 the effective phi for the foundation interface without the rock knobs was estimated based on review of construction photos and as-built foundation profiles available from District archives. There are two main components to the sliding friction angle: basic friction angle and roughness component. The estimated basic friction angle for concrete to rock is based on the values reported in the literature for basalt. Barton and Choubey (1977) report ranges of 35 to 38 degrees under wet conditions and 31 to 36 degrees under dry conditions. The roughness component will be estimated using Barton-Bandis approach. (Barton & and Bandis, 1990) (Bandis, 1982)

In order to assess the level of risk for not meeting baseline functional requirements, it was necessary to perform a non-linear time history analysis using a suite of time histories based on a PSHA specific to the site. The District co-sponsored a PSHA for the mid-Columbia region with Chelan Co. and Douglas Co. PUDs. PSHA calculations were performed by Shannon & Wilson for three evaluation sites in the mid-Columbia region.

A period range of 0.2 to 2 seconds was targeted for the FUIs monoliths based on the estimate for their natural frequencies. The seven spectrum compatible time histories that were developed for the 0.3 second period were used in the dynamic analysis of the Wanapum FUI.

The 2D FLAC (Fast Lagrangian Analysis of Continua, 2001) model was constructed to perform stability analysis of the Wanapum Future Units Intakes (FUIs) including the concrete infill that was completed in 2010. The analysis accounts for the effects of various static and dynamic forces acting on the FUIs. The anchorage system of FUIs is modeled with the cable elements in FLAC. Figures 3 and 4 present the two different 2D FLAC models with and without a rock knob in the foundation. As-built drawings and the photographs from the original construction show that FUIs 12 and 13 were constructed on a rock knob feature while FUIs 14 through 16 were built on a relatively flat rock foundation.

The interface elements connecting the FUI and foundation allow the structure to slip or separate from the foundation. Interface elements are modeled through the rock knob so that the model is able to capture the possible shearing failure if the intact rock shear capacity were to be exceeded. Hydrodynamic loads were applied using a Westergaard subroutine in the FLAC program.

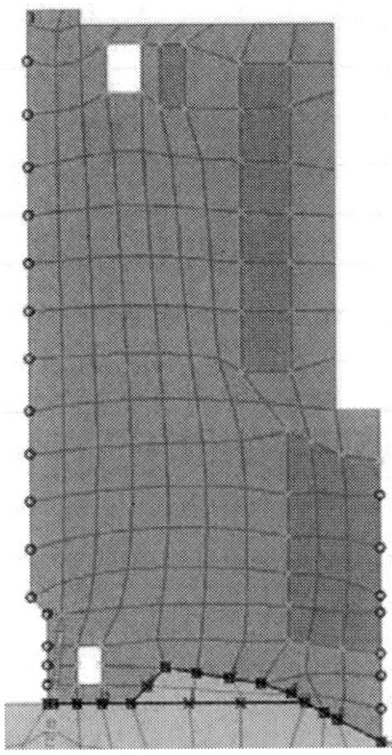

Figure 3. Wanapum FUI FLAC 2D model with Rock Knob (FUIs 12 & 13)

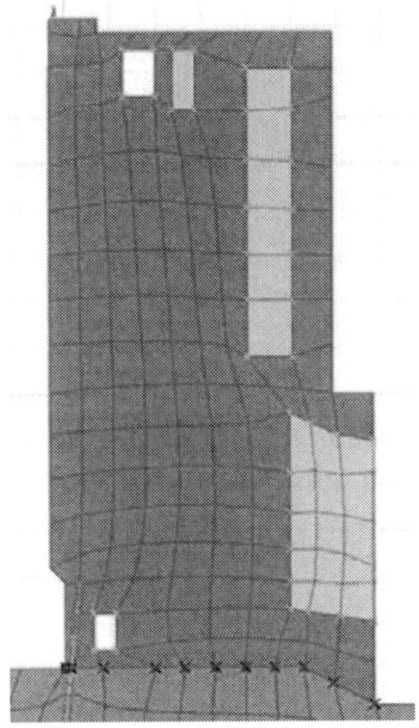

Figure 4. Wanapum FUI FLAC 2D Model (FUI 14 -16)

Numerical parameters for the cable elements were calibrated from the results of a pullout test carried out on May 10-17, 1960 at the Wanapum Dam site. The test anchor consisted of four cables, each composed of 90 ¼ inch diameter wires. A 16-inch diameter hole was drilled in rock and grouted to a height of 37.66 feet. The height of the bond zone grout placed during the test was in error placed to El. 428.77. This is 7.66 feet above the design bond zone elevation that was used in the FUI anchor installation. The numerical model constructed in FLAC for this pullout simulation produced relatively similar results compared to the field pullout test results.

MODELING OF POST TENSIONED ANCHORS

The PT anchors utilized the "BBRV" system of anchorage that was common in the 1960s. It relies on cold-formed button ends on each wire that bear on perforated head plates in each anchorage with one head plate per bundle of 90 wires. Tension is applied to the PT anchor wires by jacking the head plates. Four head plates are provided at the top and bottom of each assembly. The four lower anchor heads are secured to a common plate through which a primary grout supply pipe is fastened.

The PT anchor ultimate strength has been assumed based on a wire ultimate stress capacity of 240 ksi as noted in project design records. Anchor grout compressive strength was tested at 28 days for the test anchor. According to construction records, the grout strength exceeded 2800 psi on average. Both design and construction records show that the PT anchors were locked off at 70% of ultimate strength and they were assumed to provide a working load of 60% of ultimate strength accounting for up to 10% loss in anchor prestress over time due to relaxation and creep.

All PT anchors were grouted in two stages. Upon review of the original construction records, there was concern that a cold joint at the first-second stage grout interface was not cleaned to remove latency prior to the installation of the second grout lift. This introduced the possibility that corrosion may occur at the first-second stage grout interface located approximately 40 feet below the base of the FUI. Subsequent analysis has shown that the resistance of the rock above the hypothesized zone of corrosion could provide sufficient pullout resistance to fully develop the anchor strength.

Prior to construction of the FUI monoliths in 1960, a test PT anchor was installed and tested. The test anchor was installed in a 16-inch diameter hole drilled into the basalt bedrock in close proximity to the FUI foundations. The test anchors were installed in the hole with anchor plates at the bottom of the hole (El. 378.45). The anchor wires were jacked and locked off at approximate El. 467.45. As noted above, an error resulted in placing the top of the grout zone 7.66 feet above the design height. This was accounted for in the model.

The anchor test report provides test data showing anchor load versus elongation measurements. Based on Hook's Law that relates elongation of a material to Young's Modulus, load and material cross-section, the active stressing loss can be back-calculated. The active stressing length minus the initial stressing length gives an estimate of the debonded length. (The initial stress length is the elevation of the application of load minus the elevation of the top of the bond zone grout.)

Table 3 and Figure 5 show a comparison of the debonded lengths estimated based on measured anchor elongations during Anchor Test 3 and the debonded lengths predicted based on the 2D FLAC numerical model. The 2D FLAC predicted debonded lengths as generally about 30% higher than the debond lengths computed from the Test 3 data.

An initial debonded length of eight feet was selected for the 2D FLAC analysis based on the assumption that, if the anchor PT force is exceeded and the FUI foundation lifts relative to the base, the anchor wire bundles will debond above and below the interface almost instantaneously and the debonded length will increase quickly as the seismic event continues.

Table 3. Comparison of Debond Lengths for PT Anchor Test 3 and Numerical Model Results

Anchor Test 3 Elongation (ft)	**Anchor Test 3 Load (k)**	**FLAC 2D Model Load (k)**	**Anchor Test 3 Debonded Length (ft)**	**FLAC 2D Model Debonded Length (ft)**
0.122	1293.0	1216.9	2.18	3.19
0.209	2068.8	2020.9	5.61	7.15
0.311	2973.9	2901.4	7.44	11.11
0.400	3620.4	3614.6	10.48	14.31
0.479	3904.9	4218.3	16.74	16.95

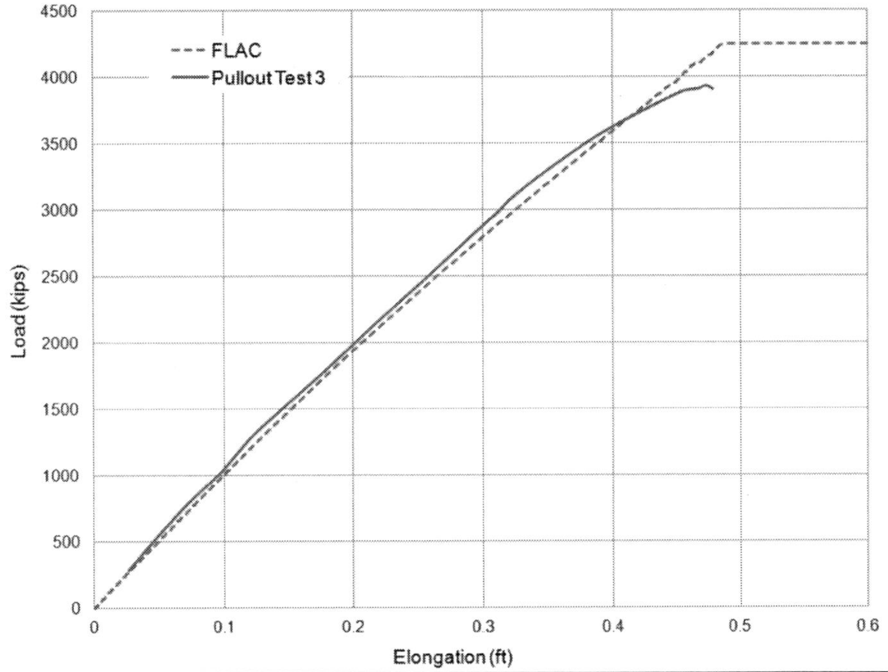

Figure 5. Comparison of load v/s elongation for PT Anchor Test-3 and FLAC numerical model

Figure 6 shows a schematic of possible PT Anchor failure mechanisms. Seismic analysis results show that the FUIs are expected to endure displacement horizontally relative to the rock foundation. This displacement forces the PT anchors to deform at the foundation interface. Each anchor is sloped upstream and is embedded in an 18-inch diameter corrugated steel pipe within the foundation. Below the foundation rock interface, the anchors are embedded in a 16-inch diameter hole drilled into rock. It appears that fracturing of the anchor grout around the anchor wire bundles will occur during a seismic event.

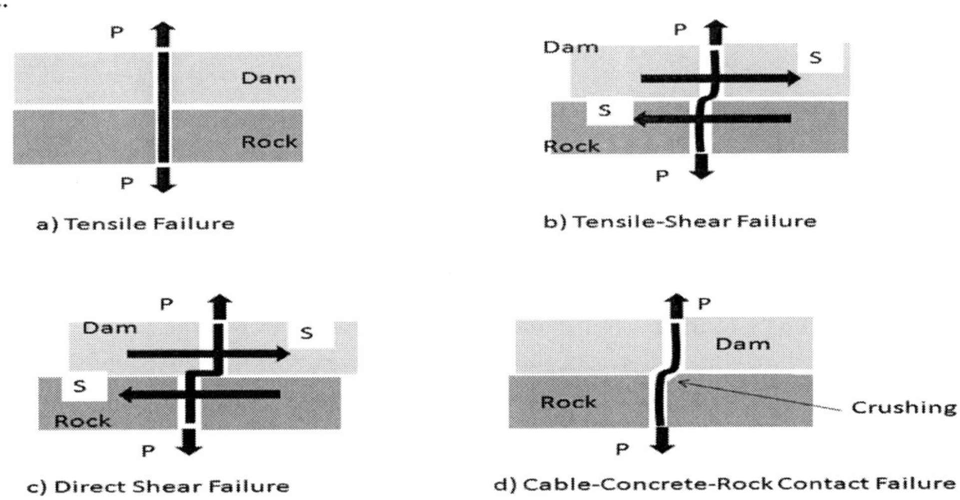

Figure 6. Schematic Showing Hypothetical PT Anchor Failure Mechanisms

Depending on the extent of the grout fracture, the wire bundles may be able to move and accommodate a substantial amount of lateral base displacement. We also observed that, if the grout between the anchor wires were to become fractured and disintegrate, up to two inches of displacement could occur if each bundle of wires is distorted and then able to compress. It appears that the base shear displacement capacity may be greater that one inch but there is not enough information available concerning grouted anchor shear displacement limits to justify setting a higher limit. We set the anchor shear deflection limit at one inch based on these observations and the analysis team's consensus. It appears that, in order to accurately estimate the base shear limit for these anchors, a laboratory model or prototype test program would be required.

The wire used in the FUI PT anchors was tested at a laboratory prior to anchor installation as part of the test program. Although the test did not continue to failure of the wire, the results show the 3% cumulative stress limit is conservative and consistent with the opinion of several experts with knowledge of this BBRV wire product that was prevalent in the late 1950s and into the 1960s.

PRELIMINARY ANALYSIS RESULTS

Seismic performance of the Wanapum FUI is evaluated based on two main criteria: 1) base displacement of the Wanapum FUI, and 2) cumulative strains in the cable anchors during the earthquake. A maximum limit of one inch is assumed for the base displacement, and a maximum limit of 3% is assumed for the cable anchors. The average results of the analyses using seven time histories are compared with the maximum limits.

Non-linear time history analysis of FUIs 12 and 13 was performed (rock knob included) for a 20,000–year return period event. Non-linear time history analysis of FUIs 14 – 16 was performed for 2,500; 5,000; 10,000; 15,000; and 20,000-year return period events. Analysis results for FUIs 12 and 13 confirmed that the rock knob provides significantly greater stability under earthquake loading than the other three FUIs without rock knobs. As discussed, the key performance parameters were base shear displacement and cumulative anchor strain. Average base shear and cumulative strain results were used to evaluate analysis results. Table 4 shows analysis results in terms of average base shear displacement and cumulative strain for each of the return periods evaluated. Seismic analysis results for the 20,000-year return period showed that the presence of rock knobs in the foundations of FUIs 12 and 13 results in significantly less shear displacement than FUIs 14 through 16 that have relative flat foundation interfaces.

Figure 7 shows time histories of axial force (kips) and strain (%) in the cable element crossing the interface of foundation and dam for a ground motion scaled for 1 in 10,000-year return period event. From Figure 7, the nonlinear behavior of FLAC cable elements (representing the PT anchors); i.e.; the accumulation of plastic strain leads to reduction in axial force in the cable element can be visualized.

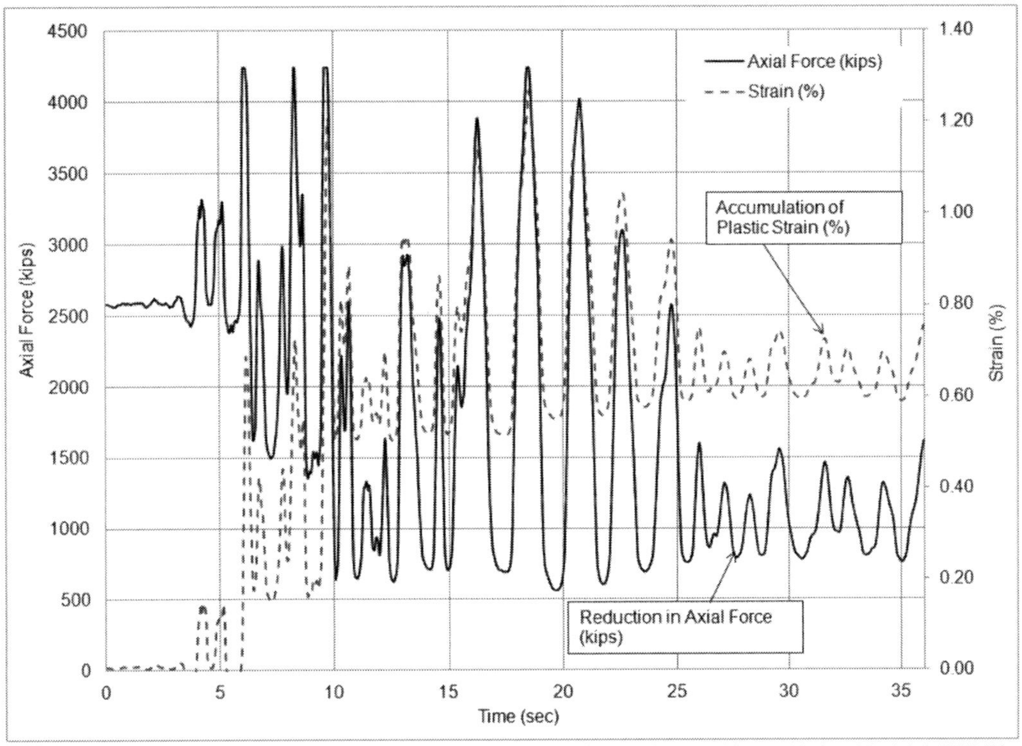

Figure 7. Anchor Axial Force (kips) and Strain (%) vs. Time (1 in 10,000 YRP)

Considerably less damage is anticipated for the FUIs with rock knob incorporated into their foundation compared to the FUIs with a relatively flat foundation. Therefore, it can be assumed that the failure of the Wanapum FUI system as a whole is governed by FUIs without a rock knob. Conservatively, a 5,000-year return interval seismic event can be assumed to be the failure event, where three out of five FUI would experience significant rotation and some uncontrolled release of pool.

There was considerable variability in the analysis results for base shear displacement, vertical (lifting) displacement, and maximum strain between each of the seven time histories analyzed. The range in base shear displacement was between approximately 0.6 and 1.15 inches for the 5000-year return period event. The base vertical displacement ranged between approximately 0.4 and 0.7 inches for the 5000-year return period event.

Table 4. Average Base Shear Displacement and Cumulative Strain Results

FUIs 12 & 13 Results

Return Period	Average Base Shear Displacement	Average Cumulative Strain
20,000	1.69	2.39

FUIs 14 – 16 Results

Return Period	Average Base Shear Displacement	Average Cumulative Strain
2,500	0.58	0.46
5,000	0.95	0.79
10,000	1.82	1.23
15,000	2.84	1.94
20,000	3.71	2.67

SOURCES OF UNCERTAINITY

Seismicity

The Mid-Columbia seismic study identifies the potential seismic risk as uniform hazard response spectra and was the basis for development of a suite of time histories for a range of return period events. However, the earthquake magnitude, frequency content, and duration are estimates that are subject to a high degree of uncertainity.

Upstream and Downstream Water Levels

Worst case condition was selected for analysis with high normal pool and low normal tailwater elevations.

PT Anchor Modeling Uncertainty

Initial debonded length: Based on the results from the pullout test, eight feet of cable is expected to debond over the course of the seismic event. However, there is a possibility that the duration of the event is short enough that the full eight feet of debonding might not take place. A shorter debonded length could result in high localized strains in the cable anchor at the dam-foundation interface. These localized strains can cause the failure of the cable anchors. Figure 8 shows a bar chart giving the variability in the maximum strain versus the assumed initial debond length used in the 2D FLAC model for Time History M4 (G994). The maximum strain is approximately twice as high for an initial debonded length of four feet than the assumed debonded length of eight feet.

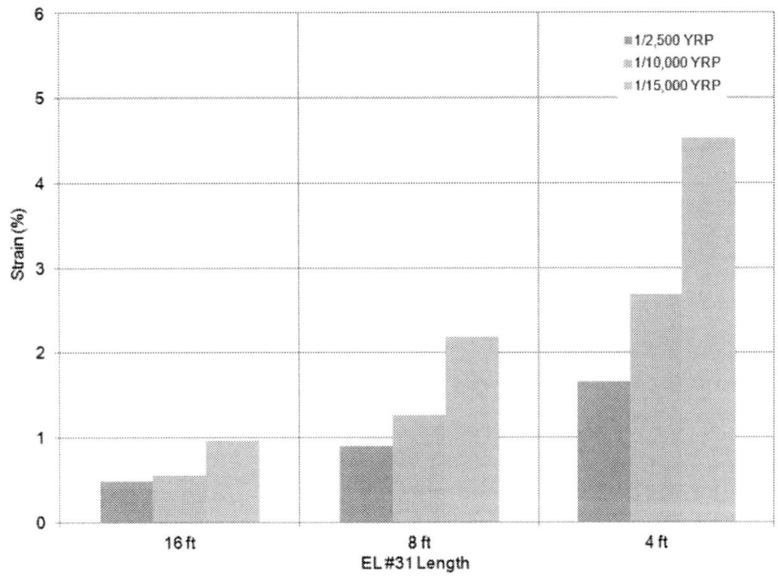

Figure 8. Sensitivity Analysis for Initial Debond Length Assumption

Three-Dimensional Effects

Three-dimensional effects such as modeling of foundation interface of roughness and irregularities that influence resultant eccentricity and side friction. Figure 9 shows a schematic of a FUI with this potential source of side friction.

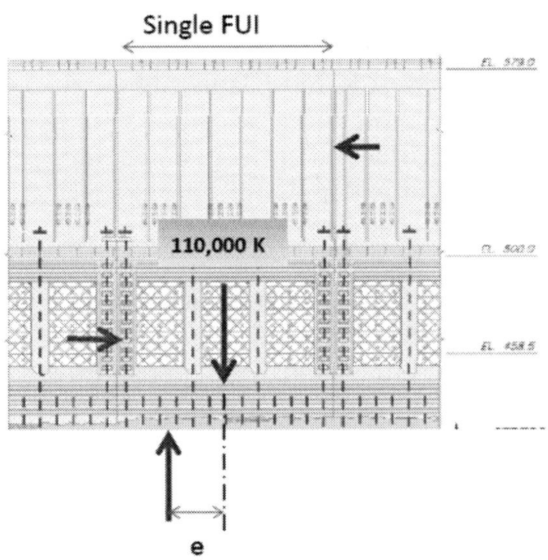

Figure 9. Schematic Showing Potential Source of Side Friction

Base Shear Displacement Capacity

An ultimate shear displacement limit of one inch was selected for this analysis based on a review of anchor displacement scenarios and estimates of combined stress and professional judgment as discussed above. The analysis output for each time history run was examined to see if the shear displacement was exceeded.

Cumulative Strain Capacity

An ultimate cumulative strain limit of 3% was selected for the 2D FLAC analysis based on anchor wire testing completed in 1960 prior to FUI construction. This limit was not incorporated into the FUI structural model. The analysis output for each time history run was examined to see if the shear displacement was exceeded.

RIDM IMPLICATIONS

In order to estimate the downstream areas that would be inundated and population at risk (PAR) from a sudden failure of the Wanapum FUIs, preliminary dam break analyses were performed. The overturning failure of all five FUI monoliths (11 – 16) would represent a very conservative assumption in terms of uncontrolled release and potential downstream impacts. For this study scenario we have assumed five 90-foot long monoliths would rotate about the toe so that the original upstream face became the top surface acting as a long, broad-crested weir. Peak discharges for this scenario were estimated for several headwater levels (El. 571 to El. 560) and for one to five monoliths failing due to overturning. The results indicate that the peak discharge over one overturned FUI monolith at reservoir El. 571 would be approximately 118,000 cfs. The peak discharge for five overturned monoliths would be approximately 590,000 cfs.

Hydraulic analysis shows that the peak discharge for the simultaneous overturning of all five FUI monoliths (590,000 cfs) would not exceed the channel capacity of the river reach between Wanapum and Priest Rapids Dams and the river reach below Priest Rapids Dam. Preliminary analysis and "worst case scenario" with five FUI monoliths failing resulted in an estimated fatality of less than 1. As noted, the presence of rock knobs in the foundations of FUIs 12 and 13 provides significantly greater shear resistance than FUIs 14 through 16 that have relative flat foundation interfaces.

Seismic return periods and annual event probabilities were used to estimate the annual seismic probabilities. Based on this estimate and the results of FUI time history analysis for different return period events, the annualized probabilities of failure could be estimated by multiplying event probabilities by the probabilities of failure for FUIs 12 and 13, and FUIs 14 through 16.

Fifty percent of the tolerable risk criterion established by the USACE was selected as a prototype basis for evaluating the FUI structural system for seismic loading. The results of this analysis are shown on Figure 10. Because of the negligible downstream impact of FUI failure it appears that the Wanapum FUIs will fall within the range of acceptable and tolerable risk even if the FUIs were judged to be more fragile than estimated.

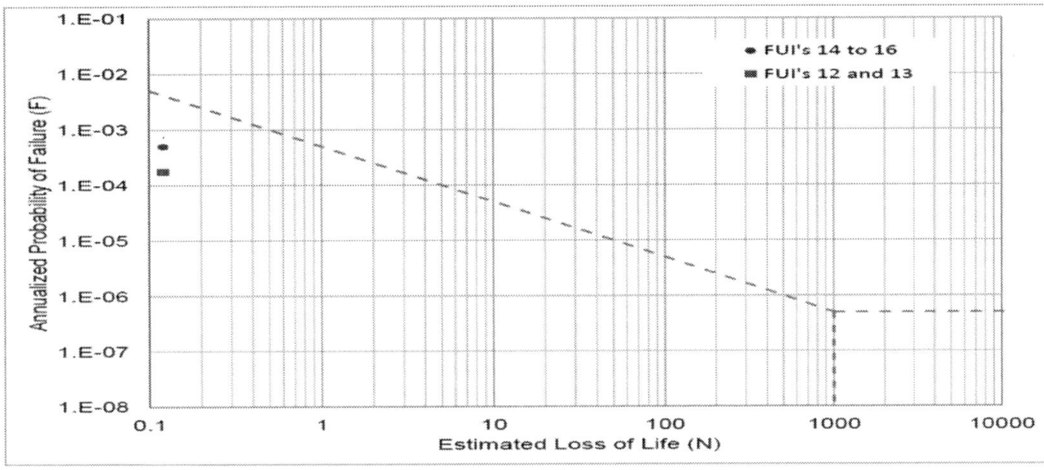

Figure 10. Estimated Seismic Life Safety Tolerable Risk

CONCLUSION

Base shear displacement appears to be the dominant performance parameter but, as discussed, the displacement limit of one inch is uncertain. More analysis or a model study would be required to refine this estimate.

FUIs 12 and 13 have significantly less fragility than FUIs 14 -16 due to the presence of the rock knob. Peak flow over a single overturned FUI is approximately 118,000 cfs so, if five monoliths overturn, total flow would be approximately 590,000 cfs. Downstream channel capacity is sufficient to handle the increase in flow without a significant change in downstream water level even if five FUIs were to overturn. This realization may represent an "off ramp" that eliminates the need to perform more in-depth risk analysis to satisfy dam safety requirements.

Consideration should be given to performing a more detailed risk analysis to assess risk-cost implications of post-seismic damage to the FUIs. Although additional risk analysis may not be required to address FERC RIDM criteria, it may be justified based on the economic exposure that failure of the FUIs may represent.

ACKNOWLEDGEMENTS

Many Grant County PUD No. 2 employees and private consultants contributed to the success of the prototype RIDM review of the Wanapum FUIs. Some of the key contributors are listed below:

Grant County	Consultants	
Rebecca Simpson	Steve Smith	Don Bowes
David Moore*	Peter Friz	Bill Perkins
Randy Nash	Dick Griffith	Ali Shahbazian
	Marty McCann	

* We would like to make special mention of the contributions of David Moore who was responsible for much of the original thinking regarding the Wanapum FUIs and the performance of the PT anchors. David was tragically struck by a car and killed while commuting home on his bicycle from the District offices in September 2011. If he were still alive, Dave would certainly have been one of the authors of this paper and participated in its presentation at USSD.

REFERENCES

Barton, N., and Bandis, S. (1990). *"Review of Predictive Capabilities of JRC-JCS Model in Engineering Practice".* Loen, Norway: Proc. Int. Symp on Rock Joints (eds N.Barton and O. Stephansson): pp 603-610.

Bandis, S. (1982). *"Effects of Block Size on the Shear Behaviour of Jointed Rock".* Berkeley, CA: 23rd U.S. Symposium on Rock Mechanics: pp 739-760.

Peck, Hanson, and Thornburn (1974). "Foundation Engineering", John Wiley & Sons, Inc.: p 100.

Rocscience 2011, Roclab Version 1.032, University of Toronto

Rutherford, J., Stump, C., Mass, G., Lehto, D., Nash, R. (2011). Wanapum Future Unit Infill Project. *U.S. Society on Dams.* San Diego, CA: USSD.

NEW INNOVATIVE TECHNIQUE FOR STRENGTHENING OLD BUTTRESS DAMS AND MULTIPLE ARCH DAMS

Thomas Konow[1]

ABSTRACT

This article presents a new method for reconstruction and strengthening of old buttress dams, where the old dam is used as a shell or formwork for construction of a new dam. The technique has been patented in Norway and has proven to be cost efficient in cases where construction of a new dam is the only solution. The solution can also be applied at a full reservoir level, which can be a major advantage in many cases.

The new technique is presently applied on a flat slab buttress dam in Norway. At this project, the solution has proven to give 40 % cost savings comparted to demolition of the existing dam and reconstruction of a new dam. In addition the solution results in a very stable and safe structure, as shown by examples in this paper.

It is assumed that the solution can be particularly relevant for slender buttress dams such as flat slab buttress dams and multiple arch dams. Examples of how the technique can be applied on these dams are also presented in this paper.

It is considered that the technique described in this paper can be the preferred solution even though the construction costs of a new dam is comparable, because the safety gained is very favourable compared other traditional dam types.

INTRODUCTION

According to International Commission On Large Dams (ICOLD), Technical dictionary of dams, there are three types of buttress dams:

- Flat slab buttress dams (or deck dam)
- Multiple-arch dams
- Solid head buttress dams

Buttress dams are characterized by a water-tight upstream side that is supported by a series of buttresses on the downstream side. The buttresses transfer the force from the water pressure to the foundation. The buttresses normally have a constant spacing along the dam. Typically, the buttresses are placed at intervals of 5 to 20 meters, depending on the size and design of the dam. Buttress dams are also sometimes referred to as hollow dams because of the open sections between the buttresses on the downstream side of the dam.

In the US, flat slab buttress dams are sometimes called Ambursen dams, since this dam type was patented in the US by the Norwegian engineer Ambjørnsen who immigrated to

[1] Thomas Konow, Consultant Engineer – Dam Safety, Norconsult, Brusetveien 73, N-1395 Hvalstad, Norway. E-mail: thomas.konow@gmail.com

the US. In Norway, the flat slab buttress dam is also called "Grøner-dam". This is a further development of the Ambursen-dam made by the engineer C.F. Grøner (former ICOLD President). Here the vertical slab joints are placed at the point of zero moment in the slab and thereby reducing the required reinforcement in the slab. Due to these two engineers, Norway and the US are considered to be the countries with the most slab buttress dams worldwide. In Norway there are a total of 240 slab buttress dams whereof 44 are defined as "large dams".

As the age of the buttress dams' increase, the need for repair and rehabilitation also increases due to general deterioration and age. In addition, many dams do not meet today's requirements for strength or over-all stability, as the required level of safety for dams in general has been increasing over the years. As a result, several of today's buttress dams are in need of rehabilitation and in some cases construction of a new dam may be the only solution.

CHALLENGES WITH NORWEGIAN BUTTRESS DAMS

With a few exceptions, all of the Norwegian buttress dams are of the type "flat slab" buttress dams.

Most of these dams were constructed during the 1950s and 1960s. The general condition of these dams depends on the age combined with the general quality of the construction works and the environment the structure has been exposed to. Mechanisms that affect the general condition of the structure can be frost and thaw effects, reinforcement corrosion and alkali aggregate reaction (AAR). Often, a combination of these different deterioration mechanisms is present at the same time.

In addition the structure need to have sufficient stability according to today's safety requirements. In Norway, the requirements for stability changed with the new regulations that came into force in 2001. The present requirements for stability of buttress dams are shown in the below table.

Table 1. Norwegian requirements for stability of slab buttress dams

Load situation	**Factor of safety for**	
	Overturning	**Sliding**
1. Ice pressure (100 kN pr. m)	1,4	1,4
2. Design flood (Q_{1000})	1,4	1,4
3. Design flood excluding rock anchor/bolts (Q_{1000})	1,1	1,1
4. Probable maximum flood (PMF)	1,3	1,1

The new regulations are also applied on existing dams. As a result, many of the existing slab buttress dams do no longer meet the stability requirements against sliding and/or overturning.

In addition, the Norwegian dam safety authorities have an intention to not allow slab buttress dams in the highest hazard class, and this also applies for existing dams.

In Norway, the above mentioned factors have resulted in an increased number of cases where slab buttress dams needs strengthening or rehabilitation. In some cases decommissioning and reconstruction of a new dam is the only option.

RECONSTRUCTION OF DAM KROKAVATN

Dam Krokavatn is 19 m high and has a total length of 100 m, where the main dam is a flat slab buttress dam about 50 m long.

The dam design is typical for Norwegian flat slab buttress dams. It is constructed with an upstream slab inclined at about 50° to the horizontal and supported by 8 concrete buttresses at 5 m spacing. Originally, the upstream deck had a thickness of 0.30 m at the top and increasing to a thickness of 0.80 m at the foundation.

At the right abutment there is a 50 m long concrete gravity spillway. Between the spillway and the buttress dam there is a corner along the dam axis. Through several years a crack had developed in the concrete wall located at the corner between the buttress dam and the spillway (see figure 2), and resulted in a large leakage through the crack at full reservoir level.

Figure 1. Dam Krokavatn. 50 m long slab buttress dam (near side) and the spillway on the far side. Notice that the dam axis has a corner on the far side between the buttress dam and the spillway.

Figure 2. Cracking of concrete at the buttress wall at the corner between the spillway and the slab buttress dam, due to15 mm expansion of the upstream slab.

Dam Safety Evaluation

In 2012, the dam owner conducted a safety evaluation of the dam, where the overall safety was assessed.

The safety evaluation identified that the dam did not meet today's requirements for stability against sliding and overturning. As the dam is a high hazard dam, the Norwegian dam safety authorities required a reconstruction of the dam.

Alkali Aggregate Reaction (AAR)

The dam safety evaluation also concluded that AAR was present in the concrete. This was not a surprise, as observations from the dam showed a relative movement of 15 mm at the buttress wall next to the spillway (see figure 2), over a length of 8.45 m.

This corresponded to an elongation of 0.18 % and indicated a tensile stress of 360 N/mm^2 in the reinforcement. As the characteristic strength of the reinforcement was 400 N/mm^2, the tension in the reinforcement was close to yielding, and it was therefore concluded that the reservoir should be restricted to maximum 2.5 m below maximum operating level. This section of the dam was only 4 m high and a collapse of the concrete slab would therefore not have any serious consequences.

Rehabilitation Works

To upgrade the dam it was decided to use the old buttress walls as formwork to cast new buttresses in every second hollow area, as shown in the figure below. The additional weight from the buttresses would then secure the required stability of the dam.

In addition, a new concrete slab on the upstream side of the old slab was necessary since the load capacity of the slab was not sufficient due to AAR.

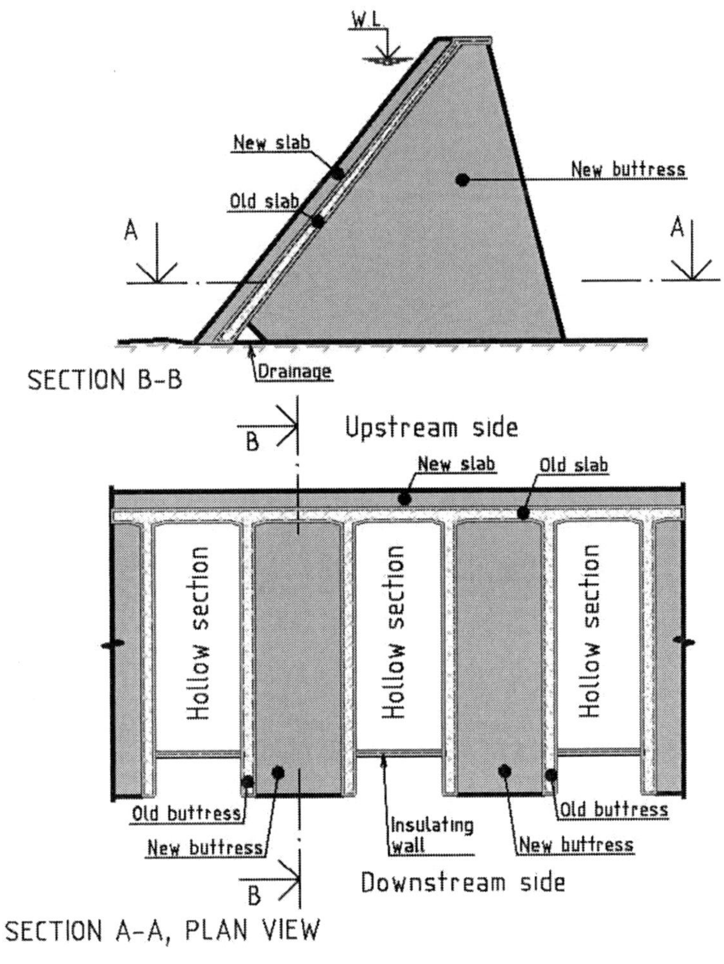

Figure 3. General arrangements for reconstruction of dam Krokavatn.

After reconstruction, the original dam is only a shell, and has no function in regard to stability or strength of the new structure.

This solution was by far considered to be the cheapest solution compared to the alternative, which would be construction of a new dam.

Cost estimates of different solutions are shown below.

Table 2. Cost estimate of different alternatives (1 NOK = 6.43 US$)

Alternative	Cost (NOK)	Cost (US$)	Difference (%)
1. Buttress dam – Reconstruction	22 000 000,-	3 420 000,-	-
2. New gravity dam in concrete	31 000 000,-	4 820 000,-	+41%
3. New rockfill dam	41 000 000,-	6 370 000,-	+86%

Evaluation of Stability

Calculations of the dam stability proved that the solution (alternative 1, table 2) was very favourable in terms of stability. This is mainly due to the additional weight from the new buttresses. In addition, the weight of the new upstream slab and a less inclined upstream face also contribute to the improved stability of the dam.

A comparison of the stability before and after reconstruction are shown in the figures below.

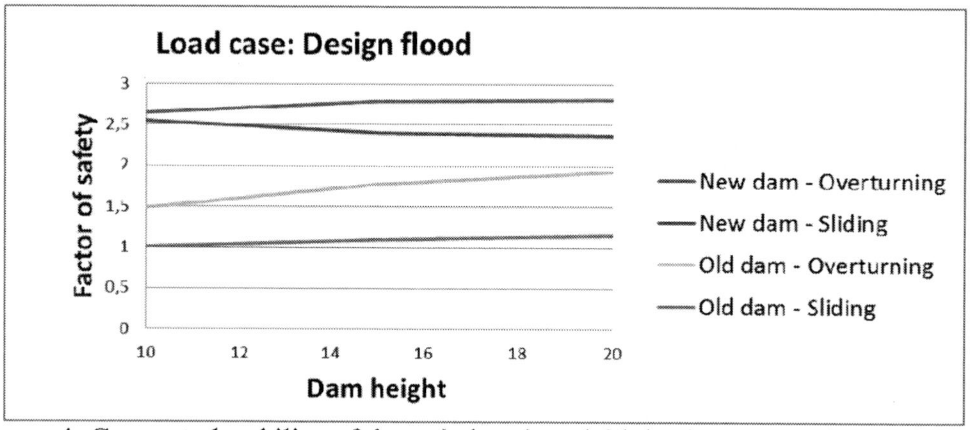

Figure 4. Computed stability of the existing dam (old dam) compared to that of the reconstructed buttress dam (new dam) for a design flood 0.5 m above normal water level.

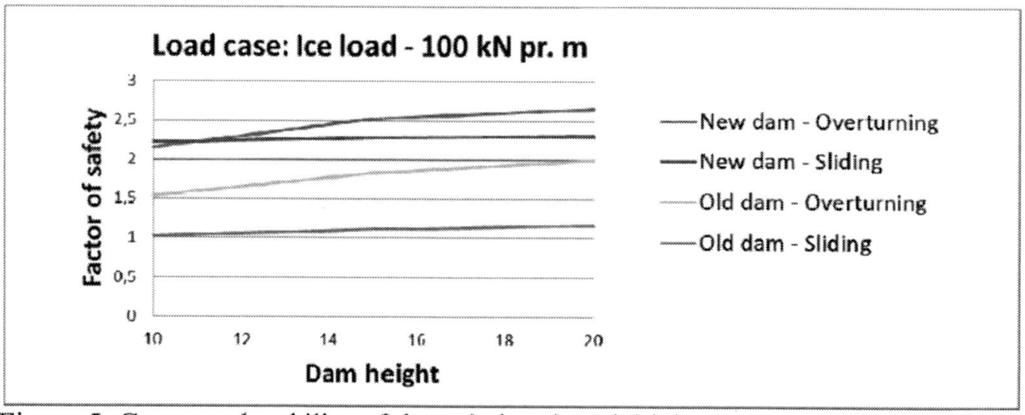

Figure 5. Computed stability of the existing dam (old dam) compared to that of the reconstructed buttress dam (new dam) for å design ice load of 100 kN pr. m length.

The calculations of stability show that the new dam has a very high level of safety. The factor of safety is higher than 2 and is much higher than the required level of safety in Norway. In general the factor of safety against sliding is increased with more than 1, compared to the existing dam.

As a simplification, it can be said that the new dam can be looked up on as a combination of a flat slab buttress dam and a solid head buttress dam, where the massive buttresses and the stabilizing weight of the water above the inclined slab together with a reduced pore pressure results in a very safe dam structure.

Pictures of the Construction Works in 2014

As mentioned, the safety evaluation in 2012 concluded that a reservoir restriction of 2,5 m below normal operating level was necessary due to the damages due to AAR. The dam owner therefore wanted the reconstruction carried out as soon as possible.

The construction works started in 2014. During the summer a new road of 7 km was constructed and the new upstream slab was cast. The new buttresses will be constructed during the summer in 2015. As the dam is located in a mountain area, it is not accessible during the winter and the work has to be carried out in the period June to October.

Some pictures from the dam reconstruction are shown below.

Figure 6. Reinforcement of two of the upstream slabs.

Figure 7. New slab on the upstream side with insulation of the concrete during hydration due to the cold climate in the late autumn 2014.

OTHER METHODS FOR STRENGTHENING FLAT SLAB BUTTRESS DAMS

The solution used at dam Krokavatn can be modified in order to suit other types of flat slab buttress dam, and some of these solutions are described with text and illustrated with figures.

New Slab on the Downstream Side (Alternative A)

This alternative solution is illustrated in the figure below.

The figure shows a cross section D-D of a new buttress where a new concrete slab is constructed on the downstream side of the existing upstream slab. Section C-C shows one possible solution for placing the construction joints and thereby connecting the new buttresses and the new concrete deck. This can be done by casting the new buttresses before cutting thorough the buttresses of the existing dam.

This solution does not require a full drawdown of the reservoir which can be a great advantage in some cases. The solution may be used on buttress dams where both stability and strength of the structure may be a problem.

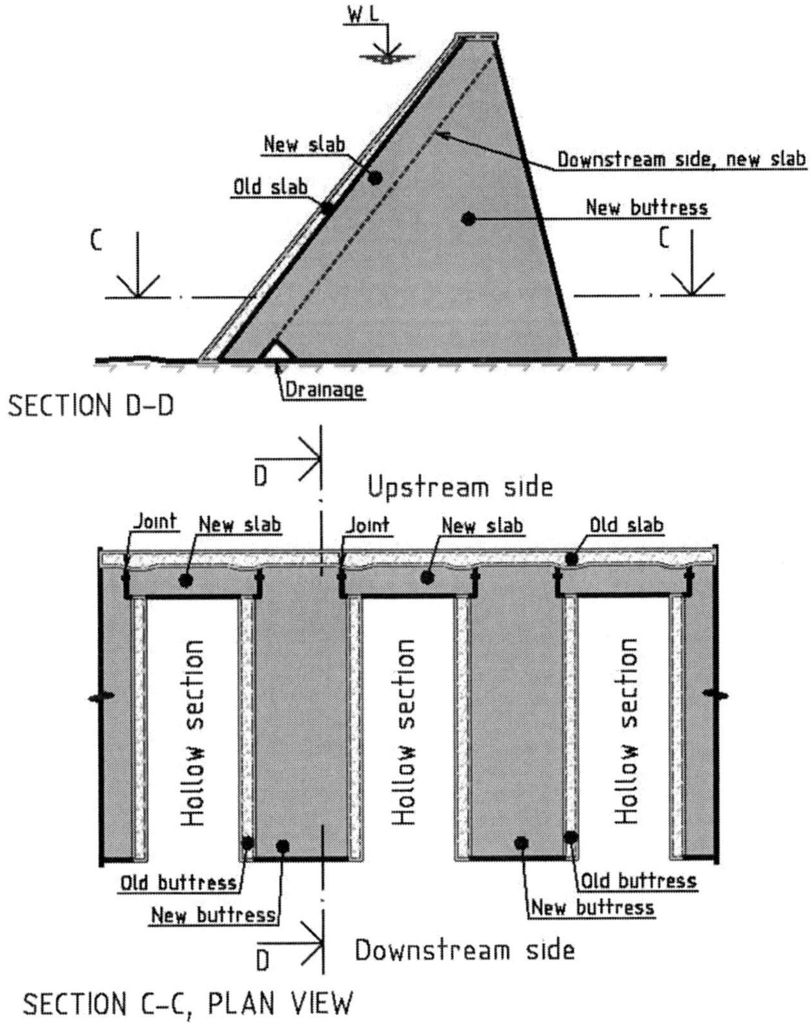

Figure 8. Reconstruction with a new slab on the downstream side.

<u>**New Buttresses Only (Alternative B)**</u>

The solution is illustrated in the figure below.

The figure shows a cross section H-H of a new buttress where a new concrete deck is not required. Section G-G shows a plan view with a possible solution for strengthening the slab in cases where this may be necessary. This can for instance be done by installing a row of beams to support the existing slab. The solution requires that the existing upstream slab do not have any major damages or leakages and can still work as an upstream watertight seal for the dam.

This solution does not require a drawdown of the reservoir and can be used where both stability and strength of the structure may be a problem. The new buttress can influence the static load distribution of the existing slab and may result in the need for strengthening the existing upstream slab.

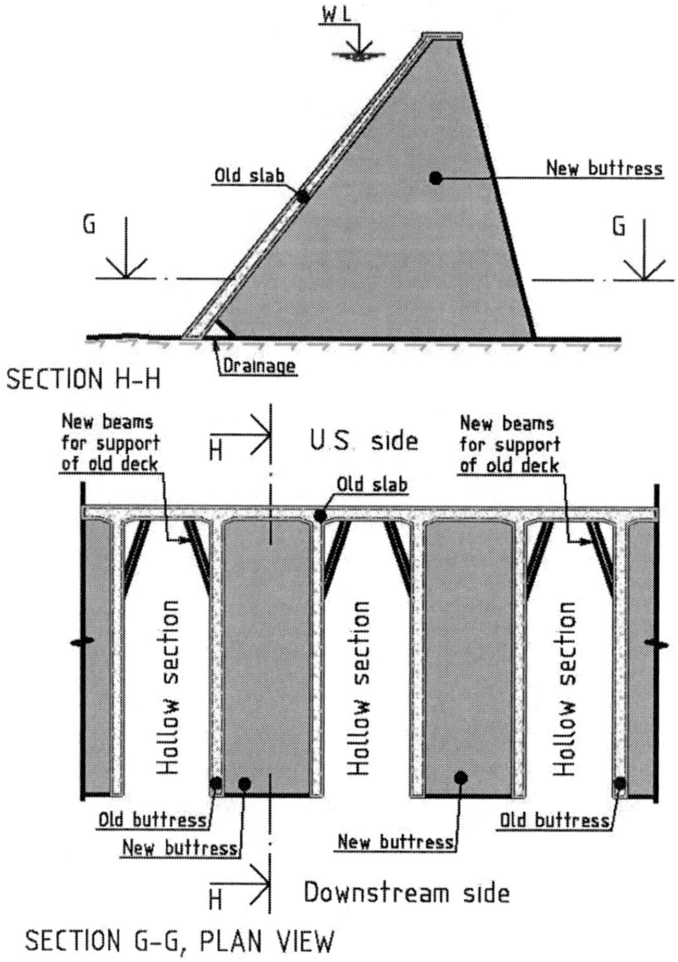

Figure 9. Reconstruction without a new upstream slab.

STRENGTHENING OF MULTIPLE ARCH DAMS

A multiple arch dam is designed with same principles as the flat slab buttress dam, with an inclined upstream face. Normally, this is also a slender dam structure where the vertical weight of the water above the slab gives a major contribution to the stability of the dam.

Multiple arch dams are generally higher than the flat slab buttress dam and have a greater spacing between the buttresses, as the arches can have a larger span than the flat slab. The stresses in the structure and on the interface to the foundation are therefore much higher compared to a flat slab buttress dams. On some older buttress dams, FEM analysis has identified unacceptable stresses in these types of dams.

In the following examples it is illustrated how multiple arch dams can be strengthened with the same method used for dam Krokavatn.

Reconstruction to a Solid Head Buttress Dam (Alternative C)

The solution is illustrated in the figure below.

The figure shows a cross section L-L of a new buttress in a reconstructed multiple-arch dam with a new concrete slab on the upstream side of the existing upstream concrete arch slab. Section K-K shows a plan view where the new buttresses are cast in every second hollow section. The new upstream concrete slab is designed as an upstream buttress head so that the structure can be defined as a solid head buttress structure. The figure show a constant thickness of the upstream concrete slab, however the thickness may vary with the dam height and water pressure.

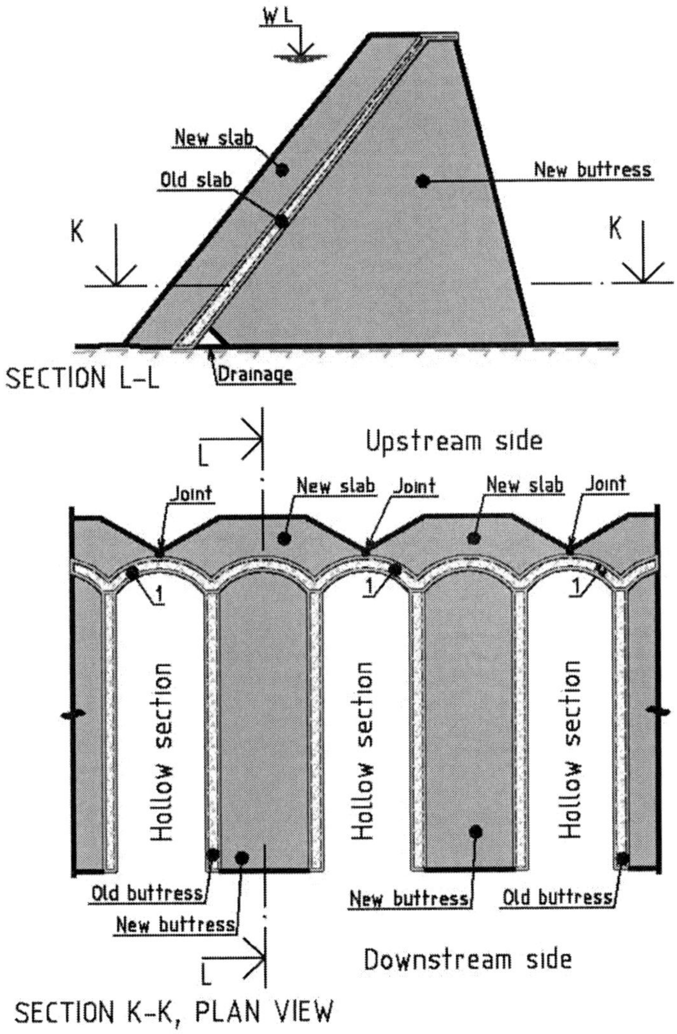

Figure 10. Reconstruction to a solid head buttress dam.

__**New Buttresses Only (Alternative D)**__

The solution is illustrated in the figure below.

The figure shows a cross section N-N of a new buttress in a reconstructed multiple-arch dam where a new upstream concrete deck is not required, as also illustrated in the plan view (Section M-M). The solution can be used to improve the overall stability of the dam, or where the existing buttresses do not have sufficient safety and require strengthening. The solution implies that the existing arches have sufficient safety.

To obtain a satisfactory waterproofing it is possible to e.g. install a watertight geo-membrane on the upstream side of the dam. It is also possible to construct a new upstream deck at a later stage, if this is considered necessary.

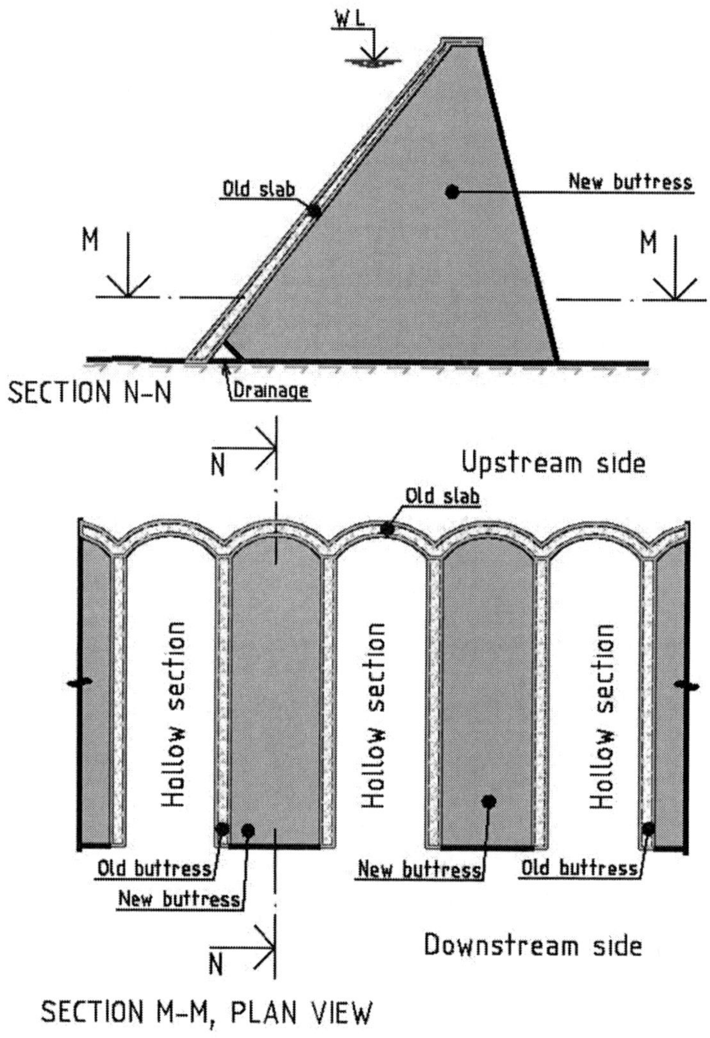

Figure 11. Reconstruction with new buttresses only.

New Slab on the Downstream Side (Alternative E)

The solution is illustrated in the figure below.

The figure shows a cross section P-P of a new buttress where a new concrete arch is constructed on the downstream side of the existing arch. Section O-O shows one possible solution for placing the construction joints and thereby connecting the new buttresses and the new concrete arch. This can be done by casting the new buttresses before cutting thorough the buttresses of the existing dam.

This solution does not require a full drawdown of the reservoir which can be a great advantage in some cases. The solution may be used on buttress dams where both stability and strength of the structure may be a problem.

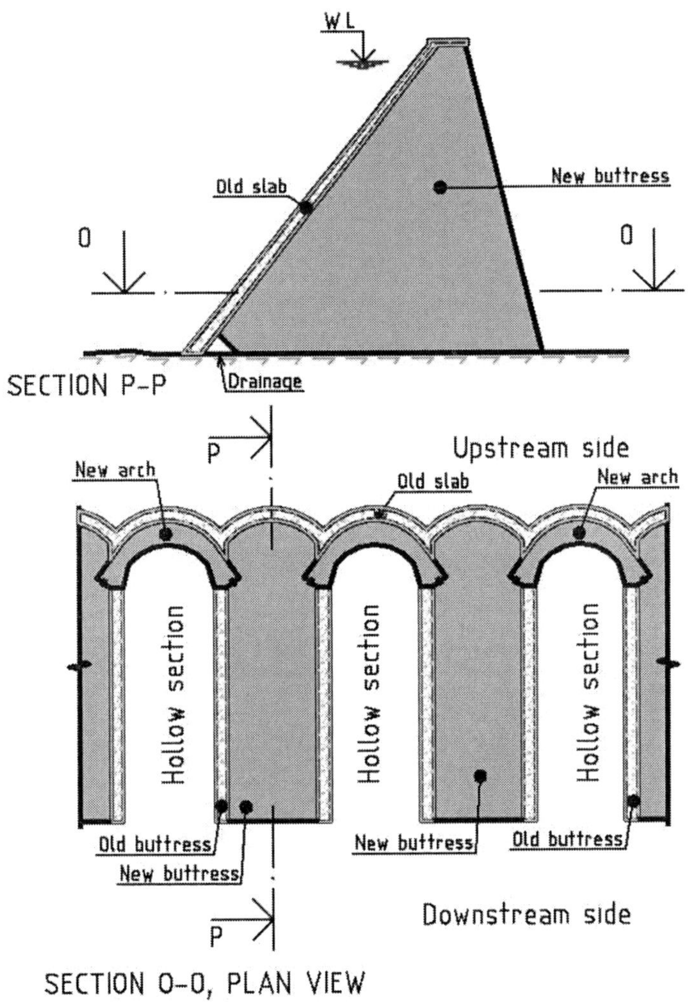

Figure 12. Reconstruction with a new arch on the downstream side.

CONCLUSION

As the ages of the buttress dams increase, the need for repair and rehabilitation also increases due to general deterioration and age. In addition, many dams do not meet today's requirements for strength or over-all stability, as the required level of safety for dams in general has been increasing over the years. As a result, many of the existing buttress dams are in need of rehabilitation and in some cases construction of a new dam may be the only solution.

This paper presents a new technique for reconstruction of buttress dams where new buttresses are cast in between the buttresses of the existing dam. The technique has been patented in Norway, and an international PCT application has been filed.

As a simplification, the solution can be illustrated as a combination of a flat slab buttress dam and a solid head buttress dam, where the massive buttresses and the stabilizing weight of the water above the inclined slab together with a reduced pore pressure results in a very safe dam structure.

The technique is presently being used on a flat slab buttress dam in Norway and the paper gives a brief description of the safety evaluation and rehabilitation works of this dam.

Design of this dam has proved that the solution is very favourable in terms of stability. This is mainly due to the additional weight from the new buttresses. In addition, the weight of the new upstream slab and a less inclined upstream face also contribute to the improved stability of the dam.

At the above mentioned project, the solution has proven to be cost efficient compared to construction of a new dam, which was considered to be the only alternative solution.

The technique described in this paper may be the preferred solution even though the construction costs of a new dam is comparable, since the safety gained by this technique is very favourable.

The paper also presents several possible different applications for the technique and illustrates different methods of how it can be applied for slender buttress dams and multiple arch dams.

EVALUATION AND MONITORING OF ARGONAUT DAM

Christopher M. Abela PE[1]

ABSTRACT

This paper presents the stability evaluation and monitoring of Argonaut Dam located in Jackson, California. Argonaut Dam, which is a concrete multiple arch dam, was designed by John S. Eastwood in 1916, for the purpose of storing tailings from the new quartz mill. The dam, which was not well inspected or monitored for the past 70 years, was stumbled upon in 2010 by the Department of Toxic Substances Control while conducting soil contamination surveys. Due to the condition of the dam, the Department of Toxic Substances Control alerted the Environmental Protection Agency (EPA) whom requested the assistance of the United States Army Corps of Engineers (USACE) to evaluate the dam's condition. Following a Phase I investigation, it was determined that the dam was susceptible to several stability failure modes, which could impact the town of Jackson should failure occur. This paper provides an overview on the dam's stability evaluation, potential repair / retrofit, and future monitoring.

INTRODUCTION

The Argonaut Dam, located in Jackson CA, is described by historic documents as being 420-450ft long and 46-50ft tall at its highest point. The dam's arches, which originally numbered at 14 (now only 13), have a tapered thickness of 18" at the base and 12" at the top. Both the arches and 15 buttress walls are reinforced with a 1" or 1 1/8" diameter steel wire rope, which daylights out the back of the walls. In addition, tie beams have been constructed between 6 of the arches to brace the tallest buttress walls; the spacing between the arches is approximately 30ft. From recent inspections of the dam, a plan and elevation view of the remaining arches and walls was developed, see Figure 1. The foundation of the dam is speculated to be composed of a weathered clayey siltstone of low plasticity based on soil borings (SB-40 & SB-41) taken upstream of the dam during a soil investigation by URS, see URS (2009). However, the exact weathering condition of the siltstone is unknown.

Following two inspections in 2013 by the US Army Corps of Engineers (USACE) and Environmental Protection Agency (EPA) concerns were raised on the dams overall stability. Given the age of the dam, condition of the tie beams, and relative close proximity of the dam to downtown Jackson both a preliminary static and dynamic assessment were recommended, see Figures 2 & 3. The assessment of the dam, detailed

[1] Civil Engineer (Structural), United States Army Corps of Engineers, Sacramento District, Christopher.M.Abela@usace.army.mil.

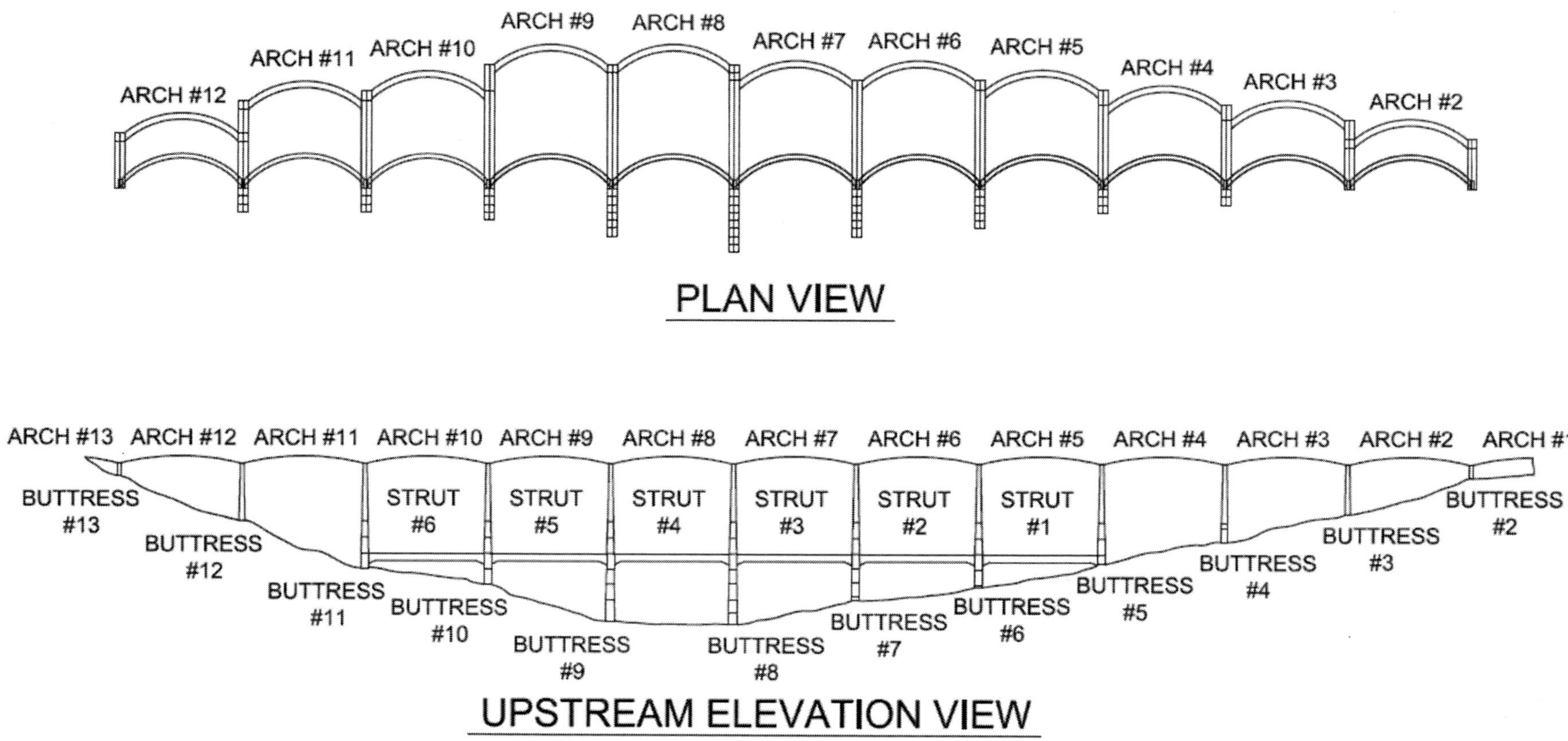

Figure 1. Plan and Elevation View of Dam Defining Arch and Buttress Numbering

herein, was performed in accordance with USACE standards ER 1110-2-1156 Safety of Dams Policies and Procedures, and EM 1110-2-2100 Stability Analysis of Concrete Structures. Please note that EM 1110-2-2201 was not used as this manual pertains to single arch dams (without buttress walls) impounding water. The inspectors included Chris Abela, PE, and Ken Pattermann, GE, of USACE, and Dan Shane of the EPA. The dam is currently privately owned and is not regulated or inspected by a State or Federal entity due to the fact that the dam impounds less than 15 acre feet of water.

Figure 2. Aerial View of Project Vicinity

Figure 3. Project Impact Areas

In regards to seismic stability of arch dams, research uncovered that multiple arch dams have an excellent track record, with no failure of an arch dam having occurred due to an earthquake (FERC, 1999). However, according to (FERC 1997), buttresses like those used in multiple arch dams, if unreinforced or unbraced, are susceptible to damage from lateral earthquakes, due to their slender design, which is better suited for in plane loads.

STABILITY ANALYSIS OF THE DAM

Chapter 7 of EM 1110-2-2100 titled, "Evaluating and Improving Stability of Existing Structures" along with ER 1100-2-1156 describe a procedure to be followed when evaluating the static and dynamic stability of concrete structures. The procedure is broken down into phases with Phase I (Preliminary Analysis and Evaluation) being the first step in analyzing the structures stability. Under this phase it is recommended that a two-dimensional (2D) analysis be performed to determined stability adequacy of the dam and to determine if the analysis should progress to Phase II, which consists of a more comprehensive analysis.

Due to the complex geometry of the dam, a three-dimensional (3D) finite element analysis (FEA) model versus a 2D FEA model was created to help evaluate the dam's stability. To simplify the analysis, Arch #8 (the tallest arch) as shown in Figure 1, was selected for the stability evaluation. The arch was analyzed for sliding, floatation, overturning, and bearing stability under both static and dynamic load cases related to usual, unusual, and extreme events. It is important to note that because USACE codified guidance does not specifically address the analysis of multiple arch dams, the stability evaluation of Argonaut Dam was supplemented with FERC documentation related to this type of structure. For the stability analysis, because of the potential for loss of life, the structure was assumed to be critical.

LOADS AND LOADING CONDITIONS

The selection of the unit weights and strength for the various materials for the project were based on historical documents of similar structures, inspection notes on the Argonaut Dam, a soil boring investigation by URS Corp (URS, 2009), and recommendations by project delivery team members.

Dead Load:

- Unit weight of concrete = 144 pcf
  (*Based on historical Argonaut Dam inspection notes*)

- Compressive strength of concrete = 2500 psi
  (*Estimated compressive strengths based on Webber Creek Dam built 1924 near city of Placerville, see Forrest, et al. 2000*)

  Creek Dam built 1924 near Placerville, see Forrest, et al.

Live Loads & Hydraulic Loads:

- Unit weight of water = 62.5 pcf
- Unit weight of moist backfill / mining tailings = 90 pcf
  (*Based on historical Argonaut Dam inspection notes and recommendations from project delivery team members*)
- Unit weight of saturated backfill soil = 115 pcf
  (*Based on recommendations from project delivery team members and URS 2009*)

From the URS geotechnical report the ground water elevation within the mining tailing was determined using the boring logs of SB-40 and SB-41, which were the closest to the dam. The boring logs estimated the ground water table to be ~5ft below the ground surface on the upstream side. During the two field inspections the ground surface was measured to be ~3ft below the top of the arches. Together, these values put the ground water table ~8ft below the top of the arches. An estimated value of 28 degrees was used for the drained peak friction angle of the soil, per the recommendations of project delivery team (PDT) members. This value was based in part on the description of the material provided within URS 2009.

From the unit weights and historical observations, the usual and unusual static load cases defined by EM 1110-2-2100 were evaluated. The usual static load case was based on the existing conditions found during the URS soil exploration with the ground water ~8ft below the top of the arches. For the unusual load case, the water elevation behind the dam was assumed to be at the top of the arches. This assumption came as a direct result of a photo taken in 2006, which showed water spilling over the arches during a large rain event. Due to the lack of information on the Argonaut Dam it was difficult to assume any value related to an extreme water event. Therefore, for Phase I of this investigation, only the usual and unusual loading conditions were considered. Uplift pressure loads under the buttress walls and arch base were applied per FERC recommendations (FERC, 1997) Chapter 10. This document provided two examples of uplift pressure distributions for buttress dams, one without a slab connecting buttress walls and one with a slab connecting buttress walls. In the case of the Argonaut Dam the selected uplift pressure distribution was that of no slab connecting the two buttress walls. This resulted in a small trapezoidal pressure distribution under the arch base, which quickly transitions to a uniform uplift pressure distribution under the buttress walls, see Figures 4 and 5.

For seismic loads, the USGS website was used to determine both the operational basis earthquake (OBE) and maximum design earthquake (MDE) peak ground acceleration (PGA) values:

Seismic Loads:

- OBE with a return period of 144 Years = 0.0685g (Unusual Event)
- MDE with a return period of 949 years = 0.1175 (Extreme Event)

Because the initial construction of the arch was to hold back mining tailings, the most appropriate loading conditions within EM 1110-2-2100 in regards to stability evaluation matched those of a retaining wall design, which has been provided in the table below.

Table 1. Description and Classification of Load Cases

Load Case	Loading Description	Classification
LC – 1	Normal Operating	U
LC – 2	Normal Operating + Flood Loads	UN
LC – 3	Normal Operating + OBE	UN
LC – 4	Normal Operating + MDE	E

The reader is directed to EM 1100-2-2100 for descriptions of the loads involved for each loading condition. Two free body diagrams illustrating some of the loads acting on the arches have been provided in Figures 4 and 5.

DEVELOPMENT OF THE FINITE ELEMENT ANALYSIS MODEL

To evaluate the stability of the dam, a FEA model of Arch #8 (the tallest arch) was modeled in SAP2000 (structural analysis computer program). Construction of the 3D model began by drawing 3D faces in AutoCAD that represented either the buttress walls or the arch. These 3D faces were then imported into SAP2000 and extruded into solids between sets of 3D faces. This method was determined to be the easiest way to create the model and maintain dimensional accuracy. However, in some cases, due to the complex geometry of both the arches and buttress walls, edge constraints were used in locations to insure a smooth stress distribution across solids whose edges did not align.

An attempt was made to maintain an aspect ratio of 3:1 for all solids throughout the model. However, due to geometrical constraints this rule may have been exceeded in some locations. The tie beams that spanned between the buttress walls were not included into the model due to their distressed state. Figure 6 below illustrates the final model created for the analysis along with a field photo of Arch #8. Figure 7 has been provided to illustrate the deteriorated condition of the tie beams.

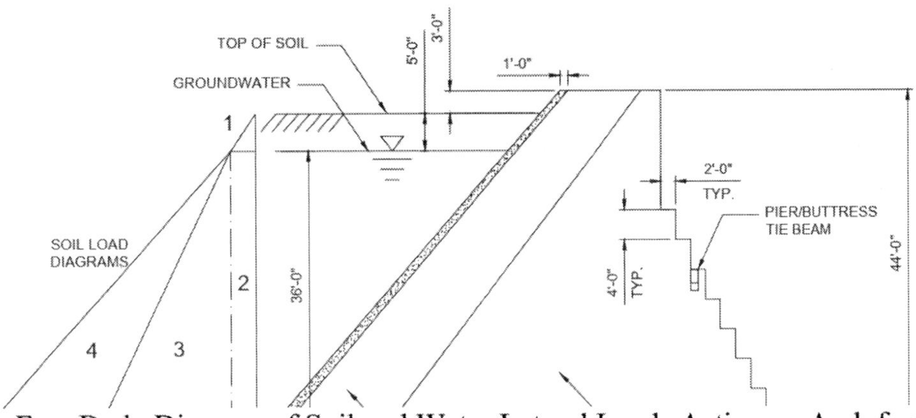

Figure 4. Free Body Diagram of Soil and Water Lateral Loads Acting on Arch for LC-1

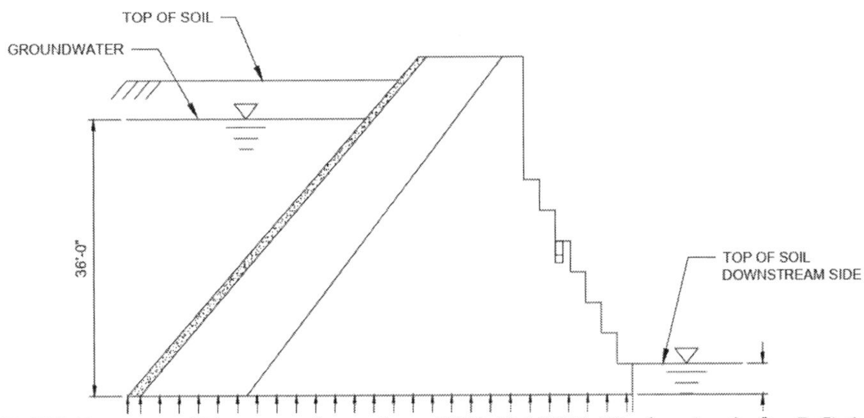

Figure 5. Uplift Pressure Diagram (based on FERC 1997) Under Arch for LC-1, LC-3, & LC-4,

The entire base of the foundation was supported on gap links in the Global Z direction with a stiffness of 1000 psi/in, which is the current assumed value for the coefficient of subgrade reaction. This assumed value will be updated following the next geotechnical investigation, discussed in detail later. In addition to the gap links, restraints were also applied in the Global X and Y direction to account for the partial embedment of the arch and buttress walls. Loads were applied to the model using surface pressures based on joint patterns for the water loads and soil loads. For the dynamic loading of the soil a joint load was applied at 0.6*height of the soil which was based on EM 1110-2-2100, EM 1110-2-2502, and the assumption that the soil was a non-yielding backfill due to the arch shape. For the inertial loading of the arch and buttress walls, a multiplier was applied into the SAP model as 2/3s*PGA for both the OBE or MDE earthquakes. For further information regarding this seismic coefficient method, please refer to EM 1110-2-2100 section 4.7b and EM 1110-2-6053 section 4.2a. It should be noted that because only the stability of a single arch and its corresponding buttress walls were checked, none of the loads within the model were factored. Figure 8 illustrate examples of the applied loads and the deflected shape of the model.

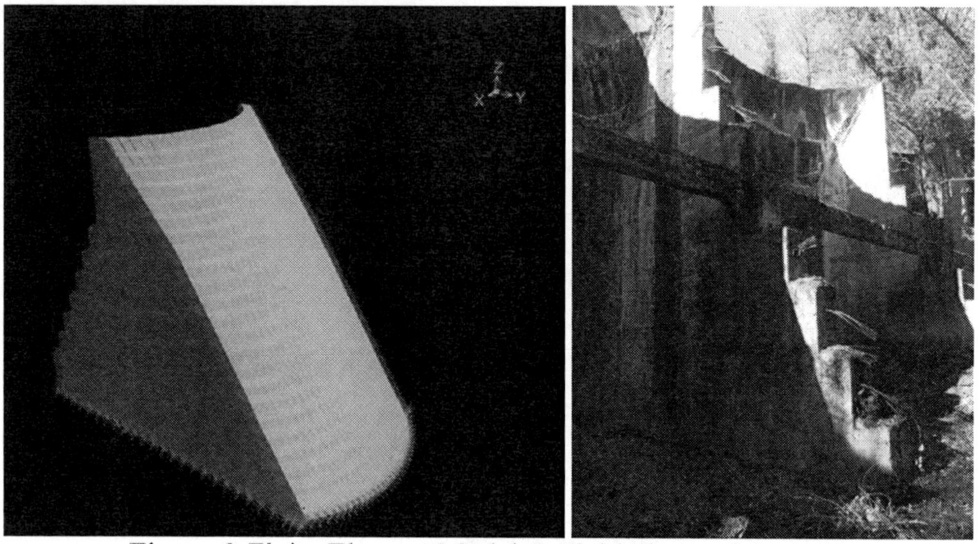

Figure 6. Finite Element Model & Field Photo of Arch #8

Figure 7. Distressed State of Tie Beams

STABILITY RESULTS

To evaluate sliding at the base, a friction coefficient (ϕ) of 0.35 and strength of cohesion (c) value of 20 ksf were assumed between the weathered rock and the dam per the recommendations of PDT members. Based on these values the following table provides the results for the sliding factor of safety for each load case based on Eq. 3-1 of EM 1110-2-2100:

$$FS = \frac{N \tan\phi + cL}{T} \tag{1}$$

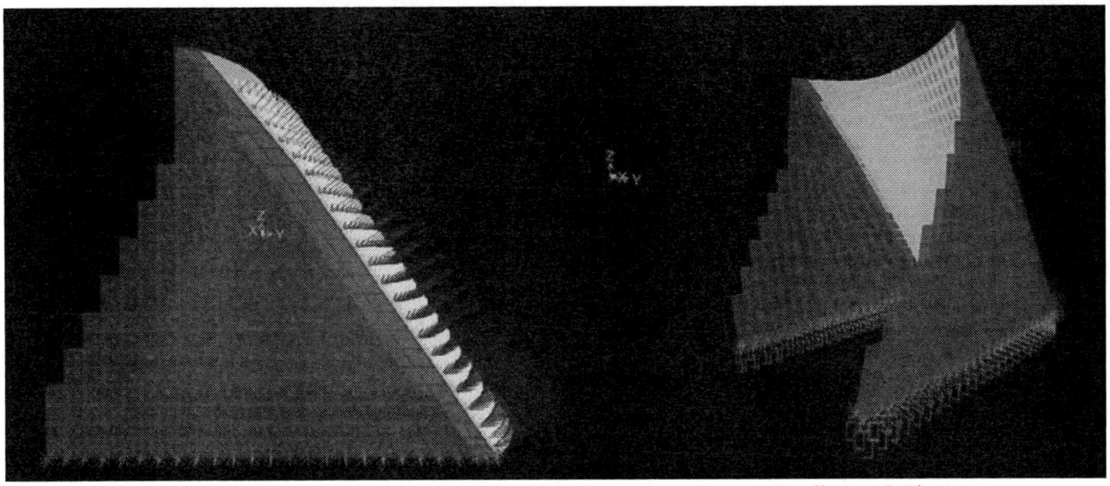

Figure 8. Applied Soil and Water Pressure Loads to Arch and Deflected Shape

Table 2. Stability Results for Sliding along Foundation

Load Case	Normal Force (N) (Kip)	Sliding Force (T) (Kip)	Calculated Factor of Safety	Required Factor of Safety	Result
LC-1	1881.94	1932.56	1.82	2.0	FAIL
LC-2	2329.04	2612.31	1.41	1.5	FAIL
LC-3	1891.14	2237.29	1.57	1.7*	FAIL
LC-4	1891.14	2445.70	1.44	1.3*	OK

*See Table 3-2 of EM 1110-2-2100 for selection of Factor of Safety value.

The normal force and sliding forces were determined from the global reactions within the SAP2000 model. The normal force was captured through the global Z direction, and the sliding force was captured through the global Y direction.

From the results it is evident that sliding stability was exceeded for LC-1, LC-2, LC-3, but not the seismic load case LC-4. It is important to note that the result of "FAIL" does not necessarily mean that failure of the dam is imminent, but rather that the dam does not currently meet safety guidance and standards.

A second sliding shear stability failure mode was also checked, which dealt with sliding shear through concrete versus sliding shear at the base of the dam. This stability check was added due to observations made in 2013, in which it was noted that the embedded cable had corroded and possibly created a weakened plane in the concrete. Because the friction coefficient of the concrete / weakened plane is unknown, a range of values were checked using ACI 318-11. For simplicity the required factor of safety for sliding shear at the base was also used to compare values for sliding shear through concrete under operating conditions or unfactored loads. The table below presents the range of safety values for the sliding shear through concrete failure mode:

Table 3. Internal Stability Results for Sliding Shear through Concrete (Operating Conditions Loads Not Factored)

Load Case	Normal Force (Kip)	Sliding Force (Kip)	Factor of Safety for m = 0.6	Factor of Safety for m = 1.4	Required Factor of Safety (Based on Table 6 Values)	Result
LC-1	1881.94	1932.56	0.58	1.36	2.0	FAIL
LC-2	2329.04	2612.31	0.62	1.44	1.5	FAIL
LC-3	1891.14	2237.29	0.51	1.18	1.5	FAIL
LC-4	1891.14	2445.70	0.46	1.08	1.1	FAIL

From the results it can be seen that neither selected friction of coefficient value reached the minimum required factor of safety. Of particular concern is how low the factor of safety is for the coefficient of friction of 0.6, which correlates with concrete cast against a hardened concrete surface. Given the amount of efflorescence and algae along some of the cracked planes it is highly unlikely that a friction coefficient of 1.4, which correlates to monolithically placed concrete, is still present. The Figure 9 below illustrates the assumed shear plane failure location within the buttress walls.

The factor of safety for floatation was also evaluated and found to greatly exceed the required minimum values. For brevity, the tabulated values have not been provided. The equation for determining the factor of safety for flotation is provided below.

$$FS = \frac{W_S + W_C + S}{U} \qquad (2)$$

To determine the percent base in compression, the FEA model was used in which the total numbers of joints in compression were compared to the total number of joints within the base. It is important to note that for this particular stability load case the percent base in compression was evaluated based on the buttress walls joints alone and not the combination of the buttress wall and connecting arch joints, which if evaluated could falsely suggest a stability issue. Under this approach the buttress walls were determined to be 100% in compression for all load cases, thus meeting the minimum requirements of the EM.

To evaluate the bearing capacity of Arch #8, an assumed allowable bearing capacity of 20 ksf was selected based on EM 1110-1-1905 Table 4-8 and the recommendations of

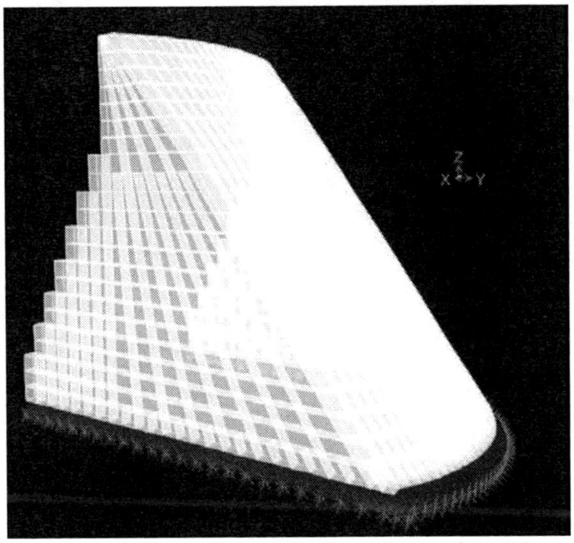

Figure 9. Assumed Shear Plane Failure Location Shown in Blue

PDT members. To determine q.max or the maximum bearing pressure, solid element 912, which has an area of 2.54 ft^2 was determined to be the element in each loading condition that had the highest gap link reaction values surrounding it. From the results it was observed that the allowable bearing capacity was exceeded in every load case. It is postulated that the exceedance of the bearing capacity could be a result of the slender footprint of the buttress walls ~ 2ft wide. In addition, these bearing capacity values are preliminary, and future geotechnical investigation results on the foundation may result in higher allowable bearing pressure values, which may be increased by 15% and 50% for unusual and extreme load cases respectively, see EM 1110-2-2100.

DISCUSSION OF STABILITY RESULTS

From the results of the various stability checks, the most concerning failure mode was sliding shear through the concrete. During inspection of the dam, no visible modern (square or circular) reinforcing bars could be seen, even in the sections that were heavily damaged. In addition, the use of ASTM A15 and ASTM A16 steel was not used as reinforcement until 1911 and 1913 respectively, see CRSI 2001. Based on this information it was assumed that no vertical reinforcement existed in the dam. Unfortunately, due to the lack of vertical reinforcement, the visible signs of corrosion (circled in yellow in Figures 10 & 11 below), and the visible horizontal cracked concrete planes (shown with arrows in Figures below), there is a high probability that this failure mode may occur. The concern of this failure mode was justified through the results, which showed failure of every load case despite the range of both high and low frictional coefficients values explored.

Figure 10. Corrosion of Cables within Arch #8 and Weakened Concrete Planes

Figure 11. Typical Observed Condition of Arches

Concern for the other stability modes that failed also exists, however the results in some cases, such as sliding at the base and bearing pressure, could improve following a more detailed geotechnical investigation. In regards to the percent base of compression, although it is not ideal for the arch to be lifting off its base, it was observed that the buttress walls were 100% in compression and were not lifting off their gap link foundation. If the buttress walls did show signs of lift off, this would underscore a potential stability issue.

REPAIR / RETROFIT OPTIONS

Several options are available to repair / retrofit multiple arch dams. Historically, many multiple arch dams have been repaired due to seismic governing load cases; however these same methods can also be used to upgrade dams for static load cases. Below are examples of similar dams that were repaired using various methods:

- Bear Valley Dam (Multiple Arch): Upgrades included filling the bays between the arches with mass concrete. (Hansen & Nuss 2013)

- Littlerock Dam (Multiple Arch): Designed also by J.S. Eastwood, used roller compacted concrete (RCC) to infill between the arch buttress to form a gravity dam section. (Hansen & Nuss 2013)

- Weber Dam (Multiple Arch): RCC was placed adjacent to the largest buttress walls to increase lateral stiffness. (Hansen & Nuss 2013)

- Stony Gorge Dam (Ambursen Slab and Buttress Dam): Simply supported diaphragm walls were place between the buttresses to support the buttresses in the cross canyon direction. (Hansen & Nuss 2013)

- Big Tujunga Dam (Thin Arch Dam): Used steel dowels grouted into the existing dam to tie to a concrete overlay, which thickened the arch. (Hansen & Nuss 2013)

The most applicable of these options to the Argonaut Dam include mass concrete, roller compacted concrete, and a concrete overlay. The simply supported diaphragm walls would not help the Argonaut Dam significantly to resist horizontal shear forces through the existing concrete.

CONCLUSION

From the stability results of the Phase I analysis it was determined that the analysis of the dam should progress to the next Phase II, a 35% retrofit design. As part of the 35% design a larger FEA model composed of a minimum of four of the tallest arches should be built to reinvestigate stability values. The FEA model and stability analysis should include updated values for the: modulus of subgrade reaction, allowable bearing pressure, friction coefficient value, and strength of cohesion value.

The aforementioned soil properties when inputted into the 35% FEA model will provide a more comprehensive analysis and evaluation of the stability results. Based on the Phase I stability results the most concerning / governing load case was not seismic, but rather LC-1 and or LC-2, which when evaluated under factored loading conditions will produce even higher static demands on the structure, especially for shear.

Options to repair / retrofit Argonaut dam against failure in the various stability load cases were briefly explored in the Phase I investigation and found suitable for use despite the controlling load case being horizontal shear. To determine the most feasible, constructible, and economical repair alternative, the various proposed options should be developed to the 35% design level and compared against one another. In addition, until a retrofit option has been implemented, the dam should be inspected on a yearly basis and during high precipitation events for any signs of dam instability / failure (i.e. significant widening of existing cracks, new horizontal cracks and or excessive deflections in the buttress walls).

REFERENCES

Concrete Reinforcing Steel Institute (CRSI), (2001) "Evaluation of Reinforcing Bars in Old Reinforced Concrete Structures". Engineering data report number 48.

ER 1110-2-1156, (2014). Engineering Regulation "*Engineering and Design -Safety of Dams - Policy and Procedures,* United States Army Corps of Engineers, Washington, DC.

EM 1110-2-2100, (2005). Engineering Manual *Stability Analysis of Concrete Structures,* United States Army Corps of Engineers, Washington, DC.

EM 1110-1-1905, (1992). Engineering Manual *Bearing Capacity of Soils,* United States Army Corps of Engineers, Washington, DC.

EM 1110-2-6053, (2007). Engineering Manual Earthquake Design and Evaluation of Concrete Hydraulic Structures, United States Army Corps of Engineers, Washington, DC.

Federal Energy Regulatory Commission, (1997) "Chapter 10 Other Dams" Engineering Guidelines for the Evaluation of Hydropower Projects.

Federal Energy Regulatory Commission, (1999) "Chapter 11 Arch Dams" Engineering Guidelines for the Evaluation of Hydropower Projects.

Forrest, M., Bureau, G., Lazarte, C., Hutchings, S., (2000) "Seismic Rehabilitation of Weber Dam" Association of State Dam Safety Officials 2000 West Region Annual Conference.

Nuss, L. K., and Hansen, K. D., (2013) "Seismic Upgrades to Concrete Dams Then and Now" ASDSO The Journal of Dam Safety Volume 11 Issue 4, pp. 21-35.

URS (2009) "Argonaut Mine Tailings Site", Site Investigation Report, Agreement No. 07-T337.5 for California Environmental Protection Agency.

THE DAMAGE RATING INDEX: ASSESSING THE SEVERITY OF ALKALI-AGGREGATE REACTION IN CONCRETE DAMS

Fred Shrimer, P.Geo., LG/LEG (WA)[1]

ABSTRACT

The use of Petrographic examination, along with various staining techniques to identify Alkali-Silica Reaction (ASR) gel products, is well known within the concrete rehabilitation and management fields as methods to identify the nature of concrete condition and deterioration. For structures such as dams, where management strategies are critically dependent upon accurate and reliable assessments, a new method, called the 'Damage Rating Index' (DRI), has been used to assess not only the degree of severity of deterioration in concrete due to ASR, but also to provide a means of relative degrees of damage in various components of the structure.

A description of the method is provided that illustrates how the DRI procedure measures such damage, how it can be used to distinguish ASR damage from other mechanisms of deterioration, and how the DRI values relate to damage observed in the structure. A number of case examples are provided, along with some discussion and thoughts for further consideration.

INTRODUCTION

The effects of concrete deterioration in hydraulic dams can be a major cost and engineering consideration in the management of these structures. Maintenance costs, dam safety, service life and environmental effects represent major elements of managing dam infrastructure.

When dams are constructed of concrete, the expectations for longevity, durability and troublefree service are increased, due in part to the costs and efforts associated with this construction approach. Concrete construction is capable of providing greater longevity and service life for dams, when compared with earthen or rockfill dams.

Often, due the locations of dams, producing concrete at the site provides challenges associated with sourcing of cement and aggregates, as well as equipment and crews; thus, "getting it right" is all the more important. Most dams typically source the construction materials (rock and/or sand and gravel aggregates) locally, but for earthen and rockfill dams, the criticality of material sourcing is considerably lower.

Once construction of a concrete dam has been completed, the service life of the concrete elements begins. The ingredients and the workmanship are set, and environmental and service conditions are the only variables that will continue to affect the 'life' of the concrete. It is therefore critically important that the concrete ingredients are compatible with the expected service life of the structure, and that the execution of the concrete

[1] Senior Geoscientist, Golder Associates, Vancouver, BC, Canada, fshrimer@golder.com

construction has utilized good practices typical for production and the placement, consolidation, finishing and curing of the concrete. In common practice, concrete dams are expected to fulfill service life requirements of one hundred years or more.

Thus, changes in the concrete that result from chemical reactions involving the aggregates used to manufacture the concrete are considered unacceptable, and should be diligently avoided.

In practice, however, dams may frequently develop alkali-aggregate reactions (AAR) that affect the concrete, for a variety of reasons. The primary among these is that reactive aggregates may have been used. This in turn may be because the technology available at the time of construction was not capable of identifying the aggregates as “reactive”.

Tremendous developments in the field of correctly characterizing the type and the degree of potential for AAR have occurred over the past thirty to forty years; thus, dams that were constructed prior to that period may have been unknowingly constructed using aggregates and cement combinations that are reactive, albeit to varying degrees.

Photograph 1. A dam in British Columbia in which ASR expansion in the concrete prevented the safe operation of the radial gate.

For concrete dams, the need to deal with or manage various mechanisms that may affect the intended service life of the structure is of paramount importance. If, for example, a dam constructed thirty five years ago is expected to provide an additional ninety years of service, the determination that AAR is or may be affecting the concrete is important, if it appears that the AAR may reduce the structure's serviceability or safety to an unacceptable degree.

To adequately understand the nature and extent of AAR in a dam, a suitably-designed and executed AAR investigation is required. For example, it may be that the concrete used in the dam's construction utilized different aggregates, with differing reaction potentials, such that some concrete elements exhibit differing AAR responses, potentially ranging from "nil" to "highly reactive". Thus, while one concrete element may be in very good condition, exhibiting no signs of deterioration attributable to AAR, adjacent elements may show significant cracking, ASR gel, distortion and distress. Distinguishing between variations in concretes that were used in the construction, and distinguishing the relative degree of AAR in those concretes is important if proper characterization of the dam's concrete elements is to be relied upon in making decisions about the continued management of the structure.

The best methods for such characterization are (1) visual assessment and examination by a qualified AAR investigator of the concrete during field survey supported by (2) detailed petrography-based examination of samples of the concrete in the laboratory. This paper describes the Damage Rating Index, which is a petrography-based procedure that has been developed and has been used for such characterization work.

DAMAGE RATING INDEX METHOD — A DESCRIPTION

The Damage Rating Index (or "DRI") procedure utilizes Petrographic observations made during microscopic examination of the concrete. Samples of concrete are first prepared by sawcutting and polishing the sawn surface. The sample is then marked into a grid of 1 cm^2 squares. Examination of the prepared sample is done using a stereomicroscope, ideally also equipped with a means of photographing relevant features that are observed in the sample.

A number of prescribed features are recorded if observed; the features which are recorded are those which would be related to the occurrence of AAR in concrete. These include (a) cracks in aggregate, (b) cracks in aggregate that contain ASR gel, (c) cracks in paste, (d) cracks in paste accompanied by ASR gel, (e) air voids containing ASR gel, (f) reaction rims and (g) debonded aggregate. Recently, other features have been to the list, on the basis that they have been observed by different analysts performing the method: (h) ASR gel-infused paste and (i) corroded aggregate particles. In addition, some discussion about the role and importance of fine aggregate features has also occurred. The features listed above are each assigned a factor, based on their relative importance in the production of "damaging effects" of ASR.

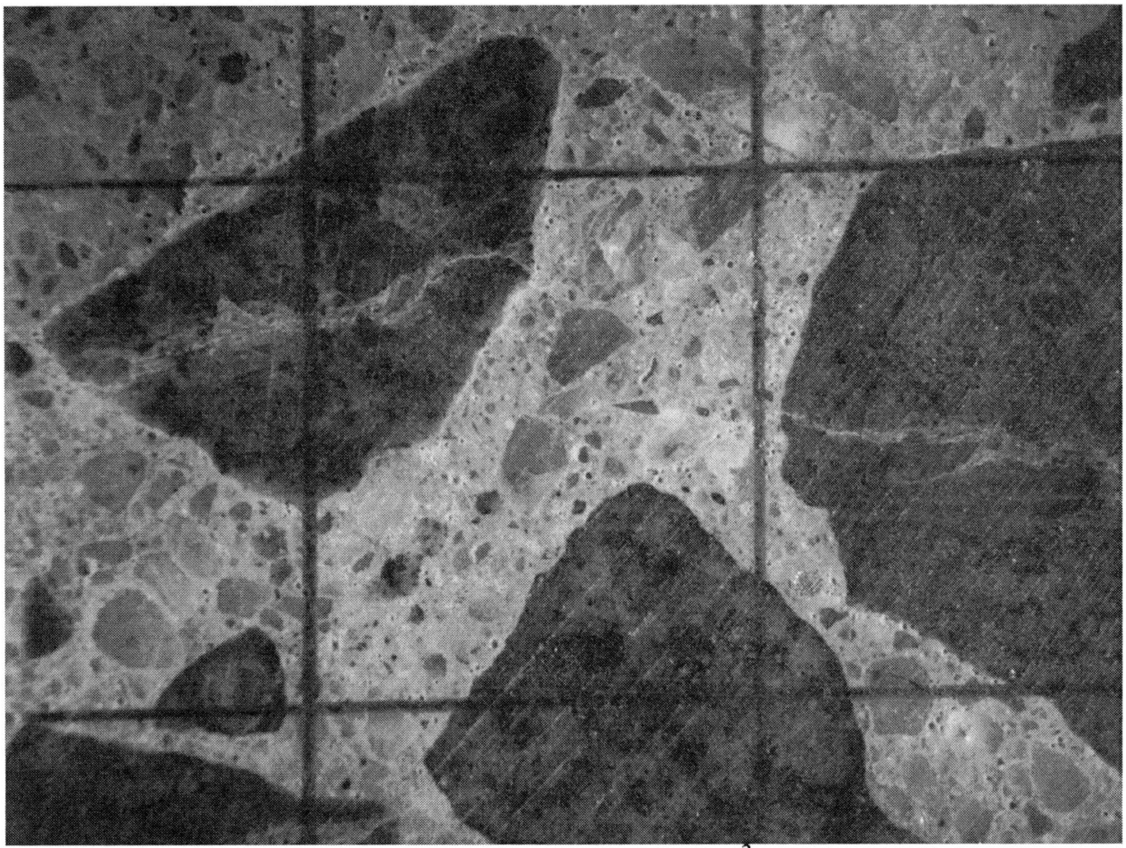

Photograph 2. Polished concrete surface scribed with 1 cm^2 grid, illustrating an ASR-gel-filled crack passing through two aggregate particles and the paste matrix of concrete

The factors currently used by the author for these features are as follows:

Feature	Multiplier
Crack in aggregate	0.25
Crack in aggregate with ASR gel	2
Debonded aggregate	3
Reaction rim	0.5
ASR gel in air void	2
Crack in matrix	0.5
Crack in matrix with ASR gel	4
Corroded aggregate	3

In the examination, the number of each of these features is counted for each square centimeter of the grid. After these features have been tabulated for the sample, the total is multiplied by the appropriate factor, to produce a gross DRI value. This is then normalized to a standard area of 100 cm^2, to produce a final DRI for the sample.

USE OF THE DAMAGE RATING INDEX IN DAM INVESTIGATIONS

Originally developed in the 1990s specifically for use in assessing ASR in concrete dams (Dunbar and Grattan-Bellew, 1995[i]), the DRI has been applied in evaluation of a variety of structure types. Powers and Shrimer (2007)[ii] demonstrated the successful use of the DRI method in a variety of ASR investigations, including concrete pavements at airports, concrete roadway pavements, bridges, tunnels, reservoirs, mine and industrial facilities, railway structures, and others. In addition, it has been used to characterize laboratory concrete test specimens.

Often it has been found to be of great assistance in the following:

- Distinguishing between the extents of ASR effects in concrete from different parts of the same structure or project
- Distinguishing between the effects of ASR and other deterioration mechanisms (e.g., freeze-thaw attack, reinforcing steel corrosion, sulphate attack, impact/excessive loading). This is augmented by classical petrographic methods.
- Characterizing the amount of ASR in concrete and projecting whether additional effects of the reaction might be anticipated
- Relating the serviceability of the concrete to its service life expectancy
- Determining whether the ASR that is known or thought to be present in concrete has reached a stage that is or may be of concern
- Identifying whether ASR in concrete is due to fine aggregate or coarse aggregate reaction
- Determining whether ASR in concrete is producing/has produced damage of significance or merely cosmetic (non-structural) effects
- Identifying the progress of ASR in concrete by reviewing current-stage DRI values and comparing them with previous DRIs obtained in the past

ADVANTAGES/BENEFITS OF THE DRI

There are several positive benefits associated with the use of the DRI as part of an investigation of AAR in a dam.

Chief among them is the ability to convey qualitative Petrographic information to users who need to know answers to questions such as:

- Is the damage due to AAR or something else?
    - The reason this is important is that different deterioration mechanisms should be repaired using different methods and strategies. If the repair approach that would be used to address freeze-thaw damage were used to repair AAR-affected concrete, it would only be a matter of time before the effects of AAR were once again observed.
- Is the AAR fairly minor and superficial, or is it quite serious and in need of immediate repair / mitigation?
    - This is important because if the AAR is of a minor nature, then it is likely that repairs will be minimal, or may not be required at all. If the AAR is severe and

  - advanced, then budgeting for appropriate measures will be of significant importance in management of the dam
  - If the safe operation of the dam is affected by AAR, such as the jamming of radial gates due to expansion of concrete, then this will/should be of prominent importance in the overall management of the dam
- How far along is the AAR? Does it appear to be done, is it about to stop, or is it just beginning and can there be more expected?

The best method for characterizing AAR in concrete has been (and continues to be) Petrographic Examination, using the Guide in ASTM C 856. However, the data produced in this method is highly descriptive, utilizing the language of petrography: it provides qualitative information of a type that is not readily grasped by engineering staff, managers, operations and construction staff, and the like.

In essence, the DRI "translates" the qualitative/descriptive Petrographic data into a numeric value so that the output can be readily understood.

The currently used DRI scale is easy to understand: the higher the number, the worse the effects of the reaction in that sample of concrete.

CASE EXAMPLES

Dam A

A dam constructed in the mid-1950s was experiencing difficulties with operation of its single radial gate. The dam was constructed using locally-derived sand-and-gravel materials for concrete aggregates. Although no records were kept at the time of construction, regional cement supplies had been producing cements with Na_2O-equivalents of 0.6% or lower. No evaluation of the aggregates was conducted at the time that construction was undertaken.

Photograph 3. Spillway wall detail, showing advanced map cracking and efflorescence in concrete

The dam exhibited considerable damage from ASR, including cracking, differential movement and displacement of concrete elements, jamming of the single radial gate due to expansion of concrete in the gravity sections and walls, staining, and loss of concrete strength.

Following a detailed field inspection, sites for drilling cores were selected. A series of core samples were extracted from various elements of the dam to represent various condition of the concrete, from slight to serious ASR effects.

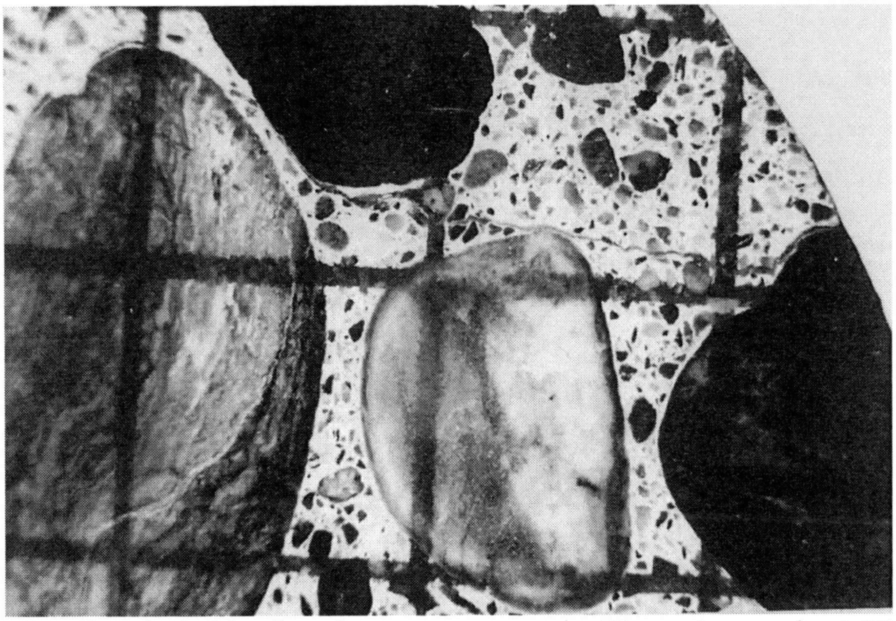

Photograph 4. Polished slab taken from a concrete core, illustrating cracks, ASR gel and reaction rims.

Photograph. Core fractured, exposing fresh ASR gel rims on aggregate in concrete.

Petrographic examination of the cores confirmed that the concrete had been affected by ASR, albeit to different degrees. In order to evaluate more precisely the relative effects of ASR in the concrete, we performed DRI analyses on several cores from visually more affected and less affected concrete in the dam. The purpose of the DRI testing was to evaluate whether the samples from the more affected concrete could be demonstrated to have more ASR damage in them relative to the samples from less affected concrete elements.

Approximately twelve cores were subjected to the DRI method analysis. The results of the DRI testing are given in Table 1, below.

Table 1. DRI Results from Concrete Cores, Dam A

CONCRETE CONDITION	DRI VALUES	DAMAGE RATING
Significant damage: cracking is frequent, efflorescence common, staining and gel exudation observed	552; 459; 609; 520	Significant – Serious
Moderate damage: Cracking observed but not extensive, some efflorescence, modest staining	179; 273; 225; 303	Moderate
Minor damage: Light/faint and infrequent to no cracking, minor staining and efflorescence	46; 70; 104; 115	Infrequent / minor

On the basis of the DRI analyses, we found that there was a good correlation between the DRI data and the condition of the concrete in the structure, in that, for concrete that was in poor condition in the dam produced correspondingly high DRIs. The opposite was also found, that concrete exhibiting minor or light indications of damage in the field had DRIs that were typically less than 100.

Of significant note was that the manifestations of damaging effects of ASR were more apparent in the concrete samples with higher DRIs: ASR gel was more common, cracks that contained ASR gel were more widespread and common, and they were longer/wider. These features correspond with greater damaging effects in concrete, such as expansion and cracking, distortion and loss of strength.

In this particular case, the safe operation of the dam was being compromised by a radial gate that was not operating properly, and could not be fully opened. After having conducted a detailed site investigation, including (1) a concrete condition survey and (2) completion of ranking exercise for concrete damage in various elements of the dam, (3) laboratory evaluations including (3a) Petrographic Examination (3b) Damage Rating Index and (3c) compression strength tests, it was concluded that the gravity sections and wall of the spillway have been expanding due to ASR and were 'squeezing' the portion of the dam that housed the radial gate.

Dam B

This dam, located in central/eastern Canada in the Canadian Shield region, was constructed in the 1940s, with some additional construction in the 1960s. Portions of the dam structures were suspected to be affected by the effects not only of ASR but also of freeze-thaw cycles and degradation.

In order to assess these and other possible degradation mechanisms, and to attempt to gauge the relative effects that could be ascribed to ASR, we conducted a series of analyses that began with Petrographic examination and were supplemented by DRI analyses.

Initial Petrographic examinations, conducted in accordance with ASTM C856 (Petrographic Examination of Hardened Concrete), were completed on the samples, on both fractured surfaces and on slab-cut polished faces. Supplementally, thin-section mounts were also prepared to enable examination under polarized light.

These examinations indicated that the concrete had been affected by minor carbonation, some freeze-thaw activity, and ASR.

DRI tests were conducted on six cores, selected to represent various conditions of concrete, ranging from “good” to “fair” to “poor” – these designations were selected by the Owner of the dam, in conjunction with the lead petrographer on the project.

Table 2. DRI Results from Concrete Cores, Dam B

CONCRETE CONDITION	DRI VALUES	DAMAGE RATING
Poor Condition: Cracks are noted with great frequency, some displacement and distortion is evident, efflorescence common, staining and gel exudation common	516; 632	Significant – Serious
Fair Condition: Cracking is observed but is not pervasive, some efflorescence, modest staining	236; 347	Moderate
Good Condition: Cracking is not significant, occasional, cracks are discontinuous and not wide/deep. Minor staining and efflorescence	31; 84	Infrequent / minor

It was found that there was a good correlation between the DRI values obtained for the concrete and the observed condition of the concrete in the dam.

Where cracking and evidence of expansive disruption of the concrete was pronounced in the concrete elements of the dam, the DRI numbers were high. The opposite was also true, that cores sampled from areas of the dam that exhibited little damage had DRI values that were low.

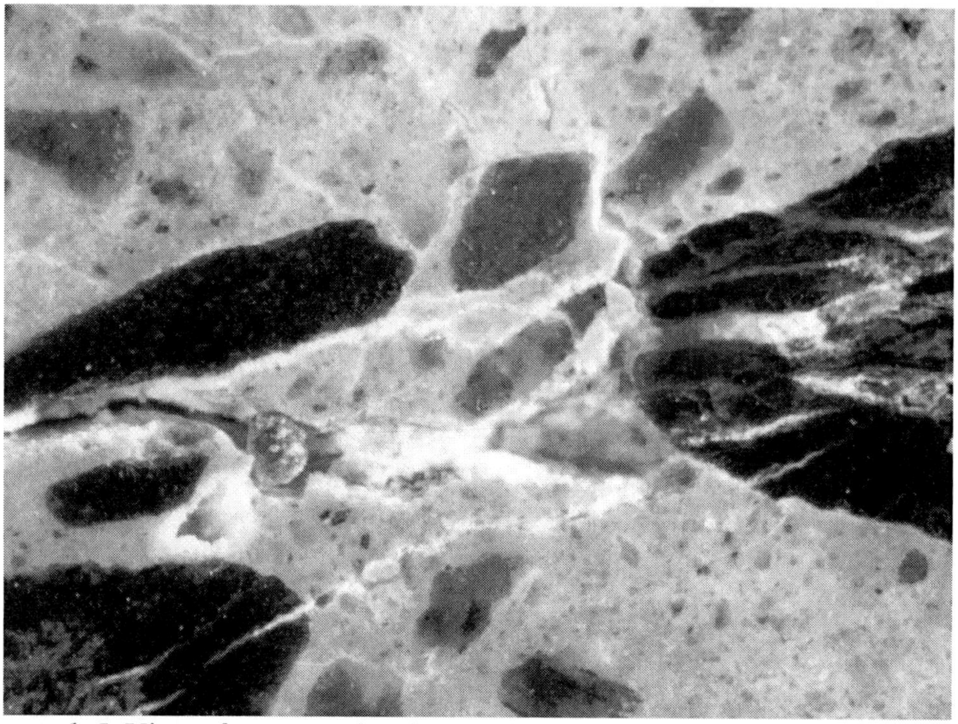

Photograph 5. View of concrete showing cracks in paste, containing ASR gel, passing through and around aggregates.

Using the results of the DRI assessments, it was possible to augment the information developed in the Petrographic examination. While Petrographic work alone had identified that both ASR and freeze-thaw had affected the concrete, the DRI was able to further the assessment and assign weighting to these mechanisms.

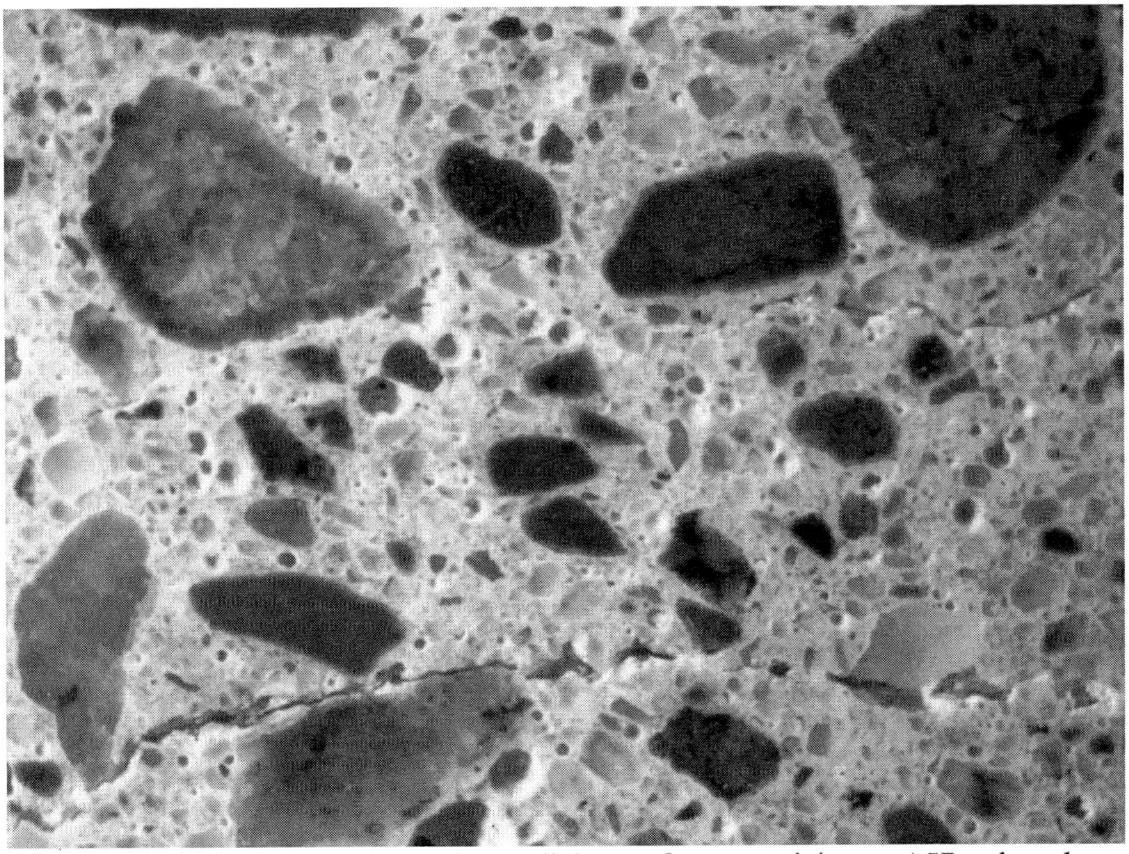

Photograph 6. Cracks in paste matrix parallel to surface, containing no ASR gel product. These cracks were inferred to be due to the effects of freeze-thaw damage rather than the effects of ASR.

The DRI analyses indicated that the concrete had been more damaged by the effects of ASR than by freeze-thaw. This was based upon observations such as the following:

- Cracks that contained ASR gel were far more frequent than cracks that were devoid of gel, the latter being attributable to the effects of freeze-thaw.
- The paste-aggregate interfacial zones of debonded aggregates were typically noted to contain vestigial ASR gel in these zones, indicating that ASR had been involved in the debonding process rather than freeze-thaw processes
- Cracking occurred at significant depths within the concrete, rather than being more common in surficial concrete and decreasing with depth
- Reaction rims and ASR gel were present throughout the concrete and were proportional in number to the DRI

Additionally, it was inferred that the relative amounts of ASR-induced damage, as expressed in the various cores, potentially also reflected relative stages of ASR in the concrete, with low DRI values representing concrete that had only undergone a low amount of the reaction, while high DRI values represented concretes that had undergone considerably stronger reaction.

Further to this, the low DRI concretes – given that the were made using the same ingredients (i.e., cement and aggregates) and a similar proportioning of those ingredients (cement content, aggregate proportions, no supplementary cementing materials) – were seen as having the potential to develop the same level of ASR damage that was observed in the high-DRI concretes. It was thought that these concretes may have had a lower/different opportunity to react, perhaps due to specific exposure conditions. In some cases, the low-DRI concretes had been less directly exposed to humid air, for example, while most of the high-DRI concretes were exposed on a sometimes continuous basis to humid conditions.

DISCUSSION

Development and use of a method that 'translates' qualitative petrographic data into numeric data that is readily understood by engineers and management staff, and that provides a sense of relative importance of ASR in concrete, has been the objective of the DRI method since its inception.

Review and improvements for this procedure have been discussed in Rivard, Fournier and Ballivy (2002)[iii] , in Shrimer (2006)[iv], and in Sanchez et al. (2014)[v]. Some suggestions for revisions, improvements or additions of/to the elements of the DRI method have been presented in these and other papers. These have included discussion around the weighting factors associated with the DRI features, and centering on the relevance of various features in terms of (1) the reaction itself and (2) the damage stemming from the reaction. These are both linked with the (a) potential for the concrete to undergo continued ASR and (b) the present and future performance requirements of the concrete.

Some of the difficulties that have been noted in these and other papers relate to the relationship between the condition of the concrete samples that are used for DRI analyses and the following:

- Current condition of the concrete in the structural elements
- The amount of reaction that the concrete has undergone
- The amount of reaction still possible (e.g., the 'Potential for Future Expansion') in the site concrete
- The amount and types of effects of the reaction that are likely to produce damage in the concrete
- Distinguishing between features that are (or may be) linked with ASR vs other mechanisms such as freeze-thaw, corrosion, etc.
- Apportioning causality to the various features in terms of relevance/importance

Some workers have expressed that the choice of weighting factors to be used will need to be agreed upon within specific projects and for specific needs, but should be used consistently, once selected. Additionally, a scale that assists with interpretation as to the meaning of the DRI should be provided, in order to convey meaning to the output of the method.

The citations noted above have posited varying weighting factors, and have advanced arguments to support the use of the values to be assigned for these factors. There have been discussions that center on whether some features are of significance in determining DRIs, and considerations have related to the relationship of the feature with observed damage or deterioration of the concrete.

Differences in the interpretation of ASR-induced damage, as expressed in the DRI value, are illustrated in the table below which provides DRI scales correlating DRI values with severity of ASR in concrete. The compared DRI ranges and rankings are based on the author's reference system and that given in Fournier et al. (2009)[vi]

DRI	Degree of ASR (after Shrimer [2006])	DRI	Degree of ASR (after Fournier et al. [2009])
0 – 40	Negligible	0 – 250	Non to trace
40 – 125	Minor		
125 – 300	Moderate	250 – 500	Fair to moderate
300 – 500	Significant	800 – 800/1000	Moderate to severe
500 – 650	Serious		
>650	Very serious	≥800/1000	Severe to very severe

Review of the DRI ranges and the ratings assigned to those ranges, presented above, indicates the following:

- Agreement that higher DRI values correspond with higher damage and low DRIs reflect comparatively little damage
- Slight to significant disagreement regarding where the threshold points fall for various condition/severity ratings

It is apparent that additional work is required to bring various DRI schemes and methods into better alignment. Some initial round-robin studies have been undertaken, and a workshop to introduce practitioners who have interest in the method was held in October 2008.

At this time, a standardized approach has not been adopted by workers involved in these assessments. The implications are that use of the DRI should recognize that repeatability associated with the results should be undertaken with a degree of caution. DRI analysts should provide a guide to interpret their results, as well as a synopsis of the rationale for the selection/application of DRI factors.

CONCLUSIONS

Despite the comments above concerning differences observed in various DRI approaches, it is concluded that the method offers a significant benefit beyond that provided in the

Petrographic Examination (C856) procedure in that it produces numeric data that can be appropriately scaled and understood by non-petrographic persons.

Interpreting the DRI data and utilizing it properly can add significantly to the optimal understanding of what can often be difficult and complex problems in managing expensive concrete infrastructure. At this time, the utilization of the DRI tool is considered to be a significant benefit to and enhancement in concrete condition survey work.

Working towards a unified system for DRI analysis is a goal that is achievable and would benefit the industry that deals with management of concrete infrastructure.

REFERENCES

[i] Dunbar, P. and Grattan-Bellew, P. (1995) Results of damage rating evaluation of condition of concrete from a number of structures affected by AAR, *Proceedings, CANMET/ACI Workshop on Alkali-Aggregate Reactions in Concrete*, Dartmouth NS, pp. 257 – 265.

[ii] Powers, L.J. and Shrimer, F.H.(2007) Quantification of ASR in Concrete: An Introduction to the Damage-Rating Index Method, International Cement Microscopy Association meeting, Montreal, QC.

[iii] Rivard, P., Fournier, B. and Ballivy, G. (2002) The Damage Rating Index Method for ASR Affected Concrete – A Critical Review of Petrographic Features of Deterioration and Evaluation Criteria. In Cement, Concrete and Aggregates, ASTM, Vol 24 – pp. 81-91.

[iv] Shrimer, F., Development of the Damage Rating Index Method as a Tool in the Assessment of Alkali-Aggregate Reaction in Concrete: A Critical Review", in *Proceedings*, Marc-André Bérubé Symposium on Alkali-Aggregate Reactivity in Concrete, Montréal, Canada, 2006, pp. 391 – 412.

[v] Sanchez L.F.M., Fournier, B., Jolin, M., Duchesne, J. (2014) Reliable quantification of AAR damage through assessment of the Damage Rating Index (DRI). Cement and Concrete Research (67) 74-92.

[vi] Fournier, B., Tremblay, S. and Frenette, J. 2009. "Results of the Petrographic and Stiffness Damage Testing of Concrete Cores from the Princess Margaret Bridge, Fredericton, NB." Gemtec – Consulting Engineers and Scientists, 71p.

AUTHOR INDEX

AUTHOR INDEX

AUTHOR INDEX

United States Society on Dams (USSD)
1616 Seventeenth Street #483
Denver, CO 80202
USA

CURRAN ASSOCIATES INC.
proceedings.com

ISBN 978-1-5108-2941-1
9 781510 829411

35th Annual USSD Conference 2015

Managing Risk and Uncertainty – Ensuring a Sustainable Future for Dams, Levees and Reservoirs

Louisville, Kentucky, USA
13-17 April 2015

Volume 2 of 2

35th Annual USSD Conference 2015

Managing Risk and Uncertainty – Ensuring a Sustainable Future for Dams, Levees and Reservoirs

Louisville, Kentucky, USA
13-17 April 2015

Volume 2 of 2

ISBN: 978-1-5108-2941-1

Printed from e-media with permission by:

Curran Associates, Inc.
57 Morehouse Lane
Red Hook, NY 12571

Some format issues inherent in the e-media version may also appear in this print version.

Printed by Curran Associates, Inc. (2016)

For permission requests, please contact United States Society on Dams (USSD)
at the address below.

United States Society on Dams (USSD)
1616 Seventeenth Street #483
Denver, CO 80202
USA

Phone: 303-628-5430
Fax: 303-628-5431

gene@ussdams.org

Additional copies of this publication are available from:

Curran Associates, Inc.
57 Morehouse Lane
Red Hook, NY 12571 USA
Phone: 845-758-0400
Fax: 845-758-2633
Email: curran@proceedings.com
Web: www.proceedings.com

USSD

Managing Risk and Uncertainty — Ensuring a Sustainable Future for Dams, Levees and Reservoirs

35th Annual USSD Conference
Louisville, Kentucky, April 13-17, 2015

Hosted by Corps of Engineers

On the Cover

Aerial view of the Cannelton Hydroelectric Project — one of American Municipal Power's four run-of-river hydroelectric projects currently under construction on the Ohio River. The powerhouse is a three-unit, single monolith, reinforced concrete, combination dam/hydroelectric powerhouse founded on alluvial river deposits and designed to be submerged in floods. The powerhouse is equipped with three horizontal bulb turbines each with a capacity of 28 MW and is capable of generating a total of 84 MW at a gross head of 25 feet. The powerhouse is provided with an exterior-type 175-ton capacity permanent bridge crane that operates on concrete runway frames above the roof in a direction parallel to flow and is an unique application for hydroelectric projects on the Ohio River, first implemented at the Belleville Hydroelectric Project.

U.S. Society on Dams

Vision

A world class organization dedicated to advancing the role of dam and levee systems and building the community of practice.

Mission

USSD, as the United States member of the International Commission on Large Dams, is dedicated to:

ADVOCATE:	Champion the role of dam and levee systems in society.
EDUCATE:	Be the premier source for technical information about dam and levee systems.
COLLABORATE:	Build networks and relationships to strengthen the community of practice.
CULTIVATE:	Nurture the growth of the community of practice.

FOREWORD

Included in these Proceedings are papers presented during the **35th USSD Annual Meeting and Conference**, held April 13-17, 2015, in Louisville, Kentucky. A separate book includes abstracts for each paper.

The theme of the 35th USSD Conference was ***Managing Risk and Uncertainty — Ensuring a Sustainable Future for Dams, Levees and Reservoirs.*** The Conference technical program was organized by several USSD Committees under the leadership of **B. Alex Grenoble**, HDR. The theme highlights that, while dams and levees are critical components of the nation's infrastructure, there are always risks and uncertainties associated with ownership and operation of these facilities. Recognizing and effectively managing these risks and uncertainties is critical to ensuring that these structures will continue to provide the societal benefits that we depend on without posing an unacceptable hazard to life, property or the environment.

Technical Sessions during the Conference focused on best practices and new technologies related to concrete dams, embankment dams and levees; dam safety and security; construction and rehabilitation; hydraulics; foundations, monitoring; and seismic and environmental issues.

The papers in the Proceedings were selected from abstracts submitted in response to a Call for Papers, and include both oral and poster presentations. Authors are specialists with broad experience from government agencies, utilities, academia, water districts, consulting firms and private industry.

The Conference Organizing Committee extends thanks and appreciation to the **Corps of Engineers**, Host of the 35th Annual Meeting and Conference.

Special thanks are also extended to the Committee Members who selected the abstracts and reviewed the technical papers, and to the authors who prepared the papers included in the Proceedings.

CONTENTS

Embankment Dams

Concrete Dams

Levees

Tailings and Ash Impoundments

Dam Safety and Security

Monitoring

Foundations

Public Awareness

Dam Decommissioning

AMP Hydro Projects

OLD DAMS, NEW FAULTS — NOW WHAT?

John C. Stoessel, P.E.[1]
C. Mike Knarr, S.E.[2]
Nicholas Von Gersdorf, P.E.[3]
Matt Muto, PhD.[4]
Zee Duron, PhD.[5]

ABSTRACT

Decades of freeze-thaw have taken a toll on the structural capacity of the three concrete dams associated with the Rush Creek Hydroelectric Project in the high Sierra Nevada Mountains of California. Signs of deterioration include weeping along the downstream face, concrete spalling, and disintegration of the cement/sand/rock matrix. The level of deterioration has led to concerns about the ability of the dams to maintain full reservoir under extreme load conditions. To further influence this concern, a seismic fault, located in close proximity to the three dams and thought to be inactive, was recently determined to be active, with the capability to produce high levels of ground shaking. After a thorough evaluation of each of the three dams, including nondestructive testing, core samples, and linear and nonlinear finite element analysis, it was concluded that the three dams could not be operated under full reservoir conditions. Agnew Lake was drained; Gem Lake was reduced to 60% level; and Waugh Lake was reduced to 50% level. At these reduced levels, the dams would continue operate safely under a seismic event with a probability of occurrence of 1 in 10,000 (0.0001) years until permanent structural changes could be implemented.

The project owner, Southern California Edison (SCE), has developed extensive retrofit designs for the three dams in order to get these projects back to full operation. These designs incorporate unique components to address the logistics of operating not only in a National Forest, but, in the case of two of the dams, operating in a designated Federal Wilderness Area. With the projects located in the high Sierra Mountains, transportation, utilization of modern construction equipment, and workforce support are complicated by physical and regulatory restrictions, requiring detailed plans and close coordination with the United States Forest Service.

This paper addresses the uniqueness and complexities of designing, permitting, and constructing the retrofits for these remote dams, which are at 8,000 to 10,000 feet elevation and have no access by vehicles, since they are several miles from paved roads.

---

[1] Senior Engineer, Southern California Edison Co., 1515 Walnut Grove Ave., Rosemead, CA 91770, (626) 302-0381, john.stoessel@sce.com
[2] Consulting Civil/Structural Engineer to Southern California Edison Co., 1515 Walnut Grove Ave., Rosemead, CA 91770, (626) 302-0521, mike.knarr@sce.com
[3] Engineer, Southern California Edison Co., 1515 Walnut Grove Ave., Rosemead, CA 91770 (626) 302-0332 nicholas.von@sce.com
[4] Engineer, Southern California Edison Co., 1515 Walnut Grove Ave., Rosemead, CA 91770 (626) 302-0473 matthew.muto@sce.com
[5] Professor, Harvey Mudd School of Engineering, 301 Platt Blvd., Claremont, CA 91711 (951) 440-4735

INTRODUCTION

The Rush Creek Project, licensed under the Federal Energy Regulatory Commission (FERC) Project 1389-CA, is comprised of three dams with associated reservoirs, two penstocks, one powerhouse containing two Pelton-Doble impulse turbines (13 megawatts total capacity), and two transportation railroad trams. The project is located in the Eastern Sierra Nevada mountains, approximately 300 miles north of Los Angeles, California, USA. The project has operated continuously since 1916 with generation occurring for six months out of the year, as a result of annual snow melt.

A plan has been developed to address the impact that environment deterioration has had on the structural integrity of each dam. This plan accounts for current extreme load conditions (seismic and flood), which are significantly higher than had been previously accounted for in analyses.

AGNEW LAKE DAM

The smallest of the three dams, Agnew Lake Dam is a concrete, multiple arch dam comprised of five complete and two partial arches. The dam is 30 feet high and 278 feet long. It is located at an elevation of 8500 feet and is subject to extreme weather conditions, including below freezing temperatures, high winds, and substantial snowfall. Construction was completed in 1916. The dam is located on United States Forest Service (USFS) land, but does not reside in a designated wilderness area. Figure 1 contains a photo of Agnew Lake Dam.

Figure 1. Agnew Lake Dam

Generation is possible directly off Agnew Lake. A flowline extends downstream of the dam several hundred feet, where it ties into the Gem Lake penstock. Generation using Agnew Lake as a water source is typically done when the lake is being drained or when the Gem Lake penstock is out of service. Generation is somewhat reduced, due to the lower head at Agnew Lake.

GEM LAKE DAM

The largest of the three dams, Gem Lake Dam is a concrete, multiple arch dam comprised of 16 complete and two partial arches. The dam is 80 feet high and 688 feet long. It is located at an elevation of 9058 feet and, like Agnew Lake Dam, is subject to extreme weather conditions. Construction of the dam was completed in 1916, but after being in service for seven years, a gravity section was added to the downstream side of the dam to address severe freeze/thaw damage that had occurred to several of the arches. This gravity section extended from the base of each arch to 30 feet below the crest. The crest of Gem Lake Dam is the boundary of the Ansel Adams Wilderness Area. Figure 2 contains a photo of Gem Lake Dam.

Figure 2. Gem Lake Dam

The primary water source for generation at Rush Creek Powerhouse is Gem Lake.

RUSH MEADOWS DAM

Rush Meadows Dam is a concrete single arch dam that is 50 feet high and 463 feet long. It is located at an elevation of 9419 feet. The dam is 2.5 feet thick at the top and 11.4 feet thick at the bottom. The spillway is a 3 foot notch in a wingwall at the right abutment. Construction of the dam was completed in 1925. The dam is located entirely in the Ansel Adams Wilderness Area. Figure 3 contains a photograph of Rush Meadows Dam.

Figure 3. Rush Meadows Dam

Generation directly from the reservoir (Waugh Lake) at Rush Meadows Dam is not possible, as there is no flowline connecting Waugh Lake to Rush Creek Powerhouse. Water released from the low level outlets at Rush Meadows Dam flows down Rush Creek and empties into Gem Lake.

Figure 4 contains a schematic that illustrates the interdependence of the three reservoirs and generation.

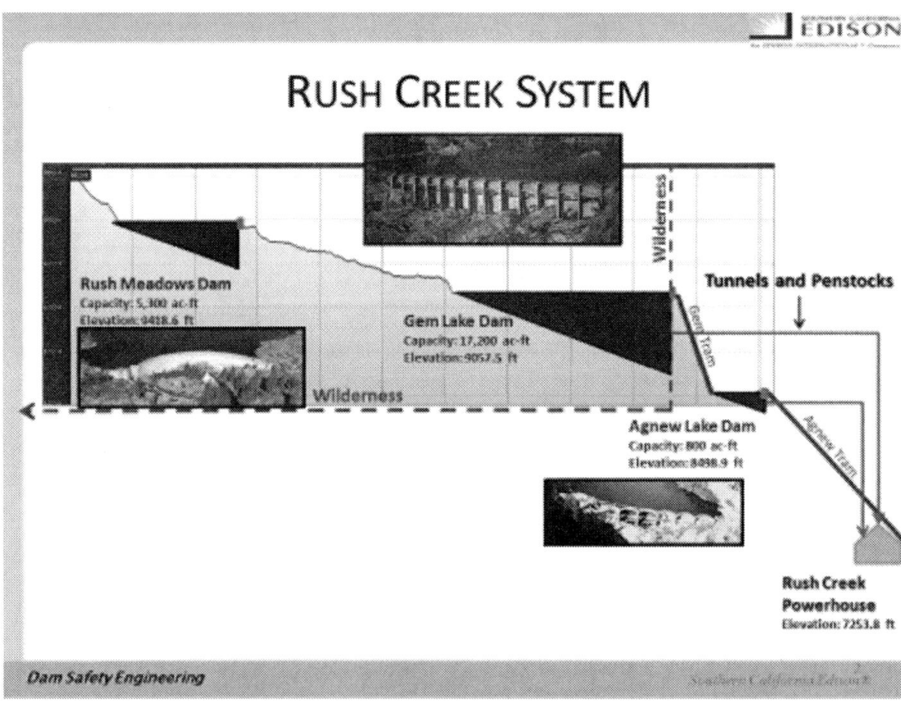

Figure 4. Schematic of Rush Meadows Project Hydraulics

CURRENT CONDITION OF RUSH MEADOWS, GEM LAKE, AND AGNEW LAKE DAMS

Given the extremely cold temperatures typical to these dam locations, the concrete has been damaged by recurring freeze-thaw. That condition is exacerbated by the relatively poor concrete mix that was typically used in early 1900's construction and was not developed to resist multiple years of freeze-thaw cycles. Because of the remote location of these dams and the difficulty of transporting materials to the jobsites, local sources of construction materials were used whenever possible. Most, if not all, of the aggregate used in the concrete during original construction was from rock adjacent to the jobsite. Much of this rock was flat and angular, which resulted in a relatively porous concrete that allowed migration of water into the dam. In addition, there is evidence of areas where large sized rock was used (cyclopean concrete), which also created a porous matrix. When temperatures dropped below freezing, the destructive forces of expansion took their toll in the form of cracking and spalling. As the concrete deteriorated, paths for water to easily flow through the concrete developed, eventually exiting the downstream face in the form of weeping or, in extreme cases, exiting in streamflow.

After decades of freeze-thaw cycles, pockets of very loose material resembling rubble occur. The overall strength of the structure degrades and extreme load conditions (seismic and flood) can lead to localized or global failures.

DESIGN CONDITIONS THAT LIMIT STRUCTURAL CAPACITY OF DAMS

Each of the three dams has design conditions that create challenges when analyzed for high seismic load conditions. Each of these is discussed below.

Agnew Lake Dam

Multiple arch dams were developed by John S. Eastwood as a design alternative for locations where construction materials were scarce or difficult to transport to the jobsite [Ref. 2]. Agnew Lake Dam, at the time of construction, was well suited for a multiple arch configuration. However, as time went by and seismic design became better understood, the deficiencies of a multiple arch design became apparent. While strong in the streamflow direction (perpendicular to the dam face), multiple arch dams are inherently weak in the cross-stream direction (parallel to the dam face). Without substantial strengthening of the buttresses and/or significant bracing between buttresses, the arches are prone to failure under moderate to large earthquakes (return periods greater than 1,000 years). With the added cost and complexity of buttress strengthening or bracing, the benefits of a multiple arch design diminish.

Agnew Lake Dam was constructed with struts that were designed to transfer the cross-stream loading predicted at the time of construction. The struts, when analyzed to modern seismic requirements or, lack sufficient structural capacity to resist the cross-stream load condition at Agnew Lake Dam.

One of the buttresses at Agnew Lake Dam sits directly on and at the edge of an elevated rock outcropping. Under seismic loading, this rock could fracture, causing the buttress to lose support and collapse or slide away, resulting in a failure of the arch.

Gem Lake Dam

As with Agnew Lake Dam, the multiple arch design of Gem Lake Dam makes it relatively weak in the cross-stream direction. In addition, the arches are much taller. The struts included in the design are of little value, as they would buckle under even moderate seismic loads. Even if the struts were capable of supporting the induced loads, they do not provide an adequate system to resist the full cross-stream loading.

At Gem Lake Dam, the buttresses on the tall arches are relatively thin. Analysis shows that these buttresses will be overstressed in shear under the design seismic loading.

Rush Meadows Dam

Rush Meadows Dam was originally designed to be 30 feet high but was revised to 50 feet during construction. By raising the height, the right end of the dam extended above the existing rock abutment by 20 feet. In response to this, a concrete block was added to the right end to simulate an abutment and the spillway was constructed to the right of that concrete block (see Figure 5).

Figure 5. Concrete Block at Right End of Rush Meadows Dam

The result of this design is that the right end of the dam tends to “float” rather than rigidly resist thrust loading. The concrete block does not provide the rigidity of a natural rock abutment, as exists on the left end of the dam. Under high seismic loading, the concrete block will be unable to resist the thrust loads imposed and will allow excessive deformation, leading to failure of the upper 20 feet of the dam.

The dam profile (shallow arch) makes the dam act more like a cantilever dam in the center portion, becoming more “arch-like” as the abutments are approached. Coupled with the relative thin cross-section of the upper 20 feet, high seismic loads will overstress the center portion as it resists the inertial loads through cantilever action.

SEISMICITY AND THE SILVER LAKE FAULT

The Rush Creek Project dams are located in a region where high seismic events are a distinct possibility. Prior to 2007, the dams were analyzed using peak ground accelerations approaching 0.4g, corresponding to earthquakes in the M6.0 range. In 2007, the Silver Lake Fault, considered inactive, was determined to be active. The Silver Lake Fault is located within one mile of Agnew Lake Dam and has the potential to generate seismic events in the M6.0 range. Prompted by this new fault, SCE conducted a thorough evaluation of several nearby fault to develop appropriate seismic source models [Ref. 1]. The models were then used in a seismic hazard evaluation for the area to develop new seismic response spectra for the Rush Creek dams [Ref. 3]. The outcome of this evaluation was that the peak ground acceleration was only slightly larger (0.42g) but the spectral acceleration at a period of 0.1 second, which greatly influences the response of all three dams, increased by 50%. When these new seismic response spectra were used to analyze the three dams, the results indicate the potential for failure under seismic loading (shear failure of the buttresses at Gem and Agnew and buckling of the arch at Rush Meadows dam) [4].

Figure 6 shows the previous response spectrum (designated MCE) and the new response spectra for various return periods. Utilizing risk informed decision making (RIDM), SCE selected a 5,000 year return seismic event for the Rush Meadows dams. This results in a 1 in 10,000 probability of dam failure occurring from a seismic event, as the dams are drained each year for one-half of the year. The 5,000 year earthquake level is used to design retrofits that will keep stresses and deformations low enough to prevent, with high confidence, a failure of the dam and subsequent uncontrolled rapid release of water (URRW), thus protecting the public below the dams.

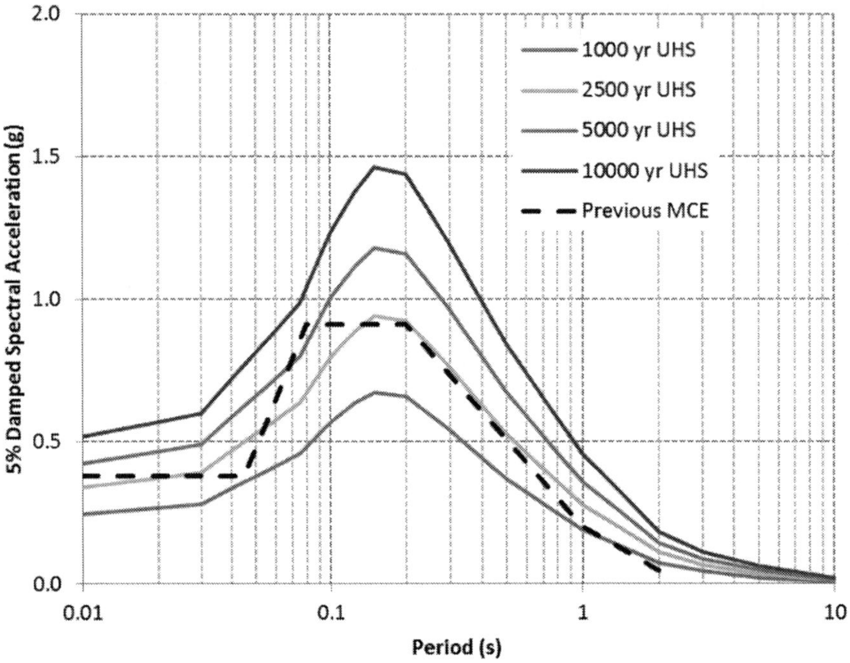

Figure 6. Current vs. Previous Dam Site Response Spectrum

PUTTING THINGS TOGETHER

In summary, three facts are driving the necessity to upgrade each of the three Rush Meadows Project dams.

1. Significant increases in the seismicity at the dam sites
    - As the site seismicity increases, measured by peak ground acceleration and energy content of the ground motion, the dam is subjected to higher inertial loads, which result in increased deformation and increased stresses.
2. Deteriorated concrete in each of the dams
    - Decades of freeze/thaw cycles have weakened the concrete
    - Decreases in concrete compressive strength of as much as 50%.
3. Inherent design inadequacies
    - Rush Meadows behaves in part as an arch dam and in part as a cantilevered structure, due in large part to the lack of an adequate abutment at the right end.
    - Multiple arch designs are vulnerable to cross-stream loading and require additional detailing to address this vulnerability.

SHORT TERM — ADDRESSING DEFICIENCIES THROUGH OPERATIONAL RESTRICTIONS

The three dams are classified as high hazard dams. In order to continue to operate the project, the system needed to be reconfigured to ensure public safety. Reservoir levels were analytically determined that would allow the dams to be safely operated under the current seismic environment. This analysis lead to the following restrictions:

- Agnew Lake – completely drained
- Gem Lake – lowered to the top of the gravity section of the dam, or about 25 feet below the spillway crest
- Waugh Lake – lowered to one-half the height of the dam, or about 22 feet below the spillway crest

LONG TERM - ADDRESSING DEFICIENCIES THROUGH SEISMIC RETROFITS

Agnew Lake Dam

Figure 7 illustrates the conceptual retrofit design for Agnew Lake Dam. The design involves adding concrete on the downstream side to thicken the upper portion of the arch and to create a gravity section for the lower portion of the arch. Thickening is done to the upper ten feet of each arch, increasing thickness from 1.5 feet to 3.5 feet. The gravity section then decends from that point to the base. Additional concrete is placed around each buttress to increase the section shear capacity.

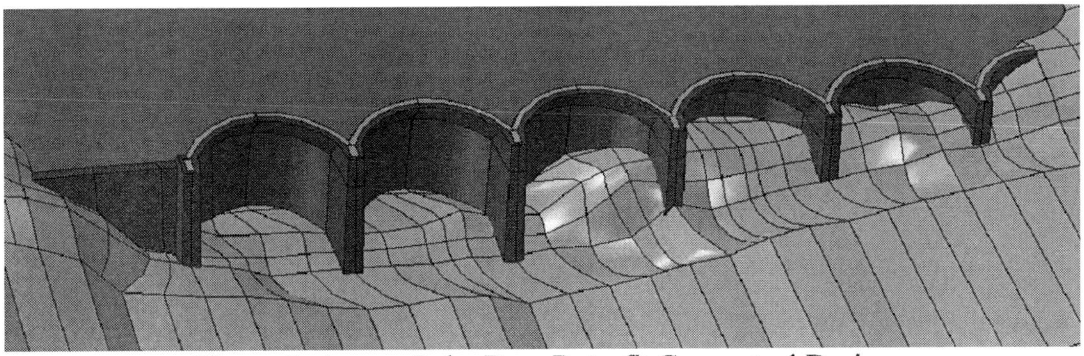

Figure 7. Agnew Lake Dam Retrofit Conceptual Design

Gem Lake Dam

Figure 8 illustrates the conceptual retrofit design for Gem Lake Dam. Chevron steel bracing is applied between buttresses to strengthen the dam for cross-stream loading. Bracing is also applied to the top of the arches in the horizontal plane in order to strengthen the arch. Additional concrete is placed around each buttress to increase the shear capacity.

Figure 8. Gem Lake Dam Retrofit Conceptual Design

Rush Meadows Dam

Figures 9 and 10 show the two conceptual designs for retrofitting Rush Meadows Dam. Conceptual Design No. 1 involves adding buttresses at intervals along the upstream face of the dam, increasing the concrete block at the right abutment, and placing a "mini arch" section along the lower half of the middle section of the dam. Conceptual Design No. 2 is converting the dam to a rockfill structure. Though the preferred concept, Design No. 2 would be very difficult to implement, as Rush Meadows Dam is located in the Ansel Adams Wilderness Area, and the Wilderness Act of 1964 explicitly prohibits mining activities in designated wilderness areas. A rockfill option is only viable if local rock is used, as the cost to import the estimated quantity of rock would be prohibitive. Mining local rock would require an amendment to the Wilderness Act.

Figure 9. Rush Meadows Dam Retrofit Conceptual Design No. 1

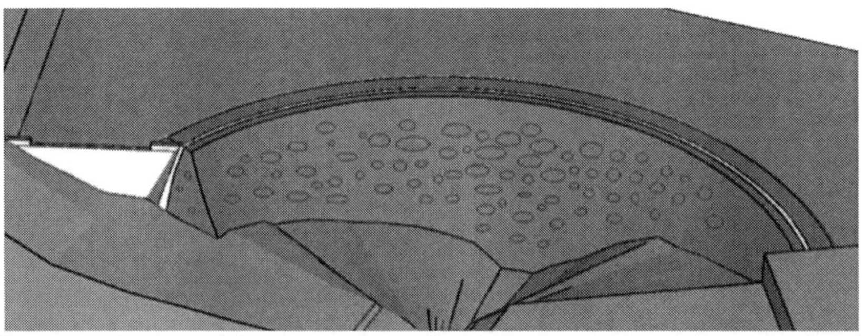

Figure 10. Rush Meadows Dam Retrofit Conceptual Design No. 2

COST-BENEFIT ANALYSIS

In order to determine the financial viability of the proposed retrofits, a cost-benefit analysis was undertaken. Table 1 shows the results of this analysis (normalized to $1 to illustrate relative values). Compared to decommissioning, the retrofit costs, which include capital investments and Operation and Maintenance funding, show a substantially lower present value (PV) cost. Decommissioning is very costly, given the wilderness locations and the need to restore the land to its natural state. Decommissioning provides no benefit, as the lost in-house generation needs to be replaced with purchased power. However, the present value of partial decommissioning is very close to that of the full retrofit, while maintaining the benefit of in-house generation.

Table 1. Cost/Benefit Analysis Results

DESCRIPTION	PV Cost	PV Benefit	Benefit/Cost
Current Configuration – Not Sustainable	$1.00	$0.94	0.94
Full System Retrofit	$1.88	$0.94	0.50
Gem, Rush Retrofit; Agnew Decommission	$1.88	$0.94	0.50
Gem, Agnew Retrofit; Gem Decommission	$1.94	$0.94	0.48
Full System Decommissioning	$4.56	N/A	N/A

RELICENSING — THE BIG UNKNOWN

The operating license for the Rush Meadows Project expires in 2027, which is only 12 years away. Owners are discovering that there are many uncertainties when negotiating the relicensing with interested parties. Changes to the operating parameters, such as minimum releases, can require significant modifications and can impact generation. Any large capital investment needs to evaluate the remaining time on the operating license when evaluating cost recovery. With an amortization of 12 years, cost recovery becomes much more difficult than with the typical 25 to 50 year license term.

Contemporary relicensing of hydropower projects have shown a propensity toward increased releases that bypass generation. These are mainly driven by environmental agendas that are spearheaded by outside agencies. In many instances, these may result in reduced generation and/or increased Operation and Maintenance costs. In extreme cases, improvements need to be made to dams and water conveyance systems in order to achieve those increased releases. Taken to an extreme, these changes could render a project financially unacceptable and lead to license forfeiture and decommissioning.

The uncertainty associated with the relicensing process, coupled with the 12 years remaining on the current license, lead to a change in strategy regarding the retrofitting of the dams. Before beginning retrofit construction, early relicensing would be pursued. This would allow negotiating of the new license conditions to move forward without the complexities a large capital outlay for retrofitting could introduce. As the relicensing process proceeds, the cost/benefit analysis for the retrofits can be revised to incorporate any expected changes that might influence continued operation.

CONCLUSIONS

The Rush Meadows Project provides value as a renewable energy generator, helping Southern California Edison meet California's goal of a statewide 20% renewable energy portfolio. The project also supports local recreation and water management.

Inherent design conditions combined with environmental factors result in a complex and costly program of retrofits that are necessary to allow the dams to operate at full reservoir capacities. Conceptual retrofit designs address the necessary target areas. When compared to the alternative of decommissioning, retrofit costs are significantly lower.

Agnew Lake and Gem Lake Dams are both multiple arch designs, but the retrofits are very different. For Agnew Lake Dam, retrofitting with mass concrete (gravity/thickened arch) is less costly than Chevron steel bracing. Utilizing mass concrete at Gem Lake Dam would be significantly more costly than the Chevron bracing approach.

The relatively short period of time remaining on the license (12 years till 2027) means that a significant capital investment needs to be recovered in a very short period of time. The relicensing process has the potential, through interveners, to lead to operational changes that make the project less financially attractive. Since this is a negotiation process between the owner and the interested parties, a large capital investment so close to the expiration of the license could compromise negotiations, to the detriment of the owner. An early relicensing, in advance of any retrofit investments, establishes any new operation conditions that can then be factored into the cost/benefit analysis.

REFERENCES

AMEC Geomatrix (2011), *Silver Lake Fault Evaluation, Mono County, California.* Oakland, California.

Jackson, Donald C. (2005), *Building the Ultimate Dam: John S. Eastwood and the Control of Water in the West.* Norman: University of Oklahoma Press, 2005.

SCE Dam Safety (2012), *Preliminary Seismic Hazard Evaluation for Lee Vining Creek, Mill Creek, and Rush Creek Dams.* San Dimas, California.

SCE Dam Safety (2012), *Rush Creek Dams, Report for Event Risk Assessment Workshop on September 24 and 25, 2012.* San Dimas, California.

GATE TRUNNION ANCHOR REMOVAL AND ANALYSIS

Jesse Kropelnicki[1]
Robert F. Lindyberg[2]
Tony Plizga[2]

ABSTRACT

The hydroelectric dam presented in this paper was constructed in 1920-21, and consists of a west earth embankment, a powerhouse, a decommissioned log sluice and fishway housing, six Tainter gate controlled spillway bays, an uncontrolled overflow spillway, a roller-compacted concrete fuse plug spillway, an east earth embankment, and a detached embankment.

In 2008, during the FERC Part 12 Inspection, it was discovered that one of the four anchors installed on Pier No. 6 had separated from the pier. The anchors are used to prevent movement of the Tainter gate pins by providing extra compressive resistance within the concrete pier. In 2013, an analysis was completed by BARR Engineering and Environmental Consultants confirming that the three unaffected anchors could still provide the tension capacity necessary. In 2014, Parsons Brinckerhoff (PB) acted as Owner's Engineer during the anchor removal, field testing, and borescope inspection of the failed anchor. The anchor was sent out for laboratory testing to determine the cause of the failure and forensic analysis. PB and FDH Engineering (FDH) performed the field inspection, Dispersive Wave Impact-Echo Testing, and Bending Wave Testing on all of the remaining anchors to determine their condition and load carrying capacity.

This paper discusses the methodology that PB and FDH used to determine the cause of failure of the anchor and the methodology to determine the current condition of the remaining gate trunnion anchors.

[1] Principal Engineer, Parsons Brinckerhoff, 75 Arlington St., Boston, MA 02116, (617) 960-4975 Kropelnicki@pbworld.com
[2] Vice President, FDH Engineering, 6521 Meridien Dr., Raleigh, NC 27616, (919) 367-5280, RLindyberg@fdh-inc.com
[3] Senior Principal Engineer, Parsons Brinckerhoff, 75 Arlington Street, Boston, MA 02116, (617) 960-4972, Plizga@pbworld.com

Figure 1. Downstream view of the spillway section.

INTRODUCTION

In 2008, during the FERC Part 12 Inspection, it was discovered that one of the four gate trunnion anchors installed at one of the Piers had separated, about one foot, from the pier. The post-tensioned trunnion anchors were installed in the spillway piers in 1984 to provide anchorage resistance within the concrete piers against the gate trunnion loads and hydrostatic loading on the spillway gates. In 2013, an analysis was completed by BARR Engineering and Environmental Consultants (BARR) confirming that the three remaining anchors in Pier could provide the necessary anchorage to resist the gate trunnion loads. On March 25, 2014 the Federal Energy Regulatory Commission (FERC) sent a letter to the Owner requesting an investigation into the cause of the trunnion anchor failure, verification that the remaining anchors are intact, and visual inspection of the remaining trunnion anchors be done on a monthly basis. This paper discusses the subsequent investigation and analysis.

BACKGROUND

The project consists of a earth embankment, a non-overflow concrete section, a powerhouse, a decommissioned log sluice and fish way housing the low-level outlet works, six Tainter gate bays, an uncontrolled overflow spillway, a roller-compacted concrete fuse plug spillway, an east earth embankment.

The controlled spillway portion of the dam consists of six Tainter gate bays between seven piers. The Tainter gate trunnions, installed from 1983 to 1985, are located at the nose of the pier with approximately a four foot clear distance to the edge of concrete. In each pier there are four trunnion anchors designed to provide the tensile force to counteract the hydrostatic and gate forces. During the 2008 Seventh FERC Part 12 Inspection the independent consultant noticed that one of the anchor rods on Pier 6 had displaced.

In March 2014 FERC sent a letter to the owner requesting an investigation be done to determine the cause of failure, deicing measures be taken to prevent ice build up on the face of the dam, and a visual inspection of the piers for concrete cracking and any signs of distress for the anchors. The owner was instructed to perform the visual inspections on a monthly basis.

In response to the FERC directive, BARR performed an analysis on whether the remaining three gate trunnion anchors provided adequate resistance for the forces acting on the gate. BARR used conventional code based methodology with gravity load, hydrostatic load, ice pressure load, gate lifting load, moving gate load, earthquake and hydrodynamic load, wind load, and wave load using the load combinations defined in U.S. Army Corps of Engineers, "Design of Spillway Tainter Gates" (EM2702). A sensitivity analysis was performed to accommodate relative unknowns like the weight of the gate, location of gate centroid, ice/mud gravity load and centroid, and machinery loads which ensured that fluctuations in the values would not be significant enough to cause a failure of the pier. The analysis ultimately found that the utilization ratio of the steel was highest (0.89 to 0.97) for a load case where the earthquake and ice pressure loads from the Potential Failure Mode (PFM) were present. The earthquake load from the PFM is unlikely and conservative while the ice pressure loading can be mitigated by keeping ice off of the face of the dam.

The decision to use Non-Destructive Testing (NDT) instead of lift off testing was chosen as NDT allows for the condition of the trunnion anchors to be examined without increasing the risk of failure. Lift off testing, which elastically deforms the anchor records the tension on the anchor to validate the in-situ capacity. However, this test risks overloading the installed anchor capacity and causing the steel to enter the plastic deformation zone, or even rupturing the bar causing failure. Therefore, NDT was used for the engineering/field evaluation because it is much safer for staff to run and it poses no long term risks for the dam as lift off testing often does.

ENGINEERING EVALUATION

Inspection of Failed Anchor

PB performed a visual inspection of the failed anchor prior to its removal and found it was displaced 27-7/8" (see Figure 2) from the face of the pier indicating the anchor was most likely under load when it failed (causing the displacement). After removal, an inspection of the rupture point led to the discovery of what appeared to be a circular cut through the anchor (See Figure 3). Following this discovery, the top of Pier 6 at the deck level was inspected for evidence of a concrete core boring that could have struck the anchor. As part of this inspection, a previously grouted-in core drill hole from the deck was found directly above the anchor location and very likely was the cause of the anchor failure. Pier 6 also has an existing monitoring well installed just lateral to the grouted-in core hole, that given its orientation at the deck level, would not intercept the gate anchorage system. A similar monitoring well was observed in the same location as Pier 6 at Pier 2.

Figure 2. Displacement of failed anchor before removal.

Figure 3. View of rupture point of failed anchor.

After the failed anchor was removed from Pier 6 by the Owner's contractor, FDH attempted to perform a bore hole inspection of the anchor hole. Since the anchor sheath had not leaked or lost any grease the borescope's camera and light were coated with grease before the borescope could reach the area of the anchor's rupture. Removal of the grease would cause waste concerns and require an indefinite amount of time to sufficiently clear the grease from the sheath. In light of corrosion not being the cause of the rod failure, the decision was made to abandon the borescope inspection.

Following the initial site inspection, additional drawings were provided by the Owner showing the monitoring wells in Pier 2 and Pier 6 (see Figure 4). The drawings show monitoring wells BF W 33 and BF W 31 cored in 1988, located in Piers 2 and 6, respectively. The core boring logs for these locations, also provided by the Owner, do not indicate that a coring was attempted and then aborted during initial drilling at Pier 6. However, based on PB's evaluation and discussions with the Owner, it is believed that a coring was attempted at Pier 6 that intercepted the anchor and partially drilled thru it. Once the boring struck the anchor, the coring location was aborted and grouted in. This monitoring well location was then moved laterally toward the powerhouse and drilled without interference over its full vertical depth. This successful coring is the coring that was documented within the drill logs.

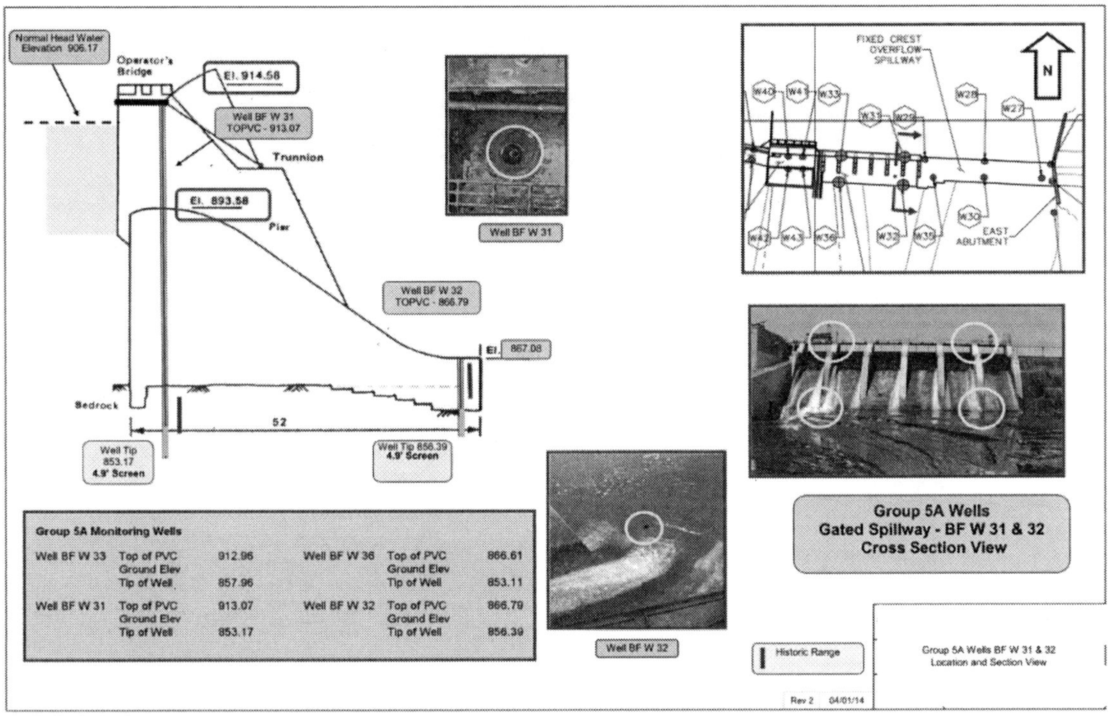

Figure 4. Drawing showing the locations of monitoring wells at Pier 2 and Pier 6.

Inspection of Piers

PB performed a visual inspection of the remaining pier noses and remaining trunnion anchors. The piers and remaining anchors were all in excellent condition with only minor surface rust on the exposed anchor ends. During the visual inspection it was observed that Pier 7 had 1-1/4" diameter trunnion anchors while the trunnion anchor for Piers 1-6 were 1-3/8" diameter.

Based on PB's inspection, all piers were in excellent condition with no signs of overstress or deterioration. Piers 1, 2, 6, and 7 did have minor hairline cracking observed, but none of this cracking presents structural concerns at this time.

Figure 5. Side view of typical pier.

Figure 6. Upstream view of typical pier nose.

<u>Lab Evaluation of Failed Anchor</u>

Altran was the consultant for the laboratory evaluation of the failed anchor at Pier 6. Through a visual inspection at their lab, Altran was able to determine that a cut 3/4" deep by 5/8" wide occurred on the anchor. The cut had concentric grooves 1/8" to 1/16" apart that are consistent with common hole-saw/boring hole tools. High-resolution examination of the remaining fracture surface at the failed end of the rod found the fracture morphology is consistent with overload failures of high strength materials.

Once the anchor was partially cut by the concrete core bit, a significantly reduced cross sectional area remained. This reduced cross section resulted in significant overstress of the material and eventually resulted in tensile failure of the rod. Although the core boring was likely done in 1988, the failed anchor was not observed or detected until 2008 when it was observed to be displaced downstream.

The mechanical, hardness, and chemical evaluation of the anchor found similar ultimate tensile strength and alloy composition as SAE 1500 series steel. The tensile test found the ultimate strength to be 159 ksi and the yield strength to be 141 ksi. The full rod net diameter was measured to be 1.25 in. Using a nominal ultimate strength of 150 ksi and the fracture surface area of 0.59 in^2 would require 88,500 lbs to rupture the anchor.

Scanning Electron Microscopy (SEM) imaging was performed on the fracture surface of the rod. SEM images use focused beams of electrons across a specimen surface to reflect on the surface of an object to create an image with fine spatial resolution. The images can reveal whether the anchor rod failed due to a manufacturing defect within the steel. The SEM imaging performed on the failed anchor rod showed a typical steel cleavage fracture which proves the failure was not caused by an abnormality in manufacturing (see Figure 7).

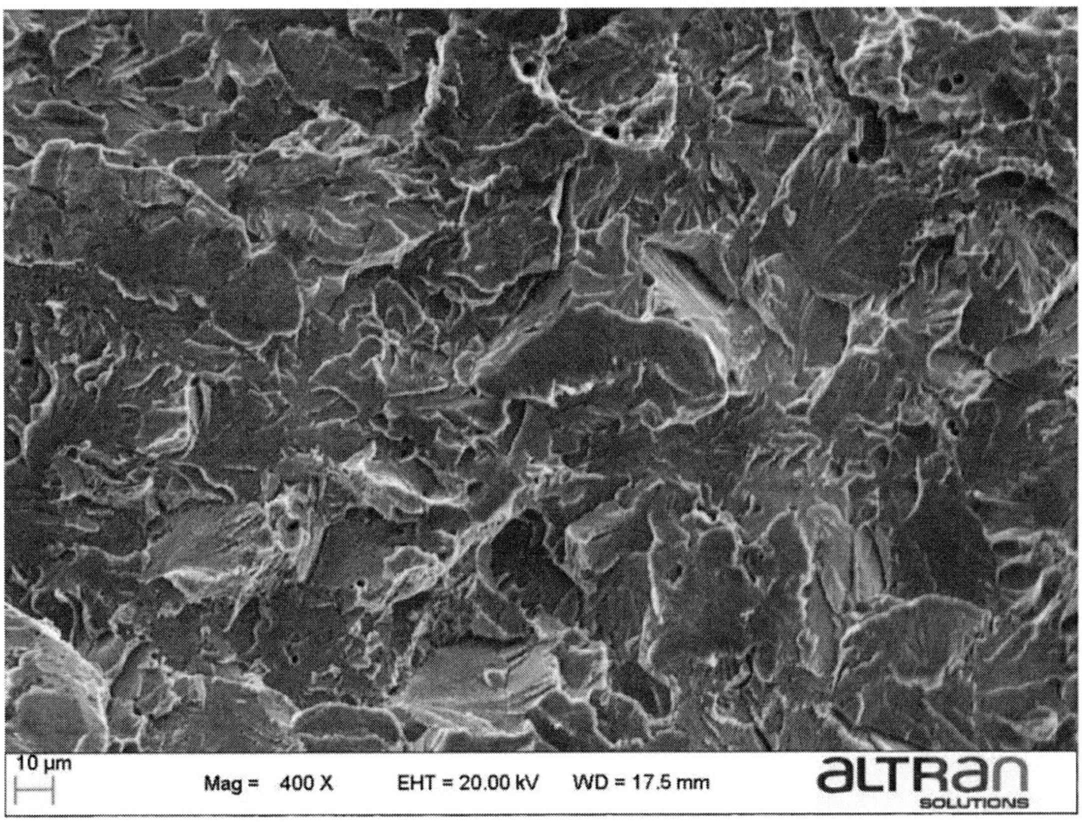

Figure 7. SEM Image of the Rod Fracture Surface Showing Typical Steel Cleavage Fracture.

Non-Destructive Testing of Anchors

FDH was the consultant for the Non-destructive Dispersive Wave investigation of the remaining trunnion anchors. FDH personnel performed Impact Echo Tests and Bending-Wave Tests to determine the condition of the remaining anchors. Non-destructive testing was chosen to use for this project because it allowed for the condition of the anchors to be checked with the material that is currently exposed. The Impact Echo Test records how long it takes for dispersive waves, produced by a small device with accelerometer ages that are mounted adjacent to the impact area, to propagate the length of the rod. The data is then is analyzed and used to compute the length of the anchor rod. The Bending Wave Testing is conducted by temporarily mounting accelerometer gages to the side of the anchor and striking the anchor with a multiple different impact devices. The propagations of the impact waves are recorded by the accelerometers. The data recorded can then be analyzed to compute excitation frequencies. The excitation frequencies and rod length can be combined with the rod diameter to create an empirical relationship with the rod tension.

The Impact Echo Test results showed only two anchor rod lengths that were outside of the tolerance range (+/- 5%) of 18'-20'. In both instances, the rod lengths were slightly longer than 20'. The failed anchor was tested before removal and found to share no signal characteristics to the intact anchors. The data collected from the failed anchor shows the amplitude of the impact oscillating throughout the entire length of the anchor while the data collected from intact anchors show the amplitude of the impact become progressively weaker throughout the length. No defects or structurally concerning anomalies were found within the intact anchors.

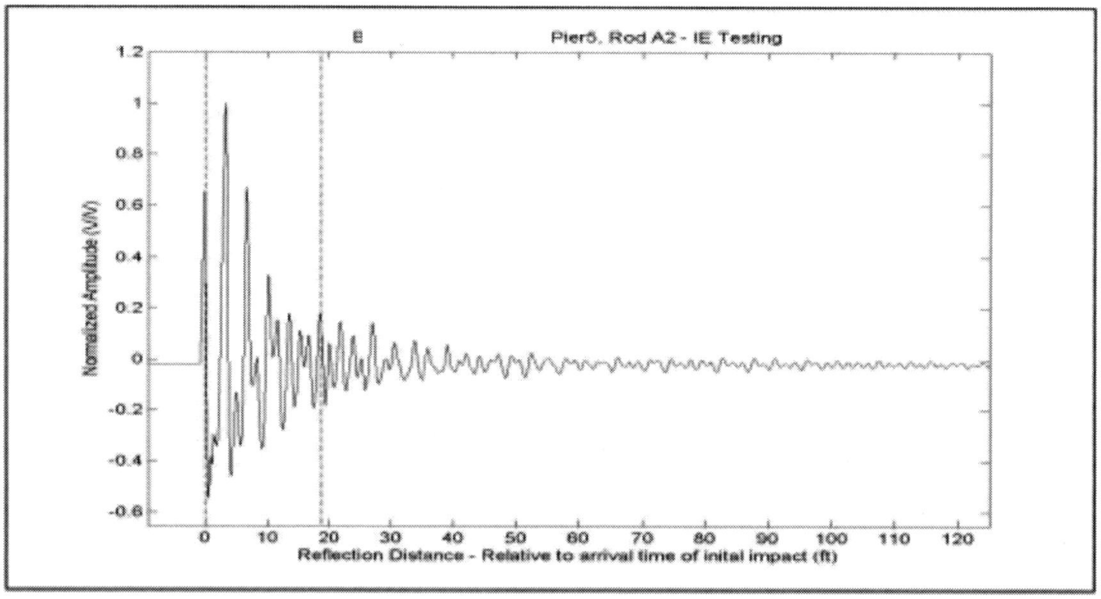

Figure 8. Typical plot of measured signal from Impact Echo Testing for intact anchor rod.

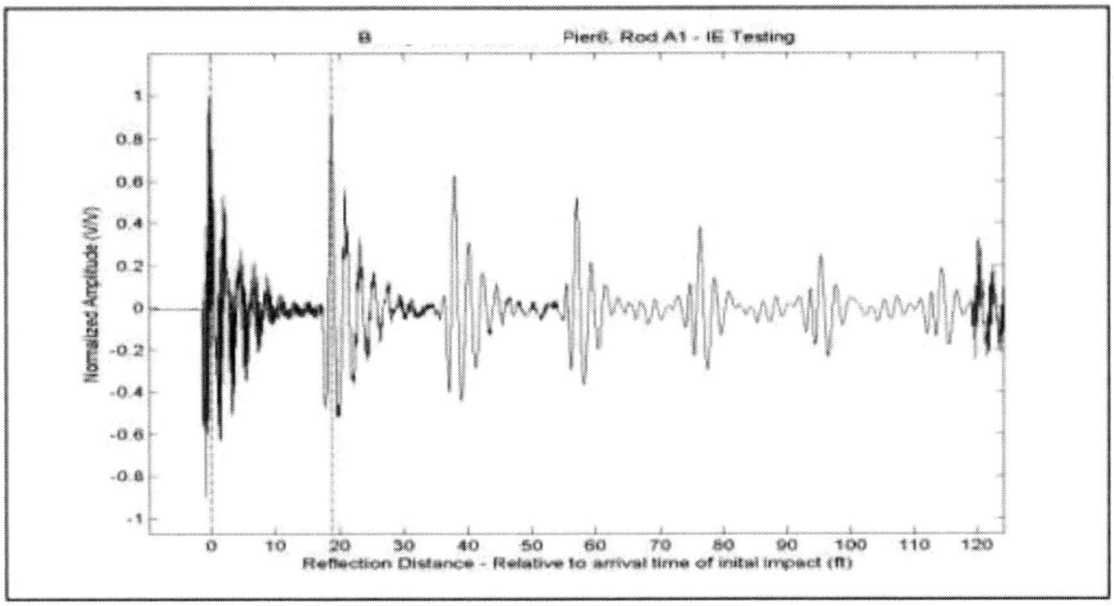

Figure 9. Plot of measured signal from Impact Echo Test for Pier 6 failed anchor rod.

Performing the Bending-Wave Test resulted in data that could be used later to develop approximate anchor loads. The prototype would allow FDH to develop an empirical model from the bending wave data to estimate the actual tensions for the trunnion anchors.

Table 1 and Table 2 show the results found by FDH's Impact Echo Test for the 1-3/8" and 1-1/4" trunnion anchor rods, respectively.

Table 1. Computed Trunnion Anchor Rod Lengths – 1-3/8" Diameter Rods

Pier	Column (1)	Row (1)	Impact Echo Test Length (ft)	Pier	Column (1)	Row (1)	Impact Echo Test Length (ft)
1	A	1	20.3	4	A	1	19.9
1	A	2	19.7	4	A	2	19.9
1	B	1	19.9	4	B	1	19.5
1	B	2	20	4	B	2	19.4
2	A	1	20.1	5	A	1	18.7
2	A	2	19.5	5	A	2	19
2	B	1	19.8	5	B	1	19.1
2	B	2	19.4	5	B	2	19.8
3	A	1	19.9	6	A	1	19.1*
3	A	2	18.9	6	A	2	19.4
3	B	1	20	6	B	1	19.2
3	B	2	19.2	6	B	2	19.8

* Broken Rod

Table 2. Computed Trunnion Anchor Rod Lengths – 1-1/4" Diameter Rods

Pier	Column	Row	Impact Echo Test Length (ft)
7	A	1	18.4
7	A	2	18.7
7	B	1	18.4
7	B	2	18.6

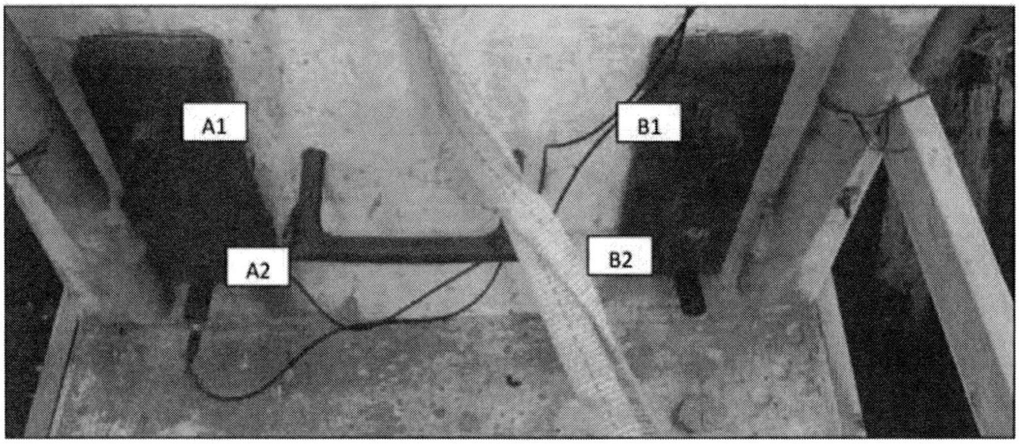

Figure 10. Diagram showing column/row legend.

CONCLUSIONS/LESSONS

An engineering team led by PB provided a field inspection and subsequent evaluation of the concrete piers and steel anchors, addressing concerns about their safety and structural integrity, in light of the failure of a post-tensioned anchor at Pier 6, causing the anchor to displace from the downstream concrete face. The PB team provided an assessment of all anchors and piers at the project to identify short term and long term risks to the continued safe operations of the facility.

The conclusions are as follows:

- The failure of the upper, powerhouse side anchor at Pier 6 was caused by a concrete core boring through a portion of the anchor, leading to its' tensile fracture. This core boring was likely associated with the drilling of a monitoring well from the deck.

- Presently the project is considered safe, with no immediate modifications required. Under current conditions, the trunnion anchorage system and piers are safe.

- The Pier 7 anchors were found to have a smaller diameter (1-1/4") than anchors on Piers 1 through 6 (1-3/8"). When evaluating the strength of anchors at a project be sure to visually inspect all anchors. The size or the condition of the anchors could vary with location.

- Thoroughly inspect the site and original construction drawings before determining the location of new bore holes.
- Previous construction or instrumentation work can cause issues as this work was not included in the original planning of the dam.
- Monitor contractors working on dams closely. Contractors may be unaware of the consequences of mistakes made or may try to cover mistakes up while working on site.
- Non-Destructive Testing is a safer alternative to lift off testing that is able to evaluate anchor tension without risking damage or failure of anchors.

METOLONG DAM AND RAW WATER PUMP STATION — CHALLENGES, SUCCESSES, AND LESSONS LEARNED

Stephen L. Whiteside, P.E.[1]
David Bosshart[2]
Peter P. Macy, P.E.[3]

ABSTRACT

For the Metolong Dam and Water Supply Program (MDWSP) in Lesotho, Africa, CDM Smith is serving as the program manager. The MDWSP includes the following components: an 80-meter-high roller compacted concrete (RCC) dam with multi-stage raw water intake and pump station; a 75-million-liter-per-day water treatment plant, including raw water transmission main and high-lift pump station; downstream conveyance system; advanced infrastructure including access roads and bridges, power supply, water and sanitation systems, and telecommunications; and environmental and social management.

Lesotho is a small, mountainous country that is totally surrounded by South Africa. Lesotho has abundant water resources that are so important in much of Africa and the world. Lesotho has two very large dams (Katse Dam completed in 1997 and Mohale Dam completed in 2002) that are part of the Lesotho Highlands Water Project (LHWP) Phase I that exports water to South Africa. Phase II of the LHWP commenced in 2013 and is planned to include a 165-meter tall CFRD concrete-faced rockfill dam. The MDWSP is an important project for Lesotho as it provides a reliable supply of potable water for the people of Lesotho. The USA Millennium Challenge Corporation was a key donor to the program.

This paper includes an overview of the design and construction of the Metolong Dam and Raw Water Pump Station. Challenges, successes, and lessons learned are highlighted. Construction commenced January 12, 2012, impoundment of the reservoir began February 17, 2014, and first water was delivered to the Water Treatment Works on May 26, 2014. The project was substantially complete in December 2014.

INTRODUCTION

Two major milestones were achieved in th e first half of 2014 on the Metolong Da m Project in Lesotho. The first was the co mmencement of impounding the reservoir on February 17, 2014 and the second was the first water delivery from the dam to the Water Treatment Works (WTW) on May 26, 2014.

[1] CDM Smith, 5400 Glenwood Avenue, Suite 400, Raleigh, NC 27612, whitesidesl@cdmsmith.com
[2] Metolong Program Management Unit, Red Cross Building, 29 Mabile Rd., Old Europa, Private Bag 9137, Maseru 100, Lesotho, david.bosshart@mpmu.org.ls.
[3] CDM Smith, 891 Duncan Street, Brooklyn, Pretoria, South Africa, 0181, macypp.@cdmsmith.com.

It was critical that impounding begin early enough in the rainy season (typically October – April) to allow capturing sufficient water so that final commissioning of the WTW and the Downstream Conveyance System (DCS) could begin. Otherwise, there would likely have been a delay of approximately 6 months, or more, waiting for the next rainy season to fill the reservoir above the intakes to the Raw Water Pump Station (RWPS). This would have delayed the overall Metolong Dam and Water Supply Program (MDWSP) with significant impacts and costs.

There have been numerous challenges and delays on the project. To achieve these two critical milestones during the first half of 2014, as needed, it was necessary to impound the reservoir well before the dam was complete and likewise deliver the first water to the WTW well before the RWPS was complete. Figuring out how to do this was not easy and required changing attitudes and forging a partnership of cooperation followed by intense teamwork and hard work.

Based on these milestones, the MDWSP was available for commissioning and operations training to commence the second half of 2014 along with the first potable water delivery. Final project completion and handover to the Operator, the Water and Sewerage Company Lesotho (WASCO) will follow in 2015.

This paper focuses the following items for Metolong Dam:

- Background information related to the overall MDWSP with focus on funding and stakeholders.
- An update on the construction progress including steps taken to achieve the critical early impoundment and water delivery.
- Various challenges encountered along with successes.
- Lastly, lessons learned.

Figure 1. Upstream view of Metolong Dam on February 17, 2014 shortly before closing the gate to the diversion tunnel to start filling the reservoir.

BACKGROUND

The Government of Lesotho (GOL) established the Metolong Authority (MA) to oversee the implementation of MDWSP. The Metolong Authority is supported by the Metolong Program Management Unit (MPMU) that manages the implementation of the entire program. MPMU was formerly established by the Millennium Challenge Corporation (MCC).

MDWSP Composition

The MDWSP is a multi-donor[4] funded program aimed at improving the supply of water to Maseru and nearby towns in support of the GOL's efforts to sustain socio-economic growth and development. The MDWSP will deliver potable bulk water to Maseru and the peri-urban communities of Roma, Mazenod, Morija, and Teyateyaneng.
The MDWSP comprises the following five major inter-related components:

1. Environmental and Social Programs (E&S)
2. Advance Infrastructure (AI)

[4] Government of Lesotho (GOL); Kuwait Fund for Arab Economic Development (KFAFD), the Saudi Fund for Development (SF), the OPEC Fund (OFID), the Arab Bank for Economic Development in Africa (BADEA), and the Abu Dhabi Fund for Development (ADFD), collectively termed KBOSA; Millennium Challenge Corporation (MCC); European International Bank (EIB); International Development Association (IDA); and the Republic of South Africa (RSA)

Metolong Dam and Raw Water Pump Station

3. Metolong Dam (MD) and raw water pump station (RWPS) being constructed on the South Phuthiatsana River about 35 km from Maseru;
4. Water Treatment Works (WTW) with 75 Ml/day average capacity and peak output of 93.4 Ml/day.
5. Downstream Conveyance System (DCS) of approximately 160 kilometers of varying diameter pipelines being constructed under three contracts:
   i. Contract No. 1 – Primary pipeline to Maseru
   ii. Contract No. 2 – Secondary pipeline to Mazenod, Morija, and Roma
   iii. Contract No. 3 – Secondary Pipeline to Teyateyaneng

The MDWSP includes two independent panels of experts:

- A two-member Environmental and Social Issues Panel
- A five-member Panel of Experts (POE) to review dam safety and compliance with World Bank OP/BP 4.37[5]

In August 2009, the Metolong Authority secured consulting services for both the detailed design and construction supervision of the Metolong Dam and RWPS. Contracts were signed with a joint venture of Arcus Gibb Holdings from South Africa and Consolidated Consultants for Engineering and Environment from Jordan (Gibb-CC JV).

MDWSP Composition

Beginning in the year 2003, the MDWSP was subject to the preparation of a number of comprehensive feasibility level cost estimates. Based on the recommendations of these previous studies, the Millennium Challenge Corporation (MCC) began to take serious interest in developing the MDWSP. An update of the 2003 feasibility study was subsequently requested by MCC, and a supplementary study entitled "Additional Feasibility Study," was completed in 2007[6]. This study formed the basis for the initial 2008 funding arrangement reproduced below which amounted to approximately US$285 million.

[5] Operational Policy and Bank Policy 4.37, the World Bank, October 2001 (revised April 2013)
[6] In 2006, MCC retained Mott MacDonald to update the 2003 Feasibility Study for which an interim report was submitted on 14 February 2007.

Table 1. Summary of Original 2008 MDWSP Funding Commitments (Given in USx10^6$)

Program Component	GOL	BADEA	OFID	KFAED	SF	RSA	IDA	MCC
Advance Infrastructure	34.30					6.00		
Metolong Dam & Raw Water Pump Station		10.00	6.00	18.36	15.00			
Water Treatment Works								55.20
Conveyance Systems	68.53						13.00	18.97
Environmental - Social	4.60						10.00	
Institutional	9.70						2.00	12.60
Total	117.13	10.00	6.00	18.36	15.00	6.00	25.00	86.77

In September 2010, the European Investment Bank (EIB) joined with the other funding groups/donors indicated above, and picked up a substantial shortfall that had already become identified with the major DCS components. Unfortunately, timing of the MDWSP tenders coincided with the infrastructure development necessary to host the 2010 World Cup games being held in the surrounding Republic of South Africa. The influence of this event on the prices of services and materials had a dramatic impact on the previously estimated feasibility costs. By mid-2011, all the major components comprising MDWSP had been tendered for construction; consequently, the major construction cost of the MDWSP was well identified. What could be readily concluded was that the cost of all the components comprising the MDWSP had more than doubled over the 2003 feasibility cost estimate and were more than 50% over the 2008 revised cost estimate.

The construction cost of the dam was particularly hard hit. Feasibility cost estimates placed this amount at approximately US$ 36 million. Revised estimates in late 2008 raised this estimate to US$ 55 million. In response to this potential cost increase, additional supplementary funding for the dam had to be sourced, and a donor's conference was held in early 2009. At this conference the Arab Funders committed to funding the tendered price for the dam.

The Invitation to Bid was issued December 2010. Six bids were received for the construction of the dam and raw water pump station on April 11, 2011. The six bids ranged from US$80 million to US$ 194 million, and the mean was US$ 143 million. The low bid which was submitted by Sinohydro Corporation Limited was US$47 million below the second lowest bid. The construction contract was awarded on August 8, 2011 to Sinohydro.

In early July 2011, the consortium of Arab funders, which now included the Abu Dhabi Fund as an additional member, confirmed their earlier verbal commitment to fund the full tendered price of Metolong Dam. However, the Arab Funders did not commit to funding any contingency amounts associated with the construction of the dam, and it was agreed that GOL would source any additional funding for contingencies, if later required. The very fact that the Arab Funders agreed to finance the US$80 million tender, based on an original commitment of only US$36-million, was most gratifying.

Thus, MDWSP currently has ten (10) funders/donors. Contributions provided by each of these ten have been allocated on a concessionary basis to specific MDWSP components. The following table su mmarizes the current status of the MDW SP funding agreem ents which amounts to approximately US$ 470 million:

Table 2. Summary of Current MDWSP Funding (Given in US$x10^6)

Component	GOL	EIB	BADEA	OFID	KFAED	SF A	DFD	RSA	IDA	MCC
AI	34.30	16.08						6.00		
MD & RWPS			15.00	9.00	23.36	25.00	21.00			
WTW										66.17
DCS		150.06							18.00	8.00
E&S	3.00	28.64							13.00	
Institutional	8.00	5.22							2.00	12.60
Total	45.30	200.00	15.00	9.00	23.36	25.00	21.00	6.00	38.00	86.77

Metolong Dam Commencement Delayed

Delay transpired in issuing the instruction to Commence the Works, as it was necessary for the Arab Funders to put the necessary supplementary loans in place. This resulted in an overall delay of 157 days. The Commencement Date was finally established as January 12, 2012 when the final portion of the ten-percent advance payment was received by the Contractor.

Claims, Cash Flow and Formulation of a Supplementary Agreement

Roller compacted concrete (RCC) placement in the dam began on August 5, 2013 which was in itself a significant construction achievement. However, despite this advancement, an accumulation of challenging issues had severely impacted the Contractor's cash flow (Actual cumulative monthly cash flow is compared against the original tendered projection of cash flow, graphically below.). By August 2013, the actual cash flow in earnings was running less than 50 percent of projected revenue. This situation was severely impacting the ability of the Contractor to resource the work appropriately with his suppliers. In early August, the Contractor's Project Manager approached MA/MPMU representatives and suggested a global settlement to resolve the many outstanding claims that were affecting progress of the work. It was also anticipated that this would help alleviate the Contractor's deteriorating cash flow situation.

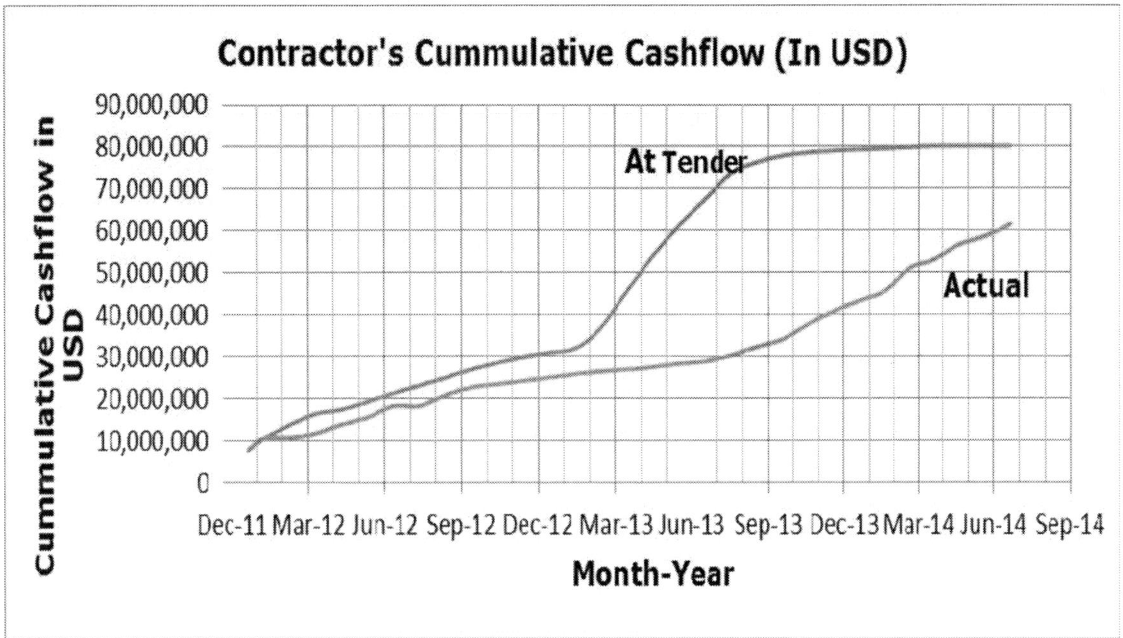

Figure 2. Contractor's Cash Flow

By August 2013, the Contractor had submitted more than 50 notices of claims and variations with a value exceeding US$15 million. It was likely that the regular claim notices would continue as previous problems concerning labor unrest, obtaining visas and work permits for expatriate staff, border clearances, and late payments continued and some seemed to be getting worse. Issues relating to differing conditions needing effective responses from the Engineer were intensifying.

Metolong Authority was extremely interested in the Contractor's global settlement suggestions because despite numerous appeals for the Contractor and Engineer to work together, the relationship remained confrontational which was detrimental to the project. This was largely due to numerous claims, with significant value, and associated contractual maneuvering. Too much effort was being spent fighting over claims and contractual issues and not enough towards constructing the works to enable impounding during the upcoming rainy season. The settlement also was beneficial to MA in so much as the negotiated amount was less than the Engineer's determination and significantly less than what the Contractor had originally claimed.

Metolong Authority (the Employer) and Contractor began informal discussions regarding the proposed global settlement in September 2013 which continued on through November. These discussions eased some of the tensions on the project, and progress was good even as new challenges emerged. Metolong Authority management requested approval from its MA Board to enter into formal negotiations with the Contractor and received approval in mid-December 2013. The approval allowed for a global settlement to resolve all outstanding claims for an amount not to exceed US$ 9 million. Payments would be made when new "milestone" key dates such as first impoundment and first water delivery were achieved.

In January 2014 the Employer, Engineer, and Contractor began formal negotiations and Supplemental Agreement No. 1 was finally reached on January 27, 2014. This agreement increased the contract amount to US$89,991,160. Comparatively, the second lowest bid for the project was US$ 127.5 million so even with this increase of US$ 9 million, the contract price was still more than US$ 37 million below the second lowest bid. The agreement included new milestone dates for a) commencement of impoundment; b) water delivery of 25, 35, and 93 Ml/day; c) completion of the RWPS; and d) completion of the entire project. It also included significant delay damages for not meeting the agreed dates.

Institutional Stakeholders

Various loan covenants with the Funders required GOL to address the formation of appropriate institutional structures that would ensure long-term sustainability of the MDWSP. Thus, it was not until October 31, 2011 that the eventual operator of the MDWSP facilities was finally identified. Up until that time, creation of several alternative bulk water supply entities had been envisioned. The Water and Sewerage Company (WASCO), was selected in this capacity. Complicating this selection process was the fact that existing water utilities in Lesotho did not have significant experience in the operation of large dams. Even at the time of being designated as the operator of the MDWSP facilities, WASCO remained resistant to taking on dam safety and operation responsibilities.

Under the present contractual arrangements for the major MDWSP components, WASCO is to participate as observers during the basic commissioning processes, and subsequently, the contractors will provide assisted operations training. During the latter period, the WASCO personnel assigned to MDWSP will be expected to physically run the MDWSP facilities. The current situation is such that WASCO needs to expand on available resources and engage the necessary personnel to fulfil adequately these contractual obligations.

Through the guidance mainly of the World Bank, Lesotho has also developed the formation of an Asset Management Authority (AMA) as the owner of the MDWSP assets. The AMA is in the process of compiling a comprehensive MDWSP asset register. The AMA functions directly under the Commissioner of Water. In its present conception, the AMA will not service the associated debt. This function remains with the Lesotho Ministry of Finance until such time as the revenue basis can support alternative arrangements.

In the preparation of the required Emergency Preparedness' Plan (EPP) pursuant to compliance with the World Bank OP 4.37, the Lesotho Disaster Management Authority (DMA) has been of valuable assistance. The DMA has integrated its capabilities into the EPP to work with WASCO in alerting the communities residing downstream of the dam in case of pending emergency associated with the sudden release of water into the South Phuthiatsana River.

A number of other institutional stakeholders are involved directly and indirectly with the MDWSP. Among these is the Department of Water Affairs which monitors and records stream flow data, and which in the future will monitor compliance with instream flow release requirements.

METOLONG DAM CONSTRUCTION PROGRESS

The Metolong Dam, when completed, will be 83 m high (Elevation. 1595 to EL. 1678), with a crest length of 278m and crest width of 7 m. The full supply level will be at El 1671. It will impound a reservoir approximately 15.5 km long having a storage capacity of 64 million m^3. The mean annual precipitation in the catchment is 850 mm and the mean annual runoff is 66 million m^3. The RCC gravity dam will contain approximately 270,000 m^3 of concrete. The upstream face of the dam is vertical and the downstream face is stepped with a slope of 0.8:1. The uncontrolled spillway will be 75 m long and has been designed to pass a flood of 2,000 cubic meters per second.

As of September 30, 2014, 227,000 m^3 of RCC had been placed in the dam which is approximately 84% of the total amount. The height of the dam was 66 m out of 83 m or 80%. The RCC placements were completed late 2014, and all concrete placements in the dam are expected to be completed early 2015.

Figure 3. Downstream view of the dam and pump station September 30, 2014 (also visitor's center upper left)

EARLY IMPOUNDMENT AND ITS CHALLENGES

Challenges Associated with Availability of Water

The five major components of MDWSP are all being constructed under separate contracts; however, commissioning of these individual contracts is inter-dependent upon the availability of water from the dam. Whereas the MCC Compact tenure of five years had provided an initial constraint/stimulus in coordinating component implementation, it was recognized that any delay in water availability could have severe cost implications with the other components. At inception, it had been envisioned that the hydrologic cycle could experience drought conditions, making it necessary to defer impounding to any great extent until the next occurrence of a wet season. Thus, provision for staged impoundment at the dam was incorporated into the contract documents. Because of the complexities that eventually transpired in funding for the dam, delay materialized, and it was necessary to revert to a comparable staged impoundment strategy. Most of the success of achieving component completion deadlines did however rest upon initially achieving early impoundment.

Challenges to Early Impoundment

Following the initial delay in commencement of the project there were numerous challenges and other significant setbacks as outlined below:

- The mobilization and site establishment was very slow due to the fact that the delay in commencement resulted in the Contractor diverting key personnel and equipment planned for Metolong to other worldwide projects.
- The Contractor lacked familiarity with local conditions and requirements.
- Both the Contractor and Engineer faced problems in obtaining properly qualified staff and replacing staff that have left the project.
- Excavation for the dam and diversion tunnel did not start until June 2012.
- Just as critical excavation works started, there was an extremely violent strike and the Contractor threatened to leave the job unless security measures were put in place. A project shutdown of a month resulted while agreements were reached and security ensured. The Lesotho Mounted Police established an outpost on-site to assist with security.
- Issues have developed with the nearby villagers over employment, blasting, dust, and noise control. Some issues have led to work slowdowns and stoppages imposed by local inhabitants.
- Labor issues continued through 2013 with several work slowdowns and two short stoppages due to illegal strikes. Earlier, Metolong Authority had engaged the services of an industrial relations specialist. Through the efforts of both MA and the Specialist, the Contractor was dissuaded from imposing strict counter measures to these illegal strikes.
- There have been significant delays in obtaining visas and work permits for expatriate staff which delayed commissioning of critical aggregate processing and concrete production plants.

The many delays severely impacted the Contractor's expected cash flow as discussed previously. This was also compounded at times by slow payments from the funders. The cash flow constraints resulted in several work slowdowns by the Contractor.

In mid-2013 the probability of impounding during the upcoming rainy season starting in late 2013 seemed unlikely as neither the RCC placements in the dam nor the concrete placements in the intake tower had commenced. Furthermore, the Engineer advised in June 2013 that they were of the opinion that the intake tower had to be complete to full height before commencing impoundment.

It was later agreed that the intake tower did not need to be complete before the start of impounding. It only needed to be constructed to an elevation higher than the dam with sufficient free-board and above the lowest intake so water could be passed through the dam. In November and December 2013, after delays in the delivery of the intake piping, the Engineer's opinion was that the earliest possible impoundment was mid-March 2014. This would have greatly reduced the probability of capturing enough runoff before the dry season.

Measures Taken to Achieve Early Impoundment

In October 2013 the Contractor and Engineer each developed small teams whose main focus was the achievement of early impounding. Regular meetings were held and the minimum requirements for commencing impoundment were jointly determined. The meeting frequency was increased to weekly in 2014 and checklists were created to monitor progress.

The RCC works were progressing well and by October 2013 the lower gallery was completed which allowed the start of grouting. The first intake pipes were expected late October but were delayed due to fabrication and lining / coating issues. By the end of October it was becoming clear that the critical path for impounding was moving from the RCC dam and grouting works to the intake tower.

Figure 4. RCC being placed August 7, 2013 - note the concrete is covered on the right side to protect against cold night time temperatures

By mid-November the critical intake pipes had still not been released for delivery to the site. The Engineer advised that the pipes had been rejected because either the pipes had been lined without prior inspection or because the lining was not compliant with the specifications. In order to progress the critical works, the Employer reviewed the implications of the risk involved and requested an immediate release of the 1.2-meter-diameter pipes. It was readily recognized that repairs to the lining could be accomplished on site along with the reinstatement of the lining / coating at the locations of the field welds. The first pipes were delivered to site on November 19. After installing the pipes, the Contractor was ready to begin field welding in early December, but was prevented from doing so because the welding certification / procedure paperwork for his staff was not complete. On December 6, the Employer authorized the welding to commence, at the Contractor's risk, prior to receipt of all paperwork, with the understanding that all welds would be 100% x-ray tested and any defects would be identified and corrected.

Installation of the critical intake tower piping work was now able to proceed and the Contractor made very good progress through the year-end holiday period. By late December, the two parallel intake/outlet pipes had been installed through the dam, and the lowest intake tee and bell mouth had been installed. Unfortunately, the straight vertical pipe for the second intake stack had not yet been delivered and would not be delivered until the second or third week of January 2014 because the supplier was closed for the year-end holidays. On December 22, the Contractor proposed a blockout for the missing pipe so that concrete placing for the intake tower could continue. The Engineer rejected the Contractor's proposal. The Employer did not understand the reason for the rejection and consulted the POE who likewise could not see significant reasons why the proposed blockout should not be adopted. On December 31, the Employer consulted with the Engineer, and the proposed blockout approach was adopted.

In January 2014, it was finally becoming clear that impounding could be achieved in February and at last everyone was on board to make it happen. The Contractor proposed impounding on Friday, February 14, 2014, but the Employer requested a delay to Monday 17 February to accommodate a GOL presence. The closure of the diversion tunnel gate and Impoundment Ceremony were successfully completed on February 17, exactly one year after the river was diverted.

Figure 5. Closure of diversion gate February 17, 2014

Figure 6. Impoundment Ceremony

Unexpected Fast Filling of Reservoir and Overtopping

It was expected that the reservoir would fill the 25 m to the lowest intake within 1 – 2 months and then possibly overtop the dam. Because of this potential, a low spot was constructed in the RCC, near the center of the dam, for a planned overtopping (see Figure 1 – the depression is just to the left of the dump truck on the dam surface). Unusually heavy rainfall immediately followed the closure of the diversion gate, and the reservoir started to overtop only 10 days later on February 27, 2014.

Figure 7. Preparing for overtopping February 26, 2014; start of overtopping February 27

Heavy rains continued into March leading to a large overtopping, estimated at approximately 120 m^3/sec, on March 11 which damaged formwork and flooded the lower gallery. The rains then subsided, the overtopping ended March 13, and RCC placements resumed March 14.

Figure 8. Gallery inspection during overtopping on March 8, 2014 with Contractor, Engineer, POE and WASCO representatives; large overtopping March 11

Although the overtopping, of approximately 2 weeks, caused delay and added costs, there were also benefits as it demonstrated good performance of the dam and also provided WASCO, who has little experience in dams, with added confidence regarding the operation and safety of the dam. Additionally, it provided valuable training and practice regarding implementation of the Emergency Preparedness Plan.

According to the Lesotho Meteorological Services, the rainfall during February and March 2014 was 141% and 417% of long-term averages, respectively. Furthermore, following the very heavy rains in early March it was unusually dry.

The push to impound in February 2014 was crucial because had impoundment been delayed to mid-March the reservoir would not have filled to the lowest intake and pumping to the water treatment works from the RWPS would not have been possible. This would have delayed the commissioning of the water treatment works and downstream conveyance pipelines for many months.

EARLY WATER DELIVERY

The delivery of water from the RWPS to the water treatment works (WTW) was also challenging due to various delays in the construction of the pump station.

Delays in Construction of RWPS

The excavation of the platform for the pump station just downstream of the dam was in massive sandstone and was complicated due to numerous stress relief joints and bedding planes. The upper portion of the excavation did not go as planned due to a combination of the challenging geology and aggressive blasting techniques.

In November 2012, the MA authorized the Engineer to engage both a blasting specialist and a slope stability specialist to evaluate the situation and provide recommendations on how best to proceed. The specialists visited the site in January 2013 and provided valuable guidance for stabilizing the rock mass and improving blasting techniques for the remaining excavation. It then took considerable time for the Engineer to design the remedial works. The instruction for additional slope support was issued mid-May 2013 and the instruction for a considerable amount of additional excavation was issued mid-June. The excavation and slope support work then continued and was mostly completed in August 2013.

Then there were delays in finalizing the foundation design with construction details provided mid-September. The first blinding concrete was placed in October, and the pump room foundation / floor was finally placed December 2013. It was now clear that the entire pump station concrete structure could not be completed by the desired first water delivery date of May 2014 and alternative approaches needed to be considered.

Measures Taken to Achieve Early Water Delivery

Early impoundment of the reservoir was the highest priority because it was realized that if impoundment did not occur during the wet season, there would be a significant delay to the availability of water. Even if the RWPS was complete, it would not have water to pump to the WTW. For this reason the RWPS became the critical focus after impoundment on February 17, 2014.

The Engineer's Resident Engineer, who had been key in orchestrating the early impoundment, resigned and left the project in late February, which was a major setback. The huge challenge of being able to deliver water in May was becoming clearer to the project team.

On February 21, 2014 the Contractor called a meeting with the Engineer, the Employer, and its major mechanical and electrical subcontractor KSB. The Contractor presented his plan to achieve early water delivery in May by installing one pump and associated piping in the partially completed pump station utilizing temporary electrical installations. The plan was well thought out and was agreed to by the project participants.

The Contractor and Engineer again developed teams whose focus became the early delivery of water. The teams met weekly and developed checklists that were used to monitor progress.

KSB mobilized and set the first 600 l/s pump April 23, 2014; however the suction and discharge piping expected April 15 was delayed and did not arrive until April 30. Once the installation of the pipes began it was discovered that the concrete pipe supports did not properly align and significant field modifications were required. This led to a stressful period as the Contractor was pushing hard to make required field modifications to progress the works without full concurrence by the Engineer.

The Contractor successfully pumped water to the WTW on May 26, 2014 but without full acceptance of the works by the Engineer. During the pump shutdown the variable speed drive did not ramp down properly which resulted in a significant surge on the discharge piping. Several cracks were observed in the concrete pipe supports so pumping was suspended while the Contractor and Engineer agreed on remedial safety measures. These measures were implemented, and pumping resumed June 4, 2014 and continued until June 7 when an unforeseen problem was discovered at the WTW requiring remedial action. This effectively took the RWPS off the critical path of the overall program and the WTW took center stage. On August 14, the problem was rectified and pumping resumed. The project was back on track to deliver the first potable water during the last quarter 2014 as scheduled.

Figure 9. RWPS pumping first water to the WTW May 26, 2014 – note excavation in the massive sandstone and also the status of construction with one pump installed and operational.

Figure 10. WTW receiving first water from RWPS May 26, 2014 – concrete plant for the dam in background of left photo along with visitor's center

CONCLUSION AND LESSONS LEARNED

During the implementation of the MDWSP there have been many challenges and lessons learned. Following are some of the most significant lessons:

- A major lesson learned revolved around sourcing additional funding to service identified shortfalls. To put additional funding in place for the dam required a considerable amount of time as entirely new financing contracts needed to be prepared. This involved funders going back to their respective Board of Directors subsequent to conducting fully fledged appraisal missions. The process is lengthy.
- For the dam, it took 157 days from the date of signing the construction contract (August 11, 2011) to issuance of the Date of Commencement (January 12, 2012). The Metolong Authority believed that the full advance payment to the Contractor could be drawn from the initial loan commitments prior to having supplementary funding in place. The funders initially did not agree to this manner of disbursement resulting in significant delay and associated ramifications. A disbursement plan needs to be in place as soon as possible to avoid delays.
- One ramification resulting from the above lessons is to carefully consider mobilizing the construction supervision until supplementary funding is in place, as this can be costly. In the case of the dam, construction supervision was mobilized congruent with signing the construction contract. This was undertaken on the assumption that the dedicated personnel would be retained throughout the job. As it turned out, the personnel that had been retained through early mobilization later left the job for personal reasons.
- Although not mentioned in this paper, the Arab Funders did not establish a common basket of funds to draw from. Instead, each of the five Arab funders is individually invoiced on each Interim Pay Certificate. This has resulted in a complex accounting effort, particularly as each makes disbursements accordingly. The lesson learned is to try an induce coalescing multiple funders into one common basket to simplify accounting practices.
- There were many stakeholder's involved in the MDWSP. Besides the 10 Funders and various Lesotho agencies directly involved as detailed in section 2.5, there

were several other GOL agencies directly involved with the implementation of the project. These agencies included:

a) Lesotho Revenue Authority – customs, taxes, tax refunds
b) Immigration – visas, work permits, residence permits
c) Labor – safety, health and industrial relations
d) Department of Environment
e) Ministry of Mining
f) Lesotho Mounted Police and Lesotho Security Services
g) Metolong Authority Board and Sub-Committees

- Coordination with the many stakeholders was time consuming and at times challenging. A dedicated person focused on stakeholder alignment and management would have likely proved valuable.
- There were risks in accepting the lowest bidder as their bid was significantly below both the Engineer's estimate and the second bid. The risks were evaluated by the Employer and Engineer. It was understood that additional management would be required and that there would likely be many challenges to overcome because the low bidder would have little or no contingencies built into their bid and would likely aggressively pursue claims. There were many challenges and claims as detailed in this paper. Nevertheless, it appears the challenges and claims will be worked through and the project will end up costing several tens of million dollars less than the second bid.
- Active involvement by the Employer was key to achieving program objectives.
- An experienced MA/MPMU team was crucial for driving the multiple components, making critical decisions, and taking necessary risks to advance the project.

The figure below shows the status of the dam on December 31, 2014.

Figure 11. Metolong Dam on December 31, 2014

REFERENCES

Reynolds, K., Rafoneke, B., Havenga, B. (2013). An overview of the institutional arrangements for the implementation of Metolong Dam. SANCOLD Conference 2013, 30 – 44.

Bosshart, D., Neumann, MJE. (2013). Metolong Dam – An overview of the construction and challenges faced to date. SANCOLD Conference 2013, 136- 151.

Ministry of Energy, Meteorology and Water Affairs; Lesotho Meteorological Services; Press Release Issued: 13-08-2014; Winter Weather Performance (May-June-July 2014)

HIGHLIGHTS AND CHALLENGES OF CONSTRUCTING THE PORTUGUES THICK-ARCH RCC DAM IN PUERTO RICO

Rafael Ibáñez-de-Aldecoa[1]
David Hernández[2]
Eskil Carlsson, PE[3]

ABSTRACT

At the end of 2013 the construction of the Portugues Dam near Ponce, Puerto Rico, was completed by Dragados USA Construction Company. The project was designed and procured by the U.S. Army Corps of Engineers, and is owned by the USACE and the Puerto Rican Department of Natural and Environmental Resources. Upon completion, Ponce residents have now much needed flood damage protection during the intense rainy season that often includes tropical storms and hurricanes.

The work included construction of the following elements:

- **Roller-Compacted Concrete (RCC) Dam:** a 220-foot high single centered thick-arch dam structure with a crest length of 1,230 feet and 355,000 yd^3 of RCC
- **Appurtenant structures:** integral spillway, intake structure, and a control and valve house located on top of a 24,000 yd^3 RCC buttress

The project also included foundation rock excavation, foundation treatment work, developing an onsite quarry for aggregate production, and an access road connecting the crest of the dam to the valve house.

The Portugues Dam is the first arch-gravity (or thick-arch, term preferred by USACE's designers) RCC dam built in the United States, and the first RCC dam in Puerto Rico. As such, it presented unique design and construction challenges.

Focusing primarily on the construction process, this paper addresses particular constraints of very different nature that had to be overcome, among which include: very limited availability of local personnel with experience in RCC; limited availability of certain materials, equipment, subcontracting and managerial required resources; very seasonally hot and wet climate with frequent river floods during the wet season (reason for which only winter windows were allowed for RCC placement, preventing a steady workforce and requiring retraining of personnel throughout the entire dam structure construction period); high temperatures for placing concrete even during the winter windows; complex geology at the dam foundation leading to extensive foundation treatments; etc.

[1] Dragados USA, Inc. 701 5th Avenue, Suite 7170, Seattle, WA 98104, ribanez@Dragados-USA.com
[2] Dragados USA, Inc. 810 Seventh Ave., New York, NY 10019, dhernandezg@Dragados-USA.com
[3] Dragados USA, Inc. 810 Seventh Ave., New York, NY 10019, ecarlsson@Dragados-USA.com

INTRODUCTION

The Portugues Dam is the first Roller-Compacted Concrete (RCC) dam ever built in Puerto Rico and the first single-centered thick-arched RCC dam in the United States. The project was awarded to Dragados USA in 2008, was completed in December 2013 and the dedication ceremony was conducted February 2014.

This dam completes the last phase of the Portugués and Bucaná Rivers Project to protect the city of Ponce from floods. The city of Ponce is subject to frequent and serious flooding from recurrent torrential rains, tropical storms, and hurricanes, especially when these rains occur in the mountainous terrain to the north of the city. This complex climatology also greatly affected the construction of the dam.

Being the first RCC project in Puerto Rico, local experience was not available and an extensive training program on RCC was implemented to be able to build the dam, meet the multiple requirements and achieve the project objectives.

The fact of the location of the project in an island, greatly conditioned supplies of all types (materials, spare parts, etc.) that not being available on the island had to come from the mainland.

A summary of the main site specific challenges are presented in the document below.

BACKGROUND

Project Location

The Portugues Dam is located in the central region of Puerto Rico near the southern Caribbean cost, less than eight miles from the mouth of the Portugues River and approximately three miles northwest of the city of Ponce. Ponce is the second most populous city in Puerto Rico.

Project History

The mountainous terrain north of Ponce is subject to sudden and rapid runoff of the existing rivers that caused frequent floods in the downstream urban areas. These floods often resulted in extensive property damages and even loss of life in the city of Ponce.

The Portugues Dam was originally designed by the United States Army Corps of Engineers (USACE) to be its first three-centered double curvature thin-arch dam, as a traditional arch concrete dam. In the early 2000s, the cost of the single construction bidder, significantly exceeded the government's cost estimate and USACE decided to revise its original design for a more cost effective alternatives, finally selecting the current RCC design.

The Portugues Dam, completed in December 2013, is the last of the two dam structures built as part of the Portugués and Bucaná Rivers flood control project that was authorized in 1970. The first dam, the Cerrillo Dam was completed in 1992. The entire project consists of two multipurpose reservoirs with downs-stream channel improvements from the dams to the river outflows as showed in Figure 1. The Puerto Rican Department of Natural and Environmental Resources (DNER) is the ultimate owner of the project.

Figure 1. Project location map

Project Description

The Portugues Dam was built as a flood protection. The Portugues Dam is a 220-ft high single centered thick-arched RCC dam with a crest length of 1,230 ft. The project was built with over 355,000 yd^3 of RCC for the dam, over 10,000 yd^3 of conventional concrete on the spillway, intake and top of dam, nearly 20,000 yd^3 of mass concrete for foundation and dental concrete and over 24,000 yd^3 of RCC for the foundation of the Valve House. The dam has a vertical upstream face and a stepped downstream face. The downstream steps are 4 ft high (four 1 ft thick lifts) resulting in equivalent downstream slope of 0.35H:1V. Even though the RCC lifts were built concurrently, the body of the dam had 18 sections created by installing contractions joints spaced 70 ft apart. An internal gallery (7 ft wide and 12 ft in height) was built in the RCC approximately 20 ft above the design foundation level. An upstream slopped grout curtain and vertical foundation drain holes were drilled from inside the gallery. The dam has a 140 ft wide un-gated ogee-crested and flip bucket spillway which was constructed with conventional concrete in the central section of the dam, above the original river channel.

The permanent outlet works consists of two 60-inch diameter regulating outlet pipes traversing the dam through a conventional concrete block built between the dam foundation and the RCC structure. The outlet pipes extend from the sluice gates at the intake structure just upstream of the Dam to the fixed-cone valves located at the Valve House just downstream of the Dam.

Figure 2. Aerial view of the dam under construction and RCC batch plant facilities

The slow strength development pozzolan-cement used in the dam was designed to achieve a compressive strength of 4500 psi after one year, but generally exceeded 6000 psi. The slow strength development nature of the RCC allowed continuous RCC placement on "hot" surface joints (i.e., relatively "fresh" concrete surfaces) for a period up to a maximum of 24 hrs between lifts. Partially set concrete surface or "warm" joint surfaces (up to 72 hrs) and "cold" joints surfaces (over 72 hrs), required special joint surface treatments before the following lift was placed.

Construction Facilities

Figure 3. Aerial view of the RCC batch plant and pre-cooling facilities

The project required the installation of several facilities to support the construction of the dam. The principal support structures included:

- An aggregate production plant to produce three types of coarse aggregates (NMAS: 2-in, ¾-in, and 3/8-in) and fine aggregate (sand). The different size aggregates allowed the production of a wide variety of different concrete
- A sand washing plant to remove fine of a portion of the processed sand
- An on-site double concrete plant able to produce up to 5 yd^3 per batch each of RCC, conventional concrete, or mortar, supported with:
  - Two 1,000 ton cement silos
  - One 1,000 ton flyash silo

  - Three coarse aggregate silos of 2,000 tons each
  - Two fine aggregate silos of 1,000 tons each (one for washed sand and another for unwashed sand)
  - A plant for fabricating ice flakes capable of producing 200 tons/day
  - A water cooling plant for the "wet belts" tunnel and chilled mixing water
  - Two 400 ft long wet belts located inside an aggregate cooling tunnel to lower the temperature of the coarse aggregate to below 50° F and help meet the maximum average RCC placement temperature of 60° F
- An off-site portable batch concrete plant to produce conventional concrete and mortar
- A conveyor system, with a length of up to 978 ft, to transport the concrete from the plant to the central portion of the dam
- Miscellaneous support facilities such as the mechanical, electrical, welding and carpentry shops

The construction of the Portugues Dam did not have access to the electrical grid system and all the on-site facilities were powered with generators.

CONSTRUCTION CHALLENGES

Site and Subsurface Conditions

The project is located in heavily fractured and in areas highly weathered sandstone, siltstones and metamorphic volcanic conglomerates intersected by shears and diorite dikes. The highly fractured foundation geology made it difficult to obtain a regular foundation surface. The aggregate material was excavated from a quarry located just over 1 mile north from the dam site.

Heavy excavation equipment was permitted to be used by contract up to the theoretical foundation level, leaving a significant portion of the final cleanup to be performed by extensive use of hand tools and occasional very light excavation equipment (see Fig. 4). This resulted in a very labor intensive and time consuming dental treatment process that was often impacted by heavy rain and sudden floods.

The mountainous terrain of the site created difficulties for the layout of the support facilities and forced the division of the aggregate stockpiles in multiple locations. Likewise, the required temporary construction accesses to different levels of the dam were difficult to deploy.

Figure 4. Labour intensive foundation preparation

RCC Placement Requirements and Supplies Constraints

The Portugues Dam was constructed with a series of very restrictive requirements. These requirements, although not unusual for many RCC projects and many are common USACE and USA standards, presented special challenges during construction of the project. As indicated earlier, this project is the first RCC structure in Puerto Rico, hence many of these requirements were not previously used in Puerto Rico. This impacted particularly services, materials and equipment. Because of these requirements, many of the local companies in the island could not provide services, materials and equipment to the project. Many of the materials and services had to be provided either by selected qualified local suppliers or by off-island companies.

Producing RCC is basically an industrial process, where continuous RCC plant operation is vital for a consistent product. Minor problems and small breakdowns can cause costly disruptions and even stoppages of the production operation where hundreds of people would be sitting idle due to spare part delivery dependencies.

Puerto Rico is an island that does not have similar facilities to the ones used for the construction of this dam, so the vast majority of the specialized spare parts and other equipment had to be brought from off-island suppliers, resulting in a critical logistical challenge to minimize the impact of delivery, maintenance and repair delays. To mitigate these high risks, Dragados USA established a special program to stockpile on site a significant amount of spare parts and long lead items, not commonly stockpiled in other projects, to reduce the potential impact on the schedule. Despite this, on several occasions, spare parts and materials had to be shipped by air from the US. In addition, Dragados USA had on its payroll a larger than normal team of mechanical and electrical personnel working 24 hrs per day, 7 days per week. These measures increases the cost of a project, but are essential to maintaining the project on track.

Some RCC placement contract requirements included:

- Place RCC at an average temperature below 60° F
- RCC placement windows allowed only during the cooler months of the year
- Cover placed RCC with new layer within 24 hrs (maximum exposure time between layers); otherwise special joint surface treatment and placement of mortar required between lifts (warm joint treatment)
- Produce, transport, place and compact RCC within a 45 minute window
- Extensive and labor intensive RCC surface cleaning requirements between lifts

Figure 5. General view of the RCC placement

Limited Access to the Crest of both Abutments

The only permanent access road to the dam crest is from the downstream side towards the right abutment. But to connect this road with the dam crest there is a bridge which logically could not be built until the end of the upper part of the right abutment.

Therefore, to have access for the auxiliary equipment to the proximity of the upper portion of the right abutment while constructing the RCC, and to remove the RCC placement equipment from the top of the dam, and move it to build the upper part of the left abutment, a difficult temporary road was arranged at the upstream side of the right abutment.

Likewise and due to the left abutment does not have a permanent access road, to place the RCC placement equipment into the left abutment in order to finish the upper part of it, and to remove all the equipment from the dam after completing the portion of the dam left of the central spillway, another very difficult temporary road, because of the steep natural slope conditions of the left bank of the valley, had to be built at the downstream side.

Both accesses, as were no part of the project design, were removed upon completion of their role.

Weather

Puerto Rico is one of the islands that separate the Caribbean Sea from the Atlantic Ocean. It is located in an area that is prone to heavy tropical storms and hurricanes. As indicated earlier, the project was built to protect the city of Ponce from frequent and dangerous floods. The project itself is located in a narrow valley with steep slopes in the mountainous terrain just north of the city. The watershed area of the Portugues River receives on average a precipitation exceeding 80 inches per year. The frequent tropical rains often come down heavily and suddenly, giving little time to protect the work areas, removing equipment and safely relocate personnel from the incoming river floods. The steep terrain required the temporary diversion system to be designed for overtopping conditions. The work also had to stop until the water in the inundated access road and low laying work areas receded. Floods were a constant threat during the early construction stages, requiring a vigilant eye on the weather. When rain occurred during RCC placement, often there was not sufficient time to protect the work in progress. This resulted in extensive repairs and stoppages that exceeded the allowed time between lifts for regular placement and required extensive joint surface treatment before resuming construction.

Overtopping of the diversion cofferdam during heavy rain events, divided the construction site and limited access to the quarry, RCC plant, and right abutment area.

The frequent rains also created continuously varying and non-homogeneous moisture content of the fine aggregates. RCC is very sensitive to moisture variations and slight changes in moisture had an impact on the consistency of the RCC. This required that a portion of the sand be stockpiled and handled two or three times into covered sheds to maintain a more consistent moisture content for optimum RCC production.

The contract required that RCC be placed below a daily average temperature of 60° F. The average temperature in Ponce, even during the winter months, is above 76° F. To meet the project requirements, all coarse aggregates used for RCC production were chilled to a temperature below 50° F and ice chips were used in lieu of most of the batch water to lower the RCC temperature. The relatively high moisture content of the fine aggregates caused by the frequent rains, in combination with the fact that the coarse aggregate leaving the cooling tunnel was wet, limited the amount of ice that could be added to the batches to cool down the RCC, so maintaining the fine aggregates dry was very important to achieve the low placement temperatures.

Figure 6. One of the several floods experienced during the dam construction

Work Force Availability and Experience

The construction of RCC dams is a continuous operation and is essential to maintain a 24 hours-7 days per week work schedule. This required the availability of a large pool of skilled construction labor force. During peak construction periods, the Portugues Dam employed up to 800 personnel. The city of Ponce and its vicinity could not provide all the needed experienced field personnel for the project, so a significant portion of the labor force had limited or no construction background.

Additionally and as indicated earlier, the Portugues Dam was the first RCC project constructed in Puerto Rico. Therefore, finding local technical personnel with experience in RCC technology was not possible and bringing experienced personnel from outside the island had to be limited due to cost constraints. Therefore, most personnel had to be locally trained in RCC placement technology.

Most of the activities in the construction of RCC are very time sensitive. This required that all activities be carefully coordinated with specialized crews for the success of the project; from the time between fabrication and placement of the RCC, to the maximum time allowed between placement of consecutive lifts, beyond which special joint surface treatment is required. Some of the specialized activities that the different crews had to accomplish include:

- Dental concrete treatment
- Drilling and Grouting
- RCC fabrication
- RCC conveyance
- RCC placement and consolidation
- Cleaning and preparation or RCC surface
- Moisture control of RCC surface
- Raising and preparation of formwork
- Grout Enriched RCC consolidation near exposed surfaces
- Contraction joint cutting
- Instrumentation installation and monitoring
- Inspection and testing
- Equipment maintenance

Poor execution or delays in any of the above activities posed a significant risk of stopping RCC placement that required costly and time consuming remedial work to resume construction after "warm" or "cold" joint treatment. So a reliable and well trained team is essential.

A sizeable portion of the available resources did not have the required construction experience and could not easily adjust to the continuous shift work. This shift work is generally very demanding and resulted in a recurrent turnover of the work force that had to be replaced and retrained.

The fact that RCC placement was limited to a six month window in the winter months, complicated the experience retention even further, since the majority of the RCC workforce had to be dismissed during the summer months. For the next placement window, an attempt was made to rehire the same personnel; however, a significant portion of the workforce did not returned and the previously acquired local knowledge was lost and new personnel had to be re-trained and a second learning curve of the field personnel had to be absorbed by the project.

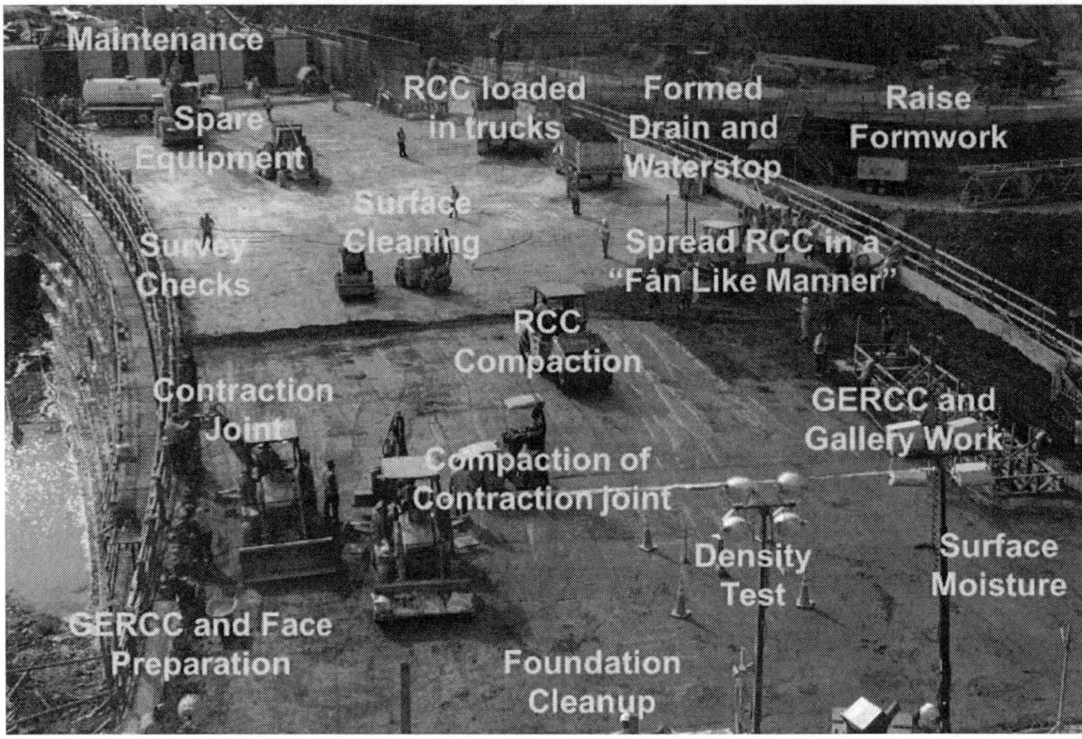

Figure 7. Typical RCC placement activities

Safety

Safety is of paramount importance to USACE and Dragados USA. An extensive safety program was implemented in the project with dedicated safety personnel available in the project at all times during the construction activities.

The time sensitive nature of RCC construction requires that equipment and personnel work fast and in close proximity posing a serious safety risk. Furthermore, RCC dam construction uses concrete conveying systems that have to be raised regularly as the dam height increases, demanding personnel to work from ever increasing elevations.

The continuous changing nature of the workforce, the limited construction experience of some of the available personnel hired for the project and the construction requirements described above, also posed considerable risks of serious injury to the field personnel that had to be mitigated throughout the construction period.

The safety requirements and expectations from USACE and Dragados USA for the project were significantly more stringent than the typical safety requirements for construction projects in the region. This made the implementation of the safety plan more challenging and prompted Dragados USA to implement an extensive safety awareness and safety culture program through continuous safety training. It is important to note that even though the project experienced a series of minor incidents, it did not suffer a fatality or other serious accident.

Quality Program

The overall safety of a dam is greatly dependent on the quality of its construction. The Portugues Dam was built to the highest quality standards. Quality Control (QC) was the responsibility of Dragados USA and Quality Assurance (QA) the responsibility of USACE. During peak construction periods, the project quality personnel had a staff of over 30 QC and 20 QA personnel.

A wealth of information was collected and analyzed during the construction process. Some highlights of the quality checks are presented below:

- Submittal process – All activities were thoroughly documented through the submittal process. Over 4,500 submittals were prepared for this project
- Construction tolerances – Construction tolerances were very demanding. For example the spillway required that the finish grade of the 140 ft wide ogee, chute and flip bucket be built to within 1/8-in
- Testing frequencies – Vebe consistency, RCC placement temperatures, RCC density, compressive strength, aggregate gradation, were just a few of the numerous tests and other quality data points collected during the construction of the dam. Many of these tests were performed at high frequencies, for example, most of the over 2,000 conventional concrete truck loads delivered to the project were tested for slump, air content, temperature, and unit weight

The Dam was constructed using the USACE's three step Quality Control and Quality Assurance Process strengthened by Dragados USA's extensive RCC dam experience. First, all definable features of work of the project required to a formal planning session prior to its execution, where its execution plan, as well as all safety, quality, and environmental requirements were thoroughly reviewed and planned by the Dragados USA's construction and quality teams and approved by the USACE's QA team. Then, before any activity was started, the supervisor reviewed all safety, quality and construction requirements with the construction crew to ensure alignment. Finally, the execution was continuously inspected and verified to ensure compliance with the requirements.

Before RCC placement begun, Dragados USA used the initial RCC test section and other RCC test strips to train the different field crews, select best RCC placement equipment and optimize placement methodologies. For the second RCC placement window, Dragados USA constructed a second test section to refresh the knowledge of existing employees, capture lessons learned from the previous placement session and help train new employees. This second test section was critical to achieve the anticipated quality of the RCC placement during the second window.

RCC TECHNOLOGY KNOWLEDGE TRANSFER

As indicated earlier, the Portugues Dam resulted in the first single-centered, thick-arch RCC dam designed and constructed in the United States and the first RCC dam in Puerto Rico.

The dam has served as training ground for the next generation of USACE dam engineers through the *Dam Safety University* program. Using this program USACE trained many of their engineers in workshops and temporary field assignments in dam engineering. The project was also used to transfer RCC technology knowledge to the local engineering community through workshops and temporary employment.

CONCLUSION

The Portugues Dam was successfully completed in December 2013. However, the combination of the project being located in an area where delivery of materials and availability of spare parts for specialized equipment is limited, with lack of experienced personnel in RCC technology, can present a significant challenge to the project without sufficient planning and a strong team experienced in the RCC construction technology.

To mitigate these difficulties, an extensive training program must be implemented to ensure all personnel can work safely and effectively with the numerous time sensitive activities that must be coordinated to achieve a quality RCC dam structure.

Finally, it is very important that the owner, designer, and contractor teams work cooperatively to resolve the difficulties that are sure to occur during construction of complicated projects like the Portugues Dam.

Figure 8. USACE's and Dragados USA's personnel celebrating the completion of the RCC placement

ACKNOWLEDGEMENTS

Grateful thanks must be accorded to the USACE, Jacksonville District, the designer and temporary owner of the Portugues Dam project, for permission to publish this paper. Nevertheless USACE cannot be held responsible for the views expressed in this paper, they being solely those of the authors.

REFERENCES

USSD, 2013, Achievements and Advancements in U.S. Dam Engineering.

Vazquez, P. and Gonzalez, A., Moving Successfully from a Conventional Concrete into an RCC Design for Portugues Dam. Proceedings of the 6th International Symposium on RCC Dams, Zaragoza, Spain, October 2012.

Figure 9. Aerial view of the completed Portugues Dam

ANALYSIS OF SHIP GROUNDING FOR DAMS OF THE PANAMA CANAL EXPANSION PROJECT

Lelio Mejia[1]
Ethan Dawson[2]
Rajendrum Arulnathan[3]
Maximiliano De Puy[4]

ABSTRACT

The Panama Canal Expansion Project includes construction of four new dams, known as the New Borinquen Dams, to provide navigation access from the new Post-Panamax locks to the existing waterway of the Panama Canal. The dams are critical components of the canal expansion and must withstand their design loads with a high level of reliability. Because the dams could be subjected to grounding of Post-Panamax-size ships, their design requires that the dams be capable of withstanding impact loads from ship grounding without loss of function.

Two sets of analyses were performed to estimate the potential damage to the dams due to ship grounding. Initial analyses were performed with the computer program SHIPCOL, a semi-empirical program in which grounding forces are represented using empirical constants calibrated on the basis of scale model tests. Penetration of a ship bow into the dams and maximum grounding forces were computed for various grounding scenarios. Once the critical grounding scenarios had been identified, three-dimensional (3D) elasto-plastic numerical simulations were performed using the program FLAC3D, to evaluate the stresses and strains induced in the dam core by the ship impact. The FLAC3D analyses also provided an independent evaluation of the penetration distance into the dam and the induced grounding forces.

The collision scenario resulting in penetration closest to the dam core involved a head-on collision from a ballasted ship at an approach velocity of 8 knots. In this scenario the ship comes to a stop about 22.5 m from the core. Analysis of collision scenarios producing the greatest grounding forces indicated that the maximum induced compressive and tensile stresses within the dam core are less than 10% of the initial long-term effective stresses. On this basis, it was concluded that the dams can resist the design ship grounding without major disruption of their function.

INTRODUCTION

The Borinquen Dams (referred to as Dams 1E, 2E, 1W and 2W) are part of the new Pacific Access Chanel (PAC), which will connect the Gaillard Cut section of the Panama

[1] URS Corporation, Oakland, California ,USA, lelio.mejia@urs.com
[2] URS Corporation, Los Angeles, California ,USA, ethan.dawson@urs.com
[3] URS Corporation, Oakland, California ,USA, rajendram.arulnathan@urs.com
[4] Senior Earthquake Engineer, Geotechnical Section, Autoridad del Canal de Panama, Panama, AAbrego@pancanal.com

Canal to the new Pacific Post-Panamax Locks. The channel will be 6.7 km long and 218 m wide at its bottom. The Borinquen Dams 1E and 2E extend along the east bank of the new channel, and retain the channel water 10.7 m above the nominal elevation of Miraflores Lake. Dams 1W and 2W form the west bank of the channel in low-lying areas located just north of the new locks entrance.

The dams are critical components of the Panama Canal and must be built to withstand their design loads with a high level of reliability. Because the dams could be subjected to grounding of Post-Panamax-size ships, the design criteria requires that the dams be capable of withstanding the potential impact loads from ship grounding without loss of function.

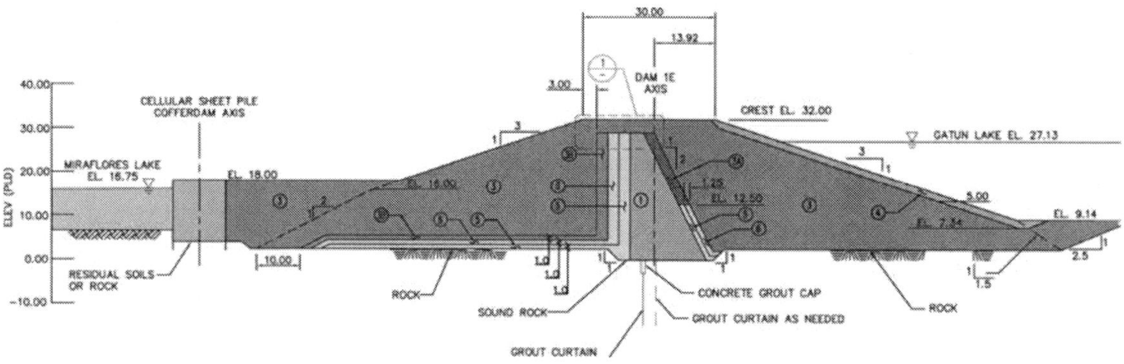

Figure 1. Typical cross section of Dam 1E.

A typical cross section for Dam 1E is shown in Figure 1. The dam consists of upstream and downstream rockfill shells and a clay core founded on bedrock. The crest elevation is 32 m, and the crest width is 30 m. The inboard and outboards slopes are generally 3:1 (H:V), except for a portion of the outboard slope below elevation 16 m where the slope is 2:1. The face of the inboard slope is covered with riprap for erosion protection.

Two sets of analyses were performed to estimate potential damage to the dams due to ship grounding. Initial exploratory analyses were performed with the program SHIPCOL, a semi-empirical program in which grounding forces are represented using empirical constants calibrated on the basis of scale model tests (Brink-Kjær et al., 1983; Denver, 1983; Havnø and Knott, 1986; Simonsen, 1997). From these analyses, penetration of the ship bow into the dams and maximum grounding forces were computed for various grounding scenarios (Ben Gerwick Inc., 2008). The SHIPCOL analyses were used to identify the critical grounding scenarios and provided estimates for ship penetration distance into the dam. However, they did not directly provide any information on the potential stresses and strains that would be induced in the dam core by the ship grounding.

To estimate grounding-induced stresses and strains within the dam core, 3D elasto-plastic numerical simulations were performed using the program FLAC3D (Itasca, 2005). These analyses considered a head-on grounding scenario, which had been determined to be the critical case from the SHIPCOL analyses. The 3D elasto-plastic analyses not only allowed an evaluation of the stresses and strains induced in the dam core by a grounding event, but also provided an independent check on the penetration distance and grounding forces.

SHIP CHARACTERISTICS AND GROUNDING SCENARIOS

Various ship grounding scenarios were considered in the design criteria for the Borinquen Dams (URS, 2008a). For the Lake Miraflores side of the dams, a Panamax container ship with a displacement of 75,000 metric tons was considered. A single scenario consisting of a 4 knot approach velocity at an angle of incidence of zero (head on impact) was specified.

For the Pacific Access Channel side of the dams, grounding of a 12,000 TEU (twenty-foot equivalent units) Post-Panamax container ship with at displacement of 160,000 metric tons was specified. Ship dimensions and particulars are listed in Table 1.

Table 1. Post-Panamax Container Ship Specifications for Grounding Analyses

Beam	49.0 m
Length overall	366.0 m
Draft	15.2 m
Displacement	160,000 metric tons
Bow Shape	Bulbous

The following grounding scenarios were specified:

- 8 knots approach velocity at an angle of incidence of 55°
- 4 knots approach velocity at an angle of incidence of 45°

The angle of incidence is measured from the perpendicular to the dam axis.

Since, the overall dimensions and displacement of the Panamax size ships are significantly smaller than those of the Post-Panamax ships, grounding on the Lake Miraflores side of the dams was considered to be less critical. Furthermore, because of the presence of cofferdams and planned backfilling on the Miraflores Lake side, the dams are unlikely to be affected by ship grounding from this direction. Therefore, the ship grounding analyses focused on the Pacific Access Chanel side.

While the design criteria called for ship grounding with an angle of incidence oblique to the dam axis, analyses were also performed for an angle of incidence normal to the dam axis (head on collision). This hypothetical scenario is believed to provide an upper bound for the effects of ship impact loading on the dam. This scenario is also simpler to understand and simulate than the more complex oblique grounding scenarios.

ANALYSES WITH SHIPCOL PROGRAM

The SHIPCOL program was developed by the Danish Hydraulics Institute (DHI) for the design of artificial islands around bridge piers. It computes the distance and path of penetration of a ship bow into an earth embankment, along with the grounding forces generated by the collision. SHIPCOL models the ship as a rigid body with six degrees of freedom (surge, heave, sway, yaw, roll, pitch), with the ship bow assumed to be non-deformable. The equations of motion for this rigid body are integrated numerically in a time-stepping procedure to track the position and velocity of the ship throughout the collision process. Embankment (soil or rock fill) forces acting on the ship bow were developed from theoretical soil yield mechanisms, combined with empirical constants, calibrated on the basis of model tests.

For the analyses of the Borinquen Dams, the dam geometry was approximated as a plane surface with a 3:1 slope (Ben Gerwick Inc., 2008). The rockfill shell was assigned a friction angle of 42 degrees, a moist density of 2,162 kg/m^3, and a saturated density of 2,243 kg/m^3. The riprap layer on the dam was assumed to have the same material properties as the dam shell. A friction coefficient of 0.44 was assigned between the ship hull and the rockfill. The shape of the ship's bulbous bow was approximated as an ellipse with vertical sides and a flat bottom. Analyses were performed for both 'fully laden' cases (vessel loaded to its maximum draft) and for 'ballasted' cases (unloaded, with water ballast to improve stability).

Output from the SHIPCOL program includes the trajectories of the ship bow into the dam, and time histories of horizontal and vertical grounding forces acting on the ship. Figure 2 shows the penetration trace for Case 5: a fully laden ship with approach velocity of 8 knots and angle of incidence of zero degrees (Ben Gerwick Inc., 2008). This case was found to exert the highest loads on the dam core. The ship approach is from left to the right in Figure 2, with the initial impact point (at abscissa -60 m) located near the toe. The penetration into the dam shell is approximately 30 meters, with the plot showing the ship bow riding up slightly toward the end of penetration. The computed grounding forces acting on the ship bow for Case 5, plotted in Figure 3, can be shown to reach their maximum values at the deepest penetration.

Computed penetrations and resting distances from the bow to the dam core are summarized in the Table 2, for the various cases analyzed. The distance from the bow to the dam core is defined as the horizontal distance between the core and the final position of the bow. Not surprisingly, the results show that penetration increases with approach velocity, and also with vessel mass (fully laden versus ballasted). The oblique approach angles typically show more run-up on the dam face with deflection of the ship, rather than boring into the dam. This corresponds to longer stopping distances, but greater resting distances from the bow to the dam core.

While the penetration distances for the ballasted cases are smaller than those for the fully laden cases, they result in closer approaches of the ship to the dam core. With a shallower draft, contact for the ballasted cases occurs farther up the dam face, closer to the core.

Case 6 (head-on impact of a ballasted vessel with approach velocity of 8 knots) resulted in the closest proximity to the dam core: 22.5 m. Thus, none of the analysis cases indicate penetration of the dam core.

Although in the ballasted cases the ship comes closest to the dam core, the maximum grounding forces are significantly higher for the fully laden cases. Case 5 (head-on impact of a fully laden vessel with approach velocity of 8 knots) produced a horizontal grounding force of about 100 MN (see Figure 3), the maximum force calculated in the analyses.

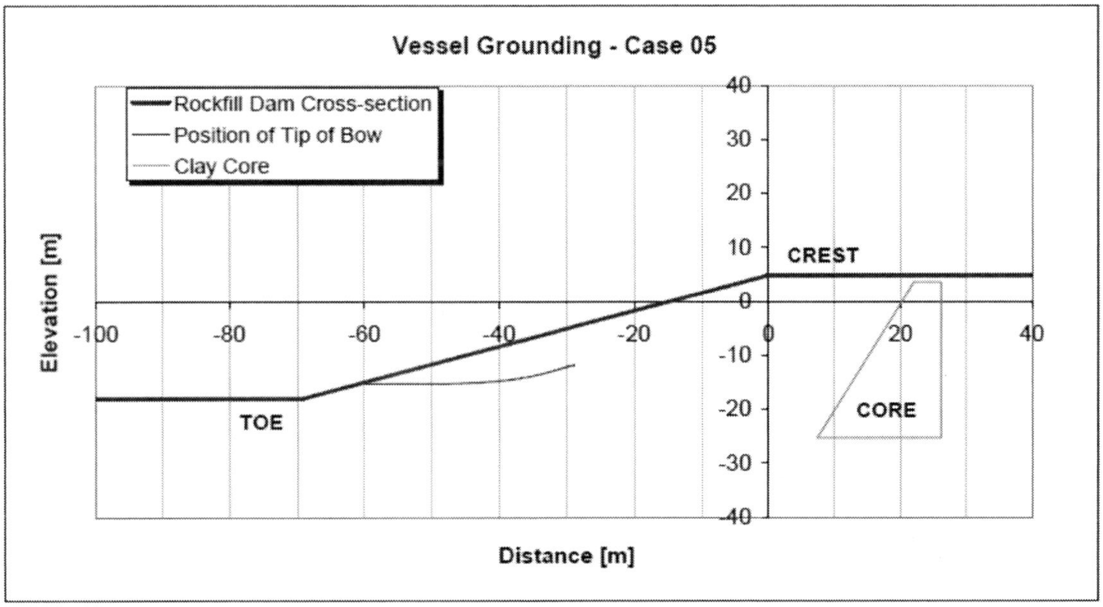

Figure 2. Computed Trace of Bow Penetration into Dam Shell from SHIPCOL Program

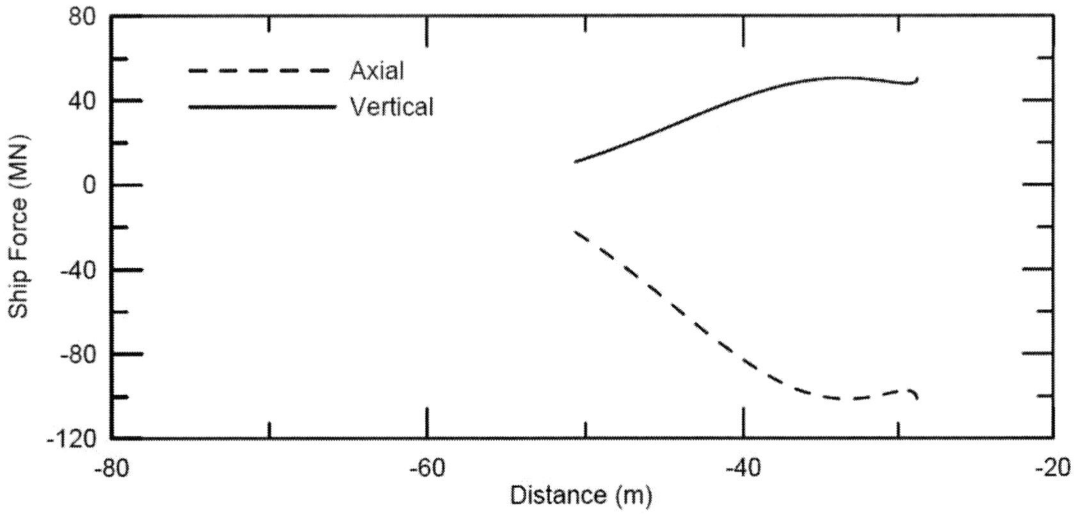

Figure 3. Computed Grounding Forces Acting on Ship Bow from SHIPCOL Program

Table 2. Summary of SHIPCOL Analysis Cases and Results

Case	Approach Velocity (knots)	Angle of Incidence (deg.)	Loading State	Penetration (m)	Distance from Clay Core (m)
1	6	0	Laden	25.3	41.0
2	6	0	Ballasted	16.7	25.6
3	4	0	Laden	19.2	46.7
4	4	0	Ballasted	13.2	29.0
5	8	0	Laden	31.2	35.5
6	8	0	Ballasted	20.2	22.5
7	6	55	Laden	87.6	69.5
8	6	55	Ballasted	50.8	50.3
9	6	45	Laden	63.3	57.2
10	6	45	Ballasted	37.1	40.7

3D ELASTO-PLASTIC SIMULATIONS

The SHIPCOL analyses helped to identify the critical ship grounding scenarios and provided estimates for ship penetration distance into the dam, and total grounding forces. However, they did not directly provide information on the stresses and strains induced in the dam core by ship grounding. To address this issue, 3D elasto-plastic numerical simulations were performed, focusing on the simple head-on grounding scenario, which was determined to be the critical case from the SHIPCOL analyses. The 3D elasto-plastic analyses not only allowed an evaluation of the stresses and strains induced in the dam core by a grounding event, but also provided an independent check on the penetration distance and grounding forces. In addition, with the elasto-plastic analyses it was possible to investigate the sensitivity of the results to various soil strength and deformation parameters.

Analysis of the ship grounding problem involves consideration of at least the following:

1. Soil forces acting on the ship bow during the collision
2. Deceleration and eventual stopping of the ship after some penetration into the embankment

3. Forces transmitted to the dam and the corresponding stresses and strains in the dam.

Each of these aspects was considered in the 3D simulations.

It was not possible to simulate the grounding process in a single, continuous numerical analysis, as the problem involves extremely large strains and drastic rearrangement of the embankment geometry. Instead, the problem was modeled approximately through several separate small-strain analyses, each for a particular distance of ship penetration into the dam.

An initial set of analyses focused on the soil reaction forces acting on the ship as a function of penetration distance. For these analyses, the numerical meshes included only small areas of the dam around the bow. A typical numerical mesh for these analyses is presented in Figure 4, which shows the model for the analyses corresponding to 15 m of ship penetration. Various parametric studies were performed with these models to investigate the effects of embankment soil properties and ship-soil interface properties on reaction forces. Then, using best-estimate material properties, the relationship between reaction force and penetration distance was developed by performing analyses for several penetration distances (Figure 5).

The relationship between reaction force and penetration distance was then used to estimate the stopping distance for various approach velocities. The distance at which the ship comes to a halt was computed by integrating the work done by the soil reaction force and equating this work to the initial kinetic energy of the ship. Time histories of ship velocity were also obtained from this calculation.

Finally, the forces transmitted to the dam and the resulting stresses and strains within the dam core were analyzed for a penetration distance of 25 m. This distance was roughly the maximum penetration distance computed from the energy balance analysis, and produced the maximum grounding force. For this analysis, the numerical mesh used for modeling the area of impact was inset into a coarser mesh of the entire dam cross as shown in Figures 6 and 7.

For the elasto-plastic analyses, only horizontal movement of the ship was considered. Upward movement of the bow, as the ship rides up onto the dam, was neglected. The SHIPCOL analyses indicate that vertical movement is relatively small and only occurs in the final stages of the collision process. Neglecting vertical forces and displacements allowed for a simple 1D computation of penetration distance based on horizontal momentum only. Allowing for vertical movement would require accounting for vertical as well as rotational momentum terms.

Numerical analyses were performed with the computer code FLAC3D, Version 3.1 (Itasca, 2005). The dam materials and the basalt bedrock were modeled with a linear-elastic, perfectly plastic Mohr-Coulomb constitutive model. In this model, the soil shear strength is defined in terms of a friction angle and cohesion. For stresses states below the shear strength, the behavior of the model is linear elastic. For stresses at the shear strength, the model behavior is perfectly plastic.

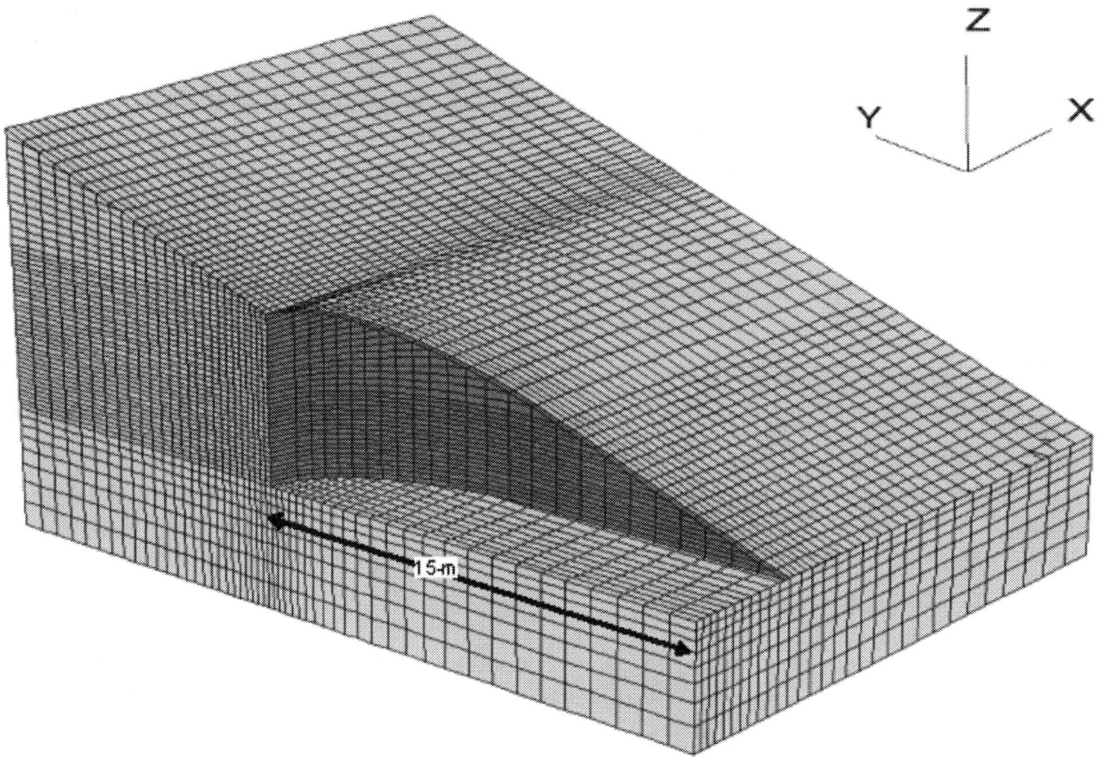

Figure 4. Fine Inset Mesh around Ship Bow

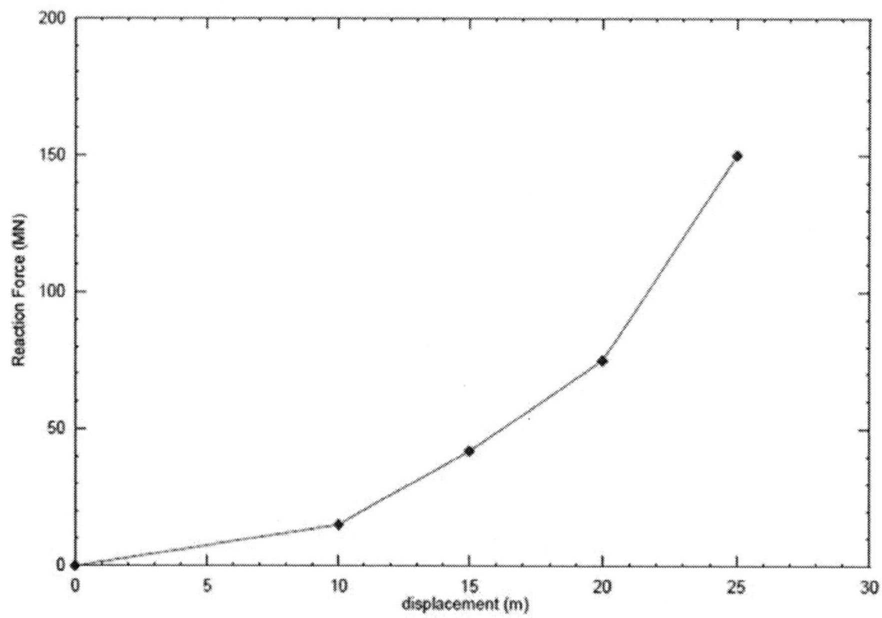

Figure 5. Reaction Force vs Penetration Distance

The soil properties used for the ship grounding analyses are listed in Table 3. These were taken from the dam design Geotechnical Interpretive Report (URS, 2008b) and URS (2008c).

Table 3. Soil Properties for Elasto-Plastic Analysis

Soil Type	Moist Unit Weight (kN/m^3)	Saturated Unit Weight (kN/m^3)	Cohesion (kPa)	Friction Angle (degrees)	Young's Modulus (kPa)	Bulk Modulus (kPa)
Dam Core	15.7	17.0	20	26	stress dependent	stress dependent
Rockfill Shell	21.0	22.8	0	47	stress dependent	stress dependent
Bedrock	21.6	21.6	200	32	2.05E6	4.7E5

The elastic stiffness properties for the dam core and rockfill shell were assumed to be stress dependent, using the formulation of Duncan et al. (1980) where Young's modulus and bulk modulus are functions of the minimum principal stress according to the following formulae:

$$E = Kp_a\left(\frac{\sigma_3}{p_a}\right)^n \quad (1)$$

and

$$B = K_b p_a\left(\frac{\sigma_3}{p_a}\right)^m \quad (2)$$

where E is Young's modulus, B is the bulk modulus, p_a is atmospheric pressure, σ_3 is the minimum effective principal stress, and K, K_b, n and m are dimensionless parameters. Values for the dimensionless parameters are listed in Table 4.

Table 4. Stiffness Parameters for Core and Rockfill

Soil Type	Young's Modulus		Bulk Modulus	
	K	n	K_b	m
Dam Core	150	0.45	140	0.20
Rockfill Shell	450	0.40	125	0.20

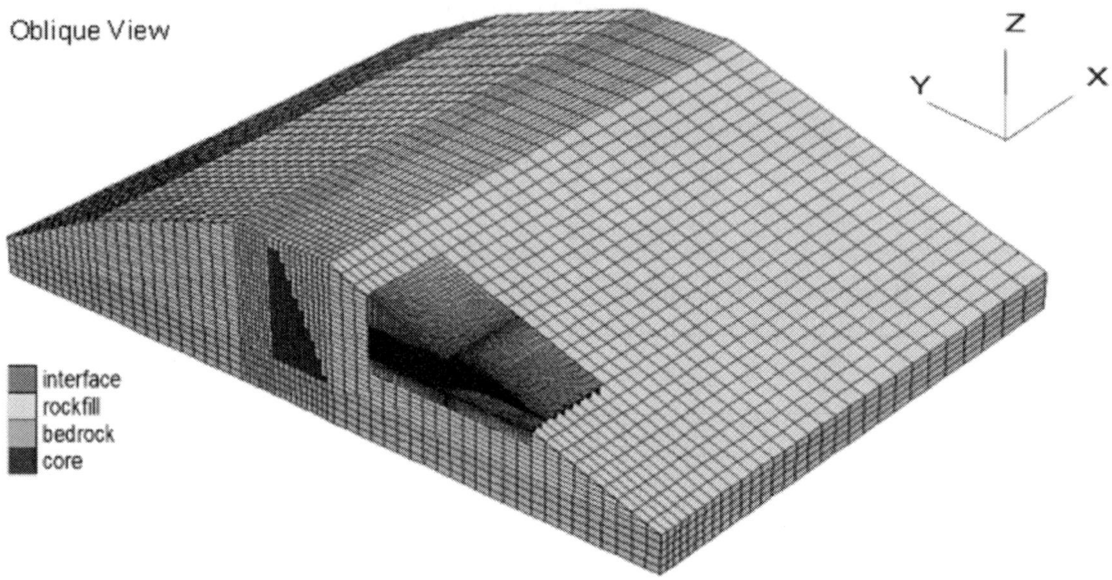

Figure 6. Numerical Mesh for the Dam with Fine Inset Mesh around Ship Bow

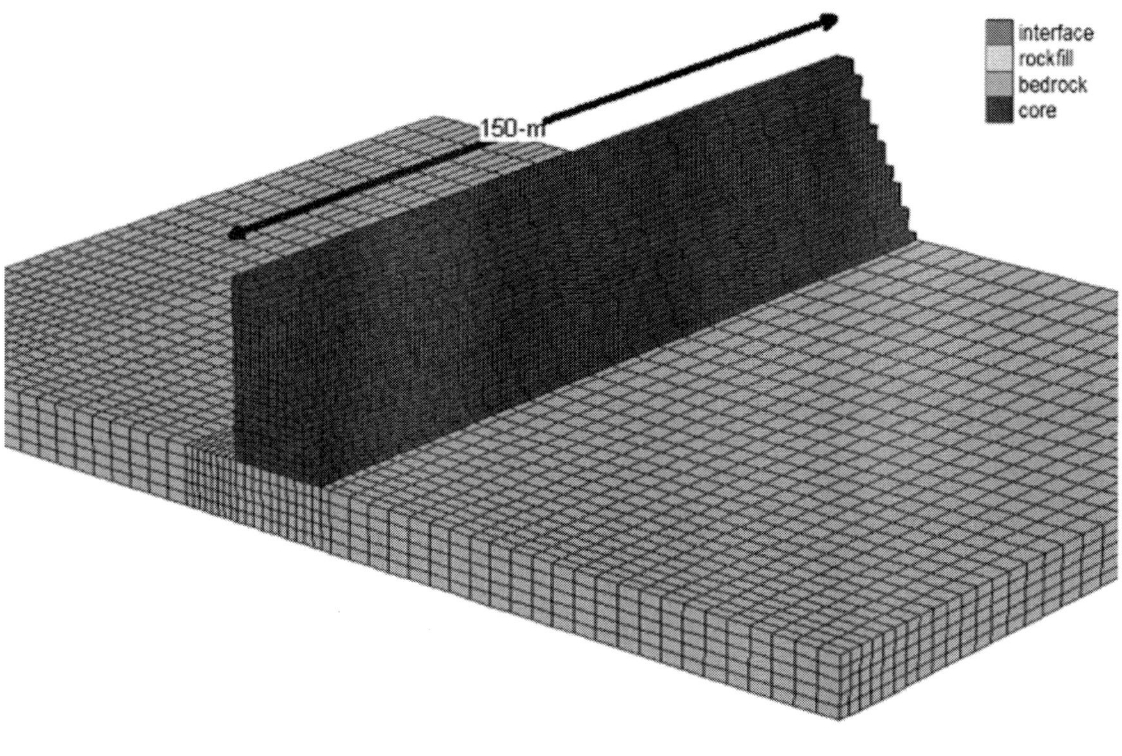

Figure 7. Numerical Mesh for the Dam Core

The penetration distances computed for approach velocities of 6, 4 and 8 knots are presented in Table 5, along with the corresponding penetration distances computed with the SHIPCOL program. The penetration distances estimated using the two different analysis approaches agree reasonably well, with the distances computed from SHIPCOL being roughly 10% to 15% higher than the distances computed from the FLAC3D analysis.

Table 5 Penetration Distanced Computed with FLAC3D and SHIPCOL

Case	Approach Velocity (knots)	Angle of Incidence (degrees)	Loading State	SHIPCOL Penetration Distance (m)	FLAC3D Penetration Distance (m)
1	6	0	Laden	25.3	22.7
3	4	0	Laden	19.2	17.5
5	8	0	Laden	31.2	26.9

Forces transmitted to the dam and the resulting stresses and strains within the dam core were analyzed for a penetration distance of 25 m. The ship-grounding-induced displacements of the dam core are shown in Figure 8. The maximum induced displacements in the y-direction (the direction of ship motion) are 8 mm, while the maximum displacement in the x-direction (along the dam axis) are 2 to 3 mm. The grounding-induced normal strains ε_{yy} and ε_{xx} are plotted in Figure 9. The maximum increment in ε_{yy} is 0.06% (compression) and the maximum increment in ε_{xx} is 0.01% (extension). A maximum induced shear strain of 0.07% (Figure 10) is calculated at the toe of the core. The maximum induced stresses in the x- and y-directions (not shown) are about 18 kPa (compression) and 8 kPa (tension), respectively. The induced principal stresses reach a maximum value of 40 kPa at the toe of the core. These induced stresses are less than 10% of the initial effective stresses.

CONCLUSION

Two sets of analyses were performed to estimate the potential damage to the Borinquen Dams due to ship grounding. Initial exploratory analyses performed with the program SHIPCOL, focused on estimating the penetration of a Post-Panamax ship into the dams for various analysis cases. The results show that for all the cases considered, the design ship does not penetrate into the core of the dam. The collision scenario resulting in the closest approach to the core involved a head-on collision from a ballasted ship at an approach velocity of 8 knots, which comes to a stop about 22.5 m from the core.

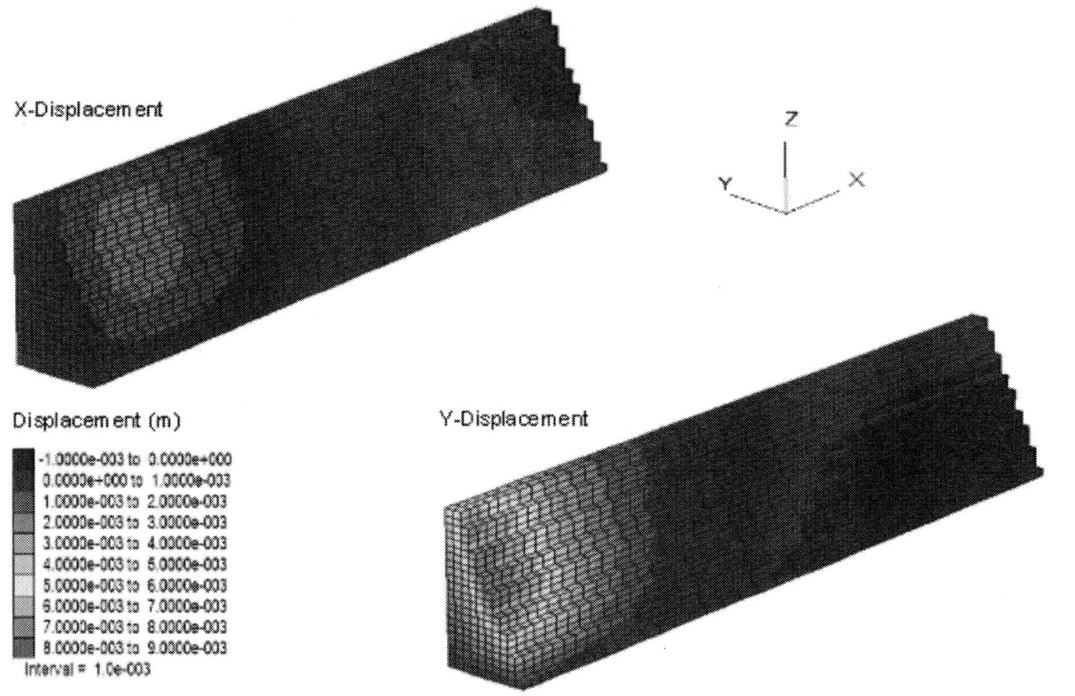

Figure 8. Contours of Induced Displacements in the Dam Core

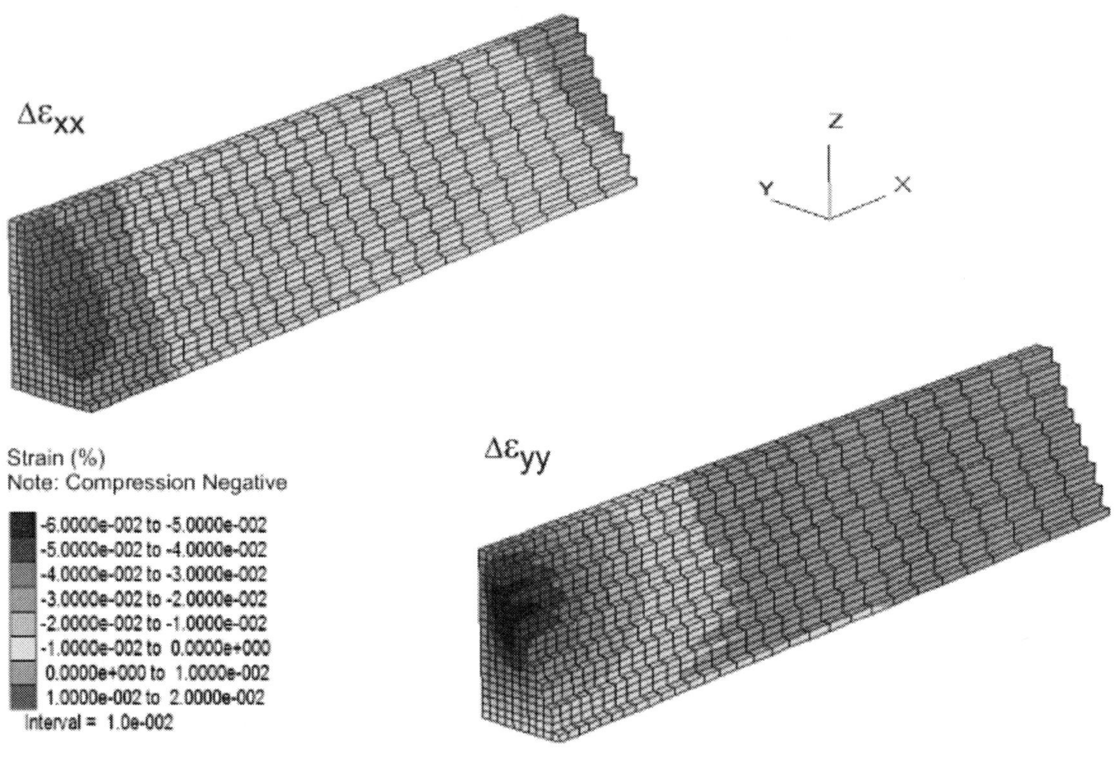

Figure 9. Contours of Induced Normal Strains in the Dam Core

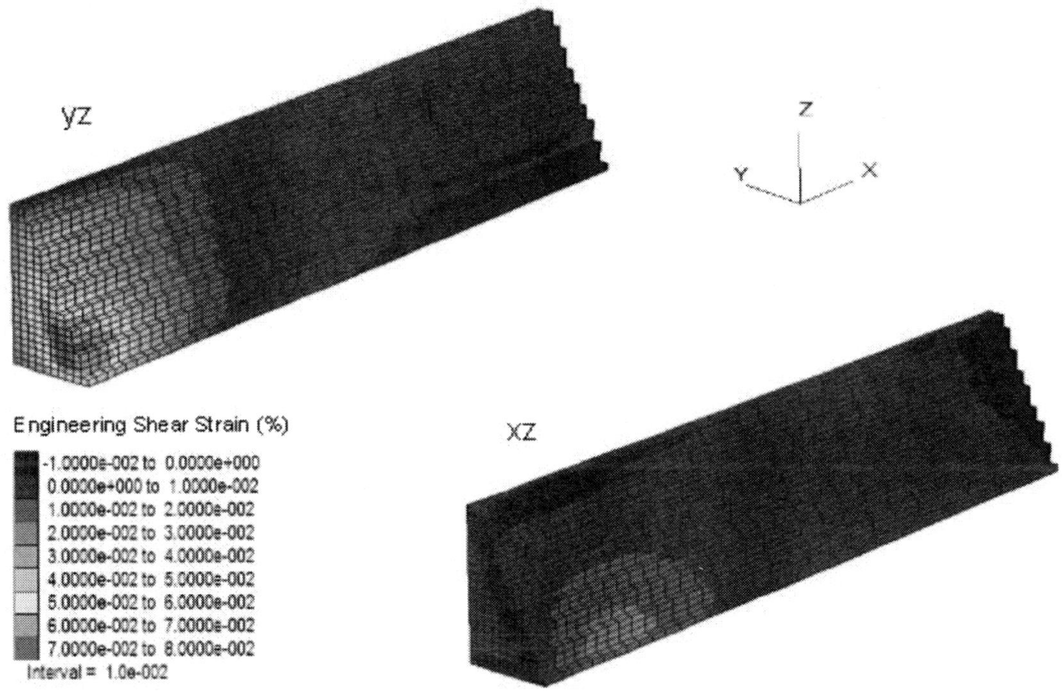

Figure 10. Contours of Induced Shear Strains in the Dam Core

Elasto-plastic analyses with the program FLAC3D focused on estimating the stresses and strains induced by the ship impact for a case where the ship penetrates25 m into the dam. The analysis results show that the maximum induced displacement within the core in the direction of ship motion is less than 8 mm, whereas it is only 3 mm along the axis of the dam. The maximum induced compressive and tensile stresses within the core are about 18 kPa and 8 kPa, respectively. These stresses are less than 10% of the initial effective stresses.

The analysis results demonstrate that, even under conservative grounding scenarios, deformations and induced stresses within the dam core are quite small. Thus, the dams would be relatively unaffected by the design ship grounding impacts and would not suffer significant disruption of their function.

REFERENCES

Ben Gerwick Inc. (2008). *Panama Canal Design of Borinquen Dams, Vessel Grounding Analysis*, Report to URS Corporation, March 3, 2008.

Brink-Kjær, O, Brodersen, F P and Nielsen, A.H. (1983). Modelling of ship collisions against protected structures. *Proceedings of the International Association for Bridge and Structural Engineering (IABSE) colloquium.* Copenhagen, Denmark.

Denver, H (1983). Geotechnical model tests for the design of protective islands. *Proceedings of the International Association for Bridge and Structural Engineering (IABSE) colloquium. Copenhagen, Denmark.* IABSE Reports - Volume 42.

Duncan, J. M., Byrne, P., Wong, K. S., and Mabry, P. (1980). *Strength, stress-strain and bulk modulus parameters for finite element analyses of stresses and movements of soil masses*, *Report No. UCB/GT/80-01*, University of California Berkeley, Department of Civil Engineering.

Havnø, K and Knott, M (1986). Risk analysis and protective island design for ship collisions. *Proceedings of the International Association for Bridge and Structural Engineering (IABSE) colloquium,* IABSE Reports - Volume 51, Tokyo, Japan.

Itasca Consulting Group (2005). *FLAC, Fast Lagrangian Analysis of Continua in 3-Dimensions, Version 3.1.* Minneapolis, Minnesota, USA.

Simonsen, B.C. (1997). *Mechanics of Ship Grounding*, Ph.D. thesis, D epartment of Naval Architecture and offshore Engineering, Technical University of Denmar k, Lyngby, Denmark.

Sterndorff. M.J. and Peders en, P.T. (1996), Groundi ng Experiments on Soft Bot toms, *Journal of Marine Science and Technology*, pp 174-181.

URS. (2008a). *Dam Design Criteria (Task Order 4, Task A.4),* Draft, prepared for ACP contract no. CMC-172538, February.

URS (2008b). *Foundation Materials for Dams 2E, 1W, and 2W (Task Order 4, Task A.1.4),* Geotechnical Interpretive Report, Final Technical Memorandum, prepared for ACP contract no. CMC-172538, May.

URS (2008c). *Insitu Construction Materials for dams 2E, 1W, and 2W (Task Order 4, Task A.2.3),* Geotechnical Interpretive Report, Final Technical Memorandum, prepared for ACP contract no. CMC-172538, May,

MODELING GEOMORPHIC FEATURES IN LEVEE RELIABILITY ANALYSES

Lourdes Polanco-Boulware[1]
John Rice PE, GE, PhD[2]

ABSTRACT

It has long been understood that subsurface geology affects the potential for internal erosion due to underseepage in levees. Geomorphic features such as abandoned meanders, cross channels, point bars, and meander scrolls have been known to concentrate or block seepage, resulting in sand boils and internal erosion. However, due to the complexity of the flow regime that occurs when geomorphic features dominate the subsurface geometry, incorporating the effects of geomorphic features into either deterministic or probabilistic analysis has remained a challenge. Based on the assumption that the preponderance of the underseepage risk to a levee reach is due to the geomorphic features along that reach, methodology is being developed to incorporate the combined effects of these geomorphic features into reliability analyses of underseepage risk. The methodology consists of a Response Surface-Monte Carlo analysis that takes into account the uncertainty in the subsurface geometry and the soil properties in assessing the flow regime associated with each geomorphic feature. Three-dimensional finite-element seepage analyses will be used to develop the response surface in order to take into account the inherent three-dimensional aspects of the features. A failure modes and event tree analysis will then be performed for each geomorphic feature to assess the likelihood of the resulting hydraulic conditions developing into a levee failure. The probabilities of failure for the geomorphic features located along a levee reach will then be combined to assess the total probability of failure of the levee reach due to underseepage.

INTRODUCTION

Reliability assessments of levees founded on variable geologic conditions have often struggled with quantifying the probability of unsatisfactory performance or failure due to localized adverse foundation conditions. With levees, a failure at any one location along the perimeter of a protected basin often leads to failure of the entire system. The analogy of a weak link in a chain is often used to illustrate this situation. Natural geologic variation along the alignment of a given levee affects the overall reliability of the system and in some cases can even affect the critical mode of failure. As the length of levee protecting a basin increases, the probability of encountering unfavorable subsurface conditions increases, and thus, the probability of unsatisfactory performance or failure increases. This phenomenon has often been referred to as length effects.

---

[1]Graduate Student, Department of Civil and Environmental Engineering, College of Engineering, Utah State University, 4110 Old Main Hill, Logan, UT 84321-4110; phone: (435) 881-2223; E-mail: lourdes.polanco@aggiemail.usu.edu.
[2] Assistant Professor, (same organization); phone: 435-797-8611; E-mail: john.rice@ usu.edu.

Adverse foundation conditions along a levee's alignment are often associated with localized geomorphic features. For instance, an abandoned, sand-filled creek channel crossing beneath the levee may represent an increased probability of a piping failure and may even change the critical mechanism of internal erosion for that levee reach. In many cases, the locations of highest risk along a structure's alignment can be tied to the occurrence of one or more problematic geomorphic features. By quantifying the risk to the levee of the individual geomorphic features and combining those risks for the entire length of the levee, the total probability of unsatisfactory performance for the levee can be quantitatively assessed.

Of course, not all geomorphic features of the same type will represent the same level of risk. Variation in geometric configuration and soil properties will lead to different probabilities of internal erosion due to underseepage. Thus, the models for each feature must have flexibility to account for variations in geometry, soil properties, depth, and other relevant parameters. Furthermore, there may be interdependence of closely situated features. This is especially true for levees where seepage affects the internal erosion failure mechanism, as adjacent geomorphic features and even the alignment of the structure itself may affect the hydraulic pressure and flow regime on a feature.

The goal of this research is to develop methodology for assessing the risk of internal erosion related failure due to underseepage in a system of levees protecting a basin.

BACKGROUND

Underseepage and Internal Erosion

Underseepage is a phenomenon that occurs when a differential hydraulic head forces water through the foundation soils or bedrock beneath a structure. In geotechnical engineering practice, underseepage is commonly encountered in dams, levees, construction excavations, detention ponds, road embankments and other structures capable of maintaining a differential hydraulic head. In many cases, underseepage can occur without damage to the foundation or overlying structure. However, in cases where the forces imposed on the soil by the seeping water exceed the forces retaining the soil, the soils may begin to erode by one of several mechanisms, a phenomenon which is commonly referred to as "internal erosion". Over time, this type of internal erosion often progresses, further reducing the resistance to seepage flow and accelerating the erosion process, eventually leading to failure of the structure.

Terminology describing the various mechanisms of internal erosion has evolved over recent years, as our understanding of the mechanisms of erosion has developed. Nomenclature for these mechanisms has been and continues to be inconsistent in practice and in available technical literature on the subject. In a recent publication, the International Committee on Large Dams (ICOLD 2013) has developed a system of nomenclature describing the various mechanisms of internal erosion. This paper has essentially adopted the ICOLD nomenclature. The term "internal erosion" has been accepted as a generic term to describe erosion of particles by water passing through a body of soil or rock. Other terms described below refer to specific mechanisms by which soil particles are eroded by the seeping water.

The term backward erosion piping (BEP) describes the erosion of soil due to seepage water through a soil mass that initiates when particles of soil are dislodged from the soil matrix at an unprotected seepage exit point. As BEP continues, a pathway or "pipe" is formed by progressive erosion at the upstream end of the erosion void. Heave is the movement of mass of soil due to underlying hydraulic pressure or seepage forces. Both heave and BEP mechanisms can contribute to the potential for internal erosion beneath a levee and therefore both mechanisms are included in the assessments presented herein. Concentrated leak erosion occurs along a concentrated flow path due to scour caused by shear forces imposed by the flowing water. Concentrated leak, contact erosion, and suffusion, while possible under a levee, are considered much less likely forms of internal erosion and thus are not included in this assessment.

Geomorphic Features

Geomorphic features underlying a structure often have a significant effect on the underseepage behavior and the mechanisms of internal erosion. While the methodology presented in this paper can be applied to a broad array of geomorphic settings, this paper will present a case study where the foundation consists of alluvial deposits that are commonly encountered in a meandering river flood plain. As such, our discussion here will focus on the geomorphic features that are typically encountered in meandering river deposits.

The typical depositional environment of a meandering river is characterized by river channel deposits and flood (overbank) deposits (Leopold and Wolman 1957; Brierley and Fryirs 2005). Channel deposits in an active river channel are typically granular and tend to become finer in the upward direction (Walling et al. 1997; Walling et al. 2004; Filgueira-Rivera et al. 2007; Smith and Pérez-Arlucea 2008; Ritter et al. 2011). Overbank deposits are typically located above the channel deposits, and are deposited when flood waters exit the river channel and deposit the finer grained materials due to decreasing flow velocity (Walling et al. 2004; Ritter et al. 2011). The thickness of overbank deposits is generally greatest adjacent to the river bank, resulting in low ridges adjacent to the active channel known as natural levees (Brierley et al. 1997; Cazanacli and Smith 1998; Ferguson and Brierley 1999; Hudson and Heitmuller 2003; Adams et al. 2004; Temmermana 2004; Smith and Pérez-Arlucea 2008; Hudson 2011; Ritter et al. 2011). Channel deposits are often referred to as the *foundation* layer and the overbank deposits are often referred to as the *blanket* layer.

Figure 1 is a model illustrating the simplified meandering river stratigraphy described above. This pattern of deposition has a significant effect on the underseepage behavior and associated internal erosion mechanisms. Due to the high hydraulic conductivity of the foundation and low conductivity of the blanket, only a small percentage of the total head loss occurs in the foundation. This results in large pressures beneath the blanket that can lead to heave and cracking of the blanket. In some cases, heave may result in a catastrophic failure (blowout), but more commonly the blanket cracking leads to a concentrated flow through the blanket, high gradients, and the initiation of piping in the foundation. This process initially results in sand boils, but can progress to full piping failures of the structures (USACE 1956; Glynn and Kuszmaul 2010; Polanco and Rice 2011; Rice and Polanco 2012).

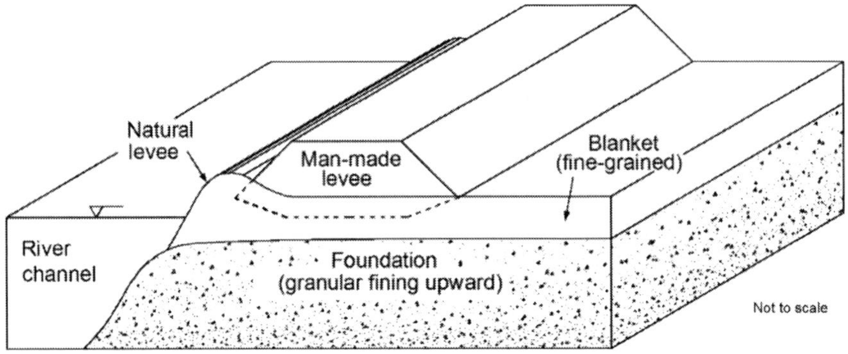

Figure 1. Typical subsurface cross section of a meandering river depositional environment.

In addition to the general meandering river geomorphic profile described above, a number of local geomorphic features are common within the meandering river environment including: abandoned channels, point bars, meander scrolls, and crevasse splays. These features are shown schematically on Figure 2. Where these features intersect the alignment of a man-made levee, flow concentration or localized blanket thinning may occur, resulting in a significantly higher potential for internal erosion than in the surrounding areas (Glynn and Kuszmaul 2010). Abandoned channels (paleo-channels) are river or tributary channels that were once active. They are often infilled by overbank deposits, but in some cases (i.e., chute and neck cutoffs and tributary channels) they may be partially infilled with coarse-grained soil (Bridge 2003; Brierley and Fryirs 2005; Fryirs and Brierley 2013). These layers may represent a point of concentrated seepage flow or local blanket thinning that locally increases the potential for internal erosion. Point bars consist of sand and gravel bars that commonly form on the inner (convex) meandering bank of a river (Nanson 1980; Brierley and Fryirs 2005; Ritter et al. 2011). They can represent both concentrated flow due to their granular makeup and localized thinning of the blanket layer. Meander scrolls (Figure 3) are a complex depositional feature associated with successive point bar formation and infilling between point-bar ridges with fine-grained soils (Nanson 1981; Saucier 1994; Woolfe and Purdon 1996; Brierley and Fryirs 2005; William Lettis & Associates 2008; Glynn and Kuszmaul 2010). The low hydraulic conductivity of the fine-grained infilling has potential to block or concentrate seepage into the point bars resulting in higher hydraulic gradient than surrounding areas. Crevasse splays (Figure 4) form in association with the natural levee. During paleo flood events, flood waters breach the natural levee forming a crevasse through it (Bristow et al. 1999; Bridge 2003; Brierley and Fryirs 2005; Ritter et al. 2011). As the water passes through the crevasse, it splays out on the landside of the natural levee and deposits coarse-grained sediment. The crevasse splay deposit represents a concentrated pathway to a layer of sand within the blanket, resulting in elevated potential for the heave and piping mechanisms.

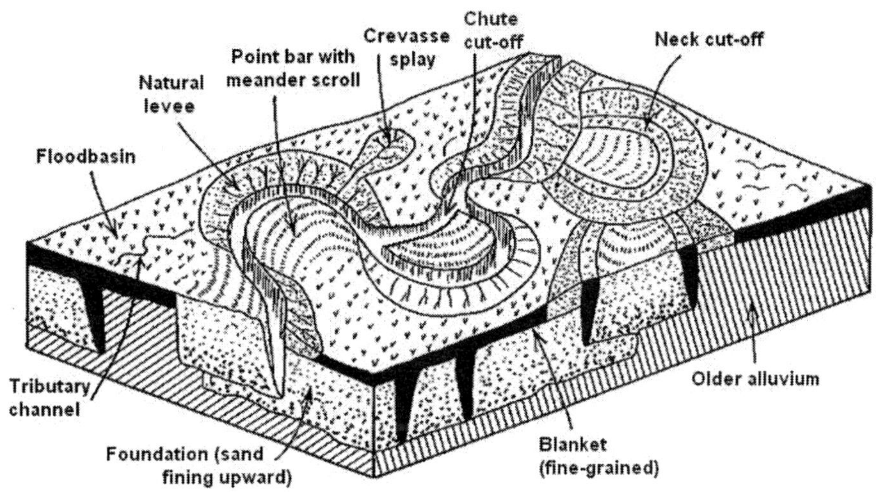

Figure 2. Common geomorphic features within the meandering river environment. (Modified from Allen 1970).

Analysis of Internal Erosion

Some of the earliest work in deterministic underseepage analysis was performed by Bligh (1910, 1913, 1915) who developed an empirical relationship between piping potential and the shortest flow path length beneath a water-retaining structure. Lane (1934) later recognized a distinction between flow along the base of the structure, vertical flow along vertical barriers, and flow through a granular media and modified Bligh's work to develop the weighted creep ratio method. Lane's empirical method also took into account the varied erosion resistance of different soil types (i.e., average gradient versus concentrated flow).

Later efforts by the U.S. Army Corps of Engineers (USACE) led to theoretical relationships for the computation of seepage flow beneath levees (USACE 1956). These relationships provided the basis for the "blanket theory" equations that are presented in the USACE Engineering Manual "Design and Construction of Levees" (USACE 2000). The presented equations assume two soil layers in the levee foundation with constant thickness and horizontal boundaries (the *blanket* and *foundation* layers described above) (USACE 2000, 2005). Flow is assumed to be horizontal for the foundation layer and vertical for the blanket layer. The blanket theory equations calculate the head at the base of the blanket layer that can be used to calculate either a gradient through the blanket or an uplift pressure beneath the blanket. Thus, the factor of safety can be calculated for either piping or heave.

The main limitation of the "blanket theory" equations is that the subsurface geometry is restricted to two homogenous layers of constant thickness. This limitation was partially resolved by the finite difference computer program LEVEEMSU (Wolff 1989; Gabr et al. 1995) that numerically solves the blanket theory equations to allow for nonuniform thickness of layers and varied soil hydraulic conductivity. The development of the finite-element seepage analysis program CSEEP by the USACE (Knowles 1992; Tracy 1994) has enabled more complex geometries to be modeled. Numerous commercial finite-element seepage analyses programs have become available since the development of CSEEP. Finite element analyses have enabled the modeling of complex subsurface geometries where converging and diverging seepage flows can affect the magnitude of exit gradients.

Probabilistic methods for assessing underseepage have been developed by coupling the deterministic methods discussed above with probability models such as the Taylor series method (Harr 1987; Duncan 2000; Baecher and Christian 2003; USACE 2004; Sleep and Duncan 2008). One common approach that is well documented by the USACE consists of using the "blanket theory" equations as the performance function and the first order-second moment (FOSM) Taylor series method as the probability model (Wolff 1994; Crum 1996; Wolff et al. 1996; Wolff 2008). This methodology is subject to the same limitations as the blanket theory, that is, it works well for simple profiles but is characteristically a two-dimensional analysis and cannot model the complex subsurface conditions and three-dimensional aspects often associated with the fluvial depositional environment.

Basin Behavior and Length Effects.

Linear earth structures having significant length (levees, canals, detention basins, road embankments, pipelines, and dams of considerable length) pose a challenge in terms of evaluating the overall reliability of the system. Failure at any point of the structure often represents failure of the entire system, allowing the water to escape from impoundment or render transportation along the structure impossible. As a result, they can be functionally compared to a chain, in that if any link were to break the entire chain would fail (Steenbergen et al. 2004; Vrouwenvelder 2006; Vrouwenvelder et al. 2010; Kanning 2012). In current practice, the overall reliability of these structures is typically evaluated by dividing the structure into reaches and evaluating the reliability of the individual reaches using a characteristic subsurface profile representing generic conditions within the reach. It is recognized that within this reach there are perturbations away from the generic conditions and that failure will have the highest probability of occurrence where these perturbations create conditions most susceptible to one or several modes of failure (Vrouwenvelder 2006; Wolff 2008; Bernitt and Lynett 2010; Vrouwenvelder et al. 2010; Shannon & Wilson 2011; Kanning 2012). It is also recognized that the longer the reach being characterized is, the greater the likelihood is that the perturbations will create a condition susceptible to failure; a phenomenon generally referred to as *length effects*. The increased hazard due to length effects can be reduced in some situations by limiting the length of structure protecting an area. This is a common practice with levees in the Netherlands where levee basins are often divided into smaller sub-basins or polders to reduce the length of levee protecting any one area and also reduce the affected area at risk from failure of any one reach of levee.

Compared to other failure mechanisms, internal erosion is particularly sensitive to length effects (Wolff 2008; Bowles et al. 2012; Kanning 2012). The practice of dividing the systems into reaches with a common profile and then analyzing each reach to calculate a separate measure of reliability, such as factor of safety or probability of unsatisfactory performance, has been common practice (Wolff 2008). However, due to a combination of aleatory uncertainty (due to natural variability) and epistemic uncertainty (due to a lack of data) with respect to the variation of the subsurface profile and soil conditions within the reach, there is uncertainty as to when the variation in conditions may result in a condition or combination of conditions that results in unsatisfactory performance. While it is understood that the chances of encountering a location where the conditions result in failure increase with the length of the reach being analyzed, quantifying the increase in risk with reach length has proven to be difficult.

METHODOLOGY

The methodology presented herein is based on the concept that the majority of the underseepage hazard along a levee comes not from the background hazard due to a characteristic subsurface profile but from discrete geomorphic features that interrupt the characteristic profile, providing locations for concentration of seepage flow and/or buildup of hydraulic pressure. The interactive hydraulic behavior between these features, the structure, and the surrounding characteristic subsurface profile is definitely three-dimensional, concentrating seepage flow into the geomorphic feature and decreasing flow in the surrounding area (or vice versa if the feature represents a seepage block). Thus, by identifying and analyzing the geomorphic features along the alignment of a structure, the vast majority of the underseepage hazard can be evaluated. This is not to say that the remaining portion of the characteristic subsurface profile does not contribute to the overall hazard. In fact, numerous substructures within the characteristic profile that cannot feasibly be identified by normal means of exploration pose a level of background hazard that will still need to be considered.

The series of steps for calculating the total failure probability of a levee-protected basin using the proposed methodology are outlined in Figure 3 and the analysis tools for performing these steps are listed. Details of these steps are presented below. This paper describes the methodology for Step 2 in detail while the remaining steps, which are the subject of further research and implementation, are briefly described to provide a context for how the Step 2 results will ultimately be used.

Step 1: Identify Geomorphic Features. This step consists of identifying the locations and properties of geomorphic features along the levee reach being analyzed. It is anticipated that this will generally be accomplished using existing geologic information such as maps prepared along the Sacramento River Levee System (William Lettis & Associates 2008; Pearce et al 2009, 2010). In other cases the likelihood of encountering geomorphic features will be estimated based on data from similar depositional environments.

Step 2 - CADFs for Geomorphic Features. The assessment of the hazard that each geomorphic feature poses to the levee system begins with developing a cumulative ascending distribution function (CADF) for either the exit gradient or uplift pressure of each geomorphic feature. Depending on the failure mechanism (i.e., piping, heave, etc.) either the exit gradient or the uplift pressure on the land side of the levee will be the key parameter dictating the likelihood of initiating internal erosion. A CADF is a plot of increasing values of the exit gradient or uplift pressure versus the probability of being less than that value. The probability of exceeding a limit value of the exit gradient or uplift pressure can be obtained directly from the CADF, such as the anticipated gradient at which erosion will initiate (i.e., at a factor of safety of 1.0). However, this probability is not the probability of levee failure because other factors will need to be considered in the assessment of the probability of the internal erosion progressing to a levee failure. These additional factors will be included using the event tree analysis presented as Step 5 in Figure 3.

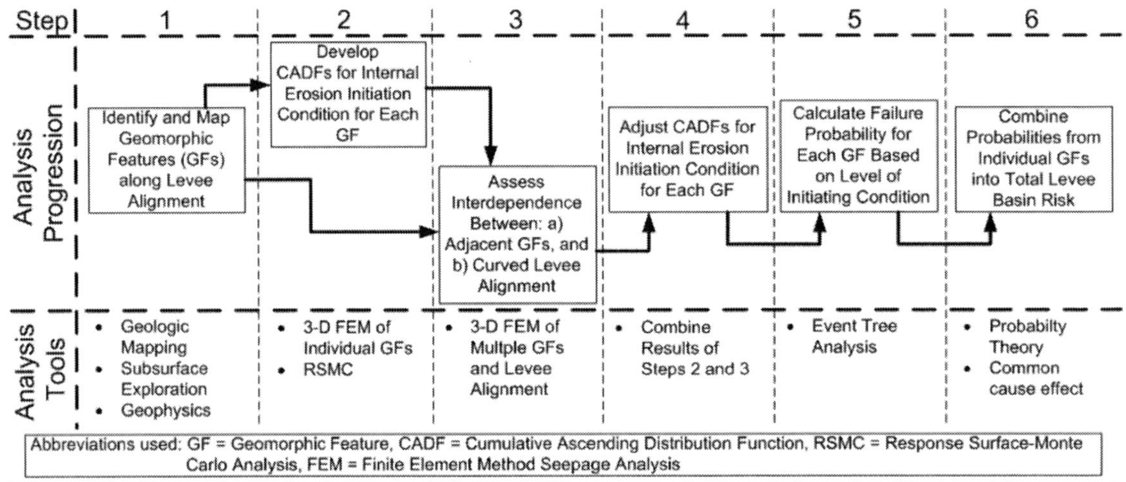

Figure 3. Steps for calculating reliability of levee reaches and length effects with respect to underseepage.

The CADFs will be evaluated for each of the mapped geomorphic features using a probabilistic assessment of hydraulic conditions model developed using a Response Surface-Monte Carlo (RSMC) simulation methodology. Rice and Polanco (2011, 2012), Polanco and Rice 2013) have developed two-dimensional models for assessing hydraulic conditions in geometrically complex levee profiles. The proposed methodology uses multiple finite element analyses to develop a relationship between the key input parameters (hydraulic conductivity of the soil layers and subsurface geometry of the soil layers) and the probability of reaching critical hydraulic conditions (uplift pressures or hydraulic exit gradients). The analyses are conditional on the river reaching a certain level (such as the level associated with a flood with an annual exceedance probability of 1% or the 100-year flood level). This relationship, generally called a *Response Surface* (Xu and Low 2006; Low 2008), is used to perform a Monte Carlo simulation that results in the CADF. The range and probability distribution for each soil or geometric input parameters are represented using a probability density function (PDF) that describes the relative likelihood for this parameter to take on a given value. For this study Rice and Polanco's (2012) RSMC methodology has been expanded using three-dimensional finite element analyses to account for the three dimensional seepage aspects of the individual geomorphic features.

The RSMC analyses will produce CADFs of key hydraulic parameters (i.e. uplift pressure or exit gradient) for each geomorphic feature mapped along the levee alignment in Step 1. To accomplish this, a general response surface will be developed for each geomorphic feature type (e.g. crevasse splay, cross channel, etc.) that describes the relationship for calculating the key hydraulic parameters from parameters describing the feature geometry and the soil parameters. The CADFs for the individual geomorphic features will then be produced by performing the Monte Carlo simulations using PDFs for the input parameters that are specific to the individual geomorphic features. The feature-specific PDFs will be based on mapped geometry, subsurface explorations, and knowledge of the geomorphic processes creating the features. Each PDF will represent a combination of the aleatory and epistemic uncertainty associated with the respective parameter for the geomorphic feature in question. One benefit of developing a response surface for each feature type is that the resulting response surfaces can be used beyond the context of the current study, for other studies having similar geomorphic features.

The development of the various response surfaces that are needed is not trivial. The response surface development consists of a parametric analysis involving each of the uncertain input parameters. The PDF distributions for each parameter are discretized into five or six values (or more if needed) that represent the range of values in the PDFs. Finite element seepage analyses are then run for every combination of values for each variable using a 3-D finite element model (for instance SoilVision Systems Ltd.- SVFlux, 2009) representing the respective geomorphic feature. The results of the finite element analyses are plotted to develop curves or surfaces describing the relationship between the variables and the exit gradient and the uplift pressure. While there are many soil and subsurface geometry parameters that may affect the results of the finite element analysis, part of the challenge of performing the RSMC analysis is reducing the number of parameters to those that have a significant effect on the analysis outcome when varied over the parameter's possible range of values. For example, consider the crevasse splay deposit model depicted on Figure 4 that is modeled using a finite element model similar to that presented on Figure 5. The geometric parameters that may be considered for this feature include the dimensions of the crevasse channel, the dimensions of the splay, the location of the splay with respect to the man-made levee, and the depth that the deposit is buried in the blanket. Soil parameters include the hydraulic conductivities of the channel, splay, blanket, and underlying foundation deposits as well as the unit weight of the blanket soils. If nine variables are considered in the above analysis with the potential range of each represented by five discretized values, the total number of finite element analyses run will be at least 5^9 or nearly 2 million runs; an impossible number to complete. Therefore, it is necessary to reduce the number of parameters in the finite element analyses by identifying parameters that have insignificant effect on the outcome of the analysis and combining some of the remaining parameters to reduce the number of parameters defining the response surface.

The process of combining parameters is illustrated in the example presented in the following section of this paper. In essence, the parameters listed above are combined into three parameters controlling the hydrauic behavior of the crevasse-splay depsit: the conductance of the crevasse channel, the transmisivity of the splay, and the conductance of the blanket overlying the splay. The theory being that the total hydraulic head in the splay is controlled by the flow capacity of the three elements. Parametric analyses are performed to assess the effects the simplification has on the results; that is, the difference between the analysis results using all the true parameter values and the results using the simplified model. The simplified model is then used to create a response surface for each type of geomorphic feature encountered along the levee alignment.

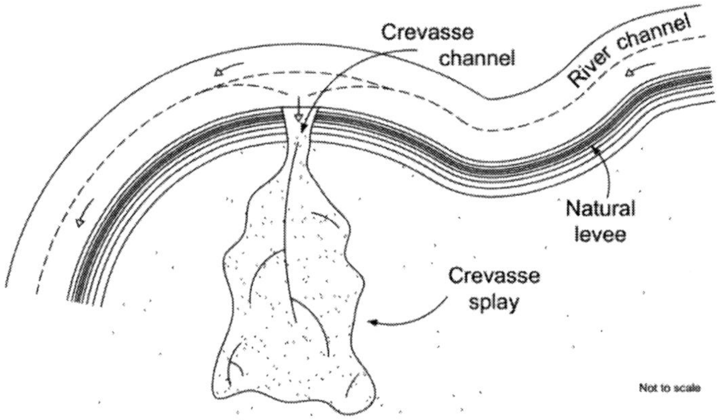

Figure 4. Schematic illustration of a crevasse-splay deposit.

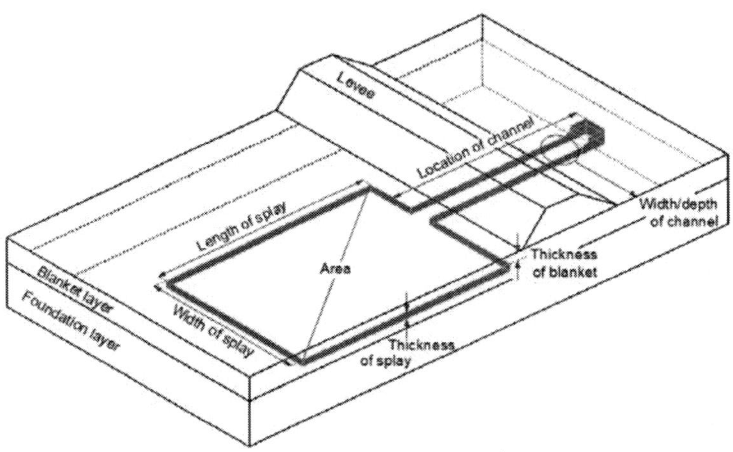

Figure 5. Three-dimensional finite element model of a crevasse-splay deposit.

Once the response surface has been developed, a CADF is produced for each geomorphic feature with a Monte Carlo simulation (enhanced with Latin Hypercube sampling) using the response surface and the computer program @Risk (Palisade 2013). PDFs for the input variables that are specific to each mapped geomorphic feature based on the mapped size, subsurface information, and knowledge of the depositional processes are input into the analysis. Although some of the input parameters have been combined for the response surface development, individual PDFs for each of the combined parameters will be input into the analysis. The parameters are then combined into the three combined parameters discussed above by computer coding before calculating the exit gradient or uplift pressure using the response surface.

In locations where the subsurface profile is not dominated by the geomorphic features discussed above, there will still be a "background hazard" of underseepage failure. In the absence of the geomorphic features the near surface soil profile in these locations is dominated by overbank deposits and can be characterized with a relatively simple profile similar to that presented on Figure 1. Due to the simplicity of the profile, reliability assessment of this type of profile has been shown to be accurately performed using a FOSM Blanket Theory analysis discussed in the analysis of internal erosion section (Wolff 1994; Wolff et al. 1996; Crum 1996; Wolff 2008; Polanco and Rice 2011; Rice and Polanco 2012; Polanco and Rice 2013).

<u>Step 3 - Assess Interdependence of Features and Levee Alignment.</u> In addition to the inherent three-dimensional character of the seepage analysis for the geomorphic features, three-dimensional seepage flow affects the underseepage potential in two additional ways: 1) seepage interaction between adjacent geomorphic features and, 2) converging or diverging seepage in areas where the levee makes a tight bend. The interdependence of adjacent features will be assessed using three-dimensional seepage analyses modeling multiple features in one model. The effects of river bends on the background hazard will be assessed using methods developed by Benjasupattananan and Meehan (2012a, 2012b, 2013a, 2013b; Meehan and Benjasupattananan 2012, 2013) using modifications of the USACE's Blanket Theory equations modified for three-dimensional effects.

<u>Step 4 - Event Tree Analysis.</u> The probabilities of achieving the uplift pressure or exit gradient levels calculated in Steps 2 and 3 are contingent on the water level rising to a specified level (such as the 100 year flood). The various levels of pressure or gradient calculated in Step 2 represent conditions at which piping or heave have varying levels of potential to initiate. However, several other conditions are necessary for the initiation condition to propagate into failure. For example, for a certain geomorphic feature, the additional conditions may include: 1) the water level must be sustained for an adequate length of time, 2) the erosion pipe must form during the duration of the high water, 3) the pipe must stay open without collapse (i.e., a stable roof supports the pipe), and 4) efforts of intervention (flood fight) are unsuccessful. To assess the potential for levee failure to occur in this situation, an event tree analysis will be performed.

<u>Step 5 - Combine the Probabilities of Failure.</u> Once the probabilities of underseepage failure for the mapped geomorphic features and the background hazard have been assessed in Steps 2 through 4, these probabilities must be combined to calculate the probability of failure for the entire levee system. Because failure at any one location along the alignment constitutes failure of the entire system, the system is considered to fail when any one of the sources of underseepage hazard (geomorphic features or background hazard) progresses to failure. In this sense, the total probability of failure would be the sum of the probabilities from all of the hazard sources. However, because of the interdependence of the hazard sources, a *common cause* correction is likely needed for the analysis results (e.g., Bowles et al. 2001, 2012). The *common cause effect* occurs because failure of any one of the hazards would result in flooding of the basin, so once failure occurs through one of the hazards the probability of the other hazards propagating

to failure is essentially zero. (I.e., if the probabilities of failure are simply summed, the total probability of failure will be overestimated) (Bowles et al. 2001, 2012).

EXAMPLE — CREVASSE SPLAY DEPOSIT

The following section presents the development of a response surface for the crevasse splay geomorphic feature. The response surface is then used to produce two CADFs for a hypothetical crevasse splay deposit; one to represent the probably of exceeding given values of hydraulic gradient and one to represent probability of exceeding values of factor of safety against blanket heave.

Model Description

The three-dimensional finite element seepage model is based on the simplified crevasse-splay model presented in Figure 5. It consists of 1) a crevasse channel leading from the river water to the splay on the land side of the levee, 2) a splay area on the land side of the levee, and 3) a low permeably blanket on top of the splay. The geometry and seepage characteristics for the entire feature are represented by the 10 parameters described in Table 1.

Table 1. Parameters describing the geometry and seepage characteristics of the crevasse-splay model.

Parameter	Description	Parameter	Description
W_{ch}	Width of the crevasse channel	L_s	Length of the splay area
L_{ch}	Length of the crevasse channel	t_s	Thickness of the splay
t_{ch}	Thickness of the crevasse channel	K_s	Horizontal hydraulic conductivity of the splay area
K_{ch}	Horizontal hydraulic conductivity of the crevasse channel	t_b	Thickness of blanket above the splay
W_s	Width of the splay area	K_b	Vertical hydraulic conductivity of the blanket above the splay

The range and probability distribution for each of the 10 parameters can be represented using a probability density function (PDF) that describes the relative likelihood for this parameter to take on given values. However, as previously described, creating a response surface with this many parameters would require millions of finite element analyses. Therefore, in order to reduce the number of analyses to a reasonable number, the 10 parameters are combined into three parameters that adequately describe the flow regime in the crevasse splay deposit. The conductance of the crevasse channel, $C_{channel}$, describes the resistance to flow from the river to the splay. The transmisivity of the splay, T_s, describes how easily the flow reaching the splay is distributed througouth the splay. The conductance of the blanket, $C_{blanket}$, defines the ease at which the pressures in the splay can be dissipated though the blanket. The combined parameters are calculated as presented in the following equations:

$$C_{channel} = \frac{K_{ch}W_{ch}t_{ch}}{L_{ch}} \quad (1)$$

$$T_s = K_s t_s \quad (2)$$

$$C_{blanket} = \frac{K_b W_s L_s}{t_b} \quad (3)$$

Verification of the Simplified Model

Table 2 presents the results of the parametric analyses used to confirm the validity of the simplified model. The combined parameters that make up the simplified models are presented in the first column. The second columns presents the dependent parameters (those that are constituents of the respective combined variable) that are held constant during the respective parametric analysis. The second column presents the independent parameters (those that are not constituents of the respective combined variable) that are held constant during the analysis. Columns four and five respectively present the dependent and independent variables that are varied in the respective parametric analyses. Finally, the maximum amount of variation resulting from each analysis is presented in the final column. The numbers in the final column represent the largest variation that occurs with variation of the parameters listed in columns 4 and 5. Thus, if the simplified model perfectly matched the model with all parameters used the values in the final column would all be zero.

As can be observed from Table 2, the maximum variation resulting from using the simplified model is less than one foot of total head in all but one case. Furthermore, most of the variation that is observed occurs when using values at the ends of PDF distributions where the probability of occurrence is very low relative to the values in the center of the distribution. Therefore, it can be concluded that the simplification of the model imposes insignificant errors to the results.

Developing the Response Surface

Using the combined parameters identified above (T_s, $C_{blanket}$, $C_{channel}$), the response surface was generated for the crevasse-splay geomorphic feature using multiple runs of a three-dimensional finite element analysis. The response surface was generated over ranges of values for the three combined parameters that cover all of the possible values resulting from variation of the 10 original parameters of the model. The possible range of each combined parameter was discretized into six to eight values to represent the variation of the parameter. Finite element analyses were then performed on every possible combination of the discretized values of the combined parameter. The results of the analyses were plotted on a "family of curves" that together represent a three-dimensional surface that defines the relationship between the three combined parameters and the maximum total head in the splay. The family of curves is presented in Figure 6. Table 2. Results of parametric analysis to assess the validity of the simplified flow model for crevasse-splay response surface

Combined Parameter for Response Surface	Constant Dependent Parameters	Constant Independent Parameters	Varied Dependent Parameters	Varied Independent Parameters	Maximum variation in resulting head (ft)
Transmissivity of Splay, T_s	N/A	W_s, L_s, W_{ch}, L_{ch}, t_b, K_b, K_f	K_s, t_s	K_{ch}	0.25
		W_s, L_s, W_{ch}, L_{ch}, t_b, K_{ch}, K_f		K_b	0.69
				t_b	0.61
Conductance of Blanket, $C_{blanket}$	W_s, L_s	W_{ch}, L_{ch}, t_s, t_{ch}, K_{ch}, K_f	K_b, t_b	K_{ch}	0.25
	W_s, L_s	W_{ch}, L_{ch}, t_s, t_{ch}, K_s, K_f	K_b, t_b	K_s	0.69
	t_b, L_s		K_b, W_s		0.02
	t_b, W_s		K_b, L_s		0.51
	K_b, L_s		t_b, W_s		0.02
	K_b, W_s		t_b, L_s		0.06
	K_b, t_b		W_s, L_s		0.08
Conductance of Channel, $C_{channel}$	W_{ch}, L_{ch}	W_s, L_s, t_s, t_b, K_b, K_f	K_{ch}, t_{ch}	K_s	1.27
	t_{ch}, L_{ch}		K_{ch}, W_{ch}		0.22
	t_{ch}, W_{ch}		K_{ch}, L_{ch}		0.41
	K_{ch}, L_{ch}		t_{ch}, W_{ch}		0.38
	K_{ch}, W_{ch}		t_{ch}, L_{ch}		0.36
	K_{ch}, t_{ch}		W_{ch}, L_{ch}		0.39
	W_{ch}, L_{ch}		K_{ch}, t_{ch}	K_b	0.97

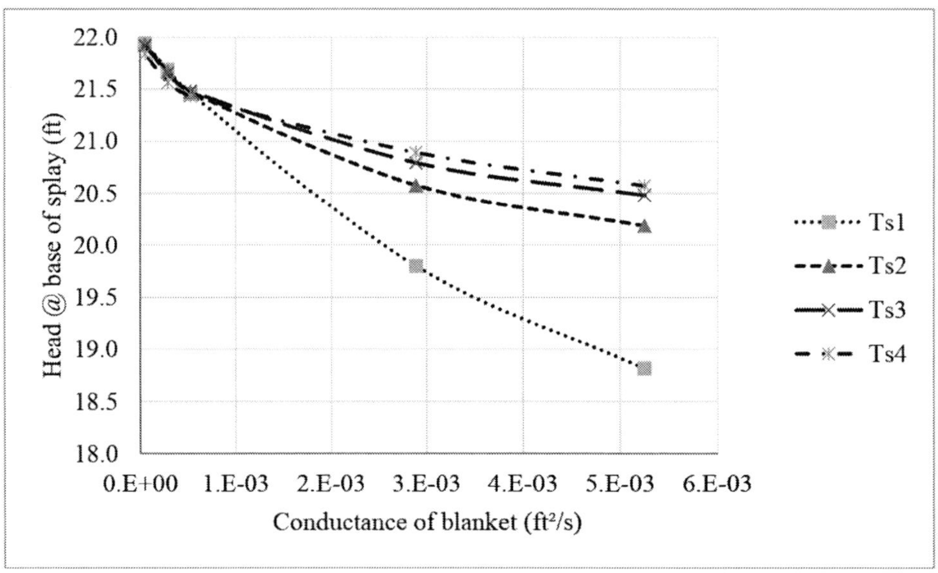

Figure 6. Family of curves for the crevasse-splay model for one constant $C_{channel}$ and different ranges of T_s and $C_{blanket}$.

Thus, given values of each of the combined parameters, the maximum total head in the splay can be calculated. Equations were developed through regression analysis to fit the curves of the response surfaces to facilitate computer coding the response surface into a spreadsheet linked with the computer program @Risk which runs the Monte Carlo Simulation.

RSMC Analysis

A Monte Carlo analysis was performed using a program written in an Excel spreadsheet and linked with the computer program @Risk. PDFs for each of the 10 original model parameters plus one additional parameter (the unit weight of the blanket, γ_b) are input into the program. Details of the PDFs are presented in Table 3. For each iteration of the Monte Carlo analysis the gradient through the blanket, $i_{blanket}$, and the factor of safety against heave, F_{heave}, are calculated using the following sequence:

- Values of each of the 10 parameters are randomly selected based on the PDF distributions.
- These parameter values are then combined into the three combined parameters using Equations 1 through 3.
- The three combined parameters are then used with the response surface to calculate the maximum total head in the splay h_{max} (the zero head datum is set as the land side ground surface elevation).
- Using the maximum head value the gradient through the blanket and the factor of safety against heave are calculated using Equations 4 and 5:

$$i_{blanket} = \frac{\Delta h}{t_b} = \frac{h_{max}}{t_b} \tag{4}$$

$$F_{heave} = \frac{t_b \gamma_b}{u_{splay}} = \frac{t_b(\gamma_b)}{(h_{max}+t_b)\gamma_w} \tag{5}$$

The sequence above was repeated for 10,000 iterations in order to produce CADFs representing the probability of various values of $i_{blanket}$ and F_{heave}. These CADFs are presented in Figures 7 and 8. It should be noted that Figure 7 is actually presented as a CDDF (Cumulative Descending Density Function) to make comparison with Figure 8 easier (i.e. failure probability on the right side on both figures).

The probabilities of exceeding and exit gradient of 1.0 or a factor of safety of 1.0 against heave are indicated in Figures 7 and 8. The figures can also be used to calculate the probabilities of exceeding lesser or greater values of these parameters that may indicate greater or lesser probabilities of initiating a BEP failure of the levee. Thus, the CADFs may be used in conjunction with the event tree analyses described in the previous section to assess the likelihood of levee failure due to an individual geomorphic feature along the alignment. The procedure could be repeated for all of the geomorphic features (of various types) located along the levee alignment and the probabilities of failure for each feature combined to produce the total probability of failure for the entire levee.

Modeling Geomorphic Features

Table 3. PDFs for the 10 Input Parameters

Variable	Type of PDF distribution	Most likely value	Standard deviation	Truncated minimum value	Truncated maximum value
Thickness of the splay (t_s)	Normal	3	3	1	10
Thickness of the channel (t_{ch})	Normal	3	3	1	10
Thickness of the blanket (t_b)	Normal	4	3	1.5	10
Width of the splay (W_s)	Normal	200	86	50	300
Width of the channel (W_{ch})	Normal	150	86	75	300
Length of splay (L_s)	Normal	700	347	700	1500
Length of channel (L_{ch})	Normal	95	3	95	105
Log of channel hydraulic conductivity (log K_{ch})	Normal	-4	1	-5	-3
Log of hyd. cond. of channel to splay ratio (log K_{ch}/K_s)	Normal	300	100	1	1000
Log of blanket hydraulic conductivity (log K_b)	Normal	-8	1	-9	-7
Unit weight of the blanket (γ_b)	Normal	120	5	110	130

The hydraulic exit gradients and pore pressures calculated for this crevasse splay model are generally higher than those computed by a two-dimensional or blanket theory model. For comparison, a gradient calculated using an open-seepage-exit blanket-theory scenario model using most likely values (MLVs) and a blanket thickness equal to t_b is about 1.1. In the actual levee model, the differential head is 22ft. with blanket thicknesses as low as 4ft. and as high as 20ft. and, consequently, gradients greater than 1.0 are expected. That said, these results illustrate to some extent the three-dimensional concentration effect that this geomorphic feature has on the seepage flow. It is also important to note that these results represent a condition that may initiate internal erosion and that the probability of failure would require further assessment though the event-tree analyses that are the subject of further research and development.

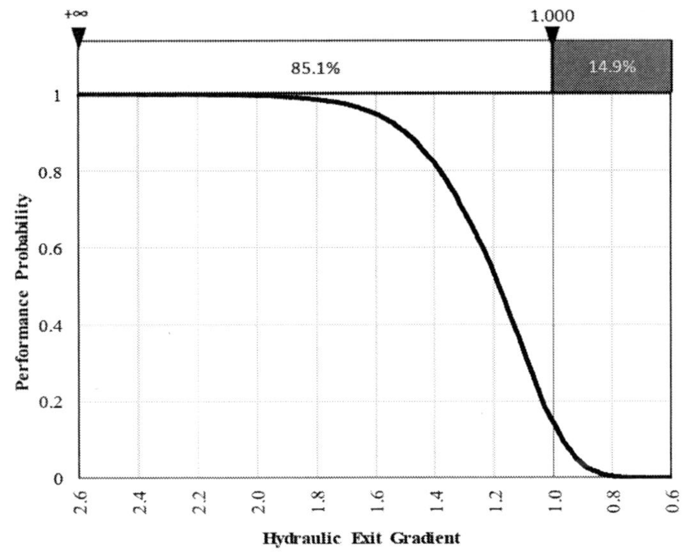

Figure 7. CDDF for gradient through the blanket, $i_{blanket}$.

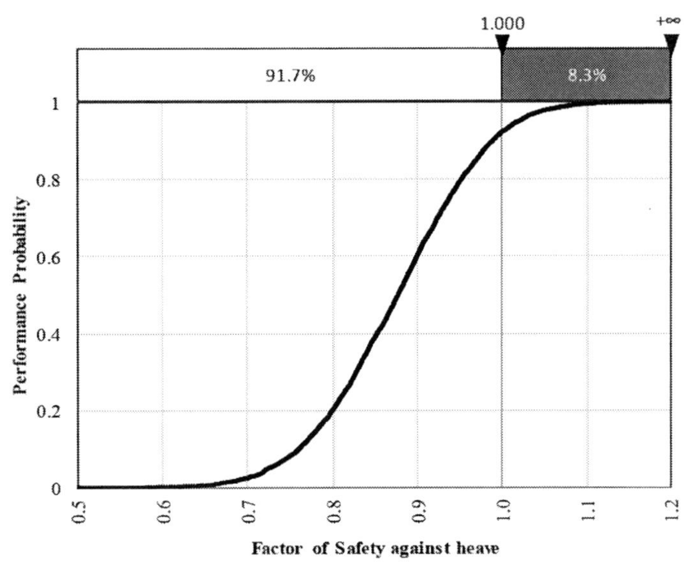

Figure 8. CADF for factor of safety against heave, F_{heave}.

CONCLUSION

The paper presents methodology for calculating the probability of failure of a reach of levee by calculating the individual failure probabilities due to geomorphic features that intersect the levee alignment. The paper outlines the total procedure that includes identifying the geomorphic features, producing CADFs for the probabilities of various heave or BEP initiation, calculating the interaction between geomorphic features, calculating failure probabilities using event tree analyses, and calculating the total probability of failure to a levee reach by combining the probabilities from all of the geomorphic features.

Modeling Geomorphic Features

Details of the RSMC procedure for producing CADFs for the geomorphic features are presented. An example is presented where CADFs for heave and exit gradient are calculated for a crevasse splay deposit. Procedures for the remaining steps in the outlined procedure are currently under development with ongoing research.

ACKNOWLEDGEMENTS

This material is based upon work supported by the National Science Foundation (NSF) under Grant CMMI 1400640. Any opinions, findings, and conclusions or recommendations expressed in this material are those of the authors and not necessarily the views of NSF.

REFERENCES

Adams, Peter N., Slingerland, Rudy L. and Smith, Norman D. (2004). Variations in natural levee morphology in anastomosed channel flood plain complexes. *Geomorphology*, 61, 127–142.

Allen, J. R. (1970). *Physical Processes of Sedimentation*. London: George Allen & Unwin.

Baecher, G. B. and Christian, J. T. (2003). *Reliability and statistics in geo-technical engineering*. Chichester, England: Wiley.

Benjasupattananan, S. and Meehan, C. L. (2012a). Deterministic and Probabilistic Approaches for Two- and Three-Dimensional Levee Underseepage Analyses. *Proceedings of Dam Safety 2012*: Annual Conference of the Association of State Dam Safety Officials, Denver, CO, September 16-21, 2012, ASDSO, Lexington, KY, Published on CD-ROM.

Benjasupattananan, S. and Meehan, C. L. (2012b). An Investigation of Three Probabilistic Approaches for Levee Underseepage Analysis. *Proceedings of Geo-Congress 2012*: State of the Art and Practice in Geotechnical Engineering, Geotechnical Special Publication No. 225, Oakland, CA, March 25-29, 2012, ASCE, Reston, VA, 2912-2921. (DOI:10.1061/9780784412121.298)

Benjasupattananan, S. and Meehan, C. L. (2013a). Probability-Based Design for Levee Underseepage: Heaving vs. Piping Phenomena. Proc., *Geo-Congress 2013*: Stability and Performance of Slopes and Embankments III, Geotechnical Special Publication No. 231, San Diego, CA, March 3-7, 2013, ASCE, Reston, VA. (Accepted, In Press)

Benjasupattananan, S. and Meehan, C. L. (2013b). Analytical Solutions for Levee Underseepage Analysis: Straight and Curved Levee Sections with an Infinite Blanket. *Proceedings of Geo-Congress 2013*: Stability and Performance of Slopes and Embankments III, Geotechnical Special Publication No. 231, San Diego, CA, March 3-7, 2013, ASCE, Reston, VA. (Accepted, In Press)

Bernitt, Laurits and Lynett, Patrick (2010). Breaching of sea dikes. *Proceedings of 32nd Conference on Coastal Engineering*, Shanghai, China, 1(32), 1-10.

Bligh, W. G. (1910). Dams, barrages, and weirs on porous foundations. *Engineering News*-Rec., 64(26), 708–710.

Bligh, W. G. (1913). Lessons from the failure of a weir and sluices on porous foundations. *Engineering News*-Rec., 69(6), 266–270.

Bligh, W. G. (1915). Submerged weirs founded on sand. In *Dams and Weirs* (pp. 151-179). Chicago, American technical society.

Bowles, D. S., Anderson, L. R., and Chauhan, S. S. (2001). Approaches to the common cause adjustment in event tree used in dam safety risk analysis. An IDSRM-USU Working Paper, Logan, Utah: Institute for Dam Safety Risk Management;

Bowles, D. S., Chauhan, S. S., Anderson, L. R. and Grove, R. C. (2012). Baseline Risk Assessment for Herbert Hoover Dike. *ANCOLD Conference on Dams*, Perth, Western Australia, Australia. 27-28 October 2012.

Bridge, John (2003). *Rivers and floodplains: Forms, Processes, and Sedimentary Record.* Oxford, UK: Blackwell Publishing.

Brierley, G. J. and Fryirs, K. A. (2005). *Geomorphology and river management: applications of the river styles framework.* Oxford, UK: Blackwell Publishing.

Brierley, G. J., Ferguson, Rob J. and, Woolfe, Ken J. (1997). What is a fluvial levee? *Sedimentary Geology*, 114, 1-9.

Bristow, C. S. Skelly, R. L. and Ethridge, F. G. (1999). Crevasse splays from the rapidly aggrading, sand-bed, braided Niobrara River, Nebraska: effect of base-level rise. *Sedimentology*, 46,1029-1047.

Cazanacli, Dan and Smith, Norman D. (1998). A study of morphology and texture of natural levees–Cumberland Marshes, Saskatchewan, Canada. *Geomorphology*, 25, 43–55.

Crum, D. A. (1996). Reliability applied to levee seepage analysis. *Probabilistic Mechanics and Structural Reliability*: ASCE Proceedings of 7th Specialty Conference, Worcester, MA, 1, 946–949.

Duncan, M. J. (2000). Factors of safety and reliability in geotechnical engineering. *Journal of Geotechnical and Geoenvironmental Engineering*, ASCE, 126(4), 307-316.

Ferguson, Rob J. and Brierley, Gary J. (1999). Levee morphology and sedimentology along the lower Tuross River, south-eastern Australia. *Sedimentology*, 46, 627-648.

Filgueira-Rivera, Manuel, Smith, Norman D. and Slingerland, Rudy L. (2007). Controls on natural levée development in the Columbia River, British Columbia, Canada. *Sedimentology*, 54, 905–919.

Fryirs, K.A. and Brierley, G.J. (2013). *Geomorphic analysis of river systems*. Oxford, UK: Wiley-Blackwell.

Gabr, M. A., Taylor, H. M.Jr., Brizendine, A. L., and Wolff, T. F. (1995). LEVEEMSU: Analysis software for levee underseepage and rehabili-tation. *Technical Rep. GL-95-9*, U.S. Army Engineer Waterways Experiment Station, Vicksburg, MS.

Glynn, M. E. and Kuszmaul, Joel (2010). Prediction of Piping Erosion along Middle Mississippi River Levees-An Empirical Model. USACE, Technologies and Operational Innovations for Urban Watershed Networks Research Program, *ERDC/GSL TR-04-12*. Originally published in 2004 and revised in 2010.

Harr, M. E. (1987). *Reliability-based design in civil engineering*. New York: McGraw-Hill.

Hudson, Paul F. (2011). Natural levees. In *Encyclopedia of water science*, (second edition, pp. 763-767). Taylor Francis.

Hudson, Paul F. and Heitmuller, Franklin T. (2003). Local and watershed scale controls on the spatial variability of natural levee deposits in a large fine-grained floodplain: Lower Pánuco Basin, Mexico. *Geomorphology*, 56, 255–269.

ICOLD (2013). Internal erosion of existing dams, levees, and Dikes, and their foundations - International Glossary. *CIGB-ICOLD Bulletin*, (1).

Kanning, Willem (2012). *The weakest link*. Doctoral dissertation, Delft Technical University. Available online at: http://repository.tudelft.nl/search/ir/?q=isbn%3A%229789461860880%22

Knowles, V. R. (1992). Applications of the finite element seepage analysis Corps program CSEEP (X8202). *Technical Rep. ITL-92-6*, U.S. Army Engineer Waterways Experiment Station, Vicksburg, MS.

Lane, E. W. (1934). Security from under-seepage: Masonry dams on earth foundation. *Proceedings, ASCE*, 60, 1235-1272.

Leopold, L. B. and Wolman, M. G. (1957). River channel patterns: braided, meandering and straight. *U.S. Geological Survey Professional Paper*, 282B, 39-85.

Low, B. K. (2008). Reliability of levee systems, p. 134-168. in Kok-Kwang Phoon (Ed.). *Reliability-based design in geotechnical engineering*. Taylor & Francis Group, New York.

Meehan, C. L. and Benjasupattananan, S. (2012). An Analytical Approach for Levee Underseepage Analysis. *Journal of Hydrology*, Elsevier, 470-471, 201-211. (DOI:10.1016/j.jhydrol.2012.08.050)

Meehan, C. L. and Benjasupattananan, S. (2013). An Analytical Approach for Modeling Axisymmetric Levee Underseepage. *Journal of Geotechnical and Geoenvironmental Engineering*, ASCE. (In Review).

Nanson, Gerald C. (1980). Point bar and floodplain formation of the meandering Beatton River, northeastern British Columbia, Canada. *Sedimentology*, 27, 3-29.

Nanson, Gerald C. (1981). New evidence of scroll-bar formation on the Beatton River. *Sedimentology*, 28(6), 889-891.

Palisade Corparation (2013). *@Risk 6.1*. Integrated Excel program that performs risk analysis using Monte Carlo simulation.

Pearce, J.T., Marlow, D., Avila, C. and Selvamohan, S., (2009). Use of geomorphic and airborne geophysical data for analysis of levees and floodplain processes for levee evaluations in California. The Association of State Floodplain Managers, *ASFPM Annual Conference*, Orlando.

Polanco, L. and Rice, J. (2011). Reliability based underseepage analysis in levees using Monte Carlo simulation. *2011 GeoFrontiers Proceedings*, Paper #2978.

Polanco, L. and Rice, J. (2013). The effects of geometry in levee underseepage reliability. *Proceedings of Dam Safety 2013*: Annual Conference of the Association of State Dam Safety Officials, Providence RI, September 8-13, 2013, Published on CD-ROM.

Rice, John D. and Polanco, Lourdes (2012). Reliability-based underseepage analysis in levees using a response surface–monte carlo simulation method. *Journal of Geotechnical and Geoenvironmental Engineering*, ASCE, 138 (7), 821-830.

Ritter Dale F., Kochel R. Craig and Miller Jerry R. (2011). *Process Geomorphology*, 5th Edition., United States: Waveland Press.

Saucier, R.T. (1994). *Geomorphology and Quaternary Geologic History of the Lower Mississippi Valley*. U.S. Army Engineer Waterways Experiment Station, Vicksburg, Mississippi, Volume I, 414 p.

Shannon & Wilson, Inc. (2011). Skagit River Levee General Investigation-Levee Risk and Reliability Analysis. *Report submitted to USAGE*, Seattle District, Skagit County, Washington, USA.

Sleep, M., and Duncan, M. J. (2008). Manual for geotechnical engineering reliability. *Report by Virginia Tech Center for Geotechnical Practice and Research*, Blacksburg, VA.

Smith, Norman D. and Pérez-Arlucea, Marta (2008). Natural levee deposition during the 2005 flood of the Saskatchewan River. *Geomorphology,* 101, 583–594.

Steenbergen H. M. G. M., Lassing B. L., Vrouwenvelder A. C. W. M., Waarts P. H. (2004). Reliability analysis of flood defence systems, *Heron*, 49 (1), 51–73.

SoilVision Systems Ltd. -SVFlux (2009). Groundwater Seepage Modeling Software.

Temmermana, S., Goversa, G., Meireb, P. and Wartel, S. (2004). Simulating the long-term development of levee–basin topography on tidal marshes. *Geomorphology*, 63, 39–55.

Tracy, F. T. (1994). Seepage Package. *CSEEP Micro Version, X8202*, U.S. Army Engineer Waterways Experiment Station, Vicksburg, MS.

USACE (1956). Investigation of underseepage and its control. *Technical Memorandum No. 3-424*, U.S. Army Engineer Waterways Experiment Station, Vicksburg, MS.

USACE (2000). Design and construction of levees. *Engineering Manual EM 1110-2-1913*, U.S. Army Core of Engineers, Washington, DC.

USACE (2004). Geotechnical reliability of dam and levee embankments. *ERDC/GSL CR-04-1*, Geotechnical and Structural Laboratory, U.S. Army Core of Engineers, Washington, DC.

USACE (2005). Engineering and design: Design guidance for levee underseepage. *ETL 1110-2-569*, U.S. Army Core of Engineers, Washington, DC.

Vrouwenvelder, Ton (2006). Spatial effects in reliability analysis of flood protection systems. *Second International Forum on Engineering Decision Making*, Lake Louise, Canada, April 26-29.

Vrouwenvelder, A. C. W. M, Van Mierlo, M. C. L. M., Calle, E. O. F., Markus, A.A., Schweckendiek, T. and Courage, W. M. G. (2010). Risk analysis for flood protection systems. *Deltares, Main Report*, The Netherlands.

Walling, D. E., Owens, P. N. and Leeks, G. J. L. (1997). The characteristics of overbank deposits associated with a major flood event in the catchment of the River Ouse, Yorkshire, UK. *Catena,* 31(1-2), 53-75.

Walling, D. E., Fang, D., Nicholas, A. P. and Sweet, R. J. (2004). The grain size characteristics of overbank deposits on the floodplains of British lowland rivers. *Proceedings of Sediment Transfer through the Fluvial System*, Moscow, 288, 226–234.

William Lettis & Associates, Inc. (2008). Surficial Geologic Map and Initial Geomorphic Assessment, Sacramento River (East Side), Sacramento County, California. *Unpublished consultant report prepared for URS Corporation.*

Wolff, D. (2008). Reliability of levee systems. In *Reliability-based design in geotechnical engineering* (pp. 448-496). New York: Taylor & Francis Group.

Wolff, T. F. (1989). LEVEEMSU: A software package designed for levee underseepage analysis. *Technical Report GL-89-13*, U.S. Army Engineer Waterways Experiment Station, Vicksburg, MS.

Wolff, T. F. (1994). Evaluating the reliability of existing levees. *Research Report Prepared for U.S. Army Engineer Waterways Experiment Station*, Michigan State University, East Lansing MI.

Wolff, T. F., Demsky, E. C., Schauer, J., and Perry, E. (1996). Reliability assessment of dike and levee embankments. *ASCE Geotechnical Special Publication*, 1, 636–650.

Woolfe, Ken J. and Purdon, Richard G. (1996). Deposits of a rapidly eroding meandering river: terrace cut and fill in the Taupo Volcanic Zone. *New Zealand Journal of Geology and Geophysics*, 39 (2), 243–249.

Xu, B. and Low, B. K. (2006). Probabilistic stability analyses of embankments based on finite-element method. *Journal of Geotechnical and Geoenvironmental engineering*, ASCE, 132(11): 1444-1454.

WHEN YOUR LEVEE CANNOT BE ACCREDITED . . . THEN WHAT?

Maureen P. Cissel, CFM[1]

ABSTRACT

Local officials in Maryland received notification from the federal government declaring their local levee system could not be accredited. The levees were designed and constructed by the federal government (albeit in the 1950s), no major problems had been identified during the annual inspections, and there had never been a levee failure. What had changed?

Post Hurricane Katrina the National Levee Safety Committee was formed to evaluate the standards for levee accreditation. This levee, like so many in other jurisdictions throughout the nation, needed to be upgraded to meet the new standards for FEMA accreditation. There were a *flood* of questions that followed:

- Are our constituents in danger? (Answer: not immediate danger but let's get to work)
- How can we explain this to our residents and property owners? (Though flood insurance is always recommended, will it be required?)
- How do we fix this?

The first step towards accreditation was a thorough system review; the next step was a complex analysis of the system components. The analysis, design and construction were completed over the next five years. The project team included the local agency/levee sponsor, federal regulatory agencies, design team, and construction contractors.

A systematic management and design approach to identify deficiencies, design, construct, and ultimately obtain levee accreditation. The system includes almost five miles of earthen levee, pumping stations, pressure conduit, and various levee penetrations, sluice gates and flapper valves. The project reduced overall flood risk and improved safety and welfare of more than 1,350 properties and their residents and owners.

INTRODUCTION

Once every four years the America Society of Civil Engineers (ASCE) conducts a comprehensive assessment of the nation's major infrastructure using a simple A to F grading format. The ASCE website provides the details related to the 2013 Report Card for America's Infrastructure (http://www.infrastructurereportcard.org/) which demonstrates that the overall infrastructure Grade Point Average rose slightly. The levee systems grade is below the overall grade average; levees received an average D- grade. This is a very scary finding considering that there are an estimated 100,000 miles of

[1] Project Manager, McCormick Taylor, 509 Exeter Street, 4th Floor, Baltimore, MD 21202
mpcissel@mccormicktaylor.com – 410.662.7400

levees nationwide, in each of the 50 states and the District of Columbia, providing some level of protection to millions of people living or working in these levee protected areas.

Levees are designed to protect against a certain level of flooding. The National Flood Insurance Program stresses on its website (https://www.floodsmart.gov/floodsmart/pages/flooding_flood_risks/levees.jsp) that no levee provides full protection from flooding–even the best flood control structure cannot completely eliminate the risk of flooding. Levees can and do decay over time, making maintenance a serious challenge. Many levees throughout the country are over 50 years old and were originally used to protect farmland but are now protecting developed communities. The reliability of these aging structures is unknown in many cases, increasing the risk to public health and safety. Repairs or rehabilitation is needed to maintain the functionality of flood control systems that have deteriorated over time and/or systems that do not meet current design standards.

The US Army Corps of Engineers (USACE) has established the National Levee Database that accounts for most of the levees designed by the USACE. This database is available to the public at http://nld.usace.army.mil. Many of the flood protection systems in place today were not designed by the USACE. Federal Emergency Management Agency (FEMA) has developed a Midterm Levee Inventory which includes some of these systems not constructed by USACE. The plan is that eventually these two databases will be combined into one system to provide for a comprehensive list of all levees. The USACE and FEMA are collaborating on efforts to share information and data, synchronize work efforts and align program requirements while involving local and state agencies and other stakeholders.

Who is responsible for the levees and what this responsibility means is not always apparent. While the USACE inspects and evaluates some levees to determine if they meet accreditation requirements, FEMA identifies and maps risk behind the levees – most levees are owned, operated and maintained by local communities. In many cases the local jurisdictions may not have the knowledge or experience to identify and correct deficiencies and rely on other outside resources for assistance.

WRRA 2014 put in place the requirement for several things that will change the future of levee projects:

- A one time inventory and review of all levees
- Comprehensive National Levee Safety Guidelines
- Establishment of Levee Hazard Potential Classification system

These initiatives have not yet been funded, full funding may take years.

This paper will discuss a federally (USACE) designed and constructed flood protection system, which is important because there are distinctions between the responsibilities and roles for federal verses non-federal constructed systems. We can consider how WRRA will change similar projects in the future.

PROJECT DESCRIPTION

The Anacostia Flood Risk Management Project, located in Prince George's County, Maryland, was authorized by Congress in 1950. The USACE completed the design and construction and the project was formally transferred to the local project sponsor, The Washington Suburban Sanitary Commission (later transferred to Prince George's County Government), in 1959 for operation and maintenance. The flood risk management project consists of 2.7 miles of flood control channels and 5.3 miles of levee embankment along the Anacostia River main stem, the Northeast Branch and the Northwest Branch. The project consists of four separate systems: Brentwood, Riverdale-Hyattsville, Edmonston-Bladensburg and Colmar Manor levee systems. Four flood control pumping stations connecting storm sewers for interior drainage, a pressure conduit and gravity outlet pipe for interior drainage are also part of the flood risk management project. The project was designed to protect seven Maryland communities–the incorporated towns of Bladensburg, Edmonston, Riverdale, Hyattsville, Brentwood, Cottage City and Colmar Manor - from high flows. The figure below shows the Anacostia Flood Risk Management Project, each of the individual features identified above and the highly developed, densely populated, areas located behind these levees.

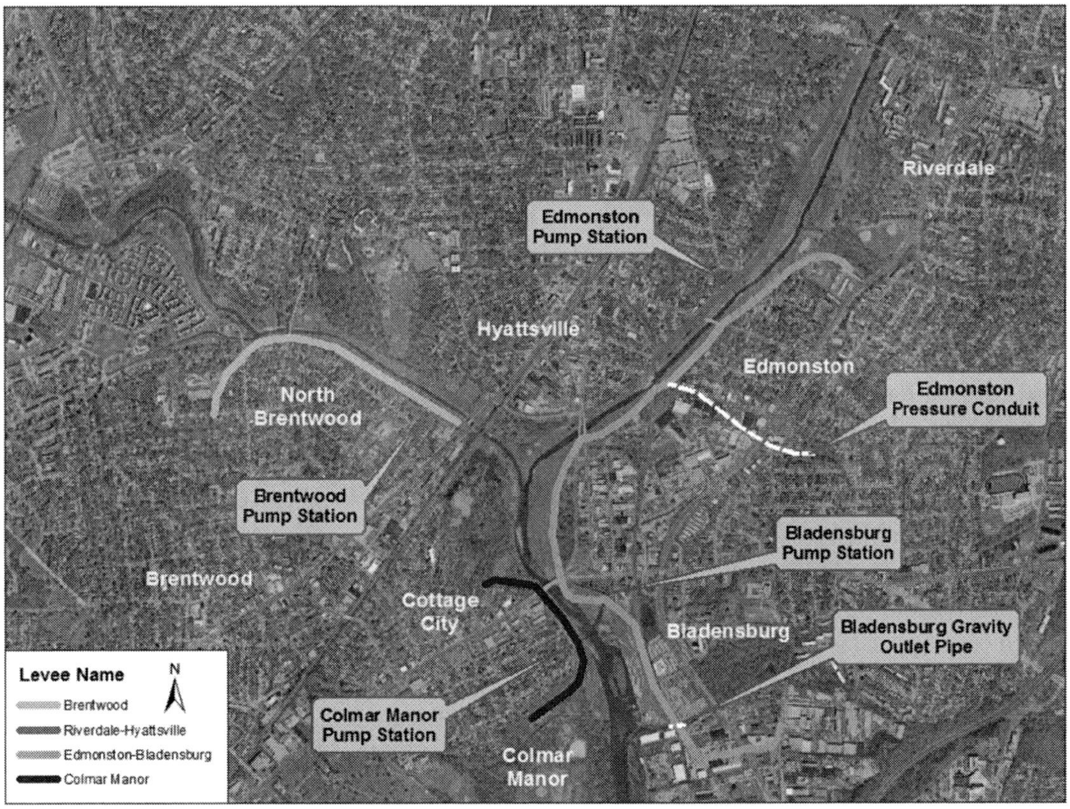

Figure 1. Project Location Map.

In 2010, as part of the National Flood Insurance Program, FEMA was in the process of updating the Flood Insurance Rate Maps (FIRM) in this jurisdiction. The local sponsor realized that the existing levee would not meet the current adequacy requirements to contain the 100-year event. This would result in the updated FIRM showing the areas behind the levee systems as special flood hazard area zone AE which is an area designated as an area subject to inundation with mandatory flood insurance purchase requirements and floodplain management standards The local sponsor was shocked; the levees were designed and constructed by the USACE (albeit in the 1950s), and there had never been a levee failure. The USACE conducted annual inspections and the local sponsor was not aware of any major problems. What had changed? Katrina and more. As stated in the National Levee Safety Committee website (http://www.leveesafety.org), the nation's attention has been refocused on the role of levees as critical infrastructure after the Mississippi River floods in 1993 and 2008, California floods in 1986 and 1997 and the tragedy of Hurricane Katrina in 2005.

LEVEE ACCREDITATION

As a result of the nationwide focus on levees, many levees have been identified to need rehabilitation in order to meet the standards for FEMA accreditation. For the local sponsor there are a *flood* of questions associated with that finding:

- Are our constituents in danger? Answer: Not immediate danger, but let's get to work.
  We must always be mindful that levees reduce, not eliminate, flood risk.
- How can we explain this to our residents and property owners? Though flood insurance is always recommended, will it be required? Answer: Over time infrastructure needs repair/rehabilitation and regulations/standards change or are improved.
- Can we fix this and if so how? Answer: One deficiency at a time.

In the case of the Anacostia Flood Risk Management Project the local levee sponsor started working towards FEMA levee accreditation. The first step was a thorough system review which considered historical documentation, condition, operation, maintenance and repairs of the entire project. The next step was a complex analysis of each of the system components: hydrology and hydraulics, interior flooding, structural, geotechnical, mechanical, electrical, Operations & Maintenance (O&M) manuals and an emergency action/evacuation plan.

The analysis, design and construction for the Anacostia Flood Risk Management Project were completed over the next four years. The project team included the local agency/levee sponsor, USACE and FEMA, consultant design team and construction contractors. The team forged strong partnerships to develop a systematic management and design approach to identify deficiencies, finalize design, complete construction and ultimately obtain levee accreditation. In the case of the Anacostia systems, the USACE was able to provide a Positive Levee System Evaluation for each of the systems that was submitted for consideration as part of the National Flood Insurance Program. The

USACE was only able to serve this role because there was an ongoing study and active project regarding the H&H work effort for this flood risk management project. The collaboration and success of this project resulted in these levees being one of the first reaccredited systems in FEMA's Region III and the approach will serve as a model for the remaining 140 systems. The project reduced overall flood risk and improved safety and welfare of more than 1,350 properties, as well as their residents and owners.

LEVEE ACCREDITATION PROCESS

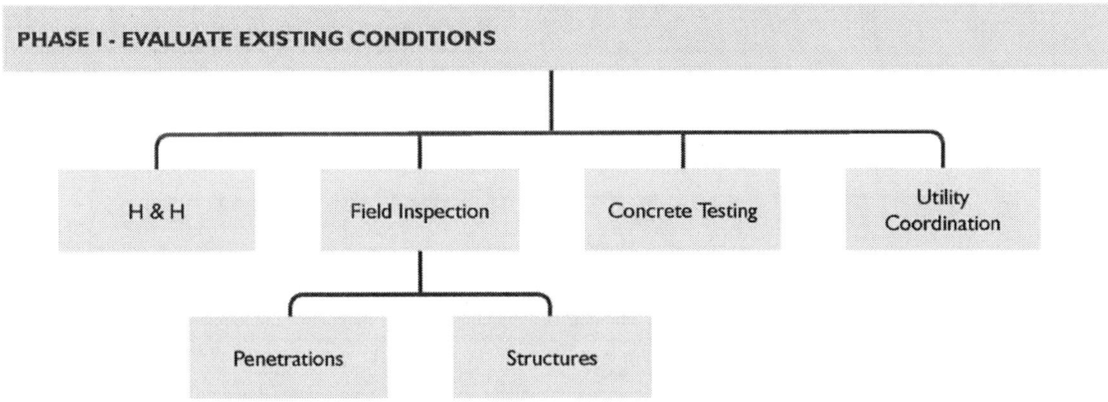

Figure 2. Phase I – Evaluate Existing Conditions.

Hydraulic and Hydrologic Engineering

USACE's Hydrologic Engineering Center report served as the initial step in the risk and uncertainty analysis. Subsequently, the local sponsor contracted with the McCormick Taylor team to perform the engineering design and prepare the plans and specifications for the rehabilitation projects. The local sponsor, the consultant team and the USACE district work closely together to ensure that the work met the USACE guidelines and requirements.

Field Inspection

Levee Penetration Evaluation. Prepared contract specifications and oversaw the clean out and inspection of each levee penetration. This effort required the removal of sediment and debris; work in confined spaces and detailed field evaluation and inspection. Several of the penetrations were located within areas of tidal influence, making the inspection and ongoing flood control efforts complex.

Structure Evaluation. Inspected and exercised all existing sluice gates, flapper valves and trash racks. This was followed by a comprehensive maintenance effort to clean, lubricate, paint and perform other maintenance as required.

Concrete Testing. Concrete coring, testing and chemical analysis were completed. Evaluation of findings in advance of conducting detailed structural analysis of the pump station structural deficiencies was performed to complete the structural repair design efforts.

Utility Coordination. Identified, inspected and evaluated utility crossings located on and through the levee system to include water, sewer, poles with telephone, cable and power lines. Utility line relocation/raising was required in some cases to provide adequate construction access and clearances for flood control.

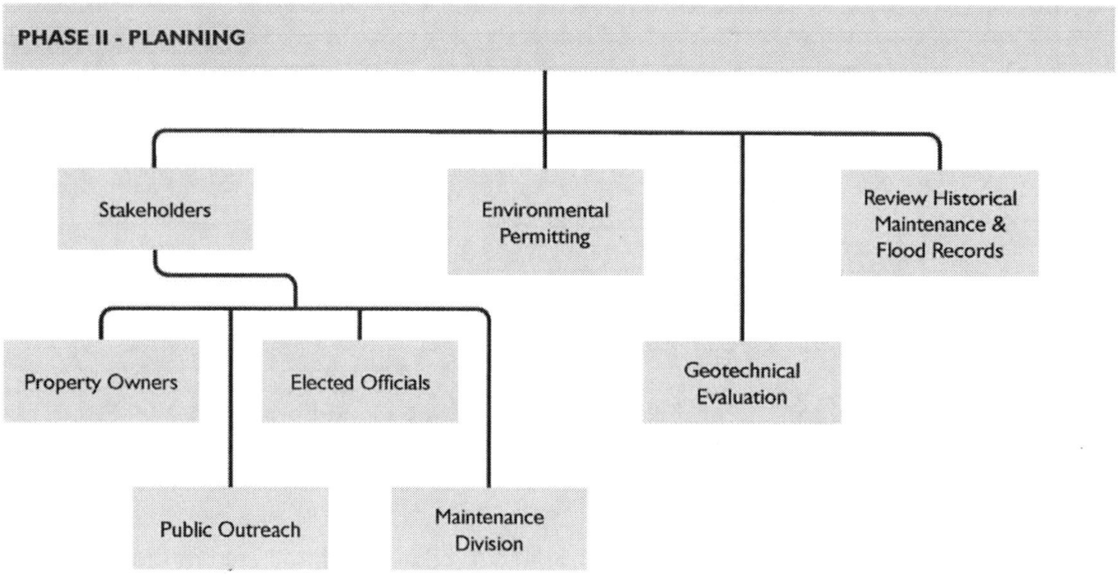

Figure 3. Phase II – Planning.

Stakeholders

Property Owner Coordination. Assembled and verified right of way and evaluated and refined impacts of the construction effort. The flood control project is located on property that was deeded to the local park authority, not the local levee sponsor. The park uses parts of the flood control project area for bike trails and other recreational purposes. The local sponsor needed to establish the limits of the flood control project and review jurisdictional priorities and requirements.

Public Outreach. Attended community meetings and provided press releases and website updates to keep the community informed of the project progress and construction requirements.

Elected Officials. Provided briefings to the elected officials and the municipal leaders to ensure understanding of the requirements for accreditation and construction. The local

County was the levee sponsor, but local municipalities were protected by the levee system so it was required all officials had a clear understanding of the impacts to their constituents. Aerial and engineering display maps, graphics and presentation slides were prepared and presentations to various stakeholders were conducted.

Maintenance Division – Engaged the levee sponsor maintenance staff to ensure all constructed improvements were clearly understood and appropriate maintenance activities could and would be performed going forward.

Environmental Permitting

Local, state and federal permits for work within the floodplain, waters of the US, tidal and non-tidal waterways were required. Coordination was completed with USACE and the Maryland Department of the Environment as part of the Joint Permit Application process. Approval was obtained by the Dam Safety Division, and a Waterway Construction Permit was issued. Delegation of Engineer-In-Charge responsibility was approved and storm water management requirements were satisfied. Coordination was performed with and approval obtained by Critical Area Commission, Natural Resources and Historic Trust divisions.

Geotechnical Evaluation

Soil borings were performed on the levee systems. An evaluation was performed that confirmed the historic construction file documents and existing conditions were acceptable.

Historical Maintenance and Flood Records

Reviewed historical records to identify any significant or re-occurring maintenance concerns. In addition any historical flooding events were reviewed to identify and analysis any system failures.

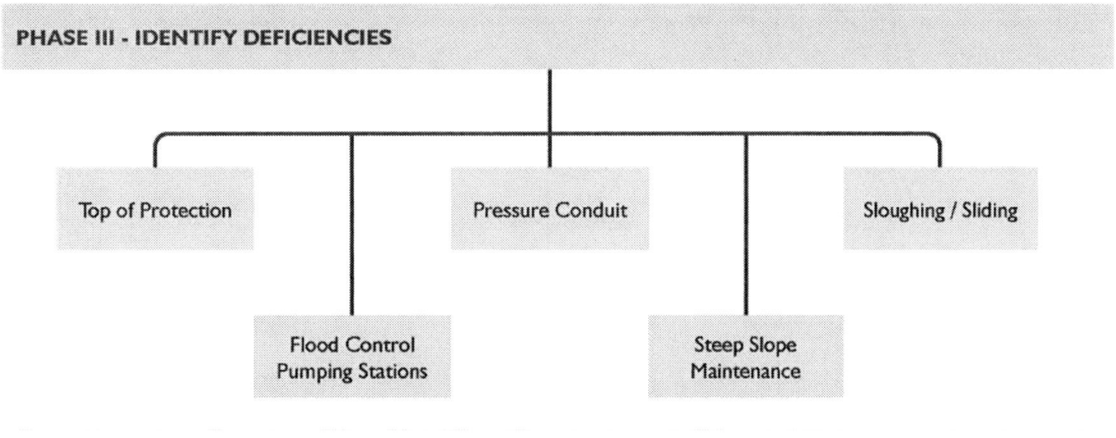

Figure 4. Phase III – Identify Deficiencies.

Top of Protection

Analyzed settlement, erosion and rutting of the levee crest to identify the cause and if adequate/consistent elevation for flood protection was available. The required top of levee protection was identified based on the H&H and risk and uncertainty analysis. A factor of safety was added for constructability concerns and the constructed top of protection elevation for the levee crest was established.

Flood Control Pumping Stations

Conducted detailed inspections performed by structural, electrical, mechanical and civil engineering teams to identify deficiencies at four flood control pumping stations. Deficient items included a full range of elements from sluice gate frames, concrete deterioration, to non-compliant access ladders.

Pressure Conduit

Determined adequate access needs for the pressure conduit to allow for safe inspection and maintenance. Field verified and tested existing access vaults to reveal opportunities for improvement.

Steep Slope Maintenance

Performed critical slope failure analysis. Developed alternative slope treatments to ensure safe maintenance of steep slope within the levee system. The alternatives reviewed included rip rap, gabions and the combination of both. Final design included a tiered rip rap slope treatment with structural elements.

Sloughing/Sliding on Embankment – Examined and investigated depressions on the levee slopes. Performed geotechnical investigation and engineering to determine that repair of slope failures required excavation past the failure arc. Reconstruction of the levee slope along the reach was required to correct localized depressions and rutting found on the landside of the embankment.

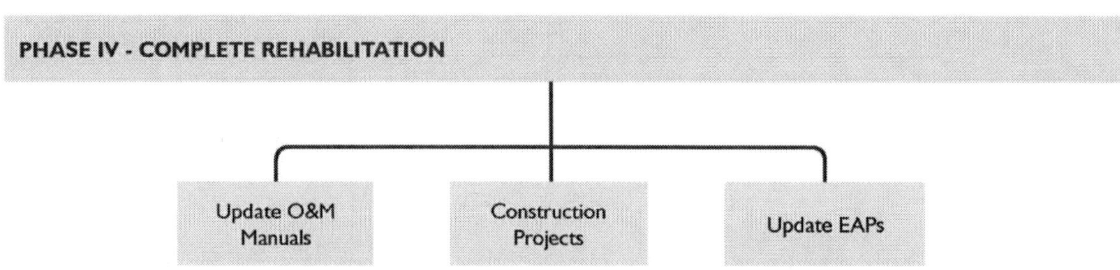

Figure 5. Phase IV – Complete Rehabilitation.

Operation and Maintenance Manuals

Reviewed the original O&M manuals for each system and developed updated manuals. The new manuals identified the location and description of each element of the levee system and pumping station. The O&Ms included recommended operating and maintenance procedures, as well as troubleshooting recommendations and flow charts to be used by field personnel. Detailed responsibilities for management, field staff and operators, electrical and mechanical contractors were provided and procedures for record keeping and safety requirements were detailed.

Construction Projects

The USACE District Office issued letters of no objection for each phase of the rehabilitation work prior to construction. The overall approach was to address each deficiency item. This entailed extensive coordination and separate construction contracts.

Levee Rehabilitation. Completed design and construction to raise the levee in areas of insufficient protection. The contract plans showed the specific limits of raising to the closest six inches (rounding up for constructability) and tie-ins to the undisturbed levee crest. The specifications required a Test Strip to ensure the contractor had acceptable material and construction methods as the initial step in the construction. Strict quality control and quality assurance measures were required by the contractor and confirmed by the owner's field representative.

Pumping Station Upgrades. Prepared detailed contract specifications and plans that included replacement and repair of existing sluice gates, installation of new sluice gates, and repair to the damaged concrete were all elements included. This effort required an elaborate dewatering sequencing plan to control tidal influence, perform the construction and provide ongoing flood protection. Other pumping station enhancements included electrical, mechanical and SCADA systems as part of the overall upgrade effort.

Pressure Conduit Rehabilitation and Access Project. Developed construction plans and specifications to replace existing access vaults and install new access vaults along an existing 2,200 linear foot pressure conduit. The work effort included cleaning and video inspection, concrete repair.

Emergency Action Plans.

Updated Emergency Action Plans (EAPs) were developed for each levee system and distributed to the appropriate action teams to include fire, police, homeland security and others. These plans identified evacuation routes and structures of particular concern, such as schools and senior citizen facilities. Specific flood levels were defined and safety precautions and protocols required for each level were developed

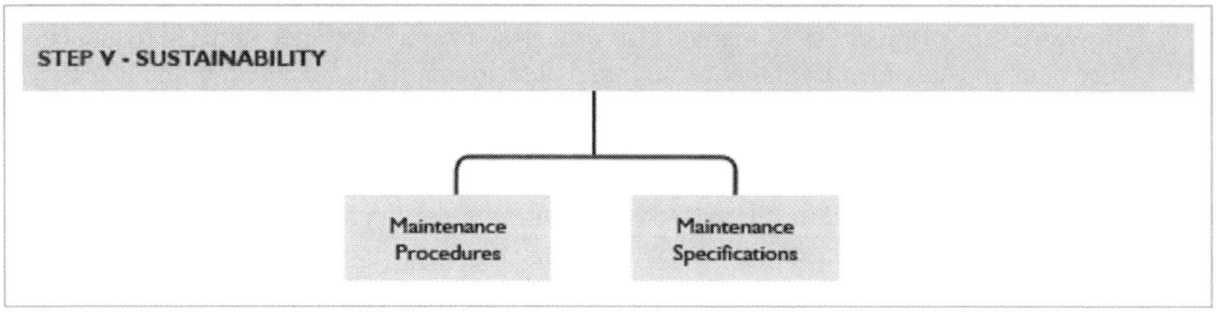

Figure 6. Phase V – Sustainability.

Maintenance Procedures

Established maintenance procedures to insure all inspection, maintenance efforts are performed and documented in accordance with USACE guidelines and recommendations.

Maintenance Specifications

Developed the requirements for contracted maintenance services needed to maintain and document inspection and testing performed on the levee, pumping stations, sluice gates, flapper valves and trash racks in accordance with the updated USACE requirements. This included the development of detailed exhibits, maps and forms to support the maintenance, testing and inspections.

OTHER RELATED ACTIVITIES

Maintenance of Traffic

Designed plans for construction access points, flagging operations and trail closures for the many access points to provide safe maintenance of traffic operations.

Automated Trash Rack

Evaluated the existing manual bar screen structure and initiate the development of structural design alternatives to replace the existing manual trash rack with an automated monorail system to automate this function. Review included electrical and structural design requirements, access and maintenance needs.

Timber Pile Deflector

Developed a feature to provide protection to outfall sluice gates located within the tidal flood plain during high water events. This feature was also used to provide for a redundant system to field verify and provide a visual confirmation of the flood level during flood operations. Flood Level signs were developed to be installed at appropriate elevations to serve this dual purpose.

Anacostia Flood Warning Network

Coordinated the existing flood gauge system with the operation and emergency action plan to improve the interactions and reactions.

Construction Support Services

Attended pre-bid and pre-construction meetings, reviewed product material lists and shop drawings, responded to contractor Requests for Information, and attended field and progress meetings as requested. Prepared final as-built plans to confirm top of protection elevation was obtained during construction.

CONCLUSSION

Clearly the effort to evaluate, plan, identify and evaluate deficiencies followed by the extensive collaborative effort to correct the deficiencies and complete the rehabilitation and provide for long term sustainability was a huge effort that took five years to complete.

Most levee sponsors would agree that at least some part of this effort is currently needed for their flood risk reduction systems. Although it could easily be overwhelming, the message I would hope to convey is that if you take levee accreditation one step at a time, it is attainable and will most likely be easier than expected. We think it begins with education and information; once the levee owner and the elected officials understand the potential risk, the required support for the efforts needed for the levee evaluation is attainable. Providing the necessary resources to take corrective and preventive action can be accomplished in smaller, more manageable pieces to allow for project programing. The process will not happen overnight, but as the needs are identified they can be funded, resourced and implemented.

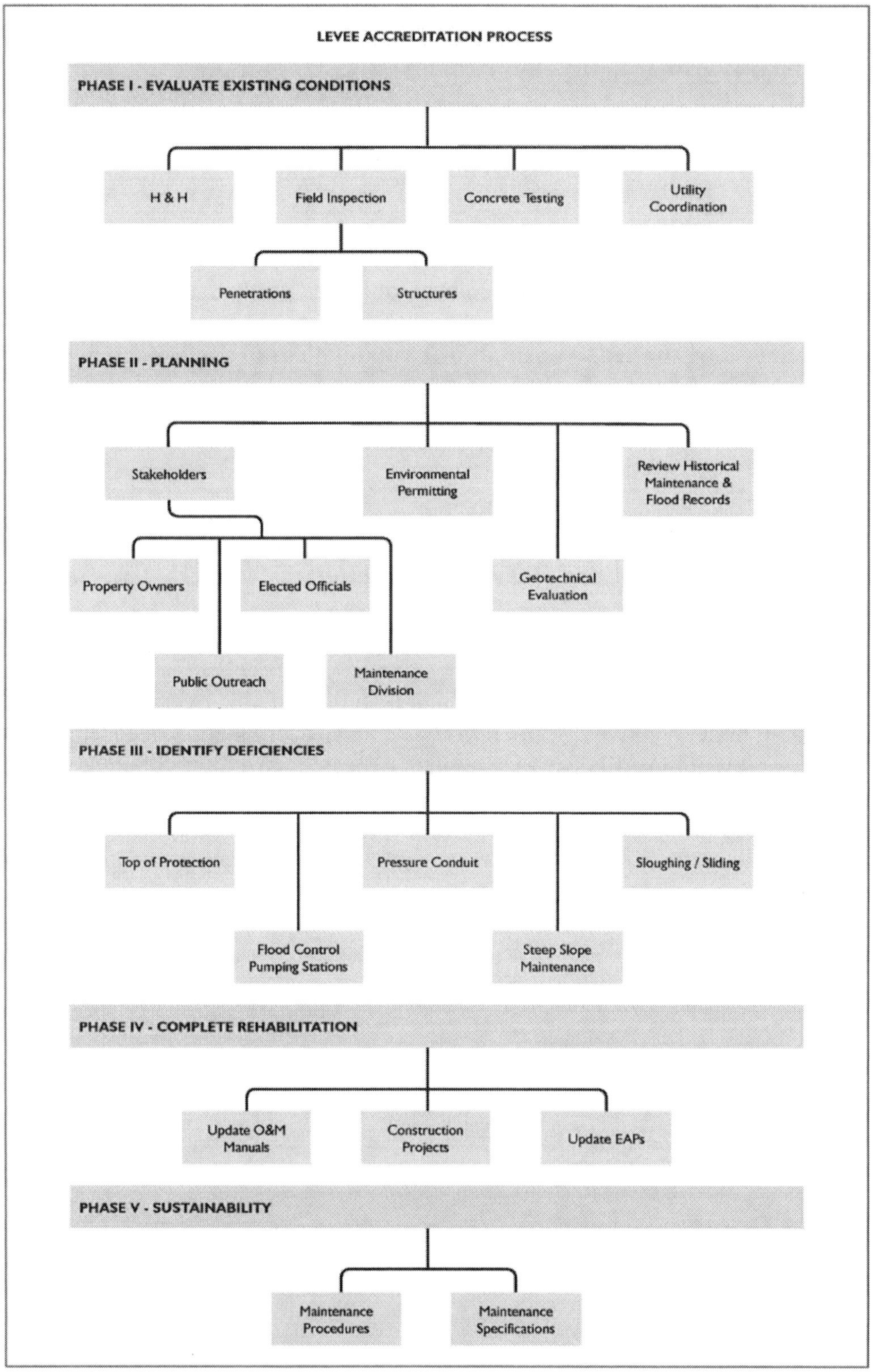

Figure 7. Levee Accreditation Process.

BEST PRACTICES FOR SEEPAGE DATA COLLECTION DURING LEVEE FLOOD FIGHTS

Cory Williams, P.E.[1]
Noah Vroman, P.E.[2]

ABSTRACT

Inspections are key elements to assessing the physical condition of levee systems. Performing inspections during flood events and non flood events provide insight to the performance of the levee system under different loading conditions. For most levee systems, through seepage and under seepage related levee failure modes are significant risk drivers. Thus, performing inspections and recording seepage issues during flood events is essential for making informed engineering assessments of levees. During the historic flood of 2011 on the lower Mississippi River, a significant effort was made to accurately record performance data such as quantity of seepage and location of sand boils. Several methods for the collection of seepage data were utilized during the flood event with various levels of success. The resulting reports of seepage observations provided a wealth of information about the event, allowing the US Army Corps of Engineers in the Mississippi Valley Division to prioritize levee repairs and providing insight to future levee performance. This paper summarizes lessons learned for recording seepage issues during the Mississippi River flood of 2011. In addition, this paper discusses best practices of field collection and report preparation such as using standard terminology for seepage and sand boils, utilizing air and land inspection techniques and collecting data with a standardized format that is compatible with GIS applications.

INTRODUCTION

Routine and periodic inspections are key elements to assessing the physical condition of levee systems. However, these inspections are normally performed during low water or "dry" conditions. For most levee systems, through seepage and under seepage related levee failure modes are significant risk drivers. Inspections during flood events allow for observations of seepage conditions. Proper recording of these observations are key elements in making informed assessments of the levee system.

During flood events, inspections may be performed concurrently with flood fighting activities. Those of us with the responsibility for flood fighting levees do a good job recognizing issues during flood events and resolving them the field. However, once the flood is over and the crisis has ended, the notes and observations either stay in our desk drawer or in our head, never to be shared with others. Without written documentation and a system to keep it, these precious data are lost over time.

[1] Chief, Geotechnical Engineering Branch, U.S. Army Corps of Engineers, Memphis District. 167 N. Main St. Memphis, TN, 38103, cory.h.williams@usace.army.mil
[2] Director, MVD Dam and Levee Safety Production Center, Vicksburg District, 4155 East Clay Street Vicksburg, MS 39183, noah.d.vroman@usace.army.mil

Without the benefit of recorded observations, engineers must rely exclusively on calculations and judgment to make decisions. Although our understanding of levee performance is increasing, calculations alone do not provide enough information to predict future problems or assess the physical condition of the levee. The best indicator of future levee performance is the levee's past performance during flood events.

Levee performance data gathered during a flood event informs future levee inspections. Because the condition of levees changes over time due to erosion, animal activity and flood loadings, it is necessary to perform regular inspections to identify deficiencies, verify that the levee is properly maintained and inform the stakeholders about the overall levee condition. Although most levee deficiencies can be observed at any time during the year, some issues such as seepage in levees and floodwalls and sand boils can be observed only during a flood event. Having a written flood fight record is the only way to accurately capture seepage and sand boil issues.

Flood fight data also informs levee risk assessments. A levee risk assessment evaluates the likelihood of all of the potential failure modes for the levee system. The risk assessments are used to inform interim risk reduction measures and communicate risk to stakeholders and the public. Generally, the assessment considers three components of risk; (1) the hazard (flood loading), (2) the system response (how the levee is expected to perform), and (3) the consequences (people and property in the leveed area). Although the hazard and consequence can be determined using other means, the system response cannot be estimated accurately without a record of flood performance data. Without a thorough assessment of all three components of risk, the risk assessment is incomplete.

During the historic flood of 2011 on the lower Mississippi River, a significant effort was made to accurately record performance data such as seepage/sand boil locations and severity. Many different methods were used to record these issues observed during the event. This paper will discuss these methods and evaluate which were the most effective.

STANDARD TERMINOLOGY

One of the most important methods of communicating a problem is using standard terminology to describe the issue. If the language used to describe problem does not have a clear definition, then listener can only use his own experience to imagine the scope of the issue. For example, if someone discovered a sand boil and described it as "large", the observer and the listener may have two different ideas of what a "large" sand boil looks like. Most likely, both are comparing that description to the largest sand boil that each one has ever seen in their own respective experiences. When preparing to document issues during a flood event, all inspectors must use consistent flood fight terminology. The three most important definitions that must be established before any flood event include seepage and sand boils.

Seepage

The amount of seepage observed during an event is an excellent indicator of the potential for the development of sand boils in the future. The entire length of levee should be evaluated just after the flood crest. Seepage is generally measured using an estimate of the quantity of seepage water emerging from the ground. These definitions are recommended for use in future events so that consistent comparisons can be made, which will provide insight if conditions are changing over time.

- Impounded water – Area covered in water but unable to determine source.
- No seepage - No evidence of saturated areas
- Very light seepage - Saturated areas only (no flow)
- Light Seepage - Ground surface is saturated with seepage flowing slowly out of the ground surface as a uniform thin sheet seepage flow.
- Medium Seepage - Seepage over large areas flowing into collecting streams on the surface (No sand boils)
- Heavy Seepage - Seepage over large areas flowing into collecting streams with sand boil activity

Figure 1, Example of Heavy Seepage.

Almost as important as recording the quantity of seepage in a given levee reach is recording the absence of seepage in a given reach. In some cases, areas have no written documentation that indicates the amount of seepage observed. To future readers, the lack of description either means there was no seepage observed in the area or the area was not evaluated at during the flood. So that there is no doubt about what was observed, the words "no seepage" should be used in areas where no saturated areas or seepage is present.

Seepage Data Collection

Sand Boils

A sand boil is a cone of sand and water that is the result of internal erosion of foundation sand due to uncontrolled and filtered seepage emerging from the foundation. Because internal erosion can degrade the foundation of levees, sand boil activity should be considered very dangerous. Thus, all sand boils should be documented. In order to document properly, sand boils should be differentiated by their size. The tendency in the field is often to describe the severity of a sand boil by measuring and recording the size of the sand cone formed by the sand boil (measured as the diameter of the sand boil foot print) and the diameter of the flow channel through the deposited sand cone. Although the size of the sand cone is an important indicator of the amount of material that has been removed from the levee foundation to date, the future risk of internal erosion and levee breach can best be described by accurate measurements of the size of the channel extending beneath the sand cone, through the subgrade soils and consistent estimates of the rate of the flow emerging from the sand boil. The flow channel that extends through the subgrade soils is termed the sand boil throat. Sand boils with larger throat diameters may have a higher potential to cause internal erosion of the foundation or represent a more matured state of the development of the failure mode. The size of the boil may be consistently measured in the field by inserting one's hand in the boil and comparing the diameter to the number of fingers that can be inserted. The following definitions were used during the 2011 flood event and recommended for use in future events. (See Figure 2)

- Pin Boil – Throat size less than 1/2-inch in diameter (less than 1 finger width wide)
- Small – Throat size 1/2-inch to 2-inches in diameter (up to 3 finger widths)
- Medium – Throat size 2-inch to 4-inches in diameter (3 fingers to Fist)
- Large – Throat size 4-inch to 6-inches in diameter (Fist and extended thumb)
- Very Large - Throat size greater than 6-inches in diameter (Larger than fist and extended thumb)

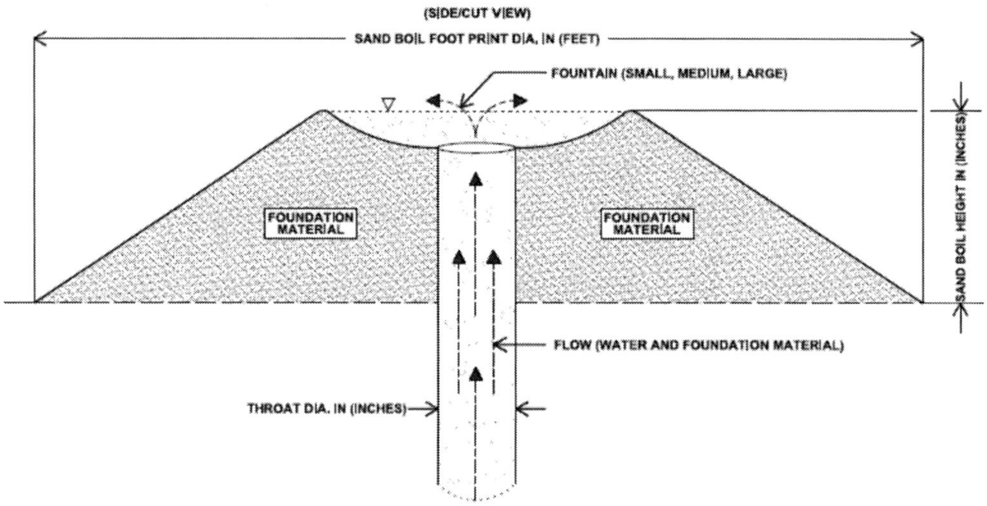

Figure 2. Sand Boil

COLLECTION METHODS

There are several methods of collecting seepage data and their usefulness varies depending on a number of factors including the size of the levee system impacted by the flood event and the duration of flood event. This paper focuses on the collection methods explored during the 2011 Mississippi River flood and includes aerial, land based, and mobile based methods. Advantages and disadvantages of each method are discussed below.

Aerial Documentation Methods

A number of air based video systems are available to monitor and document flood activity. In general, these cameras attach to aircraft and are monitored and controlled in the cockpit. The main advantage of using this equipment is that it allows us to document hundreds of miles of levee per day. Additionally, seepage may be observed in inaccessible areas such as areas of impounded water and areas surrounded by thick vegetation. Most of these cameras also record the geographic position of the aircraft using global positioning systems (GPS). Because of the relative high cost of acquiring aircraft and the large data sets resulting in the video collected, the number of flights during a flood event is generally limited. Ideally, the flights should be performed just at or with 24 hours of the crest of the flood event.

Land Based Methods

Land based collection methods are the most important method of collecting data. Individual sand boils are easily overlooked from the air or from a vehicle riding along the levee crown. Walking areas exhibiting medium to heavy seepage or near areas of impounded water is essential to discovering sand boils.

The most obvious sign of sand boil activity is flowing water near the landward side of a levee. Although flowing water is most obvious in areas where surface waters are well drained, sand boils often occur in areas of impounded water and may be observed as a rippling surface or cloudy/discolored area surrounded by clear water. Additionally, sand boils often produce tan/light brown foam on the surface of the water. This foam is generally the result of the movement of trapped air in the ground mixing with soil particles as it is pushed out by seepage pressures. This foam typically lingers for days during a flood.

Generally, potential seepage areas should be patrolled just before sand boils develop. The day, time, location and hydraulic loading at the time of discovery should be recorded for all sand boils. Generally, land based methods can generate large amounts of data and this data is normally stored in a database. As the flood progresses, changes in size of the sand boil, amount of sand transported to the surface and flood fight technique (i.e. sand bagging) performed should be documented. This information assists the geotechnical engineer in establishing appropriate boundary conditions needed for initiation of a sand boil, confirms geologic conditions at a site and is pertinent information needed for predicting future levee performance.

MICA (Mobile Information Collection Application)

In the past, many method of documenting levee performance issues have been employed. Before the widespread use of GPS equipment, topographic maps were marked with notes indicating the location of problems. Occasionally, field notes were maintained by individuals, but these notes were not always provided to those preparing routine levee inspections or passed down to the next generation of first responders. In the last few years, first responders are sent out with paper forms, a regular digital camera, and a handheld GPS device. They are asked to "write down" the coordinates of the location. Once they get back to hotel or the regional field office (RFO) ,the recorded data required a large amount of effort to transfer to an appropriate database.

MICA stands for Mobile Information Collection Application (MICA). This software is an android based application that incorporates the mobile device's gps, camera and internet capability to capture data in the field and communicate it with the command center. With the MICA data collection system, tablets or phones are deployed with the mission specific categories of data already pre-loaded into the application. There is no need for other equipment since the tablet and phone all have built-in GPS and camera capabilities.

MICA was developed and is maintained by the U. S. Army Corps of Engineers (USACE) Engineer Research and Development Center (ERDC), Information Technology Laboratory in Vicksburg, Mississippi. Because of the flexibility of the application, MICA can be used in many different kinds of first responder missions within USACE. MICA has assisted the U. S. Army Corps of Engineers' (USACE) response to debris and blue roof missions, bridge evaluations and flood response.

The general procedures for using MICA are that the user must "sync" their mobile device with the server by selecting the mission from the mission list. Once the device synchronizes, all the folders to store specific data that were created via the MICA website are automatically added to the tablet or phone. Field personnel then go into the field and if they come across a levee performance issue, they simply press a button on their phone or tablet. This will turn on the GPS and capture the latitude and longitude of the issue (where they are standing) and then open a screen that allows them to take photos, videos, text notes, sound or voice, a sketch, or fill the previously created form. Once they collect this information, that point is saved on the device and synced back to the MICA website for immediate viewing by the mission commander. Near-real-time access of the field data is available to the decision makers.

MICA can be used in remote areas where cell phone service coverage is limited. No internet access is required to collect data. Field personnel can collect data without internet access (4G or Wifi). The data is stored locally on the tablet or phone, and once an internet connection is established the data can then be synced to the MICA server. Users can collect as many photos, videos, forms, etc for each point as necessary. Obviously the more data the user collects, the longer it will take to sync.

Memphis Districts used the MICA system and paper forms side by side during the 2011 Mississippi River flood event. It has been our experience that paper forms have an increased risk of errors in transcribing the GPS position from the device to the paper and also again from the paper form to a database. Other weakness noted in paper forms is the issues with associating the correct photograph to the data point. By comparison, the use of MICA system allowed quicker and more accurate data collection than the use of paper forms. MICA saves a significant amount of time during collection by automatically associating photographs and GPS position when collecting points. This is important not only because it makes the job of the first responder easier, but also we discovered that notes taken were more complete. It is our experience that more effort required to record information, the less detailed the information becomes.

LEVEE PERFOMANCE REPORTS

The air mounted video equipment technology and MICA discussed above provides us with extremely valuable tools that will allow us to better to document individual problems during a flood event. However, the data from these tools are not sufficient to fully document the overall performance of the levee. There must be a concerted effort to pull all of the data together into a single report that summarizes all activities and problems encountered during the flood event. This report should not only include all data points collected but more importantly the observations and conclusions of those who experienced the flood. The summary report should contain the following key pieces of information contained in the following paragraphs.

Hydrographs & Flood History

Since all data collected will be tagged with a date and time stamp, the report should contain a hydrograph within the report that relates the river stage with the date and time of the observation. Graphs like the one shown in Figure 3 satisfy this purpose in a very simple and concise way. Because gages may be located miles from specific levee sites, equations should also be provided that relate river stage readings to specific levee miles.

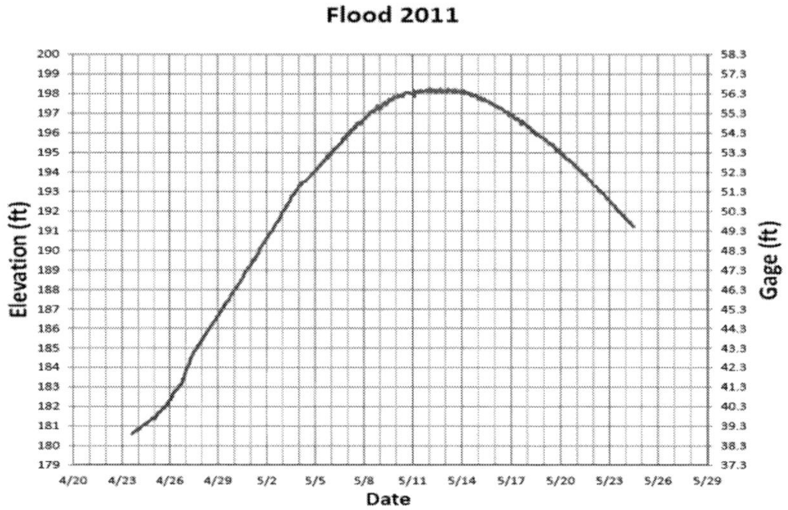

Figure 3. Example Hydrograph during a Flood Event.

Additionally, the historic loading of the levee may be useful to estimate the levee's future performance. Illustrations like the one in Figure 4 below are also useful in comparing the magnitude of past floods.

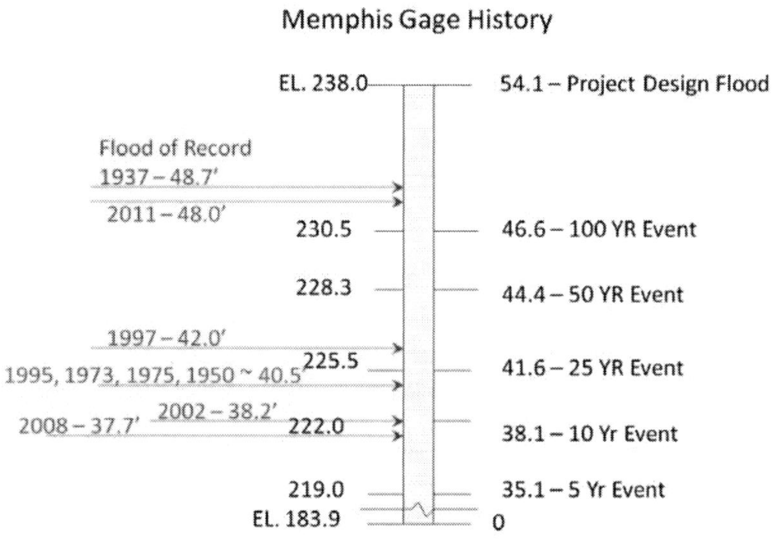

Figure 4. Example River Gage History Illustration

Instrumentation Data

Instrumentation such as piezometers for levees, flood walls and pump station are installed to indicate the pressures that develop in the foundation when the levee is loaded by a flood. Because there are many activities occurring during a flood, the people placed in charge of patrolling levees are sometimes reluctant to read and record piezometer levels. Although it is important to read piezometers regularly throughout the year, the most important time to read a piezometer is during the flood event. As shown in Figure 5 below, the most significant points on the graph are at the highest levee loadings.

Some planning is required prior to event in order to make the piezometer readings accurate. First, all piezometers should be cleared of vegetation and clearly marked by number on first day of the flood fight. All supplies and equipment such as piezometer readers, connectors and extension pipe should be inspected. Upon the first reading, pertinent data such as bottom depth of piezometer, top elevation of piezometer, geographic coordinate should be verified and compared to the installation records.

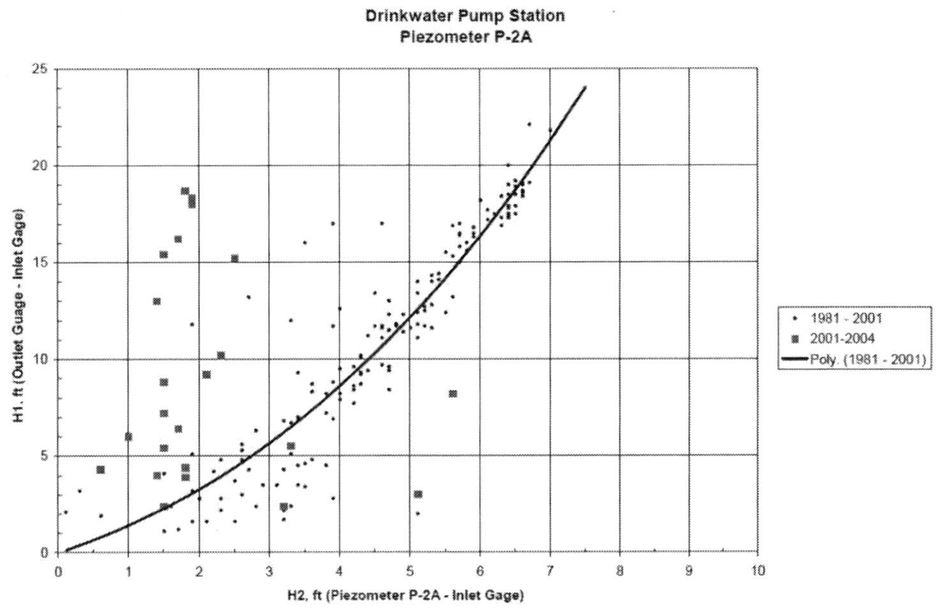

Figure 5. Example Piezometer data.

Photographs

Photographs and videos are essential to any flood report. All photographs should be accompanied with a short description of the subject in the photo. Although we have all heard that a picture is worth a thousand words, without an accompanying description, it may also be interpreted a thousand different ways to those who were not present when the picture was taken. The flood fight team is responsible for adequately describing the photographs and their significance in the final report.

Brief training sessions on taking photos should be given before flood fighting activities. The quality of the photographs is extremely important to the quality of the subsequent report. The examples in Figure 6 and Figure 7 below are good and bad examples of photographs taken during a flood. Figure 6 is similar to hundreds of photographs taken during the 2011 flood event. Some of the problems with this photograph are that there is no date stamp or GPS position on the image itself and is therefore indistinguishable between the other photographs just like this one. Additionally, there are no other objects in the picture to indicate the scale or relative size of the sand boil. A ruler, boot, hand or any other common object should be inserted in order to provide scale. Figure 7 is a better example of a photograph of a sand boil. This picture includes a time, date and GPS position stamped on the image. Additionally, a person and sandbags are in the image to provide scale.

Figure 6. Sand Boil Photograph (Bad Example)

Figure 7. Sand boil Photograph (Good Example).

Maps

As noted in Figure 8 below, when individual sand boils locations are shown on a map, patterns become more evident. These patterns help identify the scope and scale of the problem and provide hints about the geologic conditions that exist near problem areas. For example, the map shown in Figure 8 indicates a distinct line of sand boil activity extending at a slight angle perpendicular to the levee. Although this area was well know to have seepage problems in past flood fights, the line of sand boils were obscured by the trees in the wooded areas and houses in the residential areas. Only after each observed sand boil was collected and shown on a map, was the pattern evident.

Figure 8. Example Flood Map with Seepage and Sand Boils (USACE-MVM 2011)

Descriptions of Issues by Levee Station

Photographs, maps and water data are essential to creating a useful flood fight report. However, the most important piece of information that can be provided in a flood report is the written description of the levee performance by those who observed the event. Geologic conditions beneath the levee can change gradually along a levee system, and the seepage and related remediation activity can be consistent within reaches of hundreds or thousands of feet at a time.

Seepage observations should be reported and grouped into reaches along the levee.. Integrated tools that display levee stations with GPS positions allow the observer to identify their location as they patrol areas of levee. The activity and performance of every foot of levee should be described using standard terminology as described in previous paragraphs.

In areas where no problems were observed, the words "no seepage observed" should be used to describe that reach. Likewise if the area was inundated with water, but no problems were observed, then that reach should be described as "area inundated with water, no issues were discovered during or after the event". The absence of such a statement can be interpreted in several ways by future readers. Obviously, a future reader may interpret the lack of data to mean that there were no significant issues in that reach. However, the lack of any statement leaves some doubt that the area was patrolled at all or if any findings were properly recorded. Including a definitive statement that "no seepage observed" leaves no doubt about the levee performance in that reach.

In the days following the historic 2011 flood event on the Mississippi River the geotechnical engineers assigned to flood fight the Mississippi River levees prepared summary reports as described above. The reports included all of the key elements such as photos, hydrographs, and maps that have been described in the preceding paragraphs. Almost immediately after the completion of the reports, requests for assistance came from all levee districts affected by the flood. These flood fight reports became the basis for prioritizing repairs. Because the reports included standard terminology regarding the quantity of seepage and size of sand boils, the severity of the problem areas were easy to compare between levee districts no matter who authored the flood report. Photographs and related descriptions allowed ample justification for proposed repairs. An example of this narrative is included as Figure 9 below.

<u>Levee Mile 2 - (Sta. 2/00+00 to 2/52+97):</u>

- **2/00+00 to 2/24+00** -No evidence of sand boils was observed in this area. Water was stacked up behind the Mounds Creek Culverts to elev. 318.5 ft. submerging most of the levee toe and making it very difficult to see smaller boils. It was impossible to differentiate between seepage and backwater.
- **2/24+00 to 2/45+60** – Heavy seepage was observed in the sump area for the north pump station in Mound City. One large high-energy boil found on May 9. The boil was located between the Ohio River levee toe and a railroad embankment maintained by ADM Grain. A roughly 50 ft. long section of the railroad embankment had settled about 1 ft. on the side facing the sand boil. ADM Grain said that they had experienced a similar failure roughly 1 year before and had placed large riprap at the toe of the railroad embankment to try to correct the problem. The pump station was shut off to try to raise the water level in the sump and slow the boil down, but there was no gage in the sump to determine the exact water elevation. This boil is especially significant because this area is protected by the Mound City Slurry Trench Parcel 1. An inspection should be conducted during the low water period to evaluate the integrity of the slurry trench and clay cap in this area.

Sand boil in Mound City north pump station sump area. Riprap from previous repair visible behind boil. LAT 37.090832, LONG -89.155116

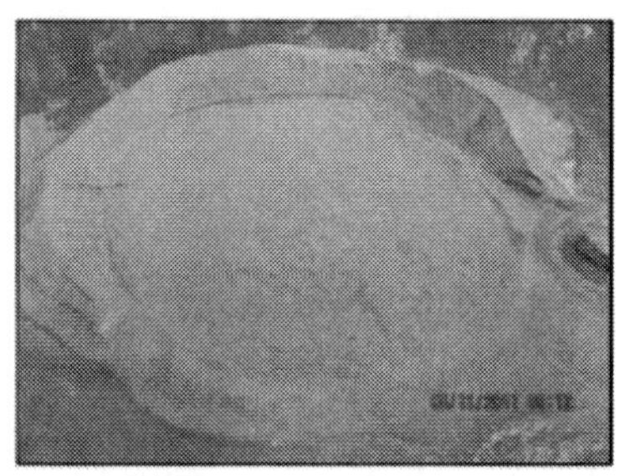

Close up view of sand cone. Throat was roughly 12 in. in diameter and piping coarse sand. Water in sump was about 3 ft. deep when photos were taken.

Figure 9. Example Report Narrative with Photographs from the 2011 Mississippi River Flood Report. (USACE-MVM 2011)

DATA STORAGE

Once all of the data has been collected, it must be accessible for anyone interested for years to come. Multiple electronic storage systems exist that may be used to retain the data. The National Levee Database (NLD) is the single source for the stakeholders to become informed and evaluate the risk of living and working behind levees. For that reason, the NLD should be primary repository of all performance data related to levees. Other systems such as HP trim and Project wise is also used by the Corps of Engineers to retain data and documents. Although all of the above electronic systems are quite suffocated, in the author's opinion, few of these systems can replace the reliability of a well organized hard copy file system. For that reason, it is recommended that at least one complete hard copy of flood fight records be maintained by each organization that is responsible for the design and maintenance of a levee system.

CONCLUSION

The U.S. Army Corps of Engineers has a long history of providing assistance to levee owners and taking decisive action during flood events to maintain the integrity of the levee systems that provide flood risk reduction to the public. U.S. Army Corps of Engineers first responders receive excellent training for flood fighting techniques and have access to state of the art technology to support our work. The USACE Levee Safety Program has developed excellent tools to ensure consistent inspection procedures, consistent terms to describe levee maintenance deficiencies, and a system for recording and organizing inspection observations. These same types of tools and consistencies are needed during flood monitoring operations. Developing a disciplined and consistent approach to document, describe, and record seepage performance issues during flood events would enhance the Levee Safety Program.

Many tools, including the MICA system, have been developed recently that allow efficient and effective collection and documentation of seepage performance data. Considering that the Levee Safety Program relies on observations of levee performance to inform levee risk screenings, higher level risk assessments, routine inspections and periodic inspections; there is an increased responsibility to document the levee performance when it is tested the most. To engage in effective risk management principles across the entire nation, a consistent language is imperative to ensure that "significant seepage" recorded in California means the same thing as "significant seepage" recorded in Mississippi. The best indicator of future levee seepage performance is the levee's past seepage performance during flood events. Documentation of past seepage performance is only meaningful if it is recorded in a commonly understood and standardized language.

ACKNOWLEDGEMENTS

We gratefully thank Melissa Mullen for her valuable guidance during the preparation of this paper.

REFERENCES

USACE-MVM (Memphis District U.S. Army Corps of Engineers) (2011) Flood Fight Report for flood of 2011

EVALUATION OF MINE BREAKTHROUGH AT COAL TAILINGS IMPOUNDMENTS

Aaron J. Antell, P.E.[1]
Erik R. Schuller, E.I.T.[2]
Robert M. Shusko, P.E.[3]

ABSTRACT

The evaluation of a mine breakthrough event and potential for impounded coal tailings to flow into mine workings was brought to the attention of the coal mining industry as a result of the Martin County, KY breakthrough event in October 2000. The far reaching effects of this event and other lesser known breakthrough events resulted in design professionals in the coal industry and regulatory agencies, including the Office of Surface Mining Reclamation and Enforcement (OSM), to review the state of practice with regard to evaluating potential for mine breakthrough and flow potential of impounded coal slurry. Recent publications evaluating these issues have addressed identifying potential failure modes and their reliability, generalizing soil behavior issues related to strength of coal tailings, simplified approaches to identifying flow characteristics, and the effects of the impoundment's operational status on in-place coal tailings.

The authors propose that mine breakthrough and the potential of subsequent uncontrolled flow of coal tailings into mine workings can be initiated by seepage and hydrostatic pressure, overburden failure, or a combination of both. Given these generalized scenarios, there are existing methods of geotechnical analyses to estimate the level of safety against uncontrolled flow occurring as a result of mine breakthrough. The state of practice for evaluating flow potential of impounded coal tailings (fine coal refuse) into underground mines is also discussed.

INTRODUCTION

Coal refuse slurry impoundments are some of the largest manmade structures and are prevalent throughout the Appalachia region. These impoundments are generally designed embankments constructed of coarse coal refuse (CCR) that impound slurry consisting of water and fine coal refuse (FCR). Prior to 2000, the primary focus during geotechnical design of these facilities was the stability of the embankment dam and its ability to safely control water from mining operations and surface run off during storm events. The failure of the Big Branch Slurry Impoundment in Martin County, KY on October 11, 2000, in which the tailings impoundment broke into underground mine

[1] Senior Project Engineer, D'Appolonia Engineering Division of Ground Technology, Inc., 275 Center Rd., Monroeville, PA 15146, ajantell@dappolonia.com
[2] Engineer, D'Appolonia Engineering Division of Ground Technology, Inc., 275 Center Rd., Monroeville, PA 15146, erschuller@dappolonia.com
[3] Senior Principal Engineer, D'Appolonia Engineering Division of Ground Technology, Inc., 275 Center Rd., Monroeville, PA 15146, rmshusko@dappolonia.com

workings and ultimately released between 250 to 300 million gallons of slurry into nearby streams, brought mine breakthrough failures into the national spotlight. Identifying and accounting for the potential breakthrough of tailings impoundments into mine workings now receive greater attention during the design, construction, and reclamation of these facilities. The purpose of this paper is to expose the larger dam engineering community to the potential of breakthrough of impounded coal tailings and slurry into underground mine workings. This paper provides a background of documented mine breakthrough events, identifies potential breakthrough exposures and locations within impoundments, summarizes failure modes that could result in breakthroughs, and discusses the authors' view on state of practice for evaluating flow of coal tailings during a mine breakthrough event.

BACKGROUND

The Big Branch Impoundment release in 2000 brought attention to the issue of breakthrough into mine workings, but it was not the first such documented incident to occur at a coal refuse slurry impoundment in the Appalachia region. A prior breakthrough event occurred at the Big Branch impoundment in 1994, but it did not have the environmental impact that resulted from the 2000 event (Michael et al., 2010). Michalek, et al. (1996) summarized coal refuse slurry releases as a result of breakthroughs that occurred at two different facilities in 1996. The breakthroughs that occurred at the Miller Cove Impoundment and the Buchanan #1 Impoundment in Lee County and Buchanan County, VA, respectively, resulted in discharge of FCR slurry into streams located downstream of the underground mine workings (Michalek et al, 1996).

Due to the significant environmental effects stemming from the breakthrough failure at the Big Branch Impoundment, the evaluation of the breakthrough into underground coal mines has become a forefront issue for the Office of Surface Mining Reclamation and Enforcement (OSM). The purpose of the OSM is to ensure that coal mines are operated in a manner that protects citizens and the environment during mining operations, assures that the land is restored to beneficial use following mining, and to mitigate the effects of past mining through reclamation of abandoned coal mines and their support facilities. Recent publications by OSM and their representatives have evaluated mine breakthrough and identify potential failure modes, the reliability of mine maps, generalizing soil behavior issues related to strength of coal tailings, simplified approaches to identifying flow characteristics, and the effects of the impoundment's operational status on in-place coal tailings. Soil behavior issues that are briefly discussed in the publications include rate of consolidation, susceptibility to liquefaction, and thixotropic strength. The nature and type of the loading on coal tailings and the potential effect of seepage conditions during a breakthrough event has received less discussion.

The OSM also recently completed its third oversight study in West Virginia relating to the risk of coal slurry impoundment breakthrough into underground mines (OSM, 2013). The study encompassed the assessment of 15 permitted coal slurry and mining related freshwater impoundments throughout the state and reiterates the ongoing importance regulatory agencies in the Greater Appalachia region have placed on the identification, evaluation, and remediation of potential impoundment breakthroughs into underground mine workings.

EXPOSURE AND LOCATIONS OF POTENTIAL BREAKTHROUGH

The geologic and topographic characteristics within Greater Appalachia, a region known for deep valleys with steep-slopes, often results in multiple locations within a coal slurry impoundment that a potential breakthrough could occur. The presence of multiple mineable coal seams at varying elevations is common and the seams can either outcrop along the valley sides or be present below the valley bottom. Mine configurations within a valley used for construction of a coal slurry impoundment can be identified as one of three configurations: 1) mine workings in a coal seam below the impoundment, 2) mine workings in a coal seam that outcrops at the hillside within the limits of the impoundment, and 3) mine workings in a coal seam that outcrops at the hillside above the limits of the impoundment. Figure 1 presents a schematic of these three scenarios.

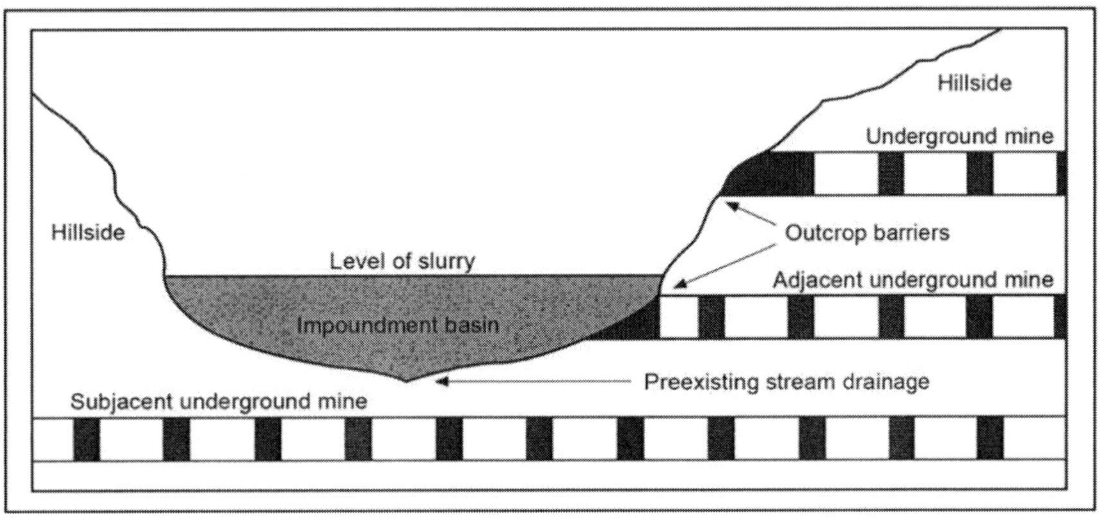

Figure 1. Schematic cross section of impoundment basin and proximate underground mines. (Michael et al., 2013).

Mine Workings Below Impoundment

Mine workings and potential future mine workings in coal seams identified below an impoundment can be susceptible to breakthrough depending on several variables including depth to the mine workings, rock strata between the impoundment and mine workings, and type of mining completed within the coal seam. In general, the likelihood of breakthrough decreases as the depth to mine working increases. The type, thicknesses, and conditions of the rock strata below the impoundment affect the likelihood of a breakthrough into underlying mine workings. Highly fractured rock strata can provide seepage paths for water to travel from the impoundment to the mine workings, potentially initiating uncontrolled seepage into the mine. The use of modern longwall mining techniques versus room and pillar mining affects how the potential for breakthrough is analyzed. For room and pillar mining analyses such as pillar failure, floor failure (pillar punching), and roof failure should be considered. For longwall mining methods, concerns regarding strain and fracturing of bedrock from the induced subsidence should be considered.

Mine Workings Adjacent to Impoundment

Mine workings that outcrop within the impoundment are historically the most likely to present a potential breakthrough due to the proximity and sometimes direct contact with the impoundment. The Big Branch, Miller Cove, and Buchanan #1 slurry impoundment breakthroughs all occurred through mine workings that outcropped within the respective impoundments. Even with updated mine maps that are believed to be reliable, it is imperative to complete a comprehensive site reconnaissance to identify all potential mine workings that will likely become inundated or that will require a barrier as a result of impoundment construction. These workings include outcrops of coal seams that have been mined, coal seams that could be mined in the future, auger mining holes, adits, and any faults in the surrounding strata that may hydraulically connect the impoundment to mine workings below. Mine workings outcropping in the impoundment are generally classified as sealed or unsealed.

Unsealed Mine Workings. Mine workings in coal seams outcropping within the inundation limits of the impoundment that have not been actively mitigated to ensure a proper seal are considered unsealed. Unsealed mine workings can include coal outcrops covered by colluvium, unfilled or reclaimed auger mine holes and portals, and workings backfilled with mine spoil. The lack of designed barriers makes these unsealed mine workings susceptible to seepage and internal erosion as well as an overburden failure as deposition of the water and slurry covers the workings. Especially challenging are any unidentified mine workings adjacent to an impoundment. Even if they do not directly daylight into the impoundment, they may not have a sufficient barrier between the mine workings and the impoundment's inundation limits to prevent a release into the mine. Minimum widths of coal barriers between mine workings and the coal seam outcrops have been previously reviewed and summarized (Kohli & Block, 2007).

Sealed Mine Workings. Following the identification of openings and other discontinuities that may lead to mine workings within the inundation limits of an impoundment, a sealing plan should be developed that is specific to the impoundment configuration, mine workings, openings, and other discontinuities to reduce the potential of breakthrough. There are many possible configurations for post mining sealing of outcrops to prevent failure during exterior loading, such as inundation from an impoundment, as discussed further in MSHA (2009) and Pearson et al. (1981).

Below or Above Drainage Mine Workings

An important distinction that must be made throughout the identification process of mine workings at risk of potential breakthroughs is whether the mine working is "below drainage" or "above drainage". Below drainage is used to describe mine workings that are in a region of a coal seam that does not daylight at the ground surface, and, therefore, does not provide a conduit through which released tailings can exit the mine. Conversely, above drainage refers to a mine in a coal seam that daylights at the ground surface and provides an exit through which released tailings can be discharged above ground. Generally, mine workings that are adjacent to the impoundment are more likely to be "above drainage" than mine workings located below the impoundment bottom. Below drainage mine workings generally will not result in the release of material out of the mine during a breakthrough event, but pose a workers safety risk if the mine is active at the time of the breakthrough.

INITIATION OF BREAKTHROUGH EVENTS

Based on the authors' review of documented impoundments where breakthroughs occurred and other studies completed on mine collapses in the Appalachia region, the authors have identified two generalized modes that can initiate a breakthrough event. These failure modes are seepage failures and overburden failures and are discussed in detail herein.

Seepage and Hydrostatic Pressures

One of the major engineering concerns with mine workings in close proximity to impounded tailings and free water is the potential for uncontrolled seepage into the mine workings. For this discussion, uncontrolled seepage is considered to be water/slurry emanating from an unfiltered exit or with seepage forces sufficient to cause soil migration. Depending on the location of the mine workings in relation to the impoundment elevation, there can be a significant difference in the total head at the surface of the impoundment and the mine void. Hydraulic pathways that may be present between the impoundment and the mine void can potentially lead to internal erosion of the pathway and ultimately uncontrolled seepage into the mine void. This failure mode is generally more likely to occur when the mine void is adjacent to the impoundment valley, but is theoretically possible in mine voids beneath the impoundment as well.

The potential for a breakthrough caused by internal erosion, as was the case in the three breakthrough incidents previously discussed, depends on the following three major factors; the impoundment pool level with respect to the mine workings, the type and condition of material between the impoundment and the mine workings, and the distance between the impoundment and the mine workings. These factors should be identified and addressed when evaluating the potential inundation of new mine workings adjacent to an impoundment as well as for analyses of old or existing mine workings that are already below the pool level of an adjacent impoundment.

From the standpoint of minimizing potential for breakthrough, it is preferred to leave a thick barrier of undisturbed coal between the edge of the mine and the valley outcrop. Since an undisturbed coal seam can be more pervious than surrounding rock, a barrier of undisturbed coal may not be considered adequate to provide a complete cut off for water from the impoundment to the mine workings, but can provide good protection from internal erosion relative to an engineered barrier. This situation, however, represents one possibility and is not always the case. More often than not, there are mine workings, portals, auger mining holes, or adits that extend through the valley sides and daylight into the proposed impoundment limits. For these cases, a barrier between the impoundment and mine working must be constructed to reduce or eliminate negative effects of seepage.

Engineered barriers to prevent seepage from infiltrating are common structures where open access to mine workings are found within impoundment limits and can consist of a combination of compacted coarse coal refuse, engineered soil fills, geotextile fabrics, geomembrane liners, and low permeability clay fills. Engineered barriers are designed to mitigate uncontrolled seepage into the mine and potential internal erosion by reducing the hydraulic gradient between the impoundment and the mine workings and providing a filtered exit to resist particle movement resulting from seepage.

As summarized in Michael (2005), there is debate regarding the mode of failure that resulted in the mine breakthrough at Martin County. Studies of the event indicate the breakthrough is the result of seepage degrading the mine seal. Other studies based on FEM analyses indicate that the scenario of mine seal degradation could not occur and the breakthrough was the result of increased hydrostatic pressure that caused the barrier to fail. Although there are differing opinions regarding the mode of failure, the important point is that both modes of failure can be evaluated using existing methods and tools available to the designer. FEM computer programs are useful tools to calculate pressure heads and seepage flows at various locations. These programs, such as SLIDE, Seep/W, etc., allow the user to easily change parameters such as geometry and material properties, which is useful in analyzing different configurations for use in comparison (Thacker, 2002).

Overburden Failure

Overburden failures consist of the physical failure of the mine roof or the outcrop barriers of the mine under a given vertical loading. The loading on the underlying mine workings consists of the deposition of FCR and water into the impoundment as well as accumulated water as a result of runoff following a rainfall event. Under these loads, an overburden failure can occur in mine workings adjacent to an impoundment and mine workings below an impoundment. Although none of the documented breakthrough events were directly attributed to overburden failures, overburden failures are possible and should be considered during the evaluation of a potential break through event.

Mine Workings Adjacent to Impoundments. When analyzing mine workings located adjacent to an impoundment, the three primary modes of overburden failure are uplift, sliding wedge, and subsidence. Each of these failure modes can be analyzed by identifying the forces involved for a given scenario and a factor of safety against failure can be calculated. Newman (2003) provides a more comprehensive breakdown of the safety factor calculation and the identification of forces involved in uplift and sliding wedge failures. A typical diagram of a mine working outcrop adjacent to an impoundment is shown in Figure 2.

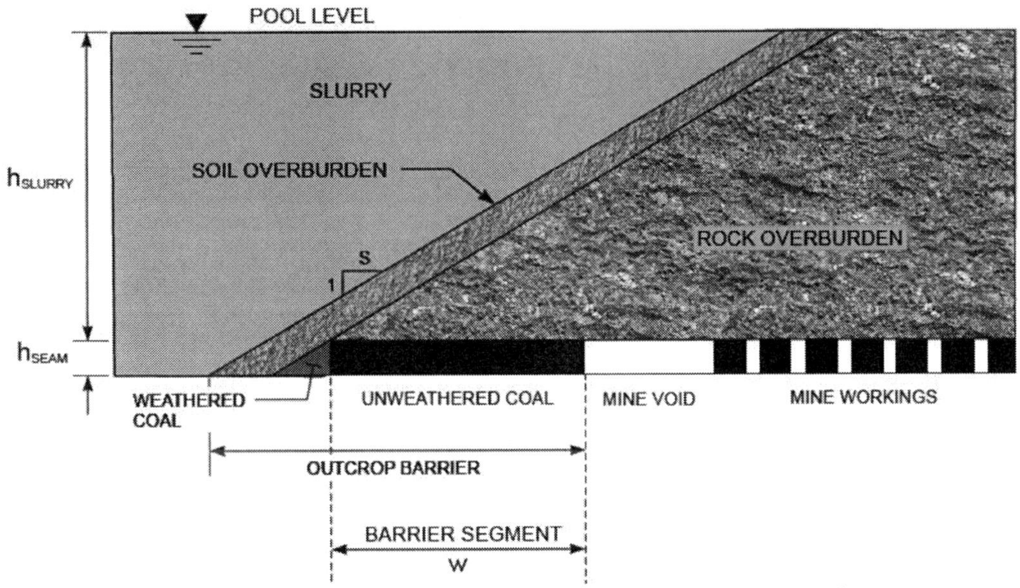

Figure 2. Outcrop Barrier Diagram (MSHA, 2009)

Mine Workings Below Impoundment. Overburden failures of mine workings below an impoundment are geometrically constrained to subsidence type failures. These type of failures include roof failures, pillar failures, and floor failures (pillar punching). Roof stability of room and pillar mine workings are analyzed by assuming a span between two pillars will act similar to that of an elastic beam and, as such, will fail in tension and is based on the strength of the roof material, the overburden, and the length of the roof span. Pillar collapse involves determining the strength of the pillar and comparing it to the overburden pressure present. Provided the pillar is stable, the floor of the coal seam must also be checked against pillar punching. Pillar punching, also considered a floor failure, occurs when the floor of a coal seam cannot resist the overburden transferred through the pillar resulting in the intact pillar punching through the floor slab. Pillar punching failures result in deformations of the coal seam floor and lowering of the mine roof, which results in subsidence. A subsidence failure that occurs in mine workings either adjacent to or beneath an impoundment can result in either a direct connection between the impoundment and the mine workings or the creation of discontinuities that can lead to uncontrolled seepage.

Sudden Failure vs. Progressive Failure. Depending on the depth of mine working in relation to the impoundment and the size of the failure, a failure in mine workings below an impoundment could result in either a direct opening or a discontinuity in overburden between the impoundment and mine. A large failure with a shallow overburden is likely to create a direct connection between the impoundment and the mine workings below such that a release of water or tailings into the mine can occur. A small failure with a thicker overburden that results in subsidence may not result in a direct connection, but provide a weakened zone or discontinuity that may provide cracks through which internal erosion and piping can occur. An impoundment with mine workings only a short distance beneath it is more likely to experience a failure that will directly connect the impoundment and mine workings whereas mine workings that are deep compared to the impoundment are more likely to experience subsidence that will not directly connect the impoundment and mine workings.

Operational Conditions vs. Reclamation Conditions

During the operation of the impoundment, as discussed previously, analyses must account for hydrostatic pressure based on the pool level as well as the overburden due to the settled FCR and water. Following the operational life of an impoundment, the facility must be capped and reclaimed to reduce environmental impacts of the impoundment in its completed state. The load conditions present during reclamation must be taken into account and be analyzed for breakthrough potential before proceeding. General procedures for reclaiming impoundments consists of dewatering the impoundment of free water, placement of cover material over the settled fine coal refuse, and vegetation of the cover material to reduce sediment runoff during rainfall events.

The two major differences in the loading conditions between operational and reclamation configurations are the reduction of water in the impoundment and the additional overburden due to the cover. Because the impounded free water is generally eliminated during reclamation, pore pressures within the FCR are reduced and seepage rates at mine barriers or coal outcrops are reduced. Accordingly, the potential for mine breakthroughs due to seepage and pore pressures should be reduced. As discussed in Michael & Chavel (2008), a big concern during reclamation regarding breakthrough in to mine workings is the increased overburden pressure from the material required to cap the impoundment. In their discussion of environmental risks involved with the reclamation of impoundments, Michael & Chavel (2008) discuss the use of a capping layer of coarse coal refuse or stripping material with a thickness of two to 30 feet and a soil layer with a thickness ranging from one to five feet to support vegetative growth.

In the experience of the authors, the effective thickness of the capping layer can potentially be greater than the planned thickness due to mixing and sink-in of the cap material into underlying fine coal refuse. The mixing and sink-in can be limited by dewatering impounded water prior to construction of the cap; however, the impoundments are often still in operation during much of the cap construction. During the incremental filling or covering of the impoundment cap, material is either immediately supported by the finer coal tailings or displaces and mixes with the underlying coal tailings until the cap material is supported. This achieves some level of strength gain in the upper reaches of the coal tailings under the increased confinement caused by the cap construction. The increased confinement and related strength gain will tend to increase resistance of the FCR to breakthrough.

FLOW CHARACTERISTICS OF COAL TAILINGS

Supplemental to the potential for breakthrough into mine workings is the flowability of impounded FCR. Based on the limited data on FCR flowability and documentation of breakthrough events, there is not an established procedure to determine if coal tailings will flow into a mine given a potential breakthrough event.

State of Practice in Coal Mining Community

The notion and evaluation of in-situ soil, including natural or industrial waste deposits, reaching a static condition that initiates flow type behavior is not new to the conventional and tailings dam communities. However, the evaluation of flow characteristics of coal refuse in a mine breakthrough event has received little focus in the community of civil and geotechnical engineers involved with design and operation of coal tailings impoundments. As discussed previously, the design of mine barriers to prevent breakthrough received some attention prior the Big Branch failure and is included in the Design Manual for Coal Refuse Disposal Facilities (MSHA 2009). The evaluation of subsidence and surface ground movement related to underground mining, although highly empirical, has established procedures and approaches in the coal mining community. However, the evaluation of potential flow characteristics of impounded coal slurry, particularly as it relates to a mine breakthrough event, is new in the coal mining community.

Michael et al. (2005) presents a review of the stability and potential flow characteristics of impounded coal refuse tailings. Based on their review, there is no empirical data regarding the potential flow characteristics of coal refuse. Their review includes discussion of issues that are viewed to relate to potential flow of impounded fine refuse, including consolidation strength, matrix suction, thixotropic strength, and the effects of additives on impounded refuse and tailings. In their discussion of consolidation strength the authors address the effects of permeability, moisture content, rate of consolidation, rate of impoundment construction, apparent low shear strength of coal tailings in existing impoundments, uncertainty in the development of shear strength, and susceptibility of refuse and tailings to liquefaction. Their discussion of susceptibility of refuse to liquefaction includes the potential of static liquefaction resulting from the sudden shear stress induced by mine-barrier breakthrough and mine subsidence. Michael et al. (2013) point out that the liquefaction potential for a material is governed by the relative values of the material's "in-situ density (or void ratio) and effective stress." Michael et al. (2013) also reviewed the use of non-Newtonian flow model Bingham Plastic flow model as potentially applicable to coal tailings.

The subsequent publication by Michael et al. (2010) reiterates similar discussion of factors affecting the potential flow behavior of coal tailings. Of note, Michael et al. (2010) briefly mentions the use of the "collapse surface" approach for evaluation of liquefaction and the use of in-situ testing, including cone penetration testing, standard penetration testing, and vane shear testing for evaluation of seismic behavior of coal tailings. Overall, both Michael et al. (2005) and Michael et al. (2010) provide valuable overview of engineering principles and established methods that relate to the potential flow behavior of impounded coal tailings, but these publications do not recommend a method of analysis or specific criteria for evaluation of flow potential.

In OSM (2011), it is suggested that comparing the measured in-situ moisture content to the liquid limit of the coal tailings can be used as one method to determine if impounded coal tailings are flowable. The concept of using the liquid limit as the boundary between flowable and non-flowable material is subsequently discussed and recommended by Michael et al. (2013). The procedure indicates that the lowest test result for liquid limit can be used as the in-situ dividing line between flowable and non-flowable for samples obtained from the impoundment.

Suitability of Perceived State of Practice

Considering the limited documentation of empirical data and discussion of evaluating the flowability of in-place coal tailings, it is the authors' view that the liquid limit criteria set forth by OSM (2011) can be perceived as the state of practice for evaluating the potential flowability of impounded coal tailings. The basis for this view is that the criterion is set forth in a technical guidance document by OSM, who has significant regulatory stakes in the permitting, operation, and abandonment of coal tailings impoundments. Although OSM (2011) briefly indicates the potential use of "other relevant test" to establish that a material will not flow, it is the opinion of the authors of this paper that when a performance criterion is put forth by a regulatory entity, it is sometimes difficult to use other "undefined" methods to establish suitable performance for plan approval.

As defined in ASTM D4318-10, the liquid limit is the "arbitrarily defined boundary between the semi-liquid and plastic states." Although the liquid limit is somewhat arbitrary, it has been shown to be a strong indicator of mechanical and hydraulic properties of a soil and, accordingly, it is fundamentally part of how the geotechnical community characterizes soils. For example, the liquid limit of naturally occurring soils has been indicated to correspond to a remolded shear strength ranging from 50 to 27 psf with a mean of 36 psf (Wroth & Wood, 1978 and Youssef et al., 1965), where remolded shear strength implies a man-made specimen.

For naturally occurring soils, the liquid limit is generally proportional to the amount of clay particles and indicates the clay particles effect on the overall behavior of the soil (Seed et al, 1964). It is also only applicable to plastic materials. Materials with enough non-plastic particles (silt and sand) do not yield results for liquid limit and are determined to be non-plastic. Therefore, the use of liquid limit and moisture content to determine the flowablity of in-place coal tailings does not encompass all gradations and types of material potentially encountered in a coal tailings impoundment. More specifically, impounded material with larger amounts of fine-grained sand or non-plastic silt cannot be screened for flow potential using the liquid limit test. Due to typical slurry discharge procedures and the depositional environment in the coal tailings impoundments, the coarser non-plastic materials tend to be located near the embankment where mine breakthrough is less applicable and finer materials that are more plastic tend to be in the back of the impoundments where mine breakthrough is more applicable. However, during the early stages of the impoundment when the surface area is typically smaller the coarser, non-plastic materials tend to be more distributed throughout the impoundment and are potentially near areas with exposure to mine breakthrough.

Based on the experience and in the opinion of the authors, the use of liquid limit and moisture content to determine the flowablity of in-place coal tailings is potentially conservative. Table 1 summarizes field vane shear test results, corresponding liquid limits, moisture contents, and calculated void ratios for in-place impounded coal tailings at an impoundment in the Appalachian coal fields. As shown in the table, the water contents range from 4% dry to 5% wet of the liquid limit and field vane shear strengths are notably higher than the mean shear strength values at the liquid limit reported by Wroth & Wood (1978) and Youssef et al. (1965). Also note that the calculated void ratios based on the water contents are approximately two-thirds of that for a naturally occurring soil because of the lower specific gravity of solids. Given the calculated void ratio based on water content near the liquid limit is lower than naturally occurring soils, it is reasonable to expect the shear strength to be higher. A full discussion of the potential behavioral differences of coal tailings and natural soils is not part of the scope of this paper. The results in Table 1 are intended to illustrate the potential for the liquid limit criteria for flowability of impounded coal tailings being conservative.

Table 1. Field Vane Shear Test Results for Impounded Coal Slurry

Boring	Sample ID	Depth (ft)	USCS Symbol	Water Content (%)	LL (%)	PI (%)	Specific Gravity	Void Ratio	Peak FVS, Su (psf)
B-11	OS-25	84.0	CL	34	38	16	1.79	0.61	1050 1150
	OS-26	89.0	CL	35	37	13	1.71	0.60	1060 1280
	OS-27	94.0	ML	34	33	8	1.66	0.56	1370 1330
B-12	S-6	36.0	CL	38	33	13	1.90	0.72	110 150 200
	S-9	52.3	CL	31	37	18	-	-	930 1510 570
	S-11	62.3	CL	41	38	18	-	-	480 320 290

CONCLUSIONS

The Big Branch release in 2000 brought greater awareness to mine operators, design professionals, and regulatory stakeholders of factors affecting the potential breakthrough of coal tailings into adjacent underground mines. Given the potential exposures to mine breakthough, the greatest risks for breakthrough are mines in coal seams that daylight within the inundation limits of a coal slurry impoundment. Mine breakthroughs can initiate due to seepage and hydrostatic pressures at barriers (natural or engineered) between underground coal mines and impounded slurry, or due to overburden failures such as subsidence and related sinkholes or cracks in overburden material.

Based on the authors' review of publications by OSM, the perceived state of practice for evaluation of flow potential of impounded coal tailings is comparing the liquid limit to in place water content, such that material with a water content over the liquid limit is considered flowable. Based on the authors' experience, this criterion may be conservative and cannot be applied to all material found in coal tailings impoundments. The authors consider the liquid limit criterion a screening level criterion that should be used with other established methods for evaluating flow and static liquefaction potential. The authors understands that OSM has funded a project that will perform standard engineering tests and employ small-scale flow models to help assess flow behavior of impounded coal tailings under static and dynamic loading. This project is scheduled for completion in 2015 and will provide valuable data to help determine if the liquid limit criterion is applicable and what established methods can be employed for evaluation of flow potential of impounded coal tailings in a breakthrough situation.

REFERENCES

Pearson, M. L., Preber, T., & Conroy, P. J. (1981). *Outcrop barrier design guidelines for Appalachian coal mines. Open file report (final) Jul 79-Feb 81*(No. PB-82-113838). Dames and Moore, Park Ridge, IL (USA).

Kohli, K.K., & Block, F. (2007). *Guidance Manual: Outcrop barrier design for above drainage Coal Mines.* U.S. Department of the Interior, Office of Surface Mining and Reclamation and Enforcement, Washington, DC.

Michael, P.R., & Chavel, L. (2008). *Environmental risks associated with coal refuse impoundment reclamation: An assessment of the possibility of an underground mine breakthrough occurring as a result of the impoundment reclamation process.* U.S. Office of Surface Mining. Office of Surface Mining Appalachian Region Management Council Open-File Report, Pittsburgh, PA.

Michael, P.R., Murguia, R., & Kosareo, L. (2005). *The Flowability of Impounded Coal Refuse: A Review of Recent Work and Current Ideas in the Engineering Profession.* US Office of Surface Mining Appalachian Region Management Council Open-File Report, Pittsburgh, PA, 25.

Michael, P.R., Richmond, M., Lane, D., & Superfesky, M. (2013). *Preventing Breakthroughs of Impounded-Coal-Waste-Slurry into Underground Mines.* 2013 West Virginia Mine Drainage Task Force Symposium Papers, Morgantown WV, March 26-27, 2013.

Michael, P.R., Richmond, M.W., Superfesky, M.J., Stump, D.E., & Chavel, L.K. (2010). *Potential breakthroughs of impounded coal refuse slurry into underground mines.* Journal of Environmental and Engineering Geoscience, 26(3), 299-314.

Michalek, S.J., Gardner, G.H., & Wu, K.K. (1996). *Accidental Releases of Slurry and Water from Coal Impoundments Through Abandoned Underground Coal Mines.* U.S. Department of Labor, Mine Safety and Health Administration. Safety and Health Technology Center, Pittsburgh, PA.

Mine Safety and Health Administration. (2009). *Engineering and Design Manual: Coal Refuse Disposal Facilities.* U.S. Department of Labor. Prepared by D'Appolonia Engineering Division of Ground Technology, Inc.

Newman, D.A. (2003). *Rock Mechanics and the Analysis of Underground Mine Stability Adjacent to Coal Refuse Impoundments.* Proceedings of the 22nd International Conference on Ground Control in Mining. West Virginia University, Morgantown, WV.

Office of Surface Mining Reclamation and Enforcement. (2011). *Technical Position Paper: Potential of impounded-coal-waste-slurry breakthrough into underground mines.*

Office of Surface Mining Reclamation and Enforcement. (2013). *Oversight Report: Coal Slurry Impoundment Breakthrough Potential (Phase III).* Charleston, WV.

Sames, G. P., & Moebs, N. N. (1989). *Hillseam Geology and Roof Instability Near Outcrop in Eastern Kentucky Drift Mines.* Report of Investigations 9267. U.S. Department of the Interior, Bureau of Mines.

Seed, H.B., Woodward, R.J., & Lundgren, R. (1964). *Fundamental aspects of the Atterberg Limits.* Journal of Soil Mechanics and Foundation Divisison, ASCE, 90(6), 75-105.

Thacker, B. K. (2002). *Big Branch slurry impoundment breakthrough: Why it happened and lessons learned.* Proceedings, Tailings Dams, 47-58.

Wroth, C. P., & Wood, D. M. (1978). *The correlation of index properties with some basic engineering properties of soils.* Canadian Geotechnical Journal. 15(2), 137-145.

Youssef, M. S., El Ramli, A. H., & El Demery, M. (1965). *Relationships between shear strength, consolidation, liquid limit and plastic limit for remoulded clays.* In Proceedings of the 6th International Conference on Soil Mechanics and Foundation Engineering. Montréal. Vol. 1, 126-129.

FACTORS AFFECTING UPSTREAM CONSTRUCTION AT COAL TAILINGS IMPOUNDMENTS

Robert M. Shusko, P.E.[1]
Aaron J. Antell, P.E.[2]
Andrew J. Antell, E.I.T.[3]

ABSTRACT

The evaluation of staged embankment construction overtop weak foundation conditions has long been part of geotechnical and dam engineering practice, but has historically not been a substantial component of the design and analysis of coal tailings impoundments. Due to operational and production related constraints of coal mines and preparation plants, the construction of coal tailings dams and impoundments often require the use of some variation of the upstream construction method to meet freeboard and disposal capacity requirements. In light of recent notable upstream failures at coal tailings impoundments such as the Nolan Run Slurry Impoundment failure, the undrained analysis of upstream embankments constructed over settled, fine coal tailings is receiving more attention as part of the evaluation of such impoundments (MSHA, 2014). Although the methods and approaches used for staged-construction analyses of upstream embankments at impoundments is similar to conventional situations, there are unique features that exist at coal tailings impoundments that affect the staged-construction analyses of upstream embankments.

There are several features of the operation and design of coal tailings impoundments that affect the construction of upstream embankments and the stage-construction analyses. Recent environmental regulations has led to more restrictive requirements including the use of geosynthetic lining systems. Facilities operate to limit discharge of coal contact water, which has led to the use of tailings impoundments for storage of water resulting in an increase of free water level above the tailings. There is limited availability of typical and site specific characterization of undrained shear strength and rate of pore pressure dissipation characteristics of coal tailings. The potential exists for the specific gravity of coal tailings to be 30% to 40% lower than the specific gravity of naturally occurring soils.

The features and issues listed above are being addressed through design features and implementation of additional exploration and material characterization programs. The results of these programs serve to provide a better understanding of the behavior of coal tailings during staged construction loading associated with upstream embankment construction.

---

[1] Senior Principal Engineer/President, D'Appolonia Engineering Division of Ground Technology, Inc., 275 Center Road, Monroeville, PA 15146, (412) 856-9440, rmshusko@dappolonia.com
[2] Senior Project Engineer, D'Appolonia Engineering Division of Ground Technology, Inc., 275 Center Road, Monroeville, PA 15146, (412) 856-9440, ajantell@dappolonia.com
[3] Assistant Project Engineer, D'Appolonia Engineering Division of Ground Technology, Inc., 275 Center Road, Monroeville, PA 15146, (412) 856-9440, antellaj@dappolonia.com

INTRODUCTION

Coal tailings impoundments are constructed to dispose of waste materials produced at preparation plants that process and clean raw coal to generate a coal product meeting the customer's specifications. Coal preparation plants produce two types of waste material; coarse coal refuse (CCR) and fine coal refuse (FCR). CCR typically consists of a well graded mixture of coarse sand and gravel size particles with some fine-grained particles and cobble sized particles. CCR is normally generated from the mechanical processing of the raw coal to extract the coarser fraction of coal and consists of the more durable portion of the refuse that does not readily break down. CCR's specific gravity is normally much higher than that of the coal and can be separated from the coal fraction with less processing. FCR typically consists of fine sand, silt, and clay size particles and is normally generated by processing of the finer portion of the raw coal. In most instances, the finer portion of the raw coal also contains less durable rock such as claystone and carbonaceous shale that have lower specific gravities as compared to the coarser portion. Because the specific gravity of the finer portion is closer to the specific gravity of coal, additional processing is required to extract the finer portion of the coal. The refuse generated by the additional processing comprises the majority of the FCR has a lower specific gravity as compared to the CCR.

Coal tailings impoundments are often staged embankments constructed of CCR with hydraulically placed FCR pumped upstream of the CCR embankment. The CCR staged embankments are constructed using either upstream, centerline, or downstream construction methods. The upstream construction method, or some variation thereof, is often required to accommodate operational and production related constraints of coal mines and coal preparation plants or geometric constraints of the site.

During upstream construction of coal refuse staged embankments, CCR is pushed out over top of hydraulically placed FCR. Prior to CCR push outs, the hydraulically placed FCR is under to normally consolidated under its self-weight and has high water contents. When CCR push-outs are constructed in vertical increments using low ground pressure bulldozers, excess pore pressure in the FCR can develop from the loading of the CCR push outs. The excess pore pressures would result in undrained conditions in the FCR push out foundation. If the push out is constructed quicker than the development of undrained shear strength there is risk of embankment failure and potential unsafe work conditions for construction equipment operators. The evaluation of staged embankment construction overtop potentially weak foundation soils in undrained conditions has long been a part of geotechnical and dam engineering, but has not historically been a substantial component in the design of coal tailings impoundments. Due to recent notable upstream failures at coal tailings impoundments, the undrained analysis of upstream embankments constructed over FCR is receiving more attention as part of the evaluation and design of coal tailings impoundments.

Some efforts have been made to understand the mechanical properties of FCR prior to, during, and after CCR push out construction, but the availability of geotechnical data for FCR in undrained conditions under varying loading conditions is limited due to access restraints when sampling or performing in-situ testing of the soft or loose FCR. Evaluating the behavior of FCR is also site specific and depends greatly on the operational functions of the coal mine and preparation plant as well as the geologic lithology surrounding the coal being mined. While FCR can generally be described as fine sand, silt, and clay sized particles in terms of grain size, the particle shape, specific gravity of solids, mineralogical composition, and depositional nature is potentially different than fine-grained soils typically associated with staged embankment construction.

BACKGROUND

Historically, undrained staged construction analyses of upstream CCR push outs have not been a substantial component in the design of coal tailings impoundments. Until recently upstream CCR push outs were not given a lot of attention because upstream push outs could normally be performed at a relatively slow rate. For upstream CCR push outs, a slow rate of construction is preferred so that pore pressures in the FCR can dissipate as construction is being performed that allows for the development of shear strength. Due to the slow rates of construction that allow pore pressures to dissipate, undrained staged construction analyses of upstream embankments at coal tailings impoundments were generally not applicable or required.

With new and more restrictive environmental regulations, the costs and time associated with both permitting and constructing new tailings facilities has greatly increased. The increased time and costs to permit and construct new tailings facilities have resulted in coal producers constructing existing tailings impoundments to greater heights using upstream push-out methods. Also, with advances in coal preparation processes that increase plant capacity and also to reduce operating costs, operators are consolidating coal preparation activities resulting in a greater amount of refuse generation for disposal at existing tailings impoundments that are not designed for the increased refuse production. In many cases the development of an upstream constructed embankment allows for a substantial increase in impoundment storage volume at an existing facility and allows continued operations without developing a new site. To reduce the potential of interrupting coal preparation, upstream push outs may need to be constructed at a more rapid pace that inherently increases the risk of undrained failure of weak foundation material.

Recent upstream failures at coal refuse slurry impoundments have occurred suddenly during construction and with large portions of the upstream embankment breaking off into the impoundment leaving near vertical high walls. Construction equipment, such as bulldozers and haul trucks, have been on the portions of the upstream embankment that broke off. The recent failures generally indicate undrained conditions due to the failures occurring abruptly during or shortly after construction.

The most notable failure that occurred was in 2012 at the Nolan Run Slurry Impoundment, which is part of the Robinson Run Mine No. 95 Mine, located in Harrison County, West Virginia. At the Nolan Run Slurry Impoundment severe bubbling occurred in the impoundment along the upstream edge of a CCR push out. Minutes after the severe bubbling began, a portion of the upstream embankment broke off into the impoundment leaving a near vertical, 20 foot tall high wall of CCR. The portion of the upstream embankment that broke off was in the shape of an eye with its greatest dimensions being 700 feet long, longitudinal to the embankment, and 200 feet wide, transverse to the embankment (MSHA, 2014). This failure resulted in the fatality of a bulldozer operator who was operating on the push-out embankment.

Procedures and requirements for analyzing the stability of coal tailings impoundment embankments for long-term, drained and seismic loading conditions are outlined in regulatory agency manuals, but there is little specific guidance and requirements for undrained analyses. The lack of historical requirements for undrained staged analysis of CCR embankments creates uncertainty for consultants and coal companies relative to what is considered sufficient analysis when submitting plans and calculations to regulatory agencies.

OPERATIONAL FEATURES OF COAL TAILINGS IMPOUNDMENTS

Although the method and approach used for staged construction analyses of upstream embankments at coal tailings impoundments is similar to conventional staged construction over weak natural soils, there are unique features that exist at coal tailings impoundments that affect the staged construction analyses of upstream embankments.

Foundation Stresses and Pore Pressures

During upstream construction at coal tailings impoundments, FCR acts as the foundation soil and the resisting forces for CCR upstream push-outs. FCR is pumped from the preparation plant, in the form of slurry through piping systems and is hydraulically placed in the coal tailings impoundment. FCR consolidates under self-weight and, prior to CCR upstream push-outs, is normally consolidated with high water contents. The presence of carbonaceous shale, clay and coal in FCR can affect the specific gravity of the solids, resulting in values that are approximately 30% to 40% lower than naturally occurring soils, such that the in-situ unit weight and effective confining stress prior to the upstream construction is less than that for naturally occurring soils. The specific gravity of solids comprising CCR, although potentially lower than naturally occurring soils, is in most cases higher than the FCR which results in higher theoretical unit weights than the FCR. The result of the unit weights in the CCR being inherently higher than the FCR (due to the differing specific gravity of solids) is a greater stress increase in the FCR relative to the initial stress prior to CCR upstream push outs.

Recent Environmental Regulations

Recent environmental regulations for the development and construction of coal tailings impoundments has led to more restrictive requirements with respect to the protection of ground water. Most notably is the use of geosynthetic materials such as geosynthetic clay liners (GCL) or flexible membrane liners (FML) on the existing ground surface within the footprint of coal tailings impoundments. The regulatory requirement to use geosynthetic liners is becoming more common to prevent contact water from seeping into the existing ground and surrounding natural waterways.

Geosynthetic liners have low permeabilities relative to the CCR and FCR that comprise coal tailings impoundments. In some situations, the geosynthetic liner's low permeability provides a single drained case when considering the rate of consolidation for FCR beneath CCR upstream push-outs. The rate of consolidation for FCR is significant because pore pressures need to dissipate for the FCR to gain strength. The rate of construction of CCR upstream push-outs is important to monitor so that excess pore pressures do not develop in the FCR foundation soils.

The rate of construction for CCR upstream push outs is often controlled by the need to raise the embankment to meet freeboard requirements or increase slurry disposal capacity. Due to environmental regulations, coal tailings impoundments operate to limit discharge of coal contact water into adjacent waterways. Along with the FCR slurry being pumped into the impoundment from the preparation plant, additional water is sometimes pumped into the impoundment from underground mine workings or other operational features associated with coal processing. The additional water pumped into the impoundment has led to increased free water levels over the settled FCR. The increased free water levels create a less desirable condition for construction of CCR upstream push outs because it makes it more difficult to develop FCR deltas around the upstream perimeter of the impoundment and limits consolidation of the FCR.

From an upstream construction and stability standpoint, it is preferred to operate the impoundments with limited free water and to move the slurry discharge point around the upstream perimeter of the impoundment where future CCR upstream push outs are planned. When operated with limited free water and proper movement of the slurry discharge point around the perimeter of the impoundment, the FCR can flow over a previously developed delta and cause the sand and silt size particles in the FCR slurry to be deposited over a longer distance from the upstream slope. In this ideal situation, an obvious and broad delta or beach is created that provides a foundation for CCR upstream push-outs that will dissipate pore pressure and gain strength quicker, in comparison to the finer clay-size particles in the FCR slurry which typically stay suspended longer in the slurry and have a greater affinity for water. With the increased free water in the impoundment, the ability to develop a delta prior to upstream construction features is reduced causing less desirable conditions for initiating upstream construction procedures.

There is also limited data on the true geometry of upstream embankments at coal tailings impoundments. Design sections used for analyses typically include the proposed grading on the upstream side of the embankment, and assume a depth of CCR submersion into underlying FCR. As CCR upstream push-outs are constructed, the initial CCR that is pushed out into the impoundment sinks into the FCR, potentially due to consolidation of the FCR, mixing with the FCR, or displacing the FCR into the impoundment, or a combination of all. Since this sometimes occurs below the impoundment's free water level there is a degree of uncertainty of the true embankment geometry. Since there can be substantially more CCR or mixed coal refuse (CCR and FCR) below upstream CCR push-outs than anticipated, the design sections used in analyses can sometimes provide a conservative case for the analyses.

Limited Typical and Site Specific Characterization

Limited published typical or site specific geotechnical data, including normalized undrained strengths and time rate of consolidation or pore-pressure dissipation information, is available for coal refuse relative to undrained analyses of staged upstream construction at coal tailings impoundments. Because rigorous undrained analysis of upstream construction features at coal tailings impoundment has not historically been a key feature of design or regulatory requirements, there are little published values of key parameters.

Common issues encountered when attempting to collect geotechnical data on coal refuse for use in upstream analyses are access constraints, difficulties in sampling, and restrictions related to safety precautions. The FCR that will act as the foundation soil can normally only be accessed from the edge of or within the impoundment, such that there is limited opportunity to safely collect physical samples of FCR and/or perform in-situ testing on FCR prior to a CCR upstream push out. It is somewhat common for the collection of FCR samples and to perform in-situ testing to construct a small upstream push out for the development of a platform over top of FCR for access. A shortcoming of performing a subsurface exploration from the small CCR push out is that the samples of FCR collected are not representative of the FCR prior to starting actual upstream embankment construction due to increased vertical stress in the FCR. The increased stress results in the start of consolidation in the FCR and increased difficulty establishing the preconsolidation stress history. It is much less common to work from a barge within a coal tailings impoundment. The difficulty of obtaining physical samples of FCR and/or performing in-situ testing on FCR limits the information available on characterizing FCR with respect to the rate of pore pressure dissipation and undrained shear strength.

DESIGN FEATURES AND APPROACHES TO MITIGATE ISSUES

The features and issues that affect the performance of undrained staged construction analysis of coal tailings impoundments are being addressed through design features and implementation of additional exploration and material characterization programs. The results of these programs serve to provide a better understanding of the behavior of coal tailings during staged construction loading associated with upstream embankments.

Foundation Stresses and Pore Pressures

Piezometers, typically pneumatic or vibrating wire, are installed during earlier stages of upstream CCR embankment construction to monitor pore pressures during construction of later stages and provide a warning system for potentially unsafe conditions. The piezometers are typically installed at a depth that is below any mix or sink-in zone that occurs during the construction of the push out. Due to the increased attention in the industry relative to upstream construction, it is becoming more common, and to some degree, a requirement for plan approval to include piezometers at multiple depths in a particular installation. The use of piezometers at multiple depths presumably allows for better characterization of pore pressure behavior during upstream construction.

As is common with piezometers installed within a dam, an action or "trigger" level is selected for each piezometer. The action levels are used as a threshold to warn operators of a potential increase in pore pressures during the construction of the embankment. If unacceptable pore pressures are measured in the piezometers during construction, then construction is stopped until the pore pressures reduce to an acceptable level or stability analyses demonstrate acceptable conditions for the site specific geometry. The action levels are generally selected by the designer and generally specified as excess pressure above the impoundment pool level. The action levels should be based on slope stability analyses and the pore pressure conditions required to maintain suitable factors of safety during the staged construction.

Dissipation of Pore Pressure

The placement of subsurface drainage features either below or immediately abutting upstream push-out embankments may be used to mitigate the effect from geosynthetic liners within the impoundments that limit pore pressure dissipation. As discussed previously, the presence of a liner limits drainage at the interface between the impoundment and natural ground and can promote the development of pore pressures during upstream construction.

Aggregate drains or wick drains in a horizontal orientation can be used to promote drainage and mitigate pore pressure development. Aggregate drains can be considered for installation at the interface between the push out and the embankment; however, it is difficult to maintain the integrity of the drainage aggregate due to the location and nature of the upstream construction. If the configuration of the embankment and push outs allow, the placement of aggregate beneath the push out, and on top of the geosynthetic liner is more desirable. In some configurations, the aggregate can be used as protective cover over the geosynthetic liner and serve the purpose of both protection and drainage. The use of a drainage layer as cover material over the geosynthetic liner provides the advantage of relying on double drainage for the analysis and construction of the push outs.

Horizontal wick drains have been used at coal tailings impoundments to mitigate the development of pore pressures during upstream construction. The wick drains should be installed well before the start of upstream construction and are normally placed over the FCR delta and sometimes extend into the pool area. The downstream portion of the wick drains are normally connected to an aggregate drain on the main portion of the embankment that daylights to a surface drainage control.

The authors have proposed using CCR cover over the liner system to provide a means for dissipating pore pressures. By installing a layer of CCR, which in most cases has a higher permeability than the FCR, there is a means for pore pressure to dissipate laterally in areas that are being loaded by upstream methods. The use of CCR in this capacity is advantageous and acceptable from the mine operator's perspective because the material is widely available onsite. If CCR is considered for use as a drainage layer, it is recommended that an evaluation be performed regarding the suitability of the CCR with respect to permeability and durability.

Limited Typical and Site Specific Characterization

The increased requirements for undrained staged construction analyses of upstream embankments at coal tailings impoundments will likely require collection of site specific data. The use of laboratory shear strength testing, including anisotropically consolidated triaxial testing and direct simple shear testing, on high quality undisturbed samples to develop normalized undrained shear strength using the SHANSEP method coupled with in-situ shear strength testing to confirm lab results is generally the "best" state of practice (LADD, 1991). When suitable data cannot be acquired, typical values (for conventional soils) are often assumed. Typical values of peak undrained strength normalized for consolidation stress (S_u/σ'_{vc}) for normally consolidated soils range from 0.2 to 0.3. Based on the authors' experience this range is applicable for normally consolidated FCR and potentially conservative. Results from anisotropically consolidated triaxial compression and extension tests and direct simple shear tests performed on fixed piston samples of FCR collected from an impoundment in the Appalachian Coal Fields are shown in Table 1.

Table 1. Peak Undrained Shear Strength Ratio of FCR
Coal Tailings Impoundment Appalachian Coal Fields

Depth (ft)	USCS	<#200 (%)	LL (%)	PI (%)	G_s	Wc (%)	Test Type (1)	$\sigma'_{vc}/\sigma'_{In\ Situ}$ (2)(3)	S_u/σ'_{vc} (2)(4)
80.0-82.0	CL	87	37	13	1.67	36.3	DSS	1.2	0.30
						27.4	CK_oU-C	1.2	0.49
						29.2	CK_oU-E	1.2	0.24
87.0-89.0	CL	88	40	16	2.12	33.6	DSS	1.0	0.36
						37.0	CK_oU-C	1.0	0.53
92.0-94.0	CL	59	36	13	1.68	38.1	DSS	1.6	0.31
						16.2	CK_oU-C	1.5	0.44
97.0-99.0	ML	90	42	11	1.51	34.9	DSS	2.3	0.38
						44.5	CK_oU-C	2.3	0.54
33.3-35.3	ML	98	42	16	1.76	49.0	DSS	3.0	0.27
						49.2	CK_oU-C	3.0	0.47
						46.2	CK_oU-E	3.0	0.17
64.3-66.3	CL	92	37	15	1.72	37.3	DSS	3.1	0.29
						23.3	CK_oU-C	3.1	0.35

Test Type Abbreviations: DSS = Direct Simple Shear, CK_0U-C = Anisotropic Consolidated Undrained Triaxial in Compression, CK_0U-E = Anisotropic Consolidated Triaxial in Extension.

σ'_{vc} = Laboratory Effective Vertical Consolidation Stress

$\sigma'_{In\ Situ}$ = In Situ Effective Vertical Stress

S_u = Peak Undrained Shear Strength

Table 1 indicates a normalized undrained strength ratio (S_u/σ'_{vc}) ranging from 0.17 to 0.54, with an average value for all testing of 0.37. The average normalized undrained strength ratios for each test type are summarized in Table 2.

Table 2. Peak Undrained Shear Strength Ratio of FCR Based on Test Type
Coal Tailings Impoundment Appalachian Coal Fields

Test Type	Average S_u/σ'_{vc}
K_0 Consolidation Triaxial Compression	0.47
K_0 Consolidation Triaxial Extension	0.21
Direct Simple Shear	0.32

CONCLUSIONS

Due to the increased attention being given to staged construction of upstream embankments, undrained analyses of coal tailings impoundments will become more common, particularly in active impoundments that require future expansion using upstream construction methods. There is also increased environmental regulatory pressure that is resulting in new design challenges that affect the analysis and construction of upstream embankments in coal tailings impoundments. The use of geosynthetic liners has well intended basis from the standpoint of water and environmental protection, but the liners can potentially limit pore pressure dissipation and rate of strength development in FCR. Additional design features can be used to provide a means to dissipate pore pressures; however, the additional free water being impounded due to environmental regulations may limit the development of and positive effects resulting from a delta in the impoundment. The additional water also needs to be accounted for in the analyses and when performing subsurface explorations to support the analyses.

Although there is limited available documentation for typical parameters for coal tailings and limited ability to collect site specific data for undrained analyses of staged construction at coal tailings impoundments, it is the experience of the authors that typical values of normalized peak undrained shear strength (S_u/σ'_{vc}) for natural soils are suitable for impounded FCR. Lab testing on FCR from an impoundment located in the Appalachian Coal Fields indicates the normalized peak undrained shear strength ratio for FCR is within or potentially higher than the typical range of 0.2 to 0.3 for normally consolidated natural soils.

REFERENCES

Ladd, C. C. (1991). *Stability Evaluation During Staged Construction. Journal of Geotechnical Engineering*, 117(4), 540-615.

MSHA (2014). *Report of Investigation: Dam Failure Accident (Drowning), Nolan Run Slurry Impoundment.* Report CAI-2012-19. U.S. Department of Labor, Mine Safety and Health Administration.

SEISMIC EVALUATIONS FOR COAL COMBUSTION RESIDUAL IMPOUNDMENTS IN THE EASTERN SEISMIC ZONE

W. Shin, Ph.D., P.E.[1]
James McNash, P.E.[2]
Ramachandran Kulasingam, Ph.D., P.E.[3]

ABSTRACT

An extensive field investigation and laboratory test program was developed and implemented to characterize the dike, subsurface soils, and impounded Coal Combustion Residuals (CCRs) for two CCR ponds. Seismic hazard, site response, and liquefaction analyses were performed. These analysis results indicated, in general, that liquefiable zones are predominately located along the downstream toe of the pond perimeter dike. The areas exhibiting liquefaction potential were evaluated for the consequences of liquefaction and the results indicated that mitigation measures need to be applied along a section of the dike toe to reduce potential for liquefaction-induced hazards. The option of lowering the water level within the ponds was selected because the plant owner anticipates proposed United States Environmental Protection Agency (USEPA) regulations will require pond closures. The selected option is a risk reduction measure that mitigates the flow potential of the contained CCR materials in the event of a dike failure. The analysis work presented here illustrates the effect of using region specific modulus reduction and damping curves and age scaling factor to account for the Pleistocene origin of the foundation soils on the final results and mitigation decision. The use of these region specific correlations results in significant reduction in the calculated extent of liquefiable zones. While it is important to incorporate emerging region specific methodologies in the eastern seismic practice, parameter studies to evaluate the effect of these methodologies when compared to standard generic methodologies should be considered as part of the evaluation as one done for this project.

INTRODUCTION

The December 2008 failure of a 65 ft high ash disposal cell in Tennessee heightened the awareness of the environmental regulatory community regarding the stability of CCR impoundments across the United States. In response to the 2008 incident in Tennessee, the USEPA initiated a program requiring owners of CCR impoundments to assess the stability of storage pond dikes across the country. For an electric generating station (Site) located in the Eastern United States, the results of USEPA review of historical engineering documents warranted additional seismic stability and liquefaction potential evaluations for the perimeter dikes of a pond storing CCRs. The remaining sections of

[1] Engineer, Geosyntec Consultants, 1300 South Mint St,, Ste. 410, Charlotte, NC, 28203, 704-227-0840, WShin@geosyntec.com.
[2] Senior Staff Engineer, Geosyntec Consultants, 5420 Corporate Blvd. Ste. 202, Baton Rouge, LA 70808, 225-9299-7333, JMcNash@geosyntec.com.
[3] Associate, Geosyntec Consultants, 1300 South Mint St., Ste. 410, Charlotte, NC, 28203, 704-227-0840, RKulasingam@geosyntec.com.

this paper describe: (i) general site background; (ii) geotechnical characterization; (iii) seismic hazard; (iv) site response analyses; (v) liquefaction analyses; (iv) liquefaction consequence evaluation and mitigation; and (iv) conclusions.

GENERAL SITE BACKGROUND

The Site is a coal-fired, electric generating station located in the Eastern US, and was constructed in the early 1970's. Impoundments were designed and constructed in a phased approach, as additional storage was required to manage CCRs at the Site. The USEPA review indicated that one of the impoundments, denoted as "Pond 1", required additional stability analysis for seismic and liquefaction failure mechanisms, since the Site is influenced by the Charleston Seismic Zone. The remaining impoundments were excluded from these further evaluations based on the potential downstream consequences of failure.

Pond 1 currently receives sluiced Flue Gas Desulfurization (FGD) residuals that do not meet the specifications for beneficial reuse in residential wallboard production. The perimeter dikes received a "High Hazard Classification" in the USEPA review, because the structures are situated adjacent to a creek. Pond 1 shares the perimeter dikes with a fly ash impoundment (Pond 2), and the two ponds are separated by a divider dike. Since these impoundments are located in close proximity, the owner decided to extend these evaluations to Pond 2 as well. The geotechnical investigation locations and representative dike cross section A (discussed in subsequent sections) are provided in Figure 1.

The perimeter dikes were constructed primarily with clayey sands excavated from the interior area of these ponds. The length of the perimeter dikes is approximately 11,000 ft (not including the length of the divider dike). Based on review of boring logs performed prior to the perimeter dike construction, the pre-construction grade sloped down

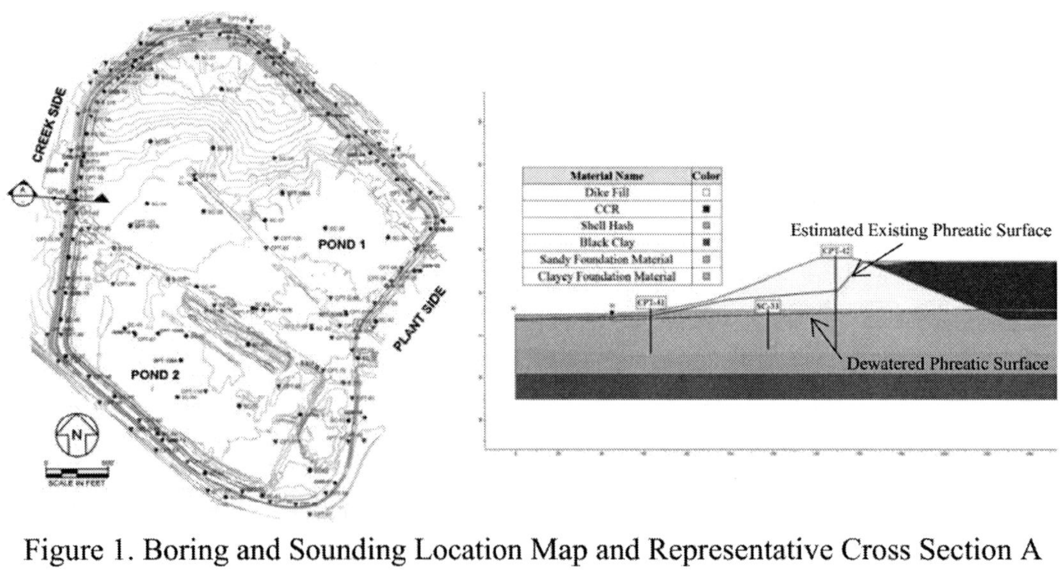

Figure 1. Boring and Sounding Location Map and Representative Cross Section A

generally toward the northwest and the adjacent creek. The crest of the perimeter dike is relatively flat. Ponds 1 and 2 have surface areas of approximately 100 acres and 62 acres, respectively. Pond 1 contains free water above the settled FGD residuals approximately over one-half of the pond surface area. Water levels in Pond 1 are controlled via a culvert and interior channels diverting surface water to a pump station in Pond 2, where it is ultimately pumped to an adjacent pond. The perimeter dike is about 15 to 30 ft high from the outside toe, with the northwestern section along the creek side being the highest. The upstream and downstream slopes of the perimeter dikes range from 2 horizontal to 1 vertical (2H:1V) to 3H:1V.

GEOTECHNICAL CHARACTERIZATION

General Subsurface Stratigraphy

Historical engineering reports and boring logs were reviewed at the onset of the project in order to evaluate the available data in the vicinity of Ponds 1 and 2. During the design of these two ponds, twenty borings (typically beneath the dike structures) with Standard Penetration Tests (SPTs) and fifteen test pits within the pond interiors were conducted. During a subsequent hydrogeological investigation, eighteen borings were advanced through the perimeter and divider dikes, generally with SPTs until the top of the Shell Hash stratum was encountered. However, a number of these borings were drilled without sampling and were utilized to measure the depth to Shell Hash strata. The historical information indicated that the subsurface soils primarily consist of four geotechnical units, at the depths most relevant for this evaluation. A brief description on each unit is presented as follows:

- Dike Fill: Materials within the dike consist predominantly of medium dense, brown to gray, silty/clayey fine to medium sands, and stiff, sandy clays to low to medium plasticity clays.

- Foundation Soils: Foundation soils consist typically of brown to gray, clayey sands, silty sands, and poorly graded fine to medium sands with varying amounts of shells. Clay lenses were occasionally encountered in the more sandy foundation soils. The relative density of the foundation soils ranged from very loose to medium dense, and were observed to generally be 10 to 30 ft thick. In the majority of borings, the lower 5 to 10 ft of foundation soils consist heavily of shell fragments and shells, and was described typically as loose to medium dense, clayey fine sand with interbedded shells. This geotechnical unit is of the Pleistocene age.

- Shell Hash: A layer of dense to very dense, partially cemented to heavily cemented shells was encountered beneath the foundations. Blow counts in this layer exceeded 50 blows for less than 6 in. penetration with minimal sample recovery. Historical data indicate that the thickness of this layer typically ranges from 1 to 10 ft, and was found to be the thinnest in the eastern corner of Pond 1.

- Dark Gray / Black Clay: A dark gray to black clay was encountered beneath the Shell Hash stratum. This stratum was described as stiff to hard, dark gray to black, medium to high plasticity, over-consolidated clay with some sand. Borings performed in the site indicated that this material has SPT blow counts in excess of 50 blows per foot, in general.

Supplementary Investigation

A focused geotechnical investigation of the perimeter dike and perimeter dike toe was warranted as preliminary liquefaction potential and post-liquefaction seismic slope stability analyses indicated the presence of potentially liquefiable zones and associated slope stability concerns during the design earthquake. However, these analyses were based on historical boring logs where the energy ratio (ER) of the hammer system was not reported. Therefore, the objectives of the supplementary subsurface investigation included: (i) the characterization of material properties of dike fill, foundation soils, and CCRs; (ii) the evaluation of potentially liquefiable soils with significantly improved spatial and depth resolution; and (iii) the calibration of historical borings to borings with a known hammer calibration.

Cone Penetration Test (CPT) soundings with pore pressure measurements were selected as the primary investigation tool since they provide a fine resolution of in-situ measurements, and the ability to collect shear wave velocity data for selected soundings. Thirty five CPTs were advanced at both the crest and outside toe of the perimeter dikes (seventy soundings); while, seven soundings were advanced into CCR materials. Eleven and four soil test borings with SPTs were advanced at the dike crest and outside dike toe, respectively, by mud-rotary wash methodology in accordance with the procedures recommended by Idriss and Boulanger [2008]. Split spoons and undisturbed samples were collected for laboratory testing at each boring location.

Track mounted equipment was used while accessing the interior of each Pond. Access to Pond 2 (fly ash) with equipment was found to be easier than access to Pond 1 (FGD residuals). This may be due to the difference in material properties and water levels. The surface condition of Pond 1 limited access to the interior of the pond by higher ground pressure rigs. Low ground pressure rigs were able to traverse the interior of Pond 2 without significant rutting or pumping of water to the surface. For relatively soft areas, the rig was connected to a winch mounted dozer. Borings and soundings located along the dike crest and dike toe were advanced to the top of the Shell Hash stratum; while, borings and soundings located within the pond interior were terminated at the bottom of CCR material. The automatic hammer utilized was found to have an ER of 78%. Investigation locations were spaced 200 to 300 ft. A boring and a CPT sounding were located within 5 ft of a historical boring from each of the previous two subsurface investigations, and the resulting data were interpreted to estimate the hammer calibration for the historical soil borings. Figure 1 provides the location of each investigation point within the vicinity of the Ponds 1 and 2.

An extensive laboratory testing program including index testing (natural moisture content, grain size distribution, Atterberg Limits, fines content), specific gravity, in-situ unit weight, and shear strength tests was conducted to evaluate the material properties of the subsurface dike fill and foundation soils. Index testing was particularly focused in samples classified in the field as loose, clean to slightly silty sands in support of the liquefaction potential evaluation.

Cone Penetration Test Sounding Interpretation

The Soil Behavior Index (Ic) for each CPT sounding was calculated using the methodology presented by Robertson and Cabal [2012], and utilized to develop the subsurface model at each sounding location. Additionally, shear wave velocity (Vs) measurements were collected during CPT soundings in 5 ft depth intervals at several locations along the perimeter dike crest, dike toes, and CCR materials within the ponds. The soil measurements were supplemented by a correlation (Equation 1) developed by Mayne [2006] using sleeve friction (fs) measurements to develop representative Vs profiles for the materials overlying the Shell Hash stratum.

$$V_S = (118.8 * \log(f_S/20.885) + 18.5)/3.2808 \quad (1)$$

where:

Vs = shear wave velocity (ft/s); and
fs = sleeve friction (psf).

Representative Vs-profiles were developed for three soil profiles for the perimeter dike and three free-field (dike toe) soil profiles, and were subsequently utilized to model the site response at the Site. Shear wave velocity profiles were combined with soil index test results for shallower soils and site-specific historical and regional shear wave velocity measurements for deeper soils to develop representative soil columns for the site response analysis. An example of dike and free field Vs profiles and soil columns are provided in Figure 2.

Phreatic Surface

The phreatic surface through the perimeter dikes was developed based on CPT pore pressure dissipation tests and borehole water level measurements collected between 48 and 72 hours after boring completion. Since water levels in boreholes advanced by mud rotary wash in clayey soils are often unreliable, three additional borings were advanced without sampling by means of a hollow stem auger. Water levels in these borings were collected 24 hours after termination and utilized to select the phreatic surface through the perimeter dikes.

CCR Materials

The CCR materials generally classified as non-plastic silts. Vs measurements collected within CCRs typically ranged between 200 and 300 ft per second and were found not to increase with depth. Undrained triaxial shear testing conducted in the laboratory for both

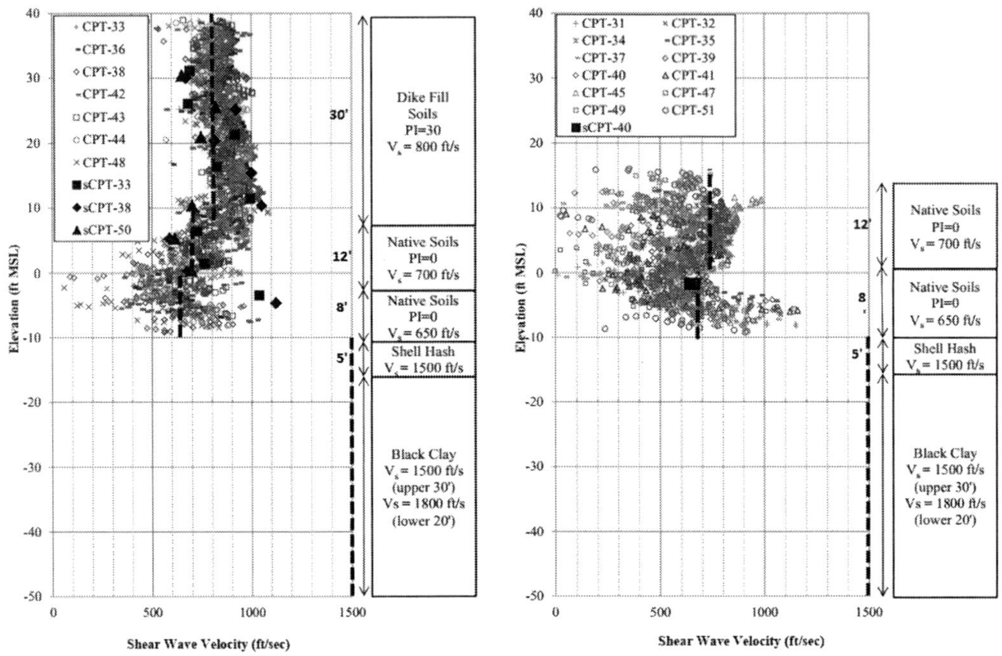

Figure 2. Representative Vs Profiles (Left –Dike Centerline; Right - Free Field)

undisturbed piston samples and reconstituted samples prepared by slurry deposition techniques showed strain hardening behavior despite the relatively low Vs values. It is noted that only Piston samples were tested for the fly ash. The test results from the slurry deposition technique were found to be similar to the undisturbed piston sample test results for the FGD residuals. Friction angles values of 33 and 40 degrees were interpreted for the fly ash and FGD residuals, respectively, from the test results. However, friction angle values were limited to a maximum of 33 degrees for both materials until more testing could be carried out. It is noted that the analyses and conclusions reached focused on the dike stability, and hence not dependent on the CCR properties.

SEISMIC HAZARD

Peak Ground Acceleration (PGA) was selected to be consistent with a 98 percent or greater probability that the PGA will not be exceeded in 50 years. While the USGS seismic hazard maps indicated a higher PGA, region specific work by Chapman and Talwani [2006] for geologically realistic site conditions for this coastal plain site indicated a PGA of 0.16g. The moment magnitude of the earthquake was selected to be 7.3 in the Charleston seismic zone. Five ground motions were selected to reasonably envelope the target design spectra developed based on regional seismic hazard maps developed by Chapman and Talwani [2006]. These include design motions downloaded from the USGS website based on site location which provided good coverage for the low period region specific target spectra and synthetic motions developed for the eastern US to provide better coverage for the long period region.

SITE RESPONSE ANALYSES

Analysis Methodology and Input Parameters

Site response analyses were conducted using SHAKE2000 [Ordonez, 2013], a one-dimensional equivalent-linear site response analysis computer program. One of the main goals of the site response analyses was to calculate cyclic shear stresses within the soil profiles selected, which were used for the liquefaction analysis. For the SHAKE analyses, a total of six representative profiles (three profiles at the dike centerline and three profiles at free-field) were developed based on shear wave velocity profiles. In addition, transitional dike profiles within the dike slope were also considered to evaluate two-dimensional effect of the dike geometry on ground response, although the results are not presented in this paper. The five ground motions, described in the previous section, were used as input for the SHAKE analyses.

In addition to the input ground motions and the soil columns, the site response analyses require dynamic shear modulus reduction and damping curves. Two cases were analyzed to evaluate the effects of regional soil characteristics on site response: (i) case with generic curves (Case I); and (ii) case with region-specific curves (Case II). Several researchers proposed dynamic shear modulus reduction and damping curves based on laboratory test results [Vucetic and Dobry, 1991; Ishibashi and Zhang, 1993; Darendeli, 2001; Andrus et. al, 2003; Zhang et. al., 2005]. Among them, those proposed by Ishibashi and Zhang [1993] were selected for the analyses of Case I because they can account for the effects of effective mean pressure and plasticity of soils. Andrus et. al. [2003] and Zhang et. al. [2005] presented equations to estimate dynamic shear modulus reduction and damping curves based on existing resonant column and torsional shear test results for relatively undisturbed soil samples collected from the Southeastern U.S. The equations were developed for different geologic units and soil plasticities. The geologic units include Holocene, Pleistocene, Tertiary, and Saprolite (Residual Soil). For the analyses of Case II, the equations for Holocene-age and Pleistocene-age soils proposed by Andrus et. al. [2003] were used to develop dynamic shear modulus reduction and damping curves for Dike Fill and Native Foundation Soils, respectively. The Holocene curves were developed using compacted dike fill and natural soil samples. The dynamic shear modulus reduction and damping curves used for the analyses of Cases I and II are presented in Figure 3.

Analysis Results

Average values of the site response analysis results for five input ground motions were calculated for each profile. Average maximum horizontal equivalent accelerations and average maximum cyclic shear stresses are presented in Figures 4 and 5, respectively. Here the horizontal equivalent acceleration was calculated by

$$\text{Horizontal Equivalent Acceleration} = \tau_h / (\rho\, z) \quad (2)$$

where:

τ_h = cyclic shear stress at depth z

ρ = mass density of material above z
z = depth to top of layer

The results shown in Figures 4 and 5 indicate that calculated average maximum horizontal accelerations and average maximum cyclic shear stresses for Case I are higher than those for Case II by approximately 30 to 50 percent. Such increase has direct impact on the calculated cyclic shear stress ratios (CSRs) for the liquefaction analysis. Relatively lower maximum horizontal equivalent accelerations and cyclic shear stresses for Case II seem to be the result of using Pleistocene age region-specific modulus reduction curves for native soils. Those modulus reduction curves exhibit larger nonlinearity even at low amplitudes of shear strain.

LIQUEFACTION ANALYSIS

An initial step in performing a liquefaction potential evaluation is the application of screening criteria based on the geotechnical index properties of the soil to evaluate whether subsurface materials are potentially liquefiable. In general, soils that show "clay-like" behavior are not susceptible to liquefaction. The Modified Chinese criteria [Finn et. al 1994] and Idriss and Boulanger [2008] criteria were applied to screen soils generally not considered to be liquefiable. Soils classified as Organic Peat, Silt and Clay, or a combination of these materials, were not considered to be susceptible to liquefaction during this evaluation. Additionally, soils that exhibit "clay-like" behavior according to CPT soundings were also screened from liquefaction potential calculations. Soils that are "clay-like", which was defined as soils with a Soil Behavior Index (I_c) greater than 2.60, were assumed to be non-liquefiable.

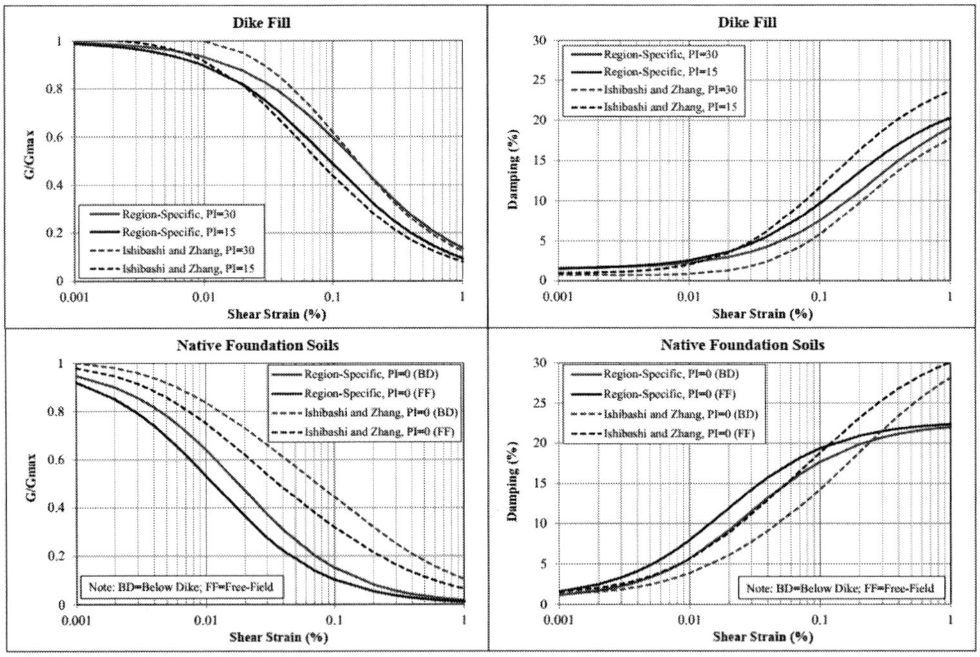

Figure 3. Modulus Reduction and Damping Curves

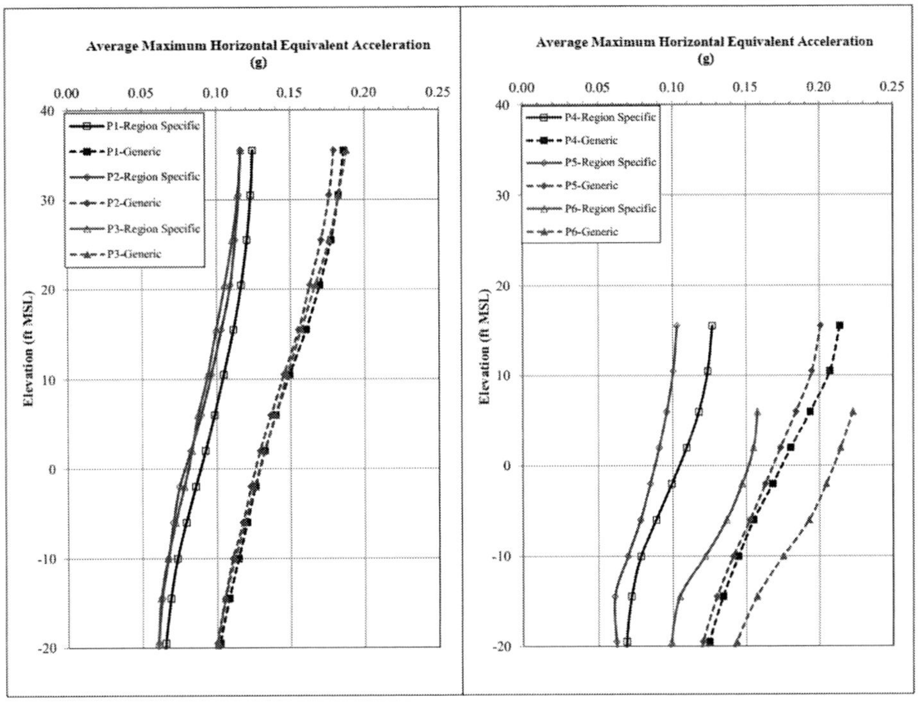

Figure 4. Average Maximum Horizontal Equivalent Acceleration Profiles (Left –Dike Centerline; Right - Free Field)

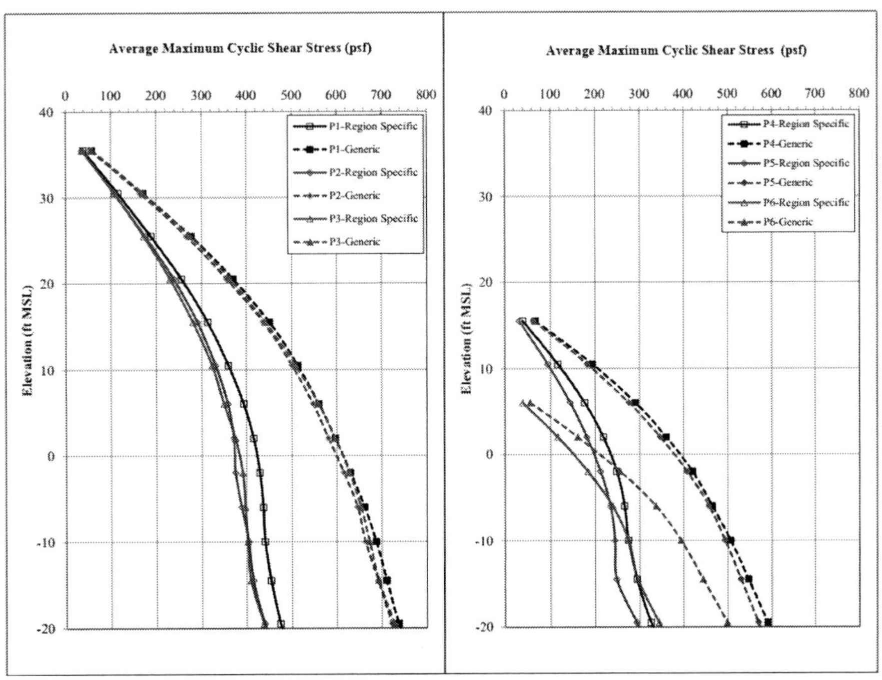

Figure 5. Average Maximum Cyclic Shear Stress Profiles (Left –Dike Centerline; Right - Free Field)

Liquefaction analyses were performed based on the simplified procedure recommended by Seed and Idriss [1971] and later updated by Idriss and Boulanger [2008]. This approach is based on comparing in-situ test results with case histories of occurrences and non-occurrences of liquefaction from historical records of earthquakes. The Idriss and Boulanger [2008] methodology interprets in-situ data to evaluate the cyclic resistance ratio (CRR) of a soil that has been corrected for earthquake magnitude and overburden effects.

Age Scaling Factor

Correlations associated with liquefaction potential analysis were developed based on case histories of relatively young soil deposits (i.e. Holocene Era). Youd and Perkins [1978] identified that younger, loose soils are more susceptible to liquefaction than older soils. Arango et al. [2009] identified that the influence of soil age on the cyclic resistance is significant, and should not be ignored. Geologic maps indicated that the surface soils at the site are of Pleistocene origin. Therefore the CRR was adjusted by an Age Scaling Factor (ASF) to account for the influence of soil age. It is noted that "geologic" age differs from "geotechnical" age. Geologic age refers to the overall age of the soil since deposition. Geotechnical age in this sense refers to the age of the soil since the last instance of liquefaction. The geotechnical age was considered in the selection of ASF as recommended by Andrus et al. [2008].

Regional correlations relating the soil age to ASF developed by Andrus et al. [2008] were applied to calculate the ASF values shown in Table 1. Work by Leon at al. [2005] indicates the geologic age of soils near the Site is 450,000 years and geotechnical age based on evidence of paleoliquefaction events to be between 500 and 450,000 years. Their work did not find evidences of recent occurrences of liquefaction near the Site from the 1886 Charleston earthquake. Based on this information, an ASF of 1.3 was selected for the Pleistocene age soils found beneath the perimeter dikes at the Site. ASF of 1.0 was used for all dike soils.

Table 1. Computed ASF values using an equation proposed by Andrus et al. [2008]

Age, t (years)	**ASF**
126	1.08
546	1.20
5,038	1.38
10,000	1.44
450,000	1.75

Cases

The factor of safety against liquefaction (FS_{Liq}) was computed at every interval where data were collected for each soil test boring (in 2 ft or 5 ft intervals) and each CPT sounding (in 0.16 ft intervals) for the Generic (Case I) and Region-Specific (Case II) modulus reduction and damping curves. For each case, FS_{Liq} were computed considering an ASF = 1.0 (Cases IA and Case IIA) and an ASF = 1.3 (Cases IB and IIB) for the foundation soils.

Results

The FS_{Liq} for each Case was evaluated at each boring or sounding location. Results indicated that the use of Generic dynamic modulus and damping curves result in larger computed liquefiable zones for ASF = 1.0 and 1.3. Relatively smaller liquefiable zones were computed under Case IB conditions (Generic reduction and damping curves and an ASF=1.3). Region-Specific curves (Case II) resulted in smaller computed zones of liquefaction. Results from each individual boring and sounding were compared spatially along the crest and toe of the perimeter dikes and along numerous cross sections. One such cross section is shown for a section located along the creek side in Figure 6, where SPT and CPT based methods appear to show consistent results, in general. The computed liquefiable zones were not found to be continuous from the downstream toe of the perimeter dike to the dike centerline. However, the computed liquefiable zones were considered to be laterally continuous in most areas along the creek side dikes and limited areas along other sections of the downstream toe of the perimeter dike. These liquefiable zones are predominantly located along the downstream toe of the perimeter dike of Ponds 1 and 2. Significant liquefiable zones were not calculated to be present under the dikes for Case II or Case IB.

To visualize FS_{Liq} and the risks associated with the extent of liquefiable zones during the design earthquake, each Case was viewed using a Geographic Information System (GIS) model as shown in Figure 7. This model can be spun around to view the liquefiable zones along the toe as well as along the dike centerline. A linear kriging algorithm was implemented to predict the extent of liquefiable zones in between each calculation point. While these models were utilized to visualize and to aid in the selection of critical areas for further evaluations with respect to the potential consequences of liquefaction, it was assumed that calculated liquefiable zones at the downstream toe of the perimeter dikes were laterally continuous in areas adjacent to the creek. The evaluation of potential consequences of liquefaction is discussed in the following section.

LIQUEFACTION CONSEQUENCE EVALUATION AND MITIGATION

Evaluation of Liquefaction Consequences

As discussed in the previous section, the liquefaction analysis results indicated that some areas were identified as potentially liquefiable and that the potentially liquefiable areas were significantly influenced by the selected modulus reduction and damping curves and age scaling factor, especially related to the use of generic versus region specific properties. Based on the liquefaction analysis results, the consequences of liquefaction were evaluated for the Creek side perimeter dikes site conditions. As part of the evaluation, post-liquefaction static slope stability analyses were performed for selected critical sections. Two slip modes were considered in the analyses: (i) global circular slip surface mode; and (ii) non-circular (planar) slip surface mode through potentially liquefiable soil zones. Residual shear strengths were assigned to the potentially liquefiable soil zones based on the semi-empirical approach presented in Idriss and Boulanger [2008]. Post-liquefaction static slope stability analysis results indicated that a portion of the perimeter dike of Ponds 1 and 2 along the creek side does not meet the target FS of 1.1 for toe slip surfaces. Therefore, mitigation measures needed to be considered for the referenced area to reduce liquefaction-induced hazards.

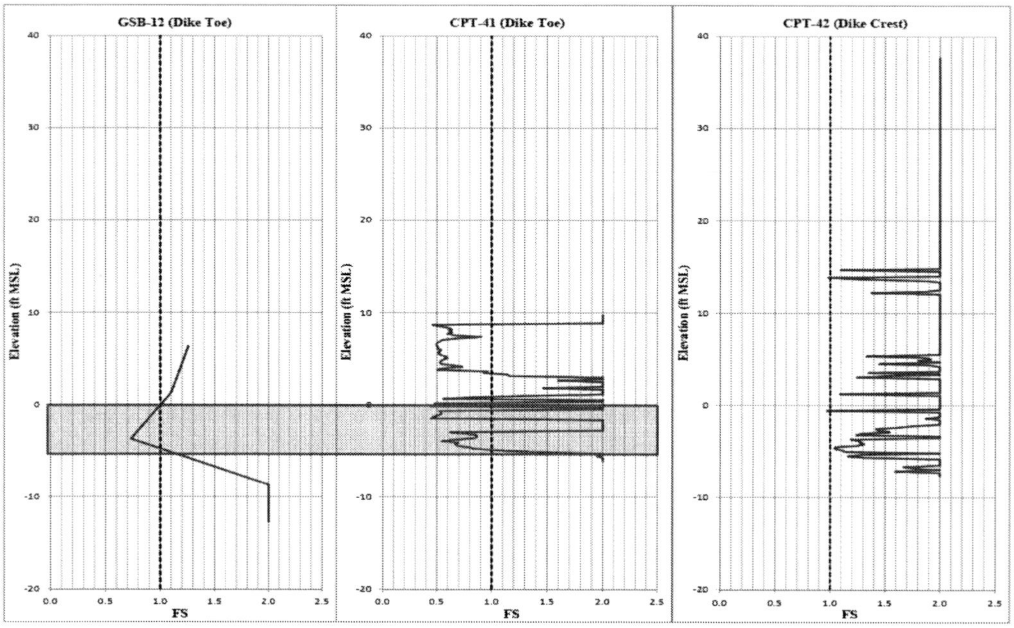

Figure 6. CPT and SPT Liquefaction Results for a Cross Section (FSLiq = 1 shown as dashed line).

Seismic Mitigation

By screening several applicable seismic mitigation options, two options were considered for further evaluation: (i) construction of a toe buttress berm; and (ii) lowering the water level within the ponds. Although the construction of mitigation toe berm can achieve the required mitigation at a relatively low cost, the option of lowering the water level within the ponds was selected because the plant owner plans to excavate ponded CCR materials in a few years as part of closure of the ponds. The target level for lowering would almost take the water level down to the bottom of the dikes. This water level lowering within the ponds would also provide the additional benefit of mitigating the flow potential of the contained CCR materials in the event of a dike failure.

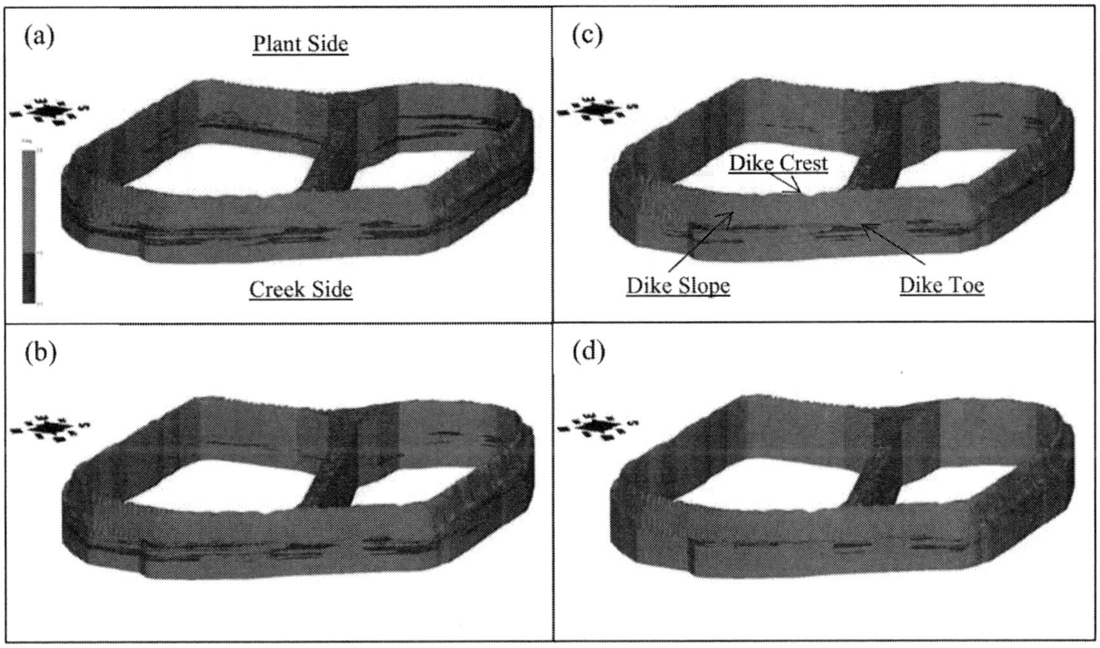

Figure 7. GIS Models of FS_{Liq} Results for the (a) Case IA; (b) Case IB; (c) Case IIA; (d) Case IIB. (10X Vertical Exaggeration - Viewed from the Creek Side. FSLiq < 1.0 is shown in blue; FSLiq > 1.0 is shown in green.)

CONCLUSION

An extensive field investigation and laboratory test program was developed and implemented to characterize the dike, subsurface soils, and impounded CCRs for an FGD residual and ash pond. Seismic hazard, site response, and liquefaction analyses were performed. These analysis results indicated, in general, that liquefiable zones are predominately located along the downstream toe of the perimeter dike, but not under the containment dikes themselves. The areas exhibiting liquefaction potential were evaluated for the consequences of liquefaction and the results indicated that mitigation measures need to be applied for the dike area along a section adjacent to a creek to reduce potential for liquefaction-induced hazards. The option of lowering the water level within the ponds was selected because the plant owner anticipates proposed USEPA regulations will require pond closures. The selected option is a risk reduction measure that mitigates the flow potential of the contained CCR materials in the event of a dike failure.

The analysis work presented here illustrates the effect of using region specific modulus reduction and damping curves and ASF to account for the Pleistocene origin of the foundation soils on the final results and mitigation decision. The use of these region specific correlations results in significant reduction in the calculated extent of liquefiable zones. While it is important to incorporate emerging region specific methodologies in the eastern seismic practice, parameter studies to evaluate the effect of these methodologies when compared to standard generic methodologies should be considered as part of the evaluation. GIS based visualization tools can be used to present the effect of different assumptions and input parameters on the overall risk to the owner and make a reasonable informed risk-based decision in terms of the need and extent of mitigation measures.

REFERENCES

Andrus, R.D., Zhang, J., Ellis, B.S., and Juang, C.H., "Guide for Estimating the Dynamic Properties of South Carolina Soils for Ground Response Analysis", Report No. FHWA-SC-03-07, South Carolina Department of Transportation, Columbia, South Carolina, 2003.

Andrus, R. Gassman, S. L., Talwani, P., Hasek, M., Camp, W. Hayati, H. and Boller, R., "Characterization of Liquefaction Resistance of Aged Soils: Summary of Selected First Year Findings", Proceedings of 2008 NSF Engineering Research and Innovation Conference, Knoxville, Tennessee, NSF Grant # CMS-0556006, January 2008.

Arango, I., Lewis, M. R., and McHood, M. D., "Site Characterization Philosophy and Liquefaction Evaluation of Aged Sands", Bechtel Technology Journal, Vol. 2, No. 1, December 2009.

Chapman, M.C. and Talwani, P., "Seismic Hazard Mapping for Bridge and Highway Design in South Carolina", South Carolina Department of Transportation, FHWA-SC-06-09, 2006.

Darendeli, M.B, "Development of New Family of Normalized Modulus Reduction and Material Damping Curves," Ph.D. thesis, University of Texas at Austin, Austin, Texas, 2001.

Finn, W. D.L., Ledbetter, R.H., and Wu, G. "Liquefaction in Silty Soils: Design and Analysis", Ground Failures under Seismic Conditions, Geotechnical Special Publication 44, ASCE, New York, pp. 51 - 76. 1994.

Idriss, I. M., and Boulanger, R. W., "Soil Liquefaction During Earthquakes", Earthquake Engineering Research Institute, EERI Publication MNO-12, 2008.

Ishibashi, I. and Zhang, X.J., "Unified Dynamic Shear Moduli and Damping Ratios of Sand and Clay", Soils and Foundations, Vol. 33, No. 1, pp. 182-191, 1993.

Leon, E., Gassman, S. L., and Talwani, P., "Effect of Soil Aging on Assessing Magnitudes and Accelerations of Prehistoric Earthquakes", Earthquake Spectra, Vol. 21, No. 3 pg. 737-759, August 2005.

Mayne, P.W., "The 2nd James K. Mitchell Lecture: Undisturbed Sand Strength from Seismic Cone Tests," Geomechanics and Geoengineering, Vol. 1, No. 4, pp. 239–247, 2006.

Ordonez, G.A., "SHAKE 2000: A computer program for the 1-D analysis of geotechnical earthquake engineering problems", www.geomotions.com, 2013

Robertson, P, K, and Cabal, K. L., "Guide to Cone Penetration Testing for Geotechnical Engineering", 5th Edition, November 2012.

Seed, H.B, Idriss, I.M., "Simplified Procedure for Evaluation Soil Liquefaction Potential", Journal of the Soil Mechanics and Foundations Division, ASCE, Vol. 107, NO. SM9, 1971.

Vucetic, M. and Dobry, R., "Effect of Soil Plasticity on Cyclic Response", Journal of Geotechnical Engineering, Vol. 117, No. 1, January, ASCE, pp 89-107, 1991.

Youd, T. L., and Perkins, M., "Mapping liquefaction-induced ground failure potential", J. Geotechnical Eng. Div., ASCE 104(GT4), 433-46, 1978.

Zhang, J., Andrus, R.D., and Juang C.H., "Normalized Shear Modulus and Material Damping Ratio Relationships", Journal of Geotechnical and Geoenvironmental Engineering, Vol. 131, No. 4, April, pp. 453-464, 2005.

CONDITIONS OF COAL ASH EMBANKMENTS

John Seymour, P.E.[1]
Omer Bozok[2]
Amanda Hughes, Ph.D.[3]
Brad Bodine, P.E., CHMM[4]

ABSTRACT

There are nearly 500 coal ash disposal/treatment ponds contained by earthen and coal ash embankments up to 400 ft high in the United States. They are not consistently regulated among states and many are not regulated under dam safety programs, instead relying on treatment pond (National Pollutant Discharge Elimination System) or disposal area (solid waste) regulatory programs. In the wake of the dredge cell failure at the Tennessee Valley Authority (TVA) Kingston Fossil Generating Plant in 2008, the United States Environmental Protection Agency (USEPA) inspected 180 power plant facilities and more than 429 impoundments from 2009 through 2011.

This paper presents a statistical summary of the findings of the inspections for the following eight key characteristics: embankment hazard classification, height, downstream slope, evidence of failure, construction materials, evidence of seepage or sloughing, presence of woody vegetation, and major erosion or slope deterioration.

A summary of past failures is presented with an assessment of whether the key characteristics observed during the inspections are sufficient indicators of future failure. The statistical data and the evaluation of past failures indicate that while embankment inspections are standard practice, they should not be solely relied upon to assess embankment safety. Further, the Potential Failure Mode Assessment (PFMA) is offered as an additional method to assess the safety of embankments.

INTRODUCTION

After the dredge cell failure at the TVA Kingston Fossil Generating Plant, the USEPA mandated inspections of disposal unit embankments at power plants between 2009 and 2011 as a part of their national effort to assess the management of CCRs. Findings were summarized in USEPA Coal Combustion Residuals Impoundment Assessment Reports (Reports). The Reports include a checklist for one or more disposal units at each facility.

---

[1] John Seymour, P.E., Geosyntec Consultants, 134 N. LaSalle Street, Chicago, IL 60602, 312-416-3919, jseymour@geosyntec.com
[2] Omer Bozok, Geosyntec Consultants, 134 N. LaSalle Street, Chicago, IL 60602, 312-416-3924, obozok@geosyntec.com
[3] Amanda Hughes, Ph.D., Geosyntec Consultants, 134 N. LaSalle Street, Chicago, IL 60602, 312-416-3931, ahughes@geosyntec.com
[4] Brad Bodine, P.E., CHMM Geosyntec Consultants, 134 N. LaSalle Street, Chicago, IL 60602, 312-416-3909, bbodine@geosyntec.com

The Reports for 180 facilities available on the USEPA website were reviewed.[5] The Reports contained 429 ash disposal unit inspection checklists. The number of disposal units in the data base is slightly more than 429 checklists because some disposal units shared an embankment and were therefore assessed in a single checklist. The checklists were completed by contractors with expertise in the area of embankment integrity and are stamped by a professional engineer.

This paper presents a statistical summary and discussion of checklist responses to key characteristics that are generally considered to be good indicators of embankment conditions and therefore potential failure of the disposal unit. In total, the inspection checklists requested information on 25 topics. Eight key characteristics of embankments were selected for use in this study because they are most likely to indicate potential for failure. The characteristics have discrete (e.g. a simple "yes" or "no") or numeric responses.

This paper then provides a summary of six previous disposal unit failures, which are then compared to the key characteristics to evaluate whether the characteristics can be used as predictors of future failure. Finally, the paper addresses the ability of key characteristics and other inspection data to adequately assess the safety of CCR embankments.

METHOD

The checklists enabled extraction of CCR disposal unit information in a consistent method despite a number of different authors and diverse facility site settings. Additionally, aerial imagery was acquired to aid in the review of facility layout and design characteristics. General information on facility location provided in the Reports was used to obtain accurate aerial imagery.

Key Characteristics

A brief discussion of the eight key characteristics extracted from the inspection checklist follows.

Hazard potential classification. The hazard potential classification is reported as one of the following four categories in the inspection checklist: high, significant, low or less than low. Embankments assigned a high hazard potential classification will likely cause the loss of human life in the event of a failure or misoperation. A significant potential classification describes embankments that are unlikely to cause the loss of human life in the event of a failure or misoperation but may result in economic loss, environmental damage, or disruption of lifeline facilities. Embankments whose failure or misoperation will not result in the loss of human life and have low economic or environmental losses are classified as having a low hazard potential. A less than low hazard potential classification rating is assigned to embankments resulting in no loss of human life, or

[5] The electronic USEPA Coal Combustion Residuals Impoundment Assessment Reports repository used for this evaluation can be found at the following internet address: http://www.USEPA.gov/osw/nonhaz/industrial/special/fossil/surveys2/.

economic or environmental losses in the event of failure or misoperation. Note that the hazard potential classification does not provide an indication of the potential for failure.

What is the dike/embankment height (feet)? The maximum dike/embankment height for the embankment was typically provided in the inspection checklist or in the text of the Report. For facilities that contained two or more tiers, the total height of the disposal unit's dike/embankment was entered measured from the downstream toe to the crest at the highest point.

What is the dike/embankment downstream/outboard slope? The downstream slope of the dike/embankment was obtained from text descriptions of the facilities or from review of drawings. In instances where a range was provided for the slope, the steepest slope was reported.

Is there evidence of failures (Y/N/Minor)? The inspection checklist contains a section to discuss failures. If this section contained a discussion of past failures (generally defined as a slope stability failure on the outside slope or a failure resulting in the release of CCR materials through the embankment or overtopping the embankment to outside of the containment area) the question was answered "Yes". An answer of "Minor" indicated that failures were not discussed in the inspection checklist but the report identified minor historic or current failures.

What are the construction materials? Descriptions of the dike/embankment construction materials are recorded in the questionnaires as described in the Report. Responses to the dike/embankment construction materials query were reduced to two possible categories: "Ash" or "Other" during data processing. Materials classified as "Ash" include bottom ash, fly ash or flue gas desulphurization (FGD) waste. Embankments constructed with a mixture of ash and other materials were classified as "Ash". Embankments that did not include ash as a construction material were classified as "Other."

Is there evidence of seepage and/or sloughing (Y/N)? Information on seepage and sloughing was obtained from the inspection checklist. The inspection checklist question regarding sloughing (checklist question #18) only provides information on conditions observed during the inspection. Seepage is addressed by two parts of the inspection checklist: (i) checklist question #21 addresses seepage observed during the inspection, and (ii) a supplement page is provided that requests information on past seepage and subsequent mitigation measures. A "Y" answer indicates sloughing and/or historic or current seepage has occurred at the facility.

Are trees growing on embankment (Y/N)? Checklist question #9 relates the presence or absence of woody vegetation on slopes of the embankment.

Is there major erosion or slope deterioration (Y/N)? Checklist question #19 of the inspection checklist relates the presence of major erosion or slope deterioration on the embankment.

Database Development

At an initial stage, the type of impoundment was identified based on the criteria in the USEPA inspection checklist. Incised impoundments were excluded from the analyses because they do not have the ability to fail like a diked impoundment. Cross–valley, embankment, side-hill impoundments, or any combination impoundment that included one of these three types, were retained in the analysis. Additionally, embankments were excluded if they were considered landfills or did not receive CCR.

Generally, the review was limited to the report text, tables, and figures in the Report, including the checklist. If data could not be found in the report after a reasonable examination, "unknown" was entered into the database. A "reasonable examination" included reading applicable sections of the text, tables, figures and the inspection checklist in the Report.

A Report was not prepared for the Kingston Dredge Cell Area as it existed in 2008. Therefore, to permit comparison to other facilities, information on physical and operational features for the Kingston Dredge Cell Area was obtained from other existing reports and interviews with knowledgeable personnel. The primary source of data was the 2009 "Root Cause Analysis of TVA Kingston Dredge Pond Failure on December 22, 2008" (AECOM June 25, 2009).

The authors did not verify the accuracy of the Reports. However, many of the Reports contain more than one value for the same key characteristic. For example, an impoundment's dike/embankment height may be assigned differing values on site drawings, in the main body of text, and in the inspection checklist. While these differences may represent reporting errors, discrepancies may also be attributed to actual variations in the dike/embankment height or to embankment alternations occurring over time. While such discrepancies were observed intermittently in the Reports, differences between the reported values were generally small.

Responses to the eight key characteristics are compiled in a Microsoft Access® database. From the database, trends and ranges in embankment characteristics are quantified.

Approximately 30% of the facilities were randomly selected for quality control (QC). Database output for 54 facilities (and all associated impoundments) was evaluated against the original Report to verify accuracy. Any errors identified during the QC review were evaluated to determine if disagreements were because of data entry error, differing report interpretation methods, or systemic data handling procedures. No systemic errors were identified in the QC review process.

Graphics were then created in Microsoft Excel®. Histograms relate the frequencies of numeric responses by bins.[6] Summaries of characteristics having discrete responses are related in pie charts. Findings are discussed in the Results section.

---

[6] Bin sizes are discrete intervals appearing on the x-axis of a histogram.

RESULTS

The results of the data assessment of all impoundments (greater than 429) described in the 180 Reports are summarized as follows:

1. <u>Hazard Potential Classification.</u>

The hazard potential classifications of embankments described in the Reports are as follows: 45 (10%) high, 198 (46%) significant, 160 (37%) low, and 25 (6%) less than low (Figure 1). One embankment was not rated. Three of the embankments that had previously failed were classified as posing a significant hazard and three were classified as high hazard.

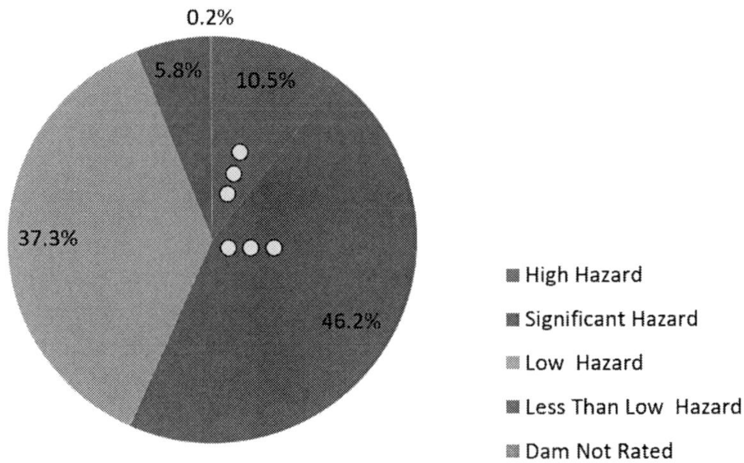

Figure 1. Hazard potential classifications of embankments. Yellow circles indicate the category of failed embankments.

2. Embankment Height.

Histograms for embankment heights for all impoundments are provided in Figure 2. The tallest embankment (400+ feet) is at the FirstEnergy Little Blue Run impoundment. The results indicate the majority of the embankments had a height greater than 2 feet and less than or equal to 25 feet (227 embankments, 53%), and greater than 25 feet and less than or equal to 50 feet (132 embankments, 31%). The heights of the six embankments which have previously failed ranged from 28 feet to 85 feet, with four embankments in the 25 to 50 foot range, one embankment in the 50 to 75 foot range, and one embankment in the 75 to 100 foot range.

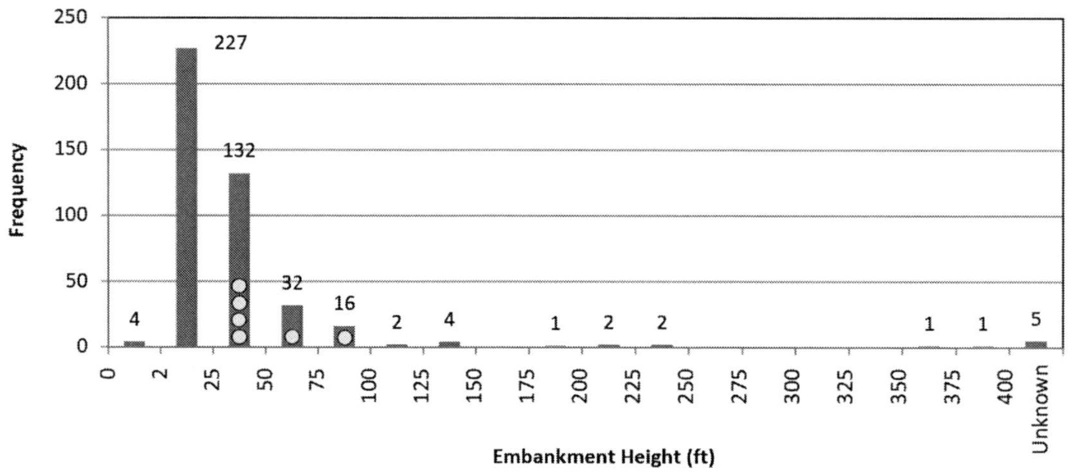

Figure 2. Embankment height. Yellow circles indicate the category of failed embankments. Bins are inclusive of their maximum value (e.g. 227 embankments have heights that are greater than 2 and less than or equal to 25 feet).

3. Downstream Slope.

Reported downstream slopes of embankments ranged from 0.25H: 1V to 30H: 1V (Figure 3). The downstream slopes of 54 embankments were not described in the Reports. The median slope was 2.5H: 1V, the average slope was 2.67H: 1V, with a standard deviation of 2.3. Of the six embankments that previously failed, three had slopes of 2H: 1V, two had slopes of 3H: 1V, and one embankment had a slope of 4H: 1V.

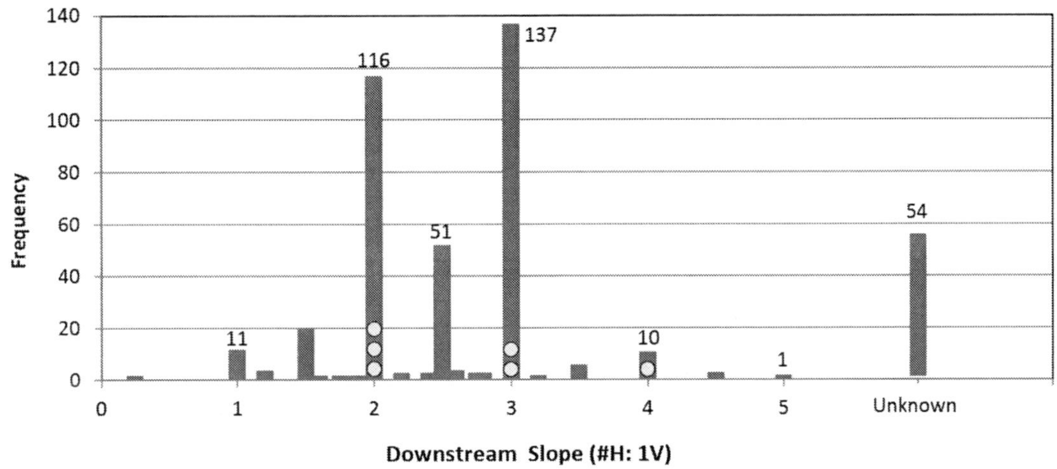

Figure 3. Histogram of downstream slope (#H:1V) of embankments.
Yellow circles indicate the category of failed embankments.
Three embankments with slopes greater than 5H:1V are not shown.

An example of the height of a 145-ft high, fly ash pond, earthen embankment and downstream slope are shown in Figure 4.

Figure 4. An example downstream slope of a fly ash pond earthen embankment.
Source: Lockheed Martin, 2009

4. Evidence of Failures.

Only *evidence of failures* that may lead to failure is included; the data exclude the six failed embankments or facilities if a feature other than the embankment failed. Four of the failed facilities occurred before the inspection and are excluded. The results indicate that a minor or significant failure has occurred at 73 (17%) of the reviewed CCR impoundments (Figure 5). The remaining 83% of embankments are reported to have no evidence of failures. One embankment (Primary Ash Storage Basin at the Dan River Steam Station) failed after the inspection. USEPA inspections for the five remaining embankments that failed prior to the USEPA inspections had evidence of failures that was not related to the more significant previous failure.

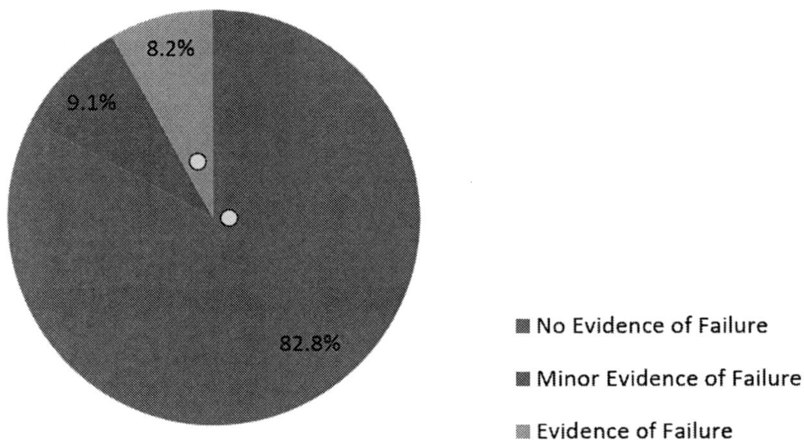

Figure 5. Evidence of failure of embankments.
Yellow circles indicate the category of failed embankments.

5. **Construction Materials.**

The construction materials of 62 (14.5%) of the embankments reviewed are either entirely or partially of ash (Figure 6). The remaining 367 embankments reviewed (86.5%) are earth embankments and did not include ash. Three of the six embankments that previously failed contained ash.

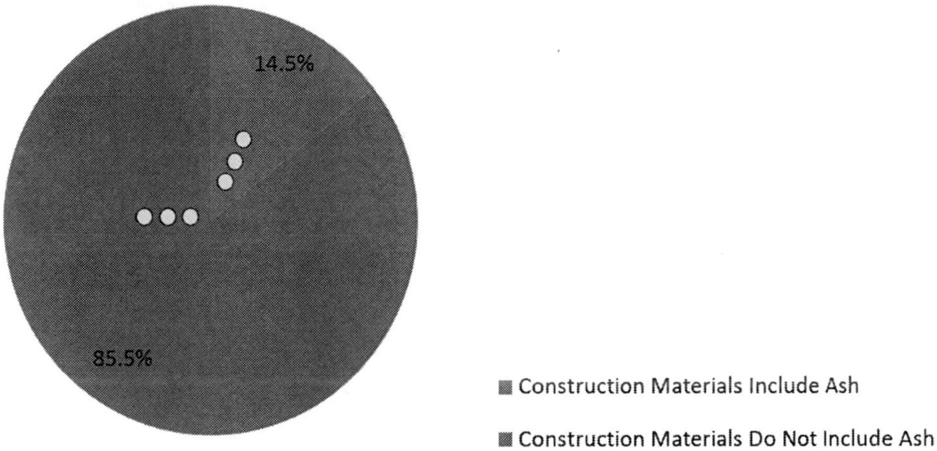

Figure 6. Embankment dike construction materials.
Yellow circles indicate the category of failed embankments.

6. **Evidence of Seepage or Sloughing**.

Evidence of seepage was present at approximately 40% (171 embankments) of the reviewed impoundments, including five embankments that previously failed (Figure 7). The remaining 60% of embankments (258) had no evidence of seepage or sloughing.

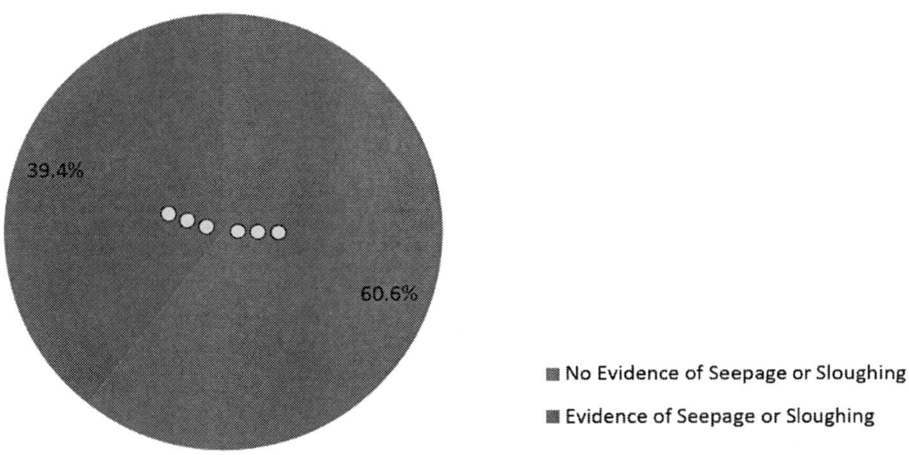

Figure 7. Evidence of seepage or sloughing observed on embankments.
Yellow circles indicate the category of failed embankments.

An example of evidence of seepage is shown in Figure 8 and a slough is shown on Figure 9.

Figure 8. An example of perennially wet ground and cattails from seepage.
Source: Lockheed Martin, 2009

Figure 9. An example of a slough/scarp.
Source: GZA, 2012

7. Presence of Woody Vegetation.

Woody vegetation was present on 167 (39%) of the embankments having Reports, including two embankments (Secondary Fly Ash Pond at Baldwin Power Station and the Gypsum Stack Embankment at Widows Creek Fossil Plant) that had previously failed (Figure 10). The remaining 262 embankments reviewed (61%) were reported to not have woody vegetation present on the slopes.

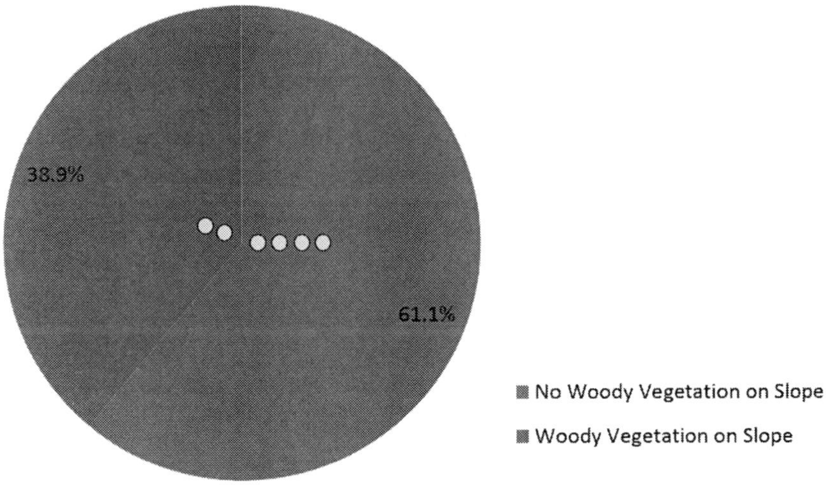

Figure 10. Woody vegetation on embankment slope.
Yellow circles indicate the category of failed embankments.

An example of woody vegetation on embankment slope is shown on Figure 11.

Figure 11. An example of woody vegetation on a downstream slope.
Source: Dewberry, 2011

8. **Major erosion or slope deterioration.**

The majority (364, 85%) of the embankments described in the Reports were reported to not have major erosion or slope deterioration (Figure 12). Five of the six embankments that previously failed reported no major erosion or slope deterioration; we have made the assumption that this condition was reflective of the failed facilities at the time of failure. The remaining 65 (15%) of the embankments reviewed reported major erosion, including one of the embankments that had previously failed.

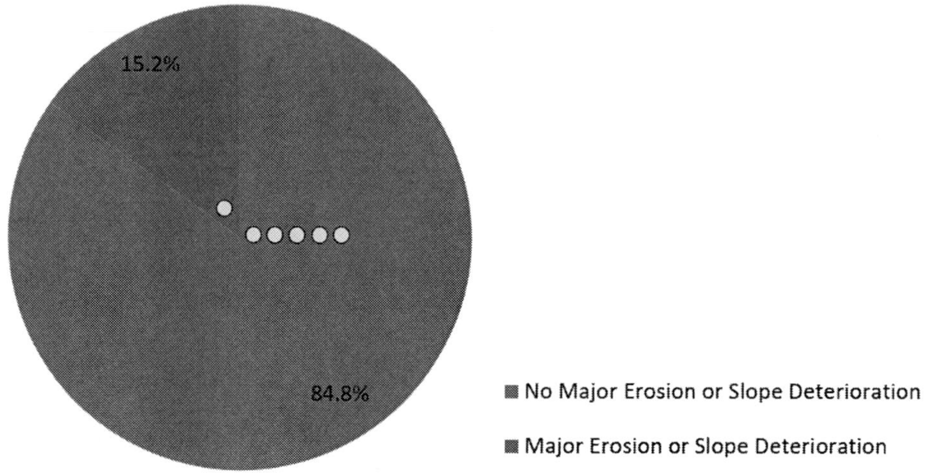

Figure 12. Major erosion or slope deterioration on embankment. Yellow circles indicate the category of failed embankments.

PREVIOUS FAILURES

This section presents a summary of documented failures. Failure is described as a loss of containment of CCR stored behind the embankment. The failures are presented chronologically from the oldest to most recent. Of the six failures, only one failure, that was located at the Dan River Facility, occurred after the inspection was conducted.

This section also examines: (i) whether the failure could have been predicted based upon information that was obtained during a visual inspection - for the case where failure occurred after the inspection; and (ii) whether the failure could have been predicted based upon information that would have been obtained if the inspection was conducted before the failure.

A summary of the eight key characteristics for six facilities with previous failures is presented in Table 1.

Table 1. Report Data for Failed Facilities.

Impoundment Name	Hazard Classification	Embankment Height (ft)	Approximate Downstream Slope (#H:1V)	Evidence of Failure	Embankment Construction Material	Evidence of Seepage or Sloughing	Woody Vegetation on Slope	Major Erosion or Slope Deterioration
Dan River Steam Station Primary Ash Storage Basin	Significant	40	2	No	Other	Yes	No	No
Baldwin Power Station Secondary Fly Ash Pond	Significant	55	3	Yes	Other	No	Yes	No
Eagle Valley Generating Station Pond D	High	38	3	Yes	Ash	No	No	Yes
Eagle Valley Generating Station Pond E	Significant	28	3	Yes	Ash	No	No	Yes
Martin's Creek Electric Station Ash Basin Number 4	High	45	2	Yes	Other	No	No	No
Tennessee Valley Authority Kingston Dredge Cell	High	85	4	Yes	Ash	Yes	No	No
Widows Creek Fossil Plant Gypsum Stack	Significant	35	2	Yes	Other	Yes	Yes	No

Baldwin Facility

The Baldwin Ash Pond Facility had a deep-seated embankment failure in 1995. The facility is located in Baldwin, Illinois. An aerial view is provided in Figure 13. The embankment was reported to be constructed using "earth fill" and "impervious fill" that is presumed to be locally obtained loess deposits and low plastic clay, respectively. The embankment was originally constructed to be 35-ft high and later was vertically extended with a 20-ft dam rising. The raised embankment was constructed out of clayey soil over bottom ash and with no offset from the original embankment (GZA 2012).

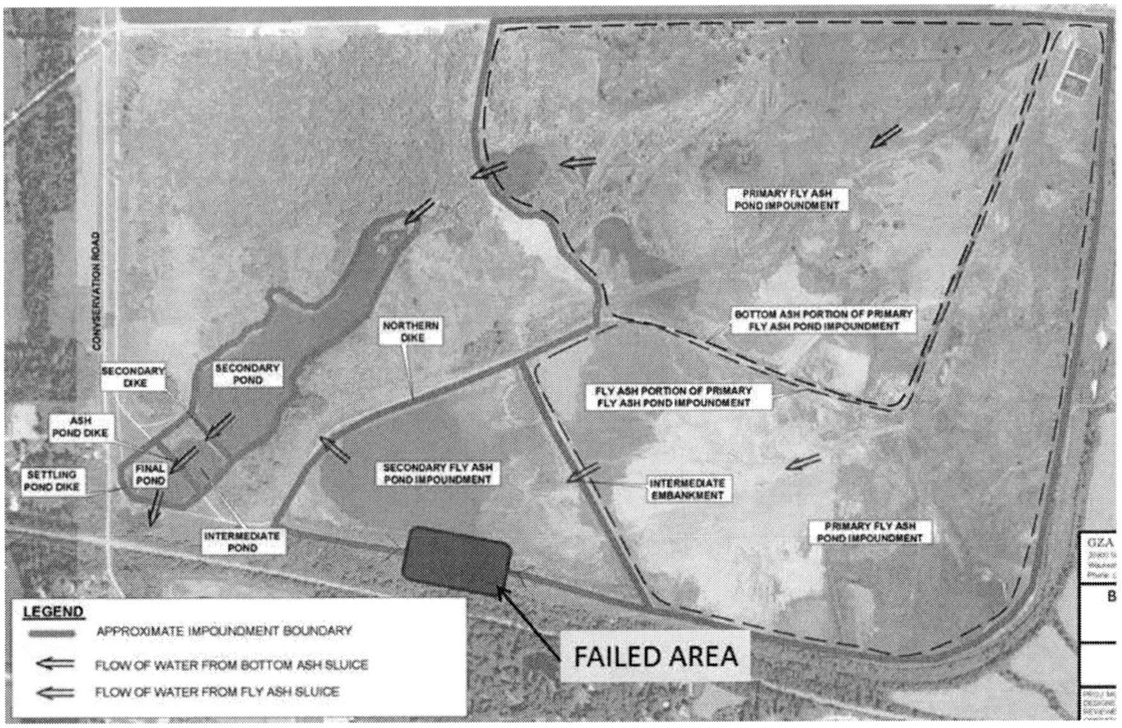

Figure 13. Baldwin Facility Site Plan
Source: GZA, 2012, with highlights by the authors

The downstream side slope of the embankment where the failure occurred was 3H:1V. Failure analysis indicated that the failed section of the embankment had a factor of safety less than the generally excepted value of 1.5. No further information was provided.

Of the key characteristics, "evidence of seepage or sloughing" and "major erosion or slope deterioration" were identified at the facility in the USEPA inspection. If those features existed before the failure, they would have been an indication of a higher potential for failure.

Martins Creek Facility

The Martins Creek Steam Electric Station Facility is located in Martins Creek, Pennsylvania. An aerial view is provided in Figure 14. The facility had four CCR treatment and disposal ponds that were utilized for disposal of CCRs. In 2005, an incident occurred where a discharge water system failed causing a significant release of water and fly ash. The mixture of water and fly ash spread into surrounding fields, into Oughoughton Creek and the Delaware River. The coal fired units have been retired since 2007 and the ponds were decommissioned.

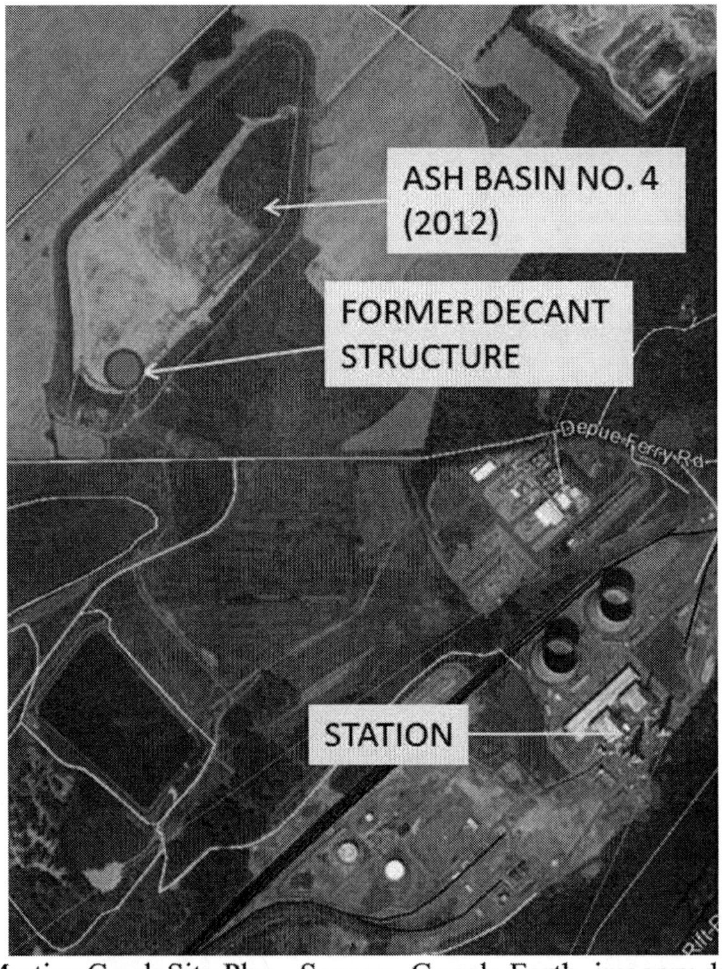

Figure 14. Martins Creek Site Plan. Source: Google Earth, imagery date 5/19/2012

The cause of the failure was reported to be a failed wooden stop log, which was reported to have had a fabrication defect. The stop log was the second (out of forty) from the bottom and is presumed to normally have been submerged. Once the stop log failed, there was no shutoff valve in the discharge pipe or other secondary barrier to prevent the spill or slow the release. The failure of the stop log caused an uncontrolled release of approximately 100 million gallons of fly ash slurry over three day period (D&D 2009). USEPA considered this incident as a proven damage case based on the results of an environmental study conducted after the failure. It is reported the response action cost approximately $37 million (USEPA 2010).

It appears that the failed stop log could only have been visually inspected using a remote underwater video camera. If it could have been observed, the defect would likely not have been detected through visual observation by remote camera inspection. It appears that none of the typical inspection characteristics would have identified a manufacturing defect and therefore the visual inspections could not have predicted this failure.

Eagle Valley Facility

The Eagle Valley Generating Station Facility is located north of Martinsville, Indiana. An aerial view is provided in Figure 15. It has five ponds (Ponds A through E) that receive various slurried CCR products for treatment. These ponds are separated with internal divider embankments.

Figure 15. Eagle Valley Site Plan.
Source: CDM, 2010, with highlights by the authors

The embankment was designed and constructed with generally 3H:1V slopes. Embankment failures occurred at the facility in 2007 and 2008. Both failures occurred in the same embankment separating Pond D and Pond E. The Reports indicate that the failed embankment was made of fly ash. The containment embankments were reported to be constructed of "compacted fill & ash" or "earthen/ash". These two failure events are considered one failure because they occurred at the same location.

In 2007, the embankment between Ponds D and E failed causing Pond D to flow into Pond E causing the northern embankment of Pond E to overtop, releasing 30 million gallons of treated pond water into the discharge canal. A similar failure occurred in 2008 at the same location causing another 30 million gallons of treated pond water to spill into the discharge canal.

The cause of the 2007 failure was reported by BT Squared to be slope instability combined with piping (seepage that carries or erodes soil particles from within the embankment) through the embankment (CDM 2010). The cause of the 2008 failure was reported by BT Squared to have resulted from an overestimation of the shear strength of fly ash by the designer and the redesigned embankment was inadequate to support the loading conditions on the embankment between Pond D and Pond E (CDM 2010). For an embankment failure of this type the steepness of the slope would be generally reflective of the design. For example, steeper slopes would be reflective that the embankment materials and subgrade are stronger than for an embankment with a flatter slope given an analysis yielding the same factor of safety for both slopes. Therefore, in both failure instances, the embankment slope, which is approximately 3H:1V, would not have been a good indicator of potential instability because all of the slopes for Pond D and Pond E were reported to be constructed at slopes of approximately 3H:1V and are constructed of similar materials and only the divider embankment failed.

Further, a visual inspection would not have been predictive of failure if the cause is a design parameter that was not reflective of the strength of the embankment. However, if piping would have been observed during an inspection, it could have been an indicator of potential future instability, or at a minimum, an indication that further analysis would be merited.

Kingston Facility

The Kingston Fossil Plant Facility is located in Harriman, Tennessee. An aerial view is provided in Figure 16. It had two main CCR handling areas: (i) Ash Pond C, used for treating and storing fly ash and bottom ash; and (ii) the FGD Landfill, used for storing FGD sludge. In addition to Ash Pond C and the FGD Landfill, the plant had a dredge cell to store fly ash (currently decommissioned due to the failure in 2008). The dredge cell had received CCR in the form of slurry until it reached its capacity in the early 1980s, after which the dredge cell was vertically expanded by constructing elevated cell containment embankments primarily out of ash. The expansion embankment was set back a minimum of 200 ft from the main embankment surrounding the dredge cell. The facility underwent a number of subsequent raisings of the containment embankment using the upstream construction method.

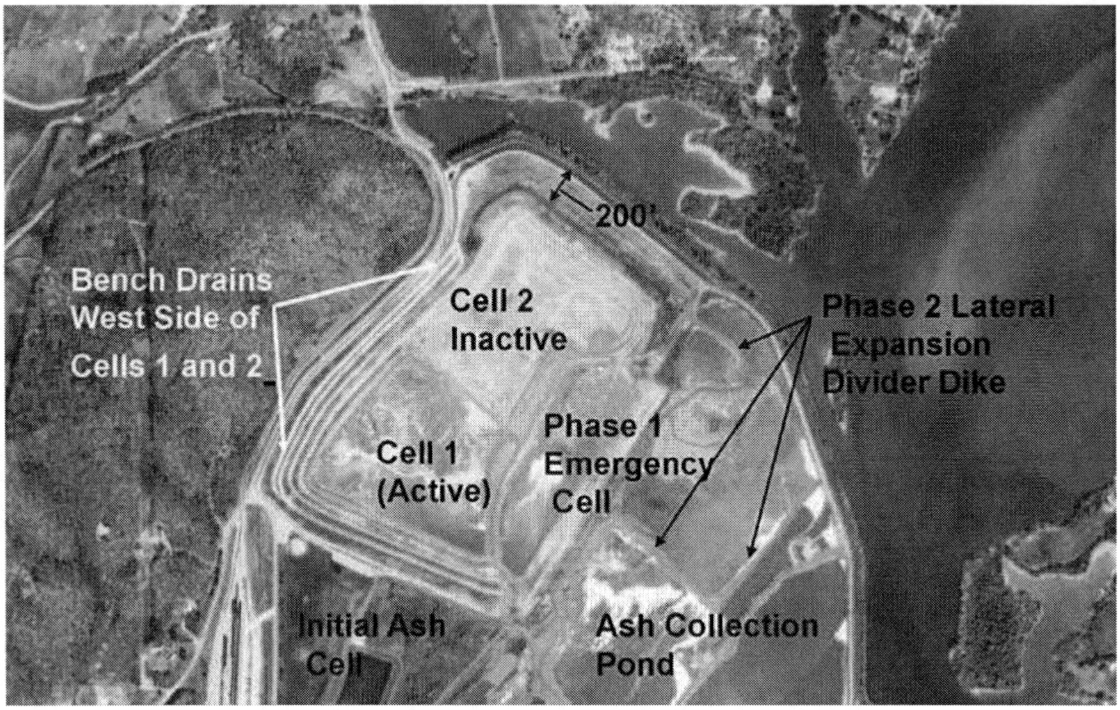

Figure 16. Kingston Dredge Cell Site Plan (2008).
Source: AECOM, 2009A

In 2008, the dredge cell had a major failure in the northern portion of the site causing an uncontrolled release of approximately five million cubic yards of ash slurry that flooded more than 300 acres of land, impacted the Emory River and Clinch River, disrupted power, ruptured a gas line, knocked one home off its foundation and damaged others. Fortunately, there were no casualties. The failure caused the largest ash release in the United States.

An aerial view of the facility after the failure is shown on Figure 17.

Figure 17. An aerial view of Kingston Dredge Cell 2 Failure.
Source: AECOM, 2009A

The release occurred as a result of a series of progressive failures of dredge cell containment embankments that were constructed as part of vertical expansions of the dredge cell. The dredge cell containment embankments failed over a period of hours. It was reported that several factors lead to failure, including: i) the relatively fast rate of vertical expansion; ii) the nature of loose wet ash foundation materials below the containment embankments; iii) fill setback and geometry, and iv) unusually weak slimes layer located under the raised dredge cell containment embankments (AECOM, 2009). The cost of failure is estimated to be approximately $3 billion including the cleanup cost and ecological and socio-economic damages (USEPA 2010).

The USEPA could not conduct an inspection of the facility as it had previously failed and was not rebuilt. However, the key characteristics from the USEPA inspection checklist were recreated by Geosyntec personnel that had worked at the dredge cell before the failure and from the Root Cause Analysis (AECOM, 2009B).

The following key characteristics were identified in the recreated inspection sheet:

- Seepage was identified in areas that were addressed along Swan Pond Road by improvement of drainage (pore pressure relief) from behind the embankment. Because seepage mitigation did not occur on the failed embankment, it is assumed that seepage was not identified in TVA inspections along the failed embankment.
- Sloughing was identified along Swan Pond Road. Because sloughing mitigation did not occur on the failed embankment, it is assumed that sloughing was not identified in TVA inspections along the failed embankment.

- The slopes of the embankment were approximately 4H:1V which is significantly flatter than the majority of the data which indicates 42% are in the range of 2.5H:1 to 3H:1V. For comparison, the Eagle Valley divider dike was sloped at 3H:1V.
- Evidence of failure could be considered to have been identified along Swan Pond Road but that "failure" was not a loss of containment from behind the containment berm. However, the "failure" was relatively small (several dump trucks) from the surface of the embankment that was primarily made out of fly ash. This failure is more accurately considered a "slough".
- The embankment was constructed primarily of fly ash. The data indicate that 14% of the embankments evaluated for this study were made primarily of ash.

The inspection data collected does not appear to have been a good predictor of failure. The slopes were approximately 4H: 1V, flatter than other failed slopes constructed primarily of fly ash (failed embankment at the Eagle Valley facility had approximately 3H:1V slopes). Seepage and more importantly piping and significant erosion were not reported[7] on prior periodic TVA inspections in the failed embankment. Further, one of the causes of failure was the slimes layer, which was not comprehended prior to the failure.

Widows Creek Facility

Widows Creek Fossil Plant Facility of Tennessee Valley Authority, located in Stevenson, Alabama, has two main CCR collection areas: (i) the Main Ash Complex – used for impounding fly ash and bottom ash; and (ii) the Gypsum Stack – used for storing gypsum. An aerial view is provided in Figure 18. The Gypsum Stack received sluiced gypsum prior to failure. TVA would then mechanically collect the wet gypsum, spread it out within the dredge cell for drying, and when dried, placed the material onto the stack.

[7] Significant erosion was documented along Swan Pond Road, but that area was repaired and internal drainage was constructed and this portion of the embankment was not the main failed embankment.

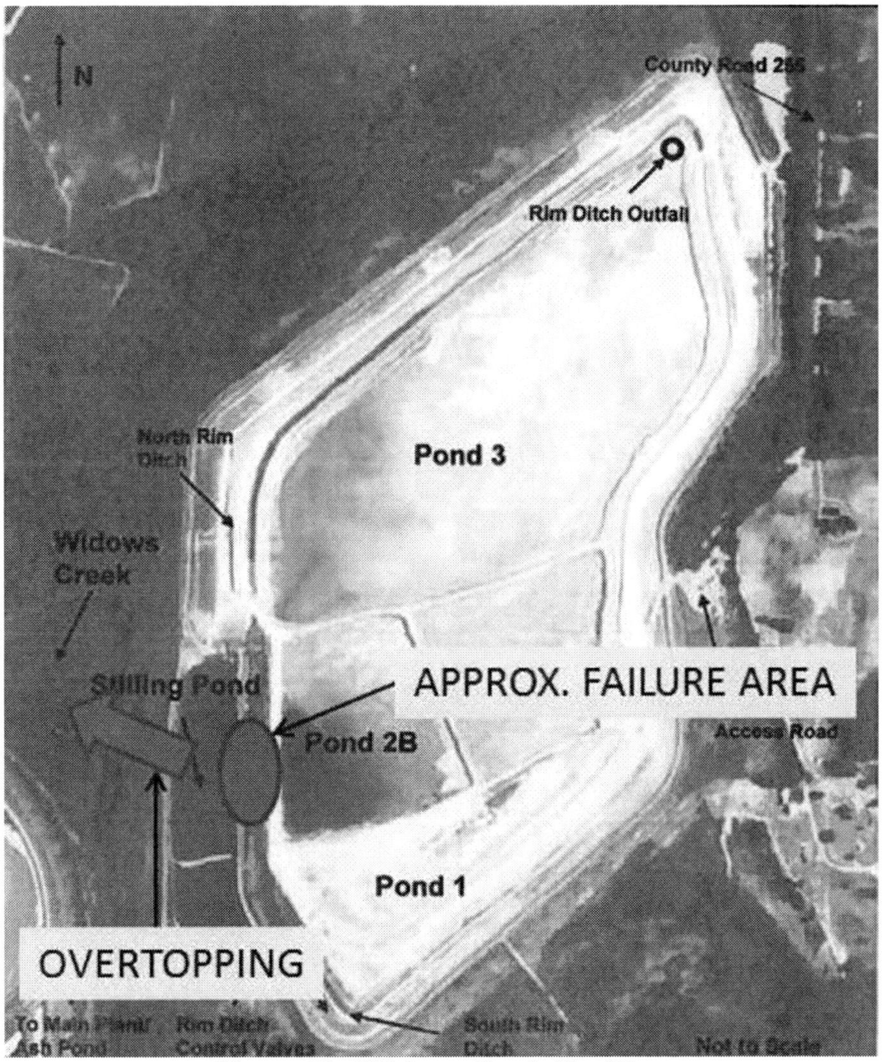

Figure 18. Widows Creek Site Plan.
Source: Dewberry & Davis, 2012 (Appended Stantec February 2010 Report of Phase 2 Geotechnical Exploration) with highlights by the authors

There were two incidents that occurred at the facility, one in 2009 and one in 2010, each causing an uncontrolled release of gypsum. In 2009, an abandoned weir from an internal drainage system failed releasing gypsum into the gypsum stack stilling pond and into Widows Creek (Dewberry & Davis 2012). Approximately 6.1 million gallons of slurry carrying approximately 5,000 cubic yards of gypsum and some fly ash was released into Widows Creek which flows into the Tennessee River. The cleanup cost for the 2009 incident was reported to be approximately $9.2 million (USEPA 2010). USEPA did not classify this incident as a damage case, because samples at relevant points of potential exposure did not exceed applicable standards (USEPA 2010).

In 2010, a newly constructed spillway pipe from a stormwater pond in the Gypsum Stack washed out causing gypsum to slide down the stack and into the stilling pond. It was reported that spill in 2010 was contained in the stilling pond (D&D 2012) and was not released into nearby Widows Creek and therefore is not a "failure" as defined here.

It appears that none of the typical inspection parameters included in this analysis would have been a precursor to failure. However, checklist question 21 provides related information on seepage, including whether seepage is coming from "Around the outside of the decant pipe". While it was an "abandoned weir" from an internal drainage system that failed and not a decant pipe, it could be presumed that an experienced inspector would have noted seepage from around the weir if it was on the inspection checklist.

Dan River Plant

The Dan River Plant is located in Eden, North Carolina. It has two ash ponds, the Primary Pond and Secondary Pond, both of which were utilized for impounding CCR slurry. An aerial view is provided in Figure 19. These ponds were constructed over 36-in. and 48-in. diameter stormwater pipes that had been installed prior to construction of the ash ponds. It was reported that a minimum of three feet of soil cover was maintained between the top of the stormwater pipes and the bottom of the ponds (PCR 2009).

Figure 19. Dan River Site Plan.
Source: Paul C. Rizzo & Associates, 2009 with highlights by the authors

In February 2014, a portion of the 48-in stormwater pipe under the Primary Pond collapsed (NCDENR 2014), causing approximately 30,000 to 39,000 tons of ash to drain into the Dan River. A timeline of events is presented here:

http://portal.ncdenr.org/web/guest/dan-river-spill

Visual inspection of the collapsed pipe performed after the failure revealed that portion of the pipe where the pipe collapsed was corrugated metal pipe. The rest of the pipe was concrete (NCDENR 2014).

It appears that it is unlikely that any of the typical inspection characteristics would have identified a precursor to failure of a buried stormwater pipe.

CONCLUSIONS

The review of the selected data from the USEPA inspection sheets indicated that four of the six facilities did not inspect the features that contributed to failure. Of the other two, one (Eagle Valley) could have identified piping if the divider dike was on the inspection regime, and the other (Baldwin) did not have enough information to draw a conclusion.

Therefore, the inspection data do not appear to be reliable to predict failure based upon the six observed failures (which is a relatively small sample size) out of hundreds of facilities. However, inspections are required by the standards of the industry and are effective methods of maintaining embankments that reduces the potential for failure.

There are other tools available that could be used to assess the potential for an embankment to fail and develop a monitoring program to detect and mitigate signs of potential failure. These tools are the standard of practice for water retention dam safety in the United States. The Federal Energy Regulatory Commission (FERC) utilizes dam safety performance monitoring plan tools for hydropower dams. They can be found at:

http://www.ferc.gov/industries/hydropower/safety/guidelines/dspmp/background/per-plan.asp

In summary, FERC states:

> *"The Engineering Guidelines, Draft Chapter 14 titled, "Dam Safety Performance Monitoring Program" provides recommended procedures and criteria to develop a Performance Monitoring Program based upon "potential failure mode thinking" which assists in reviewing and evaluating the safety and performance of water retaining structures regulated by FERC".*

This approach can be used for any water retention structure, including embankments that retain water and wet CCRs. Essentially, a Potential Failure Mode Analysis (PFMA) is conducted and a Performance Monitoring Plan (PMP) is customized for the facility based on the PFMA.

Per FERC,

"The PFMA is conducted jointly by the licensee, Independent Consultant and FERC staff. This process is guided by a facilitator. For the most part the PFMA is a one-time exercise. Due to its "question asking approach" the potential failure mode examination process:

- *Enhances the dam safety inspection process*
- *Enhances and focus the visual surveillance and instrumented monitoring program*
- *Identifies shortcomings or oversights in data, information or analyses necessary to evaluate dam safety and a potential failure mode*
- *Helps identify the most effective dam safety risk reduction measures.*

Based upon the results of the PFMA, the Performance Monitoring Program is developed. The PMP defines the appropriate monitoring for the water retaining structures based upon the PFMA."

The integration of a PFMA with a PMP results in an efficient and effective dam safety program. The following are examples of the author's experience where a PFMA was conducted to identify and mitigate potential failure modes.

1. An abandoned spillway structure that had not been filled with concrete or grout under the embankment that eventually could collapse, similar to the Dan River facility, was identified and subsequently filled with grout to reduce the potential for such failure.
2. Deteriorating stop logs were identified and subsequently repaired, which reduced the potential for uncontrolled discharge, similar to the Martins Creek facility.

Additional potential failure modes (PFMs) were identified, prioritized and addressed such that the potential for failure was reduced. The keys to the success of the PFMA process was the openness of all Owner stakeholders (plant operations, safety, environmental and legal personnel) to identify the PFM, initiate and maintain a performance monitoring program, and prioritize mitigation to address the highest priorities first followed by lower priorities.

QUALIFICATIONS

This paper presented the raw data from 180 inspection reports completed by many consulting engineering firms and took that data and compared it to six CCR facility failures. This paper did not provide an independent nor exhaustive analysis of all of the failures. Recognizing that most if not all of the six facilities also involve litigation and/or third party damage assessments, the conclusions are not intended to provide any new causes of failure or responsibility for failure, but relied upon statements in the literature that were available.

Therefore, conclusions have been drawn to provide general guidance that there are more thorough methods to assess embankment stability than uniform visual inspections. Those other methods are the PFMA and PMP processes implemented at FERC facilities.

REFERENCES

AECOM, "Kingston RCA, Kingston Dredge Cell Failure Root Cause Failure Analysis", public presentation, June 2009A

AECOM, "Root Cause Analysis of TVA Kingston Dredge Pond Failure from December 22, 2008", June 2009B

CDM. "Assessment of Dam Safety of Coal Combustion Surface Impoundments – Indianapolis Power & Light Company, Eagle Valley Generating Station, Martinsville, IN" Final, November 2010

Dewberry & Davis, LLC. "Coal Combustion Waste Impoundment Dam Assessment Report – Martins Creek Steam Electric Station – PPL Generation, Bangor, PA" Final, December 2009

Dewberry, LLC, "Coal Combustion Waste Impoundment Round 7- Dam Assessment Report, Lansing Generating Station, Interstate Power and Light Company (IPL), Lansing, Iowa, March 2011.

Dewberry Consultants, LLC. "Coal Combustion Residue Impoundment Round 11 – Dam Assessment Report - Kingston Fossil Plant, Tennessee Valley Authority, Harriman, TN" Final, April 2013

Dewberry & Davis, LLC. "Coal Combustion Residue Impoundment Round 11 – Dam Assessment Report - Widows Creek Fossil Plant, Tennessee Valley Authority, Stevenson, AL" Final, December 2012

GZA GeoEnvironmental, Inc. (GZA) "Round 10 Dam Assessment, Dynegy Midwest Generation, LLC – Baldwin Energy Complex Primary Fly Ash Pond, Secondary Fly Ash Pond, Secondary Pond, Intermediate Pond, Final Pond" Final, December 2012

Lockheed Martin, Final Report, Assessment of Dam Safety of Coal Combustion Surface Impoundments, American Electric Power General James Gavin Power Plant, Cheshire, OH, September 2009

N.C. Department of Environmental and Natural Resources "Dan River Coal Ash Spill – Time Line of Events", May 2014, URL: http://portal.ncdenr.org/web/guest/dan-river-spill

Paul C. Rizzo Associates, Inc. "CCW Impoundments Inspection Report - Dan River Steam Station, Rockingham County, NC" Final, September 2009

U.S. Environmental Protection Agency (U.S. EPA). "Hazardous and Solid Waste Management System; Identification and Listing of Special Wastes; Disposal of Coal Combustion Residuals from Electric Utilities" Federal Register, June 2010

GEOTECHNICAL PROPERTIES AND DIAGENESIS OF PONDED FLY ASH

Robert C. Bachus, PhD., P.E., D.GE[1]
J. Carlos Santamarina, PhD[2]

ABSTRACT

Ponded fly ash is a unique sediment, exhibiting distinct particle and bulk-scale characteristics compared to natural sediments. However, these materials are not unlike materials retained by mine tailings dams and are often managed similar to mine tailings. A recently completed EPRI-funded laboratory study focused on assessing the fundamental geotechnical engineering properties of fly ash and its behavior in ash disposal ponds. This information is important to understand from the perspective of stability of the impounded material, as the ponded fly ash serves as the foundation for the cover system used to close a decommissioned ash pond. For the EPRI study, samples of ponded and dry fly ash were obtained from several electric utilities and subjected to extensive laboratory characterization. Laboratory test results helped discern the distinguishing physical and chemical characteristics of ponded fly ash, and led to the identification of potential pitfalls in using standard geotechnical correlations and analyses when dealing with these materials, particularly those correlations that include reference to specific gravity, void ratio, and unit weight. Importantly, short-term aging was observed in some fly ash specimens. This early diagenesis is readily detected by shear wave velocity and has a marked effect on self-weight consolidation, which alters the contractive/dilative tendency of fly ash subjected to shear. In most cases, laboratory results indicate that "water pluviated and immediately loaded" fly ash specimens would tend to dilate upon shear in ponds of moderate depth. These tendencies are contrary to those of materials prone to static liquefaction, a phenomenon that has been attributed to ponded fly ash and potentially to loose low-plasticity tailings. Results from this testing program help guide the design of proper field characterization studies and are relevant to the selection of strategies for "closing" existing fly ash ponds. These results may prove beneficial in understanding the behavior of ponded tailings materials. This paper and presentation provides a summary of the engineering properties of ponded fly ash, examines the mechanism of diagenesis of these materials, and assess the potentially significant effects of diagenesis on the engineering performance and the flow potential of hydraulically placed materials in fly ash ponds and behind tailings dams.

INTRODUCTION

The sluicing and wet disposal of coal combustion residuals (CCRs) including fly ash in unlined basins or ponds is one of the more common disposal techniques in the Eastern U.S. A 2008 accident at the Tennessee Valley Authority (TVA) Kingston facility in Harriman, TN resulted in the failure of the containment system for the ponded materials

[1] Principal, Geosyntec Consultants, 1255 Roberts Blvd NW, Kennesaw, GA, 30144, rbachus@geosyntec.com.
[2], Professor, Georgia Institute of Technology, School of Civil and Environmental Engineering, 790 Atlantic Dr., Atlanta, GA, 30332, jcs@gatech.edu

and a significant release of water and water-laden ash to the surrounding environment. The TVA accident involved a dredge cell used for the permanent disposal of fly ash that was hydraulically excavated (i.e., dredged) from an adjacent ash pond and allowed to settle in the new ash pond. After an increment of dredging and settling in the new pond, the perimeter of the dredge cell (i.e., the dike) at the Kingston facility was raised by placing a relatively narrow compacted earth berm over a portion of the previously placed earth berm and sluiced ash in an "upstream construction" manner common to the mine tailings industry. Photographs from the TVA accident confirm that some of the released fly ash comprised intact blocks of material that appeared to float within a liquid-like slurry of fly ash, contributing to the appearance that the fly ash tended to "flow" from the impacted dredge cell.

A phenomenon referenced as "static liquefaction" has been hypothesized as one of the potential contributory mechanisms of the dredge cell failure. Static liquefaction is a phenomenon in which the granular skeleton of a saturated loose soil matrix collapses during undrained shear, resulting in a significant reduction in effective stress and the flow of the granular mass if unconstrained. The TVA accident raised concerns within the electric power industry regarding the geotechnical stability of existing CCR wet disposal facilities and the United States Environmental Protection Agency (USEPA) commissioned investigations and inspections that focused on assessing the geotechnical stability of ash disposal facilities (see http://www.epa.gov/osw/nonhaz/industrial/special/fossil/surveys2/). The industry is currently awaiting a decision from USEPA regarding recommended/required procedures and timelines for managing and closing existing wet-disposal ash ponds. To assess the stability of CCR ponds and to provide appropriate designs for closure systems, the engineering properties of the hydraulically placed (i.e., ponded) in situ fly ash need to be established. While the engineering properties of fly ash have been extensively studied in the past, most of the previous work focused on the beneficial re-use of dry fly ash or ash that had been excavated from wet-disposal ponds. There has been little information provided to date regarding the in situ properties of ponded fly ash. Furthermore, there was even less information available to understand flow potential and static liquefaction of these materials. Personnel from Geosyntec Consultants and the Georgia Institute of Technology recently collaborated on a laboratory study sponsored by the Electric Power Research Institute (EPRI) to assess the engineering characteristics of these materials. This paper was prepared to provide a summary of the findings from the EPRI-funded study and the implications that the study results have on closing existing wet-disposal ash ponds.

OBJECTIVES AND STRATEGY FOR EPRI STUDY

The four primary objectives of the EPRI (2012) study were to: (i) characterize the engineering properties of ponded fly ash; (ii) understand the self-weight consolidation mechanism and static (i.e., non-seismic) shear response of ponded fly ash that could lead to static liquefaction; (iii) develop testing techniques and/or protocols to assess flow potential and the potential for skeletal collapse and static liquefaction of ponded fly ash; and (iv) develop suggested guidelines regarding the site- and material-specific testing and assessment of ponded fly ash. The primary focus of this study was related to ponded fly ash, as opposed to the wider family of CCRs that include dry fly ash, bottom ash, and flue gas desulfurization (FGD) by-products, primarily because: (i) fly ash is anticipated to be the primary by-product in most existing CCR wet-disposal facilities; and (ii) when compared to the other CCRs, fly ash is anticipated to be most susceptible to flow and static liquefaction. The following four-phase testing strategy was followed to accomplish each of the four stated objectives.

- Characterization of Ponded Fly Ash – To develop an understanding of the similarities and differences between ponded fly ash from utilities across the country, bulk samples of fly ash from five participating utilities and a total of 22 coal-fired generating units were obtained for testing. Fly ash was recovered from existing ash ponds. Characterization tests included: specific gravity of solids (G_s), Atterberg limits, particle size distribution, and scanning electron microscopy (SEM). Tests were also performed on selected dry ash samples from the storage silos (i.e., hoppers) adjacent to the coal-fired generating units. Using simple characterization testing concepts, techniques were developed during this study to form specimens at the loose and dense states and to establish the range of "realizable" densities (i.e., the practical minimum and maximum densities that could be achieved under actual field conditions) for the ponded fly ash from a specific disposal facility. Results obtained from this study were compared whenever possible to the historical results compiled from the literature regarding ponded ash from the U.S., Europe, and Asia.

- Self-weight Consolidation and Static Shear Response of Ponded Fly Ash – To complement the characterization tests, specimens were formed in a loose condition and submitted for conventional one-dimensional consolidation (i.e., oedometer) testing. Some of these specimens were tested using end caps modified to include bender elements to assess: (i) shear wave velocity of the consolidated specimens, and (ii) change in shear wave velocity over time to assess the potential for diagenetic aging of the specimens. Similarly, conventional laboratory static triaxial compression tests were performed on specimens to assess strength under both drained and undrained conditions.

- Potential for Flow and Static Liquefaction of Ponded Fly Ash – The primary objective of this study was to assess the potential skeletal collapse of ponded fly ash, which would lead to a loss of strength by static liquefaction and flow. To accomplish this, both laboratory and in situ characterization techniques were investigated. In the laboratory, the characterization, consolidation, and strength tests results were used to assess this potential. To assess in situ conditions, an electrical needle probe and a model cone penetrometer (CPT) were developed to assess stratification and strength variability. Finally, a sedimentation simulation and flow simulation test apparatus was developed to understand fly ash sedimentation and to assess the ponded fly ash flow potential after removal of the containment structure.

- Guidelines for Site- and Material-specific Testing and Assessment of Ponded Fly Ash – The final phase of the study was focused on reviewing the compiled information and extracting guidelines for assessing pond fly ash. These guidelines include standard tests and new procedures to identify mechanical signatures of specific ponded materials in view of their potential to statically liquefy upon containment loss. These guidelines also include comments regarding the importance of the ash containment facility design; however, the authors note that this study did not investigate the engineering properties or design associated with the containment facility.

SUMMARY OF TEST RESULTS

Details regarding the laboratory testing program and a comprehensive presentation of testing results are beyond the scope of this paper but are documented in ERPI (2012). In general, however, most of the tested materials were characterized as Class F fly ash and were obtained from coal-fired electric power utilities located in the Midwest and Southeast U.S. This section is organized to provide a summary of the results and select graphics.

Characterization Test Results

Grain Size. Fly ash is typically referenced as a low-plasticity, silt-sized material. Experimental results from this study are consistent with this characterization and the tested fly ash specimens had a median grain size of approximately 7μm to 50μm. The materials were relatively well graded with a coefficient of uniformity (Cu) between 4 and 20 in most cases. A representative set of individual grain size curves is presented in Figure 1. The horizontal "bar" at 50 percent passing corresponds to the range in particle sizes as previously reported in the literature.

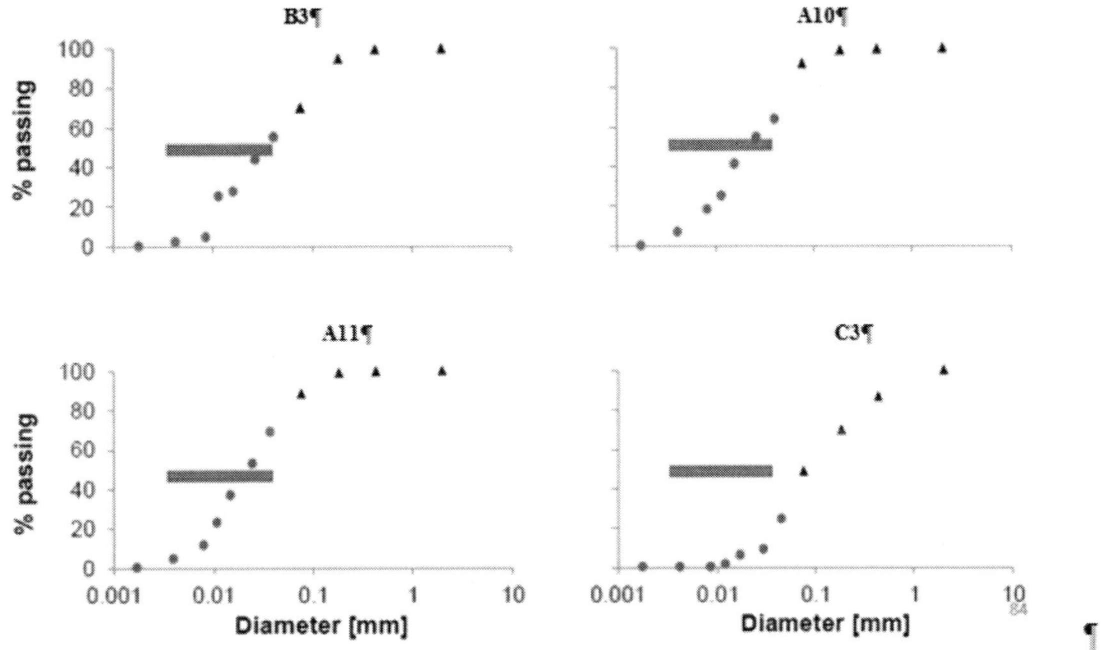

Figure 1. Representative Particle Size Distribution Curves
(Note: The horizontal "bar" at 50 percent passing corresponds to the range in particle sizes as previously reported for fly ash in the literature)

Plasticity and Classification. For the EPRI study, all but one of the tested fly ash specimens exhibited low plasticity and if considered as a "soil" would be classified in the Unified Soil Classification System (USCS) as ML. Typically the liquid limit (LL) was <45 and the plasticity index (PI) was <10.

pH. In the presence of water, fly ash typically produces a mildly acidic to very basic condition. The pH for the nine tested specimens in this study ranged from 6 to 12, within the typical fly ash range reported in the literature. The potential significance of pH will be addressed subsequently.

Contact Angle. All specimens tested as part of this study showed high contact angle when using both water and non-polar fluids, in part due to the inherent micron-scale roughness exhibited the granular surfaces, as will be discussed later in this paper. These results have important implications to the determination of specific gravity, as discussed later.

Ferromagnetic Properties. The mass fraction of ferromagnetic grains was measured using magnetic separation and weighting. The separated phases were investigated using x-ray diffraction (XRD) and results show the presence of hematite Fe_2O_3 (weakly magnetic), magnetite Fe_3O_4 (strongly magnetic), and maghemite Fe_2O_3 (strongly magnetic), in significant (i.e., as high as 17 percent) mass fraction. It is known that such a high ferromagnetism affects the interpretation of electromagnetic geophysical measurements and needs to be recognized when assessing potential in situ characterization techniques that include electromagnetics.

Particle Shape, Crushing Potential, and Mineral Precipitation. Scanning electron microscope (SEM) microphotographs were gathered at different scales to explore the grain morphology. Figure 2 shows the prevalent spherical shape and the possible occurrence of cenospheres, which indicate long exposure to high temperature (i.e., melting and ensuing surface tension effects). Sintered conglomerates and particles exhibiting significant surface roughness were also apparent. Most grains in the micrographs were in the range of 4 to 40μm and many grains exhibited the shell discontinuities and surface "holes" that reveal the internal porosity. The comparison of SEM micrographs in Figure 3 shows no evidence of grain crushing in fly ash specimens that are subjected to 10 MPa vertical effective stress in the oedometer cell. Some preliminary SEM evidence points to the precipitation of new species on the mineral surface leading to the observed changes in stiffness, as will be discussed later.

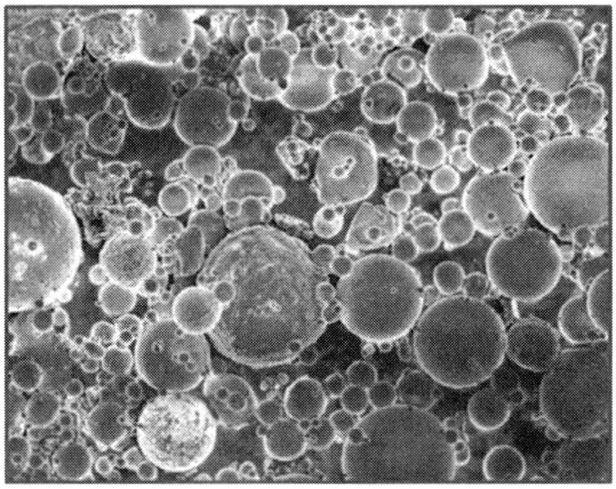

Figure 2. SEM image (1 cm ~10 micrometers)

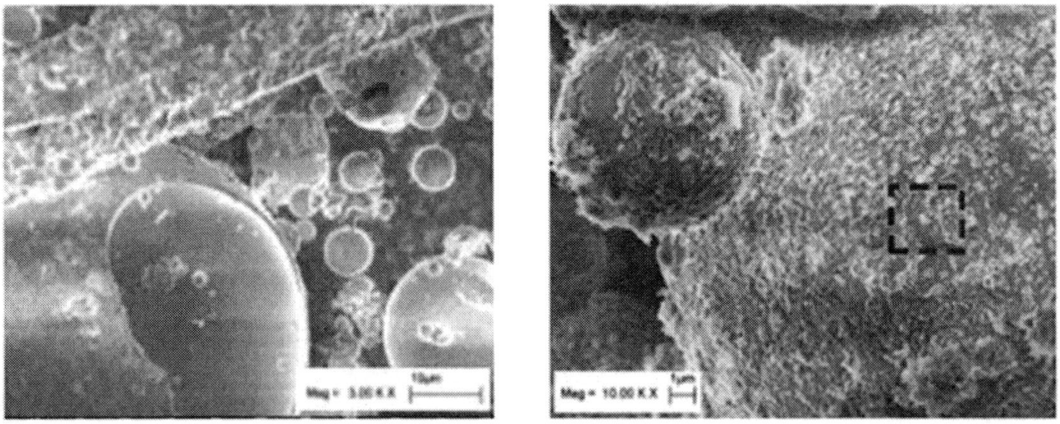

Figure 3. SEM Micrographs Before Testing (left) and After Wetting, Aging, and Testing under High Normal Stress (right)

Specific Gravity. The measured specific gravity of solids (Gs) showed a wide range that varied from 1.96 to 2.66, compared to values of most natural sediments (i.e., Gs=2.7±0.1). However, the range reported in this study was smaller than the historical range reported in the literature, as shown in Figure 4. A comparison of the calculated Gs using water (polar liquid) and alcohol (non-polar liquid) was also performed and the differences between calculated Gs values measured using water and alcohol could not be readily explained by wetting differences. For this study, the specific gravity was computed using water was termed the "effective" specific gravity of solids. Geotechnical engineers typically assume that "void ratio" comprises interparticle (i.e., Type 3) voids, while for fly ash, the conventional calculation of void ratio explicitly includes the occluded (i.e., Type 1) and internally connected (i.e., Type 2) voids. Therefore, for fly ash, properties that are referenced or correlated by comparing the measured void ratio to a typical value in soil can expectedly be misleading.

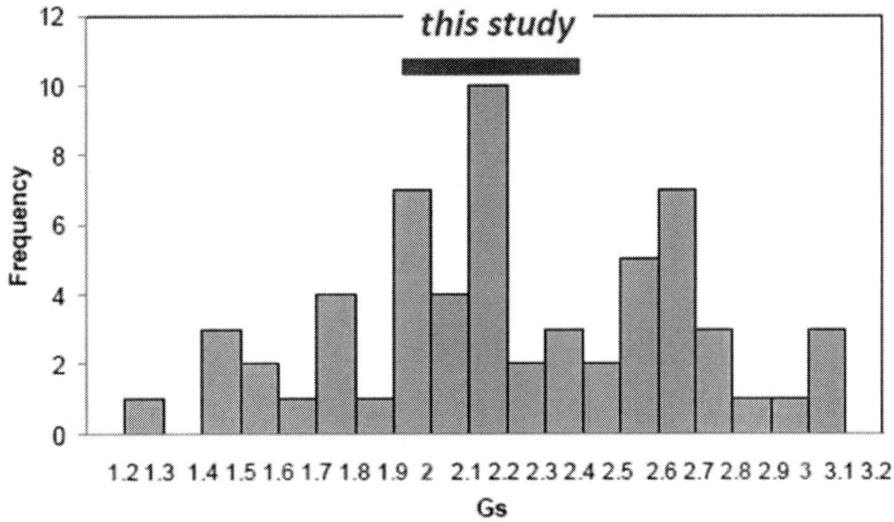

Figure 4: Comparison of Gs Measurements from this Study to Historical Values

Realizable Density. The results of the tests performed as part of this study clearly indicate that void ratio and porosity are not adequate state-parameters for the characterization of fly ash deposits due to the wide range in specific gravity and the presence of both interparticle voids as well as internally connected pores. Consequently, analysis of fly ash deposits should avoid using previous guidelines/correlations from geotechnical engineering that explicitly use void ratios (or porosity), as these guidelines were developed for natural soils made of solid mineral grains. An alternative interpretation strategy in the form of credible or realizable densities is introduced. The authors note that procedures for the determination of minimum and maximum void ratios developed for sands that utilize dry bulk specimens were not intended (and are not adequate) for fine-grained fly ash, particularly those that may be hydraulically deposited. Therefore, efforts were made to explore the assessment of minimum and maximum credible (i.e., realizable) densities or unit weights using the six formation methods identified in Table 1. These include both wet and dry specimen preparation procedures.

Details of these six methods and the calculated nominal energy associated with each method are beyond the scope of this paper, but are included in EPRI (2012).

Table 1. Specimen Formation Techniques to Achieve the Maximum and Minimum Realizable Density of Fly Ash

Method	Nominal Energy Density [J/m^3]
Water Pluviation using a Sieve	10^{-5}
Suspension Settlement	10^{-5}
Formation under Shear	0.15
Free Fall through a Funnel	500
Tapping Compaction	4×10^{4}
Standard Proctor Compaction	6×10^{5}

Ultimately, based on the consistency of test results from the EPRI study, the "standard Proctor compaction test" was selected to achieve the maximum realizable density and the "formation under shear using a dry specimen" was selected to achieve the lowest realizable density. Results of standard Proctor compaction tests for six samples provided in Figure 5 clearly support the previous observation/conclusion that fly ash from different sources exhibit different properties, in this case different compaction characteristics. Figure 6 presents the results of a full suite of realizable density test results for fly ash from one site. A summary of the results for six fly ash bulk samples is presented in Table 2. These results suggest that the minimum and maximum realizable densities are specific to a given ponded ash. Importantly, these test results provide a frame of reference (i.e., extremes to assess relative realizable density) to compare in situ dry density values obtained on undisturbed samples of ponded fly ash and to assess the as-tested relative density for the subsequently referenced performance tests. Based on very limited field information, the measured insitu unit weight of ponded ash appears to be achieve realizable density values that exceed 70 percent, although much more work is needed to assess these conditions in the ponds. .

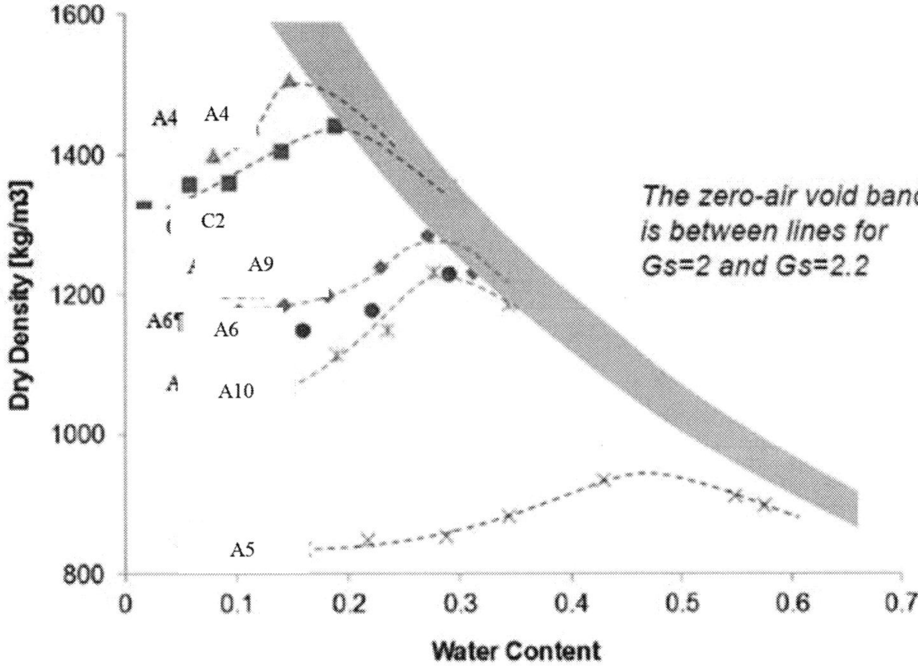

Figure 5. Standard Proctor Compaction Test Results for Ponded Fly Ash from Six Sources

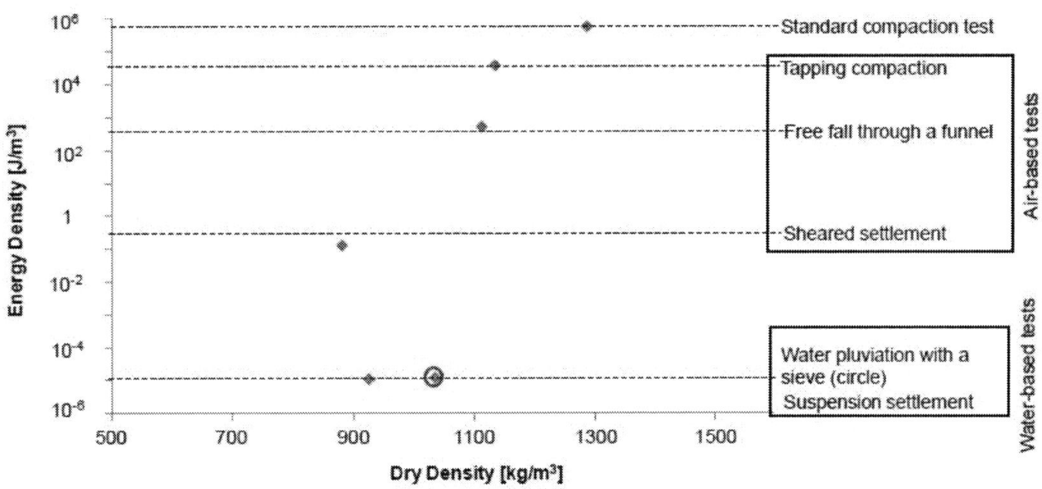

Figure 6. Summary of Realizable Density Tests for One Fly Ash Bulk Sample

Table 2. Summary of Realizable Density Test Results for Fly Ash from Six Sources

Methods	Energy density level [J/m³]	Density (kg/m³)					
		A-9[1]¶ G_s: 2.26	C-2[1]¶ G_s: 2.28	A-5[1]¶ G_s: 2.05	A-6[1]¶ G_s: 2.28	A-3[2]¶ G_s: 2.36	A-11[2]¶ G_s: 2.28
Water pluviation with a sieve	1 10^{-5}	1034	1182	688	--	1071	946
Suspension settlement	1 10^{-5}	924	1130	603	864	1126	853
Formation under shear	0.15	880	1099	605	943	1061	559
Free fall through a funnel	300 ~600	1111	1145	626	945	1084	639
Tapping compaction	3 10^4 ~ 4.5 10^4	1133	1312	767	1211	1290	908
Standard compaction test	6 10^5	1285	1442	950	1250	1508	1233

[1] Initially saturated sample from the site
[2] Initially dry sample from the site

16

Segregation Potential. If ponded fly ash is assumed to exhibit properties of a fine-grained silt, the behavior of ponded fly ash will be affected by the in situ structure of the material, which is controlled by the method of deposition in the ash pond. There are no standard methods for assessing the depositional mechanism or the characteristics of sedimented materials in the laboratory. As part of this study, an assessment of the settling/sedimentation characteristics of fly ash was made to show that grain size variation and possible differences in effective specific gravity between different types of fly ash particles (i.e., spherical solids, cenospheres, ferromagnetic grains, etc.) combine to cause segregation during deposition in the ash pond. Tests were then performed to evaluate the degree and significance of particle segregation. The experimental results show that fly ash readily segregates during sedimentation, albeit, different sources of fly ash exhibit different segregation trends. The result of the segregation is a horizontally layered sediment as documented by visual assessment, x-ray examination, and sounding using a needle probe (see Figure 7). These results suggest that the sedimentation may lead to pronounced hydraulic conductivity and strength anisotropy in the pond, observations that were later corroborated in column flow tests. The results of the relatively simple tests to assess potential segregation have potentially significant implications regarding the behavior of ponded fly ash with respect to the vertical and horizontal hydraulic permeability, rapid lateral dissipation of excess pore pressure in the coarser components of the horizontally stratified stratum, and preferential planar shear failure surfaces through the fine stratum.

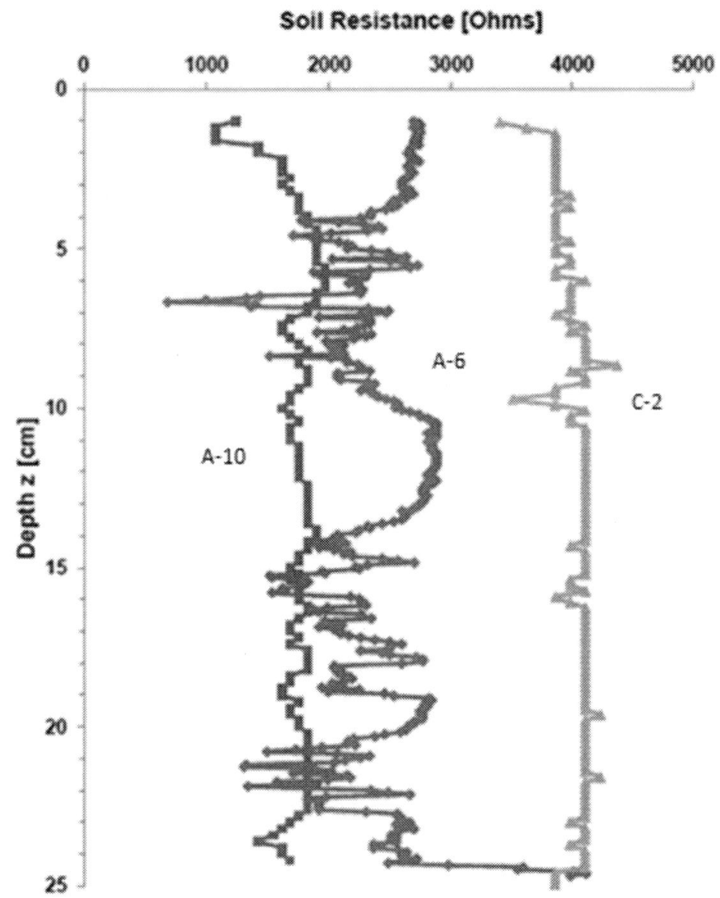

Figure 7. Comparison of Electrical Needle Probe Soundings in Three Fly Ash Sediments

Consolidation, Diagenesis, and Static Shear Strength Test Results

Consolidation: One of the questions regarding ponded fly ash relates to the in situ compressibility, given the potential loose structure that could result from hydraulic deposition. A series of relatively standard and non-standard one-dimensional (1-D) consolidation tests were conducted on specimens formed by different sample preparation techniques. Results are presented for the specimen formed using water pluviation to simulate hydraulic placement conditions in the pond. Bulk samples of dry fly ash were used for these tests. Therefore, the mixing with water prior to sample formation represented the initial contact between fly ash and water, conditions that occur immediately before hydraulic deposition in the ash pond at an electric power generating plant. Results in Figure 8 (left) show a relatively "normally consolidated" response when the specimen is loaded shortly after its initial contact with water using a load increment ratio (LIR) of <1.0. These results show that the granular skeleton formed by fly ash is relatively stiff (i.e., relatively small strain when loaded) when specimens are formed and tested under these conditions. For all of the fly ash specimens tested in this program the compressibility, permeability, and consolidation coefficient of fly ash were compared to values anticipated for "conventional" sand, low plasticity silt and kaolinite clay. In many cases, the stiffness for the fly ash specimen is actually greater than that anticipated for sand, an unexpected response for a loose, silt-sized matrix. In addition, the stiffer-than-expected response was observed for many specimens that were allowed to age before being tested. A typical result is shown in Figure 8 (right). In this latter case, the formed specimen was allowed to "cure" or "age" for a period of about 5.5 days before loading. The changes in the log vertical stress versus strain plots in Figure 8 due simply to specimen aging are dramatic. Post-depositional physical and chemical changes in ponded fly ash appear as a possible reason for the anomalous response shown in Figure 8 (right). Because the consolidation test occurs over an extended period of time, any potential diagenetic effects due to the interaction of the fly ash and water would be inherently "built into" these time-dependent data. These initial test results were used as a basis for performing a series of tests to assess the potential processes and implications of diagenesis, as will be discussed subsequently.

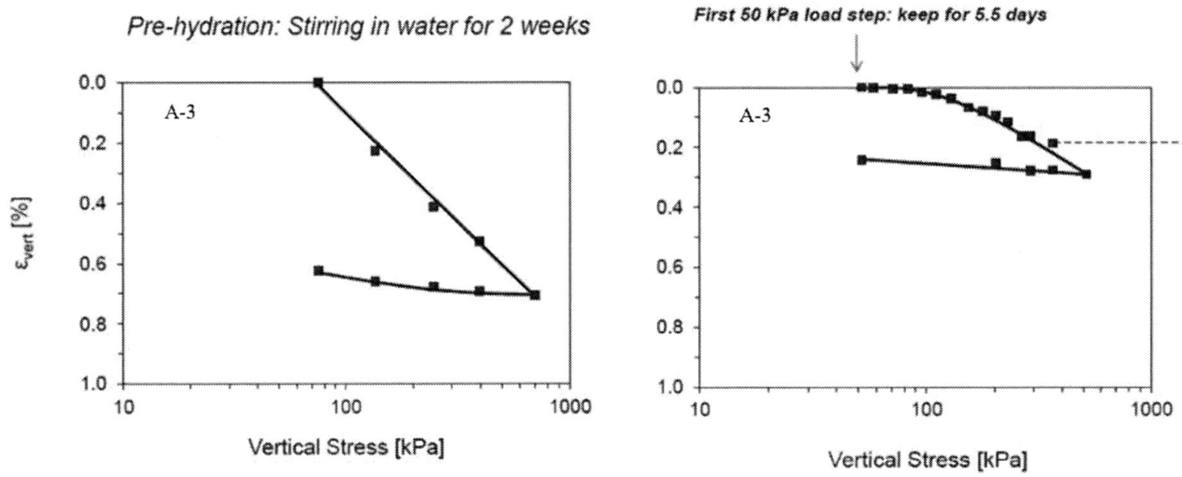

Figure 8. One-dimensional Consolidation Tests Results on Freshly Loaded Specimens (left) and Specimens that were Aged before Testing (right)

Diagenesis. Early qualitative observations indicated that some of the ash samples subjected to "first wetting" (i.e., when dry fly ash from the hopper was mixed for the first time with water), experienced stiffness increase over time. This suggests that some types of ash experience diagenesis through geochemical reactions, although the reactions may be "ash-specific." While extensive diagenesis may potentially (and beneficially) stiffen and strengthen the mechanical response of ponded ash (e.g., compare results in Figure 8), a detrimental aspect of the process could be that early light diagenesis may potentially hinder consolidation and preserve an open, brittle structure in the ash. This could, in turn, cause the specimen to contract/collapse upon further normal or shear loading. Because of the potential implication of this phenomenon, the following two different test procedures were implemented to assess early diagenesis and its potential effects: (i) assess potential for diagenesis to "lock in" loose/dense in situ structure by loading only after the onset of diagenesis; and (ii) assess and contrast potential quantitative effects of diagenesis by measuring shear wave velocity during formation diagenesis. A discussion of the results of these tests is presented below.

With regards to the potential mechanism to lock-in the structure, bulk samples of dry fly ash from each of two sites were collected. From each of these bulk samples, loose and dense specimens were formed under controlled conditions, for a total of four test specimens. Each specimen was loaded gradually in the consolidometer to 100 kPa in the absence of water to avoid any diagenesis during this initial phase. Water was subsequently added to each consolidometer to allow wetting and saturation under a constant vertical stress for a time period of 120 hours. After this aging period, the 1-D loading was continued for all four specimens. Results of the four tests are shown in Figure 9. All of the specimens expectedly contracted when loaded in the dry condition. This was followed by an abrupt vertical strain, immediately after introduction of water as a result of interparticle capillary forces and possibly augmented by contact dissolution. Test results for both specimens in Figure 9 (right) show similar consolidation characteristics before and after "curing" except for the abrupt volume change caused by the introduction of water. This behavior is contrasted by the response shown in Figure 9 (left). Both of these fly ash specimens show a relatively flat load-deformation response after wetting/curing of the specimen, as well as contrasting slopes of the load-deformation curve in the pre- and post-wetting condition. This behavior strongly suggests that the development of diagenetic cementation during saturation/hydration periods strengthens/stiffens the specimens and hinders normal consolidation. To further demonstrate the likely development of cementing diagenetic bonds, a complementary test was performed. After completion of the 1-D consolidation tests, the dense specimen that experienced diagenesis was recovered, dried, "rodded" to break up cementing bonds and repacked in the oedometer. The wetting/curing/loading test was repeated following identical procedures to the initial consolidation test. However, the results captioned "remold" in Figure 9 (left) show that digenetic tendencies are now largely non-existent. The specimen shown in Figure 9 (right) does not exhibit any signs of diagenetic cementation.

To further quantify the degree and rate of diagenesis, a consolidation test was completed in which shear waves were propagated through a wet fly ash specimen to identify changes in shear stiffness over time at constant effective stress. The small strains induced in this test would not be expected to disturb any diagenetic process in the specimen. Experimental results presented in Figure 10 clearly confirm that some fly ash specimens experience very marked early diagenesis (i.e., within 10 to 20 days) accompanied by a 20- to 50-times increase in stiffness. Other specimens tested under identical conditions exhibited little, if any, change in shear wave velocity as the specimen aged. Preliminary SEM evidence points to the precipitation of new species on the particle surface that may lead to the observed stiffness changes (see Figure 3 (right)).

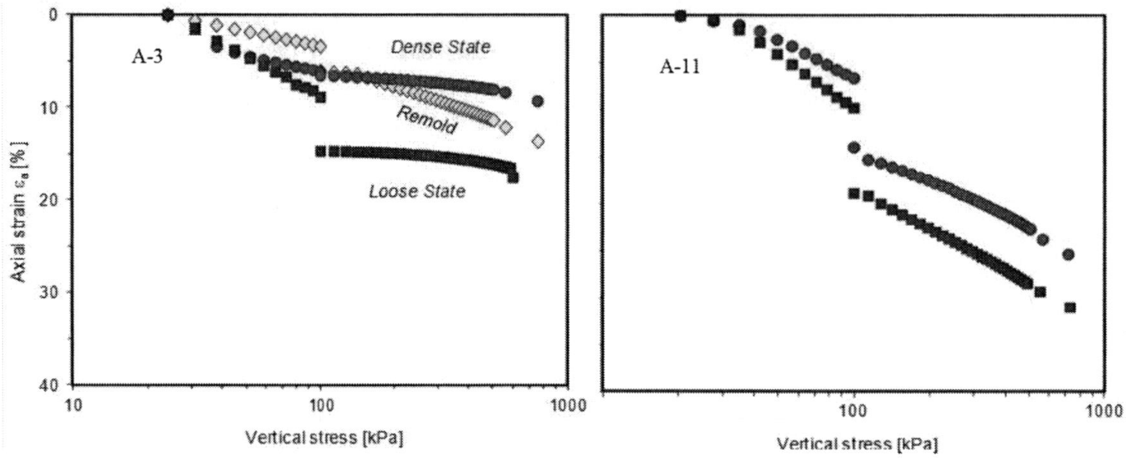

Figure 9. Diagenesis in Initially Dry Specimen that is Wetted

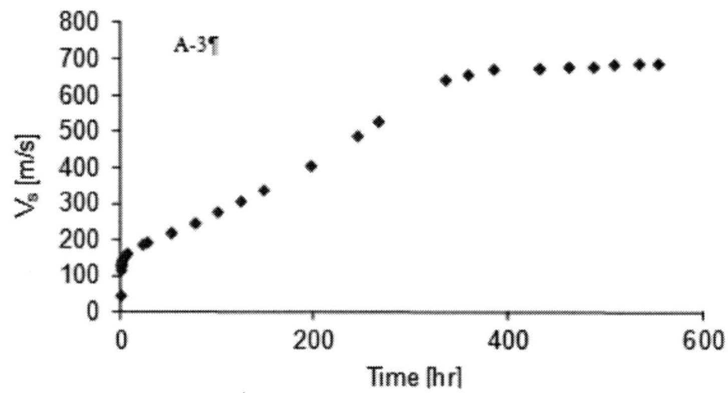

Figure 10. Diagenesis Detection Using Shear Wave Propagation

<u>Static Shear Strength</u>. Triaxial tests were conducted on six different fly ash specimens that were selected for their wide range of index properties. In one series, dry specimens were tested to guarantee drained behavior and to prevent the potential consequences of short-term diagenesis. In the second series, saturated specimens were tested undrained to corroborate the drained results, specifically to assess changes in pore pressure during shear in the transition from contractive to dilative behavior. The drained and undrained triaxial responses were evaluated from the context and perspective of critical state soil mechanics (CSSM).

Both types of tests present their own challenges with regards to test procedures. Volume change in dry specimens must be monitored using the fluid that flows into and out of the test chamber. Volume change measured in this manner must be carefully compensated for the volume displaced by the piston itself. On the other hand, it is particularly challenging to attain the high degree of saturation needed for undrained tests given the internally connected voids within the fly ash. Both drained and undrained triaxial test series were designed to represent field densities and in situ stress conditions. Specimens were prepared to represent relatively loose conditions by either air or water pluviation. Tests were conducted at three different effective confining pressures, the highest effective confining pressure was limited to ~200 kPa as this initial effective confining stress represents stress conditions within fly ash ponds that are <25 m deep. A discussion of specific observations follows:

Results of both test series indicate a significant compression (i.e., densification) during the application of the confining stress. This indicates that fly ash will undergo self-weight consolidation in the absence of diagenesis. Drained tests were characterized by early contraction that extends until the axial strain exceeds ~1 to 2 percent, followed by a quasi-steady state condition at a mobilized friction angle of ~25°, which marked the onset of dilation. Peak friction angles for both drained and undrained specimens are mobilized at an axial strain between 5 and 10 percent; these deformation levels are significantly higher than those typically measured in sands. Test results indicate that the critical state friction angle at constant volume drained shear varies between 29° and 36°, while the critical strength friction angle for the undrained tests varies between 35° and 36°. Typical test results from one of the drained test series is shown in Figure 11. There was systematic shear localization observed in virtually all of the drained and undrained specimens initiating just before the peak strength. The development of localization restricts the ability to accurately assess the critical state density for the drained tests. In any case, test results point to a relatively flat critical state line in the density-p' projection, within the stress range of interest for typical ponds, as shown in the lower portion of Figure 11..

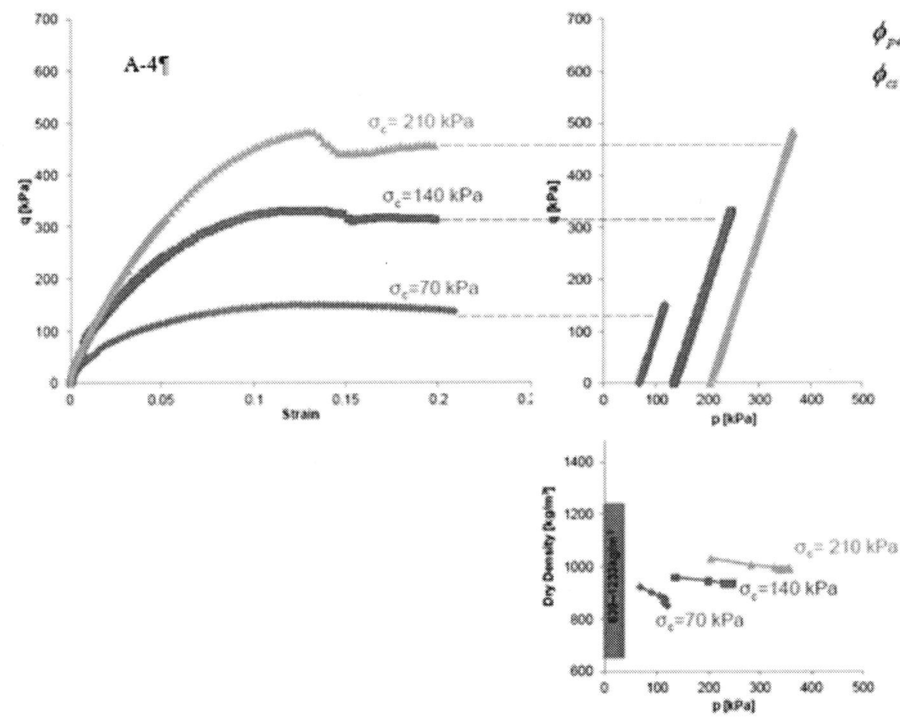

Figure 11. Results of Drained Triaxial Loading

Potential for Flow and Static Liquefaction Results

Flow Potential. The release of fly ash from an ash pond in a manner similar to the previously referenced "flow" failure can only take place after the loss (i.e., failure) of the perimeter ash containment structure. The opportunity for essentially uncontrolled release of fly ash was qualitatively modeled in the laboratory using fly ash specimens that were formed by sedimentation in transparent acrylic cylinders. To simulate the loss of containment, the acrylic cylinder is suddenly lifted from its base to release the confinement of the previously sedimented fly ash. Representative distinctive and contrasting trends regarding "failure" modes and behavior are shown in Figure 12.

- Figure 12a shows a failure by "slumping" and limited 'flow" of some layers;
- Figure 12b shows that the presence of free water above the sediment column when confinement is released favors "flow" when the boundary/containment is removed;
- Figure 12c shows a "dilative" tendency in some sediments/layers, as characterized by the development of a dull-dry surface and the sediment-hindered mobility until it is either wetted or vibrated; and
- Figure 12d shows a notably "stable" response in sediments prone to early diagenesis (subsequently described and studied) when specimens were left in the tube beyond the sedimentation time prior to having the confinement released.

a b c d

Figure 12. Representative and Contrasting Behavior and Failure Modes upon Containment Loss

While all of the test specimens were formed using identical water pluviation procedures, they exhibited clear differences in "flow" behavior when the containment is removed. Although this is not a completely representative simulation of the actual mechanisms that may occur in the field in which flow may commence at the top of the deposit and be exacerbated by free water, these results demonstrate the difficulties in drawing general characterization about the behavior of ponded fly ash as the results reflect the consequences of particle size, segregation, layer unit weight, availability of supernatant water, and early diagenesis prior to the release of containment.

Static Liquefaction Results:

It is important to acknowledge that static liquefaction is attributed historically to strongly contractive sediments subjected to undrained loading. A summary of the dilative and contractive tendencies is presented in Figure 13 for all of the triaxial test specimens of pluviated ash as a function of the post-consolidation relative realizable density. In the absence of diagenetic cementation, all these freshly remolded specimens densified during drained confinement and it was not possible to form specimens with a post-consolidation relative realizable density of <50 percent. Based on study of the various sample forming procedures identified previously in this paper and observations from field sampling of fly ash from active ponds, the authors believe that achieving a realizable density of <50 percent in the field is highly unlikely *unless there is early diagenesis* of the fly ash in the field. Even the loosest undrained specimens from this study (i.e., Dr≈60 percent) still show a tendency to develop negative pore pressure and to approach cavitation despite the use of back-pressure during the tests. This is a clear indicator of dilative tendency for the loose fly ash, a tendency that will oppose static liquefaction. Drained tests developed localized shear surfaces; the contractive tendencies are considered to be "incipient," as the tendencies were all attributed to the early stages of loading.

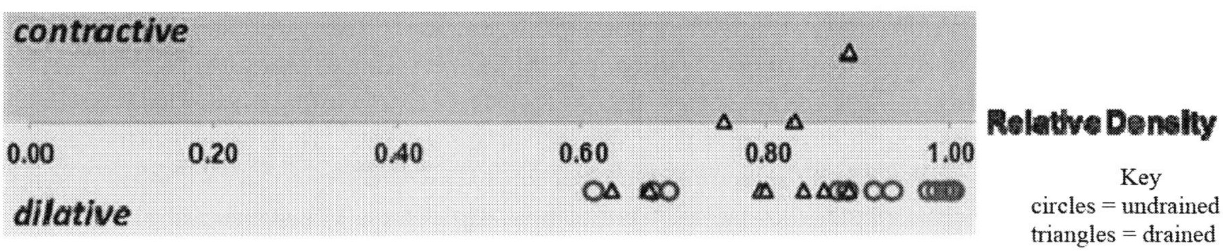

Figure 13. Prevalent Dilative Tendency for Specimens with Dr> 65%

For reference, the test results from this study indicate that the fly ash response resembles natural sands, except that: (i) higher strain levels are required in fly ash compared to sand to reach peak and quasi-steady state conditions; (ii) fly ash exhibits pronounced difficulties in forming decisively contractive specimens for the nominal effective confinement of interest (e.g., approximately <200 to 300 kPa to simulate conditions in a fly ash pond; and (iii) fly ash exhibits high internally connected voids (i.e., Type 2) that are difficult to completely evacuate of air. With regards to this last point, these internal compressible air "pockets" in the fly ash may limit the generation of positive excess pore pressure during undrained loading, although the effect of long-term submergence on the existence of these air pockets is unknown.

The ability to load ponded fly ash under undrained conditions in the field is related to the rate of load application compared to the rate of pore pressure dissipation. The rate of pore pressure dissipation is governed by the coefficient of consolidation, which is a function of hydraulic conductivity and matrix compressibility, and the rate of load application. The time for pore pressure dissipation depends on the coefficient of consolidation (c_v) for the particular fly ash material and the largest distance from the center of the compressible layer to a drainage boundary (H), assuming the areal extent of the load (L) is much greater than the distance H, a condition to be expected in an ash pond. While the finest fraction particle size of fly ash may be indicative of low permeability materials (i.e., silt- and clay-sized particles), the hydraulic depositional environment leads to an anisotropic layered deposit, where coarse fly ash layers drain more effectively than the more compressible and less conductive fine-grained fly ash layers. These observations suggest that excess pore pressure is unlikely to build up during ponding, except for a transient condition related to a sudden layer collapse under some critical load. In the absence of confirmatory field evidence, the authors believe that it is prudent to monitor pore pressure during field construction events until sufficient field data are obtained that confirm these analysis results based on laboratory observations, although these results suggest that the potential for ponded fly ash to statically liquefy and flow is considered to be small unless the materials are affected by early diagenesis.

IMPLICATIONS OF TEST RESULTS TO ASH POND CLOSURE AND CONSIDERATION OF IMPOUNDMENT AS A DAM

The implications of these recent laboratory testing and characterization results have potentially significant effects on the site- and plant-specific assessments of ash disposal ponds. First and foremost is the importance of the containment system for the ponded ash. The authors note that some fly ash cells are constructed using the upstream construction techniques described in this paper that are similar to techniques common to mine tailings disposal operations, while the majority of the existing ash disposal facilities are constructed as "earth dams" and then the ash is sluiced into the dam as "tailwater." For facilities constructed as dams, the potential for a flow failure is relatively low because of the inherent stability of the containment structure. The second most significant implication is the availability of "free water" in the pond. For the referenced TN accident, the impoundment contained free water on top of the ash. The authors further note that several ash disposal operations utilize active and passive dewatering techniques to minimize the amount of free water in the impoundment. These activities are anticipated to contribute significantly to the stability of the impounded ash. As noted in the introduction for the TN accident, large intact blocks of "dewatered ash" were observed to be floating on the water-laden ash. At this facility, one of the cells that did not fail had been "dewatered" during the previous year by simply allowing water to infiltrate and/or evaporate. When the results regarding the containment structure and dewatering are considered in the context of the characterization results from the recent EPRI study, the authors believe that it may be demonstrated that impounded fly ash in properly designed facilities that are undergoing active (or potentially passive) dewatering can be demonstrated to exhibit no tendencies to flow or statically liquefy. Therefore, these facilities will not actively impound "free water" in a manner common to a "conventional" dam.

CONCLUSIONS

Occluded and internally connected voids in fly ash add significant difficulty to the accurate and consistent determination of specific gravity of solids (Gs) for fly ash. This limitation has a direct bearing on the calculation of void ratio, porosity, and degree of saturation. Failure to recognize occluded and internally connected voids can lead to the erroneous utilization of pre-existing soil correlations/experience when characterizing the geotechnical behavior of ponded fly ash. These correlations, in turn, can lead to incorrect conclusions regarding the behavior of fly ash. In particular, void ratio is the central parameter used to assess the tendency for volume change (i.e., contractive or dilative) in natural soils, while the degree of saturation determines whether excess pore pressure (i.e., positive or negative) will be generated during undrained loading. Therefore, alternative characterization approaches… that do not rely on void ratio… are needed to more accurately characterize ponded fly ash.

The terms "minimum and maximum realizable densities", ρ_{min} and ρ_{max}, are proposed in this study (or γ_{min} and γ_{max} when considering minimum and maximum realizable unit weights). Test procedures are suggested to determine these parameters. Specifically, the maximum realizable density uses the conventional standard Proctor compaction test common to most geotechnical engineers. The minimum realizable density uses a technique referenced as "formation under shear." These two density limits can be consistently obtained to provide a robust reference range that can be used to assess the in situ density (ρ_o) in terms of relative realizable density (Dr) where $Dr=(\rho_o-\rho_{min})/(\rho_{max}-\rho_{min})$. The classical geotechnical engineering framework used for granular materials can be extended to include fly ash when the fly ash is characterized in terms of relative realizable density. It is explicitly recognized that threshold values (and techniques) used for assessing the relative density of sands do not necessarily translate to fly ash.

Test results in the laboratory that include pluviation through water anticipate that significant segregation will occur during the sedimentation in a fly ash pond. The resulting layered structure will exhibit pronounced anisotropic properties, including higher hydraulic conductivity in the horizontal direction than in the vertical direction. The presence of layered fine zones can lead to preferential weak planes, although this study did not address this potential consequence.

Early diagenesis during the initial fly ash wetting and deposition was observed in most of the "dry" fly ash specimens included in this study, particularly for materials that exhibit relatively high pH. The high pH may reflect high calcium oxide content relative to sulfur, as noted by others, but a systematic assessment was not performed as part of this study. Early diagenetic cementation prior to self-weight consolidation potentially can lock-in high interparticle porosity during deposition and increase the contractive tendency in the ponded fly ash. In a pond setting, locked-in high porosity may manifest as collapse when a critical vertical load is reached, or sediment flow when shear is applied (e.g., after failure of the containment structure). Therefore, ponded fly ash materials that are prone to early diagenesis must be assessed for high in situ porosity. In the absence of diagenetic cementation (and for some specimens that exhibited diagenesis), laboratory test results show that normal wet deposition leads to relative realizable densities that support dilative behavior for ponded ash in the upper 20 to 25 m in a pond when subjected to monotonic shear loading, thus hindering static liquefaction. Even if fly ash does not statically liquefy, it is possible for the material to "flow" only if free water is present and the containment structure fails. If containment is assured, then a fly ash flow failure is unexpected.

In conclusion, results from this study indicate that the most important processes/parameters that affect the behavior of ponded fly ash are: (i) occluded and internally connected voids that result in a low specific gravity and a high potential for trapping air in the structure of the fly ash; (ii) segregation-induced heterogeneity and anisotropy that impact permeability, pore water dissipation, and strength; (iii) early diagenesis that increases stiffness and strength, but also may restrict self-weight consolidation and lock-in high porosity; and (iv) presence of supernatant ponded water that may favor flow conditions if the pond containment system is compromised. Water-deposited fly ash is inherently anisotropic, and it generally exhibits a dilative load-deformation behavior that hinders static liquefaction in ponds shallower than 20 to 25m.

ACKNOWLEDGEMENTS

The EPRI study was a highly interactive collaboration between Geosyntec Consultants, the Georgia Institute of Technology (Georgia Tech), EPRI, and several EPRI members. The support and interaction is greatly appreciated. Most importantly, the authors note that the study would not have been possible without the dedication and commitment of the following team of student researchers at Georgia Tech under the direction of Dr. Santamarina: M. Cha, S. Chong, S. Dai, J. Jang, S. Kim, A. Mezencevova, E. Papadopoulos, C. Pasten, X. Ren, S. Roshankhah, F. Santamarina, A. Sivaram, and M. Terzariol. Each student member took individual responsibility and ownership for a specific aspect of the project. The authors are indebted to this group for their tireless efforts to execute a very demanding laboratory testing program.

REFERENCES

Bachus and Santamarina, (2012), Final Report, Geotechnical Properties of Fly Ash and Potential for Static Liquefaction, Volume I – Summary and Conclusions, November, 2012.

Bachus and Santamarina, (2012), Final Report, Geotechnical Properties of Fly Ash and Potential for Static Liquefaction, Volume II - Data Summary, November, 2012.

PRESERVATION OF PIEZOMETERS AND ASSOCIATED AUTOMATED DATA ACQUISITION SYSTEMS IN TAILINGS DAMS

Dennis J. Fela, P.E.[1]
David R. Welters, P.E.[2]

ABSTRACT

Tailings impoundment dams represent some of the most severe environments for piezometers and associated automated data acquisition systems (ADAS), which are currently the most actively constructed dams across the United States with new permits regularly being granted by state and federal agencies. Agencies currently require the monitoring of phreatic levels within the dams throughout the staged operation of the impoundment. This is commonly being performed with piezometers, typically employing vibrating wire (VW) technology alone or in conjunction with other piezometers, installed as part of initial construction or at key stages, and maintained during operation as new stages are built for facility expansion. The installation and operation of VW piezometers on tailings impoundments presents a challenge due to the constantly changing embankment and pool conditions, as well as the aggressive environment. The failure rate of VW piezometers in active tailings impoundments has been greater than at traditional dam installations due to factors relating to the construction and material composition of the dams. VW piezometers typically require the installation of cable which is typically buried underground and routed to an enclosure. Active disposal sites where the dam is being raised present challenges to avoid the cables being damaged or severed during construction operations. Lightning strikes near enclosures or buried cables can impact instruments. VW piezometers are sometimes installed within tailings above the phreatic level and sit idle for months or years prior to the pool levels rising to influence the instrument. Instrument failure is common in these installations, particularly where the phreatic surface fluctuates around the piezometer tip. Preservation of instrumentation in these environments is dependent on selection of piezometer materials and methods to achieve life expectancy, reliability and accuracy. Planning for ADAS enclosures is also required to account for expansion, exposure to the elements, lightning, etc.

INTRODUCTION

Coal refuse disposal facilities use impoundments for disposal of slurried materials produced in the process of separating coal from soil and rock and recovering process water. The dam embankments are typically constructed from soil and rock, called coarse coal refuse (CCR), which is separated from coal at the mines' preparation plants. These dams can exceed 300 feet in height. Slurry containing fine coal refuse (FCR) particles (fine sand, silt, and clay) is pumped into the impoundments as they are simultaneously

[1] Senior Project Engineer, D'Appolonia Engineering Division of Ground Technology, Inc., 275 Center Rd., Monroeville, PA 15146, djfela@dappolonia.com
[2] Project Engineer, D'Appolonia Engineering Division of Ground Technology, Inc., 275 Center Rd., Monroeville, PA 15146, drwelters@dappolonia.com

being constructed using the coarser soil and rock from the coal preparation plant. The embankments can be constructed in stages using the upstream method of construction, where the new stages of the embankments are being constructed partially on top of existing stages of the embankments and partially on top of deposited FCR that has settled out of the slurry; or the downstream method where the new stages and subsequent crests of the embankment extend progressively downstream.

Piezometers, standpipe (sometimes called Casagrande piezometers), vibrating wire, and less frequently pneumatic, are used at facilities for long term monitoring of the piezometric levels within tailings dam embankments. During upstream construction, for stages being built over settled fine coal refuse deposits, piezometers are also being installed in the upstream portion of the embankments to monitor pore water pressures in the underlying fine coal refuse. Pore-water pressures in the upstream portion of the embankment and piezometric levels in the downstream portion of the embankments are important factors in assessing the structural stability of the embankment. Many regulatory agencies require the regular monitoring and reporting of dam embankment phreatic levels for comparison against predetermined action levels.

A case study of monitoring two long-term impoundment sites summarized here provides insight into the frequency and potential causes of instrumentation failure with a focus on vibrating wire piezometers. The VW wire piezometers were susceptible to environmental conditions which subsequently damaged or destroyed the instrumentation. A look at the history of failure can provide an indication of the possible pattern of failure. Preventative measures, some of which have already been implemented, are further evaluated.

CASE HISTORY

Two tailings impoundment dams located in southwestern Pennsylvania were instrumented and monitored for a period exceeding ten years during staged construction. The dams were outfitted with vibrating wire piezometers installed throughout the period at differing elevations and locations within the dams. The piezometers were integrated into an ADAS system through enclosures outfitted with radios to transmit the data to a central computer at each facility.

The VW piezometers installed at both impoundments are embedded in either sand or cement bentonite grout usually within boreholes at varying elevations within the embankment. The VW piezometers cables extend up from the boreholes and run along shallow trenches to ADAS enclosures typically located near the dam crest, downstream benches or the abutments. In some locations the boreholes and trenches have been subsequently covered by embankment construction. The ADAS enclosures contain equipment which reads the VW piezometers and transmits data to a host computer for further reduction or submittal.

Tailings Impoundment No. 1

Instrumentation has been present at Tailings Impoundment No. 1 since six pneumatic piezometers were initially installed in May 2001. The lines to these pneumatic piezometers were severed in October 2002, and the instrumentation was replaced with VW piezometers PZ-1 through PZ-6 in May 2003. Since that time all piezometers installed at the facility have been vibrating wire piezometers. As shown in Figure 1, the total number of VW piezometers gradually has increased to twenty-eight through 2014. There have been sixteen instances of piezometer malfunction over the 11 year monitoring period. In nine cases the damage to the piezometers was irreparable; in seven cases the damage was identified and the piezometer continued to function after being repaired.

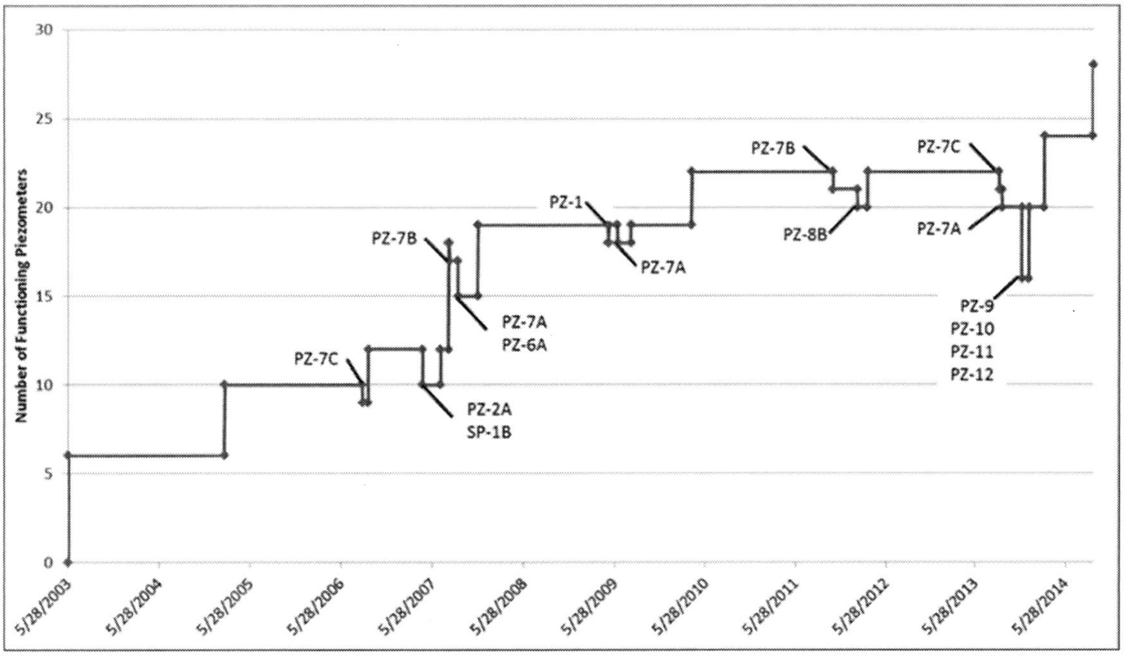

Figure 1. Total Number of VW Piezometers and History of Failure at Tailings Impoundments No. 1.

Of the nine cases of irreparable damage to piezometers at the site, eight failures have occurred at the VW piezometers PZ-6A, PZ-7A, PZ-7B and PZ-7C (including replacements), which were installed in the upstream embankment in February 2005. These installation locations have failed two or three times each, with the exception of PZ-6A, which has failed once. Piezometers PZ-7A, PZ-7B and PZ-7C were installed during upstream construction at the same elevation, but in different locations of the upstream embankment (Refer to Figure 2 for locations). The initial failure of these three piezometers occurred at different times while their tip elevations were still above the elevation of the pool. The cables of these piezometers were buried along with the cable from piezometer PZ-6A in a 3-foot deep trench down the middle of the embankment face to an ADAS enclosure. No rain events were observed on the days that these piezometers failed, but widely fluctuating readings were observed at three of these four piezometers in the period prior to failure.

Replacements for piezometers PZ-6A, PZ-7A, PZ-7B and PZ-7C were installed in November 2007. The cables of these piezometers were trenched to the middle of the embankment crest and then down the face of the embankment about 350 feet to the ADAS enclosure. Before PZ-7A failed for the second time in June 2009, the crest of the embankment was built up about 30 feet over the portion of the trench installed on the crest. An electrical storm event was recorded on the day that PZ-7A failed. A replacement of PZ-7A was installed in August 2009 and the cable was connected to the closed ADAS enclosure.

Figure 2 presents a general plan showing the locations of most of the VW piezometers installed at Impoundment No. 1 (some of the VW piezometers lie in adjacent saddle dams). The locations marked in red signify piezometers which malfunctioned and were replaced and the locations marked in blue signify lost connections due to cable damage (which was subsequently repaired).

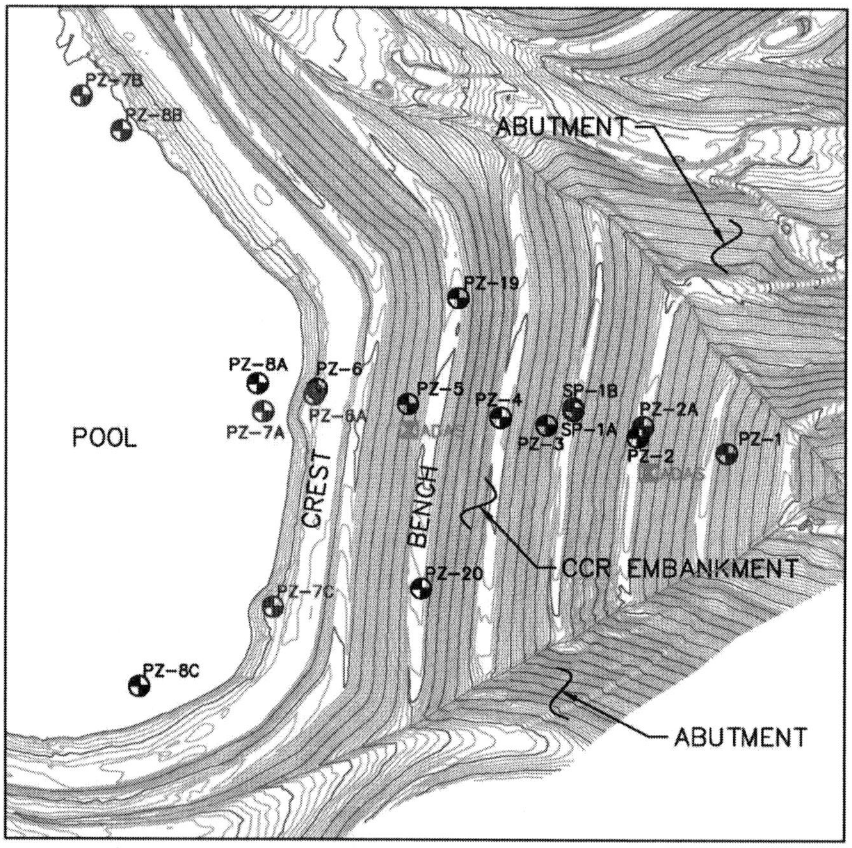

Figure 2. Instrumentation Plan - Impoundment No. 1

In April 2010, piezometers PZ-8A, PZ-8B and PZ-8C were installed in the upstream portion of the embankment. Of these three piezometers, one has failed (PZ-8B). In December 2011, readings taken at PZ-8B indicated that the water level in the embankment began rising at rates inconsistent with the pool level and other instrumentation. The piezometer or cables were assumed to be damaged and the piezometer was replaced in March 2012, after which readings returned to normal levels.

Between October 2011 and September 2013, piezometers PZ-7A, PZ-7B and PZ-7C failed again. Two of the three experienced fluctuating readings in the period prior to failure.

As shown on the centerline cross section in Figure 3, the majority of the piezometer failures have occurred in the upstream push-out portion of the embankment. These piezometers have some of the longest cable runs due to the enclosure location and are located nearest to the impoundment pool.

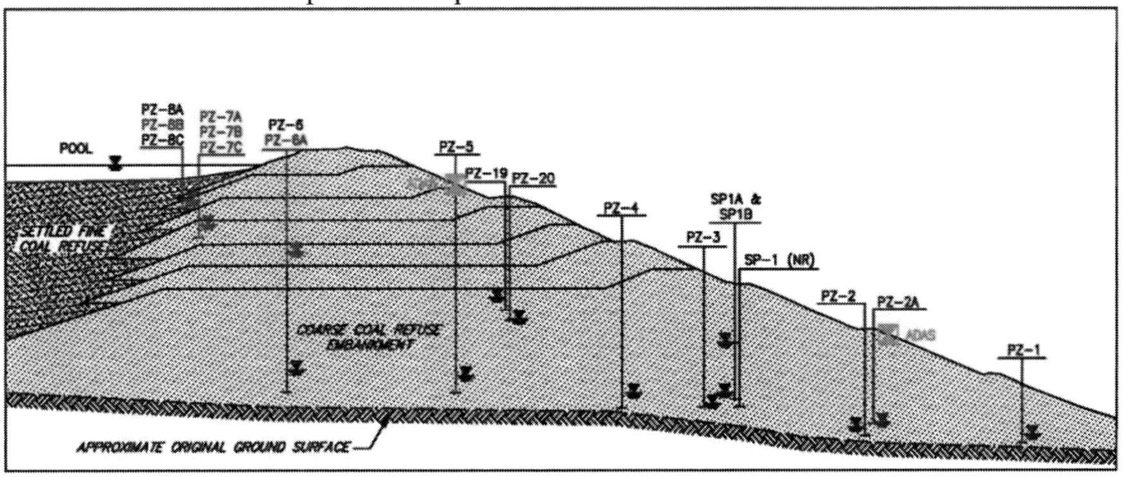

Figure 3. Centerline Cross Section - Impoundment No. 1

The causes of the piezometer failures at Impoundment No. 1 were not often completely understood since the instruments were typically buried deep within the embankment and could not be recovered. However examining the location of the piezometers and cable trenches with respect to elevations and locations within the embankment, possible failure modes were determined to be embankment displacement damaging cable or electrical damage to either cables or VW piezometers from lightning strikes. Four of the five piezometers experiencing irreparable damage were installed in the upstream push-out partially over FCR which was hydraulically placed. These areas experienced substantial deformation as the FCR consolidated. The deformations could have damaged the cables near their installation point. One indication that embankment deformation may be causal is the observation that several piezometer failures occurred after a period of wide fluctuations in the readings which followed stage construction where push-outs were made into the impoundment.

In other cases, instrument failures have occurred on days that experienced precipitation events and electrical damage from lightning strikes was suspected. The piezometer installations at this site generally feature cables buried in shallow trenches located down the center of the embankment running to an ADAS enclosure. It is plausible that lightning struck somewhere along the length of cable or, due to the upstream locations, the pool near the piezometers.

As mentioned previously seven of the VW piezometer and six of the pneumatic piezometer failures have been associated with cutting of the cables/lines, usually due to excavation by various subcontractors. In the case of the seven cables of the VW piezometers, the cables were repaired and the piezometers continued to provide readings.

Tailings Impoundment No. 2

At the second refuse facility, fifteen VW piezometers, named P1 through P15, have been installed from December 2005 to the present. As with the first facility, the piezometers were installed in both the upstream embankment and the main part of the embankment for long term monitoring of the phreatic levels within the embankment. As shown in Figure 4, the total number of VW piezometers gradually increased to seventeen, corresponding to the staged construction; a total of four piezometers were rendered out of service over the 9 year monitoring period.

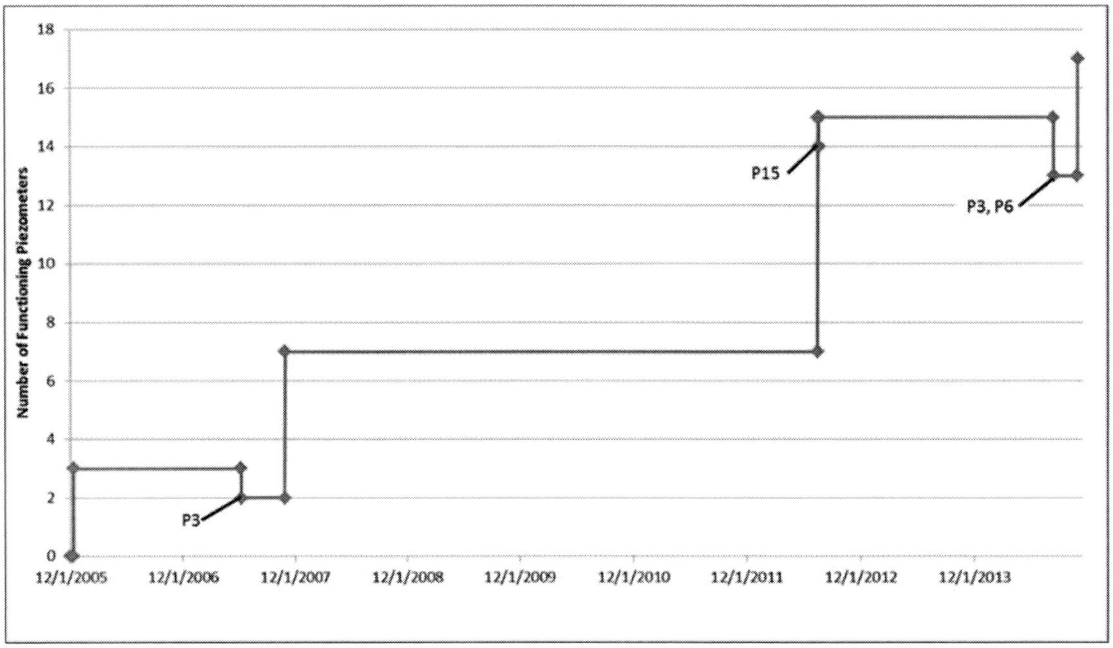

Figure 4 - Total Number of VW Piezometers and History of Failure at Tailings Impoundments No. 2.

Piezometers P1, P2, P3 were installed in the dam embankment in December 2005. Piezometer P3 failed in June 2007 and was replaced in October 2007. At that time, four additional VW piezometers, P4, P5, P6 and P13, were installed. In July 2012, VW piezometers P7 through P12 were installed in the main portion of the embankment and P14 and P15 were installed in the upstream construction area of the embankment. Within days of installation, readings from P15 ceased. It was discovered that a bulldozer had severed the cable near the piezometer borehole and that the sensing unit functioned properly once the cable was repaired.

In August 2014, P3 failed a second time, together with P6 during an electrical storm. The last reading recorded by the ADAS system coincided with the time that the storm occurred. The timing along with the sudden loss of data support the idea that these instruments suffered electrical damage during the storm. Unlike Impoundment No. 1, most of the instrumentation cable at Impoundment No. 2 is buried deep beneath the embankment having been previously covered by staged construction. An exception however is, Piezometers P3 and P6 as shown on Figure 5. These piezometers were installed in the embankment and the piezometer cables were buried in a shallow trench running across the embankment to an ADAS enclosure located on an abutment next to the dam. The location where piezometers P3 and P6 are installed is near an outslope corner of the embankment away from the abutments, far from the enclosure. The cable trenches were not subsequently covered by staged construction. Elsewhere at Impoundment No. 2, the length of trench from the embankment to the ADAS enclosures is relatively short. Additionally the ADAS enclosures are located near the relatively sheltered groin area of the dam.

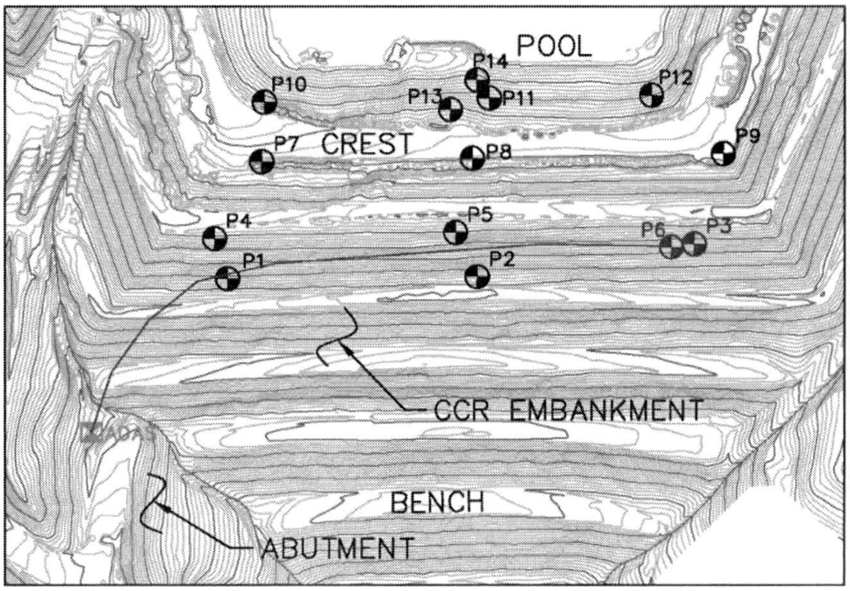

Figure 5. Instrumentation Plan - Impoundment No. 2

ADAS Interruptions of Service

In addition to the failure of the VW piezometers the ADAS system has been interrupted at both facilities for various reasons including:

- Damage to cables entering ADAS enclosures. In one instance the cables were severed by a weed trimmer. Conduit was subsequently installed from the ground to enclosure.
- Battery voltage dropping below operable thresholds was common. Usually this occurred during the winter months and with improperly aligned solar panels or undersized batteries/solar panels. In one instance the solar panel fell off the mount, was simply placed flat on the enclosure, and became covered with snow. The solar panels were subsequently upgraded from 10 watt (W) to 20W. The power supplies would also occasionally fail and were replaced.
- A power outage cut communications to one of the radios using A/C power.

- Radio communications would occasionally be lost. This occurred for various reasons including damaged coaxial cable, damage radios, and damaged or misaligned antennae. In one instance the dam embankment was raised in elevation and blocked line of sight between radio relays requiring a complete redesign of the radio network. In another instance moisture entered the enclosure and shorted the radio.

INSTRUMENTATION FAILURE MODES

The modes of failure of instrumentation are often related to the environments in which they are installed and operate. Dunnicliff et al. (1993) compiled a list of potential environments/conditions which could impact instrumentation. The list consists of: deformations, high pressures, corrosive conditions, temperature extremes, electric shock, vandalism, dirt/dust, humidity/water, erratic power supplies, and loss of accessibility. VW piezometers can be impacted by many of these including those suspected in the case study tailings impoundment. Many of these environments were present in the two case study facilities with the most common being electric shock from lightning strikes and loss of accessibility associated with construction damage.

Identification of Failure

The first step in identifying faulty VW piezometers or instrumentation is the recognition of questionable data. This often includes piezometer readings when the dam threshold levels (critical state, allowable, alert, action, etc.) are reached (Crum 2011). Threshold levels should be defined as to their specific meaning and source.

Crum (2011) states that piezometic readings are potentially influenced by precipitation/infiltration, tail water, barometric pressure, groundwater temperatures, excess pore pressure from surcharge loads or other construction induced pressures, liquefaction or cyclic loading, drainage features, or pipe/conduit/structure leakage, However the most potential to be indicative of a failure mode is the impoundment pool level.

Unfortunately VW piezometers cannot be checked like standpipe piezometers which can be inspected by use of a down-hole camera. Crum (2011) recommends response testing, consisting of either bailing or adding water to a Casegrande piezometer, to confirm that the instrument reading corresponds to the piezometric levels in the dam. For VW piezometers the resistance of the coils can be checked using and ohmmeter connected across the gage terminals. The measured resistance should be in the range of 200 to 400 ohms depending on the manufacturer and cable length. If the resistance is very high or infinite the cable is likely broken or cut. If the resistance is very low the gage conductors may be shorted.

If the fault is suspected to be a cable break, a time-domain reflectometer (TDR) may be used to locate the cable break. A TDR is an electronic instrument that uses time-domain reflectometry to characterize and locate faults in metallic cables. The TDR is typically connected to the enclosure end of the cable and can accurately locate the length of cable to the break or fault. A Velocity of Propagation (VOP) is required for the specific cable to accurately determine length. For instance, Geokon, Inc. recommends using a VOP of 0.63 for their cable using a Riser Bond model 1550 TDR.

Lightning Strikes

Lightning strikes can generate surge voltages in the signal cables and VW piezometers. The surges are readily capable of burning out individual signal channels, damaging insulation, or in some instances damaging piezometers, data loggers and associated equipment.

The risk of lightning strikes occurring at a facility is dependent on its geographical location. Klebba (2003) has asserted that the risk of lightning strikes to instrumentation systems varies based on geographic location, with lightning being "attracted" to Florida, the Himalayas, and central Africa. Klebba (2003) referenced the average yearly lightning strikes compiled by NASA and provided in Figure 6. Some of the piezometers evaluated at the two case studies, located in western Pennsylvania and a zone of moderate exposure to lightning strikes, were nonetheless impacted.

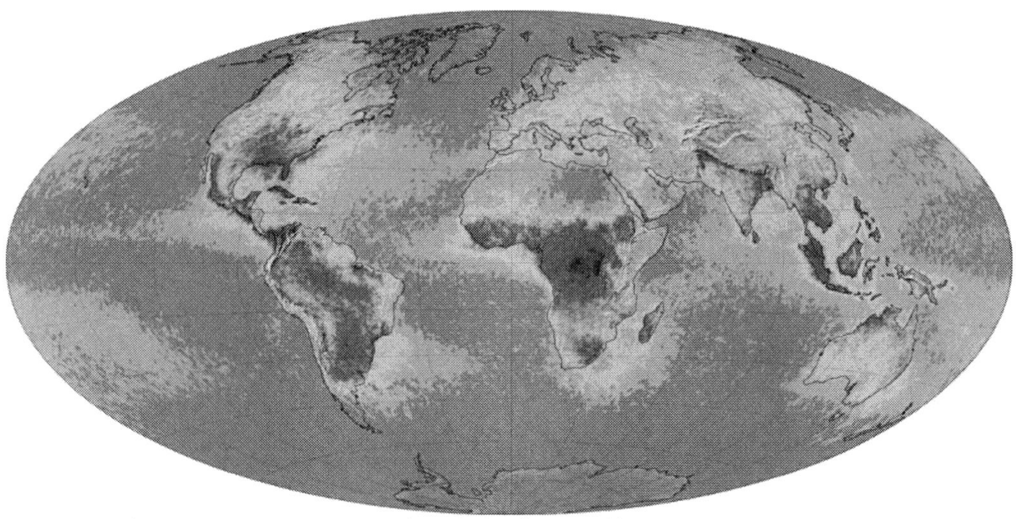

Figure 6. Average Yearly Lightning Strikes (NASA 1995-2002)

The majority of lightning current responds horizontally when striking the ground or being transferred to ground. Klebba (2003) speculated that less than 15% of the energy from lightning strikes penetrated the earth. High voltage transients due to lightning strikes tend to travel near the ground surface away from the point of impact of the strike. Long horizontal cable runs will experience a difference of potential along their length which increases with the length of the cable. High current surges can occur both at the sensor end and the datalogger end. Therefore, the risk of damages due to lightning is increased when designing ADAS network with long instrument cables (Choquet et al. 2014).

Moulds et al. (2000) also noted that the long cable runs increase the risk of lightning damage. Each line in the following Figure 7 represents a voltage equipotential line in 50 kV increments. Any cable not following the iso-voltage path may be exposed to large voltage differences along its length. Therefore if the VW piezometer is buried at depth and the lightning strike occurs near the cable, the difference between voltage at the sensor and strike location could exceed 10 kV or more. The same potential voltage differential is also mentioned by Shoup (1992) in relation to buried sensors and signal cable. A lightning strike which occurs some distance away from a long run of cable can still produce a voltage differential of similar magnitudes. Routing cables into a common trench increases the risk to all sensors since arcing can occur between adjacent cables as well as between ground and cables.

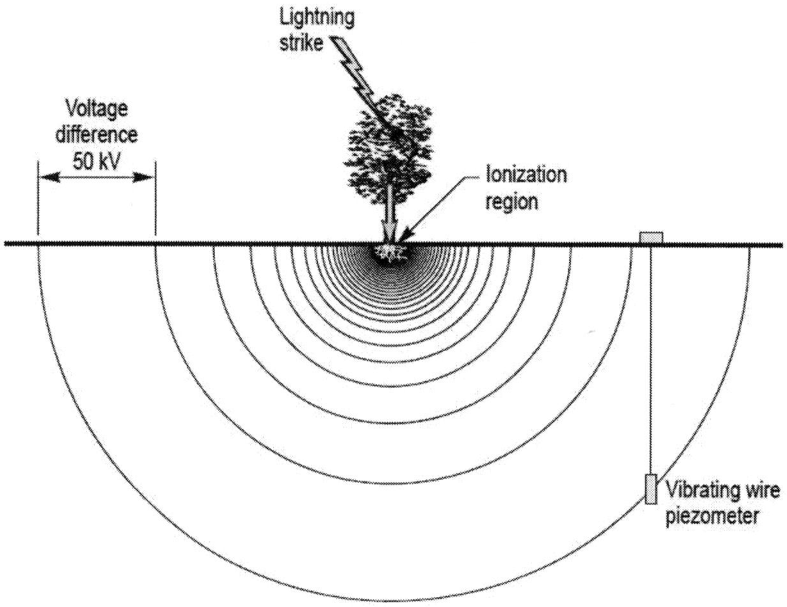

Figure 7. Lightning Strike Iso-Voltage Gradient (Moulds et al. 2000)

Clairmont et al. (1999) stated that damage to equipment can be caused by improper grounding including using the wrong type of electrical cable, not following the manufacturers installation recommendations, inadequate temporary protection, or improper shielding and surge protection devices.

Construction Damage

Construction damage commonly occurs to ADAS equipment, particularly on active tailings impoundments. Any exposed component is susceptible to damage from equipment and shallow buried cables are often severed from excavation as occurred with the case study presented here. Many facilities often do not survey nor mark the locations of buried ADAS cables or they do not make their locations/markings known to excavation subcontractors. Damage to piezometer installations occurred in a facility studied by Quaranta et al. (2008). VW piezometers were installed within standpipes along the crest of the dam. In this instance heavy machinery was a constant threat to the casing. New casing segments regularly needed to be installed as the dam crest was raised.

Power Loss

Due to their remote locations on tailings facilities, ADAS enclosures are often powered by batteries charged with solar panels. Impoundments Nos. 1 and 2 both had enclosures that lost power, halting data communication. The causes ranged from undersized solar panels, faulty power supplies, or extended periods of little sunlight. AC power was also lost in radio communication relays. Quaranta et al. (2007 & 2008) experienced battery power loss during the instrumentation of a coal waste impoundment. Rapid loss of power was experienced in the systems 12V, 7.5 AH batteries. The power drain from the instrumentation was more than the 5 watt solar panels could handle. The solution to the problem included wiring an additional 5W solar panel in parallel and changing the batteries from 7.5 AH to 18AH based on the assumption that the 7.5AH batteries could not fully store all the available energy from the two solar panels.

Communications

Communications loss can be common in tailings facilities with ADAS that use radios. Antennae need to be appropriate for the system and often need line of sight. Many factors can interrupt the communication, including damage to the cable, antennae, radios, or blockage of line of sight. Communication loss was experienced in the facility studied by Quaranta el al. (2007 & 2008). The loss of data was connected to a bad datalogger, which had to be replaced with a new unit. There were also line-of-site issues which created problems for data communications.

Other Factors

Consistent with the environmental conditions identified by Dunicliff et al. (1993) other contributing factors have been encountered:

- Ground deformations, settlement, or slope failures typically affect buried ADAS cables by straining or breaking buried cables and causing inaccurate readings or loss of data. Liquefaction has also been shown to affect VW piezometers instrumentation (Youd et al. 1994). Landslides can destroy entire clusters of instruments and enclosures.
- Instrumentation sensors at tailings impoundments are often installed in mine tailings, which can be acidic depending on their composition. Corrosive environments can damage VW piezometers, although the extent is uncertain since the instruments are rarely recovered.
- VW piezometers are calibrated for a specific pressure and/or installation at varying pressures. For instruments installed at differing piezometric levels, varying pressures can result in inaccurate readings or even failure of the instrument.

- Sometimes ADAS data cannot be collected due to loss of accessibility as an impoundment is constructed. This is more common for non-ADAS systems and components such as Casagrande piezometers.

PRESERVATION OF INSTRUMENTATION

Proper ADAS planning and layout can significantly reduce the risk of damage and failure. The risk of lightning strikes should be assessed based on geographic location. Historic lightning strike information is available from NASA and NOAA. If lightning strikes are anticipated, Klebba (2003) recommended the use of "lightning protection zones" varying in protection from areas receiving direct lightning strikes to those not subjected to strikes but where some instruments could see reduced current and attenuated electromagnet fields. The recommendations are to avoid power and signal interconnections of equipment that cross equipotential zones, or to construct a lightning protection zone within which it is safe for the equipment to be installed. At the case study facilities lightning protection zones have been or will be established for enclosures and radio relay stations.

Alternatives to avoiding damage to piezometers and ADAS dataloggers include keeping the cable runs to the sensors short, increasing the number of dataloggers in the network and, if possible, using radio communications in order to shorten cable lengths (Choquet et al. 2014). Transient protection and grounding of all instruments is done in the enclosure boxes: all incoming wires from instruments are connected to the multiplexers that incorporate transient protection, and all shield wires of instruments are connected to a common ground connector. A heavy gage copper cable can be utilized as a separate grounding network which is independent from other grounding circuits (Choquet et al. 2014). Transient protection can also be used at the ground surface above the VW piezometer to protect the piezometer from lightning strikes affecting the cables. The cable run locations should be accurately documented and marked. Often the locations are staked on the surface and a marker tape is placed in the trench above the cables.

The protective ground system needs to be at least as deep as the deepest sensor, as elaborated by Shoup (1992). An optical isolator for RS-232 is highly recommended to further isolate the computer's serial port from the data acquisition system. Isolation should be rated at a minimum of 500 Volts. Primary lightning protection devices, such as gas discharge tubes, may be used to protect sensor connections to data acquisition modules, but only if an appropriate protective ground system is installed first. If a protective ground system is not installed, such lightning protectors could actually increase the severity of damage to sensors and associated cables. For AC powered ADAS equipment, AC arrestors should be wired inline (Clairmont et al. 1999). For battery powered systems, a minimum of 10 or 20 watt solar panels are preferred to charge 12V batteries.

For ADAS using radios, the system should be laid out considering current layout and future expansion. A good line of site from enclosures for radio communications as well as a network of multiple repeaters is recommended for ADAS systems (Quaranta et al. 2008).

The correct VW piezometers should be chosen for the applications. This includes the correct pressure range, corrosion resistance, and/or direct burial heavy duty applications. The VW piezometers should also be installed using the manufacturer's recommendations (i.e. soaking the tip, grout mix, etc.) with care taken do avoid damage to the instruments and cables.

CONCLUSION

The case study cited here revealed a history of failed VW piezometers at two tailings impoundment facilities. Many of the failures have been attributed to embankment deformation, broken buried cables or electrical damage from lightning strikes. Vulnerability to problems caused by lightning and ground faults increases dramatically with cable length. The two impoundments feature some VW locations stationed far from the ADAS enclosures with cables often grouped in the same trench. Common failures occurred in the piezometers with long cable runs located near the top or upstream side of the dam. Another possible cause of failures is electrical damage from lightning strikes to the impoundment pool damaging piezometers buried near the FCR in the impoundment.

Troublshooting errors in an ADAS system begins with the data. When determining the problem, it is better to have no data than misleading data (Crum 2011). VW Piezometers can be investigated to determine whether the instrument or cable is damaged. Unfortunately, the only solution is to often re-drill and install another VW piezometer. VW piezometers can be suspended in larger diameter Casagrande piezometers enabling their replacement. However this negates the advantage of short hydrodynamic time lag and ease of installation associated with grouted-in VW piezometers (Contreras et al. 2012).

Proper ADAS planning is recommended for tailings impoundments. Preventative measures can be implemented to reduce the risk of electrical damage from lightning strikes, construction damage, communications loss, etc. The planning should consider past, present and future conditions on the impoundment. Facilities should reduce the number of long cable runs and consider placing more enclosures with radio or cable modem communication devices.

ACKNOWLEDGEMENTS

Special thanks to Geokon, Inc. for reference and TDR information.

REFERENCES

Contreras, I.A., Grosser, A.T., and Ver Strate, R.H. "Update of the Fully-Grouted Method for Piezometer Installation", *Geotechnical News*, June 2012.

Choquet, P., and Taylor, R.M., "Automatic Data Acquisition Systems (ADAS) for Dam and Levee Monitoring", *Geo-Congress 2014*, ASCE, 2014.

Clairmont, J., Dupuis, M., Choquet, P., and Budin, R., "Lightning Protection and Shielding for Geotechnical Instruments and Data Acquisitions Systems", *Geotechnical News*, September 1999.

Crum, D. A., "Rogue Piezometers", *21st Century Dam Design – Advances and Adaptations,* 31st Annual USSD Conference, April 11-15, 2011.

Dunicliff, J., and Green, G.E., "Geotechnical Instrumentation for Monitoring Field Performance", John Wiley & Sons, Inc., 1993.

Klebba, J.M., "Eliminating Lightning Damage for Field Data Acquisition Systems", *Conference Proceedings of 6th International Symposium on Field Measurements in GeoMechanics*, September 2003.

Moulds, A., and Watson, A. M., "Lightning Protection for Dam Instrumentation a Case Study of the New Victoria Dam." *Ancold Bulletin,* 2000.

Quaranta, J.D., Banta, L.E., and Altobello, J.A., "Remote Monitoring of a High Hazard Coal Waste Impoundment in Mountainous Terrain Case Study", *Tailings and Mine Waste '08*, 2008.

Quaranta, J.D., Banta, L.E., and Altobello, J.A., "Electronic Instrumentation of Coal Slurry Impoundments for Real-Time Data Collection to Support Monitoring and Emergency Response Actions", National Technology Transfer Center, Wheeling Jesuit University, July 2007.

Shoup, D., "Sensors in the Real World", Slope Indicator, 1992.

Youd, T. L., and Holzer, T. L., "Piezometer Performance at Wildlife Liquefaction Site, California", *Journal of Geotechnical Engineering,* Vol. 120, No. 6, June 1994.

IMPLEMENTATION OF NEW YORK STATE'S GUIDANCE FOR DAM HAZARD CLASSIFICATION

Kevin Ruswick, PE, CFM[1]
Gregory Daviero, PhD, PE[2]

ABSTRACT

New York State (NYS) promulgated new dam safety regulations in August 2009 which established specific owner activities for maintaining a safe dam. The law includes requirements to develop Inspection and Maintenance Plans and Emergency Action Plans, perform inspections and formal Engineering Assessments, and drives the need to identify and remediate deficiencies at existing dams for the betterment of public safety. Timetables for performing these activities were established based on the dam hazard classification; specifying more aggressive timetables and more rigorous design criteria for high hazard facilities than for intermediate or low hazard facilities. The appropriate hazard classification informs the State of a dam's potential risks to life and property downstream of the facility, allowing for an educated allocation of its resources; and informs the owner of the potential consequence of dam failure, potential liabilities, and appropriate design criteria. Therefore, establishing a proper hazard classification is paramount to establishing a correspondingly appropriate dam safety program. The NYS Dam Safety Section recently issued *Guidance for Dam Hazard Classification* to assist their staff, dam safety consultants, and dam owners in the consistent assignment of hazard class. This Guidance identifies several categories of population and infrastructure at risk that can drive the hazard classification of a facility including: loss of life, access for emergency services, damage to homes, damage to roads and railroads, interruption of utilities, downstream dams, and environmental damage.

The hazard class is based on the potential consequence of a dam's failure. The primary tool used to assess this is the inundation mapping associated with "sunny day" and "rainy day" dam break scenarios, and the comparison of consequences between the failure and non-failure events. Application of specific technical approaches can influence the results of these studies and should be carefully considered when determining an appropriate hazard classification or in the regulator's evaluation of these determinations. Discussions will involve topographic resolution, aerial imagery, calibration of hydrologic models, limits of analyses, downstream tributary inflows, and dam break parameters (timing and geometry), and how selection of these can influence the outcome of a hazard classification study in NYS or elsewhere. A review of the NYS guidance documentation and its application through case studies is also presented.

---

[1] Associate, Schnabel Engineering of New York, 28 Corporate Drive, Suite 104, Clifton Park, New York 12065, 518-348-8575, kruswick@schnabel-eng.com.
[2] Principal, Schnabel Engineering of New York, 28 Corporate Drive, Suite 104, Clifton Park, New York 12065, 518-348-8575, gdaviero@schnabel-eng.com.

INTRODUCTION

New York State (NYS) promulgated new dam safety regulations in August 2009 which established specific owner activities for maintaining a safe dam and is enforced by the NYS Department of Environmental Conservation (NYSDEC). The law codified the need for dam owners to investigate, identify, and remediate deficiencies at existing dams for the betterment of public safety. Tenants of responsible dam ownership, the regulations now require the development of Inspection and Maintenance Plans and Emergency Action Plans, performance of regular inspections, and conducting formal Engineering Assessments. Timetables for performing these activities were established based on the dam's hazard classification: specifying more aggressive timetables and more rigorous design criteria for high hazard facilities than for intermediate or low hazard facilities. This is a logical correlation as the hazard classification informs the State of a dam's potential risks to life and property downstream of the facility, allows for an educated allocation of its resources, and informs the owner of the potential consequence of dam failure, potential liabilities, and appropriate design criteria. Therefore, establishing a proper hazard classification is paramount to establishing a correspondingly appropriate dam safety program and for identifying owner responsibilities and timetables for compliance with the State's new dam safety regulations.

The NYSDEC Dam Safety Section (DSS) recently issued *Guidance for Dam Hazard Classification* (Guidance) to assist their staff, dam safety consultants, and dam owners in the consistent assignment of hazard class. The hazard class is based on the potential consequence of a dam's failure independent of a dam's condition. Several key categories of at risk population and infrastructure are identified that can drive the hazard classification of a facility including: loss of life, access for emergency services, damage to homes, damage to roads and railroads, interruption of utilities, downstream dams, and environmental damage. This Guidance standardizes the application of the decision matrix for typical situations encountered in the downstream impacted areas. It is particularly useful in eliminating the ambiguity of the classification process where professional opinions of impact on and importance of affected infrastructure and risk to life within structures can vary. Specific criteria is identified to standardize hazard class based on numbers and depth of flooding to residential structures (homes), and number of homes with impaired access to emergency services. For roadways and railroads, the Guidance directs that overtopping of the embankment implies damage to the infrastructure and the hazard classification can be determined based on the specific NYS Department of Transportation functional classification of overtopped roadways and the rail use identified in the NYS Rail Plan and Rail map. In each case, a quantitative assessment of impact and classification can be made through the analysis of the inundation mapping. Application of specific technical approaches may influence the results of these studies. Therefore, the series of both large and small hydrologic and hydraulic decisions made by the engineer, as well as appropriate model selection, should be carefully considered when determining an appropriate hazard classification or in the regulator's evaluation of these determinations.

REVIEW OF NYS'S GUIDANCE FOR DAM HAZARD CLASSIFICATION

The Guidance presents a quantitative way to standardize the determination of hazard classification replacing what was heretofore primarily qualitative. The Guidance follows NYS's recently promulgated dam safety regulations which established specific, hazard-based owner activities for maintaining a safe dam, and was developed to assist the NYS Dam Safety Section staff and consulting engineers in the assignment of hazard class.

NYS uses a four-tiered classification system where Classes A, B, and C dams are regulated as Low, Intermediate and High Hazard, respectively, and Class D dams are considered negligible or no hazard dams typically associated with failed or decommissioned structures. The Guidance defines Class A dams as structures where, "A dam failure is unlikely to result in damage to anything more than isolated or unoccupied buildings, undeveloped lands, minor roads such as town or county roads; is unlikely to result in the interruption of important utilities, including water supply, sewage treatment, fuel, power, cable or telephone infrastructure; and/or is otherwise unlikely to pose the threat of personal injury, substantial economic loss or substantial environmental damage." Class B dams exist where; "A dam failure may result in damage to isolated homes, main highways, and minor railroads; may result in the interruption of important utilities, including water supply, sewage treatment, fuel, power, cable or telephone infrastructure; and/or is likely to pose the threat of personal injury and/or substantial economic loss or substantial environmental damage. Loss of human life is not expected." And Class C dams exist where; "A dam failure may result in widespread or serious damage to home(s); damage to main highways, industrial or commercial buildings, railroads, and/or important utilities, including water supply, sewage treatment, fuel, power, cable or telephone infrastructure; or substantial environmental damage; such that the loss of human life or widespread economic loss is likely."

The Guidance includes a description of several factors typically considered in assigning a hazard classification. Where engineering opinions could vary with respect to the impact of dam failure, the Guidance provides specific direction in an attempt to standardize these determinations. The process relies on the analysis of the "sunny day" and "rainy day" dam break inundation mapping (and hydraulic modeling output) and the comparison of consequences between the failure and non-failure events. Specific guidance is provided for eight critical factors which are reviewed in the following paragraphs.

Loss of Human Life

The likely loss of life is identified as the most significant factor in determining if a dam is high hazard and is implied where damage to homes, regularly occupied, is "sufficiently severe or widespread." Though the Guidance does not provide new insights for determining loss of life, it references *ACER 11 (Bureau of Reclamation, 1988)* and *A Procedure for Estimating Loss of Life Caused by Dam Failure (Bureau of Reclamation, 1999)* for the engineer to consider. More easily defined quantitative criteria discussed below are linked to the likely loss of life and trigger a high hazard designation.

Secondary Consequences and Impairment of Access to Emergency Services

The occurrence of a "personal injury" which would require access to emergency services is considered a likely consequence of a dam failure. Therefore, isolation from access to emergency services is a driver to determining hazard class. The Guidance sets the isolation of 10 homes or a single hospital, nursing home, or fire and police stations as the limit for establishing a Class B designation. No specific limit on the number of homes is provided for a Class C designation, but flooding of a hospital by 1 foot is identified as a justification for a high hazard designation.

Damage to Homes

The Guidance's designation of damage to homes is one of the areas where it succeeds in introducing a quantitative measurement in lieu of engineering opinion. The process utilizes hydraulic modeling output of flood depth and velocity linked with first floor elevations. Specific limits are provided to give clarity and consistency in the designation of the hazard classification. These limits are repeated in Table 1.

Table 1. Hazard Classification Based on Numbers and Depths of Flooded Homes.

Flooding Depth	Number of Homes Impacted		
	1 to 10	11 to 99	100 or more
Up to 1 foot above lowest occupied floor	A	B	C
Greater than 1 foot above lowest occupied floor	B	B	C
Above the Low Danger Zone (per ACER 11)	C	C	C

Where the depth of flooding is utilized for the determination of hazard classification, it is assumed that the corresponding velocities are low enough such that the structure would fall in the Low Danger Zone per ACER 11. Any flood depth and velocity combination resulting in a structure within the Judgment Zone or High Danger Zone would result in a hazard Class C designation. ACER's Depth-velocity flood danger level relationship for houses built on foundations is shown in Figure 1 for clarity.

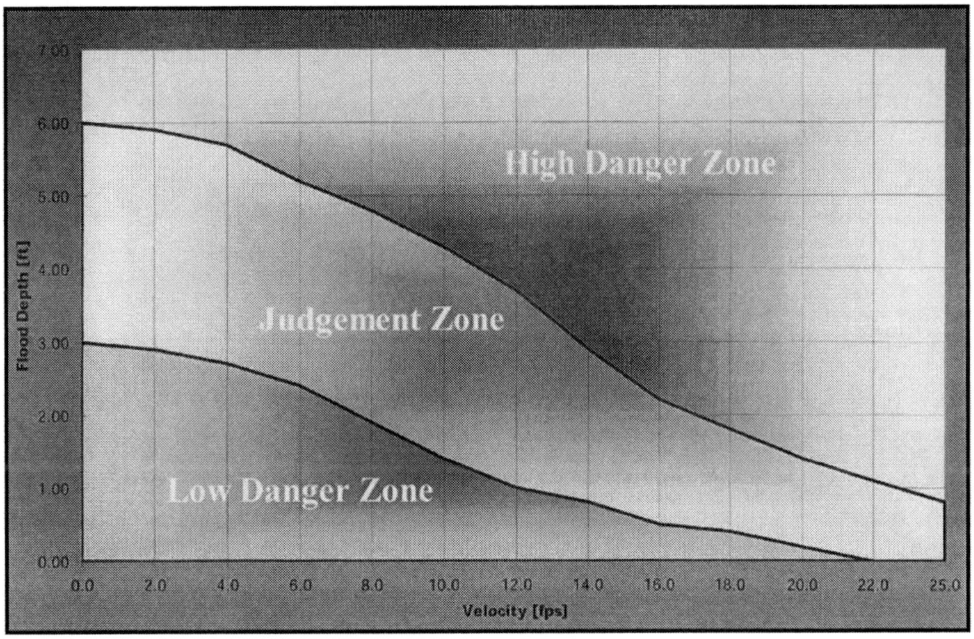

Figure 1. Depth-velocity Flood Danger Level for Houses Built on Foundations.

Main Highways versus Minor Roads

Another significant advancement is the use of the NYS Department of Transportation's functional classification as a contributor to the determination of hazard class. The NYS Department of Transportation applies a functional classification to roadways based on the type of service it provides and is an indicator of the importance of the roadway within the State's transportation system. The Guidance directs that overtopping of the embankment implies damage to the infrastructure. Specific thresholds are provided to give clarity and consistency in the designation of the hazard classification with respect to roadway overtopping. The criteria are repeated in Table 2.

Table 2. Hazard Classification Based on NYS DOT Functional Classification.

NYSDOT Functional Classification	DSS Road Type	Hazard Classification
Urban - Principal Arterial Interstate	Main Highway	C
Rural - Principal Arterial Interstate	Main Highway	C
Urban - Principal Arterial Expressway	Main Highway	C
Urban - Principal Arterial Other	Main Highway	C
Urban - Minor Arterial	Main Highway	B
Rural - Principal Arterial Other	Main Highway	B
Rural - Minor Arterial	Main Highway	B
Urban – Collector	Main Highway	B
Rural - Major Collector	Main Highway	B
Rural - Minor Collector	Minor Road	A
Rural – Local	Minor Road	A
Urban – Local	Minor Road	A

New York State Dam Hazard Classification

Railroads versus Minor Railroads

Similar to roadway embankments, the Guidance directs that overtopping of a railroad embankment implies damage to the infrastructure. The Guidance differentiates between a "railroad" and "minor railroad" with their overtopping implying a hazard class C and B, respectively. Specific discussion is provided for intercity passenger, commuter, scenic and tourist lines, and transit lines as well as freight lines.

Interruption of Important Utilities

A discussion on the linkage between a dam failure and interruption of important utilities is provided. Important utilities are defined as serving greater than 100 residences, and durations of the outages are discussed as well. Interruption of important utilities for 5 days or less generally implies Class B. Widespread or serious damage to important utilities includes but is not limited to: loss of utility services serving major populations or loss of important utility services for periods on the order of several weeks, and generally implies Class C. For wastewater treatment plants, damage to a Major or Significant facility implies Class B, while damage to a less significant plant implies Class A. Loss of a water treatment plant or water supply reservoir is generally not a factor in driving the hazard class of the dam creating the reservoir. The dam owner is encouraged to consider the benefit of the reservoir and associated infrastructure and wants to consider adopting higher design standards than those assigned by NYS DEC.

Substantial Environmental Damage

Similar to the discussion related to interruption of important utilities, the Guidance discusses the linkage between a dam failure and potential environmental damage. Clearly, a dam failure can cause loss of habitat or could release or cause the release of environmental toxins. It is stated that this factor rarely drives the hazard classification, and little direction is provided to quantify these impacts.

Downstream Dams

The Guidance clarifies that the hazard classification should consider the collective consequence of the failure of the dam in question as well as all other downstream dams which may fail as a result of the upstream failure. No detail is provided to define the basis for the determination of the sequential dam failures. The authors presume that this decision would be based on the hydrologic and hydraulic modeling of the downstream waterway, and be contingent upon whether the downstream dams were erodible and were overtopped due to the upstream dam break.

This final criterion introduces how application of specific technical approaches related to the hydrologic and hydraulic modeling can influence the results of these hazard classification studies and should be carefully considered when determining an appropriate hazard classification.

INFLUENCES OF KEY HYDROLOGIC AND HYDRAULIC ITEMS

The Guidance offers some specific and quantifiable decision matrices for determining appropriate hazard class. The hazard class is based on the potential consequence of a dam's failure. The primary tool used to assess this is the inundation mapping associated with "sunny day" and "rainy day" dam break scenarios, and the comparison of consequences between the failure and non-failure events. A quantitative assessment of impact and classification can be made through the prediction of the depth and velocity of flooding at residential structures and overtopping of roadways and railroad embankments. For many dam projects inundation mapping is developed for use in the Emergency Action Plan, and the engineer is inclined to be conservative in their selection of key hydrologic and hydraulic parameters so as to exaggerate the impacted area for the protection of public health. For the determination of hazard class, it is imperative that these parameters be representative of the site and dam break circumstances. In some cases, conservative values, rightly selected in consideration of public safety, may obscure the impact of a dam failure and, ironically, drive the apparent hazard class lower.

While this Paper is not intended to be a comprehensive evaluation of all the potentially influential hydrologic and hydraulic parameters, the remaining sections will focus on key issues which can influence the outcome of a hazard classification study. Of particular importance are the base mapping and aerial imagery resolution, selection of dam break parameters (timing and geometry), prediction of the inflow design flood, and consideration of downstream tributary inflows.

The most reliable predictor of hazard class is the sunny day dam break mapping. This mapping contains the uncertainties related to the resolution of the topographic and aerial imagery, selection of dam break parameters, and the hydraulic modeling of the downstream channel, but eliminates those associated with the prediction of the inflow design storm and contributions of downstream watersheds. The process may conclude here when the outcome of the sunny day modeling can identify impact on enough residential structures or overtopping of key roadways or railroad embankments to justify a high hazard determination. In all other cases the engineer needs to proceed to the rainy day analysis for a final determination. Also note that in NYS, any proposed reduction in hazard classification for a dam needs to be prepared by a licensed professional engineer and must be supported by detailed dam breach modeling.

Base Mapping and Aerial Imagery

The ability to reliably predict flooding of residential structures to the nearest foot and to identify and count impacted structures within the inundation area makes the application of the Guidance strongly dependent on the availability of high resolution topographic base mapping and aerial imagery. NYS has been advancing the Federal Emergency Management Agency's (FEMA) map modernization initiative throughout the state. focusing first on its more populated regions. As of this writing, high resolution, LiDAR-based topographic mapping is available in all or portions of 42 out of 62 counties. Typically, this data is supplemented by as built or field verified dimensions of downstream hydraulic structures and instrumented or surveys of first floor elevations of key structures. Without available digital elevation models, the engineer can rely on conventional instrumented surveys to define the downstream channel dimensions as well as key first floor elevations, but this results in a significantly more costly and less accurate representation of the inundation area. Aerial imagery is now widely available through free, publically available internet based mapping tools (Google Earth ®, Bing Maps®, etc.) but should be confirmed to be representative of current features. Imagery should be confirmed with a windshield survey of the potentially impacted areas.

Dam Break Parameters

The Guidance provides some very basic information on the technical requirements of the dam break analysis suggesting the use of unsteady flow computer modeling with "piping" as the trigger for the sunny day dam break, and "overtopping" as the trigger for the wet weather dam failure. The Guidance provides no further direction for the selection of the dam break parameters or timing. Similar to the development of mapping for use in an EAP, the authors recommend that dam break parameters should be selected to maximize the peak flow of the dam break hydrograph in order to simulate a realistic worst case scenario. Though reasonable physical parameters, reflective of the dam's construction material and geometry, should be selected for the shape of the breach, the timing of the failure should be set to the smallest reasonable limit. Typical ranges for dam break parameters have been estimated by several agencies including: Bureau of Reclamation, FEMA, FERC, and the Corps of Engineers. The parameters are typically expressed as a function of the height of the dam and construction materials.

Limits of the Analyses

The Guidance suggests that the limits of the analyses be determined where the difference in flood elevations between dam failure and non-failure events is approximately two feet or less. Presumably this would apply to the rainy day dam break scenario though this criterion seems in conflict to using one foot as the basis for determining hazard class based on the numbers of impacted homes as shown in Table 1. The authors recommend careful application of this criterion so as to reduce the likelihood of missing potentially impacted structures. Though not identified within the Guidance, the authors recommend that the sunny day dam break limits be set where the dam break flood elevation converges with the regulated 100-year base flood elevation. In consideration of downstream hydraulic structures, the Guidance is strong in its assertion that these be modeled to function unobstructed and function as designed.

Inflow Design Flood

For the rainy day dam break, the investigator needs to develop a hydrologic model of the contributing watershed to predict its behavior during the required inflow design flood (IDF). NYS requires the IDF to be 50% of the Probable Maximum Flood (PMF) for existing high hazard dams, 150% of the 100-year flood for intermediate hazard dams, and the 100-year flood for low hazard dams. For large watersheds with small dams, the maximum modeled downstream flows are likely to be dominated by the wet weather flow where the IDF peak and volume are significantly greater than the dam break volume, and the incremental increase in flooding due to dam break diminishes. Conversely, in small watersheds with large dams, the maximum modeled downstream flows are likely to be dominated by the dam break flow where the IDF peak and volume are significantly less than the dam break flow. While it is easy to see the relative importance of the IDF peak and volume in these extremes, most circumstances are less self-evident and the over prediction of the IDF may obfuscate the impact of a concurrent dam failure.

In many cases, for ease of application, engineers utilize the Soil Conservation Service (SCS) unit hydrograph methodology within their hydrologic model. Based on the authors' experiences, the SCS unit hydrograph methodology tends to be overly conservative in terms of peak flow predictions, and we have found through multiple previous hydrologic model calibration efforts in the northeast that the Snyder Unit Hydrograph Method or the Clark Unit Hydrograph method can better represent the rainfall-runoff process for extreme storm events.

An over prediction of the IDF could result in an over prediction of the inundation area. While this might be an acceptable outcome for identifying residences for evacuation in an EAP, an increase in the inundation area always results in a smaller incremental rise in water surface for a given dam break volume. Note that the hazard class guidance for the rainy day breach scenario considers only the incremental increase in damage due to dam failure when comparing the breach and non-breach scenarios. The implications of this are that an overly conservative IDF could lead to a reduction in hazard classification if the addition of the dam break volume shows few additional impacted structures. This is likely an unintended consequence of the process. Regardless of the selected hydrologic transformation, the authors recommend that the hydrologic model be calibrated to the largest historical storms where rainfall and reservoir elevation data exist. In this way, the volume of the IDF flow can be validated and balanced appropriately against the dam break volume.

Downstream Tributaries

Similar to the IDF, the assumption associated with downstream tributaries can influence the outcome of the hazard class analysis. When not modeled prudently, the inclusion of downstream tributary inflows can further diminish the predicted impact of a dam failure. Again, assumptions that are appropriate for developing functional inundation maps for inclusion in an EAP can lead toward a determination of a lower hazard classification. The authors recommend developing and calibrating hydrologic models of the downstream tributary watershed to appropriately predict the timing of these flows. Contributions from downstream watersheds that peak coincidentally with breach flows may conceal the impact of the dam break. There also needs to be consideration of the size and distribution of the storm for downstream tributary areas relative to the IDF. It is not the intent of this paper to prescribe how these details should be considered, but to shed light on their relative influences on the process.

Dam owners' budgets are often constrained, which can lead to engineers utilizing simplifying assumptions in regards to downstream inflows. One common method is to use 100-year inflows for downstream tributaries based on FEMA and/or USGS StreamStats flow values. These are commonly applied as steady state inflows, while the breach hydrograph is commonly modeled as an unsteady flow hydrograph. Utilization of steady-state downstream inflows automatically results in coincident timing of the flood wave peak flows. While this may appear at first a conservative assumption, the reality is that the peak flow from downstream inflows may be significant enough to raise mainstem elevations above the channel banks into a broad floodplain where significant floodplain storage is available to attenuate the breach hydrograph. If in reality the peak timing is not coincident, this approach may overestimate the available channel and floodplain storage, thereby decreasing the estimated incremental increase in flood damages resulting from the dam breach, as well as increase travel times downstream of the confluence.

Breach Hydrograph Development and Downstream Routing

There are several methodologies available for the engineer to estimate breach outflow hydrographs, peak flows, and routing of the resultant flood wave through the downstream drainageway.

Simplified methodologies include regression based equations such as the National Weather Service's (NWS) Simple Dam Break Equation which was developed by Wetmore and Fread (1984):

$$Q_b = Q_o + 3.1B_r\left(\frac{C}{(T_f + C/\sqrt{H})}\right)^3 \quad (1)$$

Where,
Q_b = Breach Flow + Non-Breach Flow (cfs)
Q_o = Non-Breach Flow (cfs)
B_r = Average Breach Width (ft)
$C = 23.4(A_s/B_r)$
A_s = Reservoir Surface Area (ac) at Maximum Pool Level
H = Selected Failure Depth (ft) above Final Breach Elevation
T_f = Time to Failure (hrs)

Froehlich (2008) developed equations to estimate breach dimensions and failure times based only on the depth and volume of water stored by the dam. Average breach width (B_r) and failure time (T_f) are estimated by the following equations:

$$B_r = 8.239K_oV_w^{0.32}H^{0.04} \quad (2)$$

$$T_f = 3.664\sqrt{\frac{V_w}{gH^2}} \quad (3)$$

Where,
K_o = Failure Mode Factor (1.0 piping, 1.3 overtopping)
V_w = Storage Volume (ac-ft)

Note that the Simple Dam Break equation does not account for many site-specific conditions including but not limited to: submergence, downstream constrictions, or conditions where the ultimate size of the breach may be constrained by bedrock and/or narrow valley walls. Therefore, flows computed from this methodology are generally considered conservative (high end) estimates for potential breach flows and are typically recommended only for a screening level analysis. High predicted outflows from this methodology could lead to hazard classifications higher than appropriate; thus cost savings from utilizing this approach could actually lead to higher owner expenses to deal with increased regulatory compliance activities.

Other models commonly used in developing breach outflow hydrographs include the Corps of Engineer's HEC-HMS hydrologic model and HEC-RAS hydraulic model. Both models include hydrograph generation for piping and overtopping failure modes. Required input parameters include: breach bottom width, breach side slopes, breach height, and time for breach formation. For piping breach scenarios, an orifice coefficient is required to compute flow through the "pipe." For overtopping breach scenarios, a weir coefficient is required to compute flow over the breach "weir." Progression of the breach enlargement to ultimate size can be identified as linear or as a user-input progression curve. Given identical stage-versus-storage relationships and breach parameter input, both the HEC-HMS and HEC-RAS models should provide similar breach outflow hydrographs in most cases. The primary circumstance where results may differ occurs when there is a hydraulic constriction downstream of the dam. This could be a bridge or a narrow valley which results in water surface elevations that submerge the progressing breach. HEC-RAS can account for reduced breach discharges due to submergence while HEC-HMS cannot. For this reason, the authors recommend application of HEC-RAS for breach hydrograph development and unsteady flow routing through the downstream drainageway. For complex downstream areas, application of a 2-dimensional hydraulic model may be required for routing of the dam breach hydrograph. Note that a beta version of HEC-RAS that incorporates 1-D and 2-D modeling is being evaluated, and a publicly available version is anticipated in 2015. A more detailed discussion in regards to 1-D versus 2-D and even 3-D modeling is a worthy topic but beyond the scope of this paper.

CASE STUDIES

The authors have had the opportunity to apply the new Guidance on more than 25 dams throughout NYS. The scope of these past assessments varied widely from screening level hazard class assessments to formal hazard class reduction studies. We have selected a few examples which present how the new guidance was applied under varying project budgets, data availability, and project goals.

Otsego Lake Dam – Cooperstown, New York

Otsego Lake Dam is a run of river dam located on the Susquehanna River, approximately 0.5-miles downstream from its source at the south end of Otsego Lake. The dam is a concrete buttress structure with a total length of approximately 70 feet and a height of approximately 11 feet. It has a normal pool storage of 341,094 acre-feet, a maximum storage of 345,320 acre-feet, a surface area of 4,225 acres, and a drainage basin of approximately 68.34 square miles. Prior to 2013, the dam was regulated as a Class B structure.

Figure 2. Otsego Lake Dam as Viewed from Downstream Right Abutment.

A detailed hydrologic model of the watershed using the USACE's HEC-HMS was developed. Watershed parameters were refined through calibration to a historic flood in June 2006 when more than seven inches of rain fell on the watershed. The calibrated hydrologic model was then used to develop and route a storm equivalent to the regulatory SDF which is 150% of the 100-year storm for a Class B dam. Dam breach models were also developed for a sunny day breach as well as for a breach during the SDF. A detailed hydraulic model of the Susquehanna River was developed using the USACE HEC-RAS and HEC-GeoRAS programs. The model extended downstream to the confluence with Oakes Creek, a distance of approximately 5 miles. LiDAR data was not available for this portion of Otsego County; therefore, the preliminary model data was extracted from a 10m USGS Digital Elevation Model (DEM) supplemented with 5-foot contour mapping provided by the Village. Model input data for several road crossings were developed from NYSDOT bridge plans.

Dam breach hydrograph generation was performed in the HEC-RAS model, and it was found that tailwater submergence and the presence of 0.5 miles of channel with two constrictive bridges between the dam and Otsego Lake reduced potential dam breach outflows when compared to theoretical dam breach discharges computed with HEC-HMS. Downstream inflows were included for the 100-year storm based on hydrograph output from the HEC-HMS hydrologic model. The computed peak water surface elevations from the preliminary model runs were post-processed with HEC-GeoRAS in an ArcGIS platform to overlay the peak water surface elevations on a digital elevation model to develop approximate inundation boundaries. The inundation boundaries were then plotted on aerial imagery to determine whether any structures may be impacted by a failure of the dam (Figure 3).

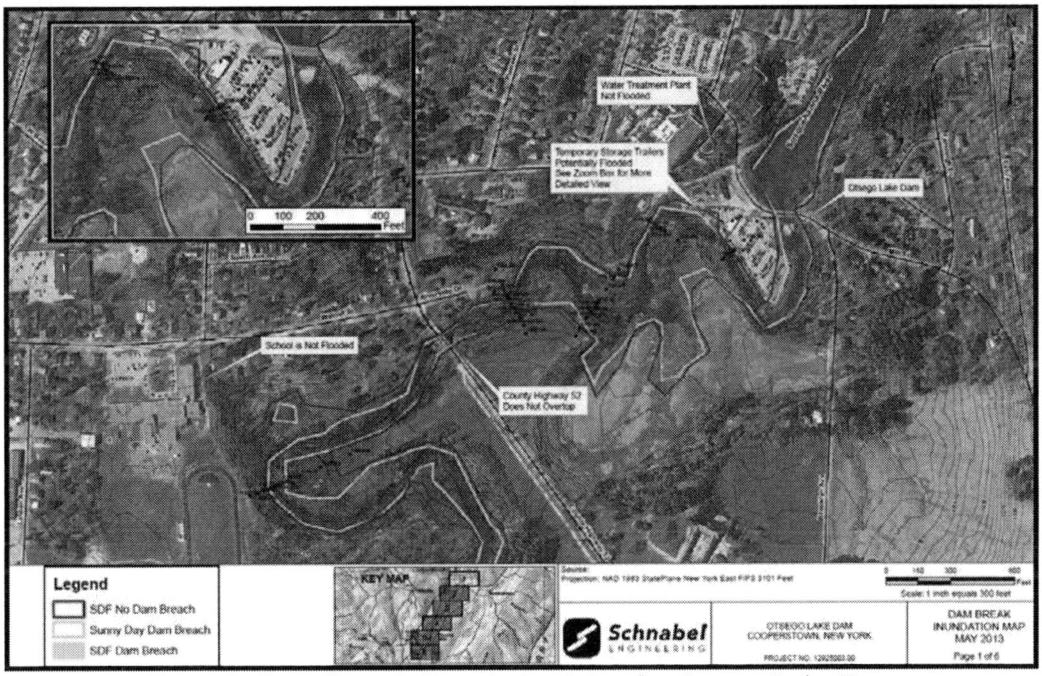

Figure 3. Dam Breach Inundation Map for Otsego Lake Dam.

County Highway 52 (Susquehanna Avenue), located approximately 3,000-feet downstream of the dam, is classified as a Rural Major Collector. Based on the Guidance, overtopping of this road would imply that the Class B designation for Otsego Lake Dam would have been appropriate. The model results indicated that this road would not be overtopped as a result of failure of Otsego Lake Dam. All other roadways within the model limits are classified as Minor roads and also were predicted to not be overtopped following a dam failure. Thus in terms of potential roadway impacts, the Otsego Lake Dam would be appropriately designated as a Class A – Low Hazard Dam.

The inundation mapping did, however, indicate that a few structures near the hospital parking lot adjacent to the dam could be flooded as a result of failure of the dam (See inset on Figure 3). As the accuracy of the USGS DEM data was estimated to be approximately five feet, the resolution was deemed inadequate for application of the Guidance to structures. Subsequently, instrumented field surveys were performed to refine the hydraulic model as well as collect key elevations including lowest occupied floor elevation. The field survey indicated that the potentially impacted structures were actually unoccupied storage sheds. Additionally, the model results indicated that the incremental increase in flooding depth between the breach and non-breach scenarios was less than one foot. Thus in terms of potential impacts to structures, Otsego Lake Dam is appropriately designated as a Class A – Low Hazard Dam.

The model output and inundation mapping from the refined HEC-RAS model were also used to evaluate the other remaining criteria contained in the Guidance, and the project team concluded that application of Guidance to the dam breach modeling results indicated that Otsego Lake Dam was consistent with a Class A – Low Hazard designation. A hazard classification reduction report was submitted to the NYSDEC who subsequently approved the change in hazard class for Otsego Lake Dam from Class B to Class A.

Kingston Reservoir No. 2 – Woodstock, New York

Reservoir No. 2 Dam is a run of river dam located on the Saw Kill near the Town of Woodstock, NY. The reservoir was originally constructed in 1897 as a water supply reservoir to serve the City of Kingston. The concrete and earthen dam is approximately 40-feet tall, and the spillway portion of the masonry dam is approximately 132-feet long at the principal overflow weir. The gravity section of the dam is abutted by embankment sections flanking each side of the spillway that bring the overall dam length to approximately 750 feet. The dam has a reported normal pool storage volume of approximately 95 acre-feet and a maximum storage volume of approximately 125 acre-feet. Prior to 2012, Reservoir No. 2 was designated as a Class C dam. At first glance, this may seem an appropriate hazard classification given the height of the dam and the presence of several roadways and residential structures along the Saw Kill downstream of the dam. However, the Saw Kill has a very steep longitudinal gradient and the storage volumes impounded by the dam are relatively small in relation to the watershed size at the dam site.

Reservoir No. 2 is located within Ulster County which has county-wide LiDAR-based topographic mapping publicly available. Additionally, the Saw Kill downstream of the dam flows through relatively populated areas of Ulster County and detailed FEMA floodplain mapping has been prepared based on HEC-RAS hydraulic modeling to delineate 100-year floodplains. The availability of this data allowed for the cost-effective development of detailed dam breach modeling for assessing the appropriate hazard classification using a comprehensive application of the Guidance.

Figure 4. Kingston Reservoir No. 2 as Viewed from Downstream Left Abutment.

The HEC-HMS model was used to generate inflow hydrographs, and the HEC-RAS model was used to simulate the breach formation and routing of the breach and non-breach floods though the downstream drainageway. The dam break modeling extended downstream to the Saw Kill's confluence with Esopus Creek. ArcMap software along with the HEC-GeoRAS application was used to develop input cross section data for the HEC-RAS dam break hydraulic model. Stream cross sections were extracted from a LiDAR-based digital elevation model provided by Ulster County. Road crossing hydraulic structure data were extracted from the existing FEMA HEC-RAS model associated with the 2007 Flood Insurance Study (FIS). Manning's roughness coefficients used in the hydraulic model ranged from 0.035-0.064 for the main channel and 0.049-0.18 for the overbank areas based on the FEMA FIS values. The hydraulic analyses were conducted using the unsteady flow computational engine in HEC-RAS, which allows for attenuation of the flood wave as it progresses downstream.

Dam break analyses were conducted for a sunny day breach as well as a breach during the SDF which is 50% of the Probable Maximum Flood (PMF) for existing Class C dams. For the wet weather analyses, four downstream tributaries were added to the HEC-RAS model to account for their inflows based on the FEMA 100-year HEC-RAS model. In this particular instance, the downstream tributaries were small in relation to the Reservoir No. 2 watershed, and had peak flows which were an order of magnitude lower than inflows and outflows for Reservoir No. 2. Subsequently, these downstream tributaries were not considered to be significant enough to impact the conclusions derived from the results of breach versus non-breach scenarios, and their inclusion as steady state inflows versus inflow hydrographs was warranted. Profile plots of the peak flood elevations from the sunny day as well as the SDF for both breach and non-breach scenarios are presented in Figure 5.

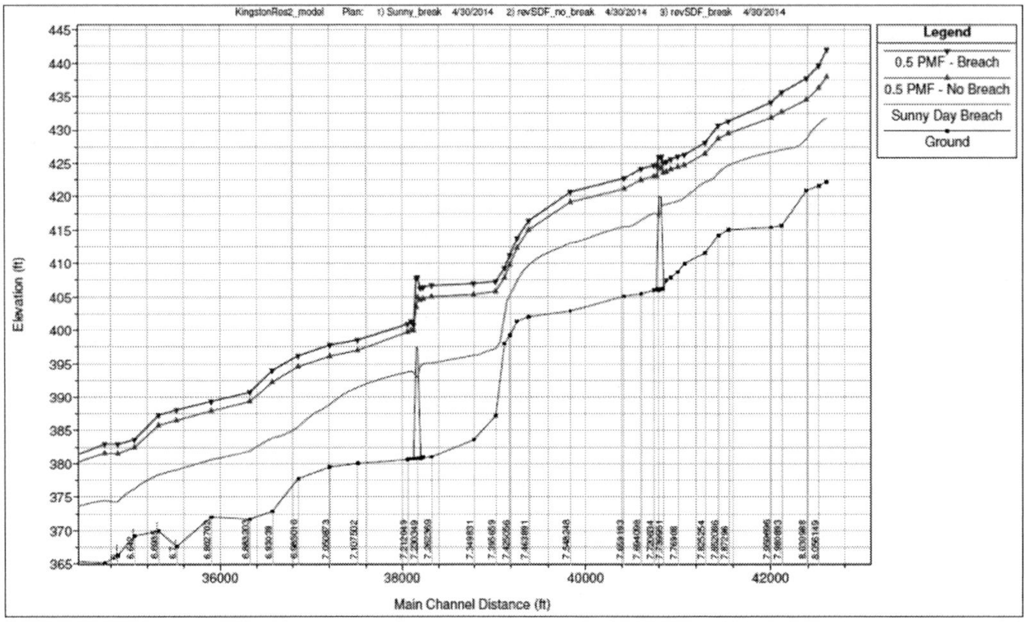

Figure 5. Flood Profiles for Kingston Reservoir No. 2 Dam Breach Analyses.

The results of the sunny day breach analysis indicated that the flood wave resulting from a normal pool failure of the dam in the absence of significant rainfall would generally remain within the channel banks, and would not overtop any downstream roadways or inundate any residential or commercial structures. The wet weather dam breach analyses were conducted for the Class C SDF as well as the Class B SDF. Note that only the 50 percent PMF results are included in Figure 5.

The HEC-RAS model results were post-processed using HEC-GeoRAS within a GIS software platform. While the sunny day dam breach results indicated that no structures, roadways, or utilities would be impacted, the wet weather inundation mapping and flood profiles indicated that numerous residential structures and roadways could be significantly impacted by a wet-weather event. As the Guidance clearly indicates that the procedure to evaluate hazard potential directly attributable to a wet weather breach of a dam should be based only on the incremental increase in flooding, it was critical to quantify the potential wet weather impacts for both the breach and non-breach scenarios. The primary impact categories consistent with application of the Guidance were associated with residential structures and roadways. Table 3 presents a summary of the number of buildings estimated to be impacted from breach and non-breach scenarios during the Class C and Class B SDF events. Table 4 provides an estimate of flooding depth and velocities for structures impacted by the breach scenario that are not predicted to be impacted during the non-breach scenario.

Table 3. Reservoir No. 2 Impacted Buildings Summary.

Flood Scenario	Number of Buildings Affected		
	Breach	**No Breach**	**Incremental Difference**
50% PMF	269	264	5
150% of the 100-Year Storm	146	145	1

Table 4. Incremental Wet-Weather Building Impact Summary.

Depth (ft)	Velocity (ft/s)	Property ID	Address	Street
0.5	< 0.5	27.4-5-16.1	1835	RT 212
< 0.1	< 0.5	27.4-5-16.1	1835	RT 212
< 0.1	< 0.5	27.4-2-29.2	127	JOHN JOY RD
< 0.1	< 0.5	27.4-2-26	119	JOHN JOY RD
< 0.1	< 0.5	38.2-1-61.1	837	ZENA RD

As depicted in Tables 3 and 4, the incremental increase in building flood impacts attributable to a wet weather failure is insignificant and based on the Guidance for impacts to homes is consistent with a Class B hazard designation.

The roadway impacts for breach and non-breach scenarios were also considered following the Guidance. While the incremental impacts to roadway overtopping are minor, the functional classification of the roadways is the overriding factor in application of the Guidance. Each potentially impacted roadway is classified as Minor roadways, consistent with a Class A hazard designation.

The availability of detailed LiDAR-based topographic mapping and a FEMA hydraulic model of the Saw Kill downstream of Kingston Reservoir No. 2 allowed for a comprehensive and cost effective application of the Guidance for a hazard class assessment of the dam. The quantitative results of the dam breach modeling, coupled with inundation mapping overlain on recent aerial imagery and detailed topographic mapping, indicated that the potential flooding impacts directly attributable to a failure of Reservoir No. 2 were consistent with a Class B hazard designation. This reduction in hazard classification for Reservoir No. 2 was formally approved by NYSDEC in 2012.

NYSDEC Dams – Various Locations, New York

The NYSDEC owns a portfolio of more than 100 dams throughout the state and is working to become self compliant with their 2009 dam safety regulations. This Paper and case study refers to 23 of these that are currently regulated as Class A and have known deficiencies and/or concerns about hazard creep. As the regulatory requirements for Class A dams are significantly less than for Class B or Class C dams, a primary goal of the owner was to determine if a change in hazard class is warranted for any of the dams. Additionally, with a fixed budget to bring all of the dams into regulatory compliance, the owner desired to conduct the hazard class assessments in a cost-effective process founded on sound engineering principles that would retain sufficient funds to conduct more detailed studies and rehabilitation of dams where required.

The 23 dams are located in state parks or wildlife preserves but within counties where detailed LiDAR-based topographic mapping has not yet been prepared. Thus, detailed dam breach modeling and inundation mapping would need to be supported by a field survey program which did not fit within the allocated budget for hazard class assessments. Therefore, a hazard screening program was developed with a primary objective to determine whether the hazard class could be confirmed through application of the Guidance without the need for detailed dam breach modeling. The general procedure of the hazard class screening process is summarized below.

Data Collection and Review. The designation of the appropriate hazard class relies upon the accuracy of the available information, particularly in regards to the dam size, storage volumes and watershed size. The initial step in the hazard class screening for each dam included a review of information provided by the NYSDEC and summarized in the NYSDEC inventory as well as a review of publicly available GIS datasets and Google Earth aerial imagery. The primary information collected for the semi-quantitative hazard class screening included:

- Height – Inventory data compared to design drawings and/or field measurements
- Length – Inventory compared to design drawings and/or field measurements
- Normal Pool and Maximum Pool Storage Volume – Inventory compared to estimated volumes computed from design drawings, topographic mapping, and aerial imagery
- Surface Area – Inventory compared to design drawings and aerial imagery
- Drainage Area – Inventory compared to USGS StreamStats value
- Construction Type – Inventory compared to design drawings as available

Screening Level Dam Breach Peak Flow Analysis. The NWS Simple Dam Break equation was used to estimate potential breach peak flows for each of the 23 dams. Breach flows were estimated for two scenarios: with the reservoir at normal pool, and with the reservoir at the crest of the dam. Peak flows from the Simple Dam Break equation provide an order of magnitude estimate that can be used to compare potential breach outflows to the estimated 100-year flood flows, and capacity estimates of downstream hydraulic structures such as roadways and dams.

USGS StreamStats Watershed Flow Analysis. StreamStats is a web-based watershed mapping and flow analysis tool developed by the USGS for many states throughout the country. It includes an interactive mapping component that allows the end user to select a point (such as a dam) along a drainageway in order to delineate the watershed based on USGS DEM data. The program can then be used to compute watershed characteristics as well as peak flow estimates for standard recurrence interval events ranging from the 1-year flood to the 500-year flood. The peak flow estimates are based on regional regression equations developed from USGS stream gage data. For the hazard screening process, StreamStats was used primarily for the efficient delineation of watershed area tributary to each dam. The peak flow estimates for extreme floods such as the 100-year and 500-year events were also compared with the dam breach peak flow estimates to ascertain the general magnitude of breach flows in relation to non-breach flood flows. These estimates helped to develop an opinion as to whether a roadway located closely downstream of the dam is likely to have adequate conveyance capacity to pass the breach outflows without overtopping. Depending on the functional class of the roadway, this may be sufficient to establish an appropriate hazard classification without detailed downstream hydraulic modeling.

Desktop Review of Dam and Downstream Areas. A qualitative review of the dam and downstream areas using Google Earth, other available aerial imagery (Bing Maps, etc.), and publicly available GIS datasets was also performed. The aerial imagery and planimetric data (roads & railroads) depicted in Google Earth were used to identify structures and infrastructure that could potentially be impacted by a dam breach. The limits of each desktop assessment were generally determined by engineering judgment factoring in watershed size, dam height and storage volume, and typically extended downstream to a point where the potential floodwave enters a large enough water body where no significant incremental increase in flooding would be anticipated. This might be a large lake in relation to dam storage volume or a waterway with a significantly greater watershed size than that contributing to the dam in question. A qualitative assessment of potential floodplain storage was also considered in terms of potential breach flood wave attenuation and establishing a reasonable termination point of the downstream screening. This was particularly important for dams that were relatively tall in height but had limited storage. The application of the Simple Dam Break equation may indicate a high peak flow in relation to 100-year storm flow estimates but superposition of the storage volume over a broad forested wetland area may indicate that overbank flood depths would likely be shallow, and significant attenuation of the flood wave is anticipated.

For any structures identified in the potential downstream inundation area, elevation data from Google Earth supplemented by site visits during dam inspections was used to estimate the elevation of the structure in relation to the stream channel, dam height, and dam crest. While the elevation data in Google Earth is not accurate enough to determine whether a residential structure is inundated by 6 inches versus 2 feet of water, if applied with engineering judgment can in many cases help establish whether a structure has the potential to be inundated by a dam failure. For instance, a house estimated to be 15 feet above the stream channel would not likely be inundated by a dam with a maximum height of 8 feet. For structures that could not logically be eliminated from potential flood impacts, USGS topographic mapping was used to estimate the adjacent channel cross section and longitudinal slope of the river to aid in the computation of the normal flow depth for the estimated breach flows. Where this procedure could not sufficiently remove a structure from the potential inundation area, additional hydrologic/hydraulic studies and field survey were recommended.

Review of Roadway Functional Class. A similar process was used to assess road and railroad crossings in the potential inundation area. For those crossings that appear likely to be overtopped from a failure of the dam, the NYS Department of Transportation (DOT) Functional Class Viewer was used to determine the road classification. As the dams are generally located in sparsely populated areas of the state, most of the roadways within the potential downstream project reach were classified as Minor roads. For these roadways, overtopping and presumed failure of the road implies a Class A hazard designation, thereby eliminating the need for detailed hydrologic and hydraulic modeling.

Downstream Hydraulic Structure Capacity Analysis. For the few dams where a roadway consistent with a Class B or Class C designation crossed the downstream drainageway within the estimated inundation zone, a cursory hydraulic capacity analysis was performed to access the likelihood that a dam failure would result in overtopping of the roadway. The capacity analysis was generally conducted using the FHWA's HY-8 hydraulic model, field measurements of bridge opening area, and NYSDOT design drawings where available. The general approach was to compute the maximum capacity of the road crossing without overtopping, and compare it to the estimated peak flows from the simplified dam breach analysis and StreamStats estimates for the 100-year flood. In some cases, it was determined that the roadway would likely be overtopped from the 100-year flood without breach of the dam; therefore, overtopping of the roadway with a failure of the dam would not constitute a significant increase in incremental flood impacts, and an increase in hazard classification attributable to roadway overtopping was not warranted. In one case, the estimated 100-year flow without a breach of the dam impinged on the low chord, and the addition of the peak flows from a dam break would indicate that the incremental increase in peak flows would be adequate to cause overtopping of the road resulting in an increase in hazard classification.

A similar approach was used for any downstream dams within the anticipated inundation impact reach. For these downstream dams, spillway capacity was estimated based upon the dam inventory data supplemented with field measurements. Where the spillway capacity was estimated to be less than the peak flow estimates from the Simplified Dam Break equation, the downstream dam was presumed to fail, and was subsequently considered in the semi-quantitative hazard class screening process.

Screening Assessment Results. The results of the hazard class screening were generally intended to determine if adequate information exists to confirm the Class A hazard designation of each of the 23 dams. The overall approach of the hazard screening process was to provide a semi-quantitative application of the NYSDEC's Guidance document in a cost efficient manner based on strong engineering principles. As of the date of this paper, the hazard screening assessment has been applied to approximately 18 of the 23 dams. Of the 18 completed, approximately 15 of the dams were able to be confirmed as Class A structures. This represents a significant cost saving approach when compared with performing detailed dam breach modeling and inundation mapping to develop an opinion of hazard classification for every dam. Additional hydrologic and hydraulic modeling is being recommended for the dams where a definitive opinion on hazard classification could not be made using the hazard screening process. Please note that the hazard screening approach using the Guidance is applicable to confirm hazard classification of an existing dam; however, detailed dam breach modeling is required to support any proposed reduction in hazard classification.

CONCLUSION

The NYS Dam Safety Section recently issued *Guidance for Dam Hazard Classification* to assist their staff, dam safety consultants, and dam owners in the consistent assignment of hazard class. This Guidance identifies several categories of population and infrastructure at risk that can drive the hazard classification of a facility including: loss of life, access for emergency services, damage to homes, damage to roads and railroads, interruption of utilities, downstream dams, and environmental damage.

While the new Guidance does provide for a more objective determination of appropriate hazard class for dams in New York State, the results of a hazard class assessment utilizing the new Guidance can be influenced by assumptions and limitations relating to key hydrologic and hydraulic items which need to be well understood and documented. It is important to note that what may appear to be conservative assumptions in developing dam breach models can actually lead to under-estimating the incremental difference between wet weather events under breach and non-breach scenarios.

Some of these nuances are captured from the case studies presented which include both detailed dam breach analyses as well as a semi-quantitative screening level application of the new Guidance. From application of the guidance to real world hazard class assessments, it was generally found that the quantitative criteria in regards to residential structure impacts and roadway impacts are the primary drivers for the majority of hazard class assessments. We believe that the case studies demonstrate that the new Guidance is a valuable tool for both the engineer as well for the regulatory agencies.

ACKNOWLEDGEMENTS

The authors would like to acknowledge the Village of Cooperstown and the Kingston Water Department for their permission to discuss their infrastructure, and CHA Consulting, Inc. and staff at the NYS Office of General Services and NYS DEC who contributed to the assessment of the NYS DEC-owned dams.

REFERENCES

New York State Department of Environmental Conservation Division of Water Program Policy, *Draft Guidance for Dam Hazard Classification*, DOW TOGS 3.1.5, 2011.

United States Department of the Interior, Bureau of Reclamation, *A Procedure for Estimating Loss of Life Caused by Dam Failure,* 1999.

United States Department of the Interior, Bureau of Reclamation, *Downstream Hazard Classification Guidelines, ACER Technical Memorandum 11*, 1988.

FROM DATA GRAVEYARDS TO SMART PHONES: LEVERAGING TECHNOLOGY TO MODERNIZE DAM HAZARD ASSESSMENTS

Benjamin Israel-Devadason, P.E., CFM[1]
Paul G. Schweiger, P.E., CFM[2]
Kate Aulenbach, E.I.T.[3]
Amanda J. Hess, P.E., CFM[4]
Karen Goff, P.E.[5]

ABSTRACT

The North Dakota State Water Commission (NDSWC) recently used National Dam Safety Program grant funds to engage Gannett Fleming to perform hazard classifications for 90 dams under their jurisdiction. The NDSWC's goal was to have consistent, up-to-date, defensible hazard classifications for these dams, and to identify those dams with the greatest potential downstream impacts to best allocate limited dam safety resources to manage risk in the most effective way.

Because of the large number of dams to be evaluated and limited budget, a technical approach was developed that provided the greatest benefits for the least cost. The final approach used two-dimensional modeling to accommodate the complex and relatively flat terrain downstream of most North Dakota dams, and programming a custom database to apply user-defined criteria for assessing consequences, including loss of life, at dwellings and road crossings. The final product made significant use of GIS tools and internet-based resources, and linked these resources to the consequence analyses through the customized database. A powerful feature of the database is the ability to quickly perform sensitivity analyses on the hazard classifications of all 90 dams concurrently. The database can also be used to monitor downstream development and perform periodic hazard classification updates without any re-analysis.

Due to the cross-platform fluidity of the technology employed, the inundation maps and other consequence analysis data were generated and published in formats that could be easily viewed, even with a smartphone so that end users with limited technical expertise can effectively use the risk data.

[1] Benjamin Israel-Devadason, P.E., CFM., Senior Hydraulic and Hydrologic Engineer, Gannett Fleming, Inc., Harrisburg, PA, bisrael@gfnet.com

[2] Paul G. Schweiger, P.E., CFM., Vice President, Gannett Fleming, Inc., Harrisburg, PA, pschweiger@gfnet.com

[3] Kate Aulenbach, E.I.T., Hydraulic and Hydrologic Designer, Gannett Fleming, Inc., Harrisburg, PA

[4] Amanda J. Hess, P.E., CFM. Hydraulic and Hydrologic Group Manager, Gannett Fleming, Inc., Harrisburg, PA

[5] Karen Goff, P.E. Dam Safety Engineer, North Dakota State Water Commission.

INTRODUCTION

A large number of the more than 80,000 dams in the United States have never undergone a formal hazard classification based on detailed hydraulic analyses. For several of these dams, the hazard classification was assigned by simply using best judgment with no or very limited supporting documentation. There are also often inconsistencies in how dams in an inventory have been classified since the classifications were made by different people over the span of decades. The hazard classification of dams can also change with time due to downstream development; a process known as "hazard creep".

The North Dakota State Water Commission (NDSWC) recently used National Dam Safety Program grant funds to engage Gannett Fleming to perform hazard classifications for 90 dams under their jurisdiction. The majority of these dams were classified as "Medium Hazard", however, it was suspected that some of the dams could be reclassified as high hazard dams or low hazard dams. The existing classifications were generally based on limited information, with little or no analyses or documentation. The NDSWC's goal was to have consistent, up-to-date, defensible hazard classifications for these dams, and to identify those dams with the greatest potential downstream impacts to best allocate limited dam safety resources to manage risk in the most effective way. A location map showing the 90 dams is provided as Figure 1.

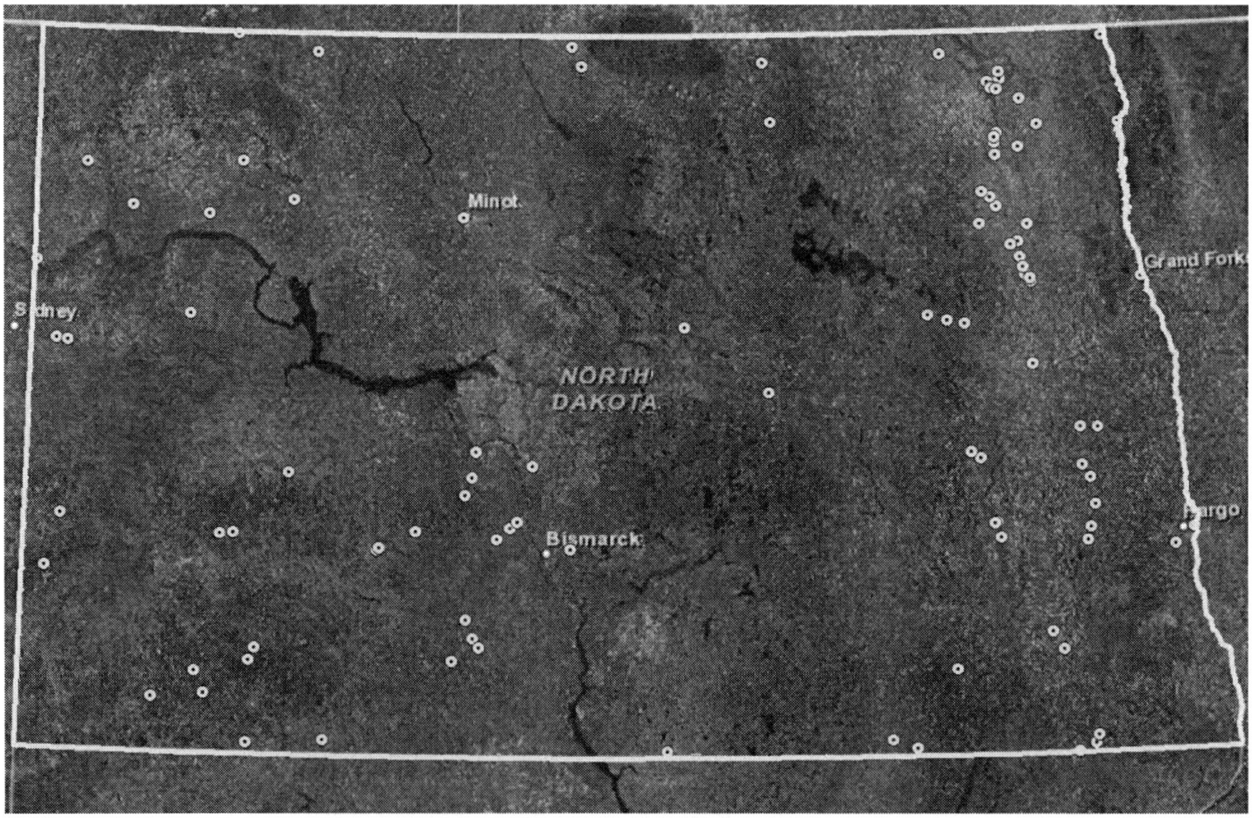

Figure 1. Location Map of 90 Dams

TECHNICAL APPROACH

Because of the large number of dams to be evaluated and limited budget, a five-step technical approach was developed that provided the greatest benefits for the least cost. A key component of the approach is the use of two-dimensional modeling to accommodate the complex and relatively flat terrain downstream of most North Dakota dams,

Step 1: ***Review Records.*** Prior to other data collection tasks, all pertinent records available for each dam from the North Dakota State Water Commission databases were reviewed. These reviewed data includes any past hydrologic and/or hydraulic studies performed on the dams, inspection reports, field photographs, and relevant site drawings. During this step, key hydraulic and hydrologic data including the watershed drainage area at the dam and pertinent dam embankment dimensional characteristics that is used in the following steps of the analysis were retrieved.

Step 2: ***Collect Data.*** At the second step, digital terrain data for the study area corresponding to each dam was obtained from the USGS National Elevation Database. This terrain was deemed to be of a resolution that is sufficient to prepare a simplified two-dimensional model of the reach downstream of each dam. This data which is of approximately 15- to 30-foot in horizontal resolution formed the basis of the Digital Elevation Models (DEM) that was used in the hydraulic analyses. A terrain map of the area downstream of Fordville Dam No.4, one of the 90 dams assessed, is provided as Figure 2. As part of this study, available current and historical aerial imagery, and other relevant photographs were also obtained to evaluate the changes over time in downstream development and to assist in the hazard evaluation process.

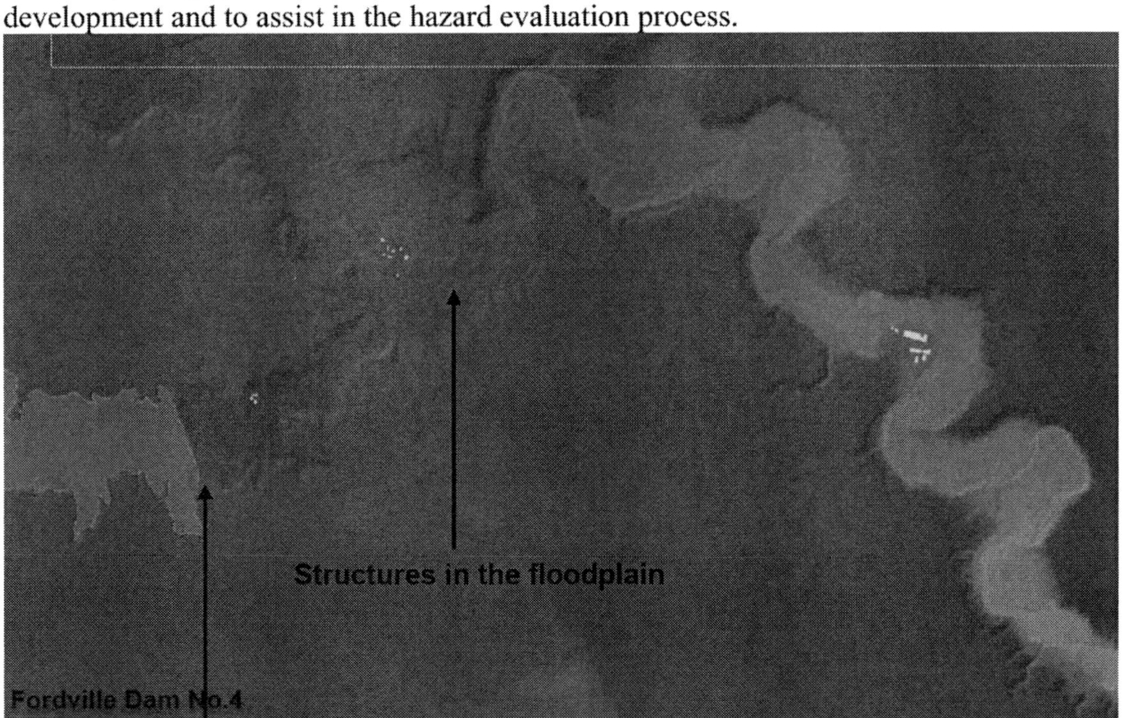

Figure 2. Terrain Map of Fordville Dam No. 4 (Yellow dots mark the location of structures)

<u>Step 3:</u> ***Develop Simplified Hydrologic Model.*** To characterize various scenarios of failure, two representative scenarios were selected namely the (1) sunny day failure scenario and the (2) failure under flood conditions scenario.

The sunny day failure scenario was conservatively simulated assuming the reservoir at the top of dam pool level with no flows through the auxiliary and principal spillway prior to the breach of the dam.

The dam failure under flood conditions scenario was simulated using a representative Probable Maximum Flood (PMF) inflow volume that was approximated as three times the maximum storage of the dam. This approach was adopted after a comparative evaluation of the drainage areas and the reservoir storages of the 90 dams. The comparative evaluation showed that assuming three times the maximum storage of the reservoir to simulate a hydrologic loading condition was a reasonable approach to achieve a simplified hydrologic characterization of the 90 watersheds studied. This simplified approach, which discounted the attenuating effects of the watershed, would also result in modeling a scenario where the downstream impacts from a dam breach are maximized.

<u>Step 4:</u> ***Develop Simplified 2D Hydraulic Model.*** A state-of-the-art, limited-detail two-dimensional hydraulic model of the reservoir, dam, and the downstream areas was prepared for each of the 90 dams to simulate the breach scenarios and to route the breach flooding downstream. The dam embankment breach was modeled within the 2D model and the reservoir storage was released dynamically through the breach opening and routed downstream. A breach time of 15 minutes was assumed for each structure. Downstream limits of modeling were selected based on a review of the terrain, downstream consequences, and a screening of the incremental flooding effects induced by a breach on downstream structures. Flow junctions and tributaries will be modeled within the 2D model to appropriately to account for the attenuating effects of storage areas.

The results from the two-dimensional model analysis include flow depths and flow velocities at various locations downstream of the dam. Maximum flow depths corresponding to the breach of Fordville Dam No.4, one of the 90 dams modeled, is presented as Figure 3.

Maps showing the extent of inundation were developed from the results of the two-dimensional model analysis. Furthermore, the results from this hydraulic analysis were used to perform a consequence analysis. The flow conditions were screened to evaluate areas that will potentially be subjected to dangerous flooding. To assess the areas that will be subjected to dangerous flooding conditions, two criteria were used, namely (1) depth of flow greater than 2 feet and (2) depth times velocity greater than 7. These criteria were developed on the basis of the U.S. Bureau of Reclamation's ACER 11 curves relating flow depths and flow velocities corresponding to high, significant, and low hazard conditions. Structures and roads located within the areas where the depth of flow exceeds 2 feet or the depth times velocity exceeds 7 were estimated to be subject to adverse consequences.

Step 5: ***Prepare Hazard Classification Spreadsheet.*** The results from the two-dimensional hydraulic analyses were processed to develop inundation shapefiles and shapefiles showing areas of consequences as described in Step 4. These shapefiles were then used to identify impacted structures and roads located within the inundation and consequence areas.

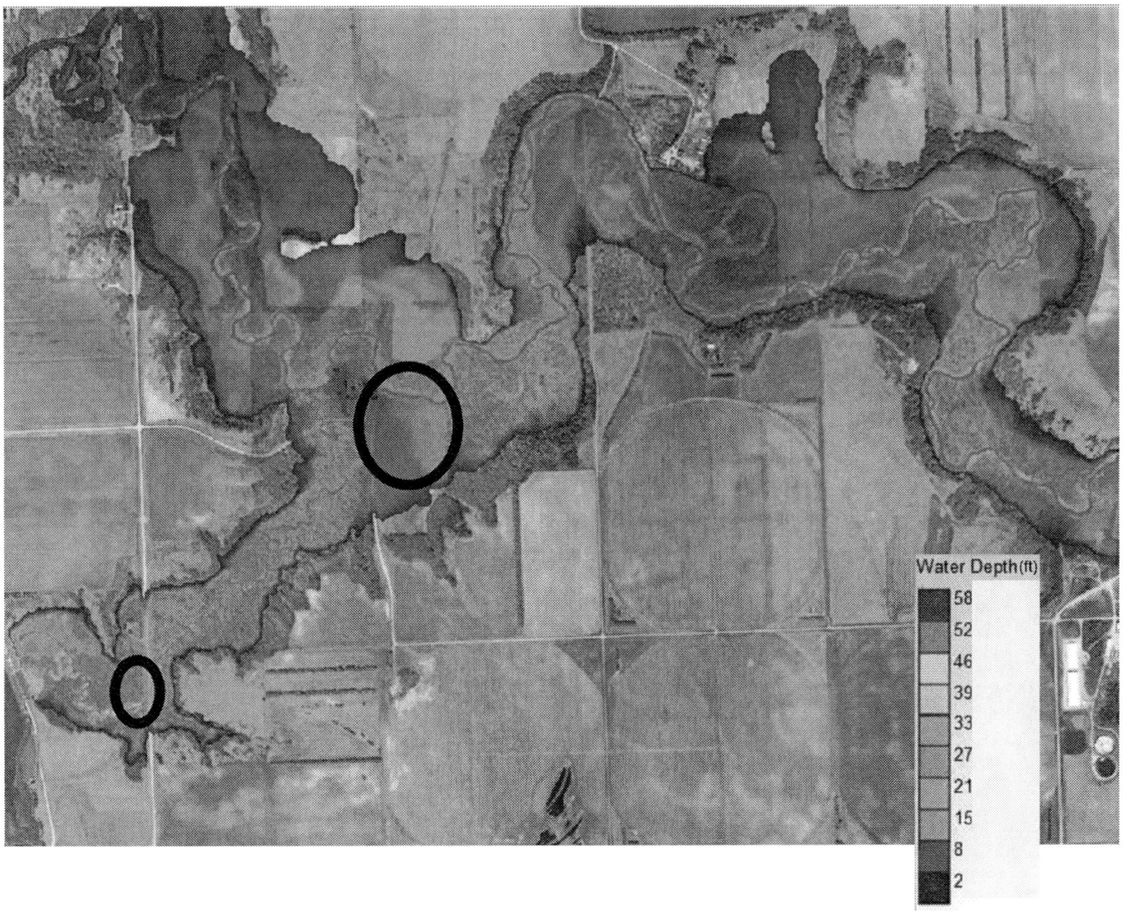

Figure 3. Maximum Breach Flow Depths from Two-dimensional model analysis of Fordville Dam No. 4. The circled areas show the location of structures in the inundation zone.

A detailed database was then developed listing the structures, roads and railroads located within the inundation area. This custom database to manage and apply user-defined criteria for assessing consequences, including loss of life, at various types of dwellings, road crossings and other structures was programmed within the spreadsheet. Figure 4 shows a map with shapefiles showing the location of impacted structures and roads within the inundation zone that was developed as part of the consequence analysis.

The final product made significant use of GIS tools and internet-based resources, and linked these resources to the consequence analyses and hydraulic analyses through the customized database. A powerful feature of the database is the ability to quickly perform sensitivity analyses on the hazard classifications of all 90 dams concurrently by simply modifying the consequence analysis criteria. The database automatically updates the hazard classifications, including the color codes used to help visualize the results. The database can also be used as a tool to monitor downstream development and perform periodic hazard classification updates without any re-analysis.

Figure 4. Map with shapefiles showing the location of impacted structures (yellow) and roads (red and black line) within the inundation zone.

Another powerful feature that was tested in this study was the use of modern data formats to view and present risk information instead of traditional reports or paper-based maps. Due to the cross-platform fluidity of the technology employed in these analyses, the inundation maps and other consequence analysis data were generated and published in formats that could be easily viewed, even with a smartphone. This innovation takes complex dam failure consequence and risk data and presents it in user friendly formats so that end users with limited technical expertise can effectively use the risk data. All of the outputs from this analysis can be loaded to mobile devices to facilitate groundtruthing.

As the final step in the study, maps showing areas of inundation and consequences were prepared to be used within the EAPs. Location of the impacted structures, roads and railroads, and flood wave arrival times were also presented in these maps. An example map is presented in Figure 5.

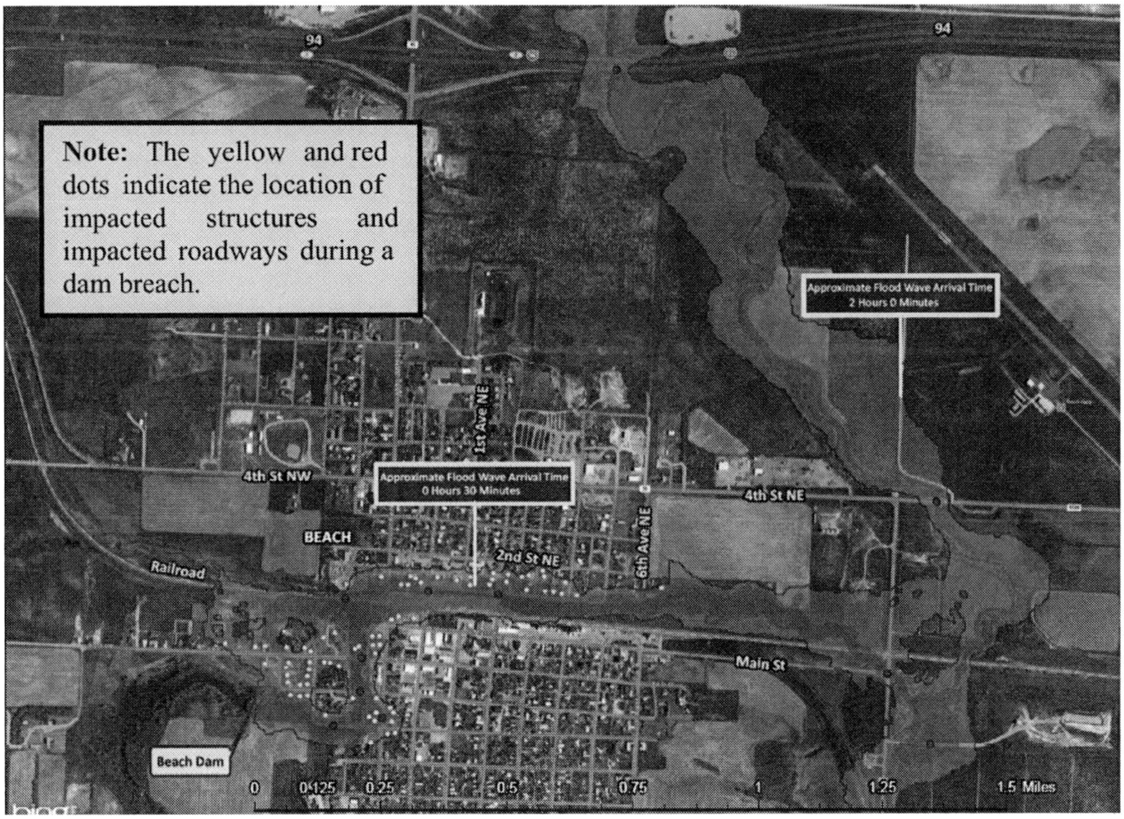

Figure 5. Example EAP inundation map

CONCLUSION

The above five-step approach developed and applied to study the breach consequences of 90 dams in North Dakota presents a promising technical approach to perform dam breach analysis with limited detail, at a significantly low cost to the dam owners and other stakeholders.

The final product made significant use of state-of-the-art GIS tools and internet-based resources, and linked these resources to the consequence analyses and hydraulic analyses through the customized database. A powerful feature of the database is the ability to quickly perform sensitivity analyses on the hazard classifications of all 90 dams concurrently by simply modifying the consequence analysis criteria at one location. The database can also be used as a tool to monitor downstream development and perform periodic hazard classification updates without any re-analysis.

Another powerful feature that was tested in this study was the use of emerging and futuristic data formats to view and present risk information instead of traditional reports or paper-based maps. Due to the cross-platform fluidity of the technology employed in these analyses, the inundation maps and other consequence analysis data were generated and published in formats that could be easily viewed even with a smartphone. This powerful simplification innovation that takes complex and detailed dam failure consequence and risk data and presents it in extremely simplified formats, is a huge benefit since even the end users with limited technical expertise can now access all of the risk data without having to go through elaborate reports. All of the outputs from this analysis approach can be loaded to mobile devices and thereby adding additional convenience.

POTENTIAL FAILURE MODES FOR A NATURAL DAM INCLUDING ESTIMATED DURATION OF FAILURE MODES

David Thompson, P.E., P.Eng.[1]
Li Yan, P.Eng.[2]
Greg Rollins, P.E., G.E.[3]
Ken Lum, P.Eng.[4]

ABSTRACT

At BC Hydro's Ruskin Dam and Powerhouse in British Columbia, Canada, the left abutment of the 58 metre high gravity dam forms an earthen "natural dam" between the reservoir and the powerhouse. The left abutment, containing power intakes/tunnels, has experienced on-going tunnel leakage, and initial engineering investigations estimated the downstream slope could undergo large deformations under seismic loading, posing a risk to the powerhouse and possibly loss of the reservoir. A Potential Failure Modes Analysis (PFMA) was carried out for the left abutment, following U.S. Government Federal Energy Regulatory Commission's (FERC's) process but modified by BC Hydro. The key modification from the FERC process is the inclusion of the anticipated duration of each step in the development of each potential failure mode (PFM) for the purpose of assessing the ability to detect, respond and intervene to arrest full development of the particular PFM, and/or identifying the needs for mitigation to prevent the occurrence of the failure mode(s). The work involved review of construction history, site geotechnical data, previous deficiency investigations, and performance history of the left abutment and associated structures. Seventy-seven PFMs were identified and classified. Following the initial PFMA, further engineering evaluations were carried out to assess potential dam safety deficiencies associated with the PFMs. Based on these analyses, the PFMs were reclassified and final engineering documents prepared to remediate and/or monitor each PFM. The PFMs of greatest significance are being remediated.

INTRODUCTION

Purpose and Scope of Potential Failure Modes Analysis

The Ruskin Dam and Powerhouse form part of BC Hydro's Stave River System. The Ruskin facility was commissioned in 1930, when the first generating unit went into service; second and third units were commissioned in 1938 and 1950, respectively. Previous left abutment studies concluded that abutment slopes above the powerhouse, which forms a "natural dam", could become locally unstable in a seismic event, but that uncontrolled release of the reservoir would be unlikely because there would be sufficient

[1] MWH Global, 2353 130th Ave NE, Suite 200, Bellevue, WA 98005, USA, 425-896-6972, David.P.Thompson@mwhglobal.com.
[2] BC Hydro, 6911 Southpoint Drive, Edmonds A02, Burnaby, BC V3N 4X8, Canada, 604-528-2491, Li.Yan@bchydro.com.
[3] MWH Global, 806 SW Broadway, Suite 200, Portland, OR 97205, USA, 503-220-5414, Gregory.C.Rollins@mwhglobal.com.
[4] BC Hydro, Edmonds A02, Burnaby, BC V3N 4X8, Canada, 604-528-2406, Ken.Lum@bchydro.com.

time to respond by lowering the reservoir. A Potential Failure Modes Analysis (PFMA) was undertaken to systematically identify potentially unstable conditions that could lead to an uncontrolled release of the reservoir through the left abutment. The analyses highlighted critical modes of potential failure for various loading conditions, identified gaps in the existing data, and proposed field investigations and further analyses necessary to either conclude that the PFMs are not a concern, or to monitor the development of the PFMs or implement remedial measures to prevent potential failures.

Project Background

Ruskin Dam is located in British Columbia, Canada approximately 60 km east of Vancouver. The dam is a 130 m long, 58 m high concrete gravity structure with seven radial spillway gates. The dam has been classified as an "Extreme" consequence facility under BC Hydro's Dam Safety Management System.

The powerhouse is located downstream of the left abutment of the dam. The powerhouse slope, which forms the downstream side of the left abutment, is located immediately above the powerhouse and was excavated in bedrock and glacial deposits. The slope is approximately 35 m high and dips steeply toward the powerhouse with a slope of approximately 1.0 horizontal to 1.0 vertical (H:V) at its steepest section. The slope has a history of seepage, sloughing and sliding. Ruskin Dam, left abutment and powerhouse are shown in Figure 1.

Figure 1. Dam, powerhouse and left abutment "natural dam"

The left abutment of Ruskin Dam is formed on a bedrock promontory with a maximum elevation of approximately 48 m (Figure 2). Bedrock drops off steeply to the south and east to below sea level forming a saddle approximately 1 km wide. The saddle is in-filled with dense glacial deposits that form a ridge between Hayward Lake and Stave River downstream of the dam. The power intake structures and three tunnels are situated at and through the left abutment, respectively.

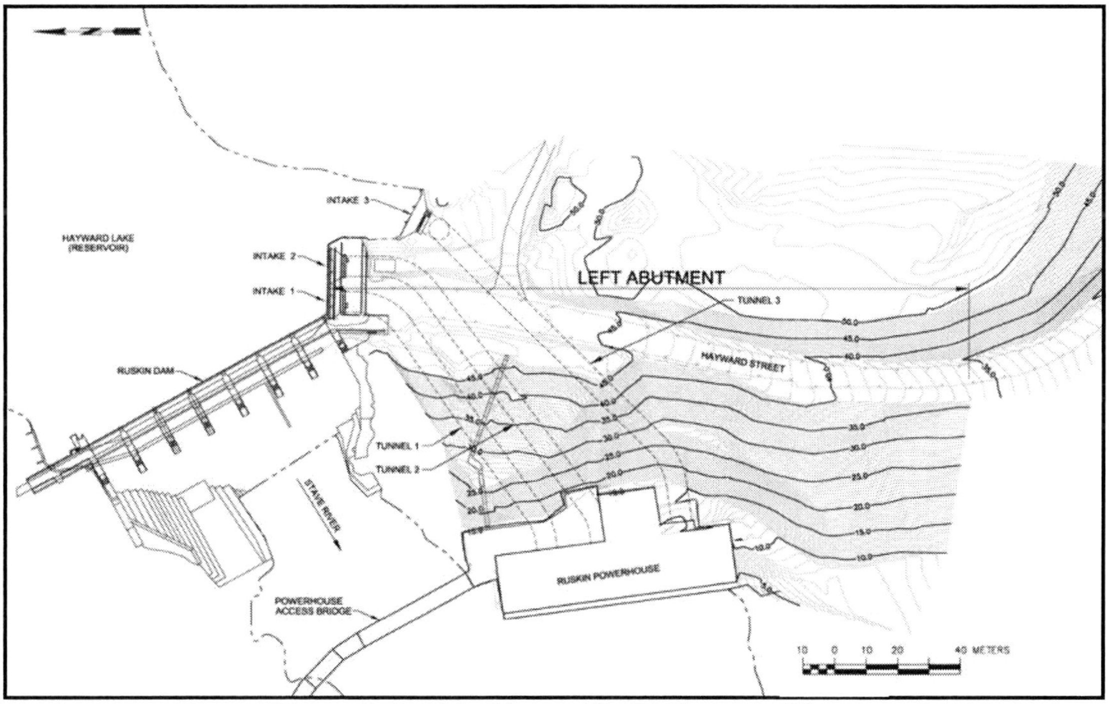

Figure 2. Ruskin Site and Left Abutment Plan

Intake 1 and 2 is a combined structure, which was constructed as part of the original dam, and conveys water to Tunnels 1 and 2. Intake 1 and 2 is approximately 25 m wide, up to 8 m in depth, and 23 m high. Intake 3 was added in 1950 with the addition of the Tunnel 3 and powerhouse Unit G3. The Intake 3 structure is approximately 17 m across the inlet face, 6 to 10 m in depth, and 15 m high. Both intake structures are founded on blasted bedrock surfaces. The structure for Intake 1 and 2 has been constructed against a near-vertical excavated rock surface. The Intake 3 structure is in contact with a blasted rock face on its right side but backfilled with loose fills at the downstream and left sides.

Tunnels 1 and 2 were constructed as part of the original dam development in 1929 and Tunnel 3 was constructed in 1950. The downstream portions of each tunnel have a 5.79 m diameter steel lining and the reminder of the tunnel lengths are concrete lined and 6.4 m in diameter. Tunnels 1 and 2 are in bedrock up to the north wall of the powerhouse; though the last 10 to 15 m immediately upstream of the powerhouse was trenched in rock and the steel liner backfilled with concrete. Most of Tunnel 3 was constructed in bedrock, but the crown and upper walls of its downstream end were constructed in overburden by cut and cover methods and encased in concrete. The downstream end of Tunnel 3 is founded on bedrock with loose granular soil backfill around the penstock pipe. There has been a history of leakage attributed to Tunnel 3, including leakage noted in 2006 and 2007, for which BC Hydro made remedial repairs to the concrete tunnel lining. In December 2008, unusually high leakage was ascribed to seasonal cold water entering the tunnels and led to the shutdown of the Unit G3 and unwatering of the tunnel for inspection.

The left abutment slope above the powerhouse consists of interlayered sequences of glacial deposits. Above approximately Elevation (El.) 35 m, the top of the slope is capped by more than 10 m of glacial till. Between El. 15 and El. 5 m there is a 20 m thick sequence of sand and silty sand. Within this deposit are two layers of silt that are 1 to 2 m thick and thought to extend continuously through the deposit at approximately El. 27 m and at the base of the till at approximately El. 35 m. The sand and silty sand sediments are underlain by either a clay/silt unit near the south end of the powerhouse or gabbroic bedrock in areas inland from the powerhouse. The top of the clay/silt unit is at approximately El. 15 m. Inland from the powerhouse, the bedrock surface rises above El. 15 m and is in direct contact with the sand and silty sand materials. Figure 3 shows Tunnel 3 and the subsurface profile of the left abutment "natural dam".

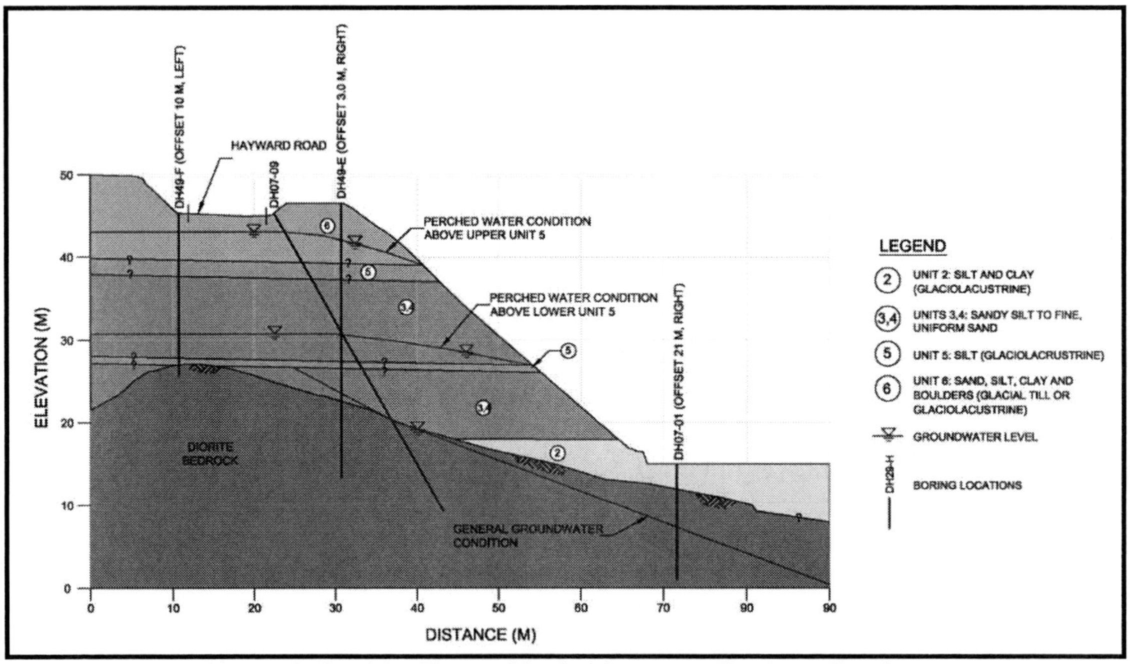

Figure 3. Ruskin Site and Left Abutment Geology Section

All of the overburden deposits have been significantly over-consolidated by Quaternary age glacier cover that was up to 2000 m thick. Consequently the materials are all very densely compacted, except for some locally disturbed loose sands. The slope above the powerhouse has experienced surficial sloughing over the years. The overall groundwater table in the slope is generally low but there is a perched water table on top of the two silt layers at El. 35 m and 27 m. Surficial seepage has been noted at the base of the slope and at mid height.

PROJECT FAILURE MODES ANALYSIS PROCESS

PFMA Identification Process

The PFMA process has been formalized in recent years by FERC to focus on the site-specific factors of greatest significance to dam safety. As part of the Part 12D Dam Safety Evaluation process, required for all regulated projects in the U.S., FERC's PFMA process requires participation of dam owners and independent consultants. The process involves first assembling and reviewing relevant project documents including design reports, construction records and post-construction analyses and reports. These documents provide a comprehensive background of the project and enable free and knowledgeable discussion. Following a review of the documents, staff of the dam owner and independent consultant identify potential failure modes associated with usual (normal static), unusual (floods) and extreme (earthquakes) loading conditions for each project water retaining feature. Each PFM is evaluated for significance and likelihood of failure under each loading condition, assigning the PFM to a category, evaluating monitoring programs relative to the PFM, and proposing risk reduction action items.

The PFMA conducted for the Ruskin Dam left abutment followed a similar process, but was modified by BC Hydro based on their experience with prior dam safety evaluation activities at other BC Hydro facilities. The key modification from the FERC process is the inclusion of the anticipated duration of each step in the development of each PFM for the purpose of assessing the ability to respond and intervene to preclude development of the particular PFM. BC Hydro staffs involved in the PFMA have a long history of studies pertaining to the left abutment. Their consultant, MWH Global, reviewed the historical documents and photographs related to the left abutment, participated in the PFMA sessions and performed several independent analyses.

Two PFMA sessions were held. At the first session the history of the left abutment was reviewed followed by development of potential failure modes for each loading condition. The process required that any PFMs identified be completely described from an initiator to failure (i.e., uncontrolled release of the reservoir). Following the first session, BC Hydro and MWH classified each potential failure mode and developed preliminary PFM flow diagrams for each PFM. At the second PFMA session which was held about one month later, the preliminary PFM flow diagrams and PFM classifications were reviewed and discussed for concurrence of the PFMA team. The second session was followed by development of a report for review and finalization by the PFMA team. Important outcomes of the report were (1) documentation of the PFMA process and PFMs and (2) development of field investigations and assessments necessary to either refute the possibility of the PFMs as being credible for the left abutment or to provide information necessary to highlight and design remedial construction or surveillance/monitoring programs.

Potential Failure Modes for a Natural Dam

PFM Classification

Each identified PFM was considered relative to its likelihood and importance and was categorized according with the FERC classification system, which involves four categories (I - IV) and descriptions as indicated in Table 1.

Table 1. Potential Failure Mode Categories

Category	Description
I. Highlighted Potential Failure Modes	Those potential failure m odes of greatest sig nificance considering need for awarenes s, potential for occ urrence, magnitude of consequence and likelihood of adverse response (physical possibility is evident, fundamental flaw or weakness is identified and events leading to failure seemed reasonable and credible) are highlighted.
II. Potential Failure Modes Considered But Not Highlighted	These are ju dged to be of lesser significance and likelihood. Note that even though these potential failure modes are considered less significant than Category I, the y are all also described and included with reasons for and against the potential occurrence of the failu re mode. The reaso n for the lesser significance is noted and summarized in the documentation report or notes.
III. More Information or Analysis are Needed in Order to Classify	These potential failure modes to s ome degree lacked information to allow a confident judgment of significance and thus a da m safety investigative action or analy sis can be recommended. Because action is req uired before resolution the need to this action may also be highlighted.
IV. Failure Mode Ruled Out	Potential failure modes may be ruled out because the physical possibility does not exist, inform ation came to light which eliminated the concern that ha d generated the development of the potential failure mode, or the failure mode is clearly so remote as to be non-credible or not reasonable to postulate.

Reference: FERC Engineering Guidelines for the Evaluation of Hydropower Projects, Chapter 14.3, Potential Failure Mode Analysis.

Potential Failure Mode Step Duration

BC Hydro has recognized that the duration of development of a PFM is an important component of the evaluation of that PFM. PFMs that have a prolonged development period may allow for the dam owner to monitor, detect, intervene and/or mitigate the PFM, as necessary, prior to an uncontrolled release of the reservoir; or to lower the reservoir in a controlled manner to minimize or eliminate the potential for loss of life or property damage prior to full development of the failure mode. As such, BC Hydro has established a system of evaluating the duration of individual failure steps ranging from Level 1 (minutes) to Level 7 (decades). The system has been captured in a color-coded table (Table 2) that identifies the duration and an example failure step for each of the failure step duration levels.

Table 2. Failure Step Duration and Response Options to Arrest PFMs

Level	Duration	Example Failure Step	Ability to Monitor and Detect Before Occurrence	Response Time	Response Options[1] to Arrest PFMs
1	Minutes (seconds to hours)	Earthquake	No	Class A: No Time to Respond	Little to no ability to respond
2	Hours (minutes to days)	Dam Overtopping	No		
3	Days (hours to weeks)	Flood	Yes	Class B: Short Time to Respond	Reactive Response: Visual inspection, monitoring of existing instruments and implementing protection system as required. Some intervention options, like lowering the reservoir, may also be available
4	Weeks (days to months)	Slope Saturation	Yes		
5	Months (weeks to years)	Piping	Yes	Class C: Long Time to Respond	Proactive Response: Visual inspection, monitoring of existing instruments, analyzing performance results and enhancing monitoring system or implementing physical remediation or intervention options as required
6	Years (months to decades)	Fines Migration	Yes		
7	Decades (years to centuries)	Corrosion	Yes		

[1] Note that the purpose of this column is not to instruct operators or others how to respond to a dam safety failure or incident. The response options provided herein are to provide a general framework for further examination of an appropriate response plan commensurate with the response time available for the specific PFM. The first response of any sign of failure should follow established emergency response procedures.

Identifying the duration of each failure step using this system allows for planning and implementing a suitable instrumentation monitoring and surveillance program to provide early detection and warning of a potential failure step within the PFM. Early detection will allow for intervention as early as possible prior to full development of a failure mode. Based on the estimated duration of each failure step, the likely total response time was identified for each failure mode, together with potential response option(s) to arrest the failure mode. Three classes of response time were introduced: Class A – no time to respond; Class B – short time to respond (reactive), and Class C – long time to respond (proactive). Thus, in the PFMA process, each PFM was assessed and classified according to FERC's categories and estimated response time classes. Response times (A through C) were assigned for Category I, II and III PFMs, and no response time was assigned for Category IV PFMs.

POTENTIAL FAILURE MODE ANALYSIS RESULTS

The PFMA team initially evaluated over 100 potential failure scenarios, identifying failure mode initiating events, failure processes, durations of failure modes, and preliminarily classifying the failure modes. A subset of the team then went through each PFM, expanding the evaluation and developed a PFM flow chart for each initiating event and major component at risk pertaining to the stability of the left abutment.

The PFM flow charts were developed by using a set of symbols to provide concise summaries of initiating events, failure mode steps, and final failure with descriptions and whether or not there had been supporting evidence. Each of the failure mode steps was color-coded based on duration of the failure steps and the symbols indicated in Figure 4. An example of the flow chart, depicting in this case PFMs resulting from an earthquake event potentially compromising the stability of the project's intake structures, is presented in Figure 5, where PFMs E.27/28 relate to potential uncontrolled release of the reservoir directly due to failure of the intake structures, thus have little to no time to respond, and E.29/30 are due to inability to timely close the intake gates when tunnel failure occurs, leading to failure of the left abutment, thus have some time to respond, and E.31/32 relate to increased seepage after failure of the intakes, and then overtopping of the left abutment "natural dam", and thus have longer time to respond.

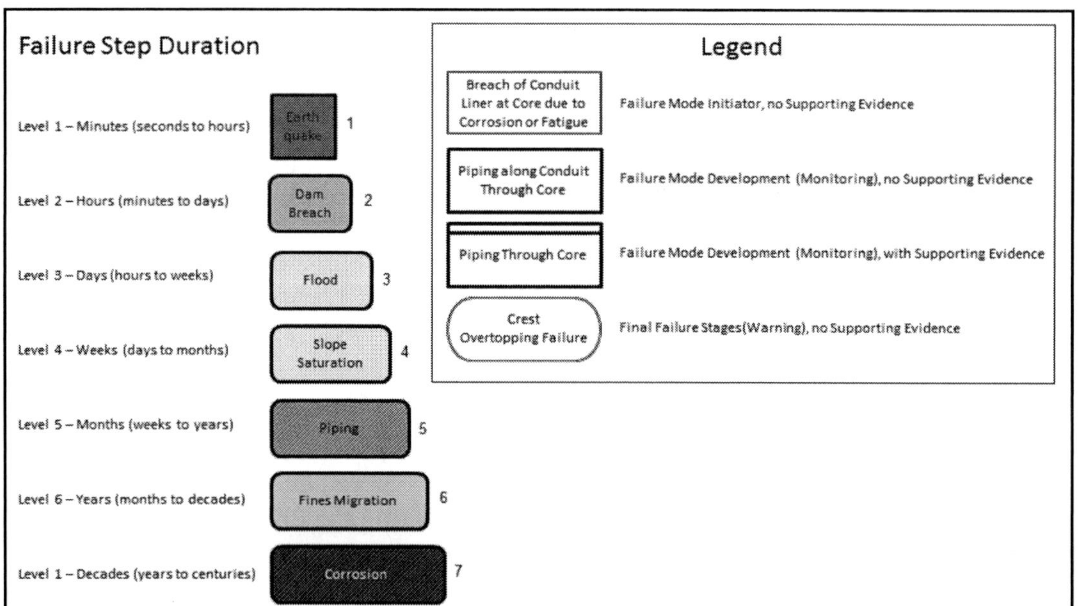

Figure 4. Failure Step Duration Color Code and PFM Flow Chart Legend

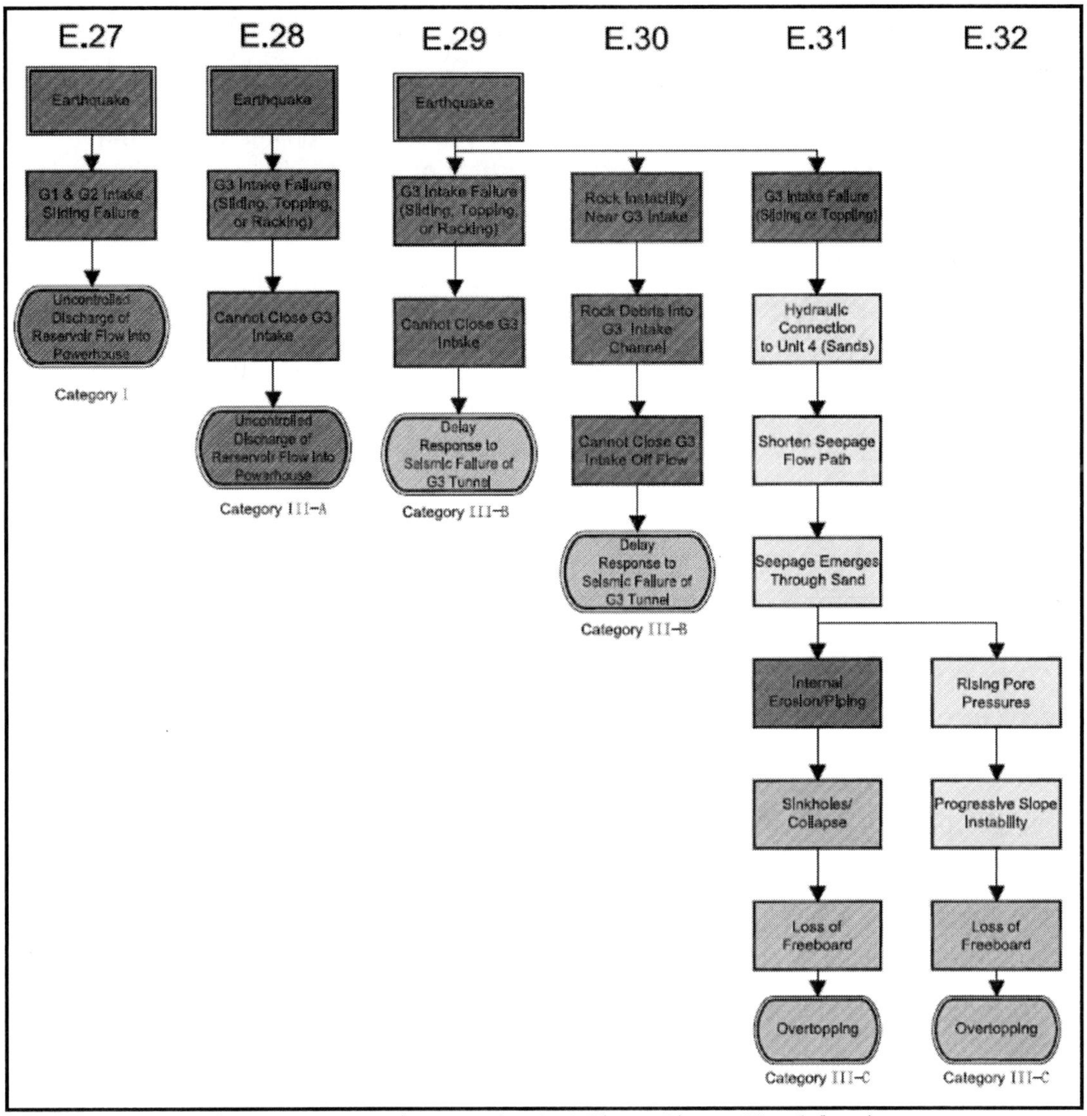

Figure 5. One of Several PFM Flow Charts Used for the Left Abutment Failure Modes (Earthquake at Intakes)

Detailed descriptions of the PFMs, the PFM category and rationale, recommended investigation and analysis, potential monitoring and actions, and the duration of the failure and the ability for BC Hydro to intervene was included in the PFMA report documenting the findings.

Based on the further evaluation stemming from development of the flow charts and appropriate combining of PFMs, the PFMA team ended with a total of 77 PFMs falling under five general failure initiators or conditions.

POTENTIAL FAILURE MODES ASSESSMENT RESULTS AND MITIGATION

The 77 PFMs were derived from five general failure mode initiators and were classified into four failure mode categories (Category I – IV) based on their likelihood and importance to dam safety using the FERC classification system and possible response time as indicated in Table 3:

Table 3. Potential Failure Modes and Classifications

PFM Initiator	Identified PFMs	FERC PFM Category				Response Time		
		I	II	III	IV	A	B	C
Static Reservoir Seepage	8	--	--	8	--	--	--	8
Earthquake	32	1	--	25	6	5	9	12
Tunnel 3 Leakage	4	2	--	1	1	--	2	1
Operational	13	--	--	9	4	--	1	8
Flooding	20	--	--	10	10	--	7	3

Of particular note, a total of three PFMs were classified as Category I, indicating that these PFMs are of greatest significance considering potential of occurrence, magnitude of consequence, and other factors. The three Category I PFMs are related to seismic stability of Intakes 1 and 2 structure and Tunnel 3 leakage issues.

Also noteworthy is that about 2/3 of the PFMs were classified as Category III, indicating that more information was required to categorize. When conducting a PFMA, it is typical to have a large number of PFMs that cannot be categorized due to insufficient information. This is not desirable as some of these Category III PFMs could be significant to dam safety. Therefore, a left abutment dam safety assessment was performed to confirm each Category I PFM and to re-classify each Category III PFM to the extent practicable. The following was concluded in the assessment report:

Static Reservoir Seepage: All 8 PFMs relating to reservoir seepage were identified as Category III-C. These PFMs could develop as a result of Hayward Lake water seeping, under normal reservoir operation, into sandy zones of native and fill material near Intake 3, causing migration of fine-grained soils, leading to downstream slope failure and possible breach of the left abutment "natural dam". Since fines migration due to reservoir seepage could not be definitively confirmed based on the currently available site investigation results and instrumentation data, these PFMs are considered credible in that they are physically possible. Therefore, these PFMs cannot be ruled out, and classification remained as Category III-C. Because each of these PFMs has a long duration (i.e., Response Time C) to reach full failure state, an enhanced instrumentation system with appropriate surveillance, monitoring and response plans has been designed to provide monitoring and early warning of these PFMs. In addition, defensive measures consisting of a downstream slope weighted filter blanket was designed as part of the powerhouse slope modification to address these PFMs. Implementation of these measures provides the required risk mitigation, rendering these PFMs to Category IV after the remediation.

Earthquake: One seismically driven PFM postulated structural damage to the Intakes 1 and 2 structure, making the gates inoperable during the Maximum Design Earthquake (MDE), was classified as Category I-A. In addition, 25 other PFMs were classified as Category III and were resolved and reclassified as follows:

- *Seismic Failure of Tunnel 3*: Four of these PFMs relate to rupture of the tunnel concrete lining in the rock and/or the rock-soil contact or rupture of the tunnel connection to the powerhouse resulting in leakage from the tunnel, instability of the left abutment slope and uncontrolled release of the reservoir. Finite element modeling was performed resulting in determination of structural capacity and displacements. It was concluded that two of the PFMs are credible and could be mitigated by re-lining/sealing portions of the concrete lined tunnel and a downstream external drainage collection and detection system. This is being included as part of the improvement works. Once the mitigations have been implemented all of these PFMs will be Category IV.

- *Seismic Failure of Upstream Slope*: These ten Category III PFMs postulated either slope failures at or near Intake 3, or soil liquefaction. Upon further study and modeling, six of these PFMs were re-categorized as I-C and the remainder as IV. Therefore, it was determined that slope regrading was required along with the installation of a reverse filter system, and a drainage collection and measurement system as indicated in Figure 6. This work is being included as part of the improvement works.

- *Seismic Failure of Downstream Slope*: All six PFMs in this set related to either shallow-seated slope failures, excessive deformation, or soil liquefaction. Upon further study and modeling, two of these PFMs were re-categorized as I-C and the remainder as Category IV. The designed regraded slope and reverse filter system as indicated in Figure 6 will mitigate these PFMs.

- *Seismic Failure of Intake 3*: These five PFMs postulated five different failure mechanisms relating to intake sliding, toppling or racking, or adjacent rock instability leading to an uncontrolled release of the reservoir. Further modeling and study was performed and determined that the structure is stable under the MDE and no mitigation is warranted. These PFMs were then re-classified as Category IV.

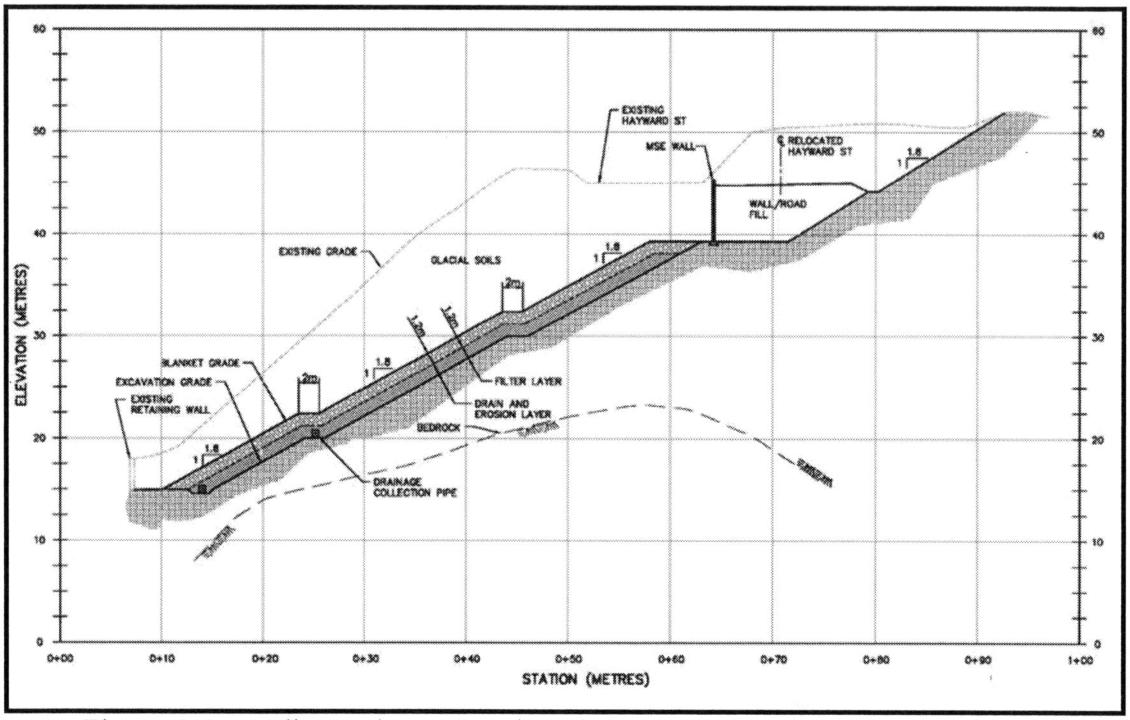

Figure 6. Regrading and Reverse Filter for Left Abutment Powerhouse Slope

Tunnel 3 Leakage: The four PFMs in this set were categorized I-B, I-C, III-B and IV due to leakage emerging from the tunnels under normal operating conditions. Upon examination it was determined that the Category III-B PFM was prevented by the presence of the existing steel liner in the tunnel and that the two Category I PFMs could be mitigated by the installation of a new liner system inside the existing concrete lined section as well as the leakage monitoring system. Upon completion of installation these Category I PFMs will be mitigated.

Operations: A total of 13 PFMs were identified that could be caused by possible detrimental operation of the facility. Of these, nine were categorized as III-C and four as IV. The Category III-C PFMs were re-assessed as follows:

- *Rapid Reservoir Drawdown*: Three PFMs proposed that rapid drawdown of Hayward Lake could cause various types of slope failures leading to compromising the reservoir rim and release of the reservoir. Slope stability analyses were carried out to assess various scenarios and concluded that such failures are not reasonable to postulate, re-classifying each as Category IV.

- *Reservoir Cycling*: Two PFMs postulated that under normal operations the variation in reservoir level could cause flushing of fines or affect seepage regime in the left abutment. Modeling performed ruled these PFMs out and they were reclassified as Category IV.

- *High Reservoir Operational Level*: Two PFMs proposed that a high reservoir level could cause leakage that could result in piping or slope instability. Seepage analyses were performed that indicated that the occasional high reservoir level would not significantly increase the seepage to cause piping or rise in pore pressure in the left abutment. These PFMs were ruled out, i.e., Category IV.

- *Rapid Dewatering of Tunnel 3*: Two PFMs postulated that rapid tunnel dewatering could either cause seepage and loss of fines through cracks in the tunnel or may result in damage to the tunnel, leading to slope failure. Further study cannot confirm or refute these PFMs, however, the planned upgrade to the tunnel will address these PFMs, either rendering them to Category IV after the upgrade or mitigating them with appropriate operational controls and instrumentation installed to monitor these long duration PFMs.

The development of the operational PFMs highlighted the importance of operational controls, visual inspections and monitoring of existing and newly-planned/installed instrumentation.

Flooding: Half of these 20 PFMs were classified as Category III-B or III-C and the other half as Category IV. The Category III PFMs were further examined and reclassified as follows:

- *Flood Spill*: One Category III-B PFM postulates potential erosion of the powerhouse backfill around Tunnel 3, causing progressive shallow slope failure, eventual loss of freeboard, and overtopping of the "natural dam". This potential does exist, but would have a long duration to reach the full failure state. Thus this PFM was reclassified as Category I-C. Remedial measures have been constructed by adding a new retaining wall and new riprap armoring of slope downstream of the powerhouse. Final categorization of this PFM is IV.

- *Probable Maximum Precipitation*: There were nine Category III-B or III-C PFMs identified relating to failures due to a high reservoir that could result in a higher seepage in sand, upstream slope instability near Intake 3, or failure caused by saturation of downstream slope from heavy precipitation during the probable maximum precipitation event. The PFMs were either ruled out (IV) by studies performed or will be mitigated by regrading the left abutment slope and installation of the reverse filter and drainage collection systems.

As a result of the PFMA and subsequent dam safety assessment, the Category I PFMs are being remediated. The remedial measures to be implemented for these highlighted PFMs include:

- Regrading and benching of the left abutment slope behind the powerhouse;

- Installation of a reverse filter system on the regraded left abutment slope, along with a drainage collection system and instrumentation, and

- Installation of a new tunnel lining system in Tunnel 3, and an exterior tunnel leakage collection and monitoring system.

CONCLUSION

The left abutment of BC Hydro's Ruskin Dam forms an earthen "natural dam" between the reservoir and the powerhouse, and has a history of sloughing and localized slope failures, and experiences on-going tunnel leakage and seepage. A comprehensive Potential Failure Modes Analysis (PFMA) was jointly carried out by BC Hydro and MWH for the left abutment and its appurtenant structures (power intakes and power tunnels) to assess issues potentially affecting the safety of the "natural dam". The PFMA performed followed a process similar to FERC's but modified by BC Hydro to include the time element of the PFMs in which the time for each failure step to develop in a given PFM was estimated and color coded in the PFM flow chart so that the total duration of each PFM can be assessed. With this approach, each PFM, in addition to being classified per FERC's Categories I to IV, was also further classified according to the availability of response time, to allow for assessment of the ability to monitor for early detection and warning, and to intervene as necessary to arrest full development of the failure modes.

The PFMA examined five PFM initiators due to reservoir seepage, earthquake, tunnel leakage, operational and flooding, and identified 77 PFMs, most of which were classified as Classes B and C for response times, indicating these PFMs will take time to develop with opportunities to monitor and intervene. The initial PFMA results also indicate a large number of Category III PFMs that required further evaluation. With the initial PFMA results, further engineering evaluations, including additional site investigations, were strategically developed and carried out to assess potential dam safety deficiencies associated with the PFMs. The further assessment confirmed the Category I PFMs, and ruled out some of the Category III PFMs and re-classified to Category IV. The PFMA and the subsequent dam safety deficiency investigations performed clarified the needs for remediation to meet dam safety requirements, and objectively identified where and types of instruments that can be installed to monitor and provide early detection and warning, together with appropriate response plans, for the target PFMs. Remedial measures are being implemented for the Category I PFMs.

ACKNOWLEDGEMENTS

The authors wish to thank the Ruskin Dam and Powerhouse Upgrade Project and BC Hydro's Dam Safety group for permission to publish this paper. The authors also wish to acknowledge the contributions of other BC Hydro staff, especially Derek Sakamoto who coordinated and compiled all previous relevant documents, and Steve Garner who introduced the time element of the PFMA in BC Hydro.

A NEAR-MISS — A HISTORIC RAINFALL LOCATED THE WEAK LINK IN UPGRADED EMBANKMENT DAM

Daniel L. Johnson, P.E.[1]
Brad Dallam, P.E.[2]
Ryan D. Schoolmeesters, P.E.[3]

ABSTRACT

The epic storm event that took place in north central Colorado in September 2013 was yet another reminder that Mother Nature will have the final say. The event tested the local infrastructure to unprecedented levels and in unanticipated ways; and weaknesses were revealed. While most of the dams in the affected area performed well, the Baseline Dams and Reservoir were among the infrastructure that was put to the test.

On September 12, the second day of the rain event, the dam owner placed a call to the Colorado Dam Safety Branch to report that a new 40-foot-long longitudinal crack had developed overnight along the downstream shoulder of the Northwest dam crest and had sloughed approximately 1 foot downslope. Adding to the complexities, several smaller scarps were cutting into the dam crest along the upstream shoulder in the same reach, and the off-channel reservoir was receiving uncontrolled inflows from South Boulder Creek well in excess of the normal inflows due to complications at the diversion structure. With the National Weather Service forecasting a stalled front over the Boulder area and continuing heavy rainfall through at least the next few days, the Emergency Action Plan for Baseline Reservoir was activated and a team of engineers was mobilized to the site to aid in the response.

Quick action by the owner and the team of engineers, digitally accessible State dam files, and 24-hour monitoring and inspection enabled the dam to "weather the storm"; however weaknesses were revealed and a near-miss was recorded. This paper endeavors to replay the events, decision-making process and response measures at Baseline Dam and Reservoir throughout the September 2013 storm event and expresses the need to fully understand safety issues prior to an upgrade.

INTRODUCTION

The epic storm event that took place in north central Colorado in September 2013 was yet another reminder that Mother Nature will have the final say. The event tested the local infrastructure to unprecedented levels and in unanticipated ways; and weaknesses were revealed. While most of the dams in the affected area performed well, the Baseline Dams and Reservoir were among the infrastructure that was put to the test.

---

[1] Daniel L. Johnson, P.E., Vice President, Tetra Tech, Inc., 4900 Pearl East Parkway, Suite 300, Boulder, Colorado 80301, 303-447-1823, dan.johnson@tetratech.com
[2] Brad Dallam, P.E., Assistant City Engineer, City of Lafayette, 1290 S. Public Rd., Lafayette, CO 80026
[3] Ryan Schoolmeesters, P.E., Dam Safety Engineer, State of Colorado, 1313 Sherman Street, Room 818, Denver, Colorado, 80203, 303-866-3581 ext. 8284, ryan.schoolmeesters@state.co.us

Baseline Reservoir is situated near the eastern limit of Boulder, Colorado and is impounded by two High- Hazard potential dams referred to as the East and Northwest dams. The Northwest Dam had undergone a major safety upgrade just two years before the storm event. The upgrading included installation of a downstream drain and flattening of the upstream slope.

On September 12, the second day of the rain event, the dam owner placed a call to the Colorado Dam Safety Branch to report that a new 40-foot-long longitudinal crack had developed overnight along the downstream shoulder of the Northwest dam crest and had sloughed approximately 1 foot downslope. Adding to the complexities, several smaller scarps were cutting into the dam crest along the upstream shoulder in the same reach, and the off-channel reservoir was receiving uncontrolled inflows from South Boulder Creek well in excess of the normal inflows due to complications at the diversion structure. With the National Weather Service forecasting a stalled front over the Boulder area and continuing heavy rainfall through at least the next few days, the Emergency Action Plan for Baseline Reservoir was activated and a team of engineers was mobilized to the site to aid in the response.

Quick action by the owner and the team of engineers, digitally accessible State dam files, and 24-hour monitoring and inspection enabled the dam to "weather the storm"; however weaknesses were revealed and a near-miss was recorded. This paper endeavors to replay the events, decision-making process and response measures at Baseline Dam and Reservoir throughout the September 2013 storm event and expresses the need to fully understand safety issues prior to an upgrade.

DAM BACKGROUND

The State of Colorado straddles the North American Continental Divide along the Rocky Mountains; a line which delineates rainfall drainage to either the Pacific Ocean to the west and the Atlantic Ocean to the east. Baseline Reservoir is situated at the base of the Eastern Front Range of the Rocky Mountains in the southeastern corner of the City of Boulder. A location map is presented in Figure 1 below.

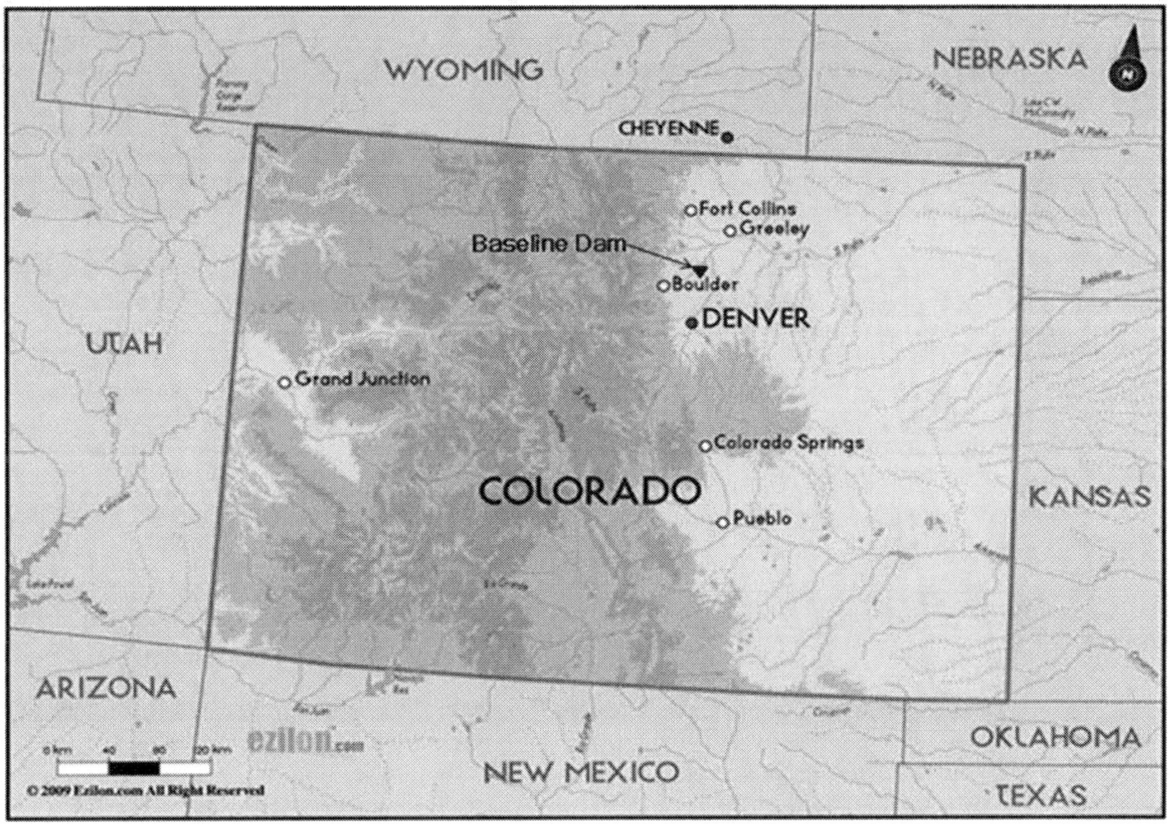

Figure 1. Baseline Location Map

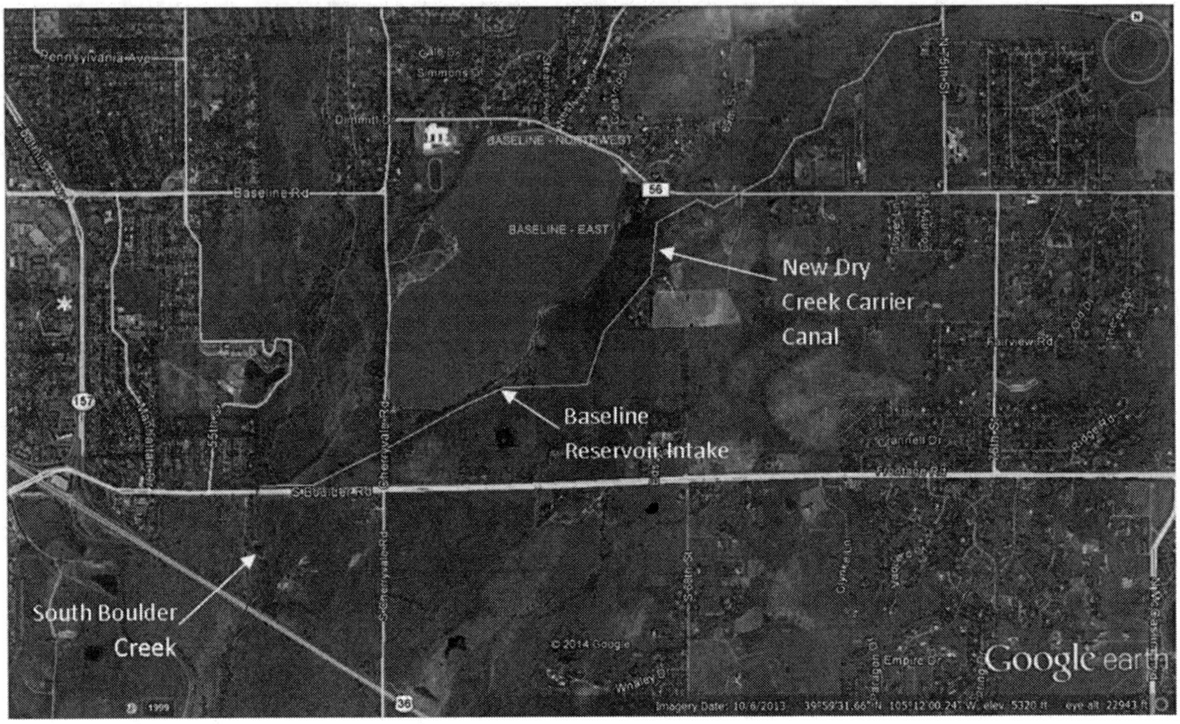

Figure 2. Baseline Reservoir Vicinity Map

In Colorado, like in many semi-arid western U.S. states, water is controlled by the Prior Appropriation Doctrine, in which the concept is "first in time, first in right" in which water rights were established before Colorado became a state. Being one of the state's most valuable resources, all administration of water is under the premise that it will not be wasted and its utilization should be maximized. Baseline Reservoir was decreed in 1906 for agricultural use, subsequently, a majority of shares have gone through a legal process and transferred to municipal use. It currently provides water for the City of Lafayette and several small agricultural users.

Baseline Reservoir is impounded by two dams commonly referred to as the East and Northwest dams. The reservoir has a normal storage volume of approximately 5,300 acre-feet and a surface area of 270 acres. Beyond the reservoir area, the precipitation tributary area is very small with a contributing watershed of approximately 108 acres. The off-channel reservoir is fed primarily by diversion flows from South Boulder Creek, which runs along the western side of the reservoir. South Boulder Creek has normal annual stream flows between 10 and 300 cubic feet per second (CFS) at the nearest stream gage several miles southwest of Baseline Reservoir as shown in Figure 3. The flows are somewhat controlled by Gross Reservoir located about 7 miles upstream of the Baseline Reservoir diversion.

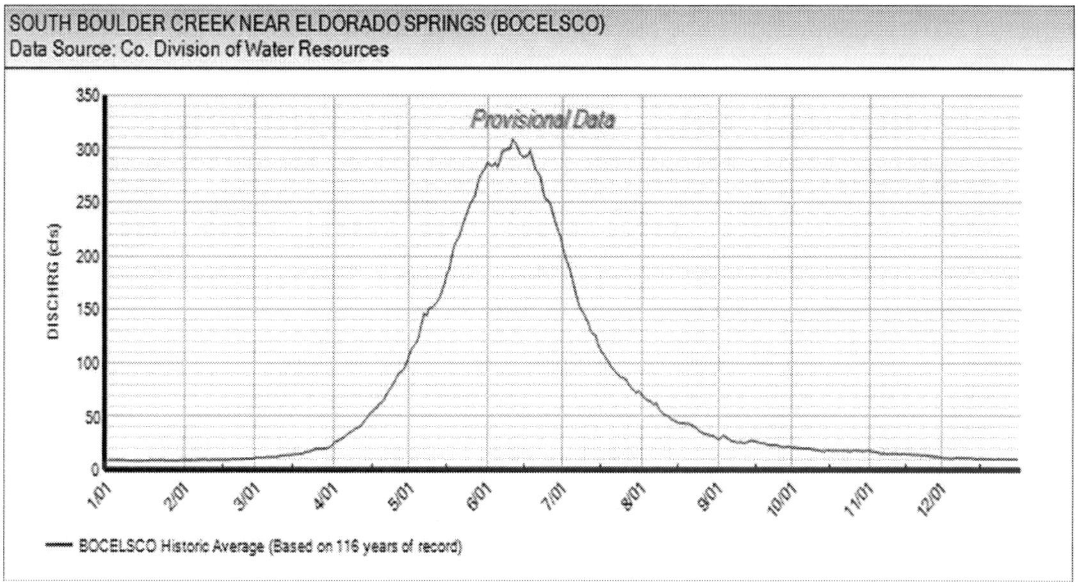

Figure 3. Historic average stream flows in South Boulder Creek

The South Boulder Creek diversion structure is located near the southwest corner of the reservoir, on the north side of South Boulder Road. South Boulder Creek stream diversions are routed through the New Dry Creek Carrier Canal along the south and east side of Baseline Reservoir. The Baseline Reservoir intake is located at the southeast corner of the reservoir, as shown in the vicinity map in Figure **2**2.

In accordance with the criteria set forth in the State of Colorado Rules and Regulations for Dam Safety and Dam Construction (2007), both dams are classified as large size, high hazard potential dams. Although recorded history of the dams is incomplete, they are believed to have been constructed around 1905 as homogenous earthen embankments with puddle cores. The dams were raised approximately 6 feet in 1925, and another 2 fee in 1936 to their present-day crest elevation of 5,315 ft. (NAVD 1988).

The East dam is approximately 40 feet high and 2,300 feet long. The upstream slope is graded to approximately 2.5H:1V with riprap erosion protection. The downstream slope has a 3H:1V grade on the upper 15 vertical feet and 4H:1V on the lower slope. The East Dam contains the outlet works at the valley bottom and the combined service and emergency spillway, excavated into bedrock on the left abutment.

The Northwest dam is approximately 20 feet high and 2,650 feet long. The upstream slope at the Northwest dam is variably graded to between 1.5H: and 2H:1V with riprap cover. Prior to 2012, the downstream slope was at 2H:1V. In 2012, a dam upgrading project was completed, which included improvement of upstream slope protection and installation of a toe drain and downstream slope flattening within the center and left sections of the dam, 2,200 feet of embankment length. Other than the right abutment (450 feet length), the downstream slope graded was graded to 2.5H:1.0V. Completion of this project in 2012 is believed to have had a significant positive impact on the events of September 2013.

The dam and reservoir are owned, operated and maintained by the Baseline Land and Reservoir Company (Company), which is a conglomerate of several municipalities and smaller entities. The City of Lafayette (City) is the majority shareholder, with approximately 70 percent of the shares. As a result, the City serves as the principal representative of the Company and has taken responsibility for the operations, maintenance, and water accounting of the facility.

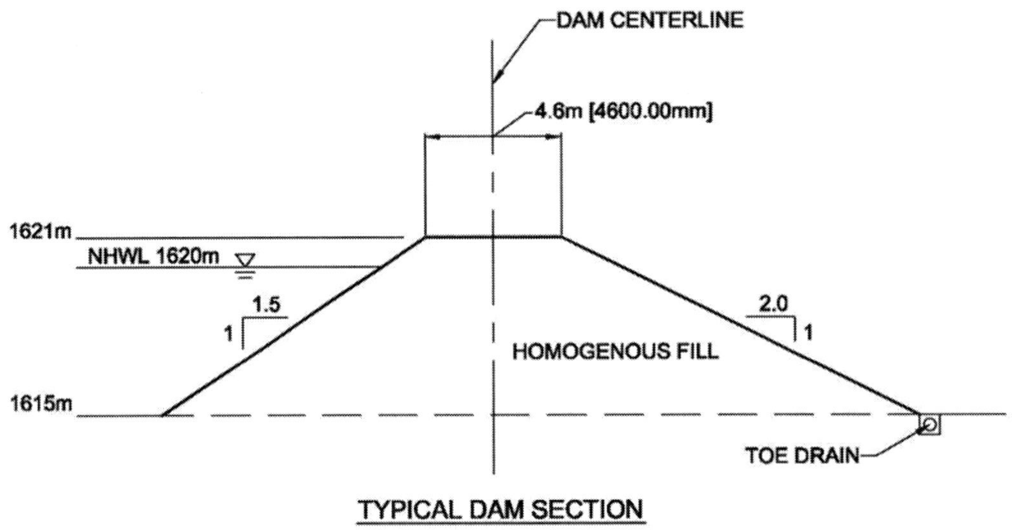

Figure 4. Baseline Dam Cross Sections

HISTORIC STORM

On September 9 and 10, 2013, unusually heavy rains along the Front Range produced around 1 to 2 inches of accumulation each day, which completely saturated the soils. Subsequently, a front over the Colorado Front Range, including heavy moisture from the Pacific Northwest and a tropical Gulf system from the southeast, was stalled for nearly 6 days. The combination of these two systems brought record amounts of "precipitable water", or the amount of moisture in the atmosphere, into the Colorado Front Range by September 12.

As it turned out, the storm provided approximately 7 days of rainfall with variable intensity levels throughout the storm duration. As can be seen in Figure 5 below, the City of Boulder and surrounding communities, the east side of Denver, and the south side of Colorado Springs were the three regions hardest hit by this storm; each receiving nearly 20 inches of rainfall over the 7 day period. For perspective, the annual average precipitation for this region is around 16 inches.

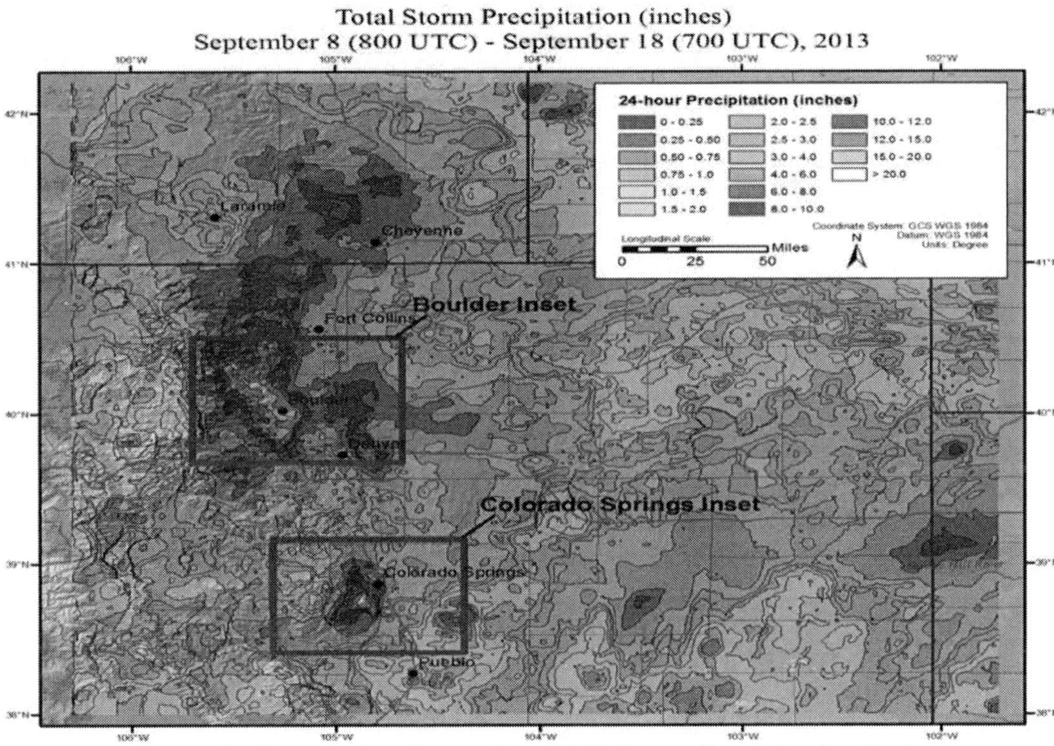

Figure 5. Cumulative September 2013 Storm Precipitation Radar (Applied Weather Associates, 2013)

DAM EMERGENCY

Northwest Dam

Around mid-day on September 12th, Brad Dallam (City of Lafayette) contacted Ryan Schoolmeesters (State of Colorado Dam Safety Branch) stating that the downstream slope of the Northwest dam of Baseline Reservoir had slumped overnight. Brad stated that he hadn't seen the slump personally; however it was described to him as approximately 30 feet long with 8 inches of drop. Brad was rightfully concerned about the safety of the dam and requested assistance. Earlier that day, Tetra Tech recognized that this was an event of epic proportions and extended an offer of assistance to the Colorado Dam Safety Branch. After a brief discussion with the State Dam Safety Branch Chief, Dan Johnson of Tetra Tech agreed to join Brad and Ryan at Baseline to provide additional technical expertise.

Initial observations found rains had saturated the embankment and pools of standing water were present on the crest. Several cracks and minor slumps had formed along the upstream shoulder of the crest for a few hundred feet left of the right abutment of the Northwest dam. Approximately 300 feet left of the right abutment, the downstream crest and slope had slid up to approximately 1-foot downward over a 40 foot length. Upon inspection, no clearly defined slip plane toe was identified, however evidence of topsoil creep was observed and a bulge was apparent on the downstream slope below the slump. Observations at the time were difficult due to the heavy precipitation and the long weed and grass vegetation. Rainwater was ponding at the dam toe, making seepage observations and measurements difficult. The crest pools were draining to both upstream and downstream slopes, and more notably, into the cracks created by the slumps.

Figure 6. Photos of Upstream and Downstream Slope Damage
(Photos were taken 2 weeks after the initial report)

Emergency Actions – Northwest Dam

Since evidence of uncontrolled, muddy seepage was not observed during the initial inspection, the slope movement on the downstream slope was attributed to saturation of the relatively steep downstream slope causing the slope to slip under its own weight. Initial response measures included the following:

- Cut shallow trenches through the crest to drain surface water to the upstream slope and prevent continued water migration into the cracks along the crest,
- Cut down vegetation along the downstream toe to facilitate observation of the toe area,
- Install a movement monitoring system along the downstream slump (two sticks, one in stable embankment and the other in the slumped mass) to monitor for continued movement,
- Cover the crest and slope with tarps to reduce continued rainfall saturation,
- Constant communication with regional emergency services and the State Dam Safety office.

Figure 7. Temporary Visqueen Cover over Damaged Embankment, Makeshift Monitoring System, and Sand Filter Berm at Toe

The City made arrangements to monitor the Northwest dam throughout the night, focusing on the downstream slope movement. Monitoring included; observation of the overall area, slope movement indicators, and change of seepage conditions along the downstream toe.

Emergency Actions – East Dam

Since the flood occurred late in the water year, the Reservoir was about 900 acre-feet below capacity (five feet of freeboard) prior to the flood enabling some flood flow retention. The outlet works has capacity of approximately 100 cfs at full reservoir and water can be taken to the City's water treatment plant. The outlet could assist in reducing the level of rise and the stress on the Northwest Dam. The outlet was fully opened on Thursday.

An inspection of the East dam was performed while the City worked on the initial mitigation measures at the Northwest dam. The inspection began at the spillway, which consisted of an open channel cut through the left abutment of the East Dam that had no history of being called into service. Subsequently the spillway channel was being used as a boat launch and for other recreational activities. Numerous lake kayaks and other obstructions were located in the spillway channel that were removed soon after the inspection. The spillway channel was not well defined downstream of the dam. The inspection showed that a portion of the spillway flows could have taken a 90-degree turn after passing the embankment and traveled along the dam toe, whereas the other portion would flow directly onto the horse property immediately east of the East Dam.

Considering that the outlet was unable to keep up with the inflow, the spillway was anticipated to begin flowing sometime in the next day. It was recommended that the private residence just downstream of the East Dam, occupied by a 90 year old woman, be evacuated. In the interest of protecting the East Dam from erosion and potential failure under spillway flow conditions, the City was requested to build up a berm at the downstream end of the spillway channel to direct flows away from the embankment toe.

Inspection of the outlet works showed discharge water that appeared notably cloudier than reservoir water. No recognizable seepage exit points were detected around the outlet pipe outfall structure; however, tail water from the heavy outlet flow was inundating the discharge channel and half of the outlet structure. To check for seepage/leakage issues in the inundated area, the caretaker closed the control gate temporarily to facilitate inspection for continued water movement around or out of the outlet pipe. Upon gate valve closure, no continued flows were observed, and standing water in the channel regained appearance of reservoir water, thereby reducing concerns about the outlet pipe.

Inspection of the toe drain pipe outfalls presented another point of interest. The drains, which had historically never operated, were each passing around 10 to 20gpm. Fortunately the water clarity was good, precluding the likelihood of a damaged drain system; however, the sudden discharge volume increase created concern.

Document Review

The State Dam Safety office had installed a digital filing system that is available over the internet (with appropriate credentials) enabling easy review from remote locations. Around 6:00 P.M., Dan and Ryan used the Tetra Tech office in Boulder to download and review record drawings, construction documents, inspection reports, and monitoring data. The primary reason for the file review was to take a step back from the immediate threats to gain a better understanding of the dams and to develop and evaluate Potential Failure Modes (PFM) at the dam and the symptoms that may provide an early warning of these critical PFMs.

During that review, several important discoveries were made. First, it had been previously stated that the 40-foot long slump had occurred in a reach of the dam that had been recently modified to address potential instability of the over-steepened slopes. During the review, however, it was found that the downstream slough occurred in an *unimproved* section of the dam. This discovery provided a better understanding of why the slump occurred in that location, and more importantly, that the dam reach that had been recently improved was probably stable.

Second, the toe drain system at the East Dam was shown to have been constructed at a shallow depth, which supported a hypothesis that the drain system was collecting surface infiltration as opposed to through-dam seepage. This hypothesis would later be confirmed by the cessation of drain flows during a break in the rainfall.

Third, a review of the spillway rating curve data revealed that the spillway had a maximum capacity of approximately 1600cfs. As mentioned previously, the peak inflows earlier that day had been estimated around 1,000cfs. On the basis of the considerable attenuation presumed to be provided by the 270 acre reservoir, it was initially decided that a dam overtopping failure was not a primary point of concern. However, subsequent review of recent topography at the dam showed a low area up to 1-foot lower than the rest of the dam centered at the location of the 40-foot slump, which reduced the spillway capacity to approximately 900cfs. This revelation heightened awareness and concern for dam overtopping and provided a better understanding of a trigger point to initiate evacuations.

Fourth, recent and historic data for in piezometers located in the downstream edge of the crest and at the dam toe indicated that the phreatic surface in the area of the slope failure typically read higher that would be expected in a homogenous dam. The City was tasked with taking piezometer readings on about 6-hour time intervals to evaluate the rise in phreatic surface with reservoir rise. This raised concern that seepage exits may form along the slope failure surface (shortened seepage paths) leading to embankment piping.

Based on the available information, the primary dam failure modes of concern were:

- Seepage through the embankment, likely expressing in the lower few feet of the embankment, resulting in piping and eventual failure of the embankment;
- Complete slope failure of the downstream slope at the present slump, followed by continued shallow slope failures, eventually leading to either piping or overtopping of the embankment and failure of the dam; or
- Some combination of these two PFM's.

After the file review, the information was relayed to the 24-hour monitoring crew along with directions for what to look for during their overnight monitoring. Cold and very wet, the inspection team retreated for a late night warm shower and dry clothes, as well as a little rest.

Complications at the Upstream Intake

As mentioned, the reservoir is an off stream reservoir with a relatively small contributing watershed. The reservoir is filled by a diversion and canal off of South Boulder Creek. Approximately 30 years ago the City of Boulder constructed a bike bridge crossing South Boulder Creek on the south side of South Boulder Road. Two box culverts beneath the bike bridge were approximately 5 feet by 5 feet, and handrails were constructed on each side of the bridge. It is believed that the intent was that during high flows, the hand rails would break away and flow would continue down South Boulder Creek.

During the flood flows of the September event, the bike bridge and hand rails completely plugged with debris, essentially stopping all flow going down South Boulder Creek. The back up on South Boulder Creek extended about 1000 feet upstream of the culvert and flows split to the east and west of the channel. The uncontrolled flood flow that went to the east crossed over South Boulder Road, putting the road out of service, and drained into the New Dry Creek Carrier canal, downstream of the canal gate structure.

The New Dry Creek Carrier is a hybrid between an irrigation ditch and natural drainage. Back in the 1870s, the farmers recognized the value of the natural drainage and its upper end proximity to South Boulder Creek. The farmers connected South Boulder Creek to Baseline Reservoir via the Dry Creek Carrier.

The flood water entered the New Dry Creek Carrier at flow rates far greater than the capacity of these facilities. The flood water overtopped the canal banks, and the water entered Base Line Reservoir as uncontrolled flood flow. Without this situation, the flows to the reservoir could have been sufficiently low to prevent reservoir level rise.

On Monday, September 16, 2013, Lafayette Staff received notification that a large reservoir upstream on South Boulder Creek (Gross Reservoir) would be spilling flood waters soon, increasing the stream discharge by up to four times. This increased flow would be routed around the bike bridge "dam" and accelerate the reservoir rise and spillway use. The State Dam Safety Engineer directed Lafayette to prevent this flood water from getting into Baseline Reservoir by whatever means necessary.

Baseline Company contacted a construction firm that arrived on site at that Monday about noon. The Construction firm removed debris from around the bike bridge to allow flow down South Boulder Creek and reduce or stop flows into the canal. It was noted that the bide bridge culverts did not have sufficient capacity to convey the flood flows and the reservoir continued to rise, uncontrollable. It was decided that the bike bridge needed to be removed and the Contractor was directed to remove the bridge. The bridge was very stout and required larger machinery than was available on Monday. On Tuesday, larger equipment arrived on site, and the bike bridge was removed but a new channel had been eroded, enabling flood flow to continue across South Boulder Road into the New Dry Creek Carrier, and to the reservoir. A channel plugging plan was developed to prevent flood flows from continuing to flow uncontrollably to the reservoir. However, there was environmentally sensitive habitat in the area, managed by the City of Boulder Open Space Program. The Open Space Program Rangers prevented initiation of any work on the Creek, with the threat of criminal justice for anybody who attempted to access to the property. At this point, the State Dam Safety Office provided a notice that all consequences would be the Open Space Program's responsibility which enabled limited access to the property. The channel was subsequently plugged and flood flow to the canal was stopped, while minimizing damage to the open space lands.

Figure 8. Image of the Pedestrian Bridge Responsible for the Uncontrolled Inflows to Baseline Reservoir

RESULTS

The period of emergency lasted over several days, involving several agencies and municipalities, dam experts, construction contractors, regulatory agency personnel, and news reporters. The dam failure was averted and downstream residents were safe. The reservoir reached its highest storage level over its nearly 90 year life, providing insight to the capabilities of the dam and spillway. The Emergency Action Plan was initiated, but not fully engaged, once the failure model and site conditions were assessed.

LESSONS LEARNED FROM THIS SPECIFIC EVENT

Floods of this magnitude are rare and test the facilities and systems that we develop. These types of tests are valuable for enhancing both the facilities and systems for the future. There are several lessons learned that are worthwhile pointing out.

Immediate Action by Owner

The personnel responsible for the Baseline Reservoir were aware of the potential for a precipitation event and flood of this magnitude to provide loading conditions that had not be previously experienced by the facility. Therefore, they were diligent in having their staff inspect the facility during the driving rain and making the appropriate contacts with the agencies that could provide support. This early awareness enabled early involvement of dam experts to assist in retarding the growing potential of a dam failure. This type of awareness and action should be exercised by all dam owners.

Involvement of Experienced Engineers

The State Dam Safety Office was prepared to have their engineer on site in a matter of hours after notice of concern was received. Coordination of local engineering firms and the State Dam Safety Office enabled immediate contact and a private dam expert to be on site to work with the State engineer. Having two engineers from different agencies enabled synergism in developing emergency measures, understanding of the construction and performance history, and the connections/authority to engage support services immediately.

Caution in Providing Direction for Emergency Repairs

As discussed in this paper, the downstream slope was covered with tall grass and week vegetation, hindering visual inspection, especially since the downstream toe was inundated with flood water. Direction was provided to clear the vegetation; however, the goal was to reduce the height of vegetation and not remove it by the roots. The owner's diligent staff initiated removal which resulted in removal and partial excavation on the slope which was not desired. Subsequent observation by the engineers enabled modification of the directions to just cutting down the vegetation. Clear direction and follow-up are required during such emergency actions as the workers likely do not have the understanding of each action taken.

IMPACT OF MULTIPLE INTEREST AGENCIES IN MANAGING OFFSITE HYDRAULIC CONDITIONS

As noted in this report, the dam has a small drainage basin and a gated canal that fills the reservoir from a significant sized stream. The ability to close the gate and lower the reservoir should have been relatively easy. However, the bicycle bridge, state highway, and multiple owner/agency interests prevented reduction in flood flow to the reservoir for several days. The lesson learned is to look well upstream of potential dam safety impacts and to understand that multiple interests may impede the ability to take emergency actions. Levels of communication need to be understood so that resolution of non-owner restraints can be understood and communicated such that dam safety actions are implemented.

CONVEYING LESSONS LEARNED FROM DAM INCIDENTS AND FAILURES

Paul Schweiger, P.E., CFM[1]
Greg Richards, P.E., CFM[2]

ABSTRACT

There is a pressing need to understand the underlying causes of dam failure to ensure such occurrences are not repeated. Historical information on dam incidents is scattered, incomplete, missing, and sometimes misleading – making it difficult for owners and practitioners to easily access meaningful information that could assist them with critical design and operational decisions. If lessons learned and best practices are not effectively communicated, there is a possibility that poor practices will be repeated and preventable incidents will not be averted.

In late 2013, the Federal Emergency Management Agency (FEMA) authorized a project to research past dam failures and incidents and select the most appropriate information to include in a compendium of lessons learned supported by case histories. The project scope includes addressing a range of failure modes, dam types, and dam safety practices, including lessons learned relating to a spectrum of dam safety topics including engineering and design, emergency planning and response, operation and maintenance, and regulatory issues. The primary objective for this project is to create a website that will allow users to access a series of lessons learned and corresponding case studies. Each lesson learned will be presented in a user-friendly, web-based format that includes access to pertinent graphics, narratives, statistics, photographs, videos, best practice resources, technical papers and other available information. The target audience includes dam safety engineers, operators, dam owners, regulators, emergency managers, academia and students.

PROJECT BACKGROUND

As dam failures and incidents occur both nationally and internationally, there is a pressing need to understand the underlying causes of failure to ensure such occurrences are not repeated. Historical information on dam incidents is scattered, incomplete, and sometimes misleading – making it extremely difficult for owners and practitioners to easily access meaningful information that could assist them with critical design and operational decisions. If lessons learned and best practices are not effectively communicated, there is a possibility that poor practices will be repeated and preventable incidents will not be averted.

In late 2013, the Federal Emergency Management Agency (FEMA) authorized a project to research past dam failures and incidents and select the most appropriate information to include in a compendium of lessons learned supported by case histories. The project scope includes addressing a range of failure modes, dam types, and dam safety practices,

[1] Vice President, Gannett Fleming, Inc., Camp Hill, PA 17011, pschweiger@gfnet.com
[2] Project Manager, Gannett Fleming, Inc., Camp Hill, PA 17011, grichards@gfnet.com

including lessons learned and best practices on emergency action plan activation and response related to dam failures and incidents. This educational information is to be conveyed in an innovative, user-friendly manner that is appealing to contemporary users. The target audience includes dam safety engineers, operators, dam owners, regulators, emergency managers, academia and students.

The primary objective of the project is to create a new website that will allow users to access a series of dynamic summaries of specific lessons learned and corresponding case studies. Many lessons learned will be specific to individual failure modes and will address the majority of generalized failure modes. Each lesson learned and case study summary will include access to pertinent graphics, narratives, statistics, photographs, videos, best practice resources, technical papers and other available information.

As of February 2015, the project team has completed an extensive review of available information on dam incidents and failures, created and prioritized a list of lessons learned, identified relevant case histories, and developed a library of best practices and other important resource materials. More than 175 different lessons learned from dam incidents and failures have been identified. These lessons learned focus on a spectrum of dam safety topics including engineering and design practices, emergency planning and response, operation and maintenance, and regulatory issues. Fifty-nine of these lessons learned have been identified as "1st Priority" and will be included in the new website as budget permits. Of the more than 625 case studies identified, 43 case studies have been selected to support the "1st Priority" lessons learned. Lists of these selected lessons learned and case studies are included as Tables 1 and 2, respectively.

The web host for the new website has not yet been finalized, but it is anticipated that the site will be deployed by late summer 2015. Once the website is live, the public will have access to both the lessons learned and case study summaries described above. The site will also allow users to submit additional lessons learned or case study information to be considered for inclusion on the website. The website will be a helpful tool for members of the dam safety community and will help convey the lessons of the past to the current generation.

Table 1. First Priority Lessons Learned

ID	Lesson Learned
1	Seepage along penetrations through embankment dams should be controlled using a filter diaphragm instead of anti-seep collars.
2	A concrete encasement or cradle around or under conduit penetrations through earthen embankments allow for better compaction of earthfill.
4	The first filling of a reservoir should be planned, controlled and monitored.
46	Uncontrolled vegetation on and around dams hinders effective dam inspection and can lead to serious structural damage, significant deferred maintenance costs and dam failure.
50	Regular operation, maintenance, and inspection of dams is important to the early detection and prevention of dam failure.
60	Emergency Action Plans can save lives and must be updated, understood, and practiced regularly to be effective.
64	Treatment of fractured foundation rock for embankment dams is important to prevent internal erosion of embankment material that is in contact with the foundation.
70	Stability of the dam foundation and other geologic features must be considered during dam design.
89	The design engineer(s) should be involved in the construction phase of dam projects.
97	High and significant hazard embankment dams should have internal filter and seepage collection systems.
102	External independent peer review of designs and decisions is an effective means of providing quality assurance and reducing the risk associated with design oversights and deficiencies.
108	Engineers performing dam inspections and assessments should have a sound knowledge of the as-built features and common design practices in use at the time of dam construction to help them identify potential failure modes and other vulnerabilities.
109	In emergency situations, lowering the reservoir can be an effective means of risk reduction.
110	Filters and drains for embankment dams must be compatible with adjacent fill or in-situ materials.
111	Many years of successful dam performance does not guarantee future successful performance.
113	Safety should not be sacrificed for cost.
119	Earth/rock cut spillways can fail by erosion and breaching and should be evaluated for integrity during passage of the design flood.
120	PMF magnitude floods do occur.
121	High and significant hazard dams should be designed to pass an appropriate design flood. Dams constructed prior to the availability of extreme rainfall data should be assessed to make sure they have adequate spillway capacity.
122	The use of corrugated metal pipes in embankment dams is discouraged.
123	Concrete gravity dams should be evaluated to accommodate full uplift.
126	Pumped storage embankment dams should have an emergency spillway or other

ID	Lesson Learned
	redundant features to prevent overfilling and overtopping of the embankment.
127	All dams need an operable reservoir drain system.
129	Warning time and rapid response is critical to saving lives during a dam failure emergency.
131	Gates and other mechanical systems at dams need to be inspected and maintained.
132	Dams need to be regulated by state or federal agencies.
135	Many dam failures are the result of a combination of events and can include human, natural, or structural factors.
136	Intervention can stop or minimize the consequences of a dam failure. Warning signs should not be ignored.
139	Dam owners need to address public safety at dams.
140	Hazardous hydraulic conditions such as hydraulic rollers can occur at dams of all sizes.
141	The consequences of intentional controlled releases from dams need to be understood and carefully executed.
143	Dams located in areas of potential seismic activity need to be evaluated for liquefaction, cracking, potential fault offsets, deformations, and settlement due to seismic loads.
144	Alkali-aggregate reactions (AAR) can cause serious concrete deterioration and other problems at concrete dams.
145	The hazard classification of a dam can change over time (hazard creep).
147	Most concrete dam failures are the result of foundation stability problems. Concrete dams founded on bedrock require subsurface investigations and testing of rock properties.
151	Outlet works and pressurized conduits in embankment dams should be provided with a means for upstream closure.
152	Dams should be inspected and monitored during and/or after large rainfall, seismic, or other unusual loading events.
165	Poor compaction, steep rock abutments, and other foundation discontinuities at embankment dams can lead to differential settlement, cracking, and failure by internal erosion.
166	Dams built from or on loose granular soil can be susceptible to liquefaction of the foundation material.
195	Settlement of earthfill embankment dams can reduce freeboard and increase the risk for overtopping. New embankment dams should be designed with allowances for settlement.
196	Discontinuities in grass-lined spillways can accelerate erosion and lead to significant damage or complete breaching of the spillway.
206	Most dam incidents and failures can be attributed to human factors.
207	Modifications to spillways can unintentionally decrease their capacity. Any modification to a spillway should be reviewed and approved by a professional engineer.
234	Geotextiles should not be used as the primary means of filtration in drainage systems for dams.
235	To the extent practicable, dispersive soils should not be used in the construction of

ID	Lesson Learned
	embankment dams.
236	For new dams, the as-built conditions of critical elements should be verified through direct observation and/or testing. This is particularly important for critical elements that are constructed "in the blind."
237	For high and significant hazard dams, elements that are critical for the safe operation of the dam should be designed using the "belts and suspenders" approach.
238	Internal erosion can occur at relatively low hydraulic gradients.
239	Internal erosion can take decades to progress.
240	Structural underdrains can lead to internal erosion under or along stilling basins and conduits.
241	Non-plastic soils are far more likely than soils with some plasticity to experience internal erosion.
242	Cutoff walls can concentrate gradients and seepage flows.
243	Karstic foundations are difficult to treat and can cause internal erosion issues.
244	Concrete dams on rock foundations are not immune to failure during flood overtopping.
245	During floods that greatly exceed the design flood, spillway failure (due to exceedance of the spillway capacity) may be a more critical potential failure mode than dam overtopping.
246	Hard rock spillway channels and hard rock concrete dam foundations may erode under the right flow conditions combined with unfavorable joint orientations and other joint characteristics.
247	Dams may overtop for floods more frequent than the design flood if the spillway capacity is reduced (due to debris plugging or gate malfunction) or if a gated spillway is not operated as assumed in design studies.
248	Dam operators may operate a gated spillway at release levels less than those specified in the SOP if there is the potential for downstream consequences.
249	Spillway gates should be tested to the maximum extent possible on a regular basis to verify the performance of the electrical/mechanical system and to ensure that the gates can travel freely without binding or being restricted.

Table 2. First Priority Case Studies

ID	Selected Case Studies	State	Country	Year of incident	Type of dam
154	Alla Sella Zerbino Dam		Italy	1935	Concrete
1	Anita Dam	Montana	USA	1997	Earthfill
115	Austin Dam (Bayless Dam)	Pennsylvania	USA	1911	Concrete
62	Austrian Dam	California	USA	1989	Earthfill
196	Balderhead Dam		Great Britain	1966	Earthfill
126	Baldwin Hills Dam	California	USA	1963	Earthfill
73	Big Bay Lake Dam	Mississippi	USA	2004	Earthfill
207	Black Creek site 53	Mississippi	USA	1983	Earthfill
72	Buffalo Creek Coal Waste Dam	West Virginia	USA	1972	Tailings
166	Canyon Lake Dam	South Dakota	USA	1972	Earthfill
24	Centennial Narrows Dam	Arizona	USA	1997	Earthfill
652	Davis Creek Dam	Nebraska	USA	2007	Earthfill
649	Deer Flat Dam	Idaho	USA	2006	Earthfill
650	El Guapo Dam		Venezuela	1999	Earthfill
89	Folsom Dam	California	USA	1995	
65	Fontenelle Dam	Wyoming	USA	1965	Earthfill
513	Fort Peck	Montana	USA	1938	Earthfill
88	Glen Canyon Dam	Colorado	USA	1983	
149	Gleno Dam		Italy	1923	Concrete
214	Hope Mills Dam	North Carolina	USA	2003	Concrete
76	Ka Loko Dam	Hawaii	USA	2006	Other

ID	Selected Case Studies	State	Country	Year of incident	Type of dam
103	Lake Delhi Dam	Iowa	USA	2010	Composite
85	Lake Needwood Dam	Maryland	USA	2006	Earthfill
116	Laurel Run Dam	Pennsylvania	USA	1977	Rockfill
108	Lawn Lake Dam	Colorado	USA	1982	Earthfill
180	Lower San Fernando Dam	California	USA	1971	Other
190	Lyell Dam		Australia	2000	Rubber
35	Malpasset		France	1959	Concrete
31	Meadow Pond Dam	New Hampshire	USA	1996	Earthfill
181	Mud Mountain	Washington	USA	1976	Composite
63	Pablo Dam	Montana	USA	1994	Earthfill
21	Penn Forest Dam	Pennsylvania	USA	1960	Earthfill
112	Percy Quin Dam (Lake Tangipahoa Dam)	Mississippi	USA	2012	Earthfill
119	Quail Creek Dike	Utah	USA	1989	Earthfill
651	Ricobayo Dam		Spain	1934	Concrete
185	Shih Kang Dam		Taiwan	1999	Concrete
114	South Fork Dam	Pennsylvania	USA	1889	Composite
117	St. Francis Dam	California	USA	1928	Concrete
75	Taum Sauk (Upper) Reservoir Dam	Missouri	USA	2005	Rockfill
64	Teton Dam	Idaho	USA	1976	Earthfill
113	Vajont Dam		Italy	1963	Concrete
155	Wister Dam	Oklahoma	USA	1949	Masonry
648	Wolf Creek Dam	Kentucky	USA	1968	Composite

EXPOSURE: A NEW DECISION METRIC FOR SELECTING EFFECTIVE SETS OF SECURITY UPGRADES

Kevin E. Burns, PhD[1]
James D. Morgeson[2]
Jason A. Dechant, PhD[3]
Yazmin Seda-Sanabria[4]
Enrique E. Matheu, PhD[5]

ABSTRACT

The United States Army Corps of Engineers (USACE) conducts Security Risk Assessments (SRAs) at its most consequential dam projects. The Common Risk Model for Dams (CRM-D) provides a mathematically rigorous and easy-to-implement way to conduct SRAs. The CRM-D quantifies risk as the product of the probability of a successful attack, given it is attempted, and consequences. Referred to as *conditional risk*, this decision metric is the expected loss given a specified attack is attempted on a particular target. A specified attack (consisting of an attacker type and an attack vector) carried out on a particular target comprises a scenario.

The CRM-D considers three attacker types and thirty-two attack vectors identified by USACE Headquarters (HQs). A dam with only a modest number of critical assets could thus have several hundred scenarios and, consequently, several hundred conditional risk estimates. This paper introduces a decision metric, *exposure*, which allows the analyst to aggregate conditional risk estimates across scenarios. The analyst can use exposure to compare risks by attack type, by target or for any useful set of scenarios. These comparisons can guide an analyst in determining a proposed set of security upgrades. A standard set of graphics and return-on-investment calculations based on exposure are introduced that summarize the current level of risk at a dam project as well as the reduced level of risk should the set of recommended security upgrades be implemented.

INTRODUCTION

In 2005, the Institute for Defense Analyses (IDA) initiated the development of the Common Risk Model (CRM) for evaluating and comparing risks associated with the Nation's critical infrastructure. This model incorporates commonly used risk metrics that are designed to be transparent, simple, and mathematically justifiable. The CRM also

[1] Research Staff Member, Strategy Forces, and Resources Division (SFRD), Institute for Defense Analyses (IDA), Alexandria, VA 22311, kburns@ida.org.
[2] Research Staff Member, SFRD, IDA, Alexandria, VA 22311, jmorgeso@ida.org.
[3] Research Staff Member, SFRD, IDAs, Alexandria, VA 22311, jdechant@ida.org.
[4] National Program Manager, Critical Infrastructure Protection & Resilience Program Office of Homeland Security, Directorate of Civil Works, U.S. Army Corps of Engineers, Washington, DC 20314, Yazmin.Seda-Sanabria@usace.army.mil.
[5] Sector Outreach and Programs Division, Office of Infrastructure Protection, National Protection and Programs Directorate, U.S. Department of Homeland Security, Washington, DC 20598, Enrique.Matheu@hq.dhs.gov.

enables comparisons of risks to critical assets both within and across critical infrastructure sectors.

Over the past few years, an extended version of the CRM has been under development by IDA in collaboration with the USACE and the U.S. Department of Homeland Security (DHS). The extended model—Common Risk Model for Dams (CRM-D)—takes into account the unique features of dams and navigation locks and provides a systematic approach for evaluating and comparing risks from terrorist threats across a large portfolio[6] (Seda-Sanabria et al., 2011a).

In general, the risk for an attack scenario is considered to be a function of three variables: *threat*—the likelihood of an attack scenario being chosen by adversaries; *vulnerability* — the likelihood of defeating the target's defenses, given that the attack is chosen; and estimated *consequences* of the attack, given the target's defenses are defeated (DHS 2013). Therefore, it can be stated that

$$R = f(T, V, C), \tag{1}$$

where R is risk, T is threat, V is vulnerability, and C is the consequences. A widely used approach is to define the risk function as the product of these three variables:

$$R = T \times V \times C. \tag{2}$$

Threat, denoted as P(A), is defined as the probability that a given attack scenario is chosen, *conditional* on one of the attack scenarios in the portfolio being chosen within a specified timeframe (usually taken to be a year). Vulnerability is defined as the conditional probability that a given attack vector will successfully defeat the target's defenses, given the attack scenario is chosen, or P(S|A). Consequences, denoted as C, are defined in terms of lives lost and economic loss given that the attack successfully breaches all of the defenses protecting the scenario target. Thus

$$R = P(A) \times P(S|A) \times C. \tag{3}$$

One can also define conditional risk for a scenario, R_C:

$$R_C = V \times C = P(S|A) \times C. \tag{4}$$

R_C is conditional on the adversary's choice of a specific scenario. Calculating conditional risks for all scenarios can be useful to an analyst at a dam project as this knowledge can inform decisions regarding how to improve the security measures at the dam's facility. The next section provides an overview of how conditional risk is estimated in CRM-D.

[6] A portfolio is a set of dam projects evaluated by a risk analyst.

BASIC CONCEPTS: THE COMMON RISK MODEL FOR DAMS

A conceptually simple model of layered defenses is used to evaluate the conditional risk of a given critical infrastructure target. As an example, Figure 1 represents the case of a target protected by three notional defensive layers.

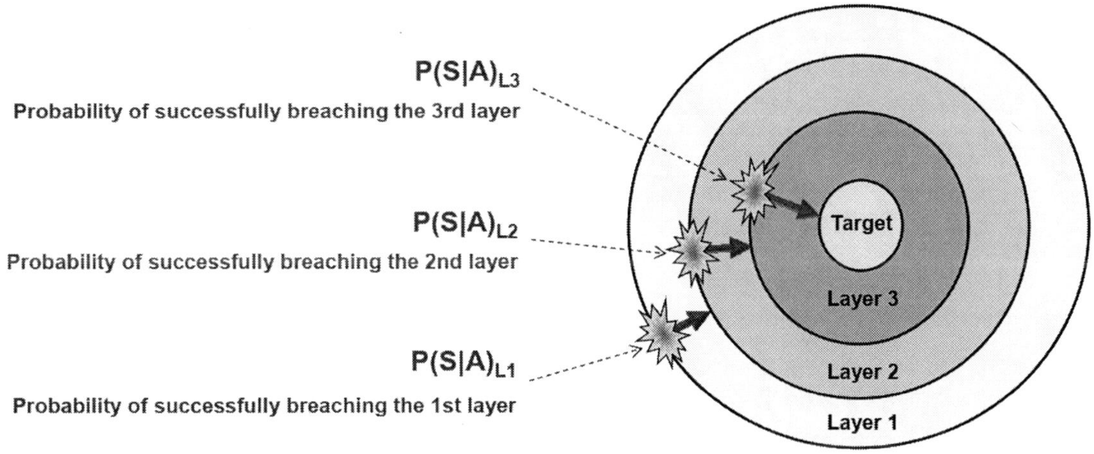

Figure 1. Conceptual Model of Layered Defenses

For the target to be successfully attacked, each defensive layer providing some form of protection would need to be successfully breached. For the case shown in Figure 1, the joint probability that a given attack, if attempted, will be successful in reaching the target, considering it being attempted (also known as the vulnerability or P(S|A)), can be determined using the following expression:[7]

$$P(S|A) = P_{L1} \times P_{L2} \times P_{L3}. \quad (5)$$

As part of the CRM-D development, a representative set of physical attack vectors was defined by USACE HQ to represent a wide spectrum of attacks that can be used to facilitate the comparison of vulnerability and conditional risk results across a large portfolio. These attack vectors,[8] which assume high-capability adversaries (well-trained attackers with significant access to resources), are listed in Table 1 and are arranged by categories known as *attack types*. The twenty-eight attack vectors listed in Table 1 are a subset of the thirty-two total attack vectors defined by USACE HQ for various adversary types.

[7] The CRM-D model explicitly accounts for time delays in the attack due to the time it takes to breach layers and the time it takes to traverse interlayer distance. Also taken into account is attack vector degradation due to armed defenders. Since these are factors normally used in the estimation of conditional probabilities, CRM-D simplifies the calculation of conditional probabilities by assuming independence among the probability estimates except for those factors explicitly accounted for. See Morgeson et al., "Incorporating Uncertainties in Estimation of Vulnerabilities for Security Risk Assessments."

[8] The attack vectors used in CRM-D can be extended to include additional attack vectors that are identified over time.

Table 1. Attack Vectors Considered for High-Capability Adversaries

Attack Mode	Attack Type	Attack Vectors			
Land	VBIED	Sedan	Cargo Van	Box Truck	Large Truck
	Assault Team	Single Attacker	Small-Size Cell	Medium-Size Cell	Large-Size Cell
	Stand-off Weapon	Large-Caliber Rifle	Rocket-Propelled Grenade	Mortar	Man-portable Guided Missile
	Sabotage	Insider	Outsider	-	-
	Excavating Attack	Mechanical	Kinetic	-	-
Water	Water-borne Improvised Explosive Device (WBIED)	Inflatable Boat	Small Boat	Large Boat	Barge
	Underwater IED	Surface Swimmer	Subsurface Swimmer	Modified Small Boat	Semi-submersible Boat
Air	Impact	Helicopter	Small Airplane	Narrow-body Airliner	Wide-body Airliner

The probability that a specific attack vector would be successful in reaching a given target depends on the probability that each of the defensive layers sequentially encountered along the attack path is successfully penetrated. These individual layer probabilities, in turn, depend on the type of adversary, the attack vector chosen, and the strength of the defenses at the particular layer (Morgeson 2013). To illustrate the concept, assume that a given target T-1 is protected by three defensive layers (L_1, L_2, and L_3). For a given attack vector denoted as AV-1, the overall probability of success for the scenario defined by attack vector AV-1 and the target T-1 is the joint probability of successfully penetrating all three layers. For this scenario, equation 5 can be written using the following notation:

$$P(S|A)_{Overall} = P(S|A)_{AV\text{-}1,\,L1} \times P(S|A)_{AV\text{-}1,\,L2} \times P(S|A)_{AV\text{-}1,\,L3}. \quad (6)$$

Equation 7 is an example of the calculation in equation 6 using notional estimates for the probabilities for each layer.

$$P(S|A)_{Overall} = 0.2 \times 0.7 \times 0.9 = 0.126. \quad (7)$$

As defined in equation 4, conditional risk is the product of the probability of successful attack (given that that attack is attempted) multiplied by the corresponding consequences.

For illustrative purposes, assume that the consequences of a successful attack on target T-1 is estimated to be \$1 million (M), and therefore

$$R_c = P(S|A) \times C = 0.126 \times \$1M = \$126{,}000. \quad (8)$$

R_C is the expected loss given a hypothetical attack on target T-1 protected by layers L_1, L_2, and L_3. Therefore, for every attack scenario (i.e., combination of an adversary type, a specific attack vector, and a given target), the CRM-D methodology provides a systematic approach to estimate conditional risk. Table 2 contains conditional risk calculations for a set of scenarios at a notional dam.

Table 2. Conditional Risk for Selected Attack Scenarios for a Notional Dam

Attack Type	Attack Vector	Impoundment					Powerhouse (Control Center)					Powerhouse (Turbines and Generators)				
		Conseq.					Conseq.					Conseq.				
		Loss of Life	Economic Loss (\$M)	$P(S\|A)_{FINAL}$	Conditional Risk (Life)	Conditional Risk (Econ. \$M)	Loss of Life	Economic Loss (\$M)	$P(S\|A)_{FINAL}$	Conditional Risk (Life)	Conditional Risk (Econ. \$M)	Loss of Life	Economic Loss (\$M)	$P(S\|A)_{FINAL}$	Conditional Risk (Life)	Conditional Risk (Econ. \$M)
VBIED	Sedan	1	1	1.0	1	1	11	1	1.0	11	1	1	22	1.0	1	22
	Van	1	3	1.0	1	3	11	3	1.0	11	3	1	22	1.0	1	22
	Box Truck	1	3	1.0	1	3	11	3	1.0	11	3	1	44	1.0	1	44
	Large Truck	100	200	1.0	100	200	11	3	1.0	11	3	1	66	1.0	1	66
Assault Team	Single Attacker	1	0	0.51	1	0	11	3	0.58	6	2	1	44	0.58	1	25
	Small Cell	1	0	0.9	1	0	11	3	0.81	9	2	1	132	0.81	1	107
	Medium Cell	1	1	1.0	1	1	11	3	1.0	11	3	1	132	1.0	1	132
	Large Cell	1	1	1.0	1	1	11	3	1.0	11	3	1	132	1.0	1	132

Source: The data in Table 2 and associated graphics throughout this paper are drawn from Michael Keleher, Steven Walser, and Samuel Himel, *The Common Risk Model for Dams Support System: A Prototype Analyst Tool* (Alexandria, VA: Institute for Defense Analyses, IDA Paper P-5220, to be published). Values in the table are rounded.

Note that Table 2 contains forty-eight conditional risk calculations (highlighted in red). Yet this represents only a fraction of conditional risk calculations that would be performed at many dams. This table includes only one adversary type, only two attack types (thus only eight attack vectors), and only three assets. The challenge facing a risk analyst is how to summarize the data contained in pages and pages of such tables in a way that informs decisions about potential security upgrades.

Security Upgrades

INTRODUCING EXPOSURE

The purpose of calculating conditional risks for many scenarios is ultimately to help the decision maker understand how to reduce such risks most effectively by investing in security improvements. The goal is to compare risks among different target assets at a dam, for different types of attack vectors, or for different dams within a portfolio of dams. To make any of these comparisons, it is necessary to aggregate risks across scenarios. Conditional risk cannot be aggregated across scenarios because by definition, each conditional risk estimate is based on the adversary choosing that scenario. A different decision metric (other than conditional risk alone) is needed.

A decision metric is needed that retains the essential information contained in conditional risk for each scenario considered, but also quantitatively summarizes conditional risk across multiple scenarios in order to inform decision makers as they contemplate potential security upgrades. Because the metric is intended to inform decision makers, the metric should ideally be intuitive. This paper introduces such a metric: *exposure*.

Defining Exposure

Exposure is defined as human lives and economic value at risk due to an attack. Exposure can be determined for any set of scenarios (including a single scenario) of interest to the analyst. Even when exposure is used as a summary metric for multiple scenarios, it is defined as human lives or economic value at risk due to *a* potential attack. Intuitively, the number of human lives at risk from a single attack at a specified dam cannot exceed the maximum number of lives lost considering all scenarios at that dam. Consequently, exposure, measured in terms of lives lost,[9] for any set of scenarios, will never exceed the maximum possible lives lost from any individual scenario in the set. For example, consider Table 3 which presents the consequences in terms of loss of life for selected scenarios at a notional dam. The largest loss of life from any single scenario is one hundred (for the scenario involving the large truck vehicle-borne improvised explosive device [VBIED] against the impoundment). Exposure for any subset of these scenarios (including exposure for *all* the scenarios in Table 3) will not exceed one hundred. The analyst can usefully compare exposure calculations by attack type or by asset or for any useful set of scenarios. The results of these comparisons guide a decision maker who is determining the most effective set of proposed security upgrades.

Calculating Exposure

The considerations discussed in the previous section suggest calculating exposure for a set of scenarios as a weighted sum of the conditional risk estimates for each scenario in

[9] Exposure is defined and calculated both in terms of human lives and economic value. The calculations for human lives and economic value are always done separately, just as conditional risk calculations are made separately. In the examples throughout the paper, either human lives or economic value will be chosen to illustrate a particular point.

the set.[10] There are numerous weighting schemes[11] that an analyst could choose. Considering the previously stated goals of a metric that (1) preserves the essential information contained in conditional risk estimates and (2) is intuitive to decision makers, the authors propose the following method for calculating exposure. Exposure—denoted R_C'—for any set of scenarios, is the sum of scaled conditional risk for each scenario so that (1) the scaling factor for each scenario is the same and (2) the exposure for the set of all scenarios at an undefended dam (i.e., when P(S|A) equals 1.0 for each scenario) is equal to the consequences of the individual scenario that equals or exceeds the consequences of all other scenarios. (This value of consequences is referred to as C_{MAX}; see example below.)

A dam (considered as an aggregation of all the critical assets at the dam) cannot be more exposed, that is, suffer more than the worst consequences a given adversary type can impose in a single attack. The exposure for each scenario is its conditional risk multiplied by a scaling constant, denoted by "k." Scaling conditional risk allows for a realistic and valid comparison of conditional risks, both at a dam and among a portfolio of dams, without sacrificing the valuable information embedded in the conditional risk estimate. Exposure allows one to compare different sets of attack vectors or different assets at a dam. It also permits estimates of the relative values of different sets of security upgrades by observing the amount of reduction in exposure following the application of each set of security upgrades. Table 3 illustrates how exposure would be calculated for a set of scenarios at a notional dam. In Table 3,

- C indicates consequences, in loss of life for an individual scenario.
- C_{TOTAL} is the sum of consequences for all scenarios considered in the table.
- C_{MAX} is the largest C for an individual scenario under consideration in the table
- $k = C_{MAX}/C_{TOTAL}$.
- Exposure is denoted R_C', and is calculated as $R_C' = k \times R_C$.
- $R_C'_{As\text{-}Is}$ indicates the exposure for a scenario, given the current security measures present at the dam project.

[10] The notion of aggregating/summarizing a collection of conditional risks, with each conditional risk tied to a distinctly unique conditioning event, may be considered analytically unsatisfying. Such a decision metric has no physical interpretation; nor does it have a stochastic interpretation, such as an expected value of loss. Nevertheless, it is a useful decision metric to decision makers because they must consider the magnitude of overall risk exposure and not isolated events.
[11] See discussion in this paper on alternative metrics.

Security Upgrades

Table 3. Exposure (Loss of Life) for Selected Scenarios from a Notional Dam

Attack Type	Attack Vector	Target Asset	C	P(S|A)	P(S|A) x C = R_C	k x R_C = $R_C{}'_{As\text{-}Is}$
$C_{TOTAL} = \Sigma\, C = 151$; $C_{MAX} = 100$; $k = C_{MAX}/C_{TOTAL} = 100/151 = 0.66$						
VBIED	Impoundment					
	Sedan		1	1.0	1.0	0.66
	Van		1	1.0	1.0	0.66
	Box Truck		1	1.0	1.0	0.66
	Large Truck		100	1.0	100	66
Assault Team	Control Center					
	Single Attacker		11	.58	6.4	4.2
	Small Cell		11	.81	8.9	5.9
	Medium Cell		11	1.0	11.0	7.3
	Large Cell		11	1.0	11.0	7.3
	Turbines and Generators					
	Single Attacker		1	.58	0.58	0.38
	Small Cell		1	.81	0.81	0.53
	Medium Cell		1	1.0	1.0	0.66
	Large Cell		1	1.0	1.0	0.66

USING EXPOSURE TO DEVELOP RISK MITIGATION OPTIONS

A Risk Mitigation Option (RMO) is a package of security upgrades intended to reduce exposure at critical assets and components of a dam project. The upgrades consist of improvements or additions to the defensive layers at the project. Options are constrained in practice by command guidance such as funding limitations or operational considerations.

The decision metric that will be used to evaluate potential RMOs is the net reduction in the exposure for the set of all scenarios considered at the project. Exposure for a dam in its current security configuration is denoted $R_C'{}_{As\text{-}Is}$. It is calculated using conditional risk estimates for each scenario based on the P(S|A) values for the current defensive configuration at each layer protecting the dam. $R_C'{}_{RMO}$ denotes exposure for the dam after implementing the proposed RMO. It is calculated using conditional risk estimates for each scenario based on revised P(S|A) values resulting from the proposed defensive configurations.

Exposure was introduced to help the analyst use conditional risk estimates to develop RMOs that effectively reduce risk. Exposure, calculated by summing scaled conditional risk values over different sets of scenarios, is particularly helpful in answering two questions from the dam owner/operator perspective:

1) Which targets are currently the most exposed?
2) Which attack vectors currently cause the most exposure?

The first question is addressed by calculating exposure for each asset at the dam. That is, calculate the sum of scaled conditional risk estimates over each scenario involving a particular asset. Similarly, the second question is addressed by calculating exposure for each attack type. That is, calculate the sum of scaled conditional risk estimates over each scenario involving a particular attack type. In the next section, graphics are introduced that help the analyst answer these two questions.

Graphical Displays to Help Develop RMOs

Each scenario contributes to the overall exposure at a dam project. In order for an analyst to recommend security upgrades that efficiently reduce that level of exposure, it would be helpful to know which scenarios contribute the most to the overall level of exposure. As previously mentioned, this can be analyzed by asset or by attack type. Pie charts are an intuitive way to display this information.

Figures 2 through 4 show how exposure is distributed over assets and attack vectors for current defenses at a notional dam. Charts such as these inform the analyst during the process of building RMOs. $R_C'{}_{AS\text{-}IS}$ is calculated for the set of all scenarios; each wedge in the pie charts represents the percentage of that total that is contributed by the set of scenarios that only involve the asset (or attack type) indicated.

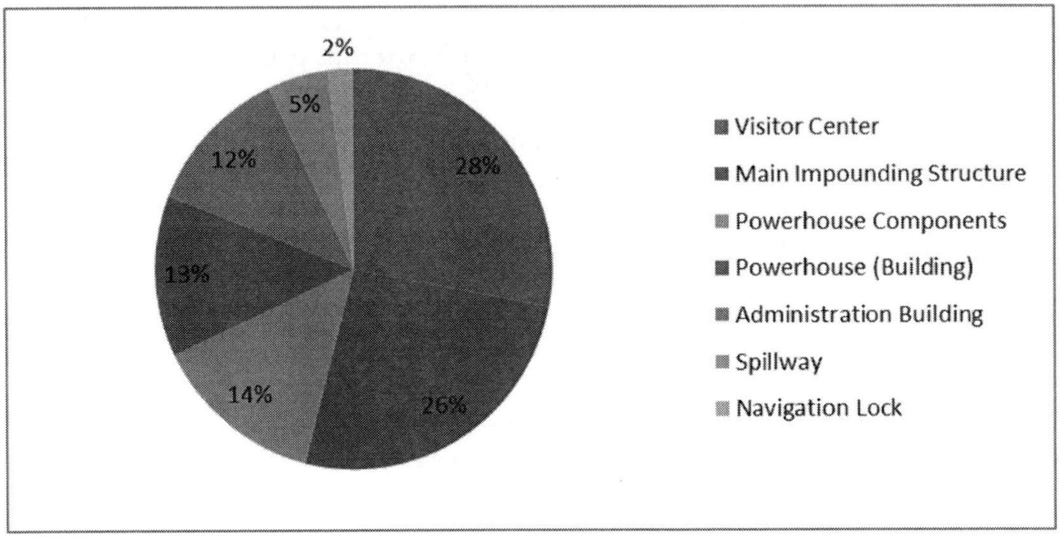

Figure 2. Distribution of Loss-of-Life Exposure Over Assets for "As-Is" Defenses

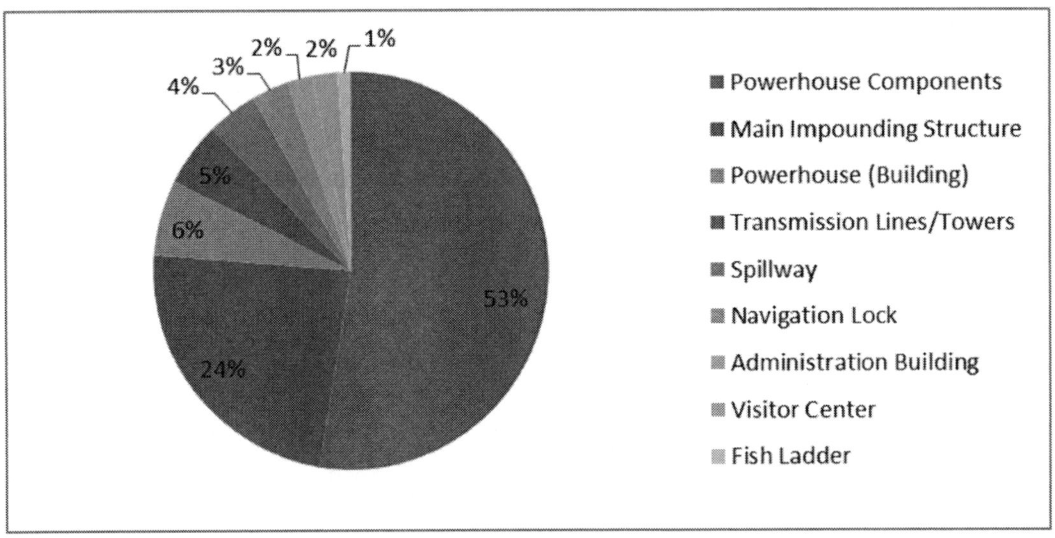

Figure 3. Distribution of Economic Exposure Over Assets for "As-Is" Defenses

Figures 2 and 3 both display relevant information to help answer the question, "Which targets are currently the most exposed?" Notice, for example, in Figure 3, the analyst and/or decision maker can see immediately that roughly three quarters of all *economic* exposure is contributed by scenarios involving only two assets—the components in the powerhouse and the main impounding structure. Figure 2, on the other hand, shows that the largest exposure in terms of loss of life is at the visitor center. These types of insights are invaluable to an analyst as he or she considers where to propose security upgrades at the dam project.

The same type of analysis can be conducted by attack type. Figure 4 helps answer the question, "Which attack vectors currently cause the most exposure?" It shows that roughly three quarters of total exposure at the project is attributed to scenarios involving VBIEDs and Assault Teams. An analyst can also readily surmise that protecting against land-based attacks is more likely to lower overall exposure than protecting against water-borne attacks.

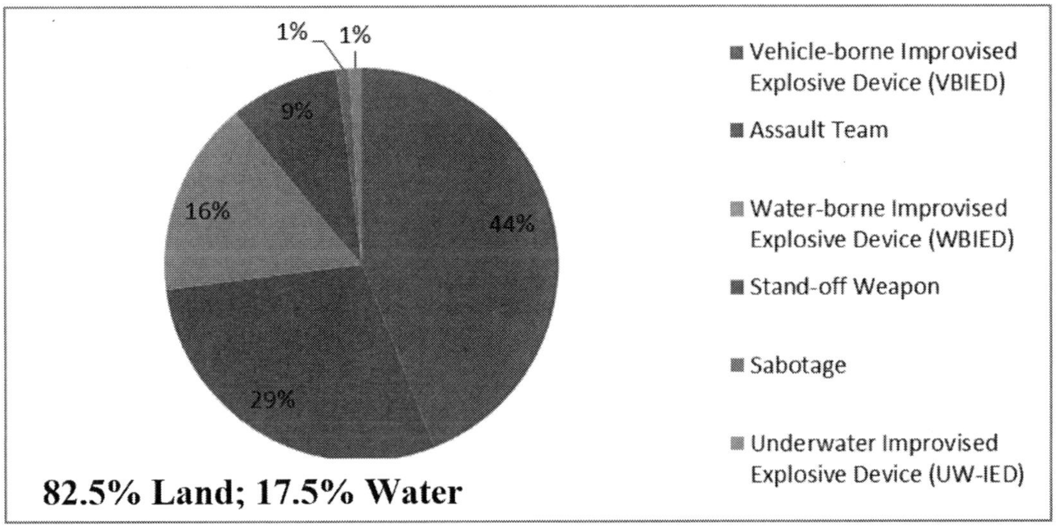

Figure 4. Distribution of Loss-of-Life Exposure Over Attack Vector Types for "As-Is" Defenses

Graphical Displays that Help Evaluate RMOs

The pie charts in Figures 2, 3, and 4 aid the analyst in creating RMOs designed to better protect certain assets or to better protect against certain types of attacks where the dam is most exposed. The bar chart in Figure 5 quantifies the reduction in exposure that would be achieved if the proposed RMO is implemented. Figure 5 presents three calculations for total exposure at the project based on three different security configurations at the dam. The red bar, denoted $R_C'_{Undefended}$, represents the level of exposure if there were no effective defenses at the dam project, that is, the P(S|A) values for every scenario is 1.0. Mathematically, this means that R_C equals consequences for every scenario. By construction, R_C', in this case, is equal to the maximum consequence level over all scenarios. Including this calculation in the graphic allows the decision maker to readily observe the current state of defenses (represented by the height of the blue bar) vis-à-vis an undefended dam. The green bar represents the level of exposure if the proposed RMO was implemented. The percent reduction expressed at the top of the chart is a comparison of the exposure levels for the RMO and the current state of defenses.

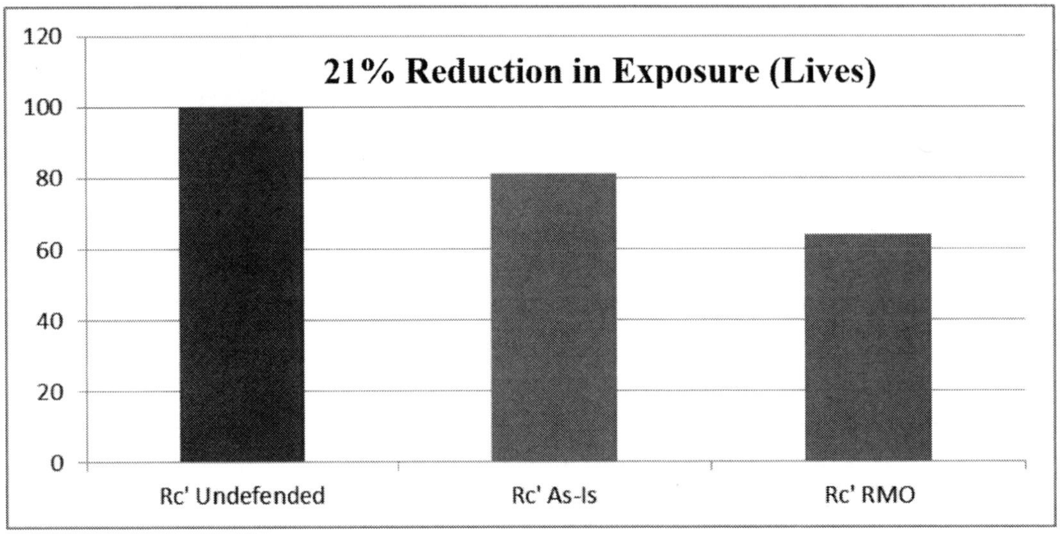

Figure 5. Reduction in Loss-of-Life Exposure for RMO

In this particular notional example, the decision maker can readily see that the proposed RMO would reduce exposure in terms of loss of human life by 9 percent.

Implementing Exposure

IDA has created a training tool known as the Common Risk Model for Dams Support System (CRM-DSS). Among other things, this tool automates the calculation of exposure and the creation of all the graphics introduced in this paper. CRM-DSS has successfully been used to train thirty-five USACE analysts to conduct a Security Risk Assessment (SRA) based on exposure calculations. Analysts have used the tool to successfully complete SRAs at multiple dams.

OTHER POTENTIAL METRICS

Exposure is a weighted sum of conditional risk estimates for a particular set of scenarios. It is not the only possible solution for the problem of summarizing conditional risk estimates across scenarios. Other weighted sums were considered; each has its own strengths and weaknesses. For developing individual SRAs at individual USACE dams, exposure was deemed to have the most desirable properties of various metrics that were considered. The following subsections present four alternative metrics along with a brief discussion of their merits and demerits.

Summing Conditional Risk Estimates

Perhaps the most intuitive way to combine conditional risk estimates across a set of scenarios is to simply add the conditional risk estimates without weighting them. This solution has two significant disadvantages. Using this metric, the measure of risk attributed to a set of scenarios will always increase as scenarios are added to the set. Thus it is theoretically possible for one dam, say Dam A, to have a higher measure of risk than say Dam B simply because Dam A is vulnerable to more scenarios than Dam B, regardless of the potential consequences from any one scenario. The second disadvantage is related to the first. Using this metric, a dam could have a measure of risk that is larger than the possible consequences from any single attack.

Maximum Possible Consequences

Another possible weighting scheme is to give the scenario with the largest consequences a weight of 1 and give all other scenarios a weight of 0. This appropriately places an emphasis on the largest consequence scenario but totally ignores all other scenarios. Imagine an RMO that recommends adding a water layer on the downstream side of the dam. This might significantly lower risk to a number of assets that can be attacked from the downstream side. However, if the largest consequence scenario can only be attacked from upstream, this RMO will not reduce the measure of risk at all if this metric is employed.

Threat Estimate: P(A)

The CRM-D contains a module (Kirpichevsky 2012) that estimates the threat parameter, P(A) in equation (3). This parameter could be considered a weight on conditional risk and has some of the desirable properties of exposure. In fact, it has the added benefit of capturing the threat component of risk. However, the P(A) module was developed for use at the portfolio level.

Average Conditional Risk

Another intuitive weighting scheme is to use the arithmetic average of the conditional risk estimates for all the scenarios in the set of interest, thus equally weighting each scenario. Initially, this metric seems to be more intuitive than exposure and might be more easily explained to decision makers. It is interesting to note that exposure is also a weighting scheme in which all the scenarios are equally weighted. The difference is that in exposure the weights do not necessarily sum to 1. The advantage of the weighting scheme used to calculate exposure is that it has a more intuitive interpretation, particularly in the case of an undefended dam. Consider for example the following notional situation depicted in Table 4. The level of exposure at the entire dam is 100, but the average conditional risk is 12.5. In this example, the average conditional risk is not a very descriptive measure of how exposed the dam is to attack, that is, how much is at risk. If more scenarios that have little or no conditional risk are added, the average conditional risk becomes a summary metric that is even less intuitive.

Table 4. Exposure (Economic) for Select Scenarios from a Notional Dam

Attack Type	Attack Vector	Target Asset	C	P(S|A)	P(S|A) x C = R_C	k x R_C = $R_C'{}_{As\text{-}Is}$
$C_{TOTAL} = \Sigma C = 200$; $C_{MAX} = 100$; $k = C_{MAX}/C_{TOTAL} = 100/200 = 0.5$						
VBIED	Impoundment					
	Sedan		20	1.0	20	10
	Van		30	1.0	30	15
	Box Truck		50	1.0	50	25
	Large Truck		100	1.0	100	50
Assault Team	Impoundment					
	Single Attacker		0	1.0	0	0
	Small Cell		0	1.0	0	0
	Medium Cell		0	1.0	0	0
	Large Cell		0	1.0	0	0

CONCLUSION

Exposure is an intuitive concept that provides risk analysts with a method for synthesizing conditional risk calculations across multiple scenarios that is easy to calculate. The synthesis of condition risk calculations at an individual dam provided by this new metric can be presented to decision makers via graphical displays that are informative and easy to comprehend. It has been successfully tested in training courses and in actual SRAs conducted at USACE dams.

REFERENCES

Dechant, Jason A., John L. Conley, Anthony M. Fainberg, David R. Goodman, Anthony C. Hermes, Michael J. Keleher, Kevin F. McCrohan, and J. Darrell Morgeson. 2012. *The Common Risk Model for Dams: Methodology and Application*. IDA Paper P-4761. Alexandria, VA: IDA.

Keleher, Michael J., Steven Walser, and Samuel Himel. To be published. *The Common Risk Model for Dams Support System: A Prototype Analyst Tool.* IDA Paper P-5220. Alexandria, VA: IDA.

Kirpichevsky, Yev, Jason A. Dechant, Victor A. Utgoff, J. Darrell Morgeson, Colin M. Doyle, and Anton V. Strezhnev. 2012. *Estimating Threat from Adaptive Adversaries: Probabilistic Decision Modeling in the Common Risk Model for Dams.* IDA Paper P-4857. Alexandria, VA: IDA.

Morgeson, J. Darrell, Yazmin Seda-Sanabria, E. E. Matheu, and Michael J. Keleher. 2013. "Incorporating Uncertainties in Estimation of Vulnerabilities for Security Risk Assessments." Proceedings 33rd USSD Annual Meeting and Conference, Phoenix, Arizona, February 11–15, 2013.

National Research Council. 2010. *Review of the Department of Homeland Security's Approach to Risk Analysis.* Washington, DC: The National Academies Press.

Seda-Sanabria, Y., M. A. Fainberg, and E. E. Matheu. 2011a. "A Consistent Approach for Vulnerability Assessment of Dams." Proceedings 31st USSD Annual Meeting and Conference, San Diego, California, April 11–15, 2011.

Seda-Sanabria, Y., M. A. Fainberg, E. E. Matheu, J. D. Tressler, and M. L. Bowen. 2011b. "Implementation of the Common Risk Model for Dams for Security Assessments of USACE Critical Infrastructure." Proceedings Dam Safety 2011 Conference, Washington, DC, September 25–29, 2011.

U.S. Department of Homeland Security. 2013. *National Infrastructure Protection Plan: Partnering for Critical Infrastructure Security and Resilience.* Washington, DC: DHS. Available at http://www.dhs.gov/sites/default/files/publications/NIPP%202013_Partnering%20for%20Critical%20Infrastructure%20Security%20and%20Resilience_508_0.pdf.

U.S. Government Accountability Office. 2005. "Risk Management." GAO-06-91. Washington, DC: GAO. Available at www.gao.gov/cgi-bin/getrpt?GAO-06-91.

BLAST IMPACTS ASSESSMENT FOR EMBANKMENT DAMS: PARAMETRIC STUDY RESULTS AND DESIGN IMPLICATIONS

James Parkes, P.E.[1]
Rodney Sedillo, P.E.[2]

ABSTRACT

Dam security and risk assessments are a growing concern in the field of dam safety. Many dams provide easily accessible, high visibility targets for those intending to commit malicious or terroristic acts. As such, the vulnerability of dams to the impact of explosive blasts should be considered in dam security and risk assessments. A recent study has been performed to evaluate the blast impacts to embankment dams, including the potential for local crater or crack formation that could lead to a devastating erosive breach. This detailed study included the use of non-linear explicit dynamics modeling of the blast impact on a trial dam computational model. Details of the development of the simplified trial dam section and the procedure for performing the blast impact assessment, including evaluating results in terms of potential crater and crack formation, are presented. A critical aspect of the analysis is the soil material model used to evaluate the dynamic impact of the blast in the numerical modeling. A parametric study of several soil models was performed to assess the overall effect on the depth of potential crater and crack formation. Soil material models were developed based on a standard software library as well as published research on landmine impacts for differing soil types. The results of the localized blast impact for different soils, as well as concrete ground cover, are discussed along with implications for design of potential mitigation measures.

INTRODUCTION

Infrastructure security and protection is a growing concern, especially for high risk and high hazard infrastructure including dams. There are over 87,000 dams in the United States, with over 26,000 classified as high or significant hazard (USACE, 2013). The consequences of a dam failure may include not only the direct loss of a roadway and the loss of a reservoir for water supply or power generation, but also the potentially devastating effect of the sudden uncontrolled release of the reservoir in terms of loss of life and severe property and infrastructure damage. Many of these dams include public roadways on the crest, making them easily accessible as well as high visibility targets for potential malicious acts or terrorist attacks.

A proper assessment of the impact of explosions on dams requires sophisticated numerical analyses that account for the size, shape, and characteristics of the structure, and the location and intensity of the explosive load. Much of the available work for analyzing blast impacts on structures has focused on structures made of manufactured

[1]Parsons Brinckerhoff, Inc. 100 S. Charles Street, Baltimore, MD 21201, 410-843-3306, parkes@pbworld.com.
[2]Parsons Brinckerhoff, Inc., 2202 N. West Shore Blvd, Tampa, FL 33607, 813-520-4389, Sedillo@pbworld.com.

materials such as steel, concrete, masonry, etc. Geomaterials, such as soil, have not been studied as extensively, and direct assessments of blast impacts on structures made of geomaterials have not been widespread.

A recent study has been undertaken to analyze the impacts of explosive blasts on an embankment dam using sophisticated numerical modeling software. A parametric study of different soil material models and explosive loads has been performed as part of this work. The results of the parametric study have implications with regard to future analyses and design of mitigation measures.

DEVELOPMENT OF A TRIAL DAM FOR ANALYSIS

A trial dam section was developed based on published industry guidelines. The trial section is intended to be representative of a common dam type and configuration. A public roadway is assumed to be present on the crest of the dam for the assumed threat scenario of an explosive load delivered by truck.

According to the National Inventory of Dams (NID) (USACE, 2013), there are over 74,000 earth dams in the United States, making up over 85 percent of the total number of dams in the country. Therefore, an earth dam was used for the trial dam section.

Development of Trial Dam Geometry

Dam Height. Over 90 percent of dams in the U.S. are 50 feet or less in height (USACE, 2013), so 50 feet was used as the dam height for the trial dam.

Dam Type. A homogeneous dam section was assumed for both simplicity in the modeling and because homogeneous dams are the most common type of embankment dam.

Slope Configurations. Published guidelines were used to develop the slope configurations for the simplified trial dam section. Guidelines presented in *Design of Small Dams* by the United States Bureau of Reclamation (USBR) (1973) were used.

For consideration in analysis, a reasonable yet minimal cross section is desired. A minimal cross section is considered to be more vulnerable to the impacts of an explosion because a larger section with additional mass will provide additional inertia and resistance to the blast impact. A minimal cross section is based on the steepest possible slope configuration, which is applicable for silty or clayey sands and gravels. For this analysis, the trial dam is assumed to be a homogeneous earth dam composed of silty or clayey sand, classified as SM or SC in accordance with the Unified Soils Classification System (USCS). Based on published criteria for SC or SM soils, the upstream and downstream slope configurations for the trial dam are 2.5H:1V and 2H:1V, respectively.

Crest Width and Freeboard. USBR (1973) provides a formula for relating the crest width to the dam height, which results in a minimum width of 20 feet for a 50-foot high dam.

USBR also notes that 20 to 25 feet is a likely minimum width if a public roadway is included on the crest. Assuming minimum travel-lanes widths of 10 to 12 feet with no shoulders and minimal area for guardrails, a crest width of 25 feet is used for the model.

Freeboard can vary considerably depending on the dam purpose and the loading conditions. USBR (1973) indicates minimum freeboard of three to seven feet above the maximum water level should be used. For the trial dam, freeboard of five feet is used.

Based on the trial dam configuration and conceptual shear strength parameters for silty or clayey sand, a steady state seepage analysis was performed. The results indicate that the trial dam has a minimum factor of safety (FS) of global stability of 1.5 and is therefore reasonable and consistent with industry guidelines.

SIZE AND LOCATION OF EXPLOSIVE LOAD

An appropriate explosive size and location were developed for the model. The explosive is assumed to be delivered using a vehicle, such as a car or truck, on the public crest roadway. The explosive is modeled as a mass of TNT. The delivery platform is not included in the model.

Size of Explosive

An explosive load consistent with a what a light truck or van can carry is used. Examples of similar events include the 1993 attack on the World Trade Center in New York City and the 1995 attack on the Oklahoma City Federal Building. An equivalent explosive size of 5,000 pounds of TNT is used as the upper end of the range for an explosive delivered by a van or light truck (Parsons Brinckerhoff, 2010). This size of the explosive is consistent with analyses that have been performed for other transportation-related infrastructure. The impact of a smaller load of 2,500 pounds of TNT is assessed as part of the parametric study. By comparison, the Oklahoma City explosive was roughly equivalent to 4,000 lbs of TNT.

Location of Explosive

The explosive location is evaluated to model cases with the vehicle in the upstream lane or in the downstream lane. Since the crest on the trial dam is 25 feet wide, the explosive will be centered approximately 6 to 6.5 feet from either the top of the upstream slope or the top of the downstream slope.

The explosive is centered above the ground surface because the vehicle platform will have ground clearance and the explosive itself will have mass. The height of the explosive above the ground surface is referred to as the "standoff distance" because it is the minimum distance from the center of the explosive to the structure being assessed. A standoff distance of 4 feet is used. The same location and standoff distance is used for the smaller 2,500-pound load used in the parametric study. The location of the explosive load and the trial dam section are presented in Figure 1.

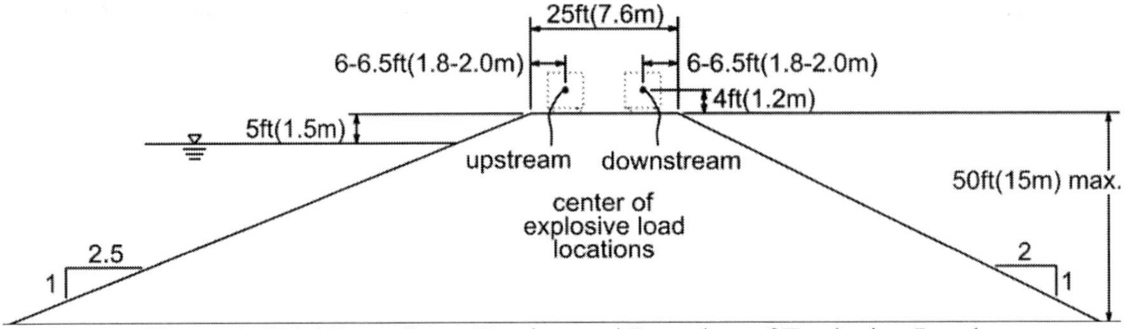

Figure 1. Trial Dam Cross Section and Location of Explosive Load

BLAST IMPACT ANALYSIS

Failure Modes

Several conceivable failure modes exist for a blast impact on the crest of a dam, including:

1. Dynamic global stability failure during the blast due to the blast shockwave.
2. Global stability failure post-blast due to impacts/alterations of the dam geometry.
3. Progressive failure of the dam post-blast due to formation of an erosive breach through a crater or cracked section.

The focus of this work is on the third failure mode, in which the blast impact creates a crater or cracked section deep enough to extend below the reservoir level. Soils used in dam construction often have some degree of cohesion, either due to fines content or apparent cohesion due to capillary action, and can therefore form tension cracks that may extend below the base of a crater. The crater or cracked section may form a preferential, uncontrolled seepage path through the embankment which can lead to an erosive breach. The time for such seepage and erosion to form may be relatively quick if the crater is deep enough to extend below the reservoir level, or relatively slow if only cracks extend below the reservoir or phreatic surface.

To properly analyze this mode of failure, it is necessary to use software that can model blast impacts on structures, including appropriate material models.

Software for Blast Impacts Assessment

Modeling the impacts of explosions requires specialized software. Conventional geotechnical analyses are based on relatively static loads of low to moderate intensity that occur over time periods ranging from days to years. Applied pressures generally within the range of conventional soil-strength parameters and deformations are relatively small.

Explosion-impact analyses, by contrast, involve high-intensity that occurs over fractions of a second. Deformations and strains can be large and load intensity (pressure) may be several times the material strengths. Transient analyses and non-linear material behavior are significant aspects of these analyses, and therefore specialized software is needed.

Analysis of explosions is typically performed using explicit-dynamics software which enables modeling of very rapid, high-intensity loads. Such analyses require complex numerical modeling software that use different solvers for fluids and solids in a coupled analysis. This is known as Euler-Lagrange coupling. The Euler processor is used to analyze large distortions of fluids and gases, including explosives and detonation products. In a Euler analysis, the fluids pass through the individual elements representing an area or space. The Lagrange processor is for modeling solid structures. In a Lagrange analysis, the elements representing the solid body deform under the loading conditions. The Euler and Lagrange analyses are depicted graphically in Figure 2. Coupled together, the impact of an explosive blast on a solid structure can be effectively modeled.

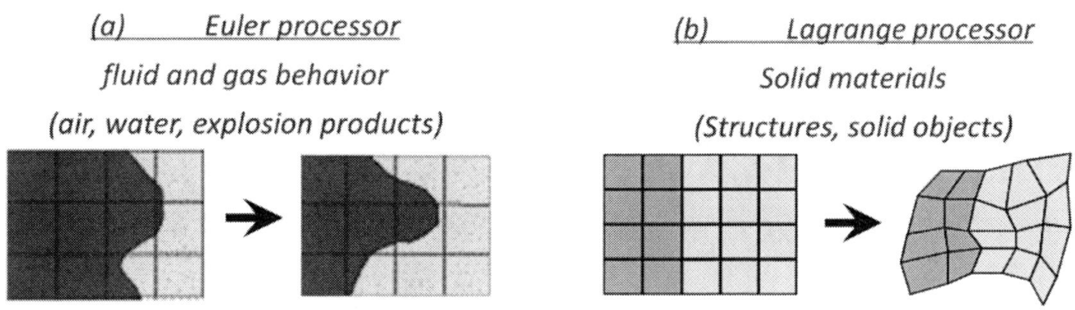

Figure 2. Euler and Lagrange solvers (Parsons Brinckerhoff, 2010)

A commonly used Euler-Lagrange software package is AUTODYN by ANSYS, Inc. AUTODYN Release 14 has been used for this work because it is widely used in practice for performing blast-impact analyses.

Soil Material Models for Blast Impacts Assessment

Most software for explicit-dynamics analyses include material libraries for fluids and structural materials. Most materials have a complex response to dynamic, high-intensity loading and these material libraries have been developed over years of practice based on evolving research. However, the majority of the material models are for manufactured materials, such as steel, concrete, masonry, etc. In most blast-impact analyses, soils are not involved or are only included as secondary components, such as confining media. This type of loading is not typical for geotechnical applications and there are relatively few structures built of soils that have required such assessments.

The default library in AUTODYN includes one material for soil, *Sand-Compaction.* This material is a compaction model that simulates deformation of dry sand in response to an explosive load. Although the type of model may be applicable (compaction model that simulates deformations), the model for a dry sand is not considered appropriate.
The failure mode under consideration is the formation of a crater or cracked section within an embankment dam. Soils for dams are generally placed wet of optimum moisture content to improve ductility and minimize post-construction settlements.

Seepage through the dam and capillary rise above the phreatic surface will lead to saturation of the dam soils if they are not already saturated as a result of the construction process. Therefore, soils models for analysis should consider a high degree of saturation.

Behavior of Soils under Explosive Loads. Research findings published by Clemson University and the Army Research Laboratory (CU-ARL) on the impact of the degree of saturation on sands and clayey sands with regard to landmine applications has provided a detailed understand of soil behavior under explosive loads (Grujicic et al., 2005, January 2007, March 2007, April 2007).

Sands consist of a soil skeleton of solid particles with void space between, as indicated in Figure 3a. The void space may be filled almost entirely with air (dry), entirely with water (fully saturated), or with some voids filled with air and some with water (partially saturated). Volume of voids, air, water and the degree of saturation are visualized using phase diagrams as shown in Figure 3b and c.

(a) **(b)** **(c)**

Figure 3: (a) Soil skeleton (b) Phase diagram of partially saturated soil, and (c) Phase diagram of fully saturated soil

Deformation of a mass of sand under an explosive blast load is believed to be a function of two mechanisms (Grujicic et al., January 2007):

1. Elastic deformations and/or fracture of the interparticle bonds /forces between the sand grains: Elastic deformations occur at low pressure levels, while fracture of the bonds occurs at higher pressure levels. The interparticle bonds/forces are a function of surface roughness and binders within the soil.
2. Elastic and plastic deformation of the constituent materials in the sand mass: The sand, air and water within the sand mass may compress in response to load.

For slow loading conditions (conventional geotechnical loading), the air or water within the pore spaces is able to drain and the deformation of the sand mass is dependent on the response of the solid particles. However, for very rapid loading conditions (explosive blasts), pore fluids do not have the opportunity to drain; therefore, the response of the soil mass is a combination of the response of the soil particles and the pore fluid.

For low levels of saturation, the soil mass contains a relatively high volume of air (Figure 3b). Interparticle bonds (friction and/or binder strength) are relatively high and the mass contains a high volume of compressible fluid (air). The compression of the air within the void space provides an energy-dissipating effect on the overall blast impact for dry or partially saturated soil.

For completely saturated sand, interparticle forces are lower because water reduces friction and buoyancy reduces normal forces. Also, the water in the void space is an incompressible fluid. Therefore, the two mechanisms of deformation for the sand mass are both adversely affected for the saturated case. The compressibility effect is the more significant. The saturated sand lacks the energy-dissipating effect of the air within the pore spaces and therefore behaves as an incompressible material (Grujicic et al., 2005). As a result, blast impacts will be more significant for saturated soils than for dry soils.

Research indicates that clay binder may coat the sand particles and reduce interparticle bonds/forces. However, the addition of up to 15% clay binder has a less-significant effect than the degree of saturation (Grujicic et al. April 2007). Other research indicates that the effect of particle size, particle distribution and presence of silt and/or clay on the blast response of sand is small compared to the degree of saturation (Grujicic et al., 2005).

There are several important implications of the landmine research with regard to blast impacts for dams. First, the degree of saturation is the most significant aspect of the soil model. As indicated previously, soil for a homogeneous dam is expected to be at or near saturation. Therefore, a soil model for a saturated soil is necessary for performing the blast-impact analyses. Second, the effect of clay or other binders is not significant compared to the degree of saturation. Therefore, it is less important whether the dam material is an SC, SM, SP, etc. and more important whether the soil is dry or saturated.

Development of an Appropriate Soil Material Models. The default *Sand-Compaction* model in the AUTODYN material library does not account for soil saturation. Therefore, saturated soil models have to be developed for this analysis.

As described above, CU-ARL performed research on saturated sand and clayey sand for landmine impacts. That research focused on modification of the AUTODYN compaction model. The CU-ARL soil-material models were developed for similar soils (saturated sand and clayey sand) using the same software (AUTODYN) for use in blast-impact analyses and have been validated using field testing (Grujicic et al. 2005, January 2007, March 2007, April 2007). Therefore, the CU-ARL models are considered appropriate for use in this analysis.

The models from the CU-ARL research were recreated by approximating the input functions from the published work. Minor changes were made to the tensile function in the soil models. The recreated models are designated *SC-Sat* and *SP-Sat* for saturated clayey sand (SC) and saturated clean sand (SP), respectively. Validation models from the research were recreated to verify that the materials had been successfully approximated.

The saturated clayey sand model (*SC-Sat*) was used for the base model. The parametric study includes use of the saturated clean sand model (*SP-Sat*) and the use of the default *Sand-Compaction* AUTODYN library model for comparison.

Trial Dam Model

Three dimensional numerical models were developed using the trial dam and explosive load shown in Figure 1 and the *SC-Sat* soil model. The analysis was performed using a full-size, 50-foot high model for primary assessment of blast impacts and a smaller, limited height model of the upper 20 feet for investigating localized impacts. The length of the model along the dam centerline was 40 feet, symmetric about the explosion. This length was necessary so that the blast impacts did not cause distortion at the model boundaries. The dam model included 183,000 elements, with smaller elements near the area of the blast and then becoming progressively larger with distance from the blast.

Euler space for the air, reservoir water and explosive was created around the dam. The element mesh size for the Euler space was comparable to the dam.

Boundary conditions around the edges of the model and base of the dam or dam section were set to allow transmission of the pressure waves out of the model and prevent reflection of the wave back into the embankment. The model as set-up is shown in Figure 4 and after analysis in Figure 5.

Figure 4. Three Dimensional Trial Dam Model, Upper 20 feet, shown without air space

Figure 5. Three Dimensional Trial Dam Model Following Analysis

ASSESSMENT OF ANALYSIS RESULTS

As indicated previously, the focus of this work is on the localized impact of the explosion and formation of a crater and local damage.

Evaluation of Localized Impacts

The models show a cratering effect which is consistent with expectations. However, it is possible that blast impacts extend below the depth of the crater itself. Reviewing the model results in terms of effective strains indicates that the impacts of the explosion extend well below the immediate cratering depth, as shown in Figure 6.

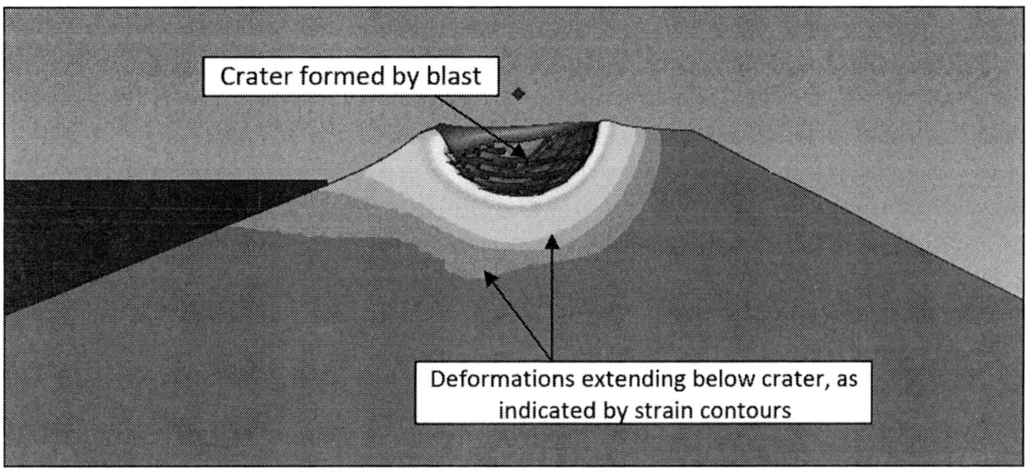

Figure 6. Model results showing effective strain contours

Most embankment dam soils include some degree of cohesion and therefore may be capable of developing cracks. Crack formation in dams is a widely recognized concern and may result from tension, differential movements, or shrinkage (USACE, 2004). Cracked soil sections are often included in slope-stability assessments (USACE 2003). In a similar fashion, it is considered possible for cracks to develop as a result of shearing and deformations caused by the blast impact. These effects may extend below the crater, and may result in internal damage that may adversely affect the overall integrity of the dam. Uncontrolled seepage into a crater or cracked section may lead to erosion or piping which could eventually result in an uncontrolled breach. It is important to be able to assess this damage potential because it may not be readily observable or apparent, and repairs for a blast impact may need to extend beyond filling of the crater.

Criteria for assessing the depth to which cracking may occur must be developed in order to assess the localized blast impact. In conventional slope stability, potential crack depth is based on soil cohesion. However, blast-loading conditions are of such high intensity and short duration that conventional shear strengths are not directly considered in the material models. Another criterion that is often considered in dam engineering is limiting strain. Limiting strain is sometimes used as criteria for interpretation of shear-strength tests, and strain compatibility is a consideration for slope stability assessments (USACE, 2003).

The use of a limiting strain criterion for developing the depth of possible cracking is proposed. In conventional geotechnical engineering practice, limiting strains for interpretation of shear-strength and strain-compatibility assessments are often in the range of 5% to 15%. For this application, strains in the range of 10% to 30% are considered. A strain of 10% is consistent with the range of conventional assessments. Blast-model results indicate that above 30% the strain contours are relatively tightly spaced, indicating high deformations that are likely indicators of soil-structure failure, excessive displacement or cratering. The range of 10% to 30% is considered a range over which the soil remains primarily intact, but may still be subject to strains large enough to produce cracks.

Development of Simplified Chart

Euler-Lagrange nonlinear dynamic analyses involve complex and time-intensive analytical skills. In order to simplify the assessments, a simplified chart is developed based on the results of the trial dam models.

Analyses were run with the blast centered on both the upstream and downstream lanes to facilitate the assessment of a blast anywhere on the crest roadway. The localized impact of the blast results in an irregular pattern of deformation due to the dam slopes and the different confining conditions that exist. Contours of strain are developed from the blast model results. As shown in Figures 6 and 7, the maximum depth of strain is offset towards the nearest slope because of the lack of external confining pressure there. Strains extend laterally and upwards from the maximum depth to the nearest slope. This pattern is evident for blasts on both the upstream and downstream sides of the crest, as shown in Figure 7. Depth of maximum strains below the crest and below the slope are comparable for blasts on the upstream and downstream sides of the crest.

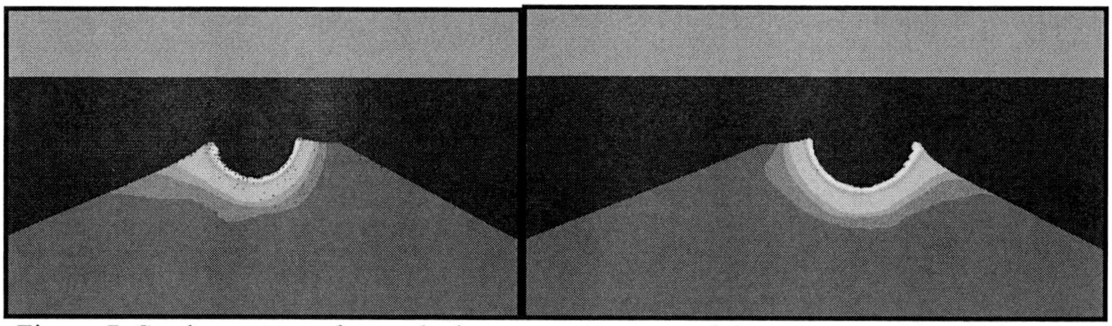

Figure 7. Strain contours for explosions on upstream and downstream sides of the crest.

The results of both the upstream and downstream blast analyses were superimposed on each other to develop a composite pattern of strain propagation for a blast anywhere across the crest. A simplified "zone of potential cracking" based on strain data was developed that encompasses the results of both analyses as shown in Figure 8. For simplicity and conservatism, the point of deepest strain is moved to directly beneath the outer edges of the crest. The deepest strain beneath the crest is used as a lower horizontal boundary. The point of deepest strain below the upstream and downstream slopes is used for both slopes. The result is a simplified "zone of potential cracking" that allows an estimate to be made of potential damage resulting from an explosive blast on the crest roadway. A similar procedure is used for developing a depth of potential crater formation. The depth of crater formation is taken from the 100% strain contour.

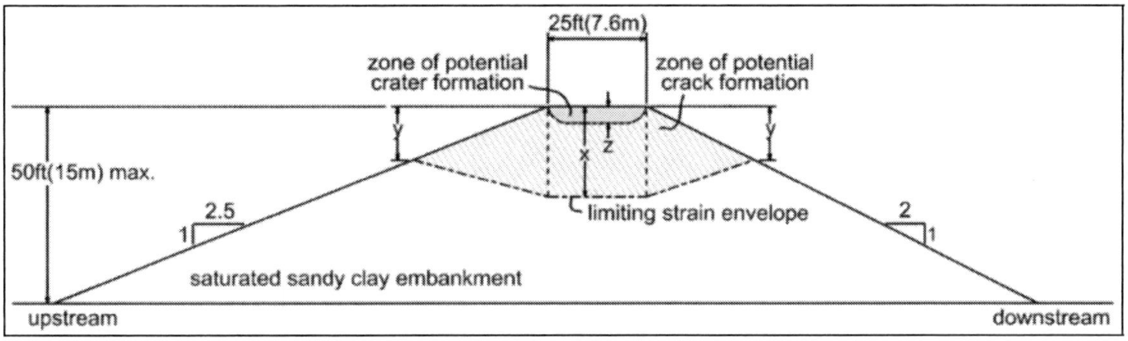

Figure 8. Simplified Depiction of Crater Depth and Zone of Potential Crack Formation Due to Explosive Blast on Dam Crest

Results of Analysis and Parametric Study

The trial dam model was analyzed for saturated clayey sand (*SC-Sat*) and an explosive size of 5,000 lb of TNT. A parametric study was performed to analyze the trial dam using saturated relatively clean sand or sand with non-plastic fines (*SP-Sat*) and the default dry sand model from the AUTODYN library (*Sand-Compaction*). The explosive size was varied to include a 2,500 lb TNT blast. A final case was run of the base model using saturated clayey sand (*SC-Sat*) and including two feet of concrete on the crest surface.

Results of Different Soil Material Models. Table 1 provides a summary of the analysis and parametric study for different material models for the two explosive sizes. Depths of the crater and potential cracked zone based on Figure 9 and strains of 10%, 20%, and 30% are presented. As indicated, the selection of the soil material model has a significant effect on the results. The saturated clayey-sand model results in the most significant impact, followed by the saturated-sand then dry-sand models. This is consistent with expectations regarding the behavior of soils under explosive loads, which indicate that the effect of saturation and clay fines would be to reduce the energy-dissipation effect of the soil material.

When evaluating potential crack formation, the selection of the strain value to be used is a matter of engineering judgment and risk tolerance. For example, if the trial dam represents a high hazard dam, a lower strain value is more appropriate because the consequences of failure are more severe.

Table 1: Crater and Potential Crack Depths for Base Case and Parametric Study

Strain Level (%)	2,500 lb. (1,135 kg) TNT Equivalent			5,000 lb. (2,270 kg) TNT Equivalent		
	Depth at Slope (Y in Fig. 8) ft (m)	Max. Depth Below Crest (X in Fig. 8) ft(m)	Crater Depth (Z in Fig 8) ft (m)	Depth at Slope (Y in Fig. 8) ft (m)	Max. Depth Below Crest (X in Fig. 8) ft(m)	Crater Depth (Z in Fig 8) ft (m)
SC-Sat	(Base Case, Saturated Clayey Sand)					
10 %	9 (2.7)	23 (7.0)	13 (4.0)	13 (4.0)	28 (8.5)	17 (5.2)
20 %	8 (2.4)	19 (5.8)		11 (3.4)	23 (7.0)	
30 %	7 (2.1)	17 (5.2)		9 (2.7)	21 (6.4)	
SP-Sat	(Modified Compaction Model, Saturated Clean Sand)					
10 %	5 (1.5)	13 (4.0)	5 (1.5)	7 (2.1)	17 (5.2)	8 (2.4)
20 %	3 (0.9)	11 (3.4)		5 (1.5)	14 (4.3)	
30 %	2 (0.6)	9 (2.7)		4 (1.2)	12 (3.7)	
Sand-Comp.	(AUTODYN Default Compaction Model, Dry Sand)					
10 %	0 (0)	6 (1.8)	2 (0.6)	0 (0)	7 (2.1)	3 (0.9)
20 %	0 (0)	2 (0.6)		0 (0)	3 (0.9)	
30 %	0 (0)	2 (0.6)		0 (0)	3 (0.9)	

Results of Model with Concrete. A model of the upper 20 feet of the dam was analyzed with two feet of concrete as a crest roadway section or structural-hardening measure. The concrete extends across the crest except for a two-foot space at each edge to accommodate guardrails, fencing, etc. as shown in Figure 9. The concrete was modelled with a structural concrete material model included in the AUTODYN material library.

The intent of analyzing concrete on the crest was to conceptually explore the potential for mitigation measures related to the crest roadway. The thickness of concrete used here would be indicative of an upgraded or hardened structure for improving blast impact resistance. The objective of the analysis was to determine what influence the structural concrete would have on the overall blast impact. As with the other parametric models, the results were evaluated in terms of the strain propagation within the embankment.

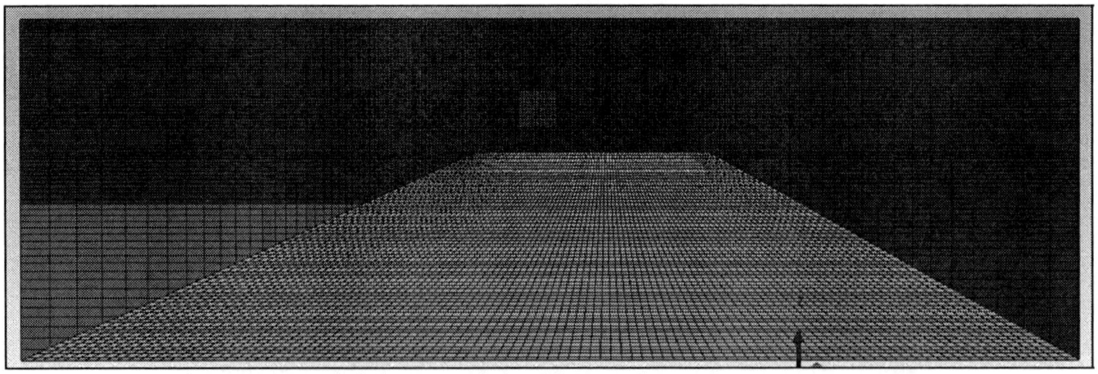

Figure 9. Partial dam model with two feet of concrete on crest

The results indicate that a crater still forms and strains still propagate into the embankment. However, the depth and diameter of the crater and the pattern of strains are different than without the concrete. The high strains in the embankment are concentrated immediately below the concrete section, while the strains above 10% are relatively tightly spaced, as can be seen in Figure 10. The strains and deformations within the concrete are not intended for evaluation; as a structural material model, this material has to be assessed with regard to the structural damage criteria in the material model. Because the intent is to assess the impact in comparison to sections without the concrete, the focus here is on the pattern of strains within the embankment.

Figure 10. Results of model (strains) with 2 feet of concrete on crest

This model result indicates that the inclusion of two feet of concrete on the crest surface may help mitigate the impacts deeper into the embankment. Specifically, the presence of the concrete reduces the size of the crater and the depths to which strains propagate within the embankment. However, this model is considered illustrative and should not be considered highly accurate.

The model uses a structural concrete material model together with a compaction model for the embankment soil. Without further research and/or validation, it cannot be definitely stated that these two different types of material models can be used together in a single structural model and still be capable of producing valid results. For example, the strain compatibility of the two materials has not been specifically accounted for. Additional research is needed to assess whether these materials can be used together or what modifications must be made in the model set-up.

Implications for Design for Blast Impact Mitigation

Although the blast-impact models presented herein have not been field validated, the results are considered conceptually correct. The parametric study results indicate the need to use appropriate soil material models that adequately consider the effects of saturation and clay fines for blast-impact assessments of dams.

The results of the model for dry sand indicate greatly reduced impact from the blast. This is unlikely to be a realistic scenario in the field because dam materials will have some degree of saturation. However, it is worthwhile to note how significant the energy-absorbing effect of the dry sand is in mitigating the blast impact. This effect results from compaction of air within the pore spaces. Although it is not possible to include dry sands in a dam or road bed because these may eventually become saturated, it may be possible to use material with entrained air, such as air-entrained flowable fill or foamed concrete, as an energy-dissipating layer. This material could be used as a capping lift on the embankment soil and/or as part of the pavement or subgrade section.

The parametric study results indicate that clean sands dissipate the blast impact better than clayey sands. With regard to design for blast mitigation, consideration should be given to using clean granular material in the upper portion of an embankment dam within the freeboard elevations where seepage control is not as critical.

The inclusion of concrete may be effective at reducing the depth of the crater as well as the depth of strain propagation. This indicates that upgrading the roadway section may be effective at reducing or mitigating blast impacts.

A combination of these measures, such as concrete with air entrainment with a foamed concrete subbase, may be the most effective upgrade measure. Using the roadway section as a potential mitigation measure may also provide the benefit of cost sharing between the dam owner and the local municipality responsible for roadway maintenance.

CONCLUSIONS

A trial dam section based on published guidelines was developed for analyzing the impacts of explosive blasts due to a truck bomb on the crest of the dam. Analyzing the impacts of explosions on embankment dams requires a complex, three dimensional numerical model using explicit-dynamics software. Appropriate material models are necessary to account for saturation and fines content of the dam materials.

Localized blast impacts may extend beyond and below the blast impact crater and may lead to preferential seepage paths and internal erosion or piping. Localized blast impacts below the crater may not be immediately apparent in the aftermath of the explosion, and seepage issues may take time to develop. Limiting strain criteria is proposed for evaluating the depth of potential cracking due to localized blast impacts.

A simplified chart has been developed based on the result of trial models. The depths of cratering and potential cracking are estimated for the trial dam section. A parametric study of two explosive sizes and three different material models has been performed and results presented with respect to the simplified chart and limiting strain criteria.

Results of the parametric study indicate that saturated soils and soils with clay content will experience more significant localized impacts compared to clean sands or dry sands. Although dry sands are unlikely to exist within a dam, the results indicate that materials with air entrainment may offer significant energy dissipating effects as mitigation measures. Analyses also indicate that structural concrete on the crest is effective at reducing the local impacts, including reduced crater size and depth of strain propagation.

Further research is needed to assess the potential mitigation effects of air-entrained materials and structural concrete. Appropriate material models for use in explicit-dynamics analyses are needed for such materials. Strain compatibility between structural concrete models and compaction models for soils also needs to be explored further.

ACKNOWLEDGEMENTS

This research was sponsored by the William Barclay Parsons Fellowship Program of Parsons Brinckerhoff. The authors would like to thank the Career Development Committee as well as mentors George Munfakh and Hugh Kelly for their support. Also, the authors would like to thank Carrie Nicholson, who prepared many of the graphics.

REFERENCES

Grujicic, M., B. Pandurangan, and B.A.Cheeseman, (2005) "The Effect of Degree of Saturation of Sand on Detonation Phenomena Associated with Shallow Buried and Ground Land Mines." *Shock and Vibration* 12: 1-21.

Grujicic, M., B. Pandurangan, B.A. Cheeseman, W. N. Roy, and R.R., Skaggs, (January 2007) "Application of the Modified Compaction Material Model to the Analysis of Landmine Detonation in Soil with Various Degrees of Water Saturation." *Shock and Vibration* 14: 1-21.

Grujicic, M., B. Pandurangan, G.M. Mocko, S.T. Hung, B.A. Cheeseman, W.N. Roy, and R.R. Skaggs, (March 2007) "A Combined Multi-material Euler/Lagrange Computational Analysis of Blast Loading Resulting from Detonation of Buried Landmines." *Multidiscipline Modeling in Materials and Structures* 4, no. 2: 105-124.

Grujicic, M., B. Pandurangan, N. Coutris, B.A. Cheeseman, W. N. Roy, and R.R. Skaggs, (April 2007) "Derivation and Validation of a Material Model for Clayey Sand for Use In Landmine Detonation Computational Analyses." *Multidiscipline Modeling in Materials and Structures* 5, no.4: 311–344.

Parsons Brinckerhoff, (2010) "Blast Protective Design Seminar." Presentation to Amtrak,, WMATA, Baltimore City, and the Maryland Transit Administration, Parsons Brinckerhoff office, Baltimore, MD.

United States Army Corps of Engineers (USACE) (2003) *Slope Stability*. Engineer Manual, no. EM 1110-2-1902. Department of the Army, Washington, D.C.

United States Army Corps of Engineers (USACE) (2004) *General Design and Construction Considerations for Earth and Rockfill*. Engineer Manual, no. EM 1110-2-2300. Department of the Army, Washington, D.C.

United States Army Corps of Engineers (USACE) (2013), *2013 National Inventory of Dams*. http://nid.usace.army.mil, NID 2013 database.

United States Bureau of Reclamation (USBR) (1973), *Design of Small Dams*. 2nd ed. Prepared by the United States Department of the Interior. Washington, D.C.

SYSTEMS ANALYSIS OF DAM SAFETY AT OPERATING FACILITIES

R. Ascila[1]
G.B. Baecher[2]
D.H.D. Hartford[3]
A.N-A. Komey[4]
R.C. Patev[5]
P.A. Zielinski[6]

ABSTRACT

This paper illustrates a new approach to assessing dam safety using systems simulation of natural, engineered, and anthropogenic processes. A case example is used to illustrate the approach and to indicate the strengths and weaknesses of viewing dam safety through the lens of systems engineering. The systems engineering approach seeks to identify emergent behaviors of dams, reservoirs, and waterways rather than focus on the simple reliability of gates, sensors, or other single components.

INTRODUCTION

The reliable performance of dams and their appurtenant systems depends on the complex interactions of a large number of natural, engineered, and human systems. Historically, such systems have been designed for extreme loads, *e.g.*, the largest flood or earthquake that might occur during some service life. But accidents and failures seldom derive from such loads: They much more often occur through uncommon combinations of mishaps that are difficult or impossible to identify during design. The challenge is develop analysis approaches that capture these emergent systems effects. Emergent behaviors are patterns or regularities that arise through the interactions among system elements which themselves do not exhibit such properties.

Adverse performance typically arises from unanticipated combinations of loadings and responses within the normal operating ranges of a dam's performance. For example, instrumentation mis-performs so pool levels are allowed to rise unbeknown to the operator, simultaneously ice forms on the spillway gates so that they cannot be opened sufficiently to accommodate necessary discharges, and finally the operator misperceives the danger and an accident follows. Failures often occur because several deviations from normal conditions occur simultaneously. These failures are caused by unanticipated properties of the systems interactions of natural, engineered, and human components.

---

[1] Vattenfall, Evenemangsgatan 13, 169 56 Stockholm SE, rommanas.ascila@vattenfall.com
[2] University of Maryland, Department of Civil Engineering, College Park, MD, USA, gbaecher@umd.edu
[3] BC Hydro, 6911 Southpoint Drive, Burnaby BC V3N 4X8, Canada, des.hartford@bchydro.bc.ca
[4] University of Maryland Department of Civil Engineering, College Park, MD, USA, adielkomey@gmail.com
[5] US Army Corps of Engineers, USACE, 696 Virginia Road, Concord MA, Robert.C.Patev@usace.army.mil
[6] Ontario Power Generation, 800 Kipling Avenue, Toronto Ontario, M8Z 5S4, Canada, pa.zielinski@opg.com

RECENT EXPERIENCE WITH ACCIDENTS AND FAILURES

Case studies of recent accidents and failures suggest that incidents occur in ways not easily summarized in traditional event tree and fault trees (DSIG 2014). As a result, there is interest in new ways of characterizing dam safety risks, what might be thought of as a "second-generation approach to dam safety risk assessment."

Among the observations made from this recent experience is that near misses of accidents and failures are surprisingly common. Although every case of a near accident or failure is unique, all share similarities from which lessons can be taken. The time-window for effectively responding to a developing situation may be narrow, and both operators and managers need to be trained ahead of time to deal with them.

Management and Corporate Memory

Among the lessons learned from recent accidents and failure are the following: At least at some organizations and at least in some accident situations, corporate management acts as if it is complacent to dam safety concerns. Reaction times in response to operator warnings or requests for direction are at times slow, taking hours or days at a moment when response time may be critical. A recurring question for operator and management reaction to emerging situations is whether the principal driver for action is protecting production or protecting safety? In many cases it appears that protecting production dominates reactions, leaving dam safety as a secondary consideration. Similarly, dam safety objectives during an emergency situation may be overridden to achieve environmental objectives.

In some recent cases, operating orders and standard operating procedures at the time of the incident have been out of date. Some procedures had not been updated for many years. This reflects in part a lack of organizational memory. Kletz (1993) has noted that people have memories but organizations do not: When people leave, their memory leaves with them and the organization forgets the lessons that led to the existing eprocedures. Sometimes organizational memory has erased the purpose of certain equipment from current understanding but such equipment can be removed or changed only at risk that it in reality serves a safety related purpose. Many case studies identify old, poorly maintained, or even antiquated equipment, sensors, and other systems as adversely affecting dam safety. Deferred maintenance and replacement of safety-related equipment seems common. The unavailability of spare parts for older systems is becoming a serious hindrance both to operations and to safety, and may cause long periods of impeded operation while spares are manufactured.

There Are Many Stakeholders

Most projects do not operate in a vacuum but reside within larger systems (*e.g.*, river basins) with both upstream and downstream projects. These others projects may be owned and operated by other organizations. There are also upstream and downstream riparians and industrial users, so that the set of stakeholders is large and varied. Good relations with the stakeholder community, and good communication to and from the stakeholders, are important to safety.

Cannot Always Blame the Operator for an Incident

Usually, the responsibility for an incident goes up and down the management chain. The performance of operators depends on many tings, including the set up of control rooms, the use or overuse of emergency warning signals, staffing levels, training, and fatigue. At one time, station operators had broad control and took deep pride in the operation of the stations under their control. Changes in corporate management, the frequent transferring of staff, and remote operation have all reduced the level of operator loyalty to the station. More and more facilities, especially small to medium sized ones, no longer have on-property operators. The operations of the facility may be directed from 100's of km away at a regional center. Impacts of remote operation on incidents need to be understood.

Human reliability research has advanced rapidly during the past two decades (Swain and Guttmann 1983). This can increasingly be brought to bear on hydropower operations. One of the observations both in this research and in hydropower experiences is that operators may not be adequately prepared to grapple with fault-conditions when things are transpiring other than expected. Event simulators or other training techniques from related technologically hazardous fields might find use in operator training.

Disturbances Are Important

Disturbances are those exogenous factors that affect operations. On a rare event level these may include earthquakes or extreme hydro-meteorological events, although such extreme disturbances may appear in traditional risk analyses and been planned for. More common but still important disturbances include such things as grid availability (or lack thereof), floating debris in the reservoir, and floating or frazil ice. Sometimes a disturbance make lead to important secondary effects, such as forest fires or avalanches leading to debris. Debris appears to be a significant disturbance to hydropower operations, and massive-debris incidents may be even more common today than in the past.

Redundancy Is Important

Lack of redundancy appeared as an important part of the chain of events in several or most of the case studies. The most common of these were lack of redundancy in backup power, access to an unmanned site during an emergency, communication systems, and sources of data (*i.e.*, can the data be believed, why don't we have other sensors for the same measurement?). The use of SCADA systems does not obviate the need for redundancy, and may make the need even more critical.

Loss of station power and grid power is common during an incident. In this event, auxiliary generator or DC power is needed to run sensor and SCADA systems, to operate gates, and to undertake many remedial actions. In many cases either these back up sources of power were not provided for, or they were in a sufficient state of disrepair as to fail.

Given the increasing use of remote operation, access to a site can be critical in an emergency. However, many emergency situations occur during storms, at night, in winter, and at other unfavorable times when roads are impassible, aircraft cannot fly, and communications to the site are interrupted. This makes lack of site access a common factor in the event chain leading to adverse outcomes.

Sensor systems in use today are often old and unreliable. Absent redundant data, it can be impossible to know whether unusual readings indicate an emerging incident or are merely wrong. Loss of site power or communications can black out all sensor data and leave a remote operator blind to developing site conditions. Autospill and other automatic response devices should parallel human decisions to in part allay the loss of sensor data at critical times.

SYSTEMS APPROACH TO DAM SAFETY

The reliable performance of a dam system depends on the varying demands placed upon it by river and reservoir hydrology, operating rules, and vagaries of human and natural factors. These systems control floods, condition flows, and filter high frequencies in the river discharge. The function of a flow-control system is to retain water volumes and to pass flows in a controlled way. Their reliability involves structural, mechanical, electrical, control systems and subsystems reliability, as well as human interactions, organization issues, policies and procedures, all of which, occurring in a broad spectrum of environmental conditions.

Adverse performance typically arises from unanticipated combinations of loadings and responses within the normal operating ranges of a dam's performance. That is, failures often do not occur under the extreme loadings considered in design, but under the uncommon combination of relatively common things happening. For example, instrumentation misperforms so pool levels rise, simultaneously ice has formed on spillway gates, which then cannot be opened to accommodate discharges, and finally the operator misperceives the danger and an accident follows. Failures occur because several and sometimes not unusual deviations from normal conditions occur simultaneously.

A new approach is required to risk analysis for dam safety, as current engineering practices do not address the character of the most probable causes of accidents and failures. Most accidents and failures occur not because an extreme event occurs (*e.g.*, a flood or earthquake), but rather because a series of more common events occurs, which in their unfortunate and unexpected combination leads to an adverse outcome: an accident or failure. Such combinations of unforeseen yet unfortunate events cannot be predicted easily—or maybe at all—ahead of time. They are what systems engineers call *emergent* behaviors.

This unfortunate combination may not be of "usual events." Accidents and failures often occur due to a combination of events that one by one are in the range of the design events. It may not be possible to get an accident or failure if all components and events are in a perfectly normal state. Some conditions must be in the range of not quite usual—for example a 50-year flood, lack of required maintenance, slightly incompetent personnel or organization, bad instructions or policies, a power blackout, or the like—but yet not extreme individually.

In a review of the performance of spillway gates and associated operating equipment reported to the National Performance of Dams Program, McCann (2013) reports that exclusive of the failure of spillway structures themselves, the performance of hydraulic systems is important even during normal operations. These systems are affected by mal-operation, control system errors, and a host of interactions among factors. McCann summarizes 65 flow-control accidents which occurred before 1995 at US dams, and grouped by apparent cause. In each case, a causal chain of contributing factors leads to an apparent final cause and thus to the subsequent accident. Although the final cause may be identified as the root cause, it need not be.

The observation that most accidents and failures occur not from extreme loads but through more common yet unforeseen sequences of events is now widely recognized. Perrow (1999) argues that ever increasing complexity in technological systems makes these failures inevitable. He argues that complex rather than linear interactions among components leads to systems performance that is difficult to understand. Flow-control in dam systems displays both sorts of interactions. Perrow also argues that tight *vs*. loose coupling among components increases the chance of accidents. Tight coupling increases the likelihood that operator intervention will make things worse rather than better, since the nature of an emerging situation may well not be understood immediately.

Many dams may be considered to be tightly coupled systems because they have little flexibility and have time-dependent operational needs. That is, a dam cannot usually wait while disturbances are attended to in an orderly way. Delays in operations may not be tolerable. Water flows into the reservoir and it must be dealt with now, not at some more convenient time in the future. Characteristics of tightly coupled systems are that delays in processing are not possible; production sequences are invariant; little slack exists in methods, equipment, or personnel; and substitutions of supplies, equipment, or personnel are limited. All of these things increase the likelihood of accidents. In tightly coupled systems, expediencies cannot be used to save a situation: Redundancy, segregation, diversity, and defence in depth must be designed into the system from the start (Rigbey 2013).

Rasmussen *et al.*, (1990) and Leveson (2012) argue that the chain of causality leading to an accident is often opaque. The causal path to an accident will be "prepared by resident conditions that are latent effects of earlier events or acts." Thus, the length of the causal path depends to a large extent on the stopping-rule for seeking contributing causes: Do the causes stop with technical faults, or with operator "errors", or do they extend back to management and design decisions? Regan (2010) makes this same argument but with specific reference to dams:

> [...] chain-of-event models, such as a typically defined failure mode or risk assessment event tree, oversimplify the causes of incidents and exclude many systemic factors and non-linear interactions. The failures of Teton, Silver Lake and Taum Sauk all had contributions from systemic factors and non-linear interactions that were unrecognized prior to the failure. Our current dam safety programs are, in essence, trying to determine the safety of the dam by examining a few components of the dam, one component at a time. [...].
>
> Analysing individual components against a prescribed standard is insufficient to assure the safety of dams. Reducing the risk associated with dams to a level that is as low as reasonably practicable requires evaluating the dam as a complex system with interactions of sub-systems that may be difficult to recognize. We must include subsystems that are not normally included in typical standards-based or risk-informed decision-making processes. A systems approach to dam safety provides the opportunity to identify places where risk can be reduced in a cost effective and timely manner.

A SECOND-GENERATION APPROACH

To achieve Regan's suggested goal of a new approach, one needs to take a systems view of the analysis of function and failure of flow control in dam systems. This is contrary to recent decades' separation of analysis according to different fields, that is, analysis being divided among different specialists. However, it is more in accordance with the approach used when most dams were planned in earlier times. The specialties and methods of analysis used today are still absolutely required, but need to be supplemented with an improved overview of how things come together and influence each other. Equally important is an improved overview and understanding of the hierarchy of the requirements (*i.e.,* the system purpose) affecting the whole system and its components.

This view of how things come together—including requirements (purpose), functions, and components—might be called a *systems view.* This is not a reversion to the old times of qualitative judgment, but demands more, not less, scientific study. Also the evolving notion of how risks arise motivates a new perspective and a new mind-set. Risk analysis is predicated on a systematic decomposition of all things that might go awry, and as a result it forces us to think about systems performance. The 'new approach' attempts to develop a systems-engineering approach into a practical methodology.

The 'new approach' needs to be in agreement with systems thinking. It needs to capture real, evidence-based causes of accidents and failures. Among other things, this means that the new approach must address instrument and control systems, system state and maintenance practices and schedules, the dynamics of the system, time histories of system states, times to repair damages, times for consequences to manifest, human reliability, and other heretofore neglected but important aspects of performance such as the dam owner's maintenance regime.

To achieve this, the entire flow control system at dams needs to be considered, including not only spillways, but also all other waterways, including low-level (bottom) outlets and power intakes. In a dam system there are special ways and arrangements for how things are connected, forms, hierarchies, delay and feedback mechanics, systems dynamics, mathematically challenging nonlinearities, and so on which lead to simulation approaches. Of course, not only the technical parts of the system are important: organization, social context and human reliability are also a major part of flow control systems (Figure 1).

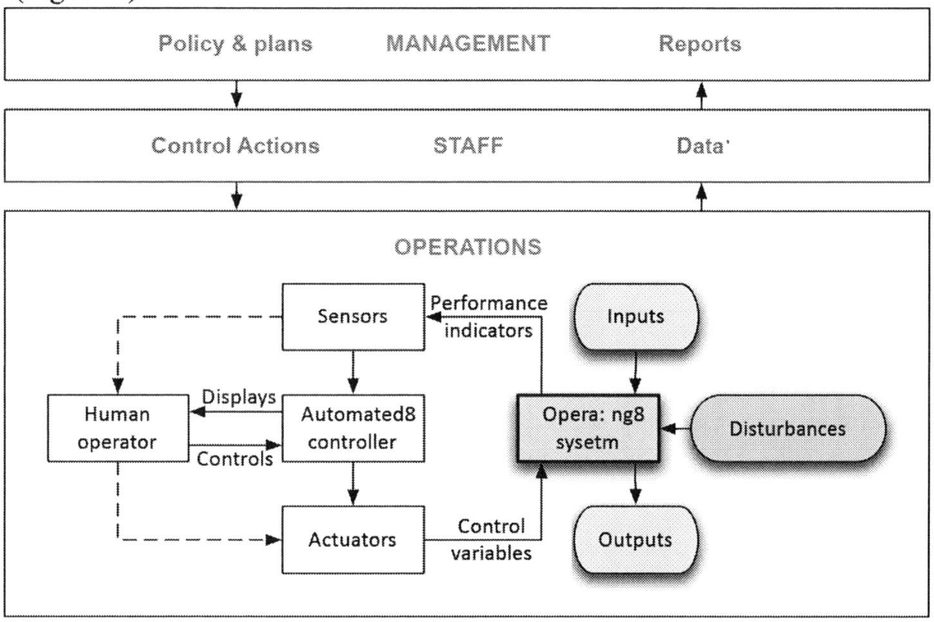

Figure 1. Systems operations of a dam (adapted from Leveson 2011).

Dam systems for flow control comprise a broad set of components such as structures, equipment, sensors, communication facilities, personnel, management arrangements and policies that enable the handling of water flows through the reservoir and past the relevant dam to the downstream reach of river (Leveson 2011). In this sense the totality created by man between the river upstream and the river downstream may be considered a flow control system. The dam body is an integral part of this system and, depending on the objectives of the modeler, it may or may not be included in the purpose-built system model for flow control being the object for certain types of system failures caused by sliding, internal or external erosion, or excessive settlements.

The term *system* is used here as it is commonly used in risk analysis, and its definition and limits in any particular instance is a matter of modeling convenience. A *system* is understood to denote a set of interacting, interrelated, and interdependent elements that form a complex whole for the purposes of analysis. Where relevant, it can include not only one dam, but a cascade of dams within a river basin. Modeling of such a system intends to simplify the reality of the complex whole in such a way that a realistic resemblance of the salient features of the system are retained and can be scientifically analyzed. Determining what these salient features are, is a crucial step in application of the method which will vary according to the specific interest of the modeler.

PROJECT EXAMPLE

This new simulation approach to systems is being applied to a project at Ontario Power Generation (OPG) in northeast Ontario. The project is a cascade of four power stations. The number of riparians in the river flood plain is few and there is no commercial riverine navigation, so potential loss of life is negligible and operational safety dominates the considerations. The problem facing the engineering analysis was to conceptualize a systems engineering model for the operation of the dams, generating stations, spillways, and other components; then to employ the model through stochastic simulation to investigate protocols for the safe operation of the project. The project comprises four hydroelectric generating stations. Together, the four stations comprise a system that came into service in 1931.The system borders two physiographic regions: the Canadian Shield which extends from the south to just north and, beyond this, the Hudson Bay Lowlands.

Generating Stations

Several hydroelectric generating stations were built in the basin during the twentieth century. In late 1989, Ontario Hydro purchased the plants and through its successor organizations has operated the facilities since.

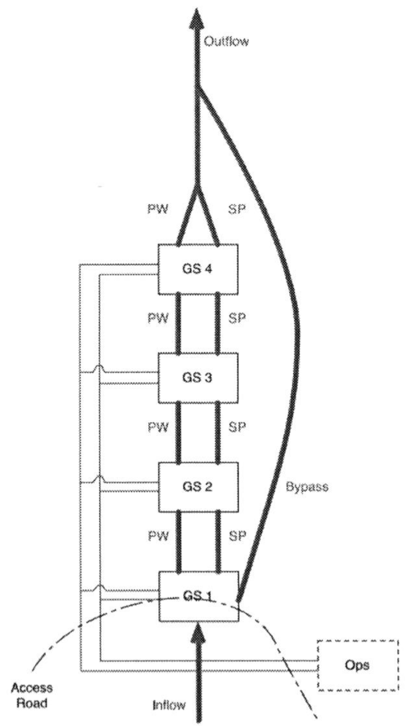

Figure 2. Project layout

Transmission of electrical energy from the four stations is provided by a 230 kV transmission line from the lowest plant through the others and to the regional grid (Figure 2).
When river flows exceed the maximum power flow of top reservior, a bypass spillway structure, located 2.5 km east of the GS, is used to pass excess water into a bypass channel. The bypass spillway comprises eight sluices with a capacity of 4870 cms. Waters bypassed flow northward and re-enter the river about 17km downstream of the lowest GS. A secondary spillway structure constructed in the former river channel has a capacity of 1217 cms, and provides for diversion flow to the downstream GSs in case of shutdown of the upper units and to augment the bypass capacity.

The second dam in the sequence is concrete and incorporates intakes for a powerhouse, a spillway structure to bypass flows in the event of a sudden unit outage, and an earth-fill retaining structure located near the spillway. The head pond extends upstream about 7km, has a surface area of about 5.3km^2 and a live storage of 106 mcm. The powerhouse contains four vertical Francis type turbine units, with the capacity to generate 52 MW at a rated flow of approximately 190cms and an operating head of 34.5m. The existing spillway structure consists of ten gated sluices, each 8.4m wide by 9.2m high, plus a 230m long overflow crest. The spillway structure was originally designed (prior to the construction of the bypass) to convey what was then the full design flood flow on the river. The third GS has a single concrete dam that incorporates the intakes for the power station and a spillway to bypass flows in the event of a plant outage. The head pond extends 4 km upstream. Its surface area is approximately 3 km^2 and live storage is about 6.9mcm. The operating head is 31m and the rated flow is 525cms. The final GS has a single concrete dam incorporating an intake structure and spillway. The head pond is 5.6 km. It has a surface area of 1.2 km^2, with live storage of 3.2mcm. The power plant is similar to those upstream but operates at 0.5m lower generating head.

Historic Hydrologic Series

Fifty year historic time series data on reservoir inflows are available and were used to generate stochastic reservoir inflows through time series forecasting (Figure 3). An autoregressive time-series model was fit to these historical data, and used to generate synthetic reservoir inflows over long periods of time.

Time-series modeling in hydrology has a long tradition and has been based on the assumption that it possible to develop a general type of models capable of representing most relevant statistical characteristics of historic stream flow series. The origins of the concept can be traced as far back as to Hazen (1914) and in more recent times to Hufschmidt (1966).

In general, a time series is understood as a series of observations that are arranged chronologically. This arrangement ensures that the order of occurrences is preserved. River flows data, unlike the data for some of the climatological phenomena, are at daily intervals and conform to the data requirements for most of the time series models. Modeling river systems by simulation has a long tradition in water and environmental management, especially for economic optimization (Simonovic 2008). The watershed level has typically been in the center of modeling attention due to its complexity in responding to external and internal forces.

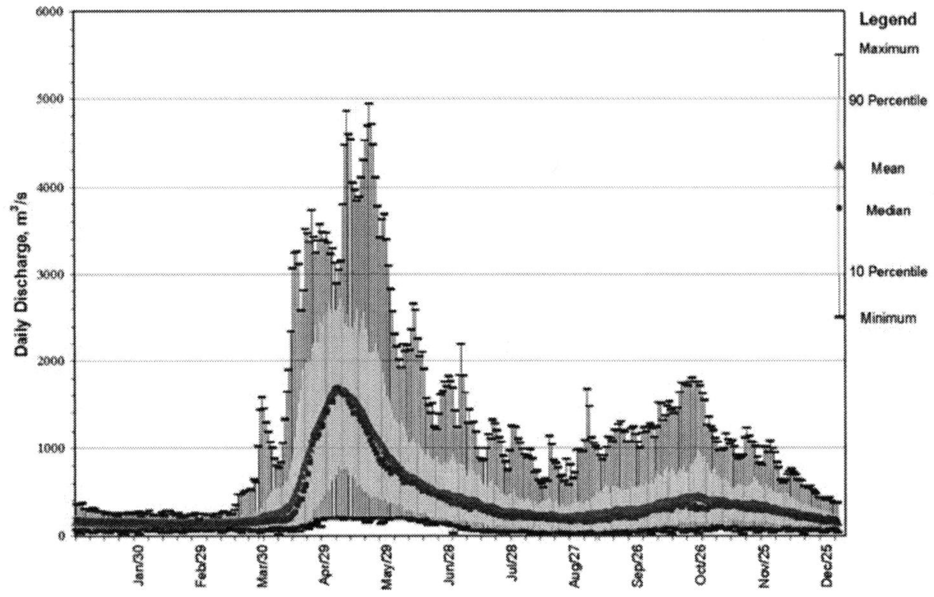

Figure 3. Historic Daily Inflow Statistics

SIMULATION MODEL

Objectives of the Case Study

The objective of the modeling was to understand the systems interactions in the cascade, and the potential for accidents and failures caused by the interactions of components of that system. This includes:

- Applying the systems modeling framework to understand the systems interactions in the cascade of four dams.
- Formulating and constructing a model to characterize the hydrodynamics of the cascade including the dynamics of transport, storage, and power generation.
- Holistically integrating river basin hydrology, routing of inflows through the reservoir system, operating rules and human factors of operating the spillway, and the dam component fragilities (structural, mechanical and electrical).
- Generating stochastic time series by using the historic inflow time series to forecast inflows for several thousand years and multiple replications to identify unforeseen chain of events that could lead to accidents and failures.
- Reviewing the current operating rules to determine whether further optimization techniques can be adopted to improve the power generation capabilities of the system.
- Modeling inherent disturbances (lightening, seismic, floating ice, grid disturbances and debris) via a probabilistic framework.

- Incorporating human reliability analysis (HRA) into the model by using expert judgment and available data to estimate human reliability probabilities on spillway gate and possibly other operations.

The System Model

The general model interface with the salient aspects of the physical system being modeled was developed. When the simulation is started, the software iterates over many years (perhaps millions) of performance and evaluates the long history of the system, saving selected results for subsequent analysis.

Initial had been based on the *Matlab*™ software platform and its associated *Simulink*™ package. However, with the proliferation of commercial off-the-shelf party simulation engines at least as well optimized for engineering reliability, other alternatives were evaluated. Ultimately, the choice was made to build the model on the *GoldSim*™ platform. While this platform is less powerful than some of the others reviewed, it had the advantage of simplicity.

Normal Operations and Power Generation

Under normal operations, water flows for the GSs are provided from the uppermost reservoir. The water level is normally within the operating headwater level range. The limit of the headwater level is the "absolute maximum operating level." The difference between the absolute maximum and maximum operating levels is the flood allowance, which is used to hold water in extreme conditions to reduce downstream flooding. The storage between the absolute minimum and minimum operating levels is used if a system energy emergency occurs. Under normal operating conditions with equivalent discharges at each station, the full operating range would rarely be utilized.

Under normal operating conditions, the outflow from the uppermost reservoir passes through all the GSs. During any outage of a GS, the spillway at the station experiencing the outage will be operated to pass the flow to the other GSs. During high river flow conditions (e.g., spring runoff) when the uppermost reservoir is near its maximum limit, the spillway into the diversion is operated in conjunction with GS to pass the full river flow.

Gate Operations

Gated spillways are designed to engineering standards, however with aging, exposure, preventative maintenance, and lack of frequent operations they become more vulnerable to non-availability. On-demand failures of gated spillways may be caused by a gate component that can be repaired in minutes to hours or a component that may cause complete failure of the gate system and unexpected release of the reservoir containment.

Gated spillways contain a complex integration of structural, mechanical, and electrical (SME) components that must operate on demand. There are a total of 10 sluices with gates. Two of the 10 sluices (Nos. 1 and 2) are alongside the generating station and open into the river, while 8 are nearly 3.2 km upstream and open into the bypass. Sluices 1, 2, 3, 4, 7 and 8 are remotely controlled. Sluices 5, 6, 9 and 10 are locally controlled by the operator agents at the gate. The sluices are numbered from right to left looking upstream at the dam.

This model includes the effects of hydraulic loads, debris, ice, and operator actions. These loads are related to gate availability through traditional fragility curve relations. The inputs to the gate operations simulation at any time step are the respective states of the input and disturbance variables. The outputs are the gate availabilities (probability of use on demand).

Gate operations are also affected by instrumentation, supervisory control and data acquisition (SCADA) systems and controls. The simulation approach is well suited to uncovering the implications of component performance assumptions on overall systems operations.

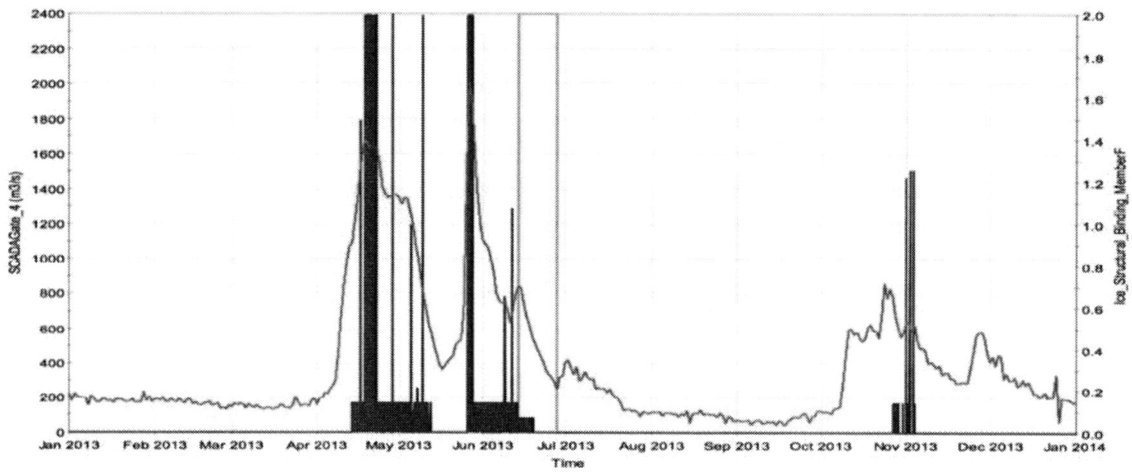

Figure 4. Simulated effect of disturbances: red is reservoir inflow, blue is SCADA controlled spill, and green is structural binding of gate.

Disturbances

Sometimes in the course of the operation of a dam system an extraordinary event occurs, such as an earthquake, an earth or rock slide into the reservoir or conveyance works, a major fire in the drainage area, or the like, at a time which cannot be anticipated, but that may affect dam operations or even dam safety. Within the simulation, such events, and their more modest siblings, are treated as exogenous disturbances. *Disturbances*, can in principle include a broad variety of phenomena, both of natural origin such as lightening strikes affecting power supplies or instrumentation, or of anthropogenic origin such as the grid being unable to accept power and thus the powerhouse waterway having diminished discharge capacity, or operational incidents or accidents such as powerhouse fires. The exact definition is a matter of modeling convenience; severe floods or droughts may for instance be considered as disturbances or alternatively just as aspects of the stochastically modeled catchment hydrology.

A disturbance of major concern in the project is ice build up. The time series in Figure 5 shows the simulated ice build up on the gates of any of the dams during one randomly simulated year. The stochastic input for this simulation is the variability in daily temperature generated from a statistical time series identified to temperature data over the past century. Daily temperatures are simulated, and Stefan's Equation (USACE 2002) is used to calculate ice thickness on the reservoir. Stefan's equation uses anticipatory degree-days below freezing with an empirical constant to forecast ice development.

The disturbances included in the present model are limited to external loss of grid availability, lightening (affecting grid availability), floating ice, icing on structures, instrument or SCADA mis-operation, and human error. Each is treated as a Poisson process of time, possibly with a corresponding probability distribution of magnitude.

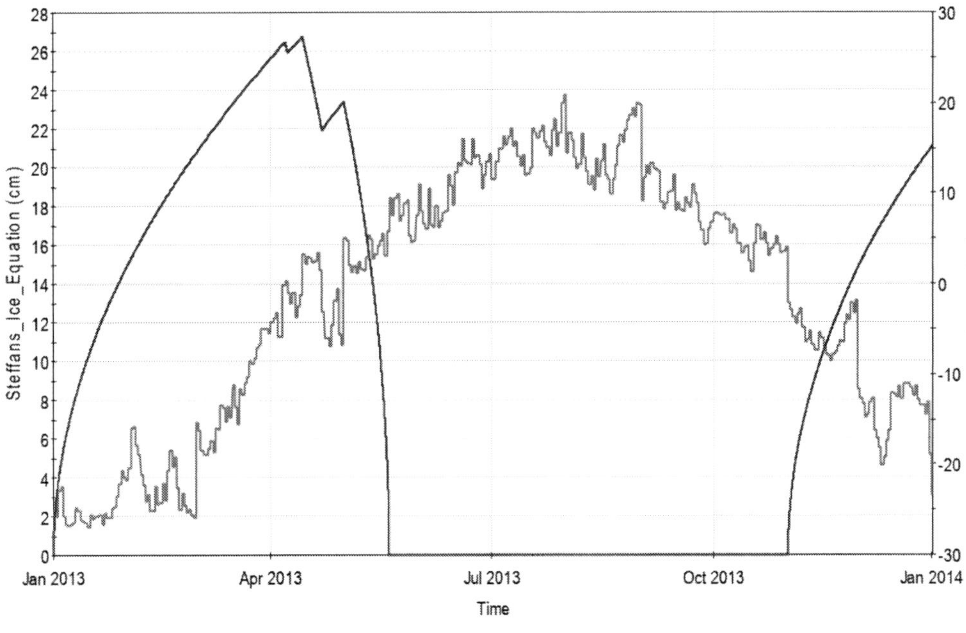

Figure 5. Ice build up in reservoir based on Stefan Equation, red curve is ice build up in cm, green curve is simulated temperature in C.

Human Reliability

It comes as no surprise that human actions are a source of vulnerability for industrial systems, giving rise to modern methods and models of *human reliability analysis* (HRA). HRA aims to deepen the examination of the human factor in the workplace, and is important due to the contributions of humans to the resilience of systems and to possible adverse consequences of human errors or oversights—especially when the human is a crucial part of the large systems.

HRA involves qualitative and quantitative methods to assess human contribution to risk. There are many and varied methods available for HRA, with some safety critical domains developing industry focused methods. Human error has been classified by Reason (1990) into three levels: behavioral, contextual, and conceptual. Reason concludes that the first two types include the different errors that can result from the same process but the conceptual level leads to understanding the causal mechanisms using more theory than observation.

The current simulation effort using the SPAR-H approach to HRA (Gertman et al. 2005). A NASA report (Chandler et al. 2006) report documents many of the HRA methodologies that have been developed to estimate human error potential. The report compares of these methods and provides a selection of four methods that could be used for the NASA risk analysis program. The report recommends for NASA HRA: THERP (Swain and Guttmann 1983), CREAM (Hollnagel 1998), NARA (Kirwan 1994), and SPAR-H. The report recommends that all the four methods be deployed and their results documented to be able to select the best suitable program for HEP estimation for future NASA endeavors. In regards to spillway systems, these HRA methods are focused to nuclear power plants and will need significant adaptations before they could be applied to the operation of spillway subsystems.

CONCLUSIONS

In the simulations to date, among the drivers to which the frequencies of accidents and failures are most sensitive appear to be: (1) loss of remote control through communications or other breakdowns, (2) incorrect operations decisions that fail to establish the appropriate spill profile for the prevailing conditions, (3) loss of power supplies or other supporting or auxiliary services, (4) loss of access to the dam in emergency, (5) excessive weather, (6) common cause disruptions such as fire, (7) failure of control or instrumentation, (8) lack of qualified personnel to provide an emergency response, and (9) failure of the gates or supporting structures. Obviously, the observation that these appear to contribute to high frequencies of accidents and failures depends on the input assumptions about rates, stochastic processes, and fragilities that the modeling has made. So, at this point no generic conclusions are warranted about the causes of operational problems from these results.

Simulation appears to be a promising approach to modeling the reliability of dam systems and supporting a second-generation evolution for dam safety risk analysis. The reasons appear to be that simulation allows for the need to account for time in failure analysis, it incorporates feedback on operating procedures and human reliability in systems function, it provides a way to fuse models across different technological and human systems, and it provides for the chaining of precursors in accident sequences.

REFERENCES

Chandler, F., Chang, J., and Mosleh, A. (2006). *Human Reliability Analysis Methods Selection Guidance for NASA*. OSMA Technical Report, NASA, Washington DC.

DSIG. (2014). "DSIG Workshop on Understanding Operational Issues and their influence on Dam Safety." CEATI, Niagara.

Gertman, D., Blackman, H. S., Marble, J., Byers, J., Haney, L. N., and Smith, C. (2005). *The SPAR-H - Human Reliability Analysis Method*. U.S. Nuclear Regulatory Commission, Washington DC.

Hazen, A. (1914). "Storage to be provided in impounding reservoirs for municipal water supply." *Trans. Amer. Soc. Civil Eng.*, 77, 1539–1569.

Hollnagel, E. (1998). *Cognitive reliability and error analysis method: CREAM*. Elsevier, Oxford ; New York.

Hufschmidt, M. M. (1966). *Simulation techniques for design of water-resource systems*. Harvard University Press, Cambridge.

Kirwan, B. (1994). *A Guide To Practical Human Reliability Assessment*. CRC Press.

Kletz, T. (1993). *Lessons from Disaster: How Organizations Have No Memory and Accidents Recur*. Gulf Professional Publishing, Houston, Tex.

Leveson, N. G. (2011). "Applying systems thinking to analyze and learn from events." *Safety Science*, 49(1), 55–64.

Leveson, N. G. (2012). *Engineering a Safer World: Systems Thinking Applied to Safety*. The MIT Press.

McCann, M. (2012). "Performance of hydraulic systesm." *National Performance of Dams Program*, National Performance of Dams Program, Montreal.

Perrow, C. (1999). *Normal Accidents: Living with High-Risk Technologies*. Princeton University Press.

Rasmussen, J. (1990). "Human Error and the Problem of Causality in Analysis of Accidents." *Philosophical Transactions of the Royal Society of London Series B-Biological Sciences*, 327(1241), 449–462.

Regan, P. (2010). "Dams as systems — a holistic approach to dam safety." US Society on Dams, Sacramento.

Rigbey, S. (2013). "BC Hydro Spillway Gate Reliability Program." HydroVision International, Denver.

Simonovic, S. P. (2008). *Managing Water Resources: Methods and Tools for a Systems Approach*. Routledge.

Swain, A. D., and Guttmann, H. E. (1983). *Handbook of Human Reliability with Emphasis on Nuclear Power Plant Applications*. Washington, DC.

USACE. (2002). *Ice Engineering Manual*. US Army Corps of Engineers, Hanover NH.

Chandler, F., Chang, J., and Mosleh, A. (2006). *Human Reliability Analysis Methods Selection Guidance for NASA*. OSMA Technical Report, NASA, Washington DC.

DSIG. (2014). "DSIG Workshop on Understanding Operational Issues and their influence on Dam Safety." CEATI, Niagara.

Gertman, D., Blackman, H. S., Marble, J., Byers, J., Haney, L. N., and Smith, C. (2005). *The SPAR-H - Human Reliability Analysis Method*. U.S. Nuclear Regulatory Commission, Washington DC.

Hazen, A. (1914). "Storage to be provided in impounding reservoirs for municipal water supply." *Trans. Amer. Soc. Civil Eng.*, 77, 1539–1569.

Hollnagel, E. (1998). *Cognitive reliability and error analysis method: CREAM*. Elsevier, Oxford ; New York.

Hufschmidt, M. M. (1966). *Simulation techniques for design of water-resource systems*. Harvard University Press, Cambridge.

Kirwan, B. (1994). *A Guide To Practical Human Reliability Assessment*. CRC Press.

Kletz, T. (1993). *Lessons from Disaster: How Organizations Have No Memory and Accidents Recur*. Gulf Professional Publishing, Houston, Tex.

Leveson, N. G. (2011). "Applying systems thinking to analyze and learn from events." *Safety Science*, 49(1), 55–64.

Leveson, N. G. (2012). *Engineering a Safer World: Systems Thinking Applied to Safety*. The MIT Press.

McCann, M. (2012). "Performance of hydraulic systesm." *National Performance of Dams Program*, National Performance of Dams Program, Montreal.

Perrow, C. (1999). *Normal Accidents: Living with High-Risk Technologies*. Princeton University Press.

Rasmussen, J. (1990). "Human Error and the Problem of Causality in Analysis of Accidents." *Philosophical Transactions of the Royal Society of London Series B-Biological Sciences*, 327(1241), 449–462.

Regan, P. (2010). "Dams as systems — a holistic approach to dam safety." US Society on Dams, Sacramento.

Rigbey, S. (2013). "BC Hydro Spillway Gate Reliability Program." HydroVision International, Denver.

Simonovic, S. P. (2008). *Managing Water Resources: Methods and Tools for a Systems Approach*. Routledge.

Swain, A. D., and Guttmann, H. E. (1983). *Handbook of Human Reliability with Emphasis on Nuclear Power Plant Applications*. Washington, DC.

USACE. (2002). *Ice Engineering Manual*. US Army Corps of Engineers, Hanover NH.

USING MONITORING PROGRAMS TO MANAGE DAM SAFETY RISKS

Jay N. Stateler, P.E.[1]
Brad Iarossi, P.E.[2]

ABSTRACT

Those involved in designing or modifying dams can use the best methods and practices available, but uncertainties will always exist regarding the completed structure. Unknowns regarding the geology of the damsite will always exist because there are practical limits relative to site exploration work. Construction unknowns will always exist to some extent since nothing is ever constructed in complete conformance with the plans and specifications. Similarly, documentation of construction work will always be somewhat less than total. And those represent uncertainties for new, state-of-the-art dams. The unknowns and uncertainties associated with existing dams, particularly older dams, obviously can be many times greater. Evaluations of existing dams can be performed using the latest techniques. However, such evaluations will always be limited by what is estimated (or can be surmised) about the material properties associated with the dam itself, as well as the dam's foundation and abutments. State-of-the-practice limitations may exist as well. Dam safety evaluations can be performed to the best of our abilities, but realism about the degree of validity and accuracy of the results is important. Hence, the need for well-designed and successfully implemented monitoring programs at dams, to address the unknowns and uncertainties, and appropriately manage the risks regarding the initiation/development of potential failure modes for dams (both those that have been identified and evaluated, and those that have not).

INTRODUCTION

The Bureau of Reclamation's Arthur V. Watkins Dam nearly failed in 2006 due to seepage flow through foundation soils that exited into an open drainage channel (the South Drain) located about 130 feet downstream of the dam. Figure 1 shows a schematic

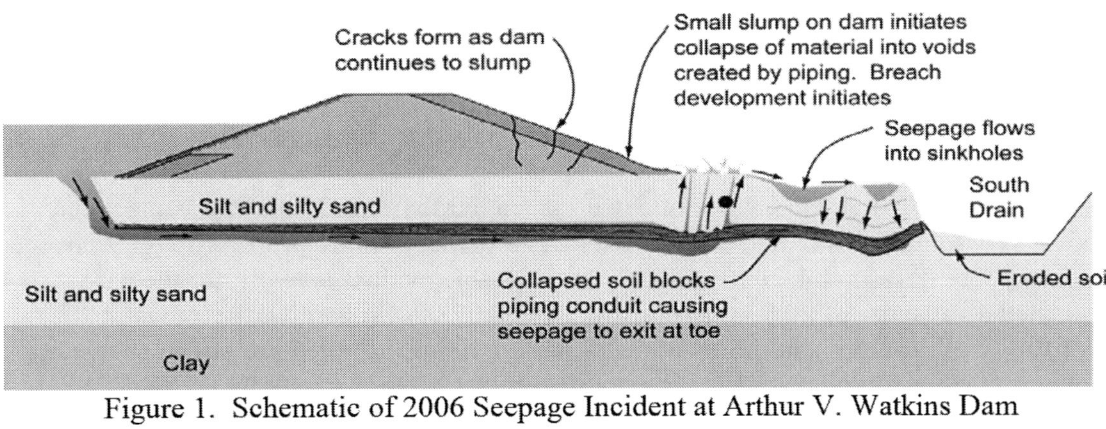

Figure 1. Schematic of 2006 Seepage Incident at Arthur V. Watkins Dam

[1]Civil Engineer, Bureau of Reclamation, Denver, Colorado, jstateler@usbr.gov
[2]Dam Safety Officer, U.S. Fish and Wildlife Service, Falls Church, Virginia, brad_iarossi@fws.gov

of the situation. The foundation seepage had eroded a large quantity of material along the seepage path, and only a rapid "last minute" emergency response prevented dam failure. This failure mode had been specifically identified in a Comprehensive Facility Review (CFR) Report prepared for the dam four years earlier. The future monitoring plan for the dam defined in that CFR Report included monthly visual inspections that were to pay "particular attention" to seepage conditions at the South Drain, looking for evidence of "any unusual turbidity or evidence of soil movement" regarding flows into the drain. The CFR included calculations that indicated that the estimated annual risk of a seepage-related failure through the foundation was 1.4×10^{-4} (or about 1 chance in 7,000 in any year), which slightly exceeds Reclamation's guideline value of 1×10^{-4}.

The developing 2006 seepage failure mode at Arthur V. Watkins Dam was not detected by the routine dam safety monitoring efforts performed by the dam's operating personnel. The only documented "monthly" visual inspection carried out at the dam between May 15, 2006 and the date of the incident, November 11, 2006, was performed on August 29, 2006, and nothing unusual was noted in the report. It is unclear whether a careful inspection of the area downstream of the toe of the dam in the incident area, including the South Drain, was performed during the August 29 visual inspection.

Rather than being detected by dam operating personnel, the unusual conditions were detected and reported by a private landowner who operates a livestock feedlot downstream of the dam at the incident area. This was very fortuitous. The dam length where the foundation conditions are similar to the incident area totals about 5 miles. The feedlot area "patrolled" by the rancher amounted to about 1 mile out of these 5 miles, and except for about another ½ mile, the remaining downstream areas are farmlands and open fields where people are not routinely present. Based solely on this information, one might conclude there was roughly a 70 percent chance that this foundation seepage situation would have developed in an area where detection was unlikely (3.5 miles of the 5 mile area did not regularly have a "human presence") and therefore there was a 70 percent chance that the incident could have proceeded (undetected) to dam breach and dam failure in 2006. Other possibilities for detection clearly would exist, which would lower the 70 percent value, perhaps significantly, but on the other hand, assuming a 100 percent detection chance in the 1.5 miles where a human presence routinely exists is clearly not realistic. In summary, based solely on this "human presence" factor, there was roughly a 30 percent chance for the opportunity for "human intervention" to prevent the seepage incident from leading to dam failure (which fortunately did occur).

If an optimistic 75 percent chance of detection (rather than the 30 percent value noted above) is assumed, this would correspond to a 25 percent failure probability (or 1 chance in 4) in 2006. Prior to the 2006 incident, the available evaluation work for the dam indicated a value of 1.4×10^{-4} (or about 1 chance in 7,000) regarding the annual probability of development of this potential failure mode, as noted previously. By early November 2006, all the risk analysis event tree nodes up to the last two nodes ("Intervention Fails" and "Dam Breaches") had actually occurred, making the actual chance of dam failure in 2006 (assumed here to be about 1 chance in 4) much higher than the pre-incident estimated value (about 1 chance in 7,000).

Before leaving this case history, some other noteworthy points bear mention:

- For the 2.5 years prior to mid-May 2006, the record of dam operating personnel performing monthly visual inspections at the dam was flawless. Having only one visual inspection performed in the 6 months preceding the incident at the dam was an aberration, not the norm.

- Performing effective routine visual inspections for a dam that is about 14 miles long is a challenge, to put it mildly. To carry out these inspections when it is hot, humid, and buggy in the summer only adds to the challenge.

Having occasional meetings at the damsite between technical personnel and dam operating personnel concerning how to actually carry out the routine visual inspections of a dam (key areas to focus on, methods that can be used for checking various areas, etc.) is beneficial to all involved parties, and helps to ensure an efficient, yet effective routine monitoring program. Such interactions could have been better for Arthur V. Watkins Dam in the years before the incident occurred.

The beneficial role played by the general public in noting and reporting unusual conditions at dams and downstream of dams obviously is not to be overlooked, as the Arthur V. Watkins Dam case history clearly demonstrates. Consideration might be given to outreach efforts or even brief training programs for landowners residing near and/or immediately downstream of a dam. Signage at a damsite can be used to provide a phone number to call if something unusual is noted by a fisherman or other recreationalist that happen to be at a damsite when something unusual has occurred or is occurring. There have been a number of incidents at Bureau of Reclamation (BOR) dams and U.S. Fish and Wildlife Service (USFWS) dams in recent years where fishermen have detected and reported important performance anomalies.

THE REALITIES OF PERFORMING DAM EVALUATONS

Those involved in designing or modifying dams can use the best methods and practices available, but uncertainties will always exist regarding the completed structure. Unknowns regarding the geology of the damsite will always exist because there are practical limits relative to site exploration work. Construction unknowns will always exist to some extent since nothing is ever constructed in complete conformance with the plans and specifications. Similarly, documentation of construction work will always be somewhat less than total. And those represent the uncertainties or risk factors for new, state-of-the-art dams. The unknowns, uncertainties, and risk factors associated with existing dams, particularly older dams, obviously can be substantially greater.

Evaluations of dams can be performed using the latest techniques and procedures. However, such evaluations will always be limited by what is estimated (or can be surmised) about the material properties associated the dam itself, as well as the dam's foundation and abutments. Unknown geologic defects likely are present for many, if not all dams. State-of-the-practice limitations may exist as well. In short, dam safety evaluations can be performed to the best of our abilities, but realism about the degree of validity and accuracy of those assessments is important. Hence, the need for well-designed and successfully implemented monitoring programs at dams, to address the unknowns and uncertainties, and appropriately manage the risks regarding the initiation/development of potential failure modes for dams (both those that have been identified and evaluated, and those that have not).

COMMON EMBANKMENT DAM FAILURE MODES AND MONITORING

The historical record regarding embankment dam failures indicates that roughly half of the failures are seepage-related. Areas where an embankment dam may fall short of the ideal, "state-of-the-art" structure with respect to seepage risks are many:

- Use of erodible materials (e.g. silts or dispersive clays) in the dam embankment
- Inadequate embankment placement and/or compaction procedures
- Lack of filter protection for seepage paths through the dam, and its foundation and abutments
- Lack of appropriate seepage reduction measures for foundation and abutment flows
- Deficiencies in sealing fractures and openings in foundation and/or abutment rock prior to the placement of embankment material
- Potential for concentrated seepage flow along conduits and at other embankment/structure interfaces due to poor design details (e.g. use of cutoff collars, use of circular pipes without cradles, etc.) and/or poor construction work (e.g. inadequate hand compaction)
- Joint openings, cracks, perforations due to deterioration, etc. in a structure (e.g. outlet works, spillway, etc.) in contact with embankment materials that allows seepage flow to travel into or out of the structure
- Unfiltered toe drains or damaged toe drain installations that consequently do not provide filtering regarding the seepage flows collected
- Plugged or damaged toe drain pipes or other drainage facilities that do not function as intended

While it would be desirable to modify all dams having one or more of the above vulnerabilities/deficiencies to essentially eliminate the risk of seepage-related failure, the reality is that most dams have one or more of these vulnerabilities/deficiencies, and they will not be modified anytime soon to eliminate them. Consequently, appropriately designed routine dam safety monitoring programs, including visual monitoring, seepage and drain flow monitoring, and possibly other instrumentation need to be employed to address the (hopefully) low, but real risks of dam failure. The principle focus would be looking for evidence of material transport by seepage flow, but close attention also would be paid regarding changes in seepage or drain flows, the development of new seepage areas, changes in embankment or foundation water pressures, the development of sinkholes, depressions, and transverse cracks, etc. The routine dam safety monitoring program is particularly important regarding seepage failure modes since many seepage failures historically have occurred after many years of apparently successful dam operation. In some cases this was due to deterioration of materials (e.g. corrugated metal pipes) but often the failure mode just developed slowly, but persistently over time. A study by Patrick Regan published in 2009 [1] reported that approximately half of dam failures that occurred after the first five years of operation occurred after 50 years of operations. In the case of Arthur V. Watkins Dam, the 2006 near-failure incident occurred 42 years after the dam first went into operation.

For embankment dams, the risks of seepage-related failure are typically the most obvious risks to be addressed by a dam's routine dam safety monitoring program. However, other potential failure modes may need to be addressed as well. Some of the more common are noted below:

- Slope instability.- Historically not a cause of many dam failures (as opposed to incidents), but sustained elevated reservoir levels during a flood event may result in downstream slope failure which conceivably could result in a dam breach/loss of control of reservoir water in some situations. Deformation monitoring and/or water pressure monitoring may be appropriate in conjunction with visual inspections, with heightened attention when reservoir levels are unusually high.

- Seepage-related failure during a flood event.- Elevated reservoir levels, and the consequent increased water pressures and seepage gradients, may trigger initiation/development of a seepage failure mode that has not developed or could not develop under normal operating conditions. Also, areas of the dam embankment and abutments may be tested for the first time with reservoir water. The monitoring is similar to monitoring under normal operating conditions, except that the monitoring frequency is greatly increased, particularly in the event of historically high reservoir levels.

- Earthquake-induced slope instability due to liquefaction of embankment or foundation materials.- The role of routine monitoring for this failure mode typically is simply to have baseline data for comparison purposes after the earthquake to understand the impacts of the seismic shaking (deformations, cracks, changed seepage, changed water pressures, etc.). If the risks of this potential failure mode are high, some sort of early warning system regarding dam failure might be appropriate to facilitate immediate evacuation of downstream areas.

- Earthquake-caused damage which results in seepage failure through new cracks, openings, etc.- The role of routine monitoring for this failure mode is again to have baseline data for comparison purposes after the earthquake to understand the impacts of the seismic shaking (deformations, cracks, changed seepage, changed water pressures, etc.). In this case, since dam failure is not essentially instantaneous with the earthquake shaking (like the previous failure mode), if changed conditions at the damsite can be promptly recognized, it may be possible to take actions to prevent failure of the dam, as well as to take timely actions to initiate evacuation of people in downstream areas.

Monitoring Programs

As can be seen regarding the potential failure modes discussed above for embankment dams, monitoring programs can be very valuable in reducing dam safety risks.

COMMON CONCRETE DAM FAILURE MODES AND MONITORING

In the same vein as for embankment dams, monitoring programs can be used to reduce the dam failure risks at concrete dams. For concrete dams, instability of blocks of abutment rock and sliding in the dam foundation have historically proven to be the cause for the majority of concrete dam failures. Austin (Bayless) Dam, St. Francis Dam, and Malpasset Dam are just a few of the most famous examples. For Reclamation's Flaming Gorge Dam, shown in Figure 2 below, the prominent potential sliding planes regarding the shale and bentonite layers present at each abutment led to in-depth evaluations of the potential for a block of abutment rock to move under load thrust from the dam, which could result in sudden, catastrophic failure of the dam. The evaluation work indicated a

Figure 2. Aerial View of Flaming Gorge Dam

low potential for such movement, but still a noteworthy potential failure mode whose risks, though acceptably low, warranted routine monitoring to ensure that no indications of potential problems develop. Increased abutment water pressures would reduce abutment stability, so monitoring of foundation drain performance, uplift pressures, and abutment seepage is performed. Unusual deformations of the dam clearly would be of concern, and an indicator of potential abutment stability issues, so plumbline monitoring and surveyed monitoring of monuments on the dam and on the abutments is performed.

Though seismic loads for Flaming Gorge Dam are not particularly high, earthquake shaking could conceivably initiate movement of a block of abutment rock. The potential failure mode evaluation of the dam performed in 1996 [2] noted the desirability of promptly checking the twenty-one contraction joints within the dam for any offsets in the aftermath of a significant nearby earthquake (which could be an indication of abutment instability issues as well as other possible concerns). Rather than installing a large number of instruments to accomplish this, it was suggested that saw-cut scribe marks could be provided at the joints to allow quick, effective checks for possible movement. This was done at contraction joint locations in two galleries of the dam, one near the top of the dam and one paralleling the dam/foundation contact. Figure 3 shows scribe marks at one typical contraction joint location. The point of mentioning this is to note that

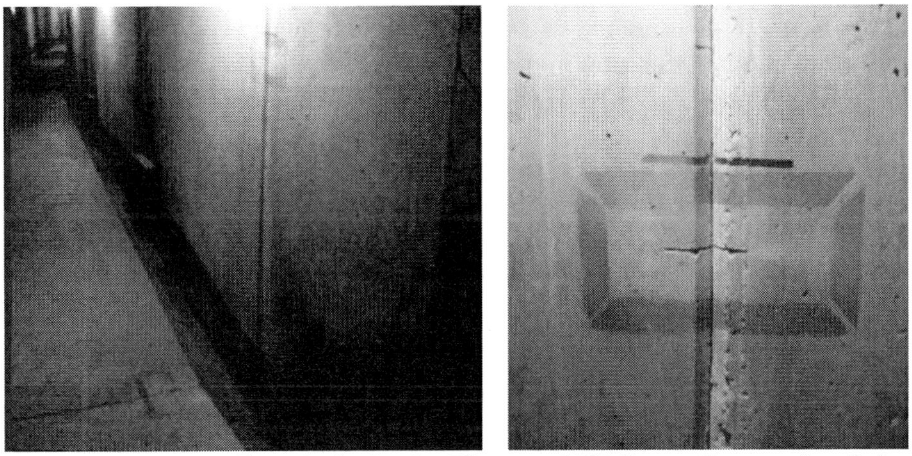

Figure 3. Saw-Cut Scribe Marks in the Floor and Wall at a Contraction Joint in a Gallery at Flaming Gorge Dam (left) and Close-Up View of a Scribe Mark (right)

effective monitoring installations do not need to be fancy, or elaborate, so long as the needed information can be obtained. In fact, simplicity is to be preferred, all other things being equal.

For Reclamation's Yellowtail Dam, shown in Figure 4 below, one of the potential failure modes is solution breccia failure of the upper left abutment. A large amount of seepage

Figure 4. Aerial View of Yellowtail Dam

is known to be traveling through the left abutment of the dam. The concern is that over time, this seepage could gradually have adverse impacts on the structural capability of the rock to support the dam. This potential failure mode is viewed to have acceptably low risks, but the monitoring program for the dam is intended to verify continuing satisfactory performance (that is expected and anticipated), looking for any changes that might develop slowly over time. Monitoring for this purpose consists of plumblines (read twice each year, at the times of maximum upstream and downstream deflection that are well-established by many years of data), and monthly visual monitoring, checking for cracking of concrete that could be structural in nature and observing scribe marks installed at contraction joints (like shown in Figure 3), looking for joint offsets.

OTHER DAM FAILURE MODES AND MONITORING CONSIDERATIONS

History tells us that the most numerous dam failures for embankment dams are seepage-related failures (discussed above) and failures due to flood overtopping that erodes and breaches the dam (where monitoring has little role, other than for awareness of the developing dam failure so that downstream evacuations can be initiated in a timely manner). For concrete dams, foundation and abutment support issues predominate, from the historical numbers standpoint. However, numerous other potential failure mode types are possible for dams, including the following:

- Overtopping of a concrete dam in a flood that leads to erosion of foundation/abutment materials, loss of foundation support for the dam, and dam failure.- The role of monitoring in this situation would be to have adequate pre-flood baseline information available so that post-flood erosion of foundation/abutment materials could be assessed and evaluated. Should the dam fail during the overtopping, the baseline information would obviously be of no benefit. However, in the event that the first overtopping event does not result in failure, monitoring information gathered before and after the flood, as well as during the flood, would allow better evaluation/assessment of the situation so that appropriate reinforcement/protection measures could be implemented in advance of the next, potentially larger flood overtopping event.

- Spillway flow erosion failure.- The spillway fails when called upon to pass large discharges, resulting in an uncontrolled release of reservoir water. Similar to the previous bulleted item, the role of monitoring in this situation would be to have adequate pre-flood baseline information available about the condition of the spillway so that flood-related damage could be effectively assessed. Also, spillway performance information would be gathered during the flood event, looking closely for hydraulic patterns at variance with expected/desired spillway performance. Again, the monitoring information gathered before, during, and after the flood event would allow better evaluation/assessment of the situation so that appropriate measures could be implemented in advance of the next, potentially larger discharge event at the spillway.

- Spillway gate failure.- A spillway gate that has water against it under normal operating conditions could fail, resulting in a dangerous, uncontrolled release of reservoir water. While this could occur under normal operating conditions, failure would be more likely during a flood, or due to earthquake shaking, when loads on the gate and gate support structures would be greater. Thorough inspections (often involving climb teams) as well as evaluations (structural as well as maintenance) would be the appropriate methods to employ regarding minimizing risks associated with this class of potential failure modes.

- Equipment malfunction or failure.- An inability to operate spillway gates as intended in a flood could result in dam overtopping failure. Many other failure scenarios are possible as well. The key is regular exercising and inspection of facilities so confidence exists that performance in an emergency will be satisfactory.

- Operator error.- Similar to the previous bullet, one typically thinks of issues that could arise in a flood event, but other possibilities exist. This general category of potential failure mode highlights the importance of an effective, regularly reoccurring training program for dam operating personnel.

Routine dam safety monitoring may not have a significant role regarding the management of the dam safety risks associated with some potential failure modes, as can be seen regarding some of the bulleted items above. However, for the potential failure modes responsible for the vast majority of dam failures historically, an important role does exist. When thinking about effectively carrying out this role, there are two key elements with respect to intervening to prevent dam failure: (1) detection of anomalous performance and (2) carrying out an effective response. The detection element has two parts. First, an appropriate monitoring program needs to be designed/developed, considering: (1) the potential failure modes that are significant for the dam, and (2) a reasonable program for "general health" monitoring (high value, low cost, typically). Then that program needs to be executed, including not only performing the routine visual inspections and reading the instruments, but also immediately checking the collected data/information to identify anomalous performance in a timely fashion. The second "key element" comes into play when anomalous performance is detected. This involves mobilizing personnel, materials, and equipment, and taking appropriate actions to prevent dam failure.

DORRIS DAM CASE HISTORY

To illustrate how adjustments to a monitoring program can be used to reduce dam safety risks, an evaluation performed regarding Dorris Dam will be discussed. Dorris Dam is a high-hazard, homogeneous embankment dam in northeastern California that is owned and operated by the U.S. Fish and Wildlife Service (USFWS) at the Modoc National Wildlife Refuge (NWR). The dam is about 2 miles long, has a maximum height of about 26 feet, and has a spillway/outlet works structure near its maximum section. A potential failure modes analysis and follow-on risk analysis were performed regarding the dam in 2013 [3] involving personnel from the USFWS and from the Bureau of Reclamation

(BOR). The highest risk potential failure mode (PFM) identified during this evaluation (which will be referred to hereafter as PFM A.4) involved seepage flow along the circular concrete pipes of the spillway/outlet works structure. Figure 5 below shows a section view of this structure, and Figure 6 below shows the downstream end of the spillway and outlet works pipes.

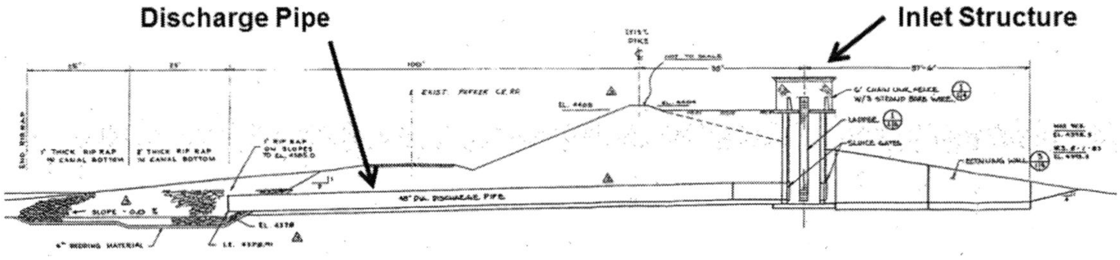

Figure 5. Section View of Dorris Dam Spillway/Outlet Works (along outlet works pipe)

Figure 6. Pipe Outfalls of Dorris Dam Spillway/Outlet Works During Irrigation Season

The center pipe shown in Figure 6 is the 4-foot-diameter outlet works pipe. Two gates are located at the inlet structure shown in Figure 5 that control the outlet works releases. The two outside pipes shown in Figure 6 are 5-foot-diameter spillway pipes. The inlet structure shown in Figure 5 has an uncontrolled overflow inlet associated with each of the two spillway discharge pipes. Figure 7 below shows a photograph taken during construction of the spillway/outlet works. Clearly, compacting material under the haunches of the circular pipes would be difficult, creating a highly vulnerable, currently unfiltered seepage path through potentially low-density material beneath a "roof" provided by the concrete pipe.

The annualized probability of failure associated with PFM A.4 was calculated to be 1.1×10^{-3}, which is an order of magnitude worse than the guideline value of 1×10^{-4}. A contributing factor to the risks associated with PFM A.4 is that from April until December of each year, substantial outlet works discharges are made that raise the water level in the downstream channel (as can be seen in Figure 6). If seepage was occurring

Figure 7. Construction of Spillway/Outlet Works at Dorris Dam (looking upstream)

along or under any of the discharge pipes during this time period, it would be very hard (or impossible) to detect under these conditions, until the flow became very large. Consequently, the dam could fail, or get so close to failing at the time the seepage is detected that failure might not be preventable.

At the other end of the spectrum is the condition that exists at the pipe outfalls from January through March. Figure 8 below shows the situation. Outlet works releases are not occurring, the water level in the downstream river channel is lower, and any significant seepage flow relating to PFM A.4 would be fairly easy to detect (at an early stage in the development of PFM A.4).

Figure 8. Pipe Outfalls of Dorris Dam Spillway/Outlet Works on February 6, 2013

The results of the evaluation of Dorris Dam indicated that it would be appropriate to perform modification work at the dam to provide filter protection around the discharge pipes of the spillway/outlet works structure (as well as around an abandoned outlet works pipe at the dam that also presented very high risks regarding a potential failure mode similar to PFM A.4). Table 1 and Figure 9 show the impact on the dam safety risks of providing this filter protection. (For reference, the Dorris Dam evaluation also noted other dam modification work that could be performed to further reduce the risks of other potential failure modes, to achieve total risk estimates for the dam that closely approach the guideline risk values indicated on Figure 9, but this other work is not relevant to the topics being discussed in this paper.)

Table 1. Dorris Dam - Reduction in Estimated Probability of Development of Potential Failure Modes Before and After Providing Filter Protection Regarding Potential Seepage Paths at the Spillway/Outlet Works Structure and at an Abandoned Outlet Works Pipe

Potential Failure Mode	Annualized Failure Probability Before Providing Filter Protection	Annualized Failure Probability After Providing Filter Protection
Static – Seepage at Spillway/Outlet Works (Red Arrows in Figure 9 below)	1.1×10^{-3}	1.1×10^{-5}
Static – Seepage at Abandoned Outlet Works (Blue Arrows in Figure 9 below)	8.6×10^{-4}	8.6×10^{-6}
Flood – Seepage-Related Failure (Green Arrows in Figure 9 below)	2.9×10^{-5}	5.2×10^{-6}

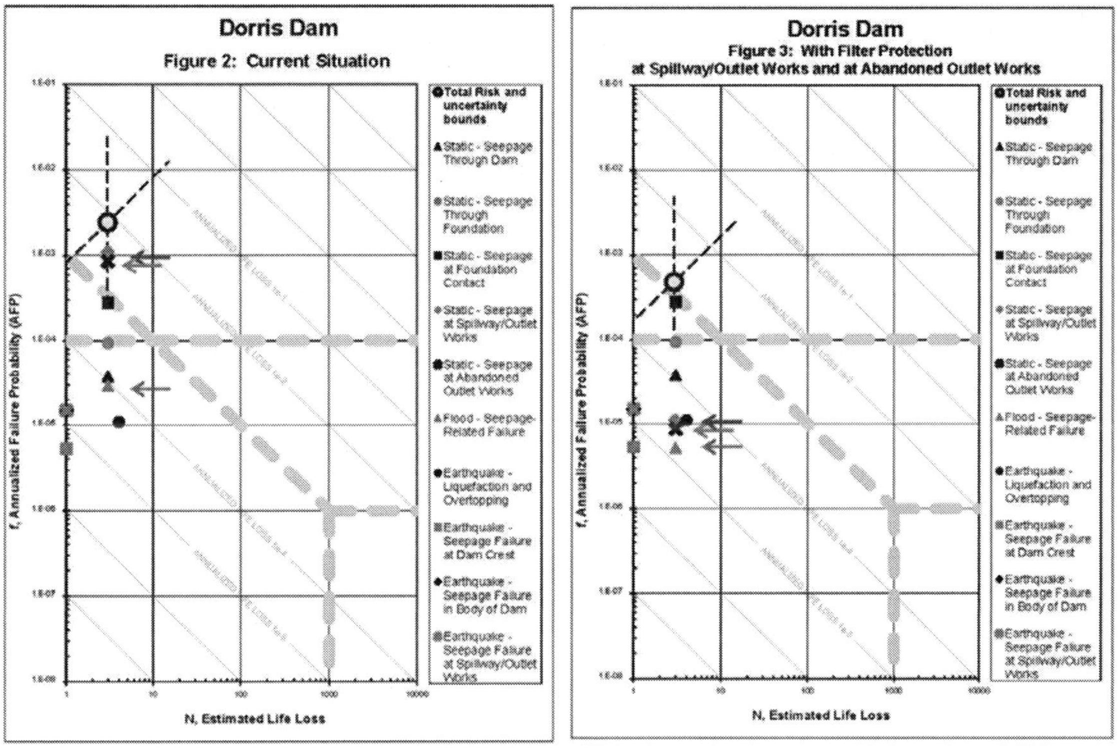

Figure 9. Presentation of Risks for Dorris Dam Before and After Providing Filter Protection Regarding Potential Seepage Paths at the Spillway/Outlet Works Structure and at an Abandoned Outlet Works Pipe

It may be a number of years before the desired structural modification work can actually be accomplished at Dorris Dam. What can be done to appropriately manage the dam safety risks with respect to PFM A.4 in the interim?

In response to this situation, the evaluation included a recommendation to: "Adjust dam operating procedures so that outlet works discharges are halted at least one day each month, so that monitoring for possible seepage flows at or near the spillway and outlet works conduit outfalls can be effectively performed at that time."

Operationally, this recommendation could be implemented without difficulty. Delivery of irrigation water could be delayed one or two days each month without any significant adverse impacts.

Clearly it would be very beneficial to monitoring effectiveness for PFM A.4 if the pool of water downstream of the conduit outfalls could be lowered (like what is shown in Figure 8), so that any seepage flow occurring at or near the conduit outfalls would be very easy to visually note. As shown in Figure 10, roughly one hundred feet downstream of the conduit outfalls is a concrete weir structure that is used to accurately determine discharges from the reservoir. This discharge data is needed for operational as well as legal reasons. The crest elevation of this weir dictates the water elevation downstream of the conduit outfalls when outlet works discharges are temporarily halted.

Monitoring Programs

Figure 10. Weir Structure Downstream of the Pipe Outfalls of Dorris Dam Spillway/Outlet Works

Installing a pipe (with a controlling valve) through the weir structure at as low an elevation as possible would allow water that is otherwise trapped between the weir structure and the conduit outfalls to drain out, which would allow monitoring for seepage flows at or near the conduit outfalls to be much more effectively performed. Providing such a gated pipe was not viewed to be difficult to accomplish, so the evaluation included a recommendation to install such a gated pipe. Less than 2 weeks after the evaluation report was completed and transmitted, the gated pipe was installed by Refuge personnel, as shown in Figure 11.

Figure 11. Gated Drain Pipe Provided at Weir Structure Downstream of the Spillway/Outlet Works Conduit Outfalls at Dorris Dam (looking upstream on left and downstream on right)

With the completion of this work, and with the operational change to halt outlet works discharges at least one day each month, it became possible to effectively monitor for possible initiation/development of PFM A.4 until the day arrives when dam modification work would directly address this PFM. Every month, outlet works flows could be halted, the pool of water upstream of the weir structure could be drained, and an effective inspection for evidence of possible initiation/development of PFM A.4 could be accomplished. The dam safety risks of PFM A.4 could be appropriately managed until the dam modification work could be accomplished, with the plan for detection and response being as indicated in the Table 2 below.

Table 2. Dorris Dam – Managing the Risks for the Development of PFM A.4 (Internal Erosion of Embankment Material by Seepage Flow Along the Alignment of the Spillway or Outlet Works Conduits that Results in Dam Failure by Breaching)

Element of Plan	Aspect	Plan for Accomplishing
Detection.- Detect the anomalous performance	Appropriate monitoring program designed/developed	Pipe outfall area dewatered once per month and checked for seepage
	Data/information collected as planned	Refuge personnel who are doing other dam monitoring work carry out this new monitoring activity
	Data/information promptly reviewed for anomalous performance	If any evidence of seepage is observed, refuge management personnel and USFWS dam safety personnel promptly notified
Response.- Carry out an effective response regarding the anomalous performance	Needed materials and equipment available	Refuge has needed heavy equipment and knows where sand, gravel, and earthfill can be obtained (near the dam)
	Needed personnel available	Several Refuge personnel live near dam. Operators for heavy equipment can be mobilized rapidly, including trained Refuge personnel
	Appropriate plan of action can be promptly developed	Emergency Action Plan has guidance. Training on dam safety response has been performed. Technical support can be promptly obtained from USFWS and BOR dam safety personnel

RISK ANALYSIS EVENT TREE FOR SEEPAGE FAILURE MODES

Thinking more generally about seepage-related potential failures, which account for a large percentage of the historical failures of dams, it is useful to look at the "standard starting-point" event tree that is typically used for them in risk analysis work:

Table 3. "Standard Starting Point" Event Tree for Seepage-Related Failure Modes for Embankment Dams

Node Number	Event Tree Node Description
1	Reservoir at or above threshold level
2	Initiation – Erosion starts
3	Continuation – Unfiltered or inadequately filtered exit exists
4	Progression – Continuous stable roof and/or sidewalls
5	Progression – Constriction or upstream zone fails to limit flows
6	Progression – No self-healing by upstream zone
7	Intervention fails
8	Dam breaches

Looking at the above event tree, where are the opportunities to lower the risks of a seepage-related potential failure mode?

One opportunity would be to lower the normal operating level of the reservoir. Commonly this is referred to as putting in place a reservoir restriction. Relative to the dam failure risks, this favorably affects the probabilities associated with both Nodes 1 and 8, but this is seldom a popular choice since it impacts the water supply benefits provided by the dam, would impact the power that could be generated (if there is a powerplant associated with the dam), and also could affect the recreational benefits provided at the reservoir. The associated lost benefits (i.e. the costs of a reservoir restriction) could be very substantial.

Another opportunity would be to structurally modify the dam to provide filter protection regarding the seepage path being studied, like was discussed above regarding Dorris Dam. Such modification work would dramatically impact the probabilities associated with Node 3. The costs of the work could be substantial, and it likely would take some time before the work could actually be carried out (due to lining up funding, obtaining needed permits, performing the design work, fitting the construction effort appropriately into the operations and activities at the dam, preparing the contract documents, and contracting for the work).

In theory, other options exist for structurally modifying a dam, besides installing filter protection regarding the seepage paths of concern. The probabilities associated with Nodes 5 and 6 could be improved by work done on the upstream face of the dam, but this would be far less cost-effective than providing filter protection on the downstream side of the dam. The probability associated with Node 2 could be improved, but this would essentially require removing and reconstructing large portions of the dam. Trying to address Node 4 is not really an option, and might be counterproductive with respect to Node 2. In short, other options exist in theory for structurally modifying dams with respect to seepage-related failure modes, but realistically, installing filter protection regarding seepage paths of concern (i.e. addressing Node 3) is typically the only truly viable option with respect to dam modification work.

One other opportunity remains. Node 7 relates to intervention efforts to prevent dam failure, which involves both monitoring for evidence of anomalous performance, and then taking action to prevent dam failure should there be such evidence. Using the BOR risk analysis methodology, if it is "very unlikely" for intervention efforts to fail, then the annualized risk estimate for dam failure can be reduced by a factor of as much as 100 (2 orders of magnitude), which is obviously very substantial. To achieve such results, going through an evaluation like is presented in Table 2 is necessary to ensure that all the "Aspects" regarding "Detection" and "Response" are addressed well. As is evident from a review of Table 2, making improvements with respect to Node 7 may not be very costly. It may amount to not much more than using as wisely as possible the time of dam operating personnel, and thinking through plans and possible actions in the event emergency conditions develop. Considering the Dorris Dam situation and the information presented in Table 2, it is viewed that the steps taken to improve the monitoring effectiveness regarding PFM A.4 (and therefore the prospects for successful intervention regarding it), have reduced the risks of dam failure due to PFM A.4 by a factor in the range of 10 to 20. A very substantial improvement.

To bring the story regarding Dorris Dam up to date, design work is currently underway to address the issues raised in the 2013 evaluation of the dam [3], which will include: (1) providing filter protection around the conduits at the spillway/outlet works structure (addressing PFM A.4), (2) providing filter protection around an abandoned outlet works pipe (noted previously), and (3) providing a filtered toe drain and drainage trench along much of the length of the dam to address concerns about seepage at the dam/foundation contact and through the foundation of the dam. The construction work will begin in 2015 and will be conducted in stages as funding allows, with the highest priority work elements (Items (1) and (2) above) proceeding first. Also noteworthy is that a one-day Emergency Action Plan (EAP) exercise at the damsite conducted in November 2014 used PFM A.4 as one of the two exercise scenarios that were carried out. Since PFM A.4 currently presents the greatest dam safety risks, it made sense to use the required EAP exercise activity to maximize the likelihood of an effective response to PFM A.4, should it develop in the time before dam modification work addressing it can be completed.

CONCLUSIONS

There are a great number of unknowns and uncertainties associated with our ability to accurately estimate the risks of dam failure, no matter how hard and carefully we work at it. Unknown geologic defects or conditions, construction unknowns and errors, errors in the estimates used for material properties (even when based on exploration work), inaccuracies in loading estimates, etc. make it important that we have a realistic view regarding the degree of validity and accuracy of our evaluations and assessments. The probability of failure of Arthur V. Watkins in 2006 was diligently estimated to be about 1 chance in 7,000, but actually was closer to about 1 chance in 4, considering the events that unfolded that year. In light of a cold, sober look at the reality of things, it is appropriate to use dam safety monitoring programs to hedge our bets, lower and manage our dam safety risks, and provide appropriate protection regarding the "known unknowns" and the "unknown unknowns."

ACKNOWLEDGEMENTS

Thanks to all the USFWS personnel who contributed to the success of the 2013 Comprehensive Review of Dorris Dam [3], including Constantine Pappas, Steve Clay, Sean Cross, Greg Albertson, Jeff Rose, and Gary Blefgen, as well as BOR personnel who contributed to the project, including Beth Boaz, Lucille Piety, Chris Wood, Feng Su, Dennis Weingardt, Sarah Derouin, and especially Scott Stevens, who peer reviewed the final report and was extensively involved in the risk analysis effort for the project.

REFERENCES

[1] Regan, Patrick J. (April 2009) An Examination of Dam Failures vs. Age of Dams. Technical paper presented at the 29th Annual United States Society on Dams Conference in Nashville, Tennessee.

[2] Stateler, Jay N., Von Thun, J. Lawrence, Scott, Gregg A. et al. (July 1996) Performance Parameters for Flaming Gorge Dam. *Technical Memorandum No. FLG-8130-CFR-TM-95-1*, Bureau of Reclamation.

[3] Stateler, Jay N. et al. (November 2013) Comprehensive Review – Dorris Dam – Modoc National Wildlife Refuge – U.S. Fish and Wildlife Service.

CONSTRUCTION AND MONITORING OF EXTRACTION WELLS AT THE BLUE RIDGE DAM EMBANKMENT

Peter Zimmerman, PE, PG[1]
Lindsay McMichen, PG[2]

ABSTRACT

Movement of the Blue Ridge Dam embankment during the seismic remediation led to a heightened dam safety concern. Numerous piezometers, observation wells, inclinometers, and survey monuments were installed and monitored following the slope movement. TVA was committed to raising the Blue Ridge reservoir to normal summer pool levels in 2013, and features designed to raise the global stability factors of safety for non-seismic loading were constructed. These features included installing extraction wells to lower the phreatic surface in the embankment, adding rock fill buttressing at the toe of the embankment to improve stability, installing gabion walls and an L-wall at the toe, and re-grading the rock fill slopes. Extraction well installation and operation at Blue Ridge Dam was begun only after careful consideration of risks and preparation of an instrumentation and monitoring program. The wells were designed to limit the risk of sediment removal from the embankment by screening the wells within the rock and weathered rock foundation and by installing the wells with a suitable sand filter. Embankment settlement was a concern and all soil piezometers and observation wells were assigned target elevations. Pumping was to be slowed or stopped if piezometric levels in the instruments fell below the target elevations. The intense monitoring that occurred following movement of the downstream slope and the observed performance of the dam during a prolonged flooding event increased confidence in the long-term performance of the dam. TVA elected not to install additional extraction wells and instead continue their robust monitoring program at the dam.

INTRODUCTION

Blue Ridge Dam, owned by the Tennessee Valley Authority (TVA), is located at River Mile 53.0 on the Toccoa River in northeast Georgia. The project was constructed in 1930 by the Toccoa Power Company, a subsidiary of the Tennessee Electric Power Company and was acquired by TVA in 1939. The Blue Ridge Project provides flood control, hydroelectric power generation, water supply, water quality, recreation and enhanced aquatic ecology benefits. At normal summer pool, the reservoir has a capacity of approximately 200,000 acre-feet and a surface area of 3,330 acres. The project consists of a large semi-hydraulic earth fill embankment, a powerhouse, a gated spillway and an ungated auxiliary spillway.

The dam embankment is approximately 175 feet high at its maximum section, 950 feet wide at its base, and approximately 1,125 feet long along its axis. The crest width is approximately 40 feet and includes a two-lane highway at approximate Elevation (El.)

[1]ARCADIS, 1210 Premier Drive, Suite 200, Chattanooga, TN 37421, peter.zimmerman@arcadis-us.com
[2] Tennessee Valley Authority, 1101 Market Street, Chattanooga, TN 37402, lcooper@tva.gov.

1710. The original design included a downstream rock fill toe that is now covered with a rock fill buttress. A small cutoff trench approximately 15 feet deep and 50 feet wide was included under the centerline of the embankment; however, the details of foundation treatment in the trench are unknown. Figure 1 shows a typical section of the downstream portion of the embankment.

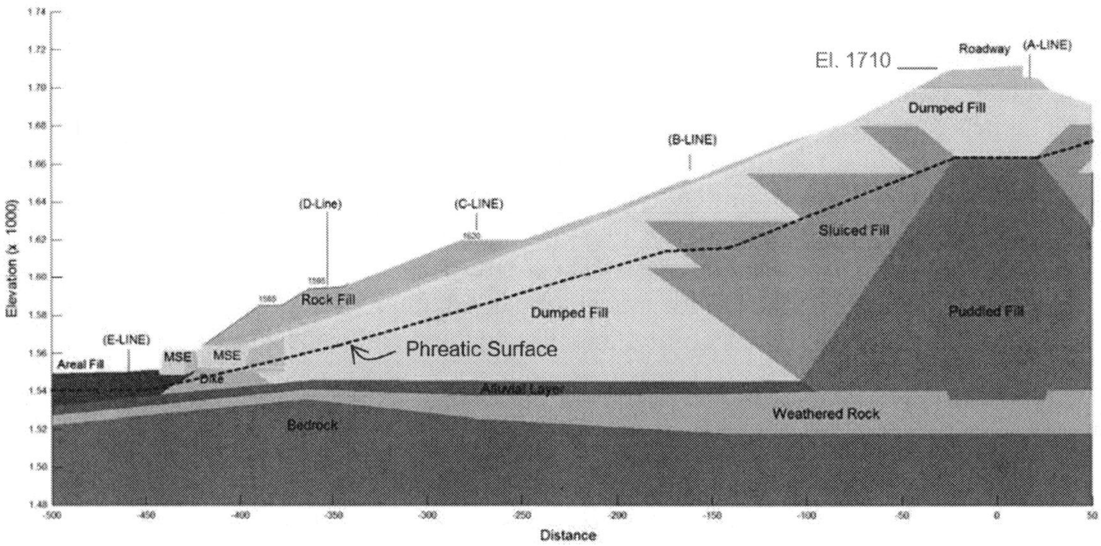

Figure 1. Typical Embankment Section

Typical of dams built in this era using semi-hydraulic construction techniques, the embankment is segregated into three general zones: outer shells of dumped fill, intermediate shells of sluiced fill, and a central puddled core. Based on historical drawings and photographs, the embankment was constructed by first placing earth fill retention dikes upstream and downstream of the embankment centerline. These dikes were raised by dumping fill from side-dump rail cars running on timber trestles constructed parallel to the embankment axis. As the dikes were raised, the dumped material was sluiced down the inner slopes by water cannons operating from floating platforms between the dikes. The sluicing action presumably created an intermediate zone of coarser material and a central puddle core of fine-grained material. As the outer dikes reached design grade, trestles were abandoned and new ones were constructed higher and closer to the dam core. This process of dumping, sluicing, and raising was repeated to eventually reach the final design grades and elevations.

The dumped fill generally ranges from micaceous silty and gravelly sand to sandy silt, the sluiced fill consists mainly of micaceous silty sand, and the puddle core includes sandy silt and silty sand. The alluvial soil located below portions of the embankment is composed of low plasticity clay and silt. The construction techniques utilized for fill placement introduced anisotropy to the sluiced fill and puddle core materials. The permeability of these materials is likely greater in the horizontal direction than in the vertical direction. In the case of the sluiced fill, the direction of the anisotropy is expected to be approximately 45 degrees from vertical since the sluiced fill layers were constructed on a sloping surface from the horizontal dumped fill lifts. Hydraulic conductivity for use in the seepage analysis was estimated based on consolidation tests, cone penetration test (CPT) dissipation tests, variable head permeability tests, and field slug tests. Hydraulic conductivity in the weathered rock and rock layers is very dependent on the presence of fractures and joints. The hydraulic conductivity of the weathered rock tends to be greater than that of both the underlying rock and the overlying alluvial or fill layers due to the prevalence of joints and fractures enlarged or created during the weathering process.

Soon after construction of the dam was completed, the penstock that runs through the dam was damaged by excess water pressure. In 2010, TVA began implementation of a plan to repair and improve the penstock, stabilize the intake tower for seismic loads, and improve the seismic stability of the embankment dam. Construction activities associated with the seismic remediation of the embankment included adding a granular filter on the downstream embankment surface, placing rock fill of varying thickness over this filter, and constructing two mechanically stabilized earth (MSE) walls at the embankment toe. At its maximum section, the rock fill was approximately 25 feet thick. Figure 2 shows construction of the filter layer and rock fill berm.

Figure 2. Photos from the Blue Ridge Dam seismic rehabilitation.

During rock fill placement in February 2012, signs of downstream slope movement developed. Cracking occurred in the upper portions of the downstream slope, the MSE walls began moving, and a small penstock connecting the surge tank to the low-flow hydropower unit began leaking. When the small penstock was disassembled at one of its mechanical joints, the pipe sprung apart by nearly 3 inches. It was apparent that slope movement had occurred and might still be occurring. Rock fill placement was quickly suspended, and interim risk reduction measures (IRRMs) were implemented. The IRRMs included lowering the reservoir, removing rock fill from the upper portions of the embankment, and placing rock fill at the toe of the dam. A comprehensive investigation of the embankment stability was simultaneously initiated.

With the interruption of construction activities, the overall schedule quickly became a significant issue. The Blue Ridge reservoir had been lowered for the penstock repairs in 2010 and had not yet been returned to normal operating levels. Because of the stability concerns and to serve as an IRRM, plans to return the reservoir to its normal operating levels were suspended. The continuation of the lowered reservoir level had substantial economic consequences for the Blue Ridge region. It was determined that risk reduction measures should be implemented as soon as possible in hope that that the reservoir could return to normal operations by February 2013.

In lieu of seismic remediation measures, efforts became focused on developing measures that improved the static stability of the embankment. The goal for the design team was to develop interim remediation measures that would increase the static stability factor of safety (FoS) of the dam to a level acceptable for resumption of normal reservoir operations by 2013. The interim condition of the dam would be required to satisfy TVA's static slope stability criteria while a permanent design was analyzed and constructed. A significant factor in selection of the interim remediation was the time constraint of having construction substantially completed for the raising of the pool in February 2013. The interim design was not intended to satisfy seismic criteria.

The primary goals for design of interim remediation included:

- Achieve and maintain an acceptable FoS for all reservoir levels up to and including normal summer pool using a combination of buttressing and control of the phreatic surface.
- Develop a constructible and operable design that improves existing conditions while doing no harm to the long-term operability of the dam.
- Provide instrumentation (monitoring wells, piezometers, inclinometers, and survey monuments) to monitor embankment behavior during construction.
- Maintain appropriate control of the phreatic surface for stability and minimize embankment deformations.

The selected interim remediation was a combination of dewatering wells, buttressing, and re-grading of the lower portions of the embankment. Major components that were considered for the interim remediation include the following items:

- Extraction wells to lower the phreatic surface in the embankment.
- Rock fill buttressing at the toe of the embankment to improve stability.
- Gabion walls at the toe near the left abutment for grade separation.
- A reinforced concrete L-wall in the vicinity of the surge tank, thrust block, small penstock, and low-flow generator building. The L-wall was necessary due to the lack of space for the conventional rock fill buttress.
- Re-grading of existing rock fill slopes

A pilot program to test the effectiveness of dewatering wells for controlling piezometric levels within the embankment was undertaken early in the construction of the interim remediation.

INSTRUMENTATION AND MONITORING

Instrumentation at Blue Ridge Dam includes numerous piezometers and observation wells, inclinometers, seepage weirs, and survey monitoring points. Much of the instrumentation was used to monitor embankment response to operation of the dewatering wells. In particular, piezometers and observation wells were closely monitored during initial pumping tests and during operation of the extraction wells.

Piezometers and instruments at the dam are generally organized into 5 lines that run parallel to the dam axis. A-Line instruments are located along the embankment crest; B-Line instruments are located in the vicinity of a bench at approximately El. 1640; C-Line instruments are located along or near the El. 1620 bench; D-line instruments are located along the El. 1594 bench; and, E-Line instruments are located along the embankment toe. The location of these lines is shown in Figure 1.

The piezometers and observation wells used at Blue Ridge Dam include those installed prior to construction, instruments installed during the initial construction, and instruments installed as part of the monitoring program for the interim static remediation. Numerous historic piezometers and wells were damaged when rock fill induced settlement caused standpipes to buckle or crack. Damage to the piezometers and wells was determined by downhole camera prior to remediation construction activities. Damaged wells and piezometers were replaced in-kind prior to commencement of construction.

The initial extraction well installation occurred on the western portion of the embankment and the piezometer and monitoring well locations in this area of the dam are shown in Figure 3. The type of instrument shown in Figure 3 can be determined by the first two letters of its name. Nested piezometers are indicated by the letters A, B, and C at the end of the instrument name. The following general types of instruments are represented:

- OW – observation wells have relatively long screens that may be screened across multiple fill or foundation layers.
- CP – Casagrande piezometers with screens installed in a specific soil layer.
- VP – vibrating wire piezometers grouted in place with cement bentonite grout.
- IP – vibrating wire piezometers attached to the outside of inclinometer casing during installation and then grouted in place.

DEWATERING WELL DESIGN AND CONSTRUCTION

The objective of the interim remediation was to produce conditions that provide a static global stability FoS of 1.5 or greater for the normal summer pool condition. The design life of the interim remediation was planned to be 3 to 5 years and a primary feature of the proposed concept was an extraction well system to lower the phreatic surface within the

Figure 3. Piezometer and observation wells on the western portion of the Blue Ridge Dam embankment

embankment. In addition, a rock fill buttress at the embankment toe was also a primary feature.

The concept of the extraction well system was to lower the phreatic surface through the embankment by installing wells with submerged pumps that would operate full time until a permanent remediation was made. Installing and operating extraction wells within the embankment introduced new potential failure modes to the project and the well system was designed to minimize the risks associated with long-term pumping. The two major risk factors that were identified are that the extraction wells could pull sediment from the embankment to form voids and increased seepage gradients, and that lowering the piezometric level would cause unacceptable settlement and deformation of the embankment.

The wells were designed with sand filter gradations that would prevent piping of embankment and foundation materials. The filter design followed U.S. Army Corps of Engineers criteria (USACE, 1986 & 1992). The final design also included settling basins for the well discharge that would allow observation of the well effluent and collect sediment that might be carried in pumped water. The well design is shown in Figure 4.

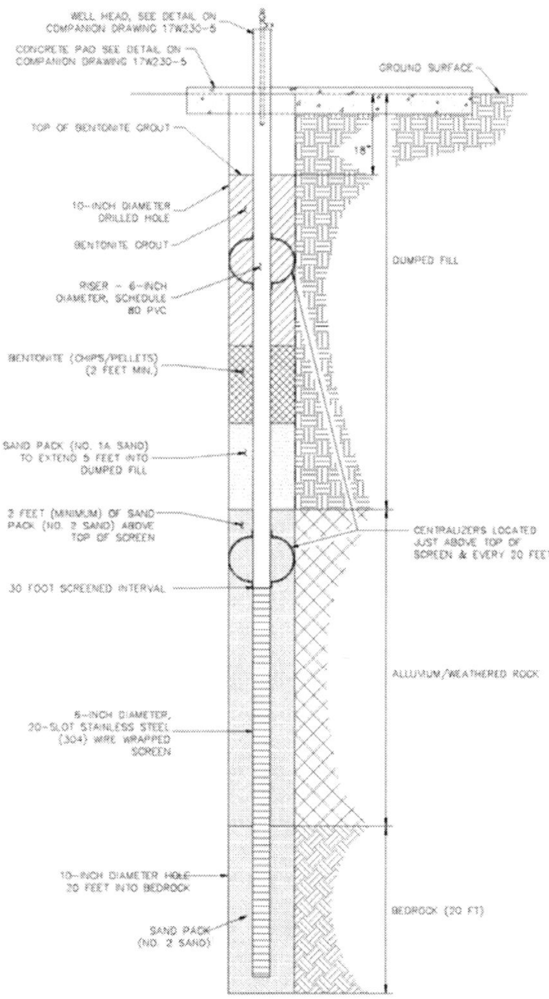

Figure 4. Pumping well design.

Given the compressible nature of the embankment soil, an increase in effective vertical stress due to lowering of the phreatic surface would likely cause settlement within the embankment and foundation soils. One way to minimize the settlement was to restrict the drawdown to elevations equal to or higher than the piezometric levels exhibited during typical winter pool drawdown.

Construction and Monitoring of Extraction Wells

During drawdown of the phreatic surface, the embankment was monitored to ensure that excessive settlement did not occur. Of particular concern was the differential settlement that might occur across the embankment. Excessive differential settlement could cause embankment cracking, which might lead to a serious dam safety problem. The amount and time of embankment settlement was estimated based on consolidation testing and observed drawdown. Threshold values (both amount and rate of settlement) were established for the tolerable differential settlement across a defined length of the embankment (for example, between adjacent settlement monuments).

Pumping was to be reduced or stopped to restrict the lowering of the phreatic surface if measured settlement began to approach threshold values. Calculations for estimated primary consolidation settlement were performed to provide a range of anticipated settlement that could occur for a variety of drawdown levels. Consolidation parameters were taken from the average of the consolidation test results for the dumped fill and the alluvium/residuum layers. Settlement was evaluated at each location for increases in vertical stress resulting from one foot increments of drawdown.

Settlement was calculated for a range of scenarios based on historic piezometric data and slope geometry. The actual amount of settlement was expected to be somewhere within the range of the high and low scenarios used in the calculations. Figure 5 presents the anticipated settlement with drawdown determined by averaging the two most likely consolidation scenarios at Sta. 22+00. Separate curves are shown for each of the well locations.

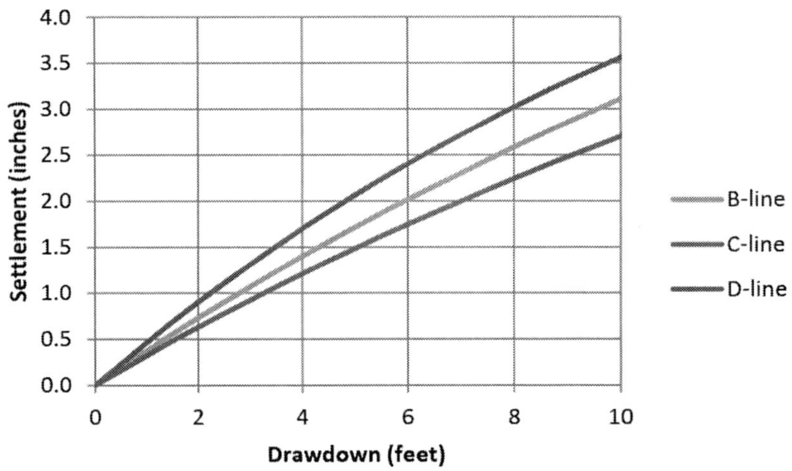

Figure 5. Anticipated Settlement vs. Drawdown

The Blue Ridge Dam operating guide calls for the winter pool to generally be lowered approximately 15 feet compared to summer pool. Little to no settlement was anticipated with the phreatic surface being maintained at or near historic winter pool operating levels. Visual monitoring for depressions near the well locations was undertaken even while attempting to maintain the current phreatic surface. Depending on the steepness of the cones of depression, differential settlement and cracking was assumed to be possible.

The likelihood of excessive or differential settlement was reduced by limiting the amount of targeted drawdown. The phreatic surface was closely monitored with observation wells and piezometers during pumping. Survey monitoring points were also observed during pumping activities. Any excessive differential settlement would be cause to stop the pumping and reassess conditions.

PILOT WELL PROGRAM

The extraction wells were designed to be screened within the weathered rock and rock foundation of the embankment. These formations provided high pumping capacity when the wells intersected joints and fractures within the rock. Five wells were installed as part of the pilot program and two of the wells, B-4 and C-2, had adequate capacity for prolonged pumping. Step-drawdown tests and short-term constant rate pumping tests were initially performed, and the two wells are still in operation as of January 2015.

The extraction well system was designed to lower piezometric pressures within the relatively permeable weathered rock and rock foundation. This would create downward gradients that would, in turn, lower piezometric pressure within the overlying fill by allowing drainage into the foundation. Pressure drops were not limited within the rock and weathered rock layers, but piezometric levels within the embankment fill were not allowed to go below the designated threshold elevations. Solinst Leveloggers were deployed in 25 piezometers and observation wells prior to commencing pumping tests. The Leveloggers recorded data at one minute intervals during the pumping tests.

Extraction Well B-4 was installed with a 30-foot screen in weathered and fractured bedrock. The step-drawdown test at B-4 was run at 10, 20, and 23.5 gallons per minute (gpm) for 120 minutes at each step. The final step was planned for 30 gpm but was reduced to 23.5 gpm due to back pressure in the discharge line. The approximate radius of influence was observed to be 220 feet in D-line piezometer OW2370D. Figure 6 shows drawdown in a number of wells monitored during the step test.

The wells at B-4 and C-2 were pumped at a constant rate from February 20 to March 7, 2013. The pumps were shut off until March 22 and then the wells were pumped at a constant rate from March 22, 2013 to September 17, 2014 when the pumps were shut off. The pumps were returned to service on October 4, 2014 and continue to be pumped. Observation wells and piezometers were monitored daily through construction of the interim remediation measures. Monitoring was moved to a monthly schedule in July 2013. The Blue Ridge reservoir was brought up to normal summer pool level through the spring of 2013 with the reservoir reaching summer pool in the beginning of May 2013. Figure 7 shows the C-line observation well water elevations from instruments screened within the rock and weathered rock foundation from January 26, 2013 to June 25, 2013. Figure 8 shows data from the C-line observation well and piezometers from instruments screened or installed within the alluvial foundation soil layer and the embankment fill.

The pumps in wells C-2 and B-4 were temporarily turned off on September 17, 2014 to monitor embankment performance and piezometer response with the wells inactive. On

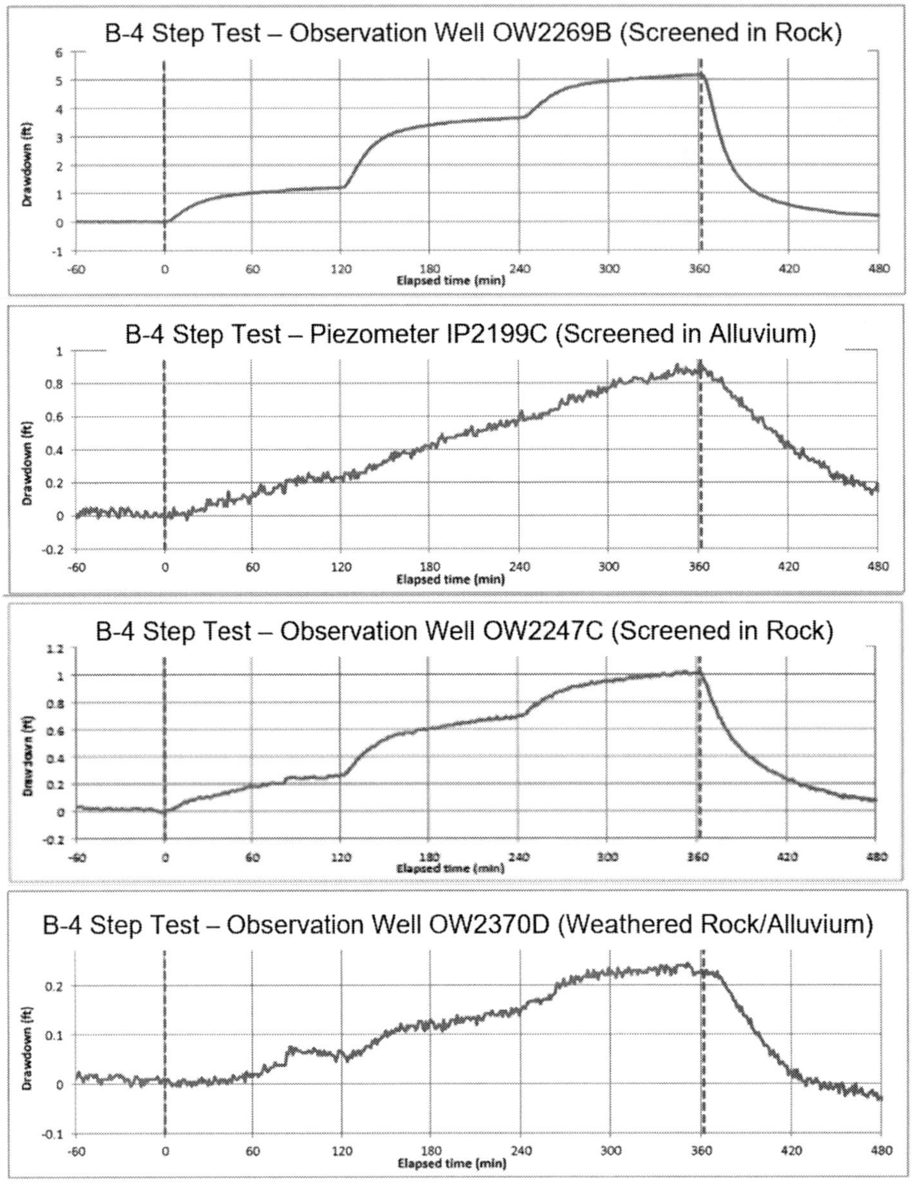

Figure 6. Piezometer and observation well measurements during step-testing at B-4

September 16, initial water levels were recorded manually at all instruments and 20 Levelogger data loggers were installed in open standpipe piezometers and observation wells with the loggers suspended 20-feet below the water elevation in each standpipe. Additionally, water level data was collected from 16 vibrating wire (VW) transducers for a total of 6 days after the pumps were turned off. C-line piezometers generally showed the biggest influence from pumping although a number of D-line and B-line instruments also were influenced. Table 1 shows the C-line instruments' maximum increase in piezometric level that was recorded within the 6 days of monitoring following termination of pumping.

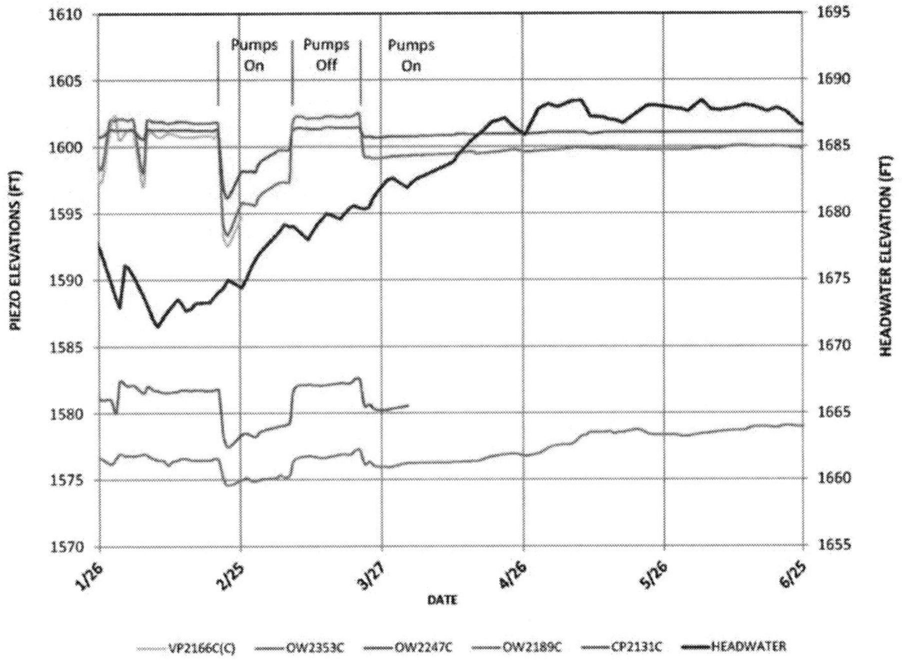

Figure 7. Observation wells and piezometers screened within the rock and weathered rock foundation

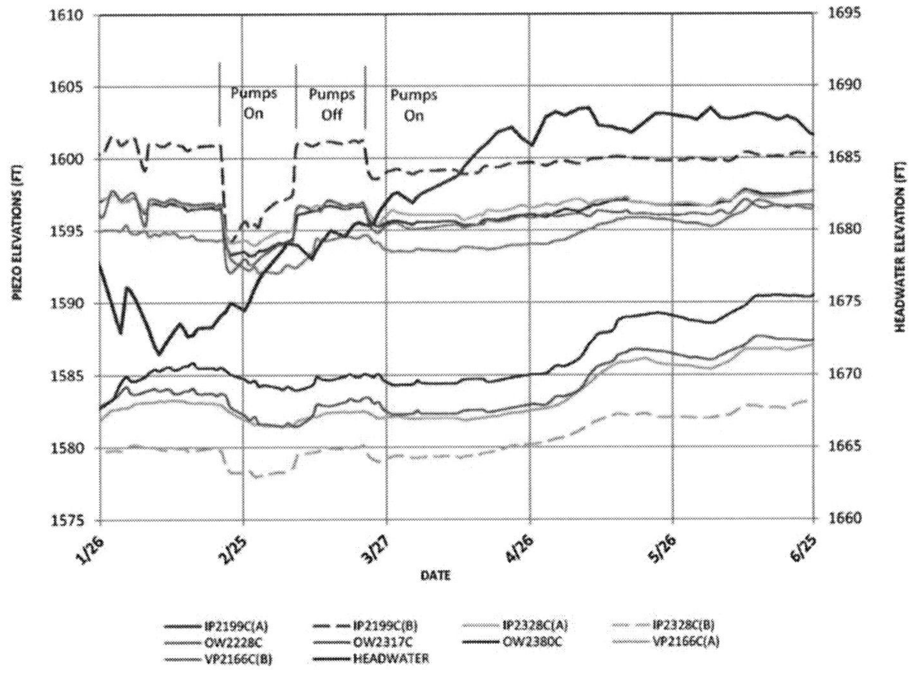

Figure 8. Observation wells and piezometers screened within the alluvial layer or embankment fill

Well ID	Distance from Well C-2 (ft)	Soil or Foundation Unit Screened	Increase in Piezometric Level After Termination of Pumping (ft)
VP2166C(A)	11	Fill	1.8
VP2166C(B)	11	Fill	1.5
VP2166C(C)	11	Alluvial	3.3
OW2189C	34	Rock	3
IP2199C(A)	44	Fill	1
IP2199C(B)	44	Alluvial	1.6
OW2247C	92	Rock	2.8
OW2317C	162	Fill	1.8
IP2338C(A)	183	Fill	0.6
IP2338C(B)	183	Alluvial	1.1
OW2353C	198	Rock	2.5

Table 1. Increase in piezometric level of the C-line instruments following termination of pumping

An evaluation of the piezometer observations collected on September 30, 2014 revealed that 22 piezometers had maximum of record readings when compared to post-construction baseline readings from July 2013. Piezometers IP2196D (A) and CP2178D (A) exhibited pore pressure elevations above the threshold limits that were established in summer of 2013, and pore pressure in each of these piezometers increased 0.4 feet when the extraction well pumps were turned off. The two extraction wells were returned to service on October 4, 2014 and continue to be operated as of January 2015.

The initial well system design included pumping wells along the entire length of the embankment, but TVA eventually elected not to install additional wells and to only operate the two productive pilot wells. This decision was based in part on performance of the embankment and on the significant instrumentation monitoring at the site.

The Blue Ridge Reservoir operating guide calls for the reservoir elevation to rise from Elevation 1670 in mid-February to Elevation (El.) 1687 in late-May, and construction of remediation features at the dam was ongoing in mid-February. The dam experienced reservoir elevations above El. 1681 in December 2012 and January 2013 due to heavy rainfall and flooding. The increased reservoir level provided an opportunity to observe the dam at a reservoir elevation not experienced since lowering the reservoir ahead of construction in July 2010. The embankment seepage and slope stability analyses were updated based on the December and January piezometer observations and found that the expected slope stability FoS at the end of construction was 1.43. Inclinometers within the dam embankment and at the downstream toe did not show movement that could be directly correlated to the rise in reservoir level.

TVA's dam safety criteria require the FoS against slope instability be equal to or greater than 1.5 during normal, long-term operations. However, smaller factors of safety, as low as 1.3 are permitted on a case-by-case basis for temporary conditions such as those which may arise during construction. The criteria may be adjusted for specific projects with the permission of TVA's Dam Safety Governance (DSG) group. DSG and the Blue Ridge Project Review Board (PRB) concurred with the recommendation to accept a post-construction stability FoS that is less than 1.5 due in part to the following site-specific exceptions:

- The target FoS for the interim design was significantly higher than the probable historical FoS for BRH dam through most of its operational history
- Thorough geotechnical investigations and monitoring programs had reduced uncertainty in the characterization of the dam (stratification, physical properties, likely failure modes, and seepage patterns)
- Geotechnical properties used in the stability analysis were considered to be conservative representations of actual soil strengths in the embankment and its foundation.
- The interim design was considered to be short-term with a design life of three to five years.
- The stability and performance of the embankment was to be verified through ongoing monitoring of instrumentation.

SURVEILLANCE AND INSTRUMENTATION MONITORING PLAN

Interim remediation features, designed to increase the slope stability FoS of the dam to a level acceptable for resumption of normal reservoir operations in 2013, were constructed between December 2012 and July 2013. The interim condition at the dam was intended to satisfy the static slope stability criteria for normal operation of the reservoir. During this interim period, TVA undertook an evaluation of the need for additional modifications to the dam to satisfy seismic stability requirements.

The instruments monitored at the site include instruments that pre-date construction as well as those added during construction. Instruments monitored in the Interim Post-Construction phase included:

- Piezometers and observation wells
- Inclinometers
- Manhole discharges
- Survey monitoring points

Threshold values and/or Action Levels were established for all instrumentation. The threshold and action levels were defined based on slope stability analysis values or on the expected dam performance depending on the type of instrument and its applicability to the analysis. Data from instruments such as piezometers and monitoring wells were compared directly to the slope stability analysis and were therefore assigned threshold limits. Inclinometers and survey data were not directly comparable to the stability analysis and the data from these instruments was assigned action levels based on comparison to historic trends and expected dam performance.

The threshold levels for piezometers and observation wells were established by first comparing the field measurements of each instrument in the vicinity of the three analysis cross sections. Next, a piezometric line was developed by assigning elevation points at the various instrument locations that were greater than the typical field measurements for those instruments. The assigned elevation points at each instrumentation location were then connected to form a piezometric surface for use in the slope stability analysis. The piezometric line was then varied upward within the slope stability model until the analysis resulted in a 1.3 slope stability FoS. Threshold Levels were assigned based on the piezometric elevation that was expected to result in a slope stability FoS of 1.3. Figure 9 shows the July 3, 2013 observed piezometric elevations compared to the threshold level piezometric line for one of the analysis cross sections.

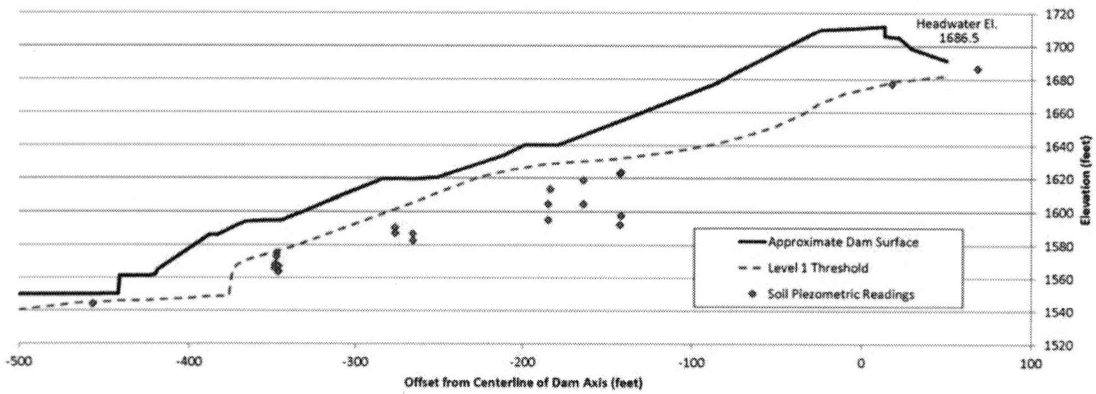

Figure 9. Field instrumentation measurement elevations compared to the threshold piezometric line.

Threshold values for evaluation of instrumentation and survey data were established for various levels of action. Exceedances of thresholds generally triggered increased monitoring and evaluation of the instrumentation, survey, and review of inspection data as a whole to identify stray data or actual trends which could be indicative of a change in the structure.

CONCLUSION

Movement of the Blue Ridge Dam embankment during the seismic remediation led to a heightened level of concern with the safety of the dam. Numerous piezometers, observation wells, inclinometers, and survey monuments were installed and monitored. Geotechnical analyses were updated using the soil strength data and other subsurface information collected during the instrument installation and other subsurface investigations.

TVA was committed to raising the Blue Ridge reservoir to normal summer pool levels in 2013, and features designed to raise the global stability factors of safety were constructed in the first half of 2013. These features included installing extraction wells to lower the phreatic surface in the embankment, adding rock fill buttressing at the toe of the embankment to improve stability, installing gabion walls and an L-wall at the toe, and minor re-grading of existing rock fill slopes.

Extraction well installation and operation at Blue Ridge Dam was begun only after careful consideration of risks and preparation of a robust instrumentation and monitoring program. The wells were designed to limit the risk of sediment removal from the embankment by screening the wells within the rock and weathered rock foundation and by installing the wells with a suitable sand filter. Embankment settlement was a concern and pore pressure heads within all soil piezometers and observation wells were assigned target elevations. Pumping was to be slowed or stopped if piezometric levels in the instruments fell below the target elevations. A limited pilot well program seemed to confirm that a system of pumping wells would be effective in lowering the piezometric elevation within the embankment.

The intense monitoring that occurred following movement of the downstream slope and the observed performance of the dam during a prolonged flooding event increased confidence in the long-term performance of the dam. TVA elected not to install additional extraction wells and to instead continue their robust monitoring program at the dam. All instruments are assigned threshold levels, and actions are defined for the event in which the thresholds are exceeded.

REFERENCES

US Army Corps of Engineers (USACE), 1992, Engineering Manual No. 1110-2-1914, Design, Construction, and Maintenance of Relief Wells

US Army Corps of Engineers (USACE), 1986, Engineering Manual No. 1110-2-1901, Seepage Analysis and Control for Dams

PERFORMANCE MONITORING TO HELP MANAGE RISK FOR DAMS AND IMPOUNDMENTS

W. Allen Marr[1]

ABSTRACT

Every dam and impoundment presents risks due to the many uncertainties about how it will perform and the large consequences of an uncontrolled release. Undesirable performance can occur unexpectedly due to sudden loads, hidden changes or unseen degradation of materials. A seemingly well performing structure based on visual inspections can suddenly display indications of poor performance and fail within hours.

Performance monitoring with visual surveillance and data from instrumentation is frequently used to provide information on the inner workings of the facility over time. An effective performance monitoring program can give indications of unacceptable performance in time for corrective measures to be taken to avoid failure. The combination of using observations and measurements to give early warning and taking action to avoid failure and/or reduce the consequences of a failure provides a powerful approach to reduce the risk of failure of a dam or impoundment.

This paper examines the ways in which performance monitoring can help reduce the risk of failure. It describes the components of a performance monitoring program that are required for it to be an effective risk management tool. It includes three cases to illustrate the use of a performance monitoring program to reduce risk. It concludes that an effective performance monitoring program can reduce the probability of failure by one to two orders of magnitude and potentially reduce consequences by similar amounts.

INTRODUCTION

Performance monitoring provides a cost effective way to reduce the risk of dam failure. For the purposes of this paper, failure is considered to be the uncontrolled release of stored contents from the dam. By indicating undesirable performance early, a performance monitoring program can provide time to perform mitigation measures to reduce the probability of a failure, or reduce the consequences of a failure, or both.

Performance monitoring consists of information from visual inspections and quantitative data from instruments. Both elements are important and the two are complementary. Visual inspection may reveal poor performance not detected by instruments. Loss of fines by piping may be detected by visual inspection while instrumentation may provide no clue that piping is occurring. Contrarily, visual inspections are limited to exposed surfaces and cannot detect conditions that may be deteriorating within the structure. Pore water pressures may be increasing within the dam and lowering the global stability but without surface manifestation until failure develops quite suddenly.

[1] Founder & CEO, Geocomp Corporation, 125 Nagog Park, Acton, MA 01720, wam@geocomp.com

Though only a small percentage of dams develop problems and even fewer of those fail, the highly indeterminate nature of each dam makes it impossible to accurately predict which dam will develop a problem. The many unknowns about the properties of the materials and the large number of possible variations in conditions can never be fully revealed. Therefore, it is prudent that any dam that may affect public safety have a performance monitoring program to monitor its vital signs.

Performance monitoring by itself does not reduce risk. It must be accompanied by an action plan that when put in effect counteracts the factors driving the dam toward failure and stops or reverses the course of deteriorating safety. The invoked action plan mitigates risk. Performance monitoring for dam safety is an instrumentation, monitoring and mitigation program, or I-M-M program in short.

Dams fail because of unknowns about their condition and performance, wrong actions by people, deterioration of materials over time, or loading conditions that exceed those used in the design. An I-M-M program is used to detect these conditions with the aim to intervene and correct them before the dam fails. Dam safety programs repeatedly detect developing problems in time for corrective actions to be completed and, as a result, serve as a highly effective risk management strategy for owners of dams. Most of these actions are never published so we don't have a record of the effectiveness of this approach. An exception is the USBR-USACE (2012) report which indicated that intervention following the detection by visual observations of the initiation of internal erosion resulted in less than 1% of the observed cases causing a dam breach.

A dam safety program aims to protect life, property, and the environment by ensuring that a dam is designed, constructed, operated, and maintained as effectively as possible. It consists of activities to continually inspect, monitor, evaluate and document the design, construction, operations, maintenance, rehabilitation, and emergency preparedness of each dam, as well as the coordination with the potentially affected public agencies having responsibility for public safety. It also includes a clear plan that defines responsibilities for each component and communicates the state of the dam safety program to all with responsibility. Visual inspections and monitoring of instruments are vital parts of many dam safety programs.

Many existing dams are not well monitored. Instrumentation installed for monitoring during construction and first filling may need to be adjusted for conditions observed during routine operations. In such cases, well designed and implemented performance indicators, such as systems to monitor deformations, seepage, and turbidity, combined with periodic visual inspections, must be in place to monitor the overall performance of the dam. Technologies exist to monitor almost any aspect of dam performance and safety. The challenge is to organize and manage the human and financial resources to apply these tools in an effective and efficient safety program and to act quickly to implement actions that restore a dam's safety when needed.

Making decisions on how to manage risk requires a systems assessment, including the following:

1. Knowledge of the current state of the risk at the dam, based on the owner's performance monitoring program.
2. The potential for adverse interactions among the numerous components and subsystems at the project – for example, seepage control system, dam operations systems, debris blockage of spillways, remote operational control, and organizational interactions.
3. The contributions of the various system failure modes to the risk.
4. Possible risk management alternatives.
5. The risk reduction each alternative would achieve at the specific site.
6. The associated cost for each risk reduction alternative.

There is no one answer to driving risk as low as reasonably practicable (ALARP). Dams differ, owner-management configurations differ, and regulatory governance models differ. In each case, a performance monitoring program, coupled with remedial actions when required, can be an effective way to enhance safety. A systematic approach to safety provides an opportunity to identify the best means and methods of instrumentation and monitoring that help manage risk in a cost-effective and timely manner.

There is widespread agreement about the safety benefits of surveillance and monitoring. Regan, Nettle and Zygaj (2008) noted that "a surveillance and monitoring plan is designed to assure that appropriate surveillance and monitoring procedures are in place to identify a developing failure mode as soon as possible so that the maximum time is available for intervention or warning and the chances of success are greatest." They advocate the development of a failure mode sequence to clarify the points where surveillance and monitoring can be successfully implemented and definition of the risk reduction steps that could be taken at each step. A USSD white Paper, "Why Include Instrumentation in Dam Monitoring Programs?" (USSD, 2008), states two important points in the foreword: "There are many historical cases of dam failures where early warning signs of failure might have been detected if a good dam safety-monitoring program had been in place," and "A good dam safety monitoring program should be a key part of every dam owner's risk management program." A recent document by USBR related to dam safety and risk also refers to the use of a monitoring program to help manage risk, (USBR, 2011) and notes that additional risk management activities are recommended when risks are of practical concern. These may include, "visual inspections, instrumentation monitoring (underlining added for emphasis), inundation mapping, exercising of Emergency Action Plans, periodic examinations and evaluations, and similar measures considered to be 'good practice.'"

In draft guidelines, FERC (2012) indicated the role of instrumented monitoring in risk management. In a graphic showing the relationship among risk analysis, risk assessment and risk management, "Monitoring" is included as a Risk Reduction option. The document also states: "Risk management encompasses activities related to making risk-informed decisions, prioritizing evaluations of risk, prioritizing risk reduction activities, and making program decisions associated with managing a portfolio of facilities….These included potential structural and non-structural actions on a given dam or project, but also include such activities as routine and special inspections, instrumented monitoring and its evaluation, (underlining added for emphasis), structural analyses, site investigations, development and testing of emergency action plans and many other activities." FERC is continuing to develop this draft document. Also FEMA is working on a document the draft of which includes the same graphic from the FERC (2012) report (FEMA, 2014).

The Corps of Engineers (USACE, 2014) has recently published comprehensive guidance on dam safety. This guidance states (underlining added for emphasis):

> *Successful risk management requires a healthy routine monitoring program, including maintenance, repair and staff who are trained in data collection and interpretation.* [...]. *in some cases where data is* (sic) *relied on for life safety risk reduction decisions, it is appropriate to utilize independent expert consultants to review instrumentation data analyses and help validate conclusions."*
>
> *Risk management for dams includes short-term Interim Risk Reduction Measures (IRRM), long-term structural risk reduction measures, and strengthening recurrent activities – such as monitoring and surveillance, emergency action planning, operations and maintenance, and staff training.*
>
> *All Civil Works water control projects must have an adequate level of instrumentation, as appropriate to address potential failure modes and risks, to enable design engineers to monitor and evaluate the safe performance of the structures during the construction period and under all operating conditions....Where it is determined that instrumentation is a necessary monitoring component, instrumentation will be utilized to enable designers and operators to verify performance is within tolerable limits relative to potential failure modes.*

This brief review of the role of surveillance and monitoring in US dam safety practice indicates a solid recognition of the benefits of performance monitoring to risk management for dams. But the obvious question arises: how much risk reduction can we achieve with a good performance monitoring program? Is there some way to quantify how much risk reduction can be achieved? A definitive answer to this question would help quantify the benefits of a good performance monitoring program to dam safety, determine the scope and dimensions of a program to help reach the desired risk goals, and demonstrate the benefits to decision makers to help justify levels of effort for a program. A search of the literature revealed almost nothing to show how others quantify the benefits of I-M-M within risk assessments.

BASELINE PROBABILITIES OF DAM FAILURE

A helpful start comes from considering the baseline rate of failure of modern dams. Baecher, Paté, and DeNeufville (1980) determined that the average annual rate of failure of a modern dam (post 1940) is about 10^{-4} (1 in 10,000) per year from all causes. A recent study by the US Nuclear Regulatory Commission has confirmed this number as a generic failure rate for modern dams (NRC, 2010).

While a failure rate of 1 in 10,000 per year might sound low, consider 100 years as a measure of "life", perhaps the dam's design life, or the approximate life expectancy of a baby born downstream of the dam at the start of its construction. There would be about a 1 in 100 chance that a generic dam would experience a failure during this 100 year life. In rough numbers 1 out of 100 babies born within the flood zones of 100 generic dams would experience a major flood sometime in its life time, or one out of 100 dams could be expected to fail sometime within a design life of 100 years. As another example, a nuclear power plant operated below a generic dam and within the flood zone would face an annual risk exposure of a major flood of 1 in 10,000 which is 100 times greater than the NRC goal for risk to containment vessels of less than 1 in 1,000,000 per year from man-made causes (NRC, 1986).

Statistics compiled by others, such as Middlebrooks (1953) and summarized in Baecher et al (1979, 1980) show that a dam is more likely to fail during construction or first filling, from which they determined an average annual probability of failure of a modern dam less than 5 years old as 10^{-3}. Once a dam has been successfully constructed and operated through its first cycle of filling then the average annual probability of failure drops to about $0.5*10^{-4}$ per year.

Baecher et al (1979, 1980) and others have also looked at the primary mechanisms of failure. Table 1 summarizes a number of these results. There is general agreement in the primary mechanisms of dam failures, probably because these references use overlapping data sets.

Table 1. Causes of Dam Failures

Failure Mode	Failure Rate, %				
	Baecher et al. (1979)	Foster et al. (2000)*	Von Thun (1985)	Regan (2010)	Rate used herein
Overtopping	23-38	37	37 43 31		
Seepage, piping	30-44	38	38 25 39		
Sliding	2-15	7	7	10	10
Spillways, misc.	9-35	18	18 22 20		
* Normalized for the inclusion of spillway failures					

Combining this information with the average annual rate of failure of dams, we can conclude that the average annual probability of failure of a modern dam more than 5 years old due to overtopping is about $0.5*10^{-4} * 0.31 = 0.15*10^{-4}$. The rate for other causes corresponding to "sunny day" events is $0.35*10^{-4}$. (A couple of clarifying details are required. Firstly, the data do not support knowing these rates to two significant digits but two significant digits are shown here to help distinguish the breakdown between failures from overtopping and failures from other causes. Secondly, technically the actual failure rates given in this section are equal to or greater than the stated rates.)

It is important to keep in mind that these are generic failure rates that do not consider the particulars of a specific dam. A dam with known defects could be shown with a site specific analysis to have a higher probability of failure, and a conservatively designed, constructed, and operated dam with a strong monitoring program could be shown to have a lower probability of failure. USBR/USACE (2012) and FERC (2012) provide much useful guidance for performing site specific risk assessments that can include considerations of the benefits of performance monitoring and risk mitigation.

STATE OF CURRENT DAM PERFORMANCE MONITORING PRACTICES

The author contends that most dams in the inventories of dams and dam failures historically did not have effective performance monitoring programs, especially during their periods of long term operation. Some of the major dams and some dams with specific challenges had strong monitoring programs during their construction and first filling but many of those programs declined over time. Based on the author's experience, many of today's performance monitoring programs are woefully inadequate as effective risk reduction tools. This is due to deficiencies in their performance monitoring programs:

- Inadequate numbers and locations of instruments to detect and monitor significant failure modes
- Poorly maintained monitoring systems so that data are not reliable or believable.
- Insufficient frequency of readings to detect rapid changes in a dam's performance.
- Poor training of staff to know what to look for and how to appropriately interpret the results of the monitoring efforts.

Regarding frequency of readings, many dam monitoring programs collect data from instruments once a quarter to once per year after the dam passed its first few years of life. This is simply inadequate for performance monitoring to provide the most value in reducing risk. Many dams fail within hours to a few days of the first warnings. Baldwin Hills failed within four hours of seepage being observed and Teton Dam failed two days after seepage was first noted (Regan et al, 2008). Walter Bouldin Dam failed less than 24 hours after a visual inspection reported no problems (FERC, 1978).

Instruments read at quarterly to annual intervals may not give adequate warning to take action to reduce risk. The author contends that the effectiveness of historical dam monitoring programs was relatively small compared to what it could be with a modern, effective surveillance, instrumented monitoring and remedial action program. As suggested by the quote above from USSD (2008) - "there are many historical cases of dam failures where early warning signs of failure might have been detected if a good dam safety-monitoring program had been in place," - supports this contention.

The relatively infrequent monitoring rate for dams was historically established as a balance between the perceived value of the data and the cost of collecting the data. But the foundations of this balance are now different. The demands of society to reduce risk from dam failures has grown enormously over the past few decades so the potential value of performance monitoring data to warn of surprising or unacceptable performance has increased considerably. The cost to collect and process information more frequently using automated equipment has dropped dramatically. The balance has clearly changed and we need to respond accordingly with modern, real-time performance monitoring programs of sufficient frequency to provide indications of an emerging risk in time to alter the outcome and reduce the risk, either by reducing the probability of failure, lowing the consequences, or a combination of both.

QUANTIFYING REDUCED RISK FROM REAL-TIME MONITORING

Consider some initiating event that could lead to a dam failure. For illustration consider a shear slide within the downstream shell of an earthen embankment dam. Assume that the slope has a probability, p, of failing in any one year from a shear slide and that such sliding would cause a failure of the dam. Next consider that a monitoring system could be implemented that would detect the precursor events to a shear slide before it resulted in a failure and remedial actions could be taken that would prevent the failure. If the monitoring system worked perfectly and if we were able to implement the remedial actions in time and scope to prevent the failure, then the probability of failure by this sliding mechanism would be reduced to zero.

But nothing is perfect. We know that monitoring programs aren't always fully successful and remedial plans don't always work, so we need to consider these facts. Suppose our monitoring program is 90% successful at detecting the development of a shear slide in time for us to take action and the remedial action has a 90% chance of success in preventing failure for an event with probability, p, of occurring. If we take these actions we will have reduced the chance of failure of the dam by shear sliding to $p*(1-0.9*0.9)$ or 0.19p which is 5 times less than the generic risk, p. From Table 1, the historical record indicates on average about 10% of dam failures result from shear slides so we are looking at reducing the generic annual failure rate from a shear slide in a modern dam more than five years old from $0.1*0.5x10^{-4} = 5x10^{-6}$ to about $1x10^{-6}$ through the use of an I-M-M program that is 90% effective.

This concept might be easier to understand when viewed as an event tree. Figure 1 shows such a simplified event tree for this example. Assuming an initiating event occurs, there is a 10% chance that the performance monitoring program does not detect the poor performance and failure occurs. There is a 0.9 x 0.1 = 9% chance that the performance monitoring program works but the corrective action does not work to prevent the failure. These two failure modes combine to create a probability of failure of 19% with the performance monitoring program, conditional on the occurrence of an initiating event that would lead to failure. The overall conditional probability of failure has been reduced from 1 to 0.19 or five times. These values need to be multiplied by the probability of the initiating event occurring to get the total probability of poor performance.

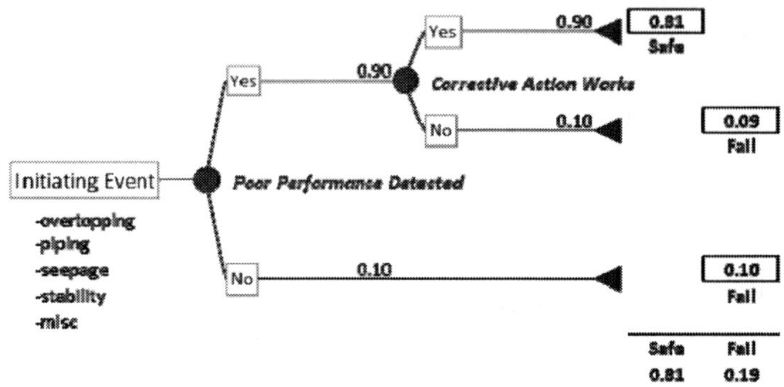

Figure 1. Simplified Event Tree to Illustrate Benefit of I-M-M

Table 2 shows the results for different probabilities of poor performance detection and corrective action effectiveness. If the performance monitoring program isn't very good and the corrective action effectiveness fifty-fifty (Case 1), this approach is not very effective. On the other hand if the performance monitoring and corrective action are very effective (Case 4), then this approach reduces the conditional probability of failure by two orders of magnitude. Cases 5 and 6 show the obvious but important point that a monitoring program that is very effective at detecting poor performance does nothing to reduce probability of failure if the corrective action is not implemented or is implemented poorly.

Table 2. Reduction in Probability of Failure with Effective Monitoring and Remedial Action

Case	Probability Detect	Probability of Remedial Success	Probability of Failure	Reduction Factor for Probability of Failure
0	No detection	None	1.0	None
1	0.5	0.5	0.75	1.33
2	0.9	0.9	0.19	5.2
3	0.95	0.95	0.098	10.2
4	0.995	0.995	0.01	100
5	0.995	0.0	1.0	1.0
6	0.995	0.5	0.50	2

Interestingly in 1992 ANSI suggested that performance monitoring might help reduce risk from the effects of dam failure on nuclear reactors for causes other than sudden events like earthquakes and extreme floods with the statement: "dam failures from other onsite causes which result from gradual changes threatening dam safety might be detected and adequate corrective measures can be taken" to reduce risk (ANSI, 1992).

CASES TO SUPPORT RISK REDUCTION ACHIEVED BY PERFORMANCE MONITORING

Case 1: Whitman's Embankment Dam

Whitman (1984) in his Terzaghi Lecture gave an example event tree for the failure of a modern earth dam by cracking which is redrawn in Figure 2(a). He assigned probabilities to the branches of the event tree based on his engineering judgment. He stressed in his paper that this was an illustrative example to stimulate discussion with the descriptors and numerical values to be considered liberally. Despite this qualification, his example seems reasonably representative of the failure mechanism of cracking because he arrived at an overall lifetime probability of failure by this mechanism of 4.5×10^{-5}. The useful point for the purpose of this paper is that Prof. Whitman included the effects of a monitoring system in his event tree with the branches "Piping Detected" and "High Pore Pressure Detected." These are events where a monitoring system indicates the development of the cracking failure mechanism in time for mitigation actions to be taken. In his assessment, he considered that good conservative design practice would have a 90-97% chance of detecting piping and a 99-100% chance of detecting high pore pressures before they caused failure. We can use this event tree to assess the value of the monitoring system by changing the probabilities of detection of piping and high pore pressure to 0 which would be the case if no monitoring was performed. Figure 2(b) shows the result. The probability of failure by cracking has increased from $4.5\text{x}10^{-5}$ to $2.5\text{x}10^{-3}$. Looked at in reverse, implementing a monitoring system for the failure mechanism of cracking reduced the probability of failure from 2×10^{-3} to 5×10^{-5}, a 57-fold reduction.

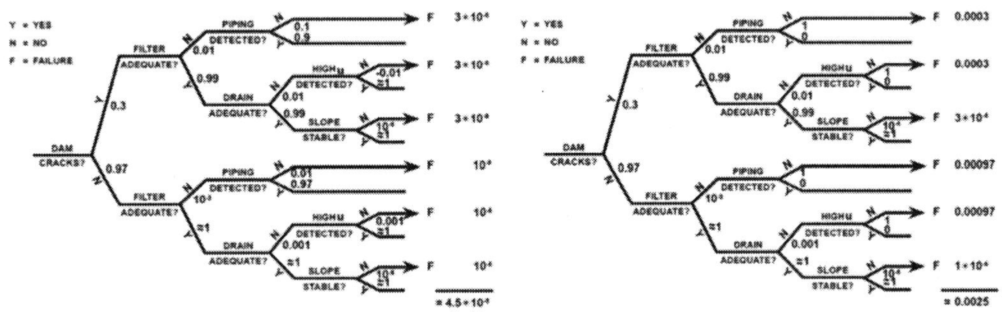

(a) (b)

Figure 2. Whitman's Event Tree for Failure of Modern Dam by Cracking

Whitman (1984) also reworked his event tree to consider the case where the same dam was constructed with an erodible core and no filter constructed against abutments containing open joints that were difficult to treat adequately with grouting and a monitoring system that did not detect piping. This is shown in Figure 3(a). He estimated a probability of failure by cracking that led to piping of 0.2. If we modify the probability of detecting piping to the values in Figure 2(a), the probability of failure by piping decreases to 0.016, as shown in Figure 3(b). This result indicates a 13-fold reduction in the probability of failure achieved with an effective monitoring and response program for this case.

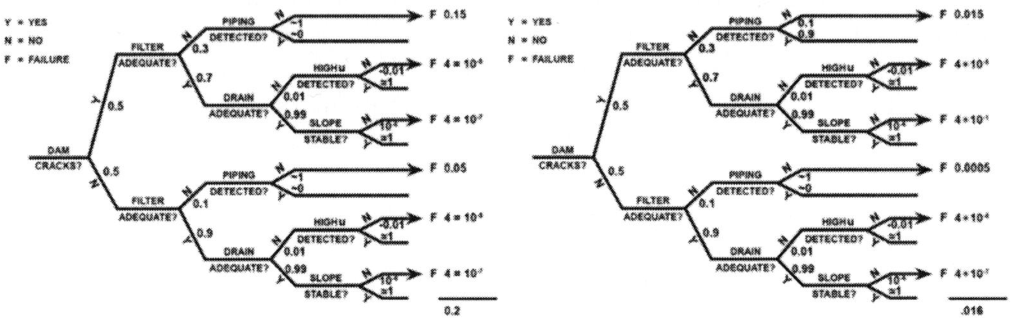

(a) (b)

Figure 3. Whitman's Event Tree for Dam with Erodible Core, No Filters and Open Joints

Case 2: Central Artery-Tunnel Project

This project is not a dam but a major highway tunnel project. However many of the potential failure modes and risks have similarity to significant failure modes for dams, i.e. global stability, piping, excessive deformations and soil-structure interface complexities. Much of the instrumentation used for the performance monitoring program was the same types as used for dams. Consequently the experience from using performance monitoring as a risk management tool on this project has direct relevance to dam safety.

The Central Artery/Tunnel Project (CA/T Project) was the largest, most complex highway project in American history at the time it was built. The $15 billion project consists of 161 lane miles of urban highway – about half underground in a 7.5-mile corridor – through the heart of one of America's oldest and most historic cities in difficult underground conditions. The main portion of the Project replaced Boston's aging, elevated Central Artery (I-93) with a modern eight-to-ten lane underground expressway. The project also extended the Massachusetts Turnpike (I-90) to Logan Airport, built a dramatic 10-lane cable-stayed bridge across the Charles River, created a state-of-the-art incident response and traffic management system, constructed five major highway interchanges and has the largest vehicular tunnel ventilation system in the world. The work for the CA/T Project was unprecedented, with the majority of the construction taking place in the middle of downtown Boston which houses the main financial and historical areas that must remain open and accessible for businesses, residents and tourists. More than 150 significant structures existed within the potential construction impact zone of the CA/T Project. These included several high-rise buildings located on mat foundations within feet of proposed deep excavations, many multi-story brick and masonry structures of historic value, major transportation infrastructure, and completed portions of the project.

Previous construction experience in Boston had shown that damage to existing structures from new construction could create significant added costs to a project. For example, in the 1980's, the Trinity Church sued the John Hancock Insurance Company for 40 million dollars in damages. They successfully argued that cracking of masonry elements in the historic church was caused by construction of a nearby excavation for the foundation of the John Hancock Building and were awarded $11.6 million. Also, other prior construction experiences in the Boston area had shown that the effects of dewatering for an excavation could extend thousands of feet from the construction (Lambe and Whitman, 1969). The potential for damage to structures adjacent to the CA/T Project was difficult to assess, but an amount of several hundred million dollars to more than a billion dollars in claims and litigation expenses seemed to be reasonable estimates of the possible damages. There was also significant financial risk from the potential for delay claims by General Construction Contractors if litigation related to damage claims halted construction at any point.

The combination of a complex geotechnical environment, numerous sensitive structures next to the work, novel methods of construction, and many involved stakeholders produced an environment where the potential costs of adverse or unexpected performance were very high. Those potential costs involved not only money but also time and political consequences. Despite an extensive geotechnical investigation and comprehensive design effort, many uncertainties remained about the likely effects of construction activities on existing facilities and performance of the newly constructed facilities. Consequently the CA/T Project team made a comprehensive geotechnical instrumentation program a key part of the Project plan.

A comprehensive geotechnical and structural monitoring program was implemented for the project. The purposes of the instrumentation program (Bechtel-Parsons Brinckerhoff, 1991) were to provide data to help the project team to:

- Prevent or minimize damage to structures by providing data to determine the source of ground deformations,
- Assess the safety of all works by comparing the observed response of ground and structures with the predicted response and allowable deformations or disturbance levels
- Develop protective and preventive measures to structures,
- Select appropriate remedial measures where required,
- Evaluate critical design assumptions where significant uncertainty exists,
- Determine the adequacy of the contractor's methods, procedures and equipment,
- Monitor the effectiveness of protective, remedial and mitigative measures,
- Provide feedback to the contractor on his performance,
- Assess the contractors' performance, contractor-initiated design changes, change orders, changed conditions and disputes,
- Provide documentation for assessing damages sustained to adjacent structures allegedly resulting from ground deformations and other construction related activities,
- Advance the state-of-the-art by providing performance data to help improve future designs.

Potential damages to adjacent structures associated with the CA/T Project construction were anticipated to be very significant if careful steps were not taken to prevent damage. To the author's knowledge, no formalized risk assessment was done during the planning and design phase of the project; note that it was not typical practice to make such estimates in the 1982-1995 time period which this project was designed. The design engineers did express the following main concerns for major problems that could have occurred:

- Collapse of one or more major structures adjacent to the excavations due to a failure of the excavation support system. Examples of major structures adjacent to the Project included the 33 floor main tower of the Federal Reserve Bank Building, the 46 floor high rise building at One Financial Center, the historic MBTA South Station and the AMTRAK South Station rail yard system.
- Major damage to structures due to a failure of the excavation support system resulting in business disruptions lasting for a significant period of time.
- Architectural damage to the many historic structures along the work alignment due to excessive deformations from excavation or ground water lowering.
- Claims from contractors for construction delays related to damage to adjacent structures, which resulted in work stoppage and delays.

As the project was approaching completion, the author led a group of the project's management team in conducting a simplified risk assessment at what could have happened if no surveillance and performance monitoring had been conducted. Marr (2007) provides more details on this assessment. The result was an estimated total risk cost to the project of $550,000,000 due to potential building collapse, major damage to existing facilities, architectural and structural damage to historic buildings, and disruption and construction delays. This estimate did not include the potential for significant loss of life and injury in the event of structural collapse.

The geotechnical and structural instrumentation and monitoring program cost approximately $60,000,000 including instruments, installation, data gathering, data evaluation, and interpretation of the results. Data were collected, evaluated, reported and interpreted within 24 hours so that near real-time control could be exercised over the project's activities. By using data from thousands of sensors, the Management Consultant was able to monitor locations and activities thought to present a high risk and to direct changes to the General Contractors' work when the measurements of actual performance indicated unsatisfactory conditions.

Overall, the approaches taken on the Geotechnical Instrumentation Program for the Central Artery/Tunnel Project were exceedingly successful. With respect to those portions of construction that could be monitored and managed with information from the instrumentation program, the following achievements occurred:

- The project paid approximately $9,000,000[2] to repair damage to adjacent structures that might possibly have been caused by the construction.

[2] There were a number of claims filed that assert damages related to CA/T construction. At the time of assessment described here, their outcomes were uncertain. No follow up study was done; however to the author's knowledge no significant additional claims were paid for damage to adjacent buildings and poor performance that would alter the conclusions reached by the assessment.

- No catastrophic failure occurred and there was no major damage to any adjacent property.
- Unexpected values and trends in the monitoring data caused further field inspections and investigations that revealed inadequate or incomplete construction means or methods for many cases. The geotechnical instrumentation helped identify these problems at an early stage, so they could be corrected before they had any serious consequences to the Project.

In simplistic terms, but terms that are meaningful from an overall facilities management perspective, it can be argued that the performance monitoring program for the CA/T project reduced the potential risk cost from geotechnical failures and associated delays from an estimated $550,000,000 to an experienced repair cost of $9,000,000. This represents a 61-fold reduction in risk. This project established the precedent for what has become standard practice for tunneling in urban areas - near real-time monitoring of geotechnical and structural features to control construction in ways that minimize damage and delays.

Case 3: Panama Canal

The Panama Canal is infamous for the many landslides that occurred during and following its construction. In the 1980s, the Panama Canal Authority implemented a program to detect incipient slides and employ appropriate remedial measures when required to stabilize slopes before they failed.

In 1988, Dr. Luis D. Alfaro of the Panama Canal Commission (PCC) performed an extensive review of the history of sliding of cut slopes in the Gaillard Cut of the Panama Canal (Alfaro, 1988). Much of this information was documented in PCC internal reports. The purpose of this study was to evaluate the risks of channel encroachments due to landslides along the entire extension of the Gaillard Cut to help allocate landslide control resources more efficiently and reduce the overall cost of landslide activity. The Gaillard Cut was divided into 24 zones based on topographic and geologic conditions. The history of distress and slides in each zone was gathered and summarized to arrive at observed failure rates. Landslides along the canal that did not potentially encroach into the canal and threaten its operational ability were not considered.

Slides in the Gaillard Cut were observed to not develop suddenly; thus the Landslide Control Program can detect incipient slides before they develop fully and give time to invoke remedial measures. This program consisted of visual observations from field inspections, surveying of monuments, monitoring of rainfall and some monitoring of pore water pressure. Dr. Alfaro concluded that this has allowed PCC to reduce the risks of slides that result in channel encroachments by at least one order of magnitude. He also concluded that a more systematic program of inspections and monitoring coupled with maintenance could reduce these risks by even another order of magnitude. Thus experience at the Panama Canal indicates that an effective performance monitoring program coupled with timely remedial/maintenance work can reduce risk from landslides into the Canal by one to two orders of magnitude.

Importantly the Landslide Control Program has two complementary components – detection through visual surveillance and instrumentation, and proactive action to stop further movements before they develop into full landslides. This case, which is entirely independent from the author's experience concluded that a proactive surveillance and instrumentation monitoring program coupled with proactive remediation/maintenance reduced the risk of landslides by roughly an order of magnitude and that with enhancements an additional order of magnitude in risk reduction could be possible. The type of landslides in the Panama Canal is similar to that which can occur in a dam or through its foundation or abutments. Therefore this experience seems applicable to dams as well.

USE OF I-M-M TO REDUCE RISK FOR VARIOUS FAILURE MODES

Principal potential failure modes for dams were summarized by Baecher et al (1980) from work by others as overtopping, seepage and piping, sliding and miscellaneous. "Miscellaneous" might include structural failure, such as gates, or damage caused by improper operation. For an I-M-M system to be effective at reducing risk it must be able to identify a developing failure mode in time for risk reduction measures to be invoked. Those measures might aim at reducing the probability that the potential failure mode will develop into complete failure, or reducing the consequences of the failure, or both reducing probability and reducing consequences.

Probability of failure would generally be reduced by taking action to alter the physical state of the dam in a way that the developing failure mode is slowed, stopped or reversed. For example berms might be hastily added to reduce the chance of overtopping, or the reservoir might be drawn down to stop the further development of piping, or stabilizing berms be constructed to stop a developing landslide. For the I-M-M approach to work, there must be actions that can be taken that will reduce the probability of failure and there must be time to complete those actions and for their benefits to be realized. This might not always be the case. For example, there might be insufficient time to prevent a large landslide caused by an earthquake or a sudden large storm that overtops the dam. However, an I-M-M program might identify an impending failure in time to take actions to reduce consequences. Potential actions with only short notice might be to move people out of the flood zone, protect expensive equipment, or hurriedly shut down a downstream facility. Very significant reductions in consequences and therefore risk can sometimes be achieved with only a few minutes of reliable warning.

A key point is that for the I-M-M approach to be effective at reducing risk the instrumentation must be capable and reliable to indicate a developing failure mode as early as possible, there must be some plan of action to slow or stop the developing failure mode that can be put in place within the available time, there must be a person to who makes the decision to deploy the action plan, and the action plan must be executed in a way that it fulfills its intended purpose. If all these components are in place, then I-M-M can be expected to reduce risk of dam failure by one to two orders of magnitude. However, one missing component will completely negate the benefit of the I-M-M program, even if the other components are all performed to the best we possibly can do.

To be an effective risk management tool, the performance monitoring program must be an effective program that includes these components done well:

- Identify potential failure modes (PFMs) and Key Risk Indicators (KRIs).
- Evaluate existing inspection protocols and instrumentation arrays for effectiveness in monitoring identified PFMs and KRIs.
- Install appropriate surveillance and/or monitoring systems to detect the development of the identified PFMs and monitor the KRIs.
- Identify action levels (threshold levels and limit levels) for all instruments and mitigation responses for each action level.
- Develop and implement policies and procedures to obtain data in a timely and accurate manner.
- Develop and implement policies and procedures to assure the data are evaluated and interpreted in a timely manner.
- Develop an action plan with mitigation measures that address the steps to be taken in the event an action level is exceeded.
- Have and use a clear communication path with assigned responsibilities to initiate the action plan when the evaluated and interpreted results of the I-M-M program indicates it necessary.
- Execute the action plan in a timely manner.

CONCLUSIONS

A review of several cases shows that risk from dam failure can be reduced by one to two orders of magnitude with an effective monitoring program coupled with workable mitigation measures. For the instrument-monitor-mitigate approach to be effective at reducing risk, the instrumentation must be capable and reliable to indicate a developing failure mode as early as possible, there must be some plan of action to slow or stop the developing failure mode that can be put in place within the available time, there must be a person who makes the decision to deploy the action plan, and the action plan must be executed in a way that it fulfills its intended purpose of mitigating risk. If all these components are in place, then I-M-M can be expected to reduce risk of dam failure by very substantial amounts. However, one missing component will completely negate the benefit of the I-M-M program, even if the other components are all performed to the best possible.

The potential benefits of an effective I-M-M program to reducing risk are not well considered in the published literature and guidance documents related to risk management for dam safety. Most of the guidance documents mention surveillance and performance monitoring in passing but fail to discuss and demonstrate how I-M-M can be used to significantly reduce risk. A search for documentation to understand how those doing risk assessments for dam, including formalized risk analysis, revealed very little treatment of the role of the I-M-M system in managing the risk for a dam. This condition needs to change. The tools are available to help quantify and demonstrate the potential benefits of I-M-M to risk management for dam safety. Branches can be added to event tree analysis that include the impact of I-M-M on the probability and consequence of failure. Subject matter experts can be used to estimate the probabilities required for those branches in the event tree. It is time that those of us engaged in dam safety make better use of these tools and determine how we better use I-M-M programs to help us quantify and manage risk.

ACKNOWLEDGEMENTS

The author expresses great thanks to Dr. Luis D. Alfaro of the Panama Canal Authority for permission to publish the Panama Canal experience. He also thanks Prof. Gregory B. Baecher and Dr. John T. Christian for their many hours of discussion and guidance on probability and risk assessment and Prof. T. W. Lambe for his many years of mentoring.

REFERENCES

Alfaro, L. D. (1988). The Risk of Landslides in Gaillard Cut," Unpublished report of the Panama Canal Commission Engineering Division Geotechnical Branch dated October 20, 1988.

ANSI (1992). "American National Standard for Determining Design Basis Flooding at Power Reactor Sites." ANSI/ANS-2.8-1992.

Baecher, G. B., Paté, M.-E., and de Neufville, R. (1979). *NED cost determination for probability of dam failure*. Report to the US Water Resources Council, Systems Analysis Inc., Cambridge MA, 65.

Baecher, G. B., Paté, E.M. and de Neufville, R. (1980). "Risk of dam failure in benefit/cost analysis." *Water Resources Research*, **16**(3): 449-456.

Bechtel-Parsons Brinkerhoff (1991). Central Artery (I-93)/Tunnel (I-90) Project, unpublished Concept Report NO. 2AB24, Geotechnical Instrumentation, March. Prepared for the Massachusetts Department of Public Works.

FEMA (2014). "Federal Guidelines for Dam Safety Risk Management – Draft," United States Federal Emergency Management Agency. May.

FERC (1978). "Walter Bouldin Dam Failure and Reconstruction," Report to the Federal Energy Regulatory Commission, Office of Electric Power Regulation, September. US Printing Office Stock No 061-002-00011-1.

FERC (2012). "Draft - Federal Guidelines for Dam Safety Risk Management", United States Federal Energy Regulatory Commission, Version 3.0, September 2012. (www.ferc.gov/industries/hydropower/safety/guidelines/draft-guidelines.pdf accessed on November 2, 2014)

Foster, M., Fell, R., and Spannagle, M. (2000). "The statistics of embankment dam failures and accidents." *Canadian Geotechnical Journal*, 37(5), 1000–1024.

Lambe, T. W. and Whitman, R.V. (1969). Soil Mechanics. John Wiley & Sons, New York.

Marr, W. A. (2007). "Why Monitor Performance?" *ASCE GSP 117*. Keynote paper presented to 7th International Symposium on Field Measurements in GeoMechanics, Boston.

Middlebrooks, T. A. (1953). "Earth-dam practice in the United States." *Transactions of the American Society of Civil Engineers*, Centennial Volume, 697–721.

NRC (1986). "Policy Statement on Safety Goals," U.S. Nuclear Regulatory Commission, Federal Register, Vol. 51, pg. 30028. August 21.

NRC (2010). "Identification of a generic external flooding issue due to potential dam failures." U.S. Nuclear Regulatory Commission memorandum Performance ML101900305.pdf dated July 19. http://pbadupws.nrc.gov/docs/ML1019/ML101900305.pdf accessed on November 2, 2014.

Regan, J.J., Nettle, J.D. and Zygaj, J.A. (2008). "Managing Dam Safety Risks Through Surveillance and Monitoring Plans and Surveillance and Monitoring Reports," *Proceedings of 28th Annual USSD Conference on the Sustainability of Experience – Investing in the Human Factor*. Portland, Oregon, April 28 - May 2, 951-965.

Regan, P. (2010). "Dams and civil structures: An examination of dam failures vs. the age of dams." *Hydropower Review*, June 1.

USACE (2014). "Safety of Dams – Policy and Procedures," United States Army Corps of Engineers ER 1110-2-1156. March 31.

USBR (2011). "Dam Safety Public Protection Guidelines: A Risk Framework to Support Dam Safety Decision-Making." U.S. Department of the Interior, Bureau of Reclamation, Dam Safety Office, Denver, CO, August.

USBR/USACE (2012). "Best Practices in Dam and Levee Safety Risk Analysis," Chapter 21, p 75. 7 Dec 2012 from usbr.gov/ssle/damsafety/risk accessed on 1/27/15.

United States Society on Dams (2008). "Why Include Instrumentation in Dam Monitoring Programs?" Version 1.00 November. ISBN 978-1-884575-45-7.Von Thun, J. L. (1985). *Application of statistical data from dam failures and accidents to risk-based decision analysis on existing dams*. U.S. Bureau of Reclamation.

Whitman, R.V. (1984). "Evaluating calculated risk in geotechnical engineering: 17th Terzaghi Lecture." *J. Geotech. Engrg*., ASCE, 110(2), 145-188.

Von Thun, J. L. (1985). *Application of statistical data from dam failures and accidents to risk-based decision analysis on existing dams*. U.S. Bureau of Reclamation.

TIME-FUNCTION METHOD TO ESTABLISH PIEZOMETER PERFORMANCE PARAMETERS

Brian Hart[1]

ABSTRACT

Dam safety monitoring is critical to the management of aging dam infrastructure. One way dam safety and risk are characterized is through the evaluation and monitoring of dam instrumentation data, including piezometer readings. Piezometer readings are gathered and analyzed, and dam instrument performance parameters are established based on this analysis. Methods for establishing performance parameters for piezometers in embankment dams that have reached pore-water pressure equilibrium have been well documented. However, methods to establish performance parameters for piezometers in embankment dams that have not reached equilibrium have not been thoroughly addressed. This paper details the assumptions and limitations of the load-response method, and proposes a time-function method of establishing performance parameters for piezometers in embankment dams that have not reached equilibrium based on time-series analysis. The time-function method is outlined through an example of its application and the results are reviewed. Piezometer data collected from two embankment dams were evaluated for this study: one embankment dam features excess pore-water pressure, which is dissipating with time, while the second embankment dam is unsaturated, with pore-water pressure that is increasing with time. In both cases, the results show the proposed time-function method adequately predicts the future performance of the piezometer readings.

INTRODUCTION

Dam safety monitoring is critical to the management of aging dam infrastructure, and the use of instrumentation to monitor the performance of dams is an important component of an effective dam safety program. A well-designed and well-executed instrumentation monitoring program can provide information that is needed to understand the ongoing performance of a dam and can help detect early indications of adverse performance. One way in which dam performance is monitored is through the evaluation of instrumentation readings, and in particular, piezometer readings. Based on evaluation work and historical data, instrumentation engineers may establish performance parameters that the readings are expected to fall within. A traditional method of establishing performance parameters is based on a load-response relationship, which assumes the pore-water pressure of an instrument in the dam directly correlates with the reservoir elevation. Effective methods of establishing performance parameters for piezometers in embankment dams that have not reached pore-water pressure equilibrium have not been established. This paper proposes a time-function method of establishing performance parameters for piezometers in embankment dams that have not reached equilibrium.

[1] Dam Safety Engineer, Bureau of Reclamation, Provo Area Office, bhart@usbr.gov

Pore-Water Pressure Equilibrium Scenarios

Pore-water pressure is the pressure retained in water held in the pore spaces of soil and exerted on its surroundings, and is important because it affects the stability of a dam. Pore-water pressure equilibrium can be described using Terzaghi's principle of effective stress (Equation 1) as a state at which the pore-water pressure in the dam is no longer changing with time during static loading conditions.

$$\sigma = \sigma' + \mu \tag{1}$$

Total Stress, σ
Effective Stress, σ'
Pore-water Pressure, μ

When equilibrium is achieved, the pore-water pressure is essentially directly related to the reservoir level. Most dams operate in quasi-equilibrium because constant reservoir fluctuations keep the pore-water pressure in flux and geophysical characteristics affect the lag response time. However, until the pore-water pressure ceases to either continuously increase or decrease with time during essentially static loading conditions, pore-water pressure equilibrium has not been achieved.

Immediately following construction of the dam and first filling of the reservoir, there are typically two scenarios for pore-water pressure before equilibrium is achieved. The first scenario involves an embankment dam with excess pore-water pressure, where high pore-water pressures are created during the construction of the dam by what ends up being excessive water relative to the voids of the compacted fill. The second scenario is one of an embankment that is not yet fully saturated, where the wetting front traveling through the embankment has yet to reach all portions of the embankment and equilibrium conditions have not been achieved.

In Scenario 1, excess pore-water pressure is created when heavy compressive forces (i.e. compaction) are applied to the dam core and excess water is present relative to the voids in the compacted material. Because water is nearly incompressible, the water tends to hold a portion of the stress applied by the compressive forces and becomes pressurized. If the pore-water pressure is high enough it can cause the embankment to lose shear strength, and threaten the stability of the dam. In most cases the structure will remain stable and water will slowly exit the embankment in the upstream and/or downstream direction. The excess water pressures will be relieved and consolidation of the embankment will occur. With sufficient time, often years, the high pore-water pressure will dissipate until the pore-water pressure approaches an equilibrium condition (Figure 1).

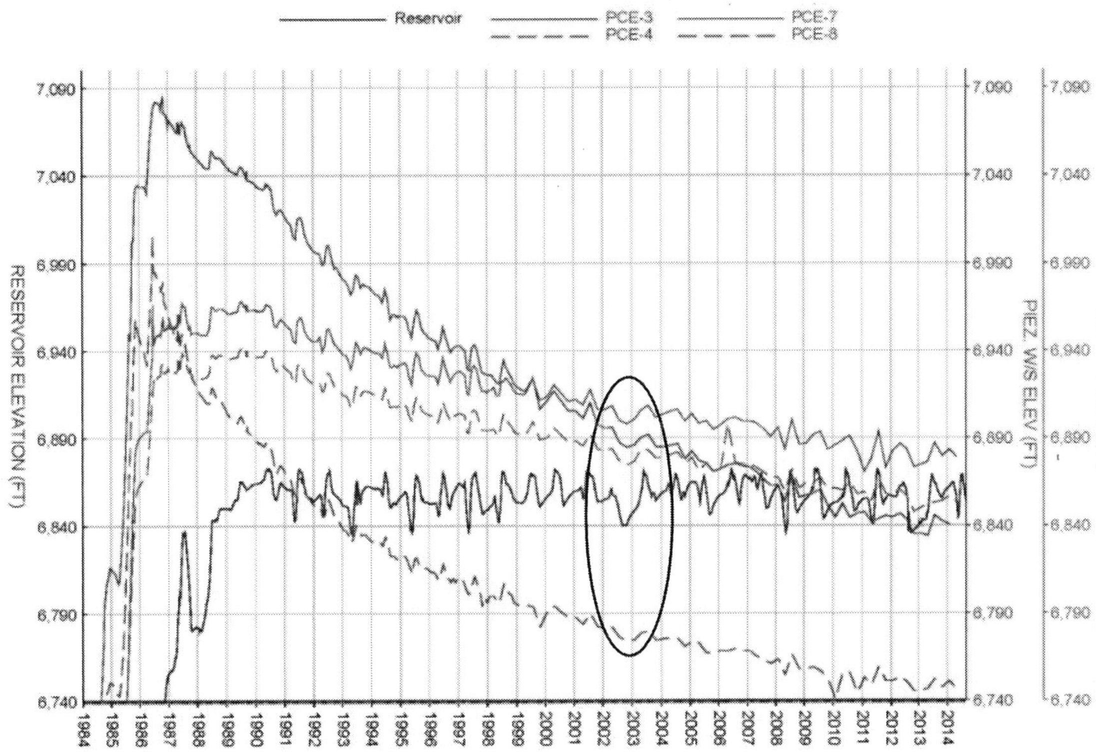

Figure 1. Scenario 1 Dam - Historic piezometer readings which are dissipating with time. Notice the piezometer readings are influenced by a logarithmic decay trend as well as the fluctuations of the reservoir elevation.

In Scenario 2, unsaturated conditions and low pore-water pressures are observed in the embankment after construction of the dam and first-filling of the reservoir. After the reservoir begins to fill, water will seep through the embankment and fill the voids in the soil material. If the soil material is well compacted and very impermeable, then the process will be very slow (Figure 2).

In both scenarios, pore-water pressure will eventually approach equilibrium given enough time. When equilibrium is achieved, the movement of water in the embankment becomes less pronounced and significant, but still occurs to some degree in response to changing reservoir levels.

What is important to note is that the manner in which pore-water pressure approaches equilibrium for both Scenario 1 and Scenario 2 can be logarithmic depending on the geophysical characteristics of the dam (e.g. cohesive clays). For instance, if a dam core was built with material that had a high percentage of fines, and a high Plasticity Index, the dam core would be very impermeable and the pore-water pressure would achieve equilibrium slowly. In situations like this it would be expected that the pore-water pressure would eventually reach equilibrium conditions in a logarithmic fashion with time, assuming a reasonably consistent driving force.

Time-Function Method

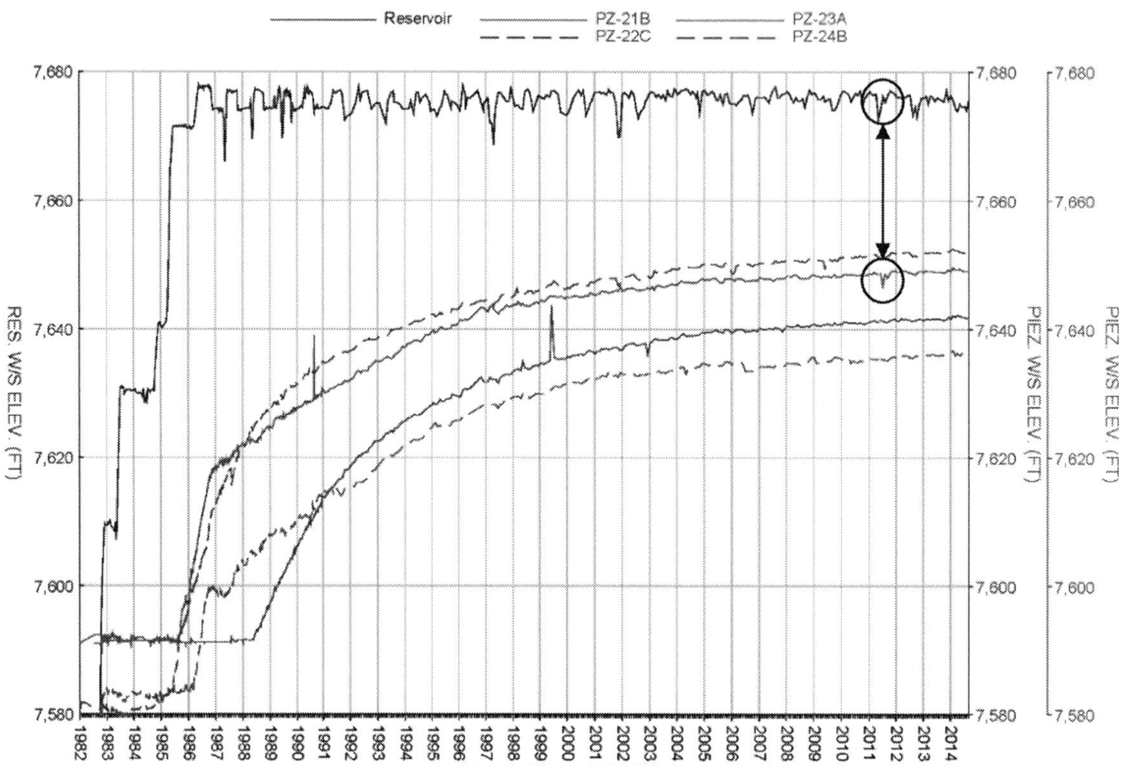

Figure 2. Scenario 2 Dam- Historic piezometer readings which are increasing with time. Notice the piezometer readings are influenced by the logarithmic trend as well as the fluctuations of the reservoir elevation.

Performance Parameters

Performance parameters are well-defined dam instrumentation limits that indicate unexpected behavior of the dam. The limits are set to bring "additional" attention to abnormal instrumentation performance, and facilitate a preemptive response to avoid the occurrence of a serious dam incident. For piezometers, performance parameters consist of an upper limit and a lower limit that the readings are expected to be within based on evaluations of historic piezometer readings and their relationship with changing reservoir levels. These bounds help to predict expected future performance of the dam instrumentation as part of the ongoing dam instrumentation monitoring program.

Performance parameters are typically established and reevaluated every 6-8 years, depending on the operating authority. A time cycle of six years, formerly used by the Bureau of Reclamation, was used for this study.

Traditional Load-Response Method

The traditional load-response method of establishing performance parameters for piezometers has been discussed in previous documents and is commonly used in industry practice (USACE Appendix E, 2004). In this approach, instrumentation data is collected at a damsite over many years and, during analysis, a load-response relationship is plotted. The resulting plot is interpreted by an engineer and judgment is used in its evaluation. This method of developing performance parameters is effective given the assumption that pore-water pressure is directly correlated with reservoir water level, as is the case in an equilibrium or near-equilibrium situation.

An example of a load-response relationship for piezometers is a piezometer level versus reservoir level correlation plot, where the piezometer readings are plotted with the reservoir level on the x-axis (Figure 3b). The correlation plot displays the historical piezometer and reservoir elevation data and illustrates the correlation between the two variables. If the piezometer level is directly influenced by the reservoir, and no significant lag effect is present, the correlation plot will show a strong correlation and a generally straight line.

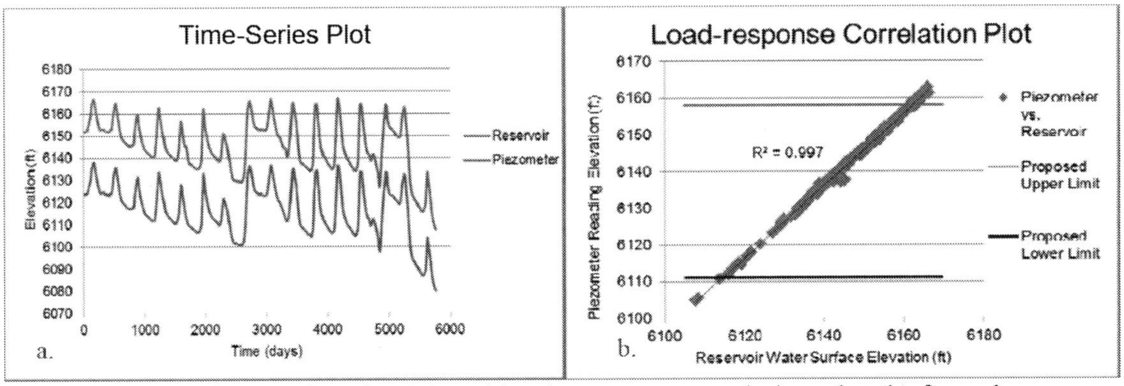

Figure 3. Time-series plot (a) and load-response correlation plot (b) for a dam core piezometer. Notice the very strong correlation, indicated by the high R^2 value, between the piezometer level and reservoir elevation.

Through use of the load-response method (plotting the reservoir elevation on the x-axis) there are multiple variations to establish performance parameters. Engineers will often look at the lowest reservoir water level and correlate that to the piezometer level observed for the related time period. Conversely, the same relationship is established for the highest reservoir level and the highest piezometer level observed for that period. These values will be designated as the upper and lower limits for the piezometer performance parameters (Figure 3b). At the discretion of the evaluating engineer, performance parameters may also be developed with a linear threshold limit equation. This equation can be used to determine the expected piezometer reading upper and lower limits based on the reservoir elevation at any given time. These variations are based on the load-response relationship; that is, the piezometer readings are plotted with the reservoir level on the x-axis and assume the piezometer level is driven only by the reservoir level at the time the readings were taken.

Limitations of the Load-Response Method

As with any applied technique or method, thorough knowledge and understanding of its applicability and limitations should be acquired before use. As discussed earlier, the load-response method works under the assumption that the piezometer level is directly related to the reservoir elevation. In situations when equilibrium has been achieved, this assumption is valid and the piezometer readings show a significant correlation with the reservoir level (Figure 3b). However, serious limitations are apparent when the load-response method is used to establish performance parameters in embankment dams that have not achieved equilibrium.

In such cases, the transient piezometer readings are influenced by two factors: both the fluctuations of the reservoir and the time dependent trending pattern of the readings (Figure 1 and Figure 2). When the piezometer readings are viewed on a load-response correlation plot, the low R^2 value indicates there is no significant correlation between the readings and reservoir level (Figure 4c & 4d). Because the readings are influenced by both the change in reservoir level and the time dependent trending pattern, the load-response method cannot adequately account for the trending pattern affecting the readings.

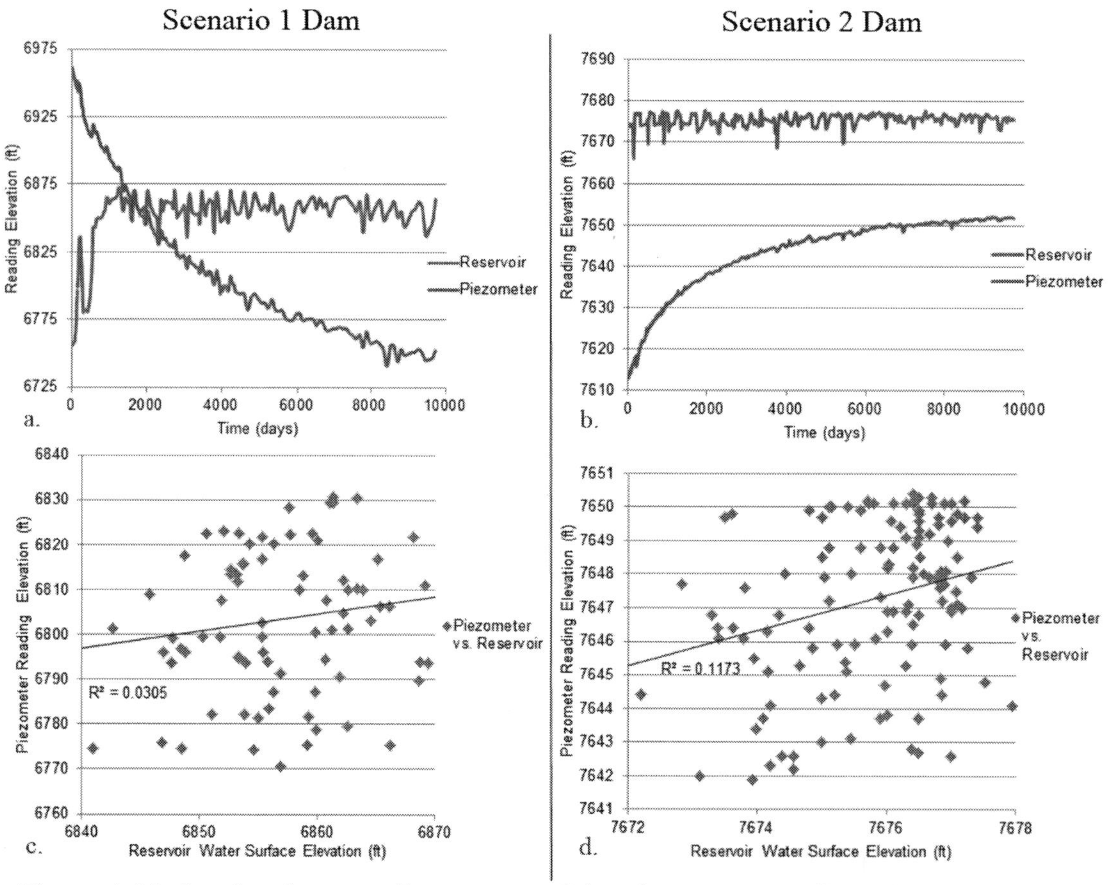

Figure 4. Notice the clear trending pattern of the piezometer reading when the piezometer data is plotted on a time-series plot (a & b), yet there is essentially no correlation when the same data is plotted on a load-response correlation plot (c & d).

This limitation has typically been accounted for by extrapolating the data, in conjunction with a time-series plot, and using judgment in determining where the readings would be in the projected six or eight year cycle. Based on this analysis, the upper and lower limits are placed near a location that would allow the transient readings to still be within the limit range near the end of the projected cycle (Figure 5). However, this common work around has limitations that must be addressed, and can actually increase the risk to the dam safety program.

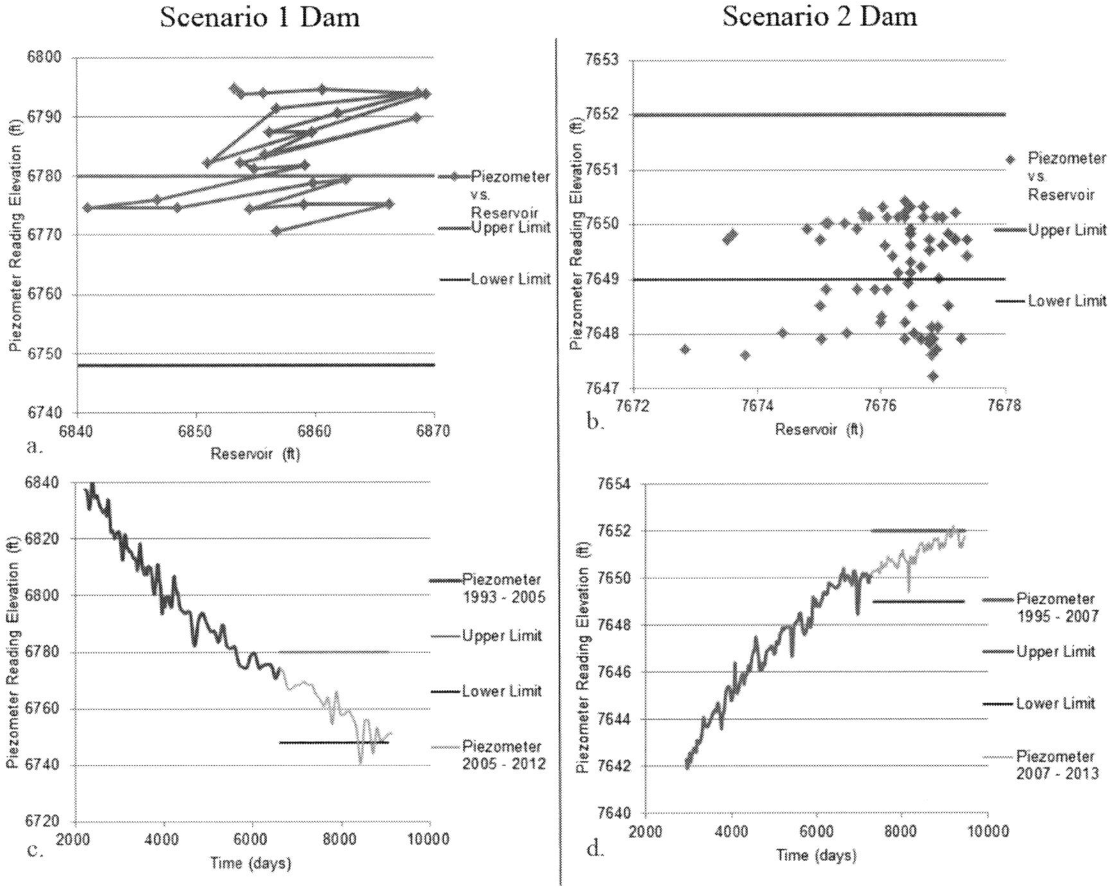

Figure 5. A plot of readings connected via time-series (a) and a scatter plot (b) are graphed. The readings connected via time-series indicate there is little correlation to only the reservoir elevation and the readings are still transient. Performance parameter limits for both piezometers are established at locations where future readings are expected to be within the projected cycle (a & b). The limits have a substantial range in which the readings can act and still be within "acceptable" range. The blue data set of readings was used to predict the performance of the orange data set of readings.

The most serious limitation of extrapolating the data on a load-response correlation plot is that the performance parameter limits are set at a location projected in the future, and are not near the current readings (Figure 5a & 5b). This ultimately creates a large and unnecessary range between the upper and lower limit that introduces risk to dam safety program (Figure 5c & 5d).

A clear example of this is the parameter limits established in Figure 5c. Once the limits are set, there is an unnecessarily large range in which the readings could drop and still be within "acceptable" behavior. Further, at the end of the cycle the readings could increase substantially and still be within the "acceptable" range. If piezometer readings have an unnecessarily large range in which they can act, then they have the capacity to dramatically fluctuate and still be within the "acceptable" performance parameter range. This diminishes the relevance of the performance parameters as they are supposed to generally predict future performance based on historical operations. Additionally, such a range does little or nothing to signal unexpected performance. In a worst case scenario, it can delay or distract engineers from adequate evaluation of the conditions in an emergency, wasting critical time.

This method also introduces an unnecessary level of subjectivity to the process of establishing performance parameters. Because there are no clear trending patterns in the dataset when viewed with the reservoir level on the x-axis (Figure 5a & 5b), engineering judgment is required to establish the limits which can lead to unnecessary subjective bias.

What is needed for transient readings are performance parameters that closely follow the readings as they adjust with time, and account for the fluctuations in the readings due to the change in reservoir elevation.

Time-Series Analysis

In statistics, serial correlation is the correlation of a dataset over time, and typically it is preliminarily screened for through a time-series analysis. Time-series analysis, plotting the dataset with time on the x-axis and observing for trends, can be a simple tool that is often used in forecasting. In the case of pore-water pressure that has not achieved equilibrium, the piezometer readings are related to both the fluctuations of the reservoir and the time dependent rate of dissipation or filling of the embankment. This temporal relationship is easily observed via a time-series plot.

This paper proposes a time-function method for establishing performance parameters in embankment dams that have not achieved equilibrium by applying a time-series analysis. The time-function method is outlined through an example and the results of its application are reviewed.

TIME-FUNCTION METHOD TO ESTABLISH PERFORMANCE PARAMETERS

Performance parameters are established for piezometers using the time-function method for example purposes. Four piezometers from two different embankment dams were chosen for this study: Scenario 1 Dam and Scenario 2 Dam. Scenario 1 Dam represents Scenario 1, where excess pore-water pressure was introduced during the construction of the dam and has been dissipating for nearly 30 years of operation. Scenario 2 Dam represents Scenario 2, where unsaturated pore-water conditions were initially present and pore-water pressure in the core of the dam continues to rise after more than 30 years of operation.

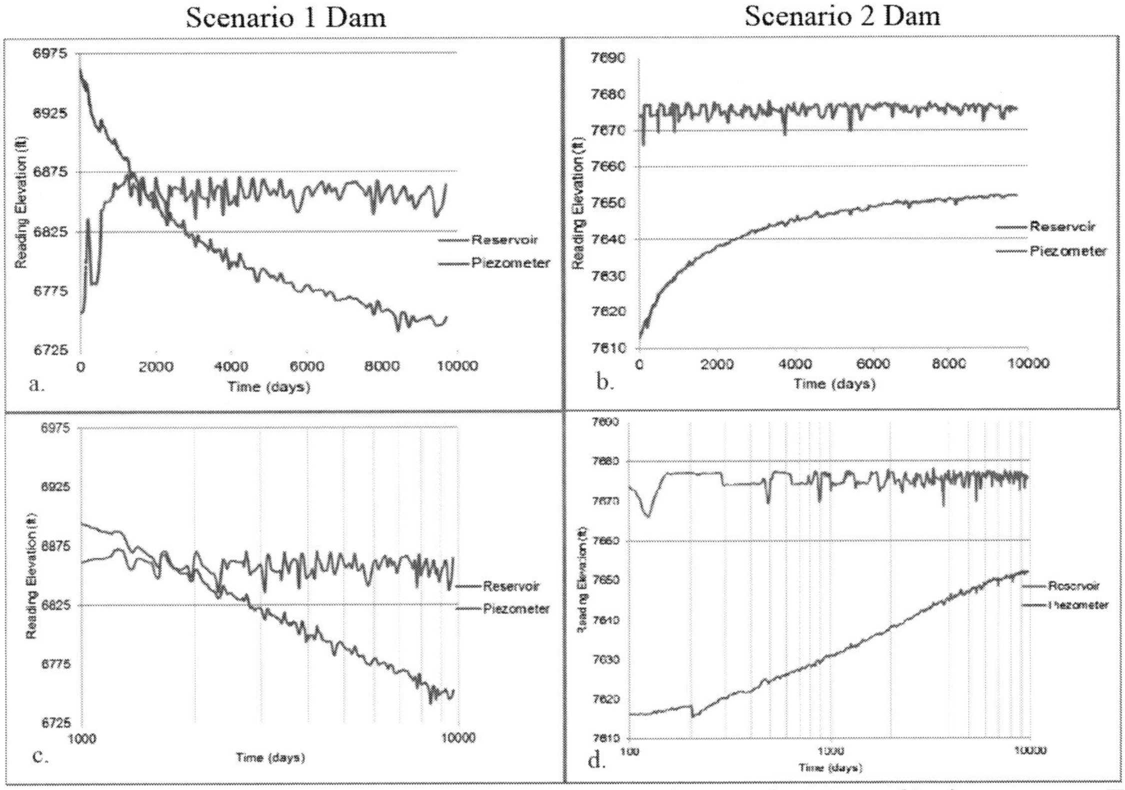

Figure 6. Time-series plot for Scenario 1 Dam (a) and Scenario 2 Dam (b) piezometers. The readings show a linear trend when plotted on a logarithmic axis (c and d), indicating the logarithmic trend of the readings.

Piezometer readings were plotted with time on the x-axis for each piezometer at each dam (Figure 6). A datum date was established for each respective dam near the start of the reservoir first fill. The piezometers all demonstrate an increasing or decreasing logarithmic trend when a time-series plot is evaluated. A logarithmic regression analysis was performed on each piezometer's historic data for a twelve year cycle (Figure 7a & 7b) and a six year cycle (Figure 7c & 7d). The twelve year cycle was evaluated to determine if using a longer time period provided better results than only the most recent six years of data.

For Scenario 1 Dam, the performance parameters for six and twelve year cycles were developed from years 1993-2005 and 1999-2005 data, respectively. For Scenario 2 Dam, the performance parameters for the six and twelve year cycles were developed from years 1995-2007 and 2001-2007 data, respectively. The instrumentation data analysis for each dam started six to ten years after first-filling of each respective reservoir to account for the close monitoring period immediately following the first-filling of the reservoir.

Time-Function Method

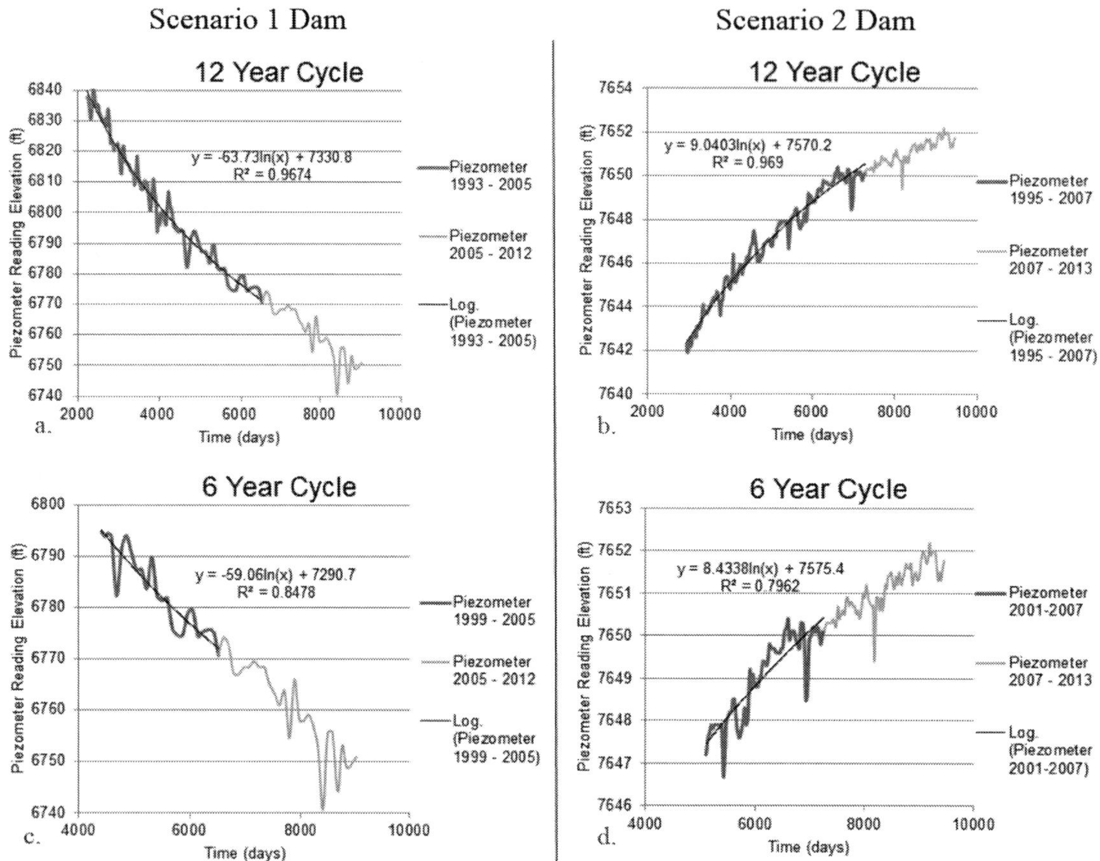

Figure 7. The regression analysis was developed using the blue set of readings and was tested against the orange set of readings. Notice the strong correlation of the blue dataset indicated by the relatively high R^2 value.

The regression analysis was performed using the Trendline Function in Microsoft Excel. All of the logarithmic regressions showed a strong correlation indicated by the relatively high R^2 value. The results from the regression analyses were used to establish logarithmic equations for how the piezometer readings function with respect to time (Figure 7).

Once the logarithmic equations were developed, the y-intercept for each equation was manipulated to establish an upper limit and lower limit by increasing or decreasing the y-intercept of the equation (Figure 8b). To establish the upper limit, the y-intercept of the equation was increased and the equation was plotted at the upper limits of the piezometer readings reached during the plotted cycle. To establish the lower limit, the same process was used except the y-intercept was decreased and plotted at the lower limits of the piezometer readings. The range between the upper and lower limit was not a significant consideration except that y-intercept values were used to provide a whole number range between the upper and lower limits.

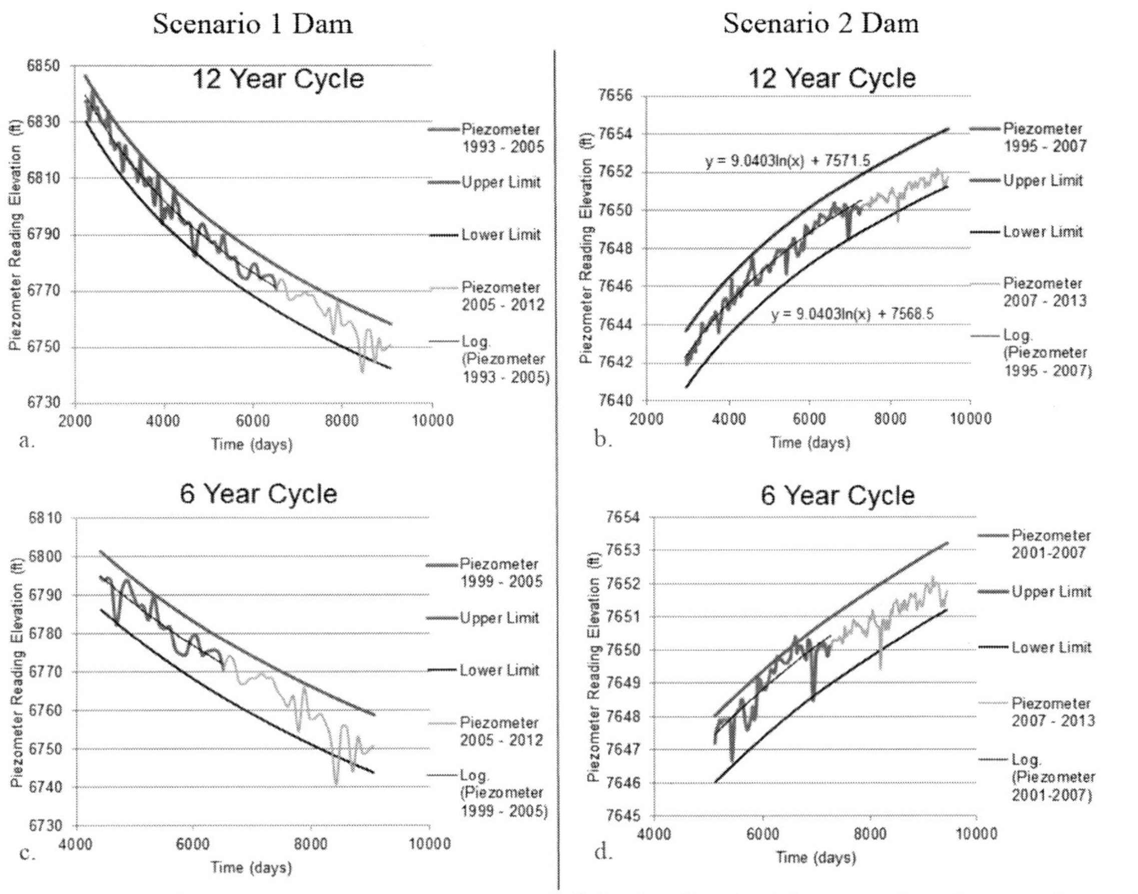

Figure 8. Performance parameters were established using the blue set of readings and tested against the orange set of readings. The upper and lower limit equations are displayed on Figure 8b as an example.

ANALYSIS

After the upper and lower performance parameter limits were established, the limits were used to predict the piezometer performance for the respective test year cycles (Figure 8). The accuracy of the performance parameters was determined by quantitatively evaluating how often the piezometer readings exceeded either the upper or lower performance parameter limits during the respective test year cycles.

For Scenario 1 Dam, the performance parameter results (Figure 8a & 8c) were compared to the 2005 - 2012 data readings. For Scenario 2 Dam, the performance parameter results (Figure 8b & 8d) were compared to the 2007 - 2013 data readings.

RESULTS

The combined results for both Scenario 1 Dam and Scenario 2 Dam show that the time-function method for the twelve year and six year cycles accurately predicted future performance 92% and 89% of the time, respectively (Table 1).

Table 1. Combined Results for Scenario 1 Dam & Scenario 2 Dam

Method	Twelve Year Time-Function Method	Six Year Time-Function Method
Average Time Within Performance Parameters	92%	89%

Also, through observation, it is clear that the limits are closely bound around the piezometer readings and follow the readings with time (Figure 8). This translates to a much tighter range as the upper and lower limit are able to progress with time following the same rate as the transient piezometer readings. Results for Scenario 1 Dam are summarized in Table 2, and results for Scenario 2 Dam are summarized in Table 3.

Table 2. Summary Table of Performance Parameter Accuracy for Scenario 1 Dam

Method	Twelve Year Time-Function Method	Six Year Time-Function Method
Average Range Between Upper and Lower Limit (ft)	17.3	15.5
Average Time Within Performance Parameters	89%	92%

Table 3. Summary Table of Performance Parameter Accuracy for Scenario 2 Dam

Method	Twelve Year Time-Function Method	Six Year Time-Function Method
Average Range Between Upper and Lower Limit (ft)	3.5	2.0
Average Time Within Performance Parameters	94%	86%

DISCUSSION

Observations

The use of engineering judgment is inherently subjective, and introduces bias to any work. The results of this paper are based on the Author's judgment to establish performance parameters for the time-function method. Further, this study is not exhaustive and only evaluates piezometers from two different dams.

Both the twelve and six year test cycles had results that accurately predicted the future performance of the piezometer level approximately 90% of the time. In light of the pore-water pressure that has not achieved equilibrium, a 90% accuracy is considered adequate (though results would likely vary depending on the piezometers). The results are inconclusive in regards to using a twelve year vs. a six year cycle for the time-function method.

Dam Safety Monitoring

Performance parameters are essentially guidelines for how instrumentation data is expected to behave based on historic performance. They are one of many tools used to monitor dam safety, and they do not represent strict rules that signal catastrophe if the instrumentation data are out of the performance parameter range.

The most critical time for a dam, based on historical dam failure rates, is within the first five years after construction, during first filling of the reservoir and the initial years of operation. Standard practice calls for close monitoring of the dam for a period of five to ten years after construction of the dam and first filling of the reservoir. Observed instrumentation data from this initial monitoring period may indicate that pore-water pressure has not reached equilibrium, and may not do so for several or many years. The pore-water pressure in embankment dam piezometers with excess or minimal pore-water pressure has been documented to act logarithmically with time before ultimately approaching equilibrium conditions. In such cases, application of the proposed time-function method is likely more appropriate than the traditional load-response method.

Applicability of the Time-Function Method

The load-response method accounts for the influence of the reservoir level on the pore-water pressure, but fails to capture the time-dependent trend of pore-water pressure prior to equilibrium. Once equilibrium in the embankment has been achieved, the reservoir level should be the primary stimulus of the pore-water pressure and the load-response method should be used. However, in some cases before equilibrium is achieved, the pore-water pressure is influenced by both a time-dependent component and a reservoir level component (2).

$$\text{Pore-water Pressure} \approx \pm f(\mathit{Time}) + f(\mathit{Reservoir\ Level}) \tag{2}$$

The time-function method is unique because by using a time-series analysis to establish the performance parameters, it accounts for the logarithmic trend of the piezometer readings as well as the influence from the fluctuations of the reservoir. The trend is accounted for by the natural logarithmic time-function of the equation, while the fluctuations of the reservoir are accounted for by the range set between the upper and lower limits (3).

Time-Dependent Component (applies to $\pm K \times Ln(t)$) | Reservoir Level Component (applies to C)

$$\text{Time-Function Method} \begin{cases} \text{Upper Limit} = \pm K \times Ln(t) + C \\ \text{Lower Limit} = \pm K \times Ln(t) + C \end{cases} \tag{3}$$

Time-Function Method

Limitations of the Time-Function Method

A significant limitation of the time-function method is the cumbersome nature of the time-series plot and equations. Developing the performance parameters is not significantly more time consuming or difficult than the load-response method, but managing the ongoing monitoring process requires more diligence. The instrumentation engineer must clearly indicate the datum starting time of each time plot and how it progresses. Further, handling specific equations that are unique for each piezometer introduces many variables, and it is foreseeable that input mistakes can occur if this method is implemented on a large scale. Thorough data management would be required to avoid mistakes if the time-function method were implemented.

Historical piezometer readings for Scenario 2 Dam did not perfectly follow one theoretical logarithmic curve. The results of this study show that the slope of the curve can change and the piezometer readings can fall out of the performance parameter range due to the changing slope of the curve. This is likely related to geophysical characteristics of the dam and how the pore-water pressure is reacting (dissipating or increasing) after construction and first fill. In such cases, the dam should receive close attention until the performance parameters are reevaluated.

CONCLUSION

Care should be taken in the application of methods used to establish performance parameters for dam instrumentation. The appropriate method is typically that which most accurately predicts the future performance of the dam. This paper has illustrated the limitations of using the load-response method to establish performance parameters for piezometers with transient readings. Conversely, the applicability of the time-function method in establishing performance parameters for piezometers in embankment dams that have not achieved equilibrium has been shown effective. Use of the time-function method can contribute to the growing body of institutional knowledge regarding the safety and performance of embankment dams.

ACKNOWLEDGEMENTS

The Author would like to express thanks to friends and colleagues who made contributions to this paper.

REFERENCES

Bureau of Reclamation. *Design of Small Dams.* 3rd ed. Denver, CO: U.S. Department of the Interior, Bureau of Reclamation, 1987. Print.

Dunnicliff, John, and Gordon E. Green.*Geotechnical Instrumentation for Monitoring* Field Performance. New York: Wiley, 1993. Print

Gall, Eugene, et al. "*Correlation Plot - A Modern Tool In Dam Safety Monitoring.*" Modernization and Optimization of Existing Dams and Reservoirs (2007): Web. 20 Sept. 2013.

Jeon, Jesung, et al. "*Development of Dam Safety Management System.*" Advances in Engineering Software 40 (2007): 554-563. Web. 20 Sept. 2013.

Terzaghi, Karl, and Ralph B. Peck. *Soil Mechanics in Engineering Practice.* 3rd ed. New Delhi: Wiley, 1996. Print

Schwartz, F.W., and Hubao Zhang. *Fundamentals of Groundwater.* New York: Wiley, 2003. Print.

U.S. Army Corps of Engineers. *General Design and Construction Considerations for Earth and Rock-Fill Dams, Appendix E, Process for Establishing Performance Parameters.* Washington, DC, DC: U.S. Army Corps of Engineers, 2004. Print.

U.S. Army Corps of Engineers. *Instrumentation of Embankment Dams and Levees.* Washington, DC, DC: U.S. Army Corps of Engineers, 1999. Print.

Time-Function Method

TVA'S INSTRUMENTATION AND MONITORING PROGRAM — HELPING MANAGE RISKS ASSOCIATED WITH WASTE IMPOUNDMENTS

W. Allen Marr, PhD, PE, NAE[1]
Nicolas McClung, PE[2]
Scott Turnbow, PE[3]
Dori Ross, PE[4]

ABSTRACT

TVA is responsible for the operation and safety of 22 waste stacks associated with its 11 coal-fired power plants located in 3 states. These stacks store various combinations of fly ash, bottom ash, and gypsum with a combined surface area of approximately 1680 acres. Fly ash comes from the removal of particulate matter from the exhaust stacks. Bottom ash is the residuals left from burning coal. Gypsum is a byproduct of the removal of particulate matter from the exhaust stacks. Historically these materials were transported to the disposal areas by pipeline and sluiced into position behind dikes. The dikes are raised by constructing perimeter embankments of local materials and in some cases using the waste materials. Most raises occurred using the upstream method in which each new dike is placed on top of the older dike but shifted inward. The dikes have typically impounded standing water for their operational life which has resulted in them being managed as dams.

In 2008, after the catastrophic failure of one of these stacks at Kingston, Tennessee, TVA undertook a major program to upgrade its performance monitoring program to a real-time monitoring system that gives advanced warning of unexpected performance. Primary potential failure modes were identified for each active waste stack and a monitoring program was developed to indicate early stage development of each significant failure mode. Instrumentation was installed and a real-time data collection and management system put into place. Currently this system monitors more than 5,400 sensors logging more than 17,000 readings per hour.

This paper describes the factors used to design the monitoring system, the key components of the system, and examples of how it is used to help manage TVA's risk for these facilities. The benefits of the instrumentation and monitoring approach as a risk management tool are discussed. Lessons learned from the implementation and operation of the system are described so that other owners of similar facilities might benefit.

[1] CEO, Geocomp, 125 Nagog Park, Acton MA 01720, wam@geocomp.com, 978-635-0012
[2] Manager, Risk Management and Quality Assurance, Tennessee Valley Authority, 1101 Market Street, LP 5G C, Chattanooga, TN 37402, 423-751-8025, namcclung@tva.gov
[3] General Manager, Strategy and Engineering, Tennessee Valley Authority, 1101 Market Street, LP 5G C, Chattanooga, TN 37402, 423-751-3031, msturnbow@tva.gov.
[4] Sr. Project Manager , Geocomp Corp., 125 Nagog Park, Acton MA 01720, dross@geocomp.com, 978-635-0012

INTRODUCTION

TVA currently has 11 coal-fired power plants with 46 active generating units located in Tennessee, Alabama, and Kentucky. These plants can produce up to 81.4 billion kilowatt-hours of electricity in one year which accounts for about 57 percent of TVA's power generating capability. Since the large failure at its Kingston facility in 2008, TVA has begun closing all wet ash systems. The closure of wet ash storage systems consists of building ash and gypsum dewatering facilities, constructing new dry storage landfills, and closing existing ash and gypsum ponds. This plan is expected to cost 1.5 to 2 billion dollars over an 8 to 10 year period.

On December 22, 2008, the north and central portions of the cell 2 ash disposal area dike failed, releasing an estimated 5.4 million cubic yards of fly ash slurry at the Kingston Fossil Plant in Roane County, Tennessee. The fly ash slurry flowed across the Emory River Channel and over the Swan Pond Creek flood plain, approximately 3200 feet beyond the limits of the original ash pond. It was estimated that 42 residential properties were affected by the failure and 22 properties were evacuated. No injuries were reported as a result of the failure. Cleanup has cost TVA more than $1.2 billon to date.

TVA contracted AECOM to perform a root cause analysis and find the cause of the ash spill at the Kingston Fossil Plant. They determined there were four key factors that lead to the failure (Walton, June 25, 2009). First, the dikes of the dredge cell were built using the "upstream method", by which each new dike added to a facility is built on loose sluiced ash. The most recently constructed dikes were built on loose, sluiced fly ash 200 feet inward from the original dike, which itself was built on more stable foundation conditions. Second, the dredge cell footprint decreased in size as ash was added and more cell height was required to accommodate the same ash volume. Consequently, the elevation of the ash was increasing faster and faster with each passing year. Third, creep deformations had developed over time due to loose slimes which were deposited early in the ash pond's history. These slimes had high water contents, high liquidity indices, and low undrained shear strengths upon which 40 to 85 feet of loose wet ash was placed. The final cause of the failure presented in AECOM's analysis was the loose wet ash itself. The initial sluiced ash had a high void ratio as a result of being deposited under water and didn't benefit from consolidation under the surcharge weight placed above these deposits. This resulted in the sluiced ash remaining very loose with a low undrained shear strength. The combination of these factors resulted in a global stability failure and the release of a large volume of fly ash. TVA then undertook a massive evaluation of the stability and safety of all of its active ash storage facilities.

One of the outcomes of this evaluation was the development of an instrumentation and monitoring program to monitor the stability of the Coal Combustion Products (CCP) disposal area structures. Specifically, the pore water pressures within the containments would be monitored by reading the depth of water in standpipe piezometers (PZ) and subsurface slope movement would be monitored by reading deflections of slope inclinometer casings at planned intervals. The CCP Instrumentation and Monitoring Program became integral to TVA's risk management program by:

- Helping demonstrate that each facility is meeting TVA's performance objectives.
- Indicating unacceptable or questionable performance to draw attention for further examination.
- Indicating unacceptable performance in sufficient time to allow remedial measures to be implemented.
- Collecting associated data and information to help interpret data from the instrumentation.
- Presenting all relevant information in an organized, expeditious manner to allow timely evaluation of the safety of a facility for any particular condition.
- Indicating where modifications to the monitoring program and operating practices are required.

DEVELOPMENT OF INSTRUMENTATION PROGRAM

Safety assessments in 2009/2010

In 2009 and 2010, TVA performed a series of geotechnical explorations and slope stability evaluations at the 11 active coal-fired power plants. The assessments were made in order to determine whether present conditions indicated an unstable condition that could cause a release of waste materials into the surrounding environments. The study of each site began with a review of historical information and site inspections, after which a geotechnical exploration plan was developed and implemented within each site's disposal areas.

Representative cross sections were developed to reflect typical conditions for each facility and a series of explorations were made that consisted of drilling borings, taking samples and installing piezometers. Subsurface investigations included Standard Penetration Testing (SPT), Cone Penetration Tests (CPT), and undisturbed soil sampling. Borings were generally positioned at the dike crest, mid-slope and toe areas of the representative cross sections. Most borings were advanced to apparent bedrock or into firm natural material.

Subsequent laboratory testing consisted of classification, moisture content, permeability, and shear strength testing. Data from the drilling and laboratory testing program were used to perform slope stability and seepage analyses in order to quantify factors of safety for current conditions. The seepage analyses helped estimate seepage gradients for the evaluation of piping potential, and pore water pressures within dikes and foundation soils for use in the slope stability analyses.

Dikes and ash stacks were assessed under static, long-term, steady state, and rapid-drawdown conditions. Factors of safety were developed using Spencer's method of analysis, circular slip surfaces, and search routines that helped identify the critical failure surface. Slope stability models were then evaluated using both pore pressures predicted with the seepage models and data from piezometer readings. The new information and analyses showed multiple cross sections with factors of safety for global stability less than 1.5, the minimum factor of safety required by TVA for long-term, static loading conditions. Evaluations also identified areas of inadequately controlled seepage. TVA undertook work to improve these locations including re-grading of perimeter ditches, slope flattening and the installation of riprap, and the installation of toe buttresses and toe drains to improve surface drainage and reduce the ponding of water. Operations and Maintenance Manuals were developed. An instrumentation and monitoring program was formulated and implemented for each facility to provide indications of developing potential failure modes so risk mitigation measures could be deployed when required.

INSTRUMENTATION PROGRAM

The purpose of the instrumentation was and is to monitor the stability of the CCP disposal area structures. It focuses on providing indications and warnings of the following potential failure mechanisms:

- Slope instability due to high pore water pressures in the interior of the slope.
- Slope instability due to lateral movement within the slope from whatever cause.
- Overtopping of the retention dikes by stored water.
- Unexpected loss of containment as indicated by an unexpected decrease in pond water level.
- Occurrence of significant precipitation.
- Occurrence of a significant earthquake.

The CCP Instrumentation and Monitoring Program is integral to TVA's risk management program by:

- Providing actionable up-to-date information on the performance of the impoundment structures.
- Indicating unacceptable performance in sufficient time to allow remedial measures to be implemented.
- Indicating unacceptable or questionable performance to draw attention for further examination.
- Providing other information, including historical records, relevant to the interpretation and understanding of performance monitoring data.
- Organizing data and information in a format that facilitates rapid and efficient interpretation.
- Helping demonstrate that the facility is meeting TVA's performance objectives.

<u>Types of instruments chosen and why</u>

The following describes the different types of instruments installed and the purpose of each.

Piezometers – Slope stability of these stacks depends directly on the nature of flow and pore water pressures within the stacks. Piezometers were installed at representative locations to define phreatic conditions and indicate pore water pressures. Standpipe-type piezometers were installed in borings advanced using 3 ¼ inch internal diameter (ID) hollow stem augers or 4¼-inch (ID) hollow stem augers. Either 5-foot or 10-foot long slotted screen piezometers were used with the boring fully grouted above the top of the screen to create a seal. The piezometers were typically installed in upper and lower dikes as part of the overall stability evaluation to provide data within the existing dikes and native foundation soils.

Inclinometers - Slope movements serve as early warning signs of impending instability. Horizontal inclinometers provide very good measurements of horizontal movement within a slope. Inclinometer borings were outfitted with 2.75-inch (ID) slope inclinometer casings which were grouted in place with bentonite-cement grout. These were generally placed at the outer edge of the top of the waste stack at representative locations.

Survey Benchmarks and Settlement Monuments – Due to construction techniques and materials used in many of the fossil impoundment structures there was concern about continued long term deformations from consolidation of the waste materials. Survey benchmarks and settlement monuments were installed as part of the monitoring program for embankment stability monitoring. These instruments are read manually by TVA's surveyors on an annual basis.

Water Level Sensors – Pore pressures and seepage conditions within the slopes depend on the pond levels and river levels. Pond and river level gauges were installed throughout the active fossil plants to measure surface water levels in each impoundment.

Typical instrumented section

A typical instrumented cross section is shown in plan in Figure 1 and in section in Figure 2. Figure 1 shows the containment berm built of natural soils with two benches. The grey-white material on the right side of the photo is stored ash.

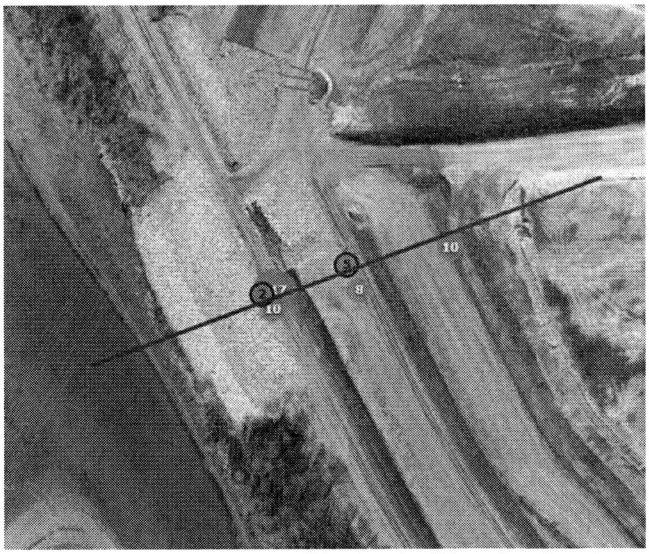

Note: Figure is a screen capture from the iSiteCentral monitoring system. Aerial photo is scaled to project coordinates. Section and instruments are geo-located in a layer overlying the aerial photo. Circles indicate locations of instruments. Number in circle indicates number of sensors at each location. Green color indicates all sensor readings are within their normal range. Light colored material in the photo is rock material added to fortify the slope.

Figure 1. Plan for Section of Cumberland

The cross section in Figure 2 has a red dashed line showing the critical failure surface for global stability. The upper orange dashed line shows the phreatic surface used by the geotechnical engineers to assess global stability in 2010-2011. The analyses calculated the pore pressures for the entire cross section using this piezometric surface. Vertical blue lines indicate pore pressure measurements made in 2012 using grouted in place piezometers. The bottom of the line indicates the elevation of the sensor. The top of the line indicates the piezometric head measured by the sensor. All piezometers in the lower part of the cross section indicate pore pressures much less than one would obtain using the shown phreatic surface. Using the measured pore pressures rather than the assumed single phreatic surface resulted in a significant increase in the calculated factor of safety for global stability for both static and earthquake loading. The data from these piezometers helped TVA avoid significant remedial work and save millions of dollars.

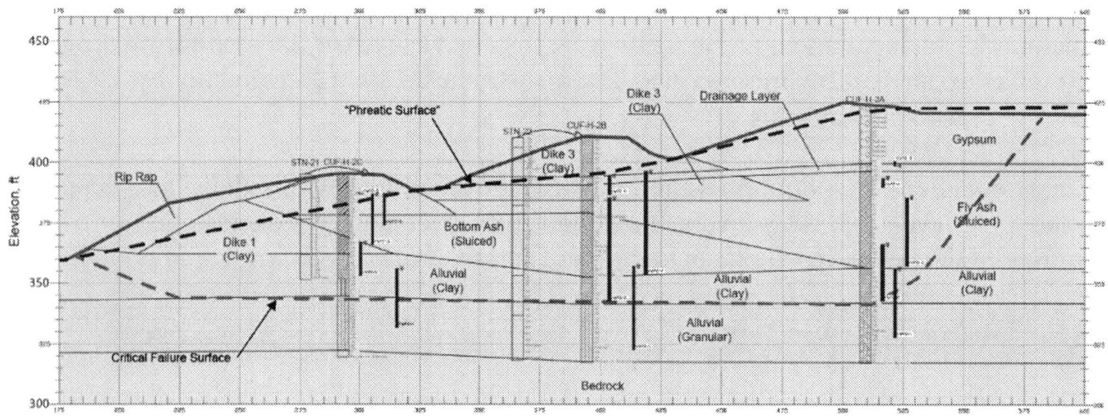

Figure 2. Cumberland Cross Section

Automation

After the Phase I instrumentation program was complete, it became apparent that conditions could change quickly within a stack and that manual readings could not be obtained quickly enough to derive maximum value from the instrumentation and monitoring program. TVA decided to implement an automated instrumentation program that would provide data every five minutes on key metrics of stack stability. Most of the standpipe piezometers were retro-fitted with Geokon 4500S Vibrating Wire Piezometer Sensors connected to data loggers. The inclinometer casings were retrofitted with Geokon In-Place Inclinometers Model 6150 connected to data loggers. The Pond and River Level sensors were likewise automated. Weather stations to measure rain, wind, temperature and barometric pressure were also installed at each site to provide supporting data.

Data Acquisition and Communication

Campbell Scientific CR1000 data loggers with AVW200 vibrating wire interface modules are used to power the sensors and convert sensor outputs to digital readings. These data are transferred to a base station at each site using spread spectrum radios. An

IP or Internet modem at each site receives the radio signals and forwards them to the central data servers at TVA's headquarters in Chattanooga, Tennessee. Figure 3 shows a typical installation of data logger, solar power supply and radio at an instrument location.

Figure 3. Typical TVA instrument station

Figure 4. Software app showing sensor status

Experience at TVA has shown that conditions can change over a few minutes so electronic instruments are read every five minutes and the readings are compared to previous readings and preset limit values. Manual readings of non-electronic instruments are taken by TVA personnel and then input into TVA's data management system while the person is still at the instrument location. This is done using a software app on a phone or tablet. Figure 4 shows one app that gives a map showing instrument locations and sensor status as well as a screen that displays the current reading of any sensor.

DATA AND INFORMATION MANAGEMENT SYSTEM

With the need and ability to collect data from more than 5,400 sensors every five minutes, TVA realized the need for a comprehensive real-time data management system to store, maintain and report these data as needed. TVA contracted with Geocomp to provide a secure and robust data management system that enables TVA users to access all data at any time to help them evaluate performance and risk of their ash storage facilities. Geocomp provided TVA with a server-based system that receives data from the field data loggers via IP connections, converts the raw data to engineering units, stores the information and reports the data in a variety of formats to TVA personnel as needed. The system uses a web browser- and GIS-based graphical user interface which allows users to navigate and access the system quickly and intuitively. Data are stored in a Microsoft SQL database. The system provides a number of useful features:

- All raw instrumentation monitoring data, both for manually and automatically read instruments are stored in the system for the complete record of data collection. This permits adjustments to calculated values if calibration settings need to be changed.

- Raw data are converted to engineering units using the associated calibration factors and/or other sensor data for correction (for example, barometric pressure corrections for piezometers). These calculated values are stored to facilitate rapid data portrayal. They are recomputed if calibration settings change.
- Customizable tables and charts can be created that present either raw or calculated data with their respective threshold and limiting value performance limits if these have been defined.
- Charts and tables are presented in customized reports which are also accessible on demand through the web interface providing TVA with an efficient and reliable reporting system. Each user can create and save his/her private charts and tables and share any of these globally.
- TVA's iSiteCentral GIS system has the functionality to send alert notifications whenever performance limits (threshold values and limit values), defined by TVA or their Engineering Design contractors, are exceeded.
- A chronological performance history is included for each site with each performance incident, such as a crack, a slip, or a pipe washout, defined in space and time with the locations of these incidents shown on a topographic plan or aerial photo.
- A collection of documents, photographs, plans, maps and aerial photographs over the life of the facility that catalogue information relevant to the evaluation of the performance of the facility and make this information readily available from within the TVA iSiteCentral platform. Historical topo maps and aerial photos can be viewed as GIS layers to show changes over time.

Description of the TVA iSiteCentral GIS system

A TVA engineer's ability to assess and manage the risks associated with their areas of responsibility is directly affected by the quality of data and observations available, the timeliness of the data being delivered, analysis of the data in context with other information, and the actions that are required as a result. As noted before, there is a significant amount of data (13,000 readings per hour every hour of each day) coming into the TVA iSiteCentral GIS system, so much that it would be overwhelming for a person or group of persons to evaluate without using an automated system.

To aid in this process, TVA's iSiteCentral GIS system provides a map-based portal to access performance data, documentation, and historic information relevant to the CCP disposal areas. It also proactively provides automated data collection, processing, and alarming services to identify field performance issues for TVA engineers so that they can focus their efforts and act as appropriate.

A schematic diagram showing the data flow and services being performed automatically by TVA's iSiteCentral GIS is shown in Figure 5. The automated instrumentation data are collected on field data loggers on pre-programmed intervals. The typical configuration on most sites is one reading every five minutes for the weather station, piezometers, pond level indicators and river level indicators; and one reading every hour for the in-place inclinometers. The data collected are uploaded into the TVA iSiteCentral system.

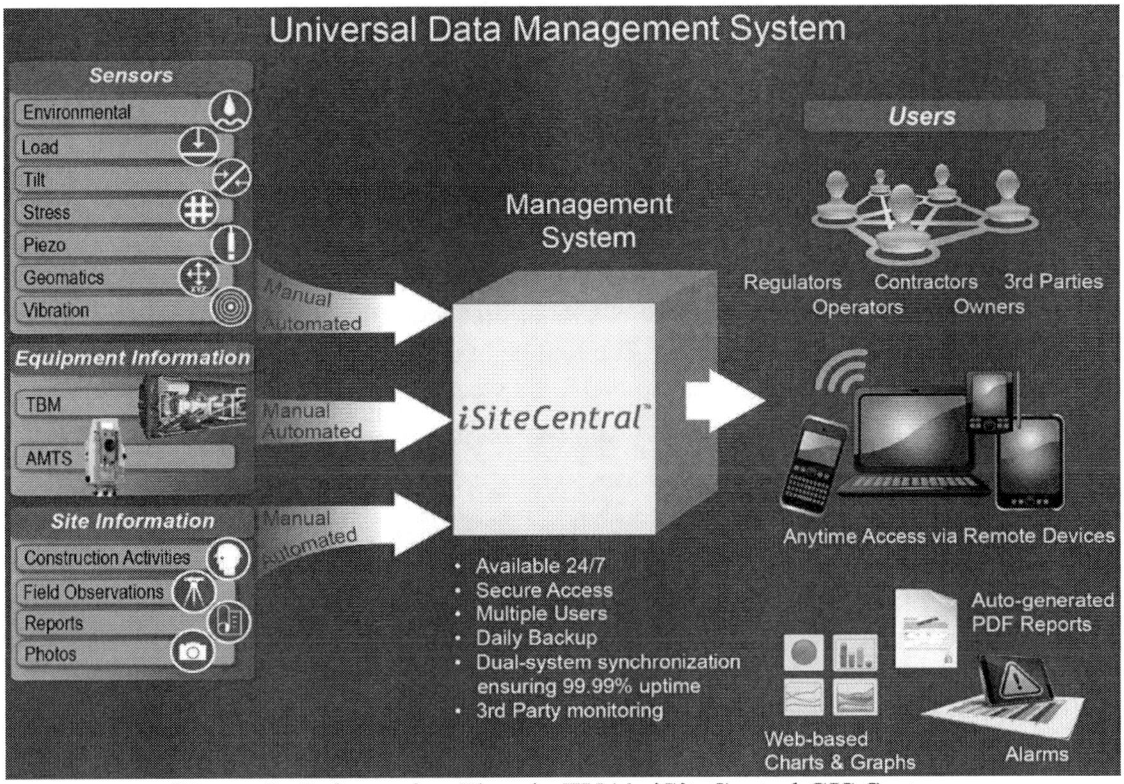

Figure 5. Information Flow in TVA's iSiteCentral GIS System

Several services are running on TVA's iSiteCentral system to identify actionable issues in the data stream. These include:

Heartbeats – the system has a user-defined threshold called the *Heartbeat Span*. When the time between the current time and the last posted reading exceeds the Heartbeat Span, an alert like that shown in Figure 6 is issued to indicate that something is not working to deliver data when expected. The Heartbeat Span is configurable for each individual sensor or as groups of sensors. Heartbeat Jobs are created and scheduled for any list of sensors and any user or user group.

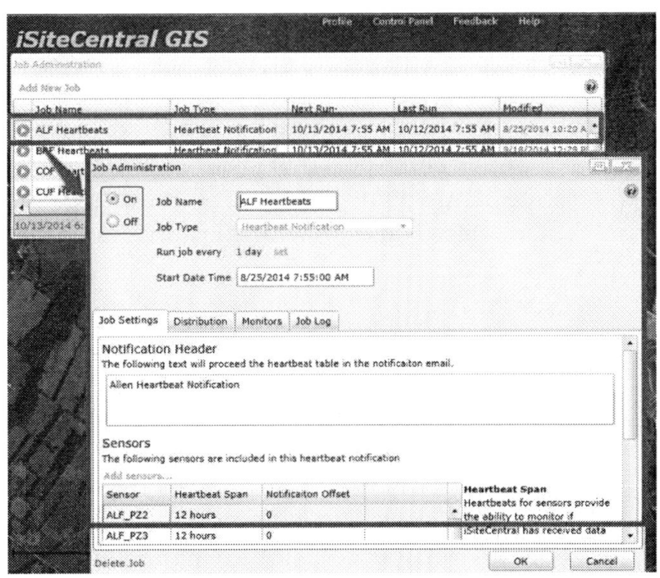

Figure 6. Heartbeat job detail

System Health alerts provide an early warning of potential data processing issues, malfunctioning equipment, or communication failures.

Invalid/Erroneous data are flagged and removed from charts, tables and reports but remain in the database for troubleshooting.

Rainfall Alerts – One driving force behind some potential failure modes at the CCP disposal facilities is heavy rainfall. TVA's iSiteCentral has two focused services to identify and track rainfall events. The first is an integrated reporting of NOAA's weather station alerts and access to NOAA's weather data. Heavy rainfall warning notices like that shown in Figure 7 are delivered to users to focus additional attention to the monitoring where and when appropriate. The second is a rainfall event tracking algorithm developed to provide information on storm severity (i.e. a 50 year storm) based on NOAA's published analyses. TVA uses this information to mobilize field staff to perform visual monitoring and inspection procedures so consequences of heavy rainfall can be identified early and managed.

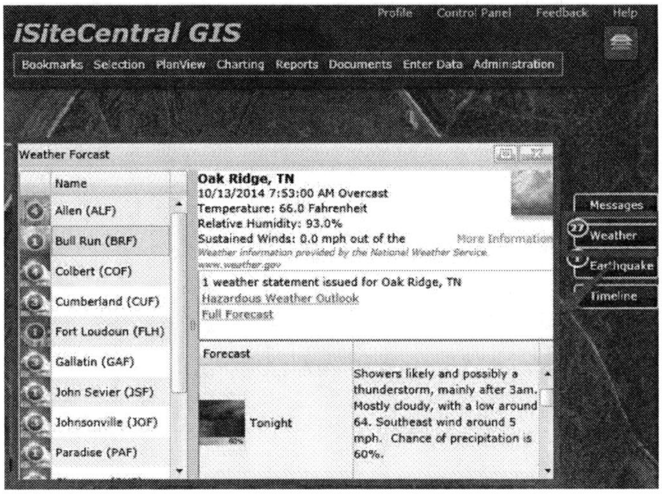

Figure 7. Rainfall Alert

Earthquake Alerts – The TVA has regulatory requirements to assess and report on any significant earthquake events that occur at their CCP Disposal facilities. These facilities are exposed to shaking primarily from the New Madrid fault system. TVA's iSiteCentral monitors the USGS earthquake's real-time reporting feeds, and issues alerts based on the specific criteria associated with TVA's regulatory obligations.

Sensor Alerts – A Limit Level reading and Alert Level reading is established for each sensor. Email messages are sent whenever these levels are exceeded and TVA initiates a pre-planned action plan. The Alert status of all sensors at a site are shown in color code on a plan of the site to provide an overall quick visual impression of the status of a site.

Each user in the system has a managed view into the system (i.e. the project administrator determines what sensors, information, and features a user can access). For TVA's iSiteCentral services listed above, a user's managed view is always enforced. This means that if a user is an active participant in a heartbeat job, but does not have permission to view the data for all sensors configured in the job, then that user will only see an alert message for the sensors it is permitted to see/view. This allows flexibility to deploy alerting services without confusing users with alerts on sensors/data for which they are not responsible. One useful application of this feature is to direct alerting messages on system performance to operations and maintenance (O&M) staff and messages on performance alerts to the Engineering staff. For example, Engineering staff don't need to be bothered by messages about a failing power supply that needs attention by O&M staff.

Typical use cases

Rain Data - Rainfall notifications are programed to alert engineers, field staff and management when threshold events are exceed. At least three events in 2013 and 2014 triggered field staff and operational staff to mobilize to several CCP impoundments and visually monitor saturated slopes and swollen ponds. These notifications helped TVA to successfully reduce risks from overtopping of dikes and slope instability.

Unusual Performance - On August 1, 2014, TVA field staff reported to TVA's engineering group a perceived potential problem in an ash pond at the Shawnee facility. At the time, unexpected bubbles were reported in the pond which could indicate water flowing out of the pond through an abandoned pipe or sinkhole. This observation concerned field staff as prior failures of this type have occurred. TVA used the iSiteCentral GIS system to evaluate the problem. The pond had been outfitted with an automated pond water level sensor. TVA accessed the data from the pond sensor (including historical information) to determine that there had been an unexpected decrease in the pond level elevation. The data were confirmed to be accurate. The change in pond level was determined to be directly related to dredging activities occurring at the site. TVA stopped the dredging activities and the pond level returned to normal.

Exceeding Threshold and Action Levels - During the last year, TVA has experienced multiple rainfall events that have produced saturated soil conditions and as a result increased pore pressures in the CCP impoundments. These increased pore pressures have caused multiple threshold levels to be exceeded which resulted in the following activities:

- Field personnel called to each area and visual inspection performed.
- Manual verification of automated readings.
- Slope stability analysis performed modeling existing conditions.
- Modification of slopes and operating procedures to increase factor of safety where warranted.
- Adjustment of threshold levels if factors of safety remained above acceptable criteria for steady state conditions.

PROJECTED BENEFITS OF THE TVA INSTRUMENTATION AND MONITORING PROGRAM

There are many projected benefits to the TVA instrumentation and monitoring program by using iSiteCentral GIS. Many of these benefits have been achieved already and are discussed below. These benefits will continue through and beyond final closure of these waste storage facilities.

Both the operations/maintenance group and the engineering group at TVA currently benefit from the early warning system that is part of TVA's iSiteCentral GIS system. Messages in the form of emails are sent to select members of TVA operations/maintenance group to notify them when a sensor needs maintenance and/or additional attention. With approximately 5400 sensors at TVA facilities, this is extremely helpful in tracking and scheduling maintenance needs. The engineering group staff also receive notifications for instrumentation readings that exceed specific threshold and action limits. This brings immediate attention to evaluate the readings and determine what action is required. Alerts also in the form of an email are sent to TVA staff including members of the engineering group and representatives at the individual facilities for a rain event that exceeds a specified amount. This information allows TVA to be alert to possible changing site conditions related to weather events. All notifications from the iSiteCentral GIS system allow for early identification and detection of unexpected or undesirable performance by TVA.

TVA's engineering group has access to vast amounts of current and historical data and documentation which is stored in one central location – TVA's iSiteCentral GIS System. The data and information management system includes material in relation to instrumentation such as boring logs, installation logs, both manual and automated data, photos, cross sections, reports, historical timelines, and site development plans. TVA staff can create charts, tables and reports directly from iSiteCentral GIS. The categorized data allows for easy access to information TVA staff need to evaluate real time issues and make informed decisions about site conditions and the possible need for additional action. If no action is necessary, TVA staff can continue to follow trends over time and look for cause and effect relationships.

LESSONS LEARNED

Advancements in logger technology and planned changes in field system configuration and data migration are being integrated to improve the overall system:

1) The field data loggers will be equipped/updated with new capability to serve data over a TCP/IP connection, and published to authorized users in XML format via a password-protected HTTP connection, and
2) The remote connections, which are currently IP modems on the local cellular networks, will be modified to transmit data via the TVA's LAN network at each site.

These changes will: reduce the lag-time between data collection and data posting/alerting in TVA's iSiteCentral GIS; make the data transmission more secure and less costly; provide a redundant means of data collection (i.e. XML data feed and conventional data files); and provide easy review of current readings by field staff without using data logger-specific software. Ultimately the monitoring systems will be more interactive and event-driven. For example, a heavy rainfall warning from NOAA, an Earthquake event alert from USGS, or an Engineer observing some event/activity in the field can automatically trigger more frequent data collection and reporting.

During the initial system development, the focus was on collecting and reporting data from the instruments as quickly as possible to support TVA's engineering assessments. It soon became clear that rapid assessment of the data required rapid access to historical information, current geometry, subsurface information and other documents to provide context to the instrumentation data. An extensive document management system was added that provides these materials using the same GIS platform. Documents in electronic form are tagged with time and location data so they can be correlated with the instrument locations. This feature allows an engineer making an assessment of the instrumentation data to readily obtain information about the subsurface conditions, sequence of development of the slope, history of performance, installation details for the instrument and other information to help determine what the measured data mean.

During deployment it became attractive to consider making it possible for on-site people to add information and data to the database and to see results from the system using mobile devices (cell phones and tablets). Specific software apps have been developed to enter manual readings, add field observations, provide inspection reports, view measurements for a specific instrument and add annotated, geo-referenced digital photos. These capabilities are greatly enhancing the effectiveness of the field activities.

CONCLUSIONS

TVA uses a proactive, near-real-time performance monitoring program to identify undesirable performance and potentially harmful events at the earliest possible time so proactive, mitigation activities can be performed. This approach is an integral part of TVA's risk management program to reduce the probability and consequences of a failure of any of its waste impoundments. This program provides several significant services to TVA personal including:

1. Providing important data and related information on facility performance in near-real-time in a presentation platform that enables efficient and accurate evaluation and interpretation.
2. Providing information on the historical development and performance of each facility, subsurface information and other useful documents to provide context to the measured performance data and aid in its interpretation.
3. Providing component status information so that the monitoring system can be kept fully operational, reliable and believable.
4. Providing alert messages on significant events that require TVA attention and action.
5. Enabling TVA to do a more comprehensive performance monitoring program with less operational resources and enabling its staff to focus on data interpretation and action rather than data gathering.

This work to obtain a central platform to store, manage and review instrumentation data and related information has been very successful in helping TVA to streamline its processes and work flow related to field monitoring of the performance of its waste impoundments. TVA is planning to expand this approach to other data intensive operations, such as groundwater and water quality monitoring.

ACKNOWLEDGEMENTS

The authors appreciate the dedicated efforts of TVA and Geocomp staff in developing and implementing the TVA iSiteCentral GIS system. We also acknowledge the significant contributions of Rob Nyren and Kris Armstrong in preparing this paper.

REFERENCES

Tennessee Valley Authority. (October 8, 2014). *Fossil Fuel Generation.* Retrieved from http://www.tva.com/power/fossil.htm

Walton, W. (June 25, 2009). *Root Cause Analysis of TVA Kingston Dredge Pond Failure on December 22, 2008.* Vernon Hills, IL: AECOM.

PREDICTING PIEZOMETRIC RESPONSE DURING HIGH LOADING EVENTS BASED ON LIMITED HISTORICAL DATA

Chun-Yi Kuo, Ph.D., P.E.[1]
Troy S. O'Neal, Ph.D, P.E.[2]
Richard B. Hockett, P.G.[3]

ABSTRACT

Piezometers are important in monitoring dam performance during routine operations and during flood events or after an earthquake event. The risk assessor is often charged with predicting ground water level for high pools when such pools have not occurred historically. Often the piezometer readings are plentiful and the data can be extrapolated to determine ground water levels during extreme events. However, some dams do not have sufficient piezometric data for meaningful regression analyses. Although over 100 working piezometers are present at Moose Creek Dam, the piezometric data record is very limited. This makes prediction of the piezometric response to future higher pools difficult. Remedial measures have been incrementally constructed, consisting of an extensive relief well system and enlarging an upstream impermeable blanket. Those remedial measures make the interpretation and extrapolation of future piezometric response even more difficult and challenging. To assist the understanding and analyses of piezometric response, a concept of "Piezometer Response Ratio" was used. This paper presents how the response ratio concept is being utilized for meaningful piezometer data analysis at Moose Creek Dam.

INTRODUCTION

The importance of monitoring programs for dam safety is widely accepted. The monitoring program provides the information to develop a better understanding of the on-going performance of the dam which provides the basis for predicting the performance of the dam under extreme loading conditions. Piezometers are generally used to measure hydrostatic uplift pressure beneath the embankment and structures, and pore pressures within the embankment and the foundation. The risk assessor is often charged with predicting the ground water level for high pools when such pools have not occurred historically, such as for a 1000-year flood or for the probable maximum flood (PMF). Often the piezometer readings are plentiful and the data can be analyzed statistically to secure a trend for the basis of extrapolation. However, some dams and levees do not have sufficient piezometer data for meaningful regression analyses. This paper presents a case study how the concept of "Piezometer Response Ratio" is utilized to assist the

[1] Geotechnical Engineer, Louisville District, US Army Corps of Engineers, Louisville, KY 40202, chun-yi.kuo@usace.army.mil
[2] Geotechnical Engineer, Louisville District, US Army Corps of Engineers, Louisville, KY 40202, troy.s.oneal@usace.army.mil
[3] Geologist, Louisville District, US Army Corps of Engineers, Louisville, KY 40202, Rick.B.Hockett@usace.army.mil

understanding and discussion of piezometric response when only limited historical data are available.

MOOSE CREEK DAM AND ITS PIEZOMETER DATA

The 7.5-mile long Moose Creek Dam(MCD) is located upstream of Fairbanks, Alaska and reduces flooding on the Chena River by temporarily impounding and diverting floodwater to the much larger Tanana River (Figure 1). The dam is constructed over a broad valley filled with mostly clean sand and gravel. Surficial material is predominantly 0 to 20 feet of nonplastic inorganic silt and sandy silt overlying the sand and gravel. The silt and sandy silt were excavated within the embankment footprint during construction. There are localized areas of permafrost which have likely degraded since construction. The depth to metamorphic schist bedrock ranges from the ground surface at the north abutment to over 600 feet under most of the dam.

In performing the risk analysis for the MCD, the embankment was divided into three distinct zones because of differences in geology, permafrost locations, project performance indicators, and downstream consequences (USACE, 2013b). The zones are defined by stationing as follows: South Zone (0+00 – 254+00), Central Zone (254+00 – 380+00) and North Zone (380+00 – 489+00).

Construction of MCD began in 1975 and the dam became operational in 1979. Initial test filling reached an elevation of 506.5 in July 1981 and the record pool occurred in June 1992, reaching an elevation of 512.7. All elevations are referenced to North American Vertical Datum of 1988 (NAVD88) unless otherwise noted. There are currently 111 open tube piezometers through the crest of the dam, along the embankment toe, and in selected downstream areas for measuring pore pressures during flood events (USACE, 2009). During a flood event, ground water levels in the open piezometers are measured twice daily. MCD is typically a dry dam and had only seven historical high pool events with dam loading longer than 3 days (Figure 2). As a result, even though there are over 100 piezometers, the data are limited. Table 1 shows a partial summary of peak piezometric response at MCD. As shown in the table, many data are missing or unavailable. This imposes great difficulties in understanding, analyzing and extrapolating piezometric response to predict future ground water levels during extreme events.

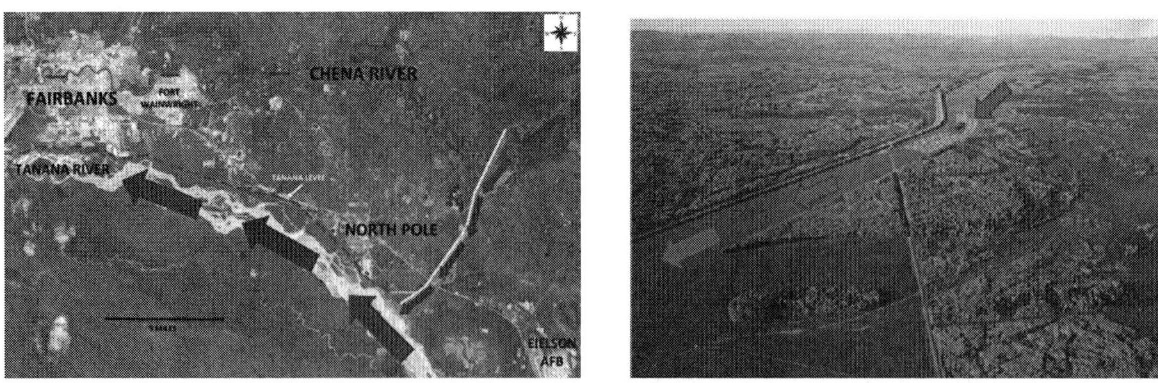

Figure 1. Operational Schematic and Aerial View of Moose Creek Dam

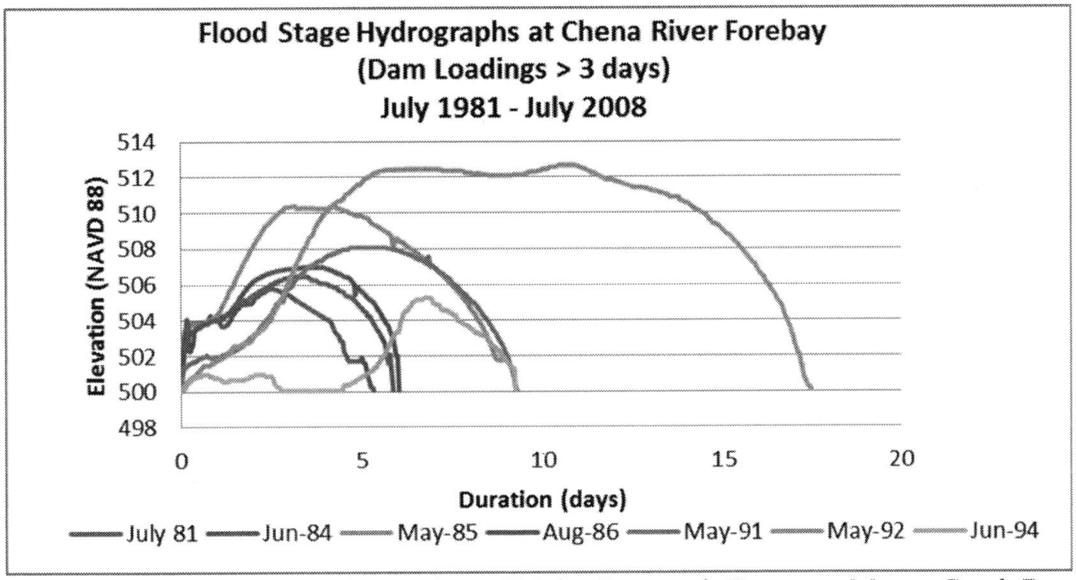

Figure 2. Pool Elevation and Duration of Major Reservoir Events at Moose Creek Dam

Table 1. Partial Summary of Peak PZ Response (after USACE, 2013a)

		Peak Response (ft, NAVD88)						
PZ No.	Location Station; Offset	1981 Test Pool 506.5'	1984 High Pool 507.0'	1985 High Pool 510.4'	1986 High Pool 505.8'	1991 High Pool 508.1'	1992 Record Pool 512.7'	1994 High Pool 505.3'
288	100+00; 11'	504.0	---	505.9	---	504.7	507.6	---
294	115+00; 11'	502.5	---	504.7	---	503.7	506.7	---
303	148+30; 11'	498.5	---	502.1	---	501.2	504.7	497.1
308	172+00; 11'	497.4	---	501.1	---	500.1	505.3	496.3
314	190+00; 11'	496.8	---	501.1	---	500.1	505.7	495.5
320	210+00; 11'	496.3	495.3	499.8	---	499.8	504.7	495.9
326	230+00; 11'	496.0	495.9	499.3	---	499.4	503.6	496.5
332	250+00; 11'	494.5	493.8	497.4	495.9	497.7	501.4	495.4
338	270+00; 11'	498.1	---	---	---	498.8	501.2	497.7
446	278+05; 11'	---	497.4	497.1	498.6	498.3	500.2	498.1
470	284+59; 10'	---	---	---	---	500.1	501.1	498.8
344	290+00; 11'	501.7	498.5	501.8	502.2	500.6	501.7	499.9
472	295+02; 11'	---	---	---	---	499.8	500.7	499.7
474	299+99; 10'	---	---	---	---	499.4	500.8	499.3
458	307+50; 11'	---	---	497.6	499.7	500.2	501.3	500.0

Moose Creek Dam can be subjected to rapid pool loadings. Sand boils have developed at the downstream toe of dam during prior loading events. Since 1981, remedial measures were constructed consisting of increasing the lateral extent of the upstream impermeable blanket and adding 188 relief wells. The original design of the embankment incorporated a constructed upstream silt blanket of varying width and thickness in areas where natural silt cover was less than three feet thick. In the 1982-1983 modification, the upstream silt blanket was extended to form a continuous section at least 600 foot wide with a minimum thickness of 3 feet along the length of the embankment dam (Figure 3). In addition to six relief wells constructed before 1981, relief wells were added during 1982-1983, 1986, 1994, and 1997 (Figure 4). The incremental installation of remedial measures complicates the interpretation and extrapolation of piezometric response due to their control on ground water levels during flood events.

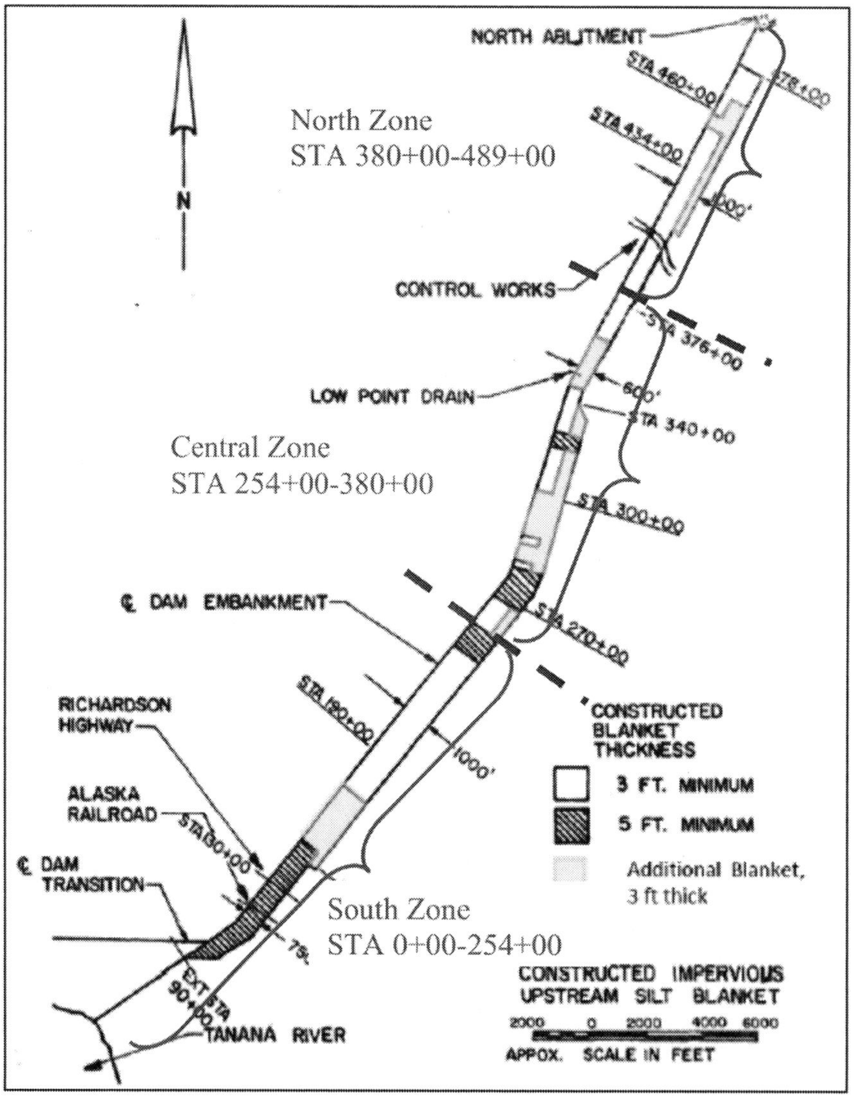

Figure 3. Constructed Upstream Silt Blanket at Moose Creek Dam

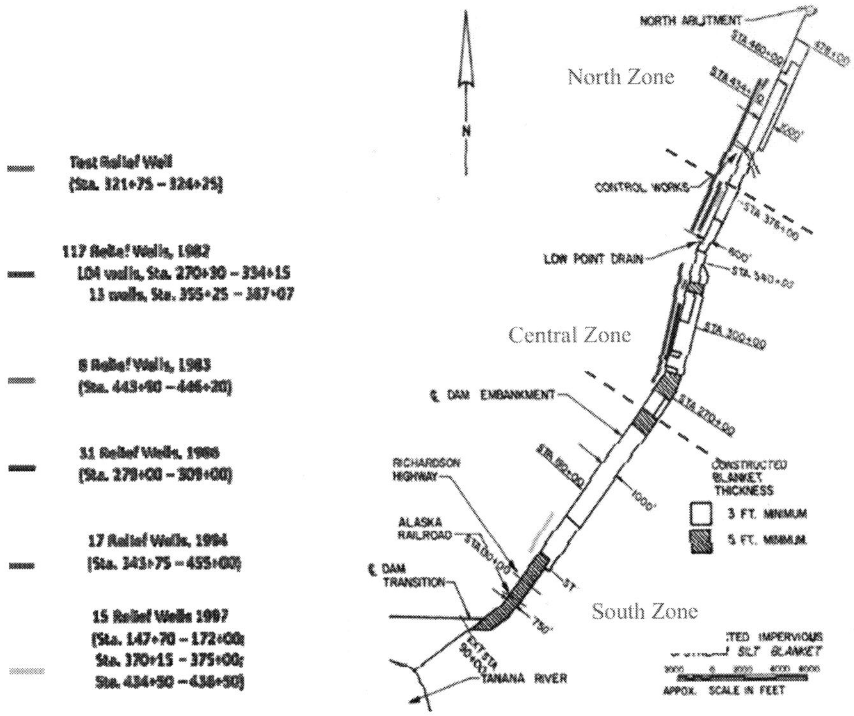

Figure 4. Relief Well Construction at Moose Creek Dam

PIEZOMETER RESPONSE RATIO

The concept of "Response Ratio" was utilized to assist the interpretation of ground water elevation response to pool loading on the dam. "Response Ratio" of a piezometer during a flood event in this paper is defined as the ratio of the peak rise of the ground water level in the piezometer to the peak rise of the flood pool, both with respect to the antecedent groundwater level (Figure 5).

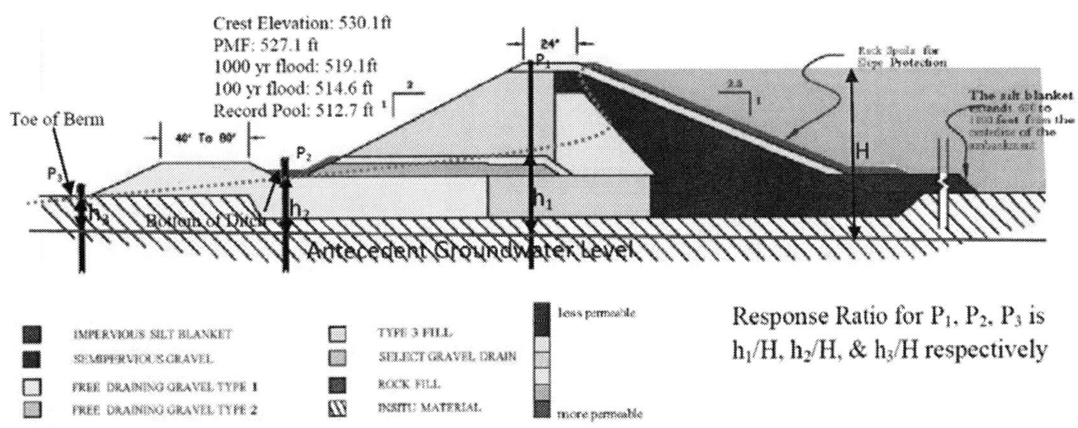

Figure 5. Illustration of Typical Dam Section & the Corresponding Piezometer Response Ratio during Flood

For piezometers located along the same seepage path, the one located further downstream has a smaller response ratio because ground water level response to pool loading decreases with increased distance from the reservoir. The response ratio is also influenced by seepage control measures such as the upstream silt blanket, and the downstream relief wells. Both the upstream silt blanket and the relief wells have the effect of reducing piezometer response during flood events. For piezometers at different locations along the dam, the ones adjacent to a wider and more competent silt blanket will have a smaller response ratio. Functioning relief wells reduce the response ratio of piezometers located nearby. For a balanced system, steady state peak response ratios for a particular piezometer during different flood levels should be fairly constant. Some minor influence will come from the antecedent ground water level which may fluctuate slightly with time.

Figure 6 shows the typical antecedent ground water level at Moose Creek Dam reported by Hydrologic Engineering Center, US Army Corps of Engineers (USACE, 1995, CRCP-PR-28). Elevations in Figure 6 are referenced to National Geodetic Vertical Datum of 1929 (NGVD29) and require an addition of 5.06 feet to convert to NAVD88 datum.

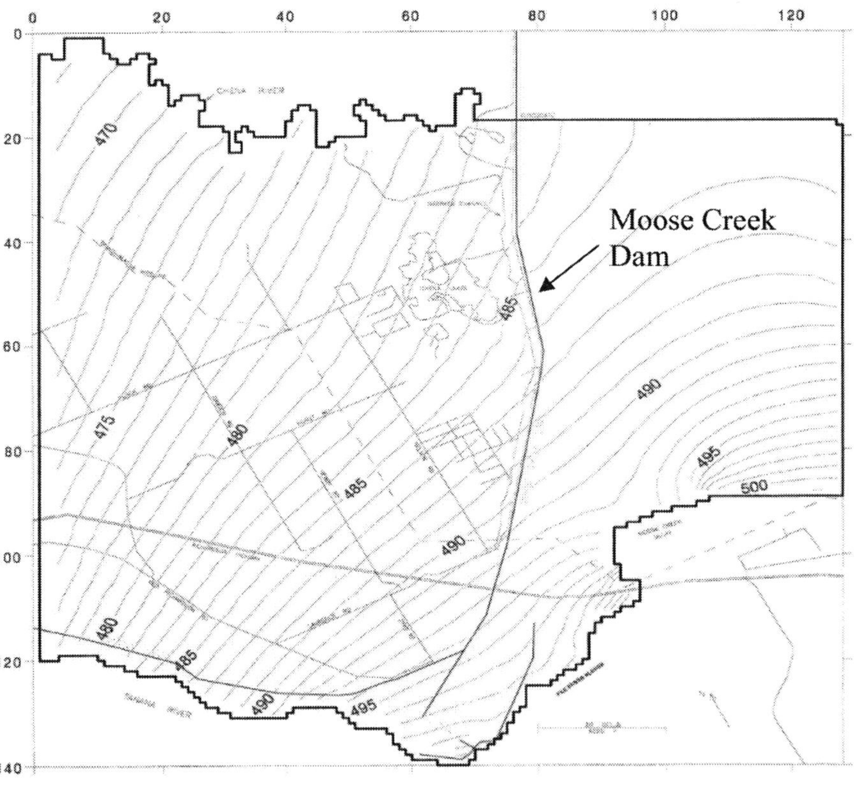

Figure 6. Antecedent Groundwater Levels at Moose Creek Dam

A similar ratio calculated using the head above tail water to total head had been used previously to extrapolate the piezometric readings (USACE, 1981 and USACE, 1987). The 1981 report utilized the piezometric measurements obtained during the 1981 test pool and extrapolated the PMF piezometric readings. The 1987 report used the May

1985 data to extrapolate the PMF phreatic surface at the toe of the downstream stability berm. In the two reports, the ratios are relative to the assumed tail waters, while in this paper; the response ratio is relative to the antecedent groundwater. Head ratio calculations have been used for a long time and are one of the many tools engineers use for evaluating instrumentation.

Such "response ratios" appear to be a useful concept. Response ratios can be used to extrapolate the PMF (or any other flood loadings) phreatic surface if they can be reasonably assessed. Response ratios are expected to be fairly constant and unusual changes of the values may indicate something has dramatically changed in the seepage system. Response ratios also provide indirect evidence regarding the effectiveness of seepage control measures. The concept of response ratios should be applied under at least near steady state conditions.

MOOSE CREEK DAM PIEZOMETRIC RESPONSE

Due to significant topographic relief in the diversion floodway, the 1992 record pool was the only event that almost completely covered the floodway. During the 1981 test filling, the dam was inundated only between approximately STA 200+00 to 430+00 (Figure 7). Thus, for the past seven major flood events, there are many piezometers at locations where the embankment has not been loaded, and therefore exhibited limited response.

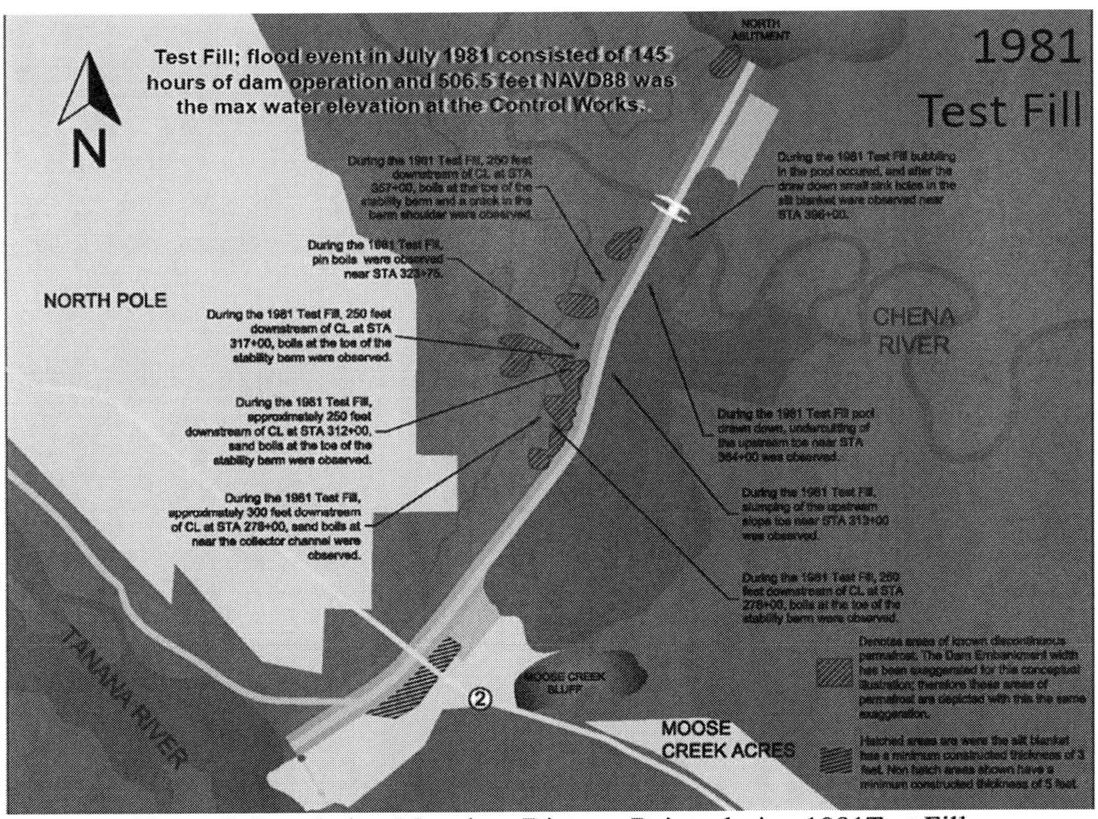

Figure 7. Inundation Mapping, Distress Points during 1981Test Fill

During the 1981 test pool, the floodway south of approximately station 220+00 was not inundated, and as expected, the piezometers located in the southern portion of the project showed very low response ratios (Figure 8). Figure 9 shows piezometric responses of P-309 and P-315 located in the South Zone. These piezometric responses are typical for most of the piezometers located in the South Zone during the 1981 test pool. They typically have low piezometric response (thus, low response ratio) and also show significant lag time in their peaks with respect to the peak of flood. For P-309 and P-315, the peaks of piezometric response occurred about ten days after the peak of flood. An exception to the generally low response ratios in the South Zone are the piezometers near STA 100+00, which had high response ratios due to the influence of the Tanana River.

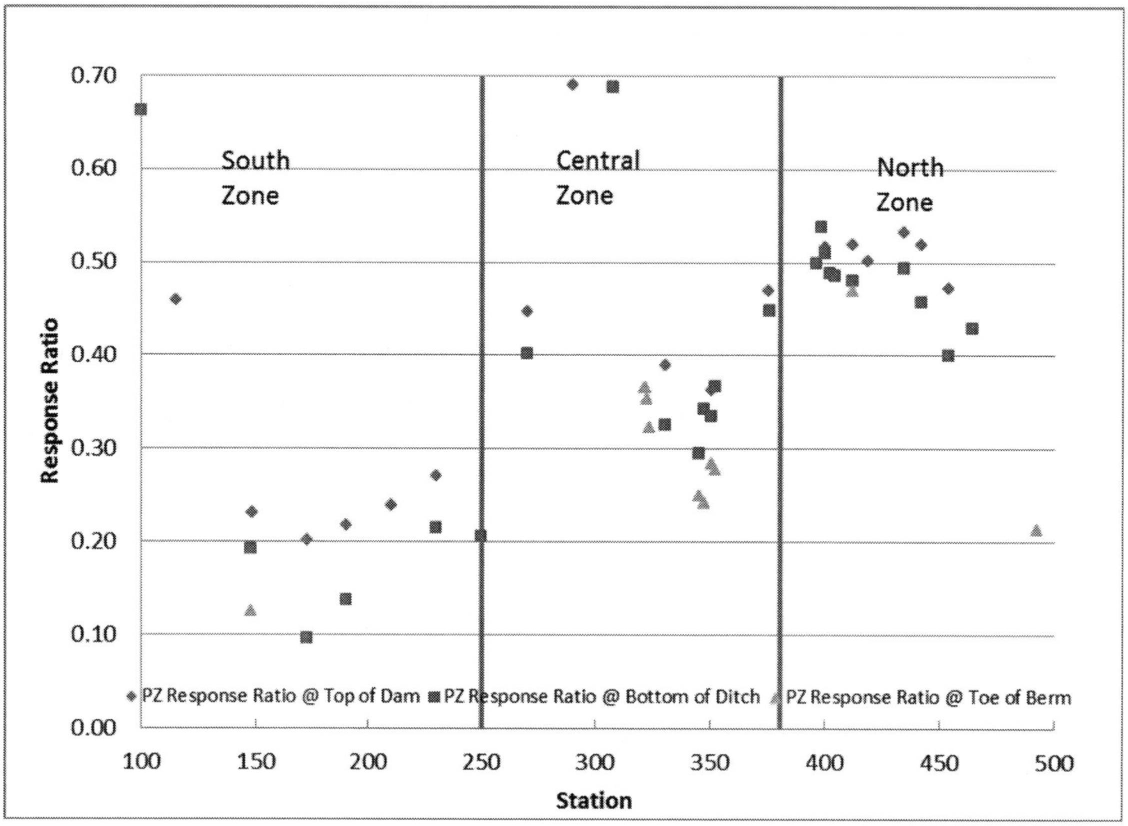

Figure 8. Piezometer Response Ratios during the 1981 Test Fill

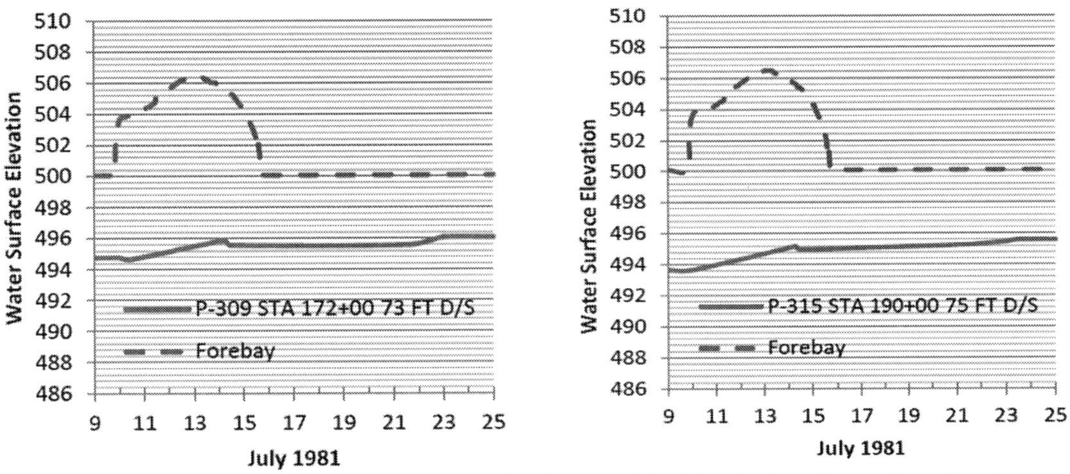

Figure 9. Piezometric Response of P-309 and P-315 during 1981 Test Pool

Figure 10 shows the piezometric response of P-338 and P-357, which exhibits the typical response for piezometers located in the Central Zone. Different than those located in the South Zone, piezometric responses in the central zone have clear peaks and show little lags time. Their peaks generally are within one day of the peak of flood.

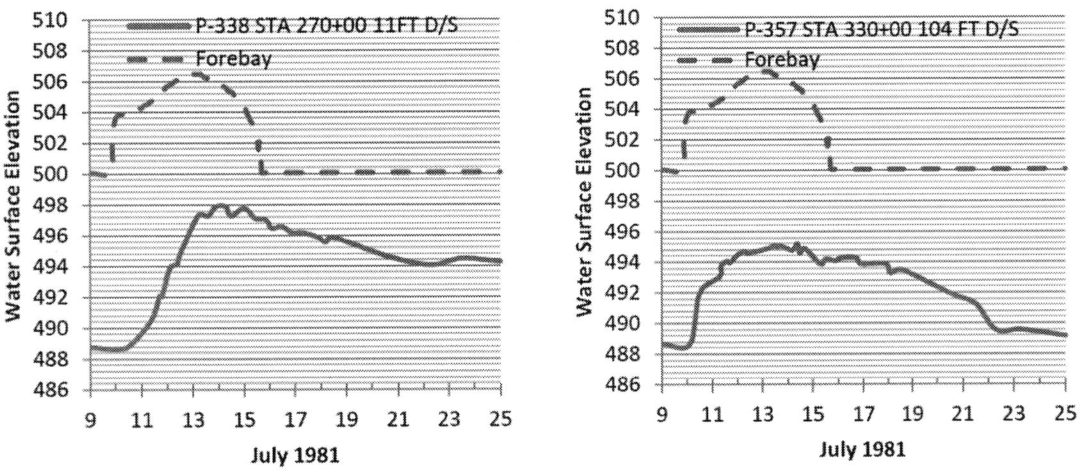

Figure 10. Piezometric Response of P-338 & P-357 during 1981 Test Pool

As seen in Figure 8, piezometers located near the Low Point Drain (Station 350+00) showed relatively low response ratios. The Low Point Drain region is characterized by the presence of extensive silt cover in this low area. The low response ratios around the Low Point Drain region are believed to relate to the extensive silt cover which effectively increases the seepage path and thus reduces the piezometer response ratios. However, the piezometers near STA 300+00 showed unexpected high response ratios. These unexpected high response ratios indicated that the natural silt cover around STA 300+00 may be not as uniform, thick, or effective as originally thought. There was originally no or little constructed silt blanket between STA 280+00 to STA 300+00 in 1981 (Figure 3).

Predicting Piezometric Response

The high response ratio region around the STA 300+00 also corresponds well with the area showing distress during the 1981 test pool (Figure 7). Near the distress points located at STA 357+00 and STA 396+00, the PZ response ratios at the bottom of the ditch appeared to be high compared to those at the top of the dam (Figure 8).

The response ratios of piezometers located in the North Zone are generally around 0.5 (Figure 8). The response ratios decrease north of STA 450+00 because the region was not inundated during the 1981 test pool (Figure 7).

Figure 11 shows the extent of floodway inundation during the 1992 record pool event. The 1992 flood almost inundated the entire floodway. Available piezometric response ratios during the 1992 record pool event are presented in Figure 12. It can be seen that the PZ response ratios mostly range between 0.5 and 0.6, except the central zone showing lower ratios. An extensive relief well system was installed in the central zone after the 1981 test pool event. The central portion of the project is known to have extensive silt cover in addition to the constructed silt blanket. Comparing the piezometric responses between 1981 and 1992 (Figure 8 vs. Figure 12), it can be seen that the unexpectedly high response ratios around STA 300+00 in 1981 no longer existed in 1992. This indicates that the 1982-1983 remediation efforts (extended upstream silt blanket and downstream relief wells) appear to be effective. Figure 12 also indicates that the extensive relief wells between STA 270+30 to 377+57 have further reduced the response ratios in this region, especially for those piezometers located at the toe of the downstream stability berm, where a blowout would be most damaging.

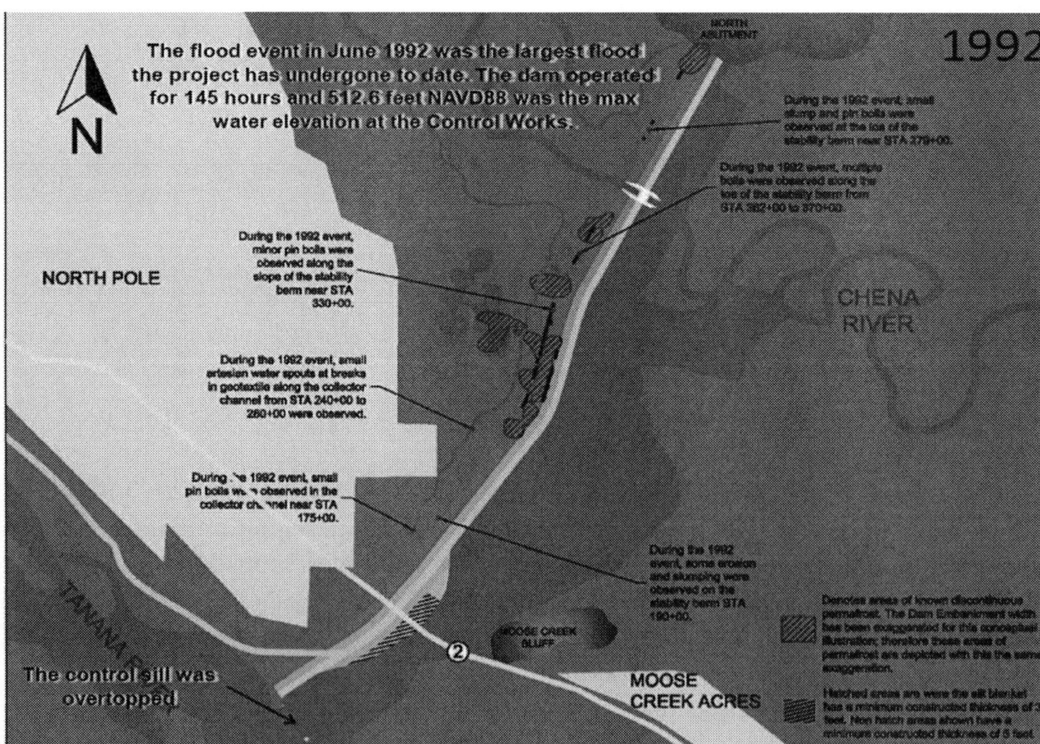

Figure 11. Inundation Mapping, Distress Points during 1992 Record Pool

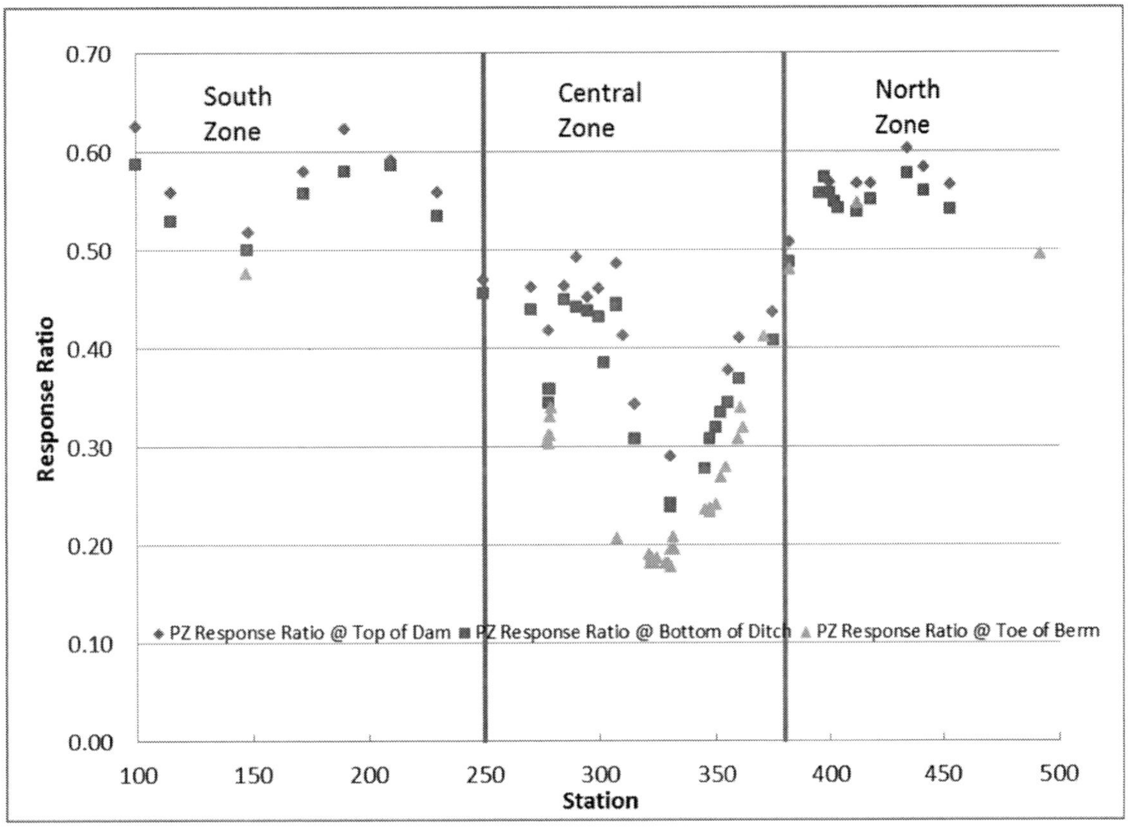

Figure 12. Piezometric Response Ratios during the 1992 Record Pool

EXTRAPOLATION OF PIEZOMETRIC RESPONSE

The response ratio concept can be used to extrapolate piezometric response. It is assumed that the response ratio for a piezometer is fairly constant for a stabilized, balanced, steady state system. Having the representative response ratios, the projected piezometric response during a PMF event or any other pool can be easily calculated by using Equation 2 shown below. Recall the definition of response ratio which is the ratio of the peak rise of the piezometer to the peak rise of the flood, both relative to the antecedent groundwater.

$$RR_{PMF} = h_{PMF}/H_{PMF} = (PZ_{PMF} - GW_A)/(PMF\text{-}GW_A) \quad (1)$$

$$PZ_{PMF} = GW_A + (PMF\text{-}GW_A) \times RR_{PMF} \quad (2)$$

where
PMF: Probable Maximum Flood
GW_A: Antecedent Ground Water Level
RR_{PMF}: Response Ratio during PMF
PZ_{PMF}: Piezometer reading during PMF
h_{PMF}: Peak rise of the piezometer during PMF
H_{PMF}: Peak rise of the pool during PMF

Figure 13 shows response ratios for piezometers near the top of dam during the seven major flood events. The 1984 flood had a similar peak and duration with 1981 test pool and the differences in the piezometer response ratios during these two events are most indicative of the effects of the 1982-1983 modifications. The 1985 and 1992 floods were the two largest floods Moose Creek Dam has experienced. Also included in the figure are the projected response ratios for the piezometers during PMF.

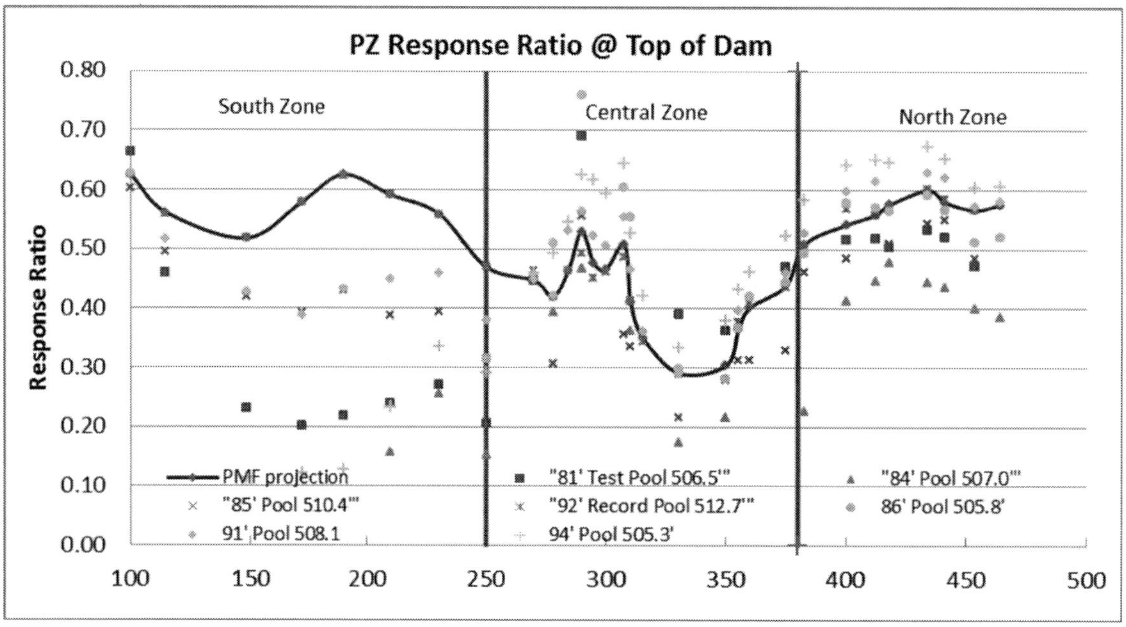

Figure 13. Piezometric Response Ratios for Piezometers near the Top of the Dam

The projected response ratios during PMF are selected, or calculated based on the following considerations.

1. The PZ response ratios during the 1981 test pool were not included in considering or calculating the PMF projections. Because major modifications have been applied after the 1981 test pool (e.g. 1982-1983 modifications and 1986-1988 modifications), the piezometric responses in 1981 were not good indicators of future piezometer response.
2. The South Zone of the dam was typically not inundated, and the PZs response ratios in these regions during 1992 record pool were the highest as compared to the other flood events. As such, for piezometers located south of station 250+00, the projected response ratios during PMF are selected to be the same as the response ratios during the 1992 record pool.
3. The response ratios for other piezometers located in the Central and North Zones are generally within a relatively narrow range as compared among different events (excluding 1981 as discussed and a few outliers), for these piezometers, the average response ratios within the narrow range are calculated and selected as the representative response ratios during PMF.

4. The June 1994 flood was composed of two closely spaced floods (see Figure 2). The first flood, peak at 501 ft, stayed for about 3 days and the second flood, occurred two days later, peak at 505.3 ft. As a result, the PZ response ratios during the 1994 floods appear to be relatively high because the antecedent ground water for the second 505.3 ft flood was higher (due to the first 501 ft flood) and are not representative.

There are a few limitations of the response ratio approach. First, the assumption of a fairly constant response ratio for a particular piezometer is only valid for a stable, balanced system under near steady state conditions. Second, piezometric levels in the vicinity of relief wells will be controlled by the elevation of the relief well outfall and will not keep rising as flood level increases. The response ratio is not a constant and using a representative response ratio to extrapolate piezometric response is not applicable to piezometers in the vicinity of relief wells.

Another possible approach for predicting piezometric responses for future floods is through modeling such as SEEP/W. Projected piezometric responses at STA 350+00 (no relief well), show very good agreement between SEEP/W and response ratio approach as shown in Figure 14. Piezometric response projections at STA 278+00 (with relief well), show departure of the two approaches as the flood level increases (Figure 15). As expected, the effect is larger for piezometer P-428 which is located at the toe of the stability berm near a relief well.

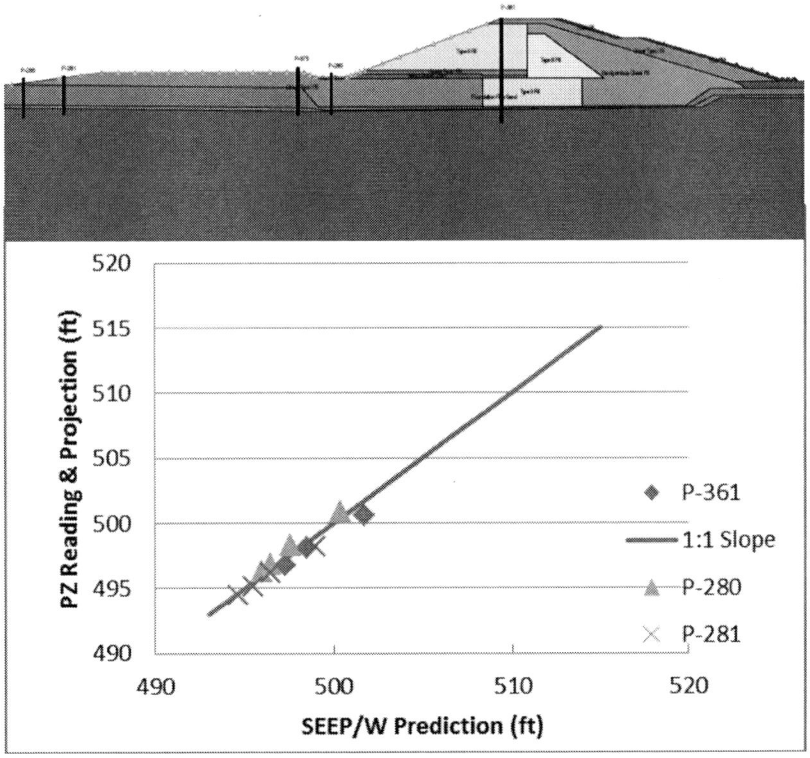

Figure 14. Comparison of PZ Projections (Station 350+00)

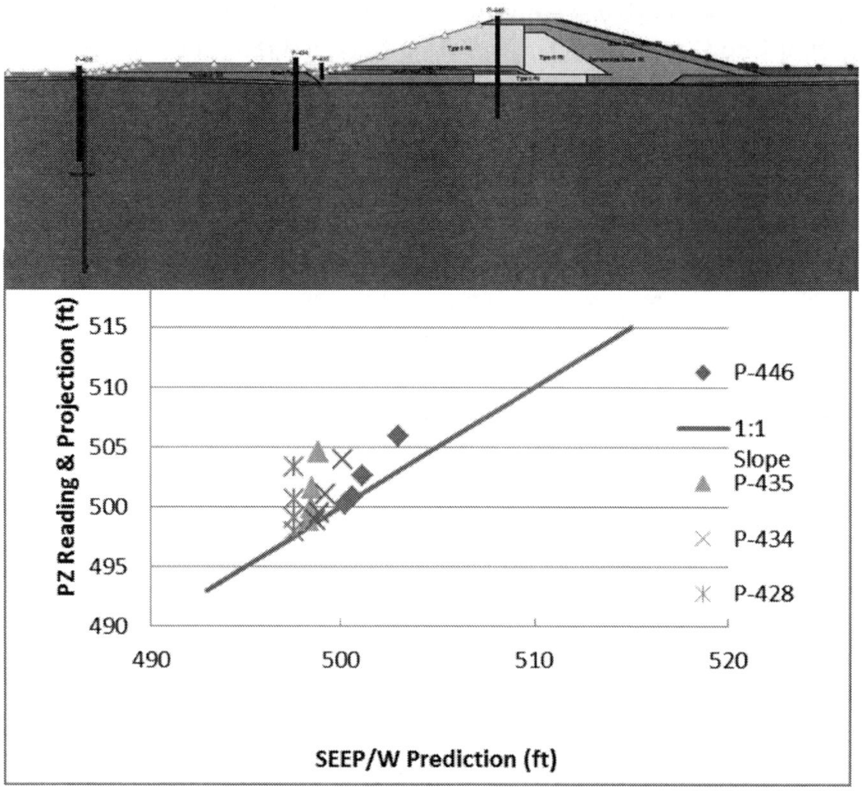

Figure 15. Comparison PZ Projections (Station 278+00)

The response ratio approach allows the extrapolation of piezometric response at Moose Creek Dam. Projection of piezometric response during extreme floods is important for assessing and forecasting the dam's performance during such floods. Existing relief wells were placed in locations that have either: 1) known boils or 2) predicted heave pressures above predicted toe resistance (thin silt layers downstream in combination with predicted tailwater conditions). Additional relief wells may be needed to improve dam safety, based on anticipated future performance, and to build robustness into the seepage control system below the toe of the dam, particularly to accommodate higher pools. A detailed profile drawing with predicted pressures by pool event, downstream silt thickness from foundation excavation data, and calculated factors of safety for uplift was generated for the Moose Creek Dam. This drawing allows for strategic relief well placement over the long term maintenance of the project, and allows for flood fight monitoring and activities to be focused. An example profile section is provided below (Figure 16).

CONCLUSION

Piezometers are important in monitoring dam and levee performance during routine operations and during flood events. This paper presents a case study for a piezometer response ratio concept being utilized for the risk assessment of Moose Creek Dam. This response ratio approach is useful in understanding, discussing, and extrapolating piezometric response under near steady state conditions when the piezometer readings are limited. The response ratio approach to extrapolation of piezometer readings does not apply to piezometers in the vicinity of relief wells.

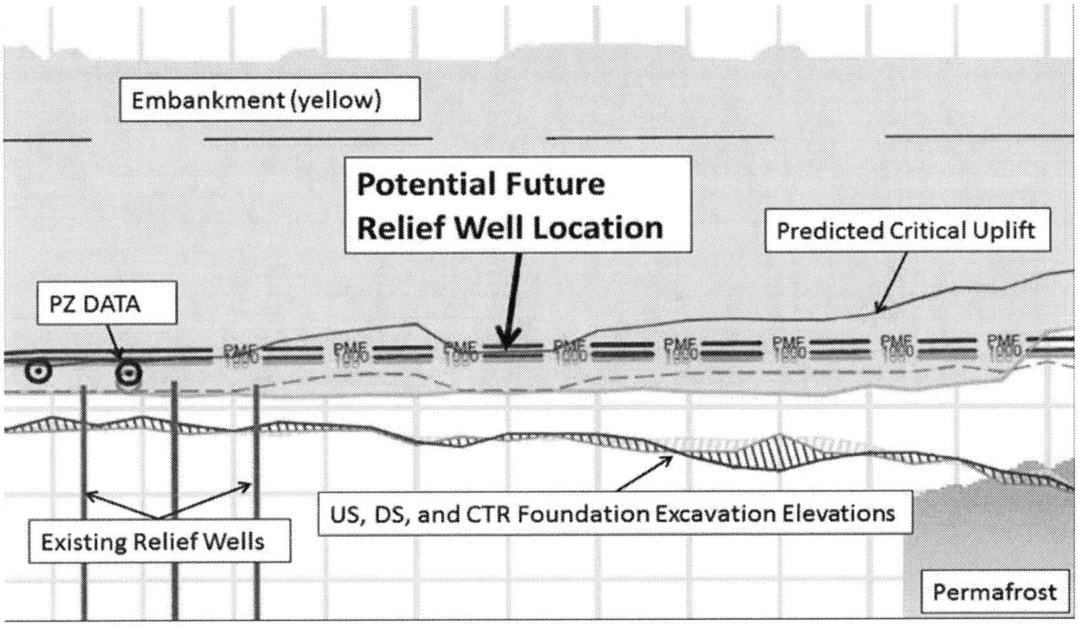

Figure 16. Developed Detailed Profile Drawing (After USACE, 2013b)

REFERENCES

US Army Corps of Engineers (USACE), 1981, After Action Report, Chena River Lake Project Test Filling, Alaska, Alaska District.

US Army Corps of Engineers (USACE), 1987, Letter Report Number 21, Moose Creek Dam, Permafrost Effects on Seepage, Chena River Lake Project Test Filling, Alaska, Alaska District.

US Army Corps of Engineers (USACE), 1995, The Development and Application of Initial Hydrogeologic Flow Model for the Chena River Lakes Flood Control Project, CRCP-PR-28, Hydrologic Engineering Center, Davis, CA.

US Army Corps of Engineers (USACE), 2009, Interim Risk Reduction Measures Plan, Moose Creek Dam, Chena River Lake Project, Alaska, Alaska District.

US Army Corps of Engineers (USACE), 2013a, Site Characterization Report, Moose Creek Dam, Chena River Lake Project, Alaska, Alaska District.

US Army Corps of Engineers (USACE), 2013b, Baseline Risk Assessment Report, Moose Creek Dam, Chena River Lake Project, Alaska, Alaska District.

THE EFFICACY OF PREDRILLING IN SEEPAGE BARRIER WALL CONSTRUCTION

John W. France, PE, D.WRE[1]
Michael F. Zoccola, PE[2]
Li Yan, Ph.D., P.Eng.[3]

ABSTRACT

In the past several years, the authors have been involved in two seepage barrier wall projects in which predrilling proved to be very effective in achieving design goals. In many cases, a construction contractor will not consider predrilling because of the added initial expense and time. However, the two cases highlighted in this paper demonstrate that the cost and time for predrilling can, in the end, be a cost and schedule benefit for a seepage barrier wall project. This paper reviews the benefits of predrilling for Wolf Creek Dam, KY and Ruskin Dam, British Columbia, Canada. The unique aspects of each project are reviewed, with emphasis on the benefits realized from predrilling.

INTRODUCTION

In the past several years, the first author has been involved as a senior technical advisor on two seepage barrier wall projects in which predrilling proved to be very effective in achieving design goals. In many cases, a construction contractor will not consider predrilling because of the added initial expense and time. However, these two cases demonstrate that the cost and time for predrilling can, in the end, be a cost and schedule benefit for a seepage barrier wall project.

The first project is the Wolf Creek Dam safety rehabilitation project in Kentucky, completed by the U.S. Army Corps of Engineers. In this project, predrilling proved very beneficial in achieving alignment of secant pile barrier wall elements to depths exceeding 250 feet. The alignment control on these elements was critical to achieving the desired barrier wall continuity requirements for the project.

The second project is the Ruskin Dam safety rehabilitation project in British Columbia, Canada. In this project, predrilling was used to control the alignment of jet grout columns which formed a slot beneath a section of the concrete dam. The slot was required for construction of an innovative asphalt membrane, which connected a panel barrier wall in the abutment to the concrete dam.

The unique aspects of each project are reviewed, with emphasis on the benefits realized from predrilling. Both projects received engineering achievement awards, and the

---

[1] AECOM, 8181 East Costilla Place, Denver, CO, john.france@aecom.com, 303-740-3812
[2] U.S. Army Corps of Engineers, Nashville, TN, michael.f.zoccola@usace.army.mil, 615-736-5693
[3] BC Hydro, 6911 Southpoint Drive, Edmonds A02, Burnaby, B.C. V3N 4X8, li.yan@bchydro.com, 604-528-2491

application of predrilling was an important contributor to the successes of these projects, which resulted in those awards.

WOLF CREEK DAM

Project Background

Wolf Creek Dam, owned and operated by the U.S. Army Corps of Engineers (USACE), Nashville District, is on the Cumberland River in South Central Kentucky. It provides flood control, hydropower, recreation, water supply, and water quality benefits for the region. The dam is a combination of a 3,940-foot-long[4], unzoned, compacted clay embankment and a 1,796 foot-long concrete gravity section. Wolf Creek Dam impounds Lake Cumberland, which, at capacity, has more than 1,200 miles of shoreline, more than 63,000 acres surface area, and stores more than 6 million acre-feet of water. Figure 1 shows an aerial view and location map.

Figure 1. Aerial View and Location Map of Wolf Creek Dam

A detailed description of the geology, design deficiencies, past problems, and past remedial efforts has been the subject of several papers (Zoccola et al, 2009; Santillan and Salas, 2012; Zoccola et al, 2014) and is not the emphasis of this paper. These factors can be briefly summarized as follows. The embankment was constructed predominately on alluvial soils that overly karstic limestone characterized by an extensive interconnected

---

[4] Design and construction for the Wolf Creek Dam project were completed using English units and those units are used in the discussion of the project in this paper.

network of solution features, varying from fist size to large caves on the order of 50 feet. The most severe solutioning is found along two well defined joint sets oriented approximately parallel and perpendicular to the dam axis. Designed in the 1930's and built in the 1940's, standards of practice for foundation treatment methods were not adequate for an embankment dam on this foundation. Figure 2 shows karst features exposed during original construction. Seepage and piping of the karst infilling and the overlying embankment is the failure mode of concern.

Figure 2. Cut-off Trench/Solution Channel in Limestone Rock Intersected by Several Large Caves. (For scale, note men at bottom, on scaffolding, and sitting atop beam.)

Sinkholes on the downstream embankment face and muddy shows in the tailrace appeared in 1967 and 1968, indicating an active internal erosion failure mode. An emergency grouting program from 1968 to 1970 arrested the internal erosion, and the first seepage barrier wall ever constructed by the USACE was constructed at Wolf Creek Dam from 1975 to 1979.

The Nashville District monitored various performance indicators after construction of the barrier wall, and, in the early 2000's, it was concluded that the wall did not go deep enough or extend laterally far enough to intercept all the significant solution features. Monitoring data indicated that seepage under and around the barrier wall was increasing.

In 2005, an evaluation conducted as part of risk screening for all dams in the USACE's portfolio rated the dam "Inadequate" for embankment and foundation seepage and internal erosion under normal reservoir levels. In 2007 an Independent External Peer Review panel concluded that there was compelling evidence that an internal erosion failure mode had re-initiated, was in an advanced "continuation" stage of development, and represented a significant risk of failure under normal operation conditions. The dam was assigned a Dam Safety Action Classification I – Urgent and Compelling (DSAC I) rating, which is the most severe USACE category of risk. A subsequent quantitative risk assessment resulted in an estimated probability of failure of around 1 in 100. As a result, it was decided to construct a new seepage barrier wall extending deeper into the foundation and further laterally than the wall constructed in the 1970's.

New Seepage Barrier Wall

In 2008, construction was begun on a new state-of-the-practice seepage barrier wall. For construction of the wall, the Nashville District partnered with a joint venture of specialty construction firms, Treviicos and Soletanche (TSJV), selected through a "Best Value, Request-for-Proposal" contract acquisition process.

The new barrier wall ties into the upstream face of the concrete monolith and extends the full length of the embankment, terminating in the right abutment. A profile is shown in Figure 3. The deepest sections of the 3,800-foot-long wall are 275 feet deep, with up to 95 feet of penetration into the limestone foundation. In most locations, the new barrier wall extends to 75 feet or more deeper than the original barrier wall. The specifications required a continuous wall with a minimum thickness of 2 feet at all locations. Grout curtains were installed upstream and downstream of the wall alignment prior to wall construction, to pre-treat the foundation and prevent slurry loss during wall element construction. The grout curtains extended to at least 50 feet below the bottom of the new barrier wall.

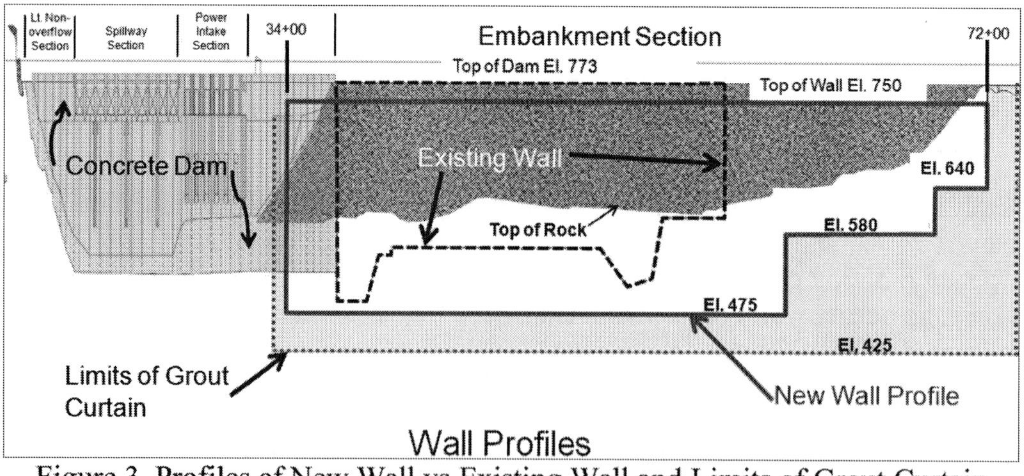

Figure 3. Profiles of New Wall vs Existing Wall and Limits of Grout Curtain

Three element configurations were considered for construction of the wall: 1) overlapping rectangular panels, 2) overlapping piles, and 3) a combination of both panels and piles. These configurations are shown in Figure 4. The wall consists of connected sequences of primary (P) and secondary (S) elements with primary elements constructed first, followed by the secondary elements to close the gaps between primary elements. For the 3,800-foot long wall, 3,176 feet of wall (84%) consisted of overlapping piles and

624 feet consisted of the combination geometry. The overlapping panel configuration was not used for the seepage barrier wall, but it was used for a protective concrete embankment wall (PCEW), discussed further below. The discussion in this paper will focus on the overlapping pile configuration, which was the primary barrier wall construction technique for the Wolf Creek project.

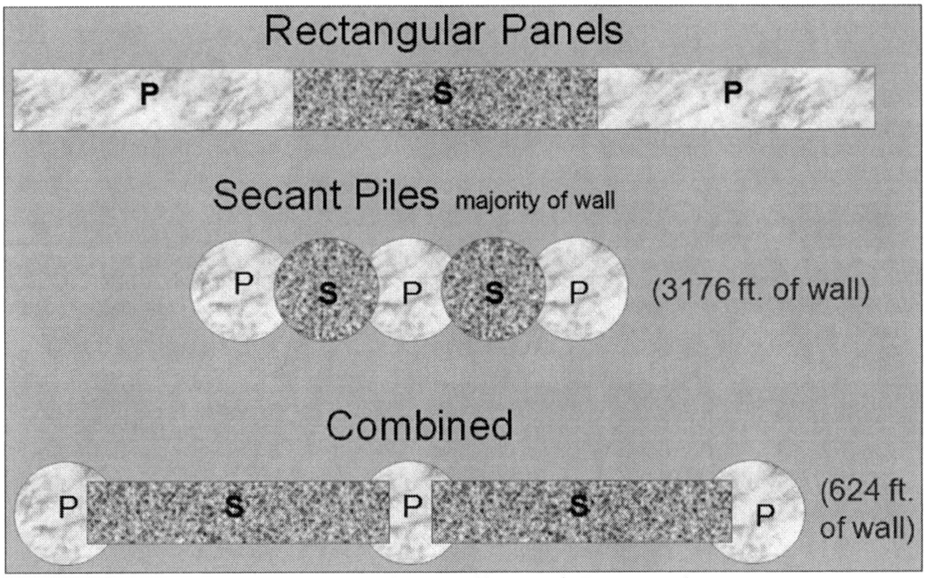

Figure 4. Barrier Wall Panel Geometries

Construction Sequence

It was anticipated that construction of each barrier wall element in the hard limestone would require an extended time period. To protect the embankment and overburden soils from fluid pressures during element construction in rock, a PCEW wall was constructed, extending to the top of rock. The PCEW wall was constructed with overlapping rectangular elements.

Each PCEW element was 6 feet wide by 9 feet-2 inches long, with a minimum required overlap of 5 inches for adjacent elements. The elements were excavated using a Soletanche hydrofraise designed and built by the TSJV specifically for the Wolf Creek project. The excavation was stabilized with polymer slurry. Upon completion of excavation, each panel was backfilled with tremie-placed concrete.

Eight-inch diameter guide or pilot holes were drilled through the PCEW and into rock at the location of each secant pile. Each pilot hole extended to about 3 feet below the design depth of the barrier wall. These holes guided the subsequent drilling of the larger overlapping piles. Maintaining verticality within tolerances for these guide/pilot holes was critical to the verticality, and thus continuity, of the pile wall elements.

The guide holes were drilled with a Wassara water-powered, down-the-hole hammer. The hammer was fitted with a slant-face bit and a bent housing to allow steering of the holes. Normally the bit was operated by rotating. If a hole began to deviate, the slant-face bit and bent housing were oriented in the hole to "steer" the hole back toward vertical, and the bit operated in hammer mode without rotation.

Figure 5 shows a schematic of the pilot holes through the PCEW and photos of the water hammer and bit.

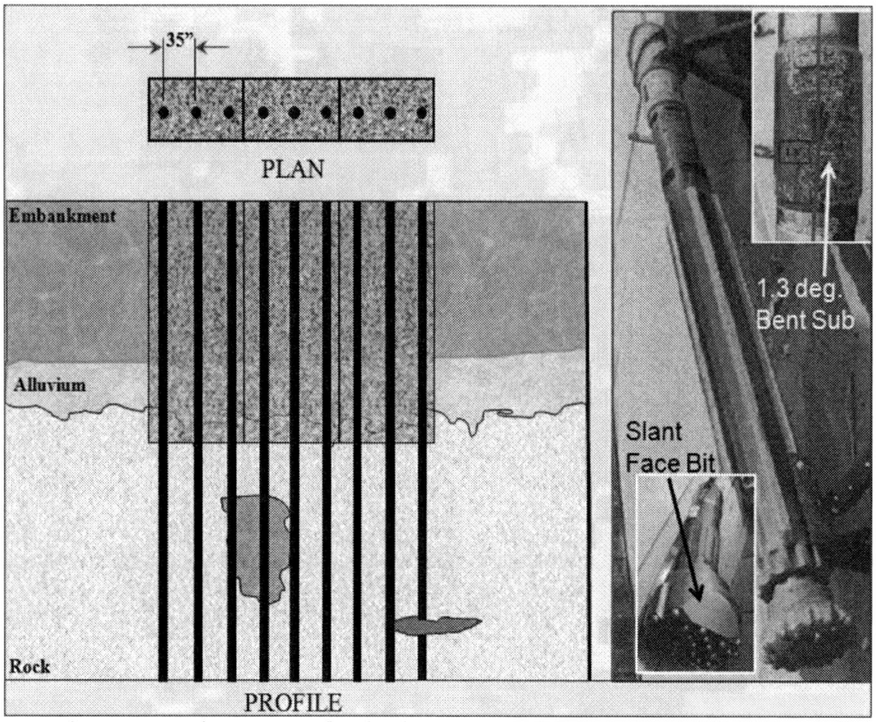

Figure 5. Schematic of 8-inch Diameter Guide/Pilot Holes and Wassara Water Hammer Bit Adapted for Steering

The pile elements were constructed using Wirth drill rigs, as shown in Figure 6. The Wirth drills include a bottom-hole assembly equipped with rock roller bit faces to cut the rock. Each drill was fitted with a stinger (see Photo 3 in Figure 6), which followed the pilot holes ahead of the main cutting face. The bottom-hole assemblies on the Wirth drills had on-board inclinometers, which provided real-time verticality data to the operators on the drills. However, the Wirth drills had no "steering" capability. The construction method depended on the stinger following the previously drilled guide holes to maintain the locations and verticality of the piles. Maintaining verticality was also aided by the length (60 feet) and dead weight (~ 70 tons) of the drill's bottom-hole assembly, which provided a pendulum effect and used gravity to advantage. The final configuration of the 50-inch diameter secant pile barrier wall elements within the PCEW is shown in Figure 7.

Figure 6. Photo 1 Shows Wirth Drilling Table Locked into Guidewall,
Photo 2 Shows the 60-ft-long, 70 Ton Bottom-Hole Assembly Being Set In Place,
Photo 3 Shows a Close-up of the Rock Roller Bit with Stinger

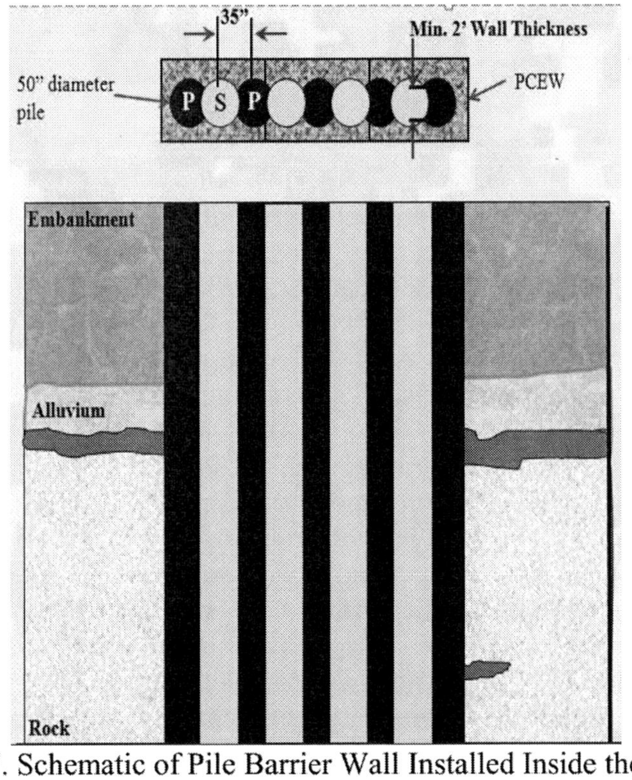

Figure 7. Schematic of Pile Barrier Wall Installed Inside the PCEW

Guide Hole Alignment Control

The TSJV chose overlapping 50-inch diameter piles on 35-inch center-to-center spacing for the secant pile wall. To meet the specified minimum 2-foot thickness, the actual deviation between adjacent pile centers could be no greater than 44.9 inches between pile centers, a total deviation of 8.86 inches from the design 35-inch spacing, over a total depth of up to 275 feet. Thus, if at any depth the center of one pile drifted right of target by 4.45 inches, the adjacent pile could drift no more than 4.45 inches left. This geometry is shown in Figure 8. For a depth of 275 ft, this results in a tolerance for this particular example of 0.13% of depth, a previously unprecedented tolerance. Of course the illustration in Figure 8 is based on no deviation in the upstream-downstream direction. If upstream-downstream deviation occurred, the left-right deviation limit would be even smaller.

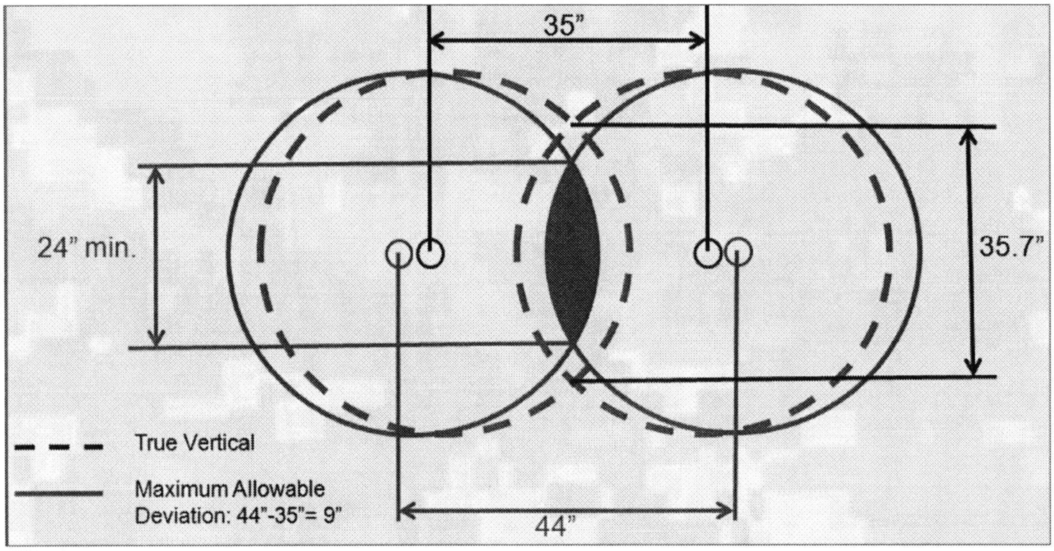

Figure 8. Pile Geometry and Tolerances

As discussed above, the 8-inch guide holes were critical to meeting verticality requirements, because the Wirth drills with stingers followed the guide hole alignments. The alignment of each guide hole was monitored during drilling and corrected as required, and then checked with a second measuring method after hole completion.

During drilling, a down-hole measuring probe called the "Paratrack 2" was used to measure hole alignment. The probe contained tri-axial accelerometers to measure the inclination and azimuth of the hole. Paratrack readings were taken every 10 feet, as each hole progressed, to measure alignment and determine whether steering was needed to bring the hole back toward true vertical. If the hole began to deviate from alignment, the slant-face bit and bent housing were engaged, as described above, to steer the hole back toward correct alignment. Once the hole was completed, an alignment check was made using a traditional inclinometer survey with inclinometer casing installed down the hole, inside the drill casing. An example plot from a completed guide hole is provided in Figure 9, showing a comparison of the alignment data in two formats: 1) a deviation versus depth plot and 2) a polar plot. The polar plot shows a trace of the direction and magnitude of the hole center deviation over its depth. Throughout the project there was generally good agreement between the two methods. The example in Figure 9 shows that at most, looking at the inclinometer readings, the hole deviated about 2 inches over its depth of 275 feet. From the Paratrack data, deviation is less than 1 inch at any depth. The results shown in Figure 9 are typical of data recorded for guide hole alignments on the Wolf Creek project.

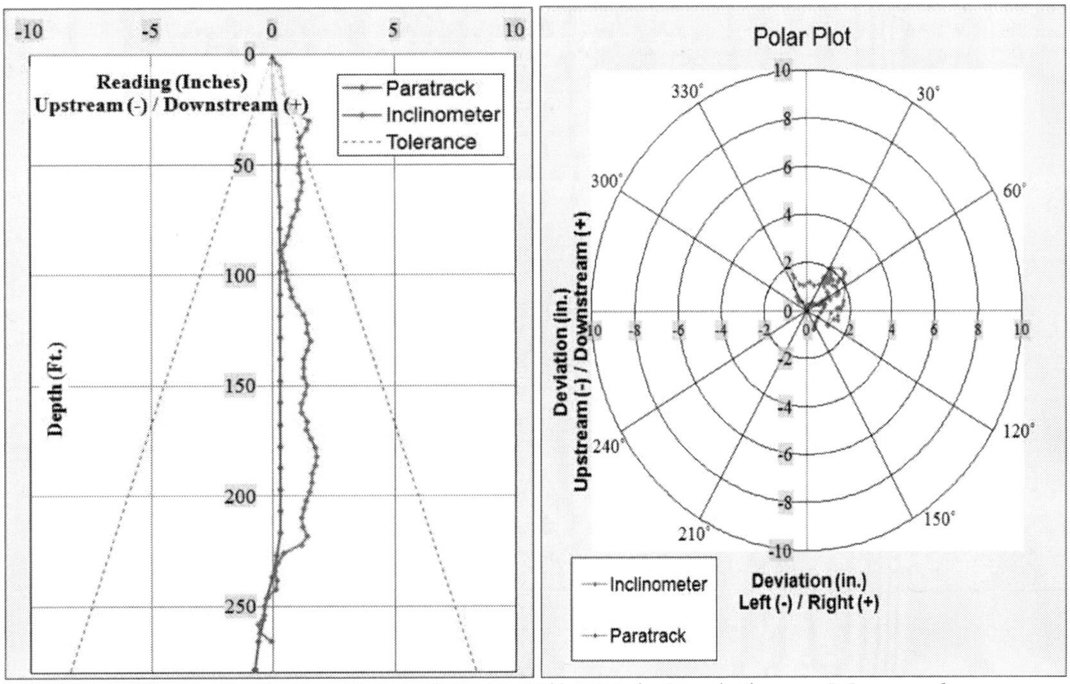

Figure 9. Comparison of Guide/Pilot Hole Deviation as Measured with Paratrack and Inclinometer. Deviation vs. Depth on the Left and Polar Plot Showing Deviation and Azimuth on the Right

Confirmation of Design Requirements

Multiple means of measuring the barrier wall elements in space were used to cross check and verify that the wall thickness as-constructed met the design requirements, i.e. a minimum thickness of 24 inches at all locations. For the secant pile barrier wall elements, four separate measurements were taken during construction, and all four were used to evaluate as-constructed pile locations over the full depth of each pile.

As described above, two location measurements were completed during the drilling of each guide hole – the Paratrack and the inclinometer. These measurements were used for two estimates of as-constructed secant pile locations, assuming perfect 50-inch diameter secant piles around the alignments indicated by these two measurements.

During the drilling of each pile, two additional measurements were made to evaluate the pile location in space. The first was the Wirth drill's on-board inclinometer system discussed above. The second measurement followed completion of the hole and was completed using a KODEN device.

Figure 10 shows a typical comparison plot of deviation vs. depth for all four methods used to measure pile verticality, and Figure 11 is a companion polar plot. These plots show the consistency that was typically achieved among the four measurements, with less than about four inches of discrepancy among them.

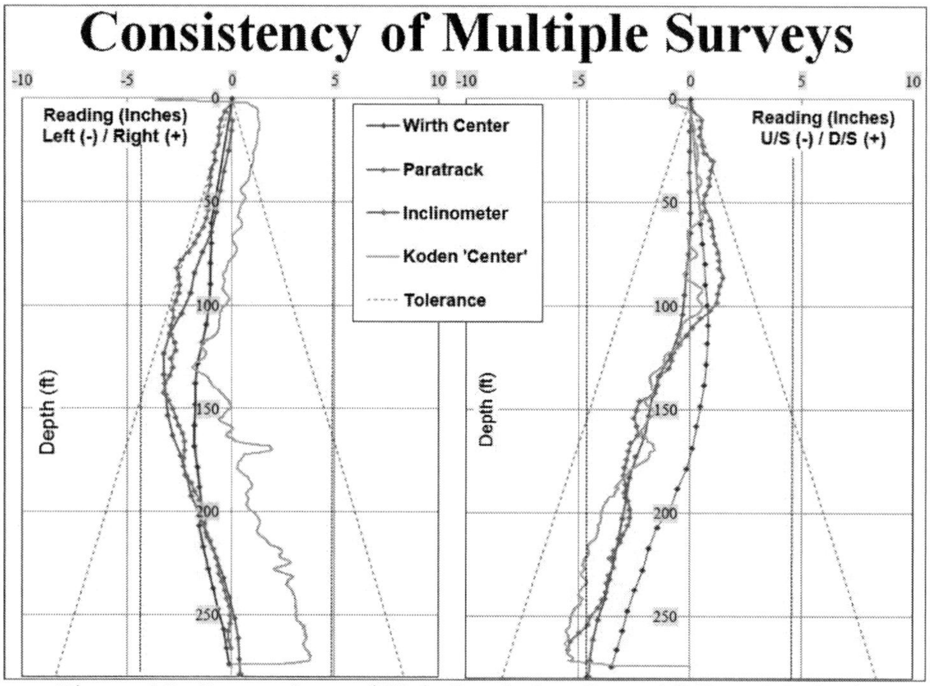

Figure 10. Comparison of Hole Deviation from All Four Methods

Using the data from all measurements, two-dimensional and three-dimensional plots of the wall were created to visualize and evaluate various combinations of measurements and the resulting wall thickness. Figure 12 is an example two-dimensional plot at 150-foot depth for five piles.

There were 1197 piles installed creating 1196 joints. This resulted in about 280,000 linear feet of joints. An analysis of data during construction resulted in only one pile being out of tolerance to an extent requiring remediation by installing a replacement pile – a truly remarkable achievement.

Wall thickness was calculated at 9,661 locations for comparison to the required 2-foot minimum. The measurements showed compliance with the minimum thickness according to all four measurements methods, except at only one location other than at the one joint requiring remediation. This was at a depth of 270 feet for one joint, where the indicated thickness was 1.95 feet according to the KODEN data. The other measurement methods indicated a wall thickness greater than 2 feet at that location. This location was ultimately accepted as meeting requirements, based on clear compliance from three out of four measurement methods and only a marginal non-compliance based on the KODEN measurements. Wall thickness statistics from all methods showed that the minimum thickness was the 1.95 feet as noted above, the maximum was 4.4 feet, and the average was 3.2 feet for the secant pile barrier wall.

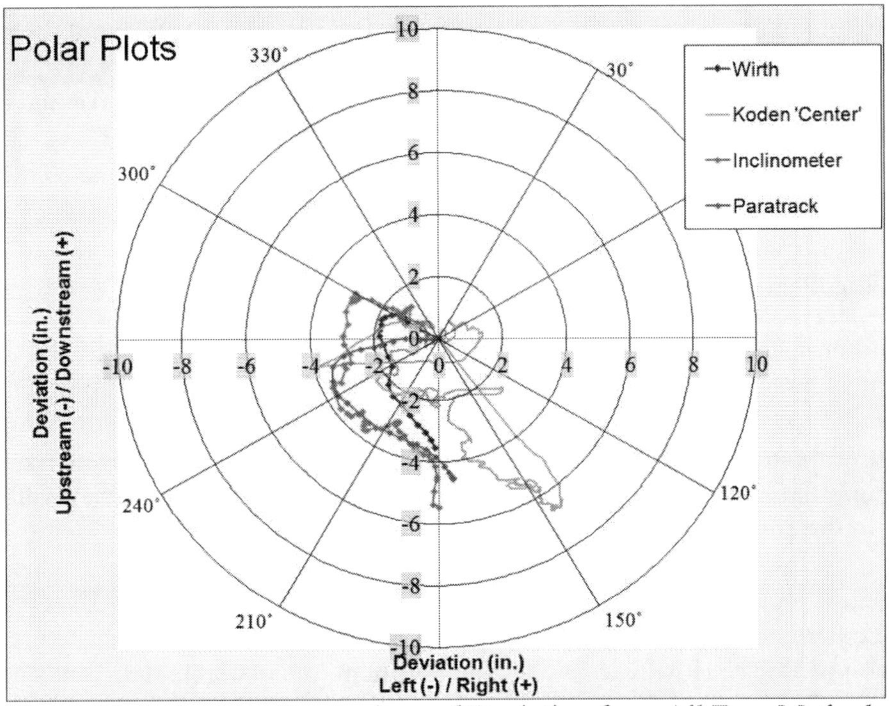

Figure 11. Polar Plot Comparison of Deviation from All Four Methods

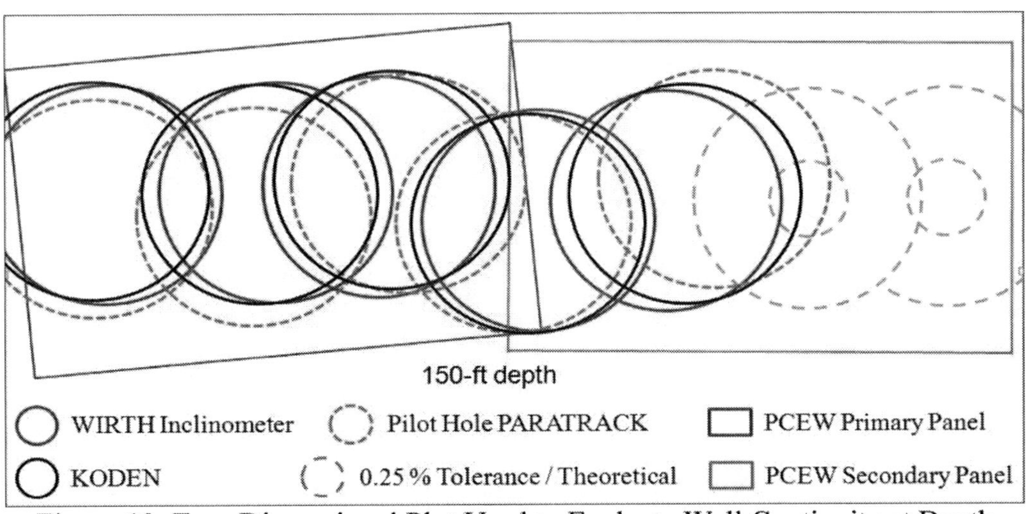

Figure 12. Two-Dimensional Plot Used to Evaluate Wall Continuity at Depth

Though not addressed in detail in this paper, the Combined Barrier Wall sections all exceeded the 2 feet required minimum thickness. The measurement statistics showed a minimum of 2.6 feet and a maximum of 3.1 feet.

These extraordinary results were due in large part to the predrilled guide holes and the alignment control of those holes. The verticality tolerances for the secant piles exceeded what would have been expected if the holes were drilled with the Wirth drill without verticality control. Since the Wirth drills had no means of verticality adjustment during drilling, the guide holes were crucial to achieving final alignment. Without the predrilled guide holes the amount of remedial work to achieve the minimum 2 foot wall thickness would like have been significantly greater, increasing ultimate project cost and extending the construction duration.

RUSKIN DAM

Project Background

The second project is the Ruskin Dam safety rehabilitation project in British Columbia. In this project, predrilling was used to 1) ensure embedment of jet grout columns into bedrock and 2) control the alignment of a series of overlapping holes in the jet grout columns, to form a slot beneath a section of the concrete dam. The slot was required for construction of an innovative asphalt membrane connecting a panel barrier wall in the abutment to the concrete dam.

Ruskin Dam is located near Mission, British Columbia and is owned by BC Hydro. The dam was constructed between 1929 and 1930 and is a concrete gravity structure founded primarily on bedrock. However, at its right abutment, the concrete dam abuts against overburden soils which consists mainly of glacially deposited but highly erodible sands and silts, and till above the bedrock. The right end of the concrete dam was actually constructed against a slope in the glacial soils, resulting in that section of the concrete dam being underlain by soil strata. The original seepage system at the right abutment consisted of an inclined concrete slab on a steep soil slope of 1.25H:1V, supported at its plinth by a concrete gravity wall, a core wall, and steel sheet piles. The concrete slab extends along the abutment about 130 m[5] upstream of the dam (see Figure 13).

Significant seepage and piping issues developed in the right abutment after first filling of the reservoir in 1930, and numerous remedial actions had been carried out to address the problems. Through extensive investigations, BC Hydro determined that the recurring abutment seepage and piping issues, as well as the low seismic withstand of the concrete slab seepage barrier, posed significant dam safety risks to the facility. A seepage control upgrade project was recently completed by BC Hydro to address the deficiencies. The upgrade work included construction of a new seepage barrier wall extending upstream of the dam, a new seepage training wall extending downstream of the dam, and a special tie-in connecting the new barrier wall to the concrete dam. The special tie-in is a seepage barrier wall which included jet grouting of the soils adjacent to and beneath the concrete dam to allow for construction of a flexible, geomembrane type cutoff connecting the slurry panel barrier wall to the dam. The tie-in was further supported by additional jet-

[5] Design and construction for the Ruskin Dam project were completed using both English and SI units and both systems of units are used in the discussion of the project in this paper.

grouting ground improvement downstream. Figure 14 shows the general arrangement of the right abutment upgrade work.

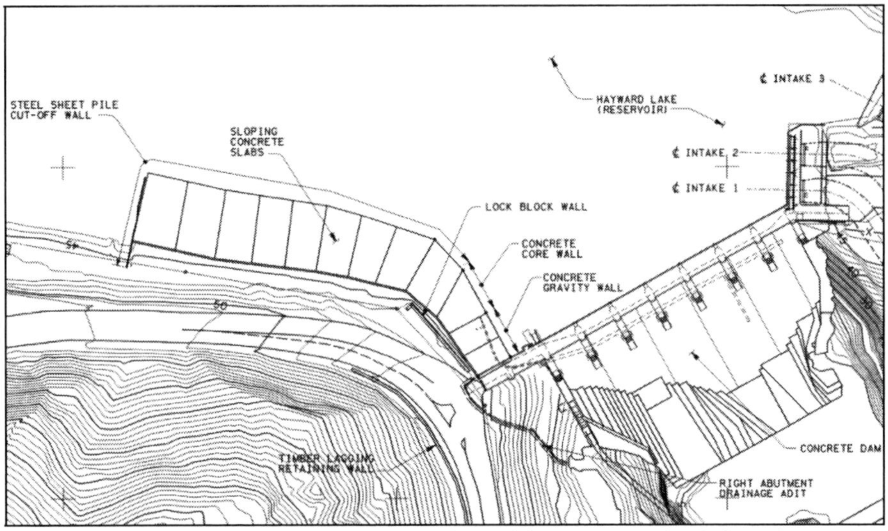

Figure 13. General Arrangement of Ruskin Dam and Right Abutment (after Sweeney and Yan, 2014)

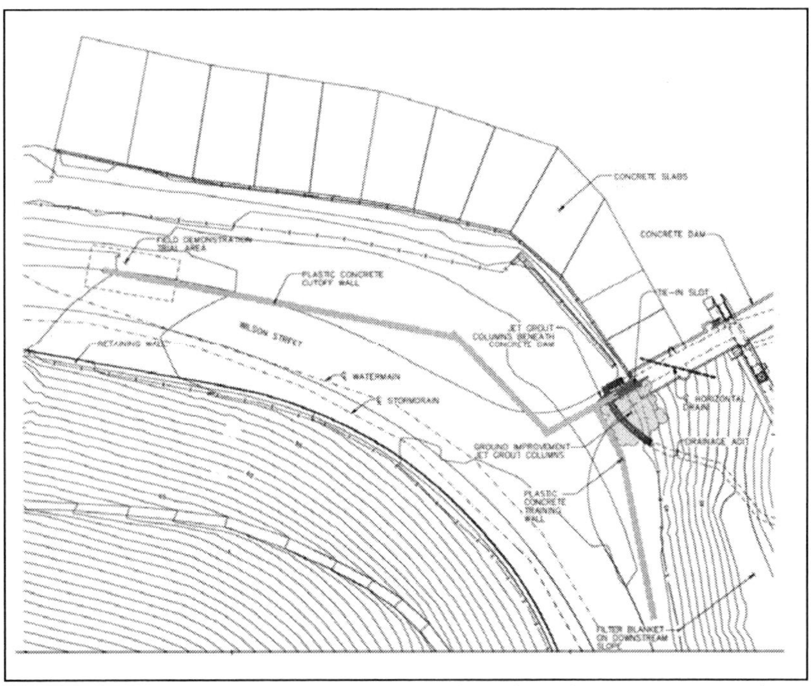

Figure 14. General Arrangement of Right Abutment Upgrades (after Sweeney and Yan 2014)

New Barrier Wall Concept

The new seepage barrier and training walls were constructed of plastic concrete to withstand the anticipated seismic ground deformations. The slurry panel wall construction method was used with a Bauer BG-40 Hydromill to construct the 1 m wide plastic concrete walls through various geologic units and with a minimum 0.5 m embedment into the bedrock.

The design for the special tie-in water barrier wall to connect to the concrete dam called for construction of two rows of overlapping stiff (> 6MPa unconfined compressive strength [UCS]) jet grout columns underneath the section of concrete dam that abuts against and overhangs the overburden. The jet grout columns were required to key into bedrock, forming a stiff wall, about 12 m long by 25 m maximum depth and 1.5 m wide, beneath the concrete dam to support the construction of a slot, within which a flexible geomembrane type water barrier wall was constructed to connect the new plastic concrete barrier wall and the section of the concrete gravity dam founded on bedrock.

A flexible water barrier wall was required for the tie-in to accommodate the anticipated differential movements (up to 50 mm) between the concrete dam and the abutment for the design earthquake event. The selected final solution for the flexible tie-in wall was placement of hot asphalt mastic material in a 7 m long by 0.17 m wide continuous slot, which extends from about 1.5 m below the dam crest to 0.5 m into the bedrock, and embeds 0.6 m into the new plastic concrete barrier wall and 2 m into the concrete gravity dam. Figure 15 shows the schematic of the flexible tie-in wall.

Predrilling for Jet Grouting

Construction of this special tie-in section of the seepage barrier wall was the most challenging part of the right abutment upgrade work, and predrilling played a key role in the successful completion of both the jet grouting and slot cutting. Initially, predrilling was not planned for the jet grouting work, except for pre-coring of the concrete to allow for jet grouting to be carried out below the overhanging concrete. However, during the field trial of the method, it became apparent that the jet grouting rig had difficulties in consistently identifying the bedrock surface to allow verification of sufficient penetration of the monitor and drill bit into the bedrock, so that the high pressure nozzles could construct the columns from the bedrock surface to ensure good contact with the bedrock, as required by the design.

To facilitate the construction and meet the design requirements, predrilling through overburden and into bedrock was carried out during production. Pre-drilling was carried out at the centroid of each planned jet grout column location, which was determined based on the jet grout field trial results for the spacing and diameter of jet grout columns. Each predrilled hole was completed using a sonic drill rig, with a 7-inch diameter core barrel and 8-inch diameter override casing for holes outside the concrete dam, and with a 10.5-inch diameter override casing for holes through the concrete dam, i.e. large enough for the jet grout monitor to pass through the concrete and into the bedrock. Predrilling with a Sonic rig also allowed retrieval of core samples of overburden soils and delineation of the bedrock surface. Once the bedrock surface was identified, the hole was drilled about 3 ft. into the bedrock, which is the offset distance of the jet grout nozzles from the bottom of the monitor and drill bit. Predrilled holes were then backfilled with sand, prior to jet grouting. In addition to providing a pilot hole and embedment into the bedrock for better alignment control and sealing of bedrock contact, predrilling with a sonic rig also allowed detailed logging of subsoils and bedrock at each location. Due to lack of the ability to detect the types of subsoils and the bedrock surface with the jet grouting rig, obtaining this information from predrilling allowed the contractor to adjust the jetting parameters for different soil units (sands, silts and till) according to the field trial results, and thus ensured all jet grout columns were overlapped and sealed against bedrock, forming a stiff, watertight wall underneath the concrete dam for subsequent construction of the tie-in slot.

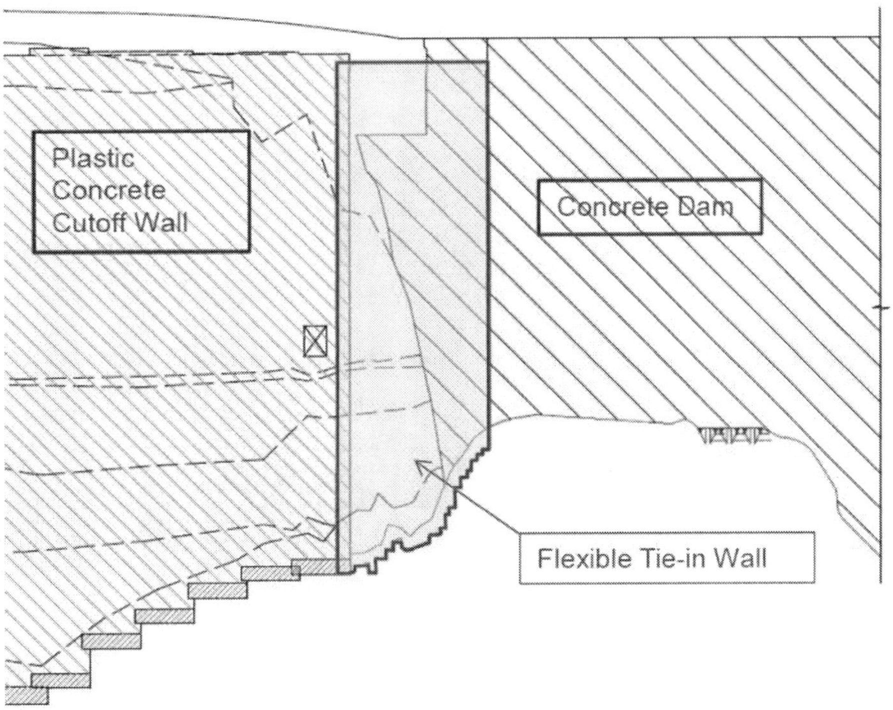

Figure 15. Schematic Showing the Flexible Tie-in Wall

Tie-in Slot Construction

Construction of the tie-in slot involved drilling a series of 12-inch diameter overlapping holes at 10-inch spacing, to create a minimum 6-5/8-inch wide continuous slot. The holes were drilled alternatively in primary and secondary sequences, extending through the concrete dam, jet grout columns, plastic concrete wall, and into the bedrock (see Figure 16). Consequently, maintaining tight verticality control of each hole was very critical for the slot construction. The general methodology developed and used by the contractor was (Wilson et al 2014):

- predrilling of a series of pilot holes to full depth at all primary hole locations;
- then drilling of all 12-inch diameter primary holes following the pilot holes; and

- finally, drilling of all secondary holes using a special guide system that used the adjacent two primary holes as a double guide to drill out the overlapping secondary hole.

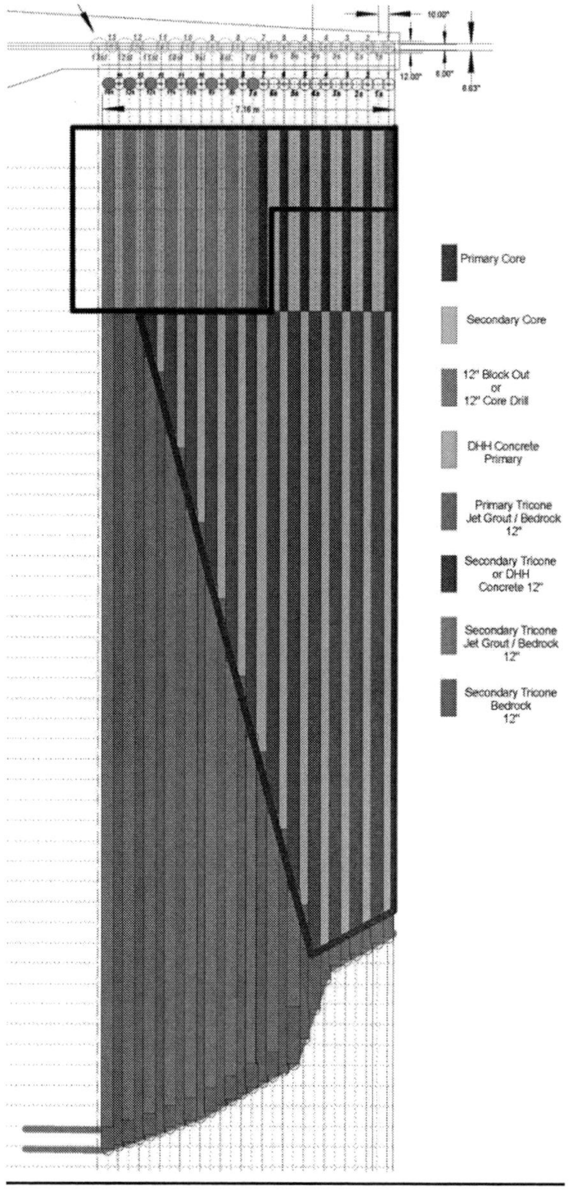

Figure 16. Drilling of the Tie-in Slot (Wilson et al. 2014)

The success of this approach relied on 1) the ability to predrill precision pilot holes from which precision primary holes were created, and 2) a special custom design guide system that followed the primary holes to drill out the overlapping secondary holes. Precision drilling of the pilot holes was achieved by coring of NQ size holes through the concrete dam and jet grout columns into the bedrock. During the NQ coring, the alignment was checked after every 5-foot run, and corrections were made if the hole deviated beyond 6 to 10 mm, using the inverted pendulum method (Wilson et al. 2014), for which BC Hydro had very good experience from the same drilling contractor at a previous project (Sherstobitoff et al 2014). The results of the NQ pilot holes using the inverted pendulum method at Ruskin Dam were also very good for the primary holes in the concrete dam and the jet grout columns, for which alignment accuracy within ±15 mm over a maximum 25 m depth was achieved. However, the NQ coring in the plastic concrete wall was not able to maintain the verticality, even with the inverted pendulum correction method, due to the relatively weak nature of the plastic concrete (1.5 MPa UCS). The two primary holes located in the plastic concrete wall were completed without the NQ pilot holes, using a mud rotary drill that used a large diameter tricone bit and heavy, stiff drill rods to achieve the required accuracy. All pilot holes were backfilled with a very soft, lean grout mixture prior to drilling of the full size holes.

To minimize dam safety risks and maintain the drilling efficiency, different drilling techniques were used in drilling of the full size holes in different materials (i.e. the concrete of the existing dam with estimated strength of 30 MPa, the new plastic concrete barrier wall with a design strength of 1.5 MPa, the jet grout columns beneath the dam with design strength of 6 MPa, and the underlying bedrock). For the primary holes, a down the hole hammer (DTH) was used to drill the large diameter holes in the concrete dam. For holes through the contact between the dam concrete and the jet grout columns, the drill was then switched at about 0.5 m above the contact to an 11.75-inch diameter custom core barrel, to drill through the transition between the concrete dam and jet grout columns, while preventing damage to the underside of the dam and/or cracking of the jet grout at the contact. The large diameter core was extracted from the hole and the drill switched to a Reversed Circulation (RC) tricone method to continue drilling in the jet grout columns and into the bedrock. During the drilling, the DTH bits, core barrels, and RC tricone bits were all equipped with a stinger, to ensure that the full size primary holes followed the NQ pilot holes formed by the inverted pendulum drilling.

After completion of the primary holes, the drill bit was equipped with a custom double guide system that followed the adjacent primary holes, to ensure the proper location of the secondary hole and provide the required uniform continuity from the overlapping holes. The guide had two offset tube guides attached to each side of the drill rods, travelling in the adjacent existing primary holes while the secondary hole was advanced (Wilson et al. 2014). Due to the variation in strength of the materials encountered in the slot trench, and the risk of cracking of the weaker materials, the DTH hammer unit assembly was utilized only in the secondary holes located within the concrete dam. The air rotary with RC tricone unit was used in the jet grout columns and plastic concrete portions of the work. Because of the uneven bedrock surface, some of the primary holes had to be drilled deep into bedrock to ensure that the secondary holes drilled with the guide could be embedded sufficiently into the bedrock. Therefore, a detailed logging of the bedrock surface was important. This logging of the bedrock surface was obtained from the predrilling of the NQ pilot holes, so that the required depth of the primary holes could be adjusted.

The final width of the slot was verified by lowering a steel template into the slot. Where the slot width was out of tolerance, a custom designed "clover" bit attached to a standard 6-5/8-inch drill string on a sonic drill head was used to "trim" the narrow portions of the slot, to ensure a minimum width was achieved throughout the slot. Figure 17 shows the drilling of the secondary overlapping holes with the guide system, and the completed slot filled with asphalt. Figure 18 shows the completed impermeable flexible asphalt mastic tie-in wall that connects the new plastic concrete barrier wall constructed by the slurry panel method and the concrete dam.

Figure 17. Guide Used in Drilling the Secondary Overlapping Holes (Left) and Completed Slot filled with Asphalt (Right)

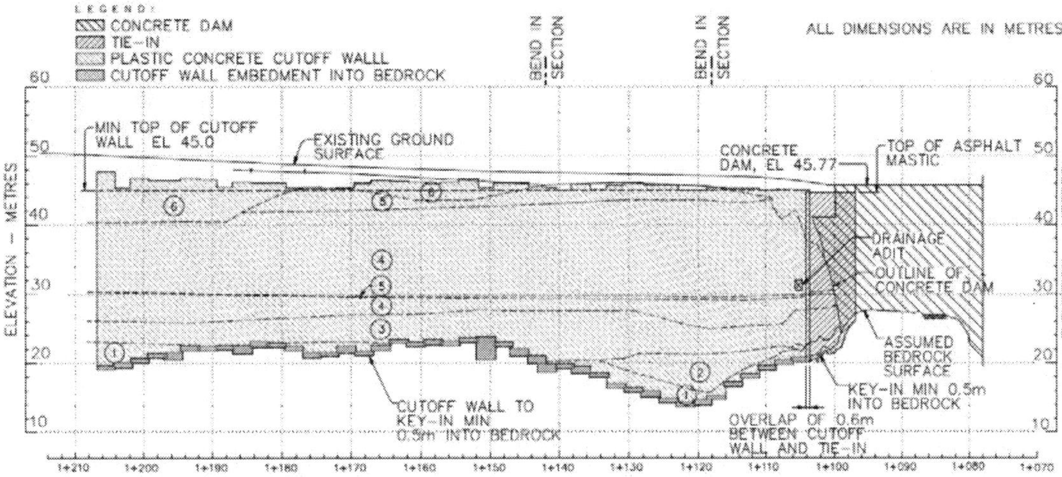

Figure 18. Tie-in Water Barrier Wall Connecting with
Plastic Concrete Cut-off and Concrete Dam

CLOSURE

In both the Wolf Creek Dam and Ruskin Dam projects, predrilling was used to substantial benefit of the construction of seepage barrier walls. In both cases, accurate alignments of vertical wall elements were required for different reasons. Predrilling of guide holes with directional control proved effective in meeting the tight location tolerances of final wall elements. The resulting accuracy of completed elements and the elimination of the need for significant remedial work more than justified the additional cost and time spent on predrilling.

REFERENCES

Santillan, Alberto Fabio and Pierre Salas, 2012, "Wolf Creek Dam Foundation Remediation – An Innovative Successful Solution," (2012), Dam Safety 2012, National Conference of the Association of State Dam Safety Officials, Denver, CO, September 2012.

Sherstobitoff, J., Yan, L. and Razavi Darbar, S. (2014), "Seismic Upgrade of Intake Tower at BC Hydro's Strathcona Dam," Canadian Dam Association 2014 Annual Conference, Banff, Alberta, October 7-9, 2014.

Sweeney, N. and Yan, L. (2014), "Dam Safety Upgrade of the Ruskin Dam Right Abutment," Canadian Dam Association 2014 Annual Conference, Banff, Alberta, October 7-9, 2014.

Wilson, B., Sweeney, N., Atkinson, M. and Diggle, R. (2014), "Construction of a very flexible, impermeable, bituminous membrane seepage barrier, to connect an existing concrete dam structure to a newly constructed abutment cut-off wal,l" Deep Foundation

Institute 39th Annual Conference on Deep Foundations, Atlanta, GA, October 21-24, 2014.

Zoccola, Michael F., Tommy A. Haskins, and Daphne M. Jackson, 2009, "Seepage, Piping, And Remediation in a Karst Foundation At Wolf Creek Dam," Managing Our Water Retention Systems: Proceedings of the 29th Annual Meeting and Conference of the U.S. Society on Dams, Nashville, Tennessee, April 2009.

Zoccola, Michael F., Vanessa Bateman, Josh Bomar, and Fabio Santillan, *Wolf Creek Dam,* Association of State Dam Safety Officials, The Journal of Dam Safety, Volume 12, Issue 1, 2014.

SEEPAGE CONTROL UPGRADE FOR RUSKIN DAM RIGHT ABUTMENT

Nathan Sweeney[1]
Li Yan[2]
Saman Vazinkhoo[3]

ABSTRACT

A new seepage control system, consisting of seepage cut-off and training walls, and tie-in of the cut-off wall into the concrete dam was constructed at Ruskin Dam, located in British Columbia, Canada. The seepage control system was part of a larger project to upgrade the right abutment to address seismic and seepage deficiencies. Due to the dam safety risks and complex nature of the site, a unique early contractor involvement (ECI) procurement process was used to encourage innovation and reduce risks by including the contractor in the development of the final design. Several specialized and innovative methodologies were used to construct the seepage control system, including slurry panel construction for the cut-off and training walls, and jet grouting, specialized drilling, and placement of an asphalt mastic for the tie-in. Jet grouting was carefully carried out at the tie-in to repair and strengthen the sands and loose fills adjacent to the dam without affecting safe operation of the dam. Successful completion of the jet grouting allowed for construction of a 7.2 m long, 0.168 m wide, and 19.5 m to 25.5 m deep slot to connect the seepage cut-off wall to the concrete dam. The slot, consisting of a series of overlapping holes, required several innovative construction techniques, including the use of the pendulum drilling method, and a custom designed guide system to complete the overlapping holes. The asphalt mastic used to fill the slot is watertight and highly flexible to accommodate potential seismic-caused deformations.

INTRODUCTION

Ruskin Dam is a concrete gravity dam, constructed between 1929 and 1930 on the Stave River (see Figure 1). The dam rests primarily on diorite bedrock, however bedrock at the right (west) abutment is located about 15 m below the dam crest, and the dam was constructed with an overhanging section placed against overburden. The overburden consists mainly of pro-glacially deposited silts and sands, which are part of the Quadra Sand deposits. The Quadra deposits are very dense, but are highly erodible and susceptible to seepage-caused piping and loosening. For this reason, a 130 m long sloping concrete slab seepage cut-off system was constructed in 1930 on the right abutment soils; however, piping issues occurred in the abutment sands immediately after the first filling of the reservoir. In addition, the dam is located in a seismically active area resulting in additional dam safety risks to the facility. To address these deficiencies, the seepage control system at the right abutment was upgraded between 2009 and 2013 to include a

[1] Senior Engineer, BC Hydro, 6911 Southpoint Dr., Burnaby, BC, Canada, Nathan.Sweeney@bchydro.com
[2] Engineering Team Lead, BC Hydro, 6911 Southpoint Dr., Burnaby, BC, Canada, Li.Yan@bchydro.com
[3] Ruskin Project Engineer, BC Hydro, 6911 Southpoint Dr., Burnaby, BC, Canada, Saman.Vazinkhoo@bchydro.com

new cut-off wall, with a special tie-in to the concrete dam, a new seepage training wall, and a reverse filter blanket and drainage system on the downstream abutment slope.

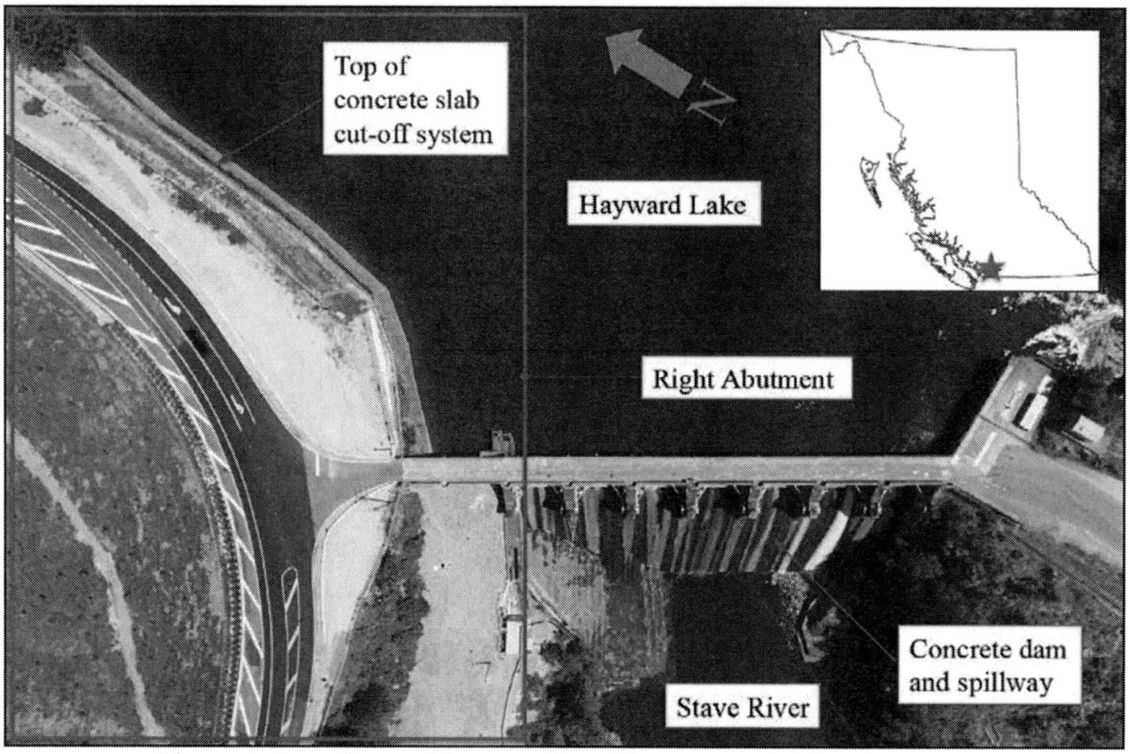

Figure 1. Location and layout of Ruskin Dam and right abutment

PROJECT DESCRIPTION

Site Geology

The right abutment overburden adjoining the concrete dam consists of five soil units above bedrock, classified based on the geologic history and engineering properties. The general stratigraphy of the abutment at the dam is shown in Figure 2. Below the foundation soils, the diorite bedrock was likely shaped by fluvial processes, which resulted in the formation of the deepest channel where the concrete dam is located. The lowest soil deposit (Unit 1) consists of silty sand and gravel, and is typically found in channels or gullies at low points in the bedrock. Unit 2 consists of glaciomarine silty clay. Unit 3 is a grey, massive, non-plastic silt layer, which was likely deposited in standing water, perhaps when the channel was blocked. Unit 4 consists of Quadra Sand deposits, which are an outwash deposit from the Fraser Glaciation, placed sub-aerially on flood plains by braided rivers. Within Unit 4 there are several silt interbeds (Unit 5), with the most prominent and continuous layer located at about El. 29-30 m. A majority of the right abutment is composed of the Unit 4 sand, which although dense due to overconsolidation from glaciation, is susceptible to loosening through seepage and piping. Further descriptions of the geology and site characterization, including other geologic units, are provided in Sweeney and Yan (2014).

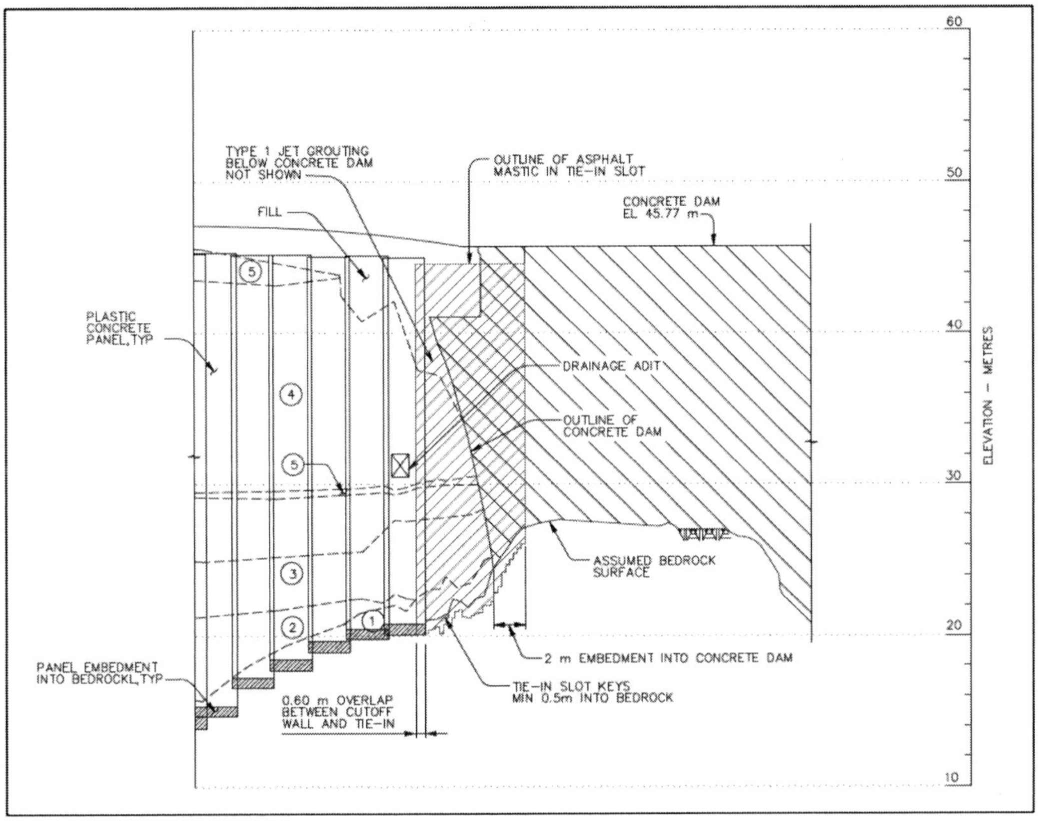

Figure 2. Foundation soil units at right abutment (along dam axis looking upstream)

New Seepage Control System

Because Ruskin Dam is an extreme consequence dam, the new seepage control system is designed to survive a maximum design earthquake (MDE) without leading to uncontrolled release of the reservoir. The MDE used for this project is a magnitude 7.5 event with a peak ground acceleration (PGA) of 0.71g, which corresponds to an annual exceedance frequency (AEF) of 1 in 10,000 years. The upgrades are also designed for continued operation of the reservoir following an earthquake with an AEF of 1 in 475 years (PGA of 0.26g).

The scope of the right abutment seepage control upgrade is described in Sweeney and Yan (2014). The general arrangement of the right abutment upgrades is shown in Figure 3. This paper focuses on the design and construction of the following components, which are the key elements of the right abutment upgrade work:

- New upstream seepage cut-off wall and downstream seepage training wall.
- Tie-in connecting the new cut-off wall with the concrete dam including jet grouting adjacent to and below the overhanging section of the concrete dam to allow construction of a continuous slot for the tie-in section. The slot was backfilled with a highly flexible, watertight material.

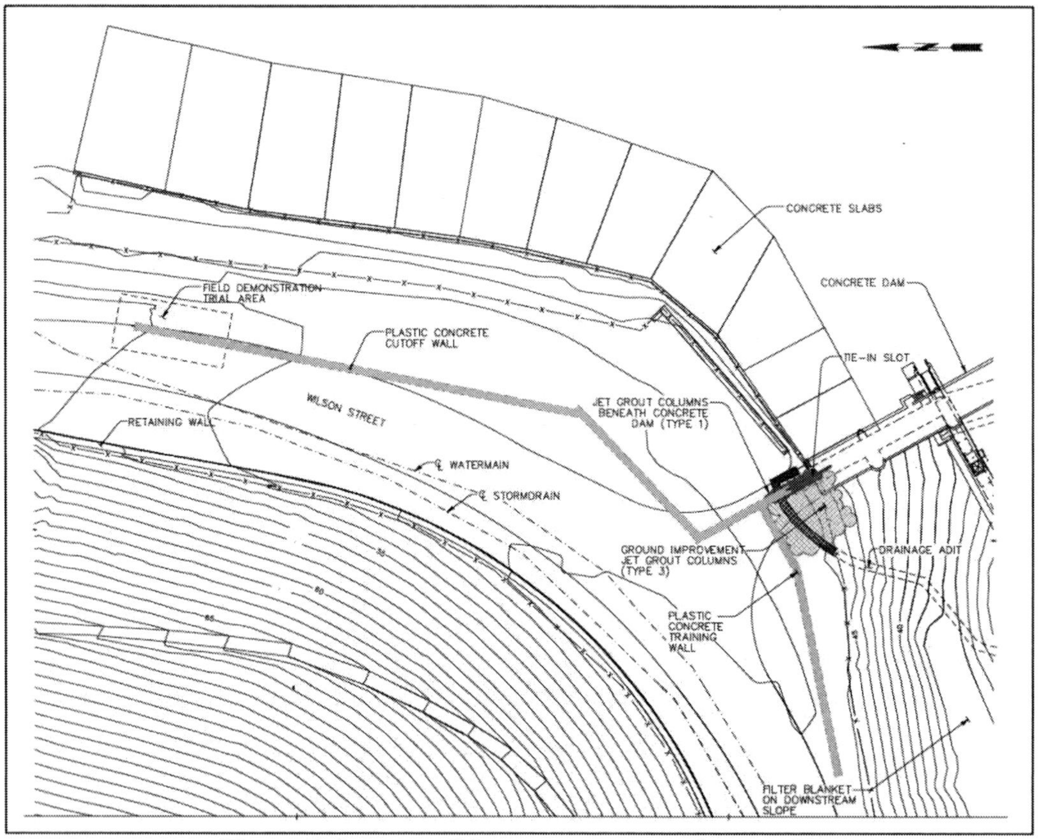

Figure 3. Layout of the seepage control upgrade

PROCUREMENT AND EARLY CONTRACTOR INVOLVEMENT

Due to significant dam safety risks during construction and the complexity associated with the work, an early contractor involvement (ECI) process was used to select a contractor for the construction of the cut-off wall, training wall, and tie-in. The process was designed to meet the following main objectives:

- Transfer knowledge of the site to the contractor,
- Identify potential risks (including dam safety risk) that could occur during construction and assess their impact on the work,
- Develop appropriate construction methodologies to minimize the identified risks,
- Develop a collaborative working relationship where ideas and experiences could be shared between the contractor and the designer to develop the best solutions, and
- Maintain cost competition between the proponents during the process.

Two contractors were selected through a public request for supplier qualification (RFSQ), and were each paid a stipend of $100,000 to participate in the ECI process. BC Hydro prepared a project requirements document, which included the performance requirements for each design item, preliminary level drawings, and submission schedule. The document was updated by BC Hydro as the design progressed and was eventually converted into the technical specifications within the construction contract.

The ECI process was organized into three stages corresponding to the 35%, 65%, and 95% design levels. Each stage consisted of an increased level of detail, and included submittals related to construction methodology, dam safety risks during construction, cost, safety, environmental protection, and drawings. The proponents were also asked to propose a site investigation to be conducted prior to construction, tailored specifically to the requirements of their design, equipment, and construction methodology. Regular meetings were held with both proponents to discuss concepts, provide feedback, and answer questions. BC Hydro conducted a formal evaluation process and provided comments to the proponents after each stage. At the end of the third stage, a preferred proponent was selected based on the evaluation process, followed by negotiation, finalization of the contract specifications, and signing of the contract. The pre-qualification stage lasted 7 months, the ECI process took about 8 months, and the negotiation phase lasted about 4 months. The contract was awarded to Golder Construction, Inc. as the general contractor. The jet grouting and tie-in slot were completed by Malcolm Drilling Company, Inc. and Foundex Explorations, Ltd., respectively, as subcontractors to Golder Construction. Once the contract was awarded, site investigations proceeded in early 2012, followed by construction of the main works between mid-2012 and mid-2013. A one-time contract price adjustment was allowed prior to commencement of construction, based on the results of the site investigation and jet grouting field trial.

The ECI process successfully encouraged innovation by allowing for the inclusion of the contractor's input during the development of the final design, while providing BC Hydro the opportunity to assess and influence the construction methodology to reduce dam safety risks during construction.

CUT-OFF WALL AND TRAINING WALL

Design

The new cut-off wall extends about 100 m upstream from the dam and is located between 18 and 20 m inland from the top of the existing concrete slab to avoid earthquake-induced ground deformation damage (see Figure 3). The cut-off wall is designed to replace the original concrete slab cut-off system and function with full reservoir levels against the abutment. The training wall is connected to the cut-off wall near the dam, and extends about 40 m downstream. The purpose of the training wall is to direct seepage away from the loose fills immediately downstream of the dam and into the filter blanket on the right abutment slope. The cut-off wall and training wall keys into a bedrock ridge at depths of about 24 m to 34 m, and the walls are embedded at least 0.5 m into bedrock. The ridge extends upstream and downstream of the dam and is approximately parallel to the river valley.

Plastic concrete was selected during the ECI phase as the wall material, due to its ability to accommodate the deformations expected at the wall during the MDE. The walls were designed with a width of 1 m to accommodate localized strain concentrations. The plastic concrete was specified to have a maximum unconfined compressive strength (UCS) of

1.5 MPa at 28 days, a minimum UCS of 200 kPa at 7 days (to ensure stability of the plastic concrete during construction), to be able to withstand 10% axial strain without cracking, and a maximum hydraulic conductivity of 1×10^{-6} cm/s. The requirements for the plastic concrete were provided to the proponents in the ECI performance requirements document.

Construction Methodology

Due to the environmental concerns with the proximity of the work to Hayward Lake (a drinking water source), and the requirement for panel excavation into hard bedrock, the successful proponent decided to use a hydromill to construct the cut-off and training walls. The hydromill and associated equipment provides a mostly closed system for slurry circulation, minimizing the environmental risks. A Bauer BG-40 rig was selected and outfitted with a BC-32 cutter, which was capable of excavating panels with dimensions of 2.8 m long by 1 m wide to a maximum depth of about 35 m. Direction control was maintained with flaps on the frame of the cutter and by adjusting the relative speed of the cutter wheels. This allowed corrections to be made during excavation, which was important for ensuring that the required panel overlap was achieved. The on-board computer in the BG-40 provided a real time display of parameters such as the rate of cutting, crowd force, pressure on the wheels, depth, and deviation in the x and y directions. During excavation, the slurry and cuttings from the trench were pumped from the cutter through a desander, which removed the fine and coarse cuttings. The slurry was then re-circulated back into the trench. Upon confirmation that both cutter wheels had encountered bedrock, the cutter was advanced at least 0.5 m into bedrock. Fresh slurry was then pumped into the top of the trench and removed from the bottom by the pump in the cutter. Placement of plastic concrete was carried out using the tremie method, once the sand content in the slurry was acceptable and the panel geometry was confirmed. The contractor prepared laboratory trial mixes of plastic concrete, followed by field trial mixes to confirm compliance with specifications. The final plastic concrete mix consisted of 762.9 kg/m^3 of sand (fine aggregate), 867.5 kg/m^3 of gravel (coarse aggregate), 140 kg/m^3 cement, 33.2 kg/m^3 bentonite, and 336.4 kg/m^3 water. Except for the bentonite, which was added at the site prior to mixing, the plastic concrete was produced and delivered from a commercial concrete plant which was located about 14 km from the site.

Acceptance Criteria and Results

Acceptance of the walls was based on the excavation results, including visual observation of the cuttings and the output of the hydromill's onboard instruments, the panel geometry measurements, and the test results from the plastic concrete. Acceptance of bedrock embedment and panel geometry was provided prior to placement of plastic concrete.

The determination of the bedrock surface presented a significant challenge during construction of the panels. In general, the soils at the site are very dense, and the till-like units above bedrock (Units 1 and 2) contain boulders and cobbles, which could be mistaken for bedrock. A baseline bedrock surface had been developed and accepted based on exploratory drilling conducted separately by BC Hydro in 2008 and the contractor in 2012. During panel excavation, bedrock was determined to be at elevations that were higher than the baseline at some locations, based on the response of the cutter and the type of cuttings at the desander. Several verification holes were drilled through completed panels to confirm the presence of bedrock below the panels. The results of the drilling allowed for reliance on the cutter response combined with the observation of diorite fragments at the desander for determining bedrock location.

Primary verification of the panel geometry and x-y location was based on the output from the BG-40 rig. The overlap for the panels was specified to be 300 mm, and the deviations of the panel (cross or parallel to the panel) were typically less than 50 mm. Secondary verification of the panel geometry was carried out with a KODEN tool after removal of the cutter from the trench. The KODEN tool uses ultrasonic signals to image the sidewalls of the trench and can detect collapsed trench walls.

Samples of plastic concrete were obtained directly from the ready-mix trucks for slump and temperature measurements and to cast cylinders for triaxial, UCS, and laboratory hydraulic conductivity testing. Following placement of plastic concrete, in situ hydraulic testing was carried out in several of the verification drill holes. Due to the softness of the plastic concrete, intact plastic concrete core samples could not be recovered. Hydraulic conductivity values within the panels ranged from 1×10^{-5} m/s to 6×10^{-6} m/s, which were higher than the laboratory test results by about an order of magnitude (an average of 2.4×10^{-7} cm/s was obtained from the laboratory tests). These results are consistent with other experience in plastic concrete cut-off wall construction (Singh et al., 2005, and Yan et al., 2008). Although the in situ hydraulic conductivity results were greater than allowable (1×10^{-6} cm/s), the results were considered to be acceptable, because a hydraulic conductivity of 1×10^{-5} cm/s had been used in the seepage analysis of the new seepage control system.

JET GROUTING

Design

The tie-in of the cut-off wall to the concrete dam is the most critical component of the right abutment upgrade, as leakage through this portion of the wall would result in a very short and concentrated seepage path around the end of the concrete dam. The tie-in is complicated by the geometry of the concrete dam at this location, and the potential for presence of loose sands due to historical seepage issues. The right abutment end of the dam is a narrow 2.4 m wide section founded partially on bedrock and partially on soil, where concrete was placed against steeply excavated silts and sands during the original construction (see Figure 2).

The design for the tie-in required that the soils beneath and downstream of the overhanging portion of the concrete dam be strengthened by jet grouting to form a relatively hard soilcrete material. The first panel of the plastic concrete wall was then excavated into the soilcrete to form a key. Two types of jet grouting were used (see

Figure 3). Single fluid jet grouting (Type 1) was specified in the more sensitive area beneath and adjacent to the dam to repair loosened soils, and to provide stability and allow for construction of the tie-in slot. Two fluid (with air) jet grouting (Type 3) was used for the loose fills downstream of the dam to provide additional seismic stability for the tie-in. The Type 3 jet-grouted block extends from the tie-in to 8.5 m downstream of the dam axis and about 10 m east from the training wall, extending to bedrock or till. Jetting of the Type 3 columns was carried out following completion of the Type 1 columns. The geometry and performance requirements (UCS and hydraulic conductivity) were provided to the contractor during the ECI process; the contractor then developed the detailed design for jet grouting, including the spacing, column diameter, and jetting parameters. The use of air was not allowed during drilling or jetting underneath the dam.

Jet grouting is a high pressure, high flow operation, and one of the most significant challenges was the potential that the intersection with voids and loosened sands could cause ground collapse resulting in sinkholes or depressions in the abutment. Additional risks included the potential for hydraulic fracturing of the soils if reflow blockage occurred, and the loss of grout into the rockfill beneath the concrete slabs located upstream of the dam. These risks were discussed extensively during the ECI process, and risk mitigation plans were included in the contractor's construction methodology and the dam safety risk management plan. The dam safety risk management plan was prepared jointly by the contractor and BC Hydro, and identified potential dam safety risks during jet grouting, provided instrument alarm limits, surveillance levels, and incident response requirements. A field trial was used to test the construction method, and calibrate the response of the instruments. To address the risk of grout entering the rockfill, a barrier wall, consisting of a series of overlapping holes drilled with a sonic rig, and backfilled with a shotcrete-type concrete mix, was constructed immediately upstream of the Type 1 columns. The response of pore water pressures during jetting was monitored with high frequency data collection in nearby piezometers, and at a normal rate at other piezometers and the downstream weirs. To minimize the consequences of potential sinkholes or ground collapse, the reservoir was drawn down during construction of the Type 1 columns.

Field Trial

One of the most important steps for any jet grouting project is the design and implementation of a field trial program to determine the achievable column diameters, and to select the jet grouting parameters (pressure, flow, withdrawal rate) for the desired column sizes. The jet grouting field trial was carried out about 40 m downstream of the dam on the right abutment in similar soils to those located at the dam. Four Type 1 columns and three Type 3 columns were constructed using various withdrawal rates (pressure and flow were kept constant). Following completion of jet grouting, verification drilling and coring were carried out with sonic and mud rotary drill rigs, and the top of the columns were exposed and measured. Two angled sonic holes were drilled on either side of one Type 1 and one Type 3 jet grout column to determine the column diameter near the surface and at depth. Vertical core holes were drilled with a mud rotary rig to obtain samples for UCS testing. Based on the results of the drilling and excavation, it was concluded that the diameter of the Type 1 columns ranged from about 1.5 to 2.0 m, and the diameter of the Type 3 columns ranged from about 2.0 to 2.5 m in the Unit 3 and 4 soils. However, confirmation of the diameters in the lower units (Units 1 and 2) could not be completed due to their deep location at the trial site. One additional Type 1 column was installed in the Type 3 area near the dam to assess the column diameter in the lower soil units.

Piezometric responses near the trial site (with distances ranging between about 2 and 20 m from the jet grouting) were monitored at very high sampling rates at about 1 reading per second to allow for the detection of immediate pore water pressure responses in the soils near the columns. Very high pressures (up to 8 m increase) occurred within about 4 m of the jetting, but reduced quickly to less than 0.2 m at a distance of about 9 m. The results of the groundwater pressure responses are shown in Figure 4, along with data collected during jet grouting at John Hart Dam in 2012 (Siu et al., 2013). Alarms were assigned to piezometers near the production columns based on the field trial results.

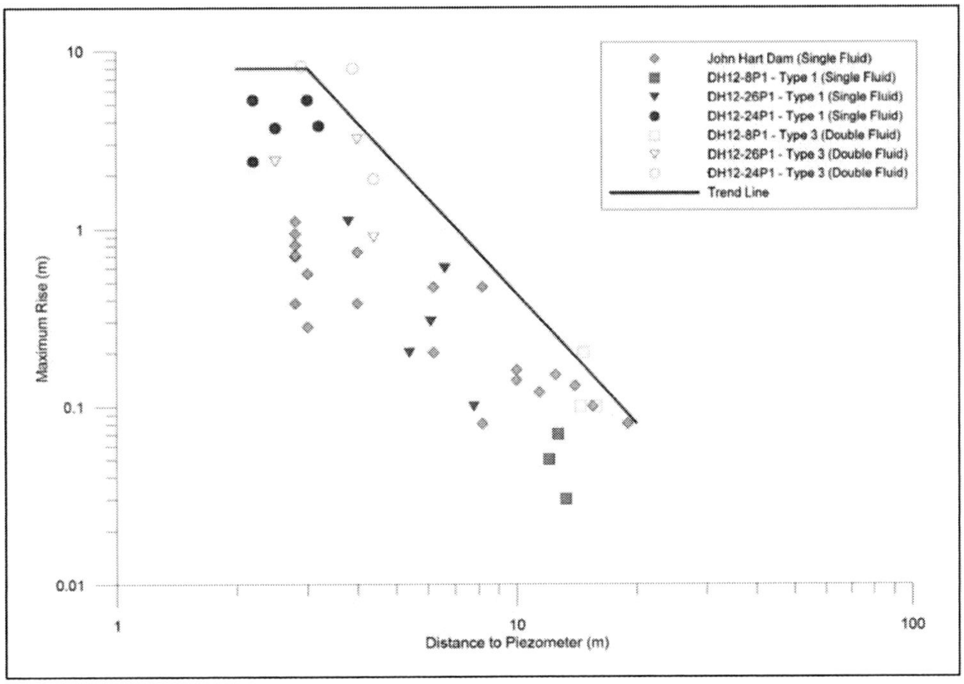

Figure 4. Piezometric response to jet grouting during field trial

Construction Methodology

Based on the field trial results, the jet grout columns were laid out using the following general parameters:

Type	Column Spacing	Column Diameter
1	2 rows spaced at 0.75 m with overlapped columns spaced at 0.85 m	Minimum 1.2 m
3	Overlapping columns spaced at 1.3 to 2.4 m	1.35, 2.0 and 2.5 m (nominal)

Achieving the required Type 1 diameter was extremely important to strengthen and stabilize the loosened sand at the tie-in area so that the slot could be constructed through the soilcrete, without any weak, collapsible sections. The Type 1 jet grouting extended through fills used to backfill the upper portion of the original excavation for the concrete dam, through the erodible Units 4 and 3, and into the stiffer Units 2 and 1 located directly above bedrock, as shown in Figure 2. All Type 1 columns were required to be sealed to bedrock. A slower withdrawal rate and a double cut (two passes of the nozzle) were used in the lower soils (Units 2 and 1) to increase the energy and ensure that the required diameter was constructed.

All jet grouting was carried out with a Bauer BG-15 rig, which uses a sophisticated on-board computer and data acquisition system to obtain data during jetting, including pressure, flow, rotation, withdrawal rate, and the inclination of the hole. The sensor for inclination measurements is located in the jet grout monitor, and when combined with GPS information obtained from sensors on the rig, provides the x-y location of the hole during drill-down. The data are transmitted through the drill rods (called E-rods) to the data acquisition system on the drill rig.

Pre-drilling was carried out for most of the Type 1 and Type 3 columns to penetrate the concrete dam, to ensure the columns were sealed against bedrock, and to prevent locking of the drill casing by cobbles and boulders. To ensure that the Type 1 columns were sealed to bedrock, it was necessary to locate the jet nozzle at the bedrock surface. The distance between the nozzle and the drill bit (either 0.76 m or 1 m depending on the bit) required that the hole extend past the contact into bedrock. Pre-drilling also allowed for the determination of the top of the double cut (i.e. top of Unit 1 or 2) and the top of bedrock. Pre-drilling was also used for the Type 3 columns to prevent cobbles and boulders in the loose fill downstream of the dam from falling against the drill casing, potentially inhibiting casing withdrawal. Pre-drilling was carried out with a sonic drill for the Type 1 columns and an air rotary drill rig for the Type 3 columns. All holes were backfilled with medium-grained sand.

Difficult ground conditions impacted the construction of both the Type 1 and Type 3 columns. The fill placed during original construction around the overhanging section of the concrete dam contains cobbles and boulders, and caused abrasion on the drill casing and locking of the drill string in some of the holes. During jetting for the Type 1 columns, the drill bit was damaged when the drilling intersected a boulder in the fill. In another case, the drill string become locked in by boulders and cobbles and had to be retrieved by over-coring with a sonic drill rig. Reflow was lost briefly for a number of Type 1 columns but no significant grout loss or ground settlement occurred during jetting of the columns.

In the Type 3 area, significant loss of reflow occurred when jetting encountered a zone of loose, open cobble and boulder fill that was placed at the bottom of the excavation during construction of the overhanging section of the dam. The Type 3 jet grout columns could not be constructed in this zone of loose and openwork fill because the reflow back to the surface could not be maintained, and it was not apparent where the grout was flowing. To continue the jet grouting, a grouting program was carried out to pre-treat the zone of the loose openwork fill with a special low mobility soft grout. The grout mix used consisted of water-cement-bentonite at a ratio of 1:7:17.5, which resulted in grout with 28-day UCS values between 250 kPa and 1,000 kPa that was soft enough for erosion by jet grouting. End-of-casing pressure grouting was used to pump a total of about 90 m^3 of the low mobility soft grout into the zone. The pre-treatment was successful in filling the openwork fill, and allowed the jet grout columns to be constructed.

During the initial Type 3 jet grouting, fracturing of the loose fills in the area occurred under the high pressure and flows of the double-fluid jetting, resulting in grout flowing through cracks in the ground, which were parallel with the original 1929 excavation surface. It was speculated that the cracks had formed as result of the difference in horizontal stresses along the boundary between the loose fill and the denser, native soils. Lower energy, single fluid jet grouting was used successfully for the first five columns to seal the area, without causing ground settlement. The remaining columns were then completed successfully with the double fluid method.

During the jet grouting work, piezometric levels around the dam were monitored in real time in an instrument control building located near the work site, and radio communication was maintained with the rig operator. Jetting pressures were immediately reduced if reflow was lost or if any of the instruments reached the alarm levels.

Acceptance Criteria and Results

The three most important factors used to determine acceptance of the results of the jet grouting were (1) visual inspection, (2) verification coring, and (3) test results from UCS and in situ hydraulic conductivity testing.

Full-time inspection was carried out during jet grouting to observe both the reflow at the collar of the columns, and the real-time display of the jetting parameters provided by the BG-15. A significant challenge with jet grouting is the confirmation of the column diameters, which can be highly variable due to the increase in confining stress with depth, and/or changes in soil type. Deviation results obtained from the BG-15 on-board computer enabled the bottom of the columns to be located relative to the top, and HQ verification core holes were completed in the Type 1 and Type 3 areas to assess the potential for gaps between columns. Cores were photographed and logged, and samples were collected for UCS testing. An optical and acoustic televiewer was used to provide high quality imaging for select core holes, and standard down-hole camera profiling was carried out where possible for the remainder of the holes. In situ hydraulic conductivity testing was also carried out at select locations within the verification holes to evaluate the water-tightness of the overlapped columns.

In general, recovery from the core holes in Type 1 area was high and very good quality soilcrete was obtained. The Unit 3 (silt) and Unit 4 (sand) materials were observed to result in very strong, homogeneous soilcrete, which was easily cored and provided high-quality samples. A sample from one of the televiewer results is shown in Figure 5 at the contact with bedrock and at the drainage adit (the location of the adit is shown on Figure 3). Good contact was observed between the soilcrete and bedrock and included penetration of the soilcrete into the rock fractures, as well as a good seal of the soilcrete to the outside of the adit formwork, both of which were also observed from the core samples. Core recovery of soilcrete from the Type 1 columns indicated that the column diameters are at least 1.2 m. In the Type 3 area, the soilcrete was observed to contain much more gravel, cobbles, and boulders, which caused poorer recovery during coring. However, the televiewer and video profiling provided confirmation that soilcrete was present throughout the Type 3 area.

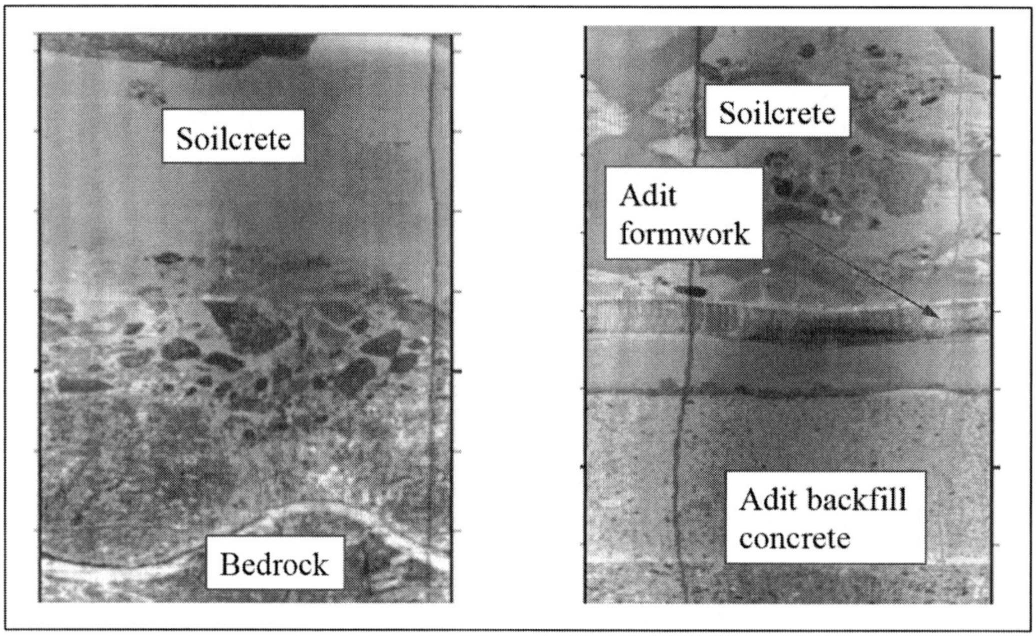

Figure 5. Televiewer results from Type 1 jet grout columns

Three different methods were used to obtain soilcrete samples for UCS testing. Samples of the reflow were collected at various times during construction of the columns and cast into cylinders. Samples were also collected from the fluid column with a hydraulic wet-grab sampler, and core samples (HQ size) were retrieved during drilling of the verification holes. The UCS results from all sampling methods are shown in Figure 6 for both the Type 1 and Type 3 columns. High strengths were observed from all methods, especially from the Type 1 columns, with most of the results falling between 8 and 20 MPa. The strengths are slightly lower for the samples from the Type 3 columns, where the quality of the recovered core samples was more variable due to the increased amount of granular particles in the soilcrete. Most of the UCS results from the Type 3 are between 4 and 16 MPa. Most of the in situ hydraulic conductivity test results were between 2×10^{-7} cm/s and 9×10^{-6} cm/s, with only one being 5×10^{-5} cm/s.

The results from the three indicators provided confidence that the Type 1 and Type 3 columns had met the design intent, and work was allowed to proceed with the construction of the slot through the Type 1 columns.

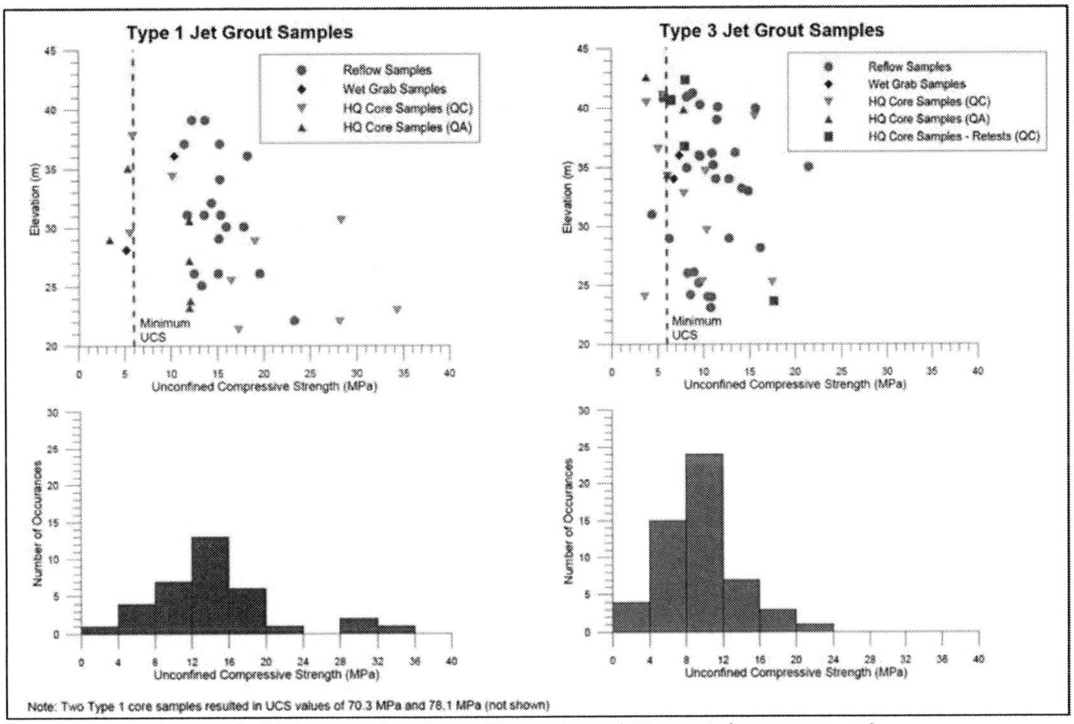

Figure 6: UCS results from Type 1 and Type 3 jet grout columns

TIE-IN SLOT AND BACKFILL

Due to the potential for seismic-caused differential deformations between the concrete dam and the right abutment, a highly flexible material was specified for the backfill of tie-in slot, to ensure that it would remain watertight following the MDE. The baseline design from BC Hydro included the use of a geomembrane surrounded by cement-bentonite grout. During the ECI process, the contractor proposed an alternative design, consisting of an asphalt-based mastic as the backfill material. This alternate design was accepted, with the requirement that it achieve the following performance requirements:

- Maximum hydraulic conductivity of 10^{-6} cm/s
- Minimum UCS of 200 kPa and maximum 300 kPa to ensure that the material remained stable and did not creep
- Minimum 10% axial strain at peak deviator stress in a consolidated undrained triaxial test with an effective confining stress of 150 kPa
- Ability to accommodate a minimum of 50 mm of lateral displacement at any location without cracking or rupturing

Due to limited precedence for the use of this material, an extensive product development and testing program was required, along with various field trials.

Development and Selection of Backfill Material

Asphalt-type products have been used for cut-off walls in Japan (Ohne et al, 1989), as water stops in an RCC dam (Hall et al. 2005), and extensively for grouting (Naudts and

Hooey, 2003). However, expertise for development and construction of asphalt mastic for cut-off walls was not readily available in North America, and so a full testing program was required to develop and demonstrate the suitability of the proposed material. The contractor carried out development of the asphalt mastic with input and review from BC Hydro. Laboratory trial mix testing, including UCS, unconsolidated undrained triaxial testing, direct shear testing and hydraulic conductivity testing, was carried out to select the appropriate asphalt content and confirm that the performance requirements could be achieved. Because the behavior of the asphalt mastic is dependent on temperature, the laboratory tests were carried out at a temperature of about 10 °C which is close to the ground temperature at Ruskin Dam. Based on the results of laboratory trial mixes, followed by field batching confirmation from a local batching plant, the mix for the asphalt mastic was chosen to contain 13.5% bitumen, 20% calcium carbonate, and 66.5% fine and coarse aggregate. The test results confirmed that the asphalt mastic has a density between about 2090 to 2160 kg/m^3, and can accommodate 50 mm of deformation without cracking while meeting the strength and hydraulic conductivity requirements. Field-testing was then carried out to develop construction methods. Photos of the mastic during direct shear testing and a UCS test are shown in Figure 7.

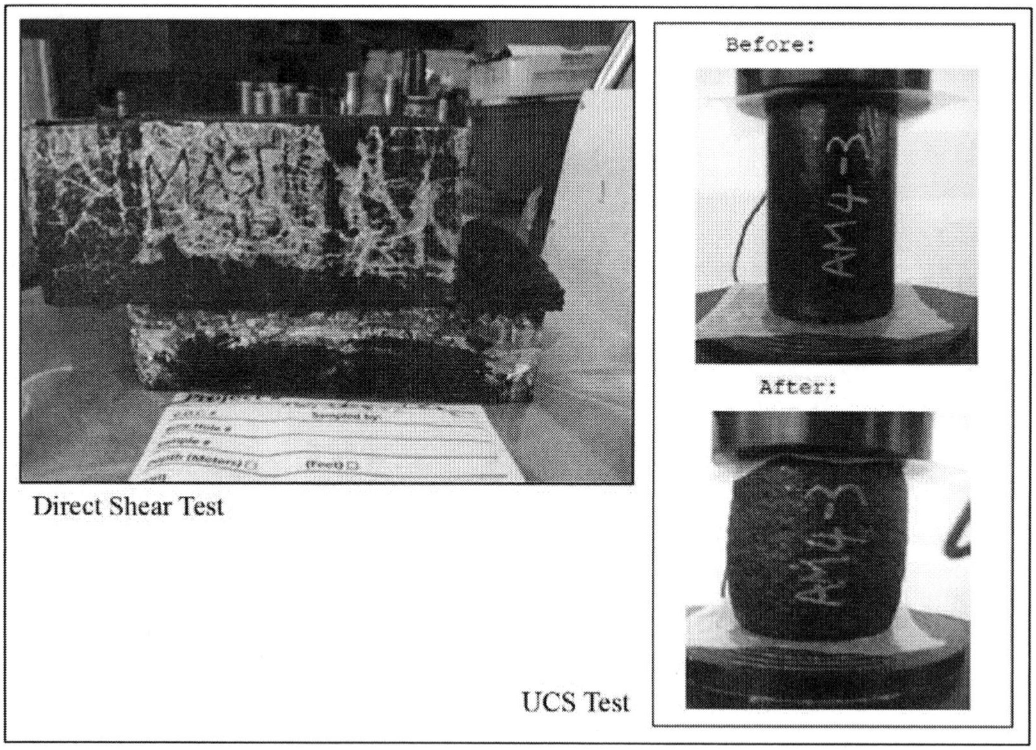

Figure 7: Direct shear test (left) and UCS test (right) on asphalt mastic

Slot Construction Methodology

The methodology chosen for construction of the tie-in slot consisted of a series of overlapping drill holes carried out in a primary and secondary sequence to create a slot between the cut-off wall and the concrete dam. The holes had to be drilled through the dam, the Type 1 jet grout columns, the plastic concrete and bedrock, requiring a highly versatile drilling methodology to accommodate the different stiffness of the materials, especially at the connection of the slot to the plastic concrete panel.

Initially, very accurate NQ primary pilot holes were drilled along the slot spaced at 0.51 m (20 in), to provide pilot holes for the large diameter primary holes. These holes were drilled with the reverse pendulum method, which BC Hydro had successfully used in a previous project at Strathcona Dam (Sherstobitoff et al. 2014). With this method, the alignment of the hole is checked every 1.52 m (5 ft.) by inserting a weight into the bottom of the hole, which is connected to a wire extending to a float at the ground surface. The offset of the wire is checked against the baseline measurement and a correction made if necessary. In general, a correction was required if the deviation was between about 6 mm and 15 mm. To correct the alignment, a small hole (about 37 mm diameter) was drilled through the bottom of the NQ hole. The alignment of the small hole was oriented with an offset template, which was placed at the bottom of the NQ hole and rotated to bring the hole back to the correct alignment. An NQ reamer was then advanced to straighten the hole by following a steel spear placed in the small hole. If the deviation was greater than 15 mm, the alignment could not be corrected, and the hole had to be grouted and re-drilled.

Figure 8: Tie-in slot with completed primary holes (top left), completed slot (lower left), and guide frame for secondary holes (right, Wilson et al., 2014)

The NQ primary holes were then opened to 0.3 m diameter with a combination of down-the-hole (DTH) hammer, coring, and reverse circulation tricone bits, each custom fit with stingers to follow the NQ primary holes. Down-the-hole (DTH) hammer drilling was used in the concrete dam portion of the slot, and reverse circulation tricone drilling was used in the soilcrete to prevent damage from the air pressure. Large diameter coring was used at the contact between the concrete and jet grout columns to ensure that the holes remained straight. The top of the slot following completion of the large diameter primary holes is shown in Figure 8. A custom-built guide system was used to drill the 0.3 m diameter secondary holes by following the primary holes on either side, resulting in a slot with a minimum width of 0.168 m. Again, DTH hammer drilling was used in the dam and reverse circulation tricone drilling was used in the soilcrete. One of the guide frames is shown in Figure 8, along with a photo of the completed top of the slot. Some difficulty was encountered during removal of the cuttings from the secondary holes. A large amount of cuttings fell to the bottom of the slot and could not be removed as the drilling progressed, requiring periodic removal of the guide system from the slot to clean out the holes.

At the connection to the cut-off wall, the NQ pilot holes could not be used in the soft plastic concrete. To complete these holes, four core barrels were stacked together to provide increased rigidity to the drill string. All the primary holes, and most of the secondary holes were drilled in the dry with air circulation; however to ensure the fully open slot remained stable, two secondary holes were left undrilled to support the slot. The slot was then filled with water to provide support and the two remaining holes were drilled using the tricone method with circulation provided by two large pumps. The water remained in the slot until it was displaced by placement of the asphalt mastic.

Mastic Placement Methodology

Initially, the contractor proposed pumping to place the hot asphalt mastic into the slot. However, after a number of field trials it became evident that the asphalt mastic was not pumpable, and its viscosity (or flowability) was highly dependent on its temperature. This led to the selection of a tremie placement methodology. To ensure flowability of the material during tremie placement, a field trial was required to establish a reasonable lower temperature limit. During this trial, the contractor used a set of 7.3 m long wood-formed slots filled with pre-heated water (to about 80 °C) to simulate the tie-in slot. Hot mastic was poured at the center of these wood-formed slots under different temperatures, and the time for the mastic to flow and reach the ends of the slots was measured (see Figure 9). The purpose of heating the water in the slot was to reduce temperature loss of the mastic during the placement. Based on these results it was determined that a minimum mastic temperature of about 110-120 °C was required to ensure the mastic would have sufficient flowability to flow horizontally in 80 °C water to the ends of the slot.

After these trials, a field tremie placement trial was carried out in a trial slot constructed in the field trial jet grout columns downstream of the dam. The trial slot was filled with water heated by hot steam to about 80 °C, as in the wood-formed slots. Several issues were encountered during this trial. Firstly, the tremie pipe was not sealed at the bottom, and was thus full of water when the 120 °C hot mastic was introduced. The mastic mixed with the water in the pipes, resulting in violent eruption of boiling water-mastic mixture due to the temperature difference. Secondly, the tremie pipe was not insulated which, together with the mixing of the mastic and water in the pipe, caused cooling of the mastic. This resulted in the plugging of the tremie pipes, making tremie placement not possible. Several modifications were then made to the tremie pipes, including the use of a cap at the bottom of the tremie pipes to avoid water entry into the pipe, and adding a second, outer wall to the tremie pipes to provide insulation between the water and the hot mastic.

Once the modifications to the installation methodology had been accepted, placement of mastic in the tie-in slot was allowed to proceed. The mastic was batched at a nearby asphalt pug mill, located about 40 km from the site, and transported to site with concrete ready mix trucks with mastic temperatures on arrival between about 140 °C and 160 °C. The water in the slot was heated to about 80 °C with the same steam heating system used during the field trial. Two sets of thermocouples were installed at each end of the slot to allow monitoring of the water and mastic temperatures. Two sets of tremie pipes were used with an inside diameter of 229 mm. The pipes were double walled (except at the joints) to provide insulation between the water and the inside of the pipe. Silicone and tape were used at the pipe joints to prevent the ingress of water. A sacrificial end cap was used to seal the bottom of the first tremie pipe to allow water to be removed by a vacuum truck prior to placement of mastic.

Figure 9: Field trial tests carried out in wood-formed slots

The two sets of tremie pipes (with staggered joints) were installed such that the slot was divided into two equal segments to allow for alternating placement. Pouring of the hot mastic started at the pipe located at the deeper end of the slot. After the pipe was filled with mastic, it was raised slightly and the weight of the mastic caused the end cap to detach from the bottom and the mastic began flowing out into the slot, and displacing the water in the slot. Placement by the tremie method was maintained throughout the pour by ensuring that the bottom of the discharge pipe was constantly embedded at least 0.5 m into the mastic. The mastic level was sounded at 4 to 5 places in the slot to confirm that mastic was flowing to the ends. Once the mastic had filled the lower portion of the slot, placement of mastic was alternated between the two tremie pipes. The hot mastic was placed by the tremie method to a depth of about 10 m, which is above the groundwater level. The water in the slot was then removed, and the remainder of the mastic was placed in the dry with the tremie pipes just above the surface of the mastic. Photos of the mastic placement are shown in Figure 10.

Figure 10: Mastic in hopper above tremie pipes (upper left), top of mastic in slot (right), and pouring of mastic from truck into hopper (lower left)

The most significant challenge with placement of the mastic was the temperature control. A drop in temperature occurred at every stage throughout the process, including material delivery, transfer of the mastic down the chute and into the tremie pipe hopper, and as it travelled down the pipes into the slot. Any mixing with the water in the slot or in the tremie pipe resulted in a violent eruption of boiling water-mastic mixture, and cooling of the mastic. It was therefore important to ensure that the tremie pipe bottom and joints were well sealed to prevent mixing with water in the pipe during the tremie process. The mastic needed to be poured as hot as possible with minimum heat loss during the placement. The goal was to achieve a mastic temperature of about 110-120 °C at the discharge point at the bottom of the tremie pipe. Thus, maintaining high temperature and minimizing heat loss with appropriate temperature monitoring was the key in quality control for the contractor, and the main criterion for BC Hydro to accept the material placed in the slot using the tremie method.

Acceptance Criteria and Results

During the placement of the mastic, the mastic levels in the slot were measured to ensure that (1) the mastic was reaching the ends of the slot, and (2) to confirm that the mastic was at least 0.5 m above the next joint of the tremie pipes before retrieving the pipes. The temperature of the mastic was also monitored by two sets of thermocouples at each end of the slot, which provided secondary confirmation that the mastic had reached the ends by showing marked temperature increases. The cap at the bottom of the tremie pipe and sealing of the joints was successful in separating the mastic from the water during the initial pour, and the insulation of the tremie pipes was key to minimizing temperature loss in the mastic. This allowed continuous placement with the tremie method, avoiding mixing with the water in the tremie pipes and therefore cooling of the mastic. Some settlement and consolidation occurred as the mastic cooled, resulting in a surface drop of about 0.3 m five days after the final placement. Samples of asphalt mastic were collected during field placement, from which laboratory specimens were formed and tested to confirm that the strength, hydraulic conductivity, and flexibility requirements had been met.

CONCLUSION

A new seepage cut-off system has been constructed at the right abutment of Ruskin Dam to address the seismic and seepage deficiencies. To properly implement the design and construction of the new cut-off wall, training wall, and the tie- in to the dam, a number of complex dam safety and construction risks were identified using a unique procurement strategy. The ECI process provided a competitive framework to communicate the risks to the contractor, to develop final design and appropriate construction methodologies by sharing ideas and experiences, and to establish risk mitigation strategies to safely carry out the work. Several innovative solutions were required for this project including pre-treatment pressure grouting of the loose openwork fill in the Type 3 area to allow for jet grouting, the use of overlapping drill holes to construct the tie-in slot, and the development of the asphalt mastic, which required significant product development, testing, and field trial demonstration. Construction of such specialized elements requires care and attention from the development of the design and specifications to monitoring and testing during construction for acceptance. Careful planning, effective communication with the contractor before and during the work, and adaptive changes during construction allowed this complex upgrade work to be completed successfully and safely, without occurrence of any significant dam safety issues.

ACKNOWLEDGEMENTS

The authors wish to acknowledge the contributions of other project team members, including S. Garner, M. Li and P. J. Ho. Permission from Dam Safety and Ruskin Dam and Powerhouse Upgrade Project to publish this paper is also gratefully acknowledged.

REFERENCES

Hall, F., Krumm, P., and Allen, J. (2005). Remediation of RCC Dam Seepage – Upper Stillwater Dam Rehabilitation – Phase 1. *Proceedings of the 2005 USSD Annual Meeting and Conference.* Salt Lake City, Utah.

Naudts, A. and Hooey, S. (2003). Hot Bitumen Grouting: The Antidote for Catastrophic Inflows. *Grouting and Ground Treatment:* pp. 1293-1304. doi: 10.1061/40663(2003)76.

Ohne, Y., Tatebe, H., Narita, K., Akutsu, T., and Okayama, T. (1988). Design and Construction of Asphalt Mixture Cut Off Wall. *Proceedings of the International Commission on Large Dams.* San Francisco, California.

Sherstobitoff, J., Yan, L., and Razavi Darbar, S. (2014). Seismic Upgrade of Intake Tower at BC Hydro's Strathcona Dam. *Proceedings of the 2014 Canadian Dam Association Annual Conference.* Banff, Alberta.

Singh, N., Douglas R., Ahlfield, S., and Grant, M. (2005). Construction of a Roller Compacted Concrete Upstream Blanket and Plastic Concrete Cutoff Wall at Cleveland Dam. *Proceedings of 2005 Association of State Dam Safety Officials Annual Conference.* New Orleans, Louisiana.

Siu, D., Gazzarrini, P., Jungaro, S., and Hunt, D. (2013). John Hart Dam – Backup Jet-Grouted Seepage Cutoff Wall Construction. *Proceedings of the International Commission on Large Dams 2013 International Symposium.* Seattle, Washington.

Sweeney, N., and Yan, L. (2014). Dam Safety Upgrade of the Ruskin Dam Right Abutment. *Proceedings of the 2014 Canadian Dam Association Annual Conference.* Banff, Alberta.

Wilson, B., Sweeney, N., Atkinson, M., and Diggle, R. (2014). Construction of a Very Flexible, Impermeable, Bituminous Membrane Seepage Barrier, to Connect an Existing Concrete Dam Structure to a Newly Constructed Abutment Cut-off Wall. *Proceedings of the 39th Annual Conference on Deep Foundations.* Atlanta, Georgia.

Yan, L., Trapp, D., and Sy, A. (2008). Construction of a Plastic Concrete Seepage Cutoff Wall for the New Coquitlam Dam. In S. Prakesh (Ed.), *Proceedings of the 6th International Conference on Case Histories in Geotechnical Engineering.* Arlington, Virginia.

MOOSE CREEK DAM — RISK ASSESSMENT OF AN 8-MILE-LONG DAM FOUNDED ON DEEP SAND AND GRAVEL DEPOSITS WITH DISCONTINUOUS PERMAFROST IN FAIRBANKS, ALASKA

Troy S. O'Neal, Ph.D., P.E.[1]
Chun-Yi Kuo, Ph.D., P.E.[2]

ABSTRACT

Moose Creek Dam is located on the Chena River upstream of Fairbanks, Alaska and reduces flooding in Fairbanks and surrounding communities by temporarily impounding and diverting floodwater to the much larger Tanana River. While classified by the US Army Corps of Engineers as a dam, it functions much as a levee.

The eight-mile long "dry" dam is constructed over a broad valley filled with stratified sands and gravels covered with a surficial silt layer of varying thickness. There are localized areas of near-surface permafrost, with some areas that have thawed since construction. Depth to bedrock ranges from being at ground surface at the north abutment, to over 600 feet along the majority of the dam axis. The southern end terminates at the Tanana River. Groundwater typically is about 6 to 8 feet below the ground surface and rises quickly during flood events but dissipates slowly.

The risk team performed an initial assessment of the existing project risk and continues to refine the risk estimate based on a recent flood event (June 2014) and future population growth trends. Many unique aspects were assessed including dam design, construction methods in cold regions, the effects of the discontinuous permafrost on ground water flow and internal erosion, uncertainty in subsurface data, and multiple breach formation locations. This paper describes some of the methods utilized to develop the existing project risk assessment.

INTRODUCTION

Moose Creek Dam construction began in 1973 and was completed in 1979. The drainage area above the Moose Creek Dam is 1,496 square miles. The project was designed to help limit flow on the Chena River at the downtown Fairbanks gage to 12,000 cfs or less. Because the Little Chena tributary flows into the Chena River below Moose Creek Dam, Fairbanks can still be potentially flooded from the Little Chena tributary depending on localized rainfall events.

The main embankment is an approximately eight-mile long earthfill structure that reaches a maximum height of 50 feet above the Chena River streambed. The embankment was constructed of semi-pervious silty gravel on the upstream side to inhibit seepage. Primary seepage control measures are a continuous upstream seepage blanket constructed of silt

[1] Geotechnical Engineer, Louisville District, US Army Corps of Engineers, Louisville, KY 40202, troy.s.oneal@usace.army.mil
[2] Geotechnical Engineer, Louisville District, US Army Corps of Engineers, Louisville, KY 40202, chun-yi.kuo@usace.army.mil

that is at least 600 feet wide with a minimum thickness of 3 feet, a toe drain system, a downstream stability berm, and seepage collection channels. Based on test and record pool performances, 158 relief wells were incrementally installed.

The outlet control structure is composed of four 25-foot-wide concrete bays divided by piers, and each bay is designed to pass Chena River flows at a maximum of 3,000 cfs, with additional flows through the fishways and fish ladder. Flow is regulated with four hydraulically operated steel vertical lift gates. A gated low point drain is used to remove trapped, stagnant water at the low point in the floodway after floodwaters have receded. Downstream seepage collection channels direct seepage and water from the main low point drain and relief wells back to the Chena River below Moose Creek Dam on both the north and south sides of the river.

The project floodway consists of an excavated and cleared channel, approximately 6.5 miles long and up to 2,400 feet wide, used to direct diverted floodwater from the Chena River into the Tanana River. The floodway control sill limits backflow into the floodway from the Tanana River. Tanana River Levee (Figure 1. Dashed purple line) connects to the south end of the Moose Creek Dam and extends approximately 22 miles west to the confluence of the Tanana and Chena rivers. It protects the greater Fairbanks area from high water on the Tanana River.

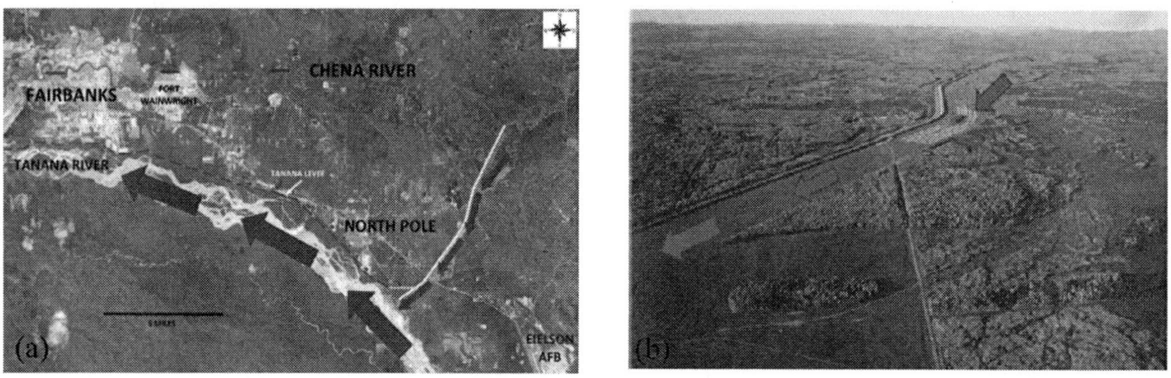

Figure 1. Operational Schematic and Aerial View of Moose Creek Dam

GEOLOGIC SETTING OF THE TANANA VALLEY

The Moose Creek Dam foundation consists of sands and gravels that vary in thickness from 0 to more than 600 feet. The right abutment is founded on bedrock and the remainder of the dam is founded on very pervious soil, which thickens in the central portion of the dam. Open work gravel and gap-graded granular materials are present. Open work gravels are relatively rare, while gap-graded gravels are common. The grain-size distribution of the granular materials varies greatly in short distances, both horizontally and vertically. The foundation can generally be considered as a sandy gravel/gravelly sand. Materials classified using the Unified Soil Classification System as GM, GP, SP, and SM are commonly present.

The granular deposits are overlain by a layer of nonplastic silty materials. The silt is present as silt and sandy silt (ML), and is often associated with silty sand (SM). The surficial silt deposits range from less than 1 foot to over 20 feet in thickness. Silt is also present in many sloughs formed by abandoned channels of the Chena River and its tributaries. Several of these sloughs pass through the Moose Creek Dam alignment.

Non-continuous areas of permafrost are primarily present under the central portion of the dam alignment which is a valley low point and former boggy area. Permafrost areas act as ground water flow blocks and may locally increase ground water gradients and divert ground water from the expected flow path.

The water table is shallow in the granular deposits, generally at 6 to 8 feet below the ground surface, and is normally in an unconfined condition when the "dry" dam is not loaded. As the pool level rises, the granular materials become semi-confined by the surficial silt layer. During flood control operation, ground water levels downstream of Moose Creek Dam respond quickly to changes in impounding pool and rise to near the ground surface as the pool develops. Figure 2 shows a conceptual model of the valley geologic conditions with the red line indicating the location of the dam axis.

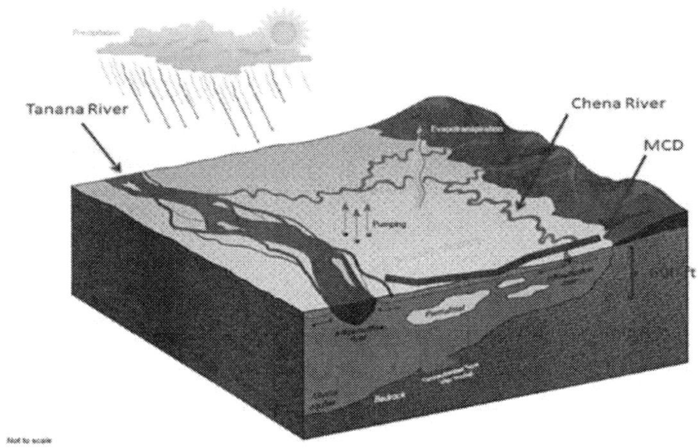

Figure 2. Conceptual Model of Valley Geologic Conditions at Moose Creek Dam

SEISMIC SETTING

Alaska is the most tectonically active region in the United States and experiences more than half of all earthquakes recorded in North America each year (Alaska Earthquake Information Center, 2013). The present tectonic framework of Alaska is dominated by subduction of the Pacific plate underneath the North American plate at an angle of about 45 degrees, with a rate of closure of about 3 inches per year.

Subduction of the Pacific plate generates earthquakes. The epicentral depth of these earthquakes traces the depth of the interface between the two plates as the Pacific plate subducts until it warms enough to lose strength and become plastic. North of the clearly demarcated subduction zone, stresses due to the plate convergence are transmitted into interior Alaska where the deformation causes substantial crustal seismicity. Fairbanks and Moose Creek Dam are located between two generally east-west trending fault zones: the northern fault zone formed by the Kaltag, Tanana, Preacher, Steven's Creek, Medicine Lake and Tintina Faults, and the southern fault zone dominated by the Denali fault.

EMBANKMENT DESIGN AND CONSTRUCTION

The main embankment of Moose Creek Dam was designed by the Alaska District project team and Harry Cedergren[3]. The design of the dam included features based on anticipated high amounts of foundation seepage. These consist of: a 5 foot thick upstream seepage blanket constructed of silt that was later augmented to widths from 600 to 1,000 feet, a toe drain system, a downstream stability berm and seepage collection channels. Extensive relief wells were later installed based on test and record pool performances.

The dam embankment above ground level was constructed in two main zones and is founded below the surficial silty soils on the underlying sands and gravels. The upstream zone is semi-pervious silty gravel that is protected with rock spalls. The remainder of the embankment is free-draining gravel with a downstream select gravel drain. Figure 3 shows a typical cross section of the embankment. Water that seeps under or through the dam is collected in an extensive collection system that parallels the downstream face of the dam and drains into the Chena River below the dam. Several granular types of material were used to construct the embankment and backfill the sloughs.

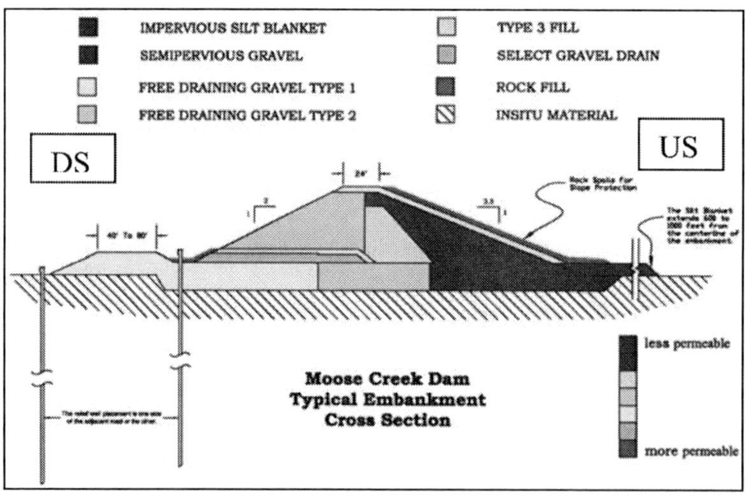

Figure 3. Typical Embankment Cross Section

[3] Harry R. Cedergren (1911-96) received his BSCE from the University of Washington in 1938 and MS in soil mechanics from Harvard in 1939, under Professor Arthur Casagrande. He worked for the Army Corps of Engineers, the California Division of Highways, and the California Department of Water Resources for about 30 years before becoming an independent consultant in 1964. His textbook on *Seepage, Drainage, and Flow Nets* is the definitive reference on seepage and drainage through soils for engineering purposes.

RISK ASSESSMENT BACKGROUND

To evaluate project risk, the team performed a site visit, prepared a site characterization report (SCR), reviewed prior risk work by others, performed a supplemental potential failure modes analysis (PFMA), developed a risk assessment strategy, held elicitation sessions in which they discussed each failure mode type in detail, and completed an existing condition risk assessment. As described in the next section, the dam was divided into north, central, and south zones based on geology, dam performance, and consequence centers to facilitate the risk assessment. Existing risks were estimated for each zone along with total project risk. Potential risk reduction measures were developed based on the findings.

In general, discontinuous permafrost within the region is known to be present under the central zone of the dam where many of the distress points occurred during the 1981 test pool. The USACE Cold Regions Research and Engineering Laboratory (CRREL) in Fairbanks, Alaska was engaged to help quantify the risks associated with seepage flow around permafrost zones. Because the dam is founded on varying granular deposits with old sloughs under the axis of the dam, "conceptual" failure modes to encompass the full spectrum of potential failure pathways were developed to bound the risk assessment.

Since construction, reasonable and prudent actions have been taken by the Alaska District to both remediate known deficiencies and increase the monitoring of the dam. Because the relief well installations and blanket extension occurred over several years, and Moose Creek Dam has no permanent pool, the ability to predict good performance with confidence is moderate. As such, continued vigilant monitoring and testing of the Emergency Actions Plan is essential and certain improvements to seepage monitoring were recommended during the course of the assessment.

The dam can be subjected to significant and rapid pool loadings. Past experience has shown that seeps and boils can develop (1981 test pool) downstream of the dam toe and in the South Seepage Collection Channel (SSCC). Since 1981, remedial measures were constructed consisting of adding 158 relief wells, augmenting the continuous constructed upstream impermeable blanket, flattening slopes and constructing filters on the upstream side of the South Seepage Collection Channel (SSCC) in discrete locations, and adding filters to the downstream locations where some sloughs passed under the dam alignment.

ZONE DELINIATION LOGIC FOR RISK ANALYSIS

The risk assessment team found that dividing the embankment into three distinct zones was necessary for estimating risk because of differences in geology, permafrost locations, project performance indicators, and downstream consequences (Figure 4).

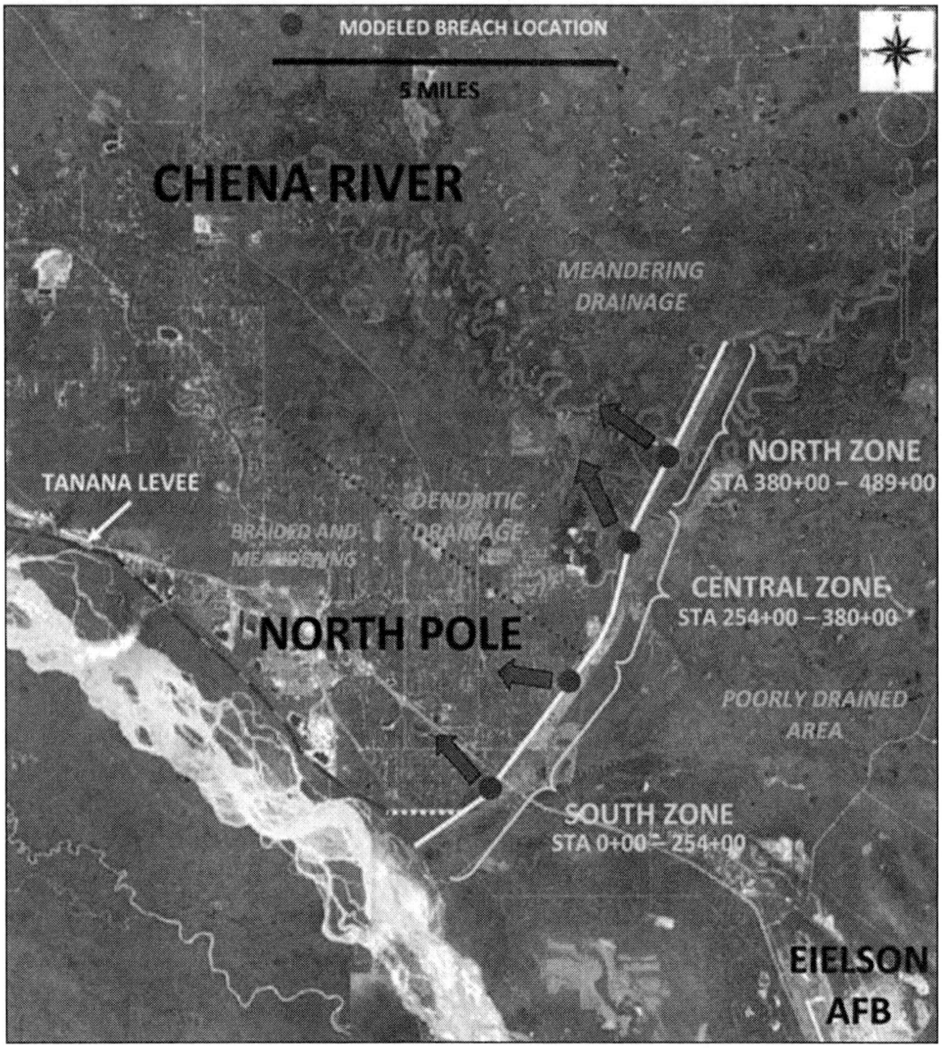

Figure 4. Plan View of Risk Zones and Modeled Breach Locations.

<u>South Zone (55+00 – 254+00)</u>

1. Geology: Primarily Tanana River braided stream, potential for isolated stream meander deposits.
2. Permafrost: No significant permafrost
3. Construction: Phase II Contract – Limited undercut (12 ft) to foundation gravels.
4. Performance: Only minor distress noted in zone.
5. Consequences: Short arrival time to the community of North Pole.

<u>Central Zone (254+00 - 380+00)</u>

1. Geology: Poorly Drained Area, transition area of Chena/Tanana Rivers.
2. Permafrost: Significant permafrost in area affecting seepage.
3. Construction: Phase I Contract – Complete undercut to foundation gravels.
4. Performance: Significant distress noted in zone.
5. Consequences: Potential short arrival time to the community of North Pole.

North Zone (380+00 – 489+00)

1. Geology: Chena River meandering area.
2. Permafrost: No significant permafrost
3. Construction: Phase I "Berg" Contract – Complete undercut to foundation gravels.
4. Performance: Only minor distress noted in zone.
5. Consequences: Long (8+ hours) flood wave arrival time to Fairbanks, North Pole not significantly inundated.

CONCEPTUAL FAILURE MODES

It was determined that regardless of the many test borings (2,000+), foundation gradations tests (2,400+) and 14 cross sections (roughly 2 per mile of dam); significant uncertainty remains in areas where borings are either sparse or shallow. Some sampling bias is also suspected for the borehole gradation data; however, based on trench and construction data and site inspections, maximum particle sizes are probably in the range of 3 to 4 inches throughout the site for the upper 30 feet of foundation material.

Evaluating only discrete locations along the dam alignment based solely on previous distress locations given the relatively low pool events to date may not fully capture the project risk, in fact, one reach of the dam has never been loaded because it is located on relatively high ground. Therefore, "conceptual" cases were brainstormed and developed for the project to attempt to bound the risk assessment while accounting for the data uncertainty in the complex outwash foundation soils.

The final risk driving conceptual cases developed for any point along the dam were:

- **Case 1** – Backward Erosion and Piping (BEP) of a continuous fine sand (SP) or silty sand (SM) layer in the foundation with vertical exit at the toe of downstream stability berm.

    - **Case 1a** – BEP of a continuous fine sand (SP) or silty sand (SM) layer in the foundation with vertical exit at the toe of downstream stability berm due to constricted flow resulting in excessive uplift pressures (previous heave distress points observed) induced by permafrost.

- **Case 2** – BEP of a continuous fine sand (SP) or silty sand (SM) layer in the foundation with horizontal exit in South Seepage Collection Channel (SSCC).

- **Case 3** – Scour of a continuous fine sand (SP) or silty sand (SM) layer through open work gravels in the foundation with horizontal exit in SSCC.

SELECTED FAILURE PATHWAYS

Each of these conceptual failure modes was evaluated at selected locations in order for sensitivity analyses to be comparable, allowing the team to apply subsurface data within the area in order to make informed judgments on the project risk. Since the foundation (and embankment) soils are primarily sand and gravel, a basic set of assumptions were developed for these internal erosion cases.

The team reviewed all available site data and concluded that the most likely failure modes were generally related to old sloughs (buried meandering stream channels), the low point drain, and permafrost "windows." Five potential failure mode locations were selected for analysis, three related to the presence of materials capable of holding a roof, one where higher than normal ground water gradients may be present, and one based on a distress point that developed during the 1981 test pool. The locations and logic are presented below.

South Zone (55+00 – 254+00)

The Chena Slough formerly connected the Tanana River and the Chena River, allowing water from the Tanana to flow into the Chena. In the early 1940s the Chena Slough was cut off near the Tanana River by the construction of Moose Creek Dike, and again in the 1970s by the Moose Creek Dam (Figure 5). Construction of the dam foundation across Chena Slough (dam stations 147+00 to 150+50) required the dewatering of a 130' x 300' area. Excavation and foundation backfill were accomplished without any problems after the foundation area had been dewatered (USACE, 1983).

Figure 5. 1971 Air Photo of Chena Slough Prior to Construction of Moose Creek Dam.

Three failure modes (Case 1, 2, and 3; no permafrost present) were evaluated at the Chena Slough: backward erosion and piping with horizontal and vertical exits, and scour.

Central Zone (254+00 - 380+00)

The low point concrete drain structure was also considered a potential roof and was evaluated by the team for the possibility of internal erosion under the floor slab. (This failure mode was later ruled out as a risk driver due to good construction practices and records).

Figure 6. Low Point Drain Construction and Downstream (DS) Dewatering.

Additionally, most of the Central Zone is dominated by discontinuous permafrost areas, which are separated by 'windows' (the term window is used to denote areas between permafrost masses). The team determined that these windows would likely experience higher than normal ground water gradients as ground water flow is funneled between the frozen areas. The window at STA 278+00 and its cross section were determined to be representative for this zone and were therefore used in the analysis. Ground water conditions and predictions at STA 294+71 were also considered because an upstream to downstream array of piezometers is also present at that location.

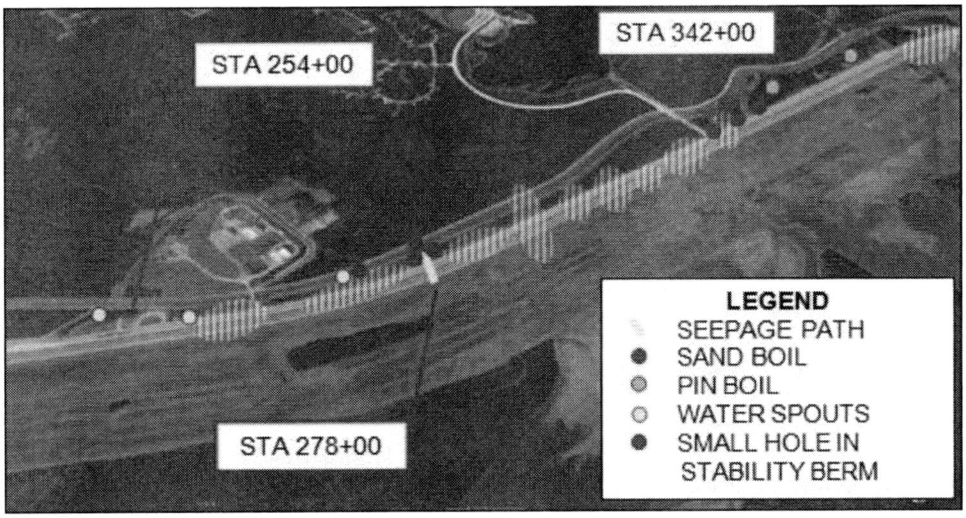

Figure 7. Permafrost Windows from STA 254+00 to 342+00. Permafrost areas are shaded blue, and are conceptual.

A sand boil developed downstream of the stability berm at STA 357+00 during the 1981 Test Pool. Because this was the first significant loading on the dam, the boil was surrounded by a large temporary dike around the distress point.

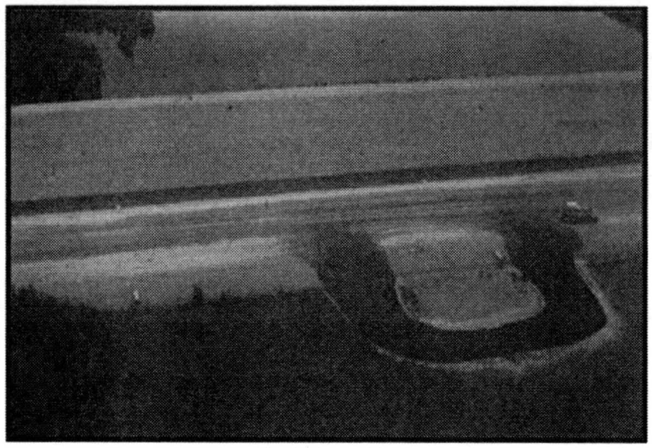

Figure 8. Dike and Boil at STA 357+50, July 16, 1981

This incident appears to be the most severe distress noted up to that point at the Moose Creek Dam, and triggered the installation of about 130 additional relief wells. The team determined it significant enough to evaluate the risk at this location.

North Zone (380+00 – 489+00)

The North Zone dominate failure mode was related to the remnant presence of the old Chena River that was routed around the control structure during construction. The Chena River was routed through the North Meander and the Control Structure was constructed in an excavated channel between the North and Central Meanders. The Central and South Meanders were then excavated and backfilled. The North Meander was backfilled after flow was routed through the Control Structure and its new approach/retreat channel.

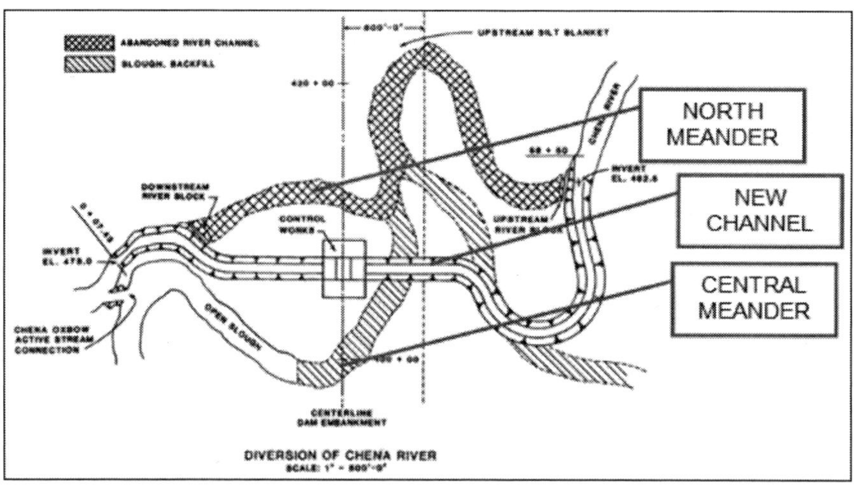

Figure 9. Backfill of North and Central Chena River Meanders, and Location of New Chena Channel at Control Structure. From USACE (1983).

The South Seepage Collection Channel (SSCC) runs parallel to the dam alignment until it reaches STA 370+00, where it turns sharply to the west and then to the northwest, until it drains into the Chena River about two miles downstream of the Control Structure. The drainage created by this design results in seepage from the South and Central meanders flowing into the old Chena Meander Channels downstream of the dam, rather than into the SSCC. This unique situation creates a different condition at the pre-construction Chena Channel area than exists at other failure mode locations.

Because drainage from the pre-construction Chena Channel meanders is not routed through the SSCC, the tailwater conditions are different at this location. During high pool events, seepage into the SSCC raises the tailwater in that channel, which partially offsets the greater head exerted by the upstream pool. At the pre-construction Chena Channel, the nearly direct connection with the Chena River effectively short-circuits the SSCC and the tailwater head is controlled by the Chena River stage. During high pool events, the Control Structure is operated to limit discharge to the Chena River to 1,000 cfs and a lower head is present on the Chena River than is present in the SSCC. This creates a larger upstream/downstream head differential at the pre-construction Chena Channel than at areas drained by the SSCC. The greater head differential in turn leads to a higher probability of initiation and progression of internal erosion.

EVENT TREE STRUCTURE

A generic event tree for internal erosion foundation backward erosion and piping failure modes is presented below. For convenience, the team placed the EXIT node immediately after the FLAW node to evaluate pool independent geologic and project design features prior to applying hydrologic load conditions. The steps highlighted in **blue** represent nodes that are pool dependent.

- ↳ FLAW - A Continuous Flaw Exists (Fine Sand Layer C_u<7)
  - ↳ EXIT - An Inadequate Filtered Exit is Present and Linked to FLAW (Continuation)
    - ↳ INTIATION – Erosion Initiates (Detachment) at Exit (BEP)
      - ↳ PROGRESSION 1 – Avg. Gradients Sufficient to Transport Materials
        - ↳ PROGRESSION 2 – Ability for Failure Pathway to Support Roof
          - ↳ PROGRESSION 3 - Upstream Zone Fails to Limit Flows
            - ↳ INTERVENTION - Unsuccessful Detection/ Intervention
              - ↳ BREACH - Uncontrolled Release of the Reservoir

The team followed Best Practices risk methodology developed by the USACE for dam risk analysis and failure probability estimation. Because of the length of Moose Creek Dam, normal pool conditions (dry), past performance, and varying consequences along the length of the dam, the risk analysis was similar to a levee risk evaluation.

In general, the team reviewed, developed, and refined project data in an iterative fashion to develop sufficient data to perform reasonable elicitations. The eight-mile length of dam required data synthesis on a relatively large scale. Additionally, past performance and iterative remediation activities (installing relief wells) required evaluations in a time-step sequence for the team to fully understand the present risks at the site.

Estimates of dam failure risks required a quantification of the probability of the loading and the probability of failure of the dam and appurtenant structures given the loading, and the adverse consequences (loss of life, property damage or other consequences) should failure occur.

CONSEQUENCE EVALUATIONS

For this analysis, impacts of 36 hydraulic events are estimated under several different human response conditions for a total of 108 simulations. Many of these simulations were aggregated to obtain 20 individual life loss estimates for a particular breach. Hydraulic events representing likely conditions for four separate breach locations, along with a non-breach scenario, along with varying breach size and development times were simulated. Best Case, Worst Case, and Most Likely simulations were made for each event by varying life loss assumptions. The risk analysis utilized the results of the "most likely" scenario.

Within the study area, up to 16,000 residential, commercial, industrial and public structures are estimated at risk of being inundated by a Maximum High Pool (MHP) event. The affected areas include the largely residential zones directly downstream of the dam in and around North Pole, Alaska; Fort Wainwright; and residential and commercial developments within Fairbanks, Alaska. The potential inundated area contains up to approximately 53,000 population at risk (PAR) during the day and 60,000 at night.

Figure 10. MHP Flood Depths Following a Breach near the Control Structure.

DAM SAFETY CASE - SELECTED POTENTIAL FAILURE MODE

A quantitative assessment of the project risk was generated for all the conceptual potential failure modes at the selected failure path locations. The total project risk was aggregated using the worst/highest risk conceptual failure mode for each zone. These failure modes were initially compared to USACE public policy guidelines. The results were presented to both a Quality Control and Consistency review panel as well as to USACE's Dam Safety Senior Oversight Group. An overview of the Dam Safety Case for one selected potential failure mode is presented below.

Potential Backward Erosion and Piping with Vertical Exit - Key Evidence

- Although the foundations soil contains lenses of highly erodible fine sand (C_u<7), the heterogeneous mix makes the ability to have a continuous erodible layer upstream to downstream of the dam (200-500 ft) unlikely.

- During construction, significant steps were taken to remove, test, and replace all highly erodible materials within the dam embankment footprint making the likelihood of a continuous erodible layer even less likely.

Figure 11. Removal of Foundation Silts below Embankment.

- Point to point piezometer data and seepage modeling indicate very low exit gradients are present and are not likely to initiate backward erosion piping at the critical exit points except for two cases and risks were tolerable.

- Blowout was evaluated and found to not to be a risk driver from the perspective of potential loss of foundation shear strength and subsequent effect on embankment stability.

- The global (average) hydraulic gradients (0.01 – 0.04) are much less than the threshold gradient (0.16) required to transport particles of an erodible layer. Sensitivity analysis of the worst case gradient included using steady state seepage conditions at the maximum high pool (MHP) load and assuming the constructed upstream silt blanket did not exist.

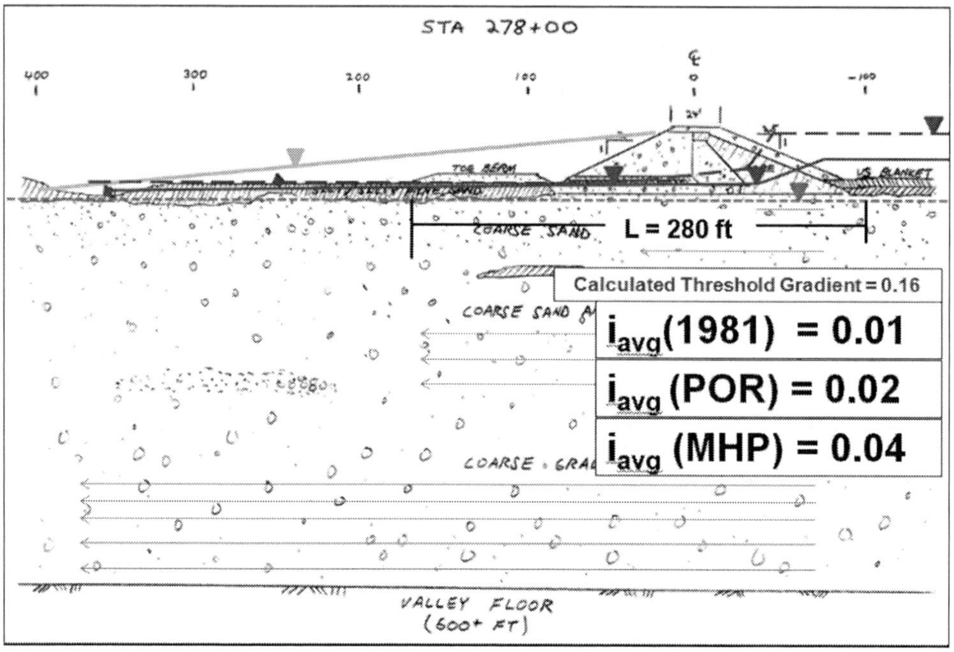

Figure 12. Average Hydraulic Gradient Calculations Compared to Threshold Gradient.

- Most past boils were numerous but small, showed no signs of progression, and were related to very thin silt cover and/or defects in the natural silt blanket near the edge of the downstream toe stability berm (except in the permafrost area, which was subsequently modified with the addition of relief wells). Subsequent upstream blanket extension and installation of 158 relief wells greatly reduced boil activity at the 1992 pool of record (512.6 ft NAVD88) than observed during the 1981 test pool (506.5 ft NAVD88).

Figure 13. Sand Boil Cones at STA 317+00, July 1981.

- More than 50% of the embankment footprint (cross section view) was built with clean granular materials (<4% fines) making it very unlikely to support a roof along a failure pathway.

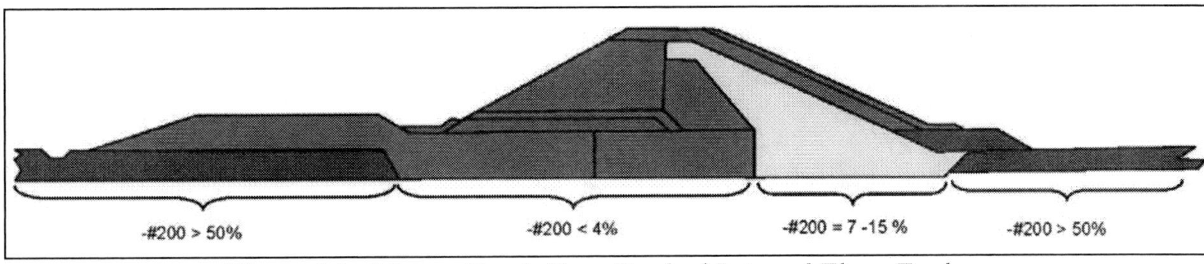

Figure 14. Moose Creek Dam Typical Internal Fines Zoning.

- Moose Creek Dam is a "dry dam." The infrequent pool events typically have relatively short durations and would inhibit the ability to progress a failure pathway except for very infrequent events.

CONCLUSIONS

The team performed an initial assessment of the existing project risk and continues to refine the risk estimate based on a recent flood event (June 2014) and future population growth trends. The team faced numerous uncertainties in evaluating the project risk that were evaluated by zoning the dam alignment based on logical parameters, selecting "worst" failure pathways, and employing conceptual failure modes to bound the risk for decision makers. The use of conceptual failure modes was successful in conveying potential risk for the project. Further evaluations continue including resolving length effects uncertainty and development of potential risk reduction measures once evaluations are complete.

REFERENCES

US Army Corps of Engineers (USACE), 1981, After Action Report, Chena River Lake Project Test Filling, Alaska, Alaska District, USACE

US Army Corps of Engineers (USACE), 1987, Letter Report Number 21, Moose Creek Dam, Permafrost Effects on Seepage, Chena River Lake Project Test Filling, Alaska, Alaska District, USACE

US Army Corps of Engineers (USACE), 2009, Interim Risk Reduction Measures Plan, Moose Creek Dam, Chena River Lake Project, Alaska, Alaska District, USACE

US Army Corps of Engineers (USACE), 2013a, Site Characterization Report, Moose Creek Dam, Chena River Lake Project, Alaska, Alaska District, USACE

US Army Corps of Engineers (USACE), 2013b, Existing condition Risk Assessment Report, Moose Creek Dam, Chena River Lake Project, Alaska, Alaska District, USACE

EVALUATING ERODIBILITY OF GRAVELLY FINE-GRAINED SOILS USING THE SUBMERGED JET EROSION TEST

Tony L. Wahl, P.E.[1]

ABSTRACT

The submerged jet erosion test is widely used to evaluate erodibility of fine-grained soils in dams, levees, stream banks and channels. Soils that contain significant quantities of gravel or coarse sand are problematic because the small ¼-inch-diameter jet is unable to transport coarse material out of the developing scour hole, so the hole becomes effectively armored during the test. To address this problem, laboratory tests were performed to determine if erodibility of gravelly fine-grained soils can be accurately determined from jet tests performed on specimens remolded from a finer control fraction of the original soil. Jet tests were performed on gravelly specimens (the parent soil) using a vertical orientation of the soil surface that prevents armoring; these results were compared to tests with a standard horizontal orientation performed on specimens prepared from the soil fractions passing the No. 4 and No. 40 sieves. Oversize corrections were applied to adjust the compaction water content and target dry density of the specimens to produce similar compaction states for the finer-grained portions of each soil. Comparison of erosion test results validated the hypothesis. Although there was scatter among individual results, averages of multiple tests of the parent and finer soils were in reasonable agreement. Results were inconclusive on whether it is preferable to remove coarse particles with the No. 4 or No. 40 sieve; screening at either size threshold appeared to yield reasonable results. For practicality, screening at the No. 4 sieve would be adequate and easiest to accomplish in most cases.

INTRODUCTION AND PURPOSE

Determining the erodibility of fine-grained soils enables modeling of erosion processes affecting embankment dams, levees, canal embankments, natural streams, and earthen spillway channels. The extreme variability of erosion thresholds and rates for typical embankment soils is well known (e.g., Wan and Fell 2004; Briaud 2008), both across soil types and for individual soils due to changes in compaction conditions (Hanson and Hunt 2007). This variability makes it imperative to measure erodibility characteristics whenever possible so that accurate modeling can be performed. When erodibility is poorly understood, unanticipated failures can take place, or the speed of erosion processes may be grossly underestimated.

A favored method for measuring soil erodibility in recent years has been the submerged jet erosion test (JET) developed by the Agricultural Research Service (ARS) at their Hydraulic Engineering Research Unit (HERU) in Stillwater, Oklahoma (Hanson and Cook 2004). The test is described in ASTM standard D5852, *Standard Test Method for Erodibility Determination of Soil in the Field or in the Laboratory by the Jet Index*

[1] Hydraulic Engineer, Bureau of Reclamation Hydraulics Laboratory, Denver, CO, twahl@usbr.gov .

Method. The Bureau of Reclamation Hydraulics Laboratory in Denver, Colorado has used a device of this type for several years (Wahl et al. 2008).

The submerged jet test simulates scour of a soil surface due to a perpendicular impinging jet. The jet is positioned above the soil surface of interest, and the depth of scour caused by the jet is recorded over time. The jet is typically produced from a ¼-inch-diameter nozzle operating under a pressure of about 1 to 8 ft of head. When the test is applied to predominately fine-grained soils that include a significant fraction of coarse sand or gravel, the strength of the jet is often too low to either detach the coarse particles or transport them out of the scour hole. This can lead to armoring of the scour hole by coarse particles. The threshold amount of coarse particles needed to create this problem is not well defined. One potential solution to the problem is to run the test with the entire apparatus tipped up on its side so that the jet is in a nearly horizontal orientation. This allows the coarse particles to fall out of the scour hole immediately after they are detached, but in both lab and field environments it creates significant logistical challenges for performing the tests.

This report presents research performed to evaluate an alternative approach to determining erodibility of fine-grained soils containing significant amounts of gravel or coarse sand (referred to hereafter as gravelly fine-grained soils). The hypothesis for this study is that the erodibility of a gravelly fine-grained soil will be primarily controlled by the erodibility of a finer-grained control fraction (i.e., the fraction of the soil that passes a pre-determined sieve size), so soil specimens could potentially be screened to eliminate the coarse fraction and jet tests could then be carried out on specimens with the modified gradation, compacted in an adjusted manner to reproduce the original compaction state of the finer-grained fraction of the parent soil. This research study tested the validity of this hypothesis and attempted to define the best methods for screening to eliminate the coarse material that impedes testing of the parent soil.

BACKGROUND

Figure 1 shows a schematic diagram of the submerged jet erosion test and an accompanying photo of the laboratory apparatus constructed by the Bureau of Reclamation from plans provided by Greg Hanson, USDA-ARS (retired). ASTM standard D5852 describes the test, although data analysis methods have evolved since publication of the standard. Data from the tests described in this report were analyzed using the methods described in Hanson and Cook (2004). Recently, other analysis procedures have also been proposed (e.g., Daley et al. 2013).

The test is typically run with a constant jet pressure that can be adjusted prior to each test. The jet is positioned over the soil surface of interest, and the initial elevation of the jet and the jet pressure are selected to apply a desired shear stress to the soil specimen. The depth of scour beneath the jet is measured over time and the recorded data are used to estimate the critical shear stress needed to initiate erosion and the detachment rate coefficient relating the rate of erosion to the applied stress in excess of the critical value. The analysis is based on a volumetric form of the excess stress erosion model:

$$\dot{\varepsilon} = k_d(\tau - \tau_c) \tag{1}$$

where $\dot{\varepsilon}$ is the volume of material removed per unit surface area per unit time ($ft^3/hr/ft^2$, or ft/hr), k_d is a detachment rate coefficient ($ft/hr/lb/ft^2$), τ is the applied stress (lb/ft^2), and τ_c is the critical shear stress (lb/ft^2).

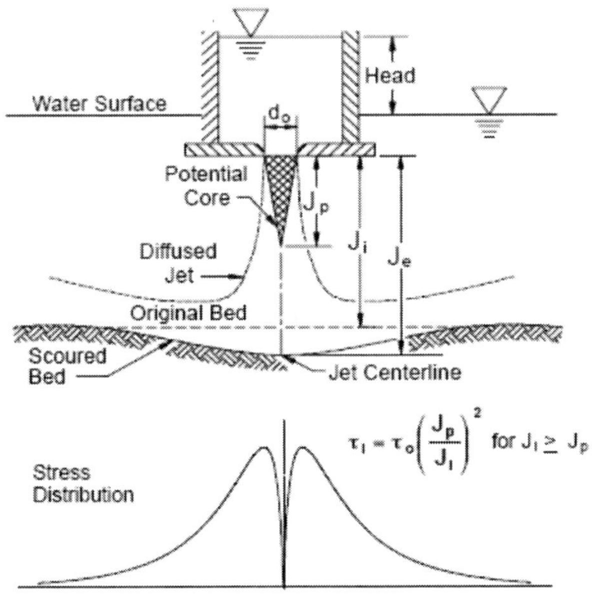

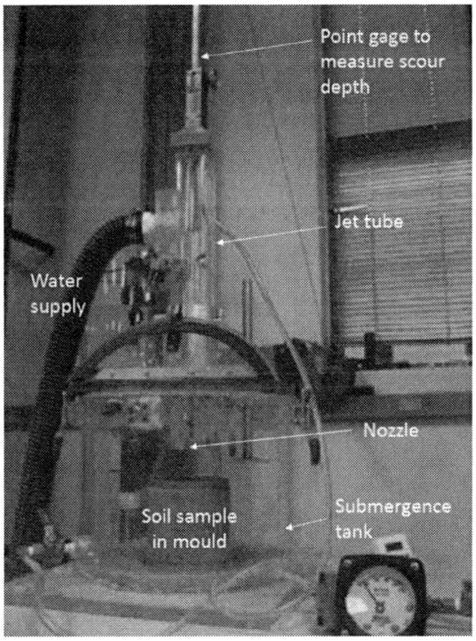

Figure 1. Jet test schematic diagram and photo of laboratory test apparatus.

The excess stress erosion model presumes that the relation between erosion rate and excess stress is linear, but this may not always be the case. For this reason, it is always desirable to perform tests at an applied shear stress level that is comparable to the stress that is expected to be applied to the soil in the situation of interest. This can be accomplished by adjusting the jet pressure and the initial distance between the nozzle and the soil surface.

THEORY

The hypothesis for this study is that the erodibility of a gravelly fine-grained soil will be controlled by the erodibility of the finer portion of the soil, which will be described as a control fraction, primarily made up of sand, silt and clay particles. To determine the erodibility of a gravelly soil, it is proposed to separate the specimen into finer and coarser fractions and to then perform submerged erosion jet testing on compacted specimens of the finer control fraction. Compaction of the finer-grained material should be adjusted so that the compaction state of the specimen is similar to the compaction state of the finer-grained material contained in the original gravelly soil specimen.

Submerged Jet Erosion Test

The ideal break point for separating the coarse and fine soil fractions is not obvious. Two potential options are the No. 4 and No. 40 U.S. Standard sieve sizes (4.76 mm and 0.42 mm, respectively). The No. 4 sieve removes all gravel-sized particles and would yield a sample containing only sand-size particles, silt, and clay. Samples passing the No. 4 sieve are typically used for compaction testing (ASTM D698, ASTM D1557). The No. 40 sieve removes all gravel, coarse and medium sand, yielding a sample that contains only fine sand, silt and clay. Samples finer than the No. 40 sieve are used for determination of the Atterberg limits that indicate soil plasticity properties (ASTM D4318). Previous work (Hanson et al. 2010) has related soil erodibility parameters to clay content, so it could be expected that erodibility parameters might correlate well with plasticity characteristics. Thus, a sample based on the minus No. 40 fraction might provide the most useful indication of erodibility characteristics. A practical consideration is that separating with the No. 40 sieve cannot readily be accomplished with moist soil and requires significantly greater effort to break down the soil specimen prior to sieving, particularly for the large specimen size required for the JET test.

SOIL AND TEST SPECIMEN PREPARATION

A test soil was created for the research by combining two soils already present in the laboratory and removing the coarse gravel fraction (larger than ¾" sieve). This soil was assigned the designation 36B-X92. Using the Unified Soil Classification System (USCS), the soil type was SC – Clayey Sand. The makeup of the soil was 47 percent fines, 39 percent sand, and 14 percent fine gravel. Atterberg limits were also determined, and the soil was plastic with liquid limit LL=39, plastic limit PL=17, and plasticity index PI=22. Specific gravities for the minus No. 40 and minus No. 4 fractions were determined, as well as the bulk specific gravity of the No. 4 to ¾" fraction in a saturated surface dry (SSD) condition. The absorption ratio for the gravel was determined to be 2.9% and the absorption ratio of the medium and coarse sand particles (+ No. 40 material) was assumed to be the same. Detailed test results are available in the appendix of Wahl (2014).

A standard Proctor compaction test (ASTM D698, Method C) was performed on soil 36B-X92, which will hereafter be described as the *parent soil.* Optimum water content was determined to be 12.2%, and the maximum dry unit weight was 115.6 lb/ft^3. A submerged jet erosion test was performed on each of the five compaction test specimens (in 6-inch diameter moulds), with the test apparatus oriented so that the jet was approximately horizontal and the soil surface was approximately vertical (Figure 2). Test results are shown in Table 1. These tests are somewhat more difficult to conduct than tests in the more typical orientation (vertical jet) because the sample must be securely fastened into the submergence tank, drainage from the test is difficult to confine, and the deflector plate that blocks the jet while scour depths are being measured was not designed to operate well in this orientation. For relatively erosion resistant soils the pattern of scour caused by the jet is similar in this orientation to the vertical jet, but for very erodible soils, accelerated erosion of the top side of the scour hole can occur because gravitational forces contribute to instability of the soil above the impinging jet. This can lead to an asymmetric scour hole shape that may contribute to accelerated erosion at the jet centerline where scour depth measurements are made (Figure 3).

Table 1. Submerged jet erosion test results from compaction test specimens.

Compaction state		**Erodibility**	
Dry density **lb/ft^3**	**Water content** **%**	k_d **$ft/hr/lb/ft^2$**	τ_c **lb/ft^2**
109.4	7.2	30.1	0.000953
113.4	8.9	7.38	0.00269
115.5	10.4	0.295	0.0594
114.5	12.9	0.17	0.1043
113.8	15.3	0.069	0.604

Figure 2. Submerged jet test performed with the jet in a horizontal orientation.

Figure 3. Asymmetric development of the scour hole in a test of a weak soil with horizontal jet orientation.

After testing of the compaction specimens was complete, the remaining soil was then split into three portions to be used for erosion tests of the parent soil, the –No. 4 fraction, and the –No. 40 fraction. Three test cylinders for each soil fraction were hand-compacted with the objective of achieving compaction equivalent to 95% of standard Proctor maximum density. For the cylinders prepared from the finer soil fractions, water contents and target densities were adjusted so that compaction of the finer-grained soils would be equivalent to the computed dry densities of the matching finer-grained fraction when the parent soil was compacted to 95% of maximum density.

The adjusted water contents and dry densities were computed using equations given in ASTM D4718, *Standard Practice for Correction of Unit Weight and Water Content for Soils Containing Oversize Particles*. The adjustments are based on three assumptions:

- the coarse-grained particles (oversize material) being excluded (either +No. 4 or +No. 40 material) are floating in a matrix of the finer particles, with all voids between the coarse particles filled by the finer soil, or stated in another way, there is no interference between the coarse-grained particles that affects the compacted unit weight of the finer fraction;

- the coarse-grained particles are internally saturated (saturated, surface dry = SSD) during compaction; and

- the coarse sand and medium sand particles in soil 36B-X92 have the same absorption ratio as the fine gravel particles.

ASTM D4718 states:

> "This practice is based on tests performed on soils and soil-rock mixtures in which the portion considered oversize is that fraction of the material retained on the No. 4 sieve. Based on these tests, this practice is applicable to soils and soil-rock mixtures in which up to 40% of the material is retained on the No. 4 sieve. The practice also is considered valid when the oversize fraction is that portion retained on some other sieve, such as the 3/4-in. sieve, but the limiting percentage of oversize particles for which the correction is valid may be lower. However, the practice is considered valid for materials having up to 30% oversize particles when the oversize fraction is that portion retained on the 3/4-in. sieve."

Since the limiting percentage of oversize particles is believed to decrease with an increase in the size of the oversize material, it can be inferred that when the oversize particle division point is finer than the No. 4 sieve, the maximum amount of oversize particles should increase to something greater than 40%. In the case of soil 36B-X92, the percentage of particles retained on the No. 4 sieve is only 14 percent and the percentage retained on the No. 40 sieve is 33 percent. Thus, the practice described in ASTM D4718 should be applicable when we choose either the No. 4 or No. 40 sieve to define oversize particles.

The equations used to compute the water content and dry unit weights are:

$$w_F = \frac{(100w - w_C P_C)}{P_F} \tag{2}$$

$$\gamma_{d,F} = \frac{\gamma_d G_C \gamma_w P_F}{100(G_C \gamma_w) - \gamma_d P_C} \tag{3}$$

where:

w_F = water content of control fraction, (F = finer fraction in this application)
w_C = water content of oversize (coarse) fraction, (C=coarse)
w = water content of total sample,
$\gamma_{d,F}$ = dry unit weight of control fraction,
γ_d = dry unit weight of total sample,
γ_w = unit weight of water,
G_C = bulk specific gravity of coarse fraction,
P_F = percent of control fraction by weight,
P_C = percent of coarse fraction by weight.

The equations above can be derived from relations that compute the water content and dry unit volume (volume per unit weight) of the total sample as the weighted average of the associated quantities for the fine and coarse fractions. To apply the equations to this soil, the water content of the coarse fraction was assumed to be the absorption ratio (2.9%), and the bulk specific gravity was calculated as a weighted average of the specific gravities determined for the gravel and sand fractions represented in the oversize component. Table 2 shows the computed target water contents and dry densities for each series of erosion test specimens. The objective was to conduct each series of tests with the finer-grained control fraction soil compacted to the same state that would occur if the parent soil were compacted to 95% of standard Proctor maximum density. The table also shows the actual water content and dry densities achieved for each specimen. Densities were within about 1% of the target values and water contents were within 0.3%.

Table 2. Hand-compacted soil specimens for jet erosion testing.

Specimen Description	Target compaction state		Actual compaction and comparison to target value	
	Dry density, lb/ft^3	Water content, %	Dry density, lb/ft^3	Water content, %
Parent soil	109.8 (95% of 115.6)	12.2	109.7 (-0.1)	12.0 (-0.2)
			109.7 (-0.1)	11.9 (-0.3)
			110.4 (+0.6)	12.1 (-0.1)
–No. 4	104.3	13.7	104.7 (+0.4)	13.4 (-0.3)
			104.7 (+0.4)	
			104.7 (+0.4)	
–No. 40	94.3	16.8	95.3 (+1.0)	16.7 (-0.1)
			95.3 (+1.0)	
			95.3 (+1.0)	

EROSION TESTING AND ANALYSIS

Submerged jet erosion tests were conducted on the three hand-compacted specimens of the parent soil (in 6-inch moulds) with the soil surface nearly vertical, permitting gravel to fall out of the scour hole as erosion occurred. The –No. 4 and –No. 40 specimens in 4-inch moulds were tested in the traditional orientation, with the soil surface horizontal and the jet impinging vertically down on the soil surface. Two of the –No. 40 specimens were tested twice, once on the top surface of the specimen and a second time with the specimen inverted. Table 3 shows the jet erosion test results.

Table 3. Submerged jet erosion test results.

Specimen Description	Compaction state		Erodibility	
	Dry density lb/ft^3	Water content %	k_d ft/hr/lb/ft^2	τ_c lb/ft^2
Parent soil, hand-compacted to 95%	109.7	12.0	0.088	0.299
	109.7	11.9	2.485	0.072
	110.4	12.1	1.152	0.010
Parent soil from compaction test (approx. 99-100% density)	115.5	10.4	0.295	0.0594
	114.5	12.9	0.17	0.1043
–No. 4,	104.7	13.4	0.202	0.110

Specimen Description	Compaction state		Erodibility	
	Dry density lb/ft^3	Water content %	k_d ft/hr/lb/ft^2	τ_c lb/ft^2
hand-compacted	104.7		0.246	0.058
	104.7		0.253	0.125
–No. 40, hand-compacted	95.3	16.7	0.827	0.012
	95.3		0.027	0.340
			0.136	0.070
	95.3		0.262	0.291
			0.005	0.298

Figure 4 shows the results plotted in relation to erodibility categories proposed by Hanson and Simon (2001). These categories span the range of erodibilities observed in a study of natural cohesive streambed deposits in loess areas of eastern Nebraska, western Iowa, and northern Mississippi. They also represent typical ranges of erodibility measured in compacted soils used in civil engineering infrastructure, such as dams and levees (Wahl et al. 2008; Hanson et al. 2010). Five erodibility classes are recognized: very resistant, resistant, moderately resistant, erodible, and very erodible. Hanson and Simon (2001) also proposed a relation between the detachment rate coefficient and the critical shear stress, and this relation is shown on the figure. (Subsequent researchers have proposed other relationships in recent years).

Figure 4 shows that there is large variability of the erosion test results, between and even within the test series. Results for the –No. 4 tests are tightly clustered, but the –No. 40 tests exhibit detachment rate coefficients that vary more than ±1 order of magnitude and the tests on the parent soil also vary by about ±0.75 orders of magnitude. Considered together, the nine tests on hand-compacted specimens exhibit detachment rate coefficients that vary by ±1.5 orders of magnitude. Erodibility rates are expected to exhibit order-of-magnitude variation, so the range observed here is noticeably large, but not extreme. The range of critical shear stress values is about ±0.75 orders of magnitude. This is somewhat unusual, as the variability of the critical shear stress is often greater than the variability of the detachment rate coefficient.

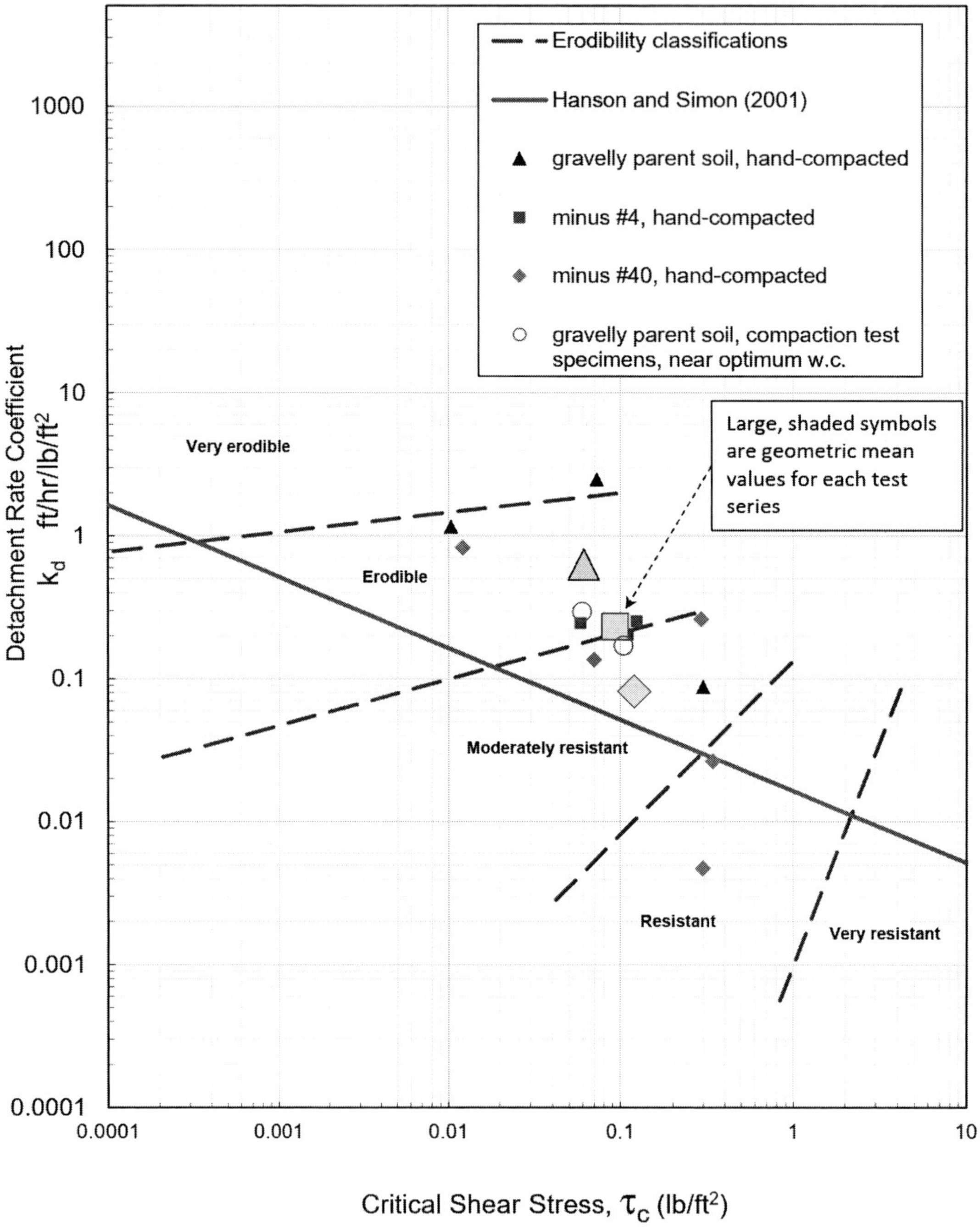

Figure 4. Erodibility test results.

To reduce the impact of variability on interpretation of the results, averages of the data from each test series were computed, and these are also shown in Figure 4. Results for two of the compaction test specimens bracketing the optimum water content value are also included. These 5 data points cluster tightly together. The –No. 4 and –No. 40 soils exhibit slightly more erosion resistance on average than the parent soil; this is consistent with the slight overcompaction of those test specimens that was shown in Table 2.

Unintentionally, there was significant delay between the time of specimen compaction and erosion testing for some specimens, and even though the specimens were sealed to prevent moisture loss, this may have had an impact on the results. Figure 5 shows the detachment rate coefficients plotted versus the number of days of sample curing that took place prior to testing. There is a suggestion of a trend toward greater erosion resistance as curing time increases, although there is also greater variability in the results from those specimens that cured for a long period of time.

CONCLUSIONS

This research study tested the hypothesis that the erodibility of a gravelly fine-grained soil could be determined by evaluating the erodibility of a finer-grained control fraction. Specimens of the parent material and two different finer-grained derivative soils were evaluated with submerged jet erosion tests, and the results appear to validate the hypothesis. Although there was significant scatter among individual test results, averages of detachment rate coefficients and critical shear stresses from multiple tests of the parent and derivative soils plotted within ±0.5 orders of magnitude of a central value. The testing also sought to determine whether the gravelly sample should be screened at the No. 4 or No. 40 sieve, but the results were inconclusive on this question. It appears that screening at either size threshold would yield reasonable results. For practicality, screening at the No. 4 sieve would be adequate and easiest to accomplish in most cases. Considering the inherent variability of soil erodibility and the fact that all important parameters could not be fully controlled in this study, further testing is recommended to confirm the findings of this study. Future research could evaluate additional soils, including soils that push the limits for applicability of the water content and density correction formulas given in ASTM D4718. Future testing could also evaluate the effects of curing time and control the curing time in order to limit its potential effects on erosion test results.

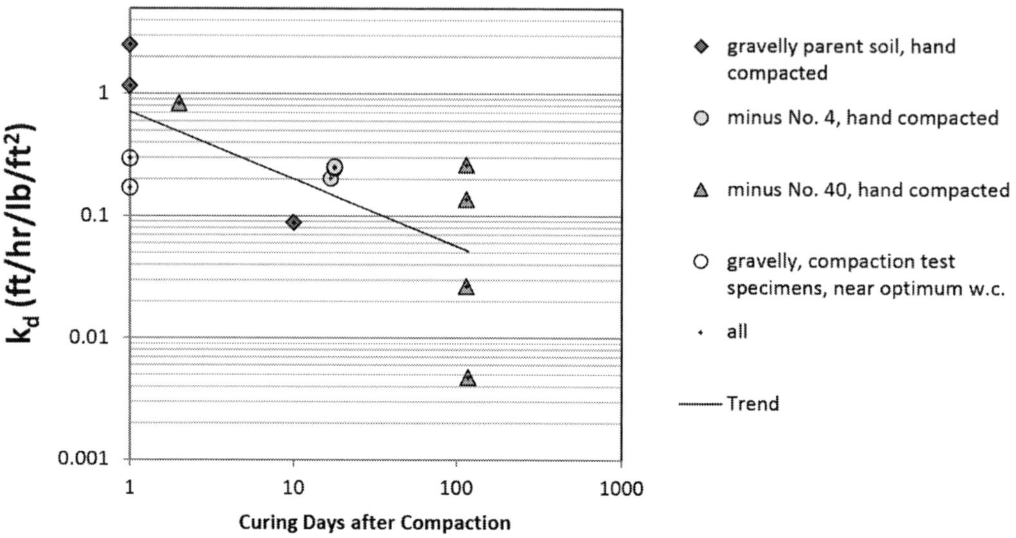

Figure 5. Jet erodibility test results versus specimen curing time after compaction.

A typical application scenario for this research is a situation in which one needs to evaluate the erodibility of a gravelly soil encountered in the field. To accomplish this, the in situ dry density of the soil and water content would be measured in the field, perhaps using a sand cone test or nuclear density gauge. If the soil was mechanically compacted, an estimate of the water content at the time of compaction could be made, if it is believed to differ from the in situ water content. A sample of the soil would then be obtained for laboratory testing. The soil sample would be sieved to remove +No.4 particles and the remaining soil would be used to create erosion test specimens. The absorption ratio of the gravel particles and the specific gravity of the coarse and fine fractions would be determined through laboratory testing. The adjusted dry density and water content for the –No. 4 soil would be calculated using the equations from ASTM D4718, and specimens would then be hand-compacted at the appropriate water content to achieve the desired dry density. Submerged jet erosion tests could then be conducted on these specimens in the typical vertical jet orientation. Tests should be conducted at an applied shear stress level that is comparable to the stress that is expected to be applied to the soil in the situation of interest.

ACKNOWLEDGEMENTS

Dr. Robert Rinehart and Bethany Jackson in Reclamation's Materials Engineering Research Laboratory created the soil used for this research, performed soil properties analysis and compaction testing, computed oversize corrections, and prepared all soil erosion test specimens.

REFERENCES

ASTM Standard D5852, 2007. Standard test method for erodibility determination of soil in the field or in the laboratory by the jet index method. American Society for Testing and Materials.

ASTM Standard D4718, 2007. Standard practice for correction of unit weight and water content for soils containing oversize particles. American Society for Testing and Materials.

ASTM Standard D698, 2007. Standard test methods for laboratory compaction characteristics of soil using standard effort. American Society for Testing and Materials.

ASTM Standard D1557, 2009. Standard test methods for laboratory compaction characteristics of soil using modified effort. American Society for Testing and Materials.

ASTM Standard D4318, 2010. 10 standard test methods for liquid limit, plastic limit, and plasticity index of soils. American Society for Testing and Materials.
Bureau of Reclamation, *Earth Manual*, Part 2, 3rd Edition, Denver, CO, 1990.

Briaud, J.-L., 2008. Case histories in soil and rock erosion: Woodrow Wilson Bridge, Brazos River meander, Normandy cliffs, and New Orleans levees. 9th Ralph B. Peck Lecture. *Journal of Geotechnical and Geoenvironmental Engineering*, 134(10):1425-1447.

Daley, E.R., G.A. Fox, A.T. Al-Madhhachi, and R.B. Miller, 2013. A scour depth approach for deriving erodibility parameters from jet erosion tests. *Transactions of the ASABE*, 56(6): 1343-1351 ISSN 2151-0032 DOI 10.13031/trans.56.10350

Hanson, G.J., and A. Simon, 2001. Erodibility of cohesive streambeds in the loess area of the midwestern USA. *Hydrological Processes*, Vol. 15, pp. 23-38.

Hanson, G.J., and K.R. Cook, 2004. Apparatus, test procedures, and analytical methods to measure soil erodibility in situ. *Applied Engineering in Agriculture*, 20(4):455-462.

Hanson, G.J., and S.L. Hunt, 2007. Lessons learned using laboratory jet method to measure soil erodibility of compacted soils. *Applied Engineering in Agriculture*, 23(3):305-312.

Hanson, G.J., T.L. Wahl, D.M. Temple, S.L. Hunt, and R.D. Tejral, 2010. Erodibility characteristics of embankment materials. In: *Dam Safety 2010*. Proceedings of the Association of State Dam Safety Officials Annual Conference, September 19-23, 2010, Seattle, WA. (CDROM).

Wahl, T.L., P.L. Regazzoni, and Z. Erdogan, 2008. *Determining Erosion Indices of Cohesive Soils with the Hole Erosion Test and Jet Erosion Test*, Dam Safety Technology Development Report DSO-08-05, U.S. Dept. of the Interior, Bureau of Reclamation, Denver, Colorado, 45 pp.

Wahl, T.L., 2014. *Measuring erodibility of gravelly fine-grained soils.* Hydraulic Laboratory Report HL-2014-05, U.S. Dept. of the Interior, Bureau of Reclamation, Denver, Colorado, September 2014.

Wan, C.F., and R. Fell. 2004. Investigation of rate of erosion of soils in embankment dams. *Journal of Geotechnical and Geoenvironmental Engineering*, 130(4):373-380.

USING COMBINED FIELD GEOLOGIC MAPPING AND PHOTOGRAMMETRY TO DOCUMENT FOUNDATION STRUCTURAL GEOLOGY AND WEATHERING, FOLSOM AUXILIARY SPILLWAY JOINT FEDERAL PROJECT, FOLSOM, CA

Tatia R. Taylor, M.S., Ph.D.[1]
F. James Martin[2]
Joshua E. Domme[3]

ABSTRACT

A combination of traditional field geologic mapping and geospatially referenced photogrammetric images are being used to create a precise 3-dimensional geologic map of the Folsom Auxiliary Spillway Joint Federal Project (JFP) foundation. Geologic documentation of project foundations for permanent historical record is a key objective during construction of civil works projects by the US Army Corps of Engineers (USACE). The Folsom JFP is a USACE megaproject consisting of an approach channel, control structure, spillway chute and stilling basin, the foundation of which is being geologically mapped by USACE geologists with access facilitated by the project prime contractor, Kiewit Infrastructure West (Kiewit). A methodology for geologic mapping has been developed wherein the excavated and cleaned foundation is photographed using photogrammetric techniques and subsequent geologic field mapping is completed using the photogrammetry images as a map base. Photogrammetry images are processed by the geologists using proprietary software which combines them into geospatially oriented 3-dimensional projects. On completion of field mapping, the geologists then synthesize the field collected data with the photogrammetric projects, resulting in geospatially oriented 3-dimensional geologic maps. Photogrammetry images are captured both on and above ground, with elevation for invert photographs accomplished using a professional aerial platform model S800 EVO, manufactured by DJI, and operated by Kiewit. The foundation rock at Folsom JFP consists of pervasively jointed and locally sheared granodiorite of the Rocklin pluton that is preferentially weathered and altered at depth. Geologic mapping consists primarily of structural geology and weathering grades, with details including mineralogy, joint fillings, and rind thicknesses. The project permanent record will include both the 3-dimensional data set and traditional 2-dimensional geologic maps and cross sections. The combined field mapping and photogrammetry approach, enhanced by the aerial platform access method, is resulting in significant improvements in field flexibility and safety, efficiency, and reduced expense.

---

[1]U.S. Army Corps of Engineers, Sacramento District, 1325 J Street, Sacramento CA, 95814
Ph. 916 402 0495, tatia.r.taylor@usace.army.mil
[2]U.S. Army Corps of Engineers, Sacramento District, 1325 J Street, Sacramento CA, 95814
Ph. 916 557 6679, f.james.martin@usace.army.mil
[3]Kiewit Infrastructure West Co., 2200 Columbia House Blvd., Vancouver, WA, 98661 Ph. 360 693 1478, josh.domme@kiewit.com

INTRODUCTION

The Folsom Auxiliary Spillway Joint Federal Project (JFP) is a US Army Corps of Engineers (USACE) civil works megaproject in Folsom, CA, consisting of an approach channel, control structure, spillway chute and stilling basin. The underlying project foundation is being geologically mapped by USACE geologists with field access facilitated by the project's Phase IV prime contractor, Kiewit Infrastructure West (Kiewit). Project authorizations and phase overview are detailed in APPENDIX A. Geologic documentation of project foundations for permanent historical record is a key objective during construction of civil works projects by USACE in accordance with USACE Engineer Manual 1110-1-2908: *Engineering and Design, Rock Foundations*, and USACE Engineer Manual 1110-1-1804: *Engineering and Design, Geotechnical Investigations*. Elements of required permanent documentation of the foundation include detailed geologic conditions, collection of unobstructed foundation photographs, and a survey-controlled topographic record of the foundation. This data is crucial during excavation and construction, for example, to determine field-fit slope stabilization methods and foundation preparation and treatment prior to concrete placement, and serves as the first reference for any future post-construction issues, such as seepage or slope stability.

A geologic mapping approach combining traditional field mapping and photogrammetry was developed and adapted for the project site and conditions, and has realized a number of advantages over other methods. Flexibility of the approach, speed of processing, and innovative field site access support for relatively low startup cost has expanded the USACE geologists' capabilities and efficiency in support of Kiewit's Phase IV schedule and construction goals.

Site Geology

The JFP auxiliary spillway project area is underlain by the early Cretaceous Rocklin pluton quartz diorite (locally referred to as granodiorite). The Rocklin pluton is intrusive to the adjacent late Jurassic Penryn pluton, and both occur as part of a limited population of small Mesozoic plutons found within the larger Foothills Metamorphic Belt (FMB; Swanson, 1978). The FMB includes metavolcanic and plutonic rocks and is believed to be a suture zone associated with accretionary tectonics of the Late Jurassic Nevadan orogeny (Schweickert and Cowan, 1975; Snow and Scherer, 2006). In the project area, the intrusive rocks were initially overlain by deposits of the Mehrten Formation, which was removed early in the Phase I and Phase II excavations (see APPENDIX A for phase details). Foundation rock consists of pervasively jointed and locally sheared granodiorite with a complex preferential weathering profile that extends at depth throughout the rock mass. Observational evidence suggests that the primary joint sets were likely original cooling/unroofing features within the pluton, which were subsequently reactivated by local geologic events on multiple occasions. Cross-cutting relations, secondary mineralization, and regional fault history indicate a complex deformational sequence.

METHODS

A combination of traditional field geologic mapping and geospatially referenced photogrammetric images are being used to create a precise 3-dimensional geologic map of the JFP foundation. A methodology for geologic mapping was developed wherein the excavated and cleaned foundation is photographed using photogrammetric techniques and corresponding geologic field mapping is completed using the photogrammetry images as a map base.

Geologic Mapping

Documentation of the structural geology and degree of weathering of the foundation are the key observations forming the basis of the geologic hand mapping of the JFP foundation. These characteristics of the rock informed the rock bolting and slope stabilization approaches both during the design stage of the project, and during construction which required extensive field-fit slope stabilization methods and spot bolting of excavated rock slopes. Pervasive joints and shear zones are the primary structural features occurring in the granodiorite (Figure 1). Table 1 describes the four primary joint sets found within the project area, which bound scaled, variably oriented removable blocks. Where shearing occurs, it is localized on and trends along preexisting joints, primarily on the D- and B-joint sets. Interaction between joint-bounded rock blocks and final design slope orientation was critical in understanding and addressing slope stability challenges such as overhangs and block egress, and resulted in irregular final cutslopes characterized by exposed joint faces and block casts (Figure 2).

Figure 1. Geologists performing field mapping of wall and invert surfaces. Foundation characterized by pervasive jointing and shearing. Stilling Basin, JFP.

Table 1

Main Joint Set Average Orientations and General Orientation Ranges Based on Corps' Compilation of Plotted Field Measurements by Reclamation (2010), Left Cutslope (Stas. 10+00 to 27+00) and Right Cutslope (Stas. 10+00 to 31+00), Phase II Excavation, Upper Chute

Joint Set	Average Orientation			General Range		Number of Measurements
	Strike (degrees)	Dip Direction (degrees)	Dip (degrees)	Dip Direction (degrees)	Dip (degrees)	
D	223	313	33NW	280-340	15-65	798
B	082	172	77SE	140-200	60-90	992
A	336	066	73NE	035-090	45-90	677
C	221	311	79NW	295-325	70-90	98

Figure 2. Irregular final cutslopes characterized by exposed joint faces and block casts. Stilling Basin, JFP.

The granodiorite has a well developed but variable weathering profile that grades downward with decreasing weathering at depth. The weathering degree was divided into five zones during core logging and early surface mapping by USACE, based on Engineer Manual 1110-1-2908. From more weathered to less weathered, the zones are as follows: Decomposed, highly weathered, moderately weathered, slightly weathered and unweathered. The depth of weathering depends on the spacing and permeability of joints, fractures, and shear zones. Below the top of highly weathered bedrock, the degree of weathering and the overall fracture densities generally decrease rapidly with depth. Individual discontinuity-controlled weathering becomes increasingly prevalent. Narrow zones of decomposed to highly weathered rock extend down along discontinuities well below the general base of the highly weathered rock and presents a 3-dimensional framework of heterogeneous, preferentially weathered rock throughout the excavation (Figure 3). Although rock quality generally increases with depth, zones of highly weathered and decomposed granodiorite have contributed substantially to blasting, excavation, slope stabilization, and foundation grouting issues during all phases of the project.

Figure 3. Preferential weathering along joints. Stilling Basin, JFP.

Field geologic hand mapping of these features is performed by USACE geologists (Figure 1) using Brunton compasses on map bases which are created by capturing, preliminarily processing, and printing out photogrammetric images of the map areas. The distance from the rock face or invert required to shoot photogrammetry pictures ensures a consistent scale of observation, and determines the number of overlapping photo sheets printed out for field mapping of an area. To increase the efficiency of the geologic mapping, geologists take measurements of joints and fractures and note them on the rock faces while waiting for the map bases to be pre-processed and printed. These pre-measurements are then transferred to the map bases when they are returned to the field area, and hand mapping continues. Other significant features included on the field maps are springs and seepage, presence of gouge and mineral infilling on joints, thickness of weathering rinds, and characterization of shear zones and sheared joints.

Photogrammetry – Rationale and Advantages

USACE geologists chose photogrammetry for documenting the Phase III and Phase IV JFP foundation for a number of reasons. In comparing photogrammetry to Lidar, which was used by the US Bureau of Reclamation (USBR) on earlier project phases, several advantages of photogrammetry over Lidar were noted based mainly on lessons learned by USBR. The initial startup costs were an order of magnitude smaller than the cost of using Lidar: equipment for photography, software program, licensing, and technical support for photogrammetry, computer equipment, and survey support were acquired for approximately $150,000, with recurrent contracts for software technical and survey support. Whereas using Lidar would have required an external contractor to perform the field work, processing, and final map production (exceeding $1 million), photogrammetry can be performed and processed in-house by USACE geologists. Other beneficial features of photogrammetry included the field flexibility and improved synthesis between the imagery and the geologic data. Because they are produced and processed in-house, photogrammetric images could be printed and returned to the field immediately for mapping, and there is no angular disparity between the hand mapping perspective and the photogrammetry data because mapping is done on the photogrammetry. A shortcoming of using Lidar was that the scanning, image capture, and processing were not timely enough to use the Lidar imagery as a base map, requiring separate photos to be taken which were not consistently from the same perspective or scale, and this introduced challenges during synthesis with geologic data due to angular disparities.

The use of photogrammetry also satisfies USACE Engineer Manual guidance for required elements necessary to have complete documentation of a civil works project foundation. These include detailed geologic conditions, collection of unobstructed foundation photographs, and a survey-controlled topographic record of the foundation. Photogrammetry synthesized with the geologic hand mapping accomplishes all of these directives without the need for separate contract actions or field activities. The purpose of permanent foundation records is for future reference, but some advantages of having the maps and photogrammetry are already being realized while construction is ongoing. Although the final report-grade AutoCAD drawings have not been completed, geologic maps and photogrammetry are available for examining 3D models and geology of the foundation beneath areas already covered by project structures, such as the control structure and spillway slabs. The real-time availability of this data allows analysis of a range of features such as joint and block spacing for understanding curtain grouting processes, projections for slope stability, and items as small as making oriented measurements in 3D space.

Limitations of using photogrammetry have been learned on the job and are primarily associated with site conditions. Lighting is key to taking high quality photos. On an active construction site, providing the amount and control of artificial lighting is prohibitive, thus photogrammetry must be performed during daylight hours. Weather also impacts photogrammetric images. Due to the high quality of the images, rain drops can be captured in the photos. This distorts the spatial distance to the map surface, resulting in unsuccessful processing of the photogrammetry. In adapting the use of photogrammetry to the JFP which is in California, the long days, long summers, and intermittent precipitation have not made these limitations prohibitive, and the advantages described previously easily outweigh the disadvantages.

Photogrammetry – Field Approach

Typical photogrammetry and field mapping projects range in size from 50 to 200 feet long by 5 to 10 feet high for walls, and 50 by 50 foot squares for inverts, and are performed by geologists at the completion of excavation, scaling, and preliminary cleanup. These activities are performed prior to drilling for rock reinforcement and any concrete placements.

Figure 4. Laying out targets, invert. Stilling Basin, JFP.

Survey target placement and map boundary painting for photogrammetry are the first steps in the process (Figure 4). Next, photogrammetry photos are taken and the targets are surveyed by the USACE survey contractor. The geologist takes photos of the area twice, usually having a second geologist take the backup photos if it is a wall area (Figure 5). For an invert area the geologist would take multiple photos at each camera position so that there are backup photos. The photos require between 60-75% overlap from one photo to the next and must be in a straight line. For inverts, the photos need to maintain approximately the same height above the surface in addition to proper overlap and being in a straight line. Geospatial survey data (station, offset, and elevation) for targets marking the project limits is collected in the field from the surveyor and the full survey data is sent electronically for subsequent photogrammetry processing. All survey target data is sent in UTM coordinates following the 1983 State Plane Coordinate System.

Figure 5. Photogrammetry photos being taken, upstream headwall, Stilling Basin, JFP.

Figure 6. 3DM CalibCam, random points generation (red) and survey control points (green).

Photogrammetry is performed using a Sony NEX-7 camera with a 20mm lens utilizing the aerial camera platform. Previously a Nikon D700 with 20mm lens and a remote shutter on a pole was used by a geologist in a man lift for inverts and walls (see Field Access Techniques below). Processing of photogrammetry images and data is performed by the geologists using proprietary software programs by AdamTechnology including 3DM CalibCam, DTM generator, and 3DMAnalyst. This process starts with 3DM CalibCam where all the images for an area are loaded using the calibration file for the camera and lens used to take the photos. By using the calibration file, 3DM CalibCam generates relative points on the photos and assigns those points numbers (Figure 6). These points are on features the program identifies from image to image, so if it identifies the same feature in adjacent photos it will assign the same point number to it. When the points are viewed on two photos at the same time, the yellow points are shared by both photos, the red points are not shared by those two photos, and the green points are the surveyed control points manually assigned by the geologist (Figure 7).

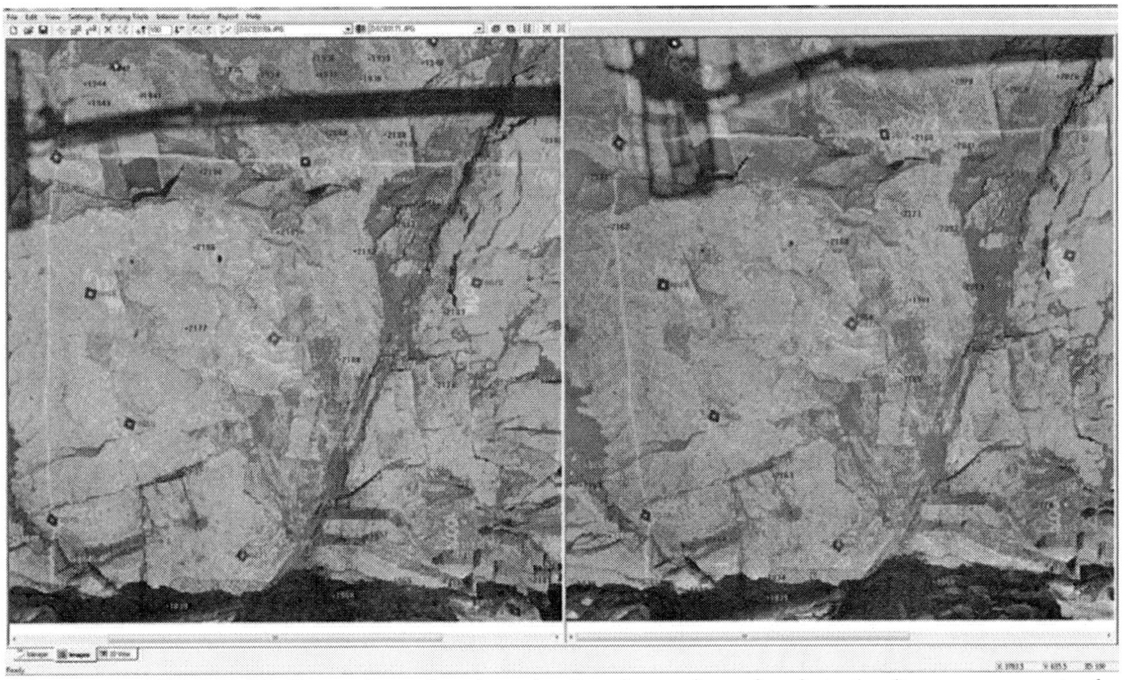

Figure 7. 3DM CalibCam, shared points (yellow), unshared points (red), survey control points (green).

The control points are not needed for the initial processing right after taking the photos in the field. The program can process the data/photos without control data to ensure the project worked, and it will be properlsy geospatially referenced once the control data file is loaded and the control points are added. Once the points are generated the geologist runs the exterior orientation process in the program, either as a free network if the control data has not been added or a control network if the control data has been added. The geologist will set the image accuracy X and Y, which are known values from when the camera and lens calibration file was created. Once the image accuracy is set, then the image resection process followed by the bundle adjust process is run (Figure 8, left). This produces the 3DM CalibCam Adjustment Results report (Figure 8, right). The report contains information on how well the generated points relate to each other 3-dimensionally and when those values are in the acceptable range the geologist moves on to the next step. If the values in the Adjustment Results report are not acceptable the geologist either adds or removes points until acceptable values are reached, or uses the backup set of photos for that area. Once acceptable values in the adjustment results are achieved, the geologist prints the photogrammetry images to be used for the geologic field mapping (Figure 9). The final step in 3DM CalibCam is generation of the digital terrain models (DTMs) after the control data has been added. The geologist picks the image pairs for the desired DTMs then has the program generate the DTMs (Figure 10). After the DTMs are generated the geologist uses 3DM Analyst to finalize the DTMs and synthesize the field geologic data with the photogrammetry model.

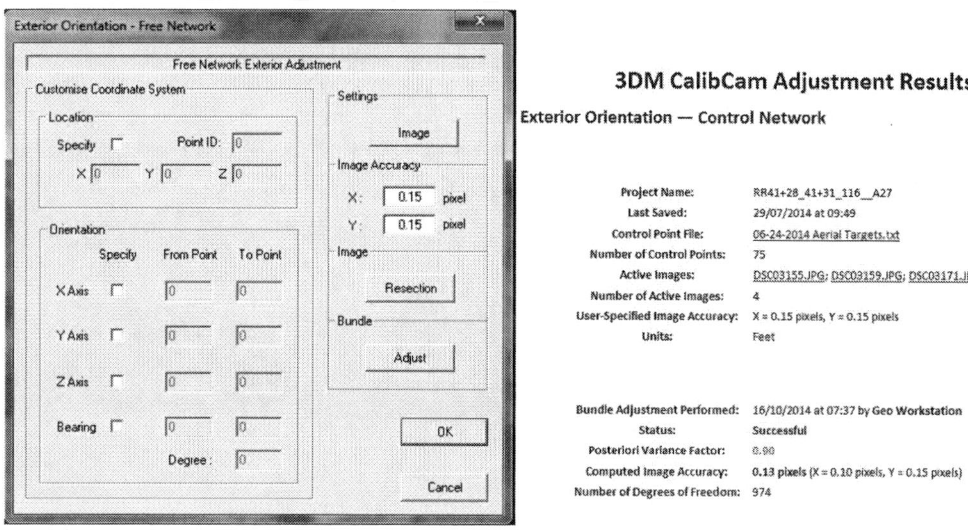

Figure 8. 3DM CalibCam image resection and bundle adjustment screen (left) and Adjustment Results report (right).

Figure 9. Geologic field mapping performed on photogrammetric image base. Stilling Basin, JFP.

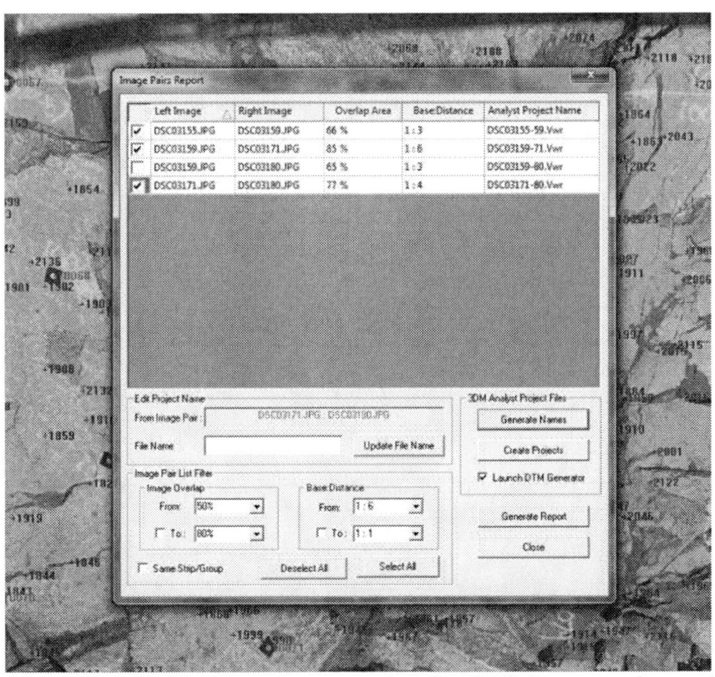

Figure 10. 3DM CalibCam, image pairs for DTM generation selection.

Synthesis

Using 3DM Analyst, the geologist loads the DTMs for an area (Figure 11), then crops them down to the area desired in order to remove excess overlap and to utilize the DTM parts with the most data. Once the DTMs are cropped (Figures 12 and 13) to the geologist's satisfaction, they are saved with a different file name so that all the finalized DTMs can be identified and located together. With all the finalized DTMs loaded in 3DM Analyst, the geologist takes the field maps and starts incorporating the field data into the photogrammetry. The feature types that can be added are user defined (Figure 14) and for these maps include elevation contour lines, joint planes, weathering contact lines, shear zone boundary lines, control points, seepage zone boundary lines and other feature boundary lines. The program will generate the elevation contour lines at the interval set by the geologist (Figure 15). Using the field maps the geologist digitizes the weathering contact lines, shear zone boundary lines, reference control points, and any other feature boundary lines. Once all features have been digitized from the field maps by the geologist they are saved as a 3DM Analyst feature file. The geologist also exports those features as .dxf files for use in creating the final record geologic map. The geologist has the program create a high resolution georeferenced .tif image file also used for the final record geologic map. Next the geologist digitizes all the field joint measurements (Figure 16). Once all the field joint measurements have been added, the program plots a stereonet containing all the joints, which the geologist can take an image of for inclusion on the final record geologic map. An Excel .csv file is exported containing all the joint planes generated, including X, Y, Z of plane center, dip, dip direction, and any other user defined information. The geologist saves this as a 3DM Analyst feature file. These saved 3DM Analyst feature files along with the finalized DTMs are the 3-dimensional geospatially referenced record of the foundation. The exported data and images are imported into 3D AutoCad for final annotation and formatting of the final record geologic map (Figure 17).

Figure 11. 3DM Analyst, uncropped DTM.

Figure 12. 3DM Analyst, cropped DTM.

Figure 13. 3DM Analyst, cropped DTM, 3D view.

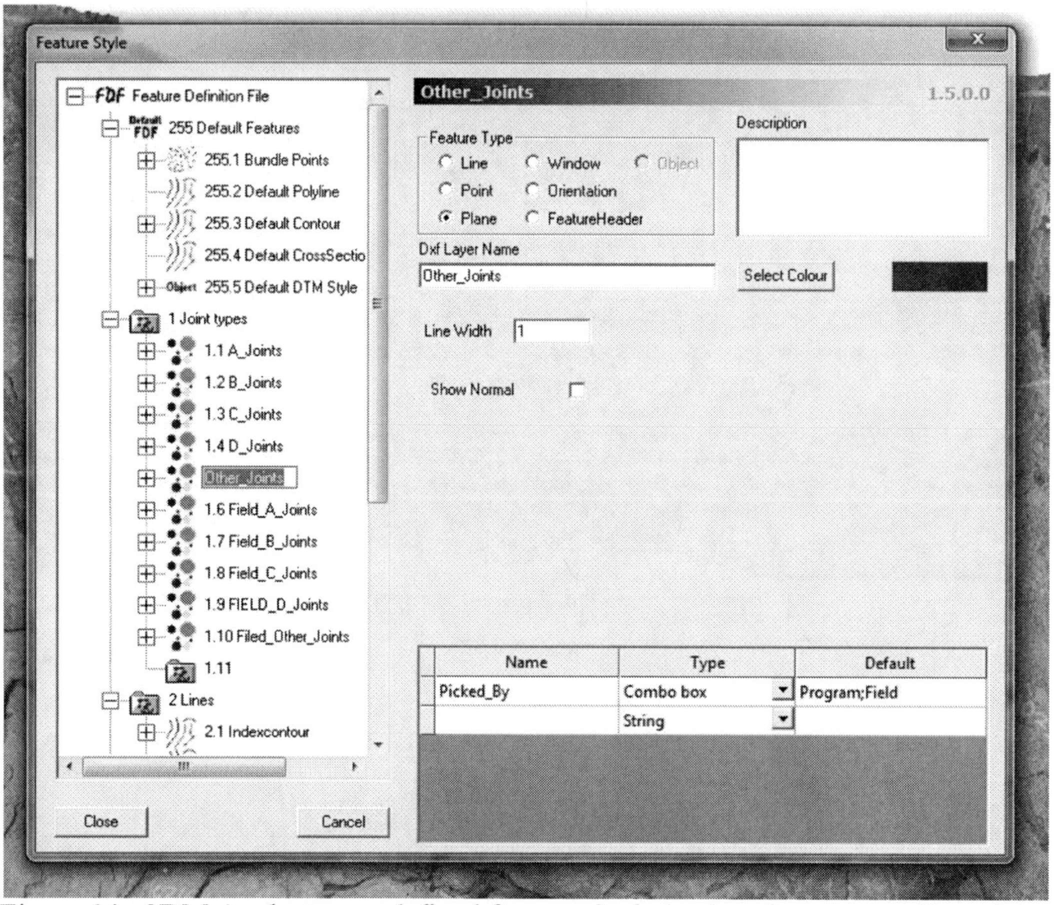

Figure 14. 3DM Analyst, user-defined features include elevation contour lines, joint planes, weathering contact lines, shear zone boundary lines, control points, and other features.

Figure 15. 3DM Analyst, adding features, 1-ft contour.

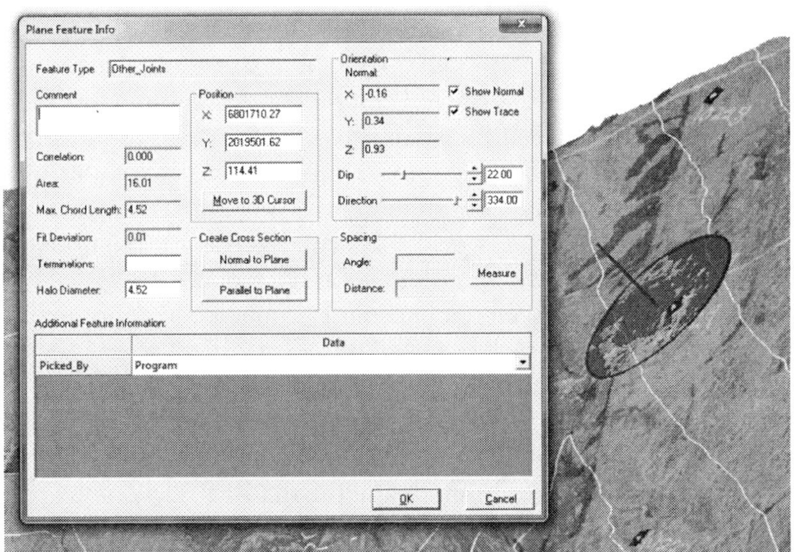

Figure 16. 3DM Analyst, digitizing joint from geologic mapping. User-defined features include dip and dip direction.

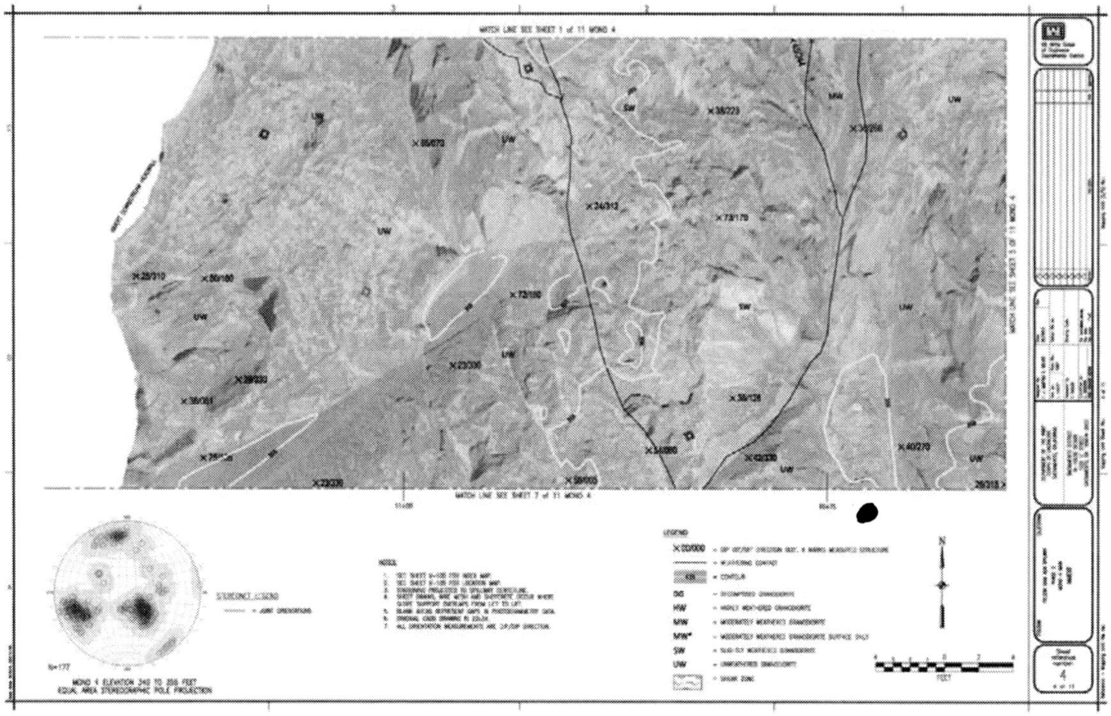

Figure 17. DRAFT finalized foundation map: 3DM Analyst feature files and finalized DTMs are imported into 3D AutoCad for final annotation and formatting of the final record geologic map.

Field Access Techniques

Photogrammetry images are captured both on and above ground, depending on whether a wall is being mapped, or an invert surface. Contract specifications for both Phase III and Phase IV of the JFP require the Contractor to provide the access to the slopes and inverts needed by USACE geologists to shoot photogrammetry images and perform geologic mapping. Wall images may be shot from the ground, using the proper distance to the face, horizontal spacing of the images, and associated vertical limitations. For walls that exceed the vertical limits of ground shooting, contractor man-lifts have been utilized to take a second row of overlapping photos in order to capture the entire map area. For invert surfaces, the photographer was elevated to the specified height above the invert, either in a man-lift or in a crane basket (Figures 18 and 19). This approach was time consuming, with the use of a crane basket being even more costly than use of a man-lift, and falling under critical pick crane lift rules. The invert areas to be mapped on the earlier Phase III of the project were limited, and the man-lift and crane approach was used throughout that phase.

Figure 18. Geologist shooting invert photos from a man basket. Left, camera setup. Right, taking photographs.

Figure 19. Geologist shooting invert photos from a crane basket.

A significant and innovative advance in the approach to field access for photogrammetry has been achieved by Kiewit on the Phase IV JFP work. Invert geologic mapping and photogrammetry on Phase IV is spatially extensive (approximately 649,000 square feet) and includes the inverts of the Approach Channel, the entire Spillway Chute and the Stilling Basin. The Approach Channel walls add approximately 48,000 square feet, for a total of approximately 697,000 square feet of geologic mapping. The lower chute is created with a spiral vertical curve generating a gradually steepening slope. The slope extends 600 linear feet and changes in elevation of 200 feet. The introduction of a ground-controlled aerial camera platform model S800 EVO, manufactured by DJI and provided by Kiewit (Figure 20), allows controlled elevation of the photogrammetry camera for capturing images, with USACE geologists designing the shots, Kiewit operators running the aerial camera platform, and together directing and capturing the photography. The S800 EVO is a remote controlled, battery-operated, multi-rotor propeller craft made of carbon fiber and designed for aerial photography. It is equipped with an independently operated camera gimbal that can move in any direction. The propeller craft is capable of a stationary hover and can also move in any direction. The S800 EVO is spatially located by GPS orientation, and operated by two separate remote controls. The first controls the propeller craft, while the second remote controls the on-board camera gimbal. Safety features include a control tether, a wireless ground-based viewing screen that sees what the aerial camera sees, and built in GPS safety features that return it to the point of take off if signal is lost or the battery is low. Map areas are photographed according to spacing, overlap, and elevation/distance determined by the geologists. Kiewit personnel operate the remote control of the S800 EVO, and USACE geologists operate the camera gimbal remote control while utilizing the viewing screen to precisely locate and orient each image (Figure 21).

Figure 20. Aerial camera platform model S800 EVO manufactured by DJI. Stilling Basin, JFP.

Figure 21. Tethered aerial camera platform S800 EVO. Kiewit personnel operate the S800 EVO remote control, and USACE geologists operate the camera gimbal remote control while utilizing the viewing screen to precisely locate and orient each image. Approach Channel, JFP.

Advantages of using the S800 EVO include reduced startup cost, increased efficiency, flexibility of field operations, improved personnel safety, and adaptability of the platform to site conditions. The initial investment for the S800 EVO was on the order of $10,000-$20,000 and when combined with the reduction in time to complete the capture of photogrammetry images, is a far preferable technique than previous approaches. Field flexibility for photogrammetry is further improved using the aerial camera platform because the photos can be taken much faster, thus allowing initial processing of the photogrammetry and production of the map base images to be completed and returned to the field for hand mapping much sooner. Photos taken from a man-lift or crane basket can take an hour or hours depending on conditions, while photos taken with the S800 EVO can be completed in minutes. Personnel safety is enhanced by ground based operation of the aerial camera platform, which reduces risk by putting the camera apparatus in the air rather than the camera operator. The aerial camera platform has also been adapted for elevation access in areas where flying the equipment is not feasible due to lack of direct GPS or other unfavorable conditions. The camera platform has been hung from a rope and pulley system stretched across the map area at the proper elevation and then conveyed from one side to the other to capture the images (Figures 22 and 23). In areas where GPS is obstructed or variable, it has been suspended from the basket of a man-lift (Figure 24) or from a crane (Figure 25). All of these adaptations produce the same high quality photogrammetry results as flying the camera platform.

Foundation Structural Geology

Figure 22. Securing S800 EVO to rope in preparation for shooting photogrammetry pictures. Stepped Chute, JFP.

Figure 23. Photogrammetry with S800 EVO suspended from rope over invert. Stepped Chute, JFP.

Figure 24. S800 EVO suspended from man basket over invert. Stilling Basin, JFP.

Figure 25. S800 EVO suspended from crane over invert. Approach Channel, JFP.

CONCLUSION

Geologic physical and spatial documentation of project foundations for permanent historical record is a key objective during construction of civil works projects by USACE. USACE geologists have developed and applied a methodology for geologic mapping wherein the excavated and cleaned foundation is documented by a combination of traditional field geologic mapping and geospatially referenced photogrammetric images. Mapping and photogrammetry are then synthesized to create a precise 3-dimensional geologic map of the Folsom JFP foundation. This approach has several advantages over other methods such as Lidar, including reduced startup investment, in-house utilization, increased field efficiency, flexibility, and comprehensive data which satisfies USACE technical requirements for foundation records, and is available immediately for use as project construction is ongoing. The limitations of photogrammetry on this project, including daylight constraints and weather impacts, have been out-distanced by these advantages. The innovative addition of an aerial camera platform to the photogrammetry technique has proven superior to previous approaches. Use of the aerial camera platform has further enhanced flexibility and efficiency of field operations, improved personnel safety, and has proven to be an adaptable tool to use in changing project and site conditions for a reasonable investment. On the Folsom JFP, this increased flexibility, speed of processing, innovative field site access support, and relatively low startup cost has expanded the USACE geologists' capabilities and efficiency in support of USACE civil works foundation documentation requirements and Kiewit's Phase IV schedule and construction goals.

ACKNOWLEDGEMENTS

The authors would like to thank the US Army Corps of Engineers, Sacramento District, and Kiewit Infrastructure West for supporting the JFP geologic mapping and photogrammetry program. The authors also would like to thank each other for the positive working relationship needed to develop and successfully continue this joint effort. We also wish to thank Carmen Williams and Mike Knarr for their helpful and insightful reviews of this manuscript.

REFERENCES

Schweickert, R. A.; Cowan, D. S., 1975: "Early Mesozoic tectonic evolution of the western Sierra Nevada, California" Geological Society of America Bulletin, v. 86, no. 10, p. 1329-1336.

Snow, C. A.; Scherer, H., 2006: "Terranes of the Western Sierra Nevada Foothills Metamorphic Belt, California; A Critical Review". International Geology Review, 48(1): pp. 46-61.

Swanson, S. E., 1978: "Petrology of the Rocklin Pluton and Associated Rocks, Western Sierra Nevada, California". Geological Society of America Bulletin, v. 89: pp. 679-686.

USACE Engineer Manual 1110-1-2908: Engineering and Design, Rock Foundations, November, 1994

USACE Engineer Manual 1110-1-1804: Engineering and Design, Geotechnical Investigations, January, 2001

APPENDIX A

Project Authorizations and Summary

Under Section 101(a) (6) of the Water Resources Development Act (WRDA) of 1999 (PL 106-53), the United States Congress authorized physical modifications at Folsom Dam for improved flood damage reduction measures. The modifications were to be consistent with the descriptions provided in the U.S. Army Corps of Engineers' (USACE) *Supplemental Information Report for the American River Watershed Project, California*, dated March 1996, as modified by the *Folsom Dam Modification Report, New Outlets Plan*, dated March 1998, prepared by the Sacramento Area Flood Control Agency (SAFCA).

In 2006, the Energy and Water Development Appropriations Act (PL 109-103) provided further direction regarding actions at Folsom Dam. Specifically, Section 128 directed the Secretary of the Army and the Secretary of the Interior to collaborate on authorized activities to maximize flood damage reduction improvements and address dam safety needs at Folsom Dam and Reservoir. The Act included a potential auxiliary spillway as a reasonable alternative under the previous authorizations.

The Folsom Auxiliary Spillway Joint Federal Project (JFP) is being constructed in lieu of two single-purpose projects meeting the objectives of both the U.S. Bureau of Reclamation (Reclamation) and the USACE (i.e., dam safety and flood damage risk reduction, respectively) and allowing both agencies to save costs. Both agencies are participating in the design and construction of the project features. The JFP is being developed in five phases, as follows;

- Phase I: Preliminary excavation for Spillway Chute designed and executed by Reclamation (completed in 2008)
- Phase II: Final excavation for Spillway Chute and Stilling Basin, designed and executed by Reclamation (completed in 2010)
- Phase III: Final excavation for and construction of Control Structure, designed jointly and executed by the USACE (awarded September 2010, scheduled completion March 2015)
- Phase IV: Final excavation for Approach Channel and construction of concrete features in Approach Channel, Spillway Chute, and Stilling Basin, designed and executed by the USACE (awarded May 2013, scheduled completion May 2017)
- Phase V: Final restoration of haul roads, site grading, and miscellaneous features, designed and executed by the USACE (scheduled completion October 2017)

Foundation Structural Geology

BUILDING A FLOODWALL IN MANHATTAN

Matthew Redington, PE[1]

ABSTRACT

Superstorm Sandy came onshore in late October of 2012, causing widespread destruction on the mid-Atlantic coast. With tropical storm-force winds spanning over 940 miles in diameter, the storm was the largest Atlantic storm on record. Effects of the storm were felt in seven countries and twenty-four states in the U.S.

A coastal surge, high tides, strong winds, and heavy rainfall combined to cause extensive flooding in low lying areas of New York City. The Veterans Affairs Medical Center, located in downtown Manhattan just two blocks from the East River, was evacuated prior to landfall of the storm, but was heavily damaged by flooding. Although no patients or staff members were injured as a result of the evacuation or the storm, the damaged facility remained out of service for months, making it especially challenging to provide healthcare services to veterans in the New York area.

This paper tells the story of a hospital's path to recovery from initial landfall and damages, to completion of final design. Temporary flood protection measures were installed in the aftermath of the storm to allow resumption of veteran services and protection of the facility during interim construction conditions. Upon beginning design for a permanent flood protection system, it became critical to build consensus among multiple stakeholders for the project. This included meeting with multiple layers of staff at the U.S. Department of Veterans Affairs to ensure the needs of the agency, local staff, and patients could be met by the project.

Insight is provided on how to effectively communicate design risks, post-project residual risks, and project uncertainty with a diverse set of stakeholders. The public awareness component of the project, through multi-agency coordination and completion of an Environmental Assessment, is discussed. Final design of the floodwall system and incorporation of stakeholder preferences is also discussed.

INTRODUCTION

In the fall of 2012, conditions in the warm waters of the Caribbean were optimal for development of a hurricane. In mid October of 2012, multiple U.S. maintained climate models were indicating that a hurricane would likely develop and follow a typical path. On October 23, 2012, a developing storm was officially declared a hurricane, and the weather model maintained by the Europeans indicated a startling development. The European model was indicating the potential for the hurricane (named Hurricane Sandy) to travel northward into the Atlantic, and then veer westward due to the presence of a high pressure system in the northern Atlantic region. This westward turn was a

[1] Civil Works Technical Advisor, HDR Engineering, Inc. 701 Xenia Avenue South, Suite 600, Minneapolis, MN 55416. Matthew.Redington@hdrinc.com

significant deviation from typical storm tracks, and would put at risk the most densely populated areas of the U.S.

The storm path prediction by the European model turned out to be accurate. Hurricane Sandy traveled northward across Cuba and into the Atlantic Ocean. While cooler waters caused the storm to eventually lose official hurricane status, the combination of storm strength, size, and the path was to make it particularly devastating. Superstorm Sandy, as it ultimately came to be known, made landfall near Brick, New Jersey, on October 31, 2012. In addition to exposing the mid-Atlantic coastal region to 80 mph winds and rainfall depths up to 13 inches, the storm made landfall during a full moon and at high tide. The low lying nature of the densely populated areas, and the vulnerability of infrastructure, homes, and commercial properties resulted in lives lost and extensive damages.

The Veterans Affairs Medical Center (VAMC) in Manhattan, New York, was severely damaged by Superstorm Sandy. These damages resulted in the hospital remaining closed for six months after landfall. This paper provides a description of the damages to the facility, and the hospital's path to recovery through completion of the design phase.

Temporary flood protection measures were installed in the aftermath of the storm to allow resumption of veteran services and protection of the facility during interim construction conditions. Upon beginning design for a permanent flood protection system, it became critical to build consensus among multiple stakeholders for the project. This included meeting with multiple layers of staff at the U.S. Department of Veterans Affairs to ensure the needs of the agency, local staff, and patients could be met by the project.

Insight is provided on how to effectively communicate design risks, post-project residual risks, and project uncertainty with a diverse set of stakeholders. The public awareness component of the project, through multi-agency coordination and completion of an Environmental Assessment, is discussed. Final design of the floodwall system and incorporation of stakeholder preferences is also discussed.

LANDFALL AND RESULTING VAMC DAMAGES

The path of the hurricane, showing landfall in the mid-Atlantic region is shown in Figure 1. The VAMC is located on the eastern side of Manhattan Island on the block between East 23rd Street, East 25th Street, 1st Avenue, and Asser Levy Place. The VAMC consists of multiple buildings and occupies most of the city block (a separate school is located at the southwest corner of the block). Figure 2 shows the limits of the VAMC owned property. As seen in the figure, the VAMC is located in very close proximity to the East River. The east edge of the VAMC property is approximately 300-400 feet from the East River. The East River drains to Lower Bay and the Atlantic Ocean. The close proximity to the East River and low elevation of the site and finished floor elevations make the facility susceptible to damage during coastal flooding events.

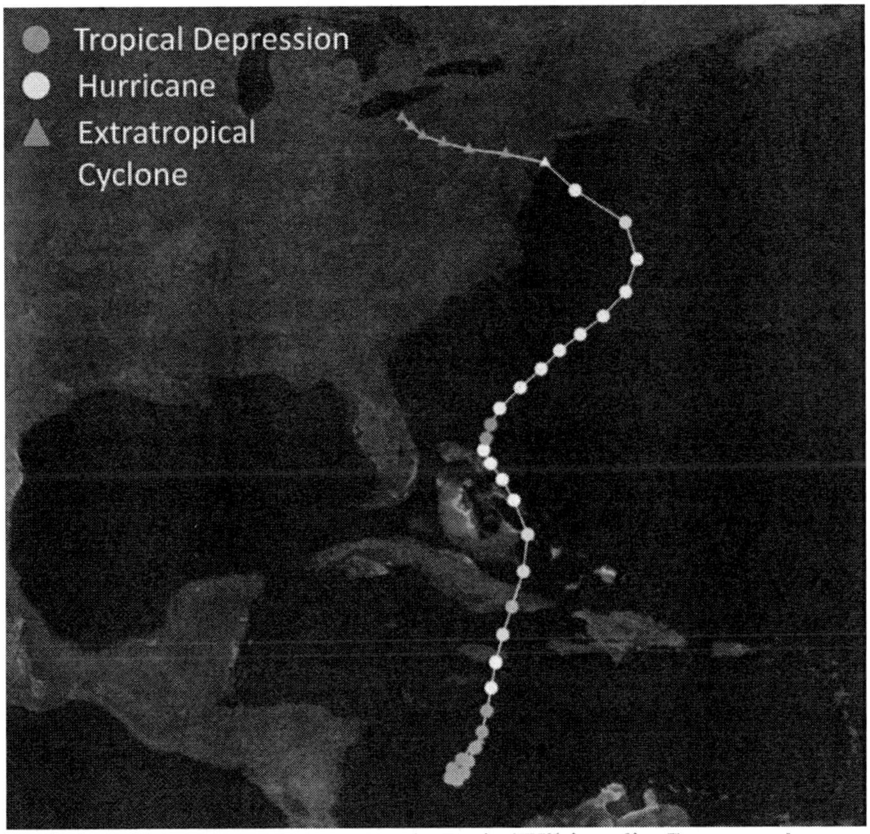

Figure 1. Superstorm Sandy Path (Wikimedia Commons)

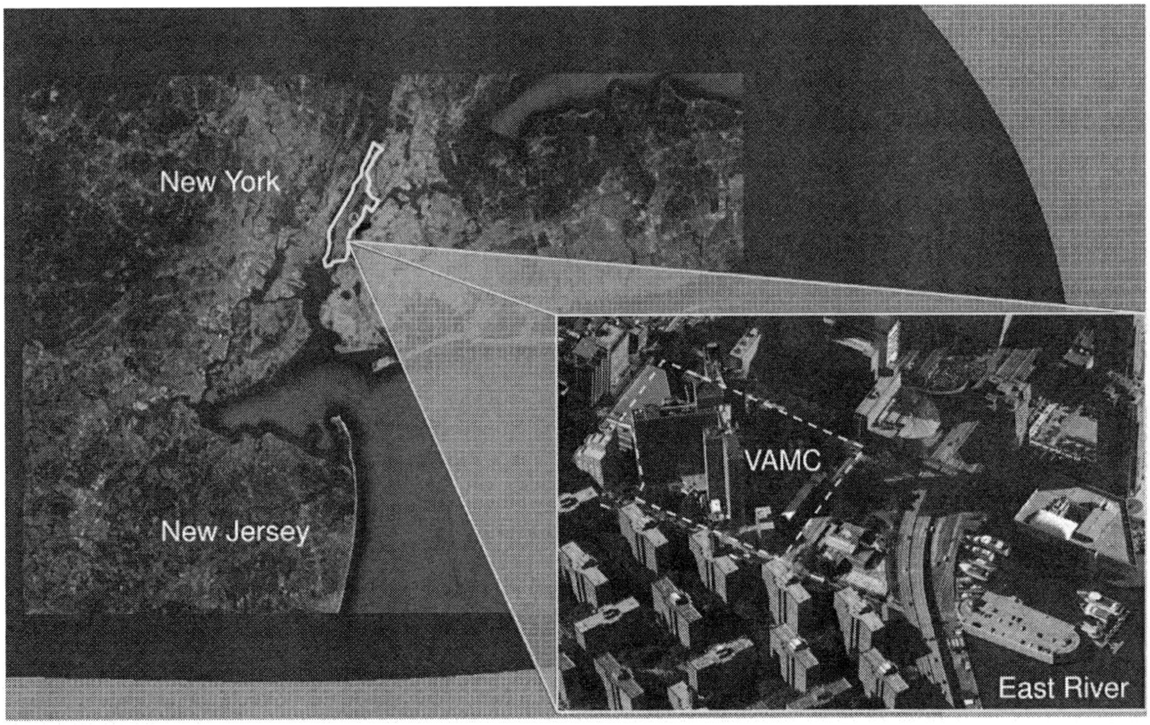

Figure 2. Project Location

The grade elevations of the VAMC parcel vary from between 2.5 feet and 21 feet above sea level. The finished floor elevation of the main hospital is located at 5 feet above sea level. Superstorm Sandy caused coastal inundation reaching elevations of approximately 10 feet above sea level at the VAMC. As a result of the inundation, the following critical hospital features were damaged:

- Outpatient clinic on ground floor
- Morgue
- Chapel
- Warehouse
- Sterile processing
- Domestic water pumps
- Hot water pumps
- Fire pumps
- Emergency generators
- Normal power and emergency power switch gear
- MRI

The extensive damages to key hospital departments and facility infrastructure resulted in the facility being shut down for 6 months. Due to the widespread destruction caused by Superstorm Sandy, it was particularly challenging for the VA to get replacement switchgear for the electrical system which was required to restore power to the entire hospital. Many other facilities also needed new switchgear and the demand for new equipment greatly exceeded inventories and production capacities.

TEMPORARY FLOOD PROTECTION

Establishing Temporary Protection

The closure of the VAMC facility due to flooding damages and lack of power resulted in significant modifications to the approaches used by the VA to provide health care to veterans in Manhattan. These temporary measures to sustain care provided significant challenges to the VA and caused inconveniences to some veterans who were forced to travel to alternate locations. It was imperative to get the facility back to standard operating capacity as quickly as possible. The following actions were required to get the facility up and running:

- Removal of water
- Removal of immediate hazards
- Demolition of damaged equipment, walls, and furniture
- Reestablishment of electricity to the hospital
- Reestablishment of mechanical equipment
- Reestablishment of key operations or departments, or outsourcing to new locations

Due to the critical nature of the facility, the VA established a line of temporary flood protection. This temporary protection consisted of installation of HESCO baskets around the perimeter of the site (to prevent inundation from coastal flooding) and installation of backflow prevention valves in sanitary and storm sewer lines. This temporary flood protection will provide protection over a 2 year time period covering the design and construction phases of the floodwall protection project. The temporary protection was built at the beginning of the design phase for the new floodwall project. The temporary protection measures will be systematically dismantled as permanent flood protection measures are constructed so that the facility can be protected from flooding during interim construction conditions. Figure 3 shows a portion of the temporary line of HESCO basket defense.

Figure 3. Temporary Flood Protection Measures

BUILDING CONSENSUS

Identifying Stakeholder Needs

There were multiple layers of the U.S. Department of Veterans Affairs organization that had an interest in shaping the project to make sure the needs of the agency, the staff, and patients were met. The organizational layers and their primary interests included:

- Local Engineering Staff: wall alignment, traffic patterns through site entrances/exits, connections to adjacent streets, compliance with New York City agency requirements
- Local Administration Staff: parking, impacts to hospital operations during construction, coordination with other ongoing projects, conflicts with City park project located adjacent to the VAMC project, aesthetics, and affects on operations (post construction)

- Eastern Regional Administration Staff: meeting Federal agency perimeter security requirements, schedule, and budget

In-person meetings were held at the VAMC Manhattan site to talk through each party's interests, and what would be required in order to make the project a success. Each meeting included walking on the grounds of the facility, photo documentation, and meeting minutes.

Wall Height and Alignment Presentation

A key part of the process of building consensus was educating the staff at each level of the VA organization as to the information needs of the design team, potential options for design features, degrees of protection, relevant design standards, and areas of uncertainty or risk. At the same time, it was critical for the design team to ask questions about the needs of each agency layer, and potential operational impacts to the facility. It was only through a collaborative process of presenting, listening, and revising, that a design could be developed that met the needs of all parties.

Design presentations proved instrumental in the early phase of design work to assist with development of design criteria for the project. A presentation was first given to design team members to confirm the approach and develop design recommendations for the VA. The presentation was subsequently given to VA Administration Staff, regional staff, and local VA engineering staff to emphasize the link between operations and flood protection criteria, discuss recommendations, refine criteria, and gain support for a mutually agreed upon set of criteria. Multiple presentations required significant time and budget, but it was an essential process that was required to build consensus and it ultimately made implementation of the design more efficient at later stages of the project.

The presentations were used to provide clear documentation of calculation methods and visual depiction of how criteria could potentially impact facility operations. Figure 4 was shown as part of the presentation in order to show the VA what considerations enter into a top of wall height determination. Figure 5 was shown in order to illustrate the effect of wave height (which varies depending on location on site) on top of wall elevation calculations. Several images were shown, similar to Figure 6, to explain where water elevations were in relation to various entrance/exit points at the site perimeter in order to explain what points would not be accessible during a flood.

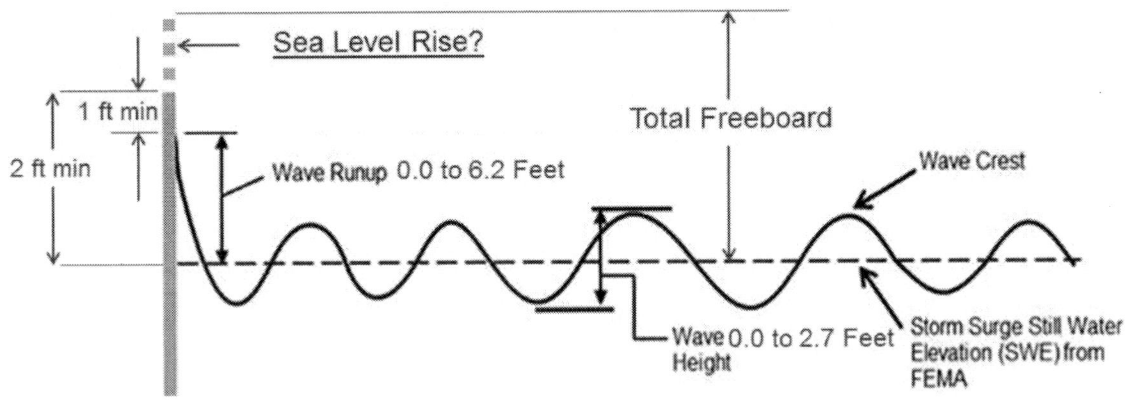

Figure 4. Determination of Top of Wall Elevation (HDR, October 2013)

Stillwater Elevations

Return Period, yr	SWE, ft (Manhattan Datum) (from FEMA)	
	2007	2013
10	4.5	5.1
50	6.1	8.1
100	6.9	9.4
500	9.1	12.5

Total Freeboard

Wave Run-up: 6.2 feet
FEMA additional 1.0 feet
Sea Level Rise: 2.0 feet
9.2 feet

Max Protection Elevation (east side) = 12.5 + 9.2 = 21.7 feet
Max wall height = 21.7 – 4.5 = 17.2 feet +/- above grade

Figure 5. Elevation Calculation for Worst-Case Design Scenario of Meeting 500-year Wave Conditions (HDR, October 2013)

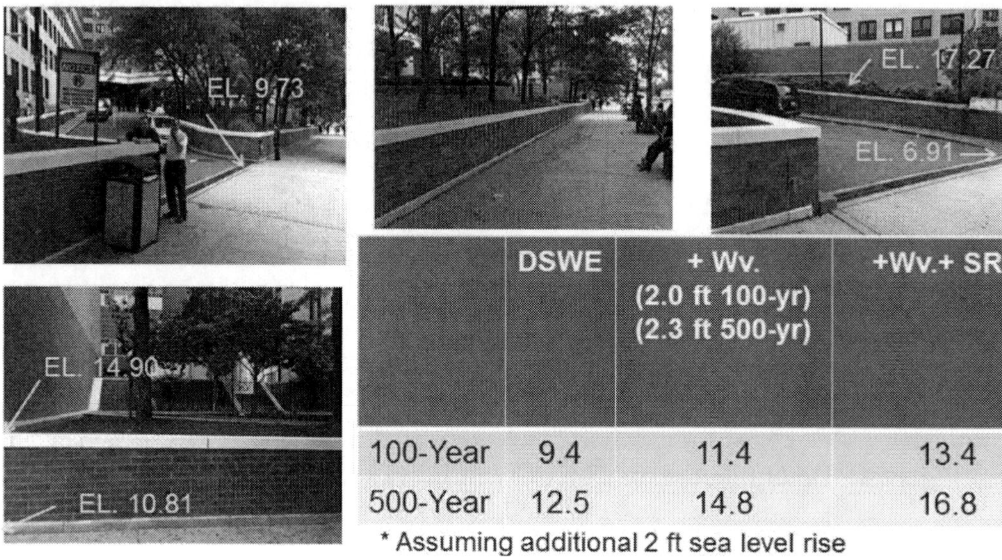

	DSWE	+ Wv. (2.0 ft 100-yr) (2.3 ft 500-yr)	+Wv.+ SR *
100-Year	9.4	11.4	13.4
500-Year	12.5	14.8	16.8

* Assuming additional 2 ft sea level rise

Figure 6. Linking Flood Protection Criteria to Facility Operations (HDR, October 2013)

Selecting and Documenting Criteria

A criteria memorandum was created to provide a concise summary of key criteria and guidance documents that would be a guide for the development of construction plans and specifications. This document was created early in the design phase of the project in order to communicate alternate design approaches and achieve agreement among project stakeholders. The existence of this document, and the use of the document as the first deliverable to be reviewed and approved by the VA, encouraged timely agreement on criteria among the design team and project stakeholders. This deliverable requirement created a measurable goal to keep the project design moving, and helped mitigate the risk of rework later in the design process due to confusion over criteria or misunderstandings on objectives achieved by the design.

Basis of Design Report

While the criteria memorandum was used to document the direction of the project, a Basis of Design Report (BODR) was used to provide documentation that criteria were being met, and describe the decision making process during design. The development of a BODR (or similar document) is especially important when designing flood protection for critical facilities such as hospitals. The use of a BODR for such projects includes the following benefits:

- Provides documentation of the basis for design
- Provides documentation of client participation in selecting criteria to prevent undue liability to the designer
- Prevents misunderstandings in regards to what the flood protection system can accomplish

- Provides the background necessary for the owner to establish operational protocols (such as closure or evacuation of facility)

Due to the fact that the VAMC is a critical medical facility, additional sections addressing residual risks and engineering during construction were added to the BODR.

COMMUNICATING RISK AND UNCERTAINTY TO STAKEHOLDERS

Residual Risks

The section of the BODR related to residual risks provides clarity to the owner that no system is bullet-proof. Conditions which exceed criteria and affect hospital operations and the safety of patients or staff could occur. A brainstormed list of potential system failures of the system was provided in the BODR. Examples of potential failure modes include items such as the following:

- Overtopping of the wall due to inundation exceeding design criteria
- Overtopping would be most likely at the south east corner of the site due to wave exposure
- Asphalt overlays covering manhole vents could affect the ability of internal drainage area systems from performing as intended
- Failure of check valves could require manual activation of shut-off of valves on utility lines in order to prevent interior flooding
- Exposure of vertical building faces to wind driven rainfall could increase the size of contributing drainage areas and the resulting storm runoff in the internal drainage area

Engineering During Construction

While high quality design services, construction plans, construction specifications, and documentation are important to project stakeholders, it is only a high quality constructed product that results in achieving flood protection. To minimize risk of system failures and ensure the owner gets the system they believe they are paying for, it is especially important for the designer to have a role during the construction phase of the project. This involvement is especially important when the design protects a critical facility.

The VA has assembled a construction management team (from existing VA staff) for the construction phase of the project. The BODR has been used as a tool to document concerns related to areas of particular risk and uncertainty to the VA and make recommendations on involvement by the designer. Special attention is given to areas of specialized construction or with greater uncertainty. The recommendations included:

- Proper installation is required in order for the design to function as intended.
- Soil conditions are highly variable because Manhattan was once a landfill. Changes to site conditions along the alignment could warrant changes to construction techniques or foundation design.

- The depth to rock on site is highly variable. Use of rock anchors, when required to counter uplift, should be observed by the Engineer because installation quality is critical.
- The potential for unidentified utilities is high. Site observation is recommended to help identify and mitigate for newly discovered conflicts during construction.
- The close proximity of construction to existing VA site features and City utilities warrants careful observation of techniques by the Engineer.
- Subsurface soil mixing is a specialized construction technique. This work should be observed in order to gain confidence that the installed product will provide the protection that is intended by the Engineer.

PUBLIC AWARENESS

An Environmental Assessment was completed for this project. The Mayor's Office of Environmental Coordination served as the lead agency. Additional interested agencies that participated in the review included New York Department of City Planning, Department of Environmental Protection, Landmarks Preservation Commission, and the Department of Parks and Recreation.

The primary interests identified during the development of the Environmental Assessment included the following items:

- Noise which could disturb hospital staff, patients, and people on adjacent streets or buildings
- Vibration due to installation of project features having the potential to damage adjacent utilities and the nearby Asser Levy Public Baths Building that is on the National Register of Historic Places
- The aesthetics of the wall fitting in with the architecture of the existing VAMC, the historic bathhouse building, and the future park that was planned for construction along the existing Asser Levy Place

A secant pile wall was incorporated into the design for seepage control underneath the floodwall in order to address noise and vibration issues. Although a secant pile wall is typically slower to build and more expensive than use of alternatives such as steel sheet pile, the secant pile wall is constructed through relatively quiet auguring methods. Auger cast piles were selected for the pile supports for the floodwall for the same reason. Any changes to this design approved by the Engineer and implemented by the contractor would need to be supported by a construction protection plan. This plan would require monitoring to ensure that changes do not result in detrimental effects on adjacent utilities or the historic bathhouse.

Aesthetic concerns resulted in an architectural review of alternative floodwall designs. Alternatives considered included a basic unfinished concrete wall, a concrete wall with formliner, and use of masonry on the exterior a concrete core wall. The concrete core masonry faced design with column representations was ultimately selected in order to best match the character of the bathhouse, the existing VAMC landscaping walls, and the future park setting. Renderings were developed to assist in the selection process. Figure 7 shows a rendering of the selected wall type running along the future Asser Levy Place Park.

Figure 7. Floodwall Rendering Along the Future Asser Levy Park (HDR, April 2014)

COMPLETED DESIGN

Permanent Flood Protection System Design

The permanent flood protection system includes construction of a floodwall, seepage cutoff wall, interior drainage improvements, four sanitary sewer pumping stations, interior site paving, utility service reconstruction, construction of manholes, installation of backflow prevention valves, and landscaping. The floodwall will be an L-wall configuration in order to locate the floodwall as close to the property line as possible. The floodwall will traverse from the main building entrance at 23rd Street, east to Asser Levy Place, north along Asser Levy Place, and then west along 25th Street. The total length of the floodwall will be approximately 1,570 feet. The wall will be approximately 10.5 feet high above existing grade along Asser Levy Place. Along 23rd and 25th Streets, the floodwall will maintain a height of at least 8 feet above grade as the ground elevations rise toward 1st Avenue. Five flood gates will be provided around the perimeter of the site which will allow vehicle and pedestrian access to the site, while allowing closure of openings in the event of a coastal flooding event.

There are four locations where pump stations will be installed in order to allow pumping of building sewage and roof drainage off site and into the combined sewer lines in the adjacent city streets during times at which perimeter backflow prevention valves are closed. Utility services will be replaced across the property lines in order to protect against seepage failures, and to have new piping beneath the new floodwall which reduces the likelihood pipe ruptures and the likelihood of required maintenance.

Structural Evaluation

The presence of multiple obstructions such as key utilities, utility vaults, building foundation corners, gates, and alignment turns at property corners required numerous modifications to the floodwall footing in the form of notches out of the footing, decreases in footing length, rearrangement of pile spacing, and footing elevation changes. As a result of the high number of unique features, finite element analysis was performed (using STAAD software) for the entire floodwall alignment rather than using a typical approach of design based on a typical section that represents the most severe loading condition. STADD analysis was performed on 11 separate monoliths (from expansion joint to expansion joint). Figure 8 shows the modeling results for a single monolith with footing modification and footing elevation change.

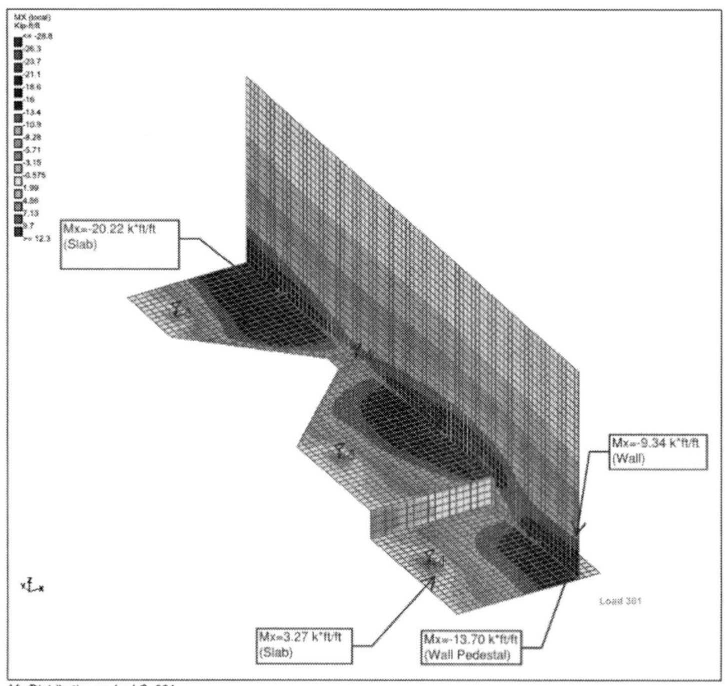

Figure 8. Finite Element Analysis (HDR, February 2014)

Seepage Countermeasures

Portions of the City of Manhattan (including the VA site) were built on a landfill. As a result, subsurface conditions are highly variable. Although much of the area surrounding the project is, or will ultimately become, paved as part of the project, the presence of pavement cracks, numerous utilities under pressure flow during flood events, and scattered vegetation areas necessitated use of a seepage cutoff wall. Multiple options were considered for seepage cutoff including use of sheet pile, soil bentonite cutoff trench, and secant wall construction. Ultimately, for reasons discussed in the text addressing the environmental assessment, a secant wall methodology was selected for installation of a soil-bentonite-cement cutoff wall.

Utility Conflicts

Utility conflicts were numerous along the alignment. Two engineers worked full time for 5 months to complete the utility plans and associated agency coordination. Civil 3D software was used to 3-dimensionally model the floodwall and utilities. The use of visualization tools allowed the design team to observe the project from multiple angles in order to check for conflicts. Figure 9 provides a view from underneath the floodwall looking up. This figure shows the floodwall being supported by piles (grey and black), utility vaults/manholes in green, and utility lines in yellow. As shown in the figure, there is a conflict between an outgoing storm sewer line and one of the supporting wall footings. Visualization was a valuable time saving tool for confirming and resolving conflicts, and communicating the nature of conflicts to stakeholders.

Figure 9. Utility Conflict Visualization

Facility Access

Designing a flood protection system for a hospital introduces unique operational design requirements. Hospital access points are critical for maintenance of facility operations. At all times of the day, during all days of the year, staff, equipment, and supplies must be able to come and go from the site. Any closure of entrances/exits from the site can have a significant impact to services. The designed floodwall travels along approximately three fourths of the parcel boundary. This alignment crosses the majority of the access points to the facility.

Manually activated gates would need to be activated prior to landfall of a coastal storm. This manual activation would sever access to the facility. In order to maximize the duration of time for which entrances/exits from the facility can be open, a floating flood gate was used at entry points. Gates operated by buoyancy remain flush with grade (allowing site access) until the moment that floodwaters arrive on site. Flood waters spill into a grate adjacent to the gate, allowing the gate to rotate upwards into place. Figure 10 shows a test activation for a similar floodgate at an alternate site.

Figure 10. Flotation Activate Gate (FloodBreak)

CONCLUSION

The devastation caused by Superstorm Sandy highlights the importance of protecting critical medical facilities from coastal inundation. Although the VAMC was badly damaged as a result of coastal flooding during the storm, the VAMC is undertaking construction of a flood protection system to mitigate for risks from future flooding events.

Key to the success of this project was considering the needs of multiple stakeholders. Consensus building was undertaken among multiple layers of the U.S. Department of Veterans Affairs in order to refine and select criteria, and design a system that could meet the operational needs of the agency, its local staff, and its patients. Presentations were made by the design team to VA staff in order to communicate design risks, post-project residual risks, and project uncertainty. An Environmental Assessment was conducted to evaluate the needs of the local community and environment. This flood protection system, once built, will mitigate for exposure of the facility to coastal surges from the East River and allow the VAMC to continue meeting its mission of providing health services for veterans.

ACKNOWLEDGEMENTS

The author would like to express appreciation to the U.S. Department of Veterans Affairs for the opportunity to provide design services on this critical infrastructure project and to submit a paper on behalf of the project team. Although it is not possible to thank everyone involved in the project, the authors would like to recognize the following HDR staff for their efforts on this project: Bill Nathans (Project Manager-Architecture), Srikanth Gorugantula (Project Manager-Engineering), Scott Anderson and Ron Jefferys (Geotechnical Engineers), Tony Falcone and Rajayra Diaz (Civil Engineers – Utilities and Agency Coordination), Kathleen Thompson and Ted Shannon (Interior Drainage), Dave Haugen (Pump Stations), Brian Cheng (Architectural Renderings), Dylan Coonrad and Erin Disbrow (Graphics Designers), Desiree Mervau (Graphics Artist), Anthony Cortese (Landscape Architect), Sean Crane and Richard Prakopcyk (Architects), Jennifer Curran, David Davis, Debra Peer, and Rebecca Hartless (Environmental Assessment Specialists), Paul Van Hagen, Ken Standig, Jesse Freeberg, Tan Enhui, John Larson, and Jerry Mulvihill (Structural Engineers), Ed Taraba (Security Designer), Dan Heilman (Wave Evaluation), John Hess (Technical Advisor), Kris Hodne (CAD Design), Werner Mueller (Project Principal), Nancy Brandenburg (Specifications), Francisco Brilhante (GIS), and Dale Peterson (Electrical Engineer).

REFERENCES

HDR. October 2013. Presentation to U.S. Department of Veteran's Affairs. On file with HDR.

HDR. April 2014. Final Environmental Assessment for Flood Mitigation Measures, *Design and Construction of Perimeter Floodwall, Manhattan, NY.* On file with HDR.

HDR. February 2014. *Basis of Design Report.* On file with HDR.

Wikimedia Commons, http://commons.wikimedia.org/wiki/File:Sandy_2012_track.png

Floodbreak. File photograph provided by Victor Althoff of FloodBreak.

JUGGLING CONFLICTING NEEDS AT TIMS FORD DAM FOR WARM AND COLD-WATER SPECIES, RECREATION, AND FLOOD CONTROL

Colleen R Montgomery, M.S., P.E[1].
John T Baxter, M.S.[2]

ABSTRACT

The 40 mile reach on the Elk River between Tims Ford Dam and Fayetteville, Tennessee has conflicting water temperature needs for an endangered fish species and game fish. Releases from the dam are being managed in a unique way to accommodate both species, and to also provide wade fishing flow needs and downstream flood damage protection.

Tennessee Valley Authority (TVA) is involved in a consultation with the US Fish and Wildlife Service on this reach to protect several mussels and the Boulder Darter, a small endangered warm water fish that inhabits the Elk River near Fayetteville, about 40 miles downstream of Tims Ford dam.

TVA cannot just release warm water from the spillway to accommodate the darters because the State stocks trout immediately below the dam, and the tailwater is a popular fishery. Trout are cold water fish, requiring water below 68°F.

The dam's outflow structures include a hydro turbine and sluice using a low-level intake, and the emergency spillway, whose crest is 100 feet higher than the turbine intake. There can be a 25°F difference between the spillway and turbine release temperatures in the summer due to reservoir stratification. Warm spillway releases alone would kill the trout, and cold turbine releases alone stress the darters.

Numerical modeling predicted water temperatures from using different blends of turbine, sluice, and spillway releases to determine the best release combinations to provide warmer water to the Fayetteville area, while still maintaining the trout fishery near the dam, in lieu of adding a costly selective withdrawal structure in the reservoir.

The preferred solution was to abandon use of the hydro turbine from May to November, unless there is a flood or other emergency. Blended sluice and spillway releases provide flow downstream. These operations have allowed growth of the darter population and fish habitat improvement in the entire Elk River.

INTRODUCTION AND BACKGROUND INFORMATION

Tims Ford Dam (Figure 1) creates a multipurpose reservoir located on the Elk River in Franklin County, Tennessee, in south-central part of the state. The dam is located at Elk River Mile (ERM) 133.3, and is 36.7 river miles downstream of Elk River Dam (Woods Reservoir). Tims Ford dam was completed on December 1, 1970, and its single

[1] Tennessee Valley Authority, crmontgomery@tva.gov
[2] Tennessee Valley Authority, jtbaxter@tva.gov

hydroturbine started producing power in March, 1972. Downstream environmental protection has been a goal of TVA's since the mid 1980's, and toward that end, a sluice was added in 1986 to help provide a continuous minimum flow to the tailwater reach. This eliminated drying of the channel when the turbine was idle. The sluice discharges between 80 and 240 cfs. The releases are usually less than 50°F year-round.

Figure 1. Tims Ford Dam with Outlet Structures Identified

Water released from Tims Ford Dam during normal hydropower operations passes through the large hydroelectric turbine which generates approximately 38 Megawatts (MW) and discharges about 3800 cfs to the tailwater. When the turbine is not in use, water is discharged via the sluice. The emergency spillway conveys water through the dam for flood control, and is sometimes used to convey very low flows for environmental flow maintenance in the downstream channel. The turbine intake centerline is at elevation 751 ft-msl. The spillway crest is approximately 100 feet higher in the dam than the turbine or sluice intake. Additionally, the spillway crest is 35 feet below the normal summer pool elevation, and therefore releases water that is cooler than the water on the surface of the reservoir.

The original purposes of Tims Ford Dam were flood control and recreation. To accommodate the recreation objective, a summer pool between elevation 888 and 883 ft msl is maintained from May 15 through October 15 in most years, when inflows are adequate. In drought conditions, however, the reservoir pool may be drawn lower, to fulfill downstream water requirements. Figure 2 shows the Tims Ford Operating Guide, along with pool levels for the last two years.

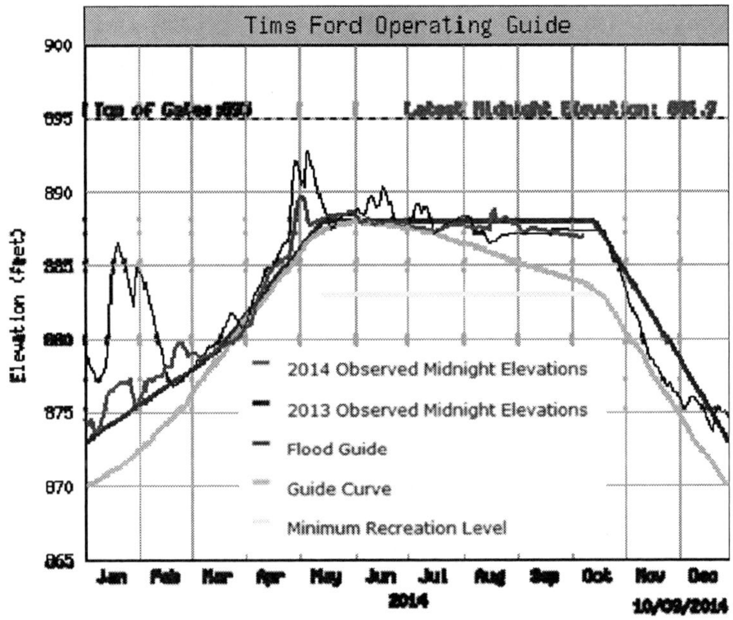

Figure 2. Tims Ford Reservoir Operating Guides with 2013 and 2014 Pool Levels

Recreation in the reservoir involves boating, swimming and water skiing, as well as passive enjoyment by lakefront owners. Most lakefront landowners live toward the upstream end of the reservoir in the outskirts of Winchester, TN. Many docks on the reservoir are built at a fixed elevation of 888 ft msl, and these would become submerged if the reservoir pool were to exceed this level. Likewise homeowners' boats would be beached if the pool level dropped too low.

The other original purpose of Tims Ford Dam is flood control. Fayetteville, TN, located 40 miles downstream of the dam, is one of the state's biggest flood damage centers, and Tims Ford Dam has helped avert many millions of dollars in flood damages at Fayetteville . The sizeable natural drainage area between the dam and Fayetteville requires that the dam holds back the water until the natural flood crest passes by Fayetteville, at which time turbine and/or spillway releases can be filled in behind the flood crest.

Before TVA's Reservoir Release Improvements (RRI) Program of 1991, the first five to ten miles of the Tims Ford tailwater were inhospitable to fish, because the dissolved oxygen (DO) content of the releases was near zero for much of the year.

As part of the RRI program, a bubble plume DO augmentation system was installed inside the penstock. A forebay aeration system was installed in 2006 to replace the ailing penstock aeration system, which had exceeded its expected lifespan at that time. The forebay system allows for repairs without having a unit outage. The spillway releases are well-aerated due to the rocky nature of the downstream end of the spillway chute. Newer to the area is tailwater recreation, which has quickly become economically important to the region, and people are drawn to the area for trout fishing and floating down the river in kayaks or rafts. Several outfitters run float trips in the tailwater. The trout fishery below the dam is regularly stocked with brown and rainbow trout by Tennessee Wildlife Resources Agency (TWRA) and can be maintained because of the cold temperature of the Tims Ford dam releases. The trout fishery extends about 14 miles from the dam, ending at Beans Creek, a warm tributary. Wade fishing is the most common fishing method in the tailwater, and water depths generally require fishermen to wear chest waders.

Other aquatic species that inhabit the tailwater downstream of Beans Creek are less tolerant of the cold water than trout are, and even more intolerant of the large temperature fluctuations that come with alternating turbine use and minimum flow. These species include as many as 49 different warm water fish species, the most sensitive of which is the endangered boulder barter (*Etheostoma wapiti*), a small fish which dwells in moderately flowing pools and among large, flat rocks in areas that are relatively clean of silt. Several endangered mussel species also inhabit the tailwater and have similar sensitivities to those of the boulder darter.

Managing flows for endangered species' needs, a game fishery, flood control, recreation, and water quality has posed some unique challenges for the operation of Tims Ford Dam in recent years, but TVA has modeled, analyzed and adjusted the releases and has come up with solutions that have made all stakeholders reasonably happy.

ENDANGERED SPECIES PROTECTION BELOW TIMS FORD DAM

Even though the dam releases now contain adequate oxygen, and provide a continuously wetted channel, they are too cold for endangered species such as the boulder darter and the crackling pearlymussel. They are even too cold to be ideal for some trout species, with releases as cold as 45°F.

TVA has been involved in a consultation with the U.S. Fish and Wildlife Service (USFWS) since 2004, involving protection of endangered species around 31 of its dams.

Even before the consultation began, alternatives were being explored to provide warmer water farther downstream for the darters. Steadier releases also help reduce bank erosion that can impair spawning of fish and mussel species.

A Biological Opinion issued by the USFWS in 2006 established a number of protective measures to be implemented at Tims Ford, including:

- Establishing a multi-agency working group to oversee the implementation of alternative dam release strategies and to document their impacts for at least 10 years.
- Modification of operations at Tims Ford Dam to warm tailwater temperatures and lessen water level fluctuations year-round, especially at times critical to the reproduction of the boulder darter, attempting to simulate natural flow regimes
- Deployment of a water temperature monitoring network in the tailwater
- Financial assistance to support repopulation of the boulder darter

- Conducting fish and mussel surveys to track the status of all listed species in the Elk River (USFWS, 2006)

Since 2006, the Tims Ford tailwater has continued to have the majority of the focus of the consultation, and has been studied a number of times to determine ways to alter dam operations to help protect the federally endangered fish and mussel species in the tailwater and maintain the trout fishery, while fulfilling the project's other purposes.

The intent of this and other studies have been to evaluate options for warming Tims Ford dam releases, while utilizing all of the available outlet structures in different ways than they had been used in the past.

Boulder Darter Water Temperature and Flow Criteria

Because boulder darters live only two or three years, successful reproduction ***every year*** is important for survival of the species. And because boulder darters live in rocky river-bottom terrain, extended field observation is difficult. Electrofishing surveys are discouraged because the species is so small and delicate, with a typical fish length of less than 3 inches. Therefore, only limited information from brief fish surveys are generally available. From laboratory observations of fish behavior and health at Conservation Fisheries, Inc. (CFI), it is thought that the darters prefer less variable flows than they have been exposed to in the tailwater, and temperatures between 16-21 degrees C (60°- 70°F) for spawning, followed by warmer temperatures for growth of fry(very young fish). Spawning season is typically May to June.

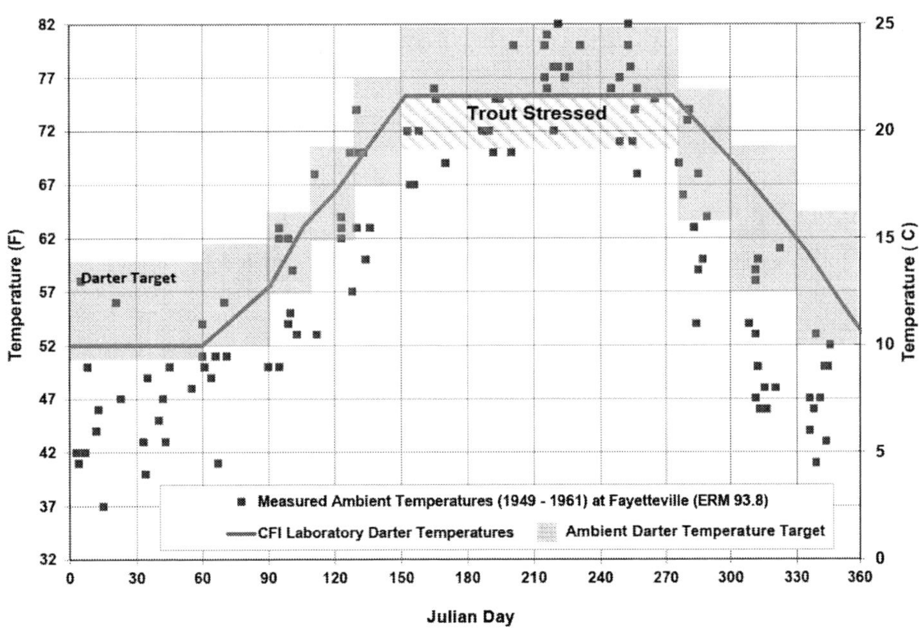

Figure 3. The Boulder Darter Water Temperature Criteria, and Pre-Impoundment River Temperatures

Target water temperatures for Boulder Darters were pulled from fishery studies by Rakes et. al at Conservation Fisheries, Inc. (CFI), in 2004 and 2005. Figure 3 shows the temperatures at which the Boulder Darters were maintained in the fishery control group with good spawning success rates and growth of fry. These laboratory temperatures were simplified to monthly target ranges shown in block shading in the plot, and they have become the target water temperatures for the Elk River at Fayetteville.

CFI also performed temperature change tests on laboratory darter populations mimicking changes that occur from turbine releases. These tests confirmed that the sudden water temperature changes like those caused by turbine releases can cause slowed development and significantly increased mortality of the darter embryos and fry related directly to the water temperature changes, compared to a control population for which the water temperatures were not decreased (Petty, 2013).

Changes to water temperatures from turbine use can be significant and can impact the entire 100+ mile reach of the non-impounded portion of the river, as shown in Figure 4.

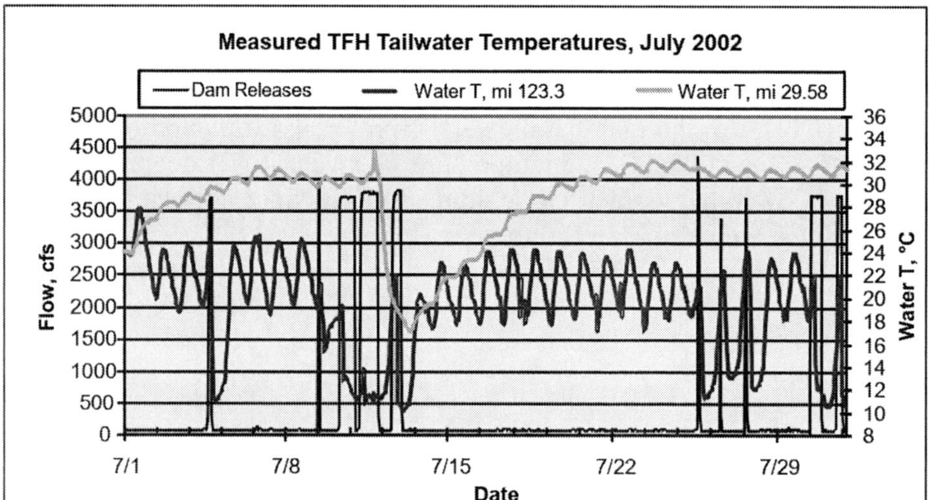

Large generation blocks on 7/9-12/2002 decreased water temps 14°C and impacts carried all the way to the downstream reservoir boundary

Figure 4. Measured Elk River water temperatures near upstream and downstream ends of riverine segment, July 2002

NUMERICAL MODELING TO EVALUATE DAM RELEASE ALTERNATIVES

One of the primary goals of the proposed operational changes at Tims Ford Hydro (TFH) is to reduce the sudden and significant changes in water temperatures from turbine use during the spring, summer, and fall. Any improvements in the upstream reaches of the Elk River would also benefit boulder darters and other sensitive fish and mussels in the lower reaches of the Elk River downstream of Fayetteville.

Numerical modeling was utilized to allow assessment of a wide variety of dam release alternatives, to help screen options methodically to more logically come up with reasonable options to test in the field, with minimal stress to endangered species. Field testing of the most promising alternatives could then be conducted and evaluated through a network of water temperature monitors and species surveys over a longer time period.

Two separate areas of study were evaluated in different models: First, the entire tailwater reach from Tims Ford Dam at Elk River Mile (ERM) 133.3 to the boundary of Wheeler Reservoir (ERM 29.8) was modeled using a one-dimensional unsteady flow and water quality model with water temperature simulation capabilities. Second, the entire reservoir, extending from Tims Ford Dam at ERM 133.3 upstream to ERM 165.3, was modeled, using a 2-dimensional reservoir water quality model to determine the impacts of operating changes on the reservoir water temperatures and water quality.

Model Application for Tims Ford Tailwater

The Tims Ford tailwater model utilized in this study is a one-dimensional TVARMS (TVA River Modeling System) one-dimensional hydrodynamic model (Hauser, 2003) which performs complete hydrodynamic and water quality computations for the modeled reach. The tailwater model utilizes hourly dam release data and hourly release temperature data, as well as hourly meteorology data including dry bulb and wet bulb temperature, wind speed, and solar radiation. Numerous cross sections are included in the model to depict changes in the river geometry along the reach. At the downstream end of the model, Wheeler reservoir elevations are the downstream boundary condition.

Previous Model Results

Many water temperature model simulations had been run for previous studies on the Tims Ford tailwater (Montgomery, 2005, 2007, 2008), in which different turbine release patterns were evaluated to assess their temperature impacts at Fayetteville and other locations. The most important results from these studies indicated that to reduce water temperature fluctuations:

- Turbine use should be extended throughout a 24-hour period (rather than scheduling the releases in a large block around the peak demand hours) to minimize water temperature impacts in the tailwater.
- The longer the time duration of turbine generation, the farther downstream the temperature decreases extended.
- The spillway *must* be used to help warm the releases.
- The spillway cannot be used simultaneously with the turbine because of the narrow (non-vibrational) operating range of the turbine unit. The turbine cannot run at a lower flow rate or efficiency. The discharge combination of the two would cause minor flooding in many places.

More recently evaluated tailwater warming alternatives include using the spillway to release various percentages of the total flow to the tailwater from April 15 to late October. The spillway draws water from about 100 feet above the turbine withdrawal zone and consequently releases warmer water than the turbines during the warm half of the year. Generally the temperature difference between turbine and spillway releases is 8° to 12°C, but it can be greater than that, depending on the month and the antecedent conditions.

Evaluated Spillway Release Alternatives

Initial spillway release alternatives were geared toward meeting reservoir pool elevation goals, and included:

- For normal to wet years, spill 50% of the weekly average flow, release the remainder through the turbine as evenly spaced as possible, OR
- Spill 75% of the weekly average flow, release the remainder through the turbine as evenly spaced as possible, OR
- Spill 100% of the weekly average flow
- Utilize a blend of sluice and spillway releases. From mid-April to late October when the flood guide is not being exceeded, run water through the sluice up to its maximum capacity of 240 cfs, and supplement this flow rate with spill if additional flow needs to be released for reservoir pool control. Do not use the hydro turbine.

A graphical inflow estimation tool allows easy estimation of future inflows to the watershed from a unit hydrograph (Parsly, 2006). A spill rate equal to the weekly average inflow rate can be calculated using the inflow estimator, and dam operators only need to make gate changes once a week, or less often in drier periods.

Results from 50% Spill Alternative

Initial results indicated that the 50% spill option warms the releases considerably, and reduces the large temperature fluctuations the tailwater experiences by nearly 50% when the turbine is turned on, compared to actual conditions.

However this alternative still had water temperatures that were too cool for the boulder darter in the late spring and early summer, especially during June and July. Figure 6 shows these water temperature results compared with actual water temperatures for 2001 just below Beans Creek at ERM 119.3. Target water temperatures for the boulder darter by CFI are superimposed on the graph as well (shaded blocks). Water temperature results from this alternative for other locations in the tailwater are found in an earlier report (Montgomery, 2006). Water temperature fluctuations at Beans Creek for the 50% spill option are in the 7-8°C range, which still has potential to stress the fish.

75% Spill Alternative

Results from the 75% spill alternative did not look significantly different than the 50% spill alternative in terms of reducing water temperature fluctuations, so it was eliminated.

100% Spill Alternative

This option eliminates use of the turbine completely, and would utilize the emergency spillway at very low gate settings. It had essentially the same upper limits on the water temperature as the other alternatives but it reduced the temperature fluctuations even more. Water temperature variations in this alternative are mainly due to diurnal cycles and meteorological events.

Unfortunately, the 100% spill option produces water that is way too warm for trout in the reach near the dam, even at only 240 cfs (Figure 7). This was true for all years modeled.

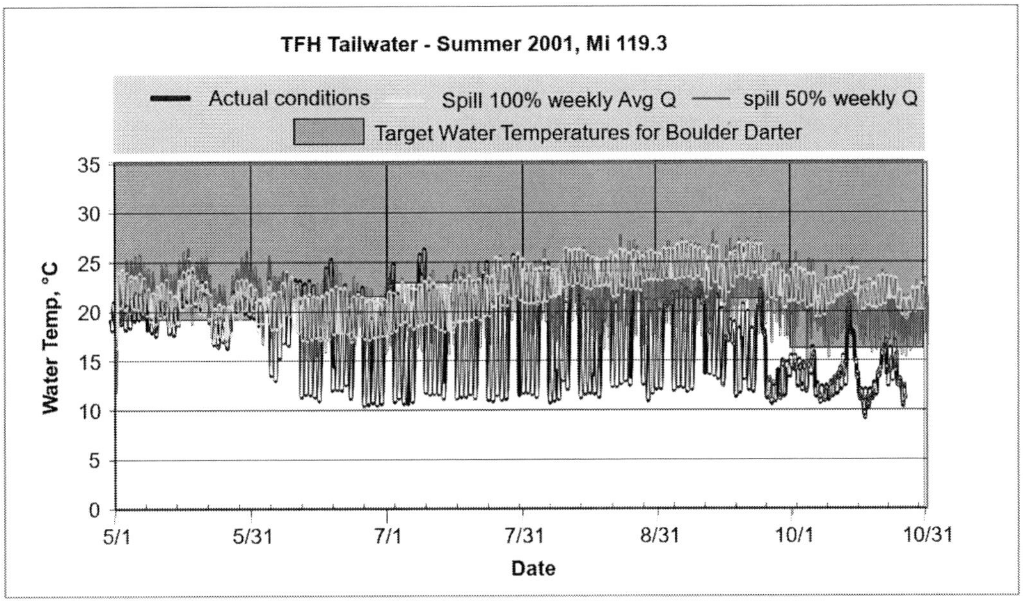

Figure 5. Water Temperatures at Beans Creek, (ERM 119.3) for spill alternatives and actual conditions

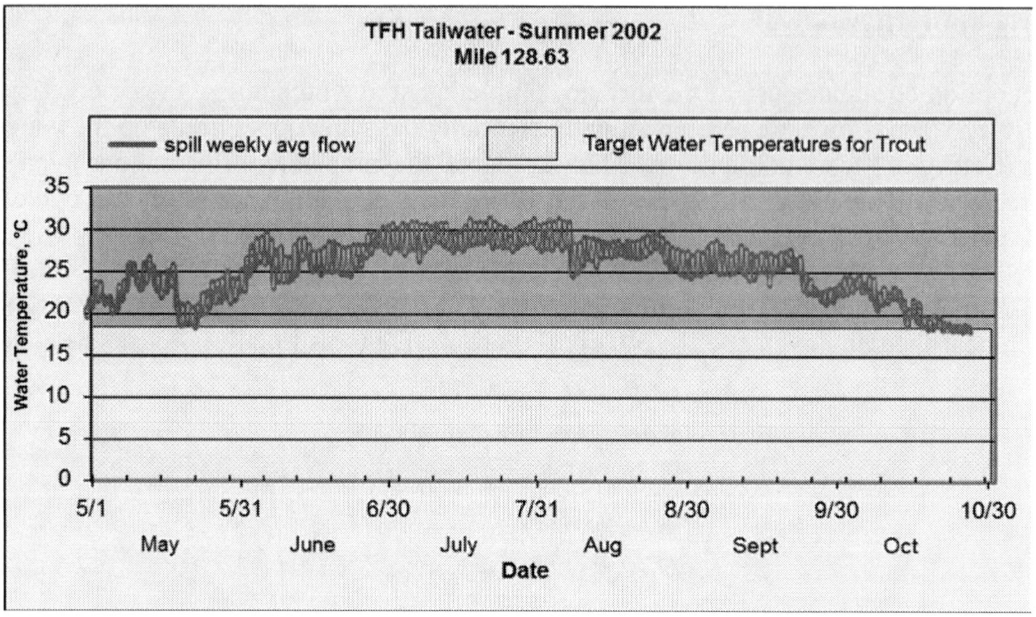

Figure 6. 2002 Water Temperatures at ERM 128.33 from the 100% spill option releasing 240 cfs

Blended Sluice and Spill Alternative

Because the 100% spill option releases water that is too warm for trout for much of the spring and summer, and the spill + turbine options produce significant water temperature fluctuations which can stress the darters, a blend of the sluice and the spillway may do a better job of meeting the desired water temperature range both upstream and downstream of Beans Creek more consistently. In this alternative, the turbine would be idle from mid-April to late October. Dam releases would come primarily from the sluice during dry periods and from both the sluice and spillway during periods where more than 240 cfs needs to be released for reservoir pool control.

IMPLEMENTATION OF THE BLENDED ALTERNATIVE

Reservoir and river water temperatures and reservoir pool levels are carefully watched starting in early April, to determine whether or not it is feasible to cease turbine use on April 15. During a wet year, turbine use may need to continue a bit longer to prevent flooding of docks in the reservoir. Sluice or blended sluice and spillway releases occur 24 hours a day between April and October.

Although the target date to resume turbine use is October 15, the ambient water temperatures need to be similar to the turbine release temperatures so the flow transition occurs without causing shock to the darters. In some years, TVA has not resumed turbine use until early November to allow better water temperature alignment with target temperature criteria. Communication occurs between TVA, USFWS, and TWRA when turbine use ends in April, for flood events that occur between April and October, and again in late October, when turbine generation resumes.

Positive Results: Fish Surveys Showed Increasing Populations in Recent Years

Since eliminating use of the turbine use for the warm half of the year, the numbers of fish species and their population counts have been gradually increasing downstream of Tims Ford Dam. Figure 8 shows the Index of Biotic Integrity (IBI) scores at different Elk River miles since 2007. Most tailwater locations show increased IBI scores, which includes increases in numbers of native fish speciesas well as several other indicators.

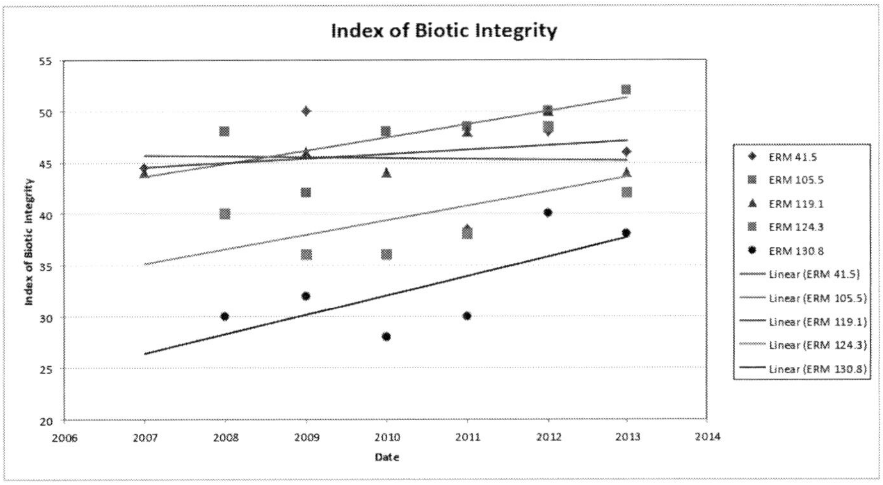

Figure 7. Fish IBI trends at five locations in the Tims Ford Tailwater, 2007-2013 (Consultation Compliance Report 2013)

Water Temperature Monitoring Demonstrates Success of Operational Changes

Water temperatures are being monitored at seven locations in the Tims Ford tailwater with Hobo® water temperature monitors, and at two other locations with real-time monitors. The Hobo® data collectors are small and self-contained and collect data at regular user-specified intervals but do not transmit data by themselves. An optical port allows quick downloading of the data for analysis. These units have been deployed every 15 to 20 river miles from ERM 29.8 (Wheeler Reservoir boundary) to ERM 119 (downstream of Beans Creek, just outside of the trout zone). At ERM 124, TVA placed a Hobo Weather Station real-time satellite-transmitting temperature monitor , sending hourly water temperature data from this remote location to the TVA water information database. This monitor provides hourly water temperature readings inside the trout zone. A USGS stream gauge with flow and water temperature measurement capabilities exists at Fayetteville, at ERM 93. The water temperature data collection is jointly funded.

Water temperature monitors have been installed at Tims Ford Dam at four depths near the spillway crest, as well as at the sluice release elevation, so water temperatures can be tracked in the reservoir and for about 100 miles on the Elk River. Most of these monitors have been in place since 2007 and can be used to track the water temperature impacts of any change in Tims Ford dam releases. Results of this monitoring are reported to USFWS and TWRA yearly.

Juggling Conflicting Needs

The Elk River water temperatures at Fayetteville are also compared to those on unregulated, benchmark streams in the region to ensure that the water temperature patterns mimic those on a natural stream whenever possible. Figure 9 shows a comparison between the Elk River and the Buffalo River water temperatures in 2014. The temperature patterns on both streams are similar, but the regulated stream is a few degrees cooler in the summer.

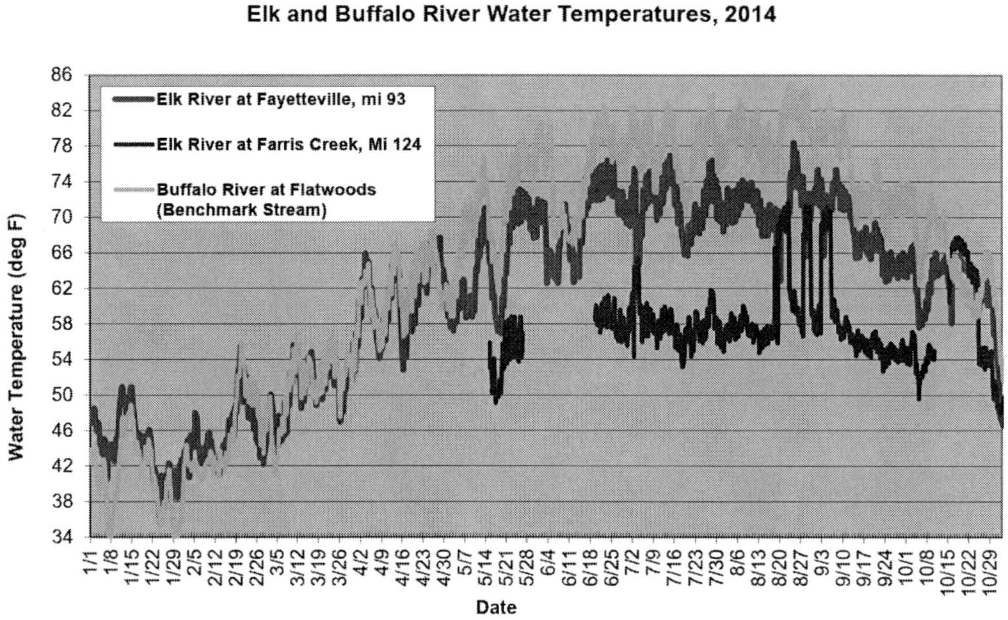

Figure 8. Elk River Water Temperatures compared to an unregulated Benchmark Stream

Sacrificing Hydro Power Revenue for more than half of the year

The preferred release alternative utilizing only the sluice and spillway from April 15 to at least October 15 (or later) generally releases a combined flow of less than 600 cfs to the tailwater, but more often the flow is 240 cfs or less. This alternative idles the hydro turbine for about 6 1/2 months of the year. The lack of hydropower revenue at Tims Ford Dam comes with a financial loss ranging from about $200K to $600K each year since 2008 according to TVA Hydro Value Analysis software that analyzes the efficiency of our hydro operations (TVA HVA, 2014).

Tims Ford was TVA's first hydroelectric facility retrofitted with a small generating unit utilizing the sluice outlet for the purpose of being able to generate some power while maintaining tailwater minimum flows for aquatic habitat improvement during periods when the turbine is not generating. The small turbine unit produced 0.5 MW of power and released about 60 cfs to the tailwater. Unfortunately, the small hydro unit is not presently in use due to damage incurred in the May, 2004 flood that requires overhaul of the entire unit to repair, because the small hydro turbine would only help TVA regain about 1/70th of the lost revenue from not using the large unit, therefore the $300K+ that would fix it does not have a favorable benefit/cost ratio.

Although the loss of hydropower revenue seems significant, Tims Ford is one of TVA's smallest power producing dams in terms of Megawatts, and the sacrifice of the power revenue in this location has been received positively by the USFWS in the protection of the endangered species. This tradeoff has allowed TVA to continue to have a positive working relationship with the USFWS and perhaps has helped retain greater operational flexibility at other hydropower projects involved in the consultation.

RESERVOIR IMPACTS FROM CHANGING WATER WITHDRAWAL ELEVATION

Changing *where* the water is removed from the reservoir does have some impacts on water temperatures in the reservoir, and reservoir water quality. Figure 10 shows two animation snapshot sequences of reservoir water temperatures and DO at different times during the summer for actual conditions in 2001 versus the 100% spill option for the same year, produced from Ce-Qual-W2 modeling (Cole, 1995). The Ce-Qual-W2 model breaks the reservoir down into vertical layers and horizontal segments, forming a two-dimensional mesh. Using a graphical post-processor, model results can be displayed in various ways, enhancing the interpretation of the results.

The reservoir model was utilized to show how changes in the location of the dam release (turbine vs. spillway) would impact water temperatures and water quality in the reservoir. In the 100% spill option, there is more cool water left in the bottom of the reservoir during the summer, and because more of the upper level water is being pulled through the spillway, there is less warm water near the top of the reservoir. In terms of water quality, the spill only option provides better DO in the reservoir for longer in the year than the actual conditions did.

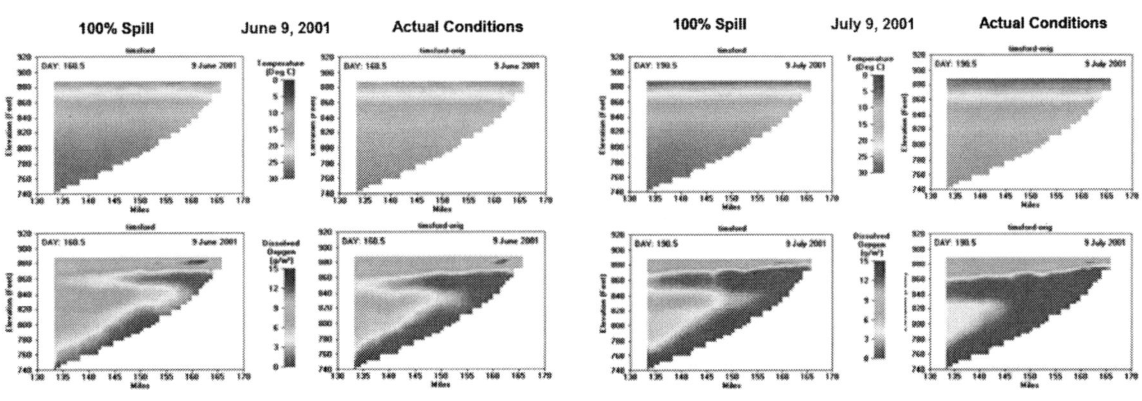

Figure 9. Reservoir Temperature and DO conditions in August 2001 for 100% spill and Actual conditions

BENEFITS OF ALTERED FLOWS FOR TAILWATER RECREATION

Changing dam releases can enhance or impede tailwater recreation activities. There are presently two major forms of recreation that have developed in the Tims Ford tailwater. These are trout fishing and float trips via kayak, canoe or raft.

More Float Trip and Wade Fishing Opportunities

Float trip outfitters cannot operate when the large hydro unit is releasing water at Tims Ford, because the 3800 cfs release is too deep and swift. Because of this, TVA had previously been running an 80 cfs minimum flow from the sluice on weekends to allow float trips, but still ran the turbine during the week. With complete cessation of turbine use for 6+ months, the outfitters can now schedule weekday float trips too. Flow rates over 600 cfs produce a velocity that is too swift to make floating enjoyable.

A low, steady flow is also desirable for wade fishing. The river is not generally comfortable or safe to wade in at flow rates greater than 350 cfs. For safe wading, the product of depth and velocity is generally of concern. TVA considers a product of depth and velocity of **four or less** to be safely wadeable, and depths of more than four feet are generally not wadeable for most people, no matter what the velocity is. Model-computed velocities were graphed for a variety of flow rates in the trout fishing reach, and are shown in Figure 11.

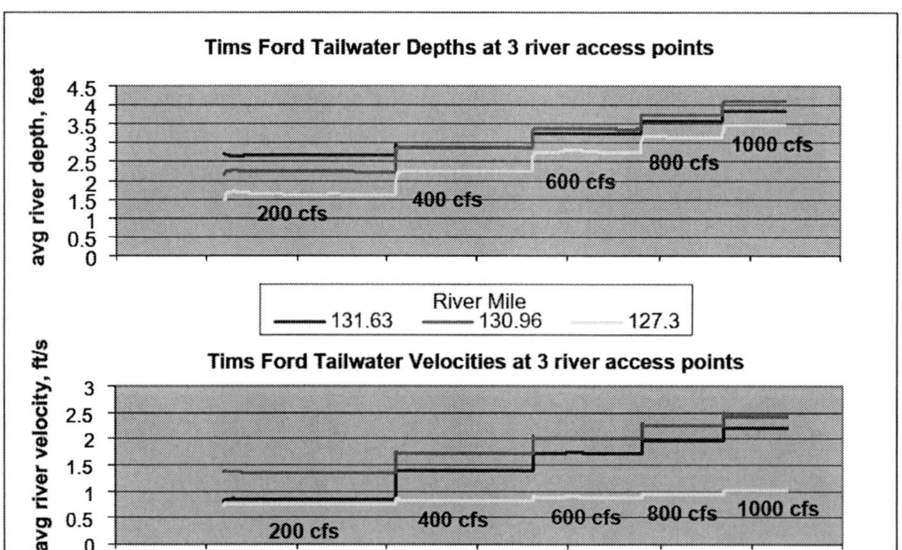

Figure 10. Average Depths and Velocities in Tims Ford Tailwater at 3 access points

Since changing the warm-season hydro operations, there is more time available for better quality tailwater recreation. The flow-duration curve presented in Figure 12 shows the percent of time the flow was greater than or equal to different values. If the safe limit for floating the river is 600 cfs, as an example, people could comfortably float the river about

68% of the time now, whereas when the turbine was used, it was about 45% of the time (subtracting out the very low-flow and the high-flow ends of the curves in the Figure 12).

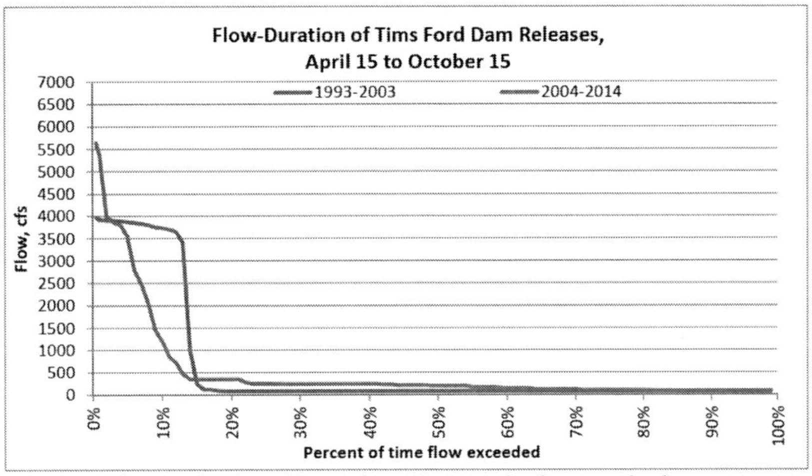

Figure 11. Comparison of Flow-Duration Curves before and after warm season flow modification

There is also a greater amount of time where the flow is between 80 and 350 cfs, and a smaller amount of time where the flow is greater than 500 cfs. The flows are generally also steady flows, released 24 hours a day. This intermediate flow rate provides a better experience for float trips, and still allows for a pleasant wade fishing experience.

Rather than sacrifice recreation in the tailwater for weeks at a time when higher flows need to be released, if the water temperature permits it, TVA would choose to spill a larger amount of water than the weekly average flow, but for a shorter amount of time, to pull the reservoir level back down to an acceptable level faster and allow tailwater recreation to resume as soon as possible. The only time of year this may not be feasible is in August or September, when reservoir water temperatures are at their warmest in the reservoir and a large spill release may cause water in the trout zone to reach the low 70's.

CONCLUSION

A number of dam release options will warm the tailwater temperatures below Tims Ford Dam significantly above the temperature of water from the turbine to make the tailwater a more suitable environment for endangered fish and other species. But only some of these release options are suitable to maintain both trout habitat near the dam and boulder darter habitat further downstream from the dam.

Eliminating use of the turbine during the warm half of the year has allowed water temperatures on the Elk River to better follow the water temperature patterns of an unregulated stream, which has benefited not only the endangered species in the river, but all of the surveyed fish species.

In addition, there have been benefits to tailwater recreation from the elimination of turbine use during the warm part of the year. This has allowed weekday float trips and more daylight hours in which the river can be accessed for wade fishing.

The economic losses from foregone hydro power generation at Tims Ford are significant, and have added up to several million dollars over the last few years; however the species and habitat benefits, as well as the increased recreation opportunities all have benefitted the region, and the losses may be greater if we had a less cooperative relationship with the USFWS.

Monitoring of water temperatures and aquatic species will continue so that all interested parties can continue to observe the effects of the river operations on the downstream environment, as every year brings different conditions and often, new challenges.

REFERENCES

Montgomery, 2006. *Tims Ford Tailwater Temperature Adjustment Study*, Montgomery, C. R., TVA, April 2005.

Parsly, 2006. I*nflow Adjustment Tool program*, Parsly, J. A., TVA, May 2006.

Petty, 2013. EFFECTS OF TEMPERATURE ON REPRODUCTION AND EARLY LIFE HISTORY OF BOULDER DARTERS, *ETHEOSTOMA WAPITI*, Petty, Missy, Pat Rakes, J.R. Shute, and Crystal Ruble, Conservation Fisheries, Inc. December 9, 2013

TDEC, 2003. *Upper Elk River Watershed (C0603003) of the Tennessee River Basin water quality Management plan. Nashville, Tennessee.* Available: http://www.state.tn.us/environment/wpc/watershed/wsmplans/upelk/

TVA, 1965, *The Tims Ford Project on the Elk River*, Tennessee Valley Authority, Planning Report No. 54-100, Knoxville TN, March 1965.

TVA HVA, 2001-2014. *TVA Hydro Value Analysis program*, Parsly, James A.

USFWS 2006. *Biological Opinion.* Section 7 Consultation - Routine Operations and Maintenance of TVA's Water Control Structures in the Tennessee River Basin, FWS Log No. 42430-2006-F-0146, October, 2006

Hauser, 2003. TVARMS ADYN/RQUAL, TVA River Modeling System one-dimensional model, Hauser, Gary E, January, 2003

Cole, 1995. Cole, T. and Buchak, E. "CE-QUAL-W2: A Two-Dimensional, Laterally Averaged, Hydrodynamic and Water Quality Model, Version 2.0," Tech. Rpt. EL-95-May 1995, Waterways Experiments Station, Vicksburg, MS, 1995.

ROCC 2013. Henry, T.H, et. al. "River Operations Consultation Compliance FY2012 Annual Report", Tennessee Valley Authority, April 2013.

DEMOLITION OF DAMS AND POTENTIAL DEVELOPMENT OF RUN-OF-RIVER PROJECTS USING ALTERNATIVE FISH SCREEN INTAKES

Jon Y. Kaneshiro, PE, GE, PG, CEG[1]
Don Dodds, PE[2]

ABSTRACT

Run of River (ROR) conduit hydroelectric projects with no or limited dams can be a significant additional energy resource. Many existing dam sites slated for demolition as well as many undeveloped quality sites still exist for conduit only (or with only limited or very small dams) projects. These sites are generally located in, but not limited to, upland forested areas with significant rainfall, low run off coefficients, and moderate stream gradients. A topographic map survey of northwestern California, western Oregon and Washington, and Southeast Alaska, show 85 potential sites with a total capacity greater than 2,000 MW that could be built for less than half the cost of equivalent wind power and 9 times more efficient than solar power. Plant output ranges from 150 MW to 4 MW with an average of 40 MW. In a world where green power needs are great, and when small ROR conduit projects are considered green, why are these resources not taken more advantage of, and what can be done to optimize or enhance these opportunities? The US Department of Energy website states that those issues are site and technology dependent.

This paper examines and proposes one part of the possible solution to the main environmental challenge – protection of even the tiniest of fish. This part of the solution provides for a very economical intake approach for dams, reservoirs, and ROR conduit sites that differ from more costly approaches to protect fish. It is fish friendly, and builds on available water and waste water technology. To understand this somewhat innovative solution for an intake, one must understand the behavior of the fish. Fish follow the flow and are attracted to horizontal flow, but not vertical flow, a priori! The proposed intake consists of a series of horizontal wells buried in engineered graded coarse filter beds fitted with back flushing elements for routine maintenance as required to remove even the finest clays. The intake of water is in the vertical direction not in the horizontal direction. This paper provides examples of sites that could be developed as well as engineering details of the layout and sizes of the intakes (and outlets). As it turns out, the proposed outlet designs are similar in layout to the intakes and provide for beneficial grounds for fish egg laying. Moreover could such an approach to intake technology be used for even larger scale intakes for major water supply projects? Some of the technology discussed in this paper may be protected. This paper is based on the Dodds and Kaneshiro's Hydrovision 2011 paper with only slight modifications to reach out to a broad based audience and disseminate the concepts and ideas, some of which may be worth incorporating into designs of intake and outlet systems for consideration by planners, designers, and owners.

---

[1] Senior Engineering Manager, Parsons Corporation, 110 West A Street, Suite 110, San Diego, CA 92101, 619-515-5122, jon.y.kaneshiro@parsons.com
[2] President, North Pacific Research, 9676 SE Chula Vista Court, Happy Valley, Oregon 97086, don@northpacificresearch.com

---

Alternative Fish Screen Intakes

DAMS, RESERVOIRS AND FREE RUNNING RIVERS

Currently in this country, all good sites for building dams to produce power are already taken. Many dams are being removed because they block "free running" rivers. Strangely enough, this technology can make dam removal more green and efficient without the loss of the power. However many good sites as well as sites slated for dam removal are still available for ROR Hydro projects. Currently according to the DOE, hydro power is renewable but not green because they have negative effects on the environment. These effects are broken down into two categories: land use issues, and water use issues.

Typical Land Use Issues

A search of environmental publications produced the following list of land use objections to hydropower without specific examples of the damage.

1. Submerges land,
2. Damages riparian zones of the rivers,
3. Impacts recreation,
4. Holds back debris,
5. Slows stream current.

These examples appear to be rather broad. Any inquiry for more definite criteria was refused. These issues must be of concern because of surface structures, like the dam and the reservoir. Since ROR projects do not use a dam or reservoir, it is reasonable to assume that all these land use issues do not apply to ROR hydro small or large. No specific land issues are listed that were caused by the intake, penstock, powerhouse or outlet. Further, penstocks and some powerhouses are presently buried and, as such would have little surface impact. Thus, by elimination, the intake and outlet, no surface features remain to cause land use problems.

Typical Water Use Issues

The same search produced the following common water use issues.

1. Disrupts fish migration upstream and down,
2. Alters the chemical and physical characteristics of the water,
3. Changes the silt and nutrient loads of the streams,
4. Temporarily reduces stream flow effecting fish passage,
5. Some smolt killed when they pass through the turbine

Again, without a dam or reservoir, fish passage up and down stream is unaffected, and there is no change to the silt and nutrient load, nor any change in the water temperature or chemical properties. Thus, by elimination, that leaves only maintaining minimum stream flow and smolt kill as the issues that make large ROR currently unacceptable as green power. Can these issues be removed or reduced significantly?

FISH PASSAGE FLOW

Fish passage water is handled successfully and routinely for typical small run of the river hydro, or otherwise it would not be categorized as Green. Adequate fish passage water is a function of many factors, only one of which is how much water is sent to the turbines. The percentages of the total water, needed for fish passage, or used for power need to be assessed rather than limiting green power plants by some arbitrary power output. In some instances, the power companies seem to stay away from run of the river plants because the licensing requirements are more stringent and difficult.
There are many rivers in upland areas with significant rainfall and steep stream gradients that produce flows considerably larger than the minimum fish flow for a large portion of the year. It is more reasonable to assume that fish passage issues are site dependent and they should be treated so, and each power plant small or large should be evaluated on the site parameters. Potential sites should not be eliminated, simply, because they are above some arbitrarily power production limit. The bottom line is that currently fish flow can and is routinely managed.

Smolt Kill

This problem is isolated to the intake structure because the smolt are only killed when they pass through the intake and into the turbines. Whereas smolt kill has been reduced at great expense, by modifying the turbines and intakes screens it has not been eliminated. Presently, it is impossible to build a hydro plant, normal impoundment or ROR (small or large) without killing some smolt. For years, the solution to making hydro green revolves around ways to keep the smolt out of the penstock or improve their passage through the turbine using various improvements to the equipment. There is always another side to every problem.

Smolt Behavior

No one to date has recognized that smolt itself is not an innocent bystander, but part of the problem. In order to keep the smolt out of the turbines, we need to think like a smolt. Consider that as a fry changes to a smolt it consumes its egg sack, which provides it with sustenance. When it is empty, they get hungry and since there is little food in a gravel bed, they are driven by their instinct to exit the safety of gravel and move into the river to look for food.

Once into the river, they are confronted with current that comes in two major categories, horizontal and vertical. Vertical current produces a relatively short trip in most rivers, eliminating vertical movement as a direction for the food search. So they chose to swim horizontally. The smolt senses any vertical movement of the water and adjusts its course instinctively to maintain its horizontal path through the water without continuously bouncing off the bottom. This still leaves the smolt with two choices; look for food against the current or with the current. The smolt are dumb, but they choose to go with the flow because it takes less energy. It is, however, just a coincident that "with the current" is also downstream. Unfortunately, this is not as romantic as the notion that their instinct magically leads them downstream to the mother ocean, but it is infinitely more useful. Thus for thousands of years smolt have followed their instinct to swim with the horizontal current to lead them to food and coincidentally to the sea. This simple system worked until someone built a dam and changed the rules, whereby now some horizontal currents lead to turbines and fish deaths.

Mitigation of Smolt Behavior by Design of Intake

Consider that presently all water that enters a penstock is taken from the rivers horizontally simply because it is convenient for humans. Suppose we focus on the smolt rather than on the turbine for a change, and the solution to smolt kill becomes obvious. Since neither smolt nor fish follow vertical current anywhere, the solution is to extract the turbine water out of the river vertically. Then the smolt's will become part of the team and their instinct will work with us to avoid turbine blades.

MINIMUM IMPACT INTAKE (MII)

There are several different ways to design a Minimum Impact Intake (MII); however, the most efficient is shown in Figure 1. It is essentially a super-sized modified French Drain that extracts the penstock water out of the river bottom, vertically through an engineered porous stream bottom collection system. Similar to the French Drain, the collection system uses clean coarse gravel and a perforated pipe to collect the water and subsidiary pipe to carry the water to the penstock. This is not a sand filter but a gravel screen made with coarse gravel that relies on the instincts of the smolt for success. Figure 2 shows the size of the small grain portion of the screen. (The silver circle in the lower left is a Quarter). The smolt's instinct to not swim 6 ft vertically down through gravel. The filter is generally processed from insitu riverbed, thus it looks and acts like the natural stream. Estimates of the cost of a typical surface concrete intake and screen system compared to the MII show the MII to be simpler to build and considerably less expensive.

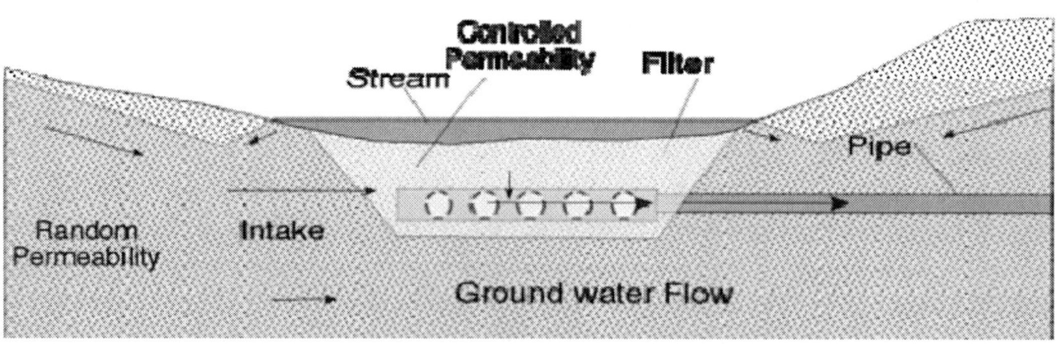

Figure 1. MII Intake Schematic

Figure 2. Fine grain portion of MII Filter

The vertical velocity through the filter is controlled by the power intake valve, the pore size and the overall size of the filter. Even though the pore size is about 2 cm in diameter, the velocity through the filter is generally less than 2 cm/sec because the intake regulation valve controls amount of the water extracted and the velocity of the water through the filter. The extreme low velocity is to protect terrestrial life and reduces filter maintenance. For example, a gravel filter that is 30 meters wide and 150 meters long can provide a streambed interface of 4500 meter^2 (about a football field in area) and, with a controlled vertical velocity of 1.5 cm/sec (typical permeability of clean gravel is about 10 times that value) can provide a flow of about 70 cubic m/s. Dropping that quantity of water 100 meters through a penstock can generate about 120 MW of power with no smolt kill and no disruption of terrestrial life.

Potential Green ROR Hydro and Intake Sites

The typical site for the optimum use of the MII system is commonly in upland forested areas with high rainfall, low runoff coefficient, steep terrain, V-shaped valleys, and a small flood plain with a relatively stable streambed. The average stream gradient for example, is about 0.011 for pipeline projects, which parallel the streambed and about .009 for tunnel projects, which can take a shorter route. A quick topographic map survey of north western California, western Oregon and Washington and SE Alaska, show 85 potential sites with a total capacity greater that 2,000 MW that could be built for less than half the cost of equivalent wind power. Plant output ranges from 125 MW to 6 MW with an average of 40 MW. The smaller plants are pipeline penstock projects and the larger are tunnel penstock projects. In a world where green power needs are great, this potential should not be ignored.

Technology Behind the Minimum Impact Intakes (MII)

Whereas the MII uses no new technology that requires billions to test and prefect, keep in mind, that it uses the old technology differently. It is important to emphasize here that the MII is definitely not a standard sand filter or infiltration gallery, and as such does not have the maintenance problems associated with those filters. The MII is composed of 4 to 15-cm gravel with pores size about 2.5 cm in diameter. The troublesome silt and clay, (carried load), that routinely plugs sand filters and infiltration galleries, flows through the MII filter and for that matter through the turbines as well. In order to understand better how the MII works we need to review some basic principles of river and groundwater hydrology.

River Hydrology and MII

Several terms commonly used in river hydrology are important to proper location and design of MII. Terms of interest are Influent and Effluent Streams, Surface load, Carried load and Bed load. Streams are separated into two distinct types depending on whether the water flows out of the stream, Influent, into the ground, or into the stream, from the ground, Effluent streams. The groundwater scientists applied these designations before river hydrologists, thus the apparent illogical nomenclature. Effluent streams are commonly but not always found near the upper reaches of the river with steep gradients, and are the ones best suited for MII. A site visit can supply information on the rate of stream erosion or deposition and general information on the river's local bed load. ROR hydro cannot be successfully built anywhere, for that matter neither can dams, wind turbines or many other forms of power generation.

MATERIAL TRANSPORTATION

Stream hydrology affects the design because besides the water, the river generally carries three different types of materials called surface load, suspended load, and bed load. Each of these zones has their problems.

The surface load is the material carried along on the surface of the stream, like algae, leaves, branches, leaves, beer cans other floating debris and even terrestrial life that enters the stream large or small. The stream surface is a moving target, rising and falling with the flow carried by the stream.

The suspended load is those particles, which circulate between the surface and the bottom. Commonly, considered to consist of clay and silt because the turbulent flow constantly circulates this material and keeps it afloat. However, technically smolt and other aquatic life, are also a part of the suspended load. All of these particles are not dense enough to sink to the bottom, and others have control of their position and a small intellect that directs their path. This fact is often ignored. But it is important if we are to prevent smolt kill associated with hydropower generation.

The bed load is the cobbles to coarse-grained sand size particles that are bounced along the river bottom, and moved erratically downstream by the current. This load is heavily dependent on the stream velocity, the higher the velocity the large the particles that can be moved. Thus, as the velocity increases the smaller particles are swept downstream. Further increase will move gravel and eventually high velocity flow will move the cobbles. This material is transported during peak flows when the water velocity is high, and deposited when the river velocity slows down. It is this bed load that controls the process of erosion and deposition. In a stable stream the bed, where both erosion and deposition are minimal, the bed load is transient, what moves in one year moves out the next and the river bed is fairly stable.

Standard Horizontal Surface Intake Problems

A surface intake must be designed to extract the minimum necessary water during low flows and exclude the entire surface load. Thus during normal flows the intake is over sized. Excluding, all of the surface load require a trash rack that extends above the high flow and below the low flow to keep the surface load material out of the turbines. Trash racks require constant maintenance to keep the floating debris, from blocking the water flow. Further, since part of the water is taken from the middle of the stream it also has to deal with suspended load. The clay and silt particles can be passed through the turbines without a problem. However, the aquatic life needs to be excluded. In order to prevent this from happening it is necessary to forcibly exclude them from turbine water by using subsurface screens because they have no way of knowing why they cannot swim downstream. Presently that is not fool proof. Occasionally aquatic or terrestrial life can be trapped against the screen and injured or killed.

SUBSURFACE VERTICAL INTAKE, MII, PROBLEMS

One of the important factors for locating the MII is a stable streambed. The bed load for a stable stream by definition is the bed load left in place during the last high water episode and will be transported further downstream with the next high water episode. Thus, the problem of the deposition properly located MII is minimized because the upper level of the filter constantly being changed by Mother Nature every high water period. Thus, in a perfect world, the surface load, the bed load, the smolt and other aquatic and terrestrial life all move downstream, and the suspended load passes through the turbines. Any medium sand or small gravel that can lodge in the pores of the filter will reduce the permeability of the filter. However, since the filter is relatively inexpensive to construct it can be oversized, and also the original permeability of the filter is much higher than needed because the turbine valves regulate the flow. The need for a low rate of flow through the filter is controlled by the safety concerns for terrestrial life and a major effect on how much groundwater is captured by the intake. If any sand or gravel particles that make it through intake that might damage the turbines can be trapped by a gravity separator (sedimentation basin or sandbox) in the valve house or other location before those particles can enter the penstock. Like any filter it can be back washed or in serious flow impediment, the top foot or so can be super suckered off, washed and replaced, with little to no down time. Site selection of the intake should consider sites which would minimize sedimentation of fine particles but also consider the possibility that foliage (autumn leaves falling) so as to minimize the amount of maintenance of the gravel beds. Ironically, foliage may be more of concern in uplands area that can be heavily vegetated, while downstream sites with wide rivers may be more conducive to large MII style water supply intakes in terms of clogging due to foliage.

Groundwater Hydrology

Permeability of the local ground and the streambed, and the location and gradient of the groundwater table are all important design factors. See Figure 3. The water flows out of the page and left and right into the stream, but always parallel to the phreatic surface. Natural stream permeability varies greatly from place to place. Typical streambed material at an MII site would be classified between a poorly graded silt sand and gravel mixture with a permeability of about 0.000001 cm/sec and a well-graded gravel, (GW) 0.001 cm/sec (USBR, 1979). The permeability of a typical filter would be similar to clean gravel between 15 and 1.5 cm/sec (Leonard, 1962). The differences in the permeabilities control the design of the MII.

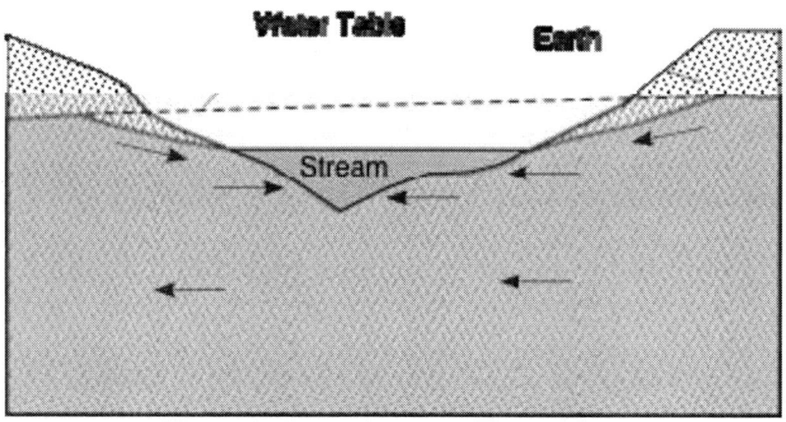

Figure 3. Groundwater Effluent Flow

In order to understand the process we need, step up in complexity slowly and the next step is a look at the typical groundwater conditions around a normal well, because in this case the water flow is symmetrical parallel and perpendicular to the page. The dotted line again shows the original ground water surface before pumping the well. See Figure 4. The arrows show the water flow. Note that all the water again flows parallel to the phreatic surface, i.e., the ground water table. If the water is withdrawn from the well faster than the ground water can recharge the well, a cone of depression is formed around the well but the water still flows parallel to that phreatic surface. If more is pumped out than can flow in, the cone of depression becomes deeper and eventually the well will be pumped dry. The volume of the water that can be pumped from the well and the slope of the cone are dependent on the diameter, the length of the well and the permeability of the surrounding ground. The first two properties of the well can be controlled by the design and the permeability can be improved by a process called hydrofracing, which increase the permeability.

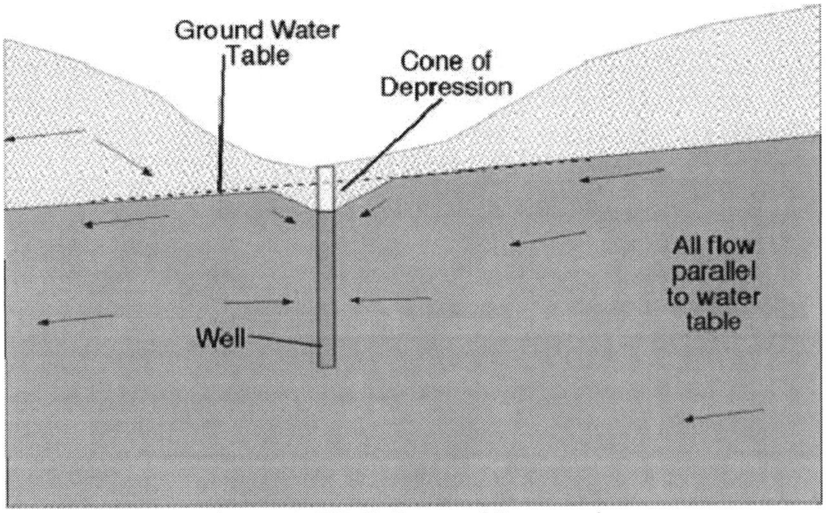

Figure 4. Well Cone of Depression

In the case of the MII, the stream and the system become more complicated. Again, the MII alters the phreatic surface, but in this case, by increasing the length, width and permeability of the intake, the designer can alter the flow into the intake by several orders of magnitude. Since the phreatic surface under the stream is horizontal, all water must flow horizontally in the stream and directly under the stream. See Figure 5. The ground water flows both ways into the stream and also flows perpendicular to the page. In this case, the MII appears to draw no water out of the stream and only water out of the ground. Again, like in the well we can enlarge the interface area and alter the permeability of the filter. However, the water entering from the ground is controlled by the ground permeability outside of the MII.

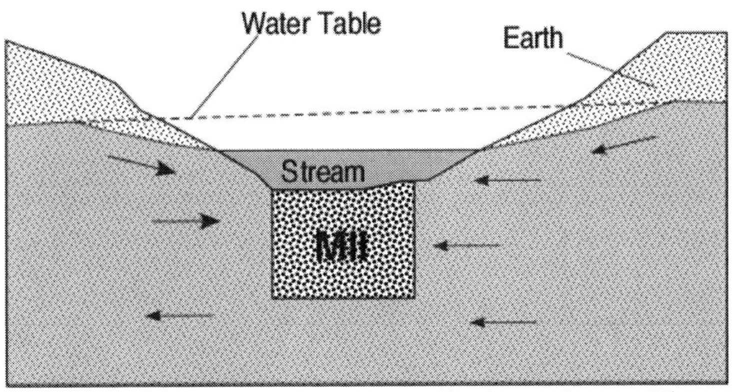

Figure 5. MII Depression Perpendicular to Stream Flow

However, in truth some of the water is still supplied by the stream, because unlike the well, the presence of the stream overhead affects the inflow. This can be seen if we draw a section parallel to the stream flow. See Figure 6. In this section the surface water is drawn down upstream of the intake, because water is being extracted, and remains down for the length of the intake. This situation is independent of how the water is extracted, i.e. vertical with an MII or horizontal with a traditional intake. Finally, the actual slope of the surface is not readily apparent on the ground, but it occurs. Since the stream surface is being drawn down, the phreatic surface is now sloping down near the upstream end of the filter and the water flowing parallel to this surface has a downward component that will supply a portion of the water to the MII. Downstream of the intake, the ground water replenishes the stream and the stream surface again becomes parallel to the stream bottom.

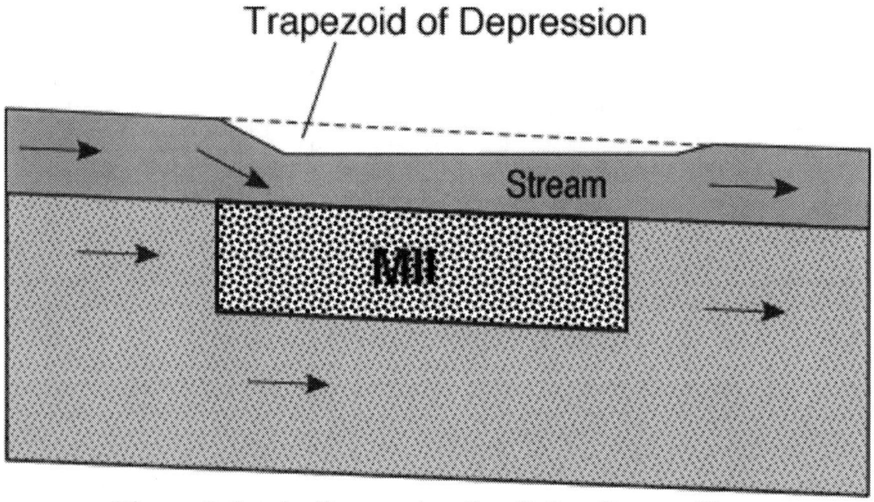

Figure 6. Intake Depression Parallel to Stream Flow

The Result

Ninety seven percent of the world's potable water supply is ground water (McWhorter and Sunada, 1977). Seventy five to eighty percent of that water flows into the ocean without ever reaching the surface. If 15% of the turbine water comes from the ground then the plant operator has two choices, generate 15% more power, while using the same surface flow or reduce the amount of water extracted from the surface flow by 15% and produce the same power. In either case the stream flow below the outlet is increase by 15% and the system provides an economical way to recapture large amounts of ground water for municipal, agricultural, or environmental applications. On the troubled Klamath River where dams are being removed, there are seven possible sites for this technology. Because the flow is increase by ground water at each site, if all were replace by MII power plants, surface fish flow could be increased by as much as 1000 cfs.

The MII Technology's Impact on Dam Removal

This technology can also be used to remove a dam in an orderly manner the same way it was built. If the stream water is by passed around the reservoir and dam, then the entire dam structure can be removed in an orderly manner and the reservoir can be clean and restored to its pre dam condition with no silt impact downstream. If during the dam removal an MII is installed near the upstream end of the old reservoir then after the dam is removed, the water can be diverted back into the river, and a power plant can be installed in the diversion tunnel. Once that the powerhouse is operational then the MII can be activated and a run of the river power plant will provide more green and renewable energy than the original dam and provide income from the power to pay for the removal. The result is the dam is gone, without power lose, smolt are protected, no surface structures, and the stream flow is increased.

The Maximum Impact Outlet

So far, we have been only talking about the intake, but this technology can also provide environmental value for the outlet other than no surface impact. Most of the time, the outlet water is just spilled into the stream, which is inexpensive and has minimal land use issues. However, a significant benefit can be obtained by returning the water to stream in a manner similar to the intake filter system only this time the flow is reversed and water flows upward into the stream. This occurs in nature from time to time and is referred to as artesian flow.

Interestingly enough, research shows that salmon prefer artesian conditions for spawning because the upward flow makes it easier for them to excavate their spawning pits and also slightly opens the pores in the gravel and cleans the fine material out of the pores, making it easier for roe to hide in the river gravel until they are more capable of protecting themselves from predators. The most prolific spawning ground in the Pacific NW is the Hanford reach on the Columbia (where co-author Dodds managed the groundwater department for Rockwell on the Basalt Waste isolation project, in charge with mapping the groundwater at the Hanford site). It is not a coincident that the ground water in that area is artesian.

Neither the salmon nor the roe care what causes the upward flow. Designing an outlet gravel filter screen is similar to the intake gravel filter screen. Thus, the spawning salmon would instinctively believe it to be artesian water and chose the gravel screen area for spawning. Thus, every ROR hydro plant would come with its own wild salmon spawning ground and increase the population of the prized wild salmon.

Construction & Maintenance

The construction of the filter is quite simple, because it is essentially building a French Drain by digging a hole in the streambed, installing perforated HDPE pipe, and backfilling the hole with selected porous gravel. Much of the gravel can be screened from the existing riverbed. Whereas the construction most likely will require diversion of the stream, that diversion is short and can be simplified by using the penstock to divert the water around the construction site during the short time the intake and outlet are constructed. The cost of the MII intake is about half that of a standard intake due to the simplicity of the construction. According to Russell (2011), wind power costs about $2,000/kwh. The construction cost of a MII Hydro power plant costs about $800/kwh. The maintenance of any filter depends directly on the size of the openings in the filter. Obviously, a filter with a 2.5 cm opening will plug less often that one with a 0.003 cm opening. The savings in maintenance alone will pay for the construction of the intake. Further, the filter construction is so inexpensive that it can be sized larger than needed so that some degradation can occur before power generation capacity is affected. Next, in the unlikely event that small particles should clog the filter, it can be back washed like any other filter using compressed air adapted from wastewater technology.

Alternative Fish Screen Intakes

CONCLUSIONS

Using this MII system, we can extract large amounts water sufficient to generate hundreds of Mega Watts of inexpensive, green, ROR hydro without land issues, while significantly reducing fish issues, increasing river flow and adding wild salmon spawning habitat. These benefits are significant and certainly deserve close consideration and study. Smaller pilot scale ROR MII intakes would be recommended not unlike those for small municipalities or for irrigation for study.

The only new technology is the location (sub river bottom) of the intake and the outlet. If the MII fails, the risk involved in this technology consists of the cost of a normal intake to replace the MII, since the penstock and powerhouse are in place. Further, based on contractors estimates, the MII costs about half of the standard intake. The reward of opening up many green, fish and environmentally friendly ROR hydropower plants to a world that needs green power is certainly worth a small risk.

The Big Picture

Finally, it is of interest to look at the global point of view. The system proposed in this paper would look schematically like Figure 7. On the Left of the figure, we have the Run of the River hydro project with the MII extracting water and sending it to a generator. The water is then put back into the river and it flows to the ocean. There the water is evaporated by solar power, and rises into the sky as clouds. The clouds are then blown over the mountains by wind power, where the water condenses into rain and falls on the mountains. It then flows down and through the ground into the river where it passes into the MII and starts the journey over. That is as close to perpetual motion as humans can currently get and without a doubt the most efficient use of wind and solar power. We can't build these hydro plants everywhere, but we should be building them where ever possible because the efficiency of Hydro is greater than any other energy technology by a factor between two and nine depending on the other technology.

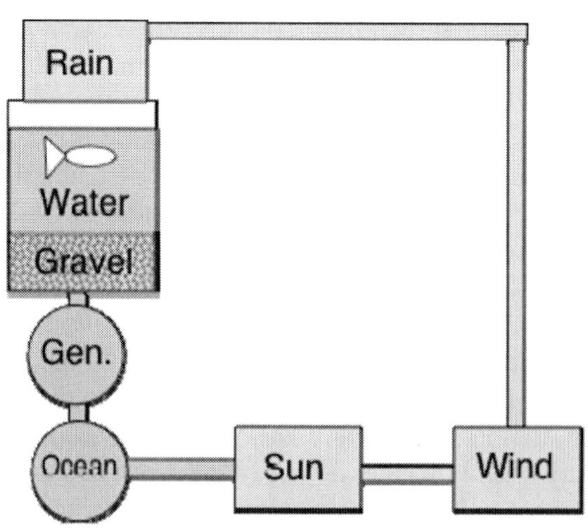

Figure 7. The Big Picture

Environment Impact

Finally, the MII's presence is significant because of its absence. Figure 8 shows the surface impact of the MII on a typical stream. Note no MII actually exists at this site, but the yellow line indicates where one could be built. However, the site is a typical site for optimum performance of the MII. That is, the riverbed is composed of gravel and cobbles, the normal bed load for this site, and there is also no indication of severe erosion or deposition indicating that the riverbed is stable. Maybe more important, this is a streambed that children can play in, rafters can camp on, and fisherman can seek out the elusive cutthroat, all unaware of the intake's presence.

Figure 8. The MII as it might look in Operation

SUMMATION

In summation, this technology provides the following benefits:

1. Works with the smolt's instinct to go horizontal downstream.
2. Opens thousands of potential ROR hydro river sites.
3. Simplifies permitting because surface features and fish concerns are eliminated.
4. Supplies water for Up to 300 MW plants.
5. Eliminates visual impact.
6. Requires no trash rack or screens maintenance.
7. Encourages wild spawning.
8. Recaptures groundwater.
9. Improves dams removal while still providing green renewable power.
10. No trash racks, or fish s creens to clean. Over the lifetim e of the project the M II will more than pay for itself in savings on maintenance cost alone.
11. Proven technology. Requires little resear ch and developmental expense relative to other new power generation systems.

ACKNOWLEDGEMENTS

Thanks to Dave Driscoll CEO of Geotechnical Resources Inc., Beaverton, OR, for review of the 2011 paper and concepts.

REFERENCES

Dodds, D. and J.Y Kaneshiro, 2011, Greening up large run-of-the-river hydro power, Hydrovision Proceedings, Sacramento.

DOE [United States Department of Energy] website accessed February 2011.

Leonard, G.A., ed., 1962, Foundation Engineering, McGraw-Hill.

McWhorter, D. and D.K. Sunada, 1977, Groundwater Hydrology and Hydraulics

Radtke Russell, P, 2011, Engineering News Record, April , 2011; pg 13

USBR [United States Bureau of Reclamation], 1979, Design of small dams.

GEOTECHNICAL MEETS GEOLOGY — INVESTIGATING THE CARMEL RIVER REROUTE AND DAM REMOVAL PROJECT

Kami Deputy, PE[1]

ABSTRACT

San Clemente Dam was constructed between 1920 and 1921 and was built to impound drinking water for the Monterey Peninsula. It is a 106 feet high, 300 foot wide concrete arch dam with the original capacity to store 1425 acre-feet of water. Since the dam has been commissioned, the reservoir has silted in with over 2.5 million cubic yards of sediment and impounds less than 100 acre-feet of water.

During the 1990s the California Department of Water Resources – Division of Safety of Dams issued a safety order, finding that the San Clemente Dam could fail during an earthquake or probable maximum flood. Following an environmental impact study, California American Water presented the option to remove the dam. Concurrent with this option, geotechnical investigations were performed which led to the development of an indicative design approach for removal of the dam, rerouting the Carmel River, and stabilizing the reservoir sediments in place.

The first year of this multi-year project included a geotechnical investigation program consisting of geologic mapping, soil/rock core borings, in situ permeability testing, test pits, cone penetration testing, and geophysical logging of select boreholes. The investigation was performed during a two month window before the site was closed for winter. The project team had to acquire permits, perform surveys of the existing site conditions including animal/aquatic life, and develop and obtain approval of a Field Investigation Plan. Obstacles onsite included accessing drill sites that varied from soft surface soils to steep slopes, methane gases within the strata, collecting soft soils below the water table, and drilling and recovering highly fractured rock.

INTRODUCTION

San Clemente Dam (SCD) is a 106-foot high concrete arch dam located approximately 18.5 miles from the Pacific Ocean on the Carmel River in Monterey County, California (Figure 1). The confluence of the Carmel River and San Clemente Creek is located just upstream of the dam. When the dam was constructed in 1921, it had a reservoir storage capacity of approximately 1,425 acre-feet. Today the reservoir has been filled by more than 2.5 million cubic yards of sediment, leaving a reservoir storage capacity of approximately 70 acre-feet as of 2008 (MPWMD).

[1] Project Professional, Kleinfelder, Denver, CO 80202, kdeputy@kleinfelder.com.

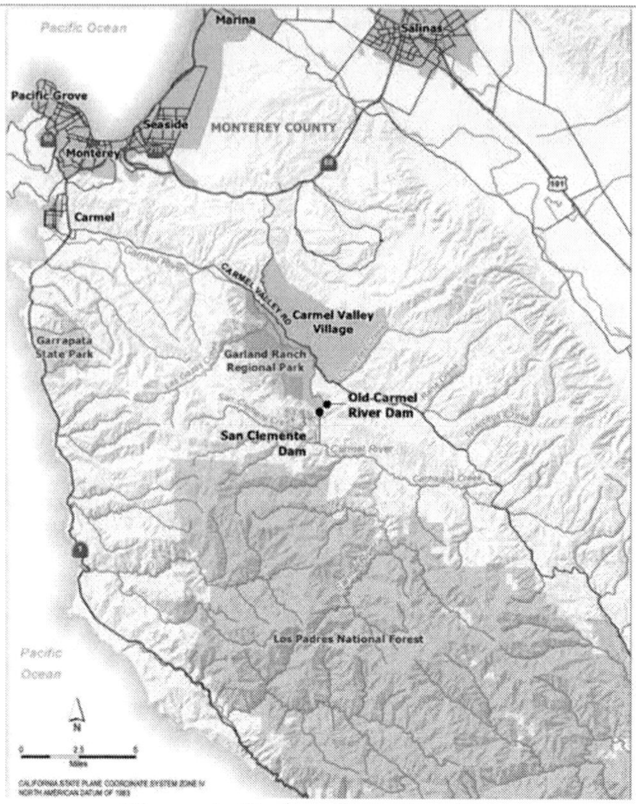

Figure 1. Project Location Map.

PROJECT DESCRIPTION

California American Water (CAW) developed an approach with State Coastal Conservancy, National Marine Fisheries Service, and the Planning and Conservation League Foundation to remove San Clemente Dam and implement the objectives of the Carmel River Reroute and Dam Removal (CRRDR) Project to address the public safety and ecological issues posed by the dam. In December 2007, the California Division of Water Resources (CDWR) Division of Safety of Dams (DSOD) certified the Final environmental impact report, and in February 2008 CDWR indicated that the dam safety issue could be addressed through implementation of the CRRDR.

The CRRDR accomplishes dam removal while reducing the volume of accumulated sediment requiring excavation, placement and stabilization. The concept involves rerouting the Carmel River into San Clemente Creek at a location approximately 3,000 feet upstream of SCD, bypassing the majority of accumulated sediment in the Carmel River. An overview of the key CRRDR elements is shown on Figure 2.

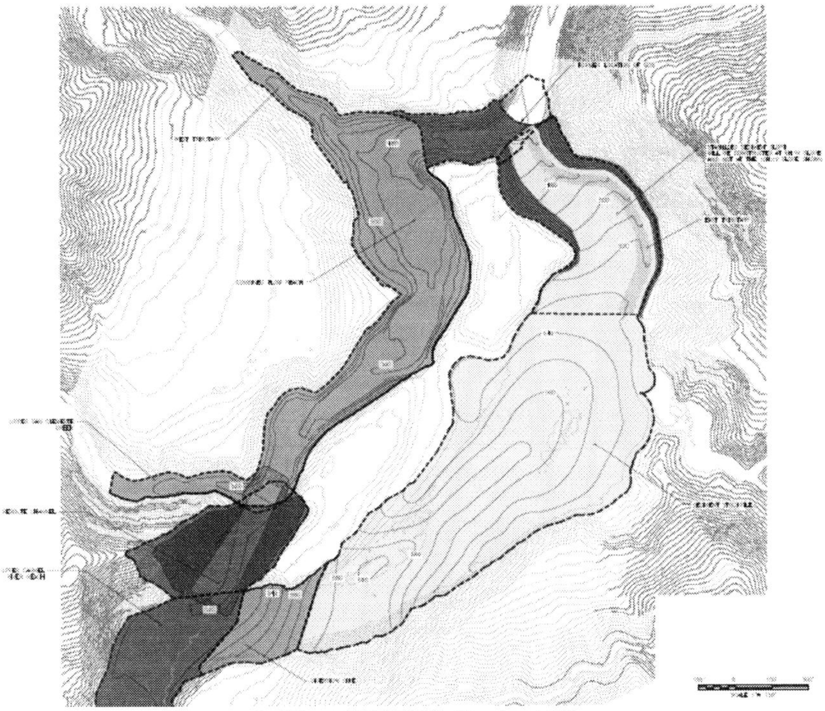

Figure 2. Key CRRDR Project Elements.

The bypass (Reroute Channel) will be excavated through the drainage divide between the Carmel River and San Clemente Creek. Upstream of the Reroute Channel, the Upper Carmel River will be excavated and restored to transition to existing channel grades at the upstream end of the project limit. The new Combined Flow Reach (formerly the San Clemente Creek arm of the reservoir) between the Reroute Channel and the present SCD location will be restored with a focus on steelhead passage. Upstream of the Combined Flow Reach, the Upper San Clemente Creek will also be excavated and restored to transition to existing channel grades at the upstream end of the project limit. Sediment excavated from the Combined Flow Reach (CFR) will be relocated to the abandoned Carmel River arm, and the sediment in the abandoned Carmel River arm will be stabilized in place to form the Sediment Stockpile (abandoned reservoir).

A Diversion Dike will be constructed upstream of the Sediment Stockpile to divert the Carmel River flow into the San Clemente Creek drainage through the Reroute Channel and to prevent flow from entering the upstream end of the Sediment Stockpile. The Stabilized Sediment Slope will be constructed at the downstream end of the abandoned Carmel River arm to retain sediments in the Sediment Stockpile area. A new channel will be constructed to convey flows from the East Tributary across the surface of the Sediment Stockpile and down the Stabilized Sediment Slope. The new channel will be designed to fit in with the natural setting and character of the site.

After the construction of the major structures described above, the entire site will be revegetated with native grasses, brush, plants, and trees as part of the habitat restoration of the site. The Old Carmel River Dam, located downstream of SCD, will be removed as part of the project and finally, the removal of SCD will complete the CRRDR. After project completion, the CAW property around SCD will be conveyed to the U.S. Department of the Interior Bureau of Land Management.

GEOLOGIC SETTING

The CRRDR project is located on the southwest margin of the Carmel Valley, within the Coast Ranges Geomorphic Province of Central California. This province is comprised of a discontinuous series of northwest-southeast trending mountain ranges, ridges and intervening valleys characterized by complex folding and faulting and the dominant structural regime in the region. Geologic structure within the Coast Ranges Province is generally controlled by the San Andreas fault system, which is a major tectonic transform plate boundary. This right-lateral, strike-slip fault system extends from the Gulf of California in Mexico to Cape Mendocino in northern California and forms a portion of the boundary between two tectonic plates. In this portion of the Coast Ranges Province, the Pacific Plate (located west of the transform boundary) moves north relative to the North American Plate (located east of the transform boundary). Deformation along this plate boundary occurs across a wide zone that is referred to as the San Andreas fault system.

The project is located within the Salinian Block, which is one of the distinct continental terranes of the central Coast Ranges. In the region, the Salinian Block is bounded by the San Andreas fault on the east and the Sur-Nacimiento fault zone on the west (Page, 1966). The basement rock of this block is composed of Cretaceous age (about 145 to 65 million years old) granitic and Jurassic age (about 200 to 145 million years old) pre-batholithic metamorphic rocks. Major orogenic features within the Salinian Block include the Gabilan Range and Sierra de Salinas to the east/northeast of the site, and the Santa Lucia Range (within which the site is located).

Overlying the granitic and metamorphic basement rocks of the Salinian Block are Cretaceous, Tertiary and early Quaternary (approximately 65 to 1.4 million years old), marine and continental sedimentary rocks and occasional Tertiary volcanic rocks. These Cretaceous and Tertiary age rocks are typically folded and faulted into a series of generally northwest-southeast trending blocks, largely as a result of stresses related to movement along the San Andreas fault system. In the vicinity of the project, the Cretaceous and Tertiary age rocks are overlain by younger Quaternary age (1.8 million years old to present) alluvial, colluvial, and landslide deposits.

The project site has been mapped by Dibblee and Minch (2007), among others. Dibblee and Minch (2007) indicate the northwest-trending Cachagua fault transects the site, extending onto the site along the West Tributary, across the main saddle area, to the East Tributary and beyond. Two geologic units are in contact at the site along the fault. The reference indicates the area northeast of the fault (including San Clemente Dam) is underlain by Jurassic to Cretaceous age quartz monzonite of the Salinian Block. The project area southwest of the fault is mapped as being underlain by a complex of granitic and meta-sedimentary rock. Dibblee and Minch (2007) identify a third bedrock unit, consisting dominantly of granodiorite, which underlies the project site south of the proposed Reroute Channel. The level areas to the west of the dam and north of San Clemente Creek are mapped as being underlain by Pleistocene age (1.8 million to 11,000 years old) dissected alluvium.

The project site and surrounding vicinity were included in the Multi-Jurisdictional Hazard Mitigation Plan, Monterey County, California (MCOES, 2007). The plan's assessment of the seismically induced landslide hazard designates the slope southwest of the proposed Reroute Channel and the slope adjacent to the proposed right (east) abutment of the Stabilized Sediment Slope as highly susceptible to landsliding during a seismic event. The two locations approximately coincide with debris slide scar areas identified by URS (2012). According to the hazard mitigation plan, the remaining slope areas at the site have low susceptibility to seismically induced landsliding. Note that the hazard mitigation plan was created for planning purposes, and given its low resolution, is intended to provide guidance as to where site-specific investigation may be warranted; it is not intended to replace or supersede data acquired from a site specific investigation.

Understanding the geology of the CRRDR site provides important upfront knowledge about potential issues that may be encountered during exploration and construction. For instance, the Cachagua Fault is mapped as crossing through the site, making it important to identify the trace of the fault and understand the seismic implications posed to the new features being designed. Additionally, given the location and amount of tectonic activity in the area, we can expect to see highly fractured and sheared rock and the metamorphic/igneous rock types indicate that a variation in weathering and strength can be expected along contact zones, shears, and within the rock types. Knowing the geology also aids in logging of rock core borings and being able to identify when a different rock type is encountered that may not be mapped at the site.

GEOTECHNICAL INVESTIGATION

The purpose of the geotechnical investigation was to collect information to supplement existing data for use in the final design of the CRRDR. The investigations for Final Design were performed in accordance with the Field Investigation Plan prepared by Kleinfelder (2013). The field investigation was conducted during the months of September and October, 2013, with laboratory testing and geologic mapping extending through January 2014. The investigations included the following:

- Geologic and Rock Mass Mapping
- Soil and Rock Core Borings
- Cone Penetration Tests Test Pits
- In-situ Hydraulic Conductivity Testing (LeFranc and Lugeon Testing)
- Downhole Rock Core Imaging
- Downhole Seismic Velocity Measurements
- Geotechnical Laboratory Testing of Soil and Rock

KEY GEOTECHNICAL DESIGN CONSIDERATIONS AND APPROACH

The geotechnical investigations for final design were performed to characterize the geotechnical and hydrogeological properties of the geologic units and establish their distribution (see Table 1 below). Key geologic and geotechnical design considerations include:

- Geotechnical characterization and hydrogeological characteristics of soil and rock conditions that will affect the design and construction of major project components including the Diversion Dike, Stabilized Sediment Slope, Sediment Stockpile, and Reroute Channel. Geotechnical characterization of the Combined Flow Reach were performed primarily in 2014 and reported separately.
- Type, quality, and estimated particle size range of rock to be excavated from the Reroute Channel.
- Distribution and quantity of sediment to be excavated from structure foundations.
- Potential for re-use of on-site materials in construction.
- Site-specific design earthquake (ground motion) parameters and time-histories.
- Identification of geologic hazards that could potentially impact the project.

Table 1. Approach for Addressing Key Geotechnical Design Considerations

Technical Objective	Activity
Geotechnical Characterization	• Surface Geologic Mapping • Rock Mass Mapping • Exploratory Borings, CPTs and Test Pits • Downhole Geophysical Surveys • Laboratory Testing
Hydrogeological Characterization	• Exploratory Borings • In-situ Permeability and Pump Testing • Laboratory Testing • Piezometer Installations
Distribution of Geologic Units	• Surface Geologic Mapping • Exploratory Borings • Preparation of Geologic Map and Sections

CHALLENGES

Access to the site was delayed by nearly 4 months, requiring the geotechnical investigation to be performed during a 2 month window. The delay was incurred because of the numerous permits that needed to be processed as well as biological surveys and monitoring of the site needed to be performed before site clearing and grubbing could occur. Additionally, a Field Investigation Plan (FIP) had to be developed and approved by the project team including CAW and DSOD. The FIP posed a unique problem in that it was reviewed by numerous parties and the specificity that was detailed in the FIP created a very lengthy plan. Comments received during reviews of the FIP eventually led to separating the plan into the feature being investigated (i.e., Reroute Channel, Diversion

Dike, Stabilized Sediment Slope, etc), the specific drilling type, sampling type and interval, the anticipated soil or rock condition to be sampled, as well as the laboratory testing likely to occur on the specific sample. Each field person had a copy of the FIP detailing exactly what to do for each interval, and sampling type and depth was only changed if a change in the lithology was observed. Detailing the information in the FIP, led to time delays from CAW and DSOD in approving the start of the exploration program.

Once clearing and grubbing began, we were able to mobilize the first of four drilling crews to the site and begin drilling. It was projected that one drilling crew would be able to complete all of the soil borings and a separate drilling crew would be brought onsite to perform the deeper rock core borings in the difficult to access area of the Reroute Channel. Because the contractor was able to clear the site faster than originally projected and because of the delay in accessing the site, two additional drill rigs were brought on to the site to complete the drilling before the site had to be shut down for the upcoming winter. Finding drillers that could access the site and collect the samples detailed in the FIP made for a very small pool of drillers to select from and created an increase in the original cost for completing the field work. Accessing the site also required that personnel attend a site specific health, safety, and environmental sensitivity training that required additional coordination with the contractor and mandated that no one was allowed onsite unless they had attended training.

Drilling had to be performed around continuous aquatic surveys, biology sweeps and monitoring, and collection of aquatic and land animals. Surveys were performed before each work shift by a biological monitor and again before a drill rig was moved to a new site. Additional surveys were performed if hydraulic hoses or oil leaks were observed. To mitigate shutdowns for environmental concerns, the drill rigs were setup on plastic sheeting so leaks and drilling fluid could be better contained. When drilling near the existing Carmel River, drill rigs would have to be moved at night to a higher elevation in the event that a storm developed and created flood conditions. The extra time for moving the drill rigs on and off holes was not accounted for and increased the time and cost for completing drilling. Lastly, because of the environmental sensitivities to the river and surround biology, all of the drilling fluid and cuttings had to be drummed and removed to a settling pond area, also extending the time to complete each boring.

The CRRDR site has steep slopes, soft surface soils, poison oak, and loose colluvial and debris slide material. Access to the steep slopes was performed with a small track rig after the areas were cleared by biologists and grubbed of vegetation. During this process, the loose surface soils were also removed or compacted under the weight of the grubbing equipment. The soft surface soils did not pose a problem to the tracked drill rigs; however, they did pose a problem to passenger vehicles attempting to drive on them. For many of the locations, support vehicles with tracks were used to transport materials directly to the drill rig or the drilling crew transported materials to the site on foot.

Drilling encountered methane gases within the strata, difficulty collecting soft soils below the water table, and drilling and recovering highly fractured and highly weathered rock.

Methane gases were encountered in the infilled reservoir sediments and were observed when the drilling fluid was forcibly pushed back up out of the drill hole. The FIP and field protocols were revised to monitor for methane gases. Each drilling crew and field logger carried multi-gas meters to detect when a methane pocket might be encountered. When methane was detected, the crew was directed to stop drilling until bubbling from the escaping gases dissipated, or the methane readings were deemed to be within an acceptable level that work could continue.

To collect soils below the ground water table, catchers at the base of the sampler were used to help retain the coarse grained material. For fine grained material, piston samplers were used. Drilling and recovering the highly fractured and highly weathered rock was performed by a specialty driller with reliable experience in rock core drilling. Because of the environmental sensitivities, drilling fluid additives aside from bentonite were not allowed. This requirement extended the time of drilling to ensure that the rock was being recovered and not damaged during the drilling process.

CONCLUSION

The CRRDR project is a large, complex, and multi-faceted project. Because the project is operating under a Design-Build contract, there are very short deadlines that, if missed, can delay the project schedule. While access to the site was delayed for multiple reasons, a concerted effort was made by Kleinfelder and the Design-Build team to perform the geotechnical investigation and collect the information required to complete the design phase. Without the information, the project could have been delayed by up to a year.

The geology of the site was similar to previous studies with the main differences noted in the mapped extents and loading of debris slides. The previous work performed was not as intensive in the investigation but were not required to be as intensive as the investigation performed for the actual design and construction of the CRRDR.

The challenges experienced during the geotechnical investigation provided some lessons learned for the instrumentation phase of the project. Instrumentation was installed after the designs were completed and before construction and excavation started. The biology and environmental requirements were easier to prepare for as well as coordinating site access and movements onsite for the drilling crew and the drill rig.

The most pertinent lesson learned was flexibility. Design-Build projects move fast and often have changes based on the information being obtained, schedule, funding, and team decisions. Being flexible and addressing the challenges as they arose assisted in a successful completion of this phase of the CRRDR project.

ACKNOWLEDGEMENTS

I would like to acknowledge and thank Craig Hall, Steve Brown, Pete Stroud, Kellie Stratton and all of the other Kleinfelder personnel that spent many days, nights, and weekends working on the project through all of its phases and on every aspect from project administration to design and field work.

REFERENCES

Dibblee, T.W. and Minch, J.A. (2007). Geologic Map of the Carmel Valley Quadrangle, Monterey County, California. Dibblee Geological Foundation, Dibblee Foundation Map DF-357, scale 1:24,000.

California Geological Survey (CGS) (2010). Digital Images of Official Maps of Alquist-Priolo Earthquake Fault Zones of California; CD 2000-003; updated through December 2010 at: http://www.quake.ca.gov/gmaps/ap/ap_maps.htm.

International Society for Rock Mechanics ISRM (1981). Rock Characterization, Testing and Monitoring: Suggested Method for Determination of Uniaxial Compressive Strength of Rock Materials. Pergamon Press, London, U.K.

Kleinfelder (2002). Sediment Characterization Study: San Clemente Reservoir, Monterey County, California, prepared for Mussetter Engineering, Inc., November 2002.

Kleinfelder (2013). Field Investigation Plan, Carmel River Reroute and San Clemente Dam Removal Project, Monterey County, California, dated September 11, 2013.

Montgomery Watson Harza (2008). Final Basis of Design Report, Carmel River Reroute and San Clemente Dam Removal, Monterey County, California, prepared for California State Coastal Conservancy, February 15, 2008.

Monterey County Office of Emergency Services (2007). Multi-Jurisdictional Hazard Mitigation Plan, Monterey County, California. Prepared by URS Corporation on behalf of the Monterey County Office of Emergency Services. September 2007.

Monterey Peninsula Water Management District. http://www.mpwmd.dst.ca.us/

Page, B.M. (1966). Geology of the Coast Ranges of California, in Bailey, E.H., editor, Geology of northern California: California Division of Mines and Geology Bulletin No. 190, p. 255-276.

URS (2012). Geotechnical Data Report, San Clemente Dam Removal Project, prepared for State Coastal Conservancy, revised March 19, 2012.

U. S. Department of the Interior Bureau of Reclamation (1998). Engineering Geology Field Manual. Vol. 1-2.

U.S. Geological Survey (USGS) (2010). Quaternary Fault and Fold Database for the United States, from web site: http//earthquakes.usgs.gov/hazards/qfaults/.

Woodward-Clyde Consultants (1998). Subsurface Investigation, San Clemente Dam, Seismic Retrofit San Clemente Dam, prepared for California-American Water Company, August 6, 1998.

MANAGING UNCERTAINTIES DURING HYDROPOWER DAM REMOVALS

Peter E. Haug, PE[1]

ABSTRACT

Dam removal design teams should be aware of inherent uncertainties in both costs and schedules for dam removal design, deconstruction, and post-removal monitoring. Case studies of three dam removals, in addition to the author's experience with six other dam removals, are used to illustrate ways engineers can reduce owner and regulatory risk for dam removal costs, schedules, and expectations. These risk reduction measures include upfront questions about watershed contamination, sediment volume, lakefront property owners, local fish management strategies, and status of project licensure and ownership.

INTRODUCTION

The author's employer (Ayres Associates) entered the dam removal industry in the 1980s and 1990s. With few exceptions, early dam removals included just a few drawings showing existing and proposed contours with few staging, dewatering, or restoration details. Dam removals proceeded without much legal opposition, except perhaps where a dam was forcibly removed against the owner's wishes. In general, the removal process was designed as a simple earthwork project, included little stream restoration effort, and cost as little as one third of the inflation-adjusted costs to remove a dam today. The design effort and construction costs for these early removals were predictably structured with few surprises for the designer.

As with many other firms, removal project managers at Ayres Associates experienced remarkable differences between 20th and 21st century removals. As environmental advocacy groups increased public awareness of dam removal permits and promoted public involvement, project costs and durations became more uncertain. Although these advocacy groups funded and supported dam removals in the interest of habitat and stream connectivity, their methods focused on constant publicizing of the negative impacts of dams and encouraging public participating in dam removal hearings. Unfortunately for dam removal advocates, not all the public participants wanted their favorite fishing hole or view modified by dam removal. Over the years of public participation in dam removal hearings, the author has noticed increased public outcry against dam removal and less support of dam removal. Ironically, as environmental advocacy groups publicized their support of dam removals, residents living near dam removals realized that they could also publicize their opposition to dam removals. As a result, it seems that the public perception has changed from "dam removal benefits outweigh the costs" to "dam removal costs may outweigh the benefits." The advocacy groups promoted more public involvement as a means to force dam removals, but now the public involvement ironically produces more uncertainty for prospective dam removal projects – uncertainty with regard to permit lawsuits, schedule delays, and costs of dam deconstruction.

[1] Water Resources Engineer at Ayres Associates, 3433 Oakwood Hills Parkway, Eau Claire, Wisconsin 54701, haugp@ayresassociates.com.

While public involvement and project accountability improve both designers and regulators, this article focuses on how project uncertainties arise and attempts to make engineers better at communicating dam removal uncertainties to owners and regulators. Three case studies are presented, including the 2008 Woodley Dam Removal, the 2012 Grimh Dam Removal, and the in-progress 2015 Gordon Dam Removal.

Case 1: Woodley Dam, Polk County, Wisconsin

In 2001 Woodley Dam nearly overtopped during a large flood event, and the Department of Natural Resources (DNR) soon ordered a mandatory drawdown until repairs could be completed. Citing excessive repair expenses, the grist mill building (last producing hydroelectric power in 1994) was removed in 2003 with little opposition (Figure 1), but the County's subsequent application for dam removal was met with multiple unexpected oppositions.

Figure 1. Woodley Dam Pre-Removal (Powerhouse Already Removed).

Figure 2. Woodley Dam Post-Removal (Same View as Figure 1).

The permit for the Woodley Dam removal was contested because the local river canoe industry teamed with Trout Unlimited to oppose artificial riprap on geotextile, even though this sediment stabilization measure was designed to preserve the trout stream's clarity and restore trout migration across the four feet of stream grade between upstream and downstream bed elevations. Arguing that the riprap fish weirs and boulders (designed to restore fish passage) prevented a "free-flowing, natural stream," the constituents requested the DNR order a wider dam removal and removal of more reservoir sediments. Additionally, a local snowmobile advocacy group contested the dam removal permit, arguing that loss of the dam prevented them from using the dam as their river crossing.

While the dam owner argued the right to remove the dam in court, the plaintiffs disagreed on whether the restoration plans were in the public interest. The competing restoration goals halted the project as the case proceeded to an administrative law judge. As a measure to progress the removal permit, the dam owner conceded to dredge an extra 1,300 cubic yards, limit riprap use to protection of the bridge structure, and replace the in-stream armor with a new snowmobile bridge (an additional cost of $114,000). In addition to the construction cost increases, this change to the dam removal plan and permits required an additional 25% in engineering fees. Once the contested case hearing was discontinued, the plans and permit application were revised to include the above concessions, and the dam was removed as shown in Figure 2. While the actual deconstruction required only five months, the engineering design and permits had taken five years to complete.

Case 2: Grimh Dam, Sawyer County, Wisconsin

When North Central Power Company decided to remove the 30 foot high, 1100 foot long Grimh Dam in a sparsely populated region of northern Wisconsin, few anticipated the legal, environmental, and public safety challenges ahead. Grimh Dam originally supplied power to a local shingle factory, and the resulting Village of Radisson grew around the Grimh Dam impoundment. As the shingle business faded, the dam was raised nine more feet to produce hydroelectricity for the region's wood-products industries and surrounding communities. By 1997, the dam required significant repairs to meet Federal Energy Regulatory Commission (FERC) standards, so the owner filed a legal notice to abandon the dam. In what might have been a short dam removal process in the late 1990s, the removal process was delayed during attempts to sell the dam to the community and other hydropower producers.

In 2000, the DNR required a drawdown to address unrepaired safety concerns. This drawdown initiated a civil lawsuit from local property owners. Local residents claimed their shallow wells were losing water, the once pristine fishing area was destroyed, and property values were lowered because of the unsightly mud flats where the reservoir once flowed. Because objecting to the DNR drawdown was not successful, residents quickly turned to contesting the dam abandonment permit, and an administrative law judge was appointed to oversee eight months of legal arguments about dam removal. Eventually the judge ruled that the owner and DNR had sufficient rights to proceed with removal, but afterwards the DNR requested additional justification for the sediment management plan to handle 100,000 cubic yards of silt behind the dam plus revised plans to accommodate native species and stream restoration concerns. For each adjacent private landowner bordering the impoundment the owner should budget litigation expenses to cover uncertainty in securing a dam removal permit through the contested case process.

Figure 3. Grimh Dam Pre-Removal.

Figure 4. Grimh Dam Post-Removal.

Figures 3 and 4 show, respectively, the before and after removal photographs (note the same house under the red arrow, though 15 years later).

Environmental hazards encountered during deconstruction of Grimh Dam included a large flood event, contaminated soil discoveries, asbestos mitigation, and constant concerns about the 100,000 cubic yards of very soft sediment. Cofferdams have inherent risks of overtopping, and designing a cofferdam for a ten-year flood (a non-codified but generally-accepted regional design standard) creates a 10, 19, or 27% risk of overtopping in a one, two, or three year construction period, respectively. As luck (or unluckiness) would have it, two rainfall events (4.0 inches of rain in one day and a separate event of 5.5 inches of rain in two days) both produced ten-year floods. The first flood overtopped the first stage cofferdam in early September and the second flood nearly overtopped the second stage cofferdam in early October (a normally low-flow period). The takeaway from this experience is to make sure annually-compounded risk and overtopping uncertainty are fully understood by the contractor, DNR, and owner.

While the project's asbestos items were fairly inexpensive (approximately $11,000) to mitigate in the first weeks of construction, fuel-contaminated soil was discovered during the second year of construction in soil lenses below the concrete slab of the powerhouse's old diesel generator building. The unexpected contamination required extensive remediation of the west bank, including an additional $100,000 to remove contaminated soils and complete the required DNR testing and site deed restriction paperwork.

Over 75 years of project operation, the reservoir had amassed 100,000 cubic yards of extremely soft sediments. The DNR required the owner to post "danger" signage all around the flowage, and the author can confirm by personal experience that one small step off of the upstream cofferdam and onto grassed-over reservoir sediments required an hour to extract one's lower body from the soft sediments. While sand drains easily, it became apparent through the project that the reservoir silts and clays were going to take many years to stabilize. Through frequent, purposeful communication with the DNR and owner, Ayres Associates was able to inform them about risks of sediment mobilization during each removal phase. While the DNR always retained the rights to stop the removal process at any time, the DNR did acknowledge the following risk reduction measures: 1) sediment "dilution" since the larger Chippewa River was only 1.5 miles downstream of the dam, 2) partial sediment "consolidation/stabilization" since the reservoir had been half way drawn down for 10 years prior to removal, and 3) natural sediment "accommodation" as a slow release of the reservoir's remaining volume over two growing seasons would give time for the downstream reaches to adjust.

Deconstruction proceeded tenuously with this inherent uncertainty of not knowing when or if the DNR would halt the construction activities due to sediment flux. While the DNR did not require upfront dredging at Grimh Dam, the design team and owner were always prepared for the DNR to require additional sediment stabilization measures, even up to the last month of deconstruction. In fact, at the end of the project (Figure 4) all reservoir sediments had been exposed to at least two growing seasons, had vegetation growing on them, and yet were still easily liquefied by passing construction traffic. The lack of reservoir sediment consolidation and drainage required an additional 132 cubic yard berm of riprap to lessen cutting into the final western channel bank. The DNR also accepted three additional years of slow head-cutting upstream before the head-cuts halted in a reach of large boulders. In short, sediment removal could have added $500,000 to $800,000 in construction costs, but the above risk reduction measures allowed the DNR to stand by their plan to 1) slowly release sediments at the rate at which the natural environment could accept them and 2) undertake only reactive repairs to discrete areas of bank instabilities.

For Grimh Dam, the total dam decommissioning to removal process elapsed 15 years between the first newspaper publication of impending dam removal and the removal contractor's demobilization.

Case 3: Gordon Dam, Douglas County, Wisconsin

After a protracted search for a buyer, Dahlberg Light & Power decided to remove the 33 foot high, 1550 foot long Gordon Dam in 2010. An earlier FERC Petition for Declaratory Order and attempts to relicense the dam were stalled by requirements to increase hydraulic capacity and fix embankment seepage issues. In terms of dam removal uncertainty, Gordon Dam offered a unique situation where the dam owner, the State of Wisconsin, and a large timber company own 95% of the flowage perimeter. Therefore, the risks and uncertainties for this dam were much different than for Grimh Dam or Woodley Dam. First, with only one private landowner on the upstream impoundment, the project had a low risk of proceeding to a contested case permit hearing. Second, the impoundment had no sediment near the dam, mainly because sediments dropped out about a mile upstream of the dam and remained stationary once the dam was half drawn down in 2010. While there was some mercury in the local soils and a few leaking underground storage tanks within the watershed, the sediments in the upstream delta were not likely to contain heavy metals or other contamination sufficient to require dredging. Therefore, no dredging was budgeted for this project. Third, since lake sturgeon was the main species of interest, the stream restoration did not require complex design procedures to accommodate fish passage other than mimicking the natural streambed downstream of the dam removal site.

The DNR and Corps of Engineers staff were involved early in the Gordon Dam removal process. DNR dam safety staff considered a cofferdam sized for the 10-year event to be sufficient but desired the remaining reservoir head to be drawn down over two construction phases. During a preliminary consultation, regulatory staff revealed a design preference to remove all of the dam's concrete and shape the remaining embankments to 1) not cause any backwater during the 100-year flood and 2) include at least some flat area as an overbank to mimic the downstream channel. These preferences were accepted early in the removal design phase and allowed both the owner and engineer to proceed with less uncertainty about project costs and review schedule.

The design uncertainty was reduced by following the regulatory suggestions as far as the owner's budget allowed, but other methods also helped to reduce design risk. First, Gordon Dam has no record of diesel generators onsite, so the DNR did not make the owner test for contaminated embankment soils. However, the DNR did claim that an upstream prison facility justified a metals check for reservoir sediments. Therefore, 25 holes were drilled through the ice to probe for sediment depths and collect samples for metals testing. This reduced design risk by informing both the owner and DNR that the bed sediments were primarily coarse sands and not soft silts. Since silts tend to carry more contaminants and mobilize easier than coarse sands, the design team could advocate for larger drawdowns and less sediment excavation during construction. Second, native species seeding is generally recommended for all dam removals, especially for areas below the Ordinary High Water Mark because any seed sown on the exposed reservoir bed have a good chance of being transported during floods to environmentally sensitive areas downstream. Therefore, a three step seeding plan (dry, mesic, and wet) was specified for this project to reduce the DNR's concerns about impacting the downstream wild rice and wetland areas. Third, construction costs were more certain without the unknowns of impoundment stabilization (no sediment is adjacent to the dam), contaminated soils (not found), or protracted delays due to contested case litigation (there is only one private landowner). The project so far has been ahead of schedule and the construction costs are expected to be much lower than for the Grimh Dam project.

Figure 5. Gordon Dam Pre-Removal.

CONCLUSIONS

This article presented a challenge for designers to reduce uncertainties in estimating dredge extents, bank stabilization costs, contaminated soil mitigation costs, cofferdam overtopping events, and other construction issues. Some lessons learned include the following:

1) Even if the dam is apparently an eyesore to the community, citizens may not agree on the restoration goals (Woodley Dam) or the removal itself (Grimh Dam). Each private home or business bordering the flowage increases the chances of a contested permit. (The author budgets about $2000 per parcel for litigation.)

2) Dredging of sediments, especially contaminated sediment, can be the primary construction cost for dam removals. The owner needs to know early in the process if dredging will be required.

3) Stream restoration goals (fish passage improvements, rock arch rapids, etc.) are another large cost for removals in some states (Minnesota, for instance). Designers should inform the DNR about dam removal examples where natural slow release of sediment did not adversely impact the environment (Grimh Dam, Marmot Dam).

4) Upfront communication and risk education are key in most engineering projects, but especially critical in dam removals. Engineers owe dam owners an upfront list of project uncertainties. A line item in an early removal cost opinions might read "$100,000 for contaminated soil removals" if diesel generators have been used within the removal extents. A line item of "$140,000 for contested case litigation" when no one has yet objected to the project could be justified if six dozen vacation homes surround the lake. Let the owner know what can reasonably happen during the dam removal permitting process so that they are not surprised later.

To facilitate owners' awareness of dam removal design and construction cost uncertainties, Ayres Associates has developed cost estimating regressions based on dam height, reservoir area, embankment length, maximum flow capacity, project location (state), and other physical parameters. This provides the owner with a Class 5 "Concept Screening" cost opinion without much additional analysis. Before a Class 4 "Feasibility Study" cost opinion is developed, the owner and DNR should answer the following questions:

____ Has the FERC license been surrendered already?
____ Have diesel generators been used onsite frequently? Any known spills?
____ Do local DNR fish biologists favor fish passage modifications post-removal?
____ Does the watershed have a history of soil contamination (mills, mines, etc.)?
____ How many privately-owned parcels can view the reservoir impoundment?
____ How much has the impoundment filled with sediment?
____ Does the dam form an artificial barrier to invasive species migration upstream?
____ Does the dam create wild rice, wetlands, or other protected habitat upstream?

In answering those questions upfront, designers can relay to owners the ways in which their removal project compares with other dams in terms of project complexity.

Ayres Associates has reduced some environmental hazard uncertainties and resulting permit review times by pushing owners to accept upfront (prior to regulatory agency involvement) sediment retention strategies like staged drawdown, multiple season removal efforts, and native species restoration. Upper Midwestern dam removal permits almost never allow a drawdown more than 15 feet in one year, and most drawdown volumes are also used as a storage reduction in the inflow hydrograph for cofferdam height design. Multiple growing seasons allow the engineer to correct deficiencies in the initial native species mix, ascertain how fast reservoir sediments can consolidate, and reduce the risk of a single mid-construction flow event eroding all the reservoir sediments.

As a concluding statement, dam removal projects are generally unique with individual challenges that make prediction of design effort and construction costs difficult. However, by recognizing the common sources of design uncertainty (contested case litigation potential, sediment handling issues, and contaminated soil presence), offering regulatory reviewers low cost consolations (native species seeding, removal sequences spread over multiple growing seasons, and making the grading plan mimic downstream channel shapes), and educating owners and contractors about risks (cofferdam overtopping, sediment transport risk, and experiences on other similar projects), the overall uncertainty with regard to design times, permit schedules, and construction costs can be reduced.

TIME TO GENERATE POWER NEW HYDRO DEVELOPMENT ON THE OHIO

Paul R. Blaszczyk[1]
Kevin Griebenow[2]
Philip E. Meier[3]

ABSTRACT

American Municipal Power, Inc. (AMP) is currently constructing four new hydroelectric plants on the Ohio River. The projects are jointly the largest new hydroelectric development in the United States and are an excellent example of how renewable hydropower can be a viable part of our nation's energy future. The power plants are all regulated by the Federal Energy Regulatory Committee (FERC) and are located at existing U.S. Corps of Engineer (USACE) locks and dams fulfilling the long standing purpose of the USACE to add renewable energy resources to the Sites. MWH is serving as that Owner's Engineer for the projects and has worked with AMP from the screening study phase through licensing, design and now construction. The projects entail reinforced concrete powerhouses designed to be overtopped in flood conditions as well as closure structures constructed of hardfill and concrete that tie the powerhouses into the existing dams and the newly formed landside abutments. Though the plants are similar from site to site, the foundation conditions varied greatly from alluvium atop karstic limestone to slickensided claystone. This paper provides an overview of the projects and goes into detail regarding the coordination and collaboration between the Owner, Engineer and regulators throughout design and construction including the implementation of a Board of Consultants (BOC), design improvements resulting from the Potential Failure Mode Analysis (PFMA) process, and reporting and addressing issues during construction.

INTRODUCTION

The Cannelton, Smithland, Willow Island, and Meldahl Hydroelectric Projects are being developed by American Municipal Power, Inc. (AMP), which is a nonprofit corporation organized in Ohio in 1971 for the purpose of owning and operating electric facilities or otherwise providing for the generation, transmission and distribution of electric power and energy to its members. AMP has 130 members – 129 member communities in Indiana, Kentucky, Michigan, Ohio, Pennsylvania, Virginia and West Virginia, and one joint action agency in Delaware.

[1] Vice Preside nt, MWH Americas, Inc., 175 W. Jackson Blvd., Suite 1900, Chicago, IL 60604, paul.r.blaszczyk@mwhglobal.com
[2] Civil Engineer, Federal Energy Regulatory Commission, 230 South Dearborn St, Room 3130, Chicago, IL 60604, kevin.griebenow@ferc.gov
[3] Vice President Hydro Development and Operations, American Municipal Power, Inc., 1111 Schrock Road, Suite 100, Columbus, OH 43229, pmeier@amppartners.org

AMP is developing these run-of-the-river hydroelectric facilities that will add nearly 300 MW of new, renewable generation at existing dams on the Ohio River, which extends 981 miles from the confluence of the Allegheny and the Monongahela Rivers in Pittsburgh, Pennsylvania, to its confluence with the Mississippi River at Cairo, Illinois, dropping approximately 430 feet over this distance. Along the way, it flows through or borders six states: Illinois, Indiana, Kentucky, Ohio, Pennsylvania, and West Virginia. It is the largest tributary of the Mississippi River and is about 1 mile wide at the Smithland Locks and Dam.

THE OHIO RIVER

The drainage basin for the Ohio River is about 200,000 square miles, including nearly all of Ohio and Kentucky, and portions of 12 other states. The Ohio River basin is one of the most populated and industrialized regions of the United States. More than 25 million people, almost 10 percent of the U.S. population, live in the Ohio River Basin. The Ohio River is also the second most heavily used commercial river in the United States, due in large part to river transportation being an efficient mode of transportation. A typical tow consisting of 12 barges and a tug uses only about 10 percent of the fuel that truck transportation would for the same amount of cargo. Currently, over 200 million tons of cargo is transported on the Ohio River each year. Coal comprises about one-half of the cargo, with the balance comprised of goods such as agricultural products, aggregates, fertilizers, chemicals, and steel.

Early navigation on the Ohio River was dependent on the seasonal flow and hampered by sandbars. From 1874 to 1929, the USACE constructed 53 locks and dams that turned the river into a series of pools that could be navigated safely year-round. Beginning in the 1950s, the USACE began replacing the original locks and dams with higher locks and dams and reducing the number of locks and dams from 53 to 19. The last remaining two original locks and dams (numbers 52 and 53) will be replaced in the coming years with the completion of the 19th project, the Olmsted Locks and Dam. Figure 1 shows the current locks and dams on the Ohio River.

The locks and dams were designed with the idea that hydropower might be added in the future. Figure 2 shows a typical layout of the locks and dam at each site (Figure 2 is an aerial photo of the Cannelton Locks and Dam). The photograph is oriented such that the river flow is from left to right. The set of locks, one 600 feet long and the other 1,200 feet, are visible at the bottom of Figure 2. The gated spillway bays are located adjacent to the locks and vary in number from site to site. At Cannelton, there are 12 bays, each equipped with a radial gate. A fixed weir comprised of steel sheet piling cells connects the last gate bay to the shore, as shown in the upper portion of the photo. These weirs were added for additional spill capacity in high flow periods and as a possible location for the development of future hydropower projects.

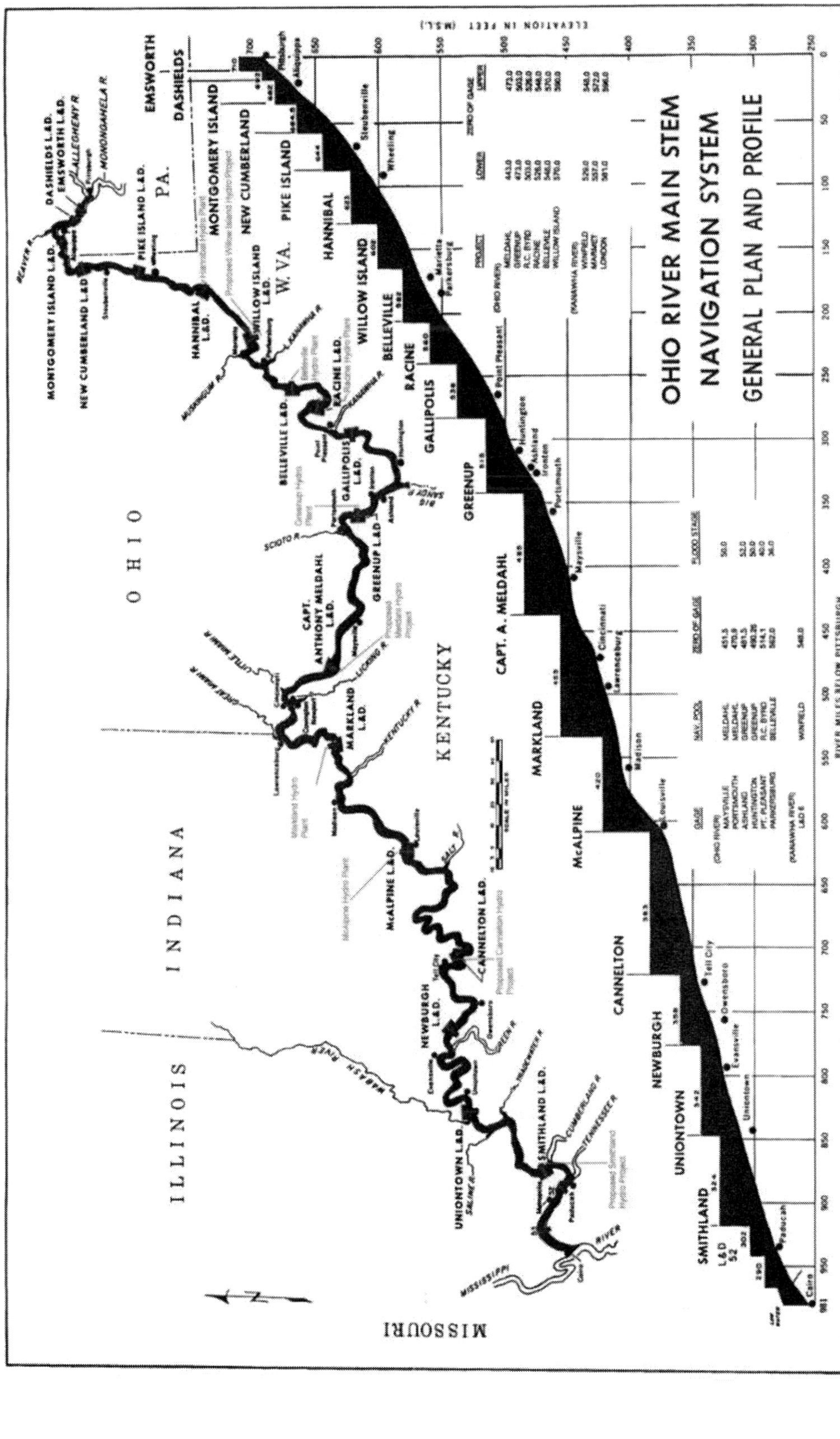

Figure 1. Locks and Dams on the Ohio River.

Figure 2. Aerial pre-construction photograph of the existing Cannelton Locks and Dam showing the typical layout of the USACE locks and dams along the Ohio River.

The flow in the Ohio River varies seasonally and can exceed one million cubic feet per second (cfs) at the confluence with the Mississippi River. This leads to large flow variances at each dam along the Ohio River. For example, the flow in the Ohio River at the Cannelton Locks and Dam varies from about 3,000 cfs to just over 600,000 cfs, with flows exceeding 100,000 cfs almost 50 percent of the time. Although the flow in the Ohio River can be quite large, it is not feasible to take advantage of the larger flows for hydropower due the variations in the pool elevations. Hydropower potential is not only related directly to the flow, but also the pressure head differential at the site (i.e., the difference in elevation between the normal upper pool and lower pool). Because of the way the USACE manages the pools on the Ohio River, there is an inverse relationship between flow and head. When the flow is high, the head is low and not suitable for economic hydropower generation. Conversely, when the head is at its highest, the flow is too low. Therefore, the challenge for hydropower development on the Ohio River is to take advantage of the flow ranges where there is significant head at the sites.

The USACE typically operates the projects to maintain the upper pool (headwater) at a constant elevation by gradually opening the gates while flows increase. As the gates progressively open, the lower pool (tailwater) rises with the increased flow. When the tailwater is nearly equal to the upper pool level, the gates are opened fully and the river is free flowing. Figure 3 shows a typical rating curve, which for this case is for the Cannelton Locks and Dam. As shown, the headwater is held constant at elevation 383 feet for flows up to about 320,000 cfs (all elevations in this paper are in Ohio River Datum). The minimum tailwater is elevation 358 feet and the tailwater rises with increasing flow until at 320,000 cfs, where it is only about one foot below the headwater elevation (due to the head loss across the dam structures). With further increasing river flow, the headwater and tailwater levels rise together.

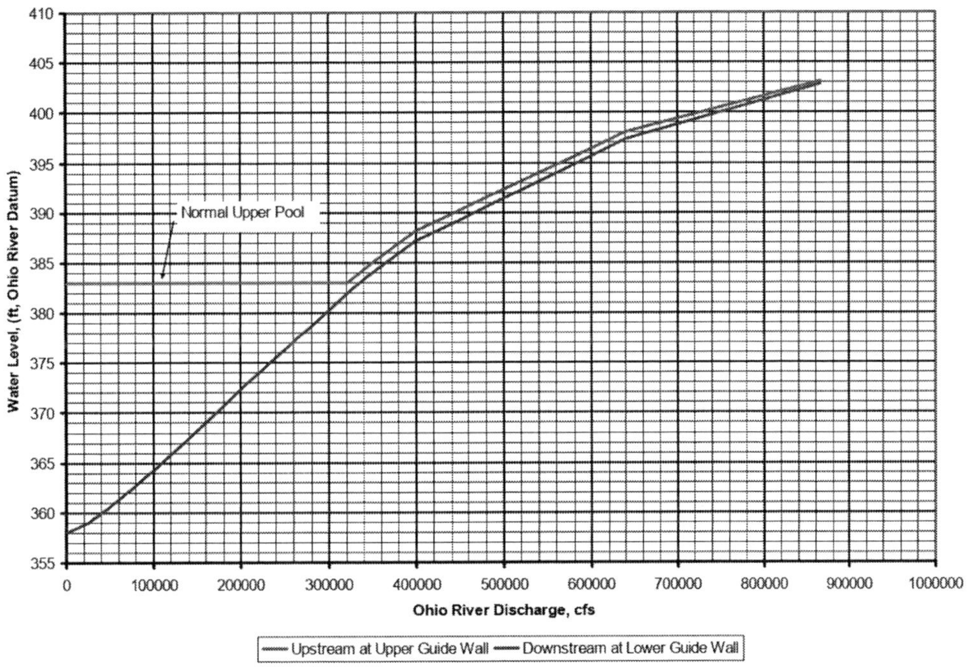

Figure 3. Headwater and Tailwater rating curves for Cannelton Dam.

THE HYDROELECTRIC PROJECTS

In 2006, AMP commissioned MWH Americas, Inc. (MWH) to perform a Phase 1 screening study of undeveloped hydroelectric sites on the Ohio River adjacent to USACE dams. This was quickly followed by a Phase 2 study that ultimately identified the top four sites for development on the Ohio River. Based on the results of that study, AMP sought and obtained the Federal Energy Regulatory Commission (FERC) licenses for the four projects.

In April 2007, AMP selected MWH as its engineer for the Willow Island, Cannelton and Smithland Hydroelectric Projects as it recognized that it would take an engineering firm with significant hydro experience, expertise and resources like MWH to develop these projects simultaneously. Since then, MWH has assisted AMP in all aspects of the projects including project management; planning; permitting; financing and regulatory support; preliminary and final design; preparation of bidding documents for the cofferdams, equipment procurements, transmission lines, and the general construction contracts; negotiations and contract awards; engineering services during construction; and resident engineering. In 2008, when AMP decided to proceed with concurrently development of the Meldahl Hydroelectric Project with the other three projects, AMP increased MWH scope to include Meldahl.

AMP developed and operates the Belleville Hydroelectric Facility, a 42 MW run-of-the-river hydroelectric generation facility located on the Ohio River at the Belleville (W.Va.) Locks and Dam. The facility, which went online in April 1999, was also designed by MWH and served as the template for the current projects; thus the hydroelectric projects are similar in many ways. Similar to Belleville, the new reinforced concrete powerhouses will be constructed in the abutments of the locks and dam structures on the opposite side of the river from the locks. Water will be supplied to the powerhouses through open headrace channels excavated through the existing river banks in the upstream pools and will be discharged back to the downstream pools through excavated tailrace channels in the existing river banks. The projects will use the head differential created by the locks and dams along with the flows in the river to produce power.

However, each plant has its unique geological, hydrological, and environmental site constraints that require adaptation of the common design. For example, two of the projects are excavated into and founded on rock, while two are founded on alluvium requiring different designs. Two of the projects are also located in areas of higher seismicity that require special measures. The turbine-generator units, which are all nearly identical horizontal bulb type units with 25.3 feet diameter runners, are actually designed for the specific head and flow conditions for each site. Finally, regulatory and environmental concerns such as threatened and endangered species, cultural resources, and recreation require site specific adoptions and solutions that are regulated by different state agencies (Kentucky for Cannelton, Smithland and Meldahl and West Virginia for Willow Island), USACE Districts (Louisville for Cannelton and Smithland and Huntington for Meldahl and Willow Island), and FERC Regional offices (New York for Willow Island and Chicago for the others).

The Cannelton Hydroelectric Project was the first of AMP's four run-of-the-river hydroelectric projects currently under construction to break ground, in 2009. The 84-MW Cannelton plant is located on the Ohio River near Hawesville, Kentucky. The Cannelton Project will divert water from the existing USACE Cannelton Locks and Dam and is expected to generate an average gross annual output of about 458 million kilowatt-hours (kWh). The powerhouse will house three horizontal 28-MW bulb-type turbine and generating units with an estimated total rated capacity of 84 MW at a gross head of 25 feet. Figure 4 is an aerial photo from December 2014 illustrating the layout of the Cannelton project, which is generally representative of the Meldahl and Smithland projects as well.

The Smithland hydroelectric facility, located near Smithland, Kentucky, will add 72 MW of new, renewable generation to the region. Similar to Cannelton, the Smithland project will divert water from the existing USACE Smithland Locks and Dam through three horizontal 24-MW bulb-type turbine and generating units with an estimated total rated capacity of 72 MW at a gross head of 22 feet to generate an average gross annual output of approximately 379 million kilowatt-hours (kWh). Smithland is located approximately 62 river miles upstream of the confluence of the Ohio and Mississippi rivers, in Livingston County, Ky.

Figure 4. Aerial Photo of Cannelton from December 2014 looking upstream with the newly constructed hydroelectric facilities on the right bank.

Located near St. Marys, West Virginia, the Willow Island hydroelectric facility will divert water from the existing USACE Willow Island Locks and Dam through two horizontal 17.5-MW bulb-type turbine and generating units with an estimated total capacity of 35 MW at a gross head of 20 feet to generate an average annual output of approximately 239 million kilowatt-hours (kWh).

Finally, the Meldahl Hydroelectric facility will become the largest hydroelectric power plant on the Ohio River with an estimated capacity of 105 MW. This run-of-the-river project will divert water from the existing Captain Anthony Meldahl Locks and Dam located near Maysville, Kentucky, approximately an hour southeast of Cincinnati. The Meldahl powerhouse will house three horizontal 35-MW bulb turbines that will generate an average gross annual output of approximately 558 million kilowatt-hours (kWh) and an estimated total rated capacity of 105 MW. Figure 5 is an aerial photo of the Willow Island project from December 2014, which had to be located much closer to the existing structures due to site constraints.

Figure 5. Willow Island Hydroelectric Project – Aerial Photo from December 2014. This view is looking from the landside towards the river with flow going from left to right.

All of the sites utilize a similar powerhouse layout built around the horizontal bulb-type generating units. Due to the hydrology of the river and the fact that the hydroelectric projects could not affect upstream water elevations during any river flow, the powerhouses and dams constructed on either side of the powerhouses (referred to as closure structures), had to be designed to be overtopped during higher flow events. This lead to the powerhouse layout shown in Figure 6, which allows for the powerhouses to be overtopped by as much as 25 feet during peak flood events. Further discussion on the design and details of the powerhouses is provided in in Bodily et al (2015).

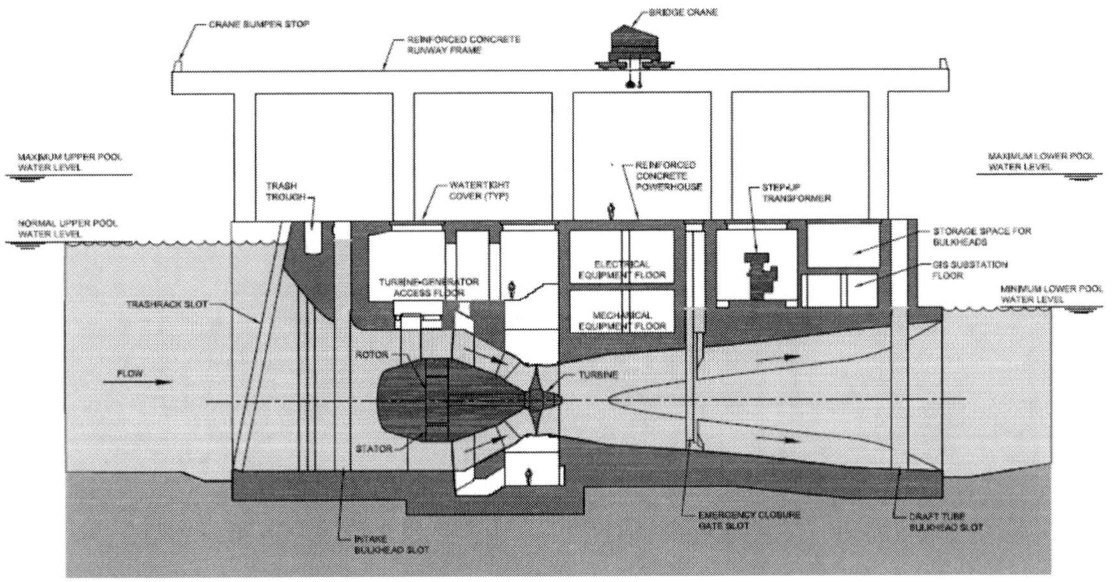

Figure 6. Typical Powerhouse Cross Section

Design of the projects was a truly collaborative effort amongst the owner, the engineer, the regulators, and in some cases, the contractors. It did not start that way however. To understand this, one needs to understand the history of the sites; namely, the numerous starts and stops of both licensing and construction efforts at these sites over the last several decades.

DEVELOPMENT AND LICENSING HISTORY

A FERC license was originally issued for an 80 MW development at the Cannelton Locks and Dam on June 21, 1991. The project was to include a powerhouse with three 26.67 MW bulb-type turbine/generators located at the left abutment of the USACE's dam. Fortunately for AMP, construction never started.

Then, an amendment was issued by FERC on October 8, 1997 for a development that was to include eight power modules with a total capacity of 79.2 MW. The amendment with the power modules, shown in Figure 7, was a radical change in the design from the previously submitted and licensed, which was to consist of a typical low head installation, to a new concept where the power modules were to be installed in the spillway gate bays. The license was then amended a second time on October 18, 2000 due to the fact that the design had evolved to a 46.2 MW development using a matrix of 140 bulb type units, each with a capacity of 330 Kw, installed in the gate slots.

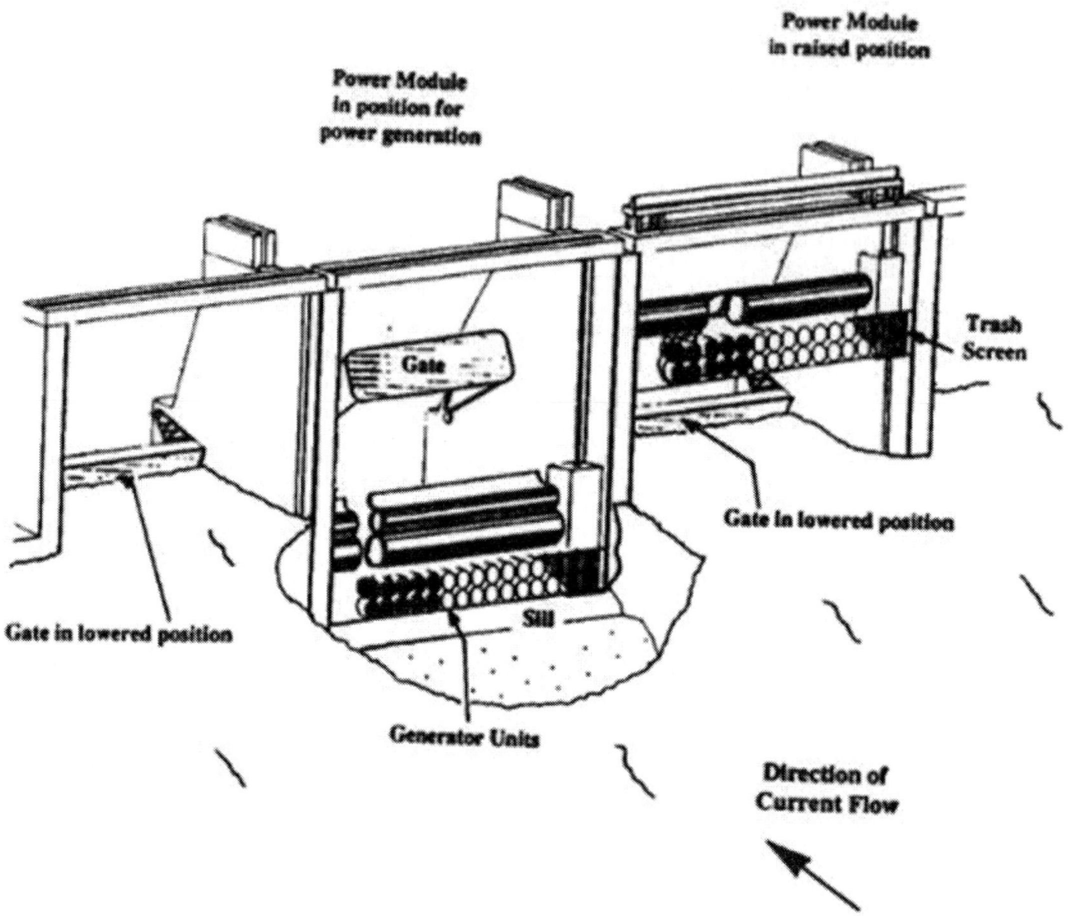

Figure 7. Schematic diagram of 1997 licensee's proposed power generation system for the Cannelton and Smithland sites (Source: modified by the FERC staff from W.V. Hydro, 1994).

The start of construction was approved by a FERC order issued on August 8, 2001. By 2004, draft tubes and other equipment were being fabricated along with plans to install a prototype module to demonstrate the concept. The prototype was to be installed at the Smithland Dam. However, although interconnection agreements with the Midwest ISO for both projects were completed, long-term power purchase agreements that were critical to financing the projects had not been reached with four utilities. Based on this and the fact that current and projected project costs had increased greatly, notice of the fabrication of the units being suspended was provided to FERC in February 2004.

Then, on May 5, 2005, an application was submitted to FERC to change the design from the 140 matrix unit layout to three large vertical turbines. This layout, which was very similar to what was originally authorized in the 1991 license, involved 24.6 feet diameter turbines that would have a siphon intake with the powerhouse located next to the last spillway gate at Cannelton. Again however, construction did not start.

Based upon the results of the Phase 1 and 2 studies mentioned earlier, AMP filed an application with FERC to transfer the license from the previous licensee on December 21, 2006. With no objections raised by the resource agencies, the third major amendment to the original license was issued on March 22, 2007, just three short months after filing the transfer request. This amendment was basically a new license due to the revised project design with only several original articles reinstated.

Similarly, the FERC license for Willow Island was issued in 1989, but through a series of FERC and Congressional extensions, remained active until AMP acquired it from the City of New Martinsville, West Virginia (an AMP Member) in 2007. The Meldahl Hydroelectric Project was a bit different. The licensee for this project is the City of Hamilton, Ohio, a member of AMP. This project was also previously licensed however, when the previous licensee failed to develop the project in a timely manner, the license was terminated by FERC. As a result, the City prepared a new license application, which was subsequently granted in the summer of 2008. The project is now being developed jointly between the City of Hamilton and AMP.

COLLABORATION MAKES IT HAPPEN

Given that all of the sites had been previously licensed with multiple amendments issued thereafter but with construction never really starting at any of the sites, the regulatory agencies were extremely skeptical that AMP would actually build the projects. But build they did. In fact, the regulatory agencies, particularly FERC's Chicago Regional Office (CRO), became acutely aware of AMP's intentions at Cannelton, Smithland and Meldahl at a coordination meeting on November 2, 2007, where it was disclosed that an extensive subsurface exploration program was being conducted to understand the local geology. Coordination, communication, and reporting protocols were established amongst the parties at that time as were design requirements such as the need to conduct Potential Failure Modes Analyses (PFMA) for all projects, for both the cofferdams and powerhouses.

It was at this meeting where FERC realized the scope of the foundation issues at the Smithland and Cannelton sites. While Andrews et al. (2015) and Mirza et al. (2015) discuss these issues in detail, bedrock elevations at both of these sites were well below the powerhouse foundation levels, versus Meldahl and Willow Island, which are founded on rock. The foundation rock at Smithland is composed of lightly weathered limestone bedrock with voids and solution cavities (karst) up to 15 feet deep within the upper 25 feet of rock, and both projects are also located in areas of higher seismicity that required special measures. Given these facts, FERC notified AMP that a Board of Consultants (BOC) would need to be convened and also brought in their own consultant, Dr. Alfred Hendron, Jr. to assist in the overall review of the project designs and construction By December 2007, FERC accepted AMP's recommended BOC participants, which included Mr. Joseph Ehasz, Mr. James W. Irwin, and Mr. Eric B. Kollgaard. Meetings with the USACE and FERC as well as a site visit to the Cannelton site were also conducted at that time during which AMP and MWH reported the status of progress to date and made a detailed presentation of the proposed project scheme and construction sequencing to suit the aggressive schedule. The site inspection involved walking the majority of the proposed construction footprint, the potential borrow and spoil areas and tie-in with the existing USACE Cannelton lock and dam cellular cofferdam structure. No unfavorable site conditions or major scheduling discrepancy was discovered over the course of the day.

The first BOC meeting was held towards the end of February 2008. Prior to the meeting, preliminary design documents for the Cannelton project had been prepared and filed for review. It was at this meeting where the collaboration between the owner, the engineer, and the regulators truly began. For example, initial design concepts for the powerhouse included the use of large diameter caissons drilled into the underlying rock formations. However, as discussions were held and foundation information reviewed, the BOC ultimately recommended that a soil supported mat foundations be considered for the powerhouses at both Smithland and Cannelton. This was ultimately incorporated into the design and although other foundation treatment was required (see Mirza et al., 2015), the collaborative approach ultimately saved the project time and money through the elimination of the drilled caissons.

Similar discussions and collaborative results were achieved at subsequent BOC meetings, which occurred at intervals ranging from three months to a year depending upon the stage of the projects. All aspects of the projects were reviewed including:

- conceptual cofferdam designs by MWH;
- final cofferdam designs by design-build contractors, which included record setting cutoff walls;
- 100-ft high embankments constructed in the wet;
- dewatering systems designed for up to 35,000 gpm of flow; and
- final powerhouse and closure structure designs by MWH.

Not all parties agreed on all items; there were many instances where the parties very much disagreed, or where for example, the BOC and MWH agreed, but the FERC and its consultant did not or vice versa. However, discussions remained respectful and positive, always focusing on dam safety and the best approach for the projects to achieve AMP's objectives.

The BOC's involvement was also very useful in expediting design reviews. For each major design package, the package was prepared and submitted a week or two in advance of the BOC meeting. At the meeting, which generally lasted two or three days and was typically attended by the BOC, FERC, FERC's consultant, USACE, AMP and MWH. MWH would then present the design including the design criteria and cases used, design assumptions and analysis techniques employed, the results and interpretations of the analyses, and ultimately the implementation of the results via the design drawings and specifications. Questions and discussions would occur throughout until such point that all parties were satisfied. The BOC would then deliberate amongst them and prepare a report that would be published and presented to all parties at the conclusion of the meetings. The BOC report would discuss their overall observations, agreement or concerns with design concepts, and recommendations regarding design criteria, cases, assumptions and analysis techniques employed as well as results and the design drawings and specifications. Participation in the meetings and the quick conclusion to the BOC process greatly assisted the approval process by both the FERC and USACE.

This also proved effective when designs required modifications due to discoveries made during construction or unforeseen events. One example of this is the Five-Foot [high] Flood Fighting Dike (FFFFD) that was required during the construction of the cofferdam at the Cannelton Project. On January 25, 2010, a river stage prediction was made that was higher than the current elevation of the cofferdam, which was under construction. About 2,400 feet of the cofferdam along the riverside portion was predicted to be overtopped, potentially causing damage to the partially constructed cofferdam and flooding of the excavation that had commenced with the cofferdam's interior. The contractor proposed to build a temporary FFFFD upstream of the cofferdam's concrete-bentonite slurry cutoff wall by dumping clay fill material and moving and shaping it with a Caterpillar D6R dozer. No additional compaction would be done. The USACE, FERC and BOC were provided design information, which was reviewed and ultimately approved via a series of conference calls.

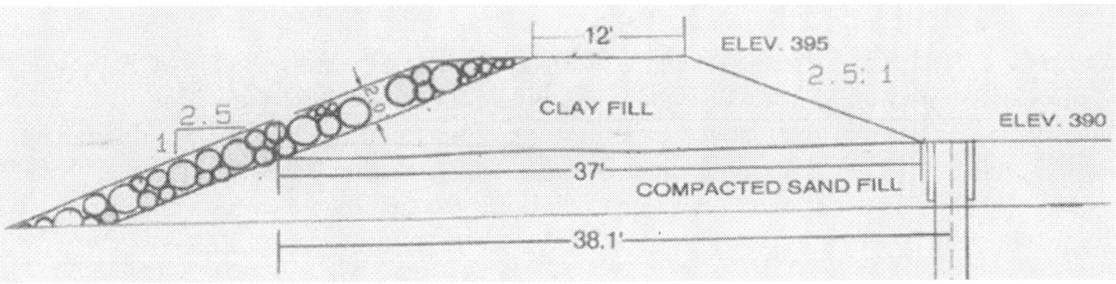

Figure 8. Five-Foot [high] Flood Fighting Dike (FFFFD) at Cannelton. The permanent clay cap would extend up and over to Elev. 405 feet and behind it for about 40 feet. The permanent cap would also tie in to the cement bentonite slurry wall at the toe of the FFFFD. (Source: FERC Construction Report 10 and 11, April 30, 2010.)

The FFFFD took about three days to complete with the placement of riprap taking several more days. Due to the urgency of the potential flood situation, the material was placed even though it was wetter than optimum due to the rain that was leading to the high river flows. The temporary dike held and resisted scour on the riverside slope, mostly providing freeboard against the flood. After the event, the contractor questioned whether the FFFFD could be left in place and worked into the permanent cofferdam. Density tests found it to not meet the compaction specification for the remainder of the project, so the FERC required a slope stability analysis based on the reduced soil density and corresponding strength. After lengthy discussions with the BOC, USACE, FERC, and MWH, it was agreed that the five-foot high, 37-foot wide dike could be incorporated in to the rest of the clay cap.

Another example of excellent collaboration in a time-of-need was when the Smithland cofferdam had to be flooded. In May 2011, a record flood occurred on the Ohio River at the Smithland project location. On May 2, 2011, the river was predicted to exceed the crest of the cofferdam by ½ foot. At that time, the USACE was contemplating demolition of portions of an existing levee to open the Birds Point-New Madrid Floodway to protect the city of Cairo, Illinois, but had not made a decision as to which portions or if any would be demolished. Therefore, direction was given by the USACE on May 2, 2011 to flood the Smithland cofferdam to prevent an uncontrolled overtopping of the cofferdam.

As discussed in Andrews et al. (2015), the cofferdams at all four projects are equipped with Emergency Flood Structures (EFS) to accommodate this scenario or other emergency situation that would require flooding of the cofferdam. However, the EFS at Smithland was not complete at the time the high river flows were anticipated and the controlled flooding was ordered. Therefore, an interim spillway was quickly designed incorporating the partially constructed EFS chute and the materials on site since the access road was flooded. The spillway was cut into the downstream embankment slope with the interior portions cut first and a make-shift stilling basin was constructed at the base of the chute. The design, review, approval, and implementation process was completed within three days – a true testament to the collaboration of all project stakeholders.

On May 2, 2011, the cofferdam began to fill and that same evening the USACE demolished a portion of its levee system to activate the Birds Point-New Madrid Floodway. While this attenuated the flood wave in the Ohio River at the Smithland site, river levels continued to rise and filled the cofferdam to within 2.6 feet of the crest by May 5, 2011. The peak river levels exceeded the 100-year flood elevation by approximately 0.4 feet.

The interim spillway performed as expected and resulted in a controlled filling of the cofferdam with no observed damage to the cofferdam or adjacent USACE facility. Figure 9 shows the cofferdam being filled with water and the Ohio River in flood stage all around the cofferdam.

Figure 9. Aerial photos taken on May 3, 2011 during the flooding of the Smithland cofferdam. Left photo is close up of the cofferdam with the right photo illustrating the extent of the flooding at the location of the Smithland Lock and Dam.

CURRENT STATUS

Construction on all four of the projects continues. As of January 2015, the powerhouse concrete and closure structures are complete at the Cannelton project and the cofferdam has been removed. Final excavation and placement of riprap within both the approach and tailrace channels are also complete. The transmission line has been energized and is currently supplying backfeed power to the power plant. Construction efforts are focusing on mechanical completion of the turbine generator components for all three units as well as completion of terminations and input/output checkouts for all of the electrical equipment within the powerhouse. Commercial operation in currently anticipated to occur in mid-2015.

The Meldahl project is also nearing completion. Mechanical completion activities for the turbine generators are slightly ahead of the progress at Cannelton, but completion and turnover of the unit and plant systems lags a bit behind. The cofferdam at Meldahl has been removed and final grading of the approach and tailrace channels in nearing completion. Commercial operation at Meldahl is also currently anticipated to occur in mid-2015.

At Willow Island, the powerhouse structure has just completed along with the closure structures. With the dewatering systems also complete, the powerhouse has been deemed watertight allowing the initial stages of the cofferdam removal process to commence. All of the major turbine generator components have been lowered into the powerhouse and assembly is well underway. Installation efforts within the powerhouse structure are focused on mechanical piping as well as electrical conduits and cable trays. Commercial operation of this two-unit plant is also currently scheduled for mid-2015.

Construction at the Smithland project is underway as well. As of January 2015, the powerhouse concrete is approximately 90% complete. The closure structures are also nearly complete, with only facing concrete left to place. Excavation outside the footprint of the cofferdam has been completed and riprap placement is nearly complete. The powerhouse crane has been erected and commissioned for limited use. Commercial operation is currently envisioned to be sometime in early to mid-2016.

While these projects have been and continue to be challenging, we believe the successes at Cannelton and Smithland are representative of the true collaboration that can exist amongst owners, regulators, and designers. Early on, project participants put away their personal agendas and each put in the effort and time required to find the best solution for the project – and all in a timely manner. The implementation of an experienced and practical BOC coupled with an experienced owner and dedicated engineering and regulatory team greatly assisted the design review process, ultimately resulting in a better project. While these two projects serve as reminders of how true collaboration can work, we believe all four of the projects will help to build a better world by producing almost 300 MW of renewable, clean energy and are sincerely appreciative of the efforts of the many people involved in making these projects a success.

ACKNOWLEDGEMENTS

The authors would like to acknowledge AMP and their municipal members for their initiative to develop the innovative new hydroelectric projects on the Ohio River and for their vision of renewable energy.

The authors would also like to thank the Board of Consultants for the Cannelton and Smithland Hydroelectric Projects, which was comprised of Eric Kollgaard, Jim Erwin and Joe Ehasz, the many individuals from the USACE and FERC, and FERC's consultant, Dr. Alfred J. Hendron Jr., for their collaborative engagement on these new hydroelectric projects.

REFERENCES

Andrews, T.G., L. Zhao, J.M. Hynes and P.E. Meier (2015). "Holding Back the Ohio – Cofferdam and Cut-Off Wall Design and Construction," USSD 2015 Annual Conference, Louisville, KY.

Bodily, J.A., C. Konstantellos and P.E. Meier (2015). "Upgraded – New Powerhouses at Existing USACE Dams on the Ohio," USSD 2015 Annual Conference, Louisville, KY.

Mirza, A.A., L. Zhao, J.M. Hynes, T.G. Andrews and P.E. Meier (2015). "From the Ground Up – Ground Improvement for Hydro Development on the Ohio River," USSD 2015 Annual Conference, Louisville, KY.

HOLDING BACK THE OHIO — COFFERDAM AND CUT-OFF DESIGN AND CONSTRUCTION

Thomas G. Andrews[1]
Lin Zhao[2]
John M. Hynes[3]
Philip E. Meier[4]

ABSTRACT

Four hydroelectric projects are currently under construction on the Ohio River adjoining existing U.S. Army Corps of Engineers (USACE) locks and dams. These projects, Willow Island, Meldahl, Cannelton and Smithland, which are being developed by American Municipal Power (AMP), required large cofferdams up to 3,800 feet in length to encompass excavations up to 100-feet deep and enable powerhouse construction to proceed in the dry. Since it was critical to dam safety that the integrity of the existing dams and navigation pools be maintained during all phases of construction, detailed construction staging requirements were developed along with mitigation measures to address possible contingencies. The conditions at each site were unique and required site-specific designs and extensive coordination between design-build contractors, the USACE, the Federal Energy Regulatory Commission (FERC), an external Board of Consultants (BOC), and MWH as the Owner's Engineer. Weak rocks, including slickensided claystone, and a thrust fault were present at Willow Island requiring 2,000 kip rock anchors; soft clays required support during excavation at Meldahl; a very large scour hole required construction of a 100-foot-high zoned embankment dam in the Ohio River at Cannelton; and the foundation conditions at Smithland consisted of thick alluvium over karstic limestone – all of which complicated cofferdam design and construction. Each cofferdam included an integral cut-off wall constructed using slurry trench methods, the largest of which was the 3,800-foot-long soil-cement-bentonite cut-off wall at Smithland, which had a maximum depth of 187 feet. Each cofferdam also required an extensive array of deep dewatering wells to provide a dry excavation and maintain stability during construction. This paper discusses the design, construction and performance of these four large cofferdams and the lessons learned.

INTRODUCTION

Four hydroelectric projects (Smithland Hydroelectric Project, Cannelton Hydroelectric Project, Meldahl Hydroelectric Project, and Willow Island Hydroelectric Project) are currently under construction on the Ohio River adjacent to existing U.S. Army Corps of

[1] Principal Geotechnical Engineer, MWH Americas, Inc., 175 W. Jackson Blvd., Suite 1900, Chicago, Illinois. thomas.g.andrews@mwhglobal.com
[2] Lead Geotechnical Engineer, MWH Americas, Inc., 175 W. Jackson Blvd., Suite 1900, Chicago, Illinois. lin.zhao@mwhglobal.com
[3] Senior Geotechnical Engineer, MWH Americas, Inc., 175 W. Jackson Blvd., Suite 1900, Chicago, Illinois. john.m.hynes@mwhglobal.com
[4] Vice President – Hydroelectric Development and Operations, American Municipal Power, Inc., 1111 Schrock Road, Suite 100, Columbus, Ohio, pmeier@amppartners.org

Engineers (USACE) locks and dams (Figure 1). These projects are being developed by American Municipal Power (AMP), a nonprofit wholesale power supplier for municipal electric systems. AMP serves 130 members – 129 member communities in Indiana, Kentucky, Michigan, Ohio, Pennsylvania, Virginia and West Virginia, and one joint action agency in Delaware. Combined, these publicly owned utilities serve approximately 625,000 customers. MWH Americas, Inc. (MWH) was selected by AMP to serve as the Owner's Engineer for the design and supervision of construction for these new hydroelectric projects.

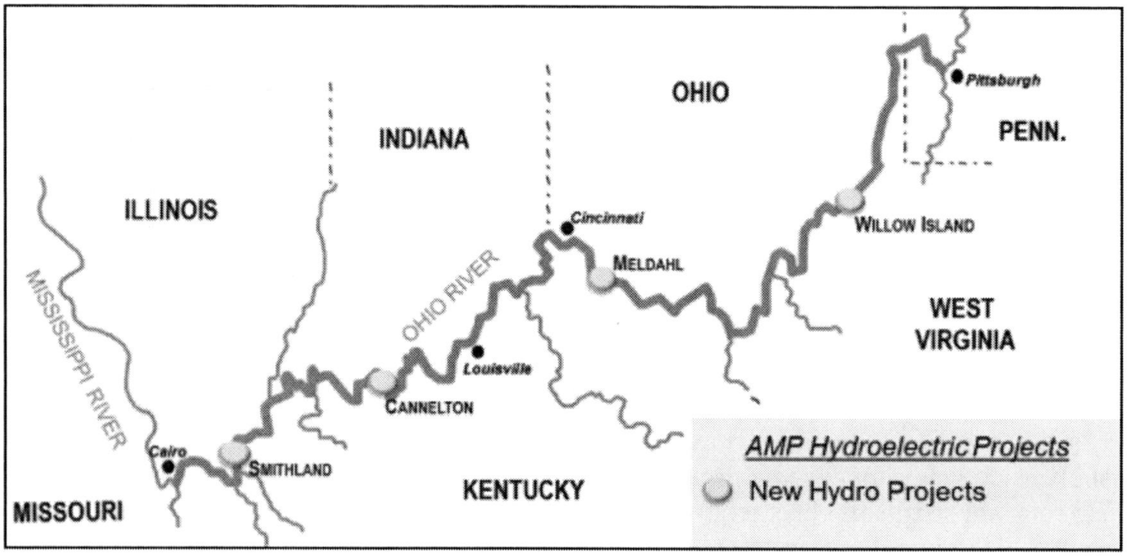

Figure 1. Project Location Map

The construction of these hydroelectric projects required large cofferdams with integral cut-off walls and robust dewatering systems to enable powerhouse construction to proceed in the dry. These cofferdams ranged between about 2,500 and 3,800 feet in circumference. Excavations in excess of 100 feet were required to reach the foundation levels for the powerhouses.

The cofferdams and powerhouse excavations were designed and constructed to provide a stable dry excavation, provide safe working conditions, optimize costs, preserve navigation on the Ohio River, and maintain the integrity of existing dams and new hydroelectric works throughout construction.

Prior to finalizing the cofferdam designs, a Potential Failure Mode Analysis (PFMA) was conducted for each cofferdam.

PROJECT DESCRIPTIONS AND CONSTRUCTION STAGES

Each of the four hydroelectric projects required the construction of an approach channel, tailrace channel, powerhouse (2 or 3 units per project with 25-foot diameter horizontal bulb turbines), riverside closure structures connecting to the existing dam (e.g., 72-foot-high concrete gravity dam at Meldahl) and landside closure structures (e.g., 90-foot-high hardfill dam at Meldahl). The reinforced concrete powerhouses are designed for overtopping during flood events. The generation capacity of each project is as follows:

- Willow Island Hydroelectric Project: 35 MW capacity (20-foot gross head)
- Meldahl Hydroelectric Project: 105 MW capacity (30-foot gross head)
- Cannelton Hydroelectric Project: 84 MW capacity (25-foot gross head)
- Smithland Hydroelectric Project: 72 MW capacity (22-foot gross head)

The design of each project was tailored to address site-specific conditions including:

- weak rocks, including slickensided indurated clays and a thrust fault present at Willow Island;
- soft clays required support during excavation at Meldahl;
- proximity of Cannelton and Smithland to the New Madrid and Wabash Valley Seismic Zones with potentially liquefiable alluvial soils; and
- karstic limestone which further complicated the design at Smithland.

Unique project-specific design features included:

- 2,000-kip rock anchors at Willow Island to stabilize the foundation below the cellular cofferdam and significant dental excavation and rock bolting to treat the fault where it crossed the excavation.
- Steel piles to stabilize excavation slopes at Meldahl.
- An embankment cofferdam at Cannelton with a 100-foot-high zoned section constructed in the Ohio River and stone columns to improve the foundations below the closure structures by mitigating against liquefaction potential and minimizing immediate and earthquake-induced settlement.
- An embankment cofferdam at Smithland with a 150- to 187-foot-deep, 4000-foot-long soil-cement-bentonite cut-off wall and a robust dewatering system. Stone columns were installed beneath the closure structures. Compaction and consolidation grouting was used to treat the foundation soils and karstic rock below the powerhouse.
- Of eight (8) closure structures (dams) required for the project, six (6) were constructed of hardfill (roller-compacted soil-cement) to maximize the use of on-site materials and minimize the importation of aggregate.

All four projects have similar project layouts and consist of an intake approach channel, a reinforced concrete powerhouse, a landside closure structure, a riverside closure structure and a tailrace channel (Figure 2).

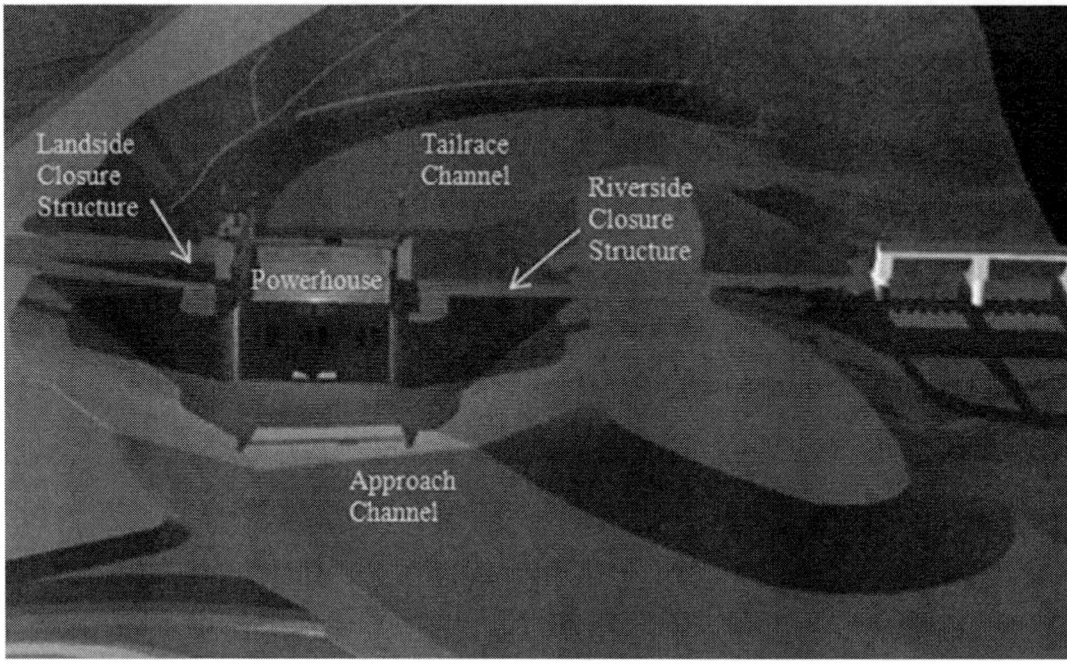

Figure 2. Typical Layout of Project Components (rendering of Cannelton shown)

Photos of the Willow Island, Meldahl, Cannelton and Smithland projects are shown in Figure 3. Further details regarding the development of these hydroelectric projects can be found in Blaszczyk et al. (2015).

Figure 3. Aerial Photos of Construction at Smithland (top left), Cannelton (top right), Meldahl (bottom left), and Willow Island (bottom right) taken during the fall of 2014.

COFFERDAM CONSTRUCTION CONTRACTS

Separate standalone contracts for cofferdam construction and powerhouse excavation were established for each hydroelectric project. This isolated potential differing site condition risks associated with the required geotechnical and marine construction from the larger follow-on general construction contracts for the powerhouses and appurtenant work. This was also done to minimize potential consequential delay claims if differing site conditions were encountered.

Design-Build Contracts for Willow Island, Cannelton and Smithland Cofferdams

The cofferdam and powerhouse excavation contracts for the Willow Island, Cannelton and Smithland projects were each bid as design-build contracts. Design-build contracts were used to enable the projects to go into construction prior to the expiration of the Federal Energy Regulatory Commission (FERC) licenses, which AMP had acquired from the prior licensees for Willow Island, Cannelton, and Smithland. Prior to bidding the cofferdam and powerhouse excavation contracts for these three projects, MWH prepared conceptual designs for review by the external Board of Consultants (BOC) for Cannelton and Smithland and by the USACE and FERC for all three projects to obtain approvals for the general cofferdam layouts for each project. MWH then developed Owner's Requirements, which presented design and construction requirements for each design-build cofferdam and powerhouse excavation contract.

Cofferdam and Cutoff Design and Construction

The design-build contractors and their engineering consultants for the cofferdam and powerhouse excavation contracts for Willow Island, Cannelton and Smithland, which are listed in Table 1, were responsible for preparing the final cofferdam and excavation designs.

Table 1. Design-Build Cofferdam Contractors

Project	Design-Build Cofferdam Contractor	Contractor's Engineer(s)
Willow Island	The Ruhlin Company	Mueser Rutledge / Geocomp / E.L. Robinson
Cannelton	Kiewit-Traylor Joint Venture	Mueser Rutledge
Smithland	C.J. Mahan	Paul Rizzo Assoc. / Florence & Hutcheson

The design-build contractors developed the necessary construction details, including seepage control systems (cut-off walls and dewatering systems) required to support cofferdam and excavation construction, and staging requirements required to maintain the integrity of the USACE dams and upper and lower navigation pools throughout construction.

MWH, as the Owner's Engineer, reviewed the final designs developed by the design-build contractors. These designs also required approval by the BOC (Cannelton and Smithland only) and the USACE and FERC. The Owner's Requirements allowed for phased approvals of the contractor's design. This enabled the contractors to commence with the early stages of construction while they worked to obtain final approval of all aspects of their design.

Design-Bid-Build Contract for Meldahl Cofferdams

Cofferdam construction and powerhouse excavation for the Meldahl project was performed under a traditional design-bid-build contract. Since this project was constructed under a newly-issued FERC license, time was available to complete the entire design and obtain approvals from the USACE and FERC for the cofferdam prior to bidding. MWH prepared the final cofferdam and powerhouse excavation design for Meldahl. Since bidders did not need to pair with engineering consultants or accept design responsibility, a wider pool of bidders submitted bids on the Meldahl cofferdam and powerhouse excavation contract relative to what was received for the design-build contracts used for the other three projects. The deeper pool of bidders and inclusion of a geotechnical baseline report, which defined and limited the contractor's subsurface risk, resulted in lower bid prices relative to what had been received for the design-build contracts. The Meldahl cofferdam and powerhouse excavation contract was ultimately awarded to Angelo Iafrate Construction Company on the basis of best value.

FUNCTIONAL REQUIREMENTS FOR COFFERDAMS

Each cofferdam was designed for river levels up to and including the 100-year flood. The decision to have the cofferdams designed and constructed to provide protection against 100-year floods was driven by AMP's experiences during the construction of the Belleville Hydroelectric Project, which nearly had to be flooded during the middle of powerhouse construction due to rising flood waters. By designing the cofferdams to provide protection against 100-year flood events, AMP greatly reduced the possibility of delays and potential re-work resulting from flooding of powerhouse construction. This also reduced the costs of AMP's Owner Controlled Insurance Policy (OCIP).

Emergency flooding systems were incorporated in the designs for each cofferdam to enable controlled flooding of the interior of each cofferdam and avoid uncontrolled overtopping in the event of a flood greater than a 100-year event.

The excavations within the interior of the cofferdams that were required to construct the powerhouse foundations generally extended to about 100 feet below the pre-existing ground surface. At Willow Island and Meldahl, the powerhouse excavation extended about 50 and 40 feet into bedrock, respectively. The powerhouse excavations at Cannelton and Smithland were in soil for their entire depth. The cofferdams were sized so that large portions of the approach and tailrace channels could also be constructed in the dry. This enabled better grade control and improved hydraulic flow conditions upstream and downstream of the powerhouse, which is an important factor affecting generation in low-head hydroelectric projects. This also enabled large portions of the riprap in the approach and tailrace channels to be placed and inspected in the dry.

Cut-off walls were required to minimize seepage quantities through the cofferdams and their foundations for river levels up to and including those associated with the 100-year flood. The cut-off walls were continuous around the circumference of the cofferdams and extended to the top of bedrock.

Dewatering systems incorporating deep dewatering wells, sumps and drainage ditches to collect surface water and groundwater were required to control water entering the excavation and to enable work within the cofferdam to be accomplished in the dry. The dewatering systems were required to have sufficient capacity such that during and subsequent to the excavation, the groundwater level at any point inside the cofferdam was maintained at or below the levels required to satisfy stability requirements. Piezometers were installed prior to commencing with powerhouse excavation to evaluate the performance of the cut-off walls and dewatering systems. It was required that dewatering be performed continuously, without interruption. Redundant and standby dewatering equipment was required for maintenance and emergencies. Backup power equipment and fuel was also required.

Slope protection was required within the interior of the cofferdams and powerhouse excavations. Where seepage emerged from soils within the powerhouse excavations (generally near the top of rock), filtered drains were installed to protect against piping and erosion.

Comprehensive instrumentation and monitoring systems were required at each site to monitor the performance and safety of the cofferdam and excavated slopes throughout all phases of construction. The cofferdam instrumentation utilized piezometers, inclinometers, survey monuments and tiltmeters, although the specific instrumentation varied from project to project based on site-specific geology and project features. Instrumentation and monitoring for the cofferdams, powerhouse excavations, and permanent works is discussed in detail in Hynes et al. (2015).

All designs were in accordance with applicable USACE guidelines, manuals and regulations, and FERC guidelines. The minimum required factors of safety for embankment and cellular cofferdams were taken from USACE Engineering Manuals EM 1110-2-1902 and EM 1110-2-2503, respectively. Where not specifically addressed in those documents, one or more internationally recognized standards of practice were used. The stability cases presented in Table 2 were analyzed for each cofferdam and powerhouse excavation.

Table 2. Cofferdam Analysis Cases

Conditions	Exterior River Level		Strength Parameters	
	Upstream	Downstream	Coarse Grained Soil	Fine Grained Soil
End of Construction	Normal Pool	Minimum Pool / Ordinary High Water**	Drained	Undrained
Normal	Normal Pool	Minimum Pool / Ordinary High Water**	Drained	Drained
Seismic (Construction Basis Earthquake*)	Normal Pool	Minimum Pool / Ordinary High Water**	Drained***	Drained***
Construction Flood	100-year Flood	100-year Flood	Drained	Drained
Post-Flood (after flood recedes, cofferdam interior remains filled)	Normal Pool	Minimum Pool	Drained	Drained
Interior Rapid Drawdown (after unwatering of cofferdam interior following flooding)	Normal	Ordinary High Water	Drained	Duncan, Wright and Wong (1990)
Exterior Rapid Drawdown (loss of downstream pool)	---	Water level associated with loss of lower pool	Drained	Duncan, Wright and Wong (1990)

* The Construction Basis Earthquake (CBE) was established for cofferdam design using 3% exceedance probability over the maximum service life of the cofferdam.

** The minimum pool and ordinary high water levels bracket normal lower pool fluctuations. The most conservative level was used depending on direction of sliding being evaluated.

*** Liquefaction and cyclic softening analyses using the methods recommended by Idriss and Boulanger (2008) were performed to determine that the CBE would not induce a loss of soil strength.

COFFERDAM LAYOUTS

The cofferdam and excavation layouts were based upon AMP's and MWH's experience with similar projects, including the Belleville Hydroelectric Project, and geometric information for the powerhouse water passages and unit settings obtained from Voith, the turbine-generator manufacturer. MWH compiled this information and developed the maximum anticipated powerhouse dimensions and minimum unit submergence. This information was used to locate the powerhouses and to develop the conceptual layouts of the cofferdams and powerhouse excavations. The layout and type of cofferdam for each project were then finalized based on site-specific conditions, particularly space constraints and economics.

The space available at Willow Island was particularly tight because of the adjoining railroad and industrial property. This required that the riverside of the powerhouse be located within 60 feet of the existing dam's nearest spillway pier (Pier No. 9) and forced the riverside of the cofferdam to extend out in the river so that it encompassed the existing dam's nearest (leftmost) spillway bay. This necessitated that a cellular cofferdam be used at Willow Island so that the cofferdam could connect to the upstream and downstream sides of Pier No. 8 of the existing dam and not extend further out into the Ohio River.

Embankment cofferdams were constructed at Meldahl, Cannelton and Smithland, because the greater depth to bedrock made cellular cofferdams cost prohibitive. Since space was available, these embankment cofferdams were shifted landward to reduce the amount of marine construction and material to be imported than would otherwise have been required for construction.

COFFERDAM DESIGN AND CONSTRUCTION

Cellular Cofferdam at Willow Island

The riverside cofferdam at Willow Island was 1,300 feet in length and composed of sixteen sheet pile cells founded on bedrock and connected to the upstream and downstream sides of Pier No. 8 of the existing dam. The existing spillway Bay No. 8 and Pier No. 9 were on the interior of the cofferdam enclosure adjacent to the powerhouse excavation. The cellular cofferdam alignment was laid out so that no part protruded into the river beyond the riverside face of the Pier No. 8 of the USACE dam.

The cofferdam cells at Willow Island were approximately 63 feet in diameter with thirty degree wyes and about 70-foot-tall. A weighting berm was constructed along the interior side of the cells to obtain the minimum required factor of safety against sliding for the 100-year design flood when the cofferdam's interior was unwatered. The cellular cofferdam with interior weighting berm is shown in Figure 4.

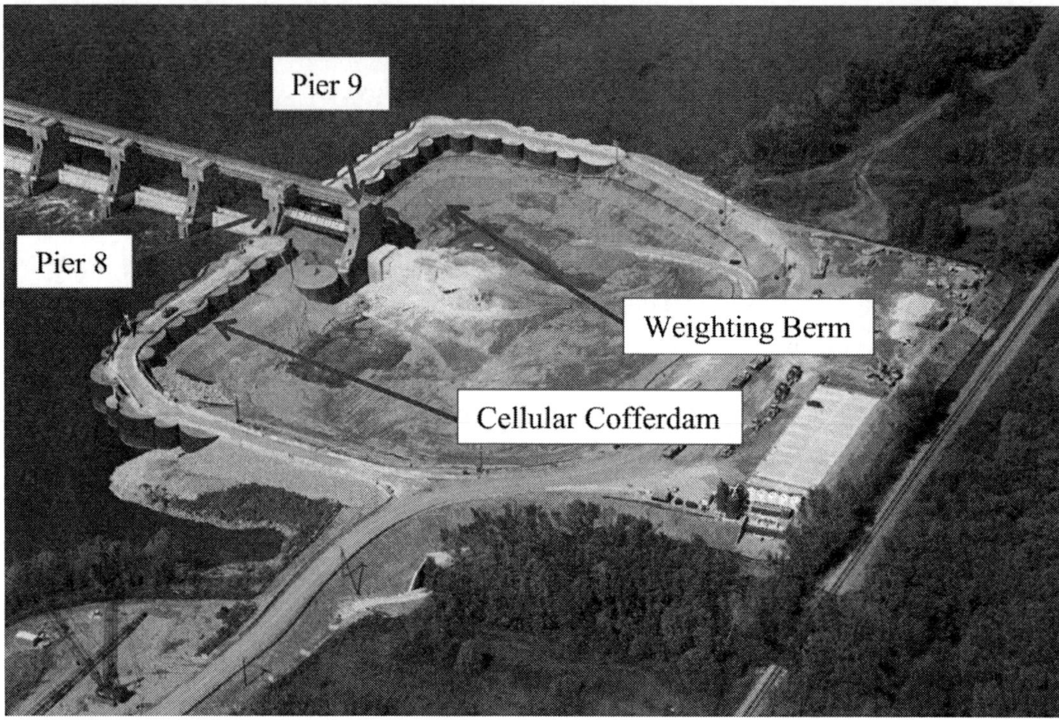

Figure 4. Willow Island Cofferdam Prior to Start of Rock Excavation

The design of the cellular cofferdam was analyzed for overturning, bearing capacity, bursting, internal shear, and pullout (pile rise). The stability of Pier Nos. 8 and 9 of the Willow Island Dam was also evaluated with the cofferdam in place and excavation dewatered to verify their ability to meet the required factors of safety when the cofferdam's interior was unwatered. These analyses were performed in accordance with the applicable USACE Engineering Manuals (EM 1110-2-2503 and EM 1110-2-2200) and the FERC Engineering Guidelines for Evaluation of Hydropower Projects (FERC, 2002) using the minimum required factors of safety designated therein. Adequate factors of safety were found for all applicable analysis cases.

The cellular cofferdam for Willow Island was constructed using the general sequence as described below:

1. Remove unsuitable foundation materials, as required, by conventional landside excavation or dredging from the footprint of the cellular cofferdam.
2. Remove riprap under the cofferdam footprint and stockpile for future use.
3. Construct cofferdam cells using marine equipment simultaneously on the upstream and downstream sides of the dam.
4. Fill cofferdam cells with free-draining materials, and place fill within the existing USACE spillway Bay No. 8 to maintain stability of the dam piers when the interior of the cofferdam is unwatered.
5. Construct watertight connections between cellular cofferdam and Pier No. 8 of the existing dam.

6. Place free-draining materials in the "wet" below water on the interior side of the cofferdam cells to construct weighting berms prior to unwatering the area enclosed by the cofferdam.

Embankment Cofferdams at Meldahl, Cannelton and Smithland

The river and land sides of the cofferdams at Meldahl, Cannelton and Smithland and the landside cofferdam at Willow Island were composed of earth fill materials. The riverside embankment cofferdams consisted of zoned fills that allowed construction to proceed in the "wet" below river levels using rockfill and granular material and provided for the construction a cut-off wall through the center of the constructed embankment.

The general zonation consisted of outer rockfill sections with interior sand fill sections that were overlain by impervious clay fill, as shown in Figure 5. The rockfill sections generally had 1.75 horizontal to 1.0 vertical exterior side slopes. The top of the rockfill sections were set to elevations that were above typical river levels during the summer months, when the initial phase of marine cofferdam construction was generally performed. The slopes of the clay fills were typically 2.5 horizontal to 1.0 vertical. The exterior face of the clay fill was protected by riprap.

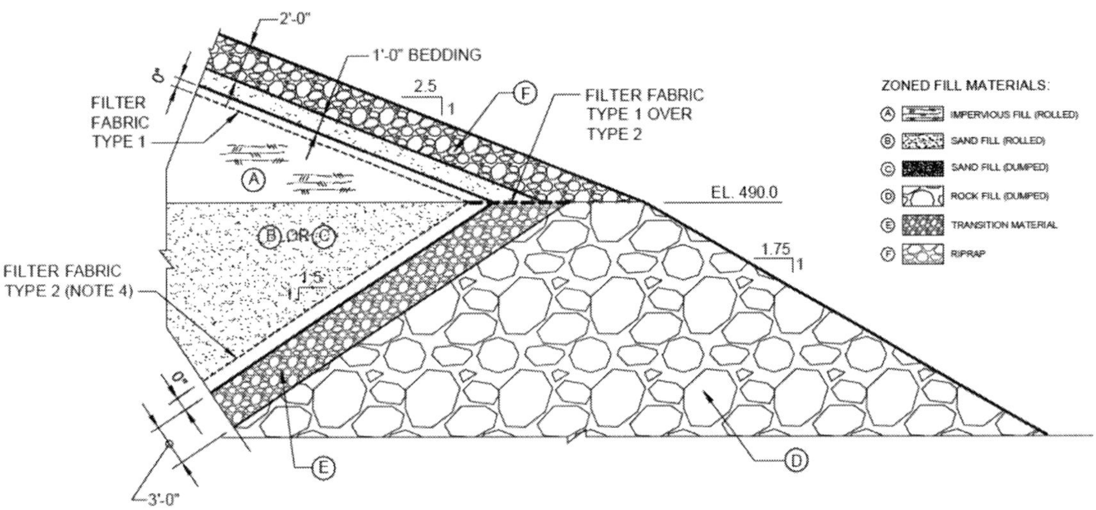

Figure 5. Typical Embankment Cofferdam Zonation (Meldahl shown).

The dimensions of the embankment cofferdams for the Willow Island, Meldahl, Cannelton and Smithland Hydroelectric Projects are summarized in Table 3.

Table 3. Embankment Cofferdam Dimensions

Project	Embankment Length (ft.)	Maximum Height (ft.)	Embankment Alignment
Willow Island	1,300	25	Landside only
Meldahl	3,600	85	River and Land Side
Cannelton	3,600	135	River and Land Side
Smithland	3,800	50	River and Land Side

The designs of the embankment cofferdams were analyzed to assess stability for all applicable design cases including, normal, seismic, construction flood, post-flood and rapid drawdown. The 100-year flood was used for the construction flood case. Analyses were performed in accordance with the applicable USACE Engineering Manual (EM 1110-2-1902) and the FERC Engineering Guidelines for Evaluation of Hydropower Projects (FERC, 2006).

The embankment cofferdams for Meldahl, Cannelton and Smithland were constructed using the general sequence as described below:

1. Perform clearing and grubbing and topsoil removal and stockpiling.
2. Remove unsuitable foundation materials, as required, both on land and in the area where cofferdam extends into the river.
3. Perform limited powerhouse excavation in areas upstream and downstream of the existing dam's sheet pile cut-wall, which extends outward from the end of dam.
4. Construct a work platform (using impervious fill obtained from the limited powerhouse excavation) for cut-off wall construction along the landside of the cofferdam.
5. Construct cut-off wall along the centerline of the landside of the cofferdam using the slurry trench method.
6. Construct rockfill berm(s) along the outer portion(s) of the cofferdam's footprint where it extends into the Ohio River.
7. Place a transition zone of gravel and cobbles along the interior side of the rockfill berms.
8. Place a heavy geotextile over the transition zone material.
9. Place sand fill in the "wet" below river level in the area between the rockfill section and the existing river bank (or between the inner and outer rockfill sections where the entire footprint of the cofferdam extends into the river).
10. To increase strength and limit deformation, vibro-compact thick zones of sand fill that were placed underwater, while making sure to maintain a buffer between vibro-compaction probe and heavy filter fabric.
11. Place and compact sand fill in lifts at elevations where the cofferdam is above river level.

12. Construct a work platform of impervious fill for construction of the cut-off wall along the riverside of the cofferdam.
13. Construct cut-off wall along centerline of riverside of cofferdam using slurry trench methods.
14. Place and compact impervious fill in lifts to construct the upper elevations of the cofferdam atop the fill material placed on the riverside and atop the existing ground on the cofferdam's landside.
15. Place riprap to armor the outer face of the cofferdam fill sections above the rockfill section.

The landside of the Willow Island cofferdam was constructed using similar methods as used to construct the landsides of the cofferdams at the other three projects.

Details of cofferdam construction are presented in Marold (2012).

Flood Events during Cofferdam Construction

In late January 2010, a flood event was forecast for Cannelton while construction of the cofferdam was still on-going. The contractor and MWH decided to stop placing impervious fill across the entire width of the cofferdam and to focus efforts on constructing a 5-foot-wide dike of impervious fill in a low area along the cofferdam's riverside, as shown in Figure 6. The contractor was able to construct this dike in advance of rising river levels and avoided having to flood the partially completed cofferdam. Additional details are presented in Blaszczyk et al. (2015).

Figure 6. Construction of Temporary Impervious Fill Dike at Cannelton in Advance of Rising River Levels

In early May 2011, the USACE predicted a flood in excess of the 100-year event at Smithland during the initial stages of powerhouse excavation. Unfortunately, the contractor had yet to construct the cofferdam's emergency flooding structure since its construction would have required excavation below common spring flood levels. MWH worked with the contractor to have a temporary flooding chute constructed from stockpiled riprap, which enabled flooding of the cofferdam interior through a small cut that was made into the crest of cofferdam embankment. This allowed for the gradual flooding of the cofferdam's interior to proceed without erosion, which can be seen in Figure 7.

Figure 7. Emergency Flooding of Smithland Cofferdam

At the time when it was decided to commence with flooding of the Smithland cofferdam, the USACE forecast that Ohio River levels would rise to the 500-year level. The USACE subsequently demolition of portions of an existing levee to open the Birds Point-New Madrid Floodway to protect the city of Cairo, Illinois on May 2, 2011 to reduce flooding. The flood waters at the Smithland ultimately peaked at just above the cofferdam's 100-year design level. Fortunately, this very large flood event occurred prior to completing the cofferdam and powerhouse excavation and, thus, occurred prior to beginning powerhouse construction when the financial and schedule impacts of such flood would have been substantial.

Cut-off Walls

Cut-off walls were constructed through the embankment cofferdams to control seepage into the interior of the cofferdam and to enable the construction of a dewatered excavation suitable for construction of the powerhouses.

The cofferdam cut-off walls were generally located along the centerline of the cofferdams and were continuous and extended to the bedrock around the entire cofferdam alignment.

With the exception of the riverside of the Willow Island cofferdam, where steel piling was used to construct a cellular cofferdam, cut-off walls along the land and river sides of the cofferdams were generally constructed using slurry trench techniques to form soil-bentonite, soil-cement-bentonite or cement-bentonite walls. Soil-bentonite, soil-cement-bentonite, and cement-bentonites slurry walls were constructed at the various sites, as presented in Table 4. Each slurry wall had a minimum thickness of 30 inches.

Table 4. Cofferdam Cut-Off Walls

Project	Type of Cut-off Wall	Cut-off Wall Length (ft.)	Maximum Depth of Cut-Off Wall (ft.)
Willow Island	Soil-Bentonite	1,300	55
Meldahl	Cement-Bentonite	3,600	80
Cannelton	Cement-Bentonite	3,600	150
Smithland	Soil-Cement-Bentonite	3,800	187

Cemented slurry walls (cement bentonite or soil-cement-bentonite) were used at Meldahl, Cannelton, and Smithland to minimize the potential for erosion of the wall into the adjoining pervious soils under the relatively high gradients across the cut-off walls that would be induced at these sites following the dewatering of the area enclosed within the cofferdams.

A key into bedrock was not required for the cofferdam cut-off walls, since the corresponding reduction in seepage inflow was judged to not be worth the significant cost increase associated with excavating a key into rock. However, the cut-off walls were required to be taken to competent rock, as defined as rock that could not be removed by a hydraulic clamshell or a long-reach excavator. Loose or highly weathered rock at the top of bedrock was to be removed, but chiseling or grinding was not required.

At Meldahl, the construction of the cement-bentonite cut-off wall required special attention where it crossed the existing dam's concrete weir. In this area, the slurry trench was widened from 30 inches to 6 feet to minimize seepage along the interface between the cement-bentonite wall and the upstream and downstream faces of the concrete weir. Similar steps were taken at Cannelton and Smithland to obtain good contact between the cofferdam slurry walls and the existing permanent sheet pile walls that extended landward from the ends of the existing dams.

Details of cofferdam cut-off wall construction for Cannelton are presented by Ressi di Cervia (2010); the details are similar to the other projects.

Powerhouse Excavation

The cofferdams and powerhouse excavations were analyzed under applicable design cases. Slope stability analyses were used to identify excavations that would need to be supported and to determine maximum allowable phreatic surfaces (and minimum amount of drawdown) needed to maintain the required factors of safety. Figure 8 shows the final allowable phreatic surface for Meldahl.

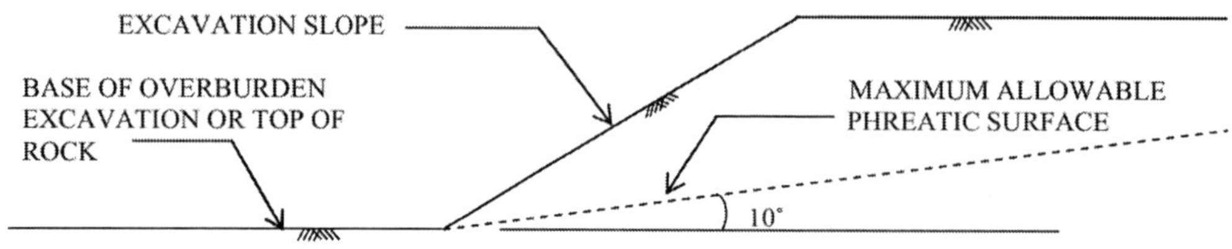

Figure 8. Maximum Allowable Phreatic Surface from Stability Analysis for Meldahl

Dewatering Systems

The cofferdams and integral cut-off walls controlled groundwater inflows into the powerhouse excavations, but were not designed to completely eliminate seepage inflows.

Therefore, dewatering systems were required to intercept and control seepage through the cofferdams and their cut-off walls and underlying bedrock foundations. The dewatering systems were also needed to remove groundwater trapped within the cofferdams after construction and due to precipitation.

The dewatering systems included deep wells, sumps, and drainage ditches to control ground and surface water flows into each cofferdam, maintain drawdowns required for slope and excavation stability, and to allow work within the cofferdams to be accomplished in the dry. Extensive arrays of deep dewatering wells were installed in rings on the interior of each cofferdam's cut-off wall.

The number of deep wells installed ranged from 16 to 65 depending on the amount of seepage anticipated at each site based on seepage analyses. The design capacity of the dewatering systems varied from 3,000 to 35,000 gpm depending on the subsurface conditions. Each dewatering well was installed with a Rossum Sand Tester port to enable monitoring of pumped sand and a flow meter to monitor changes in pumping rates. Header pipes collected the water from the dewatering wells and discharged into sediment ponds, which then discharged into the river.

A supplemental wellpoint system was installed by the contractor at Meldahl (Angelo Iafrate Construction Company) to drawdown groundwater near the top of rock so that toe drains could be installed along the base of the excavation slopes in soil.

Since cofferdam and excavation slope stability depended on maintaining a minimum drawdown of the phreatic surface, standby replacement pumps were kept on-site and backup generators maintained in-place in the event of a power loss. Hynes et al (2015) discusses the use of piezometers to monitor and identify potential malfunction of the pumps and dewatering system overall.

Emergency Flooding Systems

Since uncontrolled overtopping of the cofferdams would likely cause the cofferdams to fail, each cofferdam had to be constructed with an emergency flooding system that would enable controlled flooding of the cofferdam's interior in case river levels were predicted to exceed the cofferdam's 100-year design flood.

The discharge capacities of the emergency flooding systems were calculated based on anticipated entrance conditions, energy losses, and river pool levels. The emergency flood systems were sized so that the cofferdam's interior, including the completed powerhouse excavation, could be filled within about 12 hours. The ability to perform controlled filling of the cofferdam interior in a short period of time would provide additional time to evaluate the need to perform emergency flooding and reduced the possibility of unnecessary flooding of the cofferdam.

Emergency flooding structures were constructed at the downstream sides of the cofferdams to pass river water into the area enclosed by the cofferdams (see Figure 9). These cast-in-place reinforced concrete structures typically consisted of four adjoining 8-foot-wide by 8-foot-high box culverts. The emergency flooding structures were located near the top of the cofferdams to enable easy access and provide low velocities (weir flow) instead of high velocity (orifice flow) into the excavation. To address the potential for piping to occur along the sides of the emergency flooding structures during high river events, diaphragm filters were installed around the emergency flooding structures in accordance with the Federal Emergency Management Agency technical manual on conduits through embankment dams (FEMA, 2005).

Figure 9. Emergency Flooding Structure and Grouted Riprap Chute at Cannelton

The upstream side of the emergency flooding structures were generally sealed with a row of wood "needle" beams placed side-by-side with one "king pin" steel pile in the middle of the wood beams. The exterior face of the beams was covered by a membrane to minimize seepage during times of high water. If required, the emergency flood structure would be operated by removal of the king pin and subsequent removal of each wood beam until the entire area in front of the box structure was opened to the river. The king pin and wood beams would be removed using a small crane, which was required to be maintained on-site. If needed, it was anticipated that the needle beams could all be removed in about 2 hours. At Smithland, the design-build contractor (C.J. Mahan) opted to install slide gates on the exterior side of the emergency flooding structure in lieu of needle beams.

Grouted riprap spillway chutes were constructed to transfer the flow from the emergency flooding structures safely down the excavation slopes to the base of the powerhouse excavations. Grouted riprap was chosen for the channel surface because of its durable rough surface, which would reduce terminal velocities to the maximum acceptable velocity and the grout would prevent riprap from being plucked from the chute's surface.

Emergency Flooding Plans (EFP) were developed for each site and included in the Temporary Construction Emergency Action Plan (TCEAP) for each project. The EFP included pre-flood monitoring and preparation for flooding, flooding procedures, and post-flood unwatering procedures. If the river was predicted to rise above the 100-year flood event, the emergency flooding system would be operated at the direction of the USACE. The timing of commencement would depend on the river level and rate of rise.

As of January 2015, the operation of the emergency flood structures has not been required. During a flood event at Cannelton in April 2011, Ohio River levels rose to over 8 feet above the invert of the emergency flooding structures box culverts prior to receding, as shown in Figure 10. Some water seeped through the needle beams, but this reduced with time, possibly due to swelling of the wood beams. The needle beams and cofferdam itself performed as intended throughout this event.

Figure 10. Flood Water against Emergency Flooding Structure at Cannelton

SITE-SPECIFIC CHALLENGES

Conditions present at each project site required the development of project-specific engineered solutions to enable the construction of stable cofferdams and excavations.

Weak Rock and a Reverse Fault at Willow Island

Bedrock at Willow Island consists of interbedded siltstone, claystone/indurated clay, sandstone, shale and limestone of the Conemaugh Group, which exhibits frequent facies changes with thinning and thickening of beds. The bedrock stratigraphy at the site is extremely variable in character and composition, both horizontally and vertically. Bedding layers at the site are generally nearly horizontal to sub-horizontal. The joint pattern at the site consists of a set of joints trending north-northeast and another set perpendicular to the first.

While the siltstone, sandstone and limestone beds are medium hard to hard, the claystone/indurated clay and shale are very weak to weak and slickensided surfaces are common within claystone/indurated clay.

Deep-seated sliding into the powerhouse excavation along weak rock horizons or slickensided surfaces was identified as a potential mode of failure for the Willow Island cofferdam. Based on an extensive subsurface exploration and laboratory testing program, the weak rock horizons and slickensided surfaces were assessed to have a strength envelope with a cohesion intercept of 0 and a friction angle of 14°.

The stability of the Willow Island cofferdam was further complicated by the presence of a reverse fault that the USACE found when excavating for construction of the foundation of Pier No. 9 during the original construction of the Willow Island Locks and Dam in the 1970's. Based on USACE documentation and an assessment of borehole data, the fault was assessed to have a strike of about N75-80°W and dip 30° to the south. This indicated that the fault would cross under the downstream side of the cellular cofferdam and that the fault's hanging wall would be undercut by the powerhouse excavation in rock.

To address the weak rock horizons and reverse fault, the design-build contractor at Willow Island (The Ruhlin Company) installed high-capacity rock anchors around the riverside of the powerhouse rock excavation to obtain the minimum required factors of safety for the 100-year design flood. The 59-strand rock anchors, which extended to depths of more than 150 feet into rock and provided up to 2,000 kips of anchor capacity, were used. The rock anchor installation is shown in Figure 11.

Figure 11. Installation of High-Capacity Rock Anchors at Willow Island

Due to the importance of the cofferdam to maintain safe conditions and the known presence of weak rock horizons and the reverse fault, an extensive array of instrumentation was installed at Willow Island to monitor cofferdam performance during and following rock excavation. The instrumentation included an automated total station to monitor the displacement to survey targets, inclinometers, tiltmeters, piezometers, and load monitoring sensors on rock anchor strands, as discussed in Hynes et al. (2015). While minor deformations were recorded, no significant or sudden movements were noted and the rock anchor loads generally remained constant. Therefore, the rock anchors were assessed to have performed as intended.

When the powerhouse rock excavation exposed the reverse fault, it was found to have an apparent width of in excess of 10 feet. The fault zone was generally composed of broken overturned beds of sandstone and claystone/indurated clay with observed dips up to 80°. Immediately beyond this zone, the rock was relatively undisturbed and bedding dipped approximately 3° towards the south. The broken rock in the fault's shear zone consisted of interlocking slabs of rock that were generally up to 3-foot-long and 6-inches-thick along with smaller crushed rock fragments. The broken rock was generally free of clay gouge. No seepage or free water was observed within the faulted area.

The fractured rock within the exposed fault zone initially held up well along the face of the rock excavation, but once this zone was subjected to rain, it began to ravel back into the face of the excavation as shown in Figure 12.

Figure 12. Reverse Fault in Riverside Wall of Willow Island Powerhouse Excavation

To minimize the potential for material degradation and further raveling in the fault's shear zone and the slabbing off of overhanging competent beds, the open area of the exposed fault zone was covered with 4 inches of fiber-reinforced shotcrete to control raveling. Untensioned Grade 150 ksi all-thread No. 8 bars were installed through the shotcrete at least 8 feet into sound rock and grouted in place on a typical spacing of about 5 feet in each direction. The anchor bars were extended to the face of the rock excavation to provide support for forms that were used to place concrete within the opened fault zone. Concreting was performed in lifts to minimize pressure against the forms. After removal of the forms, bearing plates with anchor nuts were installed on the anchor bars where they extend through the face of the concrete. The fault stabilization is shown in Figure 13.

Figure 13. Stabilization of Reverse Fault in Riverside Wall of Willow Island Powerhouse Excavation

Very Soft Clay at Meldahl

Very soft clay with Standard Penetration Test (SPT) N-values ranging from weight of hammer to 2 was present at depths of about 20 to 30 feet below the pre-construction ground surface across portions of the Meldahl site. While this material was discontinuous across the site, it was found across a significant fraction of the site.

To obtain the minimum required factor of safety for the end-of-construction (undrained) case for sliding stability during powerhouse excavation, wide flange steel piles were installed at select mid-slope locations where the very soft clay zones were present to stabilize the cofferdam and powerhouse excavation slopes. Steel piles were chosen because they enabled powerhouse excavation to proceed as scheduled; there was not sufficient time to improve clay strength via pre-loading or other means.

Supplemental Cone Penetration Testing (CPT) was performed to identify the areas that needed additional reinforcement. An analysis was performed to determine the section, length, and spacing of piles to achieve the required factor of safety for the end-of-construction (undrained) case. Pile lengths varied from 35 to 45 feet with spacings of 3 to 4 feet. The reinforcing piles are shown in Figure 14.

Figure 14. Installed Slope Reinforcing Piles at Meldahl

Scour Hole at Cannelton

A large scour hole was present downstream of the existing dam at Cannelton, which required the design-build contractor (Kiewit-Traylor Joint Venture) to construct a large zoned embankment cofferdam in the Ohio River that was founded on the alluvial deposits that form the river bottom. The maximum height of this cofferdam from the exterior riverside toe to the cofferdam crest was about 100 feet. The portion of the riverside cofferdam that extended into the Ohio River was about 1,250 feet long and required extensive marine construction.

Successive stages of marine cofferdam construction for the riverside cofferdam at Cannelton are shown in Figures 15, 16 and 17.

Figure 15. Marine Construction of Rock Berms for the Cofferdam at Cannelton

Figure 16. Vibro-Compaction of Sand Fill for the Cofferdam at Cannelton

Figure 17. Cannelton Cofferdam and Powerhouse Excavation upon Completion

It was recognized that river bottom conditions could vary from season to season at Cannelton. To address this variability, a contract allowance was established for supplemental rockfill and removal of unsuitable foundation material in the event that the river bottom topography was found to be lower and/or softer than anticipated. A side-scan sonar survey conducted at the start of marine construction at Cannelton indicated that the river bottom contours were in fact lower than the contours surveyed 2 years earlier prior to bidding. Therefore, part of the allowance was used to cover the cost of additional cofferdam fill.

Karstic Limestone at Smithland

Karstic limestone was present at the Smithland site below approximately 140 to 160 feet of alluvium. Voids and solution cavities with apparent depths of up in excess of 15 feet were found within the upper 25 feet of bedrock; such solution features were particularly common in the top 20 feet of bedrock. Some voids appeared to be open, while others appeared to be partially filled with soil. The average Rock Quality Designation (RQD) of the upper 25 feet of bedrock was 69 with an average core recovery of 80%.

Limestone recovered from rock coring from the geotechnical exploration for the project, which was managed by MWH and performed by Terracon, exhibited features ranging from pinpoint vugs to large voids. The weathered condition of the rock core, associated iron oxide staining, and crystal formation observed along most open joints and walls of voids indicated the movement of groundwater through the rock mass.

Overburden at the site is generally composed approximately 20 feet of clay overlying about 120 to 140 feet of loose to medium dense sand and gravel, with occasional clay layers and lenses.

The possibility that the karstic limestone strata and overlying pervious sands and gravels could potentially convey high flows through open channels in the karstic limestone and result in the development of sand boils and/or sinkholes during powerhouse excavation within the cofferdam was identified as a potential failure mode during preliminary cofferdam design. This concern was heightened by the knowledge that the USACE experienced high sudden inflows during construction of the Smithland Locks and Dam, which on at least one occasion was accompanied by the formation of a vortex in the Ohio River.

Due to the cost associated with trying to obtain a positive cut-off through the karstic limestone to depths of over 160 feet below the pre-construction ground surface, the design-build cofferdam contractor at Smithland (C.J. Mahan) chose to construct a soil-cement-bentonite cut-off wall to top of rock and, recognizing that the solution features in rock would make this wall effectively act as a "hanging" wall, decided to install over sixty (60) deep dewatering wells to control seepage. By using this extensive array of deep dewatering wells to drawdown the phreatic surface below the base of the powerhouse excavation, the contractor was able to mitigate potential issues with heave or sand boils in the event that a sudden increase in inflows were to occur under the cofferdam's cut-off wall. The dewatering system at Smithland was also designed to provide redundant capacity to address the possibility of increased inflows over time.

Possible slurry trench instability during cut-off wall construction, due to a loss of slurry from the cut-off wall excavation trench into open channels, was also a concern considered during preliminary design. To address this, the contractor kept more than 100 feet of slurry trench open at the ground surface, but only excavated a small portion all the way down to the top of rock at any one time prior to backfilling the bottom of the trench with soil-cement-bentonite backfill. Since the large open trench was filled with bentonite slurry, a very large amount of slurry would need to be lost into an open feature in rock to affect slurry levels within the trench or trench stability.

The bottom of the 3800-foot-long soil-cement-bentonite cut-off wall at Smithland stepped up and down along the irregular top of rock surface. The maximum depth of this cut-off wall was 187 feet. The construction of this deep cut-off wall is shown in Figure 18.

Figure 18. Soil-Cement-Bentonite Cut-Off Wall Construction at Smithland

Slurry losses from within the cut-off wall trench did occur on a few occasions during slurry trench excavation at Smithland, but the contractor was able to plug them with excavation spoils and continue advancing slurry wall construction.

While the seepage inflows intercepted by the dewatering system at Smithland were considerably more than pumped out at the other sites, the dewatering system has had sufficient capacity to handle this amount of seepage and maintain the lowered phreatic surface. As of January 2015, there have been no occurrences of sand boils or sinkholes.

PERFORMANCE OF CUT-OFF WALLS

The performance of the cofferdam cut-off walls varied significantly based on geologic conditions at each site. The dewatering pumping rates at all sites decreased with time as pore water trapped within the cofferdam dam was removed. The initial decrease in pumping rates at Cannelton is shown in Figure 19. The performance of the cutoff walls was also monitored using piezometers and by visual surveillance for potential seepage on the cofferdam's interior slopes, as described in Hynes et al. (2015).

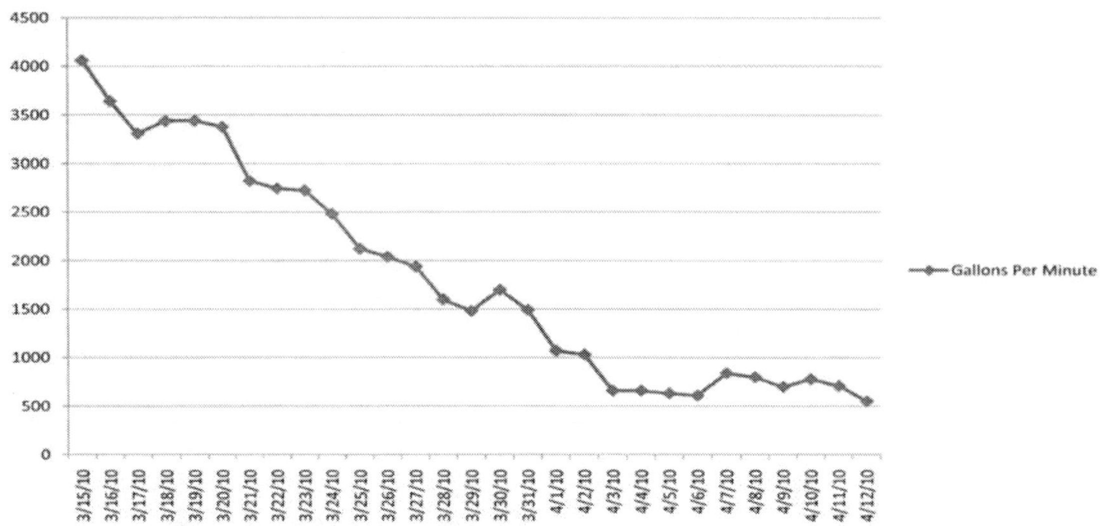

Figure 19. Pumping Rate at Cannelton during Initial Dewatering

The steady-state dewatering pumping rates after completion of initial dewatering are as follows:

- Cannelton – 100 gpm
- Willow Island – 500 gpm
- Meldahl – 2,500 gpm
- Smithland – 12,000 gpm

The effectiveness of the cofferdam cut-off wall at Cannelton exceeded expectations. The cut-off walls at the other three sites generally performed as expected based on pre-construction seepage modeling.

LESSONS LEARNED

Lessons learned from the construction and performance of the cofferdams and cut-off walls for the Willow Island, Meldahl, Cannelton, and Smithland Hydroelectric Projects are described below.

- The project sites had each previously served as staging areas during construction of the adjacent USACE's locks and dams. During cofferdam construction, sheet piling, concrete, pipes and other buried debris was uncovered. To avoid delays in the cut-off wall construction, pre-probing was conducted at Meldahl and Cannelton so that obstructions could be removed from the upper 20 feet of soil along the cofferdam's cut-off wall alignment.
- Atmospheric conditions prevented the moisture conditioning (drying) of impervious fill from being accomplished during winter months. To avoid construction delays, fly ash was mixed with the fill material at Cannelton and lime treatment was used at Meldahl so that fill placement and compaction could proceed on schedule.

- It can be difficult to identify top of rock during slurry trench excavation for cut-off walls, particularly when cobbles or dense material are present above the top of rock. At Meldahl, closely spaced borings along the cut-off wall alignment were needed to better define the top of rock profile.
- The cofferdams enclosed large areas, which formed large catchment areas during precipitation events. When the slopes were disturbed by construction activities, rain events could results in large amounts of erosion along the slopes and sediment deposition at the base of the excavation. The ongoing maintenance of erosion protection on interior slopes helped contractors avoid having to clean sediment from work areas at the base of the excavation and from needing to re-grade the excavation slopes.
- Issues with the dewatering systems, while not frequent, were not uncommon. Therefore, standby pumps and other dewatering accessories were kept on-site and projects were staffed 24-hours a day to minimize any outages.
- PVC header pipes for the dewatering system at Meldahl broke on two occasions as a result of construction activity and caused erosion in the underlying cofferdam and excavation slopes. HDPE was found to preferable over PVC for header pipes, because it is less brittle and not as prone to damage when disturbed by construction activities.
- The piezometer readings were generally quite responsive to changes in dewatering pumping rates and times. Therefore, piezometers can provide timely notice if a dewatering well goes offline and should be checked regularly; this is discussed in greater detail in Hynes et al (2015).

CONCLUSIONS

While the cofferdams and integral cut-off wall constructed for the Willow Island, Meldahl, Cannelton and Smithland Hydroelectric Projects are temporary structures that were solely constructed to facilitate construction of the projects' permanent works, these structures constitute large, complicated civil works in their own right, which required site-specific design and regular coordination among the owner, contractors, engineers and regulators throughout design and construction. Dam safety considerations necessitated that detailed construction staging requirements be implemented and mitigation measures be developed to address potential dam safety issues. Ultimately, the cofferdam and cut-off wall for each project performed as intended and enabled powerhouse construction to proceed as planned within a dry, stable excavation.

ACKNOWLEDGEMENTS

The authors would like to acknowledge AMP and their member municipalities for their initiative to develop these innovative hydroelectric projects and the cofferdam contractors and their engineering consultants for their respective contributions to the success of the Willow Island, Meldahl, Cannelton and Smithland Hydroelectric Projects.

The authors would also like to thank the Board of Consultants for the Cannelton and Smithland Hydroelectric Projects, which was comprised of Eric Kollgaard, Jim Erwin and Joe Ehasz, FERC's consultant, Dr. Alfred J. Hendron Jr., and the many individuals from the USACE and FERC, and FERC who provided constructive recommendations and feedback during cofferdam design and construction.

The aerial photographs of the Smithland, Cannelton, Meldahl and Willow Island projects were provided to AMP by CJ Mahan Construction Company, Kiewit-Traylor Joint Venture, Walsh Construction Company, Alberici-Baker Joint Venture, The Ruhlin Company, and Voith.

REFERENCES

Blaszczyk, P.R., K. Griebenow, P.E. Meier (2015). "Time to Generate Power – New Hydro Development on the Ohio", USSD 2015 Annual Conference, Louisville, KY.

Duncan, J.M., Wright S.G. and Wong, K.S. (1990). "Slope Stability during Rapid Drawdown," Proceedings of H. Bolton Seed Memorial Symposium. Vol. 2.

Federal Emergency Management Agency (FEMA) (2005). *Technical Manual: Conduits through Embankment Dams*, September.

Federal Energy Regulatory Commission (FERC) (2002). *Engineering Guidelines for Evaluation of Hydropower Projects*, Chapter 3, "Gravity Dams," October.

Federal Energy Regulatory Commission (FERC) (2006). *Engineering Guidelines for Evaluation of Hydropower Projects*, Chapter 3, "Embankment Dams," September.

Hynes, J.M., L. Zhao, A.A. Mirza, T.G. Andrews, P.E. Meier (2015). "Better Safe than Sorry – Surveillance and Monitoring of Hydro Construction on the Ohio," USSD 2015 Annual Conference, Louisville, KY.

Idriss, I.M. and R.W. Boulanger (2008). "Soil Liquefaction During Earthquakes", MNO-12, Earthquake Engineering Research Institute.

Marold, W.J. (2012). "Construction of a 100-foot Deep Cofferdam in the Ohio River," Proceedings of USSD 2012 Annual Conference, p. 1129-1143.

Ressi di Cervia, A.L. (2010). "Construction of a Deep Cut-off Wall for a New Powerhouse," Proceedings of USSD 2010 Annual Conference, p. 857-869.U.S. Army Corps of Engineers (USACE) (1989). "Engineering and Design – Design of Sheet Pile Cellular Structures Cofferdams and Retaining Structures," EM 1110-2-2503.

U.S. Army Corps of Engineers (USACE) (2003). "Engineering and Design – Slope Stability," EM 1110-2-1902.

FROM THE GROUND UP — GROUND IMPROVEMENT FOR HYDRO DEVELOPMENT ON THE OHIO RIVER

Abid A. Mirza[1]
Lin Zhao[2]
John M. Hynes[3]
Thomas G. Andrews[4]
Philip E. Meier[5]

ABSTRACT

Of the four new hydroelectric projects that American Municipal Power (AMP) is developing adjacent to existing U.S. Army Corps of Engineers (USACE) locks and dams on the Ohio River, the Cannelton and Smithland projects required extensive ground improvement to create suitable foundation conditions for the projects' permanent water-retaining structures. Both projects are located near the New Madrid and Wabash Valley Seismic Zones and founded on deep alluvial foundations that were potentially liquefiable, if left untreated. The foundation conditions at Smithland were further complicated by the presence of karstic limestone beneath the alluvium, where voids were present in bedrock and ground loss into solution features was believed to have created "chimneys" of loosened soil. At Cannelton and Smithland, vibro replacement (stone column installation) was performed within and adjacent to the foundations of the permanent closure structures (hardfill dams) on the riversides and landsides of the powerhouses to mitigate against potential liquefaction and minimize immediate and earthquake-induced settlement. At Smithland, compaction grouting was employed to treat the foundation soils and karstic bedrock below the powerhouse. Consolidation grouting was also performed at Smithland along the powerhouse's cut-off wall to reduce seepage through the karstic bedrock below the cut-off wall. This paper discusses the selection of the ground improvement techniques employed along with the results of the ground improvement programs and lessons learned during construction.

INTRODUCTION

Four hydroelectric projects (Smithland Hydroelectric Project, Cannelton Hydroelectric Project, Meldahl Hydroelectric Project, and Willow Island Hydroelectric Project) are currently under construction on the Ohio River adjacent to existing U.S. Army Corps of Engineers (USACE) locks and dams (Figure 1). These projects are being developed by

[1] Geological Engineer, MWH Americas, Inc. 175 W. Jackson Blvd., Suite 1900, Chicago, IL 60604, abid.a.mirza@mwhglobal.com
[2] Lead Geotechnical Engineer MWH Americas, Inc. 175 W. Jackson Blvd., Suite 1900, Chicago, IL 60604, lin.zhao@mwhglobal.com
[3] Senior Geotechnical Engineer MWH Americas, Inc. 175 W. Jackson Blvd., Suite 1900, Chicago, IL 60604, john.m.hynes@mwhglobal.com
[4] Principal Geotechnical Engineer, MWH Americas, Inc. 175 W. Jackson Blvd., Suite 1900, Chicago, IL 60604, thomas.g.andrews@mwhglobal.com
[5] Vice President – Hydroelectric Development and Operations, American Municipal Power, Inc. 1111 Schrock Road, Suite 100, Columbus, OH 43229, pmeier@amppartners.org

American Municipal Power (AMP), a nonprofit wholesale power supplier for municipal electric systems. AMP serves 130 members – 129 member communities in Indiana, Kentucky, Michigan, Ohio, Pennsylvania, Virginia and West Virginia, and one joint action agency in Delaware. Combined, these publicly owned utilities serve approximately 625,000 customers. MWH Americas, Inc. (MWH) was selected by AMP to serve as the Owner's Engineer for the design and supervision of construction for these new hydroelectric projects.

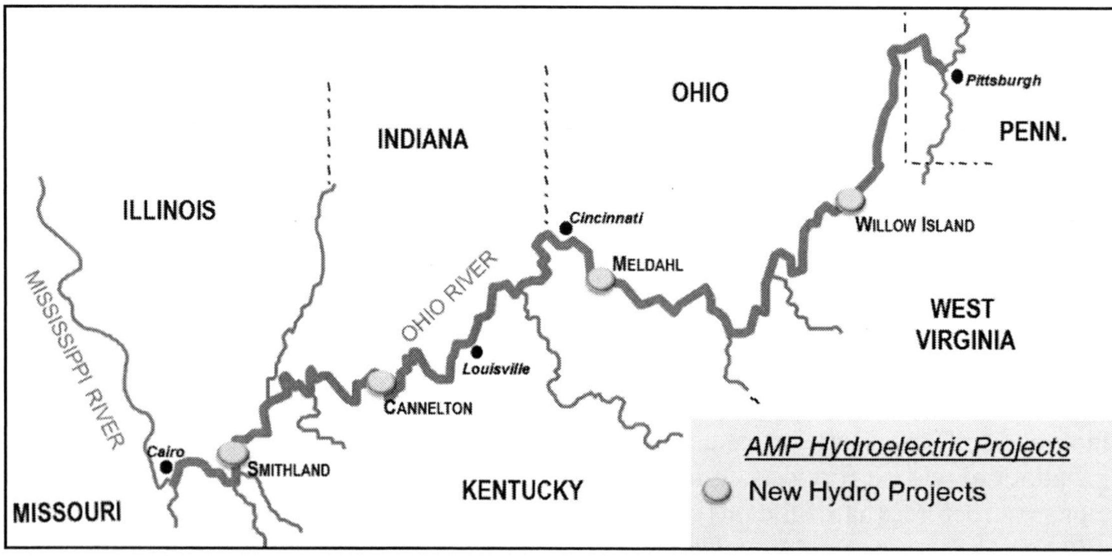

Figure 1. Project Location Map

The performance and stability of the powerhouse and closure structures foundations for these hydroelectric projects is critical for dam safety and project performance. Project- and site-specific foundation preparation and ground improvement methods were implemented to provide suitable foundations for these water-retaining structures. MWH was responsible for ground improvement design and evaluation.

PROJECT DESCRIPTIONS AND CONSTRUCTION STAGES

Each of the four hydroelectric projects required the construction of an approach channel, tailrace channel, powerhouse (2 or 3 units per project with 25-foot diameter horizontal bulb turbines), riverside closure structures connecting to the existing dam (e.g., 72-foot-high concrete gravity dam at Meldahl) and landside closure structures (e.g., 90-foot-high hardfill dam at Meldahl). The reinforced concrete powerhouses are designed for overtopping during flood events. The generation capacity of each project is as follows:

- Willow Island Hydroelectric Project: 35 MW capacity (20-foot gross head)
- Meldahl Hydroelectric Project: 105 MW capacity (30-foot gross head)
- Cannelton Hydroelectric Project: 84 MW capacity (25-foot gross head)
- Smithland Hydroelectric Project: 72 MW capacity (22-foot gross head)

The design of each project was tailored to address site-specific conditions including:

- weak rocks, including slickensided indurated clays and a thrust fault present at Willow Island;
- soft clays required support during excavation at Meldahl;
- proximity of Cannelton and Smithland to the New Madrid and Wabash Valley Seismic Zones with potentially liquefiable alluvial soils; and
- karstic limestone which further complicated the design at Smithland.

Unique project-specific design features included:

- 2,000-kip rock anchors at Willow Island to stabilize the foundation below the cellular cofferdam and significant dental excavation and rock bolting to treat the fault where it crossed the excavation.
- Steel piles to stabilize excavation slopes at Meldahl.
- An embankment cofferdam at Cannelton with a 100-foot-high zoned section constructed in the Ohio River and stone columns to improve the foundations below the closure structures by mitigating against liquefaction potential and minimizing immediate and earthquake-induced settlement.
- An embankment cofferdam at Smithland with a 150- to 187-foot-deep, 4000-foot-long soil-cement-bentonite cut-off wall and a robust dewatering system. Stone columns were installed beneath the closure structures. Compaction and consolidation grouting was used to treat the foundation soils and karstic rock below the powerhouse.
- Of eight (8) closure structures (dams) required for the project, six (6) were constructed of hardfill (roller-compacted soil-cement) to maximize the use of on-site materials and minimize the importation of aggregate.

All four projects have similar project layouts and consist of an intake approach channel, a reinforced concrete powerhouse, a landside closure structure, a riverside closure structure and a tailrace channel (Figure 2).

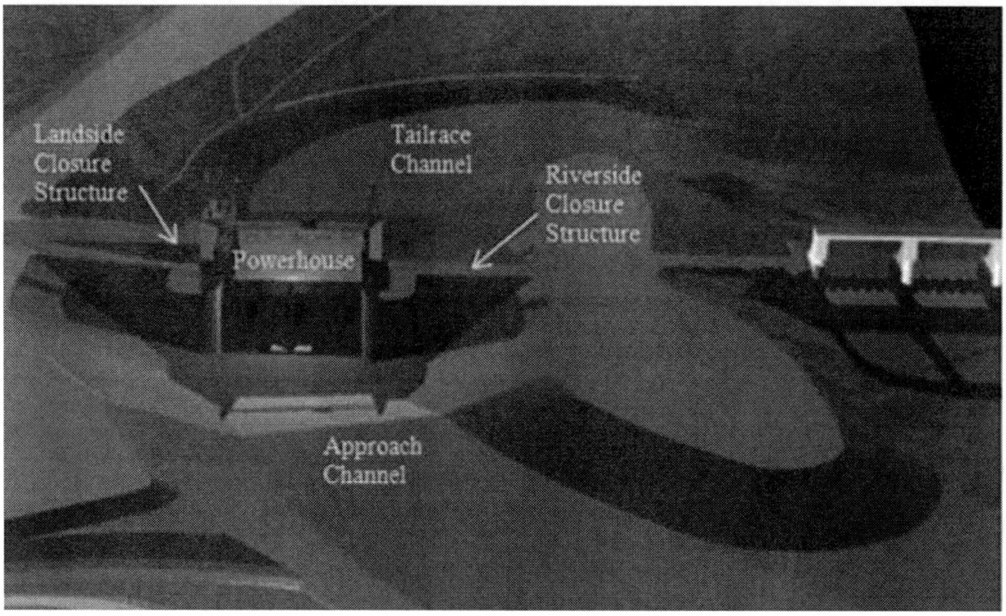

Figure 2. Typical Layout of Project Components (rendering of Cannelton shown)

Photos of the Willow Island, Meldahl, Cannelton and Smithland projects are shown in Figure 3. Further details regarding the development of these hydroelectric projects can be found in Blaszczyk et al. (2015).

Figure 3. Aerial Photos of Construction at Smithland (top left), Cannelton (top right), Meldahl (bottom left), and Willow Island (bottom right) taken during the fall of 2014.

GENERAL FOUNDATION CONDITIONS

Willow Island and Meldahl Project Sites

The water-retaining structures for the Willow Island and Meldahl Hydroelectric Projects are both generally founded on bedrock with the exception of the outer portions of the landside closure structures that are founded on alluvial soils. The following conventional foundation preparation techniques were used for the water-retaining structures that are founded on rock at these two projects:

- Curtain grouting was performed along the upstream faces of the riverside closure structure, powerhouse and landside closure structure at Willow Island and along the upstream side of the riverside closure structure at Meldahl. Curtain grouting was not performed along the upstream face of the powerhouse or landside closure structure at Meldahl due to the length of the seepage path and tightness of rock under these structures. A stable grout mix with 1.1:1 water to cement ratio, 1.5% bentonite by weight of water, and superplasticizer (approximately 2% by volume of water or 0.375% by weight of cement) was typically used.
- The rock foundations below the powerhouses and closure structures at Willow Island and Meldahl were each prepared by removal of severely or moderately weathered rock blocks; loose, semi-detached and unsound rock fragments; and overburden material and debris to provide a stable and competent foundation for the powerhouses and closure structures. This was performed by scaling, barring, hand working, pressure washing, and vacuuming. Areas of blast-damaged rock that might affect stability of the foundation were removed and backfilled with slush grout. In addition, slush grout was locally applied to seal gapped joints and to backfill over-drilled blast holes. The treated foundation was inspected and mapped by MWH prior to placement of concrete over the rock foundation.
- A reverse fault that was found when the USACE constructed the Willow Island Locks and Dam in the 1970's was known to cross the footprint of the hydroelectric project. As discussed in Andrews et al. (2015), the area where the fault was exposed in the riverside wall of the powerhouse excavation was stabilized with untensioned rock dowels and shotcrete and backfilled with concrete. The reverse fault was also exposed across the right upstream corner of the powerhouse foundation and in a portion of the riverside closure structure foundation. Dental excavation and dental concrete was performed where the fault was exposed in the foundation. Loose or broken rock, sand and clay were removed from the fault zone using an excavator, hand working, water jetting, and vacuuming. The excavated fault trench ranged from about 2 to 8 feet in depth and from a few inches to about 2 feet in width. Dental concrete was placed in the excavated fault trench after loose material was removed to expose sound rock surfaces (Figure 4). Approximately 250 yd^3 of dental concrete was placed in the powerhouse footprint and an additional 50 yd^3 of dental concrete was placed in the riverside closure structure footprint.

Figure 4. Dental Concrete Placement in Fault Trench at Willow Island

Cannelton and Smithland Project Sites

Unlike Willow Island and Meldahl, the powerhouses and closure structures at Cannelton and Smithland were founded entirely on alluvial soils and both sites required extensive ground improvement to provide suitable foundations for the powerhouses and closure structures. These projects are both located near the New Madrid Seismic Zone (NMSZ) and Wabash Valley Seismic Zone (WVSZ) (see Figure 5) and their sandy alluvial foundations were potentially liquefiable, if left untreated. The foundation conditions at Smithland were further complicated by the presence of karstic limestone beneath the alluvium, which raised concerns of the potential for high flows through solution features in the rock below the powerhouse's soil foundation with possible sinkhole formation and ground loss into voids in rock.

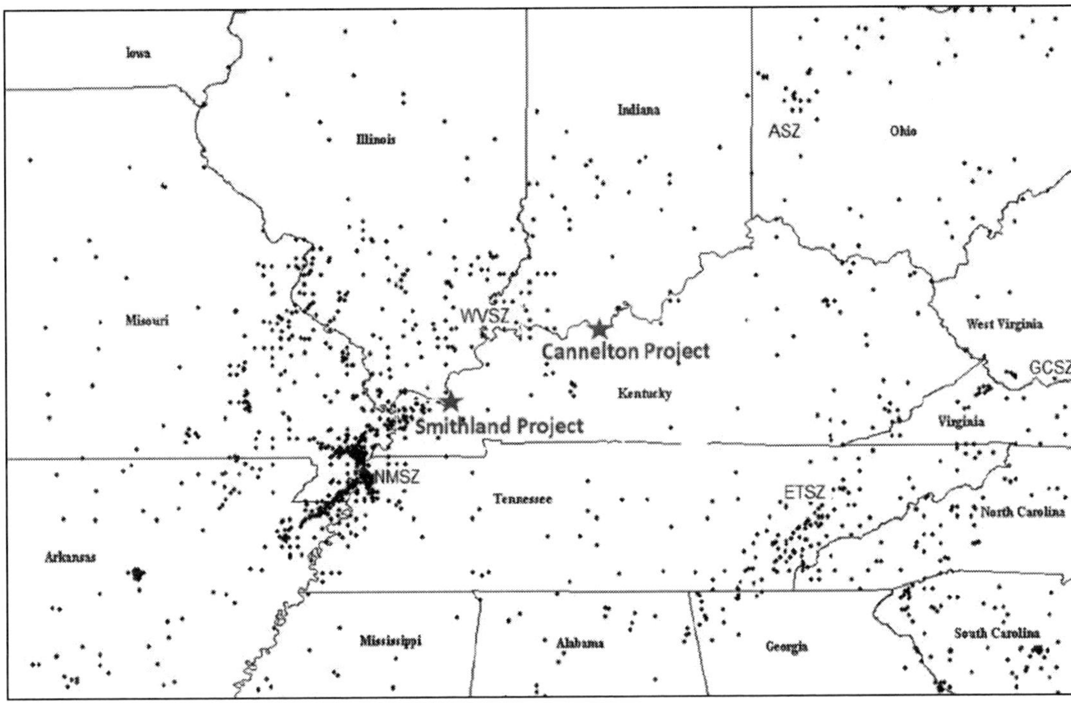

Figure 5. Seismicity near Cannelton and Smithland (1699 to 2009)

At Cannelton and Smithland, stone columns were installed to densify the areas within and adjacent to the riverside and landside closure structure foundations to mitigate against potential liquefaction and minimize settlement during construction and earthquake-induced settlement. At Smithland, compaction grouting was employed to treat the foundation soils and karstic rock below the powerhouse foundation. Consolidation grouting was also performed at Smithland along the permanent seepage cut-off wall to address the potential for soil to migrate into solution features in rock below the powerhouse. The selection of suitable ground improvement techniques for Cannelton and Smithland required detailed study of the foundation conditions at each project site.

GROUND IMPROVEMENT AT CANNELTON

Subsurface Conditions at Cannelton

Prior to commencing with powerhouse excavation, the subsurface materials at Cannelton consisted of approximately 100 to 140 feet of alluvium overlying sedimentary bedrock. Overburden was generally composed of approximately 10 feet of loose to medium dense sand overlying about 20 feet of stiff clay that overlies loose sand with localized lenses of stiff sandy silt, which in turn overlies medium dense to very dense sand. Underlying the alluvial deposits is the Buffalo Wallow Formation, featuring a succession of alternating strata of limestone, sandstone and shale.

The Cannelton powerhouse is founded directly on dense to very dense sand at about 100 feet below the pre-construction ground surface (the required powerhouse excavation is described in Andrews et al., 2015). The foundations of the closure structures on the riverside and landside of the powerhouse slope upward at 2.5 horizontal to 1.0 vertical from the powerhouse foundation level to the pre-construction ground surface. The closure structures, which were constructed of hardfill (see Arnold et al., 2015, and Truschke et al., 2015, for details), are founded directly on alluvial soils, generally loose to medium dense sands at the upper elevations transitioning to medium dense to dense sands at lower elevations.

Foundation Issues Identified at Cannelton

Seismicity. Since Cannelton located is near the New Madrid and Wabash Valley Seismic Zones, both sources of large historic earthquakes, seismic hazards such as strong ground motions, liquefaction, earthquake-induced settlement and post-earthquake instability were required to be addressed during foundation design for the powerhouse and closure structures. A site-specific seismic hazard assessment was performed to develop design earthquakes using probabilistic and deterministic methods.

The Cannelton Project is classified as having Significant Hazard potential based on USACE and FERC criteria because of potential loss of life and major economic losses that could be associated with an uncontrolled release of the upper navigation pool. The Maximum Credible Earthquake (MCE) was established as the Maximum Design Earthquake (MDE) for the permanent water-retaining structures. The assessed seismic design parameters for Cannelton's MCE/MDE are summarized below:

- Peak Ground Acceleration: 0.15 g (rock outcrop)
- Design Earthquake Magnitude: 7.5 (Mw) at 85 km

Ground Motion Amplification. As seismic waves propagate from rock into overlying soil, ground motions in soil can be amplified. To assess potential ground motion amplification in the soil at Cannelton, a site response analysis was performed by MWH using SHAKE91 (Idriss and Sun, 1992). A total of four sets of recorded rock ground motions were used to perform site response analysis. These rock ground motions were selected from recorded rock acceleration time histories that were generally consistent with the design earthquake (magnitude, source-to-site distance, type of faulting and duration of strong shaking). The peak ground acceleration presented above represents the ground motions for a rock outcrop, if present at the project site. Since alluvial soil is present above bedrock at Cannelton, a deconvolution analysis of the assessed rock outcrop ground motions was performed using SHAKE91 to obtain ground motions for the actual top of bedrock. The deconvoluted time histories were then used to perform site response analysis and evaluate ground motion amplification in soil. Figure 6 presents peak horizontal ground acceleration profile obtained from one subsurface profile. The site response analysis results indicated that the peak ground accelerations at the ground surface for the MCE would range from 0.21 g to 0.31 g, corresponding to amplification factors of 1.5 to 2.2.

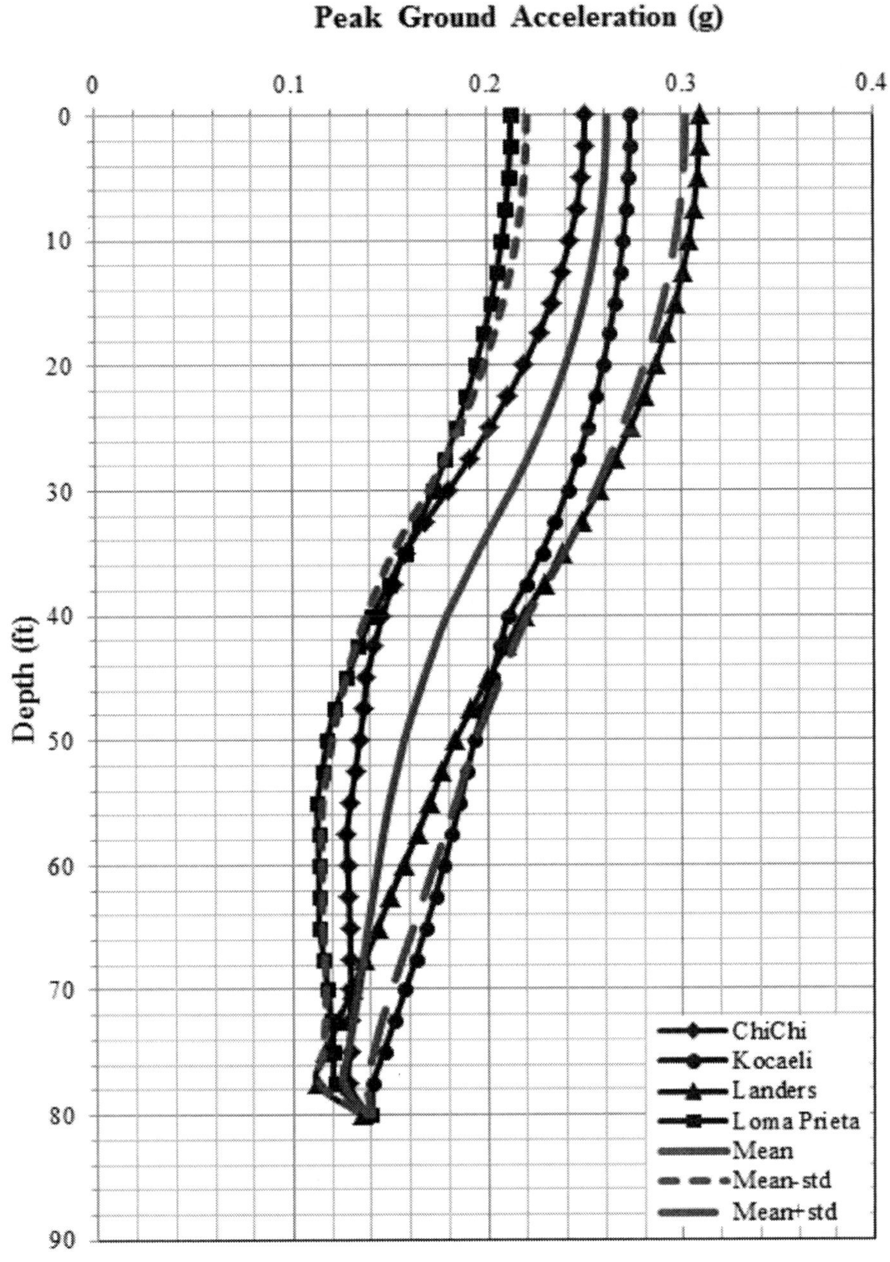

Figure 6. Typical Peak Ground Acceleration of Soil Profile obtained from Site Response Analysis at Cannelton

Liquefaction Potential. The analysis of liquefaction and cyclic softening potential was conducted using the approaches developed by Idriss and Boulanger (2008) for the various confining pressures under and adjacent to the powerhouse and closure structures. These analyses used maximum cyclic shear stresses assessed by the site response analysis. The liquefaction and cyclic softening analyses indicted that the soil under the powerhouse at Cannelton will not experience liquefaction/cyclic softening under the MCE, but did indicate that the sandy soils under the closure structures and slopes adjacent to the powerhouse would likely experience liquefaction during the MCE, if left untreated.

Earthquake-Induced Settlement. The potential liquefaction-induced settlement due to the MCE at Cannelton was estimated using the simplified method developed by Tokimatsu and Seed (1987). The estimated liquefaction-induced settlement of the closure structures, if the foundations were not improved, was estimated to vary between 19 inches of settlement at the abutments to less than an inch in the areas near the powerhouse. If such differential settlements were to occur following an earthquake, the closure structure's hardfill and concrete facing could potentially fracture and water-stopped connections between the closure structures and powerhouse could potentially rupture.

Dynamic Deformation and Post-Earthquake Instability. If the foundations below the closure structures were left untreated, ground motions associated with the MCE could induce unacceptable permanent deformations that could damage the water-stopped connections between the closure structure and the sheet pile cut-off wall and powerhouse. Similarly, the presence of liquefied soils and excess pore pressures in non-liquefied soils could lead to post-earthquake instability in the closure structure foundations following the MCE, if these foundations were not treated.

Ground Improvement Program at Cannelton

Selection of Ground Improvement Method. As noted above, liquefaction, stability and deformation analyses indicated that ground improvement of the closure structure foundations and adjacent approach and tailrace channel slopes at Cannelton was required to mitigate against liquefaction potential, earthquake and post-earthquake instability, and earthquake-induced settlement for the MCE. Various ground improvement alternatives were considered including: deep dynamic compaction, jet grouting, soil mixing, compaction grouting, vibro compaction, wet top-feed vibro replacement (stone columns) and dry bottom-feed vibro replacement (stone columns). The installation of stone columns by the dry bottom-feed method was assessed to be the best suited, most economic ground improvement method for Cannelton for the following technical and commercial reasons.

- Stone columns provide ground densification and reinforcement and have been widely and successfully used for liquefaction mitigation.
- Stone columns can be used to treat soils to depths in excess of 80 feet.
- Stone columns are generally more economic than the other applicable ground improvement methods, with the exception of deep dynamic compaction.
- Unlike deep dynamic compaction, the installation of stone columns can be performed on slopes without increasing the potential for slope instability.
- There are sufficient U.S.-based specialty contractors that install stone columns to provide adequate price competition.

- The dry bottom-feed method eliminates the main problem with wet top-feed vibro replacement, which is the management of large quantities of water and spoils, which can be expensive and disruptive to other construction operations.
- Multiple vibro rigs on multiple shifts can be used to address schedule concerns.

When installing stone columns by the vibro-replacement using the dry bottom-feed method, pre-drilling is commonly performed to loosen soils in a pilot "hole", which enables the vibrator (also called a "vibroflot") to penetrate to depth with minimal water jetting. This method eliminates the generation of excess water and spoils during penetration. Once the vibrator has penetrated to the planned bottom of the column, stone (gravel) is delivered to the bottom of the vibrator using a hopper and supply tube through the center of the vibrator. The vibrator is then raised and plunged to compact lifts of stone into the surrounding soil. This method typically constructs fairly uniform stone columns in the ground, although the column diameter varies depending on the original soil density. The constructed stone columns and in-situ soil form an integrated system having low compressibility and high shear strength.

<u>Ground Improvement Design.</u> The limits of stone column treatment was designed so that the powerhouse and closure structures, which compose the permanent water-retaining structures at Cannelton, and the approach and tailrace channels immediately adjacent to these structures remain stable and would not undergo excessive deformation if subjected to the MCE.

Based on the liquefaction, slope and global stability, and deformation analyses described above, ground improvement was not required below the footprint of the powerhouse due to the presence of dense and very dense sand below the foundation level elevation with Standard Penetration Test (SPT) N-values generally in excess of 30. However, these analyses did indicate that ground improvement was needed in the soils located at upper elevations upon which the closure structures were founded to densify and reinforce the in-situ soils to mitigate against potential earthquake-induced settlement and potential instability of the closure structures.

It was determined that stone columns would be installed within the footprints of the landside and riverside closure structure foundations and in portions of the adjacent approach and tailrace channel slopes, as shown in Figure 7. The maximum depth of the stone columns at Cannelton was 60 feet.

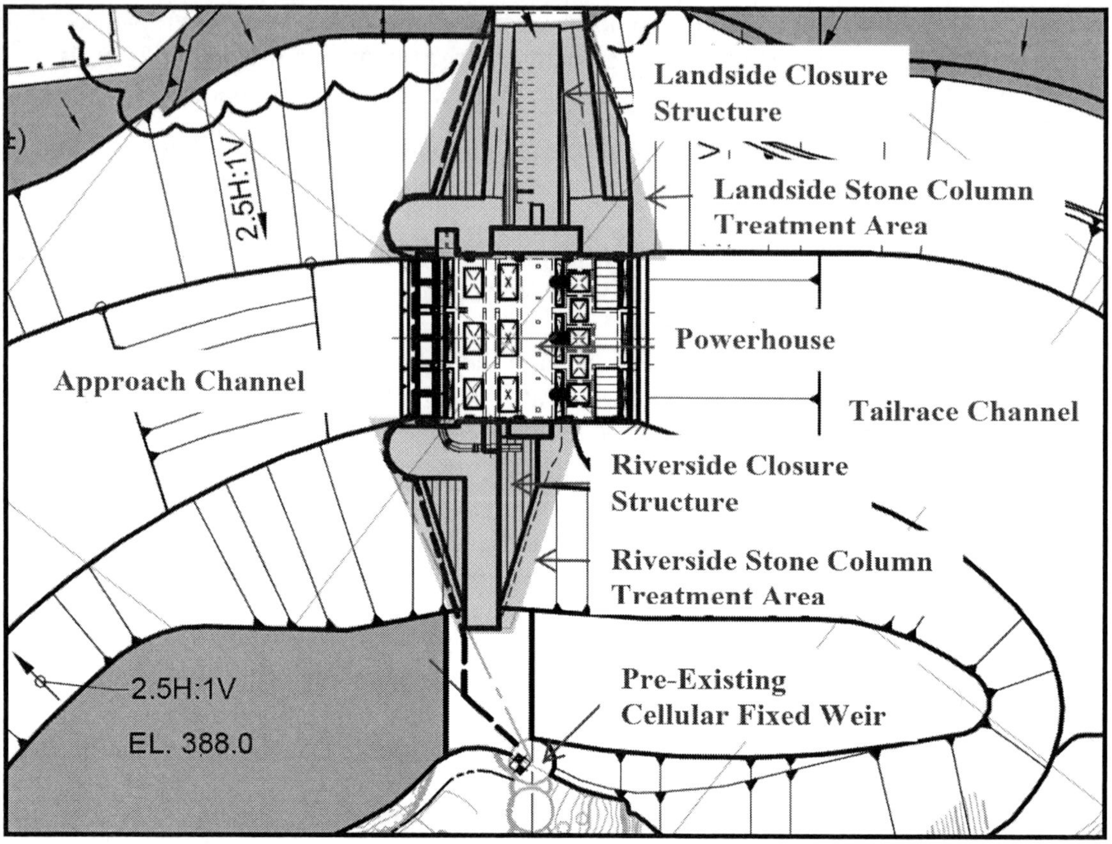

Figure 7. Landside and Riverside Stone Column Treatment Areas at Cannelton (treatment areas are shown in red)

The following constructability issues were identified for stone column installation at Cannelton.

- To minimize impacts of ground improvement work on the overall project schedule, the stone column installation work was performed concurrently with other general construction activities, including the start of powerhouse construction. Production stone column work needed to begin near the base of the powerhouse excavation slopes so that this area could be opened up for construction of the powerhouse foundation.
- To install stone columns in the sloping closure structure foundations, a series of temporary work benches needed to be constructed to facilitate stone column installation work.
- Stone column work was required to proceed upwards from the bottom to the top of the slopes to mitigate against potential slope stability issues associated with construction surcharge loads and temporary cuts and fills. The contractor was required to perform stability analyses to demonstrate the suitability of its proposed construction sequence and to establish minimum set-back distances from the edges of the temporary work benches described above.

- The stone columns at Cannelton were installed in soils that were largely unsaturated, since they were located within the unwatered cofferdam that was required for concurrent powerhouse construction. While the closures structure foundations would be saturated following filling and decommission of the cofferdam, the unsaturated conditions at the time of stone column installation is unusual and could potentially reduce the amount of densification achieved by an individual stone column.
- The dry bottom-feed method pushes the stone laterally into the in-situ soils and induces considerable mixing on the outer boundary of the stone column, but the inner portion of each column generally consists of relatively clean gravel. Therefore, the introduction of stone columns can increase the vertical permeability of the soil mass. While the maximum upstream-to-downstream river level differential across the permanent water-retaining structures at Cannelton is only 25 feet, the permanent sheet pile cut-off along the upstream face of the closure structures and powerhouse, which was installed in advance of stone column construction, was driven to a depth at least 15 feet below the bottom of the adjacent stone columns or to refusal on the top of bedrock to increase the seepage gradient and reduce the quantity of seepage under the permanent water-retaining structures.
- Since it is difficult to achieve significant densification of soils around the upper few feet of a stone column due to the lack of confinement, the design required that the upper 3 feet of the subgrade below the closure structure be over-excavated, proof rolled and compacted in lifts using a large vibratory roller following completion of stone column installation.

Stone Column Installation. The stone column installation work was included in the general construction contract for the Cannelton powerhouse and appurtenances. The general contractor was Walsh Construction Company, who subcontracted the stone column installation work to Moretrench. The stone column work at Cannelton is shown in Figure 8.

Figure 8. Stone Column Installation from Temporary Work Bench at Cannelton (adjacent to initial powerhouse construction)

Prior to undertaking production work, the ground improvement specialty contractor was required to undertake a stone column test program at pre-selected locations where ground improvement was required. The test program was used to assess the vibratory probe spacing needed to obtain the required densification. The test program included both pre- and post-treatment Cone Penetration Test (CPT) soundings and SPT borings.

A triangular pattern for stone column installations was used at Cannelton. The center-to-center spacing between stone columns was determined from the stone column test program, which consisted of one test section on the landside slope of the powerhouse excavation and one on the riverside slope. Each test section included 37 stone columns installed with center-to-center spacings of 8 and 9 feet.

The stone column installation methods employed in some portions of the test sections failed to meet the acceptance criterion at Cannelton. Subsequent stone column installation a supplemental test section and within the production areas utilized modified installation procedures that (1) reduced the quantity of water jetted during initial vibrator penetration, (2) reduced the height of vibrator lifts between re-penetrations, and (3) increased the target maximum re-penetration amperage, which improved densification.

Based on pre- and post-treatment CPT and SPT results in the test sections, an 8-foot center-to-center spacing was initially established for production stone columns. The center-to-center spacing was later increased to 9 feet following an additional test section conducted on the riverside that demonstrated a larger spacing could be used with the contractor's refined installation procedures. Typical pre- and post-treatments test results from the test section are presented in Figure 9.

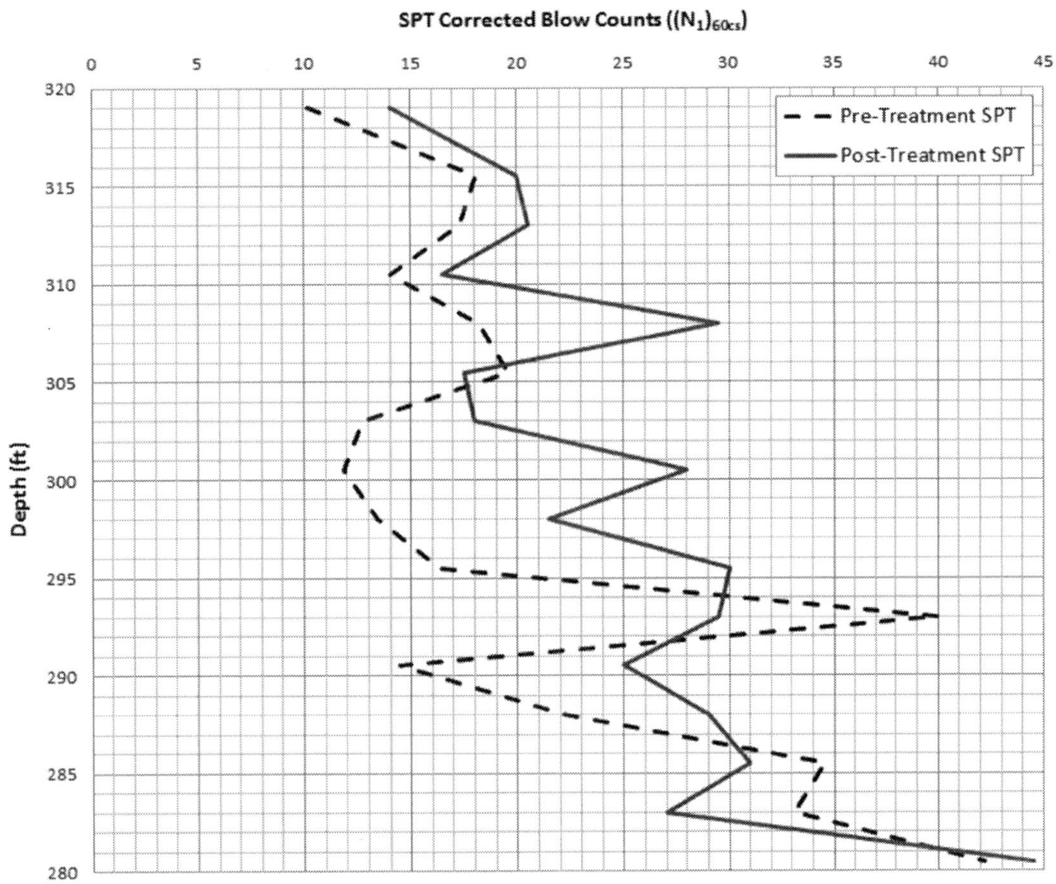

Figure 9. Typical Post-Treatment SPT Results for Stone Columns at Cannelton

A total of 1325 stone columns were installed in the landside treatment area at Cannelton and 668 stone columns were installed in the riverside treatment area, including the stone columns installed in test sections and supplemental split-space columns, as described below.

Ground Improvement Verification. To prevent liquefaction, improve dynamic stability and reduce earthquake-induced settlements and deformations, the contractor was required to construct stone columns such that a relative density of approximately 65% or more was achieved within the zone of ground improvement at Cannelton. CPT soundings were used to perform verification testing during production work.

Rather than establish multiple acceptance criteria to represent the varying confining pressures at different locations under and adjacent to the closure structures, one CPT acceptance criterion was developed to envelope all areas regardless of location or confinement. Since the soils in areas along the slopes adjacent to the closure structure are generally unaffected by the closure structure surcharge, these areas would be more susceptible to liquefaction than equally dense soil under the closure structures and thus were used to develop the CPT acceptance criterion. A similar alternative SPT acceptance criterion was also developed to evaluate SPT results from the stone column test sections.

The acceptance criteria were developed for a factor of safety of 1.0 against liquefaction for the MCE at mid-spacing between stone columns in the slopes adjacent to the closure structures. Soils located closer to the stone columns would be expected to have higher factors of safety as a result of being exposed to a greater compactive effort. Furthermore, soils within the footprint of the closure structures meeting the acceptance criteria would be expected to have even greater factors of safety due to the confining pressures associated with the closure structures.

The post-treatment CPT and SPT results indicated that most of the area treated by the production columns met established the acceptance criteria with the exception of a few zones. One zone was related to areas where stone columns penetrated a thin layer of indurated clay cobbles and gravel that was found across portions of the Cannelton project site. In some areas where this "hard" layer was present, the acceptance criteria were not satisfied in the horizon immediately below the "hard" layer. The zones that failed to meet the acceptance criteria at Cannelton were discontinuous across the project site and generally located several feet below the closure structure foundations. Subsequent liquefaction and post-earthquake stability analyses of the Cannelton closure structures indicated that sufficient ground improvement was achieved in the strata immediately below the "hard" layer to maintain stability if it were subjected to the MCE. A Newmark (1965) sliding block analysis, which conservatively assumed that the maximum developed pore pressures occur at the start of earthquake shaking, was performed to assess potential dynamic displacements for the MCE using analysis sections that incorporated the relatively loose horizon below the "hard" layer. The Newmark analysis indicated that permanent displacement would not occur under the MCE, since the yield accelerations were greater than the peak ground motions calculated for the critical sliding surface. Therefore, the amount of ground improvement achieved immediately below the "hard" layer was sufficient to satisfy the design intent to provide suitable foundations for the design earthquake.

Additionally, the upper area of the landside treatment area failed to meet the acceptance criteria after the initial installation of stone columns on a 9 foot center-to-center spacing. This area was treated with supplemental split-spaced stone columns. A small area of the riverside treatment area also required supplemental stone columns, as did portions of the initial tests sections. A total of 236 supplemental stone columns were ultimately installed on the landside and 41 supplemental stone columns were installed on the riverside. Subsequent CPT soundings indicated that the acceptance criterion was met in the areas where supplemental stone columns were installed.

Evaluation of Ground Improvement at Cannelton

Analyses were performed to evaluate the closure structure foundation after installation of stone columns based on the results of the post-treatment CPT soundings. These include liquefaction analysis, settlement analysis, Newmark sliding block analysis, and post-earthquake stability analysis. The results of these analyses indicated that the post-improvement closure structure foundations would be stable and earthquake-induced settlement and deformation would be negligible under the MCE. Therefore, the stone column installation program at Cannelton was assessed to have satisfied the design intent to prevent liquefaction, improve dynamic stability, and reduce earthquake-induced settlements and deformations.

GROUND IMPROVEMENT AT SMITHLAND

Subsurface Conditions at Smithland

Prior to powerhouse excavation, the subsurface materials at Smithland consisted of approximately 140 to 160 feet of alluvium overlying karstic limestone bedrock. Overburden was generally composed of approximately 20 feet of stiff clay that overlies loose to medium dense sand, which in turn overlies medium dense sand and gravel with occasional clay layers and lenses. Underlying the alluvial deposits is limestone of the Ste. Genevieve Formation. This formation consists of oolitic to cherty dolomitic limestone which occurs in generally flat-lying, slightly undulating thin to massive beds at Smithland.

Voids and solution cavities with apparent depths in excess of 15 feet were found within the upper 25 feet of bedrock. Some solution features appeared to be open, while others appeared to be partially filled with soil. The average Rock Quality Designation (RQD) of the upper 25 feet of bedrock was 69 with an average core recovery of 80%.

The powerhouse and closure structures at Smithland are founded directly founded on soil, generally composed of sand and gravel that was characterized as medium dense prior to ground improvement.

Foundation Issues Identified at Smithland

Seismicity. Similar to Cannelton, Smithland located is near the New Madrid and Wabash Valley Seismic Zones and seismic hazards needed to be addressed by the foundation design for the powerhouse and closure structures. A site-specific seismic hazard assessment using probabilistic and deterministic methods was performed to develop design earthquakes. Since the project is classified as having Significant Hazard potential based on USACE and FERC criteria, the MDE was taken as the MCE. The assessed seismic design parameters for Smithland's MCE/MDE are summarized below:

- Peak Ground Acceleration: 0.29g (rock outcrop)
- Design Earthquake Magnitude: 7.5 (Mw) at 43 km

Ground Motion Amplification. To account for the potential ground motion amplification in the soil profile above rock at Smithland, a site response analysis was performed, which indicated that the peak ground accelerations on the ground surface for the MCE would range from 0.34 g to 0.63 g, corresponding to amplification factors of 1.2 to 2.2.

Liquefaction Potential. Liquefaction and cyclic softening analysis using the methods developed by Idriss and Boulanger (2008) for the MCE indicated that most granular soil layers beneath both the powerhouse and closure structures at Smithland would likely liquefy, if it were left untreated and subjected to the MCE.

Dynamic Deformation and Post-Earthquake Instability. If the foundations below the powerhouse and closure structures were left untreated, ground motions associated with the MCE could induce unacceptable permanent deformations and potentially excess pore pressures generation and liquefaction that could destabilize the powerhouse and closure structures. Therefore, treatment of the powerhouse and closure structure foundations was required.

Karstic Limestone Bedrock. Loss of ground could occur under static and/or seismic conditions if the foundation soils were to migrate into the karstic limestone bedrock at Smithland, which contains numerous open and partially filled solution channels and voids. Ground loss into solution features over time appears to have created "chimneys" of loosened soil across the project site, which may have caused the in-situ soil to be relative loose at depth.

Significant ground loss below the powerhouse foundation could cause differential settlements, which could possibly result in structural damage to the powerhouse. Solution features in the karstic limestone could also result in large amounts of underseepage and the possible formation of sinkholes, if left untreated.

Ground Improvement Program at Smithland

To address the general foundation issues and seismic hazards at Smithland, the project's ground improvement program needed to satisfy the following general performance requirements.

- Powerhouse foundation: ground improvement measures needed to reduce liquefaction potential, maintain earthquake and post-earthquake stability, reduce earthquake-induced settlement and deformation (including differential settlements), and prevent loss of ground into the underlying karstic solution features under static and seismic conditions.
- Closure structure foundations: ground improvement measures needed to reduce liquefaction potential, maintain earthquake and post-earthquake stability, and reduce earthquake-induced settlement and deformation (including differential settlements).
- Seepage cut-off along the upstream side of the powerhouse and closure structures: ground improvement measures needed to reduce seepage and control exit

gradients, prevent piping in soil and rock, and provide a durable seepage cut-off. Of particular concern is the potential presence of open and partially filled solution channels within the bedrock (karst foundation).

Selection of Ground Improvement Method for Closure Structure Foundations. The available techniques for liquefaction mitigation assessed for the closure structure foundations included deep dynamic compaction, deep soil mixing, jet grouting, compaction grouting, and vibro compaction. Similar to Cannelton, the installation of stone columns by the dry bottom-feed method was assessed to be the best suited and most economic method for improving the foundations of the closure structures at Smithland.

Selection of Ground Improvement Method for Powerhouse Foundation. Compaction grouting was selected to mitigate liquefaction potential of the overburden under the powerhouse and to treat solution features in the underling karstic bedrock at Smithland. Compaction grouting was performed in areas within and immediately adjacent to the powerhouse footprint to (1) fill open voids in the karstic limestone bedrock and densify loose material within solution features in the rock to mitigate against future migration of overburden soils into voids and (2) densify granular soils above the top of rock to mitigate against potential earthquake-induced settlement and strength loss of granular foundation soils.

Compaction grouting is performed by injecting a low slump grout into soil and rock under large pressures to fill voids and push surrounding soils into a denser state. This technique has been used for liquefaction mitigation on other projects, but the results within 20 to 30 feet of the ground surface are often mixed since pressurized grout can cause upward heaving rather outward displacement under reduced confining pressures. To provide confinement for compaction grouting, the powerhouse excavation was stopped approximately 30 feet above the general foundation level of the powerhouse so that compaction grouting could be performed from a work platform at that elevation.

Consolidation grouting was performed along the permanent sheet pile cut-off wall that runs along the upstream side of the powerhouse and closure structures to perform additional filling of voids in rock to reduce underseepage and the potential for soil migration through solution features in rock below the cut-off wall.

Figure 10 presents the areas for the ground improvement in the powerhouse and closure structure footprints and surrounding area (about 70 feet from the limits of footprint of the powerhouse and the closure structures). Figure 11 shows the cross section of the ground improvement through the landside closure structure, powerhouse and riverside closure structure.

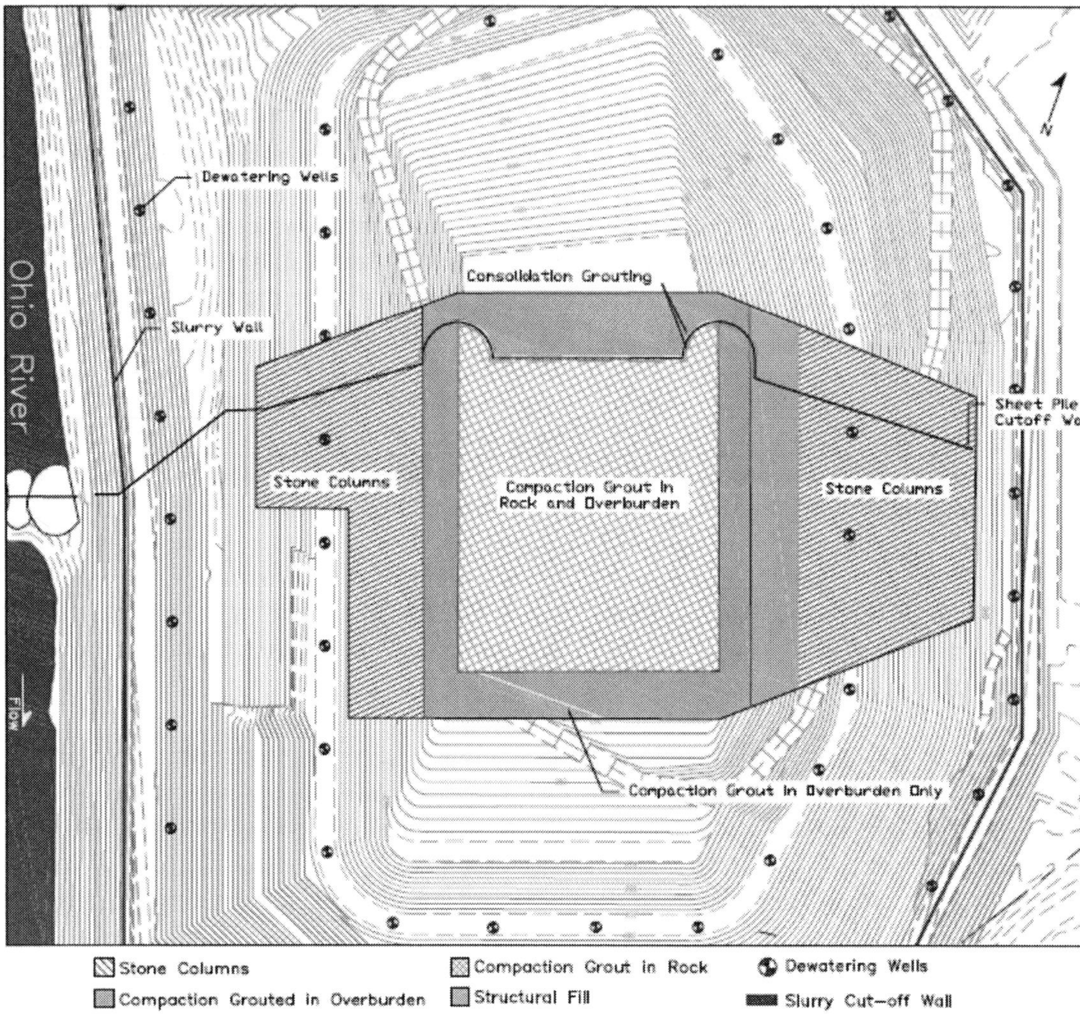

Figure 10. Limits of Ground Improvement at Smithland

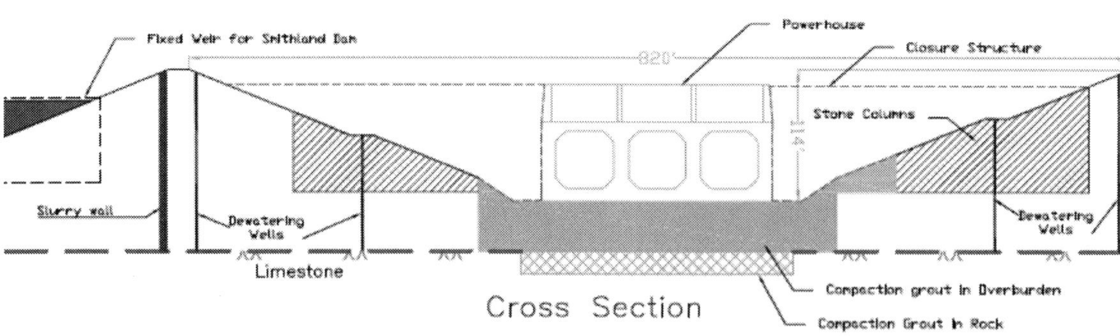

Figure 11. Cross Section of Ground Improvement at Smithland

<u>Stone Column Installation in Closure Structure Foundations</u>. The stone column installation work at Smithland was performed as change order to the cofferdam and powerhouse excavation contract. The cofferdam was C.J. Mahan Construction.

Moretrench was the contractor for the installation of the initial stone columns and Hayward Baker completed the stone column work.

An equilateral pattern of stone columns was used for stone column installations at Smithland. Before stone column production, two stone column test sections in the footprints of the landside and riverside closure structures were evaluated to determine the column-to-column spacing. A 9-foot center-to-center spacing was established for the landside stone columns and 8-foot spacing was established for the riverside stone columns.

Using the established spacing, 653 production stone columns were installed in the landside stone column treatment area at Smithland and 556 production stone columns were installed in the riverside treatment area, including the stone columns installed in the test section and supplemental stone columns, as described below.

A verification testing program was performed to determine if the required densification was achieved. Acceptance criteria were developed to evaluate post-treatment CPT soundings and SPT blow counts of ground treated by stone columns.

Figure 12 presents a typical comparison of pre-treatment and post-treatment CPT tip resistances, which indicates notable densification. However, some areas within the footprint of the landside closure structure experienced only minimal improvement near the ground surface (typically the top 20 to 30 feet) and did not meet the acceptance criteria. To improve post-earthquake stability and minimize possible earthquake-induced settlement, supplemental split-spaced stone columns were installed in the areas that did not initially meet the acceptance criteria. The acceptance criteria were ultimately met at all locations within the stone column treatment area.

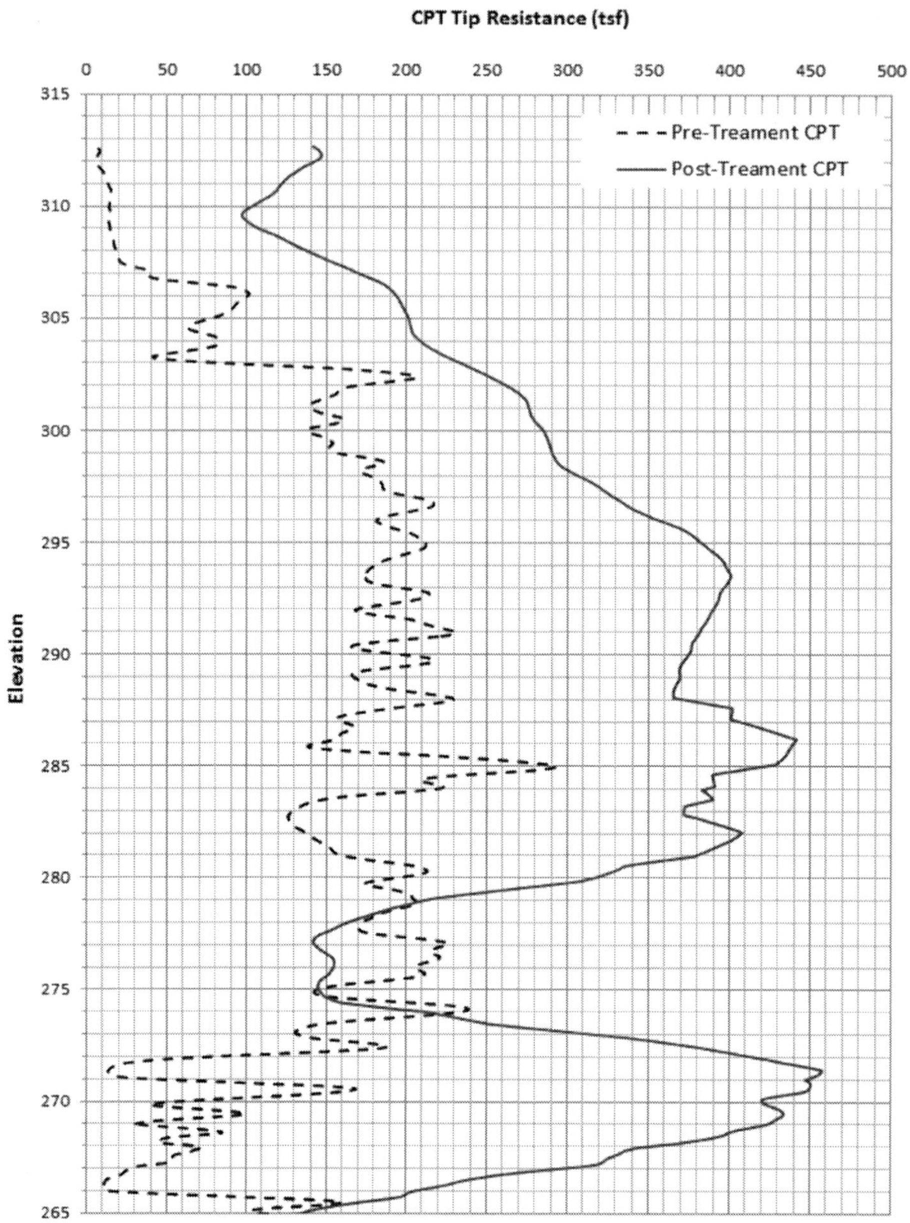

Figure 12. Typical Pre-Treatment and Post-Treatment CPT Tip Resistance for Stone Column Treated Ground at Smithland

Compaction Grouting in Powerhouse Foundation. The compaction grouting work at Smithland was also performed as a change order to the cofferdam and powerhouse excavation contract. Hayward Baker was the subcontractor for the compaction grouting work.

The highly congested work area available for compaction grouting in the vicinity of the powerhouse foundation required a high level of coordination amongst work crews by Hayward Baker to meet the tight schedule allotted for this work, which enabled powerhouse construction to begin as scheduled. The compaction grouting work at Smithland is shown in Figure 13. A three-dimensional view of compaction grouting holes with visualized grout take quantity is shown in Figure 14.

Figure 13. Compaction Grouting Operations at Smithland

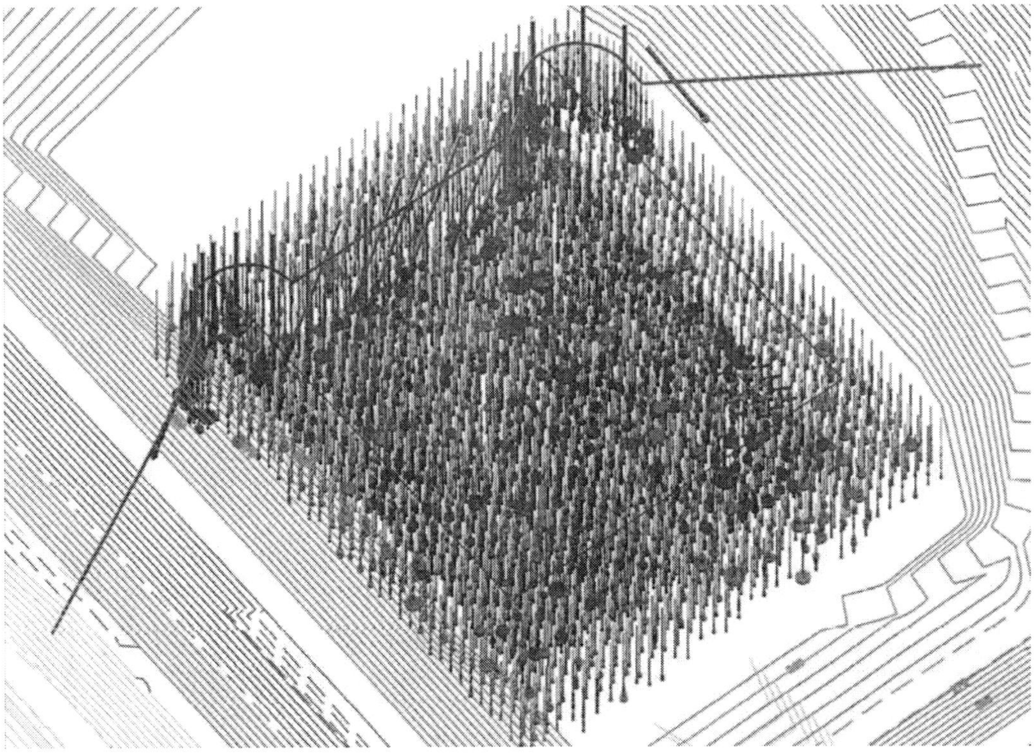

Figure 14. 3-D Visualization of Grout Takes in Compaction Grouting Holes at Smithland

A compaction grouting test program at Smithland was conducted to establish the spacing between primary grout holes and grouting procedures, including target (maximum) pressures, grout take limits and split-spacing criteria to achieved optimal densification, while also preventing heave at the ground surface. Based on the results of the test program, the following grouting procedures and split-spacing criteria were adopted for the production work:

- All drilling and grouting was performed from a work platform approximately 30 feet above the general foundation level of the powerhouse.
- A 10-foot × 10-foot center-to-center rectangular grid was used to layout all primary grout holes.
- All primary and secondary holes were grouted in 2-foot vertical stages using the target pressures and grout take limits presented in Table 1.

Table 1. Target Compaction Grouting Pressures and Grout Take Limits at Smithland

Material	Zone	Target Pressure	Maximum Grout Take per 2 ft. Stage	Other Criteria
Bedrock	Top 25 ft. of bedrock.	450 psi	405 ft^3 (15 yd^3)	If volumes exceed 405 ft^3 (15yd^3) in a 2 ft. stage, then pump 13.5 ft^3 (0.5 yd^3) per stage for remaining rock profile. Maximum total volume of 540 ft^3 (20 yd^3) per hole.
Overburden	Top of bedrock to 20 ft. below general foundation level of powerhouse.	600 psi	30 ft^3	The 30 ft^3/stage limit is not applicable in the first 2 ft. stage above apparent top of bedrock.
	20 ft. below general foundation level of powerhouse.	400 psi	6 ft^3	---
	Powerhouse foundation level to compaction grouting work platform.	---	---	Backfill / non-pressure grouting only.

- Compaction grouting in overburden was performed from the general foundation level of the powerhouse to top of bedrock (a depth that ranged from about 25 to 50 feet).
- All primary holes within the limits of rock grouting were also drilled and grouted to a depth of 25 feet below the apparent top of rock (the depth in which large solution features was encountered by the investigation borings).

- Secondary holes were taken 25 feet into rock if (1) one or more holes in a quadrant of four adjoining holes had a total grout take in rock exceeding 500 ft^3 per hole or (2) two or more holes in a quadrant of four adjoining holes had a total take in rock exceeding 250 ft^3 per hole.
- Where secondary holes were required, they were drilled at the centroid of the square grid, so as to split-space the primary holes. The secondary holes were grouted with the same criteria as was used for primary holes.
- In grout holes where drill tool rotation was arduous (high rotation hydraulic pressures) during grouting in rock or overburden, a lower target pressure was used to avoid seizing up the casing.

A summary of the quantity of compaction grout injected into bedrock and overburden soil is presented in Table 2.

Table 2. Compaction Grout Takes at Smithland

Grout Holes	Take in Bedrock (ft^3)	Average Take in Bedrock (ft^3/ft.)	Take in Overburden (ft^3)	Average Take in Overburden (ft^3/ft.)	Total Grout Take (ft^3)
Primary	48,589	2.6	76,868	0.91	125,457
Secondary	2,874	2.0	12,362	0.85	15,236
Total	51,464	2.3	89,230	0.90	140,693

The post-treatment verification SPT borings in overburden were conducted at center points between quadrants of four adjoining grout holes and thus, assessed areas receiving relatively reduced compactive effort. The results of the verification SPT borings indicated that the SPT acceptance criterion was generally exceeded at all depths within granular foundation materials.

Secondary compaction grout holes were drilled in overburden in the locations of the few SPT borings that indicated that additional densification was required. Overall, the post-treatment SPT borings show that the compaction grouting program significantly densified the powerhouse foundation. Representative pre- and post-treatment SPT results for the compaction grouting at Smithland are shown in Figure 15.

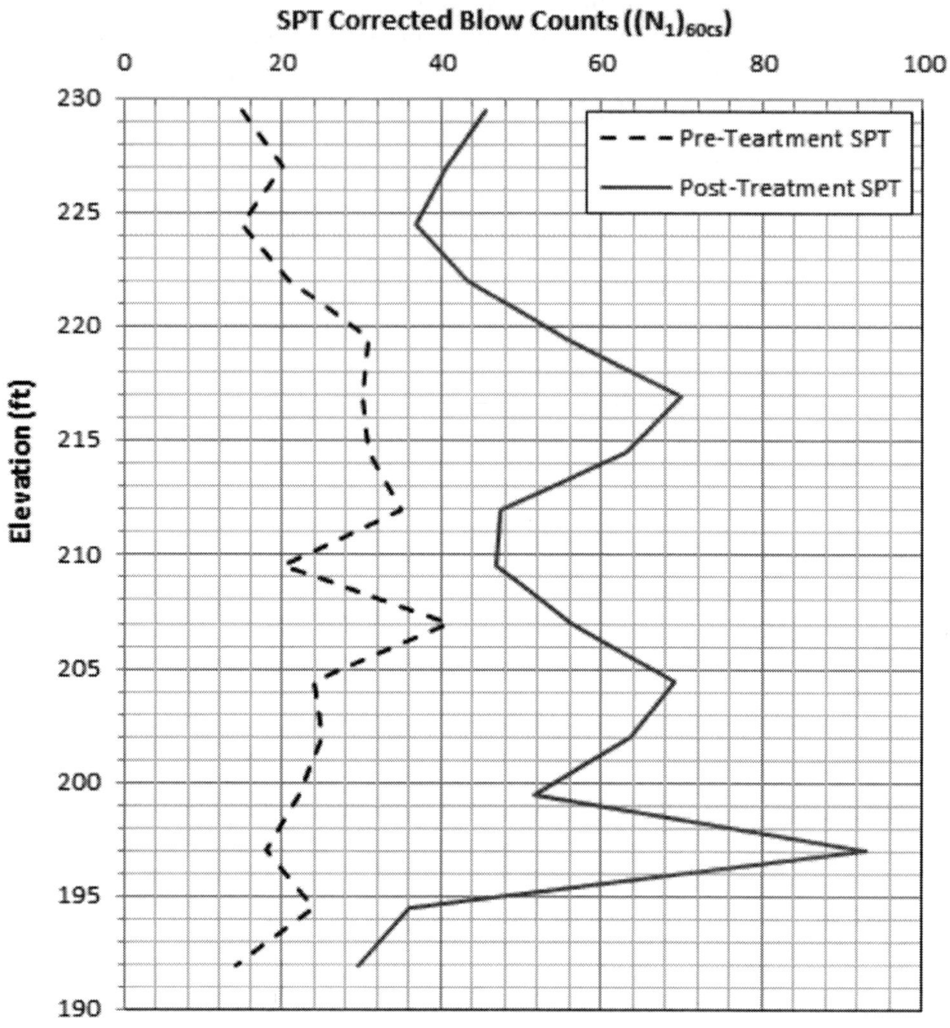

Figure 15. Pre-Treatment and Post-Treatment SPT Results for Compaction Grouting at Smithland

Consolidation Grouting along Cut-off Wall. Consolidation grouting was designed to improve soils and rock in areas adjacent to the permanent sheet pile cut-off wall along the upstream face of the powerhouse and adjacent areas along the upstream faces of the closure structures where the sheet piling extends to the top of bedrock. Consolidation grouting in rock at Smithland was designed to fill open voids and fissures in the bedrock and minimize the overall amount of under-seepage through rock.

Primary consolidation grout holes and mandatory secondary consolidation grout holes were located upstream and downstream of the sheet pile wall, respectively. Primary and secondary holes were split-spaced by tertiary holes and in turn quaternary holes in areas where the grout take into the rock measured exceeded 1.0 ft^3 per foot in the preceding hole(s). The spacing between the mandatory primary and secondary grout holes, as measured along the cut-off wall, was typically 5 feet. Where tertiary and quaternary holes were used, the spacing between grout holes was typically 2.5 feet and 1.3 feet, respectively.

Three sonic cores were drilled at the locations along the periphery of the upstream sheet pile cut-off wall in an attempt to obtain a continuous sample of soil and rock after completion of consolidation grouting. While intact grout (compaction or consolidation) was not observed in the sonic cores, pea gravels that could have been from the compaction grout mix were present in the core samples. The grout matrix may not have obtained adequate strength to withstand the vibrations associated with sonic coring. The sonic coring operations were observed to have steady penetration rates in soil and rock and did not appear to encounter any loose zones or voids.

Evaluation of Ground Improvement at Smithland

The effectiveness of ground improvement was evaluated based on liquefaction potential and earthquake-induced settlement, and Newmark permanent displacement assessed based on post-treatment CPT and SPT results. The results of these analyses indicate that post-improvement powerhouse and closure structure foundations would be stable under the MCE and that the stone column and compaction grouting programs achieved the design intent to mitigate liquefaction potential and reduce earthquake-induced settlements and deformations.

LESSONS LEARNED

Lessons learned from the ground improvement programs at the Cannelton and Smithland Hydroelectric Projects are described below.

- When planning a ground improvement program, it is important to have contingency plans and to recognize that the project's overall schedule can be impacted by supplemental ground improvement work.
- The engagement of a suitably experienced geotechnical specialty contractor with trained, knowledgeable personnel is very important to the success and efficiency of a ground improvement program. When ground improvement work is included in a larger general construction contract, it is recommended that specialty subcontractors be pre-qualified.
- Test sections are useful to evaluate and refine a ground improvement contractor's methods. Since ground conditions can vary across a project site, the contractor may need to modify procedures.
- The amount of densification achieved by vibro-replacement depends on the amount of energy input into the ground by the vibrating probe, which is operator dependent. The amount of energy can be assessed based on the amount of amperage build-up, if an electric powered vibrator is used, and the duration of effort at each elevation.
- The amount of densification achieved by vibro replacement below a hard layer can be less than is obtained in other horizons. This is possibly due to the removal

of material by water-jetting during vibrator penetration through a hard layer and/or constrained vibrator motion due to confinement in a hard layer.

- The densification achieved by compaction grouting at Smithland was generally better than achieved by stone columns, even though stone columns were performed on a closer spacing. This result was unexpected based on available case histories and may have been due to the higher amount of confinement during compaction grouting or can possibly be related to the specific installation methods used by the respective specialty ground improvement contractors.

CONCLUSIONS

The powerhouse and closure structures for the four hydroelectric projects currently being constructed by AMP along the Ohio River require sound foundations under both static and seismic conditions. Therefore, various efforts were made to identify foundation issues that may affect the performance of these critical energy infrastructure projects during site investigation, design, and construction.

Conventional methods for rock foundation preparation were employed at Willow Island and Meldahl to prepare suitable foundations for permanent water-retaining structures founded on rock. The powerhouses and closure structures at Cannelton and Smithland are founded on soil and required extensive ground improvement to mitigate liquefaction potential and reduce earthquake-induced settlements and deformations. At Smithland, ground improvement was also needed to address potential ground loss and excessive under seepage in the solution features through the underlying karstic bedrock.

The effectiveness of ground improvement activities performed at the Cannelton and Smithland was evaluated based on post-treatment SPT and CPT results, sonic coring, installation logs, and engineering analyses. The results indicate that the desired ground improvement objectives were satisfied and that the improved foundations are suitable for all evaluated loadings conditions, including those associated with the MCE.

ACKNOWLEDGEMENTS

The authors would like to acknowledge AMP and their municipal members for their initiative to develop the innovative new hydroelectric projects on the Ohio River and for their vision of renewable energy.

The authors would also like to thank the Board of Consultants for the Cannelton and Smithland Hydroelectric Projects, which was comprised of Eric Kollgaard, Jim Erwin and Joe Ehasz, FERC's consultant, Dr. Alfred J. Hendron Jr., and the many individuals from the USACE and FERC who provided constructive recommendations and valuable feedback during ground improvement design and foundation evaluation.

The aerial photographs of the Smithland, Cannelton, Meldahl and Willow Island projects were provided to AMP by C.J. Mahan Construction Company, Walsh Construction Company, Alberici-Baker Joint Venture, and The Ruhlin Company, respectively.

REFERENCES

Andrews, T.G., L. Zhao, J.M. Hynes and P.E. Meier (2015). "Holding Back the Ohio – Cofferdam and Cut-Off Wall Design and Construction," USSD 2015 Annual Conference, Louisville, KY.

Arnold, T.E., A. Brandi, P.E. Meier and T.M. Scott (2015.) "Optimized – Hardfill Mix Design and Testing." USSD 2015 Annual Conference, Louisville, KY.

Blaszczyk, P.R., K. Griebenow and P.E. Meier (2015). "Time to Generate Power – New Hydro Development on the Ohio", USSD 2015 Annual Conference, Louisville, KY.

Idriss, I.M. and J.I. Sun (1992). "User's Manual for SHAKE91, A computer Program for Conducting Equivalent Linear Seismic Response Analyses of Horizontally Layered Soil Deposits Program Modified based on the Original SHAKE Program Published in December 1972 by Schnabel, Lysmer and Seed."

Idriss, I.M. and R.W. Boulanger (2008). "Soil Liquefaction During Earthquakes", MNO-12, Earthquake Engineering Research Institute.

Newmark, N. M. (1965). "Effects of earthquakes on dams and embankments," Geotechnique, Vol. 15, No. 2, pp. 139-160.

Tokimatsu, K. and H.B. Seed (1987). "Evaluation of settlements in sand due to earthquake shaking." Journal of Geotechnical Engineering, ASCE, Vol. 23, No. 4, pp. 861-878.

Truschke, S.M., T.E. Arnold, T.M. Scott, P.E. Meier (2015). "Built to Last – Hardfill Construction." USSD 2015 Annual Conference, Louisville, KY.

OPTIMIZED — HARDFILL MIX DESIGN AND TESTING

Terrence (Terry) Arnold, P.E.[1]
Brandi Antal[2]
Philip E. Meier[3]
Taylor M. Scott[4]

ABSTRACT

Each of the four hydroelectric projects that American Municipal Power, Inc. (AMP) is developing on the Ohio River require water-retaining structures to be constructed within the large powerhouse excavations: providing closure between the powerhouse and existing dam (riverside) and between the powerhouse and excavation in the bank of the Ohio River (landside). Trapezoidal dams composed of hardfill were selected to construct the landside closure structures for the Willow Island and Meldahl Projects. The Cannelton and Smithland sites used trapezoidal hardfill dams for both the landside and riverside closure structures. These structures are founded directly on a wide range of foundation conditions including in-situ alluvial soils. Hardfill is a variation of roller-compacted concrete that utilizes locally available soil aggregate in lieu of conventional concrete aggregate. Project-specific hardfill mixes were developed using a combination of on-site (or nearby) soil aggregate sources, fly ash and cement. Extensive mix design programs were performed to evaluate the soil aggregates available for each project and develop individual hardfill mixes. This enabled the selection of constructible hardfill mixes, including incorporation of fly ash to reduce construction costs while still providing the required strength and durability. This paper presents the trial mix design programs that were conducted to develop hardfill mixes for each project site and the modifications required to analyze the effect of fly ash usage due to limited published historical data.

PROJECT DESCRIPTION

American Municipal Power, Inc. (AMP) is developing four hydroelectric projects at existing US Army Corps of Engineers (USACE) locks and dams along the Ohio River. The USACE provided for future hydropower construction when the dams were originally constructed. The locations of the four hydroelectric projects (Smithland, Cannelton, Meldahl, and Willow Island) under construction on the Ohio River are shown on Figure 1.

---

[1] Senior Project Manager, MWH Americas, Inc., 1560 Broadway St., Suite 1800, Denver, CO 80202, terrence.e.arnold@mwhglobal.com
[2] Civil Engineer, MWH Americas, Inc., brandi.antal@mwhglobal.com.
[3] Vice President – Hydroelectric Development and Operations, American Municipal Power, Inc. 1111 Schrock Road, Suite 100, Columbus, OH 43229, pmeier@amppartners.org.
[4] Associate Civil Engineer, MWH Americas, Inc., 1560 Broadway St., Suite 1800, Denver, CO 80202, taylor.scott@mwhglobal.com.

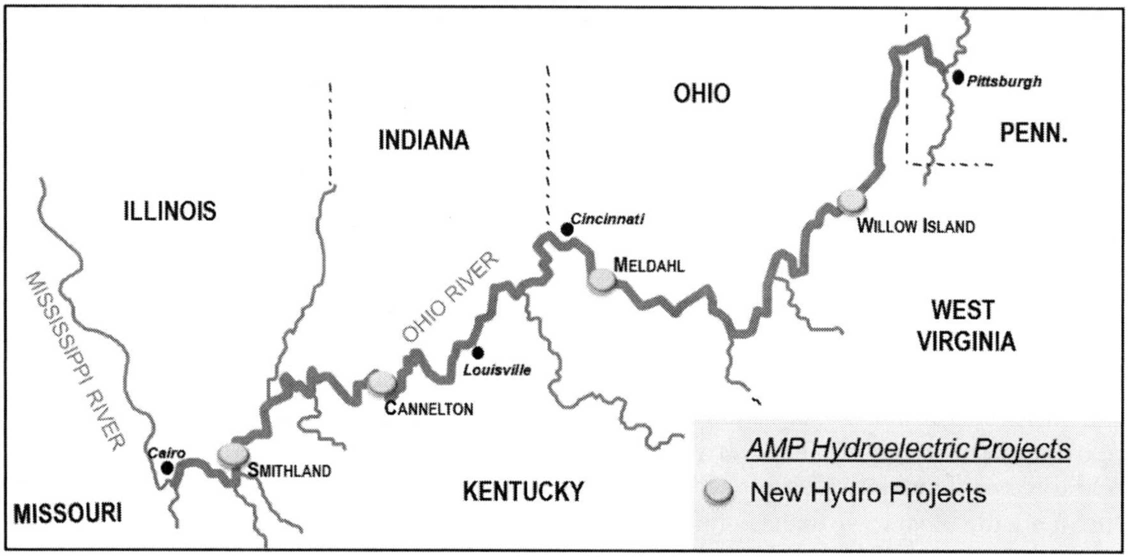

Figure 1. Project Location Map.

The design of each project was tailored to site-specific conditions. Foundation conditions varied wildly ranging from weak rock, including slickensided indurated clays and a thrust fault at Willow Island to soft clays which required support during excavation at Meldahl. The Cannelton and Smithland Projects are located near the New Madrid Seismic Zone and involve potentially liquefiable alluvial soils while the design at the Smithland Project was further complicated due to karstic limestone. Unique project-specific design features included:

- 2,000-kip rock anchors at Willow Island to stabilize the foundation below the cellular cofferdam and significant dental excavation and rock bolting to treat the fault where it crossed the excavation. Steel sheet piles to stabilize excavation slopes at Meldahl.
- A cofferdam at Cannelton with a 100-foot-high zoned embankment constructed in the Ohio River and stone columns to improve the foundations below the closure structures to mitigate against liquefaction potential and minimizing earthquake-induced settlement.
- The embankment cofferdam at Smithland with a 150- to 187-foot-deep, 4000-foot-long soil-cement-bentonite cut-off wall and dewatering system. Stone columns were also installed beneath the closure structure. Compaction and consolidation grouting was used to treat the foundation soils and karstic rock below the powerhouse.
- Of eight (8) closure structures (dams) required for the project, six (6) were constructed of hardfill (roller-compacted soil-cement) to maximize the use of on-site materials and minimizes the importation of aggregate.

Additional details regarding the foundation conditions and details regarding project-specific ground improvement are provided in Mirza et al. (2015).

Relevant details for the closure structures at each project are summarized below. The descriptions, LSCS refers to landside closure structures and RSCS refers to riverside closure structures. The projects are presented in the order that the mix design programs were conducted. The hardfill mix design programs were modified based on the results of the previous programs.

Cannelton

The Cannelton Hydroelectric project is located about 720.7 river miles below Point Bridge, Pittsburg, PA.

- The maximum height of the LSCS is 116 ft and is founded on in-situ soil.
- The maximum height of the RSCS 99 ft and is founded on in-situ soil.

Figure 2: Cannelton Aerial View, 31-Mar-2013

Smithland

The Smithland Hydroelectric Project is located in Kentucky about 918.5 river miles below Point Bridge, Pittsburgh PA, and 62.5 river miles upstream of the confluence of the Ohio and Mississippi Rivers.

- The LSCS will have a maximum height of about 114ft and is founded on in-situ soil.
- The RSCS will have a maximum height of about 93ft and is founded on in-situ soil.

Figure 3. Smithland Overall Aerial View, 24-Oct-2014

Meldahl

The Meldahl Hydroelectric Project is located northeast shore of the Ohio River about 436.2 river miles below Point Bridge, Pittsburgh, PA.

- The LSCS will have a maximum height of about 90ft and is founded on rock and in-situ soil.
- The RSCS will have a maximum height of about 82ft and is founded on competent rock.

Figure 4. Meldahl Overall Aerial View, 8-Oct-2014

Willow Island

The Willow Island Hydroelectric Project is located on the West Virginia shore about 161.7 river miles below Point Bridge, Pittsburgh PA.

- The LSCS will have a maximum height of about 75ft and is founded on rock and in-situ soil.
- The RSCS will connect the powerhouse to the existing spillway pier will have a maximum height of about 58ft and is founded on competent rock.

Figure 5. Willow Island Overall Aerial View, 24-Oct-2014

CLOSURE STRUCTURE GENERAL DESIGN

Hardfill was selected for the design of the powerhouse closure structures (the area between powerhouse and excavation slopes) to meet the primary design criteria including a design that would provide: 1) a water retaining structure between the approach channel and tailwater, upstream and downstream, of the powerhouse; 2) resist applied loads; and 3) ability to withstand overtopping since the powerhouse structures are located within the Ohio river floodplain and are expected to be flooded during the operating life of the hydropower plants.

Hardfill is a relatively new technology used in dam construction in the U.S. Hardfill is a concrete construction product that typically consists of aggregate, water and cement. ICOLD Bulletin 117 (ICOLD, 2000) defines hardfill as a lean RCC or a low strength cemented sand/gravel product, placed by roller compaction methods in a symmetrical cross section. Hardfill structures are adaptable to a wide range of foundation conditions and can be constructed using a variety of aggregate materials. This was key in developing project design concepts that could be applicable for each hydroelectric project, while minimizing customization of the design concept at each site.

Structural and stability analysis were performed for the closure structures for construction, normal operating, and earthquake loading cases at each of the four sites. A summary of the target engineering criteria for the hardfill mix design programs for the closure structures included:

Hardfill Properties

- Friction angle (φ) = 35°
- Apparent cohesion (c) = 20 psi
- Compressive strength = 1,000 psi
- Modulus of elasticity = 100,000 psi

Foundation Interface Properties

- Friction Angle (φ) = 34°
- Cohesion (c) = 0 psi

Hardfill Construction Material

Due to the relatively low strength requirements for symmetrically placed hardfill, a wide range of aggregate types are suitable for hardfill mixes. The aggregate used for hardfill is typically locally sourced and varies between a well graded sand and gravel (roller compacted concrete type material) to a clean to silty sand (soil-cement type material). Alluvium (silty sands) is a common material available from the required excavations for the project construction at the existing locks and dams in the Ohio River. A standard borrow assessment program to evaluate and select suitable borrow material for hardfill, and hardfill mix design testing was developed and then customized for each project. Geotechnical investigations performed during the preliminary phases of the project confirmed that sandy soil was the predominant soil type locally available at each of the project sites.

Since the predominant soil consisted generally of sand, soil-cement mix design procedures and testing methods were used for the mix design program. Soil-cement has historically been used primarily for slope protection for reservoirs and for lining of channels. Consequently, minimal performance history is available for use of a soil-cement based hardfill material to construct a large water retaining structure. An additional consideration for these projects was the ready availability of fly ash in the Ohio River Valley, with the unit cost of fly ash less than half of the cost of cement.

Design of the hardfill mixes started with assessment of potential borrow materials followed by testing to develop hardfill mix proportions and engineering properties. The borrow assessment and hardfill mix design program generally consisted of assessment of on-site material available from required excavations, and a survey of available nearby, commercial borrow materials. A hardfill mix design program to investigate the potential borrow materials, develop mix proportions and engineering properties of the design mix was developed and consisted of four general phases as described below.

- Phase 1 – Evaluation of potential borrow material. Borrow sources from on-site required excavations and nearby commercial source were identified and testing of basic engineering properties for comparison with published hardfill data was performed,
- Phase 2 – Initial evaluation of the suitability of select borrow sources for hardfill at a standard cement content,
- Phase 3 – Testing to develop a design cementitious content using the selected aggregate, and
- Phase 4 – Confirmation testing of engineering properties for the selected hardfill mix proportions.

HARDFILL MIX DESIGN PROGRAM

The objective of the hardfill mix design program was to develop a hardfill mix that would provide engineering properties to meet structural, water retention and stability criteria, use locally available material, and provide an economical and constructible mix. Historically, sandy soils and soil cement mixes have used cement only. The use of fly ash in the hardfill mixtures presented a unique challenge. There is limited published history[5] of using fly ash replacement in sandy soil-cement/fly ash mixes available.

Fly ash contributes to the long term strength of concrete/RCC mixtures but can delay/retard the early strength gain of mixes. Therefore, the typical soil-cement testing program was modified to include the strength gain due to fly ash over a longer time period. Additionally, accelerated cure procedures were used to evaluate later phases of laboratory testing in a quick manner to meet the construction schedule. Modifications to a typical soil-cement testing program included:

- Adding standard compressive strength tests at 90-, 180-, and 365-days age.
- Adding accelerated cure compressive strength tests.
- Adding (F-T/W-D) durability testing at later ages to allow some strength gain of the fly ash to occur prior to the test.

The phased testing program and test results for the four projects are summarized below.

Phase 1: Evaluation of Potential Borrow Material

Hardfill Aggregate Evaluation. On-site borrow material available at the Cannelton and Smithland sites consisted of a clean to silty sand with some scattered zones of silty, clayey sand. Selective stockpiling from the Cannelton powerhouse excavation limited the quantity of clayey soil in the stockpile. Material from the powerhouse excavation at Smithland contained seams and layers of clayey material and the stockpile was highly variable ranging from a clean to silty or clayey sand as well as containing seams of clay. Required excavations at the Meldahl and Willow Island sites were insufficient to provide

[5] See Antal et al. (2013) for a summary of published information on fly ash replacement methods.

the required closure structure volume. Therefore, off-site commercial sources of soil aggregate were further investigated at those sites. Based on the predominant soil type available at the four sites, the following criteria was specified to identify the soil aggregate for hardfill:

- Soil aggregate shall be classified as SP, SW, SP-SM, and/or SM (ASTM D2487),
- No particle larger than 1 ½ in. in any dimension,
- No more than 20 percent passing the No. 200 sieve size (ASTM D422),
- Plasticity index (ASTM D4318) of no more than 6,
- Soil pH (ASTM D4972) greater than 5.3, and
- Organic content (ASTM D2974) of less than 1 percent.

Soil Aggregate Availability. On-site borrow material varied widely over the 750 river miles from the upper most Willow Island Project to the Smithland Project, the furthest down river of the four hydro-electric projects. Material testing of the on-site stockpiles was performed to develop basic engineering and soil aggregate properties of the borrow material, and confirm that the material was suitable for a hardfill. The materials available on site, or from nearby commercial sources for each site, are described below.

Cannelton. Sampling and testing of the on-site stockpile at Cannelton showed that the borrow material consisted of silty sands with a fines content of about 10-15% (SP-SM). Scattered pockets or zones of silty/clayey soil aggregate (SC-SM) with a fines content of about 33%, and silty soil aggregate (SM) with a fines content of about 20% also occurred in the stockpile. Soils containing elevated fines content and/or plasticity are known to affect the cementitious content requirements, compressive strength, and durability of soil cement mixtures. Therefore, hardfill mix testing of the range of soil aggregate types was performed. The range of gradations encountered in the on-site stockpile, as well as an average "composite" gradation are shown on Figure 6.

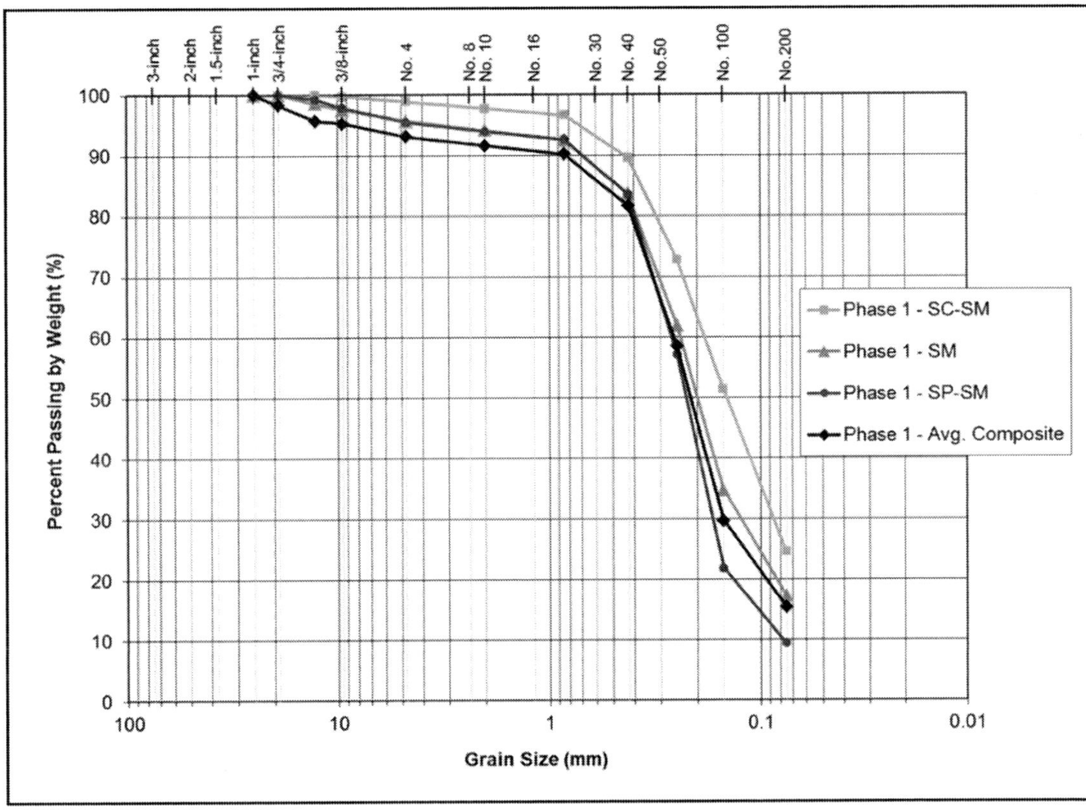

Figure 6. Range of Aggregate Type from Required Powerhouse Excavation at Cannelton

Smithland. Geotechnical investigations performed during the preliminary phases of the project confirmed that sandy soil was locally available at the Smithland site. Different material types identified in the onsite stockpile included: 1) clean sand (SP); 2) slightly silty to silty sand (SP-SM); 3) silty sand (SM) and 4) clay with some clayey sand pockets (SC). Based on the results of the engineering properties, four samples were tested. The three material types identified (SP, SP-SM, and SM) were tested in addition to an "average" composite sample (composited by mixing together a weighted average of the other samples to represent an average of what is expected from the stockpile). The range and average gradation of samples are shown in Figure 7.

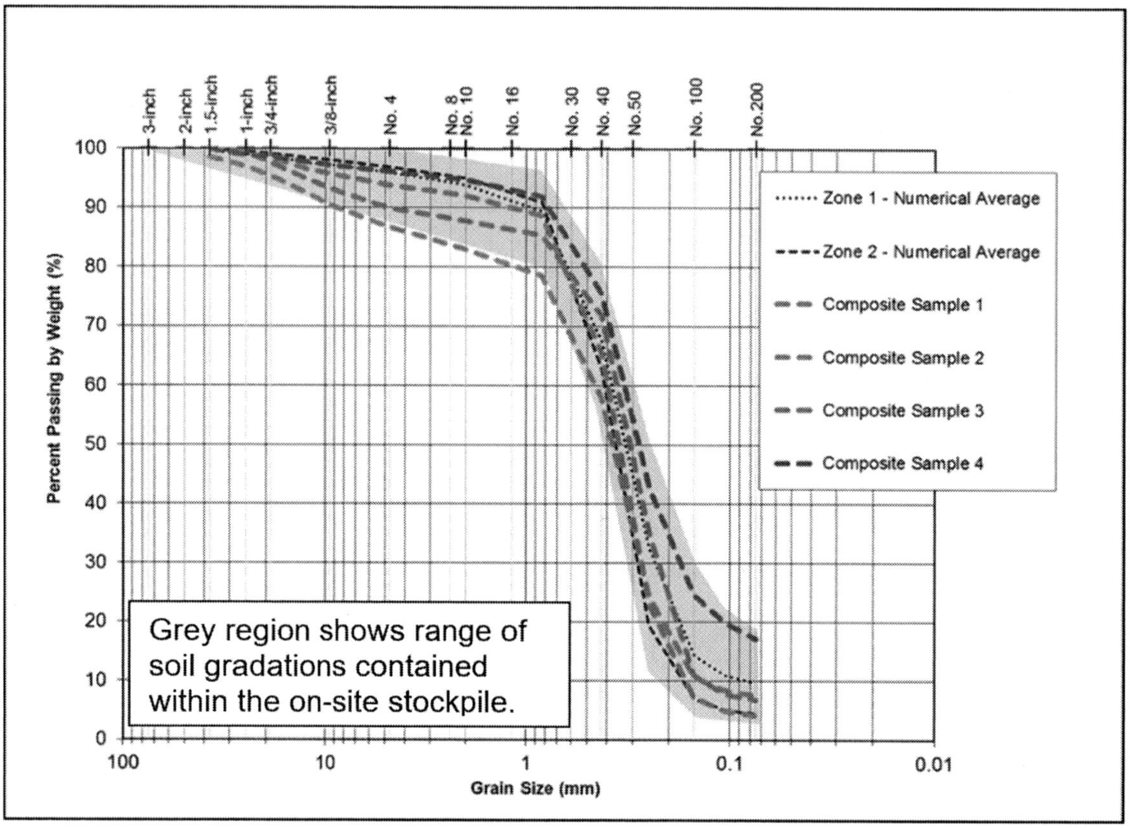

Figure 7. Smithland - Range of Soil Gradations Contained within the On-Site Stockpile

Meldahl. Geotechnical investigations performed during the preliminary phases of the Meldahl project confirmed that sandy soil was locally available; however, during powerhouse and cofferdam excavation, borrow quantities stockpiled from selective excavation were insufficient for construction of the closure structures. Consequently, it was decided that an off-site soil aggregate material meeting project soil aggregate specifications would be used.

Local aggregate suppliers were investigated and quarries capable of producing the type and quantity of soil aggregate material needed for the Meldahl Project were identified. The Hilltop Companies Butler Quarry in Butler, Kentucky, approximately 13 miles from the project site, was selected as the primary soil aggregate source and three (3) potential soil aggregates were identified: Hilltop – L910, Hilltop – L801, and Hilltop L803 – for hardfill evaluation. The gradations of the soil aggregates selected for further testing are presented on Figure 8.

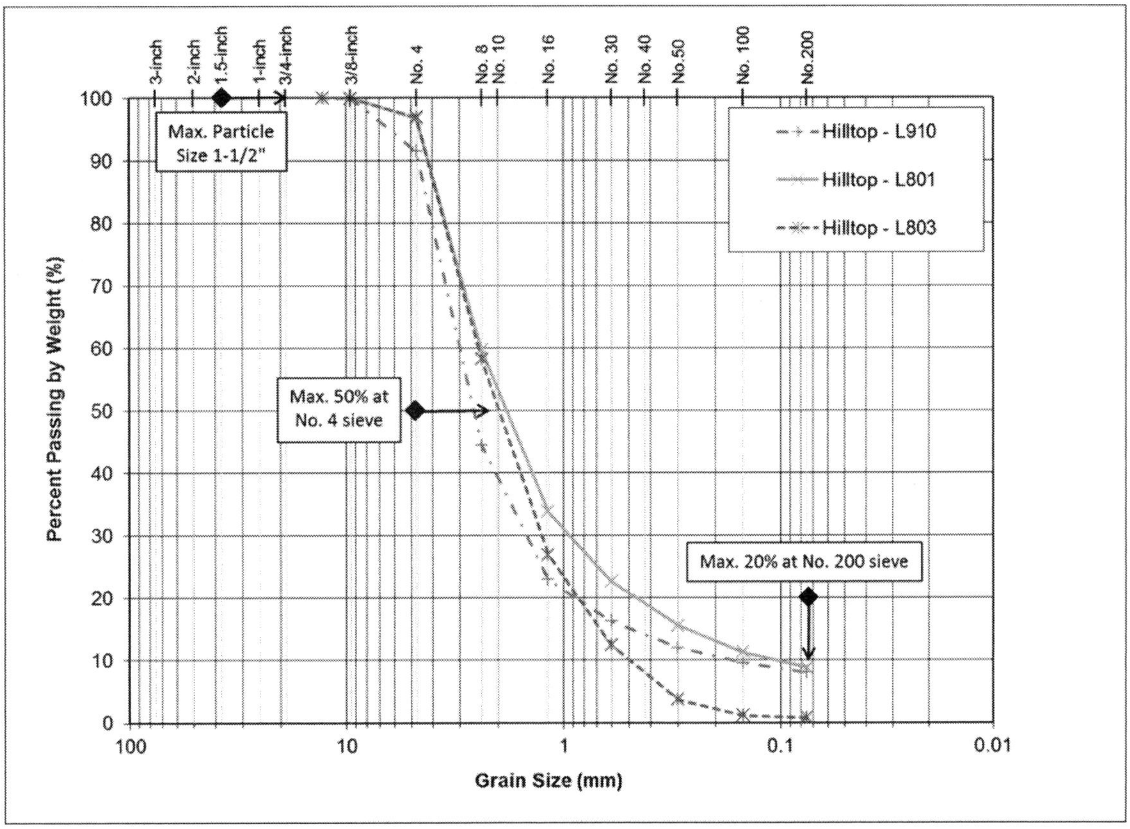

Figure 8. Meldahl – Imported Soil Aggregate Gradations

Willow Island. Geotechnical investigations performed during the preliminary phases of the project confirmed that sandy soil was locally available. However, geotechnical investigations of the site showed that quantities of hardfill aggregate from required excavations were insufficient to meet the required closure structure hardfill volume. Therefore, similar to the Meldahl Project, off-site sources were considered. The Hilltop Companies Big Bend Quarry in Battletown, KY and the Belpre Sand and Gravel Company in Hocking, OH were identified as local aggregate supplies that that were capable of producing adequate material suitable for the Willow Island Project. One potentially suitable aggregate type from each quarry was identified and selected for further testing as shown on Figure 9.

Hardfill Mix Design and Testing

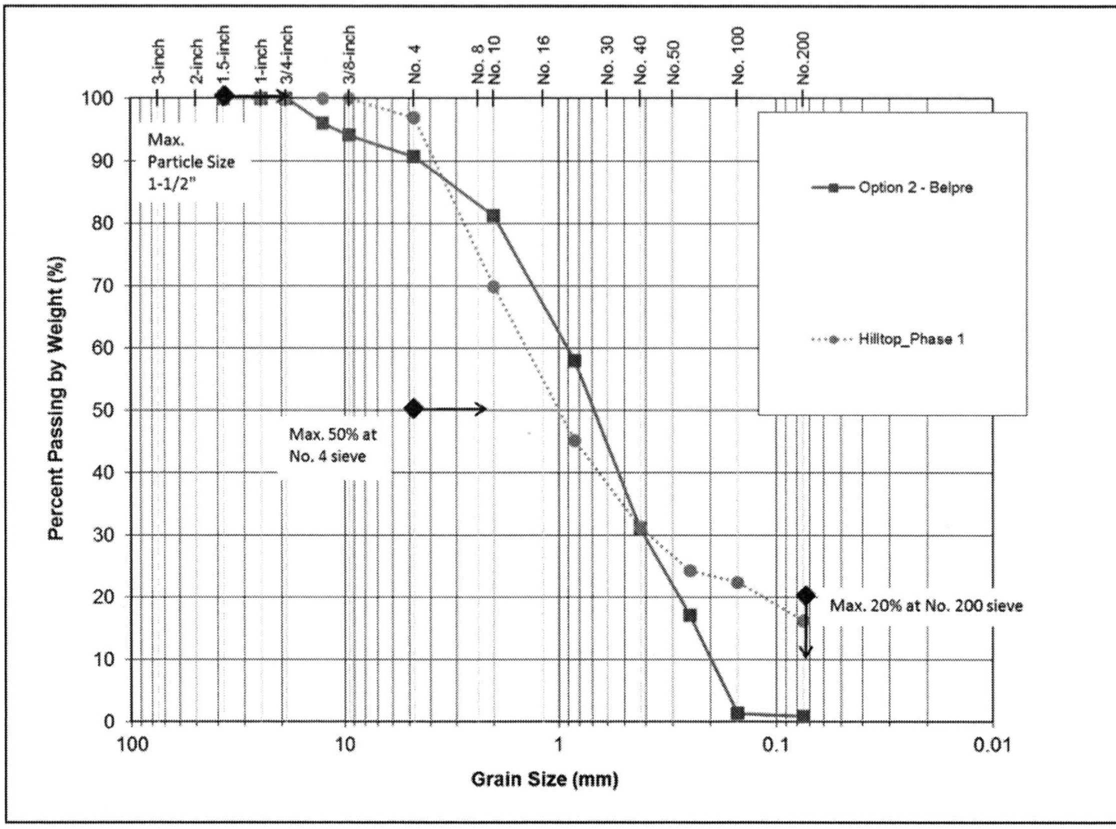

Figure 9. Willow Island –Range of Soil Gradations Contained from Nearby Available Commercial Sources

Hardfill Aggregate Selection

The characteristics of the various hardfill aggregates were generally compared at a constant cement content, typically 8% (with the exception of the Willow Island project where the cement content was lowered based on the testing results from the previous three projects), to evaluate the comparative strength gain of each aggregate. Tables 1 and 2 provide a comparison between the soil aggregate properties for each soil type and the associated compressive strengths at the Cannelton and Smithland Projects, respectively. The SP-SM material at the Cannelton Project provided the lowest compressive strength (20 to 40% lower than the other soil aggregate types); a wide range of compressive strengths in the various soil aggregate sources was noted. The compressive strengths for the four composite soil aggregate sources at the Smithland Project were closer in value with Composite aggregate 4 showing the lowest compressive strength as shown in Table 2.

Table 1. Cannelton - Soil Aggregate Type vs. Hardfill Compressive Strengths at 8% Cement Only Content

Soil Aggregate	AASHTO Soil Classification (ASTM D3282)	Atterberg Limit D4318			Fines Content (%)	Compressive Strength – Standard Cure (ASTM D1633) 28-days age (psi)
		LL	PL	PI		
SC-SM	A-2-4	17	14	3	24.5	1145
SP-SM	A-3	NP	NP	NP	9.3	645
SM	A-2-4	NP	NP	NP	17.2	900
Composite Average	A-2-4	NP	NP	NP	15.3	875

Table 2. Smithland - Soil Aggregate Type vs. Hardfill Compressive Strengths at 8% Cement Only Content

Soil Aggregate	AASHTO Soil Classification (ASTM D3282)	Atterberg Limit D4318			Fines Content (%)	Compressive Strength – Standard Cure (ASTM D1633) 28-days age (psi)
		LL	PL	PI		
Composite 1	A-3	NP	NP	NP	4.2	490
Composite 2	A-3	NP	NP	NP	4.3	525
Composite 3	A-3	NP	NP	NP	6.9	495
Composite 4	A-3	18	14	4	17.1	435

The results for the hardfill materials tested for the Meldahl Project are presented in Table 3. The Hilltop - L801 soil aggregate was selected for the Meldahl project based on the initial comparison of the compressive strength of each sample with 8 percent cement content as shown in Table 3.

Table 3. Meldahl - Soil Aggregate Type vs. Hardfill Compressive Strengths at 8% Cement Only Content

Soil Aggregate	AASHTO Soil Classification (ASTM D3282)	Atterberg Limit D4318			Fines Content (%)	Compressive Strength Standard Cure (ASTM D1633) 28-days age (psi)
		LL	PL	PI		
Hilltop L801	A-1-b	NP	NP	NP	8.8	2505
Hilltop L901	A-1-a	NP	NP	NP	8.0	1980

Hardfill Mix Design and Testing

Hilltop L803	A-1-a	NP	NP	NP	0.8	1080

The compressive strength of borrow sources at Willow Island were generally the highest of all the materials tested in equivalent cement content basis as shown when comparing the test results in Tables 1 through 4.

Table 4. Willow Island - Soil Aggregate Type vs. Hardfill Compressive Strengths at Variable Cement Only Contents

Soil Aggregate	AASHTO Soil Classification (ASTM D3282)	Atterberg Limit D4318			Fines Content (%)	Cement Content	Compressive Strength Standard Cure (ASTM D1633) 28-days age (psi)
		LL	PL	PI			
Hilltop Sand	A-1	NP	NP	NP	16.2	6.4%	2450
Belpre Sand	A-1	NP	NP	NP	3.1	7.4%	1645
Belpre Sand	A-1	NP	NP	NP	4.1	5.7%	1465

Generally, based on the test results, the Smithland soil aggregate and the Cannelton soil aggregate were similar, and the Meldahl and Willow Island soil aggregates are a coarser material obtained from commercial aggregate sources were similar. The Willow Island soil aggregate also consisted of a relatively low percentage of material passing the No. 200 sieve. Hardfill gradations for each project are presented on Figure 10.

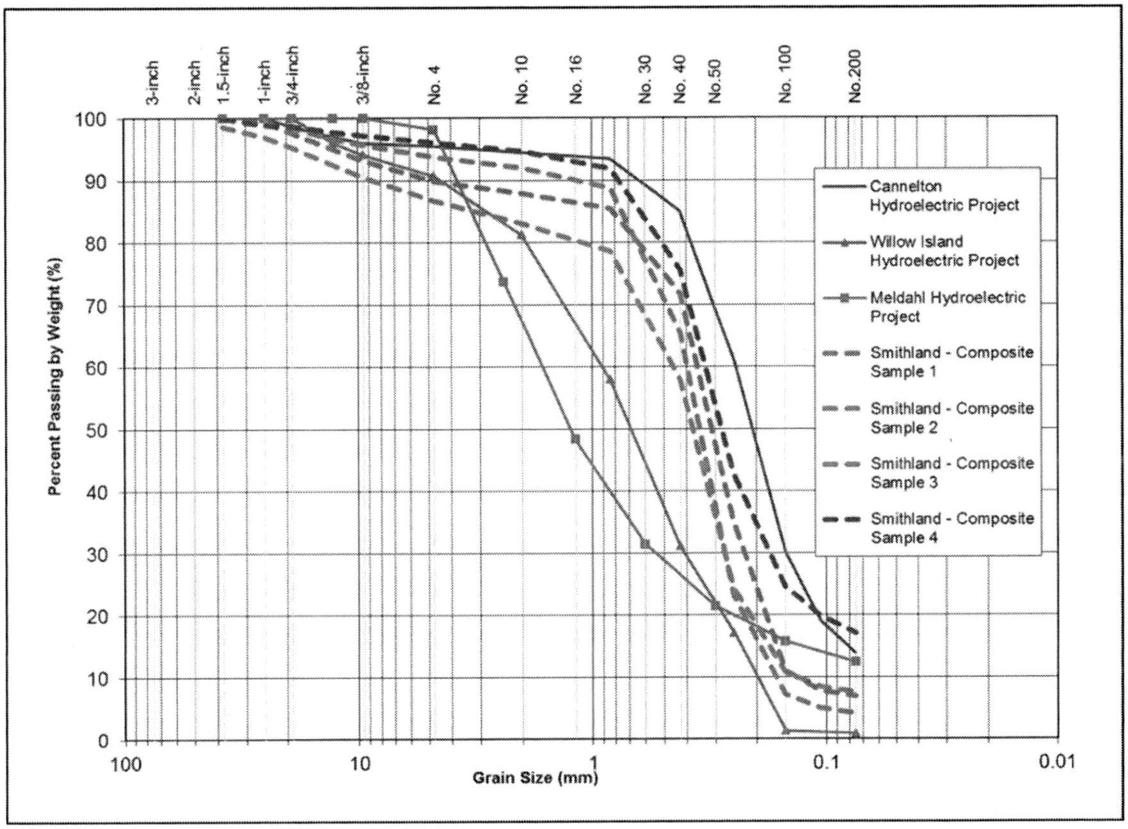

Figure 10. Summary of Hardfill Gradations for Each Project

Hardfill Test Methods

To aid in the selection of mix design proportions, a range of cement and fly ash proportions were evaluated to develop index values for strength development, and durability based on cementitious content. The tests performed are discussed below.

Compressive Strength. Compressive strength testing was performed for both cement-only and cement plus fly ash mixes. Cement-only mixes were tested to allow better evaluation of the strength gain properties associated with the addition of fly ash. Compressive strength cylinders were prepared following ASTM D1633, "*Standard Test Methods for Compressive Strength of Molded Soil-Cement Cylinders*." Compressive strength testing of cement-only mixes was performed at 2-, 7-, and 28-days age, and cement plus fly ash mixes were tested at 2-, 7-, 28-, 90-, 180-, and 365-days age. In addition, testing of some cement-only mixes at 90 to 180-days age were also performed to allow for analysis of the incremental strength gain provided by the fly ash.

Accelerated Cure Compressive Strength. Accelerated cure compressive strength testing was performed to provide an early indication of the ultimate strength of a hardfill mixture. Mixtures that contain both cement and fly ash would be expected to reach ultimate strengths between about 180- and 365-days. Therefore performance of the accelerated cure test was important to provide early projection of compressive strengths to allow for acceleration of the overall mix design program, as well as for produced material during the construction phase.

Initial phases of the hardfill mix design program considered ASTM C684, "*Standard Test Method for Making, Accelerated Curing, and Testing Concrete Compressive Test Specimens*" Method A and Method B accelerated cure methods. Test results using the ASTM C684 procedures provided accelerated cure strengths generally lower than conventional compressive strength cylinders at equivalent ages. This is not consistent with typical accelerated cure results for conventional concrete; therefore, an alternate accelerated cure testing procedure developed for RCC mix design programs on previous projects, including the Olivenhain Dam (Holderbaum et al. 2002), was selected for use in the Cannelton mix design program. The accelerated cure cylinder testing procedure generally included:

1. Standard cure cylinders for 7 days,
2. After the 7-day standard cure, submerge cylinders in sealed water container with pressure relief hole on top of container,
3. Place submerged cylinder in 160-deg-F oven for 7 and 14 days,
4. Remove the cylinder from the oven after the desired accelerated cure age (either 14- or 21-days age) and allow cylinder to cool down to room temperature while in the container without placing cylinder in cold water, and
5. Cap and test cylinder.

The results of the warm water bath accelerated cure procedure generally correlated to the ultimate strength of the hardfill (Antal et al. 2013).

Durability. Soil-cement testing programs typically include freeze-thaw and wet-dry durability testing at 7-days age. Freeze-thaw (F-T) and wet-dry (W-D) durability testing at later ages was also tested for this project to allow some strength gain of the fly ash to occur prior to testing. Freeze-thaw testing was performed following ASTM D560, "*Standard Test Methods for Freezing and Thawing Compacted Soil-Cement Mixtures*", and wet-dry testing was performed following ASTM D559, "*Standard Test Methods for Wetting and Drying Compacted Soil-Cement Mixtures*".

Compressive strength testing was performed for both cement-only and cement plus fly ash mixes. As above, cement-only mixes were tested to allow better evaluation of the strength gain properties associated with the addition of fly ash.

Modulus of Elasticity. Compressive strength testing also included measurement of stress-strain data on select samples in order to estimate the modulus of elasticity of the hardfill material. Testing and calculation of the modulus of elasticity was performed in accordance with ASTM C469. The modulus of elasticity was calculated as the ratio of the stress and strain corresponding with 40% of the ultimate compressive strength and a strain of 0.000050 (in/in).

Direct Shear. Direct shear testing (ASTM D5607) was performed to evaluate the hardfill material properties of cohesion and friction angle. Direct shear testing was performed for the parent (intact) material and an untreated cold joint to simulate the range of conditions that could be encountered during construction. Testing was performed at 28-days age for the cement-only mix and at 180-days age for the cement and fly ash mix.

Test Results: Phases 2 and 3

Phase 2 and 3 testing consisted of compressive strength and F-T/W-D testing at various ratios of cement plus fly ash. Cement and fly ash proportion index values were developed to aid in the selection of final mix design proportions, as well as selecting proportions for testing in subsequent mix design programs. The indices calculated included the strength gain per pound of cement and the strength gain per pound of fly ash at 90- and 180-days age. These indices provided a better understanding of the incremental strength gain for projecting strength gain at other cement contents.

The strength gain for cement-only mixes ranged from 1.7 to 7.5 psi/lb at 7-days age and from 2.2 to 9.3 psi/lb at 28-days for all of the sites. The strength gain index for cement-only mixes did not change significantly for 90- and 180-days age tests, as shown in Table 5. There was a noticeable difference in the strength gain index between the pit run site materials and the processed commercial soil aggregates. Pit run materials generally showed a strength gain index ranging between 1.7 and 4.6 psi/lb at 7-days age and 2.2 to 5.5 psi/lb at 28-days age, while the offsite commercial sources showed strength gain indices ranging from 4 to 7.2 psi/lb at 7-days age to 6 to 9.3 psi/lb at 28-days age.

In addition to the strength contribution from cement, the contribution from fly ash was also evaluated. Different combinations of cement plus fly ash were expected to provide similar compressive strength results., The compressive strength gain through 28 days for cement-only mixes is primarily due to cement content and is essentially complete at 28-days age. Fly ash generally begins to hydrate and contribute to strength gain between about 56- and 90-days age. Therefore, index values were calculated to evaluate the compressive strength gain due to fly ash at 90- and 180-days age. The resulting index values ranged between 3.2 and 5.1 psi/lb of cementitious material (cement plus fly ash) at 90-days age and between 2.7 and 8.1 psi/lb of cementitious material at 180-days age. The difference between on-site pit run material and offsite commercial soil aggregates was not as distinct as was observed for the cement-only mixes. The strength gains for the cement-only and cement plus fly ash mixes for each site are summarized in Table 5. Results of compressive strength testing for cement-only and cement plus fly ash mixes for each site are plotted with age on Figures 11 through 14. The Willow Island mix design program was optimized based on the testing results from the other three sites, therefore only cement plus fly ash mixes were tested.

Table 5. Compressive Strength Index Values: Cement and Cement plus Fly ash

Project	Hardfill Mix	Age (Days)				
		7	28	90	180	365
CEMENT ONLY Compressive Strength (psi/lb of cementitious material)						
Cannelton	5.6 -14%	2.6 – 4.6	3.9 – 5.5	4.2 – 6.0	4.5 – 6.5	N/A
Smithland	6.1 – 8%	1.7 -2.0	2.2 – 2.7	2.7 – 3.4	2.6 – 3.6	N/A
Meldahl	4.9 - 8%	6.9 – 7.2	7.0 – 8.9	7.5 – 10.1	7.6 – 9.9	N/A
Willow Island	5.7 - 6.4%	4.0 – 7.5	6.0 – 10.7	7.0 – 10.7	7.2 – 10.7	N/A
CEMENT + FLY ASH Compressive Strength (psi/lb of cementitious material)						
Cannelton	C 5.2 – 6.8% F 1.2 – 2.8%	N/A	N/A	3.2 – 3.9	5.4 – 7.8	7.1 – 12.0
Smithland	C 4.9 – 6.7% F 2.7 – 3.6%	N/A	N/A	1.7 – 3.2	2.7 – 5.3	N/A
Meldahl	C 4.9 - 7% F 2.2 – 3.7%	N/A	N/A	1.5 – 5.1	3.1 – 9.0	6.3 – 10.1
Willow Island	C 4.9 - 7% F 2.2 – 3.7%	N/A	N/A	1.9 – 4.5	4.6 – 8.1	N/A

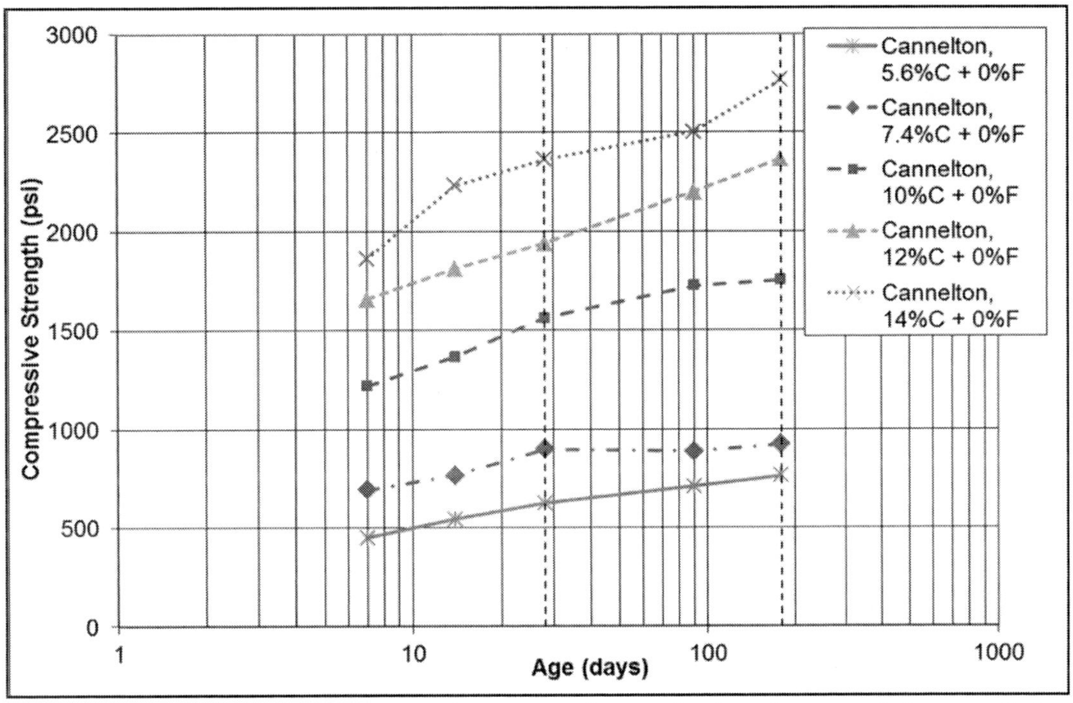

Figure 11a. Cannelton - Compressive Strength Results for a Range of Cement Only Proportions

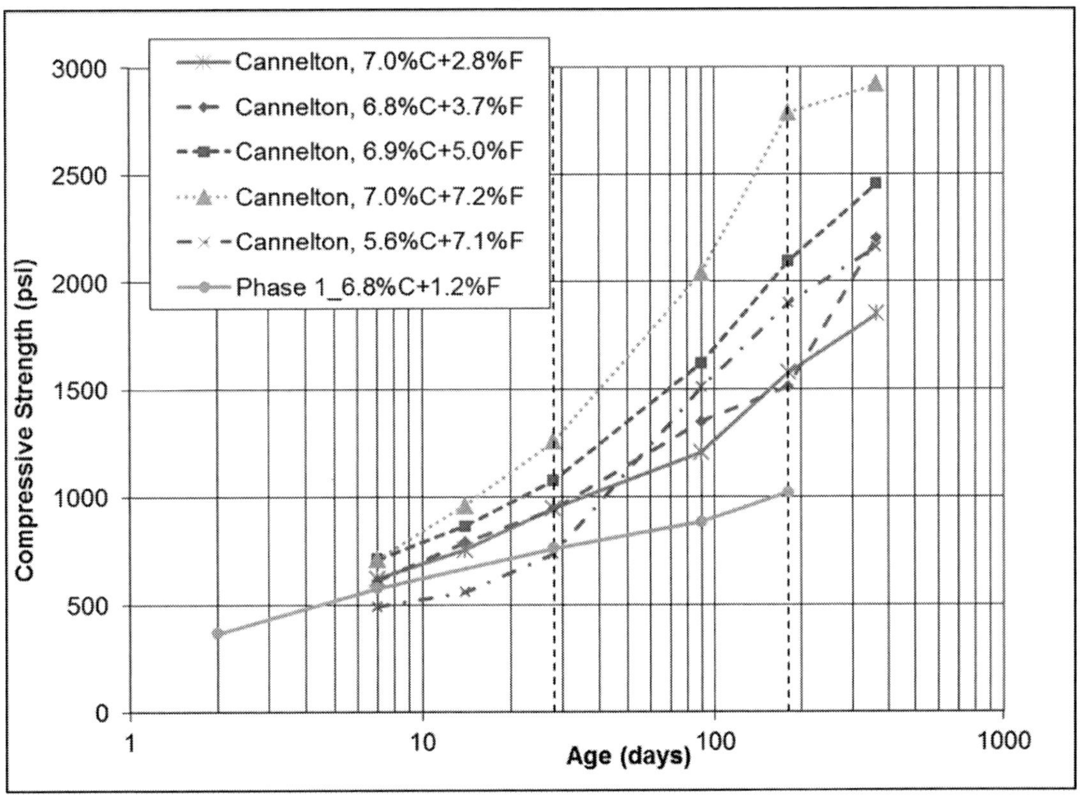

Figure 11b. Cannelton - Compressive Strength Results for a Range of Cement Plus Fly ash Proportions

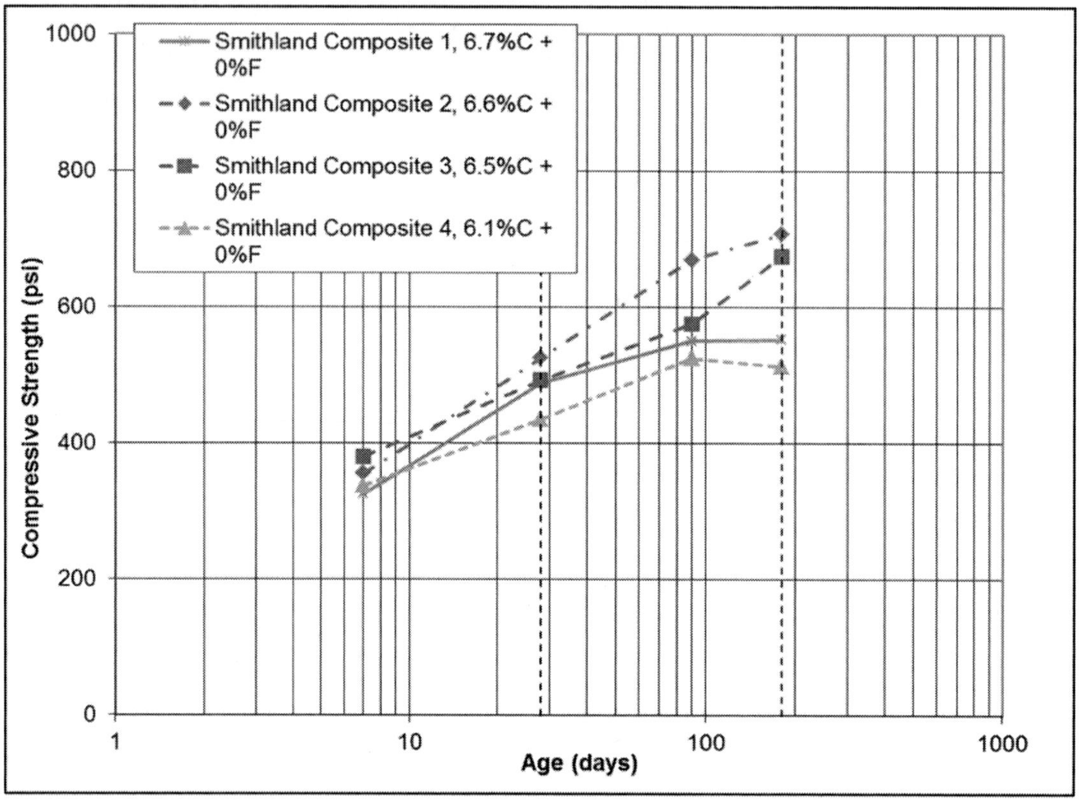

Figure 12a. Smithland - Compressive Strength Results for a Range of Cement Only Proportions

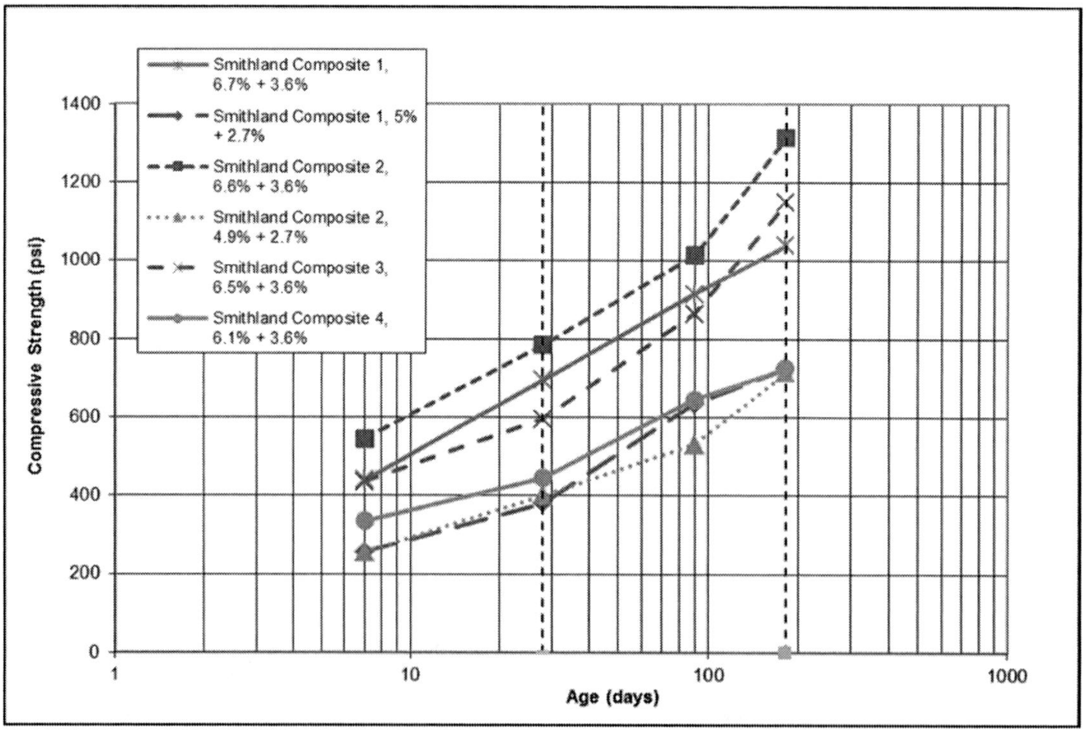

Figure 12b. Smithland - Compressive Strength Results for a Range of Cement Plus Fly ash Proportions

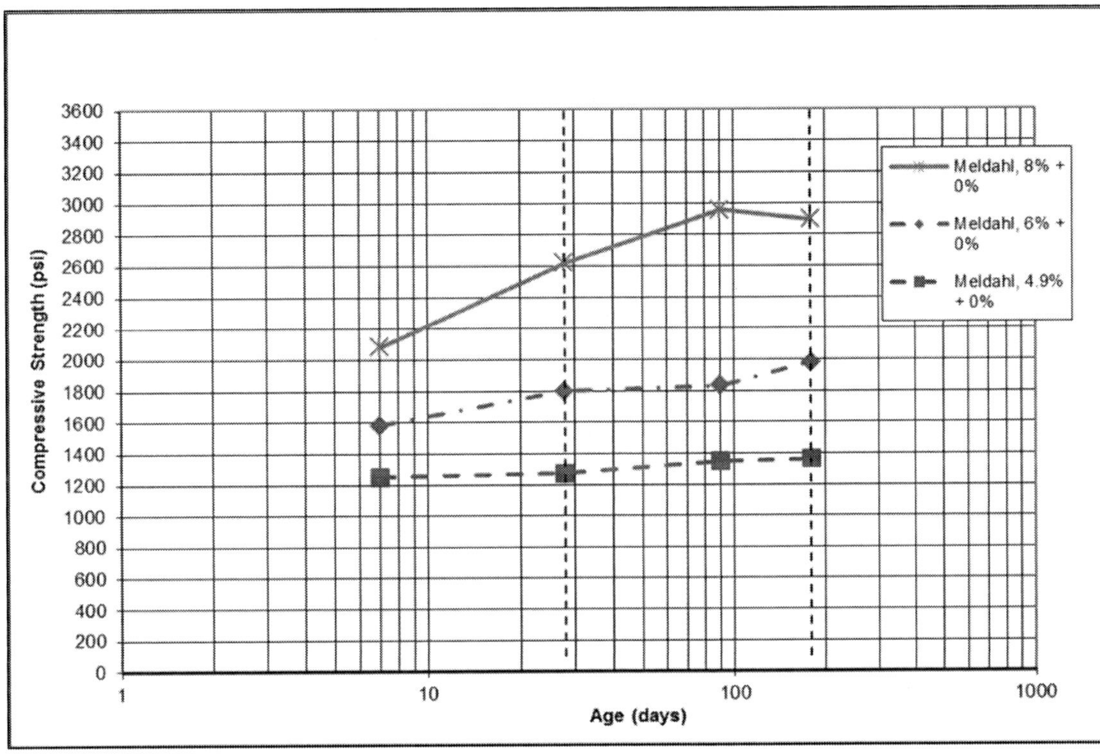

Figure 13a: Meldahl - Compressive Strength Results for a Range of Cement Only Proportions

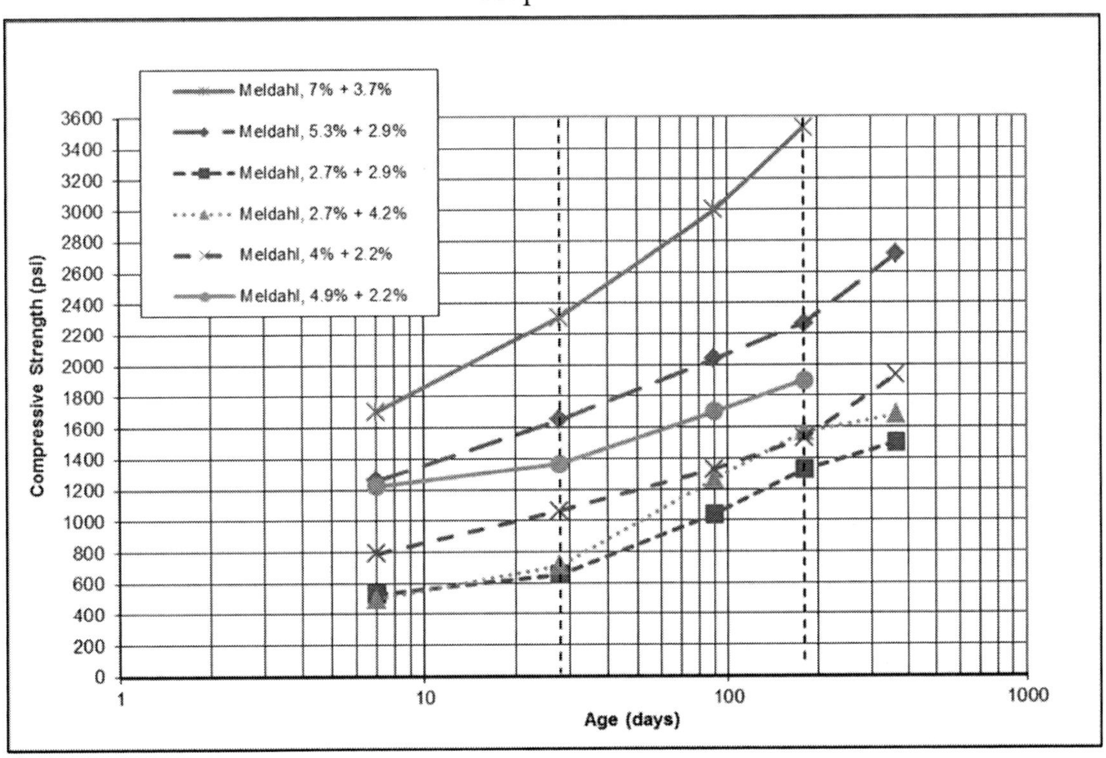

Figure 13b: Meldahl - Compressive Strength Results for a Range of Cement Plus Fly ash Proportions

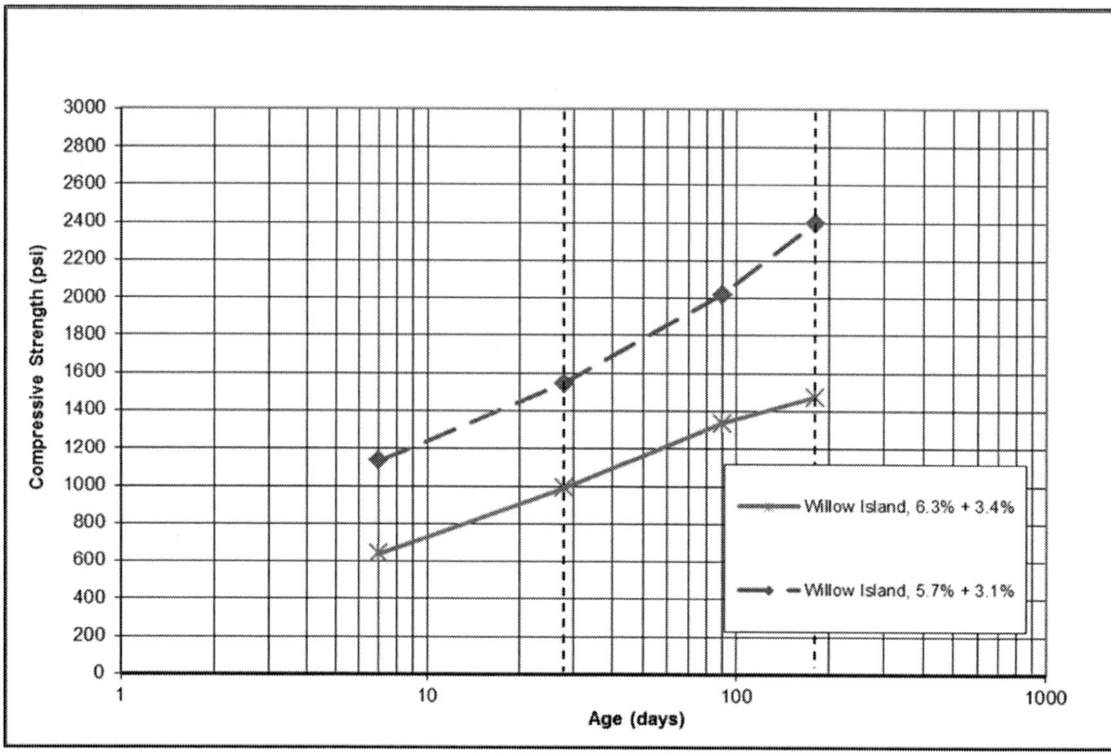

Figure 14. Willow Island - Compressive Strength Results for a Range of Cement Plus Fly ash Proportions

Durability. Evaluation of the durability of soil-cement mixes for slope protection generally consists of (F-T) and (W-D) testing for deterioration, and measurement of sample breakdown following multiple cycles of each test. The Portland Cement Association recommends a maximum soil-cement loss following F-T or W-D is 14% for a well hardened soil-cement mixture (PCA 2006). Criteria developed by the U.S. Bureau of Reclamation (USBR) is more stringent than the PCA guidelines; USBR recommends a maximum loss of 6% for the freeze-thaw test and a maximum of 6% for the wet-dry tests (USBR. 1990). The standard test procedure starts the freeze-thaw/wet-dry cycles at 7-days age. Cement plus fly ash mixes have not typically been used for soil-cement; therefore, there is a lack of performance history with this type of mix. In addition, samples containing low cement content but with fly ash that would provide higher compressive strength, and therefore higher durability, at later ages would not be expected to provide representative resistance to F-T and W-D deterioration testing samples at 7-days age. Therefore, the hardfill mix testing program included freeze-thaw and wet-dry testing at 28- , 90- and 180-days age to evaluate the performance of hardfill mixes with cement and fly ash.

Due to the lack of published performance history of cement plus fly ash mixes, the USBR criteria for samples at 90-days age were adopted for the mix design analysis. The mix design test results for durability with cement only and cement plus fly ash hardfill mixes are presented in Table 6.

Table 6. Cement and Fly ash Compressive Strength Index Values for Freeze-Thaw and Wet-Dry Tests

Project	Hardfill Mix	Durability Tests - Freeze-Thaw (%) ASTM D560				Durability Tests - Wet-Dry (%) ASTM D559			
		Age (Days)				Age (Days)			
		7	28	90	180	7	28	90	180
CEMENT ONLY Compressive Strength									
Cannelton	5.6 -14%	3.3	2.8	2.5	2.5	2.8	2	2	1.8
Meldahl	4.9 - 8 %	0.8 – 1.3	N/A	0.6 – 1.2	N/A	N/A	N/A	N/A	N/A
Willow Island	5.7 - 6.4%	1.0 – 1.1	0.9	0.8 – 0.9	N/A	0.5 – 0.7	0.8	0.1 – 0.4	N/A
CEMENT + FLY ASH Compressive Strength									
Cannelton	C 5.2 – 6.8% F 1.2 – 2.8%	4.4 – 7.4	4.6	5.3	3.5	3.5 – 7.6	3.1	1.8	2.8
Smithland	C 4.9 – 6.7% F 2.7 – 3.6%	N/A	1.0 – 3.1	0.9 – 2.1	N/A	N/A	0.3 – 1.5	0.1 – 1.0	N/A
Meldahl	C 4.9 - 7% F 2.2 – 3.7%	1.3 – 2.1	N/A	0.6 – 2.8	N/A	0.7 – 0.8	0	0.4	N/A
	C 2.7 - 4.4% F 2.2 – 2.9	17 - 100	N/A	0.6 – 4.6	2.5	0.8 – 2.6	N/A	0.3 – 0.9	N/A
Willow Island	C 4.9 - 7% F 2.2 – 3.7%	1.3 – 3.4	0.5	0.8 – 0.9	N/A	0.7 – 0.8	0	0.4	N/A

Accelerated Cure vs. Standard Cure. Accelerated cure testing is used to provide an early indication of the ultimate strength of concrete cylinders. Testing and correlation has been performed with conventional concrete cylinders to establish appropriate procedures (ASTM C684). However, testing of the hardfill samples using ASTM C684 did not correlate with standard cure samples, and provided significantly lower strengths. Therefore, the effectiveness of the use of an alternative method was evaluated. A trial test method using accelerated cure procedures used for RCC in previous projects was followed. Relative comparison of the standard cure and accelerated cure results varied between mixes with cement-only and mixes with cement and fly ash. Using the accelerated cure testing, the 21-day accelerated cure results were roughly equivalent to the 180- to 365-day standard cure results. Based on the test results, the 21-day accelerated cure results obtained using the alternate RCC method generally showed good correlation with the "ultimate" compressive strengths. The results of the accelerated cure versus standard cure testing for both the cement-only and cement plus fly ash samples are shown on Figures 15 and 16, respectively.

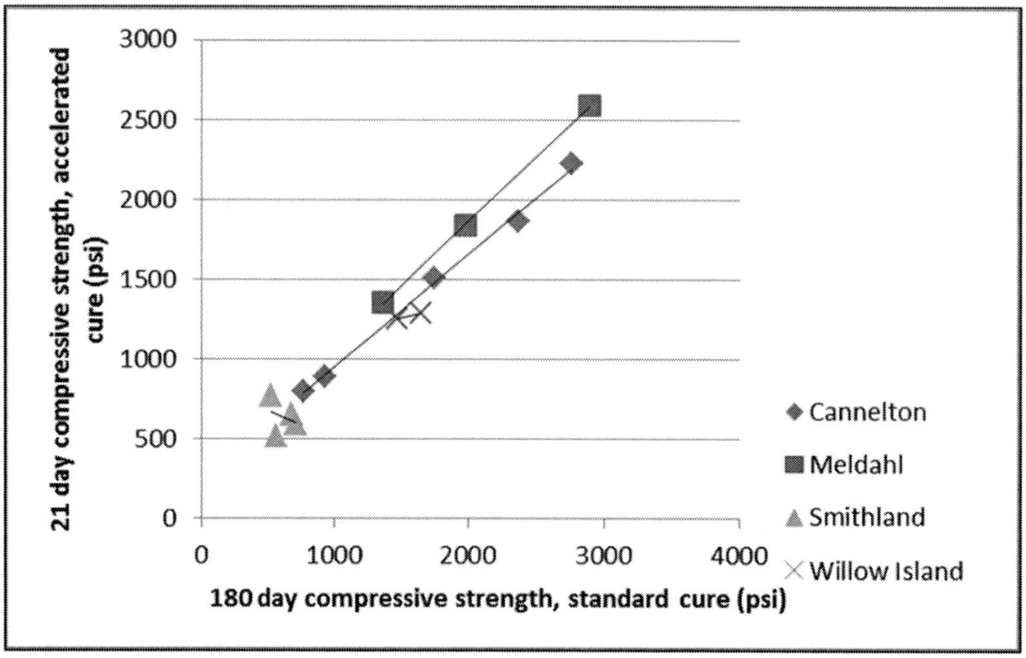

Figure 15. Accelerated 21-day Versus Standard 180-day Cure, Cement-Only Mixes

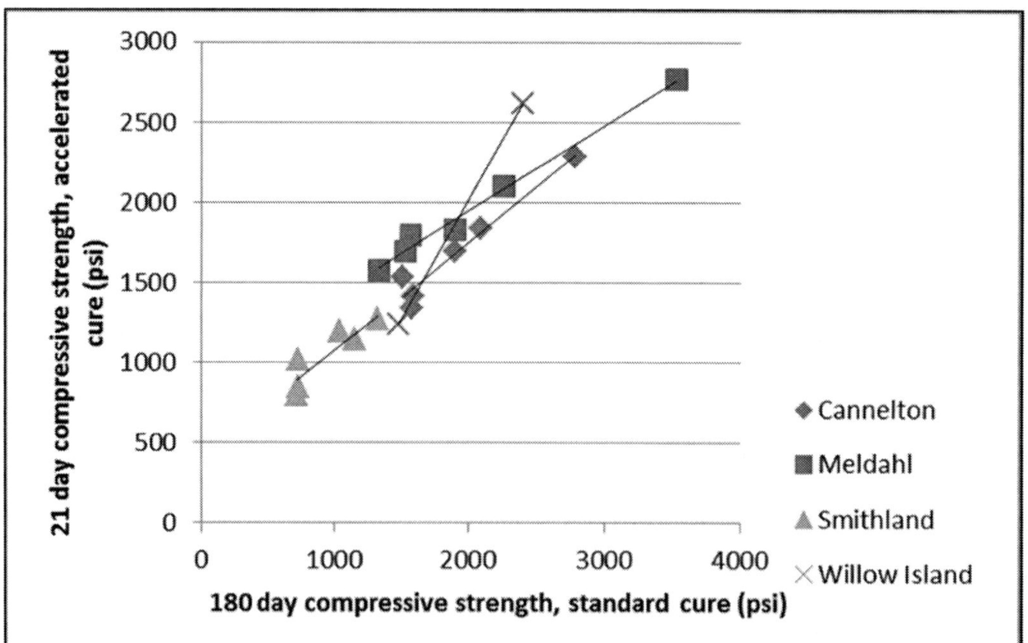

Figure 16. Accelerated 21-day Versus Standard 28-day Cure, Cement Plus Fly Ash Mixes

Test Results: Phase 4

The Phase 4 program was used to confirm the engineering properties of the final hardfill mixture proportions. The modulus of elasticity for the hardfill was estimated as 1.5 x 10^6 psi at 180-days age. The hardfill attained a parent material cohesion of 26.8 ksf and a friction angle of 60.4 degrees. The mix proportions selected met the varied material property requirements for the closure structure. The results of various engineering properties are summarized below.

Modulus of Elasticity. Modulus of elasticity test results are summarized in Table 7. The cement and fly ash mixture meets the design strength of 1,000 psi at 180-days and the modulus of elasticity closely matches the assumed design value of 1.0x10^6 psi.

Table 7. Modulus of Elasticity Results

Project/Soil Aggregate	Cement Content % (lbs/cy)	Fly ash Content % (lbs/cy)	Compressive Strength ASTM C469 (psi) Age (Days)		Modulus of Elasticity ASTM C469 (psi) Age (Days)	
			28	180	28	180
Cannelton/Average Composite	5.6% (171)	0% (0)	640	N/A	0.83x10^6	N/A
	5.6% (171)	3.7% (113)	N/A	1585	N/A	1.05x10^6
Meldahl/L801	4.0% (145)	2.2% (80)	1060	1525	N/A	1.70x10^6
	4.9% (180)	0% (0)	1270	1360	1.25x10^6	N/A
	4.9% (180)	2.2% (80)	1365	1900	1.50x10^6	1.96x10^6
Willow Island/Belpre	5.7% (195)	0% (0)	1285	1465	2.00x10^6	N/A
	5.7% (195)	0% (0)	1545	2400	2.85x10^6	3.10x10^6

Direct Shear. Direct shear results for the parent (intact) material and for un-treated cold joint are presented in Table 8. The peak shear stress properties exceed the target design criteria of cohesion (C) equal to 20 psi (2.9 ksf) and friction angle (phi) of 35 degrees for all mixes except the cold joint cement-only mix. Based on the shear strength results, acceptable performance of the hardfill is expected, however, preparation of the cold joints with a method such as cleaning and/or power brooming the surface was recommended for construction. Details concerning the preparation of joints during construction can be found in the paper focusing on the construction of the AMP closure structures (Truschke et al. 2015).

Table 8. Direct Shear Strength Results

Project Soil/Aggregate	Cement Content % (lb/cy)	Fly Ash Content % (lb/cy)	Testing Age Days	Parent Material Peak Shear Strength ASTM D5607 Phi (deg)	Parent Material Peak Shear Strength ASTM D5607 C (ksf)	Unbonded Lift Shear Strength ASTM D5607 Phi (deg)	Unbonded Lift Shear Strength ASTM D5607 C (ksf)
Cannelton (Avg. Composite)	5.6 (171)	N/A	28	55.2	6.5	31.1	0.8
	5.6 (171)	3.7 (113)	90	31.9	28.5	32.3	1.4
	5.6 (171)	3.7 (113)	180	60.4	26.8	31.2	1.1
	5.6 (171)	3.7 (113)	360	54.3	44.7	46.9	0.0
Smithland (Composite 1)	6.7 (195)	3.6 (105)	28	68.5	0.0	N/A	N/A
	6.7 (195)	3.6 (105)	90	33.9	13.7	N/A	N/A
	5.0 (146)	2.7 (79)	28	49.8	34.0	N/A	N/A
	5.0 (146)	2.7 (79)	90	55.4	7.8	N/A	N/A
Smithland (Composite 2)	6.6 (195)	3.6 (105)	28	64.2	0.0	N/A	N/A
	6.6 (195)	3.6 (105)	90	59.4	9.8	N/A	N/A
	6.6 (195)	3.6 (105)	180	60.4	18.6	N/A	N/A
	5.0 (146)	2.7 (79)	28	58.7	0.0	N/A	N/A
	5.0 (146)	2.7 (79)	90	48.1	6.4	N/A	N/A
	5.0 (146)	2.7 (79)	180	48.5	15.2	N/A	N/A
Meldahl (L801)	4.9 (180)	N/A	28	48.8	18.8	28.1	1.9
	4.9 (180)	2.2 (80)	90	64.9	20.7	53.2	0.0
	4.9 (180)	2.2 (80)	180	N/A	N/A	29.7	2.1
Willow Island (Hilltop Sand)	5.7 (195)	N/A	28	64.1	152	N/A	N/A
	5.7 (195)	3.1 (105)	90	N/A	N/A	29.6	1.9
	5.7 (195)	3.1 (105)	180	N/A	N/A	39.6	0.0

CONCLUSION

The use of hardfill with concrete facing for the powerhouse closure structures provides a water retaining structure that can be safely overtopped during periodic river flooding events. The hardfill mix design was proportioned and tested following procedures that are representative of the material type used for construction. Due to the relatively low strength for symmetrically placed hardfill, a wide range of aggregate types are suitable for a hardfill mixture. The laboratory testing program for the AMP Ohio project shows that a wide range of materials can be used to design a hardfill water retaining structure.

Engineering properties required for a hardfill water retaining dam can be obtained using sandy soil, both with cement only, and with partial replacement of cement with fly ash. There is limited performance data for sandy soil used to build tall water retaining structures, and for the use of fly ash as a partial replacement for cement in sandy soil mixes. To develop engineering properties including the contribution of fly ash, testing up to 1-years age were performed. Accelerated cure testing using warm water bath procedures developed for RCC projects, provided good correlation with "ultimate" compressive strength of both cement only and cement plus fly ash mixes. Durability testing (consisting of freeze-thaw and wet-dry tests) with cement plus fly ash showed similar results at 90-days age that were obtained at 7-days age for cement only samples.

REFERENCES

Antal, Brandi; Arnold, Terry; Konstantellos, Christ; Meier, Phil; and Truschke, Stefano. 2013. Hardfill on the Ohio River – Optimizing the Use of Construction Materials.

ASTM C469 / C469M-14, Standard Test Method for Static Modulus of Elasticity and Poisson's Ratio of Concrete in Compression, ASTM International, West Conshohocken, PA, 2014, www.astm.org

ASTM C684-99(2003), Standard Test Method for Making, Accelerated Curing, and Testing Concrete Compression Test Specimens (Withdrawn 2012), ASTM International, West Conshohocken, PA, 1999, www.astm.org

ASTM D422-63(2007)e2, Standard Test Method for Particle-Size Analysis of Soils, ASTM International, West Conshohocken, PA, 2007, www.astm.org

ASTM D559-03, Standard Test Methods for Wetting and Drying Compacted Soil-Cement Mixtures (Withdrawn 2012), ASTM International, West Conshohocken, PA, 2003, www.astm.org

ASTM D560-03, Standard Test Methods for Freezing and Thawing Compacted Soil-Cement Mixtures (Withdrawn 2012), ASTM International, West Conshohocken, PA, 2003, www.astm.org

ASTM D1633-00(2007), Standard Test Methods for Compressive Strength of Molded Soil-Cement Cylinders, ASTM International, West Conshohocken, PA, 2007, www.astm.org

ASTM D2487-11, Standard Practice for Classification of Soils for Engineering Purposes (Unified Soil Classification System), ASTM International, West Conshohocken, PA, 2011, www.astm.org

ASTM D2974-14, Standard Test Methods for Moisture, Ash, and Organic Matter of Peat and Other Organic Soils, ASTM International, West Conshohocken, PA, 2014, www.astm.org

ASTM D3282-09, Standard Practice for Classification of Soils and Soil-Aggregate Mixtures for Highway Construction Purposes, ASTM International, West Conshohocken, PA, 2009, www.astm.org

ASTM D4318-10e1, Standard Test Methods for Liquid Limit, Plastic Limit, and Plasticity Index of Soils, ASTM International, West Conshohocken, PA, 2010, www.astm.org

ASTM D4972-13, Standard Test Method for pH of Soils, ASTM International, West Conshohocken, PA, 2013, www.astm.org

ASTM D5607-08, Standard Test Method for Performing Laboratory Direct Shear Strength Tests of Rock Specimens Under Constant Normal Force, ASTM International, West Conshohocken, PA, 2008, www.astm.org

Holderbaum, Rodney; Kline, Robert; Rogers, Michael; Hartman, Randall; and Grant, Russell. 2002. Design of Roller-Compacted Concrete Materials for the Olivenhain Dam.

ICOLD. 2000. Bulletin 117: The Gravity Dam: A Dam for the Future.

Mirza, A.A., L. Zhao, J.M. Hynes, T.G. Andrews and P.E. Meier (2015). "From the Ground Up – Ground Improvement for Hydro Development on the Ohio River," USSD 2015 Annual Conference, Louisville, KY.

P.J. Mason and R.A.N. Hughes. 2008. "The Design and Construction of a Faced Symmetrical Hardfill Dam", MWH Ltd, UK and J.D. Molyneux, Black & Veatch Inc., USA – Hydropower & Dams, Issue Three.

Portland Cement Association. 2006. Soil-Cement Guide for Water Resources Applications.
Portland Cement Association. 1992. Soil-Cement Laboratory Handbook.

Truschke, Stefano; Arnold, Terry; Scott, Taylor; and Meier, Phil. 2015. "Built to Last – Hardfill Construction," USSD 2015 Annual Conference, Louisville, KY.

USBR. "Concrete Manual – A Water Resources Technical Publication". Eighth Edition – Revised. Pages 27-29, and 40.

USBR. 1990. Design Standard No. 13 – Chapter 17 – Soil-cement Slope Protection.

BUILT TO LAST — HARDFILL CONSTRUCTION

Stefano M. Truschke, P.E.[1]
Terrence (Terry) E. Arnold, P.E.[2]
Taylor M. Scott[3]
Phillip E. Meier[4]

ABSTRACT

American Municipal Power (AMP) is developing four new hydroelectric projects adjacent to existing U.S. Army Corps of Engineers (USACE) locks and dams on the Ohio River. Each include closure structures between the powerhouses and existing structures; which are largely constructed using hardfill dams. These trapezoidal gravity-type closure structures were designed to retain water, resist erosion, maintain differential water levels between headwater and tailwater, and to be overtopped and completely submerged during flood events. The hardfill used to construct these structures consisted of locally available soil aggregates, fly ash, and cement, mixed on-site and placed and compacted similar to roller compacted concrete. Cast-in-place reinforced concrete toe walls, precast concrete blocks for vertical and sloping hardfill faces, and anchor bars embedded in the hardfill were incorporated to facilitate construction of these hardfill structures. The hardfill mix designs were prepared based on an extensive laboratory testing program undertaken by MWH, the design engineer. The construction contractor for each project was required to perform full-scale trial hardfill placements to demonstrate and refine their procedures prior to constructing the permanent closure structures. MWH monitored hardfill construction, including foundation preparation, lift surface preparation, hardfill mixing, hardfill delivery, spreading and compaction, and quality control testing. Construction at the four AMP sites encountered a wide range of weather conditions at placement ranging from a low of 20 degrees Fahrenheit to a high of 108 degrees Fahrenheit. Field observations and field adjustments to mixes and placement techniques as well as lessons learned are described for the hardfill closure structures.

PROJECT DESCRIPTION

American Municipal Power (AMP) is developing four hydroelectric projects at existing US Army Corps of Engineers (USACE) locks and dams along the Ohio River. The USACE provided for future hydropower construction when the dams were originally constructed. The location of the four hydroelectric projects (Smithland, Cannelton, Meldahl, and Willow Island) under construction on the Ohio River is shown on Figure 1.

---

[1] Senior Civil Engineer; MWH Americas, Inc., 1560 Broadway St., Suite 1800, Denver, CO 80202; 303-291-2222; Stefano.M.Truschke@mwhglobal.com
[2] Senior Project Manager; MWH Americas, Inc., 1560 Broadway St., Suite 1800, Denver, CO 80202; 303-291-2222; Terrence.E.Arnold@mwhglobal.com
[3] Associate Civil Engineer; MWH Americas, Inc., 1560 Broadway St., Suite 1800, Denver, CO 80202; 303-291-2136; Taylor.Scott@mwhglobal.com
[4] Vice President – Hydroelectric Development and Operations; American Municipal Power, Inc.; 1111 Schrock Road, Suite 100, Columbus, OH 43229; 614.540.1111;pmeier@amppartners.org

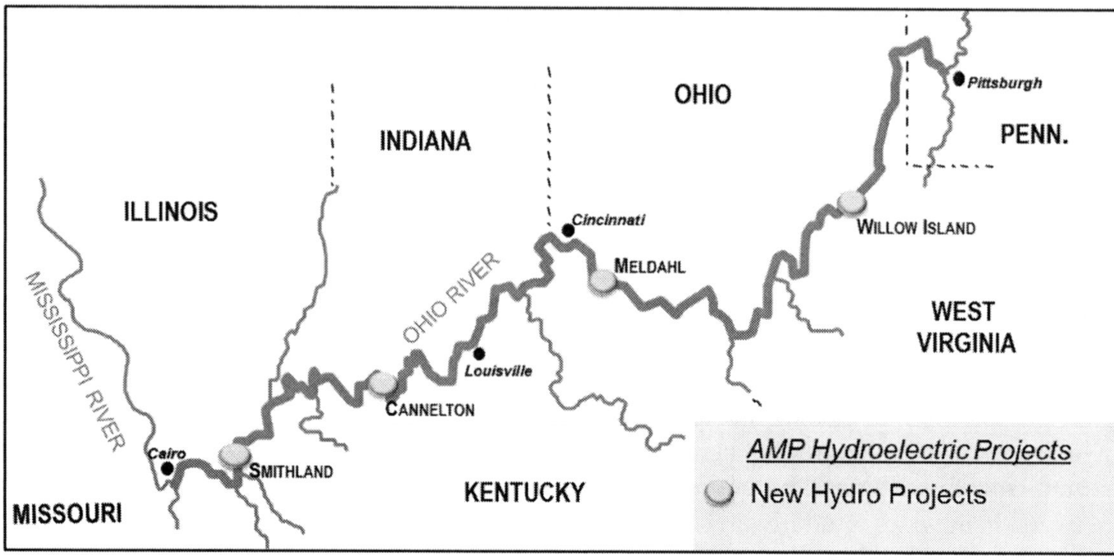

Figure 1. Project Location Map

The design of each project was tailored to address site-specific conditions. Foundation conditions varied wildly ranging from weak rock, including slickensided indurated clays and a thrust fault at Willow Island; soft clays required support during excavation at Meldahl; Cannelton and Smithland are located near the New Madrid Seismic Zone and had potentially liquefiable alluvial soils; and Smithland also had karstic limestone which further complicated the design. Multiple structure types were evaluated which would meet the project requirements, and are described further in the paper titled *Hardfill on the Ohio – Optimizing the Use of Construction Materials (Antal et al 2013).* Ultimately, six (6) out of eight (8) were constructed of hardfill (roller-compacted soil-cement) to maximize the use of on-site materials and minimize the importation of aggregate. The general layout for the Cannelton Project is shown on Figure 2, with the remaining sites having a similar layout. The structure type, height, and foundation type for each project, listed in order of construction, is summarized in Table 1.

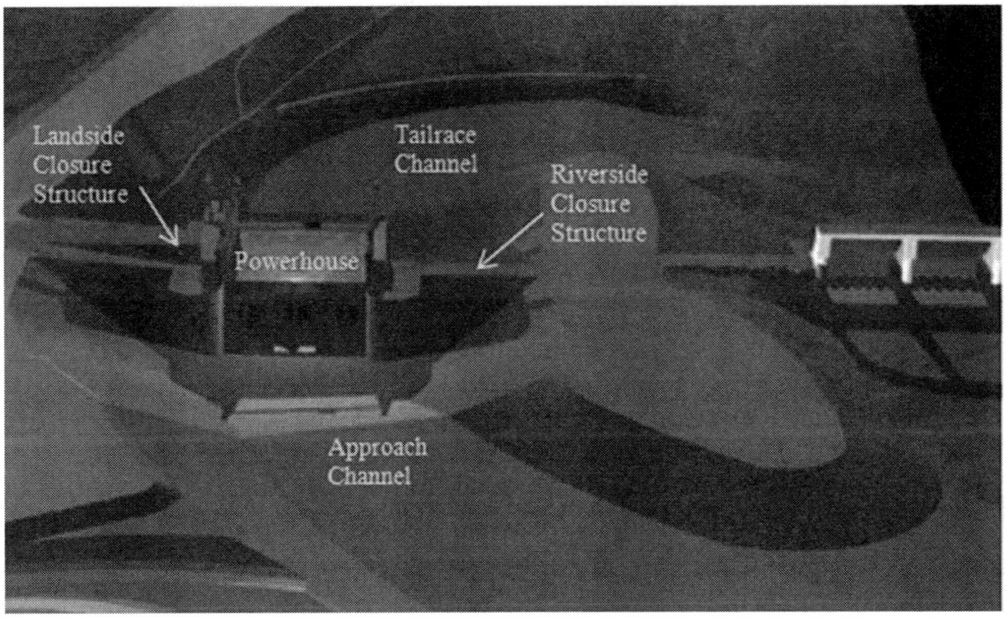

Figure 2. Typical Layout of Project Components (rendering of Cannelton shown)

Table 1. Closure Structure Properties

Project Site	LSCS			RSCS		
	Type	Height (ft)	Foundation Type	Type	Height (ft)	Foundation Type
Cannelton	Hardfill	116	Soil	Hardfill	99	Soil
Meldahl	Hardfill	90	Soil/Rock	Concrete	82	Rock
Willow Island	Hardfill	75	Soil/Rock	Concrete	58	Rock
Smithland	Hardfill	114	Soil	Hardfill	93	Soil

CLOSURE STRUCTURE DESIGN CRITERIA

Hardfill is a relatively new technology used in dam construction in the United States. ICOLD Bulletin 117 defines hardfill as a concrete construction product that typically consists of cemented sand or gravel aggregate in a trapezoidal gravity section and encased in conventional concrete and is also referred to as a Cemented Materials Dam. Due to the relatively low strength requirements for symmetrically placed hardfill, a wide range of aggregate types can be accommodated and lower strength foundation than rock foundations for gravity dams can also be used.

For the AMP hydroelectric projects, the closure structures fill the area between the existing dam and the powerhouse, and between the powerhouse and the excavated river abutment, to retain the headwater pool. As such, these structures were designed to be watertight and structurally capable of resisting lateral and applied loads. Based on previous project experience with similar materials, target hardfill properties were selected for the structural and stability analysis. An ultimate compressive strength of 1,000 psi at 180 days age was selected for design, with an initial compressive strength of 500 psi at 28 days age. In order to provide the water-tight barrier, the design included an upstream conventional concrete barrier. Since the water-tightness is primarily controlled by the upstream control barrier, individual lift surfaces were not required to be water-tight. Therefore, individual lifts required a relatively low cohesion. The typical plan and section of the closure structures are shown on Figure 3 and Figure 4, respectively.

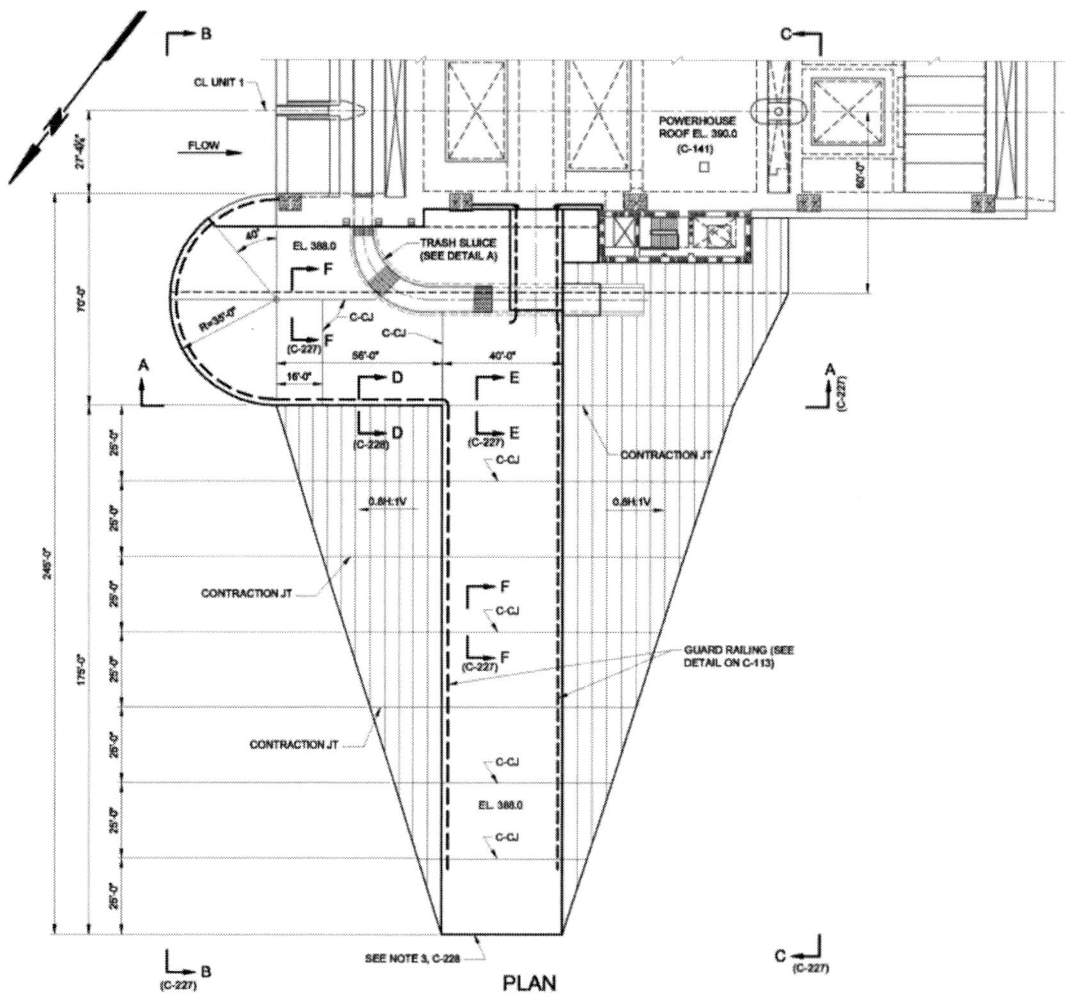

Figure 3. Typical Closure Structure – Plan View

Hardfill Construction

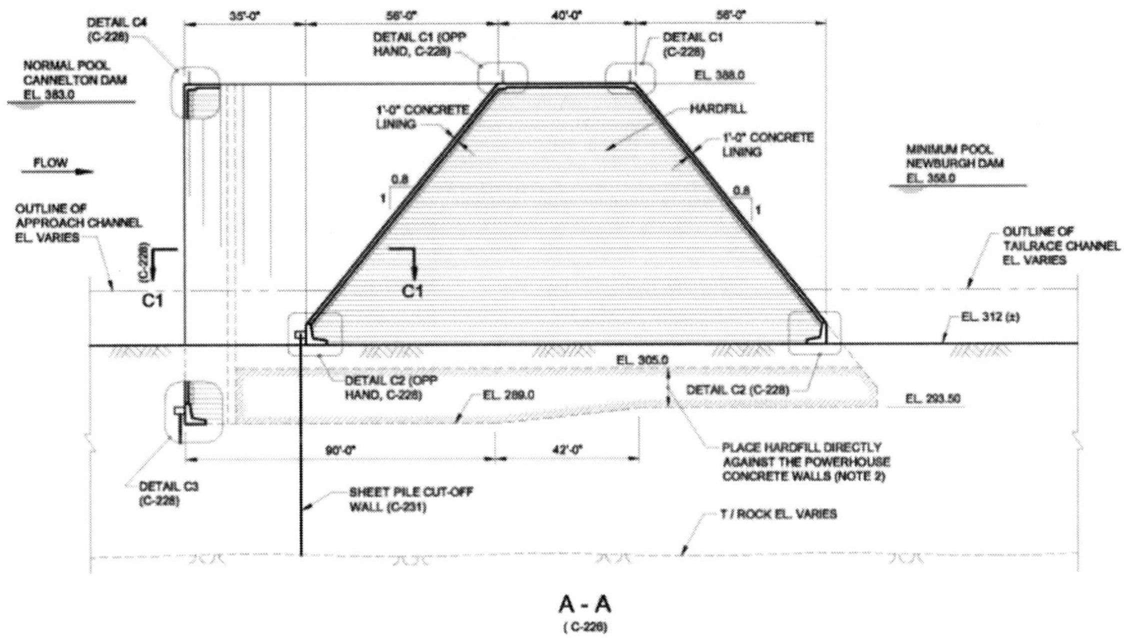

Figure 4. Typical Closure Structure – Cross Section

HARDFILL MIX DESIGN PROGRAM – TEST RESULTS

In order to develop the required mix proportions to use for construction, a laboratory testing program was developed and implemented during the cofferdam and powerhouse excavation phases. Generally, potentially suitable materials, such as alluvial soils with low plasticity, were selectively stockpiled for use as hardfill aggregate during excavation of the powerhouse. Representative samples were obtained from the on-site stockpiles, tested for evaluation as and analyzed for potential hardfill materials.

Several testing methods were used to evaluate the cement and fly ash mixtures. Testing methods included compressive strength testing, freeze-thaw, wet-dry cycles, and direct shear. The testing results were analyzed with respect to the range of mix proportions considered. More detail on the mix design development and analysis is contained in the paper entitled *Optimized – Hardfill Mix Design and Testing (Arnold et al, 2015).*

Mix proportions for testing at similar cementitious contents were performed as a baseline for comparison between project materials. The "baseline" testing primarily consisted of testing at 195 pcy cement and 105 pcy fly ash, the results of which are summarized in Table 2. Mix proportions were also tested at a generally lower proportion of 146 pcy cement and 79 pcy fly ash and are summarized in Table 3 for Meldahl, Willow Island, and Smithland. The lowest proportion tested for the Cannelton site of 213 pcy cement and 80 pcy fly ash is provided for comparison of mix designs and initial mix properties. For the trial placements and the first 30 days of hardfill placement, a strength overdesign factor of 25% was included to account for potential differences between laboratory and in-field placement and for uncertainty due to lack of published history of fly ash usage for sandy hardfill mixtures. The baseline mix design was ultimately used at the beginning of production, with the potential to reduce mix proportions at a later date depending on strength trends during actual placement.

Table 2. Preliminary Mix Design Results at 195 pcy Cement plus 105 pcy Fly Ash (about 8% total cementitious content by dry weight of aggregate) for Cannelton, Meldahl, Willow Island and Smithland Projects

Project Name	Standard Cure Compressive Strength (psi) ASTM D1633					Accelerated Cure Compressive Strength (psi)		Water Factor
	Age (Days)					Age (Days)		
	7	14	28	90	180	14	21	
Cannelton	610	790	945	1350	1510	1415	1535	1.66
Meldahl	1260	1440	1650	2040	2060	1930	2100	0.94
Willow Island	1550	1775	1985	2420	2440	1880	2120	0.92
Smithland	440	N/T	695	913	1040	N/T	1195	1.15

N/T = Not Tested

Table 3. Preliminary Mix Design Results at 146 pcy Cement plus 79 pcy Fly Ash for, Meldahl, Willow Island and Smithland Projects and at 213 pcy Cement plus 80 pcy Fly Ash for the Cannelton Project

Project Name	Standard Cure Compressive Strength (psi) ASTM D1633					Accelerated Cure Compressive Strength (psi)		Water Factor
	Age (Days)					Age (Days)		
	7	14	28	90	180	14	21	
Cannelton (213C+80F)	620	755	950	1205	1575	1145	1335	1.14
Meldahl (146C+79F)	790	900	1060	1330	1525	1400	1690	1.30
Willow Island (146C+79F)	1270	1340	1480	1640	1900	1600	1610	1.23
Smithland (146C+79F)	255	N/T	380	633	725	N/T	845	1.54

N/T = Not Tested

HARDFILL TRIAL PLACEMENT

An important consideration in specialized construction procedures for roller compacted concrete, soil cement, and hardfill dams is to perform a trial placement. A trial placement is an invaluable tool to evaluate the workability of the laboratory mix design, the contractor's means and methods, the ability to uniformly mix hardfill, evaluate the number of passes required to achieve the specified degree of compaction, and other construction procedures. The trial placement allows the contractor to demonstrate and experiment with different techniques of placement, without impacting the quality of the final product. With an emerging technology such as hardfill, the trial placement is an even more important component of successful completion of the project. At each of the AMP sites, a different contractor was performing the hardfill construction, and each contactor had widely varying prior experience, equipment, and ideas for means and methods in the placement of hardfill. The trial placement at each site yielded invaluable knowledge for both the contractor and the engineer. With the exception of the anticipated delayed strength gain of the fly ash hardfill mixture, little impact to the overall construction means and methods were observed as opposed to a cement only hardfill mixture.

The trial placement specified for each project required a demonstration of a minimum of 6 lifts of hardfill to evaluate the complete process. The components of the project demonstrated during the trial placement included:

- hardfill batching and mixing;
- delivery, spreading, and compaction;
- placement of the outside facing precast concrete blocks;
- lift surface preparation and curing;
- interface with the alluvial abutment; and
- quality control activities.

At the completion of the hardfill trial placement, hardfill cores were obtained to evaluate the properties of the placed hardfill and a sawcut through the trial placement was completed for visual evaluation of the placement.

Coring and saw cutting were performed following a minimum of 14 days of field curing. Both 4-inch and 6-inch diameter cores extended through multiple lifts in order to evaluate the lift interface and the variability in bonding between lift surfaces. Laboratory testing provided an opportunity to confirm field test data when tests could be conducted in similar location.

Laboratory density testing of cores was performed generally at locations that corresponded with field density tests locations of the nuclear gauge and/or sand cone test procedures. Density of the cores was measured in general conformance with ASTM C642. The field testing program identified erroneous readings by the nuclear gauge at one site based on a comparison to a sand cone test and the laboratory measured core density. Evaluation of the difference between the nuclear density gauge included performing sand cone tests adjacent to locations of nuclear gauge readings were taken. Cores were later obtained at the same location. The sand cone and core density results compared favorably; and, the nuclear gauge was replaced with a new gauge.

The sawcut through each trial placement section was used to inspect the hardfill lifts on a more continuous basis. The sawcut at each site confirmed that a well-compacted mass was achieved with good uniformity. In general, the lifts were well compacted, were uniform in appearance throughout each lift, and the lift interface was difficult to identify in most cases. Field observations indicated that a moderate amount of bonding was present. The sawcuts for each site are shown on Figure 5 to Figure 8, with apparent lift lines indicated by a dashed line.

Figure 5. Cannelton Site Trial Placement Saw Cut

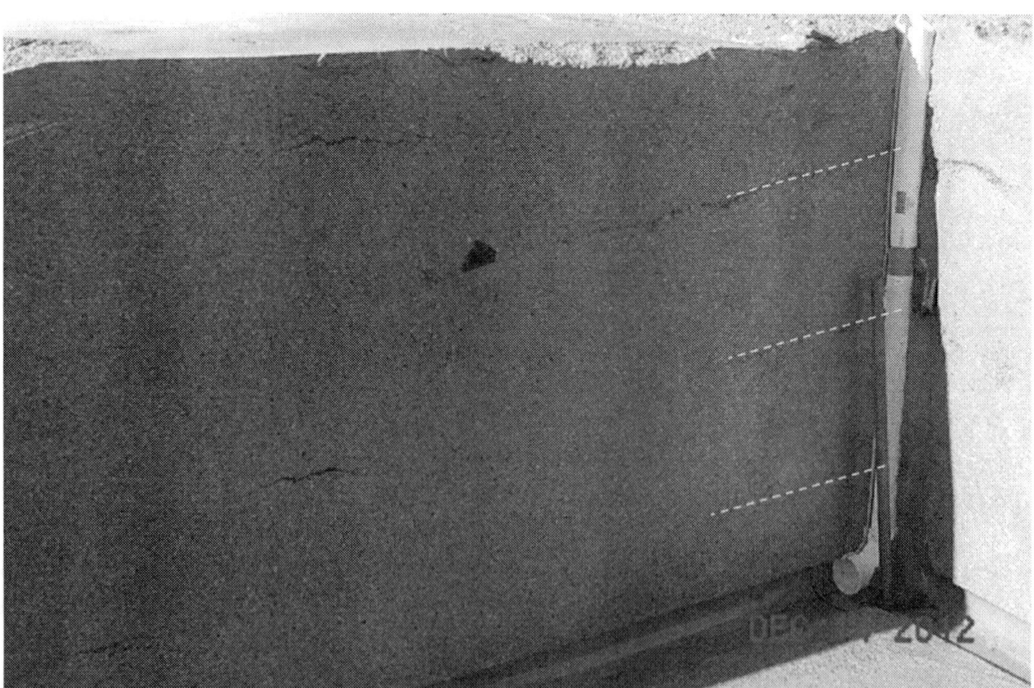

Figure 6. Meldahl Site Trial Placement Saw Cut

Figure 7. Willow Island Site Trial Placement Saw Cut

Figure 8. Smithland Site Trial Placement Saw Cut

CONSTRUCTION METHODS

Hardfill Mixing

One of the most important components of a successful hardfill project is the mixing equipment and providing a uniformly mixed hardfill at the desired proportions. There are several differences between conventional concrete and hardfill that require more detail in the selection of an appropriate mixing plant. As hardfill is a "zero slump" product, there is a relatively low moisture content in the final product and the quantity of "add water" is often very low. Therefore, it is important that the mixer has a vigorous mixing action, relative to what is required for a conventional concrete mixture. A compulsory pugmill type mixer has a favorable performance history for low cementitious content, low water content mixes. Additionally, hardfill needs to be spread in a continuous and rapid manner, which is best accomplished with an on-site mixing plant. An on-site, twin shaft compulsory style mixer was specified for use for the hardfill for the AMP projects. A twin shaft compulsory mixer operates with two shafts with paddles rotating in opposing directions to thoroughly mix the hardfill raw materials. At two of the sites, a continuous mix compulsory mixer was used, and at a third, a batch style compulsory mixer was used.

An alternative mixer style, a horizontal spiral blade batch mixer, was proposed at one site. The horizontal spiral blade mixer is a batch type mixer and operates as a single shaft within a cylinder with angled paddles rotating along the circumference of the mixer cylinder.

Limited previous published experience/historical records of mixing hardfill, soil cement or RCC mixes for water resources applications with a horizontal spiral blade mixer were available. In order to evaluate the effectiveness of the horizontal spiral bladed mixer and the required mixing times to achieve a uniform mixture; a mixer uniformity test was performed.

Mixer Uniformity Test. A Mixer Uniformity Test can be useful for evaluating the effectiveness of a mixer and selecting an appropriate mixing time. The test involves batching multiple batches of hardfill and performing testing on each batch. The uniformity tests are repeated at a range of mixing times to select the minimum required mixing time. The mixer uniformity testing procedure was adapted from U.S. Army Corps of Engineers *CRD 55 – Test Method for Within Batch Uniformity for Freshly Batched Concrete*. Uniformity testing was performed to evaluate the alternative proposed horizontal spiral bladed mixer.

Testing performed for the mixer uniformity test included compressive strength, cement content (as measured by the Heat of Neutralization method), moisture content, fines content (passing No. 325 sieve), and visual observations of uniformity. Individual batches were separated into 3 segments and testing performed on each. Compressive strength testing was performed at 7 days age following standard cure procedures and at 14 days age following accelerated cure procedures. The method for accelerated cure testing is a method used among some hardfill and RCC practitioners consisting of 7 days of standard cure followed by 7 days of accelerated cure in a 160 degree Fahrenheit water bath. The methodology for the accelerated cure testing is described further in *Arnold et al. 2015*. The fines content at the No. 325 sieve is typically used to evaluate the cement content of the mixture. The cement content was measured using the Heat of Neutralization Method – ASTM D5982. The results of a Mixer Uniformity test are presented in Table 4. Mixer Uniformity Test Results.

Table 4. Mixer Uniformity Test Results

TEST	WITHIN BATCH 1 AVERAGE	WITHIN BATCH 1 MAX DIFF	WITHIN BATCH 2 AVERAGE	WITHIN BATCH 2 MAX DIFF	BATCH TO BATCH MAX DIFF	MAX ALLOWED DIFF	PASS/ FAIL
MAXIMUM DRY DENSITY (PCF)	123.7	0.40%	124	0.60%	0.30%	2%	PASS
WATER CONTENT OF MIX (%)	6.90%	1.40%	6.60%	6.10%	4.90%	15%	PASS
HARDFILL MIXTURE FINES CONTENT – SIEVE NO. 325 (%)	18.60%	1.10%	18.90%	2.60%	1.80%	10%	PASS
CEMENT CONTENT (%)	7.20%	6.90%	8.10%	1.20%	11.70%	10%	**FAIL (Batch to Batch)**
COMPRESSIVE STRENGTH (7-DAY)	854	5.20%	894	2.30%	4.60%	15% (Within Batch) 25% (Batch-to Batch)	PASS
ACCELERATED CURE COMPRESSIVE STRENGTH (14-DAY)	1571	9.50%	1730	8.20%	9.60%	25%	PASS
VISUAL OBSERVATIONS	Appeared thoroughly mixed, uniform in color, with consistent moisture content based on grab testing. No obvious dry spots or clods.		Mix generally appeared thoroughly mixed, uniform in color, with consistent moisture content based on grab testing. Infrequent cementitious clods		Cementitious clods were observed in the second batch. Overall, appeared uniformly mixed.		

The mixer uniformity test indicated that a relatively uniform mixture would be provided by the horizontal spiral blade mixer and was used for the production of hardfill for the site. The mixer generally performed acceptably, providing a relatively uniform mixture. Some areas of concern noted in the mixing included that the limestone aggregate tended to agglomerate (bind/aggregate particles together) during stockpiling and were not completely broken up by the mixer. Observations of various mixers and aggregates indicated that a compulsory pugmill would have deaggregated the agglomerated particles, at least partially in the mixing, and would have provided a more uniform hardfill mixture.

<u>Batching Accuracy.</u> In a typical hardfill mixture, the amount of cementitious materials and water required is typically much lower than in a conventional concrete mixture. This presents unique challenges when using a concrete plant that has not been configured

specifically to address these potential issues. The typical mixture for the AMP sites included approximately 6% cement and 3% fly ash by dry weight of aggregate, with a typical optimum moisture content of about 8%. The resulting maximum dry density of aggregate of 131 pounds per cubic yard (pcy) and an optimum moisture content of approximately 8%, required batching 210 pcy of cement and 105 pcy of fly ash. In a typical batch size of 5 cy, requires the weighing of 1,050 pounds cement, 525 pounds fly ash. More importantly, the total water content of hardfill requires only 283 pounds of water. Depending on the moisture content of the stockpiled aggregate, typically in the range of 3-5%, the required add water can be extremely low, on the order of 100 pounds or 11 gallons per cubic yard. In order to increase the accuracy of the add water, it was necessary to outfit the batch plant with a smaller diameter water meter to facilitate accurate measurement of these add water quantities.

Electronic Batch Records. With the advance of computer batch plant software, it has become easier to monitor hardfill mixture properties. With over 1,000 cubic yards of hardfill placed in a typical production shift during full scale placement, there were numerous "batch" proportions or frequent "timed" readings of mix proportions to evaluate the batch records. Therefore, batch plant records were exported to a spreadsheet for further evaluation, to check the batch proportions and calculate values that may not be provided on the batch tickets. The spreadsheet was set up to automatically highlight cells that exceed established tolerances. In addition, statistics for the daily batches, such as the average batch variance and total amount of material used were compiled to verify the daily material usage at the plant and project future usage of materials. Additionally, the water content of the batched hardfill was compared to in-place hardfill moistures contents and compressive strength results.

Lift Surface Preparation

Since the lift surfaces were not required to be water-tight or act as a homogenous concrete mass, a relatively low cohesion value was required for the hardfill lift surface. In order to obtain some cohesion between lifts, each lift was required to be cleaned of loose debris and unbonded hardfill prior to placement of subsequent lifts.

Several methods were employed to enhance lift bonding. When placing a subsequent lift within approximately 4 to 6 hours, high velocity, wheeled directional blowers, capable of propelling air at a velocity of approximately 200 miles per hour, were utilized to remove loose material, along with hand cleaning with shovels to remove small areas of unbonded hardfill, as shown in Figure 9. The wheeled directional blower provided a low angle cleaning action that did not damage/loosen compacted hardfill that had not yet achieved initial set of the cement. If hardfill placement was suspended for a significant amount of time, higher energy cleaning was required to remove unbonded material. At many of the sites, a hydraulic power broom with steel and polypropylene bristles was employed to successfully remove loose material from the lift. After precipitation, additional cleaning was required to remove laitance and accumulated debris.

Figure 9. Lift Surface Preparation with Directional Blower and Hand Tools

Delivery and Spreading

Hardfill is a cementitious product and the chemical reaction which ultimately results in the bonding of the aggregate with the cementitious materials begins at the introduction of water to the cement. Thus, the time from mixing to compaction is important for quality hardfill. Cement begins to react when water (from the aggregate or add water) is available and mixing, transport, spreading and compaction must be completed in a timely matter. Typically, it is important to complete final compaction of the hardfill including lane shoulders within 90 minutes of introduction of cement to the mixture. Stiffening of the mix from evaporation and/or initial set is affected by weather, which in turn affects the time to complete compaction (less time in hot weather, and more time in cold weather).Therefore, the rapid and continuous delivery and spreading of hardfill is critical.

Different hardfill delivery methods were used by the different contractors at the AMP sites ranging between truck delivery to conveyors. A transfer unit was used to transport hardfill to the placing surface when trucks hauled material from the mixing plant to the placement area. Ultimately, the transfer points of hardfill were noted to be a critical part of the delivery process. The selection of equipment that can efficiently transport hardfill is essential to enhance the quality of the hardfill placement.

Transfer Conveyor Delivery. At multiple sites, material was delivered to the placement site by a haul truck and transferred to a haul truck that remained on the hardfill surface. The transfer of material was accomplished using a material transfer placer, similar to equipment utilized for pavement construction. The sizing of the material transfer placer was critical to the efficient transfer of hardfill. Transfer conveyors ranged from 18" to 36" wide belts with varying degrees of performance. Ultimately, the 18" wide conveyor belt material transfer placers were replaced with a 36" wide conveyor, which greatly increased the production rate. At one site, the material transfer placer was replaced with direct truck delivery of the hardfill to the lift surface to facilitate an increase in the production rate, as shown on Figure 10. However, this delivery method had different challenges to providing a quality hardfill product and lift surface.

Haul Truck Delivery. The delivery of hardfill directly to the lift surface by the haul truck was utilized for a portion of the hardfill delivery at some sites. Direct delivery of hardfill by truck to the lift surface required additional activities to control contamination including placement of a coarse gravel mud pad at the access ramp and the cleaning of haul truck tires prior to entry to the lift surface. Additionally, the access point for the haul truck needed to be raised continuously for each subsequent lift.

Spreading. Spreading of the hardfill was performed with dozers with elevation controls. At most of the sites, level control was maintained using laser mounted receivers. However, GPS mounted elevation control installed on the dozer blade was also successfully implemented at one of the sites. Typically a 10-inch loose lift was placed and compacted to achieve a 9-inch compacted lift. A tolerance of individual lifts of +/- 1-inch was achieved.

Figure 10. Delivery and Transfer of Hardfill to Placement Surface

Contraction Joint Installation

One major feature of a hardfill gravity structure is the installation of contraction joints perpendicular to the axis of the structure to provide controlled locations for shrinkage and cracking of the hardfill. Contraction joint design consisted of installation of plastic sheeting within each lift of fresh hardfill. Installation methods used at the various AMP sites included:

- The hardfill was initially spread in a loose lift, the plastic sheet was inserted into the loose hardfill using a metal template pressed by the dozer blade. The metal template was removed and an initial pass by the roller was made in the static mode to "lock" the sheet in place, followed by standard compaction. This method typically required two to three laborers.
- The contraction joint was installed by digging a trench in the loosely spread hardfill. The trench was dug using narrow bladed shovels by hand. The plastic sheet was then placed in the trench and the trench backfilled with loose hardfill. The joint was also compacted first with the static mode of the vibratory roller, following by the typical compaction procedure. This method typically required a team of three to six laborers.
- Another site employed the use of a specially modified miniature excavator with a vibratory plate attachment to drive the plastic sheeting in place in the loosely placed hardfill. The joint was then compacted by the standard roller in static mode followed by normal passes in vibratory mode. This method typically required two laborers and equipment operator.

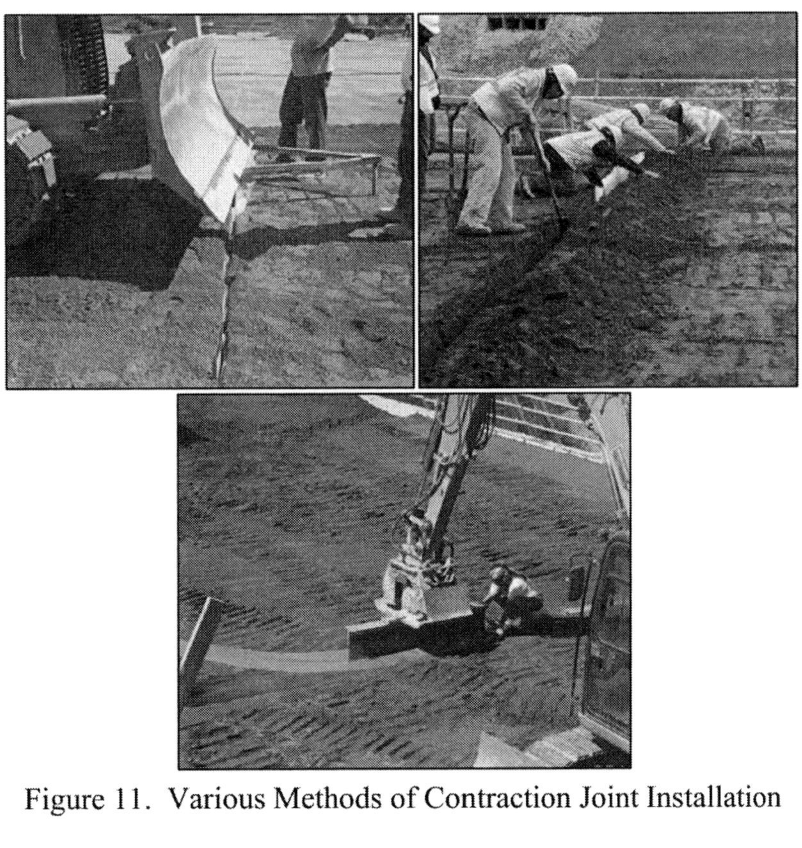

Figure 11. Various Methods of Contraction Joint Installation

Precast Concrete Blocks and Cast-in-Place Concrete Facing

The upstream and downstream faces of the closure structures were formed with an anchored precast concrete block system and the cast-in-place reinforced concrete facing was anchored to the precast blocks. The precast blocks were formed with a slight batter on the interior (hardfill) side, and either a vertical or inclined face on the exterior side. The anchoring system was designed to provide structural stability of the closure structure faces during placement of the hardfill, and also retain the concrete facing against potential unbalanced hydrostatic head during drawdown conditions after flood events. The anchors consisted of reinforcing steel bars of differing length with a square plate anchored within the hardfill to provide pullout resistance.

The precast concrete blocks were designed to stack one block directly on top of another. Individual precast blocks were outfitted with removable hooks to facilitate lifting and placement. A bed of non-shrink grout was placed under the lip of the hardfill block. The lip of the precast block was included to provide resistance of the precast block to the freshly placed hardfill. A photo of the anchored precast concrete block system being constructed is shown on Figure 12.

Figure 12. Installation of Vertical Face, Anchored Precast Concrete Blocks

One challenge experienced at the AMP sites during the placement of the precast blocks was the smoothness of the top and bottom surfaces of the blocks. The precast blocks were intended to be placed directly on top of one another with minimal gap between blocks. However, the variability in the "smoothness" and achieving level placement of the hundreds of blocks required grinding of blocks or placing shims to provide a level block surface. It was important that the placement of the precast blocks advance at the same rate as the hardfill placement. The custom inclined and curved precast concrete blocks provided a vertical surface to place hardfill against which facilitated adequate compaction of hardfill at the edges.

The concrete facing was typically placed at a later date following the precast block and hardfill placement. The use of the precast blocks as a forming system allowed for the concrete facing to be placed independently of the hardfill operation. This was advantageous at the AMP hydroelectric projects as the powerhouse consisted of multiple massive conventional concrete pours, which allowed other activities to occur simultaneously with construction of the closure structures.

Compaction

Selection of equipment to achieve the desired amount of compaction is critical to the successful completion of hardfill. The target density for the projects was 98% of the maximum dry density as determined by the standard proctor test (ASTM D558) of the hardfill mixture including cementitious materials. In order to allow for occasional/infrequent outlying test results, some variance in the compaction test results were allowed. The specifications required that the compaction be maintained at a running average of the previous 10 test results of at least 98%, with no individual test below 95% of maximum dry density.

A smooth steel single drum vibratory compactor was utilized for primary compaction at each of the sites. The overall gross weight of the compactor exceeded 21,000 pounds and the steel drum provided a dynamic force of a minimum of 120 pounds per inch of drum width. In order to provide a smoother hardfill surface, the vibratory passes were preceded by a single pass in static mode to "lock" material into place. With the moisture content of the hardfill material ranging between 0.5%-1.0% below optimum moisture content by dry weight of hardfill), the desired target density was achieved with a minimum of 8 passes of the vibratory roller with a back-and-forth roundtrip by the compactor defined as 2 passes.

Compactors with variable amplitude and frequency settings allow for adjustment to achieve the desired final compaction density and lift surface in a more efficient manner. Often even a small change in the moisture of the material resulted in a change in compaction effort or appearance of the compacted product. At majority of the sites, the most efficient compaction was achieved with the vibratory compactor in a relatively high amplitude and low frequency. A typical, well compacted lift surface is shown on Figure 13.

One common issue that occurs with hardfill/RCC/soil-cement is a less compacted surface exhibited by signs of "peeling". The compactor settings for frequency and amplitude can be adjusted and typically avoid the shearing/tearing of the hardfill surface which creates an undesirable zone at the lift interface. An example of the shearing of the surface of the hardfill when operating at low amplitude and high frequency is shown on Figure 14. The best settings for material type and compactor settings varied for each material.

In order to achieve sufficient compaction along the abutments and precast block formed surfaces, a secondary, smaller compactor was required to fit in more confined and curved areas. A variety of smaller double drum rollers were used, both operator ridden (refer to Figure 15) and remote controlled.

Figure 13. Example of Typical Hardfill Lift after Compaction

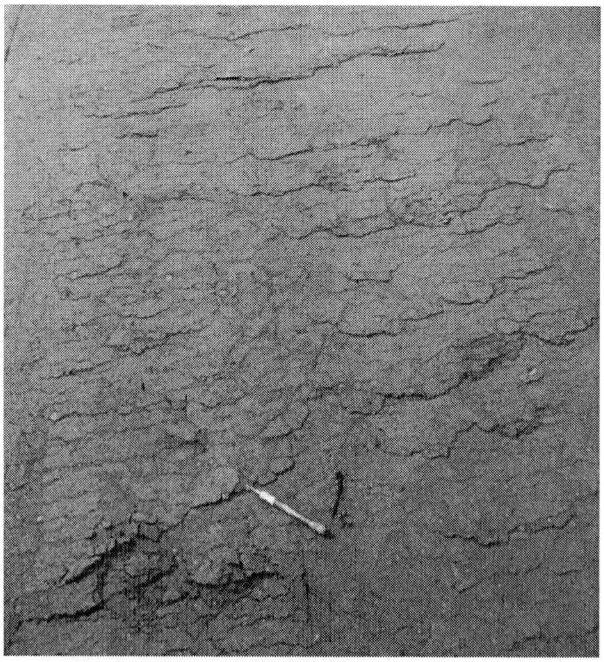

Figure 14. Example of Peeling and shearing resulting from the single drum roller on low amplitude and high frequency after compaction

Figure 15. Secondary compaction along curved radius

Hot Weather Placement

Weather has a significant effect on hardfill placement and compaction. Due to the relatively low moisture content of the hardfill and the relatively narrow moisture range to achieve optimum compaction, the control of moisture loss during hot weather, in particular, is critical. Temperature of the hardfill also needs to be controlled during placement to limit the amount of induced thermal stresses in the hardfill.

Air temperatures during placement and curing across the various sites ranged from a high of about 108 degrees Fahrenheit to a low of near zero degrees Fahrenheit. Moisture loss occurs during the transport from the plant to the placement area, as well as in the hardfill after dumping. Typically, a moisture loss of about 1-2% by weight of the hardfill mixture was noted between batching and delivery during very hot weather with both conveyor and truck delivery. The mix moisture content for batching was selected to account for evaporation losses with a target of placement at about optimum moisture content. Moisture control of the hardfill prior to compaction included modifications to typical placement procedures to keep the hardfill from losing excessive moisture in the placement area. The rapid delivery of hardfill during hot weather placement was essential to limit loss of moisture prior to spreading. Once spread, initial compaction was commenced as rapidly as practical. After initial compaction, the hardfill material was much less susceptible to moisture loss, as the compaction helps seal the surface and retain moisture. Typically, the hard fill remained "alive" for initial compaction up to about 45 minutes. During hot weather, especially when air temperatures exceeded about 85 degrees Fahrenheit, the hardfill mix tended to stiffen and became more difficult to compact, the time to start compaction was reduced to approximately 30 minutes to limit the loss of significant moisture as observed by the "workability" of the hardfill mix.

Project specifications required limiting the hardfill temperature at time of placement to less than 95 degrees Fahrenheit would limit the development of thermal stresses. Temperatures exceeding 80-90 degrees were encountered. However, the hardfill remained below the required 95 degrees without further temperature controls. This was primarily due to the low temperature of the aggregate stockpile material. Alluvial soils from the required excavation were used at the site that experienced the highest temperatures during placement. The alluvium stockpiled generally retained cooler than ambient temperatures. The aggregate in the stockpile was generally in the range of 60 degrees Fahrenheit, and the hardfill did not gain significant heat during stockpiling, excavation and mixing of the hardfill.

The most significant impact to the hardfill placement procedures was the need for vigilant control of the lift surface curing. During the highest temperatures and when the hardfill was exposed to full sun, constant moistening of the surface after compaction was required to maintain a moist surface, while not flooding the lift surface and potentially increasing the water content of the hardfill before reaching initial set. The most common method to maintain surface moisture was utilizing garden hoses and nozzles in a misting or light spray mode. This method required multiple laborers full time to cover the entire lift surface. During extended shutdowns, the lift surface was covered with plastic sheeting to limit evaporation. Supplemental moisture was provided using sprinklers set on the hardfill surface, which could be monitored by a single person.

Cold Weather Placement

Cold weather occurred during a portion of the hardfill placement at most sites. In order to maintain the surface temperature above the required 40 degrees Fahrenheit to prevent freezing, concrete insulation blankets were typically spread over the surface. The hydration of the hardfill created sufficient heat to maintain the required temperature and moisture of the hardfill except for one site where hardfill placement primarily occurred during the winter months. Blanket removal was sequenced to remove just before placement was scheduled to start, to maintain the required surface temperature.

Hardfill placement was performed primarily during the winter months at one site. The contractor utilized glycol ground heaters with coiled hose that were placed under the blankets to maintain the surface temperature during nights and non-placement days. The combined system of glycol heaters and concrete blankets typically maintained a temperature between 40 and 50 degrees Fahrenheit, in ambient air temperatures that were well below freezing. In addition, the aggregate was stockpiled during warmer temperatures and was generally at a temperature of about 60 degrees, which aided in meeting the temperature requirement.

QUALITY CONTROL

Construction quality control (QC) included monitoring of batch plant records, compressive strength testing, aggregate gradation and quality testing, compacted density, and field observations such as placement times. In order to monitor the hardfill placement a full time inspector and lab technician were utilized during hardfill placement.

Aggregate Gradation

Aggregate was monitored for conformance with the specifications at an interval of once per 5,000 cubic yards. The average and range of gradations at each of the AMP sites is plotted on Figure 16. The aggregates at Cannelton and Smithland consisted of selectively excavated material (alluvial soils with low plasticity), while hardfill aggregate material at the Meldahl and Willow Island was locally sourced from commercial quarries.

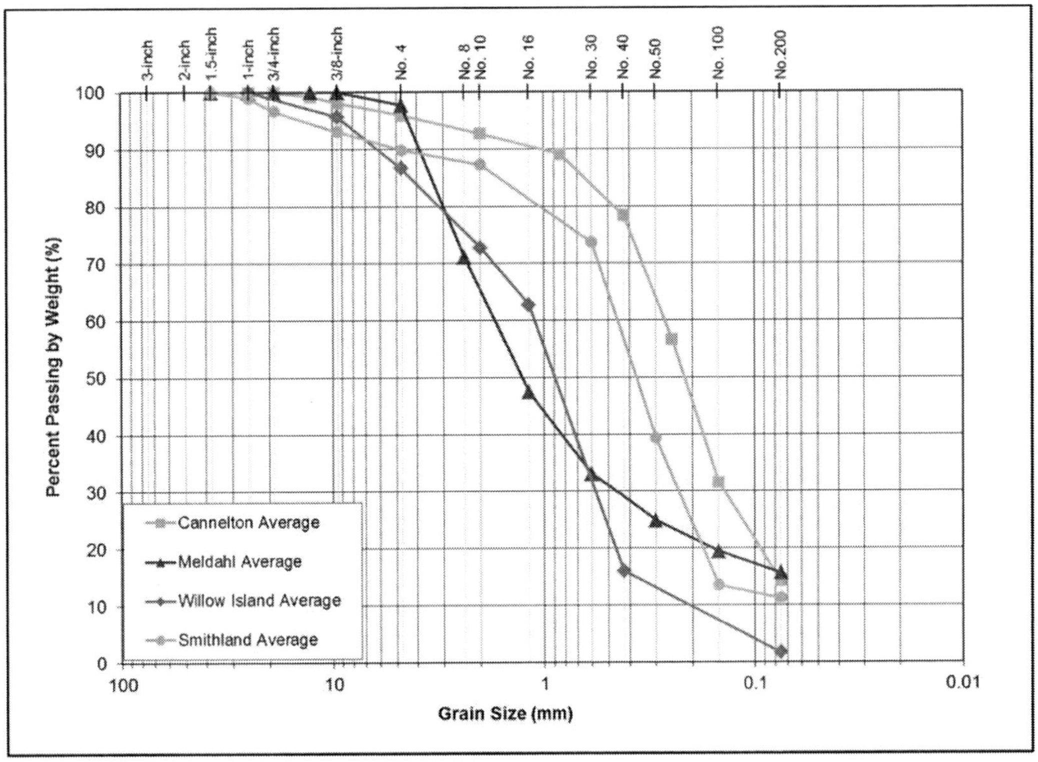

Figure 16. Average Gradation Results at the Four AMP Sites

Cement Content

One method used to measure the cement content of the freshly mixed hardfill was the Heat of Neutralization Method (ASTM D5982). The heat of neutralization test consists of generating a calibration curve of the hardfill material with specific cement contents in the laboratory over a range of cement contents. The calibration curve represents a linear relationship between the heat rise and the cement content. The same test procedures are then applied to a field sample and then compared to the calibration curve to estimate the cement content of the field sample.

Three separate calibration curves were developed during the initial project development, the complete hardfill mixture, cement only, and fly ash only. The heat of neutralization test was developed for cement only mixtures. The heat rise measured in tests with fly ash only was negligible. Based on the comparison of the various calibration curves, it was assumed that the heat rise generated in the neutralization reaction measured the cement content only. Therefore, the test was utilized to confirm the cement content of the freshly mixed hardfill during placement. The calibration curves based on the total cementitious content are shown on Figure 17.

The results of the heat of neutralization test method generally reflected the actual batched cement content within approximately plus or minus 1 percent. The heat of neutralization results are plotted versus the measured batch amount at the Meldahl site on Figure 18. The heat of neutralization test provided a method to confirm that cement was being batched and general trends and was a valuable QA tool when cylinder test results indicated lower than average test results. However, the test is not accurate enough to use for quality control and/or acceptance of the hardfill in-place.

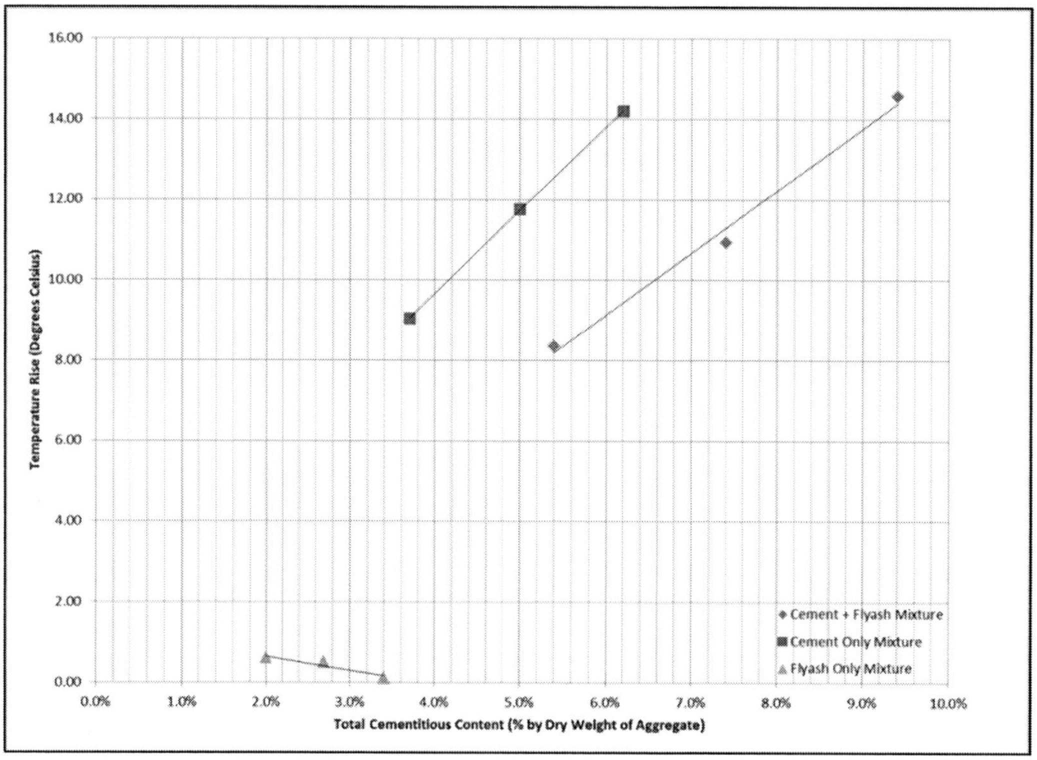

Figure 17. Heat of Neutralization Calibration Curve for Cement and Fly Ash, Cement Only, and Fly Ash Only Hardfill Mixtures. (Cement Content and Fly Ash Content Held Constant in Combined Hardfill Mixture)

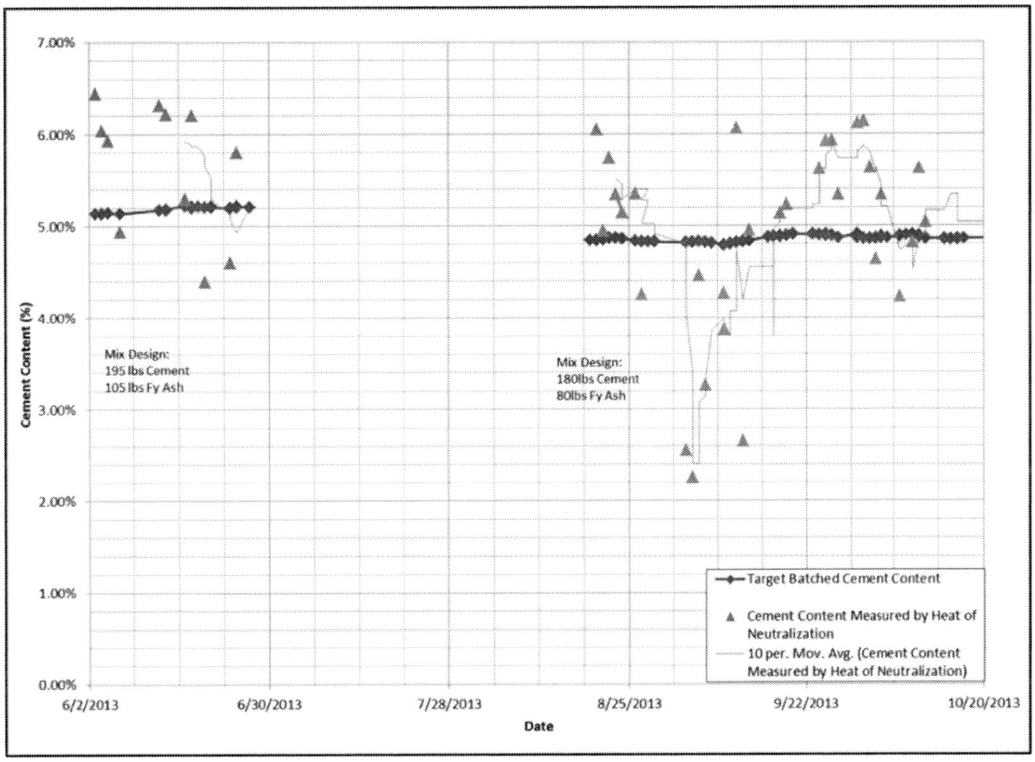

Figure 18. Cement Content as Measured by the Heat of Neutralization Method Compared to Batched Cement Content

Density

The percent compaction was determined relative to the Standard Proctor test following procedures for soil cement mixtures (ASTM D558). The in-place dry density of the hardfill was measured using a nuclear gauge at a frequency of at least once per lift or once per 1,000 cubic yards placed. The in place dry density of the hardfill was compared to the maximum dry density for the material. Compaction curves were prepared on at least a weekly basis to account for changes in the soil aggregate at each of the project sites. A summary of the in-place dry density and proctor maximum dry density for each site are shown on Figure 19 through Figure 22. The variation in maximum dry density and in-place compacted density showed trends in parallel. In general, the in place dry density of the hardfill equaled or exceeded the maximum dry density as determined by ASTM D558. Both the in-place dry density and proctor density test results at the Cannelton and Smithland sites showed gradual changes over time. Soil aggregates from these two sites consisted of on-site material from required excavations. The soil aggregate source of Meldahl and Willow Island were local commercial quarry material. Some variation over time was also noted in the in-place and proctor density results. Frequent proctor compaction testing provided the necessary adjustment in the maximum dry density value as the soil aggregate properties varied for both the on-site and quarry supplied soil aggregate.

Hardfill Construction

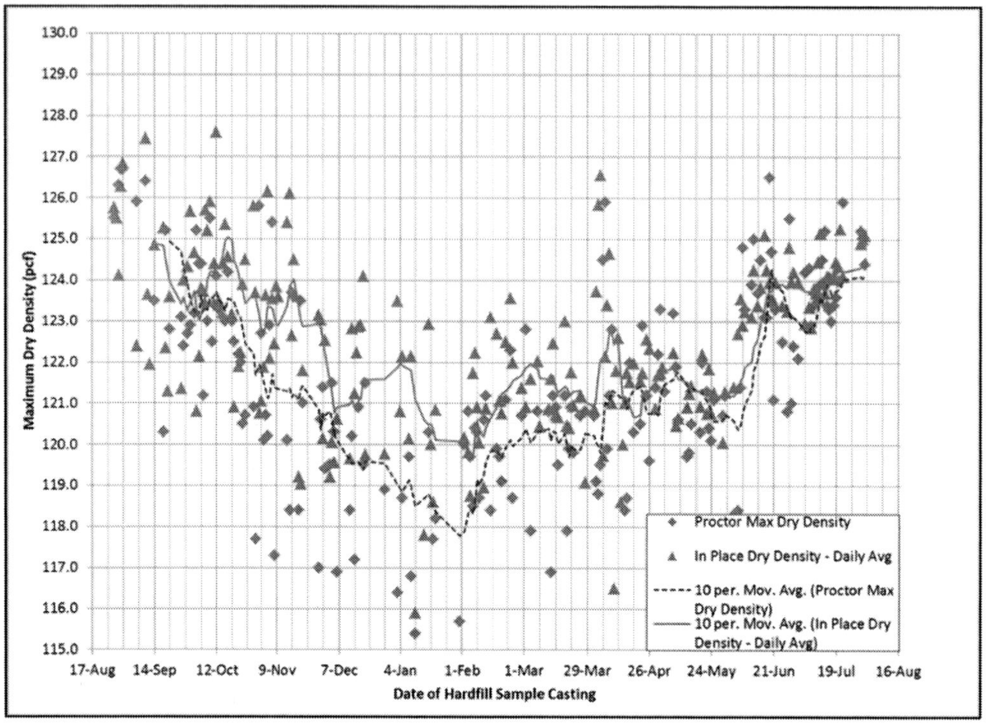

Figure 19. Comparison of In-Place Dry Density by Nuclear Gauge and Standard Proctor Maximum Dry Density Results for Cannelton Site

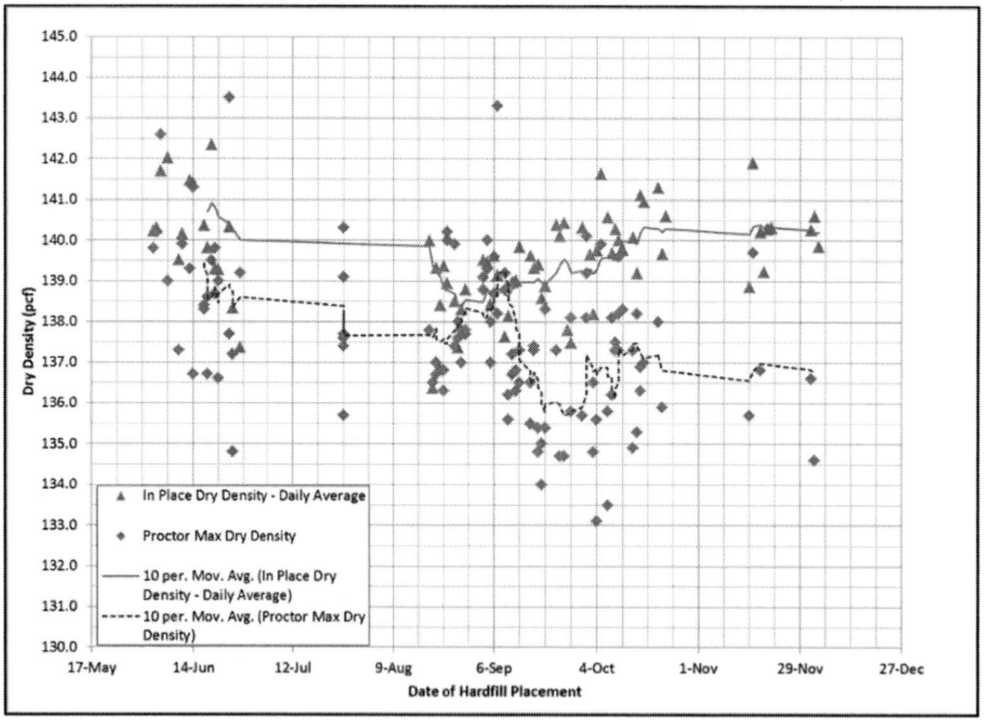

Figure 20. Comparison of In-Place Dry Density by Nuclear Gauge and Standard Proctor Maximum Dry Density Results for Meldahl Site

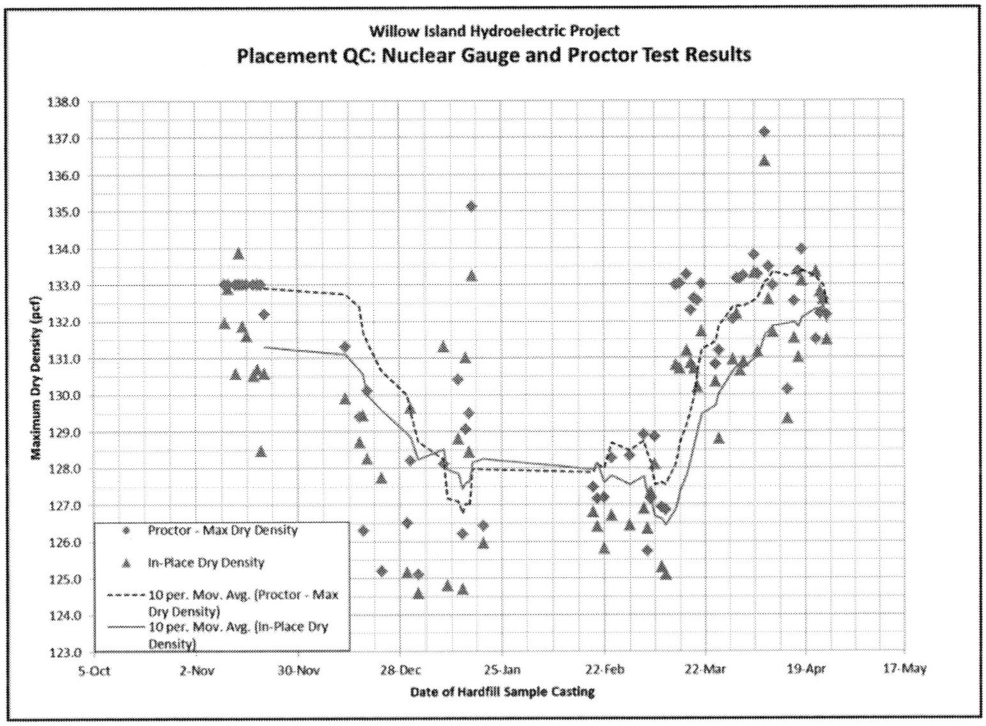

Figure 21. Comparison of In-Place Dry Density by Nuclear Gauge and Standard Proctor Maximum Dry Density Results for Willow Island Site

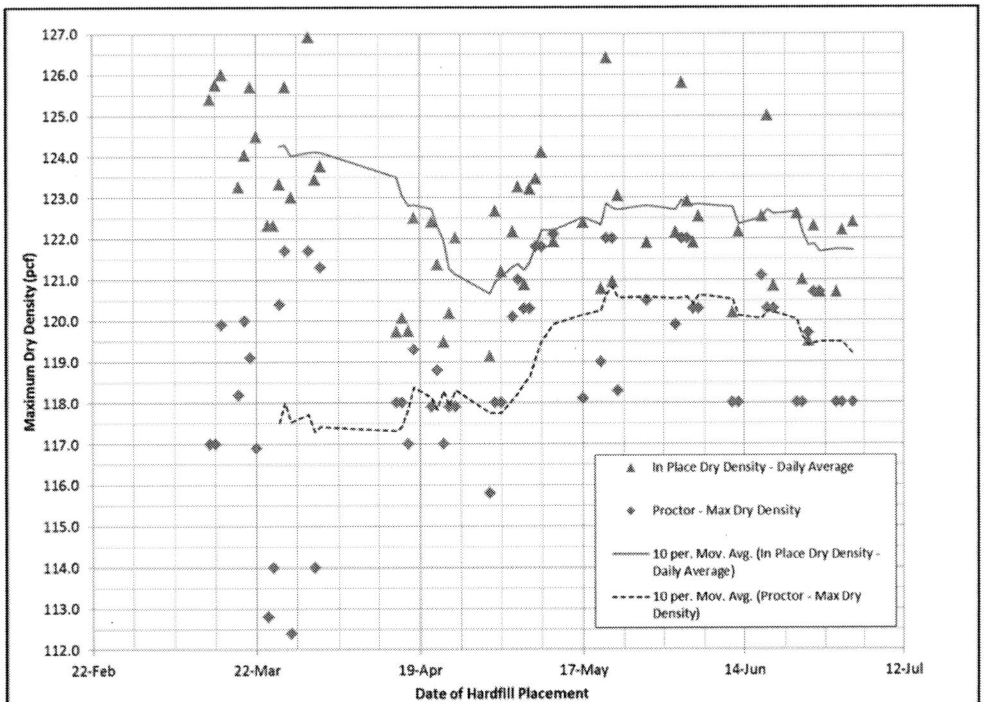

Figure 22. Comparison of In-Place Dry Density by Nuclear Gauge and Standard Proctor Maximum Dry Density Results for Smithland Site

Compressive Strength

One of the key indicators for the development of the required hardfill properties is the compressive strength. Compacted hardfill samples (ASTM D558) were prepared for testing at 7, 28, and 180 days age at a frequency of at least once per lift or once per 500 cubic yards.

The mix design ultimately used at each site was very similar, although the gradation of the soil aggregate borrow varied significantly depending on the source of soil aggregate ranging from on-site alluvium to quarry produced clean to silty sands. The mix design utilized as the baseline was 195 pounds of cement and 105 pounds of fly ash for a 1 cubic yard batch.

A plot of the compressive strength versus time was developed for the four sites at 28 day and 180 day age. Following the initial 30 days of placement at Meldahl, a reduction in mix proportions from the baseline to 180 pounds of cement and 80 pounds of fly ash was implemented. The cement and fly ash content at 3 sites primarily used a mix similar to the baseline mix design, although early test results and the laboratory test program indicated that the cementitious content could be decreased.

Compressive strength test results are summarized on Figure 23 and Figure 24. Compressive strength test results at the Meldahl and Willow Island (using quarry produced soil aggregate) sites were higher than the two sites using on-site alluvium. The target compressive strength requirements were met at both the 28 day and 180 day test ages at all of the four sites, with the exception of a short period during winter placement at the Cannelton site. At the Willow Island site, 180 day age cylinders were not tested since the 28 day test results exceeded the 180 day strength requirement.

At the Cannelton site, a decreasing trend in compressive strength was observed beginning in December, as shown in Figure 25. The frequency of the heat of neutralization tests was increased when the trend for decreasing compressive strength was observed. The heat of neutralization tests indicated a lower cement content (as low as 2.6% cement by dry weight of hardfill) in contrast to the target mix proportion of 6.7% by dry weight, recorded by the batch plant. Investigation of the potential causes of the decreasing strength included re-calibration of the plant and evaluation of the cement and fly ash measurement in the plant. The initial plant setup included an impact plate measuring device for cement and fly ash measurement instead of direct weigh measurement devices. Ultimately, the measurement device was changed to a weight method on April 13 and the 10 test average trended upward after the measurement device change.

As noted above, compressive strength at all of the sites met the target design strength with the exception of a short term period at Cannelton when issues were encountered in the cement measuring equipment. The distribution of the strength data within the structure was evaluated. The scattered test cylinders that did not meet the target strength at 365 day age were not concentrated in specific areas of the closure structure and were judged not to impact the performance of the structure.

Hardfill Construction

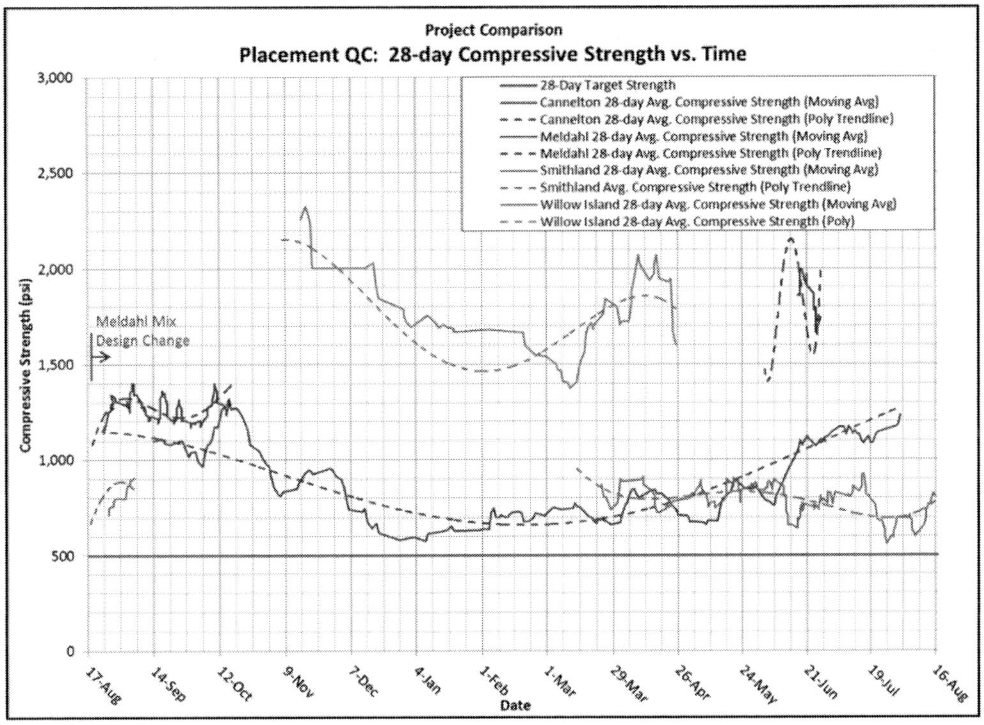

Figure 23. 28 Day Compressive Strength for AMP Hydroelectric Projects by day/month

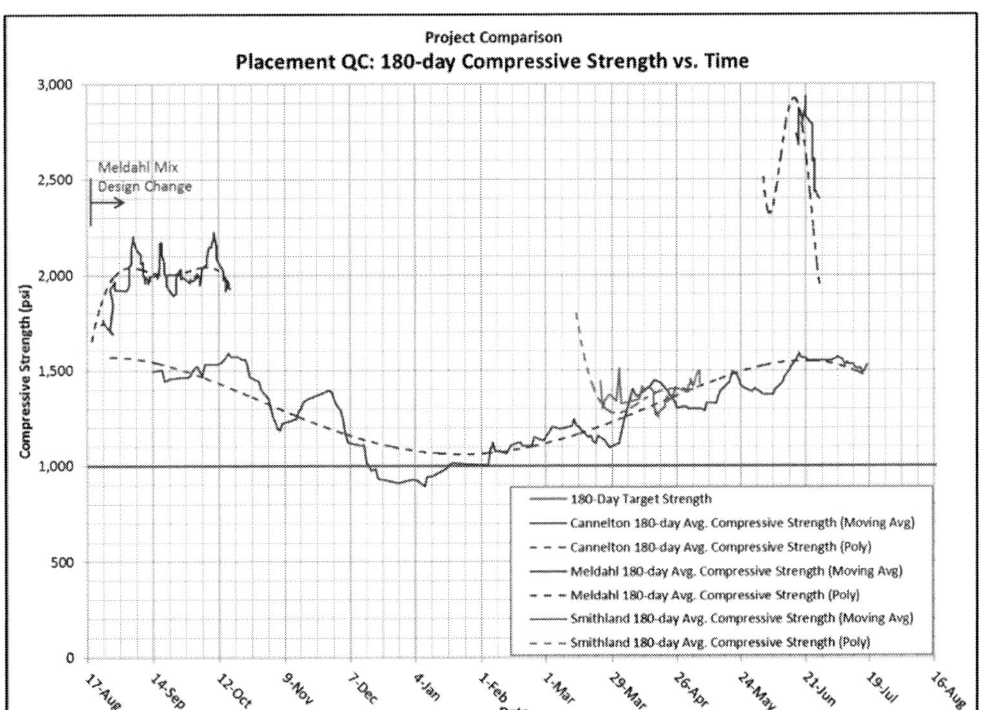

Figure 24. 180 Day Compressive Strength for AMP Hydroelectric Projects by day/month

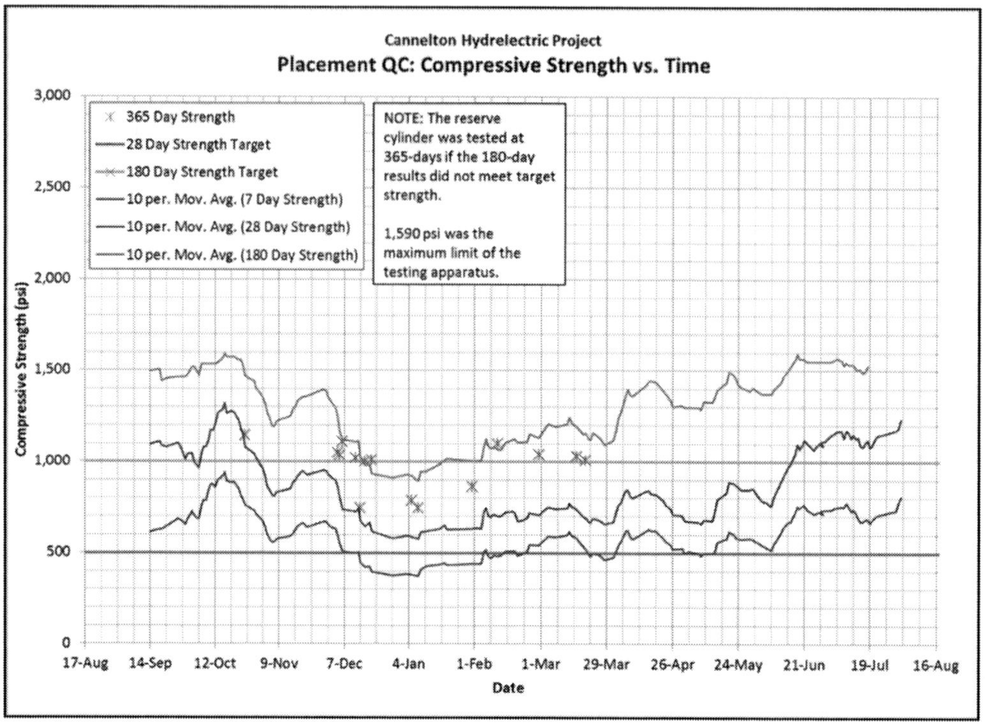

Figure 25. Cannelton 7, 28, and 180 Day Age Compressive Strength Test Results.

CONCLUSION

The use of hardfill with reinforced concrete facing for the AMP powerhouse closure structures provided a water retaining structure, as well as a structure that can be safely overtopped during periodic river flooding stages. The hardfill material was placed in relatively rapid construction and at a lower cost than a conventional concrete structure, and utilizing a hardfill mixture with fly ash replacement. Both on-site alluvium and locally available quarried clean to silty sands were used, demonstrating the adaptability of hardfill for hydraulic structure design and construction. Quarry processed soil aggregate provided higher compressive strength (or lower cementitious content was required to meet the design properties), but was higher in cost than on-site material. Construction was performed by four different contractors (one at each site), utilizing a variety of means and methods specific to each site, and encountering varied weather conditions throughout placement. Each contractor adapted methods that successfully achieved the intended design.

REFERENCES

ASTM C642-13, Standard Test Method for Density, Absorption, and Voids in Hardened Concrete, ASTM International, West Conshohocken, PA, 2013, www.astm.org

ASTM D558-11, Standard Test Methods for Moisture-Density (Unit Weight) Relations of Soil-Cement Mixtures, ASTM International, West Conshohocken, PA, 2011, www.astm.org

ASTM D1633-00(2007), Standard Test Methods for Compressive Strength of Molded Soil-Cement Cylinders, ASTM International, West Conshohocken, PA, 2007, www.astm.org

ASTM D5607-08, Standard Test Method for Performing Laboratory Direct Shear Strength Tests of Rock Specimens Under Constant Normal Force, ASTM International, West Conshohocken, PA, 2008, www.astm.org

ASTM D5982-07, Standard Test Method for Determining Cement Content of Fresh Soil-Cement (Heat of Neutralization Method), ASTM International, West Conshohocken, PA, 2007, www.astm.org

Antal, Brandi; Arnold, Terry; Konstantellos, Christ; Meier, Phil; and Truschke, Stefano. 2013. Hardfill on the Ohio River – Optimizing the Use of Construction Materials.

Arnold, Terry; Antal, Brandi; Meier, Phil; and Scott, Taylor. 2015. Optimized – Hardfill Mix Design and Testing.

U.S. Army Corps of Engineers. 1992. CRD 55 – Test Method for Within Batch Uniformity for Freshly Batched Concrete.

UPGRADED — NEW POWERHOUSES AT EXISTING USACE DAMS ON THE OHIO

Jason A. Bodily, P.E.[1]
Christ Konstantellos, P.E., S.E.[2]
Philip E. Meier[3]

ABSTRACT

American Municipal Power, Inc. (AMP) is currently constructing four concrete combination dams/hydroelectric powerhouses – Willow Island, Meldahl, Cannelton and Smithland – on the Ohio River adjacent to existing U.S. Army Corps of Engineers (USACE) locks and dams. The projects are jointly the largest new hydroelectric development in the United States and are an excellent example of how renewable hydropower can be added to existing dams and can be a viable part of our nation's energy future. Designs for all four concrete combination dams/powerhouses have been completed by MWH Americas, Inc. (MWH). The powerhouses at Willow Island and Meldahl are founded on rock, and the powerhouses at Cannelton and Smithland are founded on alluvial river deposits. This paper discusses the general layout of the powerhouses and presents the methodologies employed in the analysis and design of the cast-in-place reinforced concrete powerhouses, which will act as dams and have been designed to be fully submerged during flood conditions. This paper also discusses state-of-the-art computer programs and techniques utilized to model and structurally analyze and design the ten-story tall monolithic structures, which will house a combined total of eleven horizontal bulb-type turbine-generator units with runner diameters of (7.7 m) 25.3 ft.

INTRODUCTION

The USACE provided for future hydropower development at its Ohio River locks and dams when they were originally constructed. Dam abutments opposite the locks generally consist of overflow structures that in some cases could be replaced by a hydropower plant. AMP is currently constructing four hydropower plants - Willow Island, Meldahl, Cannelton, and Smithland - on the Ohio River adjacent to existing USACE locks and dams. AMP is upgrading the existing installations by developing these renewable energy sources for the benefit of its member communities and the nation. The river water levels for the four projects are summarized in Table 1.

[1] Senior Structural Engineer, MWH Americas, Inc., 175 W. Jackson Blvd., Suite 1900, Chicago, IL 60604, (312) 831-3000, jason.bodily@mwhglobal.com
[2] Vice President, MWH Americas, Inc., 175 W. Jackson Blvd., Suite 1900, Chicago, IL 60604, (312) 831-3000, christ.konstantellos@mwhglobal.com
[3] Vice President – Hydroelectric Development and Operations, American Municipal Power, Inc., 1111 Schrock Road, Suite 100, Columbus, OH, pmeier@amppartners.org

Table 1. Ohio River Water Levels (feet).

Project	Upper Pool					Lower Pool				
	Normal	Flood Levels				Min.	Flood Levels			
		10-yr	25-yr	100-yr	500-yr		10-yr	25-yr	100-yr	500-yr
Willow Island	602.0	613.5	616.9	621.8	626.9	582.0	612.0	615.4	620.3	625.4
Meldahl	485.0	497.1	501.7	507.9	514.5	455.0	496.4	501.0	507.4	514.0
Cannelton	383.0	398.0	399.9	403.2	406.8	358.0	396.7	399.4	403.0	406.6
Smithland	324.0	340.0	342.2	345.6	349.5	302.0	338.7	341.1	344.5	348.2

Note: All elevations are based on the Ohio River Datum (ORD).

WILLOW ISLAND HYDROELECTRIC PROJECT

The Willow Island Hydroelectric Project (FERC Project Number 6902) is located on the West Virginia (left) bank adjacent to the existing Willow Island Locks and Dam on the Ohio River, about 161.7 river miles below Point Bridge, Pittsburgh PA. The existing navigable structure consists of a 1,200-ft-long by 110-ft-wide main lock and a 600-ft-long by 110-ft-wide auxiliary lock located adjacent to each other on the Ohio (right) bank. The dam is a non-navigable gated structure consisting of eight tainter gates, each with a clear span of 110 ft between piers and a height of 26 ft above the sills. The total length of the dam is 1,128 ft including the 111-ft-long fixed cellular weir.

The hydroelectric project will utilize the 20-ft maximum gross head at the site, and the centerline of the powerhouse is located approximately 137 ft from the landward face of the existing pier 9, which is the leftmost pier of the dam. The existing cellular weir structure on the landward side of pier 9 was removed and replaced with a new concrete (riverside) closure structure, which connects the powerhouse to the existing concrete weir adjacent to pier 9. The normal upstream pool is maintained during non-flood events at El. 602.0. The downstream pool fluctuates above the minimum pool level of El. 582.0 and is regulated by the USACE's Belleville Locks and Dam facility. Figure 1 shows the general layout of the Willow Island Hydroelectric Project which includes the following principal features:

- a two unit, single monolith, reinforced concrete, combination dam/hydroelectric powerhouse founded on rock and designed to be submerged in floods;
- a closure structure on the landside of the powerhouse constructed of hardfill (roller compacted soil-cement) and capped with reinforced concrete. The top of the landside closure structure is at El. 625.0 and it is founded on rock and in-situ materials;
- a closure structure on the riverside of the powerhouse constructed of concrete connects the powerhouse to the existing spillway pier 9. The top of this structure is at about El. 604.0 and it is founded on competent rock;
- a hydraulic (J-wall) structure on the right side of the approach channel immediately upstream of the riverside closure structure. The top of the J-wall is at El. 604.0 and it is founded on and anchored to competent rock; and

- an approach channel and a tailrace channel, both of which will be lined with riprap.

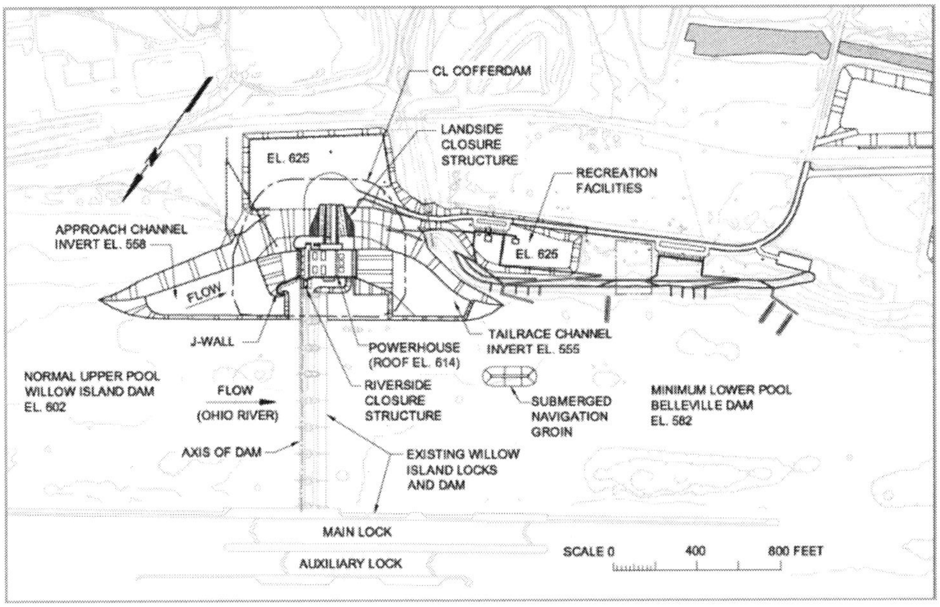

Figure 1. Willow Island, General Project Layout.

Figure 2. Willow Island Aerial View, January 23, 2015.

MELDAHL HYDROELECTRIC PROJECT

The Meldahl Hydroelectric Project (FERC Project Number 12667) is located on the Kentucky (left) bank adjacent to the existing Captain Anthony Meldahl Locks and Dam on the Ohio River, approximately 436.2 river miles below Point Bridge, Pittsburgh PA. The existing navigable structure consists of a 1,200-ft-long by 110-ft-wide main lock and a 600-ft-long and 110-ft-wide auxiliary lock. The locks are located on the Ohio (right) side of the river. The 1,384-ft-long spillway consists of twelve tainter gates, each 100 ft long and 37 ft high. Five of the twelve spillway gates were originally intended for either normal (raised) operation or lowered (submergible) operation to pass water over the top of the gates. The other seven gates were constructed for normal operation. At the south (Kentucky) end of the spillway, there is a concrete gravity overflow weir with a 360-ft-long crest at El. 487.0. The normal upper pool is maintained during non-flood events at El. 485.0. The downstream pool fluctuates above the minimum pool level of El. 455.0 and is regulated by the USACE's Markland Locks and Dam facility.

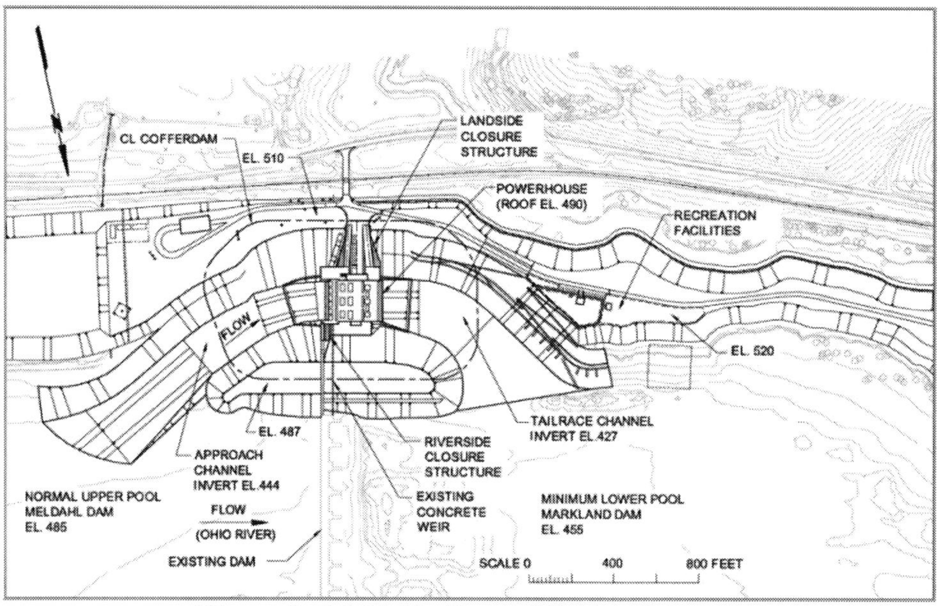

Figure 3. Meldahl, General Project Layout.

The Meldahl Hydroelectric Project will utilize the 30-ft maximum gross head at the site, and the centerline of the powerhouse is located approximately 560 ft from the landward face of the existing pier 13, which is the leftmost pier of the dam. A new concrete (riverside) closure structure connects the powerhouse to the existing concrete gravity overflow weir.

Figure 3 shows the general layout of the project which includes the following principal features:

- a three unit, single monolith, reinforced concrete, combination dam/hydroelectric powerhouse founded on rock and designed to be submerged in floods;
- a closure structure on the landside of the powerhouse constructed of hardfill (roller compacted soil-cement) and capped with reinforced concrete. The top of the landside closure structure is at El. 508.0, and it is founded on rock and in-situ materials;

- a closure structure on the riverside of the powerhouse constructed of concrete connecting the powerhouse to the existing overflow weir. The top of this structure is at El. 487.0, and it is founded on competent rock; and
- an approach channel and a tailrace channel, both of which are lined with riprap above the top of rock.

Figure 4. Meldahl Aerial View, January 5, 2015.

CANNELTON HYDROELECTRIC PROJECT

The Cannelton Hydroelectric Project (FERC Project Number 10228) is located on the Kentucky (left) bank adjacent to the existing Cannelton Locks and Dam on the Ohio River, approximately 720.7 river miles below Point Bridge, Pittsburgh PA. The existing navigable structure consists of a 1,200-ft-long by 110-ft-wide main lock and a 600-ft-long by 110-ft-wide auxiliary lock. Both navigation locks are located on the Indiana (right) bank. The dam is about 1,412 ft long and includes a concrete fixed weir with 12 tainter gates, each 100 ft long and 42 ft high. At the southern end of the gated section of the dam a concrete capped cellular overflow weir extends to the south (Kentucky) bank of the Ohio River.

The hydroelectric project will utilize the 25-ft maximum gross head at the site and is constructed adjacent to the existing cellular overflow weir. The centerline of the powerhouse is located approximately 677 ft from the landward face of the existing pier 13, which is the leftmost pier of the dam. Figure 5 shows the general layout of the project which includes the following principal features:

- a three unit, single monolith, reinforced concrete, combination dam/hydroelectric powerhouse founded on alluvial river deposits and designed to be submerged in floods;

- a closure structure on the landside of the powerhouse constructed of hardfill (roller compacted soil-cement) and capped with reinforced concrete. The top of this structure is at El. 404.5 and it is founded on in-situ materials;
- a closure structure on the riverside of the powerhouse also constructed of hardfill and capped with reinforced concrete. The top of this structure is at El. 388.0 and it is founded on in-situ materials; and
- an approach channel and a tailrace channel, both of which are lined with riprap for the majority of their length.

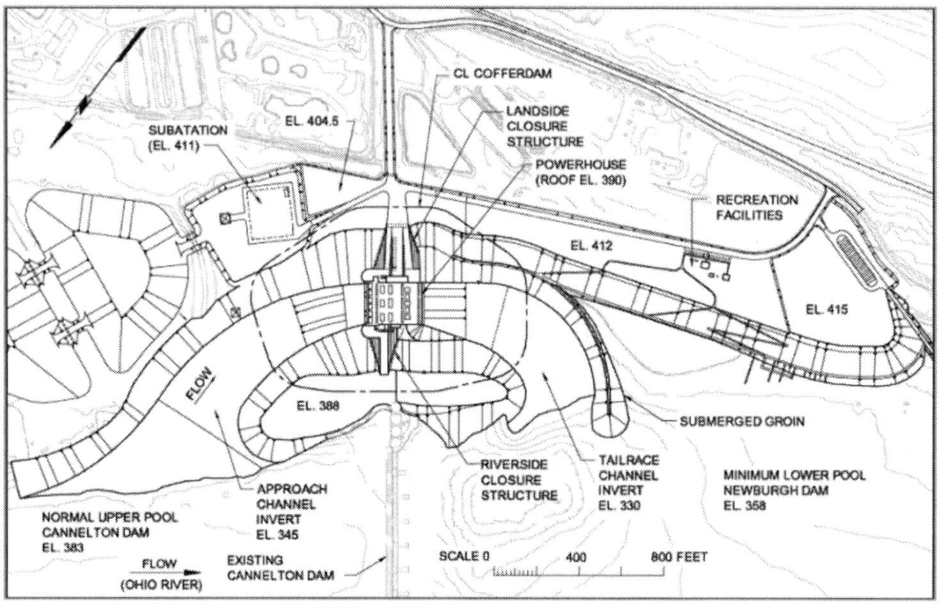

Figure 5. Cannelton, General Project Layout.

Figure 6. Cannelton Aerial View, December 29, 2014.

SMITHLAND HYDROELECTRIC PROJECT

The Smithland Hydroelectric Project (FERC Project Number 6641) is located on the Kentucky (left) bank adjacent to the existing Smithland Locks and Dam on the Ohio River, about 918.5 river miles below Point Bridge, Pittsburgh PA, and 62.5 river miles upstream of the confluence of the Ohio and Mississippi Rivers. The existing navigable structure consists of twin 1,200-ft-long by 110-ft-wide navigation locks adjacent to each other and located on the Illinois (right) bank. The existing dam is a concrete fixed weir with 11 tainter gates, each 110 ft long and 36 ft high. The dam is approximately 2,954 ft long, and at the eastern end of the gated section of the dam, a concrete capped cellular overflow weir extends to the east (Kentucky) bank of the Ohio River.

The hydroelectric project will utilize the 22-ft maximum gross head at the site and is constructed adjacent to the existing cellular overflow weir. Figure 7 shows the general layout of the project which includes the following principal features:

- a three unit, single monolith, reinforced concrete, combination dam/hydroelectric powerhouse founded on alluvial river deposits and designed to be submerged in floods;
- a closure structure on the landside of the powerhouse constructed of hardfill (roller compacted soil-cement) and capped with reinforced concrete. The top of this structure is at El. 347.0, and it is founded on in-situ materials;
- a closure structure on the riverside of the powerhouse also constructed of hardfill and capped with reinforced concrete. The top of this structure is at El. 326.0, and it is founded on in-situ materials; and
- an approach channel and a tailrace channel, both of which will be lined with riprap for the majority of their length.

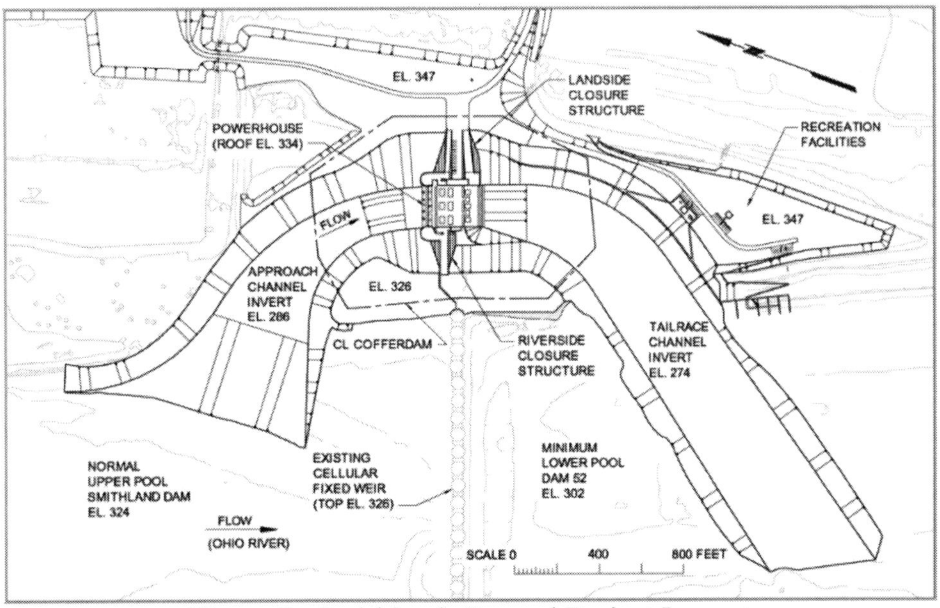

Figure 7. Smithland, General Project Layout.

Figure 8. Smithland Aerial View, December 26, 2014.

POWERHOUSE FEATURES

Water will be conveyed to the powerhouse by an approach channel excavated into the left river bank. Water will then be discharged back into the river through a tailrace channel also excavated into the left river bank. As shown on Figures 1, 3, 5 and 7, this arrangement is typical for all four projects. A typical longitudinal powerhouse section is shown in Figure 9.

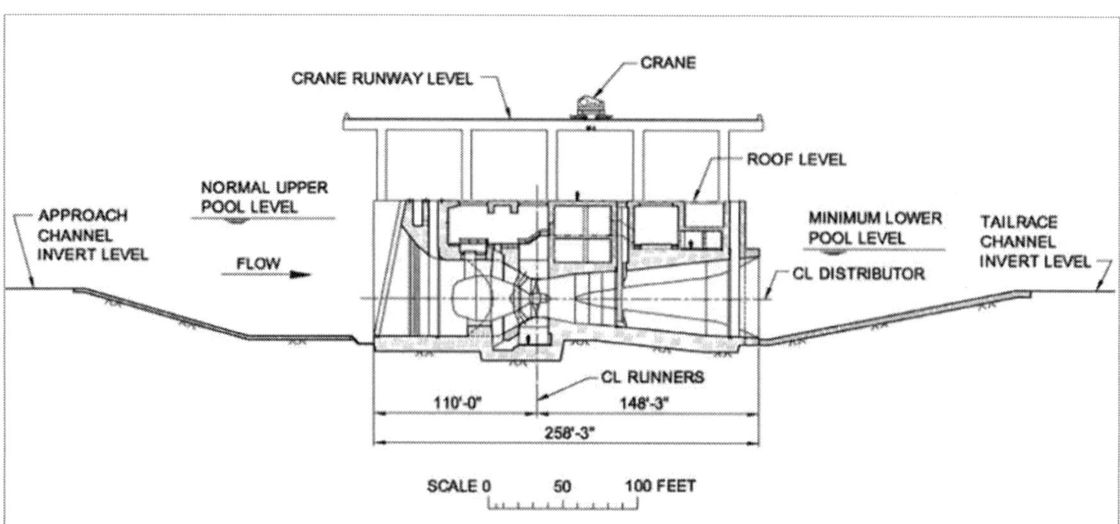

Figure 9. Powerhouse, Typical Longitudinal Section.

All four combination dam/hydroelectric powerhouses are constructed of conventional cast-in-place reinforced concrete, and have been designed to be submerged during flood events, similar to the existing Belleville Hydroelectric Powerhouse (designed by Harza Engineering Company - now MWH - and developed by OMEGA JV5 / AMP in the mid 1990's) at the USACE's Belleville Locks and Dam facility. As a result, there are no ventilation openings below the 500-yr flood level, and all other access openings below that level are provided with watertight covers, doors, or stoplogs. The principal powerhouse data is summarized in Table 2.

Table 2. Powerhouse Data.

Project	Approx. Structure Size (ft) L x W x H	Number of Units	Maximum Head (ft)	Power (MW)	Average Annual Energy (GWh)	Centerline Distributor Elev. (ft)	Roof Elev. (ft)	Main Entrance Elev. (ft)	PH Crane Runway Elev. (ft)
Willow Island	258 x 140 x 107	2	20	35	195	549.0	614.0	625.5	668.0
Meldahl	258 x 210 x 112	3	30	105	522	420.0	490.0	508.5	544.0
Cannelton	258 x 210 x 107	3	25	84	390	325.0	390.0	405.0	444.0
Smithland	258 x 210 x 107	3	22	72	350	269.0	334.0	347.5	388.0

Each powerhouse roof will have several equipment hatches, covered with watertight concrete covers as large as 840 square feet, to provide access to interior powerhouse spaces. The primary function of the hatches is to facilitate the initial installation as well as future maintenance of the generators, turbines, and main power transformers. To increase the water-tightness of the roof slab and hatch covers, the entire top surface of the roof will be protected with a spray-on waterproof membrane.

Each powerhouse is provided with an exterior type 175-ton capacity permanent bridge crane that operates on concrete runway frames above the roof in a direction parallel to flow. This unique application for hydroelectric projects on the Ohio River, which was first implemented at the Belleville Hydroelectric Project, covers the entire powerhouse structure and provides for installation and maintenance of equipment, from the intake trashracks to the draft tube bulkheads. Elevated 54 ft above the roof to remain above the 500-yr flood level, the crane was sized to handle all equipment located inside and outside the powerhouse, including the watertight roof hatch covers.

Main access to the powerhouse interior is above the 100-yr flood level and is provided at the entrance building located on the landside of the powerhouse. All levels are accessible by stairs and an elevator complying with the Americans with Disabilities Act (ADA). An equipment access shaft equipped with an overhead l0-ton crane is located in the entrance building. A second stair shaft is also provided on the riverside of the powerhouse, also ADA compliant. Both stairs have openings for regular doors at the roof level, but these are also equipped with steel doors that during floods can be closed and locked either from inside or outside of the stair shafts.

POWERHOUSE STRUCTURAL DESIGN CRITERIA

This and the following sections present the methodologies employed in the structural analysis and design of the four reinforced concrete, combination dam/hydroelectric powerhouses. State-of-the-art computer programs were utilized to model, structurally analyze and design the monolithic structures. Since the layout of the four powerhouses, and the forces to which they are subjected to, are very similar to one another, the analyses and designs are also very similar to one another, except for the foundation parameters because Willow Island and Meldahl are founded on rock, and Cannelton and Smithland are founded on alluvial river deposits.

Figure 10 shows a typical longitudinal powerhouse section cut at the centerline of one of the units and Figure 11 shows a typical transverse powerhouse section cut through the steel-lined portion of the draft tubes. These figures show the five color-coded regions, A through E, which were separately structurally analyzed and designed. For discussion purposes, the five regions are identified as follows:

- Region A, highlighted in blue, is the concrete runway support frames for the 175-ton powerhouse bridge crane (Figures 10 and 11).
- Region B, highlighted in grey, is the precast concrete covers for the powerhouse roof hatches (Figure 10).
- Region C, highlighted in red, is principally the powerhouse roof, but also includes several walls and columns that support the powerhouse roof and an interior slab (Figures 10 and 11).
- Region D, highlighted in magenta, is the entire cast-in-place reinforced concrete powerhouse minus regions A, C and E (Figures 10 and 11).
- Region E, highlighted in green, is the entrance building and riverside access building (Figure 11).

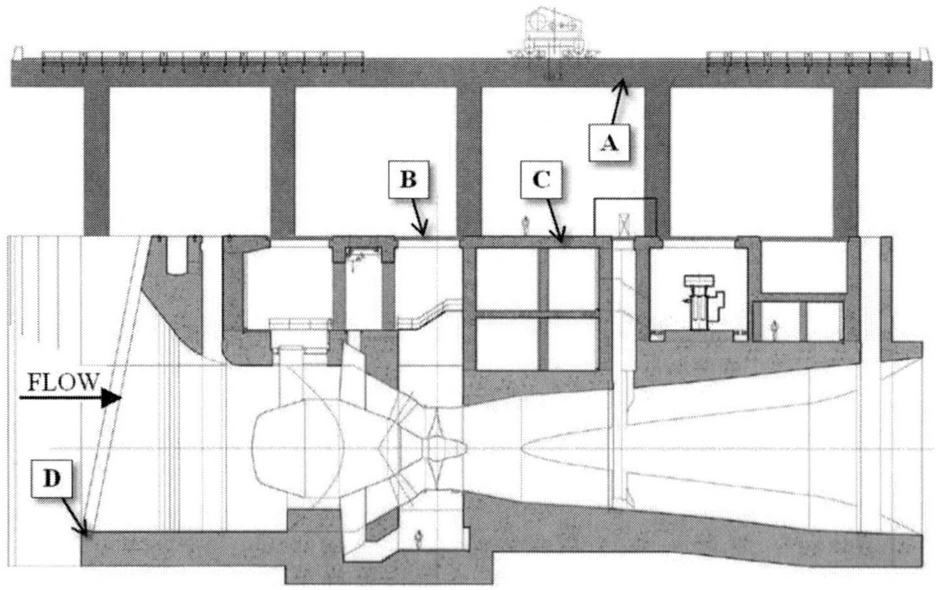

Figure 10. Longitudinal Powerhouse Section with Color-Coded Regions.

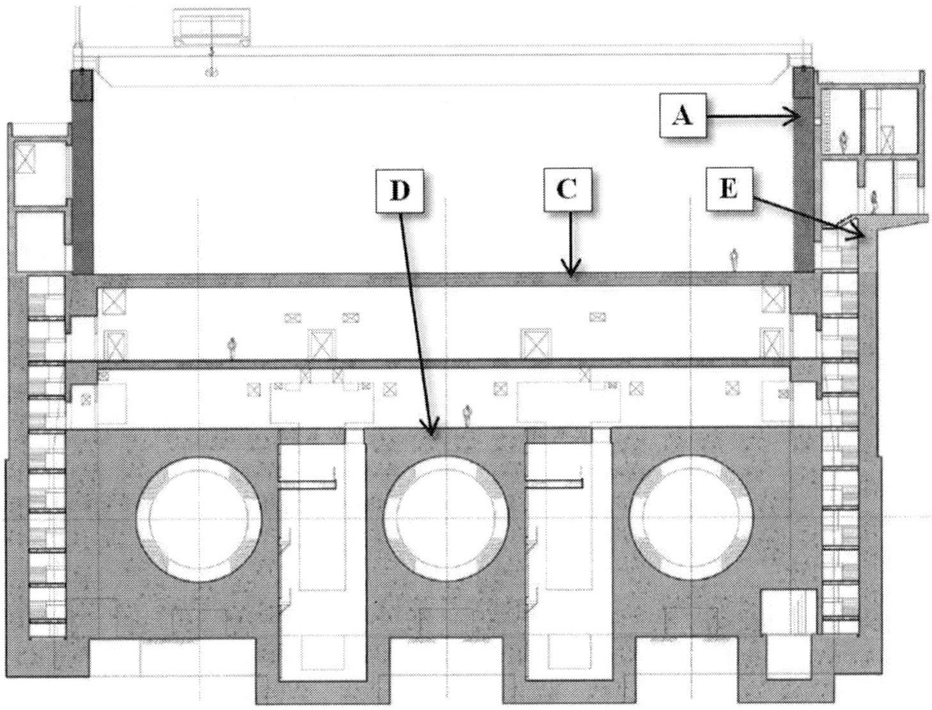

Figure 11. Transverse Powerhouse Section with Color-Coded Regions.

All structures were structurally analyzed and designed in accordance with the following:

- ACI 318-08, Building Code Requirements for Structural Concrete and Commentary
- USACE EM 1110-2-2100, Stability Analysis of Concrete Structures
- USACE EM 1110-2-2104, Strength Design for Reinforced Concrete Hydraulic Structures
- USACE EM 1110-2-3001, Planning and Design of Hydroelectric Power Plant Structures
- ASCE 7-05, Minimum Design Loads for Buildings and Other Structures

The following computer programs were used to perform the structural analysis and reinforced concrete design:

- SAP2000 by Computers and Structures, Inc.
- STAAD.Pro by Bentley Systems, Inc.
- Abaqus/CAE by Dassault Systèmes Simulia Corp.
- CSI Column by Computers and Structures, Inc.
- Mathcad by Parametric Technology Corp.
- Microsoft Excel by Microsoft Corp.
- AutoCAD by Autodesk, Inc.

The powerhouses were designed using load factors for hydraulic structures in accordance with USACE EM 1110-2-2104. The following load cases were considered for the structural design of the powerhouses. Table 1 lists the project specific headwater and tailwater elevations. Categories of usual, unusual and extreme are as defined in USACE EM 1110-2-2100:

- **Case 1** – Construction Condition: Structure completed within dewatered cofferdam. Backfill in place with construction surcharge load where applicable. Wind load. No water. (Unusual Case)
- **Case 2** – Normal Operating Condition: Normal headwater and tailwater plus ice load where applicable. Water passages of powerhouse filled. Hydraulic thrust due to operating units load rejection. (Usual Case)
- **Case 3** – Loss of Downstream Pool: Same as Case 2, but tailwater at the tailrace channel invert level and no ice load. (Extreme Case)
- **Case 4** – Inspection Condition: Same as Case 2, but water passage (between the intake bulkheads and the draft tube bulkheads) of any or all three units of the powerhouse empty. (Unusual Case)
- **Case 5** – Flood Condition (100-yr): 100-yr flood headwater and tailwater. Water passages of powerhouse filled. (Unusual Case)
- **Case 6** – Maximum Flood Condition (500-yr): 500-yr flood headwater and tailwater. Water passages of powerhouse filled. (Extreme Case)
- **Case 7A** – Earthquake Condition: Same as Case 1, plus seismic loads due to horizontal acceleration. (Extreme Case)
- **Case 7B** – Earthquake Condition: Same as Case 2 but without ice load, plus seismic loads due to horizontal acceleration. (Extreme Case)
- **Case 8** – Bulb Breach: Normal headwater and tailwater. Water passages of powerhouse filled, emergency gate is closed, but all interior spaces of the powerhouse are filled with water to normal headwater elevation. (Extreme Case)

USACE and FERC criteria were adopted for the structural stability analysis of each powerhouse. Each powerhouse satisfied the established factors of safety against sliding, overturning and flotation for all loading conditions, including the 500-yr flood.

POWERHOUSE CRANE RUNWAY SUPPORT FRAMES DESIGN

The powerhouse crane runway support frames are identified as Region A and highlighted in blue on Figures 10 and 11. Each frame consists of five rectangular columns and one rectangular beam, and they have been sized as follows:

- For Meldahl, Cannelton and Smithland: Column = 7.5-ft-deep x 5.5-ft-wide; Beam = 8.0-ft-deep x 5.5-ft-wide
- For Willow Island: Column = 6.5-ft-deep x 4.5-ft-wide; Beam = 7.0-ft-deep x 4.5-ft-wide

Each frame was constructed with 5,000 psi concrete. The frames were modeled and analyzed using STAAD.Pro software. One of the reasons STAAD.Pro was used is that once a moving wheel load is defined, STAAD.Pro automatically generates and analyzes numerous wheel load cases, which makes it simple to determine which wheel loads cause the maximum internal forces in the support frames. Frame elements were used to model the two-dimensional beam and columns, which were subjected to both in-plane and out-of-plane loads. Dead, live, seismic and wind loads were considered. Static elastic p-delta analyses were performed using STAAD.Pro to determine the non-seismic load combination design forces. Using site specific response spectra, STAAD.Pro was also used to perform response spectrum analyses of the frame. The modal responses were combined using the SRSS (square root of the sum of squares) method. The results of the response spectrum analyses were appropriately combined with the results of the static analyses to determine the seismic load combination design forces.

The design forces from STAAD.Pro and Mathcad were used to design the longitudinal and transverse reinforcement of the support frame beam and the transverse reinforcement of the support frame columns. The design forces from STAAD.Pro and CSI Column were used to create column interaction diagrams and design the longitudinal reinforcement of the support frame columns.

PRECAST CONCRETE COVERS DESIGN

The precast concrete covers are identified as Region B and highlighted in grey on Figure 10. Figure 12 is a typical key plan view identifying the covers (labeled MK-1 through MK-7) for all four powerhouses. Since the powerhouses are designed to be submerged periodically during floods, the precast concrete covers have been designed to withstand the hydrostatic flood loads acting on them.

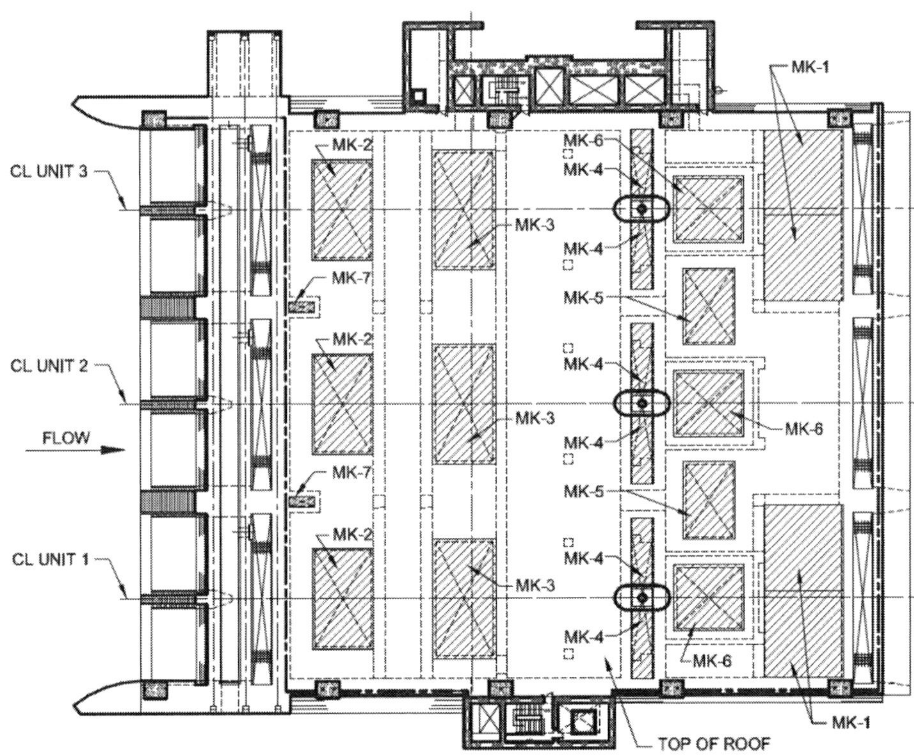

Figure 12. Powerhouse Roof, Precast Concrete Covers.

All covers were constructed with 6,000 psi concrete and lightweight aggregate. The covers are reinforced with a rebar mat in their top and bottom faces. All covers have been provided with heavy duty lifting devices. Each cover has two or four lifting devices, depending on its size.

The precast concrete covers were modeled and analyzed using SAP2000 and rectangular shell elements. Depending on the width to length ratio, the covers were analyzed and designed as two-way reinforced concrete slabs simply supported on all four edges or as one-way reinforced concrete slabs simply supported by two opposite edges. The following loads were considered:

- Self-weight of the cover
- Uniform live load of 1,000 psf
- Hydrostatic load from the 100-yr flood event (normal stress condition)
- Hydrostatic load from the 500-yr flood event (overstress condition)
- HS 20-44 wheel load

Static linear elastic analyses of the covers were performed using SAP2000 to determine the design forces. It is important to note that the hydraulic factor, H_f, is applied to the overall load combination instead of performing an additional serviceability analysis per USACE EM 1110-2-2104 when designing structural members of a hydraulic structure. Since the covers are removable, they are not permanent structural components of the hydraulic structure (i.e. the powerhouse) and can be replaced if they are damaged. Therefore, the hydraulic factor was not used to analyze and design the precast concrete covers for the powerhouse roof hatches. The design forces from SAP2000 and Mathcad were used to design the reinforcement of the covers and perform the necessary shear capacity checks.

POWERHOUSE ROOF DESIGN

Region C, highlighted in red on Figures 10 and 11, is principally the 3.5-ft-thick reinforced concrete roof slab, but also includes several walls and columns that support the powerhouse roof and an interior slab. The powerhouse roof and its supporting members were modeled and structurally analyzed using SAP2000. Shell elements were used to model the slabs and walls and frame elements were used to model the beams and columns. The powerhouse roof slab, the interior slab, most of the walls and the beams were designed as flexural members using Mathcad and some of the walls and the columns were designed as compression members using interaction diagrams created in CSI Column.

Like the precast concrete covers, the 3.5-ft-thick reinforced concrete powerhouse roof slab was designed to withstand a uniform live load of 1,000 psf, hydrostatic load from the 100-yr flood event (normal stress condition), hydrostatic load from the 500-yr flood event (overstress condition), and HS 20-44 wheel loads, in addition to its self-weight.

During construction of the powerhouses, the contractors have placed large concentrated loads (e.g. crawler cranes, gantry cranes, turbine-generator components, etc.) on the partially completed and completed powerhouse roof slabs. MWH has carefully reviewed the contractors' plans to verify that the powerhouse roof slabs have adequate structural capacity before permitting the contractors to implement their plans (i.e. place large concentrated loads on the powerhouse roof slabs). Large crawler cranes and gantry cranes have been driven onto the powerhouse roof slabs and have made significant picks while supported by the powerhouse roof slabs.

POWERHOUSE DESIGN

Region D, highlighted in magenta on Figures 10 and 11, is the entire cast-in-place reinforced concrete powerhouse minus regions A, C and E. The Region D reinforcement, which is the vast majority of the powerhouse reinforcement, was designed using the structural analysis and reinforced concrete design methodologies described in this section.

Using the powerhouse general arrangement drawings and the Voith (turbine manufacturer) final prototype hydraulic neatline drawings, a three-dimensional solid model of the powerhouse was created using AutoCAD.

Using project specific geotechnical data of the powerhouse foundation, a three-dimensional solid model of the powerhouse foundation was created using AutoCAD. The powerhouse and powerhouse foundation solid models were then exported from AutoCAD and imported into Abaqus/CAE (Abaqus), which is a three-dimensional finite element method (FEM) software program. This program was used to model and structurally analyze the entire powerhouse, excluding the two powerhouse crane support frames, the entrance building and the riverside access building (see Figures 10, 11, 13, 14 and 15).

Figure 13 shows the basic geometry and materials of the Cannelton and Smithland Abaqus models, Figure 14 shows the basic geometry and materials of the Meldahl Abaqus model, and Figure 15 shows the basic geometry and materials of the Willow Island Abaqus model.

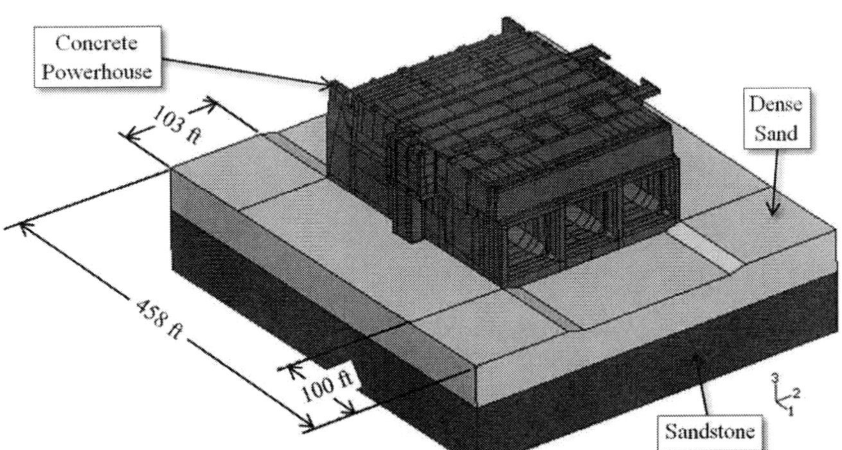

Figure 13. Abaqus Model for Cannelton and Smithland.

The model was then subdivided and meshed within Abaqus. The powerhouse was meshed using quadratic brick elements. Next, the loads for the relevant load cases and the boundary conditions were applied to the model within Abaqus. Subsequently, the linear elastic finite element analyses were run, which produced extremely large quantities of results data to be reviewed within Abaqus and used to design the powerhouse reinforcement.

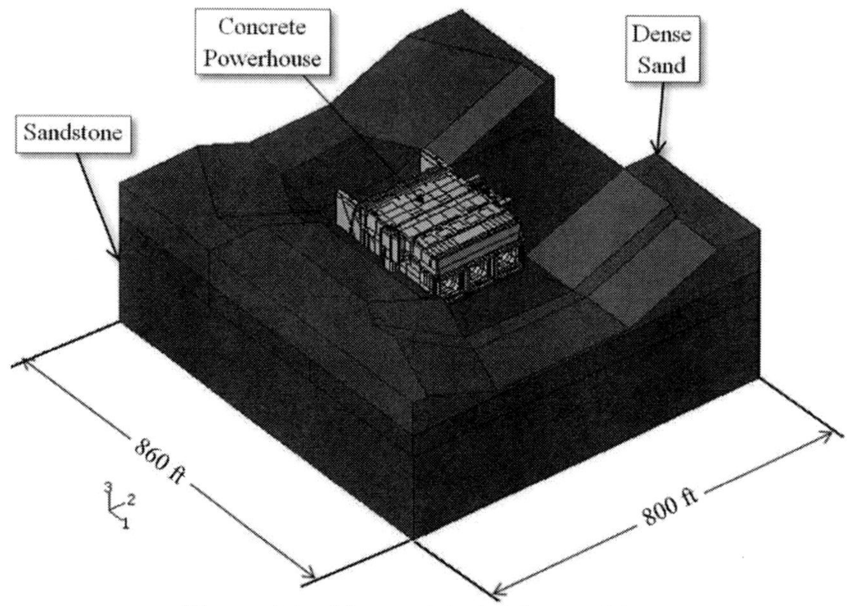

Figure 14. Abaqus Model for Meldahl.

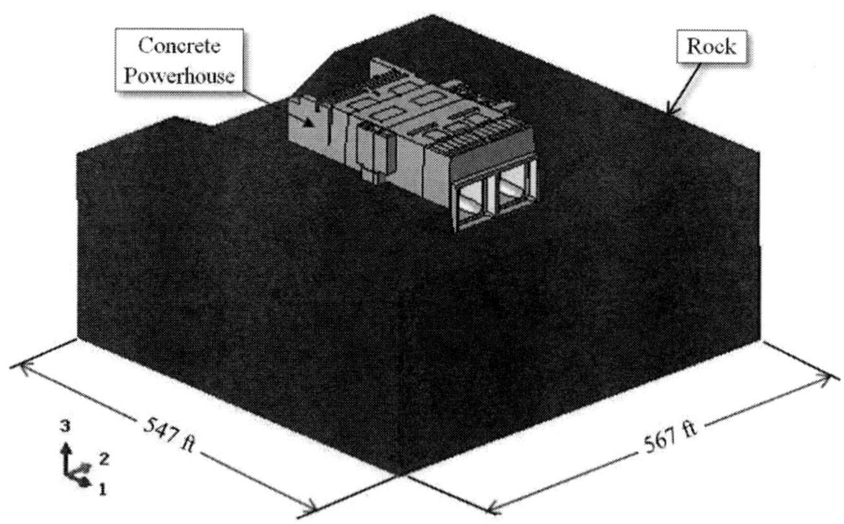

Figure 15. Abaqus Model for Willow Island.

All of the powerhouse models used 3,000 psi concrete with a 450,000 ksf modulus of elasticity. The Cannelton, Smithland and Meldahl models used dense sand (500 ksf modulus of elasticity) and sandstone (76,500 ksf modulus of elasticity). The Willow

Island model used rock with a 7,800 ksf modulus of elasticity. These soil and rock moduli were based on interpretation of in-situ geological information.

The Cannelton and Smithland powerhouses are founded on alluvial river deposits, dense sand, and the Meldahl and Willow Island powerhouses are founded on rock. For the Cannelton and Smithland powerhouses the purpose of the Abaqus analyses, in addition to providing stress results that were integrated for the powerhouse reinforcement design, was to gain an understanding of how the mat foundations on soil, with its potential for differential settlement, would affect the global design of the powerhouse.

The Abaqus models used a right-handed Cartesian coordinate system. The positive x-axis is parallel to the centerline of the units and points downstream, in the direction of flow. The positive y-axis lies in the horizontal plane perpendicular to the x-axis and points towards the left side (landside) of the powerhouse. The positive z-axis is vertical and points up. Normal stress S11 is parallel to the x-axis, normal stress S22 is parallel to the y-axis and normal stress S33 is parallel to the z-axis.

For example, for the powerhouse base slab reinforcement design, S11 normal stresses were integrated to design the longitudinal horizontal reinforcement (parallel to flow) and S22 normal stresses were integrated to design the transverse horizontal reinforcement (perpendicular to flow).

Another example, for the powerhouse landside exterior wall reinforcement design, S11 normal stresses were integrated to design the horizontal reinforcement (parallel to flow) and S33 normal stresses were integrated to design the vertical reinforcement.

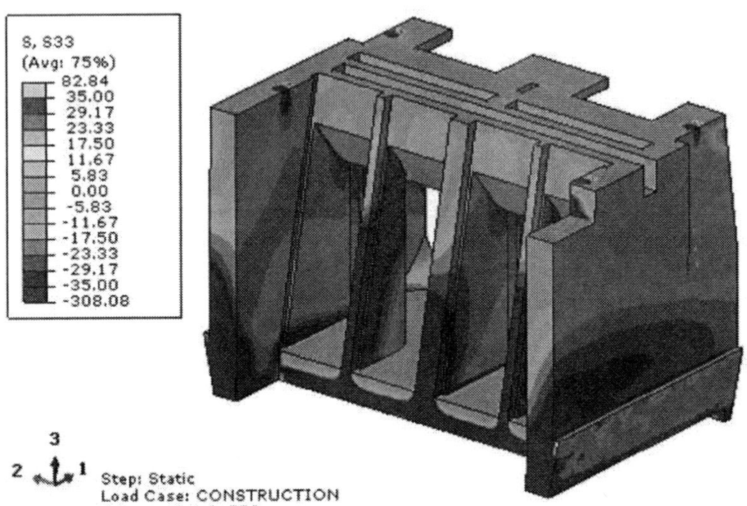

Figure 16. Stress Plot at the Intake Portion of the Powerhouse.

Stresses could be obtained in any orientation and at any point in the structure from the Abaqus FEM analysis results. The following procedure is the implemented powerhouse reinforcement design methodology, which is based on stress integration:

1. Select a structural member of the powerhouse (e.g. landside intake wall or intake base slab) and the orientation of the reinforcement (e.g. horizontal or vertical) to be designed. In this case the vertical reinforcement of the landside intake wall is going to be designed, which requires integration of S33 stresses (see Figure 16).
2. Select an area of interest to design the selected powerhouse structural member based on a review of the relevant stress plots. Define a path in Abaqus in the area of interest. Figure 17 is a view of the exterior face of the landside intake wall and shows several paths (e.g. USW-5, USW-6 and UW-2) that have been defined through the landside intake wall. The defined paths shown on Figure 17 were selected based on a review of the S33 stress plot (see Figure 16). Note that USW and UW stand for upstream wall.
3. Extract the path stresses for the various load cases from Abaqus and integrate them to calculate the unfactored internal stress resultants (axial force, P, and moment, M) at the path. Figure 18 shows the equations used to determine the unfactored internal stress resultants.
4. Factor the internal stress resultants in accordance with USACE EM 1110-2-2104 and then use the factored axial forces, P_u, and factored moments, M_u, to design the reinforcement utilizing interaction diagrams created in CSI Column.

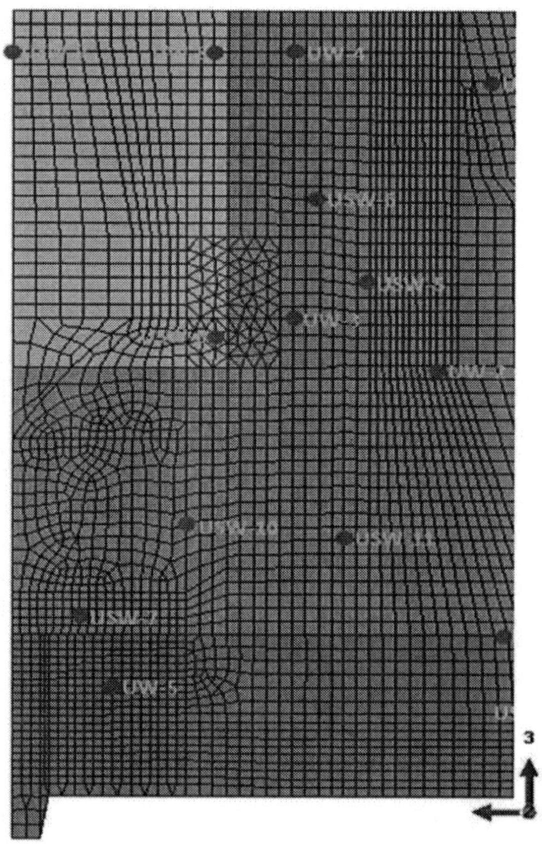

Figure 17. View of the Exterior Face of the Landside Intake Wall Showing Defined Paths.

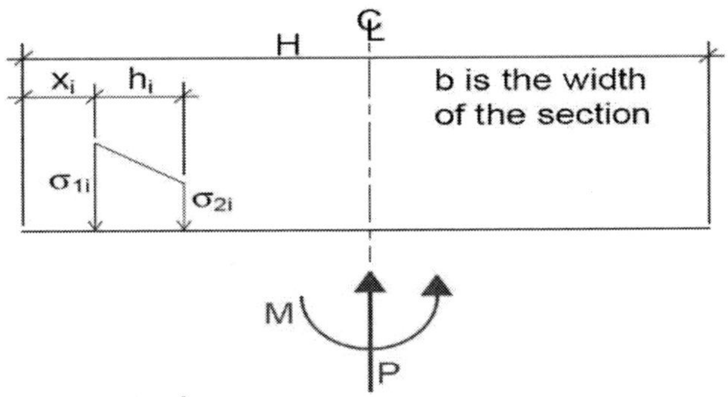

$$P = \sum_i \frac{\sigma_{1i} + \sigma_{2i}}{2} \cdot b \cdot h_i$$

$$M = \sum_i \frac{\sigma_{1i} + \sigma_{2i}}{2} \cdot b \cdot h_i \cdot \left(x_i + \frac{h_i(2\sigma_{2i} + \sigma_{1i})}{3(\sigma_{2i} + \sigma_{1i})} - H/2\right)$$

Figure 18. Equations Used to Calculate the Unfactored Internal Stress Resultants of the Path.

ENTRANCE BUILDING & RIVERSIDE ACCESS BUILDING DESIGN

Region E, highlighted in green on Figure 11, is the entrance building and riverside access building. The entrance building and riverside access building are rectangular reinforced concrete buildings that form a portion of the powerhouse superstructure. These buildings are essentially composed of up to 2 ft thick reinforced concrete slabs and walls, and they were modeled using shell elements in SAP2000, which was used to perform linear elastic static and dynamic structural analyses of these buildings. The slabs were designed as flexural members using Mathcad and most of the walls were designed as compression members using interaction diagrams created in CSI Column.

CONCLUSION

AMP is taking advantage of the USACE's foresight and is upgrading four existing USACE lock and dam installations (Willow Island, Meldahl, Cannelton, and Smithland) on the Ohio River by adding renewable energy sources to these installations for the benefit of its member communities and the nation. The general layout of these hydroelectric projects and the methodologies employed in the analysis and design of the cast-in-place reinforced concrete, combination dams/hydroelectric powerhouses have been presented. The state-of-the-art computer programs and techniques utilized to model and structurally analyze and design the ten-story tall powerhouses, which will house a combined total of eleven horizontal bulb-type turbine-generator units with runner diameters of 25.3 ft (7.7 m), have also been presented.

ACKNOWLEDGEMENTS

The authors would like to acknowledge AMP and their municipal members for their initiative to develop the innovative new hydroelectric projects on the Ohio River and for their vision of renewable energy.

REFERENCES

Bodily, Jason; Konstantellos, Christ; and Meier, Phil; *"The Layout and Design of Four Hydroelectric Powerhouses on the Ohio River"*, HydroVision 2012, Louisville, KY, July 17-20, 2012.

Harris, Craig; Meier, Phil; and Blaszczyk, Paul; *"New Hydropower Development on the Ohio River"*, HydroVision 2008, Sacramento, CA, July 14-18, 2008.

Hughes, Brian; Sims, Nancy; Gemperline, Eugene; Konstantellos, Christ; and Meier, Phil; *"Hydropower Developments at Existing Lock and Dam Projects"*, Waterpower XVI, 2009.

Meier, Phil; Harris, Craig; and Blaszczyk, Paul; *"Powering the Ohio River"*, HydroVision 2010, Charlotte, NC, July 27-30, 2010.

BETTER SAFE THAN SORRY — SURVEILLANCE AND MONITORING OF HYDRO CONSTRUCTION ON THE OHIO

John M. Hynes[1]
Lin Zhao[2]
Abid A. Mirza[3]
Thomas G. Andrews[4]
Philip E. Meier[5]

ABSTRACT

The four new hydroelectric projects that American Municipal Power (AMP) is developing adjacent to existing U.S. Army Corps of Engineer (USACE) locks and dams along the Ohio River require project-specific dam safety surveillance and monitoring programs to be developed, implemented and modified throughout the various phases of project construction, including cofferdam construction, foundation excavation, powerhouse and closure structure excavation, and controlled filling and removal of the cofferdams. The primary goal of these projects is maintaining or improving dam safety and preventing an uncontrolled loss of the navigation pools during all phases of construction and project operations while adding electrical generation capacity to the existing USACE dams. Since the cofferdams required for construction extend into both the upstream and downstream pools and encompass excavations over 100 feet below normal pool levels, robust programs to monitor cofferdam, excavation and dewatering system performance were developed that utilized regular visual inspection along with extensive instrumentation, much of which was automated to provide timely information. Separate programs were developed to monitor the permanent works during first filling of the cofferdam interiors and during cofferdam removal. Since the conditions change throughout construction, frequent assessment and reporting was required. This paper provides an overview of the surveillance and monitoring program throughout construction and discusses lessons learned.

INTRODUCTION

Four hydroelectric projects (Smithland Hydroelectric Project, Cannelton Hydroelectric Project, Meldahl Hydroelectric Project, and Willow Island Hydroelectric Project) are currently under construction on the Ohio River adjacent to existing U.S. Army Corps of Engineers (USACE) locks and dams (Figure 1). These projects are being developed by American Municipal Power (AMP), a nonprofit wholesale power supplier for municipal electric systems. AMP serves 130 members – 129 member communities in Indiana, Kentucky, Michigan, Ohio, Pennsylvania, Virginia and West Virginia, and one joint

---

[1] Senior Geotechnical Engineer, MWH Americas, Inc., 175 W. Jackson Blvd., Suite 1900, Chicago, IL 60604, john.m.hynes@mwhglobal.com.
[2] Lead Geotechnical Engineer, MWH Americas, Inc., lin.zhao@mwhglobal.com.
[3] Professional Geotechnical Engineer, MWH Americas, Inc., abid.a.mirza@mwhglobal.com.
[4] Principal Geotechnical Engineer, MWH Americas, Inc., thomas.g.andrews@mwhglobal.com.
[5] Vice President – Hydroelectric Development and Operations, American Municipal Power, Inc. 1111 Schrock Road, Suite 100, Columbus, OH 43229, pmeier@amppartners.org.

action agency in Delaware. Combined, these publicly owned utilities serve approximately 625,000 customers. MWH Americas, Inc. (MWH) was selected by AMP to serve as the Owner's Engineer for the design and supervision of construction for these new hydroelectric projects.

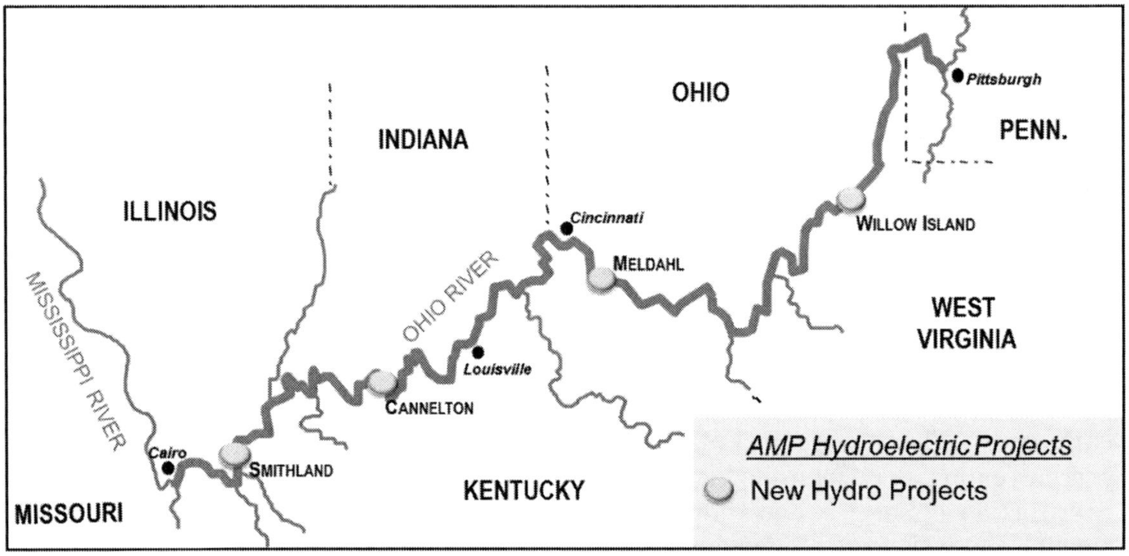

Figure 1. Project Location Map.

Programs were developed for each project to monitor the cofferdam and permanent works (i.e., powerhouse, closure structures, and approach and tailrace channels). Surveillance and monitoring programs included visual observations as well as both automatic and manually-read instrumentation.

PROJECT DESCRIPTIONS AND CONSTRUCTION STAGES

Each of the four hydroelectric projects required the construction of an approach channel, tailrace channel, powerhouse (2 or 3 units per project with 25-foot diameter horizontal bulb turbines), riverside closure structures connecting to the existing dam (e.g., 72-foot-high concrete gravity dam at Meldahl) and landside closure structures (e.g., 90-foot-high hardfill dam at Meldahl). The reinforced concrete powerhouses are designed for overtopping during flood events. The generation capacity of each project is as follows:

- Willow Island Hydroelectric Project: 35 MW capacity (20-foot gross head)
- Meldahl Hydroelectric Project: 105 MW capacity (30-foot gross head)
- Cannelton Hydroelectric Project: 84 MW capacity (25-foot gross head)
- Smithland Hydroelectric Project: 72 MW capacity (22-foot gross head)

The design of each project was tailored to address site-specific conditions including:

- weak rocks, including slickensided indurated clays and a thrust fault present at Willow Island;
- soft clays required support during excavation at Meldahl;
- proximity of Cannelton and Smithland to the New Madrid and Wabash Valley Seismic Zones with potentially liquefiable alluvial soils; and
- karstic limestone which further complicated the design at Smithland.

Unique project-specific design features included:

- 2,000-kip rock anchors at Willow Island to stabilize the foundation below the cellular cofferdam and significant dental excavation and rock bolting to treat the fault where it crossed the excavation.
- Steel piles to stabilize excavation slopes at Meldahl.
- An embankment cofferdam at Cannelton with a 100-foot-high zoned section constructed in the Ohio River and stone columns to improve the foundations below the closure structures by mitigating against liquefaction potential and minimizing immediate and earthquake-induced settlement.
- An embankment cofferdam at Smithland with a 150- to 187-foot-deep, 4000-foot-long soil-cement-bentonite cut-off wall and a robust dewatering system. Stone columns were installed beneath the closure structures. Compaction and consolidation grouting was used to treat the foundation soils and karstic rock below the powerhouse.
- Of eight (8) closure structures (dams) required for the project, six (6) were constructed of hardfill (roller-compacted soil-cement) to maximize the use of on-site materials and minimize the importation of aggregate.

All four projects have similar project layouts and consist of an intake approach channel, a reinforced concrete powerhouse, a landside closure structure, a riverside closure structure and a tailrace channel (Figure 2).

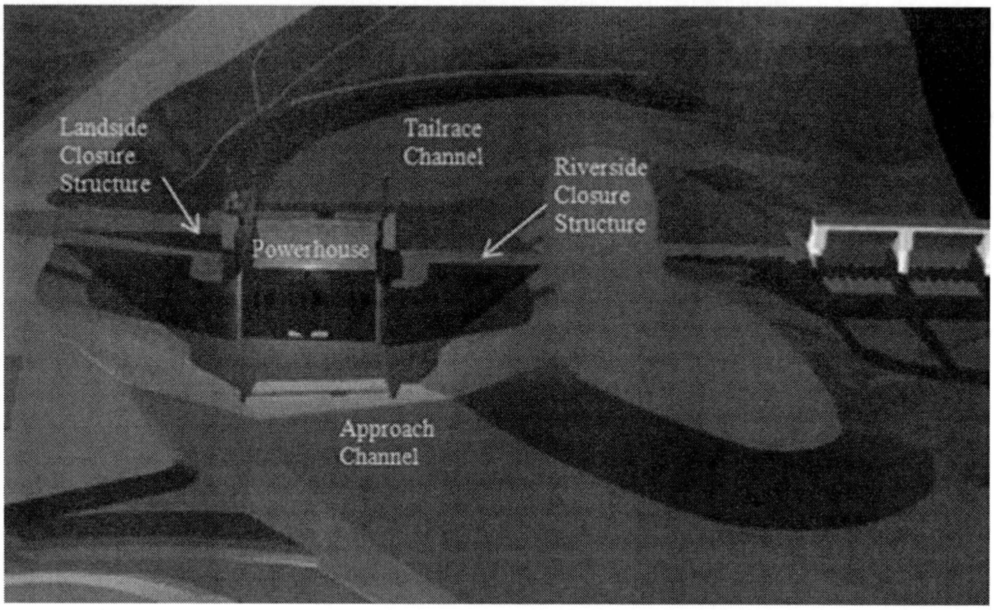

Figure 2. Typical Layout of Project Components (rendering of Cannelton shown)

Photos of the Willow Island, Meldahl, Cannelton and Smithland projects are shown in Figure 3. Further details regarding the development of these hydroelectric projects can be found in Blaszczyk et al. (2015).

Figure 3. Aerial Photos of Construction at Smithland (top left), Cannelton (top right), Meldahl (bottom left), and Willow Island (bottom right) taken during the fall of 2014.

For the purpose of surveillance and monitoring, the projects were divided into three phases – Cofferdam Construction and Powerhouse Excavation; Powerhouse Construction and Cofferdam Decommissioning; and Commercial Operation – which are described below.

Cofferdam Construction and Powerhouse Excavation Phase

This phase consisted of the construction of the cofferdam to provide protection for up to a 100-year flood event and subsequent excavation to the powerhouse foundation level within the interior of the cofferdam.

Construction of cofferdams was required to allow powerhouse construction to be performed under dry conditions. The cofferdams at Smithland, Cannelton, and Meldahl consisted of an earthen embankment with an integral slurry trench cutoff wall. The cofferdam at Willow Island consisted of a cellular cofferdam on the riverside and an earthen embankment with an integral slurry trench cutoff wall on the landside. Additional details of the cofferdams can be found in Andrews et al. (2015).

Powerhouse excavation at Smithland and Cannelton consisted of excavation through alluvial soils. Excavation at Meldahl and Willow Island was through both alluvial soils and bedrock.

Powerhouse Construction and Cofferdam Decommissioning Phase

This phase consisted of construction of the permanent works (i.e., powerhouse, closure structures, and approach and tailraces channels). During this phase, after the powerhouse and closure structures were deemed watertight to the 100-year flood level, decommissioning of the cofferdam began. Cofferdam decommissioning consisted of controlled filling of the cofferdam's interior, and subsequent removal of the cofferdam to form the final configuration of the approach and tailrace channels. As of January 2015, controlled filling of the cofferdam has been completed for Cannelton and Meldahl, and has yet to begin for Willow Island and Smithland.

Commercial Operation Phase

This phase starts after substantial completion of construction when the bulb turbine generating units are commissioned for commercial operation. Units for Cannelton, Meldahl, and Willow Island are to be commissioned in 2015 and it is anticipated that Smithland will be commissioned in 2016.

SURVEILLANCE AND MONITORING PROGRAMS

Surveillance and monitoring plans (SMP) were developed for each project for all three of the listed phases. The SMPs were developed in accordance with Chapter 14 of the Federal Energy Regulatory Commission's (FERC) Dam Safety Surveillance and Monitoring Plan guidelines dated January 2008. Temporary SMPs were developed for the cofferdam and powerhouse construction phases, while long-term SMPs were developed for the commercial operation phase.

Cofferdam Construction and Powerhouse Excavation Phase

The SMPs for the cofferdam construction and powerhouse excavation phase were developed to assess the performance of the cofferdam, temporary cutoff wall, and dewatering system for each project during construction.

Instrumentation installed for the projects included:

- piezometers to monitor piezometric levels both inside and outside the cofferdam,
- flow meters to measure dewatering rates,
- Rossum sand testers to measure sand content of flow from dewatering wells, and
- inclinometers, tilt meters and survey monuments to measure potential movement of the cofferdam and existing structures.

Willow Island, which included rock anchors to support the rock excavation, also utilized DYNA Force elasto-magnetic sensors to measure the tension in the anchor strands.

All instrumentation installed for the cofferdam construction and powerhouse excavation phase was temporary, and was decommissioned during or immediately after controlled filling of the cofferdam for each project.

The Willow Island, Cannelton, and Smithland cofferdams were constructed using design-build contracts, so the contractor was responsible for overseeing the surveillance and monitoring during cofferdam construction at those sites. The Meldahl cofferdam was a design-bid-build contract. Therefore MWH, as AMP's engineer, was responsible for surveillance and monitoring for this phase at Meldahl.

A summary table showing the number of each type of instruments is provided in Table 1.

Table 1. Number of Cofferdam (Temporary) Instruments at Each AMP Site

Instrument	Smithland	Cannelton	Meldahl	Willow Island
Vibrating Wire Piezometers	19	18	23	44
In-Place Inclinometers (IPI)	9	5	7	7
Manual Inclinometers	1	0	3	1
Survey Monuments	32	14	21	64
Tilt Meters	0	0	0	4
Flow Meters	65	20	20	32
DYNA Force Sensors	0	0	0	45

In general, the inclinometers were in-place inclinometers (IPI) and were read automatically via an automated data acquisition system (ADAS). Manually read inclinometers were also utilized to verify readings provided by the IPIs. Inclinometer depths varied from 70 to 100 feet.

All cofferdam piezometers utilized vibrating wire sensors that were read via an ADAS. A portion of the piezometers were fully grouted vibrating wire piezometers (i.e., vibrating wire piezometers are installed in borehole which is then backfilled entirely with cement-bentonite grout without the use of a standpipe), while others consisted of vibrating wire piezometers in an open standpipe, which allowed for manual confirmation of the vibrating wire readings. Piezometer depths varied from 40 to 150 feet.

Two of the project sites (Smithland and Willow Island) utilized automated motorized total stations (AMTS), which were connected to the ADAS, to read the survey monuments. At the other project sites, prisms were surveyed on a regular basis by a licensed professional surveyor. Flow meters and Rossum sand tests were read on a regular basis by site personnel. Each of the four projects utilized separate websites which allowed remote access to all instrumentation data collected by ADAS.

Threshold and Action Levels. Threshold and action levels were developed to aid in immediate field-verification of instrumentation readings and to assist in determining if readings were approaching a level that would cause concern regarding the stability of the cofferdam and powerhouse excavation. Readings exceeding threshold levels prompted further investigation into the situation and the consideration of potential corrective actions. Readings exceeding action levels prompted the implementation of a corrective action (e.g., check of the reading/instrumentation, increased surveillance and monitoring frequency, review of stability analysis assumptions, or potentially emergency action).

Threshold and action levels for the piezometers were developed from the results of slope stability analyses. Threshold values and action levels for the inclinometers, crack gauges, survey monuments, flow meters, and Rossum Sand Testers were established based on stability analysis, numerical analysis, engineering judgment, and experience at other projects. Typical threshold and action levels across the four projects are provided in Table 2. Figure 4 shows the layout of cofferdam instrumentation at Meldahl.

Table 2. Typical Threshold and Action Levels Used For Cofferdam Instrumentation

Instrument	Threshold Values	Action Levels
Piezometers	Based on Allowable Phreatic Surface Calculated by Slope Stability Analyses	
Inclinometers in Soil	4 inches	6 inches
Inclinometers in Rock	0.25 inches	0.50 inches
Embankment Survey Monuments	3 inches	6 inches
Survey Monuments on Concrete Structures Founded on Rock	0.15 to 0.25 inches	0.30 to 0.50 inches
Tilt Meters on Concrete Structures Founded on Rock	0.06 degrees	0.08 degrees
Flow Meters	15% increase from steady state rates	30% increase from steady state rates
Rossum Sand Testers	3 ppm	5 ppm

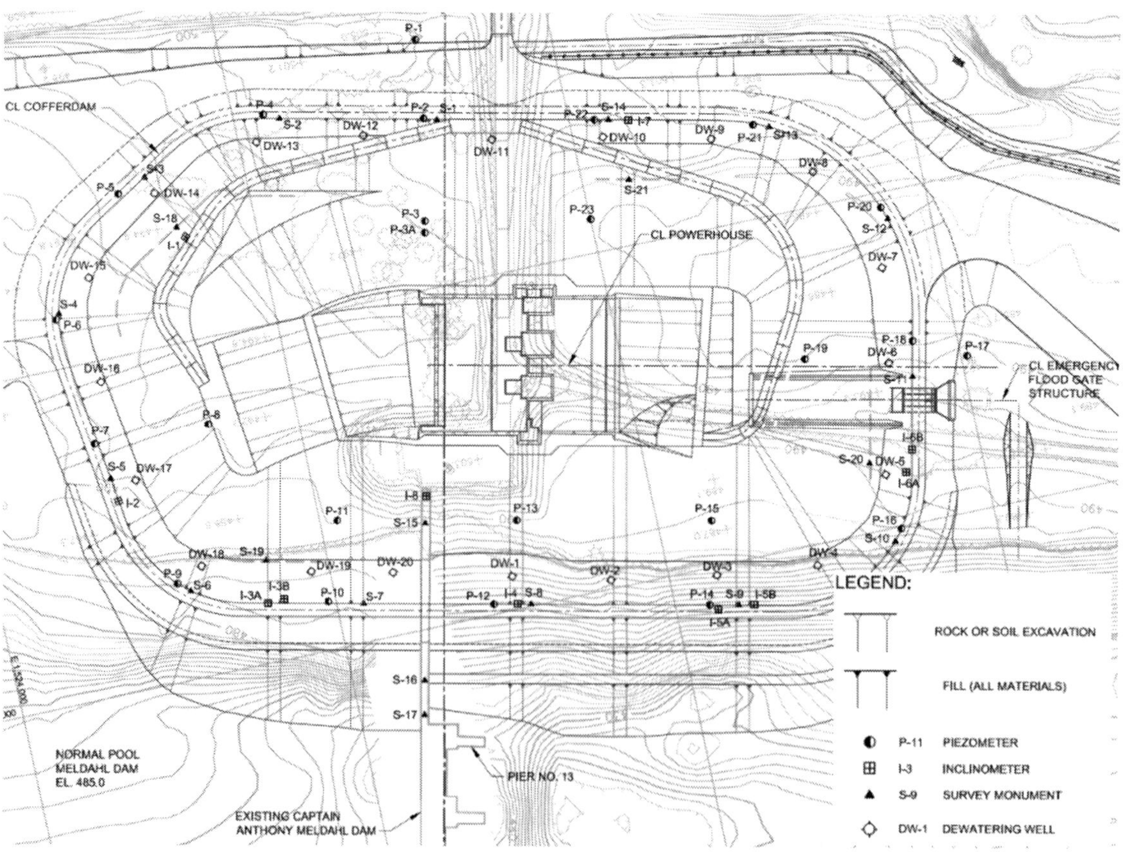

Figure 4. Cofferdam Instrumentation Layout at Meldahl

Readings for instruments connected to the ADAS were collected typically once every 1 to 4 hours; these readings were posted on the individual project instrumentation websites. Reading frequencies for instruments not automatically read by the ADAS generally ranged from weekly to monthly, with higher frequencies during excavation.

Formal visual surveillance was performed on a weekly to bi-weekly basis with informal observations also noted by field staff on a daily basis. Surveillance items included:

- Checking the crest for signs of subsidence or cracking
- Checking for slope deformation.
- Noting any seeps or springs, as well as turbidity of seepage.
- Noting any signs of erosion.
- Checking the functionality of the instrumentation and dewatering systems.
- Checking riprap for any signs of deterioration or subsidence.

Powerhouse Construction and Cofferdam Decommissioning Phase

The SMPs for the powerhouse construction and cofferdam decommissioning phase were developed for the purpose of continuing to monitor the performance of the cofferdam during construction while transitioning to monitoring of the permanent works as coordinated with controlled filling of the cofferdam and cofferdam removal.

Powerhouse construction was performed under a design-bid-build contract for all four projects. Therefore MWH, as AMP's engineer, was responsible for all surveillance and monitoring during the powerhouse construction and cofferdam decommissioning phase.

Reading frequencies and threshold and action levels for cofferdam instrumentation remained the same for this phase. Cofferdam instrumentation continued to be read up until the point when controlled filling of the cofferdam commenced, at which point this temporary instrumentation was decommissioned.

All permanent instrumentation was installed as a part of this phase and baseline readings were taken as a prerequisite for controlled filling of the cofferdam. Permanent instrumentation includes:

- Powerhouse piezometers to measure uplift pressures at the powerhouse foundation and assess effectiveness of the permanent cutoff wall. The powerhouse piezometers were installed in the lowermost powerhouse gallery and consist of a pressure gauge mounted on a stainless steel standpipe.
- Closure structure piezometers to measure uplift pressures at the closure structure foundations and assess effectiveness of the permanent cutoff wall. The closure structure piezometers were installed through the closure structure into the foundation and consist of a vibrating wire piezometer in an open standpipe.

- Survey monuments to measure potential movement.
- Strong motion accelerometer to record any potential ground motions.

Figure 5 shows the layout of permanent instrumentation at Meldahl.

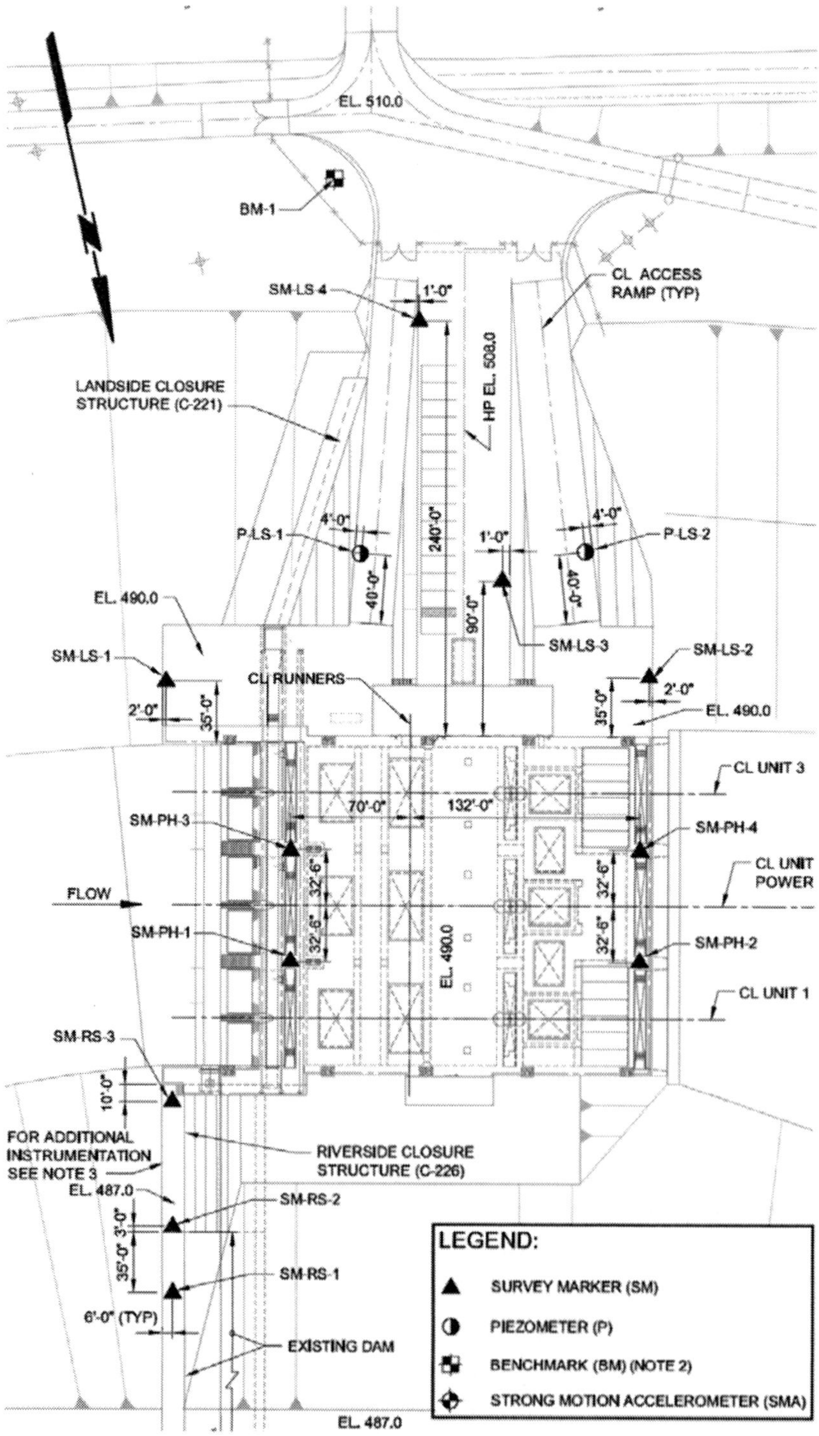

Figure 5. Permanent Instrumentation Layout at Meldahl

Readings of permanent survey monuments at each project site began approximately one month prior to the start of cofferdam decommissioning in order establish a baseline. At the start of cofferdam filling, permanent instrumentation was read daily to assess the response of the structures under initial loading. After cofferdam filling was complete, the instrument reading frequency was gradually reduced to monthly as readings continued to indicate that the structures were performing as expected.

Threshold and action levels for survey monuments were established based on previous experience at other projects. In general, threshold and action levels for structures founded on soils were 1 inch and 2 inches, respectively, while threshold and action levels for structures founded on rock were 0.4 and 0.8 inches, respectively. Threshold and action levels for piezometers were established based on stability analyses of the structures.

Formal and documented visual surveillance also continued for this phase, but with added surveillance of the permanent works, including:

- Checking the powerhouse and closure structures for signs of cracking or movement.
- Checking the contacts between powerhouse and closure structures for signs of movement or damage.
- Checking for signs of movement in the approach and tailrace channels.

Commercial Operation Phase

The SMPs for the commercial operation phase were developed to monitor permanent instrumentation over the long-term operation of the project. These plans will go into effect at the start of commercial operation of the generating units, several months following the completion of controlled filling and removal of the cofferdam.

Monitoring permanent instrumentation and visual surveillance of permanent works will continue but at a reduced frequency (monthly to yearly). Instrumentation readings taken after the completion of controlled filling but prior to the start of commercial operation will be used to establish long-term baselines. Threshold and action levels will be reduced slightly for survey monuments as little movement is anticipated after initial loading of the structures during controlled filling.

NOTABLE OBSERVATIONS AND LESSONS LEARNED

The following are notable observations of surveillance and monitoring data along with lessons learned from project experience.

Piezometer Data Used to Identify Well Breakdowns

Monitoring of piezometer data at the Meldahl site aided in identification of dewatering wells requiring maintenance. Once a spike in piezometer data was identified, field personnel were able to investigate nearby wells to that piezometer, and the contractor was able to repair the well. The response was then immediately reflected in the piezometer data. Figure 6 shows an example of piezometer data that indicated a well breakdown.

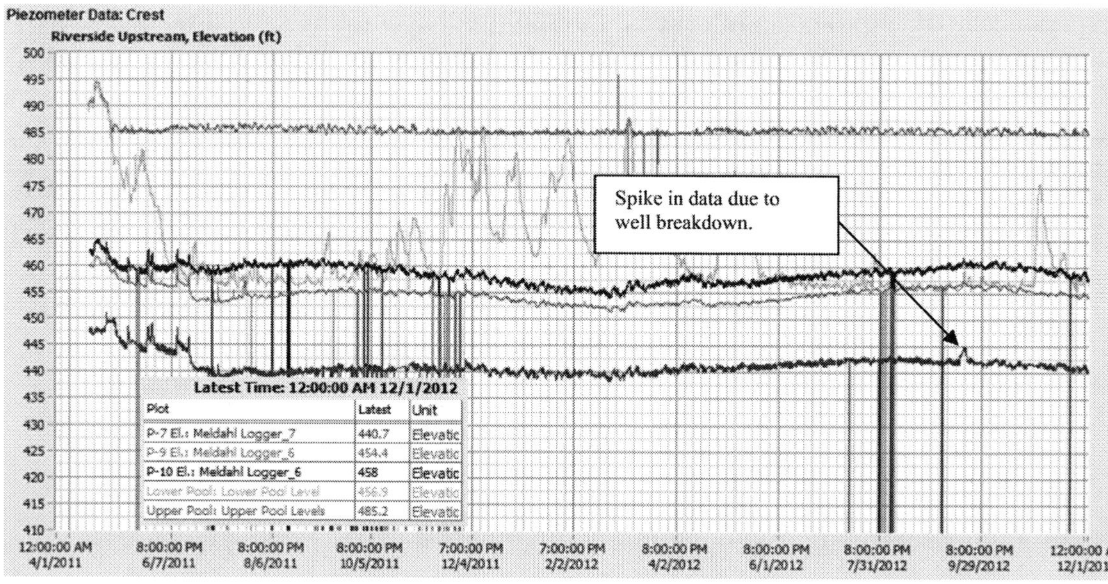

Figure 6. Meldahl Piezometer Data Showing Spike Indicating Well Breakdown

Observations during Filling

As of January 2015, controlled filling of the cofferdam's interior has occurred at Cannelton and Meldahl. Controlled filling of the cofferdams generally occurred over a period of approximately one month. Figure 7 shows photographs of the Cannelton and Meldahl projects during the controlled filling process.

Figure 7. Photos of Controlled Filling at Cannelton (left) and Meldahl (right)

Figure 8 is a plot of piezometer data recorded during and after the controlled filling process at Cannelton. Note that the approach and tailrace channel elevations represent the water levels inside the cofferdam, while the upper and lower pool levels represent the river levels outside of the cofferdam (after the cofferdam is excavated below river levels, these two levels are equal). The piezometers plotted in green (U1 to U3) measure piezometric pressures immediately below the powerhouse. The piezometers plotted in blue and red (LSUS, LSDS, RSUS, RSDS) measure pressures immediately below the closure structures.

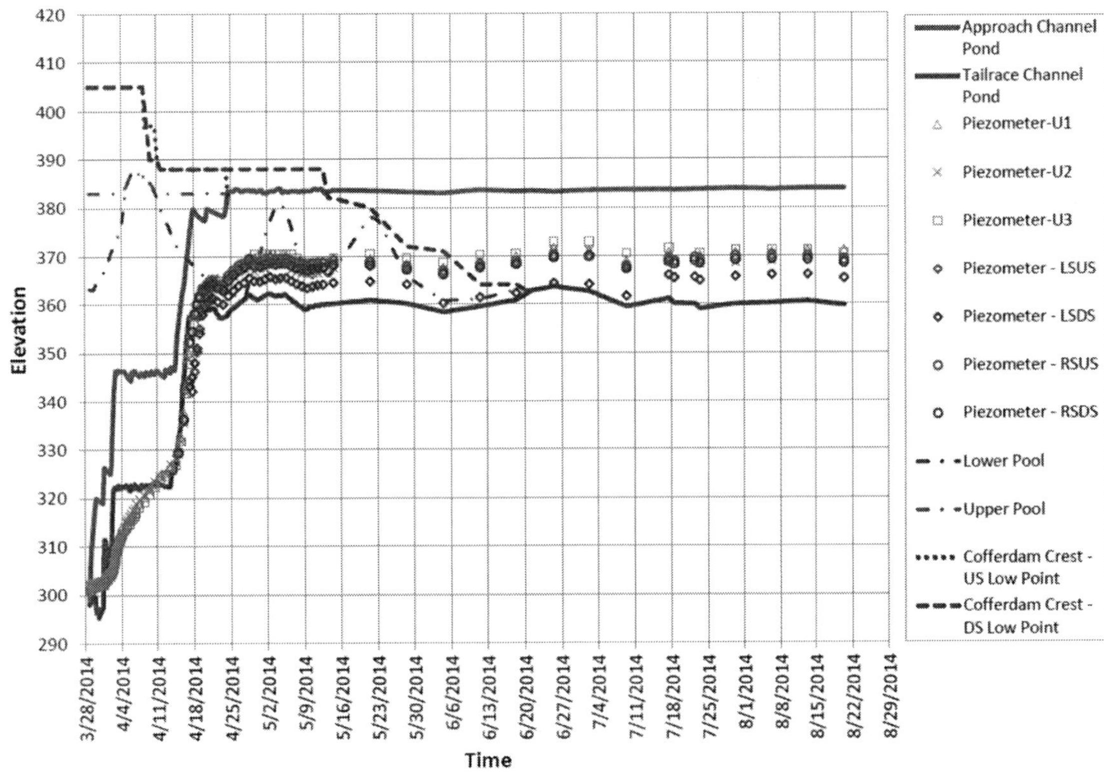

Figure 8. Cannelton Piezometer Readings during and after Controlled Filling

At Cannelton where the powerhouse is founded on soil, piezometric levels measured in the powerhouse and closure structure piezometers initially lagged behind the tailrace levels as the foundation soils saturated, but readings eventually stabilized 5 to 10 feet above tailrace levels.

At Meldahl, where the powerhouse is founded approximately 30 feet below the top of bedrock, powerhouse piezometer readings were initially higher than the elevations of the ponded water in the approach and tailrace channels, presumably due to the higher piezometric levels in the rock immediately adjacent to either side of the powerhouse. Interestingly, piezometric levels ultimately stabilized slightly below tailwater levels. It is believed that this reduction in pressure is caused by minor amounts of leakage (on the order of 5 GPM) into the powerhouse coupled with the relatively low permeability of the limestone bedrock causing a localized decrease in pressure at the powerhouse foundation.

Survey Accuracy and Survey Control

Survey accuracy and survey control was a significant challenge for the projects during all phases. With relatively tight threshold levels (as low as 0.15 inches), it was often difficult to achieve survey accuracy within those limits. One of the causes of this difficulty was the fact that cofferdam construction required placement of large amounts of fill and active dewatering. Thus, it was difficult to establish benchmarks and survey pedestals in locations that had line of sight to the monuments, were not placed on fill material, and were in a location that would not be disturbed by construction activities. Benchmarks were generally located on soil (often fill) and several were found to have moved during the course of construction. Thus, when planning the locations of survey monuments, it is recommended to also plan the location of benchmarks and survey pedestals considering line of sight as well as construction sequencing. Also, it should be recognized that benchmarks founded on fill are more likely to be susceptible to movement than benchmarks constructed on a properly treated soil or rock foundation. Therefore, in cases where benchmarks are to be placed on fill, multiple benchmarks should be used so that benchmark displacement can be evaluated.

ADAS Maintenance

Use of ADAS to collect instrumentation data allows real-time monitoring, processing, and ability for off-site viewing data. However, an ADAS also requires a significant amount of maintenance. The ADAS systems for the AMP projects used radio frequency communications to transmit data, which would occasionally encounter interference from construction activity and/or barge traffic communications. In addition to communication issues, instruments also went offline due to dead batteries, malfunctioning control boards, and lightning strikes. Lightning also caused damage to the on-site server hosting the monitoring website, which resulted in the site being down for several days on multiple occasions. Instruments read by the ADAS also showed occasional dropouts in data (returning null values) and anomalous spikes (unrealistic jumps in data that returned to normal on the next reading).

All of these items require the attention of an individual with knowledge of the system. Thus, if using an ADAS, it is recommended to plan to have a person on-site full time to manage and maintain the system. It is also recommended to have a backup server off-site in case of problems with the on-site server. It is also recommended that an appropriate number of instruments have manually-read backups that can help provide confirmation when instruments read by the ADAS are malfunctioning. Manually read instruments also provide an additional layer of backup in the event of an ADAS outage.

Collaboration with Regulatory Agencies

The regulatory agencies (FERC and USACE) have had a high level of involvement in surveillance and monitoring during construction of the AMP hydroelectric projects. All SMPs were submitted to FERC and USACE for approval. In several instances, threshold and action levels were developed based on collaboration between MWH and the agencies. Regular surveillance and monitoring updates were provided to the USACE and FERC in the form of monthly Surveillance and Monitoring Reports (SMR), weekly teleconferences, as well as regular email updates. During cofferdam filling, daily email updates were sent to the agencies. It is recommended to keep an open line of communication with regulatory agencies both when developing SMPs and throughout the process of collection and interpreting surveillance and monitoring data.

Quantity of Instrumentation and Redundancy

When an instrument reading exceeds a threshold or action value or otherwise displays a notable trend, it is important to have some redundancy in the instrumentation to evaluate such readings. At Meldahl, inclinometer and tilt meter readings and observations of sawcut tell-tales (i.e., epoxy filled sawcuts in concrete across construction joints that serve as a visual reference for potential displacements) helped to show that an anomalous jump in survey readings was caused by issues with survey methods and not actual movement of the structure. At both Meldahl and Willow Island, manual inclinometer readings were taken to verify readings from IPIs. Manual piezometer readings were taken at all sites to verify vibrating wire piezometer readings. Additionally, all sites used survey monuments and inclinometers to monitor potential displacement of the cofferdam.

While it is important to ensure redundancy in the system to allow for reading verification, it is also important to limit the number of instruments to what is needed and manageable. Every piece of instrumentation should have a specific purpose for monitoring, because with every added piece of instrumentation comes an additional requirement for maintenance and additional data to reduce, analyze, and report. With nearly 200 instruments at Willow Island, the monthly surveillance and monitoring report is approximately 650 pages. This level of instrumentation was needed for the Willow Island project to assess site specific challenges including a known reverse fault and slickensided claystone. However, with that amount of data, it can be a challenge to efficiently and effectively sift through and identify which instruments, if any, are showing potential problems. For Willow Island, daily automated email updates were sent out to flag any readings that were beyond threshold or action levels.

Notable Observations of Visual Surveillance

The most common observation noted in the formal and informal visual surveillance was erosion of the slopes of the powerhouse excavation. Due to long slopes in soil (up to approximately 100 feet high), erosion was often noted after high intensity rain events, particularly in instances where the grass cover had been removed due to work in that area and not yet re-established. Erosion rills and gullies were tracked through visual surveillance activities and were prioritized for repair.

Additionally, seeps or springs in the powerhouse excavation were mapped and documented through visual surveillance. The seeps and springs were regularly monitored for potential sediment transport and visible changes in flow rate.

Another notable observation was the ability to observe the effectiveness of the cutoff wall and downstream graded filter during controlled filling. At Cannelton and Meldahl, the upstream portion of the cofferdam was partially filled (up to approximately 25 feet) prior to beginning filling on the downstream side. This allowed for observation of the seepage rate from upstream to downstream beneath the powerhouse and to observe whether the seepage carried sediment. Visual observations showed that the filters functioned properly and that the seepage water was free of sediment.

CONCLUSION

Four new hydroelectric projects are being developed by AMP adjacent to existing USACE locks and dams. Due to the importance of maintaining dam safety and preventing the uncontrolled loss of the navigation pools during construction and operation, robust surveillance and monitoring plans were developed. The plans covered surveillance and monitoring activities starting with cofferdam construction and continuing all the way through commercial operation of the plants. Important lessons learned included establishing a detailed plan for survey control and accuracy, planning for continual maintenance of the ADAS system, and including redundancy in instrumentation to provide verification for readings.

ACKNOWLEDGMENTS

The authors would like to acknowledge AMP and their municipal members for their initiative to develop the innovative new hydroelectric projects on the Ohio River and for their vision of renewable energy.

The aerial photographs of the Smithland, Cannelton, Meldahl and Willow Island projects were provided to AMP by CJ Mahan Construction Company, Walsh Construction Company, Alberici-Baker Joint Venture, and The Ruhlin Company, respectively.

REFERENCES

Andrews, T.G., L. Zhao, J.M. Hynes and P.E. Meier (2015). "Holding Back the Ohio – Cofferdam and Cutoff Design and Construction," USSD 2015, Louisville, KY.

Blaszczyk, P.R., K. Griebenow, P.E. Meier (2015). "Time to Generate Power – New Hydro Development on the Ohio", USSD 2015, Louisville, KY.

Federal Energy Regulatory Commission (2008). *Engineering Guidelines for Evaluation of Hydropower Projects*, Chapter 14, "Dam Safety Performance Monitoring Program," Appendix J, "Dam Safety Surveillance and Monitoring Plan Outline," rev. 2. FERC 0119-2, January 15.

AUTHOR INDEX

AUTHOR INDEX

AUTHOR INDEX

United States Society on Dams (USSD)
1616 Seventeenth Street #483
Denver, CO 80202
USA

CURRAN ASSOCIATES INC.
proceedings.com

ISBN 978-1-5108-2941-1

9 781510 829411